¡Sí hay en español!

HOLT CIENCIAS Y TECNOLOGÍA

Holt Science and Technology

Now our popular middle school science program is available in Spanish!

The closer you look... the better we look!

Everything Your Students

Vocabulary and **Objectives** appear at the beginning of each lesson, so students know what to expect.

Visuals are highlighted and integrated into the narrative so students can easily make connections between the text and the illustrations.

Diagrams are clearly labeled and demonstrate processes while providing visual reference to help students fully grasp the concepts.

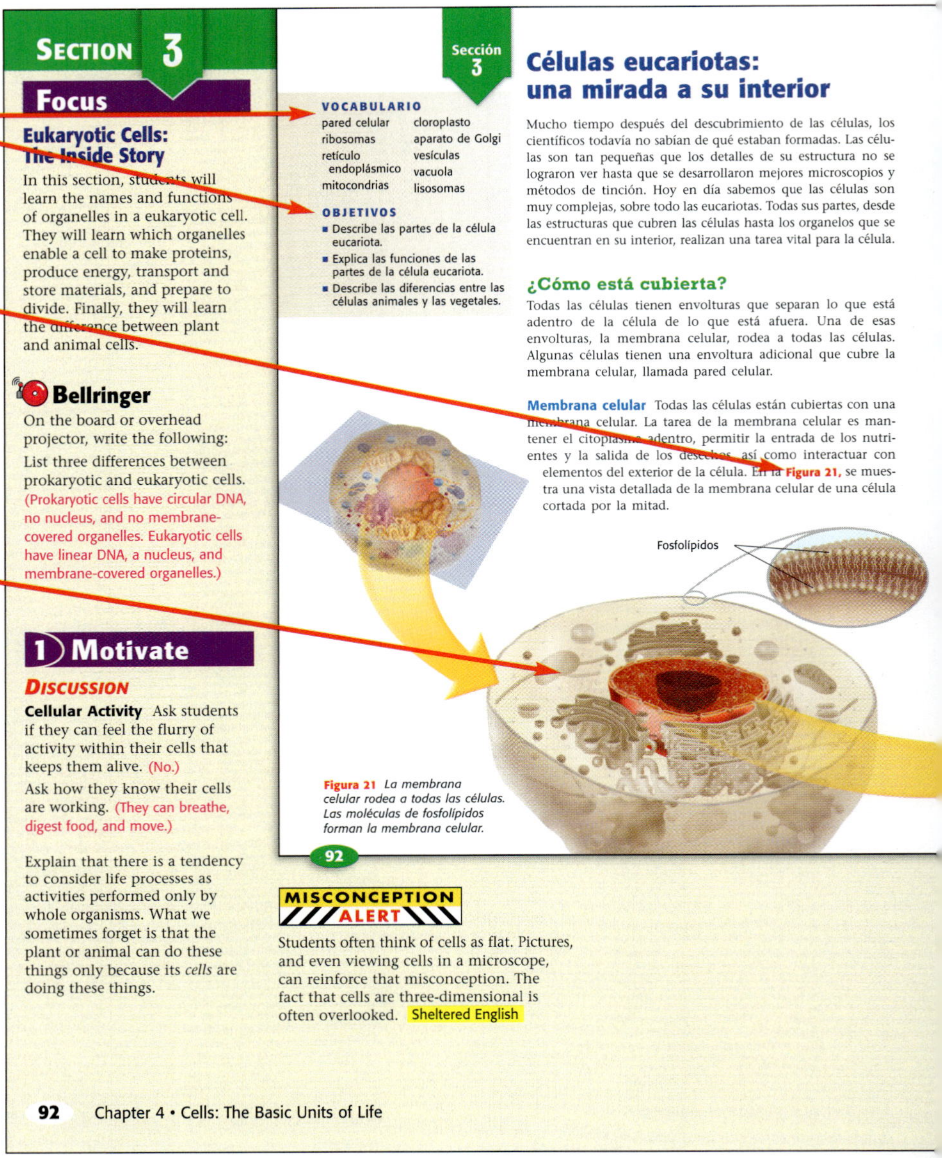

The following shows a reproduced textbook page:

EDICIÓN DEL ESTUDIANTE (Pupil's Edition)

A Complete Spanish Translation

The *Pupil's Edition* provides complete content coverage for life, earth, and physical science while making learning fun. The text keeps students interested and motivated with an easy-to-read and engaging narrative, colorful scientific art, and hands-on activities. An in-text **LabBook** also provides flexible and creative labs for every chapter.

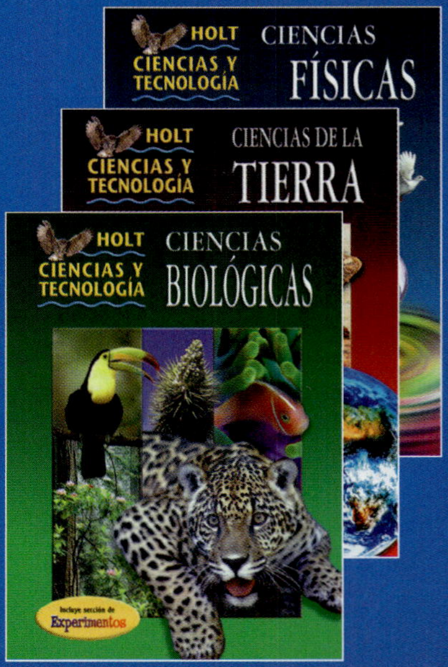

Need in Spanish

Pared celular Las células de las plantas y algas tienen una pared celular dura hecha de celulosa. La **pared celular** rodea y ayuda a la membrana celular. Cuando entra o sale mucha agua de una célula vegetal, la pared celular impide que la membrana se rompa. La fuerza de mil millones de paredes celulares en las plantas permite que un árbol no se caiga y que sus ramas desafíen la gravedad. Cuando observas heno seco, palos y tablas de madera, estás viendo las paredes celulares de células vegetales muertas. Las células de hongos, como champiñones, hongos venenosos, moho y levaduras tienen células hechas de una substancia química parecida a la que se encuentra en las conchas de los insectos. La **Figura 22** muestra una célula vegetal típica y una imagen detallada de la pared celular.

La biblioteca de la célula

El organelo más grande y visible de una célula es el núcleo. La palabra *núcleo* significa "centro" o "nuez" (se ve como una nuez dentro de un dulce). Observa la **Figura 23;** el núcleo está cubierto con una membrana por la que pasan los materiales.

El núcleo también es conocido como el centro de control de la célula. Como sabes, almacena el ADN, que contiene la información genética de las células. Casi todas las reacciones químicas vitales para la célula se relacionan con algún tipo de proteína. A veces, se puede ver una mancha negra dentro del núcleo. Esta mancha se llama *nucléolo,* y es como una especie de núcleo pequeño dentro de un núcleo grande. El nucléolo almacena los materiales que se usan para crear ribosomas en el citoplasma.

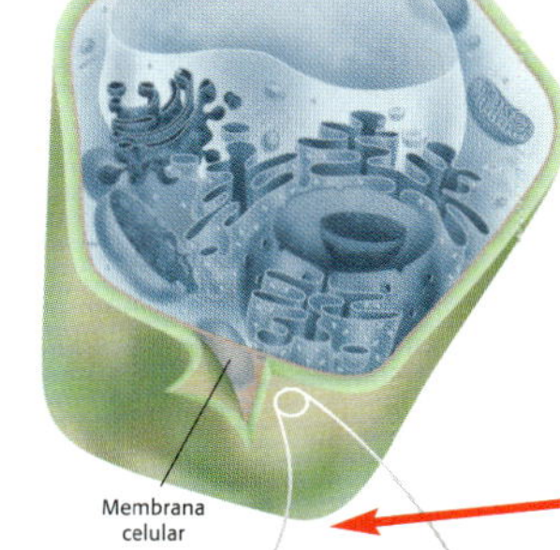

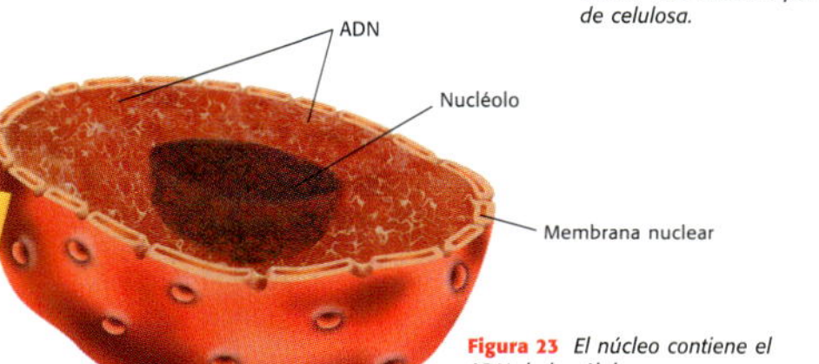

Figura 22 *La pared celular rodea la membrana celular. En las células vegetales, la pared celular está formada por fibras de celulosa.*

Figura 23 *El núcleo contiene el ADN de la célula.*

93

2) Teach

DEMONSTRATION

Cell Walls With a stick, a mushroom, and your own hand, you can illustrate the difference between a rigid cell wall and a flexible cell membrane like that found in human skin cells. Bend the stick, and it will break. Bend the mushroom, and it will come apart. Make a fist, and your skin stretches to accommodate the flexing of muscles and bone joints. If we had rigid cell walls like plants, we would find it extremely difficult to move. **Sheltered English**

MEETING INDIVIDUAL NEEDS

Writing **Advanced Learners** Because all multicellular plants and animals are composed of eukaryotic cells, stress that the eukaryotic cell can be an entity in itself, not just a component of a larger organism. Have students research one-celled eukaryotic organisms, like a yeast or a one-celled protist, and compare them with eukaryotic cells that are part of a multicellular plant or animal. Students should include drawings, and record their findings in their ScienceLog. **PORTFOLIO**

Directed Reading Worksheet 4 Section 3

Homework

PORTFOLIO **Poster Project** Have students investigate red blood cells and create a poster comparing the red blood cell with the cheek skin cell. (RBCs are the only cells in the human body that do not have a nucleus or mitochondria. Without a nucleus they cannot divide and reproduce. They live for only about 120 days, but new ones are made by bone marrow at the rate of up to 200 billion per day.)

Spanish Resources

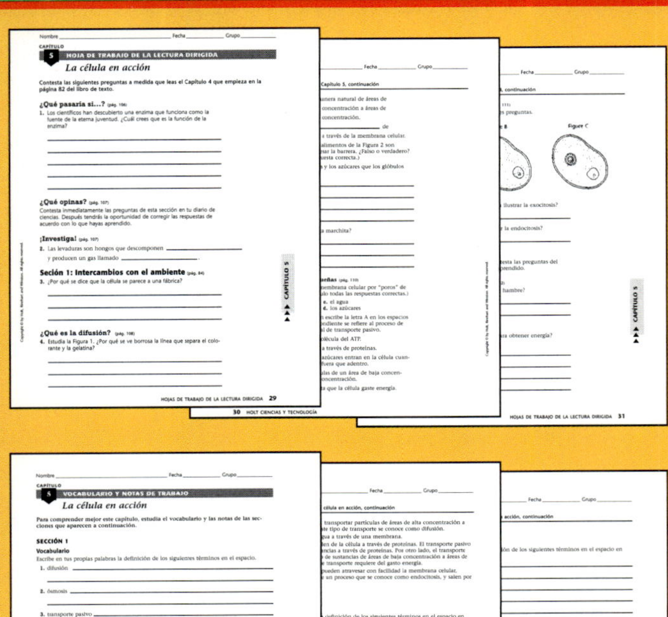

HOJAS DE TRABAJO DE LA LECTURA DIRIGIDA (Directed Reading Worksheets)

These worksheets break lessons into small sections, providing multiple-choice or fill-in-the-blank questions and engaging activities to help your students stay focused.

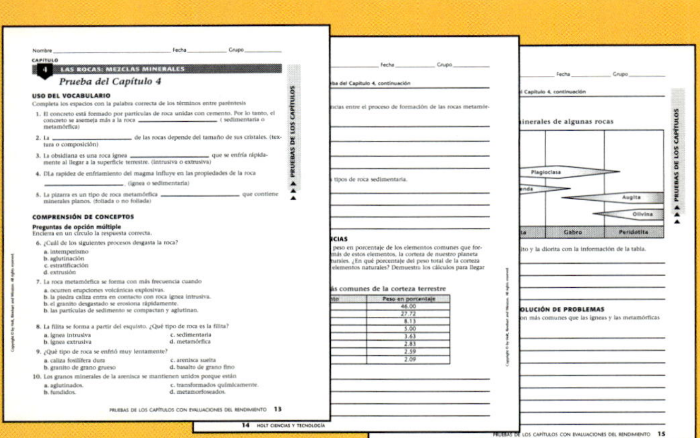

GUÍA DE ESTUDIO (Study Guide)

This resource helps your students study for tests and quizzes. The guide includes blackline master worksheets of **Chapter Highlights** and **Chapter Reviews** from the *Pupil's Edition*.

EXÁMENES DE LOS CAPÍTULOS Y EVALUACIONES BASADAS EN EL RENDIMIENTO (Chapter Tests with Performance-Based Assessment)

These tests and assessment materials accurately measure your students' understanding of concepts and skills. This resource includes multiple-choice, short-answer, concept-mapping, critical-thinking, performance-based questions, and much more!

PROGRAMA DE DISCOS COMPACTOS DE LA LECTURA DIRIGIDA (Guided Reading Audio CD Program in Spanish!)

Bridge the Gap Between Spanish and English

This CD program in life, earth, and physical science provides a direct reading of each chapter in Spanish. This is a powerful tool for helping your English-language learners read and review content. The program includes audio instructional hints, lively performances, and music to help learners stay on task.

When your Spanish-speaking students are ready to take the next step, they can use the **Guided Reading Audio CD Program** in English.

Regardless of skill levels, language abilities, or learning styles, students who use these audio CDs gain a deeper understanding of science and the world around them.

5 Easy Steps to Better Student Comprehension

Using these steps, your Spanish-speaking students have an equal opportunity to learn science concepts.

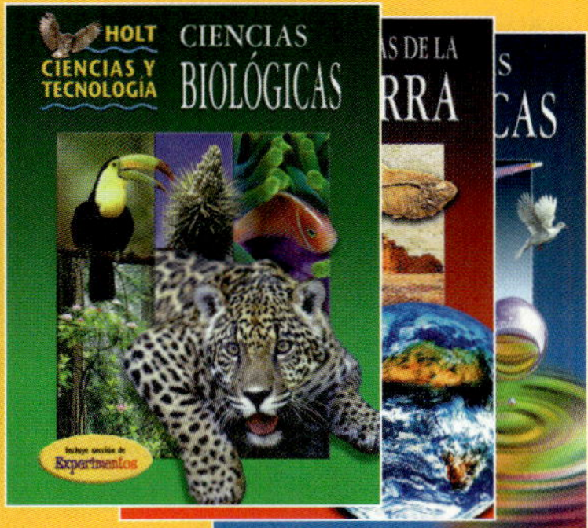

1 EDICIÓN DEL ESTUDIANTE (Pupil's Edition)

Start with the text to explore important science concepts in Spanish. If students have difficulty understanding the text or staying focused, move on to...

2 PROGRAMA DE DISCOS COMPACTOS DE LA LECTURA DIRIGIDA (Guided Reading Audio CD Program in Spanish)

Help your students gain a deeper understanding of science by listening to a direct reading of each chapter in Spanish. To get your Spanish-speaking students moving toward proficiency, this program is also available in English. To reinforce what your students have just read and heard, you can assign...

3 HOJAS DE TRABAJO DE LA LECTURA DIRIGIDA (Directed Reading Worksheets)

These worksheets break each lesson into smaller, more manageable parts. This resource is a complete Spanish translation of a variety of strategies and fun activities that further develop your students' reading and comprehension skills. At the end of each chapter, students can review with...

4 GUÍA DE ESTUDIO (Study Guide)

To help your students study for assessment, we have translated these worksheets to reinforce science concepts. Your students are now ready for...

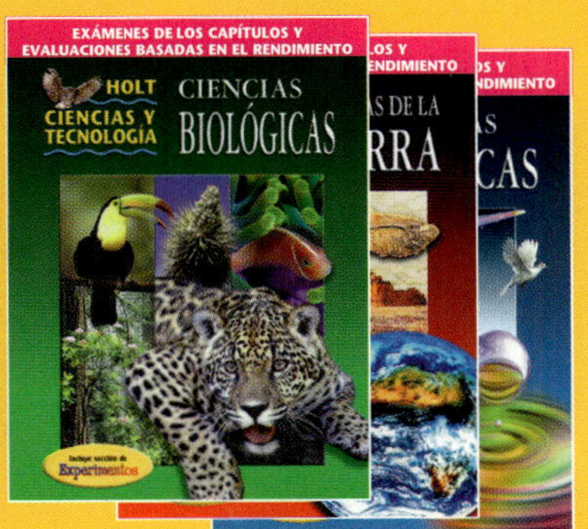

5 EXÁMENES DE LOS CAPÍTULOS Y EVALUACIONES BASADAS EN EL RENDIMIENTO (Chapter Tests with Performance-Based Assessment)

Spanish versions of the end-of-chapter tests are included in this resource.

Components Available in Spanish

HOLT CIENCIAS Y TECNOLOGÍA: CIENCIAS BIOLÓGICAS

EDICIÓN DEL ESTUDIANTE..H64749-5

EDICIÓN ANOTADA DEL MAESTRO H64751-7

GUÍA DE ESTUDIO H64752-5

HOJAS DE TRABAJO DE LA LECTURA DIRIGIDA H64753-3

EXÁMENES DE LOS CAPÍTULOS Y EVALUACIONES BASADAS
EN EL RENDIMIENTO H64754-1

PROGRAMA DE DISCOS COMPACTOS DE LA
LECTURA DIRIGIDA H65348-7

HOLT CIENCIAS Y TECNOLOGÍA: CIENCIAS DE LA TIERRA

EDICIÓN DEL ESTUDIANTE..H64756-8

EDICIÓN ANOTADA DEL MAESTRO H64757-6

GUÍA DE ESTUDIO H64758-4

HOJAS DE TRABAJO DE LA LECTURA DIRIGIDA H64759-2

EXÁMENES DE LOS CAPÍTULOS Y EVALUACIONES BASADAS
EN EL RENDIMIENTO H64761-4

PROGRAMA DE DISCOS COMPACTOS DE LA
LECTURA DIRIGIDA H65349-5

HOLT CIENCIAS Y TECNOLOGÍA: CIENCIAS FÍSICAS

EDICIÓN DEL ESTUDIANTE..H64762-2

EDICIÓN ANOTADA DEL MAESTRO H64763-0

GUÍA DE ESTUDIO H64764-9

HOJAS DE TRABAJO DE LA LECTURA DIRIGIDA H64766-5

EXÁMENES DE LOS CAPÍTULOS Y EVALUACIONES BASADAS
EN EL RENDIMIENTO H64767-3

PROGRAMA DE DISCOS COMPACTOS DE LA
LECTURA DIRIGIDA H65351-7

For more information on
HOLT CIENCIAS Y TECNOLOGÍA or
Holt Science and Technology (English version)

Call or Click.
1-800-HRW-9799
www.hrw.com

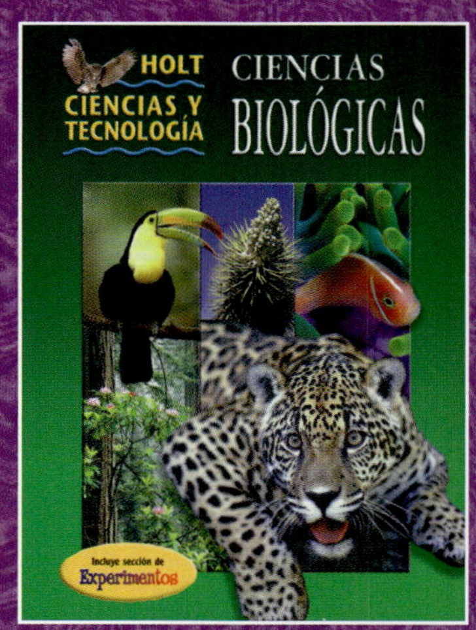

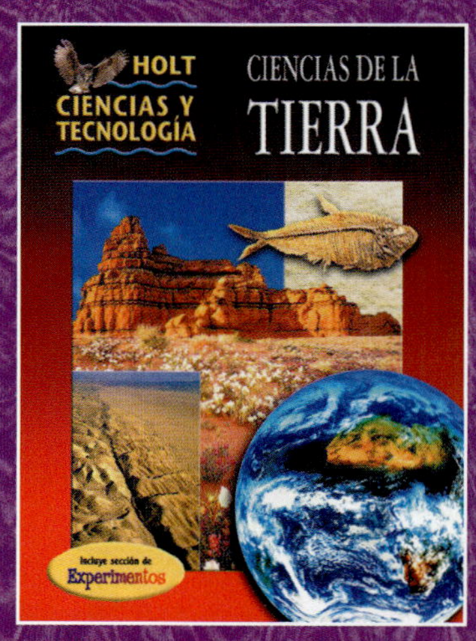

Welcome to
HOLT SCIENCE & TECHNOLOGY
Life
Science
EXPECT EXCITEMENT!
EXPECT RESULTS!

A Text that Grabs and Holds Your Students' Attention

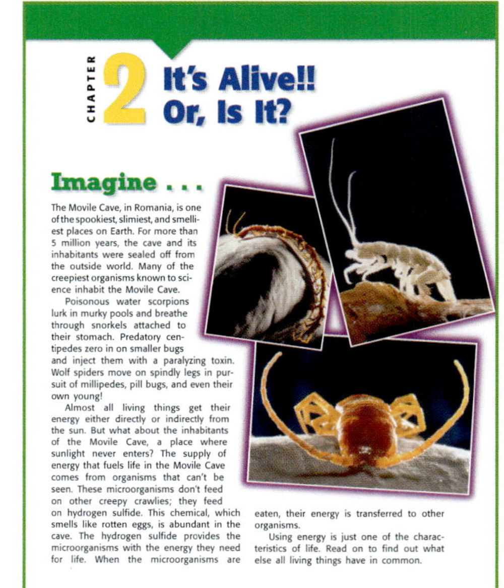

Begins with a bang!

Each chapter begins with a brief introduction designed to pique your students' interest. Here they may encounter a true story or a hypothetical situation that poses prereading questions, such as **Imagine. . . .**

A text that motivates

Holt Science and Technology motivates your students in a variety of ways.

Visuals

- are integrated into the narrative
- clearly reveal macro-to-micro relationships
- support English-language learners and reluctant readers
- are functional, accurate, and understandable

Narrative

- contains concise, outline-style headings to help students find information easily
- presents content in a clear, logical sequence
- contains friendly language to make reading accessible and enjoyable
- incorporates analogies to help students relate concepts to the real world

Pupil's Edition

Applies to real life

Some of your students may ask you why they are studying science. *Holt Science and Technology* provides answers with motivating features:

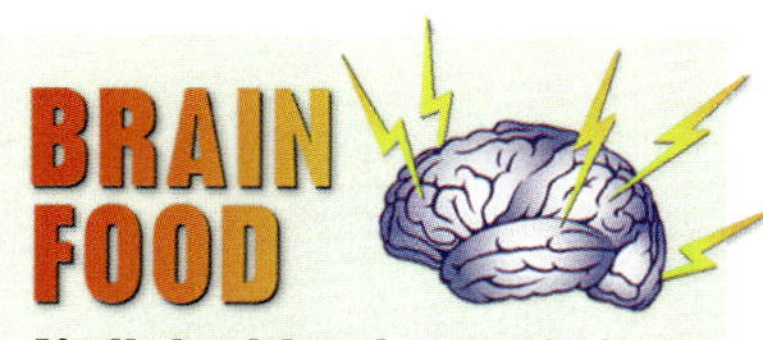

- **Investigate!** stimulates your students' curiosity about upcoming chapter concepts with a hands-on activity.
- Tidbits to feed the mind are presented in small captions called **Brain Food**. There's nothing better than a fun fact to captivate a young audience!
- **Apply** poses real-world questions and asks your students to answer them by applying what they have just learned.
- **Activity** gives your students the opportunity to use their imaginations and to expand their learning.

Brings focus to the Internet

***sci*LINKS**—a National Science Teachers Association-sponsored Web service—links you and your students to interactive activities and current information directly related to chapter content. (see page T23)

Ends with enrichment

Weird Science, **Health Watch**, **Careers**, **Scientific Debate**, and other end-of-chapter features extend chapter content with real-world examples, articles, and motivating activities.

Focus on Reading and Understanding

Holt Science and Technology makes instruction accessible to all your students—English-language learners, special needs students, those having difficulty mastering content, students who need more practice or hands-on experience, and advanced learners.

Read for understanding

Each lesson gives you suggestions to help your students read for understanding.

- **What Do You Think?** assesses your students' prior knowledge and serves as a reading warm-up.

- **NOW What Do You Think?** allows students to see how their understanding has changed.

- **Sheltered English** highlights activities that help English-language learners grasp content.

- **READING STRATEGY** emphasizes key concepts in order to guide reading and ensure comprehension.

- ***Directed Reading Worksheets*** makes reading an active process. A variety of strategies and fun activities help your students identify the main idea, then organize and synthesize supporting information.

- ***Reinforcement & Vocabulary Review Worksheets*** makes reviewing and reinforcing chapter content easy. Students have opportunities to examine issues in each section from different perspectives or to benefit from a different instructional approach.

READING STRATEGY

Prediction Guide Have students respond to the following statement before reading this section:
cells
bon
to e
answ
thei

READING STRATEGY

Prediction Guide Before reading this section, ask students to answer the following question in their ScienceLog. *Amphibian* means "double life." Why would we give this class of animals such a name? After reading this section, ask students to evaluate their answers.

NOW What Do You Think?

Take a minute to review your answers to the ScienceLog questions found on page 303. Have your answers

What Do You Think?

In your ScienceLog, try to answer the following questions based on what you already know:

1. Why don't all humans look exactly alike?
2. What determines whether a human baby will be a boy or a girl?

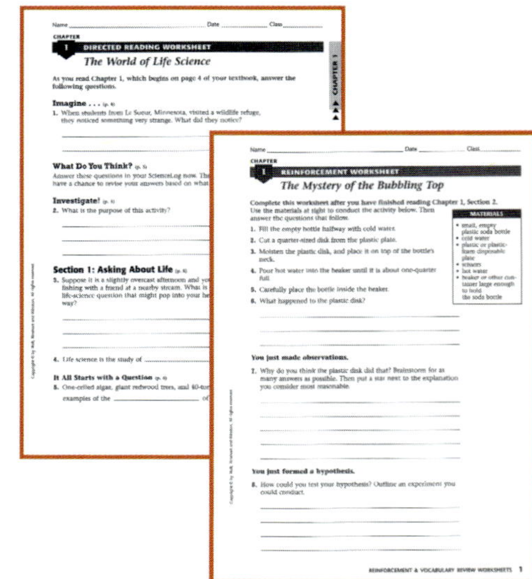

*"I need a program that helps me teach **today's students.**"*

Guided Reading Audio CD Program

This audio program provides students with a direct reading of each chapter using instructional visuals as guideposts. Auditory learners, students with limited reading proficiency, and Spanish-speaking students receive the explanation they need from this alternative text format. **Available in both English and Spanish.**

Provide universal access

Holt Science and Technology helps all your students learn science.

- **Meeting Individual Needs** in the teacher's wrap provides engaging demonstrations and hands-on activities to help learners of all levels.
- **Reteaching** gives you alternate methods of instruction for those students who need it.
- **Homework** options use a variety of teaching strategies to complement diverse learning styles.
- ***Critical Thinking & Problem Solving Worksheets*** provides challenging activities connecting science concepts to the "real world." Your students learn to think through a problem and to use the scientific method and other strategies to find a solution.

MEETING INDIVIDUAL NEEDS

Advanced Learners

In a given temperate region, different animals have different...

MEETING INDIVIDUAL NEEDS

Learners Having Difficulty

In a given temperate region, different animals have different ways of surviving cold winters. Have students list as many behaviors or adaptations for winter survival that they can think of.

Sheltered English

Approach learning from different angles

with these in-text features and ancillaries. You can make sure your students understand science concepts no matter what their learning styles.

Science CONNECTION

Environment CONNECTION

GROUP ACTIVITY

HOLT ANTHOLOGY OF SCIENCE FICTION

CROSS-DISCIPLINARY FOCUS

Multicultural CONNECTION

REAL-WORLD CONNECTION

COOPERATIVE LEARNING

HOLT SCIENCE POSTERS

Teaching Transparency

Science Puzzlers, Twisters & Teasers

Labs to Make Learning Active and Meaningful

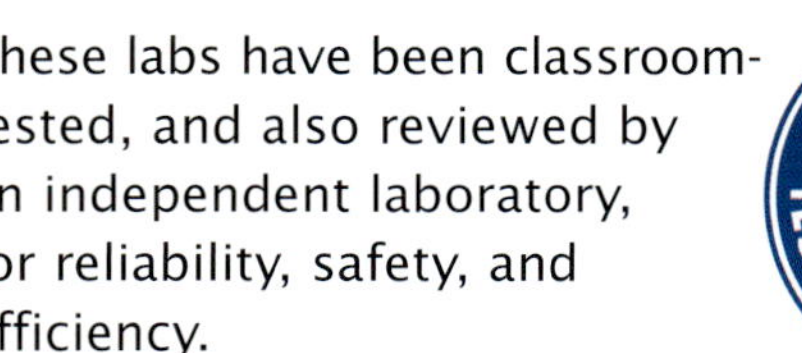

Carbon Dioxide Breath

Plants take in carbon dioxide and give off oxygen as a byproduct of photosynthesis. Animals, including you, use this oxygen and release carbon dioxide as a byproduct of respiration.

In this activity, you will explore your own carbon dioxide exhalation. Phenol red turns yellow in the presence of carbon dioxide. You will use it to detect carbon dioxide in your breath.

Materials

- 150 mL graduated cylinder
- 100 mL of water
- 150 mL Erlenmeyer flask
- eyedropper
- phenol red indicator solution
- plastic drinking straw
- paper towel
- clock with a second hand or a stopwatch
- protective gloves

Procedure

1. Place 100 mL of water into a 150 mL flask. Using an eyedropper, carefully place four drops of phenol red indicator solution into the water. The water should turn orange.

2. Place a plastic drinking straw into the solution of phenol red and water. Drape a paper towel over the beaker to prevent splashing. Carefully blow through the straw into the solution. **Caution:** Do not inhale through the straw. Do not drink the solution, and do not share a straw with anyone.

3. Have your lab partner time how long it takes for the solution to change color. Begin timing when you start blowing. Record the time in your ScienceLog. What color does the solution become?

Analysis

4. Compare your data with those of your classmates. What was the longest length of time it took to see a color change? the shortest? How do you account for the difference?

5. Is there a relationship between the length of time it takes to change the solution from orange to yellow and the person's physical characteristics, such as gender or whether the tester has an athletic build?

Going Further

Help solve The Perfect Taters Mystery on page 698 of the LabBook!

Holt Science and Technology provides a strong yet flexible lab program that meets lab science requirements, regardless of limited lab equipment or time restrictions. Labs include clear procedures, demonstrate scientific concepts, and develop students' understanding of scientific methods. **Using Scientific Methods**

These labs have been classroom-tested, and also reviewed by an independent laboratory, for reliability, safety, and efficiency.

Terry Rakes
Elmwood Junior High School
Rogers, Arkansas

In-text LabBook

LabBook, in the back of the *Pupil's Edition*, allows for

- more labs and activities,
- greater flexibility in lesson planning,
- a wider variety of labs,
- more detailed lab procedures and explanations,
- an uninterrupted chapter narrative, and includes
- separate *Datasheets for LabBook*.

Additional, in-text labs and activities

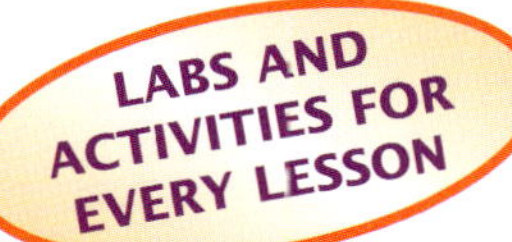

- **QuickLabs** are easy to execute and require minimal time and materials—great for quick in-class activities, teacher demonstrations, or group presentations.
- **Investigate!** stimulates your students' curiosity about scientific concepts in the upcoming chapter.
- **Activity** gives your students the opportunity to use their imaginations and expand their learning.
- **Apply** poses real-world questions and asks your students to answer them by applying what they have just learned.
- **Demonstration** and **Activity** in the teacher's edition give you options to demonstrate labs and procedures to the whole class or provide fun, hands-on activities.

Lab Ratings make choosing labs easy

Lab Ratings, for all labs, make it easy for you to determine at a glance which labs are most appropriate for your class.

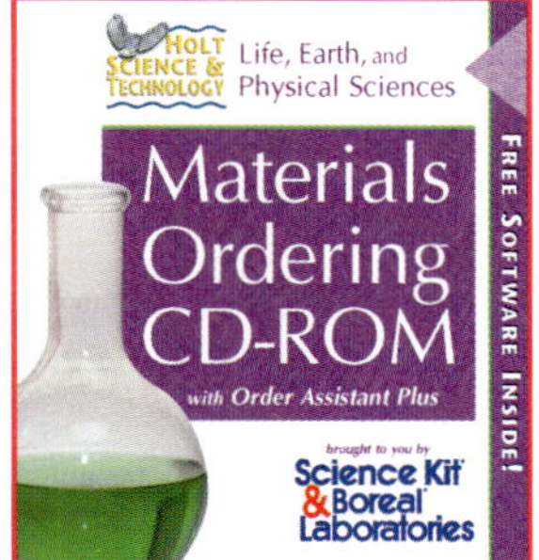

Easy-to-order lab materials

Ordering lab materials is more efficient than ever with the *Holt Science and Technology Materials Ordering Software*. This software, developed by Science Kit®, creates a "shopping list" of materials and their costs. The program also lists required materials for every lab investigation in the program, including consumable and non-consumable kits.

For a complete materials list, see page xxiv in the Annotated Teacher's Edition

Lab Manuals Extend Your Options

Lab Manuals

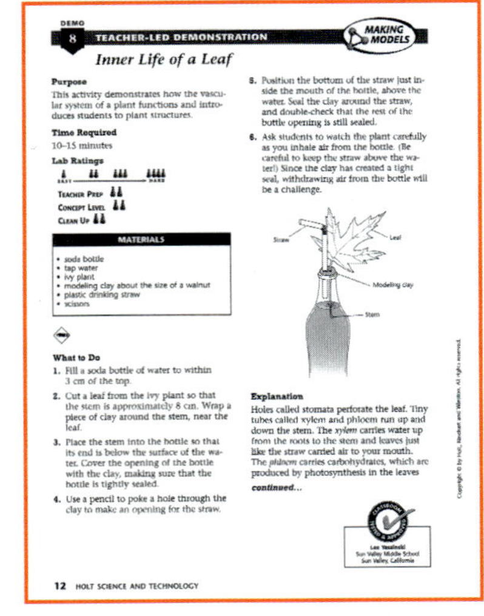

Whiz-Bang Demonstrations gives you a rousing way to get your students' attention at the beginning of a lesson. **65 labs in all!**

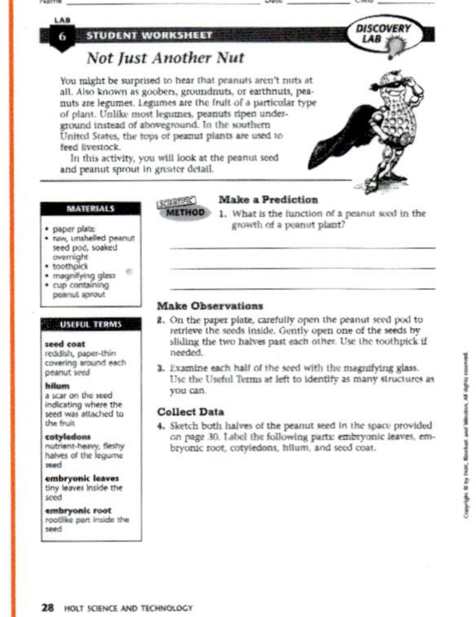

Labs You Can Eat safely incorporates food into the classroom to provide a fun inquiry-based learning tool. **25 labs in all!**

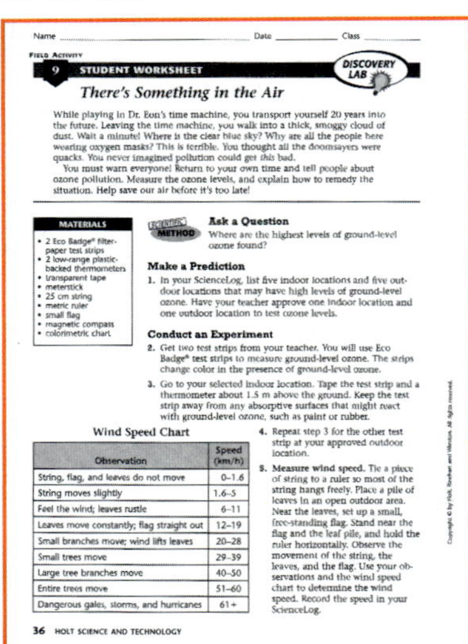

EcoLabs & Field Activities includes activities that address specific ecological questions and increase environmental awareness. **23 labs in all!**

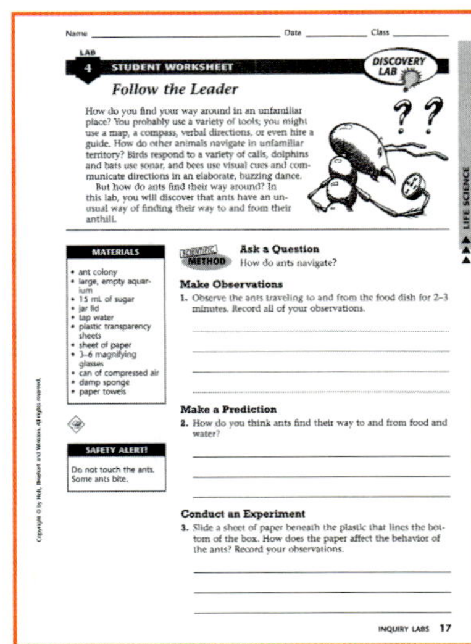

Inquiry Labs encourages your students to ask questions and investigate problems in order to find solutions. **23 labs in all!**

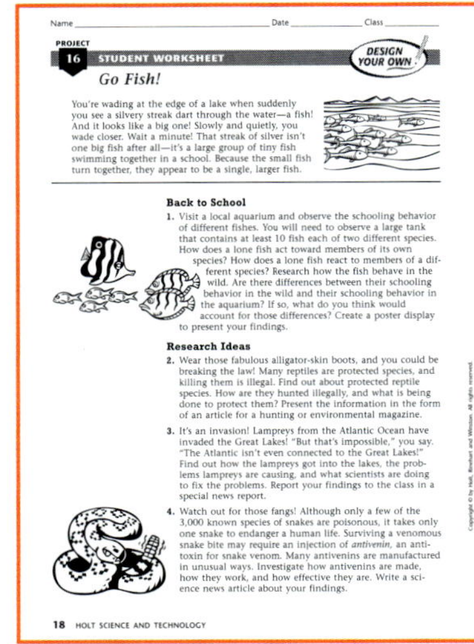

Long-Term Projects & Research Ideas extends and enriches chapter content with experiments, activities, inquiry-based projects, and Internet and library research. **2 for every chapter!**

Comprehensive Skill Development

Science skills ensure future success

Holt Science and Technology gives your students ample opportunities to master the skills necessary for future success in science. Science skills are developed in a variety of ways:

- The in-text **LabBook**, activities, and lab booklets provide lots of practice using scientific methods.
- The **Appendix** helps your students refresh their measuring and data-analysis skills, as well as their understanding of the **scientific method**.
- **Apply** and **Activity** allow your students to develop science skills in fun and motivating ways.

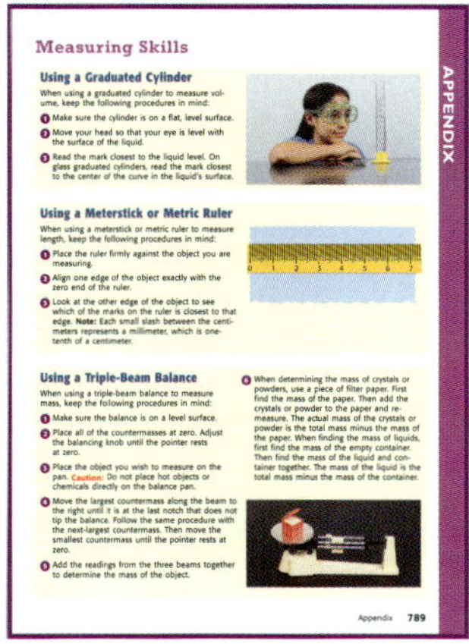

In addition, your students can find plenty of skill-building practice in *Science Skills Worksheets*. These worksheets help your students hone important science skills, such as thinking objectively, conducting research, designing investigations, keeping accurate records, and creating and analyzing graphs.

Math is covered everywhere you look!

Holt Science and Technology also strengthens your students' math skills. From practice problems to reviews, math skills are continually developed:

- **MathBreak** provides practice with direct application to the science concepts being taught.
- **Math Concepts** reviews the math lessons presented in the chapter.
- **Math Refresher**, found in the Appendix, reviews basic math skills, such as averages, ratios, percentages, and more.

Surface-to-Volume Ratio

The shape of a cell can affect its surface-to-volume ratio. Examine the cells below, and answer the questions that follow:

Math Skills for Science helps your students develop and apply basic math skills to scientific problems with two types of worksheets.

- **Math Skills Worksheets** provide a brief introduction to a relevant math skill, a step-by-step explanation of the math process, and example and practice problems.
- **Math in Science Worksheets** give your students practice using math in real-life science situations.

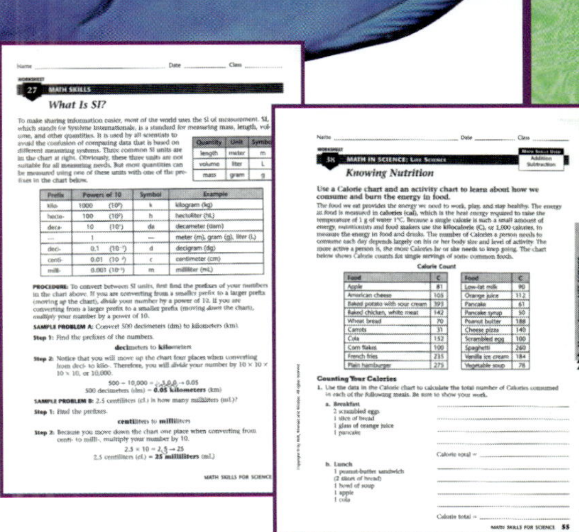

HOLT SCIENCE & TECHNOLOGY

A Versatile Teacher's Edition that is Easy-to-Use

The Chapter Organizer—
your easy-to-follow road map

With such a wealth of program resources, you'll be glad to know we've included a convenient, timesaving guide suggesting how and when to use them. The **Chapter Organizer**

- integrates all labs, technology, and print resources
- is organized according to time requirements

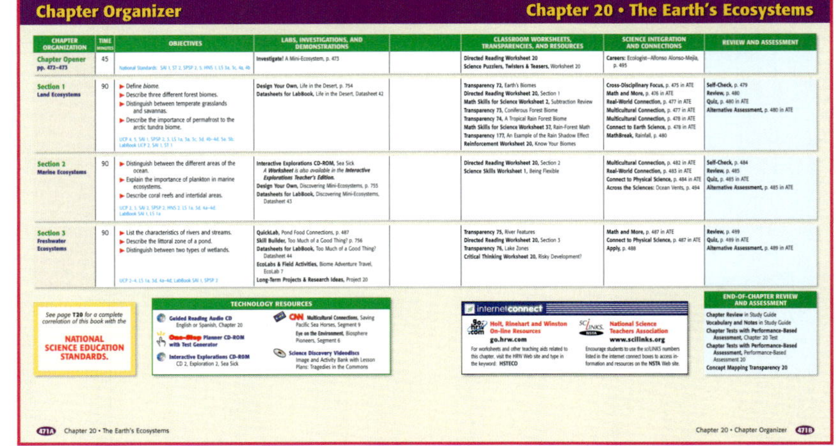

Chapter Resources & Worksheets makes choosing teaching resources easy by showing them as reduced pages. Available resources are categorized by

- Visual Resources
- Meeting Individual Needs
- Review and Assessment
- Lab Worksheets
- Applications & Extensions

The **Chapter Background** provides additional information to help you enrich upcoming lessons.

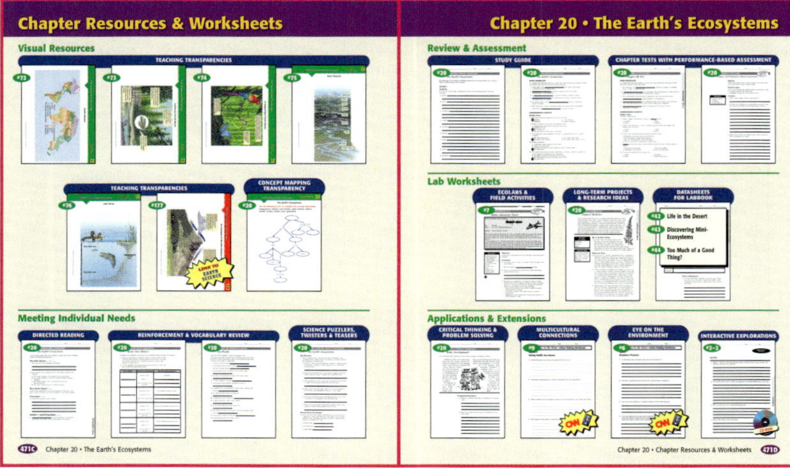

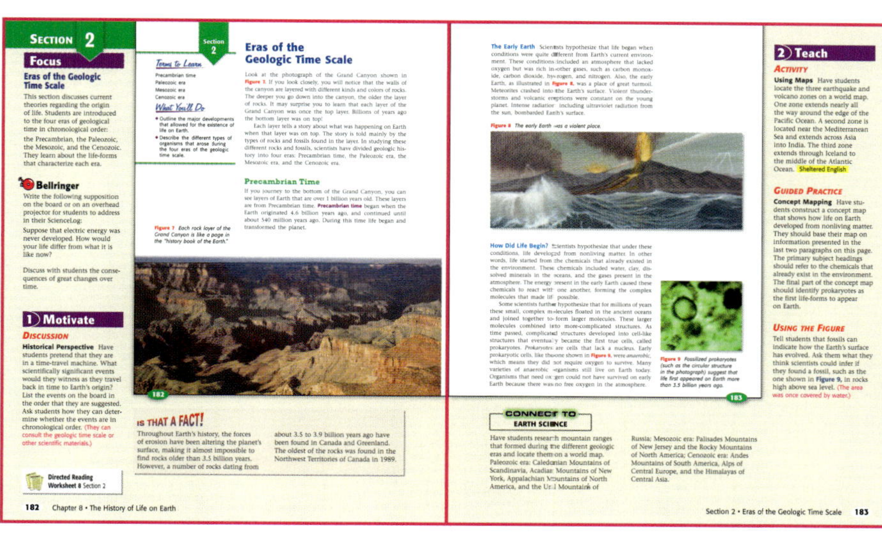

Keep the focus on the lesson

The complete lesson cycle helps you keep your students interested and involved. An array of both traditional and new teaching strategies, creative reinforcement, and thought-provoking extensions help you teach to a wide variety of learning styles, ability levels, and interests.

Fuel your presentation

Found on almost every page, fun features and intriguing stories ignite class discussion and get your students thinking.

Q: Why didn't the skeleton cross the road?

A: It didn't have the guts.

IS THAT A FACT!

Research shows that baleen-whale sounds, which scientists believe are produced by the larynx, may be the loudest sounds produced by any animal on land or in the sea. Such sounds may carry hundreds of kilometers underwater.

MISCONCEPTION ALERT

Can ancient DNA be used to produce dinosaurs as seen in the movies Jurassic Park and The Lost World? In these movies, scientists make dinosaurs by combining fragments of ancient DNA with DNA from modern-day frogs.

Science Bloopers

In 1798, when English scholars first observed the duck-billed platypus that had been sent to them by a scientist in Australia, they were convinced they were the victims of a joke. Surely, they thought, some prankster had pieced together parts of various animals. The English scholars cut and sliced the dead animal for signs of stitches holding the bill and webbed feet to its mammal-like body. It took a lot of convincing, but eventually they came to the conclusion that the animal was indeed real.

WEIRD SCIENCE

Environmental stimuli can sometimes affect flower color. Hydrangeas growing in acidic soil produce blue flowers. If the soil is made alkaline, the same plants will produce pink flowers.

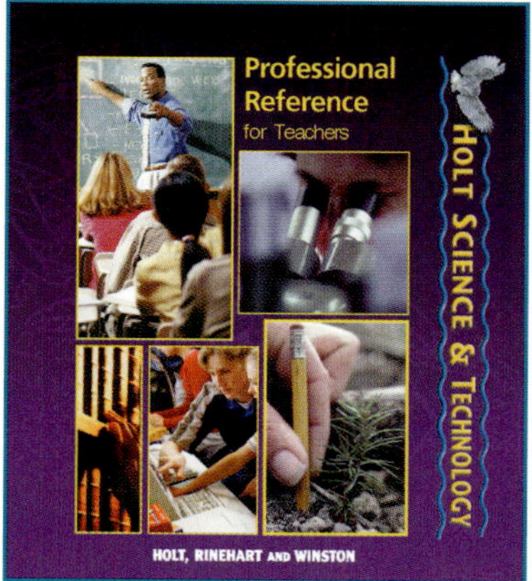

Sharpen your saw

The *Professional Reference for Teachers* provides current information about pertinent issues in science education today. In professional articles, you can learn more about the National Science Education Standards, block scheduling, classroom management, and more. A bibliography of books, lectures, magazines, and Web sites is included.

Visualize science concepts

Teaching Transparencies, many with images taken directly from the text, reinforce important science concepts and processes.

Two *Concept Mapping Transparencies* are included for each chapter—a partial map transparency to use with your students as they progress through the chapter and a completed concept map to serve as an answer key.

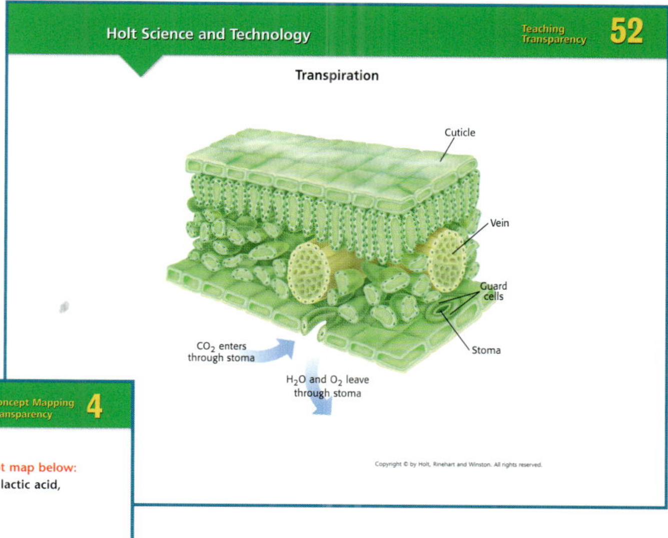

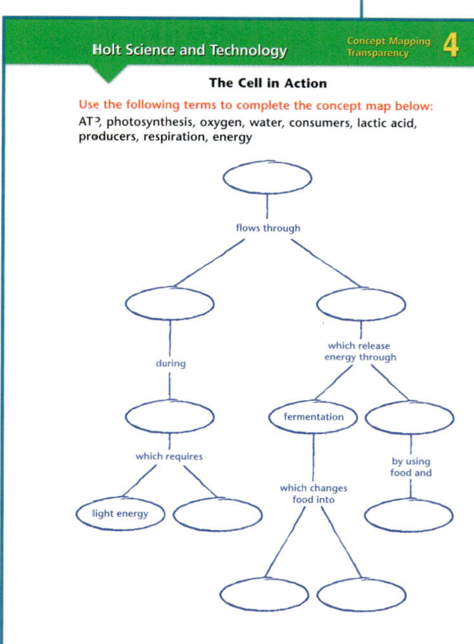

Bellringer Transparency Masters (part of the *One-Stop Planner CD-ROM*) help you focus your students' attention quickly at the beginning of class while you are dealing with administrative demands.

Assessment

Assessment that Accurately Measures Mastery of Content

Check progress

Self-Check encourages your students to evaluate their own learning by answering questions found intermittently within the chapter. A page reference allows them to check their own answers. After reading each lesson, your students explore, evaluate, and extend what they've learned by answering questions in the section **Review**. In the teacher's wrap, a **Quiz** provides an objective assessment of each lesson.

> ✓ **Self-Check**
>
> How are the structures of arteries and veins related to their functions? *(See page 782 to check your answer.)*

Chapter Highlights lists vocabulary and provides content summaries in a concise, visual format. This helps your students organize their thoughts and synthesize information.

Study Guide contains blackline masters for Chapter Highlights and Chapter Reviews that help your students gear up for tests and quizzes.

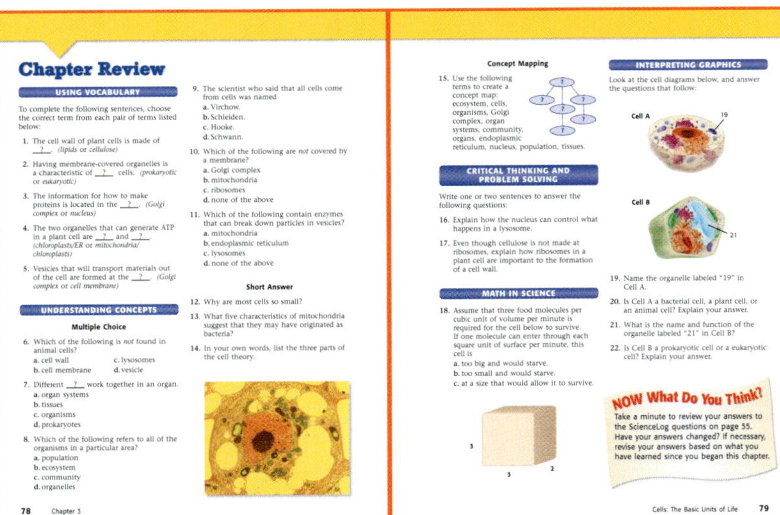

Chapter Review question types are identical to those found on the chapter tests, making the Chapter Review an excellent resource for pretest practice.

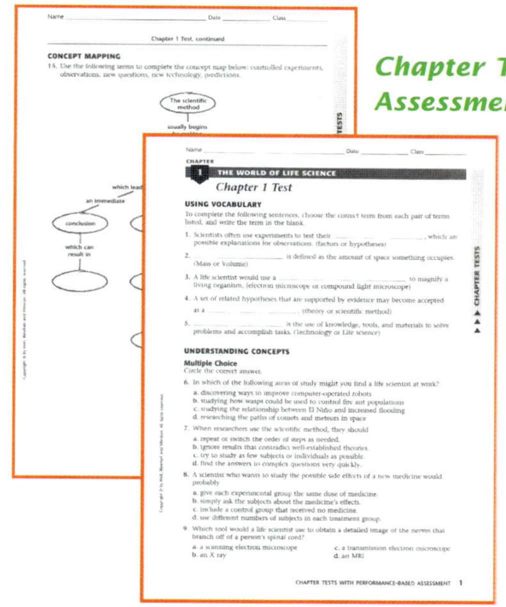

Chapter Tests with Performance-Based Assessment includes multiple-choice, concept-mapping, critical-thinking, interpreting graphics, math-in-science, and alternative assessment questions, to name a few.

Alternative Assessment in the teacher's wrap provides you with different evaluation options, such as expository writing and concept mapping, to ensure a thorough assessment.

Create your own assessments

One-Stop Planner CD-ROM

With the *One-Stop Planner CD-ROM with Test Generator*, you can create, revise, and edit quizzes, section and chapter reviews, and chapter tests, drawing from thousands of questions organized by chapter and linked to chapter objectives. Performance-based assessment is also included.

The *Test Generator: Test Item Listing* provides a printed copy of thousands of assessment items on the *One-Stop Planner CD-ROM*. This handy guide allows you to preview test items before making selections.

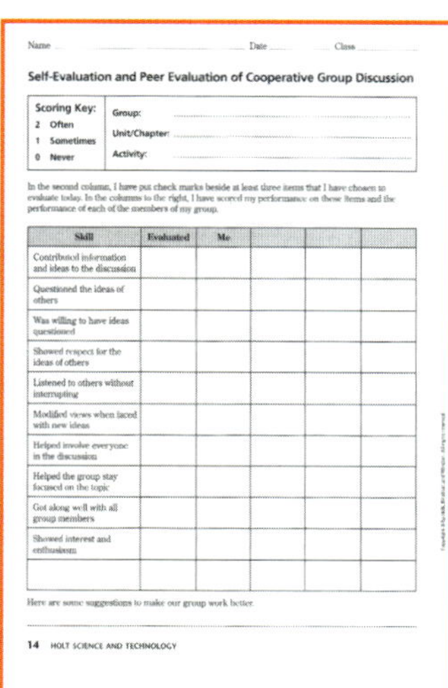

Assessment Checklists & Rubrics, available on the *One-Stop Planner CD-ROM* or as blackline masters, gives you guidelines for evaluating your students' progress, including performance and portfolio assessment tools. You can also create a customized checklist for each class, helping you gather daily scores and determine grades.

Technology

Technology that Meets Your Goals and Expands Your Options

Holt Science and Technology provides the right combination of fully integrated technology resources—including videos, CD-ROMs, and Internet connections—to make your teaching more effective, efficient, and creative.

Teacher resources

Finding, printing, and editing teaching resources is easy with the *One-Stop Planner CD-ROM with Test Generator*. You can use this cutting-edge technology to sort through thousands of pages of resources, including

- hundreds of printable worksheets
- customizable lesson plans
- a powerful test generator
  (see pages T10 and T15)

Classroom resources

The *CNN Presents Science in the News: Video Library* helps your students see the impact of science on their everyday lives. The CNN and CNN NEWSROOM Team brings to the classroom actual news coverage in *Scientists in Action*; *Multicultural Connections*; *Science, Technology & Society*; and *Eye on the Environment*. Each Teacher's Guide offers background information, a description of each segment, viewing questions, and teaching suggestions. In addition, a critical-thinking worksheet for each news segment enhances skill development.

Science Discovery Videodisc Programs will excite your students with visually stunning scientific images as well as provide a fun means for solving scientific problems in real-world situations. This package includes the popular **Science Sleuths** series and the **Image and Activity Bank**.

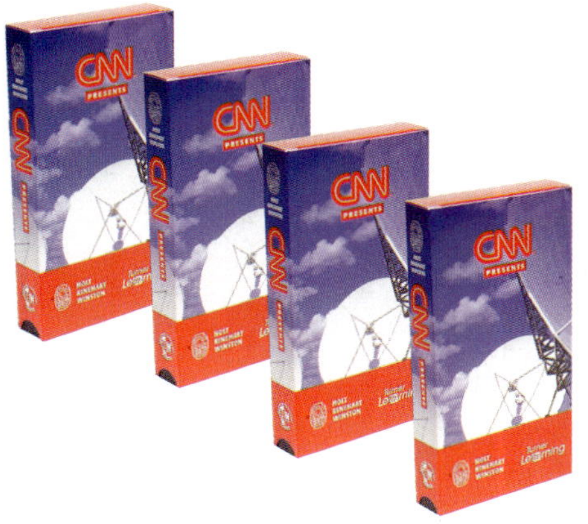
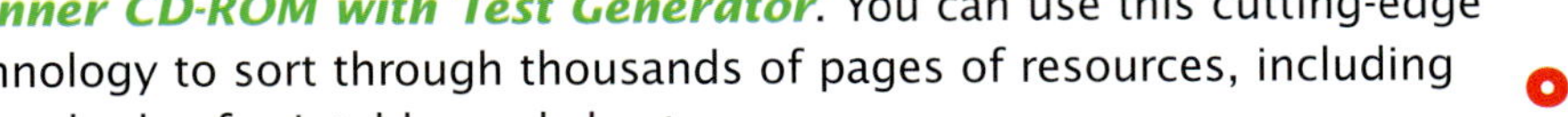

Technology

Student interactive resources

*sci*LINKS is a Web service developed and maintained by the National Science Teachers Association that links you and your students to online educational resources directly related to chapter topics in *Holt Science and Technology*.

*sci*LINKS saves you time searching for relevant Web sites. The *sci*LINKS staff, consisting of science educators and content experts, identifies, reviews, and monitors featured sites—so you can be assured that they contain appropriate and up-to-date information. In addition, you and your students never have to worry about a site disappearing. *sci*LINKS will replace an expired site with another using the same keyword.

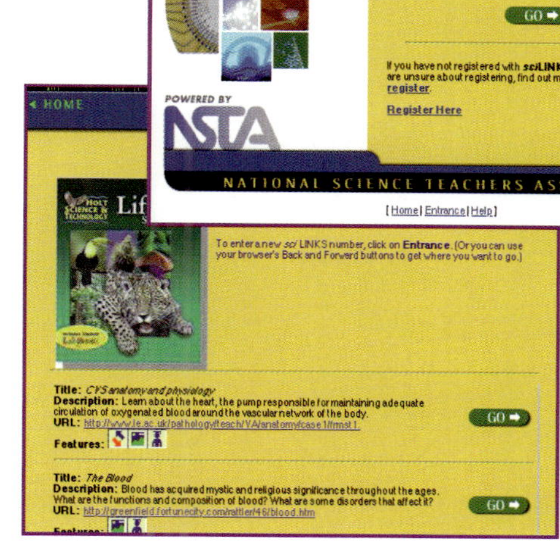

Your students can also enrich their knowledge by exploring the Internet through the **go.hrw.com** site, which links them to online chapter activities and resources.

The **Smithsonian Institution** maintains special Web sites for use with *Holt Science and Technology*. These sites include interactive exhibits, classroom activities, interviews with scientists, and a variety of application and extension topics.

Interactive Explorations CD-ROM Program turns a computer into a virtual laboratory where your students act as lab assistants in solving real-world problems—developing their inquiry, analysis, and decision-making skills.

The *Interactive Science Encyclopedia CD-ROM* gives your students instant access to more than 3,000 cross-referenced science entries, in-depth articles, science fair project ideas, interactive activities, and more.

EXPECT EXCITEMENT!
EXPECT RESULTS!

Components Listing

Pupil's Edition

Annotated Teacher's Edition

Life Science Teaching Resources

- Study Guide
- Study Guide Answer Key
- Critical Thinking & Problem Solving Worksheets
- Reinforcement & Vocabulary Review Worksheets
- Science Puzzlers, Twisters & Teasers
- Chapter Tests with Performance-Based Assessment
- Directed Reading Worksheets
- Directed Reading Worksheets Answer Key
- Datasheets for LabBook
- Datasheets for LabBook Answer Key
- Test Generator: Test Item Listing

LabBank

- Labs You Can Eat
- Whiz-Bang Demonstrations
- Inquiry Labs
- EcoLabs & Field Activities
- Long-Term Projects & Research Ideas

Program Teaching Resources

- Science Skills Worksheets
- Math Skills for Science
- Professional Reference for Teachers
- Holt Anthology of Science Fiction
- Assessment Checklists & Rubrics
- Science Fair Guide

Holt Science Posters

Technology Resources

Teaching Transparencies with Concept Mapping Transparencies

Guided Reading Audio CD Program

Guided Reading Audio CD Program, Spanish

CNN Presents Science in the News: Video Library

- Scientists in Action
- Multicultural Connections
- Science, Technology & Society
- Eye on the Environment

One-Stop Planner CD-ROM with Test Generator for Macintosh® and Windows®

Materials Ordering Software CD-ROM for Macintosh® and Windows®

Interactive Explorations CD-ROM Program for Macintosh® and Windows®

Science Discovery Videodisc Programs

Interactive Science Encyclopedia CD-ROM

- Macintosh®
- Windows®

HOLT
CIENCIAS Y TECNOLOGÍA

CIENCIAS
BIOLÓGICAS

EDICIÓN ANOTADA DEL MAESTRO

HOLT, RINEHART AND WINSTON

A Harcourt Classroom Education Company

Austin • New York • Orlando • Atlanta • San Francisco • Boston • Dallas • Toronto • London

Staff Credits

Editorial

Robert W. Todd, Executive Editor
David F. Bowman, Managing Editor
Barbara Howell, Senior Editor
Charlotte Luongo, Annette Ratliff, Tracy Schagen, Laura Zapanta, Robin Goodman (Feature Articles)

ANNOTATED TEACHER'S EDITION

David Westerberg, Bill Burnside, Kelly Graham

ANCILLARIES

Jennifer Childers, Senior Editor
Erin Bao, Kristen Karns, Andrew Strickler, Clay Crenshaw, Wayne Duncan, Molly Frohlich, Amy James, Monique Mayer, Traci Maxwell

COPYEDITORS

Steve Oelenberger, Copyediting Supervisor
Brooke Fugitt, Tania Hannan, Denise Nowotny

EDITORIAL SUPPORT STAFF

Christy Bear, Jeanne Graham, Rose Segrest, Tanu'e White

EDITORIAL PERMISSIONS

Cathy Paré, Permissions Manager
Jan Harrington, Permissions Editor

Art, Design, and Photo

BOOK DESIGN

Richard Metzger, Art Director
Marc Cooper, Senior Designer
Sonya Mendeke, Designer
Alicia Sullivan, Designer (ATE)
Cristina Bowerman (ATE)
Eric Rupprath (Ancillaries)

IMAGE SERVICES

Elaine Tate, Art Buyer Supervisor
Kim Baker, Art Buyer

PHOTO RESEARCH

Tim Taylor, Senior Photo Researcher
Stephanie Friedman, Assistant Photo Researcher

PHOTO STUDIO

Sam Dudgeon, Senior Staff Photographer
Victoria Smith, Photo Specialist

DESIGN NEW MEDIA

Susan Michael, Art Director

DESIGN MEDIA

Joe Melomo, Art Director
Shawn McKinney, Designer

Production

Mimi Stockdell, Senior Production Manager
Beth Sample, Production Coordinator
Suzanne Brooks, Sara Carroll-Downs

Media Production

Kim A. Scott, Senior Production Manager
Nancy Hargis, Production Supervisor
Adriana Bardin, Production Coordinator

New Media

Jim Bruno, Senior Project Manager
Lydia Doty, Senior Project Manager
Jessica Bega, Project Manager
Armin Gutzmer, Manager Training and Technical Support
Cathy Kuhles, Nina Degollado

Design Implementation and Production

Mazer Corporation

Acknowledgments

Chapter Writers

Katy Z. Allen
*Science Writer and Former Biology
    Teacher*
Wayland, Massachusetts

Linda Ruth Berg, Ph.D.
*Adjunct Professor–Natural
    Sciences*
St. Petersburg Junior College
St. Petersburg, Florida

Leila Dumas
Former Physics Teacher
LBJ Science Academy
Austin, Texas

Jennie Dusheck
Science Writer
Santa Cruz, California

Mark F. Taylor, Ph.D.
Associate Professor of Biology
Baylor University
Waco, Texas

Lab Writers

Diana Scheidle Bartos
Science Consultant and Educator
Diana Scheidle Bartos, L.L.C.
Lakewood, Colorado

Carl Benson
Technology Coordinator
Plains High School
Plains, Montana

Charlotte Blassingame
Science Teacher and Dept. Chair
White Station Middle School
Memphis, Tennessee

Marsha Carver
Science Teacher and Dept. Chair
McLean County High School
Calhoun, Kentucky

Kenneth E. Creese
Science Teacher
White Mountain Junior High
    School
Rock Springs, Wyoming

Linda Culp
Science Teacher and Dept. Chair
Thorndale High School
Thorndale, Texas

James Deaver
Science Teacher and Dept. Chair
West Point High School
West Point, Nebraska

Frank McKinney, Ph.D.
Professor of Geology
Appalachian State University
Boone, North Carolina

Alyson Mike
Science Teacher
East Valley Middle School
East Helena, Montana

C. Ford Morishita
Biology Teacher
Clackamas High School
Milwaukie, Oregon

Patricia Morrell, Ph.D.
*Assistant Professor, School of
    Education*
University of Portland
Portland, Oregon

Hilary C. Olson, Ph.D.
Research Associate
Institute for Geophysics
The University of Texas
Austin, Texas

James B. Pulley
Science Teacher
Liberty High School
Liberty, Missouri

Denice Lee Sandefur
Science Teacher
Nucla High School
Nucla, Colorado

Patti Soderberg
Science Writer
The BioQUEST Curriculum
    Consortium
Beloit College
Beloit, Wisconsin

Phillip Vavala
Science Teacher and Dept. Chair
Salesianum School
Wilmington, Delaware

Albert C. Wartski
Biology Teacher
Chapel Hill High School
Chapel Hill, North Carolina

Lynn Marie Wartski
*Science Writer and Former Science
    Teacher*
Hillsborough, North Carolina

Ivora D. Washington
Science Teacher
Hyattsville Middle School
Hyattsville, Maryland

Academic Reviewers

David M. Armstrong, Ph.D.
Professor of Biology
University of Colorado
Boulder, Colorado

Alissa Arp, Ph.D.
*Director and Professor of
    Environmental Studies*
Romberg Tiburon Center
San Francisco State University
Tiburon, California

Russell M. Brengelman
Professor of Physics
Morehead State University
Morehead, Kentucky

Linda K. Butler, Ph.D.
Lecturer of Biological Sciences
The University of Texas
Austin, Texas

Barry Chernoff, Ph.D.
Associate Curator and Head
Division of Fishes
The Field Museum of Natural
    History
Chicago, Illinois

**Donna Greenwood
    Crenshaw, Ph.D.**
Research Associate
Department of Biology
Duke University
Durham, North Carolina

Hugh Crenshaw, Ph.D.
Assistant Professor of Zoology
Duke University
Durham, North Carolina

Joe W. Crim, Ph.D.
Professor of Biology
University of Georgia
Athens, Georgia

Andrew J. Davis, Ph.D.
Manager of ACE Science Center
Department of Physics
California Institute of
    Technology
Pasadena, California

Peter Demmin, Ed.D.
Former Science Teacher and Chair
Amherst Central High School
Amherst, New York

Gabriele F. Giuliani, Ph.D.
Professor of Physics
Purdue University
West Lafayette, Indiana

Joseph L. Graves, Jr., Ph.D.
*Associate Professor of Life
    Sciences*
Arizona State University West
Phoenix, Arizona

**Laurie Jackson-Grusby,
    Ph.D.**
*Research Scientist and Doctoral
    Associate*
Whitehead Institute for
    Biomedical Research
Massachusetts Institute of
    Technology
Cambridge, Massachusetts

William B. Guggino, Ph.D.
*Professor of Physiology and
    Pediatrics*
The Johns Hopkins University
    School of Medicine
Baltimore, Maryland

David Haig, Ph.D.
Assistant Professor of Biology
Department of Organismic and
    Evolutionary Biology
Harvard University
Cambridge, Massachusetts

John E. Hoover, Ph.D.
Associate Professor of Biology
Millersville University
Millersville, Pennsylvania

Joan E. N. Hudson, Ph.D.
Professor of Biology
Sam Houston State University
Huntsville, Texas

George M. Langford, Ph.D.
Professor of Biological Sciences
Dartmouth College
Hanover, New Hampshire

V. Patrick Lombardi, Ph.D.
Professor of Biology
Department of Biology
University of Oregon
Eugene, Oregon

William F. McComas, Ph.D.
*Director of the Center to Advance
    Science Education*
University of Southern
    California
Los Angeles, California

LaMoine L. Motz, Ph.D.
Coordinator of Science Education
Oakland County Schools
Waterford, Michigan

Acknowledgments (cont.)

Nancy Parker, Ph.D.
Associate Professor of Biology
Southern Illinois University
Edwardsville, Illinois

Barron S. Rector, Ph.D.
Associate Professor
Department of Rangeland
   Ecology and Management
Texas A&M University
College Station, Texas

John Rigden, Ph.D.
Director of Special Projects
American Institute of Physics
Colchester, Vermont

Miles R. Silman, Ph.D.
Assistant Professor of Biology
Wake Forest University
Winston-Salem, North Carolina

Robert G. Steen, Ph.D.
Manager, Rat Genome Project
Whitehead Institute–Center for
   Genome Research
Massachusetts Institute of
   Technology
Cambridge, Massachusetts

Jack B. Swift, Ph.D.
Professor of Physics
The University of Texas
Austin, Texas

Martin VanDyke, Ph.D.
Professor of Chemistry Emeritus
Front Range Community
   College
Westminister, Colorado

E. Peter Volpe, Ph.D.
Professor of Medical Genetics
Mercer University School of
   Medicine
Macon, Georgia

Harold K. Voris, Ph.D.
Curator and Head
Division of Amphibians and
   Reptiles
The Field Museum of Natural
   History
Chicago, Illinois

Peter Wetherwax, Ph.D.
Professor of Biology
Department of Education
University of Oregon
Eugene, Oregon

Mary Wicksten, Ph.D.
Professor of Biology
Texas A&M University
College Station, Texas

R. Stimson Wilcox, Ph.D.
Professor of Biology
Behavioral Ecology &
Communication of Animals
Binghamton University
Binghamton, New York

Conrad Zapanta, Ph.D.
Research Engineer
Sulzer Carbomedics, Inc.
Austin, Texas

Safety Reviewer

Jack Gerlovich, Ph.D.
Associate Professor
School of Education
Drake University
Des Moines, Iowa

Teacher Reviewers

Barry L. Bishop
Science Teacher and Dept. Chair
San Rafael Junior High School
Ferron, Utah

Carol A. Bornhorst
Science Teacher and Dept. Chair
Bonita Vista Middle School
Chula Vista, California

Paul Boyle
Science Teacher
Perry Heights Middle School
Evansville, Indiana

Yvonne Brannum
Science Teacher and Dept. Chair
Hine Junior High School
Washington, D.C.

Gladys Cherniak
Science Teacher
St. Paul's Episcopal School
Mobile, Alabama

James Chin
Science Teacher
Frank A. Day Middle School
Newtonville, Massachusetts

Randy Christian
Science Teacher
Stovall Junior High School
Houston, Texas

Kenneth Creese
Science Teacher
White Mountain Junior High
   School
Rock Springs, Wyoming

Linda A. Culp
Science Teacher and Dept. Chair
Thorndale High School
Thorndale, Texas

Georgiann Delgadillo
Science Teacher
East Valley Continuous
   Curriculum School
Spokane, Washington

Alonda Droege
Biology Teacher
Evergreen High School
Seattle, Washington

Michael J. DuPré
Curriculum Specialist
Rush Henrietta Junior-Senior
   High School
Henrietta, New York

Rebecca Ferguson
Science Teacher
North Ridge Middle School
North Richland Hills, Texas

Susan Gorman
Science Teacher
North Ridge Middle School
North Richland Hills, Texas

Karma Houston-Hughes
Science Teacher
Kyrene Middle School
Tempe, Arizona

Kerry A. Johnson
Science Teacher
Isbell Middle School
Santa Paula, California

Martha R. Kisiah
Science Teacher
Fairview Middle School
Tallahassee, Florida

Kathy LaRoe
Science Teacher
East Valley Middle School
East Helena, Montana

Jane M. Lemons
Science Teacher
Western Rockingham Middle
   School
Madison, North Carolina

Scott Mandel, Ph.D.
*Director and Educational
   Consultant*
Teachers Helping Teachers
Los Angeles, California

Maurine O. Marchani
Science Teacher and Dept. Chair
Raymond Park Middle School
Indianapolis, Indiana

Jason P. Marsh
Biology Teacher
Montevideo High School and
   Montevideo Country School
Montevideo, Minnesota

Edith C. McAlanis
Science Teacher and Dept. Chair
Socorro Middle School
El Paso, Texas

Kevin McCurdy, Ph.D.
Science Teacher
Elmwood Junior High School
Rogers, Arkansas

Alyson Mike
Science Teacher
East Valley Middle School
East Helena, Montana

Gabriell DeBear Paye
Biology Teacher
West Roxbury High School
West Roxbury, Massachusetts

James B. Pulley
Former Science Teacher
Liberty High School
Liberty, Missouri

Terry J. Rakes
Science Teacher
Elmwood Junior High School
Rogers, Arkansas

Debra Sampson
Science Teacher
Booker T. Washington Middle
   School
Elgin, Texas

Charles Schindler
Curriculum Advisor
San Bernardino City Unified
   Schools
San Bernardino, California

**Acknowledgments
continue on page 690.**

Contents in Brief

Contents

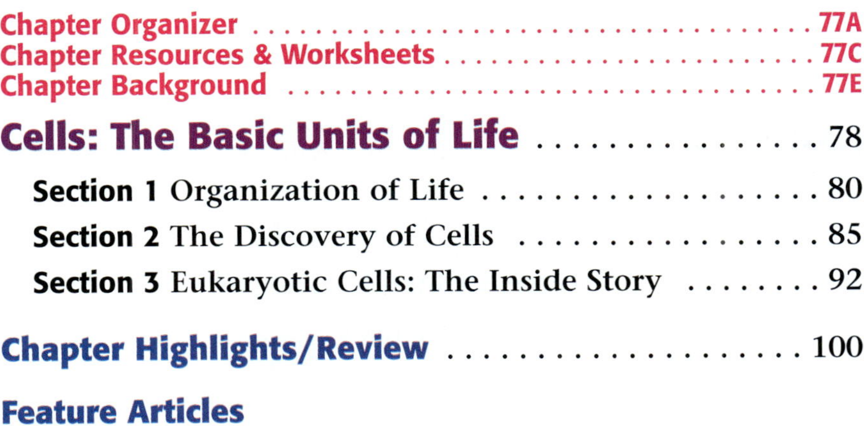

Unit 2 … **Cells**

Unit 3 · · · Heredity, Evolution, and Classification

Contents

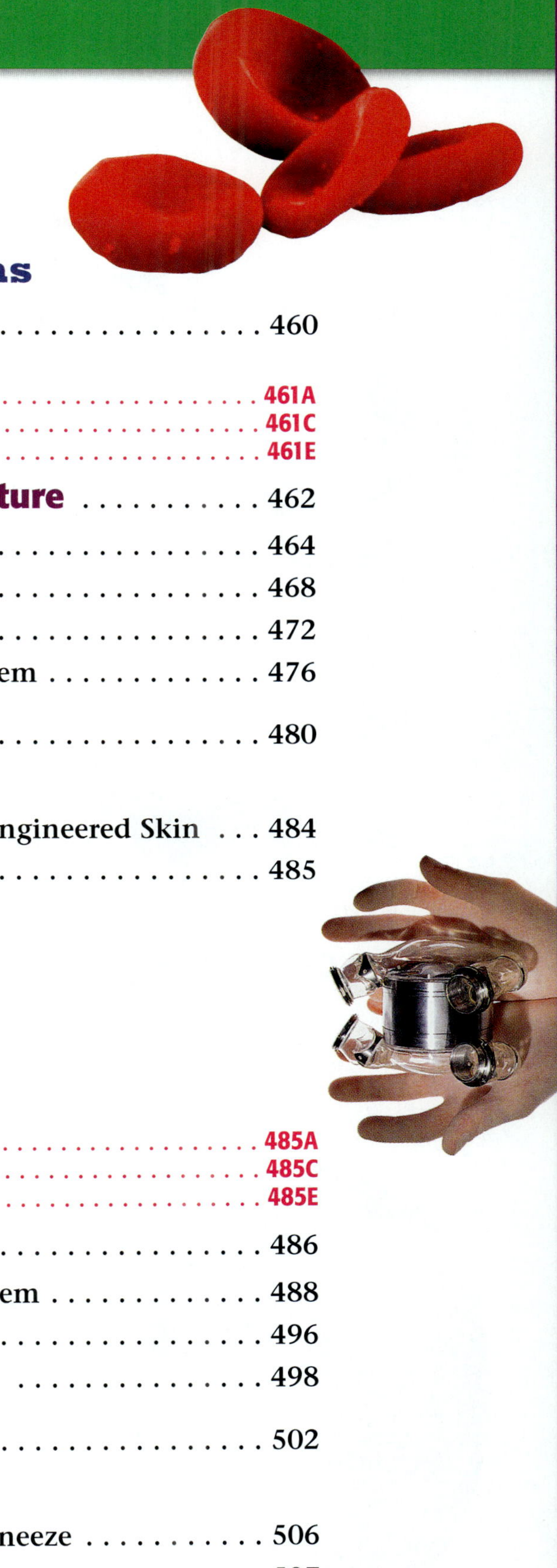

Contents

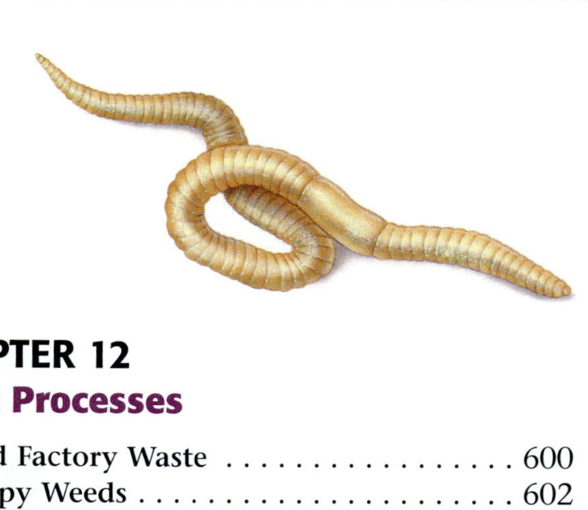

Investigate!

Now is the time to Investigate!

Science is a process in which investigation leads to information and understanding. The **Investigate!** at the beginning of each chapter helps you gain scientific understanding of the topic through hands-on experience.

QuickLab

Not all laboratory investigations have to be long and involved.

The **QuickLabs** found throughout the chapters of this textbook require only a small amount of time and limited equipment. But just because they are quick, don't skimp on the safety.

MATH BREAK

Science and math go hand in hand.

The **MathBreaks** in the margins of the chapters show you many ways that math applies directly to science and vice versa.

APPLY

Science can be very useful in the real world.

It is interesting to learn how scientific information is being used in the real world. You can see for yourself in the **Apply** features. You will also be asked to apply your own knowledge. This is a good way to learn!

Connections

One science leads to another.
You may not realize it at first, but different areas of science are related to each other in many ways. Each **Connection** explores a topic from the viewpoint of another science discipline. In this way, areas of science merge to improve your understanding of the world around you.

oceanography CONNECTION

astronomy CONNECTION

chemistry CONNECTION

earth science CONNECTION

environmental science CONNECTION

physical science CONNECTION

physics CONNECTION

Feature Articles

Feature articles for any appetite!

Science and technology affect us all in many ways. The following articles will give you an idea of just how interesting, strange, helpful, and action-packed science and technology are. At the end of each chapter, you will find two feature articles. Read them and you will be surprised at what you learn.

CAREERS

ACROSS THE SCIENCES

Science, Technology, and Society

EYE ON THE ENVIRONMENT

Eureka!

Health WATCH

SCIENTIFIC DEBATE

Science Fiction

WEIRD SCIENCE

Master Materials List

The following chart provides a comprehensive list of all the materials you would need in order to teach all of the labs and investigations in *Holt Science and Technology, Life Science.*

For added convenience, Science Kit® provides materials-ordering software on CD-ROM designed specifically for *Holt Science and Technology.* This software allows you to create an electronic materials list, complete with item numbers. Using this software, you can order complete kits or individual items, quickly and efficiently.

For more information about this software, contact your HRW representative, call Science Kit® directly at 1-800-828-7777, or visit the Web site: www.sciencekit.com.

MATERIALS AND EQUIPMENT		*Quick*Lab	Investigate!	LabBook
CONSUMABLE	**AMOUNT***	PAGE NO.	PAGE NO.	PAGE NO.
Algae	1 sample			572
Aluminum foil	1 sheet		279	609
Apple	1			609
Bag, heavy-plastic sealable, 9 × 12 in.	2			609
Bag, plastic sealable	2			616, 634
Bag, plastic trash	1			630
Baking soda	10 g			600
Balloon, round	2	363		
Balloon, slender	1			612
Balloon, small	1			630
Birdseed	1/8 lb			616
Bone, chicken	1	469		
Bottle, soda, 2 L	1	61	279, 415	630
Box, large	3		129	
Card, index, 3 × 5 in	1			599
Card, index, 3 × 5 in	20		429	
Cardboard, 2 × 2 ft	1	288		
Carton, egg	2	187		
Celery leaves	1			604
Chalk	1 stick			564
Charcoal briquette	1		151	
Chocolate, candy-coated	75			587
Clay, modeling	2 sticks		175	
Clay, water-resistant modeling	2 sticks			630
Cotton ball	6	90		

* Amount is for one group of students.

| MATERIALS AND EQUIPMENT | | **QuickLab** | **Investigate!** | **LabBook** |
CONSUMABLE *(CONTINUED)*	AMOUNT *	PAGE NO.	PAGE NO.	PAGE NO.
Cotton ball	30			634
Cricket, live	2			609
Cricket, small live	4			614
Cup, clear plastic, 300 mL	1			568
Cup, paper, 300 mL	1		175	
Cup, plastic	3		107	566, 574
Cup, plastic or paper	2			568
Earthworm	1			604
Egg, chicken	1		355	
Egg, chicken	2			634
Elodea sprig, 20 cm long	3			600
Fertilizer	1/4 cup			620
Flour	1/2 cup			566
Food samples, various types	5	43		
Gloves, protective	1 pair			560, 564, 596, 600, 614, 620, 631
Gravel	6 oz		415	616
Gumdrop, black	10			578
Gumdrop, green	10			578
Iodine solution	20 mL	43		
Isopod	4			564
Leaf, fresh	1		175	
Leaf, fresh, various kinds	5			596
Magazine	2			536, 624
Marker, black nonpermanent	1			560
Marker, black permanent	1		227	534, 602, 627
Marker, fine-point washable	1			632
Marker, various colors	1 pack			536, 573, 588, 604, 617
Marshmallow, large	15			578
Marshmallow, small colored	50			586
Marshmallow, small white	50			586
Milk	2 mL	61		
Moss, dry sphagnum	1 sample	255		
Newspaper	1		439	563, 624
Oil, cooking	1 cup	342		634
Oil, cooking	70 mL	363		
Paint, watercolor	1 palette			568

* Amount is for one group of students.

Master Materials List

MATERIALS AND EQUIPMENT		QuickLab PAGE NO.	Investigate! PAGE NO.	LabBook PAGE NO.
CONSUMABLE (CONTINUED)	AMOUNT*			
Paper, adding machine	1 roll		201	
Paper, black construction	1 sheet		53	
Paper, graph	1 sheet			562, 602, 627, 632
Paper, graph	2 sheets			635
Paper, tracing	1 sheet		151	
Paper, wax, 15 × 15 cm	1 sheet	501		
Paper, white	1 sheet		151	
Paper, white	2 sheets			568, 573
Paper towel	1 roll			604
Paper towel tube	1		53	
Pencil, assorted colored	4			635
Pencil, wax	1	282		620
Petroleum jelly	1 oz		175	
Phenol (carbolic acid) solution	4 drops			631
Pipe cleaner	3			578, 599
Plant, various species	5			596
Plant, potted	3	288		
Plant stem cutting	1			602
Plaster of Paris	300 mL		175	
Plastic wrap, clear, 3 × 10 cm	1			563
Plastic wrap, clear, approx. 1 × 2 ft	1 sheet	108		609, 620
Plate, paper	1		175	
Pond water with living organisms	300 mL			620
Poster board	1			536, 570, 588, 604
Poster board, 3 × 10 cm	1			563
Poster board, colored	1			582
Poster board, white	1			582, 617
Potato	3			574
Potato, small slice	1			564
Rice	1/4 lb	187		
Rubbing alcohol	25 mL			560
Salt	1 box			574
Seed, bean	1			598
Seed, bean	12	282		
Seed, kidney bean	4		279	
Shoe box	1			564, 604
Shell	1		175	

* Amount is for one group of students.

MATERIALS AND EQUIPMENT		*Quick*Lab	Investigate!	Lab Book
CONSUMABLE *(CONTINUED)*	AMOUNT*	PAGE NO.	PAGE NO.	PAGE NO.
Soil	8 oz		415	604
Soil, potting	4 cups		279	
Stain, methylene blue	1 drop		79	
Straw, drinking	1	386		616, 630, 631
String (or yarn)	3 m	429		616
Sugar, granulated	1/4 cup		107	566
Swab, cotton	1	90		
Tape, duct	1 m	288		
Tape, masking	25 cm	136		568, 609
Tape, transparent	4 cm	342	53, 151	563
Tape, transparent	1 roll			570, 616
Toothpick	1		79	
Toothpick, green	6			578
Toothpick, red	6			578
Vinegar	2 cups	469		
Yeast, active dry baking	1 packet		107	566
Yeast, inactive dry baking	1 packet			566
Yeast, active or inactive dry baking	1 packet			566
Yogurt	1 cup	90		

MATERIALS AND EQUIPMENT		*Quick*Lab	Investigate!	Lab Book
NONCONSUMABLE	AMOUNT*	PAGE NO.	PAGE NO.	PAGE NO.
Bag, large paper	1			582
Bag, medium paper or plastic	7			578
Bag, small paper	1			534
Bead, set of 3 different colored	5 sets	108		
Beaker, 150 mL	1			631
Beaker, 250 mL	1			566
Beaker, 400 mL	1		255, 439	562
Beaker, 600 mL	1			600, 609, 614
Bean, pinto	40			534
Binoculars	1		303	
Blender	1		439	
Board, flat, 0.5 × 0.5 m	1		439	
Bottle, spray	1			604
Bowl, large plastic	1	108, 363	463	608
Box, sand, 2 × 3 m	1			584

* Amount is for one group of students.

Master Materials List

MATERIALS AND EQUIPMENT		*QuickLab*	Investigate!	LabBook
NONCONSUMABLE (CONTINUED)	AMOUNT*	PAGE NO.	PAGE NO.	PAGE NO.
Calculator	1			534, 608
Cloth, colored, 50 × 50 cm	1			586
Container, plastic with lid	1			564, 591
Cork, small	1			612, 632
Coverslip, plastic	1	90	79	572, 620
Diffraction grating	1		53	
Eyedropper	1			563, 604, 620, 631, 632
Fishbowl, medium-sized	1			612, 614
Flashlight	1	61	35	604
Flashlight	3			568
Frog, live	1			614
Funnel, glass	1			600
Gloves, heat-resistant	1 pair			562
Gloves, various styles	5 pairs		129	
Graduated cylinder, 50 mL	3			560
Graduated cylinder, 100 mL	1			566, 608, 620, 631
Hat, various styles	5		129	
Hole punch	1			563
Hot plate	1			562, 566
Jar, 1 qt	3			620
Jar, wide-mouthed	1	469	415	
Knife, plastic	1		355	
Lamp, goose-neck	1			609
Light filter, blue	1			568
Light filter, green	1			568
Light filter, red	1			568
Magnifying lens	1	20	151, 303	566, 609
Meterstick	1		509, 533	570
Microscope, compound	1	90	79	572, 620
Microscope slide, plastic	1	90	79	572, 620
Paintbrush	1			568
Pan, dissecting	1			604
Pan, square	1		439	
Penny	100			591
Petri dish	2	282		
Pin, straight	1	386		
Pin, dissecting	1			632

* Amount is for one group of students.

| MATERIALS AND EQUIPMENT | | *Quick*Lab | Investigate! | Lab Book |
NONCONSUMABLE *(continued)*	AMOUNT*	PAGE NO.	PAGE NO.	PAGE NO.
Pin, map	15			578
Pipe, PVC, 3/4 in. diam, 12 cm	1			612
Plant guidebook	1			596
Probe	1			604
Pushpin, blue	10			578
Pushpin, green	10			578
Rock, large, approx. 3 lb	1			614
Rubber band	1			612
Rubber band	2			630
Ruler, metric	1	201, 547	107	563, 564, 568, 584, 600, 602, 604, 627, 632
Sand	approx. 20 lb			584
Sand, fine	1 lb			570
Scarf, various styles	5		129	
Scales (or balance)	1			570, 608, 618
Scales, bathroom	1			576
Scissors	1		5, 381	536, 570, 578, 582, 598, 599, 616
Scoopula	1			566
Shoe, various styles	10		227	
Spatula	1		439	
Sponge	2			618
Sponge, natural	1			608
Stirring rod	1		107	566, 602
Stirring stick, wooden	1			566
Stopwatch	1	312		562, 564, 586, 626, 631
Tape measure	1		533	576
Test tube	1			566, 600
Test tube	2		107	602
Test-tube rack	1		107	566, 602
Thermometer, Celsius	1			560, 562, 566, 626
Thermometer clip	1			562, 566
Wire screen, approx. 2 × 2 ft	1		439	

* Amount is for one group of students.

Science & Math Skills Worksheets

The *Holt Science and Technology* program helps you meet the needs of a wide variety of students, regardless of their skill level. The following pages provide examples of the worksheets available to improve your students' science and math skills whether they already have a strong science and math background or are weak in these areas. Samples of assessment checklists and rubrics are also provided.

In addition to the skills worksheets represented here, *Holt Science and Technology* provides a variety of worksheets that are correlated directly with each chapter of the program. Representations of these worksheets are found at the beginning of each chapter in this Annotated Teacher's Edition.

Many worksheets are also available on the HRW Web site. The address is **go.hrw.com.**

Science Skills Worksheets: Thinking Skills

BEING FLEXIBLE

#1

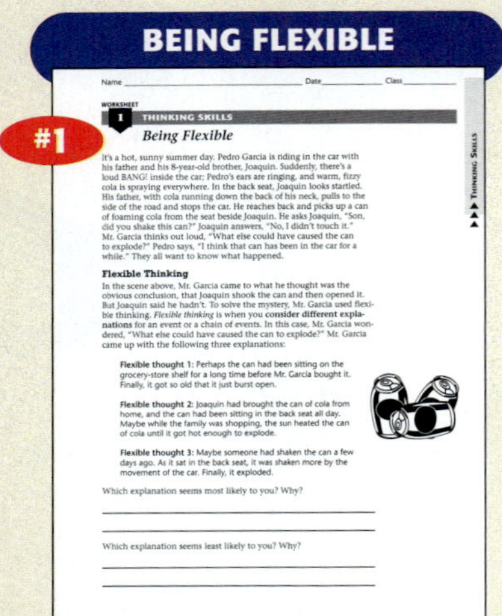

USING YOUR SENSES

#2

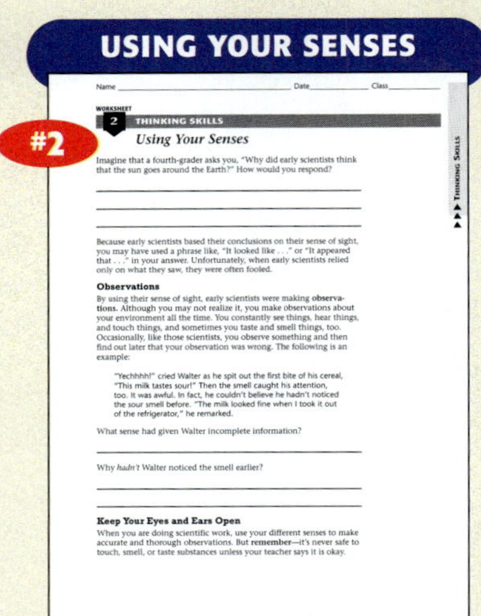

THINKING OBJECTIVELY

#3

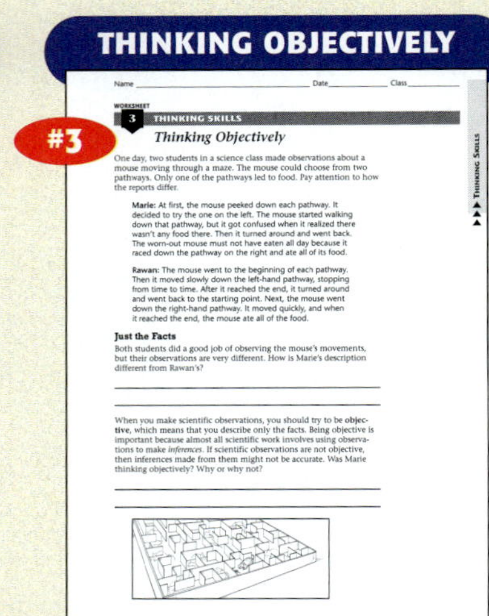

UNDERSTANDING BIAS

#4

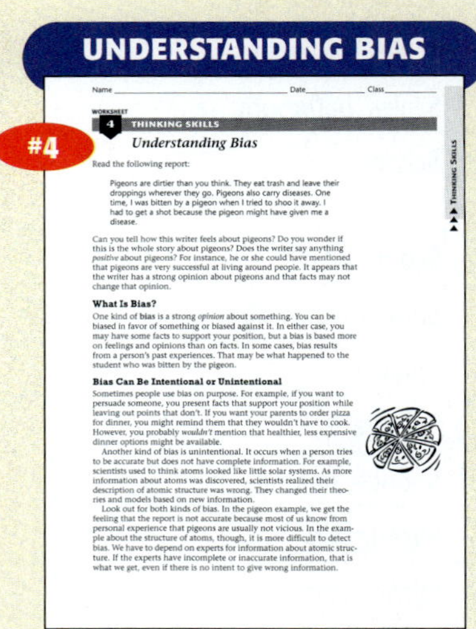

USING LOGIC

#5

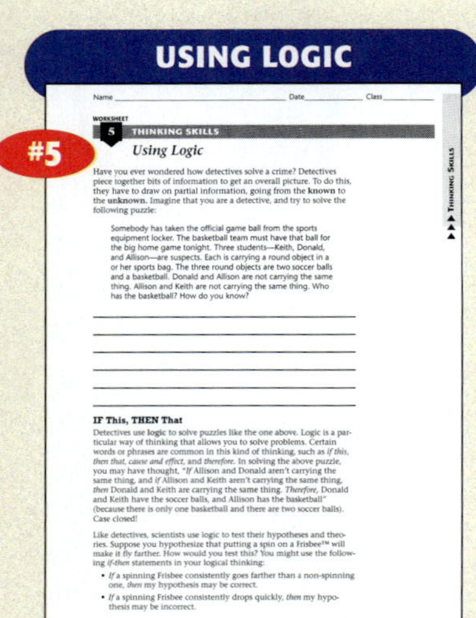

BOOSTING YOUR MEMORY

#6

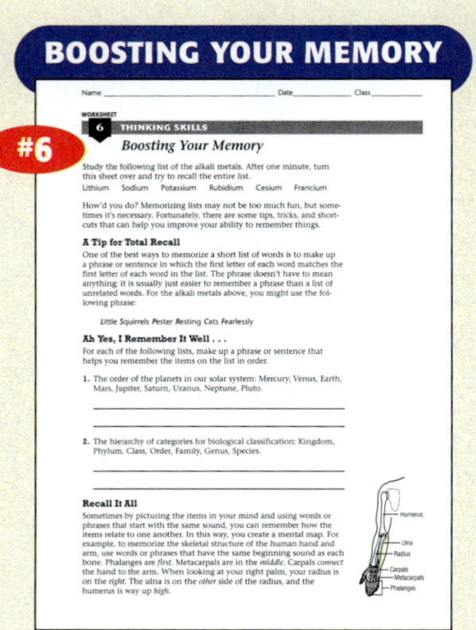

IMPROVING YOUR STUDY HABITS

#7

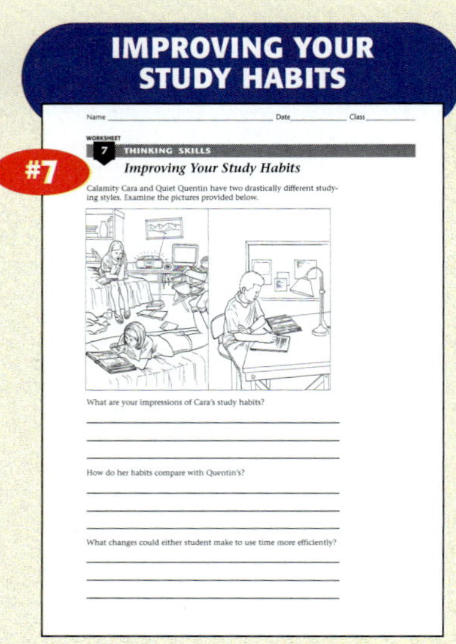

READING A SCIENCE TEXTBOOK

#8

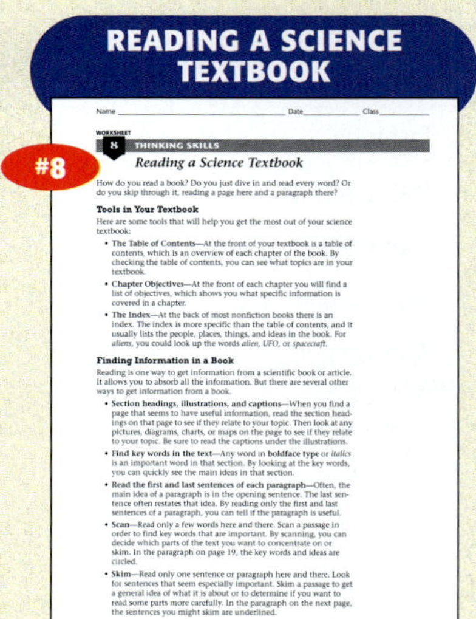

Science Skills Worksheets: Experimenting Skills

SAFETY RULES!

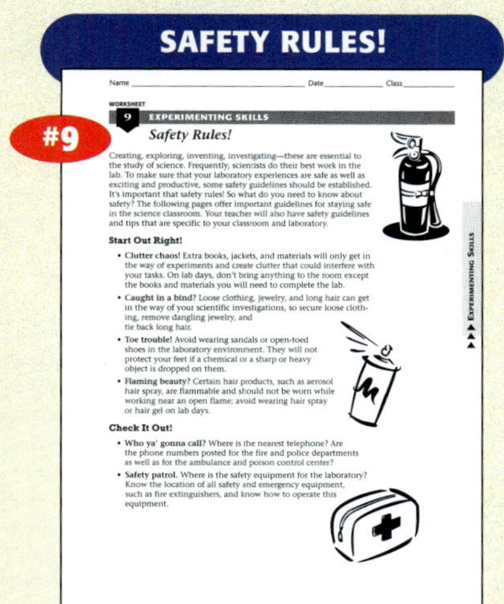

DOING A LAB WRITE-UP

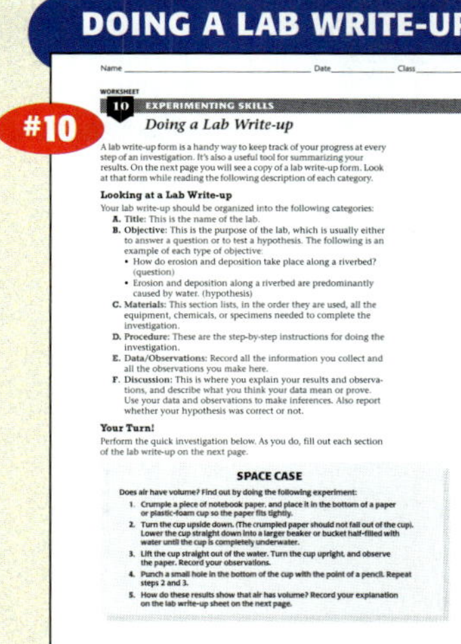

UNDERSTANDING VARIABLES

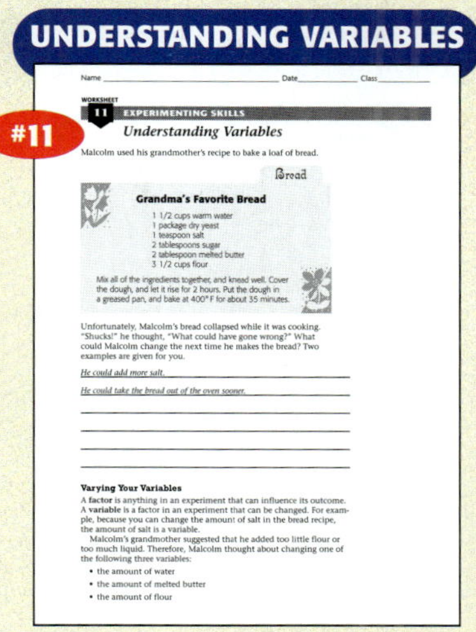

WORKING WITH HYPOTHESES

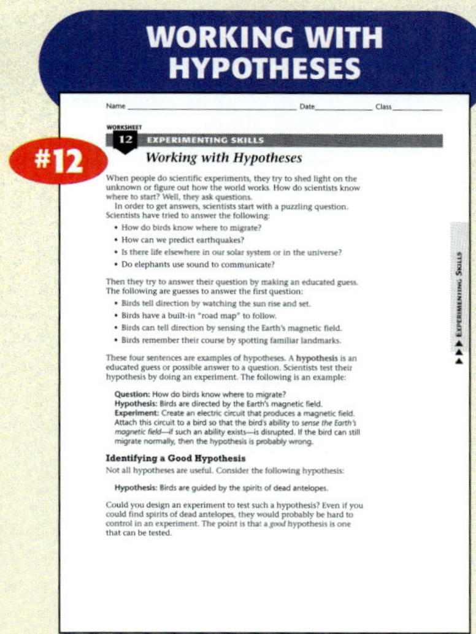

DESIGNING AN EXPERIMENT

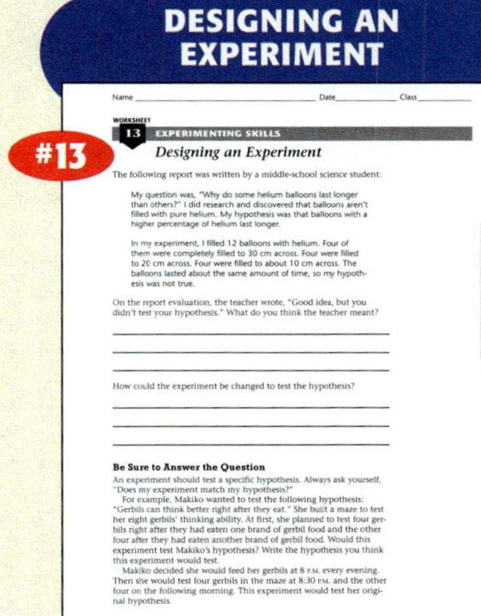

USING THE INTERNATIONAL SYSTEM OF UNITS (SI)

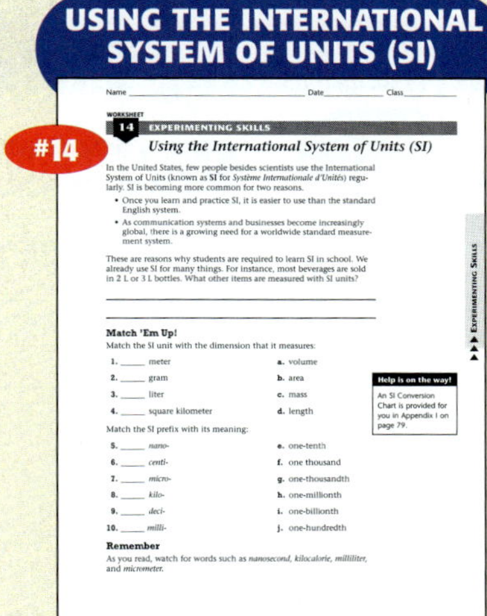

MEASURING

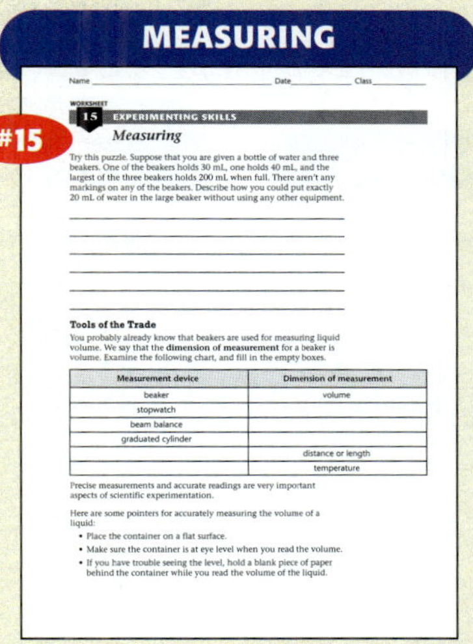

Science Skills Worksheets: Researching Skills

CHOOSING YOUR TOPIC

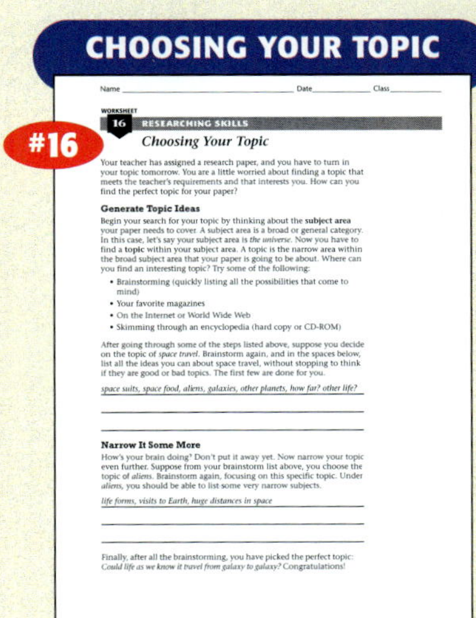

ORGANIZING YOUR RESEARCH

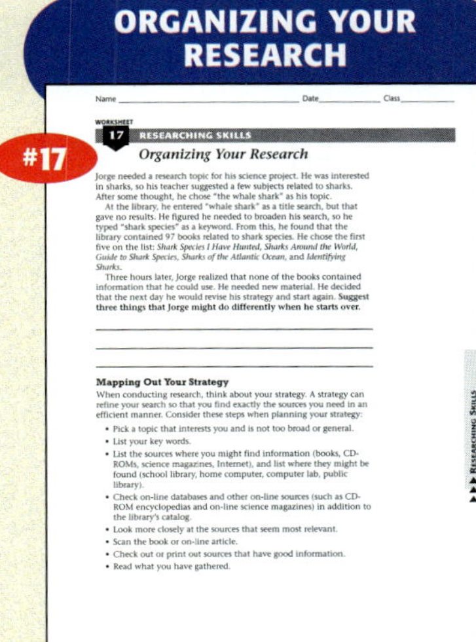

FINDING USEFUL SOURCES

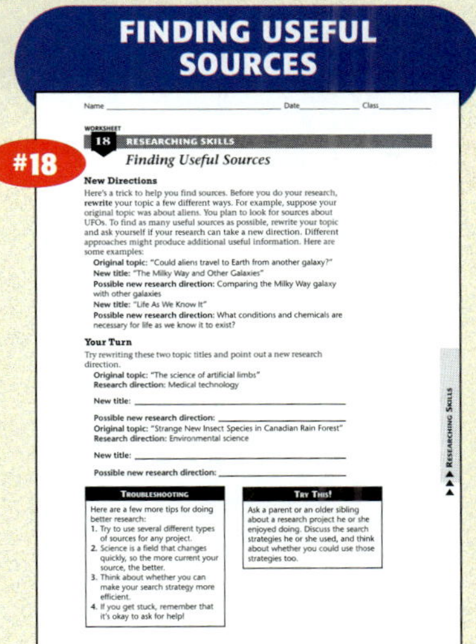

RESEARCHING ON THE WEB

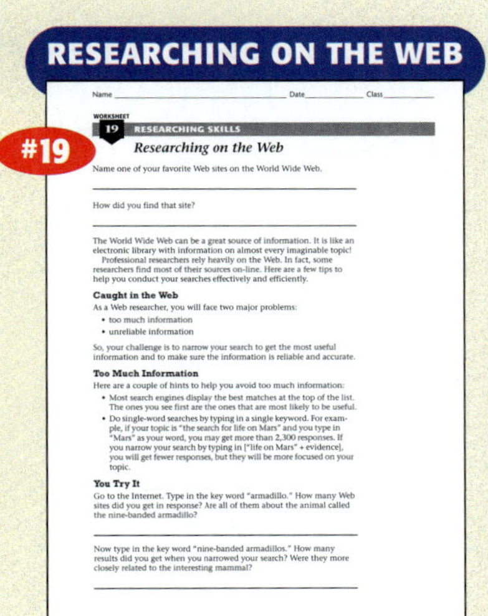

Science Skills Worksheets: Researching Skills (continued)

IDENTIFYING BIAS

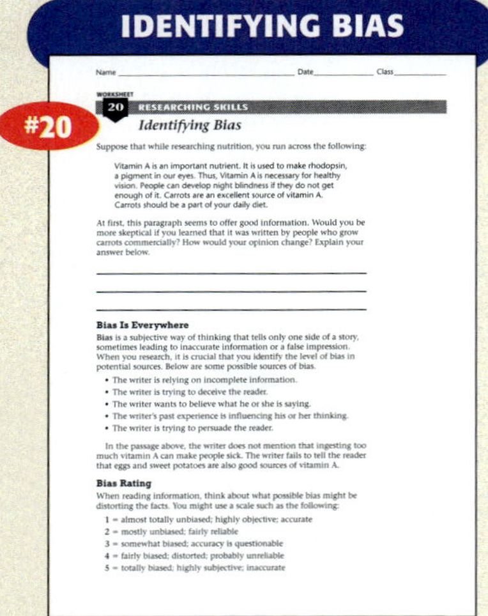

#20

TAKING NOTES

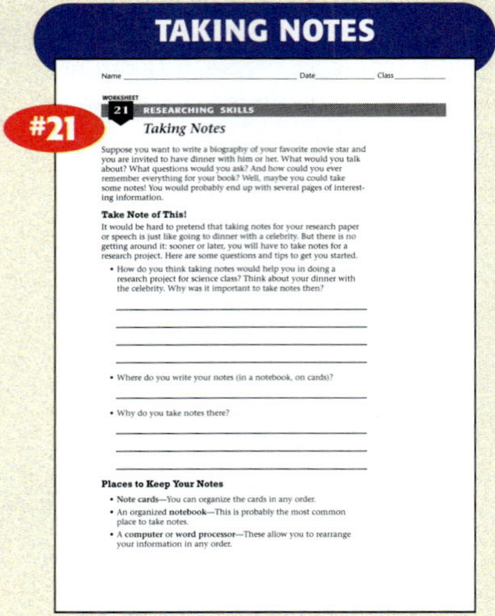

#21

SCIENCE WRITING

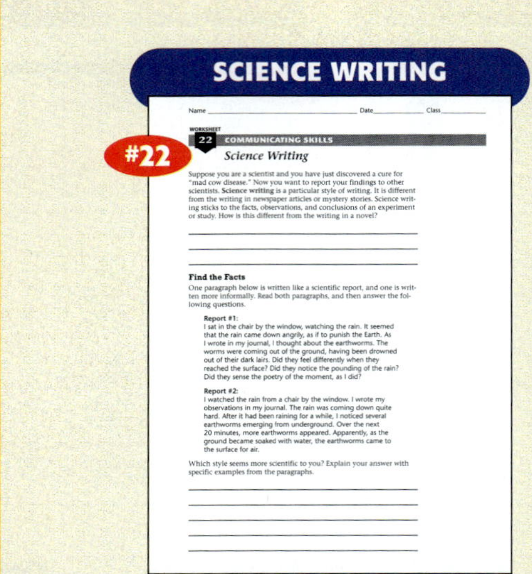

#22

Science Skills Worksheets: Communicating Skills

SCIENCE DRAWING

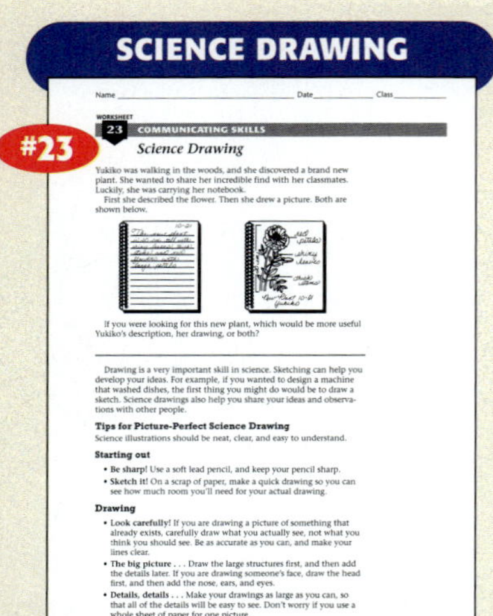

#23

USING MODELS TO COMMUNICATE

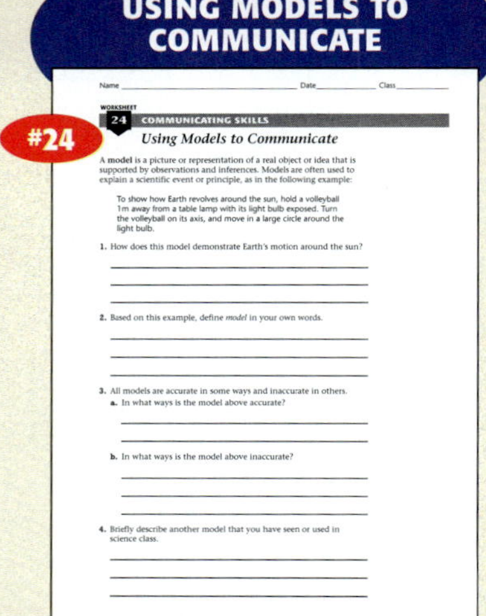

#24

INTRODUCTION TO GRAPHS

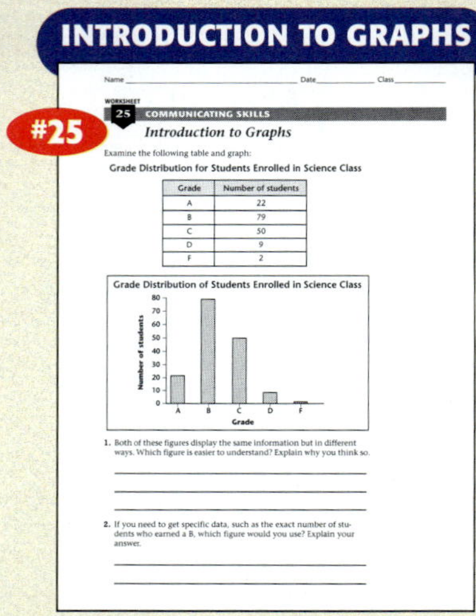

#25

GRASPING GRAPHING

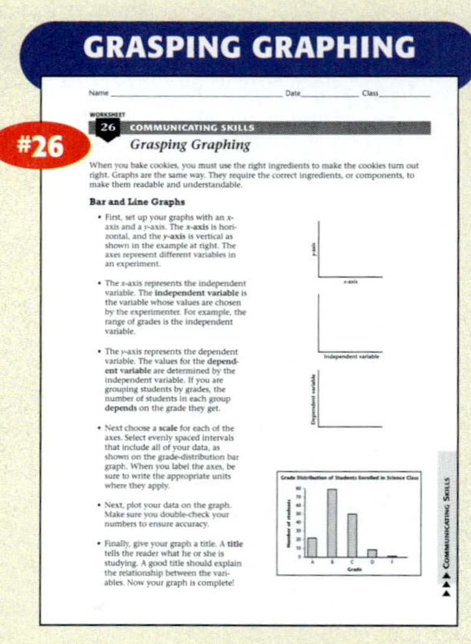

#26

INTERPRETING YOUR DATA

#27

RECOGNIZING BIAS IN GRAPHS

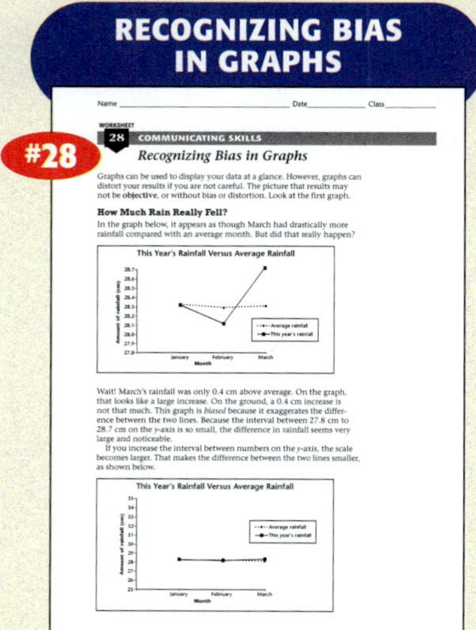

#28

MAKING DATA MEANINGFUL

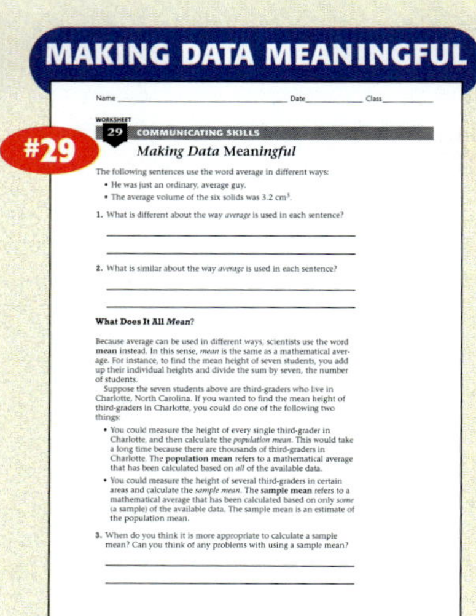

#29

HINTS FOR ORAL PRESENTATIONS

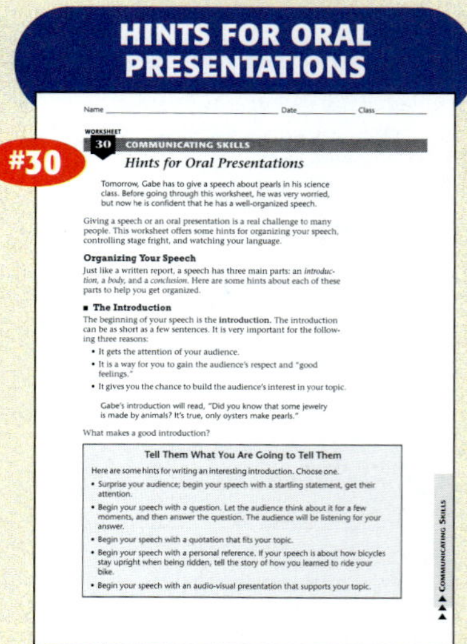

#30

Math Skills for Science

ADDITION AND SUBTRACTION

#1 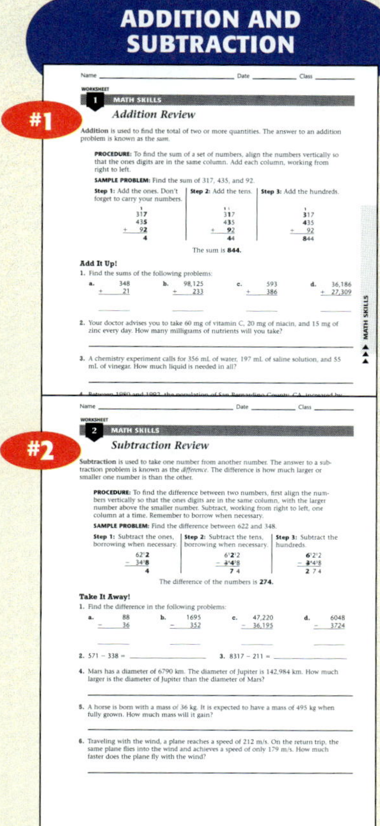

#2

MULTIPLICATION

#3 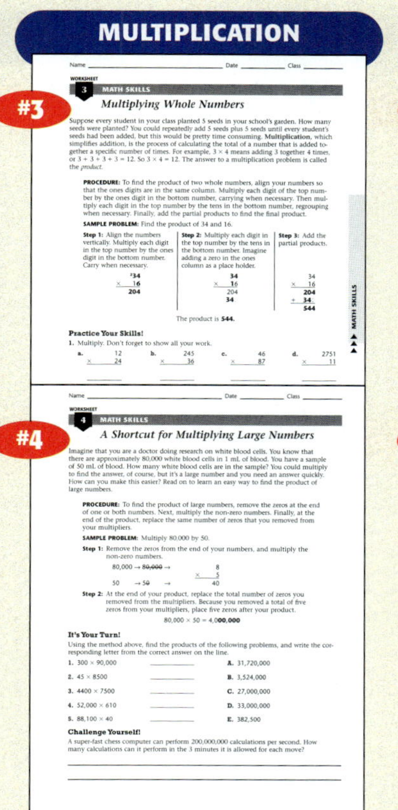

#4

DIVISION

#5 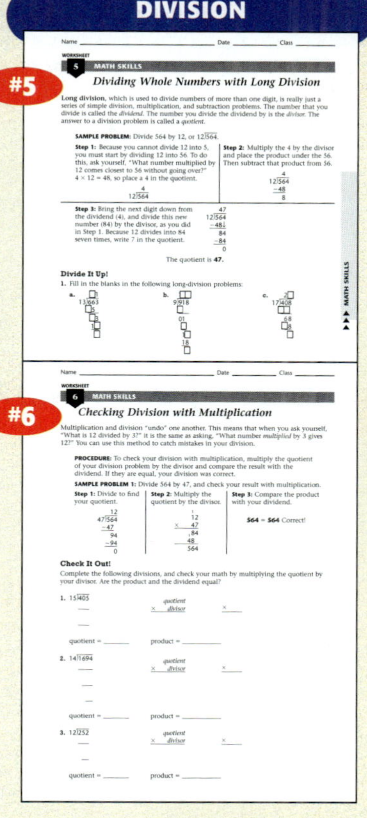

#6

AVERAGES

#7 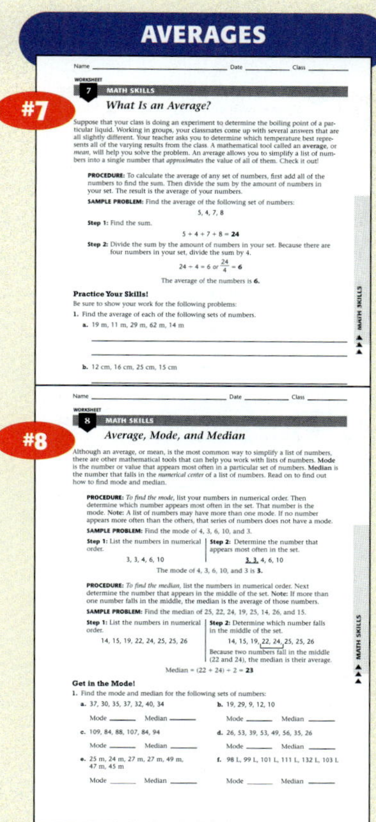

#8

POSITIVE AND NEGATIVE NUMBERS

#9 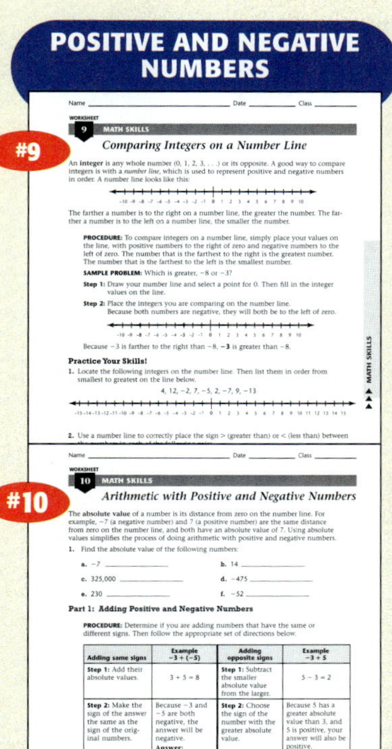

#10

FRACTIONS

#11 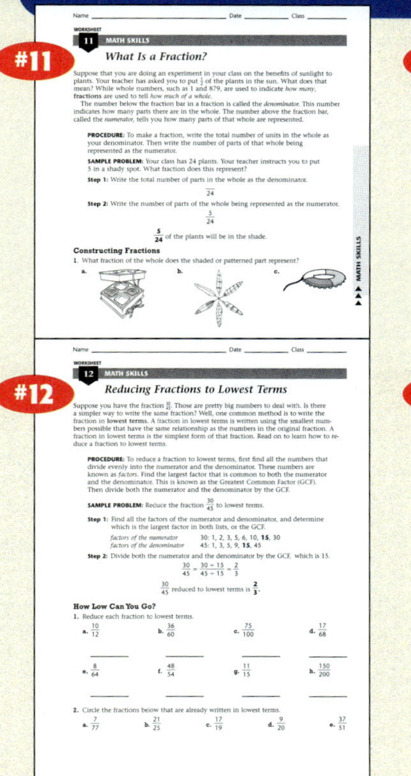

#12

#13 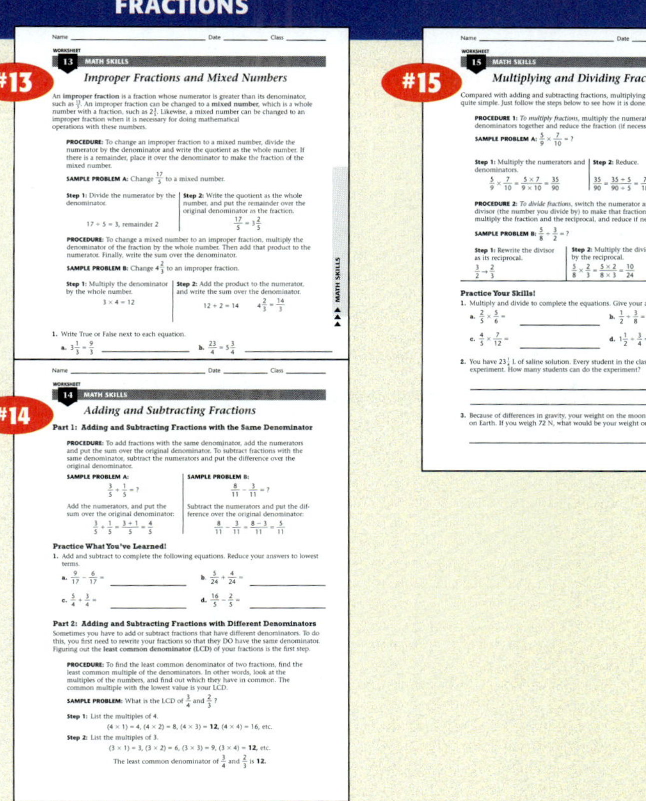

#14

#15

Math Skills for Science (continued)

RATIOS AND PROPORTIONS

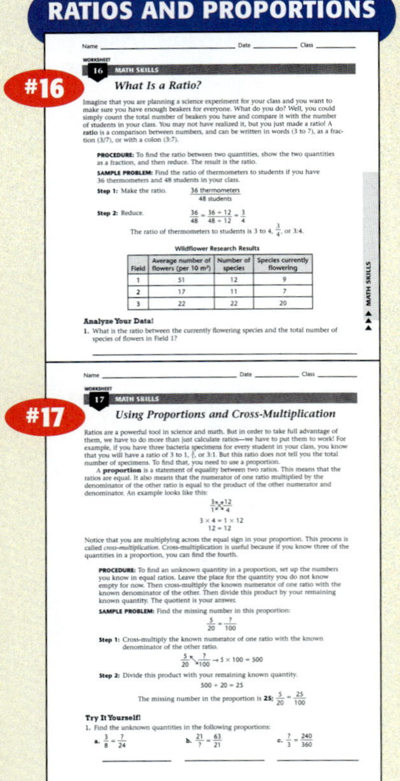

DECIMALS

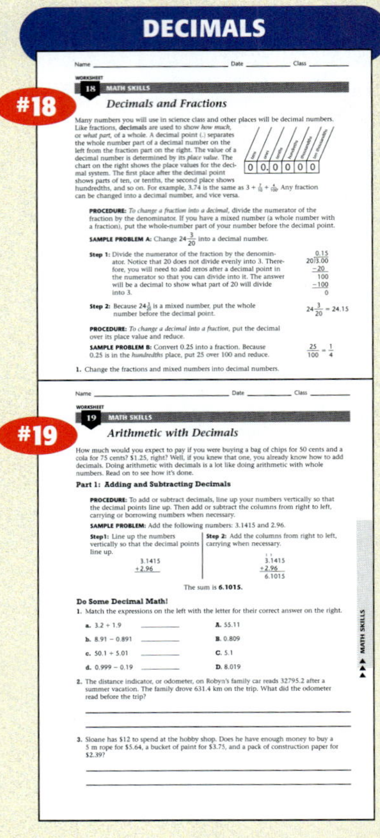

PERCENTAGES

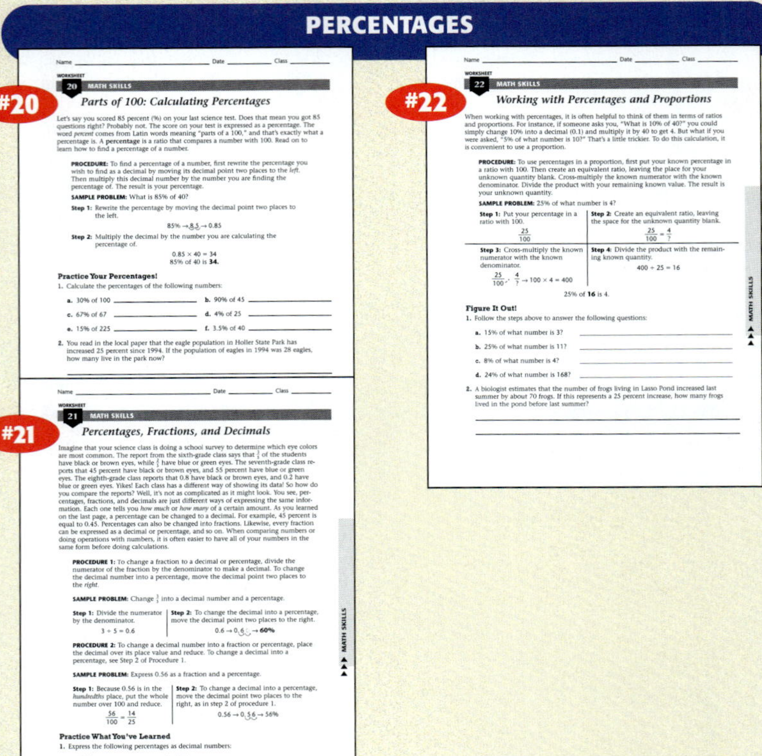

POWERS OF 10

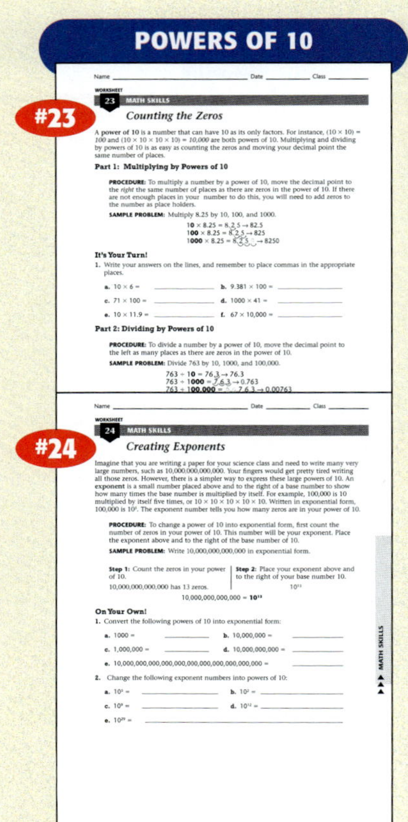

SCIENTIFIC NOTATION

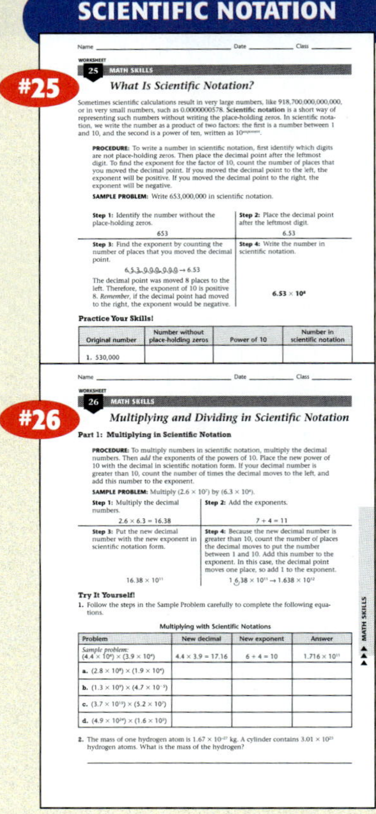

SI MEASUREMENT AND CONVERSION

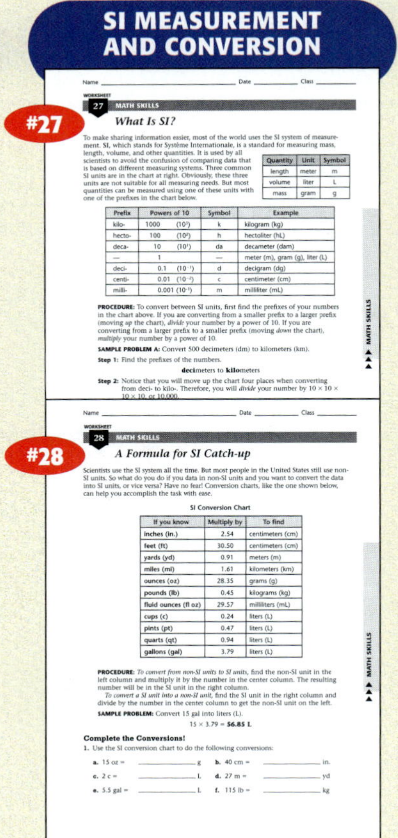

Math Skills for Science (continued)

GEOMETRY

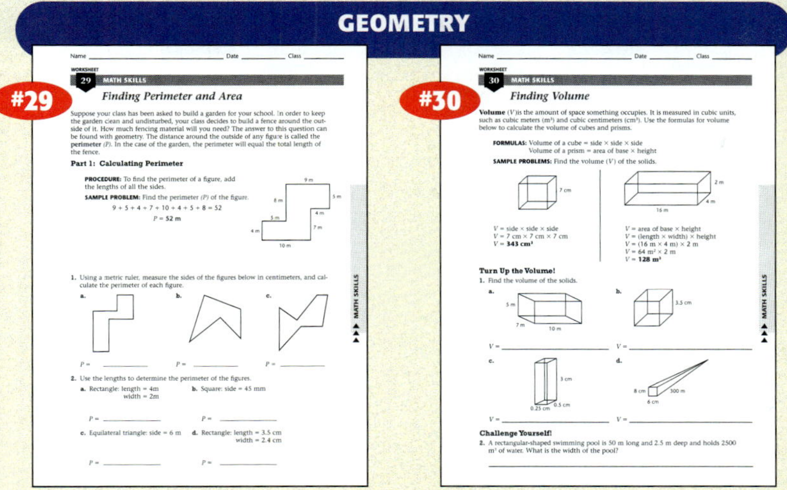

THE UNIT FACTOR AND DIMENSIONAL ANALYSIS

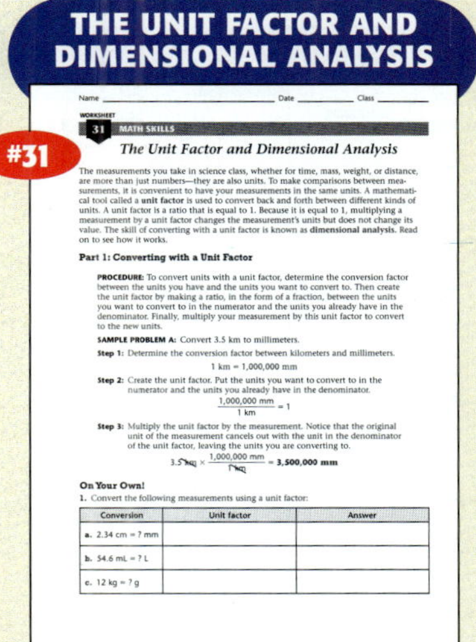

#29   Finding Perimeter and Area

#30   Finding Volume

#31   The Unit Factor and Dimensional Analysis

MATH IN SCIENCE: INTEGRATED SCIENCE

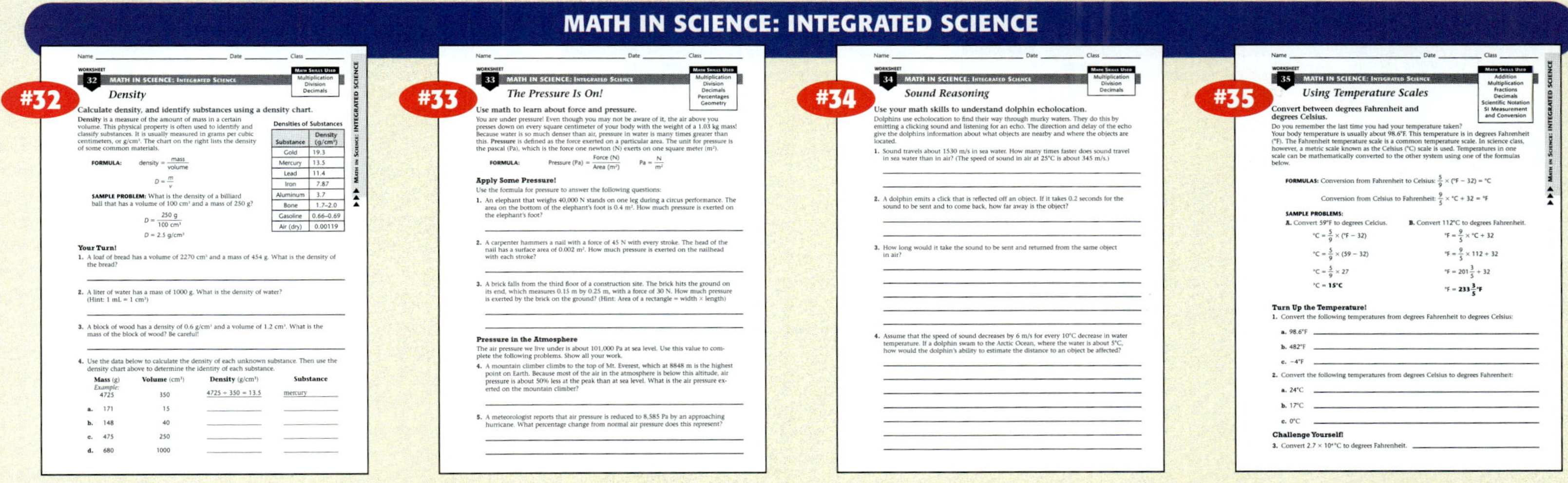

#32   Density

#33   The Pressure Is On!

#34   Sound Reasoning

#35   Using Temperature Scales

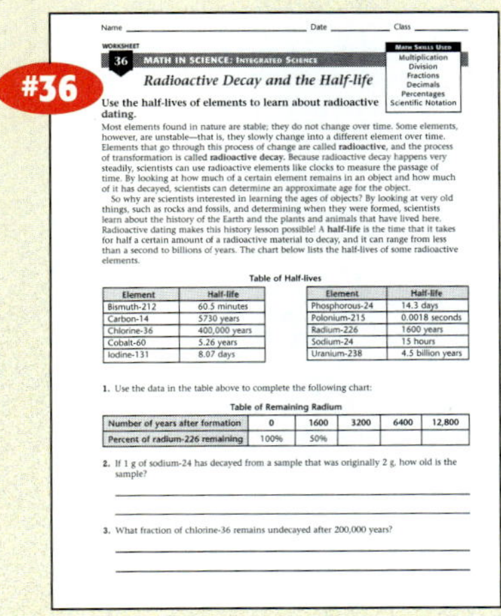

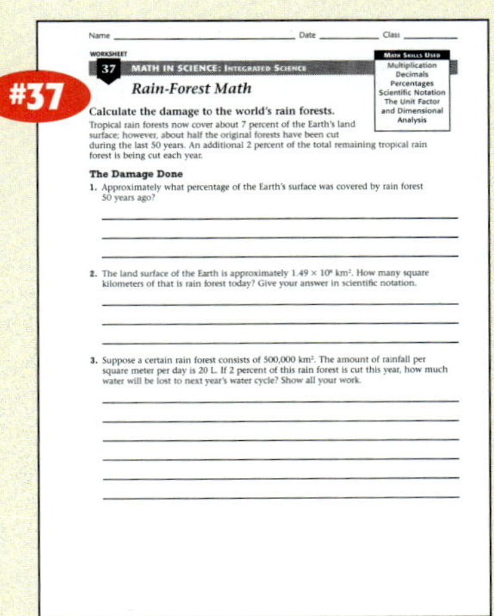

#36   Radioactive Decay and the Half-life

#37   Rain-Forest Math

Math Skills for Science (continued)

MATH IN SCIENCE: LIFE SCIENCE

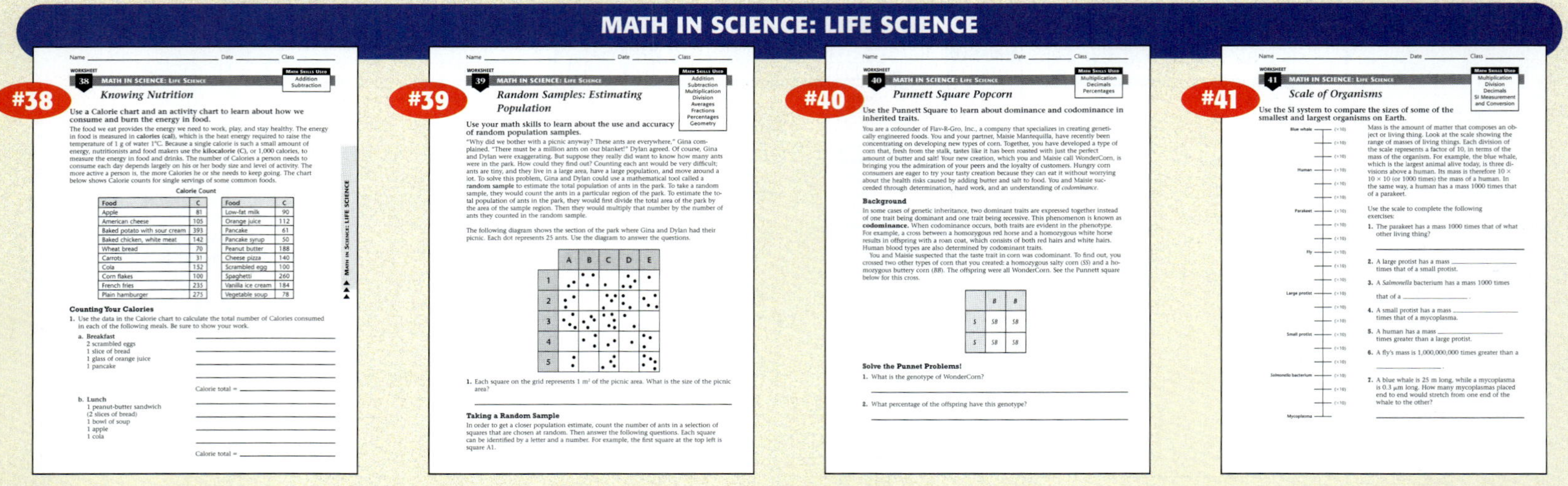

MATH IN SCIENCE: EARTH SCIENCE

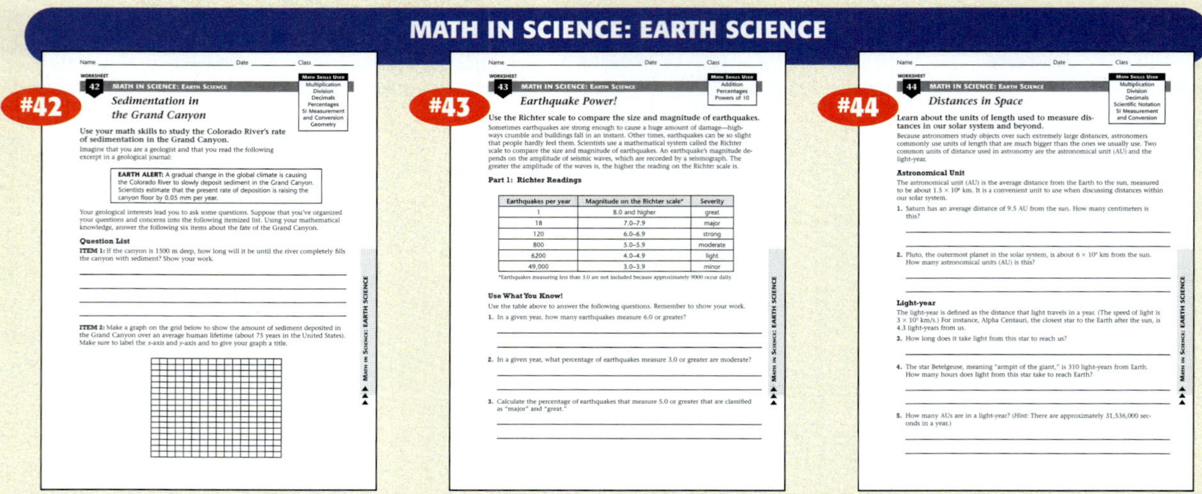

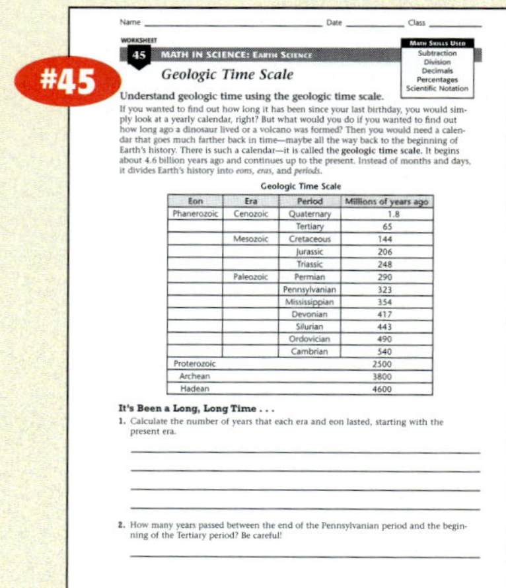

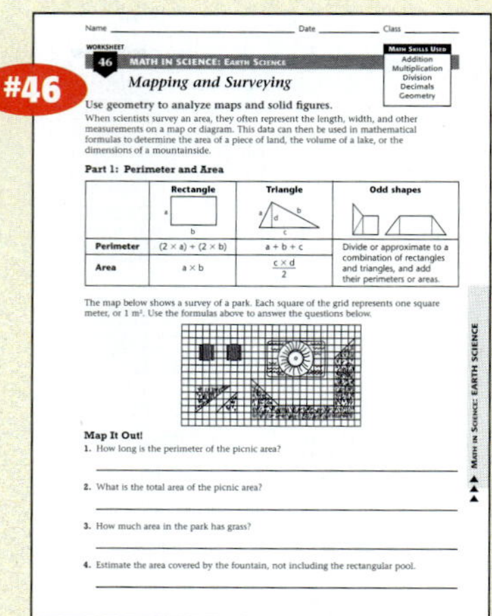

Math Skills for Science (continued)

MATH IN SCIENCE: PHYSICAL SCIENCE

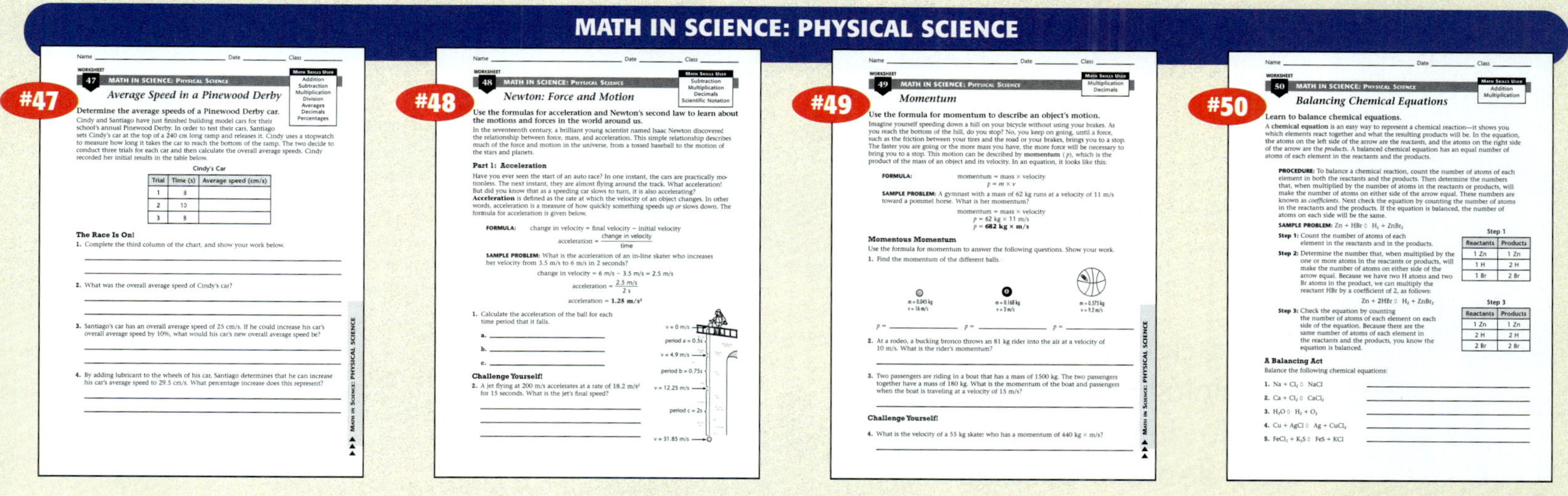

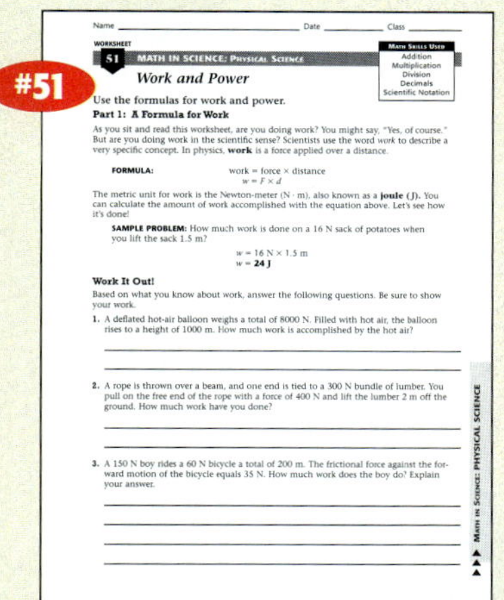

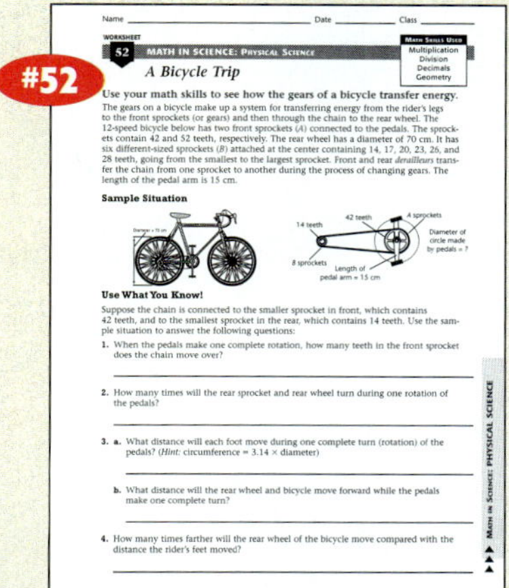

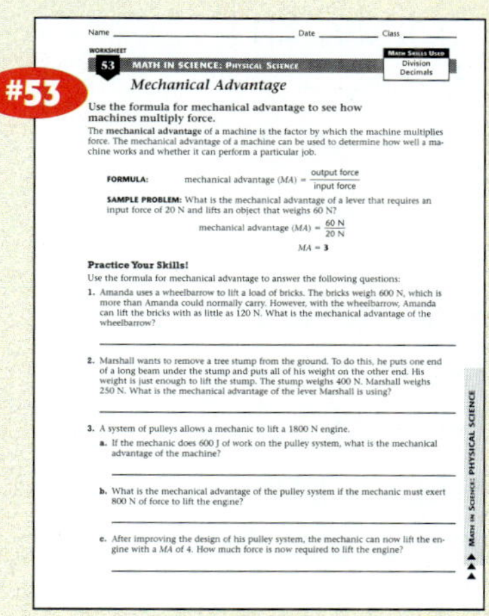

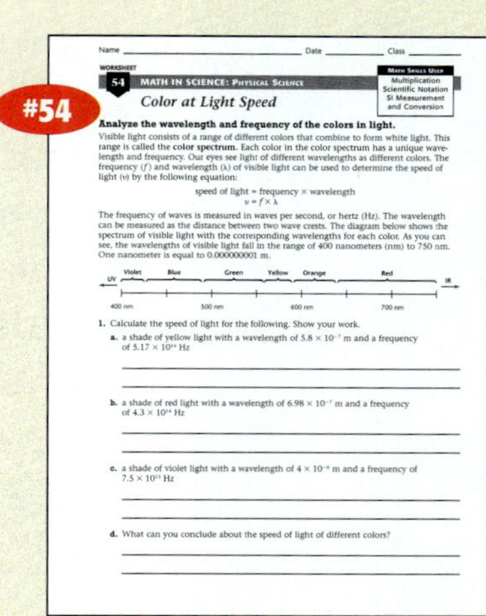

Assessment Checklist & Rubrics

The following is just a sample of over 50 checklists and rubrics contained in this booklet.

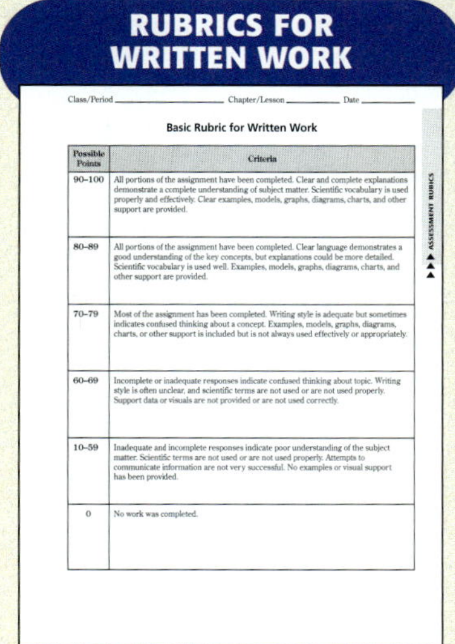

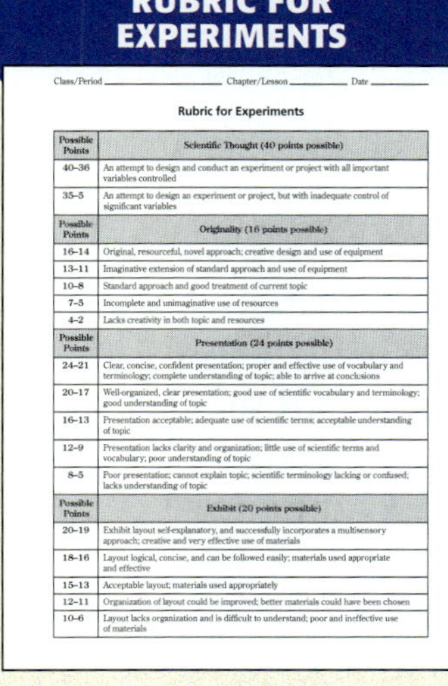

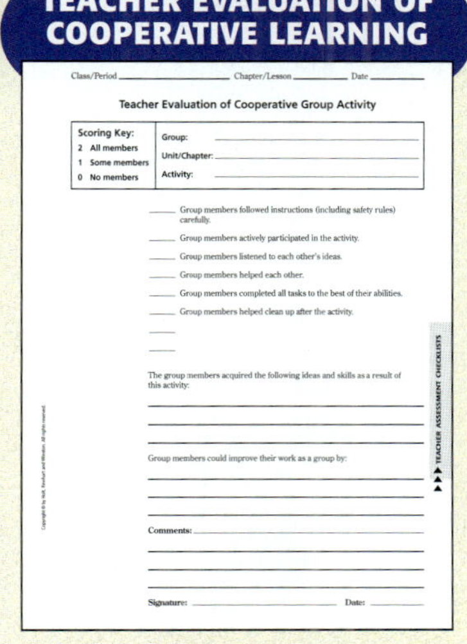

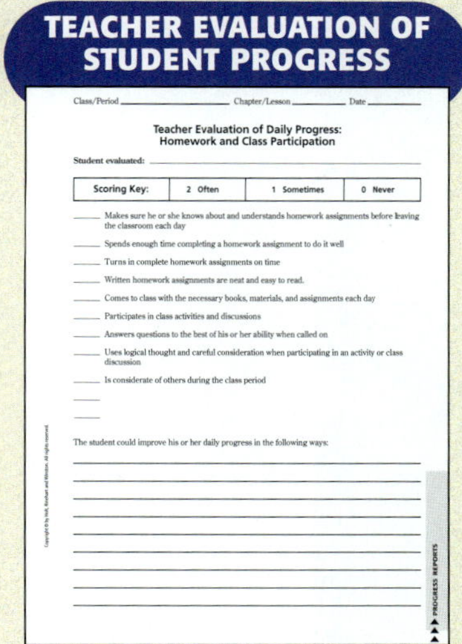

1

El estudio de los seres vivos

Las ciencias biológicas son el estudio de los seres vivos, desde la bacteria más pequeña hasta el árbol más grande. En esta unidad, descubrirás en qué se parecen todos los seres vivos. Aprenderás sobre las herramientas que usan los científicos, así como a formular preguntas sobre lo que te rodea.

Los seres humanos siempre hemos buscado respuestas sobre la vida. Esta cronología incluye algunas de las personas que han estudiado los seres vivos.

2640 a.C.

Si Ling-Chi, emperatriz de China, observa los gusanos de seda de su jardín y desarrolla un proceso para cultivarlos y producir seda.

1944

Oswald T. Avery propone que el ADN es el material que contiene las propiedades genéticas de los organismos vivos.

1946

Se construye la primera computadora completamente eléctrica, ENIAC. Pesa 30 toneladas y ocupa 450 m².

1967

El Dr. Christian Barnard realiza con éxito el primer trasplante de corazón humano.

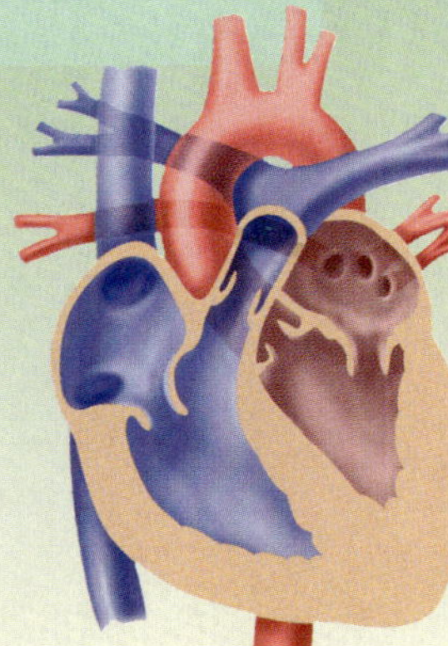

1970

Se inventan los discos flexibles para almacenar datos en las computadoras.

1010

El físico árabe Ibn al Haytham descubre que la visión es causada por la reflexión de la luz desde los objetos a los ojos.

1590

Zacharius Jansen construye el primer microscopio, el cual estaba compuesto por dos lentes y un tubo.

1685

Los adelantos en los microscopios permiten observar los glóbulos rojos por primera vez.

1914

Los estudios de George Washington Carver sobre la conservación de la agricultura y el suelo condujeron a la investigación sobre el cacahuate.

1934

Dorothy Crowfoot Hodgkin usa técnicas de rayos X para determinar la estructura de las proteínas.

1931

Se desarrolla el primer microscopio electrónico.

1983

Dian Fossey escribe *Gorilas en la bruma*, un libro sobre su investigación de los gorilas beringei de África y sus esfuerzos por protegerlos de los cazadores furtivos.

1984

Alec Jeffries desarrolla un proceso conocido como huellas de ADN.

1998

En China, los científicos descubren un fósil de dinosaurio con plumas.

Chapter Organizer

CHAPTER ORGANIZATION	TIME MINUTES	OBJECTIVES	LABS, INVESTIGATIONS, AND DEMONSTRATIONS
Chapter Opener pp. 4–5	45		**Investigate!** Figure It Out, p. 5
Section 1 Asking About Life	45	▶ Explain the importance of asking questions in life science. ▶ Give three reasons why life science is beneficial to living things.	
Section 2 Thinking Like a Life Scientist	90	▶ Describe the scientific method. ▶ Evaluate the designs of experiments. ▶ Interpret the information in tables and graphs. ▶ Explain how scientific knowledge can change.	**Demonstration,** Making Models, p. 12 in ATE **Skill Builder,** Does It All Add Up? p. 560 **Datasheets for LabBook,** Does It All Add Up? Datasheet 1 **Skill Builder,** Graphing Data, p. 562 **Datasheets for LabBook,** Graphing Data, Datasheet 2 **Inquiry Labs,** One Side or Two? Lab 1 **Whiz-Bang Demonstrations,** Air Ball, Demo 1 **Whiz-Bang Demonstrations,** Getting to the Point, Demo 2
Section 3 Tools of Life Scientists	90	▶ Describe the tools life scientists use for seeing. ▶ Explain how life scientists use computers. ▶ Explain the importance of the International System of Units.	**Demonstration,** p. 19 in ATE **QuickLab,** See for Yourself, p. 20 **Demonstration,** p. 20 in ATE **Demonstration,** Relative Size, p. 22 in ATE **Demonstration,** Displacement, p. 24 in ATE **Interactive Explorations CD-ROM,** Something's Fishy! A *Worksheet* is also available in the *Interactive Explorations Teacher's Guide.* **Making Models,** A Window to a Hidden World, p. 563 **Datasheets for LabBook,** A Window to a Hidden World, Datasheet 3 **Long-Term Projects & Research Ideas,** Project 1

TECHNOLOGY RESOURCES

Guided Reading Audio CD
English or Spanish, Chapter 1

One-Stop Planner CD-ROM with Test Generator

Science Discovery Videodiscs
Image and Activity Bank with Lesson Plans: Science and the Constitution, Models and Predictions, Through the Microscope, Making Sense of the Census

Science Sleuths: The Traffic Accident

CNN **Multicultural Connections,** Hopi Science, Segment 1

Scientists in Action, A Biologist's Dolphin Investigation, Segment 1

Interactive Explorations CD-ROM
CD 1, Exploration 1, Something's Fishy!

CLASSROOM WORKSHEETS, TRANSPARENCIES, AND RESOURCES	SCIENCE INTEGRATION AND CONNECTIONS	REVIEW AND ASSESSMENT
Directed Reading Worksheet 1 **Science Puzzlers, Twisters & Teasers,** Worksheet 1 **Science Skills Worksheet 8,** Reading a Science Textbook		
Directed Reading Worksheet 1, Section 1	**Multicultural Connection,** p. 8 in ATE **Real-World Connection,** p. 8 in ATE **Connect to Environmental Science,** p. 8 in ATE **Career:** Zoologist—Eric Pianka, p. 32	**Homework,** p. 9 in ATE **Review,** p. 9 **Quiz,** p. 9 in ATE **Alternative Assessment,** p. 9 in ATE
Transparency 1, The Scientific Method **Directed Reading Worksheet 1,** Section 2 **Math Skills for Science Worksheet 7,** What Is an Average? **Reinforcement Worksheet 1,** The Mystery of the Bubbling Top **Problem Solving Worksheet 1,** The Case of the Bulge	**Cross-Disciplinary Focus,** p. 11 in ATE **Apply,** p. 13 **MathBreak,** Averages, p. 16 **Cross-Disciplinary Focus,** p. 16 in ATE **Real-World Connection,** p. 17 in ATE	**Self-Check,** p. 12 **Homework,** pp. 12, 13, 15, 16 in ATE **Self-Check,** p. 15 **Review,** p. 18 **Quiz,** p. 18 in ATE **Alternative Assessment,** p. 18 in ATE
Directed Reading Worksheet 1, Section 3 **Math Skills for Science Worksheet 19,** Arithmetic with Decimals **Transparency 2,** Common SI Units **Math Skills Worksheet 28,** A Formula for SI Catch-up **Math Skills for Science Worksheet 27,** What Is SI? **Transparency 3,** Scale of Sizes **Math Skills for Science Worksheet 29,** Finding Perimeter and Area **Math Skills for Science Worksheet 30,** Finding Volume **Transparency 116,** Three Temperature Scales **Science Skills Worksheet 9,** Safety Rules!	**MathBreak,** Magnification, p. 19 **Multicultural Connection,** p. 20 in ATE **Multicultural Connection,** p. 21 in ATE **Math and More,** p. 23 in ATE **MathBreak,** Finding Area, p. 24 **Math and More,** p. 25 in ATE **Real-World Connection,** p. 26 in ATE **Connect to Earth Science,** p. 26 in ATE **Holt Anthology of Science Fiction,** *The Homesick Chicken*	**Homework,** p. 25 in ATE **Review,** p. 27 **Quiz,** p. 27 in ATE **Alternative Assessment,** p. 27 in ATE

END-OF-CHAPTER REVIEW AND ASSESSMENT

Chapter Review in Study Guide
Vocabulary and Notes in Study Guide
Chapter Tests with Performance-Based Assessment, Chapter 1 Test
Chapter Tests with Performance-Based Assessment, Performance-Based Assessment 1
Concept Mapping Transparency 1

internet connect

Holt, Rinehart and Winston On-line Resources

go.hrw.com

For worksheets and other teaching aids related to this chapter, visit the HRW Web site and type in the keyword: **HSTLIV**

National Science Teachers Association

www.scilinks.org

Encourage students to use the *sci*LINKS numbers listed with the Chapter Highlights to access information and resources on the **NSTA** Web site.

Chapter Resources & Worksheets

Visual Resources

TEACHING TRANSPARENCIES

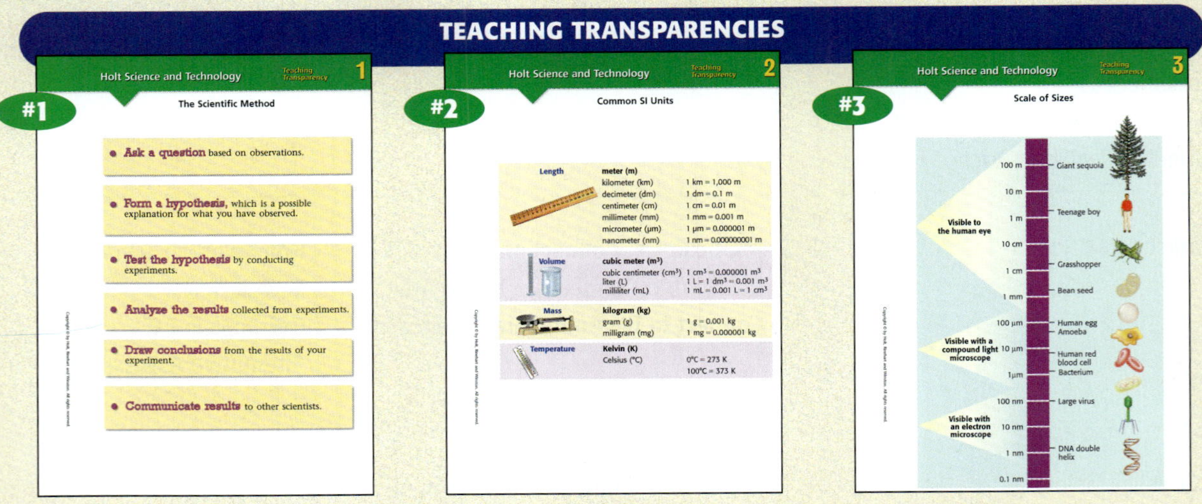

TEACHING TRANSPARENCIES   &   CONCEPT MAPPING TRANSPARENCY

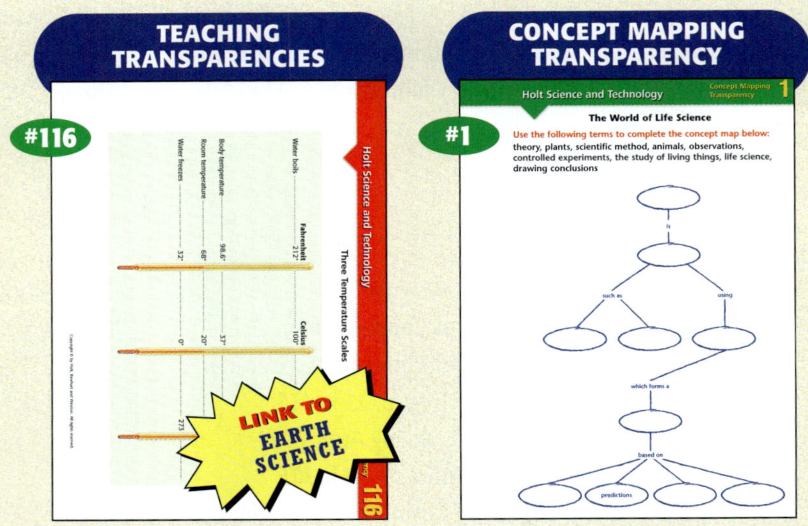

Meeting Individual Needs

DIRECTED READING | REINFORCEMENT & VOCABULARY REVIEW | SCIENCE PUZZLERS, TWISTERS & TEASERS

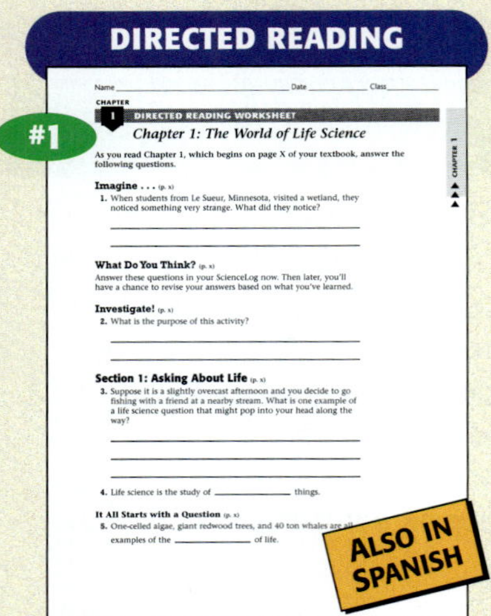

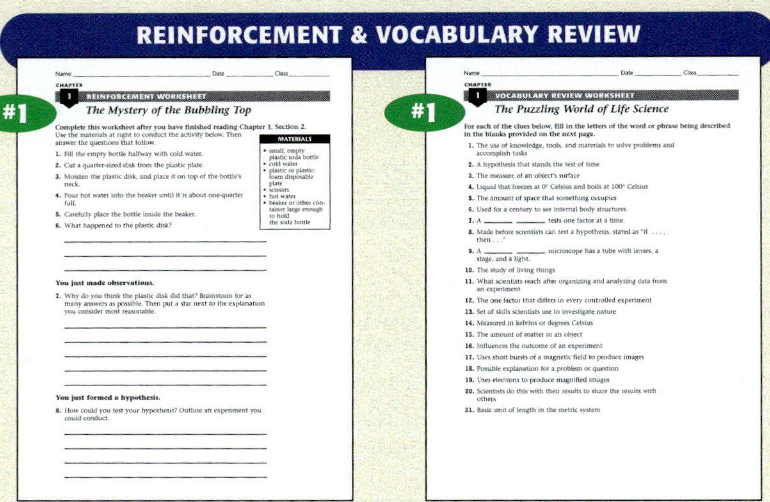

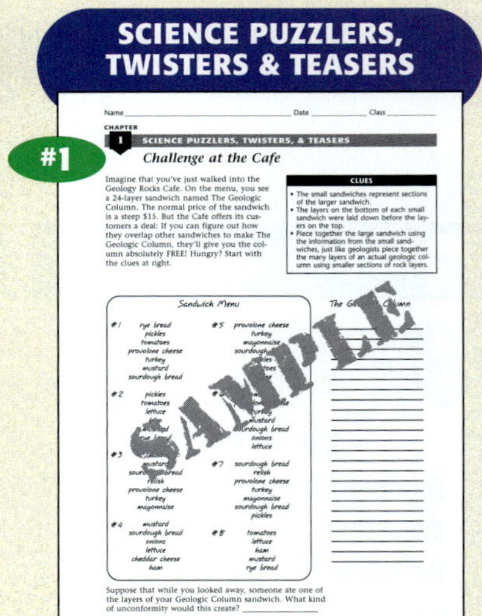

Review & Assessment

STUDY GUIDE

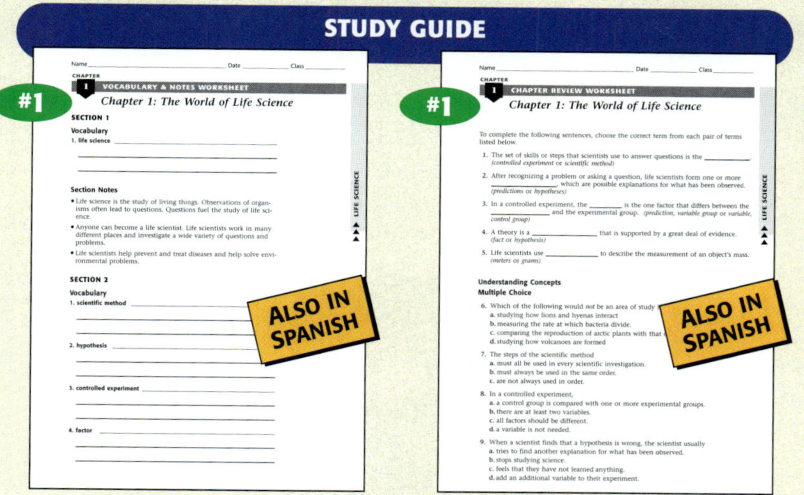

#1 (Vocabulary & Notes Worksheet — Chapter 1: The World of Life Science)

#1 (Chapter Review Worksheet — Chapter 1: The World of Life Science)

ALSO IN SPANISH

CHAPTER TESTS WITH PERFORMANCE-BASED ASSESSMENT

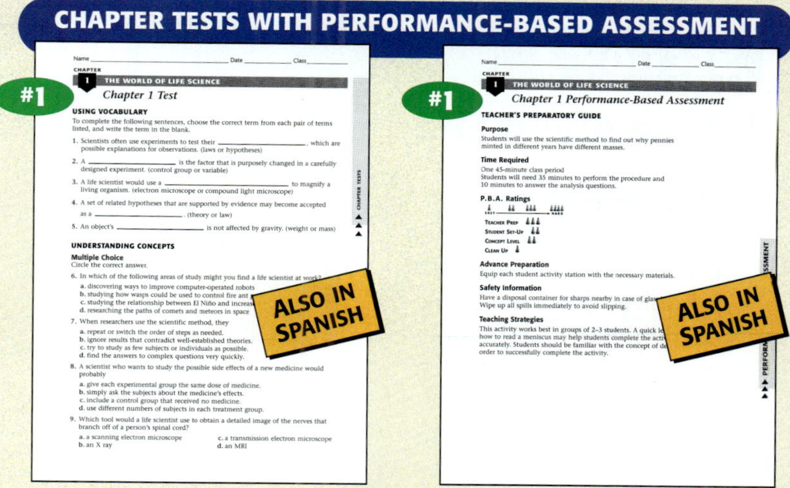

#1 (Chapter 1 Test — The World of Life Science)

#1 (Chapter 1 Performance-Based Assessment — The World of Life Science)

ALSO IN SPANISH

Lab Worksheets

INQUIRY LABS

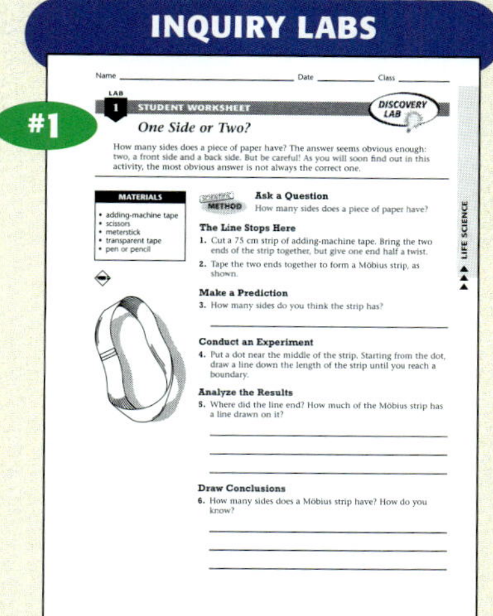

#1 (Student Worksheet — One Side or Two?)

WHIZ-BANG DEMONSTRATIONS

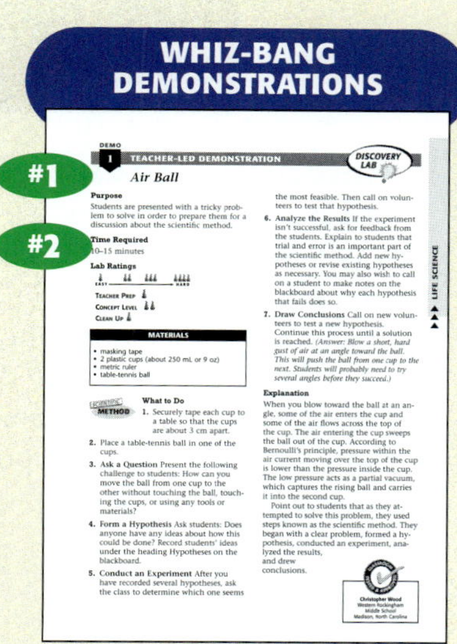

#1 #2 (Teacher-Led Demonstration — Air Ball)

LONG-TERM PROJECTS & RESEARCH IDEAS

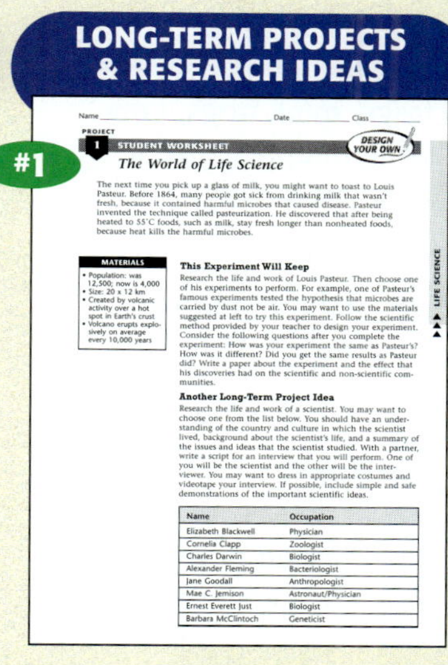

#1 (Student Worksheet — The World of Life Science)

DATASHEETS FOR LABBOOK

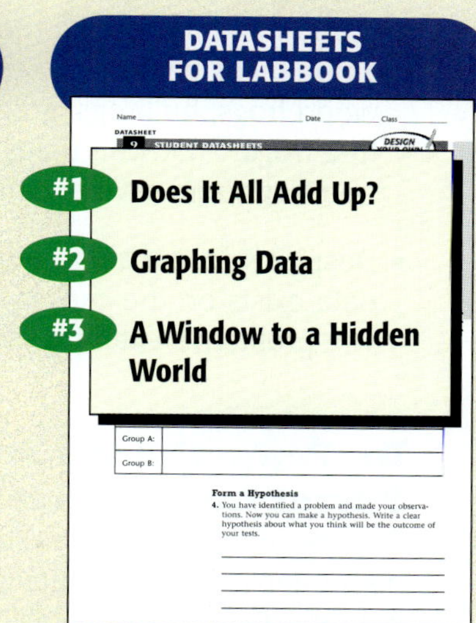

Applications & Extensions

CRITICAL THINKING & PROBLEM SOLVING

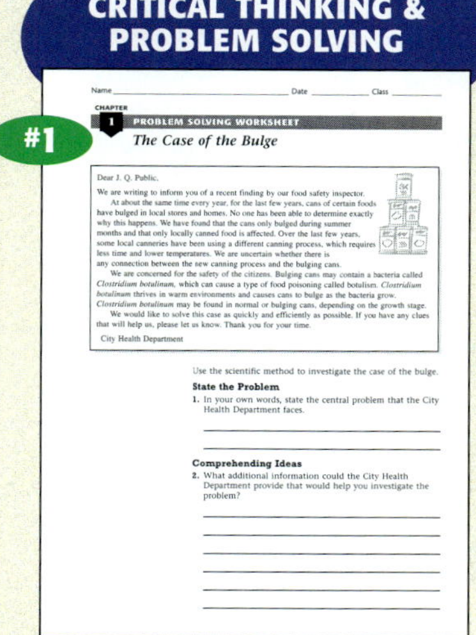

#1 (Problem Solving Worksheet — The Case of the Bulge)

MULTICULTURAL CONNECTIONS

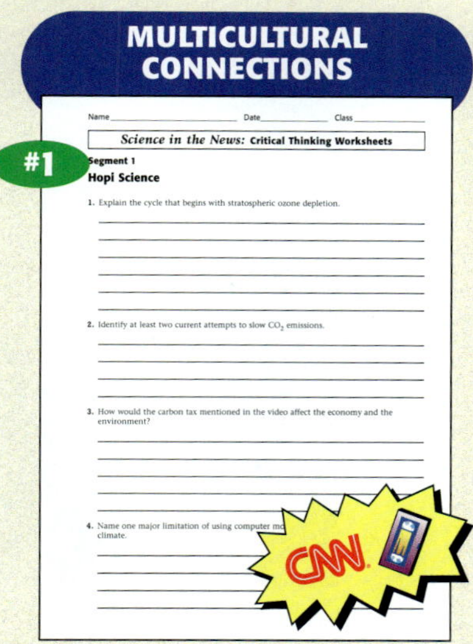

#1 (Science in the News: Critical Thinking Worksheets — Segment 1: Hopi Science)

SCIENTISTS IN ACTION

#1 (Science in the News: Critical Thinking Worksheets — Segment 1: A Biologist's Dolphin Investigation)

INTERACTIVE EXPLORATIONS

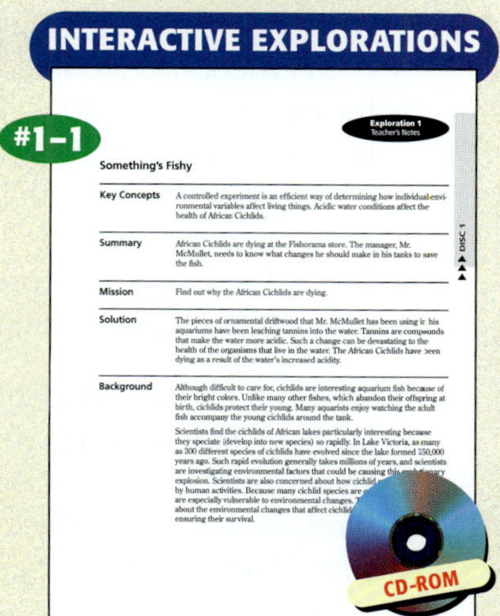

#1–1 (Exploration 1 — Something's Fishy)

CD-ROM

SECTION 1

Asking About Life

▶ Deformed Frogs

The discovery of deformed frogs by Minnesota middle-school students in 1995 sparked much attention around the country. Since that summer, reports of amphibian malformations have poured into agencies from many parts of the continent.

- Besides northern leopard frogs, the type of frog found in Minnesota by the students, other deformed amphibians have been reported, including deformed wood frogs, bullfrogs, green frogs, mink frogs, gray tree frogs, American toads, spring peepers, Pacific tree frogs, and several types of salamanders.

- Despite a recent surge of research, little solid data exists on the causes of the malformations. The problem's extent is also not known.

▶ Dr. Pepperberg's Studies on Parrot Intelligence

Parrots, or psittacids, are rarely mentioned during discussions of animal intelligence, but recent studies indicate that they are intelligent animals. Dr. Irene Pepperberg, an associate professor at the University of Arizona's Department of Ecology and Evolutionary Biology, has demonstrated that African grey parrots can process information and make decisions.

- Pepperberg has studied Alex, an African grey parrot, for more than 20 years. Pepperberg says that she has used a variety of techniques to establish a form of interspecies communication with Alex. "The existence of such behavior," she says, "demonstrates that at least one avian species is capable of interactive, referential communication."

- Alex can count and identify more than 35 objects, such as paper, a key, wood, and grain; recognize seven colors; identify five different shapes; and combine names to identify, request, refuse, and categorize more than 100 objects; and even learned to boss around lab assistants in order to modify his environment.

- Pepperberg and her assistants are busy teaching Alex phonics, and evidence suggests that Alex's communications skills will improve to the point that he may one day be able to read.

▶ Scientists Seek to Improve Polio Vaccines

The Salk and Sabin polio vaccines have nearly eliminated from the Americas the virus that causes poliomyelitis. Although the disease persists in some corners of the world, the World Health Organization is working to ensure that poliomyelitis is eventually eradicated. This would make polio the second disease in history to be erased by vaccination. The first was smallpox.

IS THAT A FACT!

- Only about 430 Siberian, or Amur, tigers still exist in the wild. Most of them can be found in eastern Russia. Another 490 or so Siberian tigers live in captivity.

SECTION 2

Thinking Like a Life Scientist

▶ Vanishing Amphibians

Scientists are perplexed by the steady decline in the world's amphibian population since the mid-1980s. Although population reductions or eliminations in the past were easily explained by the activities of people in specific areas, populations have been dropping even in protected wilderness areas. What's the cause?

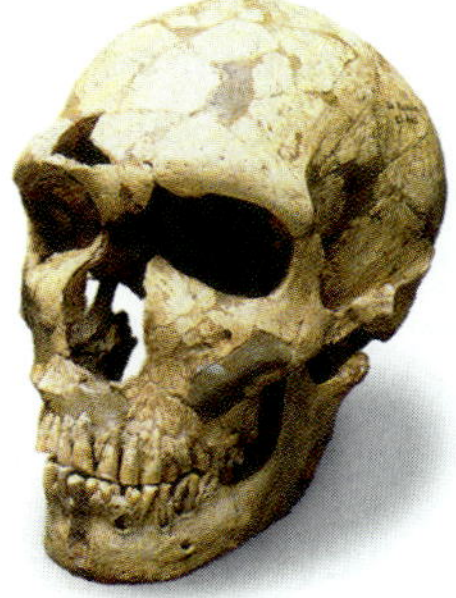

- Some scientists point to the thinning of the ozone layer, which has resulted in an increase of ionizing radiation (UV-B). A second possible cause is pollution by chemicals, either by acid rain or by fertilizers and pesticides. Two other possible causes are the introduction of nonnative competitors and predators into amphibian environments and attacks from new or existing pathogens.

- Frog populations in Australia and Central America have been decimated by outbreaks of chytrid skin fungi. Recent studies have suggested that the fungus may be responsible for mysterious frog deaths in the United States also.

▶ Neanderthal DNA Study

Some scientists believe that Neanderthals are our ancestors; some don't. Researchers who analyzed DNA extracted from a Neanderthal tooth discovered a significant difference between the mitochondrial DNA of *Homo sapiens* and that of Neanderthals. This supports the hypothesis that Neanderthals are not our ancestors. Some scientists, however, believe that the conclusion is a bit premature. They point out that the DNA findings were based on a very small sequence of DNA from one individual and that the data are still compatible with what is known about Neanderthal ancestry. The debate goes on.

SECTION 3

Tools of Life Scientists

▶ The Amazing Electron

The electron microscope, particle accelerators, nuclear energy, radar, and lasers are technological developments that resulted from the discovery of the electron by physicist Joseph John Thomson in 1897. Around the home, technological marvels, such as the microwave oven and the television, owe their beginnings to Thomson.

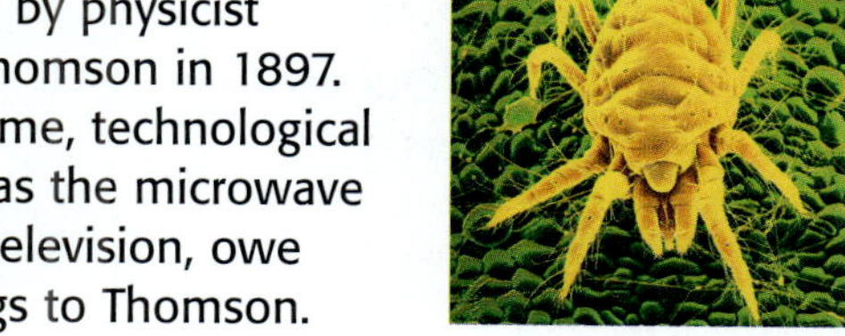

- Scientists still do not understand why the electron appears not to take up any space, even though it has mass.

▶ How Does MRI Work?

Magnetic resonance imaging (MRI) utilizes large magnets, radio-frequency signals, and computers to capture images of internal human structures. When a body is placed in a magnet, hydrogen protons in the body (the body is mostly water) align themselves with the magnetic field. A radio-frequency signal is then transmitted through the body. An interaction between the newly aligned protons and the radio-frequency signal produces a new signal, which is then received by a computer. The computer uses the data to produce detailed magnetic resonance images.

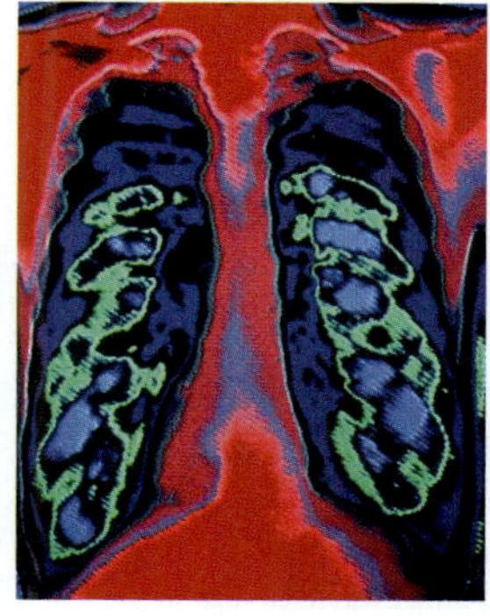

> **For background information about teaching strategies and issues, refer to the *Professional Reference for Teachers*.**

Chapter Preview

Section 1
Asking About Life
- It All Starts with a Question
- Looking for Answers
- Why Ask Why?

Section 2
Thinking Like a Life Scientist
- The Scientific Method
- Scientific Knowledge Changes

Section 3
Tools of Life Scientists
- Tools for Seeing
- Computers
- Systems of Measurement
- Safety Rules!

Directed Reading Worksheet 1

Science Puzzlers, Twisters & Teasers Worksheet 1

Guided Reading Audio CD
English or Spanish, Chapter 1

CAPÍTULO

1

El mundo de las ciencias biológicas

Imagínate. . .

Vas caminando por el campo con unos compañeros y, de pronto, ¡te das cuenta de que hay ranas saltando por todos lados! Tus compañeros y tú se ponen a atraparlas con una red. Cuando sacas la primera rana de la red, descubres algo raro: parece que tiene las patas rotas. Miras la rana que tiene tu amiga y también parece estar herida. ¡Y así están todas las demás! Te acercas un poco más y ves que una de las ranas no tiene ojos. Un momento… Estas ranas no están heridas, ¡están deformes! ¿Serán ranas extraterrestres? ¿O una especie nueva?

Aunque parezca imposible, esto de verdad le sucedió a un grupo de estudiantes de Le Sueur, Minnesota, mientras visitaban una reserva natural. Casi la mitad de las ranas que reunieron estaban deformes. Los estudiantes y su maestra estaban impresionados por su descubrimiento. ¿Qué causaría esas deformidades? ¿Podría ser un extraño fenómeno natural o el resultado de la exposición de las ranas a alguna substancia química creada por los seres humanos? Los estudiantes reunieron más información sobre las ranas y les avisaron a los científicos locales sobre su hallazgo. Ahora, estudiantes y científicos de todo el país se han unido para resolver el misterio de las ranas raras.

Los estudiantes, como harían la mayoría de los científicos, comenzaron su investigación observando algo en la naturaleza y haciéndose preguntas sobre ello. En este capítulo aprenderás cómo las preguntas estimulan el estudio de las ciencias y cómo los científicos encuentran las respuestas.

Imagine . . .

The mystery of the freaky frogs began during a field trip to a wildlife refuge in the summer of 1995. Middle-school students discovered that about half the northern leopard frogs they caught that day were malformed. Since that day, reports of amphibian malformations have become increasingly common in many other parts of North America. In Minnesota alone, scientists from the Minnesota Pollution Control Agency confirmed deformities in 27 counties in 1997.

Usa tus conocimientos para responder a las siguientes preguntas en tu cuaderno de ciencias:

1. ¿Qué herramientas usan los científicos de las ciencias biológicas?

2. ¿Qué métodos siguen los científicos para estudiar las ciencias biológicas?

3. ¿Puede cualquiera ser un científico?

Resuélvelo

En esta actividad probarás tu habilidad para realizar algunas actividades que los científicos realizan: hacer observaciones y utilizarlas para resolver un acertijo.

Procedimiento 

1. Dobla una **hoja cuadrada de papel** como se muestra en la siguiente figura. Primero, dobla el cuadrado por la mitad. Dóblalo de nuevo por la mitad. Después, dobla las esquinas de A hacia B y de A hacia C. Desdobla la hoja. Con unas **tijeras,** corta por los dobleces indicados por las líneas obscuras en la siguiente figura. Así, tendrás cinco figuras geométricas.

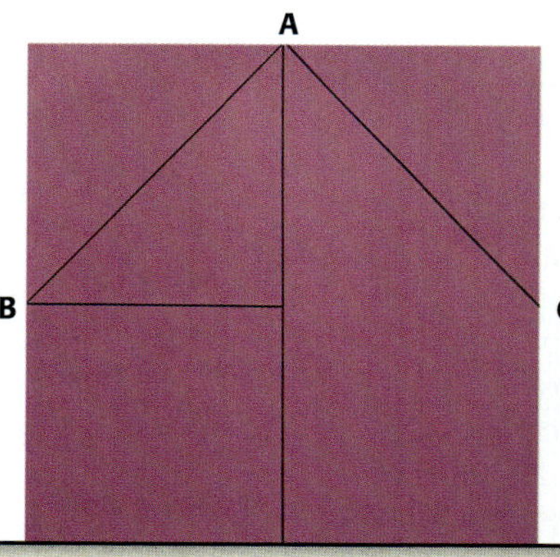

2. Observa el dibujo del pez que se muestra abajo. Trata de imaginar cómo ordenar las cinco figuras pequeñas para formar el pez. Prueba tu idea e inténtalo varias veces si es necesario. Pista: Es posible que tengas que darle vuelta a una o más figuras. Anota la solución en tu cuaderno de ciencias.

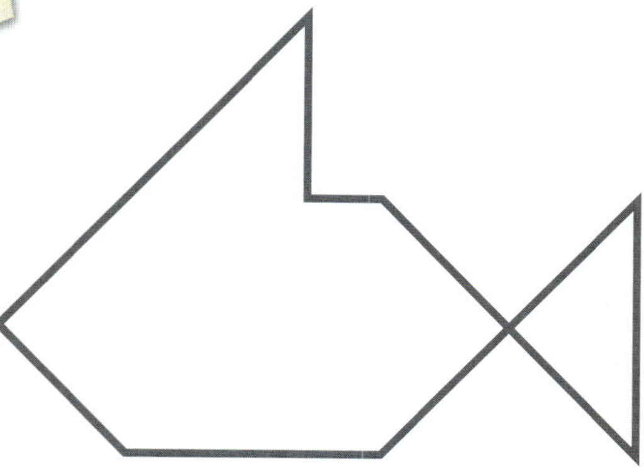

Análisis

3. ¿Pudiste ordenar correctamente las figuras sólo mediante la observación? ¿Qué observaciones te sirvieron de pista?

4. ¿De qué te sirvió poner a prueba tus ideas para resolver el acertijo?

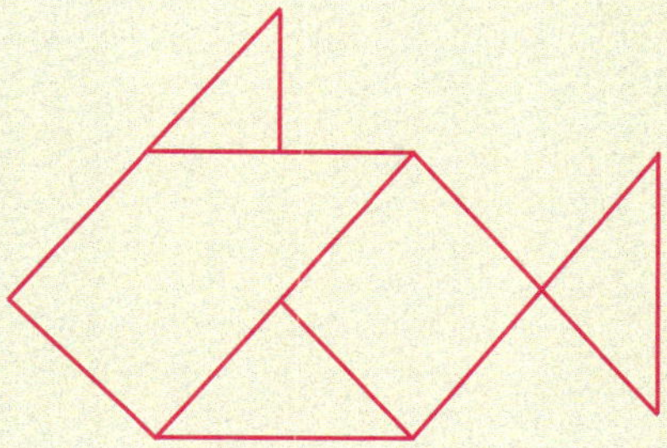

Focus

Asking About Life

This section provides students with an introduction to the life sciences. Students discover that the first step in learning about the living world around us is to ask plenty of questions. Students also learn about various areas of study in the life sciences and meet several scientists who are conducting important studies in their fields.

Bellringer

Have students write five questions about the natural world in their ScienceLog. (Examples might be: How do homing pigeons find their way home? How do plants grow new leaves? Why are dinosaurs extinct?)

Ask several students to share their questions with the class.

1 Motivate

COOPERATIVE LEARNING

Divide the class into groups of four or five. Have the groups brainstorm about what types of life scientists (botanists, bacteriologists, physiologists, etc.) might be working in their community. Have each group choose a professional in the Life Sciences field to interview by phone or E-mail. Encourage them to consider health-care workers, physicians, professors, environmentalists, science teachers, and agricultural experts, as well as the more traditional zoologists, botanists, microbiologists, and so on. Students can present reports and perhaps invite a scientist to visit the class.

VOCABULARIO
ciencias biológicas

OBJETIVOS
- Explica la importancia de formular preguntas en las ciencias biológicas.
- Da tres razones del porqué las ciencias biológicas son beneficiosas para los seres vivos.

Preguntarse sobre la vida

Es verano. Estás recostado sobre el pasto en el parque, observando tu entorno con indiferencia. Hay tres perros jugando a tu izquierda y un pajarito está posado en el árbol a tu derecha. Algunos abejorros vuelan aldredor de las flores cercanas mientras una hormiga se roba una miga de tu emparedado. De repente, se te ocurre una pregunta: "¿Por qué las hormigas no crecen tanto como las personas?" Luego se te ocurre otra pregunta: "¿Por qué los abejorros visitan las flores púrpuras y no las blancas?" Finalmente te preguntas: "¿Por qué no habrá perros azules?"

¡Felicidades! Has dado los primeros pasos para convertirte en un científico de las ciencias biológicas. ¿Cómo lo hiciste? Observaste el mundo viviente a tu alrededor y, por curiosidad, formulaste preguntas sobre tus observaciones. Después de todo, de eso se tratan las ciencias. Las **ciencias biológicas** son el estudio de los seres vivos.

Todo empieza con una pregunta

El mundo a tu alrededor está lleno de vida. En los estanques flotan algas unicelulares invisibles. Árboles de secoya gigantes parecen tocar el cielo. Ballenas de cuarenta toneladas nadan en los océanos. Dentro de tus intestinos hay bacterias diminutas alimentándose. ¿Habrías podido inventar todos los seres vivos u organismos que existen en la Tierra? Podrías hacer una infinidad de preguntas sobre cada organismo que ha existido, por ejemplo: "¿Cómo se alimenta?" "¿Dónde vive?" y "¿Por qué se comporta de una forma particular?"

En tu propio jardín Es fácil encontrar preguntas. Observa tu dormitorio, tu hogar y tu vecindario. ¿Qué preguntas sobre las ciencias biológicas se te ocurren? El muchacho de la izquierda no tuvo que ir muy lejos para darse cuenta de que tenía preguntas sobre algunos organismos muy conocidos. ¿Sabes la respuesta a alguna de ellas?

Recorrer el mundo Tu vecindario te puede dar una idea de algunas de las interrogantes que existen en el mundo. El mundo tiene muchos lugares diferentes para vivir, por ejemplo, desiertos, océanos, bosques, arrecifes de coral y charcas de marea. En casi todos los rincones de la Tierra encontrarás algún tipo de organismo vivo.

Q: What's the difference between a friendly dog and a marine biologist?

A: One's a tail wagger; the other's a whale tagger.

Buscar respuestas

Cierra los ojos por un momento e imagínate a un científico de las ciencias biológicas. ¿Qué ves? ¿Es alguien en un laboratorio observando a través de un microscopio? ¿Qué personas de la **Figura 1** parecen científicos?

Figura 1 *Los científicos son personas que formulan diversos tipos de preguntas.*

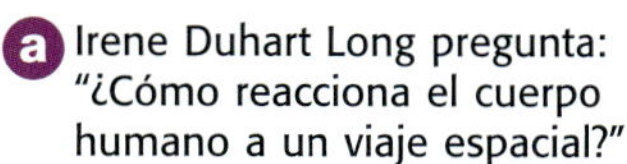

(a) Irene Duhart Long pregunta: "¿Cómo reacciona el cuerpo humano a un viaje espacial?"

(b) Geerat Vermeij, quien es ciego, pregunta: "¿Cómo han cambiado las conchas con el tiempo?"

(c) Irene Pepperberg pregunta: "¿Pueden los loros aprender un lenguaje?"

¿Quién? Si pensaste que todas las personas de la Figura 1 son científicos, acertaste. Cualquier persona puede investigar el mundo que nos rodea. Hombres y mujeres de distinta cultura y origen étnico pueden ser científicos.

¿Dónde? La investigación en el laboratorio es una parte importante de las ciencias biológicas, pero éstas también pueden estudiarse en muchos otros lugares. Los científicos llevan a cabo investigaciones en granjas, bosques, el fondo del mar e incluso en el espacio. Trabajan para empresas, hospitales, organismos gubernamentales y universidades. Muchos de ellos también son maestros.

¿Qué? La curiosidad de un científico es lo que determina qué investigará. Los científicos se especializan en diversas áreas de las ciencias biológicas. Estudian cómo funcionan y se comportan los organismos y cómo éstos interactúan entre ellos y con su medio ambiente. Estos científicos exploran cómo se reproducen los organismos y cómo transmiten características de una generación a otra. Algunos científicos de las ciencias biológicas investigan los orígenes de los organismos y cómo éstos cambian con el tiempo.

¿Te gustaría inventar una profesión que investigue los lagartos? Pasa a la página 32 para leer sobre alguien que lo hizo.

IS THAT A FACT!

Although the word *science* has been used in English since the 1300s, the term *scientist* wasn't coined until about 1840. Investigators such as Benjamin Franklin and Chevalier de Lamarck, therefore, were not known as scientists until long after their deaths.

🌐 Multicultural CONNECTION

Shisaburo Kitasato, born in 1852, was a Japanese life scientist who searched for bacteria and found ways to fight diseases. Kitasato was one of the first scientists to discover the bacteria for tetanus, diphtheria, and the bubonic plague. A graduate of the University of Tokyo Medical School, Kitasato accomplished what many thought was impossible: He developed a procedure to grow pure tetanus bacteria. This enabled him to develop effective treatments for tetanus infections and led him to discover the techniques and materials he would use to fight diphtheria and the plague.

REAL-WORLD CONNECTION

Dr. Neal Haskell is a forensic entomologist—a scientist trained in getting information about crimes from insects. Insects develop and grow at very regular rates. When Dr. Haskell finds particular larvae on a corpse, he can calculate exactly how long ago a person must have died. When a man was murdered and left in a junk pile in Oklahoma, Dr. Haskell was asked to establish the time of death using photographs, case reports, and a few vials of fly larvae. Using this information, Dr. Haskell determined exactly when the murder took place. This information was crucial to the team that solved the crime.

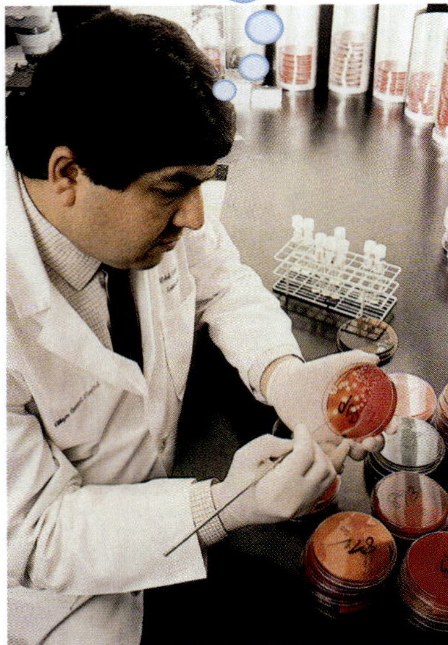

Figura 2 *Abdul Lakhani estudia el SIDA para encontrar una cura para la enfermedad.*

8

To help students understand how pollution is spread from one area to the next, squeeze or peel a lemon or an orange in front of the class. Tell students to raise their hand when they can smell the fragrance.
`Sheltered English`

¿Por qué hay que preguntar por qué?

¿Cuál es el objetivo de formular todas estas preguntas? Los científicos encuentran respuestas interesantes pero, ¿en realidad importan estas respuestas? ¿Afectarán *tu* vida? ¡Por supuesto! Al estudiar las ciencias biológicas, verás cómo te afectan a ti y a lo que existe a tu alrededor.

Combatir enfermedades La poliomielitis es una enfermedad que afecta el cerebro y los nervios, y que causa parálisis. ¿Conoces a alguien que haya tenido poliomielitis? Probablemente no. El virus de la poliomielitis se ha eliminado en muchas regiones del mundo. Pero hace algún tiempo era mucho más común. En la década de 1950, antes de que los científicos descubrieran formas de prevenir el virus de la poliomielitis, éste infectó a 1 de cada 3,000 estadounidenses.

Hoy en día, los científicos siguen investigando formas de combatir enfermedades que nos han atacado por mucho tiempo, como la tuberculosis, así como también enfermedades que han aparecido recientemente, como el síndrome de inmunodeficiencia adquirida, o SIDA. El científico de la **Figura 2** trata de aprender más sobre el SIDA, enfermedad mortal para miles de personas cada año. Los científicos de las ciencias biológicas han descubierto cómo se transmite el virus del SIDA de una persona a otra y cómo afecta al cuerpo. Al aprender más sobre el SIDA, los científicos podrían encontrar una cura para esta enfermedad mortal.

Algunas enfermedades, como la fibrosis cística, son hereditarias, es decir, se transmiten de padres a hijos. Susumu Tonegawa, en la **Figura 3,** es uno de los muchos científicos del mundo que estudian el genoma humano. El genoma de un organismo es heredado de los padres. Es información contenida en las células de un organismo y controla las actividades de las células. Los errores en algunas partes del genoma de un organismo pueden provocar que éste nazca con ciertas enfermedades o que las desarrolle con el tiempo. Al aprender sobre el genoma humano, los científicos esperan encontrar formas de curar o prevenir enfermedades hereditarias.

Figura 3 *El trabajo de Susumu Tonegawa puede ayudar a combatir las enfermedades hereditarias.*

Proteger el medio ambiente ¿Conoces algunos problemas del medio ambiente? La mayoría se debe al mal uso y a la eliminación inadecuada de los recursos naturales por parte de las personas. El primer paso para solucionar los problemas de contaminación, deforestación y extinción de la vida silvestre es entender cómo afectamos el mundo que nos rodea.

¿Por qué hay que disminuir la contaminación del aire y del agua? La contaminación afecta nuestra salud y la de los animales y las plantas. La contaminación del agua podría ser una de las causas de las deformidades de las ranas observadas en Minnesota y en otros estados. La contaminación por petróleo en los océanos mata mamíferos marinos, aves y peces. Si encontramos formas de disminuir la cantidad de contaminación que producimos, haremos que nuestro mundo sea más saludable.

Cuando cortamos árboles para cultivar la tierra o para construir casas, alteramos y destruimos el hábitat de otras criaturas. Dale Miquelle, en la **Figura 4,** pertenece a un equipo de científicos rusos y estadounidenses que estudian el tigre siberiano. La caza y la desforestación han estado a punto de causar su extinción. Al aprender sobre las necesidades de los tigres, los científicos desarrollan un plan de conservación para asegurar su supervivencia.

Figura 4 *Para aprender cuánto territorio abarca un tigre siberiano, Dale Miquelle rastrea un tigre que tiene un collar radiotransmisor.*

REPASO

1. Describe las ciencias biológicas.

2. ¿Qué beneficios pueden aportar las ciencias biológicas a los seres vivos?

3. ¿Dónde trabajan los científicos de las ciencias biológicas? ¿Qué estudian?

4. ¿Estás de acuerdo con la siguiente afirmación? *Los conocimientos de los profesionales de las ciencias biológicas no son muy importantes.* Explica tu respuesta.

5. **Aplicar** Observa la escena de la derecha y haz cinco preguntas acerca de los organismos que ves. Averigua si alguno de tus compañeros o compañeras sabe las respuestas. Anota las preguntas y respuestas en tu cuaderno de ciencias.

3) Extend

Homework

Writing Have students create an imaginary animal and an environment in which it can live. Then have them write a story about it. Students should consider what the animal will eat, which other animals and plants will live there, and how the lifeforms will coexist.

4) Close

Quiz

1. Who can be a life scientist?
   (Anyone from any background can learn to be a life scientist.)

2. Why is polio no longer a significant health concern?
   (Life scientists developed a vaccine.)

ALTERNATIVE ASSESSMENT

Writing Encourage students to develop ideas about what they can do to help preserve animal habitats. Encourage them to contact local nature organizations to find out if there are specific things they can do to promote animal preservation in their area. Have the students write a report and prepare a brief presentation on what they learned.

Answers to Review

1. Life science is the study of living things and their relationships to each other and to their environment.

2. combating disease and protecting the environment

3. Life scientists work in laboratories and offices, outdoors, underwater, and in space. Life scientists study anything dealing with living things.

4. Answers may vary but should reflect an understanding that knowledge of living things has a direct impact on the students' own lives.

5. Questions and answers will vary. Example questions are: What does the lion eat? Why aren't the other animals running away?

Focus

Thinking Like a Life Scientist

This section introduces the scientific method. Students see how the method's six steps are used in an actual investigation of deformed frogs. Students will also discover how new technologies lead to new answers about old questions.

Bellringer

As an exercise in observation, write 15–20 words on the overhead projector. Allow the students to look at the words for 10 seconds. Turn the projector off, and have the students spend a few minutes writing down as many of the words as they can remember in their ScienceLog.

1) Motivate

DEBATE

Write this question on the board:

"Which is more important, imagination or knowledge?"

Ask students to debate this question. Stress to them that creativity, flexible and original thinking, and openness to new and even wild ideas are important qualities for a good scientist to possess.

Teaching Transparency 1 "The Scientific Method"

Directed Reading Worksheet 1 Section 2

VOCABULARIO

método científico   teoría
hipótesis   tecnología
factor
experimento controlado
variable

OBJETIVOS

- Describe el método científico.
- Evalúa los diseños de experimentos.
- Interpreta la información de tablas y gráficas.
- Explica cómo puede cambiar el conocimiento científico.

Pensar como un científico de las ciencias biológicas

Sin importar dónde trabajen los científicos de las ciencias biológicas o qué preguntas traten de responder, todos tienen dos cosas en común: sienten curiosidad por la naturaleza y usan métodos similares para investigarla. Imagina que eres uno de los estudiantes que descubrieron las ranas deformes de las que hablamos al principio de este capítulo. Si quisieras investigar la causa de las deformidades, ¿por dónde comenzarías? En realidad, ya has dado el primer paso del método científico. Has observado y has hecho preguntas. ¿Qué sigue?

El método científico

La base de las investigaciones de la mayoría de los científicos es el método científico. El **método científico** es una serie de pasos que se utilizan para responder una pregunta o resolver un problema. La **Figura 5** resume los seis pasos que lo componen.

- **Haz una pregunta** basada en observaciones.
- **Formula una hipótesis** que sea una explicación posible de lo que has observado.
- **Comprueba la hipótesis** mediante experimentos.
- **Analiza los resultados** reunidos a partir de los experimentos.
- **Saca conclusiones** basadas en los resultados de tu experimento.
- **Comunícale los resultados** a otros científicos.

Figura 5 *Los científicos generalmente usan el método científico para resolver problemas y responder preguntas.*

Las ciencias son un proceso creativo Después de leer los pasos del método científico, es posible que pienses que si sigues cada paso en orden encontrarás automáticamente la respuesta correcta a tu pregunta o problema. Pero muchas preguntas no se pueden responder tan fácilmente. Los científicos de las ciencias biológicas deben usar su imaginación para obtener explicaciones sobre lo que observaron y para diseñar experimentos que comprueben sus explicaciones. Para encontrar soluciones, a veces tienen que repetir los pasos del método científico o aplicarlos en un orden diferente. Además, no todas las preguntas necesitan otro experimento. A veces sólo se necesitan más observaciones para encontrar una respuesta y otras veces no se puede encontrar ninguna respuesta.

En las próximas páginas volverás a estudiar el misterio de las ranas raras para ver cómo se siguieron los seis pasos del método científico en una investigación real.

10

IS THAT A FACT!

North American bullfrogs use an amazing method to make their distinctive call. The sound travels up their throat and blasts out from their large, flat eardrum.

Hacer una pregunta ¿Has observado alguna vez algo fuera de lo normal o que no tenga una explicación simple? Las observaciones suelen generar preguntas. Si eres un científico, las preguntas exigen respuestas. Esto a menudo implica observar más. Cuando los estudiantes de Le Sueur se dieron cuenta de que algo les pasaba a las ranas, decidieron que continuarían sus observaciones recopilando más información. Comenzaron a observar y anotar sus observaciones, como se muestra en la **Figura 6.** Contaron cuántas ranas deformes y cuántas normales habían atrapado. Describieron las deformidades y otras características de las ranas, como el color. También tomaron fotografías, hicieron mediciones y dieron una descripción precisa de cada rana.

Figura 6 *Recopilar información fue el primer paso en la investigación de los estudiantes.*

Además de la información sobre las ranas, los estudiantes recopilaron información sobre los otros organismos del estanque. También le hicieron muchas pruebas al agua: midieron el nivel de acidez y anotaron cuidadosamente la información y las observaciones. Después de todo, eran científicos haciendo su trabajo.

Las observaciones pueden ser de varios tipos, por ejemplo, mediciones de longitud, volumen, temperatura, tiempo o velocidad. Pueden describir el volumen de un sonido o el color o forma de un organismo. Pueden indicar la cantidad de organismos de un área, su estado de salud o su patrón de comportamiento. El tipo de observaciones que los científicos pueden hacer es infinito. Pero sin importar lo que indiquen, las observaciones son útiles si se hacen y registran en forma precisa. Algunas de las herramientas que los científicos usan para observar se muestran en la **Figura 7.**

Figura 7 *Los científicos usan herramientas como microscopios, reglas y termómetros para recopilar información.*

READING STRATEGY

Mnemonics Have students develop a mnemonic device that will remind them of the six steps in the scientific method: **A**sk a question, **H**ypothesize, **T**est, **A**nalyze results, **D**raw conclusions, and **C**ommunicate results. An example is, "**A**nn **H**as **T**wenty **A**ngry **D**ogs and **C**ats."

CROSS-DISCIPLINARY FOCUS

Social Studies To help students become aware of how science and technology affect their lives, have them work together to develop a timeline of discoveries that have occurred during their lifetime. Divide the class into small groups, and assign each group a year. Have them use library resources to research their year. Each student will be responsible for selecting two events and preparing a notecard for each event. The cards should include the year, what the discovery was, and a few descriptive sentences about the discovery and its significance. Have several volunteers assemble the timeline on a bulletin board. Several other volunteers could then be called on to illustrate the timeline.

Does It All Add Up?

WEIRD SCIENCE

In 1786, Luigi Galvani noted that frog legs jerked when he touched them with a metal probe. When he touched a nerve with the probe, the nerve, the metal, and the liquid around the nerve created a small electrical current, which stimulated the nerve. The nerve in turn stimulated the muscle. By observing the physiological result of an electrical current, Galvani helped to establish the study of neurophysiology and clinical neurology.

ACTIVITY

Investigate Your Area *(Writing)* Have students observe the daily activities in and around a local pond over a period of several weeks. Tell them to record their observations in a journal.

MEETING INDIVIDUAL NEEDS

Advanced Learners Encourage students to come up with a hypothesis and test it themselves. Questions that might lead to hypotheses they can test include:

- Does soda pop help plants grow?
- Does soft music help plants grow?
- Which degrades faster in compost—natural or human-made fibers?

Students should check the design of their experiment with their teacher before they begin.

DEMONSTRATION

Making Models The northern leopard frog, which inhabits the pond that was studied by the Minnesota students, has quite a peculiar call—a mixture of grunts, snores, and squeaks that sound like a wet palm being rubbed across an inflated balloon. Ask a volunteer to demonstrate the frog's call for the class.

`Sheltered English`

Answer to Self-Check

2. Insecticides and fertilizers caused the frog deformities.

Formular una hipótesis Después de observar, los científicos formulan una o más hipótesis. Una **hipótesis** es una explicación o respuesta posible a una pregunta. Cuando los científicos formulan una hipótesis, piensan en forma lógica y creativa, y consideran lo que ya saben.

Una hipótesis debe ser comprobable mediante experimentos. Una hipótesis no es verificable si no se pueden reunir observaciones o información, o si no se puede diseñar un experimento para comprobarla. Sólo porque una hipótesis no sea comprobable no significa que no sea correcta. Sólo quiere decir que no hay forma de corroborarla o refutarla.

Es posible que diferentes científicos tengan diferentes hipótesis para un mismo problema. En el caso de las ranas, se formularon las hipótesis que ves abajo. ¿Cuál de las explicaciones será la correcta? Para averiguarlo, las hipótesis deben comprobarse.

Hipótesis 1: Las deformidades fueron causadas por uno o más contaminantes químicos del agua.

Hipótesis 2: Las deformidades fueron causadas por ataques de parásitos o de otras ranas.

Hipótesis 3: Las deformidades fueron causadas por una mayor exposición a los rayos ultravioleta del Sol.

✔ Autoevaluación

¿Cuál de las siguientes afirmaciones es una hipótesis?

1. Hay ranas deformes en los Estados Unidos y Canadá.
2. Los insecticidas y fertilizantes causaron las deformidades de las ranas.
3. Las ranas absorben fácilmente los contaminantes a través de la piel.

(Consulta la página 636 para comprobar tus respuestas.)

Homework

Testing Hypotheses Answers to questions in many areas of life can be found by forming and testing a hypothesis. Write a hypothesis based on the following observation:

The blacktop feels hotter than the sidewalk.

Ask students how such a hypothesis might be tested.

Predicciones

Antes de que los científicos puedan comprobar una hipótesis, deben hacer predicciones. Una predicción es una afirmación de causa y efecto que se usa para comprobar una hipótesis. Generalmente se expresan como: "Si…, entonces…" Por ejemplo, se formuló la siguiente predicción para la Hipótesis 3:

> **Hipótesis 3**
>
> Predicción: Si un aumento en la exposición a la luz ultravioleta causa las deformidades, entonces algunos huevos de ranas expuestos a la luz ultravioleta en el laboratorio se convertirán en ranas deformes.

Se puede hacer más de una predicción para cada hipótesis. Después de hacer las predicciones, los científicos pueden diseñar experimentos para determinar si alguna de las predicciones es verdadera y corrobora la hipótesis. En las próximas páginas verás cómo se comprobó la Hipótesis 3.

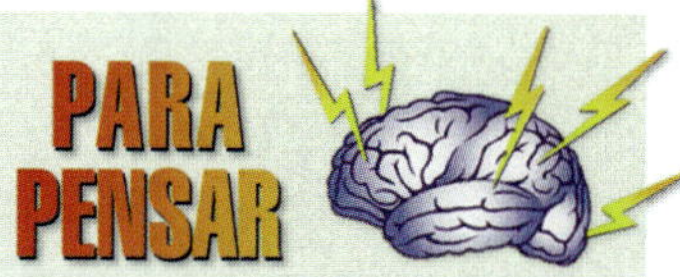

PARA PENSAR

Para cualquier conjunto de hechos hay una cantidad infinita de hipótesis que podrían explicarlos. Con respecto a este tema, el filósofo inglés William de Occam estableció un principio que se conoce como el "Principio de Occam", el cual establece que, casi siempre, la explicación más simple es la mejor.

APLICA

Vas caminando por un parque lleno de árboles con una amiga. De repente, llegan a un área pequeña donde todos los árboles están derribados en el suelo. ¿Qué sería lo que los tumbó o los hizo caer? Tu amiga piensa que fueron extraterrestres. ¿Se puede comprobar su hipótesis? Explica tu respuesta en el cuaderno de ciencias. ¿Qué otras hipótesis se te ocurren?

Answer to APPLY

The hypothesis that extraterrestrials caused the trees to fall is not testable because there is no way to support or disprove the hypothesis. A viable hypothesis is that a volcanic eruption knocked the trees down.

Writing **Researching Galileo**

The Italian astronomer and physicist Galileo (1564–1642) was the first scientist to be credited with using a scientific method to solve problems. He recognized the importance of carefully controlled experiments in developing scientific theories. The steps that he followed became the foundation of the scientific method used by scientists today. Have students research some of Galileo's discoveries and present their findings in an oral report to the class.

MEETING INDIVIDUAL NEEDS

Learners Having Difficulty

Have students use a dictionary to find the definitions and origins of the words *thesis* and *hypothesis*. Have volunteers write the information they find on the board. As an alternative or additional activity, small groups of students could then compile lists of words that contain *hypo-* or *-thesis*—for example, *hypodermic, synthesis,* and *photosynthesis*.

MISCONCEPTION ALERT

Students may believe that an experiment is a failure if their hypothesis is not supported by the data gathered. Remind them that the point of conducting experiments is to investigate the hypothesis and to learn from the data. Often, more can be learned from experiments that do not support a hypothesis than from experiments that do.

RETEACHING

Have students propose other possible experiments that scientists could use to test the effect of UV light on frogs. Ask them if conducting the experiments in the frogs' natural environment would be a good idea. Discuss how such an experiment might be set up.

USING THE TABLE

Have students study the table on this page. Ask them the following questions:

What is the only factor that differs between the control group and the experimental groups? (the variable, which is UV light exposure time)

What would happen if the number of eggs was different for each of the groups? (The experiment would not be controlled; the results would be invalid.)

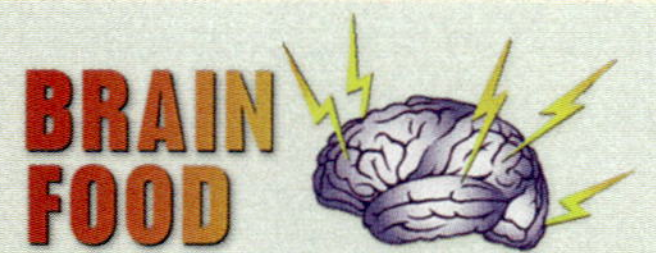
BRAIN FOOD

At the forefront of the deformed-frog situation is this concern:

Is human health at risk? Do the malformed frogs signal a widespread environmental problem?

Discuss this concern with students, and pose this question:

What steps can scientists take to find out whether humans are also at risk?

Comprobar la hipótesis Después de que los científicos hacen una predicción, comprueban la hipótesis. Tratan de diseñar experimentos que demuestren claramente si un factor en particular fue la causa del efecto que se observó. Un **factor** es cualquier cosa que influye en el resultado de un experimento. Los factores pueden ser cualquier cosa, desde la temperatura hasta el tipo de organismo que se estudia.

Los científicos se esfuerzan por realizar experimentos controlados. Un **experimento controlado** prueba sólo un factor a la vez. En este tipo de experimentos hay un grupo de control y uno o más grupos experimentales. Todos los factores para el grupo de control y los grupos experimentales son los mismos, excepto uno. El único factor diferente se llama **variable.** Debido a que la variable es el único factor diferente entre el grupo de control y los experimentales, los científicos pueden estar más seguros de que las variables son las causas de cualquier diferencia observada en el resultado del experimento. El diseño de un buen experimento requiere de mucha reflexión y planificación. Veamos cómo montar un experimento para comprobar la predicción de la Hipótesis 3: *Si un aumento en la exposición a la luz ultravioleta causa las deformidades, entonces algunos huevos de ranas expuestos a la luz ultravioleta en el laboratorio se convertirán en ranas deformes.*

La primera cosa que hay que identificar es la variable. En este caso, la variable es la cantidad de luz ultravioleta (UV) a la que se exponen las ranas. El único factor diferente entre el grupo de control y los experimentales es el tiempo que los huevos de rana están expuestos a la luz UV. Esto se muestra en la siguiente tabla.

Los otros factores, como el tipo de rana, la cantidad de huevos en cada acuario y la temperatura del agua, deben ser iguales tanto en el grupo de control como en los experimentales.

Diseño del experimento para comprobar el efecto de la luz UV en las ranas				
Grupo	**Factores**			
	Tipo de rana	Número de huevos	Temperatura del agua	Variable: exposición a luz UV
#1 Control	rana leopardo	100	25°C	0 días
#2 Experimental	rana leopardo	100	25°C	15 días
#3 Experimental	rana leopardo	100	25°C	24 días

14

Students often assume that worthwhile experiments should be dazzling—something must explode or change color or sizzle with electricity. However, as the frog-egg example shows, most good experiments are not so dramatic. In addition, they often take awhile to yield results. The upside is that anyone can perform experiments; they aren't just for "mad scientists."

Como ves en la tabla, cada grupo del experimento contiene 100 huevos. ¿Para qué incluir tantos individuos en cada grupo? Los científicos siempre tratan de tener muchos individuos en el grupo de control y en el experimental. Mientras más organismos prueban, más seguros están de que las diferencias en el resultado de un experimento realmente se deben a las diferencias en la variable y no a las diferencias naturales entre los individuos. Repetir experimentos es otra forma en que los científicos corroboran sus conclusiones acerca de los efectos de una variable. Si un experimento produce los mismos resultados una y otra vez, los científicos pueden asegurarse del efecto que la variable produce en el resultado del experimento.

En la **Figura 8** se ilustra el experimento que se hizo para comprobar la predicción de la Hipótesis 3.

Figura 8 *Este experimento controlado se diseñó para averiguar si la luz UV provocó las deformidades de las ranas.*

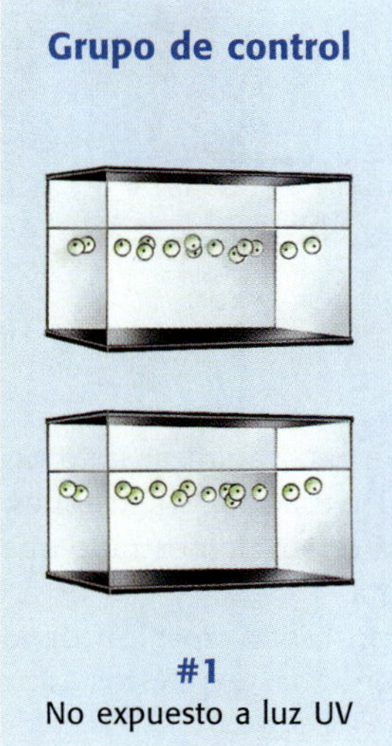

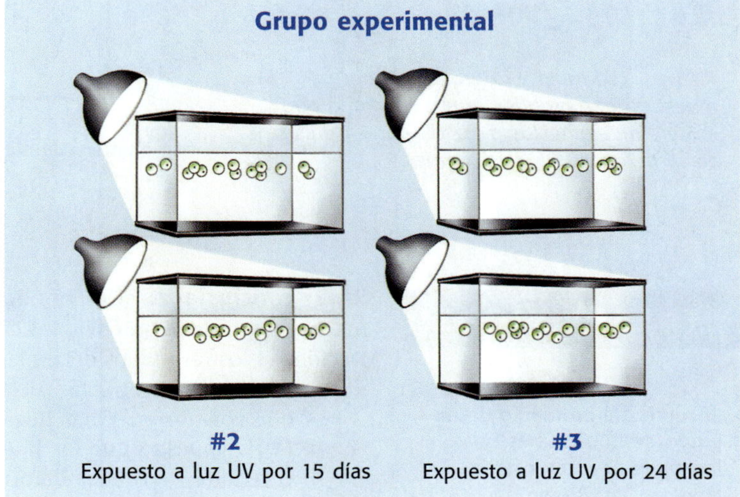

✔ Autoevaluación

Enrique está probando los efectos de diferentes jabones antibacteriales en el desarrollo de bacterias. Utiliza varios frascos de la misma cepa de bacterias. ¿Cuál frasco es el grupo de control?

1. Al frasco A, Enrique le agrega dos gotas de Superjabón.
2. Al frasco B, le agrega dos gotas de Antiburbujas-B.
3. Al frasco C, no le agrega jabón.

(Consulta la página 636 para comprobar tu respuesta.)

DISCUSSION

Experimental Conditions

Discuss the environment of each of the aquariums in the experiment. Ask students if they think that any special precautions would need to be implemented before the frogs are introduced to the aquariums. Consider such factors as flooring material (rocks, grass), the general cleanliness of each aquarium (were they all washed with similar ingredients?), the food that is introduced, and so on. Ask students:

How might the experiment be ruined if the environments varied in any significant way? (A new variable might be accidentally introduced, and the experiment would therefore not be valid.)

MEETING INDIVIDUAL NEEDS

Writing **Learners Having Difficulty** Assemble a group of objects for students to investigate, observe, and take apart if they wish. Possible objects include a flashlight, a pen, and a stapler. After students examine each object, ask them to speculate about how each object works. To get them started, suggest that they make a list or draw a diagram of the different parts in each object. Then they can proceed by writing down the possible function of each part. Ask students to summarize the ways in which they behaved like scientists during this activity.

Answer to Self-Check

Jar C is the control group.

Homework

Concept Mapping Have students research the criminal investigation process and identify the steps in the method police use to solve a crime. Students could create a poster-size concept map to illustrate the similarities and differences between a police investigation and the scientific method.

USING THE FIGURE

Table and Bar Graph Have students carefully compare the data presented in both graphics. Ask students which method of organizing the data is clearer to them? Does one have an advantage over the other? Ask if there are other ways that the scientists could have presented their findings. If students have not already done so, point out that the data could also have been presented in paragraph form. Ask what the possible advantages and disadvantages of a written summary might be.

Homework

Point out to students that according to the data in **Figure 9,** the deformities began quite abruptly after 24 days of exposure to UV light. Ask students to speculate why there were no deformities found before then. Shouldn't a shorter period of exposure have resulted in at least a few mutations? (Some scientists have determined that the end of the 24-day period of exposure corresponds to the same period that leg buds begin to form in tadpoles. No deformities are seen before then because the tadpoles don't form legs until approximately the 24th day.)

Answer to MATHBREAK

$$\frac{(6 + 5 + 4)}{3} = \frac{15}{3} = 5 \text{ days}$$

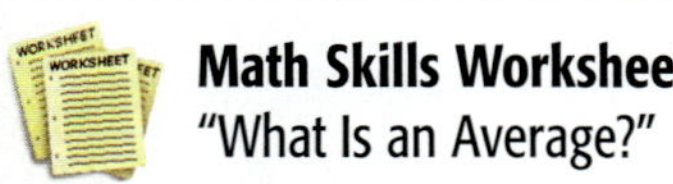

Math Skills Worksheet 7
"What Is an Average?"

Analizar los resultados El trabajo de un científico no termina cuando se completa un experimento. Los científicos deben organizar la información obtenida para poder analizarla. La información se puede organizar de varias formas, por ejemplo en tablas, gráficas o diagramas. La información que se recopiló en el experimento de luz UV se muestra en la **Figura 9.**

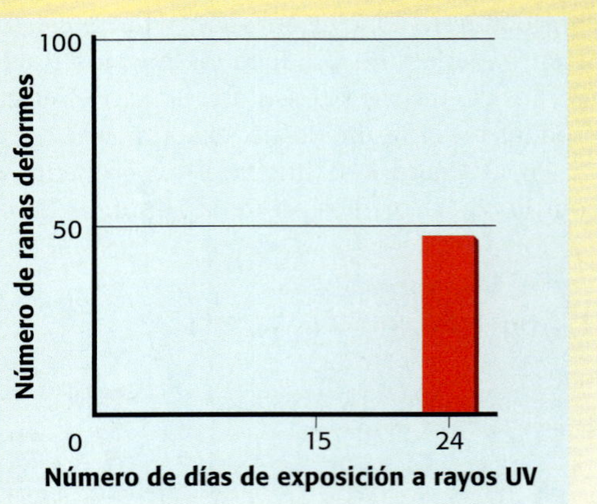

Grupo	Duración de exposición a luz UV	Número de ranas deformes
#1	0 días	0
#2	15 días	0
#3	24 días	47

Figura 9 *La tabla y la gráfica de barras muestran que algunos huevos de rana que estuvieron expuestos a luz UV por 24 días se convirtieron en ranas deformes.*

¡MATEMÁTICAS!

Promedios

Encontrar el promedio de un grupo de números es una forma de analizar información. La Dra. Brown descubrió que 3 semillas a 25°C brotaron en 8, 8 y 5 días. Para encontrar el promedio de días que las semillas tardaron en brotar, sumó 8, 8 y 5 y dividió la suma entre 3, la cantidad de sujetos (semillas) del grupo.

$$\frac{(8 + 8 + 5)}{3} = \frac{21}{3} = 7 \text{ días}$$

La Dra. Brown descubrió que 3 semillas a 30°C tardaron 6, 5 y 4 días en brotar. ¿Cuál es el promedio de días que se tardaron en brotar estas semillas?

Sacar conclusiones Después de organizar y analizar la información de un experimento, los científicos sacan conclusiones. Deciden si los resultados del experimento han demostrado que una predicción era correcta o incorrecta.

¿A qué conclusiones has llegado con la información de la **Figura 9?** ¿Demuestra que los huevos de ranas expuestos a luz UV se convierten en ranas deformes?

Los científicos que llevaron a cabo el experimento concluyeron que las pruebas corroboraron la Hipótesis 3. Efectivamente, la exposición a la luz UV en el laboratorio produjo deformidades. La información también demostró algo más: la cantidad de días que los huevos estuvieron expuestos a luz UV fue un factor importante en el desarrollo de las deformidades.

Cuando los científicos averiguan que una hipótesis no se ha corroborado con las pruebas, deben tratar de encontrar otra explicación para lo que observaron. Comprobar que una hipótesis es incorrecta es tan útil como corroborarla, porque, de cualquier manera, los científicos han aprendido algo. Y ése es el propósito del método científico en la investigación o en la búsqueda de la respuesta a una pregunta.

16

PG 562

Graphing Data

CROSS-DISCIPLINARY FOCUS

Statistics Scientists often use mathematical techniques and formulas to determine whether their data are meaningful or are just the result of chance or coincidence. The discipline called statistics deals with the development of techniques to analyze and interpret data.

El experimento de la luz UV corroboró la hipótesis de que las deformidades de las ranas se deben a la exposición a la luz UV. ¿Significa esto que la luz UV definitivamente causó la deformidad de las ranas que vivían en el pantano de Minnesota? No, no es así. ¿Qué demuestra este experimento acerca de las otras dos hipótesis? ¿Comprueba que las deformidades no se debieron a parásitos u otra substancia del agua del estanque? No, no es así. Lo único que demuestra este experimento es que la luz UV puede ser la causa de las deformidades.

Las otras hipótesis no se han descartado y muchos científicos piensan que dos o más factores pueden estar involucrados. Algunos científicos incluso especulan que los diferentes tipos de deformidades podrían deberse a varios factores.

Un acertijo tan complejo como el misterio de las ranas generalmente no se resuelve con un sólo experimento. La búsqueda de una solución puede seguir por años o hasta décadas. Como ves, una investigación no siempre termina al encontrar una respuesta. A menudo, esa respuesta da origen a otra investigación.

Comunicar los resultados En el mundo actual, los científicos forman parte de una comunidad global. Después de terminar sus investigaciones, hacen un informe y comunican sus resultados a otros científicos, como el estudiante de la **Figura 10**. Quizás otros científicos repitan los experimentos para ver si obtienen los mismos resultados. El informe también puede ayudarles a descubrir otras preguntas y respuestas. Es posible que otras respuestas fortalezcan las teorías científicas o demuestren que deben cambiarse. El paso desde las observaciones y preguntas a la comunicación de los resultados se muestra en la **Figura 11.**

Figura 10 *Este estudiante de ciencias está presentando los resultados de su investigación en una feria de ciencias.*

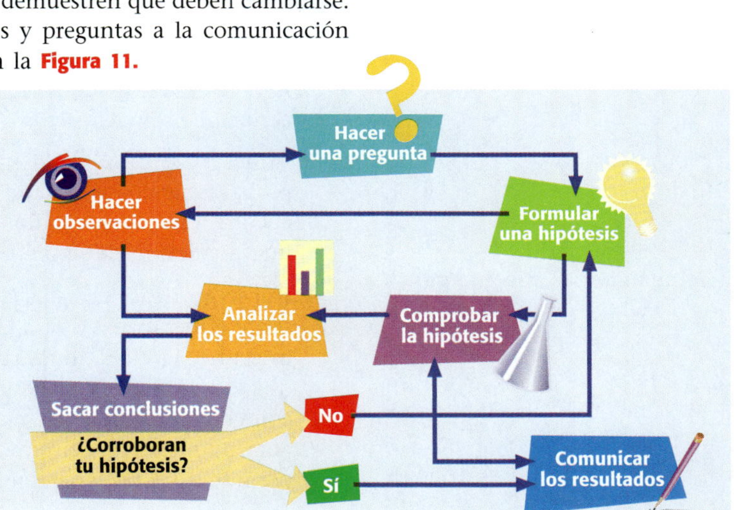

Figura 11 *Las investigaciones científicas no siempre van de un paso al siguiente. A veces los pasos se pueden omitir y otras veces se deben repetir.*

IS THAT A FACT!

Evidence suggests that UV light levels are highest in late spring and early summer, the time of year when Minnesota's frog population is laying eggs. Such evidence points to a need for UV experiments in natural environments at various times of the year.

3) Extend

GOING FURTHER

Have students read articles in scientific or nature magazines and find experiments that illustrate the scientific method. Students might notice that scientists don't always rigidly employ the steps in the same order. Have them outline as many steps as they can in each experiment and copy them into their ScienceLog.

REAL-WORLD CONNECTION

Suggest that students find out what it takes to become a data collector. Some scientific studies require massive collections of data over a huge area or over long time periods. Have students investigate organizations that are searching for sightings or facts about animals in their community. As a result of the Le Sueur students' discovery, a program called "A Thousand Friends of Frogs" was started to encourage students and other citizens to collect data about frogs and toads. The data collected can then be used to help scientists in their research.

MISCONCEPTION ALERT

Students should be reminded about the difference between a scientific theory and a hypothesis. Emphasize that theories need to be supported by a convincing body of evidence. Explain that it often takes many years to develop enough evidence to push a set of hypotheses into the realm of theory.

Quiz

1. Why is proving a hypothesis wrong just as helpful as supporting it? (You learn something either way.)

2. What was the variable in the frog experiment? (the amount of UV light exposure)

ALTERNATIVE ASSESSMENT

Classifying is an important element in any scientific investigation. Have small groups of students practice categorizing and differentiating between random items based on shared characteristics. Provide students with items such as buttons, paper clips, screws, rubber bands, and so on, and have them group them in as many ways as possible. Afterward, you might like to introduce students to dichotomous keys and explain how they can use them to help differentiate items.

REINFORCEMENT

Have students provide written answers to the Section Review. (Answers are provided at the bottom of the page.) Then have students make lists of technologies they are familiar with and a short description of how scientists might make use of such technologies.

Reinforcement Worksheet 1
"The Mystery of the Bubbling Top"

Problem Solving Worksheet 1
"The Case of the Bulge"

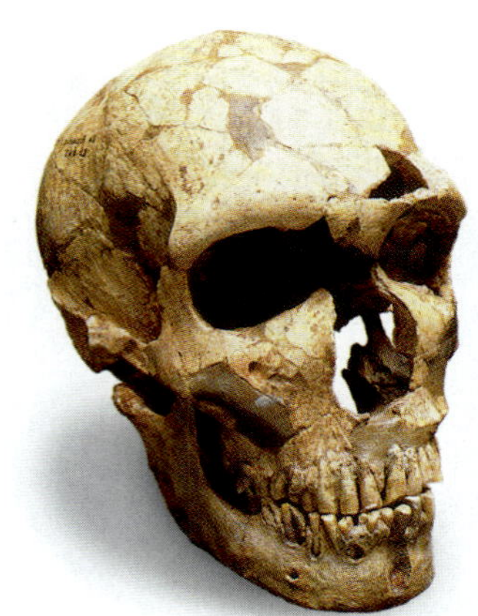

Figura 12 *El Dr. Mark Stoneking usa técnicas modernas para averiguar si los hombres de Neandertal son nuestros ancestros.*

El conocimiento científico cambia

Puede haber más de una predicción para una hipótesis. Cada vez que se comprueba que una predicción es verdadera, la hipótesis se fortalece. Una explicación que unifica varias hipótesis y observaciones corroboradas es una **teoría.**

A veces, cuando los científicos vuelven a examinar la información, llegan a conclusiones distintas. Otras veces, las observaciones posteriores demuestran que las conclusiones eran incorrectas, o que se necesita más investigación.

Nuevas tecnologías conducen a respuestas. La **tecnología** es el uso del conocimiento, las herramientas y los materiales para resolver problemas y realizar tareas. También les permite a los científicos obtener información que no estaba disponible antes. Por ejemplo, los científicos se han preguntado si los hombres de Neandertal son los ancestros de los seres humanos. Ni sus fósiles ni las técnicas de datación han dado suficiente información para responder la pregunta. Recientemente, se ha desarrollado la tecnología para comparar la información genética. El científico de la **Figura 12** la usa para buscar pruebas de que los hombres de Neandertal son nuestros ancestros.

Los científicos siempre formulan nuevas preguntas o se plantean preguntas antiguas desde un ángulo distinto. Al encontrar respuestas, el conocimiento científico sigue creciendo y cambiando.

Temperatura (°C)	Tiempo para duplicarse (min)
10	130
20	60
25	40
30	29
37	17
40	19
45	32
50	no hay crecimiento

REPASO

1. ¿Cuáles son los seis pasos básicos del método científico?

2. ¿Qué elementos producen el cambio y el desarrollo del conocimiento científico?

3. Aplicar Conceptos Diseña un experimento para comprobar la siguiente hipótesis: "La temperatura afecta la frecuencia en que los grillos cantan". Comienza haciendo una predicción del tipo: "Si…, entonces…".

4. Hacer gráficas La tabla de la izquierda da información sobre el desarrollo de un tipo de bacteria que vive en tu intestino. La tabla muestra cuánto tarda una bacteria en dividirse en dos. Traza esta información en una gráfica, con la temperatura en el eje de las *x* y el tiempo requerido para duplicarse en el eje de las *y*. ¿Qué temperatura les permite a las bacterias multiplicarse en menos tiempo?

▼ **Answers to Review**

1. Ask a question, form a hypothesis, test the hypothesis, analyze the results, draw conclusions, and communicate results.

2. the discovery of new information, the use of new technologies, the asking of new questions, and the reexamination of old questions and ideas

3. Answers will vary, but the design of the experiment should include a control group. The variable should be temperature. All other factors should be identical in the control and experimental groups. Sample prediction: If the temperature is raised, then the crickets will chirp more frequently.

4. The graph will slope down and to the right and then rise up again; 37°C.

Herramientas de las ciencias biológicas

Los científicos de las ciencias biológicas usan herramientas que los ayudan a hacer observaciones y analizar información.

Herramientas para observar

Si miras una jarra de agua de estanque, es posible que veas suciedad y algunas criaturas nadando. Pero si examinas el agua con un microscopio o una lente de mano, de repente aparece una comunidad compleja de organismos.

Para observar en forma precisa los organismos y sus partes cuando son muy pequeños y no se pueden ver a simple vista, los científicos usan herramientas de aumento. El vidrio se ha usado para aumentar el tamaño de las imágenes por casi 3,000 años. Las herramientas de aumento que hoy en día están disponibles para un científico son, entre otras, las lentes de mano y los microscopios.

Microscopio compuesto Un tipo de microscopio que se usa comúnmente hoy en día es el microscopio compuesto, que se muestra en la **Figura 13.** El **microscopio compuesto** está formado por tres partes: un tubo con lentes, una platina y una luz. Los especímenes que se ven por un microscopio compuesto a veces están coloreados con tintes especiales para poder observarlos más claramente.

Los especímenes se ubican en la platina para que la luz los atraviese. Las lentes, que están en cada extremo del tubo, aumentan la imagen del espécimen, mostrándolo más grande de lo que es en realidad.

¡MATEMÁTICAS!

Aumento

Si usas un microscopio para observar un objeto que mide 0.2 mm bajo un aumento de 100×, ¿de qué tamaño se verá?

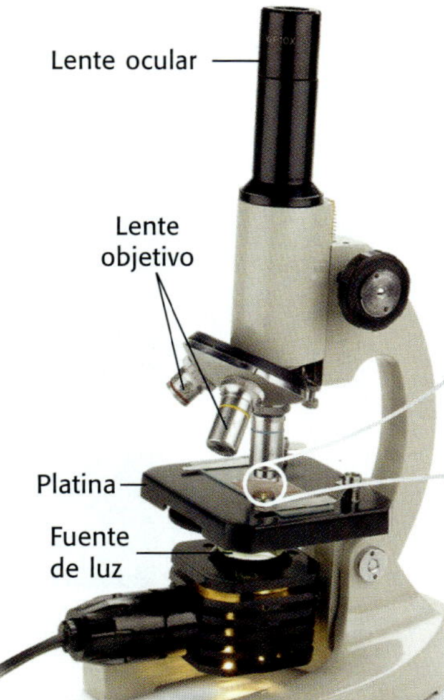

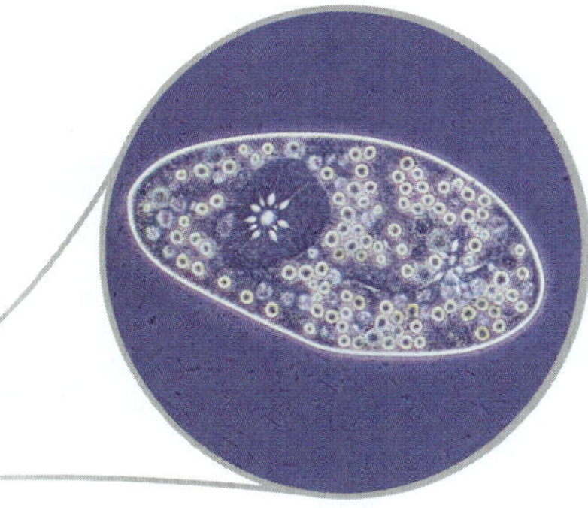

Figura 13 *Un microscopio compuesto puede producir una imagen 1,000 veces (1,000×) más grande que el espécimen real. Este paramecio se ha aumentado 200×.*

19

Math Skills Worksheet 19
"Arithmetic with Decimals"

Directed Reading Worksheet 1 Section 3

SECTION 3

Focus

Tools of Life Scientists

In this section students will explore the variety of tools that scientists use in their jobs. They will discover how scientists use computers to help them solve complex problems. Students will also learn why scientists use the International System of Units (SI).

Bellringer

As you are taking attendance, tell students that researchers recently built the world's smallest guitar—about 10 micrometers long, the size of a single cell—to demonstrate a new technology for making small mechanical devices. Ask students to think of everyday devices that they regularly use that might be better if they were smaller. Have them write these answers in their ScienceLog.

1 Motivate

DEMONSTRATION

Bring to class a collection of microscopic images of ordinary objects. On an overhead projector, show students the images. Ask students to identify the objects. Sheltered English

Answer to MATHBREAK

The object will appear to be 20 mm long.

2 Teach

LabBook PG 563

A Window to a Hidden World

QuickLab

MATERIALS

FOR EACH STUDENT:
• hand lens

Safety Caution: Tell the students to be careful with the lens. It is a delicate piece of scientific equipment and can break. Remind them not to use the lens to focus sunlight.

Have students note five new observations they can make using the lens.

Answers to QuickLab

The lens magnifies objects. Student drawings will vary.

DEMONSTRATION

Using a Microscope Help students become familiar with the parts of a compound microscope and its functions. Encourage students to ask questions as you are demonstrating. Then prepare a wet-mount slide with a drop of pond water. Allow each student to view the slide through the microscope and turn the focus knob to view the water at different depths of focus. Discuss what the students observed.

Experimentos

¿Un mundo oculto a tu alrededor? Averigua de qué se trata en la página 563.

Figura 14 *El microscopio electrónico de transmisión produce una imagen muy aumentada. El microscopio electrónico de barrido proporciona una vista clara de las características de la superficie.*

Laboratorio

Compruébalo

Observa una de tus uñas. En el cuaderno de ciencias dibuja y describe lo que ves. Después, mira la uña con una lupa. ¿Cómo cambia lo que ves al observarlo a través de una lente de mano? Dibuja y describe cómo se ve la uña cuando su imagen aumenta de tamaño.

Microscopio electrónico Los **microscopios electrónicos** usan partículas diminutas de materia que se llaman electrones para producir imágenes aumentadas. El proceso que prepara los especímenes para ser observados mata a los especímenes vivos. Por eso, no se pueden examinar seres vivos con un microscopio electrónico. Hay dos tipos de microscopios electrónicos que se usan en las ciencias biológicas: el microscopio electrónico de transmisión y el microscopio electrónico de barrido.

Los microscopios electrónicos de transmisión pueden aumentar especímenes hasta 200,000 veces (200,000×) su tamaño real. El microscopio electrónico de barrido puede producir imágenes de hasta 100,000 veces su tamaño real. Las imágenes que producen los microscopios electrónicos son más nítidas y detalladas que las de los microscopios compuestos. En la **Figura 14** se muestran los tipos de microscopios electrónicos, junto con una descripción de su propósito especializado y un ejemplo de las imágenes que pueden producir.

Microscopio electrónico de transmisión

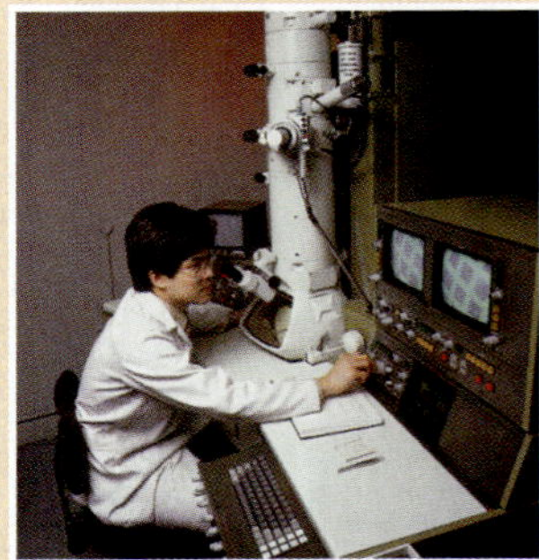

• Los electrones atraviesan el espécimen.
• Se produce una imagen plana.

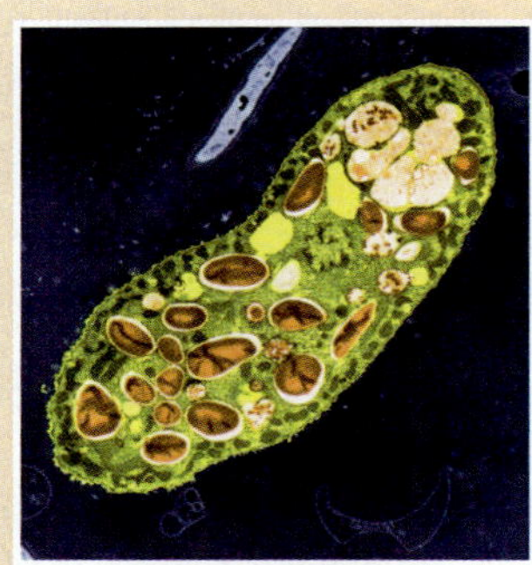

Paramecio (15,000×)

Microscopio electrónico de barrido

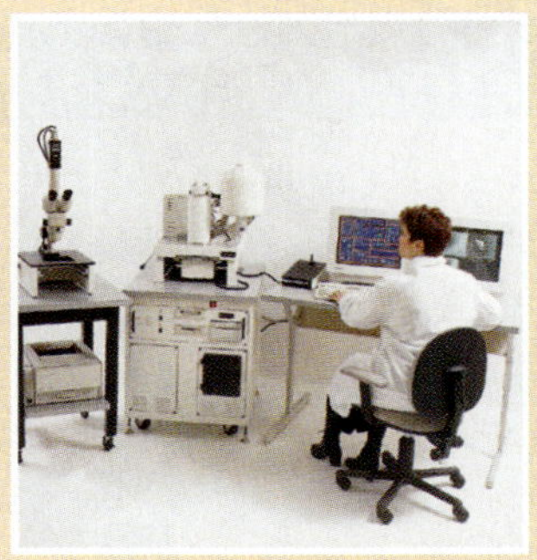

• Los electrones rebotan en la superficie del espécimen.
• Se produce una imagen tridimensional.

Paramecio (1500×)

20

Multicultural CONNECTION

Writing Have students research and write reports about Alhazen, who was born in A.D. 965 in what is now Iraq and who is considered one of the greatest scientists of the medieval period. In his book, *Optics,* he challenged Euclid's and Ptolemy's doctrines on visual rays and formulated his own.

Observar estructuras internas Los científicos de las ciencias biológicas usan herramientas para ver las estructuras internas de los organismos. Estas herramientas les ayudan a ver el interior de los organismos y a entender la composición química de los materiales.

Los rayos X proporcionan imágenes de las estructuras internas del cuerpo como lo ilustra la **Figura 15.** También se han usado para ayudarles a los científicos a aprender acerca de la estructura de las proteínas.

Las imágenes de tomografía computarizada y de resonancia magnética proporcionan imágenes más detalladas de los tejidos internos que los rayos X. En una imagen de tomografía computarizada como la de la **Figura 16,** los rayos X atraviesan el cuerpo en diferentes ángulos. Generalmente se inyecta un medio de contraste para destacar los tejidos. Las imágenes de resonancia magnética usan ráfagas pequeñas de un campo magnético y producen imágenes como la de la **Figura 17.** Con las imágenes de una tomografía computarizada o de una resonancia magnética se transfiere información a una computadora, la cual crea una imagen que un experto puede interpretar. Ambas sirven para estudiar el cerebro y el tejido espinal.

Computadoras

Desde que se construyó la primera computadora electrónica en 1946, los adelantos tecnológicos han producido computadoras más poderosas, las cuales pueden reunir, almacenar, organizar y analizar grandes cantidades de información. Hoy en día realizan miles de millones de cálculos en el mismo tiempo que las computadoras antiguas realizaban sólo miles. Con la ayuda de las computadoras, los científicos resuelven problemas que antes no podían resolver.

Las computadoras se pueden usar para crear gráficas y resolver problemas matemáticos complejos. Los científicos también usan computadoras para decidir si las diferencias en la información experimental son importantes. Además, las computadoras les facilitan la tarea de compartir información e ideas y preparar informes y artículos sobre su investigación.

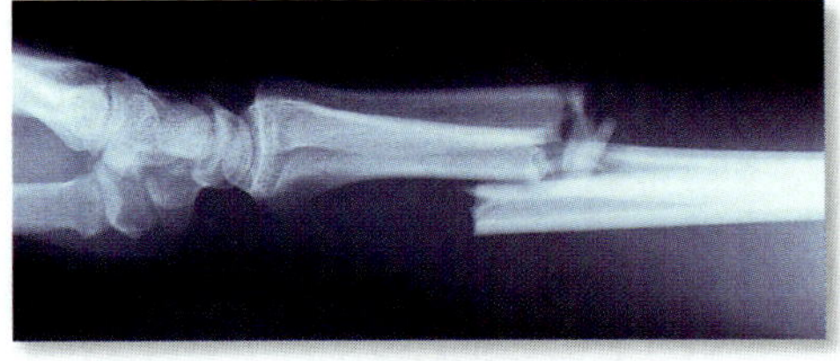

Figura 15 *Esta imagen de rayos X muestra un antebrazo quebrado.*

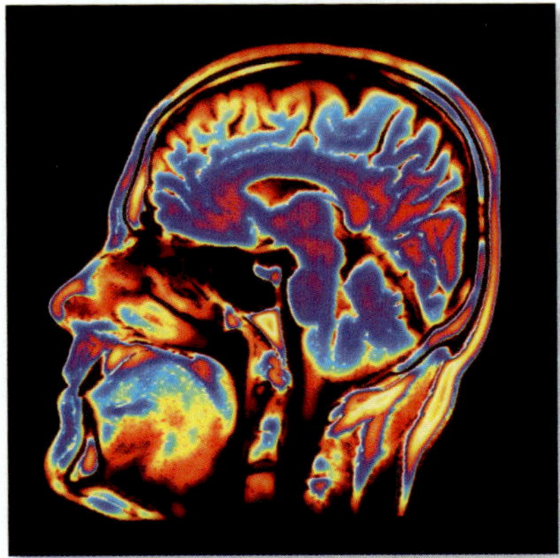

Figura 16 *Los tejidos internos del cerebro se muestran en esta imagen de tomografía computarizada.*

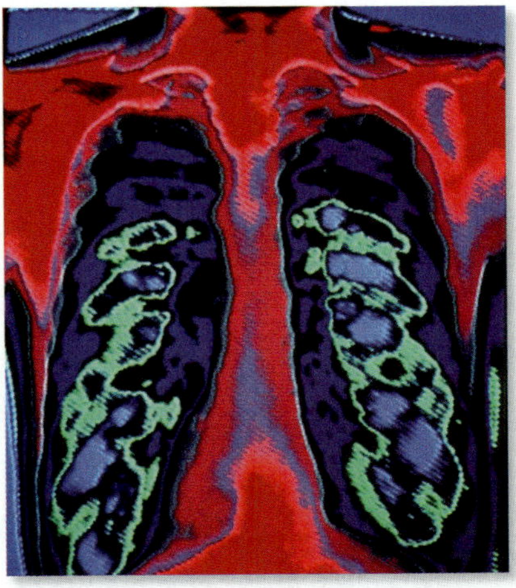

Figura 17 *Esta imagen, que muestra la circulación sanguínea por los pulmones, se obtuvo por imágenes de resonancia magnética.*

WEIRD SCIENCE

Back in the 1940s and 1950s, many shoe stores had a shoe-fitting X-ray unit. Customers inserted their feet into the device and then peeked into a port to glimpse the fluorescent image of their toes. Although the machine was popular, it leaked radiation like a sieve! Bans and heavy regulations removed most of the machines by 1970. A few might still be in existence. One was discovered in a department store in 1981 and was promptly removed.

DEMONSTRATION

Relative Size Show students just how small a nanometer is! Draw a 1 m line on the chalkboard. Tell students that a nanometer is one-billionth of that meter. Ask a student to try to draw a 1 nm line beneath the 1 m line. (Any length drawn will be too long.) Then ask a student volunteer to give you a strand of hair. As you hold the strand of hair, tell students that the diameter of the hair is about 200,000 nm—huge compared with a single nanometer!

USING THE CHART

Writing Tell students to note the prefixes for the common SI units. Ask students what they think each of the prefixes mean. (*kilo-* means 1,000; *centi-* means 0.01; *milli-* means 0.001; and so on)

Have students use their dictionaries to discover more words that use each of these prefixes. Students could then write the definitions of these words and record them in their ScienceLog.

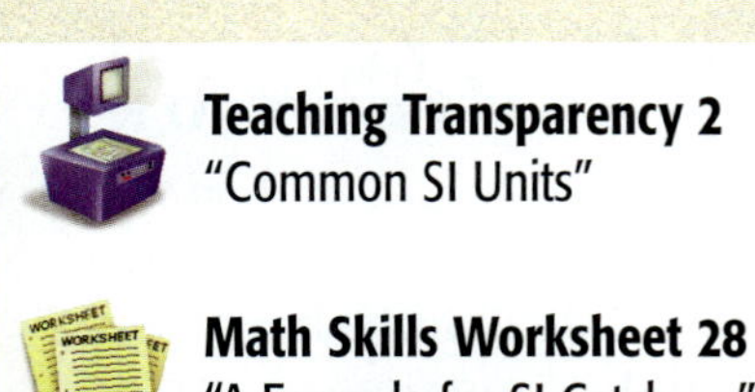
Teaching Transparency 2
"Common SI Units"

Math Skills Worksheet 28
"A Formula for SI Catch-up"

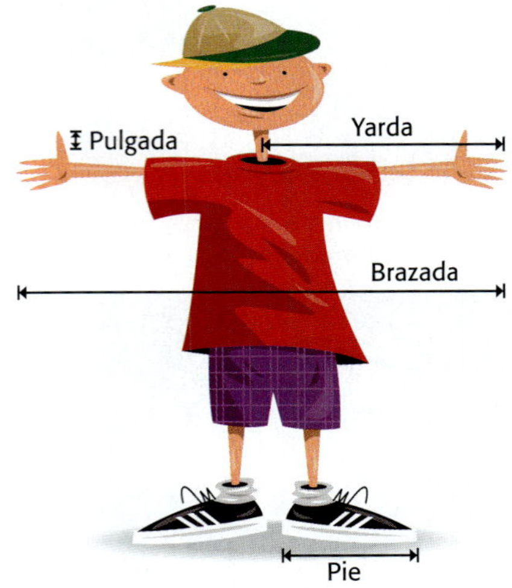

Figura 18 *El sistema inglés moderno se usa mucho en los Estados Unidos. Las unidades, que ahora están estandarizadas, se basaban originalmente en las partes del cuerpo humano.*

Sistemas de medición

Hacer mediciones precisas y confiables es una herramienta importante de las ciencias. Hace cientos de años, cada país usaba un sistema de medición distinto. Cada sistema usaba diferentes patrones para comparar las propiedades de las cosas. Por ejemplo, en Inglaterra, el patrón para una pulgada era tres granos de cebada puestos uno al lado del otro. Incluso las medidas estandarizadas del sistema inglés moderno que se usan en los Estados Unidos, en algún momento se basaron en partes del cuerpo, como se muestra en la **Figura 18.** Estas medidas no eran muy confiables porque se basaban en objetos cuyo tamaño variaba.

A fines del siglo XVIII, la Academia Francesa de las Ciencias comenzó a desarrollar un sistema global de medidas, que ahora se llama Sistema Internacional de Unidades, o SI. Hoy en día, todos los científicos y casi todos los países usan este sistema. Las medidas del SI permiten compartir y comparar observaciones y resultados.

La siguiente tabla contiene las unidades del SI de uso común de longitud, volumen, masa y temperatura. Se usan prefijos junto a las unidades SI para convertirlas en unidades más grandes o más pequeñas. Por ejemplo, *kilo* indica 1,000 veces y *mili* indica 1/1,000 veces. El prefijo que se usa depende del tamaño del objeto que se va a medir. Todas las unidades se basan en el número 10, que facilita la conversión de una unidad a otra.

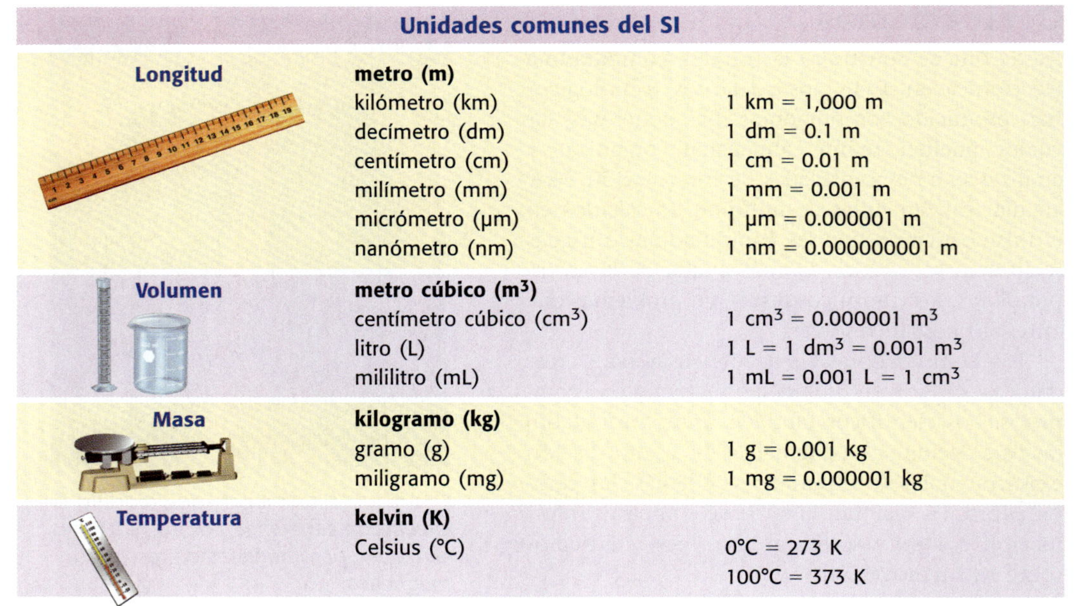

Unidades comunes del SI

Longitud	**metro (m)**	
	kilómetro (km)	1 km = 1,000 m
	decímetro (dm)	1 dm = 0.1 m
	centímetro (cm)	1 cm = 0.01 m
	milímetro (mm)	1 mm = 0.001 m
	micrómetro (µm)	1 µm = 0.000001 m
	nanómetro (nm)	1 nm = 0.000000001 m
Volumen	**metro cúbico (m³)**	
	centímetro cúbico (cm³)	$1\ cm^3 = 0.000001\ m^3$
	litro (L)	$1\ L = 1\ dm^3 = 0.001\ m^3$
	mililitro (mL)	$1\ mL = 0.001\ L = 1\ cm^3$
Masa	**kilogramo (kg)**	
	gramo (g)	1 g = 0.001 kg
	miligramo (mg)	1 mg = 0.000001 kg
Temperatura	**kelvin (K)**	
	Celsius (°C)	0°C = 273 K
		100°C = 373 K

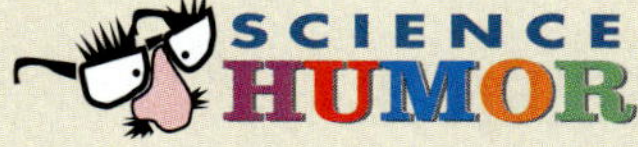
SCIENCE HUMOR

There are three kinds of mathematicians: those who can count and those who cannot.

Tamaños ¿Cuánto mide una lagartija? Para describir cuánto mide una lagartija pequeña, un científico probablemente usaría los centímetros (cm). La unidad básica de longitud del SI es el **metro** (m). Se agregan prefijos del SI a la unidad básica para expresar números muy grandes o muy pequeños. Por ejemplo, 1 kilómetro (km) es igual a 1,000 metros (m). Un metro es igual a 100 centímetros o 1,000 milímetros. Entonces, si divides 1 m en 1,000 partes, cada parte es igual a 1 mm. Esto significa que 1 mm es la milésima parte de un metro. Aunque suena muy pequeño, algunos organismos y estructuras son tan pequeños que se deben usar incluso unidades más pequeñas. Para describir la longitud de objetos microscópicos se usan los micrómetros (µm) o los nanómetros (nm). En la **Figura 20,** la escala compara los tamaños de diferentes organismos.

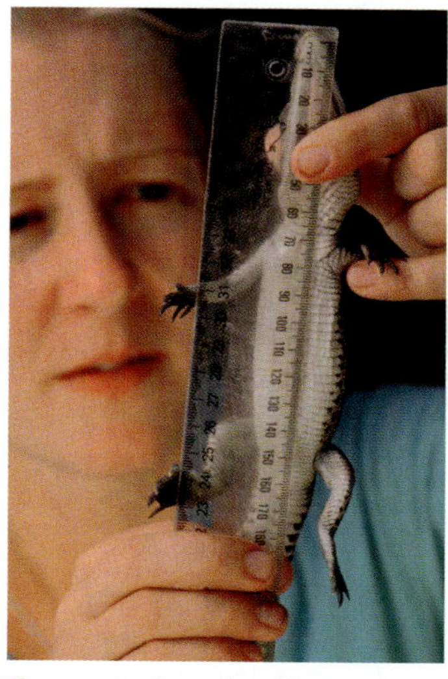
Figura 19 Este científico está midiendo la longitud de una lagartija con una regla métrica.

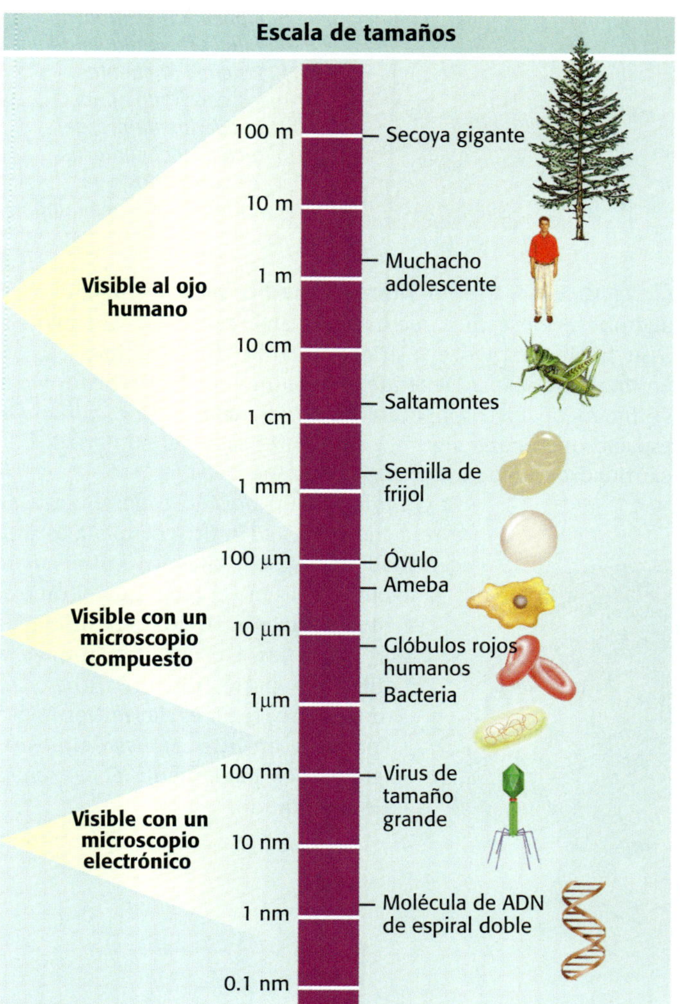

Explora

Mide el ancho de tu escritorio, pero no uses una regla ni una cinta métrica. Elige un objeto que usarás como medida, como un lápiz, tu mano o cualquier otra cosa. Averigua cuántas unidades mide el ancho de tu escritorio y compara tu medición con las de tus compañeros. En tu cuaderno de ciencias explica por qué es importante usar medidas estándar.

Figura 20 Esta escala compara organismos que se pueden ver a simple vista con organismos y estructuras microscópicas.

MATH and MORE

International System of Units   Write the following chart on the board:

1 km = 1 m × 10 × 10 × 10
1 hm = 1 m × 10 × 10
1 dam = 1 m × 10
1 m = 1 m
1 dm = 1 m ÷ 10
1 cm = 1 m ÷ 10 ÷ 10
1 mm = 1 m ÷ 10 ÷ 10 ÷ 10

Have students copy this chart in their ScienceLog, and then ask them to describe any patterns they see. (All units are structured in multiples of 10, using meters as the base unit.)

To help students understand SI conversions, have them answer the following questions:

1. How many meters are there in a kilometer? (1,000 m)

2. How many centimeters are in a meter? (100 cm)

3. What is 20 hm in millimeters? (2,000,000 mm)

Math Skills Worksheet 27 "What Is SI?"

Teaching Transparency 3 "Scale of Sizes"

IS THAT A FACT!

The International System of Units is abbreviated SI because it stands for the French *Système International d'Unités.*

Answer to Explore

Standard units allow people to communicate data to each other easily.

DEMONSTRATION

Displacement Use a graduated beaker or cylinder to demonstrate to students that a piece of clay will always displace the same amount of water, no matter how the clay is shaped. (Be sure to keep your piece of clay constant. Any additions, losses, or air pockets will change your results.)

ACTIVITY

Calculating Volume Have students bring in empty cereal boxes and fill them with dried beans, rice, or other inexpensive material. Have them predict the volume of the box. Then have them pour the contents into a large graduated cylinder or small square box of predetermined volume, and note the volume of the box itself. Ask the students why cereal manufacturers might use tall, narrow boxes for their packaging. (Tall, thin boxes appear to hold more than small square ones, they offer a large front space for eye-catching graphics, and they are easier to pour from.)

Answers to MATHBREAK

1. 25 m²
2. Answers will vary.
3. 4 cm

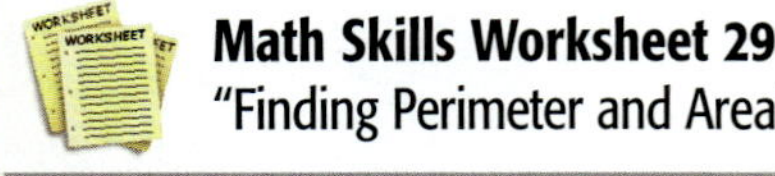

Math Skills Worksheet 29
"Finding Perimeter and Area"

¡MATEMÁTICAS!

Averiguar el área

Puedes usar la ecuación a la derecha para averiguar el área de cualquier superficie rectangular.

1. ¿Cuál es el área de un cuadrado con lados de 5 m?
2. ¿Cuál es el área de la parte superior de tu escritorio?
3. Un rectángulo tiene un área de 36 cm² y una longitud de 9 cm. ¿Cuál es su ancho?

Área ¿Qué cantidad de alfombra se necesitaría para cubrir el piso de tu salón de clases? Para responder esta pregunta es necesario averiguar el área del piso. El **área** es la medida de la superficie de un objeto.

Algunas cantidades, como el área, no se pueden expresar con una medida sino con dos o más medidas. Para calcular el área, mide el ancho y la longitud, y aplica esta ecuación:

$$\text{Área} = \text{longitud} \times \text{ancho}$$

Las unidades de área se llaman unidades cuadradas, como m², cm² y km². La **Figura 21** te ayudará a entender las unidades cuadradas.

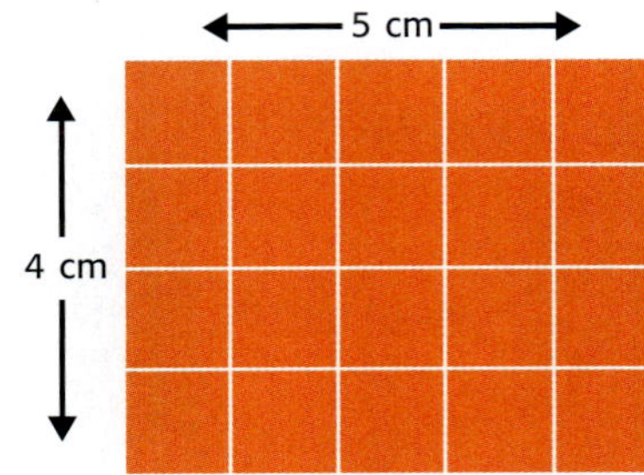

Figura 21 *El área de este rectángulo es de 20 cm². Si cuentas los cuadrados más pequeños dentro del rectángulo, contarás 20 cuadrados que miden 1 cm² cada uno.*

¿Está el vaso medio lleno o medio vacío? Imagina que algunos hipopótamos que nacieron en el zoológico van a volver a su hábitat nativo en África. ¿Cuántos hipopótamos cabrán en un contenedor? Depende del volumen del contenedor y del volumen de los hipopótamos. El **volumen** es la cantidad de espacio que ocupa algo o, como en el caso del contenedor, la cantidad de espacio que algo contiene.

El volumen de un líquido generalmente se expresa en litros (L). Un litro ocupa la misma cantidad de espacio que un cubo cuyos lados miden 1 cm. Al igual que el metro, el litro se puede dividir en unidades más pequeñas. Un mililitro (mL) es la milésima parte de un litro. Un microlitro (µL) es la millonésima parte de un litro. Se usan cilindros graduados para medir el volumen de los líquidos.

Q: What has holes but holds water?

A: a sponge

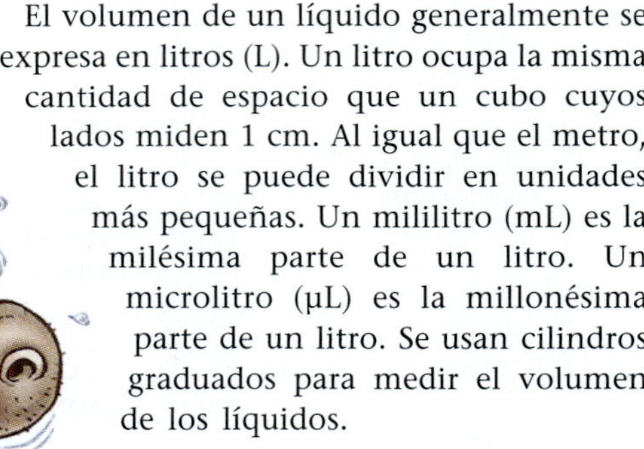

IS THAT A FACT!

While he was taking a bath, Archimedes, a famous Greek mathematician, discovered that an object's volume can be calculated by measuring the amount of water that the object displaces. In this case, the object was himself!

El volumen de un objeto sólido, como un contenedor, se expresa en metros cúbicos (m^3). Los objetos más pequeños se pueden medir en centímetros cúbicos (cm^3) o milímetros cúbicos (mm^3). Un cm^3 es igual a un mL. Para calcular el volumen de un cubo (o cualquier otra forma rectangular), multiplica la longitud por el ancho y por la altura. En la **Figura 22,** puedes tratar de calcular el volumen de un acuario.

Algunas cosas como los hipopótamos o las rocas tienen forma irregular. Si sólo multiplicaras su longitud, ancho y altura, no obtendrías una medida muy precisa de su volumen. Una forma de averiguar el volumen de un objeto de forma irregular es medir cuánto líquido desplaza.

La estudiante de la **Figura 23** está midiendo el volumen de una roca poniéndola en un cilindro graduado que contiene una cantidad de agua específica. La roca desplaza un poco de agua, lo que provoca que el nivel de agua suba. La estudiante puede averiguar el volumen de la roca restando el volumen del agua sola al volumen del agua con la roca. Luego, el volumen del agua en mililitros que desplazó la roca se debe convertir a cm^3.

Figura 22 *Para encontrar el volumen de este acuario, multiplica su longitud por su ancho y por su altura. ¿Qué volumen tiene?*

Figura 23 *Este cilindro graduado contiene 70 mL de agua. Después de agregar la roca, el nivel de agua subió a 80 mL. Debido a que la roca desplazó 10 mL de agua y debido a que 1 mL = 1 cm^3, sabemos que el volumen de la roca es de 10 cm^3.*

REAL-WORLD CONNECTION

The carat is a unit of measure used only for expressing the mass of diamonds and other precious gems. It was originally based on the weight of local grains or seeds and varied widely from place to place. The carat is now a standard unit, with one metric carat equal to 0.200 g. The masses of diamonds, rubies, emeralds, sapphires, aquamarines, zircon, spinel, opals, and pearls are all properly expressed in carats. Ask students:

Why is it important for the mass of a diamond to be expressed in a universally understood measurement?

CONNECT TO
EARTH SCIENCE

If temperature can be defined as a measure of the energy, or movement, of atoms, then atoms that are very still must be very cold. Using this logic, physicist Steven Chu chilled sodium atoms to nearly absolute zero (–273°C, or –523°F) by using laser beams aimed at different angles to restrict the atoms' movement. The result was a glowing pea-sized blob of chilled atoms that seemed to hang in the air at the intersection of the laser beams. Use the following Teaching Transparency to demonstrate the difference between absolute zero and the freezing point of water.

Teaching Transparency 116
"Three Temperature Scales"

Un compromiso masivo La **masa** es la cantidad de materia de que está hecho un objeto. El kilogramo (kg) es la unidad básica de masa. La masa de un objeto muy grande, como un hipopótamo, se expresa en kilogramos (kg) o en toneladas métricas. Un kilogramo es igual a 1,000 g; por lo tanto, un gramo es la milésima parte de un kilogramo. Una tonelada métrica es igual a 1,000 kg. Se usan gramos para expresar la masa de objetos pequeños. Una manzana de tamaño mediano tiene una masa de unos 100 g. Como se muestra en la **Figura 24,** la masa se puede medir con una balanza.

Figura 24 La masa de la manzana es igual a la masa de las pesas.

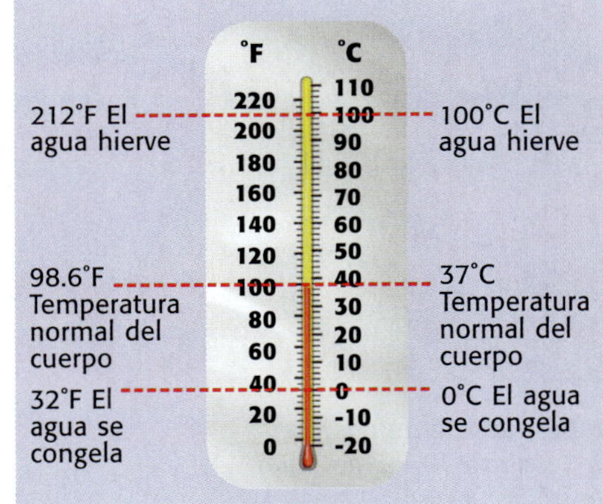

Figura 25 El agua se congela a 0°C y hierve a 100°C. La temperatura normal de tu cuerpo es de 37°C, que es igual a 98.6°F.

¿Está caliente? ¿Qué temperatura se necesita para eliminar las bacterias? Para responder esto, un científico mediría la temperatura a la que las bacterias mueren. La **temperatura** mide qué tan caliente (o frío) está algo. Aunque no te des cuenta, las moléculas que conforman la materia se mueven continuamente. Cuando se transfiere energía a estas moléculas, se mueven todavía más, lo cual hace que la temperatura aumente. De acuerdo con esto, la temperatura también se puede definir como una medida del promedio de energía que tienen las moléculas de una substancia.

Probablemente estás acostumbrado a describir la temperatura en grados Fahrenheit (°F). Los científicos por lo general usan grados celsius (°C), a pesar de que los grados kelvin son las unidades oficiales de temperatura del SI. Usaremos °C en este libro. El termómetro de la **Figura 25** muestra la relación entre °F y °C.

MISCONCEPTION ALERT

The terms *mass* and *weight* do not mean the same thing. Mass is a fundamental property of an object, while weight is a relative property—a measure of the gravitational force experienced by an object.

For example, a rocket has the same mass whether it is standing on Earth or on the moon, but it will weigh less on the moon because the moon exerts a smaller gravitational force on it.

¡La seguridad manda!

Las ciencias biológicas son emocionantes, pero también pueden ser peligrosas. Así que, ¡no te arriesgues! Siempre sigue las instrucciones de tu maestro o maestra y no intentes saltarte pasos, ni siquiera cuando creas que no hay peligro.

Antes de comenzar un experimento, pídele permiso a tu maestro y lee los procedimientos de laboratorio atentamente. Presta atención a la información de seguridad y a los avisos de precaución. La siguiente tabla muestra los símbolos de seguridad que se utilizan en este libro. Apréndete estos símbolos y su significado. Lee la información de seguridad que comienza en la página 556. ¡Esto es muy importante! Pregúntale a tu maestro o maestra si no estás seguro de lo que significa un símbolo.

Experimentos

Para no correr ningún riesgo, lee la información de seguridad que aparece en la página 556.

¡Debes hacer esto antes de realizar un experimento!

SÍMBOLOS DE SEGURIDAD

Protección de los ojos	Protección de la ropa	Protección de las manos
Cuidado con el calor	Cuidado con la electricidad	Cuidado con los objetos punzantes
Cuidado con las substancias químicas	Seguridad de los animales	Seguridad de las plantas

REPASO

1. ¿Cómo se relaciona la temperatura con la energía?

2. Si midieras la masa de una mosca, ¿qué unidad métrica sería más apropiada?

3. Menciona dos ventajas de usar el Sistema Internacional de Unidades.

4. **Comprender la tecnología** ¿Qué herramienta se usó para producir la imagen de la derecha? ¿Cómo lo sabes?

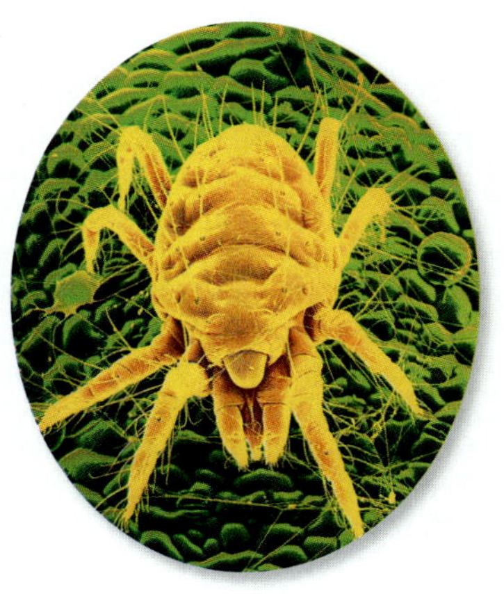

27

4) Close

  **PG 556**

Safety First!

Quiz

1. What are the three main parts of a compound microscope? (tube with lenses, stage, light)

2. What are CT scans and MRI especially useful for studying? (brain and spinal tissue)

ALTERNATIVE ASSESSMENT

Writing Have students create an illustrated dictionary of vocabulary terms that were used in this chapter. Students can share their dictionaries with the rest of the class. Sheltered English

REINFORCEMENT

Writing Have students provide written answers to the Review. Answers are provided at the bottom of this page. Then have students write a letter to a scientist, dated well into the future, that describes some of the tools that scientists use today.

 **Science Skills Worksheet 9** "Safety Rules!"

 **Interactive Explorations CD-ROM** "Something's Fishy!"

▼ **Answers to Review**

1. The higher the temperature is, the greater the energy.

2. milligram

3. All scientists use the International System of Units so that they can share and compare their results. Another benefit is that calculations became easier because SI units are based on powers of 10.

4. A scanning electron microscope. The image is three-dimensional.

VOCABULARY DEFINITIONS

SECTION 1

life science the study of living things

SECTION 2

scientific method a series of steps that scientists use to answer questions and solve problems

hypothesis a possible explanation or answer to a question

factor anything in an experiment that can influence the experiment's outcome

controlled experiment an experiment that tests only one factor at a time

variable the factor in a controlled experiment that changes

theory a unifying explanation for a broad range of hypotheses and observations that have been supported by testing

technology the application of knowledge, tools, and materials to solve problems and accomplish tasks; technology can also refer to the objects used to accomplish tasks

Resumen del capítulo

SECCIÓN 1

Vocabulario

ciencias biológicas (*pág. 6*)

Notas de la sección

- Las ciencias biológicas son el estudio de los seres vivos. La observación de los organismos generalmente conduce a preguntas. Las preguntas estimulan el estudio de las ciencias biológicas.

- Todos podemos llegar a ser científicos de las ciencias biológicas. Estos científicos trabajan en muchos lugares diferentes e investigan una gran variedad de preguntas y problemas.

- Los científicos de las ciencias biológicas ayudan a prevenir y curar enfermedades y también a resolver problemas ambientales.

SECCIÓN 2

Vocabulario

método científico (*pág. 10*)
hipótesis (*pág. 12*)
factor (*pág. 14*)
experimento controlado (*pág. 14*)
variable (*pág. 14*)
teoría (*pág. 18*)
tecnología (*pág. 18*)

Notas de la sección

- El método científico es una serie de pasos que los científicos usan para responder una pregunta o resolver un problema.

- Los pasos del método científico no siempre se siguen en el mismo orden; a veces se omiten o se repiten.

- Los científicos realizan experimentos controlados para investigar los efectos de un factor a la vez.

- Los científicos deben observar cuidadosamente, registrar información en forma precisa y ser creativos en la búsqueda de respuestas y el diseño de experimentos.

- Una hipótesis es una explicación posible y comprobable de lo que se ha observado.

✓ Comprobar destrezas

Conceptos de matemáticas

CONVERSIÓN DE UNIDADES Imagina que escribes un informe sobre el *Empire State Building*, de la ciudad de Nueva York. Un libro de la biblioteca dice que la altura del edificio es de 381,000,000 micrómetros. Este es un número muy grande, así que probablemente querrás convertirlo a un número más manejable, en metros en lugar de micrómetros. Como ves en la tabla de la página 22, 1 μm = 0.000001 m. Para convertir micrómetros a metros, debes multiplicar por 0.000001.

$$381{,}000{,}000 \times 0.000001 = 381 \text{ m}$$

Comprensión visual

¿DE QUÉ TAMAÑO ES? Para repasar los tamaños de las cosas en función de lo que se observa con diferentes tipos de microscopios, pasa a las páginas 19 y 20. La escala de tamaños de la página 23 también te ayudará a visualizar los tamaños descritos con el sistema métrico.

¡SEGURIDAD ANTE TODO! Asegúrate de conocer y comprender los diferentes símbolos de seguridad de la página 27.

28

Lab and Activity Highlights

Does It All Add Up? `PG 560`

Graphing Data `PG 562`

A Window to a Hidden World `PG 563`

Datasheets for LabBook
(blackline masters for these labs)

SECCIÓN 2

- Una teoría es una explicación que unifica un amplio rango de hipótesis y observaciones que han sido corroboradas por pruebas.

- El conocimiento científico cambia y crece constantemente con las preguntas que los científicos se hacen, las diferentes respuestas que obtienen y el uso de tecnologías que les permiten reunir información de formas novedosas.

Experimentos

¿Tiene sentido?
(pág. 560)

Hacer gráficas
(pág. 562)

SECCIÓN 3

Vocabulario

microscopio compuesto
(pág. 19)

microscopio electrónico
(pág. 20)

metro (pág. 23)

área (pág. 24)

volumen (pág. 24)

masa (pág. 26)

temperatura (pág. 26)

Notas de la sección

- Generalmente, los científicos usan microscopios compuestos y electrónicos para observar organismos o partes de organismos que no se pueden ver a simple vista debido a su tamaño.

- Los rayos X, las tomografías computarizadas y las imágenes de resonancia magnética se utilizan para ver las estructuras internas de los organismos.

- La información se recopila, almacena, organiza, analiza y comparte por medio de computadoras.

- El Sistema Internacional de Unidades (SI), es un sistema simple y confiable de medición que todos los científicos utilizan.

Experimentos

Una ventana a un mundo oculto
(pág. 563)

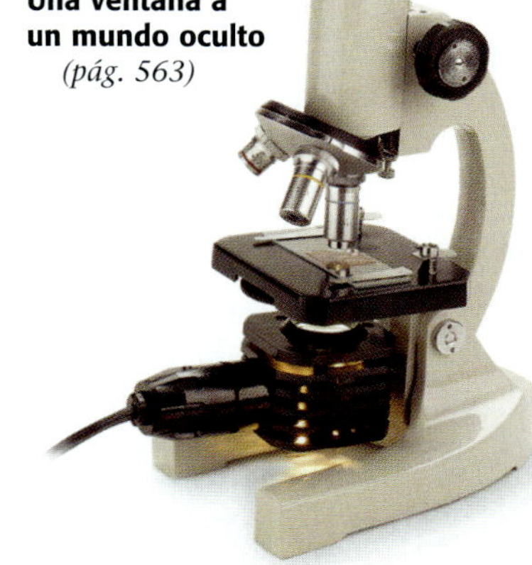

SECTION 3

compound light microscope a microscope that consists of a tube with lenses, a stage, and a light source

electron microscope a microscope that uses tiny particles of matter to produce magnified images

meter the basic unit of length in the SI system

area the measure of how much surface an object has

volume the amount of space that something occupies or the amount of space that something contains

mass the amount of matter that something is made of; its value does not change with the object's location

temperature a measure of how hot (or cold) something is; specifically, a measure of the average energy of the molecules of a substance

---

 **internet**

 **VISITA:** go.hrw.com

Visita el sitio web de HRW para encontrar una serie de herramientas de aprendizaje relacionadas con este capítulo. Sólo tienes que escribir la palabra clave:

PALABRA CLAVE: HSTLIV

*SCI**LINKS***
N S T A

 **VISITA:** www.scilinks.org

Visita el sitio web de la **Asociación Nacional de Maestros de Ciencias** *(National Science Teachers Association)* para encontrar recursos de Internet relacionados con este capítulo. Sólo escribe el **ENLACE DE CIENCIAS** para obtener más información sobre el tema:

TEMA	ENLACE
TEMA: Ranas deformes	**ENLACE:** HSTL005
TEMA: Profesiones de las ciencias biológicas	**ENLACE:** HSTL010
TEMA: Herra mientas de las ciencias biológicas	**ENLACE:** HSTL015
TEMA: Unidades del SI	**ENLACE:** HSTL020

 **Vocabulary Review Worksheet 1**

 Blackline masters of these Chapter Highlights can be found in the **Study Guide.**

29

Lab and Activity Highlights

LabBank

 **Whiz-Bang Demonstrations**
- Air Ball, Demo 1
- Getting to the Point, Demo 2

Inquiry Labs, One Side or Two? Lab 1

Long-Term Projects & Research Ideas, Project 1

Interactive Explorations CD-ROM

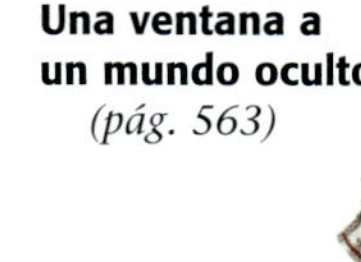 **CD 1, Exploration 1,** "Something's Fishy!"

Chapter Review
Answers

USING VOCABULARY

1. scientific method
2. hypotheses
3. variable, control group
4. area
5. grams

UNDERSTANDING CONCEPTS

Multiple Choice

6. d
7. c
8. a
9. a
10. c
11. b

Short Answer

12. A hypothesis must be testable by experimentation in order for scientists to disprove the hypothesis or support it.
13. A prediction is a statement of cause and effect used to set up an experiment.
14. The SI units for measuring volume are liters (L), units based on the liter, and units based on the cubic meter (m^3, cm^3, mm^3). The mass of an object is measured in kilograms (grams, milligrams, etc.).

Science Skills Worksheet 7
"Improving Your Study Habits"

Repaso del capítulo

UTILIZAR EL VOCABULARIO

Escoge el término correcto para completar las siguientes oraciones:

1. El conjunto de destrezas o pasos que los científicos siguen para responder preguntas es el ____.
(experimento controlado o *método científico)*

2. Después de identificar un problema o hacer una pregunta, los científicos formulan una o más ____, que son posibles explicaciones de lo que se ha observado. *(predicciones* o *hipótesis)*

3. En un experimento controlado, la ____ es el factor diferente entre el ___ y el grupo experimental. *(predicción, grupo variable* o *variable, grupo de control)*

4. El ____ es la medida de la superficie de un objeto. *(área* o *volumen)*

5. Los científicos usan ____ para medir la masa de un objeto. *(metros* o *gramos)*

COMPRENDER CONCEPTOS

Opción múltiple

6. ¿Cuál de las siguientes *no* sería un área de estudio de las ciencias biológicas?
a. investigar cómo interactúan los leones y las hienas
b. medir la velocidad a la cual se divide una bacteria
c. comparar la reproducción de plantas árticas con la de las plantas del desierto
d. investigar cómo se forman los volcanes

7. Los pasos del método científico
a. se deben usar en todas las investigaciones científicas.
b. siempre se deben usar en el mismo orden.
c. no siempre se usan en orden.
d. comienzan con el desarrollo de una teoría.

8. En un experimento controlado
a. se compara un grupo de control con uno o más grupos experimentales.
b. hay por lo menos dos variables.
c. todos los factores deben ser diferentes.
d. no se necesita una variable.

9. Si una científica descubre que una hipótesis es errónea, generalmente ella
a. tratará de encontrar otra explicación de lo que ha observado.
b. dejará de estudiar ciencias.
c. sentirá que no aprendió nada valioso.
d. agregará una variable adicional a su experimento.

10. ¿Qué herramienta usaría un científico para obtener una imagen tridimensional de un organismo microscópico?
a. una tomografía computarizada
b. rayos X
c. un microscopio electrónico de barrido
d. una lente manual

11. El Sistema Internacional de Unidades
a. se basa en un conjunto estandarizado de medidas.
b. contiene unidades que se basan en el número 10.
c. sólo sirve para medir longitudes.
d. es un mecanismo para medir el volumen.

Respuesta breve

12. ¿Por qué las hipótesis tienen que ser comprobables?

13. ¿Qué es una predicción?

14. ¿Qué unidades del SI se pueden usar para medir el volumen? ¿Y la masa?

Organizar conceptos

15. Usa los siguientes términos para crear un mapa de ideas: observaciones, predicciones, preguntas, experimentos controlados, variable, hipótesis.

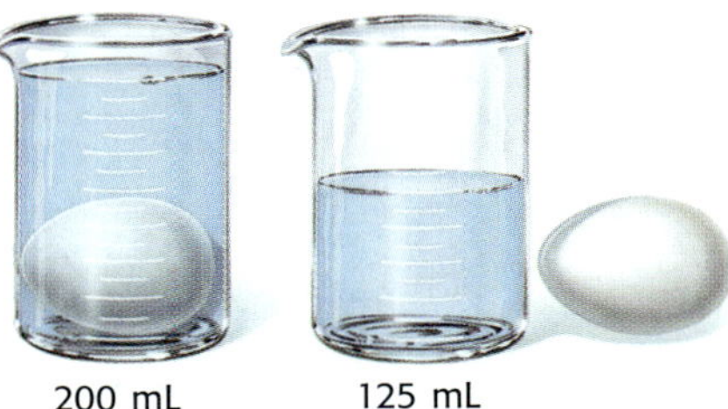

RAZONAMIENTO CRÍTICO Y RESOLUCIÓN DE PROBLEMAS

Escribe una o dos oraciones para responder a las siguientes preguntas:

16. En un experimento controlado, ¿por qué debe haber varios individuos en el grupo de control y en cada uno de los grupos experimentales?

17. Un científico que estudia ratones observa que el día que les da vitaminas con la comida, los ratones se mueven mejor en los laberintos. ¿Qué hipótesis formularías para explicar este fenómeno?

18. El volumen de un huevo con agua en un vaso graduado es de 200 mL. Después de retirar el huevo, el volumen del agua es de 125 mL. ¿Qué volumen tiene el huevo en cm^3?

200 mL   125 mL

LAS MATEMÁTICAS EN LAS CIENCIAS

19. Si aumentas 1,000 × la imagen de un organismo que mide 5 µm de largo, ¿qué longitud parecería tener el organismo en milímetros (mm)?

INTERPRETAR GRÁFICAS

Examina la siguiente ilustración de un experimento montado para comprobar esta predicción: **Si** las abejas se sienten más atraídas a las flores amarillas que a las rojas, **entonces** las abejas visitarán las flores amarillas más a menudo que las flores rojas.

11 visitas   3 visitas   11 visitas

2 visitas   9 visitas   4 visitas

20. ¿Cuántas visitas recibieron las flores amarillas en total? ¿Cuántas visitas recibieron las flores rojas en total?

21. ¿Cuál es el número promedio de visitas para las flores amarillas? ¿Cuál es el número promedio de visitas para las flores rojas?

22. ¿Por qué crees que este montaje sería una prueba no confiable de la predicción?

AHORA, ¿qué piensas?

Revisa tus respuestas a las preguntas de la página 5 que escribiste en el cuaderno de ciencias. ¿Han cambiado tus respuestas? Si es necesario, corrige tus respuestas basándote en lo que has aprendido en este capítulo.

Concept Mapping

15.  An answer to this exercise can be found at the end of this book.

CRITICAL THINKING AND PROBLEM SOLVING

16. The more individuals there are in the groups, the more confident scientists can be that differences between the groups were caused by the variable and not by natural differences between individual organisms.

17. Possible answer: Vitamins increase the mice's spatial reasoning capabilities.

18. 75 mL = 75 cm^3

MATH IN SCIENCE

19. 5 mm

INTERPRETING GRAPHICS

20. 20; 20

21. 10; 5

22. There is more than one variable. There are different kinds of flowers used in the experiment. If flower color is being tested, then flower color should be the only difference between the flowers. There should also be the same number of flowers of each color.

Concept Mapping Transparency 1

Blackline masters of this Chapter Review can be found in the **Study Guide.**

NOW WHAT DO YOU THINK

1. Answers could include microscopes, computers, tables and graphs, X rays, and SI measurements.

2. Scientists use the steps of the scientific method to study life science. They ask questions, form hypotheses, test hypotheses, analyze results, draw conclusions, and communicate results.

3. yes

Background

Zoology dates back more than 2,300 years, to ancient Greece, where the philosopher Aristotle observed and theorized about animal behavior. About 200 years later, Galen, a Greek physician, began dissecting and experimenting with animals. However, there were few advances in zoology until the 1700s and 1800s. During this period, the Swedish naturalist Carolus Linnaeus developed a classification system for animals, and Charles Darwin published his theory on natural selection and evolution.

Today, zoology is divided into many specialized areas. These areas include comparative anatomy (the study of anatomical structures among many different animals), physiology (the study of how bodies function), genetics (the study of heredity), embryology (the study of embryo development), entomology (the study of insects), ichthyology (the study of fish), and herpetology (the study of reptiles and amphibians).

PROFESIONES

ZOÓLOGO

Eric Pianka se interesó en los lagartos por primera vez cuando tenía 6 años. "Durante un viaje por el campo con mi familia, vi un gran lagarto verde en un parque que quedaba por el camino", explica Pianka. "Traté de atraparlo, pero lo único que pude agarrar fue la cola. En ese momento, supe que tenía que averiguar todo lo posible sobre su estilo de vida". Pianka es ahora un profesor de zoología mundialmente famoso de la Universidad de Texas, en Austin.

Una de las cosas que a Eric Pianka le gusta más de su trabajo es estar en un terreno virgen y ver cosas que pocas personas han visto. "¡He visitado muchísimos lugares! He pasado mucho tiempo investigando los desiertos del oeste de los Estados Unidos y también he estado en los desiertos del sur de África, India y Chile. Lo que más me interesa en la actualidad (y desde hace mucho tiempo) son los desiertos de Australia. No he tenido la oportunidad de investigar la zona amazónica del Brasil pero, ¡es mi meta para el futuro!"

La ecología de los lagartos del desierto

En sus investigaciones como zoólogo, Pianka se ha concentrado en la ecología de los lagartos del desierto. Va al desierto, reúne lagartos, los examina y los clasifica. Después, recopila información y la interpreta en libros o informes. Pianka nos dice: "Trato de responder preguntas como: ¿Por qué hay más lagartos en un lugar que en otro? ¿Cómo reaccionan entre ellos o con otras especies? ¿Cómo se han adaptado a su entorno?"

Recientemente, Pianka realizó un estudio sobre los efectos de los incendios forestales en la ecología y las diversas especies de lagartos. Espera que este trabajo muestre cómo las especies de lagartos se adaptaron a los grandes incendios que hace tiempo ocurrían regularmente en las áreas desérticas y que hoy en día son controlados por los seres humanos.

Aprender sobre la fauna silvestre

Pianka cree que con su investigación sobre los lagartos y otros animales puede ayudar a proteger el medio ambiente. "Todos me preguntan: '¿Por qué estudias lagartos?' Y yo les pregunto a ellos: '¿Por qué no?' El sentimiento general es que todo en la Tierra tiene que estar a disposición de los seres humanos. Observar cómo han vivido, muerto y evolucionado otras especies a través de millones de años nos permite entender mejor el mundo en que vivimos".

▶ *El Crotaphytus collaris vive en las regiones rocosas del sudoeste de los Estados Unidos.*

Zoólogo por un día

▶ Selecciona un animal común que viva en tu área y que se pueda observar fácilmente. Durante un par de horas, observa lo que come, lo que hace y a dónde va. Registra cuidadosamente todo lo que observes. ¿Descubriste algo que no sabías? Presenta tus descubrimientos a la clase.

32

Answer to Be a Zoologist for a Day

Student observations will vary according to the choice of animal and the duration of observation. However, students should demonstrate clear observational skills and organize the records of what they observed.

"El pollo nostálgico"

de Edward D. Hoch

¿Por qué el pollo cruzó el camino? Seguro sabes la respuesta a este acertijo, ¿cierto? O quizá crees que la sabes. Pero "El pollo nostálgico", de Edward D. Hoch puede sorprenderte. Es posible que ese viejo pollo no sea exactamente lo que parece…

Verás, uno de los pollos de las Granjas de Investigación Tangaway se ha escapado, no sólo del gallinero, sino de la granja. Hizo un orificio en una cerca de alta seguridad y, después, cruzó una carretera de ocho vías para fugarse. Pero después de todo ese esfuerzo, ¡simplemente se detuvo! Se le encontró en un lote vacío al otro lado de la carretera de Tangaway, picoteando tranquilamente.

Se le pidió a Barnabus Rex, un especialista en resolver acertijos científicos, que trabajara en el misterio. Él está intrigado por la huída del pollo. ¿Por qué se habrá tomado el trabajo de hacer un orificio en la cerca de seguridad, arriesgar su vida en la autopista y después detenerse al llegar al otro lado?

La historia nos da algunas pistas. Al leerla, quizás puedas ver lo que ve el Sr. Rex. Si sabes algo sobre los pollos, podrías resolver el acertijo. Escápate a la *Antología Holt de Ciencia Ficción* y lee "El pollo nostálgico".

33

Chapter Organizer

CHAPTER ORGANIZATION	TIME MINUTES	OBJECTIVES	LABS, INVESTIGATIONS, AND DEMONSTRATIONS
Chapter Opener pp. 34–35	45		**Investigate!** Lights On! p. 35
Section 1 **Characteristics of Living Things**	90	▶ List the characteristics of living things. ▶ Distinguish between asexual reproduction and sexual reproduction. ▶ Define and describe homeostasis.	**Discovery Lab,** Roly-Poly Races, p. 564 **Datasheets for LabBook,** Roly-Poly Races, Datasheet 4
Section 2 **The Simple Bare Necessities of Life**	90	▶ Explain why organisms need food, water, air, and living space. ▶ Discuss how living things obtain what they need to live.	**Demonstration,** Fire and Life, p. 41 in ATE
Section 3 **The Chemistry of Life**	90	▶ Compare and contrast the chemical building blocks of cells. ▶ Explain the importance of ATP.	**Demonstration,** Protein Model, p. 42 in ATE **QuickLab,** Starch Search, p. 43 **Discovery Lab,** The Best-Bread Bakery Dilemma, p. 566 **Datasheets for LabBook,** The Best-Bread Bakery Dilemma, Datasheet 5 **Labs You Can Eat,** Say Cheese! Lab 1 **Long-Term Projects & Research Ideas,** Project 2

TECHNOLOGY RESOURCES

Guided Reading Audio CD English or Spanish, Chapter 2

One-Stop Planner CD-ROM with Test Generator

 **Science, Technology & Society,** Tapping into Yellowstone's Hot Springs, Segment 1

CLASSROOM WORKSHEETS, TRANSPARENCIES, AND RESOURCES	SCIENCE INTEGRATION AND CONNECTIONS	REVIEW AND ASSESSMENT
Directed Reading Worksheet 2 **Science Puzzlers, Twisters & Teasers,** Worksheet 2		
Directed Reading Worksheet 2, Section 1 **Math Skills for Science Worksheet 4,** A Shortcut for Multiplying Large Numbers **Math Skills for Science Worksheet 26,** Multiplying and Dividing in Scientific Notation **Critical Thinking Worksheet 2,** Intergalactic Planetary Mission **Math Skills for Science Worksheet 18,** Decimals and Fractions **Math Skills for Science Worksheet 21,** Percentages, Fractions, and Decimals	**Oceanography Connection,** p. 37 **Math and More,** p. 37 in ATE **Math and More,** p. 38 in ATE **MathBreak,** Body Proportions, p. 39 **Apply,** p. 39	**Self-Check,** p. 37 **Review,** p. 39 **Quiz,** p. 39 in ATE **Alternative Assessment,** p. 39 in ATE
Directed Reading Worksheet 2, Section 2 **Transparency 154,** Climate Zones of the Earth **Reinforcement Worksheet 2,** Amazing Discovery	**Connect to Earth Science,** p. 40 in ATE **Scientific Debate:** Life on Mars? p. 50	**Review,** p. 41 **Quiz,** p. 41 in ATE **Alternative Assessment,** p. 41 in ATE
Directed Reading Worksheet 2, Section 3 **Transparency 4,** Phospholipid Molecule and Cell Membrane **Transparency 5,** Energy for Cells **Reinforcement Worksheet 2,** Building Blocks	**Multicultural Connection,** p. 42 in ATE **Holt Anthology of Science Fiction,** *They're Made Out of Meat*	**Homework,** p. 44 in ATE **Review,** p. 45 **Quiz,** p. 45 in ATE **Alternative Assessment,** p. 45 in ATE

internet connect

go.hrw.com

Holt, Rinehart and Winston On-line Resources

go.hrw.com

For worksheets and other teaching aids related to this chapter, visit the HRW Web site and type in the keyword: **HSTALV**

National Science Teachers Association

www.scilinks.org

Encourage students to use the *sci*LINKS numbers listed with the Chapter Highlights to access information and resources on the **NSTA** Web site.

END-OF-CHAPTER REVIEW AND ASSESSMENT

Chapter Review in Study Guide
Vocabulary and Notes in Study Guide
Chapter Tests with Performance-Based Assessment, Chapter 2 Test
Chapter Tests with Performance-Based Assessment, Performance-Based Assessment 2
Concept Mapping Transparency 2

Chapter Resources & Worksheets

Visual Resources

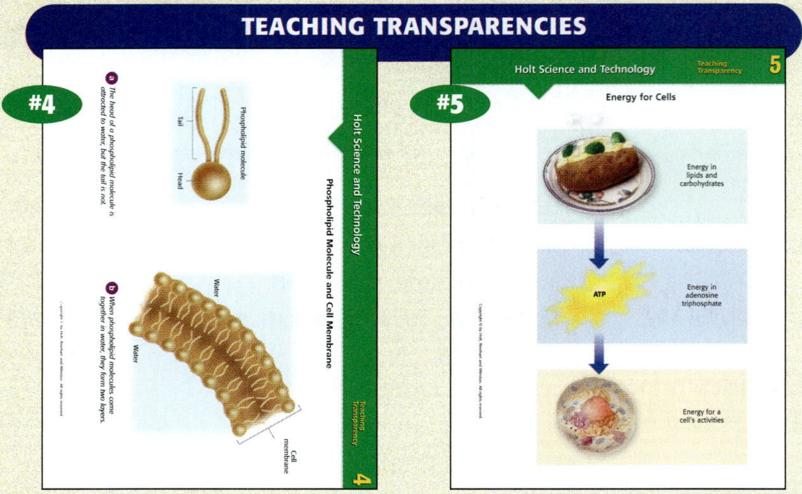

TEACHING TRANSPARENCIES

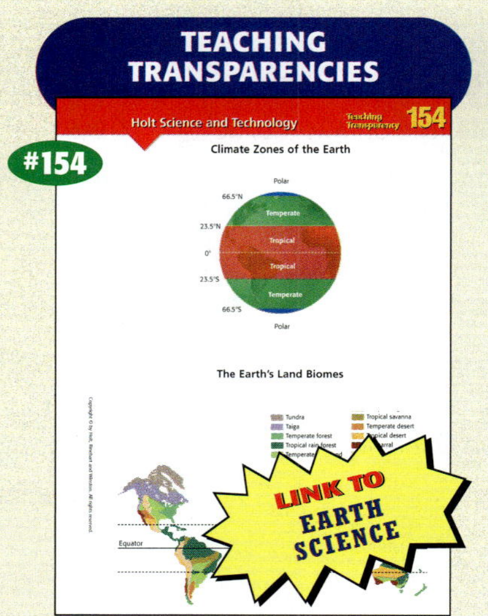

TEACHING TRANSPARENCIES

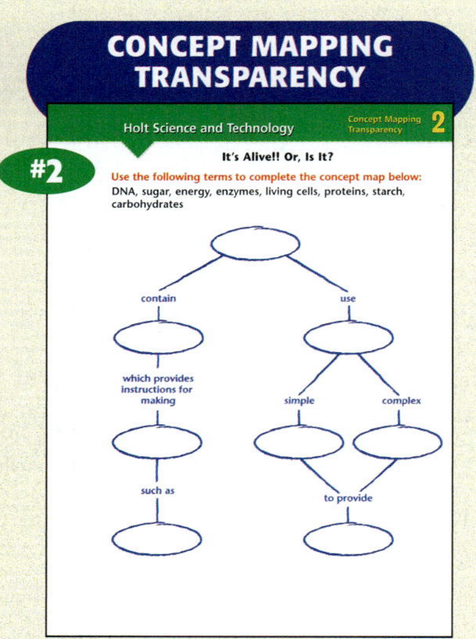

CONCEPT MAPPING TRANSPARENCY

Meeting Individual Needs

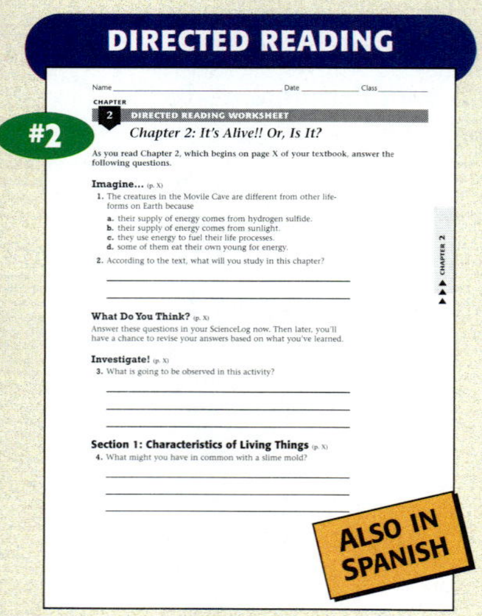

DIRECTED READING

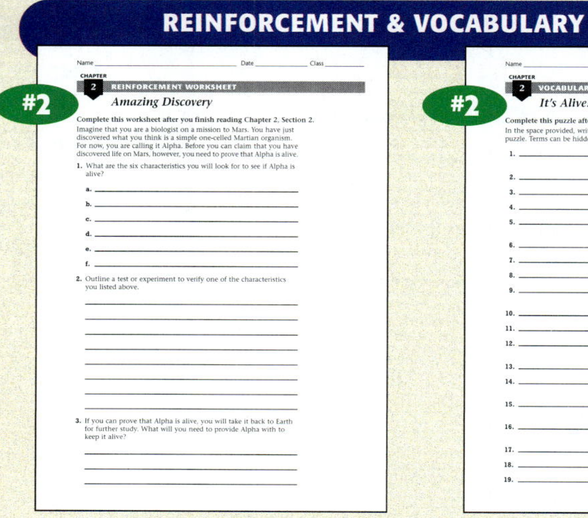

REINFORCEMENT & VOCABULARY REVIEW

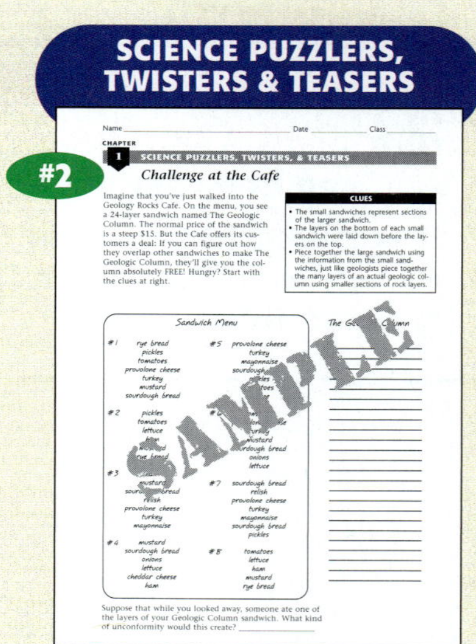

SCIENCE PUZZLERS, TWISTERS & TEASERS

Review & Assessment

STUDY GUIDE

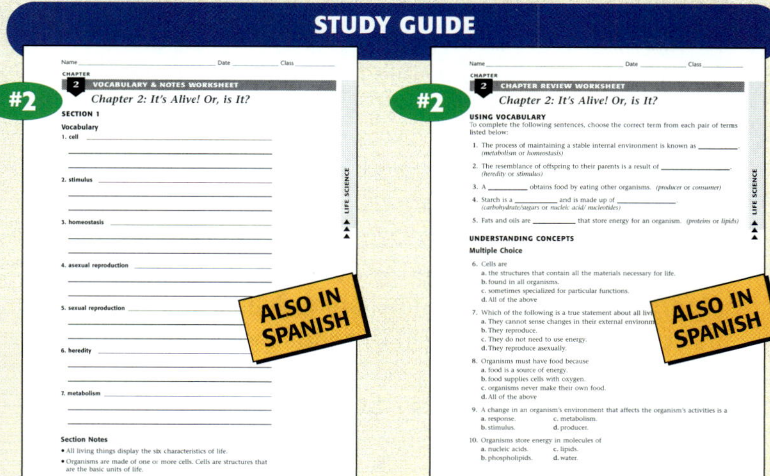

#2 — VOCABULARY & NOTES WORKSHEET
Chapter 2: It's Alive! Or, is It?

#2 — CHAPTER REVIEW WORKSHEET
Chapter 2: It's Alive! Or, is It?

ALSO IN SPANISH

CHAPTER TESTS WITH PERFORMANCE-BASED ASSESSMENT

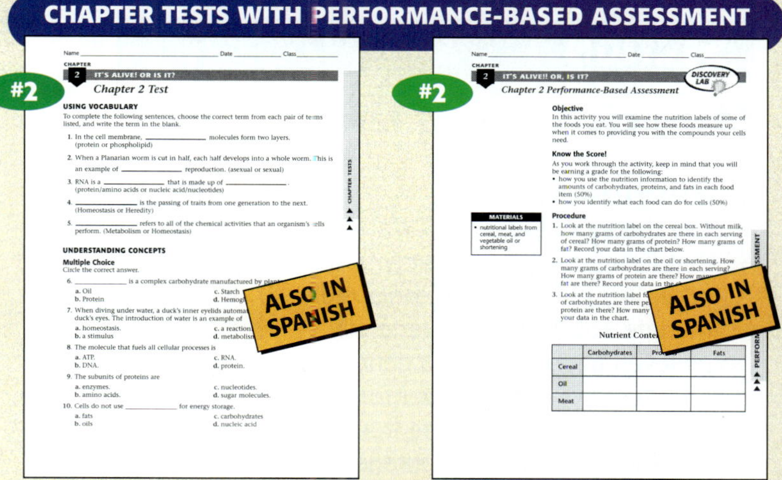

#2 — IT'S ALIVE! OR IS IT?
Chapter 2 Test

#2 — IT'S ALIVE! OR, IS IT?
Chapter 2 Performance-Based Assessment

ALSO IN SPANISH

Lab Worksheets

LABS YOU CAN EAT

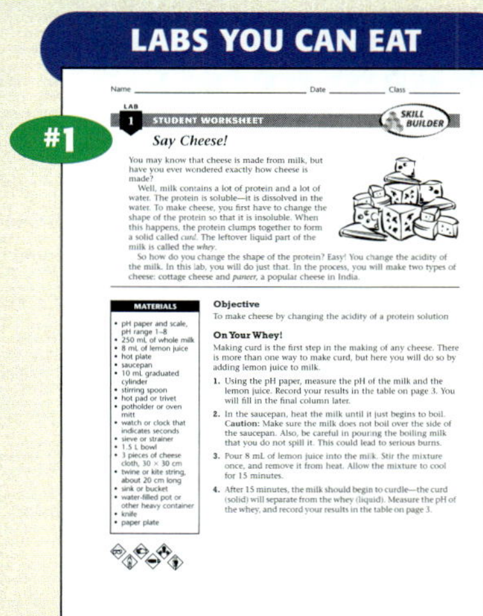

#1 — STUDENT WORKSHEET — SKILL BUILDER
Say Cheese!

LONG-TERM PROJECTS & RESEARCH IDEAS

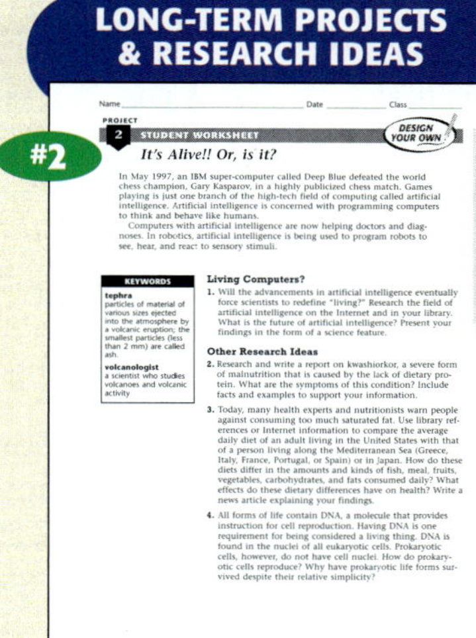

#2 — STUDENT WORKSHEET — DESIGN YOUR OWN
It's Alive!! Or, is it?

DATASHEETS FOR LABBOOK

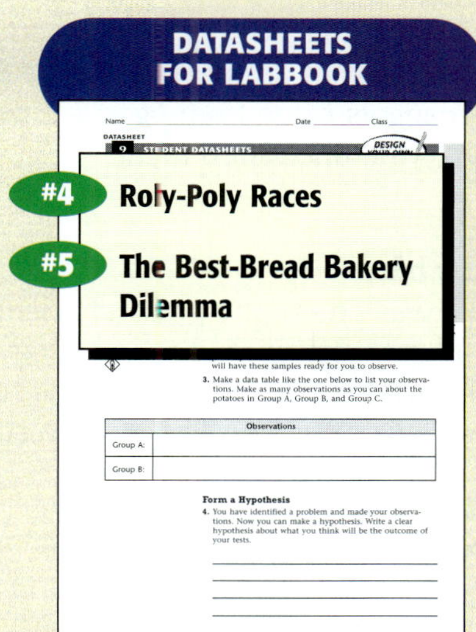

#4 Roly-Poly Races

#5 The Best-Bread Bakery Dilemma

Applications & Extensions

CRITICAL THINKING & PROBLEM SOLVING

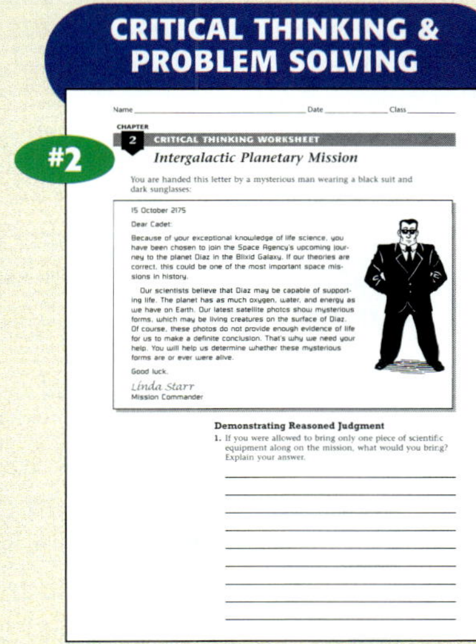

#2 — CRITICAL THINKING WORKSHEET
Intergalactic Planetary Mission

SCIENCE TECHNOLOGY

#1 — Science in the News: Critical Thinking Worksheets
Segment 1
Tapping Into Yellowston'e Hot Springs

SECTION 1

Characteristics of Living Things

▶ Biogenesis

The theory of biogenesis states that living things come only from other living things. However, until the late 1600s, people generally believed in spontaneous generation, the theory that lower forms of life, such as insects, come from nonliving things. The first evidence to disprove spontaneous generation came from controlled experiments conducted in 1667 by Italian Francesco Redi.

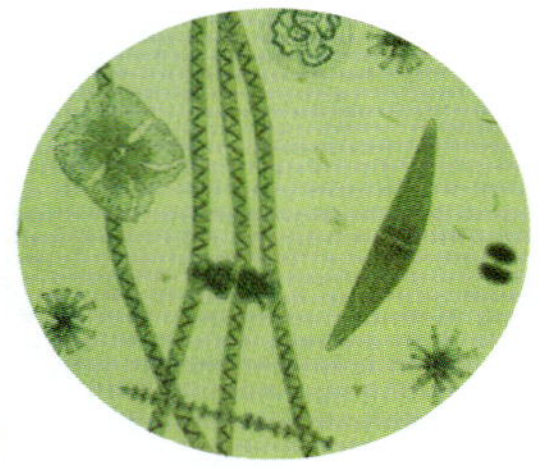

- Redi showed that maggots will appear on meat in an uncovered jar but not on meat in a closed container. Why? The maggots came from eggs laid by flies that had access to the uncovered meat.

▶ Robert Hooke

Robert Hooke was one of the greatest scientists of his time. In addition to his studies in biology with the compound microscope, he was also involved in physics, astronomy, chemistry, geology, and architecture.

- Hooke applied his discovery of the law of elasticity (that the stretching of a solid material is proportional to the force applied to it) to the design of balance springs for watches and clocks. His sketches of Mars were used 200 years later to determine that planet's rate of rotation. He studied the crystal structure of snowflakes. In 1672, he developed the wave theory of light to explain diffraction, which he had also discovered. Hooke was the first person to examine fossils with a microscope and to recognize, 200 years before Charles Darwin was born, that fossils are evidence of changes in organisms on Earth over millions of years.

IS THAT A FACT!

- Different cells in the human body have different life spans, ranging from a few days for intestinal cells to about 120 days for red blood cells and years for brain cells. Scientists have recently discovered that the brain does make new cells. However, they only mature in the hippocampus, which is responsible for learning and memory.

▶ Anton van Leeuwenhoek

Anton van Leeuwenhoek was born in 1632 into a family of tradespeople in Delft, Holland. He was poor and received no advanced education. In 1648, he was apprenticed to a fabric merchant to learn this trade. On his own initiative, he began to grind lenses to experiment with magnification. Historians think he may have been inspired by Robert Hooke's book, *Micrographia* (Small Drawings).

SECTION 2

The Simple Bare Necessities of Life

▶ Cold

Antarctica is a harsh and unforgiving environment. Most of the subantarctic islands are solid rock, and 98 percent of the continent is covered with ice. But consider the following:

- It is home to more than 400 types of lichens and 85 mosses. Lichens, a symbiotic association of algae and fungi, can tolerate low temperatures and minimal moisture.

- Green algae thrive near penguin colonies.

- Moss grows on the few small patches of soil that exist.

- Only two species of flowering plants live in Antarctica.

▶ Heat

In Death Valley National Park, summertime temperatures routinely reach 50°C, and rainfall averages 3.8 cm per year. Yet consider the following:

- More than 900 types of plants live there.

- More than 400 animal species live in this region including bats, kangaroo rats, gophers, bighorn sheep, lizards, tortoises, snakes, spiders, scorpions, beetles, turkey vultures, and roadrunners.

▶ Pressure

Three to four kilometers deep in the ocean, where the pressure is 275 times that at sea level and it is cold and dark, lurk huge yellow jellyfish, giant clams, blind fish, and red worms that are 2 m long. These animals live near deep-sea vents. Water that is trapped beneath the ocean floor and is heated by volcanic activity to as much as 300°C escapes through these vents. The cold ocean cools the water around the vents to about 13°C. Because there is no sunlight at this depth, the animals use the chemicals in the water for energy through a process called chemosynthesis. However, the amount of energy available through chemosynthesis is far less than that obtained through photosynthesis on Earth's surface.

IS THAT A FACT!

- The sand grouse of Chad, in northern Africa, builds its nest many miles from a water source. When the chicks hatch, the parents fly to Lake Chad, where they soak their breast feathers before flying back to their chicks. The chicks then drink the water from their parents' feathers; this both feeds and cools them.

- Some animals that hunt at night, such as cats and some dogs, have a special eye structure that helps them see in the dark. The *tapetum lucidum* "bright carpet," a mirrorlike layer of cells, enhances the eye's ability to see in low light. It makes a cat's eyes appear to glow in the dark.

The Chemistry of Life

▶ "Chemical" Menu

Our bodies need certain kinds of chemicals to live.

- Carbohydrates are the body's primary source of energy. Simple carbohydrates, or sugars, are found in fruits, some vegetables, and milk. Complex carbohydrates, which include starches, are obtained from pasta, seeds, nuts, and vegetables such as peas, beans, and potatoes.

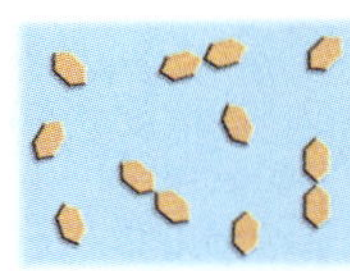

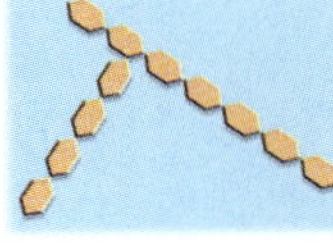

- Nucleic acids, which may contain thousands of components called nucleotides, contain essential information for the construction of proteins.

- Proteins are made of 20 different amino acids. Our cells arrange these amino acids in different combinations to make all the proteins in our body.

- Lipids include saturated and polyunsaturated fats. Saturated fats are present in greater amounts in animal products. Vegetable-based oils have more polyunsaturated fats.

IS THAT A FACT!

- If stretched out end to end, the DNA in an average human body would measure 20 billion kilometers.

- For about 100 years, beginning in the late 1700s, sperm whales were a major source of oil for lubricants and fuel for lamps. These huge animals grow to 18 m long. Sperm oil came from the whale's blubber and unusually large head. Whaling nearly made sperm whales extinct.

▶ Metabolism

Biochemical reactions that take place within a cell are collectively known as metabolism. Enzymes, which are themselves proteins, catalyze or accelerate most of the chemical reactions within a cell. Each type of reaction is catalyzed by a specific enzyme.

- A metabolic pathway is the sequence of chemical reactions needed to make a particular biological molecule. If a disruption occurs somewhere along the pathway, then the organism might develop an illness or suffer a deficiency.

IS THAT A FACT!

- When bears sleep in their dens during winter, their body temperature decreases several degrees. We know this because scientists crawled into the dens and took their temperatures. The lower body temperature reduces energy requirements, so bears can sleep for weeks or months without eating.

It's Alive!! Or, Is It?

Chapter Preview

Section 1
Characteristics of Living Things
- All Living Things Have Cells
- All Living Things Sense and Respond to Change
- All Living Things Reproduce
- All Living Things Have DNA
- All Living Things Use Energy
- All Living Things Grow and Develop

Section 2
The Simple Bare Necessities of Life
- Food
- Water
- Air
- A Place to Live

Section 3
The Chemistry of Life
- Proteins
- Carbohydrates
- Lipids
- Nucleic Acids
- The Cell's Fuel

Directed Reading Worksheet 2

Science Puzzlers, Twisters & Teasers Worksheet 2

Guided Reading Audio CD
English or Spanish, Chapter 2

¡Está vivo! ¿O no...?

Imagínate . . .

La cueva Movile, en Rumania, es uno de los lugares más aterradores y apestosos de la Tierra. Durante más de 5 millones de años, la cueva y sus habitantes han estado aislados del mundo exterior. Ahí viven algunos de los organismos más espeluznantes conocidos por el mundo científico. Escorpiones acuáticos venenosos rondan en pozos oscuros y respiran a través de tubos que les salen del estómago. Centípedos depredadores con antenas gigantes acechan a los insectos más pequeños y les inyectan una toxina paralizadora. Arañas feroces se mueven como relámpagos persiguiendo ciempiés, cochinillas iy hasta a sus propias crías!

Todos los seres vivos necesitan energía para realizar sus actividades. La mayoría la obtienen directa o indirectamente del Sol. Pero, ¿qué sucede con los habitantes de la cueva Movile, que jamás reciben la luz del Sol? En la cueva Movile, la energía proviene de organismos que no podemos ver. Estos microorganismos no se alimentan de otros insectos, sino de sulfuro de hidrógeno. Este compuesto químico abunda en la cueva y deja un horrible olor a huevos podridos en el aire. La energía que los microorganismos obtienen del sulfuro de hidrógeno les sirve de combustible para llevar a cabo sus procesos vitales.

Cuando los microorganismos son comidos por otros organismos, su energía se transfiere.

El uso de la energía es sólo una de las características de la vida. Sigue leyendo para averiguar qué tienen en común todos los organismos.

34

Imagine . . .

A natural sponge, as opposed to the rectangular colorful ones purchased at a store, is usually tan or brown colored, irregularly shaped, and extremely porous. Few people would think that it had ever been a living animal. It may seem odd, but natural sponges are, in fact, the skeletons of marine animals. What an animal or plant looks like can sometimes fool us into thinking that it is a nonliving thing, but if something has the characteristics of living things as described in this chapter, it is alive!

Usa tus conocimientos para responder a las siguientes preguntas en tu cuaderno de ciencias:

1. ¿Qué características comparten todos los seres vivos?

2. ¿Qué necesitan los organismos para vivir?

¡Enciendan las luces!

Reaccionar ante un cambio es una característica de todos los seres vivos. En esta actividad, trabajarás con un compañero para averiguar cómo responden los ojos a los cambios de luz.

Procedimiento

1. Observa los ojos de tu compañero o compañera en un salón con luz normal. Ubica la pupila, el área negra de la parte coloreada del ojo. La luz entra en el ojo a través de la pupila. Observa su tamaño.

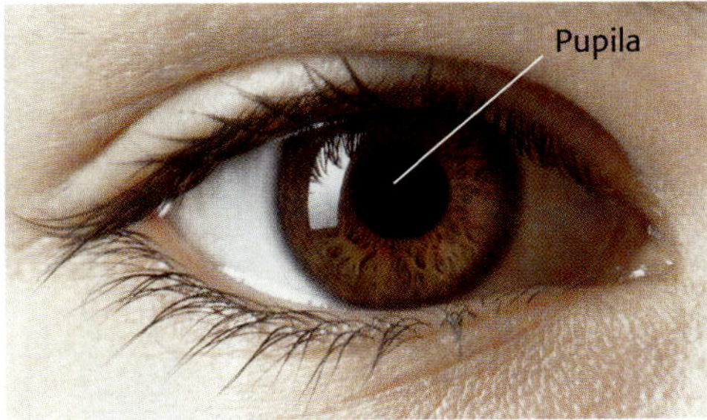

2. Haz que tu compañero mantenga los ojos abiertos y que se los tape con la mano de modo que no les llegue la luz. Espera más o menos 1 minuto.

3. Dile a tu compañero o compañera que retire las manos rápidamente. Observa sus pupilas inmediatamente. Registra en tu cuaderno de ciencias lo que les pase.

4. Ahora, enciende una **linterna** y dirígela brevemente a los ojos de tu compañero o compañera. Registra en tu cuaderno de ciencias los efectos que esto tiene sobre sus pupilas. **Cuidado:** No utilices el Sol como fuente de luz.

5. Cambia de lugar con tu compañero o compañera y repite todo el procedimiento de modo que él o ella pueda observar cómo respondieron tus ojos a los cambios de luz.

Análisis

6. ¿Cómo respondieron los ojos de tu compañero o compañera a los cambios de luz?

7. ¿Cómo afectó tu visión el cambio en el tamaño de tus pupilas? ¿Qué puedes deducir respecto a por qué cambia el tamaño de la pupila?

What Do You Think?

Accept all reasonable responses.

Students will have a chance to revise their answers in the Chapter Review under NOW What Do You Think?

Investigate!

MATERIALS

FOR EACH GROUP:
• flashlight

Safety Caution: Students must not use the sun as a source of light.

Answers to Investigate!

3. The pupils were enlarged.

4. The pupils became smaller.

6. The pupils enlarge when there is little light and become smaller when there is a lot of light.

7. The larger the pupils are, the brighter things appear. In a dark environment, pupils become larger, so more light enters the eye. The surroundings appear brighter and can be more easily seen. In a bright environment, the pupils become smaller, and less light enters the eye. The surroundings are clearly visible without extra light entering the eye.

 **SCIENCE**

On the sea floor of the Gulf of Mexico, a chunk of ice and gas, called a methane hydrate, is home for 2–5 cm worms that live on and in the hydrate. Scientists are investigating whether the worms feed on bacteria or get energy directly from the gas.

Focus

Characteristics of Living Things

This section explains the characteristics that describe living things. Students will learn that living things have cells; that they sense and respond to stimuli; that they reproduce, have DNA, and use energy; and that they grow and develop. Based on this information, students will be able to identify things as either living or nonliving.

Bellringer

Display this question on the board or an overhead projector:

What are four living and four nonliving things that you interact with or see everyday? (living: family members, pets, house plants and trees, birds, insects; nonliving: clothes, books, automobile, furniture, radio, sidewalk) Sheltered English

1) Motivate

DISCUSSION

Stimuli Ask students what they do when they go outside and it is cold. (They put on a jacket or go back inside, where it is warmer.)

Explain that feeling cold is a stimulus and that their reaction to the cold is a response. Ask students how people use technology to improve their ability to respond to environmental stimuli. (Examples are furnaces and wood stoves to heat buildings, air conditioners to cool buildings, sunglasses to shield eyes from bright sunlight, greenhouses to extend plant growing seasons.)

Technology enables us to respond to stimuli and live in more places.

VOCABULARIO

célula	reproducción sexual
estímulo	ADN
homeostasis	herencia
reproducción asexual	metabolismo

OBJETIVOS

- Enumera las características de los seres vivos.
- Distingue entre la reproducción asexual y la reproducción sexual.
- Define y describe la homeostasis.

Características de los seres vivos

Un día, notas algo raro en el pasto del patio. Es una cosa viscosa, de color amarillo y del tamaño de una moneda de diez centavos. ¿Será parte de una planta que se cayó de un árbol? ¿Estará vivo?

Aunque existe una cantidad impresionante de seres vivos en nuestro planeta, todos tienen ciertas cosas en común. ¿En qué se parecen un perro y un árbol? ¿En qué se parecen un pez y un hongo? Y, ¿en qué te pareces *tú* a una criatura viscosa (también conocida como moho viscoso)? Sigue leyendo para conocer las seis características que todos los organismos comparten.

Moho viscoso

1 Todos los seres vivos tienen células

Todos los seres vivos están compuestos de una o más células. Una **célula** es una estructura cubierta por una membrana que contiene todos los materiales esenciales para la vida. La membrana separa el contenido de la célula de su ambiente externo.

Muchos organismos, como los de la **Figura 1,** están compuestos de una sola célula. Otros, como los monos y los árboles de la **Figura 2** están compuestos de billones de células. La mayoría de las células son tan pequeñas que no pueden verse a simple vista.

Cada una realiza las funciones fundamentales de la vida. En un organismo con muchas células, éstas pueden realizar funciones específicas. Por ejemplo, las células nerviosas están especializadas en el transporte de señales y las células musculares están especializadas en realizar movimientos.

Figura 1 *Todos estos organismos están compuestos de una sola célula.*

Figura 2 *Estos organismos están compuestos de billones de células.*

36

Directed Reading Worksheet 2 Section 1

internetconnect

SCILINKS
NSTA

TOPIC: Characteristics of Living Things
GO TO: www.scilinks.org
*sci*LINKS NUMBER: HSTL030

Science BlOOpers

Leonardo da Vinci made many scientific discoveries and observations in the fifteenth and sixteenth centuries, but he mistakenly believed that the eye emitted a ray that struck and then rebounded from the observed object.

---

2 Todos los seres vivos detectan y responden a los cambios

Todos los organismos tienen la capacidad de detectar los cambios del medio ambiente y de responder a los mismos. Cuando tus pupilas se exponen a la luz su respuesta es contraerse. Un cambio en el medio ambiente de un organismo que afecta su actividad se denomina **estímulo.**

Los estímulos pueden ser substancias químicas, la gravedad, la obscuridad, la luz, sonidos o sabores. Son factores que hacen que los organismos respondan de alguna manera. Un leve roce provoca una respuesta en la planta de la **Figura 3.**

✔ Autoevaluación

¿Es tu reloj despertador un estímulo? Explica. *(Consulta la página 636 para comprobar tu respuesta.)*

Homeostasis Aunque el medio ambiente externo cambie, los organismos deben mantener un medio ambiente interno estable para sobrevivir. Esto se debe a que los procesos vitales de un organismo comprenden distintos tipos de reacciones químicas que sólo pueden ocurrir en ambientes equilibrados. Mantener un medio ambiente interno estable se denomina **homeostasis**.

Tu cuerpo mantiene una temperatura de alrededor de 37° C. Cuando sientes calor, tu cuerpo responde mediante la transpiración. Cuando sientes frío, tus músculos se sacuden bruscamente con el objeto de generar calor y eso te hace tiritar. Ya sea transpirando o tiritando, tu cuerpo trata de volver a la normalidad.

Otro ejemplo de homeostasis es el proceso para mantener estable el nivel de azúcar en la sangre. Después de comer, el nivel de azúcar en la sangre se eleva. El cuerpo responde liberando un compuesto que elimina el azúcar de la sangre y la almacena en las células musculares y hepáticas (del hígado). Cuando has pasado un buen tiempo sin comer, tu nivel de azúcar sanguínea comienza a disminuir. Entonces, tu cuerpo produce otro compuesto que libera el azúcar de los músculos y el hígado.

Figura 3 *El roce de un insecto hace que la* Dionea muscípula *cierre sus hojas rápidamente.*

oceanografía
CONEXIÓN

Los peces de las aguas heladas de la Antártida producen un anticongelante natural que los protege del congelamiento.

37

WEIRD SCIENCE

The first indication that the pancreas was the organ that secreted insulin, the compound that removes sugar from the blood, came when flies were noticed swarming over the urine of a dog whose pancreas was damaged. The flies were attracted to the excess sugar in the urine.

Critical Thinking Worksheet 2
"Intergalactic Planetary Mission"

GUIDED PRACTICE

Writing Have students list stimuli that they experience in their lives. Write them on the board. (heat, cold, a red traffic light)

Ask each student to write this list on a sheet of paper and then to write how a person would respond to each stimulus. `Sheltered English`

MEETING INDIVIDUAL NEEDS

Learners Having Difficulty Have students use pictures from magazines to create a poster that shows three living and three non-living things. Underneath each picture, have students write a brief paragraph explaining the characteristics that identify each thing as either living or nonliving. `Sheltered English`

Answer to Self-Check

Your alarm clock is a stimulus. It rings, and you respond by shutting it off and getting out of bed.

MATH and MORE

Red blood cells are very small. Healthy humans have about 5 million red blood cells per milliliter (mL) of blood. If there are 1,000 mL per liter, how many cells are there in a liter? (5 million × 1,000 = 5 billion, or 5,000,000,000)

Math Skills Worksheet 4
"A Shortcut for Multiplying Large Numbers"

Math Skills Worksheet 26
"Multiplying and Dividing in Scientific Notation"

 **PG 564**

Roly-Poly Races

DEBATE

Nature Versus Nurture
Scientists have proven that we inherit our physical characteristics from our parents (nature). They continue to research whether we inherit our personalities from our parents. Some scientists say that where we live and how we are raised are more important (nurture). What do students think is the critical factor, nurture (care) or nature (heredity)? Why?

USING THE FIGURE

Ask the question posed in **Figure 4.** (Students can find two buds forming close to the base of the hydra.)

MATH and MORE

The red kangaroo can cover 12 m in a single jump. The African sharp-nosed frog can leap 5.35 m. What percentage of the kangaroo's jump is the frog's leap?

(5.35 ÷ 12 = 0.4458, or 0.45
0.45 × 100 = 45%)

The common flea can leap 19 cm. What percentage of the frog's leap is the flea's?

(5.35 × 100 = 535 cm
19 ÷ 535 = 0.0355, or 0.04
0.04 × 100 = 4%)

 **Math Skills Worksheet 18**
"Decimals and Fractions"

 **Math Skills Worksheet 21**
"Percentages, Fractions, and Decimals"

Figura 4 *La hidra es un animal que se reproduce asexualmente mediante la formación de brotes que luego se desprenden para crecer como nuevos individuos. ¿Puedes encontrar los brotes de esta hidra?*

Figura 5 *Como la mayoría de los animales, los osos producen crías mediante la reproducción sexual.*

3 Todos los seres vivos se reproducen

Todos los organismos producen otros semejantes a ellos. Esto se realiza por medio de la reproducción asexual o de la reproducción sexual. En la **reproducción asexual** un solo progenitor puede producir crías idénticas a sí mismo. La **Figura 4** muestra un organismo que se reproduce asexualmente. La mayoría de los organismos unicelulares se reproducen así. La **reproducción sexual** generalmente requiere dos progenitores, cuyas crías tendrán características de ambos. La mayoría de los animales y plantas se reproducen así. Los ositos de la **Figura 5** fueron producidos sexualmente por sus padres.

4 Todos los seres vivos tienen ADN

Las células de los seres vivos contienen una molécula denominada **ADN** (**á**cido **d**esoxirribo **n**ucleico). El ADN tiene instrucciones para fabricar *proteínas*. Las proteínas participan en casi todas las actividades de las células de un organismo, y determinan muchas de sus características.

Cuando los organismos se reproducen, traspasan copias de ADN a sus crías. El traspaso de características de una generación a otra se denomina **herencia.** Las crías, como los niños de la **Figura 6,** se parecen a sus padres debido a la herencia.

Figura 6 *Los niños se parecen a sus padres debido a la herencia.*

5 Todos los seres vivos utilizan energía

Los organismos usan energía para realizar los procesos químicos de la vida, como la producción y descomposición de alimentos, la transferencia de material al interior y al exterior de las células y la producción de nuevas células. El **metabolismo** de un organismo es el conjunto de sus actividades químicas.

38

MISCONCEPTION ALERT

Though very much alive, mules and most other hybrids cannot reproduce. Hybrids are the result of mating organisms from different species. A mule is the offspring of a mare (a female horse) and a jack (a male donkey). Mules often live long, healthy lives, but they never have babies.

IS THAT A FACT!

Lichens, which dominate the flora of Antarctica, have an extremely slow growth rate. Certain species grow only 1 mm every 100 years. Scientists estimate that some lichens may be more than 5,000 years old.

Las computadoras realizan muchas funciones, como almacenar información y hacer cálculos complejos. Algunas computadoras han sido programadas para aprender, es decir, para resolver problemas más rápida y eficientemente con el tiempo. ¿Crees que las computadoras lleguen a ser tan avanzadas que se les deba considerar como vivas? ¿Por qué?

¡MATEMÁTICAS!

Proporciones corporales

Al crecer, tus proporciones corporales cambian. Cuando naciste, tu cabeza era el 25% de tu altura. ¿Qué porcentaje de tu altura es tu cabeza ahora? Para averiguarlo, mide tu estatura y el largo de tu cabeza. Luego, divide el largo de tu cabeza entre tu estatura. Finalmente, multiplica el resultado por 100 para calcular qué porcentaje de tu altura corresponde a tu cabeza.

6 Todos los seres vivos crecen y se desarrollan

Todos los seres vivos crecen durante ciertos períodos de su vida. Un organismo unicelular crece a medida que la célula crece. Los organismos que están compuestos de muchas células crecen al aumentar el número de células.

Además de aumentar en tamaño, los seres vivos también se desarrollan y cambian a medida que crecen. Al igual que los organismos de la **Figura 7,** tú también pasarás por muchas etapas en tu vida a medida que te conviertas en adulto.

Figura 7 *Con el tiempo, las bellotas se convierten en plántulas que, a su vez, llegarán a ser robles.*

1. ¿Qué características de los seres vivos tiene un río? ¿Está vivo el río?

2. ¿Qué tiene que ver el pelaje de un oso con la homeostasis?

3. ¿Cómo se relaciona la reproducción con la herencia?

4. **Aplicar conceptos** Nombra algunos estímulos presentes en tu medio ambiente. ¿Cómo respondes a estos estímulos?

Explora

Imagínate que tienes la habilidad de crear un nuevo organismo. ¿Cómo sería ese organismo? ¿De qué manera presentaría todas las características de un ser vivo?

---

4 Close

Answer to APPLY

Answers will vary. Students should include the six characteristics of living things in their answer and explain why a computer could or could not do those things and be considered alive.

Answer to MATHBREAK

Answers will vary, but the length of a student's head should be about one-fifth (or 20 percent of) the student's height.

Quiz

1. An apple tree is a living thing. Can it make oranges? Why or why not? (No; living things reproduce only themselves, not different living things.)

2. What is the difference between growth and development? (Growth is an increase in size. Development is the change of form of an organism.)

3. Name three activities of an organism that require energy. (Organisms need energy to make food, to break down food, to move materials into and out of cells, and to build cell parts.)

ALTERNATIVE ASSESSMENT

Writing Have students read a story of their choice and find five examples of stimuli and responses. Then have students write an explanation of why it is important to be able to respond to all of these stimuli.

Answer to Explore

Answers will vary. Students should explain how their organism displays each of the six characteristics of living things.

---

1. A river has energy (it moves) and can grow larger (after rain, or when snow melts). But it is not alive because it is not made of cells, cannot respond to stimuli, has no DNA, and cannot reproduce.

2. Homeostasis is the maintenance of a stable internal environment. The fur coat of a bear helps it keep a stable body temperature.

3. Heredity is the passing of characteristics from parents to offspring. When organisms reproduce, offspring inherit copies of their parents' DNA.

4. Answers will vary. Examples can include such things as the way something tasted, smelled, felt, sounded, or looked. Responses to the stimuli should describe the action or effect the stimuli produced in the student.

Focus

The Simple Bare Necessities of Life

This section identifies the things that an organism needs to live. Students will learn the roles that food, water, and air play in an organism's survival. They will also learn that where an organism lives is related to its ability to obtain the necessities of life.

 **Bellringer**

Pose the following question to students:

What do you think your mass would be if there were no water in your body? Write your answer in your ScienceLog. (The human body is approximately 70 percent water. If a student has a mass of 40 kg, the water's mass is 40 kg × 0.70 = 28 kg. The student's mass without water would be 40 kg − 28 kg = 12 kg.)

1) Motivate

DISCUSSION

Adaptation Show students pictures of a desert, the canopy of a rain forest, the Arctic, and seaside cliffs. Ask students to describe animals that could live in these places and to explain how each is adapted to obtain its necessities.

 **Directed Reading Worksheet 2** Section 2

Sección 2

Las necesidades básicas de la vida

VOCABULARIO
productores
consumidores
descomponedores

OBJETIVOS
- Explica por qué los organismos necesitan alimento, agua, aire y un espacio en donde vivir.
- Analiza cómo los seres vivos obtienen lo que necesitan para vivir.

¿Sabías que tú tienes las mismas necesidades básicas que un árbol o un sapo? De hecho, casi todos los organismos tienen las mismas necesidades básicas: alimento, agua, aire y un espacio en donde vivir.

Alimento

Los seres vivos necesitan alimento. Los alimentos entregan la energía y las materias primas necesarias para que los organismos realicen sus procesos vitales y construyan y reparen células y partes del cuerpo. Sin embargo, no todos los organismos obtienen alimento de la misma manera. Los organismos se agrupan en tres categorías en función de la forma en que obtienen alimento.

Algunos organismos se denominan **productores,** ya que producen su propio alimento. Las plantas de la **Figura 8** son un ejemplo. Las plantas utilizan la energía del Sol para producir alimento a partir del agua y dióxido de carbono. Algunos productores obtienen energía y alimento a partir de productos químicos del medio ambiente.

Otros organismos se denominan **consumidores,** ya que se comen (consumen) otros organismos para obtener alimento. La salamandra de la **Figura 8** es un ejemplo de un consumidor. Obtiene la energía mediante el consumo de otros organismos.

Algunos consumidores son descomponedores. Los **descomponedores** obtienen su alimento mediante la descomposición de nutrientes provenientes de organismos muertos o desechos animales. El hongo de la **Figura 8** está absorbiendo nutrientes de plantas muertas.

Figura 8 *La salamandra y el hongo sobre el cual se arrastra son consumidores. Las plantas son productores.*

Agua

El cuerpo humano está compuesto en su mayoría de agua. Las células de todos los organismos están compuestas de aproximadamente 70 por ciento de agua. El metabolismo de los organismos depende del agua ya que la mayoría de las reacciones químicas que ocurren en ellos necesitan del agua para llevarse a cabo.

Los organismos se diferencian según la cantidad de agua que requieren y cómo la obtienen. Sin agua, podrías sobrevivir sólo 3 días. Obtienes agua de los líquidos y alimentos que consumes. La rata canguro del desierto la obtiene de sus alimentos.

40

CONNECT TO EARTH SCIENCE

Use Teaching Transparency 154 to encourage student discussion of how animals thrive in so many different climates and biomes.

 **Teaching Transparency 154** "Climate Zones of the Earth"

TOPIC: The Necessities of Life
GO TO: www.scilinks.org
*sci*LINKS **NUMBER:** HSTL035

Aire

El aire es una mezcla de gases, como oxígeno y dióxido de carbono. Los seres vivos necesitan oxígeno para vivir. El oxígeno se utiliza en la *respiración,* el proceso químico que libera energía de los alimentos. Los organismos que habitan sobre la tierra obtienen oxígeno del aire. Los que viven en agua dulce y salada obtienen oxígeno disuelto del agua o suben a la superficie para obtenerlo del aire. Otros, como la araña acuática de la **Figura 9,** llegan a grandes extremos para obtener oxígeno.

Además de oxígeno, las plantas verdes, las algas y algunas bacterias también requieren dióxido de carbono. Dichos organismos producen su alimento a partir del dióxido de carbono y el agua mediante la *fotosíntesis,* el proceso que convierte la energía del Sol en energía que se almacena en los alimentos.

Figura 9 *Esta araña se encierra en una burbuja de aire para obtener oxígeno debajo del agua.*

Un lugar donde vivir

Todos los organismos necesitan un espacio que tenga todo lo que requieren para vivir. Algunos organismos, como los elefantes, requieren de un espacio amplio. Otros, como las bacterias, pueden pasar toda su vida en un poro en la punta de tu nariz.

Como el espacio sobre la Tierra es limitado, muchas veces los organismos compiten con otros para obtener alimento y agua. Muchos animales, como la curruca de la **Figura 10,** reclaman un espacio y tratan de alejar a los demás animales. Las plantas también compiten entre sí para obtener espacio y acceso al agua y a la luz del Sol.

Figura 10 *El canto de una curruca es más que una bella melodía. La curruca canta un mensaje para alejar a las demás currucas y proteger así su hogar.*

REPASO

1. ¿Por qué se clasifica a los descomponedores como consumidores? ¿En qué se diferencian de los productores?

2. ¿Por qué la mayoría de las células están compuestas de un 70 por ciento de agua?

3. **Hacer deducciones** ¿Podría existir vida en la Tierra si el aire sólo contuviera oxígeno? Explica.

4. **Relacionar** ¿Cómo pueden una cueva, una hormiga y un lago satisfacer las necesidades de un organismo?

¿Hay vida en Marte? Averígualo en la página 50.

41

---

2 Teach

DEMONSTRATION

Fire and Life Demonstrate that both a human and a burning candle share some qualities of life. Briefly hold a cold drinking glass that is inverted over a candle flame. The glass will be fogged with water droplets. Now, breathe into a cold glass. The same will happen. Besides giving off water, both use oxygen and fuel (food or wax) and give off carbon dioxide and energy.
Sheltered English

3 Close

Quiz

1. Give an example of a producer, consumer, and decomposer. (producer: any plant; consumer: any animal; decomposer: fungi)

2. What factors affect where a plant or animal lives? (competition with other organisms and the availability of water and food that is sufficient to meet the organism's needs)

ALTERNATIVE ASSESSMENT

Writing Have students collect pictures from magazines and create a poster that shows the home of a plant and an animal with all the necessities of life that were discussed in this section. Then have students write a script for a nature documentary to accompany the poster.
Sheltered English

Reinforcement Worksheet 2
"Amazing Discovery"

Answers to Review

1. Decomposers are consumers because they must obtain the food they need from other organisms. Unlike producers, decomposers cannot produce their own food.

2. Most of the chemical reactions that occur in cells depend on the presence of water.

3. Life could not exist as we know it. Green plants, algae, and some bacteria need carbon dioxide gas as well as oxygen. Without the carbon dioxide, they could not survive, and other organisms could not rely on them as a food source.

4. Answers will vary. A cave could be a place to live. An ant could be food. A lake could be a place to live as well as a source of water.

Focus

The Chemistry of Life

In this section, students will learn about life on a cellular level. They will learn about proteins, carbohydrates, lipids, phospholipids, nucleic acids, and ATP and why they are essential to sustain life. Students will learn how these substances are alike and how they are different in both form and function.

Bellringer

Have students unscramble the following words and then use all four of them in a single sentence:

cdporesru (producers)
gnreey (energy)
dofo (food)
rwtea (water)

(Sample sentence: Producers use energy from the sun to make food from carbon dioxide and water.)

1 Motivate

DEMONSTRATION

Protein Model On each of 10–15 small self-adhesive notes, write a single letter of the alphabet. The letters should spell a few simple words when placed side by side. Place the note papers on two or three of the colored papers. Tell students that the binder contents represent proteins. The letters are the "amino acids." Each word is a "protein." Dismantle the "proteins," and use some of the "amino acids" to assemble a new "protein." Sheltered English

---

VOCABULARIO

proteínas
enzimas
carbohidratos
lípidos
fosfolípidos
ácidos nucleicos
ATP

OBJETIVOS

- Compara los elementos químicos fundamentales de una célula.
- Explica la importancia del ATP.

La química de la vida

Los seres vivos están compuestos de células, pero... ¿de qué están compuestas las células? Todo lo que existe está compuesto de pequeñas unidades denominadas *átomos*. Existen alrededor de 100 tipos de átomos que se combinan para crear todo lo que existe.

Una substancia compuesta de un solo tipo de átomo es un *elemento*. Cuando dos o más átomos se unen, forman una *molécula*. Una molécula puede estar compuesta de un solo elemento o de dos o más elementos distintos. Las moléculas de los seres vivos generalmente están compuestas de distintas combinaciones de seis elementos: carbono, hidrógeno, nitrógeno, oxígeno, fósforo y azufre.

Las proteínas, carbohidratos, lípidos, ácidos nucleicos y ATP son compuestos que se encuentran en las células.

Proteínas

Casi todos los procesos vitales de una célula requieren de las proteínas para llevarse a cabo. Después del agua, las proteínas son los materiales más abundantes en las células. Las **proteínas** son moléculas grandes que están compuestas de subunidades denominadas *aminoácidos*.

Los organismos descomponen las proteínas de los alimentos para proveer aminoácidos a sus células. Luego, estos aminoácidos se unen para formar nuevas proteínas. Algunas proteínas están compuestas de pocos aminoácidos, mientras que otras contienen más de 10,000.

Las proteínas cumplen muchas funciones distintas. Algunas forman estructuras fáciles de ver, como las de la **Figura 11,** mientras que otras funcionan a nivel celular. La proteína *hemoglobina,* que se encuentra en los glóbulos rojos, se une al oxígeno para distribuirlo por todo el cuerpo. Algunas proteínas protegen a las células de los materiales extraños. Las **enzimas** son proteínas que aceleran las reacciones químicas en las células. Las células las utilizan para fabricar los compuestos químicos que requieren y para descomponerlos con el fin de obtener energía.

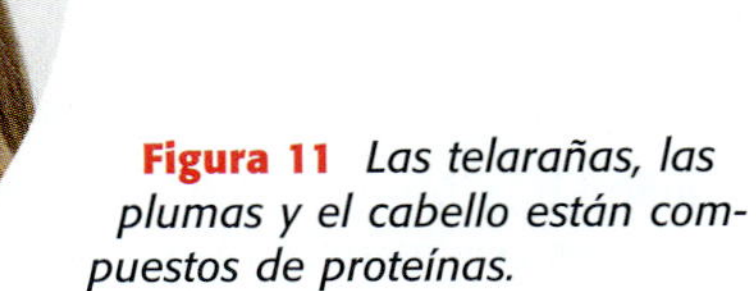

Figura 11 *Las telarañas, las plumas y el cabello están compuestos de proteínas.*

42

Multicultural CONNECTION

Hunters and gatherers of all races historically required very large areas of land to sustain them. Most cultures later developed farming and herding techniques that made possible higher population densities. Staple crops vary around the world, but the millet and sorghum grains of Africans, the wheat and barley of Europeans, the corn and squash of Native Americans, and the rice and soybeans of Asians, in correct amounts and supplemented with other foods, all are equally nutritious.

Carbohidratos

Los **carbohidratos** son un grupo de compuestos formados por azúcares. Las células los utilizan como fuente de energía y para almacenarla. Cuando un organismo requiere energía, sus células descomponen carbohidratos para liberar la energía que está almacenada en ellos.

Existen dos tipos de carbohidratos: simples y complejos. Los carbohidratos simples están compuestos de una molécula de azúcar o de unas pocas moléculas de azúcar enlazadas. El azúcar de las frutas y el del cereal son carbohidratos simples.

Cuando un organismo obtiene más azúcar de la que necesita, el azúcar adicional puede almacenarse como carbohidratos complejos para ser utilizada en el futuro. Los carbohidratos complejos están compuestos de cientos o miles de moléculas de azúcar enlazadas. El cuerpo produce carbohidratos complejos y los almacena en el hígado. Sin embargo, las plantas producen los carbohidratos más complejos. *El almidón* es un carbohidrato complejo producido por las plantas. Como se muestra en la **Figura 12,** una planta de papa almacena el azúcar adicional como almidón. Cuando comes puré de papas o papas fritas, estás comiendo el almidón que la planta almacenó. Tu cuerpo puede descomponer este carbohidrato complejo y liberar la energía que contiene.

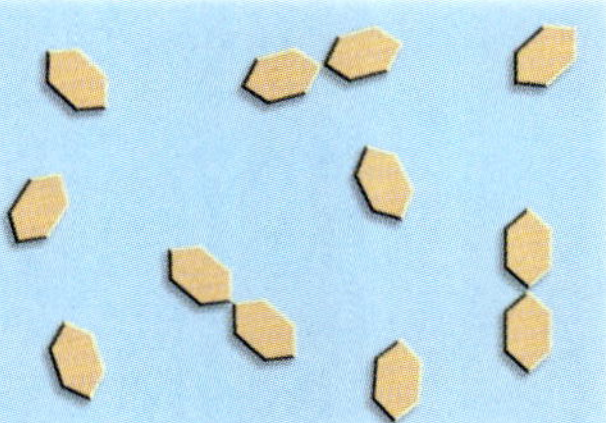

Azúcares

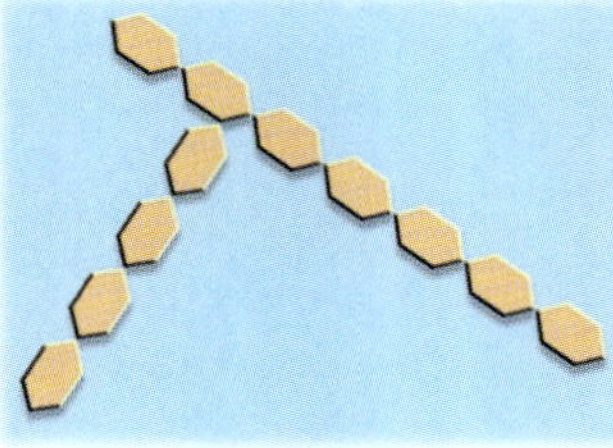

Almidón

Figura 12 *La mayoría de los azúcares son carbohidratos simples. El azúcar que se extrae de la planta de la papa se almacena en la papa en forma de almidón, un carbohidrato complejo*

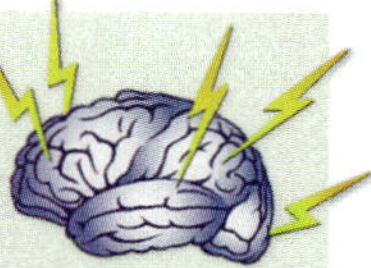

Los carbohidratos especiales que se unen a las proteínas de la superficie de los glóbulos rojos determinan tu tipo de sangre.

Laboratorio

Cómo detectar el almidón

Cuando el **yodo** entra en contacto con el almidón, se vuelve negro. Utiliza esta característica para averiguar qué **muestra del alimento** que te dé tu maestro o maestra contiene almidón.

Cuidado: El yodo puede manchar la ropa. Usa guantes protectores, gafas de seguridad y un delantal.

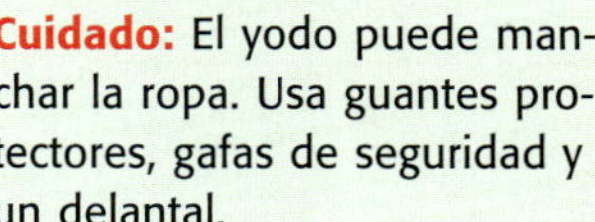

43

Fireflies produce their flashing light by a chemical reaction. The enzyme luciferase acts on the chemical luciferin in the presence of ATP to create the light. Scientists now use luciferase in the laboratory to study everything from heart disease to muscular dystrophy.

TOPIC: The Chemistry of Life
GO TO: www.scilinks.org
*sci***LINKS NUMBER:** HSTL040

---

② Teach

Quick Lab

MATERIALS

FOR EACH GROUP:
- iodine solution in small bottle
- plastic eyedropper
- 25 × 25 cm piece of aluminum foil to hold food samples
- cracker, small piece of bread, potato, chocolate, apple, broccoli, celery, piece of hot dog

Safety Caution: Remind students to review all safety cautions and icons before beginning this lab activity.

Students should not eat any of the food samples. Iodine will stain and can be toxic. Dilute the iodine to prevent injury. Have a functioning eyewash available. Each student should wear safety goggles, an apron, and protective gloves.

Instruct students to use only a few drops of iodine on each sample.

DISCUSSION

Food Choices Ask students which snack foods they prefer to eat before playing or participating in sports. Write their responses on the board. Ask students why they prefer these foods. Taste? Simply to eliminate hunger? Do they think these foods give them more energy? Explain that this section may change their opinions about the foods they eat.

Directed Reading Worksheet 2 Section 3

MEETING INDIVIDUAL NEEDS

Advanced Learners
When the human body is unable to make a necessary protein, the result is often a disease. Hemophilia and diabetes are two conditions caused by missing or defective proteins. Have students prepare a report on the cause of and the treatment for one of these conditions. Their information should include the specific protein that is lacking, a brief description of how the condition affects the body, and the role of DNA in the disease.

LabBook · **PG 566**

The Best-Bread Bakery Dilemma

Homework

Concept Mapping Have students collect the nutrition labels from five food items in their home and examine the number of grams of carbohydrates, fats, and proteins in each item. Tell students to construct a concept map that best relates the items with the headings *carbohydrates*, *lipids*, and *proteins*, based on the nutrient content. If an item belongs in two categories, the map should reflect that information.

Teaching Transparency 4
"Phospholipid Molecule and Cell Membrane"

Figura 13 *Estos son dos lípidos comunes utilizados en la cocina. La manteca proviene de la grasa de los animales, mientras que el aceite proviene del maíz.*

Figura 14 *El interior de la célula está rodeado por una membrana de moléculas de fosfolípidos.*

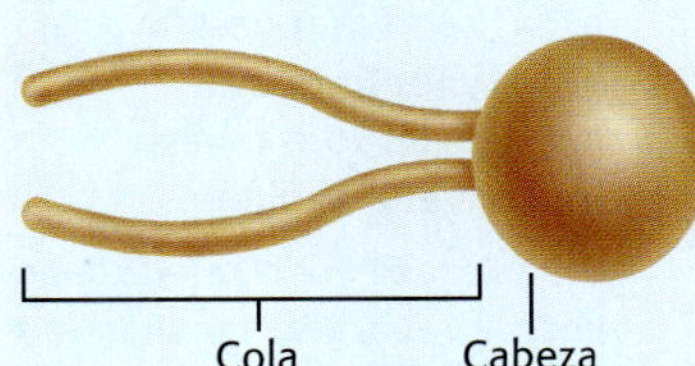

a *La cabeza de una molécula de fosfolípidos es atraída por el agua, pero la cola no.*

Lípidos

Los **lípidos** son compuestos que no se mezclan con agua. Cumplen funciones importantes en la célula, y algunos almacenan energía. Otro lípidos forman las membranas de las células.

Grasas y aceites Las grasas y los aceites son lípidos que almacenan energía. Cuando un organismo ha utilizado casi todos sus carbohidratos, puede obtener energía de sus lípidos. Las estructuras de las grasas y los aceites son casi idénticas, pero a temperatura ambiente, la mayoría de las grasas son sólidos y los aceites son líquidos. La mayoría de los lípidos almacenados en las plantas son aceites, mientras que la mayoría de los lípidos de los animales son grasas. Algunas de las fuentes de grasas y aceites que consumes aparecen en la **Figura 13.**

Fosfolípidos Todas las células están rodeadas por una *membrana celular*. Los **fosfolípidos** son moléculas que forman la mayor parte de la membrana celular. Cuando los fosfolípidos están en el agua, las colas se unen y las cabezas se orientan hacia el agua. La cabeza de la molécula de un fosfolípido es atraída por el agua, pero la cola no. La **Figura 14** muestra cómo las moléculas de fosfolípidos forman dos capas cuando están en el agua.

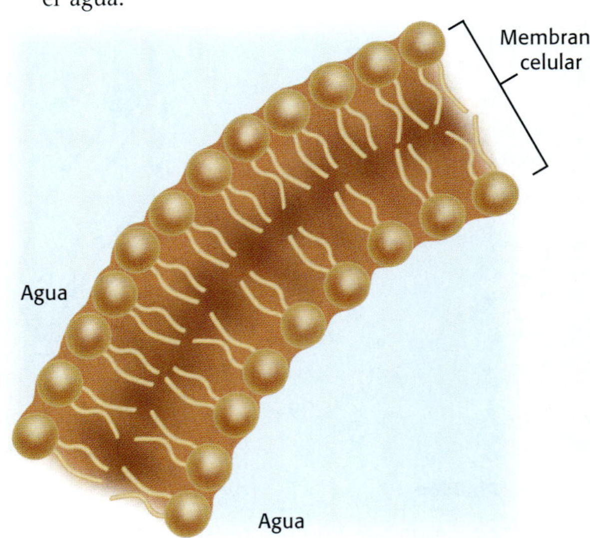

b *Cuando las moléculas de fosfolípidos se unen en el agua, forman dos capas.*

Ácidos nucleicos

Los **ácidos nucleicos** son compuestos que contienen subunidades denominadas *nucleótidos*. Un ácido nucleico puede contener miles de nucleótidos. A menudo, los ácidos nucleicos se denominan los planos de la vida porque contienen toda la información necesaria para que una célula produzca todas sus proteínas.

El ADN es un ácido nucleico. Como ya sabes, una molécula de ADN es como un libro de recetas titulado *Cómo hacer proteínas.* Cuando la célula necesita producir una proteína, obtiene información del ADN para saber cómo enlazar los aminoácidos para obtener esa proteína. En las siguientes páginas aprenderás más sobre el ADN.

El combustible de las células

Otra molécula importante para las células es el ATP (adenosin trifosfato). El **ATP** es el combustible principal para todas las actividades de la célula que requieren energía.

Cuando las moléculas de los alimentos se descomponen, como los carbohidratos y las grasas, una porción de la energía liberada se transfiere a las moléculas de ATP, como se muestra en la **Figura 15**. La energía de los carbohidratos y los lípidos debe ser transferida al ATP antes de que pueda ser utilizada por las células como energía almacenada para realizar sus procesos vitales.

Figura 15 *La energía de los carbohidratos y los lípidos debe ser transferida a las moléculas de ATP antes de poder ser utilizada por las células.*

REPASO

1. ¿Cuáles son las subunidades de las proteínas? ¿del almidón? ¿del ADN?

2. ¿Qué tienen en común los carbohidratos, las grasas y los aceites?

3. ¿Son enzimas todas las proteínas? Explica tu respuesta.

4. **Hacer predicciones** ¿Qué ocurriría con el ATP de tus células si no comieras suficientes carbohidratos? ¿Cómo afectaría esto a tus células?

4 Close

Quiz

1. Explain the difference between simple and complex carbohydrates. (Simple carbohydrates are made of one or two sugar molecules. Complex carbohydrates are made of many sugar molecules that are linked together.)

2. Name two functions of lipids. (Some lipids store energy, and others form the cell membrane.)

3. How are proteins used by an organism? (An organism breaks down proteins and uses their amino acids to build other proteins that are used to carry out chemical reactions in cells, transport materials, and protect the cell.)

ALTERNATIVE ASSESSMENT

 Writing — Have students write a job description for the cell's basic chemical building blocks. Tell students to write a classified ad that describes the required job responsibilities and physical qualifications. Have them include a description of the expected workload by explaining whether the building block will have to work constantly or sporadically. Finally, they should indicate whether the building block will work independently or with other cell components.

Answers to Review

1. The subunits of proteins are amino acids. Sugar molecules are the subunits of starch, and nucleotides are the subunits of DNA.

2. All three compounds store energy.

3. Not all proteins are enzymes. Enzymes are a special type of protein that speeds up certain chemical reactions in the cell.

4. The supply of ATP would decrease. A decrease in ATP would cause a cell to have less energy than it needs to carry out its activities. Your body would have to get ATP from other sources, like lipids.

Teaching Transparency 5
"Energy for Cells"

Reinforcement Worksheet 2
"Building Blocks"

Chapter Highlights

SECTION 1

cell a membrane-covered structure that contains all of the materials necessary for life

stimulus anything that affects the activity of an organism, organ, or tissue

homeostasis the maintenance of a stable internal environment

asexual reproduction reproduction in which a single parent produces offspring that are genetically identical to the parent

sexual reproduction reproduction in which two parents are required to produce offspring that will share characteristics of both parents

DNA deoxyribonucleic acid; hereditary material that controls all the activities of a cell, contains the information to make new cells, and provides instructions for making proteins

heredity the passing of traits from parent to offspring

metabolism the combined chemical processes that occur in a cell or living organism

SECTION 2

producers organisms that use sunlight directly to make sugar

consumers organisms that eat producers or other organisms for energy

decomposers organisms that get energy by breaking down the remains of dead organisms and consuming or absorbing the nutrients

Resumen del capítulo

SECCIÓN 1

Vocabulario

célula (*pág. 36*)

estímulo (*pág. 37*)

homeostasis (*pág. 37*)

reproducción asexual (*pág. 38*)

reproducción sexual (*pág. 38*)

ADN (*pág. 38*)

herencia (*pág. 38*)

metabolismo (*pág. 38*)

Notas de la sección

- Los seres vivos comparten las seis características de la vida.

- Los organismos están compuestos de una o más células. Las células son las unidades fundamentales de la vida.

- Los organismos detectan y responden a estímulos.

- Los organismos tratan de mantener su ambiente interno estable para no interrumpir las actividades químicas de sus células. Esto se denomina homeostasis.

- Los organismos se reproducen, en forma sexual o asexual, para procrear más organismos semejantes.

- Las crías se parecen a los padres. El traspaso de estas características se denomina herencia.

- Los organismos crecen y cambian durante su vida.

- Los organismos utilizan energía para llevar a cabo sus procesos químicos vitales. El metabolismo es el conjunto de estos procesos.

SECCIÓN 2

Experimentos

Carreras de cochinillas (*pág. 564*)

Vocabulario

productores (*pág. 40*)

consumidores (*pág. 40*)

descomponedores (*pág. 40*)

Notas de la sección

- Los organismos deben alimentarse. Los productores fabrican su propio alimento, mientras que los consumidores se alimentan de otros organismos. Los descomponedores descomponen los nutrientes de los organismos muertos y de los desechos animales.

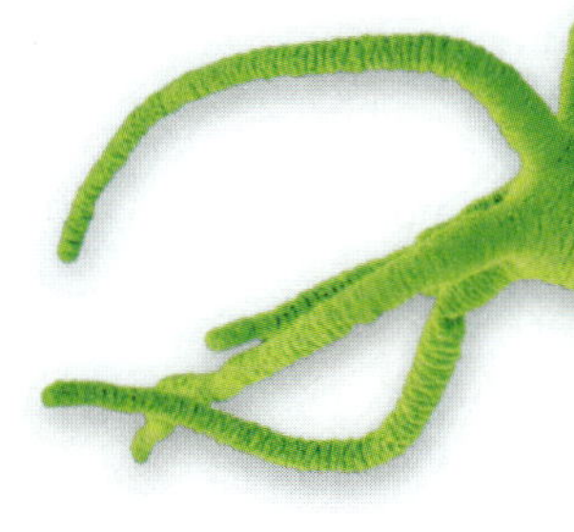
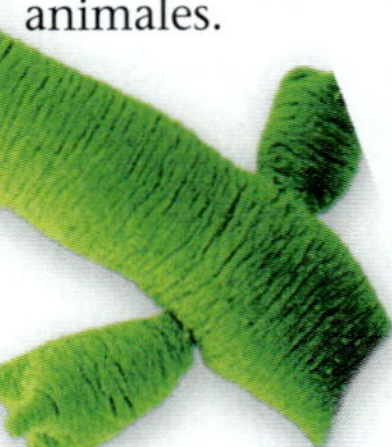

✓ Comprobar destrezas

Conceptos de matemáticas

¿QUÉ PORCENTAJE ES? En la sección ¡Matemáticas! de la página 39 calculaste el porcentaje de tu altura que ocupa a tu cabeza.

$$\frac{\text{largo de la cabeza}}{\text{altura total}} \times 100\% = \%\text{ de la altura que ocupa la cabeza}$$

Si una mujer tiene una estatura de 160 cm y su cabeza mide 20 cm significa que su cabeza representa el 12.5 por ciento de su estatura.

$$\frac{20 \text{ cm}}{160 \text{ cm}} \times 100\% = 12.5\%$$

Comprensión visual

AZÚCAR Y ALMIDÓN En la página 43 puedes encontrar una ilustración del azúcar y el almidón. Los hexágonos de cada ilustración representan moléculas de azúcar. Las azúcares son carbohidratos simples compuestos de una o dos moléculas de azúcar. El almidón es un carbohidrato complejo compuesto de muchas moléculas enlazadas. Las ilustraciones te ayudan a visualizar las diferencias en la estructura de los carbohidratos simples y complejos.

46

Lab and Activity Highlights

Roly-Poly Races `PG 564`

The Best-Bread Bakery Dilemma `PG 566`

Datasheets for LabBook (blackline masters for these labs)

proteins biochemicals that are composed of amino acids; their functions include regulating chemical reactions, transporting and storing materials, and providing support

enzymes proteins that make it possible for certain chemical reactions to occur quickly

carbohydrates biochemicals composed of one or more simple sugars bonded together that are used as a source of energy and to store energy

lipids biochemicals that do not dissolve in water, including fats and oils; their functions include storing energy and making up cell membranes

phospholipids molecules that form much of a cell membrane

nucleic acids biochemicals that store information needed to build proteins and other nucleic acids; made up of subunits called nucleotides

ATP adenosine triphosphate; molecule that provides energy for a cell's activities

SECCIÓN 2

- Los organismos necesitan agua para mantener el metabolismo.

- Los organismos necesitan oxígeno para liberar la energía que contienen los alimentos. Las plantas, las algas y algunas bacterias también requieren dióxido de carbono.

- Los organismos deben vivir en lugares en donde puedan obtener lo que necesitan.

SECCIÓN 3

Vocabulario

proteínas (*pág. 42*)
enzimas (*pág. 42*)
carbohidratos (*pág. 43*)
lípidos (*pág. 44*)
fosfolípidos (*pág. 44*)
ácidos nucleicos (*pág. 45*)
ATP (*pág. 45*)

Notas de la sección

- Los compuestos más importantes para la vida son las proteínas, los carbohidratos, los lípidos, los ácidos nucleicos y el adenosin trifosfato (ATP).

- Las células utilizan los carbohidratos para almacenar energía. Los carbohidratos están compuestos de azúcares.

- Los lípidos, como la grasa y los aceites, almacenan energía. Los fosfolípidos son los lípidos que forman parte de las membranas celulares.

- Las proteínas están compuestas de aminoácidos y cumplen muchas funciones importantes. Las enzimas son proteínas que permiten que las reacciones químicas ocurran rápidamente.

- Los ácidos nucleicos están compuestos de nucleótidos. El ADN es un ácido nucleico que contiene la información para producir proteínas.

- Las células utilizan moléculas de ATP como combustible para realizar sus actividades.

Experimentos

El dilema de la panadería "Rico Pan" (*pág. 566*)

 **internet**

HRW **VISITA:** go.hrw.com

Visita el sitio web de HRW para encontrar una serie de herramientas de aprendizaje relacionadas con este capítulo. Sólo tienes que escribir la palabra clave:

PALABRA CLAVE: HSTALV

SC/LINKS **NSTA** **VISITA:** www.scilinks.org

Visita el sitio web de la **Asociación Nacional de Maestros de Ciencias** (*National Science Teachers Association*) para encontrar recursos de Internet relacionados con este capítulo. Sólo escribe el **ENLACE DE CIENCIAS** para obtener más información sobre el tema:

TEMA: Las características de la los seres vivos	**ENLACE:** HSTL030
TEMA: Las necesidades de la vida	**ENLACE:** HSTL035
TEMA: La química de la vida	**ENLACE:** HSTL040
TEMA: ¿Hay vida en otros planetas?	**ENLACE:** HSTL045

47

 **Vocabulary Review Worksheet 2**

 Blackline masters of these Chapter Highlights can be found in the **Study Guide.**

Lab and Activity Highlights

LabBank

 **Labs You Can Eat,** Say Cheese! Lab 1

Long-Term Projects & Research Ideas, Project 2

USING VOCABULARY

1. homeostasis
2. heredity
3. consumer
4. carbohydrate/sugars
5. lipids

UNDERSTANDING CONCEPTS

Multiple Choice

6. d
7. b
8. a
9. b
10. c
11. c
12. a

Short Answer

13. Asexual reproduction can occur with just one parent, and offspring are identical to the parent. Two parents are usually required for sexual reproduction, and offspring share characteristics of both parents.
14. Living things must have air because both plants and animals need oxygen, which is one component of air. Producers also need carbon dioxide to make food.
15. ATP is the energy-containing molecule in a cell. It is the major fuel for all cellular activities.

Concept Mapping

16. An answer to this exercise can be found at the end of this book.

Concept Mapping
Transparency 2

Repaso del capítulo

UTILIZAR EL VOCABULARIO

Escoge el término correcto para completar las siguientes oraciones:

1. El proceso de mantener un ambiente interno estable se conoce como ____. (*metabolismo* u *homeostasis*)

2. Las crías se parecen a sus padres debido a ____. (*la herencia* o *el estímulo*)

3. Un ____ se alimenta de otros organismos. (*productor* o *consumidor*)

4. El almidón es un ____ que está compuesto de ____. (*carbohidrato/azúcares* o *ácido nucleico/nucleótidos*)

5. Las grasas y los aceites son ____ que almacenan energía en los organismos. (*proteínas* o *lípidos*)

COMPRENDER CONCEPTOS

Opción múltiple

6. Las células
   a. son las estructuras que contienen todos los materiales esenciales para la vida.
   b. están presentes en todos los organismos.
   c. a veces se especializan para realizar ciertas funciones.
   d. Todas las anteriores

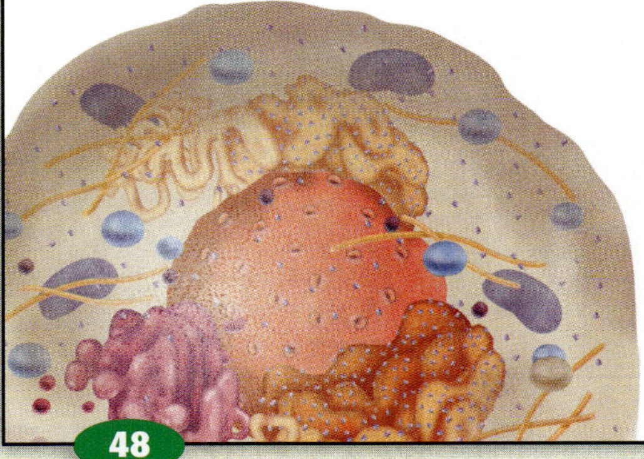

7. ¿Cuál de las siguientes oraciones acerca de los seres vivos es verdadera?
   a. No pueden detectar los cambios en su ambiente externo.
   b. Se reproducen.
   c. No necesitan utilizar energía.
   b. Se reproducen en forma asexual.

8. Los organismos deben alimentarse porque
   a. el alimento es una fuente de energía.
   b. el alimento provee oxígeno a las células.
   c. los organismos nunca producen su propio alimento.
   d. Todas las anteriores

9. Un cambio en el medio ambiente de un organismo que afecta sus actividades es
   a. una respuesta.   c. el metabolismo.
   b. un estímulo.   d. un productor.

10. Los organismos almacenan energía en moléculas de
    a. ácidos nucleicos.   c. lípidos.
    b. fosfolípidos.   d. agua.

11. La molécula que contiene la información para producir proteínas es
    a. el ATP.
    b. un carbohidrato.
    c. el ADN.
    d. un fosfolípido.

12. Las subunidades de los ácidos nucleicos son los
    a. nucleótidos.   c. azúcares.
    b. aceites.   d. aminoácidos.

Respuesta breve

13. ¿Qué diferencia existe entre la reproducción asexual y la sexual?

14. Explica en una o dos oraciones por qué los seres vivos necesitan aire.

15. ¿Qué es el ATP y por qué es importante para las células?

Organizar conceptos

16. Usa los siguientes términos para crear un mapa de ideas: célula, carbohidratos, proteína, enzimas, ADN, azúcares, lípidos, nucelótidos, aminoácidos, ácidos nucleicos.

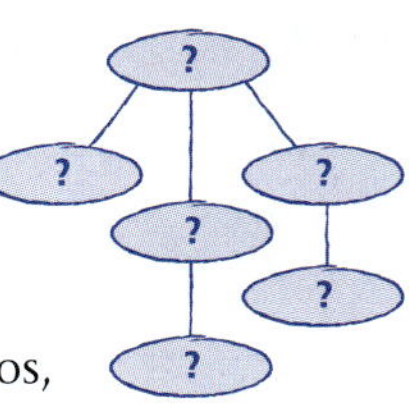

RAZONAMIENTO CRÍTICO Y RESOLUCIÓN DE PROBLEMAS

Escribe una o dos oraciones para responder las siguientes preguntas:

17. El fuego puede moverse, crecer en tamaño y dar calor. ¿Está vivo? Explica.

18. Según lo que sabes acerca de los carbohidratos, los lípidos y las proteínas, ¿por qué es importante tener una dieta balanceada?

19. Un amigo te cuenta que el estímulo de la música hace que su pez nade más rápido. ¿Cómo podrías diseñar un experimento controlado para comprobar esa afirmación?

LAS MATEMÁTICAS EN LAS CIENCIAS

20. Un elefante tiene una masa de 3,900 kg. Si el 70 por ciento de su masa es agua, ¿cuántos kilogramos de agua contiene el elefante?

INTERPRETAR GRÁFICAS

Observa el siguiente dibujo que muestra la misma planta durante 3 días.

21. ¿Qué hace la planta?

22. ¿Qué características de un ser vivo presenta la planta?

AHORA, ¿qué piensas?

Revisa tus respuestas a las preguntas de la página 5 que escribiste en el cuaderno de ciencias. ¿Han cambiado tus respuestas? Si es necesario, corrige tus respuestas basándote en lo que has aprendido en este capítulo.

49

Background

Scientists have always been intrigued by the possibility of life on other planets. Only recently have advanced biological and chemical techniques been sufficient to begin to investigate such claims. Science is not always black and white, and scientists often have heated debates over scientific evidence. Students should appreciate that scientists often have very different ways of interpreting the same evidence and that science does not always provide a clear-cut answer.

The scientists who question the studies of the ALH84001 meteorite argue that each component of the evidence can be accounted for by an inorganic explanation. Supporters argue that if all of the evidence is considered collectively, they point toward an organic explanation. One point of debate is whether Mars was cool enough to support life. Many scientists believe that Mars was much cooler at the time the meteorite left Mars. Although the Martian crust appears very dry, research has indicated that water, a necessity for life, was present in low concentrations. Scientists believe that the Martian atmosphere was much thicker at one time and that many of the atmospheric components necessary for life, including carbon dioxide and oxygen, were probably much more abundant then.

DEBATE**CIENTÍFICO**

¿Hay vida en Marte?

A fines de 1996 los titulares decían: "Evidencia de vida en Marte". ¿Qué tipo de vida? ¿Serían extraterrestres? No exactamente, pero la historia detrás de los titulares no deja de ser fascinante.

Una nave espacial muy especial

En 1996, un grupo de investigadores encabezados por científicos de la Administración Nacional de Aeronáutica y el Espacio, NASA, estudiaron un meteorito denominado ALH84001 que tenía 3,800 millones de años. Los científicos concuerdan en que es un pedazo del planeta Marte del tamaño de una papa. También están de acuerdo en que cayó a la Tierra hace 13,000 años. Se descubrió en la Antártida en 1984 y, de acuerdo con el equipo de la NASA, trajo pruebas de que existió vida en Marte.

Restos de formas de vida

Sobre la superficie del ALH84001, los científicos encontraron *moléculas orgánicas* (moléculas de compuestos que contienen carbono). Dichas moléculas son similares a las que quedan después de que un ser vivo descompone substancias para obtener alimento. En

▲ *Se piensa que esta imagen de micrografía electrónica de una estructura tubular encontrada en el meteorito ALH84001 es una prueba de que hubo vida en Marte.*

el interior del meteorito, encontraron las mismas moléculas. Debido a que se encontraban esparcidas a través del meteorito, concluyeron que no correspondían a contaminantes de la Tierra. El equipo de la NASA cree que estos restos son una prueba de que organismos diminutos, similares a las bacterias, vivieron y murieron en Marte hace millones de años.

Agua sucia o polvo de estrellas. . .

Muchos científicos no creen que el ALH84001 contenga pruebas de que haya habido vida en Marte. Algunos sostienen que los compuestos orgánicos son contaminantes de las aguas de la Antártida que se infiltraron en el meteorito.

Otros dicen que las moléculas de carbonatos fueron creadas mediante procesos a temperaturas muy elevadas. Piensan que los compuestos se formaron durante la creación de las estrellas y permanecieron en Marte cuando éste se convirtió en planeta. Otros creen que se crearon durante la formación de las rocas en Marte. Sostienen que ninguna forma de vida podría existir a temperaturas tan elevadas.

El debate continúa

Los científicos siguen debatiendo sobre las pruebas que contiene el ALH84001. Buscan pruebas de vida biológica, como proteínas, ácidos nucleicos y paredes celulares. Otros buscan pruebas en el planeta mismo: esperan encontrar aguas subterráneas que podrían haber albergado vida, o reunir muestras de tierra y roca que prueben que Marte fue alguna vez un planeta con vida.

¡Piénsalo!

▶ Si fueras a Marte, ¿qué pruebas reunirías para probar que hubo vida en ese planeta? ¿Cómo podría el descubrimiento de ácidos nucleicos o de aminoácidos probar la existencia de vida en Marte?

50

Answer to Think About It

Answers will vary. The discovery of nucleic acids, amino acids, and cell walls would be strong evidence of life on Mars because they are components of living organisms.

"Están hechos de carne"

por Terry Bisson

Dos viajeros espaciales se encuentran a millones de años luz de su hogar. Visitan sectores desconocidos para encontrar señales de vida. Su misión es contactar, saludar y registrar a cualquier ser en este cuadrante del universo. Una vez que descubran un ser vivo, deben comunicarse con él.

Durante su misión se encuentran con una forma de vida distinta de las que han visto hasta ahora. Estos seres extraños pueden pensar y comunicarse. Incluso han fabricado algunas máquinas simples, de modo que no son sólo suciedad de un estanque.

No obstante, los exploradores no saben si deben agregar esta nueva especie a su lista de formas de vida desconocidas del universo. Las criaturas son demasiado extrañas. Además, al tener habilidades limitadas, es poco probable que puedan establecer contacto con otras formas de vida en otros lugares del universo.

Quizás sea mejor que los exploradores finjan que nunca se encontraron con estos seres. Sin embargo, la tarea oficial de los viajeros es contactar y darles la bienvenida a todas las formas de vida sin importar cómo sean ni de qué estén compuestas. ¿Podrán seguir adelante con su tarea oficial? ¿Quién les creará si lo hacen?

Averígualo al leer el cuento corto de Terry Bisson, "Están hechos de carne". Esta historia se encuentra en la *Antología Holt de Ciencia Ficción*.

51

SCIENCE FICTION
"They're Made Out of Meat"
by Terry Bisson

A remarkable and intelligent life-form has been discovered at the far reaches of the universe, but it's tough to get excited about this find . . .

Teaching Strategy
Reading Level This is a relatively short story that should not be difficult for the average student to read and comprehend.

Background
About the Author Terry Bisson has written everything from comic books to short stories, novels, plays, how-to articles about writing, and news editorials. He has taken the scripts of several popular movies and converted them to novels. Some of Bisson's works have appeared in digital-audio format on the World Wide Web. "They're Made Out of Meat" is just one of several stories featured in the SciFi Channel's *Seeing Ear Theater*. In 1991, Bisson's short story "Bears Discover Fire" received the highest honors possible for science fiction writers—both the Nebula Award and the Hugo Award.

In addition to being a writer, Bisson has been an automobile mechanic, an editor, a publisher's consultant, and a teacher. Bisson teaches writing at Clarion University, in Pennsylvania, and at the New School for Social Research, in New York City. He also maintains a personal Web site full of interesting information, works by guest writers, and excerpts from his novels and stories.

Further Reading If students like Terry Bisson's style, suggest more of his stories to students. Some of his works include the following:

Bears Discover Fire and Other Stories, Tor Books, New York City, 1993

"10:07:24," *Absolute Magazine,* 1995

"First Fire," *Science Fiction Age,* Sept 1998

"The Player," *Fantasy and Science Fiction Magazine,* Oct/Nov 1997

Chapter Organizer

CHAPTER ORGANIZATION	TIME MINUTES	OBJECTIVES	LABS, INVESTIGATIONS, AND DEMONSTRATIONS
Chapter Opener pp. 52–53	45		**Investigate!** Colors of Light, p. 53
Section 1 The Electromagnetic Spectrum	90	▶ Explain how electromagnetic waves differ from other waves. ▶ Describe the relationship between a wave's wavelength and its frequency. ▶ Describe the relationship between the energy of a wave and its wavelength and frequency. ▶ Identify ways visible light and ultraviolet light are helpful or harmful.	**Demonstration,** p. 54 in ATE **Interactive Explorations CD-ROM,** In the Spotlight *A **Worksheet** is also available in the **Interactive Explorations Teacher's Edition.***
Section 2 Reflection, Absorption, and Scattering	90	▶ Compare regular reflection with diffuse reflection. ▶ Describe absorption and scattering of light. ▶ Explain how the color of an object is determined. ▶ Compare the primary colors of light with the primary pigments.	**QuickLab,** Scattering Milk, p. 61 **Demonstration,** p. 62 in ATE **Skill Builder,** Mixing Colors, p. 568 **Datasheets for LabBook,** Mixing Colors, Datasheet 6 **Inquiry Labs,** Eye Spy, Lab 23 **Labs You Can Eat,** Fiber-Optic Fun, Lab 25
Section 3 Refraction	90	▶ Define *refraction.* ▶ Explain how refraction can separate white light into different colors of light. ▶ Describe the differences between convex lenses and concave lenses. ▶ Identify examples of lenses used in your everyday life.	**Demonstration,** What Are Light Rays? p. 66 in ATE

TECHNOLOGY RESOURCES

 **Guided Reading Audio CD**
English or Spanish, Chapter 3

 **CNN.** Science, Technology & Society, Correcting Color Blindness, Segment 2

 **One-Stop Planner CD-ROM with Test Generator**

 **Interactive Explorations CD-ROM**
CD 3, Exploration 7, In the Spotlight

CLASSROOM WORKSHEETS, TRANSPARENCIES, AND RESOURCES	SCIENCE INTEGRATION AND CONNECTIONS	REVIEW AND ASSESSMENT
Directed Reading Worksheet 3 **Science Puzzlers, Twisters & Teasers,** Worksheet 3		
Directed Reading Worksheet 3, Section 1 **Transparency 6,** The Electromagnetic Spectrum **Transparency 195,** Wave Speed, Wavelength, and Frequency **Math Skills for Science Worksheet 54,** Color at Light Speed	**Cross-Disciplinary Focus,** p. 56 in ATE **Astronomy Connection,** p. 58 **Science, Technology, and Society:** Fireflies Light the Way, p. 74	**Homework,** pp. 55, 57 in ATE **Review,** p. 58 **Quiz,** p. 58 in ATE **Alternative Assessment,** p. 58 in ATE
Transparency 7, The Law of Reflection **Directed Reading Worksheet 3,** Section 2 **Transparency 7,** Regular Reflection Versus Diffuse Reflection **Reinforcement Worksheet 3,** Light Interactions	**Multicultural Connection,** p. 60 in ATE **Connect to Earth Science,** p. 60 in ATE **Apply,** p. 63 **Cross-Disciplinary Focus,** p. 62 in ATE **Cross-Disciplinary Focus,** p. 63 in ATE	**Homework,** p. 61 in ATE **Self-Check,** p. 62 **Review,** p. 64 **Quiz,** p. 64 in ATE **Alternative Assessment,** p. 64 in ATE
Directed Reading Worksheet 3, Section 3 **Transparency 8,** White Light Is Separated by a Prism **Transparency 9,** Thick and Thin Convex Lenses **Transparency 9,** A Concave Lens **Transparency 10,** Comparing a Camera to Your Eye **Transparency 11,** How Telescopes Work **Critical Thinking Worksheet 3,** Now You See It, Now You Don't!	**Connect to Physical Science,** p. 66 in ATE **Connect to Earth Science,** p. 67 in ATE **Cross-Disciplinary Focus,** p. 67 in ATE **Real-World Connection,** p. 67 in ATE **Connect to Physical Science,** p. 68 in ATE **Eye on the Environment:** Light Pollution, p. 75	**Self-Check,** p. 68 **Review,** p. 69 **Quiz,** p. 69 in ATE **Alternative Assessment,** p. 69 in ATE

Holt, Rinehart and Winston On-line Resources

go.hrw.com

For worksheets and other teaching aids related to this chapter, visit the HRW Web site and type in the keyword: **HSTLLT**

National Science Teachers Association

www.scilinks.org

Encourage students to use the *sci*LINKS numbers listed with the Chapter Highlights to access information and resources on the **NSTA** Web site.

END-OF-CHAPTER REVIEW AND ASSESSMENT

Chapter Review in Study Guide
Vocabulary and Notes in Study Guide
Chapter Tests with Performance-Based Assessment, Chapter 3 Test
Chapter Tests with Performance-Based Assessment, Performance-Based Assessment 3
Concept Mapping Transparency 3

Chapter Resources & Worksheets

Visual Resources

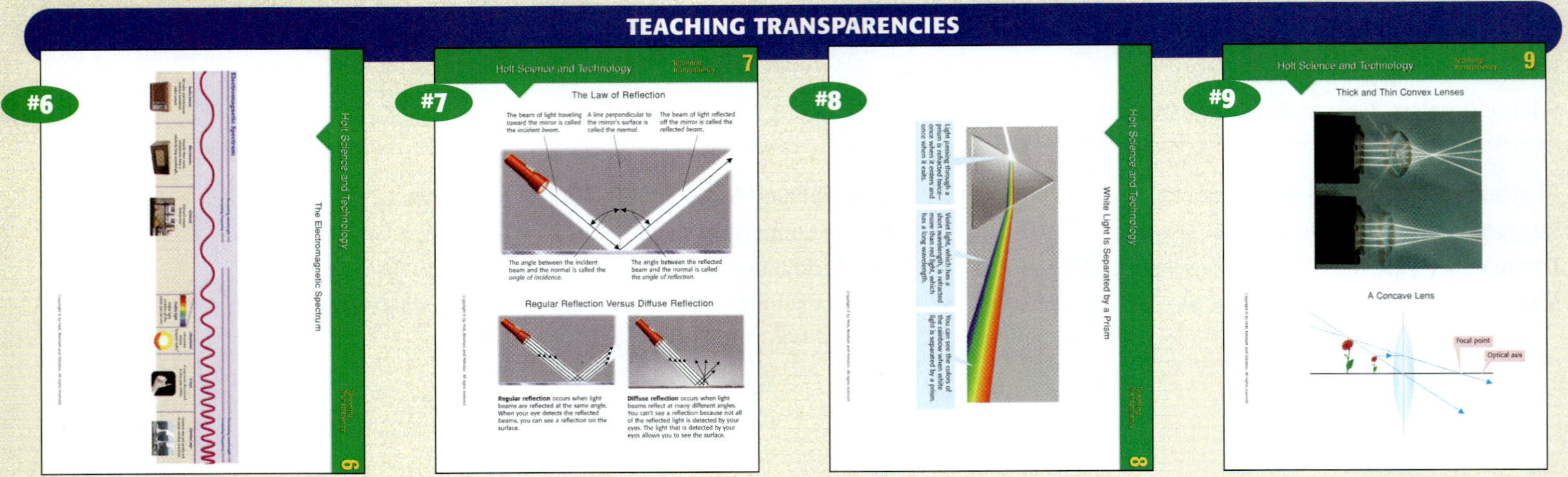

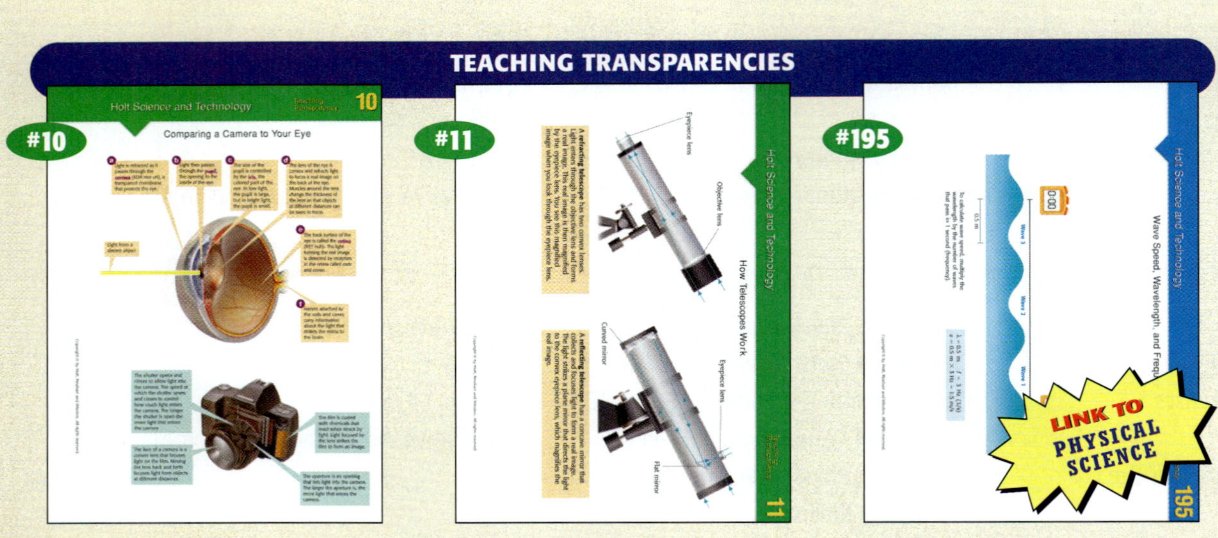

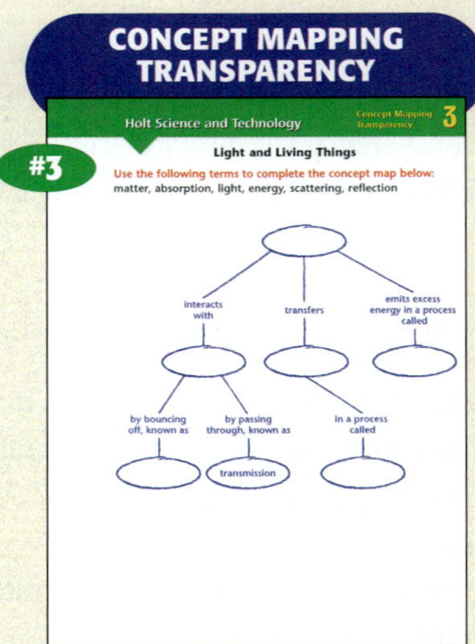

Meeting Individual Needs

DIRECTED READING

REINFORCEMENT & VOCABULARY REVIEW

SCIENCE PUZZLERS, TWISTERS & TEASERS

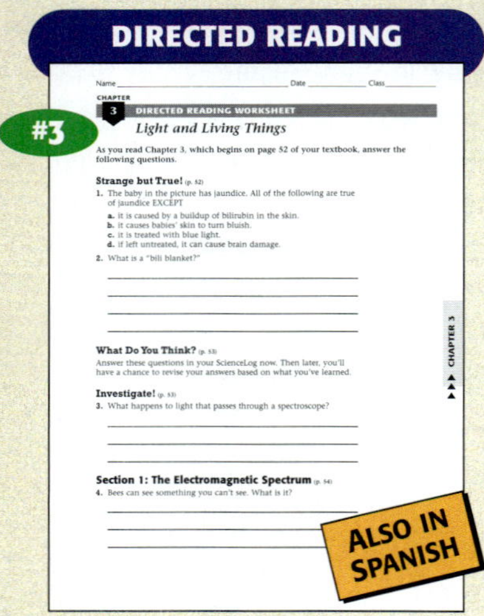

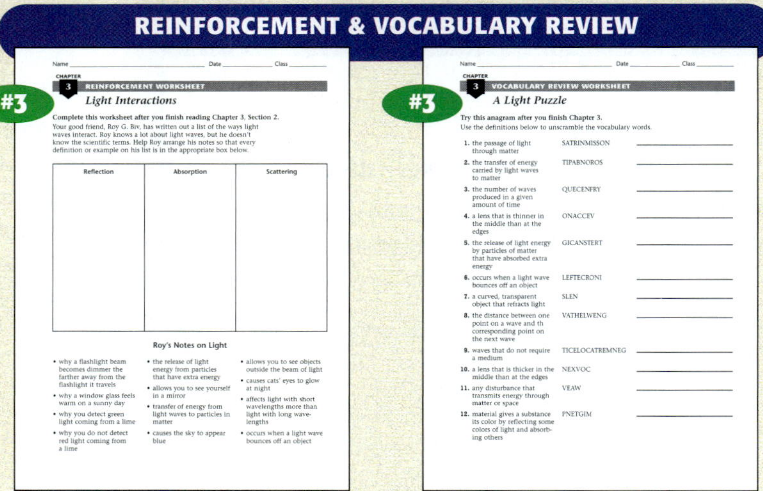

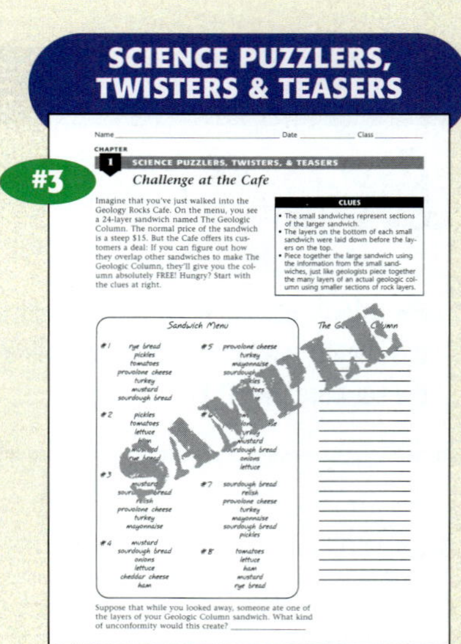

 Chapter 3 • Light and Living Things

Review & Assessment

STUDY GUIDE

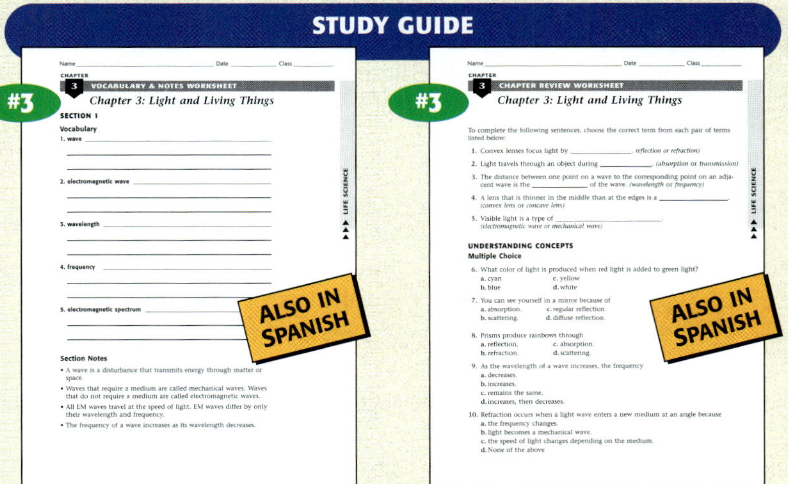

CHAPTER TESTS WITH PERFORMANCE-BASED ASSESSMENT

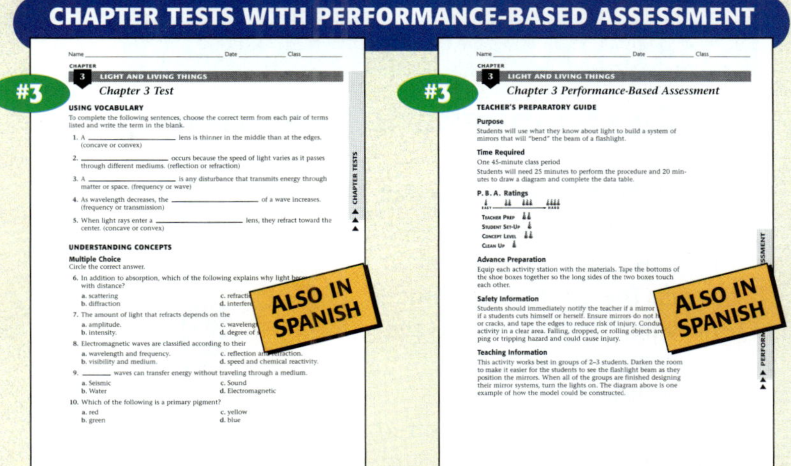

Lab Worksheets

INQUIRY LABS

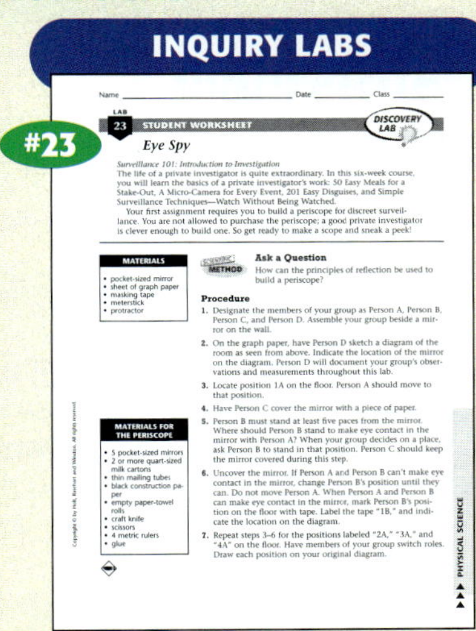

LABS YOU CAN EAT

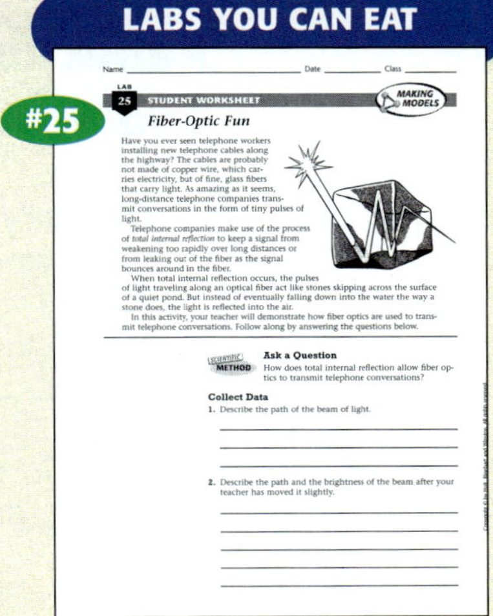

DATASHEETS FOR LABBOOK

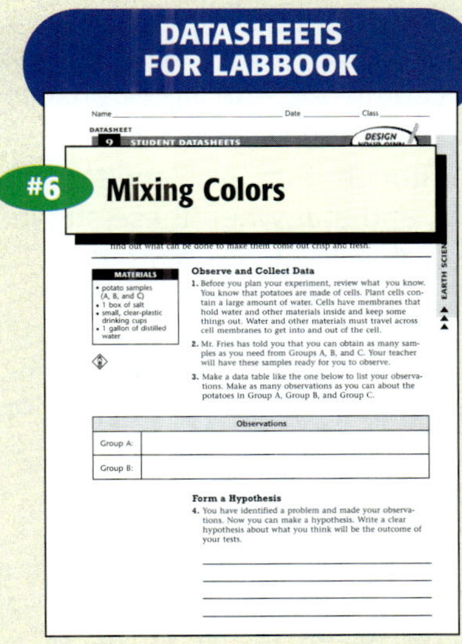

Applications & Extensions

CRITICAL THINKING & PROBLEM SOLVING

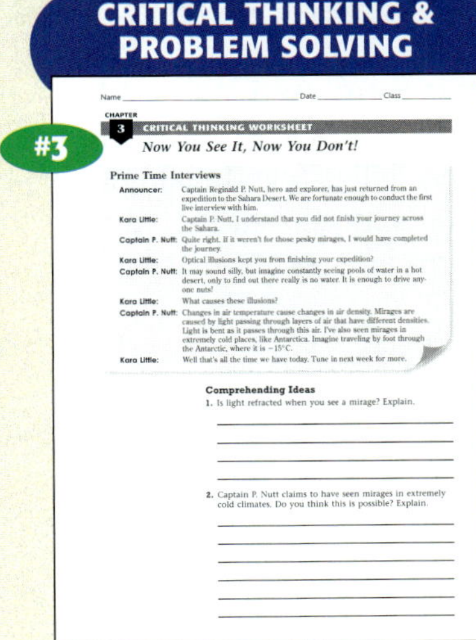

SCIENCE TECHNOLOGY

INTERACTIVE EXPLORATIONS

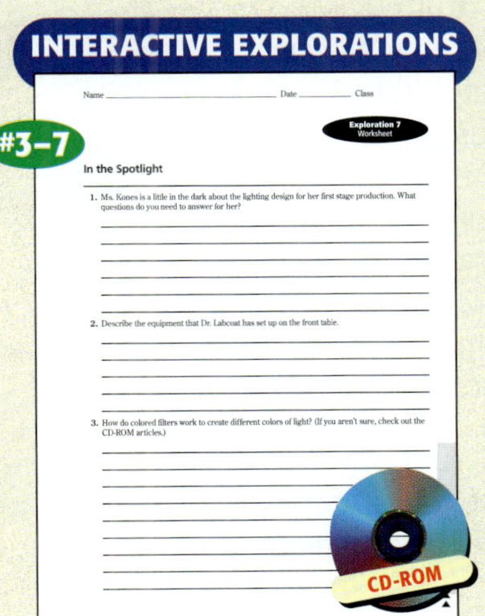

Chapter Background

SECTION 1

The Electromagnetic Spectrum

▶ **The Nature of Light**

The nature of light has been debated for thousands of years. The argument about how light travels—as a wave or as a particle—began when the Pythagoreans believed that light is emitted from a source in the form of tiny particles.

- However, Empedocles (c. 490–c. 430 B.C.) taught that light travels from its source as waves.

- In the fifth century B.C., the Greek philosophers Socrates (c. 470–399 B.C.) and Plato (c. 428–c. 348 B.C.) thought that eyes emitted streamers, or filaments, and that when these streamers made contact with objects, sight occurred.

- Even as late as the 1500s, René Descartes (1596–1650), a great French mathematician and philosopher, had beliefs similar to Plato's.

IS THAT A FACT!

- Galileo Galilei (1564–1642) once tried to measure the speed of light from one hilltop to another using a lantern. He soon realized that the light traveled very fast.

- The infrared part of the spectrum was discovered by William Herschel (1738–1822), a famous British astronomer. In 1800, he was investigating the thermal energy produced by certain waves. The waves were just below the red part of the visible spectrum. He named the waves *infrared. Infra* is Latin for "below."

- In 1865, James Maxwell (1831–1879), a Scottish physicist, developed a theory predicting the existence of electromagnetic waves. His equations predicted that such waves propagated through space at the speed of light. Maxwell's results led him to believe that light was an electromagnetic wave.

- In 1886, Heinrich Hertz (1859–1894) was trying to prove Maxwell's equations experimentally. His experimentation was very fruitful: Not only did Hertz prove Maxwell correct, but he also was the first to detect electromagnetic waves and to generate them experimentally. The unit for frequency was named in his honor.

SECTION 2

Reflection, Absorption, and Scattering

▶ **History of Mirrors**

Natural mirrors made of obsidian were used in Turkey 7,500 years ago. Later, polished mirrors of copper, brass, bronze, tin, and silver were used. The Venetians found a way to use polished silver to make mirrors in the 1200s. The silvering process used on mirrors today was begun in 1835 by a German chemist named Justus von Liebig (1803–1873).

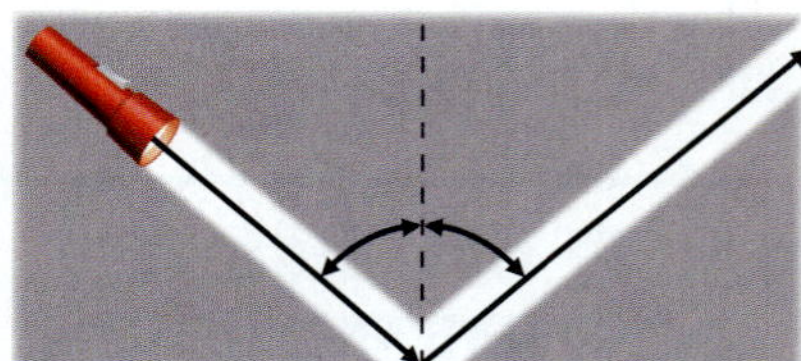

- Bronze mirrors were used in Egypt from as early as 3500 to 3000 B.C. Metal mirrors were luxury items because of the difficulty in making a flat, highly polished metal surface that would reflect light well enough to form images.

IS THAT A FACT!

- The Keck and Keck II telescopes, in Mauna Kea, Hawaii, are 10 m in diameter. They are the largest reflecting telescopes in the world today. Each telescope uses 36 mirror segments fitted together so seamlessly that they act as one large curved mirror. The segments are realigned about 1,000 times a second by computers to counteract the effect of gravity and other distortions.

▶ Light, Color, and Vision

In humans, light rays pass through the lens and strike the retina, the back part of the eye. The retina contains two types of cells, *rods* and *cones,* that react to light energy.

- Rods have a visual pigment that is very sensitive to light energy. Rods absorb all wavelengths of light, but the brain translates their signals as shades of gray.

- Cones come in three types and are sensitive to different wavelengths of light. One type of cone is triggered by light energy at the blue end of the visible spectrum, the second responds to the red end of the visible spectrum, and the third type is stimulated by the middle of the visible spectrum, or the greens.

- When light energy stimulates the cones, they send signals to the brain. The brain interprets these signals as colors, depending on how many of each type of cone has been stimulated. (See Chapter 21, "Communication and Control," for more information about the eye.)

SECTION 3

Refraction

▶ Early Uses of Lenses

The Greeks and Romans made the earliest known lenses. Their lenses consisted of spheres of glass that were filled with water. They were called burning glasses because they were used to focus sunlight into a fine point that could start a fire.

- Another early use of lenses is known from references to the Roman Emperor Nero (A.D. 37–68). When he watched performances in the arena, he used a piece of emerald to correct his poor eyesight. Convex lenses in eyeglasses came into use in Italy in about 1287.

▶ The First Telescope

- A Dutch eyeglass maker named Hans Lipperhey invented the telescope in 1608. He put a convex and concave lens in a tube after he discovered that holding the two lenses in front of each other made distant objects appear larger. Most telescopes that were made during Lipperhey's time were not used for astronomy. Instead, they were most often used by military officers to see distant armies or ships.

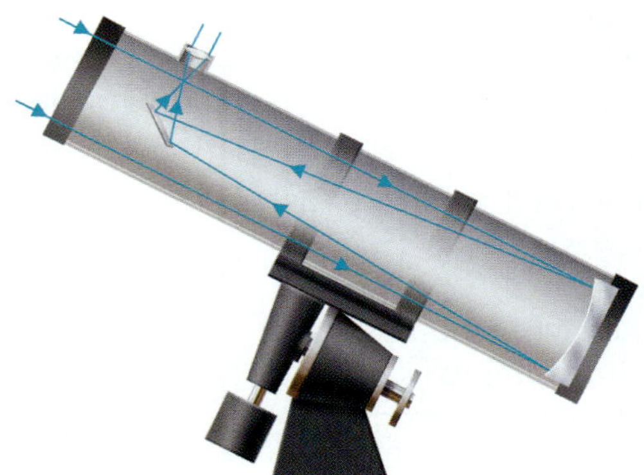

IS THAT A FACT!

- Glass lenses are made from high-quality glass known as optical glass. First a lens blank is cut from a block of optical glass with a diamond-edged saw. The edges are then ground into the rough shape of the lens. Grinding also forms the curved surfaces of the lens. A convex lens is ground on a concave surface, and a concave lens is ground on a convex surface. Finally, the lens is polished and finished with a bit more grinding of the edge.

*For additional background resources, please refer to the **HST Reference Library.***

Directed Reading Worksheet 3

Science Puzzlers, Twisters & Teasers Worksheet 3

Guided Reading Audio CD
English or Spanish, Chapter 3

Increíble…¡pero cierto!

¿Qué pensarías si entraras en un hospital y vieras al bebé de la fotografía anterior? ¡Parece que estuviera en una cabina de bronceado! Sin embargo, no se está bronceando: es un tratamiento para una enfermedad llamada ictericia.

Esta enfermedad aparece en algunos bebés debido a una acumulación de bilirrubina (un pigmento de los glóbulos rojos sanos) en el flujo sanguíneo cuando los glóbulos rojos se descomponen. El exceso de bilirrubina se deposita en la piel, dándole un color amarillento. La ictericia no es peligrosa si se actúa rápidamente. Pero si no se trata, puede dañar el cerebro.

La mejor forma de eliminar el exceso de bilirrubina en la piel es con luces azules claras. Por eso, los hospitales ponen tubos fluorescentes azules especiales sobre las cunas de los recién nacidos que requieren tratamiento. A veces se equilibra la luz azul con luces de otros colores para que los doctores y enfermeras puedan asegurarse de que el bebé no esté azul por falta de oxígeno.

Una forma más conveniente de tratamiento es la "manta bili", una almohadilla suave hecha de materiales de fibra óptica conectados a una caja lumínica que produce una luz azul. Esta manta puede envolverse alrededor del bebé y hasta se puede tomar en brazos al recién nacido y abrigarlo durante el tratamiento.

El tratamiento con luz para los bebés que sufren de ictericia es sólo uno de los usos importantes de la luz. En este capítulo, aprenderás la naturaleza de la luz, cómo interactúan las ondas luminosas y otras formas en que la luz juega un papel importante en tu vida.

52

Strange but True!

Newborn babies have extra red blood cells to provide them with plenty of oxygen during the strenuous birth process. After birth, the baby's body begins to break down the extra red blood cells, which results in a higher than normal level of bilirubin in the blood. The liver is responsible for removing excess bilirubin from the blood. Frequently, however, an infant's liver cannot remove the bilirubin fast enough, leading to jaundice. Pediatricians can perform tests to determine whether an infant's bilirubin level is high enough for light treatment.

Colores de la luz

La luz azul se usa para tratar a los bebés con ictericia. Pero cuando un bebé está bajo luz azul, es difícil saber si se ve azul por falta de oxígeno o por el reflejo de la luz azul. Para resolver este problema, se mezcla la luz azul con otros colores de luces para producir luz blanca. Cuando esta luz blanca alumbra al bebé, los doctores y enfermeras saben su color verdadero. En esta actividad estudiarás dos tipos de luz blanca.

Procedimiento

1. Enciende un **foco incandescente.** Pon una **rejilla de difracción** frente a uno de tus ojos y observa el foco a través de ésta. Gira la rejilla hasta que veas colores a ambos lados de la luz. Con un **trozo de cinta** adhesiva marca la parte superior de la rejilla.

Usa tus conocimientos para responder a las siguientes preguntas en tu cuaderno de ciencias:

1. ¿Qué son las ondas electromagnéticas?
2. ¿Qué determina el color de un objeto?
3. ¿Cómo se forma un arco iris?

2. Pega un trozo de **cartulina negra** a un extremo de un **tubo de toallas de papel** de modo que el papel cubra el orificio del tubo. Tu maestro o maestra cortará una ranura angosta en el centro de la cartulina.

3. Pega la rejilla en el extremo opuesto del tubo para que la parte superior de la rejilla se alinee verticalmente con la ranura. Has construido tu propio espectroscopio. Los espectroscopios se usan para dividir la luz en sus colores componentes. Cuando mires por el espectroscopio, verás la luz dividida a los lados del tubo.

4. Sostén el espectroscopio frente a uno de tus ojos y observa el foco eléctrico. Describe lo que ves.

5. Repite el paso 4 con un **foco fluorescente**. ¿Cómo se compara con lo que viste cuando miraste el foco incandescente?

Análisis

6. Tanto el foco eléctrico incandescente como el fluorescente producen luz blanca. ¿Qué aprendiste sobre la luz blanca con el espectroscopio?

What Do You Think?

Accept all reasonable responses.

Students will have a chance to revise their answers in the Chapter Review under NOW What Do You Think?

Investigate!

MATERIALS

FOR EACH GROUP:
• clear incandescent light bulb
• diffraction grating
• tape
• black construction paper
• paper-towel tube
• fluorescent light bulb

Safety Caution: If students are going to be close to the light bulbs, remind them to be very careful. Light bulbs get hot very quickly. Students must avoid touching or handling the bulbs.

Teacher Notes:

• Cut a slit about the thickness of an index card in each piece of construction paper.
• Toilet-paper tubes will also work for this experiment.
• Diffraction gratings are often sold as small squares held in plastic slide holders (like slides for a slide projector).

Answer to Investigate!

4. Students should see a continuous spectrum of colors (rainbow colors) on both sides of the tube. All colors have about the same brightness.
5. Students should see a spectrum of colors again. However, there will be bright bands and faint bands within the spectrum.
6. White light is made up of all the different colors of light.

IS THAT A FACT!

Diffraction gratings are pieces of glass, metal, or plastic with many slits (sometimes thousands per centimeter). When lightwaves pass through the slits, they are bent in a wave interaction called diffraction. The amount a wave bends depends on its wavelength. Light waves of different wavelengths are bent different amounts. Thus, white light passing through a grating is separated into the colors of the entire visible spectrum.

Focus

The Electromagnetic Spectrum

This section introduces electromagnetic waves and their properties. Students will learn how electromagnetic waves are different from other waves and how different types of electromagnetic waves are different from each other.

Bellringer

Draw a transverse wave with a wavelength of about 1 m on the board. Label each part of the wave, such as the wavelength, crest, and trough. Have students copy the wave in their ScienceLog and label their drawings carefully. Ask them to explain in their own words what they think wavelength and frequency are.

1 Motivate

DEMONSTRATION

Using tongs to hold one end of a small piece of copper wire, place the other end into the flame of a Bunsen burner. (Any source of thermal energy, such as a lighter or alcohol burner, will work.) A bright green luminous glow will be produced. Ask students to explain the source of the green glow. Guide the discussion to help students realize that something (atoms) in the copper wire emits a green light after absorbing thermal energy.

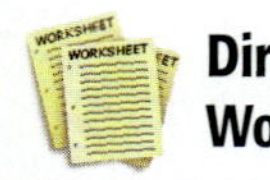

Directed Reading Worksheet 3 Section 1

El espectro electromagnético

VOCABULARIO
onda
onda electromagnética
longitud de onda
frecuencia
espectro electromagnético

OBJETIVOS
- Explica cómo las ondas electromagnéticas se diferencian de las demás ondas.
- Describe la relación entre la longitud de onda y la frecuencia de una onda.
- Describe la relación entre la energía de una onda, su longitud de onda y su frecuencia.
- Identifica las formas en que la luz visible y la luz ultravioleta son útiles o perjudiciales.

Puedes ver los objetos a tu alrededor debido a que la luz se refleja en ellos. Sin embargo, si una abeja mira los mismos objetos los vería distintos, tal como se muestra en la **Figura 1.** Esto se debe a que las abejas pueden ver un tipo de luz que tú no puedes ver: la luz ultravioleta.

Quizás te parezca extraño llamar *luz* a algo que no puedes ver, ya que la luz con la cual estás familiarizado es la luz visible. Sin embargo, la luz ultravioleta es muy similar a la luz visible. Ambas son tipos de ondas electromagnéticas. En esta sección, aprenderás acerca de las ondas electromagnéticas y cómo se distinguen de los otros tipos de ondas. También aprenderás qué es lo que distingue a una onda electromágnetica de las demás.

Figura 1 *Para ti, los pétalos de la flor a la derecha son de un color amarillo fuerte. Sin embargo, una abeja que observa la misma flor puede ver manchas ultravioleta que la dirigen hacia el centro de la flor.*

Las ondas transportan energía

Cuando piensas en ondas, quizás te imaginas las olas del mar o de un lago. Sin embargo, existen distintos tipos de ondas. Por ejemplo, todo lo que oyes viaja a través de ondas sonoras, los sismos causan ondas sísmicas y la luz que te permite ver es una onda electromagnética. Una **onda** es una perturbación que transmite energía a través de la materia o el espacio.

Cuando una onda viaja, su energía puede mover materia. Por ejemplo, la energía de las olas en una laguna hace que el agua se mueva hacia arriba y hacia abajo. La energía de las olas también mueve cualquier objeto que esté flotando sobre la superficie del agua. Por ejemplo, los botes y los patos se mueven con las olas, pero no se mueven en la misma dirección que las olas.

54

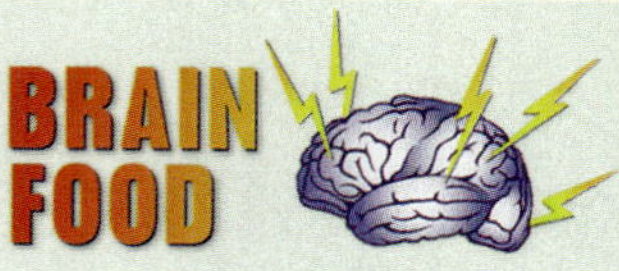

Did you know that scorpions can have up to 12 eyes and that marine flatworms can have more than 100 eyes? Bees, like most insects, have compound eyes—arrays of hundreds of single eyes. Each single eye has its own lens and looks in a different direction. And bees can see ultraviolet light, a part of the electromagnetic spectrum that humans cannot see.

Una onda transporta la energía desde su fuente, pero el material a través del que viaja no se mueve con la energía. Por ejemplo, las ondas sonoras viajan a través del aire, pero el aire no viaja con el sonido. Si fuera así, ¡sentirías un soplo de brisa cada vez que sonara el timbre del teléfono! La **Figura 2** ilustra cómo las ondas transportan energía pero no materia.

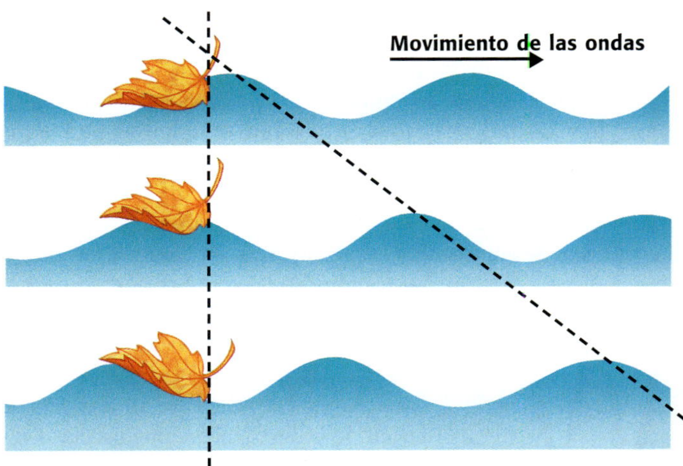

Figura 2 *La línea diagonal muestra el movimiento de una ola a medida que viaja a través de una laguna. La línea vertical muestra que la hoja que flota sobre la superficie no se mueve con la ola.*

Transferencia de energía a través de un medio Algunas ondas transfieren energía mediante la vibración de las partículas de un medio. Un *medio* es una substancia a través de la cual puede viajar una onda y puede ser un sólido, un líquido o un gas.

Cuando una partícula vibra, posee energía y puede transmitirla a la partícula de al lado. Ya que la energía fue es transferida a otra partícula, esa partícula vibra y transmite la energía a una tercera. Así, la energía se transmite a través de un medio. Las ondas que requieren un medio se denominan *ondas mecánicas.* Las olas del agua, las ondas sonoras y las ondas sísmicas son ejemplos de ondas mecánicas.

Transferencia de energía sin un medio Algunas ondas pueden transferir energía sin viajar a través de un medio. La luz visible es un ejemplo de una onda que no requiere de un medio. Otros ejemplos son las microondas, utilizadas en los hornos de microondas; las ondas de radio que transmiten señales de televisión y de radio; y los rayos X, utilizados por dentistas y doctores. Las ondas que no requieren un medio se denominan **ondas electromagnéticas** u ondas EM. Aunque las ondas electromagnéticas no requieren un medio, viajan a través de substancias como el aire, el agua y el vidrio; pero viajan más rápidamente a través del espacio vacío.

55

DISCUSSION

Show students photographs or a video of people surfing the giant waves off the coast of California, Hawaii, or Australia. Explain that this section covers some of the basic concepts about waves, especially electromagnetic waves, but that the waves they see "breaking" as surfers ride them are not covered—scientists do not fully understand why waves break. Remind them that there are still natural phenomena that scientists do not understand and that science is a process for answering questions about natural phenomena.

USING THE FIGURE

Sitting Ducks Discuss **Figure 2** with students. Place students in groups. Give each group a dishpan or some other large container partially filled with water. Explain that their first task is to create and observe waves of different sizes in the container. Students may use their hands, pencils, or other items to create waves gently. Next provide each group with one or more corks or floating bathtub toys. Have students place these objects near the center of the container. Now ask them to determine if they can move the objects to the side of the container just by using waves. Have them write their observations in their ScienceLog. Discuss their results.

Homework

Describing Waves Have students write in their ScienceLog a brief description of what they think a wave is. Ask them to describe a time they might have experienced waves. Discuss students' responses and help them understand that a wave carries energy away from its source through matter or space.

internet**connect**

SC*i*LINKS
NSTA

TOPIC: The Electromagnetic Spectrum
GO TO: www.scilinks.org
*sci*LINKS NUMBER: HSTL705

CROSS-DISCIPLINARY FOCUS

History Galileo was the first scientist to predict that light travels at a very high speed. The Danish astronomer Ole Roemer (1644–1710) was the first to demonstrate that the speed of light was finite, in 1675, by observing the eclipses of Jupiter's satellites.

USING THE FIGURE

Figure 3 illustrates wavelength and frequency for a type of wave called a *transverse wave* (a wave in which the particles of the wave's medium vibrate perpendicular to the direction the wave is traveling). Discuss with students this type of wave and the other common type of wave, the *longitudinal wave* (a wave in which the particles of the medium vibrate back and forth along the path that the wave travels). Ask students if they can think of examples of each kind of wave. (transverse—a wave traveling in a rope; longitudinal—a sound wave)

BRAIN FOOD

AM radio waves can bounce off the ionosphere layer of Earth's atmosphere and come back to Earth. That is why you can sometimes hear an AM station broadcasting from halfway across the country. You cannot tune in to an FM station broadcast from very far away because FM waves pass through the ionosphere and are not reflected back to Earth.

Ondas electromagnéticas

La luz viaja más rápido que cualquier otra cosa del universo. Pero, ¿sabías que todas las ondas electromagnéticas viajan más rápido que la luz? En el espacio casi vacío, la velocidad de la luz (y de todas las demás ondas electromagnéticas) es de aproximadamente 300,000,000 m/s. Las ondas electromagnéticas viajan más lento en el aire, el cristal y otros tipos de materia.

Todas las ondas electromagnéticas son esencialmente iguales. Sin embargo, están clasificadas en distintos grupos según su longitud y su frecuencia. Como se muestra en la **Figura 3,** la **longitud de onda** es la distancia entre un punto sobre una onda y el punto correspondiente sobre otra onda adyacente. La **frecuencia** de una onda es el número de ondas que se producen dentro de un lapso de tiempo definido. La longitud y la frecuencia de una onda determinan parcialmente la cantidad de energía que ésta transporta. En general, las ondas con longitudes de onda cortas y frecuencias elevadas transportan más energía que las ondas con longitudes de onda largas y frecuencias bajas.

La gama completa de ondas EM se denomina **espectro electromagnético.** Este espectro está organizado de longitudes de onda largas a cortas y de frecuencias bajas a elevadas. El diagrama muestra las distintas categorías de las ondas en el espectro electromagnético. Dos categorías de ondas EM que son esenciales para la vida en la Tierra son la luz visible y la luz ultravioleta.

Figura 3 *A una velocidad determinada, la frecuencia aumenta a medida que la longitud de onda disminuye. A la inversa, la frecuencia disminuye a medida que la longitud de onda aumenta.*

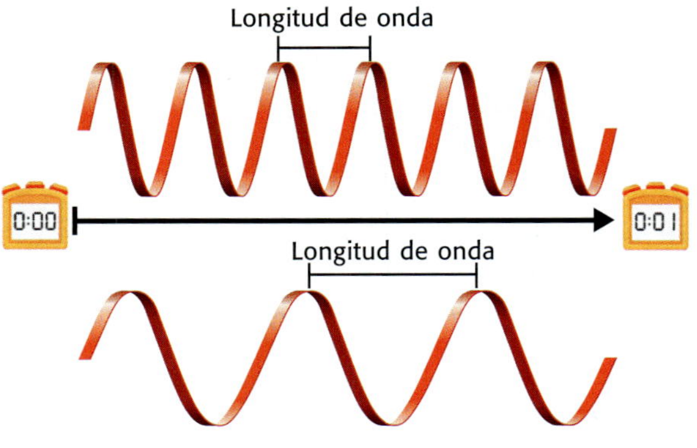

Espectro electromagnético

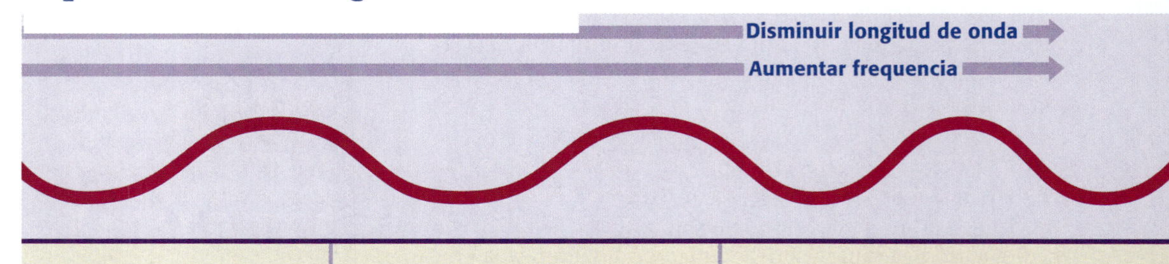

Ondas de radio
Todas las estaciones de radio y de televisión transmiten ondas de radio.

Microondas
A pesar de su nombre, las microondas tienen una longitud de onda relativamente larga.

Ondas infrarrojas
Infrarrojo significa "más abajo del rojo".

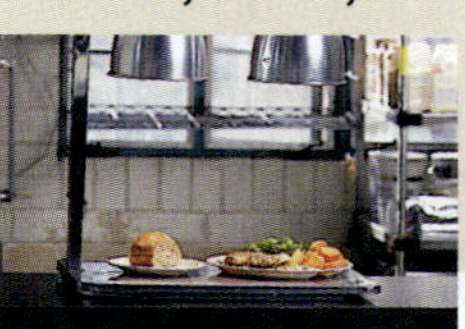

56

Teaching Transparency 6 "The Electromagnetic Spectrum"

IS THAT A FACT!

Wave speed is the speed at which a wave travels. Wavelength and frequency are related by the equation for wave speed:

$$v = \lambda \times f$$

where v = wave speed
λ = wavelength
f = frequency

Luz visible La *luz visible* es la limitada gama de longitudes de onda y frecuencias del espectro magnético que puede ser vista por el ser humano. Podemos ver las longitudes de onda como colores distintos, tal como se muestra en la **Figura 4.** Vemos las longitudes de onda más largas como una luz roja y las longitudes de ondas más cortas como una luz violeta. Debido a que la luz violeta tiene la longitud de onda más corta, transporta la mayor cantidad de energía de las ondas de la luz visible. La luz azul transporta energía suficiente para descomponer la bilirrubina en la piel de un recién nacido.

La luz visible entrega la energía necesaria para las reacciones químicas denominadas fotosíntesis, el proceso mediante el cual las plantas producen su propio alimento. La fotosíntesis es importante debido a que el alimento producido por las plantas provee energía a casi todos los organismos vivos de la Tierra.

La gama de colores que compone la luz visible se denomina el *espectro visible*. Los colores de la luz visible son: **r**ojo, **n**aranja, **a**marillo, **v**erde, **a**zul, **í**ndigo y **v**ioleta. Cuando se combinan todos los colores de la luz visible, puedes ver la luz como una luz blanca. La luz del Sol y la luz de los focos incandescentes y fluorescentes son ejemplos de luz blanca.

Figura 4 *La luz blanca, al igual que la luz solar, es luz visible de distintas longitudes de onda combinadas. Puedes ver todos los colores de la luz visible en un arco iris.*

ACTIVITY

Making Models Place the class into small groups. Give each group a piece of paper at least 4 m long. Explain that the wavelength of radio waves broadcast by their favorite FM radio station is about 3 m. Have students use a meterstick to measure and draw an FM radio wave on their paper. By comparison, the wavelength of waves produced by the typical microwave oven is about 12 cm. Have students draw waves with a 12 cm wavelength on their paper. Be sure they label their waves. Challenge students to imagine how small the wavelength of visible light is, which ranges between 400 nm and 700 nm (0.0000004 and 0.0000007 m)! `Sheltered English`

GOING FURTHER

Obtain and display a microwave oven, exposed X-ray film, and a radio (or photographs of these items). Use Teaching Transparency 195 to help students understand the electromagnetic spectrum. Discuss each item or photo with students and ask whether they have ever used or seen the items. Then discuss with students the part of the EM spectrum involved with each item. (The X-ray film involves two parts of the EM spectrum: X rays to expose the film and visible light to read it.) `Sheltered English`

Homework

Challenge students to think of mnemonic devices similar to "Roy G. Biv" for the parts of the electromagnetic spectrum shown in the diagram on pages 56 and 57 (radio waves, microwaves, infrared waves, visible light, ultraviolet light, X rays, and gamma rays).

MISCONCEPTION ALERT

The "visible spectrum" is called visible because humans can see it. However, other animals can see electromagnetic radiation with wavelengths that humans cannot see.

Teaching Transparency 195 "Wave Speed, Wavelength, and Frequency"

LINK TO PHYSICAL SCIENCE

Quiz

1. What do all waves have in common? (They carry energy.)

2. At a given wave speed, what generally happens to the energy carried by waves if their frequency decreases and their wavelength increases? (The waves will carry less energy.)

3. How is visible light different from other parts of the electromagnetic spectrum? (It isn't really, except that humans can see that narrow band of wavelengths and frequencies.)

4. If you could see some other part of the EM spectrum in addition to visible light, which part would you choose? Why? (Answers will vary.)

ALTERNATIVE ASSESSMENT

Concept Mapping Have students refer to the diagram of the electromagnetic spectrum and then prepare a concept map of the properties of the different types of EM waves. Their concept map should include the vocabulary terms from this section and an example for each term.

Math Skills Worksheet 54
"Color at Light Speed"

Interactive Explorations CD-ROM "In the Spotlight"

Luz ultravioleta La *luz ultravioleta* es la categoría de las ondas electromagnéticas que se encuentra más allá de la luz visible. Aproximadamente el 10 por ciento de la energía solar llega en forma de luz ultravioleta. Las ondas ultravioletas tienen longitudes de onda más cortas y frecuencias más elevadas que las ondas de luz visible. Por lo tanto, las ondas ultravioleta transportan más energía que la luz visible. Esta gran cantidad de energía nos afecta tanto en forma positiva como negativa.

Desde el punto de vista positivo, la luz ultravioleta se utiliza para destruir bacterias en los alimentos y en los instrumentos quirúrgicos. Además, la exposición moderada a la luz ultravioleta es beneficiosa para tu cuerpo. Al exponernos a la luz ultravioleta, las células de la piel producen vitamina D, una substancia que les permite a los intestinos absorber el calcio. Sin calcio, tus dientes y tus huesos serían muy débiles.

Desde el punto de vista negativo, una sobreexposición a la luz ultravioleta puede causar quemaduras de sol, cáncer de piel, lesiones de los ojos y aparición de arrugas. Afortunadamente, gran parte de la luz ultravioleta proveniente del Sol no llega a la superficie de la Tierra. Sin embargo, debes protegerte de la luz ultravioleta que te alcanza. Para hacer esto, debes usar un bloqueador solar con un FPS alto (**f**actor de **p**rotección **s**olar) y utilizar anteojos de sol para bloquear la luz ultravioleta, tal como la persona en el lado derecho de la **Figura 5.** Necesitas esta protección incluso en los días nublados, ya que la luz ultravioleta puede atravesar las nubes.

Figura 5 *Las quemaduras de sol no sólo son dolorosas sino también pueden causar arrugas y cáncer en la piel.*

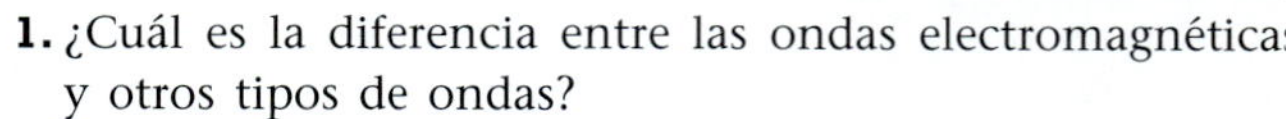

astronomía
CONEXIÓN

Los rayos gamma se producen mediante las reacciones nucleares que ocurren en el centro del Sol. Sin embargo, no debes preocuparte de la sobreexposición a los rayos gamma aquí en la Tierra. Los gases que rodean el centro del Sol absorben los rayos gamma y estos ni siquiera alcanzan la superficie de este astro.

REPASO

1. ¿Cuál es la diferencia entre las ondas electromagnéticas y otros tipos de ondas?

2. ¿Qué relación hay entre la frecuencia de una onda y la cantidad de energía que transporta?

3. Si la frecuencia de una onda disminuye, ¿qué sucede con su longitud de onda?

4. Explica cómo la luz ultravioleta puede ser tanto beneficiosa como perjudicial para el ser humano.

5. **Utilizar gráficas** Estudia el espectro electromagnético en las páginas 56 y 57 para determinar qué tiene más energía: la luz visible o las microondas. Explica tu respuesta.

58

Answers to Review

1. EM waves do not require a medium through which to travel. All other waves require a medium.

2. The amount of energy a wave carries increases as its frequency increases.

3. If the frequency of a wave decreases, the wavelength increases.

4. UV waves are useful because they can kill bacteria and because they help the body produce vitamin D. However, overexposure to ultraviolet waves can cause sunburn, wrinkles, and skin cancer.

5. Visible light has more energy than microwaves because visible light waves have a higher frequency and shorter wavelength than microwaves have.

Reflexión, absorción y dispersión

Caminas en la obscuridad y oyes algo que se mueve entre los arbustos. Diriges tu linterna hacia el sonido y puedes ver dos ojos que brillan. Al principio te asustas, pero después te das cuenta de que es sólo un gato. ¿Por qué será que los ojos de los gatos brillan en la noche? Los gatos tienen una capa especial de células en el fondo de sus ojos que refleja la luz. Dicha capa les permite ver mejor pues les da la capacidad de detectar la luz. Los cocodrilos, los venados y los tiburones también tienen este tipo de capa de reflexión en sus ojos. La reflexión es sólo una de las formas en que interactúan las ondas luminosas. En esta sección aprenderás los tres tipos de interacciones de las ondas luminosas: reflexión, absorción y dispersión.

Reflexión

La **reflexión** ocurre cuando la luz o cualquier otro tipo de onda rebota sobre un objeto. Cuando te ves en un espejo, en verdad estás viendo luz que ha sido reflejada dos veces: primero se ha reflejado en ti y luego en el espejo. La reflexión te permite ver objetos que no producen su propia luz. Cuando la luz alcanza un objeto, un poco de esa luz se refleja en él y tus ojos lo detectan. Si esa luz se refleja en ti y en los objetos que te rodean, ¿por qué no puedes ver tu reflejo en el muro? Para contestar esta pregunta, primero debes aprender la ley de la reflexión.

Figura 6 La ley de reflexión

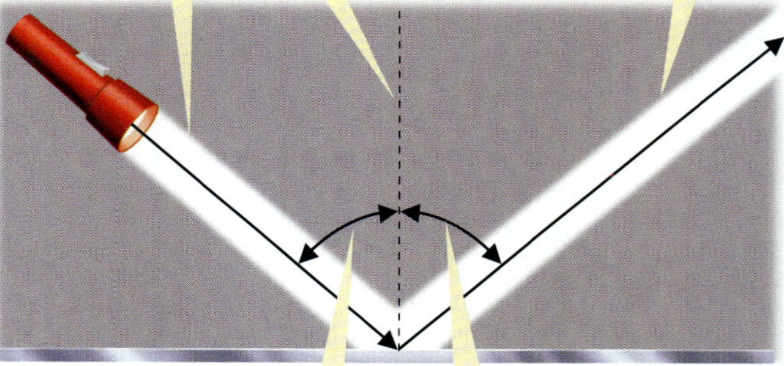

La ley de reflexión La luz rebota sobre los objetos de la misma manera que una pelota rebota en el suelo. Si tiras una pelota directamente hacia abajo rebotará directamente hacia arriba. Si la tiras en un ángulo, también rebotará en un ángulo. Esto es un ejemplo de la ley de la reflexión. La **ley de reflexión** sostiene que el ángulo de incidencia es igual al ángulo de reflexión. La incidencia es la caída de un haz de luz sobre una superficie. La **Figura 6** ilustra esta ley.

59

Teaching Transparency 7 "The Law of Reflection"

Directed Reading Worksheet 3 Section 2

Focus

Reflection, Absorption, and Scattering

In this section, students learn how light waves are reflected, absorbed, or scattered when they come in contact with objects. Students also explore how the color of an object is determined and the difference between the primary colors of light and the primary pigments.

Bellringer

Give students the following:

People use mirrors every day. From your experience with mirrors, explain how they work and what they do to light waves. Write your answer in your ScienceLog.

1) Motivate

ACTIVITY

Making a Periscope Place students in groups. Give each group a shoe box, two small hand mirrors, some modeling clay, and a pair of scissors. The mirrors must be small enough to stand upright inside the shoe box. Have students cut a 3 cm hole on the left side of each end of the box (so the holes will not be directly opposite each other). Then tell students that their job is to arrange the mirrors inside of the box with the modeling clay. The mirrors must be arranged in such a way that someone can look straight into one hole and see out of the other hole. Ask students to explain what property light must have that allows this experiment to work.

USING THE FIGURE

Concept Mapping Have students study **Figure 7.** Then ask them to make a concept map that explains the difference between regular reflection and diffuse reflection.

`Sheltered English`

Multicultural CONNECTION

Greek philosophers Hero (first century A.D.) and Ptolemy (second century A.D.) believed that eyes emitted rays that were reflected back by objects. However, an Iraqi scientist named Abu 'Ali al-Hasan ibn al-Haytham (c. 965–c. 1039) believed differently. His theory of vision was very similar to today's: Objects are seen because of the light they reflect or emit.

CONNECT TO
EARTH SCIENCE

"Red sky at night . . ." As the sun sets in the evening, light from the sun appears redder because the shorter blue wavelengths have been scattered away. When there is dust in the air, lower frequencies of light scatter more.

Teaching Transparency 7 "Regular Reflection Versus Diffuse Reflection"

Tipos de reflexión Volvamos a la pregunta inicial: "¿Por qué puedes ver tu reflexión en el espejo y no en un muro?" La respuesta se relaciona con lo lisa que sea la superficie sobre la cual se refleje la luz. Si la superficie es lisa, como un espejo o un metal pulido, el haz de luz podrá reflejarse en todos los puntos de la superficie en el mismo ángulo. Esto se conoce como la *reflexión regular*. Si la superficie es áspera, como la de un muro, los haces de luz se reflejan en él en distintos ángulos. Esto se conoce como *reflexión difusa*. La **Figura 7** ilustra la diferencia entre estos dos tipos de reflexión.

Figura 7 Reflexión regular y reflexión difusa

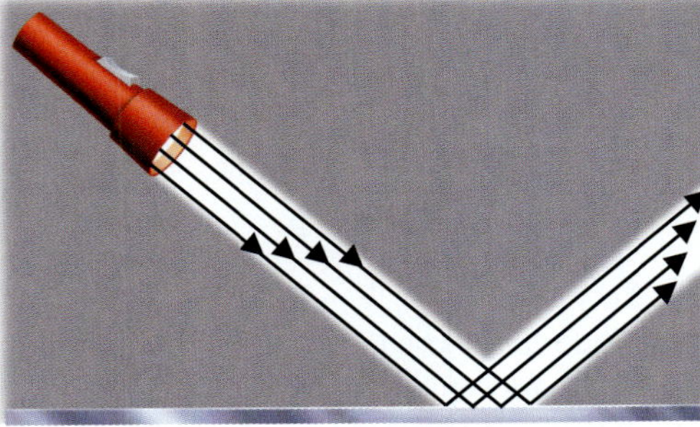

La **reflexión regular** ocurre cuando los haces de luz se reflejan en el mismo ángulo. Cuando tus ojos detectan los haces reflejados, puedes ver un reflejo en la superficie.

La **reflexión difusa** ocurre cuando los haces de luz se reflejan en distintos ángulos. No puedes ver una reflexión porque tus ojos no detectan toda la luz reflejada. La luz que tus ojos pueden detectar te permite ver la superficie.

Absorción y dispersión

Quizás te hayas dado cuenta de que cuando utilizas una linterna, los objetos que están más cercanos a ti se ven más brillantes que aquellos que están más lejos. La luz parece debilitarse mientras más se aleja de la linterna. Esto ocurre, en parte, porque el haz de luz se extiende y, en parte, por la absorción y la dispersión.

La **absorción** es la transferencia de energía transportada por las ondas luminosas a las partículas de materia. Cuando enciendes una linterna en el aire, las partículas del aire absorben un poco de la energía proveniente de la luz. Esto hace que la luz se debilite, como se muestra en la **Figura 8.** Mientras más lejos se traslade la luz de la linterna, más se absorbe en las partículas del aire. Por eso, la luz se debilita cuando viaja grandes distancias.

Figura 8 Un haz de luz se debilita debido a la absorción y a la dispersión.

WEIRD SCIENCE

The *fenestraria*, or window plant, is a succulent that grows in the deserts of southern Africa. Sunlight enters the plant's leaves and is reflected down its underground stem to its roots. There the light energy is absorbed by chlorophyll-containing cells.

IS THAT A FACT!

A blue jay's feathers are not really blue. The air molecules on the surface barbs of the feathers scatter or absorb the red and green light of the visible spectrum, leaving blue light to reflect to our eyes.

La **dispersión** es la liberación de la energía de la luz por partículas de materia que han absorbido energía adicional. Cuando la luz se libera se dispersa en todas direcciones. La luz de una linterna se dispersa debido a las partículas de aire. La luz dispersa te permite ver objetos fuera del haz, pero como la luz del haz se dispersa, el haz se debilita.

La dispersión hace el cielo azul. La luz con longitudes de onda más cortas se dispersa más que la luz con longitudes de onda larga. La luz solar está compuesta de distintos colores, pero la luz azul (que tiene una longitud de onda muy corta) se dispersa más que cualquier otro color. De modo que cuando ves el cielo, puedes ver un fondo azul. Puedes aprender más de la dispersión al realizar el experimento a la izquierda.

Luz y color

¿Alguna vez te has preguntado qué es lo que le da a un objeto su color? Ya sabes que la luz blanca está compuesta de todos los colores de la luz. Sin embargo, cuando ves una fruta bajo la luz blanca, vez un color. Las fresas son rojas, los limones son verdes y los plátanos son amarillos. ¿Por qué no se ven blancos? Para responder a esta pregunta, primero debes aprender cómo interactúa la luz con la materia.

La luz y la materia Cuando la luz alcanza cualquier forma de materia, puede interactuar con ella en tres formas distintas: se puede reflejar, absorber o transmitir. Ya has aprendido la reflexión y la absorción. **Transmisión** es el paso de luz a través de la materia. Puedes ver la transmisión de luz en todo momento. Toda la luz que llega a tus ojos es transmitida por el aire. La luz puede interactuar con la materia en distintas formas al mismo tiempo, como se muestra en la **Figura 9.**

Figura 9 *La luz se transmite, se refleja y se absorbe cuando alcanza el vidrio de una ventana.*

Laboratorio

Leche dispersa

1. Llena una **botella transparente de 2 litros** con **agua.**

2. Apaga las luces y enciende una **linterna** a través del agua. Observa el agua de todas partes de la botella. Describe lo que ves en tu cuaderno de ciencias.

3. Agrega unas gotas de **leche** al agua y agita la botella para mezclarla.

4. Repite el paso 2. Describe cualquier cambio de color. Si no ves ningún tipo de cambio, agrega más leche hasta que veas algún cambio.

5. ¿En qué se asemeja la mezcla de agua y leche con las partículas de aire de la atmósfera? Escribe tu respuesta en tu cuaderno de ciencias.

READING STRATEGY

Predicting Before students read page 61, ask them if they can explain why the sky appears blue. Write their explanations on the board. After students have read the section on scattering, discuss with them how air molecules scatter the light waves to make the sky look blue.

QuickLab

MATERIALS
FOR EACH GROUP:
• clear 2 L plastic bottle
• water
• flashlight
• milk

Answers to QuickLab

2. Students should see the flashlight beam shining straight through the water. They can also see the beam from the sides because some of the light is scattered to the side.

4. Viewing the bottle from the side, students should see a bluish color from blue light being scattered to the sides. Looking straight on at the flashlight, the light will appear reddish.

5. Milk particles scatter the light traveling through the water just as air particles scatter sunlight as it travels through the air.

ACTIVITY

Writing Ask students: What is your favorite color? It is a simple question, but have you ever thought about why you like a particular color more than others? In a short paragraph, explain why you like your favorite color. Also explain how certain colors may affect your mood.

Homework

Concept Mapping The cones in our eyes react to ranges of wavelengths of light around what we call red, green, and blue. Because of this, humans can detect a wide range of colors, hues, tints, and shades. Have students look around at home and make a concept map of all the different colors they see. The main idea of the map is colors, and the first four subcategories are red, green, blue, and others. If students run out of room on their map for all the colors they find, have them list the rest.

DEMONSTRATION

Cover a high-intensity flashlight with a green filter and a second flashlight with a red filter. In a darkened room, turn the "green" light on, and shine it on a white wall or overhead screen. Turn the "red" light on, and shine it on a different area of the screen. Ask the students what colors they see. (green and red)

Ask them to predict what color they will see when the green and red light overlap. (Yellow will appear.) (Sometimes a reddish yellow or a greenish yellow results because the filters are not a true red or green.)

`Sheltered English`

CROSS-DISCIPLINARY FOCUS

Art Encourage students to find information about pointillist artists Georges Seurat, Henri Edmond Cross, and Paul Signac. In art, *pointillism* is the theory or practice of applying small strokes or dots of color to a surface so that when they are viewed from a distance, they blend together to form an image. Distribute to small groups of students magnifying lenses and pages from the Sunday comics section of the newspaper. Instruct students to look at the comics through the lens. Discuss what they see. (For this lesson, the only point is that the larger image is made up of many small dots. You may want to repeat this activity when discussing color subtraction, which is covered on page 64.)

Answer to Self-Check

The paper will appear blue because only blue light is reflected off the paper.

Los colores de los objetos ¿De qué manera determina el color de un objeto la interacción de la luz con la materia? Ya sabes que el color de la luz está determinado por la longitud de onda de la onda luminosa. El rojo tiene la longitud de onda más larga, el color violeta el más corto y los demás colores tienen longitudes intermedias.

El color de un objeto está determinado por el color de la luz que llega a los ojos. La luz puede alcanzar tus ojos después de rebotar de un objeto o después de haber sido transmitido a través de un objeto.

Cuando la luz blanca alcanza un objeto de color, algunos colores de la luz son absorbidos y otros reflejados. Solamente la luz reflejada alcanza tus ojos y puede ser detectada. Por ende, los colores de la luz que se reflejan por un objeto determinan el color que tú ves. Por ejemplo, si tu suéter refleja la luz azul y absorbe los demás colores, verás que el suéter es azul. Otro ejemplo se muestra en la **Figura 10.**

Figura 10 *Cuando la luz blanca brilla sobre una fresa, solamente se refleja la luz roja. Todos los demás colores de la luz son absorbidos. Por lo tanto, para ti la fresa es roja.*

Si los objetos verdes reflejan luz verde y los objetos rojos reflejan luz roja, ¿qué colores de luz se reflejan en la vaca que se muestra a la izquierda? Recuerda, la luz blanca contiene todos los colores de la luz. De modo que los objetos blancos, como el pelaje blanco de la vaca, aparecen blancos porque todos los colores de la luz se reflejan. Por otra parte, el negro es la ausencia de color. Cuando la luz brilla sobre un objeto negro, todos los colores son absorbidos.

✓ Autoevaluación

Si la luz azul brillara sobre una hoja de papel blanco, ¿de qué color se vería el papel? *(Consulta la página 636 para revisar tu respuesta.)*

IS THAT A FACT!

The human eye can distinguish millions of different colors and hues.

Mezclar los colores de la luz

Para obtener una luz blanca, debes combinar todos los colores de la luz ¿verdad? Bueno, esa es una forma de obtenerla. También puedes obtener la luz blanca simplemente combinando tres colores de la luz: rojo, azul y verde, tal como se muestra en la **Figura 11.** De hecho estos tres colores pueden ser combinados en distintas proporciones para producir todos los colores de la luz visible. Por lo tanto, el rojo, el azul y el verde se denominan los colores primarios de la luz.

Cuando se combinan los colores de la luz, existen más longitudes de onda. Por ende, la combinación de los colores de la luz se denomina adición de colores. Cuando se combinan dos colores primarios, se produce un color secundario. Los colores secundarios son el cian (azul y verde), el magenta (azul y rojo) y el amarillo (rojo y verde). En la Figura 11, los colores secundarios aparecen entre los colores primarios.

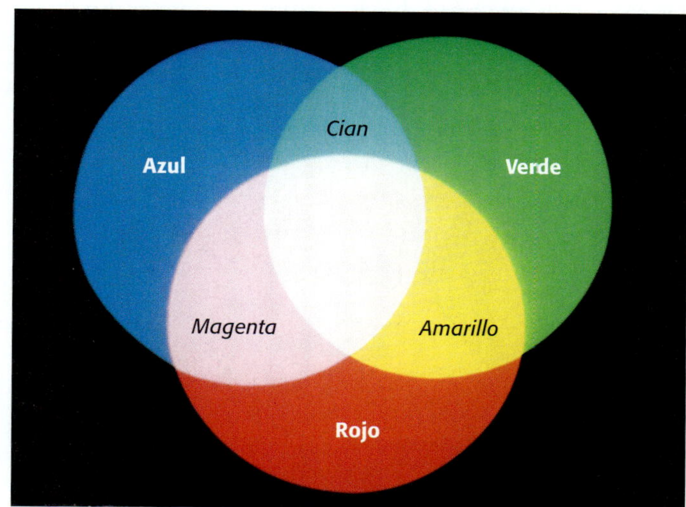

Figura 11 *Los colores primarios, escritos en negrilla, se combinan para producir la luz blanca. Los colores secundarios, escritos en cursiva, son el resultado de dos colores primarios que se combinan.*

Los colores que ves en la televisión se producen por la adición de los colores primarios de la luz. Una pantalla de televisión está compuesta de pequeños puntos que emiten luz, denominadas píxel. Cada píxel puede emitir el rojo, el azul y el verde. Los colores emitidos por los píxel se combinan para producir los distintos colores que ves en la pantalla.

APLICA

Magdalena está encargada de las luces para la obra de teatro de la escuela. Ella ha revisado la sala del equipo y ha encontrado reflectores de luz que producen tres colores de luz distintos: rojo, azul y verde. Ella sabe que también necesitará reflectores de luz de color cian, magenta, amarillo y blanco. Sin embargo, no hay un presupuesto destinado para comprar proyectores de luz. ¿Qué puede hacer para obtener los colores que necesita?

Cross-Disciplinary Focus

Art A good photographer must understand the science behind photography, such as how light will interact with the subject of a photograph, what the camera lens will do to the light, and how the light will affect the film. Invite a professional photographer to address students and to demonstrate the art and science of taking and printing photographs.

---

3 Extend

Going Further

Writing Have students research how a television works, especially how a color image is broadcast over the airwaves as an electromagnetic signal and ends up being a brightly colored picture in their home. Ask them to explain the difference between analog television signals and digital signals. Encourage them to be creative in presenting their results.

Activity

Seeing Pixels Refer students to the Brain Food on this page. If you have access to computer monitors and televisions, allow students to sprinkle one or two drops of water on each screen. The drops of water will magnify the picture elements, or "pixels," in the screen. Give students an opportunity to see the colored pixels on the screen. Discuss their observations. Ask them to predict how mixing colored dots on paper (the Sunday comics or pointillist artworks) might be different.

Answer to APPLY

Magda can use color addition of the three spotlights that she has to create light spots of all the other colors she needs. Cyan light can be produced by adding blue light and green light. Magenta light can be produced by adding red light and blue light. Yellow light can be produced by adding red light and green light. Finally, white light can be produced by adding red light, blue light, and green light.

Reinforcement Worksheet 3 "Light Interactions"

Quiz

1. Why can you see your reflection in a mirror but not on a wall? (Refer to the explanation of the law of reflection on page 59 and of types of reflections on page 60.)

2. List some things that would be different if all matter transmitted light. (People could see through walls; plants could not absorb the light energy they need; people could see right through your body.)

3. Explain how you can tell the color of an object. (Light of various wavelengths strikes the object. Some light is absorbed, and some is reflected and reaches my eyes. The wavelength of the light reaching my eyes determines the color.)

4. For the school play, why is it important for the lighting director and the scenery director to work together? (to make sure that all the colors of light and colors of pigments that are used lead to the colors the directors want for the play)

ALTERNATIVE ASSESSMENT

Concept Mapping Have students make a concept map that shows the difference between mixing colors of light and mixing colors of pigment.

LabBook **PG 568**
Mixing Colors

Mezclar los colores de los pigmentos

Los procesos para mezclar la pintura y los procesos para mezclar la luz son distintos porque la pintura contiene pigmentos. Por eso no se puede obtener blanco mezclando pintura roja, azul y verde. Un **pigmento** es un material que le da color a una substancia mediante la absorción de algunos colores de la luz y la reflexión de otros colores.

Los pigmentos le dan color a casi todo, mediante la reflexión y la absorción de la luz. La clorofila y la melanina son pigmentos. La clorofila le da el color verde a las plantas y la melanina le da color a tu piel.

Cada pigmento absorbe al menos un color de la luz. Cuando los mezclas, se absorben o sustraen más colores de la luz. La mezcla de colores de pigmentos se llama substracción de colores. Los *pigmentos primarios* son el amarillo, el cian y el magenta. Se pueden combinar para producir otro color. Todos los colores en este texto se obtuvieron mediante el uso de los pigmentos primarios y tinta negra. La tinta negra se utilizó para dar contraste a las imágenes. La **Figura 12** muestra cómo los cuatro pigmentos se combinan para producir colores distintos.

Figura 12 *El dibujo del globo de la izquierda se realizó superponiendo tintas de color cian, magenta, amarillo y negro.*

Experimentos

¿Cuándo se parece la clase de ciencias a la de arte? Cuando utilizas pinturas para mezclar los colores en la página 568.

REPASO

1. Explica la diferencia entre la absorción y la dispersión.

2. Las plantas deben obtener la energía que requieren para vivir del Sol. ¿Por qué es una ventaja el hecho de que la células de las plantas contengan clorofila?

3. **Aplicar conceptos** Explica qué sucede con los distintos colores de la luz cuando la luz brilla sobre un objeto de color violeta.

4. **Hacer inferencias** Explica por qué puedes ver tu reflejo sobre una cuchara y no sobre un pedazo de tela.

64

▼ **Answers to Review**

1. absorption: the transfer of energy carried by light waves to particles of matter; scattering: the release of light energy by particles that have absorbed extra energy

2. Chlorophyll, a green pigment, absorbs all colors of light except green. The light energy absorbed by chlorophyll is used by plants to live.

3. When white light shines on a violet object, violet light is reflected. All other colors of light are absorbed.

4. Light reflecting off a spoon is regular reflection; you can see your image in the spoon. Light reflecting off a piece of cloth is diffuse reflection. Thus, you can see the cloth but not your image in it.

Refracción

VOCABULARIO

refracción　　lente convexa
lente　　lente cóncava

OBJETIVOS

- Define *refracción*
- Explica cómo la refracción puede dividir la luz blanca en diferentes colores de luz.
- Describe las diferencias entre las lentes convexas y las lentes cóncavas.
- Identifica ejemplos de lentes que se usan en la vida cotidiana.

¿Conoces a alguien que use anteojos o lentes de contacto? ¿Los usas tú? Aproximadamente el 37 por ciento de las personas de los Estados Unidos usan anteojos o lentes de contacto para corregir problemas de la vista. Si examinas diferentes pares de anteojos, podrás ver diferencias entre las lentes dentro de los marcos. Por ejemplo, algunas lentes son más gruesas en el medio que en los bordes y otras son más delgadas en el medio que en los bordes.

Tus ojos tienen lentes también. Cuando ves algo, la luz que se produjo o se reflejó en un objeto, entra en tus ojos por la pupila, atraviesa la córnea y la lente y forma una imagen en la superficie posterior del ojo, la retina. ¿Por qué la córnea y el lente del ojo son importantes en la formación de imágenes en la retina? Para encontrar la respuesta, necesitas aprender cómo viaja la luz y lo que sucede cuando pasa de un tipo de materia a otro.

Los rayos muestran la ruta de las ondas luminosas

Como todas las ondas, las ondas luminosas viajan desde su fuente en todas direcciones. Si pudieras rastrear la ruta de una onda al alejarse de una fuente luminosa, descubrirías que la ruta es una línea recta. Como las ondas luminosas viajan en línea recta, puedes mostrar la ruta y dirección de una onda luminosa con una flecha llamada *rayo*. La **Figura 13** muestra algunos rayos provenientes de un foco.

Aunque las ondas luminosas normalmente viajan en línea recta, pueden cambiar de dirección al interactuar con la materia. Por ejemplo, las ondas luminosas cambian de dirección cuando se reflejan en un objeto. Las ondas luminosas también cambian de dirección y velocidad cuando pasan de un medio, como el aire, a otro, como el vidrio.

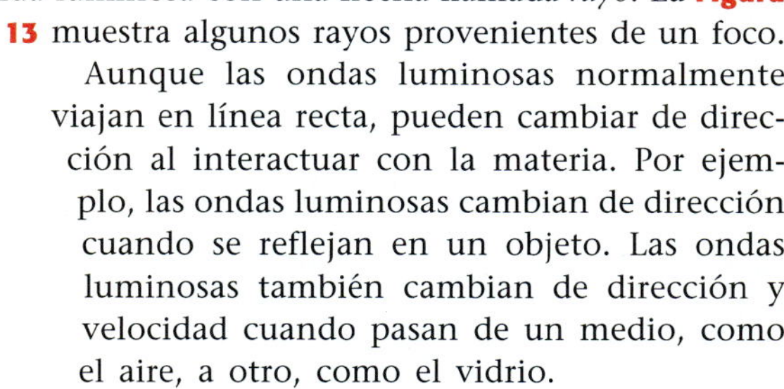

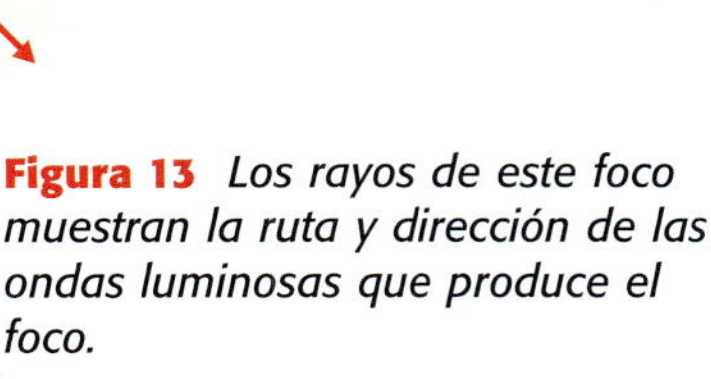

Figura 13 *Los rayos de este foco muestran la ruta y dirección de las ondas luminosas que produce el foco.*

65

Focus

Refraction

In this section, students learn what *refraction* is and how it can separate white light into different colors. Students also learn about concave and convex lenses and about some optical instruments that use lenses.

Bellringer

Display a prism; if possible, show it separating light into a spectrum of colors. Ask students to predict, in their ScienceLog, how a prism works. (It separates white light into the spectrum because of refraction. It does not change light or create colors.)

1) Motivate

DISCUSSION

Ask students why a pencil standing in a glass of water appears to be broken right at the place where the air and water meet. Challenge them to explain this phenomenon. (Light is bent, or refracted, when it passes from one medium to another at an angle.)

Directed Reading Worksheet 3 Section 3

WEÏRD SCIENCE

Natural rainbows form only when the sun is lower than 42° (about the width of four fists) above the horizon. This is due to the way light waves refract through raindrops. Thus, you can be skeptical if someone tells you that he or she saw a rainbow at noon! The formation of a rainbow also requires both reflection and refraction.

DEMONSTRATION

What Are Light Rays? A good model of a light ray is a beam of light from a bright flashlight. (A laser pointer is even better if you have access to one.) Darken the room, clap two chalkboard erasers together to create some dust, and turn on the flashlight or pointer. As you shine the light around the room, remind students that light waves travel away from a light source in all directions. Each individual point on a wave can be represented by a straight line, which is called a light ray. So a ray is an imaginary line that gives the direction of the wave. On the board, draw some ray diagrams showing what happens when light waves interact with matter.

MEETING INDIVIDUAL NEEDS

Advanced Learners Have students research the archerfish, a species of fish that hunts by using a jet of water to knock insects sitting on leaves above the surface into the water. Ask students to explain how the archerfish overcomes refraction in order to be a successful hunter. Encourage them to be creative in presenting their results.

Teaching Transparency 8 "White Light Is Separated by a Prism"

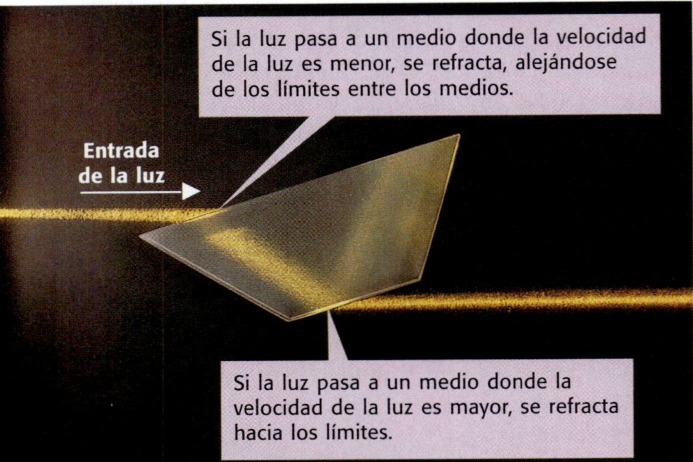

Figura 14 *La luz viaja más lentamente a través del vidrio que del aire. Entonces, la luz se refracta al pasar en ángulo desde el aire al vidrio o desde el vidrio al aire.*

Figura 15 *Un prisma es un trozo de vidrio que divide la luz blanca en los colores de la luz visible por refracción.*

Refracción

Refracción es la acción de curvarse una onda cuando pasa de un medio a otro en un ángulo. La refracción de ondas luminosas ocurre porque la velocidad de la luz varía según el material por el cual viajan las ondas. En un vacío (un lugar donde no hay materia) la luz viaja a 300,000,000 m/s, pero viaja más lento a través de la materia. Cuando una onda entra en un nuevo medio formando un ángulo, la parte de la onda que entra primero empieza a viajar a una velocidad diferente del resto de la onda. La **Figura 14** muestra cómo se curva un haz de luz por refracción.

Refracción y división de colores Cuando la luz se refracta, el grado en que la luz se curva depende de su longitud de onda. Las ondas luminosas que tienen longitudes de onda cortas se curvan más que las que tienen longitudes de onda largas. Ya has aprendido que la luz blanca está compuesta por todos los colores de la luz visible. También sabes que los diferentes colores corresponden a diferentes longitudes de onda. Por esta razón, la luz blanca se puede dividir en diferentes colores durante la refracción, como se ve en la **Figura 15.** La división de colores durante la refracción es parcialmente responsable de la formación de los arco iris. Los arco iris se forman cuando la luz solar se refracta en las gotas de agua.

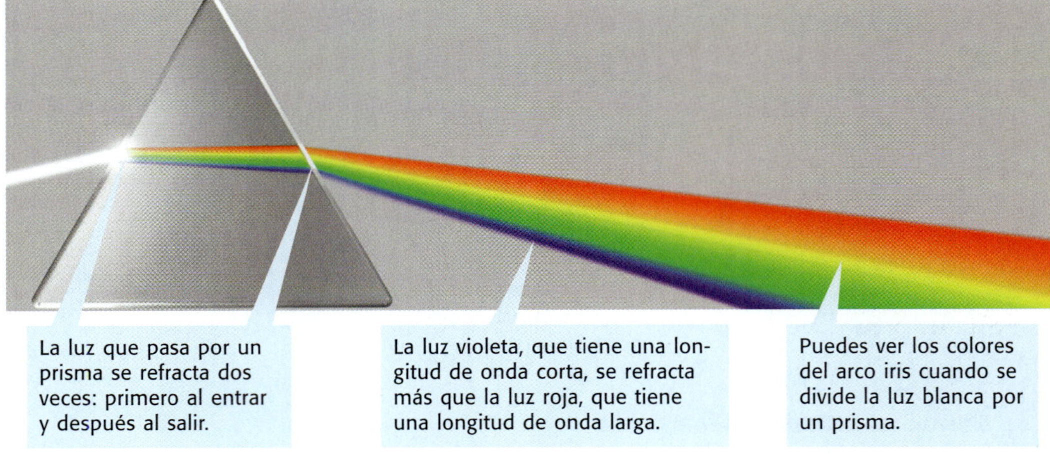

CONNECT TO PHYSICAL SCIENCE

The speed of light in a vacuum divided by the speed of light in some other medium is a useful ratio called the *index of refraction*. The index of refraction (n) for air is nearly 1. In other materials, n is greater than 1. The larger n is, the more light refracts (the greater the angle) when it enters the medium. The path is reversible. The angles will be same whether the light is going from the air into the medium or from the medium into air. Refraction is responsible for interesting phenomena such as items underwater not being where they appear to be and people being able to see the sun after it has set.

Las lentes refractan la luz

¿Qué tienen en común las cámaras, los binoculares, los telescopios, los proyectores y tus ojos? Todos tienen lentes. Una **lente** es un objeto curvo y transparente que refracta la luz. Las lentes se clasifican según su forma. Las lentes son convexas y cóncavas.

Lente convexa Una lente que es más gruesa en medio que en los bordes es una **lente convexa.** Cuando los rayos luminosos entran en una lente convexa, siempre se refractan hacia el centro. La cantidad de refracción depende de la curvatura de la lente, como se muestra en la **Figura 16.** Los rayos luminosos que pasan por el centro de un lente no se refractan.

Una lente convexa puede concentrar la luz que entra en ella. Por esta razón, las lentes convexas se usan en cámaras, telescopios, proyectores y en anteojos para las personas con hipermetropía. Tu ojo tiene una lente convexa que enfoca la luz cambiando su forma, como se muestra en la **Figura 17.**

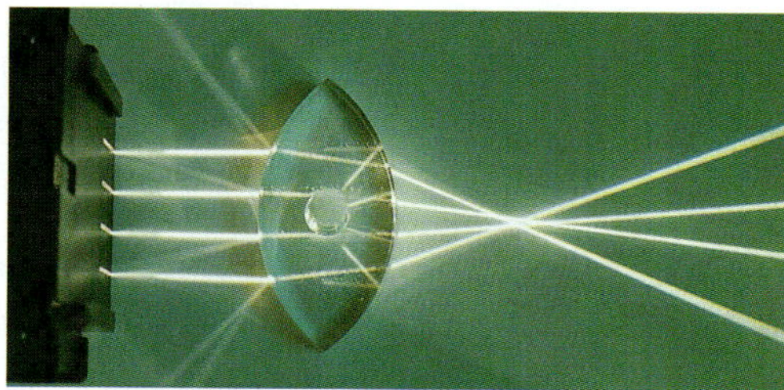
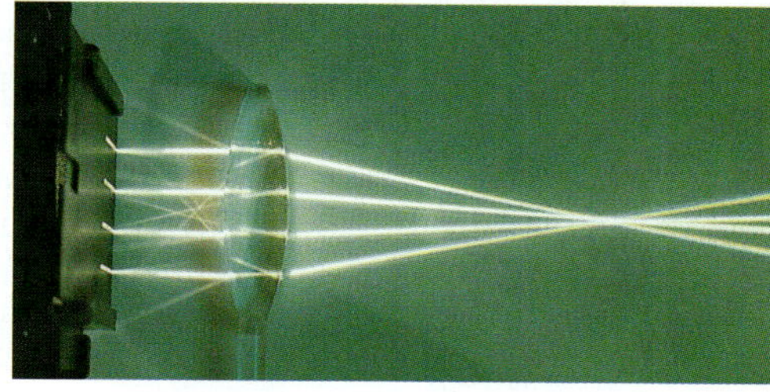

Figura 16 *La refracción de los rayos luminosos en una lente convexa depende de la curvatura de la lente.*

Figura 17 *Los rayos luminosos que entran se refractan en la córnea y las lentes. El tamaño de la pupila determina la cantidad de luz que entra al ojo. Los rayos luminosos convergen en la retina, formando una imagen.*

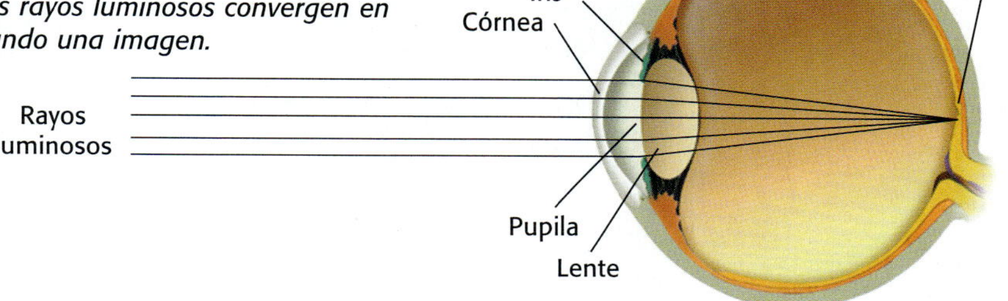

Lentes cóncavas Una **lente cóncava** es más delgada en el medio que en los bordes. Los rayos luminosos que atraviesan las lentes cóncavas siempre se curvan alejándose entre sí y acercándose a los bordes de las lentes, como se ve en la **Figura 18.** Las lentes cóncavas se usan generalmente en anteojos que corrigen la vista de las personas que tienen miopía. Las lentes cóncavas también se pueden usar en combinación con otras lentes en telescopios. La combinación de lentes ayuda a producir imágenes más nítidas de objetos distantes.

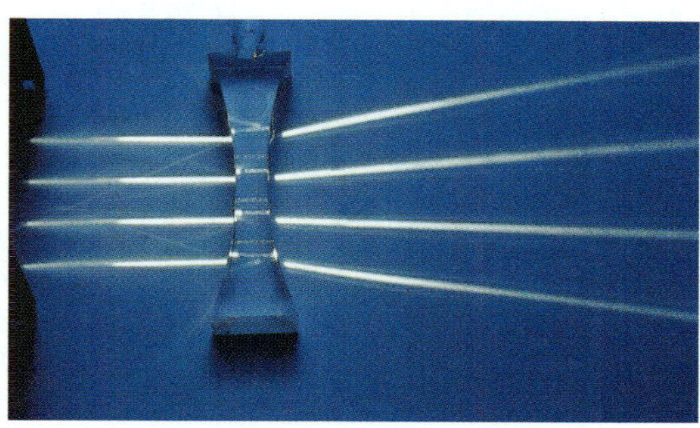

Figura 18 *La luz que atraviesa una lente cóncava se refracta hacia afuera.*

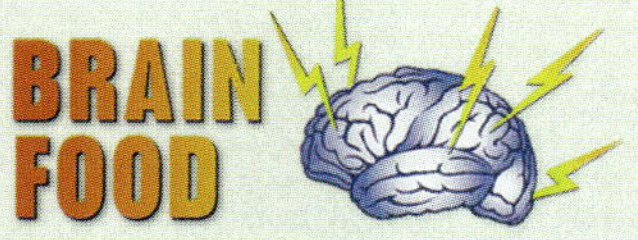

Did you know that some chickens wear red contact lenses? But the lenses don't improve the chickens' vision—they just make the chickens see everything in red! For some reason, chickens that see in red are less aggressive and produce more eggs.

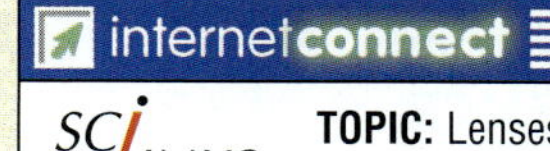

internet**connect**

TOPIC: Lenses
GO TO: www.scilinks.org
*sci***LINKS NUMBER:** HSTL720

CROSS-DISCIPLINARY FOCUS

History The process of grinding glass to make lenses has not changed much since the Middle Ages, when the first glass lenses were made. The equipment has been modernized, but lenses are still cut from blocks of glass and then ground into shape.

REAL-WORLD CONNECTION

Today, most contact lenses are thin, delicate disks worn directly on the eye to correct vision problems. They are held in place by a natural layer of tears that covers the cornea of the eye. Contacts can be shaped to correct most vision problems.

Teaching Transparency 9
"Thick and Thin Convex Lenses"
"A Concave Lens"

GOING FURTHER

Most students are familiar with point-and-shoot or single-lens reflex cameras that use 35 mm film. Now there are digital cameras that download images directly into a computer. Have students research the changes in film and camera technology since 1990. They can compare the different kinds of film available to photographers, or they can compare photographs on film with digital images. Encourage them to be creative in presenting their results.

INDEPENDENT PRACTICE

If possible, acquire some inexpensive cameras for students to use. Encourage those who are interested to do a photo essay on a topic of their choosing. Pictures can be in black and white or in color. Encourage students to experiment with different light conditions, and if possible with different shutter speeds and apertures, to see how these factors affect the images.

Answers to Self-Check

a. the pupil

b. the cornea and the lens

c. the retina

Teaching Transparency 10 "Comparing a Camera to Your Eye"

Instrumentos ópticos

Los *instrumentos ópticos* son artefactos tecnológicos que usan espejos y lentes para ayudar a las personas a observar. Algunos de estos instrumentos ayudan a las personas a ver objetos que están muy lejos y otros ayudan a ver objetos que son muy pequeños. El instrumento óptico que probablemente conoces mejor es la cámara.

¡Sonríe! La forma en que funciona una cámara es similar a la forma en que funcionan los ojos. Una cámara tiene una lente que enfoca la luz y un orificio que permite la entrada de la luz. La principal diferencia entre una cámara y tus ojos es que la lente de una cámara se puede mover hacia adelante y atrás para enfocar la luz mientras que la lente de tu ojo cambia de forma cuando necesita enfocar. La **Figura 19** muestra las partes de una cámara y sus funciones.

Figura 19 *Las cámaras son útiles para registrar imágenes en forma permanente en fotografías.*

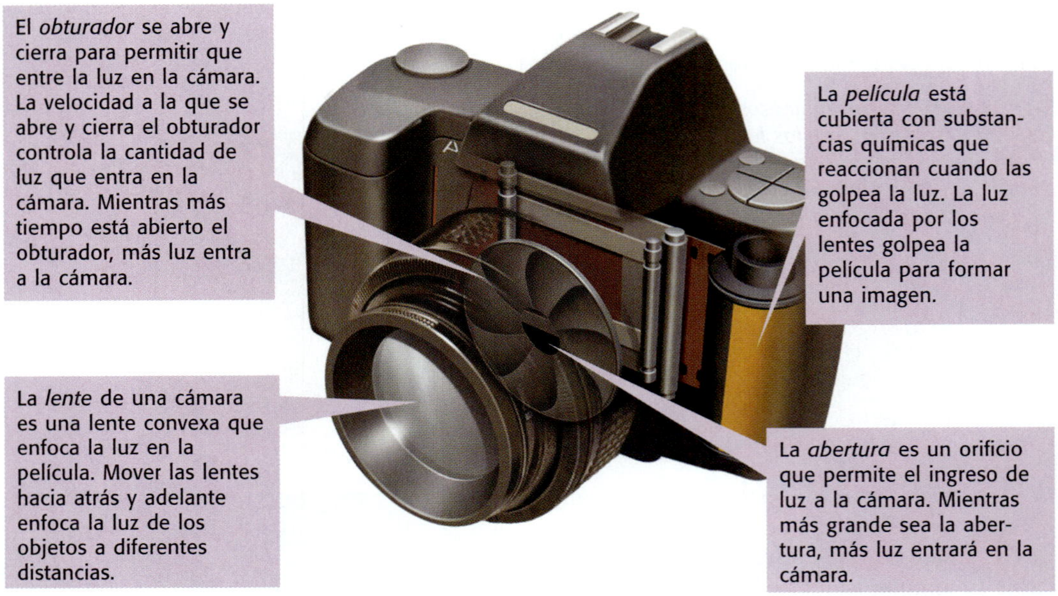

Autoevaluación

¿Qué partes del ojo humano corresponden a las siguientes partes de una cámara?

a. la abertura **b.** la lente **c.** la película

(Consulta la página 636 para comprobar tu respuesta.)

68

CONNECT TO PHYSICAL SCIENCE

Imagine a glass thread thinner than a human hair, stronger than steel, and able to carry more than 1,000 telephone messages at one time. These strands, which range from 0.0005 to 0.01 cm in diameter, are called optical fibers. Basically, light travels from one end of an optical fiber to the other, reflecting back and forth from wall to wall along the length of the fiber. Optical fibers are used in medicine to detect tumors and injuries and to transmit images of organs and lesions to a camera outside the body. Fibers can even be used to transmit light into narrow places in the body.

Telescopios Los astrónomos usan telescopios para estudiar los objetos del espacio, como la Luna, los planetas y las estrellas. Los telescopios se clasifican como refractores o reflectores. Los telescopios refractores usan lentes para capturar la luz y los reflectores usan espejos. La **Figura 20** ilustra cómo funcionan los telescopios refractores y reflectores simples.

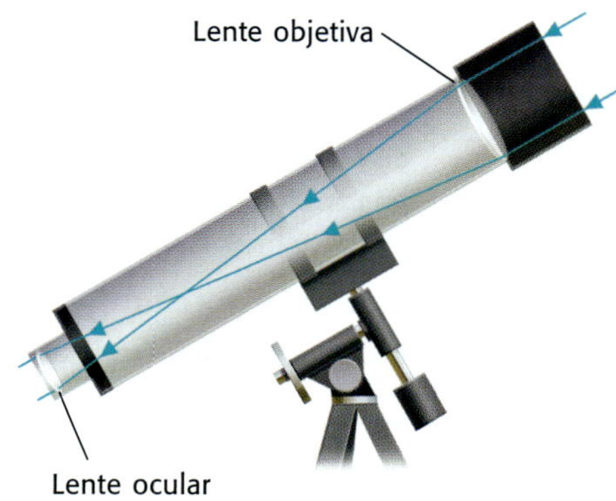

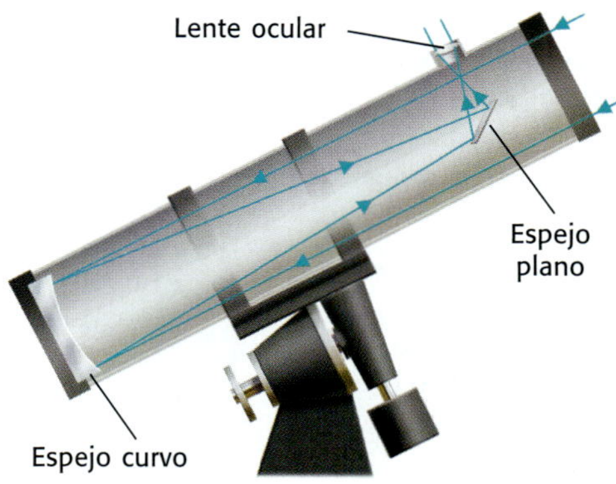

Figura 20 *Tanto los telescopios refractores como los reflectores se usan para ver objetos lejanos.*

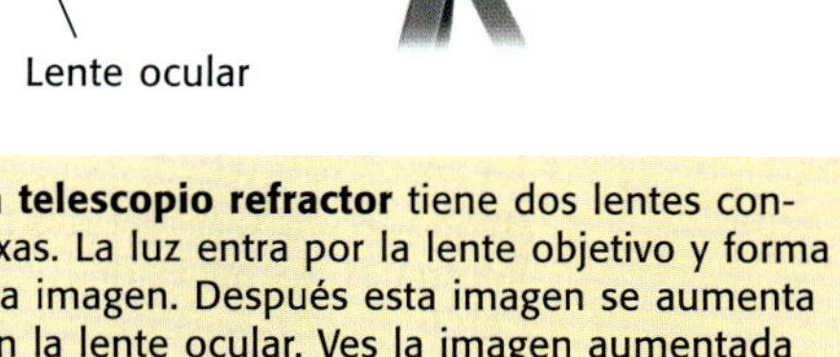

Un **telescopio refractor** tiene dos lentes convexas. La luz entra por la lente objetivo y forma una imagen. Después esta imagen se aumenta con la lente ocular. Ves la imagen aumentada cuando miras por la lente ocular.

Un **telescopio reflector** tiene un espejo curvo que captura y enfoca la luz para formar una imagen. La luz golpea un espejo plano que dirige la luz al lente ocular convexo, que aumenta la imagen.

Microscopios ópticos Los microscopios ópticos simples son similares a los telescopios refractores. Tienen dos lentes convexas: una lente objetiva, que está cerca del objeto que se estudia y una lente ocular, por la que tú miras. La diferencia entre los microscopios y los telescopios es que los primeros se usan para aumentar imágenes de objetos diminutos y cercanos y no imágenes de objetos grandes y lejanos. Los microscopios ópticos más complejos tienen varias lentes para un mayor aumento y para reducir la distorsión.

REPASO

1. ¿Qué es la refracción?

2. ¿Por qué puedes ver los colores del arco iris cuando la luz blanca se refracta a través de un prisma?

3. ¿En qué se diferencian las lentes convexas de los cóncavas?

4. **Comparar conceptos** Explica por qué un microscopio óptico simple es similar a un telescopio refractor.

69

Chapter Highlights

SECTION 1

wave a disturbance that transmits energy through matter or space

electromagnetic wave a wave that does not require a medium

wavelength the distance between one point on a wave and the corresponding point on an adjacent wave

frequency the number of waves produced in a given amount of time

electromagnetic spectrum the entire range of electromagnetic waves

SECTION 2

reflection the bouncing back of a wave after it strikes a barrier or object

law of reflection law that states that the angle of incidence is equal to the angle of reflection

absorption the transfer of energy carried by light waves to particles in matter

scattering the release of light energy by particles of matter that have absorbed extra energy

transmission the passing of light through matter

pigment a material that gives a substance its color by absorbing some colors of light and reflecting others

Resumen del capítulo

SECCIÓN 1

Vocabulario

onda (pág. 54)
onda electromagnética (pág. 55)
longitud de onda (pág. 56)
frecuencia (pág. 56)
espectro electromagnético (pág. 56)

Notas de la sección

- Una onda es una perturbación que transmite energía por la materia o el espacio.

- Las ondas que necesitan un medio se llaman ondas mecánicas. Las que no necesitan un medio se llaman ondas electromagnéticas.

- Todas las ondas electromagnéticas viajan a la velocidad de la luz. Estas ondas se diferencian sólo en su longitud de onda y frecuencia.

- La frecuencia de una onda aumenta al disminuir su longitud de onda.

- La energía de una onda se determina por su longitud de onda y frecuencia.

- La serie de ondas electromagnéticas se llama espectro electromagnético.

- La luz visible es el rango de longitudes de onda que los humanos podemos ver. Las longitudes de onda se ven en colores diferentes.

- La radiación ultravioleta elimina bacterias y produce vitamina D en el cuerpo. La sobreexposición puede producir problemas de salud.

SECCIÓN 2

Vocabulario

reflexión (pág. 59)
ley de reflexión (pág. 59)
absorción (pág. 60)
dispersión (pág. 61)
transmisión (pág. 61)
pigmento (pág. 64)

Notas de la sección

- Puedes ver una imagen reflejada desde una superficie durante la reflexión regular pero no durante la reflexión difusa.

- La absorción y la dispersión de la luz son dos razones por las cuales los haces de luz se debilitan con la distancia.

- La luz puede interactuar con la materia por reflexión, absorción o transmisión.

✓ Comprobar destrezas

Comprensión visual

EL ESPECTRO ELECTROMAGNÉTICO Toda la serie de ondas electromagnéticas se llama espectro electromagnético. Las ondas electromagnéticas se ordenan desde las longitudes de onda largas a las cortas o desde la frecuencia baja a la alta en el espectro electromagnético. Estudia el diagrama en la parte inferior de las páginas 56 y 57 para repasar las categorías de las ondas electromagnéticas en el espectro electromagnético.

LUZ Y MATERIA La luz puede interactuar con la materia de tres formas diferentes: se puede reflejar, absorber o transmitir. Estas tres formas se ilustran en la Figura 9 de la página 61.

70

Lab and Activity Highlights

Mixing Colors `PG 568`

Datasheets for LabBooks
(blackline masters for this lab)

SECCIÓN 2

- El color de un objeto se determina por el color de la luz que refleja.

- Los colores primarios de la luz son el rojo, el azul y el verde. Todos los colores primarios de la luz se combinan por adición de color para producir la luz blanca.

- Los pigmentos les dan color a los objetos. Los pigmentos se combinan por substracción de color.

- Los pigmentos primarios son amarillo, azul obscuro y magenta, y se pueden combinar para producir cualquier otro color.

Experimentos

Mezclar colores (*pág. 568*)

SECCIÓN 3

Vocabulario

refracción (*pág. 66*)
lente (*pág. 67*)
lente convexa (*pág. 67*)
lente cóncava (*pág. 67*)

Notas de la sección

- Los haces de luz se curvan durante la refracción.

- La refracción depende de la longitud de onda de la luz. Por esto, la luz se puede separar en diferentes colores por refracción.

- Una lente es un objeto curvo y transparente que refracta la luz. Las lentes pueden ser convexas o cóncavas.

- Una lente convexa es más gruesa en el medio que en los bordes. Este tipo de lente puede concentrar la luz que entra en ella.

- Una lente cóncava es más delgada en el medio que en los bordes. La luz que pasa por una lente cóncava se curva hacia los bordes de la lente.

- Las cámaras, telescopios y microscopios son instrumentos ópticos que tienen lentes.

SECTION 3

refraction the bending of a wave as it passes at an angle from one medium to another

lens a curved, transparent object that refracts light

convex lens a lens that is thicker in the middle than at the edges

concave lens a lens that is thinner in the middle than at the edges

 **Vocabulary Review Worksheet 3**

 Blackline masters of these Chapter Highlights can be found in the **Study Guide.**

---

internet

 **VISITA:** go.hrw.com

Visita el sitio web de HRW para encontrar una serie de herramientas de aprendizaje relacionadas con este capítulo. Sólo tienes que escribir la palabra clave:

PALABRA CLAVE: HSTLLT

 **VISITA:** www.scilinks.org

Visita el sitio web de la **Asociación Nacional de Maestros de Ciencias** (*National Science Teachers Association*) para encontrar recursos de Internet relacionados con este capítulo. Sólo escribe el **ENLACE DE CIENCIAS** para obtener más información sobre el tema:

TEMA: El espectro electromagnético	**ENLACE:** HSTL705
TEMA: Luz ultravioleta	**ENLACE:** HSTL710
TEMA: Colores	**ENLACE:** HSTL715
TEMA: Lentes	**ENLACE:** HSTL720

71

Lab and Activity Highlights

LabBank

 **Labs You Can Eat,** Fiber-Optic Fun, Lab 25

Inquiry Labs, Eye Spy, Lab 23

Interactive Explorations CD-ROM

 **CD 3, Exploration 7,** "In the Spotlight"

Repaso del capítulo

UTILIZAR EL VOCABULARIO

Escoge el término correcto para completar las siguientes oraciones:

1. Las lentes convexas enfocan la luz por __?__. *(reflexión o refracción)*

2. La luz viaja a través de un objeto durante __?__. *(la absorción o la transmisión)*

3. La distancia entre un punto de una onda y el punto correspondiente de una onda adyacente es la __?__. *(longitud de onda o frecuencia)*

4. Una lente que es más delgada en el medio que en los bordes es una __?__ *(lente convexa o lente cóncava)*

5. La luz visible es un tipo de __?__. *(onda electromagnética u onda mecánica)*

COMPRENDER CONCEPTOS

Opción múltiple

6. ¿De qué color es la luz que se produce cuando se agrega luz roja a una luz verde?
   a. azul obscuro
   b. azul
   c. amarillo
   d. blanco

7. Puedes verte en un espejo debido a la
   a. absorción.
   b. dispersión.
   c. reflexión regular.
   d. reflexión difusa.

8. Los prismas producen los arco iris por
   a. reflexión.
   b. refracción.
   c. absorción.
   d. dispersión.

9. Al aumentar la longitud de onda, la frecuencia
   a. disminuye.
   b. aumenta.
   c. permanece igual.
   d. aumenta y luego disminuye.

10. La refracción ocurre cuando una onda luminosa entra a un nuevo medio formando un ángulo porque
    a. la frecuencia cambia.
    b. la luz se convierte en una onda mecánica.
    c. la velocidad de la luz cambia según el medio.
    d. ninguna de las anteriores

11. Las ondas transfieren
    a. materia.
    b. energía.
    c. partículas.
    d. agua.

12. Un telescopio refractor simple tiene
    a. una lente convexa y una cóncava.
    b. un espejo y una lente cóncava.
    c. dos lentes convexas.
    d. dos lentes cóncavas.

Respuesta breve

13. Menciona dos formas en que las ondas electromagnéticas se diferencian entre ellas.

14. Describe la ley de la reflexión.

15. Describe la diferencia entre las lentes convexas y los cóncavas.

Organizar conceptos

16. Utiliza los siguientes términos para crear un mapa de ideas: luz, color, absorción, materia, reflexión, transmisión, dispersión.

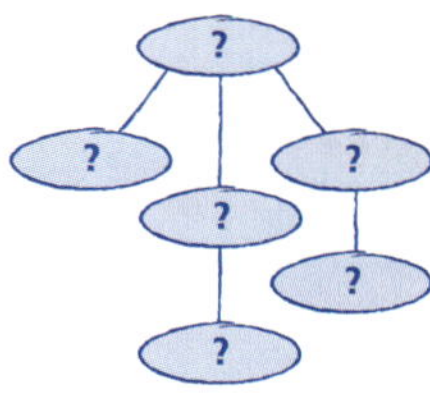

Escribe una o dos oraciones para responder las siguientes preguntas:

17. Las ondas de radio y los rayos gamma son tipos de ondas electromagnéticas. La exposición a las ondas de radio no afecta el cuerpo humano; en cambio, la exposición a los rayos gamma puede ser muy peligrosa. ¿Qué diferencia hay entre estos tipos de ondas electromagnéticas? ¿Por qué los rayos gamma son más peligrosos?

18. Si comparas la ropa que usas en verano con la que usas en invierno, observarás que la mayor parte de la ropa de verano es blanca o de colores claros pero la ropa de invierno es negra o de colores oscuros. Según tus conocimientos de cómo se determinan los colores de los objetos, trata de explicar esta tendencia.

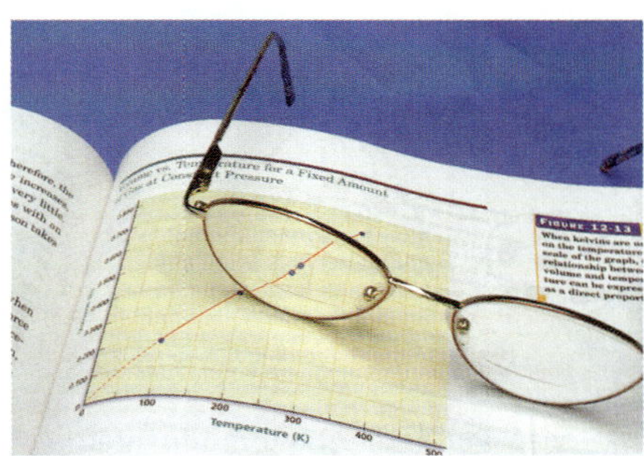

19. Algunas personas sufren de miopía e hipermetropía. Los anteojos que usan se llaman bifocales y contienen dos tipos diferentes de lentes. ¿Por qué se necesitan dos tipos de lentes?

INTERPRETAR GRÁFICAS

20. ¿Qué hay dentro de cada caja sorpresa? La única pista que tienes son los rayos luminosos que entran y salen de la caja.

a

b

c

?

d

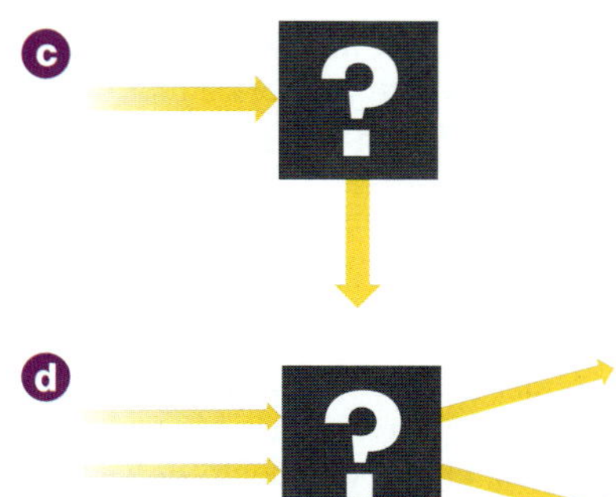

AHORA, ¿qué piensas?

Revisa tus respuestas a las preguntas de la página 53 que escribiste en el cuaderno de ciencias. ¿Han cambiado tus respuestas? Si es necesario, corrige tus respuestas basándote en lo que has aprendido en este capítulo.

73

Concept Mapping

16. An answer to this exercise can be found at the end of this book.

CRITICAL THINKING AND PROBLEM SOLVING

17. Radio waves carry a lot less energy than gamma rays, so radio waves are not dangerous. However, the energy carried by gamma rays is so high that gamma rays can kill healthy living cells in your body.
18. Summer clothes are often white or light-colored because they reflect more colors of light than dark-colored clothes. This helps people stay cooler in the summer. Winter clothes are black or dark-colored because they absorb most colors of light. This helps people stay warmer in the winter.
19. Two lenses are necessary because a concave lens is needed to correct nearsightedness and a convex lens is needed to correct farsightedness.

INTERPRETING GRAPHICS

20. a. a prism
    b. a convex lens
    c. a mirror (at a 45° angle to the side of the box)
    d. a concave lens

Concept Mapping Transparency 3

Blackline masters of this Chapter Review can be found in the **Study Guide.**

NOW WHAT DO YOU THINK?

1. Electromagnetic waves are waves that do not require a medium through which to travel. Visible light, ultraviolet light, and microwaves are all examples of electromagnetic waves.
2. An object's color depends on which colors of light are absorbed and which are reflected.
3. A rainbow is formed when white light from the sun is refracted through drops of water. As the light is refracted, it is separated into its component colors.

Background

Luciferase assays have a number of relevant uses in medicine as well. Bacterial infections can be deadly. Researchers have developed techniques to use luciferase assays to measure bacterial ATP in the urine and blood. Luciferase requires the substrate ATP to emit light and is an effective measure of the amount of ATP present. Sometimes antibiotics, the medicines used to treat bacterial infections, are not effective. Luciferase tests can be used to evaluate the effectiveness of antibiotic therapy on particular patients.

ATP is one of the most significant sources of energy in biological systems. ATP has three phosphate bonds, each capable of giving off energy. One phosphate cleaved off the ATP molecule yields an adenosine diphosphate, or ADP. With the help of luciferase, this energy is converted from chemical energy into light energy by exciting electrons in luciferins.

If you have entrepreneurs in your class, they may be interested in finding out that some companies pay people to collect fireflies. However, students may be discouraged to find out that the firefly business pays only about a penny a beetle.

Las luciérnagas iluminan el camino

Del mismo modo que la luz de un faro advierte a los barcos del peligro, los científicos están utilizando la luz de una fuente poco común, las luciérnagas, para advertirles a los inspectores de alimentos la presencia de contaminación bacteriana peligrosa. Miles de personas mueren cada año por la contaminación bacteriana de la carne. Se está utilizando la luz de la luciérnagas para estudiar enfermedades, tratamientos de aguas residuales y la protección del medio ambiente.

La luz que guía la naturaleza

Muchos organismos emiten luz, como algunos peces, calamares, escarabajos y bacterias. Las luciérnagas usan esta luz para atraer una pareja. ¿Cómo usan la energía estos organismos para emitir luz?

Todos estos organismos usan una substancia llamada *luciferina* y una enzima llamada *luciferasa* para producir luz. La luciferasa utiliza la energía del *adenosin trifosfato* (ATP) para descomponer la luciferina y producir luz como brillo o destello. Las luciérnagas son focos muy efectivos. Casi el 100 por ciento de la energía que reciben del ATP se utiliza para producir luz. Esto es más del 10 por ciento de la energía que emite un foco eléctrico como luz; ¡el 90 por ciento restante es calor!

Cómo dominar la luz de la vida

¿Cómo han podido los científicos dominar la habilidad de la luciérnaga para producir luz y así detectar bacterias? Encontraron el gen responsable de la producción de luciferasa y se lo insertaron a un virus que se alimenta de bacterias. Este virus, que no daña al ser humano, se mezcla en la carne. Cuando el virus infecta las bacterias, les transfiere el gen, el cual se une a la maquinaria genética bacteriana. Estas bacterias comienzan a producir luciferasa y brillan en presencia de la luciferina.

▲ *Es probable que veas luciérnagas agrupadas alrededor de un árbol.*

Este proceso se está utilizando para encontrar numerosas bacterias peligrosas que contaminan los alimentos, como la *Salmonella* y la *E. coli* y que causan la muerte de miles de personas todos los años. La prueba no sólo es efectiva sino rápida. Antes de que los investigadores la desarrollaran, tardaban hasta 3 días en determinar si un alimento estaba contaminado por bacterias. ¡Para entonces el alimento ya estaba en el almacén!

¡Piénsalo!

▶ Las luciérnagas usan el ATP como fuente de energía para emitir luz. Las plantas usan la energía de la luz para producir ATP. Después, las plantas pueden usar el ATP como fuente de energía. Crea una hipótesis para explicar qué color de luz les da más energía a las plantas. Investiga y averigua si tu hipótesis es correcta.

74

Answer to Think About It!

Given the information on light energy in this chapter, students will likely hypothesize that light toward the blue end of the spectrum will provide the most energy to plants. Students will find that photosynthesis is much more complex than this and requires a variety of light.

VENTANA AL MEDIO AMBIENTE

Contaminación lumínica

Las ciudades se pueden ver desde lejos durante la noche. Las luces de las ventanas perfilan los edificios. Las luces de los estadios brillan como faros. ¡La vista es asombrosa!

Desgraciadamente, los astrónomos consideran estas luces una forma de contaminación. La contaminación lumínica reduce nuestra capacidad de ver más allá de la atmósfera. Los astrónomos están perdiendo la capacidad de ver el espacio a través de la atmósfera.

Resplandor de las ciudades

Hace veinte años, las estrellas se podían ver, hasta sobre las ciudades grandes. Aún están ahí, pero ahora están ocultas detrás de las luces de las ciudades. Este brillo, denominado resplandor de las ciudades, se produce cuando la luz se refleja en el polvo y otras partículas suspendidas en la atmósfera. Todos los lugares se ven afectados por la contaminación lumínica.

La contaminación lumínica proviene principalmente de luces exteriores como faros, semáforos y luces de estadios. Otras fuentes incluyen los incendios forestales y las combustiones de gas en yacimientos petrolíferos. Al agregar más partículas al aire, la contaminación del aire empeora las cosas, pues causa una mayor reflexión.

Una luz de esperanza

La contaminación lumínica tiene soluciones simples. De hecho, se puede eliminar en el mismo tiempo que demora apagar una luz. No es práctico apagar todas las luces de la ciudad, pero hay estrategias simples que pueden lograr cambios. Por ejemplo, el uso de luces exteriores cubiertas mantiene el ángulo de la luz hacia abajo, y esto evita que la mayor cantidad de luz alcance las partículas del cielo. Además, el uso de luces sensibles al movimiento y luces programadas ayuda a eliminar la luz innecesaria. Muchas de estas estrategias ahorran energía, y también dinero.

Los astrónomos esperan que la conciencia pública ayude a mejorar la visibilidad del cielo nocturno. Algunas ciudades, como Boston y Tucson, ya han hecho avances. Los científicos han pronosticado que si no nos preocupamos, la contaminación lumínica afectará a todos los observatorios de la Tierra en la próxima década.

Compruébalo

▶ Con el permiso de tus padres, sal en la noche y cuenta las estrellas que puedas ver. Luego, enciende una linterna o una luz de la terraza. ¿Cuántas estrellas ves? Compara tus resultados. ¿Cuánto se redujo tu visibilidad?

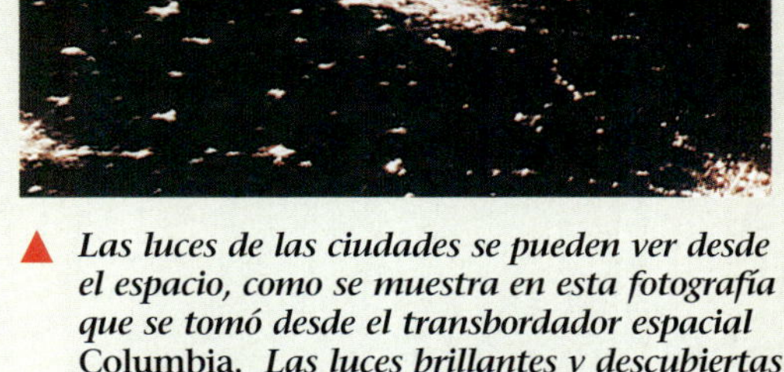

▲ *Las luces de las ciudades se pueden ver desde el espacio, como se muestra en esta fotografía que se tomó desde el transbordador espacial Columbia.* *Las luces brillantes y descubiertas (recuadro) producen neblina luminosa en el cielo nocturno sobre la mayoría de las ciudades de los Estados Unidos.*

75

La célula

UNIDAD 2

Las células están en todas partes. Aunque la mayoría de ellas no se pueden observar a simple vista, forman parte de todo organismo viviente. Tu cuerpo contiene billones de células.

En esta unidad, estudiarás las celúlas y aprenderás a distinguir entre las células animales, las vegetales y las bacterianas. Aprenderás también las diferentes partes de una célula y sus funciones. Las células fueron descubiertas en 1665 y, desde entonces, hemos aprendido mucho sobre ellas y sobre la forma en que trabajan. Esta cronología muestra algunos de los descubrimientos que se han hecho a lo largo de la historia, pero todavía hay mucho que aprender sobre el fascinante mundo de las células.

1620

Los peregrinos establecen la colonia de Plymouth.

1665

Robert Hooke descubre las células después de observar una muestra de corcho con un microscopio.

1937

El puente Golden Gate en San Francisco se abre por primera vez.

1873

Anton Schneider observa y describe con precisión la mitosis.

1952

Martha Chase y Alfred Hershey demuestran que el ADN, que se encuentra en el núcleo de las células, es el material de la herencia.

1941

George Beadle y Edward Tatum descubren que los genes controlan las reacciones químicas en las células al dirigir la síntesis de proteínas.

1831

Robert Brown descubre el núcleo de una célula vegetal.

1838

Matthias Schleiden descubre que todos los tejidos de las plantas están formados por células.

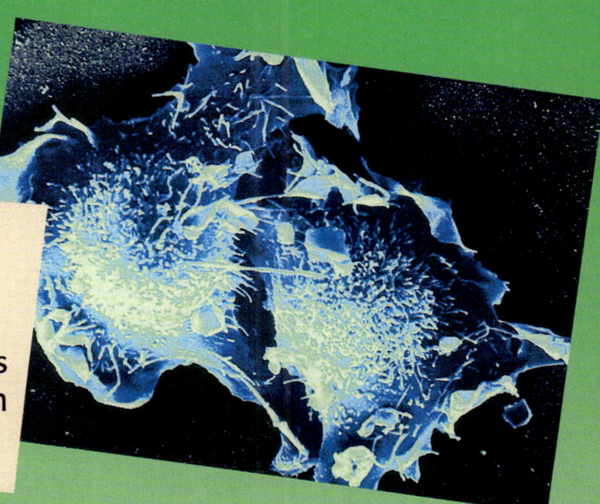

1839

Theodor Schwann demuestra que todos los tejidos animales están formados por células.

1861

Empieza la Guerra Civil estadounidense.

1858

Rudolf Virchow determina que todas las células provienen de otras células.

1956

Se descubre que la síntesis de proteínas en la célula ocurre en los ribosomas.

1971

Lynn Margulis propone una teoría sobre el origen de los organelos celulares.

1997

Una oveja llamada Dolly se convierte en el primer animal clonado a partir de una sola célula.

Chapter Organizer

CHAPTER ORGANIZATION	TIME MINUTES	OBJECTIVES	LABS, INVESTIGATIONS, AND DEMONSTRATIONS
Chapter Opener pp. 78–79	45		**Investigate!** What Are You Made Of? p. 79
Section 1 Organization of Life	90	▶ Explain how life is organized, from a single cell to an ecosystem. ▶ Describe the difference between unicellular organisms and multicellular organisms.	
Section 2 The Discovery of Cells	90	▶ State the parts of the cell theory. ▶ Explain why cells are so small. ▶ Calculate a cell's surface-to-volume ratio. ▶ List the advantages of being multicellular. ▶ Explain the difference between prokaryotic cells and eukaryotic cells.	**Demonstration,** Membranes, p. 87 in ATE **QuickLab,** Do Bacteria Taste Good? p. 90 **Making Models,** Elephant-Sized Amoebas? p. 570 **Datasheets for LabBook,** Elephant-Sized Amoebas? Datasheet 7
Section 3 Eukaryotic Cells: The Inside Story	90	▶ Describe each part of a eukaryotic cell. ▶ Explain the function of each part of a eukaryotic cell. ▶ Describe the differences between animal cells and plant cells.	**Demonstration,** Cell Walls, p. 93 in ATE **Discovery Lab,** Cells Alive! p. 572 **Datasheets for LabBook,** Cells Alive! Datasheet 8 **Skill Builder,** Name That Part! p. 573 **Datasheets for LabBook,** Name That Part! Datasheet 9 **Labs You Can Eat,** The Incredible Edible Cell, Lab 2 **Whiz-Bang Demonstrations,** Grand Strand, Demo 3 **Long-Term Projects & Research Ideas,** Project 3

TECHNOLOGY RESOURCES

 **Guided Reading Audio CD**
English or Spanish, Chapter 4

 **One-Stop Planner CD-ROM with Test Generator**

 **CNN** **Science, Technology & Society,** Flavor Cells, Segment 3
Treating Pets with Laser Light, Segment 4

CLASSROOM WORKSHEETS, TRANSPARENCIES, AND RESOURCES	SCIENCE INTEGRATION AND CONNECTIONS	REVIEW AND ASSESSMENT
Directed Reading Worksheet 4 **Science Puzzlers, Twisters & Teasers,** Worksheet 4		
Directed Reading Worksheet 4, Section 1 **Transparency 12,** From Cell to Organism **Reinforcement Worksheet 4,** An Ecosystem	**Chemistry Connection,** p. 82 **Real-World Connection,** p. 83 in ATE **Connect to Earth Science,** p. 83 in ATE	**Homework,** p. 81 in ATE **Review,** p. 84 **Quiz,** p. 84 in ATE **Alternative Assessment,** p. 84 in ATE
Directed Reading Worksheet 4, Section 2 **Transparency 13,** Surface-to-Volume Ratio **Math Skills for Science Worksheet 16,** What Is a Ratio? **Math Skills for Science Worksheet 29,** Finding Perimeter and Area **Math Skills for Science Worksheet 30,** Finding Volume	**Connect to Astronomy,** p. 86 in ATE **Real-World Connection,** p. 86 in ATE **Math and More,** Surface Area, p. 88 in ATE **MathBreak,** Surface-to-Volume Ratio, p. 89 **Apply,** p. 89 **Connect to Earth Science,** p. 89 in ATE **Connect to Earth Science,** p. 90 in ATE **Health Watch:** The Scrape of the Future, p. 105	**Self-Check,** p. 87 **Self-Check,** p. 90 **Review,** p. 91 **Quiz,** p. 91 in ATE **Alternative Assessment,** p. 91 in ATE
Directed Reading Worksheet 4, Section 3 **Transparency 214,** Structural Formulas **Critical Thinking Worksheet 4,** Cellular Construction **Transparency 14,** Organelles and Their Functions **Transparency 15,** Comparing Animal and Plant Cells **Reinforcement Worksheet 4,** Building a Eukaryotic Cell	**Connect to Physical Science,** p. 94 in ATE **Multicultural Connection,** p. 96 in ATE **Multicultural Connection,** p. 97 in ATE **Across the Sciences:** Battling Cancer with Pigs' Blood and Laser Light, p. 104	**Homework,** pp. 93, 95, 98 in ATE **Self-Check,** p. 94 **Review,** p. 99 **Quiz,** p. 99 in ATE **Alternative Assessment,** p. 99 in ATE

Holt, Rinehart and Winston On-line Resources

go.hrw.com

For worksheets and other teaching aids related to this chapter, visit the HRW Web site and type in the keyword: **HSTCEL**

National Science Teachers Association

www.scilinks.org

Encourage students to use the *sci*LINKS numbers listed with the Chapter Highlights to access information and resources on the **NSTA** Web site.

END-OF-CHAPTER REVIEW AND ASSESSMENT

Chapter Review in Study Guide
Vocabulary and Notes in Study Guide
Chapter Tests with Performance-Based Assessment, Chapter 4 Test
Chapter Tests with Performance-Based Assessment, Performance-Based Assessment 4
Concept Mapping Transparency 4

Chapter Resources & Worksheets

Visual Resources

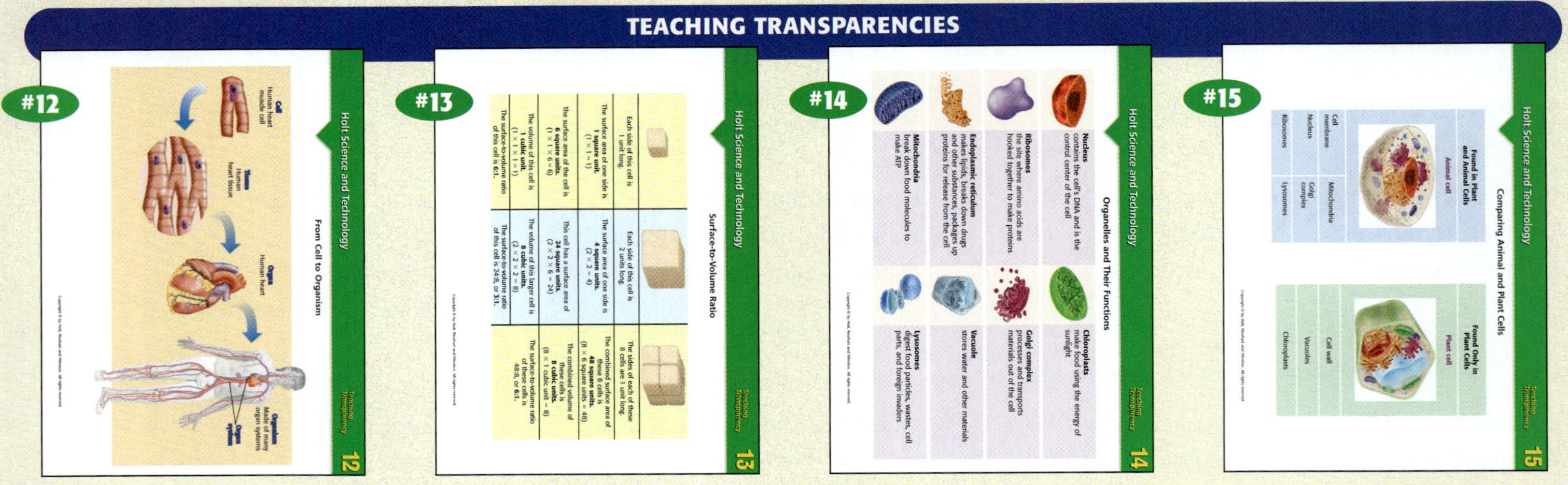

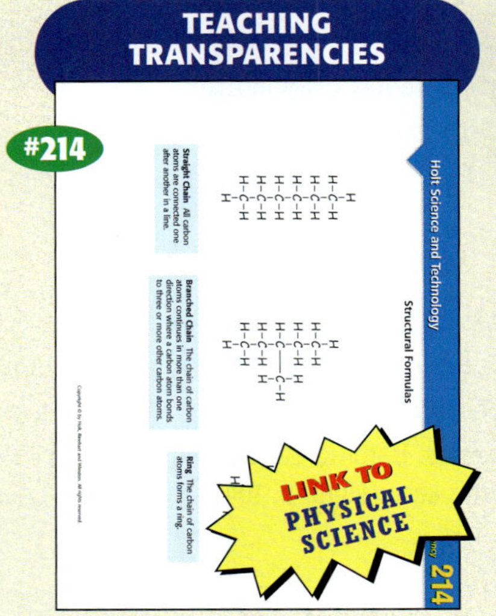

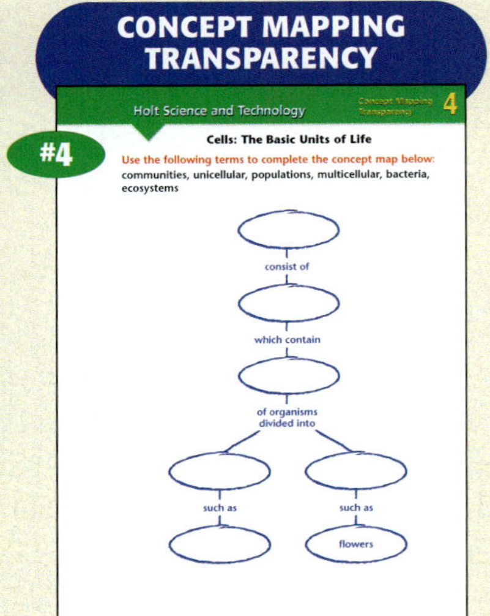

Meeting Individual Needs

DIRECTED READING

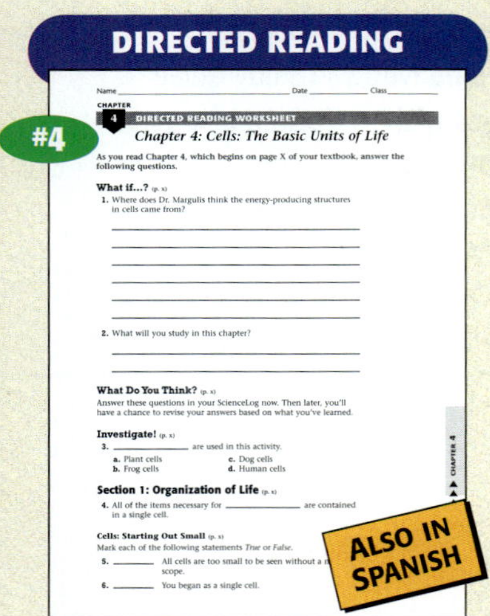

REINFORCEMENT & VOCABULARY REVIEW

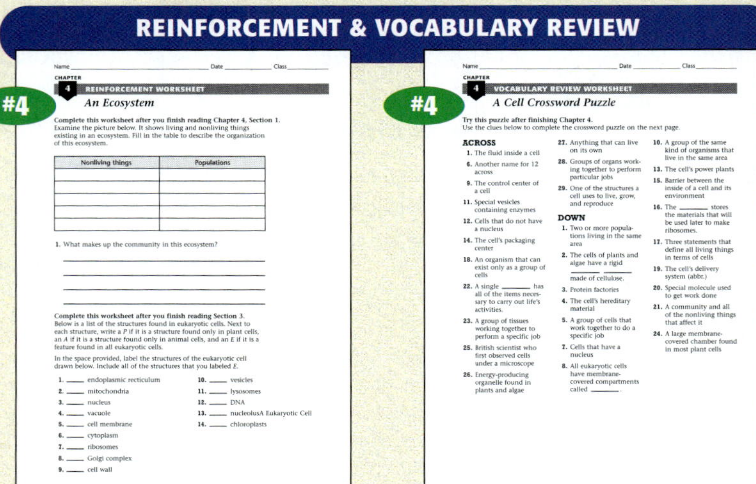

SCIENCE PUZZLERS, TWISTERS & TEASERS

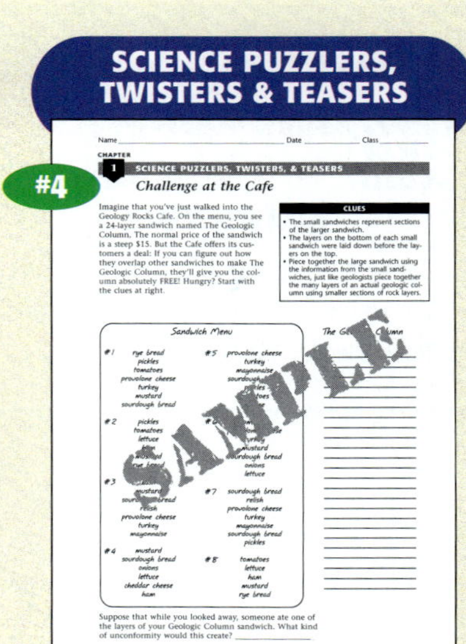

Review & Assessment

STUDY GUIDE

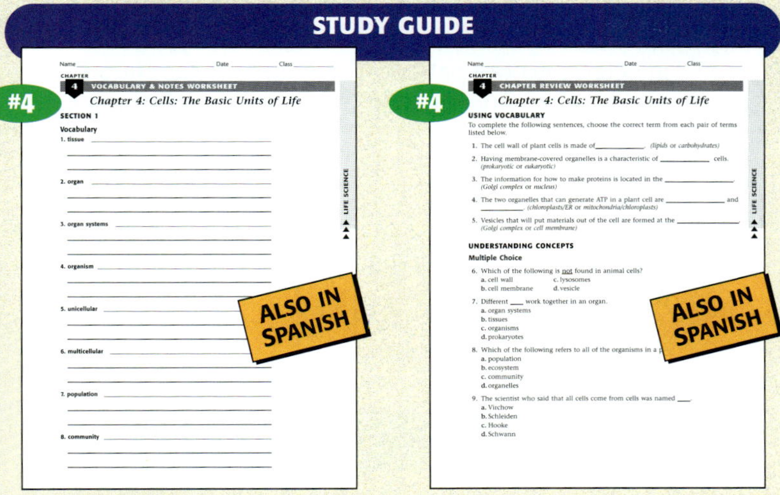

#4 — Chapter 4: Cells: The Basic Units of Life (Vocabulary & Notes Worksheet)

#4 — Chapter 4: Cells: The Basic Units of Life (Chapter Review Worksheet)

CHAPTER TESTS WITH PERFORMANCE-BASED ASSESSMENT

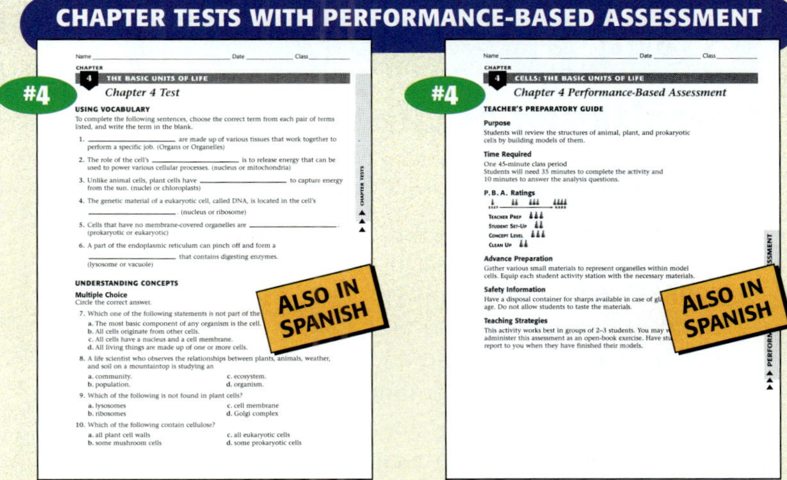

#4 — Chapter 4 Test

#4 — Chapter 4 Performance-Based Assessment

Lab Worksheets

LABS YOU CAN EAT

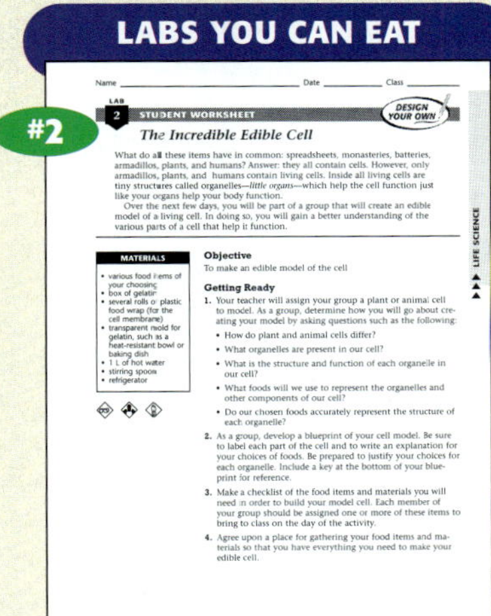

#2 — The Incredible Edible Cell (Student Worksheet)

WHIZ-BANG DEMONSTRATIONS

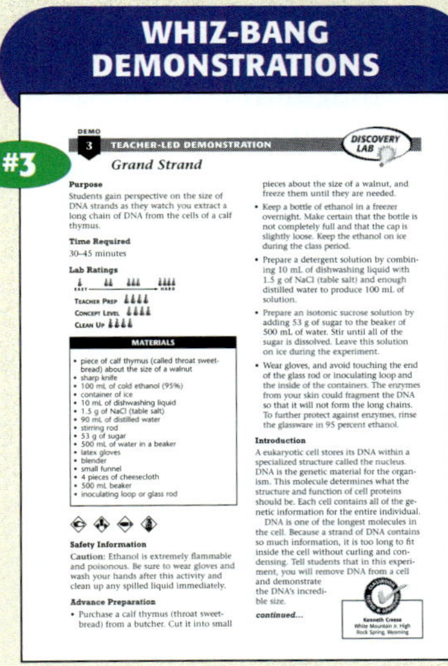

#3 — Grand Strand (Teacher-Led Demonstration)

LONG-TERM PROJECTS & RESEARCH IDEAS

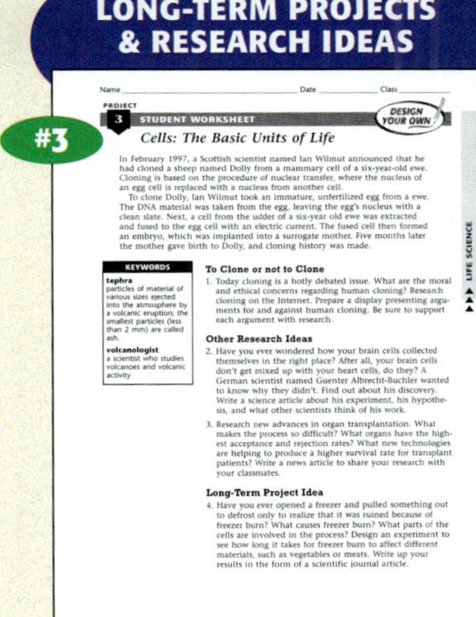

#3 — Cells: The Basic Units of Life (Student Worksheet)

DATASHEETS FOR LABBOOK

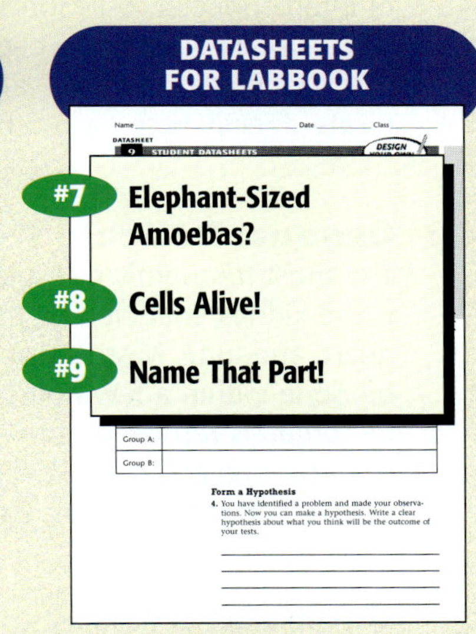

#7 Elephant-Sized Amoebas?

#8 Cells Alive!

#9 Name That Part!

Applications & Extensions

CRITICAL THINKING & PROBLEM SOLVING

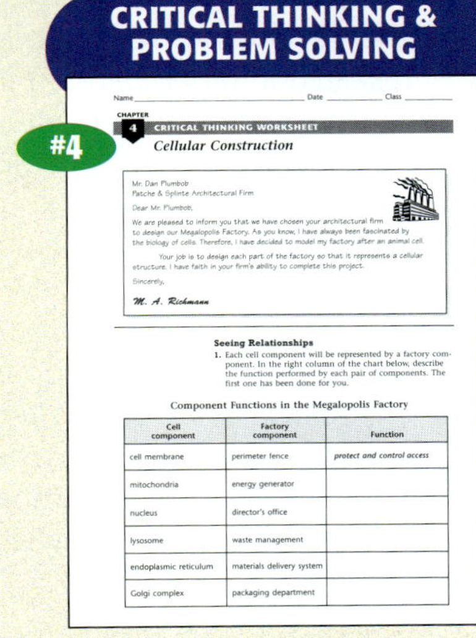

#4 — Cellular Construction (Critical Thinking Worksheet)

SCIENCE TECHNOLOGY

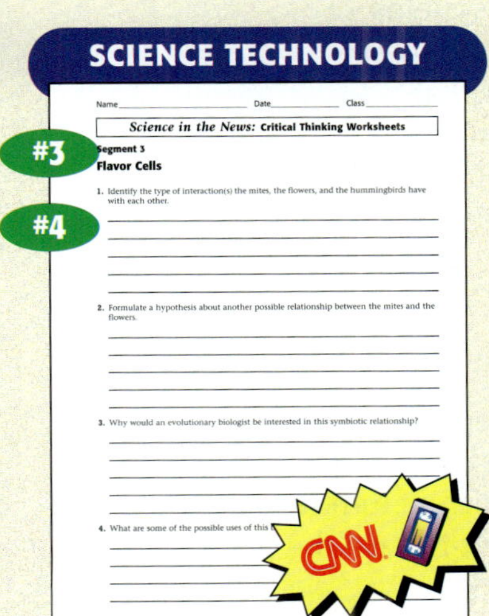

#3 — Science in the News: Critical Thinking Worksheets, Segment 3, Flavor Cells

#4

Organization of Life

▶ In a Heartbeat

The heart will function properly only if the cells that form the connective tissue and muscle perform their jobs in coordination. Scientists can use an enzyme to dissolve an embryonic heart into its individual cells. When placed in a dish, these cells, called myocytes, will continue to beat, although out of sync with each other. After a couple of days, sheets of interconnected cells form and they beat in unison. Why? Openings develop between cells that touch, and their cytoplasms connect, allowing the cells to communicate directly with each other.

▶ Organs: Delicate Workhorses

The most frequently transplanted organ is the kidney, followed by the liver, heart, and lung. Most transplants must be done within a few hours after the organ is removed from a donor because organs are too delicate to survive current long-term storage procedures.

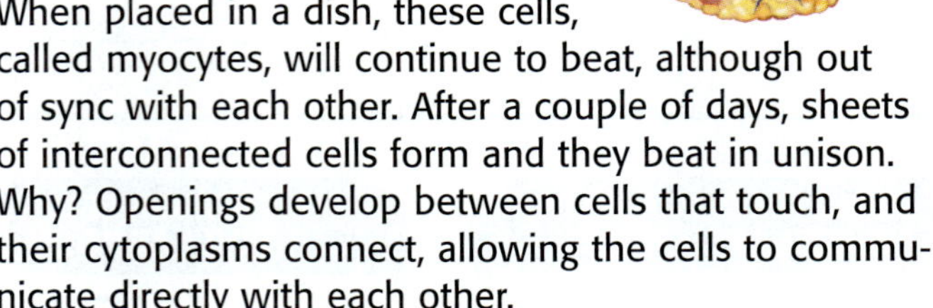

- Cryobiologists, scientists who study how life systems tolerate low temperatures, are studying the possibility of storing organs and organ systems at subfreezing temperatures. They are investigating the fluids that keep insects alive during subfreezing temperatures, hoping this knowledge can be applied to human organs.

IS THAT A FACT!

- In 1931, a doctor removed a patient's parathyroid glands in error. These glands control the amount of calcium in the blood, which in turn regulates the heart. As a last-ditch effort to save the patient, a cow's parathyroid glands were ground up and injected into the patient. The patient recuperated and lived another 30 years with similar treatments.

- Researchers are studying hypnosis as a weapon against cancer. Individuals with cancer are hypnotized to help them manage their pain. Data suggest that this treatment also helps patients live considerably longer.

The Discovery of Cells

▶ Microtomy

The development of high-magnification microscopes required that the preparation of specimens for viewing also become more sophisticated. Microtomy used to refer only to specimen cutting because a microtome is the instrument used to slice tissue sections. Today microtomy refers collectively to the art of preparing specimens by any number of techniques. When microscopic organisms are viewed as whole-mounts, they are preserved, stained, dried (alcohol removes the water), and made transparent with clove or cedar oil. Then the organism is mounted in a drop of resin on a glass slide and covered with a piece of glass only 0.005 mm thick.

▶ Physiology and the Cell Theory

The development of the cell theory aided research in other fields. In the mid-1800s, French physiologist Claude Bernard proposed that plants and animals are composed of sets of control mechanisms that work to maintain the internal conditions necessary for life. He recognized that a mammal can sustain a constant body temperature regardless of the outside temperature. Today we recognize this ability as homeostasis. But at the time no one knew what the "organized sets of control mechanisms" were. The discovery of cells, and how their many components function to sustain life in an organism, gave credence to Bernard's position.

IS THAT A FACT!

- Aeolid nudibranchs are mollusks that eat hydroids, small polyps that have protective stinging cells. The nudibranchs' digestive systems carefully sort out the hydroids' stinging cells and send them to the protective tentacles on their own backs.

SECTION 3

Eukaryotic Cells: The Inside Story

▶ "Protein" Therapy

As scientists delved deeper into the cell, they moved past the initial stages of merely identifying its structures to asking, "What do these organelles do?" and "How do they do it?" Decades of investigation have produced "gene therapy," which refers to the use of a cell's genetic material to cure disease. It might be more appropriate to call this rapidly expanding field of science "protein therapy."

- The gene can be thought of as a blueprint for the proteins essential to life. People with Duchenne's muscular dystrophy lack dystrophin, an essen- 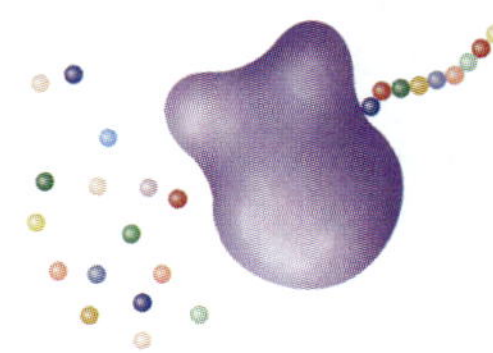 tial muscle protein that maintains the structure of muscle cells. Researchers have been able to remove the harmful genetic components of a virus and replace them with the gene for dystrophin. Their plan is to inject the dystrophin gene (the gene that codes for the dystrophin protein) directly into the muscles of Duchenne's patients and trick the body into maintaining healthy muscle.

▶ Cell Scientists

Microbiologists study the characteristics of bacteria and other microorganisms to understand how they interact with people, plants, and animals. Virologists investigate viruses, which are active only inside a living cell. Mycologists study fungi, which include molds, and yeast. Environmental microbiologists inspect the water in rivers and lakes. Those in agriculture are concerned with organisms that affect soil quality.

▶ Mitochondrial Diseases

Mitochondrial diseases are a group of illnesses caused by malfunctioning mitochondria. The problem can be with either the genes of the mitochondria or the genes of the cell. Any activity or organ that requires energy is affected by these diseases. Because the brain requires huge amounts of energy to function, it often suffers in people who have a mitochondrial disease. Other commonly affected areas are muscles, including the heart; organs, such as the kidneys; and bone marrow.

▶ Cell Walls

Every culture in the world relies in some way on the cell walls of dead organisms. Materials such as thatch, reed, and wood are composed of cell walls that remain after an organism has died.

For background information about teaching strategies and issues, refer to the *Professional Reference for Teachers.*

Chapter Preview

Guided Reading Audio CD
English or Spanish, Chapter 4

CAPÍTULO

4 La célula: unidad fundamental de la vida

¿Qué tal si...?

Imagínate esta escena de una película de horror: un muchacho se sienta a comer y se da cuenta de que su mamá hizo otra vez calabacitas. Como no tiene escapatoria, se come esas espantosas cosas verdes. Luego, descubre que, en lugar de ser digerida, una de las calabacitas se instaló dentro de su cuerpo, y lo peor de todo es que, ¡está viva! ¿No se te hace espantoso? ¿Qué tal que calabacita empezara a hacer cosas maravillosas por el muchacho, como por ejemplo darle la energía que él nunca había imaginado tener? Lynn Margulis, una científica, cree que algo parecido pasó con ciertos organismos unicelulares que vivieron hace más de mil millones de años y que crearon los tipos de células de las que estamos hechos.

Según la teoría de Margulis, hace alrededor de 1200 millones de años, algunas células más grandes empezaron a comerse a las células más chicas. Como los glóbulos blancos que se muestran en esta página, las células más grandes atrapaban a las células más chicas con prolongaciones de su cuerpo celular. Pero algunas de las células se resistían a ser digeridas. En realidad, la mayoría de ellas empezaron a sentirse muy bien en sus nuevos hogares. Las células más grandes también se beneficiaron de sus nuevos huéspedes, pues las células más chicas liberaban enormes cantidades de energía del alimento que ingerían las células grandes. Otros tipos de células pequeñas usaban la energía solar para crear suficiente comida para ellas y las más grandes. Se cree que las estructuras que producen energía en la mayoría de las células, inclusive las tuyas, descienden de estas células más chicas. En este capítulo, aprenderás más sobre las células y sus estructuras.

78

What If ...?

The energy-producing structures referred to are mitochondria. One reason scientists believe that mitochondria were once separate organisms is that they are not coded for in our genes. Though they live within our cells, they have their own genetic material and independently make 90 percent of the proteins they need to function.

If mitochondria are not coded for in our DNA, how do we get them? They exist in the cytoplasm of egg cells and so are passed on to us by our mothers. They multiply independently within our cells, so that as our cells divide, mitochondria are available in the cytoplasm of the new cells. With the exception of red blood cells, mitochondria exist in every cell of our body.

¿Tú qué piensas?

Usa tus conocimientos para responder las siguientes preguntas en tu cuaderno de ciencias:

1. ¿Qué son las células y dónde se encuentran?

2. ¿Por qué hay células y por qué son tan pequeñas?

¿De qué estás hecho?

Como ya sabes, todos los organismos vivientes están hechos de células. ¡Tú también! ¿Cómo son algunas de tus células? Haz esta actividad para descubrirlo.

Procedimiento

1. Durante este experimento usa lentes protectores, guantes y una bata. Pídele a tu maestra que ponga una gota de **azul de metileno en un portaobjetos de plástico. Cuidado:** No derrames este colorante sobre tu piel o ropa. Si esto sucede, lava la parte manchada de inmediato y avísale a tu maestra o maestro.

2. *Raspa* muy suavemente la parte interior de tu mejilla con el extremo romo de un **palillo.**

3. Revuelve el material que raspaste con la gota de colorante. **Cuidado:** ¡Ya no te metas ese palillo en la boca! Deséchalo como te lo indica la maestra.

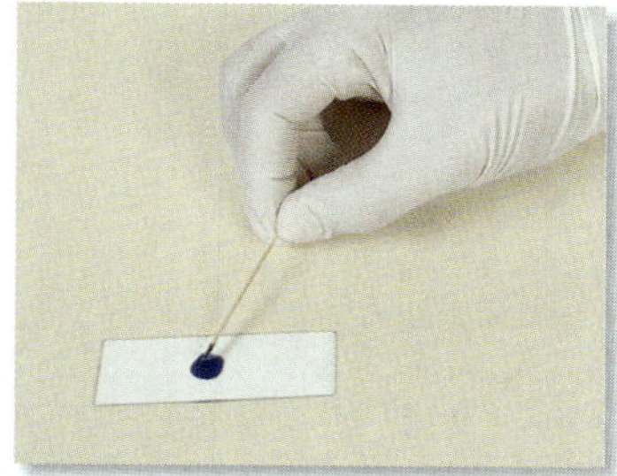

4. Coloca un **cubreobjetos de plástico** sobre la gota, como se muestra en la ilustración.

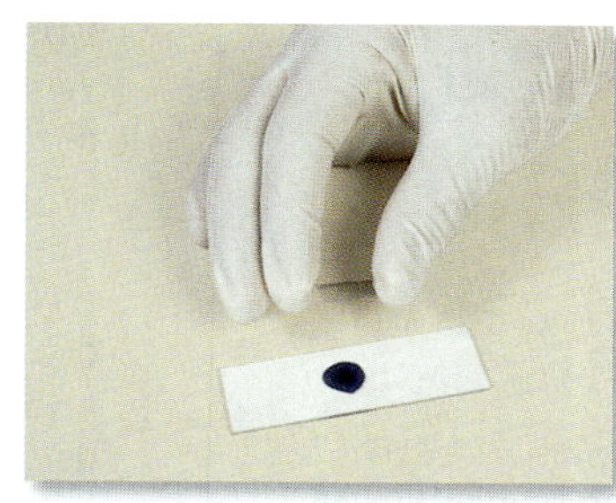

5. Coloca el portaobjetos en el **microscopio** y trata de encontrar las células. En tu cuaderno de ciencias, haz un dibujo de lo que ves.

Análisis

6. ¿Cómo son tus células? ¿Son todas iguales?

7. ¿Crees que todas las células de tu cuerpo son como éstas? Explica tu respuesta.

What Do You Think?

Accept all reasonable responses.

Students will have a chance to revise their answers in the Chapter Review under NOW What Do You Think?

Investigate!

MATERIALS

FOR EACH STUDENT:
- methylene blue
- plastic microscope slide
- coverslip
- flat toothpick or cotton swab

Safety Caution: Remind students to review all safety cautions and icons before beginning this lab activity. To prevent the spread of disease, students should not share toothpicks and should immediately dispose of their toothpick in a beaker of 10 percent bleach solution. The toothpicks can then be thrown into the garbage, and the bleach can be poured down the sink with plenty of running water. Toothpicks are sharp and could cause injury. Cotton swabs are a safer alternative for collecting cells. Remind students to review all safety cautions and icons before beginning this lab activity. Methylene blue will stain skin and clothing.

 **Directed Reading Worksheet 4**

 **Science Puzzlers, Twisters & Teasers Worksheet 4**

Answers to Investigate!

6. By comparing their drawings, students should see that all the cells share similar structures but are not exactly the same.

7. Cells vary widely, but accept all reasonable responses.

Focus

Organization of Life

In this section, students will learn that a cell is the smallest unit of life. In most multicellular organisms, groups of cells form tissues that compose organs. Two or more organs can interact to form an organ system. Students will also learn that organisms can be further organized into populations, communities, and ecosystems.

Bellringer

On the board or overhead viewer, write the following questions:

Why can't you use your teeth to breathe? Why can't you use your arm muscles to digest food?

Have students answer these questions in their ScienceLog.

1 Motivate

ACTIVITY

Concept Mapping Divide the class into small groups. Provide each group with pictures of tissues, organs, and organ systems. Have the students arrange the pictures into concept maps. Encourage them to notice unusual relationships between organs. For example, the stomach and the heart may seem very different, but both are made of muscle tissue, and both function by holding and moving substances through their cavities.

VOCABULARIO

tejido
órgano
sistemas
organismo
unicelular
multicelular
población
comunidad
ecosistema

OBJETIVOS

- Explica cómo está organizada la vida, desde una célula sencilla hasta un ecosistema.
- Describe la diferencia entre organismos unicelulares y multicelulares.

Figura 1 *La primera célula de un pollo es una de las células más grandes del mundo.*

24 horas

40 horas

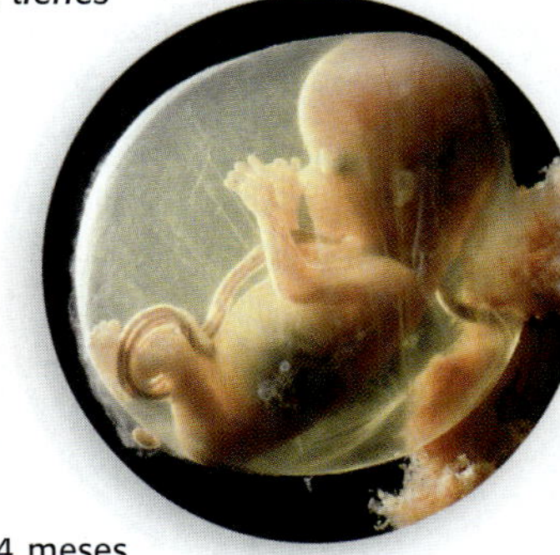

6 días

4 meses

Directed Reading Worksheet 4 Section 1

Organización de la vida

Imagínate que vas hacer un viaje a Marte. Debes empacar en tu maleta todo lo que necesitas para sobrevivir. ¿Qué empacarías? Para empezar, vas a necesitar comida, oxígeno y agua. Y esto es sólo el principio. Quizá vas a necesitar una maleta grande y bonita, ¿no? En realidad, tienes todo esto dentro de las células de tu cuerpo. La célula es más pequeña que el punto final de esta oración, pero tiene todo lo necesario para cumplir con todas las actividades de la vida.

Todo ser vivo tiene por lo menos una célula. Muchos organismos con vida están formados por una sola célula, mientras que otros tienen billones de ellas. Para que te des una idea de cómo es un ser vivo con casi 100 billones de células, ¡mírate al espejo!

Células: todo empieza con una célula

La mayoría de las células son muy pequeñas y no se pueden ver sin la ayuda de un microscopio, pero es posible que en tu refrigerador tengas una de las más grandes. ¿Quieres saber qué es? Observa la **Figura 1.** La primera célula de un pollo es amarilla con un puntito blanco y está rodeada por un fluido gelatinoso que se le conoce como la clara del huevo. El punto se divide una y otra vez hasta formar un pollito. La yema (de la primera célula) y la clara les dan los nutrientes a las células del pollito que se está formando. Igual que un pollo, tú empezaste siendo una sola célula. Observa la **Figura 2** y descubre algunas de las primeras etapas de tu desarrollo.

No todas tus células se ven y actúan igual. Tienes cerca de 200 tipos diferentes de células y cada uno realiza una función determinada. Hay células óseas, sanguíneas y cutáneas. Cuando alguien ve todas esas células juntas, te ve a ti.

Figura 2 *Empezaste como una célula, pero después de muchas divisiones, tienes alrededor de 100 billones de ellas.*

Q: Why did the chicken cross the playground?

A: to get to the other slide

Tejidos: las células trabajan en equipo

Cuando observes de cerca tu ropa, vas a ver que los hilos han sido agrupados, o tejidos, para que la ropa pueda cumplir su función. De la misma forma, las células se agrupan para formar un tejido que tiene una tarea. Un **tejido** es un grupo de células que trabajan juntas para cumplir una función específica en el cuerpo. El material que está alrededor y entre las células también es parte del tejido. En la **Figura 3** se muestran algunos ejemplos de tejidos de tu cuerpo.

Órganos: equipos que trabajan juntos

Cuando dos o más tejidos trabajan juntos para hacer una tarea específica, el grupo de tejidos recibe el nombre de **órgano.** Algunos órganos son el estómago, los intestinos, el corazón, los pulmones y la piel. Así es: hasta tu piel es un órgano, ya que contiene diferentes tipos de tejidos. Para conocerlos más de cerca, observa la **Figura 4.**

Las plantas también tienen varios tipos de tejidos que trabajan juntos. Una hoja es un órgano con tejidos que atrapan la energía de la luz para crear alimentos. Otros ejemplos de órganos vegetales son los tallos y las raíces.

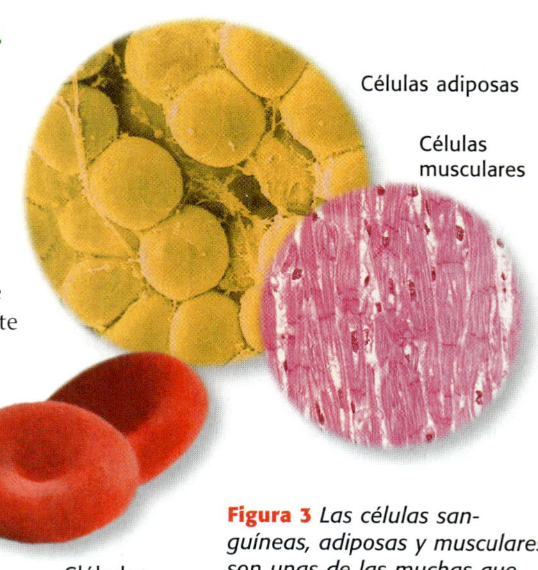

Figura 3 Las células sanguíneas, adiposas y musculares son unas de las muchas que forman los tejidos de tu cuerpo.

Figura 4 La piel es el órgano más grande del cuerpo. La piel de una persona adulta tiene una masa de casi 4.5 kg.

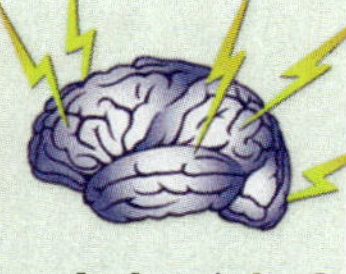

La parte que ves de la piel, el cabello y las uñas es tejido muerto. ¿No te parece raro que nos esforcemos tanto en hacer que nuestras células muertas se vean bonitas?

IS THAT A FACT!

In your lifetime you will shed about 18 kg (almost 40 lb) of dead skin.

2 Teach

DISCUSSION

Muscles Ask students to list all the ways they use their muscles. Responses will probably include walking, riding a bike, swimming, and throwing or kicking a ball. Explain that muscles are also involved in swallowing food (tongue), digestion (stomach), and blinking eyes (eyelids). Sometimes muscles act voluntarily (jumping), and sometimes they act involuntarily (the heart beating).

MISCONCEPTION ALERT

Hair and fingernails grow out of specialized skin cells. Even though they grow continuously, both are composed of dead cells along with a protein called *keratin*. If they were alive and contained nerve cells, like the deep skin layers, haircuts and manicures would be quite painful.

Homework

Writing Not all living things have the same kinds of tissues and organs. Yet all must perform similar life processes. Have students research and compare the structures a fish uses to breathe with those that a human uses. Their report should also answer the question, "What parts of the human and fish respiratory systems are similar?" (Even though a fish has gills and a human has lungs, both have cells that exchange and transport oxygen and carbon dioxide.)

MEETING INDIVIDUAL NEEDS

Learners Having Difficulty

To help students understand the levels of organization within an organism, instruct them to write the following headings on the board:

> Cell, Tissue, Organ, Organ system

Tell students to write these headings across the top of their paper and list at least two examples of each under the headings. Then ask students to share their information with their classmates.

Answer to Explore

Accept all reasonable responses.

USING THE FIGURE

Writing Have students write an expanded caption for **Figure 5.** They should read additional references for information describing the functions of the leaf, stem, and root organ systems. Then have them identify systems in the human body, if any, that perform similar functions.

RETEACHING

Writing A great way to learn something is to teach it to someone else. Have students write a letter to a friend explaining how cells, tissues, organs, and organ systems are all related.

Sobre las células de todo ser humano hay proteínas que actúan como tarjetas de identificación. Cuando a una persona se le hace un transplante de órgano, las células del nuevo órgano deben tener casi el mismo "código" que el anterior. Si las células son muy diferentes, el cuerpo del receptor rechazará el órgano.

Explora

¿Cómo crees que se organiza una escuela? La tarea del sistema escolar es educar a los alumnos en todos los niveles. Las diversas escuelas (primaria, secundaria, preparatoria) son como los diferentes órganos de un sistema. Los grupos de maestros de cada escuela trabajan juntos para enseñarle a un grupo específico de estudiantes. Si consideráramos a cada grupo de maestros de una escuela como el tejido que cumple una función particular, ¿qué sería cada maestro? ¿Qué otros ejemplos puedes usar para representar las partes de un sistema? Explica tu respuesta.

Sistemas: una gran combinación

Los órganos trabajan en grupos para hacer tareas específicas. Estos grupos se llaman **sistemas.** Cada sistema realiza una tarea específica en el cuerpo. Por ejemplo, la tarea del aparato digestivo es descomponer la comida en partículas para que las células del cuerpo la utilicen. La tarea del sistema nervioso es transmitir información entre el cerebro y el resto del cuerpo. Entre los sistemas de las plantas están los sistemas de hojas, de raíces y de tallos, como se muestra en la **Figura 5.**

Tu cuerpo tiene varios sistemas. El aparato digestivo se muestra en la **Figura 6.** Cada órgano del aparato digestivo cumple una tarea. Un órgano determinado puede realizar su tarea gracias a los diferentes tejidos que tiene. Los órganos de un sistema dependen el uno del otro; si cualquier parte del sistema falla todo el sistema es afectado, y la falla de un sistema puede afectar otros sistemas. Imagínate qué pasaría si tu aparato digestivo dejara de convertir la comida en energía. ¡Ninguno de los otros sistemas tendría energía para funcionar!

Figura 5 *Las plantas también tienen sistemas. Por ejemplo, el sistema de tallos incluye los tallos y las ramas.*

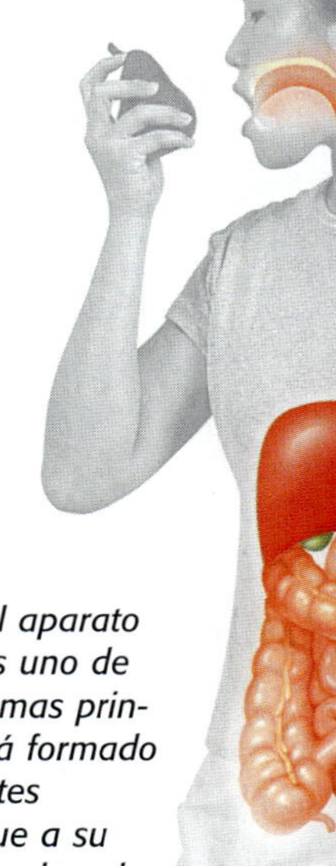

Figura 6 *El aparato digestivo es uno de los 11 sistemas principales. Está formado por diferentes órganos, que a su vez están hechos de varios tejidos.*

IS THAT A FACT!

An elephant's trunk is constructed of 135 kg (300 lb) of hair, skin, connective tissue, nerves, and muscles. The muscle tissue is composed of 150,000 tiny subunits of muscle, each coordinated with the others to enable an elephant to drink, breathe, grab, and greet its friends.

Organismos: vida independiente

Un organismo es todo ser viviente que puede sobrevivir por su cuenta. Todos los organismos están hechos de por lo menos una célula. Si una célula vive por su cuenta, es un **organismo** unicelular. La mayoría de los organismos unicelulares son tan pequeños que necesitas un microscopio para verlos. En la **Figura 7** se muestran algunos tipos diferentes de organismos unicelulares.

Tú eres un **organismo** multicelular. Esto significa que sólo puedes existir como un grupo de células y que la mayoría de tus células sobreviven si siguen siendo parte de tu cuerpo. Cuando te caes en la calle y te raspas la rodilla, las células que dejaste en la calle no pueden vivir por su cuenta. La **Figura 8** muestra cómo tus células trabajan juntas para formar un organismo multicelular.

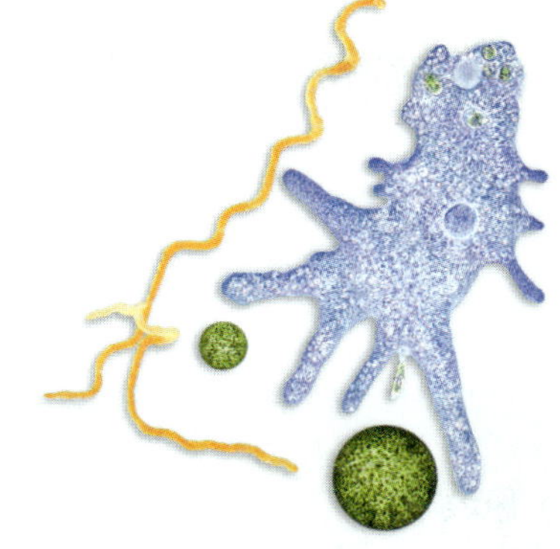

Figura 7 *Los organismos unicelulares tienen una gran variedad de formas y tamaños.*

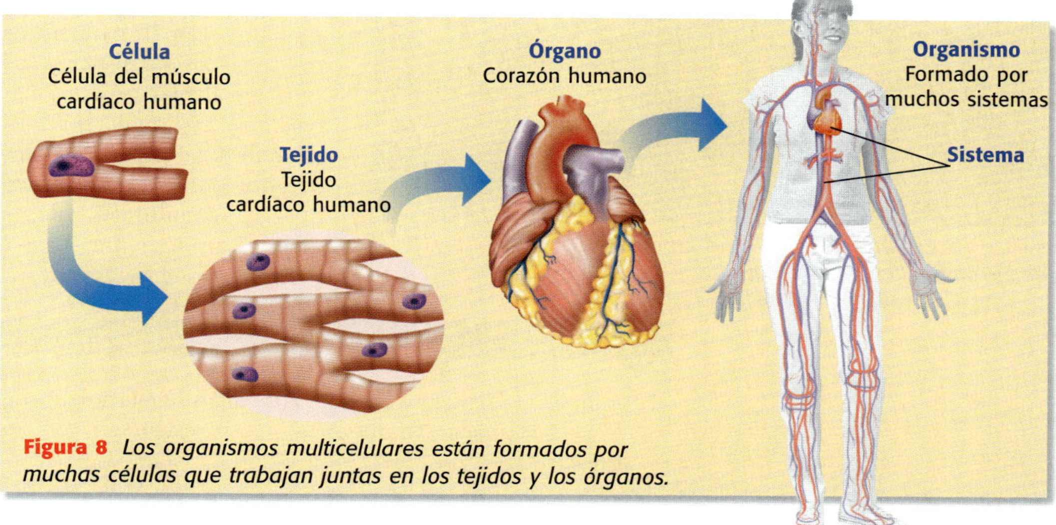

Figura 8 *Los organismos multicelulares están formados por muchas células que trabajan juntas en los tejidos y los órganos.*

El panorama

Aunque los organismos unicelulares y multicelulares pueden vivir por su cuenta, por lo general no viven solos. Los organismos interactúan con los demás en diferentes formas.

Poblaciones Un grupo de organismos del mismo tipo que viven en la misma área forman una **población.** Todas las catarinas que viven en el bosque, como las de la **Figura 9,** forman la población de catarinas de ese bosque. Todos los robles rojos forman la población de robles rojos del bosque.

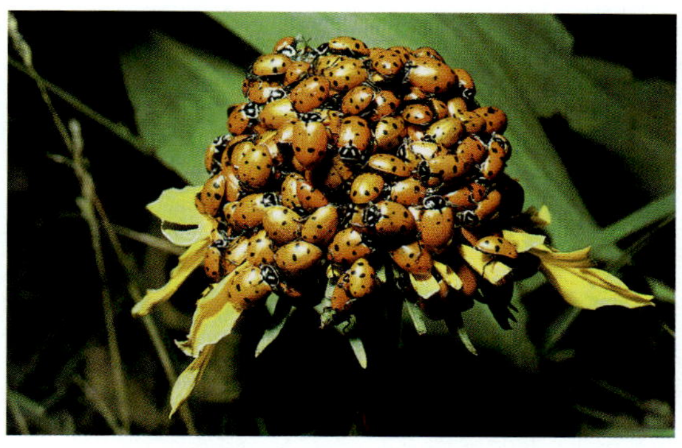

Figura 9 *Una población está formada por todos los individuos del mismo tipo que viven en la misma área.*

83

Your skin cells depend on other body cells for nutrients and oxygen, which is why they cannot live independently of your body. When someone suffers a serious burn and no longer has a protective covering of skin cells, scientists can grow skin for them in a laboratory. They take healthy skin cells, mix them with collagen proteins from cow skin, and suspend them in a nutrient solution. These new cells are applied to the burned skin as a bandage while the damaged skin heals.

CONNECT TO EARTH SCIENCE

Populations are sometimes joined or divided by Earth's physical features, such as mountains, rivers, and islands. The islands of Madagascar and Comoro, off the east coast of Africa, are home to 50 species of lemurs, a type of primate. The lemurs are geographically isolated, so while they are quite diverse on Madagascar and Comoro, there are no other wild populations of lemurs anywhere else in the world. Speciation as a result of geographic isolation is not limited to islands. Species can form when populations are divided for thousands of years by mountains, rivers, and other barriers.

WEIRD SCIENCE

Living on your own doesn't mean you can't cooperate with others. On coral reefs, a group of striped fish called cleaner wrasses attract fish that are infested with parasites to an area on the reef. The infested fish lie still while wrasses eat the parasites and fungi off their body.

Teaching Transparency 12
"From Cell to Organism"

Reinforcement Worksheet 4
"An Ecosystem"

GOING FURTHER

Writing Have students write promotional materials that encourage cells to join an organ or organ system or to remain a single cell. For example, a single cell might have independence but lack the help it needs to do its work. A tissue provides a cell with helpers that have similar interests, but it requires cooperation and good "intercellular skills."

Quiz

1. What is the relationship between your digestive system, stomach, and intestines? (The digestive system is an organ system. The stomach and intestines are organs that are parts of the digestive system.)

2. The blood cells in your body are alive, are genetically similar to each other, and can be found in the same location (inside your body). Are these cells a population? (No. A population consists of organisms that are capable of living independently. Blood cells cannot live outside the body.)

ALTERNATIVE ASSESSMENT

Concept Mapping Have students choose a human organ system and identify its component organs. Then ask students to describe the function of these organs and their relationship to one another in a concept map.

Comunidades Dos o más poblaciones que viven en la misma área forman una **comunidad.** Las poblaciones de zorros, robles, lagartijas, flores y otros organismos de un bosque son parte de la comunidad de ese bosque, como se muestra en la **Figura 10.** El lugar donde vives es una comunidad que incluye a gente, perros, gatos y otros organismos que viven ahí.

Figura 10 *El zorro, las flores y los árboles son parte de la comunidad de un bosque.*

Ecosistemas La comunidad y todas las cosas sin vida que la afectan, como el agua, la tierra, las rocas, la temperatura y la luz, forman un **ecosistema.** Los ecosistemas de tierra se llaman ecosistemas *terrestres,* como los bosques, los desiertos, las praderas e incluso el patio trasero de tu casa. Los ecosistemas de agua se llaman ecosistemas *acuáticos,* como los ríos, los estanques, los lagos, los océanos e incluso los acuarios. La comunidad que se muestra en la Figura 10 vive en un ecosistema terrestre.

REPASO

1. Completa la siguiente oración: *Las células* se relacionan con __?__ de la misma forma que __?__ se relacionan con los *sistemas.*

2. ¿Cuál es la diferencia entre las células de los organismos unicelulares y las células de los organismos multicelulares?

3. Aplicar conceptos Usa la foto del acuario de la derecha para responder las siguientes preguntas:

a. ¿Cuántas clases *diferentes* de organismos puedes ver?

b. ¿Cuántas poblaciones identificas?

c. ¿Cuántas comunidades puedes ver?

▼ **Answers to Review**

1. tissues, organs

2. The cells of unicellular organisms can survive on their own, but the cells of multicellular organisms must remain a part of the organism's body to survive.

3. a. 8; don't forget to count the plants.
b. 8
c. 1; the community includes all the living things in the aquarium.

El descubrimiento de las células

VOCABULARIO

teoría celular   núcleo
membrana celular   procariota
organelos   eucariota
citoplasma   bacterias
relación de
   superficie a volumen

OBJETIVOS

- Enuncia las partes de la teoría celular.
- Explica por qué las células son tan pequeñas.
- Calcula la relación de superficie a volumen de una célula.
- Enuncia las ventajas de ser un organismo multicelular.
- Explica la diferencia entre células procariotas y células eucariotas.

La mayoría de las células son tan pequeñas que no se pueden observar a simple vista. ¿Cómo descubrimos que las células son la unidad básica de los organismos con vida? ¿Qué te hace pensar que un conejo, un árbol o una persona están hechos de pequeñas partes que no se puden ver? En realidad, la persona que descubrió las células lo hizo por accidente ya que las vio cuando ni siquiera las estaba buscando.

Primer encuentro con las células

En 1665, un científico inglés llamado Robert Hooke estaba buscando algo interesante que pudiera mostrarles a otros científicos en una reunión. Ya había construido un microscopio compuesto con varios juegos de lentes que le permitía observar objetos diminutos. Un día decidió observar un pedacito de corcho, el tejido suave que se encuentra en la corteza de los árboles como los de la **Figura 11.** Para su sorpresa, el corcho se veía como si tuviera cientos de cajitas, como un panal. A estas cajitas les puso el nombre de *células,* que en latín significa "cuartitos".

Aunque Hooke no se dio cuenta, estas cajas eran en realidad las capas exteriores de las células de corcho que se quedaron ahí al morir las células. Luego, observó laminillas de zanahorias, helechos y otras plantas y vio que también estaban hechos de células. Vio también que las células vivas estaban llenas de "jugo". En la **Figura 12** se muestra el microscopio de Hooke y los dibujos de las células de corcho.

Hooke también observó plumas de aves, escamas de peces y ojos de moscas, pero la mayor parte del tiempo la pasó observando plantas y hongos. Como las paredes celulares de las plantas y de los hongos son más fáciles de ver, Hooke pensó que sólo había células en esos tipos de organismos y no en los animales.

Figura 11 *El corcho es un material suave que se encuentra en algunos árboles. Las células de corcho fueron las primeras que se vieron con un microscopio.*

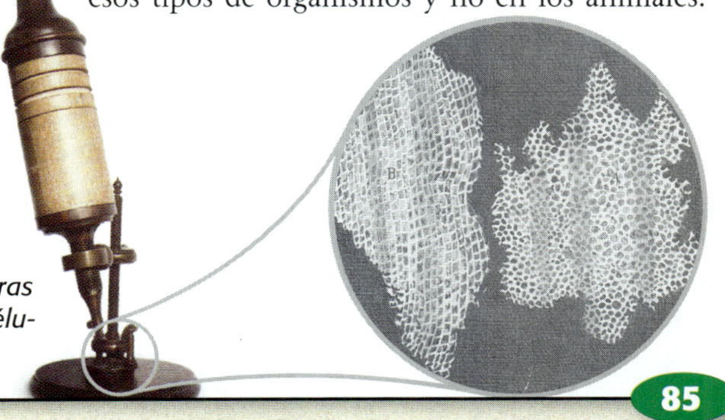

Figura 12 *Éste es el microscopio compuesto con el que Hooke vio las primeras células. Hooke hizo un dibujo de las células de corcho que vio por primera vez.*

Q: What did Robert Hooke say to his stockbroker?

A: Cell! Cell!

The Discovery of Cells

This section introduces students to the internal structures of a cell. They will learn about the parts of a cell and why cells are so small. Finally, they will learn the differences and similarities between prokaryotic and eukaryotic cells.

Bellringer

Write the following on the board or overhead projector:

Why weren't cells discovered until 1665? What invention made their discovery possible? Write your answers in your ScienceLog. (Cells weren't discovered until 1665 because most cells are too small to be seen with the naked eye. The microscope is the invention that made their discovery possible.)

1) Motivate

ACTIVITY

Modeling Cell Discovery
Before beginning this section, have students model Robert Hooke's discovery. Divide the class into small groups, and provide each group with a microscope and a prepared slide of cork cells. Have them describe and sketch their observations in their ScienceLog.

Directed Reading Worksheet 4 Section 2

REAL-WORLD CONNECTION

The yeast used in baking is a relative of single-celled eukaryotes living among lots of other organisms in the air around us. It was probably an accidental discovery that wheat dough exposed to air for a length of time would rise and, when baked, produce bread. Sourdough breads require the right combination of yeast and bacteria in the dough, and strains of native yeast vary regionally the same way many other organisms do. Sourdough from San Francisco has its own characteristic taste because bakers there use a yeast that is most common in the air around that city.

Figura 13 *Leeuwenhoek vio organismos unicelulares parecidos a éstos, que se encuentran en la lama de los estanques.*

Figura 14
Anton van Leeuwenhoek

Las células en otras formas de vida

En 1673, un comerciante holandés llamado Anton van Leeuwenhoek usó un microscopio que había fabricado para observar espuma de un estanque, y vio algo parecido a lo que se muestra en la **Figura 13.** Se sorprendió de ver muchas criaturitas nadando en la lama pegajosa; les puso el nombre de *animálculos,* que significa "animalitos".

Leeuwenhoek, que aparece en la **Figura 14,** también observó la sangre y el sarro de los dientes. Observó que las células de la sangre de los peces, los pájaros y las ranas tienen forma de óvalo, y las de los humanos y de los perros son más planas. Fue el primero en ver las bacterias y descubrió que la levadura usada para hacer que el pan se esponje está formada por organismos unicelulares.

La teoría celular

Después de que Hooke vio las células de corcho, pasaron casi dos siglos para que se dieran cuenta de que las células están presentes en *todos* los seres vivos. Matthias Schleiden, un científico alemán, observó muchas laminillas de tejidos de plantas y leyó sobre lo que otros científicos habían visto con el microscopio. En 1838, llegó a la conclusión que todas las partes de las plantas están hechas de células.

El año siguiente, Theodor Schwann, un científico alemán que estudiaba los animales, afirmó que todos los tejidos animales estaban hechos de células. Escribió las primeras dos partes de lo que ahora se conoce como la **teoría celular:**

• **Todos los organismos están compuestos de una o más células.**

• **La célula es la unidad fundamental de todos los organismos vivientes.**

Casi 20 años después, en 1858, Rudolf Virchow, un médico alemán, vio que las células sólo se podían desarrollar de otras células. Entonces escribió la tercera parte de la teoría celular:

• **Todas las células provienen de células preexistentes.**

Similitudes entre las células

Las células tienen distintas formas, tamaños y funciones, pero todas tienen estas cosas en común:

Membrana celular Todas las células están rodeadas por una **membrana celular.** Esta membrana actúa como una barrera entre el interior de la célula y su medio ambiente. También controla el paso de los materiales dentro y fuera de la célula. La **Figura 15** muestra la parte exterior de una célula.

Material hereditario Parte de la teoría celular establece que las células están hechas de células preexistentes. Cuando se forman nuevas células, reciben una copia del material hereditario de las células originales. Este material es el *ADN* (ácido desoxirribonucleico), que controla las actividades de una célula y contiene la información necesaria para que forme células nuevas.

Figura 15 *La membrana celular guarda el contenido de toda la célula.*

Citoplasma y organelos Las células tienen substancias químicas y estructuras que les permiten comer, crecer y reproducirse, las cuales se llaman **organelos.** No todos los organelos son del mismo tipo. Algunos están rodeados de membranas, pero otros no. La célula de la **Figura 16** tiene organelos cubiertos por una membrana. Las substancias químicas y las estructuras de una célula están rodeadas por un fluido. Este fluido y casi todo lo que está en él se conoce comúnmente como **citoplasma.**

De tamaño pequeño La mayoría de las células son tan pequeñas que no se pueden ver a simple vista. Tú estás formado por 100 billones de células y para cubrir el punto de la letra *i* se necesitan unas 50.

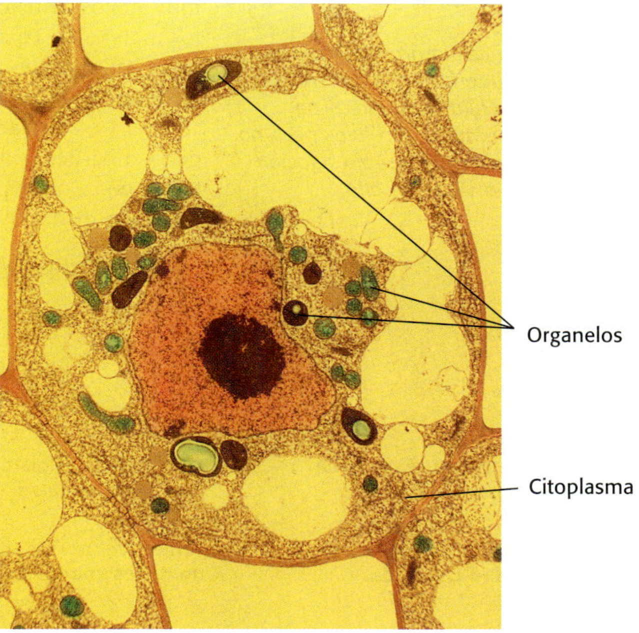

Figura 16 *Esta célula tiene muchos organelos, algunos de ellos rodeados por una membrana.*

✔ Autoevaluación

¿Por qué las células necesitan ADN? *(Consulta la página 636 para comprobar tu respuesta.)*

87

IS THAT A FACT!

The largest cell in the world is the yolk of an ostrich egg. It's the size of a baseball.

Answer to Self-Check

Cells need DNA to control cell processes and to make new cells.

Prediction Guide Have students respond to the following statement before reading this page: Blood cells are completely different from bone cells (true/false). Ask students to explain the reasons for their answer. Have students evaluate their answer after they read the page.

DEMONSTRATION

Membranes

MATERIALS
• wire mesh food strainer
• 250 mL of sand
• 250 mL of water
• 250 mL of gravel similar to that used to line fish tanks
• 250 mL of marbles or large pebbles
• pan to place under strainer

Place each material in the strainer, and have students observe and explain the results. Tell students that the cell membrane functions somewhat like the strainer. It lets some materials pass through, but not others. Explain also that the process works in both directions.

MISCONCEPTION ALERT

The physical relationship between molecules and cells is often confusing to students. Molecules are not alive; they are much smaller and fit inside of cells. Remind students of the discussion in Chapter 2 about the ATP molecules that provide energy to cells.

2 Teach, continued

READING STRATEGY

Prediction Guide Before reading this page, ask students to choose one of the following reasons for why they think cells are so small:

1. There isn't enough microscopic food available for them.

2. There isn't enough room in a multicellular organism.

3. another reason (ask for suggestions)

Have students evaluate their answer after they read the page.

MATH and MORE

Surface Area The following problem simplifies the principle of surface-to-volume ratio because it concerns a one-dimensional figure. Give each student a sheet of $\frac{1}{4}$ in. grid paper. Tell them to outline a rectangle 4 squares wide and 5 squares long. What is the area of this rectangle? (4 × 5 = 20 squares)

Now tell students to outline another rectangle that is twice as big, 8 squares wide and 10 squares long. What is the area of the second rectangle? (8 × 10 = 80 squares)

Next ask students to calculate the perimeter, or "surface area," of each rectangle. (5 + 5 + 4 + 4 = 18) (10 + 10 + 8 + 8 = 36)

How did doubling the size of the rectangle affect its internal area? How did it affect its surface area? (The internal area quadrupled: 80 ÷ 20 = 4. The surface area only doubled: 36 ÷ 18 = 2.)

Amiba gigante se come la ciudad de Nueva York

Este no es un encabezado que vayas a leer en algún periódico. ¿Por qué no? Porque las amibas están formadas por una sola célula. Es más, la mayoría de las amibas sólo se pueden ver con un microscopio. Es por eso que cuando una célula crece, más materiales deben moverse hacia adentro y hacia afuera a través de la membrana celular. Entre más grande sea la célula, necesitará más comida y oxígeno, y producirá más desechos.

Para mantener estas demandas, una célula en crecimiento necesita un área más grande en donde pueda intercambiar desechos y nutrientes. Cuando el volumen de la célula aumenta, también crece su superficie exterior. Pero hay un problema: el volumen de una célula aumenta a una velocidad mayor que el área de su superficie. Por lo tanto, si una célula crece mucho, su superficie tendrá muy pocos orificios para permitir la entrada y salida de los materiales. Para que entiendas por qué el volumen de una célula aumenta más rápido que su superficie, observa la relación de superficie a volumen de las células de la **Figura 17**. La **relación de superficie a volumen** es el área de la superficie exterior de una célula en relación con su volumen. Como puedes ver, la relación de superficie a volumen disminuye cuando la célula aumenta.

Figura 17 *La célula grande tiene una relación de superficie a volumen más baja que la célula pequeña. Si se aumenta el número de células pero no su tamaño, se mantiene una relación de superficie a volumen alta.*

Relación de superficie a volumen

		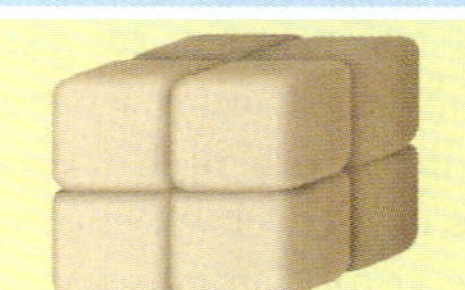
Cada lado de esta célula tiene 1 unidad de largo.	Cada lado de esta célula tiene 2 unidades de largo.	Los lados de cada una de estas 8 células tienen 1 unidad de largo.
La superficie de un lado es de **1 unidad cuadrada.** (1 × 1 = 1)	La superficie de un lado es de **4 unidades cuadradas.** (2×2=4)	La superficie combinada de estas 8 células es de **48 unidades cúbicas.** (8 × 6 unidades cúbicas = 48)
La superficie de la célula es de **6 unidades cuadradas.** (1 × 1 × 6 = 6)	Esta célula tiene una superficie de **24 unidades cuadradas.** (2 × 2 × 6 = 24)	El volumen combinado de estas células es de **8 unidades cúbicas.** (8 × 1 unidad cúbica = 8)
El volumen de esta célula es de **1 unidad cúbica.** (1 × 1 × 1 = 1)	El volumen de esta célula más grande es de **8 unidades cúbicas.** (2 × 2 × 2 = 8)	La relación de superficie a volumen de estas células es 48:8 ó **6:1.**
La relación de superficie a volumen de esta célula es **6:1.**	La relación de superficie a volumen de esta célula es 24:8 ó **3:1.**	

88

Teaching Transparency 13 "Surface-to-Volume Ratio"

Q: Define *bacteria.*

A: the rear entrance to a cafeteria

Las ventajas de ser multiceluar ¿Ahora entiendes por qué estás hecho de tantas células diminutas en lugar de tener sólo una o varias células grandes? Una sóla célula de tu tamaño y con tu forma tendría una relación de superficie a volumen demasiado baja. La célula no podría sobrevivir ya que su superficie sería tan pequeña que no permitiría la entrada de los materiales necesarios. Los organismos multicelulares crecen al producir células más pequeñas, no más grandes. Las células de un elefante son del mismo tamaño que las tuyas, sólo que él tiene más.

Además de tener la capacidad de crecer más, los organismos multicelulares pueden hacer muchas otras cosas ya que tienen diferentes tipos de células. Así como hay maestras que se especializan en enseñar, doctores y enfermeras que ayudan a enfermos y mecánicos que reparan coches, las células se especializan en realizar diferentes tareas. Una sóla célula no puede hacer todas las cosas que muchas células diferentes hacen. Al tener una gran variedad de células que se especializan en tareas específicas, los organismos multicelulares realizan más funciones que los unicelulares. Los diferentes tipos de células forman tejidos y órganos con varias funciones. Gracias a que el ser humano tiene células especializadas, como células óseas, musculares, oculares y cerebrales, puede sentarse, caminar, correr, ver una película, reflexionar sobre lo que ve y hacer otras cosas. Si te gusta hacer muchas cosas, entonces alégrate de no tener una sóla célula.

Figura 18 *Un elefante es más grande que un ser humano porque tiene más células y no porque sus células son más grandes.*

¡MATEMÁTICAS!

Relación de superficie a volumen

La forma de una célula afecta su relación de superficie a volumen. Examina estas células y responde las siguientes preguntas:

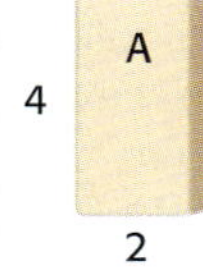
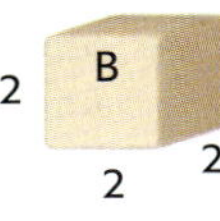

1. ¿Cuál es la superficie de la célula A y de la célula B?
2. ¿Cuál es el volumen de la célula A y de la célula B?
3. ¿Cuál de las dos células tiene la relación de superficie a volumen mayor?

APLICA

Imagínate que eres el afortunado merecedor de un paramecio, un tipo de organismo unicelular. Para cuidar bien a tu nueva mascota, tienes que calcular cuánto necesitas para alimentarla. Las dimensiones de tu paramecio son aproximadamente 125 µm × 50 µm × 20 µm. Si siete moléculas de comida pueden entrar en cada micrómetro cuadrado de superficie cada minuto, ¿cuántas moléculas puede comer tu mascota en 1 minuto? Si tu mascota necesita para sobrevivir una molécula de comida por micrómetro cúbico de volumen cada minuto, ¿cuánta comida tendrías que darle por minuto?

Answers to MATHBREAK

1. 28 square units; 24 square units
2. 8 cubic units; 8 cubic units
3. Cell *A*'s surface-to-volume ratio is the largest, at 28:8. (Even though Cell *A* has a larger surface area, the two cells have the same volume.)

 **Math Skills Worksheet 16** "What Is a Ratio?"

 **Math Skills Worksheet 29** "Finding Perimeter and Area"

 **Math Skills Worksheet 30** "Finding Volume"

Answers to APPLY

- The surface area of the pet is 19,500 μm^2, so it can eat 136,500 molecules of food every minute.
- The volume of the pet is 125,000 μm^3. If it needs one food molecule for every cubic micrometer, then it needs to be fed 125,000 food molecules every minute to survive.

Answers to Self-Check

1. The surface-to-volume ratio decreases as the cell size increases.

2. A eukaryotic cell has a nucleus and membrane-covered organelles.

CONNECT TO
EARTH SCIENCE

Astronomers are interested in the work of scientists who investigate bacteria and other microscopic organisms in Earth's crust. Microbiologists have drilled deep into the crust and found microbes nearly 3 km below the surface, where the temperature is 75°C (167°F). Because other planets have surface conditions similar to the harsh environment within the Earth's crust, astronomers believe it may be possible for microbes to live elsewhere in the solar system. Have students research and write a brief report on the conditions in Earth's crust and learn about the organisms that live there.

Quick Lab

MATERIALS

FOR EACH STUDENT:
- cotton swab
- yogurt with active culture
- plastic microscope slide
- plastic coverslip

Answer to QuickLab

Drawings should depict rod-shaped bacteria.

✓ Autoevaluación

1. Cuando una célula crece demasiado, ¿qué pasa con su relación de superficie a volumen?

2. ¿Qué tiene una célula eucariota que una procariota no tenga? *(Consulta la página 636 para comprobar tu respuesta.)*

Laboratorio

¿Son sabrosas las bacterias?

Un tipo de bacterias se encuentra en el yogurt, y... ¡son deliciosas! Con un **bastoncillo de algodón,** pon un pequeño punto de **yogurt** en un **portaobjetos.** Agrégale una gota de **agua** al yogurt y revuélvela con el bastoncillo. Ponle el **cubreobjetos** y examina la laminilla con un **microscopio compuesto.** Enfoca la muestra con el lente de corto alcance, luego con el de mediano alcance y por último con el de alto alcance. Dibuja lo que ves.

Los montones de células en forma de bastoncitos se llaman *Lactobacillus.* Estas bacterias se alimentan del azúcar de la leche (lactosa) y la convierten en ácido láctico. El ácido láctico hace que la leche se espese, ¡y de esta manera se forma el yogurt!

90

Dos tipos de células

Los diferentes tipos de células se dividen en dos grupos. Como ya sabes, todas las células tienen ADN. En un grupo, las células tienen un **núcleo,** que es un organelo cubierto con una membrana que guarda el ADN de las células. En el otro grupo, el ADN de las células no se guarda en un núcleo. Las células que no tienen un núcleo son **procariotas,** y las que sí lo tienen son **eucariotas.**

Células procariotas Las células procariotas (también conocidas como **bacterias**) son las más pequeñas del mundo. Se llaman *procariotas* porque no tienen un núcleo (*procariota* significa en griego "sin núcleo"). El ADN de una célula procariota es una molécula larga circular, en forma de liga.

Las bacterias no tienen organelos cubiertos por una membrana, sino pequeños organelos redondos llamados ribosomas. Estos organelos trabajan como pequeñas fábricas para sintetizar proteínas. La mayoría de las bacterias están cubiertas por una pared celular dura que cubre a una membrana celular más suave. Imagínate la membrana que presiona la pared como si fuera un globo que presiona la parte interior de una jarra de vidrio. Pero a diferencia del globo y la jarra, la membrana y la pared permiten que pasen las moléculas de comida y de desechos. La **Figura 19** muestra un esquema general de una célula procariota. Se cree que las bacterias fueron el primer tipo de células en la Tierra. Algunos científicos piensan que han existido desde hace unos 3800 millones de años. Se calcula que los fósiles más viejos alguna vez encontrados tienen cerca de 3500 millones de años.

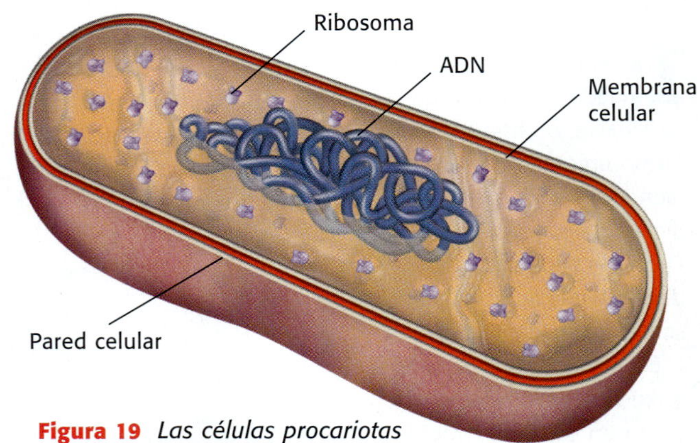

Figura 19 *Las células procariotas no tienen un núcleo ni ningún otro organelo cubierto con una membrana. El ADN circular se agrupa en el citoplasma.*

WEIRD SCIENCE

In 1969, the *Apollo 12* crew retrieved a space probe from the moon that had been launched nearly three years earlier. In the probe's camera, NASA scientists found a stowaway. The bacterium *Streptococcus mitis* had traveled to the moon and back. Despite the rigors of space travel, more than two and a half years of radiation exposure, and freezing temperatures, the *Streptococcus mitis* was successfully reconstituted.

Células eucariotas Las células eucariotas son más complejas que las procariotas. Aunque son casi 10 veces más grandes que las procariotas, tienen una relación de superficie a volumen lo bastante alta para sobrevivir. La evidencia fósil sugiere que las células eucariotas aparecieron por primera vez hace 2 mil millones de años. Todos los organismos vivientes que no son bacterias están hechos de una o más células eucariotas, es decir, las plantas, los animales, los hongos y los protistas.

Las células eucariotas tienen un núcleo (*eucariota* significa en griego "núcleo verdadero") y otros organelos cubiertos con una membrana. Una ventaja de que las células se dividan en compartimientos es que permite que muchos procesos químicos ocurran al mismo tiempo. En la **Figura 20** se muestra el esquema de una célula eucariota.

Las células eucariotas tienen más ADN que las procariotas y éste se almacena en el núcleo. Las moléculas de ADN de las células eucariotas son lineales, en lugar de ser circulares. Todas las células eucariotas tienen una membrana celular y algunas de ellas tienen pared celular. Las que tienen pared celular se encuentran en plantas, hongos y algunos organismos unicelulares. Las paredes celulares de las eucariotas son químicamente diferentes de las paredes de las bacterias. La siguiente tabla resume las diferencias entre las células eucariotas y procariotas.

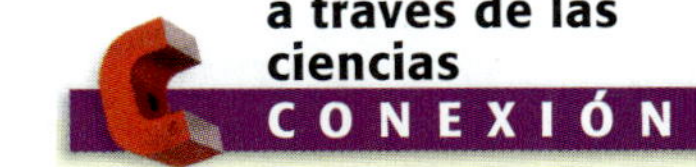

Figura 20 *Las células eucariotas tienen un núcleo y otros organelos.*

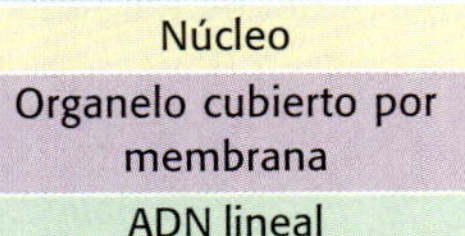

CÉLULAS PROCARIOTAS	CÉLULAS EUCARIOTAS
Sin núcleo	Núcleo
Organelos no cubiertas por membrana	Organelo cubierto por membrana
ADN circular	ADN lineal
Bacteria	Todas las demás células

a través de las ciencias

C O N E X I Ó N

¿Una nueva forma de curar las células enfermas? Consulta la página 104.

REPASO

1. ¿Cuáles son las tres partes de la teoría celular?

2. ¿En qué se parecen todas las células?

3. Menciona dos ventajas de ser un organismo multicelular.

4. Si un organismo unicelular tiene una pared celular, ribosomas y ADN circular, ¿es eucariota o procariota?

5. **Aplicar conceptos** ¿Cuál de las dos tiene una relación de superficie a volumen más alta, una pelota de tenis o una de baloncesto? Explica tu respuesta. ¿Qué se podría hacer para aumentar la relación de superficie a volumen de las dos?

91

Focus

Eukaryotic Cells: The Inside Story

In this section, students will learn the names and functions of organelles in a eukaryotic cell. They will learn which organelles enable a cell to make proteins, produce energy, transport and store materials, and prepare to divide. Finally, they will learn the difference between plant and animal cells.

Bellringer

On the board or overhead projector, write the following:

List three differences between prokaryotic and eukaryotic cells. (Prokaryotic cells have circular DNA, no nucleus, and no membrane-covered organelles. Eukaryotic cells have linear DNA, a nucleus, and membrane-covered organelles.)

1 Motivate

DISCUSSION

Cellular Activity Ask students if they can feel the flurry of activity within their cells that keeps them alive. (No.)

Ask how they know their cells are working. (They can breathe, digest food, and move.)

Explain that there is a tendency to consider life processes as activities performed only by whole organisms. What we sometimes forget is that the plant or animal can do these things only because its *cells* are doing these things.

VOCABULARIO

pared celular	cloroplasto
ribosomas	aparato de Golgi
retículo endoplásmico	vesículas
mitocondrias	vacuola
	lisosomas

OBJETIVOS

- Describe las partes de la célula eucariota.
- Explica las funciones de las partes de la célula eucariota.
- Describe las diferencias entre las células animales y las vegetales.

Células eucariotas: una mirada a su interior

Mucho tiempo después del descubrimiento de las células, los científicos todavía no sabían de qué estaban formadas. Las células son tan pequeñas que los detalles de su estructura no se lograron ver hasta que se desarrollaron mejores microscopios y métodos de tinción. Hoy en día sabemos que las células son muy complejas, sobre todo las eucariotas. Todas sus partes, desde las estructuras que cubren las células hasta los organelos que se encuentran en su interior, realizan una tarea vital para la célula.

¿Cómo está cubierta?

Todas las células tienen envolturas que separan lo que está adentro de la célula de lo que está afuera. Una de esas envolturas, la membrana celular, rodea a todas las células. Algunas células tienen una envoltura adicional que cubre la membrana celular, llamada pared celular.

Membrana celular Todas las células están cubiertas con una membrana celular. La tarea de la membrana celular es mantener el citoplasma adentro, permitir la entrada de los nutrientes y la salida de los desechos, así como interactuar con elementos del exterior de la célula. En la **Figura 21,** se muestra una vista detallada de la membrana celular de una célula cortada por la mitad.

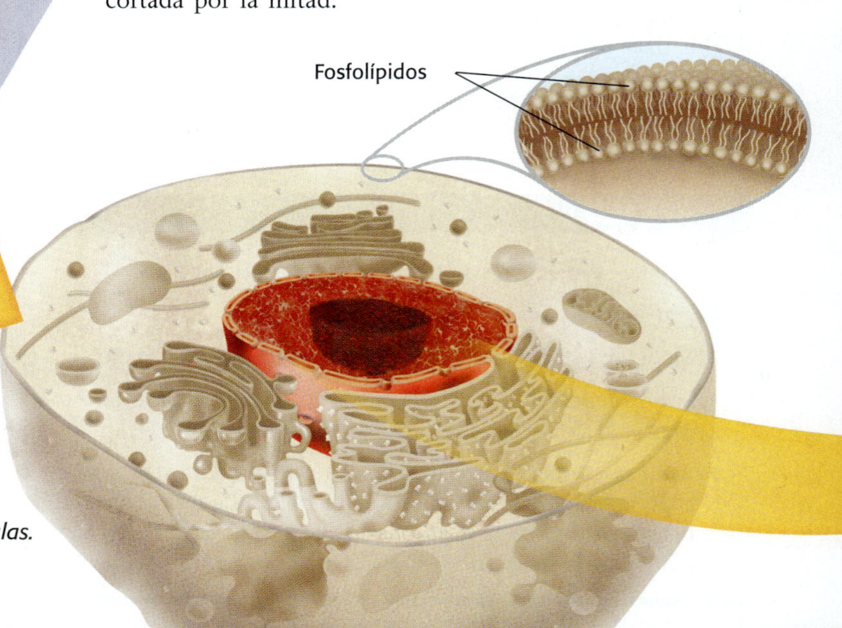

Figura 21 *La membrana celular rodea a todas las células. Las moléculas de fosfolípidos forman la membrana celular.*

92

Pared celular Las células de las plantas y algas tienen una pared celular dura hecha de celulosa. La **pared celular** fortalece y ayuda a la membrana celular. Cuando entra o sale mucha agua de una célula vegetal, la pared celular impide que la membrana se rompa. La fuerza de mil millones de paredes celulares en las plantas permite que un árbol no se caiga y que sus ramas desafíen la gravedad. Cuando observas heno seco, palos y tablas de madera, estás viendo las paredes celulares de células vegetales muertas. Las células de hongos, como champiñones, hongos venenosos, moho y levaduras tienen células hechas de una substancia química parecida a la que se encuentra en las conchas de los insectos. La **Figura 22** muestra una célula vegetal típica y una imagen detallada de la pared celular.

La biblioteca de la célula

El organelo más grande y visible de una célula es el núcleo. La palabra *núcleo* significa "centro" o "nuez" (se ve como una nuez dentro de un dulce). Observa la **Figura 23;** el núcleo está cubierto con una membrana por la que pasan los materiales.

El núcleo también es conocido como el centro de control de la célula. Como sabes, almacena el ADN, que contiene la información genética de las células. Casi todas las reacciones químicas vitales para la célula se relacionan con algún tipo de proteína. A veces, se puede ver una mancha negra dentro del núcleo. Esta mancha se llama *nucléolo,* y es como una especie de núcleo pequeño dentro de un núcleo grande. El nucléolo almacena los materiales que se usan para crear ribosomas en el citoplasma.

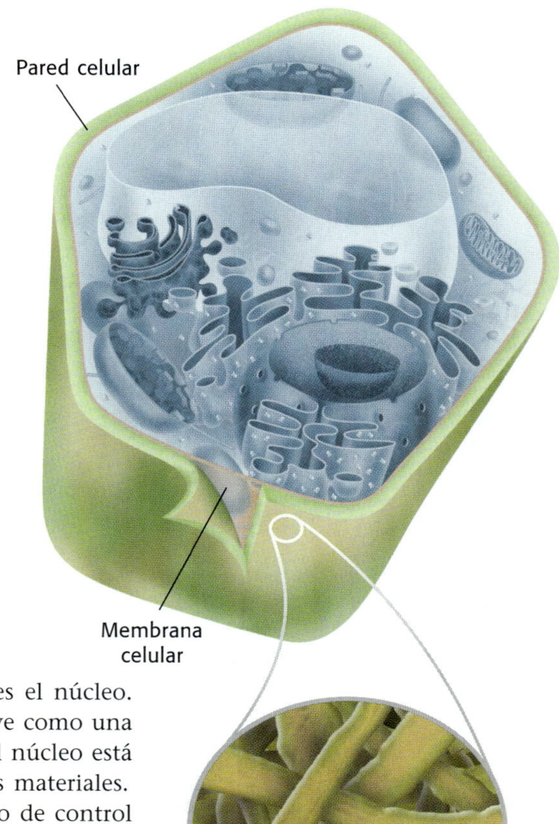

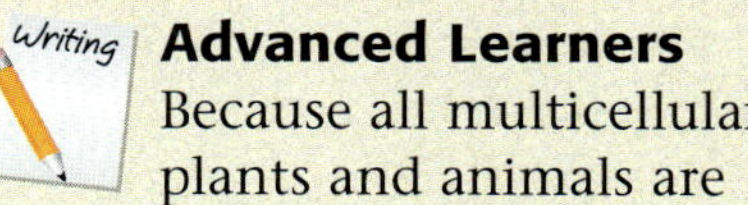

Figura 22 *La pared celular rodea la membrana celular. En las células vegetales, la pared celular está formada por fibras de celulosa.*

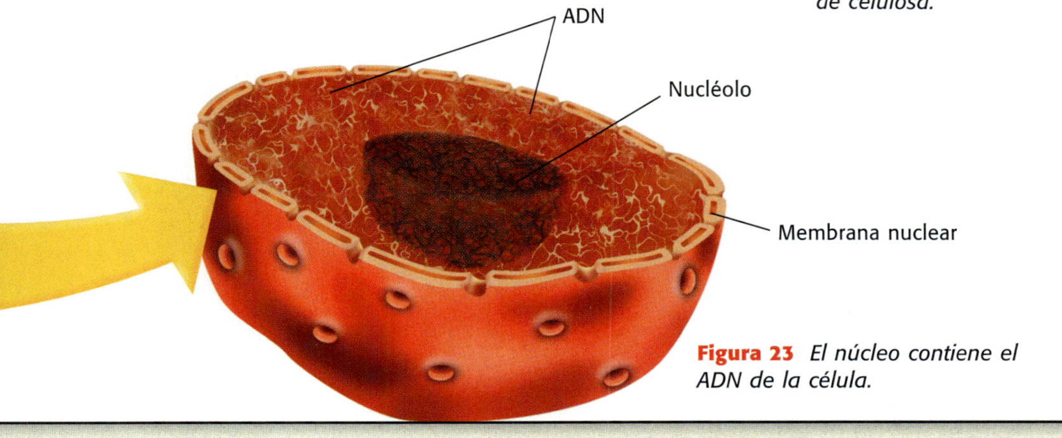

Figura 23 *El núcleo contiene el ADN de la célula.*

2) Teach

DEMONSTRATION

Cell Walls With a stick, a mushroom, and your own hand, you can illustrate the difference between a rigid cell wall and a flexible cell membrane like that found in human skin cells. Bend the stick, and it will break. Bend the mushroom, and it will come apart. Make a fist, and your skin stretches to accommodate the flexing of muscles and bone joints. If we had rigid cell walls like plants, we would find it extremely difficult to move. Sheltered English

MEETING INDIVIDUAL NEEDS

Writing **Advanced Learners**
Because all multicellular plants and animals are composed of eukaryotic cells, stress that the eukaryotic cell can be an entity in itself, not just a component of a larger organism. Have students research one-celled eukaryotic organisms, like a yeast or a one-celled protist, and compare them with eukaryotic cells that are part of a multicellular plant or animal. Students should include drawings, and record their findings in their ScienceLog. 

Directed Reading Worksheet 4 Section 3

Homework

Poster Project Have students investigate red blood cells and create a poster comparing the red blood cell with the cheek skin cell. (RBCs are the only cells in the human body that do not have a nucleus or mitochondria. Without a nucleus they cannot divide and reproduce. They live for only about 120 days, but new ones are made by bone marrow at the rate of up to 200 billion per day.)

MEETING INDIVIDUAL NEEDS

Writing **Advanced Learners**
Have students write a science fiction story about an animal whose cells are invaded by chloroplasts. Students should describe how that animal's life processes would be affected and how that animal would use this unusual occurrence to its advantage. Encourage students to write about an animal other than a mammal.

Answer to Self-Check

Cell walls surround the cell membranes of some cells. All cells have cell membranes, but not all cells have cell walls. Cell walls give structure to some cells.

CONNECT TO
PHYSICAL SCIENCE

Biophysics uses tools and techniques of physics to study the life processes of cells. Biophysicists are interested in the relationship between a molecule's structure and its function. Sophisticated techniques, such as electron microscopy, X-ray diffraction, magnetic resonance spectroscopy, and electrophoresis, allow them to study the structure of proteins, nucleic acids, and even parts of cells, such as ribosomes. Use the following Teaching Transparency to illustrate molecular structure.

Teaching Transparency 214
"Structural Formulas"

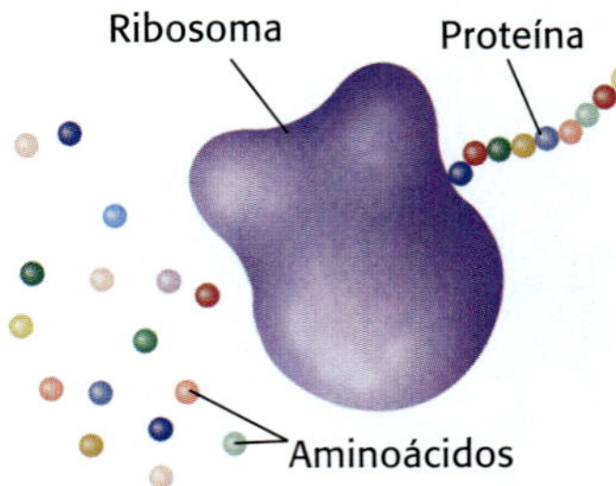

Figura 24 *Los aminoácidos se ensamblan en los ribosomas para sintetizar proteínas.*

✓ Autoevaluación

¿Cuál es la diferencia entre una pared y una membrana celular?
(Consulta la página 636 comprobar tu respuesta.)

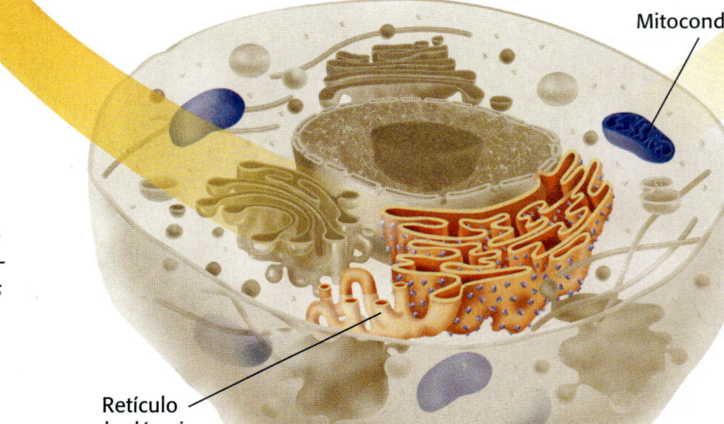

Figura 25 *El RE está hecho de compartimientos aplanados y de túbulos. Los ribosomas están unidos a algunos de los RE.*

Fábricas de proteína

Las proteínas, los "cimentos" de todas las células, están hechas de substanciaas químicas conocidas como *aminoácidos*. Estos aminoácidos se ensamblan para sintetizar proteínas en organelos muy pequeños llamados **ribosomas.** Los ribosomas son los organelos más pequeños, pero los más abundantes, y se muestran en la **Figura 24.** *Todas* las células tienen ribosomas porque todas necesitan proteínas para vivir. A diferencia de la mayoría de los organelos, los ribosomas no están cubiertos con una membrana.

El sistema de distribución de la célula

Las células eucariotas tienen un organelo llamado retículo endoplásmico, que se muestra en la **Figura 25.** El **retículo endoplásmico** o RE, es un compartimiento cubierto con una membrana que fabrica lípidos y otros materiales que se usan dentro y fuera de la célula. Procesa además los medicamentos y otras substancias químicas que pueden perjudicar la célula. El RE también es el sistema de distribución interno de una célula. Las substancias del RE se mueven de un lado a otro a través de muchos túbulos, de la misma manera en que los autos circulan por los túneles.

El RE tiene la forma de un costal aplanado o de una tela doblada sobre sí misma varias veces. Algunos están cubiertos con ribosomas, lo que hace que su superficie se vea áspera. Las proteínas sintetizadas en esos ribosomas entran al RE y luego se liberan para ser utilizadas en otra parte.

☢ WEIRD SCIENCE

Scientists at the University of New Mexico have created artificial muscles that are twice as strong as the real thing. Pending government approval, these artificial muscles may be implanted to replace paralyzed tissue in people with serious injuries.

La fuente de poder de la célula

Hoy en día, usamos muchas fuentes de energía, como petróleo, gasolina, carbón, energía nuclear e incluso basura. Necesitamos energía para hacer muchas cosas: iluminar la casa, ponerle gasolina al auto, o cocinar. Las células también necesitan energía para funcionar y vivir. ¿De dónde la obtienen?

Mitocondrias Dentro de las células, las moléculas de comida "se queman" (se desintegran) para liberar energía. Ésta se transfiere a una molécula que la célula utiliza para realizar sus tareas. Esta molécula se llama ATP.

El ATP se forma en varios lugares dentro de las células eucariotas, pero la mayoría se produce en organelos en forma de frijol llamados **mitocondrias,** que se muestran en la **Figura 26.** Estos organelos están rodeados por dos membranas. En la membrana interior, que tiene muchos dobleces, es donde se forma la mayoría de ATP. Las mitocondrias sólo trabajan si tienen oxígeno. La razón por la que respiras aire es para asegurar que tus mitocondrias tengan el oxígeno necesario para formar el ATP. Las células altamente activas, como las del corazón y las del hígado, tienen miles de mitocondrias, en cambio las otras células sólo tienen unas cuantas.

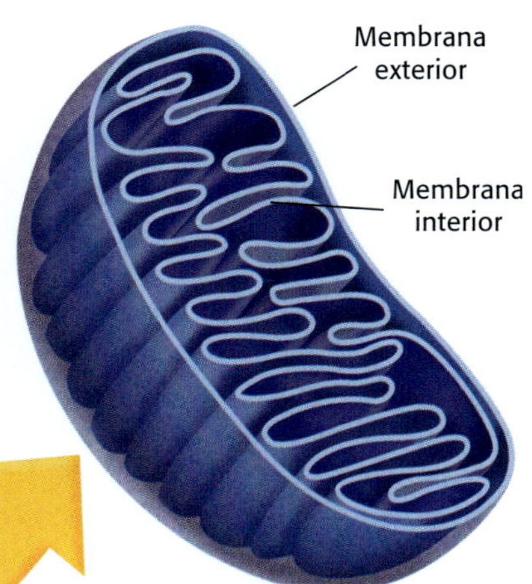

Figura 26 *Las mitocondrias tienen dos membranas. La membrana interior tiene varios dobleces.*

Cloroplastos Las plantas y las algas tienen un tipo adicional de organelo convertidor de energía, el **cloroplasto,** que se muestra en la **Figura 27.** La palabra *cloroplasto* significa "estructura verde". Como ves, también tiene dos membranas y estructuras que parecen pilas de monedas. Estos son sacos aplanados cubiertos con una membrana, y contienen una substancia llamada clorofila. La clorofila es la que hace que los cloroplastos sean verdes. Gracias a ella, el cloroplasto es la fuente de poder de la célula. La clorofila atrapa la energía de la luz del Sol y la utiliza para hacer azúcar. Este proceso se llama *fotosíntesis.* Las mitocondrias utilizan el azúcar producida para fabricar ATP. En otro capítulo descubrirás más sobre la fotosíntesis.

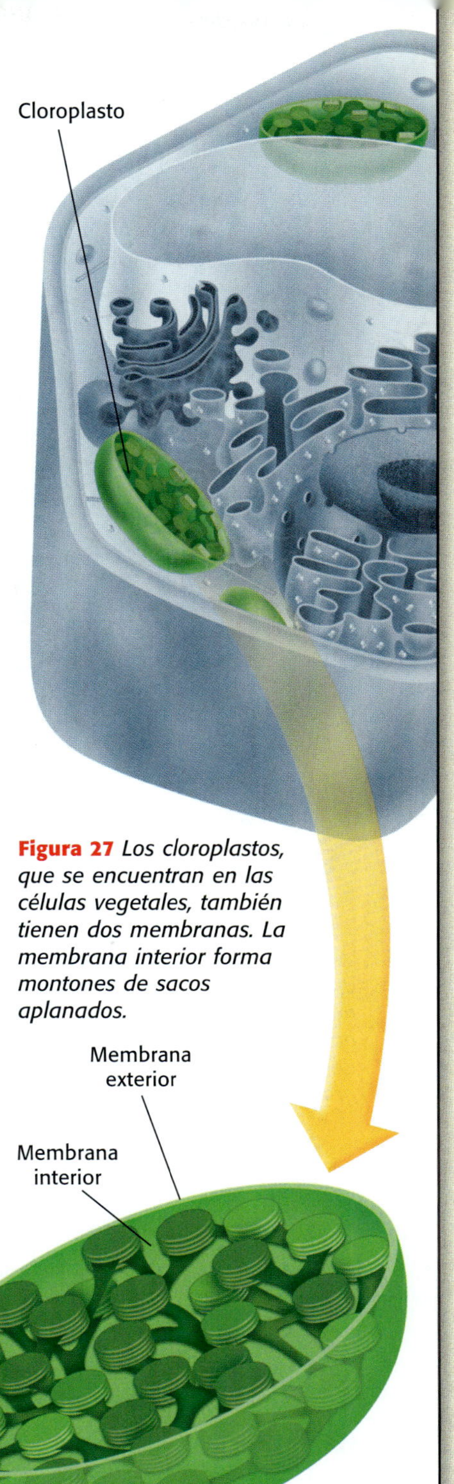

Figura 27 *Los cloroplastos, que se encuentran en las células vegetales, también tienen dos membranas. La membrana interior forma montones de sacos aplanados.*

IS THAT A FACT!
Nerve cells, called *neurons,* are the longest cells in the human body. They can be more than 1 m long.

ACTIVITY

Making Models
Have the students create a cell using a shoe box for the cell wall. Any material readily available can be used for the cell parts. Parts can be hung on a string, glued to the box, or attached by any method the student chooses. The boxes can be displayed in the classroom. **Sheltered English**

Multicultural CONNECTION

Camillo Golgi was born in 1843 in a small town called Corteno, in the northern Italian province of Lombardy. He received a medical degree in 1865. Although he was a psychiatrist, he was interested in the microscopic nature of the nervous system. He discovered the "black reaction," which is a method of staining tissue with silver nitrate so it can be viewed clearly with a microscope. In 1897, using this method, he saw the cellular structures we know today as the Golgi complex. Golgi died at the age of 83.

LabBook   PG 573

Name That Part!

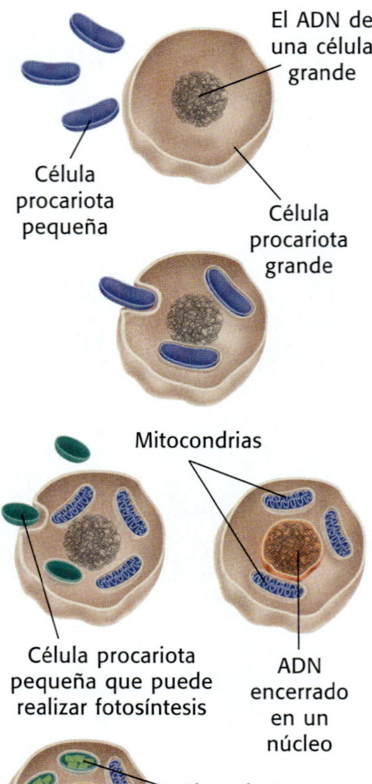

Figura 28 *Tal vez las mitocondrias y los cloroplastos se originaron de ancestros que producían energía y que fueron devorados por células más grandes.*

Teoría endosimbiótica Muchos científicos piensan que las mitocondrias y los cloroplastos se originaron como células procariotas que fueron ingeridas por células más grandes. En lugar de ser digeridas, las bacterias sobrevivieron. Esta teoría sobre el origen de las mitocondrias y los cloroplastos se conoce como "teoría endosimbiótica".

¿Qué evidencias tienen los científicos para afirmar que fue así como se formaron estos organelos? La primera evidencia es que las mitocondrias y los cloroplastos son del mismo tamaño de una bacteria. La segunda es que ambos están rodeados por *dos membranas*. Si la teoría es correcta, la membrana exterior se creó cuando las bacterias fueron devoradas por las células más grandes. Otra evidencia que apoya la teoría es que las mitocondrias y los cloroplastos tienen el mismo tipo de ribosomas y de ADN circular que las bacterias. También se dividen como las bacterias. Si las bacterias son los ancestros de las mitocondrias y de los cloroplastos, la **Figura 28** te muestra cómo hubiera sido el proceso.

El centro de procesamiento de la célula

Cuando las proteínas y otros materiales necesitan ser procesados y transportados fuera de una célula eucariota, la tarea se dirige a un organelo llamado **aparato de Golgi.** Esta estructura se llama así por Camilo Golgi, el científico italiano que la identificó por primera vez.

El aparato de Golgi se parece al RE, pero está más cerca de la membrana celular. En la **Figura 29** puedes ver el aparato de Golgi de una célula. Los lípidos y las proteínas del RE son distribuidos al aparato de Golgi, donde se modifican para realizar diferentes funciones. Los productos finales se encierran luego en una prolongación de la membrana del aparato de Golgi que luego se desprende como un brote pequeño. Este compartimiento transporta su contenido a otras partes de la célula o fuera de ella.

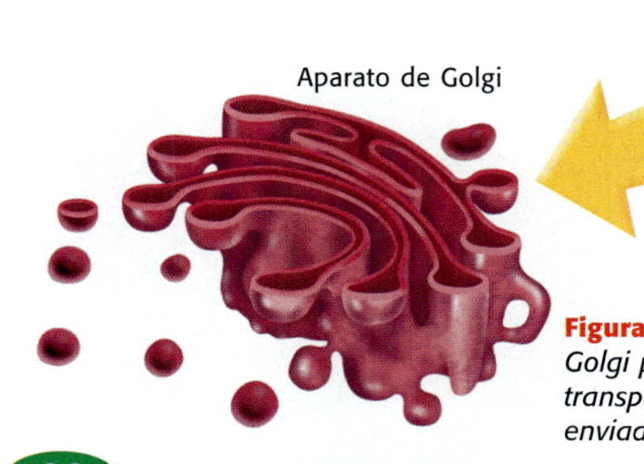
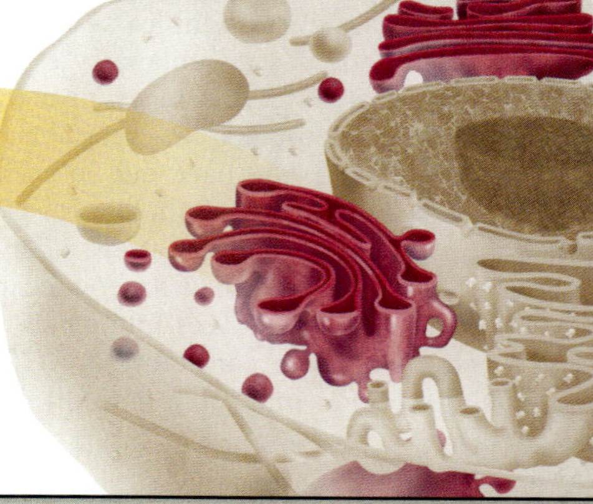

Figura 29 *El aparato de Golgi procesa, empaca y transporta los materiales enviados desde el RE.*

SCIENTISTS AT ODDS

Many scientists did not believe Golgi's claims about the Golgi complex. They thought he just saw tiny globs of the staining material. The existence of the Golgi complex was finally confirmed in the mid-1950s with the aid of the electron microscope.

Los centros de almacenamiento de la célula

Las células eucariotas tienen compartimentos cubiertos con una membrana llamados **vesículas.** Algunos se forman cuando parte de la membrana del RE o del aparato de Golgi se separan como brotes. Otras se forman cuando parte de la membrana de la célula rodea un objeto fuera de ésta. Así, los glóbulos blancos devoran otras células en tu cuerpo, como en la **Figura 30.**

Vacuolas La mayoría de las células vegetales tienen un compartimento grande cubierto con una membrana, llamado **vacuola,** como la de la **Figura 31.** Las vacuolas son recipientes que almacenan agua y otros líquidos. Las que están llenas de agua contribuyen al mantenimiento de la célula. Algunas plantas se marchitan cuando sus vacuolas pierden agua. Si quieres lechuga fresca, sólo llena las vacuolas con agua: deja remojando la lechuga en agua limpia toda la noche. ¿Te has preguntado por qué las rosas son rojas y las violetas son azules? Esto sucede gracias a un líquido de colores almacenado dentro de las vacuolas. Las vacuolas también contienen los jugos amargos y dulces que asocias con el limón, la naranja y otras frutas.

Algunos organismos unicelulares que viven en ambientes de agua dulce tienen problemas con la gran cantidad de agua que entra a las células. Tienen una estructura especial llamada "vacuola contráctil" que puede exprimir el exceso de agua de la célula. Funciona como una bomba que extrae el agua del interior de un barco.

Figura 30 *La célula más pequeña es una levadura que está siendo devorada por un glóbulo blanco.*

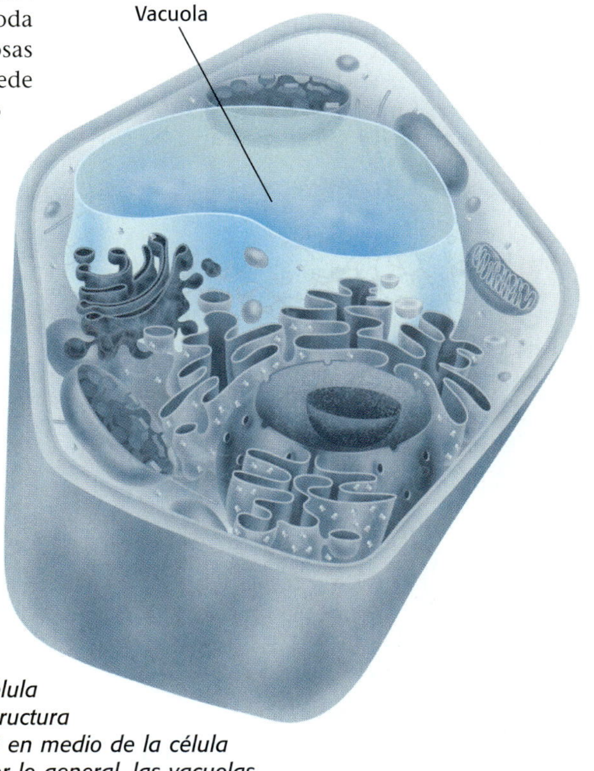

Figura 31 *La vacuola de la célula vegetal es la estructura grande que está en medio de la célula de color azul. Por lo general, las vacuolas son los organelos más grandes de una célula vegetal.*

Dr. Jewell Plummer Cobb, a cell biologist, was the first African American woman to be named president of a university in the California university system. Before she was named president of California State University at Fullerton in 1981, Dr. Cobb researched and helped explain how pigment cells normally function and how they change when they become cancerous. Dr. Cobb also taught at Sarah Lawrence College in New York and later at Rutgers University in New Jersey, where she also served as Dean. Throughout her career as a research biologist and university administrator, Dr. Cobb has demonstrated a deep commitment to research and education.

Meeting Individual Needs

Learners Having Difficulty
You can make a demonstration model of a vacuole within a cell by filling a balloon with water or air and putting it inside a clear-plastic storage bag. Show that the vacuole is part of the cell and is distinct within the cell. You can demonstrate that cells are full of motion and activity by moving the balloon around inside the bag.
Sheltered English

Critical Thinking Worksheet 4
"Cellular Construction"

IS THAT A FACT!

The vacuoles in grapes hold so much juice that they must dry in the sun for several weeks before they become raisins. A grape loses about three-fourths of its original weight in the process.

USING THE FIGURE

Have students refer to **Figure 32** and the chart below it to create their own drawing of a cell. But instead of drawing realistic images, tell them to draw an object that provides a visual clue about the organelle's job. For example, the Golgi complex, which transports materials, might be a car or a bus.

Sheltered English

PG 572

Cells Alive!

Homework

The organelles in a cell are rebelling against the nucleus. They all think they work too hard and want to take a vacation. Have students write a dialog between the nucleus and the other organelles. Tell them to help each organelle present a case for why it needs a rest and then have the nucleus explain what would happen if even one of them took two weeks off.

Teaching Transparency 14
"Organelles and Their Functions"

Teaching Transparency 15
"Comparing Animal and Plant Cells"

Reinforcement Worksheet 4
"Building a Eukaryotic Cell"

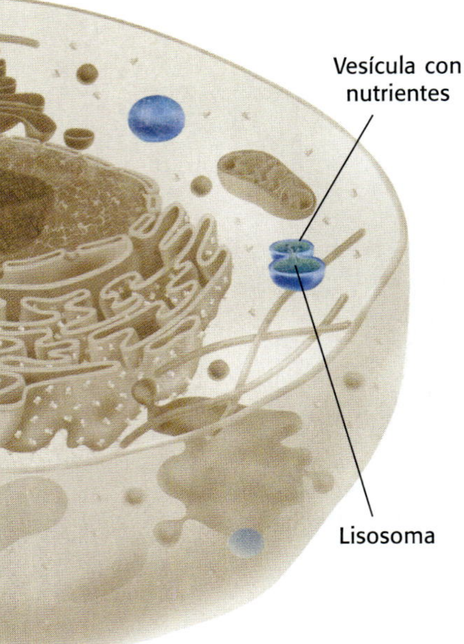

Figura 32 *Este lisosoma vierte enzimas en una vesícula que contiene partículas de alimento. Una vez digeridas, las moléculas de alimento se liberan en el citoplasma para que la célula las use.*

Paquetes de destrucción

¿Por qué la mayoría de las células de una oruga se disuelven en el material viscoso del capullo? ¿Por qué la cola de un renacuajo se contrae y luego desaparece? ¡Todo por los lisososmas!

Los **lisosomas** son vesículas especiales de las células animales que contienen enzimas. Cuando una célula devora una partícula y la guarda en una vesícula, los lisosomas se meten a estas vesículas y vacían las enzimas. Esto se muestra en la **Figura 32.** Los enzimas digieren las partículas que se encuentran en las vesículas. Los lisosomas destruyen a los organelos que están gastados o dañados. También se deshacen de los desechos y protegen a la célula de invasores. A veces, las membranas del lisosoma se rompen y se derraman las enzimas, matando la célula. Seguramente esto es lo que le pasa a un renacuajo cuando se convierte en rana. Los lisosomas hacen que las células de la cola de un renacuajo mueran y se disuelvan cuando el renacuajo se convierte en rana. ¿Sabías que los lisosomas jugaron un papel similar en tu desarrollo? Antes de que nacieras, los lisosomas provocaron la destrucción de las células que formaban una pequeña membrana entre tus dedos. La destrucción de las células provocada por los lisosomas también puede ser uno de los factores que contribuye al proceso de envejecimiento en el ser humano.

Los organelos y sus funciones

Núcleo
contiene el ADN de la célula y es su centro de control

Cloroplastos
producen alimento con la energía de la luz solar

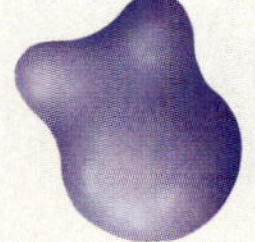
Ribosomas
lugar donde se ensamblan los aminoácidos para la síntesis de proteínas

aparato de Golgi
procesa y transporta materiales fuera de la célula

Retículo endoplásmico
fabrica lípidos, procesa medicamentos y otras substancias, empaca las proteínas para liberarlas desde la célula

Vacuola
almacena agua y otros materiales

Mitocondrias
descompone las moléculas de alimento para producir ATP

Lisosomas
digieren partículas de alimento, desechos, partes celulares e invasores

internet connect

SciLINKS
NSTA

TOPIC: Eukaryotic Cells
GO TO: www.scilinks.org
sciLINKS NUMBER: HSTL070

IS THAT A FACT!

Some tadpoles are three years old before they become frogs. The tadpole of the American bullfrog may take as long as 36 months before leaping to adulthood.

¿Planta o animal?

¿Qué diferencia crees que hay entre una célula vegetal y una animal? Ambas tienen una membrana celular, las dos tienen núcleo, ribosomas, mitocondrias, retículo endoplásmico, aparatos de Golgi y lisosomas. Pero las células vegetales tienen cosas que las células animales no tienen: una pared celular, cloroplastos y una vacuola grande. En la **Figura 33** puedes encontrar las diferencias entre las células vegetales y las animales.

Se encuentran en las células vegetales y animales	
Célula animal	
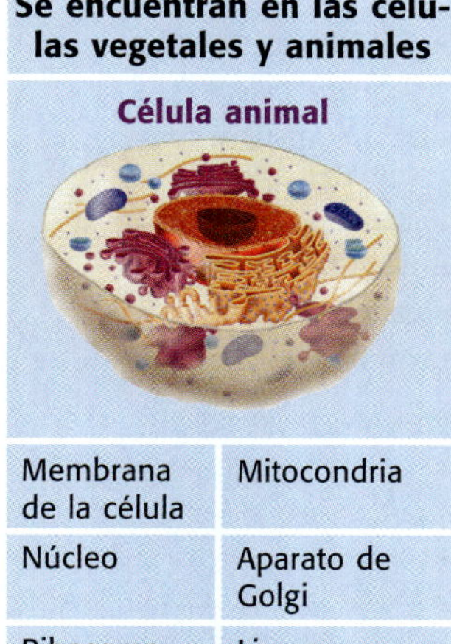	
Membrana de la célula	Mitocondria
Núcleo	Aparato de Golgi
Ribosoma	Lisosomas

Sólo se encuentran en las las células vegetales
Célula vegetal

Pared celular
Vacuolas
Cloroplastos

Figura 33 *Las células animales y vegetales tienen algunas estructuras en común, pero también tienen algo que las hace únicas.*

REPASO

1. ¿Cómo controla el núcleo las actividades de la célula?

2. ¿Cuál de los siguientes elementos no se encuentra en una célula animal: mitocondrias, pared celular, cloroplasto, ribosoma, retículo endoplásmico, aparato de Golgi , vacuola, ADN, clorofila?

3. Forma una oración con las siguientes palabras: oxígeno, ATP, respiración y mitocondrias.

4. **Aplicar conceptos** Imagina que te asignan la tarea de ponerle nuevos nombres a varias cosas de una ciudad, pero estos nombres deben ser las partes de una célula eucariota. Escribe algunas cosas que te gustaría ver en una ciudad y ponles el nombre de la parte de una célula que mejor se adapte a su función. Explica tus respuestas.

99

▼ Answers to Review

1. It contains the information on how to make the cell's proteins, which are involved in just about all the chemical activities of a cell.

2. cell wall, choroplast, vacuole, chlorophyll

3. Possible answer: Breathing supplies oxygen, which is needed by mitochondria to make ATP.

4. Answers will vary. Accept all reasonable responses.

GOING FURTHER

Writing Students now know that plant cells are a bit different from animal cells because of things such as cell walls and chloroplasts. But scientists group plant-like and animal-like organisms together and call them *protists*. Some protists are single-celled, like bacteria. Have students research and report on the differences between bacteria and protists. Encourage them to use additional resources to find information about protists and the two kingdoms of bacteria. Have them prepare a report on their findings.

PORTFOLIO

4) Close

Quiz

1. Every cell needs a membrane. Why? (A membrane is needed to keep the cytoplasm inside, to help transport nutrients and waste products, and to interact with things outside the cell.)

2. What would be the negative effects of a cell's not having lysosomes? (The cell would not be able to break down food molecules, get rid of wastes, get rid of damaged cells, or protect itself from invasion by foreign matter.)

ALTERNATIVE ASSESSMENT

Have students draw two cells, one plant and one animal. They should use one color to label all the structures common to both types of cells and another color to label all the structures that are different. ==Sheltered English==

Chapter Highlights

VOCABULARY DEFINITIONS

SECTION 1

tissue a group of similar cells that work together to perform a specific job in the body

organ a combination of two or more tissues that work together to perform a specific function in the body

organ systems groups of organs working together to perform body functions

organism anything that can live on its own

unicellular made of a single cell

multicellular made of many cells

population a group of individuals of the same species that live together in the same area at the same time

community all of the populations of different species that live and interact in an area

ecosystem a community of organisms and their nonliving environment

SECTION 2

cell theory the three-part theory about cells that states: (1) All organisms are composed of one or more cells, (2) the cell is the basic unit of life in all living things, and (3) all cells come from existing cells

cell membrane a phospholipid layer that covers a cell's surface and acts as a barrier between the inside of a cell and the cell's environment

organelles structures within a cell, sometimes surrounded by a membrane

cytoplasm cellular fluid surrounding a cell's organelles

surface-to-volume ratio the amount of a cell's outer surface in relationship to its volume

nucleus membrane-covered organelle found in eukaryotic cells that contains a cell's DNA and serves as a control center for the cell

Resumen del capítulo

SECCIÓN 1

Vocabulario

tejido (*pág. 81*)

órgano (*pág. 81*)

sistemas (*pág. 82*)

organismo (*pág. 83*)

unicelular (*pág. 83*)

multicelular (*pág. 83*)

población (*pág. 83*)

comunidad (*pág. 84*)

ecosistema (*pág. 84*)

Notas de la sección

- La célula es la unidad de vida más pequeña en la Tierra. Los organismos se componen de una o más células.

- En los organismos multicelulares, existen grupos de células que trabajan juntas y forman los tejidos. Los órganos están formados por diferentes tejidos y trabajan con otros órganos en los sistemas.

- El grupo de organismos que conviven al mismo tiempo en un área determinada forman una población. Una comunidad está compuesta por diferentes poblaciones que viven en la misma área. Un ecosistema abarca la comunidad y los componentes sin vida del área, como el agua y la tierra.

SECCIÓN 2

Vocabulario

teoría celular (*pág. 86*)

membrana celular (*pág. 87*)

organelos (*pág. 87*)

citoplasma (*pág. 87*)

relación de superficie a volumen (*pág. 88*)

núcleo (*pág. 90*)

procariota (*pág. 90*)

eucariota (*pág. 90*)

bacterias (*pág. 90*)

Notas de la sección

- La teoría celular establece que todos los organismos están compuestos por células, que la célula es la unidad básica de la vida y que cada célula proviene de otras.

- Las células tienen una membrana celular, ADN, citoplasma y organelos. La mayoría no se ven a simple vista porque son muy pequeñas.

☑ Comprobar destrezas

Conceptos de matemáticas

RELACIÓN DE SUPERFICIE A VOLUMEN La relación de superficie a volumen de una célula o de un objeto se calcula dividiendo el área superficial entre el volumen. Para determinarla primero debes determinar el área superficial.

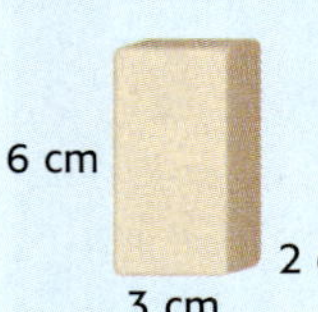

El área superficial es el área total de los lados de una figura. Esta figura tiene dos lados con un área de 6 cm × 3 cm, dos lados con un área de 3 cm × 2 cm y dos con un área de 6 cm × 2 cm.

$$\text{área superficial} = 2\,(6\text{ cm} \times 3\text{ cm}) + 2\,(3\text{ cm} \times 2\text{ cm}) + 2\,(6\text{ cm} \times 2\text{ cm}) = 72\text{ cm}^2$$

Luego debes calcular el volumen, multiplicando la longitud de los tres lados.

$$\text{volumen} = 6\text{ cm} \times 3\text{ cm} \times 2\text{ cm} = 36\text{ cm}^3$$

Para que encuentres la relación de superficie a volumen, divide el área superficial entre el volumen:

$$\frac{72}{36} = 2$$

La relación de superficie a volumen de esta figura es 2:1.

Lab and Activity Highlights

Elephant-Sized Amoebas? `PG 570`

Name That Part! `PG 573`

Cells Alive! `PG 572`

Datasheets for LabBook
(blackline masters for these labs)

prokaryotic describes a cell that does not have a nucleus or any other membrane-covered organelles; also called bacteria

eukaryotic describes a cell that has a nucleus

bacteria extremely small single-celled organisms without a nucleus; prokaryotic cells

SECTION 3

cell wall a structure that surrounds the cell membrane of some cells and provides strength and support to the cell membrane

ribosomes small organelles in cells where proteins are made from amino acids

endoplasmic reticulum a membrane-covered cell organelle that produces lipids, breaks down drugs and other substances, and packages proteins for delivery out of the cell

mitochondria cell organelles surrounded by two membranes that break down food molecules to make ATP

chloroplast an organelle found in plant and algae cells where photosynthesis occurs

Golgi complex the cell organelle that modifies, packages, and transports materials out of the cell

vesicles membrane-covered compartments in a eukaryotic cell that form when part of the cell membrane surrounds an object and pinches off

vacuole a large membrane-covered structure found in plant cells that serve as storage containers for water and other liquids

lysosomes special vesicles in cells that digest food particles, wastes, and foreign invaders

---

SECCIÓN 2

- Los materiales que las células aceptan o liberan deben pasar por la membrana celular.

  La relación de superficie a volumen es una comparación de la superficie exterior de la célula con su volumen. Esta relación disminuye cuando la célula crece.

- Las células eucariotas tienen un núcleo que contiene ADN lineal y poseen además varios organelos cubiertos con una membrana. Las procariotas tienen ADN circular y organelos que no están cubiertos con membranas.

Experimentos

¿Amibas del tamaño de un elefante? (*pág. 570*)

SECCIÓN 3

Vocabulario

pared celular (*pág. 93*)
ribosomas (*pág. 94*)
retículo endoplásmico (*pág. 94*)
mitocondrias (*pág. 95*)
cloroplasto (*pág. 95*)
aparato de Golgi (*pág. 96*)
vesículas (*pág. 97*)
vacuola (*pág. 97*)
lisosomas (*pág. 98*)

Notas de la sección

- Todas las células tienen una membrana celular que rodea el contenido de la célula. Algunas tienen una pared celular afuera de la membrana.

- El núcleo es el centro de control de la célula eucariota y contiene el ADN de la célula.

- Los ribosomas son los lugares donde los aminoácidos se enlazan para formar las proteínas. Estos organelos no están cubiertos con una membrana.

- El retículo endoplásmico (RE) y el aparato de Golgi son compartimentos cubiertos con una membrana en los que se forman y se procesan los materiales antes de ser transportados a otras partes de la célula o fuera de ella.

- Las mitocondrias y los cloroplastos son organelos que producen energía.

- Las vesículas y las vacuolas son secciones cubiertas con una membrana que almacenan material. Las vacuolas se encuentran en las células vegetales y los lisosomas en las células animales.

Experimentos

¡Células vivas! (*pág. 572*)
¿Cómo se llama esa parte? (*pág. 573*)

internet

VISITA: go.hrw.com

Visita el sitio web de HRW para encontrar una serie de herramientas de aprendizaje relacionadas con este capítulo. Sólo tienes que escribir la palabra clave:

PALABRA CLAVE: HSTCEL

VISITA: www.scilinks.org

Visita el sitio web de la **Asociación Nacional de Maestros de Ciencias** (*National Science Teachers Association*) para encontrar recursos de Internet relacionados con este capítulo. Sólo escribe el **ENLACE DE CIENCIAS** para obtener más información sobre el tema:

TEMA: Organización de la vida	**ENLACE:** HSTL055
TEMA: Poblaciones, comunidades y ecosistemas	**ENLACE:** HSTL060
TEMA: Células procariotas	**ENLACE:** HSTL065
TEMA: Células eucariotas	**ENLACE:** HSTL070

101

Lab and Activity Highlights

LabBank

Labs You Can Eat, The Incredible Edible Cell, Lab 2

Whiz-Bang Demonstrations, Grand Strand, Demo 3

Long-Term Projects & Research Ideas, Project 3

Vocabulary Review Worksheet 4

Blackline masters of these Chapter Highlights can be found in the **Study Guide.**

Using Vocabulary

1. cellulose
2. eukaryotic
3. nucleus
4. mitochondria/chloroplasts
5. Golgi complex

Understanding Concepts

Multiple Choice

6. a
7. b
8. c
9. a
10. c
11. c

Short Answer

12. Cells must be small to have a large surface-to-volume ratio.
13. Mitochondria have a double membrane, possess their own ribosomes, contain circular DNA, divide like bacteria, and are about the same size as bacteria.
14. Answers should paraphrase the following points: all organisms are composed of one or more cells; the cell is the basic unit of life in all living things; all cells come from existing cells.

Concept Mapping

15. 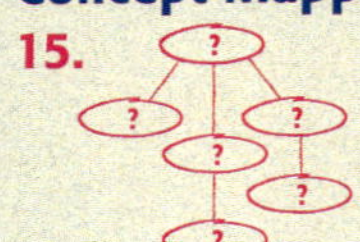 An answer to this exercise can be found at the end of this book.

**Concept Mapping
Transparency 4**

Repaso del capítulo

Escoge el término correcto para completar las siguientes oraciones:

1. La pared celular de las células vegetales está hecha de __?__. (*lípidos* o *celulosa*)

2. Tener organelos cubiertos con una membrana es una característica de las células __?__. (*procariotas* o *eucariotas*)

3. La información para hacer proteínas se encuentra en el __?__. (*aparato de Golgi* o *núcleo*)

4. Los dos organelos que pueden generar ATP en una célula vegetal son __?__ y __?__. (*cloroplastos/RE* o *mitocondrias/cloroplastos*)

5. Las vesículas que transportan los materiales fuera de la célula se forman en __?__. (*el aparato de Golgi* o *la membrana celular*)

COMPRENDER CONCEPTOS

Opción múltiple

6. Las células animales *no* tienen:
   a. pared celular.
   b. membrana celular.
   c. lisosomas.
   d. vesícula.

7. Diferentes __?__ trabajan de manera conjunta en un órgano.
   a. sistemas
   b. tejidos
   c. organismos
   d. procariotas

8. ¿Qué término se refiere a todos los organismos de un área determinada?
   a. población
   b. ecosistema
   c. comunidad
   d. organelos

9. ¿Cómo se llama el científico que dijo que todas las células provienen de otras células?
   a. Virchow
   b. Schleiden
   c. Hooke
   d. Schwann

10. ¿Cuáles de estos elementos *no* están cubiertos con una membrana?
    a. el aparato de Golgi
    b. las mitocondrias
    c. los ribosomas
    d. ninguno de los anteriores

11. ¿Cuál de estos órganos tiene enzimas que disuelven las partículas en las vesículas?
    a. las mitocondrias
    b. el retículo endoplásmico
    c. los lisosomas
    d. ninguno de los anteriores

Respuesta breve

12. ¿Por qué la mayoría de las células son tan pequeñas?

13. Menciona cinco características que sugieran que las mitocondrias se originaron como las bacterias.

14. Con tus propias palabras, menciona tres partes de la teoría celular.

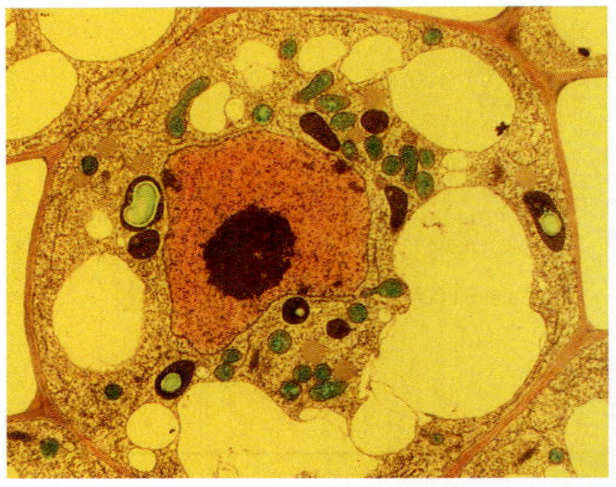

Organizar conceptos

15. Usa los siguientes términos para crear un mapa de ideas: ecosistema, células, organismos, aparato de Golgi, sistemas, comunidad, órganos, retículo endoplásmico, núcleo, población, tejidos.

Escribe una o dos oraciones para responder las siguientes preguntas:

16. Explica de qué manera el núcleo puede controlar lo que pasa en un lisosoma.

17. Aunque la celulosa no se forma en los ribosomas, explica por qué estos son importantes para la formación de la pared celular en una célula vegetal.

LAS MATEMÁTICAS EN LAS CIENCIAS

18. Supongamos que para que esta célula sobreviva, se necesitan tres moléculas de comida por unidad cúbica de volumen por minuto. Si una molécula puede entrar en cada unidad cuadrada de superficie por minuto, esta célula:

a. es muy grande y se moriría por falta de nutrientes.

b. es muy pequeña y se moriría por falta de nutrientes.

c. tiene un tamaño que le permite sobrevivir.

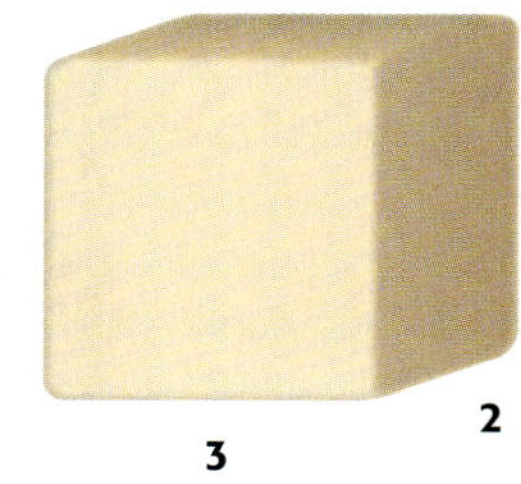

INTERPRETAR GRÁFICAS

Observa estos esquemas y responde las siguientes preguntas:

Célula A

Célula B

19. ¿Cómo se llama el organelo marcado con el número 19 en la célula A?

20. ¿Es la célula A una célula bacteriana, vegetal o animal? Explica tu respuesta.

21. ¿Cómo se llama y cuál es la función del organelo marcado con el número 21 en la célula B?

22. ¿Es la célula B una célula procariota o eucariota? Explica tu respuesta.

AHORA, ¿qué piensas?

Revisa tus respuestas a las preguntas de la página 79 que escribiste en el cuaderno de ciencias. ¿Han cambiado tus respuestas? Si es necesario, corrige tus respuestas basándote en lo que has aprendido en este capítulo.

103

CRITICAL THINKING AND PROBLEM SOLVING

16. The nucleus controls the production of proteins. The enzymes needed to break down materials in a lysosome are proteins.

17. Enzymes are used to make cellulose, and enzymes are proteins. All proteins are made at ribosomes.

MATH IN SCIENCE

18. a. too big and would starve

INTERPRETING GRAPHICS

19. mitochondrion

20. Animal; it is not a bacteria because it has a nucleus, and it is not a plant because it has no cell wall.

21. vacuole; storage of water and other materials

22. eukaryotic, because it has a nucleus

NOW WHAT DO YOU THINK?

1. A cell is a membrane-covered structure that contains the items necessary to carry out life processes. Cells are found in all living things.

2. Cells allow life processes to occur. Cells are small because a large surface-to-volume ratio allows them to exchange food and wastes with the environment more easily.

Blackline masters of this Chapter Review can be found in the **Study Guide.**

Battling Cancer with Pigs' Blood and Laser Light

Background

Students may want to know more about the traditional forms of cancer treatment. Some common therapies are chemotherapy, radiation treatment, and surgery. These are often used in combination to combat a single cancer. Although these treatments are highly effective on many forms of cancer, there are side effects that accompany them. For instance, chemotherapy and radiation treatments are generally toxic to the human body, causing side effects that can even be life threatening. Side effects include nausea, diarrhea, hair loss, organ failure, and bone marrow toxicity. Also, many cancer treatments lower a patient's ability to fight other diseases.

Chemotherapy is usually administered intravenously and is designed to interrupt the growth of cancer cells at different points in their development. For instance, one drug is effective on cancer cells during their division stage, while another attacks the cancer in its growth phase. Radiation treatments use high-energy beams of X rays and gamma rays to destroy cancerous tissues. The radiation impairs the cells' ability to multiply. Unfortunately, many forms of cancer are resistant to these treatments, and doctors continually seek new and better ways to cure cancer.

A TRAVÉS DE LAS CIENCIAS

CIENCIAS BIOLÓGICAS • CIENCIAS FÍSICAS

En la lucha contra el cáncer usando sangre de cerdos y rayos láser

¿Qué pasa cuando atraviesas con rayos láser la sangre de un cerdo? ¿Crees que hay un tratamiento para el cáncer? Los investigadores han desarrollado un tratamiento para el cáncer llamado *terapia fotodinámica.* Combina rayos láser muy fuertes con un medicamento sensible a la luz derivado de la sangre de los cerdos.

▲ *La sangre de los cerdos produce substancias que ayduan a tratar el cancer.*

¡Cerdos al rescate!

En el primer paso de esta terapia se usa una substancia sensible a la luz llamada *porfirina.* Las porfirinas son substancias químicas naturales que están en los glóbulos rojos y se unen a las lipoproteínas, las cuales transportan el colesterol en la sangre. Las porfirinas son importantes ya que absorben la energía de la luz. Las membranas celulares contienen lipoproteínas. Las células que se dividen rápidamente, como las cancerosas, fabrican membranas celulares a mayor velocidad que las normales. Como usan más lipoproteínas, acumulan más porfirinas. Los científicos desarrollaron una porfirina sintética, la Fotofirina®, hecha de porfirinas naturales de la sangre de los cerdos. Cuando se le inyecta a un paciente, actúa como las porfirinas naturales y se convierte en parte de las membranas celulares formadas por las células cancerosas. El siguiente paso consiste en la eliminación del tejido canceroso con rayos láser.

Tiro al blanco

El cirujano inserta un tubo largo y delgado con punta de láser en el área cancerosa donde se acumuló la Fotofirina. Cuando el rayo láser toca el tejido canceroso, la Fotofirina absorbe la energía de la luz. Luego, libera oxígeno el cual lesiona proteínas, lípidos, ácidos nucleicos y otros componentes de las células cancerosas. Esto mata a las células cancerosas, pero no a las sanas. La Fotofirina es más sensible a ciertas longitudes de onda de luz que las porfirinas naturales. Además, el intenso rayo del láser se puede centrar en el tejido canceroso sin afectar el tejido sano cercano.

¿Alguna alternativa?

La terapia fotodinámica mata las células cancerosas sin muchos de los efectos secundarios causados por otros métodos, como la quimioterapia. Sin embargo, el paciente puede sufrir quemaduras de Sol hasta que su cuerpo haya desechado el medicamento, lo cual tarda unos 30 días. Se está desarrollando un medicamento, llamado benzoporfirina, que tiene menos efectos secundarios y responde a varias longitudes de onda de rayos láser; también se está probando para atacar algunas enfermedades de los ojos, como la psoriasis.

¡Descúbrelo!

▶ Investiga por qué los científicos usaron sangre de cerdo para crear la fotofirina.

104

Answer to Find Out for Yourself

Researchers used pigs' blood because it is similar to human blood and is available in the large quantities necessary for research.

Una muestra del futuro

¿Qué hiciste la última vez que te raspaste la rodilla? Tal vez te pusiste una venda y ni te diste cuenta en qué momento tu rodilla sanó. Las vendas sirven como barreras que previenen infecciones y más daños. Pero, ¿y si existiera una venda viviente que cure tu cuerpo cada vez que esté lastimado? Suena como ciencia ficción, ¡pero no es así!

▲ *El Dr. Daniel Smith sostiene la "venda biológica fabricada genéticamente" que diseñó.*

El factor principal

Cuando hay una herida en la piel, como cuando te raspas la rodilla, las células producen y liberan un flujo constante de proteínas que curan la herida. Estas proteínas, que se forman de manera natural, se llaman *factores de crecimiento,* y se especializan en la reconstrucción del cuerpo. Unos reconstituyen el tejido conjuntivo que forma la piel nueva, otros reconstruyen los vasos sanguíneos del área lastimada y otros estimulan el sistema inmunológico del cuerpo. Gracias a los factores de crecimiento, la piel raspada sana en pocos días.

La venda viviente

Por desgracia, el proceso de curación no siempre es natural y sencillo. Alguien con un sistema inmunológico débil no puede producir suficientes factores de crecimiento para sanar una herida. Por ejemplo, tal vez alguien con quemaduras severas perdió la habilidad en el área quemada de producir proteínas necesarias para reconstruir tejidos sanos. En estos casos, los factores sintéticos de crecimiento pueden ser muy útiles en el proceso de curación.

Algunos avances en bioingeniería pueden ayudar a personas cuyo sistema inmunológico les impide sanar de manera natural. Existe una especie de "venda biológica", fabricada por procesos de ingeniería genética, que está hecha de células vivas de piel de donadores. El ADN de las células se manipula para producir factores humanos de crecimiento. Esta venda tiene 1 cm de grueso y tres capas: una capa de gasa delgada, una membrana artificial permeable y una bolsa de silicona en forma de cúpula que contiene los factores de crecimiento. La herida se cubre con la venda, con la capa de gasa en contacto con la lesión. Los factores de crecimiento salen de la bolsa de silicona por la membrana y pasan a través de la gasa a la herida. Ahí, actúan sobre la herida igual como lo hacen los factores de crecimiento naturales.

Fórmula de alivio rápido

La "venda biológica" también cura las heridas más rápido, pues aumenta la eficacia de las hormonas de crecimiento liberándolas a una velocidad constante en un período de tres a cinco días.

Debido a que imita los procesos de curación del cuerpo, es posible que en el futuro se usen otras versiones de esta venda para tratar una variedad de heridas y problemas de la piel, como el acné severo.

¿Tú qué piensas?

▶ Piensa en otros avances de la tecnología médica, como los anteojos o los audífonos, que imiten o mejoren las funciones que el cuerpo humano hace de manera natural.

Answers to Think About It

Answers will vary, but there are mechanical voices; artificial muscles; wheelchairs that respond to breath commands; hearing aids; glasses; and artificial hips, hands, legs, and even hearts.

HEALTH WATCH
The Scrape of the Future

Background

Skin is a naturally healing tissue helped by growth factors. Bones, muscles, blood, and many major organs, including the liver and lungs, can also heal and repair major damage with the help of the body's own growth factors. Nerve cells, including those in the spinal cord and brain, are less able to recover from damage.

Scientists are studying how nerves grow and what growth factors are involved in hopes of discovering a way to promote healing in people who suffer from nerve cell damage.

Chapter Organizer

CHAPTER ORGANIZATION	TIME MINUTES	OBJECTIVES	LABS, INVESTIGATIONS, AND DEMONSTRATIONS
Chapter Opener pp. 106–107	45		**Investigate!** Cells in Action, p. 107
Section 1 Exchange with the Environment	120	▶ Explain the process of diffusion. ▶ Describe how osmosis occurs. ▶ Compare and contrast passive transport and active transport. ▶ Explain how large particles get into and out of cells.	**QuickLab,** Bead Diffusion, p. 108 **Demonstration,** Membrane Model, p. 108 in ATE **Demonstration,** Crossing Membranes, p. 109 in ATE **Interactive Explorations CD-ROM,** The Nose Knows A *Worksheet* is also available in the *Interactive Explorations Teacher's Edition.* **Design Your Own,** The Perfect Taters Mystery, p. 574 **Datasheets for LabBook,** The Perfect Taters Mystery, Datasheet 10 **Inquiry Labs,** Fish Farms in Space, Lab 2 **Whiz-Bang Demonstrations,** It's in the Bag! Demo 5
Section 2 Cell Energy	120	▶ Describe the processes of photosynthesis and cellular respiration. ▶ Compare and contrast cellular respiration and fermentation.	**Demonstration,** Light Response, p. 112 in ATE **Skill Builder,** Stayin' Alive! p. 576 **Datasheets for LabBook,** Stayin' Alive! Datasheet 11
Section 3 The Cell Cycle	120	▶ Explain how cells produce more cells. ▶ Discuss the importance of mitosis. ▶ Explain how cell division differs in animals and plants.	**Labs You Can Eat,** The Mystery of the Runny Gelatin, Lab 3 **Whiz-Bang Demonstration,** Stop Picking on My Enzyme, Demo 4 **Long-Term Projects & Research Ideas,** Project 4

TECHNOLOGY RESOURCES

 **Guided Reading Audio CD**
English or Spanish, Chapter 5

 **One-Stop Planner CD-ROM with Test Generator**

 **Science Discovery Videodiscs**
Image and Activity Bank with Lesson Plans:
Outside and Inside

 **CNN** Science, Technology & Society,
Radioactive Medicine, Segment 5

 **Interactive Explorations CD-ROM**
CD 3, Exploration 1, The Nose Knows

CLASSROOM WORKSHEETS, TRANSPARENCIES, AND RESOURCES	SCIENCE INTEGRATION AND CONNECTIONS	REVIEW AND ASSESSMENT
Directed Reading Worksheet 5 **Science Puzzlers, Twisters & Teasers,** Worksheet 5		
Directed Reading Worksheet 5, Section 1 **Transparency 16,** Passive and Active Transport **Transparency 17,** Endocytosis **Transparency 17,** Exocytosis **Reinforcement Worksheet 5,** Into and Out of the Cell	**Math and More,** p. 109 in ATE	**Self-Check,** p. 109 **Homework,** p. 110 in ATE **Review,** p. 111 **Quiz,** p. 111 in ATE **Alternative Assessment,** p. 111 in ATE
Directed Reading Worksheet 5, Section 2 **Transparency 121,** Solar Heating Systems **Transparency 18,** Photosynthesis and Respiration: What's the Connection? **Reinforcement Worksheet 5,** Activities of the Cell **Critical Thinking Worksheet 5,** A Celluloid Thriller	**Connect to Earth Science,** p. 113 in ATE **Connect to Physical Science,** p. 114 in ATE **Apply,** p. 115 **Earth Science Connection,** p. 115 **Across the Sciences:** Electrifying News About Microbes, p. 124	**Homework,** p. 114 in ATE **Review,** p. 115 **Quiz,** p. 115 in ATE **Alternative Assessment,** p. 115 in ATE
Directed Reading Worksheet 5, Section 3 **Math Skills for Science Worksheet 24,** Creating Exponents **Science Skills Worksheet 24,** Using Models to Communicate **Transparency 19,** The Cell Cycle: Phases of Mitosis **Reinforcement Worksheet 5,** This Is Radio KCEL	**MathBreak,** Cell Multiplication, p. 116 **Math and More,** p. 117 in ATE **Holt Anthology of Science Fiction,** *Contagion*	**Self-Check,** p. 117 **Homework,** p. 118 in ATE **Review,** p. 119 **Quiz,** p. 119 in ATE **Alternative Assessment,** p. 119 in ATE

Holt, Rinehart and Winston On-line Resources

go.hrw.com

For worksheets and other teaching aids related to this chapter, visit the HRW Web site and type in the keyword: **HSTACT**

National Science Teachers Association

www.scilinks.org

Encourage students to use the *sci*LINKS numbers listed with the Chapter Highlights to access information and resources on the **NSTA** Web site.

END-OF-CHAPTER REVIEW AND ASSESSMENT

Chapter Review in Study Guide
Vocabulary and Notes in Study Guide
Chapter Tests with Performance-Based Assessment, Chapter 5 Test
Chapter Tests with Performance-Based Assessment, Performance-Based Assessment 5
Concept Mapping Transparency 5

Chapter Resources & Worksheets

Visual Resources

TEACHING TRANSPARENCIES

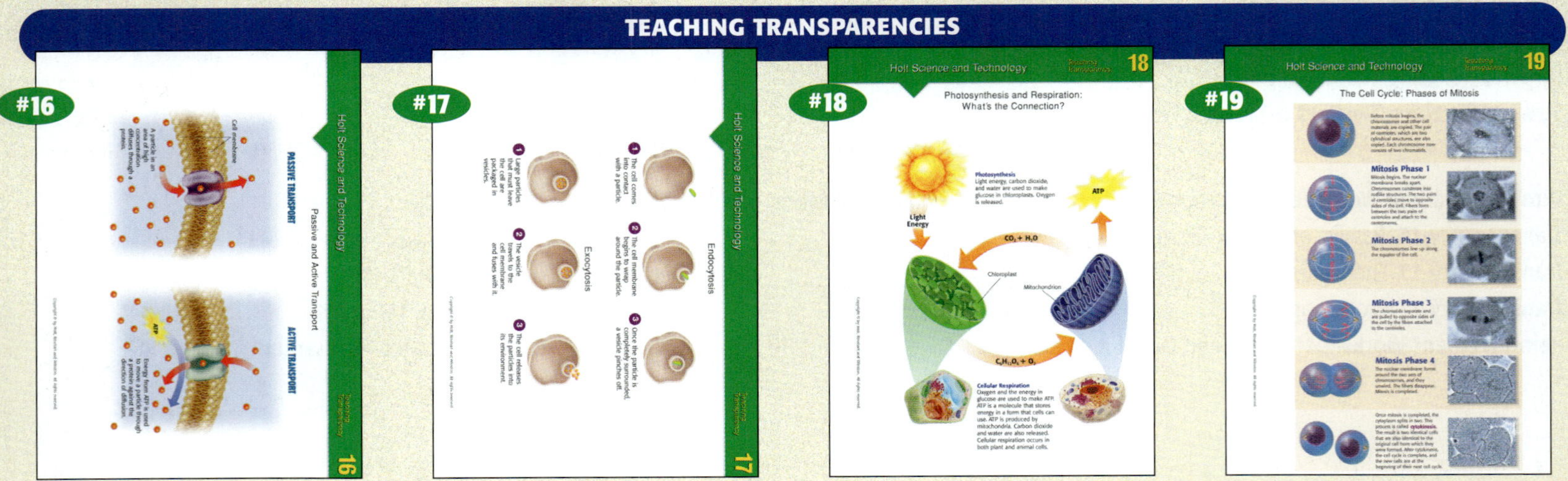

#16 · #17 · #18 · #19

TEACHING TRANSPARENCIES

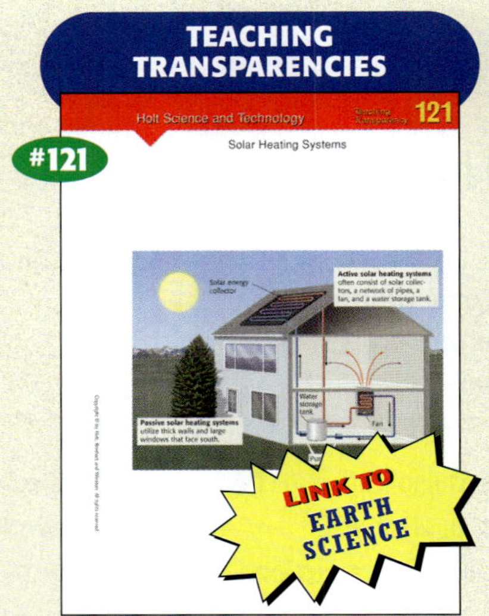

#121

CONCEPT MAPPING TRANSPARENCY

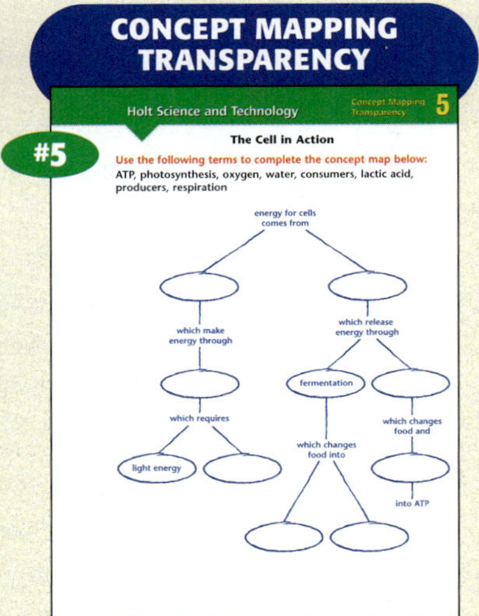

#5

Meeting Individual Needs

DIRECTED READING

#5

REINFORCEMENT & VOCABULARY REVIEW

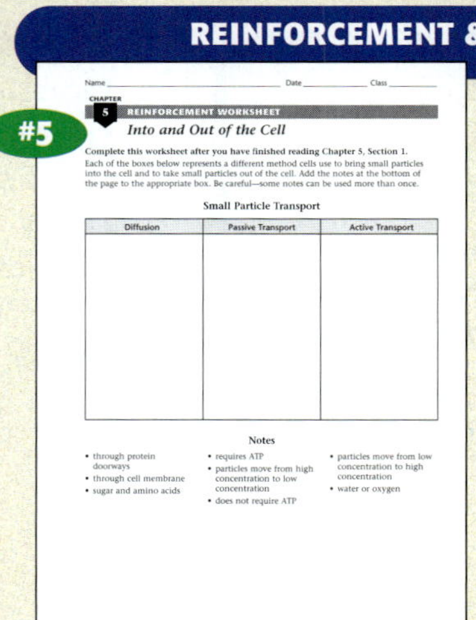

#5 · #5

SCIENCE PUZZLERS, TWISTERS & TEASERS

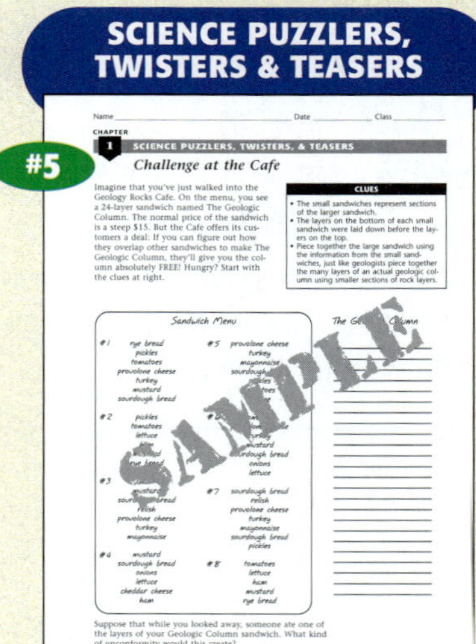

#5

Review & Assessment

STUDY GUIDE

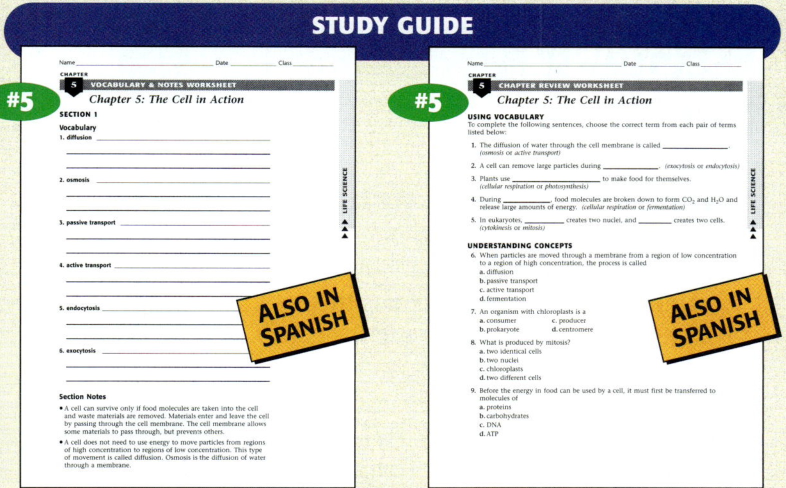

#5 — VOCABULARY & NOTES WORKSHEET — *Chapter 5: The Cell in Action*
ALSO IN SPANISH

#5 — CHAPTER REVIEW WORKSHEET — *Chapter 5: The Cell in Action*
ALSO IN SPANISH

CHAPTER TESTS WITH PERFORMANCE-BASED ASSESSMENT

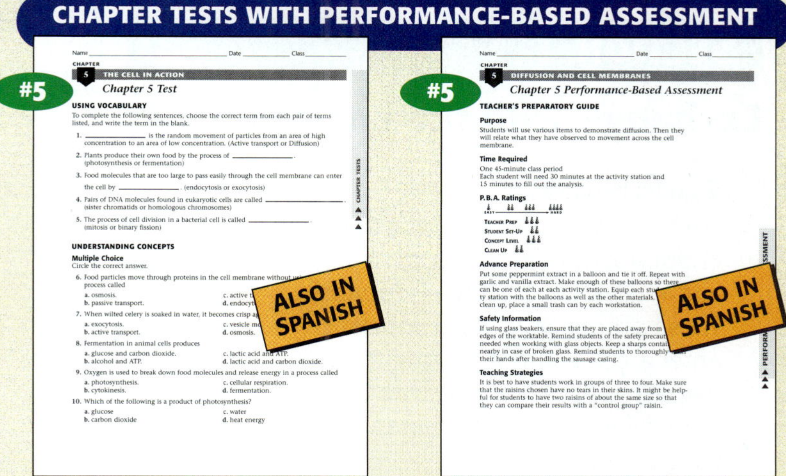

#5 — THE CELL IN ACTION — *Chapter 5 Test*
ALSO IN SPANISH

#5 — DIFFUSION AND CELL MEMBRANES — *Chapter 5 Performance-Based Assessment*
ALSO IN SPANISH

Lab Worksheets

INQUIRY LABS

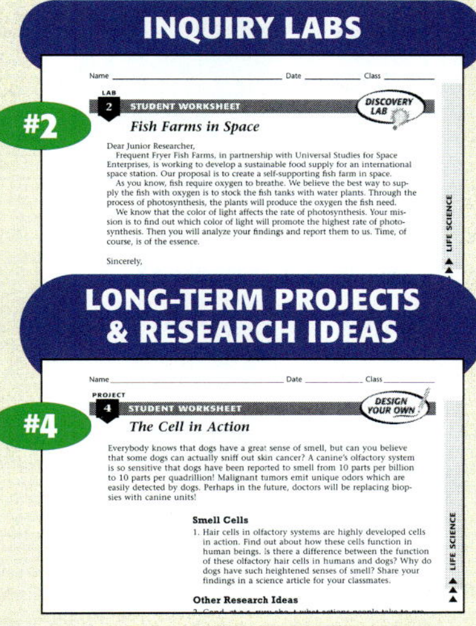

#2 — STUDENT WORKSHEET — *Fish Farms in Space*

LONG-TERM PROJECTS & RESEARCH IDEAS

#4 — STUDENT WORKSHEET — *The Cell in Action*

LABS YOU CAN EAT

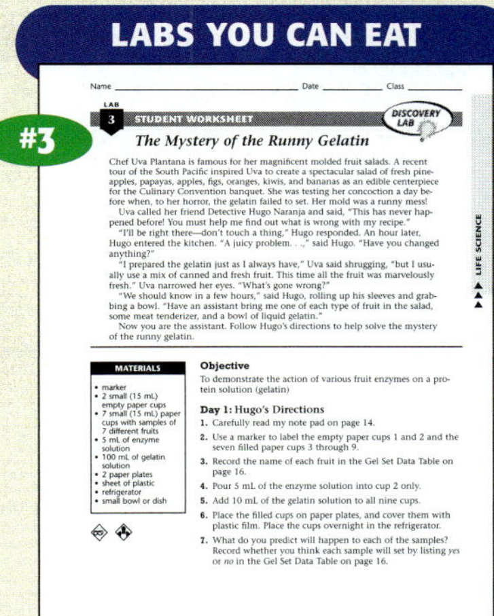

#3 — STUDENT WORKSHEET — *The Mystery of the Runny Gelatin*

WHIZ-BANG DEMONSTRATIONS

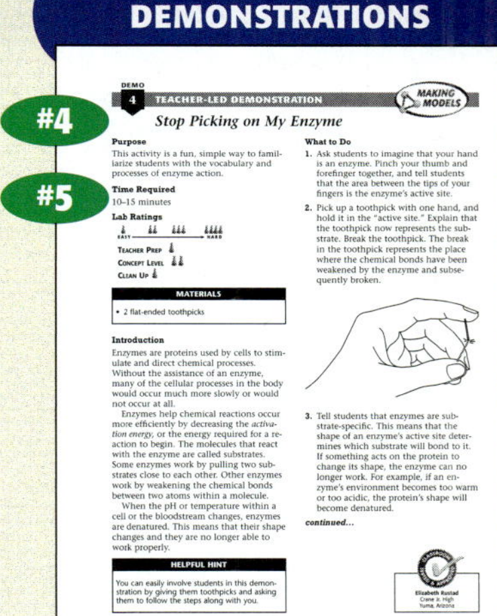

#4 **#5** — TEACHER-LED DEMONSTRATION — *Stop Picking on My Enzyme*

DATASHEETS FOR LABBOOK

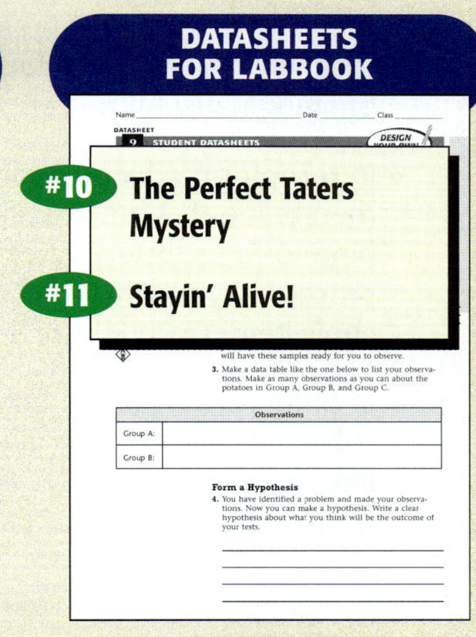

#10 **The Perfect Taters Mystery**

#11 **Stayin' Alive!**

Applications & Extensions

CRITICAL THINKING & PROBLEM SOLVING

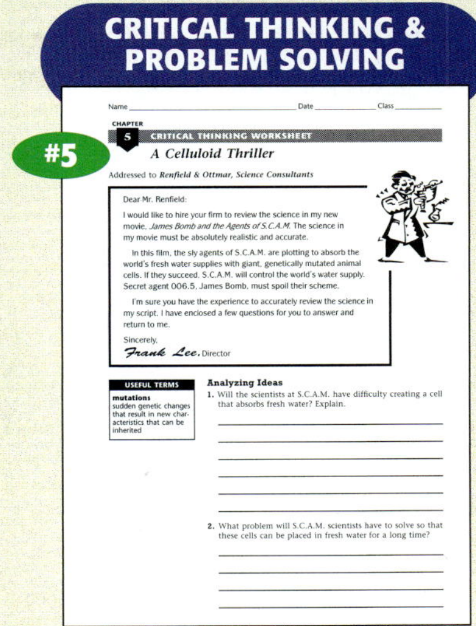

#5 — CRITICAL THINKING WORKSHEET — *A Celluloid Thriller*

SCIENCE TECHNOLOGY

#5 — *Science in the News: Critical Thinking Worksheets* — Segment 5: Radioactive Medicine

INTERACTIVE EXPLORATIONS

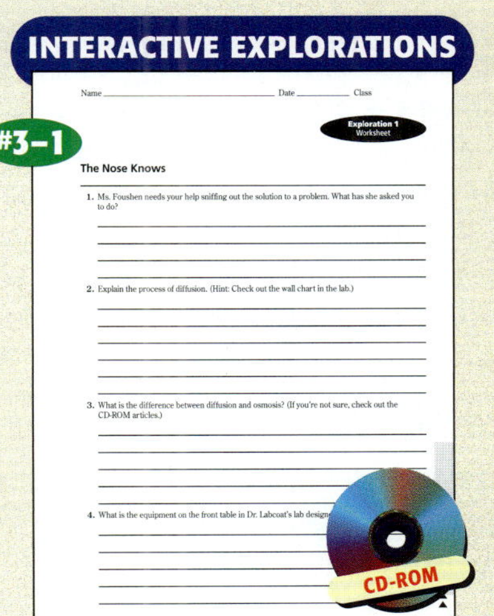

#3–1 — Exploration 1 Worksheet — The Nose Knows

SECTION 1

Exchange with the Environment

▶ Endocytosis

There are three different mechanisms of endocytosis: pinocytosis, phagocytosis, and receptor-mediated endocytosis. These processes allow a substance to enter a cell without passing through the cell membrane. The type of material involved determines which method is used.

- Large particles such as bacteria enter the cell by phagocytosis. The host cell changes shape, and the membrane sends out projections called pseudopods, meaning "false feet," which surround the particle, bringing it inside the cell.

- In receptor-mediated endocytosis, receptors on the membrane that are specific for a given substance bind to the substance before the endocytotic process begins. This method is used during cholesterol metabolism.

- In pinocytosis, the cell membrane surrounds the substance and forms a vesicle to bring the material into the cell. Pinocytosis usually involves material that is dissolved in water.

  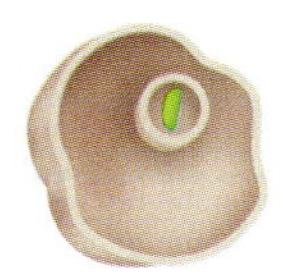

▶ Reverse Osmosis

Reverse osmosis is a process that forces water across semipermeable membranes under high pressure. The high pressure reverses the natural tendency of the solutes on the concentrated side of the membrane to pass through to the less-concentrated side. In this way water passing through the membrane is purified.

IS THAT A FACT!

- ◆ The largest single-celled organism that ever lived was a protozoa that measured 20 cm in diameter. It is now extinct.

SECTION 2

Cell Energy

▶ Jan Baptista van Helmont (1580–1644)

Van Helmont was a Belgian chemist, physiologist, and physician who coined the word *gas.* He was the first scientist to comprehend the existence of gases separate from the atmospheric air. Although he didn't know that it was carbon dioxide, van Helmont stated that the *spiritus sylvestre,* or "wild spirit," emitted by burning charcoal was the same as that given off by fermenting grape juice. He applied chemistry to the study of physiological processes, and for this he is known as the "father of biochemistry."

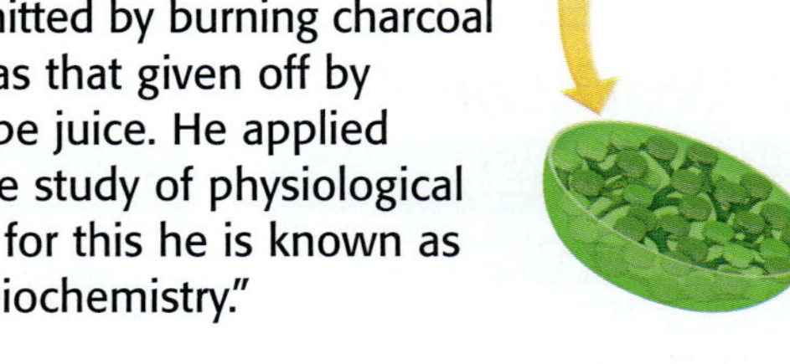

- Joseph Priestly (1733–1804) was an English clergy-man and physical scientist who was one of the discoverers of oxygen. He also observed that light was vital for plant growth and that green leaves released oxygen.

- Jan Ingenhousz (1730–1799), a Dutch-born British physician and scientist, discovered photosynthesis.

▶ Carotenoids and Photosynthesis

Carotenoids are responsible for the orange colors in plants. Their presence is usually masked by chlorophyll. They are sensitive to wavelengths of light to which chlorophyll cannot respond. Carotenoids can absorb the light waves and transfer the energy to chlorophyll, which then incorporates that energy into the photo-synthetic pathway.

IS THAT A FACT!

- Cellular respiration was discovered in 1937 by German biochemist Hans Krebs (1900–1981). The Krebs cycle is a series of chemical reactions that are essential to metabolic activities in all living organisms.

SECTION 3

The Cell Cycle

▶ Cytogenetics

Cytogeneticists study the role of human chromosomes in health and disease. Chromosome studies can reveal abnormalities such as whether a person is carrying the genetic material for a genetically linked disease.

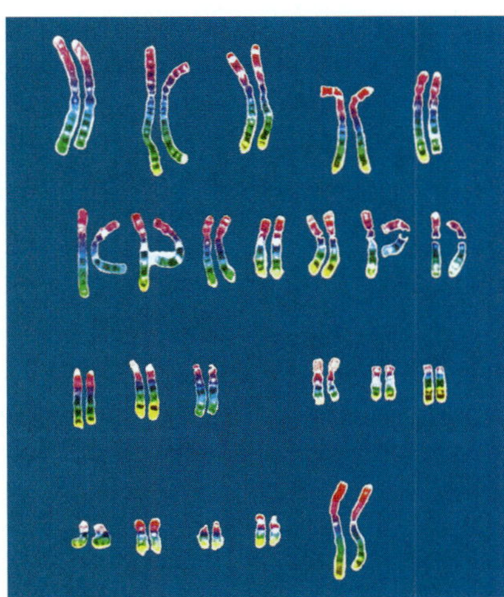

▶ Cell Division

The frequency of cell division varies a great deal. Fruit-fly embryo cells divide about every 8 minutes. Human liver cells may not divide for up to 1 year. Scientists are still trying to determine what orchestrates growth and regulates cell division. This information would help scientists understand diseases of unregulated cell division, such as cancer.

IS THAT A FACT!

- Cell division occurs at least 10 million times every second in an adult human body.

- DNA and chromosomes are related but are not the same thing. A chromosome is made up of DNA that has been wound up and organized with proteins that hold it all together. For much of the cell cycle, DNA is loose and not very visible.

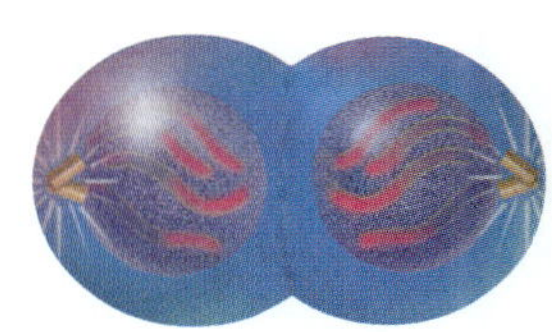
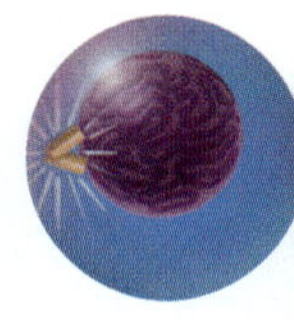
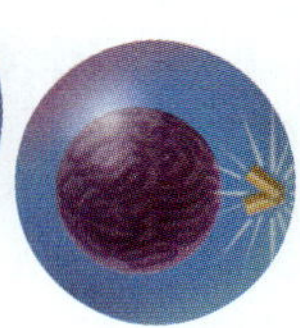

▶ Cell Adhesion

Blood cells exist individually in the body, but most other cells are connected to each other. Usually this involves special adhesion proteins, such as adherins, cadherins, catenins, and integrins. These proteins connect adjoining cells by physically locking the cells together, fastening one cell to the next. Sometimes these junctions are outside the cell, and sometimes they are inside. Adherence proteins can span the cell membranes and connect the inside of one cell to the inside of its neighbor cell.

IS THAT A FACT!

- In a healthy body, cells reproduce at exactly the same rate at which they die. However, some agents make cells reproduce uncontrollably, causing a disease known as cancer. One of these carcinogenic agents is ultraviolet radiation, which is emitted by the sun and ultraviolet lamps. People who spend excessive amounts of time in the sun run the risk of developing skin cancer.

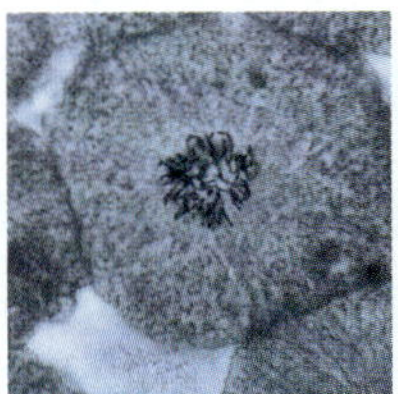
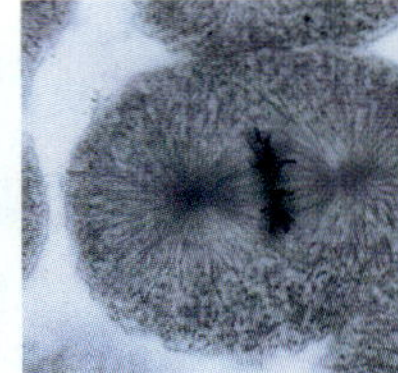
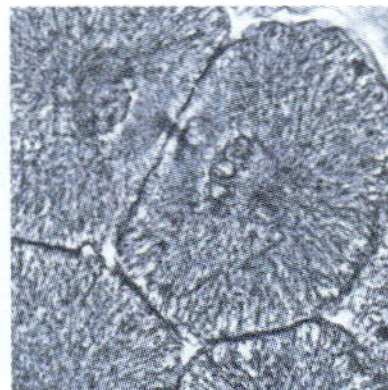

For background information about teaching strategies and issues, refer to the *Professional Reference for Teachers.*

The Cell in Action

Directed Reading Worksheet 5

Science Puzzlers, Twisters & Teasers Worksheet 5

Guided Reading Audio CD
English or Spanish, Chapter 5

La célula en acción

¿Qué tal si...?

¿Cuánto tiempo te gustaría vivir? ¿Y si llegaras a 120, 150 o más años de edad? Aún no se ha encontrado el secreto de la inmortalidad, pero recientemente los científicos hicieron un asombroso descubrimiento que podría contribuir a prolongar la vida.

En enero de 1998, unos investigadores de la Universidad de Texas informaron que habían encontrado una enzima en el cuerpo humano que funciona como una "fuente de la juventud celular". En el laboratorio, la enzima permite a las células humanas mantenerse jóvenes y multiplicarse mucho después del momento en que normalmente dejan de dividirse y mueren. Los investigadores esperan que algún día la enzima sirva para entender y tratar ciertos tipos de cáncer, así como otras enfermedades incurables. Aunque esta "enzima inmortalizante" no hará que vivamos para siempre, podría hacernos vivir más tiempo y de manera más saludable.

Todo ser vivo está hecho de células. En este capítulo aprenderás cómo las células crecen y producen más células; también entenderás cómo transportan substancias y obtienen la energía que necesitan para sobrevivir.

106

What If ...?

Most cells taken from a multicellular organism can be kept alive for days, perhaps weeks, in laboratory culture. But cancer cells can sometimes divide indefinitely. A culture of cells was taken from Henrietta Lacks, who died of cancer in 1951. The descendants of these cells continue to grow and divide today. Scientists around the world order these cells, called HeLa cells, through the mail and use them in experiments.

¿Tú qué piensas?

Usa tus conocimientos para responder a las siguientes preguntas en tu cuaderno de ciencias:

1. ¿Cómo entran y salen de la célula el agua, los alimentos y los desechos?

2. ¿Cómo utiliza la célula las moléculas de alimento?

3. ¿De qué manera puede una célula producir muchas más?

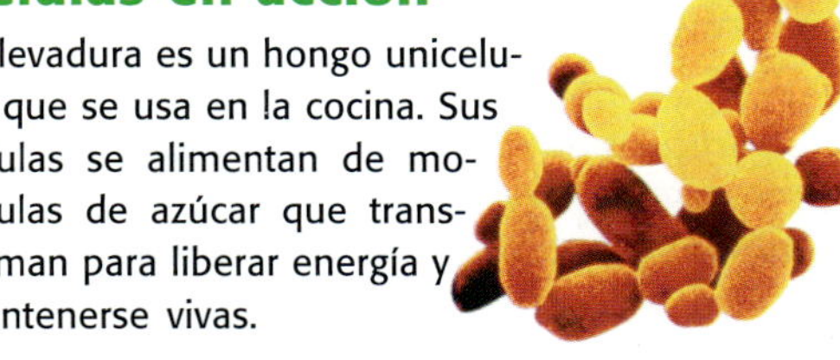

Células en acción

La levadura es un hongo unicelular que se usa en la cocina. Sus células se alimentan de moléculas de azúcar que transforman para liberar energía y mantenerse vivas.

Cuando las células de levadura liberan la energía del azúcar se produce un gas, llamado dióxido de carbono (CO_2). Las burbujas de este gas inflan la masa del pan. La cantidad de CO_2 que se produce depende de la cantidad de azúcar transformada

Procedimiento  

1. Pídele a tu maestro o maestra un **vaso de plástico pequeño** con 10 mL de una **mezcla de levadura y agua.**

2. En un **tubo de ensayo pequeño** coloca 4 mL de la **solución de azúcar** preparada por tu maestro o maestra. Después, vacíala en el vaso que contiene la mezcla de levadura y agua. Revuelve muy bien los dos líquidos con un **agitador.**

3. Regresa el contenido del vaso al tubo de ensayo pequeño.

4. Coloca un **tubo de ensayo un poco más grande** encima del tubo pequeño, tal como se observa en la figura de abajo. La boca del tubo pequeño debe tocar el fondo del tubo grande.

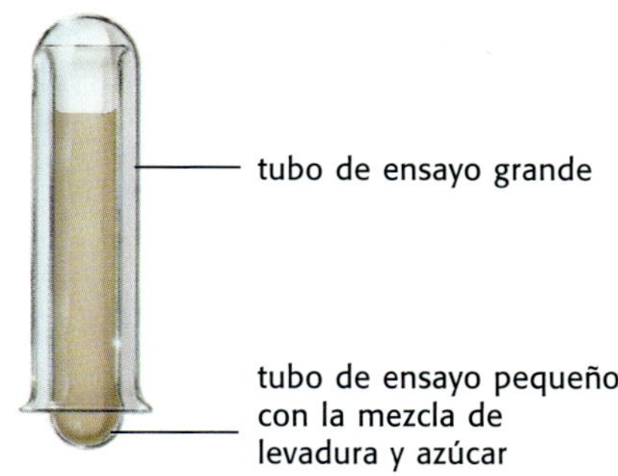

5. Voltea los tubos rápidamente. Como se muestra abajo, la mayor parte de la mezcla de levadura y agua todavía debe estar dentro del tubo pequeño, y parte de la mezcla estará en el tubo grande. Con una **regla** mide la altura del fluido contenido en el tubo grande.

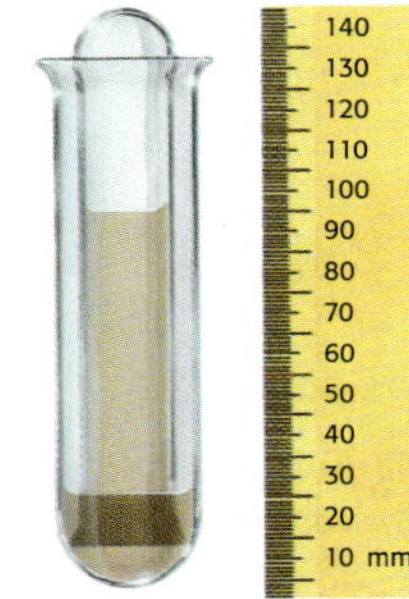

6. Coloca los tubos de ensayo en una **gradilla** y déjalos en reposo durante 20 minutos. Después, mide otra vez la altura del líquido contenido en el tubo grande.

Análisis

7. ¿Qué diferencia hay entre la primera medición y la segunda?

8. ¿A qué se debe el cambio en la altura del fluido?

What Do You Think?

Accept all reasonable responses.

Students will have a chance to revise their answers in the Chapter Review under NOW What Do You Think?

Investigate!

MATERIALS

FOR EACH STUDENT:
- small plastic cup
- yeast-and-water mixture
- small plastic test tube
- sugar solution
- stirring rod
- large plastic test tube
- test-tube stand
- ruler

Safety Caution: Remind students to review all safety cautions and icons before beginning this lab activity.

Students should wear safety goggles at all times and wash their hands when they are finished. Students should not taste the solutions.

Teacher Notes: The yeast suspension is prepared by mixing one package of dry yeast in 250 mL of water. The sugar solution is prepared by dissolving 30 mL (2 tbsp) of sugar in 100 mL of water. Some students can put their vials on ice and others can put them in a warm place, and others can leave them in a warm place overnight. In this way, students can see how temperature can affect the metabolism of yeast cells.

Answers to Investigate!

7. Answers will vary. Students should subtract the first measurement from the second measurement.

8. When the yeast cells released the energy in sugar, the CO_2 that the cells produced increased the volume of air in the smaller tube and pushed more yeast-and-sugar mixture into the larger tube, increasing the height of the liquid in the larger tube.

Focus

Exchange with the Environment

This section explains the processes involved in the exchange of materials between a cell and its environment. Students will learn about diffusion and osmosis, which is the diffusion of water across the cell membrane. Finally, students will compare and contrast active and passive transport and learn how large particles move into and out of cells.

Bellringer

Writing Write the following on the board or the overhead projector:

Which of the following best describes a living cell:

a. a building block

b. a living organism

c. a complex factory

d. all of the above

Have students write a paragraph in their ScienceLog defending their choice.

1) Motivate

DEMONSTRATION

Membrane Model Blow soap bubbles in front of the class. Explain that soap bubbles have several properties that are similar to biological membranes. One property is flexibility. The components of soap film and the cell membrane move around freely. Both soap bubbles and membranes are self-sealing. If two bubbles or membranes collide, they fuse. If one is cut in half, two smaller but whole bubbles or membranes form.

Sheltered English

VOCABULARIO

difusión   transporte activo
ósmosis   endocitosis
transporte pasivo   exocitosis

OBJETIVOS

- Explica el proceso de difusión
- Describe la ósmosis.
- Compara el transporte pasivo con el transporte activo.
- Explica cómo las partículas grandes entran y salen de las células.

Laboratorio

Difusión de canicas

Acomoda tres grupos de canicas de color en el fondo de un tazón de plástico. Cada grupo debe tener cinco canicas del mismo color. Tapa el tazón con plástico transparente y agítalo suavemente durante 10 segundos; al hacerlo, observa las canicas. ¿En qué se parece la dispersión de las canicas a la difusión de las partículas? ¿En qué se diferencia?

108

Quick Lab

MATERIALS

FOR EACH GROUP:
- three sets of colored beads
- plastic bowls
- clear plastic wrap

Intercambio con el medio ambiente

¿Qué le pasaría a una fábrica si se quedara sin energía o sin materias primas? ¿Y si no pudiera deshacerse de su propia basura? Si no tuviera energía ni materias primas y se llenara de basura dejaría de funcionar. Como la fábrica, la célula debe obtener energía y materias primas, así como eliminar sus desechos.

El intercambio de substancias entre la célula y su medio ambiente tiene lugar en la membrana celular. Antes de leer cómo entran y salen las substancias de la célula, debes saber qué es la difusión

¿Qué es la difusión?

¿Qué pasa si vacías colorante en un recipiente con gelatina sólida? Al principio, verás fácilmente dónde termina la gelatina y dónde empieza el colorante, pero con el tiempo la raya entre las dos capas se hará borrosa, como se muestra en la **Figura 1.** ¿Por qué sucede esto? El colorante y la gelatina, como toda la materia, están hechos de partículas diminutas. Las partículas de materia siempre se están moviendo y chocando entre sí. La mezcla de las distintas partículas hace que las capas se vuelvan borrosas. Esto ocurre con la materia en forma de gas, líquido o sólido.

Figura 1 *Las partículas de colorante se mezclan con las de gelatina debido a la difusión.*

Las partículas se desplazan desde áreas donde están más apretadas hasta áreas donde lo están menos. A este tipo de movimiento se le llama difusión. La **difusión** es el movimiento de las partículas de un área donde su concentración es alta a otra donde es baja. Este movimiento puede darse a través de las membranas celulares o afuera de las células. La difusión de partículas no requiere que los organismos o células utilicen energía.

Answers to QuickLab

The beads moved from areas where the colors were more concentrated to areas where they were less concentrated. Eventually, the different-colored beads were mixed somewhat evenly. The mixing of the beads required the use of the students' energy and also occurred much more quickly than diffusion normally occurs.

Difusión del agua Todos los organismos necesitan agua para vivir. Las células de los seres vivos contienen y están rodeadas por fluidos constituidos principalmente por agua. La difusión del agua a través de la membrana celular es tan importante para los procesos de la vida que se le ha dado un nombre particular: **ósmosis.**

El agua, como toda la materia, está constituida por pequeñas partículas. El agua pura tiene la concentración de partículas de agua más alta posible. Para disminuir esta concentración, sólo tienes que mezclar el agua con otra cosa, por ejemplo, con colorante de alimentos, azúcar o sal. La **Figura 2** muestra lo que sucede cuando hay ósmosis entre dos concentraciones distintas de agua.

La célula y la ósmosis Como vimos, las partículas de agua se moverán de las áreas de alta concentración a las de baja concentración. Este concepto es particularmente importante cuando lo aplicas a las células.

Por ejemplo, la **Figura 3** muestra los efectos de distintas concentraciones de agua sobre un glóbulo rojo. Como ves, la ósmosis se presenta en direcciones distintas, dependiendo de la concentración del agua que rodea a la célula. Por fortuna, los glóbulos rojos normalmente están rodeados de plasma sanguíneo, el cual está formado por agua, sales, azúcares y otras partículas. La concentración de estas substancias es igual en el interior y el exterior de estas células.

Las células de las plantas también toman y liberan agua por ósmosis. Por esta razón, una planta marchita recupera su firmeza si se la riega.

> ### ✔ Autoevaluación
>
> ¿Qué pasaría si pusieras una uva en un plato con agua pura? ¿Y si la remojaras en agua con azucár? *(Consulta la página 636 para comprobar tus respuestas)*

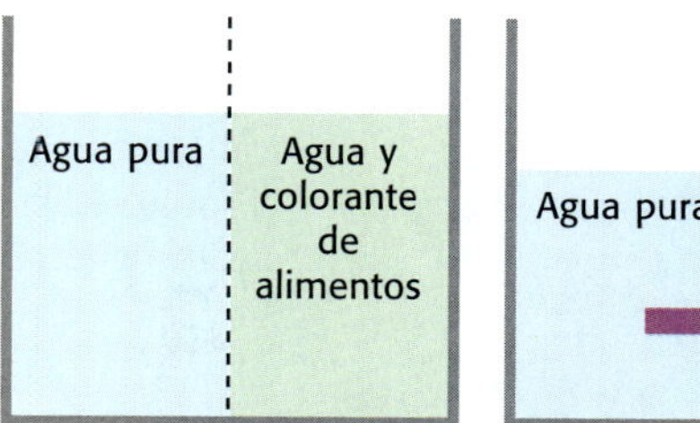

Figura 2 *Este recipiente está dividido por una barrera. Las partículas de agua son lo suficientemente pequeñas para atravesar la barrera, pero las partículas de colorante no lo son.*

1 El lado del recipiente con el agua pura tiene la concentración más alta de agua.

2 Durante la ósmosis, las partículas de agua se mueven hacia donde hay una menor concentración.

a La forma de esta célula es normal porque la concentración del agua que está en su interior es igual a la de afuera.

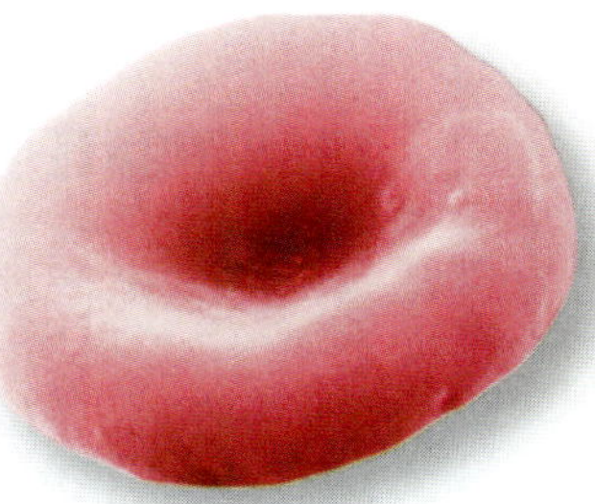

b Esta célula está en agua pura. Se llena de agua porque la concentración de partículas de agua en su interior es menor que la del exterior.

Figura 3 *La concentración de agua afuera de la célula afecta a la forma de estos glóbulos rojos.*

109

WEÏRD SCIENCE

Some salamanders don't have lungs. They get the oxygen they need right through their skin! The moist skin cells are well adapted for diffusion of gases and water.

Answers to Self-Check

In pure water, the grape would absorb water and swell up. In water mixed with a large amount of sugar, the grape would lose water and shrink.

2 Teach

MATH and MORE

Gases diffuse approximately 10,000 times faster in air than in water. If a gas diffuses to fill a room completely in 6 minutes, how long would it take the gas to fill a similar volume of still water? *(60,000 minutes)* How many hours would that be? *(1,000 hours)* How many days? *(41.67 days)*

DEMONSTRATION

Crossing Membranes

MATERIALS
• plastic sandwich bag
• twist tie
• tincture of iodine
• cornstarch
• 500 mL beakers (2)
• eyedropper
• graduated cylinder

Fill one beaker with 250 mL of water. Add 20 drops of iodine. Fill a second beaker with 250 mL water, and stir in 15 mL (1 tbsp) of cornstarch. Pour one-half of the starch-and-water mixture into the plastic bag. Secure the top of the bag with the twist tie. If any starch-and-water mixture spills onto the outside of the bag, rinse it off. Place the bag in the iodine-water mixture. Check immediately for any changes. Look again after 30 minutes. Iodine is used to test for the presence of starch and will turn the starch-and-water mixture black. The iodine particles are small enough to move through the tiny holes in the plastic bag, but the starch molecules are too large. Sheltered English

Directed Reading Worksheet 5 Section 1

LabBook PG 574

The Perfect Taters Mystery

ACTIVITY

Odor Diffusion One day prior to doing this activity, prepare the following:

MATERIALS

FOR EACH GROUP:
- small container with tight-fitting lid (one per pair of students)
- cotton ball for each container
- several strong-smelling liquids, such as vanilla, garlic oil, and eucalyptus

Soak a cotton ball in one of the liquids. Place the cotton ball in a container, and cover the container with a lid. Repeat until all containers have a soaked cotton ball.

Ask students if they can detect what is inside. If not, instruct students how to safely investigate the odors in the bottle. Have them hold the container 30 cm in front of their face and waft the air above the container toward their nose. Instruct them never to sniff directly from a container. Tell them to remove the lids, and ask them what they smell.

Explain that the cotton balls were soaked in aromatic fluids and that those fluids vaporized and diffused from the area of greatest concentration (the cotton ball) and moved to the area of lesser concentration (the air). Sheltered English

Teaching Transparency 16
"Passive and Active Transport"

Experimentos

¡Anímate a resolver el misterio de las papas perfectas! (página 574 de Experimentos).

Mover partículas pequeñas

Partículas como las de agua y oxígeno pueden pasar directamente a través de la membrana celular, la cual está formada por moléculas de fosfolípidos. Estas partículas pasan entre las moléculas de la membrana debido, en parte, a su tamaño pequeño; pero no todas las partículas que la célula necesita pueden pasar así. Por ejemplo, el azúcar y los aminoácidos no son tan pequeños como para pasar entre las moléculas de fosfolípidos de la membrana, que además los rechazan. Para entrar o salir de la célula tienen que pasar por "entradas" de proteína que están en la membrana celular.

Las partículas que pasan a través de estas proteínas lo hacen mediante transporte pasivo o activo. El **transporte pasivo,** representado en la **Figura 4,** es la difusión de las partículas a través de las proteínas. Las partículas viajan de un área de alta concentración a otra de baja concentración. La célula no utiliza ningún tipo de energía para que esto suceda.

El **transporte activo,** representado en la **Figura 5,** es el movimiento de partículas a través de proteínas en sentido opuesto al de la difusión. Las partículas pasan de un área de baja concentración a otra de alta concentración. La célula debe utilizar energía para que esto suceda. La energía proviene de la molécula de ATP, que almacena energía en una forma que la célula pueda utilizar.

TRANSPORTE PASIVO

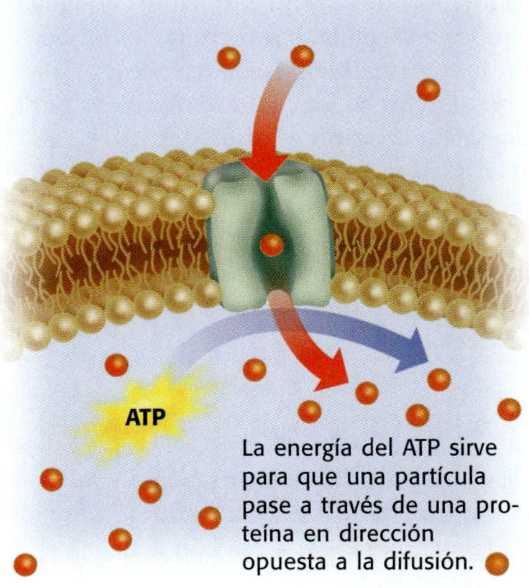

Figura 4 *En el transporte pasivo, las partículas pasan a través de la proteína de áreas de alta concentración a áreas de baja concentración.*

TRANSPORTE ACTIVO

Figura 5 *En el transporte activo, las células utilizan energía para mover partículas de áreas de baja concentración a áreas de alta concentración.*

110

Homework

Writing Ask students to describe how each of the following materials would get through a cell membrane and into a cell. Which of the materials require active transport?

a. pure water

b. sugar entering a cell that already contains a high concentration of particles

c. sugar entering a cell that has a low concentration of particles

d. a large protein

(b, d)

Mover partículas grandes

La difusión, el transporte pasivo y el transporte activo son buenos métodos para dejar entrar y salir partículas pequeñas de las células, pero ¿qué pasa con las partículas grandes? La membrana celular lleva a cabo esta tarea de dos maneras: *endocitosis y exocitosis*. En la **endocitosis,** la membrana celular rodea a la partícula y la encierra en una vesícula; de esta manera, las partículas grandes (por ejemplo, otras células) pueden entrar en la célula, como se muestra en la **Figura 6.**

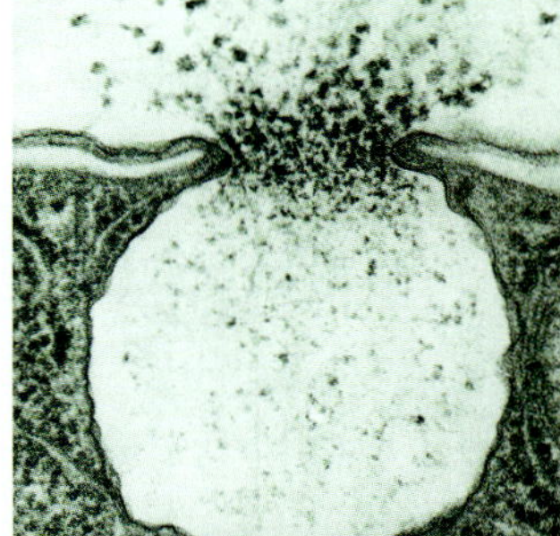

Figura 6 Endocitosis *significa "dentro de la célula".*

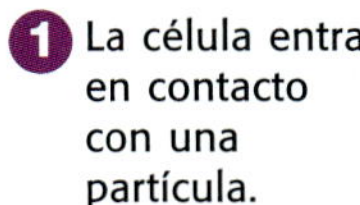

1 La célula entra en contacto con una partícula.

2 La membrana celular comienza a envolver la partícula.

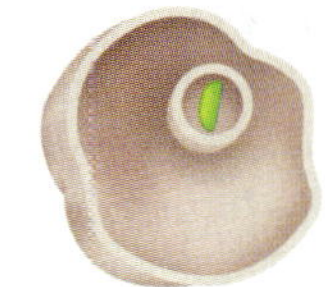

3 Cuando la partícula está completamente rodeada, la vesícula se desprende.

Si una partícula grande debe salir de la célula, el proceso es diferente. En la **exocitosis,** las vesículas se forman en el retículo endoplásmico o en el aparato de Golgi y se encargan de llevar las partículas a la membrana celular, como se muestra en la **Figura 7.**

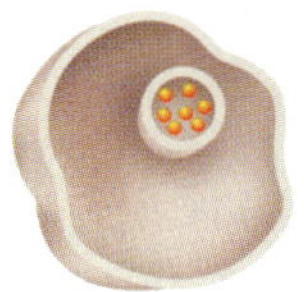

1 Las partículas grandes que deben salir de la célula son almacenadas en vesículas.

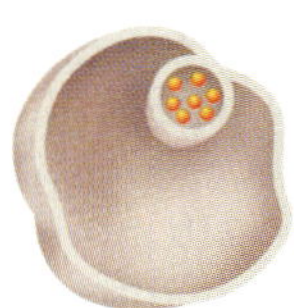

2 La vesícula va hasta la membrana celular y se fusiona con ella.

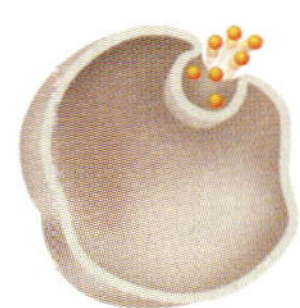

3 La célula libera las partículas en el medio ambiente.

REPASO

1. En la difusión, ¿se desplazan las partículas de las áreas de baja concentración a las áreas de alta concentración o viceversa?

2. ¿Cómo entran las partículas grandes en la célula? ¿De qué manera expulsan las células las partículas grandes?

3. **Hacer deducciones** La transferencia de glucosa al interior de la célula no requiere ATP. ¿Qué tipo de transporte proporciona glucosa a la célula? Explica tu respuesta.

RESEARCH

Writing Have students write a brief biography of Albert Claude (1898–1983), who used the electron microscope to study cells. (He shared the 1974 Nobel Prize for physiology or medicine with his student George Palade and with Christian de Duve.)

4 Close

Quiz

1. What part of the cell do materials pass through to get into and out of the cell? (the cell membrane)

2. What is osmosis? (the diffusion of water through the cell membrane)

ALTERNATIVE ASSESSMENT

Writing Have students write an instruction manual that tells a cell how to transport both a large molecule and a small molecule through the cell membrane.

Teaching Transparency 17 "Endocytosis" "Exocytosis"

Reinforcement Worksheet 5 "Into and Out of the Cell"

Interactive Explorations CD-ROM "The Nose Knows"

▼ **Answers to Review**

1. During diffusion, particles move from areas of high concentration to areas of low concentration.

2. Large particles are taken in by endocytosis and expelled by exocytosis.

3. Passive transport supplies a cell with glucose. Passive transport doesn't require the use of energy. Inform students who have trouble answering the question that glucose is a type of sugar.

Focus

Cell Energy

This section introduces energy and the cell. Students learn about solar energy and the process of photosynthesis. Finally, students learn about cellular respiration and fermentation.

Bellringer

Ask students to make a list in their ScienceLog of all the reasons why a cell might need energy. Remind students that there are many types of cells doing many different jobs.

1) Motivate

DEMONSTRATION

Light Response Cut out a square from black construction paper. Fold the square over a plant leaf of a common houseplant, such as a geranium. Affix the square with a paper clip. Be sure the leaf does not receive any sunlight. Leave the leaf covered for about 1 week. Remove the black square. The leaf will be much paler than the other leaves. In the absence of sunlight, chlorophyll is depleted and not replenished. Thus, the leaf's green color will have faded.

Sheltered English

Directed Reading Worksheet 5 Section 2

VOCABULARIO
fotosíntesis
respiración celular
fermentación

OBJETIVOS
- Describe la fotosíntesis y la respiración celular.
- Compara y contrasta la respiración celular con la fermentación.

Energía celular

¿Por qué nos da hambre? El hambre es la manera en que el cuerpo nos dice que nuestras células necesitan energía. Las células de todos los organismos utilizan energía para realizar las actividades químicas que les permiten vivir, crecer y reproducirse.

Del Sol a la célula

Casi toda la energía necesaria para la vida proviene del Sol. ¿Cómo obtienen las células esta energía? La obtienen de lo que comes. Al igual que muchos otros tipos de organismos, tienes que comer plantas u organismos que han comido plantas. Esto se debe a que las plantas son capaces de captar la luz solar y transformarla en alimento mediante un proceso llamado **fotosíntesis.** *Fotosíntesis* quiere decir "fabricado por la luz". El alimento que fabrican las plantas les da energía y es una fuente de energía para los organismos que se las comen. Si las plantas y los demás productores no existieran, los consumidores no podríamos sobrevivir.

Fotosíntesis Las células de las plantas tienen moléculas llamadas *pigmentos,* que absorben la energía de la luz. La clorofila es el principal pigmento de la fotosíntesis; gracias a ella las plantas son verdes. En las células vegetales, la clorofila se encuentra en los cloroplastos, que puedes apreciar en la **Figura 8.**

Las plantas utilizan la energía captada por la clorofila para transformar el dióxido de carbono (CO_2) y el agua (H_2O) en alimento. El alimento que producen es la glucosa, un azúcar sencillo ($C_6H_{12}O_6$). La glucosa es un carbohidrato. Al fabricarla, las plantas convierten la energía del Sol en una forma de energía que se puede almacenar. Las células de las plantas utilizan la energía contenida en la glucosa y almacenan una parte en forma de otros carbohidratos o de lípidos. La fotosíntesis también produce oxígeno (O_2). El proceso de la fotosíntesis puede resumirse en la siguiente ecuación:

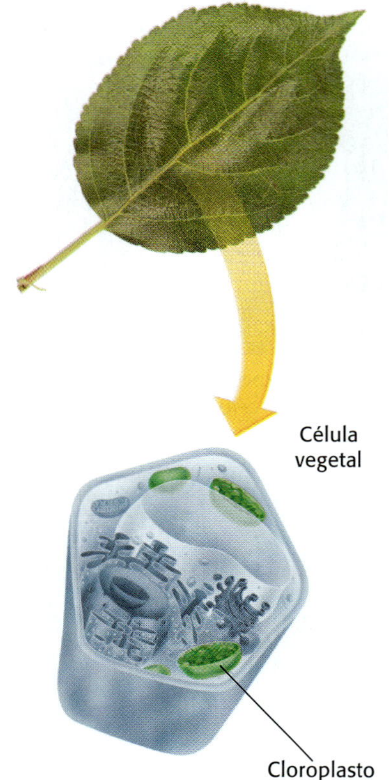

$$6CO_2 + 6H_2O + \text{energía de la luz} \longrightarrow C_6H_{12}O_6 + 6O_2$$

Dióxido de carbono   Agua   Glucosa   Oxígeno

Figura 8 *Durante la fotosíntesis, las células vegetales utilizan la energía de la luz solar para fabricar alimento (glucosa) a partir de dióxido de carbono y agua. La fotosíntesis se lleva a cabo en los cloroplastos.*

SCIENTISTS AT ODDS

In the 1800s, scientists were reluctant to change their belief that life came from a special "life force." Rudolf Virchow (1821–1902), a German scientist studying cells, believed strongly that life was a physical property of cells, not a mysterious life force. His way of thinking became known as mechanistic because life, it seemed, could be described as the sum of all the physical mechanisms of the cell. Those who clung to the life-force theory were known as vitalists. The mechanists battled the vitalists for 75 years before the cell theory became widely accepted. Today the mechanist view is still dominant in the study of biology.

Obtener energía de los alimentos

Aunque las células obtienen la energía que necesitan de los alimentos, no pueden obtenerla directamente de una manzana o de unos tacos. Los alimentos que uno come tienen que ser transformados primero para que la energía que contienen pueda ser utilizada por las células. De hecho, todos los organismos tienen que transformar las moléculas de alimento para que se libere la energía almacenada en ellas. Hay dos maneras de hacerlo. Una utiliza oxígeno y se conoce como **respiración celular;** la otra no utiliza oxígeno y se llama **fermentación.**

Respiración celular La respiración de la célula no es igual a la respiración que todos conocemos, aunque están muy relacionadas. Cuando respiramos aire, las células obtienen el oxígeno que necesitan para llevar a cabo la respiración celular y el cuerpo elimina dióxido de carbono, que es un desecho de la respiración celular.

La mayoría de los organismos, como la vaca de la **Figura 9,** obtienen la energía almacenada en los alimentos por medio de la respiración celular. Durante este proceso, el alimento (glucosa) es transformado en CO_2 y H_2O y se libera energía. Una gran cantidad de energía se almacena en forma de ATP; como sabes, el ATP es la molécula que proporciona la energía para que las células realicen sus actividades. Sin embargo, gran parte de la energía se libera en forma de calor, que en algunos organismos, como los seres humanos, ayuda a mantener la temperatura del cuerpo.

En las células de plantas, animales y otros eucariotas, la respiración celular se realiza en las mitocondrias. El proceso de la respiración celular se resume en la siguiente ecuación:

$$C_6H_{12}O_6 + 6O_2 \longrightarrow 6CO_2 + 6H_2O + \text{energía (ATP)}$$

Glucosa Oxígeno Dióxido de Carbono Agua

¿Se parece a la ecuación de la fotosíntesis? Observa el diagrama de la siguiente página para conocer la relación entre la fotosíntesis y la respiración.

Figura 9 *Las mitocondrias de las células de esta vaca utilizarán la respiración celular para liberar la energía almacenada en el pasto.*

Explora

Haz una lista de cinco alimentos que comiste en las últimas 24 horas y que originalmente eran parte de otro organismo. Escribe los nombres de esos organismos. ¿Cuáles eran productores? ¿Cuáles eran consumidores como nosotros?

a través de las ciencias
CONEXIÓN

¿Crees que la energía producida por las células pueda usarse para generar electricidad? Averígualo en la página 124.

Answers to Explore

Answers will vary. Students should note that plant foods are from producers and animal products are from consumers.

SCIENCE HUMOR

Q: How do cells communicate with each other?

A: by cellular phone

2) Teach

GROUP ACTIVITY

Writing Divide the class into groups of three or four. Have each group write the story of a carbon atom as it is used throughout time. Stories should begin with a molecule of carbon dioxide. What plant uses it for photosynthesis? What animals swallow it and use it to fuel respiration? Have students share their stories if time allows.

LabBook PG 576
Stayin' Alive!

CONNECT TO
EARTH SCIENCE

Conventional solar heating is a much simpler process than photosynthesis. The sun's energy heats either the house itself, or it heats water, which then circulates through the house. If students have ever felt the warm water from a hose that has been left in the sun, they have felt stored solar energy. Use Teaching Transparency 121 to illustrate how solar energy can be used to heat a home.

Teaching Transparency 121
"Solar Heating Systems"

USING THE FIGURE

Refer students to the diagram on this page. Ask students to answer the following questions: What happens to the ATP? Where does the ATP go? How is ATP used by the cell? How is the cell's use of CO_2 and H_2O analogous to people's recycling of paper and glass bottles?

CONNECT TO
PHYSICAL SCIENCE

Scientists have developed new solar cells that simulate photosynthesis more closely than traditional solar cells do. Just as plant cells use energy from the sun to change water and carbon dioxide into energy-rich sugars, these new solar cells use the sun's energy to convert water into energy-rich hydrogen gas, which can be used as fuel. As in plants, the byproduct of this process is clean oxygen.

Teaching Transparency 18
"Photosynthesis and Respiration: What's the Connection?"

Reinforcement Worksheet 5
"Activities of the Cell"

Critical Thinking Worksheet 5
"A Celluloid Thriller"

Fotosíntesis y respiración: ¿Qué relación hay?

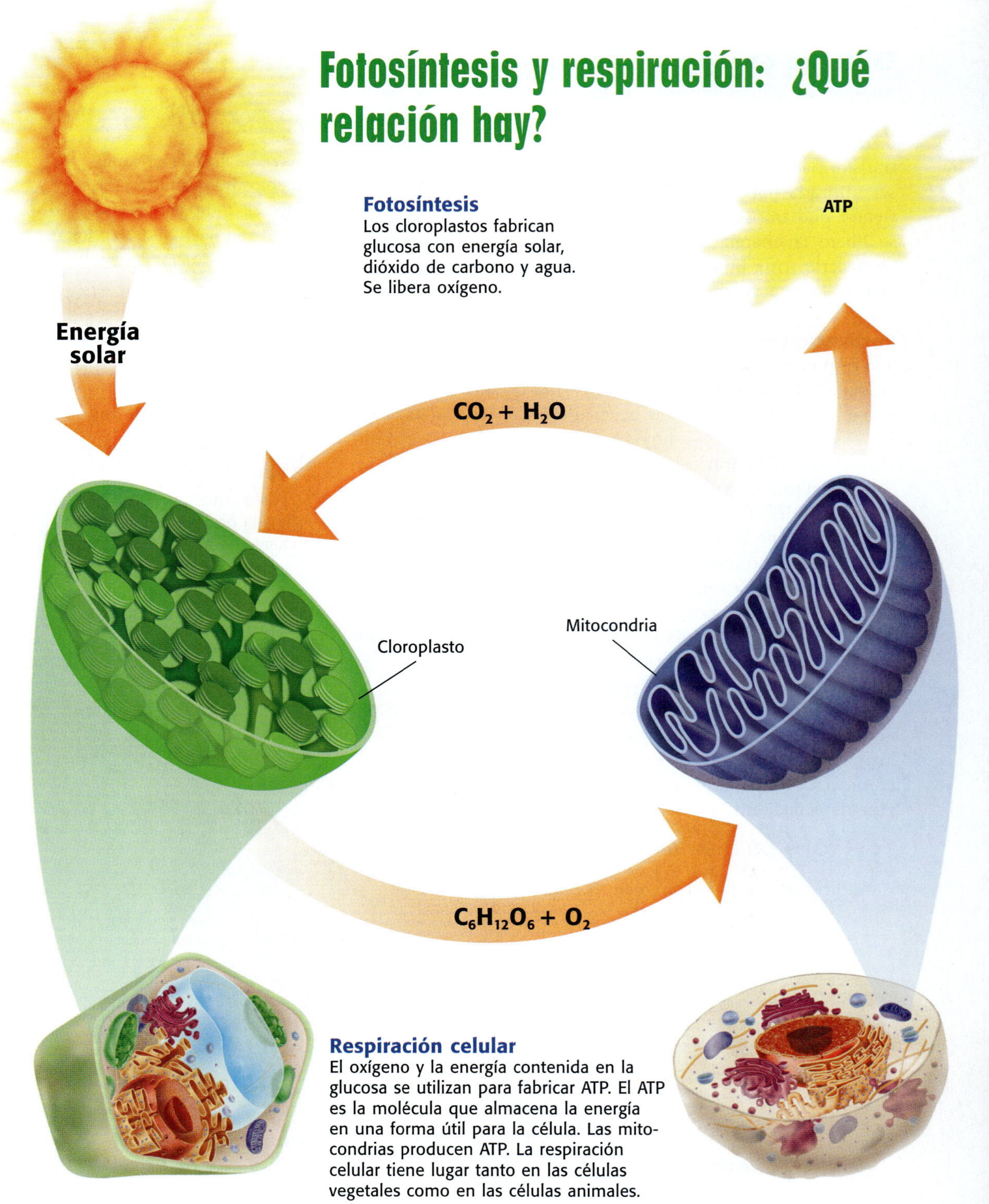

Fotosíntesis
Los cloroplastos fabrican glucosa con energía solar, dióxido de carbono y agua. Se libera oxígeno.

Respiración celular
El oxígeno y la energía contenida en la glucosa se utilizan para fabricar ATP. El ATP es la molécula que almacena la energía en una forma útil para la célula. Las mitocondrias producen ATP. La respiración celular tiene lugar tanto en las células vegetales como en las células animales.

114

Homework

Comparing Cell Processes Have students compare and contrast photosynthesis and respiration. Ask students to use diagrams to display their comparisons. Encourage students to share their diagrams with their classmates.

Se te ha asignado la misión de restaurar la vida en una isla estéril. ¿Qué tipos de organismos pondrías en la isla? Si decides poner animales, ¿qué otros organismos debe haber? Explica tu respuesta.

Fermentación ¿Alguna vez has corrido tanto que tienes la sensación de que tus músculos están ardiendo? A veces las células de los músculos no pueden obtener el oxígeno que necesitan para producir ATP mediante la respiración celular; cuando esto sucede utilizan el proceso de fermentación. El resultado de la fermentación es la producción de una pequeña cantidad de ATP y de productos provenientes de la transformación parcial de la glucosa.

Hay dos tipos principales de fermentación, los cuales se describen en las **Figuras 10** y **11.** El primero ocurre en los músculos y produce ácido láctico, que contribuye al cansancio muscular después de una actividad agotadora; también se realiza en las células musculares de otros animales y de algunas clases de hongos y bacterias. El segundo tipo de fermentación se presenta en ciertos tipos de bacterias y en las levaduras.

Figura 10 *Cuando no hay oxígeno, las células musculares usan la fermentación para fabricar ATP a partir de azúcar. También se produce ácido láctico, que hace que los músculos "ardan" durante el ejercicio.*

Figura 11 *Las levaduras fabrican dióxido de carbono y alcohol al fermentar del azúcar. El dióxido de carbono hace que el pan se infle.*

ciencias de la Tierra

C O N E X I Ó N

Cuando la Tierra era joven, su atmósfera no tenía oxígeno. Las primeras formas de vida obtenían energía mediante la fermentación. Hace aproximadamente 3 billones de años, los organismos empezaron a fotosintetizar y el oxígeno que produjeron se incorporó a la atmósfera.

REPASO

1. ¿Por qué los productores son importantes para la supervivencia del resto de los organismos?

2. ¿Cuál es la relación entre la fotosíntesis y la respiración celular?

3. ¿Qué tiene que ver la respiración normal con la respiración celular?

4. ¿En qué se parecen la respiración y la fermentación? ¿En qué se diferencian?

5. **Identificar relaciones** ¿En qué células esperarías encontrar un mayor número de mitocondrias: en las muy activas o en las que no lo son tanto? ¿Por qué?

115

---

3 Extend

GOING FURTHER

Tell students that plants are often referred to as the "lungs of the Earth." Ask students to reflect on this idea and to prepare a presentation for the class. Suggest that students research rain forests as an example of Earth's "lungs" and explain their contribution to the health of the planet.

Answer to APPLY

Answers will vary. Plants must be on the island in order to provide a source of food and energy for the animals.

4 Close

Quiz

Ask students whether the following statements are true or false.

1. Plants and animals capture their energy from the sun. (false)

2. Cellular respiration describes how a cell breathes. (false)

3. Fermentation produces ATP and lactic acid. (true)

ALTERNATIVE ASSESSMENT

Concept Mapping Have students draw a concept map of energy transfer using the following images:

sunshine; tree, for firewood; sugar cane; yeast consuming sugar, making bread rise; person chopping firewood, for baking oven; person eating bread

Students should note on their map which organisms use photosynthesis, which use respiration, and which use fermentation.
Sheltered English

▼ Answers to Review

1. Producers harness energy in sunlight to produce food (glucose). This food becomes an energy source for producers and for the organisms that consume them.

2. Photosynthesis uses light energy, carbon dioxide, and water to produce glucose and oxygen. Respiration uses the products of photosynthesis to make ATP, carbon dioxide, and water.

3. Breathing supplies the oxygen that cells need for cellular respiration.

4. Both processes release the energy stored in food. Respiration requires oxygen, but fermentation does not.

5. Active cells would have more mitochondria because they have a greater need for energy.

Focus

The Cell Cycle

This section introduces the life cycle of a cell. Students will learn how cells reproduce and the importance of mitosis. Finally, students will learn how cell division differs in plants and animals.

 **Bellringer**

On the board or an overhead projector, write the following:

> Biology is the only science in which multiplication means the same thing as division.

Have students explain this sentence in their ScienceLog. (When cells divide, they are multiplying.)

1) Motivate

ACTIVITY

Making Models Have pairs of students use string for the cell membrane and pieces of pipe cleaners for chromosomes to demonstrate the basic steps of mitosis, as described on page 117.
Sheltered English

Answer to MATHBREAK

After 24 hours, 16 cells will have formed from Cell *A*, and 8 cells will have formed from Cell *B*. Cell *A* will have formed 8 more cells than Cell *B*.

 **Directed Reading Worksheet 5** Section 3

VOCABULARIO

ciclo celular	centrómero
cromosoma	cromátidas
fisión binaria	mitosis
homólogos cromosomas	citoquinesis

OBJETIVOS

- Explica cómo las células producen otras células.
- Comenta la importancia de la mitosis.
- Explica la diferencia entre la división celular de los animales y la de las plantas.

$\div\ 5\ \div\ \Omega\ \leq\ \infty\ +\Omega\ \sqrt{}\ 9\ \infty\ \leq\ \Sigma\ 2$

¡MATEMÁTICAS!

Multiplicación celular

La célula *A* necesita 6 horas para completar su ciclo celular y producir dos células. El ciclo celular de la célula *B* requiere 8 horas. En 24 horas, ¿cuántas células más producirá la célula A en comparación con la B?

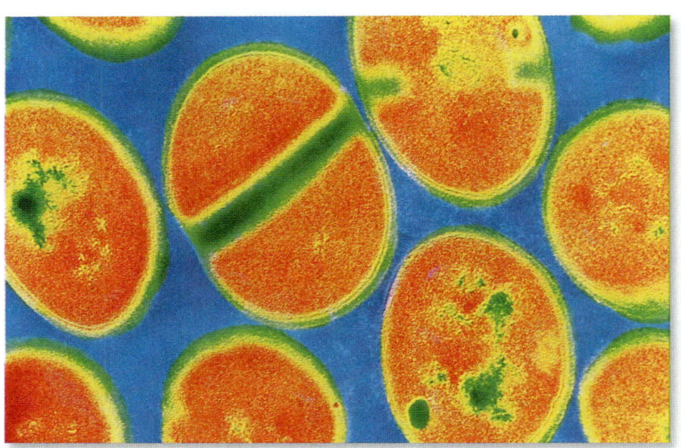

El ciclo celular

Mientras lees esta oración, ¡tu cuerpo produce millones de células nuevas! ¿Por qué necesita producir tantas células? La producción de células nuevas permite a todos los organismos pluricelulares crecer y reemplazar las células que han muerto. Por ejemplo, el entorno del estómago es tán ácido que las células que lo recubren tienen que ser reemplazadas cada semana.

La vida de la célula

De niño a adulto pasas por distintas etapas. De la misma manera, las células pasan por distintas etapas a lo largo de su ciclo de vida. El ciclo de vida de la célula se llama **ciclo celular.**

El ciclo celular comienza cuando la célula se forma y termina cuando se divide para dar origen a dos nuevas células. Antes de dividirse, la célula hace una copia de su ADN y de otras substancias que se requieren para realizar los procesos vitales. Como vimos, el ADN contiene la información que le dice a la célula cómo fabricar proteínas. El ADN de las células está organizado en estructuras llamadas **cromosomas;** en algunos organismos, los cromosomas también contienen proteínas. La copia de los cromosomas garantiza que cada célula nueva tenga todas las herramientas necesarias para sobrevivir.

¿Cómo se producen más células? Esto depende de si la célula es procariota o eucariota.

Cómo se producen las células procariotas Como vimos en capítulos anteriores, las células procariotas (bacterias) no son muy complejas; tampoco lo es su ADN. Tienen ribosomas y un único cromosoma, que es circular, pero no tienen organelos con membrana. Gracias a esto, la división de las bacterias es bastante sencilla y se conoce como **fisión binaria,** que significa "separarse en dos partes". Cada célula nueva contiene una copia del ADN. Algunas de las bacterias de la **Figura 12** están realizando el proceso de fisión binaria.

Figura 12 Las bacterias se reproducen dividiéndose en dos.

 **IS THAT A FACT!**

Before sophisticated microscopes were available, scientists could not see cells pinching and dividing. Many believed that cells came into existence spontaneously—as though crystallizing out of bodily fluids.

Las células eucariotas y su ADN   Las células eucariotas generalmente son mucho más
grandes y complejas que las procariotas. Por
esta razón, tienen mucho más ADN. Los cromosomas de los eucariotas contienen ADN y
proteínas.

El número de cromosomas varía de un
organismo a otro y no tiene nada que ver con
la complejidad del organismo. Por ejemplo, las
moscas de la fruta tienen ocho cromosomas,
las papas 48 y los seres humanos 46. En la
Figura 13 aparecen los 46 cromosomas de una
célula del cuerpo humano, alineados en pares.
Los cromosomas de cada par contienen información similar y se conocen como **cromosomas homólogos.**

Cómo se producen las células eucariotas

El ciclo celular de las eucariotas comprende tres
etapas principales. Durante la primera etapa, la
célula crece y copia sus organelos y cromosomas; las cadenas de ADN y proteínas se parecen a un trozo de cuerda que no está retorcida.
Después de la duplicación de cada cromosoma,
las dos copias se mantienen pegadas en una
región llamada **centrómero**, y cada copia se
llama **cromátida**. Como se muestra en la **Figura
14,** cada cromátida se enrolla y se condensa en
forma de X. A continuación, la célula entra en
la segunda etapa del ciclo celular.

En la segunda etapa, conocida como **mitosis,** las cromátidas se separan. Este complicado
proceso garantiza que cada célula nueva reciba
una copia de cada cromosoma. La mitosis se
divide en cuatro fases, que se muestran en las
siguientes páginas.

En la tercera etapa del ciclo celular, la célula
se divide y produce dos células idénticas a la
célula original. Comentaremos el proceso de la
división celular después de describir la mitosis.

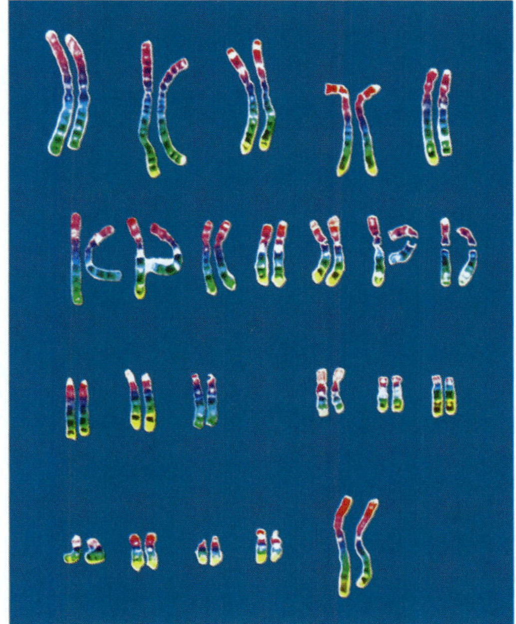

Figura 13 *Las células humanas tienen 46
cromosomas, es decir, 23 pares de cromosomas
homólogos.*

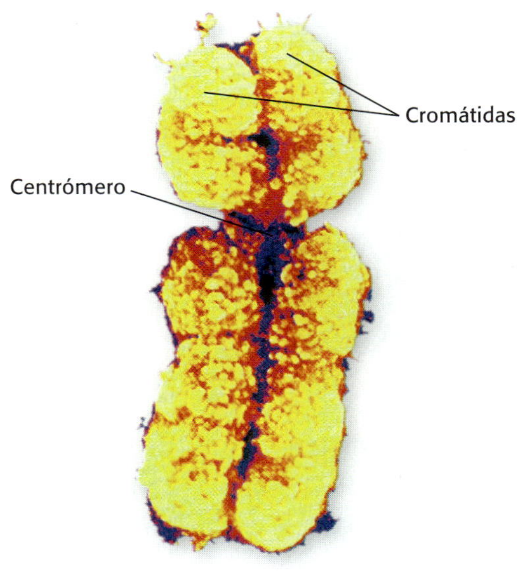

Figura 14 *Dos cadenas de ADN y proteínas se
enrollaron para formar este cromosoma duplicado, que está formado por dos cromátidas.*

✔ Autoevaluación

Tras la duplicación, ¿cuántas cromátidas hay en un par de cromosomas
homólogos? *(Consulta la página 636
para comprobar tu respuesta.)*

A fern called *Ophioglossum reticulatum*
has 1,260 chromosomes per cell, more
than any other organism.

---

② Teach

READING STRATEGY

Prediction Guide Have students guess whether the following statement is true or false
before reading this page:

> The number of chromosomes
> in eukaryotic cells differs
> because some organisms are
> more complex than others.
> (false)

Have students evaluate their
answer after they read the page.

MATH and MORE

Have students make a bar
graph to represent the
following information:

Cell Type A divides every 3
hours; Cell Type B divides
every 6 hours; and Cell Type
C divides every 8 hours. If
you start with one cell of each
type, how many A, B, and C
cells will be produced in a
24-hour period? On a graph,
compare the number of cells
produced by each type. (Cell A,
256; Cell B, 16; Cell C, 8)

 **Math Skills Worksheet 24**
"Creating Exponents"

Answer to Self-Check

After duplication, there are four
chromatids—two from each of the
homologous chromosomes.

ACTIVITY

Making Models To help students visualize cells, have them work in small groups to make three-dimensional models of plant and animal cells. Have students use different colors of clay to represent the nucleus, the chloroplasts, the mitochondria, and the cytoplasm. Students can use aluminum foil for cell walls and can use plastic wrap as cell membranes.

Next have students model mitosis. Tell them to base their model on the illustrations on pages 118 and 119. Students could use curly noodles or packaging "popcorn" to represent the four chromosomes. Sheltered English

MEETING INDIVIDUAL NEEDS

Learners Having Difficulty To reinforce the phases of mitosis, have students prepare a poster illustrating each of the four phases. It should also include brief descriptions for each phase. Have students use their poster to explain mitosis to small groups of students. Sheltered English

Homework

Cells and Diseases Most human diseases can be examined at the cellular level. Viruses take over cellular functions. Bacteria can cause infections and produce toxins that irritate or destroy our cells. Other diseases affect cellular function. Have students research and prepare an oral presentation on a disease that interests them, noting the kinds of cells affected by the disease.

Mitosis y ciclo celular

El diagrama de abajo representa el ciclo celular y las fases de la mitosis de una célula animal. Aunque la mitosis es un proceso continuo, puede dividirse en cuatro fases. Como sabes, cada tipo de organismo tiene un número distinto de cromosomas. En este diagrama sólo se muestran cuatro cromosomas para que sea más fácil ver lo que pasa.

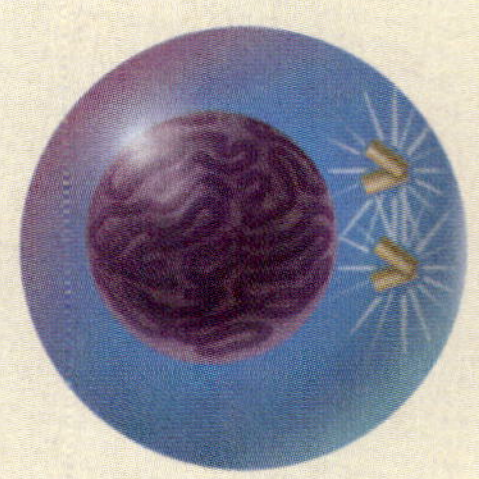

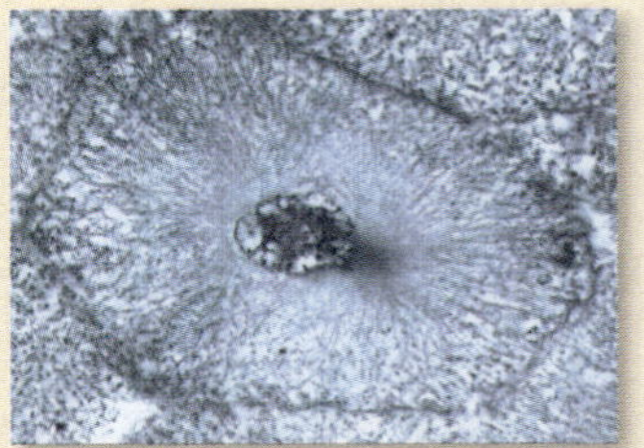

Antes del inicio de la mitosis se copian los cromosomas y otros materiales celulares, como los *centríolos,* que son dos estructuras cilíndricas. Ahora cada cromosoma está formado por dos cromátidas.

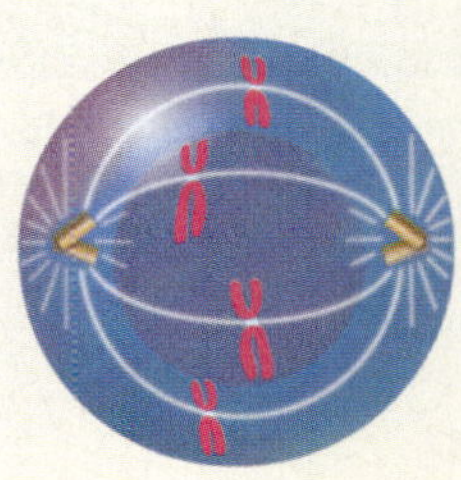

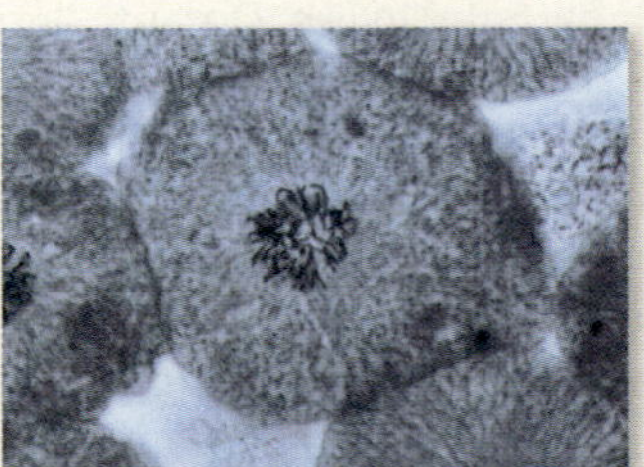

Fase 1 de la mitosis

La mitosis comienza. La membrana nuclear se rompe y los cromosomas se condensan para formar estructuras que parecen bastoncitos. Los dos pares de centríolos se mueven hacia lados opuestos de la célula. Entre ellos se forman fibras que están sujetas a los centrómeros.

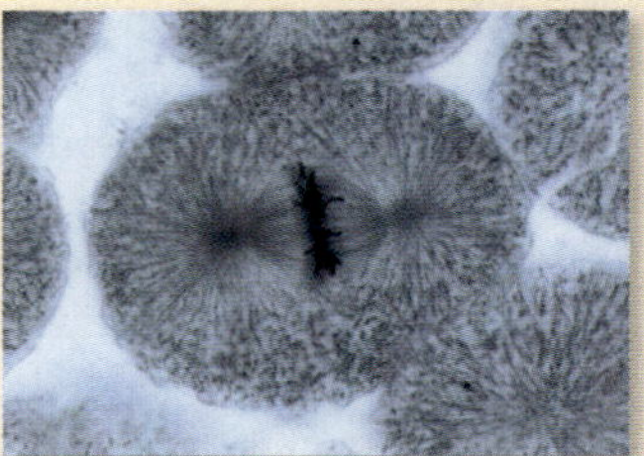

Fase 2 de la mitosis

Los cromosomas se alinean a lo largo de la parte central de la célula.

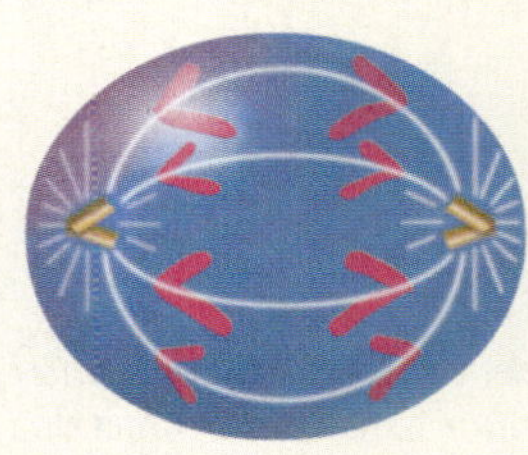

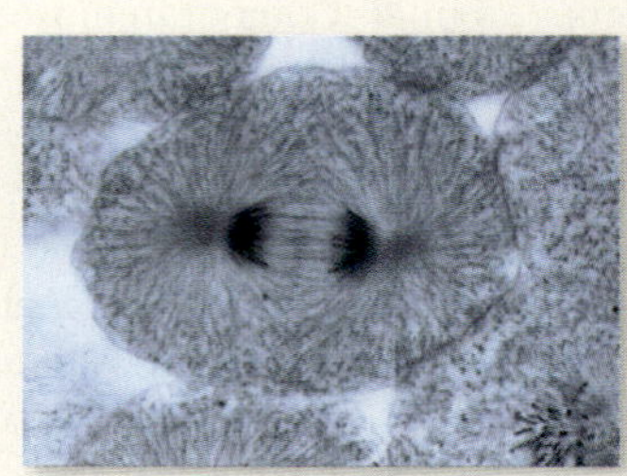

Fase 3 de la mitosis

Las cromátidas se separan y las fibras que están sujetas a los centríolos las jalan hacia lados opuestos de la célula.

118

Science Skills Worksheet 24 "Using Models to Communicate"

IS THAT A FACT!

About 1 trillion mitoses occur in an adult human every 24 hours.

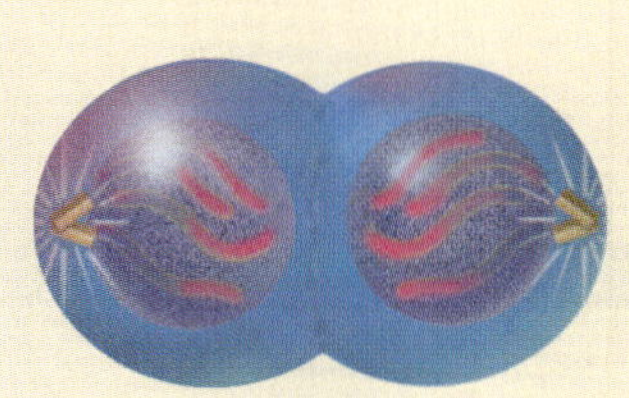

Fase 4 de la mitosis

La membrana nuclear se forma alrededor de ambos conjuntos de cromosomas y éstos se desenrollan. Las fibras desaparecen y así termina la mitosis.

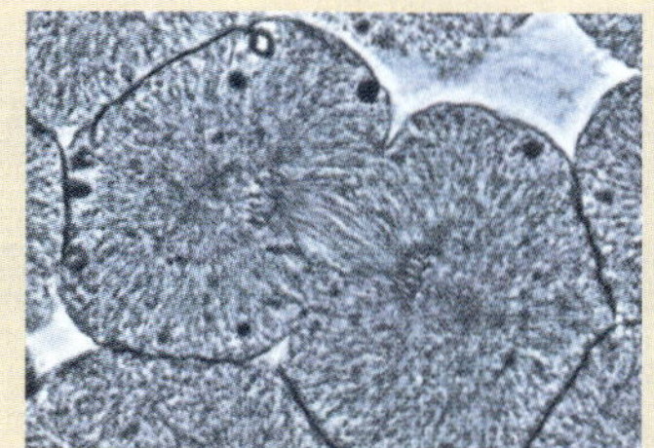

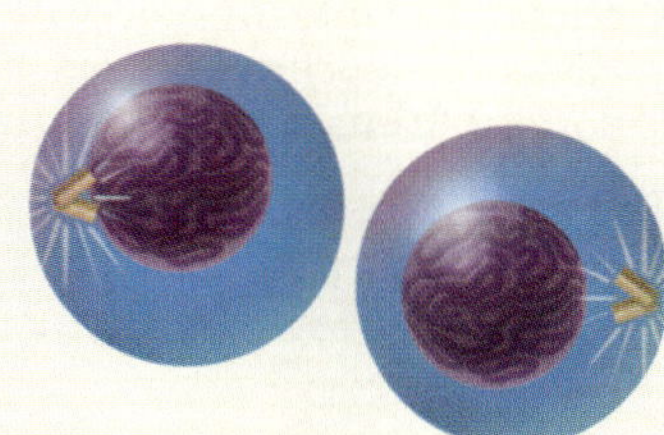

Una vez que la mitosis ha terminado, el citoplasma se divide en dos, en un proceso llamado **citoquinesis.** El resultado son dos células idénticas que también son idénticas a la célula original de la cual se formaron. Después de la citoquinesis, el ciclo celular se completa y las nuevas células están al inicio de un nuevo ciclo celular.

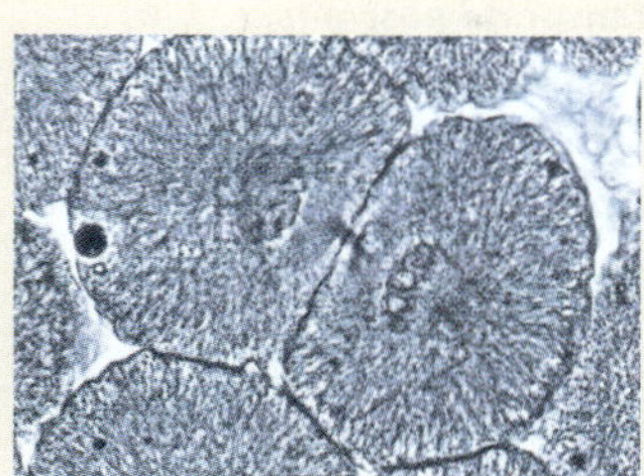

Más acerca de la citoquinesis En las células animales y en otros eucariotas sin pared celular, la membrana celular empieza a hundirse y forma una hendidura que atraviesa toda la célula; así se forman dos células hijas. Arriba se muestra la citoquinesis de una célula animal.

Las células eucariotas con pared celular, como las células vegetales, las algas y los hongos, hacen las cosas de una manera diferente. En estos organismos se forma una *placa celular* en mitad de la célula, que formará las nuevas membranas celulares que separarán a las dos células nuevas. Después de separarse, entre las dos membranas se forma una nueva pared celular. En la **Figura 15** se muestra la citoquinesis de una célula vegetal.

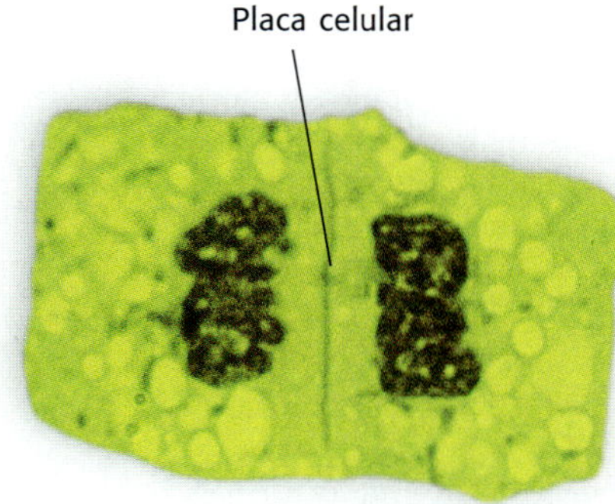

Figura 15 *Cuando las células vegetales se dividen, en la mitad se forma una placa celular que crece hacia el borde hasta que la célula se divide en dos.*

REPASO

1. ¿En qué se parecen la fisión binaria y la mitosis? ¿En qué se diferencian?

2. ¿Por qué es importante que los cromosomas se copien antes de la división celular?

3. ¿Cuál es la diferencia entre la citoquinesis de los animales y la de las plantas?

4. **Aplicar conceptos** ¿Qué pasaría si hubiera citoquinesis sin mitosis?

119

VOCABULARY DEFINITIONS

SECTION 1

diffusion the movement of particles from an area where their concentration is high to an area where their concentration is low

osmosis the diffusion of water across a cell membrane

passive transport the diffusion of particles through proteins in the cell membrane from areas where the concentration of particles is high to areas where the concentration of particles is low

active transport the movement of particles through proteins in the cell membrane against the direction of diffusion; requires cells to use energy

endocytosis the process in which a cell membrane surrounds a particle and encloses it in a vesicle to bring it into the cell

exocytosis the process used to remove large particles from a cell; during exocytosis, a vesicle containing the particles fuses with the cell membrane

SECTION 2

photosynthesis the process by which plants capture light energy from the sun and convert it into sugar

cellular respiration the process of producing ATP from oxygen and glucose; releases carbon dioxide as a waste product

fermentation the breakdown of sugars to make ATP in the absence of oxygen

Resumen del capítulo

SECCIÓN 1

Vocabulario

difusión (*pág. 108*)
ósmosis (*pág. 109*)
transporte pasivo (*pág. 110*)
transporte activo (*pág. 110*)
endocitosis (*pág. 111*)
exocitosis (*pág. 111*)

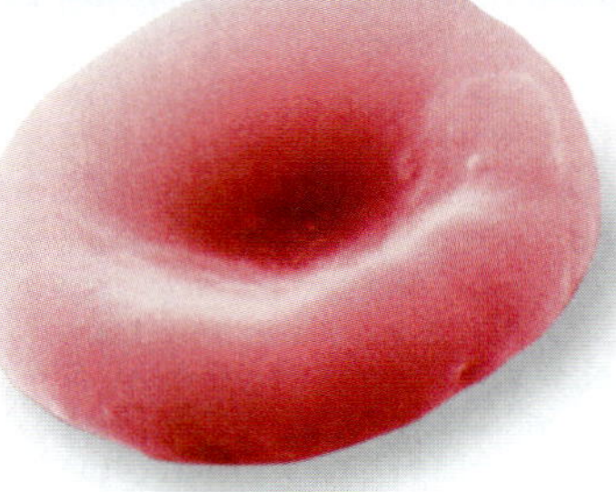

Notas de la sección

• La célula sólo puede sobrevivir si las moléculas de alimento llegan a su interior y se eliminan los desechos. Las substancias entran y salen de la célula a través de la membrana celular, la cual permite el paso de unas e impide el de otras.

• Las células no utilizan energía para mover partículas de un área de alta concentración a otra de baja concentración. A este tipo de movimiento se le llama difusión.

• La ósmosis es la difusión del agua a través de la membrana.

• Algunas substancias entran y salen de la célula pasando a través de proteínas. En el transporte pasivo las substancias se difunden a través de proteínas y en el transporte activo se mueven de un área de baja concentración a una de alta concentración. La célula debe proporcionar energía para que el transporte activo ocurra.

• Las partículas que son demasiado grandes para pasar con facilidad a través de la membrana entran a la célula mediante un proceso llamado endocitosis y salen por otro llamado exocitosis.

Experimentos

El misterio de las papas perfectas (*pág. 574*)

☑ Comprobar destrezas

Conceptos de matemáticas

CICLO CELULAR En 4 horas la célula completa su ciclo celular y produce otras dos células. ¿Cuántas células se producen en 12 horas? Primero debes calcular cuántos ciclos celulares habrá en 12 horas:

12 horas/4 horas = 3

Las células se duplican después de cada ciclo:

Ciclo 1   1 célula $\times$ 2 = 2 células
Ciclo 2   2 células $\times$ 2 = 4 células
Ciclo 3   4 células $\times$ 2 = 8 células

Por lo tanto, después de 3 ciclos celulares (12 horas) habrá 8 células.

Comprensión visual

MITOSIS La mitosis parece un proceso confuso, pero con la ayuda de estas ilustraciones podrás entenderla. En las páginas 118 y 119, observa las ilustraciones y fotografías de las fases del ciclo celular. Lee el rótulo de cada fase e identifica las estructuras celulares que describen. Dibuja el movimiento de los cromosomas en cada paso. Si estudias cuidadosamente los rótulos y las fotografías entenderás mejor la mitosis.

120

Lab and Activity Highlights

The Perfect Taters Mystery **PG 574**

Stayin' Alive! **PG 576**

Datasheets for LabBook
(blackline masters for these labs)

SECCIÓN 2

Vocabulario

fotosíntesis *(pág. 112)*
respiración celular *(pág. 113)*
fermentación *(pág. 113)*

Notas de la sección

• El Sol es la fuente de casi toda la energía necesaria para las actividades químicas de los organismos. La mayoría de los productores utilizan la energía solar para elaborar alimentos mediante la fotosíntesis. Estos alimentos se vuelven fuentes de energía para los productores y para los consumidores que se comen a los productores.

• Por medio de la respiración celular o fermentación, las células fabrican ATP. La respiración celular requiere oxígeno; la fermentación no.

Experimentos

¡Supervivencia! *(pág. 576)*

SECCIÓN 3

Vocabulario

ciclo celular *(pág. 116)*
cromosoma *(pág. 116)*
fisión binaria *(pág. 116)*
cromosomas homólogos *(pág. 117)*
centrómero *(pág. 117)*
cromátidas *(pág. 117)*
mitosis *(pág. 117)*
citoquinesis *(pág. 119)*

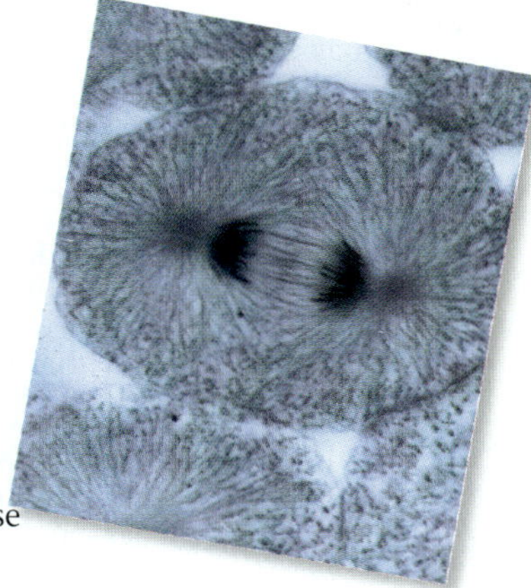

Notas de la sección

• El ciclo de vida de la célula se llama ciclo celular. El ciclo celular comienza cuando la célula se forma y termina cuando ésta se divide para dar origen a dos nuevas células. Las células procariotas dan origen a nuevas células mediante la fisión binaria, en tanto que las células eucariotas lo hacen mediante la mitosis y la citoquinesis.

• Antes de la mitosis, los cromosomas se duplican. Durante la mitosis las cromátidas se separan y se forman dos nuevos núcleos; en la citoquinesis la célula se divide.

SECTION 3

cell cycle the life cycle of a cell; in eukaryotes it consists of chromosome duplication, mitosis, and cytokinesis

chromosome coiled structure of DNA and protein that forms in the cell nucleus during cell division

binary fission the simple cell division in which one cell splits into two; used by bacteria

homologous chromosomes chromosomes with matching information

centromere the region that holds chromatids together when a chromosome is duplicated

chromatids identical copies of a chromosome

mitosis nuclear division in eukaryotic cells in which each cell receives a copy of the original chromosomes

cytokinesis the process in which cytoplasm divides after mitosis

internet

VISITA: go.hrw.com

Visita el sitio web de HRW para encontrar una serie de herramientas de aprendizaje relacionadas con este capítulo. Sólo tienes que escribir la palabra clave:

PALABRA CLAVE: HSTACT

VISITA: www.scilinks.org

Visita el sitio web de la **Asociación Nacional de Maestros de Ciencias** *(National Science Teachers Association)* para encontrar recursos de Internet relacionados con este capítulo. Sólo escribe el **ENLACE DE CIENCIAS** para obtener más información sobre el tema:

TEMA: Energía celular	**ENLACE:** HSTL080
TEMA: Fotosíntesis	**ENLACE:** HSTL085
TEMA: El ciclo celular	**ENLACE:** HSTL090
TEMA: Microbios	**ENLACE:** HSTL095

121

Vocabulary Review Worksheet 5

Blackline masters of these Chapter Highlights can be found in the **Study Guide.**

Lab and Activity Highlights

LabBank

Inquiry Labs, Fish Farms in Space, Lab 2

Whiz-Bang Demonstrations
• It's in the Bag! Demo 5
• Stop Picking on My Enzyme, Demo 4

Labs You Can Eat, The Mystery of the Runny Gelatin, Lab 3

Long-Term Projects & Research Ideas, Project 4

Interactive Explorations CD-ROM

CD 3, Exploration 1, "The Nose Knows"

Chapter Review
Answers

UNDERSTANDING VOCABULARY

1. osmosis
2. exocytosis
3. photosynthesis
4. cellular respiration
5. mitosis, cytokinesis

UNDERSTANDING CONCEPTS

Multiple Choice

6. c
7. c
8. b
9. d
10. a
11. c

Short Answer

12. Cells need chloroplasts for photosynthesis and need mitochondria for respiration.
13. At the beginning of mitosis, each chromosome consists of two chromatids.
14. The first stage is cell growth and copying of DNA (duplication). The second stage is mitosis, and the third stage is cytokinesis (cell division).

Repaso del capítulo

UTILIZAR EL VOCABULARIO

Escoge el término correcto para completar las siguientes oraciones:

1. La difusión del agua a través de la membrana celular se llama __?__. (*ósmosis* o *transporte activo*)

2. La célula elimina las partículas grandes durante la __?__. (*exocitosis* o *endocitosis*)

3. Las plantas fabrican glucosa mediante la __?__. (*respiración celular* o *fotosíntesis*)

4. Durante la __?__, las moléculas de alimento se transforman para formar CO_2 y H_2O, liberando grandes cantidades de energía. (*respiración celular* o *fermentación*)

5. En las eucariotas, se forman dos núcleos en la __?__ y dos células en la __?__. (*citoquinesis* o *mitosis*)

COMPRENDER CONCEPTOS

Opción múltiple

6. Cuando las partículas que se transportan a través de una membrana pasan de un área de baja concentración a otra de alta concentración, el proceso se llama
   a. difusión.
   b. transporte pasivo.
   c. transporte activo.
   d. fermentación.

7. Los organismos con cloroplastos son
   a. consumidores.   c. productores.
   b. procariotas.   d. centrómeros.

8. ¿Qué produce la mitosis?
   a. dos células idénticas
   b. dos núcleos
   c. cloroplastos
   a. dos células diferentes

9. Antes de que la célula pueda utilizar la energía de los alimentos debe transferirla a moléculas de
   a. proteínas.
   b. carbohidratos.
   c. ADN
   d. ATP

10. ¿Cuál de las siguientes células no lleva a cabo la mitosis?
    a. célula procariota
    b. célula del cuerpo humano
    c. célula eucariota
    d. célula vegetal

11. ¿Cuál de las siguientes células forma una placa celular durante el ciclo celular?
    a. célula humana
    b. célula procariota
    c. célula vegetal
    d. todas las anteriores

Respuesta breve

12. ¿Qué estructuras celulares participan en la fotosíntesis? ¿Y en la respiración?

13. ¿Cuántas cromátidas hay en un cromosoma al comienzo de la mitosis?

14. ¿Cuáles son las tres etapas del ciclo celular de una célula eucariota?

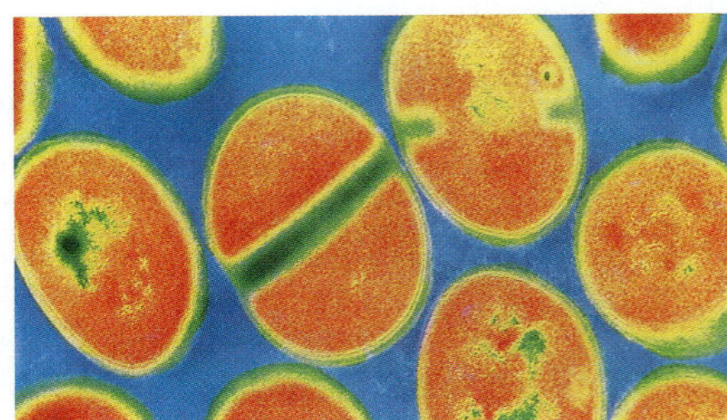

122   Chapter 5 • The Cell in Action

Organizar conceptos

15. Usa los siguientes términos para crear un mapa de ideas: duplicación de cromosomas, citoquinesis, procariota, mitosis, ciclo celular, fisión binaria, eucariota.

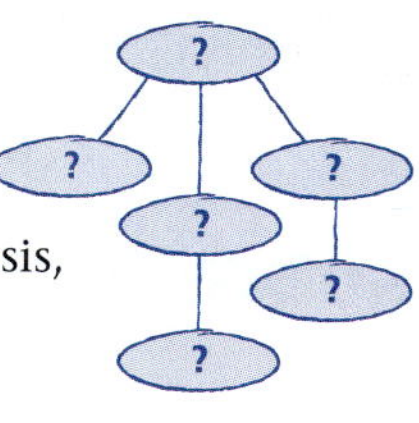

RAZONAMIENTO CRÍTICO Y RESOLUCIÓN DE PROBLEMAS

Responde a las siguientes preguntas con una o dos oraciones:

16. De las siguientes plantas, ¿a cuál se le dió agua con sal y a cuál agua pura? ¿Cómo lo sabes? Asegúrate de emplear la palabra *ósmosis* en tu respuesta.

17. ¿Por qué las células de los músculos necesitan más alimento cuando falta oxígeno que cuando lo hay en abundancia?

18. Si una célula madre tiene 10 cromosomas antes de dividirse,
    a. ¿entrará en el proceso de fisión binaria o en el de mitosis y citoquinesis para producir nuevas células?
    b. ¿cuántos cromosomas tendrá cada célula nueva al final de la división?

LAS MATEMÁTICAS EN LAS CIENCIAS

19. Si una célula tiene seis cromosomas al inicio de su ciclo celular, ¿cuántas cromátidas estarán alineadas en la parte central de la célula durante la mitosis?

INTERPRETAR GRÁFICAS

Observa la célula de abajo para responder a las siguientes preguntas:

20. ¿Es procariota o eucariota?

21. ¿En qué etapa del ciclo celular se encuentra?

22. ¿Cuántas cromátidas están presentes? ¿Cuántos pares de cromosomas homólogos hay?

23. ¿Después de dividirse, ¿cuántos cromosomas tendrá cada célula nueva?

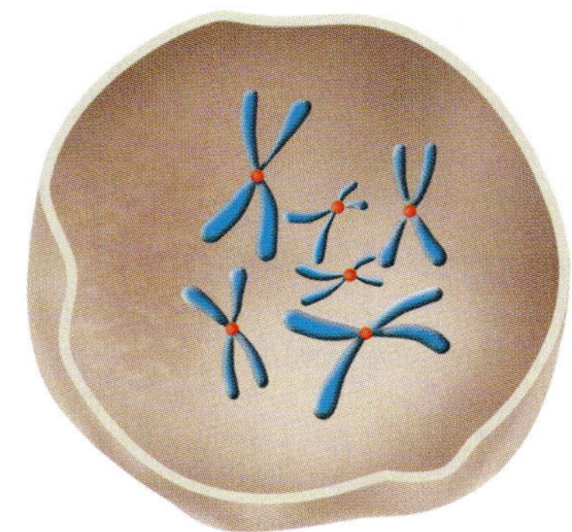

AHORA, ¿qué piensas?

Revisa tus respuestas a las preguntas de la página 107 que escribiste en el cuaderno de ciencias. ¿Han cambiado tus respuestas? Si es necesario, corrige tus respuestas basándote en lo que has aprendido en este capítulo.

123

Concept Mapping

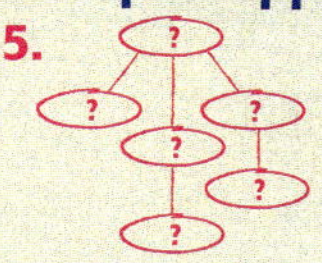

15. An answer to this exercise can be found at the end of this book.

CRITICAL THINKING AND PROBLEM SOLVING

16. The wilted plant on the right was given salt water. Osmosis occurred, and water in the plant moved into the soil, where the concentration of water was lower.

17. When there is plenty of oxygen, the cells can get energy from cellular respiration. When there is a lack of oxygen, the cell must use fermentation, which doesn't produce as much energy. For fermentation to produce more energy, more food would be required.

18. a. The cell is a eukaryotic cell and will go through mitosis and cytokinesis. Prokaryotic cells have only one chromosome.
    b. Each new cell will receive a copy of each chromosome, so each will have 10 chromosomes.

MATH IN SCIENCE

19. Each chromosome will duplicate before mitosis, so $6 \times 2 = 12$ chromatids.

INTERPRETING GRAPHICS

20. The cell is eukaryotic because it shows sister chromatids linked at centromeres.

21. The cell is in mitosis because the chromosomes have already duplicated.

22. There are 12 chromatids. There are three pairs of homologous chromosomes.

23. There will be six chromosomes in each new cell.

NOW WHAT DO YOU THINK?

1. Answers will vary but should reflect knowledge of diffusion, osmosis, endocytosis, active transport, and exocytosis.
2. Cells use food to make energy through respiration and fermentation.
3. Eukaryotic cells divide using mitosis and cytokinesis. Prokaryotic cells divide by binary fission. Every division doubles the number of cells.

Concept Mapping Transparency 5

Blackline masters of this Chapter Review can be found in the **Study Guide.**

Background

The release of energy from food is called *cellular respiration*. Cellular respiration takes place in two stages, resulting in the storage of energy in ATP *(adenosine triphosphate)* molecules.

In the microbial battery, scientists harvest some of this energy and transfer it into electricity that can be readily used.

One of the benefits of the microbial battery is its ability to make use of waste products. Ask students to consider the effect this might have on the energy demands of nations that have limited access to fossil fuels.

A TRAVÉS DE LAS CIENCIAS

CIENCIAS DE LA TIERRA • CIENCIAS FÍSICAS

Noticias electrizantes sobre los microbios

Tu auto se ha quedado sin gasolina, pero ¡no hay problema! El motor funciona con la electricidad que producen trillones de microorganismos hambrientos. Sólo tienes que meter en el tanque unos cuantos terrones de azúcar y un poco de agua fresca y ¡en marcha! Al devorar su comida, los microbios producen suficiente electricidad para llevarte sano y salvo a casa.

La batería "viviente"

¿Suena exagerado? Para Peter Bennetto y su equipo de científicos del King's College, en Londres, no lo es. Los químicos que trabajan ahí piensan que en un futuro habrá baterías "vivientes" capaces de hacer funcionar desde relojes de pulsera hasta poblados enteros. Los microorganismos pueden convertir alimentos en energía eléctrica utilizable. Una batería de prueba, de menos de 0.5 cm², mantuvo funcionando un reloj digital durante un día.

Liberación de electrones

Desde hace casi cien años, los científicos saben que los seres vivos producen y utilizan cargas eléctricas. Sin embargo, los procesos químicos que dan origen a estas pequeñísimas cargas eléctricas se descubrieron en las últimas décadas. Como parte de sus actividades normales, las células transforman almidones y azúcares; estas reacciones químicas liberan electrones. Los científicos recopilan estos electrones libres de organismos unicelulares, como las bacterias, para producir electricidad.

El Sr. Bennetto y sus colegas hicieron una lista de alimentos para relacionar los carbohidratos (como el azúcar de mesa y la melaza) con los microorganismos que los digieren de manera más eficiente. Según él, las bacterias se subdividen en flojas y eficientes. Las eficientes pueden convertir más del 90 por ciento de su alimento en compuestos que sirven como com-

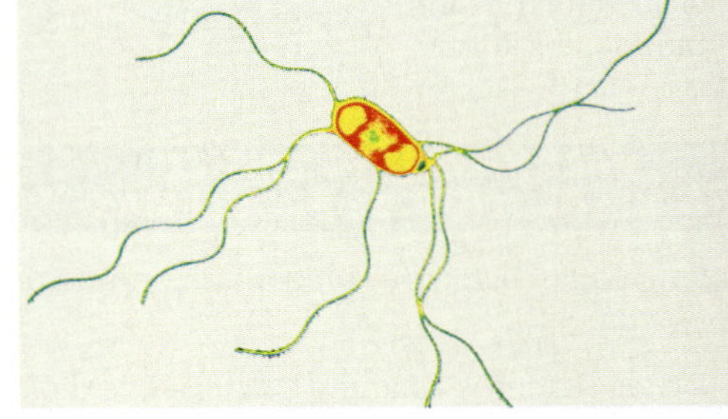

▲ *Las bacterias como éstas convierten los carbohidratos en energía eléctrica.*

bustible para una reacción eléctrica. Las menos eficientes convierten el 50 por ciento.

Se alimentan de sobras

Una de las ventajas de las baterías que funcionan con microbios es que, a diferencia de los generadores, los microbios no requieren recursos no renovables, como carbón o petróleo. Además, si consumen contaminantes (por ejemplo, subproductos de las industrias láctea y azucarera) también producen electricidad. Y, como se reproducen constantemente, no es necesario recargar la batería; sólo hay que cambiar las bacterias de vez en cuando. Para llevar a la práctica esta tecnología se necesitan otros especialistas, entre ellos, ingenieros eléctricos.

Idea de proyecto

▶ Imagínate que diriges una agencia del gobierno y que te piden dinero para una investigación sobre baterías a base de microbios. Piensa en los beneficios de desarrollar "baterías vivientes". ¿Qué problemas se te ocurren? Entre toda la clase decidan si darían dinero para la investigación.

124

Answer to Project Idea

Answers will vary, but students should recognize the two sides to investing in new discoveries. New technologies are risky, but solutions to scientific problems often begin as far-fetched ideas.

Ciencia Ficción

"Contagio"

por Katherine MacLean

A un cuarto de milla de la nave espacial *Explorer,* un equipo de médicos camina cuidadosamente por un camino angosto. Parece que están en un bosque de la Tierra en otoño; hay hojas de color verde, cobre, violeta y rojo. Pero, no es otoñ, y no están en la Tierra.

Minos se parece lo suficiente a la Tierra para ser el hogar de una colonia de seres humanos, también podría ser el hogar de organismos desconocidos capaces de causar graves enfermedades, incluso la muerte, a la tripulación del *Explorer.* Es posible que estas enfermedades sean lo bastante parecidas a las de la Tierra para ser contagiosas, aunque lo bastante diferentes para ser muy difíciles de curar.

Algo grande se mueve entre las sombras; parece un hombre. De repente, se aparece en el camino; es más alto que cualquier miembro de la tripulación, delgado, musculoso, muy bronceado y pelirrojo. Y, es increíble, habla.

"Bienvenidos a Minos, el alcalde les envía saludos desde Alexandria".

Así fue como nosotros y la tripulación del *Explorer* conocimos al pelirrojo Patrick Mead. Según él, una vez hubo una colonia de seres humanos en Minos. Aproximadamente dos años después de la llegada de la colonia, una terrible plaga mató a todo el mundo salvo a la familia Mead. Pero Patrick dice que la plaga nunca ha vuelto y que no hay más microbios contagiosos en Minos.

¿Será cierto? ¿Qué estará ocultando Patrick a la tripulación del Explorer? Si quieres saberlo, lee "Contagio" de Katherine MacLean en la *Antología Holt de Ciencia Ficción.*

125

Further Reading If students enjoyed this story, suggest some of Katherine MacLean's other works, such as the following:

The Missing Man (novella), Bart Books, 1988

The Diploids (short story collection), Gregg Press, 1981

The Man in the Bird Cage (novel), Ace Books, 1971

SCIENCE FICTION
"Contagion"
by Katherine MacLean

When they arrive on the previously unknown planet Minos, the crew of the Earth ship Explorer *immediately admire their friendly, strong, healthy, and handsome host. But could he carry a deadly disease?*

Teaching Strategy

Reading Level This is a relatively long story, containing quite a few medical terms. Students may find it challenging.

Background

About the Author Katherine MacLean's desire to write science fiction comes from an interest in combining her lifelong interests in psychology, biology, and history. In her short story collection, *The Diploids,* she applies the methods of experimentation used in physics and chemistry to anthropology and psychology. Much like "Contagion," many of these stories suggest that scientists have a choice; if they pursue science correctly, their discoveries and insights will change human interactions for the better. In one of her most famous stories, "The Snowball Effect," MacLean warns against the dangers of amateurs delving into science where only experts should venture.

MacLean's stories have appeared in many anthologies and magazines. In addition, MacLean has written several novels. In 1971 MacLean's unique blending of the sciences won her a Nebula Award for her novella *The Missing Man.*

UNIDAD 3

Herencia, evolución y clasificación

Las diferencias y las similitudes entre los seres vivos son el tema de esta unidad. Aquí aprenderás cómo las características se heredan de una generación a otra, cómo los seres vivos están clasificados de acuerdo a sus características, y cómo estas características les ayudan a sobrevivir. La ciencia no siempre ha comprendido estos temas, y aún queda mucho por aprender. Esta cronología te dará una idea de algunas de las cosas que hemos aprendido hasta ahora.

1753

Carlos Linneo publica el primero de dos volúmenes que contienen la clasificación de todas las especies conocidas.

1951

Rosalind Franklin fotografía el ADN.

1953

James Watson y Francis Crick descubren la estructura del ADN.

1960

Louis y Mary Leakey descubren los huesos fósiles del antepasado humano *Homo habilis* en Olduvai Gorge, Tanzania.

1969

El *Apolo 11* aterriza en la Luna. Neil Armstrong se convierte en la primera persona en caminar sobre la superficie lunar.

1859

Carlos Darwin sugiere que la selección natural es un mecanismo de la evolución.

1860

Abraham Lincoln es elegido presidente de los Estados Unidos. Es el dieciseisavo presidente.

1865

Gregorio Mendel publica los resultados de sus investigaciones sobre las leyes de la herencia en plantas de chícharos.

1930

Se descubre el planeta Plutón.

1905

Nettie Stevens describe cómo los cromosomas X y Y determinan el sexo.

1990

Ashanti DeSilva recibe glóbulos blancos altera-dos genéticamente para combatir su enfermedad.

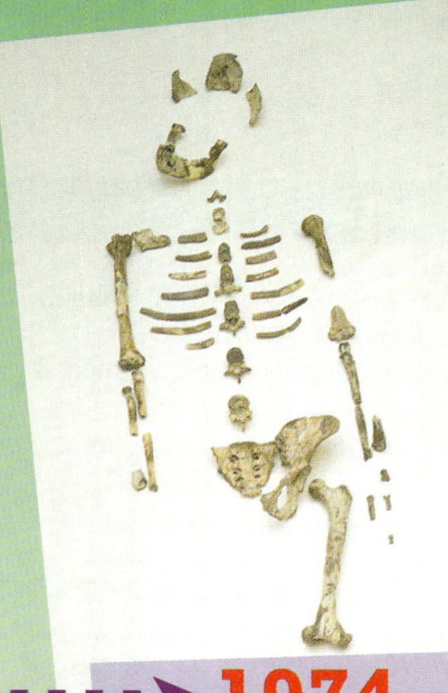

1974

Donald Johanson descubre el esqueleto fosilizado de uno de los primeros homínidos.

2000

El proyecto Genoma Humano ha identificado miles de genes humanos y tiene planeado descifrar el genoma humano completo para el año 2003.

CHAPTER ORGANIZATION	TIME MINUTES	OBJECTIVES	LABS, INVESTIGATIONS, AND DEMONSTRATIONS
Chapter Opener pp. 128–129	45		**Investigate!** Clothing Combos, p. 129
Section 1 Mendel and His Peas	90	▶ Explain the experiments of Gregor Mendel. ▶ Explain how genes and alleles are related to genotypes and phenotypes. ▶ Use the information found in a Punnett square.	**Demonstration,** Flower Dissection, p. 133 in ATE **QuickLab,** Take Your Chances, p. 136 **Making Models,** Bug Builders, Inc., p. 578 **Datasheets for LabBook,** Bug Builders, Inc., Datasheet 12 **Design Your Own,** Tracing Traits, p. 580 **Datasheets for LabBook,** Tracing Traits, Datasheet 13
Section 2 Meiosis	90	▶ Explain the difference between mitosis and meiosis. ▶ Describe how Mendel's ideas are supported by the process of meiosis. ▶ Explain the difference between male and female sex chromosomes.	**Demonstration,** Modeling Meiosis, p. 140 in ATE **QuickLab,** Round or Wrinkled, p. 142 **Long-Term Projects & Research Ideas,** Project 5

TECHNOLOGY RESOURCES

 **Guided Reading Audio CD**
English or Spanish, Chapter 6

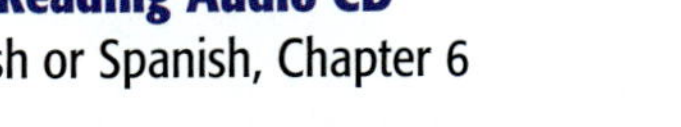 **One-Stop Planner CD-ROM with Test Generator**

 **CNN.** Science, Technology & Society,
BioDiesel, Segment 6
Bioengineered Plants, Segment 8

CLASSROOM WORKSHEETS, TRANSPARENCIES, AND RESOURCES	SCIENCE INTEGRATION AND CONNECTIONS	REVIEW AND ASSESSMENT
Directed Reading Worksheet 6 **Science Puzzlers, Twisters & Teasers,** Worksheet 6		
Directed Reading Worksheet 6, Section 1 **Science Skills Worksheet 18,** Finding Useful Sources **Math Skills for Science Worksheet 16,** What Is a Ratio? **Math Skills for Science Worksheet 40,** Punnett Square Popcorn **Transparency 20,** Punnett Square: *PP* × *pp* Cross **Transparency 20,** Punnett Square: *Pp* × *Pp* Cross **Reinforcement Worksheet 6,** Dimples and DNA **Critical Thinking Worksheet 6,** A Bittersweet Solution	**Multicultural Connection,** p. 132 in ATE **MathBreak,** Understanding Ratios, p. 134 **Math and More,** p. 134 in ATE **Apply,** p. 137 **Chemistry Connection,** p. 137 **Science, Technology, and Society:** Mapping the Human Genome, p. 148	**Homework,** pp. 130, 133, 136 in ATE **Review,** p. 137 **Quiz,** p. 137 in ATE **Alternative Assessment,** p. 137 in ATE
Directed Reading Worksheet 6, Section 2 **Transparency 21,** Meiosis in Eight Easy Steps: A **Transparency 22,** Meiosis in Eight Easy Steps: B **Transparency 23,** Meiosis and Mendel	**Cross-Disciplinary Focus,** p. 140 in ATE **Health Watch:** Lab Rats with Wings, p. 149	**Self-Check,** p. 141 **Homework,** p. 142 in ATE **Review,** p. 143 **Quiz,** p. 143 in ATE **Alternative Assessment,** p. 143 in ATE

Holt, Rinehart and Winston On-line Resources

go.hrw.com

For worksheets and other teaching aids related to this chapter, visit the HRW Web site and type in the keyword: **HSTHER**

National Science Teachers Association

www.scilinks.org

Encourage students to use the *sci*LINKS numbers listed with the Chapter Highlights to access information and resources on the **NSTA** Web site.

END-OF-CHAPTER REVIEW AND ASSESSMENT

Chapter Review in Study Guide
Vocabulary and Notes in Study Guide
Chapter Tests with Performance-Based Assessment, Chapter 6 Test
Chapter Tests with Performance-Based Assessment, Performance-Based Assessment 6
Concept Mapping Transparency 6

Chapter Resources & Worksheets

Visual Resources

TEACHING TRANSPARENCIES

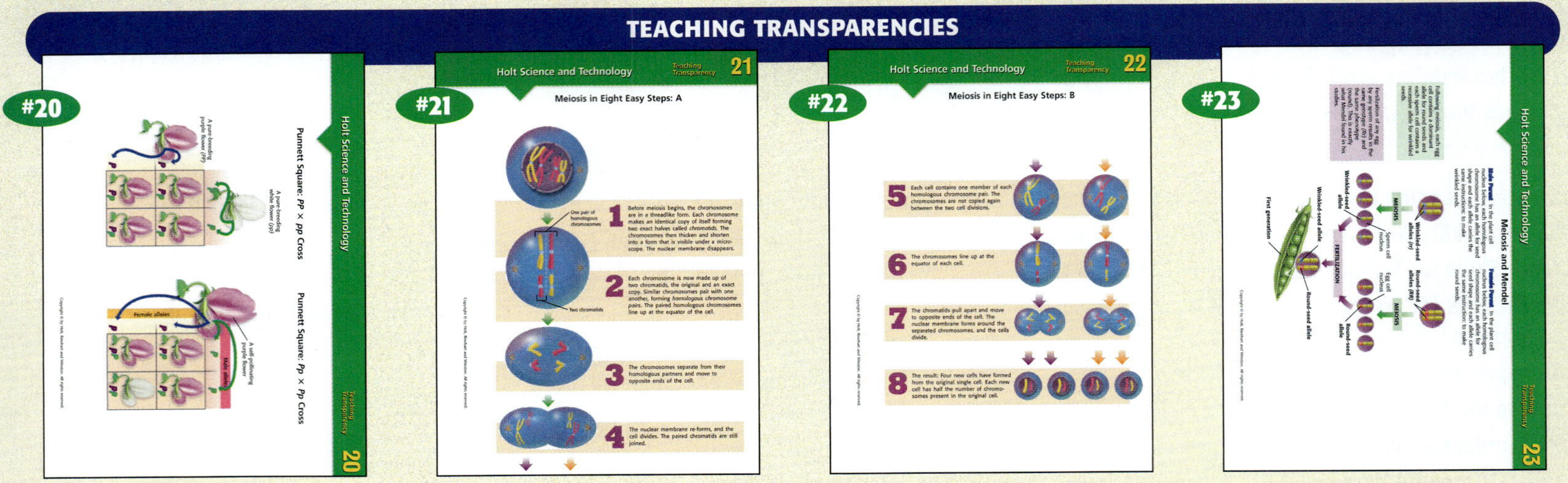

CONCEPT MAPPING TRANSPARENCY

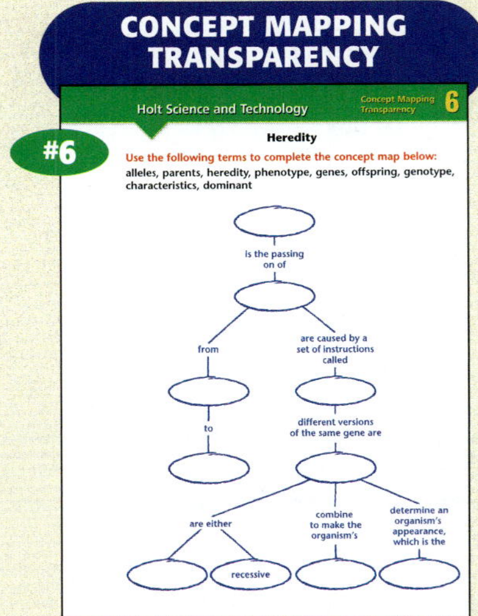

Meeting Individual Needs

DIRECTED READING

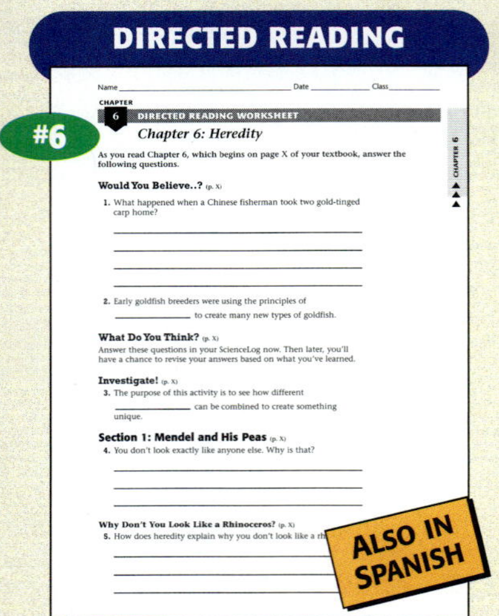

REINFORCEMENT & VOCABULARY REVIEW

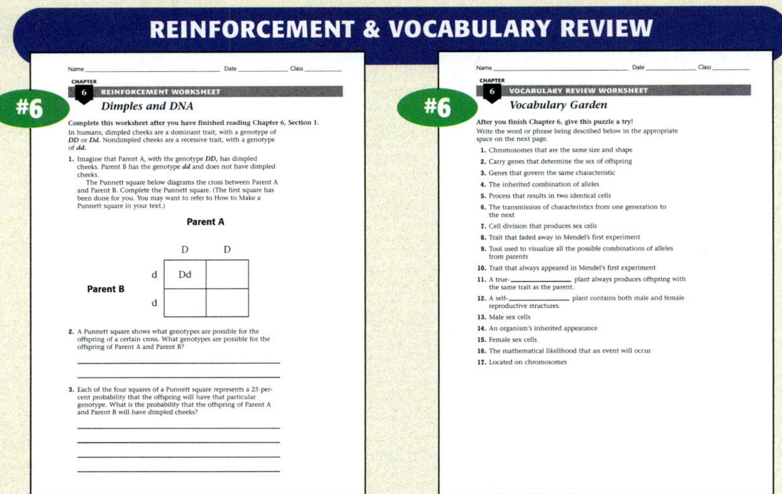

SCIENCE PUZZLERS, TWISTERS & TEASERS

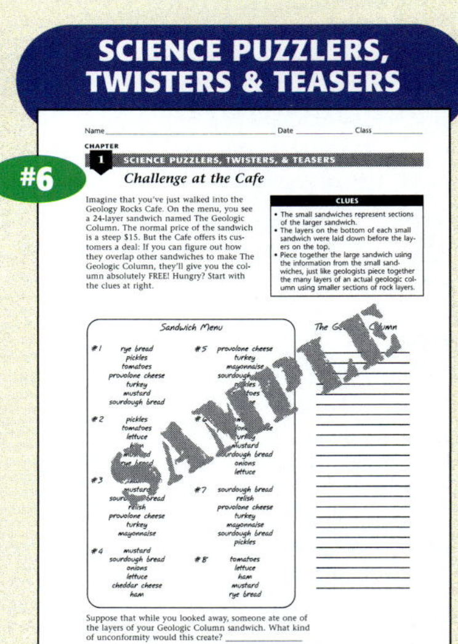

Review & Assessment

STUDY GUIDE

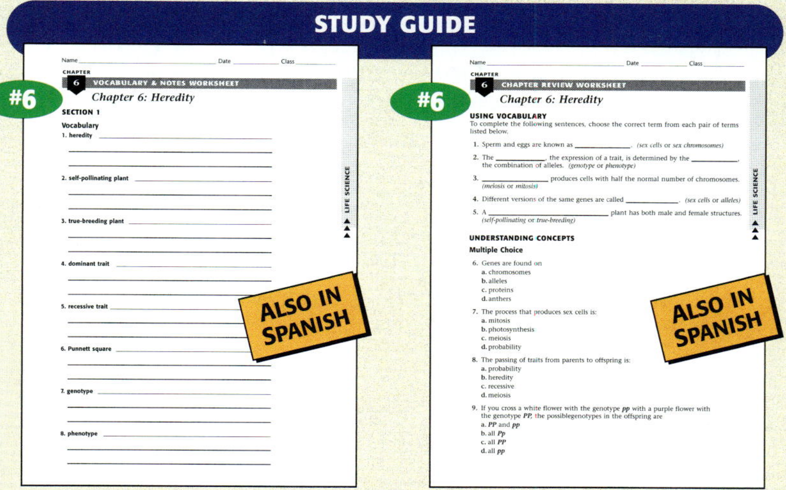

CHAPTER TESTS WITH PERFORMANCE-BASED ASSESSMENT

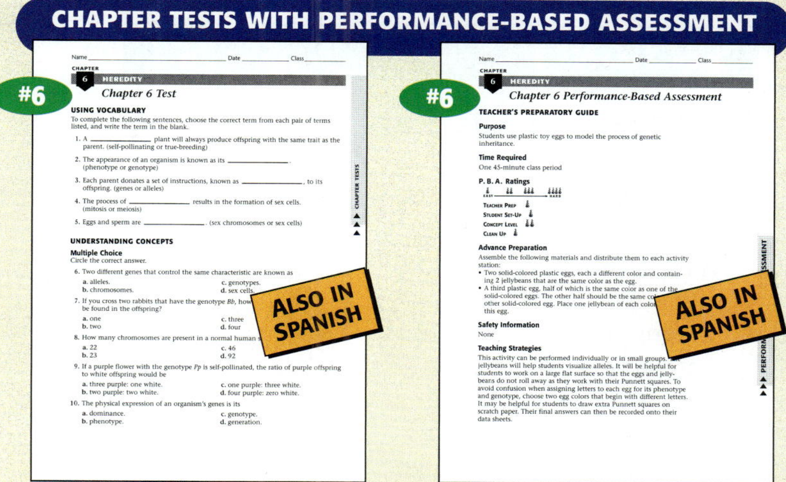

Lab Worksheets

LONG-TERM PROJECTS & RESEARCH IDEAS

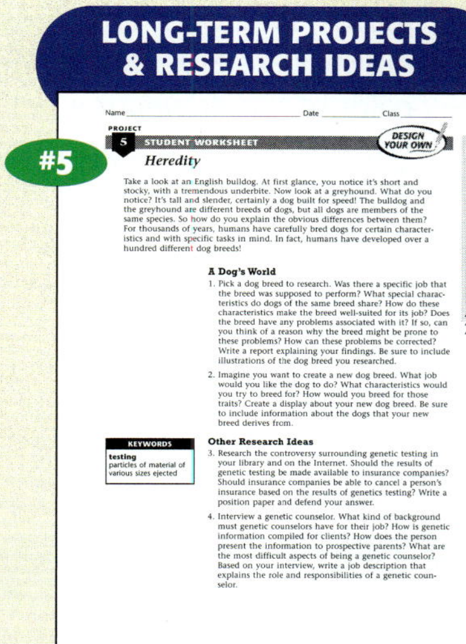

DATASHEETS FOR LABBOOK

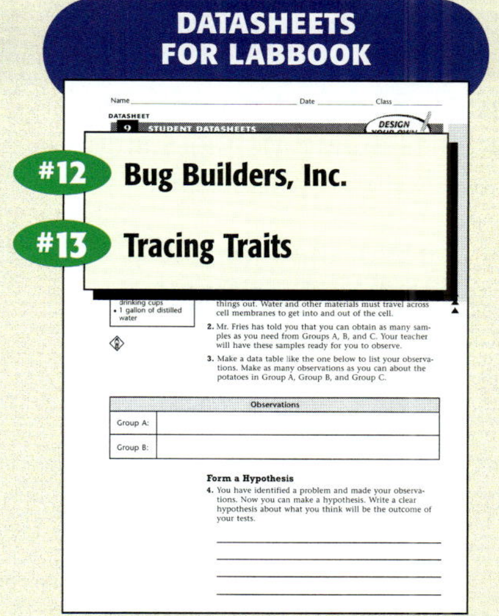

Applications & Extensions

CRITICAL THINKING & PROBLEM SOLVING

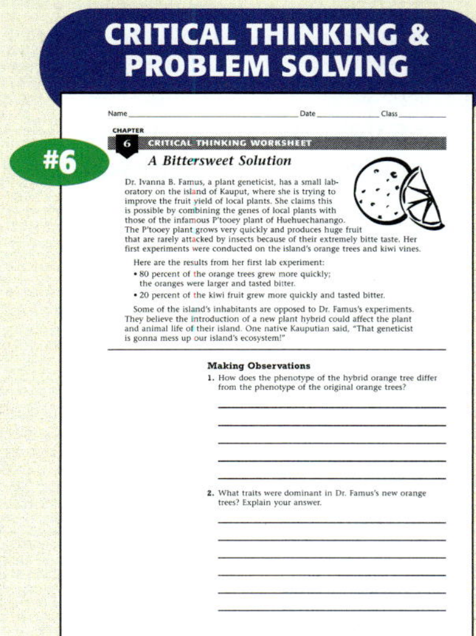

SCIENCE TECHNOLOGY

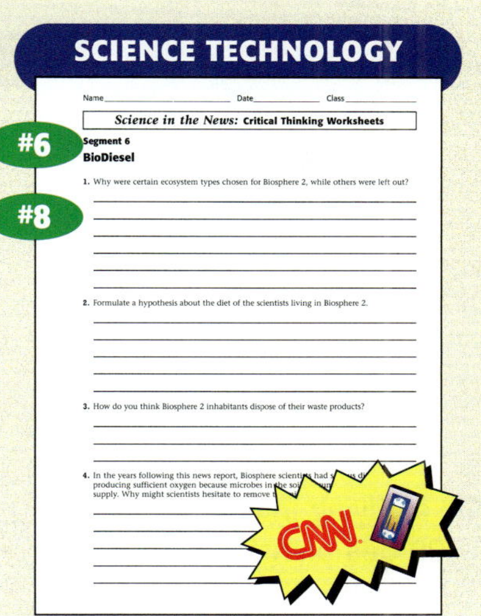

Mendel and His Peas

▶ Gregor Mendel

In 1843, in the city of Brünn, Austria (which is now Brno, a city in the Czech Republic), Gregor Mendel (1822–1884) entered a monastery. In 1865, Mendel published the results of his garden-pea experiments. Although Mendel's ideas are wide-spread today, few scientists took notice of his work during his lifetime. Mendel presented his findings in two lectures, and he had only 40 copies of his work made. Because there were no computers or photocopy machines during Mendel's time, his findings were not distributed to many scientists.

- When Mendel was elected abbot of the monastery in 1868, his duties prevented him from visiting scientists and attending conferences where he could have discussed his results. It was not until 1900, when Mendel's work was rediscovered by scientists in Holland, Germany, and Austria-Hungary, that his theories gained general acceptance in the scientific community.

- Mendel's observations were used to justify Darwin's theory of evolution by natural selection. Mendel's ideas are considered to be the foundation of modern genetics.

IS THAT A FACT!

- ◆ From 1856 to 1863, while studying inheritance, Mendel grew almost 30,000 pea plants!

- ◆ Mendel also made contributions to beekeeping, horticulture, and meteorology. In 1877, Mendel became interested in weather and began issuing weather reports to local farmers.

- ◆ Punnett squares are named after their inventor, R. C. Punnett. Punnett explored inheritance by crossing different breeds of chickens in the early 1900s, soon after Mendel's work was rediscovered.

▶ Pollination

Pollen can be transferred between plants by wind, insects, and a variety of animals. Some common pollinators are bees, butterflies, moths, flies, bats, and birds. Animals are attracted to the color of the flower, the patterns found on the petals, or the flower's fragrance. Pollen is an excellent food for some animals.

Meiosis

▶ Chromosomes

Chromosomes are composed of genes, the sequences of DNA that provide the instructions for making all the proteins in an organism. During cell division, the duplicated chromosomes separate so that one copy of each chromosome is present in the two new cells.

IS THAT A FACT!

- ◆ Male bees have only half the number of chromosomes that female bees have.

▶ Walther Flemming

Walther Flemming (1843–1905), a German physician and anatomist, was the first to use a microscope and special dyes to study cell division. Flemming used the term *mitosis* to describe the process he observed.

▶ Mitosis

In mitosis, a cell divides to form two identical cells. The steps of the process are similar in almost all living organisms. In addition to enabling growth, mitosis allows organisms to replace cells that have died or malfunctioned. Mitosis can take anywhere from a few minutes to a few hours, and it may be affected by characteristics of the environment, such as light and temperature.

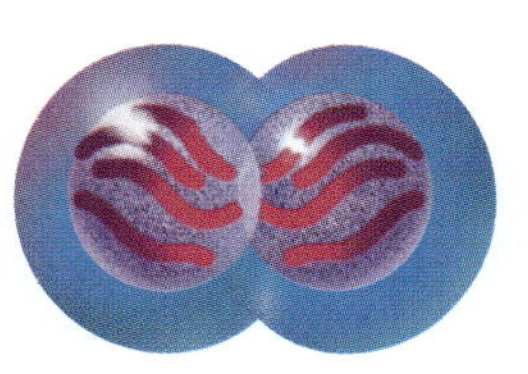
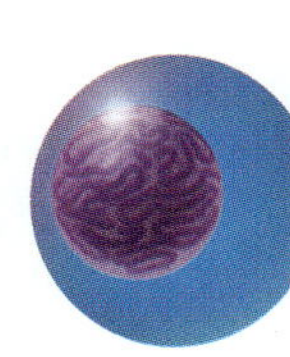
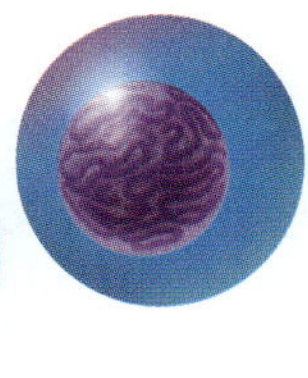

▶ Meiosis

In humans, meiosis is very different in males and females. In males, meiosis results in four similar sperm cells. In females, however, only one functional egg is produced. The other three resulting cells, which are known as *polar bodies,* contain the same amount of genetic material as the functional egg but do not mature.

▶ Genetic Disorders

A genetic disorder results from an inherited disruption in an organism's DNA. These inherited disruptions can take several forms, including a change in the number of chromosomes and the deletion or duplication of entire chromosomes or parts of chromosomes. Often the change responsible for a disorder is the alteration of  a single specific gene. However, some genetic disorders result from several of these genetic alterations occurring simultaneously. Diseases resulting from these alterations cause a wide variety of physical malfunctions and developmental problems.

- Hemophilia is an inherited blood disorder affecting about one person in 10,000. People with hemophilia are genetically unable to produce the blood proteins necessary to form blood clots. As a result, cuts and bruises considered unthreatening to anyone else can be dangerous for hemophiliacs. The clotting factors missing in people with hemophilia are now widely available and can be self-administered.

- Cystic fibrosis (CF) is a disease for which one in 31 Americans carries a recessive trait. If two of these people have children together, there is a 25 percent chance that any child born to them will have the disease. CF affects the intestinal, bronchial, and sweat glands. In people with CF, these glands secrete thick, sticky fluids that are difficult for the body to process, impeding breathing and digestion. Modern medical treatment has extended the median age for people with CF to about 30 years, an enormous improvement in just a few decades.

- Rubinstein-Taybi syndrome (RTS) is a complex genetic disorder whose characteristics include broad thumbs and toes, mental retardation, and distinctive facial features. This wide range of characteristics is believed to be linked to any one of a number of mutations in a gene responsible for providing the body with a protein called CBP. CBP is thought to be vital to the body's delicate metabolism. Because CBP greatly influences body processes, people with a problem producing CBP have a wide range of difficulties. Children with RTS can benefit from proper nutrition and early intervention with therapies and special education.

For background information about teaching strategies and issues, refer to the *Professional Reference for Teachers.*

Chapter Preview

Section 1
Mendel and His Peas
- Why Don't You Look Like a Rhinoceros?
- Who Was Gregor Mendel?
- Unraveling the Mystery
- Peas Be My Podner
- Mendel's First Experiment
- Mendel's Second Experiment
- A Different Point of View
- A Brilliant Idea
- What Are the Chances?

Section 2
Meiosis
- Two Kinds of Reproduction
- Meanwhile, Back at the Lab
- Meiosis in Eight Easy Steps
- Meiosis and Mendel
- Male or Female?

Directed Reading Worksheet 6

Science Puzzlers, Twisters & Teasers Worksheet 6

Guided Reading Audio CD
English or Spanish, Chapter 6

CAPÍTULO
6 Herencia

¿Creerías que...?

Hace mucho tiempo en China, un pescador atrapó en sus redes una carpa diferente a las demás. Estos pececitos de agua dulce son de color pardo, pero esta carpa tenía un tenue color dorado. Era demasiado hermosa para comérsela, así que el pescador se la llevó a casa como mascota.

Meses después, el pescador atrapó otra carpa dorada, y la puso con la primera. Las carpas se reprodujeron, y los nuevos peces eran de colores aún más brillantes que los de sus padres. Así fue como nacieron las primeras carpas doradas.

En los años que siguieron, la gente empezó a criar los nuevos pececitos anaranjados por toda China. Algunos dueños de estas mascotas se convirtieron en alcahuetes consumados, y escogían sólo las mejores parejas para sus peces favoritos. Cada nueva generación producía pececitos cada vez más distintos. Cuando los primeros cargamentos de carpas doradas llegaron a Japón en el año 1500 a.C., los peces ya no parecían carpas. Es más, eran peces tan aristocráticos que se prohibió que la gente común los tuviera como mascotas.

Sin saberlo, los chinos de la antigüedad utilizaron las leyes de la genética para producir nuevas variedades de carpas doradas. En este capítulo estudiarás las leyes de la herencia, los rasgos que pasan de una generación a otra. Descubrirás las leyes que hicieron posible criar hermosos peces dorados a partir de unas carpas comunes.

Would You Believe ...?

For centuries, humans have selectively bred animals for particular characteristics. Dogs, chickens, and flowers have all been selected for useful and ornamental traits. Horses have been selected for speed and strength. Crops and cattle have been selected for hardiness and productivity. Our ability to select traits is possible only because of genetic reliability.

¿Tú qué piensas?

Usa tus conocimientos para responder a las siguientes preguntas en tu cuaderno de ciencas:

1. ¿Por qué no somos todos iguales?
2. ¿Qué determina que un bebé sea niña o niño?

Combina prendas

¿Te pareces a tu mamá o a tu papá? ¿A tu hermana o a tu hermano? Puede ser que te parezcas a ellos, pero seguramente eres muy diferente. Aunque seas diferente, compartes algunas características con tus padres y hermanos.

En esta actividad vas a investigar cómo se combinan las diferentes características para crear una combinación única y diferente, como tú.

Procedimiento

1. Tu maestra te dará **tres cajas.** Una de ellas contiene **cinco sombreros,** otra contiene **cinco guantes,** y la última, **cinco bufandas.**

2. Sin ver lo que hay en las cajas, cinco compañeros sacarán un sombrero, un guante y una bufanda. Se los pondrán y los modelarán frente al resto de la clase. Luego devolverán los objetos a sus cajas para que pasen a otros cinco compañeros, y continuarán así hasta que hayan pasado por todo el salón.

3. Anota la combinación de prendas que escogiste en tu cuaderno de ciencias.

Análisis

4. ¿Hubo combinaciones iguales? ¿Cuántas combinaciones diferentes hubo en tu salón?

5. ¿Crees que se hicieron todas las combinaciones posibles? Explica tu respuesta.

6. Escoge una pareja. Con las prendas que tú y tu compañero o compañera sacaron de la caja, ¿cuántas combinaciones diferentes formarían si le dieran a otra persona un sombrero, un guante y una bufanda? Para contestar a esta pregunta, haz una gráfica como la siguiente.

	Sombrero		Guante		Bufanda	
1	X			X		X
2	X			X			X
3	X						

7. Según lo que has aprendido en esta investigación, ¿por qué crees que los hermanos de una familia son tan diferentes unos de otros?

What Do You Think?

Accept all reasonable responses.

Students will have a chance to revise their answers in the Chapter Review under NOW What Do You Think?

Investigate!

MATERIALS

FOR EACH GROUP:
- box with 5 hats
- box with 5 scarves
- box with 5 gloves

Safety Caution: Infestations of head lice are a common problem in schools. Sharing hats would, of course, be inadvisable during such a period. Jackets or sweatshirts could be substituted for hats in this exercise.

Answers to Investigate!

4. Answers will vary. There should be many different combinations.

5. Answers will vary, but students should know that the number of possible combinations is very large.

6. 8

7. Answers will vary. The number of possible genetic combinations is huge because we have so many genes.

IS THAT A FACT!

In recent decades, poachers have illegally killed thousands of elephants for their ivory tusks. Elephants without tusks are of no interest to poachers and are spared. This means that elephants lacking the genes for tusks are left to reproduce and increase the percentage of the population of elephants that can't bear ivory. In some parks in Africa, nearly 15 percent of baby elephants are tuskless. In the 1930s, only 1 percent of the elephants in these areas were tuskless.

Focus

Mendel and His Peas

This section introduces the genetic experiments of Gregor Mendel. Students explore how flowering plants are fertilized and how the offspring are affected by different crosses. Students also learn to use a Punnett square to predict the results of genetic crosses.

🔔 Bellringer

Pose the following questions to your students:

Some people have brown eyes, some have blue, and some have green. Some people have earlobes attached directly to their head, while others have earlobes that hang loose. Where do people get these different traits? How are they passed from one generation to the next? Write your thoughts in your ScienceLog.

1 Motivate

ACTIVITY

Creating Tables Ask students to notice the differences in eye color, hair color, and earlobes among their classmates. Have them count the number of students with each trait and make a data table for each trait. The tables for eye color and earlobes will each have two columns. Have students calculate the ratios of attached to unattached earlobes and brown to blue eyes. Students may have eyes that are a color other than blue or brown, and this could be noted in a third column. (Note: A class of students is not a scientific sample and may not yield statistically significant results.)

Sección 1

VOCABULARIO

herencia
planta autopolinizante
línea pura
rasgo dominante
rasgo recesivo
genes
alelos
cuadrícula de Punnett
genotipo
fenotipo
probabilidad

OBJETIVOS

- Explica los experimentos de Gregorio Mendel.
- Explica la relación de los genes y los alelos con el genotipo y el fenotipo
- Usa la información en una cuadrícula de Punnett.

Explora

Imagínate que vas al aeropuerto a recoger a un amigo al que nunca has visto en persona. ¿Qué rasgos tuyos le describirías para que te reconociera? ¿Dirías que eres alto o bajo, con pelo rizado o lacio, de ojos verdes o cafés? Haz una lista. De estos rasgos, ¿cuáles crees que heredaste? Pon una marca en los rasgos de tu lista que crees que heredaste.

Tracing Traits

Mendel y los chícharos

Si viajaras alrededor del mundo, comprobarías que no hay nadie que sea igual a ti. Eres único en el mundo. Pero ¿qué te hace diferente de los demás? Si te fijas en tus compañeros, verás que compartes muchas características con ellos. Por ejemplo, todos tienen piel en vez de escamas, pies en vez de pezuñas, y ninguno de ustedes tiene antenas; eres un ser humano, y te pareces mucho a los demás seres humanos, pero al mismo tiempo, eres distinto de los demás. A quienes más te pareces es a tus padres y hermanos, pero seguramente tampoco eres igual a ellos. ¿Sabes por qué?

¿Por qué no te pareces al rinoceronte?

La respuesta a esta pregunta es muy simple: ni tu papá ni tu mamá son rinocerontes. Pero al mismo tiempo, la respuesta no es tan simple como parece. De hecho, la **herencia**, los rasgos que se pasan de una generación a otra, es un tema muy complejo. Tu pelo, por ejemplo, puede ser rizado y el de tus padres lacio, o puede que tengas ojos azules, y tus padres tengan ojos cafés. ¿Cómo funciona esto? La gente se lo ha estado preguntando desde hace mucho tiempo. Hace unos 150 años se realizaron experimentos muy importantes, que les permitieron a los científicos encontrar algunas respuestas. Estos experimentos fueron realizados por Gregorio Mendel.

¿Quién fue Gregorio Mendel?

Gregorio Mendel nació en Heizendorf, Austria, en 1822. Su familia tenía una granja y Mendel aprendió mucho sobre el cultivo de flores y árboles desde chico; la naturaleza le fascinaba. Cuando se graduó de la universidad, Mendel entró en un monasterio. En la huerta del monasterio pudo observar las plantas para estudiar cómo se heredan rasgos de una generación a la siguiente. La **Figura 1** muestra una ilustración de Mendel en la huerta del monasterio.

Figura 1 Gregorio Mendel

Homework

What rhinoceroses look like is also genetically determined and varied. There are five species of rhinoceros, ranging in length from 2.5 m (8 ft) to 4.3 m (almost 14 ft). Have interested students research and give a written report on the rhinos alive today and the ones we know about from fossils.

El misterio se disipa

Mendel era curioso, y también era un buen observador. Se había fijado, en su trabajo en la huerta, que algunas veces los patrones de la herencia parecían sencillos, y otras no, y quería saber por qué.

Quería saber cómo se pasaban los rasgos de una generación a otra. Por ejemplo, a veces un rasgo que aparecía en una generación no aparecía en ninguna de las plantas de la siguiente generación, y en la tercera generación aparecía otra vez. Mendel había observado patrones como éste en plantas, personas y en muchos otros seres vivos.

Para simplificar su investigación, decidió estudiar un solo tipo de organismo. Como ya había usado la planta de chícharos en otros experimentos, decidió usarla también en su nueva investigación.

¿Te gustan los chícharos? Los chícharos resultaron ser una buena elección por varias razones: crecen muy rápido, son autopolinizantes y existen muchas variedades. Las **plantas autopolinizantes** tienen órganos reproductores masculinos y femeninos, como la flor de la **Figura 2.** Así, el polen de una planta o una flor puede fertilizar los óvulos de la misma flor, o los de otra flor de la misma planta. Para entender los experimentos de Mendel, necesitas primero conocer las partes de la flor y entender cómo se lleva a cabo la fertilización en las plantas. La **Figura 3** ilustra la fertilización.

Figura 2 *Esta fotografía muestra los órganos reproductores femeninos y masculinos de una flor.*

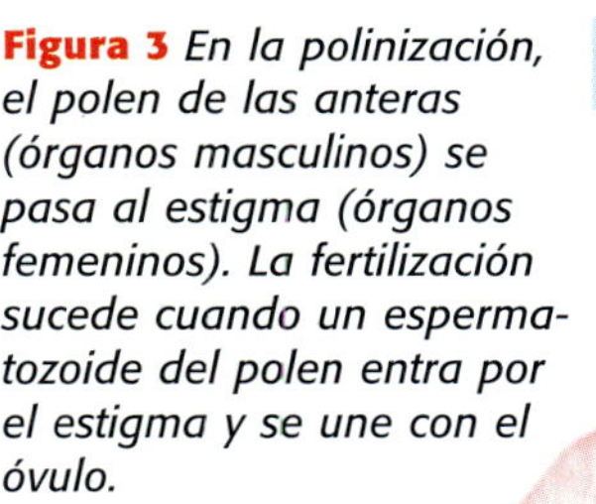

Figura 3 *En la polinización, el polen de las anteras (órganos masculinos) se pasa al estigma (órganos femeninos). La fertilización sucede cuando un espermatozoide del polen entra por el estigma y se une con el óvulo.*

IS THAT A FACT!

Although Mendel was brilliant, he had difficulty learning from scientific texts. In the monastery gardens, Mendel explored the scientific ideas he had trouble with in school. While trying to grow better peas, he discovered genetics, an entirely new field of science!

---

2 Teach

USING THE FIGURES

Discuss the physical processes involved in the fertilization of the flowers illustrated in **Figure 3.** The flower on the right can be fertilized by another flower or can fertilize itself. Compare this figure with **Figure 5,** and point out that removing the anthers from the flower makes it impossible for the plant to self-pollinate.

Sheltered English

DISCUSSION

Scientific Method Have students identify the steps of the scientific method in Mendel's work.

- **Ask a question**—How are traits inherited?
- **Form a hypothesis**—Inheritance has a pattern.
- **Test the hypothesis**—Cross true-breeding plants and offspring.
- **Analyze the results**—Identify patterns in inherited traits.
- **Draw conclusions**—Traits are inherited in predictable patterns.

What step did Mendel omit? (Hint: Why was his work overlooked for so long?) (communicate the results)

Directed Reading Worksheet 6 Section 1

READING STRATEGY

Prediction Guide Before students read this page, ask them the following question:

If a true-breeding pea plant that has purple flowers is crossed with a true-breeding pea plant that has white flowers, what will the offspring look like?

a. all purple flowers.

b. all white flowers.

c. some purple flowers and some white flowers.

d. all light-purple flowers.

(a)

Have students evaluate their answer after they read about Mendel's experiments.

MEETING INDIVIDUAL NEEDS

Learners Having Difficulty
Ask students the following questions: What other traits might vary among flowers of the same species? Why do traits vary among individuals of the same species? Can any traits of a plant's offspring be predicted?
Sheltered English

Multicultural CONNECTION

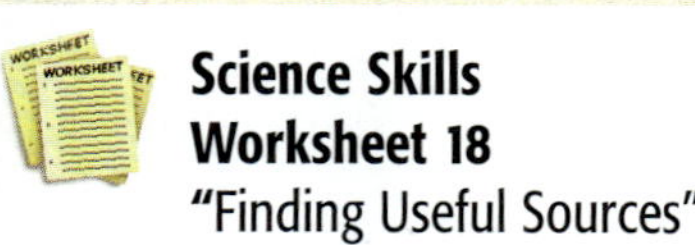

Writing Throughout history, cultures have developed different interpretations of how traits are inherited between generations. Social prohibitions have helped cultures avoid genetic interbreeding. Have interested students write a research report to share with the class.

Science Skills Worksheet 18
"Finding Useful Sources"

Chícharos y chácharas

Para simplificar las cosas, Mendel decidió estudiar sólo una característica a la vez, como la altura de la planta o el color del chícharo. De esta manera podría entender mejor los resultados. Escogió plantas que tenían dos formas distintas de cada característica que quería estudiar. Por ejemplo, para la altura, había una forma que siempre producía plantas altas y otra que siempre producía plantas bajas. Para la característica de color de la flor, Mendel escogió una forma de la planta que siempre producía flores moradas y otra que siempre producía flores blancas. El color morado y el blanco eran dos rasgos que correspondían a la característica de color de la flor. En la **Figura 4** puedes ver algunas de las características que Mendel estudió. También puedes ver los dos rasgos distintos de cada característica.

Plantas de línea pura Mendel eligió cuidadosamente sólo aquellas plantas que eran de línea pura. Cuando una planta de **línea pura** se autopoliniza, siempre produce nuevas plantas que tienen los mismos rasgos de la planta madre. Por ejemplo, una planta alta de línea pura siempre producirá plantas que también son altas.

Mendel decidió investigar qué pasaría al cruzar plantas con diferentes formas de la misma característica. Para realizar este experimento, usó un método conocido como *polinización cruzada*, en el que se cortan los estambres de una planta para que no pueda autopolinizarse y se usa polen de otra planta para fecundarla. Con este método, Mendel podía elegir el polen con el que iba a fecundar la planta. Esto se muestra en la **Figura 5.**

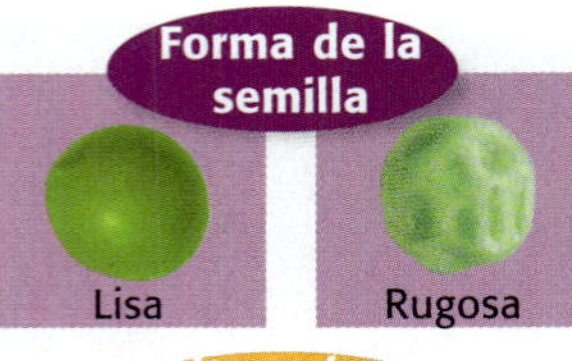

Figura 4 *Estas son algunas de las características que estudió Mendel.*

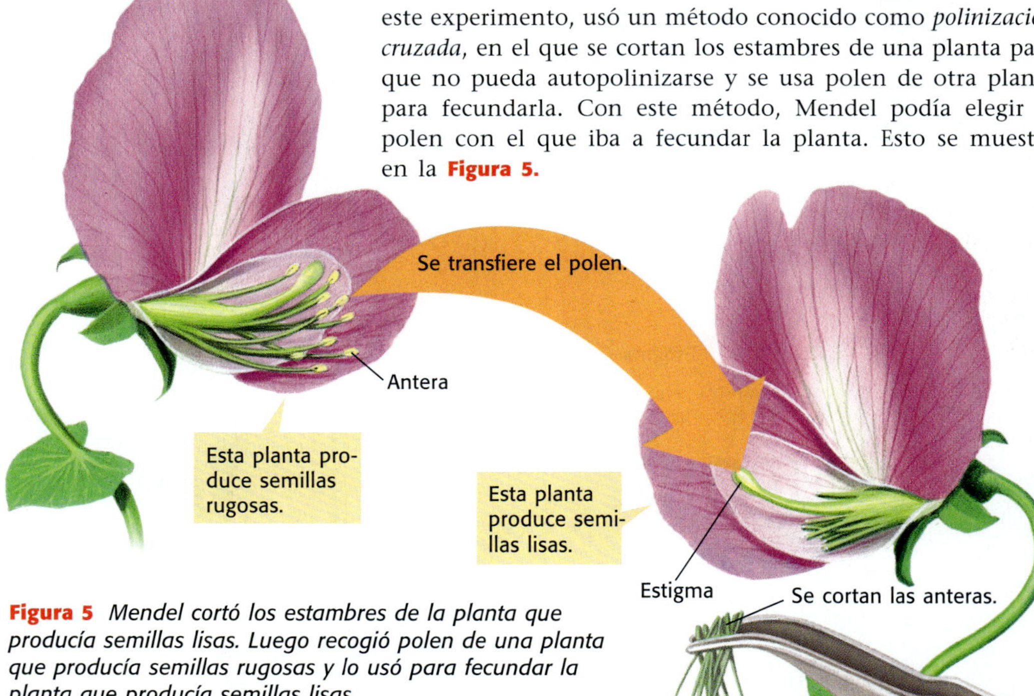

Figura 5 *Mendel cortó los estambres de la planta que producía semillas lisas. Luego recogió polen de una planta que producía semillas rugosas y lo usó para fecundar la planta que producía semillas lisas.*

132

internet connect

SCILINKS
NSTA

TOPIC: Heredity
GO TO: www.scilinks.org
sciLINKS NUMBER: HSTL110

WEIRD SCIENCE

Environmental stimuli can sometimes affect flower color. Hydrangeas growing in acidic soil produce blue flowers. If the soil is made alkaline, the same plants will produce pink flowers.

El primer experimento de Mendel

En su primer experimento, Mendel realizó cruces para estudiar siete características distintas. Cada uno de ellos unía las dos formas de la misma característica. Los resultados del cruce entre plantas que producían semillas lisas y plantas que producían semillas rugosas se muestran en la **Figura 6.** Las plantas que nacen de este cruce se conocen como *primera generación.* ¿Te sorprenden los resultados? ¿Qué crees que pasó con el rasgo de semillas rugosas?

Los resultados que Mendel obtuvo en los demás cruces fueron parecidos a los del primero: un rasgo se manifestaba, y el otro se esfumaba. Mendel decidió llamar **rasgo dominante** al rasgo que se manifestaba y **rasgo recesivo** al que se ocultaba. Para averiguar lo que había pasado con el rasgo recesivo, Mendel decidió hacer otro experimento.

El segundo experimento de Mendel

Mendel dejó que la primera generación de plantas de cada uno de los siete cruces se autopolinizara, como se muestra también en la Figura 6. Esta vez la planta con el rasgo dominante de forma de semilla (lisa) se autopolinizó. Como puedes ver, el rasgo recesivo de semillas rugosas volvió a aparecer.

Mendel repitió este experimento con los dos rasgos de cada una de las siete características. Sin importar qué característica fuera, siempre que la primera generación se autopolinizaba, el rasgo recesivo aparecía de nuevo.

Figura 6 *Una planta que produce semillas rugosas se fertiliza con polen de una planta que produce semillas lisas.*

133

GROUP ACTIVITY

Mendelian Crosses Give each student a purple bead and a white bead, and ask them to perform a Mendelian cross. Tell students to begin with the first generation with the allele combination *Pp*. Have students randomly "pollinate" with 10 other members of the class. To "pollinate," students should pair up and shake all four beads together. Without looking, each student should choose one bead. The two beads selected will determine the genotype of the offspring. Have them do this 10 times. Students should record the genotype for each pollination. Have students tally the results and determine the ratio of white-flowering plants to purple-flowering plants resulting from the matches.

DEMONSTRATION

Flower Dissection Obtain a flower that has anthers and a stigma, such as a pea flower, a tulip, or a lily. Be careful: Pollen can stain clothing and cause allergic reactions. Dissect the flower, and show students the anthers and the stigma. Could this flower self-pollinate? (Yes; it has both anthers and a stigma.)

Demonstrate how Mendel removed the anthers of his flowers and then used a small brush to transfer pollen from plant to plant. Sheltered English

Homework

Poster Project Have students create posters to illustrate Mendel's first and second experiments. Have each student demonstrate one of the seven traits Mendel studied. Encourage students to use materials such as flowers, yellow and green seeds, or wrinkled and round peas. Each project should clearly identify the parents, the first generation, and the second generation. (This activity can also be done with seven groups, one for each trait.) Sheltered English

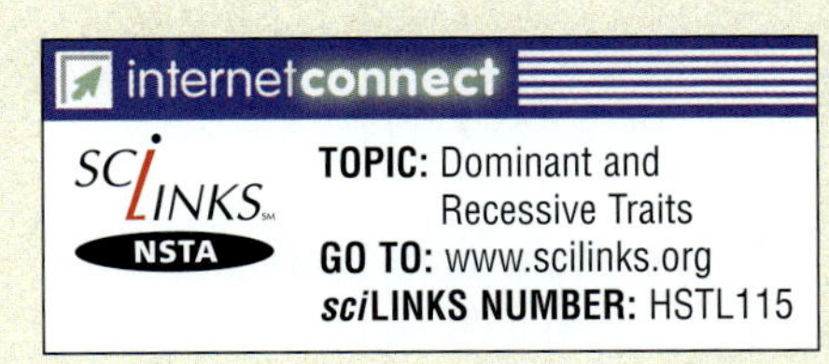

Bug Builders, Inc.
PG 578

USING THE TABLE

As seen in the table on this page, Mendel used at least 580 plants for each trait he studied. Why did Mendel work with such large samples? (Because Mendel was studying probability, larger samples increased the accuracy of his results.)

Would his data have been different if he had used much smaller samples? (Yes; the data would probably have been less accurate.)

MATH and MORE

Ratios are commonly used to compare two values. For instance, ratios are often used to express speeds, such as 55 km/h, and prices, such as 79 cents/kg. In these examples, the ratios are 55:1 and 79:1.

Math Skills Worksheet 16
"What Is a Ratio?"

Answer to MATHBREAK

The ratio of nougat-filled chocolates to caramel-filled chocolates is 18:6, or $\frac{18}{6}$, which can be reduced to $\frac{3}{1}$. This can be rewritten as 3:1 or 3 to 1.

Math Skills Worksheet 40
"Punnett Square Popcorn"

Experimentos

¿Quieres diseñar nuevos bichos? Mira las recetas de la página 578.

¡MATEMÁTICAS!

¿Qué son las proporciones?

Una proporción es un modo de comparar dos números utilizando la división. En los resultados de Mendel, la proporción entre plantas con flores moradas y plantas con flores blancas se puede expresar como 705 a 224, o 705:224. También se puede expresar como una fracción:

$$\frac{705}{224}$$

Si divides, puedes reducir la fracción, y el resultado será:

$$\frac{3.15}{1}$$

Esto quiere decir que por cada tres plantas que dan flores moradas, hay aproximadamente una planta que da flores blancas. He aquí otro problema para ti:

En una caja de chocolates hay 18 rellenos de turrón y 6 rellenos de caramelo. ¿Qué proporción hay entre los chocolates de turrón y los de caramelo?

134

Un punto de vista diferente

Luego, Mendel hizo algo que a nadie se le había ocurrido antes: contó el número de plantas con rasgos diferentes que nacieron en la segunda generación. Pensó que eso le ayudaría a entender sus resultados. Mira los resultados que obtuvo en la tabla que se muestra a continuación.

Los resultados de Mendel			
Característica	**Característica dominante**	**Característica recesiva**	**Proporción**
Color de la flor	705 morada	224 blanca	?
Color de la semilla	6,002 amarilla	2,001 verde	?
Forma de la semilla	5,474 lisa	1,850 rugosa	?
Color de la vaina	428 verde	152 amarilla	?
Forma de la vaina	882 lisa	299 abultada	?
Posición de la flor	651 en el tallo	207 en la punta	?
Altura de la planta	787 alta	277 baja	?

Como puedes ver, los rasgos recesivos aparecen otra vez, pero no tan frecuentemente como los dominantes. Mendel decidió calcular la *proporción* entre rasgos dominantes y rasgos recesivos para cada característica. Calcula la proporción entre rasgos dominantes y recesivos para cada característica como si fueras Mendel. (Si necesitas ayuda, consulta el apartado de ¡Matemáticas! a la izquierda) ¿Qué patrón encuentras si comparas las proporciones?

Ratios for Mendel's Results

The reduced ratios of dominant to recessive traits in Mendel's specimens are as follows:

Flower color	3.15:1
Seed color	3.00:1
Seed shape	2.96:1
Pod color	2.82:1
Pod shape	2.95:1
Flower position	3.14:1
Plant height	2.84:1

Answer to the question at the bottom of the student page

All the ratios can be rounded off to 3:1.

Una idea brillante

Mendel se dio cuenta de que sus resultados sólo se explicaban si cada planta tenía dos juegos de instrucciones para cada característica, y cada progenitor aportaba un juego de instrucciones a la siguiente generación. Los juegos de instrucciones se conocen como **genes.** Así, el óvulo fecundado tendría dos genes por cada característica; uno de cada planta. Los dos genes que determinan la misma característica se llaman **alelos**.

Compruébalo en la cuadrícula de Punnett Para entender cómo sacó Mendel estas conclusiones, usaremos un diagrama llamado cuadrícula de Punnett. La **cuadrícula de Punnett** sirve para visualizar las posibles combinaciones de alelos aportadas por las plantas progenitoras. Los alelos dominantes se representan con letras mayúsculas y los recesivos con minúsculas. Así, los alelos de una planta de línea pura con flores moradas se escriben *PP,* y los de una planta de línea pura que da flores blancas se escriben *pp.* El cruzamiento de estas plantas, se escribe *PP* × *pp,* como se muestra en la **Figura 7.** En la cuadrícula están las combinaciones de alelos en las plantas producto del cruzamiento. La combinación de alelos que hereda la planta se llama **genotipo.**

En la cuadrícula de Punnett de la **Figura 7,** las plantas hijas tenían el mismo genotipo: *Pp.* ¿Qué aspecto tenían estas plantas? El aspecto que hereda un organismo se llama **fenotipo.** El alelo dominante en cada genotipo, *P,* determina que las plantas hijas tendrán el mismo aspecto para esa característica; es decir, tendrán flores moradas. El alelo recesivo, *p,* determina que las instrucciones para producir flores blancas pasen a la siguiente generación.

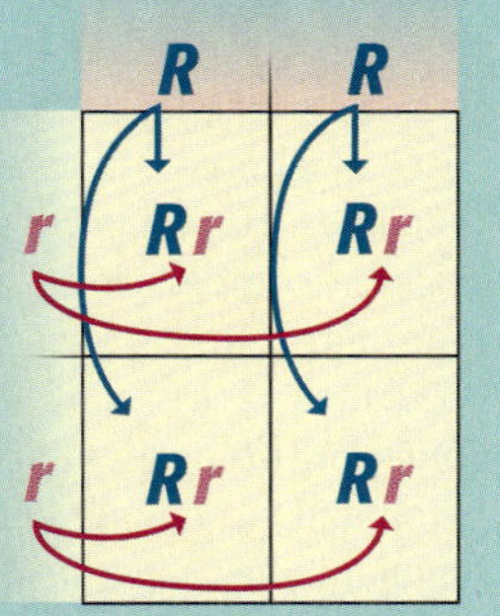

Planta de cruce puro de flores blancas *(pp)*

Planta de cruce puro de flores moradas *(PP)*

Figura 7 *Cuando un organismo es de línea pura para cierta característica, cada alelo debe llevar las mismas instrucciones. Las posibles combinaciones de alelos en las plantas hijas de este entrecruzamiento son iguales:* Pp.

Cómo hacer una cuadrícula de Punnett

Para hacer una cuadrícula de Punnett, traza un cuadrado y divídelo en cuatro secciones. Luego, escribe las letras que representan los alelos de uno de los padres en la parte de arriba. Escribe las letras que representan los alelos del otro padre a un costado.

El cruce que se muestra en esta cuadrícula de Punnett es entre una planta que sólo produce semillas lisas *RR,* y una planta que sólo produce semillas rugosas, *rr.* Sigue las flechas para llenar la parte de adentro de la cuadrícula. Las combinaciones de alelos de la cuadrícula muestran todos los genotipos de este cruzamiento. ¿Cómo serían los fenotipos de estas plantas?

Q: What do you get when you cross a bridge with a bicycle?

A: to the other side

Answer to How to Make a Punnett Square

All the phenotypes would be round.

---

3 Extend, *continued*

Quick Lab

MATERIALS

FOR EACH GROUP:
- masking tape
- 2 quarters

Answers to QuickLab

Students should find that they get the *bb* combination about $\frac{1}{4}$, or 25 percent, of the time. The more trials there are, the closer the probability will be to 25 percent. Note that each coin represents one parent's alleles. For example, the female parent has the alleles *B* and *b*. The probability that the offspring will inherit either of these alleles from the female parent is $\frac{1}{2}$, or 50 percent. The same is true for the male parent. The probability of throwing two *b* alleles in a row is calculated as follows: $\frac{1}{2} \times \frac{1}{2} = \frac{1}{4}$, and $\frac{1}{4} \times 100 = 25$ percent.

Homework

Punnett Squares Have students create Punnett squares for the different crosses in Mendel's experiments. Students should include the phenotype and genotype of the parents and offspring.

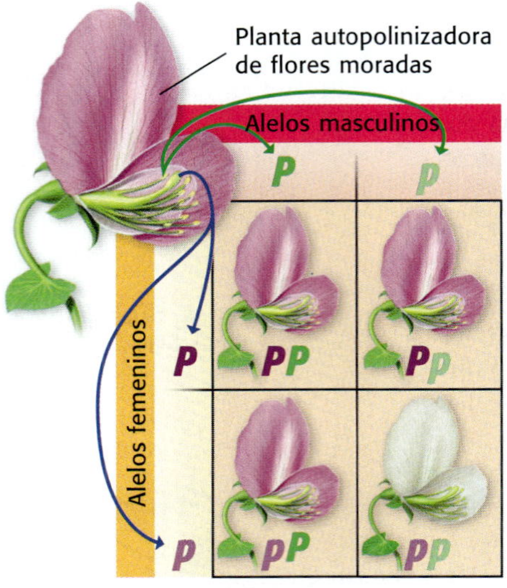

Figura 8 *Esta cuadrícula de Punnett muestra los posibles resultados del cruce entre Pp y Pp.*

Laboratorio

A cara o cruz

Imagínate que tienes dos conejillos de indias que quieres cruzar. Son de color café y tienen el genotipo **Bb.** ¿Qué posibilidades hay de que las crías sean blancas con el genotipo **bb?** Realiza este experimento para averiguarlo. Pega una tira de **cinta de papel** a ambos lados de **dos monedas de veinticinco centavos.** En un lado de la moneda escribe una **B** mayúscula y del otro lado una **b** minúscula. Lanza las dos monedas al aire 50 veces y apunta tus resultados cada vez. ¿Cuántas veces salió la combinación **bb**? Lánzalas otras 50 veces. ¿Cuántas veces salió la combinación **bb** esta vez? ¿Qué probabilidad hay de que la siguiente vez salga **bb**?

Más pruebas En su segundo experimento, Mendel dejó que las plantas de la primera generación se autopolinizaran. Los resultados de este experimento también pueden representarse en una cuadrícula de Punnett. La **Figura 8** muestra el cruzamiento por autopolinización de una planta con genotipo *Pp.* Los alelos de las plantas cruzadas indican que sus óvulos y espermatozoides pueden contener un alelo *P* o uno *p.*

¿Cómo serían los genotipos de la nueva generación? Algunos cuadros muestran la combinación de alelos *Pp* y otros la combinación *pP.* Estos genotipos son iguales, aunque el orden de las letras sea distinto. Los otros genotipos posibles en esta generación son *PP* y *pp.* Las combinaciones *PP, Pp,* y *pP* tienen el mismo fenotipo, flores moradas, porque cada una tiene por lo menos un alelo dominante para esta característica, el alelo *P.* Sólo la combinación *pp* produce plantas con flores blancas. La proporción entre rasgos dominantes y recesivos es de 3:1, como Mendel calculó.

¿Que probabilidades hay?

La nueva generación tiene las mismas probabilidades de heredar cualquier alelo de sus padres. Imagínate que lanzas una moneda. Hay un 50 por ciento de posibilidades de que salga cara y un 50 por ciento de que salga cruz. Como al lanzar la moneda, las posibilidades de heredar un alelo u otro son al azar. Para predecir la posibilidad de que cierto genotipo se herede, debemos considerar las leyes de la probabilidad.

Probabilidad La **probabilidad** es la posibilidad matemática de que ocurra un evento. Comúnmente, se expresa como una fracción o un porcentaje. Si lanzas una moneda, la probabilidad de que salga cara es de $\frac{1}{2}$. Esto significa que la mitad de las veces que lances una moneda, va a salir cara. Para expresar la probabilidad como un porcentaje, divide el numerador de la fracción entre el denominador y multiplica por 100.

$$\frac{1}{2} \times 100 = 50\%$$

Para encontrar la probabilidad de que salga cruz dos veces seguidas, multiplica la probabilidad de los dos eventos.

$$\frac{1}{2} \times \frac{1}{2} = \frac{1}{4}$$

El porcentaje sería $1 \div 4 \times 100$, que es igual al 25 por ciento.

WEIRD SCIENCE

Many ordinary fruits and vegetables carry recessive genes for bizarre traits. For instance, a recessive gene in tomatoes causes the skin to be covered with fuzzy hair!

Un gato de orejas gachas, como el de la derecha, se cruzó con un gato de orejas normales. Si la mitad de los gatitos tienen el genotipo *Cc* y orejas gachas, y la otra mitad tienen el genotipo *cc* y orejas normales, ¿qué alelo es el que determina las orejas gachas? ¿Qué genotipo tenían los padres? (Pista: haz una cuadrícula de Punnett con el genotipo de los gatitos, y deduce de ahí el genotipo de los progenitores.)

Probabilidad de los genotipos El mismo método se utiliza para calcular la probabilidad de que una cría herede cierto genotipo. Para que una planta de chícharos herede el rasgo de flores blancas, tiene que recibir un alelo *p* de cada progenitor. Las probabilidades de heredar un alelo u otro es del 50 por ciento. Entonces, la probabilidad de heredar dos alelos *p* es $\frac{1}{2} \times \frac{1}{2}$, lo que es igual a $\frac{1}{4}$, o $1 \div 4 \times 100$, que es igual a 25 por ciento.

Gregorio Mendel: las personas mueren, pero las ideas perduran Muchas veces las ideas nuevas no son aceptadas en su tiempo. Así fue con las ideas de Gregorio Mendel. En 1865 publicó sus descubrimientos ante la comunidad científica, pero no recibieron mucha atención. Sólo después de su muerte, casi 30 años más tarde, las ideas de Mendel recibieron la atención que merecían. Cuando sus ideas salieron de nuevo a la luz, se abrieron las puertas de la genética moderna.

química
CONEXIÓN

Las semillas lisas son más bonitas, pero las rugosas saben más dulces. El alelo dominante para la forma de la semilla, *R,* hace que la semilla almacene almidones (que son moléculas que guardan azúcar). Esto hace que la semilla sea gorda y lisa. Las semillas con el genotipo *rr* no producen ni almacenan almidón y por eso la semilla está rugosa, pero como su azúcar no ha sido convertida en almidón, la semilla es más dulce.

REPASO

1. El alelo que determina la barbilla partida, *C*, es dominante en los seres humanos. ¿Cómo sería el bebé de una mujer con el geontipo *Cc* y un hombre con el genotipo *cc*? En tu cuaderno de ciencias, traza una cuadrícula de Punnett que muestre este cruzamiento.

2. De las combinaciones posibles que encontraste en la pregunta 1, ¿qué proporción hay entre el número de bebés con la barbilla partida y el de bebés con la barbilla normal?

3. **Aplicar conceptos** La cuadrícula de Punnett a la derecha muestra las combinaciones de los alelos de color de pelo en los conejos. Negro, *B*, es dominante sobre blanco, *b*. ¿Qué genotipos tienen los padres?

▼ Answers to Review

1.

	c	*c*
C	*Cc*	*Cc*
c	*cc*	*cc*

2. 1:1

3. *BB, bb*

Answers to APPLY

The dominant allele is the allele for curly ears, *C*. The parents' genotypes were *Cc* (curly ears) and *cc* (normal ears).

Quiz

For rabbits, the allele for black fur, *B*, is dominant over the allele for white fur, *b*. Suppose two black parents have four bunnies— three black and one white.

1. What are the genotypes of the parents? (The parents both have the recessive allele, so they are both genotype *Bb*.)

2. What are the possible genotypes of all four siblings? (The white bunny has genotype *bb*, and the black bunnies may be *BB* or *Bb*.)

ALTERNATIVE ASSESSMENT

Ask students to imagine two animal parents with different genetic traits. Have them assign three characteristics to each parent, such as tall or short and red-nosed or blue-nosed. For each pair of characteristics, have students choose one as dominant and the other as recessive. Students should use Punnett squares to determine the possible first-generation genotypes and phenotypes for each trait in a cross between the two parents. Then have students choose two genotypes for each characteristic from the first generation and create a Punnett square showing the possible second-generation genotypes and phenotypes resulting from that cross.

Focus

Meiosis

This section discusses chromosomes, describes the process of meiosis, and explains the difference between meiosis and mitosis. The section explains how meiosis supports Mendel's findings and concludes with a discussion of sex chromosomes and how sex is determined.

Bellringer

Ask students to write a sentence for each of the following terms:

    heredity, genotype, and
    phenotype

(**Heredity** is the passing on of traits from parents to offspring.

The combination of an organism's alleles is its **genotype.**

The way that an organism looks is known as its **phenotype.**)

1 Motivate

DISCUSSION

Inherited Traits Lead a class discussion about traits that are passed from parents to their children. Have the students list examples of traits that "run in families" and that could be genetically determined. (Answers may include traits such as hair color; a tendency to develop diseases, such as diabetes and some forms of cancer; or personality traits, such as shyness.)

Ask students to think about how traits are inherited. For example, some traits are carried by one sex, and some diseases are said to "skip a generation." Explain that this section will introduce the physical processes that determine genetic inheritance.

VOCABULARIO

gametos
cromosomas homólogos
meiosis
cromosomas sexuales

OBJETIVOS

- Explica la diferencia entre mitosis y meiosis.
- Explica por qué el proceso de meiosis respalda las ideas de Mendel.
- Explica la diferencia entre cromosomas sexuales masculinos y femeninos.

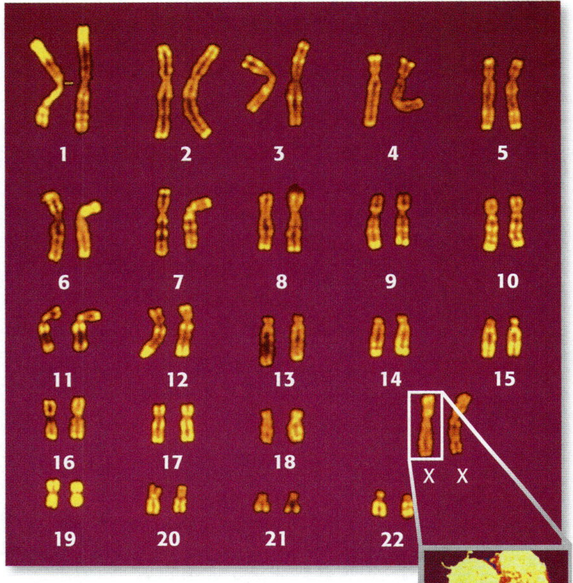

Figura 9 *Las células del cuerpo humano tienen 46 cromosomas, es decir, 23 pares de cromosomas homólogos. A la derecha puedes ver uno de los miembros de un par de cromosomas homólogos. Los cromosomas homólogos normalmente tienen la misma forma y tamaño.*

138

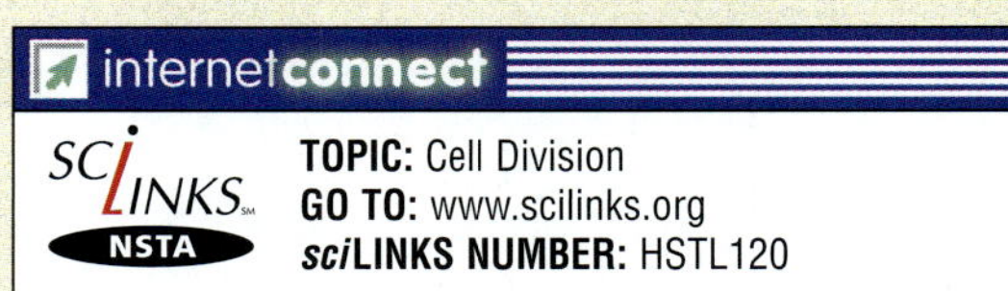

Meiosis

A principios del siglo XX, algunos científicos empezaron a realizar experimentos como los de Mendel. Los resultados eran muy emocionantes, y los científicos investigaron para ver si alguien había obtenido resultados parecidos. Así fue como encontraron el artículo de Mendel y vieron que sus resultados no eran nuevos; Mendel había observado lo mismo 35 años antes. Pero los genes seguían siendo un misterio: ¿dónde estaban exactamente?, ¿cómo pasaban su información de una célula a otra? El primer paso para contestar estas preguntas era entender el proceso de reproducción.

Dos tipos de reproducción

En lecciones anteriores, aprendiste que hay dos tipos de reproducción: asexual y sexual. La *reproducción asexual* se puede realizar con una sola célula madre. Primero, los órganos internos de la célula se duplican en un proceso llamado mitosis. Luego, la célula madre se divide y produce otras células que son sus copias exactas. La mayoría de los organismos unicelulares se reproducen así. Casi todas las células de tu cuerpo también se reproducen asexualmente.

Para formar un nuevo ser humano o una planta de chícharos, hay otro tipo de reproducción. En la *reproducción sexual*, dos células se unen para formar un nuevo individuo. Las células sexuales, conocidas como **gametos**, son diferentes del resto de las células del cuerpo. Las células del cuerpo humano, por ejemplo, tienen normalmente 46 cromosomas (o 23 pares), como puedes ver en la **Figura 9.** Estos pares de cromosomas se llaman **cromosomas homólogos.** Pero los gametos humanos tienen sólo 23 cromosomas, es decir, la mitad del número normal. Los gametos masculinos se llaman espermatozoides, y los femeninos, óvulos. Cada espermatozoide y cada óvulo tienen sólo un cromosoma de cada par de cromosomas homólogos.

Science BlOOperS

In 1918, a prominent scientist miscounted the number of chromosomes in a human cell. He counted 48. For almost 40 years, scientists thought this number was correct. In fact, it wasn't until 1956 that chromosomes were correctly counted and found to be only 46.

Menos es más ¿Por qué es tan importante que los gametos tengan la mitad del número normal de cromosomas? Cuando un óvulo y un espermatozoide se unen para formar un nuevo individuo, cada uno de los padres aporta un cromosoma para formar cada par de cromosomas homólogos. Así, se asegura que la cría tenga un número normal de cromosomas en cada célula de su cuerpo. Para funcionar bien, cada célula del cuerpo debe tener un juego completo de 46 cromosomas.

La meiosis al rescate Los gametos se crean en la meiosis, un proceso de división celular diferente de la mitosis. La **meiosis** produce células nuevas que sólo tienen la mitad del número normal de cromosomas. Para crear un gameto, los cromosomas se duplican una vez y el núcleo de la célula se divide dos veces. Los espermatozoides y óvulos que resultan de este proceso tienen la mitad del número de cromosomas de una célula normal.

Y mientras tanto, en el laboratorio

¿Qué tiene que ver todo esto con la búsqueda de los genes? Poco después de que se encontrara el artículo olvidado de Mendel, Walter Sutton, un joven estudiante de posgrado, descubrió algo importante. Sutton, que estaba estudiando los espermatozoides del saltamontes, conocía las investigaciones de Mendel, que mostraban que el óvulo y el espermatozoide aportan la misma cantidad de información genética a la cría. Sólo así se podía justificar la proporción de 3:1 que aparecía en la segunda generación. Sutton también se había dado cuenta de que, aunque los óvulos y los espermatozoides eran diferentes, tenían algo en común: sus cromosomas estaban dentro del núcleo. Con lo que había observado de la meiosis, su conocimiento de los experimentos de Mendel y su creatividad, Sutton propuso algo decisivo:

¡Los genes están en los cromosomas!

Como se comprobó más tarde, Sutton tenía razón. En las dos páginas siguientes encontrarás un resumen de la meiosis. Primero, vamos a repasar la mitosis para que puedas comparar los dos procesos.

¡A recordar la mitosis!

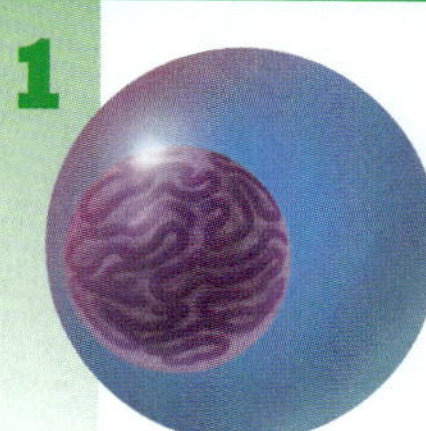
1 ¿Qué pasa adentro de una célula típica? Cada una de las cadenas largas (los cromosomas) se duplica.

2 Cada cromosoma está formado por dos mitades iguales llamadas cromátidas. Los cromosomas se vuelven más cortos y gruesos.

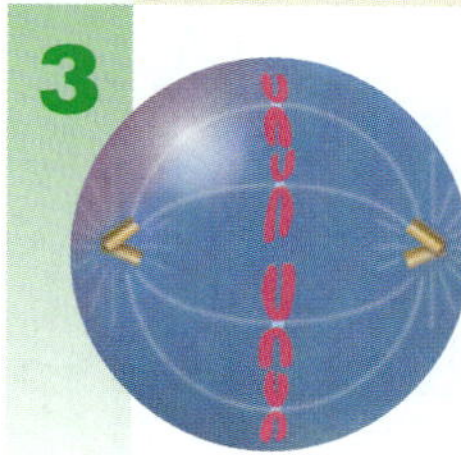
3 La membrana del núcleo se disuelve. Los cromosomas se alinean a lo largo de la parte central de la célula.

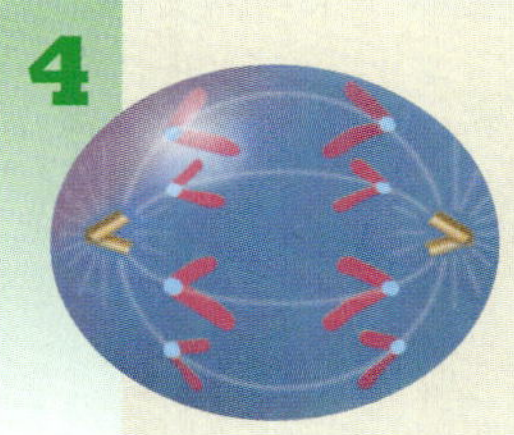
4 Las cromátidas se separan.

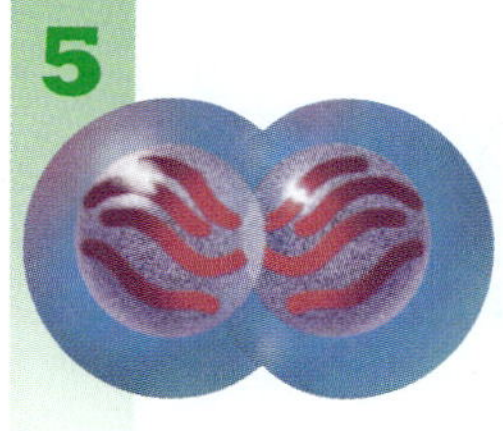
5 Una membrana nuclear rodea las cromátidas separadas. Los cromosomas se separan, y la célula se divide.

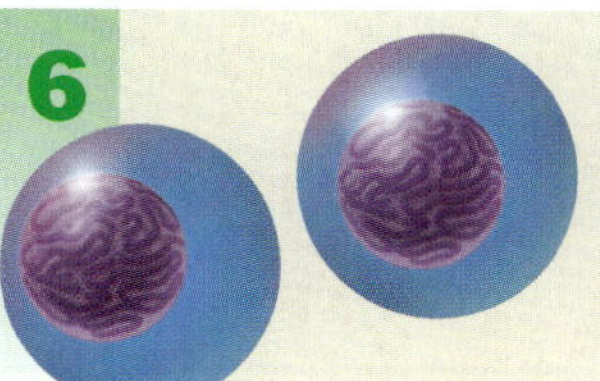
6 El resultado: dos copias idénticas de la célula madre.

139

IS THAT A FACT!

In human males, meiosis and sperm production take about 9 weeks. It's a continuous process that begins at puberty. In females, meiosis and egg production begin before birth. The process stops abruptly, however, and does not begin again until a girl enters puberty. Until a woman reaches menopause, one egg each month resumes meiosis and finishes its development. Therefore, the meiosis of a single egg may take up to 50 years to complete!

Directed Reading Worksheet 6 Section 2

DEMONSTRATION

Modeling Meiosis Select all cards that are the same color, and shuffle them. Each card will represent one chromosome. Tell students that two cards of the same number or face—for example, a queen of spades and a queen of clubs—represent homologous chromosomes. Next pair up the cards according to face or number. Make two piles; each pile should contain one of a pair. Ask students what this division represents. (the separation of homologous chromosomes into two cells)

Emphasize that this activity models only the first cell division of meiosis. Stress that after the second cell division of meiosis, each new cell will have half the number of chromosomes.

CROSS-DISCIPLINARY FOCUS

Theater Students may enjoy watching a choreographed dance or skit that walks them methodically through the steps of meiosis. Have interested students design the skit or dance, and allow them time to perform it. This can be a very effective way to present this difficult material.

Teaching Transparency 21 "Meiosis in Eight Easy Steps: A"

La meiosis en ocho pasos fáciles

El diagrama que te presentamos en estas dos páginas muestra cada etapa de la meiosis. Lee lo que dice sobre cada etapa y observa el diagrama. Los seres vivos tienen cantidades diferentes de cromosomas. En este diagrama mostraremos sólo cuatro cromosomas.

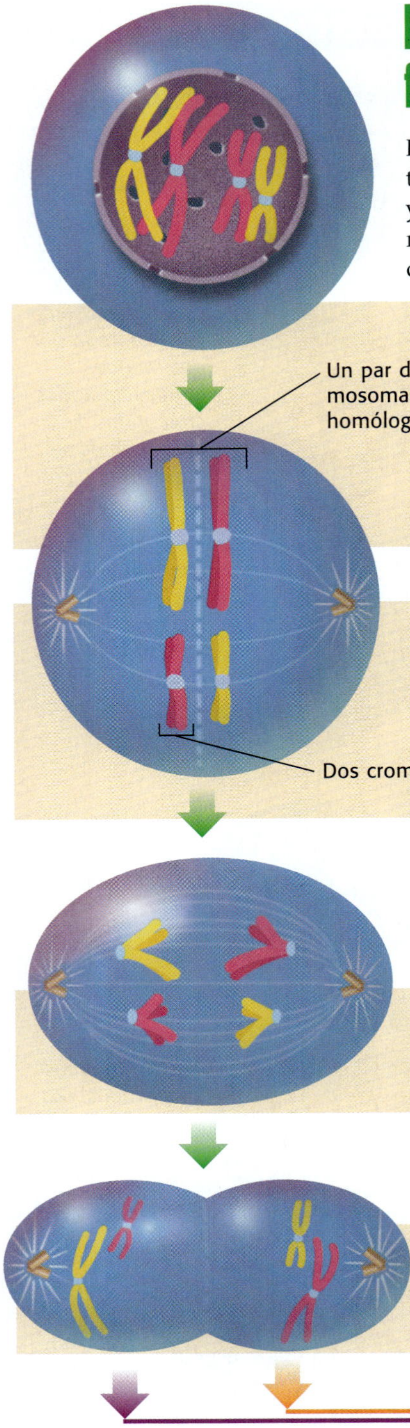

1 Antes de que comience la meiosis, los cromosomas tienen forma de filamentos. Cada cromosoma se duplica, forma dos mitades idénticas llamadas *cromátidas.* Luego, los cromosomas se vuelven más gruesos y cortos, y se pueden ver con un microscopio. La membrana del núcleo se disuelve.

2 Ahora, cada cromosoma está compuesto por dos cromátidas: la original y la copia idéntica. Los cromosomas que se parecen se juntan para formar *pares homólogos.* Estos pares de cromosomas se alinean a lo largo de la parte central de la célula.

3 Los cromosomas se separan de sus pares homólogos y se van a lados opuestos de la célula.

4 La membrana nuclear se vuelve a formar, y la célula se divide. Los pares de cromátidas siguen unidos.

140

SCIENCE HUMOR

Q: If human sex cells are created by meiosis, how are cat sex cells produced?

A: by meowsis

WEIRD SCIENCE

In some species of animals, there is only one gender! For instance, all desert whiptail lizards (*Cnemidophorus neomexicanus*) are female. Eggs are produced through parthenogenesis, and the unfertilized eggs develop into females that are genetically identical to their mother.

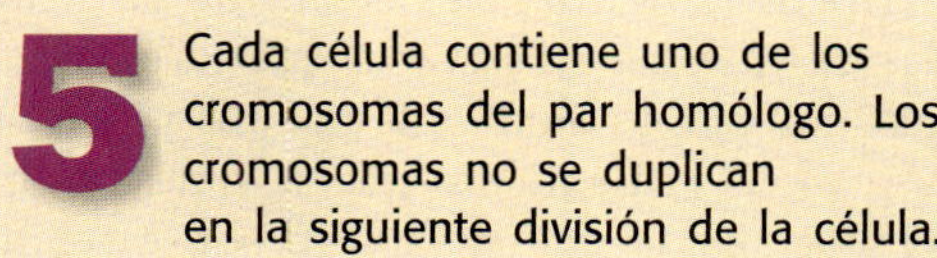

5 Cada célula contiene uno de los cromosomas del par homólogo. Los cromosomas no se duplican en la siguiente división de la célula.

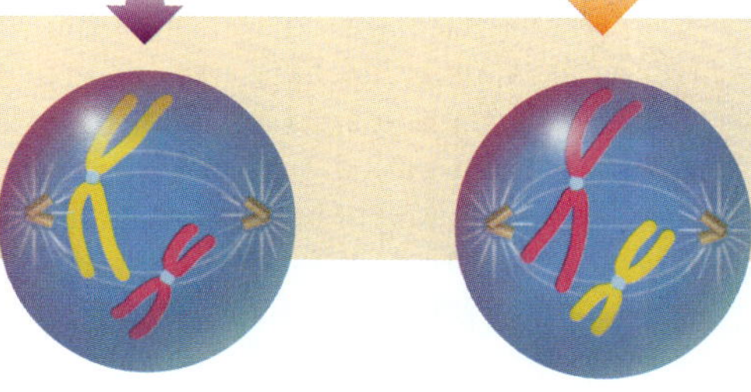

6 Los cromosomas se alinean a lo largo de la parte central de la célula.

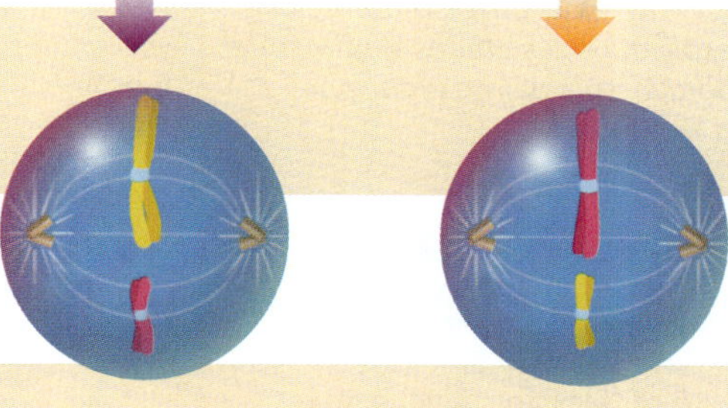

7 Las cromátidas se separan y se van a lados opuestos de la célula. Una membrana nuclear rodea los cromosomas separados y la célula se divide.

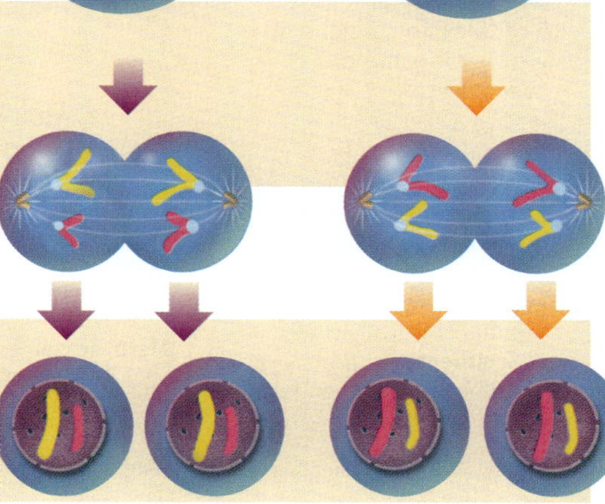

8 El resultado: cuatro nuevas células formadas de una sola célula madre. Cada nueva célula tiene la mitad del número de cromosomas de la célula madre.

✔ Autoevaluación

1. ¿Cuántos cromosomas hay en la célula original?
2. ¿Cuántos pares homólogos ves?
3. ¿Cuántas veces se duplican los cromosomas? En la meiosis, ¿cuántas veces se divide la célula?
4. ¿Cuántos cromosomas hay en cada célula al final de la mitosis?
5. ¿Qué se separa primero, las cromátidas o los cromosomas homólogos?

(Consulta la página 636 para comprobar tus respuestas.)

141

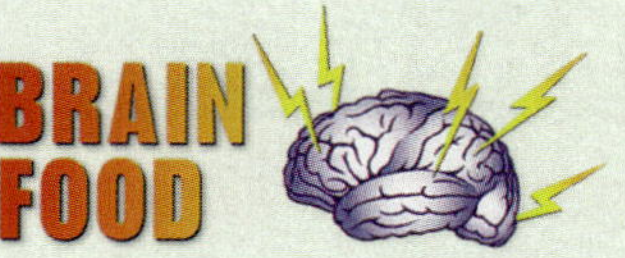

Scientists have bred watermelons that have an extra set of chromosomes. These watermelons are sterile, and their seeds do not develop. The result is seedless watermelons.

MEETING INDIVIDUAL NEEDS

Learners Having Difficulty
Visual learners and students with limited English proficiency may benefit from making a flip book that animates the phases of meiosis. First have students draw the events of meiosis in at least 50 sketches on sturdy cards about 6 in.^{2} each. Explain that each drawing should vary only slightly from the one before it. When the book is flipped through quickly, the images should appear to be in motion, and students will be able to watch meiosis in action. This activity could be repeated to demonstrate mitosis. Sheltered English

ACTIVITY

Concept Mapping Have students use the new terms in this section to create a concept map.

Teaching Transparency 22 "Meiosis in Eight Easy Steps: B"

Answers to Self-Check

1. four
2. two
3. They make copies of themselves once. They divide twice.
4. Two, or half the number of chromosomes in the parent are present at the end of meiosis. After mitosis, there would be four chromosomes, the same number as in the parent cell.
5. The homologous chromosomes separate first.

INDEPENDENT PRACTICE

Writing Have each student write a ScienceLog entry to chronicle the events of a chromosome containing an allele for a specific trait. Have the student describe the chromosome's role in the parent organism, the first-generation offspring, and the second-generation offspring. Descriptions should define whether the trait is dominant or recessive and should include an analysis of the factors that determine the genotype and phenotype of the parent and the offspring.

MISCONCEPTION ALERT

A common misconception is that many types of cells undergo meiosis. Make sure students understand that meiosis occurs *only* during sex-cell formation.

Homework

Making Models
Have students use markers, yarn, glue, and poster board to make a poster illustrating the process of meiosis. Each step of the process should include the sex chromosomes. The posters should demonstrate an understanding of meiosis, of sex cells, of sex chromosomes, and of sex determination. Sheltered English

Answers to QuickLab

One genotype is possible from the first cross, *Rr*. Three genotypes are possible for the second cross: *RR*, *Rr* (*rR*), and *rr*. The two possible phenotypes are wrinkled seeds, *rr* and round seeds, *RR, Rr*.

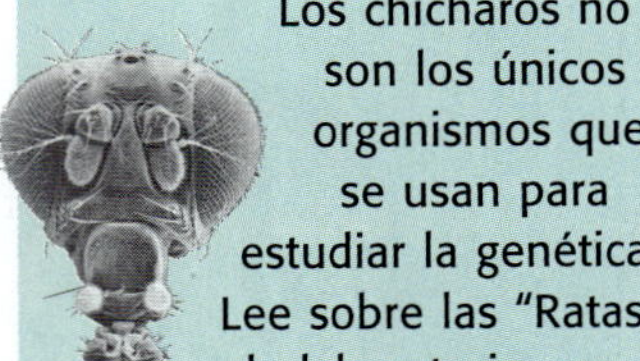
Los chícharos no son los únicos organismos que se usan para estudiar la genética. Lee sobre las "Ratas de laboratorio con alas" en la página 149.

Figura 10 *El proceso de meiosis explica los resultados que Mendel obtuvo en sus investigaciones.*

Después de la meiosis, cada espermatozoide contiene un alelo recesivo de semillas rugosas y cada óvulo contiene un alelo dominante de semillas lisas.

La fecundación de un óvulo por un espermatozoide resulta en el mismo genotipo, *Rr*, y el mismo fenotipo (lisa). Esto es exactamente lo que Mendel descubrió en sus experimentos.

Laboratorio

Lisa o rugosa

Traza una cuadrícula de Punnett para el cruce **RR** X **rr,** de la Figura 10. Luego, traza otra para el cruce de la primera generación, entre **Rr** X **Rr.** ¿Cuántos genotipos resultan del primer cruzamiento? ¿Cuántos del segundo? ¿Cuáles son? ¿Cuáles son los fenotipos del segundo cruzamiento?

Mendel y la meiosis

Tal como lo había pensado Sutton, las etapas de la meiosis respaldan las ideas de Mendel. Observa lo que le pasa a un par de cromosomas homólogos durante la meiosis y la fecundación en la **Figura 10**. El cruzamiento que se muestra en el diagrama es entre una planta que siempre produce semillas lisas y una planta que siempre produce semillas rugosas.

Padre En el núcleo de la célula vegetal que ves abajo, cada cromosoma homólogo tiene un alelo de forma de semilla y cada alelo lleva las mismas instrucciones: producir semillas rugosas.

Madre En el núcleo de la célula vegetal que ves abajo, cada cromosoma homólogo tiene un alelo de forma de semilla y cada alelo lleva las mismas instrucciones: producir semillas lisas.

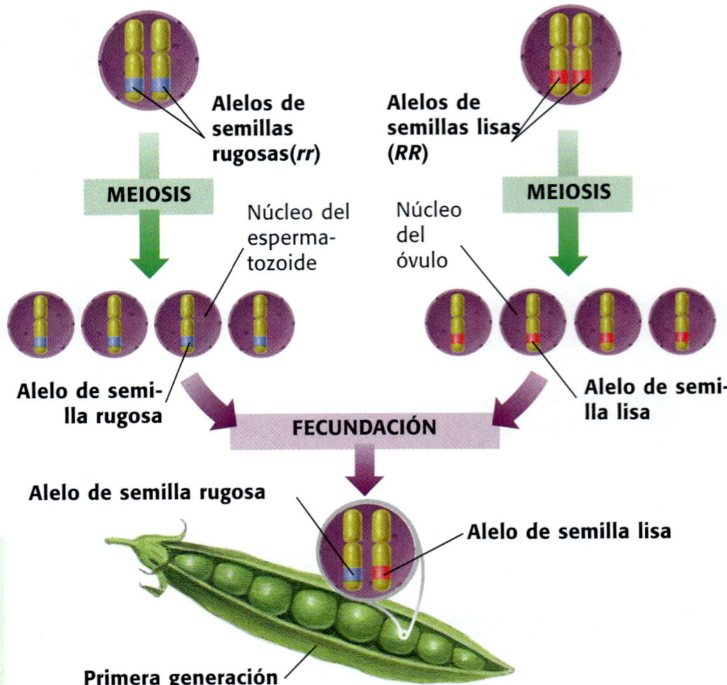

El óvulo fecundado en la primera generación contenía un alelo dominante y uno recesivo para el tipo de semilla. Sólo había un genotipo posible, porque todos los espermatozoides producidos en la meiosis contenían el alelo de semilla rugosa y todos los óvulos el de semilla lisa. Cuando la primera generación se autopolinizó, los posibles genotipos cambiaron. Hay tres genotipos posibles en este cruzamiento. ¿Cuáles son? Realiza la actividad de laboratorio de la izquierda para averiguarlo.

Teaching Transparency 23
"Meiosis and Mendel"

WEIRD SCIENCE

Theoretically, there should be exactly the same number of boy babies and girl babies. The statistics in North America, however, indicate that the birth ratio is about 104 males to every 100 females. Scientists are not sure why this happens.

¿Macho o hembra?

Los organismos tienen maneras diferentes de determinar su sexo. Para ver cómo sucede este proceso en los seres humanos, observa la **Figura 11,** y luego vuelve a mirar la Figura 9, en la página 138. Las dos fotografías muestran los cromosomas de las células de un ser humano. ¿Cuál foto es de una mujer y cuál de un hombre? Pista: Las mujeres tienen 23 pares de cromosomas semejantes, y los hombres tienen 22 pares semejantes y uno distinto.

Cromosomas sexuales El último par de cromosomas que se muestra en la Figura 11 es el de los cromosomas sexuales. Los **cromosomas sexuales** contienen genes que determinan una característica muy importante: el sexo de la cría. En los seres humanos, las mujeres tienen dos cromosomas X (el par de cromosomas semejantes), y los hombres tienen un cromosoma X y uno Y (el par de cromosomas distintos) Los cromosomas de la Figura 11 son de un hombre y los de la Figura 9 son de una mujer.

En la meiosis, uno de cada par de cromosomas va a dar origen a un gameto. Esto también es cierto en los cromosomas X y Y. Las mujeres, por ejemplo, tienen dos cromosomas X en cada célula del cuerpo. Cuando se producen óvulos en la meiosis, cada uno contiene un cromosoma X. Los hombres tienen un cromosoma X y uno Y en cada célula del cuerpo. Cuando estos cromosomas se separan en la meiosis, cada espermatozoide contiene un cromosoma X o uno Y. El óvulo y el espermatozoide se unen para formar una combinación XX o una combinación XY. Esto se muestra en la **Figura 12.**

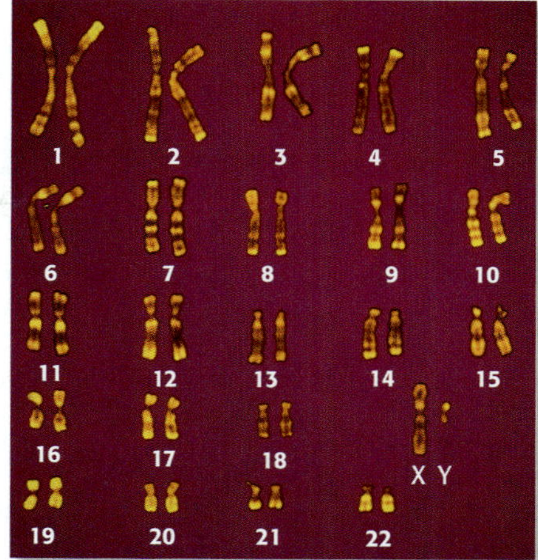

Figura 11 *¿Son estos cromosomas de un hombre o de una mujer? ¿Cómo lo sabes?*

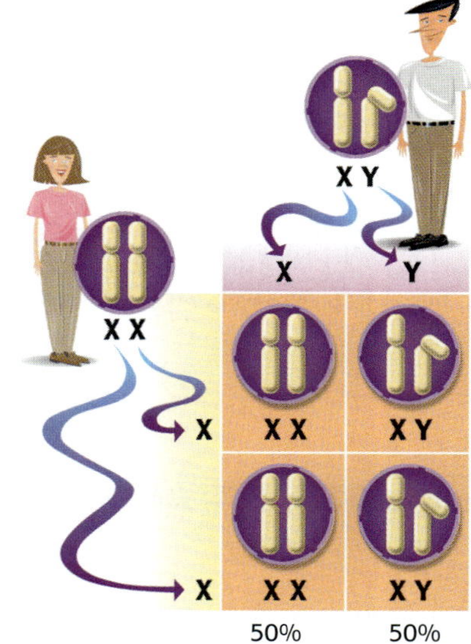

Figura 12 *Si un espermatozoide que contiene un cromosoma X fecunda un óvulo, la cría será de sexo femenino. Si el espermatozoide contiene un cromosoma Y, la cría será de sexo masculino.*

REPASO

1. Explica la diferencia entre cromosomas sexuales y gametos.

2. Si hay 14 cromosomas en las células de la planta del chícharo, ¿cuántos hay en un gameto de chícharo?

3. **Interpretar ilustraciones**
   Observa la ilustración de la derecha. ¿Muestra una etapa de la mitosis o de la meiosis? ¿Cómo lo sabes?

143

VOCABULARY DEFINITIONS

SECTION 1

heredity the passing of traits from parent to offspring

self-pollinating plant a plant that contains both male and female reproductive structures

true-breeding plant a plant that always produces offspring with the same trait as the parent(s)

dominant trait a trait observed when at least one dominant allele for a characteristic is inherited

recessive trait a trait that is apparent only when two recessive alleles for the characteristic are inherited

genes segments of DNA that carry hereditary instructions located on chromosomes and passed from parent to offspring

alleles alternative forms of a gene that govern the same characteristics

Punnett square a tool used to visualize all the possible combinations of alleles from parents

genotype the inherited combination of alleles

phenotype an organism's inherited appearance

probability the mathematical chance that an event will occur

Resumen del capítulo

SECCIÓN 1

Vocabulario

herencia (*pág. 130*)
planta autopolinizante (*pág. 131*)
planta de línea pura (*pág. 132*)
rasgo dominante (*pág. 133*)
rasgo recesivo (*pág. 133*)
genes (*pág. 135*)
alelos (*pág. 135*)
cuadrícula de Punnet (*pág. 135*)
genotipo (*pág. 135*)
fenotipo (*pág. 135*)
probabilidad (*pág. 136*)

Notas de la sección

- La herencia son los rasgos que se pasan de una generación a otra.

- Los rasgos son características hereditarias.

- Gregorio Mendel estudió las leyes de la herencia en las plantas de chícharos.

- Las plantas de chícharo de Mendel eran autopolinizantes y por lo tanto, contenían órganos reproductivos masculinos y femeninos. Además eran plantas de línea pura, o sea que siempre producían una nueva generación con los mismos rasgos de la primera.

- La nueva planta hereda dos grupos de instrucciones para cada rasgo, uno de cada progenitor.

- Los grupos de instrucciones se conocen cono genes.

- Las diferentes versiones del mismo gen se llaman alelos.

- Si se heredan los alelos dominante y recesivo para cierta característica, sólo se expresa el alelo dominante.

- Los rasgos recesivos sólo se manifiestan cuando se han heredado dos alelos recesivos para esa característica.

- El genotipo es la combinación de alelos que se da en un rasgo en particular.

- El fenotipo es la expresión física del genotipo.

- La probabilidad es la posibilidad matemática de que suceda un evento. Comúnmente se expresa como una fracción o un porcentaje.

Experimentos

Constructora de Bichos, S.A. (*pág. 578*)
Sigue la pista de los rasgos (*pág. 580*)

☑ Comprobar destrezas

Conceptos de matemáticas

PROPORCIONES Un frasco contiene 24 canicas verdes y 96 canicas rojas. Hay 4 canicas rojas por cada canica verde.

$$\frac{96}{24} = \frac{4}{1}$$

Esta proporción también se puede expresar así:

$$4:1$$

Comprensión visual

CUADRÍCULA DE PUNNETT La cuadrícula de Punett te ayuda a visualizar todas las combinaciones posibles de alelos heredados de una generación a otra. Consulta la página 135 para repasar cómo se hace una cuadrícula de Punett.

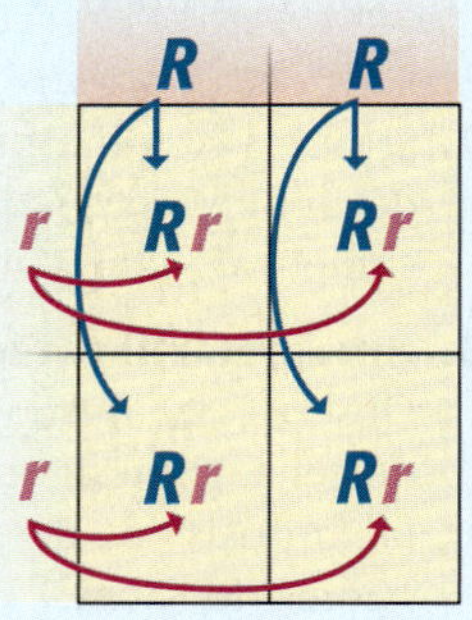

144

Lab and Activity Highlights

Tracing Traits `PG 580`

Bug Builders, Inc. `PG 578`

Datasheets for LabBook
(blackline masters for these labs)

sex cells eggs or sperm; a sex cell carries half the number of chromosomes found in other body cells

homologous chromosomes chromosomes with matching information

meiosis cell division that produces sex cells

sex chromosomes chromosomes that carry genes that determine the sex of offspring

Vocabulary Review Worksheet 6

Blackline masters of these Chapter Highlights can be found in the **Study Guide.**

SECCIÓN 2

Vocabulario

gametos *(pág. 138)*

cromosomas homólogos *(pág. 138)*

meiosis *(pág. 139)*

cromosomas sexuales *(pág. 143)*

Notas de la sección

- Los genes se encuentran en los cromosomas.

- La mayoría de las células humanas contiene 46 cromosomas, es decir, 23 pares.

- Cada par contiene un cromosoma donado por la madre y otro por el padre. Estos pares se llaman cromosomas homólogos.

- El proceso de meiosis produce gametos: óvulos y espermatozoides.

- Los gametos tienen la mitad del número normal de cromosomas.

- Los gametos contienen genes que determinan el sexo del bebé.

- Los genes femeninos contienen dos cromosomas X, y los masculinos contienen un cromosoma X y uno Y.

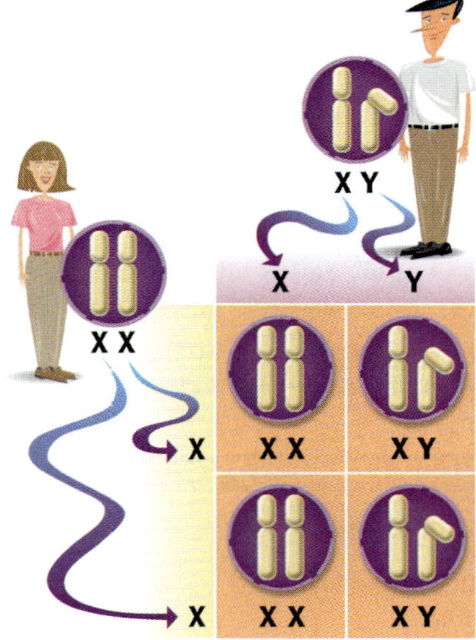

 internet

HRW **VISITA:** go.hrw.com

Visita el sitio web de HRW para encontrar una serie de herramientas de aprendizaje relacionadas con este capítulo. Sólo tienes que escribir la palabra clave:

PALABRA CLAVE: HSTHER

 **SC**i**LINKS** **NSTA** **VISITA:** www.scilinks.org

Visita el sitio web de la **Asociación Nacional de Maestros de Ciencias** *(National Science Teachers Association)* para encontrar recursos de Internet relacionados con este capítulo. Sólo escribe el **ENLACE DE CIENCIAS** para obtener más información sobre el tema:

TEMA: Gregorio Mendel	**ENLACE:** HSTL105
TEMA: Herencia	**ENLACE:** HSTL110
TEMA: Rasgos dominantes y rasgos recesivos	**ENLACE:** HSTL115
TEMA: División celular	**ENLACE:** HSTL120

145

Lab and Activity Highlights

LabBank

Long-Term Projects & Research Ideas, Project 5

USING VOCABULARY

1. sex cells
2. phenotype, genotype
3. Meiosis
4. alleles
5. self-pollinating

UNDERSTANDING CONCEPTS

Multiple Choice

6. a
7. c
8. b
9. b
10. c
11. c
12. c

Short Answer

13. Females have two X chromosomes. Males have one X and one Y chromosome.
14. Sample answer: A recessive trait is a genetic trait that is expressed only if there are two recessive alleles for the gene. A recessive trait is not expressed if an allele for a dominant trait is present.
15. Sex cells have half the number of chromosomes as other body cells.

Concept Mapping

16. 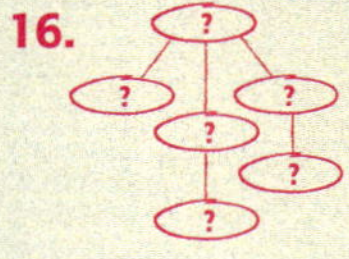 An answer to this exercise can be found at the end of this book.

CRITICAL THINKING AND PROBLEM SOLVING

17. The trait for blue eyes must be determined by a recessive allele. Brown must be dominant to blue. So both parents must carry the blue recessive allele, and a blue recessive allele must have been passed on to the child from both parents.
18. Meiosis is important for reproduction because it ensures that the characteristic number of chromosomes will not be doubled when an egg and a sperm come together.
19. No; true-breeding plants always produce the same trait. If Mendel had not used true-breeding plants, then he would have seen much more variation in traits. He would not have been able to tell whether traits were dominant or recessive because there would not have been reproducible ratios of dominant to recessive traits.

Repaso del capítulo

UTILIZAR EL VOCABULARIO

Escoge el término correcto para completar las siguientes oraciones:

1. Los espermatozoides y los óvulos son___?___ (*gametos* o *cromosomas sexuales*)

2. El ___?___ es la expresión de un rasgo, y está determinado por el ___?___, que es la combinación de los alelos. (*genotipo* o *fenotipo*)

3. La ___?___ produce células con la mitad del número normal de cromosomas. (*meiosis* o *mitosis*)

4. Las versiones distintas de los mismos genes se llaman ___?___. (*gametos* o *alelos*)

5. Una planta ___?___ puede polinizar sus propios óvulos. (*autopolinizante* o *de línea pura*)

COMPRENDER CONCEPTOS

Opción múltiple

6. Los genes se encuentran en
   a. los cromosomas.
   b. los alelos.
   c. las proteínas.
   d. los estambres.

7. El proceso que produce gametos se llama
   a. mitosis.
   b. fotosíntesis.
   c. meiosis.
   d. probabilidad.

8. La transferencia de rasgos de una generación a otra constituye
   a. la probabilidad.
   b. la herencia.
   c. los genes recesivos.
   d. la meiosis.

9. Si cruzas una flor blanca (con el genotipo *pp*) con una flor morada (con el genotipo *PP*), los genotipos posibles en la siguiente generación son:
   a. *PP* y *pp*.
   b. todos *Pp*.
   c. todos *PP*.
   b. todos *pp*.

10. ¿Cuáles serían los fenotipos del cruce mencionado arriba?
    a. todas blancas
    b. todas altas
    c. todas moradas
    d. $\frac{1}{2}$ blancas, $\frac{1}{2}$ moradas

11. En la meiosis,
    a. los cromosomas se duplican dos veces.
    b. el núcleo se divide una vez.
    c. se crean cuatro células a partir de la célula original.
    d. todas las anteriores

12. La probabilidad
    a. siempre se expresa como una proporción.
    b. es un 50% de posibilidades de que algo ocurra.
    c. es la posibilidad matemática de que ocurra un evento.
    d. es la proporción 3:1 de que ocurra un evento.

Respuesta breve

13. ¿Qué cromosomas sexuales tienen las mujeres? ¿Qué cromosomas sexuales tienen los hombres?

14. En tus propias palabras, y en una o dos oraciones, explica qué es un *rasgo recesivo*.

15. ¿En qué se diferencian los gametos de las células del cuerpo?

Concept Mapping Transparency 6

Organizar conceptos

16. Utiliza los siguientes términos para crear un mapa de ideas: meiosis, óvulos, división celular, cromosoma X, gametos espermatozoides, mitosis cromosoma Y.

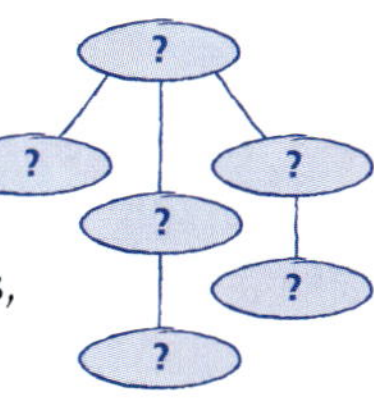

RAZONAMIENTO CRÍTICO Y RESOLUCIÓN

Escribe una o dos oraciones para responder a las siguientes preguntas:

17. Si una niña tiene ojos azules y su papá y mamá tienen ojos cafés, ¿qué puedes deducir sobre el rasgo de ojos azules? Explica tu respuesta.

18. ¿Qué importancia tiene la meiosis en la reproducción sexual?

19. Gregorio Mendel trabajó sólo con plantas de línea pura. Si no hubiera usado plantas de línea pura, ¿crees que su descubrimiento de los rasgos dominante y recesivo hubiera ocurrido? ¿Por qué?

LAS MATEMÁTICAS EN LAS CIENCIAS

20. Si *R* fuera el alelo dominante de semillas amarillas, y *r* el alelo recesivo de semillas verdes, ¿qué probabilidad habría de que el entrecruzamiento entre una planta de chícharos con el genotipo *Rr* y una con el genotipo *rr* produzca plantas hijas con genotipo *rr*?

INTERPRETAR GRÁFICAS

Observa la cuadrícula de Punnett que se muestra abajo y contesta las siguientes preguntas:

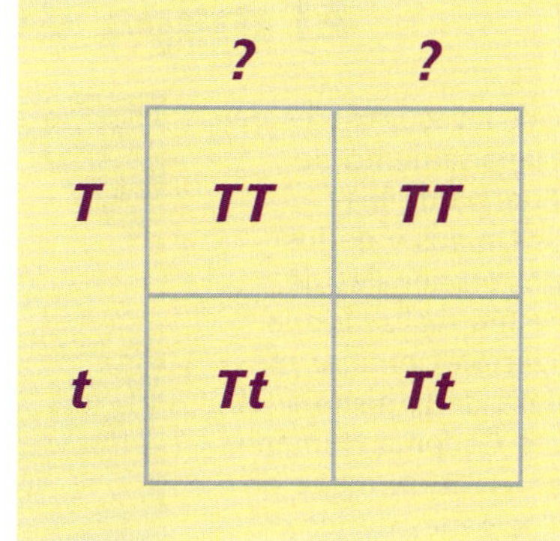

21. ¿Cuál es el genotipo desconocido?

22. Si *T* representa el alelo de plantas altas y *t* representa el alelo de plantas bajas, ¿cuál es el fenotipo de cada padre y de los hijos?

23. Si cada uno de los hijos se autopolinizara, ¿cuáles son los genotipos posibles en la siguiente generación?

24. ¿Qué probabilidad existe para cada genotipo mencionado en la pregunta 23?

AHORA, ¿qué piensas?

Revisa tus respuestas a las preguntas de la página 129 que escribiste en el cuaderno de ciencias. ¿Han cambiado tus respuestas? Si es necesario, corrige tus respuestas basándote en lo que has aprendido en este capítulo.

MATH IN SCIENCE

20. The probability of inheriting a *y* allele from the *Yy* plant is $\frac{1}{2}$. The probability of inheriting a *y* allele for the *yy* plant is $\frac{1}{1}$. Therefore, the probability that the offspring will inherit the *yy* genotype is $\frac{1}{1} \times \frac{1}{2}$, or 50 percent. These results can be visualized by using a Punnett square.

INTERPRETING GRAPHICS

21. *TT*
22. All the parents and offspring are tall pea plants.
23. Students should make two new Punnett squares. Self fertilization of *TT (TT × TT)* will yield offspring of all *TT*. Self fertilization of *Tt (Tt × Tt)* will yield offspring of *TT, Tt,* and *tt.*
24. *TT* has a 100 percent probability with a *TT* parent and a 25 percent probability with a *Tt* parent. *Tt* has a 50 percent probability with a *Tt* parent, and a 0 percent probability with a *TT* parent. The genotype *tt* has a 25 percent probability with a *Tt* parent, and a 0 percent probability with a *TT* parent.

Blackline masters of this Chapter Review can be found in the **Study Guide.**

NOW WHAT DO YOU THINK?

1. Humans do not all look alike because we have a huge variety of genes that are combined differently in each of us.
2. A baby will be a boy if he has one X sex chromosome and one Y sex chromosome. A baby will be a girl if she has two X sex chromosomes.

Ciencia, Tecnología y Sociedad

El mapa del genoma humano

Los investigadores del Departamento de Energía de los Estados Unidos y de los Institutos Nacionales de Salud están realizando la investigación científica más ambiciosa de la historia: el proyecto Genoma Humano. Quieren crear un mapa de todos los genes de los cromosomas humanos. El genoma son las instrucciones genéticas de un organismo. Esta información sería muy valiosa para prevenir y hasta curar enfermedades genéticas.

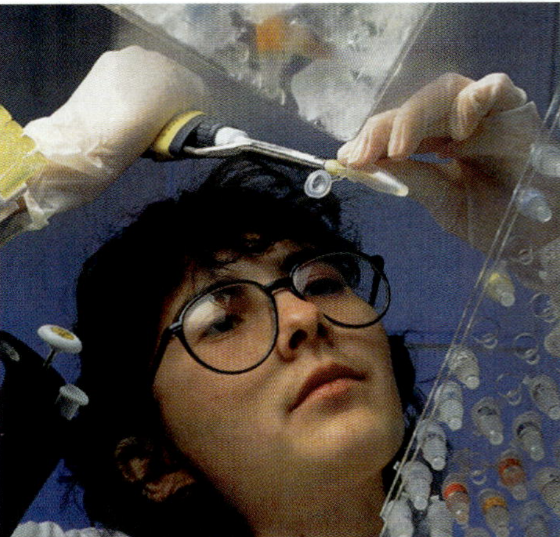

Esta investigadora está realizando una de las muchas tareas que se llevan a cabo en el proyecto Genoma Humano.

¿De quién son estos genes?

¿De quién son los genes que los investigadores están descifrando? No importa, porque el material genético sólo varía un 1 por ciento, aproximadamente. El objetivo es identificar cómo ese 1 por ciento de ADN hace que cada uno sea único y entender por qué algunos cambios genéticos causan enfermedades.

Medicina genética

Las mutaciones son cambios pequeñísimos en el material genético. Pueden ser hereditarias y causar enfermedades. Cuando los investigadores sepan el orden de nuestros genes, podrán ayudar a los médicos a detectar mutaciones en sus pacientes y así los médicos podrían avisarles del riesgo de una enfermedad ¡antes de que se enfermen! Si el médico te dijera a tiempo que tienes un riesgo genético de desarrollar niveles altos de colesterol, podrías empezar a hacer una dieta sana y más ejercicio para evitar problemas.

La tecnología avanza

La tecnología del proyecto Genoma Humano cambia constantemente. Se hacen miles de copias de un gen para analizar los cromosomas.

Los investigadores esperan completar la secuencia precisa del genoma humano (que tiene entre 50,000 y 100,000 genes) para el 2003. Algún día, una persona podrá recibir un gen sano para reemplazar uno que haya mutado. El uso de la genética en la medicina, o terapia genética, podría curar muchas enfermedades genéticas en el futuro.

¿Tú qué crees?

Aunque la investigación del genoma humano ofrece avances para la medicina, se siguen cuestionando los aspectos éticos, sociales y legales de este proyecto. Investiga y comparte tu punto de vista con tus compañeros.

148

Ratas de laboratorio con alas

Adivina: ¿qué mide menos de 1 mm de largo, zumba en la cocina y a veces tiene patas en los ojos? ¿Te rindes? La respuesta es *Drosophila melanogaster,* mejor conocida como la mosca de la fruta, porque de eso se alimenta. Este insecto ha ayudado a estudiar muchas enfermedades, en especial las que se originan en ciertas etapas del desarrollo humano. Los científicos han aprendido mucho sobre el cáncer, la enfermedad de Alzheimer, la distrofia muscular y el síndrome de Down gracias a la mosca de la fruta.

¿Por qué una mosca?

Las moscas de la fruta son los animales de laboratorio preferidos por muchos científicos. En pocos meses pueden criar varias generaciones y como tienen un ciclo de vida de sólo dos semanas, se pueden alterar sus genes y no hay que esperar mucho para ver los resultados.

Otra ventaja de las moscas es su tamaño: miles de ellas caben en un espacio reducido, y los investigadores pueden comparar y mantener una variedad de moscas para sus experimentos.

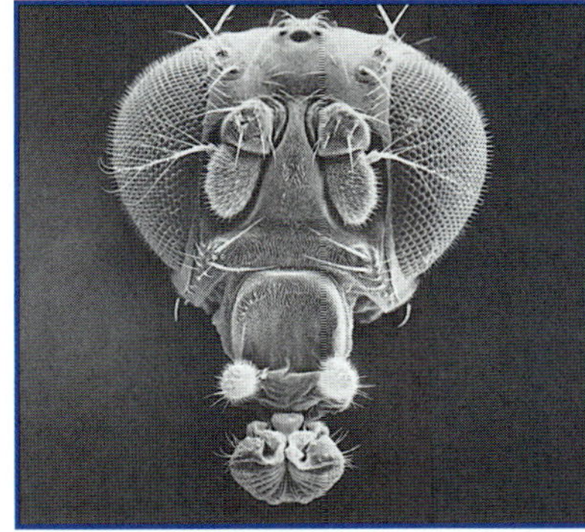

Así se ve una mosca de la fruta normal en un microscopio electrónico.

¡A esta mosca le crecieron patas en los ojos!

Comparación de códigos

Otra razón por la que estas "ratas de laboratorio con alas" son útiles es que su código genético es simple, y los investigadores lo conocen bien. Las moscas de la fruta tienen 12,000 genes, en cambio, los humanos tenemos más de 70,000. Muchos de los genes de la mosca de la fruta tienen funciones parecidas a las de los genes humanos y los científicos han aprendido a manipularlos para producir mutaciones genéticas. El estudio de estas mutaciones es importante para entender las mutaciones genéticas en los seres humanos. Sin las moscas de la fruta, la información que tenemos sobre algunos problemas genéticos de los seres humanos, como el cáncer de las células basales, podría haber costado mucho más tiempo y dinero.

¿Dónde está el límite?

¿Te parece bien que los científicos usen moscas, ratas, ratones y conejos en sus experimentos? ¿Qué opinas? Comparte tu punto de vista con tus compañeros.

149

Answers to Where Do You Draw The Line?

Answers will vary, but students should be aware that animals are used in biological research. Scientists have learned a great deal using animal models. Many scientists who work with animals take great care to minimize animal suffering during the experiments. In recent years, other models—such as computer models and cultured cells—have become available and can be used as substitutes for animals in some experiments.

Teaching Strategy

This activity has two objectives. First, it will allow the students to obtain a better grasp of the relationship a gene has to other body structures, such as cells or organs. Second, it will give students a better understanding of what is meant by the word *mutation* in the fruit-fly article.

For this activity to work, students must understand what blueprints are. Explain that blueprints are sheets of paper that have sketches of all the aspects of the house to be built. These plans are followed precisely in order to build the house. Show students sample picture of blueprints, and point out all the dimensional information they contain.

Next present an analogy to the students using the following representations:

body = city
organ = neighborhood
cell of organ = house in
   neighborhood
chromosomes in cell =
   blueprints for house in
   neighborhood
gene on chromosome = part of
   blueprints that gives instructions for certain aspect of
   house (like dimensions of the
   master bedroom or direction
   in which a door will open)

Chapter Organizer

CHAPTER ORGANIZATION	TIME MINUTES	OBJECTIVES	LABS, INVESTIGATIONS, AND DEMONSTRATIONS
Chapter Opener pp. 150–151	45		**Investigate!** Fingerprint Your Friends, p. 151
Section 1 **What Do Genes Look Like?**	90	▶ Describe the basic structure of the DNA molecule. ▶ Explain how DNA molecules can be copied. ▶ Explain some of the exceptions to Mendel's heredity principles.	**Demonstration,** p. 157 in ATE **Making Models,** Base-Pair Basics, p. 582 **Datasheets for LabBook,** Base-Pair Basics, Datasheet 14
Section 2 **How DNA Works**	90	▶ Explain the relationship between genes and proteins. ▶ Outline the basic steps in making a protein. ▶ Define *mutation,* and give an example of it. ▶ Evaluate the information given in a pedigree.	**QuickLab,** Mutations, p. 163
Section 3 **Applied Genetics**	90	▶ Define *genetic engineering.* ▶ List some of the benefits of combining the DNA of different organisms.	**Interactive Explorations CD-ROM,** DNA Pawprints *A* **Worksheet** *is also available in the* **Interactive Explorations Teacher's Edition.** **Long-Term Projects & Research Ideas,** Project 6

TECHNOLOGY RESOURCES

 **Guided Reading Audio CD**
English or Spanish, Chapter 7

 **One-Stop Planner CD-ROM with Test Generator**

 **Interactive Explorations CD-ROM**
CD 3, Exploration 8, DNA Pawprints

 **CNN** Science, Technology & Society,
Developing the Perfect Pepper, Segment 9

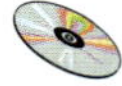 **Science Discovery Videodiscs**
Science Sleuths: Twins or Not?

CLASSROOM WORKSHEETS, TRANSPARENCIES, AND RESOURCES	SCIENCE INTEGRATION AND CONNECTIONS	REVIEW AND ASSESSMENT
Directed Reading Worksheet 7 **Science Puzzlers, Twisters & Teasers,** Worksheet 7		
Directed Reading Worksheet 7, Section 1 **Transparency 24,** DNA Structure **Math Skills for Science Worksheet 4,** A Shortcut for Multiplying Large Numbers **Science Skills Worksheet 23,** Science Drawing	**Real-World Connection,** p. 153 in ATE **MathBreak,** Genes and Bases, p. 155 **Math and More,** p. 155 in ATE **Cross-Disciplinary Focus,** p. 156 in ATE **Scientific Debate:** DNA on Trial, p. 172	**Homework,** pp. 152, 157 in ATE **Self-Check,** p. 155 **Review,** p. 159 **Quiz,** p. 159 in ATE **Alternative Assessment,** p. 159 in ATE
Directed Reading Worksheet 7, Section 2 **Transparency 25,** The Making of a Protein **Transparency 146,** The Formation of Smog **Transparency 26,** An Example of Substitution **Transparency 27,** Pedigree **Reinforcement Worksheet 7,** DNA Mutations	**Math and More,** p. 161 in ATE **Earth Science Connection,** p. 162 **Connect to Earth Science,** p. 162 in ATE **Cross-Disciplinary Focus,** p. 163 in ATE **Apply,** p. 164	**Self-Check,** p. 161 **Review,** p. 164 **Quiz,** p. 164 in ATE **Alternative Assessment,** p. 164 in ATE
Directed Reading Worksheet 7, Section 3 **Transparency 28,** Recombining DNA **Critical Thinking Worksheet 7,** The Perfect Parrot	**Cross-Disciplinary Focus,** p. 166 in ATE **Holt Anthology of Science Fiction,** *Moby James*	**Self-Check,** p. 166 **Review,** p. 167 **Quiz,** p. 167 in ATE **Alternative Assessment,** p. 167 in ATE

Holt, Rinehart and Winston On-line Resources

go.hrw.com

For worksheets and other teaching aids related to this chapter, visit the HRW Web site and type in the keyword: **HSTDNA**

National Science Teachers Association

www.scilinks.org

Encourage students to use the *sci*LINKS numbers listed with the Chapter Highlights to access information and resources on the **NSTA** Web site.

END-OF-CHAPTER REVIEW AND ASSESSMENT

Chapter Review in Study Guide
Vocabulary and Notes in Study Guide
Chapter Tests with Performance-Based Assessment, Chapter 7 Test
Chapter Tests with Performance-Based Assessment, Performance-Based Assessment 7
Concept Mapping Transparency 7

Visual Resources

TEACHING TRANSPARENCIES

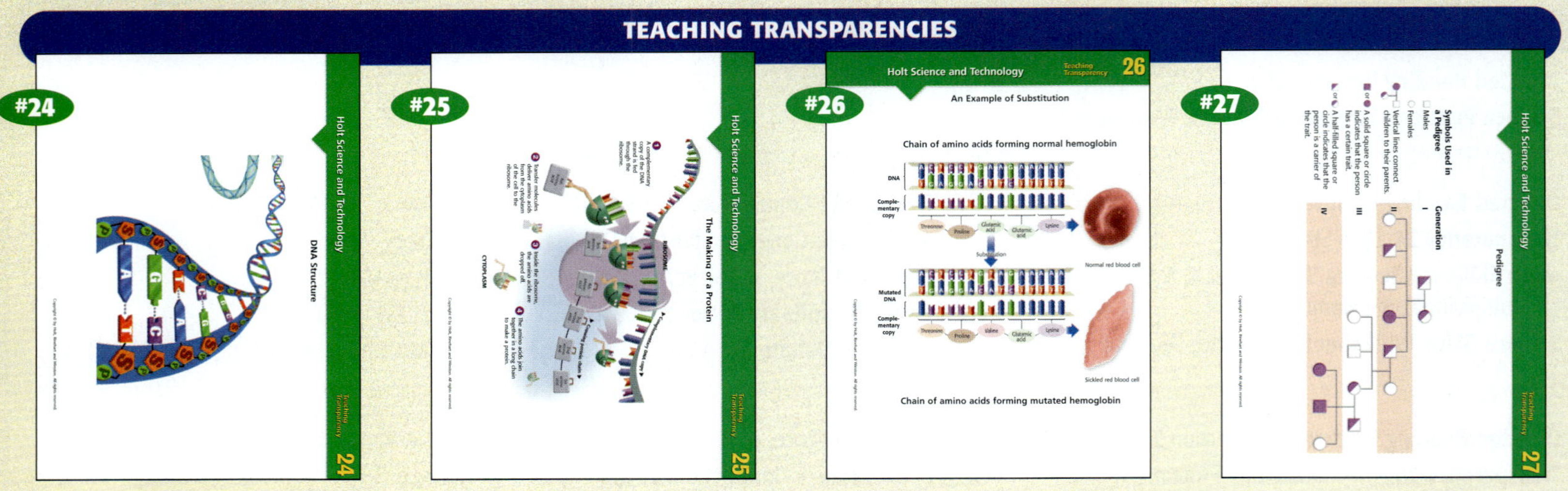

TEACHING TRANSPARENCIES

CONCEPT MAPPING TRANSPARENCY

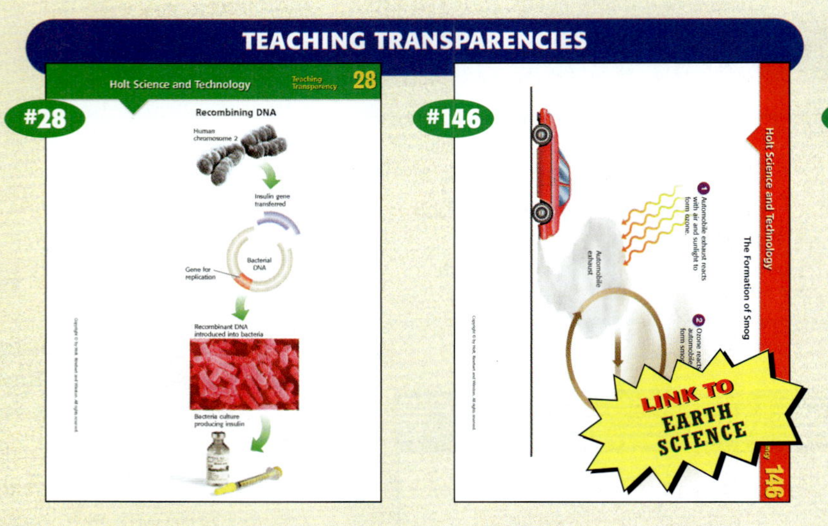

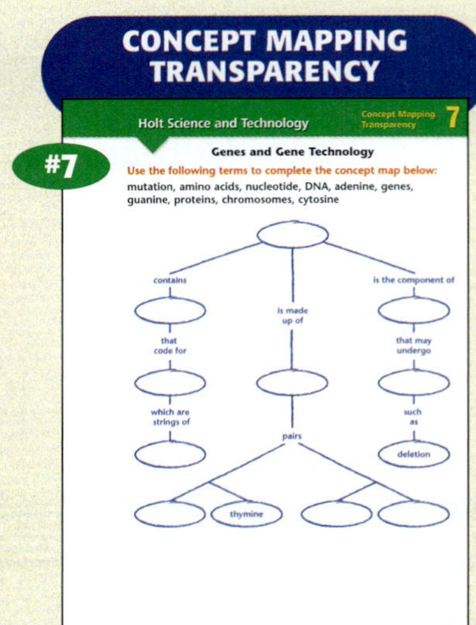

Meeting Individual Needs

DIRECTED READING

REINFORCEMENT & VOCABULARY REVIEW

SCIENCE PUZZLERS, TWISTERS & TEASERS

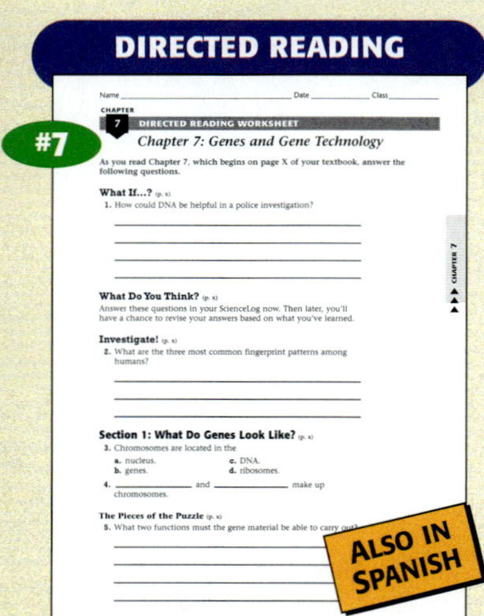

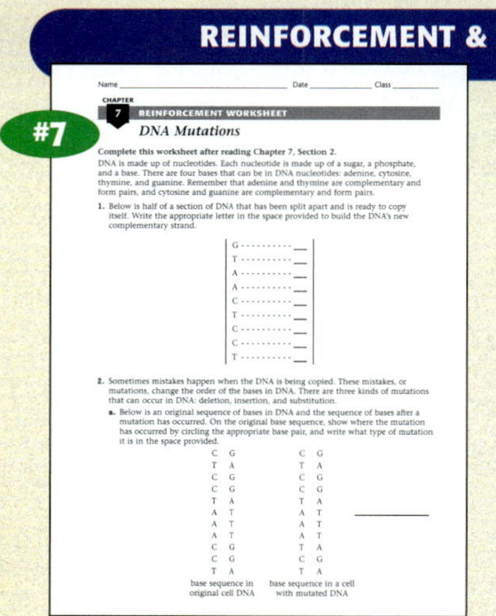

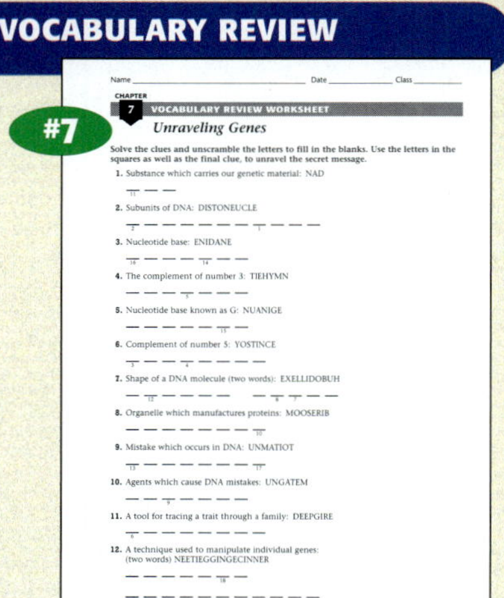

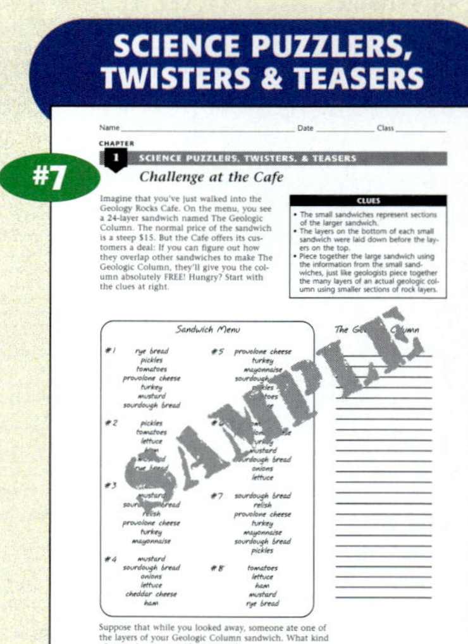

Review & Assessment

STUDY GUIDE

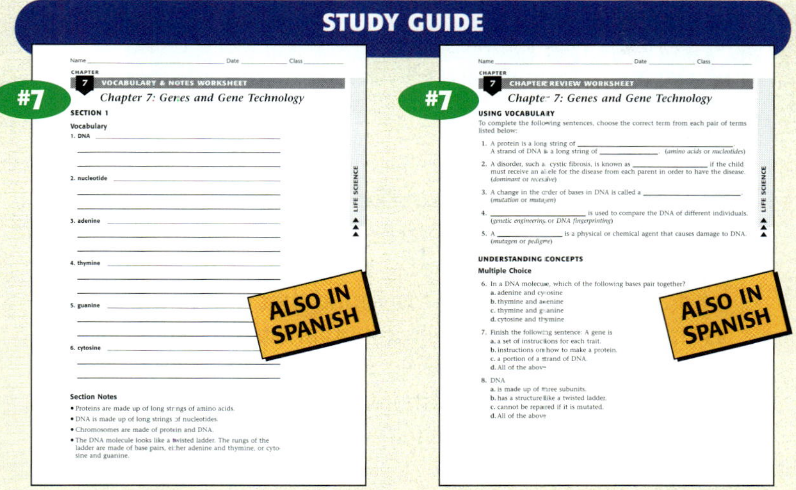

#7 — VOCABULARY & NOTES WORKSHEET — Chapter 7: Genes and Gene Technology

#7 — CHAPTER REVIEW WORKSHEET — Chapter 7: Genes and Gene Technology

CHAPTER TESTS WITH PERFORMANCE-BASED ASSESSMENT

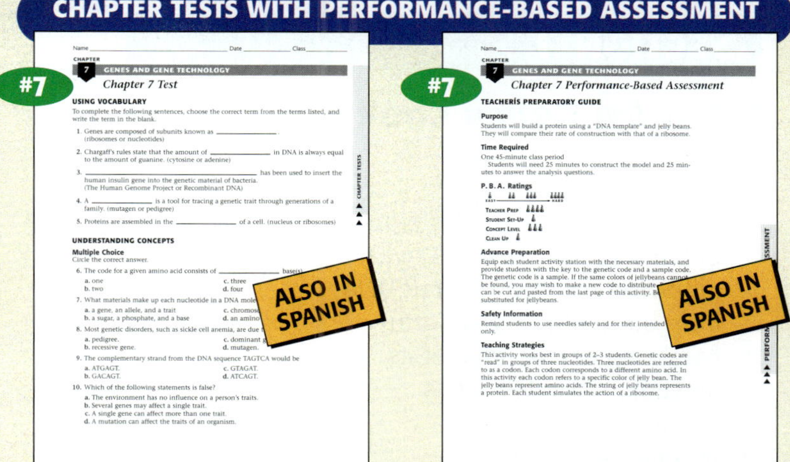

#7 — GENES AND GENE TECHNOLOGY — Chapter 7 Test

#7 — GENES AND GENE TECHNOLOGY — Chapter 7 Performance-Based Assessment

Lab Worksheets

LONG-TERM PROJECTS & RESEARCH IDEAS

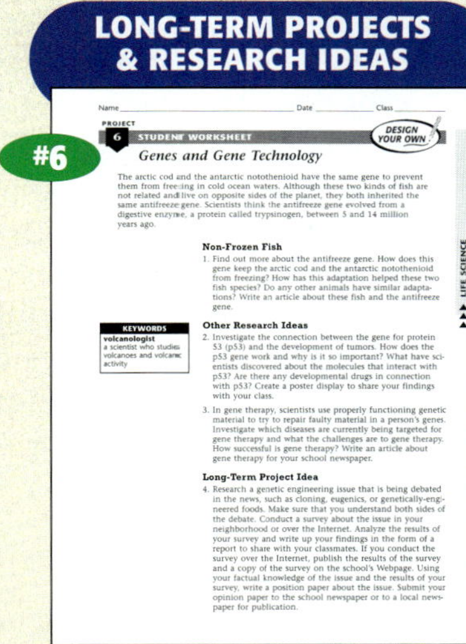

#6 — STUDENT WORKSHEET — Genes and Gene Technology — DESIGN YOUR OWN

DATASHEETS FOR LABBOOK

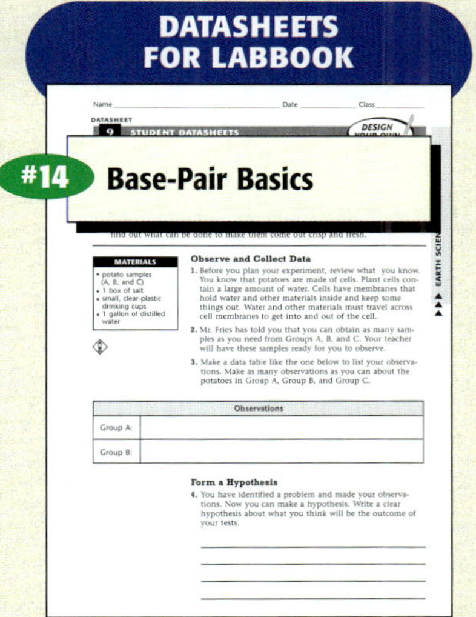

#14 — STUDENT DATASHEETS — Base-Pair Basics — DESIGN YOUR OWN

Applications & Extensions

CRITICAL THINKING & PROBLEM SOLVING

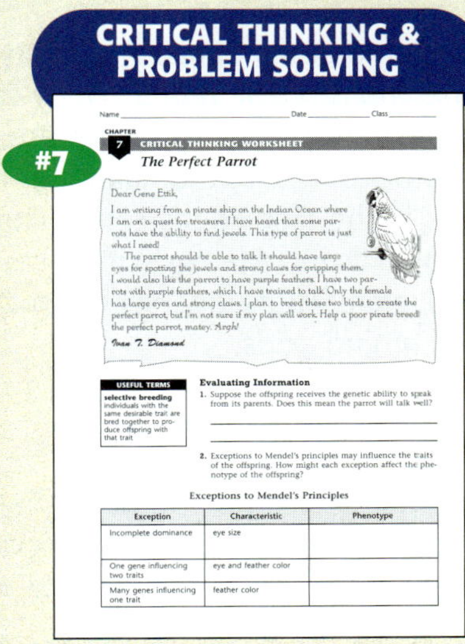

#7 — CRITICAL THINKING WORKSHEET — The Perfect Parrot

SCIENCE TECHNOLOGY

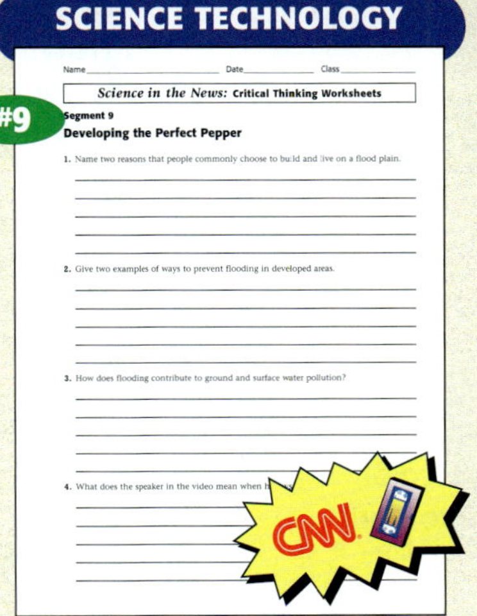

#9 — Science in the News: Critical Thinking Worksheets — Segment 9 — Developing the Perfect Pepper

INTERACTIVE EXPLORATIONS

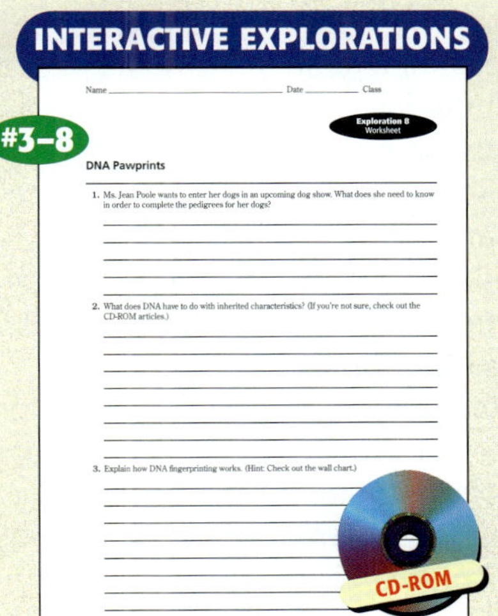

#3–8 — Exploration B Worksheet — DNA Pawprints

SECTION 1

What Do Genes Look Like?

▶ DNA

In 1869, long before the time of Watson and Crick, a 22-year-old Swiss scientist isolated DNA from a cell nucleus. Unfortunately, he had no idea of its function, much less of its role in inheritance. It was not until 75 years later, in 1944, that an American geneticist named Oswald T. Avery found evidence that DNA is the carrier of genetic information.

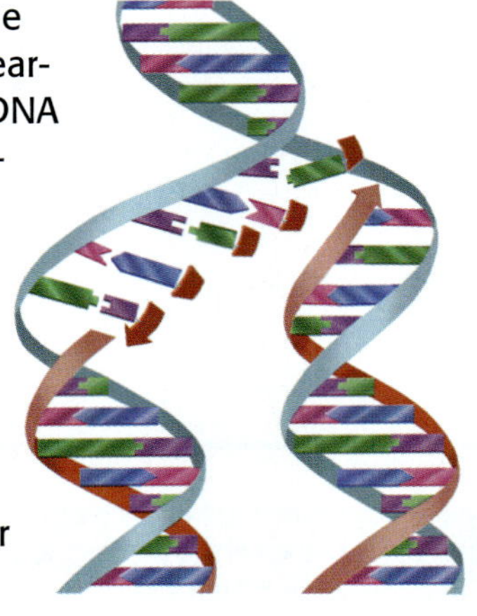

IS THAT A FACT!

- ◆ Human DNA consists of about 3 billion base pairs.

- ◆ If you could print a book with all the genetic information carried in just one human cell, it would be 500,000 pages long.

- ◆ The uncoiled DNA in the nucleus of a human body cell is about 2 m long. The DNA in chromosomes is so tightly coiled that the 46 chromosomes from a human body cell would be only about 0.00032 cm long if they were lined up end to end!

▶ Eye Color

Eye color is a trait that is influenced by the presence of melanin, a dark pigment, in the iris. People whose eyes have the least pigment have blue eyes, and people whose eyes have the most pigment have brown eyes. Scientists 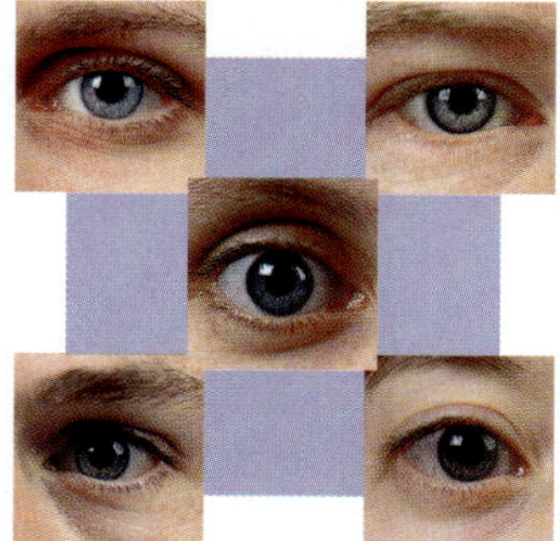 think that the number of alleles present for eye pigment control eye color. For example, one allele is thought to contribute to a medium blue eye color, while as many as eight alleles may contribute to dark brown eye color.

SECTION 2

How DNA Works

▶ Cracking the Genetic Code

In the 1960s, scientists cracked the genetic code—the translation between codons, the three base sequences and amino acids. They have found that the genetic code is universal in almost all living organisms. If a codon aligns with a particular amino acid in humans, the same codon aligns with the same amino acid in bacteria. This similarity suggests that all life-forms have a common evolutionary ancestor.

▶ Mutations and Natural Selection

The discovery that changes in DNA can lead to new traits supports Darwin's theory of evolution by natural selection. When an environment changes, mutations may enhance an organism's chances of survival in the new conditions. For example, a certain mutation makes houseflies resistant to DDT but also reduces their growth rate. Originally, this mutation was harmful, but after DDT became part of the environment of many houseflies, the mutation helped the flies survive. By the process of natural selection, the mutation was spread throughout the fly population.

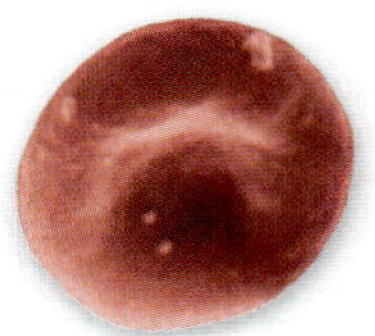

▶ Amino Acids

Of the known amino acids, 20 are necessary for human growth and metabolism. The human body can manufacture 10 of these. The other 10, which are called the essential amino acids, must be obtained from plant or animal proteins in the diet. Foods containing all the essential amino acids include eggs, milk, seafood, and meat. Legumes, grains, nuts, and seeds contain many of the essential amino acids.

▶ Protein Synthesis

It took many years for scientists to determine how protein is synthesized in the cell. The discovery that DNA's nucleotide sequence corresponds to a certain amino acid sequence was a key step in unlocking this mystery. This link was conclusively proven by Charles Yanofsky and Sydney Brenner in 1964.

- The genetic sequences used to make proteins can be compared to sentences. Where each three-letter "word" in the genetic "sentence" starts and stops is very important for constructing a protein. For example, suppose the sentence to code for a particular protein read "PAT SAW THE FAT CAT." If you start just one base pair too late, the sentence would read "ATS AWT HEF ATC AT," which is meaningless.

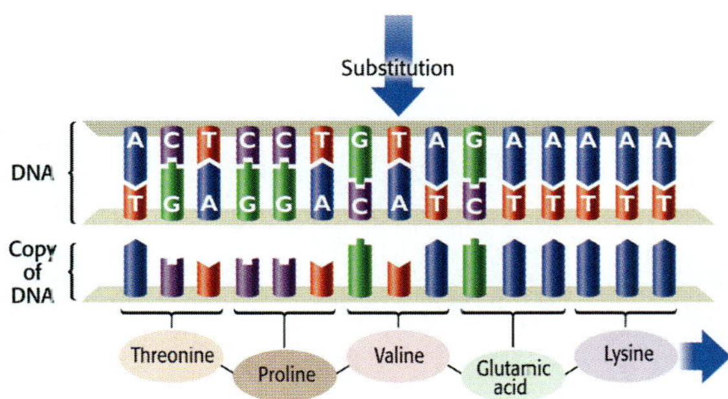

▶ Huntington's Disease

Huntington's disease is a hereditary degenerative brain disease caused by a dominant allele. People affected by Huntington's often display peculiar, dancelike movements. In the United States, many cases of Huntington's disease can be traced back to two brothers. The two men immigrated to North America from England in the 1600s because their family was accused of witchcraft. Apparently, they were persecuted because of their strange behaviors, which are now understood to be symptoms of Huntington's disease.

Applied Genetics

▶ Chimeras

Genetically engineered hybrid creatures are often called chimeras. The word *chimera* comes from Greek mythology; the Chimera was a fire-breathing she-monster, usually depicted as a composite of a lion, a goat, and a serpent.

▶ Ethical Issues

Scientists disagree about both the ethics of genetic engineering and the safety risks involved.

- Dr. Maxine Frank Singer was one of the first scientists to alert the National Academy of Science to the potential hazards of genetic engineering. Due to the efforts of Dr. Singer and her colleagues, the National Institutes of Health developed specific guidelines for genetic research in 1973. These guidelines regulate the production and use of genetically engineered DNA and organisms.

▶ Worm Work

Caenorhabditis elegans, a tiny worm used in genetic experiments, has many genes in common with humans. In fact, scientists have given these worms human anti-depressants to study the genetic causes of depression.

▶ DNA Fingerprints

DNA fingerprints are frequently used in criminal investigations. The DNA can come from hair, skin cells, blood, or other body fluids left at the crime scene by the perpetrator. Scientists use enzymes to make many copies of the DNA and then use other enzymes to cut the DNA into fragments. The fragments are separated by size and other characteristics on a specially treated plate. A photograph of the plate is taken, showing a unique set of dark bands. This set of bands is known as a DNA fingerprint. The fingerprint is then compared with the DNA fingerprint of the suspect to help determine innocence or guilt.

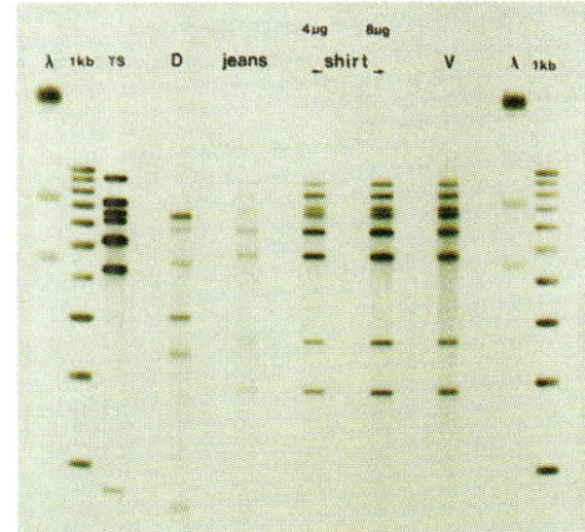

For background information about teaching strategies and issues, refer to the *Professional Reference for Teachers.*

Chapter Preview

Section 1
What Do Genes Look Like?
- The Pieces of the Puzzle
- A Picture of DNA
- Eureka!
- DNA Structure
- Making Copies of DNA
- From Trait to Gene
- More News About Traits

Section 2
How DNA Works
- Genes and Proteins
- The Making of a Protein
- Changes in Genes
- An Example of Substitution
- Genetic Counseling

Section 3
Applied Genetics
- Designer Genes
- Living Factories
- DNA Fingerprints
- The Big Picture

Directed Reading Worksheet 7

Science Puzzlers, Twisters & Teasers Worksheet 7

Guided Reading Audio CD
English or Spanish, Chapter 7

CAPÍTULO

7 Los genes y la tecnología genética

¿Qué tal si...?

Te acusan de un crimen que no cometiste. Un testigo te señala como culpable. Tu tipo de sangre corresponde al que se encontró en la escena del crimen. En realidad estabas en tu casa cuando ocurrió, pero no tienes forma de probarlo. Sin embargo, hay un testigo que puede ayudarte: tu ADN. El ADN es una substancia que se encuentra en todas tus células.

Una técnica conocida como "impresiones de ADN" puede producir una imagen de los patrones hechos por tu ADN. A menos que tengas un gemelo idéntico, tu ADN es único y puede utilizarse para identificarte. Pero, ¿qué es el ADN? ¿Por qué es único en cada persona? ¿Qué tiene que ver el ADN con lo que eres? En este capítulo lo sabrás.

150

What If . . . ?

Headlines like these are becoming more common. As scientists gain a better understanding of DNA, they are able to use it to solve crimes.

All individuals except for identical twins have a unique DNA fingerprint. That is why DNA fingerprints can be used for identification just as ordinary fingerprints can.

¿Tú qué piensas?

Usa tus conocimientos para responder a las siguientes preguntas en tu cuaderno de ciencias:

1. ¿Por qué no somos todos iguales?

2. ¿Qué son los genes? ¿Dónde están?

3. ¿Cómo se usa el conocimiento científico del ADN para tratar enfermedades?

Toma las huellas digitales de tus amigos

Un método de identificación es la dactilografía (tomar las huellas digitales). ¿Funciona? ¿Todas las huellas digitales son diferentes? Haz esta actividad para descubrirlo.

Procedimiento

1. Frota un **carboncillo** sobre **una hoja de papel de calco.** Pon la yema de un dedo sobre la marca del carboncillo y pon un **pedazo de cinta adhesiva transparente** en el dedo manchado. Quita la cinta y pégala en una **hoja de papel blanca.** Haz lo mismo con los demás dedos.

2. Observa las huellas con una **lupa.** ¿Qué tipo de huellas ves? Las de la derecha, son las más comunes entre los humanos. ¿Tienen tus dedos el mismo patrón?

Análisis

3. Compara tus huellas con las de tus compañeros. ¿Cuántas huellas iguales ves? ¿Tienen dos de tus compañeros las mismas huellas? Explica.

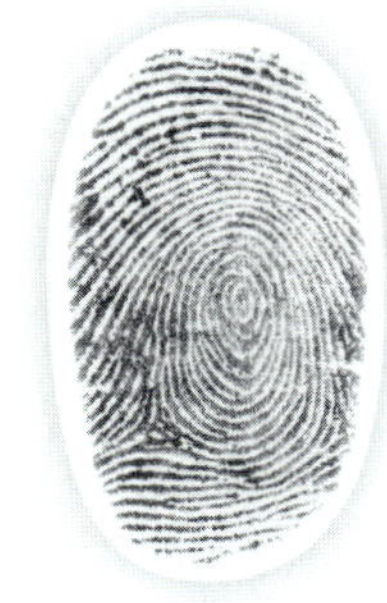

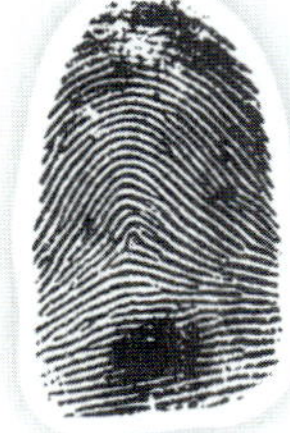

151

What Do You Think?

Accept all reasonable responses.

Students will have a chance to revise their answers in the Chapter Review under NOW What Do You Think?

Investigate!

MATERIALS

FOR EACH GROUP:
- piece of charcoal
- sheet of tracing paper
- transparent tape
- white paper
- magnifying lens

Safety Caution: Remind students to review all safety cautions and icons before beginning this lab activity. Charcoal is nontoxic, but it can stain clothes.

The loop pattern is found in about 65 percent of the population, the whorl in about 30 percent, and the arch in about 5 percent.

Answers to Investigate!

3. The number of fingerprint types will vary for each class. No two students should have the same fingerprint (unless there are identical twins in the class). Any logical, reasonable explanation for this is acceptable, but students should discuss genetic variety in their explanation.

WEIRD SCIENCE

Some people don't have fingerprints at all. For example, some scientists have lost their fingerprints by handling toxic substances, like strong acids and bases. Through frequent exposure to a strong base, such as sodium hydroxide, fingerprints can be worn away. Some people whose hands have been severely burned have also lost their fingerprints. The resulting scar tissue, however, can be just as unique as the original fingerprints.

Focus

What Do Genes Look Like?

This section introduces students to the structure and function of DNA, the process of DNA replication, and the relationship of DNA to traits. The section concludes with a discussion of some exceptions to Mendelian genetics.

Bellringer

Have students unscramble the following words and use them in a sentence:

NDA (DNA)

etcutsurr (structure)

(Sample sentence: DNA has a complex structure.)

1 Motivate

ACTIVITY

Modeling Genetic Code

Create a code by pairing each letter of the alphabet with a numeral. For example, the numeral 1 could represent the letter *a*. Have students encode a brief message. Then have students exchange and decode the message. Compare the process of encoding messages with the encoding of genetic information in DNA. Explain that the genetic code is based on the sequence of the four nucleotide bases.

Directed Reading Worksheet 7 Section 1

VOCABULARIO

ADN	timina
nucleótido	guanina
adenina	citosina

OBJETIVOS

- Describe la estructura básica de la molécula de ADN.
- Explica cómo se copian las moléculas de ADN.
- Explica algunas de las excepciones de las leyes básicas de la herencia de Mendel.

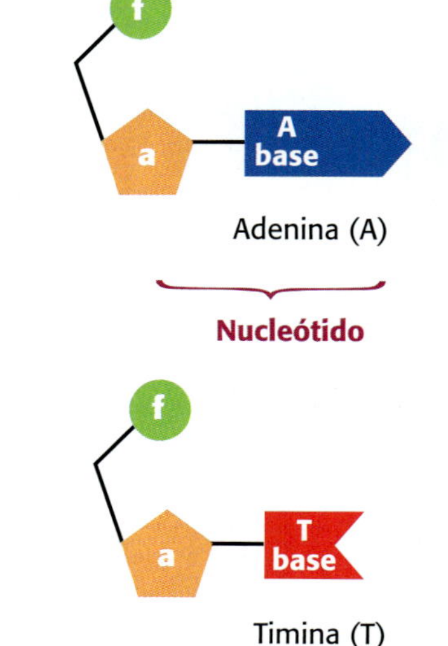

Figura 1 *Cada nucleótido está hecho de un azúcar, un fosfato y una base. Hay cuatro bases diferentes.*

152

¿Cómo son los genes?

Se sabe que los genes determinan los rasgos que pasan de una generación a otra, y también que se encuentran en los cromosomas, unas estructuras que se localizan en el núcleo de la mayoría de las células. Los cromosomas están hechos de **ADN** (ácido desoxirribonucleico) y proteínas. Pero, ¿de qué están hechos los genes? Los científicos llevan discutiendo este tema desde hace 50 años.

Las piezas del rompecabezas

El material genético desempeña dos funciones. Primero, proporciona instrucciones complejas en los procesos celulares y en la construcción de estructuras celulares. Segundo, se copia cada vez que una célula se divide, para que cada célula tenga un grupo idéntico de genes. Esto asegura que la información hereditaria se transmita a la siguiente generación. Los primeros estudios del ADN sugerían que éste era una molécula sencilla. Por esto, se pensaba que las proteínas eran las que transportaban la información hereditaria. Después de todo, las proteínas son moléculas complejas.

Sin embargo, en la década de 1940 dos experimentos sorprendentes mostraron que los genes de bacterias y virus estaban hechos de ADN. ¿Cómo algo tan sencillo podía tener la clave para crear y dirigir un organismo viviente? Para encontrar la respuesta, echemos un vistazo a las subunidades que forman una molécula de ADN.

Nucleótidos: las subunidades del ADN El ADN está formado por cuatro subunidades que se conocen como **nucleótidos.** Cada nucleótido que forma una molécula de ADN tiene tres tipos diferentes de material: un azúcar, un fosfato y una base. Los nucleótidos son idénticos, excepto por el tipo de base. Las cuatro bases son **adenina, timina, guanina** y **citosina,** cada una de las cuales tiene una forma un poco diferente. Por lo general, las bases se identifican con las primeras letras de su nombre: **A, T, G** y **C.** La **Figura 1** muestra los diagramas de los cuatro nucleótidos. Imagínate cómo encajan.

Homework

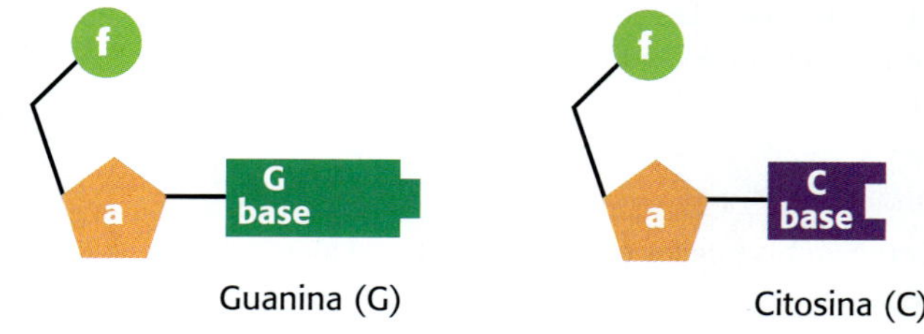

Research Have students collect information on the use of amino acids to gain muscle. Suggest that they look at ads for amino acid supplements in health-food or fitness magazines or at labels of amino acid powdered drinks at the supermarket.

Use these materials to discuss issues such as the expense of such supplements and how they might be used by the body. Discuss how amino acids might be acquired in a balanced diet.

La regla de Chargaff En la década de 1950, el bioquímico Erwin Chargaff, estudiando muestras de ADN de diferentes organismos, descubrió que la cantidad de adenina en el ADN siempre es igual a la de timina, y que la cantidad de guanina siempre es igual a la de citosina. Su descubrimiento, conocido como la regla de Chargaff, se representa así:

$$A = T \text{ y } G = C$$

En ese tiempo, nadie sabía qué hacer con el descubrimiento de Chargaff. ¿Cómo ayudaría la regla de Chargaff a resolver los misterios de la estructura del ADN? Sigue leyendo para averiguarlo.

Una fotografía del ADN

Del laboratorio del científico británico Maurice Wilkins llegaron nuevas pistas. Allí, la química Rosalind Franklin, que aparece en la **Figura 2,** creó imágenes de las moléculas de ADN. El proceso que utilizó para crear la imagen que ves en la **Figura 3** se conoce como difracción de rayos X. En este proceso, los rayos X bombardean la molécula de ADN; al golpear una partícula dentro de la molécula, el rayo rebota. Esto crea un patrón que se registra en una película. Las imágenes que Rosalind Franklin captó sugieren que el ADN tiene forma de espiral.

¡Lo encontré!

Luego los jóvenes científicos, James Watson y Francis Crick, que se ven en la **Figura 4,** basándose en el trabajo de otros, construyeron modelos de ADN con materiales muy simples, como piezas de cartón marcadas. Después de ver las imágenes de rayos X del ADN tomadas por Rosalind Franklin, Watson y Crick armaron el rompecabezas. En un momento de genialidad, descubrieron que el ADN tiene la forma de una escalera de caracol, conocida como *espiral doble*. Watson y Crick usaron su modelo de ADN para predecir cómo se copia el ADN. Cuando hicieron este descubrimiento, se dice que Crick exclamó: "¡Hemos descubierto el secreto de la vida!"

Figura 2 Rosalind Franklin, 1920–1958

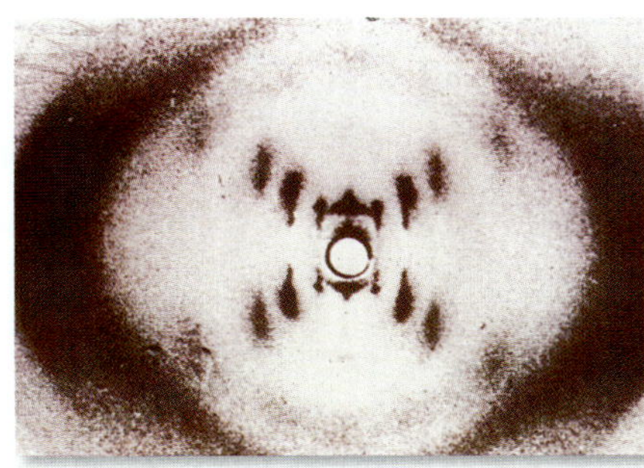

Figura 3 Esta fotografía sacada por difracción de rayos X del ADN fue tomada por Rosalind Franklin.

Figura 4 Esta fotografía muestra a James Watson (a la izquierda) y a Francis Crick (a la derecha) con su modelo de ADN.

153

READING STRATEGY

Mnemonics Have students create a mnemonic device that will remind them of the names of the bases and how they form pairs. Examples such as "**A**toms are **T**iny" or "**A**toms are **T**errific" might help remind students that **a**denine pairs with **t**hymine. "**C**athy is **G**reat" or "**C**andy is **G**ross" might remind them that **c**ytosine pairs with **g**uanine.

REAL-WORLD CONNECTION

Bipolar disorder, which is characterized by extreme behavioral changes, from deep depression to mania, is being investigated for a genetic component. One 19-year study published in 1987 followed an extended family, many members of which suffered from bipolar disorder. The family members who suffered from the disorder all had the same genetic marker (an active section of DNA), and members without the marker did not exhibit the disorder. Later studies on different populations, however, noted that people without the disorder sometimes also have that marker. Furthermore, scientists point out that the rate of bipolar disorder among Americans has risen dramatically during the last 50 years and that the average age of onset has plummeted from 32 to 19—suggesting that inherited genes alone could not possibly be responsible for bipolar disorder. These scientists suggest that environmental causes and personal experiences are more likely causes of bipolar disorder.

SCIENTISTS AT ODDS

In 1951, Rosalind Franklin began working with Maurice Wilkins in the lab at King's College, in London, but she and Wilkins never got along. Both Wilkins and James Watson belittled Franklin's abilities and accomplishments. Francis Crick, however, respected her work. Franklin's study suggesting the helical structure of DNA was instrumental in Watson and Crick's discovery of DNA's double-stranded shape. If Franklin had not died in 1958, she almost certainly would have shared the Nobel Prize awarded to Watson, Crick, and Wilkins in 1962 for their discovery of the structure of the DNA molecule.

ACTIVITY

Making a DNA Model

MATERIALS

For Each Group:
- 2 licorice whips (60 cm long)
- 50–60 toothpicks
- 50–60 gumdrops in four different colors

Have students work in groups to construct a candy model of DNA. The licorice represents the sides of the DNA molecule, and the gumdrops represent the base pairs. Each group should stretch out two licorice whips parallel to each other. Tell them to slide two gumdrops onto the middle of a toothpick to resemble a complementary base pair and to then insert each end of the toothpick into the licorice whips, connecting them the way rungs connect the two sides of a ladder. After the ladder of toothpicks grows to 20–30 base pairs, have students shape the helix by spiraling it around a tube or pole.

MEETING INDIVIDUAL NEEDS

Learners Having Difficulty

To help students better understand how the term *complementary* relates to the structure of DNA, point out that the term means "completing." Using **Figure 5,** explain that complementary base pairs join together to *complete* each rung on the spiral-staircase structure of DNA. Then point out that complementary strands of DNA join together to complete one DNA molecule. Sheltered English

Teaching Transparency 24
"DNA Structure"

Figura 5 *La estructura del ADN se puede comparar con una escalera de caracol.*

Estructura del ADN

En la parte izquierda de la **Figura 5** se muestra la escalera de caracol o espiral doble. Como puedes ver en la **Figura 6,** los dos lados de la escalera están hechos de moléculas alternadas de azúcar y de fosfato. Los peldaños de la escalera están formados por un par de bases nucleótidas. La adenina que está en un lado siempre forma un par con la timina del otro lado. De la misma manera, la guanina siempre forma un par con la citosina. ¿Cómo puede esta estructura explicar los hallazgos de Chargaff?

Figura 6 *En una molécula de ADN, las bases deben formar pares en una forma determinada. Si hay un error y las bases no se forman en pares correctamente, el gen no va a transportar la información adecuada.*

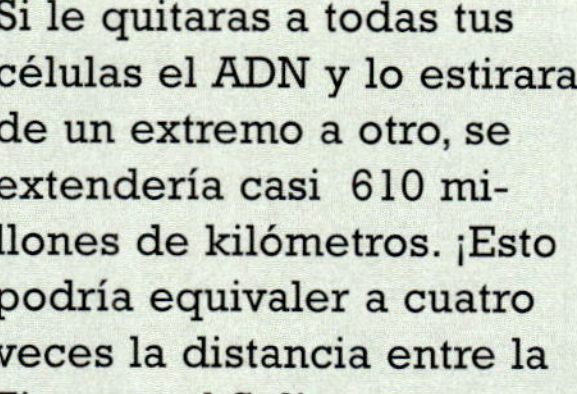

Si le quitaras a todas tus células el ADN y lo estiraras de un extremo a otro, se extendería casi 610 millones de kilómetros. ¡Esto podría equivaler a cuatro veces la distancia entre la Tierra y el Sol!

154

WEÏRD SCIENCE

Do werewolves really exist? No, but there is a gene believed to be located on the X chromosome that causes thick and abundant hair to grow on the upper body and face, including the ears, nose, cheeks, forehead, and even eyelids of affected males. This condition is sometimes called the werewolf syndrome because people with this gene resemble werewolves depicted in movies. This condition only affects people's appearance, however, not their behavior.

Copiar el ADN

¿Por qué es tan maravilloso el ADN? El modelo de Rosalind Franklin y de Watson y Crick explica cómo se copian las moléculas de ADN. Como la adenina siempre se une con la timina y la guanina siempre se une con la citosina, las bases de un lado de la molécula se vuelven una plantilla o modelo para el otro lado. Esto quiere decir que un lado es *complementario* del otro. Por ejemplo, una secuencia como ACCG tendría que tener la secuencia TGGC para formar un nuevo lado complementario. El otro lado se copia de la misma forma. Esto crea dos moléculas de ADN idénticas.

Como se muestra en la **Figura 7,** la molécula de ADN se divide en dos en el punto de enlace de las dos bases y cada lado de la hélice puede entonces aparearse con nucleótidos adicionales dentro del núcleo.

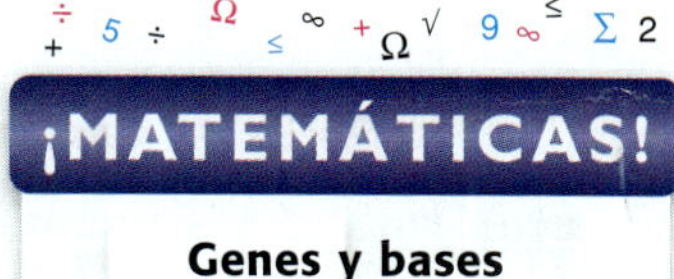

¡MATEMÁTICAS!

Genes y bases

Un ser humano tiene casi 100,000 genes. Si hay aproximadamente 30,000 bases en cada gen humano, ¿cuántas bases hay en 100,000 genes?

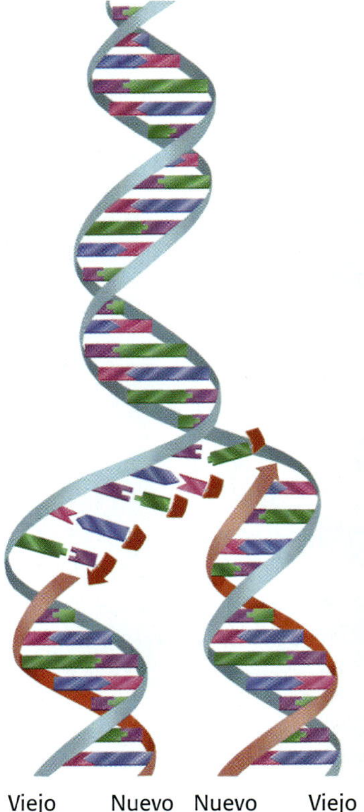

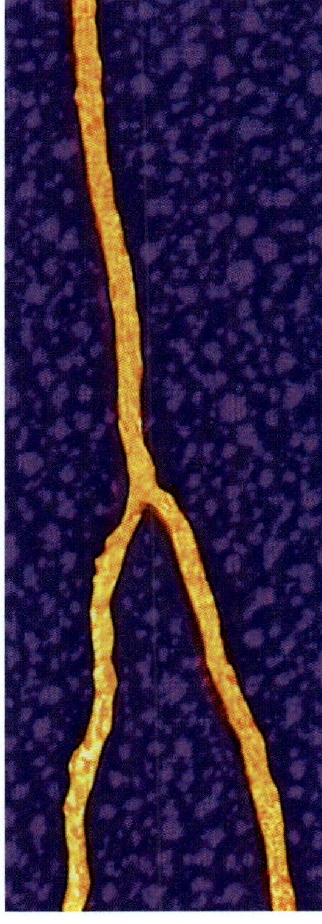

Figura 7 *La ilustración de la izquierda muestra la molécula de ADN dividiéndose para hacer una copia de sí misma. Cada mitad de la molécula original sirve como plantilla en la que se forma una nueva cadena complementaria. La fotografía de la derecha muestra una molécula de ADN que se ha separado. Este ADN está amplificado casi 1 millón de veces.*

Autoevaluación

¿Cuál sería la cadena complementaria de ADN para la siguiente secuencia de bases? ACCTAGTTG *(Consulta la página 636 para comprobar tu respuesta.)*

155

MISCONCEPTION ALERT

Can ancient DNA be used to produce dinosaurs as seen in the movies *Jurassic Park* and *The Lost World*? In these movies, scientists make dinosaurs by combining fragments of ancient DNA with DNA from modern-day frogs. In reality, fragments of ancient DNA have indeed been found, but a fragment of DNA does not provide enough information to make an entire organism. Moreover, there is no way to know whether the DNA fragments are even from a dinosaur. In addition, a frog's DNA will grow a frog, which is an amphibian, not a reptile. Most scientists agree that dinosaurs were reptiles.

USING THE FIGURE

Have students compare the illustration with the photograph of DNA replication in **Figure 7.** Have them identify the fork, or upside-down V, in both images. Point out that this is the point where the two strands of DNA have separated. Ask the following question:

How many molecules of DNA will there be when the process is complete? (two)

Are the new DNA molecules complementary or identical to the original DNA molecule? (identical)

Answer to MATHBREAK

3,000,000,000

Answer to Self-Check

TGGATCAAC

MATH and MORE

By causing DNA to replicate in a test tube, scientists can make as many as a billion copies of the original DNA template in very little time. This is a chain reaction in which DNA multiplies **exponentially** with time. This process is extremely important in biotechnology and in genetic research. To help students understand the concept of exponential growth, calculate how much money you will have in 30 days if you start with a penny on day 1 and the amount of money you have doubles every day. Using a calculator, write the products of each multiplication on the board.

(In 30 days you will have $5,368,709.12!)

Math Skills Worksheet 4 "A Shortcut for Multiplying Large Numbers"

2 Teach, continued

CROSS-DISCIPLINARY FOCUS

Music Researchers at the University of California, San Francisco, are trying to find out if a gene is responsible for "perfect pitch," the ability to immediately determine any musical note upon hearing it. It is a rare ability—possessed by perhaps only one in every 2,000 people—found most often among musicians. People with perfect pitch can easily determine the musical note of a dial tone, the hum of a refrigerator, or any sound they hear. Preliminary findings indicate that people with perfect pitch may inherit the ability, but that an early education in music may also be necessary for the trait to be fully expressed.

MISCONCEPTION ALERT

Recessive traits are sometimes referred to in a negative way, as if having a recessive trait is a bad thing. Emphasize that some recessive traits are beautiful, such as the light blue eyes of the white tiger.

internetconnect

SCLINKS. NSTA
TOPIC: Genes and Traits
GO TO: www.scilinks.org
***sci*LINKS NUMBER:** HSTL135

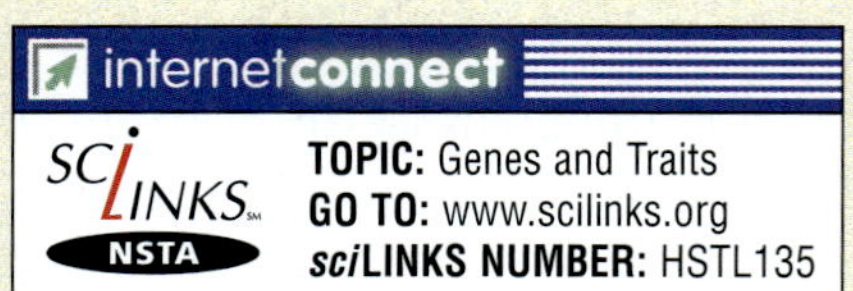

De rasgo a gen

El modelo de Watson y Crick explica porqué el ADN contiene tanta información. Las bases de la molécula se pueden poner en cualquier orden. Esto permite codificar una gran variedad de genes. Cada gen está formado por una cadena de bases. El orden de las bases da la información celular para la expresión de cada rasgo.

Panorama general El ADN funciona igual en todos los organismos: en las bacterias, los mosquitos, las ballenas y los seres humanos, pero también nos hace únicos. En el siguiente diagrama verás la relación que hay entre un rasgo y una base del ADN.

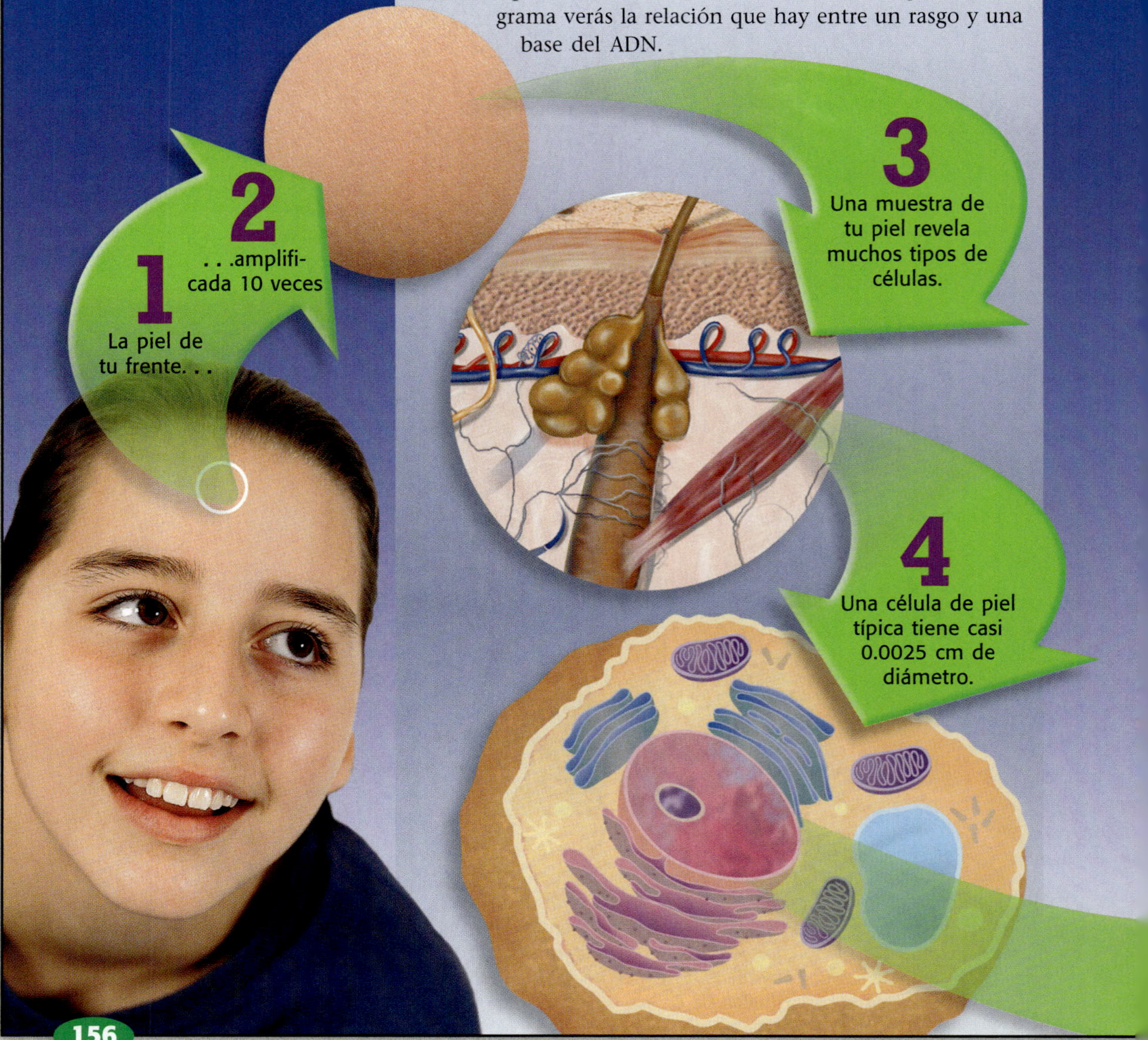

156

Science Bloopers

James Watson disliked Rosalind Franklin so much that at a 1951 lecture where she gave information about the size and possible shape of the DNA molecule, Watson refused to take notes. It took Watson and Crick 2 more years to discover some of the same information on their own. When they did, their great discovery of the shape of the DNA molecule came within 2 weeks!

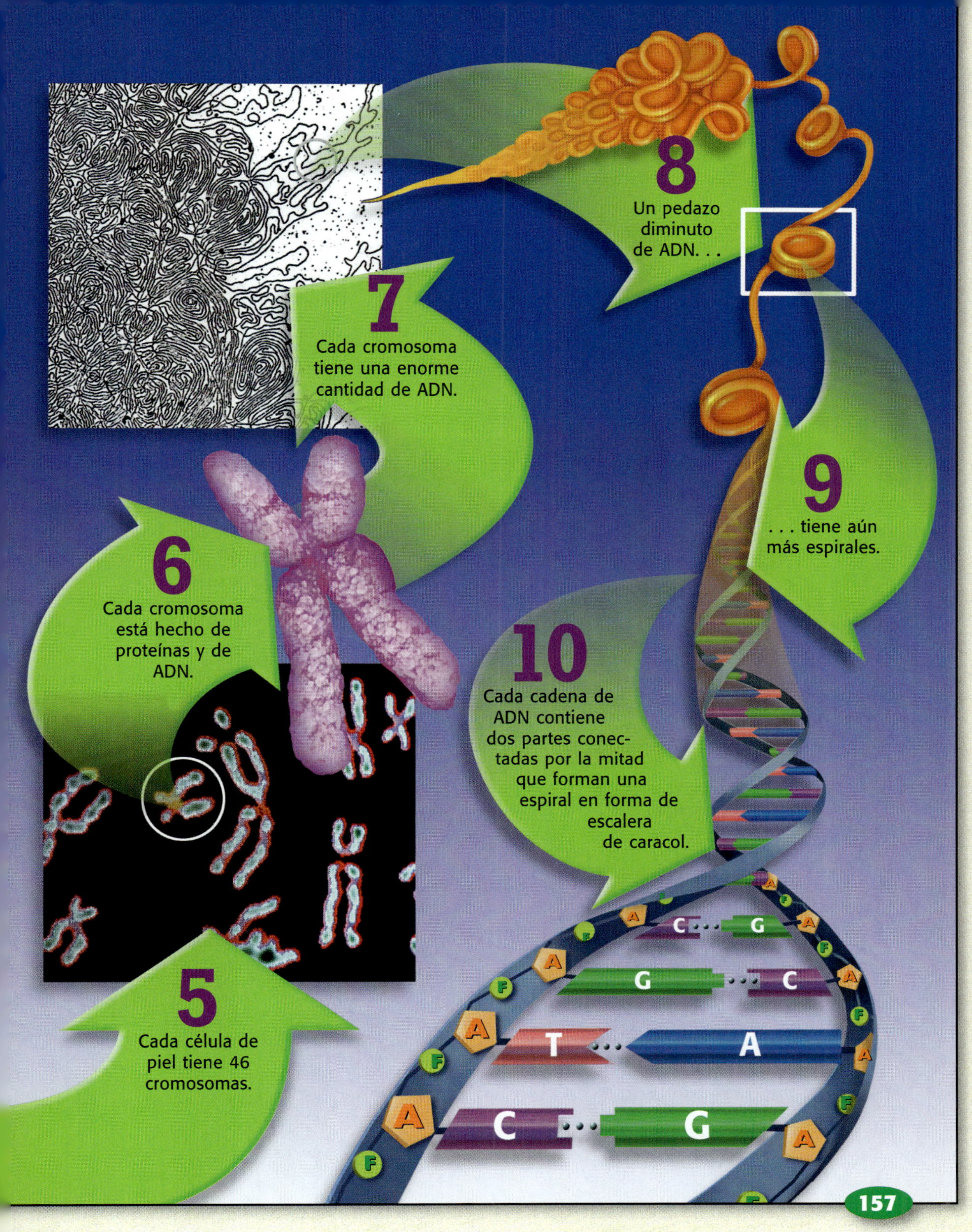

3 Extend

Demonstration

To illustrate the difficulty of fitting all of the DNA into a cell, challenge one or more students to stuff 2 m of embroidery floss into a size "0" Vegicap™. Vegicaps are available at most health-food stores. Sheltered English

Group Activity

Have students imagine that they have just discovered the structure of DNA and must present their findings to a group of scientists. Have small groups of students use a model of DNA, a poster, or another visual aid to briefly describe the structure of DNA to their classmates.
Sheltered English

Homework

Poster Project Have students use colored pencils to draw a portion of a DNA molecule. The drawing should reflect the correct structure of DNA and should contain a legend for identifying the subunits. You may also want students to outline the steps involved in copying DNA.

Science Skills Worksheet 23
"Science Drawing"

internetconnect

SCILINKS NSTA
TOPIC: DNA
GO TO: www.scilinks.org
***sci*LINKS NUMBER:** HSTL130

IS THAT A FACT!

Genetic disorders are a serious health problem for humans. Scientists know that faulty or missing genes cause diseases such as cystic fibrosis and sickle cell anemia. Scientists hope to be able to treat genetic disorders someday by altering genes within body cells.

READING STRATEGY

Prediction Guide Before students read this page, ask them if they agree with the following three statements. Students will discover the answers as they explore Section 1.

- Tigers with white fur are likely to have blue eyes. (true)
- There are four possible shades of blue among people with blue eyes. (false)
- If a person inherits genes for tallness, that person will grow tall no matter what. (false)

USING THE FIGURE

Emphasize that not all phenotypes result from completely dominant or completely recessive genes. Ask students what the flowers in **Figure 8** would look like if the gene for red flower color were completely dominant over the gene for white flower color. (All the offspring would be red.)

RETEACHING

Writing Have students describe three exceptions to Mendel's heredity principles in their ScienceLog.

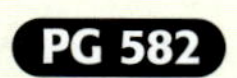

Base-Pair Basics PG 582

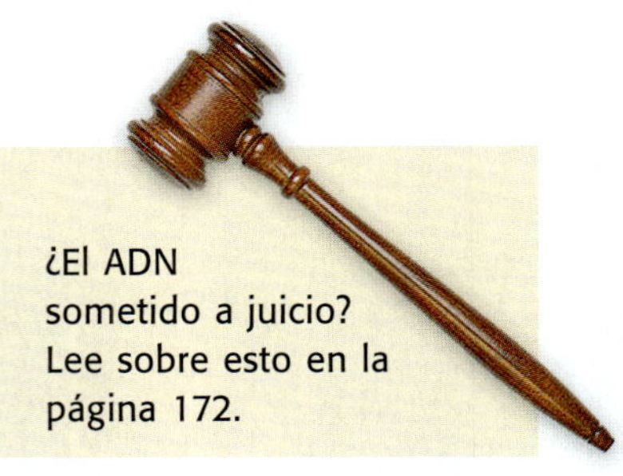

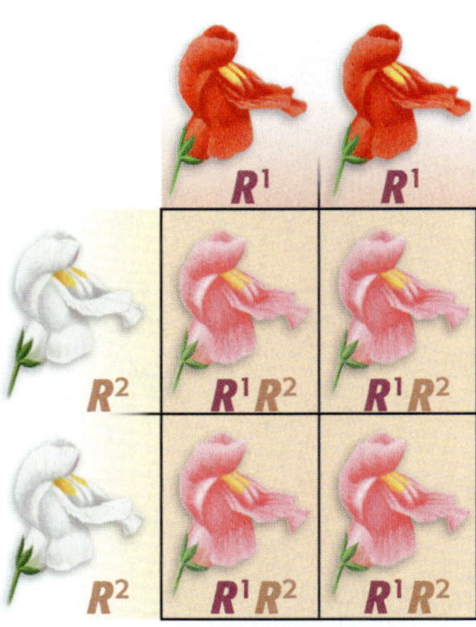

Figura 8 *La boca de dragón es un buen ejemplo de dominancia incompleta.*

Experimentos

A C G T

Aprende el alfabeto del ADN en la página 582.

Más información sobre los rasgos

Como ya habrás descubierto, las cosas son más complicadas de lo que parecen. Esto sucedió con los descubrimientos de Mendel, que ya estudiaste en el capítulo anterior. Mendel reveló los principios básicos de la transmisión de genes de una generación a otra. Pero como se ha estudiado más sobre la herencia, se han encontrado excepciones a los principios de Mendel. Algunas de estas excepciones se explican en los siguientes párrafos.

Dominancia incompleta En sus estudios con chícharos, Mendel descubrió que los diferentes rasgos no se mezclaban para producir una forma intermedia. Desde entonces, los investigadores han descubierto que algunas veces un rasgo no domina a otro. Estos rasgos no se mezclan, pero cada alelo tiene su propio grado de influencia. Esto se conoce como *dominancia incompleta*. Un ejemplo de esto es la flor "boca de dragón". La **Figura 8** muestra un entrecruzamiento entre una boca de dragón roja (**R¹**) y una boca de dragón blanca (**R²**). Como puedes ver, todos los fenotipos posibles para su descendencia son color rosa porque ambos alelos del gen tienen cierto grado de influencia.

Un gen puede influir en muchos rasgos Algunas veces un solo gen influye en varios rasgos. Un ejemplo de este fenómeno es el tigre blanco que se muestra a la derecha. El pelaje blanco es provocado por un solo gen, pero este gen no sólo influye en el color del pelaje. ¿Ves otra cosa rara en el tigre? Si lo observas bien, te vas a dar cuenta de que el tigre tiene ojos azules. El mismo gen que influye en el color del pelaje también influye en el color de los ojos.

IS THAT A FACT!

One of the many traits Mendel studied in pea plants was seed shape. He found that round seeds were dominant over wrinkled seeds. Seen under a microscope, the *RR* seeds have many starch grains that give them a full, round shape. The *rr* seeds have few starch grains, so they have a wrinkled shape. *Rr* seeds have an intermediate number of starch grains; they have enough starch to be full and round, but they have fewer starch grains than *RR* seeds. This is an example of incomplete dominance.

Muchos genes pueden influir en un solo rasgo Algunos rasgos, como el color de la piel, del cabello y de los ojos, son el resultado de algunos genes que actúan juntos. Por esta razón, es difícil decir si un rasgo es el resultado de un gen dominante o de uno recesivo. Como se muestra en la **Figura 9,** tal vez el color café de tus ojos sea de un tono ligeramente diferente al café de los ojos de alguno de tus compañeros. Las diferentes combinaciones de alelos tienen como resultado ligeras diferencias en la cantidad del pigmento.

La importancia del medio ambiente Recuerda que los genes no son los únicos que intervienen en tu crecimiento y desarrollo, también influye el medio ambiente. Considera la importancia de una dieta saludable, el ejercicio y los buenos ejemplos que te dan tu familia o tus amigos. Por ejemplo, tus genes pueden determinar que vas a ser alto, pero durante tu desarrollo debes recibir los nutrientes adecuados para alcanzar tu altura máxima. Tal vez has heredado un talento especial, pero necesitas practicarlo. La bailarina Tatum Harmon, que aquí se muestra, practica para desarrollar su talento.

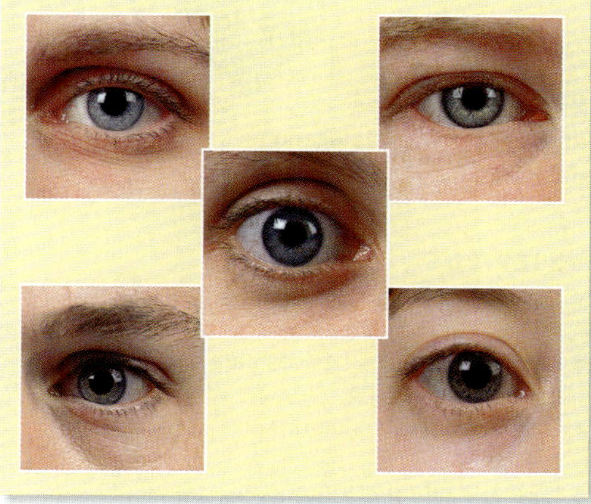

Figura 9 *Por lo menos dos genes, que muestran dominancia incompleta, determinan el color de los ojos del ser humano. Por esta razón, hay varios tonos de un mismo color de ojos.*

Tatum Harmon practica en un salón de la Compañía de Danza Alvin Ailey (Alvin Ailey American Dance Theater), donde estudia.

REPASO

1. Enumera y describe las partes de un nucleótido.

2. ¿Qué bases forman pares en una molécula de ADN?

3. ¿Qué formas sugerían las imágenes de rayos X captadas por Rosalind Franklin?

4. Explica qué quiere decir el enunciado: "El ADN une a todos los organismos".

5. **Hacer cálculos** Si una muestra de ADN tiene el 20 por ciento de citosina, ¿qué porcentaje de guanina tiene esta muestra? ¿Por qué?

159

Focus

How DNA Works

This section shows how DNA is used as a template for making proteins and how errors in DNA can lead to mutations and genetic disorders. Finally, students learn about pedigrees and how they are interpreted.

Bellringer

Have students unscramble the following words and use them both in a sentence:

tpsoneir (proteins)

neesg (genes)

(Genes contain instructions for making proteins.)

1) Motivate

GROUP ACTIVITY

Ask students to work in small groups to come up with as many different three-letter codes as possible using the four different bases. Give each group four pieces of paper, with one of the following four letters printed on each piece: *A, T, C,* or *G.* Tell students that each piece of paper represents a different amino acid. (There are 64 possible three-letter codes—the number of codons in the genetic code responsible for making proteins.) `Sheltered English`

Directed Reading Worksheet 7 Section 2

VOCABULARIO

ribosomas   mutágeno
mutación   árbol genealógico

OBJETIVOS

- Explica la relación entre genes y proteínas.
- Resume los pasos básicos para hacer una proteína.
- Define *mutación* y da un ejemplo.
- Evalúa la información que da un árbol genealógico.

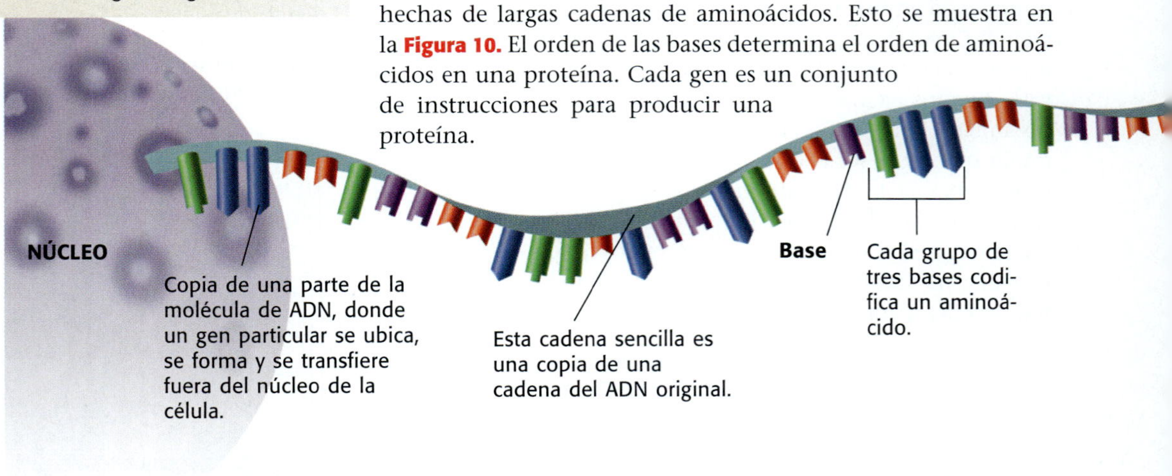

Figura 10 *Un gen es una sección de ADN que contiene instrucciones para formarse en línea y crear una proteína.*

¿Cómo trabaja el ADN?

Los científicos saben que el orden de las bases constituye un código que de alguna manera le dice a cada célula lo que debe hacer. El siguiente paso para entender el ADN consiste en descifrar este código.

Genes y proteínas

Ahora se sabe que las bases del ADN se leen como un libro, de un extremo al otro y en una sola dirección. Las bases **A**, **T**, **G** y **C** forman el alfabeto del código. Los grupos de tres bases codifican un aminoácido específico. Por ejemplo, las tres bases **CCA** codifican el aminoácido prolina y las bases **AGC** codifican el aminoácido serina. Como sabes, las proteínas están hechas de largas cadenas de aminoácidos. Esto se muestra en la **Figura 10.** El orden de las bases determina el orden de aminoácidos en una proteína. Cada gen es un conjunto de instrucciones para producir una proteína.

¿Por qué proteínas? Tal vez te estés preguntando, "¿qué tienen que ver las proteínas conmigo y con mi apariencia?" Las proteínas se encuentran en todas partes de las células. Actúan como mensajeros químicos, ayudan a determinar tu estatura, los colores que puedes ver y si tu cabello es rizado o lacio. Las células humanas tienen aproximadamente 100,000 genes, y cada uno deletrea las secuencias de aminoácidos para proteínas específicas. Existe una variedad casi ilimitada de proteínas. El cuerpo humano tiene alrededor de 50,000 tipos. Las proteínas son las responsables de la gran cantidad de diferentes formas, tamaños, colores y texturas en los organismos vivientes, como los cuernos, las garras, el pelo y la piel.

160

Q: Why did the mutant chromosome go to the tailor?

A: because it had a hole in its genes

Cómo se forma una proteína

Como ya se dijo, el primer paso para crear una proteína es copiar la sección de la cadena de ADN que contiene un gen. La copia de esta sección se hace con ayuda de enzimas copiadoras. Las moléculas mensajeras extraen la información genética de las secciones de ADN del núcleo y la llevan al citoplasma.

En el citoplasma, la copia de ADN pasa por una especie de línea de ensamblaje de proteínas. Esta "fábrica" se conoce como **ribosoma.** Las bases de la copia pasan a través del ribosoma de tres en tres. Las moléculas de transferencia actúan como traductores del mensaje de la copia de ADN. Cada molécula de transferencia recoge un aminoácido específico del citoplasma, definido por el orden de las bases de la molécula de transferencia. Como piezas de un rompecabezas, las bases de las moléculas de transferencia se unen a las bases de la copia de ADN en el ribosoma. Luego, las moléculas de transferencia sueltan sus "maletas" de aminoácidos, que se unen en cadena para formar una proteína. Este proceso se ilustra en la **Figura 11.**

Figura 11 *Este diagrama muestra cómo se hace una proteína. El orden de las bases en la copia de ADN determina qué aminoácidos se deben transferir al ribosoma.*

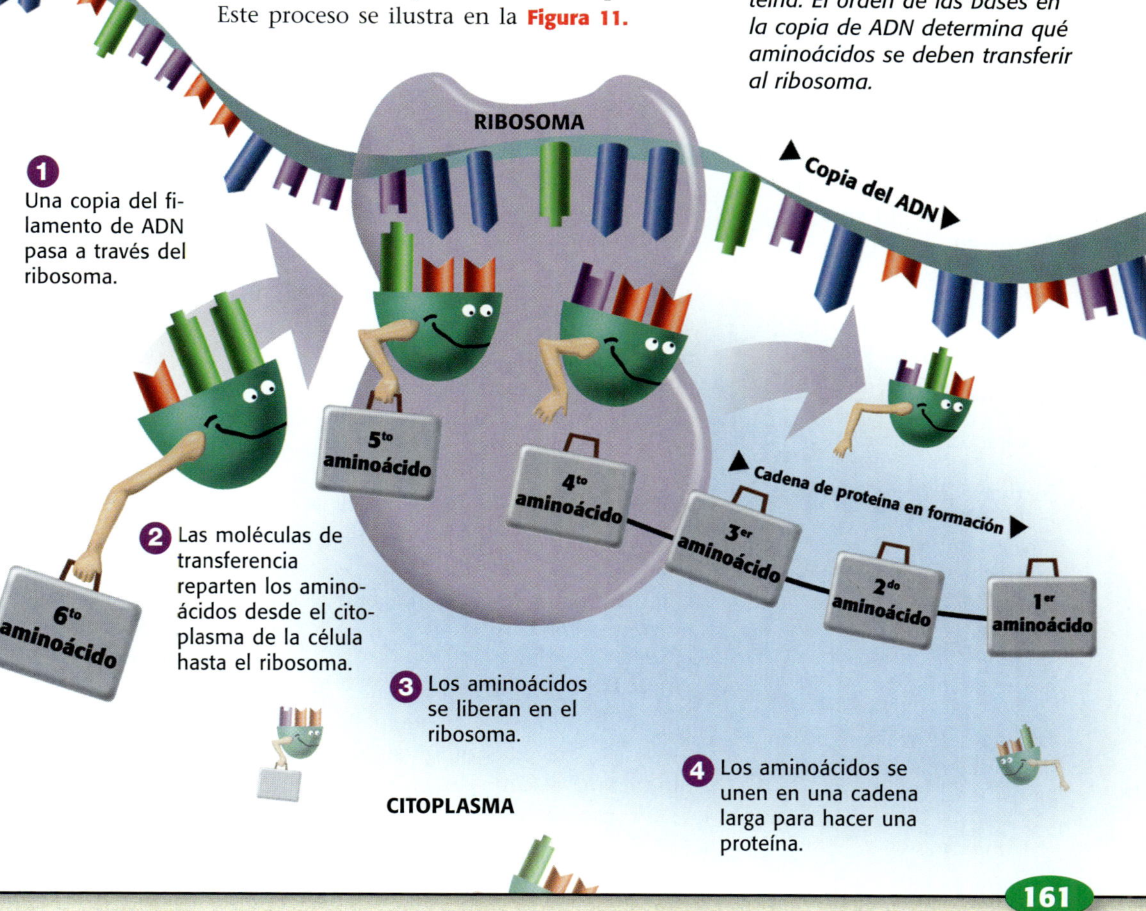

2) Teach

Answers to Self-Check

1. 1,000
2. DNA codes for proteins. Your flesh is composed of proteins, and the way those proteins are constructed and combined influences much about the way you look.

MATH and MORE

Use mathematics to discuss how DNA codes for amino acids. With four possible nucleotides in three possible positions, there are $4 \times 4 \times 4$, or 64, possible combinations. (for example, AAA, AAT, AAG, and AAC)

These combinations are called codons. DNA produces only 20 amino acids; thus, most amino acids have several corresponding codons.

COOPERATIVE LEARNING

Skit Have groups of students write and perform a short skit to demonstrate the formation of a protein. For instance, students could play the roles of a ribosome, an amino acid, a transfer enzyme, and a DNA copy.

Teaching Transparency 25 "The Making of a Protein"

GUIDED PRACTICE

Write a sequence of DNA, such as AACTACGGT, on the chalkboard. Ask students to write the sequence for a copy of the DNA using base-pairing rules. (TTGATGCCA)

Then ask students to give examples of deletion, insertion, and substitution mutations to the DNA.

MISCONCEPTION ALERT

Are mutations rare? Scientists estimate that we inherit hundreds of mutations from our parents. In addition, new mutations can happen due to environmental factors and mistakes made during DNA replication and cell division. Mistakes are made during DNA replication in approximately one out of every 1,000 base pairs. But thanks to repair enzymes and other proofing mechanisms, the final error rate is much lower—somewhere between one in a million and one in a billion.

CONNECT TO EARTH SCIENCE

Ozone is a gas made of three oxygen atoms. High in the atmosphere, ozone absorbs dangerous ultraviolet radiation (the high-energy light that can cause cancer). When produced near the surface of the Earth, however, ozone is a pollutant that affects plant growth and makes breathing more difficult. Use Teaching Transparency 146 to illustrate the process of ozone production.

Teaching Transparency 146
"The Formation of Smog"

LINK TO EARTH SCIENCE

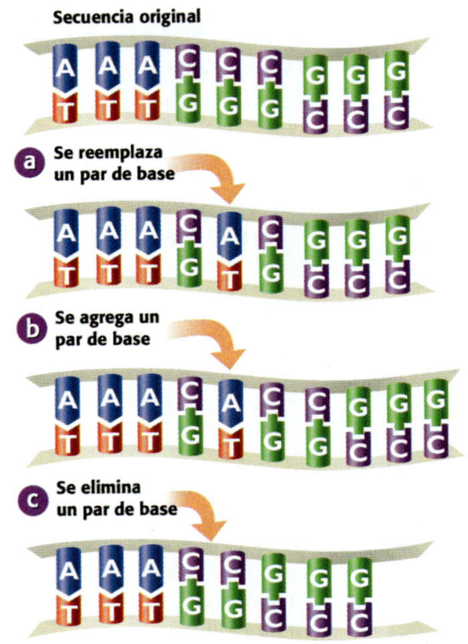

Figura 12 *La secuencia original de par de bases de arriba se ha modificado para explicar la (a) substitución, (b) inserción y (c) deleción.*

ciencias de la Tierra
CONEXIÓN

La capa de ozono de la atmósfera de la Tierra protege la superficie del planeta de la radiación ultravioleta. La radiación ultravioleta causa mutaciones en las células de la piel que provocan cáncer de piel. Cada año, más de 750,000 personas contraen alguna forma de cáncer de piel. Los científicos temen que el daño de la capa de ozono aumente el número anual de casos de cáncer de piel.

Cambios en los genes

Imagínate que te subes a la montaña rusa más nueva de un parque de diversiones. Justo antes de que te subas al carrito, te dicen que algunas partes de metal de la montaña fueron cambiadas por partes hechas de otra substancia. ¿Todavía querrías subirte?

Tal vez las partes nuevas estén hechas de un metal mejor y más fuerte, o quizá se usó otro metal con las mismas propiedades. La tercera posibilidad es que se haya usado un material que no es adecuado. ¡Imagínate lo qué pasaría si en lugar de usar metal se usara cartón! Errores como estos ocurren continuamente en el ADN y se conocen como **mutaciones.** Las mutaciones suceden cuando hay un cambio en el orden de las bases del ADN de un organismo. Algunas veces se excluye una base; esto se conoce como *deleción*. Otras veces se agrega una base; esto se conoce como *inserción*. El error más común es cuando una base incorrecta reemplaza a una correcta. Esto se conoce como *substitución*. La **Figura 12** explica estos tres tipos de mutaciones.

Los errores son posibles Por suerte, las enzimas reparadoras siempre están revisando si hay errores en la molécula de ADN. Cuando se detecta una falla, por lo general se arregla. Pero, a veces, las reparaciones no son exactas y los errores se vuelven parte del mensaje genético. Como en la montaña rusa, los cambios en el ADN pueden tener tres consecuencias: una mejoría, ningún cambio o un cambio grave. Si no se fabrica la proteína correcta, los resultados pueden ser fatales. Si el daño ocurre en las células sexuales, el error puede pasar de una generación a otra.

¿Cómo se puede dañar el ADN? Además de los errores accidentales que suceden cuando se copia el ADN, el daño puede ser causado por agentes físicos y químicos conocidos como **mutágenos.** Un mutágeno es cualquier cosa que causa una mutación en el ADN. Algunos ejemplos son la radiación de rayos X de alta energía y la radiación ultravioleta. La radiación ultravioleta es la energía de la luz solar responsable del bronceado y de las quemaduras de la piel. Algunos ejemplos de mutágenos químicos son el asbesto y las substancias químicas del humo del cigarrillo.

WEIRD SCIENCE

A human cell contains between 50,000 and 100,000 genes. Human DNA is about 3 billion base pairs long. Only about 3 percent of those base pairs are used in making proteins; the other 97 percent are regulatory sequences and nonfunctioning genes.

Ejemplo de una substitución

Supongamos que la secuencia de ADN tiene las tres bases **GAA**. **GAA** son las tres letras que dan la instrucción: "Pon ácido glutámico aquí". Si hay un error y la secuencia se cambia a **GTA**, se envía un mensaje completamente diferente: "Pon valina aquí".

Este cambio tan sencillo puede causar la *anemia drepanocítica,* enfermedad que afecta a los glóbulos rojos. Cuando la valina substituye al ácido glutámico en una proteína de la sangre, como se muestra en la **Figura 13,** los glóbulos rojos se deforman.

Los drepanocitos no pueden transportar oxígeno a todo el cuerpo, como lo hacen los glóbulos rojos normales. Por el contrario, se pegan en los vasos sanguíneos, y causan coágulos dolorosos y peligrosos.

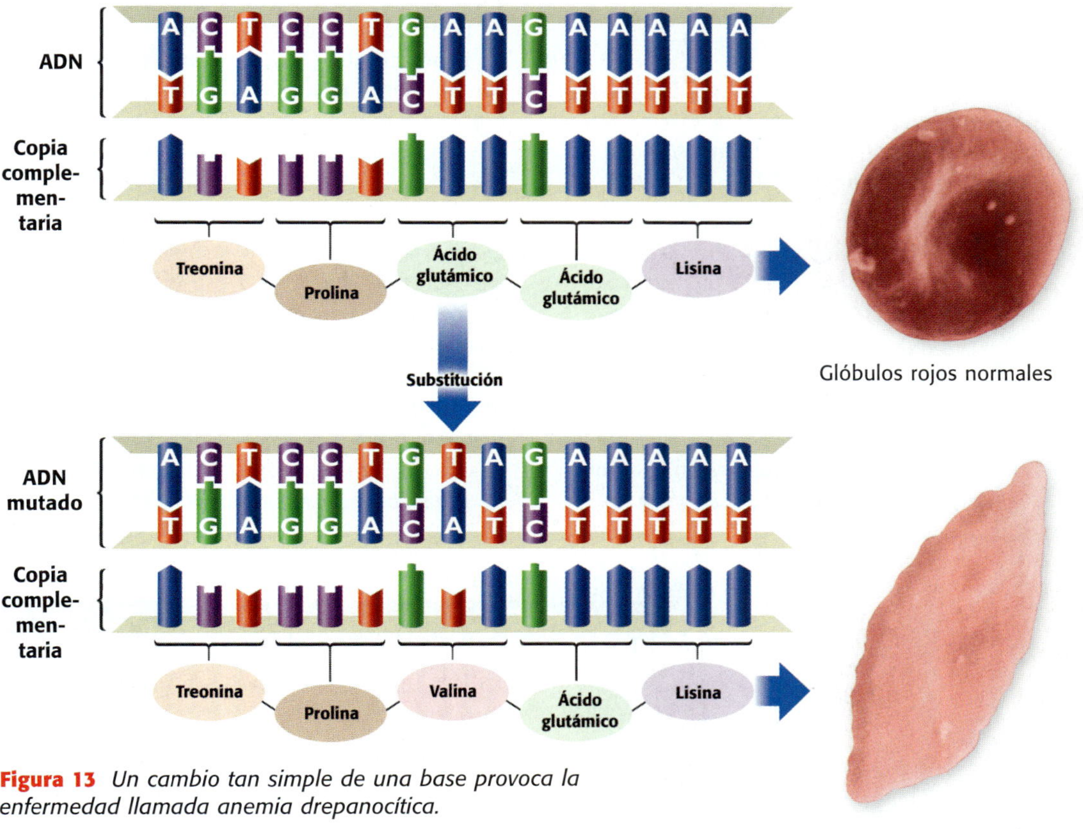

Figura 13 *Un cambio tan simple de una base provoca la enfermedad llamada anemia drepanocítica.*

Laboratorio

Mutaciones

La siguiente oración tiene palabras de tres letras, pero ha sufrido una mutación. Trata de encontrar la mutación y corrígela.

ELI RTG ATO ICO MEH MRA TON AT ¿Qué tipo de mutación encontraste? Ahora, ¿qué dice la oración?

Science Bloopers

A person who carries a single allele for sickle cell anemia is said to have *sickle cell trait*. In the past, many people did not understand that people who have sickle cell trait do not pose a risk to the community. In some areas, children with the trait were banned from public schools because people feared the condition was contagious.

---

3 Extend

RETEACHING

Write a sequence of DNA on the chalkboard, and invite students to come up to the board and change the sequence. Then ask the class to discuss the possible consequences of such a mutation in DNA. (It might cause a different amino acid to be substituted in a protein. This could result in a genetic disorder, death, or no change at all.)

Ask how the mutation could be corrected. (Enzymes may find and repair the error.) Finally, ask what could have caused the mutation. (Mutations are caused by random errors in the copying of DNA, by physical and chemical damage to the DNA, or by X rays and other mutagens.) Sheltered English

Answers to QuickLab

The mutation is a deletion. THE BIG RED CAT ATE THE BIG BAD RAT

CROSS-DISCIPLINARY FOCUS

Writing **History** Have students conduct Internet or library research and write a report on the history of Queen Victoria and her descendants. The queen was a carrier of hemophilia, a recessive disorder in which a person lacks a protein needed to form blood clots. The hemophilia gene passed from Queen Victoria to her daughters and from them to the Russian, Spanish, and German families they married into. 

Teaching Transparency 26 "An Example of Substitution"

Quiz

1. **What is the function of the ribosome?** (In the ribosome, the DNA code is translated into proteins.)

2. **List some causes of DNA mutations.** (UV radiation, cigarette smoke, or X rays)

ALTERNATIVE ASSESSMENT

Writing Have students prepare an instruction manual for their DNA. The manual should include instructions for copying their DNA and translating it into proteins. It should also include information about protecting their DNA from mutations by avoiding mutagens and correcting any mutations that occur.

Answer to APPLY

Individuals I_2 and III_2 are nearsighted. Both of Jane's parents are *Nn*. Jane's genotype can be either *NN* or *Nn*. It can't be *nn* because Jane is not nearsighted.

Jane's Fiancé

		N	n
		N	n
Jane	N	NN	Nn
	N	NN	Nn

Jane's Fiancé

		N	n
Jane	N	NN	Nn
	n	Nn	nn

Teaching Transparency 27
"Pedigree"

Reinforcement Worksheet 7
"DNA Mutations"

Generación

I   1   2
II 1   2   3   4   5   6
III 1   2   3   4
IV 1   2   3

☐ Hombres   ○ Mujeres

Las líneas verticales relacionan los hijos con sus padres.

■ o ● Un cuadrado o círculo sólido indica que la persona tiene un rasgo característico.

◪ o ◖ Un cuadrado o círculo a medio llenar indica que la persona es portadora del rasgo.

Figura 14 *La fibrosis cística es una enfermedad hereditaria recesiva que afecta el sistema respiratorio. Este árbol genealógico muestra una familia con esta enfermedad.*

El árbol genealógico de la derecha muestra el rasgo recesivo de la miopía en la familia de Rita. Ella, sus padres y su hermano tienen visión normal. ¿Quiénes en el árbol genealógico son miopes? ¿Cuáles son los genotipos posibles de los padres de Rita? ¿Cuáles son los genotipos posibles de Rita? Rita va a casarse, su novio tiene visión normal pero porta el rasgo de miopía. Haz dos cuadrículas de Punnett para los genotipos de los futuros hijos de Rita.

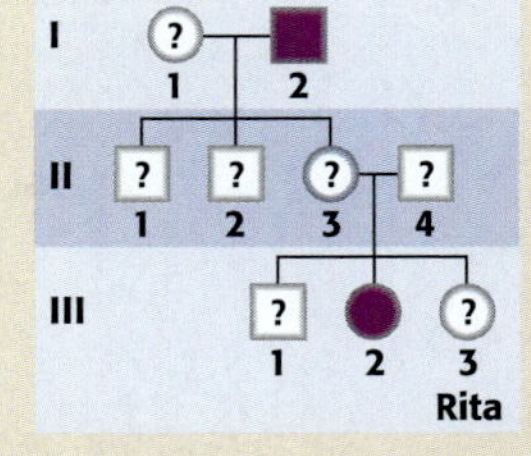

Asesoría genética

La mayoría de los transtornos hereditarios, como la anemia drepanocítica, son recesivos. Esto significa que la enfermedad sólo ocurre cuando un niño o una niña hereda un gen defectuoso del padre y de la madre. Algunas personas, llamadas portadores, sólo tienen un alelo para esta enfermedad, y pueden pasar el gen defectuoso a sus hijos sin saberlo.

La asesoría genética proporciona información a las parejas que desean tener hijos pero tienen miedo de transmitirles alguna enfermedad. Por lo general, los asesores genéticos usan un diagrama conocido como **árbol genealógico,** que es una herramienta para buscar un rasgo a través de generaciones de una familia. Al hacer un árbol genealógico, se puede predecir si una persona es portadora de una enfermedad hereditaria. En la **Figura 14,** el rasgo de la fibrosis cística se rastreó a través de cuatro generaciones. Cada generación está marcada con números romanos y cada individuo con números arábigos. En la Figura 14, se explican los símbolos de árbol genealógico.

REPASO

1. Enumera los tres tipos de mutación. ¿En qué se diferencian?

2. ¿Qué tipo de mutación causa la anemia drepanocítica?

3. **Aplicar conceptos** Las mutaciones suceden en las células sexuales y en las del cuerpo. ¿Qué células transmiten una mutación de una generación a otra? Explica por qué.

164

▼ **Answers to Review**

1. insertion: an extra base is inserted; deletion: a base is deleted; substitution: one base is substituted for another

2. substitution

3. Mutations can only be passed from generation to generation if they occur in sex cells. This is because the genes in an individual's sex cells are used to reproduce a new individual.

Genética aplicada

VOCABULARIO

ingeniería genética
ADN recombinante
impresiones de ADN
poyecto Genoma Humano

OBJETIVOS

- Define la *ingeniería genética*.
- Enumera algunos de los beneficios de combinar el ADN de diferentes organismos.

Durante miles de años, los científicos han sido conscientes de las ventajas de la reproducción selectiva. Por medio de ésta se cruzan organismos con determinadas características. Probablemente ya has disfrutado de estas ventajas sin darte cuenta. Quizá ya te comiste un huevo de una gallina que fue cruzada para producir muchos huevos. O tal vez, ya has probado el trigo reproducido de manera selectiva para resistir plagas y enfermedades. Incluso tu perro puede ser un resultado de reproducción selectiva. Algunos tipos de perros, por ejemplo, tienen un pelaje grueso para poder atrapar a sus presas en el agua helada.

Diseño de genes

Ahora, los científicos pueden manipular genes individuales con una técnica conocida como ingeniería genética. Como todos los tipos de ingeniería, la ingeniería genética pone en práctica los conocimientos científicos. Básicamente, la **ingeniería genética** permite a los científicos transferir genes de un organismo a otro. Hoy en día, se usa para fabricar proteínas, reparar genes dañados e identificar a individuos que pueden ser portadores de un alelo para una enfermedad. Algunos otros usos se muestran en la **Figura 15.**

Figura 15 *Éstos son algunos de los muchos ejemplos de ingeniería genética.*

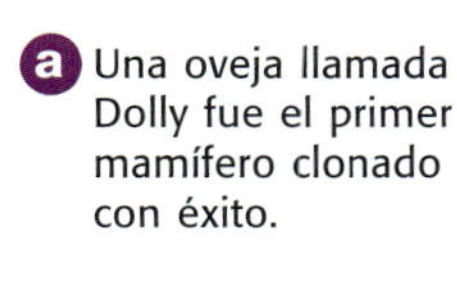

a Una oveja llamada Dolly fue el primer mamífero clonado con éxito.

b La fotografía de arriba muestra dos copos de algodón de dos plantas diferentes. El de la derecha ha sido alterado genéticamente para resistir plagas.

c Esta micrografía electrónica muestra gránulos de plástico producidos dentro de una célula vegetal.

d Los científicos agregaron un gen de luciérnaga a esta planta de tabaco. Ahora, la planta produce una enzima que hace que la planta brille.

165

Q: What has orange hair and lives in a test tube?

A: Bozo the Clone

SCILINKS
NSTA

TOPIC: Genetic Engineering
GO TO: www.scilinks.org
sciLINKS NUMBER: HSTL140

Focus

Applied Genetics

In this section, students are introduced to genetic engineering. Students learn how scientists combine DNA from different organisms to make useful products.

Bellringer

Have students pretend that they have been hired to genetically engineer the ideal fruit. Have them answer the following questions about this new fruit:

- What will the fruit look like?
- What will it taste like?
- How will it grow?
- How will it reproduce?
- What strategy will it use to spread its seeds?

1 Motivate

DISCUSSION

After introducing and explaining the concept of genetic engineering, ask students to list possible advantages and disadvantages of gene manipulation. Guide a discussion of the pros and cons of cloning and gene manipulation. (possible advantages: better crops and domestic animals, plants that grow quickly to restore destroyed forests, the eradication of some diseases; possible disadvantages: unplanned dangers to health, damaging environmental side effects, threats to human rights)

Directed Reading
Worksheet 7 Section 3

CROSS-DISCIPLINARY FOCUS

Social Studies Czar Nicholas II ruled Russia from 1894 to 1917. Nicholas, his wife, and their children were shot to death by the Bolsheviks in 1918 in the aftermath of the Russian Revolution. Their bodies were not found until 1991, when Russian anthropologists discovered their bones. To be certain of the identity of the remains, DNA samples were taken from the remains and from living members of the family. The DNA fingerprints of the deceased were similar to those of the surviving family members. The remains were determined to be Czar Nicholas II and his family.

Answer to Self-Check

The human gene for a particular protein can be inserted into bacteria that can use the gene to produce the needed protein very rapidly and in great quantities. In this way, bacteria are living factories.

Teaching Transparency 28 "Recombining DNA"

Critical Thinking Worksheet 7 "The Perfect Parrot"

Interactive Explorations CD-ROM "DNA Pawprints"

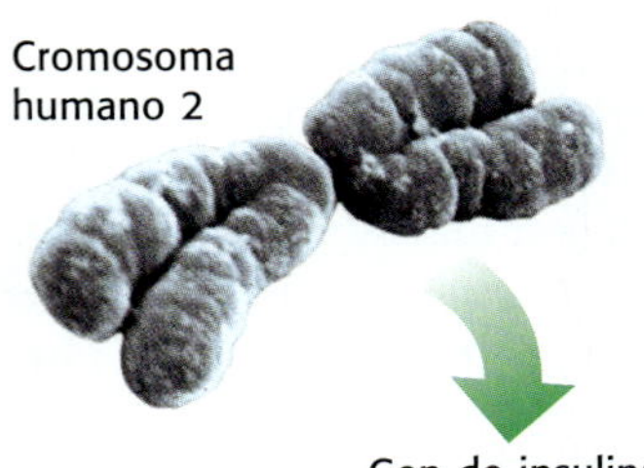

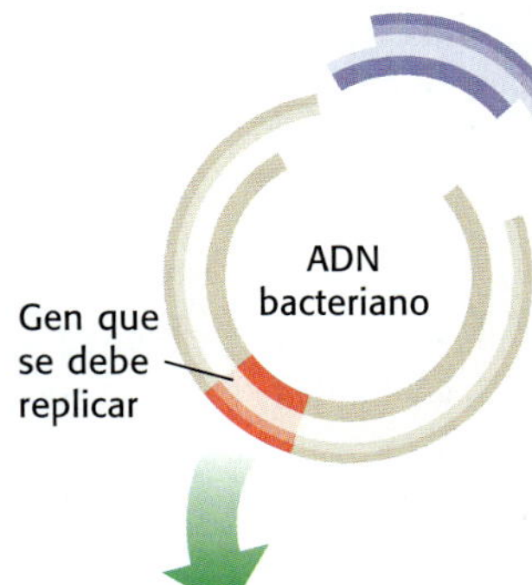

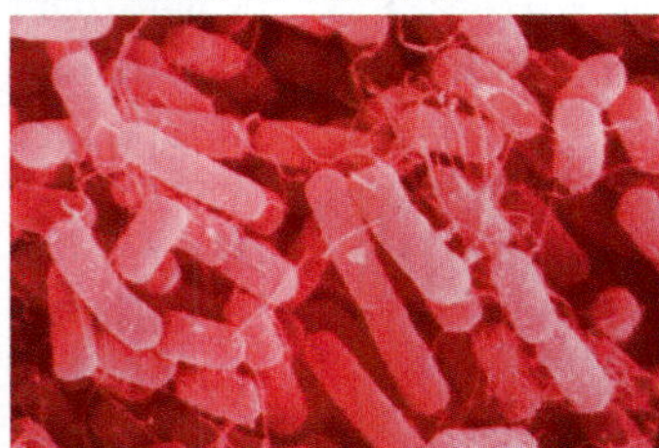

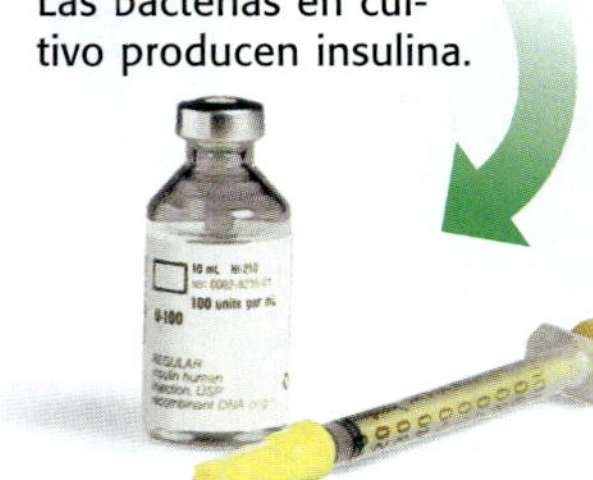

Figura 16 *El gen que produce la insulina en los humanos se encuentra en el cromosoma 2. Una copia se transfiere a un fragmento de ADN bacteriano. Las bacterias que tienen el ADN recombinante producen insulina.*

Fábricas vivientes

La estructura del ADN es la misma en todos los organismos vivientes. De hecho, la uniformidad de la estructura del ADN permite que los genes de un organismo funcionen en otro. Cuando se ponen los genes de un organismo en otro usando la ingeniería genética, el ADN resultante se conoce como **ADN recombinante.** Pero, ¿por qué hacen esto los científicos?

Una de las respuestas a esta pregunta es que el ADN recombinante se usa para tratar enfermedades. La diabetes, por ejemplo, se trata con un producto hecho con ADN recombinante. Las personas con diabetes no pueden producir suficiente insulina (una proteína). Sin suficiente insulina, la mayoría de los diabéticos morirían. Para producir ADN recombinante, los científicos insertan un gen de insulina humana normal en el ADN de determinadas bacterias. Las bacterias se vuelven fábricas de insulina, y producen grandes cantidades de esta hormona. Este proceso se muestra en la **Figura 16.**

Impresiones de ADN

Al principio de esta unidad, se te pidió que imaginaras que te culpaban de un crimen que no cometiste. El único "testigo" para tu defensa era el ADN. Como ya sabes, el ADN de cada persona es único.

En el laboratorio, los fragmentos de ADN se separan según su tamaño. De esta manera, los patrones formados por los fragmentos se pueden comparar. Puedes ver un ejemplo de estos fragmentos en la **Figura 17.** La comparación de los fragmentos de diferentes personas se conoce como análisis del ADN o **impresiones de ADN.** Es imposible que dos personas tengan las mismas impresiones de ADN a menos que sean gemelos idénticos.

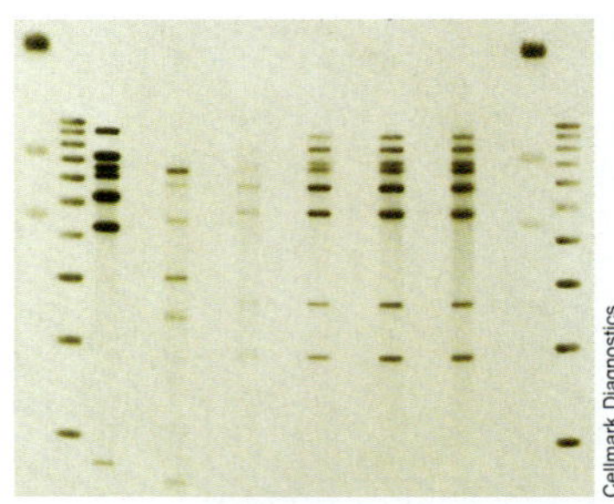

Figura 17 *Las franjas negras son fragmentos de ADN. La ubicación de las franjas negras es diferente en todas las personas.*

> ✔ **Autoevaluación**
>
> Explica qué quiere decir "fábricas vivientes" cuando se habla del ADN recombinante. *(Consulta la página 636 para comprobar tu respuesta.)*

⚛ WEIRD SCIENCE

Gene therapy is an experimental field of medical research in which defective genes are replaced with healthy genes. One way to insert healthy genes involves using a delivery system called a gene gun to inject microscopic gold bullets coated with genetic material.

Panorama general

Los científicos saben mucho acerca de algunos genes, como la ubicación y mutación del gen que causa la anemia drepanocítica. La **Figura 18** muestra uno de los resultados de la investigación del ADN. Pero todavía hay mucho que aprender sobre los genes que controlan otras enfermedades y características. La meta de un ambicioso proyecto conocido como el **Proyecto Genoma Humano** es crear un mapa que muestre la ubicación y la secuencia del ADN de todos nuestros genes.

Se espera que el conocimiento adquirido con este proyecto ayude a desarrollar terapias más eficaces para las enfermedades. Ya se han identificado genes asociados con el cáncer de colon y el glaucoma juvenil. El proyecto Genoma Humano tomará mucho tiempo y dinero para concluirse.

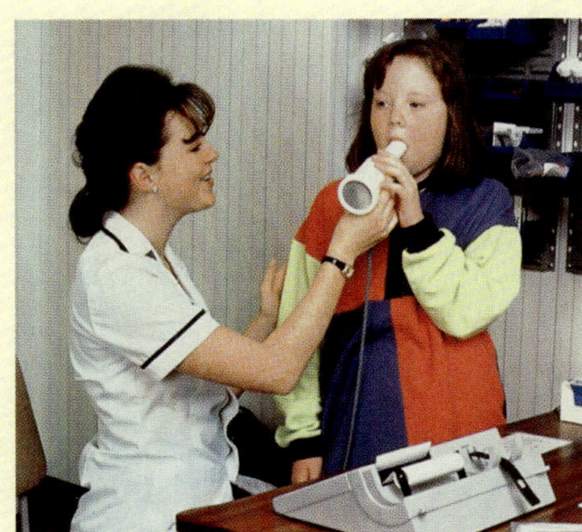

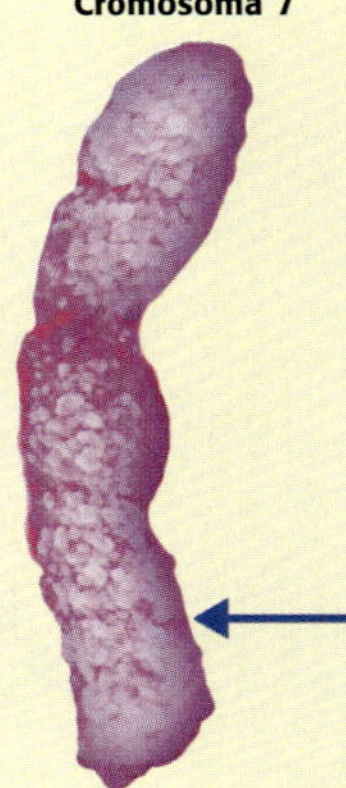

Figura 18 *La niña de esta fotografía recibe terapia para la fibrosis cística. Gracias a la tecnología del ADN se identificó el gen de esta enfermedad y se desarrolló una terapia para tratarla.*

REPASO

1. ¿Cuál es la diferencia entre ingeniería genética y reproducción selectiva?

2. Menciona algunas ventajas de la ingeniería genética.

3. ¿Por qué es una ventaja producir insulina humana dentro de una bacteria?

4. **Entender la tecnología** Describe cómo se puede usar tu ADN para identificarte.

167

SECTION 1

DNA deoxyribonucleic acid; hereditary material that controls all the activities of a cell, contains the information to make new cells, and provides instructions for making proteins

nucleotide a subunit of DNA consisting of a sugar, a phosphate, and one of four nitrogenous bases

adenine one of the four bases that combine with sugar and phosphate to form a nucleotide subunit of DNA; adenine pairs with thymine

thymine one of the four bases that combine with sugar and phosphate to form a nucleotide subunit of DNA; thymine pairs with adenine

guanine one of the four bases that combine with sugar and phosphate to form a nucleotide subunit of DNA; guanine pairs with cytosine

cytosine one of the four bases that combine with sugar and phosphate to form a nucleotide subunit of DNA; cytosine pairs with guanine

SECTION 2

ribosomes small organelles in cells where proteins are made from amino acids

mutation a change in the order of the bases in an organism's DNA; deletion, insertion, or substitution

mutagen anything that can damage or cause changes in DNA

pedigree a diagram of family history used for tracing a trait through several generations

Resumen del capítulo

SECCIÓN 1

Vocabulario

ADN *(pág. 152)*
nucleótido *(pág. 152)*
adenina *(pág. 152)*
timina *(pág. 152)*
guanina *(pág. 152)*
citosina *(pág. 152)*

Notas de la sección

- Las proteínas están hechas de cadenas de aminoácidos.
- El ADN está formado por cadenas de nucleótidos.
- Los cromosomas están hechos de proteínas y de ADN.
- La molécula de ADN parece una escalera de caracol. Los peldaños de la escalera están hechos de pares de bases: adenina y timina, o bien citosina y guanina.

- El ADN transporta información genética en el orden de las bases de nucleótidos.
- El ADN puede ser copiado ya que una cadena de la molécula sirve como plantilla para la otra.

SECCIÓN 2

Vocabulario

ribosoma *(pág. 161)*
mutación *(pág. 162)*
mutágeno *(pág. 162)*
árbol genealógico *(pág. 164)*

Notas de la sección

- Un gen es un grupo de instrucciones para formar una proteína.
- Cada grupo de tres bases en un gen codifica un aminoácido específico.
- Los genes pueden transformarse cuando cambia el orden de las bases.

Experimentos

Pares de bases *(pág. 582)*

☑ Comprobar destrezas

Conceptos de matemáticas

EL CÓDIGO GENÉTICO En la página 155 te piden que calcules el número de bases en un gen. Cada gen tiene un número específico de bases para formar una proteína determinada. Si hay un promedio de 22,000 bases en cada gen y cerca de 100,000 genes en un grupo de 46 cromosomas, ¿cuántas bases hay por cada célula humana? (Recuerda que cada célula tiene 46 cromosomas). Sólo multiplica 100,000 × 22,000 y obtendrás 2,200,000,000 bases. Es decir, ¡Dos mil doscientos millones de bases!

Comprensión visual

COPIAS DE ADN Observa la Figura 10 de la página 160 y verás el núcleo y los poros de su membrana. La copia del ADN sale por estos poros para entregarles a los ribosomas el mensaje codificado. ¿Por qué el ADN envía una copia fuera del núcleo para transmitir sus mensajes? Porque así el ADN se protege más de los factores que podrían causar una mutación si se quedara dentro del núcleo. Tal vez la copia mensajera tenga mala suerte, pero el ADN original permanece ¡sano y salvo!

168

Lab and Activity Highlights

Base-Pair Basics **PG 582**

Datasheets for LabBook
(blackline masters for this lab)

SECTION 3

genetic engineering manipulation of genes that allows scientists to put genes from one organism into another organism

DNA fingerprinting the analysis of fragments of DNA as a form of identification

Human Genome Project a worldwide scientific effort to discover the location of every gene and to create a "map" of the entire human genome

SECCIÓN 3

Vocabulario

ingeniería genética *(pág. 165)*
ADN recombinante *(pág. 166)*
impresiones de ADN *(pág. 166)*
Proyecto Genoma Humano *(pág. 167)*

Notas de la sección

- Desde hace miles de años, se ha usado la reproducción selectiva para cruzar plantas y producir mejores cosechas. También se han reproducido animales de manera selectiva con rasgos especiales durante miles de años.

- La ingeniería genética les permite a los científicos cambiar el ADN de un organismo para darle al otro organismo características nuevas.

- La meta del proyecto Genoma Humano es localizar todos los genes humanos.

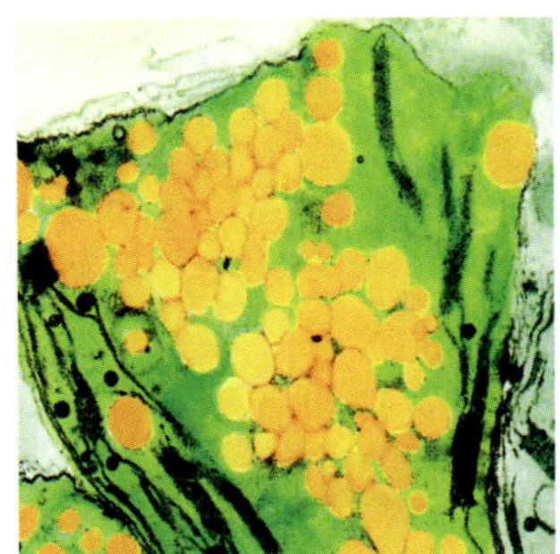

Vocabulary Review Worksheet 7

Blackline masters of these Chapter Highlights can be found in the **Study Guide.**

 internet

Visita el sitio web de HRW para encontrar una serie de herramientas de aprendizaje relacionadas con este capítulo. Sólo tienes que escribir la palabra clave:

PALABRA CLAVE: HSTDNA

Visita el sitio web de la **Asociación Nacional de Maestros de Ciencias** *(National Science Teachers Association)* para encontrar recursos de Internet relacionados con este capítulo. Sólo escribe el **ENLACE DE CIENCIAS** para obtener más información sobre el tema:

TEMA: ADN   **ENLACE:** HSTL130
TEMA: Genes y rasgos   **ENLACE:** HSTL135
TEMA: Ingeniería genética   **ENLACE:** HSTL140
TEMA: Impresiones de ADN   **ENLACE:** HSTL145

169

Lab and Activity Highlights

LabBank

Long-Term Projects & Research Ideas, Project 6

Interactive Explorations CD-ROM

CD 3, Exploration 8, "DNA Pawprints"

Chapter Review
Answers

USING VOCABULARY

1. amino acids; nucleotides
2. recessive
3. mutation
4. DNA fingerprinting
5. mutagen

UNDERSTANDING CONCEPTS

Multiple Choice

6. b
7. d
8. b
9. c
10. c
11. b

Short Answer

12. GAATCCGAATGGT
13. DNA molecules split down the middle, then each side of the molecule pairs up with an additional nucleotide. (Picture should resemble a zipper being zipped up.)
14. insertion

Repaso del capítulo

UTILIZAR EL VOCABULARIO

Escoge el término correcto para completar cada una de las siguientes oraciones:

1. Una proteína es una larga cadena de __?__. Una cadena de ADN es una larga serie de __?__. (*aminoácidos* o *nucleótidos*)

2. Un trastorno, como la fibrosis cística, se considera __?__ si el niño o la niña debe recibir un alelo de su padre y madre para sufrir la enfermedad. (*dominante* o *recesivo*)

3. El cambio en el orden de las bases del ADN se llama __?__. (*mutación* o *mutágeno*)

4. __?__ se usa para comparar el ADN de diferentes personas. (*la ingeniería genética* o *las impresiones de ADN*)

5. Un __?__ es un agente físico o químico que le causa daño al ADN. (*mutágeno* o *árbol genealógico*)

COMPRENDER CONCEPTOS

Opción múltiple

6. En una molécula de ADN, ¿cuáles de las siguientes bases forman un par?
   a. adenina y citosina
   b. timina y adenina
   c. timina y guanina
   d. citosina y timina

7. Un gen es
   a. instrucciones para cada rasgo.
   b. instrucciones para hacer una proteína.
   c. una parte de una cadena de ADN.
   d. todas las anteriores.

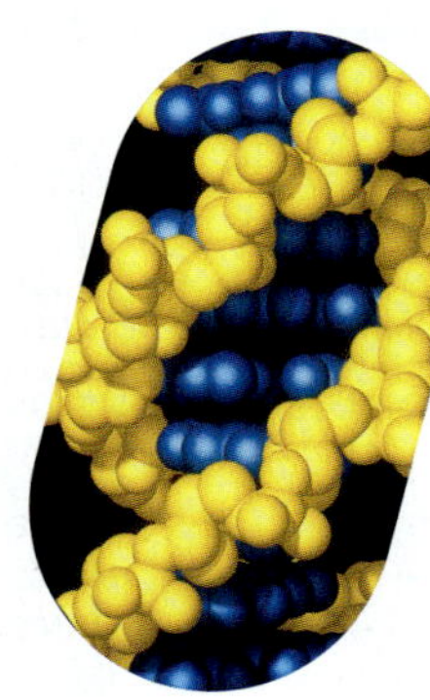

8. El ADN
   a. está formado por tres subunidades.
   b. tiene una estructura en forma de escalera de caracol.
   c. no se puede reparar si ha mutado.
   d. Todas las anteriores.

9. En una dominancia incompleta
   a. un solo gen controla muchos rasgos.
   b. todos los genes de un rasgo son recesivos.
   c. cada alelo de un rasgo tiene su propio grado de influencia.
   d. el ambiente controla los genes.

10. Watson y Crick
    a. estudiaron las cantidades de cada base de ADN.
    b. tomaron fotografías con rayos X del ADN.
    c. hicieron modelos para determinar la estructura del ADN.
    d. descubrieron que los genes se encontraban en los cromosomas.

11. ¿Cuál de los siguientes NO es un paso en la síntesis de proteínas?
    a. Las copias del ADN pasan al citoplasma.
    b. Las moléculas de transferencia llevan aminoácidos al núcleo.
    c. Los aminoácidos se ensamblan en el ribosoma para formar la proteína.
    d. Una copia del ADN pasa por el ribosoma.

Respuesta breve

12. ¿Cuál sería la cadena complementaria del ADN para la siguiente secuencia de bases?

    C T T A G G C T T A C C A

13. ¿Cómo se copia el ADN? Haz un dibujo que explique tu respuesta.

14. Si la secuencia de ADN "TGAGCCATGA" se cambia a "TGAGCACATGA", ¿qué tipo de mutación ha ocurrido?

Organizar conceptos

15. Usa los siguientes términos para crear un mapa de ideas: bases, adenina, timina, nucleótidos, guanina, ADN, citosina.

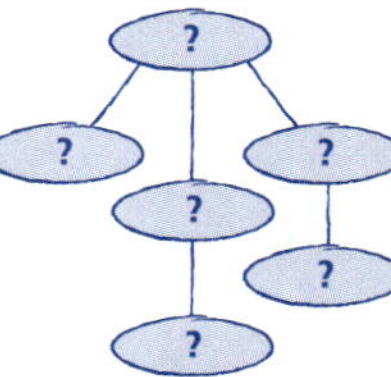

RAZONAMIENTO CRÍTICO Y RESOLUCIÓN DE PROBLEMAS

Escribe una o dos oraciones para responder a las siguientes preguntas:

16. Si el padre o la madre no muestran signos de tener anemia drepanocítica, ¿garantiza esto que su hijo o hija no va a contraer la enfermedad? Explica por qué.

17. ¿Para cuántos aminoácidos codifica esta secuencia de ADN?

    T C A G C C A C C T A T G G A

18. Las bacterias crecen con facilidad y se multiplican rápidamente. ¿Por qué son útiles para hacer proteínas fabricadas genéticamente?

LAS MATEMÁTICAS EN LAS CIENCIAS

19. La meta del Proyecto Genoma Humano es descubrir todos los genes humanos. Los científicos calculan que hay 100,000. En 1998, se habían descubierto 38,000. ¿Cuántos genes más debe descrubrir el Proyecto Genoma Humano?

20. Si los científicos encuentran 6,000 genes cada año, ¿cuántos años faltan para que termine el proyecto?

21. De los 38,000 genes descubiertos, se ha determinado la posición exacta dentro de los cromosomas de unos 7,000. ¿Qué porcentaje de los genes descubiertos ha sido ubicado en su posición real?

INTERPRETAR GRÁFICAS

Examina el siguiente árbol genealógico para el albinismo y responde luego a estas preguntas. Es posible que necesites usar una cuadrícula de Punnet. (El albinismo es un rasgo presente en individuos que no poseen pigmento en los ojos, piel o cabello).

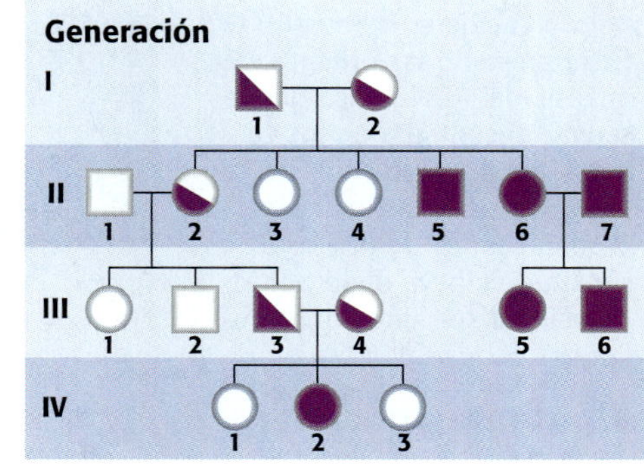

22. ¿Cuántos hombres hay en este árbol genealógico? ¿Cuántas mujeres?

23. ¿Cuántos individuos de la segunda generación tenían albinismo? ¿Cuántos eran portadores de este rasgo?

24. ¿Crees que el albinismo es un rasgo dominante o un rasgo recesivo? Explica por qué.

AHORA, ¿qué piensas?

Revisa tus respuestas a las preguntas de la página 151 que escribiste en el cuaderno de ciencias. ¿Han cambiado tus respuestas? Si es necesario, corrige tus respuestas basándote en lo que has aprendido en este capítulo.

Concept Mapping Transparency 7

Blackline masters of this Chapter Review can be found in the **Study Guide.**

Background

You may wish to point out to students that the study of DNA fingerprinting brings up some interesting issues about the use of new technologies to generate evidence for use in criminal trials. In practice, most legal bodies require that a certain technique is "generally accepted" by the scientific community before it is used as evidence in court. One problem with this requirement is determining who decides whether a technique is generally accepted or not. And if a new technique is used in court, its usefulness often depends on the ability of the judge and jury to understand it. Thus, extremely complex scientific or mathematical arguments tend to be limited in their usefulness.

The FBI and state and local crime labs are starting to establish banks of DNA fingerprints from convicted criminals. When samples are taken from crime scenes, they can be compared with those fingerprints in the bank to attempt to track down possible suspects. Opponents of such practices claim this kind of system can be abused.

Discussion

Encourage students to discuss the following question: How is DNA fingerprinting similar to traditional fingerprinting? How is it different? Accept all reasonable responses.

DEBATECIENTÍFICO

El ADN sometido a juicio

La tensión en el tribunal podía explotar en cualquier momento. El fiscal presentó las pruebas: "El análisis del ADN indica que la sangre encontrada en los zapatos del acusado coincide con la sangre de la víctima. Las probabilidades de que esta similitud se haya dado por casualidad son una en 20 millones".

Siguiente acusado: el ADN

Las batallas legales en las que se involucra el proceso de impresiones de ADN son cada vez más comunes. El proceso de impresiones digitales tradicional se ha usado por más de 100 años y ha sido una herramienta de identificación muy importante. Recientemente, muchas personas han exigido que las impresiones de ADN, o prueba de perfiles de ADN, substituya a la técnica tradicional. Esta prueba se ha usado para exonerar a miles de personas acusadas o encarceladas injustamente. La polémica empieza cuando las pruebas se usan para demostrar que un sospechoso es culpable.

Dudas razonables

Los críticos alegan que en el proceso de impresión de ADN puede haber errores humanos. El manejo de muestras de la escena de un crimen puede ser

▲ *Este médico forense junta células de piel muertas en una prenda de ropa con la esperanza de reunir muestras de ADN.*

engañoso; se pueden extraer muestras de debajo de la uña de una víctima o de la acera sucia de la calle. La contaminación por sales, químicos, mezclilla o por el estornudo de una persona en el laboratorio pueden comprometer la fiabilidad de los resultados.

Gran parte de la controversia sobre el proceso de impresiones de ADN tiene que ver con la interpretación de los resultados. Mucha gente se pregunta: "¿Es probable que alguien más, aparte del sospechoso, tenga el mismo perfil de ADN?" Las respuestas varían de una en tres a una en 20 millones, según la persona que hace la interpretación, el tamaño de la muestra y el proceso que se use.

La crítica también señala

que los resultados pueden calcularse sin considerar otros factores. Por ejemplo, es más probable que las personas de ciertos grupos étnicos compartan características de ADN con las de su grupo que con otras.

Más allá de una duda razonable

Quienes apoyan las pruebas de ADN señalan que el análisis es objetivo porque los laboratorios reciben muestras etiquetadas con códigos. La información obtenida es la que exonera o incrimina a un sospechoso. Las pruebas de ADN casi nunca se usan para condenar a alguien. Es una prueba más, como el móvil y el acceso a la escena del crimen, para llegar a un veredicto.

Los defensores del proceso de impresiones de ADN dicen que las revisiones y los balances en los laboratorios previenen errores humanos. Y algunos esfuerzos recientes para uniformar la recopilación de evidencia y la interpretación de las muestras han tenido muy buenos resultados.

¿Tú qué crees?

▶ ¿Debería admitirse el proceso de impresiones de ADN como prueba ante un tribunal? Investiga un poco más y toma una decisión.

172

Ciencia Ficción

"Moby James"

por Patricia A. McKillip

Rob Trask tiene un problema. Se trata de su hermano mayor, James. Rob está convencido de que James no es su verdadero hermano. Rob y su familia viven en una estación espacial, y él acaba de descubrir que su verdadero hermano ha sido enviado a la Tierra. Esta persona que dice ser James, es en realidad una especie de mutante de una planta o de un par de calcetines sucios.

Rob tiene otro problema, en su clase están leyendo la novela de Herman Melville *Moby Dick.* Al principio, Rob no puede concentrarse en la historia, pero cuando empieza a leerla queda fascinado con la historia del Capitán Ahab y su sed de venganza contra la gran ballena blanca Moby Dick. Moby Dick se había llevado algo de Ahab, su pierna, y Ahab quería hacer que la ballena pagara por ello. De pronto, Rob se da cuenta de que su hermano es un mutante de una enorme ballena blanca, Moby James.

A medida que Rob sigue a Ahab en su búsqueda de Moby Dick, empieza a entender qué debe hacer para que su hermano verdadero regrese. Así, observa a Moby James, esperando el momento en que cometa algún error que lo ponga en eviencia. En cuanto Rob desenmascare al James falso, podrá hacer que el verdadero regrese.

Para descubrir si Rob tiene éxito y encuentra a su hermano verdadero, lee "Moby James" en la *Antología de Holt de Ciencia Ficción.*

173

Further Reading If students liked this story, encourage them to read more of Patricia McKillip's stories, such as the following:

Fool's Run, Warner, 1987

Something Rich and Strange, Bantam, 1994

Winter Rose, Ace, 1996

SCIENCE FICTION
"Moby James"
by Patricia A. McKillip

Rob's brother is changing—but into what? a mutant robot? an evil irradiated skunk cabbage? a great white mutant whale? Whatever it is, it's making Rob very nervous . . .

Teaching Strategy

Reading Level This is a relatively short story that should not be difficult for students to read and comprehend.

Background

About the Author Patricia Anne McKillip (1948–) began her career not as a writer, but as a storyteller. As the second of six children, she often found herself in charge of looking after her younger brothers and sisters. She can't remember exactly when she first began telling her siblings stories. She does remember, however, her first attempt at writing. At age 14 she wrote a 30-page fairy tale.

Today McKillip is a full-time writer of science fiction and fantasy. In 1975, she won the World Fantasy Award for her novel *The Forgotten Beasts of Eld*. A few years later she was nominated for a Hugo Award for *Harpist in the Wind*. McKillip has written a number of other novels, and her short stories have appeared in various periodicals, including the *Science Fiction and Fantasy Review*, the *Los Angeles Times*, and the *New York Times Book Review*.

Chapter Organizer

CHAPTER ORGANIZATION	TIME MINUTES	OBJECTIVES	LABS, INVESTIGATIONS, AND DEMONSTRATIONS
Chapter Opener pp. 174–175	45		**Investigate!** Making a Fossil, p. 175
Section 1 Change Over Time	90	▶ Explain how fossils provide evidence that organisms have evolved over time. ▶ Identify three ways that organisms can be compared to support the theory of evolution.	
Section 2 How Does Evolution Happen?	90	▶ Describe the four steps of Darwin's theory of evolution by natural selection. ▶ Explain how mutations are important to evolution.	**Demonstration,** Form and Function, p. 185 in ATE **QuickLab,** Could We Run Out of Food? p. 187 **Design Your Own,** Survival of the Chocolates, p. 587 **Datasheets for LabBook,** Survival of the Chocolates, Datasheet 17
Section 3 Natural Selection in Action	90	▶ Give two examples of natural selection in action. ▶ Outline the process of speciation.	**Demonstration,** Natural Selection, p. 190 in ATE **Design Your Own,** Mystery Footprints, p. 584 **Datasheets for LabBook,** Mystery Footprints, Datasheet 15 **Discovery Lab,** Out-of-Sight Marshmallows, p. 586 **Datasheets for LabBook,** Out-of-Sight Marshmallows, Datasheet 16 **Whiz-Bang Demonstrations,** Adaptation Behooves You, Demo 6 **Long-Term Projects & Research Ideas,** Project 7

TECHNOLOGY RESOURCES

 **Guided Reading Audio CD** English or Spanish, Chapter 8

 **One-Stop Planner CD-ROM with Test Generator**

 **Multicultural Connections,** A Thailand Fossil Discovery, Segment 4

Science, Technology & Society, Deciphering Dog DNA, Segment 10

CLASSROOM WORKSHEETS, TRANSPARENCIES, AND RESOURCES	SCIENCE INTEGRATION AND CONNECTIONS	REVIEW AND ASSESSMENT
Directed Reading Worksheet 8 **Science Puzzlers, Twisters & Teasers,** Worksheet 8	**Multicultural Connection,** p. 175 in ATE	
Directed Reading Worksheet 8, Section 1 **Transparency 29,** Changes in Life over Earth's History **Transparency 96,** A Sedimentary Rock Cycle **Transparency 30,** Comparative Skeletal Structures **Transparency 31,** Vertebrate Embryos	**Earth Science Connection,** p. 178 **Multicultural Connection,** p. 178 in ATE **Connect to Earth Science,** p. 178 in ATE **Math and More,** p. 179 in ATE **Cross-Disciplinary Focus,** p. 180 in ATE	**Homework,** p. 179 in ATE **Review,** p. 183 **Quiz,** p. 183 in ATE **Alternative Assessment,** p. 183 in ATE
Directed Reading Worksheet 8, Section 2 **Math Skills for Science Worksheet 3,** Multiplying Whole Numbers **Transparency 32,** Natural Selection in Four Steps **Reinforcement Worksheet 8,** Bicentennial Celebration	**Connect to Geography,** p. 185 in ATE **Apply,** p. 189 **Eye on the Environment:** Saving at the Seed Bank, p. 198	**Homework,** p. 188 in ATE **Review,** p. 189 **Quiz,** p. 189 in ATE **Alternative Assessment,** p. 189 in ATE
Directed Reading Worksheet 8, Section 3 **Transparency 33,** Evolution of the Galápagos Finches **Critical Thinking Worksheet 8,** Taking the Earth's Pulse	**Real-World Connection,** p. 191 in ATE **Holt Anthology of Science Fiction,** *The Anatomy Lesson*	**Self-Check,** p. 191 **Review,** p. 193 **Quiz,** p. 193 in ATE **Alternative Assessment,** p. 193 in ATE

Holt, Rinehart and Winston On-line Resources

go.hrw.com

For worksheets and other teaching aids related to this chapter, visit the HRW Web site and type in the keyword: **HSTEVO**

National Science Teachers Association

www.scilinks.org

Encourage students to use the *sci*LINKS numbers listed with the Chapter Highlights to access information and resources on the **NSTA** Web site.

END-OF-CHAPTER REVIEW AND ASSESSMENT

Chapter Review in Study Guide
Vocabulary and Notes in Study Guide
Chapter Tests with Performance-Based Assessment, Chapter 8 Test
Chapter Tests with Performance-Based Assessment, Performance-Based Assessment 8
Concept Mapping Transparency 8

Chapter Resources & Worksheets

Visual Resources

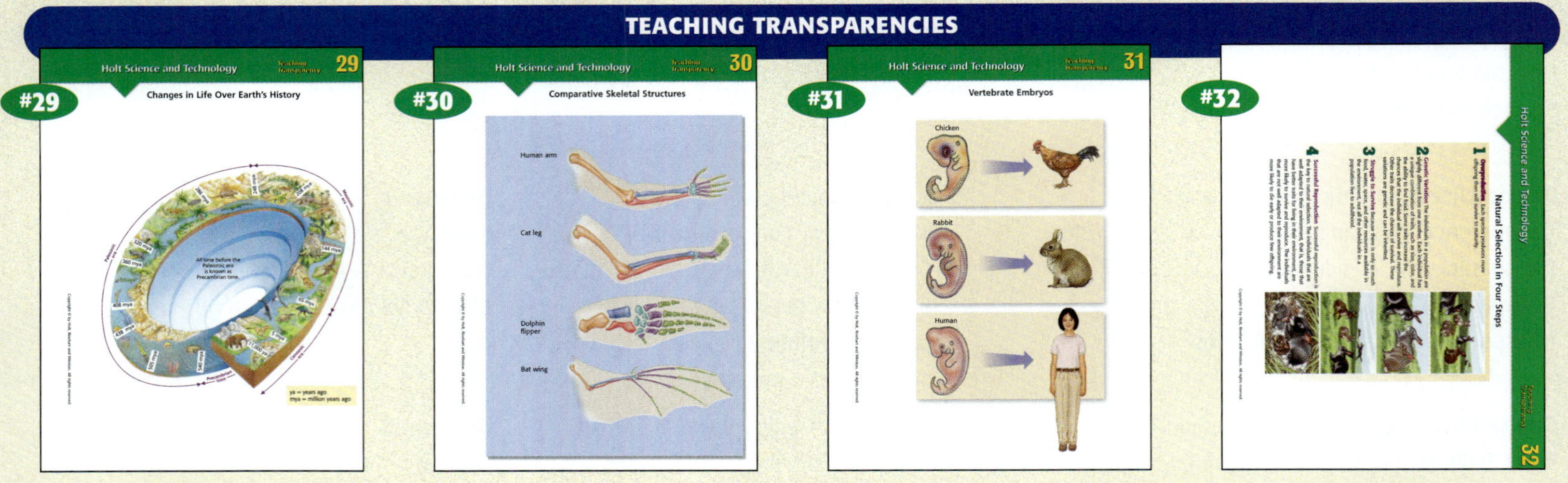

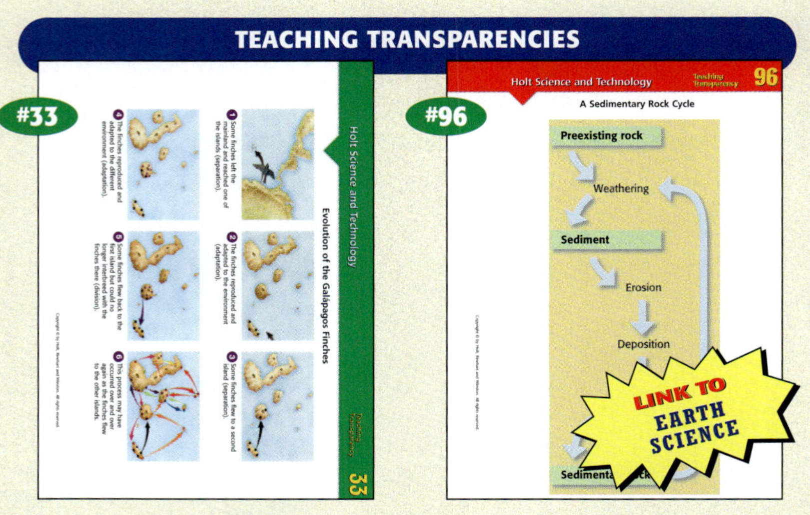

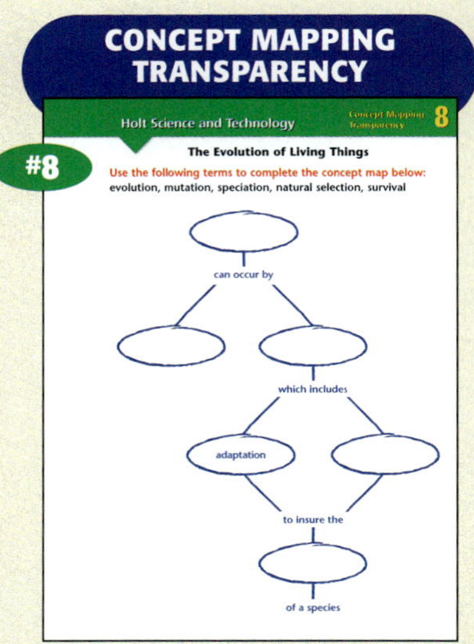

Meeting Individual Needs

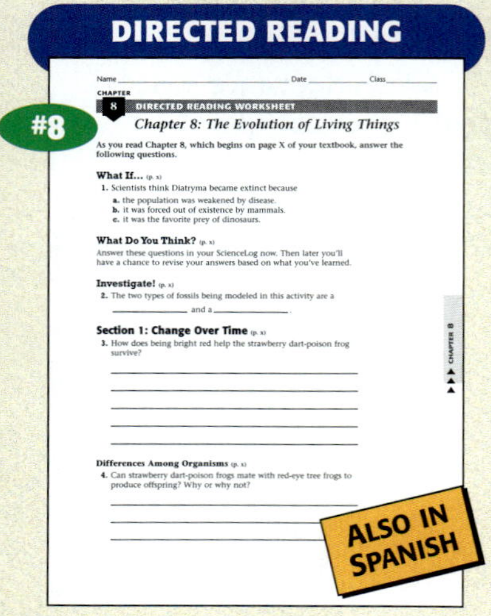

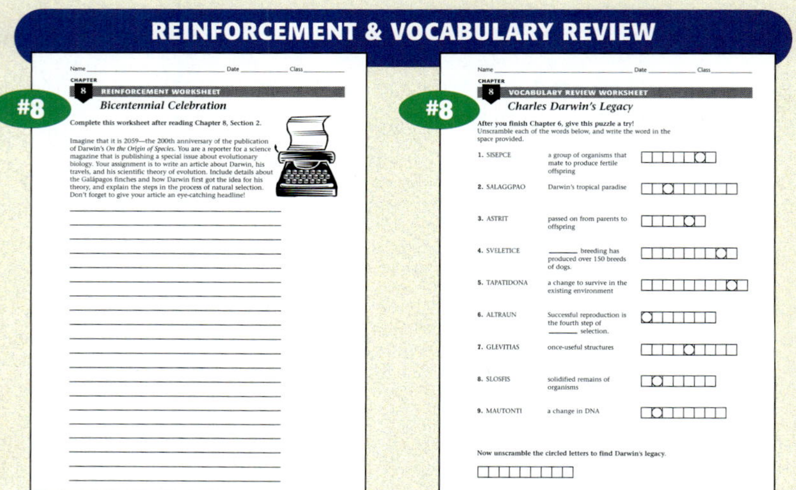

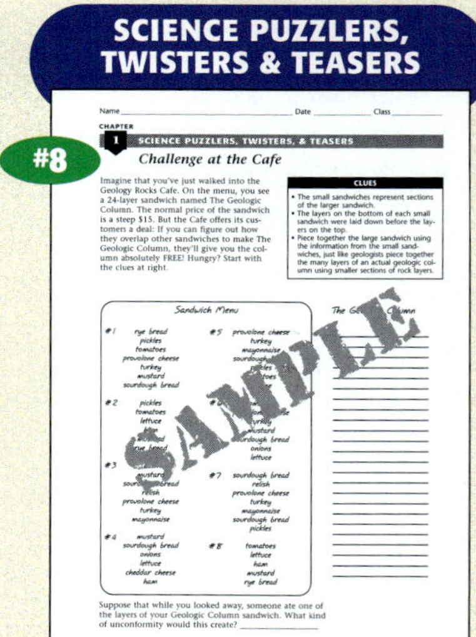

Review & Assessment

STUDY GUIDE

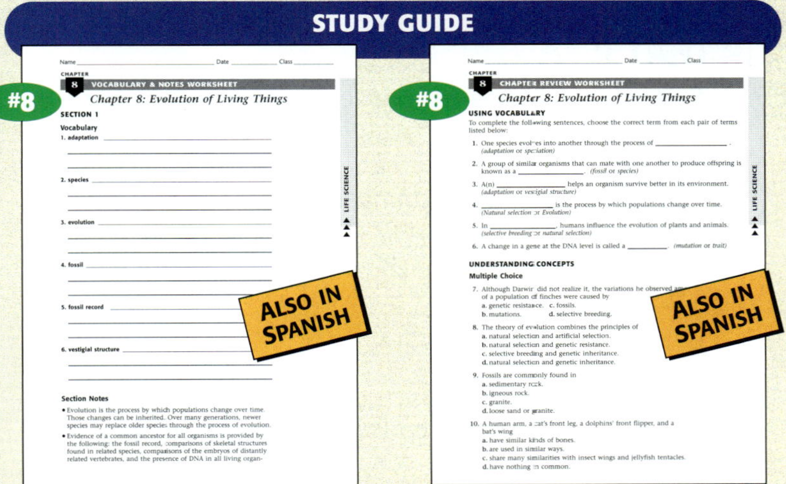

#8 Chapter 8: Evolution of Living Things — VOCABULARY & NOTES WORKSHEET

#8 Chapter 8: Evolution of Living Things — CHAPTER REVIEW WORKSHEET

ALSO IN SPANISH

CHAPTER TESTS WITH PERFORMANCE-BASED ASSESSMENT

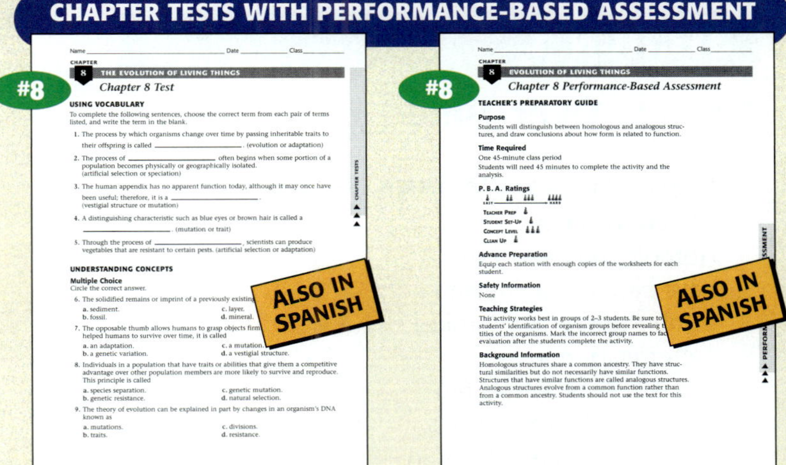

#8 Chapter 8 Test — THE EVOLUTION OF LIVING THINGS

#8 Chapter 8 Performance-Based Assessment — EVOLUTION OF LIVING THINGS

ALSO IN SPANISH

Lab Worksheets

WHIZ-BANG DEMONSTRATIONS

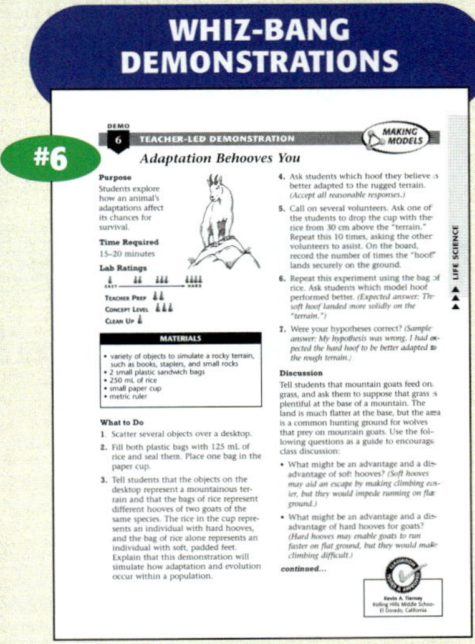

#6 *Adaptation Behooves You* — DEMO 6, TEACHER-LED DEMONSTRATION (MAKING MODELS)

LONG-TERM PROJECTS & RESEARCH IDEAS

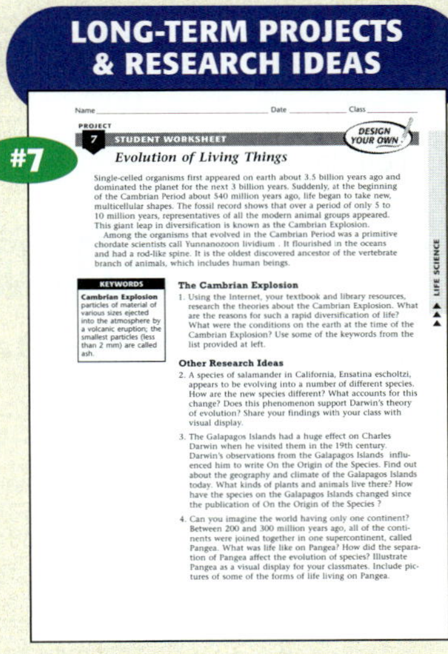

#7 *Evolution of Living Things* — PROJECT 7, STUDENT WORKSHEET (DESIGN YOUR OWN)

DATASHEETS FOR LABBOOK

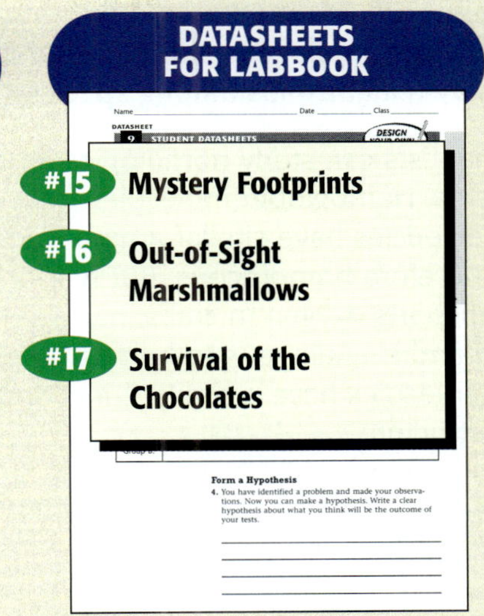

#15 Mystery Footprints

#16 Out-of-Sight Marshmallows

#17 Survival of the Chocolates

Applications & Extensions

CRITICAL THINKING & PROBLEM SOLVING

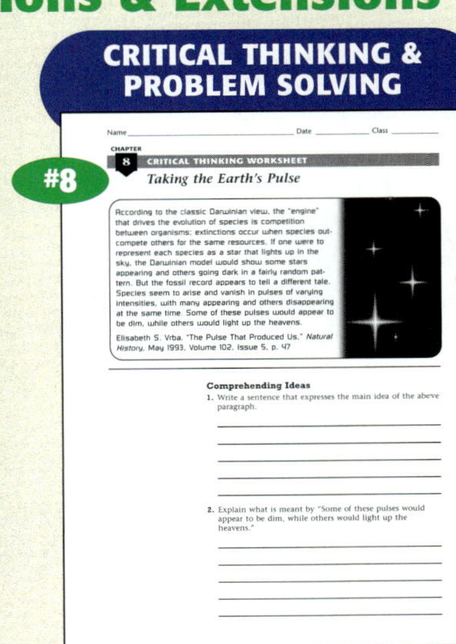

#8 *Taking the Earth's Pulse* — CRITICAL THINKING WORKSHEET

MULTICULTURAL CONNECTIONS

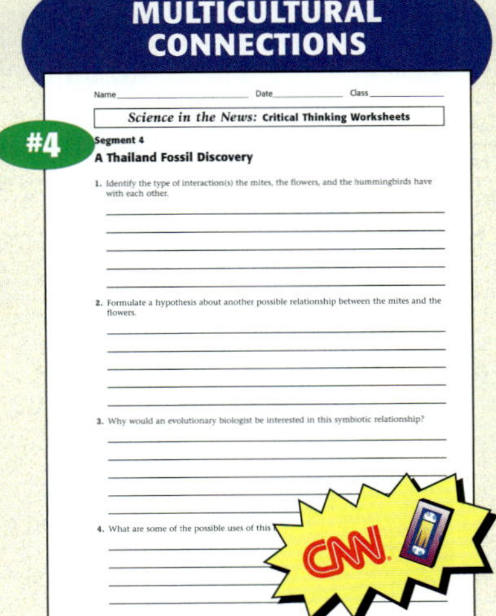

#4 Segment 4: A Thailand Fossil Discovery — Science in the News: Critical Thinking Worksheets

SCIENCE TECHNOLOGY

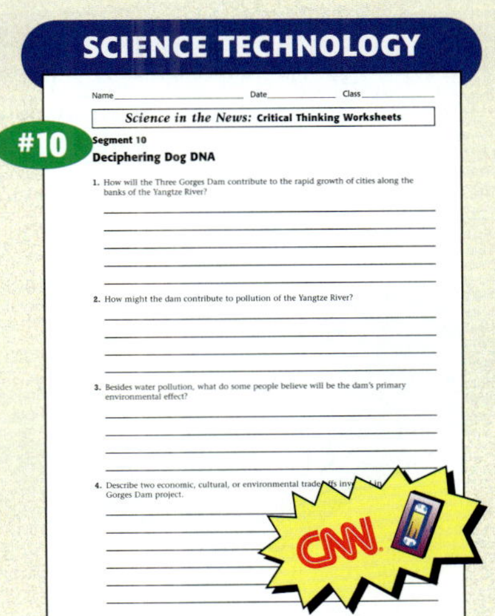

#10 Segment 10: Deciphering Dog DNA — Science in the News: Critical Thinking Worksheets

Change Over Time

▶ Homologous Versus Analogous Structures

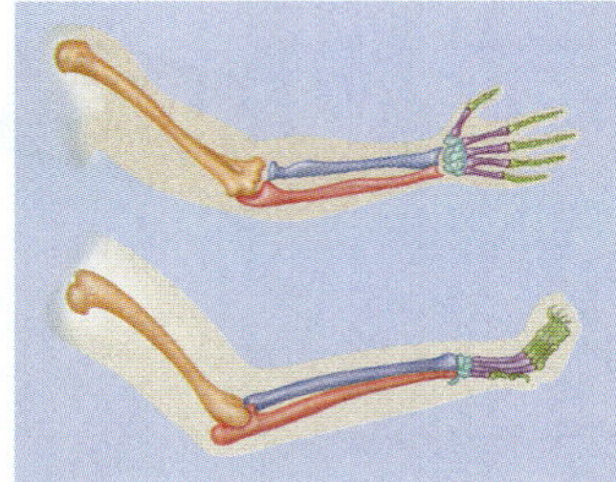

Homologous structures have similar origins and exhibit similar anatomical patterns. Bird wings, human arms, whale flippers, and deer forelimbs, for example, are similar in skeletal structure.

- In the early 1800s, French naturalist Etienne Geoffroy Saint-Hilaire studied embryos and recognized the importance of homologous structures for identifying evolutionary relationships among species.

- Today, scientists can study homologies among cellular components. Hemoglobin molecules from different vertebrate species have similar amino acid sequences and are therefore homologous. But hemocyanin, which transports oxygen in crabs, has a very different sequence and is analogous to hemoglobin; that is, the two molecules have a similar function but different structure.

- Bird wings and butterfly wings are another example of analogous structures.

▶ Convergent Evolution

When scientists study the fossils, skeletons, and DNA of species thought to be related, they sometimes find that the organisms are not related at all. For example, the jerboa and the kangaroo rat look almost identical, but they have different ancestors. Cases like this illustrate convergent evolution. The two species developed similar adaptations because they were subjected to the same environmental influences.

Is That A Fact!

- The human appendix is a vestigial organ. It's a narrow tube attached to the large intestine. It performs no function but can become infected and require surgical removal. In chimpanzees, gorillas, and orangutans the appendix is an intestinal sac that helps them digest tough plant material.

▶ Frozen Fossils

In some cases scientists can obtain DNA from ancient tissues that have not completely decomposed or fossilized. Two Japanese geneticists hope to create a mammoth-elephant hybrid by using frozen tissue from a Siberian mammoth. Critics of the project remain highly skeptical. The chances of finding intact DNA are remote, and the genetic structures of mammoths and elephants are not 100 percent compatible.

How Does Evolution Happen?

▶ Alfred Russel Wallace

Alfred Wallace (1823–1913) was born in England. He came from a poor family and had no formal scientific education. Though originally interested in botany, he began to study insects with the encouragement of British naturalist Henry Walter Bates, whom he met when he was about 20 years old. Bates and Wallace explored the Amazon from 1848 to 1852 and found much evidence to support the theory of evolution.

- From 1854 to 1862 Wallace traveled in the Malay Archipelago to find more evidence of evolution. In 1855, he published a preliminary essay, "On the Law Which Has Regulated the Introduction of New Species." Meanwhile, nearly 20 years after his voyage on the HMS *Beagle,* Charles Darwin was still mulling over his data. In 1858, Wallace mailed an essay to Darwin that explained Wallace's theory that natural selection pressures species to change.

- In July 1858, Wallace's essay was presented along with a paper by Darwin at a meeting of the Linnean Society in London. In the following year, after nearly two decades of delay (because of his doubts and repeated analyses of the data), Darwin published *On the Origin of Species by Means of Natural Selection*.

▶ Charles Lyell

Charles Lyell (1797–1875), the eldest of 10 children, was born in Scotland and raised in England. His father was a naturalist who traveled with him to collect butterflies and aquatic insects, informal research that Charles continued throughout college.

- Lyell's research led him to the belief that natural processes occurring over millions of years have shaped the Earth's features. This idea was known as uniformitarianism. Lyell's work influenced Darwin's formulation of his theory of natural selection. Darwin's proposed mechanism for evolution was plausible only if the Earth was ancient and if organisms had the requisite time for adaptation and change.

Natural Selection in Action

▶ Adaptive Coloration

Penguins, puffins, killer whales, and blue sharks are just some of the ocean animals that have white bellies and black or dark blue dorsal surfaces. This type of coloration is called countershading. When seen from below, the white underside helps the animal blend into the lighter sky above the water. When viewed from above, the dark coloration makes the animal difficult to see against the ocean depths.

IS THAT A FACT!

- Ptarmigans (chickenlike birds), Arctic foxes, and ermines change their color twice a year! All three are white in winter to blend with the snow of their northern habitat. The ptarmigan is mottled brown in summer, the fox is grayish brown, and the ermine is white below and brown on top.

- Octopuses and squid can change their color in a second. They have special cells called chromatophores that enable them to blend in with different-colored rocks and varied light conditions.

▶ The Fruitful Fruit Fly

In the mid-1800s a fruit fly that parasitized the hawthorn tree and its fruit infested apple trees in the Hudson River Valley area of New York. During the past 150 years, the apple tree variety of these flies has spread across the United States. Biologists have observed that the flies that attack apple trees do not also infest the hawthorns. Recent DNA studies revealed that the two groups are becoming isolated genetically. Scientists have concluded that speciation is occurring in these flies.

- Since the first step in speciation is separation, how did this process begin? Scientists classify this example as sympatric speciation. The flies began specializing on new host plants without geographic isolation. Separate trees were the extent of their separation.

IS THAT A FACT!

- The largest flying bird that ever lived had a wing span of more than 7 m. It was a New World vulture, and its fossil was discovered in Argentina. These birds were known as teratorns.

The Evolution of Living Things

La evolución de los seres vivos

Directed Reading Worksheet 8

Science Puzzlers, Twisters & Teasers Worksheet 8

Guided Reading Audio CD
English or Spanish, Chapter 8

¿Qué tal si...?

Imagínate que caminas por un pantano lleno de vapor, evitando culebras y arañas en un pantano de Norteamérica hace 50 millones de años. De pronto sale un gigante de 182 kg; con una cabeza enorme, cuello ancho y piernas largas y musculosas. ¿Qué bestia es? ¿Será un velociraptor o un oso perezoso gigante? ¿O quizás un oso prehistórico? No; no es ninguno de estos animales, ¡es un *Diatryma*, un ave no voladora muy común en la era Cenozoica de la Prehistoria, entre 57 y 35 millones de años atrás! Esta ave medía más de 2 m y tenía un pico enorme y garras afiladas. Se sabe que existió gracias a los fósiles que se han encontrado en Wyoming, Nuevo México y Nueva Jersey. Su desaparición está relacionada con la aparición de mamíferos de gran tamaño. Aunque esta ave se extinguió hace mucho tiempo, hay versiones más pequeñas de la misma en todo el mundo. Los fósiles indican que era uno primo distante de las gallinas actuales.

174

What If ...?

Scientists have found fossils of ancestral reptiles and mammals that are substantially larger than their modern forms. Dinosaurs, of course, are a familiar example. But there were also 1.5 m tall penguins long ago. Conversely, the ancient horses were quite small. So why did animals and plants change? Why did some organisms increase in size and others decrease? Why did they change at all? How does the altering occur? These are the questions Charles Darwin, Alfred Wallace, and others sought to answer.

Los fósiles muestran cómo eran los organismos extintos y nos proporcionan pistas del proceso de *evolución* por el cual los animales y otros seres vivos cambian a lo largo del tiempo. ¿Qué otras pruebas hay de la evolución y de cómo ocurre este proceso? Estos son algunos de los temas que estudiarás en este capítulo.

Cómo hacer un fósil

En esta actividad trabajarás con un compañero o compañera para hacer dos tipos de fósiles.

Procedimiento

1. La clase se dividirá en grupos de dos. Tu pareja y tú recibirán un **vaso** y un **plato de cartón**, **plastilina**, **vaselina** y un objeto pequeño, como **una hoja**, **una concha** o **un dinosaurio de juguete**. Marquen los vasos con su nombre.

2. Extiendan un poco de plastilina en el plato. Pongan el objeto sobre la plastilina y presionen, quitándolo después con mucho cuidado para que la impresión permanezca en la plastilina.

3. Coloquen un poco de plastilina en el fondo del vaso. Unten el objeto con vaselina y métanlo en el vaso. (Esta vez no lo presionen).

4. Pídanle a su maestra o maestro que vacíe un poco de **yeso** en el plato y en el vaso, de manera que la plastilina y el objeto queden cubiertos. Coloquénlos donde les diga la maestra o maestro y déjenlos secar toda la noche.

5. Al día siguiente, rompan el vaso y el plato con cuidado para sacar la plastilina y el objeto. Ahora tienen dos clases distintas de fósiles: un molde y una réplica. El molde es un hueco (huella) que tiene la forma de un organismo y la réplica es un molde relleno de sedimento.

¿Tú qué piensas?

Usa tus conocimientos para responder a las siguientes preguntas en tu cuaderno de ciencias:

1. ¿Qué es la evolución?

2. ¿Qué papel desempeña el medio ambiente en la supervivencia de los organismos?

3. ¿Por qué los fósiles son útiles para estudiar los cambios de los organismos?

Análisis

6. Clasifica tus fósiles en moldes o réplicas.

7. De los siguientes organismos, identifica cuáles serían buenos fósiles: una almeja, una medusa, un cangrejo y un champiñón. ¿Qué clase de fósil (molde o réplica) encontrarías de cada uno? ¿Cuáles no serían buenos fósiles? Explica tus respuestas.

175

What Do You Think?

Students will have a chance to revise their answers in the Chapter Review under NOW What Do You Think?

Investigate!

MATERIALS

FOR EACH GROUP:
- paper cups
- paper plates
- modeling clay
- petroleum jelly
- leaf, shell, or small toy dinosaur
- plaster of Paris

Teacher Notes: Some of the most famous fossilized molds are in the Laetoli region of northern Tanzania, in East Africa. In 1976, a paleontologist working with archaeologist Mary Leakey discovered thousands of animal tracks. Two years later, a team discovered human footprints. These tracks were made about 3.6 million years ago. After ash from a nearby volcano was dampened by rain, elephants, giraffes, people, and some now-extinct mammals walked across this area. Ash from another eruption of the volcano in turn covered the tracks and fossilized them.

Answers to Investigate!

6. Answers will vary.

7. The crab and clam will make the best fossils because they have hard body parts that decay slowly and leave impressions. The softer organisms—the jellyfish and mushroom—are less likely to make impressions in the sediment, and so they are less likely to make fossils.

Multicultural CONNECTION

An important set of fossilized footprints was discovered in the Laetoli region of Tanzania in 1976. Most of the people who live there are Masai, many of whom continue their traditional way of life based on maintaining herds of cattle. Their cooperation was essential to protect the site from damage by the herds. The *Loboini*, who is a traditional religious leader of the Masai, organized a meeting among the Masai and explained the tracks' significance and the need to protect them. The site remains intact because of the respect they have given it.

Focus

Change Over Time

This section introduces students to the theory of evolution. They will see how organisms change at the population level through adaptations. Students will learn about the evidence of evolution, including the fossil record and comparisons of organisms' physical structures, DNA, and embryonic structures.

Bellringer

Have the following information displayed on the chalkboard or an overhead projector when students enter:

The cockroach originated on Earth more than 250 million years ago and is thriving today all over the world. A giant deer that stood 2.1 m and had antlers up to 3.6 m evolved less than one million years ago and became extinct around 11,000 years ago.

Why do you think one animal thrived and the other perished?

Directed Reading Worksheet 8 Section 1

Sección 1

Cambios a través del tiempo

VOCABULARIO

adaptación	fósil
especie	registro fósil
evolución	vestigio

OBJETIVOS

- Explica por qué los fósiles proporcionan pruebas de que los organismos evolucionan a lo largo del tiempo.
- Identifica tres maneras de comparar los organismos para apoyar la teoría de la evolución.

Si alguien te pide que describas una rana, puedes decirle que tiene patas traseras largas, ojos saltones y la costumbre de croar de vez en cuando. Es posible que después se te ocurrieran algunas diferencias que dividen a las ranas en distintas clases. Observa las **Figuras 1, 2** y **3** de esta página. Estas ranas son diferentes entre sí, a pesar de que todas viven en un bosque tropical lluvioso. Lee la leyenda de cada figura y descubrirás que sus diferencias son más profundas.

Figura 2 *Las ranas* Leptodactylus pentadactylus *se confunden con las hojas muertas del suelo y así sobreviven.*

Figura 3 *Las ranas* Dendrobates pumilio *brincan por el suelo del bosque durante el día. El color brillante de su piel advierte a los depredadores que son venenosas.*

Figura 1 *Para sobrevivir, las ranas* Agalychin callidryas *se esconden entre las hojas de los árboles durante el día y sólo salen de noche.*

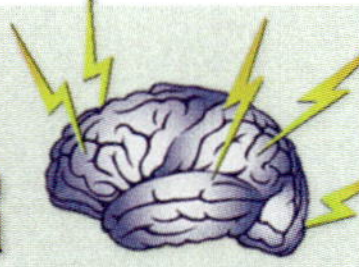

Las tribus nativas de Centroamérica untan el veneno de las ranas *Dendrobates pumilio* en las puntas de sus lanzas antes de ir de caza; el veneno paraliza a las presas.

Diferencias entre los organismos

Como ves, estas tres ranas presentan distintas adaptaciones que les permiten sobrevivir. Una **adaptación** es una característica hereditaria que permite a los organismos sobrevivir y reproducirse en su medio ambiente. Entre las adaptaciones están las estructuras y comportamientos para encontrar comida y protección, y para moverse de un lado a otro.

Los seres vivos que comparten las mismas características y adaptaciones son miembros de la misma especie. Una **especie** es un grupo de organismos que pueden aparearse y producir crías fértiles. Por ejemplo, todas las ranas *Agalychins callidryas* pertenecen a la misma especie y cuando se aparean producen ranas de la misma clase.

176

Strawberry dart-poison frogs get their toxicity from eating ants that contain the poisonous chemicals. When kept in captivity without access to their natural diet, the frogs become harmless.

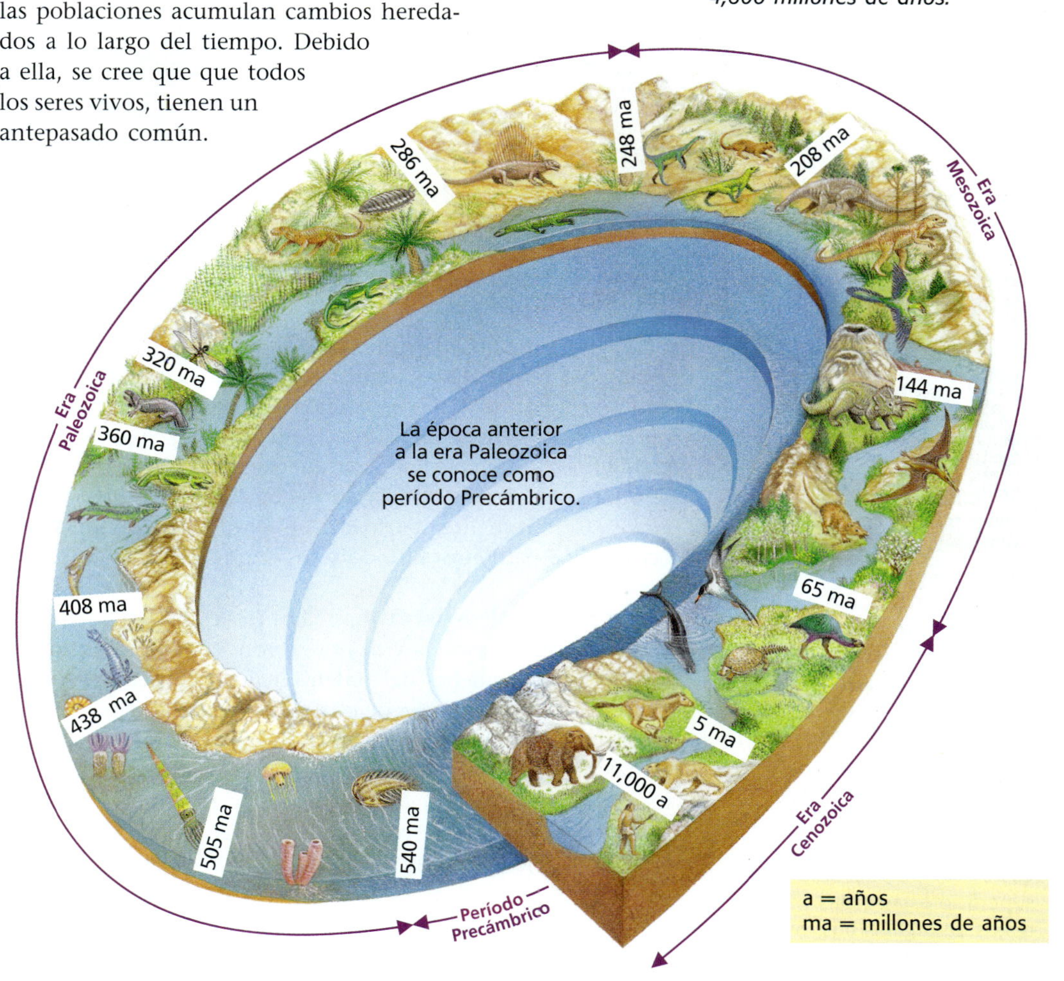

Figura 4 *Esta espiral representa los cambios en la vida desde la formación de la Tierra, hace 4,600 millones de años.*

Teaching Transparency 29 "Changes in Life over Earth's History"

1) Motivate

DISCUSSION

Adaptation Ask students if a polar bear could live comfortably in Hawaii. Ask if a fish could survive in a forest. Help students understand that each animal has characteristics that make it well-suited for its home environment. The polar bear has a thick layer of fat and dense fur to keep it warm. The fish's gills allow it to obtain oxygen from water. These are physical adaptations to specific environments.

2) Teach

USING THE FIGURE

Direct students to review **Figure 4** and explain why there are more fossils from the Cenozoic era than the earlier eras. (Fossils from earlier eras are often harder to find for two reasons. First, these fossils are located deeper in the Earth and are often covered by younger rocks. Second, many of the older rocks that once contained earlier life-forms no longer exist because rocks are recycled.)

Ask students how they think changes in the planet could have affected the appearance and disappearance of various life-forms over time. (Temperature fluctuations due to ice ages would have affected which plants and animals could survive. Climate changes would have reduced the vegetation and so limited the food available for animals. Some other possible causes for the appearance or disappearance of new species include: environmental changes caused by the impact of disintegrating meteors called bolides (BOH LIEDZ), the appearance of new predators, and the loss of a food source.)

Multicultural CONNECTION

Mary Anning (1799–1847) made some of the most important fossil discoveries of her time. She was born in Lyme Regis, in southern Great Britain, an area with many fossils. Her father, a cabinetmaker and amateur fossil collector, died when Mary was 11 years old, leaving the family in debt. Mary's fossil-finding skills provided the family with needed income. Even before she reached her teens, Mary had discovered part of the first *Ichthyosaurus* to be recognized by scientists in London. In the early 1820s, a professional fossil collector sold his private collection and gave the proceeds to the Anning family. He recognized that they had contributed many specimens for scientific investigation. Soon after, Mary took charge of the family fossil business. She later discovered the first plesiosaur. However, many of Mary Anning's finds ended up uncredited. Many scientists could not accept that a person of her financial and educational background could have acquired such expertise.

CONNECT TO EARTH SCIENCE

Using sedimentary layers as reference points, scientists can find the relative age of a fossil. Use Teaching Transparency 96, "A Sedimentary Rock Cycle," to illustrate the sedimentary rock cycle.

Teaching Transparency 96
"A Sedimentary Rock Cycle"

Pruebas de la evolución: el registro fósil

La prueba de que los seres vivos evolucionan proviene de muchas fuentes, entre ellas, los fósiles y las comparaciones entre distintos grupos de organismos.

Fósiles La capa externa de la Tierra se conoce como *corteza*. Gran parte de la corteza está formada por capas apiladas que se forman cuando el viento y el agua acarrean sedimentos, partículas de arena y polvo. Las capas más antiguas se depositan primero y están a mayor profundidad. En estas capas se encuentran los **fósiles,** que son restos solidificados o huellas de organismos que en otro tiempo estuvieron vivos. Los fósiles como los de la **Figura 5,** pueden ser de organismos completos, partes de organismos o sólo un conjunto de huellas.

Hasta el día de hoy, los científicos han descrito y nombrado a alrededor de 300,000 especies fósiles.

Figura 5 *El fósil de la izquierda es de un trilobites, un animal acuático, pariente lejano del cangrejo, que vivió hace aproximadamente 500 millones de años. Los fósiles de la derecha pertenecen a helechos con semillas que vivieron hace unos 300 millones de años.*

Por lo general, los fósiles se forman cuando una capa de sedimento cubre a un organismo muerto. Con el tiempo, se depositan más sedimentos encima del organismo, cuyos minerales se filtran al interior del organismo y lo substituyen gradualmente por piedra. Si al quedar cubierto, el organismo se pudre por completo, deja un hueco en la roca que se llama *molde*. Comúnmente los fósiles se forman en el fondo del océano, los pantanos, el lodo y los depósitos de alquitrán.

Cómo se lee el registro fósil Los fósiles proporcionan una secuencia histórica de la vida que se conoce como **registro fósil** y proporciona pruebas del orden en que ocurrieron los cambios evolutivos. Los fósiles de las capas superiores, o más nuevas, de la corteza terrestre tienden a parecerse a los organismos actuales lo que indica que eran parientes cercanos de los organismos actuales. Dos fósiles de las capas más profundas pertenecen a formas de vida muy antiguas, ahora extintas. El registro fósil también muestra los cambios en las condiciones ambientales de la Tierra a lo largo del tiempo.

ciencias de la Tierra CONEXIÓN

Por lo general, los fósiles se encuentran en capas de rocas sedimentarias, las cuales se forman cuando las rocas se rompen por acción del viento, el agua u otro factor. El viento y el agua remueven el sedimento y lo depositan. Con el tiempo, las capas de sedimento llegan a formar pilas de gran altura. Las capas más bajas se comprimen y se transforman en roca. Muchos años después, las fuerzas geológicas pueden sacar a la superficie estas capas y los restos de organismos fosilizados que contienen.

178

Espacios vacíos en el registro fósil Si cada organismo dejara su huella, el registro fósil parecería un gran árbol genealógico de la evolución. La **Figura 6** muestra un registro fósil hipotético en el que aparecen todas las relaciones entre los organismos.

Aunque se han recolectado miles de fósiles, todavía hay un gran número de espacios vacíos en el registro fósil actual, como se ve en la **Figura 7.** Esto se debe a que la formación de fósiles requiere condiciones particulares. Los organismos que no tienen partes duras generalmente no se conservan en el registro fósil. La mayoría de los organismos fosilizados tenían esqueletos o conchas que, para conservarse, quedaron enterrados bajo sedimentos muy finos y sin oxígeno. En presencia de oxígeno, gran parte del organismo no se fosiliza pues este gas ocasiona su descomposición. Como son pocos los lugares donde no hay oxígeno, los fósiles son difíciles de encontrar.

Vestigios Las ballenas son los organismos más grandes del océano. Diariamente consumen toneladas de plancton, y aunque se parecen a los peces, son mamíferos (animales de sangre caliente que respiran aire, dan a luz y alimentan a sus crías con leche). Aunque las ballenas modernas no tienen patas traseras, hay restos de huesos de estas estructuras en el interior de su cuerpo, como se ve en la **Figura 8.** Los restos de estructuras que en otro tiempo tuvieron una función se llaman **vestigios.** Se cree que a lo largo de millones de años las ballenas evolucionaron a organismos marinos a partir de organismos terrestres parecidos a los perros. Sin embargo, hasta ahora no hay pruebas fósiles para comprobar esta hipótesis. Lee el siguiente estudio sobre la evolución de las ballenas.

Figura 6 *Así se vería el registro fósil si se hubieran encontrado fósiles de cada especie.*

Figura 7 *Los científicos han logrado completar secciones pequeñas del registro fósil.*

Figura 8 *Los restos de los huesos de las patas traseras están dentro del cuerpo de la ballena.*

IS THAT A FACT!

Indricotherium, a giant land mammal that was about 4 m tall and was up to 6 m long, lived in Asia between 18 million and 32 million years ago.

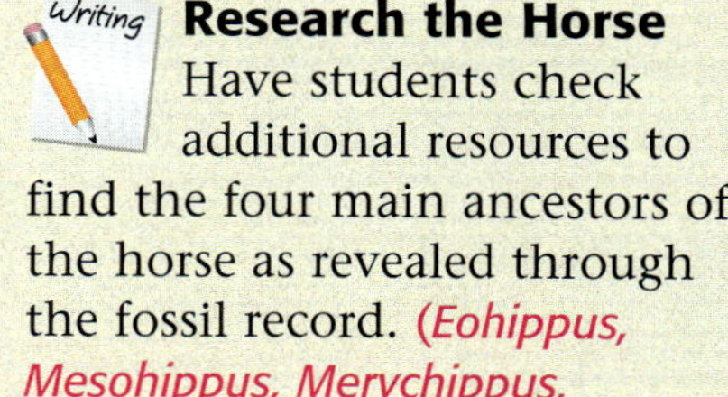

Homework

Research the Horse

Writing Have students check additional resources to find the four main ancestors of the horse as revealed through the fossil record. (*Eohippus, Mesohippus, Merychippus, Pliohippus*)

Ask students to make a poster that shows each ancestral horse in order of appearance and to write a paragraph about each one explaining its unique physical characteristics. Students should conclude their reports by answering the following question:

Are the wild horses in North America direct descendants of fossil horses found on this continent? (No; the fossil record shows that horses native to North America disappeared. Today's wild horses are descendants of horses introduced to this continent by Spanish explorers in the early 1500s.)

MATH and MORE

Say to students: Imagine that you are a scientific time traveler assigned to count the life-forms in what will become your home state 1 million years from now. You know that at present your home state contains 1,200 insect species, 550 animal species, and 600 plant species. Only 24 percent of these species are expected to survive the next million years. How many species existing now should you expect to find in 1 million years?

(insects: $1,200 \times 0.24 = 288$
animals: $550 \times 0.24 = 132$
plants: $600 \times 0.24 = 144$)

CROSS-DISCIPLINARY FOCUS

Art The role of a scientific illustrator is to make accurate pictures of organisms and things that scientists study. In the case of long-extinct species, such as dinosaurs, artists must sometimes fill in where science leaves off. Have students look for examples of illustrations of extinct animals and compare them with other illustrations of the same animal. Have students try to identify areas where artistic interpretation is used. You may wish to provide students with examples of similar illustrations from hundreds of years ago, when much less was known about these animals. **Sheltered English**

ACTIVITY

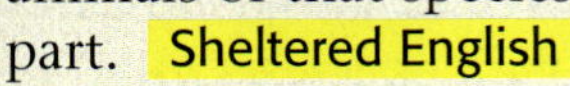

Making Posters
Have students create pictures of their own imaginary animal that is evolving from an ocean-dwelling species into a terrestrial one. Tell them to draw at least three stages of the progression and to label each significant body part with a description of how the animals of that species use that part. **Sheltered English**

internet**connect**

SC*LINKS*
NSTA

TOPIC: Species and Adaptation
GO TO: www.scilinks.org
*sci***LINKS NUMBER:** HSTL155

Estudio: la evolución de la ballena

La hipótesis de los científicos es que las ballenas evolucionaron a partir de mamíferos terrestres, como el *Mesonychid* de la ilustración de abajo, que regresaron al océano hace unos 55 millones de años. Durante las décadas de 1980 y 1990, se descubrieron varios fósiles de los antepasados de las ballenas. Estos descubrimientos sustentan la teoría de la evolución de la ballena.

El *Ambulocetus*, cuya ilustración aparece abajo, vivía en aguas costeras. Sus patas eran más cortas que las del *Mesonychid*, pero todavía tenía garras que apoyaban su peso en la Tierra. Los científicos piensan que el *Ambulocetus* usaba su cola para equilibrarse y que nadaba con sus patas como una nutria.

WEIRD SCIENCE

In 1938, some fishermen caught a live coelacanth, a primitive type of fish that was thought to have been extinct for about 65 million years.

Hace 46 millones de años, el *Rodhocetus* apareció en el registro fósil. Este animal tenía un parecido cercano con las ballenas modernas, pero conservaba las patas traseras de su antepasado terrestre. Como sus patas eran tan cortas, el *Rodhocetus* sólo podía moverse como un cocodrilo cuando estaba en tierra; no las necesitaba para nadar, como en el caso del *Ambulocetus*, sino que utilizaba su enorme cola para desplazarse en el agua. Probablemente, el *Ambulocetus* se arrastraba hasta la tierra todas las noches, mientras que el *Rodhocetus* pasaba la mayor parte del tiempo en el agua.

El *Prozeuglodon* apareció en el registro fósil 6 millones de años después que el *Rodhocetus* y estaba muy bien adaptado para la vida en el mar. A pesar de que todavía tenía un par de patas muy pequeñas, el *Prozeuglodon* vivía únicamente en el agua.

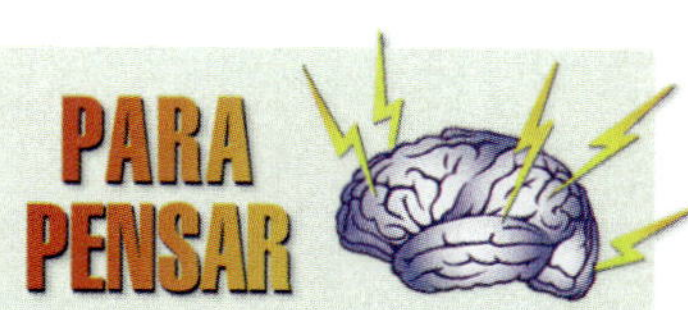

PARA PENSAR

En las primeras etapas de su desarrollo, los embriones de ballena tienen cuatro extremidades; las traseras desaparecen antes del nacimiento y las delanteras se convierten en aletas.

181

IS THAT A FACT!

Baby blue whales can weigh 9,000 kg (about 20,000 lb) at birth.

DISCUSSION

Inland Whales Explain to students that in 1849 workers constructing a railroad near the town of Charlotte, Vermont, discovered bones that were later identified as those of a beluga whale. Ask students to locate Charlotte on a map and explain why this discovery is so unusual. (It is more than 150 miles from the nearest ocean.)

Ask students what these bones tell us about the history of the land around Charlotte. (It used to be part of an ocean.)

Explain that the Champlain Sea, an extension of the ocean, existed for 2,500 years after the last glaciers retreated 12,500 years ago.

MISCONCEPTION ALERT

It is easy for students to confuse adaptation with acclimation or intentional change or adjustment. For example, a student might write that over many years, a species learned to adapt to its new environment. Explain to students that adaptation is something that happens to a species over many generations and not an activity that a species learns or chooses to do.

Acclimation is an adjustment to a condition that is within an organism's range of tolerance, such as seasonal adjustment to climate changes. This, too, is a part of an organism's evolutionary adaptations and is not something it learns to do.

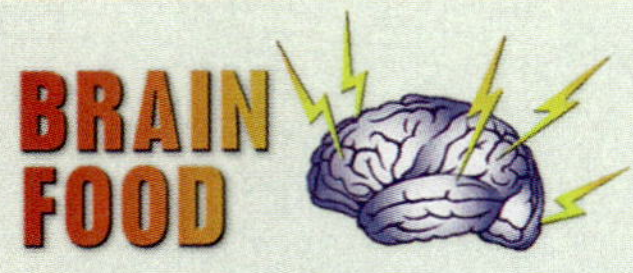

BRAIN FOOD

Over the years, some bacteria have become resistant to antibiotics that doctors use to treat or prevent diseases. Some scientists suggest that as bacteria become more and more resistant, people will be increasingly susceptible to microbes that cannot be stopped. Scientists are concerned that by using antibiotics, we are creating a larger problem for the future.

Research

Writing Scientists study animal skeletons and DNA to determine evolutionary relationships and development because merely looking at the outward appearance of a species can be misleading. The giant panda and red panda illustrate this problem. Their common names indicate that the two pandas seem closely related, but scientists now believe that the red panda is the only member of the subfamily Ailurinae of the raccoon family Procyonidae, which is quite separate from the giant panda. Have students investigate and write a report based on recent studies on the classification of these two pandas.

Teaching Transparency 30 "Comparative Skeletal Structures"

Pruebas de la evolución: comparación entre organismos

Otra prueba de que la vida evoluciona proviene de las comparaciones entre distintos grupos de organismos. En las siguientes páginas se comentan con mayor detalle las clases de pruebas que apoyan la teoría de la evolución.

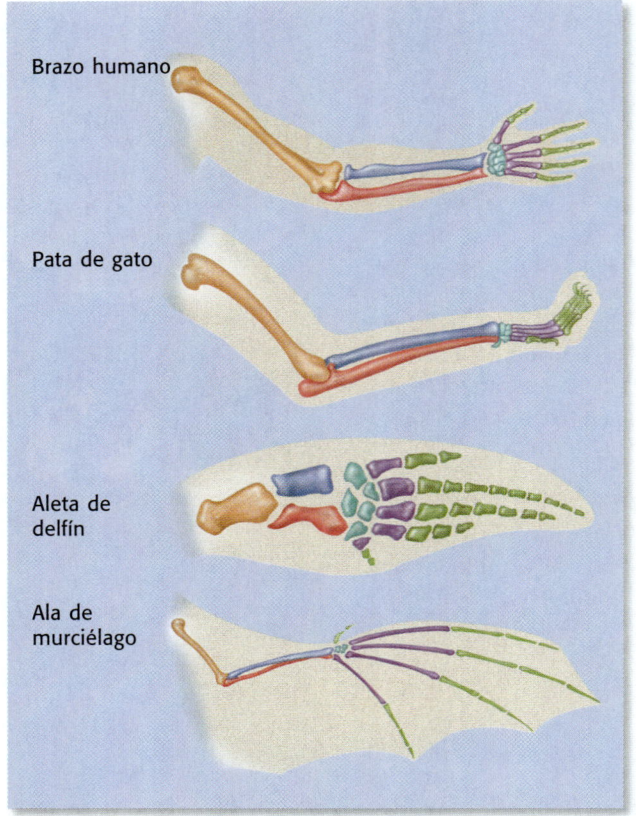

Figura 9 *Los huesos de las extremidades delanteras de estos animales son similares, a pesar de que las usan de manera distinta. Los huesos más parecidos aparecen en el mismo color.*

Comparación de estructuras óseas ¿Qué tienen en común un brazo humano, la pata delantera de un gato, la aleta frontal de un delfín y el ala de un murciélago? A primera vista, te puede parecer que tienen muy poco en común. Después de todo, estas estructuras no son muy parecidas y no se utilizan de la misma manera. Sin embargo, si las examinas con detenimiento comprobarás que la estructura y disposición de los huesos de las extremidades frontales de estos animales (**Figura 9**) son similares a las de tu brazo.

Las semejanzas indican que animales tan diferentes como gatos, delfines, murciélagos y seres humanos tenían un antepasado común. El proceso evolutivo modificó estos huesos a lo largo de millones de años para que desempeñaran funciones específicas.

Comparación del ADN de especies distintas Los científicos formularon la hipótesis de que si los organismos actuales evolucionaron a partir de un antepasado común, todos deberían tener la misma clase de material genético. Y, de hecho, la tienen. Desde las bacterias microscópicas hasta los osos polares gigantes, todos los organismos tienen el mismo material genético: el ADN.

Además, otra hipótesis dice que las especies con parentesco cercano presentan mayores semejanzas en su ADN que las especies con parentesco lejano. Por ejemplo, los chimpancés y los gorilas son parientes cercanos, pero los chimpancés y los tucanes son parientes lejanos. De hecho, el ADN de los chimpancés se parece más al ADN de los gorilas que al de los tucanes.

182

WEIRD SCIENCE

It seems as though the knee joints of birds bend backward, but they bend just like a human's knees. Birds walk on their toes. The long bone just above the toes is the foot! The first big joint above that (the one people often think is the knee) is actually a bird's ankle.

Comparación de embriones ¿Puedes distinguir entre una gallina, un conejo y un ser humano? Si comparas los adultos de cada especie es bastante fácil. Pero, ¿qué pasa si los comparas antes de su nacimiento? El lado izquierdo de la **Figura 10** ilustra los embriones de una gallina, un conejo y un ser humano.

Todos los organismos que aparecen en la figura son *vertebrados,* es decir, animales que tienen columna vertebral. Al principio, los embriones humanos y los del resto de los vertebrados son semejantes. Esto prueba que todos descienden de un antepasado común. Sin embargo, los embriones no tienen el aspecto de su forma adulta. A lo largo de millones de años, la evolución del desarrollo embrionario ha ha diversificado las estructuras embrionarias de muchas especies distintas. Estos cambios producen animales muy diferentes.

Figura 10 *Los embriones de los vertebrados se parecen mucho en las primeras etapas de su desarrollo.*

REPASO

1. ¿Por qué el registro fósil sugiere que las especies han cambiado a lo largo del tiempo?

2. ¿De qué manera las semejanzas entre los huesos de las extremidades anteriores de los humanos, gatos, delfines y murciélagos apoyan la teoría de la evolución?

3. **Interpretar Gráficas** La fotografía de la derecha muestra las capas de roca sedimentaria expuestas durante la construcción de una carretera. Imagínate que una especie que vivió hace 200 millones de años se encuentra en la capa marcada como *b*. Si su antepasado vivió hace 201 millones de años, ¿en qué capa sería más probable encontrarlo, en la *a* o en la *c*? Explica tu respuesta.

183

▼ *Answers to Review*

1. Fossils provide a historical sequence of life. The fossils found in the upper, or newer layers of the Earth's crust tend to resemble present-day organisms. The deeper in the Earth's crust fossils are found, the less they look like present-day organisms; these are fossils of earlier forms of life that are now extinct.

2. The similarities indicate that animals as different as a cat, a dolphin, a bat, and a human are all descendants of a common ancestor. They have been modified over millions of years to perform different functions.

3. c

4 Close

Quiz

1. Use the words *adaptation*, *population*, and *evolution* together in a sentence. (Sample answer: Evolution is the process by which a population changes through inherited adaptations over time.)

2. List two reasons why gaps exist in the fossil record. (Fossilization requires precise and sometimes rare conditions, including the absence of oxygen and burial in very fine sediment.)

ALTERNATIVE ASSESSMENT

Writing Charles Darwin's journals contain notes and records from his travels. Ask students to imagine that they are traveling with Darwin and keeping their own journals. Their notes and drawings should reflect what they see, the questions that arise from their observations, and the hypotheses that they form. Encourage students to write journal entries about other animals on the Galápagos Islands besides the finches, such as the Galápagos tortoise and marine iguanas.

MISCONCEPTION ALERT

Explain to students that the embryonic figures shown in **Figure 10** are not all at the same stage of development. The similarities are fleeting, but they are shown here to indicate that the vertebrate body plan is evident in early development.

Teaching Transparency 31 "Vertebrate Embryos"

Focus

How Does Evolution Happen?

This section introduces students to Charles Darwin and his voyage to the Galápagos Islands. Students will learn how artificial selection, geology, and the writings of Thomas Malthus and Charles Lyell helped Darwin formulate his theory of natural selection. Finally, students will learn that twentieth-century biologists used their knowledge of genetics to explain that species change through genetic mutation.

 **Bellringer**

On the board or on an overhead projector write the following list:

> upright walking, hair, fingerprints, binocular vision, speech

These are traits that almost all humans have in common. Ask students to list the advantages and disadvantages of each trait.

1 Motivate

DISCUSSION

Dinosaurs Ask students to describe a dinosaur. Ask them to explain why there are no dinosaurs alive today. Ask them, finally, if they think dinosaurs became extinct because they were not "evolved" enough to survive until the present. (Explain that dinosaurs were well adapted to their environment and lived over 150 million years on Earth. But a catastrophic event changed the environment faster than the dinosaurs could adapt, and they became extinct.)

VOCABULARIO

rasgo
cruza selectiva
selección natural
mutación

OBJETIVOS

- Describe los cuatro pasos de la teoría de Darwin sobre la evolución por selección natural.
- Explica por qué las mutaciones son importantes para la evolución.

Figura 11 *Este fósil de pez fue descubierto en la cima de una montaña.*

¿Cómo ocurre la evolución?

El comienzo del siglo XIX fue una época de grandes descubrimientos científicos. Los geólogos se dieron cuenta de que la Tierra era mucho más vieja de lo que se había pensado. Las pruebas indicaban que procesos graduales dieron forma a la superficie de la Tierra a lo largo de millones de años. Se hallaron restos de organismos extraños y fósiles de seres conocidos, aunque algunos estaban en lugares poco comunes. Por ejemplo, se encontraron fósiles de conchas y peces como el de la **Figura 11,** en cimas de montañas. De pronto, la Tierra parecía un lugar donde eran posibles grandes cambios. Muchos creían en la evolución, pero antes de Charles Darwin nadie había sido capaz de determinar cómo ocurría.

Charles Darwin

En 1831, a los 21 años de edad, Charles Darwin (**Figura 12,**) acababa de graduarse de la universidad. Al igual que muchos jóvenes recién graduados, Darwin no sabía qué quería hacer con su vida. Su padre quería que fuera médico, pero a Darwin le impresionaban las cirugías. Aunque terminó obteniendo un título en teología, *realmente* le interesaban las ciencias naturales: el estudio de las plantas y los animales.

Darwin decidió abandonar la carrera religiosa y le pidió a su padre permiso para hacer un viaje de 5 años alrededor del mundo, trabajando como naturalista (un científico que estudia la naturaleza) a bordo de un barco de la marina británica, el HMS *Beagle.* Durante su viaje, Darwin hizo observaciones que posteriormente dieron origen a su teoría de la evolución por selección natural.

Figura 12 *Charles Darwin, en el extremo izquierdo, navegó alrededor del mundo en un barco muy similar a éste.*

184

Directed Reading Worksheet 8 Section 2

TOPIC: The Galápagos Islands
GO TO: www.scilinks.org
*sci*LINKS NUMBER: HSTL165

The tailbone in humans is a vestigial structure that is a remnant of the tails of ancestor species.

La increíble aventura de Darwin

Durante la travesía del HMS *Beagle* alrededor del mundo, Darwin recolectó miles de muestras de plantas y animales, y tomó notas detalladas de sus observaciones. En el mapa de la **Figura 13** se señala el viaje del *Beagle*. El barcó visitó las islas Galápagos, que aparecen abajo y que están a 965 km (600 millas) al oeste de Ecuador, un país de Sudamérica.

Los pinzones de Darwin

Darwin observó que los animales y las plantas de las islas Galápagos eran muy similares, aunque no idénticos, a los que habitaban en la cercana tierra del continente. Por ejemplo, notó que los pinzones de las islas Galápagos se diferenciaban un poco de los de Ecuador. Los pinzones de las islas, además de ser diferentes a los del continente, eran diferentes entre sí. En la **Figura 14,** puedes ver que la forma de los picos y lo que comían eran las principales diferencias que había entre ellos.

Figura 13 *La línea roja indica el recorrido del HMS Beagle.*

El pinzón *Geospiza magnirostris* tiene un pico pesado y fuerte que está adaptado para romper semillas grandes y duras, como si fuera un cascanueces.

El pinzón *Geospiza conirostris* tiene un pico que le sirve para comer cactus y extraer el néctar, y funciona como un par de alicates.

El pico del pinzón *Certhidea olivacea* es pequeño y puntiagudo, y está adaptado para explorar grietas en busca de pequeños insectos. Este pico funciona como un par de pinzas.

Figura 14 *Los picos de estas tres especies de pinzón de las islas Galápagos están adaptados a las distintas maneras en que obtienen su alimento.*

185

IS THAT A FACT!

The giant tortoises of the Galápagos Islands weigh up to 270 kg and can live for over 150 years.

Survival of the Chocolates   **PG 587**

---

2 Teach

DEMONSTRATION

Form and Function Present and identify to students the following pieces of clothing:

sneaker, dress pump, loafer, necktie, scarf, anklet, knee sock

Explain that all these items are pieces of clothing but some are more closely related than others. Within each group (shoes, neckwear, socks), every item is best suited for one particular function. This relationship of similarities and differences is what Charles Darwin observed in many animal and plant species.

MEETING INDIVIDUAL NEEDS

Advanced Learners Encourage interested students to investigate Darwin's voyage and similar long-distance travel by explorers in the 1800s in greater detail. Topics for reports include the types of ships used for travel in that era, the kinds of food eaten by the explorers, and the sophistication and thoroughness of maps in the 1800s.

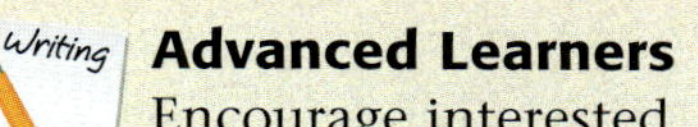

CONNECT TO GEOGRAPHY

The Galápagos Islands are administratively part of the country of Equador, though they are 1,000 km west of the mainland. They are a group of 19 volcanically formed islands. Though they have a land area of only 8,000 km^2, they are dispersed over almost 60,000 km^2 of the Pacific Ocean. It is easy to imagine how new species could arise in such a place.

2 Teach, continued

READING STRATEGY

Prediction Guide Before reading this page, have students answer the following questions:

- Why did the finches Darwin saw on the Galápagos Islands look similar to those he saw in South America?
- Why did they look a little different?

Have students evaluate their answers after they read the page.

MEETING INDIVIDUAL NEEDS

Writing **Advanced Learners** Biogeography is the study of where animals and plants are found and how they came to live in their particular location. It uses information from the fossil record and integrates ideas from biology, geology, paleontology, and chemistry. Encourage interested students to write a report about island biogeography. Have them include information about how it is used to design and manage terrestrial wildlife refuges.

RETEACHING

Writing To help students understand the process of speciation, have them write a brief paragraph about each new term in this chapter. Each paragraph should begin with a definition. Then have students write sample sentences using the term.

¿Has oído hablar de un banco que guarda semillas en lugar de dinero? Léelo en la página 198.

Las reflexiones de Darwin

Las observaciones de Darwin dieron lugar a preguntas que no se podían contestar fácilmente, por ejemplo: "¿por qué los pinzones de las islas son similares, pero no idénticos, a los del continente?" y "¿por qué los pinzones de distintas islas son diferentes entre sí?" Darwin pensó que quizá todos los pinzones de las islas Galápagos descendían de los pinzones del continente sudamericano. Es posible que una tormenta hubiera llevado a la población original de pinzones desde Sudamérica hasta las islas Galápagos y, a lo largo de muchas generaciones, los pinzones que sobrevivieron se adaptaron a distintos modos de vida en las islas.

A su regreso a Inglaterra, Darwin trabajó muchos años en una teoría que explicara la evolución. Durante este tiempo integró muchas ideas provenientes de diversas fuentes.

Darwin aprendió de los agricultores y criadores de animales En la época de Darwin, ya se habían logrado reproducir de manera selectiva muchas variedades de plantas y animales de granja. Los agricultores y criadores escogían ciertos **rasgos,** (cualidades distintivas, como granos de maíz grandes) y reproducían sólo a los individuos que los tenían. A este método se le llama **cruza selectiva** porque la selección de los rasgos que pasarán a la siguiente generación la hacen los seres humanos, no la naturaleza. La cruza selectiva de perros (**Figura 15**) ha exagerado ciertos rasgos, dando origen a más de 150 razas.

Al estudiar la genética y la herencia, aprendiste que los individuos de una misma especie presentan una gran variedad de rasgos. A Darwin le impresionó que los agricultores y criadores pudieran seleccionar estos rasgos para, en pocas generaciones, ocasionar cambios drásticos en animales y plantas. Pensó que los animales y plantas silvestres cambiaban de manera similar, pero mediante un proceso mucho más lento, ya que las variaciones se producirían por casualidad.

Figura 15 *Los perros son un buen ejemplo de cómo funciona la cruza selectiva. En los últimos 12,000 años, se los ha criado selectivamente para producir más de 150 razas.*

186

IS THAT A FACT!

As a result of selective breeding, the smallest horse is the Falabella, which is only about 76 cm tall. The largest is the Shire, originally bred in England. It can grow more than 1.73 m high at the shoulder and weigh as much as 910 kg.

Darwin aprendió de los geólogos

Los geólogos compartieron con Darwin sus pruebas de que la Tierra era mucho más vieja de lo que se había pensado. Al leer el libro *Principios de geología*, de Charles Lyell, Darwin aprendió que la Tierra se formó mediante larguísimos procesos naturales. Los datos de Lyell eran importantes porque Darwin pensaba que las poblaciones de organismos cambiaban muy lentamente.

Darwin aprendió del trabajo de Thomas Malthus En su *Ensayo sobre el principio de la población*, Malthus propuso que los seres humanos tienen el potencial para reproducirse hasta sobrepasar las reservas de alimentos. Sin embargo, también reconoció que la muerte, ya sea por hambre, enfermedad o guerra, afecta al tamaño de las poblaciones humanas. Las ideas de Malthus se representan en la **Figura 16.**

Darwin se dio cuenta de que otras especies de animales también eran capaces de reproducirse en exceso y que la falta de alimento, las enfermedades y los depredadores afectaban el tamaño de sus poblaciones. Darwin dedujo que si un número limitado de individuos sobrevivía para reproducirse era porque los supervivientes tenían algo especial. ¿Qué características los hacen sobrevivir y reproducirse? Darwin pensó que las crías de los supervivientes heredan características que les ayudan a sobrevivir en su medio ambiente.

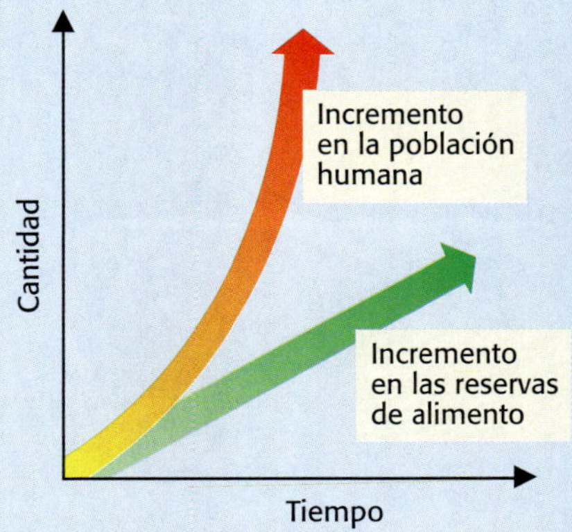

Figura 16
Malthus pensó que la población humana (representada por la línea roja) sobrepasaría las reservas de alimentos disponibles (representadas por la línea verde).

Laboratorio

¿Se puede acabar la comida?

Malthus pensaba que sí. Realiza la siguiente actividad para entender mejor la hipótesis de Malthus. Consigue **dos empaques de huevos vacíos** y **una bolsa de arroz**. En un empaque escribe "Provisiones" y en otro "Crecimiento de la población". Coloca un grano de arroz en el primer hueco del empaque "Provisiones" y cada vez que pases a otro hueco agrega un grano más. Cada grano representa una unidad de alimento. En el empaque "Crecimiento de la población", coloca un grano de arroz en el primer hueco y duplica el número de granos en cada hueco. Estos granos representan personas.

1. ¿Cuántas "personas" hay en el último hueco?
2. ¿Cuántas unidades de alimento hay en el último hueco?
3. ¿Cuál es tu conclusión?

SCIENTISTS AT ODDS

Not all scientists who study evolution agree on how the process takes place. Gradualism, the theory that Darwin supported, is based on the principle that changes in species occur slowly and steadily over thousands of years. In the 1970s, Stephen Jay Gould and others proposed the theory of punctuated equilibrium, which holds that species can remain unchanged for millions of years and then, due to dramatic environmental changes, undergo relatively rapid changes. The fossil record provides evidence that supports both sides of this debate.

Quick Lab

MATERIALS

FOR EACH STUDENT:
- 2 empty 12-egg cartons
- bag of rice

In a balanced system organisms interact so that there is maximum diversity, and population increase equals population decrease. The *Quick*Lab demonstrates an unbalanced system. There is nothing to slow the rapid increase of the rice grain population.

Answers to QuickLab

1. There are 2,048 "people."
2. There are 12 grains of rice.
3. There is not enough food to support the population.

Students should work this out mathematically before each step. They should also divide the task of counting after the first 6 cups.

Q: How did the dinosaurs listen to music?

A: on their fossil records

Math Skills Worksheet 3
"Multiplying Whole Numbers"

INDEPENDENT PRACTICE

Concept Mapping Have students make a concept map in their ScienceLog that outlines the process of change for a population of squirrels (each of which is black, red, grey or white) marooned on a treeless island of black sand that is also home to squirrel-eating foxes.

`Sheltered English`

Homework

Poster Project Have students research the natural history and current status of sea turtles (or a specific sea turtle species) to find examples for each of the four steps of natural selection. Have them construct a display to present their findings. For example, cotton balls glued to the poster board can represent eggs, and a dark plastic bag can symbolize a polluted ocean. Present the following questions as guides for their research:

1. On average, how many offspring does a sea turtle produce each year?
2. What physical adaptations have helped sea turtles survive in their environment?
3. What specific environmental factors affect their ability to survive?
4. What natural and man-made factors may be affecting their ability to survive long enough to reproduce successfully?

Teaching Transparency 32 "Natural Selection in Four Steps"

Selección natural

En 1858, alrededor de 20 años después de regresar de su viaje en el HMS *Beagle*, Darwin recibió una carta del naturalista Alfred Russel Wallace. Wallace había llegado de manera independiente a la misma teoría de la evolución en la que Darwin había trabajado por tantos años. Discutieron sus investigaciones y planearon presentar sus descubrimientos en un evento que tendría lugar ese mismo año. Después, en 1859, Darwin publicó sus propios resultados en el libro titulado *El origen de las especies por medio de la selección natural*. En su teoría, Darwin explica que la evolución ocurre por un proceso que llamó **selección natural.** Este proceso, que estudiaremos a continuación, se divide en cuatro partes.

La selección natural en cuatro pasos

1 Sobreproducción Cada especie produce más crías de las que alcanzarán la madurez.

2 Variación genética Los individuos de una población son ligeramente diferentes entre sí. Cada individuo tiene una combinación única de rasgos, como tamaño, color y capacidad para encontrar alimento. Algunos rasgos aumentan las posibilidades de que el individuo sobreviva y se reproduzca, y otros las disminuyen. Estas variaciones son genéticas y se heredan.

3 Lucha por sobrevivir El medio natural no tiene suficiente alimento, agua y otros recursos para mantener a todos los individuos que nacen; además, muchos individuos mueren a causa de otros organismos. En una población, sólo algunos individuos sobreviven hasta hacerse adultos.

4 Reproducción exitosa La reproducción exitosa es la clave de la selección natural. Los individuos que están bien adaptados a su medio, es decir, los que tienen los mejores rasgos para vivir en él, tienen mayor probabilidad de sobrevivir y reproducirse. Los individuos que no están bien adaptados a su medio tienen mayor probabilidad de morir temprano o de producir pocas crías.

188

Science Bloopers

In 1809 French naturalist Jean Baptiste Lamarck's theory of evolution stated that if an animal changed a body part through use or nonuse, that change would be inherited by its offspring. For example, larger or stronger leg muscles as a result of extensive running would be passed on to the next generation. Genetic studies in the 1930s and 1940s, however, disproved Lamarck's mechanism for inherited traits.

Imagínate que tu abuelo ha criado perros por más de 50 años, pero que nunca ha vendido uno. Los quiere mucho y los tiene en un gran corral. Al principio había seis labradores, seis terriers y seis pointers; ahora hay 76 perros y, para tu sorpresa, sólo algunos parecen pointers, labradores o terriers. Los demás se parecen entre sí, pero no se parecen a ninguna de las razas. Tu abuelo dice que en los últimos 50 años cada nueva generación se parecía menos a la precedente.

¿Por qué la mayoría de los perros se parecen entre ellos, pero no a una de las razas originales? Elabora tu respuesta en base a tus conocimientos sobre cruza selectiva.

Más pruebas de la evolución

Una de las observaciones en las que Darwin basó su teoría de la evolución por selección natural es que los padres transmiten sus características a su descendencia. Pero Darwin no sabía *cómo* ocurre la herencia ni *por qué* los individuos varían dentro de una población.

En las décadas de 1930 y 1940, los biólogos combinaron los principios de la herencia genética con la teoría de la evolución por selección natural. Esta combinación de principios explicó que las variaciones observadas por Darwin en una especie son causadas por una **mutación,** es decir, un cambio en un gen. Desde la época de Darwin, se han reunido pruebas provenientes de muchos campos científicos. Aunque los científicos admiten que otros mecanismos pueden intervenir en la evolución de una especie, la teoría de la evolución por selección natural proporciona la explicación más completa de la diversidad de la vida en la Tierra.

REPASO

1. ¿Por qué algunos animales tienen mayor probabilidad que otros de llegar a adultos?

2. **Resumir información** Según Darwin, ¿qué le pasó a la primera población de pinzones que llegó a las islas Galápagos desde Sudamérica?

3. **Calcular** Una cucaracha hembra puede producir 80 crías en una sola puesta. Si la mitad de las crías son hembras y cada una produce 80 crías, ¿cuántas cucarachas habrá en tres generaciones?

189

Answer to APPLY

The dogs were no longer being selectively bred for specific traits, so dogs with different traits were breeding together. After several generations, the dogs looked similar as the genetic mixing grew more complete.

Quiz

1. Who was Charles Lyell? (He was a British geologist.)

2. What did Darwin learn from Lyell's data about the age of Earth? (Darwin learned from Lyell that Earth was old enough for slow changes to happen in a population.)

ALTERNATIVE ASSESSMENT

Writing Locate the Rocky Mountains on a map. Explain to students that bird identification guides for North America usually classify birds into those that are east of the Rocky Mountains and those that are west of the Rocky Mountains. Have them write an explanation for why ornithologists use this system. Then tell students to research the differences and similarities of eastern and mountain bluebirds, or of blue jays and piñon jays.

Reinforcement Worksheet 8
"Bicentennial Celebration"

internet**connect**

*SCi*LINKS
NSTA

TOPIC: Darwin and Natural Selection
GO TO: www.scilinks.org
*sci*LINKS NUMBER: HSTL170

▼ Answers to Review

1. Some animals are more likely to survive because they inherit traits that enable them to find food, escape predators, or resist disease more effectively than other animals.

2. Darwin thought that the first population of finches on the Galápagos Islands gave rise to all the different species living there today.

3. The first generation = 80; 40 are female that produce 80 each. The second generation = 3,200; 1,600 are female that produce 80 each. There are 128,000 in the third generation.

Focus

Natural Selection in Action

In this section students will learn that natural selection is occurring constantly and is not just a historical relic. They will learn how a species' generation time affects its ability to adapt. Finally, students will learn the three steps of speciation: separation, adaptation, and division.

Bellringer

Display on the board or the overhead projector these instructions to students:

Write the four steps of natural selection, and create a mnemonic device to remember each step by using the first letter of each step.

1 Motivate

DEMONSTRATION

Natural Selection Place 20 black jellybeans and 20 red jellybeans on a piece of black paper, and call the display *Generation 1*. Tell students to pretend the candies are fish and ask which would most likely be eaten first by the jellybean shark. Then add 5 black jellybeans and take away 5 red ones. Call this group *Generation 2*. Ask students how many fish in this generation might survive the jellybean shark. Have students offer explanations for what happened between *Generation 1* and *Generation 2*.

Directed Reading Worksheet 8 Section 3

La selección natural en acción

VOCABULARIO
tiempo de generación
especiación

OBJETIVOS
- Da dos ejemplos de selección natural.
- Explica el proceso de especiación.

Experimentos

¡Huellas a través del tiempo! ¿A quién pertenecieron? Averígualo en la página 584.

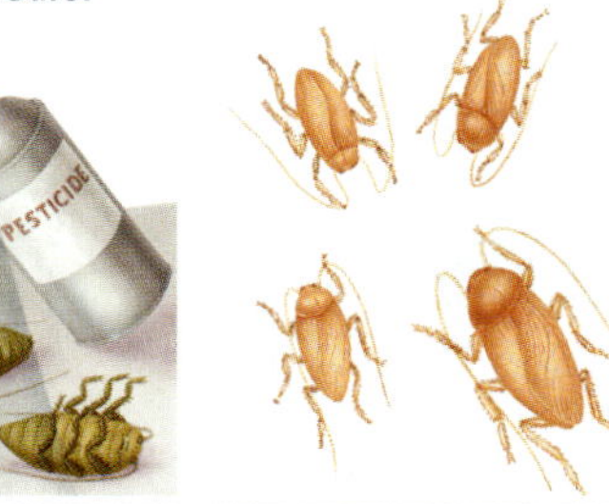

Figura 17 *La variedad de las características de una población garantiza que algunos individuos serán capaces de sobrevivir a un cambio en el medio.*

La teoría de la selección natural explica la manera en que una población cambia a través de muchas generaciones en respuesta a su medio ambiente. De hecho, los miembros de una población tienden a estar bien adaptados a su medio porque la selección natural se lleva a cabo continuamente.

Resistencia a los insecticidas Para proteger las cosechas de la acción de los insectos, algunos agricultores utilizan una gran variedad de insecticidas químicos. Sin embargo, algunos de los insecticidas que funcionaban bien en el pasado ya no son efectivos. Por ejemplo, el gorgojo del algodón es cada vez más difícil de controlar porque ha desarrollado una resistencia genética a muchos insecticidas. Y este no es el único insecto que lo ha hecho. Durante los 50 años en que se llevan utilizando insecticidas, más de 500 especies de insectos han desarrollado una resistencia a alguno de ellos.

Los insectos desarrollan rápidamente resistencia a los insecticidas porque producen muchas crías y, por lo general, sus tiempos de generación son cortos. El **tiempo de generación** es el período entre el nacimiento de una generación y el de la siguiente. Sigue el proceso de la **Figura 17** para conocer la manera en que una plaga doméstica común, las cucarachas, evolucionaron para volverse resistentes a ciertos insecticidas.

Las enfermedades, como la tuberculosis, también se han vuelven resistentes a los antibióticos que antes eran muy efectivos para combatirlas.

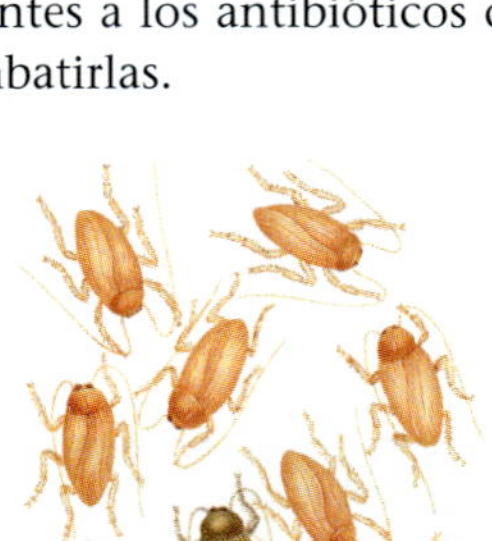

1 Cuando se aplica un insecticida efectivo a una población de insectos, la mayoría muere, pero unos cuantos sobreviven. Estos sobrevivientes tienen genes que los hacen resistentes al insecticida.

2 Después, los supervivientes se reproducen y transmiten a su descendencia los genes de resistencia al insecticida.

3 Con el tiempo, la mayor parte de la población de insectos estará formada por individuos que poseen estos genes de resistencia.

4 Finalmente, cuando se combate a los insectos con la misma clase de insecticida sólo unos cuantos mueren, ya que la población en general se ha hecho resistente.

190

Mystery Footprints **PG 584**

SCIENCE HUMOR

Q: What do you get when you cross a crocodile with an abalone?

A: a crocabaloney

Adaptación a la contaminación Las polillas *Biston betularia* que aparecen en la **Figura 18** presentan dos variaciones de color. Antes de 1850, la polilla obscura se consideraba rara y la clara era mucho más común. Pero después de 1850, las polillas obscuras se volvieron más abundantes en las zonas altamente industrializadas.

Figura 18 *En un tronco obscuro, sobresale la polilla clara (izquierda), mientras que en un tronco claro sobresale la obscura (abajo).*

¿Cuál fue la causa de este cambio en la población de polillas? Varias especies de pájaros se comen las polillas que descansan sobre los troncos. Antes de 1850, los árboles tenían una apariencia gris y las polillas claras se confundían con su entorno. Los pájaros veían más fácilmente las polillas obscuras y por eso se las comían con mayor frecuencia. Después de 1850, el hollín y el humo de las zonas industriales recién desarrolladas obscureció los árboles cercanos. Las polillas obscuras se volvieron menos visibles y las claras más, convirtiéndose en presa fácil para los pájaros. Cada vez sobrevivían más polillas obscuras, que a su vez producían descendencia de color obscuro. Por lo tanto, la población sufrió un cambio: de una mayoría de polillas claras a una mayoría obscura.

191

---

2 Teach

READING 📖 STRATEGY

After students read this page, have them draw pictures of hypothetical intermediate color variations for peppered moths in their ScienceLog. Tell them to write a caption for each picture that explains why that particular variation would have a better or worse chance for survival before 1850 and after 1850. Sheltered English

REAL-WORLD CONNECTION

Natural Pest Control The use of natural predators against insect pests provides a safe alternative to insecticides. Ladybugs, for example, which are actually beetles, are purchased in large numbers and released on crops. They are used to combat infestations of aphids, whiteflies, fruitworms, mites, the broccoli worm, and the tomato hornworm.

MAKING MODELS

Bat Houses One way to control mosquitoes is to encourage insect-eating bats to take up residence in your yard by hanging bat houses. Encourage interested students to research the design of a bat house and to construct one from cardboard. Interested students could use the model to build a bat house out of wood, hang it up outside, and see if bats move in. Sheltered English

Answers to Self-Check
1. b   3. d
2. a   4. c

Out-of-Sight Marshmallows   PG 586

RETEACHING

Ask students if they can define *subspecies*. (A subspecies is a population within a species that is different enough to be given its own name.)

Because subspecies can breed with one another, all are still members of the same species. Then tell them that all seven subspecies of a particular salamander can be found in California. Each one gradually integrates into the next subspecies. But in two locations, two of the subspecies interbreed rarely if at all. Ask students if they should be considered different species. Why or why not? (They are not yet different species because they can still interbreed.)

Ask how they would research this question. Then tell them that early DNA studies indicate that the single salamander species is becoming two different species, possibly more.

SCIENTISTS AT ODDS

Scientists at the American Ornithologists' Union are responsible for the official list of scientific bird names used in the United States. They used to recognize the Eastern towhee and the spotted towhee. Then they decided to make them one species, the rufous-sided towhee. Then they changed their minds (based on further study) and "split" the classification again. Stay tuned.

Formación de nuevas especies

El proceso de selección natural explica cómo una especie evoluciona para formar otra. Si una parte de la población de una especie se separa de la población original, con el tiempo ambas poblaciones se vuelven tan diferentes que ya no pueden cruzarse. Este proceso se conoce como **especiación**. Los siguientes pasos explican una de las maneras en que la especiación puede ocurrir:

1. Separación El proceso de especiación con frecuencia empieza cuando una parte de la población queda aislada. La **Figura 19** muestra algunas maneras en que esto puede suceder. La presencia de un cañón recién formado, una cordillera o un lago son algunas de las maneras en que las poblaciones pueden dividirse.

Figura 19 *Las poblaciones pueden separarse de diversas maneras.*

2. Adaptación Si una población ha sido dividida por alguno de los cambios que aparecen arriba, el medio también puede cambiar. Aquí es donde interviene la selección natural. Conforme el medio cambia, la población también lo hace. A través de muchas generaciones, los grupos que han sido separados se adaptan a su medio, como se observa en la **Figura 20**. Si las condiciones ambientales son distintas para cada uno de los grupos, sus adaptaciones también pueden ser diferentes.

Figura 20 *Si una sola población queda dividida, los grupos evolucionan por separado y forman especies distintas.*

IS THAT A FACT!

Some species that have adapted to live in total darkness no longer even have eyes! Just as whales have evolved into legless forms, these species have completely adapted to life without light, and some have evolved forms lacking eyes altogether. There are blind cave fish, eels, salamanders, worms, shrimp, crayfish, spiders, beetles, crickets.

3. División A lo largo de cientos, miles o incluso millones de generaciones, los grupos de una población pueden volverse muy diferentes y ya no ser capaces de cruzarse, aunque haya desaparecido la barrera geográfica. Estos grupos dejan de pertenecer a la misma especie. Los científicos piensan que los pinzones de las islas Galápagos evolucionaron por medio de estos tres pasos básicos. En la **Figura 21** se ilustra cómo pudo suceder esto.

Figura 21 *Posiblemente los pinzones de las islas Galápagos evolucionaron para formar distintas especies mediante el proceso que se describe abajo.*

❶ Algunos pinzones dejaron el continente y llegaron a una de las islas (separación).

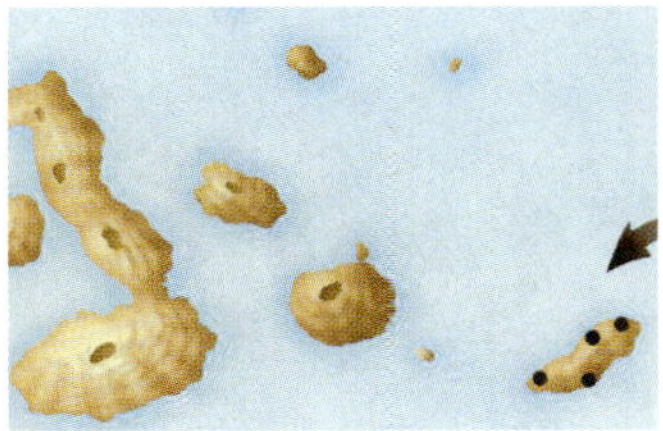

❷ Los pinzones se reprodujeron y se adaptaron al medio (adaptación).

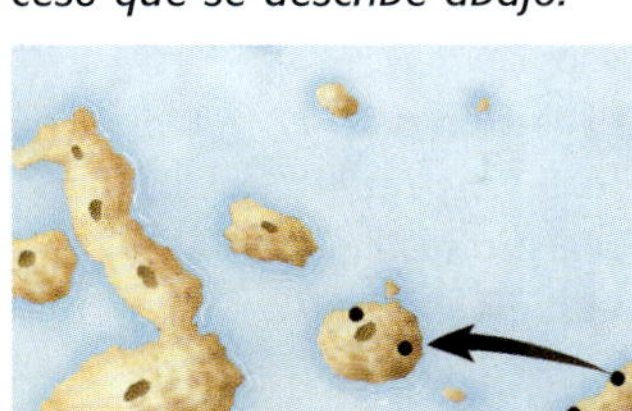

❸ Algunos pinzones volaron a una segunda isla (separación).

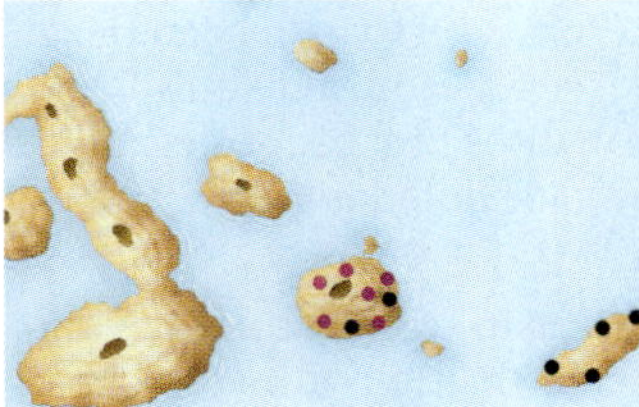

❹ Los pinzones se reprodujeron y se adaptaron al nuevo medio (adaptación).

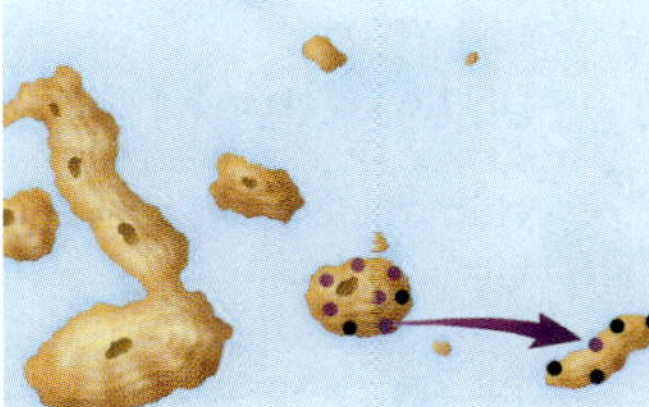

❺ Algunos pinzones regresaron a la primera isla, pero ya no pudieron cruzarse con los pinzones que había ahí (división).

❻ Es posible que este proceso haya ocurrido una y otra vez, mientras los pinzones volaban a las otras islas.

REPASO

1. ¿Por qué aumentó el número de polillas obscuras después de 1850?

2. ¿Qué factor indica que una población ha evolucionado para formar dos especies diferentes?

3. **Aplicar conceptos** La mayoría de los cactus tienen espinas (hojas modificadas para proteger la planta) que recubren el tallo jugoso, donde se almacena el agua. Explica cómo las hojas y tallos de los cactus han cambiado a través del proceso de selección natural.

193

DEBATE

People and Nature During the past several hundred years, a rapidly expanding human population has caused some species to become extinct either from habitat destruction or overhunting. Ask:

If people are as much a part of the environment as trees and birds, are their actions just another natural process?

Quiz

Concept Mapping Construct a concept map that shows how a population of mosquitoes can develop resistance to a pesticide.

ALTERNATIVE ASSESSMENT

Have each student research and give an oral presentation on how the three steps of speciation (separation, adaptation, and division) worked in providing a particular animal with a distinctive feature. For example, a student interested in giraffes might investigate how it came to have a long neck.

Teaching Transparency 33 "Evolution of the Galápagos Finches"

Critical Thinking Worksheet 8 "Taking the Earth's Pulse"

▼ **Answers to Review**

1. After the 1850s, soot and smoke from industrial areas blackened nearby trees. The dark peppered moths became less visible than the pale peppered moths on the dark tree trunks. More dark moths survived predation on the dark tree trunks and produced more dark offspring.

2. Over time, two groups of a population may become so different that they can no longer interbreed. At this point, they are no longer the same species.

3. The changes in a cactus's leaves help protect the plant and conserve water, and the stem stores water. In each generation, as plants with sharper leaves and larger stems thrived in the harsh climate, these features became more prevalent and defined in their offspring, which survived and reproduced.

VOCABULARY DEFINITIONS

SECTION 1

adaptation a hereditary characteristic that helps an organism survive and reproduce in its environment

species the most specific of the seven levels of classification; characterized by a group of organisms that can mate with one another to produce fertile offspring

evolution the process by which populations accumulate inherited changes over time

fossil the solidified remains or imprints of once-living organisms

fossil record a historical sequence of life indicated by fossils found in layers of the Earth's crust

vestigial structure the remnant of a once-useful anatomical structure

SECTION 2

trait distinguishing qualities that can be passed from one generation to another

selective breeding breeding of organisms that have a certain desired trait

natural selection the process by which organisms with favorable traits survive and reproduce at a higher rate than organisms without the favorable trait

mutation a change in the order of the bases in an organism's DNA; deletion, insertion, or substitution

Resumen del capítulo

SECCIÓN 1

Vocabulario

adaptación (*pág. 176*)
especie (*pág. 176*)
evolución (*pág. 177*)
fósil (*pág. 178*)
registro fósil (*pág. 178*)
vestigio (*pág. 179*)

Notas de la sección

- La evolución es el proceso mediante el cual las poblaciones cambian a lo largo del tiempo. Los cambios son hereditarios. A través de muchas generaciones, las especies más recientes pueden reemplazar a las más viejas por medio de la evolución.

- Las pruebas de que todos los organismos tienen un antepasado común han sido obtenidas a partir del registro de fósil, las comparaciones entre estructuras óseas encontradas en especies relacionadas, las comparaciones entre embriones de vertebrados con parentesco lejano y la presencia de ADN en todos los seres vivos.

- El ADN de las especies con parentesco cercano es más parecido que el ADN de las especies con parentesco lejano.

Experimentos

Huellas misteriosas (*pág. 584*)

SECCIÓN 2

Vocabulario

rasgo (*pág. 186*)
cruza selectiva (*pág. 186*)
selección natural (*pág. 188*)
mutación (*pág. 189*)

Notas de la sección

- Charles Darwin desarrolló una explicación de la evolución después de estudiar por muchos años a los organismos que observó durante su viaje en el *Beagle*.

- El estudio de Darwin estuvo influído por los conceptos de cruza selectiva, la edad de la Tierra y la idea de que algunos organismos están mejor dotados para sobrevivir que otros.

- Darwin explicó que la evolución ocurre por medio de la selección natural, la cual puede dividirse en cuatro partes:

☑ Comprobar destrezas

Conceptos de matemáticas

PRINCIPIO DE MALTHUS La gráfica de la página 187 muestra dos tipos de crecimiento. La línea recta representa un aumento en el que un mismo número se suma al anterior, como en 3, 4, 5, 6, ..., donde se suma 1 a cada número.

La línea curva representa un aumento en el que cada número se multiplica por el mismo factor, como en 2, 4, 8, 16, ..., donde cada número se multiplica por 2. Como se observa en la gráfica, la línea curva aumenta mucho más rápido que la línea recta.

Comprensión visual

ESTRUCTURA ÓSEA En la figura 9 de la página 182 se ilustran pruebas óseas de la evolución. La estructura ósea de las extremidades de los seres humanos, gatos, delfines y murciélagos es similar, lo cual indica que estos animales descienden de un antepasado común. En ciertas especies, la estructura ósea de los mamíferos evolucionó para llevar a cabo tareas especializadas, como volar y nadar. Al observar huesos del mismo color, entenderás cómo sucedió esto.

194

Lab and Activity Highlights

Survival of the Chocolates PG 587

Mystery Footprints PG 584

Out-of-Sight Marshmallows PG 586

Datasheets for LabBook
(blackline masters for these labs)

SECTION 3

generation time the period between the birth of one generation and the birth of the next generation

speciation the process by which two populations of the same species become so different that they can no longer interbreed

  **Vocabulary Review Worksheet 8**

Blackline masters of these Chapter Highlights can be found in the **Study Guide.**

SECCIÓN 2

(1) Cada especie produce más crías de las que podrán sobrevivir y reproducirse.

(2) Los individuos de una misma población son ligeramente diferentes entre sí.

(3) Los individuos de una población compiten entre ellos por recursos limitados.

(4) Los individuos mejor dotados para vivir en un medio determinado tienen mayor probabilidad de sobrevivir y reproducirse.

• Hoy en día, la evolución se explica combinando los principios de la selección natural con los de la herencia genética.

Experimentos

Chocolates que sobreviven *(pág. 587)*

SECCIÓN 3

Vocabulario

tiempo de generación *(pág. 190)*

especiación *(pág. 192)*

Notas de la sección

• La selección natural permite a una población adaptarse a cambios en las condiciones del medio ambiente.

• Las generaciones de organismos que han desarrollado

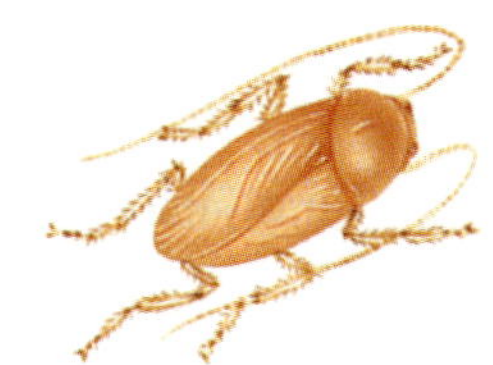

resistencia a un insecticida o a un antibiótico son una prueba de la selección natural.

• La selección natural también explica cómo una especie evoluciona a otra por medio del proceso de la especiación.

Experimentos

Malvaviscos ocultos *(pág. 586)*

 internet

 SCI*LINKS* NSTA

HRW VISITA: go.hrw.com

VISITA: www.scilinks.org

Visita el sitio web de HRW para encontrar una serie de herramientas de aprendizaje relacionadas con este capítulo. Sólo tienes que escribir la palabra clave:

PALABRA CLAVE: HSTEVO

Visita el sitio web de la **Asociación Nacional de Maestros de Ciencias** *(National Science Teachers Association)* para encontrar recursos de Internet relacionados con este capítulo. Sólo escribe el **ENLACE DE CIENCIAS** para obtener más información sobre el tema:

TEMA: Especies y adaptación	**ENLACE:** HSTL155
TEMA: El registro de fósiles	**ENLACE:** HSTL160
TEMA: Las islas Galápagos	**ENLACE:** HSTL165
TEMA: Darwin y la selección natural	**ENLACE:** HSTL170

195

Lab and Activity Highlights

LabBank

 **Whiz-Bang Demonstrations,** Adaptation Behooves You, Demo 6

Long-Term Projects & Research Ideas, Project 7

Chapter Review
Answers

Using Vocabulary

1. speciation
2. species
3. adaptation
4. Evolution
5. selective breeding
6. mutation

Understanding Concepts

Multiple Choice

7. b
8. d
9. a
10. a
11. b
12. c
13. b

Short Answer

14. 1. overproduction: Each species produces more offspring than will survive.
    2. genetic variation: Each individual has a unique combination of traits. Some traits increase the chances that the individual will survive and reproduce.
    3. struggle to survive: Individuals compete for limited resources. Some will not compete successfully and will not survive to adulthood.
    4. successful reproduction: Those individuals that are well-adapted and have traits that help them survive in their environment are more likely to survive and reproduce.

15. Fossils of the stages of whale evolution have been discovered that clearly indicate their sequence of change from land-dwelling carnivores to sea-dwelling mammals.

16. The required conditions for fossil formation are rare. The shell or bones must be completely covered in sediment in an anaerobic environment.

Repaso del capítulo

UTILIZAR EL VOCABULARIO

Escoge el término correcto para completar las siguientes oraciones:

1. Las especies evolucionan para formar otras especies mediante el proceso de ___?___. (*adaptación* o *especiación*)

2. Un grupo de organismos similares que pueden aparearse entre sí y producir crías fértiles se conoce como___?___. (*fósil* o *especie*)

3. ___?___ le ayuda a un organismo a sobrevivir mejor en su medio ambiente. (*La adaptación* o *El vestigio*)

4. ___?___ es el proceso mediante el cual las poblaciones cambian a lo largo del tiempo. (*La selección natural* o *La evolución*)

5. En la ___?___, los seres humanos seleccionan las características que pasarán de una generación a otra. (*cruza selectiva* o *selección natural*)

6. Un cambio en un gen a nivel del ADN se llama ___?___. (*mutación* o *rasgo*)

COMPRENDER CONCEPTOS

Opción múltiple

7. Las variaciones que Darwin observó entre los individuos de la población de pinzones fueron ocasionadas por
   a. la resistencia genética.
   b. las mutaciones.
   c. los fósiles.
   d. la reproducción selectiva.

8. La teoría de la evolución combina los principios de
   a. selección natural y selección artificial.
   b. selección natural y resistencia genética.
   c. reproducción selectiva y herencia genética.
   d. selección natural y herencia genética.

9. Los fósiles se encuentran comúnmente en
   a. rocas sedimentarias.
   b. rocas ígneas.
   c. granito.
   d. arena suelta o granito.

10. El brazo de un ser humano, la pata delantera de un gato, la aleta frontal de un delfín y el ala de un murciélago
    a. tienen tipos similares de huesos.
    b. tienen usos similares.
    c. comparten semejanzas con las alas de los insectos y los tentáculos de las medusas.
    d. no tienen nada en común.

11. El material genético de todos los organismos es el ADN, lo cual es prueba de que
    a. ocurrió la selección natural.
    b. descienden de un antepasado común.
    c. ocurrió la reproducción selectiva.
    d. la resistencia genética ocurre raramente.

12. Darwin pensó que el antepasado común de los pinzones de las Galápagos provenía de
    a. África.
    b. Norteamérica.
    c. Sudamérica.
    d. Australia.

13. ¿Qué parte del cuerpo de los pinzones de las Galápagos sufrió más modificaciones a través de la selección natural?
    a. las patas palmeadas
    b. los picos
    c. la estructura ósea de las alas
    d. el color de los ojos

Respuesta breve

14. Describe los cuatro pasos de la teoría de Darwin sobre la evolución.

15. ¿Por qué los fósiles de ballena prueban que evolucionaron a lo largo de millones de años?

16. ¿A qué se deben los espacios vacíos que hay en el registro de fósiles?

Concept Mapping

17. 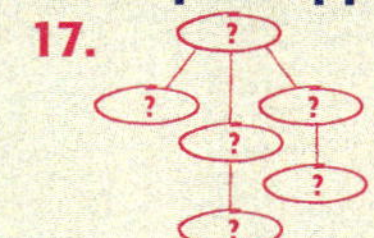 An answer to this exercise can be found at the end of this book.

Concept Mapping Transparency 8

17. Usa los siguientes términos para crear un mapa de ideas: lucha por sobrevivir, variación genética, Darwin, superproducción, selección natural y reproducción exitosa.

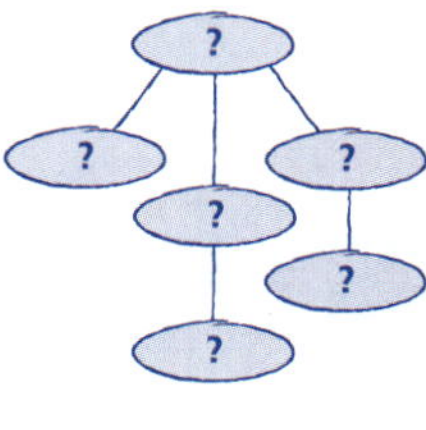

RAZONAMIENTO CRÍTICO Y RESOLUCIÓN DE PROBLEMAS

Escribe una o dos oraciones para responder a las siguientes preguntas:

18. Mediante la reproducción selectiva, los seres humanos influyen en el curso de la evolución. ¿Qué determina el curso de la evolución en la selección natural?

19. Muchas formas de bacterias desarrollan una resistencia a los antibióticos (medicamentos que matan las bacterias causantes de enfermedades). Basándote en lo que sabes sobre la resistencia de los insectos a los insecticidas, sugiere una manera en que las bacterias evolucionaron para desarrollar resistencia a los antibióticos.

20. Las dos especies de ardillas que aparecen abajo habitan en lados opuestos del Gran Cañón, en Arizona. Se parecen mucho, pero no pueden aparearse para procrear. Explica cómo una sola especie de ardilla puede formar más de dos especies.

INTERPRETAR GRÁFICAS

Con ayuda de las siguientes gráficas responde a las preguntas 21, 22 y 23:

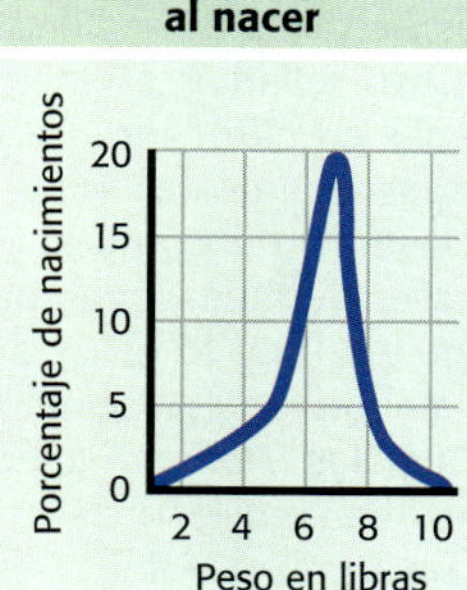

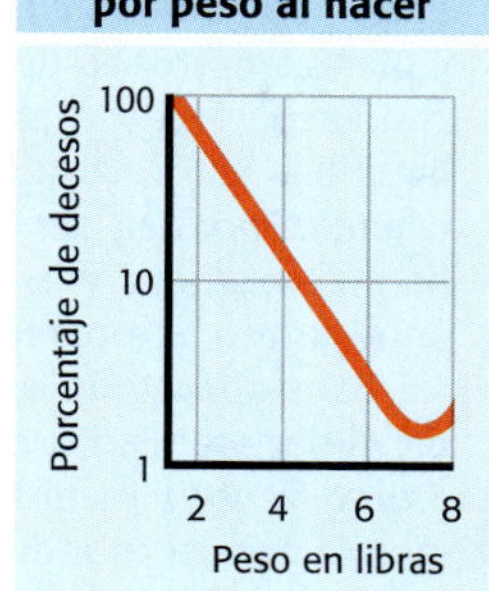

21. ¿Cuál es el peso más común al nacer?

22. ¿Cuál es el peso al nacer que tiene la mayor tasa de supervivencia?

23. ¿Cómo ayudan los principios de la selección natural a explicar por qué hay más muertes entre bebés con bajo peso al nacer que entre bebés con peso promedio?

AHORA, ¿qué piensas?

Revisa tus respuestas a las preguntas de la página 175 que escribiste en el cuaderno de ciencias. ¿Han cambiado tus respuestas? Si es necesario, corrige tus respuestas basándote en lo que has aprendido en este capítulo.

CRITICAL THINKING AND PROBLEM SOLVING

18. Conditions in the environment to which organisms must be adapted are part of the selection process in nature. Natural selection is the process of adaptation over time to changes in environmental conditions. The genetic changes that bring about adaptations in a species are thought to determine the course of evolution through natural selection.

19. A course of antibiotics may leave behind bacteria that have traits that help them survive. The survivors may reproduce, producing more individuals that have the survival trait. In this way, a strain of bacteria may become immune to an antibiotic.

20. Individual organisms in a population have their own unique set of traits. When a population becomes separated over time, each separate group may evolve in a different direction. Eventually, the groups may become different genetically and they are no longer able to interbreed. Even though they may retain a similar appearance, they have become separate species.

INTERPRETING GRAPHICS

21. 7 lb
22. 7 lb
23. Human babies are best adapted to survive at a birth weight of about 7 lb.

NOW WHAT DO YOU THINK?

1. Evolution is the process by which species change over time.
2. An organism's environment provides the selective pressures that enable some organisms to thrive and reproduce.
3. Fossils provide a physical record of organisms that lived long ago. Scientists can figure out the age of fossils. By examining fossils of similar organisms from different time periods, scientists can learn how organisms changed over time.

Blackline masters of this Chapter Review can be found in the **Study Guide.**

Background

In order to help students understand the changes in agriculture, you may wish to create a table that contrasts "traditional" and "industrial" farming practices. Under the heading *Traditional,* list words or phrases that describe traditional farming practices, such as *smaller scale, few machines or manual labor, more plant varieties,* and *mainly for sustenance.* Under the heading *Industrial,* list words or phrases that describe industrial farming practices, such as *larger scale, more mechanized, fewer plant varieties,* and *primarily for profit.* Encourage students to add their own words or phrases to contrast the types of farming practices.

In addition, point out to students that farmers in some areas, including many developing countries, practice some form of traditional agriculture. Consequently, scientists at seed banks try to visit those areas as often as necessary to obtain samples of those seeds.

VENTANA AL MEDIO AMBIENTE

Ahorros en el banco de semillas

En Fort Collins (Colorado) hay un laboratorio donde se guardan cientos de miles de semillas y plantas en tubos de ensayo sellados, cajones con llave y hasta congeladas en nitrógeno líquido a 196°C. Aunque por ahora están almacenadas, podrían ser la clave para prevenir el hambre mundial y la escasez de medicamentos en el futuro. ¿Te parece grave? Pues lo es.

Este laboratorio es el Laboratorio Nacional de Almacenamiento de Semillas *(National Seed Storage Lab)* y es el banco de semillas más grande a nivel mundial. Las semillas y los brotes de plantas que se almacenan allí representan casi todas las plantas que se cultivan para obtener alimentos, ropa y medicamentos.

▲ *Para proteger el futuro de los campos de trigo la diversidad genética debe almacenarse en bancos de semillas.*

¡Adiós burritos!

Imagínate que vas a comprar unos burritos y te encuentras con un letrero que dice "Cerrado por escasez de harina". ¿Escasez de harina? ¿Cómo es posible? ¿Crees que es una exageración? En realidad no lo es.

Si una epidemia arruinara las cosechas de trigo del mundo, habría escasez de harina. La mejor manera de combatirla, e incluso prevenirla, es cultivar nuevas variedades. A través del proceso de cruza selectiva se han mejorado muchas plantas para aumentar el rendimiento y la resistencia a enfermedades e insectos. Pero para cultivar nuevas cosechas, los agricultores requieren grandes cantidades de material genético distinto. ¿Dónde lo obtienen? En el banco de semillas, por supuesto.

Nunca lo sabremos

Pero, ¿qué pasaría si algunas plantas no llegaran nunca al banco de semillas? Si ya existen variedades nuevas y mejoradas, ¿para qué tomarse la molestia de guardar las viejas? Las variedades que se pierden muchas veces tienen característi- cas que en el fututo podrían ser útiles, como la resistencia a las enfermedades y a las sequías. Cuando se mejora una variedad de planta, la demanda de la variedad vieja puede reducirse a cero, y si deja de cultivarse puede extinguirse, a menos que se guarde en un banco de semillas. De hecho, muchas variedades ya se han perdido para siempre y nunca sabremos si alguna de ellas era capaz de resistir una sequía severa.

Todo está en el banco

Afortunadamente, los bancos de semillas llevan recolectando semillas y plantas durante más de un siglo. Conservan la diversidad genética de los cultivos al mismo tiempo que permiten a los agricultores cultivar las variedades más productivas. Mientras haya bancos de semillas en el mundo, es poco probable que nos quedemos sin harina. ¡Vamos a por unos burritos!

Profundizar

▶ Muchos bancos de semillas corren peligro. ¿Por qué? Para descubrirlo, investiga el complicado y costoso proceso necesario para su funcionamiento.

198

Answer to Going Further
Answers will vary.

Ciencia Ficción

"La lección de anatomía"

por Scott Sanders

Ya sabes cómo son estas cosas. Mañana tienes un examen importante o la entrega del proyecto final de semestre y… ¡se te olvidó el libro!, o simplemente te quedaste sin plastilina. Tú sabes lo que se siente. De pronto, las cosas parecen muy graves.

El estudiante de medicina de "La lección de anatomía", de Scott Sanders, se enfrenta a una situación como ésta. Tiene que aprenderse los huesos del cuerpo humano para un examen de anatomía al día siguiente. Cuando va a la biblioteca a sacar un modelo para armar esqueletos, se encuentra con que todos están prestados. Es imposible pasar el examen si no practica armando huesos, así que le pide a la bibliotecaria que lo verifique y, como era de esperarse, ella encuentra un modelo. Aquí es donde las cosas empiezan a ponerse raras.

Hay demasiados huesos, además, su forma no es la correcta y no encajan. ¡Debe ser una broma! Encajan de cierta manera, pero no lo suficiente para que Scott se prepare para el examen. El estudiante se queja a la bibliotecaria, pero ella no es nada comprensiva y parece tener otras cosas en la cabeza. Ahora, el estudiante está realmente preocupado.

Entérate de lo que este estudiante de medicina tiene en común con una bibliotecaria que no colabora. Averigua por qué no volverán a ser los mismos después de "La lección de anatomía". Esta historia se encuentra en la *Antología Holt de Ciencia Ficción*.

199

Further Reading If students enjoy this story, you may wish to recommend some of Sanders's other works, such as the following:

Terrarium, Indiana University Press, 1996

The Engineer of Beasts, Orchard Books, 1988

Hear the Wind Blow: American Folksongs Retold, Simon & Shuster Children's, 1985

SCIENCE FICTION
"The Anatomy Lesson"
by Scott Sanders

While studying for an exam, a medical student attempts to assemble a very unusual skeleton that may drastically change the student's future.

Teaching Strategy

Reading Level This is a relatively short story and should not be difficult for the average student to read and comprehend.

Background

About the Author Scott Sanders (1945–) writes many different kinds of stories—from folktales to science fiction. Early in life, he chose to become a writer rather than a scientist, though he has a keen interest in both writing and science. Sanders has written about a range of subjects, including folklore, physics, the naturalist John James Audubon, and settlers of Indiana. Much of his work is nonfiction. His writing has been published in many different newspapers and magazines, including the *Chicago Sun-Times, Harper's,* and *Omni.* Currently, Sanders lives and teaches in Indiana, where he belongs to writers' groups and to groups such as the Sierra Club and Friends of the Earth.

Chapter Organizer

CHAPTER ORGANIZATION	TIME MINUTES	OBJECTIVES	LABS, INVESTIGATIONS, AND DEMONSTRATIONS
Chapter Opener pp. 200–201	45		**Investigate!** Timeline of Earth's History, p. 201
Section 1 Evidence of the Past	90	▶ Describe two methods that scientists use to determine the age of fossils in sedimentary rock. ▶ Describe the geologic time scale and the information it provides scientists. ▶ Describe the possible causes of mass extinctions. ▶ Explain the theory of plate tectonics.	**Skill Builder,** Dating the Fossil Record, p. 588 **Datasheets for LabBook,** Dating the Fossil Record, Datasheet 18 **Skill Builder,** The Half-life of Pennies, p. 591 **Datasheets for LabBook,** The Half-life of Pennies, Datasheet 19
Section 2 Eras of the Geologic Time Scale	90	▶ Outline the major developments that allowed for the existence of life on Earth. ▶ Describe the different types of organisms that arose during the four eras of the geologic time scale.	**Interactive Explorations CD-ROM,** Rock On! *A **Worksheet** is also available in the **Interactive Explorations Teacher's Guide.***
Section 3 Human Evolution	90	▶ Discuss the shared characteristics of primates. ▶ Describe what is known about the differences between hominids.	**QuickLab,** Neanderthal Tools, p. 218 **Long-Term Projects & Research Ideas,** Project 8

TECHNOLOGY RESOURCES

Guided Reading Audio CD
English or Spanish, Chapter 9

One-Stop Planner CD-ROM with Test Generator

Science Discovery Videodiscs
Image and Activity Bank with Lesson Plans: The Time Machine
Science Sleuths: The Misplaced Fossil

Multicultural Connections, Protecting New Mexico's Petroglyphs, Segment 6
Scientists in Action, Creating Digital Dinos, Segment 12
Ice Age Discoveries, Segment 23

Interactive Explorations CD-ROM
CD 2, Exploration 6, Rock On!

CLASSROOM WORKSHEETS, TRANSPARENCIES, AND RESOURCES	SCIENCE INTEGRATION AND CONNECTIONS	REVIEW AND ASSESSMENT
Directed Reading Worksheet 9 **Science Puzzlers, Twisters & Teasers,** Worksheet 9		
Directed Reading Worksheet 9, Section 1 **Transparency 93,** The Rock Cycle **Transparency 34,** Unstable Atoms and the Half-life **Math Skills for Science Worksheet 36,** Radioactive Decay and the Half-life **Transparency 35,** The Geologic Time Scale and Representative Organisms **Critical Thinking Worksheet 9,** Fossil Revelations **Math Skills for Science Worksheet 45,** Geologic Time Scale **Transparency 36,** Formation of the Modern Continents **Transparency 37,** The Tectonic Plates **Reinforcement Worksheet 9,** Earth Timeline	**Connect to Earth Science,** p. 203 in ATE **Connect to Earth Science,** p. 206 in ATE **Multicultural Connection,** p. 206 in ATE **Careers:** Paleobotanist—Bonnie Jacobs, p. 225	**Homework,** p. 203 in ATE **Self-Check,** p. 205 **Review,** p. 207 **Quiz,** p. 207 in ATE **Alternative Assessment,** p. 207 in ATE
Directed Reading Worksheet 9, Section 2 **Math Skills for Science Worksheet 2,** Subtraction Review **Reinforcement Worksheet 9,** Condensed History	**Connect to Earth Science,** p. 209 in ATE **Environmental Science Connection,** p. 210 **Math and More,** p. 211 in ATE **Real-World Connection,** p. 211 in ATE **Multicultural Connection,** p. 212 in ATE **Across the Sciences:** Windows into the Past, p. 224	**Self-Check,** p. 211 **Homework,** p. 211 in ATE **Review,** p. 213 **Quiz,** p. 213 in ATE **Alternative Assessment,** p. 213 in ATE
Transparency 38, Primate Skeletal Structures **Directed Reading Worksheet 9,** Section 3	**Cross-Disciplinary Focus,** p. 215 in ATE **Cross-Disciplinary Focus,** p. 216 in ATE **Cross-Disciplinary Focus,** p. 217 in ATE	**Review,** p. 219 **Quiz,** p. 219 in ATE **Alternative Assessment,** p. 219 in ATE

END-OF-CHAPTER REVIEW AND ASSESSMENT

Chapter Review in Study Guide
Vocabulary and Notes in Study Guide
Chapter Tests with Performance-Based Assessment, Chapter 9 Test
Chapter Tests with Performance-Based Assessment, Performance-Based Assessment 9
Concept Mapping Transparency 9

internet connect

Holt, Rinehart and Winston On-line Resources
go.hrw.com

For worksheets and other teaching aids related to this chapter, visit the HRW Web site and type in the keyword: **HSTHIS**

National Science Teachers Association
www.scilinks.org

Encourage students to use the *sci*LINKS numbers listed with the Chapter Highlights to access information and resources on the **NSTA** Web site.

Chapter Resources & Worksheets

Visual Resources

TEACHING TRANSPARENCIES

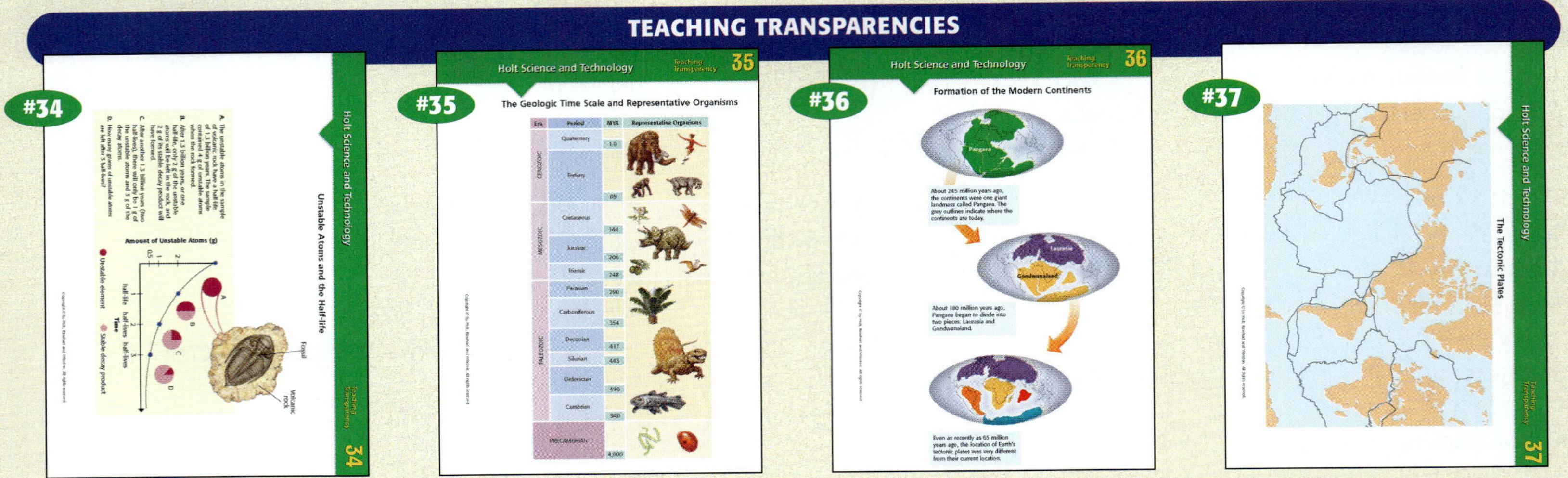

TEACHING TRANSPARENCIES

CONCEPT MAPPING TRANSPARENCY

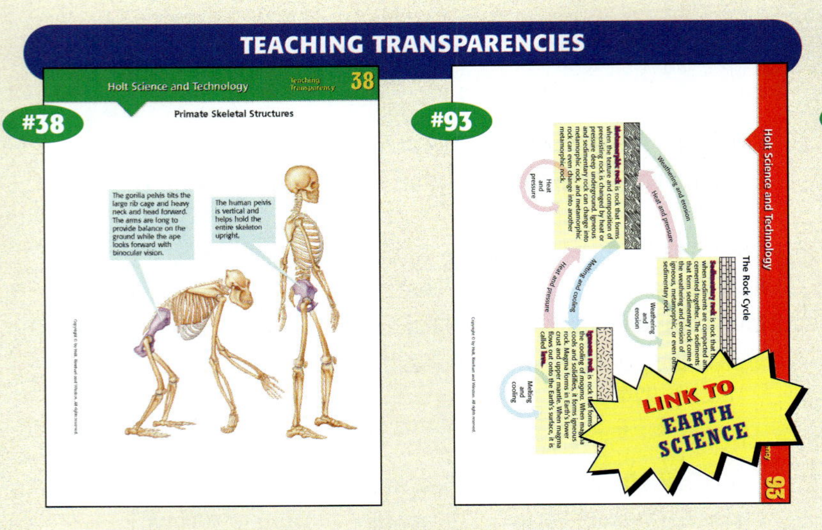

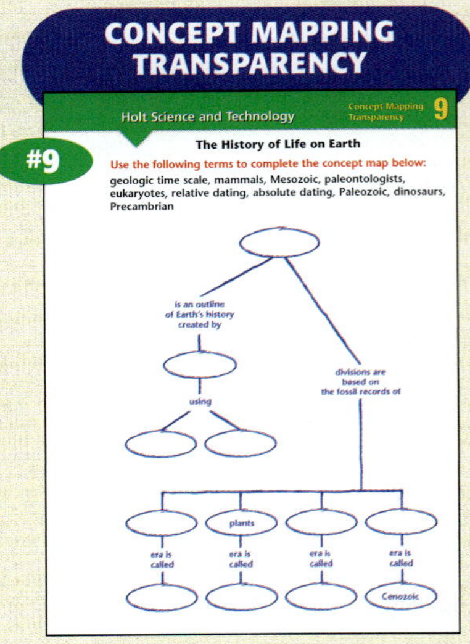

Meeting Individual Needs

DIRECTED READING

REINFORCEMENT & VOCABULARY REVIEW

SCIENCE PUZZLERS, TWISTERS & TEASERS

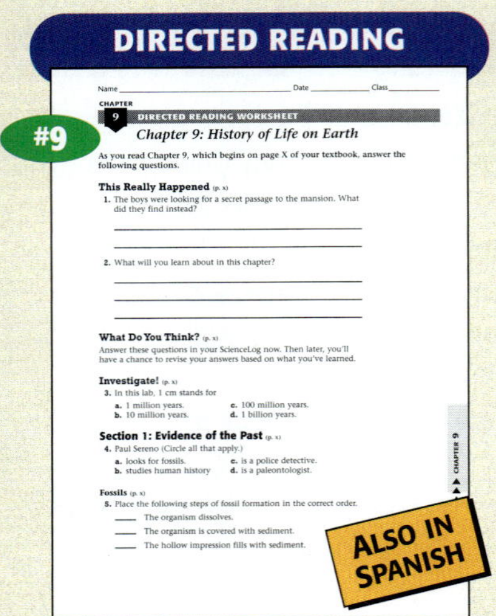

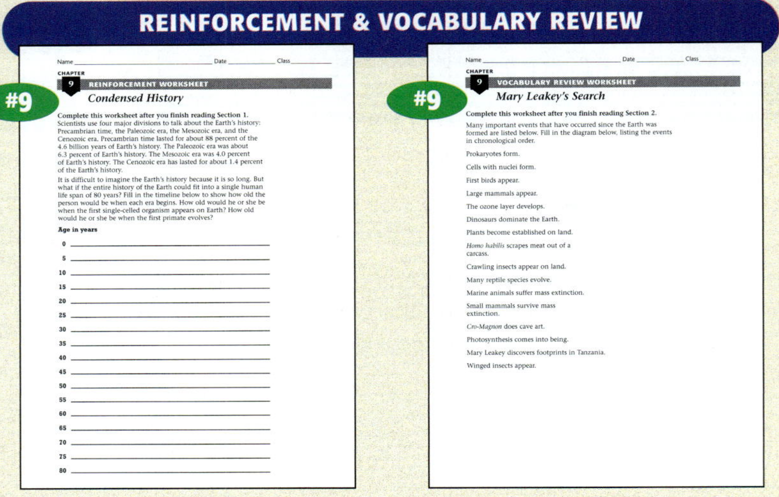

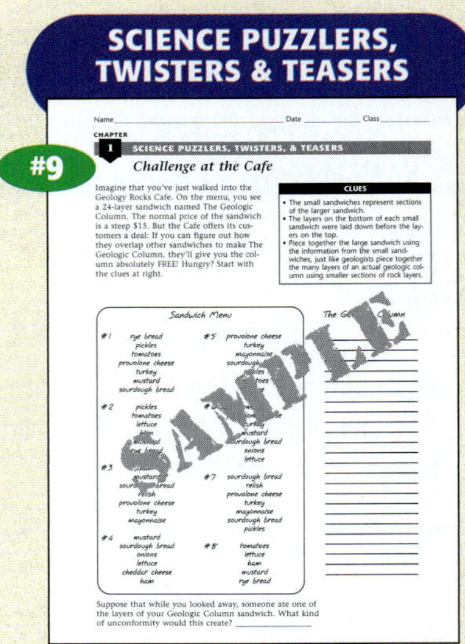

Review & Assessment

STUDY GUIDE

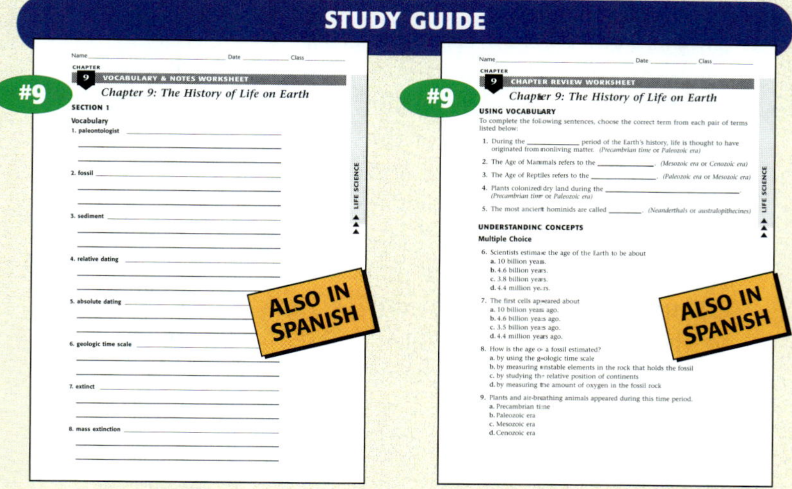

CHAPTER TESTS WITH PERFORMANCE-BASED ASSESSMENT

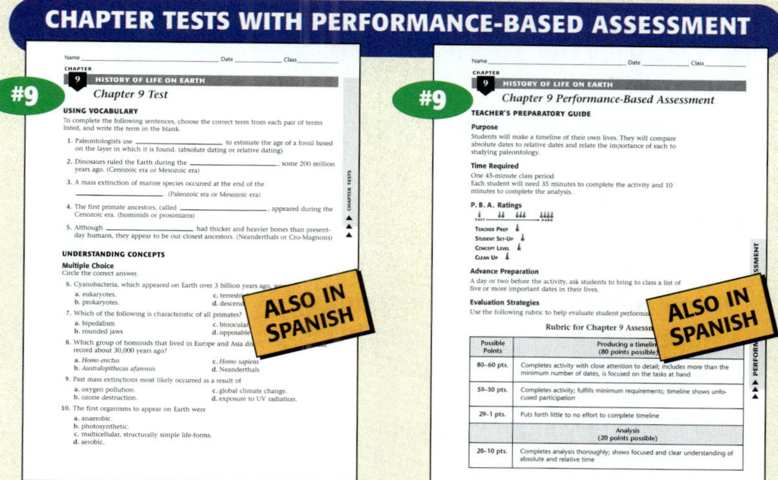

Rubric for Chapter 9 Assessment

Possible Points	Producing a timeline (80 points possible)
80–60 pts.	Completes activity with close attention to detail; includes more than the minimum number of dates; is focused on the tasks at hand
59–30 pts.	Completes activity; fulfills minimum requirements; timeline shows undiscussed participation
29–1 pts.	Puts forth little to no effort to complete timeline

Possible Points	Analysis (20 points possible)
20–10 pts.	Completes analysis thoroughly; shows focused and clear understanding of absolute and relative time

Lab Worksheets

LONG-TERM PROJECTS & RESEARCH IDEAS

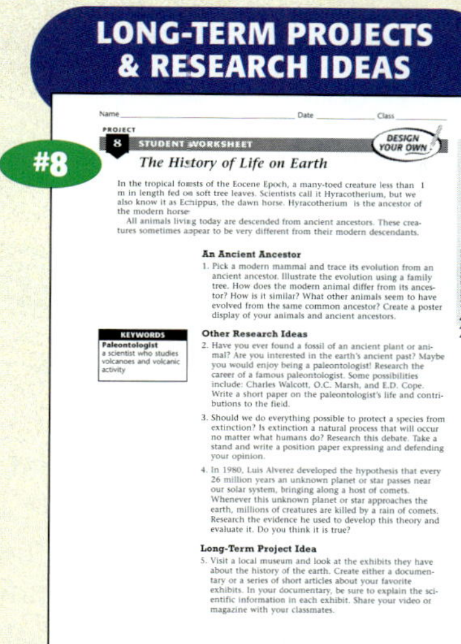

DATASHEETS FOR LABBOOK

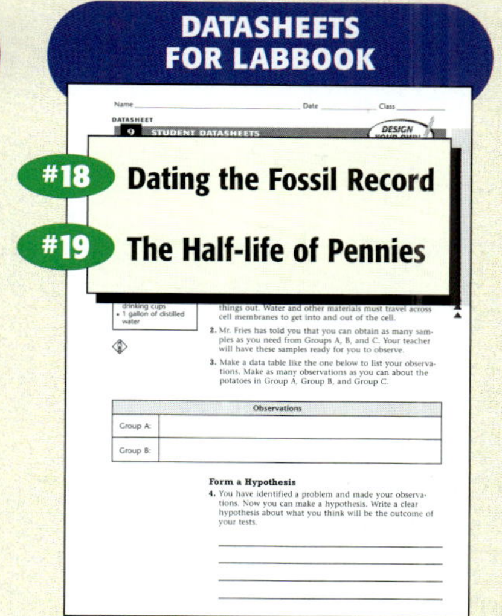

Applications & Extensions

CRITICAL THINKING & PROBLEM SOLVING

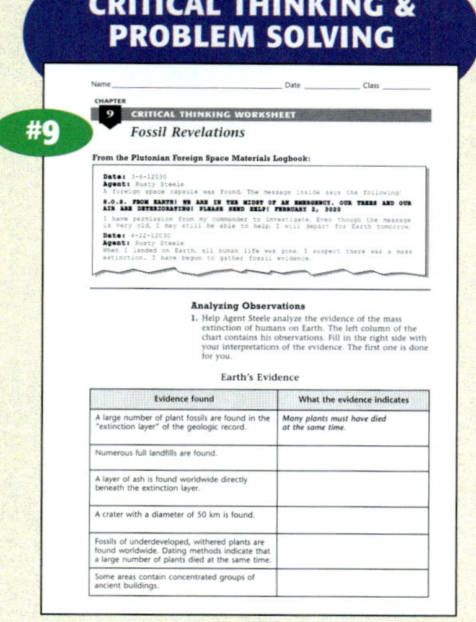

Evidence found	What the evidence indicates
A large number of plant fossils are found in the "extinction layer" of the geologic record.	Many plants must have died at the same time.
Numerous land landfills are found.	
A layer of ash is found worldwide directly beneath the extinction layer.	
A crater with a diameter of 50 km is found.	
Fossils of underdeveloped, withered plants are found worldwide. Dating methods indicate that a large number of plants died at the same time.	
Some areas contain concentrated groups of ancient buildings.	

MULTICULTURAL CONNECTIONS

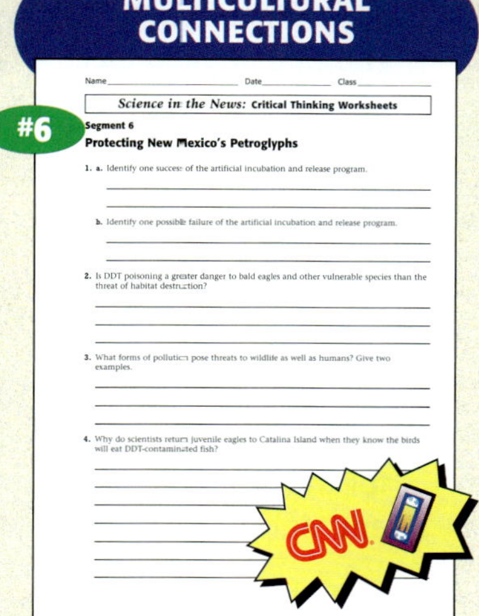

SCIENTISTS IN ACTION

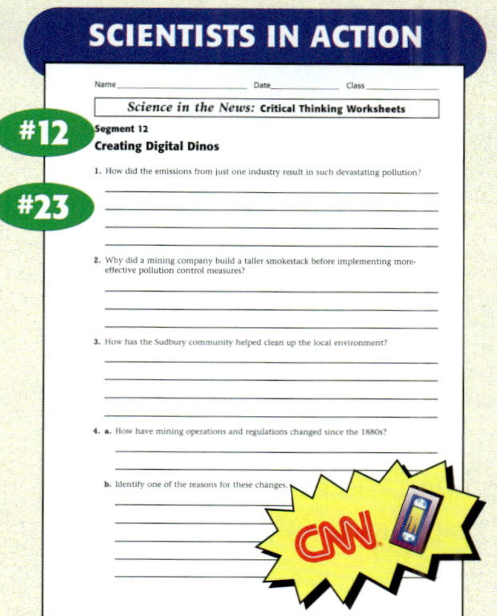

INTERACTIVE EXPLORATIONS

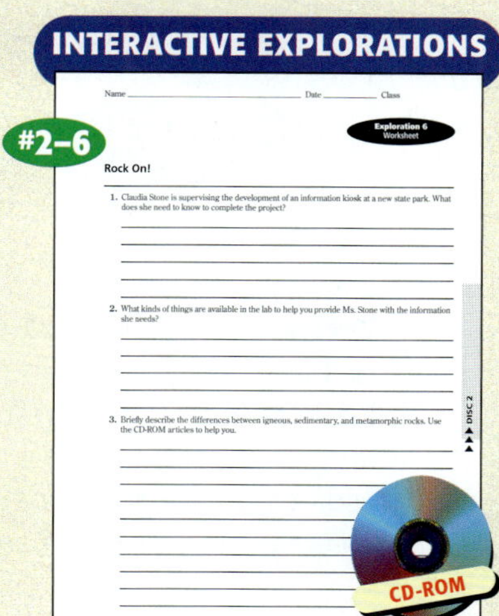

Chapter Background

SECTION 1

Evidence of the Past

▶ Fossils

Fossils are not only preserved plants and animals but also traces of plants and animals. Preserved footprints, feces, gnaw marks, and root holes can all be considered fossils.

- Despite what many people think, fossils are not particularly difficult to find. Nearly every state in the United States contains an abundance of fossils. However, scientists think that only a tiny fraction of the countless organisms that lived on Earth have been preserved as fossils. Many organisms have lived and died without leaving evidence of their existence in the fossil record.

IS THAT A FACT!

- ◆ The oldest fossils are of prokaryotes that are more than 3 billion years old.

▶ Law of Superposition

The law of superposition states that in a series of sedimentary rock layers, each layer is older than the one above it and younger than the one below it. This law is based on an observation made by Nicolaus Steno, a Danish physician, in 1669.

▶ Methods of Absolute Dating

Absolute dating determines a fossil's age in years. Usually radioisotopes are used, and the results provide a range of probable ages. One widely used radioisotope is radiocarbon. A newly developed method uses a particle accelerator and can date samples up to 60,000 years old.

▶ Modern Mass Extinction

Many scientists believe that our planet has entered another era of mass extinction and that human activities are mainly responsible for these extinctions. Urban sprawl and pollution threaten many species. During the last 200 years, more than 50 species of birds, more than 75 species of mammals, and perhaps hundreds of other species of animals and plants have become extinct.

IS THAT A FACT!

- ◆ Dinosaurs are not the biggest animals ever to live on Earth. Blue whales are bigger than the largest known dinosaur.

SECTION 2

Eras of the Geologic Time Scale

▶ Experiment About the Origin of Life

In 1953, American scientist Stanley Miller devised an experiment to simulate life-forming conditions in the early environment. He mixed together hydrogen, ammonia, and methane (to represent the early air) and water (to represent the early oceans) in a flask. Then he applied electricity to the mixture and produced amino acids. His experiment demonstrated that the building blocks of life could be created on Earth chemically. Scientists have since found amino acids in meteorites, confirming that conditions favorable for their formation exist elsewhere and not just on primeval Earth.

▶ Probing into Earth's Past and Beyond

In an effort to gain insights into Earth's past, a huge machine called an *ion microprobe* is closely examining tiny clusters of atoms in ancient rocks. When a flake of rock is placed in the machine, isotopes of certain elements in the flake can be sorted and counted. Using this technique to study the apatite from an island off Greenland, scientists were able to learn that the first signs of life on Earth came 400 million years earlier than previously thought. There weren't any fossils in the apatite, but there was chemical evidence that the 3.85-billion-year-old apatite had an organic origin.

IS THAT A FACT!

- Bristlecone pines are the oldest living trees. Some Colorado bristlecones have lived over 2,500 years. But they're young compared with the Sierra bristlecones, which have been dated as old as 4,765 years!

SECTION 3

Human Evolution

▶ Clues to Migration Route

Scientists believe that people passed through the Nile Valley of Egypt when they migrated from Africa, beginning about 100,000 years ago. Up until recently, no evidence supporting this idea existed. Now Pierre Vermeersch, an archaeologist from the Catholic University of Louvain, in Belgium, and some colleagues have found the skeleton of a child at Taramsa Hill, in the Nile Valley of southern Egypt. The skeleton of the child may be 80,000 years old

and is clearly that of a modern human. Similarities between its skull and teeth and those of equally old human remains found in East Africa and the Middle East suggest a relationship between the two populations.

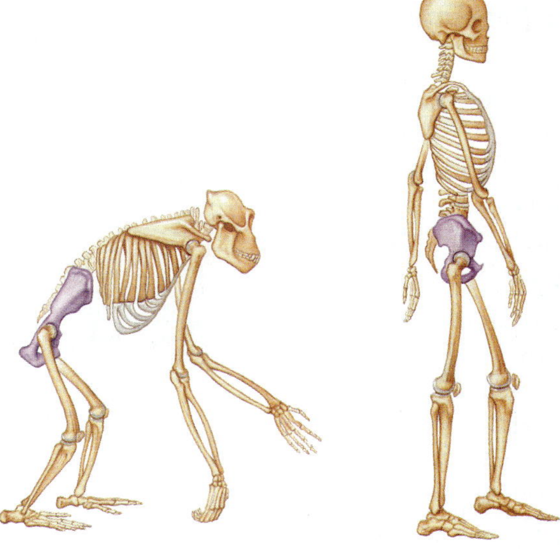

▶ Dawn of Language

Scientists Matt Cartmill and Richard Kay examined fossil hominid skulls and measured the hole through which the hypoglossal nerve passes in its course from the brain to the tongue. The hypoglossal nerve enables precise control over the tongue movements needed for speech. A large hole suggests a larger nerve. Chimpanzees have much smaller holes in their skulls than do modern humans. Because australopithecine skulls have small holes, like the skulls of chimpanzees, Cartmill and Kay think that australopithecines were unable to form words, as modern humans do.

For background information about teaching strategies and issues, refer to the *Professional Reference for Teachers*.

9

The History of Life on Earth

Directed Reading Worksheet 9

Science Puzzlers, Twisters & Teasers Worksheet 9

Guided Reading Audio CD
English or Spanish, Chapter 9

CAPÍTULO

9 La historia de la vida en la Tierra

Imagínate. . .

Un día, tus amigos y tú se enteran de la existencia de un pasadizo subterráneo que conduce a una vieja mansión abandonada, y deciden buscarlo. Cuando están en el campo buscándolo, se topan con un gran agujero entre las raíces de un árbol seco. ¿Será el pasadizo?

Uno por uno, se meten por el agujero. El agujero da a un túnel inclinado y ustedes bajan a resbalones y tropiezos hasta que caen al fondo. Mientras se sacuden el polvo, encienden sus linternas para mirar alrededor. Pero en vez de encontrar el pasadizo hacia la mansión abandonada, se encuentran en una cueva enorme. En lo alto de las paredes de la cueva hay pinturas de toros, vacas, caballos y venados. Te da la impresión de que estas imágenes llevan aquí mucho,

mucho tiempo. ¿Qué quieren decir las pinturas? ¿Por qué las pintaron en una cueva?

Esta aventura pasó de verdad en Francia, a finales de la década de 1940. Cuatro muchachos estaban buscando un pasadizo secreto a la antigua mansión de Lascaux y, en vez de encontrar el pasadizo, se toparon con un recuerdo dejado hace 17,000 años por nuestros ancestros, los hombres de Cro-Magnon. En la ilustración que ves abajo, tres de los muchachos están conversando con su maestro.

200

Imagine . . .

At first glance, the discovery that prehistoric people made paintings in a cave 17,000 years ago seems truly remarkable. However, people have been using caves much farther back in prehistory. Caves are a natural form of shelter, so it is not surprising that archaeologists have found evidence of cave occupation by early humans, Neanderthals, and *Homo erectus*. Even the first evidence of intentional burial was found in a cave. What was surprising to many scientists, though, was that the first intentional burials were done by Neanderthals, not by modern humans.

En este capítulo aprenderás más sobre las actividades, la cultura y la evolución de los seres humanos. También aprenderás sobre las pruebas que tienen los científicos del desarrollo de toda una variedad de seres vivos que poblaron la Tierra en sus primeras épocas.

Cronología de la historia de la Tierra

Haz una cronología para comprender la magnitud del tiempo que abarca la historia de la Tierra. Necesitarás un **lápiz**, una **regla métrica** y una **tira de papel de máquina sumadora** de 46 cm de largo.

Procedimiento

1. En el extremo derecho de la tira, mide secciones de 10 cm. Traza una raya con el lápiz para marcar cada sección. Divide cada sección de 10 cm en diez secciones de 1 cm. (Cada sección de 1 cm representa 100 millones de años.)

2. De arriba hacia abajo, escribe en cada marca de 10 cm lo siguiente: 1,000 ma (mil millones de años), 2,000 ma, 3,000 ma y 4,000 ma. La cronología comienza hace 4,600 ma.

3. A continuación hay una lista de algunos de los eventos más importantes en la historia de la Tierra. Anótalos en tu cronología.

   a. A la izquierda del papel (donde escribiste 4,600 ma), escribe: "El origen de la Tierra"

   b. Mide 35 cm desde el extremo derecho del papel y marca allí 3,500 ma con el lápiz. Escribe: "Aparecen las primeras células".

   c. Los dinosaurios aparecieron por primera vez hace 215 millones de años. Luego, hace unos 65 millones de años, se extinguieron. ¿Cómo medirías esto? Si 100 millones de años = 1 cm, ¿cuántos cm representarán 65 millones de años? De nuevo, desde el extremo derecho de la tira de papel, mide esta distancia y escribe: "Extinción de los dinosaurios".

   d. Hace unos 100,000 años aparecieron los primeros seres humanos con facciones modernas. Márcalo en tu cronología. (Pista: 10 millones de años = 1mm. Así que 100,000 años sería una fracción pequeñísima de 1 mm.)

   e. Sigue marcando en tu cronología los eventos que estudies en este capítulo.

Análisis

4. Las pinturas rupestres de Lascaux tienen más de 17,000 años. Compara este intervalo de tiempo con la edad de la Tierra antes de que surgiera la vida.

5. Compara el tiempo que vivieron los dinosaurios con el tiempo que los humanos han existido en el planeta.

¿Tú qué piensas?

Usa tus conocimientos para responder a las siguientes preguntas en tu cuaderno de ciencias:

1. ¿Cómo se calcula la edad de un fósil?

2. ¿Por cuánto tiempo existieron los dinosaurios?

3. Hay dos características que diferencian a los humanos de los demás animales, ¿cuáles son?

What Do You Think?

Accept all reasonable responses.

Students will have a chance to revise their answers in the Chapter Review under NOW What Do You Think?

Investigate!

MATERIALS

FOR EACH GROUP:
- pencil
- metric ruler
- strip of adding machine paper, 46 cm long

Teacher Notes: Suggest that students tape the full length of the strip of paper securely to a flat surface before they begin to record measurements.

Make sure students can relate distance to a time scale.

Answers to Investigate!

4. Accept all reasonable responses. The cave paintings are just a tiny fraction of geologic time.

5. Dinosaurs roamed for approximately 150 million years, and humans have existed for about 5 million years. Dinosaurs existed 30 times longer than humans have so far.

Q: How does a fossil get the best seat in the house?

A: It makes a preservation.

MISCONCEPTION ALERT

Movies and TV shows often depict humans interacting with dinosaurs. Students should understand this is not possible since humans and dinosaurs did not exist at the same time in history.

Focus

Evidence of the Past

This section introduces students to fossils and how they provide clues to Earth's past. Students learn how fossils form in sedimentary rock. They explore the methods scientists use to determine the age of fossils. Finally, students learn how scientists place events in the Earth's history in the correct order and what they think might have caused mass extinctions.

🔔 Bellringer

Ask students to imagine that they didn't clean their room for 30 years. After 30 years, they finally decide to sort through the 2 m pile of stuff on their floor. Ask:

"What might you find on the top of the pile? in the middle? on the bottom?"

1 Motivate

ACTIVITY

Photo Analysis Have students bring in copies (*not* originals, since they may be irreplaceable) of old photos of themselves, and tack the copies on the bulletin board. Ask students to describe how they have changed during the years since the pictures were taken. Point out that scientists use traces or imprints of living things preserved in rock in a similar way to observe how life on Earth has changed over time.

Sheltered English

Directed Reading Worksheet 9 Section 1

Pruebas del pasado

VOCABULARIO

paleontólogo
fósiles
ciclo de las rocas
edad relativa
vida media
edad absoluta
escala de tiempo geológico
extinto
extinción masiva
Pangea
tectónica de placas

OBJETIVOS

- Describe dos métodos que los científicos utilizan para determinar la edad de los fósiles en rocas sedimentarias.
- Describe la escala de tiempo geológico y la información que suministra a los científicos.
- Describe las causas posibles de las extinciones masivas.
- Explica la teoría de la tectónica de placas.

Como detectives en la escena del crimen, algunos científicos buscan pistas que les ayuden a reconstruir el pasado. Estos científicos son los paleontólogos. Los **paleontólogos,** como Paul Sereno, a quien ves en la **Figura 1,** estudian los fósiles para reconstruir cómo era la vida millones de años antes de que los seres humanos existieran. Los fósiles demuestran que las formas de vida en la Tierra han cambiado muchísimo y nos dan pistas sobre cómo sucedieron estos cambios.

Figura 1 *En 1995, el paleontólogo Paul Sereno encontró este enorme fósil del cráneo de un dinosaurio en el desierto del Sahara. Este dinosaurio fue probablemente el depredador terrestre más grande que jamás haya existido!*

Fósiles

Los **fósiles** son rastros o huellas de seres vivos, como animales, plantas, bacterias y hongos que se conservan en las rocas. Por lo general, los fósiles se forman cuando una capa de sedimento cubre a un organismo muerto. Más tarde, estos sedimentos pueden comprimirse y convertirse en rocas sedimentarias. Las rocas sedimentarias, junto con las rocas ígneas y las metamórficas, forman el ciclo de las rocas. El **ciclo de las rocas** es el proceso a través del cual un tipo de roca se transforma en otro. La **Figura 2** muestra una de las maneras en que se pueden formar fósiles en rocas sedimentarias.

Figura 2 *Mira las ilustraciones de abajo, en las que se muestra una de las maneras en que se forman los fósiles.*

❶ Un organismo muere y queda enterrado bajo una capa de sedimentos.

❷ El organismo se descompone poco a poco y deja un molde hueco, o impresión, en el sedimento.

❸ Con el tiempo, el molde se llena de sedimentos que adoptan la forma del organismo original.

202

SCIENTISTS AT ODDS

In the 1870s, two American scientists, Edward Drinker Cope and Othniel Charles Marsh, studied dinosaur fossils. They became bitter rivals and often argued. In 1878, Marsh and Cope were both excavating fossils near Como Bluff, Wyoming. They had separate excavations and didn't want to share their findings. Both groups found more fossils than they could carry. To prevent the other group from taking their fossils, each group smashed all the fossils that couldn't be carried away.

La edad de los fósiles

Los paleontólogos determinan la edad de los fósiles que encuentran mediante dos métodos: edad absoluta y edad relativa.

Edad relativa Una roca sedimentaria tiene varias capas de roca. Las más antiguas están abajo y las más recientes arriba. Si hay fósiles en una roca, el científico puede examinarla de abajo hacia arriba para ver la secuencia de fósiles en el mismo orden en el que vivieron los organismos. Este método se llama **edad relativa.**

Edad absoluta ¿Cómo calculan los científicos la edad de un fósil? La respuesta está en los *átomos*, las pequeñas partículas que forman toda la materia del universo. Los átomos están formados de otras partículas más pequeñas, unidas por fuerzas enormes. Si esas fuerzas no pueden mantener unidas las partes del átomo, se dice que ese átomo es inestable. Los átomos inestables se desintegran, dejando escapar algunas de sus partículas. Esto se llama *descomposición radioactiva*. El átomo se vuelve estable, pero al hacerlo se transforma en un tipo diferente de átomo.

Cada tipo de átomo inestable se descompone a una velocidad específica. El tiempo que tardan en descomponerse la mitad de los átomos inestables de una muestra se llama **vida media.** Algunas vidas medias duran una fracción de segundo, y otras duran miles de millones de años. Para calcular la edad de un fósil y de la roca se mide la proporción de átomos inestables y estables. Este método se llama **edad absoluta.** La **Figura 3** muestra cómo se descomponen los átomos inestables de una roca.

El *Bananabana bobana* es un fósil raro. Pasa a la página 588 para verlo.

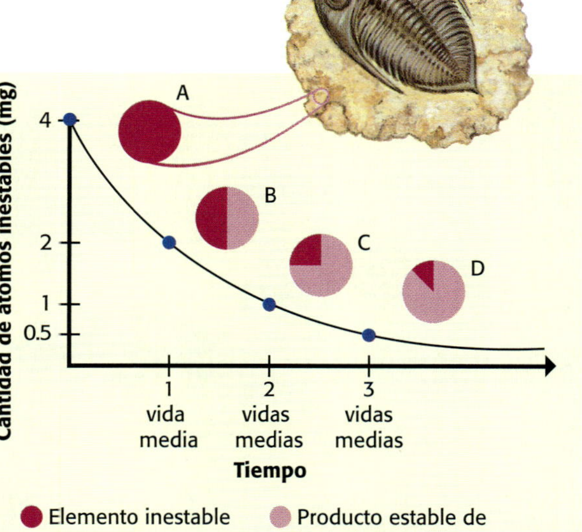

Figura 3 A veces, un organismo que está enterrado en sedimentos cerca de la superficie queda cubierto por roca volcánica. Si los científicos determinan la edad de la roca volcánica, se pueden dar una idea sobre la edad del fósil.

A. Los átomos inestables de la roca volcánica tienen una vida media de 1,300 millones de años. La roca tenía 4 mg de átomos inestables cuando se formó.

B. Después de 1,300 millones de años, o una vida media quedarán 2 mg de átomos inestables y habrá 2 mg del producto estable de su descomposición.

C. Después de 1,300 millones de años (dos vidas medias), habrá sólo 1 mg de átomos inestables y 3 mg de átomos descompuestos.

D. ¿Cuántos miligramos de átomos inestables quedarán después de 3 vidas medias?

Homework

Calculating To help students grasp the concept of half-life, have them solve these problems.

1. Thorium-232 has a half-life of 14.1 billion years. How much of an 8 mg sample will be unchanged after one half-life? (4 mg) two half-lives? (2 mg) three half-lives? (1 mg) four half-lives? (0.5 mg)

2. Carbon-14 has a half-life of 5,730 years. How much of the original sample will be left after 11,560 years? (one-fourth of the original sample) after 17,190 years? (one-eighth of the original sample)

Dating the Fossil Record

RETEACHING

Obtain a stack of old daily newspapers. Tell students that the dailies represent layers of sedimentary rock and that the pictures on the front pages represent fossils. Scramble the newspapers to make sure that they are not ordered according to date. Then have students stack the newspapers from oldest to newest, to simulate the layering of fossils over time. Sheltered English

CONNECT TO EARTH SCIENCE

The relative age of a fossil can be determined by which layer of sedimentary rock the fossil is found in. You can illustrate the formation of sedimentary rock using Teaching Transparency 93.

Teaching Transparency 93 "The Rock Cycle"

Teaching Transparency 34 "Unstable Atoms and the Half-life"

Math Skills Worksheet 36 "Radioactive Decay and the Half-life"

TOPIC: Evidence of the Past
GO TO: www.scilinks.org
*sci*LINKS NUMBER: HSTL180

MEETING INDIVIDUAL NEEDS

Advanced Learners Geologic time is the period of time that Earth has been in existence. Geologic time is divided into eras. Eras are broken into smaller divisions called periods. Periods can be divided into epochs. Challenge students to construct a geologic time line that identifies all of these divisions. Tell them they can research in the library or on the Internet for information.

RESEARCH

Students could research plant and animal species that have become extinct within the last 200 years. Many of the extinctions were caused by human activities. Extinct birds include the dodo, great auk, Labrador duck, moa, and passenger pigeon. Extinct mammals include the Steller's sea cow and the quagga.

DISCUSSION

Explain to students that *mya* means "million years ago." Likewise, *my* means "million years" and *bya* means "billion years ago." Ask students why they think geologists use this form of dating.

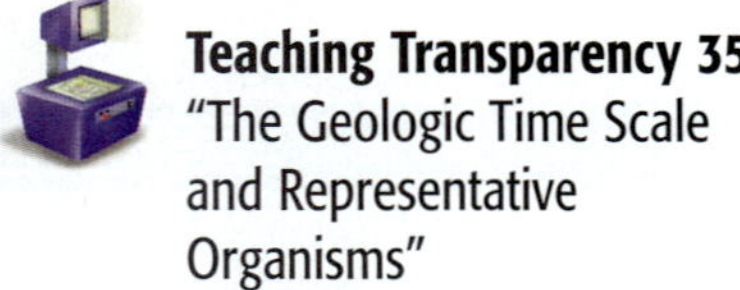

Escala del tiempo geológico

Cuando recuerdas cosas importantes de tu vida, te acuerdas del día, mes o año en que sucedieron. Estas divisiones en el tiempo te ayudan a recordar cuándo naciste, cuándo metiste el gol de la victoria o cuando pasaste a quinto grado. Como el período desde que se formó la Tierra hasta el presente es tan largo, los científicos usan un tipo de calendario para dividir la larga historia de la Tierra en grandes unidades de tiempo.

El calendario que los científicos usan para medir la historia de la vida en la Tierra se llama **escala del tiempo geológico** y se muestra a la izquierda. Después de calcular la edad de un fósil con los métodos de edad relativa y absoluta, el paleontólogo coloca el fósil en orden cronológico junto con los demás fósiles y forma una imagen del pasado que muestra cómo han ido cambiando los organismos con el tiempo.

Divisiones en la escala del tiempo geológico Los paleontólogos dividen la escala de tiempo en unidades llamadas *eras*. Cada era se subdivide en unidades más pequeñas a medida que se encuentra más información fósil.

Las eras se caracterizan por el tipo de animal predominante durante ese período. Por ejemplo, los dinosaurios y otros reptiles dominaron la era Mesozoica, y por eso se le llama la Edad de los Reptiles. El final de cada era está marcado por la extinción de ciertos organismos. La siguiente sección analiza las eras de la escala del tiempo geológico.

La escala del tiempo geológico

Era	Periodo	MA	Organismos representativos
CENOZOICA	Cuaternario	1.8	
CENOZOICA	Terciario	65	
MESOZOICA	Cretácico	144	
MESOZOICA	Jurásico	206	
MESOZOICA	Triásico	248	
PALEOZOICA	Pérmico	290	
PALEOZOICA	Carbonífero	354	
PALEOZOICA	Devónico	417	
PALEOZOICA	Silúrico	443	
PALEOZOICA	Ordóvico	490	
PALEOZOICA	Cámbrico	540	
	PRECÁMBRICO	4,600	

CONNECT TO
EARTH SCIENCE

Radioactive dating techniques have been used to show that the Earth and the solar system formed about 4.6 billion years ago. This fact was not discovered until the 1950s, when the technique for absolute dating was developed. Long before that, geologists had found a record of fossils indicating that a sequence of organisms had lived and had provided a basis for making a relative time scale.

Extinciones masivas Algunas divisiones de la escala del tiempo geológico se caracterizan por eventos que hicieron que muchas especies de plantas y animales se acabaran, es decir, se **extinguieran.** Cuando una especie se extingue, desaparece. Hay períodos en la historia de la Tierra en los que muchas especies se han extinguido al mismo tiempo. Estos períodos se llaman períodos de **extinción masiva.**

Los científicos no están seguros de la causa de las extinciones masivas. Una causa posible son grandes cambios en el clima o en la atmósfera. Algunos piensan que la extinción masiva de los dinosaurios y otras especies ocurrió después de que un gran meteorito contra la Tierra, como se ve en la **Figura 4.** El choque causó un cambio en el clima o en la atmósfera. Pudo haber otras causas de los cambios en el medio ambiente que produjeron extinciones masivas. Por ejemplo, el movimiento de los continentes pudo causar cambios climáticos.

Figura 4 *Un meteorito chocó con la Tierra hace aproximadamente 65 millones de años. Los cambios climáticos que resultaron del choque podrían ser la causa de la extinción de los dinosaurios.*

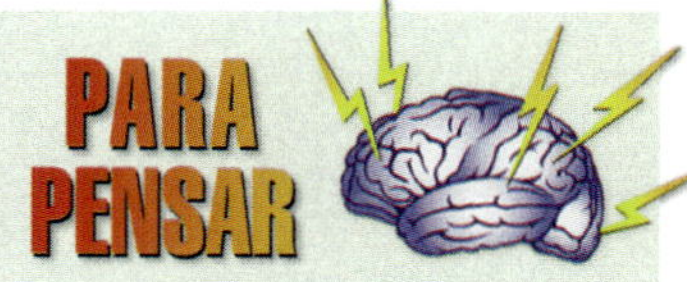

Los científicos calculan que sólo una pequeña fracción ($1/20$ de 1 por ciento) de todas las especies que han existido en la Tierra viven hoy en día. Todas las demás se han extinguido.

✔ Autoevaluación

Había diez gramos de un átomo inestable en una roca cuando se solidificó. ¿Cuántos gramos de átomos inestables quedan después de una vida media? ¿Y después de dos? *(Consulta la página 636 para comprobar tus respuestas.)*

205

WEIRD SCIENCE

One species that became extinct during the time of the dinosaurs was the insect having the largest wingspan on record. The insect belonged to the order Protodonata, and it measured an astonishing 76 cm (30 in.) from tip to tip. Its body was 46 cm (18 in.) long. It died out about 200 million years ago. Fossils of this insect have been found in Kansas.

GOING FURTHER

Poster Project Have students research what their area was like millions of years ago. They can develop a written report and a poster describing the climate, living things, and landforms at different points in time.

INDEPENDENT PRACTICE

If any students have rock collections, ask them to share these collections with the class. Allow students time to observe the different collections and to ask questions about them. Sheltered English

Answer to Self-Check

5 g, 2.5 g

Critical Thinking Worksheet 9 "Fossil Revelations"

Math Skills Worksheet 45 "Geologic Time Scale"

internet**connect**

SCI**LINKS**
NSTA

TOPIC: Mass Extinctions
GO TO: www.scilinks.org
*sci*LINKS NUMBER: HSTL185

TOPIC: The Geologic Time Scale
GO TO: www.scilinks.org
*sci*LINKS NUMBER: HSTL190

RETEACHING

Ask students to draw a picture in their ScienceLog illustrating how future plate movements might change the geography of the world. (Some predictions might include a wider Atlantic Ocean and a shift northward of Africa, Australia, and South America.)

MEETING INDIVIDUAL NEEDS

Advanced Learners Ask students: What might have happened to the animals inhabiting Pangea as it broke up? How might the changing landscape influence the evolution of living things? Have students explain how the breakup might have affected the distribution of animal species. Encourage students to use diagrams or other visual effects when sharing their ideas with the class. (Student responses should demonstrate an understanding that changes within the environment affect the ability of certain organisms to survive.)

CONNECT TO
EARTH SCIENCE

The Great Rift Valley of Africa marks a spreading center that is found on a continent. Have students identify the rift valley on a map. Ask: What kind of features mark this location? (deep valleys and lakes)

What do you think will become of this rift valley in a few million years? (It will probably continue to widen and eventually become an ocean or inland sea.)

Teaching Transparency 36
"Formation of the Modern Continents"

La Tierra cambiante

¿Sabías que se han encontrado fósiles de dinosaurios en la Antártida? La Antártida, ahora cubierta de hielo, fue en otro tiempo un lugar tibio que estos reptiles podían habitar. ¿Cómo es posible? La Antártida estaba cerca del ecuador.

Muchas especies de marsupiales (mamíferos que cargan a sus crías en una bolsa) viven en Australia. Hay varias especies de marsupiales en América del Sur y sólo una en Norteamérica. ¿Por qué hay tantos marsupiales en Australia? Australia se separó del resto de la masa continental antes de que otros mamíferos evolucionaran y empezaran a competir con los marsupiales por la comida y el espacio. Los cambios en el ambiente biológico de la Tierra fueron causados por cambios en su ambiente geológico.

Pangea Si te fijas en un mapa del mundo, verás que la forma de los continentes se parece a la de las piezas de un rompecabezas. Si movieras las piezas, verías que algunas de ellas casi encajan a la perfección. Algo así se le ocurrió al científico alemán Alfred Wegener a principios del siglo XX. Propuso la idea de que hace mucho tiempo los continentes formaban parte de una masa continental rodeada por un solo océano gigantesco. Wegener llamó a esta masa de tierra **Pangea,** que quiere decir "toda la Tierra."

Wegener pensaba que los continentes de hoy eran un solo supercontinente por tres razones. Primero, porque las formas de los continentes parecían "encajar" entre sí. Segundo, porque los fósiles de plantas y animales a ambos lados del océano Atlántico eran muy parecidos. Tercero, observó que en lugares donde ahora el clima es cálido, antes hubo glaciares. La **Figura 5** muestra cómo se formaron los continentes a partir de la Pangea.

Hace más o menos 245 millones de años, los continentes eran una única masa de tierra, llamada Pangea. El lugar que los continentes ocupan ahora se indica en gris.

Hace unos 180 millones de años, la Pangea comenzó a dividirse en dos pedazos: Laurasia and Gondwana.

Hace apenas 65 millones de años, la ubicación de las placas tectónicas de la Tierra era muy distinta a la actual.

Figura 5 *Como los continentes se mueven de 1 a 10 cm cada año, en 150 millones de años van a estar acomodados de manera muy distinta.*

Multicultural CONNECTION

Tell the class that the Mid-Atlantic Ridge rises above water at only one place, Iceland. Point to Iceland on a map, and indicate the Arctic Circle, which includes the northern part of the island. Tell students that these two features suggest that Iceland is a place of both extreme cold and extreme heat. Ask students to do some research on what life is like in Iceland. Have them focus on how geologic processes, such as glaciers, geysers, and volcanic activity, affect the lifestyle of the people who live there.

¿Se mueven los continentes? En la década de 1960, J. Tuzo Wilson pensó que las piezas que forman la corteza terrestre oscilan hacia adelante y hacia atrás empujadas por fuerzas desde el interior del planeta. Estas piezas son *placas tectónicas.* La teoría de Wilson se llama **tectónica de placas.**

Según Wilson, la corteza terrestre está fragmentada en siete grandes placas rígidas y varias pequeñas, como se ve en la **Figura 6.** Los continentes y los océanos están sobre ellas. El movimiento de las placas hace que los continentes se muevan y separa algunas especies.

Adaptación en cámara lenta

Aunque las placas tectónicas se mueven muy despacio, su movimiento afecta a los seres vivos. Normalmente, los organismos evolucionan para adaptarse a los cambios que trae el desplazamiento de los continente, pero en un mismo lugar puedes encontrar pruebas fósiles de otros seres que no sobrevivieron a los cambios. ¡Se han encontrado conchas de mar en las cimas de algunas montañas! Otra prueba de que el movimiento de las placas tectónicas sacó a la superficie rocas enterradas.

Figura 6 *Los científicos piensan que las placas tectónicas, ilustradas arriba, han reacomodado lentamente los continentes desde que se enfrió la corteza terrestre.*

REPASO

1. ¿Qué información contiene la escala del tiempo geológico y cuáles son sus principales divisiones?

2. Nombra una causa posible de las extinciones masivas.

3. Explica una de las maneras en que los cambios geológicos de la Tierra afectan a las plantas y animales.

4. ¿Qué diferencia hay entre los dos métodos (edad relativa y edad absoluta) para calcular la antigüedad de un fósil?

5. **Comprender conceptos** Los fósiles del *Mesosaurus,* el pequeño lagarto acuático de la derecha, sólo se han encontrado en África y en Sudamérica. Según lo que sabes de la tectónica de placas, ¿cómo explicas esto?

Quiz

1. What is a fossil? How are fossils usually formed? (Fossils are traces or imprints of living things that are preserved in rock. Fossils are usually formed when an organism is buried in sediments that harden into rock.)

2. What can scientists learn about Earth's past from fossils? (Fossils provide evidence that life on Earth has changed and show how those changes occurred.)

3. Why would fossils found in rock layers near Earth's surface probably be younger than those found in rock layers at the bottom of a deep canyon? (The upper layers were deposited more recently than the lower layers.)

ALTERNATIVE ASSESSMENT

Each student in a group receives three blank cards. The group looks at the geologic time scale on page 204 and thinks of questions that could be answered using information presented on the time scale. Students write a question on one side of a card. They stack the cards with the blank sides up. In turn, each student draws a card, reads the question on it, and attempts to answer the question. Group members determine if the answer is correct by consulting the geologic time scale.

▼ Answers to Review

1. The geologic time scale provides classifications for geologic time; the Precambrian, Paleozoic, Mesozoic, and Cenozoic eras.

2. perhaps a drastic change in the environment

3. Geologic changes cause animals of a single species to become separated and new species to form.

4. Relative dating is determining when a fossil was formed by examining where the fossil appears in the layers of rock. Absolute dating is measuring the percentage of radioactive material left in the rock surrounding a fossil.

5. Perhaps *Mesosaurus* evolved after Laurasia and Gondwanaland separated, but before South America and Africa separated.

Teaching Transparency 37 "The Tectonic Plates"

Reinforcement Worksheet 9 "Earth Timeline"

Focus

Eras of the Geologic Time Scale

This section discusses current theories regarding the origin of life. Students are introduced to the four eras of geological time in chronological order:

the Precambrian, the Paleozoic, the Mesozoic, and the Cenozoic. They learn about the life-forms that characterize each era.

Bellringer

Write the following supposition on the board or on an overhead projector for students to address in their ScienceLog:

Suppose that electric energy was never developed. How would your life differ from what it is like now?

Discuss with students the consequences of great changes over time.

1) Motivate

DISCUSSION

Historical Perspective Have students pretend that they are in a time-travel machine. What scientifically significant events would they witness as they travel back in time to Earth's origin? List the events on the board in the order that they are suggested. Ask students how they can determine whether the events are in chronological order. (They can consult the geologic time scale or other scientific materials.)

Directed Reading Worksheet 9 Section 2

VOCABULARIO

período Precámbrico
procariota
anaeróbico
ozono
eucariota
era Paleozoica
era Mesozoica
era Cenozoica

OBJETIVOS

- Resume los principales eventos que hicieron posible la vida en la Tierra.
- Describe los diferentes tipos de organismos que surgieron en las cuatro eras de la escala del tiempo geológico.

Las eras de la escala del tiempo geológico

Observa la fotografía del Cañón del Colorado en la **Figura 7.** Si te fijas, verás que sus paredes están formadas por capas de rocas de distintos tipos y colores. Cuanto más bajas en el cañón, más antiguas son las capas de roca. Parece increíble, pero cada capa del cañón fue en algún momento la capa superficial. Hace miles de millones de años, la actual capa inferior era la superficial. Dentro de miles de millones de años, la actual capa superficial tendrá muchas capas encima.

Cada capa nos dice algo sobre lo que sucedía en la Tierra cuando estaba arriba. El tipo de rocas y los fósiles de cada capa nos cuentan lo que pasó. Los científicos han dividido la historia geológica en cuatro eras, según sus estudios sobre las diferentes rocas y fósiles: el período Precámbrico, la era Paleozoica, la era Mesozoica y la era Cenozoica.

Figura 7 Cada capa de rocas del Cañón del Colorado es como una página del libro de la "historia de la Tierra".

Figura 8 El animal que formó este fósil precámbrico vivía en mares poco profundos.

Período Precámbrico

Si bajas al fondo del Cañón del Colorado, ¡encontrarás rocas que tienen entre 1 y 2 mil millones de años! Estas rocas se formaron en la era geológica conocida como **período Precámbrico,** hace más de 540 millones de años, cuando las únicas formas de vida que había eran organismos muy primitivos. Muchos de estos organismos no tenían partes duras en el cuerpo, de modo que hay muy pocos fósiles de esta época. La **Figura 8** muestra un fósil precámbrico.

208

IS THAT A FACT!

Throughout Earth's history, the forces of erosion have been altering the planet's surface, making it almost impossible to find rocks older than 3.5 billion years. However, a number of rocks dating from about 3.5 to 3.9 billion years ago have been found in Canada and Greenland. The oldest of the rocks was found in the Northwest Territories of Canada in 1989.

La Tierra en sus primeros tiempos Los científicos sostienen la hipótesis de que cuando comenzó la vida, el ambiente de la Tierra era muy distinto al de ahora. Ese medio ambiente tenía una atmósfera sin oxígeno, pero rica en otros gases, como monóxido de carbono, dióxido de carbono, hidrógeno y nitrógeno. La Tierra en su juventud era, como se ve en la **Figura 9,** un lugar muy turbulento. Los meteoritos chocaban contra su superficie, la agitaban violentas tormentas eléctricas y las erupciones volcánicas eran constantes. Radiaciones poderosas, como la de los rayos ultravioletas del Sol, la golpeaban.

Figura 9 *La Tierra tuvo una juventud turbulenta.*

¿Cómo empezó la vida? La hipótesis de los científicos es que, en estas condiciones, la vida surgió a partir de materia inerte. En otras palabras, la vida empezó a partir de substancias químicas que ya existían en el ambiente. Entre estas substancias había agua, arcilla, minerales disueltos en el océano y gases de la atmósfera. La energía que había en la Tierra hizo que estas substancias químicas reaccionaran entre sí y formaran las complejas moléculas que hicieron posible la vida.

Los científicos piensan que durante millones de años estas moléculas flotaron en los océanos y se unieron para formar moléculas más grandes, que reaccionaron entre sí formando estructuras más complicadas. Éstas se convirtieron en estructuras parecidas a las células y, finalmente, en las primeras células reales. La **Figura 10** muestra un fósil de este tipo de célula. Los científicos piensan que se trata de una **célula procariota,** una célula sin núcleo. Las primeras procariotas eran **anaeróbicas** que quiere decir que ni necesitaban, ni soportaban el oxígeno en su forma libre. Los organismos que necesitan oxígeno no podrían haber sobrevivido en la Tierra entonces, porque no había oxígeno libre. Muchas procariotas anaeróbicas aún viven en la Tierra, en lugares sin oxígeno libre.

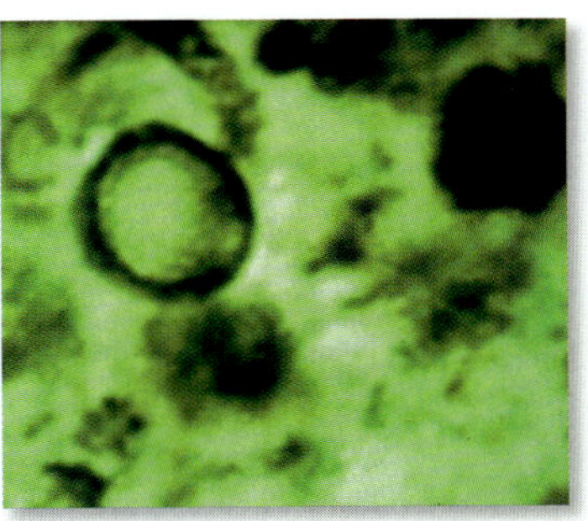

Figura 10 *Los fósiles de células procariotas (como la estructura circular de la fotografía) respaldan la hipótesis científica de que la vida comenzó en la Tierra en esta forma hace 3,500 millones de años.*

209

USING THE FIGURE

Tell students that fossils can indicate how the Earth's surface has evolved. Ask them what they think scientists could infer if they found a fossil, such as the one shown in **Figure 8,** in rocks high above sea level. (The area was once covered by ocean.)

ACTIVITY

Using Maps Have students locate the three earthquake and volcano zones on a world map. One zone extends nearly all the way around the edge of the Pacific Ocean. A second zone is located near the Mediterranean Sea and extends across Asia into India. The third zone extends through Iceland to the middle of the Atlantic Ocean. Sheltered English

GUIDED PRACTICE

Concept Mapping Have students construct a concept map that shows how life on Earth developed from nonliving matter. They should base their map on information presented in the last two paragraphs on this page. The primary subject headings should refer to the chemicals that already exist in the environment. The final part of the concept map should identify prokaryotes as the first life-forms to appear on Earth.

CONNECT TO
EARTH SCIENCE

Have students research mountain ranges that formed during the different geologic eras and locate them on a world map. Paleozoic era: Caledonian Mountains of Scandinavia, Acadian Mountains of New York, Appalachian Mountains of North America, and the Ural Mountains of Russia; Mesozoic era: Palisades Mountains of New Jersey and the Rocky Mountains of North America; Cenozoic era: Andes Mountains of South America, Alps of Central Europe, and the Himalayas of Central Asia.

DISCUSSION

Ask students how they think the emergence of cyanobacteria and their ability to photosynthesize affected the evolution of organisms on Earth. (Organisms that use oxygen were able to evolve.)

MEETING INDIVIDUAL NEEDS

Learners Having Difficulty

Help students understand the importance of photosynthesis in the evolution of life on Earth. Ask them what kind of organisms were living before the development of photosynthesis. (anaerobic prokaryotes)

Then ask them how life on Earth changed after photosynthesis. (Many aerobic organisms evolved.)

Have students write answers to the questions and accompany their answers with illustrations.

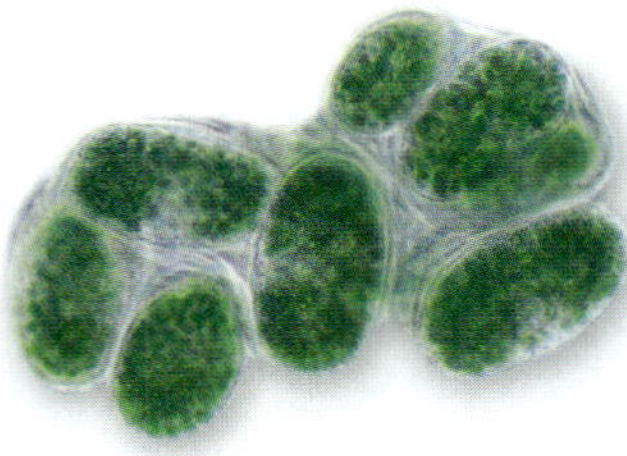

Figura 11 *Las cianobacterias o "algas azul-verdosas" son los seres vivos más simples que realizan la fotosíntesis.*

ciencias del medio ambiente
C O N E X I Ó N

El agotamiento del ozono en la capa superior de la atmósfera es un problema grave. Las substancias químicas industriales, como las que se usan en refrigeración, destruyen el ozono de la capa superior de la atmósfera. Todos los seres vivos estamos expuestos a niveles más altos de radiación, que provocan cáncer de piel. Algunos países han prohibido el uso de substancias químicas destructoras del ozono.

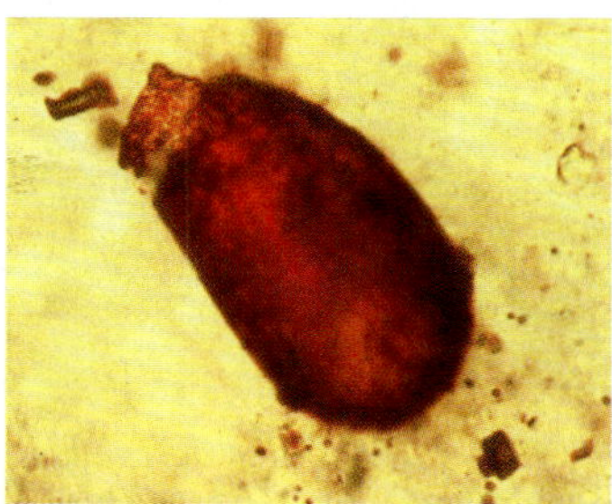

Figura 13 *Esta célula eucariota fosilizada tiene cerca de 800 millones de años.*

El primer contaminante de la Tierra: ¡el oxígeno! Hace más de 3 mil millones de años, unos organismos procarióticos llamados cianobacterias o "algas azul-verdosas" aparecieron en el registro fósil. Las cianobacterias, mostradas en la **Figura 11,** son organismos fotosintéticos: usan la energía solar para producir azúcares y almidones. Uno de los productos secundarios de la fotosíntesis es el oxígeno. Al realizar la fotosíntesis, las cianobacterias liberaban oxígeno en los océanos. Después de cientos de millones de años, habían producido tanto oxígeno que una parte escapó a la atmósfera. La fotosíntesis siguió aumentando la cantidad de oxígeno en la atmósfera, pero por 1 ó 2 mil millones de años este gas era sólo una pequeña fracción de la concentración que hay ahora.

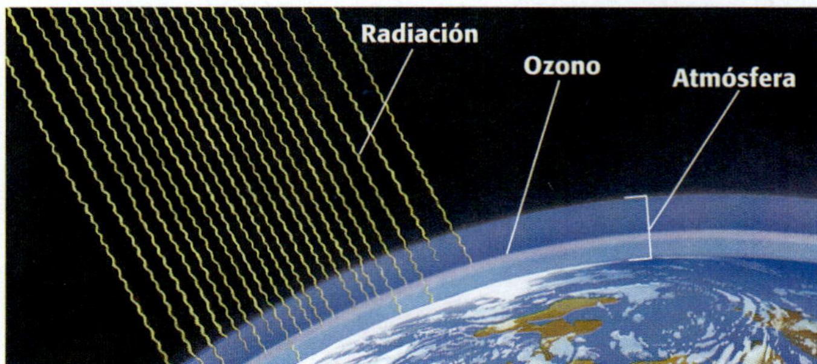

Figura 12 *El oxígeno producido en la fotosíntesis formó ozono, que absorbe la radiación ultravioleta.*

Al llenarse la atmósfera de oxígeno, algunas de sus moléculas se combinaron para formar una capa de ozono en su parte más alta, como se ve en la **Figura 12.** El **ozono** es un gas que absorbe la radiación ultravioleta del Sol. Esta radiación altera el ADN, pero se absorbe en el agua. Antes de que se formara la capa de ozono, todas las formas de vida estaban limitadas a los océanos. La capa de ozono bloqueaba la mayor parte de la radiación ultravioleta. El nivel de radiación en la superficie de la Tierra bajó lo suficiente como para permitir que los seres vivos se mudaran a tierra firme.

Después de, unos 2 mil millones de años aparecieron las primeras formas de vida complejas. Estos organismos, llamados **eucariotas,** son más grandes que los procariotas, y tienen núcleo y una estructura interna complicada. La **Figura 13** muestra el fósil de una célula eucariota. Los científicos piensan que en los últimos 1,500 millones de años estas células han evolucionado juntas para formar organismos de muchas células. Se cree que la primera célula eucariota evolucionó hace 2 mil millones de años, y es la antepasada de las plantas y animales que existen hoy. Los primeros organismos multicelulares fueron los *metazoarios,* que evolucionaron en el Precámbrico tardío.

210

MISCONCEPTION ALERT

Most of us think of Earth as an oxygen-rich planet. In actuality, only about 20 percent of Earth's present atmosphere is oxygen. The rest is made up mostly of nitrogen.

La era Paleozoica

La **era Paleozoica** comenzó hace 540 millones de años y terminó hace unos 248 millones de años. La palabra *paleozoico* viene del griego y significa "vida antigua". Comparada con el período Precámbrico, fue relativamente corta. Las rocas de esta era están llenas de fósiles de esponjas, corales, caracoles, ostras, calamares y trilobites. Los peces, que fueron los primeros animales vertebrados, aparecieron en esta era y abundaban los tiburones primitivos. La **Figura 14** muestra algunos organismos de la era Paleozoica.

La de la Tierra verde En la era Paleozoica, un período de 30 millones de años, las plantas, los hongos y los animales que respiraban aire colonizaron la tierra firme. Las plantas fueron alimento y refugio de los primeros animales terrestres. Hacia el final de esta era gran parte de la Tierra estaba cubierta por bosques de helechos gigantes, musgos, hierbas y coníferas. Todos los grupos principales de plantas aparecieron en esta era, con excepción de las plantas con flores.

Los bichos llegan a tierra firme Los fósiles indican que los insectos sin alas fueron algunos de los primeros animales en tierra firme. Les siguieron animales parecidos a las salamandras. Al final de la era Paleozoica aparecieron reptiles, insectos con alas, cucarachas y libélulas.

La mayor extinción masiva ocurrió a fines de la era Paleozoica, hace unos 248 millones de años. En ella se extinguieron el 90% de las especies marinas.

Figura 14 *Entre los organismos que aparecieron en la era Paleozoica están los primeros reptiles, anfibios, peces, gusanos y helechos.*

✔ Autoevaluación

Ordena los siguientes eventos cronológicamente:

a. Se forma la capa de ozono y los seres vivos pueblan la tierra.

b. Los gases de la atmósfera y los minerales de los océanos se combinan y forman moléculas.

c. Aparecen las células procariotas y anaeróbicas.

d. Aparecen las cianobacterias.

(Consulta la página 636 para comprobar tu respuesta.)

211

READING 📖 STRATEGY

Have students read the text on page 211. Have them write in their ScienceLog new headings for the first three paragraphs. For example, the heading for the first paragraph might be *Life Is Abundant in the Seas;* the heading for the second paragraph might be *Huge Forests of Ferns and Other Plants Develop;* and the heading for the third paragraph might be *Animals That Live on Land All the Time Appear.*

MATH and MORE

Using the premise that the Earth is 4.6 billion years old, have students calculate the length of the Earth's Precambrian and Paleozoic eras. They should use the information presented in the first paragraph on this page to determine the length of each of the eras (in millions of years). (Precambrian: 4,600 − 540 = 4,060, Paleozoic: 540 − 248 = 292)

Math Skills Worksheet 2 "Subtraction Review"

REAL-WORLD CONNECTION

The huge plants that grew in forests during the Paleozoic era later became coal. Ask students to research the locations of the world's coal deposits and mark them on a world map. (Most of the known coal reserves are in Australia, China, Germany, Poland, Great Britain, India, Russia, South Africa, the United States, and Canada.)

Homework

Comparing Organisms Have students compare the characteristics of each of the Paleozoic organisms pictured on this page with those of a descendant living today. Students can organize the information in the form of a chart.

Answer to Self-Check

b, c, d, a

READING  STRATEGY

Prediction Guide Before students read the text on these pages, ask them whether the following events occurred in the Mesozoic era, which began about 248 million years ago, or in the Cenozoic era, which began about 65 million years ago and continues today:

- Dinosaurs dominated Earth and then died out by the end of this era. (Mesozoic)
- Mammals appeared. (Mesozoic)
- Rock layers close to Earth's surface contain fossils from this era. (Cenozoic)
- The first birds appeared. (Mesozoic)
- Humans appeared. (Cenozoic)

RESEARCH

Writing Have students research reptiles and list their distinctive characteristics. (Reptiles are ectothermic; they have a distinctive type of heart; most reptiles have scales; many reptiles have claws; and most reptiles lay their eggs on land.)

Tell students that all living reptiles fall into four orders—turtles, lizards and snakes, crocodiles and related forms, and the tuatara. Have students investigate these orders and list examples of each.

Reinforcement Worksheet 9 "Condensed History"

La era Mesozoica

La **era Mesozoica** comenzó hace unos 248 millones de años y duró alrededor de 183 millones de años. La palabra *mesozoico* viene del griego y significa "vida media". Los científicos piensan que después de las extinciones de la era Paleozoica, los reptiles sobrevivientes pasaron por un período de gran evolución, que originó muchas especies diferentes. Por eso, se llama a esta era la Edad de los Reptiles.

La vida en la era Mesozoica Los dinosaurios son los reptiles más conocidos de la era Mesozoica. Los dinosaurios dominaron la Tierra durante más o menos 150 millones de años. (Imagínate: los seres humanos y sus antepasados llevamos menos de 4 millones de años en el planeta.) Los dinosaurios tenían una gran variedad de características físicas, como un pico de pato y una cresta puntiaguda en el lomo. Enormes lagartos marinos nadaban en los océanos. Las primeras aves también aparecieron en la era Mesozoica. Las plantas más importantes de la era Mesozoica temprana fueron plantas de semilla que producían conos, y que formaron grandes bosques. Las plantas con flores aparecieron más tarde. Algunos organismos de la era Mesozoica se muestran en la **Figura 15.**

Un mal rato para los dinosaurios Al final de la era Mesozoica, hace 65 millones de años, los dinosaurios y muchas otras especies de plantas y animales se extinguieron. ¿Qué les pasó a los dinosaurios? Según una de las hipótesis, un enorme meteorito chocó con la Tierra y levantó grandes nubes de polvo y suficiente calor como para provocar incendios en todo el planeta. El polvo y el humo de los incendios taparon buena parte de la luz solar y, por eso, muchas plantas murieron. Sin suficientes plantas para alimentarse, los dinosaurios herbívoros murieron, y, como resultado, los dinosaurios carnívoros que se alimentaban de los herbívoros también murieron. Puede que la temperatura global bajara durante muchos años y que sólo unos cuantos organismos, entre ellos algunos mamíferos, sobrevivieran.

Figura 15 *La era Mesozoica terminó con la extinción masiva de la muchos animales grandes, como el ankylosaurus y el plesiosaurs acuático que se ven arriba. Sobreviveron los mamíferos pequeños y el* Archaeopteryx.

Multicultural CONNECTION

Recently, a fossil plesiosaur was found in a riverside cliff in Hokkaido, Japan, with shellfish remains in its stomach. Scientists had long suspected that plesiosaurs were predators, but they had no proof. Geologist Tamaki Sato thinks that the plesiosaur swallowed its tiny prey whole because its long, sharp teeth were unsuitable for crushing hard outer shells.

La era Cenozoica

La **era Cenozoica** empezó hace 65 millones de años y aún hoy continua. La palabra *cenozoica* viene del griego y significa "vida reciente". Sabemos más de esta era que de las anteriores, porque sus fósiles están en capas de piedra cercanas a la superficie y son más fáciles de encontrar. En la era Cenozoica aparecieron muchos tipos de mamíferos, aves, insectos y plantas con flores. Algunos organismos que aparecieron en la era Cenozoica se muestran en la **Figura 16.**

El momento ideal para los grandes mamíferos A la era Cenozoica se le llama a veces la Edad de los Mamíferos. Los mamíferos dominaron la era Cenozoica como los reptiles dominaron la Mesozoica. Los mamíferos de la era Cenozoica temprana eran pequeños y vivían en los bosques. Los mamíferos grandes aparecieron más tarde. Algunos de estos últimos tenían patas largas para correr, dientes especializados según su tipo de alimentación, y cerebros grandes. Entre los mamíferos de la era Cenozoica están los mastodontes, el tigre dientes de sable, el camello, los perezosos gigantes y los caballos pequeños.

Figura 16 *Vivimos en la era Cenozoica, pero los seres humanos llevamos muy poco tiempo en la Tierra.*

REPASO

1. ¿Cuál es la principal diferencia entre la atmósfera de hace 3,500 millones de años y la atmósfera del presente?

2. ¿En qué se diferencian las células eucariotas de las procariotas?

3. Explica por qué las cianobacterias fueron tan importantes para el desarrollo de nuevas formas de vida.

4. **Identificar relaciones**   Une los organismos con el período en el que aparecieron.

   1. eucariotas
   2. dinosaurios
   3. peces
   4. plantas con flores
   5. aves

   a. período Precámbrico
   b. era Paleozoica
   c. era Mesozoica
   d. era Cenozoica

a través de las ciencias
CONEXIÓN

Normalmente, los fósiles se encuentran en rocas sedimentarias. Para ver cómo se forma esta roca, pasa a la página 224.

213

4) Close

Quiz

On index cards write the names of the organisms mentioned in the four geologic eras. Then on paper strips write the names of the geologic eras, and place the strips on a tabletop. Direct students to classify each organism named on a card by placing the card under the paper strip with the name of the appropriate geologic era.

ALTERNATIVE ASSESSMENT

Divide students into groups of four. Groups should use boxes with covers and art materials to make a diorama of each of the four geologic eras. Each group member should be responsible for a designated era. Sheltered English

GOING FURTHER

Making Models Dinosaurs varied greatly in size and appearance. Have groups of students consult reference books to find information about the many kinds of dinosaurs. Then have them use art materials to make models of different dinosaurs. Models should include flying and marine reptiles as well as land-dwelling reptiles. Sheltered English

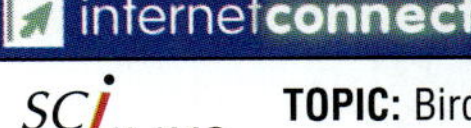

Interactive Explorations CD-ROM, "Rock On!"

internetconnect

SciLINKS
NSTA

TOPIC: Birds and Dinosaurs
GO TO: www.scilinks.org
sciLINKS NUMBER: HSTL200

▼ **Answers to Review**

1. Now we have enough atmospheric oxygen for life as we know it to exist.

2. Eukaryotes are usually much bigger, have a nucleus, and have a more complicated structure than do prokaryotes.

3. Cyanobacteria were the first significant source of atmospheric oxygen on the planet.

4. 1. a; 2. c; 3. b; 4. c ; 5. c

Focus

Human Evolution

In this section, students will learn that scientists think humans share a common ancestor and common characteristics with other primates, such as apes and monkeys. This section describes the characteristics of hominids and explains how trends in their evolution gave rise to modern humans.

Bellringer

On the board or an overhead projector, pose the following question to your students at the beginning of class:

What makes you unique among your family members? Please write the answer in your Science-Log. (Responses might include references to food preferences, physical appearances, and talents.)

Point out that understanding evolution requires recognizing similarities and differences, like those seen in families.

1 Motivate

DISCUSSION

Comparing Primates Display a picture of an ape and a picture of a human for students to compare. Have students identify characteristics that the two animals have in common. (Most likely answers will include references to common physical characteristics.)

Then ask students how the two animals are different from each other. (Most likely answers will include references to differences in intellectual ability.)

Sección 3

La evolución humana

VOCABULARIO

primate   australopiteco
prosimio   Neandertal
homínido   Cro-Magnon

OBJETIVOS

- Habla sobre las características que tienen en común los primates.
- Describe lo que conocemos sobre las diferencias entre los homínidos.

Al principio de este capítulo leíste sobre la cultura Cro-Magnon, que creó unas bellas y misteriosas pinturas rupestres hace más de 17,000 años. Gracias a los fósiles, sabemos que se parecían a nosotros. También se han encontrado otros fósiles más antiguos de seres muy parecidos a los humanos.

Tras estudiar miles de esqueletos fosilizados y otras pruebas, los científicos desarrollaron la teoría de que los seres humanos evolucionaron hace millones de años a partir de un antepasado común con el mono y los simios. Se cree que este antepasado común vivió hace más de 30 millones de años. Esta sección presenta algunas pruebas que se han reunido hasta ahora para explicar nuestra evolución.

Figura 17 Las características del orangután, el gorila y el chimpancé los hacen primates no humanos, como el dedo gordo del pie, ¡que es oponible!

Primates

Para estudiar la evolución humana, primero necesitas entender las características que nos hacen humanos. Los seres humanos están clasificados como primates. Los **primates** son un grupo de mamíferos que incluye también a los simios, monos y prosimios. Los primates tienen las características que se ilustran abajo, en la **Figura 17.**

Características de los primates

La mayoría de los primates tienen cuatro dedos flexibles y un pulgar oponible.

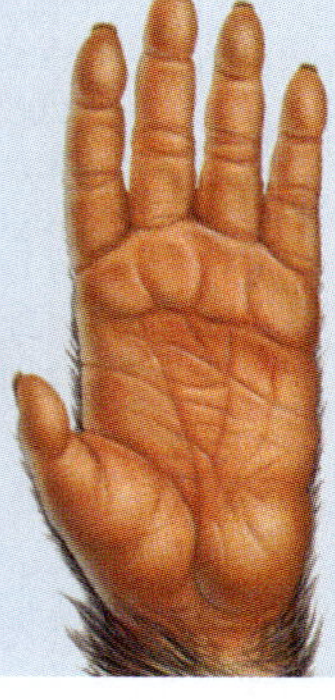

El pulgar oponible les permite agarrar objetos. Excepto el ser humano, todos tienen dedo gordo oponible en los pies.

Los ojos están en la parte frontal de la cabeza, lo que les da visión **binocular,** es decir, en tres dimensiones. Cada ojo ve una imagen ligeramente distinta de la misma escena; mientras, el cerebro mezcla las dos imágenes para crear una imagen tridimensional.

214

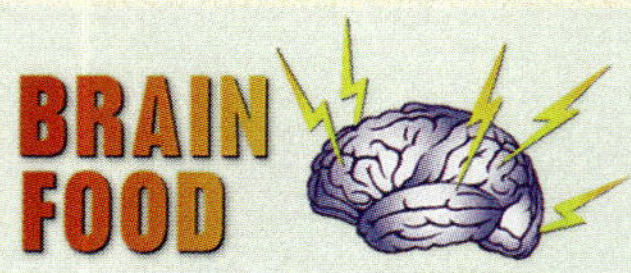

BRAIN FOOD

Although the skulls of a human and a chimpanzee appear to be very similar, there are significant differences. The cranium of a human skull is domed, whereas the chimpanzee's cranium is flatter. Also, the canine teeth in the human skull do not overlap, as they do in the chimpanzee skull. Ask students what conclusions they might make about the brain and the chewing ability of a human and of a chimpanzee based on these differences. (A human brain is larger than a chimp's brain. A chimp can't easily move its jaws from side to side when chewing.)

De acuerdo con el parecido físico y genético, el pariente vivo más cercano del ser humano es el chimpancé. Esto no quiere decir que el ser humano descienda del chimpancé. Más bien, quiere decir que los humanos y los chimpancés tienen un antepasado común. Se piensa que el antepasado de los humanos se separó del antepasado del chimpancé hace unos 7 millones de años. Desde entonces, los humanos y los chimpancés han evolucionado por caminos distintos.

A pesar del parecido entre humanos, simios y monos, los humanos se clasifican en una familia distinta llamada **homínidos.** La palabra **homínido** se refiere específicamente a los seres humanos y sus antepasados. La principal característica de los homínidos es que utilizan su capacidad de caminar erguidos sobre dos piernas como la forma principal para moverse. La capacidad de caminar sobre dos piernas se llama *bipedismo.* Examina la **Figura 18** para observar las similitudes y diferencias entre el esqueleto de los homínidos y el de otros primates. A excepción del ser humano moderno, los demás homínidos se extinguieron.

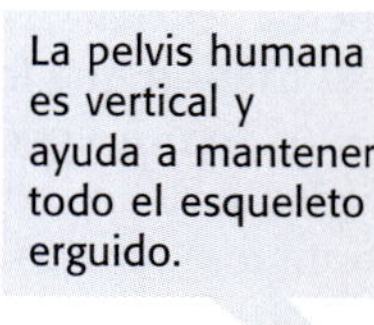

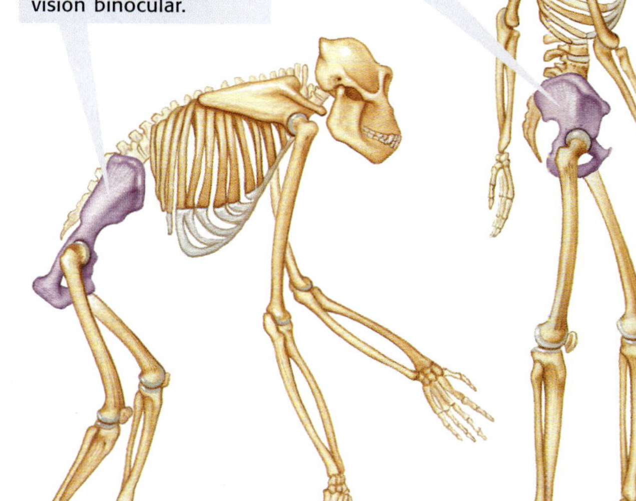

Figura 18 *Los huesos del gorila y el ser humano son básicamente iguales en forma y función, pero la pelvis de los seres humanos está diseñada para dar apoyo a los órganos internos al caminar erguidos.*

215

CROSS-DISCIPLINARY FOCUS

Anthropology The ability to walk fully upright distinguishes us from the apes. So do our large rear ends. But are they a prerequisite or a consequence of upright posture? Thomas Greiner, a physical anthropologist, believes the latter is true. Greiner developed a computer model that shows how muscle action changes in response to changes in bone shape. By means of his model, Greiner concluded that having both a smaller gluteus maximus and a larger ilium than a human hinders an ape's ability to walk upright.

History Many cartoons in the nineteenth century satirized humans' relationship to apes. In one such cartoon, Henry Bergh, the founder of the Society for the Prevention of Cruelty to Animals, chides Charles Darwin for insulting apes by suggesting that they are related to humans.

USING THE FIGURE

Discuss the characteristics of the lemur shown in **Figure 19,** How might each characteristic help the lemur survive in its environment? (Students might suggest that the lemur's fur helps keep it warm, that its tail aids in balance, and so on.) Sheltered English

MEETING INDIVIDUAL NEEDS

Learners Having Difficulty
Help students identify the characteristics that distinguish primates from other mammal groups. Show them pictures of primate and nonprimate mammals. Ask how the primates are different from the other animals. Help students determine that primates generally have flatter faces than nonprimates. Their eyes are located at the front of the head rather than at the sides, their snouts are small, and their fingers are flexible.

Figura 19 *Los prosimios, como este lémur, cazan insectos y otros animales pequeños en los árboles.*

Figura 20 *Mary Leakey aparece aquí con las huellas que descubrió, que tienen una antigüedad de 3.6 millones de años.*

216

Evolución de los homínidos

Los primeros antepasados primates aparecieron en la era Cenozoica, hace 55 millones de años, y evolucionaron en varias direcciones. Se cree que estos ancestros eran mamíferos parecidos a los ratones, que eran nocturnos, vivían en los árboles y comían insectos. Cuando los dinosaurios se extinguieron, nuestros antepasados sobrevivieron y de ellos evolucionaron los primeros primates, llamados **prosimios,** que quiere decir "antes de los monos". Hoy sólo sobreviven algunas especies, como los lémures, que puedes ver en la **Figura 19.** ¿Cuánto tiempo pasó desde que aparecieron los prosimios hasta que apareció el primer homínido? Nadie ha contestado esta pregunta, los paleontólogos han descubierto huesos fósiles de homínidos de hace 4.4 millones de años.

Australopitecos Los paleontólogos piensan que la evolución de los homínidos comenzó en África. Entre los homínidos más antiguos está el **australopiteco**. A*ustralopiteco* quiere decir "hombre mono del Sur." Estos primeros homínidos tenían brazos largos, piernas cortas y cerebros pequeños. Los rastros fósiles indican que los australopitecos eran distintos de los simios en varios puntos importantes: eran bípedos y sus cerebros eran más grandes que los de los simios (aunque eran mucho más pequeños que los de los seres humanos actuales).

En 1976, la paleontóloga Mary Leakey descubrió una serie de huellas fósiles en Tanzania. Como se muestra en la **Figura 20.** Al calcular la edad de las rocas que contenían las huellas, descrubió que tenían más de 3.6 millones de años. Las huellas indicaban que un grupo de tres homínidos había pasado caminando, con el cuerpo erguido, sobre un llano cubierto de cenizas volcánicas mojadas. Sus huellas fosilizadas son una prueba adicional del bipedismo en los antepasados del ser humano. Tras examinar las huellas que encontró en Laetoli, Tanzania, Mary Leakey observó que los caminantes se habían detenido de pronto. Mary Leakey escribió: "Este movimiento (la pausa), es muy humano. Trasciende el tiempo. Nuestro antepasado remoto había experimentado, como cualquiera de nosotros, un momento de duda."

SCIENTISTS AT ODDS

In 1975, fossils of 13 hominids were found in Ethiopia. These fossils differed in body size and jaw shape. Some anthropologists think that the larger fossils represent the males and the smaller fossils represent the females of a particular species. Other anthropologists, including Mary Leakey, believe that the differences indicate that the fossils are of two distinct species.

Lucy En 1979, se encontró un grupo de fósiles en Etiopía. En este grupo estaba el esqueleto más completo de un australopiteco que se ha encontrado hasta la fecha. Este australopiteco, apodado Lucy, vivió hace más de 2 millones de años. La **Figura 21** muestra su esqueleto. Lucy tenía un cuerpo fuerte y caminaba erguida, pero su cerebro era más o menos del tamaño del de un chimpancé. Fósiles como éste demuestran que la postura vertical evolucionó mucho antes de que el cerebro se hiciera más grande.

Un rostro como el nuestro Hace cerca de 2.3 millones de años aparecieron homínidos con facciones más humanas, que probablemente evolucionaron a partir de los australopitecos. Esta especie se conoce como *Homo habilis.* Su cráneo aparece en la **Figura 22.** Los fósiles de *Homo habilis* se han encontrado junto con toscas herramientas de piedra. Se cree que hacían y usaban herramientas. Hace 2 millones de años, *Homo habilis* fue reemplazado por su descendiente, *Homo erectus,* que tenía un cerebro más grande y aparece en la **Figura 23.** El *Homo erectus* era más grande y tenía un cráneo grueso, cejas prominentes, la frente baja y una barbilla muy pequeña.

Los fósiles indican que *el Homo erectus* probablemente vivía en cuevas, hacía fogatas y usaba ropa. Podía cazar animales grandes y descuartizarlos con herramientas hechas de hueso y piedra. Su aparición marca el principio de la expansión de los seres humanos por el mundo. *El Homo erectus* sobrevivió más de 1 millón de años, más que ninguna otra especie humana, incluídos nosotros. Estos humanos tan adaptables desaparecieron hace unos 200,000 años, cuando el ser humano moderno, u *Homo sapiens,* aparece por primera vez en el registro fósil.

Aunque el *Homo erectus* viajó por todo el mundo, se piensa que *el Homo sapiens* evolucionó en África y luego emigró a Asia y Europa.

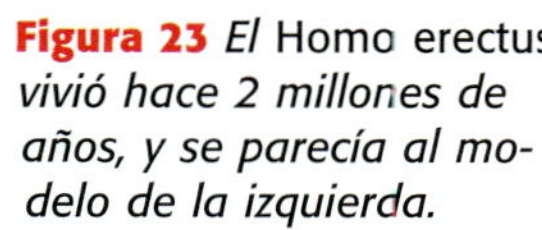

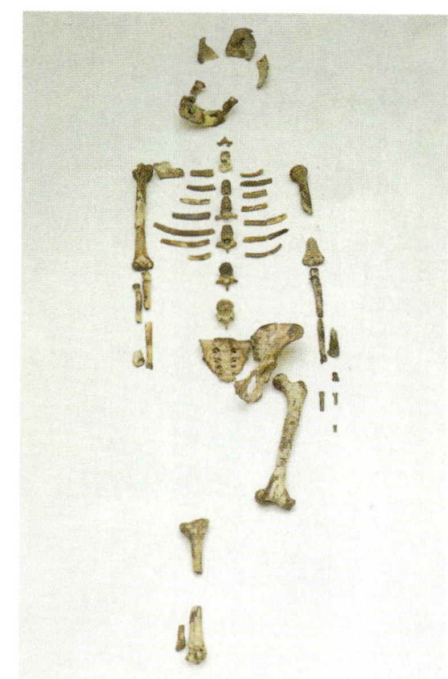

Figura 21 *La pelvis de Lucy indica que caminaba erguida.*

Figura 22 *El* Homo habilis *fabricaba herramientas de piedra.*

Figura 23 *El* Homo erectus *vivió hace 2 millones de años, y se parecía al modelo de la izquierda.*

217

3 Extend

GOING FURTHER

Have students use the description of Cro-Magnons on this page as a guide for drawing the head of a Cro-Magnon with facial features. Sheltered English

GROUP ACTIVITY

Small groups of students can work together to write a play about hunting and killing a mammoth. Tell students that Cro-Magnons hunted together and that a mammoth kill was a group endeavor. How did they do it? What weapons might they have used? How did they carve the meat? What did they do with the skin, tusks, and bones? Did the hunt include women? Have each group present its play to the class.

INDEPENDENT PRACTICE

Poster Project
Provide students with markers and poster board. Have them construct evolutionary trees of humans with whole-body sketches.

Answer to QuickLab

Answers should include spearpoints, arrowheads, choppers, skinners, or scrapers.

Figura 24 *Los Neandertales tenían cejas prominentes, como el Homo erectus,* pero un cerebro más grande que los humanos modernos.

Laboratorio

Las herramientas del Neandertal.

Los Neandertales hacían puntas de lanza y otras herramientas de piedra como las de abajo. Cada una tenía un uso especial. ¿Tienes idea sobre el uso que se le daba a cada una?

Neandertales En el valle de Neander, en Alemania, se descubrieron fósiles que pertenecían a un grupo de homínidos a los que se llamó **Neandertales.** Los Neandertales vivieron en Europa y Asia occidental hace alrededor de 230,000 años. Fuertes y de baja estatura, los Neandertales tenían cejas prominentes, como los simios, y cerebros más grandes que los de los seres humanos modernos.

Al investigar los campamentos de Neandertales, los científicos han aprendido que cazaban animales grandes, hacían fogatas y usaban ropa. Hay pruebas que indican que cuidaban a sus enfermos y ancianos y enterraban a sus muertos, colocando ofrendas de alimentos, armas y quizá flores junto a los cuerpos. La **Figura 24** es una interpretación artística de cómo era una Neandertal. Hace aproximadamente 30,000 años, los Neandertales desaparecieron, nadie sabe por qué. Algunos científicos opinan que los Neandertales son una especie aparte (*Homo neanderthalensis*) de la especie humana moderna (*Homo sapiens*). Otros piensan que los Neandertales son una raza de *Homo sapiens*. No hay suficientes pruebas para responder satisfactoriamente a esta pregunta.

Cro-Magnon En 1868, se encontraron cráneos fosilizados en unas cuevas del suroeste de Francia. Los cráneos tenían 35,000 años de antigüedad y pertenecían a un grupo de *Homo sapiens* con facciones modernas, llamados hombres de **Cro-Magnon.** Es posible que los hombres de Cro-Magnon estuvieran en África desde hace 100,000 años, emigraran de allí hace 40,000 y vivieran con los Neandertales. Comparados con ellos, los hombres de Cro-Magnon tenían una cara más pequeña y más plana, y su cabeza era más redonda, como ves en la **Figura 25.** La única diferencia significativa entre los hombres de Cro-Magnon y los humanos modernos es que los primeros tenían los huesos más gruesos y pesados.

Figura 25 *Esta es una recreación artística del aspecto de una mujer de Cro-Magnon.*

Science BlOopers

A skull of a *Homo sapiens* who had bad teeth was found in Zambia. There was a hole in one side and signs of a partially healed abscess. This skull was made famous by a writer who imagined the hole was caused by a bullet shot from an interplanetary visitor's gun 120,000 years ago.

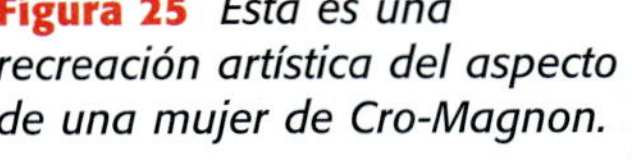

La introducción a este capítulo hablaba sobre las pinturas rupestres que hicieron los hombres de Cro-Magnon. Estas pinturas son los primeros ejemplos de arte humano: su cultura se caracteriza por la diversidad de actividades artísticas, como la pintura, la escultura y los grabados, como muestra la **Figura 26.** Las aldeas y los cementerios de los hombres de Cro-Magnon también indican que tenían una organización social compleja.

Figura 26
Los hombres de Cro-Magnon dejaron muchos tipos distintos de pinturas y esculturas, como ésta, que representa un toro.

Pruebas de la evolución humana

Aunque sabemos mucho sobre nuestros antepasados, todavía hay mucho por descubrir. Cada vez que se descubre un nuevo fósil, surgen muchas preguntas, como: "¿Dónde evolucionó el *Homo sapiens*?" Las pruebas parecen indicar que en África. "¿A partir de qué australopiteco evolucionaron los seres humanos?" Algunos científicos creen que del *Australopithecus afarensis*es, el antepasado de todos los homínidos. La **Figura 27** muestra dos interpretaciones diferentes de la evolución de los homínidos.

REPASO

1. Identifica tres características de los primates.

2. Compara el *Homo habilis* con el *Homo erectus*. ¿Qué diferencia a una especie de la otra?

3. ¿Qué pruebas existen de que los Neandertales se parecían a los seres humanos modernos?

4. **Saca conclusiones** Imagínate que estudias paleontología y que estás excavando un antiguo campamento. ¿Qué concluyes sobre los habitantes del lugar si encuentras huesos de animales grandes quemados y varios cuchillos de piedra entre los fósiles humanos?

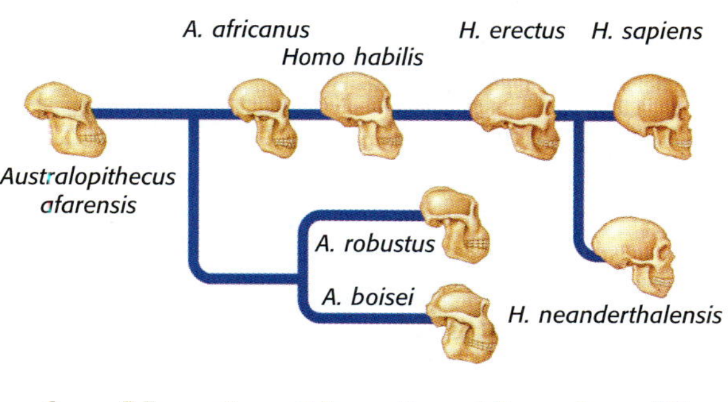

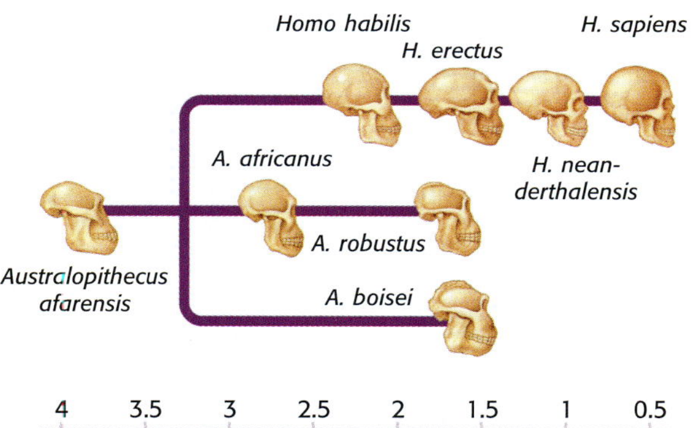

Figura 27 Estos árboles evolutivos representan dos hipótesis de cómo el Homo sapiens desciende del australopiteco.

219

Chapter Highlights

VOCABULARY DEFINITIONS

SECTION 1

paleontologist a scientist who uses fossils to reconstruct the history of life millions of years before humans existed

fossil the solidified remains or imprints of once living organisms

rock cycle the continuous process by which one rock type changes into another rock type

relative dating determining whether an event or object, such as a fossil, is older or younger than other events or objects

absolute dating estimating the age of a sample or event in years, usually by measuring the amount of unstable atoms in the sample

half-life for a particular radioactive sample, the time it takes for one-half the sample to decay

geologic time scale the division of Earth's 4.6-billion-year history into distinct intervals of time

extinct a species of organism that has died out completely

mass extinction a period in Earth's history when a large number of species died out at the same time

Pangaea the single landmass that contained all the present-day continents 200 million years ago

plate tectonics the study of the forces that drive the movement of pieces of the Earth's crust around the surface of the planet

SECTION 2

Precambrian time the period in the geologic time scale beginning when the Earth originated 4.6 billion years ago and continuing until complex organisms appear about 540 million years ago

prokaryote a cell that lacks a nucleus or any other membrane-covered organelles; also called a bacterium

anaerobic without oxygen

Resumen del capítulo

SECCIÓN 1

Vocabulario

paleontólogo *(pág. 202)*
fósil *(pág. 202)*
ciclo de las rocas *(pág. 202)*
edad relativa *(pág. 203)*
edad absoluta *(pág. 203)*
vida media *(pág. 203)*
escala del tiempo geológico *(pág. 204)*
extinto *(pág. 205)*
extinción masiva *(pág. 205)*
Pangea *(pág. 206)*
tectónica de placas *(pág. 207)*

Notas de la sección

- Los paleontólogos son científicos que estudian fósiles.
- La edad de un fósil puede calcularse a partir de la edad relativa y la edad absoluta. La edad relativa se calcula a partir de la edad de la capa de sedimentos en la que se encuentra el fósil. La edad absoluta mide la descomposición de elementos inestables que se encuentran en las rocas alrededor del fósil.
- La escala de tiempo geológico es el calendario que utilizan los científicos para describir la historia de la Tierra y de la vida en nuestro planeta.
- Muchas especies vivieron unos cuantos millones de años y después se extinguieron. En la historia de la Tierra han sucedido varias extinciones masivas.

Experimentos

La edad del registro fósil *(pág. 588)*

La vida media de un centavo *(pág. 591)*

SECCIÓN 2

Vocabulario

período Precámbrico *(pág. 208)*
procariota *(pág. 209)*
anaeróbico *(pág. 209)*
ozono *(pág. 210)*
eucariota *(pág. 210)*
era Paleozoica *(pág. 211)*
era Mesozoica *(pág. 212)*
era Cenozoica *(pág. 213)*

Notas de la sección

- El período Precámbrico abarca la formación de la Tierra, el principio de la vida y la evolución de organismos multicelulares sencillos.

✔ Comprobar destrezas

Conceptos de matemáticas

VIDA MEDIA Para que entiendas mejor lo que quiere decir vida media, imagínate que tienes $10.00 en el bolsillo y que has decidido gastar la mitad del dinero que tienes cada 30 minutos. ¿Cuánto te quedaría después de 30 minutos? ($5.00) ¿Cuánto te quedaría después de otros 30 minutos? ($2.50) ¿Cuánto te quedaría al cabo de 3 horas? (poco más de 15¢).

Comprensión visual

LA ESCALA DE TIEMPO GEOLÓGICO Seguramente has visto películas o caricaturas viejas en las que los dinosaurios y los seres humanos viven en el mismo ambiente. ¿Crees que esto es posible? Los dinosaurios y los seres humanos no vivieron en la misma época. Los dinosaurios se extinguieron hace 65 millones de años y los humanos y sus antepasados llevan menos de 4 millones de años sobre la Tierra. Mira la escala de tiempo geológico en la página 204.

Lab and Activity Highlights

Dating the Fossil Record **PG 588**

The Half-life of Pennies **PG 591**

Datasheets for LabBook (blackline masters for these labs)

SECCIÓN 2

- La edad de la Tierra es de unos 4,600 millones de años. La vida se formó a partir de materia inerte.

- Las primeras células, las procariotas, eran anaeróbicas. Luego, las cianobacterias, o "algas azul verdosas", fotosintéticas introdujeron oxígeno en la atmósfera.

- En la era Paleozoica aparecieron animales en el mar, y las plantas y los animales colonizaron la tierra.

- Los dinosaurios y otros reptiles poblaron la tierra en la era Mesozoica. Aparecieron las plantas con flores, los pájaros y los primeros mamíferos.

- Los primates evolucionaron en la era Cenozoica, que continúa hasta el presente.

SECCIÓN 3

Vocabulario

primate *(pág. 214)*
homínido *(pág. 215)*
prosimio *(pág. 216)*
australopiteco *(pág. 216)*
Neandertal *(pág. 218)*
Cro-Magnon *(pág. 218)*

Notas de la sección

- Los seres humanos y los simios monos son primates. Los primates se distinguen de otros mamíferos porque tienen pulgares oponibles y visión binocular.

- Los homínidos, un subgrupo de los primates, incluyen a los humanos y sus antepasados. El homínido conocido más antiguo es el australopiteco.

- Los homínidos más parecidos a los humanos fueron el *Homo habilis,* el *Homo erectus* y el *Homo sapiens.*

- Los Neandertales son una especie de homínido que desapareció hace unos 30,000 años.

- La cultura del Cro-Magnon era muy avanzada. No eran muy distintos de los seres humanos actuales.

ozone a gas molecule that is made up of three oxygen atoms and that absorbs ultraviolet radiation from the sun

eukaryote cells that contain a central nucleus and a complicated internal structure

Paleozoic era the period in the geologic time scale beginning about 570 million years ago and ending about 248 million years ago

Mesozoic era the period in the geologic time scale beginning about 248 million years ago and lasting about 183 million years

Cenozoic era the period in the geologic time scale beginning about 65 million years ago and continuing until the present day

SECTION 3

primate a group of mammals that includes humans, apes, and monkeys; distinguished by opposable thumbs and binocular vision

hominid a family of humans and several extinct humanlike species, some of which were human ancestors

prosimian the first primate ancestors; also a group of living primates that includes lorises and lemurs

australopithecine an early hominid that may have lived more than 3.6 million years ago

Neanderthal a species of hominid that lived in Europe and western Asia from 230,000 years ago to about 30,000 years ago, when they mysteriously went extinct

Cro-Magnon a species of humans with modern features that may have migrated out of Africa 100,000 years ago and eventually into every continent

---

 internet

HRW VISITA: go.hrw.com

Visita el sitio web de HRW para encontrar una serie de herramientas de aprendizaje relacionadas con este capítulo. Sólo tienes que escribir la palabra clave:

PALABRA CLAVE: HSTHIS

SC/LINKS NSTA   VISITA: www.scilinks.org

Visita el sitio web de la **Asociación Nacional de Maestros de Ciencias** (*National Science Teachers Association*) para encontrar recursos de Internet relacionados con este capítulo. Sólo escribe el **ENLACE DE CIENCIAS:**

TEMA: Pruebas del pasado	**ENLACE:** HSTL195
TEMA: Extinciones masivas	**ENLACE:** HSTL200
TEMA: Tiempo geologico	**ENLACE:** HSTL180
TEMA: La evolución humana	**ENLACE:** HSTL185
TEMA: Aves y dinosaurios	**ENLACE:** HSTL190

221

Lab and Activity Highlights

LabBank

Long-Term Projects & Research Ideas, Project 8

Interactive Explorations CD-ROM

CD 2, Exploration 6, "Rock On!"

Vocabulary Review Worksheet 9

Blackline masters of these Chapter Highlights can be found in the **Study Guide.**

Chapter Review Answers

USING VOCABULARY

1. Precambrian time
2. Cenozoic era
3. Mesozoic era
4. Paleozoic era
5. australopithecines

UNDERSTANDING CONCEPTS

Multiple Choice

6. b
7. c
8. b
9. b
10. c

Short Answer

11. Fossils tell us about the kinds of organisms that existed and how they changed over time.
12. Precambrian: life begins, prokaryotes and eukaryotes appear; Paleozoic: multicellular organisms, plants, insects, and amphibians appear; Mesozoic: dinosaurs and other reptiles, birds, and small mammals appear; Cenozoic: large mammals, including humans, and more-diverse birds and insects appear
13. Fossils in the Cenozoic era are closer to the surface of Earth and thus are easier to find.

Repaso del capítulo

UTILIZAR EL VOCABULARIO

Escoge el término correcto para completar las siguientes oraciones:

1. En __?__ de la historia de la Tierra la vida se originó a partir de materia inerte. *(el período Precámbrico o la era Paleozoica)*

2. La Edad de los Mamíferos se refiere a __?__. *(la era Mesozoica o la era Cenozoica)*

3. La Edad de los Reptiles se refiere a __?__. *(la era Paleozoica o la era Mesozoica)*

4. Las plantas colonizaron la Tierra en __?__. *(el período Precámbrico o la era Paleozoica)*

5. El homínido más antiguo se llama __?__. *(hombre de Neandertal o australopiteco)*

COMPRENDER CONCEPTOS

Opción múltiple

6. Se calcula que la edad de la Tierra es
   a. 10 mil millones de años.
   b. 4,600 millones de años.
   c. 3,800 millones de años.
   d. 4.4 millones de años.

7. Las células surgieron hace más de
   a. 10 mil millones de años.
   b. 4,600 millones de años.
   c. 3,500 millones de años.
   d. 4.4 millones de años.

8. ¿Cómo se calcula la edad de un fósil?
   a. usando la escala de tiempo geológico
   b. midiendo los elementos inestables en la roca en la que se encuentra el fósil
   c. estudiando la posición relativa de los continentes
   d. midiendo oxígeno de la roca fósil

9. Las plantas y los animales que respiran aire aparecieron en este período:
   a. período Precámbrico
   b. era Paleozoica
   c. era Mesozoica
   d. era Cenozoica

10. Estos homínidos fabricaban herramientas, cazaban animales grandes, usaban ropa y cuidaban a sus enfermos y ancianos. No se sabe por qué se extinguieron.
    a. los australopitecos
    b. los homínidos del género *Homo*
    c. los hombres de Neandertal
    d. los hombres de Cro-Magnon

Respuesta breve

11. ¿Qué tipo de información nos dan los fósiles sobre la evolución de la vida?

12. Menciona por lo menos un evento biológico importante que haya sucedido en estas eras geológicas: período Precámbrico, era Paleozoica, era Mesozoica y era Cenozoica.

13. ¿Por qué se encuentran más fósiles de la era Cenozoica que de cualquier otra era?

Organizar conceptos

14. Usa los siguientes términos para crear un mapa de ideas: historia de la Tierra, era Paleozoica, dinosaurios, período Precámbrico, cianobacterias, era Mesozoica, plantas terrestres, era Cenozoica.

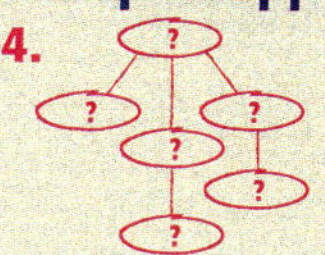

RAZONAMIENTO CRÍTICO Y RESOLUCIÓN DE PROBLEMAS

Escribe una o dos oraciones para responder a las siguientes preguntas:

15. ¿Por qué piensan los científicos que las primeras células eran anaeróbicas?

16. Nombra tres cambios evolutivos en los primeros homínidos que permitieron la evolución de los seres humanos modernos.

LAS MATEMÁTICAS EN LAS CIENCIAS

17. Tras analizar una roca en la que se ha encontrado un fósil, se descubre que contiene 5 g de una forma inestable de potasio y 5 g del elemento estable formado a partir de su desintegración. Si la vida media del potasio inestable es de 1,300 millones de años, ¿qué edad tiene la roca?, ¿qué puedes deducir de la edad del fósil?

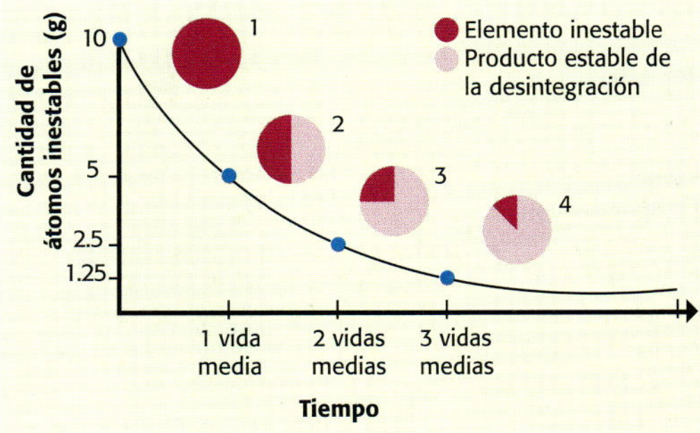

INTERPRETAR GRÁFICAS

La siguiente ilustración muestra la relación evolutiva entre distintos grupos de primates. Cuanto más abajo se separe una línea del tronco principal, más antiguo es el evento. Examina esta gráfica y responde a las siguientes preguntas.

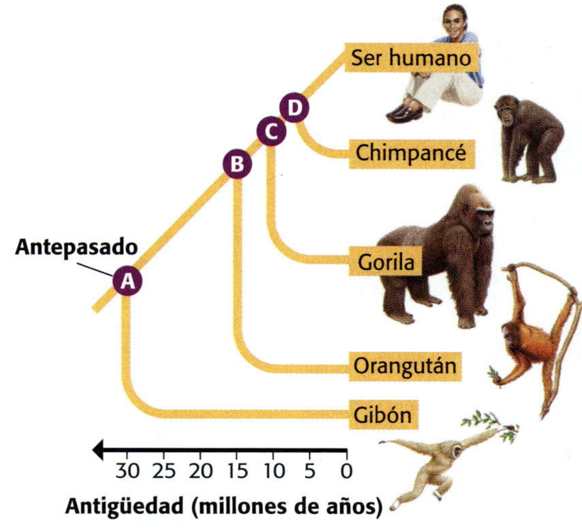

18. ¿Qué letra representa el momento en el que los seres humanos y los gorilas siguieron distintos caminos evolutivos?

19. ¿Hace aproximadamente cuántos millones de años que los orangutanes se separaron de nuestra línea evolutiva?

20. ¿Qué grupo de antropoides lleva más tiempo separado de la línea evolutiva humana?

AHORA, ¿qué piensas?

Revisa tus respuestas a las preguntas de la página 201 que escribiste en el cuaderno de ciencias. ¿Han cambiado tus respuestas? Si es necesario, corrige tus respuestas basándote en lo que has aprendido en este capítulo.

223

Concept Mapping

14.  An answer to this exercise can be found at the end of this book.

CRITICAL THINKING AND PROBLEM SOLVING

15. There was no oxygen available when they developed.
16. larger brains, longer legs, shorter arms

MATH IN SCIENCE

17. 1.3 billion years old; the fossil is also about 1.3 billion years old.

INTERPRETING GRAPHICS

18. point C
19. 15 million years ago
20. the gibbons

NOW WHAT DO YOU THINK?

1. by finding out where it was in relation to other fossils and by measuring the percentage of radioactive atoms in the sample
2. 150 million years
3. Humans walk upright and use language.

Concept Mapping Transparency 9

Blackline masters of this Chapter Review can be found in the **Study Guide**.

Background

Unlike the sun and the giant gas planets of the outer solar system, Earth and its three nearest planetary neighbors, Mercury, Venus, and Mars, are solid bodies with surfaces of rock. For this reason they are known as terrestrial planets. But even though the Earth has a solid, durable surface, it is constantly changing because of forces acting at and below its surface.

Rock is any naturally occurring solid mixture of minerals and other materials. Minerals, in turn, are any naturally occurring, inorganic chemical compound found in the Earth. Because there are so many minerals in the Earth's crust, there are also many types of rock. Rocks are broadly classified according to how they are formed.

Although all three rock types are found on the Earth's surface, most of the Earth's crust consists of igneous rock. We don't often see these igneous rocks because most of the continents are covered by a relatively thin layer of sedimentary rock. Only in a few places have natural processes brought the "basement rocks" to the Earth's surface, where we can see them.

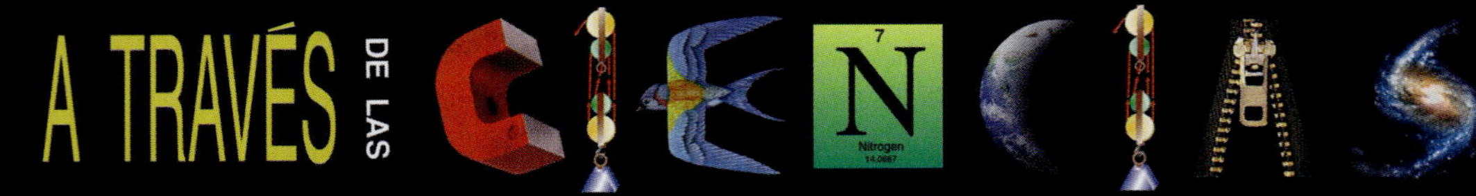

CIENCIAS BIOLÓGICAS • CIENCIAS DE LA TIERRA

Ventanas al pasado

Al pensar en la historia de la vida en la Tierra, las rocas seguramente ni te pasan por la cabeza. A fin de cuentas, no son seres vivos. ¿Qué tienen que ver con la vida en la Tierra? Te puede parecer extraño, pero mucho de lo que sabemos sobre la vida en la Tierra lo hemos aprendido de las rocas. ¿Cómo? Muy sencillo: resulta que durante millones de años posiblemente desde que la vida apareció en la Tierra los seres vivos se han quedado fosilizados entre capas de roca, y cuando encontramos uno de estos fósiles es como si encontráramos una fotografía de estos antiguos organismos.

Capas de piedra

Los fósiles casi siempre se encuentran en un tipo de roca llamado roca sedimentaria. Las rocas sedimentarias se forman cuando la superficie expuesta de la roca se desgasta por el viento, la lluvia y el hielo. Las partículas que se desprenden de la superficie de la roca se acumulan en las áreas bajas. Al amontonar las capas de partículas, su peso las compacta y se cementan gracias a reacciones químicas. Después de miles de años, las capas se convierten en roca sólida junto con los organismos que quedan atrapados entre las capas.

El ciclo de las rocas

La ilustración de la derecha muestra cómo se forman las rocas sedimentarias, las rocas ígneas y las metamórficas. Fíjate que las rocas sedimentarias y las metamórficas pueden derretirse y convertirse en rocas ígneas, lo cual normalmente ocurre en la profundidad de la Tierra. ¿Te imaginas por qué los fósiles no se encuentran normalmente en las rocas ígneas?

El ciclo de las rocas es un proceso continuo. Cualquiera de los tres tipos de roca termina convirtiéndose en otro tipo de roca. Los científicos que estudian la vida tienen suerte de que este proceso tarde millones de años. Si el proceso fuera más rápido y las rocas sedimentarias se convirtieran en metamórficas o ígneas a un ritmo más rápido, el registro fósil que tenemos sería mucho más reducido. ¡Quizás ni siquiera habríamos descubierto los dinosaurios!

¡Qué ciclo!

▶ Imagínate que encuentras varios fósiles del mismo organismo. Algunos estaban en capas profundas de rocas sedimentarias y otros en capas de rocas sedimentarias cerca de la superficie. ¿Qué te dice esto sobre el organismo?

▼ *El ciclo de las rocas*

224

Answer to Cycle This!

If fossils of the same type of organism are found in both deep and shallow layers of sedimentary rock, this probably indicates that the organism existed for the time span that the sedimentary rock containing the fossils was forming.

PROFESIONES

Cuando **Bonnie Jacobs** iba a la escuela, le fascinaban los fósiles, las culturas antiguas y la geología. "Siempre me han interesado las cosas antiguas", dice. Jacobs siguió su vocación y se convirtió en paleobotánica. "Un paleobotánico estudia los fósiles de plantas", explica Jacobs "Estudiamos fósiles de hojas, madera, polen, partes de la flor o cualquier otra parte que venga de una planta".

Bonnie Jacobs es profesora e investigadora en la Universidad Metodista del Sur en Dallas, Texas. Es paleobotánica y puede "ver" el pasado con "fotografías" especiales. Si te fijas en estas fotografías, puedes ver un pastizal antiguo, un desierto o una selva. Hasta podrías ver el lugar donde pudo haberse originado la raza humana.

Plantas fósiles y climas antiguos

Jacobs y otros paleobotánicos estudian las plantas actuales y cómo crecen en climas distintos. Es posible que las plantas que crecen en climas calientes y húmedos crecieran en el mismo clima hace millones de años. Así, cuando Jacobs encuentra el fósil de una planta que se parece a una plantas moderna, puede determinar cómo era el clima antiguo. Pero los fósiles no sólo te dan el reporte del clima.

Plantas y huesos antiguos

Algunos de estos fósiles de plantas se encuentran en capas de roca que también contienen huesos y, por lo tanto, pueden darnos pistas sobre la historia de la humanidad. "Las ideas sobre las causas de la evolución de la especie humana tienen mucho que ver con el medio ambiente", explica Jacobs. "Por ejemplo, muchos científicos piensan que justo antes de que surgiera la familia humana hubo un gran cambio en el medio ambiente, que pasó de ser boscoso a ser un espacio más abierto. Esta idea tiene que comprobarse, y el mejor modo de hacerlo es estudiando directamente las plantas".

¡A la aventura!

Para realizar sus investigaciones, Jacobs ha viajado a muchos lugares diferentes y trabajado con una gran variedad de personas. "Kiptalam Chepboi, un colega con quien trabajamos en Kenya, me llevó a una colmena de abejas mansas. Puedes tomar un trozo de panal echártelo a la boca y saborear la miel sin preocuparte de que te piquen. Fue una de las cosas más bonitas de mi visita."

Hacer un registro moderno

▶Haz tu propio fósil de una planta. Presiona parte de la hoja de una planta sobre un pedazo de arcilla. Llena con yeso el hueco que dejó la hoja. Escribe un informe sobre lo que puedes deducir a partir del fósil sobrel el medio ambiente del que proviene.

▲ *Hojas fosilizadas*

225

Background

Paleobotanists study the fossils of plants not only to learn about the plants themselves but also to search for clues about ancient climates, ecosystems, and the evolution of species.

Jacobs is one of a group of experts from all over the world who have studied in the Lake Baringo region of Kenya. Researchers studying the evolution of humans and of a wide variety of plants, mammals, fish, and microorganisms have joined forces with specialists in geology and dating techniques to gather information from the area. Human fossils from this area are among the oldest human remains ever found.

Answer to Making a Modern Record

The fossils that the students make will be representative of the environment in which the students live. Their descriptions should reflect that.

Chapter Organizer

CHAPTER ORGANIZATION	TIME MINUTES	OBJECTIVES	LABS, INVESTIGATIONS, AND DEMONSTRATIONS
Chapter Opener pp. 226–227	45		**Investigate!** Classifying Shoes, p. 227
Section 1 **Classification: Sorting It All Out**	90	▶ Explain the importance of having scientific names for species. ▶ List the seven levels of classification. ▶ Explain how scientific names are written. ▶ Describe how dichotomous keys help in identifying organisms.	**Demonstration,** Classifying Objects, p. 228 in ATE **QuickLab,** Evolutionary Diagrams, p. 231 **Skill Builder,** Shape Island, p. 592 **Datasheets for LabBook,** Shape Island, Datasheet 20 **Discovery Lab,** Voyage of the USS *Adventure,* p. 594 **Datasheets for LabBook,** Voyage of the USS *Adventure,* Datasheet 21 **EcoLabs & Field Activities,** Water Wigglers, Field Activity 1
Section 2 **The Six Kingdoms**	90	▶ Explain how classification schemes for kingdoms developed as greater numbers of different organisms became known. ▶ List the six kingdoms, and provide two characteristics of each.	**Long-Term Projects & Research Ideas,** Project 9

TECHNOLOGY RESOURCES

 **Guided Reading Audio CD**
English or Spanish, Chapter 10

 **One-Stop Planner CD-ROM with Test Generator**

CLASSROOM WORKSHEETS, TRANSPARENCIES, AND RESOURCES	SCIENCE INTEGRATION AND CONNECTIONS	REVIEW AND ASSESSMENT
Directed Reading Worksheet 10 **Science Puzzlers, Twisters & Teasers,** Worksheet 10		
Directed Reading Worksheet 10, Section 1 **Transparency 39,** Levels of Classification **Science Skills Worksheet 6,** Boosting Your Memory **Math Skills for Science Worksheet 4,** A Shortcut for Multiplying Large Numbers **Transparency 40,** Evolutionary Relationships Among Four Mammals **Transparency 40,** Evolutionary Relationships Among Four Plants **Transparency 41,** Dichotomous Key to Ten Common Mammals in the Eastern United States **Problem Solving Worksheet 10,** A Breach on Planet Biome	**Connect to Environmental Science,** p. 228 in ATE **Math and More,** p. 230 in ATE **Multicultural Connection,** p. 232 in ATE **Scientific Debate:** It's a Bird, It's a Plane, It's a Dinosaur? p. 244	**Review,** p. 233 **Quiz,** p. 233 in ATE **Alternative Assessment,** p. 233 in ATE
Transparency 95, Intrusive Igneous Rock Formations **Directed Reading Worksheet 10,** Section 2 **Math Skills for Science Worksheet 19,** Arithmetic with Decimals **Reinforcement Worksheet 10,** Keys to the Kingdom	**Connect to Environmental Science,** p. 234 in ATE **Connect to Earth Science,** p. 235 in ATE **Real-World Connection,** p. 236 in ATE **MathBreak,** Building a Human Chain Around a Giant Sequoia, p. 237 **Environmental Science Connection,** p. 237 **Multicultural Connection,** p. 237 in ATE **Apply,** p. 238 **Weird Science:** Lobster-Lip Life-Form, p. 245	**Homework,** pp. 235, 236 in ATE **Self-Check,** p. 236 **Review,** p. 239 **Quiz,** p. 239 in ATE **Alternative Assessment,** p. 239 in ATE

Holt, Rinehart and Winston On-line Resources

go.hrw.com

For worksheets and other teaching aids related to this chapter, visit the HRW Web site and type in the keyword: **HSTCLS**

National Science Teachers Association

www.scilinks.org

Encourage students to use the *sci*LINKS numbers listed with the Chapter Highlights to access information and resources on the **NSTA** Web site.

END-OF-CHAPTER REVIEW AND ASSESSMENT

Chapter Review in Study Guide
Vocabulary and Notes in Study Guide
Chapter Tests with Performance-Based Assessment, Chapter 10 Test
Chapter Tests with Performance-Based Assessment, Performance-Based Assessment 10
Concept Mapping Transparency 10

Visual Resources

TEACHING TRANSPARENCIES

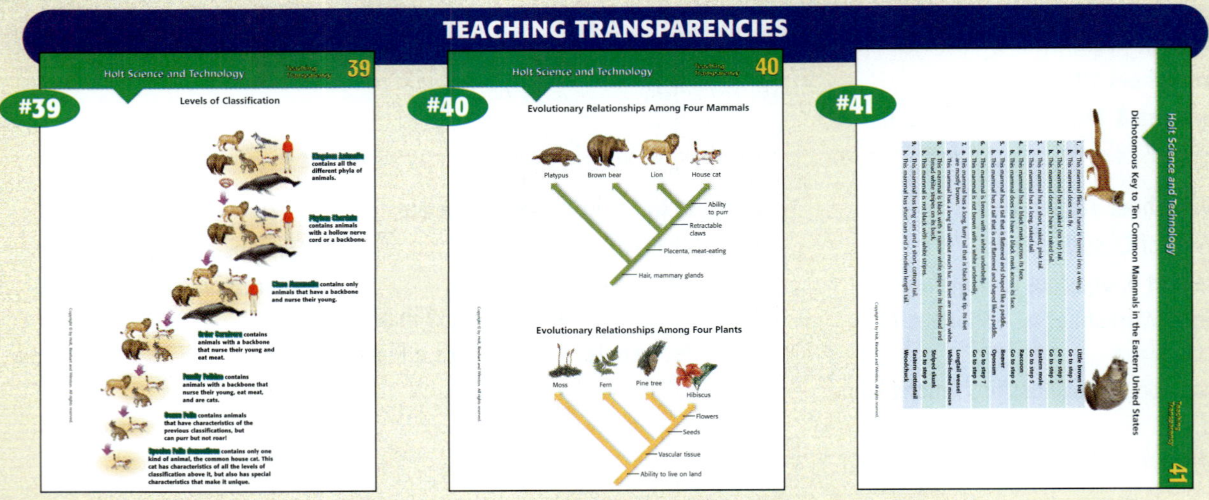

#39 Levels of Classification

#40 Evolutionary Relationships Among Four Mammals / Evolutionary Relationships Among Four Plants

#41 Dichotomous Key to Ten Common Mammals in the Eastern United States

TEACHING TRANSPARENCIES

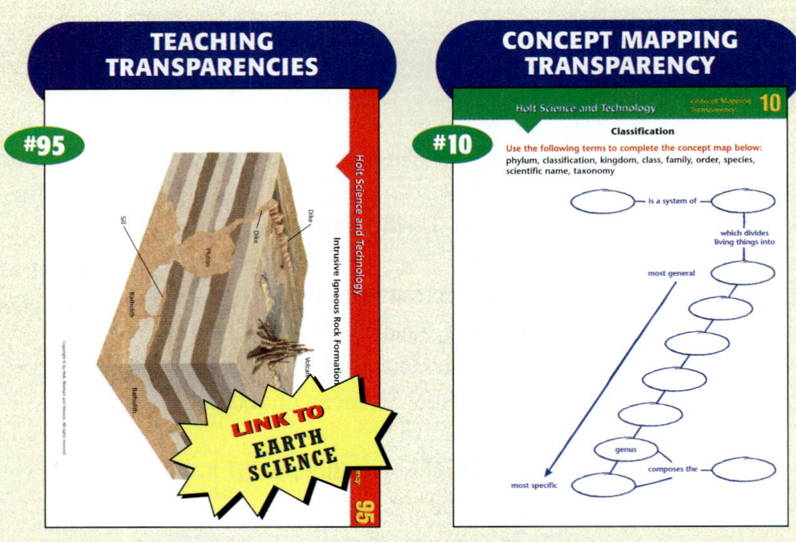

#95 Intrusive Igneous Rock Formation

CONCEPT MAPPING TRANSPARENCY

#10 Classification — Use the following terms to complete the concept map below: phylum, classification, kingdom, class, family, order, species, scientific name, taxonomy

Meeting Individual Needs

DIRECTED READING

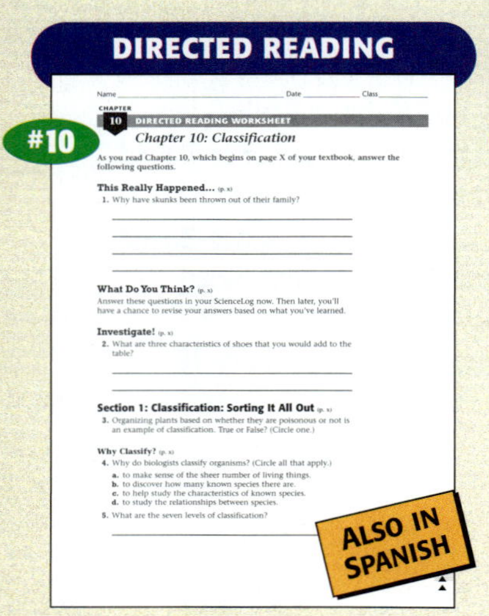

#10 DIRECTED READING WORKSHEET — Chapter 10: Classification

REINFORCEMENT & VOCABULARY REVIEW

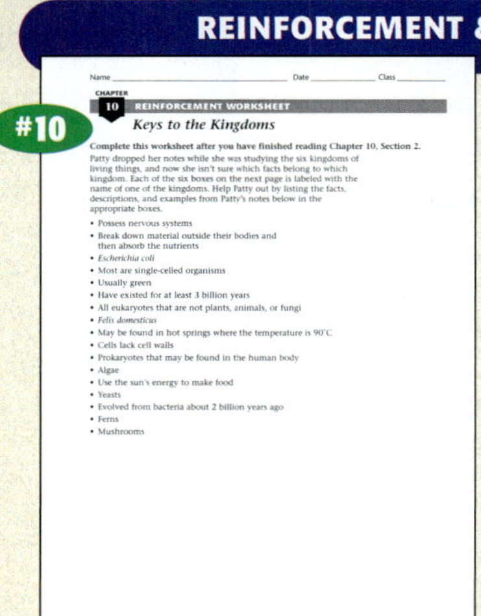

#10 REINFORCEMENT WORKSHEET — Keys to the Kingdoms

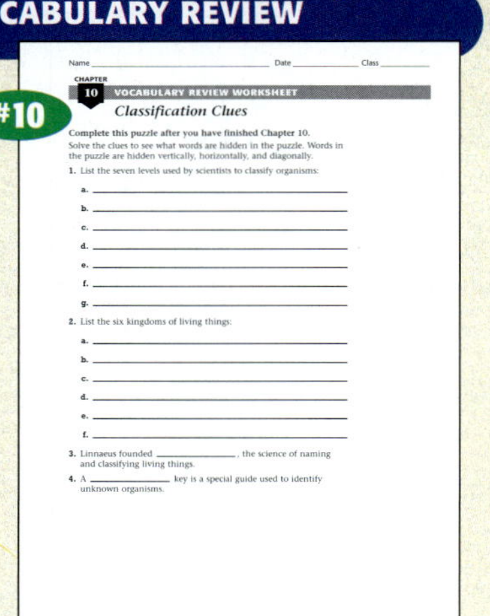

#10 VOCABULARY REVIEW WORKSHEET — Classification Clues

SCIENCE PUZZLERS, TWISTERS & TEASERS

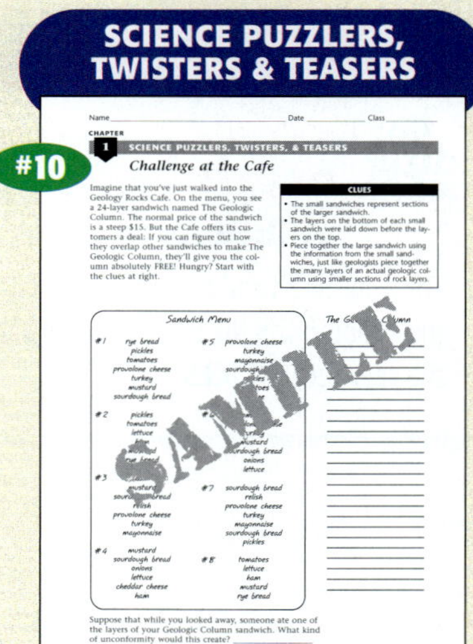

#10 SCIENCE PUZZLERS, TWISTERS, & TEASERS — Challenge at the Cafe

Review & Assessment

STUDY GUIDE

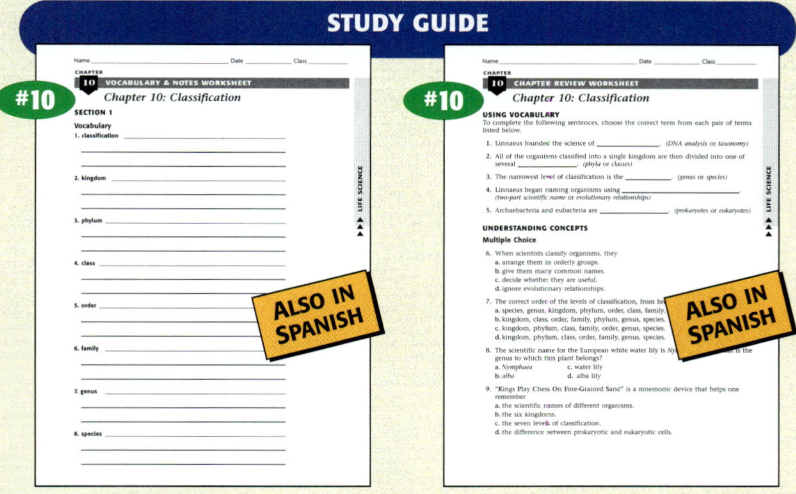

#10 — Chapter 10: Classification (Vocabulary & Notes Worksheet)

#10 — Chapter 10: Classification (Chapter Review Worksheet)

ALSO IN SPANISH

CHAPTER TESTS WITH PERFORMANCE-BASED ASSESSMENT

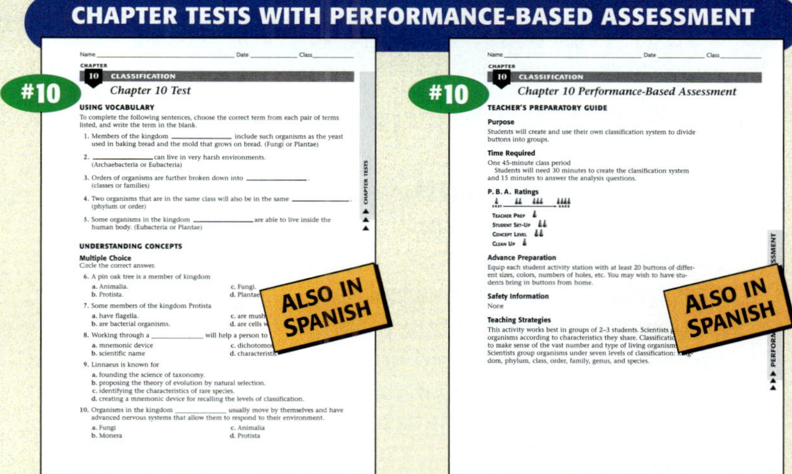

#10 — Chapter 10 Test

#10 — Chapter 10 Performance-Based Assessment

ALSO IN SPANISH

Lab Worksheets

ECOLABS & FIELD ACTIVITIES

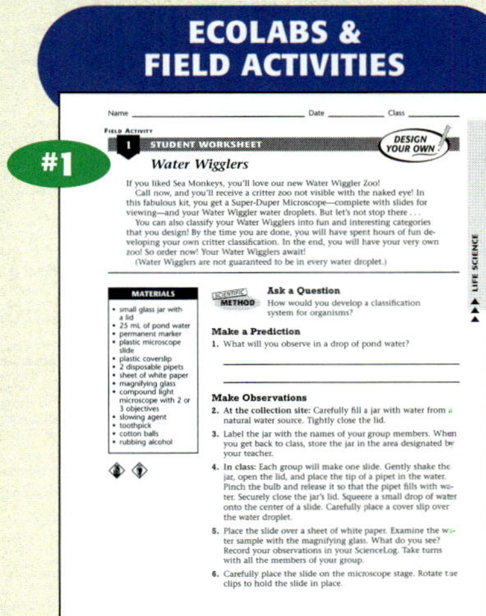

#1 — Water Wigglers

LONG-TERM PROJECTS & RESEARCH IDEAS

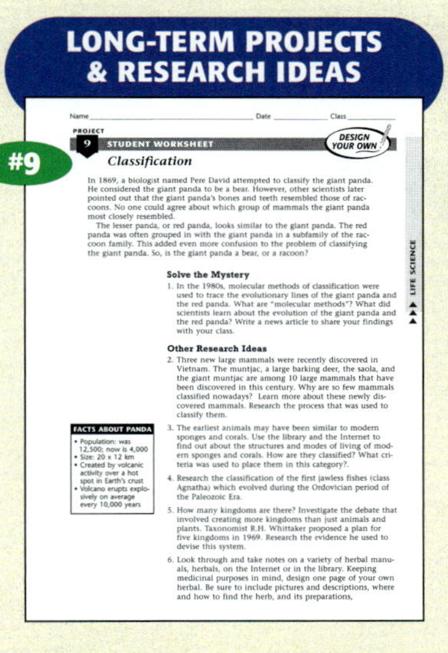

#9 — Classification

DATASHEETS FOR LABBOOK

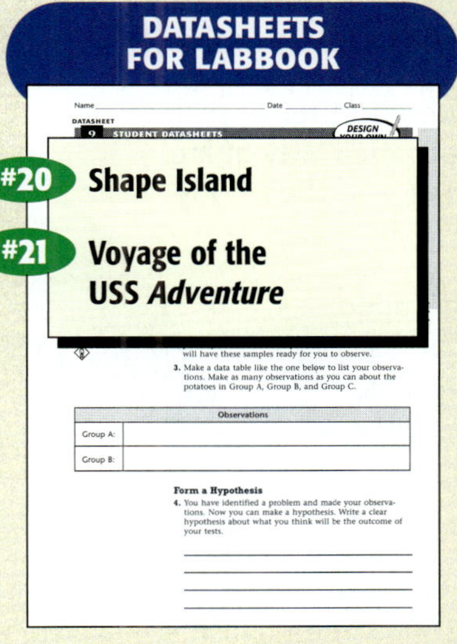

#20 — Shape Island

#21 — Voyage of the USS *Adventure*

Applications & Extensions

CRITICAL THINKING & PROBLEM SOLVING

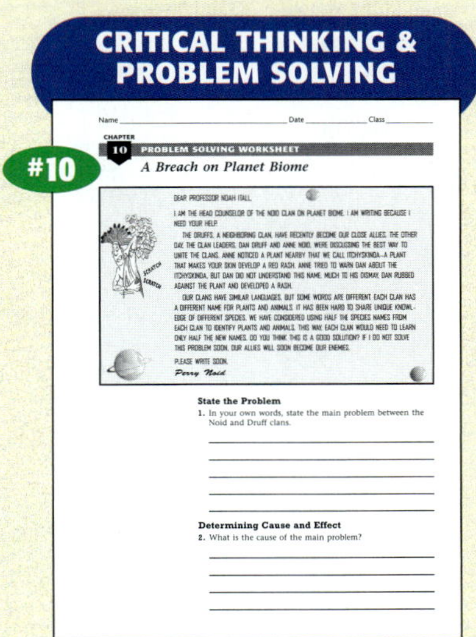

#10 — A Breach on Planet Biome

Chapter Background

Classification: Sorting It All Out

▶ Aristotle's Classification System

The great Greek philosopher and scientist Aristotle (384–322 B.C.) began classifying animals into logical groupings more than 2,000 years ago. Although Aristotle did not view different kinds of organisms as being related by descent, he arranged all living things in an ascending ladder with humans at the top.

- Animals were separated into two major groups—those with red blood and those without red blood—which correspond very closely with our modern classification of vertebrates and invertebrates.

- Animals were further classified according to their way of life, their actions, and their body parts.

- Aristotle grouped plants as herbs, shrubs, or trees, based on their size and appearance.

▶ Species in Classification

In the late 1600s, the English scientist John Ray established the species as the basic unit of classification.

▶ Basis for Modern Classification System

Our modern system of classification was introduced by Swedish scientist Carolus Linnaeus. He published a book on plant classification in 1753 and a book on animal classification in 1758.

- Organisms were classified according to their structure.

- Plants and animals were arranged into genus and species, and the categories of class and order were introduced.

- Species were given distinctive two-word names. Linnaeus's system is still in use today, although with many changes.

- Carolus Linnaeus is the Latin translation of the Swedish scientist's given name, Carl von Linné.

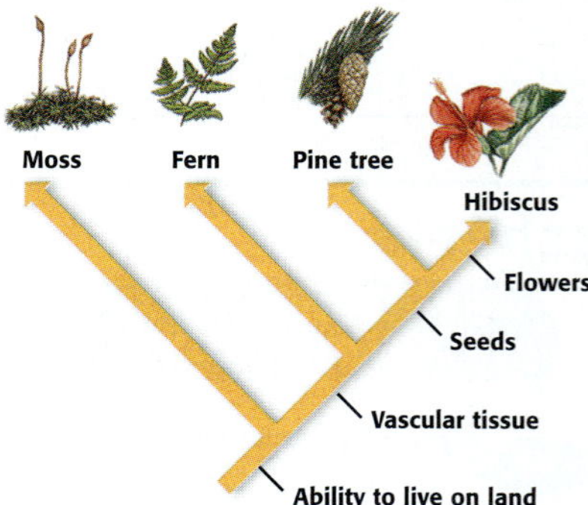

▶ Subgroups in the Animal Kingdom

Baron Cuvier Georges first divided the animal kingdom into subgroups, such as Vertebrata, Mollusca, Articulata, and Radiata, in 1817.

IS THAT A FACT!

- There are 35 known species of sea horses. Their ability to change color is superior to that of chameleons. Animals use color changes for camouflage or to express emotions. One species of sea horse, *Hippocampus hippocampus,* develops skin filaments that enable it to look like seaweed, a protist.

The Six Kingdoms

▶ Variations of the Classification System

Variations of the five-kingdom system introduced by R. H. Whittaker in 1969 are used by some modern scientists. Whittaker's system classifies organisms according to whether they are prokaryotic or eukaryotic, whether they are unicellular or multicellular, and whether they obtain food by photosynthesis, ingestion, or absorption of nutrients from their environment.

- Because studies of DNA indicate that there are significant differences between archaebacteria and eubacteria, many scientists place archaebacteria in a sixth kingdom.

▶ Different Classification Methods

Branches of biology use different methods of classification.

- The classification method microbiologists use is organized by volume, section, family, and genus.

- Botanists used to use the term *division* instead of *phylum* for plants and fungi.

▶ The Planet Within

When we organize life on Earth into categories, it is important to remember that organisms are not equally distributed throughout our classification system. Even though we often think of the Earth's living things in terms of plants and animals—organisms living above the Earth's surface—the largest kingdoms in terms of the number of species, number of individuals,

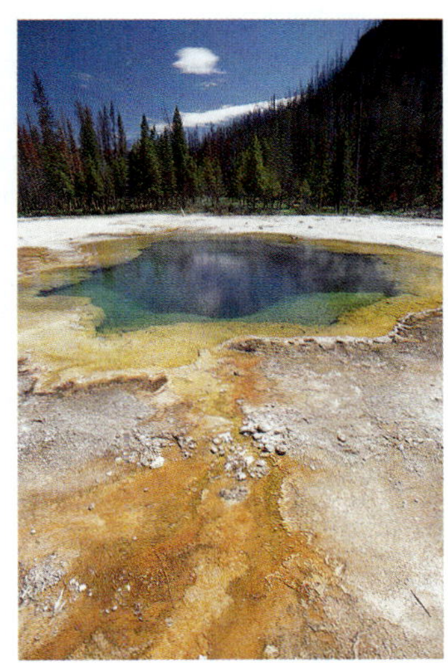

and total biomass are bacteria. And their most common home may be deep within the Earth's crust.

- Scientists have known for some time that bacteria exist all around us and that some have the ability to live in extreme environments. Some live in hot geysers, others in water with salt concentrations so high no other organisms can survive in it. Scientists have also known that many archaebacteria can thrive in anaerobic and high-pressure environments, such as those found underground. But only recently have scientists learned just how far underground they are and just how many bacteria live there.

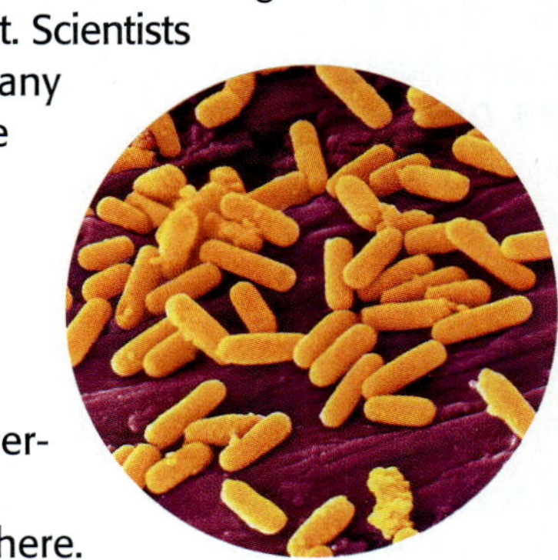

- In 1987, scientists were drilling in the rock beneath the Savannah River in South Carolina, investigating the safety of the drinking water. The cores of the rock they investigated at a depth of 500 m harbored bacteria. Other scientists found bacteria in the ocean at 750 m. A South African gold mine, as far down as 5 km, yielded other bacteria.

- Once scientists knew to look deep in the Earth for life-forms, they began looking in the sediment under the ocean and found more organisms. That sediment is 15 km deep in places, and some speculate that organisms could inhabit the sediment even to that depth. If that is the case, then the total biomass of these astonishing organisms beneath the surface of the Earth may exceed the total biomass of all the living things on the Earth's surface.

- No one knows exactly how these creatures tolerate the tremendous pressures and temperatures of their environment, but scientists have learned that these organisms are meeting their nutritional needs in a variety of ways. Some live on oxidized forms of sulfur; others on bits of organic matter found in the sediment. Some bacteria have even been found in igneous rocks, apparently subsisting on the carbon dioxide and hydrogen gas trapped in the rock.

For background information about teaching strategies and issues, refer to the *Professional Reference for Teachers.*

Directed Reading Worksheet 10

Science Puzzlers, Twisters & Teasers Worksheet 10

Guided Reading Audio CD
English or Spanish, Chapter 10

CAPÍTULO

10 Clasificación

¡Esto realmente sucedió!

Los zorrillos han sido rechazados por su familia. No por su olor, sino por su ADN. En otro tiempo se pensaba que eran los parientes más cercanos de los hurones, los visones, los tejones, las comadrejas y las nutrias. Estos carnívoros peludos de cuerpo alargado y patas cortas se agrupan en una familia llamada Mustélidos. *Mustélido* viene del latín y significa "ratón". Los zorrillos entran en el grupo de las comadrejas y los hurones porque comparten varias características físicas con los ratones, como tener las orejas chicas y redondas, y las patas cortas. Sin embargo, un investigador del Museo Suroccidental de Biología de la Universidad de Nuevo México descubrió que el ADN de los zorrillos es muy diferente del ADN de los otros Mustélidos. Los científicos pueden determinar el grado de parentesco de las especies comparando su ADN. El ADN de dos animales que son parientes cercanos, como el gato doméstico y el tigre, es más parecido que el ADN de dos animales que son parientes lejanos, como el gato doméstico y el pollo. Entonces, ¿cómo clasificamos a los zorrillos? Pues en su propia familia científica, que fue creada recientemente: los Mefítidos. *Mefítido* viene del latín y significa "mal olor".

En este capítulo aprenderás por qué los nombres científicos son tan importantes y cómo se clasifican los organismos. También aprenderás los seis reinos principales en los que se clasifican los organismos.

226

This Really Happened...

To determine evolutionary relationships, scientists are comparing the DNA not only of familiar species but also of newly discovered or rare species. Scientists on modern expeditions preserve portions of the internal tissues of specimens they collect by freezing them on-site in a tank of liquid nitrogen. Later, detailed comparative analyses of the DNA structure of the tissue samples can be done.

¿Tú qué piensas?

Usa tus conocimientos para responder a las siguientes preguntas en tu cuaderno de ciencias:

1. ¿Qué es la clasificación?

2. ¿Cómo utilizan las personas la clasificación a diario?

3. ¿Por qué los científicos clasifican los seres vivos?

Clasificación de zapatos

En esta actividad trabajarán en grupos para desarrollar un sistema de clasificación de zapatos.

Procedimiento

1. Reúne **10 zapatos diferentes** de tus compañeros. Sácalos de una tienda de segunda mano o de una venta de garaje. Marca cada zapato con un número del 1 al 10 usando un **marcador negro** y coloca los zapatos en una mesa del salón de clases.

2. Forma pequeños grupos. En tu cuaderno de ciencias, haz una tabla como la que se muestra a continuación. Úsala como ejemplo para hacer una lista de características. Completa la gráfica con la descripción de cada zapato.

	Izquierdo o derecho	Niño o niña	Con o sin cordones	Color	Talla
Características de los zapatos					
1.					
2.					
3.					
4.					
5.					

3. Usa la información de la gráfica para desarrollar una clave que sirva para identificar cada zapato. Tu clave debe ser una lista de pasos. Cada uno de ellos debe tener dos enunciados que describan los zapatos y te guíen a otro grupo de enunciados. Por ejemplo, el paso uno podría ser:

> 1. **a.** Ésta es una sandalia roja.   **Zapato #4**
>   **b.** Ésta no es una sandalia roja. **Ir al paso 2.**

Cada grupo de enunciados elimina más zapatos hasta que sólo un zapato coincide con la descripción, como en el inciso (a) de arriba. Revisa el número de la suela del zapato para ver si estás en lo cierto.

4. Cuando todos los grupos terminen, intercambia las claves con miembros de otros grupos. Trata de utilizar su clave para identificar los zapatos.

Análisis

5. ¿Para qué te sirvió hacer una lista de las características de los zapatos antes de hacer la clave?

6. ¿Pudiste identificar los zapatos con la clave de otro grupo? Si no fue así, ¿qué problemas tuviste?

227

What Do You Think?

Accept all reasonable responses.

Students will have a chance to revise their answers in the Chapter Review under NOW What Do You Think?

Investigate!

MATERIALS

FOR EACH GROUP:
- 10 different shoes (from class members, a second-hand store, or a garage sale)
- marker

Make certain students understand that the list of shoe characteristics should be unique to a particular set of 10 shoes.

Characteristics of shoes listed should be those that can easily be observed. For example, whether a shoe belongs to a boy or to a girl is not always obvious to an observer.

Answers to Investigate!

5. Making a list of characteristics is important to help you decide on the series of descriptive statements.

6. Each group may describe the shoes differently, but their descriptions should be clear enough to lead the other groups to the same conclusion.

MISCONCEPTION ALERT

Though some people think of skunks only as smelly pests, we do benefit from having them around. Skunks eat beetles, crickets, grasshoppers, mice, rats, and other insects and small rodents, helping to control the populations of those animals.

If your dog has ever been sprayed by a skunk, you know that skunks are potent adversaries. But they usually give fair warning that they are about to spray by growling, hissing, and stamping their front paws.

Focus

Classification: Sorting It All Out

In this section, students learn about the modern biological classification system. The section explains how organisms are classified based on their evolutionary relationships and how their scientific names are determined. Finally, students learn how to identify animals using a dichotomous key.

Bellringer

Ask students to think about the different ways humans classify things. Ask them to list in their ScienceLog at least five groups of things that humans classify. You may want to give them examples, such as library books, department-store merchandise, and addresses.

1 Motivate

DEMONSTRATION

Classifying Objects Display a variety of small solid objects. Ask students for their ideas on ways to put the objects into groups. For each grouping, record the defining characteristic and the objects that belong in the group. Identify objects that fit in more than one grouping. Discuss how putting objects into groups can be helpful.

Directed Reading Worksheet 10 Section 1

VOCABULARIO

clasificación	familia
reino	género
filo	especie
clase	clave dicotómica
orden	taxonomia

OBJETIVOS

- Explica la importancia de los nombres científicos de las especies.
- Menciona los siete niveles de clasificación.
- Explica cómo se escriben los nombres científicos.
- Describe cómo las claves dicotómicas ayudan a identificar los organismos.

Clasificación: División en grupos

Imagínate que vives en un bosque tropical y que tienes que conseguir tu comida, refugio y ropa en el bosque. Si quieres sobrevivir, necesitas conocer las plantas que puedes comer y las que son venenosas. Debes saber qué animales puedes comer y cuáles te pueden comer a ti si no tienes cuidado. Tienes que organizar en categorías los seres vivos que están a tu alrededor, es decir, necesitas clasificarlos. La **clasificación** es la división de los organismos en grupos según sus similitudes.

¿De qué sirve clasificar?

Durante miles de años, los seres humanos hemos clasificado los diferentes tipos de organismos en base a su utilidad. Por ejemplo, los miembros del pueblo Chácabo de Bolivia, como la familia que se muestra en la **Figura 1,** conocen 360 especies de plantas del bosque donde viven y saben para qué sirven 305 de ellas. ¿Cuántas plantas útiles en tu vida puedes mencionar?

Los biólogos también clasifican los organismos vivos y extintos. ¿Para qué hacen esto? En el mundo hay millones de especies diferentes de seres vivos. Entender el número total y la diversidad de seres vivos requiere una clasificación. Como sabes, los científicos buscan respuestas. Clasificar los seres vivos les facilita encontrar las respuestas a muchas preguntas importantes, como las siguientes:

- ¿Cuántas especies conocidas hay?
- ¿Cuáles son las características de estas especies?
- ¿Cuáles son las relaciones entre estas especies?

Para clasificar un organismo, un biólogo debe usar un sistema que agrupe los organismos según las características que tengan en común y las relaciones entre uno y otro. Los biólogos usan siete niveles de clasificación: reino, filo, clase, orden, familia, género y especie.

Figura 1 *El pueblo Chácabo conoce muy bien el medio ambiente.*

CONNECT TO
ENVIRONMENTAL SCIENCE

Some tropical rain forests are being cut down and converted into farms to feed native populations. Scientists suspect that the forests we are losing may be pharmaceutical treasure troves. One-fifth of all the world's known plant species live in tropical rain forests. Only a small percentage of the species have been studied. These plants might be sources of medicines that can be used to treat diseases. Ask students what they think could be done to ensure that rain-forest species are preserved while food needs are also met.

Niveles de clasificación

Cada organismo se clasifica en uno de varios **reinos,** que son los grupos más generales de los niveles de clasificación. Todos los organismos de un mismo reino se clasifican en varios **filos.** Los miembros de un determinado filo se parecen más entre sí que a los miembros de otro filo. Los organismos de un filo específico se agrupan en diferentes **clases.** Cada clase se subdivide en uno o más **órdenes,** los órdenes se separan en **familias,** las familias se agrupan en **géneros** y los géneros se dividen en **especies.**

Examina la **Figura 2** para seguir la clasificación del gato doméstico común desde el reino animal hasta la especie de *Felis domesticus.*

Figura 2 *El reino animal abarca todas las especies de animales, mientras que la especie Felis domesticus sólo abarca una.*

El **reino animal** abarca todos los filos de animales.

El **filo de los cordados** abarca animales con un cordón nervioso hueco (notocordio).

La **clase de los mamíferos** sólo abarca animales que tienen columna vertebral y amamantan a sus crías.

El **orden de los carnívoros** abarca animales con columna vertebral, que amamantan a sus crías y cuyos antepasados tenían dientes especiales para despedazar carne.

La **Familia de los félidos** abarca animales felinos con columna vertebral que alimentan a sus crías, provistos con garras desarrolladas y dientes especiales para despedazar carne.

El **género *Felis*** abarca animales que tienen las características de las clasificaciones anteriores, pero que no pueden rugir; sólo pueden ronronear.

Especie *Felis domesticus* sólo abarca un tipo de animal, el gato doméstico común. Este gato tiene las características de todos los niveles de clasificación anteriores, pero tiene otras más que lo hacen único.

Explora

Un recurso nemónico es una herramienta que te ayuda a recordar algo. Puede ser una palabra, una oración que tiene claves io incluso un hilo que te amarras al dedo!

Una forma de recordar los niveles de clasificación es usar un recurso nemónico como esta oración: **R**osa **F**ue **C**on **O**tra **F**eliz **G**anadora **E**mocionada.

Inventa tu propio recurso nemónico para los niveles de clasificación usando palabras que signifiquen algo para ti.

229

IS THAT A FACT!

The term *dinosaur* wasn't coined until the nineteenth century. Until then, as dinosaur bones were uncovered all over the world, the most widely accepted view was that they belonged to dragons.

Teaching Transparency 39
"Levels of Classification"

Science Skills Worksheet 6
"Boosting Your Memory"

MEETING INDIVIDUAL NEEDS

Learners Having Difficulty To help students understand what constitutes a species, genus, family, order, class, phylum, and kingdom, ask them the following questions:

What does a species contain? (organisms that have the same characteristics)

What does a genus contain? (similar species)

What does a family contain? (similar genera)

What does an order contain? (similar families)

What does a class contain? (similar orders)

What does a phylum contain? (similar classes)

What does a kingdom contain? (similar phyla) Sheltered English

USING THE FIGURE

Concept Mapping Refer students to **Figure 2.** Have them answer these questions to enhance their understanding of the sorting process used to classify the common house cat. What animals are pictured at the kingdom level? (lion, bird, human, bear, lynx, house cat, worm, whale)

Which of these pictured animals do not fit the description of a chordate? (the worm)

Which of the animals pictured at the chordate level do not fit the description of a mammal? (the bird)

Continue this questioning pattern for the remaining levels of classification. Have students put the information expressed in the diagram into a concept map.

ACTIVITY

Evolutionary Diagrams Fossils show that one difference between ancestral genera of the modern horse, *Equus*, is the number of toes:

Eohippus (55 mya*) 4 toes
Mesohippus (35 mya) 3 toes
Merychippus (26 mya) 1 large toe, 2 small toes
Pliohippus (3 mya) 1 large toe surrounded by a hoof
Equus (modern) 1 large toe, more broad and flat, surrounded by a hoof

(*mya = million years ago)

Have students use this information to construct their own evolutionary diagram. Use the diagram on page 230 as a model.

LabBook PG 592
Shape Island

MATH and MORE

Give students the following information:

There are more than 700,000 known species of insects in the world. The number of insect species accounts for about half of all known species. Have students calculate the approximate total number of Earth's known species. (over 1.4 million)

Math Skills Worksheet 4
"A Shortcut for Multiplying Large Numbers"

Teaching Transparency 40
"Evolutionary Relationships Among Four Mammals"

¿En qué se basa la clasificación?

Carolus Linnaeus, a quien ves en la **Figura 3,** fue un médico y botánico sueco que vivió en el siglo XVIII. Linnaeus fundó la **taxonomía,** que es la ciencia encargada de identificar, clasificar y darle un nombre a los seres vivos.

Linnaeus trató de clasificar todos los organismos conocidos sólo por sus características comunes. Este método cambió después de la publicación de la teoría de la selección natural de Darwin, cuando los científicos empezaron a reconocer que los cambios de evolución forman una línea descendiente a partir de un antepasado común. La taxonomía cambió para incluir estas nuevas ideas sobre las relaciones de evolución.

Los taxonomistas de hoy en día todavía clasifican los organismos en base a las relaciones de evolución establecidas. Las especies con un antepasado común reciente se pueden clasificar juntas. Por ejemplo, el ornitorrinco, el oso café, el león y el gato doméstico se relacionan porque se cree que tienen un antepasado en común: un mamífero de la antigüedad. Debido a esta relación, los cuatro animales se agrupan en la misma clase: mamíferos.

El oso café, el león y el gato doméstico se relacionan más entre sí que con el ornitorrinco. Todos son mamíferos, pero el ornitorrinco pone huevos, a diferencia de los otros tres animales. Los osos cafés, los leones y los gatos domésticos comparten un antepasado común: un carnívoro antiguo; por lo tanto, se clasifican en el mismo orden: carnívoros.

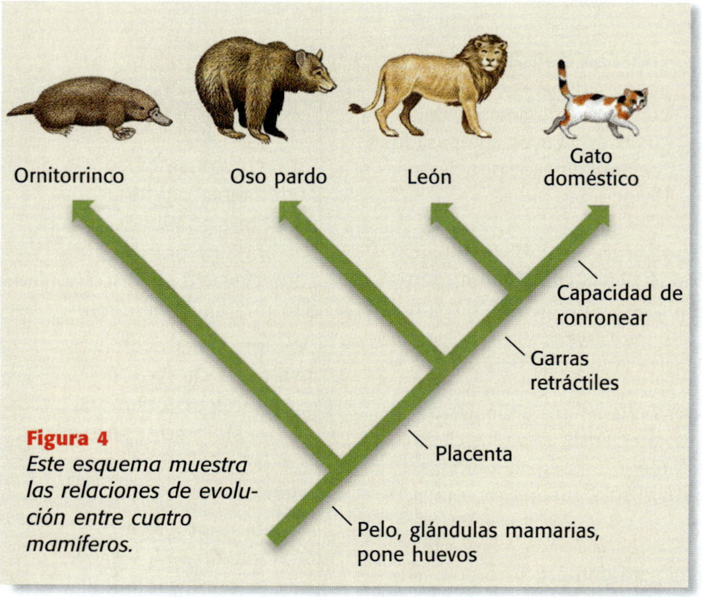

La relación de evolución cercana entre los leones y los gatos domésticos se muestra en el esquema de la **Figura 4.** Las características de la flecha que apunta hacia la derecha son las que hacen que el animal señalado sea único. El gato doméstico y el ornitorrinco comparten las características del pelo y las glándulas mamarias, pero son diferentes en otras cosas. Por ejemplo, el gato doméstico puede ronronear. La rama que conduce a los leones está más cerca de la rama que llega a los gatos domésticos. El león y el gato doméstico son parientes cercanos porque comparten el antepasado común más reciente: un gato de la antigüedad.

Figura 3 *Carolus Linnaeus clasificó más de 7,000 especies de plantas.*

Figura 4
Este esquema muestra las relaciones de evolución entre cuatro mamíferos.

SCIENTISTS AT ODDS

Chinese paleontologists have found a 121-million-year-old fossil of a dinosaur that appeared to have feathers. This, say some scientists, is further proof that birds evolved directly from dinosaurs. Other scientists refute that hypothesis because of the lack of a relationship between dinosaurs' fossilized "finger" bones and the corresponding bones in bird embryos. Additional studies by zoologists dispute the bird-dinosaur connection with comparisons of the respiratory structures of modern birds, mammals, and crocodiles and those of early bird fossils and theropod dinosaurs.

Asignación de nombres

Cuando los biólogos clasifican los organismos, les ponen nombres científicos, siempre es el mismo para un organismo dado, sin importar cuántos nombres comunes tenga.

Antes de la época de Linnaeus, los eruditos usaban nombres en latín de hasta 12 palabras para identificar las especies. Linnaeus simplificó la identificación de los organismos al darle a cada especie un nombre científico compuesto por dos partes. La primera parte del nombre identifica el género y la segunda, la especie. El nombre científico del elefante indio, por ejemplo, es *Elephas maximus*. Ninguna otra especie tiene este nombre, y todos los científicos saben que el *Elephas maximus* se refiere al elefante indio. La primera persona que describe una especie nueva es quien le da un nombre científico.

Todo me suena a griego (o latín). Los nombres científicos paracen difíciles de entender porque están en latín o en griego, pero están llenos de significado. Observa la **Figura 5;** quizá conozcas el nombre científico de este animal. ¡Claro! Es el *Tyrannosaurus rex.* La primera palabra significa "lagartija tirana" en griego y la segunda palabra significa "rey" en latín. El nombre del género siempre empieza con mayúscula y el de la especie con minúscula. Ambas palabras se subrayan o se escriben en cursiva. Tal vez has oído llamar al *Tyrannosaurus rex* como *T. rex.*

Esto es aceptable ya se ha mencionado antes el nombre del género. El nombre de la especie no está completo sin el nombre del género o su abreviatura.

Figura 5
¡Tú nunca llamarías al Tyrannosaurus rex *sólo* rex! ¿verdad?

Laboratorio

Diagramas de evolución

Un diagrama de evolución se construye para trazar el linaje de una especie aunque algunos de sus antepasados sólo sean fósiles. También se usa para mostrar las relaciones de evolución entre diferentes filos de organismos.

Haz un diagrama como el de la página 230. Usa una rana, una víbora, un canguro y un conejo. ¿Qué evolución hubo entre un organismo y el siguiente? Escríbelo en tu diagrama.

PG 594

Voyage of the USS *Adventure*

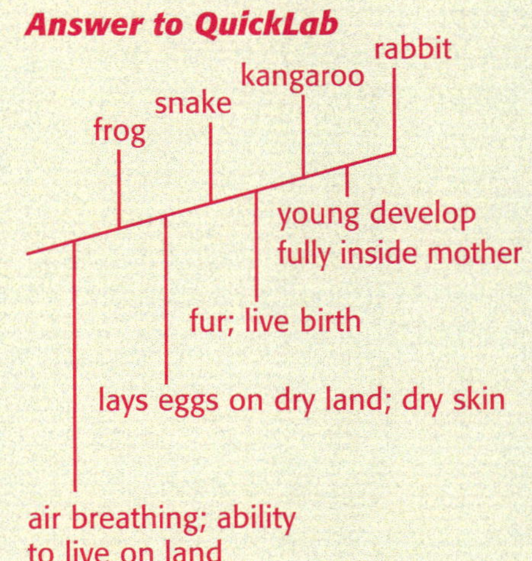

Answer to QuickLab

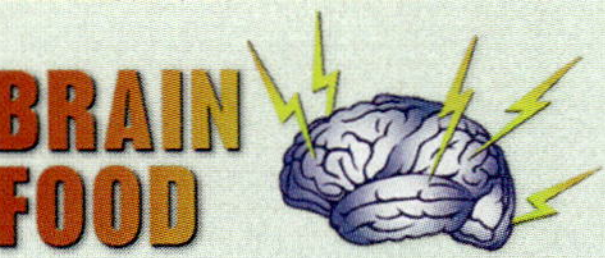

BRAIN FOOD

Have students consider the importance of classification to human thought. Ask students to try to think of something that cannot be classified in some way. Suggest that they test any item or concept they come up with by placing it in the following sentence:

(A) ______ is a type of ______.

For example, if the word is *speech*, the sentence could be filled in as follows:

Speech is a type of communication.

You may wish to hold a contest or have students share their examples in class.

IS THAT A FACT!

If you put all the insects in the world together, they would weigh more than all the people and the rest of the animals combined.

internetconnect

SCiLINKS
NSTA
TOPIC: The Basis for Classification
GO TO: www.scilinks.org
***sci*LINKS NUMBER:** HSTL205

Multicultural CONNECTION

Point out to students that some of the common names we have for animals came from other languages. For example, *burro* came from Spanish, *grebe* came from French, *macaw* came from Portuguese, and *orangutan* came from Malay. Encourage interested students to look in a dictionary for the language source of other common animal names.

RETEACHING

Display a picture of a bird whose common name is not well known to your students. Without providing any additional information about the bird, ask them to give the bird a name. Then list all the given names on the chalkboard. Help students understand that scientists around the world would have difficulty sharing information about the bird if they used more than one name for it. Sheltered English

Teaching Transparency 40
"Evolutionary Relationships Among Four Plants"

Teaching Transparency 41
"Dichotomous Key to Ten Common Mammals in the Eastern United States"

¿Por qué son tan importantes los nombres científicos?

Examina la caricatura de la **Figura 6**. ¿Cómo se llama ese animalito negro con blanco que a veces huele mal? El zorrillo tiene varios nombres comunes en inglés y por lo menos uno en cada idioma. Todos estos nombres comunes pueden provocar bastante confusión a los biólogos que quieren hablar sobre el zorrillo. Los biólogos de diferentes partes del mundo que están interesados en los zorrillos necesitan saber que están hablando del mismo animal, así que usan su nombre científico, *Mephitis mephitis*. Todos los seres vivos conocidos tienen un nombre científico compuesto.

Figura 6 *Con el nombre científico compuesto de un organismo, los científicos están seguros de que hablan del mismo organismo*

¿Cuándo y dónde vivió la primera ave? Descúbrelo en la página 244.

Claves dicotómicas

Los taxonomistas han desarrollado guías especiales que se conocen como **claves dicotómicas** como una ayuda para identificar los organismos desconocidos. Una clave dicotómica tiene varios pares de enunciados descriptivos que sólo tienen dos respuestas diferentes. De cada par de enunciados, la persona que trata de identificar los organismos desconocidos escoge el apropiado. A partir de ahí, la persona se dirige a otro par de enunciados. Si la persona trabaja con la clave usando los enunciados, logrará identificar el organismo desconocido. Trata de identificar los dos animales que se muestran en la página siguiente usando la clave dicotómica.

WEIRD SCIENCE

It's exciting when scientists find new species of plants or insects that have gone unnoticed, but it is rare when scientists find a new mammal. Between 1992 and 1998, scientists in Vietnam discovered three new species of deerlike mammals. The most recent was discovered in August 1997. This newly discovered mammal, the Truong Son muntjac, is about one-third of a meter (14 in.) tall, weighs 15.5 kg (34 lb), has a black coat and very short antlers, and barks like a dog. These mammals stay well hidden in the thick Vietnamese forest. The two other mammal species discovered recently are the Vu Quang ox and the giant muntjac.

Clave dicotómica para 10 mamíferos comunes del este de Estados Unidos

1. a. Este mamífero vuela. Sus manos tienen forma de alas.	**Murciélago café**
b. Este mamífero no vuela.	**Ir al paso 2**
2. a. Este mamífero tiene cola sin pelo.	**Ir al paso 3**
b. Este mamífero no tiene cola sin pelo.	**Ir al paso 4**
3. a. Este mamífero tiene la cola corta y sin pelo.	**Topo**
b. Este mamífero tiene la cola larga y sin pelo.	**Ir al paso 5**
4. a. Este mamífero tiene un antifaz negro.	**Mapache**
b. Este mamífero no tiene un antifaz negro.	**Ir al paso 6**
5. a. Este mamífero tiene la cola aplastada en forma de paleta.	**Castor**
b. Este mamífero no tiene la cola aplastada en forma de paleta.	**Zarigüeya**
6. a. Este mamífero es café con la panza blanca.	**Ir al paso 7**
b. Este mamífero no es café ni tiene la panza blanca.	**Ir al paso 8**
7. a. Este mamífero tiene la cola larga, peluda y con la punta blanca.	**Comadreja de cola larga**
b. Este mamífero tiene la cola larga sin mucho pelo.	**Ratón patiblanco**
8. a. Este mamífero es negro con una raya blanca delgada en la frente y rayas blancas anchas en la espalda.	**Zorrillo rayado**
b. Este mamífero no es negro con rayas blancas.	**Ir al paso 9**
9. a. Este mamífero tiene las orejas largas y la cola corta esponjada.	**Liebre de rabo blanco**
b. Este mamífero tiene las orejas cortas y la cola mediana.	**Marmota de Norteamérica**

REPASO

1. ¿Por qué usamos nombres científicos?

2. Explica las dos partes de un nombre científico.

3. Menciona los siete niveles de clasificación.

4. Describe cómo una clave dicotómica ayuda a identificar organismos desconocidos.

5. Interpretar ilustraciones Analiza la figura de la derecha. ¿Qué planta es el pariente más cercano del hibisco? ¿Y el más lejano? ¿Qué planta tiene semillas?

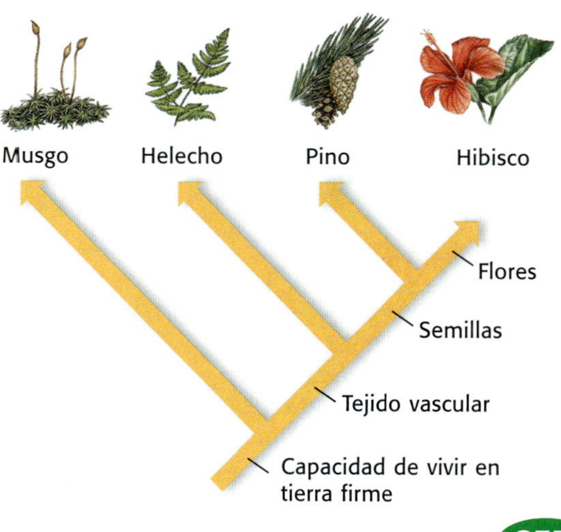

233

3 Extend

GROUP ACTIVITY

Have small groups of students work together to create an identification key that would identify common mammals in your area.

4 Close

Quiz

1. Why do scientists classify animals? (to facilitate studying them)

2. What was the basis for classification systems in the past? (shared characteristics)

3. What is the basis of modern classification systems? (evolutionary relationships)

ALTERNATIVE ASSESSMENT

Have students create a cartoon that shows how using different common names for an animal instead of its scientific name creates confusion. Students must include authentic common names and scientific names in each cartoon. **Sheltered English**

Answer to Dichotomous Key

Mammal on the top left:
1b, 2b, 4b, 6a, 7a, longtail weasel

Mammal on the top right:
1b, 2b, 4b, 6b, 8b, 9b, woodchuck

Problem Solving Worksheet 10
"A Breach on Planet Biome"

Answers to Review

1. Scientists use scientific names for organisms to be clear and precise.

2. genus and species; The genus is a broader classification, and the species name is more specific and is often given by the species' discoverer.

3. Kingdom, Phylum, Class, Order, Family, Genus, Species

4. A dichotomous key is organized into a series of pairs of questions. By working through the statements in the key, unknown organisms can be identified.

5. pine tree; moss; pine tree and hibiscus

Focus

The Six Kingdoms

This section explains how improved understanding of organisms leads to revisions in our system of biological classification. Students are introduced to the six kingdoms: Archaebacteria, Eubacteria, Protista, Plantae, Fungi, and Animalia. They learn how organisms belonging in each kingdom are distinguished.

Bellringer

Have students list seven musical artists, bands, or acts. Have them categorize the names on their lists by style of music. Ask them to describe in their ScienceLog the categories they chose and also explain which bands might fit into more than one category.

1 Motivate

DISCUSSION

Grouping Animals Ask students how zoos group animals. (Answers may include by type, by climate preferences, and by natural habitats.)

Encourage knowledgeable students to describe the layout of zoos with which they are familiar.

Have students write letters to zoos all around the country requesting a copy of the map they issue to visitors. Students could then compare the layouts of many zoos. Be sure to have students include a stamped, self-addressed envelope with their letter describing the project.

Sección 2

Los seis reinos

VOCABULARIO

arqueobacterias   reino vegetal
eubacterias   reino de los hongos
reino de los protistas   reino animal

OBJECTIVES

- Explica cómo se desarrollaron los esquemas de clasificación para los reinos cuando se descubrió un mayor número de organismos diferentes.
- Mencionar los seis reinos y dar dos características de cada uno.

Durante cientos de años, todos los seres vivos se clasificaron en plantas o animales. La clasificación entre reino vegetal y reino animal funcionó bien hasta que se descubrieron organismos como la especie *Euglena*, que se ilustra en la **Figura 7.** Si fueras un taxonomista, ¿cómo clasificarías un organismo como éste?

Figura 7 *¿Cómo clasificarías este organismo? La* Euglena, *que en esta ilustración está aumentada unas 1,000 veces, tiene características de plantas y animales.*

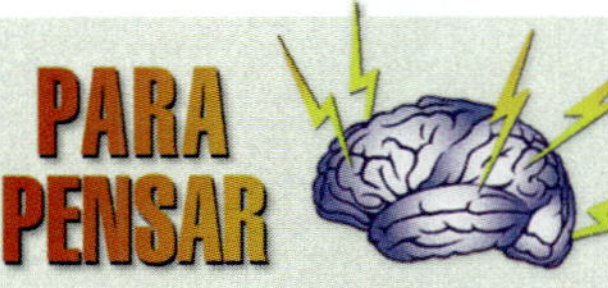

Si sus cloroplastos se mantienen retirados de la luz, la *Euglena* empieza a cazar comida como si fuera un animal. Si no reciben luz durante mucho tiempo, los cloroplastos se degeneran para siempre.

¿Qué es eso?

Como sabes, los organismos se clasifican por sus características. Como eres un excelente taxonomista, decides mencionar las características de la *Euglena*:

- Las *Euglenas* son una especie de organismos unicelulares que viven en aguas estancadas
- Las *Euglenas* son verdes y, como la mayoría de las plantas, pueden obtener su comida gracias a la fotosíntesis.

"¡Esto es fácil!", te dices a ti mismo. "Las *Euglenas* son plantas". Pero, ¡no tan rápido! Hay otras características importantes que debes considerar:

- Las *Euglenas* se desplazan de un lugar a otro moviendo sus "colas", llamadas flagelos.
- A veces, usan la comida que obtienen de otros organismos.

Ya sabes que las plantas no se mueven de un lado a otro y, por lo general, no comen otros organismos. ¿Quiere esto decir que las *Euglenas* son animales? Como puedes ver, parece que no encajan en ninguna categoría. Los científicos tuvieron el mismo problema, así que decidieron agregar otro reino para clasificar organismos como las *Euglenas*. Este reino se conoce como el reino de los protistas.

Más reinos A medida que los científicos aprendían más de los seres vivos, fueron agregando reinos para explicar las diferencias y similitudes entre los organismos. Hoy en día, la mayoría de los científicos coinciden en que el sistema de clasificación en seis reinos es el que funciona mejor. Sin embargo, todavía hay cierto desacuerdo y mucho que aprender. En las siguientes páginas aprenderás más sobre cada uno de los reinos.

CONNECT TO ENVIRONMENTAL SCIENCE

Recently, signs of salmonella infection were found in the droppings of an Antarctic gentoo penguin. The bacteria were most likely introduced from outside the Antarctic. The bacterium, *Salmonella enteritidis,* is not endemic to penguins. Scientists think that sewage dumped from passing ships or visiting albatrosses that feed on waste-contaminated squid in the oceans surrounding South America might be the sources of the bacteria. The bacteria could become infectious and pathogenic and kill the penguin chicks.

Los dos reinos de las bacterias

Las **bacterias** son organismos unicelulares muy pequeños, que se distinguen de los otros seres vivos porque son *procariotas*, es decir, organismos que no tienen núcleo. Muchos biólogos las dividen en dos reinos, **Arqueobacterias** y **Eubacterias**.

Las arqueobacterias han estado en la Tierra por lo menos desde hace 3 mil millones de años. El prefijo *arqueo* viene del griego y significa "antiguo". Hoy en día puedes encontrar arqueobacterias en lugares donde la mayoría de los organismos no pueden sobrevivir. La **Figura 8** ilustra un manantial de aguas termales en el Parque Nacional Yellowstone *(Yellowstone National Park)*. Los anillos anaranjados y amarillos alrededor de la orilla del manantial están formados por los millones de arqueobacterias que viven ahí.

La mayoría de las bacterias son eubacterias. Estos organismos microscópicos viven en la tierra, el agua ¡y hasta dentro del cuerpo humano! Por ejemplo, la eubacteria *Escherichia coli*, la de la **Figura 9**, se encuentra en grandes cantidades dentro de los intestinos humanos, donde produce la vitamina K. Otro tipo de eubacteria convierte la leche en yogurt, y otra especie causa la sinusitis, las infecciones de oído y la neumonía.

Figura 8 *El agua del gran manantial prismático del Parque Nacional Yellowstone tiene una temperatura de 90°C (194°F). El manantial es el hogar de las arqueobacterias que crecen en sus aguas termales.*

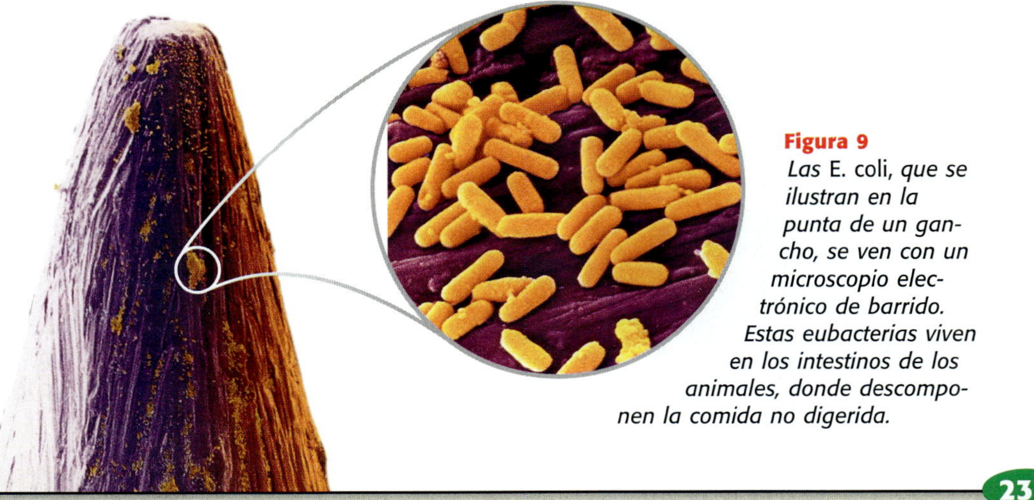

Figura 9
Las E. coli, *que se ilustran en la punta de un gancho, se ven con un microscopio electrónico de barrido. Estas eubacterias viven en los intestinos de los animales, donde descomponen la comida no digerida.*

235

REAL-WORLD CONNECTION

People in China, Japan, and other Asian countries have been practicing mariculture for thousands of years. Seaweed, shrimp, and mussels are commonly grown and harvested for food. In western countries, mariculture has experienced steady growth since the 1960s. Farmers grow kelp, fish, and shellfish for food in special farms near ocean shores or in ponds.

Homework

Writing **Researching Protists**
Have students research the protists shown in the pictures in **Figures 10, 11,** and **12** (*Paramecium*, slime mold, and giant kelp, respectively). Have them write descriptions about each of the protists, including information about its size, form, method of obtaining nutrients, method of reproduction, and in the case of the giant kelp, its commercial uses.

Answers to Self-Check

1. The two kingdoms of bacteria are different from all other kingdoms because bacteria are prokaryotes—single-celled organisms that have no nucleus.

2. The organisms in the kingdom Protista are all eukaryotes.

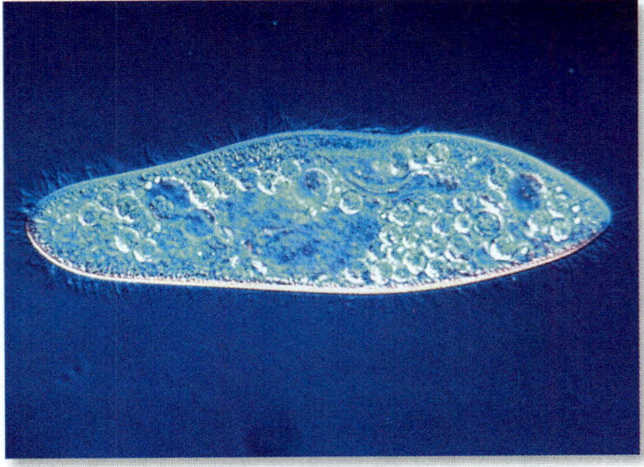

Figura 10 *Por lo general, este* Paramecium *se mueve muy rápidamente.*

Figura 11 *Este moho se extiende sobre un leño caído en el piso del bosque*

Figura 12 *Esta alga marina gigante es un protista multicelular.*

Reino de los protistas

Los miembros del **reino de los protistas,** generalmente llamado protistas, son organismos multicelulares simples. A diferencia de las bacterias, los protistas son *eucariotas,* organismos que tienen células con un núcleo y organelos rodeados por una membrana. El reino de los protistas abarca todas las eucariotas que no son plantas, animales ni hongos. Los científicos creen que los primeros protistas evolucionaron de bacterias antiguas hace unos 2 mil millones de años. Mucho después los protistas dieron lugar a las plantas, los hongos y los animales, así como a los protistas del presente.

El reino de los protistas abarca muchos tipos de organismos. Los protistas incluyen los protozoarios, parecidos a los animales; las algas, parecidos a las plantas; los hongos del limo y los hongos acuáticos, parecidos a los hongos. Las euglenas también son miembros del reino de los protistas, así como el Paramecium **(Figura 10)** y el moho de la **Figura 11.** La mayoría de los protistas son organismos unicelulares, pero algunos son multicelulares, como el alga marina gigante de la **Figura 12.**

✔ Autoevaluación

1. ¿Cuál es la diferencia entre los dos reinos de bacterias y los otros reinos?

2. ¿Cómo distingues el reino de los protistas de los otros dos reinos de bacterias?

(Respuestas en la página 636.).

SCIENTISTS AT ODDS

Is a slime mold a fungus? They were traditionally classified as Fungi because despite other differences they exhibit a similar life cycle, including the formation of spores on sporangia. But critics point out that some bacteria (myxobacteria) do this also, and those organisms are not reclassified as Fungi.

Reino vegetal

Aunque las plantas varían notablemente en tamaño y en forma, la mayoría de las personas reconocen con facilidad los miembros del **reino vegetal**. Las plantas son organismos multicelulares complejos que, por lo general, son verdes y usan la energía solar para crear azúcar a través de un proceso llamado *fotosíntesis*. Las secoyas y las plantas con flores de las **Figuras 13** y **14,** son ejemplos de los organismos del reino vegetal.

Figura 13 *Una secoya gigante puede medir 30 m de ancho y más de 91.5 m. de alto.*

Figura 14 *Las plantas como éstas se encuentran en un bosque tropical.*

¡MATEMÁTICAS!

Hacer una cadena humana alrededor del secoya gigante

¿Cuántos estudiantes necesitas para formar una cadena humana alrededor de un secoya gigante que tiene una circunferencia de 30 m? Imagínate que el estudiante promedio puede extender sus brazos unos 1.3 m. NOTA: No puedes tener una fracción de estudiante, así que asegúrate de redondear tu respuesta al número entero más cercano.

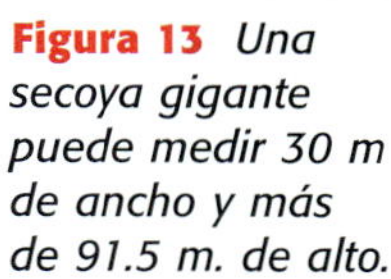

ciencias del medio ambiente
CONEXIÓN

Las secoyas son árboles muy raros. Sólo crecen en California y son una especie protegida. Algunas tienen más de 3,000 años.

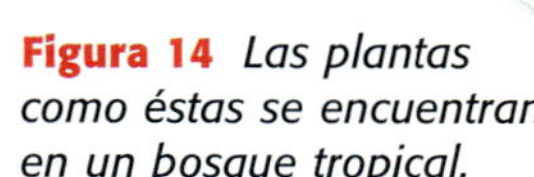

The name sequoia comes from *Sequoya,* the name of a Cherokee who is credited with developing the Cherokee written language during the 1820s.

DISCUSSION

Compare Structures Ask students what structures are common to the plants in **Figures 13** and **14.** (stems, leaves, and so on)

Ask them to assess differences and similarities between features on various kinds of plants, such as maple leaves and pine needles, tomato stems and tree trunks.

Answer to MATHBREAK

It would take 23 students to join hands and make a human chain around a 30 m sequoia.

MEETING INDIVIDUAL NEEDS

Learners Having Difficulty
Have students work together in small groups to find pictures in magazines of ferns and flowering plants that grow in North America. Provide resource books for the students to use to identify the plants. Then have students mount the plant pictures on poster board and label them. Sheltered English

MISCONCEPTION ALERT

Physical similarities are not always the best indicator of the relatedness of two organisms. For example, a small lizard, such as a skink, may look more like a salamander than a turtle, but it is more closely related to the turtle. Both the lizard and turtle are reptiles, and the salamander is an amphibian.

Math Skills Worksheet 19 "Arithmetic with Decimals"

GOING FURTHER

Writing Tell students that Pennsylvania, which has many caves, is one of the major mushroom-growing regions of the United States. Caves are ideal places in which to grow some kinds of mushrooms. Ask students to research and write a report on mushroom farming in the United States. What kinds of mushrooms are grown commercially, and what special conditions does each species require? Inexpensive kits are available for growing mushrooms, and interested students might enjoy the experience of raising their own.

RESEARCH

Students can research members of the six kingdoms that are prevalent in your immediate area. They can draw pictures of the organisms accompanied by descriptive paragraphs.

Answers to APPLY

Students are likely to think this plant is a fungus related to mushrooms because of its color and texture, but it is a plant. They will probably suggest that they need to know how it makes food because it obviously is not a green plant with leaves. Indian pipe, or *Monotropa uniflora,* is actually a wildflower member of the wintergreen family. It is a saprophyte living on the decayed roots of other plants. It gets all of its nutrients from other plants and needs neither leaves nor chlorophyll.

Figura 15 *Este hongo tan bonito del género* Amanita *es venenoso.*

Figura 16 *Este moho negro que crece en un pedazo de pan puede ser peligroso si inhalas sus esporas. Algunos mohos son muy peligrosos y otros producen antibióticos que salvan vidas, como la penicilina.*

Reino de los hongos

El moho y los hongos son ejemplos de los miembros multicelulares complejos del **reino de los hongos.** Los hongos se clasificaron originalmente como plantas, pero no obtienen nutrientes a través de la fotosíntesis. Tampoco tienen muchas de las características de los animales. Debido a sus peculiares características, se clasifican en un reino separado.

Los hongos no realizan la fotosíntesis, como las plantas, y no se alimentan de comida, como los animales. Lo que hacen es absorber nutrientes de sus alrededores después de descomponerlos con jugos digestivos. La **Figura 15** muestra un hongo muy bonito pero mortal y la **Figura 16** muestra un moho negro en un pedazo de pan. ¡Seguro que alguna vez has visto este tipo de moho en el pan!

Imagínate que caminas por un bosque con un amigo y se topan con el organismo que ves a la derecha. Piensan que es una planta, pero no se parece a ninguna de las que conocen. Tiene una flor y semillas, hojas muy chicas y raíces que crecen en un tronco podrido, y es blanco de la raíz a los pétalos. ¿A qué reino crees que pertenece? ¿En qué características se basa tu respuesta? ¿Qué información adicional necesitas para dar una respuesta más exacta?

IS THAT A FACT!

About 3,000 mushroom species grow in North America. Worldwide, about 70–80 species are poisonous.

Reino animal

Los animales son organismos multicelulares complejos que pertenecen al **reino animal.** La mayoría de los animales pueden desplazarse de un lado a otro y poseen un sistema nervioso que les permite percibir y reaccionar ante lo que sucede a su alrededor. A nivel microscópico, las células animales difieren de las células de los hongos, de las plantas, de la mayoría de los protistas y de las bacterias, en que no tienen pared celular. La **Figura 17** muestra algunos miembros del reino animal.

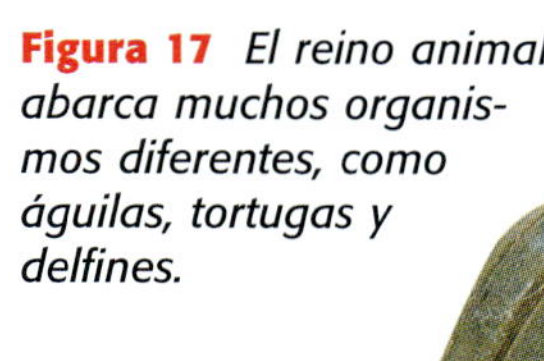

Figura 17 *El reino animal abarca muchos organismos diferentes, como águilas, tortugas y delfines.*

1. Menciona los seis reinos.

2. ¿Cuál de los seis reinos incluye los procariotas y cuál abarca los eucariotas?

3. Explica las diferentes maneras en que las plantas, los hongos y los animales obtienen nutrientes.

4. **Aplicar conceptos** Usa la información del diagrama de evolución de los primates para responder a las siguientes preguntas: ¿Qué primate comparte más rasgos con los humanos? ¿Comparten los lémures y los seres humanos las características mencionadas en el punto D? ¿Qué características tienen los mandriles que los lémures no tienen? Explica tus respuestas.

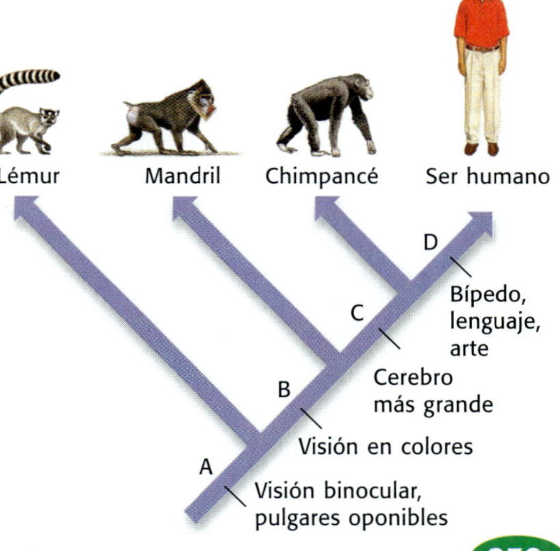

239

4 Close

Quiz

1. What causes increases in the number of kingdoms in the modern classification system? (discovery that some organisms do not fit in established kingdoms)

2. Which of the six kingdoms have single-celled organisms and which have multicellular organisms? (single-celled: Archaebacteria, Eubacteria, Protista; multicellular: Protista, Plantae, Fungi, Animalia)

ALTERNATIVE ASSESSMENT

Have students construct a chart of the six kingdoms. They should list the major characteristics of each kingdom on the chart, with a representative organism for each.

REINFORCEMENT

Have students provide written answers to the Review. (Answers are provided at the bottom of the page.)

Then have students describe and illustrate an animal in their ScienceLog that might require the formation of a seventh kingdom.

Reinforcement Worksheet 10 "Keys to the Kingdoms"

Chapter Highlights

SECTION 1

classification the arrangement of organisms into orderly groups based on their similarities and presumed evolutionary relationships

kingdom the most general of the seven levels of classification

phylum the level of classification after kingdom; the organisms from all the kingdoms are sorted into several phyla

class the level of classification after phylum; the organisms in all phyla are sorted into classes

order the level of classification after class; the organisms in all the classes are sorted into orders

family the level of classification after order; the organisms in all orders are sorted into families

genus the level of classification after family; the organisms in all families are sorted into genera

species the most specific of the seven levels of classification; characterized by a group of organisms that can mate with one another to produce fertile offspring

taxonomy the science of identifying, classifying, and naming living things

dichotomous key an aid to identifying unknown organisms that consists of several pairs of descriptive statements; of each pair of statements, only one will apply to the unknown organism, and that will lead to another set of statements, and so on, until the unknown organism can be identified

Resumen del capítulo

SECCIÓN 1

Vocabulario

clasificación *(pág. 228)*
reino *(pág. 229)*
filo *(pág. 229)*
clase *(pág. 229)*
orden *(pág. 229)*
familia *(pág. 229)*
género *(pág. 229)*
especie *(pág. 229)*
taxonomía *(pág. 230)*
clave dicotómica *(pág. 232)*

Notas de la sección

- La clasificación se refiere a la organización de los organismos según sus similitudes y relaciones de evolución.

- Los biólogos clasifican a los organismos para organizar el número y la diversidad de seres vivos, y darles nombres científicos.

- Un nombre científico siempre es el mismo para un organismo específico, sin importar cuántos nombres comunes tenga.

- El esquema de clasificación que se usa actualmente se basa en el trabajo de Carolus Linnaeus, un científico que vivió en el siglo XVIII. Linnaeus fundó la ciencia de la taxonomía, que describe, nombra y clasifica los organismos.

- Hoy en día, los organismos se clasifican en siete niveles: reino, filo, clase, orden, familia, género y especie. El género y la especie de un organismo forman su nombre científico.

- La clave dicotómica identifica los organismos.

- Los esquemas de clasificación moderna incluyen las relaciones de evolución.

Experimentos

La Isla de las formas *(pág. 592)*

☑ Comprobar destrezas

Conceptos de matemáticas

ORGANISMOS GRANDES La regla del redondeo dice que si el número que deseas redondear es mayor o igual al punto medio, debes redondear al número inmediatamente superior.

A veces, cuando trabajas con objetos en lugar de números, tienes que usar una regla diferente. En la sección "¡Matemáticas!" de la página 237, vas a redondear tu respuesta aunque incluya una fracción menor que la mitad del siguiente número. ¿Por qué? Porque si no redondeas, no tendrás suficientes alumnos para rodear el árbol.

Comprensión visual

NIVELES DE CLASIFICACIÓN Si tienes algunas dudas sobre los siete niveles de clasificación de los organismos, vuelve a la página 229. Revisa la Figura 2 y observa que el reino es el nivel más amplio y general. Por ejemplo, todos los animales están agrupados en el reino animal. A partir de ahí, los grupos se vuelven cada vez más específicos hasta llegar a un solo tipo de animal al nivel de especie. Si trabajas a partir de la especie, observa que, a medida que subes hacia al nivel de reino, cada vez hay más animales en el grupo.

240

Lab and Activity Highlights

Shape Island PG 592

Voyage of the USS *Adventure* PG 594

Datasheets for LabBook
(blackline masters for these labs)

SECTION 2

bacteria extremely small, single-celled organisms without a nucleus; prokaryotic cells

Archaebacteria a classification kingdom that contains ancient bacteria that thrive in extreme environments

Eubacteria a classification kingdom containing mostly free-living bacteria found in many varied environments

Protista a kingdom of eukaryotic single-celled or simple multicellular organisms; kingdom Protista contains all eukaryotes that are not plants, animals, or fungi

Plantae the kingdom that contains plants—complex, multicellular organisms that are usually green and use the sun's energy to make sugar by photosynthesis

Fungi a kingdom of complex organisms that obtain food by breaking down other substances in their surroundings and absorbing the nutrients

Animalia the classification kingdom containing complex, multicellular organisms that lack cell walls, are usually able to move about, and possess nervous systems that help them be aware of and react to their surroundings

---

Vocabulario

bacterias (*pág. 235*)
arqueobacterias (*pág. 235*)
eubacterias (*pág. 235*)
reino de los protistas (*pág. 236*)
reino vegetal (*pág. 237*)
reino de los hongos (*pág. 238*)
reino animal (*pág. 239*)

Notas de la sección

- Al principio, los seres vivos se clasificaban en plantas y animales. Cuando los científicos descubrieron más acerca de los seres vivos y encontraron otros organismos, se agregaron nuevos reinos más descriptivos.

- La mayoría de los biólogos reconocen seis reinos: arqueobacterias, eubacterias, reino de los protistas, reino vegetal, reino de los hongos y reino animal.

- Las bacterias son procariotas, organismos unicelulares que no tienen núcleo. Los organismos de los otros reinos son eucariotas, organismos que tienen células con núcleo.

- Las arqueobacterias habitan la Tierra desde hace unos 3 mil millones de años y viven donde la mayoría de los otros organismos no pueden.

- La mayoría de las bacterias son eubacterias y viven en casi todas partes. Algunas son dañinas y otras son útiles.

- Las plantas, la mayoría de los hongos y los animales son organismos multicelulares complejos. Las plantas realizan la fotosíntesis; los hongos descomponen el material fuera de su cuerpo y luego absorben los nutrientes; los animales se alimentan con comida, que digieren dentro de su cuerpo.

Experimentos

Viaje de la nave espacial *Aventura* (*pág. 594*)

 **internet**

HRW VISITA: go.hrw.com

Visita el sitio web de HRW para encontrar una serie de herramientas de aprendizaje relacionadas con este capítulo. Sólo tienes que escribir la palabra clave:

PALABRA CLAVE: HSTCLS

*SCI*LINKS
N S T A VISITA: www.scilinks.org

Visita el sitio web de la **Asociación Nacional de Maestros de Ciencias** (*National Science Teachers Association*) para encontrar recursos de Internet relacionados con este capítulo. Sólo escribe el **ENLACE DE CIENCIAS** para obtener más información sobre el tema:

TEMA	ENLACE
La base para la clasificación	HSTL205
Niveles de clasificación	HSTL210
Claves dicotómicas	HSTL215
Los seis reinos	HSTL220

241

 **Vocabulary Review Worksheet 10**

 Blackline masters of these Chapter Highlights can be found in the **Study Guide.**

Lab and Activity Highlights

LabBank

 **EcoLabs & Field Activities,** Water Wigglers, Field Activity 1

Long-Term Projects & Research Ideas, Project 9

Using Vocabulary

1. taxonomy
2. phyla
3. species
4. two-part scientific names
5. prokaryotes

Understanding Concepts

Multiple Choice

6. a
7. d
8. a
9. c
10. b
11. a

Short Answer

12. More than one million species have scientific names. Each of them is unique, and all scientists know specifically which organism is being discussed without the confusion of common names.

13. Two kinds of evidence of evolutionary relationships are thought to be species that have shared characteristics and species that have common ancestors.

14. A eubacterium is a prokaryote because it is always single-celled and has no nucleus or other membrane-bound organelles.

Repaso del capítulo

UTILIZAR EL VOCABULARIO

Escoge el término correcto para completar cada una de las siguientes oraciones:

1. Linnaeus fundó la ciencia de _____. (*el análisis del ADN* o *la taxonomía*)

2. Todos los organismos clasificados en un solo reino se dividen en uno de _____. (*varios filos* o *varias clases*)

3. El nivel más específico de clasificación es _____. (*el género* o *la especie*)

4. Linnaeus empezó a nombrar organismos con _____. (*nombres científicos de dos partes* o *relaciones de evolución*)

5. Las arqueobacterias y las eubacterias son _____. (*procariotas* o *eucariotas*)

COMPRENDER CONCEPTOS

Opción múltiple

6. Cuando los científicos clasifican los organismos,
   a. los ordenan en grupos.
   b. les dan varios nombres comunes.
   c. deciden si son útiles o no.
   d. ignoran las relaciones de evolución.

7. Cuando los siete niveles de clasificación se ordenan del más general al más específico, cuál se encuentra en el quinto lugar?
   a. la clase
   b. el orden
   c. el género
   d. la familia

8. El nombre científico del nenúfar blanco europeo es *Nymphaea alba*. ¿A qué género pertenece?
   a. *Nymphaea*
   b. *alba*
   c. nenúfar
   d. lirio alba

9. "Realmente Feliz, Claudia Observa Fascinada Galaxias Enormes" es un sistema nemónico que ayuda a recordar
   a. los nombres científicos de diferentes organismos.
   b. los seis reinos.
   c. los siete niveles de clasificación.
   d. la diferencia entre células procariotas y células eucariotas.

10. La mayoría de las bacterias se clasifican en el reino de:
    a. Arqueobacterias
    b. Eubacterias
    c. Protistas
    d. Hongos

11. ¿Qué tipo de organismo prospera en un medio ambiente con condiciones extremas?
    a. arqueobacterias
    b. eubacterias
    c. protistas
    d. hongos

Respuesta breve

12. ¿Por qué el uso de nombres científicos es tan importante en la biología?

13. Menciona dos tipos de evidencia que los taxonomistas modernos usan para clasificar los organismos según las relaciones de evolución.

14. ¿Es la eubacteria un tipo de eucariota? Explica tu respuesta.

Organizar conceptos

15. Usa los siguientes términos para crear un mapa de ideas: reino, helecho, lagartija, reino animal, reino de los hongos, algas, reino de los protistas, reino vegetal, champiñón.

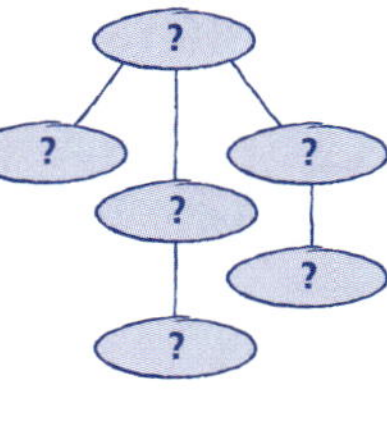

El siguiente diagrama ilustra las relaciones de evolución entre algunas plantas.

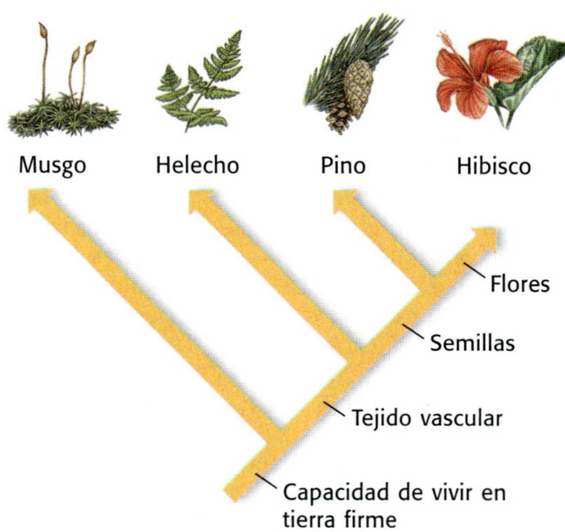

RAZONAMIENTO CRÍTICO Y RESOLUCIÓN DE PROBLEMAS

Escribe una o dos oraciones para responder a las siguientes preguntas:

16. ¿Cómo se relacionan los niveles de clasificación con las relaciones de evolución entre los organismos?

17. Explica por qué dos especies que pertenecen al mismo género, como el roble blanco *(Quercus alba)* y el alcornoque *(Quercus suber)*, también pertenecen a la misma familia.

18. ¿Qué característica tienen en común los miembros de los seis reinos?

21. ¿Qué planta es el pariente más cercano del antepasado común de todas las plantas?

22. Usa el diagrama y menciona, por lo menos, una característica que distinga cada planta de sus antepasados.

23. ¿Qué planta es pariente lejano del pino?

24. En este diagrama de evolución, ¿dónde pondrías las algas? Explica tu respuesta.

LAS MATEMÁTICAS EN LAS CIENCIAS

19. Los científicos calculan que aún hay millones de especies por descubrir y clasificar. Si sólo se han descubierto y clasificado 1.5 millones, o el 10 por ciento de las especies, ¿cuántas especies hay en la Tierra, según los científicos?

20. Las secoyas pueden crecer más de 90 m de altura. Cada metro tiene 100 cm. ¿Cuántos centímetros hay en 90 m?

AHORA, ¿qué piensas?

Revisa tus respuestas a las preguntas de la página 5 que escribiste en el cuaderno de ciencias. ¿Han cambiado tus respuestas? Si es necesario, corrige tus respuestas basándote en lo que has aprendido en este capítulo.

Concept Mapping

15. An answer to this exercise can be found at the end of this book.

CRITICAL THINKING AND PROBLEM SOLVING

16. Each level of classification groups organisms according to characteristics they share. Starting at the kingdom level, each level contains fewer organisms with characteristics in common until there is only one species at the species level.

17. The family level of classification contains more species than the genus level. All of the *Quercus* genera are in the same family because of shared characteristics.

18. All members of all six kingdoms are or once were living organisms. All living things share a common ancestor. They also have the genetic code in common, which means they all have DNA.

MATH IN SCIENCE

19. 15 million; Some scientists think that there may be as many as 15 million species on Earth today, most of which are undiscovered and unnamed.

20. 295.2 ft

INTERPRETING GRAPHICS

21. moss

22. Moss has the ability to live on land, ferns have vascular tissue, pine trees have seeds, and hibiscus has flowers.

23. moss

24. Algae would be lowest on the diagram, to the left of moss. It is a protist, not a plant, but algae share the characteristic of photosynthesis with plants.

NOW WHAT DO YOU THINK?

1. Classification is systematically organizing things or ideas.

2. People classify many things that are important in our everyday lives, such as food, addresses, and clothing.

3. Scientists classify things in order to find similarities that help us understand our world.

Concept Mapping Transparency 10

Blackline masters of this Chapter Review can be found in the **Study Guide.**

It's a Bird, It's a Plane, It's a Dinosaur?

Background

Scientists have identified many physical traits of birds that appear similar to those of dinosaurs. For example, some dinosaur skeletons contain a birdlike wishbone. Also, the two-legged upright stance of certain dinosaurs suggests that they were endothermic (birds are endothermic). All ectothermic animals are "sprawlers," meaning that they move around on all four feet.

The *Sinosauropteryx* found in northern China was first thought to support the birds-from-dinosaurs hypothesis because the fossil has featherlike features. But other paleontologists contend that rather than feathers, the structures are bristly fibers of collagen. Still other scientists believe these "feathers" are made of the same material as modern feathers but lack the organization of true feathers.

A 1997 find in Argentina gives some support to the proponents of the birds-from-dinosaurs hypothesis. A 6 ft long fossil found in Argentina shows the most birdlike dinosaur ever discovered. Its skeletal structure indicates it had arms that could flap and fold like wings. It had a birdlike pelvis as well. The sediments the dinosaur was found in suggest it is 90 million years old. But this fossil, too, has fueled the debate. Some experts say the dinosaur existed long after the development of modern birds. Birds, they argue, evolved from another line of reptiles.

DEBATE**CIENTÍFICO**

Veo, veo... ¿Qué ves?

Piensa en las aves. Pericos, palomas, buitres, emúes . . . ¡están en todas partes! Pero en otro tiempo no eran aves. ¿De dónde vinieron? ¿Cuándo evolucionaron? ¿Hace 225 millones de años?, ¿Hace 115 millones de años? Nadie está seguro. Sin embargo, éste ha sido un largo tema de debate entre científicos.

El debate empezó cuando, en 1860 y 1861, se encontraron en Alemania los restos del fósil de un dinosaurio de 150 millones de años con alas y plumas, el *Archaeopteryx*.

▲ El Archaeopteryx *fue la primera ave.*

¡Las aves son dinosaurios!

Algunos científicos piensan que las aves evolucionaron de pequeños dinosaurios carnívoros como el *Velociraptor* hace unos 115 ó 150 millones de años. Su idea se basa en las similitudes entre las aves modernas y estos pequeños dinosaurios. El tamaño, la forma y el número de uñas y "dedos" es de especial importancia, así como la ubicación y la forma del esternón y los hombros, la presencia de una estructura ósea cóncava y el desarrollo de los huesos que agitan para emprender el vuelo. Para muchos científicos, las pruebas son abrumadoras y sólo pueden llevar a una conclusión: las aves modernas descienden de los dinosaurios.

¡No es cierto!

"¡No tan rápido!", dice un grupo de científicos más pequeño pero con la misma determinación, que piensa que las aves se desarrollaron 100 millones de años antes del descubrimiento del *Velociraptor* y sus parientes. Señalan que estos dinosaurios eran terrestres y no tenían ni la forma ni el tamaño para volar. ¡Nunca se habrían despegado del suelo! Además, a estos dinosaurios les faltaba por lo menos uno de los huesos que necesitan las aves modernas para volar.

Esta idea de que las "aves descienden de los dinosaurios" se apoya en los fósiles de los *thecodonts,* pequeños reptiles trepadores que vivieron hace unos 225 millones de años. Un thecodont, un pequeño trepador de cuatro patas llamado *Megalancosaurus,* tenía los huesos y la forma del cuerpo adecuados, así como el centro de gravedad correcto para volar. Las pruebas indican, según este grupo de científicos, que las aves volaban mucho antes de que los dinosaurios existieran.

▲ *Es posible que este pequeño reptil trepador, el* Megalancosaurus, *evolucionara para dar origen a las aves que conocemos hoy en día.*

¿Quién tiene razón?

Ambos grupos están discutiendo sobre fósiles que tienen de 65 a 225 millones de años. Algunas especies dejaron muchos fósiles; mientras que otras sólo unos pocos. En los últimos años, algunos fósiles descubiertos en China, Mongolia y Argentina le han echado leña al fuego. Por lo tanto, los científicos seguirán estudiando las pruebas disponibles y haciendo conjeturas. Mientras tanto, el debate continúa.

Compara por tí mismo

▶ Busca y compara fotografías de fósiles de *Sinosauropteryx* y *Archaeopteryx*. ¿En qué se parecen? ¿En qué se diferencian? ¿Crees que las aves podrían ser dinosaurios modernos? Discute la idea con alguien que discrepe.

244

Answer to Compare for Yourself

All reasonable answers should be accepted. These are some possible responses. Similarities between *Sinosauropteryx* and *Archeopteryx* include: a reptilian skeleton, a long tail, sharp teeth, and claws of approximately the same size. The most noticeable difference between *Archeopteryx* and *Sinosauropteryx* is that *Archeopteryx* had wings and feathers and *Sinosauropteryx* had short front limbs without feathers.

Curiosidades de la CIENCIA

VIDA EN LOS LABIOS DE LA LANGOSTA

¿Alguna vez te has parado a pensar en los labios de las langostas? Es más, ¿sabías que las langostas tienen labios? Sí los tienen. Y los científicos han encontrado un animalito que vive en ellos. De eso hace ya 30 años, pero nunca lo habían estudiado de cerca. Cuando por fin lo hicieron, ¡quedaron sorprendidos! Esta diminuta forma de vida es diferente de cualquier otra cosa conocida. ¡Te presentamos a *Symbion pandora*!

Un poco extraña

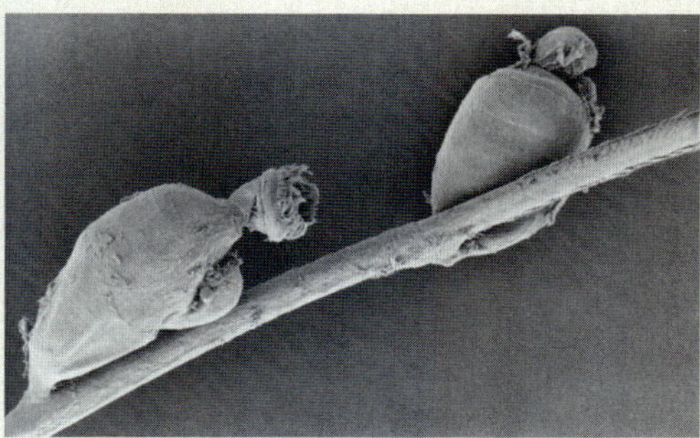

▲ *Aunque los científicos han sabido de la existencia de la* Symbion pandora *desde hace 30 años, no se habían dado cuenta de lo rara que era.*

¿Por qué la *Symbion pandora* es tan rara? Además de pasar la mayor parte de su vida en los labios de la langosta, la *S. pandora* también combina los rasgos de diferentes animales. Aquí te presentamos algunas de sus extrañas características:

- **Etapas de la vida:** El ciclo de vida de la *S. pandora* incluye varias *etapas* o formas corporales. En una época la *S. pandora* nada, mientras que en otras sólo subsiste pegada a la boca de la langosta.
- **Machos enanos:** Los *S. pandora* machos son mucho más pequeños que las hembras. Por eso se llaman *machos enanos.*
- **Costumbres alimenticias:** Los machos enanos no comen; sólo buscan una hembra, se reproducen y mueren.
- **Gemación:** Muchos organismos no tienen sexo. Estos animales se reproducen a través de un proceso llamado *gemación.* En la gemación, un nuevo animal puede brotar del adulto y asi sucesivamente.
- **Desaparición de los intestinos:** Cuando un adulto empieza a formar una gemación, su aparato digestivo y su sistema nervioso desaparecen. Parte de los intestinos contribuyen a la gemación. Luego, el adulto forma un aparato digestivo y un sistema nervioso nuevos para reemplazar a los viejos.

¿Qué tan raro es esto?

Cuando los científicos descubren una nueva planta o animal, concluyen que representa una nueva especie dentro de un género existente. En ese caso, crean un nombre para la especie nueva. Generalmente, quien encuentra el organismo lo bautiza. Si éste es *muy* raro, lo ubican en una nueva especie y un nuevo género.

La *S. pandora* es tan rara que la ubicaron en una nueva especie, un nuevo género, una nueva familia, un nuevo orden, una nueva clase ie incluso en un nuevo filo! Un descubrimiento científico como éste es excepcional. De hecho, cuando se descubrió, en 1995, se anunció en los periódicos de todo el mundo.

¿Dónde buscarías?

▶La *S. pandora* se observó por primera vez hace más de 30 años, pero nadie se dio cuenta de lo rara que era hasta que los científicos la estudiaron. Se calcula que hemos identificado menos del 10 por ciento de los organismos de la Tierra. Encuentra otras especies animales descubiertas en los últimos 10 años. ¿En qué lugares buscarías nuevas especies?

245

Answer to Where Would You Look?
Answers will vary.

UNIDAD 4

Las plantas

Probablemente estés familiarizado con las plantas, pero, ¿conoces su importancia? Al elaborar su propio alimento, las plantas proporcionan oxígeno y alimento a otros seres vivos.

A lo largo de la historia, las personas se han esforzado por entender a las plantas y en esta unidad tú también lo harás. Aprenderás cómo crecen, cómo se reproducen y cómo elaboran su propio alimento. Continúa leyendo para descubrir cosas fascinantes sobre las plantas.

250 a.c.

Los agricultores mayas construyen terrazas para controlar el agua que llega a sus cosechas.

1931

Barbara McClintock descubre que cuando los genes de una planta intercambian información se crean nuevas variedades de maíz.

1940

Se obtiene la primera imagen con el microscopio electrónico de un cloroplasto.

1580

Prospero Alpini descubre que las plantas tienen tanto estructuras masculinas como femeninas.

1763

Joseph Kohlreuter estudia la polinización de las plantas.

1776

Se firma la Declaración de Independencia.

1891

Se inventa el cierre.

1837

Se descubre que la clorofila es necesaria para la fotosíntesis.

1996

Se introduce una planta de algodón genéticamente modificada que es resistente a los insectos.

1983

Se aísla el VIH, virus responsable del SIDA.

1967

Se demuestra que el taxol, un extracto de la corteza del tejo del Pacífico, es un fármaco efectivo contra el cáncer.

Chapter Organizer

CHAPTER ORGANIZATION	TIME MINUTES	OBJECTIVES	LABS, INVESTIGATIONS, AND DEMONSTRATIONS
Chapter Opener **pp. 248–249**	45		**Investigate!** Cookie Calamity, p. 249
Section 1 **What Makes a Plant a Plant?**	90	▶ Identify the characteristics that all plants share. ▶ Discuss the origin of plants. ▶ Explain how the four main groups of plants differ.	**Demonstration,** Water Travel in Plants, p. 250 in ATE **Interactive Explorations CD-ROM,** Shut Your Trap! A **Worksheet** is also available in the **Interactive Explorations Teacher's Guide.**
Section 2 **Seedless Plants**	90	▶ Describe the features of mosses and liverworts. ▶ Describe the features of ferns, horsetails, and club mosses. ▶ Explain how plants without seeds are important to humans and to the environment.	**QuickLab,** Moss Mass, p. 255
Section 3 **Plants with Seeds**	90	▶ Compare a seed with a spore. ▶ Describe the features of gymnosperms. ▶ Describe the features of flowering plants. ▶ List the economic and environmental importance of gymnosperms and angiosperms.	**Skill Builder,** Travelin' Seeds, p. 598 **Datasheets for LabBook,** Travelin' Seeds, Datasheet 23 **EcoLabs & Field Activities,** The Case of the Ravenous Radish, EcoLab 3
Section 4 **The Structures of Seed Plants**	90	▶ Describe the functions of roots. ▶ Describe the functions of stems. ▶ Explain how the structure of leaves is related to their function. ▶ Identify the parts of a flower and their functions.	**Demonstration,** Seed Transport, p. 264 in ATE **Skill Builder,** Leaf Me Alone! p. 596 **Datasheets for LabBook,** Leaf Me Alone! Datasheet 22 **Making Models,** Build a Flower, p. 599 **Datasheets for LabBook,** Build a Flower, Datasheet 24 **Whiz-Bang Demonstrations,** Inner Life of a Leaf, Demo 8 **Long-Term Projects & Research Ideas,** Project 12

TECHNOLOGY RESOURCES

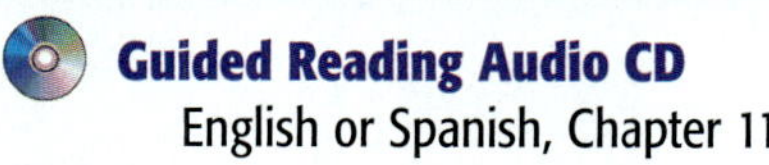

Guided Reading Audio CD
English or Spanish, Chapter 11

Eye on the Environment, Prairie Restoration, Segment 16

One-Stop Planner CD-ROM with Test Generator

Interactive Explorations CD-ROM
CD 1, Exploration 2, Shut Your Trap!

CLASSROOM WORKSHEETS, TRANSPARENCIES, AND RESOURCES	SCIENCE INTEGRATION AND CONNECTIONS	REVIEW AND ASSESSMENT
Science Puzzlers, Twisters & Teasers, Worksheet 11 **Directed Reading Worksheet 11**		
Directed Reading Worksheet 11, Section 1 **Transparency 42,** Plant Life Cycle **Transparency 43,** The Main Groups of Living Plants	**Multicultural Connection,** p. 252 in ATE **Math and More,** Percentages, p. 252 in ATE **MathBreak,** Practice with Percents, p. 253 **Careers:** Ethnobotanist—Paul Cox, p. 277	**Self-Check,** p. 252 **Review,** p. 253 **Quiz,** p. 253 in ATE **Alternative Assessment,** p. 253 in ATE
Directed Reading Worksheet 11, Section 2	**Cross-Disciplinary Focus,** p. 254 in ATE **Real-World Connection,** p. 255 in ATE	**Self-Check,** p. 256 **Review,** p. 257 **Quiz,** p. 257 in ATE **Alternative Assessment,** p. 257 in ATE
Directed Reading Worksheet 11, Section 3 **Science Skills Worksheet 22,** Science Writing **Transparency 44,** Two Classes of Angiosperms **Reinforcement Worksheet 11,** Classifying Plants **Reinforcement Worksheet 11,** Drawing Dicots	**Environmental Science Connection,** p. 259 **Apply,** p. 259 **Connect to Earth Science,** p. 260 in ATE **Cross-Disciplinary Focus,** p. 262 in ATE **Science, Technology, and Society:** Supersquash or Frankenfruit? p. 276	**Homework,** p. 261 in ATE **Review,** p. 263 **Quiz,** p. 263 in ATE **Alternative Assessment,** p. 263 in ATE
Directed Reading Worksheet 11, Section 4 **Transparency 45,** Root Structure **Science Skills Worksheet 21,** Taking Notes **Transparency 46,** Stem Structures **Transparency 47,** Leaf Structure **Transparency 190,** Photosynthesis **Transparency 47,** Flower Structure **Critical Thinking Worksheet 11,** The Voodoo Lily	**Cross-Disciplinary Focus,** p. 266 in ATE **Connect to Physical Science,** p. 269 in ATE	**Review,** p. 267 **Self-Check,** p. 268 **Homework,** p. 268 in ATE **Review,** p. 271 **Quiz,** p. 271 in ATE **Alternative Assessment,** p. 271 in ATE

Holt, Rinehart and Winston On-line Resources

go.hrw.com

For worksheets and other teaching aids related to this chapter, visit the HRW Web site and type in the keyword: **HSTPL1**

National Science Teachers Association

www.scilinks.org

Encourage students to use the *sci*LINKS numbers listed with the Chapter Highlights to access information and resources on the **NSTA** Web site.

END-OF-CHAPTER REVIEW AND ASSESSMENT

Chapter Review in Study Guide
Vocabulary and Notes in Study Guide
Chapter Tests with Performance-Based Assessment, Chapter 11 Test
Chapter Tests with Performance-Based Assessment, Performance-Based Assessment 11
Concept Mapping Transparency 11

Visual Resources

TEACHING TRANSPARENCIES

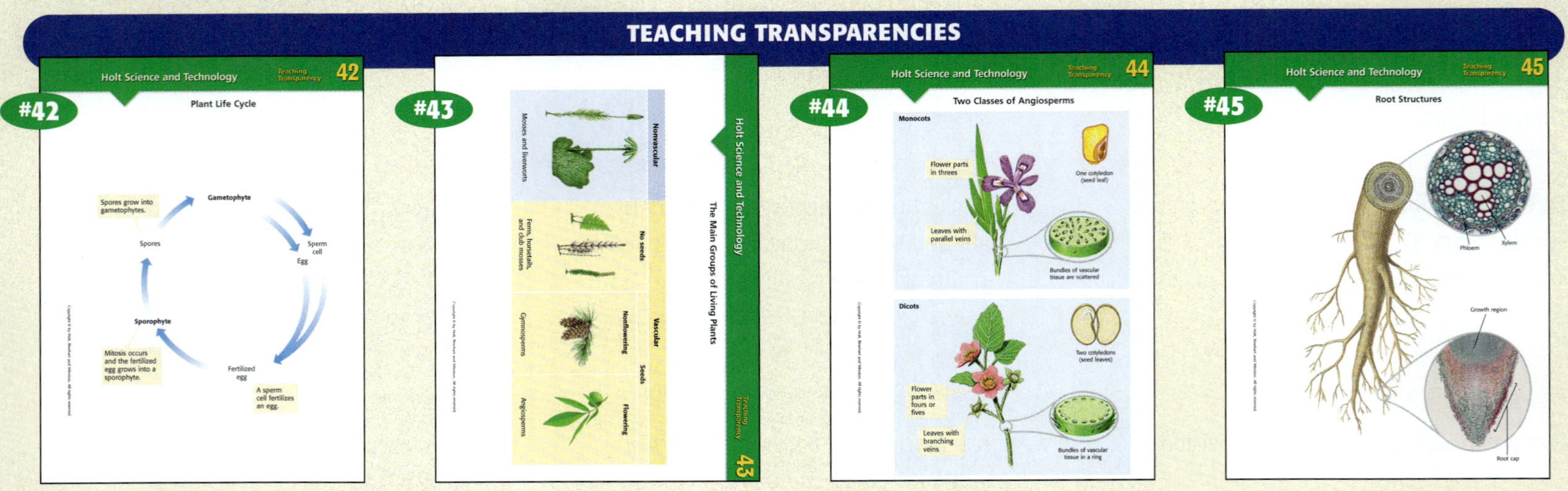

TEACHING TRANSPARENCIES

CONCEPT MAPPING TRANSPARENCY

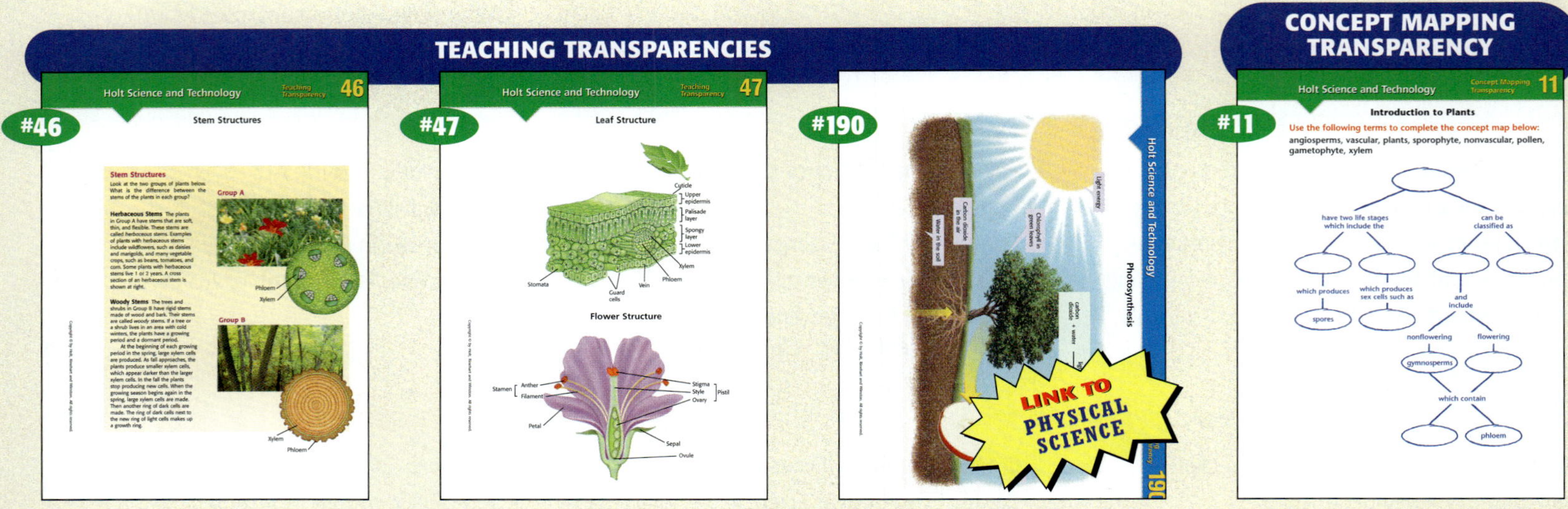

Meeting Individual Needs

DIRECTED READING

REINFORCEMENT & VOCABULARY REVIEW

SCIENCE PUZZLERS, TWISTERS & TEASERS

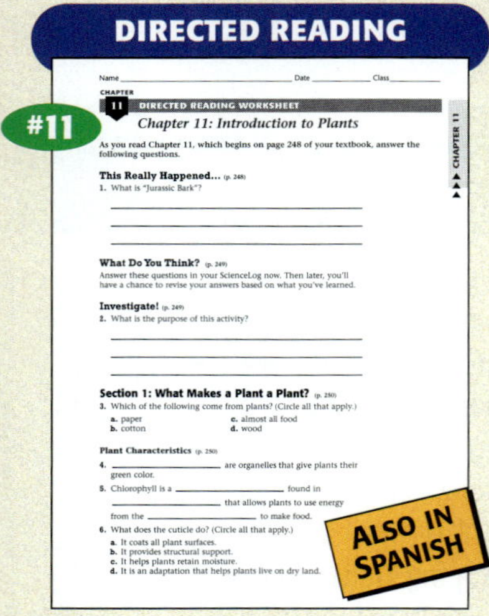

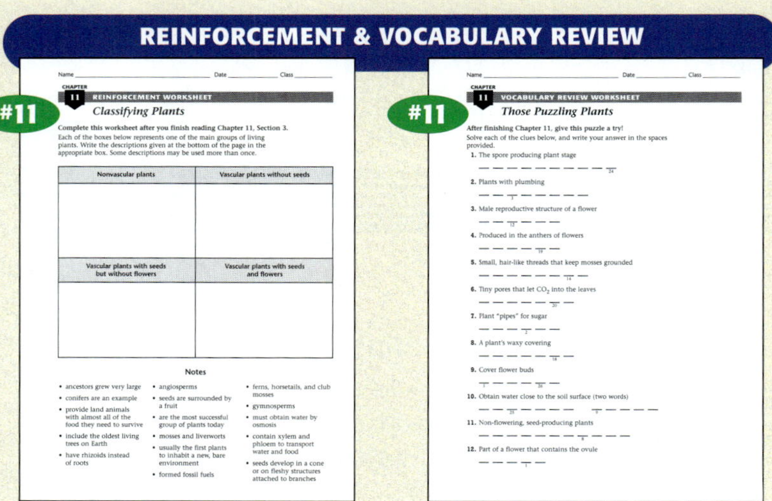

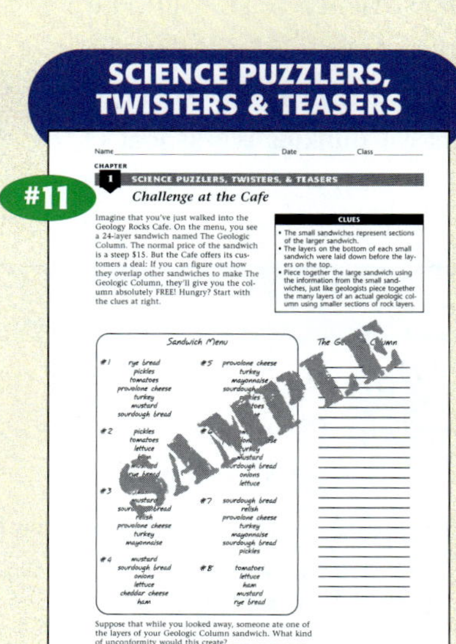

Review & Assessment

STUDY GUIDE

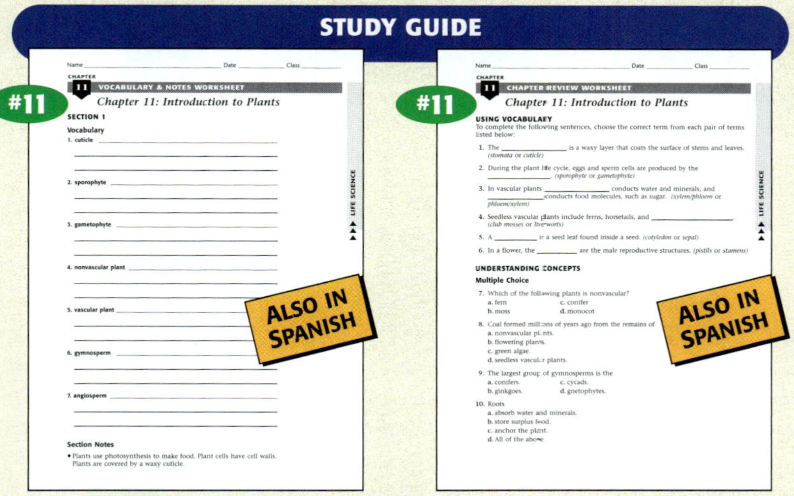

CHAPTER TESTS WITH PERFORMANCE-BASED ASSESSMENT

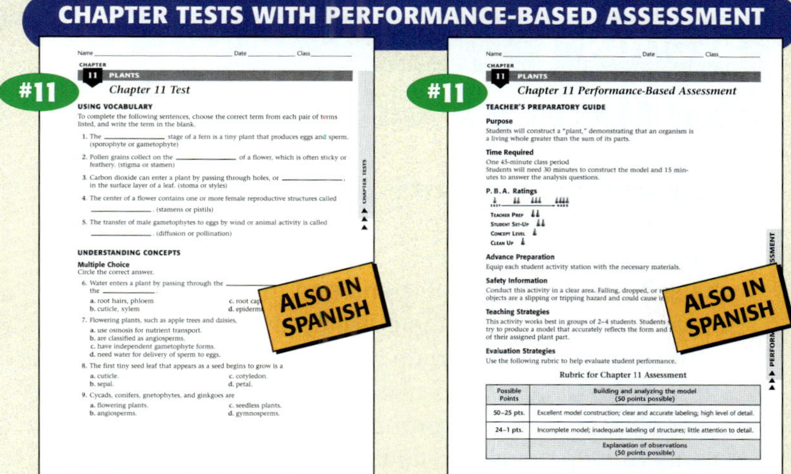

Lab Worksheets

ECOLABS & FIELD ACTIVITIES

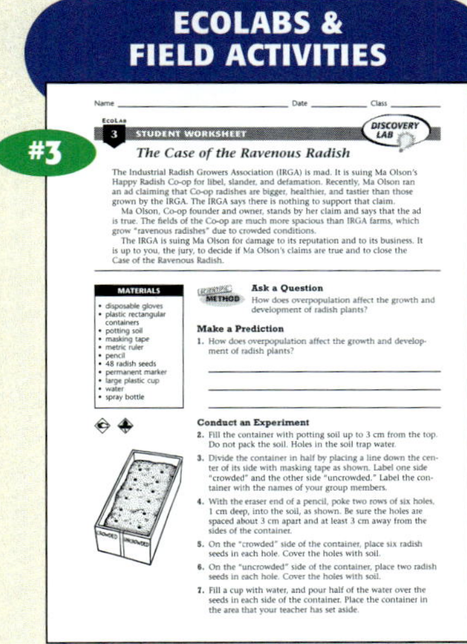

WHIZ-BANG DEMONSTRATIONS

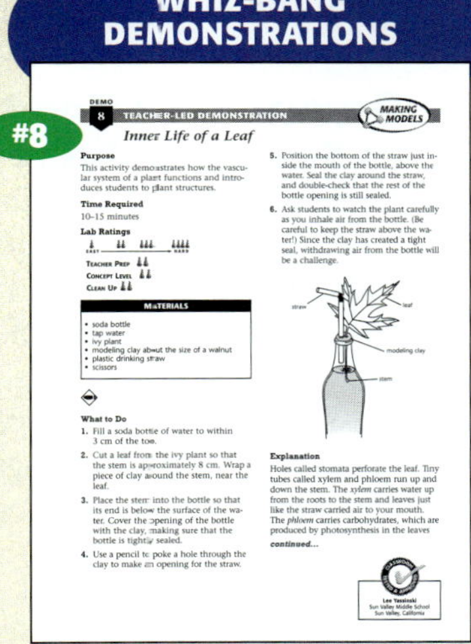

LONG-TERM PROJECTS & RESEARCH IDEAS

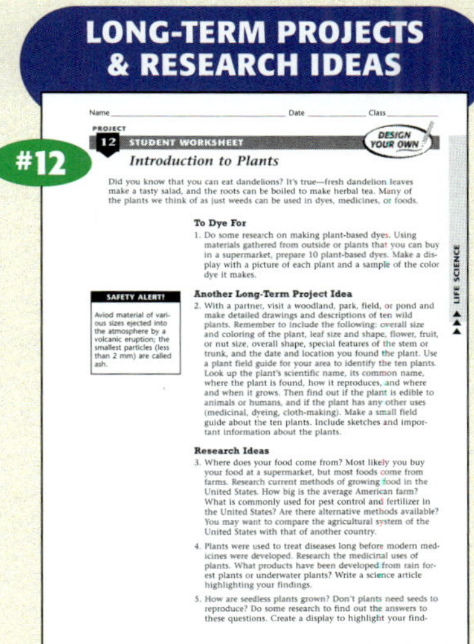

DATASHEETS FOR LABBOOK

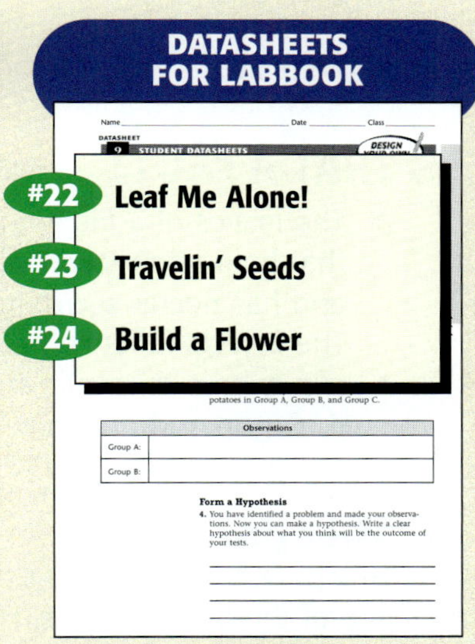

Applications & Extensions

CRITICAL THINKING & PROBLEM SOLVING

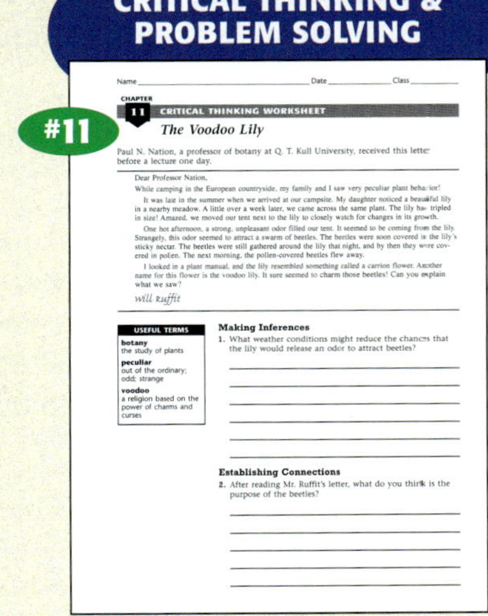

EYE ON THE ENVIRONMENT

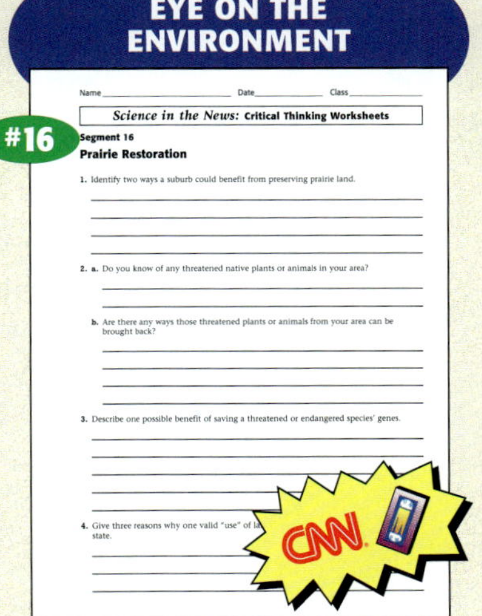

INTERACTIVE EXPLORATIONS

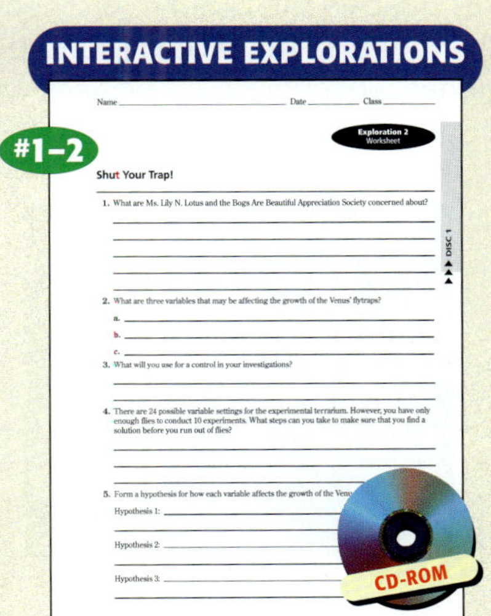

SECTION 1

What Makes a Plant a Plant?

▶ **Carnivorous Plants**

Carnivorous plants photosynthesize and are true plants. They tend to grow in bogs and marshes, where the soil is waterlogged. Bacteria and fungi cannot thrive in these soils, so there is little decomposition of organic matter to provide nutrients to plants. The small invertebrates that carnivorous plants catch provide additional nutrients, especially nitrates. The insects are digested by juices secreted by the leaves or by bacteria and fungi living in the plant.

IS THAT A FACT!

- The leaves of a pitcher plant form tall, narrow cups that hold rainwater. The tip of the plant is colorful and has nectar-secreting glands that attract insects. The insects follow a path of tiny hairs down into the cup, where the walls are smooth. The insects lose their grip and drown.

▶ **Theophrastus**

Theophrastus (372–288 B.C.), Aristotle's student, was one of the first botanists. He wrote two books, *History of Plants* and *The Causes of Plants*. He described the morphology, uses, propagation, and pollination of 500 plants and described sexual reproduction in plants. Theophrastus directed the Lyceum (a school and center of learning), in Athens. The Lyceum housed the first botanical garden. Theophrastus's writings were the standard for botanical study until the sixteenth century.

SECTION 2

Seedless Plants

▶ **Evolution of Ferns**

Ferns are an ancient group of plants with fossil records dating to the Devonian period, 408 million years ago. Nearly all of those early fern groups are now extinct. Only one or two living genera can be traced directly to Carboniferous ancestors.

▶ **Bryophytes: Good Things Come in Small Packages**

Bryophytes, which include mosses, hornworts, and liverworts, make up 15,000 described species worldwide. Various species tend to be restricted to particular environments because of sensitivities to temperature, light exposure, water availability, and chemical composition of the substrate. This makes them good indicator species for applied ecologists and conservation biologists; they can characterize an environment by identifying the bryophytes.

- Bryophytes are also useful for studies of evolution and population genetics. Their structures are complex enough to use as models for land-plant evolution but still simple enough for genetic studies. They are small and are easy to culture in the lab. Field biologists, however, can usually observe them year-round under natural conditions.

IS THAT A FACT!

- Before the invention of flashbulbs and strobe lights for indoor and low-light photography, photographers created an explosive flash of light with a powder. They spread the powder on a metal bar attached to a T-shaped device and ignited it. The powder contained millions of spores from club mosses (seedless, vascular plants that are related to ferns and horsetails and are not true mosses).

SECTION 3

Plants with Seeds

▶ The Millennium Seed Bank

The Royal Botanic Gardens, in Kew, England, has launched a project to collect seeds from 25,000 plant species around the world. One-fourth of the world's plants might become extinct in the next 50 years, but the seed bank will ensure the survival of plants vital for stabilizing soil and providing food crops, medicine, and building materials. The collected seeds are dried and stored in subzero temperatures. Scientists believe the seeds will be viable even hundreds of years in the future.

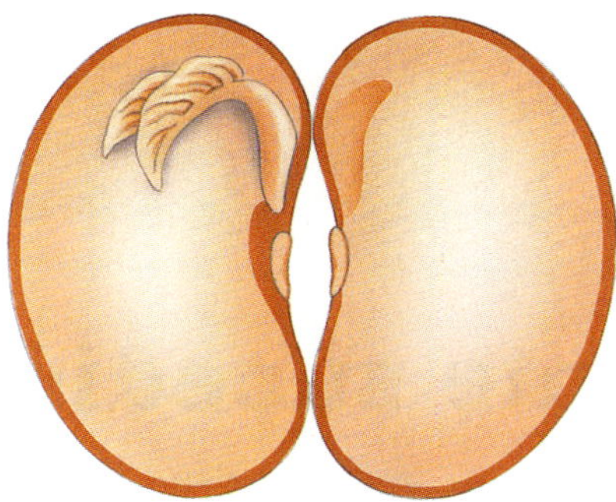

▶ The Economic and Social Importance of Plants

Farming originated thousands of years ago. Three of the earliest cultivated crops—wheat, rice, and corn—today feed more than half of the people in the world.

- Herbs and spices were valued commodities on ancient trade routes. In medieval times, explorers and merchants brought spices to Europe by camel caravan from east Asia.

- Perfume makers use essential oils from a variety of flowers, including rose, orange, lavender, and jasmine.

- The ancient Egyptians made paper from papyrus reeds. Paper can also be made from nettles, bamboo, and other plants. Today, most paper is made from wood pulp.

IS THAT A FACT!

- ▶ Alder fruits rely on water to disperse their seeds. The fruits contain oil droplets to keep the seeds afloat.

SECTION 4

The Structures of Seed Plants

▶ Can't Keep a Good Grass Down

Grass is mowed, eaten, and trampled, and still it grows. Grass buds are not on the ends of the leaf blades but are at ground level. Grazing geese and rabbits, lawn mowers, and people walking do not destroy the buds.

- People eat the seeds of grasses when they consume rice, wheat, oats, barley, rye, corn, millet, and sorghum.

▶ Inflorescence

The cluster of flowers that develops on many plants is called an inflorescence, of which there are two types. In a determinate inflorescence, the peduncle (main axis) terminates in a flower bud that prevents the peduncle from continued growth. In indeterminate inflorescence, the lower buds open first. As the peduncle continues to grow, the youngest flowers are always at the top. There are several forms of indeterminate inflorescence, including the following:

- raceme: each flower of the cluster is on a pedicel (short stem) that extends from the peduncle (snapdragon)

- spike: resembles a raceme but has no pedicels (gladiolus)

- panicle: branched raceme in which each branch has multiple flowers (lilac)

- head: short, dense spike with flowers in a circular mass (dandelion)

- umbel: all the pedicels grow from the same point at the top of the peduncle (onion)

For background information about teaching strategies and issues, refer to the **Professional Reference for Teachers.**

Science Puzzlers, Twisters & Teasers, Worksheet 11

Guided Reading Audio CD
English or Spanish, Chapter 11

CAPÍTULO 11

Introducción a las plantas

¡Esto realmente sucedió!

Un científico solitario se adentra en un lejano bosque tropical. En un estrecho cañón, nota algo fuera de lo común y descubre una especie de árbol de la época en que el *Tyrannosaurus rex* y el *Velociraptor* existían en la Tierra.

No se trata de una escena de Parque Jurásico, sino de un hecho real ocurrido en 1994, en un bosque tropical de Australia. El científico se llamaba David Noble y fue el descubridor de una especie de árbol que data del período Cretácico (hace 144 a 65 millones de años).

Los árboles, conocidos como pinos Wollemia, tienen grandes hojas en forma de cuchilla y su corteza es nudosa y café; crecen hasta 35 m de altura y sus troncos pueden alcanzar 1 m de ancho.

Desde su descubrimiento, los científicos de los Jardines Botánicos Reales, en Sidney (Australia), han sembrado semillas y cultivado plántulas de estos árboles. Pronto estarán disponibles para los amantes de la jardinería. En este capítulo, aprenderás más sobre el misterioso mundo de las plantas. Verás que son formas de vida complejas que desafían nuestro entendimiento de la naturaleza.

248

This Really Happened...

When David Noble first viewed the Wollemi Pines, he knew only that their overall appearance indicated they were gymnosperms. Then scientists used scanning electron microscopy to examine a plant's leaves and pollen. Now they have extracted DNA to identify the plant's gene sequence. We can learn many details about the plants around us even without sophisticated scientific instruments.

Usa tus conocimientos para responder a las siguientes preguntas en tu cuaderno de ciencias:

1. ¿Para qué les sirven a las plantas las flores y los frutos?

2. ¿Qué diferencias hay entre las plantas y los animales?

Galletas en peligro

Una mañana, una panadería recibe un pedido urgente de cientos de galletas con chispas de chocolate. Cuando el panadero va a por los ingredientes, se encuentra con que todos los que se derivan directa o indirectamente de las plantas se han terminado. ¿Podrá hacer las galletas?

Procedimiento

1. Examina los ingredientes de la receta de las galletas.

2. Determina qué ingredientes provienen de las plantas, ya sea directa o indirectamente.

3. Identifica las plantas relacionadas con cada ingrediente.

Análisis

4. Imagínate que la receta dijera mantequilla en lugar de margarina, ¿le serviría de algo al panadero? ¿Por qué?

5. Si quitaras de las galletas todos los ingredientes derivados de las plantas, ¿qué comerías?

What Do You Think?

Accept all reasonable responses.

Students will have a chance to revise their answers in the Chapter Review under NOW What Do You Think?

Investigate!

Teacher Notes: This is an analytical exercise that does not require experimentation.

Answers to Investigate!

4. Butter does not come directly from plants, but this would not help the baker. Students might suggest that the baker could not use the butter because it comes from cows, and cows depend on plants for food.

5. Salt and baking soda would be left because they do not come from plants (eggs come from chickens, and chickens depend on plants for food).

Directed Reading Worksheet 11

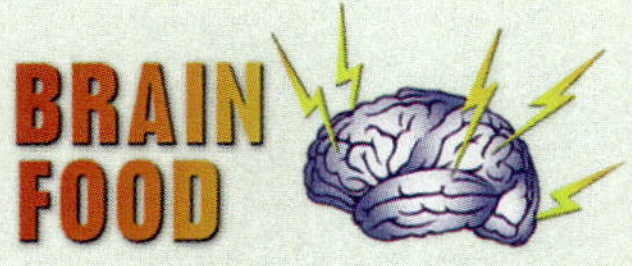
BRAIN FOOD

In 1982, the remains of a 2,000-year-old Japanese settlement were excavated by scientists. Among the findings was one seed in good condition. It was taken, planted, and watered, and it sprang into a magnificent magnolia plant. At first, it looked like the wild magnolias that grow all around Japan today. But 11 years later, when it produced its first bloom, it appeared to be different from any magnolia alive today. It had eight petals, not six. Was this just a fluke, or is it the sole survivor of an ancient species that disappeared from the face of the Earth and lay buried for 2,000 years?

Focus

What Makes a Plant a Plant?

In this section students will learn the shared characteristics of plants, such as that all plants make their own food, have a cuticle, reproduce with spores and sex cells, and have cells with cell walls. Finally, students will learn that the four main plant groups are classified as either nonvascular or vascular, depending on how materials are transported within the plant.

Bellringer

Tell students there are four major types of plants. Ask them to try to identify those types and to give at least two examples for each one. Have them review their responses when they have finished reading this section. (Students will likely respond with flowers, trees, weeds, and grasses. They might also list fruits and vegetables. They probably will not classify plants according to the chapter information.)

1) Motivate

DEMONSTRATION

Water Travel in Plants Slice a stalk of celery lengthwise to just below the leaves. Place the two halves in separate beakers, each containing a different color of water. Red and blue food coloring work best. Students should be able to see the veins in the leaves change color after the colored liquids have traveled up the stalk. Sheltered English

---

Sección 1

¿Qué distingue a una planta?

VOCABULARIO

cutícula
esporofito
gametofito
planta no vascular
planta vascular
gimnosperma
angiosperma

OBJETIVOS

- Identifica las características comunes a todas las plantas.
- Discute el origen de las plantas.
- Explica las diferencias entre los cuatro grupos de plantas principales.

Imagínate cómo sería pasar un día sin ningún derivado de las plantas. No sólo sería imposible hacer galletas con chispas de chocolate, sino muchas otras cosas, como usar pantalones vaqueros, ropa de algodón o lino, muebles de madera, lápices y el papel en todas sus formas, incluyendo el dinero. Casi todos los alimentos están elaborados a base de plantas o de animales que comen plantas, así que no se podría comer nada. Sería muy difícil sobrevivir a un día sin plantas; la vida que conocemos sería practicamente imposible si las plantas no existieran.

Características de las plantas

Como puedes observar, las plantas tienen diversos tamaños y formas. ¿Qué tienen en común los cactus, los lirios acuáticos, los helechos y las demás plantas? A pesar de que parecen muy diferentes, todas las plantas comparten ciertas características.

Las plantas elaboran su propio alimento Tal vez ya hayas notado que la mayoría de las plantas son verdes. Esto se debe a que sus células contienen cloroplastos. En capítulos anteriores, aprendiste que los cloroplastos son organelos que contienen un pigmento verde llamado *clorofila*. La clorofila absorbe la energía de la luz solar y esta energía se utiliza para fabricar moléculas de alimento, como la glucosa. Como recordarás, a este proceso se le llama fotosíntesis.

Las plantas tienen cutícula La **cutícula** es una capa cerosa que recubre la superficie de los tallos, hojas y otras partes expuestas al aire; es una adaptación que evita la desecación de las plantas, pues la mayor parte habita en tierras secas.

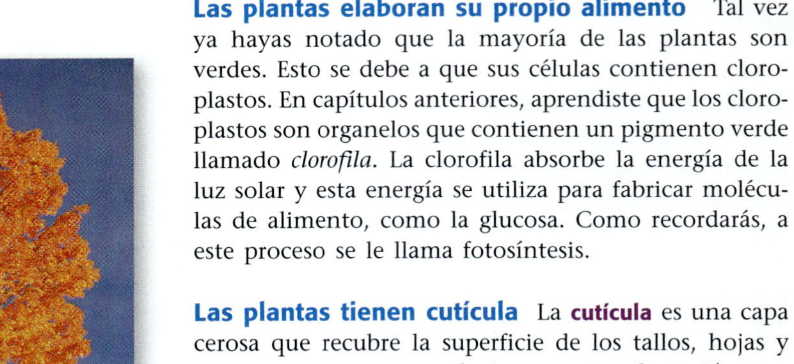

Arce

Directed Reading Worksheet 11 Section 1

internet**connect**

SCI LINKS
NSTA

TOPIC: Plant Characteristics
GO TO: www.scilinks.org
*sci*LINKS NUMBER: HSTL280

Q: How many sides are there to a tree?

A: two; inside and outside

Las células vegetales tienen paredes celulares Las células vegetales están rodeadas por una membrana celular y por una pared celular rígida, que la rodea y brinda soporte y protección a la planta gracias a los carbohidratos complejos y proteínas que contiene, y que forman un material fuerte. Cuando la célula alcanza su tamaño final, se puede desarrollar una pared celular secundaria, de consistencia resistente y leñosa; una vez formada, la célula no puede crecer más.

Las plantas se reproducen por medio de esporas y gametos El ciclo de vida de las plantas puede dividirse en dos etapas: en una se producen esporas, y en la otra gametos (óvulos y espermatozoides). La primera etapa se llama esporofito y la segunda etapa se llama gametofito. En la **Figura 2** se muestra un diagrama del ciclo vital de las plantas.

Las esporas y los gametos son células reproductoras diminutas. Las esporas que llegan a un medio adecuado, como un suelo húmedo, pueden dar origen a nuevas plantas; en cambio, los gametos no pueden hacerlo por sí solos. El gameto masculino (espermatozoide) debe unirse con el gameto femenino (óvulo) y el óvulo fertilizado que resulta de esta unión se desarrolla para formar una nueva planta.

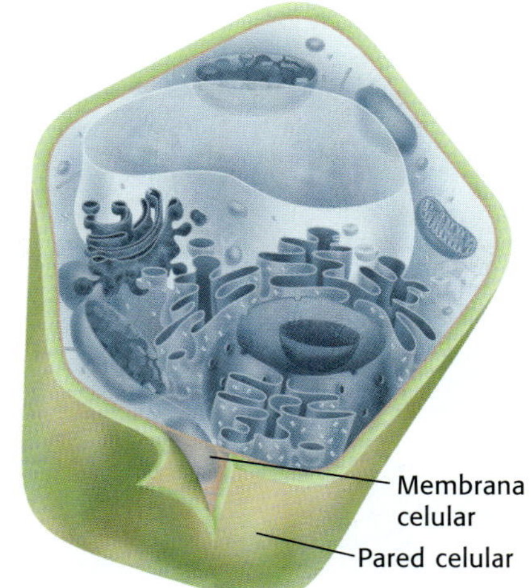

Figura 1 *Además de la membrana celular, las células vegetales están rodeadas por una pared celular.*

Figura 2 **Ciclo vital de las plantas**

Explora

Piensa de qué manera las plantas participan en tu vida diaria. Haz una lista de 10 usos de las plantas o de sus productos. Compárala con las de tus compañeros y compañeras para ver en qué se parecen y en qué se diferencian.

¿Sabías que muchos medicamentos provienen de las plantas? Pasa a la pág. 277 para aprender cómo utilizar las plantas para curar enfermedades.

251

IS THAT A FACT!

Some flowering plants have a type of surface protector that functions much like sunscreen lotion. Alpine flowers at high elevations produce purple pigments in their leaves to shield them from the damaging effects of ultraviolet light. The pigment allows sunlight in that is needed for photosynthesis but filters out harmful UV rays.

2) Teach

GROUP ACTIVITY

MATERIALS
FOR EACH GROUP:
• baby powder
• water
• eyedroppers or spoons

Organize the class into groups of 3 or 4, and give each group one set of the materials listed above. Tell students that all but one member of the group should coat the palms of their hands with the powder. Instruct the remaining member of the group to release a few drops of water on his or her classmates' hands and to record his or her observations. Explain to students that a plant's cuticle forms a similar barrier to prevent a plant from losing moisture.

Students should wash their hands immediately following this exercise. Sheltered English

Answer to Explore

Encourage students to include clothes, transportation items, furniture, writing implements, games, and food.

MEETING INDIVIDUAL NEEDS

Advanced Learners Have students compare and contrast the basic life cycle of a plant with that of bacteria and that of fungi. Tell them to create a series of illustrations that show how all three organism types reproduce and to include captions that explain similarities and differences.

Teaching Transparency 42 "Plant Life Cycle"

Multicultural CONNECTION

The ancient Maya and Aztec people made extensive use of the breadnut. It was boiled and eaten like potatoes or mashed into a gruel and sweetened. Breadnuts were ground, cooked, and mixed with corn to make tortillas. The diluted sap was fed to babies when their mother's milk was not available. The Maya and Aztecs also fed the leaves to female animals to increase their milk supply.

Answer to Self-Check

Plants need a cuticle to keep the leaves from drying out. Algae grow in a wet environment, so they do not need a cuticle.

MATH and MORE

Percentages The dandelions in your backyard produce 58,791 seeds. The germination rate is 36 percent. How many seeds will germinate?

(58,791 × 0.36 = 21,165 seeds)

Teaching Transparency 43 "The Main Groups of Living Plants"

Interactive Explorations CD-ROM "Shut Your Trap!"

Figura 3 *Las semejanzas entre las algas modernas y las plantas son prueba de que ambas surgieron de una especie de alga verde que existió hace mucho tiempo.*

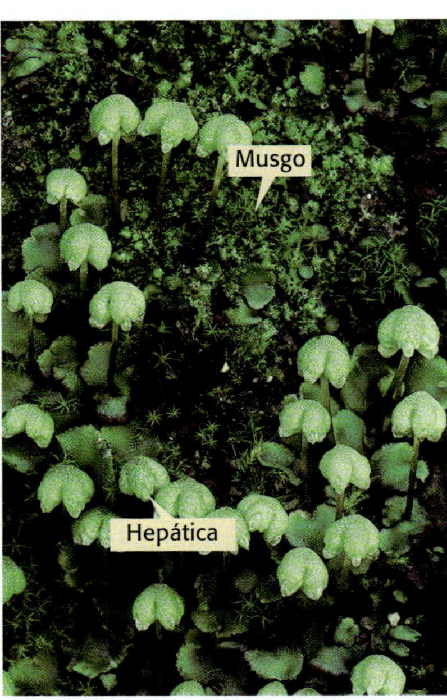

Figura 4 *Los musgos y las hepáticas no tienen tejidos conductores.*

252

El origen de las plantas

Si regresaras 440 millones de años en el tiempo, la Tierra te parecería un lugar extraño, estéril e inhóspito. La razón es que no había plantas terrestres. ¿De dónde vinieron las plantas? ¿Qué organismos fueron sus antepasados?

Observa las fotografías de la **Figura 3.** A la izquierda hay un organismo llamado alga verde. Como puedes ver, es del mismo color que las plantas, pero sus semejanzas van más allá del color. Por ejemplo, contienen la misma clase de clorofila y sus paredes celulares son similares. Además, ambas almacenan su energía en forma de almidón. Al igual que las plantas, las algas verdes tienen un ciclo de vida compuesto por dos etapas. ¿Qué te sugieren estas semejanzas? Los científicos piensan que las antiguas algas verdes que habitaron los océanos son los antepasados de todas las plantas, así como de las algas verdes actuales.

¿Cómo se clasifican las plantas?

Actualmente, existen más de 260,000 especies de plantas y, aunque todas comparten las características básicas comentadas anteriormente, se dividen en dos grupos: plantas vasculares y plantas no vasculares.

Plantas sin "tubería" Las **plantas no vasculares** (musgos y hepáticas) aparecen en la **Figura 4.** Estas plantas carecen de tubos para transportar el agua y los nutrientes, por lo que dependen de los procesos de difusión y ósmosis para transportar substancias de una parte a otra de la planta. Aunque estos procesos son lentos, proporcionan a las células de los musgos y las hepáticas las substancias que necesitan para vivir. Esto es posible porque son pequeñas; si fueran grandes (por ejemplo, del tamaño de los árboles), no habría manera de distribuir por difusión y ósmosis las substancias requeridas a todas las células.

✔ Autoevaluación

Una diferencia entre las plantas y las algas verdes es que las algas verdes no tienen cutícula. ¿Por qué las plantas necesitan cutícula y las algas verdes no? *(Consulta la página 636 para comprobar tu respuesta.)*

IS THAT A FACT!

The coconut palm sends its seed not by air but by sea. Inside the hard-shelled seed, there are supplies for a long voyage, such as plenty of food and water. The shell's exterior has a fibrous coat that helps it float. This self-contained travel package has enabled the coconut to travel miles at sea and to colonize beaches throughout the tropics.

Plantas con "tubería" Las **plantas vasculares** no dependen únicamente de la difusión y de la ósmosis para repartir las substancias a todas las células. Poseen tejidos constituidos por células que distribuyen las substancias requeridas a través de la planta, de manera similar a los tubos que llevan agua a las llaves de tu casa. Las plantas vasculares pueden alcanzar casi cualquier tamaño porque sus tejidos vasculares son capaces de transportar las substancias necesarias a grandes distancias. Algunas son muy pequeñas, pero otras son bastante grandes.

Las plantas vasculares pueden dividirse en dos grupos: las que producen semillas y las que no lo hacen. Entre las que no producen semillas están los helechos, los equisetos y los musgos. Las que producen semillas también se dividen en dos grupos: las que producen flores y las que no lo hacen. Las plantas sin flores se llaman **gimnospermas** y las plantas con flores se llaman **angiospermas**.

Los principales grupos de plantas vivientes son: (1) musgos y hepáticas; (2) helechos, equisetos y musgos; (3) gimnospermas, y (4) angiospermas. Se muestran en la **Figura 5.**

¡MATEMÁTICAS!

Practica porcentajes

En la siguiente lista aparece un cálculo del número de especies que hay en cada grupo de plantas:

Musgos y hepáticas	15,000
Helechos, equisetos y musgos	12,000
Gimnospermas	760
Angiospermas	235,000

¿Qué porcentaje de plantas no produce semillas?

Figura 5 Los principales grupos de plantas de la actualidad

REPASO

1. Menciona dos características de todas las plantas.

2. ¿Qué organismo se cree que fue el antepasado de todas las plantas? ¿Por qué?

3. ¿En qué se diferencian las angiospermas de los helechos, equisetos y licopodios?

4. **Aplicar conceptos** ¿Cómo determinarías si un organismo desconocido es un tipo de alga verde o una planta?

253

---

3 Extend

Answer to MATHBREAK

10.3 percent

GROUP ACTIVITY

Writing Arrange a visit to a local plant nursery. Have students record in their ScienceLog at least two examples, with physical descriptions, of each of the three vascular plant groups. Also encourage students to illustrate the plants.

4 Close

Quiz

1. How is a plant's size related to its method of transporting water and nutrients? (Nonvascular plants rely on osmosis and diffusion, which are efficient only in small plants. Vascular plants have conducting tissues, which enable the plant to be small or very large.)

2. What is required for a spore to grow into a new plant? (It must land in a suitable environment.)

ALTERNATIVE ASSESSMENT

Have students interview one another about the characteristics common to all plants. For example, have students ask: How do plants make their own food? Why are cell walls necessary? What is the purpose of the cuticle? What are gametophytes and sporophytes, and what roles do they play in the plant's life cycle?

▼ Answers to Review

1. Plants make their own food, have rigid cell walls, have a cuticle, and reproduce with spores and sex cells.

2. green algae, because both green algae and plants share many traits including the same kind of chlorophyll

3. Ferns, horsetails, and club mosses do not produce seeds, but angiosperms do produce seeds.

4. You should determine whether or not the organism has a cuticle. If the organism does, then it is a plant. If the organism does not, then it may be a type of alga.

Focus

Seedless Plants

In this section, students will learn that seedless plants include the nonvascular mosses and liverworts and the vascular ferns, horsetails, and club mosses. Students will learn about the features of each group and its life cycles. Finally, students will learn the importance of these plants to the environment and to humans.

Bellringer

Use the board or an overhead projector to display this question:

If plants can make their own food, why do people add fertilizer to the soil? (Fertilizers add nutrients to the soil that plants cannot make for themselves, such as minerals and nitrogen compounds.)

1 Motivate

DISCUSSION

Dandelions Ask students to describe a young dandelion flower and a mature flower that has developed seeds. (The flower's yellow head becomes white and "downy" at summer's end.)

Ask what happens to the "downy" head. (It breaks apart in the wind.)

What is blown away? Explain that although this section is about seedless plants, students can use this visual imagery of the dandelion to understand the different phases of a seedless plant's life cycle.

VOCABULARIO
rizoides
rizoma

OBJETIVOS
- Describe las características de los musgos y las hepáticas.
- Describe las características de helechos, equisetos y licopodios.
- Explica la importancia de las plantas sin semillas para los seres humanos y el medio ambiente.

Figura 6 *Los musgos nunca crecen mucho porque sus células deben obtener el agua directamente del medio o de otra célula.*

Figura 7 Ciclo vital de los musgos

254

Plantas sin semillas

Como ya sabes, hay dos grupos de plantas que no producen semillas. Uno comprende a las plantas no vasculares, los musgos y las hepáticas; y el otro, a varias plantas vasculares: helechos, equisetos y licopodios.

Musgos y hepáticas

Aunque haya plantas en todas partes, las no vasculares pasan fácilmente desapercibidas. Los musgos y las hepáticas son pequeñas y crecen en el suelo, las cortezas de los árboles y las rocas. Como no tienen sistema vascular, generalmente habitan lugares que siempre están húmedos. Cada una de sus células debe absorber agua por ósmosis directamente del medio o de una célula cercana.

Los musgos y las hepáticas no tienen tallos, raíces ni hojas verdaderas, pero poseen estructuras que llevan a cabo las funciones de estos órganos.

Rocas alfombradas De manera típica, los musgos **(Figura 6)** forman grandes grupos o colonias que recubren el suelo o las rocas con un tapete de diminutas plantas verdes. Actualmente existen alrededor de 9,000 especies de musgos.

Cada planta de musgo tiene unos hilos delgados parecidos a cabellos que se llaman **rizoides.** Al igual que las raíces, los rizoides le brindan sostén a la planta, pero no se los considera raíces porque no están formados por tejido vascular. La planta de musgo también tiene un tallo con hojas, pero como carece de tejido vascular, no se considera un tallo verdadero. El ciclo vital de los musgos alterna entre el gametofito y el esporofito, como se muestra en la **Figura 7.**

CROSS-DISCIPLINARY FOCUS

History Elizabeth Knight Britton was a botanist when female scientists faced many obstacles in their careers. She worked as the unofficial curator of mosses at Columbia University, in New York, and published 346 scientific papers between 1881 and 1930. Britton is credited with being the first person to suggest the establishment of the New York Botanical Garden, located in the Bronx, and she was a founder of the Wild Flower Preservation Society of America. One of her later achievements included helping to enact important conservation laws for the state of New York.

Hepáticas Al igual que los musgos, las hepáticas son plantas no vasculares pequeñas que generalmente habitan en lugares húmedos. Hoy en día existen alrededor de 6,000 especies de hepáticas. Su ciclo de vida es similar al de los musgos. Como se observa en la **Figura 8**, los gametofitos de las hepáticas pueden tener hojas y forma de musgo o ser anchos y aplanados. Sus rizoides se extienden hacia afuera desde el lado inferior del cuerpo de la planta y sirven para sujetarla.

La importancia de los musgos y las hepáticas

Las plantas no vasculares juegan un papel importante en el medio ambiente. Por lo general, son las primeras plantas en habitar un medio infértil, por ejemplo rocas recién expuestas. Al morir, forman una delgada capa de suelo en la que pueden crecer nuevas plantas, entre ellas más musgos y hepáticas. Al recubrir el nuevo suelo, ayudan a mantenerlo en su lugar y esto reduce la erosión. Los musgos también proporcionan a las aves materiales para hacer sus nidos.

Los musgos que forman turbas (residuos vegetales que se acumulan en pantanos y humedales, son los más importantes para los seres humanos. En ciertos lugares, como en Irlanda, hay pantanos donde se han formado gruesos depósitos de musgos muertos; esta turba se puede extraer, secar y usar como combustible.

Helechos, equisetos y licopodios

A diferencia de sus descendientes modernos, los antiguos helechos, equisetos y licopodios eran bastante altos. Los primeros bosques estuvieron formados por licopodios de 40 m, equisetos de 18 m y helechos de 8 m de altura. La **Figura 9** muestra la posible apariencia de estos bosques. Estas plantas tenían sistemas vasculares, por lo tanto podían crecer más que las plantas no vasculares.

Figura 8 Esta hepática tiene un gametofito ancho y aplanado. El esporofito parece una diminuta palmera o sombrilla.

Laboratorio

Masa de musgo

Determina la masa de una pequeña muestra de Sphagnum seco. Coloca la muestra en un vaso de precipitados grande con agua durante 10 a 15 minutos. Haz una predicción sobre cuál será la masa del musgo después de estar sumergido en el agua. Saca el musgo y determina su masa. ¿Cuánto aumentó? Compara el resultado con tu predicción. ¿Para qué se podría usar esta planta absorbente? Investiga para averiguarlo.

Figura 9 Gracias al tejido vascular, los antepasados de los modernos helechos, equisetos y licopodios alcanzaron una gran altura.

Some of the same factors that encourage the development of peat—high acidity and low oxygen—have preserved the bodies of people who died in bogs. Preserved bodies have been recovered from bogs in Europe and in Florida. Some of the bodies are 2,000 years old.

② Teach

REAL-WORLD CONNECTION

Sphagnum Moss Sphagnum moss, a primary component of peat bogs, was used during World War I as an absorbent dressing for wounds. Its hollow cells enable it to absorb up to 20 percent of its own weight in water. In earlier times, it was also used for diapers, lamp wicks, and bedding. Today gardeners often used sphagnum moss to protect fragile plants during shipment.

QuickLab

MATERIALS

FOR EACH GROUP:
• dry sphagnum moss
• large beaker of water
• balance or scale
(It may be helpful to use a dry beaker of predetermined mass to hold the wet moss on the scale or balance.)

Answers to QuickLab

Students should subtract the mass of the dry moss from the mass of the wet moss. The gained mass will be the mass of the water that the moss absorbed. See the Real-World Connection above for some uses of sphagnum moss.

MISCONCEPTION ALERT

Folklore says that moss grows on the north side of trees. But it is actually the green alga *Pleurococcus* that thrives on the moist, shaded (usually north) side of trees, stone walls, and fences.

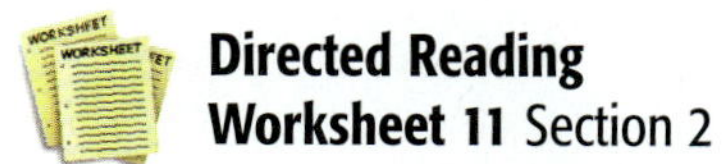

READING STRATEGY

Writing **Activity** Refer students to the diagram of the fern life cycle. Have them expand the identification of each stage with information they have learned from the text. For example, students could write a physical description of the gametophyte and of the conditions necessary for the sperm cell to fertilize the egg.

ACTIVITY

Identifying Plant Parts Before students read this page, let them view a potted fern. Tell students to draw a picture of the fern and to label as many parts as they can. Ask students if they have drawn and labeled the stem. (Answers will vary. Some students may realize that no stem was visible. Some may label a part of the fronds as a stem.)

Then remove the fern from the soil, point to the rhizome, and ask students if they know what it is. Explain that this structure is an underground stem. Stress to students that the structure and function of a plant part, not its location, is evidence of that part's identity. Sheltered English

Answers to Self-Check

Water transport is very limited for nonvascular plants. They have to stay low to the ground in order to have contact with moisture and soil nutrients. Vascular plants have xylem and phloem to transport nutrients and water throughout the plant; therefore, they can grow to be larger.

Figura 10 *Esta planta es un esporofito de helecho. Los sacos que están debajo de las frondas producen esporas. Las hojas enrolladas forman nuevas frondas.*

✓ Autoevaluación

¿Por qué algunas plantas vasculares son grandes, mientras que todas las plantas no vasculares son pequeñas? *(Consulta la página 636 para comprobar tu respuesta.)*

Helechos Los helechos crecen en muchos lugares, desde el gélido Ártico hasta los cálidos y húmedos bosques tropicales. Actualmente existen alrededor de 11,000 especies. A pesar de que la mayoría de ellos son relativamente pequeños, algunos helechos arbóreos de los trópicos pueden alcanzar 23 m de altura.

La **Figura 10** muestra un helecho típico. Casi todos los helechos tienen un tallo subterráneo llamado **rizoma,** que desarrolla raíces y hojas (*frondas*). Las frondas jóvenes están enroscadas como el extremo de un violín. Al igual que los ciclos vitales de otras plantas, el de los helechos se divide en dos etapas. Quizá la que más conoces sea la del esporofito, ilustrado en la **Figura 11.** El gametofito de un helecho es una planta diminuta del tamaño de media uña; es verde y plana, con forma de corazón. Este gametofito tiene estructuras masculinas que producen espermatozoides y estructuras femeninas que producen óvulos. Al igual que los musgos, los helechos necesitan agua para llevar los espermatozoides hasta los óvulos. Si en el suelo hay una capa delgada de agua, los espermatozoides nadarán hasta el óvulo.

Figura 11 **Ciclo vital de los helechos**

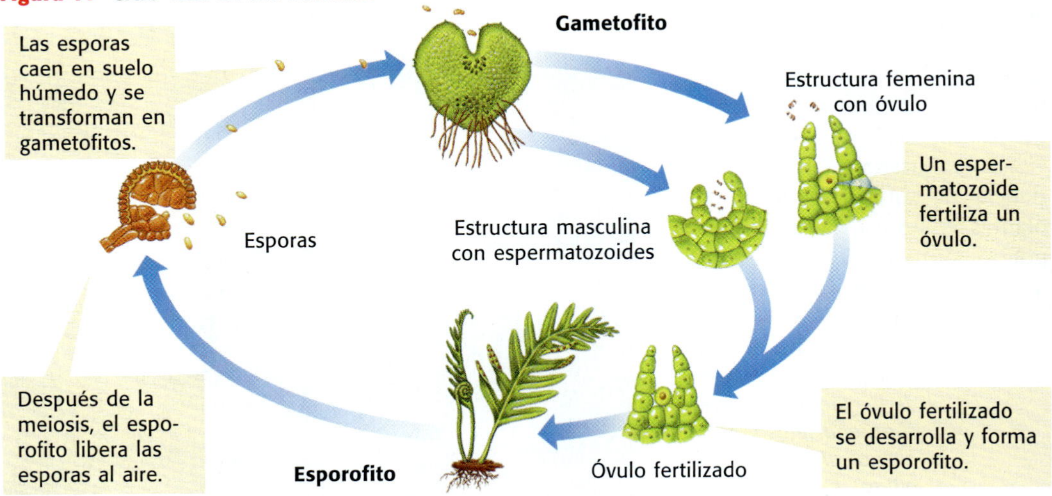

IS THAT A FACT!

The largest plant leaves in the world are those of the raffia plant of the Mascarene Islands, in the Indian Ocean, and of the bamboo palm of South America and Africa. These leaves can grow to nearly 20 m in length.

Equisetos Hace millones de años, los equisetos o colas de caballo eran plantas comunes, pero sólo unas 15 especies han sobrevivido hasta hoy. Los equisetos modernos, como los que ves en la **Figura 12,** son plantas vasculares pequeñas que generalmente miden menos de 1.3 m de altura. Crecen en lugares húmedos y pantanosos y también se les llama colas de caballo porque algunas especies tienen esa forma. Sus tallos son huecos y contienen sílice; por eso su textura es arenosa. De hecho, los primeros pobladores de los Estados Unidos les llamaron "estropajos" y los utilizaban para fregar ollas y sartenes. El ciclo vital de los equisetos es similar al de los helechos.

Licopodios Los licopodios de la **Figura 13** son plantas pequeñas que miden alrededor de 25 cm de altura y crecen en los bosques. A diferencia de los verdaderos musgos, los licopodios sí tienen tejido vascular. Hace millones de años eran plantas comunes, al igual que los equisetos. Actualmente existen unas 1,000 especies.

La importancia de las plantas vasculares sin semillas

Las plantas vasculares sin semillas juegan un papel importante en el medio ambiente. Al igual que los musgos y las plantas hepáticas, los helechos, equisetos y licopodios contribuyen a la formación del suelo y previenen su erosión al mantenerlo en su lugar.

Los helechos son populares en los hogares debido a la belleza de sus hojas. Las hojas comestibles enrolladas de algunos helechos se cosechan al inicio de la primavera.

Algunas de las plantas vasculares sin semillas más importantes para la humanidad existieron hace unos 300 millones de años. Sus restos formaron carbón, un combustible fósil que ahora se extrae de la corteza de la Tierra.

Figura 12 *Las puntas en forma de cono de los equisetos contienen esporas.*

REPASO

1. ¿Cuál es la relación entre el carbón y las plantas vasculares sin semillas?

2. ¿En qué se parecen los equisetos y licopodios a los helechos?

3. Menciona dos papeles importantes de las plantas vasculares sin semillas en relación con el medio ambiente.

4. **Hacer predicciones** Si los musgos y los helechos habitaran en un medio donde repentinamente hubiera una sequía, ¿cuáles serían los más afectados? Explica tu respuesta.

Figura 13 *Las puntas en forma de cono de los licopodios liberan esporas.*

GOING FURTHER

Writing **Botanical Timeline**
Have students construct a timeline that indicates the first appearances of each seedless plant discussed in this section and describes the animal life that existed at the same time. Tell students to include a brief report on which of those animal and plant species exist today.

4) **Close**

Quiz

1. What's the difference between a rhizoid and a rhizome? (A rhizoid is a threadlike extension of cells that anchors a moss to its substrate. A rhizome is the underground stem of a fern.)

2. Describe the ecological importance of mosses and liverworts. (They can inhabit areas that previously had no plants. When they die, they form a thin soil in which other plants can grow.)

ALTERNATIVE ASSESSMENT

Writing Have students make a chart that organizes information about the differences and similarities of the seedless nonvascular and seedless vascular plants discussed in this section. Have students present their chart to the class.

▼ *Answers to Review*

1. Coal is a fossil fuel formed from seedless vascular plants that died about 300 million years ago.

2. Like ferns, horsetails and club mosses are vascular, grow in moist environments, and do not use seeds to reproduce.

3. Seedless vascular plants help to form soil. They also help to prevent erosion.

4. Ferns would probably be able to survive longer because mosses are nonvascular and generally live in places that are always moist.

Focus

Plants with Seeds

Students will learn the characteristics that differentiate plants with seeds from those without seeds and how to compare and contrast seeds and spores. Students will also learn about the physical and reproductive features of gymnosperms and angiosperms as well as the ecological and economic importance of those plants.

Bellringer

Use the board or an overhead projector to pose this question to students:

If plants cannot move, how do they disperse their seeds?

1 Motivate

ACTIVITY

Seed Types

MATERIALS

FOR EACH CLASS:
- grey stripe sunflower seeds (not the small black oil variety)
- pumpkin seeds
- wildflower seed mix (available in nature stores and from catalogs)

Give students all of the seed varieties to examine. Tell students to compare and contrast the seeds in terms of size, shape, color, and texture. Ask students to compare this information with what they know about spores. Then ask students if they think it would be easier to introduce seed plants or seedless plants to a new plot of land. Why? Explain that this section will help them refine their answers. Sheltered English

---

VOCABULARIO

polen
polinización
cotiledón

OBJETIVOS
- Compara las semillas con las esporas.
- Describe las características de las gimnospermas.
- Describe las características de las plantas con flores.
- Conoce la importancia económica y ambiental de las gimnospermas y angiospermas.

Olmo inglés

Plantas con semillas

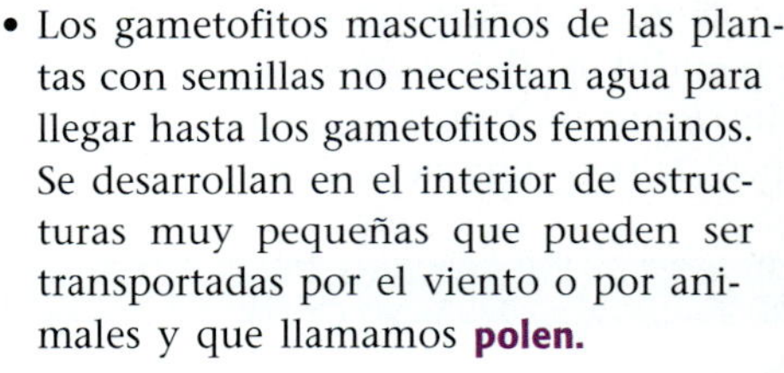
Duraznos

Probablemente estés más familiarizado con las plantas con semillas, como las que aparecen en esta página. ¿Las conoces?

Como vimos, hay dos grupos de plantas vasculares que producen semillas: las gimnospermas y las angiospermas. Las gimnospermas comprenden árboles y arbustos que tienen conos o estructuras carnosas en los tallos, donde se producen las semillas. Algunos ejemplos son el pino, el abeto, la picea y el ginkgo. Las angiospermas o plantas con flores producen sus semillas dentro de un fruto. Los árboles de durazno, los pastos, los robles, los rosales, los cactus y los ranúnculos son ejemplos de angiospermas.

Características de las plantas con semillas

Como todas las plantas, las que producen semillas alternan entre dos estadios: el esporofito y el gametofito. Sin embargo, se diferencian de las demás en lo siguiente:

- Producen semillas, estructuras dentro de las cuales se alimenta y protege a los esporofitos jóvenes.
- A diferencia de los gametofitos de las plantas sin semillas, sus gametofitos dependen del esporofito.
- Los gametofitos masculinos de las plantas con semillas no necesitan agua para llegar hasta los gametofitos femeninos. Se desarrollan en el interior de estructuras muy pequeñas que pueden ser transportadas por el viento o por animales y que llamamos **polen.**

Las características mencionadas arriba les permiten a las plantas con semillas vivir prácticamente en cualquier lugar. Por esta razón, hoy en día son las más comunes de la Tierra.

Yuca del desierto

IS THAT A FACT!

The smallest seeds in the world belong to the epiphytic orchid; 992.25 million of its seeds weigh 1 g.

¿Qué tienen de especial las semillas?

Después de la fertilización, o unión entre un óvulo y un espermatozoide, se desarrolla la semilla. Las tres partes que componen una semilla son: la planta joven (esporofito), el alimento almacenado y una cubierta resistente que rodea y protege a la planta joven. Estas partes se ilustran en la **Figura 14.**

Figura 14 *La semilla contiene alimento almacenado y una planta joven. El tegumento protege la semilla.*

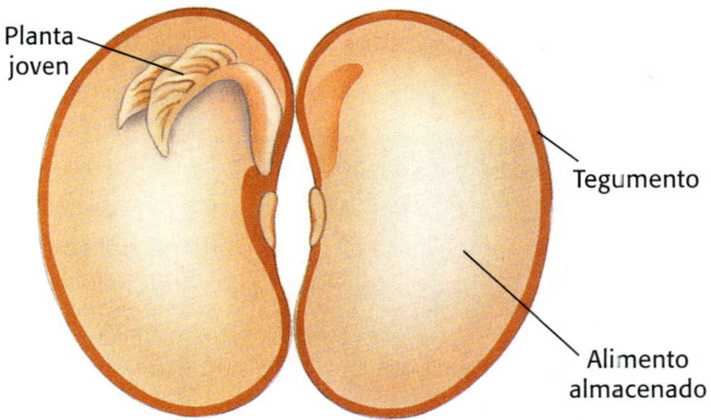

Las plantas que se reproducen por semillas presentan varias ventajas frente a las que producen esporas, como los helechos. Por ejemplo, la planta joven del interior de la semilla está formada por muchas células, está bien desarrollada y tiene raíz, tallo y hojas pequeñas. En cambio, la espora es una sola célula.

Cuando la semilla *germina,* es decir, empieza a crecer, la planta joven se alimenta de la comida almacenada en la semilla y, cuando las reservas se agotan, ya es capaz de fabricar alimento por medio de la fotosíntesis. En cambio, el gametofito que se desarrolla de la espora debe estar en un ambiente donde pueda iniciar la fotosíntesis tan pronto como empiece a crecer.

Experimentos

¿Sabías que a las semillas les gusta viajar lejos de casa? Pasa a la página 598 para ver cómo lo hacen.

ciencias del medio ambiente
CONEXIÓN

Los animales necesitan a las plantas para vivir y algunas plantas a su vez necesitan a los animales. Estas plantas producen semillas con tegumentos resistentes que no pueden empezar a desarrollarse hasta que un animal se las haya comido. Cuando la semilla se expone a los ácidos y enzimas del aparato digestivo del animal, el tegumento se destruye. Al salir del tracto digestivo del animal, la semilla es capaz de absorber agua, germinar y crecer.

A Paco y a su hermana les encanta sentarse en el porche a comer sandías jugosas durante el verano. Una vez apostaron a ver quién escupía las semillas más lejos. La primavera siguiente, Paco se dio cuenta de que unas plantas nuevas estaban creciendo en el jardín y, al examinarlas de cerca, vio unas sandías pequeñas. No tenían la menor idea de que habían iniciado una plantación de sandías. Piensa en los hábitos alimenticios de los animales silvestres, como las ardillas y los pájaros. ¿Cómo podrían comenzar una huerta?

259

GUIDED PRACTICE

Seed Dissection Give each student a lima bean that has been soaked in water to soften its seed coat. Tell students to break apart the bean with their fingers, being careful not to crush or squeeze the bean. Have students compare what they see with the information presented in **Figure 14.** Then tell students to write and illustrate their observations in their ScienceLog. Sheltered English

Answer to APPLY

Birds, squirrels, and other seed-eating animals scatter some seeds as they eat them or as they eat the fruits that contain them. Sometimes the animals bury seeds to store them, and sometimes the animals excrete the seeds after consuming them in fruit.

Directed Reading Worksheet 11 Section 3

Science Skills Worksheet 22 "Science Writing"

WEIRD SCIENCE

A large number of plants in the heathland of South Africa produce seeds with a very tasty covering called an elaiosome—tasty, that is, to ants, which carry seeds down into their underground colonies. The ants nibble off the outside covering and then leave the seed alone. The ants plant the seed at just the right depth for it to successfully germinate.

Gimnospermas: plantas con semillas y sin flores

Figura 15 *Las gimnospermas son plantas con semillas que no pro-ducen flores ni frutos.*

Entre las gimnospermas hay varias plantas ganadoras de marcas en el reino vegetal. Los árboles más antiguos que existen en la actualidad son dos especies de pinos, Pinus longaeva y P. aristata y habitan en California, Nevada y Utah. ¡Una de ellas tiene 4,900 años! Hay cuatro grupos de gimnospermas: coníferas, ginkgos, cícadas y gnetófi-tos. Sus semillas no están encerradas en un fruto. La pa-labra *gimnosperma* proviene del griego y quiere decir "semilla desnuda". Algunos ejemplos de estos grupos se muestran en la **Figura 15.**

Las **coníferas** incluyen alrededor de 550 especies y son el grupo más grande de gimnospermas. La mayoría son de hoja perenne y con forma de aguja. Sus semillas se desarrollan en conos. Los pinos, los abetos, las piceas y los cipreses son ejemplos de coníferas.

Los **ginkgos** comprenden una sola especie viva, el Ginkgo biloba. Este árbol pro-duce las semillas en estructuras carnosas que están directa-mente sujetas a las ramas.

Las **cícadas** eran comunes hace millones de años, pero actualmente sólo hay unas 140 especies. Crecen en los trópicos y sus semillas se desarrollan en conos.

Los **gnetófitos** comprenden alrededor de 70 especies muy poco comunes. Éste crece en zonas secas; sus semillas se for-man en conos.

260

Science Bloopers

When scientists compared radiocarbon dates of bristlecone pines with those obtained from tree ring patterns, they discovered that their calibrations for carbon-14 analysis were incorrect.

The new data indicated that some arti-facts found in Europe were 1,000 years older than had been previously thought. The bristlecone pines became known as the "trees that rewrote history."

Ciclo vital de las gimnospermas

Quizá las gimnospermas que más conozcas sean las coníferas. El nombre *conífera* quiere decir "llevar conos". Las coníferas tienen conos masculinos y femeninos, que se ilustran en la **Figura 16.** Las esporas masculinas se producen en los conos masculinos y las femeninas en los femeninos. Las esporas se convierten en gametofitos. Los gametofitos masculinos son el polen y producen los espermatozoides. El gametofito femenino, que está dentro de la escama del cono femenino, produce óvulos. El viento transporta el polen desde los conos masculinos hasta los femeninos de la misma planta o de otra. La transferencia del polen se llama **polinización.**

Después de ser fertilizado, el óvulo se desarrolla y forma una semilla dentro del cono femenino. Cuando la semilla está madura, el cono la libera y cae al suelo. La semilla germina y se transforma en un nuevo árbol. En la **Figura 17** se muestra el ciclo de vida de un pino.

La importancia de las gimnospermas

Desde el punto de vista económico, el grupo más importante de gimnospermas es el de las coníferas. Su madera se utiliza como material de construcción y para fabricar productos de papel. Los pinos producen resina, un líquido pegajoso que se utiliza para hacer jabón, trementina, pinturas y tinta.

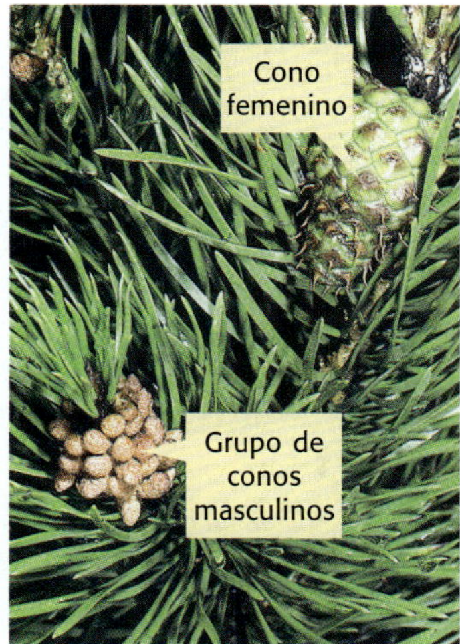

Figura 16 *Los pinos tienen conos masculinos y femeninos.*

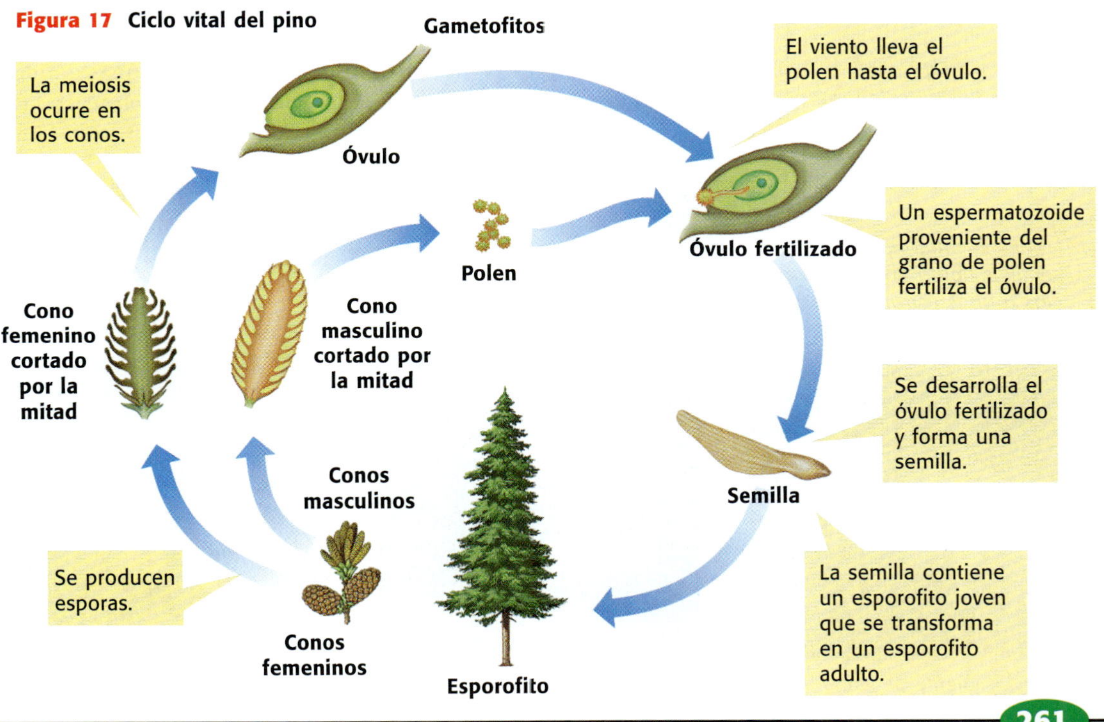

Figura 17 **Ciclo vital del pino**

261

People who buy a lot of books on gardening become good weeders.

DISCUSSION

Good Fires Ask students what environmental factors they think are necessary for a gymnosperm to germinate. (sunlight, water, soil)

Tell students that the cones of jack pines must burn to open. Explain that jack pines used to be quite numerous but have been greatly reduced in recent years. Ask students why this has happened. (There is a long-standing national policy of extinguishing forest fires, especially those that may threaten nearby towns and housing developments.)

Tell students that the Kirtland's warbler nests only in young jack pines, and ask what they think has happened to the warbler's population. Explain that conservationists now conduct controlled burns to save both the jack pines and the Kirtland's warblers.

Homework

Writing **Researching Ethnobotany** Have students research the definition of ethnobotany and write a report about ethnobotanists who have studied with indigenous South American groups to learn how they use plants.

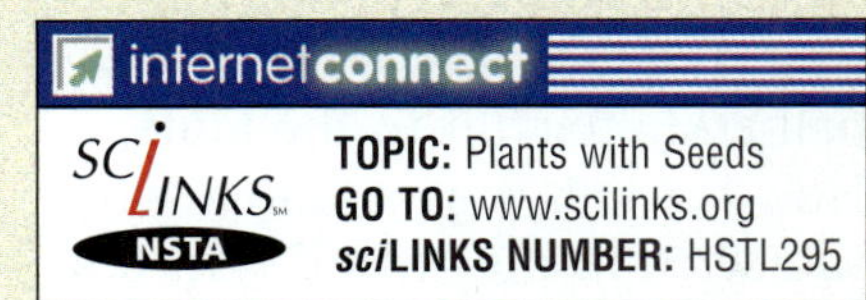

CROSS-DISCIPLINARY FOCUS

Writing **Art** Before the development of artificial dyes, artists and textile workers used solutions made from berries, roots, bark, leaves, flowers, and seeds of various plants to create colored paints and fabrics. Some of the plants they used included buckthorn, dogwood, fennel, sandalwood, and milkweed. Have students research three natural dyes that are each obtained from a different plant part. Their report should include an illustration of the plant and an explanation of how the dye is processed.

ACTIVITY

Seed Dispersal

MATERIALS
FOR EACH GROUP: • cotton balls • clear or masking tape • construction paper • scissors • fan • table

Have students shred the cotton balls and place the pieces on a table in front of the fan. Turn on the fan, and have students observe how far the cotton "seeds" travel. Have students wad a strip of tape into a marble-size ball and attach it to the table. How far does the ball move when subjected to the "wind"? Ask students how they think the seeds in this type of fruit might best be transported. (on an animal's fur)

Have students cut a "maple fruit" from construction paper. The rounded tips must be longer on one side than the other. Have students observe how this "fruit" behaves in the wind.

Sheltered English

Figura 18 *Esta abeja va camino de otra flor de calabaza, donde dejará parte del polen que lleva.*

Figura 19 *Algunos frutos y semillas flotan en el aire gracias a unas estructuras especiales.*

Diente de león

Arce

Algodoncillo

262

Angiospermas: plantas con semillas y flores

Hoy en día, las plantas con flores o angiospermas son las más exitosas. Se encuentran en casi todos los medios terrestres y hay por lo menos 235,000 especies, es decir, muchas más que todas las demás plantas juntas. Las angiospermas presentan una gran variedad de tamaños y formas, desde los dientes de león y los lirios acuáticos hasta los nopales y los olmos.

Todas son plantas vasculares que producen flores y frutos. Los tulipanes y las rosas producen flores grandes; en cambio, los pastos y los arces tienen flores pequeñas. Después de la fertilización, las angiospermas producen semillas dentro de los frutos. Entre los frutos están los duraznos, limones y uvas, así como los tomates, pepinos y muchos otros alimentos que solemos llamar verduras.

¿Para qué sirven las flores y los frutos? Las flores y los frutos son adaptaciones que permiten a las angiospermas reproducirse. Algunas angiospermas dependen del viento para la polinización y otras tienen flores que atraen a los animales. Como se muestra en la **Figura 18,** al visitar distintas flores los animales transportan el polen de flor en flor. Estas plantas no necesitan fabricar tanto polen como las que son polinizadas por el viento. ¿A qué crees que se debe esto?

Posiblemente te gusten las manzanas, tomates y calabazas. Además de un sabor agradable, también garantizan la supervivencia de las semillas al transportarlas a otras zonas donde pueden crecer nuevas plantas. Los frutos rodean y protegen las semillas. Las estructuras de algunos frutos y semillas, como los de la **Figura 19,** facilitan que el viento los transporte a distancias cortas o largas. Otros frutos atraen a los animales que se los comen y que desechan las semillas a cierta distancia de la planta progenitora. Los abrojos espinosos son frutos que van de un lugar a otro al pegarse a la piel de los animales o a la ropa y zapatos de las personas.

Teaching Transparency 44
"Two Classes of Angiosperms"

IS THAT A FACT!

The giant fan palm, which is native to Seychelles, in Africa, produces a single seed in its fruit that weighs up to 20 kg and takes up to 10 years to develop.

Monocotiledóneas y dicotiledóneas

Las angiospermas se dividen en mono-cotiledóneas y dicotiledóneas. Las semi-llas de las monocotiledóneas tienen un solo cotiledón y las semillas de las dicotiledóneas tienen dos. El **cotiledón** es una hoja que se encuentra dentro de la semilla. En la **Figura 20** se resumen otras de sus diferencias. Entre las mono-cotiledóneas están los pastos, las orquídeas, las cebollas, los lirios y las palmas; entre las dicotiledóneas están las rosas, los cactus, los girasoles, los ca-cahuetes y los chícharos.

La importancia de las angiospermas

Las plantas con flores proporcionan ali-mento a los animales. Para un venado que mordisquea pasto en una pradera, las plantas con flores son su alimento inmediato. Al comerse un ratón de campo, el búho consume indirecta-mente plantas con flores, ya que el ratón comió semillas y bayas.

Las principales cosechas de ali-mento, como el maíz, el trigo y el arroz, provienen de plantas con flores. Algunas sirven para fabricar muebles y juguetes, como los robles, y con otras, como el algodón y el lino, se fabrica ropa y cuerdas. También se usan para elaborar medicamentos, corcho, hule y aceites aromáticos.

Figura 20 Las dos clases de angiospermas

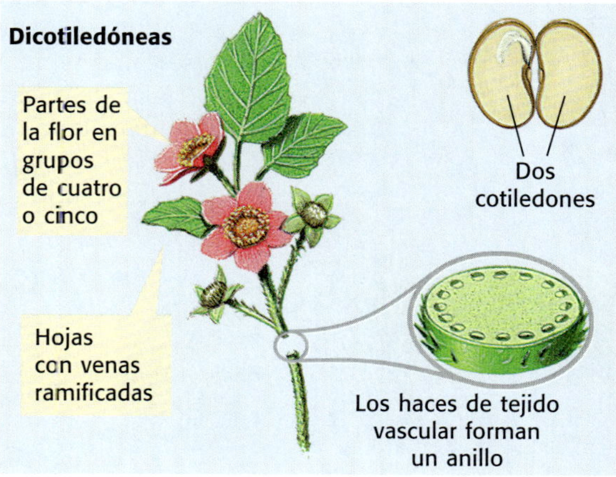

REPASO

1. Menciona dos diferencias entre las semillas y las esporas.

2. Describe brevemente los cuatro grupos de gimnospermas. ¿Qué grupo es el más grande y más importante desde el punto de vista económico?

3. ¿Qué diferencias hay entre las monocotiledóneas y las dicotiledóneas?

4. **Identificar relaciones** ¿Por qué las flores y los frutos son adaptaciones que facilitan la reproducción de las angiospermas?

---

3 Extend

DEBATE

Paper or Plastic? Pine trees are used to make paper grocery bags. Grocery stores often offer customers a choice of paper or plastic bags. Ask students to debate the merits of each.

4 Close

Quiz

1. Give one reason why angio-sperms greatly outnumber gymnosperms? (Angiosperms have adapted to almost every environment on land. In some cases, they are helped by animals that carry pollen and disperse seeds.)

2. Which plant group is more important to people, angio-sperms or gymnosperms? (Angiosperms; most plant-derived foods for people and animals come from angiosperms.)

ALTERNATIVE ASSESSMENT

Concept Mapping Organize the following terms into a concept map:

plants, gymnosperm, angiosperm, flowers, fruits, maple trees, cucumbers, conifers, cycads, monocot, dicot, orchids, lilies, peanuts, saguaro cactus, dandelion

---

▼ Answers to Review

1. Unlike seeds, spores have no stored food to nourish the new plant. Spores are single-celled. Seeds are made of many cells.

2. Conifers are the most economically impor-tant group of gymnosperms. Students should provide descriptions similar to those in **Figure 15**.

3. Students should discuss the differences illustrated in **Figure 20**.

4. The flowers help attract animals that are needed to deliver pollen to other plants. Animals eat the fruit and transport the seeds to areas where new plants can grow.

Reinforcement Worksheet 11 "Classifying Plants"

Reinforcement Worksheet 11 "Drawing Dicots"

The Structures of Seed Plants

In this section, students will learn about the physical structures and the functions of a plant's root and shoot systems. Students will also learn how those two factors are related. Finally, students will learn to identify a flower's parts and explain their functions.

Bellringer

Show students a cactus. Point out the spines, and ask students to identify them and explain their purpose in their ScienceLog. (Spines are modified leaves that help protect the plant from grazing animals.) **Sheltered English**

1 Motivate

DEMONSTRATION

Seed Transport Using a piece of Velcro™, demonstrate how some seeds become dispersed. The burdock plant served as inspiration to the inventor of Velcro. Burdocks are members of the thistle family, and they shed their seeds in burrs. These burrs are transported by animals or in people's clothing to new areas, where they take root. George de Mestral, a Swiss engineer, looked very closely at a burdock burr. He observed that the surface of a burr is covered with tiny hooks. Over the next 8 years, he worked with this design and developed a similar fastener out of nylon. **Sheltered English**

VOCABULARIO

xilema
floema
epidermis
raíz central
fibrosa
estomas

sépalos
pétalos
estambre
pistilo raíz
estigma
ovario

OBJETIVOS

- Describe las funciones de las raíces.
- Describe las funciones de los tallos.
- Explica cómo se relaciona la estructura de las hojas con su función.
- Identifica las partes de una flor y sus funciones.

Las estructuras de las plantas con semilla

¿Sabías que tenemos algo en común con las plantas? Tenemos sistemas corporales que llevan a cabo diversas funciones. Por ejemplo, el sistema cardiovascular transporta substancias a través de nuestro cuerpo y el esquelético nos proporciona sostén y protección. Del mismo modo, las plantas también tienen sistemas: el radical, el de brotes y el reproductor.

Sistemas de las plantas

El sistema radical y el de brotes le dan a la planta los recursos que necesita, tanto los que están bajo tierra como sobre ella. El sistema radical está constituido por raíces y el de brotes por tallos con hojas, conos, flores y frutos.

El sistema radical y el sistema de brotes dependen uno del otro como se muestra en la **Figura 21.** Hay dos tipos de tejido vascular: el xilema y el floema. El xilema transporta el agua y los minerales a través de la planta y el floema transporta las moléculas de azúcar. Ambos tipos de tejido se encuentran en todas las partes de las plantas vasculares.

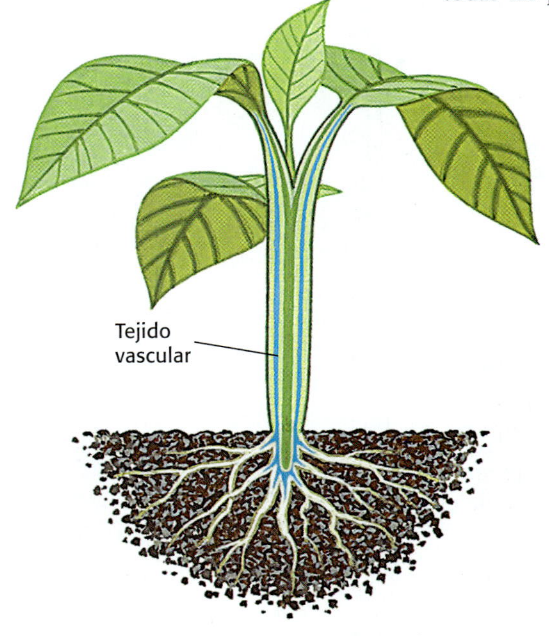

Figura 21 *Los tejidos vasculares de las raíces y brotes están conectados.*

La raíz del problema

La mayoría de las raíces son subterráneas y no nos damos cuenta de su extensión. Por ejemplo, una planta de maíz de 2.5 m de altura puede tener raíces de 2.5 m de profundidad y que estén a 1.2 m del tallo.

Funciones de la raíz Las principales funciones de las raíces son:

- **Suministran a las plantas agua y minerales que han sido absorbidos del suelo.** El xilema transporta estas substancias a través de la planta.

- **Sostener las plantas** y sujetarlas firmemente al suelo.

- **Almacenar el excedente del alimento elaborado durante la fotosíntesis.** Este alimento se produce en las hojas y el floema lo transporta en forma de azúcar hasta las raíces, donde generalmente se almacena como azúcar o almidón.

264

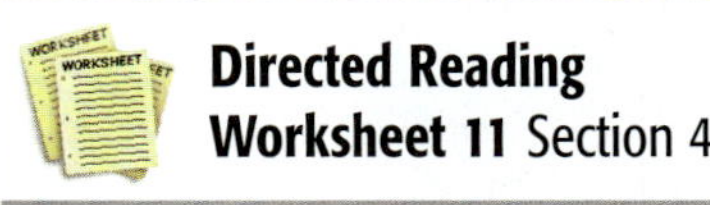
Directed Reading Worksheet 11 Section 4

IS THAT A FACT!

The deepest roots ever discovered belonged to a wild fig tree in South Africa. The roots had penetrated to a depth of 122 m.

Estructura de la raíz Las partes de la raíz se muestran en la **Figura 22.** Al igual que las células de la capa más externa de la piel humana, la capa de células que recubre la superficie de las raíces se llama **epidermis.** Algunas células de la epidermis se extienden hacia afuera de la raíz y se llaman *pelos radiculares* y su función es aumentar la superficie de absorción de agua y minerales.

Después de que la epidermis absorbe el agua y los minerales, éstos pasan por difusión al centro de la raíz, donde se localiza el tejido vascular. Las raíces crecen más en las puntas. Un grupo de células, llamado cofia, protege la punta de la raíz y produce una substancia viscosa que facilita el crecimiento en el suelo.

Tipos de raíz Hay dos tipos de raíces: centrales y fibrosas. En la **Figura 23** se ilustran ejemplos de cada una.

La **raíz central** está compuesta por una raíz principal que crece hacia abajo y de la cual salen muchas ramificaciones pequeñas. Pueden obtener agua a gran profundidad. Las dicotiledóneas y las gimnospermas tienen raíces centrales. La **raíz fibrosa** tiene varias raíces del mismo tamaño que se extienden desde la base del tallo; típicamente, absorben el agua cercana a la superficie del suelo. Las monocotiledóneas tienen raíces fibrosas.

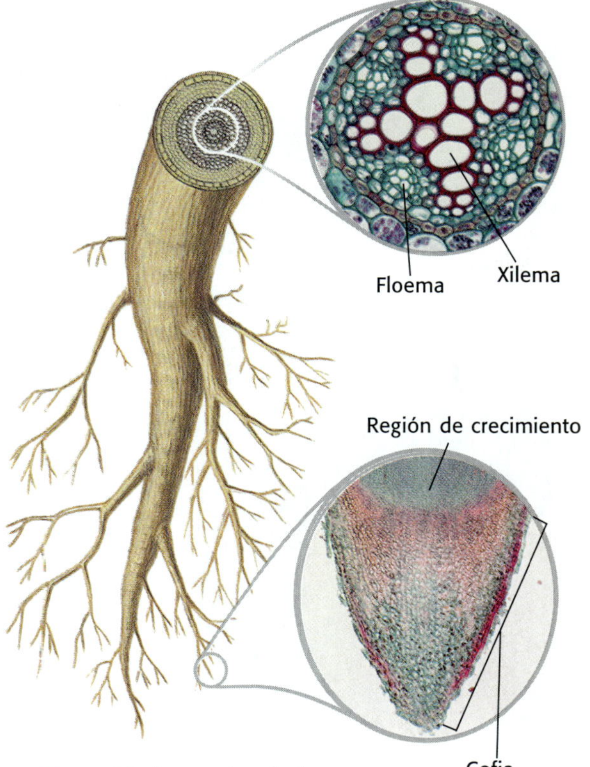

Figura 22 *Las partes de la raíz se señalan arriba.*

Raíces fibrosas

Figura 23 *La cebolla tiene una raíz fibrosa y el diente de león una raíz central. Las zanahorias tienen una raíz central que almacena alimento.*

265

2 Teach

DISCUSSION

Attracting Pollinators Discuss in class that flowers serve an important purpose for plants. They are designed to attract pollinators. They also protect pollen and fruit. Ask students to read the section on flowers and to consider how the following factors make flowers successful in the plant world: color, shape, smell, and touch. (The color of a flower may appeal to a certain insect. Bright colors are effective attractors. The shape of a flower may be specifically designed for a specific pollinator. The tubelike flower of a honeysuckle is perfect for the long beak of a hummingbird. The fragrance, or smell, of a flower is designed to attract pollinators; butterflies flock to sweet-smelling flowers, and flies are drawn to rancid-smelling flowers.)

Teaching Transparency 45 "Root Structure"

Science Skills Worksheet 21 "Taking Notes"

ACTIVITY

Have students draw pictures that illustrate the functions of stems listed on this page. Tell them to write captions for each illustration. Sheltered English

CROSS-DISCIPLINARY FOCUS

History The stems of many large aquatic grasses are called *reeds*. After reeds are harvested and dried, they can be used as a resource material to construct many useful products. For thousands of years, arrows, pens, baskets, musical instruments, furniture, and houses have been made out of reeds. Building boats from reeds is an ancient craft and is still practiced in some places where reeds are plentiful. Ancient Egyptian buildings include friezes of ocean-going ships made of reeds.

In the 1960s, a Norwegian explorer named Thor Heyerdahl wondered if a reed ship could have provided people with transportation across the Atlantic Ocean hundreds or even thousands of years ago. To demonstrate that such a journey was possible, he had Bolivian craftsmen build a traditional reed vessel, the *Ra II*, and in 1970, he sailed it across the Atlantic Ocean.

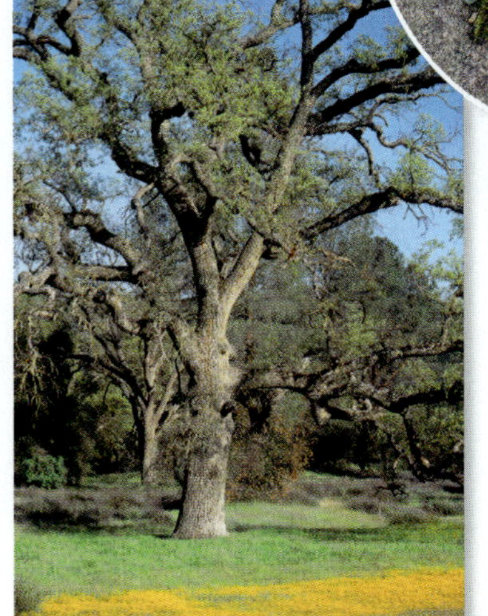

Figura 24 *Los pedúnculos de las margaritas y los troncos de los árboles son tallos.*

Margarita

Roble

¿Cuál es el sostén?

Como se observa en la **Figura 24,** las formas y tamaños de los tallos son muy variados. Por lo general, los tallos están por encima del suelo, aunque también los hay subterráneos.

Funciones del tallo El tallo conecta la raíz de la planta con sus hojas y flores; sus principales funciones son:

• **Sostener el cuerpo de la planta.** Las hojas están dispuestas en los tallos para absorber la luz solar necesaria para la fotosíntesis. Los tallos sostienen las flores y las exponen a los polinizadores.

• **Transportar substancias entre el sistema radical y el sistema de brotes.** El xilema transporta agua y minerales disueltos desde las raíces hasta las hojas y otros brotes. El floema transporta la glucosa producida durante la fotosíntesis a las raíces y otras partes de la planta.

• **Algunos tallos almacenan substancias.** Por ejemplo, los tallos de las plantas de la **Figura 25** están adaptados para almacenar agua.

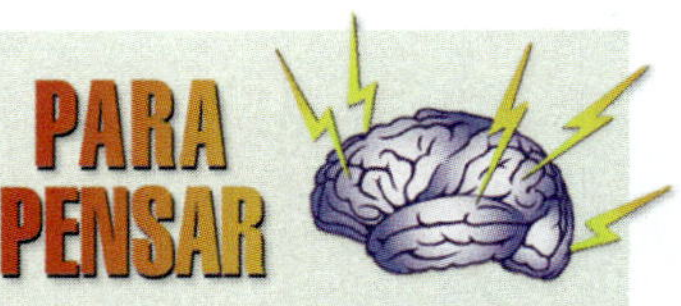

¿Raíz o brote? A pesar de que las papas crecen en la tierra, no son raíces. La papa blanca es un tallo subterráneo adaptado para almacenar almidón.

Baobab

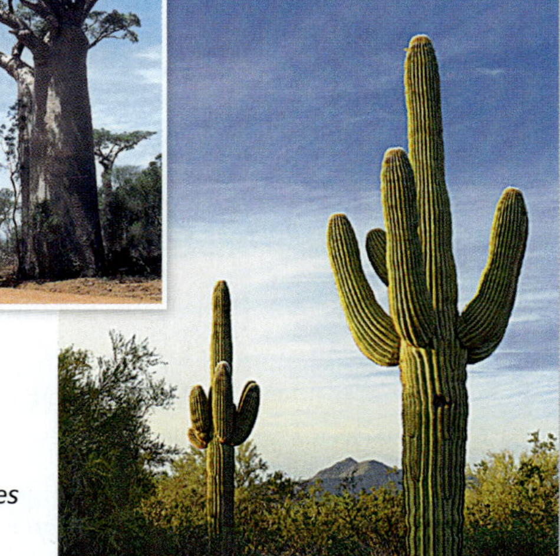

Saguaro

Figura 25 *Los baobab almacenan grandes cantidades de agua y almidón en sus enormes troncos. Los cactus almacenan agua en sus gruesos tallos verdes.*

266

WEIRD SCIENCE

Garlic bulbs are fleshy, nutrient-storing stems of garlic plants. Garlic has a pungent flavor and is often used to season foods and as a nutritional supplement. At the annual Garlic Festival, in Gilroy, California, visitors can purchase garlic ice cream and garlic candy.

Partes del tallo

Observa los dos grupos de plantas que aparecen en esta página. ¿Qué diferencia hay entre sus tallos?

Tallos herbáceos Las plantas del grupo A tienen tallos blandos, delgados y flexibles, llamados *tallos herbáceos*. Algunos ejemplos de plantas con este tipo de tallo son las flores silvestres, como los tréboles y las amapolas, y muchos cultivos, como el frijol, el tomate y el maíz. Algunas plantas con tallos herbáceos viven sólo 1 ó 2 años. A la derecha se muestra un corte transversal de un tipo de tallo herbáceo.

Tallos leñosos Los árboles y arbustos del grupo B tienen tallos rígidos de madera y corteza, llamados *tallos leñosos*. Si un árbol o arbusto habita en un área donde los inviernos son fríos, la planta tiene un período de crecimiento y un período latente.

En la primavera, al inicio de cada período de crecimiento, se producen células grandes de xilema. A medida que el otoño se aproxima, las plantas producen células de xilema más pequeñas, cuyo color es más obscuro que el de las células de xilema grandes. En el otoño, las plantas dejan de producir células nuevas. Cuando la estación de crecimiento comienza de nuevo, se producen células de xilema grandes, y de esta manera, se forma otro anillo de células obscuras. Un anillo de células obscuras que rodea a uno de células claras conforma un anillo de crecimiento.

Grupo A

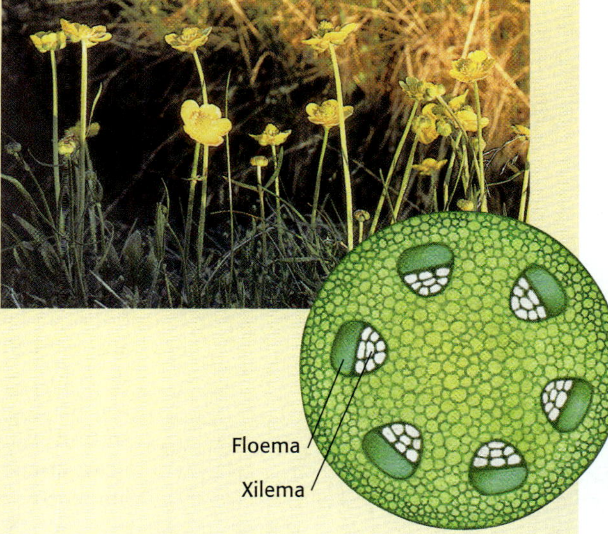

Grupo B

REPASO

1. Menciona tres funciones de las raíces.

2. Menciona tres funciones de los tallos.

3. **Aplicar conceptos** Si la sección transversal de un árbol revela 12 anillos claros y 12 obscuros, ¿cuántos años de crecimiento representa?

267

Homework

Leaf Collecting

MATERIALS

FOR EACH GROUP:
- newspapers
- heavy books
- transparent tape
- paper towels
- notebook or 4 × 6 in. index cards.
- leaves

Challenge students to see how many different types of leaves they can collect. They can mount them in a separate notebook or in their ScienceLog. Students should press the leaves a few days after collecting them. They can do this by placing each leaf between two paper towels and stacking heavy books on top. When the leaf is flat and dry, have them tape the leaves to cards or to the pages of a notebook. Have students use reference books to identify the names of the plants. Then have them label each leaf with the common name (such as red oak, honeysuckle, sugar maple), the scientific name, the date, and the location where the leaf was found. **Sheltered English**

Leaf Me Alone!

Answer to Self-Check

Stems hold up the leaves so that the leaves can get adequate sunshine for photosynthesis.

Teaching Transparency 47
"Leaf Structure"

Mimosa

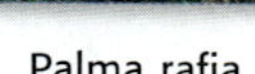
Palma rafia

Las fábricas de alimento de una planta

Hay hojas de muchas formas y tamaños: redondas, angostas, con forma de corazón o de abanico. Las hojas de la palma rafia de la **Figura 26** pueden medir seis veces más que tú, pero las de la lenteja de agua, una diminuta planta acuática, son tan pequeñas que varias cabrían en una de tus uñas.

Liquidámbar

Funciones de la hoja La principal función de las hojas es fabricar alimento. Captan la energía solar y absorben dióxido de carbono del aire. La energía de la luz, el dióxido de carbono y el agua son necesarios para llevar a cabo la fotosíntesis, durante la cual las plantas utilizan la energía luminosa para fabricar alimento.

Eucalipto

Figura 26 *Las hojas de estas plantas son muy diferentes, pero su función es la misma.*

Lenteja de agua

Experimentos

Pasa a la página 596 para averiguar más sobre las hojas.

✓ Autoevaluación

¿Cómo se relaciona la función de los tallos con la de las hojas? *(Consulta la página 636 para comprobar tus respuestas.)*

268

IS THAT A FACT!

Duckweed plants, which live on the surfaces of calm ponds, are the smallest flowering plants in the world. They can be less than 1 mm long and weigh about 150 µg, or the equivalent of two grains of table salt. It's a good thing animals like to eat duckweed, because it reproduces exponentially. One species reproduces every 30–36 hours. Left unchecked, one plant could produce 1 nonillion plants (1 followed by 30 zeros) in 4 months!

Estructura de la hoja La estructura de las hojas está relacionada con su principal función: la fotosíntesis. La **Figura 27** muestra un corte de una pequeña porción del tejido de una hoja. Las superficies superior e inferior están cubiertas por una sola capa de células, la epidermis. La luz pasa fácilmente esta delgada capa hacia el interior de la hoja. Observa los diminutos poros de la epidermis: se llaman **estomas** y su función es permitir que el dióxido de carbono entre en la hoja. Las *células oclusivas* abren y cierran los estomas.

La parte media de la hoja, donde se realiza la fotosíntesis, tiene dos capas. La capa superior es la *empalizada* y la inferior, la *capa esponjosa*. Las células de la empalizada son alargadas y contienen muchos cloroplastos, los organelos verdes que realizan la fotosíntesis. Las células de la capa esponjosa están más apartadas que las de la empalizada. Los espacios de aire que hay entre estas células permiten que el dióxido de carbono se difunda más libremente a través de la hoja. Las venas de la hoja contienen xilema y floema rodeados de tejido de apoyo. El xilema transporta el agua y los minerales a la hoja y el floema conduce el azúcar fabricada durante la fotosíntesis desde la hoja al resto de la planta.

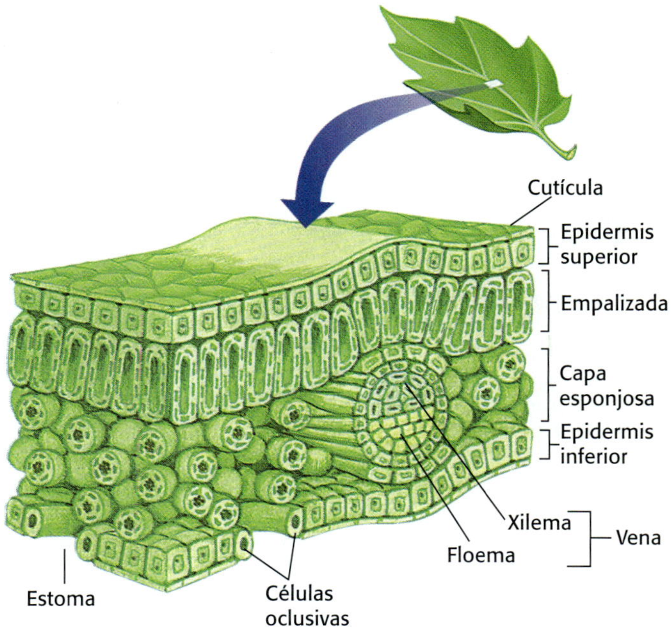

Figura 27 *Las hojas tienen varias capas de células.*

Adaptaciones de las hojas Algunas hojas realizan otras funciones además de la fotosíntesis. Por ejemplo, las espinas de los cactus son hojas modificadas, cuyas duras puntas impiden a los animales comerse el suculento tallo. Las hojas ilustradas en la **Figura 28** realizan una función poco común: atrapar insectos. Las droseras crecen en suelos que no contienen suficiente nitrógeno para cubrir sus necesidades. Sus hojas modificadas les permiten atrapar y digerir insectos que les proporcionan el nitrógeno que necesitan.

Figura 28 *Este caballito del diablo está atrapado en el líquido viscoso de una drosera.*

269

IS THAT A FACT!

The more water-repellent a leaf is, the healthier it may be, according to some scientists. Dirt, which contains disease-causing microbes, has a stronger attraction to water droplets than to a leaf's surface. When the water rolls off the leaf, so do the microbes. Water repellence thus helps prevent infection.

RESEARCH

Have students research the flower colors that are most attractive to specific pollinators, such as hummingbirds and bees. **(red and yellow, respectively)**

Tell students that some pollinators, such as bats, are active at night. Have them research what flower colors best attract nocturnal pollinators. **(white)**

GOING FURTHER

Life in a Dead Tree
Now that students have learned about live trees, encourage them to explore the life in and around a dead tree, called a snag. Once a snag has fallen down, it is called a log. Have students research the animals that use snags and logs as shelter, including songbirds, hawks, owls, snakes, bats, raccoons, and insects.

Flores

La mayoría de las personas admira la belleza de las flores, como las rosas o los lirios, sin detenerse a pensar por qué las plantas tienen flores. Como vimos, las flores son adaptaciones para la reproducción sexual. Sus diversas formas, colores y fragancias atraen a los polinizadores o permiten el paso del viento. La mayoría de las flores tienen las siguientes partes: sépalos, pétalos, estambres y uno o más pistilos. Estas partes generalmente están dispuestas en anillos alrededor de un pistilo central. La **Figura 29** muestra las partes de una flor típica.

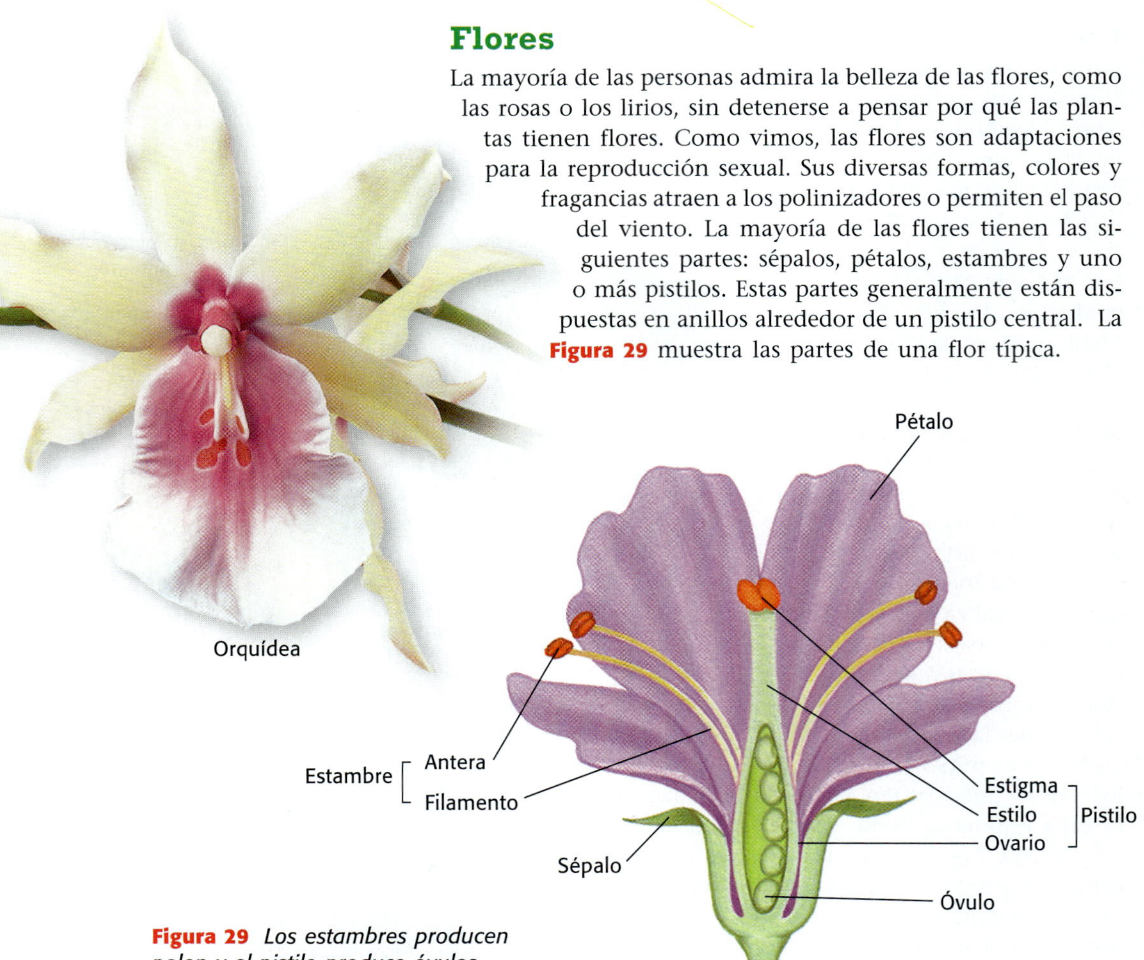

Figura 29 *Los estambres producen polen y el pistilo produce óvulos. Ambas estructuras están rodeadas por los pétalos y los sépalos.*

Los **sépalos** conforman el anillo inferior de las partes de la flor; con frecuencia son verdes como las hojas. Su principal función es cubrir y proteger a la flor inmadura cuando es un botón. Al abrirse la flor, los sépalos se repliegan para que los pétalos se agranden y se vuelvan visibles.

Los **pétalos** son anchos, planos y delgados, como los sépalos, pero su forma y color son diferentes. Los pétalos atraen insectos u otros animales a la flor; estos animales participan en la reproducción de las plantas al llevar el polen de flor en flor.

IS THAT A FACT!

The heaviest flower in the world is a species of rafflesia from Malaysia, *Rafflesia arnoldii*. The flowers can weigh 11 kg and measure 1 m in diameter.

WEIRD SCIENCE

Carrion-eating flies are probably the pollinators of *Rafflesia arnoldii*. This would explain why the flowers emit a scent reminiscent of rotting meat.

Encima de los pétalos están los **estambres,** que son las estructuras masculinas para la reproducción. Cada uno está compuesto por un *filamento* y una *antera.* Las anteras son estructuras en forma de saco que producen los granos de polen. En el centro de las flores hay uno o más **pistilos,** que son las estructuras femeninas para la reproducción. La punta del pistilo se llama **estigma** y es una estructura pegajosa donde se recolectan los granos de polen. La parte larga y delgada del pistilo se conoce como *estilo* y su base se llama **ovario.** Como se observa en la **Figura 30,** el ovario contiene uno o más óvulos. Cada óvulo contiene un huevo. En caso de fertilización, el óvulo se desarrolla y forma una semilla, y el ovario forma un fruto.

La polinización de las flores aromáticas y con pétalos de colores brillantes generalmente depende de los animales, mientras que la de las flores sin aroma ni color, como las flores del pasto de la **Figura 31,** depende del viento.

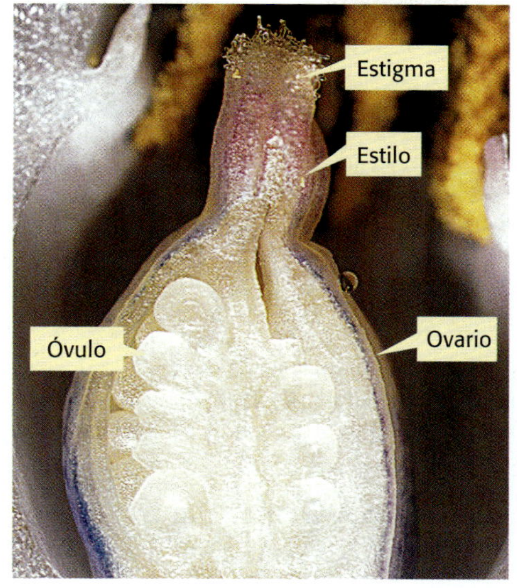

Figura 30 *El ovario de este jacinto contiene muchos óvulos.*

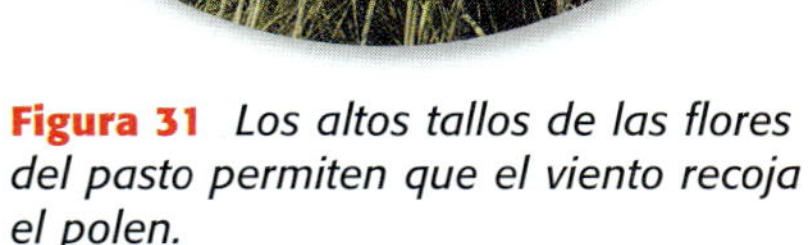

Figura 31 *Los altos tallos de las flores del pasto permiten que el viento recoja el polen.*

REPASO

1. Describe la estructura interna de una hoja típica. ¿Cómo relacionas la estructura de la hoja con su función?

2. ¿Qué parte de la flor produce el polen?

3. **Identificar relaciones** Compara las funciones del xilema y del floema en las raíces, los tallos, las hojas y las flores.

4) Close

Quiz

1. Why is it that some plants have brightly colored flowers and other plants do not? (Plants with brightly colored flowers usually need to attract insects, birds, or mammals to ensure pollination. Plants without showy flowers usually rely on the wind to assist with pollination.)

2. Describe the two types of roots, and list at least one plant that has each type. (A taproot is one main root with smaller branch roots. Dandelions and carrots have a taproot. Fibrous roots are similar in size and grow from the base of the stem. Onions have fibrous roots.)

ALTERNATIVE ASSESSMENT

Writing  Have students list ingredients for a salad. (lettuce, tomato, cucumber, carrot, mushroom, red onion, red cabbage, alfalfa sprouts)

Tell them to identify each item as a monocot, dicot, fruit, vegetable, or other. (The mushroom is a fungus.)

Then tell students to list the part of the plant that is eaten. (stem, root, leaves)

Critical Thinking Worksheet 11 "The Voodoo Lily"

▼ **Answers to Review**

1. The epidermis is transparent to allow sunlight in. Stomata allow gases to enter and leave the leaf. The palisade layer in the leaf contains chloroplasts, which carry out photosynthesis. The spongy layer contains cells that are spread out to help the diffusion of carbon dioxide. The veins transport materials to and from the leaf.

2. The anthers produce pollen.

3. Water and nutrients travel through the xylem in the roots, stems, leaves, and flowers. Sugar molecules travel through the phloem from the leaves to other parts of the plant where the sugar is used or stored.

Chapter **Highlights**

VOCABULARY DEFINITIONS

SECTION 1

cuticle a waxy layer that coats the surface of stems, leaves, and other plant parts exposed to air

sporophyte a stage in a plant's life cycle during which spores are produced

gametophyte a stage in a plant's life cycle during which eggs and sperm are produced

nonvascular plant a plant that depends on the processes of diffusion and osmosis to move materials from one part of the plant to another

vascular plant a plant that has xylem and phloem, specialized tissues that move materials from one part of the plant to another

gymnosperm a plant that produces seeds but does not produce flowers

angiosperm a plant that produces seeds in flowers

SECTION 2

rhizoids small hairlike threads of cells that help hold nonvascular plants in place

rhizome the underground stem of a fern

SECTION 3

pollen dustlike particles that carry the male gametophytes of seed plants

pollination the transfer of pollen to the female cone in conifers or to the stigma in angiosperms

cotyledon a seed leaf inside a seed

Resumen del **capítulo**

SECCIÓN 1

Vocabulario

cutícula *(pág. 250)*
esporofito *(pág. 251)*
gametofito *(pág. 251)*
planta no vascular *(pág. 252)*
planta vascular *(pág. 253)*
gimnosperma *(pág. 253)*
angiosperma *(pág. 253)*

Notas de la sección

- Las plantas elaboran su alimento mediante la fotosíntesis. Las células vegetales tienen paredes celulares. Las plantas están recubiertas por una cutícula cerosa.

- El ciclo vital de las plantas comprende una etapa de producción de esporas (el esporofito) y una de producción de células sexuales (el gametofito).

- Probablemente las plantas evolucionaron a partir de un tipo de alga verde.

- Las plantas vasculares tienen xilema, cuya función es transportar agua y minerales disueltos, y floema, que transporta las moléculas de alimento, como el azúcar. Las plantas no vasculares no tienen xilema ni floema; por lo tanto, dependen de la difusión y de la ósmosis para movilizar substancias.

SECCIÓN 2

Vocabulario

rizoides *(pág. 254)*
rizoma *(pág. 256)*

Notas de la sección

- Los musgos y las hepáticas son plantas no vasculares pequeñas. Son pequeñas porque carecen de xilema y floema. Para transportar los espermatozoides hasta los óvulos se necesita agua.

- Los helechos, equisetos y licopodios son plantas vasculares. Crecen más que las plantas no vasculares y necesitan agua para transportar los espermatozoides hasta los óvulos.

☑ Comprobar destrezas

Conceptos de matemáticas

¿QUÉ PORCENTAJE...? Si el 38 por ciento de las plantas de un bosque son plantas con flores, ¿cuál es el porcentaje de plantas sin flores? Los dos grupos dan 100 por ciento, así que réstale 38 al 100 por ciento.

100 por ciento − 38 por ciento = 62 por ciento

Consulta de nuevo la sección ¡Matemáticas! de la página 253. Puedes calcular el porcentaje de plantas que producen semillas si al 100 por ciento le restas la respuesta que obtuviste.

Visual Understanding

SEMILLAS Esta imagen muestra los dos cotiledones de una semilla dicotiledónea. La semilla ha sido separada y los dos cotiledones parecen las tapas de una hamburguesa. Estás viendo las superficies internas de los dos cotiledones. Si abres un cacahuete puedes verificar esto tú mismo, ya que en estas semillas los cotiledones se separan muy fácilmente. Hasta puedes ver la delicada planta en el interior.

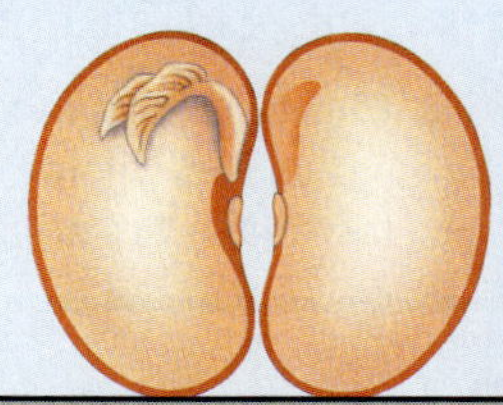

Lab and Activity **Highlights**

Travelin' Seeds PG 598

Leaf Me Alone! PG 596

Build a Flower PG 599

Datasheets for LabBook
(blackline masters for these labs)

SECCIÓN 3

Vocabulario

polen (*pág. 258*)
polinización (*pág. 261*)
cotiledón (*pág. 263*)

Notas de la sección

- Las plantas con semillas son plantas vasculares que producen semillas. Los espermatozoides de estas plantas se desarrollan dentro del polen.

- Las gimnospermas son plantas con semillas en conos o en estructuras carnosas sujetas a las ramas. Hay cuatro grupos: coniferales, ginkgoales, cicadales y gnetópsidas.

- Las angiospermas son plantas con semillas que producen sus semillas en flores. Los dos grupos de plantas con flores son las monocotiledóneas y las dicotiledóneas.

Experimentos

Semillas viajeras (*pág. 598*)

SECCIÓN 4

Vocabulario

xilema (*pág. 264*)
floema (*pág. 264*)
epidermis (*pág. 265*)
raíz central (*pág. 265*)
raíz fibrosa (*pág. 265*)
estomas (*pág. 269*)
sépalos (*pág. 270*)
pétalos (*pág. 270*)
estambre (*pág. 271*)
pistilo (*pág. 271*)
estigma (*pág. 271*)
ovario (*pág. 271*)

Notas de la sección

- Las raíces sujetan la planta al suelo, absorben agua y minerales y almacenan alimento.

- Los tallos conectan las raíces con las hojas. Sostienen las hojas y otras estructuras, transportan agua, minerales y alimento, y almacenan agua y nutrientes.

- La función principal de las hojas es la fotosíntesis.

- Las flores generalmente tienen cuatro partes: sépalos, pétalos, estambres y uno o más pistilos. Los estambres producen espermatozoides del polen. El ovario que está en el pistilo contiene óvulos; cada óvulo contiene un huevo. Los óvulos se transforman en semillas después de la fertilización.

Experimentos

¡Hojas! (*pág. 596*)
Construye una flor (*pág. 599*)

SECTION 4

xylem specialized plant tissue that transports water and minerals from one part of the plant to another

phloem specialized plant tissue that transports sugar molecules from one part of the plant to another

epidermis the outermost layer of the skin; also the outermost layer of cells covering roots, stems, leaves, and flower parts

taproot a type of root that consists of one main root that grows downward, with many smaller branch roots coming out of it

fibrous root a type of root in which there are several roots of the same size that spread out from the base of the stem

stomata openings in the epidermis of a leaf that allow carbon dioxide to enter the leaf

sepals leaflike structures that cover and protect an immature flower

petals the often colorful structures on a flower that are usually involved in attracting pollinators

stamen the male reproductive structure in the flower that consists of a filament topped by a pollen-producing anther

pistil the female reproductive structure in a flower that consists of a stigma, style, and ovary

stigma the flower part that is the tip of the pistil

ovary in flowers, the structure that contains ovules and develops into fruit following fertilization

---

internet

VISITA: go.hrw.com

Visita el sitio web de HRW para encontrar una serie de herramientas de aprendizaje relacionadas con este capítulo. Sólo tienes que escribir la palabra clave:

PALABRA CLAVE: HSTPL1

SCILINKS
NSTA

VISITA: www.scilinks.org

Visita el sitio web de la **Asociación Nacional de Maestros de Ciencias** (*National Science Teachers Association*) para encontrar recursos de Internet relacionados con este capítulo. Sólo escribe el **ENLACE DE CIENCIAS** para obtener más información sobre el tema:

TEMA	ENLACE
Características de las plantas	HSTL280
¿Cómo se clasifican las plantas?	HSTL285
Plantas sin semillas	HSTL290
Plantas con semillas	HSTL295
La estructura de las plantas con semillas	HSTL300

273

Lab and Activity Highlights

LabBank

EcoLabs & Field Activities, The Case of the Ravenous Radish, EcoLab 3

Whiz-Bang Demonstrations, Inner Life of a Leaf, Demo 8

Long-Term Projects & Research Ideas, Project 12

Interactive Explorations CD-ROM

CD 1, Exploration 2, "Shut Your Trap!"

Vocabulary Review Worksheet 11

Blackline masters of these Chapter Highlights can be found in the **Study Guide.**

USING VOCABULARY

1. cuticle
2. gametophyte
3. xylem/phloem
4. club mosses
5. cotyledon
6. stamens

UNDERSTANDING CONCEPTS

Multiple Choice

7. b
8. d
9. a
10. d
11. c
12. a
13. b
14. c

Short Answer

15. Mosses must remain relatively small plants because they are nonvascular and must absorb water directly from the environment or through diffusion.

16. The young plant in a seed is well developed and consists of many cells; a seed can survive in many environments; a seed has stored food for the new plant.

17. Water is necessary for sexual reproduction because sperm cells swim to the egg. Water also opens the spore.

Repaso del capítulo

UTILIZAR EL VOCABULARIO

Escoge el término correcto para completar las siguientes oraciones:

1. Los/la _____ son/es una capa cerosa que recubre la superficie de tallos y hojas. *(estomas* o *cutícula)*

2. Durante el ciclo vital de las plantas, los óvulos y los espermatozoides son producidos por el _____ *(esporofito* o *gametofito)*

3. En las plantas vasculares, el _____ conduce agua y minerales y el _____ conduce moléculas de alimento, como el azúcar. *(xilema/floema* o *floema/xilema)*

4. Entre las plantas vasculares sin semillas están los helechos, los equisetos y las _____. *(licopodios* o *hepáticas)*

5. Un _____ es una hoja de la semilla que se encuentra dentro de la misma semilla. *(cotiledón* o *sépalo)*

6. En una flor, los _____ son las estructuras masculinas de reproducción. *(pistilos* o *estambres)*

COMPRENDER CONCEPTOS

Opción múltiple

7. ¿Cuál de las siguientes plantas es no vascular?
   a. helecho
   b. musgo
   c. conífera
   d. monocotiledónea

8. Hace millones de años el carbón se formó a partir de restos de
   a. plantas no vasculares.
   b. plantas con flores.
   c. algas verdes.
   d. plantas vasculares sin semillas.

9. El grupo más grande de gimnospermas es el de
   a. coníferas.
   b. ginkgos.
   c. cícadas.
   d. gnetófitos.

10. Las raíces
    a. absorben agua y minerales.
    b. almacenan el excedente de alimento.
    c. sujetan la planta.
    d. todas las anteriores

11. Los tallos leñosos
    a. son blandos, verdes y flexibles.
    b. incluyen a los tallos de las margaritas.
    c. contienen madera y corteza.
    d. todas las anteriores

12. Las venas de una hoja contienen
    a. xilema y floema.
    b. estomas.
    c. epidermis y cutícula.
    d. sólo xilema.

13. En la flor, la función de los pétalos es
    a. producir óvulos.
    b. atraer a los polinizadores.
    c. proteger al botón de la flor.
    d. producir polen.

14. Las monocotiledóneas tienen
    a. partes florales en grupos de cuatro o cinco.
    b. dos cotiledones en la semilla.
    c. venas paralelas en las hojas.
    d. Todas las anteriores

Respuesta breve

15. ¿Por qué no hay musgos grandes?

16. ¿Qué ventajas tiene una semilla sobre una espora?

17. ¿Por qué es importante el agua para la reproducción de musgos y helechos?

274

Organizar conceptos

18. Usa los siguientes términos para crear un mapa de ideas: plantas no vasculares, plantas vasculares, xilema, floema, helechos, semillas en conos, plantas, gimnospermas, esporas, angiospermas, semillas en las flores.

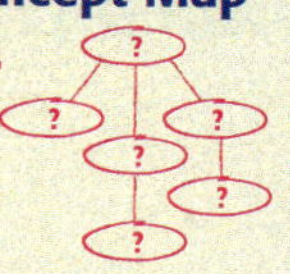

RAZONAMIENTO CRÍTICO Y RESOLUCIÓN DE PROBLEMAS

Escribe una o dos oraciones para responder a las siguientes preguntas:

19. Las plantas que son polinizadas por el viento producen mucho más polen que las plantas que son polinizadas por animales. ¿A qué crees que se debe esto?

20. Si las plantas no tuvieran cutícula, ¿dónde tendrían que vivir? ¿Por qué?

21. Los pastos no tienen aromas fuertes ni colores brillantes. ¿Qué tiene que ver esto con la manera en que son polinizados?

22. Imagina que una semilla y una espora empiezan a crecer en la grieta profunda y obscura de una roca. ¿Qué estructura reproductora, la semilla o la espora, tiene mayor probabilidad de sobrevivir y desarrollarse para formar una planta adulta? Explica tu respuesta.

LAS MATEMÁTICAS EN LAS CIENCIAS

23. En un año, un arce produjo 1,056 semillas. Si sólo el 15 por ciento de estas semillas germinaron y se transformaron en plántulas, ¿cuántas plántulas hay?

INTERPRETAR GRÁFICAS

24. Examina el corte transversal de la flor para responder a las siguientes preguntas:
    a. ¿Qué letra corresponde a la estructura en donde se produce el polen? ¿Cómo se llama esta estructura?
    b. ¿Qué letra corresponde a la estructura que contiene los óvulos? ¿Cómo se llama esta estructura?

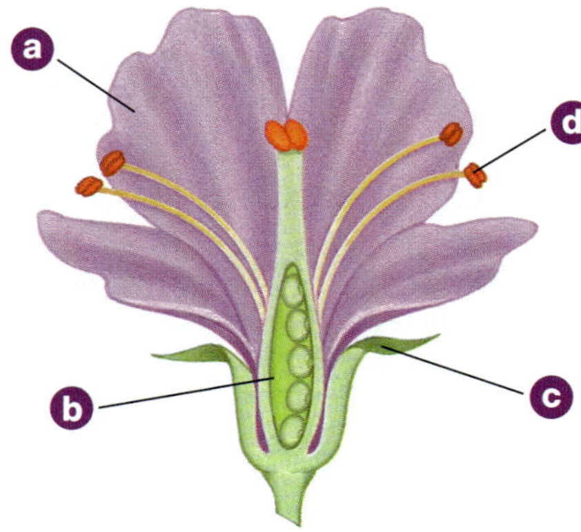

25. En los tallos leñosos, un anillo de células obscuras y uno de células claras representan un año de crecimiento. Examina la sección transversal del tronco de abajo para determinar la edad del árbol.

AHORA, ¿qué piensas?

Revisa tus respuestas a las preguntas de la página 5 que escribiste en el cuaderno de ciencias. ¿Han cambiado tus respuestas? Si es necesario, corrige tus respuestas basándote en lo que has aprendido en este capítulo.

275

NOW WHAT DO YOU THINK?

1. Answers will vary, but students should discuss the function of flowers and fruits in the reproductive cycle of flowering plants.
2. Answers may vary, but students should point out that plants make their own food and that animals have to eat the nutrients they need. Animal cells do not have chlorophyll or cell walls.

Background

The Flavr-Savr™ tomato was approved for consumer use by the Food and Drug Administration in 1994. By inserting an extra gene into the tomato, scientists were able to slow down a series of chemical reactions that cause tomatoes to rot. As a result, this genetically altered tomato can be left to ripen on the vine and can stay on the shelf much longer before becoming soft and spoiling. Many consumers think it also has a better flavor.

Consumer reaction to the tomato was not entirely positive, however. After the FDA's approval was announced, some consumer groups quickly began preparing a boycott of the product as well as planning public "tomato squashes," in which the tomatoes were crushed as a protest against genetically engineered foods.

The tomato marked a new era in commercial agriculture. Genetic engineering may eventually produce crops that can tolerate much harsher growing conditions, including the application of herbicides that are currently too strong to be used.

Ciencia, Tecnología y Sociedad

¿Supercalabaza o Frankenfruta?

Las frutas y verduras que compras en el supermercado a veces no son lo que parecen. Los científicos pueden haberlas modificado genéticamente para mejorar su apariencia y sabor, aumentar su contenido en nutrientes y prolongar su vida en el almacén.

De balas a bacterias

Con la ingeniería genética, se puede duplicar el ADN de un organismo y colocar un determinado gen en las células de otra especie. Así, los científicos pueden darles rasgos nuevos, que se transmitirán a las generaciones futuras.

Para modificar las plantas, se inserta un gen con una determinada propiedad en sus células. Hay dos métodos para insertar el ADN. Uno consiste en colocar el ADN nuevo dentro de una bacteria especial, la cual lo lleva al interior de la célula vegetal. El otro consiste en disparar partículas microscópicas de metal recubiertas con el ADN nuevo, al interior de las células vegetales. Esto se hace con una "pistola de genes" especial.

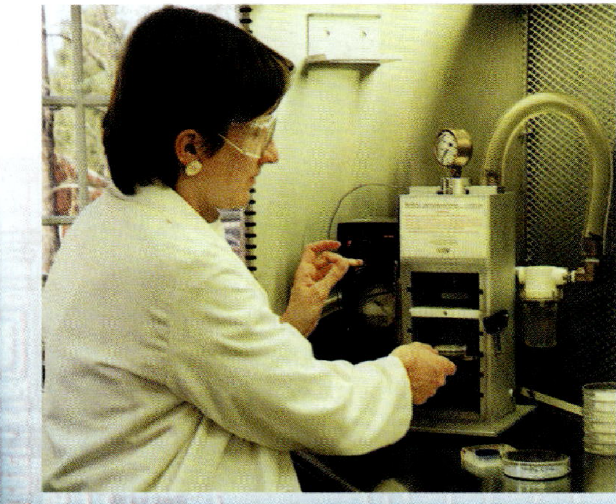

Alimentos de alta tecnología

En la última década, se han insertado genes a más de 50 clases de plantas. Generalmente, la nueva característica del gen insertado hace a las plantas más resistentes o más comerciales. Por ejemplo, a las plantas de algodón, tomate y papa se les han agregado genes de una bacteria que ataca a las orugas. Las plantas modificadas producen proteínas para matar las orugas que destruyen las cosechas. También se están desarrollando plantas de chícharo y pimiento rojo modificadas genéticamente para conservar su sabor dulce más tiempo. Y ya hay tomates modificados que duran más tiempo y saben mejor. Además, es posible que algún día existan granos de café sin cafeína.

¿Estamos listos?

Los alimentos modificados genéticamente, aunque prometedores, son motivo de controversia. Se teme que los genes introducidos a los cultivos se liberen al ambiente o que los cambios pongan en peligro la salud humana. Por ejemplo, no se sabe si la gente alérgica a los cacahuetes se enfermará si come tomates que contienen genes de cacahuetes.

¡Descúbrelo!

▶ ¿Son los alimentos modificados genéticamente un motivo de controversia en tu comunidad? Realiza una encuesta para conocer la opinión de algunas personas sobre ellos. ¿Consideran que las tiendas de comestibles deben vender estos alimentos? ¿Por qué?

◀ *Con una "pistola de genes", una científica inserta ADN dentro de unas células vegetales.*

276

Answer to Find Out for Yourself

Discuss with the class the pros and cons of genetically altered foods. Then ask students to form an opinion about the genetic engineering of animals. Is this any different from that of plants? Why or why not? Accept all reasonable responses. You may wish to divide the class into two teams and organize a debate on the topic of genetically engineered organisms.

PROFESIONES

ETNOBOTÁNICO

Paul Cox es un *etnobotánico que* viaja a lugares remotos en busca de plantas que curan enfermedades. Durante su búsqueda, consulta a chamanes y curanderos nativos. En 1984, Cox hizo un viaje a Samoa para observar a los curanderos; allí conoció a Epenesa, una curandera de 78 años de edad que podía identificar más de 200 plantas medicinales. Sus conocimientos impresionaron a Cox. Epenesa tenía un conocimiento preciso de la anatomía humana y preparaba recetas con gran cuidado y precisión.

En Samoa, el curandero es uno de los miembros más apreciados de la comunidad, pues sabe cómo tratar las enfermedades. Algunos curanderos conocen antiguos tratamientos que la medicina occidental todavía no ha descubierto. Recientemente, varios investigadores se han dirigido a ellos para hacerles preguntas sobre sus secretos médicos.

Mezcla de medicina polinesia y occidental

Después de pasar meses observando cómo Epenesa trataba a sus pacientes, ella le dio una receta para la fiebre amarilla: un té preparado con la madera de un árbol del bosque tropical. Cox trajo a los Estados Unidos el remedio contra la fiebre amarilla y, en 1986, los investigadores del Instituto Nacional del Cáncer, o NCI, *(National Cancer Institute)* empezaron a estudiar la planta. Descubrieron que contiene una substancia química que combate a los virus: la prostratina. Se ha descubierto que también tiene potencial como tratamiento para el sida.

Otro compuesto de los curanderos de Samoa sirve para tratar la inflamación. Se aplica la corteza de un árbol sobre la piel inflamada. Se están investigando las propiedades medicinales del compuesto activo de la corteza, llamado flavanona. Algún día los médicos occidentales recetarán un medicamento que contenga flavanona.

Para conservar su conocimiento

Con la muerte, en 1993, de dos de los curanderos que Cox conoció en Samoa también desaparecieron generaciones de conocimientos médicos. Esto señala la urgencia de registrar la sabiduría antigua antes de que todos los curanderos desaparezcan. Cox y otros etnobotánicos deben reunir la información de los curanderos tan rápido como puedan.

La sensación de una curación natural

La próxima vez que te pique un mosquito o tengas una quemadura leve de sol, usa un tratamiento de los curanderos nativos de América. El aloe vera es un producto vegetal que se encuentra en diversas lociones y pomadas. ¡Verás qué bien funciona!

▶ *Algún día, estas partes de una planta proveniente de Samoa podrían usarse en medicamentos para tratar diversas enfermedades.*

277

Chapter Organizer

CHAPTER ORGANIZATION	TIME MINUTES	OBJECTIVES	LABS, INVESTIGATIONS, AND DEMONSTRATIONS
Chapter Opener pp. 278–279	45		**Investigate!** Observing Plant Growth, p. 279
Section 1 The Reproduction of Flowering Plants	90	▶ Describe the roles of pollination and fertilization in sexual reproduction. ▶ Describe how fruits are formed from flowers. ▶ Explain the difference between sexual and asexual reproduction in plants.	**Demonstration,** Identify the Parts of a Flower, p. 280 in ATE **QuickLab,** Thirsty Seeds, p. 282 **Labs You Can Eat,** Not Just Another Nut, Lab 6
Section 2 The Ins and Outs of Making Food	90	▶ Describe the process of photosynthesis. ▶ Discuss the relationship between photosynthesis and cellular respiration. ▶ Explain the importance of stomata in the processes of photosynthesis and transpiration.	**Skill Builder,** Food Factory Waste, p. 600 **Datasheets for LabBook,** Food Factory Waste, Datasheet 25 **Skill Builder,** Weepy Weeds, p. 602 **Datasheets for LabBook,** Weepy Weeds, Datasheet 26
Section 3 Plant Responses to the Environment	90	▶ Describe how plants may respond to light and gravity. ▶ Explain how some plants flower in response to night length. ▶ Describe how some plants are adapted to survive cold weather.	**QuickLab,** Which Way Is Up? p. 288 **Interactive Explorations CD-ROM,** How's It Growing? *A **Worksheet** is also available in the **Interactive Explorations Teacher's Edition.*** **EcoLabs & Field Activities,** Recycle! Make Your Own Paper, EcoLab 4
Section 4 Plant Growth	90	▶ Discuss how heredity affects plant growth. ▶ Discuss how a plant's environment affects its growth. ▶ Explain what plant hormones are and what they do.	**Long-Term Projects & Research Ideas,** Project 13

TECHNOLOGY RESOURCES

 **Guided Reading Audio CD**
English or Spanish, Chapter 12

 **One-Stop Planner CD-ROM with Test Generator**

 **Science Discovery Videodiscs**
Image and Activity Bank with Lesson Plans: Plant Detectives

Science Sleuths: Green Thumb Plant Rentals #1, Green Thumb Plant Rentals #2

 **CNN. Eye on the Environment,** Tropical Reforestation, Segment 15

Scientists in Action, Growing Plants in Space, Segment 16

 **Interactive Explorations CD-ROM**
CD 2, Exploration 8, How's It Growing?

CLASSROOM WORKSHEETS, TRANSPARENCIES, AND RESOURCES	SCIENCE INTEGRATION AND CONNECTIONS	REVIEW AND ASSESSMENT
Directed Reading Worksheet 12 **Science Puzzlers, Twisters & Teasers**, Worksheet 12 **Science Skills Worksheet 2**, Using Your Senses		
Transparency 48, Pollination and Fertilization **Transparency 49**, From Flower to Fruit **Directed Reading Worksheet 12**, Section 1 **Reinforcement Worksheet 12**, Fertilizing Flowers	**Cross-Disciplinary Focus**, p. 280 in ATE **Cross-Disciplinary Focus**, p. 282 in ATE **Weird Science:** Mutant Mustard, p. 298	**Self-Check**, p. 281 **Review**, p. 283 **Quiz**, p. 283 in ATE **Alternative Assessment**, p. 283 in ATE
Transparency 50, Photosynthesis **Directed Reading Worksheet 12**, Section 2 **Reinforcement Worksheet 12**, A Leaf's Work Is Never Done **Math Skills for Science Worksheet 50**, Balancing Chemical Equations **Transparency 212**, Balancing a Chemical Equation **Transparency 51**, Transpiration	**Math and More**, p. 285 in ATE **Connect to Physical Science**, p. 285 in ATE	**Self-Check**, p. 285 **Review**, p. 286 **Quiz**, p. 286 in ATE **Alternative Assessment**, p. 286 in ATE
Transparency 52, Phototropism **Directed Reading Worksheet 12**, Section 3 **Transparency 53**, Night Length and Blooming Flowers **Transparency 53**, The Change of Seasons and Pigment Color **Reinforcement Worksheet 12**, How Plants Respond to Change	**MathBreak**, Bending by Degrees, p. 287 **Earth Science Connection**, p. 289 **Real-World Connection**, p. 289 in ATE **Apply**, p. 290	**Self-Check**, p. 288 **Homework**, p. 289 in ATE **Review**, p. 291 **Quiz**, p. 291 in ATE **Alternative Assessment**, p. 291 in ATE
Directed Reading Worksheet 12, Section 4 **Critical Thinking Worksheet 12**, Space Plants	**Eye On The Environment:** Rainbow of Cotton, p. 299	**Review**, p. 293 **Quiz**, p. 293 in ATE **Alternative Assessment**, p. 293 in ATE

END-OF-CHAPTER REVIEW AND ASSESSMENT

Chapter Review in Study Guide
Vocabulary and Notes in Study Guide
Chapter Tests with Performance-Based Assessment, Chapter 12 Test
Chapter Tests with Performance-Based Assessment, Performance-Based Assessment 12
Concept Mapping Transparency 12

Holt, Rinehart and Winston On-line Resources

go.hrw.com

For worksheets and other teaching aids related to this chapter, visit the HRW Web site and type in the keyword: **HSTPL2**

National Science Teachers Association

www.scilinks.org

Encourage students to use the *sci*LINKS numbers listed with the Chapter Highlights to access information and resources on the **NSTA** Web site.

Chapter Resources & Worksheets

Visual Resources

TEACHING TRANSPARENCIES

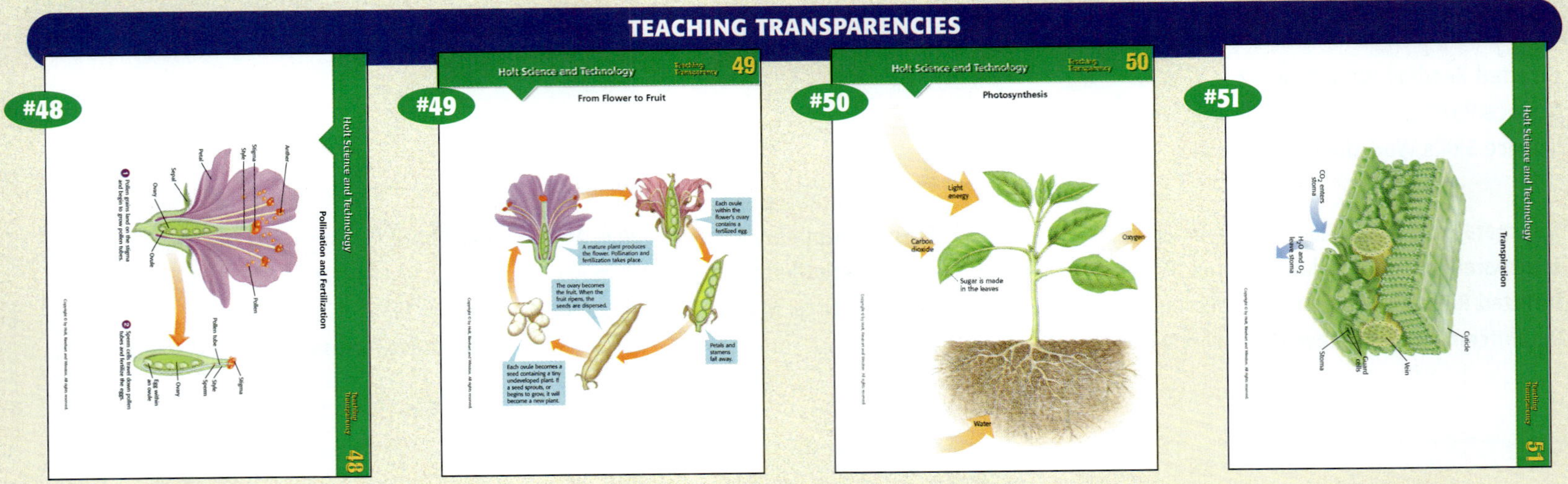

TEACHING TRANSPARENCIES

CONCEPT MAPPING TRANSPARENCY

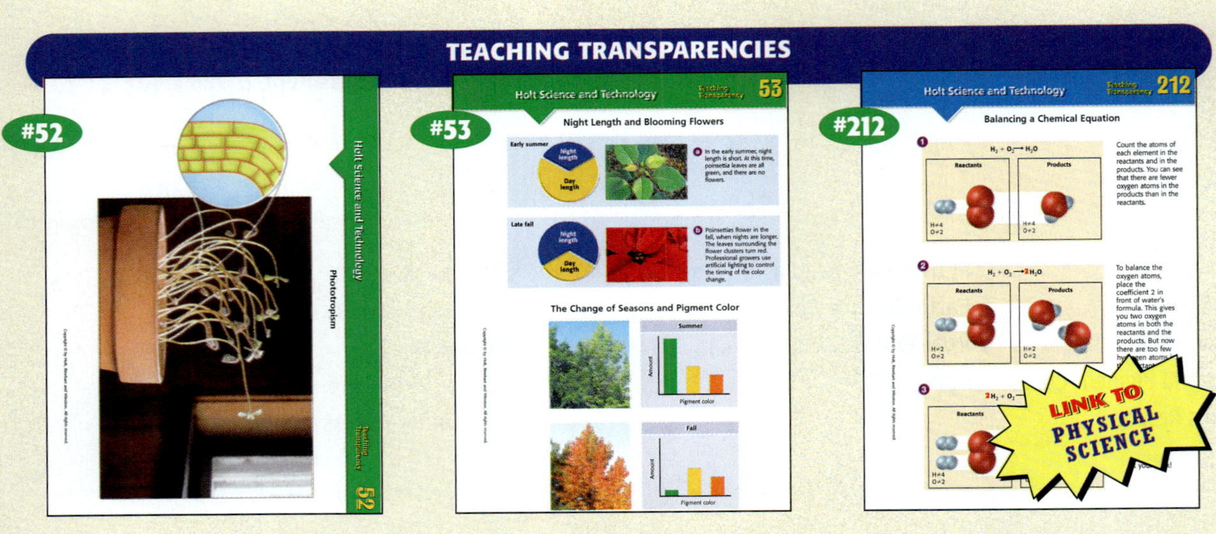

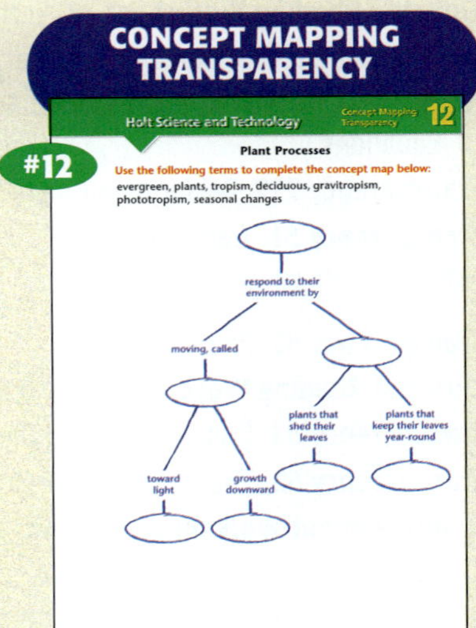

Meeting Individual Needs

DIRECTED READING

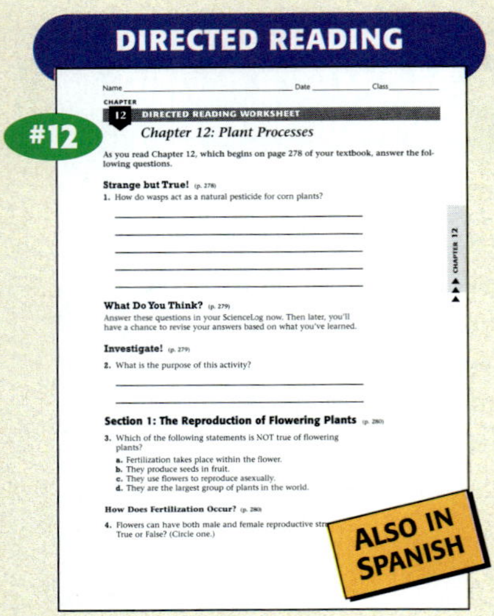

REINFORCEMENT & VOCABULARY REVIEW

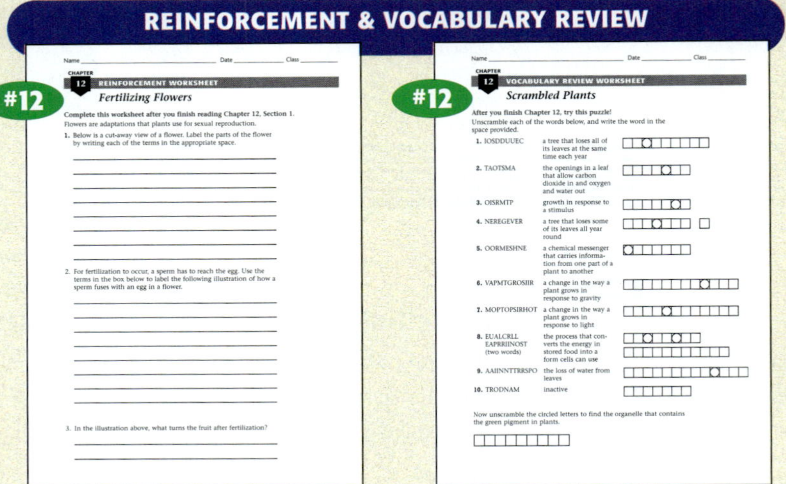

SCIENCE PUZZLERS, TWISTERS & TEASERS

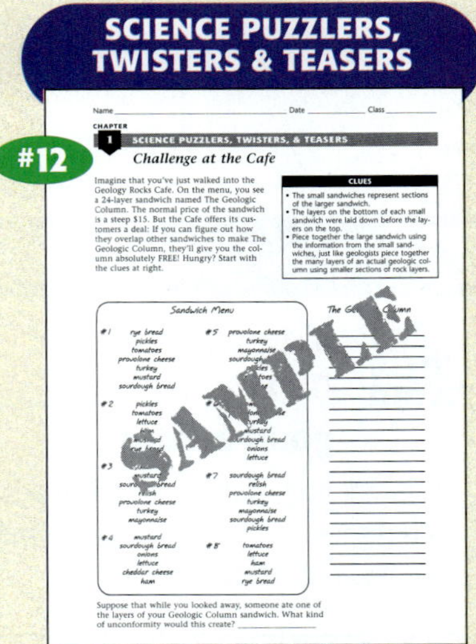

Review & Assessment

STUDY GUIDE

CHAPTER TESTS WITH PERFORMANCE-BASED ASSESSMENT

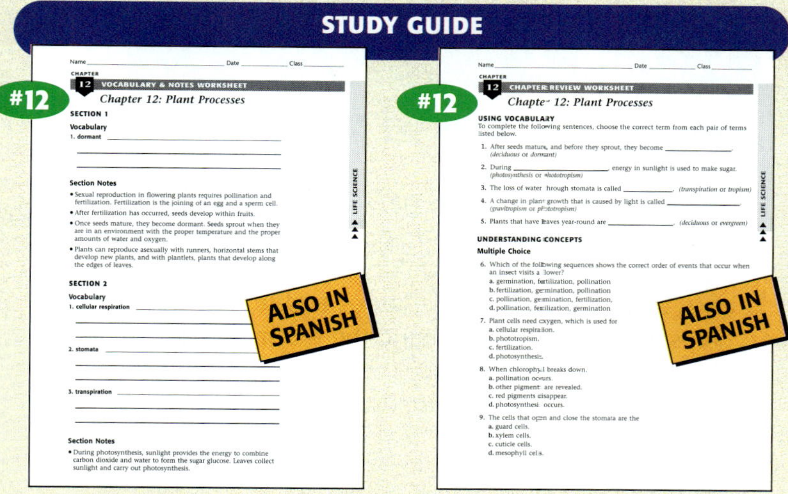

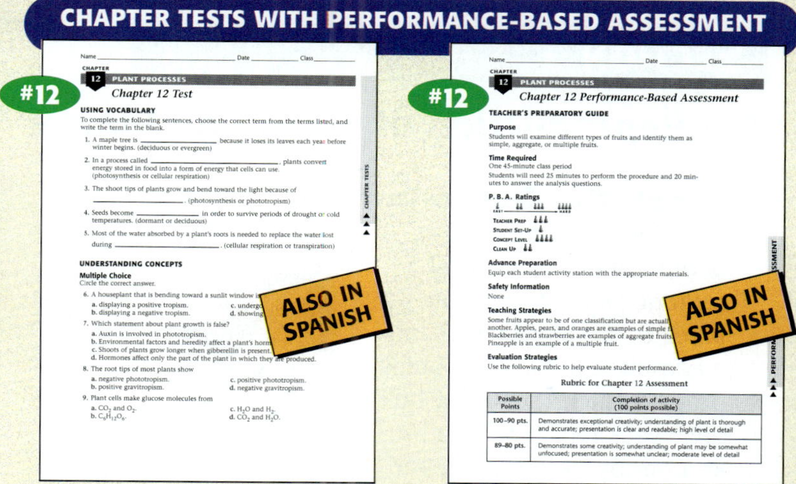

Lab Worksheets

LABS YOU CAN EAT

ECOLABS & FIELD ACTIVITIES

LONG-TERM PROJECTS & RESEARCH IDEAS

DATASHEETS FOR LABBOOK

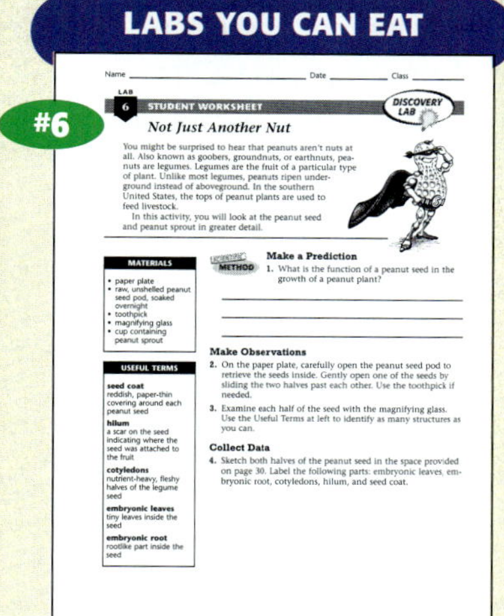

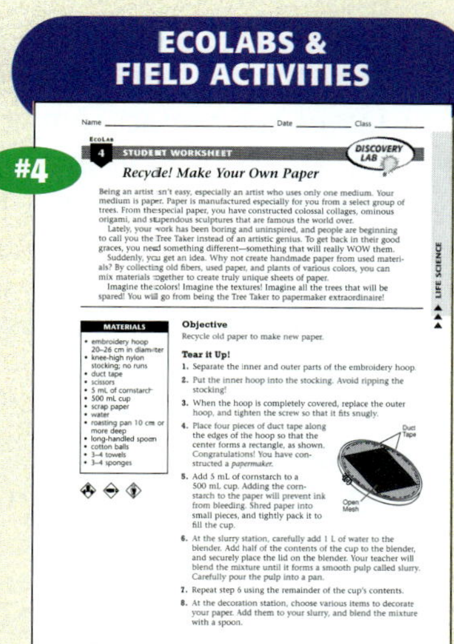

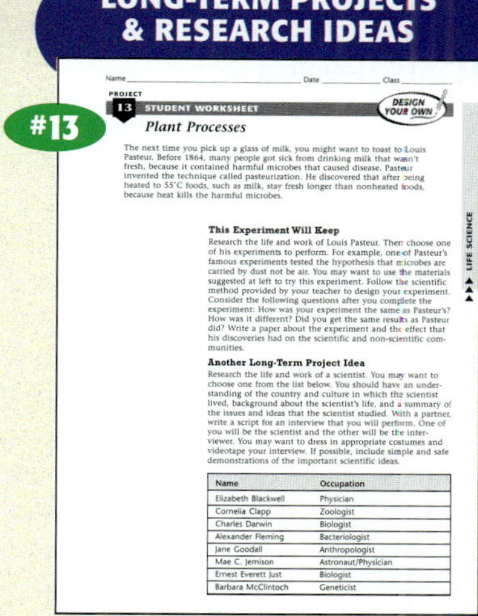

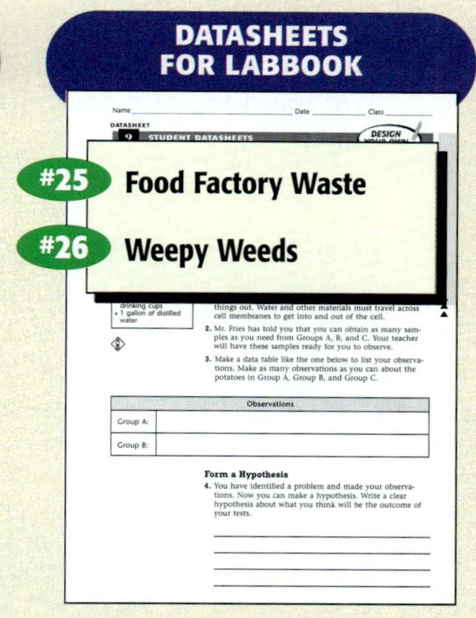

Applications & Extensions

CRITICAL THINKING & PROBLEM SOLVING

EYE ON THE ENVIRONMENT

SCIENTISTS IN ACTION

INTERACTIVE EXPLORATIONS

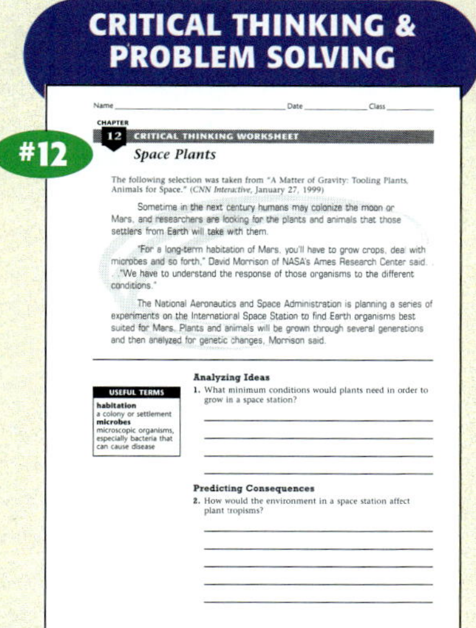

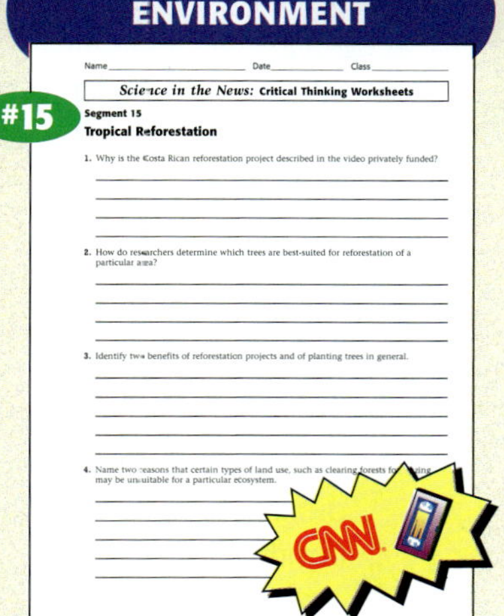

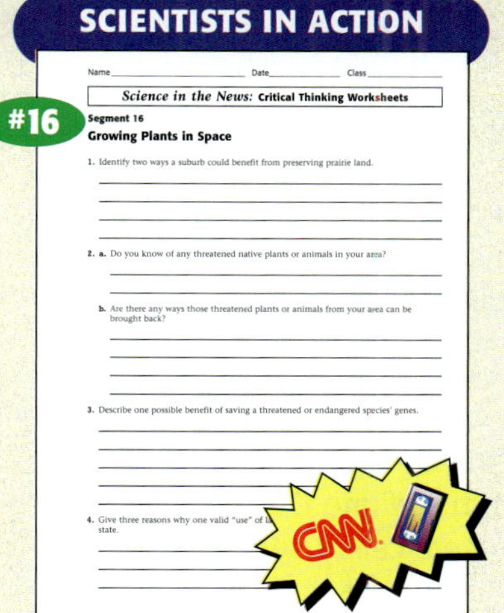

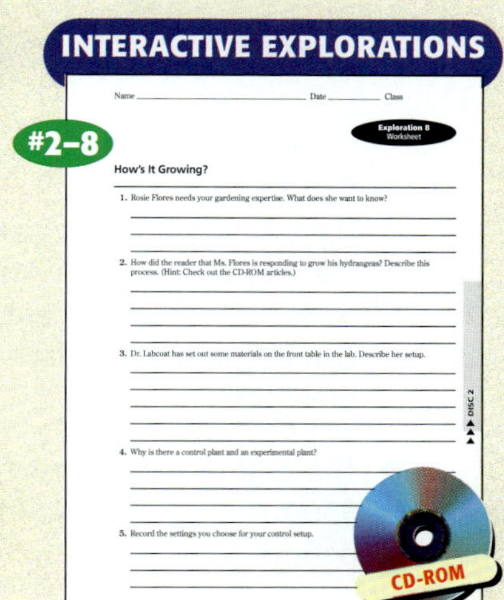

Chapter Background

The Reproduction of Flowering Plants

▶ Vegetative Reproduction

Vegetative reproduction is another term for *asexual reproduction,* in which a piece of the plant grows into a complete plant. Tulip bulbs produce one or two new bulbs each year, which can be broken off the parent plant and used to produce new plants.

- Succulent plants, such as jade plants, have fleshy leaves full of water. This water sustains the leaves if they fall off the parent plant, often long enough for them to send down roots and develop into new plants.

▶ The Perfect Flower?

Flowers can be either perfect or imperfect. Perfect flowers have both male parts (stamens) and female parts (pistils). Imperfect flowers have one or the other—stamens or pistils—but not both.

IS THAT A FACT!

- Night-blooming flowers rely on nocturnal animals, such as bats and hawkmoths, to pollinate their flowers. The flowers are usually white for increased visibility and often have a strong fragrance to attract their pollinators.

- The oldest known fossil seeds are approximately 350 million years old, from the late Devonian period. They belong to plants called seed ferns.

The Ins and Outs of Making Food

▶ Water: A Basic Ingredient

Water conservation is as important for alpine plants as it is for cactuses. At high altitudes there is little rainfall, but there are cold temperatures and high winds. In the Alps, the mountain aven and the mountain kidney vetch have hairlike coverings on their leaves to reduce water loss and provide insulation.

▶ Sunlight

Although sunlight is necessary for photosynthesis, too much sun—specifically, too much ultraviolet radiation—can damage a plant. The sun is more intense on mountains than at lower elevations. Hairlike coverings on leaves can protect some plants from excessive ultraviolet radiation. An example is the silversword, of Hawaii, which grows at altitudes up to 4,000 m.

▶ Clean Air: A Byproduct of Photosynthesis

A study sponsored by NASA has demonstrated that indoor plants in a controlled environment can extract pollutants from the air. The leaves remove low levels of pollutants, and the roots, assisted by activated carbon filters, can remove higher concentrations. The pollutants affected included formaldehyde and benzene gases.

- Scientists continue to study the effectiveness of plants at removing large particle pollutants, such as chemicals from detergents and cleaning fluids; fibers released from clothing, furniture, and insulation; and tobacco smoke.

- Photosynthesis also enables outdoor plants to cleanse the air of carbon emissions from cars and dwellings, nitrogen oxides, airborne ammonia, sulfur dioxide, and ozone. 

IS THAT A FACT!

- One 80 ft beech tree can remove the daily carbon dioxide emissions produced by two single-family homes.

SECTION 3

Plant Responses to the Environment

▶ Plant Pigments

In addition to containing the green pigment chlorophyll, plants contain other pigments that account for the colorful changes in autumn leaves. Xanthophylls are yellow, carotenes are yellowish orange, anthocyanins are red and purple, and tannins are golden yellow.

- In some years, the autumn colors of leaves—especially reds—are bright and colorful; in other years they are dull. Two factors contribute to the difference: warm, sunny days followed by cool nights (temperatures below 45°F) produce bright colors.

IS THAT A FACT!

- In 1997, the U.S. Department of Agriculture reported that nearly 60 million poinsettia (*Euphorbia pulcherrima*) plants were purchased in the United States, totaling about $222 million in sales. Poinsettias are the country's most popular potted plants.

SECTION 4

Plant Growth

▶ Discovery of Auxins

Charles Darwin (1809–1882) is credited with making the first recorded observations that led to the discovery of plant hormones. In 1881, Darwin and his son, Francis, described the occurrence of phototropism, the bending of a plant toward a light source. In an experiment, the Darwins placed lead caps on the growing tips of grass and oat seedlings and noted that the growing tips did not bend. When the lead caps were removed, the tips bent toward the light source. No additional work was done to follow up on the Darwins' experiments until the latter part of the nineteenth century.

▶ Germplasm

A plant's germplasm consists of the plant's genetic material and tissues, organs, and organisms that express the traits contained in the genetic material. The U.S. National Plant Germplasm System (NPGS) conserves and uses germplasm to improve crop plants by transferring disease resistance, improving yields, and adapting plants to grow under new conditions.

IS THAT A FACT!

- Plant hormones occur in very small quantities. In a pineapple plant, for example, only 6 µg of auxin are present for 1 kg of plant material. In terms of weight, this is equivalent to a needle in a 20-metric-ton truckload of hay.

For background information about teaching strategies and issues, refer to the *Professional Reference for Teachers*.

Chapter Preview

Directed Reading Worksheet 12

Science Puzzlers, Twisters & Teasers Worksheet 12

Guided Reading Audio CD
English or Spanish, Chapter 12

CAPÍTULO

12 Procesos de las plantas

Increíble... ¡pero cierto!

El maizal es un campo de batalla. Cuando la plaga de gusano soldado ataca el maíz, éste se defiende. De algún modo, las plantas de maíz mandan una señal de auxilio y, de repente, un comando de avispas cae sobre el maizal y ataca los gusanos ¿Cómo puede pedir auxilio una planta? Los gusanos soldado causan estragos en muchos maizales de los Estados Unidos. Cuando empiezan a devorar el maíz, una substancia de su boca hace que la planta libere otra substancia química. La hembra de la avispa parásita Aphidius colemani capta la señal del maíz y se dirige directamente hacia la planta infestada de gusanos.

Las avispas caen sobre los gusanos y depositan huevecillos bajo su piel. Cuando los huevecillos se abren, las larvas devoran el gusano.

Jim Tumlinson, el científico que descubrió esta relación entre las avispas y el maíz, dice que probablemente se dio por casualidad. Las avispas parásitas se sienten atraídas por naturaleza a la substancia que segrega el maíz.

Tumlinson quiere cultivar otras plantas para que también produzcan substancias que atraigan ciertos insectos. Las avispas parásitas podrían servir para controlar plagas, y no usar tantos plaguicidas venenosos.

Muchas plantas tienen un proceso similar al del maíz para producir y liberar substancias químicas especiales. En este capítulo estudiarás otros procesos de las plantas, como la reproducción de las plantas con flores, la fotosíntesis, la adaptación al ambiente y el crecimiento.

278

Strange but True!

The type of relationship between the corn plant and the parasitic wasps is a form of symbiosis, a close relationship between two organisms of different species. This example demonstrates mutualism, symbiosis in which both organisms benefit. The corn plant is rid of the pesky beet armyworm caterpillars, and the wasp larvae that eat the caterpillars have a source of food.

Usa tus conocimientos para responder a las siguientes preguntas en tu cuaderno de ciencias:

1. ¿Qué es un fruto?

2. ¿Cómo responden las plantas a los cambios en su medio ambiente?

3. ¿Por qué necesitan luz las plantas?

Observa el crecimiento de las plantas

Cuando plantas semillas en tu jardín, las entierras, las riegas y esperas a que las plántulas broten del suelo. Pero, ¿qué pasa cuando las semillas están bajo tierra? ¿Cómo se convierte la semilla en una planta?

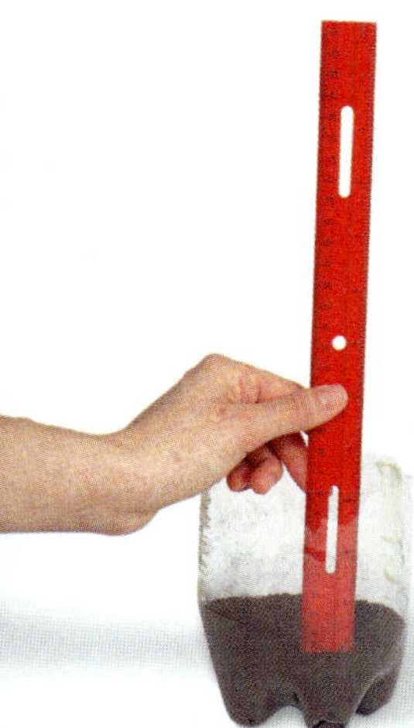

Procedimiento

1. Tu maestro te dará los materiales que necesitas: Una **botella transparente de refresco de 2 L,** con la parte de arriba recortada; **tierra para maceta,** tres o cuatro **frijoles** y **papel de aluminio.**

2. Llena la botella con tierra de maceta hasta 8 cm del borde.

3. Mete las semillas en la tierra junto a la pared de la botella, como en el ejemplo de arriba. La parte de arriba de la semilla debe estar al mismo nivel que la superficie de tierra en la botella. Añade otros 5 cm de tierra de maceta.

4. Envuelve los lados de la botella con el papel de aluminio para protegerlos de la luz.

5. Riega las semillas con unos **60 mL de agua.** Riégalas de nuevo cada vez que el primer centímetro de tierra de la superficie esté completamente seco.

6. Una vez al día, desenvuelve la botella durante unos minutos para ver cómo van tus frijoles. En tu cuaderno de ciencias, lleva un registro del crecimiento de las semillas.

Análisis

7. ¿Cuánto tardaron en germinar las semillas? ¿Cuántas germinaron?

8. ¿De dónde viene la energía que las semillas usan para empezar a crecer?

279

What Do You Think?

Accept all reasonable responses.

Students will have a chance to revise their answers in the Chapter Review under NOW What Do You Think?

Investigate!

MATERIALS

FOR EACH GROUP:
- 2 L clear-plastic soda bottle (with top half cut off)
- potting soil
- bean seeds
- aluminum foil
- water (60 mL)

Safety Caution: Some students—particularly those who suffer from allergies—may wish to wear protective gloves while handling the soil and seeds. Have students wash their hands when they are finished with the activity.

Teacher Notes: Be sure students keep seeds evenly moist to ensure germination.

Make sure students do not cover the seeds with more than 5 cm of potting soil. Doing so will delay the seedling from emerging from the surface.

Science Skills Worksheet 2 "Using Your Senses"

Answers to Investigate!

7. Answers will vary. Germination times vary depending on the seeds used. Soaking the seeds in advance will decrease the number of days to germination. Students may report that not all of their seeds germinated.

8. The seed contains stored food molecules that are used for energy.

Focus

The Reproduction of Flowering Plants

This section describes the roles of pollination and fertilization. Students will be able to explain how fruits are formed from flowers and differentiate between sexual and asexual reproduction in flowering plants.

 **Bellringer**

Ask students to list the names of all the fruits and flowers they can think of in their ScienceLog.

1 Motivate

DEMONSTRATION

Identify the Parts of a Flower

Show students a variety of fresh flowers, and ask them to describe ways the flowers are similar and different. (Student answers should focus on size, structure, fragrance, and color.)

Refer students to **Figure 1.** Point out the anthers, stigmas, petals, and sepals in each flower. Remove the petals, and shake the flower over paper. (Note: Pollen can stain skin and clothing. You may wish to wear protective gloves.)

Ask students to identify the powder on the paper. (pollen)

Explain that pollen contains the flower's male reproductive cells. Sheltered English

 **Teaching Transparency 48** "Pollination and Fertilization"

 **Teaching Transparency 49** "From Flower to Fruit"

Sección 1

La reproducción de las plantas con flores

VOCABULARIO
latente

OBJETIVOS

- Describe el papel que juegan la polinización y la fecundación en la reproducción sexual.
- Describe cómo se forman los frutos a partir de la flor.
- Explica la diferencia entre la reproducción sexual y la asexual en las plantas.

Si salieras al jardín e hicieras una lista de todas las plantas diferentes que ves, seguramente la mayoría serían plantas con flores. Las plantas con flores son el grupo de plantas más grande y diverso del mundo. Su éxito se debe en parte a sus flores, que son adaptaciones para llevar a cabo la reproducción sexual. En la reproducción sexual, el óvulo es fecundado por el espermatozoide. La fecundación en las plantas con flores se efectúa dentro de la flor y resulta en la formación de una o más semillas dentro de un fruto.

¿Cómo se lleva a cabo la fecundación?

Para que la fecundación pueda llevarse a cabo, el espermatozoide debe llegar hasta donde está el óvulo. Los espermatozoides de una planta con flores están dentro de los granos de polen. La polinización ocurre cuando los granos de polen son transportados de las anteras al estigma. Así comienza la fecundación, como se ve en la **Figura 1.** Después de que el polen cae sobre el estigma, le crece un tubo que entra a través del estilo hasta el ovario. Dentro del ovario se encuentran los óvulos. Cada óvulo contiene un núcleo.

Los espermatozoides del grano de polen bajan por el tubo hasta el óvulo y la fecundación se lleva a cabo cuando el espermatozoide se une al núcleo del óvulo.

Figura 1 *La fecundación ocurre después de la polinización.*

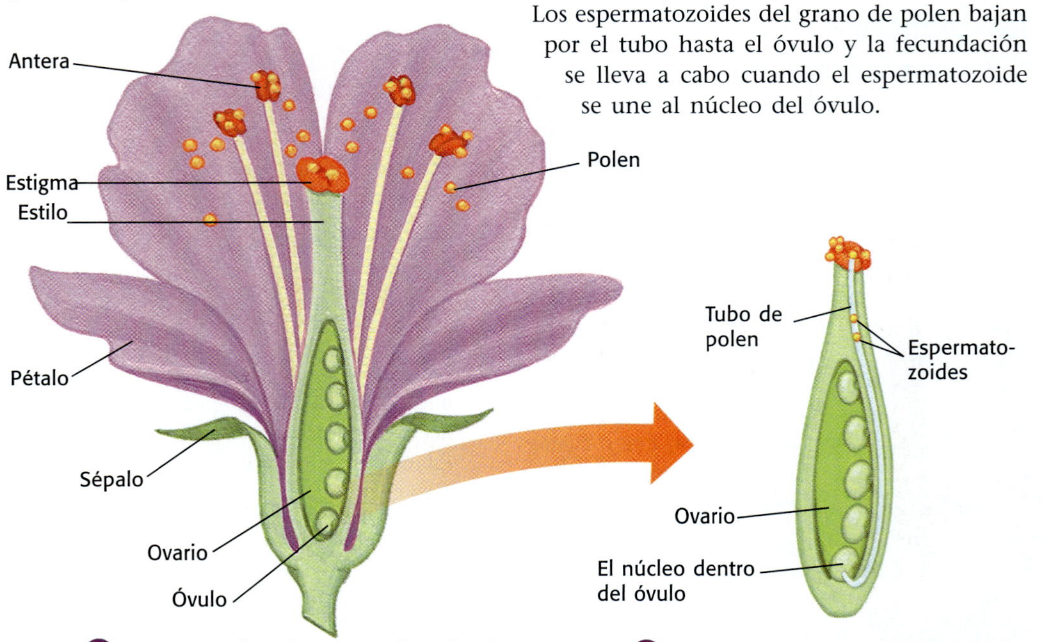

1 Los granos de polen caen sobre el estigma y comienzan a desarrollar tubos de polen.

2 Los espermatozoides bajan por el tubo de polen y fecundan los núcleos.

280

CROSS-DISCIPLINARY FOCUS

History In 1519, the explorer Hernando Cortez brought cacao beans and a recipe from Montezuma's court back to Spain. The recipe was for a new drink called chocolate made from the beans of the cacao plant. Served with honey and sugar, it greatly impressed the members of the Spanish court. In fact, they were so impressed that they kept this wonderful bean a secret for a hundred years. For the next century, Spanish monks were the only people in Europe who knew how to prepare chocolate.

De flor a fruto

Después de la fecundación, el óvulo se convierte en una semilla que contiene la pequeña planta que aún no se ha desarrollado y el ovario se convierte en el fruto que contiene la semilla. La **Figura 2** muestra cómo el ovario y el óvulo de una flor se convierten en el fruto y sus semillas.

Figura 2 *La fecundación lleva al desarrollo del fruto y las semillas.*

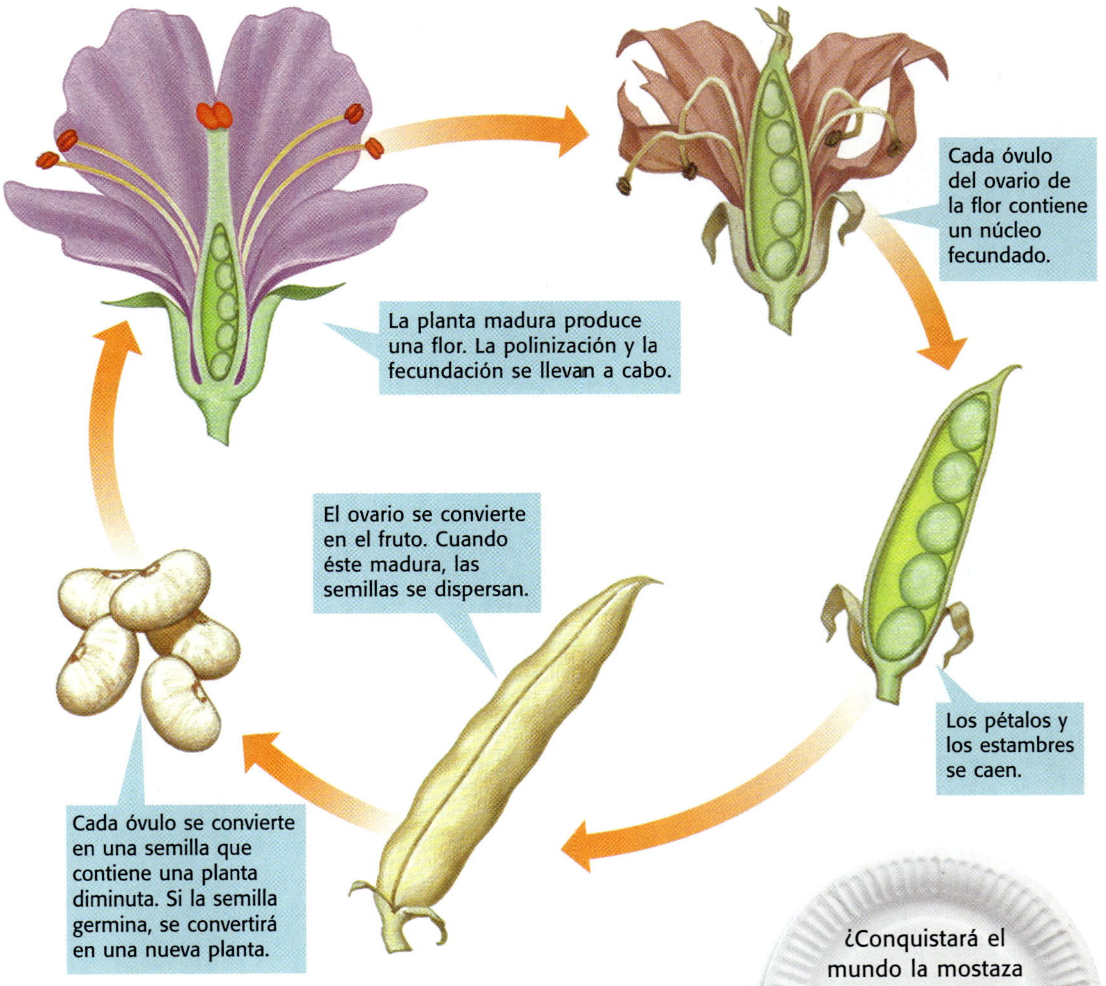

La planta madura produce una flor. La polinización y la fecundación se llevan a cabo.

Cada óvulo del ovario de la flor contiene un núcleo fecundado.

El ovario se convierte en el fruto. Cuando éste madura, las semillas se dispersan.

Los pétalos y los estambres se caen.

Cada óvulo se convierte en una semilla que contiene una planta diminuta. Si la semilla germina, se convertirá en una nueva planta.

✓ Autoevaluación

¿Cuántos frutos y semillas se pueden desarrollar de una flor que tiene un ovario con seis óvulos? *(Consulta la página 636 para comprobar tu respuesta.)*

281

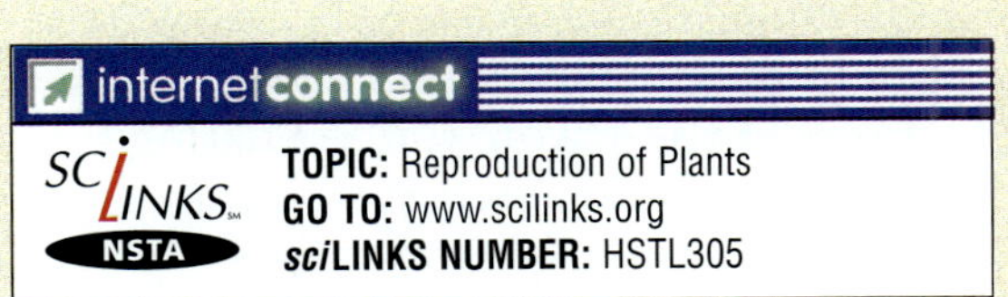

Directed Reading Worksheet 12 Section 1

---

2 Teach

ACTIVITY

Germination

MATERIALS

FOR EACH GROUP:
- packet of bean seeds
- two small, plastic containers with snap-on caps
- water

Show students that plant germination can push caps off bottles. This demonstration will take a few days. Fill a container with beans. Fill another container with beans and water. Snap the caps onto the bottles. (Don't use child-proof bottles that lock.) Place them where they can be observed. In a few days, the germinating beans will knock the caps off or even split the bottle apart. Ask students to note which beans are more powerful, the beans with water or the beans without water. (The beans with water are producing a gas, and they are also expanding as they germinate.) Sheltered English

GROUP ACTIVITY

Concept Mapping Have students work together in groups of three or four to create a concept map that details the process of sexual reproduction from the time the pollen grains reach the stigma until a seed develops inside a fruit. Encourage students to illustrate their work.

Answer to Self-Check

Fruit develops from the ovary, so the flower can have only one fruit. Seeds develop from the ovules, so there should be six seeds.

Quick Lab

MATERIALS

FOR EACH GROUP:
- 12 bean seeds
- 2 Petri dishes
- marking pen or wax pencil
- water

Students should observe that the bean seeds that soaked in water increased in size. Students might also observe that the soaked seeds have begun to crack open.

Answer to QuickLab

6. The seeds swell because they are absorbing water. The stored water can then be used by the young plant.

CROSS-DISCIPLINARY FOCUS

Art Have students use books from the library or from their home to find examples of how artists (painters, sculptors, photographers, and so on) throughout history have represented flowers and fruits in their work. You might want to show students reproductions of works by Claude Monet or Georgia O'Keeffe to encourage their interest in this activity. Interested students could create a drawing of a flower or a still life of fruit.

Sheltered English

Reinforcement Worksheet 12
"Fertilizing Flowers"

Fruto de un esfuerzo

Mientras los óvulos se convierten en semillas, el ovario se convierte en el fruto. Al crecer y madurar, contiene y protege las semillas que se están desarrollando. Observa la ilustración de abajo para ver qué partes de las frutas se desarrollaron del ovario y cuáles se desarrollaron de los óvulos.

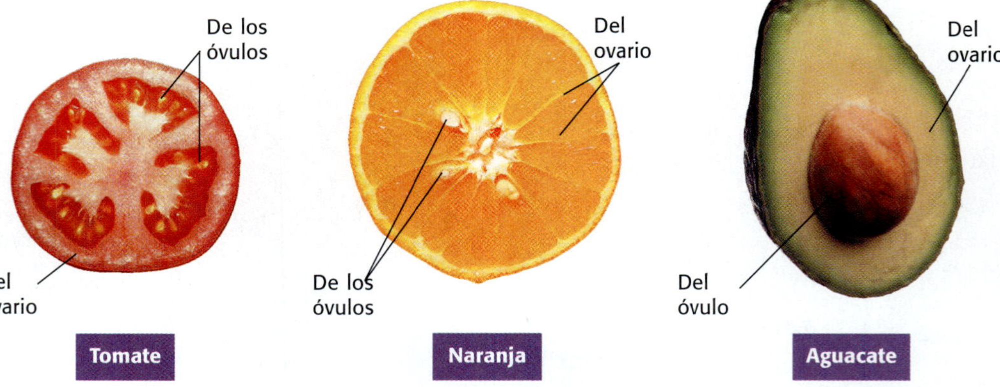

Laboratorio

Semillas sedientas

1. Pide a tu maestra o maestro **12 frijoles, 2 cajas de Petri**, **un creyón** y **agua.**

2. Llena dos terceras partes de una de las cajas de Petri con agua y añade seis frijoles. Marca la caja con tu nombre y con la palabra "Agua".

3. Pon el resto de o frijoles en la caja de Petri sin agua. Ponle tu nombre y la palabra "Control".

4. Déjalos en remojo una noche.

5. Al día siguiente, compara el tamaño de lo frijoles de los dos grupos. Anota tus observaciones en tu cuaderno de ciencias.

6. ¿Por qué cambió el tamaño de los frijoles? ¿Cómo afecta a su supervivencia?

Las semillas se convierten en plantas

Cuando la semilla ya está completamente desarrollada, la pequeña planta en su interior deja de crecer. La semilla puede quedar en estado **latente,** es decir, inactivo. Estas semillas pueden sobrevivir largas temporadas de sequía y temperaturas muy frías. Algunas semillas necesitan condiciones extremas, como un invierno helado o hasta un incendio forestal, para salir del estado latente.

Una semilla germina si cae o se siembra en un ambiente que tiene agua, oxígeno y la temperatura adecuada. Para la mayoría de las plantas la temperatura ideal para crecer es de más o menos 27 °C (80.6 °F). La **Figura 3** muestra la *germinación* de una semilla de frijol y las primeras etapas de crecimiento en una plantita de frijol.

Figura 3 *La reproducción sexual produce semillas que se convierten en nuevas plantas.*

WEIRD SCIENCE

In 1967, the seeds of the arctic tundra lupine (*Lupinus arcticus*) were discovered by a mining engineer in a frozen animal burrow in the Yukon. Carbon dating showed the animal remains in the burrow to be at least 10,000 years old. Frozen seeds of the tundra lupine were found among the animal remains. When tested, these seeds germinated within 48 hours!

Otros métodos de reproducción

Muchas plantas con flores también se pueden reproducir de manera asexual. En la reproducción asexual, las plantas no forman flores, semillas ni frutos, sino que una de sus partes, como el tallo o la raíz, produce una nueva planta. En la **Figura 4** se ven varios ejemplos de reproducción asexual.

Figura 4 *La reproducción asexual puede ocurrir de varias formas. Aquí puedes ver algunos ejemplos.*

a Los tubérculos de papa son tallos subterráneos que están henchidos de alimentos acumulados. Los "ojos" de la papa son brotes que pueden desarollarse de manera asexual y convertirse en nuevas plantas.

b La planta de la fresa produce guías, que son tallos que crecen horizontalmente por el suelo. Los retoños que crecen a lo largo de cada guía se convierten en nuevas plantas que echan raíces en la tierra.

c Esta planta del género Kalanchoe produce plántulas en los bordes de las hojas. Las plántulas finalmente caen de la planta madre y echan raíces para crecer como plantas individuales.

REPASO

1. ¿Qué diferencia hay entre la polinización y la fecundación?

2. ¿Qué parte de la flor se convierte en fruto?

3. **Relacionar conceptos** ¿Qué tienen en común las flores y las guías? ¿En qué se diferencian?

4. **Identificar relaciones** ¿Cuándo puede ser importante la reproducción asexual para la supervivencia de la planta?

---

3 Extend

RESEARCH

Writing Have students research the following different methods by which plants reproduce asexually: corm (gladiolus and crocus), tuber, (gloxinia and potato), tuberous root (dahlia), and rhizome or rootstock (iris and canna). Have students prepare an illustrated report to share with the class.

PORTFOLIO

4 Close

Quiz

1. What are the parts of a pistil? (the stigma, style, and ovary) What are the "male" parts of a flower? (the anther and pollen)

2. What is the difference between sexual and asexual reproduction in flowering plants? (Sexual reproduction involves the joining of sperm cell and egg—fertilization—to form a seed that can grow and develop into a new plant. Asexual reproduction does not involve the formation of flowers, seeds, and fruits; instead, a new plant grows from a part—such as a stem or root—of an existing plant.)

ALTERNATIVE ASSESSMENT

Have students make drawings of two common flowers in their ScienceLog and label the petals, sepals, anthers, and stigmas. Sheltered English

---

▼ **Answers to Review**

1. Pollination occurs when the pollen lands on the stigma of the female flower. Sperm cells from the pollen grain move down into the ovary. Fertilization occurs when one sperm cell fuses with the egg inside the ovule.

2. The ovary develops into the fruit.

3. They both lead to the formation of new plants, but the runner is a form of asexual reproduction while the flower is part of sexual reproduction.

4. Asexual reproduction becomes important when conditions are unfavorable for sexual reproduction.

Focus

The Ins and Outs of Making Food

This section describes photosynthesis. Students will learn about the relationship between photosynthesis and cellular respiration. In addition, they will learn about the importance of stomata in the processes of photosynthesis and transpiration.

Bellringer

Write the following question on the board or overhead projector:

Where do you get the energy you need to stay alive?

Have students write and explain their answer in their ScienceLog. (Students will probably answer that they get their energy from the foods they eat.)

Sheltered English

1 Motivate

DISCUSSION

Food and Energy Ask students the following questions before they begin reading this section.

- What kinds of foods do you eat? (Answers will vary.)

- Where do the animals you eat for food get their energy to survive? (Animals eat plants or other animals that eat plants.)

- Where do plants get their energy to survive? (Plants get their energy from sunlight.)

Explain to students that plants use energy from the sun to make their own food in a process called photosynthesis.

Sección 2

Los detalles de la fabricación de la comida

VOCABULARIO
clorofila
respiración celular
estomas
transpiración

OBJETIVOS
- Describe la fotosíntesis.
- Habla de la relación entre la fotosíntesis y la respiración celular.
- Explica la importancia de los estomas en los procesos de fotosíntesis y transpiración.

Las plantas no tienen pulmones, pero, como tú, necesitan aire para vivir. El aire es una mezcla de oxígeno, dióxido de carbono y otros gases. Las plantas necesitan dióxido de carbono para producir su alimento a través de la fotosíntesis

¿Qué pasa en la fotosíntesis?

Las plantas necesitan luz solar para producir alimentos. En la fotosíntesis, las plantas usan la energía de la luz solar para producir un tipo de azúcar ($C_6H_{12}O_6$) a partir de dióxido de carbono(CO_2) y agua (H_2O). ¿Cómo funciona esto?

Las plantas capturan la energía de la luz Las células de las plantas tienen unos organelos llamados cloroplastos, que a su vez contienen un pigmento verde llamado **clorofila**. La clorofila absorbe la energía de la luz. Aunque no se note, la luz del sol es una mezcla de todos los colores del arco iris. La **Figura 5** ilustra cómo los colores de la luz del sol se pueden separar al pasar a través de un cristal triangular llamado prisma. La clorofila absorbe casi todos los colores de la luz, menos el verde. Las plantas se ven verdes porque la clorofila refleja la luz verde.

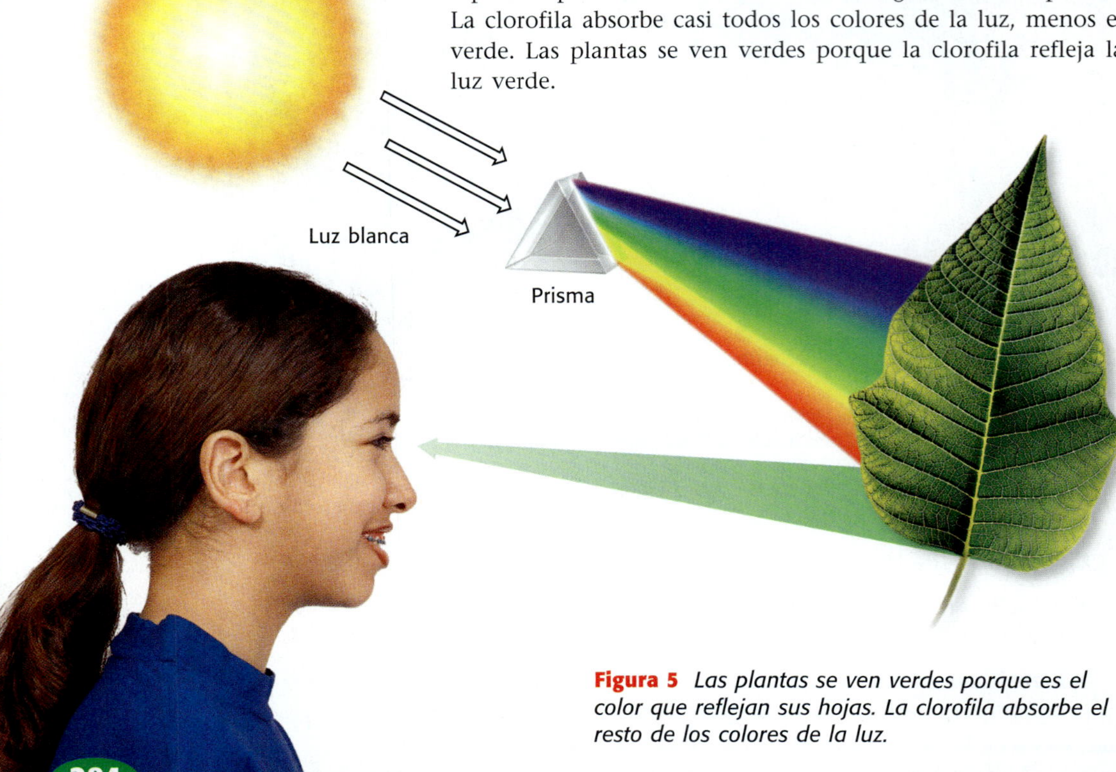

Figura 5 *Las plantas se ven verdes porque es el color que reflejan sus hojas. La clorofila absorbe el resto de los colores de la luz.*

Teaching Transparency 50
"Photosynthesis"

Directed Reading Worksheet 12 Section 2

Reinforcement Worksheet 12
"A Leaf's Work Is Never Done"

Students may think that photosynthesis occurs only in the leaves of plants. In some plants, such as aspen trees and the cactuses, photosynthesis occurs in the plant's trunks or stems.

La fábrica de azúcar   La energía de la luz que absorbe la clorofila se usa para separar la molécula del agua (H_2O) en hidrógeno (H) y oxígeno (O). El hidrógeno se combina con el dióxido de carbono (CO_2) que la planta toma del aire para formar azúcar ($C_6H_{12}O_6$). El oxígeno se libera como un producto secundario. El proceso de la fotosíntesis se resume en la siguiente ecuación:

$$6CO_2 + 6H_2O \xrightarrow{\text{Energía solar}} C_6H_{12}O_6 + 6O_2$$

La ecuación explica que se necesitan seis moléculas de dióxido de carbono y seis moléculas de agua para producir una molécula de glucosa y seis moléculas de oxígeno. Este proceso se ilustra en la **Figura 6.**

La planta usa la energía almacenada en las moléculas de alimento para realizar sus procesos vitales. Dentro de cada célula, el azúcar y otras moléculas de alimento se descomponen en un proceso llamado respiración celular. La **respiración celular** convierte la energía almacenada en el alimento en una forma de energía que las células pueden utilizar. En este proceso, la planta usa oxígeno y libera dióxido de carbono y agua.

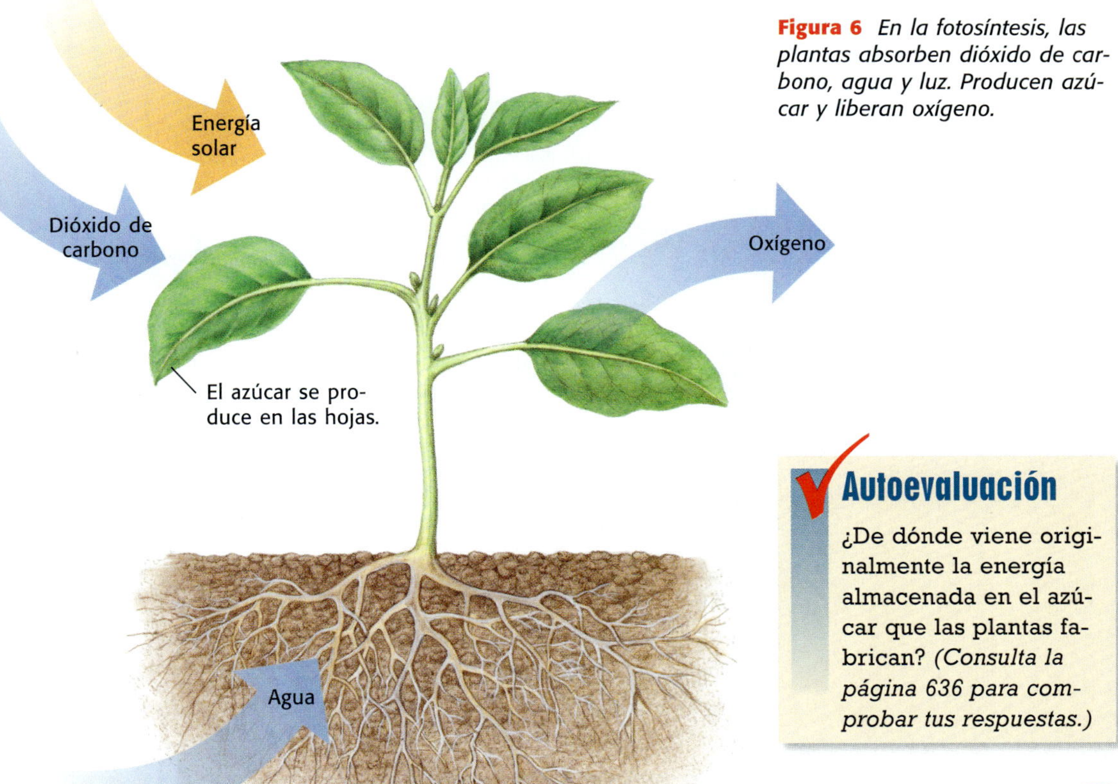

Figura 6 *En la fotosíntesis, las plantas absorben dióxido de carbono, agua y luz. Producen azúcar y liberan oxígeno.*

Autoevaluación

¿De dónde viene originalmente la energía almacenada en el azúcar que las plantas fabrican? *(Consulta la página 636 para comprobar tus respuestas.)*

285

Answer to Self-Check

The sun is the source of the energy in sugar.

Section 2 • The Ins and Outs of Making Food   **285**

2 Teach

LabBook   **PG 600**
Food Factory Waste

MEETING INDIVIDUAL NEEDS

Learners Having Difficulty
Have students with limited English proficiency read the definitions of *photosynthesis, cellular respiration,* and *transpiration* and then write a definition of each term in their own words. Sheltered English

MATH and MORE

Write the following equation— with missing numbers—for photosynthesis on the board:

$6CO_2 + \underline{}H_2O + \text{light}$
$\text{energy} \rightarrow C\underline{}H_{12}O_6 + \underline{}O_2$

Explain to students that the same number of atoms of each element must appear on each side of the equation. Have students provide the missing numbers. (6, 6, 6)

Math Skills Worksheet 50 "Balancing Chemical Equations"

CONNECT TO PHYSICAL SCIENCE

Use the following Teaching Transparency to help students understand the process of balancing chemical equations.

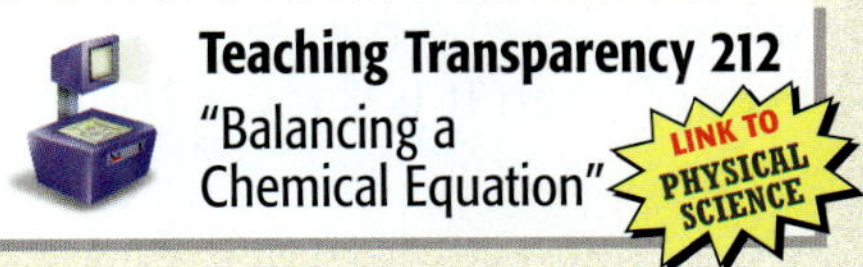
Teaching Transparency 212 "Balancing a Chemical Equation"

Weepy Weeds PG 602

GROUP ACTIVITY

Divide students into groups of four. Provide each group with 6 black marbles representing carbon atoms, 18 white marbles representing oxygen, and 12 blue marbles representing hydrogen. Have students arrange the marbles to demonstrate that six molecules of carbon dioxide and six molecules of water are needed to produce one molecule of sugar and six molecules of oxygen gas. **Sheltered English**

4 Close

Quiz

1. What molecules do plants use to make sugar? (carbon dioxide and water)

2. What substances enter and exit a leaf through the stomata? (enter: carbon dioxide; exit: oxygen and water)

ALTERNATIVE ASSESSMENT

Writing Have students write two paragraphs about stomata in their ScienceLog. The first paragraph should describe the appearance of stomata and the passage of materials through the stomata when light is available. The second paragraph should describe the same two events when it is dark.

Teaching Transparency 51 "Transpiration"

Intercambio de gases

Todas las superficies de la planta que no están bajo tierra están cubiertas por una cutícula, que es una capa cerosa que impide el paso de gases y agua. ¿Cómo recibe la planta el dióxido de carbono a través de esta barrera? El dióxido de carbono entra a las hojas de la planta a través de los **estomas**. El estoma es una abertura en la epidermis y la cutícula de la hoja. Cada estoma está rodeado por dos *células oclusivas,* que funcionan como puertas dobles para abrir y cerrar el hoyito. En la **Figura 7** puedes ver estomas abiertos y cerrados. El funcionamiento de los estomas se ilustra en la **Figura 8.**

Cuando los estomas están abiertos, el dióxido de carbono entra en la hoja. Al mismo tiempo, el oxígeno producido en la fotosíntesis sale de las células de la hoja y pasa a los espacios intercelulares, para luego salir por los estomas.

Cuando los estomas están abiertos, también escapa vapor de agua. La pérdida de agua a través de las hojas se llama **transpiración.** La mayor parte del agua que la planta absorbe por las raíces reemplaza el agua perdida en la transpiración. Cuando una planta se marchita, es porque está perdiendo más agua a través de las hojas que la que absorbe por las raíces.

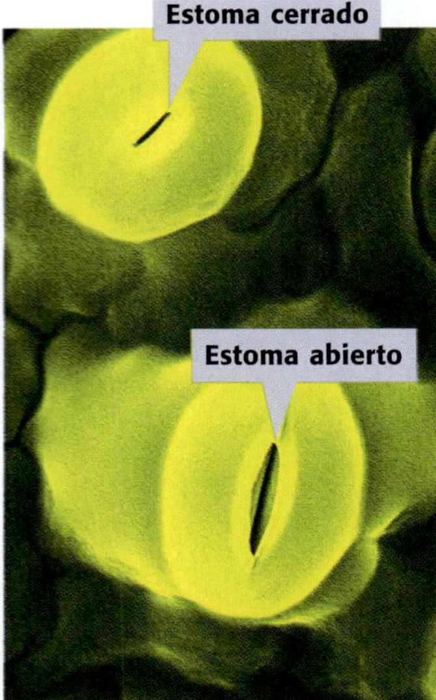

Figura 7 *Cuando la planta recibe luz y puede realizar la fotosíntesis, los estomas generalmente están abiertos. La fotosíntesis no se puede realizar en la obscuridad. El dióxido de carbono no es necesario en este momento y los estomas se cierran para conservar agua.*

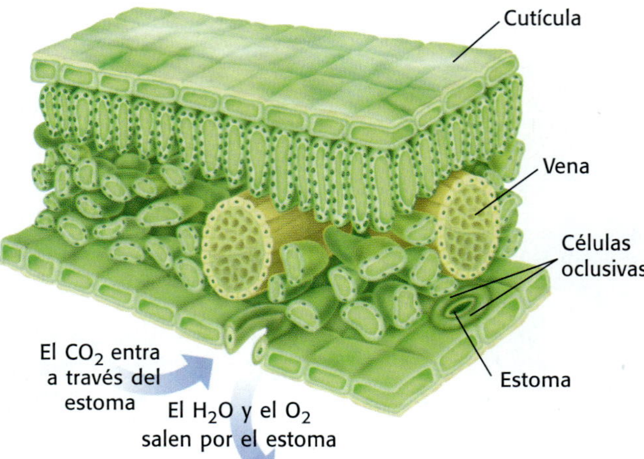

Figura 8 *Las plantas absorben dióxido de carbono y liberan oxígeno y agua a través de los estomas de las hojas.*

REPASO

1. Nombra las tres cosas que las plantas necesitan para realizar la fotosíntesis.

2. ¿Por qué necesitan respirar las células de la planta?

3. **Identificar relaciones** ¿Qué relación hay entre la función de los estomas y la transpiración? ¿Cuándo ocurre la transpiración?

286

▼ **Answers to Review**

1. Plants need light, water, and carbon dioxide for photosynthesis.

2. because the energy stored in food must be converted into a form of energy that cells can use

3. Transpiration occurs when water vapor exits the leaf through the stomata. It usually occurs when there is light available for photosynthesis and gas exchange is occurring.

Las respuestas de las plantas al medio ambiente

VOCABULARIO
tropismo
fototropismo
geotropismo
de hoja perenne
de hoja caduca

OBJETIVOS
- Describe cómo reponden las plantas a la luz y la gravedad.
- Explica cómo algunas plantas florecen según la duración de la noche.
- Describe cómo algunas plantas se han adaptado para sobrevivir en climas fríos.

$$\div \quad 5 \div \quad \Omega \quad \leq \quad \infty \quad +\Omega \quad \sqrt{} \quad 9 \underset{\infty}{\leq} \quad \Sigma \; 2$$

¡MATEMÁTICAS!

Inclinación de buen grado

Supón que una planta presenta fototropismo positivo y se inclina hacia la luz a una velocidad de 0.3 grados por minuto. ¿Cuántas horas le llevará inclinarse 90 grados?

¿Qué te pasa cuando tienes mucho frío? ¿Te castañetean los dientes? Esto sucede porque tu cerebro responde al estímulo del frío haciendo que tus músculos tiemblen para generar calor. Llamamos estímulo a cualquier cosa que provoca una reacción en un órgano o tejido. ¿También responden a los estímulos los tejidos de las plantas? ¡Claro que sí! Algunos ejemplos de estímulos que las afectan son la luz, la gravedad, los cambios de estaciones, y hasta el hecho de que alguien se las coma.

Tropismos de las plantas

La respuesta de algunas plantas a estímulos ambientales, como la luz o la gravedad, es crecer en una dirección específica. Cuando una planta crece en respuesta a un estímulo, esto se llama **tropismo.** Los tropismos pueden ser negativos o positivos, dependiendo de la dirección en que crezca la planta. Si el crecimiento es en dirección al estímulo, es positivo. Si es en dirección contraria al estímulo, es negativo. Dos ejemplos son el fototropismo y el geotropismo.

Las plantas son sensibles a la luz
Si pusieras una planta de modo que recibiera luz de una sola dirección, como por ejemplo, cerca de una ventana, las puntas de los brotes se inclinarían hacia la luz. **Fototropismo** es el cambio que la luz produce en el crecimiento de una planta. Como ves en la **Figura 9,** la inclinación de la planta ocurre porque las células de un lado del brote se hacen más largas que las células del otro lado.

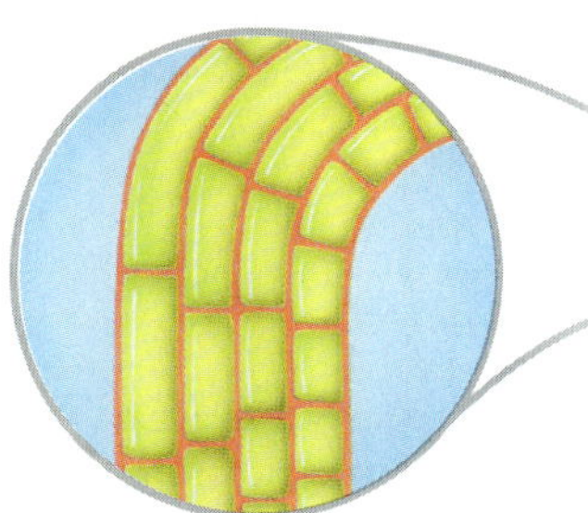

Figura 9 *Las plantas en el lado obscuro del brote se alargan más que las del otro lado. Esto hace que el brote se incline hacia la luz.*

287

Focus

Plant Responses to the Environment

This section describes how plants may respond to light and gravity and how some plants flower in response to night length. Students will be able to describe how some plants are adapted to survive in cold weather.

Bellringer

Have students write in their ScienceLog a hypothesis explaining why some leaves of some trees undergo a color change in autumn. Have volunteers read their hypothesis aloud, and let the students know they'll be learning more about such plant responses in this section.

1) Motivate

DISCUSSION

Plant Responses Bring to class a sensitive plant, such as *Mimosa pudica.* Show students how the leaves of the plant "fold up" when they are touched. You also can demonstrate plant movement with a Venus' flytrap (*Dionaea muscipula*). Have a student volunteer gently probe inside the open "trap" with a pencil, and observe how the plant responds.

Answer to MATHBREAK
5 hours

IS THAT A FACT!

The tallest tree ever measured was found in Watts River, in Victoria, Australia. The tree—an Australian eucalyptus (*Eucalyptus regnans*)—was 132 m tall in 1872.

Teaching Transparency 52
"Phototropism"

Directed Reading Worksheet 12 Section 3

USING THE FIGURE

Have students look closely at **Figure 10.** Point out that after a few days the leaves of the plant grow upward and the roots grow downward. Explain that the plant growth is in response (positive or negative) to gravity. Ask students what other stimuli the stems might be growing in response to. (Students should reason that the stems grow upward in order for the leaves to reach sunlight and roots grow downward to reach water or moisture and to anchor the plant.)

Sheltered English

Quick Lab

MATERIALS

FOR EACH GROUP:
• several potted plants
• duct tape
• cardboard

Answers to QuickLab

Observations may vary, but students should see that plant stems always try to grow perpendicular to the ground.

Gravity and light might have influenced the growth by causing the plant to grow upward.

Gravitropism benefits a plant because it directs the plant to grow so that the stems are aboveground and the roots grow downward.

¿Abajo o arriba? Llamamos **geotropismo** al cambio que la dirección de la gravedad produce en el crecimiento de una planta. La **Figura 10** muestra el efecto del geotropismo en las plantas. Pocos días después de que una planta es colocada de cabeza, se nota un cambio de dirección en el crecimiento de sus raíces y tallos. La mayoría de las puntas de los tallos muestran geotropismo negativo, es decir, crecen hacia arriba, en dirección contraria a la fuerza de gravedad. En cambio, la mayoría de las puntas de las raíces muestran geotropismo positivo, es decir, crecen hacia abajo, en la misma dirección que la fuerza de gravedad.

Figura 10 *La gravedad es un estímulo que hace que las plantas cambien la dirección en la que crecen.*

a Para crecer en dirección opuesta a la fuerza de gravedad, esta planta creció hacia arriba.

b Esta planta ha estado volteada de cabeza y sus tallos más recientes crecieron hacia abajo.

Laboratorio

¿Abajo o arriba?

¿Puede crecer de lado una planta? Vas a necesitar **varias plantas en maceta** para averiguarlo. Con **cinta** adhesiva, **sujeta un cartón** para evitar que la tierra se salga de la maceta. Voltea las plantas de lado y observa lo que sucede en los días siguientes. Anota tus observaciones en tu cuaderno de ciencias. Describe dos estímulos que creas que influyeron sobre la dirección del crecimiento de las plantas. ¿Cómo se beneficia una planta del geotropismo?

✔ Autoevaluación

1. Usa los siguientes términos para crear un mapa de ideas: tropismo, estímulo, luz, gravedad, fototropismo y geotropismo.

2. Imagínate una planta en la que la luz provoca fototropismo negativo. Si la luz le da a la planta sólo del lado derecho, ¿hacia qué lado se inclinará la planta?

(Consulta la página 636 para comprobar tus respuestas.)

288

Answers to Self-Check

1.

Tropism
is a plant's response to
stimuli
such as
gravity — light
which causes — which causes
gravitropism — phototropism

2. During negative phototropism, the plant would grow away from the stimulus (light), so it would be growing to the left.

La respuesta a los cambios de las estaciones

¿Qué pasaría si una planta que crece en un área de inviernos muy fríos floreciera en diciembre? ¿Crees que lograría producir frutos y semillas? Si dijiste que no, estás en lo correcto. Si la planta alcanzara a florecer, sus flores seguramente se congelarían y morirían sin poder producir semillas maduras. Las plantas que viven en climas fríos pueden detectar el cambio de estación. ¿Cómo lo hacen?

Tan distinto como el día y la noche Piensa en lo que pasa cuando las estaciones cambian. Por ejemplo, ¿qué pasa con la duración del día y la noche? Cuando se acercan el otoño y el invierno, los días se hacen más cortos y las noches más largas. Lo contrario sucede cuando se acercan la primavera y el verano.

La duración de los días y las noches es un estímulo ambiental muy importante para muchas plantas, pues puede hacer que comiencen a reproducirse. Algunas plantas sólo florecen al final del verano o al principio del otoño, cuando las noches son largas. Estas plantas se llaman plantas de días cortos. Algunos ejemplos de plantas de días cortos son la noche buena (que se ilustra abajo, en la **Figura 11**), la ambrosía y el crisantemo. Otras plantas florecen en la primavera o a principios del verano, cuando las noches son cortas, y se llaman plantas de días largos. El trébol, la espinaca y la lechuga son algunas de ellas.

ciencias de la Tierra
CONEXIÓN

Las estaciones son la consecuencia de la inclinación de la Tierra y de su órbita alrededor del Sol. Estamos en verano cuando el hemisferio norte está inclinado hacia el Sol y la energía del Sol nos llega más directamente. Mientras hace calor en el hemisferio norte, en el hemisferio sur hace frío. Cuando el hemisferio norte está inclinado más lejos del Sol, ocurre lo contrario.

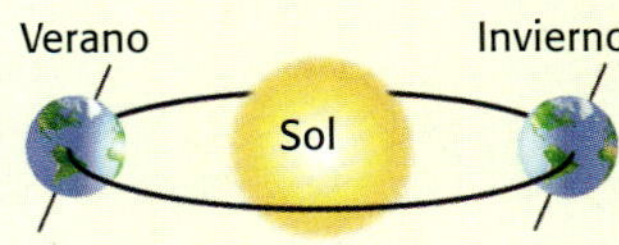

Figura 11 *La duración de la noche determina cuándo florecen las nochebuenas.*

a A principios del verano, las noches son cortas. En esta época del año, las hojas de la nochebuena son todas verdes, y no tiene flores.

b Las nochebuenas florecen en el otoño, cuando las noches son más largas. Las hojas alrededor de los botones de flores se vuelven rojas. En el cultivo comercial de nochebuenas se usan luces artificiales para controlar el momento en que las hojas cambian de color.

289

Answer to APPLY

Sample answers include the following: experiments in which entire trees were covered for part of the day to prevent them from getting sunlight; experiments in which entire trees were exposed to artificial sunlight to determine if trees could be kept from losing their leaves; experiments in which some of the leaves were protected or given artificial sunlight to determine if some leaves could be kept on the tree, even when others fell off.

READING STRATEGY

Prediction Guide Before students read this page, ask the following questions:

- What kind of tree is an evergreen? Give two examples.
  (Evergreens are trees that have leaves adapted to survive throughout the year. Examples include conifers, such as pines and firs.)
- What are deciduous trees?
  (Deciduous trees lose all their leaves at the same time each year.)

MISCONCEPTION ALERT

The colored part of the poinsettia is not the flower, as most people think it is. The colored parts are really types of leaves called bracts. The flowers are the tiny berrylike structures at the center of the colored bracts.

Teaching Transparency 53
"The Change of Seasons and Pigment Color"

Reinforcement Worksheet 12
"How Plants Respond to Change"

APLICA

Una tarde de otoño, Mónica mira su jardín y se da cuenta de que un árbol que estaba lleno de hojas la semana pasada está ahora completamente desnudo. ¿Qué hizo que se cayeran las hojas? A Mónica se le ocurrió la siguiente hipótesis:

Cada hoja del árbol podía detectar la duración de los días. Cuando los días se hicieron más cortos, cada hoja respondió a este estímulo y se dejó caer.

Inventa un experimento para comprobar la hipótesis de Mónica.

Los cambios de las estaciones en las hojas Todos los árboles pierden sus hojas en algún momento. Algunos, como los pinos y las magnolias, las pierden a lo largo del año, de modo que siempre hay algunas hojas en sus ramas. Estos árboles se llaman **de hoja perenne.** Las plantas perennes tienen hojas adaptadas para sobrevivir todo el año.

Otros árboles, como el arce de la **Figura 12,** son **de hoja caduca,** y pierden todas sus hojas en la misma época del año. Normalmente, los árboles de hoja caduca pierden sus hojas antes del invierno. En climas tropicales, que tienen una temporada seca y una de lluvias, los árboles de hoja caduca pierden sus hojas antes de la época seca. Al caer sus hojas antes del invierno o de la temporada seca, los árboles pierden menos agua en la transpiración. La pérdida de las hojas les ayuda a las plantas a sobrevivir al frío o largos períodos sin lluvia.

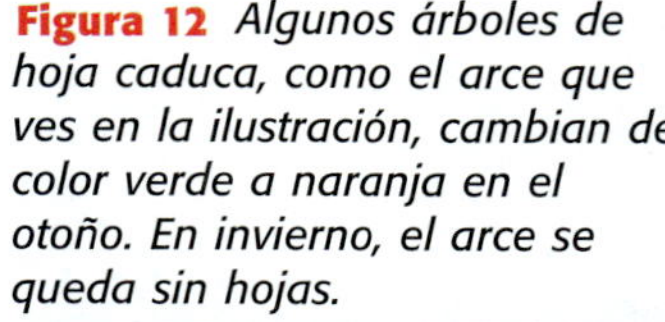

Figura 12 *Algunos árboles de hoja caduca, como el arce que ves en la ilustración, cambian de color verde a naranja en el otoño. En invierno, el arce se queda sin hojas.*

IS THAT A FACT!

When leaves change color in the fall, they make a beautiful display. Here are the autumn leaf colors of some common trees.

- flame red and orange—sugar maple, sumac
- dark red—dogwood, red maple, scarlet oak
- yellow—poplar, willow, birch
- plum purple—ash
- tan or brown—oak, beech, elm, hickory

The leaves of some trees, such as locusts, stay green until the leaves drop. The leaves of black walnut trees fall before they change color.

Como ves en la **Figura 13,** las hojas frecuentemente cambian de color antes de caerse. Cuando se acerca el otoño, la clorofila, el pigmento verde que se usa en la fotosíntesis, se descompone. Cuando las hojas pierden clorofila, se pueden ver otros pigmentos, amarillos y anaranjados. Estos pigmentos ya estaban presentes en las hojas, pero el verde de la clorofila los ocultaba. Algunas hojas también tienen pigmentos rojos, que se vuelven visibles cuando la clorofila se descompone.

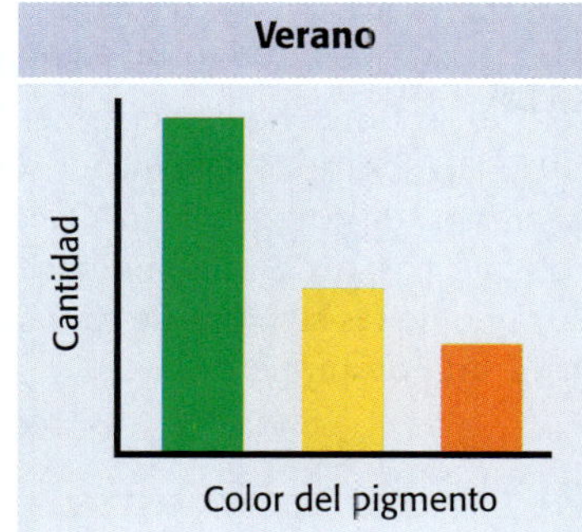

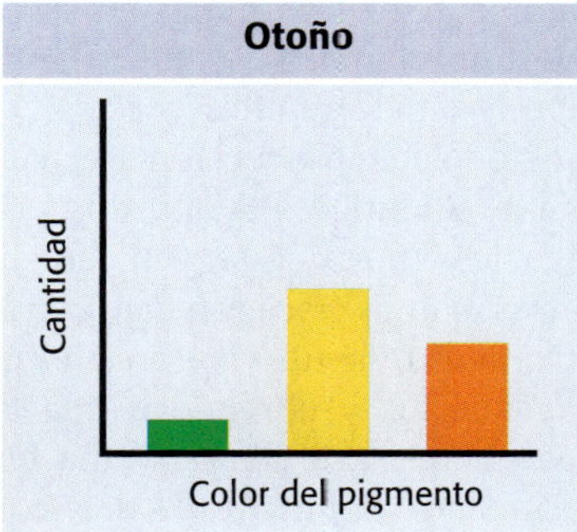

Figura 13 *La descomposición de la clorofila en el otoño es una respuesta al cambio de estación en muchos árboles. Cuando hay menos clorofila en las hojas, se pueden ver otros pigmentos.*

REPASO

1. ¿Qué efectos tienen los tropismos provocados por la luz y la gravedad?

2. ¿Qué diferencia hay entre una planta de días cortos y una de días largos?

3. ¿Por qué la pérdida de las hojas les ayuda a las plantas a sobrevivir el invierno o la temporada seca?

4. **Aplicar conceptos** Si una planta expuesta a 12 horas de luz no florece, pero lo hace cuando la exponemos a 15 horas de luz, ¿es de días largos o cortos?

Explora

Las hojas secas pueden causar un verdadero desastre sobre el pasto. El tapete de hojas secas no deja pasar el sol y el aire que el pasto necesita. Las personas dedican mucho esfuerzo a recoger las hojas, pero con los basureros casi llenos en todo el país, se recomienda hacer abono orgánico con ellas en vez de tirarlas en la basura. En muchos lugares, el gobierno local le explica a sus residentes cómo hacer el abono orgánico a partir de hojas secas y pasto cortado. Ve a la biblioteca e investiga qué beneficios para el medio ambiente aporta el abono orgánico. Prepara una exposición para la clase en la que expliques cómo se maneja en tu comunidad el problema de las hojas secas.

291

3 Extend

GOING FURTHER

Writing Have students explore other kinds of seasonal changes exhibited in plants. For example, encourage students to find out what conditions prompt bulbs to emerge in the spring and what conditions prompt seeds of flowering plants to germinate.

4 Close

Quiz

1. Why do some leaves change color in the fall? (Green chlorophyll in the leaves begins to break down, and other pigments become visible.)

2. What is the difference between an evergreen tree and a deciduous tree? (An evergreen tree has leaves adapted to survive throughout the year; they shed some of their leaves year-round. A deciduous tree loses all its leaves at the same time each year.)

3. What is a tropism in a plant? (A tropism is either a positive or negative response to an environmental stimulus, such as light or gravity.)

ALTERNATIVE ASSESSMENT

Have students make labeled drawings in their ScienceLog that illustrate plants showing positive and negative phototropism and positive and negative gravitropism.
Sheltered English

Answers to Review

1. Phototropism is caused by light. The cells on the dark side of the stem grow longer than the cells on the side facing the light, causing the stem to bend toward the light. Gravitropism is caused by gravity. Parts of the plant growing above the surface grow away from the pull of gravity. Parts below the surface grow toward it.

2. Plants that flower in only late summer or early autumn, when the night length is long, are called short-day plants. Plants that flower in the spring or early summer when night length is short, are called long-day plants.

3. Dropping its leaves prevents the tree from losing water through transpiration.

4. It is a long-day plant.

Focus

Plant Growth

This section discusses how heredity affects plant growth. Students will learn how a plant's environment affects its growth as well as what plant hormones are and what they do.

Bellringer

Write the following scrambled words on the board. Have students unscramble the words and use each word in a sentence.

mtsripo (tropism: Tropism is a change in the growth of a plant due to a stimulus.)

gnereveer (evergreen: Pine trees are evergreens.)

dseucoiud (deciduous: Maple trees are deciduous trees.)

Sheltered English

1) Motivate

DISCUSSION

Fruit Ripening Ask students if they know how to get green bananas to ripen faster. (Accept all reasonable responses.)

Tell them that bananas will ripen faster if placed in a sealed bag. Ask students why this might work. (Accept all reasonable responses.)

Tell the students that ripening fruit gives off ethylene gas, a natural plant hormone, and that exposure to this hormone prompts further ripening. By trapping the gas in the bag with the fruit, the fruit is exposed to ethylene longer and ripens faster.

VOCABULARIO
hormonas

OBJETIVOS
- Averigua cómo la herencia influye sobre el crecimiento de las plantas.
- Habla sobre cómo el medio ambiente influye sobre el crecimiento de las plantas.
- Explica qué son las hormonas de las plantas y cuál es su función.

Crecimiento de las plantas

Todos los seres vivos crecen; por ejemplo, en este mismo instante tú estás creciendo. Cuando seas mayor, alcanzarás cierta altura y luego dejarás de crecer, pero muchas plantas pueden seguir creciendo toda la vida. El crecimiento de una planta depende de sus genes, su ambiente y sus hormonas.

Herencia

Los rasgos de una planta, por ejemplo, hojas en forma de corazón o flores rojas, dependen principalmente de sus genes, o su ADN. Como el resto de los seres vivos, las plantas traspasan sus rasgos de padres a hijos. La herencia es el conjunto de rasgos que pasan de una generación a otra. La herencia explica por qué de las semillas de tomate crece una planta de tomates y no una de pimientos.

El medio ambiente

Los rasgos de una planta dependen de sus genes, pero el medio ambiente puede influir sobre la apariencia y el comportamiento de una planta. Los dos crisantemos que ves en la **Figura 14** crecieron de ramitas cortadas de la misma planta. Con este método de reproducción asexual se obtienen plantas que tienen genes idénticos. Si los dos crisantemos son genéticamente iguales, ¿por qué se ven tan diferentes? Lo que pasa es que estas plantas crecieron bajo condiciones ambientales diferentes. Crecieron en un vivero en el que se podía cambiar la cantidad de luz y obscuridad que las plantas recibían. Acuérdate de que el crisantemo es una planta de días cortos. El crisantemo que floreció tuvo días cortos y noches largas, y el que no floreció tuvo días largos y noches cortas.

La cantidad de luz y obscuridad que una planta recibe es solamente uno de los factores ambientales que pueden influir sobre su crecimiento; otros son la cantidad de agua que la planta recibe y el tipo de tierra en la que está sembrada.

Figura 14 Aunque estas plantas son idénticas genéticamente, se ven diferentes porque crecieron bajo diferentes condiciones ambientales.

292

Directed Reading Worksheet 12 Section 4

WEIRD SCIENCE

During the 1960s and 1970s, scientists conducted experiments to find out if plants grew and responded to different kinds of music. The experimenters found that plants responded best to classical music and Indian devotional music.

Las hormonas de las plantas

La producción de hormonas de una planta depende de la herencia y el ambiente. Las **hormonas** son mensajeros químicos que llevan información de una parte del organismo a otra. Se producen en cantidades pequeñas, pero su efecto sobre el organismo es muy fuerte. Las hormonas de las plantas se fabrican en partes específicas, como las puntas de los brotes. La hormona circula por la planta y provoca reacciones en todas las partes de la planta con las que tiene contacto.

Auxina   Los biólogos han identificado por lo menos cinco hormonas importantes de las plantas. Una de ellas es la auxina, una hormona que se fabrica en las puntas de los brotes y que tiene varios efectos en la planta. La auxina, por ejemplo, es la hormona responsable del fototropismo. Cuando una planta recibe luz de una dirección, la auxina se va de las puntas a la parte más sombreada de los tallos y hace que las células de ese lado se alarguen. Como viste, esto hace que los tallos se inclinen hacia la luz.

Giberelina   Otra hormona importante de las plantas es la giberelina. Como la auxina, la giberelina influye sobre el crecimiento de las plantas de varias maneras. Las plantas de la **Figura 15** son de la misma especie, pero las de la derecha recibieron aplicaciones de giberelina, que las hizo crecer muy altas. La giberelina también hace que los tallos de las flores sean más largos.

A veces, se aplican hormonas a las plantas para hacerlas crecer de cierta manera. Algunos agricultores, por ejemplo, rocían los tallos de las uvas con giberelina para que sean más largos y produzcan uvas más grandes.

¿Sabías que hay hormonas de plantas en el aire? Muchas plantas liberan una hormona gaseosa llamada etileno cuando sus frutos empiezan a madurar. El gas hace que los frutos que estén cerca liberen más etileno y maduren más rápido. Se han inventado métodos que permiten controlar el etileno, para que las frutas no maduren hasta que lleguen a la tienda.

REPASO

1. Habla sobre cómo la herencia y el ambiente influyen sobre el crecimiento de las plantas.

2. Las hormonas son mensajeros químicos. Describe uno de los "mensajes" que manda la auxina.

3. **Analizar relaciones**   ¿Qué relación hay entre el fototropismo y la producción de auxina de una planta?

4. **Aplicar conceptos**   ¿Qué puede hacer que un árbol con un tronco casi horizontal tenga ramas verticales?

Figura 15 *Las plantas de la derecha tienen tallos alargados porque las rociaron con giberelina. Las de la izquierda muestran el tamaño normal de la planta.*

293

RESEARCH

Plant Hormones Encourage interested students to research the potential advantages and disadvantages of using plant hormones, such as gibberellin, to produce larger and more productive food crops, such as grapes and beans.

Quiz

1. What are three environmental factors that affect a plant's growth? (amount of light, amount of water, type of soil in which the plant grows)

2. What are hormones? (Hormones are chemical messengers that carry information from one part of an organism to another.)

ALTERNATIVE ASSESSMENT

Writing Have students write a short science-fiction story in which an alien spacecraft inadvertently spills gibberellin over the South American rain forests. Have them describe how plant growth is affected and what can be done to correct the problem.

Critical Thinking Worksheet 12 "Space Plants"

internetconnect

TOPIC: Plant Growth
GO TO: www.scilinks.org
sciLINKS NUMBER: HSTL320

Answers to Review

1. Plant characteristics are passed from parent to offspring. When two plants are genetically identical, the environment can still affect how the plants grow and what they look like.

2. Auxin signals the cells to grow longer.

3. Auxin moves to the dark side of the plant, causing the cells on that side to grow longer. This reaction to light is called phototropism.

4. Students should demonstrate an understanding of gravitropism. If a tree has fallen over, but does not die, the new branches will extend upward, not toward the top of the tree.

SECTION 1

dormant describes an inactive state of a seed

SECTION 2

chlorophyll a green pigment in chloroplasts that absorbs light energy for photosynthesis

cellular respiration the process of producing ATP in the cell from oxygen and glucose; releases carbon dioxide and water

stomata openings in the epidermis of a leaf that allow carbon dioxide to enter the leaf and allow water and oxygen to leave the leaf

transpiration the loss of water from leaves through stomata

Resumen del capítulo

SECCIÓN 1

Vocabulario
latente (*pág. 282*)

Notas de la sección

- Para la reproducción sexual de las plantas con flores, es necesaria la polinización y la fecundación. La fecundación es la unión del óvulo y el espermatozoide.

- Cuando la fecundación se lleva a cabo, el óvulo se convierte en una semilla que contiene el embrión de la planta y el ovario se convierte en la fruta que contiene la semilla.

- Las semillas maduras se pueden quedar en estado latente. Las semillas germinan cuando tienen un ambiente con la temperatura correcta y las cantidades necesarias de agua y oxígeno.

- Muchas plantas que dan flores también pueden reproducirse de manera asexual, sin flores.

SECTION 2

Vocabulario
clorofila (*pág. 284*)
respiración celular (*pág. 285*)
estomas (*pág. 286*)
transpiración (*pág. 286*)

Notas de la sección

- En la fotosíntesis, las hojas de la planta absorben luz solar y producen glucosa a partir de dióxido de carbono y agua.

- En la respiración celular, la planta usa oxígeno y libera dióxido de carbono y agua. La glucosa se convierte en un tipo de energía que la planta puede usar.

- Las plantas absorben dióxido de carbono y liberan oxígeno por los estomas de las hojas.

Experimentos
Las sobras de la fotosíntesis (*pág. 600*)
Plantas lloronas (*pág. 602*)

☑ Comprobar destrezas

Comprensión visual

GRÁFICA CIRCULAR Las gráficas circulares sirven para visualizar fracciones sin usar números. Cada gráfica de la página 289 representa un período de 24 horas. La parte azul representa la fracción de tiempo en la que no hay luz solar, y la parte dorada la fracción de tiempo en la que sí la hay. Como se ilustra en la gráfica, al principio del verano los días ocupan $\frac{2}{3}$ del período de 24 horas, y las noches $\frac{1}{3}$.

Las **GRÁFICAS DE BARRAS** Las gráficas de barras, como las de la página 291, se usan casi siempre para comparar números. La de la derecha com- para la tasa de germinación de las semillas de plantas productoras de flores de cinco marcas distintas. Como ves, la altura de las barras indica que la tasa de germinación de la compañía D, del 88%, fue la más alta.

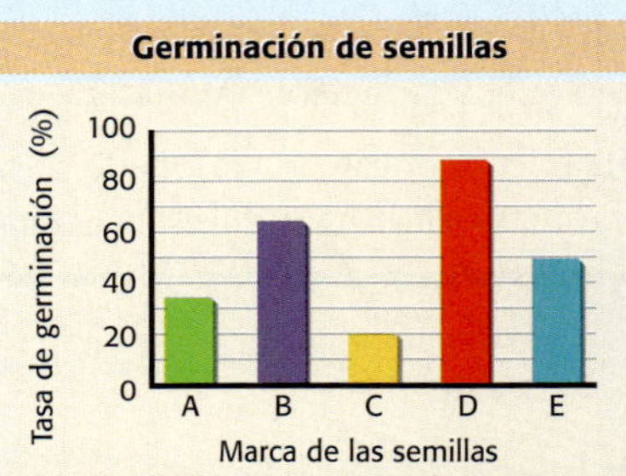

294

Lab and Activity Highlights

Food Factory Waste PG 600

Weepy Weeds PG 602

Datasheets for LabBook
(blackline masters for these labs)

SECCIÓN 3

fototropismo *(p. 287)*
geotropismo *(pág. 288)*
plantas de hoja perenne
 (pág. 290)
plantas de hoja caduca
 (pág. 290)

Notas de la sección

- El tropimo es el crecimiento de la planta que responde a un estímulo ambiental, como la luz o la gravedad. El crecimiento de la planta en dirección al estímulo se llama tropismo positivo. El crecimiento de la planta en dirección contraria al estímulo se llama tropismo negativo.

- El fototropismo es el crecimiento de la planta en dirección a la luz. El geotropismo es el crecimiento en respuesta a la fuerza de gravedad.

- Por lo general, el cambio en la cantidad de luz en las estaciones del año controla la reproducción de la planta.

- Las plantas de hoja perenne tienen hojas adaptadas para sobrevivir todo el año, pero las de hoja caduca las pierden antes de la temporada seca o fría, y eso las ayuda a sobrevivir.

SECCIÓN 4

Vocabulario

hormonas *(pág. 293)*

Notas de la sección

- La interacción de factores hereditarios y ambientales controla el crecimiento de la planta.

- Las hormonas de la planta también controlan su crecimiento. Las hormonas son mensajeros químicos que regulan el crecimiento; aunque sólo se fabrican en una parte de la planta, influyen sobre cualquier parte de la planta con la que tengan contacto. Dos hormonas importantes de las plantas son la auxina y la giberelina.

- La auxina es la hormona responsable del fototropismo, y la giberelina es la que regula la altura del tallo de las plántulas.

SECTION 3

tropism a change in the growth of a plant in response to a stimulus

phototropism a change in the growth of a plant in response to light

gravitropism a change in the growth of a plant in response to gravity

evergreen describes trees that keep their leaves year-round

deciduous describes trees that have leaves that change color in autumn and fall off in winter

SECTION 4

hormones chemical messengers that carry information from one part of an organism to another

Vocabulary Review Worksheet 12

Blackline masters of these Chapter Highlights can be found in the **Study Guide.**

 **internet**

VISITA: go.hrw.com

Visita el sitio web de HRW para encontrar una serie de herramientas de aprendizaje relacionadas con este capítulo. Sólo tienes que escribir la palabra clave:

PALABRA CLAVE: HSTPL2

SCI LINKS
NSTA

VISITA: www.scilinks.org

Visita el sitio web de la **Asociación Nacional de Maestros de Ciencias** *(National Science Teachers Association)* para encontrar recursos de Internet relacionados con este capítulo. Sólo escribe el **ENLACE DE CIENCIAS** para obtener más información sobre el tema

TEMA: La reproducción de las plantas	**ENLACE:** HSTL305
TEMA: Fotosíntesis	**ENLACE:** HSTL310
TEMA: Tropismos de las plantas	**ENLACE:** HSTL315
TEMA: Crecimiento de las plantas	**ENLACE:** HSTL320

295

Lab and Activity Highlights

LabBank

Labs You Can Eat, Not Just Another Nut, Lab 6

Ecolabs & Field Activities,
Recycle! Make Your Own Paper, EcoLab 4

Long-Term Projects & Research Ideas,
Project 13

Interactive Explorations CD-ROM

CD 2, Exploration 8, "How's It Growing?"

296

Chapter Review
Answers

USING VOCABULARY

1. dormant
2. photosynthesis
3. transpiration
4. phototropism
5. evergreen

UNDERSTANDING CONCEPTS

Multiple Choice

6. a
7. d
8. b
9. d
10. c
11. c

Short Answer

12. The cuticle is the waxy covering that prevents the leaf from losing moisture. The stomata are openings in the epidermis of the leaf that open and close, allowing only a certain amount of moisture to escape the leaf. Transpiration is the process that occurs when moisture escapes through the stomata.
13. The stimulus is light. The plant hormone auxin moves away from the light side to the dark side of the plant, causing the cells on the dark side to grow longer.
14. Phototropism is a positive tropism, and gravitropism in stems is a negative tropism.

Repaso del capítulo

UTILIZAR EL VOCABULARIO

Escoge el término correcto para completar las siguientes oraciones:

1. Cuando la semilla ya está completamente desarrollada, y antes de que germine, se puede quedar en estado __?__. (caduco o latente)

2. Durante __?__, la energía del sol se utiliza para producir azúcar. (la fotosíntesis o el fototropismo)

3. La pérdida de agua a través de los estomas se llama __?__. (transpiración o tropismo)

4. El cambio que provoca la dirección de la luz en el crecimiento de una planta se llama __?__ (geotropismo o fototropismo)

5. Las plantas que tienen hojas todo el año son __?__. (de hoja caduca o de hoja perenne)

COMPRENDER CONCEPTOS

Opción múltiple

6. Las células que abren y cierran los estomas son las
   a. células oclusivas.
   b. células del xilema.
   c. células de la cutícula.
   d. células mesófilas.

7. Las plantas necesitan el dióxido de carbono, que usan en
   a. la respiración celular.
   b. el fototropismo.
   c. la fecundación.
   d. la fotosíntesis.

8. Cuando la clorofila se descompone,
   a. ocurre la polinización.
   b. se hacen visibles otros pigmentos.
   c. desaparecen los pigmentos rojos.
   d. se lleva a cabo la fotosíntesis.

9. De las siguientes secuencias, ¿cuál muestra el orden correcto de los eventos que ocurren después de que el insecto trae polen a la flor ?
   a. germinación, fecundación, polinización
   b. fecundación, germinación, polinización
   c. polinización, germinación, fecundación
   d. polinización, fecundación, germinación

10. Cuando la cantidad de agua que la planta transpira es mayor que la cantidad de agua que las raíces absorben,
    a. la cutícula conserva el agua.
    b. el tallo muestra geotropismo positivo.
    c. la planta se marchita.
    d. la planta deja de marchitarse.

11. La hormona que hace que el tallo de la planta crezca y sea muy largo es
    a. la auxina.
    b. la glucosa.
    c. la giberelina.
    d. la clorofila.

Respuesta breve

12. ¿Qué relación hay entre la transpiración, la cutícula y los estomas?

13. ¿Qué estimula el fototropismo? ¿Cómo responde la planta a este estímulo?

14. Da un ejemplo de tropismo positivo y uno de tropismo negativo.

Organizar conceptos

15. Usa los siguientes términos para crear un mapa de ideas: plántula, flor, semilla, óvulos, reproducción de la planta, asexual, guías.

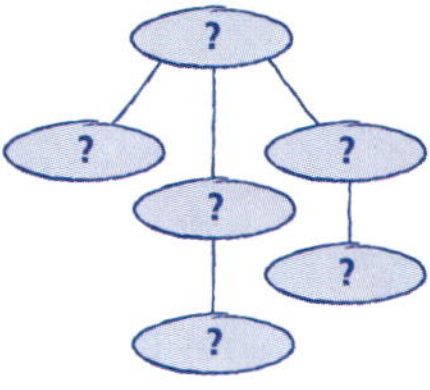

RAZONAMIENTO CRÍTICO Y RESOLUCIÓN DE PROBLEMAS

Escribe una o dos oraciones para responder a las siguientes preguntas:

16. Muchas plantas que crecen en lugares donde los inviernos son muy fríos tienen semillas que no germinan a ninguna temperatura si antes no han estado expuestas a un período largo de frío. ¿Cómo contribuye esta característica a que sobrevivan las nuevas plantas?

17. Si quisieras hacer que las nochebuenas florecieran y se pusieran rojas en el verano, ¿qué harías?

18. ¿De qué les sirve a los brotes de una planta el fototropismo positivo? ¿De qué les sirve a las raíces de una planta el geotropismo positivo?

LAS MATEMÁTICAS EN LAS CIENCIAS

19. Si la hoja de una planta tiene un área de 8 cm², ¿Qué área tiene en milímetros cuadrados? (Pista: 1 cm² = 100 mm².)

20. Las hojas de una planta tienen un promedio de 100 estomas por milímetro cuadrado de área. ¿Cuántos estomas calculas que hay en la hoja de la pregunta 19?

INTERPRETAR GRÁFICAS

Observa estos esquemas y responde a las siguientes preguntas: La ilustración muestra parte de un experimento sobre el fototropismo en plantas jóvenes. En la parte (1), las plantitas que estuvieron en la obscuridad acaban de ser colocadas donde les da la luz. La punta del tallo de una de las planta se corta y la otra punta se deja intacta. En la parte (2), las plantas se exponen a la luz proveniente de una sola dirección.

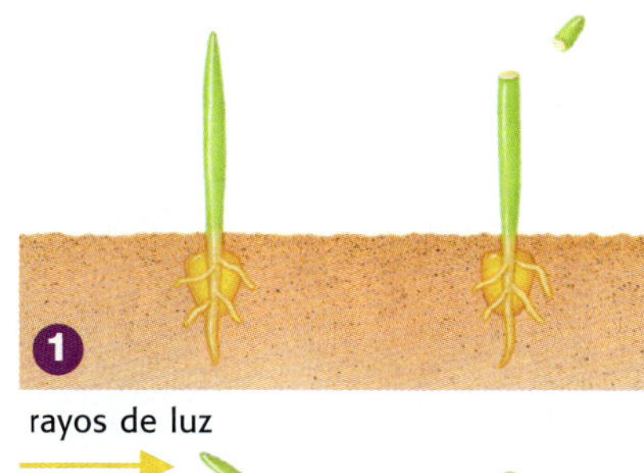

21. ¿Por qué la planta con la punta intacta se movió hacia la luz?

22. ¿Por qué la planta con la punta cortada se quedó derecha?

AHORA, ¿qué piensas?

Revisa tus respuestas a las preguntas de la página 279 que escribiste en el cuaderno de ciencias. ¿Han cambiado tus respuestas? Si es necesario, corrige tus respuestas basándote en lo que has aprendido en este capítulo.

297

Concept Mapping

15. 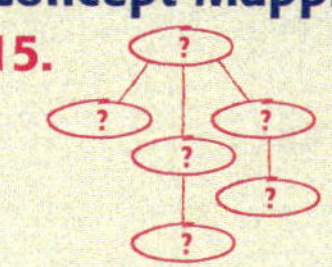 An answer to this exercise can be found at the end of this book.

CRITICAL THINKING AND PROBLEM SOLVING

16. Exposure to a long period of cold may prevent a seed from sprouting before winter. If a seed sprouted in the fall, it might not have time to produce more seeds before it dies.

17. You would have to use artificial means to provide a long period of darkness and a short period of light.

18. Positive phototropism helps the leaves get exposure to sunlight so that photosynthesis can occur. Gravitropism helps the roots grow in the correct direction so that they will have access to water and nutrients in the soil.

MATH IN SCIENCE

19. 800 mm²
20. 80,000 stomata

INTERPRETING GRAPHICS

21. The plant hormone auxin caused the shoot to bend toward the light.
22. The plant hormone auxin is produced in the shoot tip. If the tip is removed, the shoot will not bend toward the light.

Concept Mapping Transparency 12

Blackline masters of this Chapter Review can be found in the **Study Guide.**

NOW WHAT DO YOU THINK?

1. Answers may vary. Students may realize that a fruit usually contains seeds and that many fruits serve as food for other organisms, including people.

2. Answers will vary, but students may mention wilting, losing leaves, changing colors, and any other changes that show plants respond to the environment.

3. Answers will vary, but students may mention that plants need light in order to stay alive and make food.

Teaching Strategy

To help your students understand how the parts of Meyerowitz's mustard flowers are all in the wrong places, you may wish to review with your students the appropriate locations of the parts of a flower. The parts of a flower are modified leaves, usually found in four successive whorls or rings. Create a diagram that shows four concentric circles. Label the innermost circle "Pistil." Label the next circle "Stamens." (Stamens produce pollen, and each stamen has two parts, the anther and the filament.) Label the third circle "Petals." (Petals are leaflike structures that surround the carpels and stamens. Their colors and shapes give many flowers a distinctive appearance.) Label the outermost circle "Sepals." (The sepals protect the flower when it is a bud.) Explain that the inner two circles mark the locations of the reproductive organs of the flower and that the petals and the sepals are called nonessential flower parts because they are not directly involved in reproduction.

MOSTAZA MUTANTE

Las diminutas flores de mostaza que cultiva Elliot Meyerowitz están horriblemente deformadas. Parecería que han sufrido un accidente trágico, pero Meyerowitz creó estos mutantes a propósito. Es más, está orgulloso de ellas. Estas flores le pueden ayudar a resolver un importante misterio de la biología.

▲ *Elliot Meyerowitz en su laboratorio, donde ha cultivado cerca de un millón de especímenes individuales de una variedad de mostaza llamada* Arabidopsis thaliana.

Flores normales y flores anormales

Normalmente, las flores de la mostaza tienen cuatro partes bien definidas y acomodadas en lugares específicos. Pero muchas de las flores que Meyerowitz y sus compañeros cultivan no son flores normales. Algunas tienen hojas que crecen en medio de la flor; otras tienen ovarios que producen semillas donde deberían ir los pétalos. A primera vista, el orden de las partes de la flor parece accidental, pero la estructura de cada flor está determinada por unos cuantos genes.

Un modelo sencillo

Tras muchos años de investigaciones cuidadosas, Meyerowitz y sus colegas identificaron la de los genes que controlan el desarrollo de la flor de la mostaza. Con esta información, Meyerowitz descubrió ciertos patrones con los que construyó un modelo muy sencillo. El modelo indica que hay sólo tres clases de genes que determinan el desarrollo de cada parte de la flor. Meyerowitz notó que si una o más de estas clases de genes se inactiva, el resultado es una planta de mostaza mutante.

Piezas de un viejo rompecabezas

Con su investigación sobre cómo influyen los genes en el crecimiento de la flor, Meyerowitz espera añadir piezas al rompecabezas sobre el origen de las plantas con flores. Se calcula que las flores aparecieron en la Tierra hace 125 millones de años y que se convirtieron rápidamente en las plantas predominantes. Meyerowitz y sus colegas piensan que el estudio de los genes responsables del crecimiento de las flores modernas ayudará a entender la evolución de las plantas con flores.

Las plantas mutantes de Meyerowitz son ideales para entender la genética de las plantas. Pero no las encontrarás en una florería. ¡Nunca ganarían un concurso de belleza!

▲ *Meyerowitz altera los genes de una planta de mostaza para que desarrolle una flor mutante. La ilustración muestra una flor normal.*

¡Piensa!

▶ Como ves, es posible cambiar genéticamente una planta. Menciona algunos riesgos posibles de este tipo de cambios.

298

Answer to Think About It

Accept all reasonable responses. Some students may point out that the process of growing many generations of plants may be quite lengthy. In addition, it often takes many unsuccessful tests to develop a successful hypothesis. Students may also point out that genetically altering a plant may have positive or negative repercussions in terms of the usefulness of the plant.

VENTANA AL MEDIO AMBIENTE

Un arco iris de algodón

Piensa en tu camiseta favorita. Seguramente está hecha de algodón y es de colores vivos. Sin embargo, las fibras naturales de la planta de algodón son blancas. Para crear los vivos colores de las camisetas y otras telas, es necesario teñirlas con tintes que a veces son tóxicos. Sally Fox, una mujer muy ingeniosa, tuvo una idea para minimizar el uso de tintes tóxicos: ¿Por qué no cultivar *algodón de colores?*

Aprender del pasado

Las fibras del algodón vienen del mismo lugar en que crecen sus semillas, de las *cápsulas.* Las cápsulas son poco más grandes que una pelota de golf, y cuando están maduras se abren para revelar una bola de fibras y semillas. Después de quitar las semillas, las fibras se tuercen para formar hilos con los que se hacen muchos tipos de tela. Sally Fox empezó su carrera como *entomóloga,* estudiando insectos. Su primer contacto con el algodón de colores fue cuando estudiaba la resistencia de las plantas a las plagas. Aunque la mayor parte del algodón es naturalmente blanco, los indios de América han cultivado varios tonos de algodón durante siglos.

Estos tipos de algodón son más resistentes a las plagas, pero sus fibras son demasiado cortas para ser usadas en la industria textil.

En 1982, Fox comenzó el lento proceso de cruzar diferentes variedades de algodón para producir una que fuera de color y tuviera fibras largas. El tipo de algodón que Fox inventó se llama FoxFibre®y le ha ganado gran reconocimiento.

Soluciones para los problemas ambientales

La industria textil provoca dos grandes problemas ambientales: el primero es el uso de tintes para teñir las telas de algodón, y el segundo es el uso de plaguicidas en los cultivos. Ambos pueden perjudicar a los seres vivos y contaminar los recursos naturales, como el agua y la tierra.

El algodón que Fox inventó es una solución a ambos problemas. En primer lugar, como la fibra ya tiene color, no es necesario teñirla. En segundo lugar, el nuevo algodón heredó la resistencia natural a las plagas de las variedades de algodón indio que Fox utilizó en sus cruzas. Asi se necesitan menos plaguicidas para cultivar este nuevo algodón.

Los resultados que Sally Fox obtuvo demuestran que, con ingenio y paciencia, la ciencia y la agricultura pueden desarrollar juntas nuevas soluciones para los problemas ambientales.

▲ *Sally Fox en un sembradío de algodón de color.*

Trabajo de detectives

▶ Al igual que el algodón que diseñó Fox, muchos tipos de plantas y animales domésticos son producto de la selección artificial. Investiga dónde y cuándo se estableció tu fruta o tu raza de perro favorita.

299

Answers to Some Detective Work

One example is the dachshund. It was initially bred in Germany in the 1700s, where it was used to hunt badgers. Its tubular body and short legs are ideal for following prey into burrows. The word *dachshund* is German for "badger hound."

Background

Cotton has many properties that make it a superior material for fabrics. Cotton fabrics are durable, washable, and comfortable. Cotton clothing keeps you cool in the summer because it "breathes" well and allows the moisture to evaporate quickly. Some other uses for cotton include fine yarns, carpets, and blends with other fabrics.

Sally Fox grows cotton in Arizona and New Mexico. Fox is continuing to refine her crop by trying to breed fire-resistant varieties as well as new colors.

Teaching Strategy

Give students samples of different fabrics to touch and observe. Fabrics could include cotton, linen, polyester, nylon, silk, leather, or suede. Then have students classify the fabrics based on their origin—plant, animal, or synthetic. Ask students the following question:

What are the advantages of fabrics made from plants? What are the disadvantages? (Students should recognize that plant-based fabrics have different texture, come from a renewable resource, and are cheaper than animal-based fabrics. They should also recognize that plant-based fabrics may be less durable and more prone to stains than other fabrics.)

UNIDAD 5

Los animales

¿**A**lguna vez has ido a un zoológico o has visto algún programa de animales salvajes? Si es así, tienes noción de los diferentes tipos de animales que hay en la Tierra, desde pequeños insectos hasta ballenas enormes.

Los animales son fascinantes debido a su variedad de apariencia y comportamiento. También nos enseñan sobre nosotros mismos ya que los seres humanos también estamos clasificados en la categoría de animales.

En esta unidad, conocerás muchos tipos de animales, incluso algunos que quizás ignorabas que existieran. Así que ¡prepárate para una aventura animal!

1610
Galileo utiliza un microscopio compuesto para estudiar la anatomía de los insectos.

1680
El ave de la isla Mauricio, un ave no voladora, se declara en extinción.

1960
Jane Goodall, una zoóloga inglesa, empieza su investigación sobre los chimpancés en Tanzania.

1935
Francis B. Summer estudia la coloración defensiva de los peces.

1983
El cohete *Challenger* fue lanzado al espacio con Sally Ride, la primera mujer norteamericana en el espacio, como tripulante.

1987
El último cóndor salvaje de California es capturado en un esfuerzo por salvar la especie de la extinción.

1693

John Ray clasifica correctamente a las ballenas como mamíferos.

1761

La primera escuela veterinaria es fundada en Lyons, Francia.

1775

J. C. Fabricius desarrolla un sistema para la clasificación de los insectos.

1827

John James Audubon publica la primera edición de "Aves de Norteamérica" (*Birds of North America*).

1882

La investigación sobre *El Albatros* ayuda a aumentar nuestro conocimiento de la vida marina.

1839

Se construye la primera bicicleta.

1995

Catorce lobos grises canadienses son puestos en libertad en el *Yellowstone National Park*.

1998

Keiko, la orca asesina protagonista de la película "Liberen a Willy", aprende a atrapar peces para poder sobrevivir fuera del cautiverio.

Chapter Organizer

CHAPTER ORGANIZATION	TIME MINUTES	OBJECTIVES	LABS, INVESTIGATIONS, AND DEMONSTRATIONS
Chapter Opener pp. 302–303	45		**Investigate!** Go on a Safari! p. 303
Section 1 What Is an Animal?	45	▶ Understand the differences between vertebrates and invertebrates. ▶ Explain the characteristics of animals.	
Section 2 Animal Behavior	90	▶ Explain the difference between learned and innate behavior. ▶ Explain the difference between hibernation and estivation. ▶ Give examples of how a biological clock influences behavior. ▶ Describe circadian rhythms. ▶ Explain how animals navigate.	**Demonstration,** Sign Language, p. 310 in ATE **QuickLab,** How Long Is a Minute? p. 312 **Discovery Lab,** Wet, Wiggly Worms! p. 604 **Datasheets for LabBook,** Wet, Wiggly Worms! Datasheet 27 **Design Your Own,** Aunt Flossie and the Bumblebee, p. 606 **Datasheets for LabBook,** Aunt Flossie and the Bumblebee, Datasheet 28 **Inquiry Labs,** Follow the Leader, Lab 4 **Whiz-Bang Demonstrations,** Six-Legged Thermometer, Demo 9
Section 3 Living Together	90	▶ Discuss ways that animals communicate. ▶ List the advantages and disadvantages of living in groups.	**Long-Term Projects & Research Ideas,** Project 14

TECHNOLOGY RESOURCES

 **Guided Reading Audio CD**
English or Spanish, Chapter 13

 **One-Stop Planner CD-ROM with Test Generator**

 **Science Discovery Videodiscs**
Image and Activity Bank with Lesson Plans:
Signaling Animals
Science Sleuths: The Plainview Park Scandals

CNN **Eye on the Environment,** Monarch Migrations, Segment 4

Scientists in Action, Studying Dolphin Behavior, Segment 18
Learning the Language of Animals, Segment 19

CLASSROOM WORKSHEETS, TRANSPARENCIES, AND RESOURCES	SCIENCE INTEGRATION AND CONNECTIONS	REVIEW AND ASSESSMENT
Directed Reading Worksheet 13 **Science Puzzlers, Twisters & Teasers,** Worksheet 13		
Transparency 54, The Animal Kingdom **Directed Reading Worksheet 13,** Section 1 **Science Skills Worksheet 25,** Introduction to Graphs **Reinforcement Worksheet 13,** What Makes an Animal an Animal?		**Self-Check,** p. 306 **Homework,** pp. 306, 307 in ATE **Review,** p. 307 **Quiz,** p. 307 in ATE **Alternative Assessment,** p. 307 in ATE
Directed Reading Worksheet 13, Section 2 **Math Skills for Science Worksheet 21,** Percentages, Fractions, and Decimals **Math Skills for Science Worksheet 8,** Average, Mode, and Median **Transparency 87,** Finding Direction on Earth **Reinforcement Worksheet 13,** Animal Interviews **Critical Thinking Worksheet 13,** Masters of Navigation	**Math and More,** p. 309 in ATE **Math and More,** p. 311 in ATE **Connect to Earth Science,** p. 312 in ATE **Apply,** p. 313 **Physical Science Connection,** p. 313 **Eye on the Environment:** Do Not Disturb! p. 322	**Review,** p. 310 **Self-Check,** p. 312 **Review,** p. 313 **Quiz,** p. 313 in ATE **Alternative Assessment,** p. 313 in ATE
Directed Reading Worksheet 13, Section 3 **Transparency 55,** The Dance of the Bees	**Cross-Disciplinary Focus,** p. 315 in ATE **Weird Science:** Animal Cannibals, p. 323	**Homework,** p. 316 in ATE **Review,** p. 317 **Quiz,** p. 317 in ATE **Alternative Assessment,** p. 317 in ATE

END-OF-CHAPTER REVIEW AND ASSESSMENT

Chapter Review in Study Guide
Vocabulary and Notes in Study Guide
Chapter Tests with Performance-Based Assessment, Chapter 13 Test
Chapter Tests with Performance-Based Assessment, Performance-Based Assessment 13
Concept Mapping Transparency 13

Holt, Rinehart and Winston On-line Resources

go.hrw.com

For worksheets and other teaching aids related to this chapter, visit the HRW Web site and type in the keyword: **HSTANM**

National Science Teachers Association

www.scilinks.org

Encourage students to use the *sci*LINKS numbers listed with the Chapter Highlights to access information and resources on the **NSTA** Web site.

Visual Resources

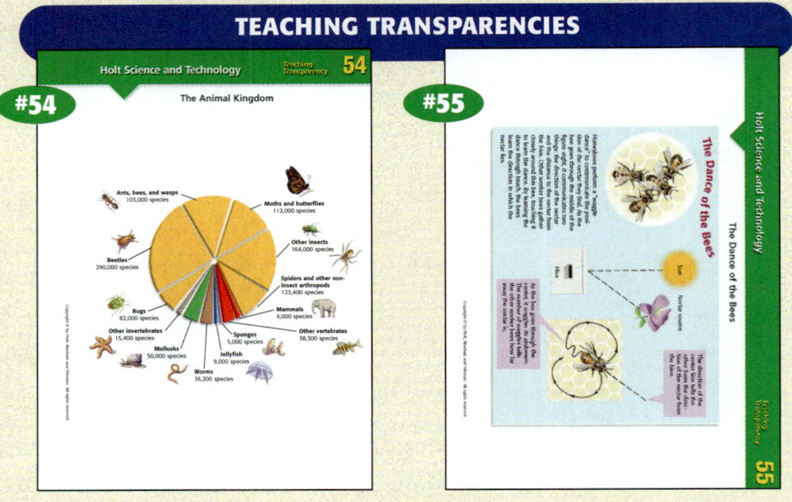

TEACHING TRANSPARENCIES

#54 · Holt Science and Technology · Teaching Transparency **54** · The Animal Kingdom

#55 · Holt Science and Technology · The Dance of the Bees · **55**

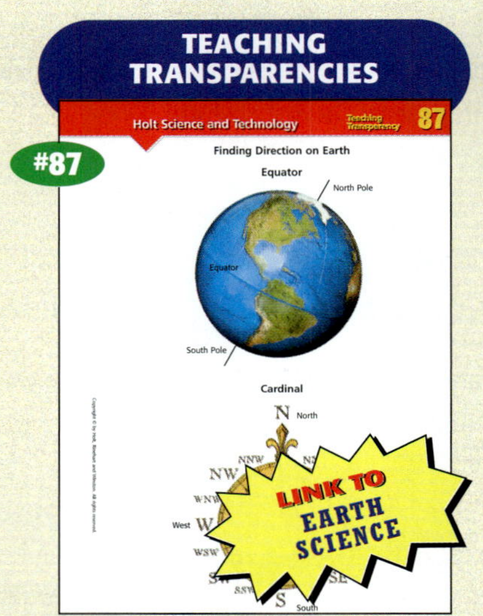

TEACHING TRANSPARENCIES

#87 · Holt Science and Technology · Teaching Transparency **87** · Finding Direction on Earth

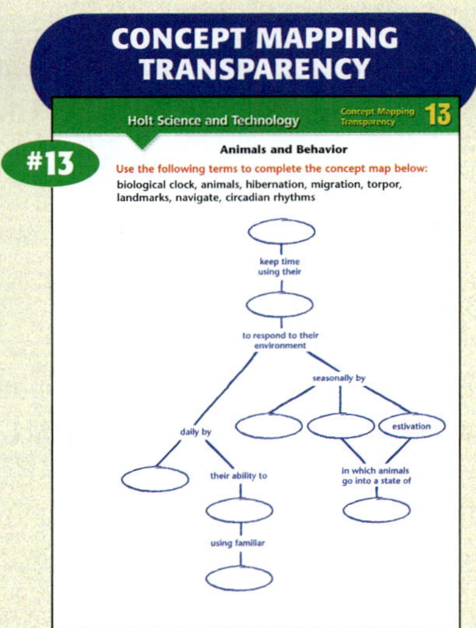

CONCEPT MAPPING TRANSPARENCY

#13 · Holt Science and Technology · Concept Mapping Transparency **13** · Animals and Behavior

---

Meeting Individual Needs

DIRECTED READING

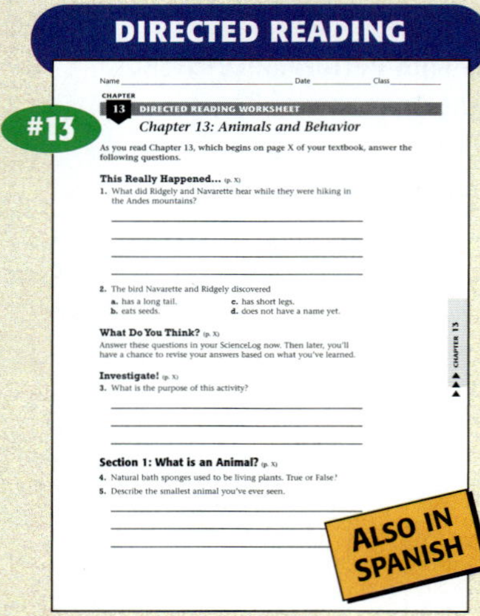

#13 — Chapter 13: Animals and Behavior

ALSO IN SPANISH

REINFORCEMENT & VOCABULARY REVIEW

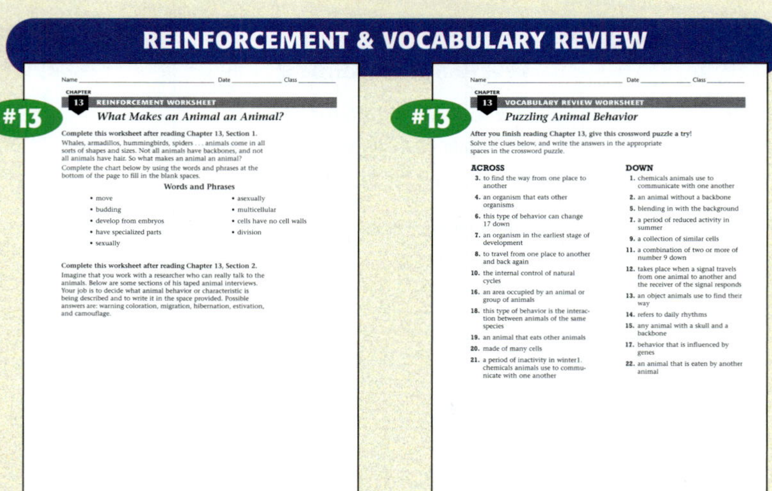

#13 — What Makes an Animal an Animal?

#13 — Puzzling Animal Behavior

SCIENCE PUZZLERS, TWISTERS & TEASERS

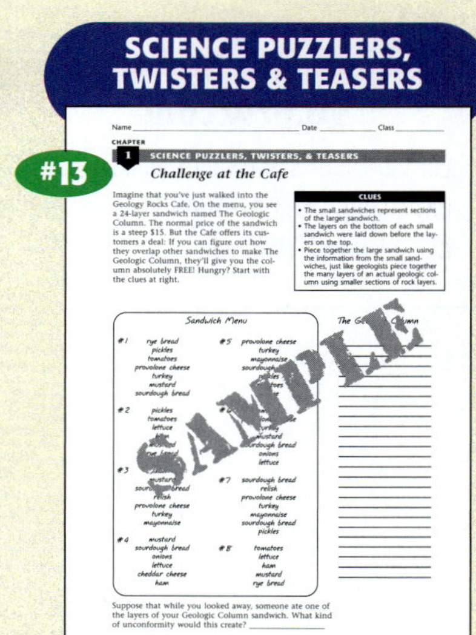

#13 — Challenge at the Cafe

Review & Assessment

STUDY GUIDE

#13 — VOCABULARY & NOTES WORKSHEET
Chapter 13: Animals and Behavior

SECTION 1
Vocabulary
1. vertebrate
2. invertebrate
3. multicellular
4. embryo
5. tissue
6. organ
7. consumer

Section Notes
• Animals with skulls and backbones are vertebrates. Animals without a backbone are invertebrates.

ALSO IN SPANISH

#13 — CHAPTER REVIEW WORKSHEET
Chapter 13: Animals and Behavior

USING VOCABULARY
To complete the following sentences, choose the correct term from each pair of terms listed below:
1. An animal with a skull and a backbone is ______________. An animal with no backbone is ______________. (an invertebrate or a vertebrate)
2. A behavior that does not depend on experience is ______________. (innate or learned)
3. In the summer, an animal enters a state of reduced activity. The animal is ______________. (estivating or hibernating)
4. Daily cycles are known as ______________. (biological clock or circadian rhythms)
5. When an egg and a sperm come together, they form an ______________. (an embryo or an organ)

UNDERSTANDING CONCEPTS
Multiple Choice
6. Which characteristic is not true of animals?
a. They are multicellular.
b. They usually reproduce sexually.
c. They make their own food.
d. They have tissues.
7. Living in groups
a. attracts predators.
b. helps prey spot predators.
c. helps animals find food.
d. All of the above
8. Warning coloration is
a. a kind of camouflage.
b. a way to warn predators away.
c. always black and white.
d. always a sign that an animal is poisonous to eat.
9. Some birds use Earth's magnetic field
a. to attract mates.
b. to navigate.
c. to set their biological clocks.
d. to defend their territory.

ALSO IN SPANISH

CHAPTER TESTS WITH PERFORMANCE-BASED ASSESSMENT

#13 — ANIMALS AND BEHAVIOR
Chapter 13 Test

USING VOCABULARY
To complete the following sentences, choose the correct term from the terms listed, and write the term in the blank.
1. All animals are ______________ and have eukaryotic cells. (vertebrates or multicellular)
2. A collection of similar cells is known as a(n) ______________. (organ or tissue)
3. ______________ is an animal's internal control of natural cycles. (Hibernation or a biological clock)
4. All animals are ______________. (predators or consumers)
5. Some animals use the Earth's magnetic field to ______________ as they travel. (communicate or navigate)

UNDERSTANDING CONCEPTS
Multiple Choice
Circle the correct answer.
6. ______________ is not an example of how animals might deal with a food shortage.
a. Migration
b. Estivation
c. Social behavior
d. Hibernation
7. Reading is an example of
a. an innate behavior.
b. a behavior controlled by genes.
c. a learned behavior.
d. an inherited behavior.
8. All ______________ lack a skull and backbone.
a. vertebrates
b. eukaryotes
c. multicellular organisms
d. invertebrates
9. Which statement about communication between animals is true?
a. Animals use it to find mates.
b. It helps animals find their food.
c. It helps animals avoid enemies.
d. It includes only friendly behavior.
10. The relationship between a worm and a robin is expressed as
a. vertebrate:invertebrate.
b. prey:predator.
c. producer:consumer.
d. prokaryote:eukaryote.
11. The use of ______________ is an example of chemical communication.
a. camouflage
b. pheromones
c. hibernation
d. estivation

ALSO IN SPANISH

CHAPTER 13 PERFORMANCE-BASED ASSESSMENT

#13 — ANIMALS AND BEHAVIOR
Chapter 13 Performance-Based Assessment

TEACHER'S PREPARATORY GUIDE

Purpose
Students will use scents and sign language to communicate ideas.

Time Required
One 45-minute class period

P.B.A. Ratings
Teacher Prep
Student Set-Up
Concept Level
Clean Up

Advance Preparation
Equip each activity station with the necessary materials. Each station should have at least five items with distinct, recognizable smells. Possibilities include vanilla, garlic, ammonia, moth balls, cloves, and orange peel.

Safety Information
Remind students not to eat anything in the lab.

Teaching Strategies
This activity works best in groups of 2 students.

ALSO IN SPANISH

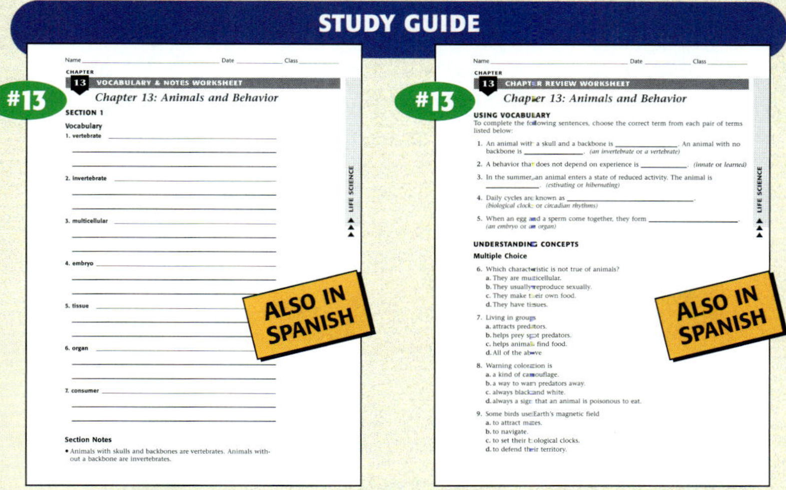
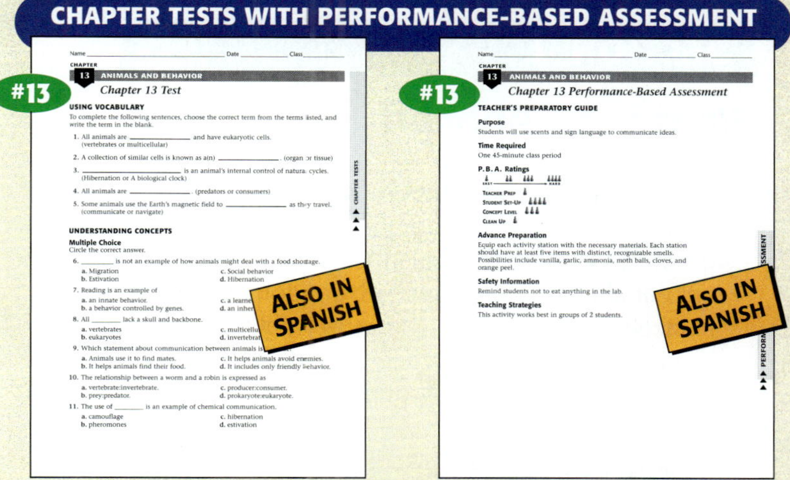

Lab Worksheets

INQUIRY LABS

#4 — LAB 4 STUDENT WORKSHEET — DISCOVERY LAB
Follow the Leader

How do you find your way around in an unfamiliar place? You probably use a variety of tools: you might use a map, a compass, verbal directions, or even hire a guide. How do other animals navigate in unfamiliar territory? Birds respond to a variety of calls, dolphins and bats use sonar, and bees use visual cues and communicate directions in an elaborate, buzzing dance.
But how do ants find their way around? In this lab, you will discover that ants have an unusual way of finding their way to and from their anthill.

MATERIALS
• ant colony
• large, empty aquarium
• 15 mL of sugar
• jar lid
• tap water
• plastic transparency sheets
• sheet of paper
• 3–6 magnifying glasses
• can of compressed air
• plastic sponge
• paper towels

SAFETY ALERT!
Do not touch the ants. Some ants bite.

Ask a Question
How do ants navigate?

Make Observations
1. Observe the ants traveling to and from the food dish for 2–3 minutes. Record all of your observations.

Make a Prediction
2. How do you think ants find their way to and from food and water?

Conduct an Experiment
3. Slide a sheet of paper beneath the plastic that lines the bottom of the box. How does the paper affect the behavior of the ants? Record your observations.

WHIZ-BANG DEMONSTRATIONS

#9 — DEMO 9 TEACHER-LED DEMONSTRATION — DISCOVERY LAB
Six-Legged Thermometer

Purpose
Students observe the effect of temperature on the frequency of a cricket's chirping and learn how an animal's behavior may be affected by its environment.

Time Required
10–15 minutes

Lab Ratings
Teacher Prep
Concept Level
Clean Up

MATERIALS
• 2 mature male crickets (available in many bait and pet stores)
• 2 glass jars
• 2 nylon stockings
• 2 rubber bands
• refrigerator
• watch or a clock that indicates seconds
• Fahrenheit thermometer

Advance Preparation
Only the mature male crickets will chirp. Keep each cricket in a separate container; male crickets are extremely territorial and have a tendency to kill other males.
You may wish to try this activity in advance. First put one cricket in each jar. Cover each jar with a nylon stocking, and secure the stocking with a rubber band. (Be sure the stocking is stretched enough to allow sufficient oxygen into the jars.) Leave one jar at room temperature. Leave the other in the refrigerator long enough to slow the chirping of the cricket significantly but not so long that the cricket is harmed. Note: Conduct this activity as soon as you remove the jar from the refrigerator. Record the temperature of the refrigerator in degrees Fahrenheit.

What to Do
1. Have students observe the cricket in the refrigerated jar. Ask them to count the number of chirps the cricket makes in 15 seconds and then to add 40 to that number. Record both numbers on the board.
2. Record the temperature of the refrigerator in degrees Fahrenheit on the board. Point out the similarity between this number and the data from step 1; they should be roughly equal.
3. Have students repeat step 1 with the other cricket.
4. Measure and record the temperature of the room in degrees Fahrenheit. Again, the temperature of the room and the data from step 3 should be almost equal.

Discussion
Use the following questions as a guide to encourage class discussion:
• What can you conclude from your observations? (Sample answer: There is a relationship between temperature and chirping frequency. A cricket can serve as a kind of thermometer.)
• If we refer to the number of chirps in 15 seconds as degrees cricket, what is the corresponding temperature in degrees cricket for 82° F? (82 − 40 = 42 degrees cricket)

LONG-TERM PROJECTS & RESEARCH IDEAS

#14 — PROJECT 14 STUDENT WORKSHEET — DESIGN YOUR OWN
Animals & Behavior

In May 1997, an IBM super-computer called Deep Blue defeated the world chess champion, Gary Kasparov, in a highly publicized chess match. Games playing is just one branch of the high-tech field of computing called artificial intelligence. Artificial intelligence is concerned with programming computers to think and behave like humans.
Computers with artificial intelligence are now helping doctors and diagnoses. In robotics, artificial intelligence is being used to program robots to see, hear, and react to sensory stimuli.

Living Computers?
1. Will the advancements in artificial intelligence eventually force scientists to redefine "living?" Research the field of artificial intelligence on the Internet and in your library. What is the future of artificial intelligence? Present your findings in the form of a science feature.

Other Research Ideas
2. Research and write a report on kwashiorkor, a severe form of malnutrition that is caused by the lack of dietary protein. What are the symptoms of this condition? Include facts and examples to support your information.
3. Today, many health experts and nutritionists warn people against consuming too much saturated fat. Use library references or Internet information to compare the average daily diet of an adult living in the United States with that of a person living along the Mediterranean Sea (Greece, Italy, France, Portugal, or Spain) or in Japan. How do these diets differ in the amounts and kinds of fish, meat, fruits, vegetables, carbohydrates, and fats consumed daily? What effects do these dietary differences have on health? Write a news article explaining your findings.
4. All forms of life contain DNA, a molecule that provides instruction for cell reproduction. Having DNA is one requirement for being considered a living thing. Prokaryotic cells, however, do not have cell nuclei. How do prokaryotic cells reproduce? Why have prokaryotic life forms survived despite their relative simplicity?

DATASHEETS FOR LABBOOK

#27 — DATASHEET 9 STUDENT DATASHEETS — DESIGN YOUR OWN
Wet, Wiggly Worms!

#28 — **Aunt Flossie and the Bumblebee**

...will have these samples ready for you to observe.
3. Make a data table like the one below to list your observations. Make as many observations as you can about the potatoes in Group A, Group B, and Group C.

Observations		
Group A:		
Group B:		

Form a Hypothesis
4. You have identified a problem and made your observations. Now you can make a hypothesis. Write a clear hypothesis about what you think will be the outcome of your tests.

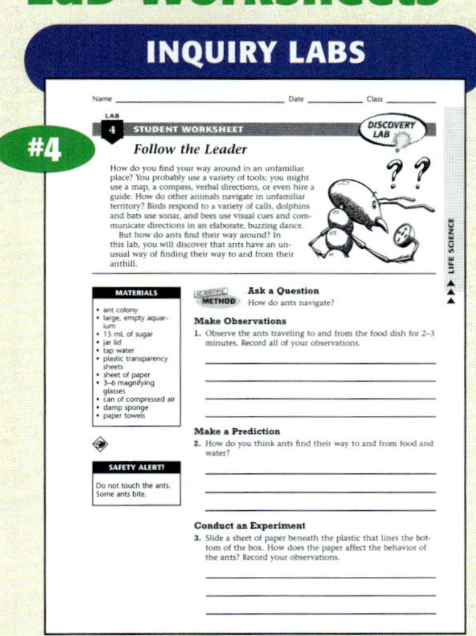
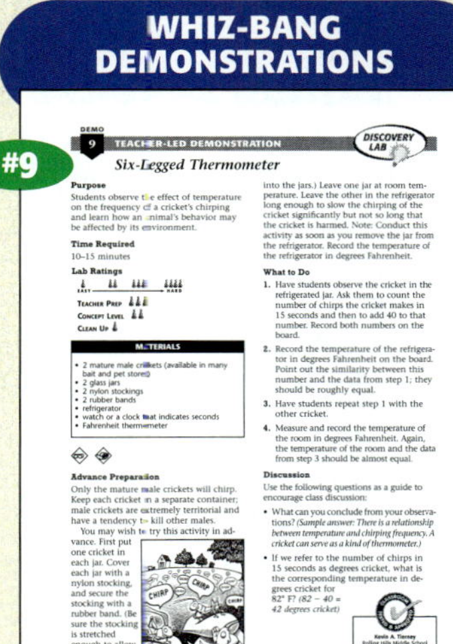
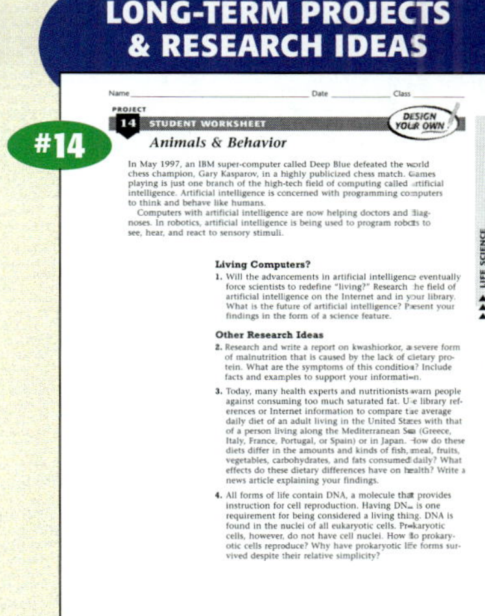
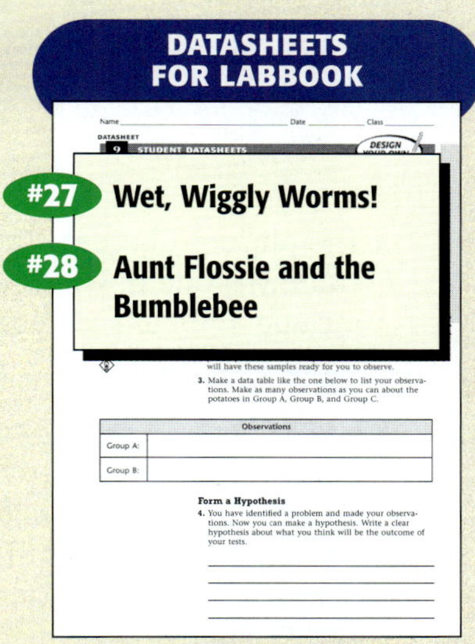

Applications & Extensions

CRITICAL THINKING & PROBLEM SOLVING

#13 — CHAPTER 13 CRITICAL THINKING WORKSHEET
Masters of Navigation

The following excerpts are taken from the article "Homing," by Eldon Greij, in Birder's World.

For many people, the most intriguing aspects of bird biology and behavior are associated with homing and migration. Several years ago, for example, a friend and I were nest-trapping Blue-winged Teal in Iowa. We captured a female that had been banded the previous year at a nest only thirty-five feet from the present one. I remember stroking her plumage and wondering if she spent the winter in Louisiana, Texas, or Mexico. It didn't make any difference as she certainly hit the bullseye on her return. While I found the accuracy of this hen teal to be remarkable, it is not extraordinary by avian standards.
A truly dramatic migration is seen in the New Zealand Bronzed Cuckoo. These birds are parasites, laying their eggs in the nests of host species that hatch and raise the young cuckoos. In the fall, about a month after their parents migrate, the young cuckoos get together and begin migrating to their wintering ground. The birds fly almost 1,220 miles west to Australia and then nearly 1,770 miles north to the Solomon and Bismarck Islands where they join their parents.

USEFUL TERMS
intriguing
exciting interest or curiosity
homing
going home
nest-trapping
trapping birds in their nests for observation
plumage
feathers
avian
of or having to do with birds
parasite
an organism that feeds on another living creature, usually without killing its host

Observing for Detail
1. a. Do you think the migration of the Blue-winged Teal is more likely innate or learned behavior? Explain.

b. Do you think the migration of the New Zealand Bronzed Cuckoo is more likely innate or learned behavior? Explain.

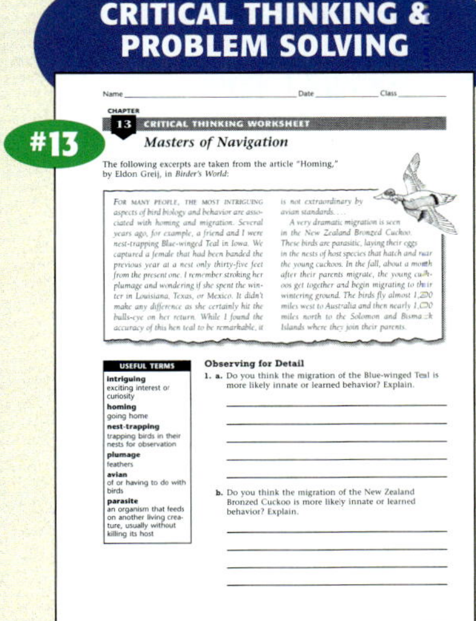

EYE ON THE ENVIRONMENT

#4 — Science in the News: Critical Thinking Worksheets
Segment 4
Monarch Migrations

1. Describe some of the methods, tools, and data used in the monarch migration research.

2. Name two unanswered questions about monarch migration that scientists hope to answer with their research.

3. a. Name one possible reason that monarch roosts are endangered.

b. How could the results of this research help solve this problem?

4. Can the methods used to study monarchs be used on other species? Explain your answer.

SCIENTISTS IN ACTION

#18 — Science in the News: Critical Thinking Worksheets
Segment 18
Studying Dolphin Behavior

1. Name two ways that exotic species may be accidentally introduced to a new ecosystem.

#19

2. Why is the source country usually not the same for both the pest and the possible biological solution?

3. Why did the researchers patent the information they learned about the fungus?

4. Describe two trade-offs between the harm the fire ants cause and the effects of pest control agents on the environment and to human health.

Chapter Background

SECTION 1

What Is an Animal?

▶ **Animal Classifications**
Zoologists classify the members of the animal kingdom on the basis of their similarities and differences. The animal kingdom has three subkingdoms (Protozoa, Parazoa, and Metazoa). The three subkingdoms are divided into phyla, and each phylum is divided into subphyla and other subgroups.

- The phylum Chordata (meaning "cord") has three subphyla, one of which is vertebrates. Vertebrates include amphibians, reptiles, birds, mammals, and three kinds of fish.

- The vast majority of animals are invertebrates. Significant invertebrate phyla include Arthropoda (meaning "jointed foot"), Porifera (meaning "hole bearers"), Nematoda (meaning "thread"), and Echinodermata (meaning "spiny skin"). Arthropoda, which includes crustaceans, spiders, millipedes, and insects, is by far the largest phylum. It includes more than 1 million living species.

▶ **Animal Reproduction**
Some animals can reproduce both sexually and asexually. The adult jellyfish, or medusa, releases sperm and eggs into the water, forming a planula. When the planula matures into a polyp, it reproduces asexually, duplicating itself and becoming ephyra. Ephyra then grow into adult jellyfish and start sexual reproduction once again.

▶ **Moving Around**
Animals use a variety of techniques to move from place to place.

- Cephalopods, which include the squid and the octopus, escape from predators by forcing a powerful jet of water from a siphon near their head. Squids have been known to swim as fast as 38 km/h (23 mph)!

- Kangaroo rats have powerful hind legs that help them quickly leap away from predators. Their tail, which acts like a rudder, enables them to change course while in midair.

SECTION 2

Animal Behavior

▶ **Animal Defense Strategies**
Animals use a variety of methods to defend themselves from predators. Many species inject or spray toxic chemicals. Other species have a foul odor or an unpleasant taste.

- Jellyfish eject poison-tipped barbs from nematocysts on their bodies. Although most jellyfish stings can't penetrate human skin, the stings of the Portuguese man-of-war and the box jellyfish are exceptions. Stings from both species can cause extreme pain, and in rare instances, even death. Don't ever step on a dead jellyfish. The nematocysts can still sting!

- Colorful ladybugs, also known as ladybirds, protect themselves from predators by virtue of their foul smell and terrible taste.

- A type of Chrysomelid beetle has such a toxic poison that the San, a tribe living mainly in Africa's Kalahari Desert, tip their arrows with it. The poison causes death by paralysis. 

- Octopuses have beaks that are used to pierce the shells of crabs and lobsters. Once the prey is pierced, the octopus injects a poison that paralyzes its prey and helps soften the meat for easy removal.

▶ Animal Migration

What mechanisms trigger animal migration? How are birds and other animals able to find their way to destinations often thousands of miles away?

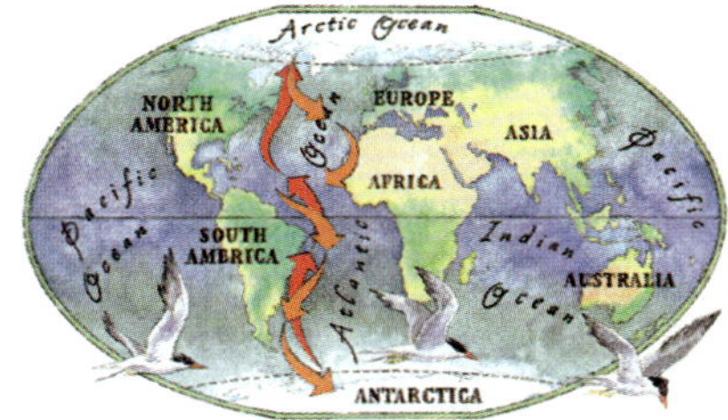

- Studies performed during the 1960s produced experiments that indicated that birds use a "celestial compass" to guide them on their migratory travels.

- Ruby-throated hummingbirds, which live as far north as Canada during the summer, fly to Central America in the fall. Their route requires them to fly 833 km (500 mi) nonstop over the Gulf of Mexico.

- Topographical features such as mountain ranges emit low-frequency sounds that some scientists believe help birds such as pigeons navigate.

▶ Human Seasonal Rhythms

Seasonal affective disorder (SAD) is a form of depression that many people, especially those in northern countries, experience during the winter. Common symptoms include fatigue, sleeping more than usual, carbohydrate craving, increased appetite, and sudden weight gain. Researchers have found that many people with SAD improve when they undergo light therapy, or phototherapy.

▶ Learned and Innate Behaviors

All behaviors, innate or learned, represent an interplay of genes and the environment. We inherit the potential for innate behavior. An innate behavior is one that appears in its fully functional form the first time it is performed. Innate behavior expresses itself without prior experience. Learned behavior depends upon experience.

SECTION 3

Living Together

▶ Animal Communication

Some kinds of animals live together, and others maintain a solitary existence. Both situations offer advantages and disadvantages. Regardless of how animals live, they communicate with each other to protect themselves, to find food, to display dominance, to find mates, and for many other reasons.

- Although octopuses and squids are both cephalopods, their social habits are quite different. Whereas the octopus is a solitary creature, squids are frequently found in schools.

- Besides using a "waggle dance" to communicate the location of nectar, honeybees also communicate information about the taste and smell of food resources. They do this through the process of trophalaxis, or the regurgitation of food into the mouths of members of the colony.

Is THAT A FACT!

- Elephants have at least 25 distinct vocal calls, including the familiar trumpet that elephants make when they are excited. Elephants can also communicate dozens of messages through low-frequency, infrasonic rumbles. Such sounds can travel up to 9.5 km.

- Antelopes raise their tail and release a warning scent to communicate danger to the herd.

For background information about teaching strategies and issues, refer to the **Professional Reference for Teachers.**

Directed Reading Worksheet 13

Science Puzzlers, Twisters & Teasers Worksheet 13

Guided Reading Audio CD
English or Spanish, Chapter 13

CAPÍTULO

13 Los animales y su conducta

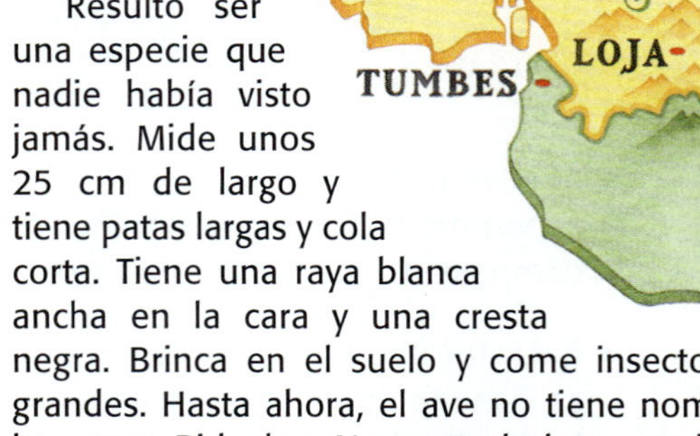

¡Esto realmente sucedió!

Robert S. Ridgely y Lelis Navarette son ornitólogos, es decir, personas que estudian las aves. En noviembre de 1997 estaban de excursión en los Andes de Ecuador, grabando el trinar de las aves. De pronto, escucharon un sonido que era una mezcla entre el ulular de un búho y el ladrido de un perro. Aunque los dos habían pasado mucho tiempo de su vida en los bosques, nunca antes habían escuchado un sonido como ése.

¿Qué había en el bosque? ¿Algún animal extraño? Siguieron caminando y casi 40 minutos después, ¡lo escucharon otra vez! Esta vez, Ridgely, quien aparece en la foto de abajo, lo grabó y lo reprodujo rápidamente. La criatura respondió y voló hacia ellos. ¡Era un ave! Pero ninguno de los expertos había visto un ave como ésa.

Resultó ser una especie que nadie había visto jamás. Mide unos 25 cm de largo y tiene patas largas y cola corta. Tiene una raya blanca ancha en la cara y una cresta negra. Brinca en el suelo y come insectos grandes. Hasta ahora, el ave no tiene nombre, pero Ridgely y Navarette la han estado estudiando y pronto le darán uno.

Durante siglos, personas como Robert Ridgely y Lelis Navarette han usado sus poderes de observación para estudiar los animales y su conducta. Hoy en día, conocemos más de 1 millón de especies y hemos aprendido mucho sobre la forma en que los animales viven e interactúan. ¡Pero siempre hay nuevos descubrimientos por hacer! Así que, la próxima vez que salgas a caminar, mantén tus ojos bien abiertos y tus oídos bien atentos. ¡Nunca sabes con qué te puedes encontrar!

302

This Really Happened...

The nearly flightless "barking" bird discovered by Ridgely, who is considered the foremost expert on the birds of South America, is a species of *Antpitta,* one of a group of very secretive, terrestrial forest birds. The bird is one of the largest to be discovered during the last 50 years. John Moore, a California real estate developer, was an amateur birdwatcher on this expedition, and his tape recorder captured the new birdcall.

¿Tu qué piensas?

Usa tus conocimientos para responder a las siguientes preguntas en tu cuaderno de ciencias:

1. ¿Qué diferencias existen entre un animal y una planta?

2. ¿Cómo saben los animales cuándo deben migrar?

¡Investiga!

¡Vete de safari!

No necesitas ir muy lejos para observar animales interesantes. Puedes ver hormigas caminando por la acera. Si buscas flores, seguramente verás abejas. Mueve un poco la tierra y tal vez encuentres lombrices. Mira hacia arriba y verás algunas aves. Si estás cerca de un arroyo, es probable que veas peces, salamandras y tal vez unos cuantos mosquitos. ¡Ah, y no te olvides de las mascotas!

Procedimiento 

1. Sal a la calle y busca **dos animales diferentes** para observar, o usa los animales que tu maestro o maestra lleve al laboratorio.

2. Sin molestar a los animales, siéntate en silencio y obsérvalos durante unos minutos desde cierta distancia. Puedes usar **binoculares** o una **lupa** para observarlos más de cerca. **Cuidado:** Ten mucho cuidado con los animales que muerden o pican.

3. Anota todo lo que observes sobre cada animal. ¿Qué tipo de animal es? ¿Cómo es? Si quieres, haz un dibujo. ¿Qué tan grande es? ¿Qué hace? ¿Por qué? ¿Se mueve? ¿Cómo lo hace? ¿Está comiendo? ¿Qué come?

Análisis

4. Compara los dos animales que estudiaste. ¿En qué se parecen? ¿En qué se diferencian?

5. ¿Cómo se mueven? ¿Puedes deducir cómo se defienden o se comunican con los demás animales si observas cómo se mueven?

6. ¿Puedes adivinar qué come cada uno? ¿Qué características de estos animales los ayudan a encontrar o atrapar comida?

303

Answers to Investigate!

4. Answers will vary, but students may refer to characteristics such as the way the animals moved, what they were eating, and their body type.

5. Answers will vary.

6. Answers will vary.

What Do You Think?

Accept all reasonable responses.

Students will have a chance to revise their answers in the Chapter Review under NOW What Do You Think?

Investigate!

MATERIALS

FOR EACH GROUP:
- binoculars or magnifying lens
- pen or pencil
- sketch pad or notebook

Safety Caution: Remind students to review all safety cautions and icons before beginning this lab activity. Students must be careful when handling the magnifying lens. Magnified sun rays should never be focused on people, animals, or flammable materials. Injuries or a fire could result. Students should observe animals from a distance and should be especially careful with animals that could bite or sting. Students allergic to insect bites or bee stings should avoid contact with these animals and receive immediate medical attention if stung.

Students should also avoid hazards such as poisonous plants, holes, cliffs, water, cars, glass, and other dangers.

Teacher Notes: This activity will work best with small groups. If possible, try to have your best science students evenly dispersed among the groups.

Focus

What Is an Animal?

This section provides students with an introduction to the animal kingdom. Students will find out the difference between vertebrates and invertebrates and will learn how scientists classify animals. Students will also discover the characteristics that set animals apart from all other living things.

Bellringer

While you are taking attendance, ask students to ponder this question:

> What is the best material for washing a car—a cotton rag, a scratch pad, or an animal skeleton?

Have them take a few moments to record their answer in their ScienceLog. Before you begin the section, call on individual students to give their answer and reasoning. (It may surprise some students to learn that genuine sponges–ones that some people use for washing cars–are animal skeletons. During the process of preparing sponges, all tissue is removed from the animals, leaving only skeletal remains. Although there are about 5,000 sponge species, fewer than 20 of them are of any commercial value.)

Teaching Transparency 54
"The Animal Kingdom"

Directed Reading Worksheet 13 Section 1

VOCABULARIO

vertebrados	tejido
invertebrados	órgano
multicelular	consumidor
embrión	

OBJETIVOS

- Entiende las diferencias entre los vertebrados y los invertebrados.
- Explica las características de los animales.

¿Qué es un animal?

¿En qué piensas cuando escuchas la palabra *animal*? Tal vez piensas en tu perro o en tu gato. A lo mejor piensas en jirafas, osos u otras criaturas que has visto en los zoológicos o en la televisión. Pero, ¿pensarías en una esponja? Las esponjas naturales de baño, como la de la **Figura 1,** ¡son los restos de un animal que vivió en el mar!

Hay animales de diferentes formas y tamaños. Algunos tienen cuatro patas y pelo, pero otros no. Unos son tan pequeños que sólo se pueden ver con un microscopio, y otros son tan grandes como un auto. Y todos son parte del fascinante mundo de los animales.

El reino animal

Los científicos han clasificado casi 1 millón de especies de animales. ¿Cuántos tipos diferentes de animales ves en la **Figura 2**? Además de las esponjas, las anémonas de mar y los corales también son animales. Y las arañas, los peces, las aves y los delfines también. Los caracoles, las ballenas, los canguros y los seres humanos también lo son. Los científicos han dividido estos animales en unos 35 filos y clases.

La mayoría de los animales no se parecen a los humanos. Pero nosotros compartimos características con un grupo de animales llamados vertebrados. Un animal que tiene cráneo y columna vertebral es un **vertebrado.** Los peces, anfibios, reptiles, aves y mamíferos son vertebrados.

Figura 1 *Esta esponja natural era un ser vivo.*

Figura 2 *Todos los seres vivos de esta fotografía se clasifican como animales. ¿Crees que parecen animales?*

A snail knocked on a man's door and asked for a donation to a snail charity. The man didn't like solicitors and kicked the snail off his porch. Ten years later, the snail knocked on the door again and said, "That wasn't a very nice thing to do!"

Aunque quizás estés más familiarizado con los vertebrados, no hay duda de que somos la minoría entre los seres vivos. Menos del cinco por ciento de las especies de animales son vertebrados. Mira la **Figura 3.** La mayoría de las especies animales son insectos, caracoles, medusas, gusanos y otros **invertebrados,** o sea, animales sin columna vertebral. De hecho, ¡la cuarta parte de todas las especies animales son escarabajos!

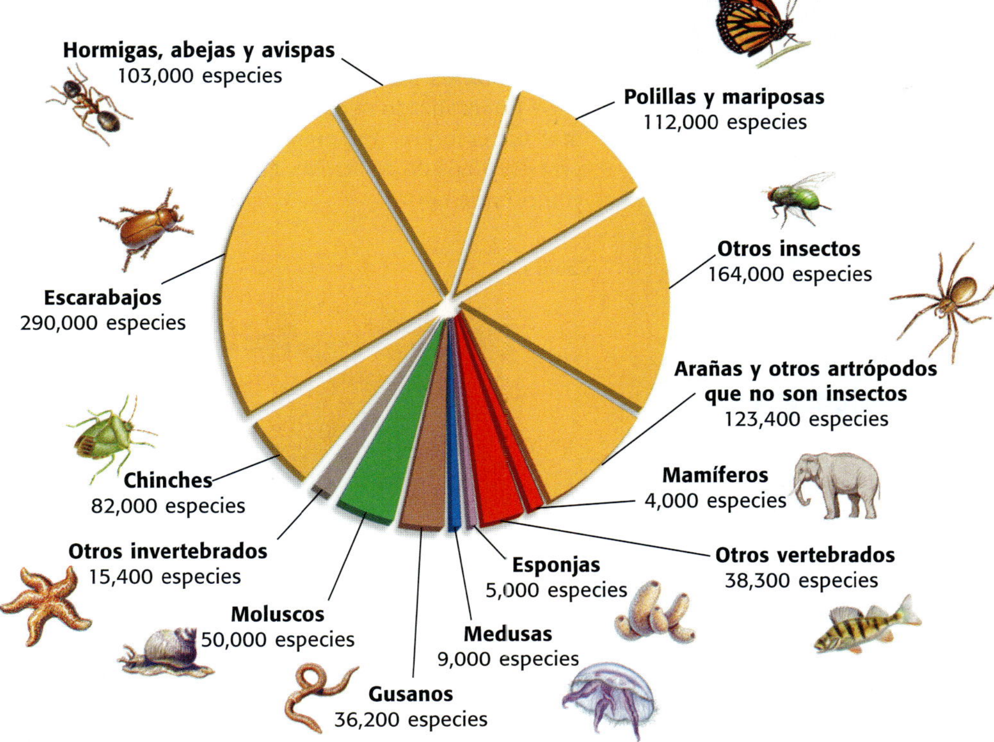

Figura 3 *Esta gráfica circular muestra los principales filos y subgrupos del reino animal. Observa que la mayoría de las especies son insectos.*

¿Eso es un animal?

La esponjas no se parecen a otros animales, hasta hace unos 200 años, casi todos pensaban que las esponjas eran plantas. Las lombrices no se parecen a los pingüinos y nadie confunde un sapo con un león. Entonces, ¿por qué decimos que todas estas cosas son animales? ¿Qué determina que algo sea un animal, una planta u otra cosa?

No hay una sola respuesta. Pero todos los animales comparten características que los distinguen de los demás seres vivos.

Los animales tienen muchas células. Todos los animales son **multicelulares,** es decir, están hechos de muchas células. Tu cuerpo tiene unos 13 billones de células. Las células animales son eucariotas sin pared celular; sólo están rodeadas por una membrana celular.

Science Bloopers

In 1798, when English scholars first observed the duck-billed platypus that had been sent to them by a scientist in Australia, they were convinced they were the victims of a joke. Surely, they thought, some prankster had pieced together parts of various animals. The English scholars cut and sliced the dead animal for signs of stitches holding the bill and webbed feet to its mammal-like body. It took a lot of convincing, but eventually they came to the conclusion that the animal was indeed real.

1 Motivate

GROUP ACTIVITY

Writing To explore the diversity of the animal kingdom, have students work in groups of four. Each group will write down examples for each of the following:

- 2 Arctic animals
  (penguin is incorrect)
- 2 Antarctic animals
  (polar bear is incorrect)
- 2 animals that crawl
- 2 animals that fly
- 2 animals with no bones
- 2 African animals
- 2 North American animals
- 2 animals that live in the soil
- 2 ocean animals
- 2 animals with more than four legs

Answers should be as specific as possible. For example, they should answer "garter snake," not "snake," or "praying mantis" instead of "insect." Have each group share its answers while other groups cross out matches on their own page. How many animals did the class think of? That's diversity!

**Science Skills
Worksheet 25
"Introduction to Graphs"**

2) Teach

MEETING INDIVIDUAL NEEDS

Learners Having Difficulty
On the board, make a chart with column headings that list the shared animal characteristics here. Label the rows:

"Tree," "Slug," "Fungus," "Coati," "Snapdragon," "Kookaburra," "Serval"

Tell students to replicate this chart in their ScienceLog. Have students check off which characteristics are found in each of these organisms. Students will probably have to check additional resources to identify some of these organisms. **Sheltered English**

Homework

Research/ Presentation
Sometimes the specialized or differentiated cells of an animal develop into a unique characteristic. For example, all mammals have hair, but only the male lion has a large, bushy mane. Have students research unusual physical characteristics of animals that are exaggerated versions of ordinary characteristics. Tell students to prepare a poster with photographs or drawings of their research results. (Suggest that students investigate a peacock's feathers, an elephant's trunk, or the "suction pad" toes of various frogs.)

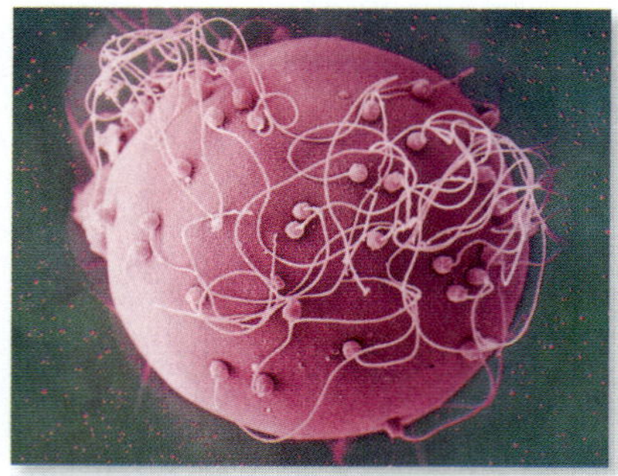

Figura 4 *Varios espermatozoides rodean un óvulo. Sólo uno de ellos puede fusionarse con el óvulo para formar un nuevo individuo.*

✓ Autoevaluación

¿Por qué los seres humanos se clasifican como vertebrados? *(Consulta la página 636 para comprobar tu respuesta.)*

Figura 6 *Al igual que la mayoría de los animales, los tiburones tienen órganos para la digestión, la circulación y la reproducción.*

Casi todos los animales se reproducen sexualmente.

Los animales producen células sexuales: óvulos o espermatozoides. Cuando un óvulo y un espermatozoide se unen en la fertilización, forman la primera célula de un nuevo individuo. La **Figura 4** muestra un óvulo rodeado por espermatozoides durante la fertilización. Algunos animales, como las esponjas y las estrellas de mar, también se reproducen asexualmente, por gemación y división.

Los animales se desarrollan a partir de embriones.

El óvulo fertilizado se divide en muchas células para formar un embrión. Un **embrión** es un organismo en la primera etapa de desarrollo. En la **Figura 5** se muestra un embrión de ratón.

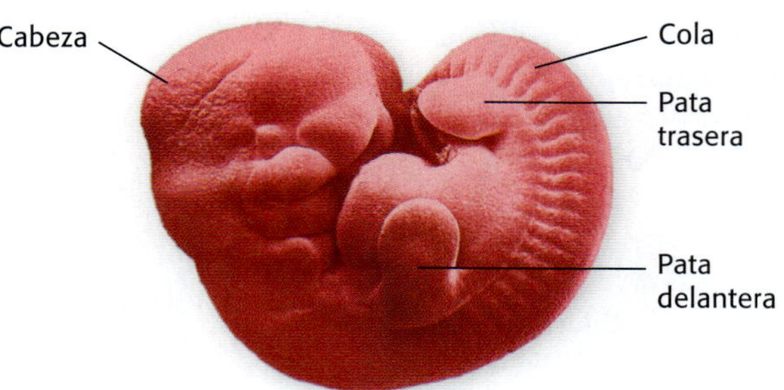

Figura 5 **Embrión de ratón**

Los animales tienen muchas partes especializadas.

El cuerpo de un animal tiene partes que hacen cosas diferentes. Cuando un óvulo fertilizado se divide en varias células para formar un embrión, las células se vuelven diferentes unas de otras. Algunas se vuelven células de la piel. Otras se vuelven células musculares, nerviosas u óseas. Estos diferentes tipos de células se organizan para formar **tejidos,** que son grupos de células similares. Por ejemplo, las células musculares forman el tejido muscular y las células nerviosas forman el tejido nervioso.

La mayoría de los animales también tienen órganos. Un **órgano** es una combinación de dos o más tejidos. El corazón, los pulmones y los riñones son órganos. Todos los animales, incluyendo el tiburón de la **Figura 6,** tienen órganos para diversas tareas.

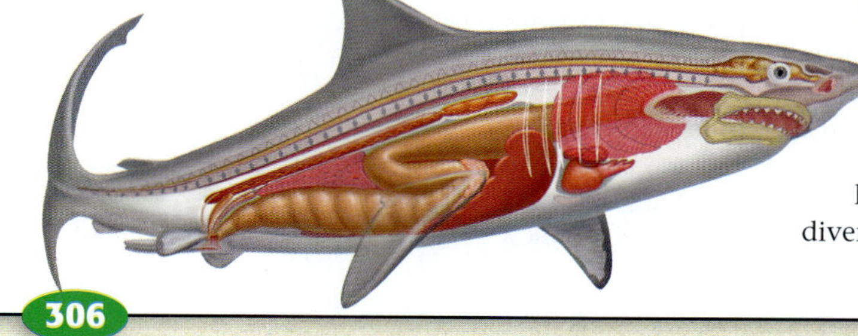

Answer to Self-Check
Like other vertebrates, humans have a skull and a backbone.

Q: What lies on the forest floor, 100 feet in the air?

A: a dead centipede

Los animales se mueven. La mayoría de los animales se mueven de un lugar a otro. Como se ve en la **Figura 7,** vuelan, corren, nadan y saltan. Es cierto que otros organismos también se mueven, pero los animales pueden hacerlo rápidamente en una sola dirección. Algunos no se mueven mucho: las anémonas de mar y las almejas se pegan a las rocas o al suelo del océano y esperan a que les llegue el alimento. Sin embargo, la mayoría de los animales son activos.

Figura 7 *Los animales se mueven de diferentes maneras.*

Los animales son consumidores. Los animales no producen su propio alimento, sino que se comen a otros organismos, o partes o productos de otros organismos. Los animales son consumidores. Un **consumidor** es un organismo que se come a otros. Este rasgo separa a los animales, como el panda de la **Figura 8,** de las plantas. Las plantas no suelen comer seres vivos, sino que producen su comida.

El alimento de los animales varía tanto como los propios animales. Los conejos y las orugas comen plantas. Los leones y las arañas comen otros animales. Los renos comen líquenes. Los mosquitos beben sangre. Las mariposas toman el néctar de las flores.

Figura 8 *Este panda gigante come hojas de bambú.*

REPASO

1. ¿Qué características separan a los animales de las plantas?

2. ¿Qué relación existe entre tejidos y órganos?

3. **Interpretar ilustraciones.** ¿Qué características del camaleón, que se ilustra a la derecha, te hacen pensar que es un animal?

307

---

1. Animals cannot make their own food; they are consumers. Plants are producers. Animal cells do not have cell walls. Plants do not usually move.

2. An organ is a combination of two or more tissues.

3. The chameleon is moving, and it is eating another organism.

Homework

Writing **Research Invertebrates**
Given that most of the animals in the world are invertebrates, encourage students to get to know them better. Have interested students choose any invertebrate and write a paragraph or two about it. Students should describe the invertebrate's range, habitat, and food sources. They should include how it obtains food, how it avoids predators, and how it affects people. Have students include one unusual fact about their subject.

4 Close

Quiz

1. What do vertebrates have that invertebrates don't? (a skull and a backbone)

2. What are collections of similar cells called? (tissues)

ALTERNATIVE ASSESSMENT

Writing Have students write a summary of the characteristics shared by all animals. Tell them not to repeat the bold-faced paragraph headings in the text. Instead, students must explain each characteristic by first giving an example. (Sample answers: Monkeys run, climb, and swing in trees. Animals move.)

Reinforcement Worksheet 13
"What Makes an Animal an Animal?"

Focus

Animal Behavior

This section introduces students to animal behavior. Students will see how animals find food and defend themselves from predators. They will discover the difference between innate and learned behavior. Finally, they will learn some of the ways animals communicate, why animal behaviors change at different times of the year, and how an animal's biological clock controls its circadian rhythms and other cycles.

Bellringer

Ask students to write a sentence for each of the following terms:

predator and *prey*

After each sentence have students list three animals that are predators and three that are prey.

• Predators hunt for their food. A list of predators might include alligators, sharks, spiders, lions, wolves, rattlesnakes, and eagles.

• Prey stay alert to avoid being eaten. A list of prey might include mice, rabbits, pigeons, flies, and deer. Sheltered English

Directed Reading Worksheet 13 Section 2

VOCABULARIO

depredador	hibernación
presa	estivación
camuflaje	reloj biológico
conducta innata	ritmo circadiano
conducta aprendida	orientarse
migrar	punto de referencia

OBJETIVOS

- Explica la diferencia entre conducta aprendida y conducta innata.
- Explica la diferencia entre hibernación y estivación.
- Da ejemplos de cómo influye el reloj biológico en la conducta.
- Describe los ritmos circadianos.
- Explica cómo se orientan los animales.

Conducta animal

Ya aprendiste las características que nos ayudan a reconocer los animales. Una es que los animales se mueven. Brincan, corren, vuelan, se lanzan, se escabullen, se arrastran y se deslizan. Pero no se mueven sólo por placer. Se mueven por una razón. Corren para alejarse de sus enemigos, se trepan para buscar alimento y construyen sus casas. Hasta la garrapata más diminuta atrapa su comida, lucha por su territorio o migra. Estas actividades se conocen como conducta.

Conducta de supervivencia

Para sobrevivir, debe encontrar comida y agua, evitar que otros animales se lo coman y tener donde vivir. Los animales tienen conductas que les ayudan a realizar estas tareas.

Buscar el almuerzo Los animales usan diferentes métodos para encontrar o atrapar su comida. Los búhos atrapan ratoncillos descuidados. Las abejas vuelan de flor en flor juntando néctar. Los koalas trepan en busca de hojas de eucalipto. Las medusas lazan sus presas con sus tentáculos. Animales como el chimpancé de la **Figura 9** usan herramientas para conseguir su cena. Sin importar su preferencia, los animales se han adaptado a su ambiente y obtienen la mayor cantidad de comida usando la menor cantidad de energía.

Cómo evitar ser comido Los animales que comen otros animales se llaman **depredadores.** El animal que es comido se conoce como **presa.** En cualquier momento, el animal que *come* se puede convertir en la *comida* de otro. Por lo tanto, los animales que buscan alimento tienen que considerar otras cosas además de la apariencia o buen sabor de la comida. Si el almuerzo es muy peligroso de conseguir, prefieren dejarlo pasar. Ser cuidadoso es sólo un método de defensa. Sigue leyendo para descubrir qué otras cosas hacen los animales para sobrevivir.

Figura 9 *Los chimpancés hacen y usan herramientas para sacar hormigas y otro tipo de alimento de lugares de difícil acceso.*

308

IS THAT A FACT!

Zoo keepers and pet owners have found that when they challenge their animals to search for food, the animals' appetites improve and their activity level and alertness increases. In addition, the animals genuinely seem to enjoy the search. At one zoo, the polar bears appeared downright depressed until keepers began to hide their food and freeze their fish in big buckets of solid ice. Working for their food simulated the hunting the bears would do in the wild, perhaps alleviating some of the stress of captivity.

A escondidas Una forma de evitar ser comido es esconderse del enemigo. Un conejo "se paraliza" de modo que su color natural se mezcla con los colores de los arbustos o el pasto. Mezclarse con los colores o formas del medio ambiente se denomina **camuflaje.** Muchos animales toman la forma o el color de ramitas, hojas, piedras, cortezas y otros materiales de su ambiente. El insecto llamado "bastoncillo" parece una ramita, algunos de ellos hasta se ladean como si los moviera el viento. En la **Figura 10** se muestra un ejemplo de camuflaje.

En tu cara Los cuernos de un toro y las espinas de un puerco espín causarían serios problemas a un depredador, pero otras defensas no son tan obvias. Algunos animales se defienden con substancias químicas. El zorrillo y el escarabajo bombardero rocían a los depredadores con substancias irritantes. Las abejas, las hormigas y las avispas les inyectan un ácido poderoso a sus atacantes. La piel de la rana sudamericana de dardos venenosos y del pájaro pitohui enmascarado de Nueva Guinea tienen una toxina mortal. Cualquier depredador que se coma o intente comerse uno de estos animales, puede morir.

Los animales que utilizan una defensa química advierten a los depredadores que busquen alimento en otra parte. Por lo general, sus armas químicas se notan por la apariencia del animal, que tiene un diseño brillante llamado *coloración de advertencia,* como se muestra en la **Figura 11.** Los depredadores evitan cualquier animal que tenga colores y diseños que asocian con dolor, enfermedad u otras experiencias desagradables. Los colores de advertencia más comunes son los tonos fuertes de rojo, amarillo, naranja, negro y blanco.

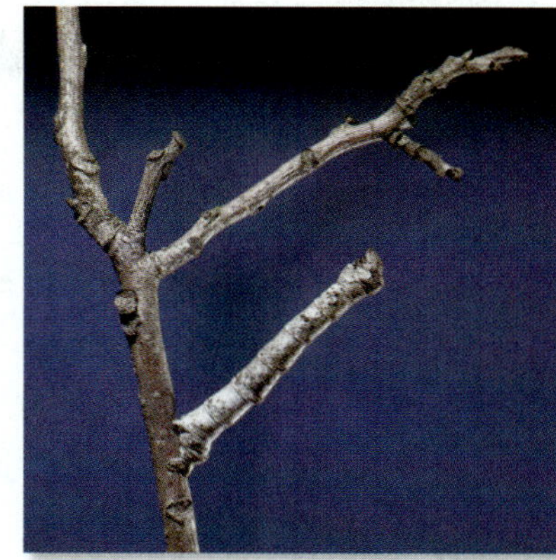

Figura 10 *Esta es una fotografía de una oruga camuflada como ramita. ¿Puedes encontrarla?*

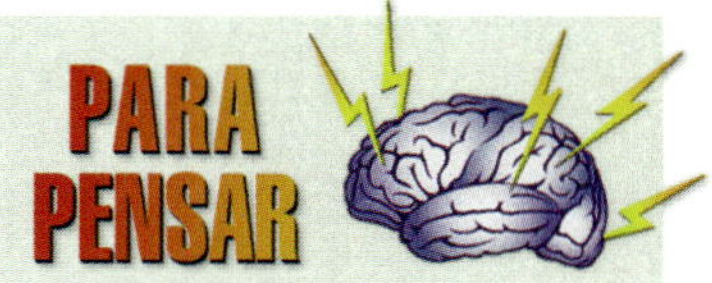

PARA PENSAR

Los pulpos son expertos en el camuflaje. Pueden cambiar el color de todo su cuerpo en menos de 1 segundo.

Figura 11 *La coloración de advertencia de los pájaros pitohui enmascarados advierte a los depredadores que son venenosos. Las rayas amarillas y negras de algunas avispas son otro ejemplo.*

309

WEIRD SCIENCE

The hooded pitohui is the only known poisonous bird. This is not the only unusual thing about the colorful, foul-smelling bird, however. The poison that the bird emits—homobatrachotoxin—is the same poison made by the New World strawberry dart-poison frog. How can two species that are so different produce the same poison? Scientists are now exploring the mystery.

Answer to Question in Figure 10

The lowest "branch" is a caterpillar.

DISCUSSION

Yawning Begin a discussion by yawning at the front of the classroom. Record how many students yawn as a result. Tell them yawning is one of the few specific behaviors shared by different animal groups. Reptiles, fish, amphibians, birds, and mammals all yawn.

Ask students: Why do we yawn?

(We yawn to rid our body of excess carbon dioxide, to speed oxygen to our brain when we are tired or nervous, or when we observe someone else yawning.)

 **PG 604**

Wet, Wiggly Worms!

MATH and MORE

Scientists have named about 1 million species of animals living in the world today, and one-fourth of those are beetles!

Based on this information, about how many different species of beetles do we know about? (250,000)

What percentage of the world's animal species are beetles? (25 percent)

 **Math Skills Worksheet 21** "Percentages, Fractions, and Decimals"

② Teach, continued

DEMONSTRATION

Sign Language Learn five basic signs in Sign Language, and demonstrate them for the class. Choose words such as *hungry, sad, angry, tired,* and *happy.*

Tell students that because of our innate language ability, it's possible to learn another language, even a nonvocal one. But perhaps humans aren't the only species with this ability. Koko, a gorilla, learned more than 500 different signs while in captivity. She even created some of her own signs, such as "finger bracelet" for "ring." Once, her trainer became frustrated because Koko was not cooperating. The trainer signed "bad gorilla," and Koko corrected her by signing "funny gorilla." Koko fibbed at times to avoid getting in trouble and insulted trainers when she was angry. She even had her own kitten, a tiny tailless cat that she named All Ball. Sheltered English

LaBo**ok**　**PG 606**

Aunt Flossie and the Bumblebee

GUIDED PRACTICE

Have students observe the eating habits of animals near their home over a period of a week or so. Have them record their observations in their ScienceLog.

Figura 12 *Comer o no comer gusanos plataneros, como éste, es una conducta innata de las culebras de agua.*

Experimentos

¿Qué le hizo el abejorro a la tía Florecita? Averígualo en la página 606.

Figura 13 *Cuando los investigadores empezaron a arrojar batatas en la playa de una isla, los macacos japoneses tuvieron un bocadillo sabroso pero arenoso. Uno de los monos enjuagó la batata y le quitó la arena. Otros vieron cómo lo hacía, y ahora todos los macacos de la isla enjuagan las batatas.*

310

¿Por qué se comportan de esa manera?

¿Cómo saben los animales cuando una situación es peligrosa? ¿Cómo saben los depredadores qué coloración de advertencia deben evitar? Algunos animales saben por instinto qué hacer, pero a veces tienen que aprender. Los biólogos llaman a estos dos tipos de conducta animal, conducta innata y conducta aprendida.

Está en los genes La conducta que no depende del aprendizaje o de la experiencia se conoce como **conducta innata.** Los genes influyen en este tipo de conducta. Como se describe en la **Figura 12,** las serpientes heredan preferencias por ciertos alimentos. Los gusanos plataneros son el alimento favorito de las culebras de agua de la costa de California, pero las culebras de otras regiones no los comen. Los humanos heredamos los genes que nos dan la capacidad de caminar. Los cachorros heredan la tendencia a masticar, las abejas la tendencia a volar y las lombrices la tendencia a cavar.

Algunas conductas innatas se presentan desde el nacimiento. Las ballenas recién nacidas tienen la capacidad innata de nadar. Otras conductas innatas se desarrollan meses o años después del nacimiento. Por ejemplo, la tendencia de un ave a cantar es innata, pero no canta hasta que está cerca de la edad adulta.

La escuela de los animales Aunque una conducta sea innata, se puede modificar. El aprendizaje puede cambiar la conducta innata. La **conducta aprendida** es la conducta que se ha aprendido a partir de una experiencia o de observar otros animales. Los seres humanos heredan la tendencia a hablar, pero el idioma que hablamos no es heredado. Podemos aprender inglés, español, chino o tagalo.

Los humanos no son los únicos que modifican las conductas heredadas con el aprendizaje. Casi todos los animales jóvenes aprenden observando a sus padres. La **Figura 13** muestra un mono que aprendió una nueva conducta observándola.

REPASO

1. ¿Cuál es la diferencia entre la conducta innata y la aprendida?

2. Aplicar conceptos ¿Cómo depende del aprendizaje la eficacia de la coloración de advertencia?

▼ **Answers to Review**

1. Innate behavior is influenced by genes and does not depend on experience. Learned behavior results from experience or observation.

2. Through experience or observation of other animals, predators learn which colors are associated with painful or undesirable consequences. Once a predator learns, it can avoid animals with these warning colors.

Conducta estacional

Muchos animales enfrentan inviernos de escasez de alimento y frío intenso. Algunos lo evitan viajando a lugares más calientes. Otros reúnen y almacenan comida. Las ranas se entierran en el lodo, los insectos se esconden en la tierra y algunos animales hibernan.

Viajeros por el mundo Cuando escasea el alimento por el invierno o las sequías, muchos animales migran. **Migrar** es ir de un lugar a otro y regresar. Los animales migran en busca de alimento, agua o lugares seguros para reproducirse. Las ballenas, el salmón, los murciélagos y hasta los chimpancés migran. En invierno, las mariposas monarca, como las de la **Figura 14,** migran de Norteamérica al centro de México en espera de la primavera. Las aves del hemisferio Norte vuelan hacia el Sur. En primavera, regresan al Norte para anidar.

Reducción de actividad Algunos animales enfrentan la escasez de alimento y agua hibernando. La **hibernación** es un período de inactividad y disminución de la temperatura corporal. Los animales que hibernan sobreviven gracias a la grasa de su cuerpo. Muchos animales hibernan, como los ratones, las ardillas, los zorrillos y los osos. Al hibernar, su temperatura, ritmo cardíaco y respiración bajan. Algunos animales bajan la temperatura de su cuerpo a sólo unos grados sobre punto de congelación y no despiertan en semanas. Otros, como los osos polares de la **Figura 15,** no entran en una hibernación profunda. La temperatura de su cuerpo no baja demasiado y duermen durante períodos más cortos.

El invierno no es la única época en la que los recursos escasean. Muchas ardillas y ratones del desierto disminuyen sus funciones vitales de manera similar en la parte más caliente del verano, cuando tienen poco alimento y agua. Este período de actividad reducida en el verano se llama **estivación.**

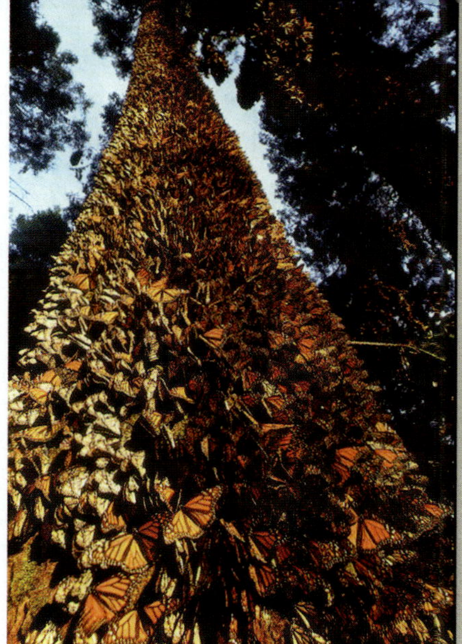

Figura 14 *Cuando las mariposas monarca se reúnen en México, puede haber hasta 4 millones por hectárea. ¡Esa sí que es una reunión familiar!*

¡Murciélagos dormidos!

Pasa a la página 322 para aprender sobre los efectos que producen los humanos en la hibernación de los murciélagos.

Figura 15 *Los osos polares y otros osos no entran a una hibernación profunda y pueden reaccionar si se encuentran en peligro. Sin embargo, tienen períodos de inactividad en los que no comen, sus funciones corporales disminuyen y se mueven muy poco.*

311

WEIRD SCIENCE

Snakes are normally solitary creatures, but they will often group together in underground pits while hibernating. Hundreds or even thousands of snakes may spend the winter together in these hibernaculums. Why do they do it? Perhaps there is some protection from the cold in a group. (The protection would be limited, though, because snakes are ectotherms.) Or maybe it makes finding a mate much easier in the spring.

GOING FURTHER

Maritime Navigation
Have students write a report about maritime navigation. Tell students to include a comparison of the instruments that were used 100 years ago and their modern counterparts. Their report should also explain how ancient people navigated.

Quick Lab

MATERIALS

FOR EACH PAIR:
• stopwatch

Answers to QuickLab

Answers will vary. Students should explain their answer using the concept of biological clocks.

Answer to Self-Check

Hibernation is not controlled by a circadian rhythm. Circadian rhythms are daily cycles. Hibernation is a seasonal behavior.

CONNECT TO
EARTH SCIENCE

Use the following Teaching Transparency to guide a student discussion about direction, mapping, and navigation.

Teaching Transparency 87
"Finding Direction on Earth"

LINK TO EARTH SCIENCE

Laboratorio

¿Qué tan largo es un minuto?

¿Tienes un reloj interno que te ayuda a cronometrar el tiempo? Haz la siguiente actividad con un compañero o compañera. Tu compañero iniciará un **cronómetro** y te dirá: "¡Ahora!" Cuando creas que ya pasó un minuto, di: "¡Alto!" Mira cuánto tiempo ha pasado. Hazlo varias veces y anota tus resultados. Luego, deja que tu compañero lo intente. ¿Se acercan los tiempos registrados al minuto? ¿Mejoraste? Inténtalo de nuevo, usando tu pulso o respiración como guía. ¿Te acercaste más al minuto esta vez? ¿Crees que tienes un reloj interno? Explica por qué.

✓ Autoevaluación

¿Controla el ritmo circadiano la hibernación? Explica. *(Consulta la página 636 para comprobar tu respuesta.)*

internetconnect

SCLINKS
NSTA

TOPIC: Rhythms of Life
GO TO: www.scilinks.org
*sci*LINKS NUMBER: HSTL340

Los ritmos de la vida

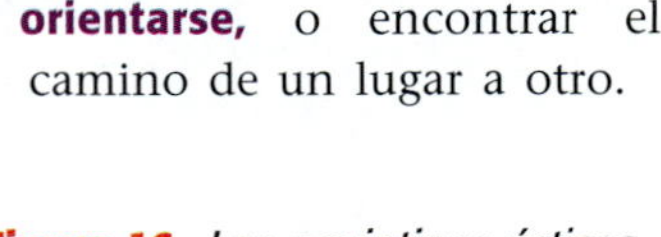

Los humanos usamos relojes y calendarios para saber a qué hora hay que levantarse para ir a la escuela, cuándo ir al cine, cuándo comprar el regalo de cumpleaños de alguien. Otros animales también necesitan saber a qué hora deben levantarse en la mañana, cuándo almacenar comida y cuándo volar hacia el sur por el invierno. Los relojes y calendarios que los animales usan se llaman relojes biológicos. Un **reloj biológico** es el control interno de los ciclos naturales. Los animales usan pistas del medio ambiente, como la duración del día y la temperatura, para fijar sus relojes.

Algunos relojes biológicos cuentan cantidades reducidas de tiempo. Otros controlan los ciclos diarios. Estos ciclos se llaman **ritmos circadianos.** *Circadiano* significa "alrededor del día." La mayoría de los animales se despiertan y se duermen a la misma hora. Éste es un ritmo circadiano.

Algunos relojes biológicos controlan ciclos más largos. Los ciclos estacionales son casi universales: los animales hibernan en ciertas épocas del año y se reproducen en otras. Saben cuándo almacenar alimento, prepararse para el invierno y migrar. Los relojes biológicos controlan todos estos ciclos.

¿Cómo encuentran los animales el camino?

Cuando planeas un viaje, quizás consultes un mapa. Si andas de excursión y no hallas el camino, te guías con una brújula o con marcas del sendero. Cuando es tiempo de migrar, ¿cómo saben los animales, como los gaviotines árticos de la **Figura 16,** qué camino seguir? Deben **orientarse,** o encontrar el camino de un lugar a otro.

Figura 16 *Los gaviotines árticos son famosos por sus largos recorridos. Cada año, hacen un viaje de 38,000 km desde el hemisferio Norte hasta la Antártida.*

MISCONCEPTION ALERT

Be sure that students understand that *hibernation* and *sleep* are not synonymous terms. Point out that animals do not hibernate continuously through the winter. They have periods that may last for days or weeks when their temperature rises to normal.

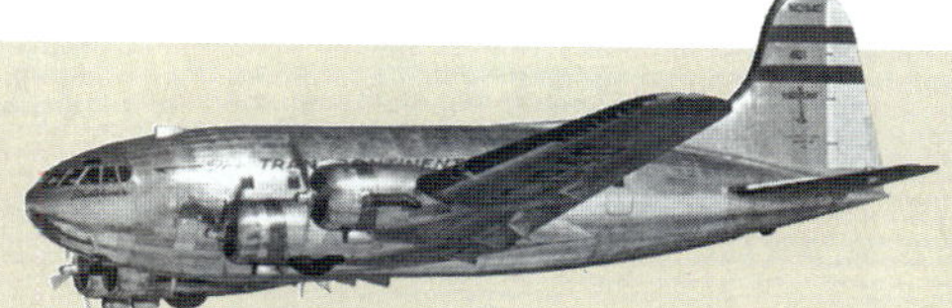

Al viajar a lugares que están en una zona de tiempo diferente, se sufre un desajuste conocido como "jet lag". Por ejemplo: el horario de Nueva York es 6 horas menos que en París. Alguien de Nueva York que va a París sufre un desajuste. Se acuesta a las 10 P. M., hora de París, pero se despierta a medianoche, y no se puede volver a dormir hasta las 6 A. M., una hora antes de que su despertador suene. ¿Cómo explican los ritmos circadianos este desfase? Si en París son las 10 P. M., ¿qué hora es en Nueva York? Si necesitas ayuda, consulta la sección "Conceptos de matemáticas" de la pág. 318.

Dobla a la izquierda al llegar al correo En viajes cortos, muchos animales se orientan con puntos de referencia. Los **puntos de referencia** son objetos fijos que sirven para hallar el camino. Cuando ves la gasolinera que está a seis cuadras de tu casa, reconoces el camino. La gasolinera es un punto de referencia para ti.

Las abejas y las palomas tienen un mapa mental de los puntos de referencia de su territorio. Las aves usan sierras, ríos y líneas costeras para hallar su camino. Los humanos y otros animales también se orientan en distancias cortas con una imagen mental del área. No todos los puntos de referencia son visuales. Los invidentes pueden orientarse en su casa porque saben dónde está todo y cuánto tiempo se tardan en atravesar un cuarto. Las palomas se orientan con el olfato y la vista.

¿Alguien tiene una brújula? Como los marineros, los animales usan la posición del Sol y las estrellas como mapa. Pero las aves migratorias tienen otros métodos para hallar su camino. Se orientan usando el campo magnético de la Tierra. Entérate de esto en la sección "Conexión: Ciencias físicas".

REPASO

1. ¿Por qué migran los animales?
2. Tres métodos que los animales usan para orientarse son:
3. ¿En qué se parecen la hibernación y la estivación? ¿En qué se diferencian?
4. **Aplicar conceptos** Algunas investigaciones sugieren que el "jet lag" se puede superar si el viajero se expone al Sol en la nueva zona horaria. ¿Por qué crees que este método funciona?

a través de las ciencias
CONEXIÓN

El núcleo de hierro de la Tierra actúa como un imán gigante, con polos magnéticos norte y sur. La fuerza y la dirección del campo magnético de la Tierra varía según el lugar y muchas aves usan esta variación como mapa. En la cabeza de las aves migratorias, sobre los orificios de la nariz, hay cristales diminutos de magnetita, el mismo tipo de material del que están hechos los imanes. Los biólogos piensan que la magnetita, que se encuentra en la roca de la ilustración, mueve o estimula los nervios del ave, indicándole su posición en la Tierra.

313

Focus

Living Together

In this section students will discover some of the ways that animals communicate with each other. They will explore how animals signal intentions and information to other animals through smell, sound, sight, and touch. Students will also learn about the advantages and disadvantages of living in a group.

Bellringer

As you are taking attendance, play a tape of humpback-whale songs. Ask students to offer suggestions related to what information the whales might be communicating. Your library might be a good source for whale tapes. Sound bytes of whale songs can also be found on the Internet.

1 Motivate

ACTIVITY

Nonverbal Communication
Play a game of charades with students to demonstrate the importance of nonverbal communication among humans and other animals. Allow the class to provide feedback after a student guesses successfully.

Sheltered English

Directed Reading Worksheet 13 Section 3

VOCABULARIO
conducta social
comunicación
territorio
feromonas

OBJETIVOS
- Discute las formas en que los animales se comunican.
- Menciona las ventajas y las desventajas de vivir en grupo.

La vida en grupo

La mayoría de los animales no viven solos; se asocian con otros animales. Cuando interactúan, lo hacen en grupos grandes o de uno en uno. Los animales pueden trabajar juntos o competir entre ellos. Toda esta conducta se llama conducta social. La **conducta social** es la interacción entre animales de la misma especie. Cualquier tipo de conducta social, ya sea hostil o amistosa, exige comunicación.

Comunicación

Imagínate cómo sería la vida si no pudiéramos hablar ni leer. No habría teléfonos, televisiones, libros ni Internet. Sin duda, el mundo sería muy diferente. El lenguaje es una forma muy importante de comunicación entre los seres humanos. En la **comunicación,** una señal debe viajar de un animal a otro y el receptor de la señal debe responder de alguna manera.

La comunicación les ayuda a los animales a vivir juntos, encontrar alimento, evadir a sus enemigos y proteger sus hogares. Los animales se comunican entre sí para alertar sobre peligros, identificar miembros de la familia, asustar a los depredadores y encontrar pareja. Uno de los usos más asombrosos de la comunicación se da cuando un animal corteja a otro. El *cortejo* es un conducta especial de animales de la misma especie, que conduce al apareamiento. En la **Figura 17** puedes ver dos grullas realizando un cortejo.

Los animales también se comunican para proteger su espacio. Muchos defienden su **territorio,** que es el área ocupada por un animal o grupo de animales, de la cual se excluyen otros miembros de la especie. Muchas especies, como los lobos de la **Figura 18,** usan sus territorios para aparearse, encontrar alimento y criar a sus cachorros.

Figura 17 *Las grullas japonesas realizan una compleja danza de cortejo.*

Figura 18 *Estos lobos aúllan para intimidar a los lobos vecinos y evitar que invadan su territorio.*

IS THAT A FACT!

Vampire bats are one of the few animals in the world that will share a meal with unrelated colony members—a truly altruistic behavior. These bats must consume 50–100 percent of their weight in blood each night (usually from cows or horses). If a bat misses a meal two nights in a row, it will die. As many as a third of the bats go without eating each night, yet few bats starve. A hungry bat will lick the wings of another bat, which in response usually regurgitates some of its meal to help the other bat survive.

¿Cómo se comunican los animales?

Los animales se comunican transmitiendo señales e información a otros a través del olfato, el sonido, la vista y el tacto. Sus señales son muy sencillas comparadas con las nuestras. Pero la señal es lo de menos, lo que importa es que transmita información específica.

¿Olores sospechosos?

Un método de comunicación es químico. Hasta los organismos unicelulares se comunican con los demás a través de substancias químicas. En los animales, estas substancias se llaman **feromonas.**

Las hormigas y otros insectos secretan una variedad de feromonas. Por ejemplo, las substancias de alarma liberadas en el aire alertan a otros miembros de la especie que hay peligro. Las substancias esparcidas dejan pistas para que los demás puedan seguirlas, encuentren alimento y regresen a sus madrigueras. Los olores de reconocimiento del cuerpo de una hormiga anuncian de qué colonia viene. Mensajes como éstos transmiten señales tanto a amigos como a enemigos y su interpretación depende del receptor del mensaje.

Muchos animales, entre ellos los vertebrados, utilizan las feromonas para atraer o influir los miembros del sexo opuesto. ¿Sabías que los elefantes y los insectos usan algunas de las mismas feromonas para atraer a sus parejas? Las mariposas reina, como la de la **Figura 19,** usan feromonas durante su cortejo.

¿Oyes lo mismo que yo?

Los animales también se comunican con ruidos. Los lobos aúllan; los delfines y las ballenas usan silbidos y ruidos complejos, para comunicarse con los demás; las aves macho cantan en la primavera para reclamar su territorio o atraer una hembra.

El sonido puede llegar a un gran número de animales en un área grande. Los elefantes se comunican con otros, a kilómetros de distancia, con ruidos de una frecuencia muy baja para el oído humano, como se describe en la **Figura 20.** Las canciones de ballenas jorobadas se pueden oír a muchos kilómetros de distancia.

Figura 19 *Las mariposas reina usan feromonas como parte de su cortejo.*

Figura 20 *Los elefantes se comunican con sonidos bajos, inaudibles para el oído humano. Cuando un elefante se comunica de esta manera, la piel de su frente tiembla.*

315

IS THAT A FACT!

Research shows that baleen-whale sounds, which scientists believe are produced by the larynx, may be the loudest sounds produced by any animal on land or in the sea. Such sounds may carry hundreds of kilometers underwater.

2 Teach

CROSS-DISCIPLINARY FOCUS

Geography Dolphins are usually thought to be ocean animals, but there is a group of small dolphins that live in fresh water in many of the world's largest rivers. River dolphins are different from their marine relatives in that they have poor eyesight and huge numbers of pointed teeth. In addition, they swim on their sides or even upside down. Have students locate on a map or globe the following rivers and coastal areas where river dolphins live: the Ganges River (India), the Indus River (Pakistan), the Yangtze River (China), the Amazon River (Brazil); the estuaries of the La Plata River (the Atlantic coast of South America) and of the Tucuxi River (the northeastern coast of South America).

MEETING INDIVIDUAL NEEDS

Learners Having Difficulty
Some students may not realize how often humans use nonverbal communication or the importance of such methods in negotiating even the mundane events of everyday life. Ask students how each of these nonverbal communication methods provides essential information:

smell (what's cooking, dangerous fumes, spoiled food)

facial expressions (friend or foe, surprise, fear, apprehension)

sound (someone approaching, a train's warning whistle, school bells, fire alarms)

Sheltered English

ACTIVITY

Investigate Grooming Tell students that grooming, or cleaning behavior, represents another way that animals communicate with each other through touch. Encourage students to do research about animals that communicate in this fashion. Examples include birds (preening), many mammals (mutual grooming), and certain fish (cleaning symbiosis). Students can use their research to create illustrated booklets that they can then share with the class.

GUIDED PRACTICE

Poster Project Tell students that wolves use body language called posturing to communicate with members of the pack. Have small groups of students research wolf posturing and then create illustrated charts showing various postures and what they mean. After posting the charts on the bulletin board, lead a discussion about wolf communication. Encourage students who have a pet dog to talk about the similarities and the differences between the body language used by their pet and the body language used by wolves. Sheltered English

Teaching Transparency 55 "The Dance of the Bees"

Figura 21 *Los perros paran las orejas y mueven la cola cuando están contentos. Cuando quieren jugar, se tienden sobre sus patas delanteras y, cuando están inquietos, agachan las orejas.*

Todo un espectáculo Muchas formas de comunicación son visuales. Al guiñarle el ojo a un amigo o fruncirle el ceño a un oponente, nos estamos comunicando con el *lenguaje corporal*. Los demás animales hacen lo mismo. La **Figura 21** muestra parte del lenguaje corporal que usa el perro.

Un animal que quiere espantar a otro hace cosas para verse más grande. Encrespa las plumas o el pelo, o abre el hocico y muestra los dientes. Las muestras visuales también son importantes en el cortejo. En la obscuridad, las luciérnagas emiten señales con luces brillantes para atraer a otras.

En contacto Los animales también pueden usar el tacto para comunicarse, como la abeja melífera. Si halla un área de flores ricas en néctar, vuelve a su colmena para decirles a sus compañeras dónde está. En la colmena, la abeja se comunica como se muestra abajo, con una danza en forma de ocho que las otras abejas aprenden por la observación y el tacto.

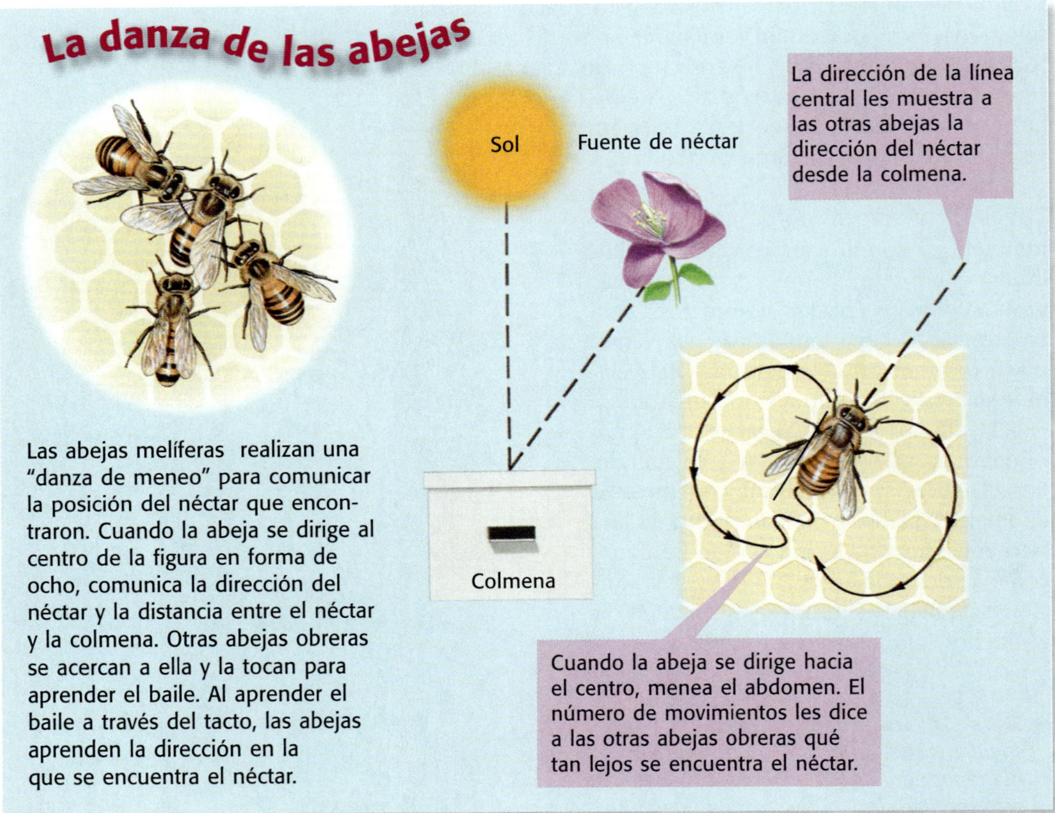

Homework

Writing **Research Ant Behavior** Have students research the behavior of ants. How do they organize the work of their colony? Several species of ants enslave other species. Have students investigate how one species of ant manages to capture and control another species.

Parte de la familia

Los tigres viven solos. Excepto durante el tiempo que una tigresa pasa con sus cachorros, el tigre rara vez está con otros tigres y, cuando lo hace, sólo es para aparearse. El león, aunque es el pariente más cercano del tigre, rara vez está solo. Los leones viven en grupos muy unidos llamados "manadas". Los miembros de una manada duermen, cazan y cuidan a sus cachorros juntos. La **Figura 22** muestra dos leonas en acción. ¿Por qué algunos animales viven en grupos y otros viven separados?

Figura 22 *Este par de leonas coopera para atrapar una gacela.*

Las ventajas de vivir en un grupo Vivir con otros animales es mucho más seguro que vivir solo. Los grupos grandes de animales localizan rápidamente depredadores u otros peligros, y pueden defenderse mejor. Por ejemplo, si un depredador amenaza una manada de bueyes almizcleros, éstos forman un círculo con las crías en el centro y los cuernos apuntando hacia afuera. Las abejas melíferas atacan en forma de enjambre cuando otro animal trata de robarse su miel. Cinco mil abejas defienden mejor un panal que una, o incluso cincuenta de ellas.

Vivir juntos también les ayuda a los animales a encontrar comida. Por lo general, los tigres y otros animales que cazan solos matan animales más pequeños que ellos. Por el contrario, las presas de los leones, los lobos, las hienas y otros depredadores que cazan en grupo son más grandes que ellos.

El lado malo de vivir en grupo Vivir en grupos también causa problemas. Los animales que viven en grupos atraen a depredadores, así que deben estar siempre atentos, como ves en la **Figura 23.** Los grupos necesitan más comida, y los animales en grupo compiten con los demás por el alimento y las hembras. Los miembros del grupo también pueden contagiarles enfermedades a los demás.

Figura 23 *Cuando se acerca un coyote, una ardilla se para y produce una fuerte llamada de alarma. Esto advierte a otras ardillas del peligro.*

REPASO

1. Los científicos han descubierto feromonas en los seres humanos. Menciona otros tres tipos de comunicación animal utilizados por los seres humanos.

2. ¿Por qué es importante la comunicación? Da tres razones.

3. **Aplicar conceptos** Con lo que aprendiste sobre la vida en grupos, menciona dos ventajas y dos desventajas de vivir en un grupo de seres humanos.

317

Chapter Highlights

SECTION 1

vertebrate an animal with a skull and a backbone; examples include mammals, birds, reptiles, amphibians, and fish

invertebrate an animal without a backbone

multicellular made of many cells

embryo an organism in the earliest stage of development

tissue a group of similar cells that work together to perform a specific job in the body

organ a combination of two or more tissues that work together to perform a specific function in the body

consumer organisms that eat producers or other organisms for energy

SECTION 2

predator an animal that eats other animals

prey an organism that is eaten by another organism

camouflage coloration and/or texture that enables an animal to blend in with the environment

innate behavior a behavior that is influenced by genes and does not depend on learning or experience

Resumen del capítulo

SECCIÓN 1

Vocabulario

vertebrados (*pág. 304*)

invertebrados (*pág. 305*)

multicelular (*pág. 305*)

embrión (*pág. 306*)

tejido (*pág. 306*)

órgano (*pág. 306*)

consumidor (*pág. 307*)

Notas de la sección

- Los animales con cráneo y columna vertebral son vertebrados. Los que no tienen columna vertebral son invertebrados.

- Los animales son multicelulares. Sus células son eucariotas y no tienen pared celular.

- La mayoría de los animales se reproducen sexualmente.

- La mayoría de los animales tienen tejidos y órganos.

- La mayoría de los animales se mueven y son consumidores.

SECCIÓN 2

Vocabulario

depredador (*pág. 308*)

presa (*pág. 308*)

camuflaje (*pág. 309*)

conducta innata (*pág. 310*)

conducta aprendida (*pág. 310*)

migrar (*pág. 311*)

hibernación (*pág. 311*)

estivación (*pág. 311*)

reloj biológico (*pág. 312*)

ritmo circadiano (*pág. 312*)

orientarse (*pág. 312*)

punto de referencia (*pág. 313*)

Notas de la sección

- Muchos animales se defienden de los depredadores por medio de camuflaje, defensas químicas o ambos.

- La conducta se clasifica como innata o aprendida. El potencial para la conducta innata se hereda. La conducta aprendida depende de la experiencia.

- Algunos animales migran para encontrar comida, agua o lugares seguros para reproducirse.

- Algunos animales hibernan en el invierno y otros estivan en el verano.

☑ Comprobar destrezas

Conceptos de matemáticas

DIFERENCIA DE HORARIOS El horario de París es 6 horas más tarde que el de Nueva York. Si en París son las 10 P.M., réstale 6 horas para obtener la hora de Nueva York.

$$10 - 6 = 4$$

En Nueva York son las 4 P.M. De la misma forma, cuando en París son las 7 A.M., en Nueva York es la 1 A.M.

Comprensión visual

LA DANZA DE LAS ABEJAS La foto de la página 316 ilustra cómo las abejas bailan la "danza del meneo" para comunicar la ubicación de una fuente de néctar. Observan la posición del Sol en relación con la colmena y la fuente de néctar. La abeja comunica esta información poniéndose en la línea central del baile, rumbo a la fuente de néctar.

Lab and Activity Highlights

Wet, Wiggly Worms! PG 604

Aunt Flossie and the Bumblebee PG 606

Datasheets for LabBook (blackline masters for these labs)

SECCIÓN 2

- Los animales tienen relojes biológicos para controlar los ciclos naturales.

- Los ciclos diarios se conocen como ritmos circadianos.

- Algunos relojes biológicos son regulados por las señales del medio ambiente de un animal.

- Los animales se orientan con puntos de referencia e imágenes mentales del área de su hogar.

- Algunos animales usan las posiciones del Sol y las estrellas o el campo magnético de la Tierra para orientarse.

Experimentos

Lombrices escurridizas *(pág. 604)*

La tía Florecita y el abejorro *(pág. 606)*

SECCIÓN 3

Vocabulario

conducta social *(pág. 314)*
comunicación *(p. 314)*
territorio *(pág. 314)*
feromona *(pág. 315)*

Notas de la sección

- La comunicación debe incluir una señal y una respuesta.

- El cortejo y el reclamo del territorio son dos tipos de comunicación importantes.

- Los animales se comunican con la vista, el sonido, el tacto y el olfato.

- Al vivir en grupo, los animales localizan sus presas y sus depredadores con más facilidad.

- Los grupos de animales son más visibles para los depredadores, y los animales en grupos deben competir con otros por la comida, las hembras y las áreas de reproducción.

learned behavior a behavior that has been learned from experience or observation

migrate to travel from one place to another in response to the seasons or environmental conditions

hibernation a period of inactivity that some animals experience in winter; allows them to survive on stored body fat

estivation a period of reduced activity that some animals experience in the summer

biological clock an internal control of natural cycles

circadian rhythm daily cycle

navigate to find one's way from one place to another

landmark a fixed object used to determine location during navigation

SECTION 3

social behavior the interaction between animals of the same species

communication a transfer of a signal from one animal to another that results in some type of response

territory an area occupied by one animal or a group of animals from which other members of the species are excluded

pheromone a chemical produced by animals for communication

 **internet**

VISITA: go.hrw.com

SC**LINKS**
N S T A

VISITA: www.scilinks.org

Visita el sitio web de HRW para encontrar una serie de herramientas de aprendizaje relacionadas con este capítulo. Sólo tienes que escribir la palabra clave:

PALABRA CLAVE: HSTANM

Visita el sitio web de la **Asociación Nacional de Maestros de Ciencias** *(National Science Teachers Association)* para encontrar recursos de Internet relacionados con este capítulo. Sólo escribe el **ENLACE DE CIENCIAS** para obtener más información sobre el tema:

TEMA: Vertebrados e invertebrados	**ENLACE:** HSTL330
TEMA: Conducta animal	**ENLACE:** HSTL335
TEMA: Los ritmos de la vida	**ENLACE:** HSTL340
TEMA: Comunicación en el reino animal	**ENLACE:** HSTL345

319

Lab and Activity Highlights

LabBank

Inquiry Labs, Follow the Leader, Lab 4

Whiz-Bang Demonstrations, Six-Legged Thermometer, Demo 9

Long-Term Projects & Research Ideas, Project 14

 **Vocabulary Review Worksheet 13**

 Blackline masters of these Chapter Highlights can be found in the **Study Guide.**

USING VOCABULARY

1. a vertebrate, an invertebrate
2. innate
3. estivating
4. circadian rhythms
5. an embryo

UNDERSTANDING CONCEPTS

Multiple Choice

6. c
7. d
8. b
9. b
10. d

Short Answer

11. Pheromones are released by one animal to send a message to another animal.
12. A territory is an area occupied by one animal or a group of animals from which other members of the species are excluded. Examples will vary but may include a bedroom, a family house, or a school.
13. Answers will vary but should be unchanging objects, such as buildings or roads.
14. Both are seasonal behaviors controlled by biological clocks.

Concept Mapping

15. 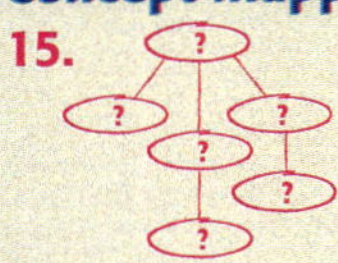 An answer to this exercise can be found at the end of this book.

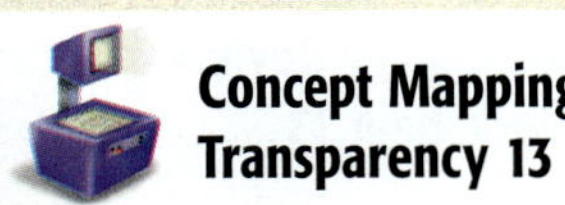

Concept Mapping Transparency 13

Repaso del capítulo

UTILIZAR EL VOCABULARIO

Escoge el término correcto para completar las siguientes oraciones:

1. Un animal con cráneo y columna vertebral es ___?___. Un animal sin columna vertebral es ___?___. *(un invertebrado* o *un vertebrado)*

2. Una conducta que no depende de la experiencia es ___?___. *(innata* o *aprendida)*

3. En el verano, un animal entra en un estado de actividad reducida. Este animal ___?___. *(estiva* o *hiberna)*

4. Los ciclos diarios se conocen como ___?___. *(relojes biológicos* o *ritmos circadianos)*

5. Cuando un óvulo y un espermatozoide se juntan, forman ___?___. *(un embrión* o *un órgano)*

COMPRENDER CONCEPTOS

Opción múltiple

6. ¿Qué característica no corresponde a los animales?
   a. Son multicelulares.
   b. Por lo general, se reproducen sexualmente.
   c. Producen su propio alimento.
   d. Tienen tejidos.

7. Vivir en grupos
   a. atrae a los depredadores.
   b. le ayuda a la presa a localizar a los depredadores.
   c. les ayuda a los animales a encontrar comida.
   d. Todas las anteriores

8. La coloración de advertencia es
   a. un tipo de camuflaje.
   b. una forma de alejar a los depredadores.
   c. siempre negra con blanco.
   d. siempre una señal de que el animal no se puede comer porque es venenoso.

9. Algunas aves usan el campo magnético de la Tierra
   a. para atraer a los machos o las hembras.
   b. para orientarse.
   c. para fijar sus relojes biológicos.
   d. para defender su territorio.

10. Para defenderse de los depredadores, un animal puede usar
    a. camuflaje.
    b. coloración de advertencia.
    c. toxinas.
    d. Todas las anteriores

Respuesta breve

11. ¿Cómo se usan las feromonas en la comunicación?

12. ¿Qué es un territorio? Da un ejemplo de un territorio de tu medio ambiente.

13. ¿Qué puntos de referencia te ayudan a orientarte cuando regresas a tu casa de la escuela?

14. ¿En qué se parecen la migración y la hibernación?

Organizar conceptos

15. Usa los siguientes términos para crear un mapa de ideas: estivación, ritmos circadianos, comportamientos estacionales, hibernación, migración, relojes biológicos.

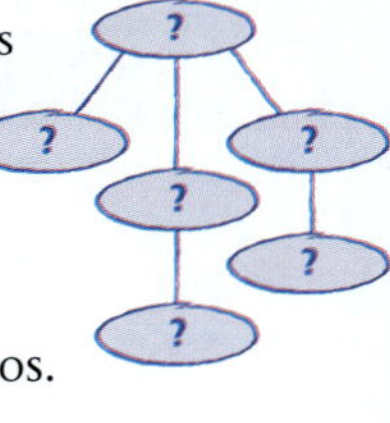

Escribe una o dos oraciones para responder a
las siguientes preguntas:

16. Si te llega el olor a zorrillo cuando vas en el
auto y cierras la ventanilla, ¿significa esto
que el zorrillo se comunicó contigo?
Explica.

17. Volar es una conducta innata de las aves.
En los humanos, ¿es una conducta innata
o aprendida? ¿Por qué?

18. Las hormigas dependen de las feromonas
y del tacto para comunicarse, pero las aves
dependen más de la vista y del sonido.
¿Por qué estos dos tipos de animales se
comunican de manera diferente?

INTERPRETAR GRÁFICAS

Esta gráfica circular muestra los diferentes filos
de las especies animales de la Tierra. Úsala
para responder a las siguientes preguntas.

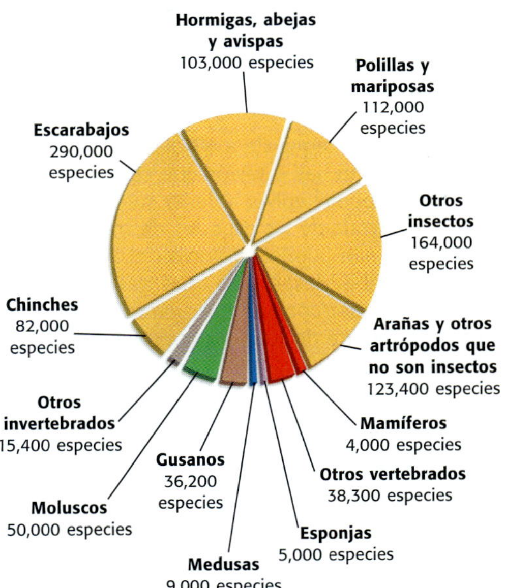

19. ¿Qué grupo de animales tiene la mayoría
de las especies? ¿Cómo se muestra en la
gráfica?

20. ¿Cuántas especies de escarabajos hay en la
Tierra? ¿Cómo se compara con el número
de especies de mamíferos?

21. ¿Cuántas especies de vertebrados se conocen?

22. Los científicos todavía están descubriendo
nuevas especies. ¿Qué sectores de la gráfica
crees que puedan aumentar? ¿Por qué?

LAS MATEMÁTICAS EN LAS CIENCIAS

Con los datos de la gráfica circular, responde a
las siguientes preguntas:

23. ¿Cuál es el número total de las especies
animales de la Tierra?

24. ¿Cuántas especies de polillas y mariposas
hay?

25. ¿Qué porcentaje del total de especies animales son polillas y mariposas?

26. ¿Qué porcentaje del total de especies animales son vertebrados?

AHORA, ¿qué piensas?

Revisa tus respuestas a las preguntas de
la página 303 que escribiste en el
cuaderno de ciencias. ¿Han cambiado tus
respuestas? Si es necesario, corrige tus
respuestas basándote en lo que has aprendido en este capítulo.

321

NOW What Do You Think?

Possible answers include the following:
1. Animals are consumers, and plants are
   producers.
2. Biological clocks control seasonal migration. Animals also migrate when food is in
   short supply.

Blackline masters of this
Chapter Review can be found
in the **Study Guide.**

CRITICAL THINKING AND PROBLEM SOLVING

16. Yes; even though you were not
    the intended recipient of the
    message, the skunk communi-
    cated with you. The skunk has
    sent a signal (its smell), and you
    have responded (by shutting the
    window).
17. It is a learned behavior. Humans
    are not born with the ability to
    fly, but we have learned to fly
    using airplanes.
18. Ants are much smaller and
    cannot see or hear over great
    distances, so they must depend
    on pheromones and touch for
    communication. Birds have bet-
    ter eyesight, and they can hear
    well and sing, so they can com-
    municate by sound and sight.

INTERPRETING GRAPHICS

19. Beetles have the greatest num-
    ber of species. Students may
    also list the group arthropods,
    which includes beetles, bugs,
    insects, and spiders—all of the
    wedges that are colored yellow.
    On the pie chart, this is shown
    by the largest wedge (or group
    of wedges) and by the label.
20. There are 290,00 species of
    beetles. There are only 4,000
    species of mammals.
21. There are 42,300 vertebrate
    species (4,000 mammals +
    38,300 other vertebrates).
22. Answers will vary, but students
    may suggest that new species
    of insects and beetles will be
    found. Because there are so
    many species in these groups,
    there is a greater chance that
    they have not all been found.

MATH IN SCIENCE

23. 1,032,300
24. 112,000
25. 10.9 percent
26. 4.1 percent

Background

Bats make up the order Chiroptera, which means "hand wing." Bats are considered the best bug control around. For example, 20 million Mexican free-tail bats live in Bracken Cave, Texas. Each night they eat 250 tons of insects. Bats are also responsible for pollinating flowers and dispersing seeds for many plants. In Arizona, nectar-feeding bats pollinate giant cacti, such as the saguaro and organ pipe. In the rain forest, many trees and shrubs depend on bats to pollinate flowers and to disperse their seeds.

There are nearly 1,000 different species of bats. The smallest bat, the bumblebee bat of Thailand *(Craseonycteris thonglongyai),* weighs less than a penny. The largest bat, the Indonesian flying fox *(Pteropus vampyrus),* boasts a wingspan of nearly 1.8 m. Only three bats are considered vampire bats—the common vampire *(Desmodus rotundus),* the hairy-legged vampire *(Diphylla ecaudata),* and the white winged vampire *(Diaemus youngi).* All three are found in Latin America.

VENTANA AL MEDIO AMBIENTE

¡No molestar!

¿Sabías que los murciélagos son los únicos mamíferos que vuelan? A diferencia de muchas aves, la mayoría de las especies de murciélagos del norte y del centro de los Estados Unidos no vuelan hacia el Sur en el invierno. En lugar de migrar, muchas especies de murciélagos hibernan. Pero si su sueño se interrumpe con frecuencia, los murciélagos pueden morir.

Siesta invernal

La mayoría de los murciélagos comen insectos, pero en el invierno los alimentos escasean. A finales del verano, muchos murciélagos de Norteamérica empiezan a almacenar grasa en su cuerpo. Estas reservas de grasa los ayudan a sobrevivir el invierno. Se dirigen: hacia cuevas donde la temperatura durante el invierno es suficientemente baja y estable para hibernar cómodamente (0°C a 9.5°C).

Durante la hibernación sufren un cambio metabólico importante. La temperatura de su cuerpo desciende a casi la misma temperatura de la cueva. La frecuencia de latidos de su corazón, normalmente de unos 400 latidos por minuto, disminuye a unos 25 latidos por minuto.

Gracias a estos cambios, la grasa almacenada dura todo el invierno, a menos que algún visitante los despierte. Si esto ocurre, ¡se pueden morir de hambre!

¡Se prohíbe la entrada!

Incluso con el cambio de su metabolismo, los murciélagos se pueden despertar de repente. Necesitan tomar agua de vez en cuando. A veces, se mueven a lugares más calientes o fríos de la cueva. Por lo general, tienen suficiente grasa almacenada para despertar unas pocas veces en el invierno y volverse a dormir.

Las personas que visitan las cuevas obligan a los murciélagos a despertar de manera innecesaria y a gastar más rápidamente la grasa que almacenaron. Cada vez que despierta, un pequeño murciélago café consume una por-

▲ *Estos murciélagos descansan en una cueva.*

ción de grasa que equivale a 67 días de alimento. Como no hay insectos que comer a su alrededor, no puede volver a llenar su reserva de grasas. La mayoría de las especies de murciélagos que hibernan pueden sobrevivir el invierno después de haber despertado unas tres veces de más, pero las interrupciones frecuentes pueden provocar la muerte de toda una colonia de murciélagos.

Aumenta tu conocimiento

▶ Investiga más sobre los murciélagos. Aprende por qué son útiles para el ambiente y qué amenaza su supervivencia. Discute algunas formas de proteger a los murciélagos y sus hábitats.

322

Answer to Increase Your Knowledge

Answers will vary. Among other services, bats help keep insect populations under control, pollinate some plants, disperse seeds, and fertilize soil. Habitat destruction and human ignorance endanger bats. Solutions might include passing laws to protect bats and educating people about what bats do and don't do.

CANÍBALES

La competencia, la super-vivencia y la reproducción son parte de la vida. En algunas especies, también lo es el *canibalismo* (alimentarse de miembros de una misma especie). Pero ¿cómo se relaciona el canibalismo con la competencia, supervivencia y reproducción? Pues, a veces, la selección de alimento se determina por la transmisión o no de ciertos genes.

▲ *Durante el apareamiento, estas arañas macho se ofrecen como alimentoa sus hembras.*

Consumidores exigentes

Las salamandras tigre empiezan su vida comiendo zooplancton, larvas de insectos acuáticos y algunos renacuajos. Si las condiciones de su pequeño estanque involucran una fuerte competencia con los miembros de su propia especie, ¡algunas de las salamandras más grandes ise vuelven caníbales!

Los científicos no están seguros de por qué las salamandras tigre se vuelven caníbales o por qué se comen animales que se son sus parientes. Tienen la hipótesis que este comportamiento elimina la competencia. Al comerse a otras salamandras, una salamandra tigre reduce la competencia por el alimento y aumenta las oportunidades de sobrevivir. Esto aumenta las oportunidades de que sus genes se transmitan a la otra generación. El no comerse a sus parientes, asegura que los genes que vienen de la misma familia se transmitan a la siguiente generación.

El último sacrificio

Las arañas de espalda roja australianas macho toman otra postura para asegurar que sus genes se transmitan. Durante el apareamiento, la araña macho rueda su cuerpo, hace un giro y mueve su panza cerca de la boca de la hembra, ofreciéndose a ella como alimento. Si tiene hambre la hembra acepta la invitación a cenar. ¡Y casi el 65 por ciento de las veces tiene hambre!

Las arañas macho quieren transmitir sus genes, así que compiten con fiereza por las hembras. Una araña de espalda roja hembra quiere asegurarse de que se fertilice la mayor parte de sus huevos, así que, por lo general, se aparea con dos. Si la hembra se come al primer macho, los estudios muestran que no se aparea con un segundo macho como lo haría si no se hubiera comida al primer pretendiente. Como comerse el macho le toma más tiempo, más huevos son fertilizados por el macho que se volvió parte del banquete. La araña macho que se ofrece como cena ha transmitido más de sus genes a la siguiente generación.

Descúbrelo

▶ Otros animales devoran miembros de su propia especie. Los científicos creen que hay una variedad de razones para este comportamiento. Con el Internet o la biblioteca, investiga el canibalismo en diferentes animales, como la mantis religiosa, el cangrejo azul, el pez espinoso, la araña viuda negra, el renacuajo y el león. Presenta tus descubrimientos a tus compañeros.

323

WEIRD SCIENCE
Animal Cannibals

Answer to On Your Own

Scientists think that cannibalism, a behavior that might appear to be detrimental to a species, may be part of a behavior adaptation called the "lifeboat strategy." Under adverse conditions, such as a scarcity of food, animals are more likely to practice cannibalism not only to feed themselves individually but also to reduce the population searching for food, which reduces competition. A good example is the blue crab, which feeds on soft-shell clams in normal circumstances. When the clams are in short supply, however, the crabs begin to eat one another. This reduces the number of crabs competing for clams and increases the chance that the species as a whole will survive.

In a related behavior, some animals attempt to improve their chances of reproductive success by interfering with the reproductive cycle of their fellow species members. For example, consider a female stickleback fish that attacks the nest of a male and eats the embryos. She is more likely to have greater reproductive success because her own offspring will have less competition for food. When a lion takes over a pride, it kills the cubs so it doesn't spend time raising and protecting young that don't carry its genes. The female lion, on the other hand, will try to protect the cubs in order to protect her genes.

Chapter Organizer

CHAPTER ORGANIZATION	TIME MINUTES	OBJECTIVES	LABS, INVESTIGATIONS, AND DEMONSTRATIONS
Chapter Opener pp. 324–325	45		**Investigate!** Classify It! p. 325
Section 1 Simple Invertebrates	90	▶ Describe the difference between radial and bilateral symmetry. ▶ Describe the function of a coelom. ▶ Explain how sponges are different from other animals. ▶ Describe the differences in the simple nervous systems of the cnidarians and the flatworms.	**Demonstration,** Sponges, p. 328 in ATE **Skill Builder,** Porifera's Porosity, p. 608 **Datasheets for LabBook,** Porifera's Porosity, Datasheet 29
Section 2 Mollusks and Annelid Worms	90	▶ Describe the body parts of a mollusk. ▶ Explain the difference between an open circulatory system and a closed circulatory system. ▶ Describe segmentation.	**Inquiry Labs,** At a Snail's Pace, Lab 5 **Labs You Can Eat,** Here's Looking at You, Squid! Lab 7
Section 3 Arthropods	90	▶ List the four main characteristics of arthropods. ▶ Describe the different body parts of the four kinds of arthropods. ▶ Explain the two types of metamorphosis in insects.	**QuickLab,** Sticky Webs, p. 342 **Discovery Lab,** The Cricket Caper, p. 609 **Datasheets for LabBook,** The Cricket Caper, Datasheet 30
Section 4 Echinoderms	90	▶ Describe three main characteristics of echinoderms. ▶ Describe the water vascular system.	**Long-Term Projects & Research Ideas,** Project 15

TECHNOLOGY RESOURCES

 **Guided Reading Audio CD** English or Spanish, Chapter 14

 **Scientists in Action,** Stopping the Termite Attack, Segment 21

 **One-Stop Planner CD-ROM with Test Generator**

CLASSROOM WORKSHEETS, TRANSPARENCIES, AND RESOURCES	SCIENCE INTEGRATION AND CONNECTIONS	REVIEW AND ASSESSMENT
Directed Reading Worksheet 14 **Science Puzzlers, Twisters & Teasers,** Worksheet 14		
Directed Reading Worksheet 14, Section 1 **Critical Thinking Worksheet 14,** A New Form of Danger in the Deep **Reinforcement Worksheet 14,** Life Without a Backbone	**Multicultural Connection,** p. 326 in ATE **Real-World Connection,** p. 328 in ATE **Environmental Science Connection,** p. 331 **Connect to Physical Science,** p. 331 in ATE **Real-World Connection,** p. 332 in ATE **Weird Science:** Water Bears, p. 352	**Review,** p. 329 **Self-Check,** p. 331 **Homework,** p. 332 in ATE **Review,** p. 333 **Quiz,** p. 333 in ATE **Alternative Assessment,** p. 333 in ATE
Directed Reading Worksheet 14, Section 2	**MathBreak,** Speeding Squid, p. 334 **Multicultural Connection,** p. 335 in ATE **Apply,** p. 337 **Eye on the Environment:** Sizable Squid, p. 353	**Review,** p. 336 **Review,** p. 338 **Quiz,** p. 338 in ATE **Alternative Assessment,** p. 338 in ATE
Directed Reading Worksheet 14, Section 3 **Math Skills for Science Worksheet 5,** Dividing Whole Numbers with Long Division **Math Skills for Science Worksheet 6,** Checking Division with Multiplication **Transparency 56,** Incomplete Metamorphosis **Transparency 57,** Changing Form—Complete Metamorphosis	**Math and More,** p. 340 in ATE **Connect to Chemistry,** p. 341 in ATE	**Self-Check,** p. 341 **Review,** p. 344 **Quiz,** p. 344 in ATE **Alternative Assessment,** p. 344 in ATE
Directed Reading Worksheet 14, Section 4 **Transparency 134,** The Three Groups of Marine Life **Transparency 58,** Water Vascular System **Reinforcement Worksheet 14,** Spineless Variety	**Connect to Earth Science,** p. 346 in ATE	**Review,** p. 347 **Quiz,** p. 347 in ATE **Alternative Assessment,** p. 347 in ATE

Holt, Rinehart and Winston On-line Resources

go.hrw.com

For worksheets and other teaching aids related to this chapter, visit the HRW Web site and type in the keyword: **HSTINV**

National Science Teachers Association

www.scilinks.org

Encourage students to use the *sci*LINKS numbers listed with the Chapter Highlights to access information and resources on the **NSTA** Web site.

END-OF-CHAPTER REVIEW AND ASSESSMENT

Chapter Review in Study Guide

Vocabulary and Notes in Study Guide

Chapter Tests with Performance-Based Assessment, Chapter 14 Test

Chapter Tests with Performance-Based Assessment, Performance-Based Assessment 14

Concept Mapping Transparency 14

Chapter Resources & Worksheets

Visual Resources

TEACHING TRANSPARENCIES

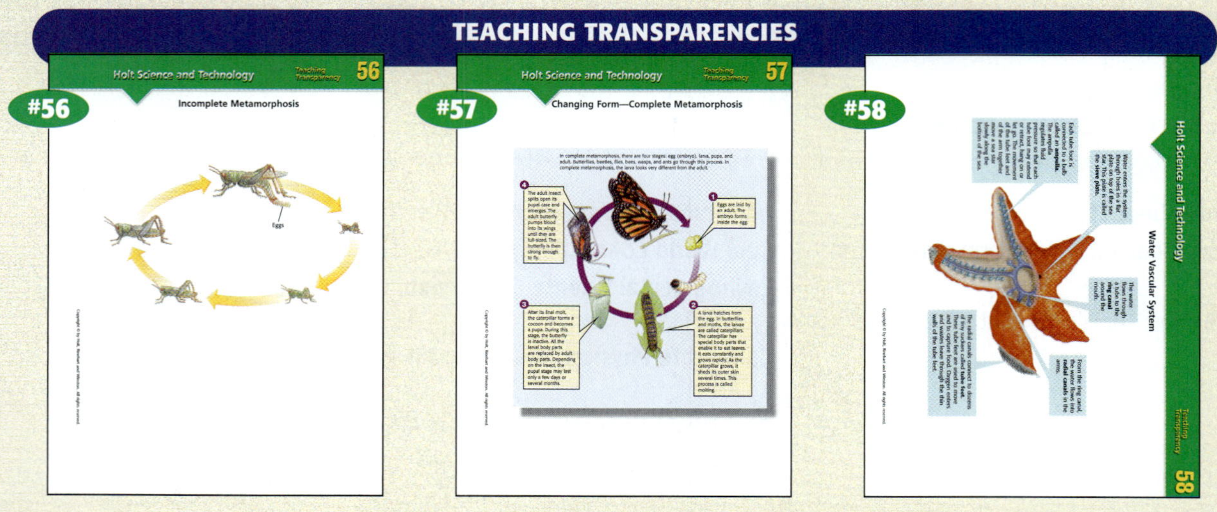

#56 — Incomplete Metamorphosis

#57 — Changing Form—Complete Metamorphosis

#58 — Water Vascular System

TEACHING TRANSPARENCIES

CONCEPT MAPPING TRANSPARENCY

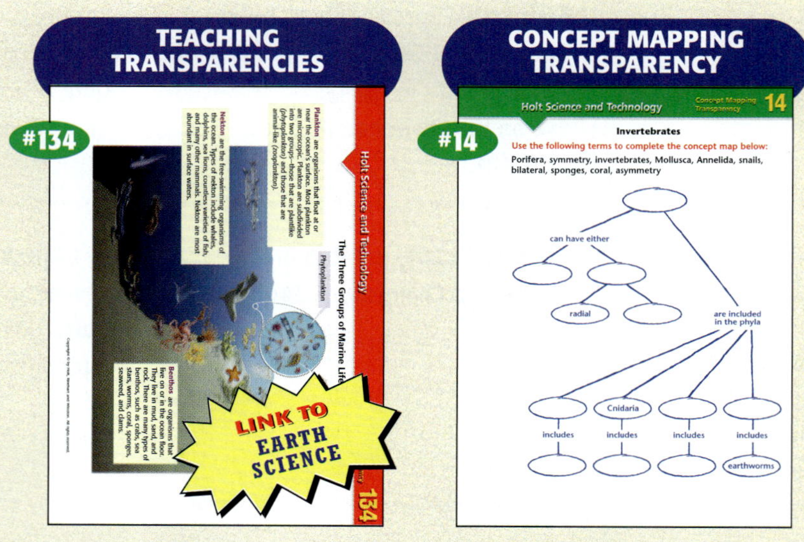

#134 — The Three Groups of Marine Life

#14 — Invertebrates

Meeting Individual Needs

DIRECTED READING

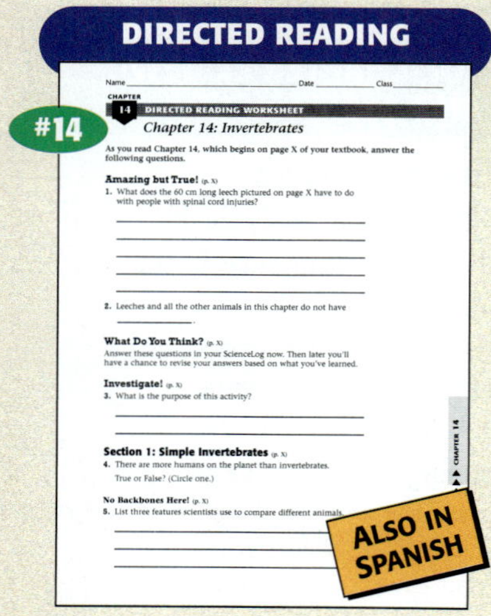

#14

REINFORCEMENT & VOCABULARY REVIEW

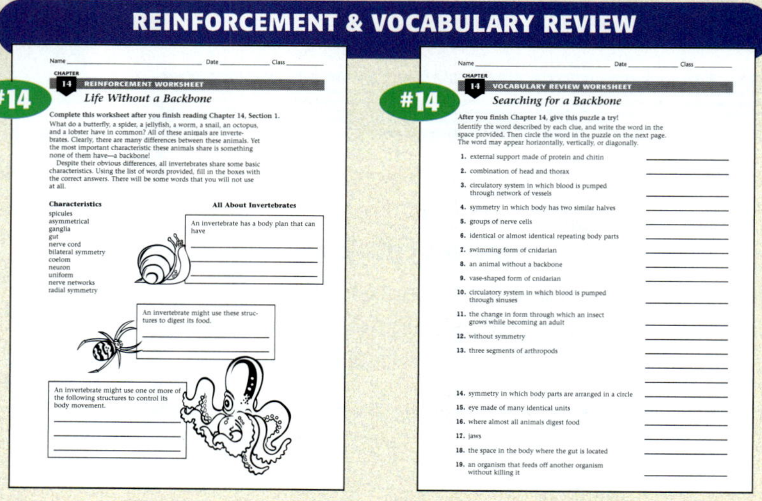

#14 — Life Without a Backbone

#14 — Searching for a Backbone

SCIENCE PUZZLERS, TWISTERS & TEASERS

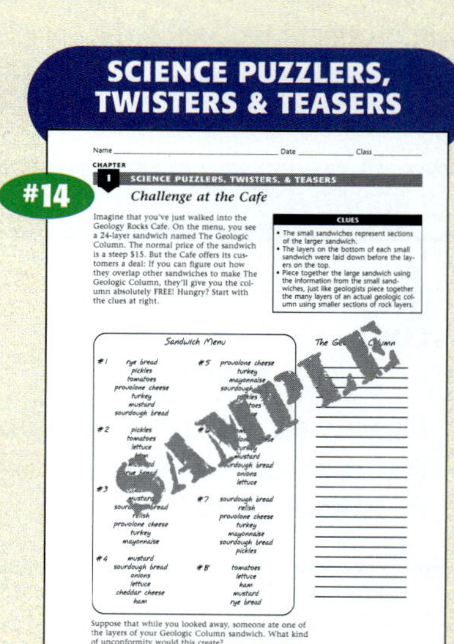

#14 — Challenge at the Cafe

Review & Assessment

STUDY GUIDE

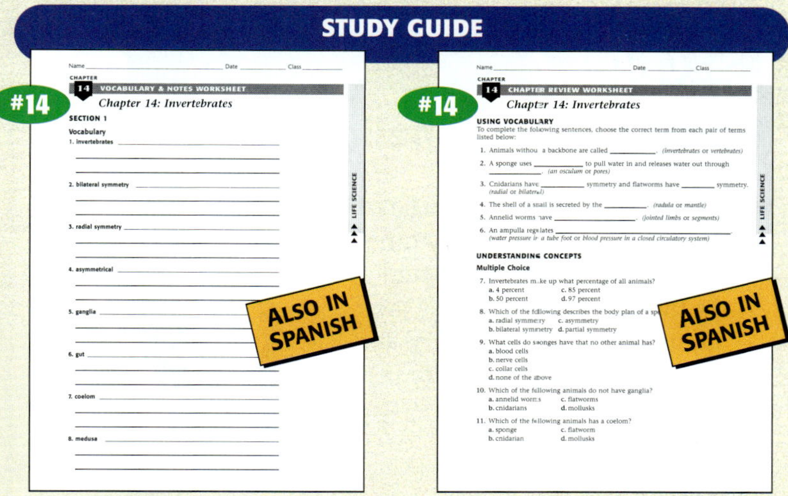

CHAPTER TESTS WITH PERFORMANCE-BASED ASSESSMENT

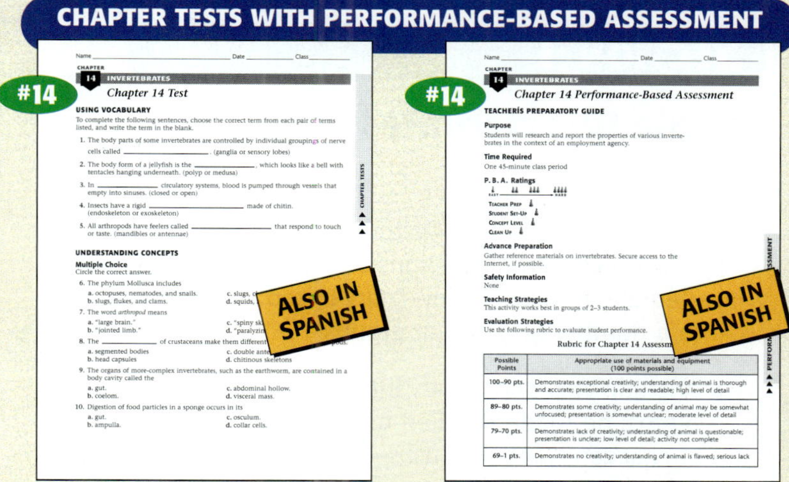

Lab Worksheets

INQUIRY LABS

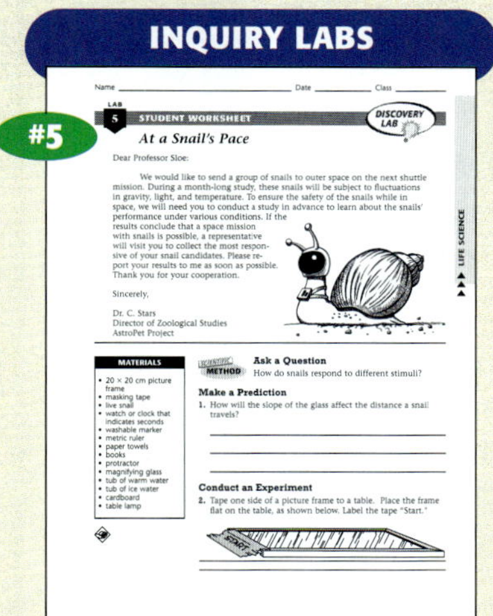

LABS YOU CAN EAT

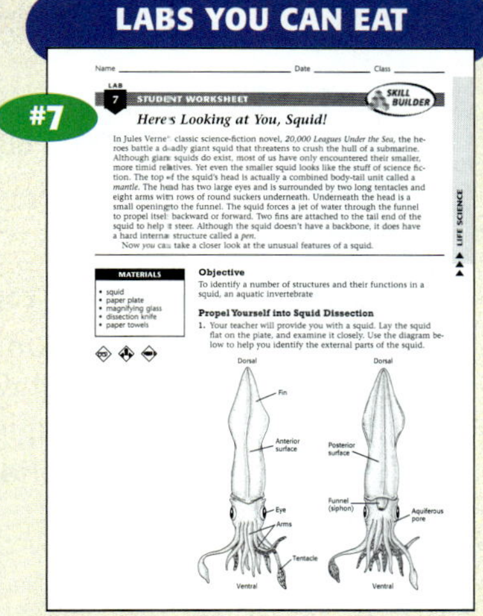

LONG-TERM PROJECTS & RESEARCH IDEAS

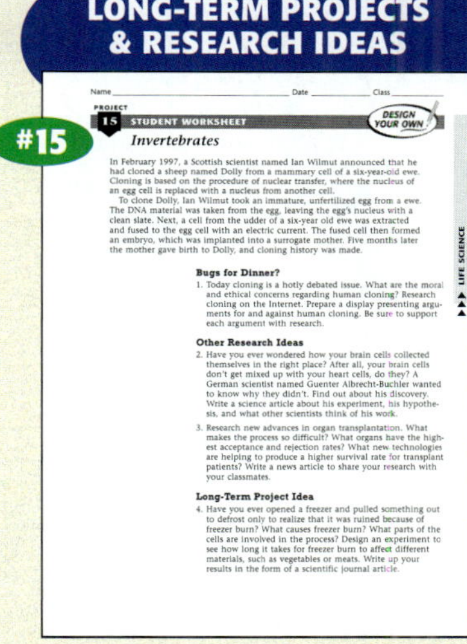

DATASHEETS FOR LABBOOK

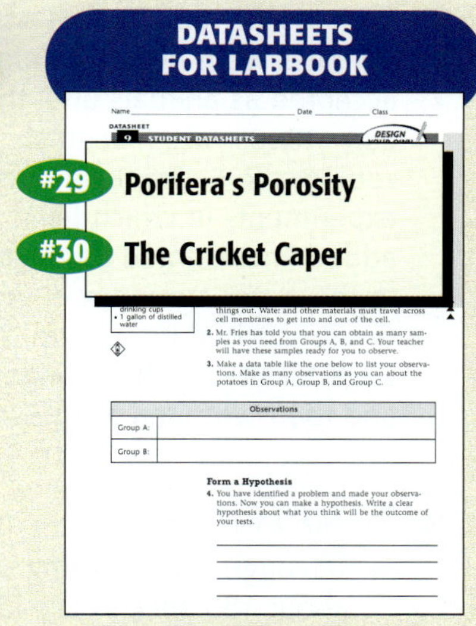

Applications & Extensions

CRITICAL THINKING & PROBLEM SOLVING

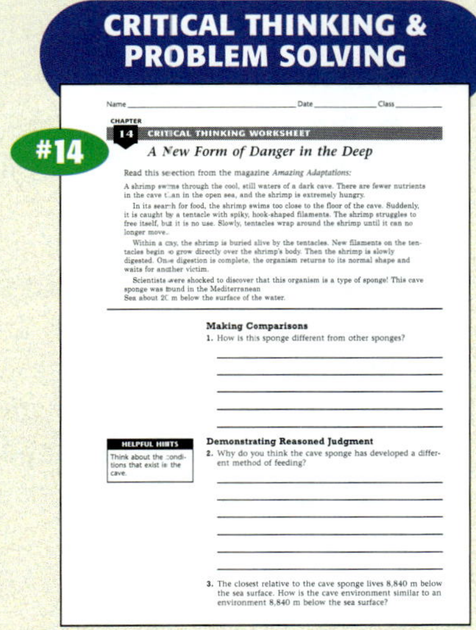

SCIENTISTS IN ACTION

SECTION 1

Simple Invertebrates

▶ **Aristotle**
The Greek philosopher Aristotle (384–322 B.C.) made the first recorded distinctions between invertebrate and vertebrate animals. He even organized invertebrates into a number of different groups.

- Aristotle was particularly interested in marine invertebrates; he made detailed observations of sea stars, crustaceans, and mollusks—especially cuttlefish, which are related to squids.

▶ **Sponges as Animals**
Sponges were not completely accepted as animals until the early 1800s, when Scottish zoologist R. E. Grant conducted experiments in which he added fine, colored particles to the water around sponges. 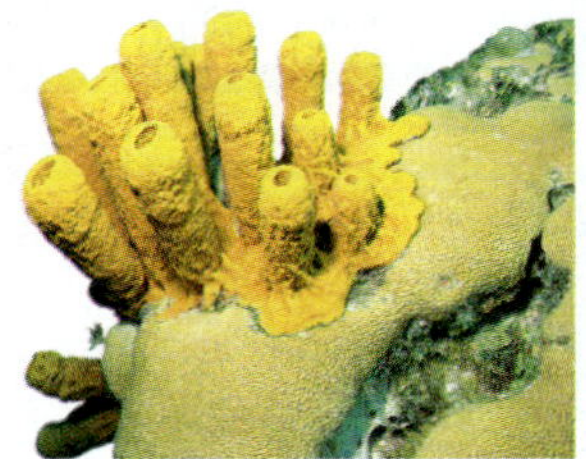 Grant watched under a microscope as the particles were taken into the sponges through microscopic pores and then "vomited forth" from the central cavity.

IS THAT A FACT!

- ◆ Of roughly 5,000 living species of sponges, only about 150 species live in fresh water. All the rest live in marine habitats.

▶ **Medusa**
The tentacle-bearing medusa form of cnidarians gets its name from Medusa, a character in Greek mythology. Medusa was a monster whose long, curly "hair" was made up of snakes.

SECTION 2

Mollusks and Annelid Worms

▶ **Useful, Edible Mollusks**
Mollusks, more than any other invertebrates, are consumed as food by people worldwide. Oysters, mussels, clams, snails, squids, and octopuses are just a few edible mollusks. 

IS THAT A FACT!

- ◆ Giant clams are the largest living bivalves. Shells of the giant clam *Tridacna gigas* can be 1.5 m long and weigh more than 225 kg.

▶ **Earth-Movers**
Charles Darwin (1809–1882) spent many years studying earthworms and calculating their remarkable earth-moving abilities.

- Scientists estimate that the amount of soil brought to the surface by earthworms each year can be as much as 90 metric tons per hectare in temperate regions and considerably more in tropical regions. This activity helps aerate the soil and improve growing conditions for plants.

IS THAT A FACT!

- ◆ Australia is home to the world's longest earthworms, which can exceed 3 m in length.

▶ **Terrestrial Leeches**
Most leeches are aquatic, but tropical rain forests are home to terrestrial leeches. The body heat of mammals attracts these tiny blood-sucking leeches. They will quickly move over vegetation to converge on any unlucky animal that stands in one place for more than a few minutes.

SECTION 3

Arthropods

▶ Diversity

Nearly 1 million species of arthropods have been identified. Scientists estimate that between 2 million and 30 million species are yet to be named. As a group, arthropods are more densely and widely distributed than members of any other animal group.

- Arthropods are found in every imaginable type of environment, from mountain peaks to deep-sea trenches, from equatorial rain forests to polar regions. Some arthropods are adapted for life on land, others for life in the air, and still others in salty, brackish, or fresh water.

▶ Beetles

The order Coleoptera (meaning "sheathed wings"), containing beetles, fireflies, and weevils, is the largest order in the animal kingdom. There are at least 250,000 known species of beetles.

IS THAT A FACT!

- All known species of spiders are predators. Their chelicerae (anterior pair of appendages) end in fangs that inject venom that kills or paralyzes their prey. When spiders bite, they also pump digestive enzymes into their victims. A spider can then suck up the resulting predigested "broth" from its prey.

SECTION 4

Echinoderms

▶ How Sea Stars Feed

The mouth of a sea star is located on the underside of its body. A short esophagus leads to a large stomach that, in many species, can be pushed out, or everted, through the sea star's mouth.

- When a sea star, such as *Asterias,* comes upon a clam, for example, it uses its tube feet and muscular arms to pull the clam's shell apart just enough so that the sea star can push its everted stomach through the opening. The stomach then wraps around the soft parts of the clam's body, and digestion begins.

▶ Class Crinoidea

The fifth class of echinoderms, Crinoidea, is less familiar to most people than are the other classes in the phylum Echinodermata. Crinoids include sea lilies and feather stars.

- Sea lilies have a stalked body topped by feathery arms that are used to snare small plankton from the water. Most sea lilies live in deep water. Feather stars  are colorful, free-moving animals with long, many-branched arms. Feather stars are common inhabitants of coral reefs.

IS THAT A FACT!

- When disturbed, many types of sea cucumbers will expel parts of their internal organs through the anus. This defense mechanism is quite effective in discouraging potential predators. The lost parts are quickly regenerated.

For background information about teaching strategies and issues, refer to the *Professional Reference for Teachers.*

Directed Reading Worksheet 14

Science Puzzlers, Twisters & Teasers Worksheet 14

Guided Reading Audio CD
English or Spanish, Chapter 14

CAPÍTULO

14 Invertebrados

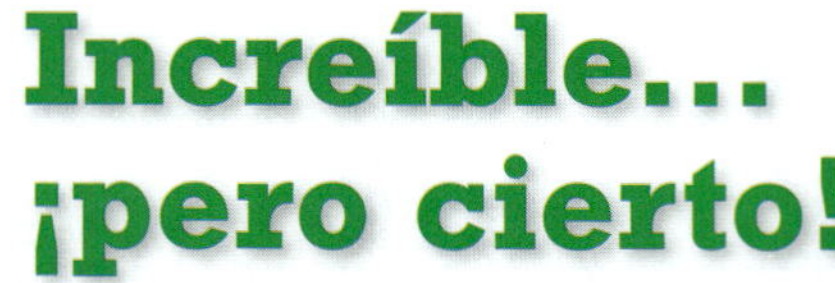

Increíble... ¡pero cierto!

En 1995, unos investigadores de Alemania fabricaron un chip de computadora que podía enviar señales a una célula nerviosa de una sanguijuela viva. Lo más increíble es que la célula podía enviarle señales de regreso al chip. ¿Qué tiene de asombroso tener un "vínculo mental" con una sanguijuela? Tan sólo en los Estados Unidos, ocurren más de 10,000 lesiones de médula espinal por año. La persona afectada puede perder el control de sus músculos, particularmente los de brazos y piernas, y casi nunca hay esperanza de que recupere su capacidad de movimiento. ¿Qué tiene que ver esto con las sanguijuelas?

Las sanguijuelas gigantes de Sudamérica, como la de la izquierda, sólo tienen unas cuantas células nerviosas, pero son iguales a las de otros organismos. Al estudiar los nervios de las sanguijuelas, los biólogos aprenden a comunicarse directamente con las células nerviosas. Esperan que la comunicación con los nervios de las sanguijuelas por medio de un chip de computadora les ayude algún día a comunicarse con células nerviosas humanas. En un futuro, las personas con lesiones de la médula espinal podrán mover los músculos comunicándose con las células nerviosas de su cuerpo mediante una computadora.

Todavía hay mucho que aprender antes de aplicar el trabajo con las sanguijuelas a los seres humanos. ¿Quién hubiera esperado un adelanto de esta magnitud con un animal que ni siquiera tiene columna vertebral? En este capítulo estudiarás los invertebrados, o animales sin columna vertebral.

324

Amazing but True!

Scientists have used many invertebrates in experiments aimed at trying to understand more about nerve cells and the nervous system. More than 50 years ago, biologists used giant axons from squids to learn how nerve impulses travel along nerve-cell processes. More recent experiments with *Aplysia*, a large sea slug, have helped scientists better understand the neurological bases of learning.

Usa tus conocimientos para responder a las siguientes preguntas en tu cuaderno de ciencias:

1. ¿Qué diferencias hay entre las esponjas y los demás invertebrados?

2. ¿Qué semejanzas y diferencias hay entre una persona y un pulpo?

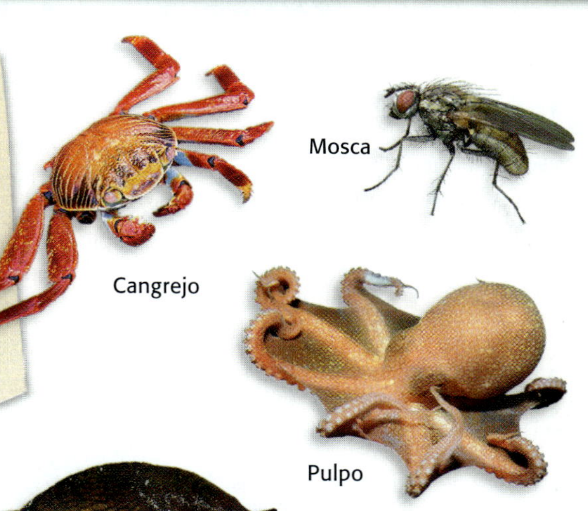

¡Clasifícalo!

El reino animal se divide en varios grupos llamados filos, que a su vez se dividen en subgrupos llamados clases. Al clasificar animales, debemos decidir qué características utilizar. Todas las características del animal, incluyendo las internas, se deben observar con cuidado. En esta actividad demostrarás tu habilidad para clasificar.

Procedimiento

1. Observa las fotografías de esta página. Estos animales están en el mismo grupo porque no tienen columna vertebral.

2. Intenta dividirlos en filos y clases. ¿Cuáles se parecen más? Agrúpalos en el mismo filo. De los animales que están en el mismo filo, determina cuáles se parecen más. Colócalos en clases dentro del mismo filo.

3. Haz un cuadro para mostrar cómo los agrupaste.

Análisis

4. ¿Qué características utilizaste para clasificar estos animales en filos diferentes? Explica por qué piensas que estas características son las más importantes.

5. ¿Qué características utilizaste para clasificar estos animales en clases?

6. ¿Qué nombres descriptivos podrías darles a estos filos y clases?

325

WEIRD SCIENCE

Coral is being used to speed up the regrowth of bone grafts. Coral has a skeletal structure that is remarkably similar to human bone.

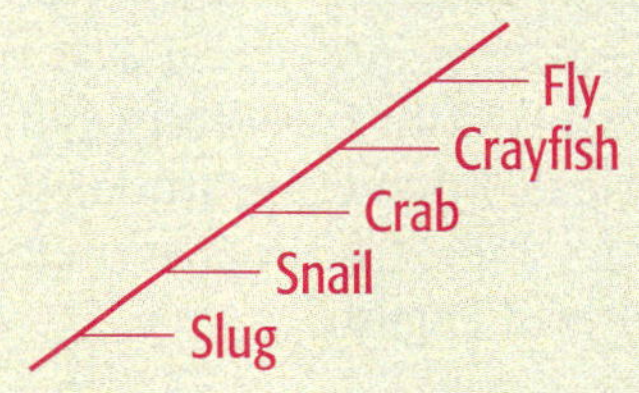

Focus

Simple Invertebrates

This section introduces students to simple invertebrates— sponges, cnidarians, flatworms, and roundworms. Students learn about the different body plans of these animals and important characteristics of their sensory, nervous, and digestive systems.

Bellringer

Pose the following questions to your students:

- What is an invertebrate? (an animal without a backbone)
- What is your favorite invertebrate?
- What special features does this invertebrate have that help it survive in its environment?

Have students write their answers in their ScienceLog.

1) Motivate

ACTIVITY

Determining Symmetry Divide students into cooperative groups of three or four. Distribute to each group copies of simple, top-view drawings of a butterfly and a sea urchin and a small, rectangular hand mirror (mirrors without frames work best). Challenge students to use the mirror to demonstrate that the butterfly is bilaterally symmetrical and that the sea urchin is radially symmetrical. Encourage students to discuss their findings as a class.
Sheltered English

VOCABULARIO

invertebrado	celoma
simetría bilateral	medusa
simetría radial	pólipo
asimétrico	parásito
ganglios	huésped
intestino	

OBJETIVOS

- Describe la diferencia entre la simetría radial y la bilateral.
- Describe la función del celoma.
- Explica las diferencias entre las esponjas y los demás animales.
- Describe las diferencias entre los sistemas nerviosos de los cnidarios y de las tenias.

Invertebrados simples

Los animales sin columna vertebral, o invertebrados, representan el 97 por ciento de todas las especies animales. Hasta ahora, se le ha dado nombre a más de un millón de invertebrados, pero la mayoría de los biólogos creen que todavía hay millones por descubrir.

Escarabajo tigre

¡Sin columna vertebral!

Hay invertebrados de distintas formas y tamaños. Los saltamontes, almejas, lombrices de tierra y medusas son invertebrados. Lo único que tienen en común es que carecen de columna vertebral. Si se examinan varias características, es posible comparar las diferencias y semejanzas entre todos los animales, incluyendo los invertebrados. Por ejemplo, el tipo de organización del cuerpo, la presencia o ausencia de cabeza y la manera como digieren y absorben el alimento.

Mariposa Morpho

Planaria

Camarón de la especie Hymenocera

Planes estructurales Los invertebrados presentan dos planes estructurales básicos o tipos de *simetría*: bilateral o radial. En la siguiente página se muestran los planes estructurales de los animales.

La mayoría de los animales tienen simetría bilateral. El cuerpo de un animal con **simetría bilateral** puede dividirse en dos mitades similares. Por ejemplo, si dibujas una línea imaginaria que divida una hormiga en una mitad derecha y una izquierda, observarás las mismas características en cada lado.

Algunos invertebrados tienen simetría radial. Las partes del cuerpo de un animal con **simetría radial** están dispuestas alrededor de un punto central. Si dibujas una línea imaginaria que atraviese una anémona de mar de arriba hacia abajo, comprobarás que las dos mitades se ven iguales; si la línea la atraviesa en cualquier dirección, las dos mitades seguirán siendo similares. Los invertebrados más sencillos, las esponjas, no tienen simetría. Los animales que no tienen simetría son **asimétricos.**

Multicultural CONNECTION

The silica spicules from many freshwater sponges are very sharp, abrasive, and strong. In Russia, dried freshwater sponges have long been used to polish silver, brass, and other metals. Indians who live along the Amazon River, in South America, add sponge spicules to clay to strengthen the pots they make from it. Have students research other uses of sponge spicules or entire sponges and create a poster based on their findings.

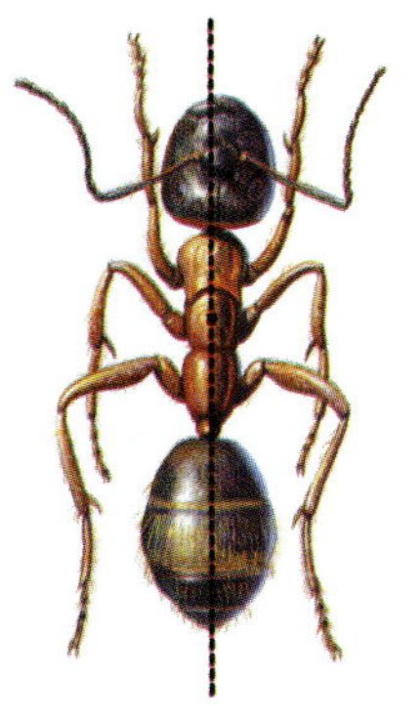

Planes estructurales de los animales

Esta hormiga tiene **simetría bilateral.** Las dos mitades de su cuerpo son idénticas; en cada lado tiene un ojo, una antena y tres patas.

Esta anémona de mar tiene **simetría radial.** El cuerpo de los animales con simetría radial está organizado alrededor del centro, como los radios de una rueda.

Esta esponja es **asimétrica.** No se puede trazar una línea recta que divida su cuerpo en dos mitades iguales.

La cabeza Todos los animales, excepto las esponjas, tienen unas fibras llamadas *nervios* que llevan las señales para controlar los movimientos del cuerpo. Los invertebrados simples tienen nervios dispuestos en redes o cordones a través del cuerpo, pero no tienen cerebro ni cabeza.

En algunos invertebrados, las células nerviosas están agrupadas en **ganglios.** Los ganglios se encuentran en todo el cuerpo y controlan distintas estructuras corporales. La **Figura 1** muestra un ganglio, el cerebro y los cordones nerviosos de una sanguijuela. Los animales más complejos tienen cerebro y cabeza; el cerebro controla los nervios.

El intestino Casi todos los animales digieren el alimento en un intestino central. El **intestino** es un tubo recubierto por células que liberan enzimas, las cuales transforman el alimento en partículas pequeñas que las células absorben. El intestino es parte del tracto digestivo.

Los animales complejos disponen de un espacio para el intestino que se conoce como **celoma** (**Figura 2**), cuya función es permitir que el intestino mueva los alimentos sin interferencia de los movimientos corporales. Además, contiene otros órganos, como el corazón y los pulmones, separados del intestino.

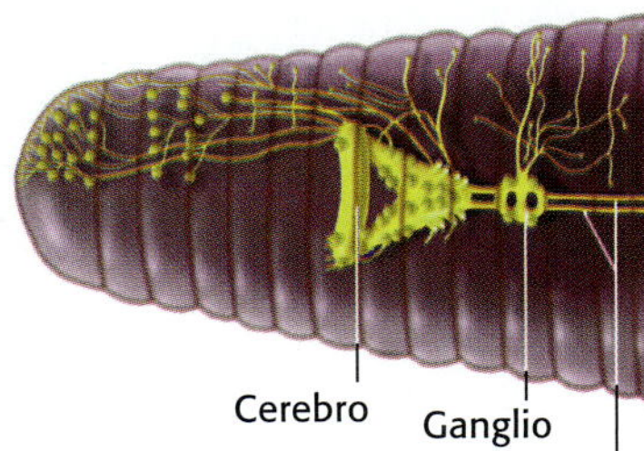

Figura 1 *Las sanguijuelas tienen un cerebro simple y ganglios. Un par de cordones nerviosos conecta el cerebro con los ganglios.*

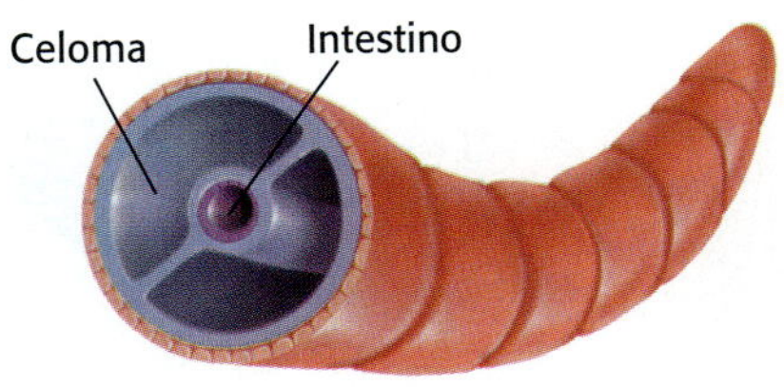

Figura 2 *Celoma de una lombriz de tierra. El intestino y los órganos se encuentran en esta cavidad.*

IS THAT A FACT!

The discovery of a new species of carnivorous sponge in the Mediterranean Sea has amazed marine biologists and confused taxonomists. A typical sponge is a filter feeder, filtering out organic matter from the ocean water. This newly discovered sponge does not eat that way. Investigators discovered that it eats crustaceans by growing thin filaments that cover its prey. The prey is dissolved and digested by the sponge in a matter of days. Now scientists are rethinking the definition of *sponge*. This newly discovered creature resembles a sponge except in its feeding habits.

2 Teach

USING THE FIGURE

Have students study the three invertebrates in the "Animal Body Plans" illustration and note their different habitats. Tell students that being bilaterally symmetrical is an advantage for animals that travel through their environment and that being radially symmetrical is an advantage for animals who live attached to a substrate and whose environment meets them on all sides. Ask students whether they think the organisms pictured in the figure support that statement. Point out to students that there is no way to divide the sponge to get two equal halves. Sheltered English

DISCUSSION

Digestion Have students discuss the advantages of having a central gut specialized for digestion.

Sample answers:

1. Powerful digestive enzymes are isolated from cells and tissues that might be harmed by those enzymes.
2. The breakdown of food particles takes place more rapidly in a centralized gut than it does in individual, unspecialized body cells.
3. Wastes are isolated from other parts of the body.

Directed Reading Worksheet 14 Section 1

REAL-WORLD CONNECTION

Sponges have few predators. The sharp spicules—or tough fibers—in their bodies discourage fish and other aquatic organisms from eating them. Many sponges also produce toxic chemicals that deter predators and keep other sponges from growing too close. A chemical isolated from a Caribbean sponge, *Cryptotethya crypta,* was one of the first marine compounds to be used in chemotherapy. Currently, many other chemical compounds produced by sponges are being tested as anticancer and antiviral drugs.

DEMONSTRATION

Sponges Place a thin, dry slice of a natural sponge under a microscope, and allow students to examine the spongin-fiber network. Next, add a few drops of water to the sponge, and have students examine the slice again. Students should be able to see clearly how the water is taken up by the fibers (the fibers will swell slightly) as well as into the spaces between the fibers.

Sheltered English

Porifera's Porosity PG 608

Esponjas

Las esponjas son los animales más sencillos: carecen de simetría, cabeza, nervios e intestino. Aunque pueden moverse, son tan lentas que su movimiento es muy difícil de apreciar. Antes se pensaba que las esponjas eran plantas, pero las plantas elaboran su alimento a partir de luz solar, agua y dióxido de carbono, mientras que las esponjas tienen que alimentarse de otros organismos. Por eso, se clasifican como animales.

Tipos de esponjas Todas las esponjas viven en el agua (la mayoría en los océanos). Como se observa en la **Figura 3,** tienen colores hermosos y formas diversas. Una de las esponjas más grandes tiene forma de almohada gigante y puede medir 2 m de ancho.

La mayoría de las esponjas tienen un esqueleto formado por astillas llamadas *espículas* (**Figura 4**). Las espículas tienen muchas formas, desde la forma sencilla de aguja hasta las formas complejas de bastón curvo y estrella. El esqueleto sostiene el cuerpo de la esponja y la protege de los depredadores.

De acuerdo con el tipo de espículas, las esponjas se dividen en dos clases. Las esponjas de la clase más extensa tienen espículas de silicato, un material que se usa para hacer vidrio. Las esponjas de baño son similares a las de silicato, pero carecen de espículas. En su lugar, tienen un esqueleto hecho de una proteína llamada *espongina,* que las hace suaves. Otro grupo de esponjas tiene espículas de carbonato de calcio, el material de las conchas de las ostras y de otros mariscos.

Rehacer y reemplazar Si el cuerpo de una esponja se rompe al ser forzado a través de un cedazo, las células separadas volverán a juntarse y reharán la misma esponja. Además, se pueden formar nuevas esponjas a partir de pedazos de otra esponja. A diferencia de la mayoría de los animales, las esponjas pueden reemplazar las partes de su cuerpo, o regenerarse.

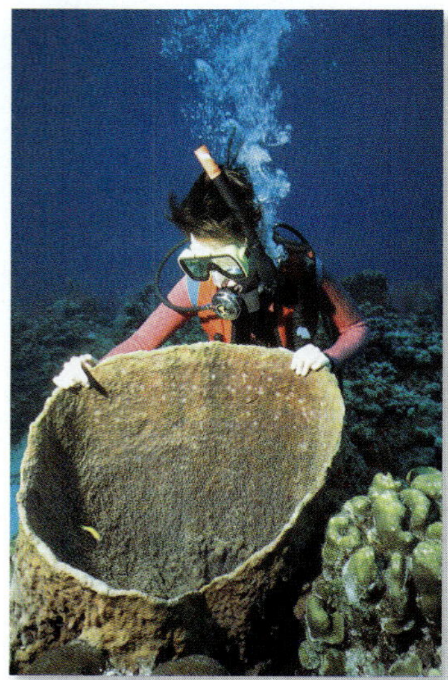
Figura 3 *Las esponjas presentan formas, tamaños y colores variados.*

Esponja gigante

Esponja tubular

Figura 4 *Esqueleto de una esponja vítrea.*

WEIRD SCIENCE

Nearly all sponges are sessile—they live attached to a surface and cannot move from place to place. *Tethya seychellensis,* which lives in the Red Sea, is an exception. Young sponges of this species can move very slowly—about 10 to 15 mm a day—by extending long, sticky projections from their body wall. The projections attach to the substrate and then contract, pulling the sponge forward.

¿Cómo se alimentan las esponjas? Las esponjas pertenecen al filo Porífera, nombre que se refiere a los miles de agujeros o poros que tienen en su exterior. A través de estos poros, las esponjas bombean agua al interior de su cuerpo, donde los coanocitos filtran las partículas de alimento y los microorganismos que están en el agua. El resto del agua fluye hacia una cavidad central y sale por un agujero que está en la parte superior de la esponja, como el humo de una chimenea. Este agujero se llama *ósculo*. La **Figura 5** ilustra este proceso.

Las esponjas no tienen intestino, sino que cada coanocito digiere sus propias partículas de alimento. Ningún otro animal tiene células parecidas a los coanocitos.

Figura 5 *Por medio de los coanocitos, las esponjas filtran el agua para obtener partículas de alimento y después la bombean hacia el exterior por el ósculo. Una esponja puede filtrar hasta 22 L de agua al día.*

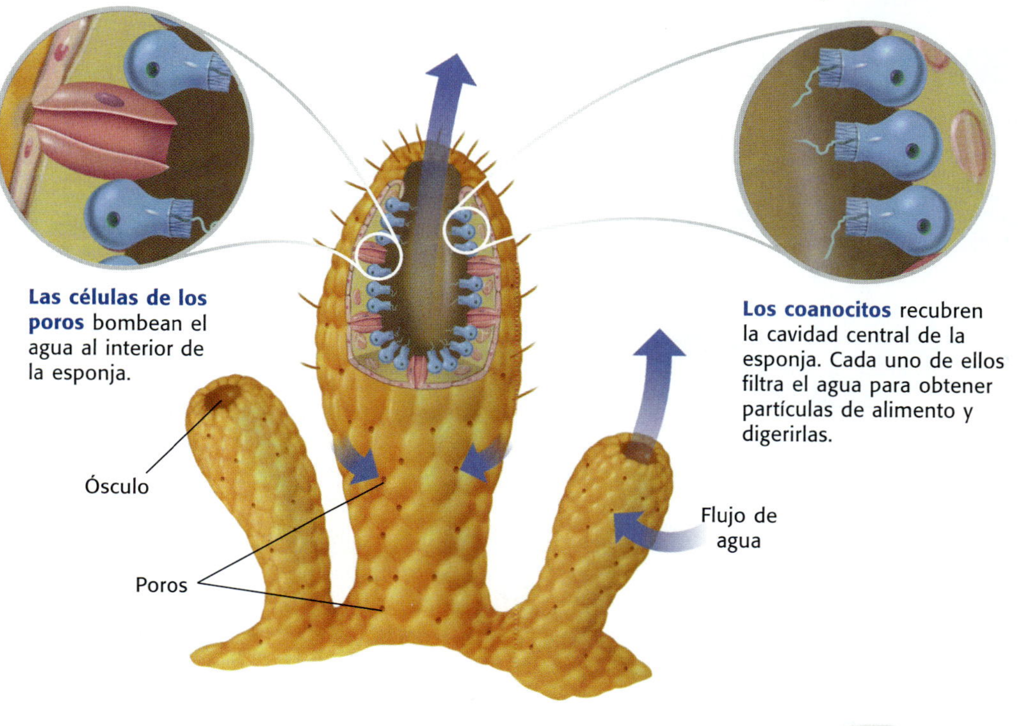

REPASO

1. ¿Por qué los coanocitos son importantes para clasificar las esponjas como animales?

2. ¿Qué es el celoma?

3. **Interpretar gráficas** ¿Qué tipo de simetría tiene el animal que se muestra a la derecha: radial, bilateral o ninguna? Explica tu respuesta.

329

READING STRATEGY

Mnemonics Write the words *medusa* and *polyp* on the chalkboard. Point out that *medusa* contains the letter *d* and that the medusa form of a cnidarian has tentacles that hang *d*own from the animal's body. By the same token, the word *polyp* contains the letter *p*, and the polyp form of a cnidarian has tentacles that project u*p* from the animal's body. Tell students they can use this letter association to remember the different body forms.

GUIDED PRACTICE

Writing Have students research the jellyfish *Aurelia*. Then have students write a story about an *Aurelia* jellyfish in which they describe the jellyfish's life cycle, how it moves, what it eats, and how it senses its environment.

ACTIVITY

Observing Hydras Obtain live hydras and water fleas *(Daphnia)* from a biological supply house. Distribute the hydras in water-filled specimen dishes, and have students work in pairs to study the hydras under a dissecting microscope. Then add several water fleas to the water in each specimen dish, and have students observe how hydras use their nematocyst-equipped tentacles to capture and subdue prey. Encourage students to make drawings of the hydras and to record their observations about how hydras move and manipulate captured prey into their mouths.

Sheltered English

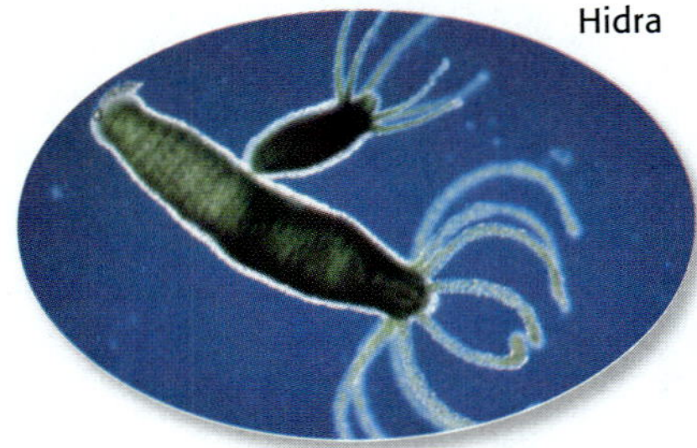

Figura 6 *Estos tres organismos son cnidarios. ¿Por qué están en el mismo filo?*

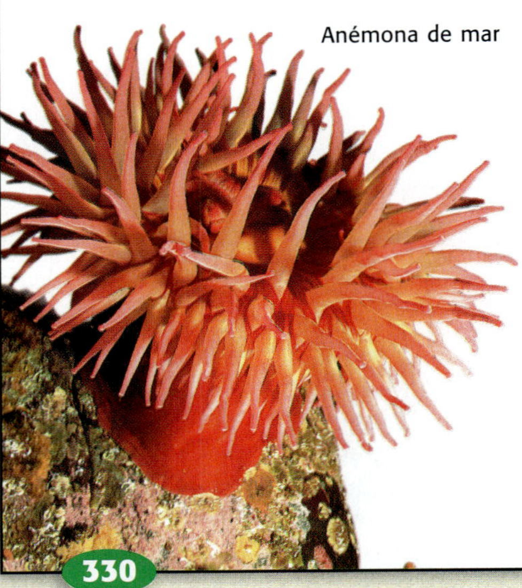

Cnidarios

Observa los organismos de la **Figura 6.** Parecen muy diferentes, pero todos pertenecen al filo Cnidaria.

Cnidaria viene de una palabra griega que significa "ortiga". Las ortigas son plantas que liberan pelos que queman la piel. Los cnidarios hacen lo mismo; todos tienen células urticantes, es decir, que pican. ¿Conoces a alguien que haya sufrido una picadura de medusa? ¡Es una experiencia muy dolorosa!

Los cnidarios son además más complejos que las esponjas, ya que tienen tejidos complejos, intestino para digerir los alimentos y sistema nervioso. Sin embargo, algunas especies comparten una característica con las esponjas: si sus células se separan, pueden juntarse y formar de nuevo el cuerpo.

La medusa y el pólipo Los cnidarios presentan dos formas: medusa y pólipo. Obsérvalas en la **Figura 7.** La **medusa** parece un champiñón con tentáculos que ondean hacia abajo. La aguamala es una medusa muy conocida. A medida que el cuerpo (o sombrilla) de la medusa se contrae y se relaja, el animal nada.

La otra forma del cuerpo de los cnidarios es el **pólipo.** Los pólipos tienen forma de florero y generalmente viven anclados a una superficie.

Algunos cnidarios son pólipos y medusas en distintas etapas de su vida, pero casi todos viven como pólipo.

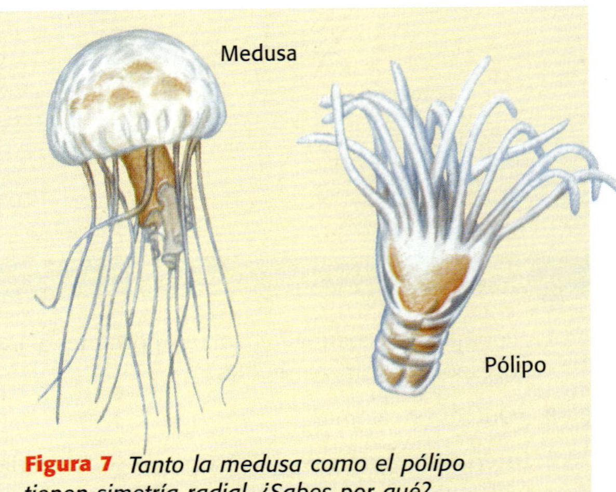

Figura 7 *Tanto la medusa como el pólipo tienen simetría radial. ¿Sabes por qué?*

WEIRD SCIENCE

The tentacles of some cnidarians are capable of stinging even if they are detached from the main body or after the animal itself is dead. *Physalia,* the Portuguese man-of-war, is a colonial cnidarian that frequently washes ashore along temperate and tropical beaches.

Barefoot beachgoers who happen to step on these stranded cnidarians quickly discover that the stinging cells can still deliver a painful sting, even if the animals themselves have been dead for several days.

Tipos de cnidarios

Tipos de cnidarios Hay tres clases de cnidarios: hidras, aguamalas, y anémonas de mar y corales. Las hidras son cnidarios comunes que viven en agua dulce y que pasan toda su vida en forma de pólipo. En cambio, el aguamala pasa la mayor parte de su vida como medusa.

Las anémonas de mar y los corales son pólipos durante toda su vida; parecen flores de colores brillantes. Los corales son cnidarios diminutos que viven en colonias, las cuales construyen enormes esqueletos de carbonato de calcio. Cada nueva generación de corales construye encima de la anterior. A lo largo de miles de años, estos diminutos animales construyen arrecifes enormes bajo el agua. Los arrecifes de coral se encuentran en las cálidas aguas tropicales de todo el mundo.

Para atrapar el almuerzo Todos los cnidarios tienen largos tentáculos cubiertos de células urticantes. Cuando un pez pequeño u otro organismo roza los tentáculos de un cnidario, cientos de células urticantes se descargan y liberan una toxina paralizadora. Cada célula urticante utiliza la presión del agua para descargar un diminuto arpón con espinas, llamado *nematocisto*. La **Figura 8** muestra un nematocisto antes y después de descargarse.

Figura 8 *Cada célula urticante contiene un nematocisto.*

Antes de la descarga Un diminuto arpón con espinas está enrollado dentro de cada célula urticante.

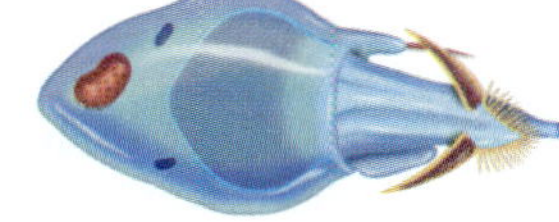

Después de la descarga Cuando el nematocisto se dispara, el largo filamento es expulsado al agua, y las espinas más largas recubren la base del filamento.

Nervios Los cnidarios poseen una red sencilla de células nerviosas, llamada *plexo nervioso*. El plexo nervioso controla los movimientos del cuerpo y de los tentáculos.

Las medusas tienen un *anillo nervioso* en el centro del plexo nervioso. Este anillo de células nerviosas coordina el nado de las aguamalas de la misma manera como nuestra médula espinal coordina la acción de caminar, pero el anillo nervioso no es un cerebro. Los cnidarios no piensan ni planean de la manera como lo hacen animales más complejos.

Los arrecifes de coral, algunos de los cuales tienen más de 2.5 millones de años, son el hogar de una cuarta parte de todas las especies de peces marinos. Desafortunadamente, los arrecifes de coral vivos están amenazados por el exceso de pesca, la minería y los daños accidentales que ocasionan los nadadores y los barcos. Los científicos buscan maneras para protegerlos.

Autoevaluación

Las medusas tienen un anillo nervioso, pero los pólipos no. ¿De qué manera el movimiento de las medusas explica la mayor complejidad de su sistema nervioso? *(Consulta la página 636 para comprobar tu respuesta.)*

Answer to Self-Check

Because medusas swim through the water by contracting their bodies, they must have a nervous system that can control these actions. Polyps move very little, so they don't need as complex a nervous system.

After students have studied **Figure 8,** perform the following demonstration. Invert the fingers of a rubber glove, and use duct tape to attach the top of the glove securely to the head of a water faucet. Tell students that the inverted fingers of the glove represent nematocysts coiled within stinging cells. Scientists have discovered that nematocysts "fire" as a result of a sudden change in water pressure inside stinging cells. When a nematocyst is stimulated to discharge, the nematocyst membrane becomes highly permeable to water. As water rushes in, the sudden increase in water pressure pushes the nematocyst out with explosive force, turning it inside out in the process. Demonstrate this phenomenon by briefly turning on the faucet. The sudden pressure increase will cause the inverted fingers of the rubber glove to evert quickly and forcefully. Sheltered English

BRAIN FOOD

Many jellyfish that capture and eat small fish have transparent or nearly transparent bodies and long, trailing tentacles that are very difficult to see in the water. Ask students to speculate about how these features are an advantage to jellyfish as predators.

3) Extend

REAL-WORLD CONNECTION

Schistosomiasis is an infectious disease caused by blood flukes of the genus *Schistosoma*. About 200 million people are afflicted by schistosomiasis worldwide, primarily in Africa, Latin America, tropical Asia, and the Middle East. Encourage students to research the *Schistosoma* blood fluke and, as a class, to create a bulletin-board display in which they describe and illustrate the fluke's complex life cycle, the symptoms of the disease, and steps that can be taken to reduce the chance of infection.

MISCONCEPTION ALERT

Students may think that tapeworm infections are relatively rare among people living in developed countries. Even in the United States, however, tapeworms from pigs and cattle can infect humans. Researchers estimate that about 1 percent of American cattle are infected with beef tapeworm. About 20 percent of all beef consumed in the United States is not federally inspected; lightly infected beef is frequently missed during inspections. As a result, if a person eats rare roast beef, hamburgers, or steaks, the chance of becoming infected with beef tapeworm is significant.

Critical Thinking Worksheet 14
"A New Form of Danger in the Deep"

Homework

Have students research the life cycle of the roundworm parasite *Trichinella spiralis* and write a persuasive paragraph in their ScienceLog on the importance of cooking pork thoroughly to prevent contracting trichinosis.

IS THAT A FACT!

The adult broad-fish tapeworm, which can infect humans, grows 10 to 20 m in length and may consist of 3,000 to 4,000 sections. A mature fish tapeworm can shed a million eggs a day.

Gusanos planos

Cuando piensas en gusanos, probablemente te vienen a la mente las lombrices de tierra o las que usas para pescar. Pero hay muchos otros tipos de gusanos y la mayoría son demasiado pequeños para verse. Los gusanos más simples son los gusanos planos.

Observa el gusano plano de la **Figura 9.** Los gusanos planos tienen simetría bilateral, a diferencia de los otros invertebrados. Casi todos tienen una cabeza definida y dos manchas oculares grandes que no parpadean. El gusano plano sabe de dónde viene la luz. Asimismo, tiene dos protuberancias con aspecto de orejas a los lados de la cabeza, llamadas *lóbulos sensoriales,* cuya función es detectar el alimento. La cabeza, las manchas oculares y los lóbulos sensoriales son indicios de que tiene un cerebro para procesar la información. En la **Figura 10** se muestra su sistema nervioso.

Figura 9 *Este gusano plano se llama planaria; tiene una cabeza bien desarrollada con ojos y lóbulos sensoriales.*

Tipos de gusanos planos Los gusanos planos se dividen en tres clases. Los gusanos planos a los que nos hemos referido se llaman *planarias*. La mayoría son pequeñas; su longitud es menor que la de una uña. Viven en el agua y en la tierra y casi todas son depredadores. Se alimentan de otros animales y pueden acechar y atacar a animales pequeños, o comérselos ya muertos. Tienen intestino, pero los alimentos y los desechos entran y salen por el mismo lugar.

Los otros dos grupos de gusanos planos son las *duelas* y las *tenias*. En la **Figura 11** se muestra una duela. Estos animales son parásitos, es decir, organismos que se alimentan de otros organismos, generalmente sin matarlos. Las víctimas se llaman huéspedes. Los parásitos pueden vivir dentro o fuera del huésped. La mayoría de las duelas y todas las tenias entran en el cuerpo de otros animales, donde viven y se reproducen. Los huevos fertilizados salen del cuerpo del huésped en sus desechos; si llegan al agua potable o a los alimentos, otro huésped puede ingerirlos, y se desarrollarán en su interior.

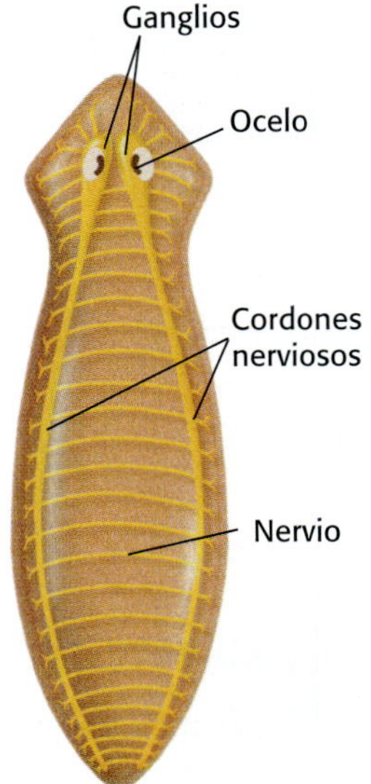

Figura 10 *En esta figura se ilustra el sistema nervioso de un gusano plano. Los dos cordones nerviosos se conectan mediante nervios delgados. Los ganglios de la cabeza constituyen un cerebro primitivo.*

Figura 11 *Las duelas se fijan al huésped con ventosas.*

El aspecto de las duelas y de las tenias es distinto del de las planarias; su cabeza es diminuta y no tienen ocelos ni lóbulos sensoriales, y se fijan al huésped por medio de ventosas y ganchos. Los gusanos planos que viven en el intestino del huésped tienen una piel especial que es resistente a la digestión enzimática del estómago del mismo. Las tenias están tan especializadas que no tienen intestino; simplemente absorben los nutrientes del intestino del huésped. La **Figura 12** muestra una tenia de los peces que se transmite a los humanos.

Gusanos cilíndricos

En un corte transversal, los gusanos cilíndricos o nematodos se ven cilíndricos, largos y delgados. Su simetría es bilateral, como la del resto de los gusanos, y el tamaño de casi todas las especies es muy pequeño. En una manzana podrida en el suelo de una huerta puede haber hasta 100,000 gusanos cilíndricos. Estas diminutas criaturas transforman los tejidos muertos de las plantas y animales, y contribuyen a enriquecer el suelo. La **Figura 13** muestra un gusano cilíndrico.

Los gusanos cilíndricos tienen un sistema nervioso sencillo, como el de los gusanos planos. El anillo de ganglios forma el cerebro primitivo y los cordones paralelos corren a lo largo del cuerpo. Su sistema digestivo es más complejo que el de otros invertebrados simples. A diferencia de los gusanos planos, que ingieren y desechan los alimentos por el mismo orificio, los gusanos cilíndricos tienen boca y ano. La mayoría son parásitos. Entre los que infectan a los seres humanos están los oxiuros y los anquilostomas. La especie *Trichinella spiralis* se transmite a los humanos por la carne de cerdo infectada y es causa de una enfermedad grave, la triquinosis. La cocción adecuada de la carne de cerdo mata los gusanos cilíndricos. Los más peligrosos viven en los trópicos.

REPASO

1. ¿Qué característica les da a los cnidarios su nombre?

2. Menciona dos características de los gusanos planos que los distinguen de los cnidarios.

3. **Analizar relaciones** Tanto los depredadores como los parásitos viven de los tejidos de otros animales. Explica la diferencia entre depredador y parásito.

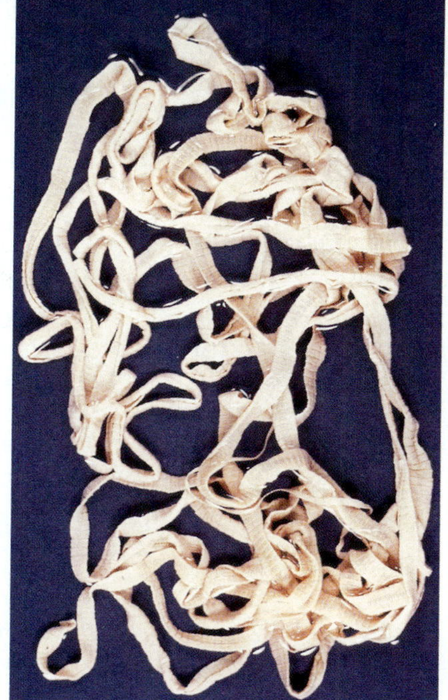

Figura 12 *Las tenias pueden alcanzar un gran tamaño. Algunas llegan a medir 13 m de largo... ¡más que un autobús escolar!*

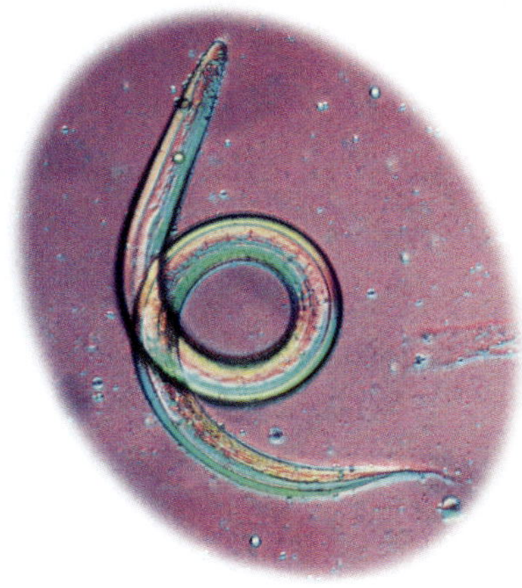

Figura 13 *Los gusanos cilíndricos tienen una cavidad corporal rellena de fluido.*

Focus

Mollusks and Annelid Worms

In this section, students are introduced to three major classes of the phylum Mollusca—gastropods, bivalves, and cephalopods. Students learn about the main parts of a mollusk's soft body, the way they feed, and the diversity in their circulatory and nervous systems. They will also explore annelid worms, including earthworms, bristle worms, and leeches.

🔔 Bellringer

Have students unscramble the following words, and write a sentence using them in their ScienceLog:

gluss (slugs)

isalns (snails)

sdusqi (squids)

klomssul (mollusks)

(Slugs, snails, and squids are all mollusks.)

1 Motivate

GROUP ACTIVITY

Writing Divide students into cooperative groups of four or five. Challenge each group to investigate how people in different countries use mollusks for food. Students can look for recipes on how to prepare and serve snails, clams, squids, and other mollusks. Ask each group to create a menu consisting of an appetizer and a main dish in which mollusks are the primary ingredient.

Answer to MATHBREAK

$$\frac{30 \text{ km/h}}{60 \text{ min/h}} = 0.5 \text{ km/min}$$

VOCABULARIO

sistema circulatorio abierto
sistema circulatorio cerrado
segmento

OBJETIVOS

- Describe las partes del cuerpo de un molusco.
- Explica la diferencia entre un aparato circulatorio abierto y uno cerrado.
- Describe la segmentación.

Figura 14 *Los caracoles, almejas y calamares son moluscos. Los caracoles son gasterópodos; las almejas, bivalvas y los calamares, cefalópodos.*

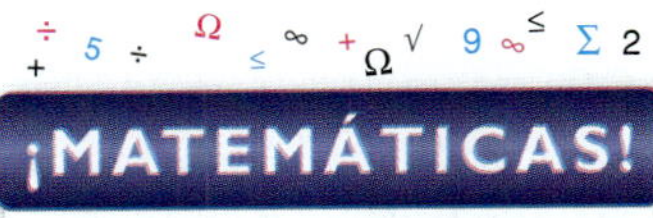

¡MATEMÁTICAS!

Calamar veloz

Si un calamar nada a 30 km/h, ¿qué distancia recorre en un minuto?

Moluscos y anélidos

¿Has probado la crema de almejas o los calamares? ¿Has visto los gusanos que andan por las aceras cuando para de llover? Si es así, ya conoces los invertebrados que estudiaremos en esta sección: los moluscos y los anélidos. Estos invertebrados son más complejos que los que has estudiado hasta ahora. Los moluscos y los anélidos tienen celoma y sistema circulatorio; su sistema nervioso es más complejo que el de los gusanos planos y los gusanos cilíndricos.

Moluscos

El filo Mollusca comprende los caracoles, las babosas, las almejas, las ostras, los calamares y los pulpos. Los moluscos constituyen el segundo filo animal más grande; la mayoría de ellos se encuentran en una de las siguientes clases: *gasterópodos* (babosas y caracoles), *bivalvos* (almejas y otros mariscos con conchas de dos valvas) y *cefalópodos* (calamares y pulpos). En la **Figura 14** se ven algunos moluscos.

La mayoría de los moluscos viven en el océano, pero algunos viven en hábitats de agua dulce; otros, como las babosas y los caracoles, se han adaptado a la vida terrestre.

El grupo de los moluscos abarca desde los caracoles de 1 mm hasta los calamares gigantes, que pueden alcanzar 18 m de largo. La mayoría de los moluscos se mueven lentamente, pero algunos calamares nadan a una velocidad de hasta 40 km/h y saltan a más de 4 m sobre el nivel del agua.

🔬 WEIRD SCIENCE

The blue-ringed octopus, found in the waters of the South Pacific, is deadly. When it is provoked, the blue rings on its skin turn so blue that they almost glow. The saliva in its bite contains a powerful toxin for which there is no known antidote! This toxin paralyzes the victim and shuts down all its life systems. If a person bitten by this octopus arrives at the hospital in time, he or she is put on a respirator for a few days until the toxin wears off.

¿Cómo identificas un molusco? El aspecto de los caracoles, almejas y calamares es bastante diferente, pero al examinarlos más de cerca sus cuerpos son casi iguales. En la **Figura 15** se describen las partes del cuerpo comunes a todos los moluscos.

Figura 15 *El cuerpo de los moluscos es blando y generalmente está cubierto por una concha. Todos los moluscos tienen un pie, una masa visceral y un manto.*

Pie La característica más evidente de los moluscos es un pie muscular ancho, que utilizan para moverse. El pie de los gasterópodos secreta un moco que les ayuda a deslizarse.

Masa visceral Encima del pie, en el celoma del molusco, se encuentra la masa visceral, que contiene las branquias y el intestino, entre otros órganos.

Manto La masa visceral, los lados del pie y la cabeza están cubiertos por una capa de tejido llamada manto, cuya función es proteger el cuerpo de los moluscos que no tienen concha.

Concha En la mayoría de los moluscos, la parte exterior del manto secreta una concha, que los protege de los depredadores y evita la desecación de los moluscos terrestres.

¿Cómo se alimentan los moluscos? Cada tipo de molusco tiene su propia forma de alimentarse. Las almejas y otros bivalvos se fijan al suelo para filtrar el agua a su alrededor y de esta manera obtener diminutas plantas, bacterias y otras partículas. Los caracoles y las babosas se alimentan por medio de la rádula, una especie de lengua en forma de cinta que está cubierta por dientes curvos. En la **Figura 16** se aprecia un detalle de la rádula de una babosa. Con la rádula, las babosas y los caracoles raspan algas de las rocas, trozos de tejido de algas marinas o pedazos de hojas de las plantas. Los caracoles y babosas depredadores tienen grandes dientes en la rádula con los que atacan a su presa. Los caracoles parásitos perforan a sus víctimas de manera muy similar a un mosquito. Con ayuda de sus tentáculos, los pulpos y calamares atrapan a sus presas y las colocan en sus poderosas mandíbulas, de manera similar a como nosotros nos llevamos el alimento a la boca.

Figura 16 *Fíjate en las filas de dientes de la rádula de una babosa. La función de la rádula es raspar el alimento de las superficies.*

335

MEETING INDIVIDUAL NEEDS

Learners Having Difficulty
Visual learners and students with limited English proficiency may have trouble understanding the difference between closed and open circulatory systems. Distribute copies of illustrations showing the closed circulatory system in a human and the open circulatory system in a clam. (A college-level biology textbook is a good source for illustrations of the clam circulatory system.) Encourage students to trace the path that blood takes in a clam. Call attention to the blood sinuses in the clam's body, and tell students that these irregular channels and spaces in the clam's tissues are filled and drained by blood vessels. Contrast this situation with the human circulatory system, in which blood is completely contained in vessels throughout the body. **Sheltered English**

COOPERATIVE LEARNING

Writing Have students work in pairs to research how cephalopods have been used in experimental studies of behavior and nerve function. Students may present their research in the form of a written report or as an oral interview, with one student acting as a magazine reporter who is gathering information for an article and the other student playing the role of a research scientist who has conducted experiments on cephalopods.

internet**connect**

*SCi*LINKS
NSTA

TOPIC: Mollusks
GO TO: www.scilinks.org
*sci*LINKS NUMBER: HSTL365

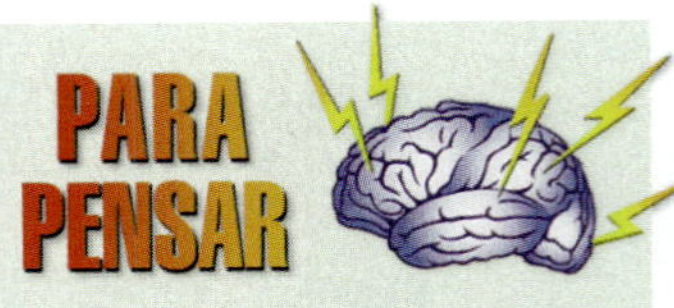

¡Los pulpos tienen tres corazones! Dos de ellos están cerca de las branquias y se llaman corazones branquiales.

Tienen corazón   A diferencia de los invertebrados más sencillos, los moluscos tienen sistema circulatorio. La mayoría de ellos tienen un **sistema circulatorio abierto,** en el que un corazón simple bombea la sangre a través de vasos sanguíneos que se vacían en espacios llamados *senos*. Este tipo de sistema circulatorio es muy diferente al nuestro, que es cerrado. En un **sistema circulatorio cerrado,** el corazón hace circular la sangre a través de una red de vasos sanguíneos que forman un circuito cerrado. El sistema circulatorio de los cefalópodos (calamares y pulpos) también es cerrado, aunque es mucho más sencillo que el nuestro.

¡Es un cerebro!   La mayoría de los moluscos tienen ganglios complejos por todo el cuerpo. Los ganglios realizan distintas funciones; unos controlan la respiración, otros mueven el pie y otros controlan la digestión.

Los cefalópodos, como el de la **Figura 17,** tienen un sistema nervioso más complejo que el del resto de los moluscos. De hecho, el sistema nervioso de los pulpos y calamares es el más avanzado de todos los invertebrados. Poseen un cerebro donde se conectan todos los ganglios. No es sorprendente que estos animales sean los invertebrados más inteligentes. Los pulpos, por ejemplo, pueden aprender a navegar en un laberinto y a distinguir formas y colores. Si se les da ladrillos o piedras, construyen una cueva para esconderse.

Figura 17 *El cerebro de los pulpos es grande. Su función es coordinar el movimiento de los ocho brazos.*

¡Sus ojos son del tamaño de unas pelotas de baloncesto! ¿Qué será? Entérate en la página 353.

REPASO

1. ¿Cuáles son las cuatro partes principales del cuerpo de los moluscos?

2. ¿Qué diferencia hay entre un sistema circulatorio abierto y uno cerrado?

3. **Analizar relaciones** ¿Cuáles son las dos características que los cefalópodos, pero no los demás moluscos, comparten con los humanos?

336

▼ **Answers to Review**

1. foot, visceral mass, mantle, and shell

2. In an open circulatory system, the heart pumps blood through vessels into sinuses. In a closed circulatory system, the blood is pumped through a closed network of vessels.

3. Like humans, cephalopods have a closed circulatory system and a brain.

Anélidos

Tal vez conozcas lombrices de tierra como la de la **Figura 18.** Estas lombrices pertenecen al filo Annelida. Los anélidos también se conocen como gusanos segmentados porque su cuerpo tiene *segmentos*. Los **segmentos** son partes del cuerpo, idénticos o casi idénticos, que se repiten. Obsérvalos en la lombriz de la Figura 18.

Los gusanos segmentados son mucho más complejos que los planos y los cilíndricos que vimos en la sección anterior. Los anélidos tienen un celoma, un sistema circulatorio cerrado, un sistema nervioso formado por ganglios en cada segmento y un cerebro. El cerebro y los ganglios están conectados por un cordón nervioso que corre a lo largo de su cuerpo.

Tipos de anélidos   Hay tres clases de anélidos: las lombrices de tierra, los gusanos con cerdas y las sanguijuelas. Viven en agua salada o dulce y en la tierra; escarban en busca de alimentos, o pueden ser depredadores o parásitos de otros organismos.

Más que una simple carnada   Las lombrices de tierra son los anélidos más comunes. Tienen de 100 a 175 segmentos, casi todos idénticos. Algunos segmentos están especializados para la alimentación o reproducción. Se alimentan de la materia orgánica del suelo y excretan desechos llamados *excrementos,* los cuales aumentan la fertilidad del suelo al proporcionar nutrientes en una forma utilizable por las plantas. También mejoran el suelo al cavar túneles, los cuales permiten que el aire y el agua lleguen a capas más profundas.

En la parte exterior de su cuerpo, las lombrices de tierra tienen cerdas duras que les ayudan a moverse. Las cerdas sostienen una parte del gusano en un lugar mientras la otra parte se abre camino por el suelo.

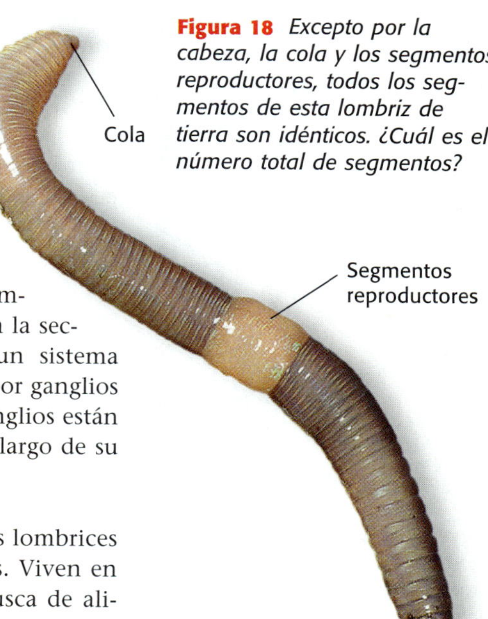

Figura 18 *Excepto por la cabeza, la cola y los segmentos reproductores, todos los segmentos de esta lombriz de tierra son idénticos. ¿Cuál es el número total de segmentos?*

Un amigo tuyo está preocupado porque su jardín está lleno de lombrices de tierra. Quiere encontrar la manera de deshacerse de ellas. ¿Crees que será buena idea? ¿Por qué? Escríbele a tu amigo una carta para explicarle lo que crees que debe hacer.

337

Q: What is worse than biting into an apple and finding a worm?

A: biting into an apple and finding half a worm

IS THAT A FACT!

The large fleshy lobe to the right of the octopus's eyes in **Figure 17** is not the head. It contains the visceral mass.

USING THE FIGURE

Writing   In their ScienceLog, have students compare and contrast the annelid worm shown in **Figure 18** with the flatworms and roundworms pictured in Section 1. Students should note that segmentation is a distinctive characteristic of the phylum Annelida. Tell students that *annelida* comes from a Latin word meaning "little ring."

GUIDED PRACTICE

To demonstrate earthworms' ability to mix soil, have the class work cooperatively to fill the bottom half of a large glass jar with sand and the top half with potting soil. Add enough water to moisten the soil and the sand, and add 5 to 10 large earthworms (available from sporting goods or hardware stores). Punch air holes in the lid and place it securely on the jar. Put the jar in a cool, dimly lit location in the classroom. Add water periodically to keep the soil moist. Encourage students to observe how the earthworms gradually mix the soil and sand during the next few weeks. Sheltered English

Answer to APPLY

It is not a good idea to get rid of the worms in his garden. Earthworms excrete wastes called castings that add nutrients to the soil that plants can use. In addition, they dig tunnels that allow air and water to get into the soil and reach the plants.

338

Quiz

Have students answer the follow-ing questions:

1. What are the three main classes of mollusks? (gas-tropods, bivalves, and cephalopods)

2. How do herbivorous snails and slugs use their radula to obtain food? (These mollusks use their radula to scrape algae off rocks, chunks of tissue from seaweed, or pieces from plant leaves.)

ALTERNATIVE ASSESSMENT

Writing Have students select a mollusk or annelid worm that interests them and research its life cycle, habitat, food, and unique structural or behavioral adaptations. Then ask students to write a rhyming or free-verse poem in their ScienceLog about the inverte-brate based on the information gathered in their research.

Figura 19 *Este gusano se ali-menta filtrando partículas del agua con sus cerdas. ¿Puedes ver sus segmentos?*

La belleza de las cerdas Si hubiera un concurso de belleza para gusanos, los gusanos con cerdas ganarían. Estos gusanos son muy diversos y de colores brillantes. La **Figura 19** muestra uno. Todos viven en el agua; algunos excavan en la arena y en el lodo, comiéndose todas las criaturas que encuentran; otros se arrastran por el fondo y comen moluscos y otros animales pequeños.

Chupadores de sangre y más Algunas sanguijuelas son parásitos que chupan la sangre de otros animales, pero otras se alimentan de animales muertos o son depredadoras de insec-tos, babosas y caracoles.

No todas las sanguijuelas son malas. Hasta el siglo XX, los doctores las usaban en tratamientos que consistían en suje-tarlas a una persona enferma para drenar la sangre "mala" del cuerpo. Esta práctica ya no se acepta, pero las sanguijuelas todavía se usan en medicina. Después de una cirugía, los médi-cos las usan para prevenir la hinchazón cerca de la herida, como se observa en la **Figura 20**. Asimismo, las sanguijuelas elaboran una substancia química que evita la formación de coágulos sanguíneos. Hoy en día, los médicos recetan medica-mentos que contienen esta substancia a pacientes que han sufrido ataques cardíacos, con el fin de evitar que los coágu-los de sangre obstruyan las arterias.

Explora

Consulta la biblioteca o la Internet para investigar el uso de las sanguijuelas en medi-cina. Haz un cartel que describa tus hallazgos.

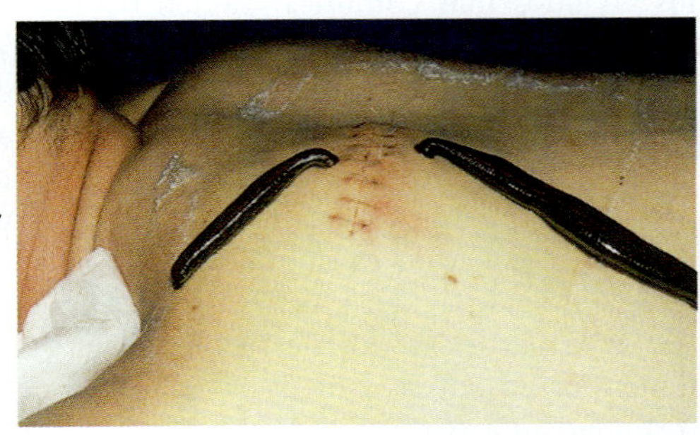

Figura 20
A veces, los doc-tores usan sangui-juelas para reducir la hinchazón después de una cirugía.

REPASO

1. Menciona las tres clases de anélidos. ¿En qué se parecen? ¿En qué se diferencian?

2. **Hacer deducciones** ¿Para qué les sirve a las sanguijue-las tener una substancia química que evita la coagulación de la sangre?

3. **Analizar relaciones** ¿En qué se diferencian los anélidos de los gusanos planos y de los gusanos cilíndricos? ¿Qué características comparten todos los gusanos?

▼ Answers to Review

1. Earthworms, bristle worms, and leeches; all are segmented, and all are annelid worms; earthworms live in soil, and bristle worms live in the water; some leeches and some bristle worms are predators, but all earth-worms are scavengers; only leeches are parasites.

2. Because they feed on the blood of fish and other animals, leeches have to keep the blood flowing (not clotting) from their hosts.

3. Annelid worms are segmented. They have a coelom and a closed circulatory system. All of the worms have bilateral symmetry.

Artrópodos

Viven aquí desde hace cientos de millones de años y están adaptados a casi todos los medios. En un acre de tierra hay millones de ellos. Se conocen con nombres más comunes, como insectos, arañas, cangrejos y ciempiés. Se trata de los artrópodos, el mayor grupo de organismos sobre la Tierra.

El setenta y cinco por ciento de todas las especies animales son artrópodos. La población de seres humanos es de unos de seis mil millones. Los biólogos estiman que la población mundial de artrópodos es de aproximadamente un trillón.

Cangrejo violinista

Características de los artrópodos

Todos los artrópodos comparten cuatro características: apéndices articulados, cuerpo segmentado con partes especializadas, exoesqueleto y sistema nervioso bien desarrollado.

Apéndices articulados Los artrópodos deben su nombre a sus apéndices articulados. *Artro* significa "articulación" y *pod*, "pie" Los apéndices articulados son las patas y otras partes similares que se doblan en las articulaciones, permitiéndoles a los artrópodos moverse fácilmente.

Segmentados y especializados Los artrópodos son *segmentados*, al igual que los anélidos. En algunos, como los ciempiés, todos los segmentos son idénticos y sólo los de la cabeza y la cola son diferentes del resto. La mayoría de las demás especies de artrópodos tienen segmentos con partes muy especializadas, como alas, antenas, branquias, pinzas y uñas. Muchas de estas partes se forman durante el desarrollo del animal, cuando dos o tres segmentos crecen juntos para formar la *cabeza*, el *tórax* y el *abdomen*. Estas partes se señalan en el saltamontes de la **Figura 21.**

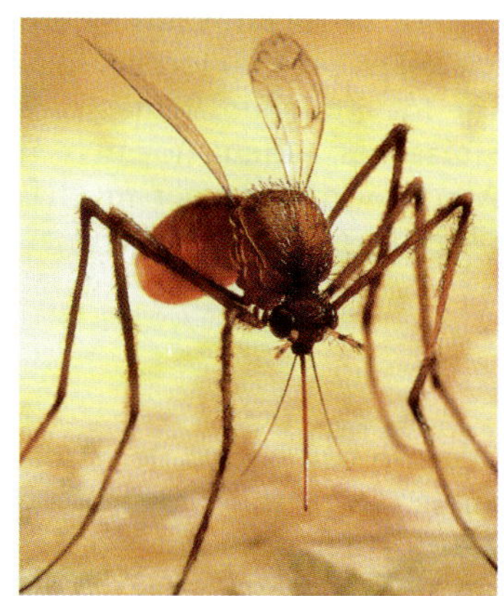
Mosquito

Tarántula

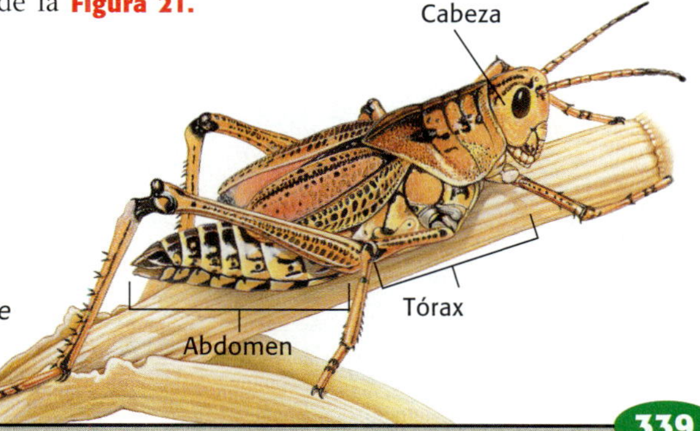

Figura 21 *Los segmentos de este saltamontes se fusionaron durante el desarrollo del embrión para formar la cabeza, el tórax y el abdomen.*

339

WEIRD SCIENCE

Mantis shrimp have powerful forelegs. They use them to smash the shells of prey such as snails and crabs. This "smasher" can deliver a blow with a force equal to that of a small-caliber bullet, which is enough to break through a glass tank!

Directed Reading Worksheet 14 Section 3

Focus

Arthropods

In this section, students learn that arthropods have the following characteristics: jointed limbs, a segmented body, a chitinous exoskeleton, and a well-developed brain and specialized sense organs. Students are introduced to four major groups of arthropods—centipedes and millipedes, crustaceans, arachnids, and insects. Insect bodies and patterns of development are examined in detail.

⏲ Bellringer

Have students pretend that, like a caterpillar, they can undergo metamorphosis and emerge from a cocoon in a new form. Ask students the following questions about their metamorphosis:

• How long will you be inside a cocoon?

• What will you look like when you emerge?

• How will you find food, and what will you eat?

• What physical or behavioral adaptations will you have after metamorphosis that you do not have now?

1 Motivate

DISCUSSION

Characteristics of Arthropods
After introducing the general characteristics of arthropods, have students discuss how these characteristics may have helped arthropods adapt to nearly all environments and to diversify to make up the largest group of animals on Earth.

The compound eyes of insects are made up of tiny bundles of light-sensitive cells called ommatidia (singular: ommatidium). The huge eyes of dragonflies contain about 28,000 ommatidia. The eyes of butterflies contain around 14,000, while those of houseflies have about 4,000. Ask students to calculate roughly how many times more ommatidia dragonflies have than butterflies or houseflies. (Dragonflies have roughly twice as many ommatidia as butterflies and seven times as many as houseflies.)

Then ask students to speculate on the relationship between the number of ommatidia and the ways these three types of arthropods get food. (Possible answer: Dragonflies are fast-flying predators and need acute vision to spot potential prey and maneuver at high speeds; many butterflies feed on flowers and must distinguish between flower types (shapes and color); houseflies rely more on odor detection than vision to find their food, which often consists of dead or stationary organisms.)

Math Skills Worksheet 5 "Dividing Whole Numbers with Long Division"

Math Skills Worksheet 6 "Checking Division with Multiplication"

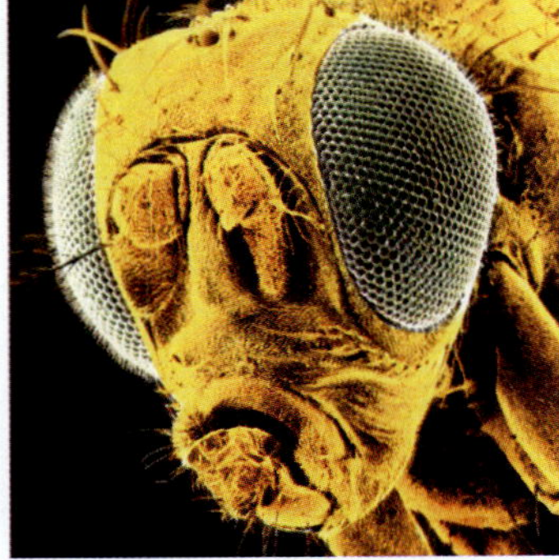

Figura 22 *Observa los ojos compuestos de una mosca de la fruta. Un ojo compuesto está formado por muchas células sensibles a la luz que funcionan juntas.*

Armadura de... ¿quitina? Los artrópodos tienen un exoesqueleto duro, o sea, un esqueleto externo hecho de proteínas y de una substancia especial llamada *quitina*. El exoesqueleto realiza algunas de las funciones de un esqueleto interno; por ejemplo, constituye un soporte rígido para el cuerpo del animal y le permite moverse. Todos los músculos están sujetos a partes distintas del esqueleto; al contraerse mueven el exoesqueleto, que a su vez mueve las partes del animal.

El exoesqueleto también hace cosas que el esqueleto interno no hace tan bien. Funciona como una armadura que protege a los órganos y músculos internos de los artrópodos, y les permite vivir en la tierra sin desecarse.

Son listos Todos los artrópodos tienen cabeza y un cerebro bien desarrollado que coordina la información proveniente de muchos órganos sensoriales, como los ojos y las cerdas del exoesqueleto. Las cerdas detectan movimiento, vibración, presión y substancias químicas. Los ojos de algunos artrópodos son muy sencillos; pueden detectar la luz, pero no forman imágenes. Sin embargo, la mayoría de los artrópodos tienen ojos compuestos que les permiten ver imágenes, aunque no tan bien como nosotros. Un **ojo compuesto** está formado por muchas células sensibles a la luz que son idénticas, como se observa en la **Figura 22.**

Tipos de artrópodos

Los artrópodos se clasifican de acuerdo con el tipo de estructuras corporales que poseen. También es posible distinguirlos contando el número de patas, ojos y antenas que tienen. Las **antenas** funcionan como órganos del tacto, del gusto y del olfato.

Ciempiés y milpiés Los ciempiés y milpiés tienen un solo par de antenas, **mandíbulas** y una cápsula cefálica dura. La manera más fácil de distinguirlos es contar el número de patas por segmento. Los ciempiés tienen un par de patas por segmento, en cambio, los milpiés tienen dos. Observa la **Figura 23;** ¿cuántas patas puedes contar?

Figura 23 *Los ciempiés tienen un par de patas en cada segmento. El número de patas varía entre 30 y 354. Los milpiés tienen dos pares de patas en cada segmento. El número récord de patas de un milpiés es 752.*

340

WEIRD SCIENCE

Did you know that compass termites from the outback of Australia are able to air-condition their mounds? Their towers are up to 2.5 m long and 3 m high but are very narrow and tall. As many as 2 million termites may be living inside. When the nest becomes too hot, worker termites rush to open a valve made of dried mud at the top of the mound. Cooler air enters the nest and sinks to the bottom of the tower. By opening and closing the mud valve in their nest, the termites have complete control over the temperature of their mound!

Crustáceos Los crustáceos incluyen a las cochinillas, camarones, percebes, cangrejos y langostas. Casi todos los crustáceos son acuáticos y tienen branquias para respirar en el agua. Todos poseen mandíbulas y dos pares de antenas. Además, tienen dos ojos compuestos, generalmente en el extremo de pedúnculos. La langosta de la **Figura 24** muestra estas características. Los crustáceos son los únicos artrópodos con dos pares de antenas.

Autoevaluación

¿Qué diferencia hay entre un gusano segmentado y un ciempiés? (Consulta la página 636 para comprobar tu respuesta.)

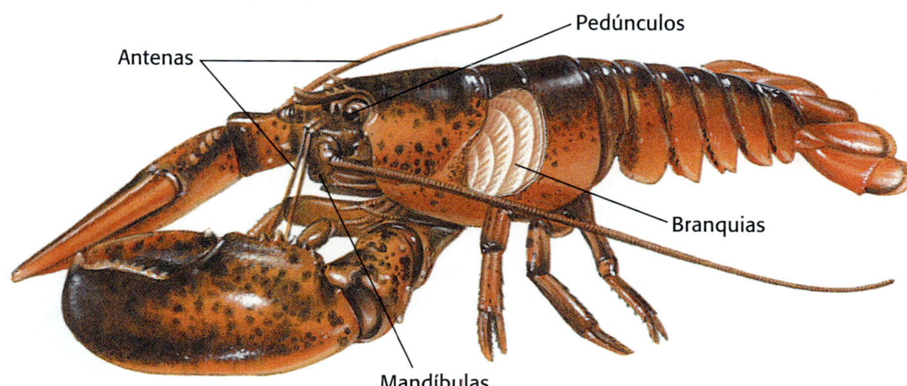

Figura 24 *Las langostas son crustáceos. Observa las branquias, las mandíbulas, dos ojos compuestos en el extremo de los tallos oculares y dos pares de antenas.*

Arácnidos Las arañas, escorpiones, ácaros, pulgas y típulas son arácnidos. La **Figura 24** muestra las partes principales de un arácnido: el cefalotórax y el abdomen. El *cefalotórax* comprende la cabeza y el tórax y, por lo general, tiene cuatro pares de patas. Los arácnidos no tienen antenas ni mandíbulas. En lugar de mandíbulas tienen unas piezas bucales especiales, llamadas quelíceros, como se observa en la **Figura 25.** Algunos quelíceros parecen pinzas o colmillos.

Los ojos de los arácnidos son característicos; a diferencia de los crustáceos e insectos, no tienen ojos compuestos. Las arañas, por ejemplo, tienen ocho *ojos simples* dispuestos en dos filas en la parte frontal de la cabeza. Cuéntalos tú mismo en la **Figura 26.**

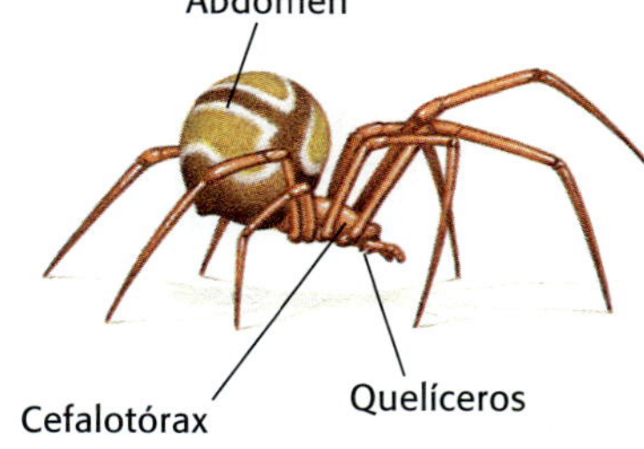

Figura 25 *Los arácnidos, como esta araña, tienen dos principales partes corporales y unas piezas bucales especiales, llamadas quelíceros.*

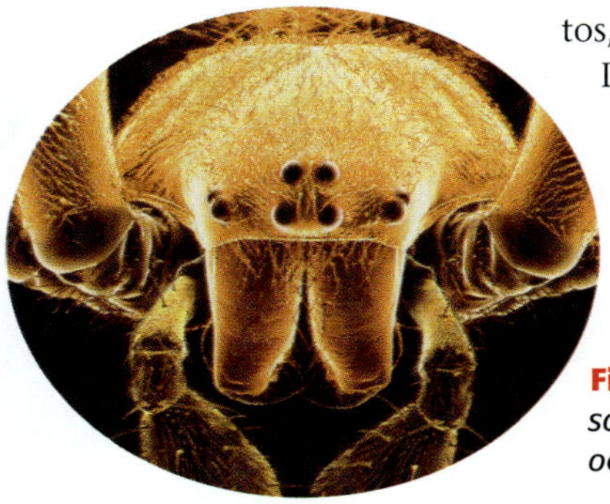

Figura 26 *La mayoría de la las personas saben que las arañas tienen ocho patas. ¡También tienen ocho ojos!*

CONNECT TO CHEMISTRY

Chitin is a strong, flexible, waterproof polysaccharide (a polymer of glucose). Chitin molecules bond readily with proteins, such as those found in the exoskeletons of arthropods. A Japanese textile and fiber manufacturing company has exploited chitin's unique chemical properties to create chitin sutures for surgery and chitin-based artificial skin. Chitin sutures don't have to be removed because they dissolve in the body; they also bind so well with proteins that they may promote healing.

Answer to Self-Check

Segmented worms belong to the phylum Annelida. Centipedes have jointed legs, antennae, and mandibles. Segmented worms have none of these characteristics.

RETEACHING

Writing In their ScienceLog, have students describe a lobster and list the characteristics it exhibits that make it a crustacean.

ACTIVITY

Making Models

MATERIALS
FOR EACH STUDENT:
• modeling clay
• pipe cleaners

It is a common misconception that spiders are insects. Challenge students to disprove this misconception by using modeling clay and pipe cleaners to create models of a spider and an insect. Have students read this section before they begin. Students' models should reflect the fact that spiders have two main body parts, four pairs of legs, and no antennae, whereas insects have three main body parts, three pairs of legs, and one pair of antennae. Sheltered English

GROUP ACTIVITY

Have students work in cooperative groups of three or four to research web-building spiders. Each group should investigate a different spider species. Using string or yarn, have each group create an example of the type of web their chosen species builds. Guide students in a class discussion in which they compare and contrast the different web shapes and designs. Sheltered English

BRAIN FOOD

Trapdoor spiders construct silk-lined burrows topped with a hinged lid that acts like a trapdoor. They lurk beneath the door, waiting for isopods, crickets, or other prey to pass by, then they bound up from below with remarkable speed. If the trapdoor is disturbed, the spiders pull their burrow doors shut with their chelicerae. Using a small spring scale, researchers have found that a trapdoor spider can exert an inward pull on its trapdoor that is 140 times its body weight. Calculate what your pulling strength would be if you could exert a force 140 times your weight.

(Example: 40 kg student × 140 = pulling strength of 5,600 kg)

Quick Lab

MATERIALS

FOR EACH STUDENT:
• transparent tape
• cooking oil

Answers to QuickLab

Fingers will not stick to tape when they have oil on them. Spiders secrete an oily substance in their legs that keeps them from sticking to their webs.

Varios arácnidos muerden o pican, pero los colmillos de las pequeñas arañas de jardín ni siquiera pueden penetrar la piel humana. En los Estados Unidos, sólo hay tres especies de arañas venenosas (la viuda negra y dos especies de *Loxosceles*), cuya picadura puede ocasionarle la muerte a un ser humano.

Las arañas no transmiten enfermedades y son muy útiles. Matan a más plagas de insectos que cualquier otro animal, incluso más que los pájaros. Se comen millones de orugas y otras plagas que destruyen cosechas, y consumen polillas y escarabajos, que destruyen la ropa y las cosechas; también comen moscas y mosquitos que transmiten enfermedades.

Las garrapatas viven en boques, malezas e inclusive en el césped. Las que pican a los seres humanos a veces son portadoras de la enfermedad de Lyme, de la fiebre manchada de las Rocallosas (**Figura 27**) o de la encefalitis. Muchas personas usan pantalones largos y sombreros cuando van a zonas donde hay garrapatas y se revisan después. Por fortuna, la mayoría de la gente que sufre picaduras de garrapatas no se enferma.

Figura 27 Garrapata de perro

Laboratorio

Telarañas pegajosas

Algunas arañas tejen sus telas con seda pegajosa para atrapar a sus presas. ¿Por qué no se quedan pegadas a su propia telaraña? Este experimento te lo explicará. Coloca un pedazo de **cinta adhesiva** en tu escritorio con el pegamento hacia arriba. Este pedazo de cinta adhesiva representa una telaraña y tus dedos representarán un insecto. Sostén el pedazo de las orillas y "camina" por él con tus dedos. ¿Qué pasa? Baña tus dedos en aceite de cocina y "camina" de nuevo. ¿Qué sucede esta vez? ¿Por qué? ¿Cómo podría explicar este experimento por qué las arañas no quedan atrapadas en sus telarañas?

342

Insectos El grupo más extenso de artrópodos es el de los insectos. Si juntáramos todos los insectos del mundo, ¡pesarían más que el resto de los animales juntos! La **Figura 28** ilustra parte de la gran variedad de insectos.

Figura 28 *Estos son algunos ejemplos de la gran diversidad de insectos. ¿Qué tienen en común?*

IS THAT A FACT!

The hind legs of the human flea, *Pulex irritans,* are especially adapted for jumping. How high can a flea jump? *Pulex* can leap 33 cm horizontally and 20 cm vertically—that's equal to an 85 m high jump for a human!

Los insectos están en casi todas partes Los insectos viven en la tierra, en el agua dulce y en las orillas del mar. El único lugar de la Tierra donde no hay insectos es el océano.

El sesenta por ciento de las especies de plantas con flores no pueden reproducirse sin los insectos; la mayoría de estas plantas dependen de insectos como abejas y mariposas para la transferencia del polen de una planta a otra. Los agricultores dependen de los insectos para polinizar cientos de cosechas de frutas, como manzanas, cerezas, tomates y calabazas.

Asimismo, muchos insectos son plagas. Las pulgas, piojos, mosquitos y moscas perforan nuestra piel, chupan nuestra sangre, o bien, transmiten enfermedades. Los insectos que se comen las plantas consumen hasta un tercio de las cosechas de este país, a pesar de la aplicación de pesticidas.

Cuerpo de los insectos El cuerpo de los insectos tiene tres partes: la cabeza, el tórax y el abdomen, como ves en la **Figura 29.** En la cabeza tienen un par de antenas, dos ojos compuestos y tres pares de piezas bucales, incluyendo un par de mandíbulas. El tórax está constituido por tres segmentos, cada uno con un par de patas.

Muchos insectos tienen un par de alas en el segundo y tercer segmento del tórax; algunos no tienen alas y otros tienen dos pares.

Desarrollo de los insectos Al pasar de huevo a la forma adulta, los insectos cambian de forma; este proceso se llama metamorfosis. Hay dos principales tipos de **metamorfosis**: la incompleta y la completa. Los insectos primitivos, como los saltamontes y las cucarachas, sufren metamorfosis incompleta. En este tipo de metamorfosis sólo hay tres estadios: huevo, ninfa y adulto, como se muestra en la **Figura 30.**

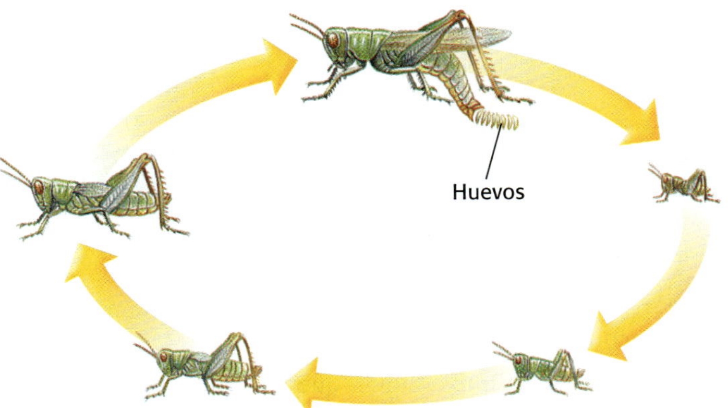

Figura 30 En la metamorfosis incompleta, las larvas (ninfas) parecen adultos pequeños.

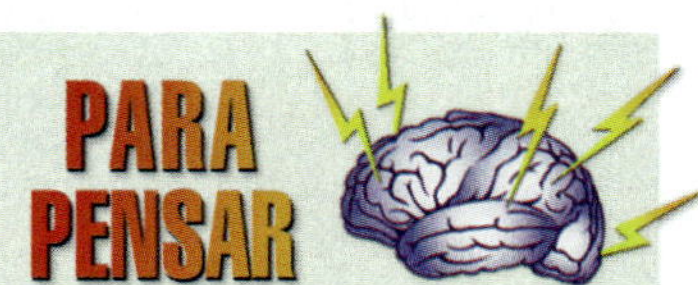

¡Una cucaracha puede vivir una semana sin cabeza! Al final, se muere de sed porque no tiene boca para tomar agua.

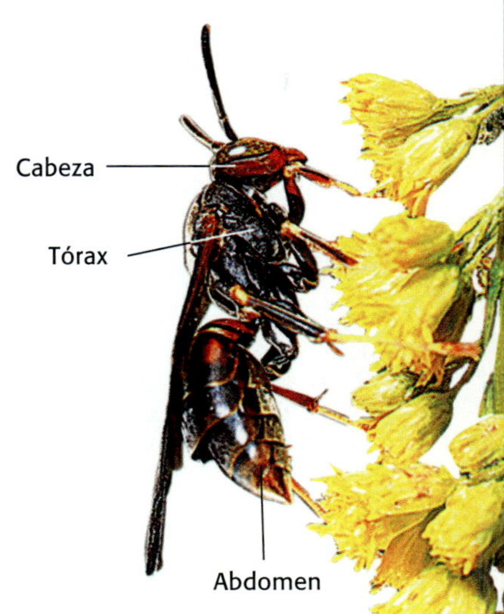

Figura 29 Las avispas tienen las mismas estructuras corporales que los demás insectos.

343

ACTIVITY

Poster Project Insecticides are routinely sprayed on lawns and gardens to kill insect pests. Unfortunately, these chemicals also kill many beneficial insects, persist in the environment, and accumulate in the bodies of animals (including people) higher up the food chain. In recent years, a variety of biological controls for insect pests have been developed that are much more environmentally safe. Have students investigate different biological controls and create a poster on the topic that could be displayed at a local garden center.

The Cricket Caper

GOING FURTHER

Encourage students with limited English proficiency to make flashcards of the vocabulary words in this chapter. You may wish to pair English-proficient students with ESL students for vocabulary practice. In addition, encourage students from other countries to share information about interesting arthropods that they remember seeing in their country. Sheltered English

Science Bloopers

In 1869, the gypsy moth was introduced into the United States in an attempt to breed a better silkworm. The results were disastrous. Some moths escaped, and the species spread throughout the northeastern part of the country. Gypsy moth caterpillars eat the leaves of deciduous trees. In years when there are especially large numbers of caterpillars, millions of acres of forest can be stripped of their leaves.

Quiz

Ask students whether the following statements are true or false:

1. The cephalothorax of a spider consists of both a head and a thorax. (true)

2. The legs of most insects arise from the abdomen. (false)

3. A few types of insects live in the ocean. (false)

4. The three stages of complete metamorphosis are egg, nymph or larva, and adult. (false)

ALTERNATIVE ASSESSMENT

Writing Have students write a narrative in which they describe a walk along a rocky ocean shore or through a tropical rain forest. Have them describe at least a dozen different arthropods that they are likely to encounter. Students should research the two different ecosystems before they begin writing. Some students may wish to create illustrations or collages to accompany their narratives.

Teaching Transparency 57
"Changing Form—Complete Metamorphosis"

Metamorfosis completa: un cambio de forma

En la metamorfosis completa hay cuatro estadios: huevo, larva, pupa y adulto. Las mariposas, escarabajos, moscas, abejas, avispas y hormigas pasan por este proceso. En la metamorfosis completa, la larva tiene un aspecto muy diferente al del adulto.

REPASO

1. Menciona las cuatro clases de artrópodos. ¿En qué se diferencian sus cuerpos?

2. ¿Qué diferencia hay entre la metamorfosis completa y la incompleta?

3. **Aplicar conceptos** Supón que encuentras un artrópodo en una alberca; tiene ojos compuestos, antenas y alas. ¿Será un crustáceo? ¿Por qué?

344

▼ Answers to Review

1. **centipedes and millipedes**—one pair of antennae, mandibles, many segments, head capsules

   crustaceans—mandibles, compound eyes on stalks, two pairs of antennae

   arachnids—two main body parts: cephalothorax and abdomen, four pairs of legs, no antennae, no mandibles, chelicerae

   insects—one pair of antennae, three pairs of legs, head, thorax, abdomen; may or may not have wings

2. incomplete metamorphosis—three stages: egg, nymph, adult; larvae (nymphs) look like smaller adults

   complete metamorphosis—four stages: egg, larva, pupa, adult; larvae look much different than adults

3. No; crustaceans do not have wings.

VOCABULARIO
endoesqueleto
sistema vascular de agua

OBJETIVOS
- Describe tres características principales de los equinodermos.
- Describe el sistema vascular de agua.

Equinodermos

El último de los filos principales de invertebrados es el Echinodermata. Todos los equinodermos son animales marinos, y el grupo incluye estrellas de mar, erizos, lirios de mar, pepinos de mar y dólares de arena. Los equinodermos más pequeños miden sólo unos cuantos milímetros y el más grande es la estrella de mar que mide 1 m.

Los equinodermos viven en el fondo de los océanos de todo el mundo. Algunos se alimentan de ostras y de otros mariscos, otros son carroñeros y otros raspan las algas de las superficies rocosas.

Estrella del género Ophiothrix

Estrella de mar

Lirio de mar

Piel con espinas

La palabra *equinodermo* significa "piel con espinas", pero las espinas no están en la superficie del animal. El cuerpo de estos animales tiene un **endoesqueleto,** que es duro y generalmente cubierto con espinas, similar al de los vertebrados. Las espinas pueden ser protuberancias filosas, como las de muchas estrellas de mar, o largas y puntiagudas, como las de los erizos. Las espinas están cubiertas por la piel externa del animal.

Bilateral o radial?

Los equinodermos tienen simetría radial, pero las estrellas de mar, erizos de mar y dólares de arena, entre otros, se desarrollan a partir de larvas con simetría bilateral. La **Figura 31** muestra una larva de erizo de mar. Ambos lados son similares.

Cuando los embriones de los equinodermos se empiezan a desarrollar, forman una boca de la misma manera que los embriones de los vertebrados. Esta es una de las razones por las que los biólogos piensan que los vertebrados están más relacionados con los equinodermos que con otros invertebrados.

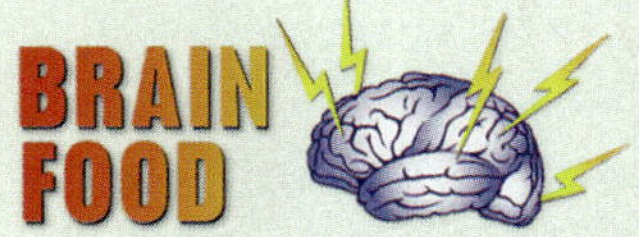

Adulto

Larva

Figura 31 *La larva del erizo de mar tiene simetría bilateral, pero el adulto tiene simetría radial.*

345

SECTION 4

Focus

Echinoderms

In this section, students are introduced to echinoderms—sea stars, brittle stars, sea urchins, sand dollars, and sea cucumbers. Students learn that echinoderms are characterized by an internal skeleton, spiny skin, radially symmetrical adults, bilaterally symmetrical larvae, a simple nervous system, and a water vascular system, which is unique to this phylum.

Bellringer

Pose the following question to your students:

Echinoderms include marine animals such as sea stars, sea urchins, and sea cucumbers. All these organisms are slow-moving bottom dwellers. How do you think they protect themselves from predators?

Have them write their thoughts in their ScienceLog.

1 Motivate

ACTIVITY

Sea Star Hypotheses Display an example of an echinoderm, such as a sea star. Have students draw it in their notebook and write a brief hypothesis describing (1) what it eats, (2) how it moves, (3) where it most likely lives. Discuss before beginning the section. Sheltered English

Directed Reading Worksheet 14 Section 4

BRAIN FOOD

Most adult echinoderms either are sessile (remain in one place) or move slowly over the sea floor. Echinoderm larvae, however, are able to swim, and with the help of ocean currents, they often travel great distances before they settle to the bottom and metamorphose into their adult form. Challenge students to speculate about why it is an advantage for echinoderm larvae to be bilaterally symmetrical and for echinoderm adults to be radially symmetrical.

READING STRATEGY

Prediction Guide Before students read this page, ask the following question:

How does a starfish move from place to place?

a. It curls up its arms and rolls across the sea floor.

b. It uses suction-cup-like tube feet that systematically attach and release to move along.

c. It uses its spines to dig into the sea floor and pull itself forward.

d. With its long arms, a starfish can swim slowly through the water.

(b)

INDEPENDENT PRACTICE

Writing **Concept Mapping** Have students make a concept map in their ScienceLog using the terms that describe echinoderms' physical characteristics and nervous and water vascular systems. Students should connect at least 12 terms, and link them with meaningful phrases. Encourage students to share their concept maps with the class.

CONNECT TO
EARTH SCIENCE

Echinoderms are members of the *benthos,* the organisms that live on the ocean floor. Use the following Teaching Transparency to illustrate the ocean context of echinoderms.

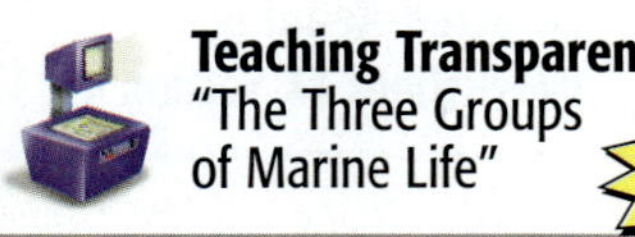

Teaching Transparency 134
"The Three Groups of Marine Life"

El sistema nervioso

Todos los equinodermos tienen un sistema nervioso similar al de las medusas. Alrededor de la boca hay un círculo de fibras nerviosas llamado anillo nervioso. En las estrellas de mar, *un nervio radial* va del anillo nervioso a la punta de cada brazo, como se observa en la **Figura 32.** Los nervios radiales controlan el movimiento de los brazos de la estrella de mar.

En la punta de cada brazo hay un ojo simple, el único órgano sensorial de la estrella de mar. El resto del cuerpo está cubierto por células sensibles al tacto y a señales químicas provenientes del agua.

Figura 32 Sistema nervioso de una estrella de mar.

Sistema vascular de agua

El **sistema vascular de agua** es exclusivo de los equinodermos. Este sistema consiste en la utilización de bombas de agua para que el animal se mueva, coma, respire y detecte estímulos del medio. La **Figura 33** ilustra el sistema vascular de agua de una estrella de mar. Obbserva cómo la presión del agua del sistema se aprovecha para realizar diversas funciones.

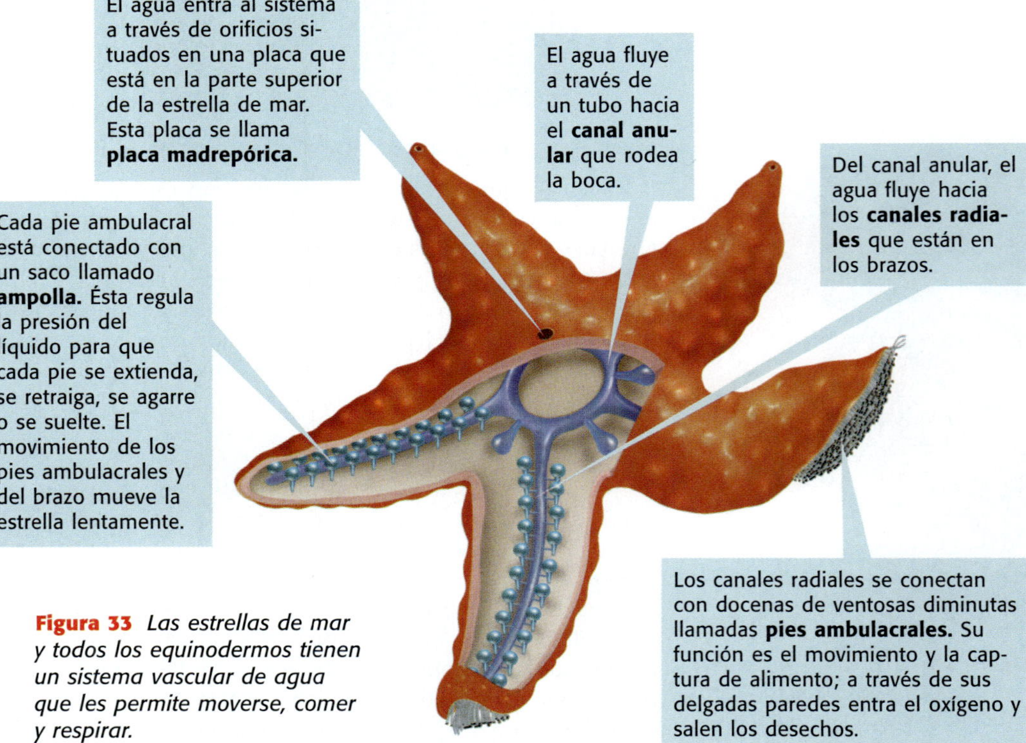

Figura 33 *Las estrellas de mar y todos los equinodermos tienen un sistema vascular de agua que les permite moverse, comer y respirar.*

346

Teaching Transparency 58
"Water Vascular System"

Reinforcement Worksheet 14
"Spineless Variety"

IS THAT A FACT!

Sea stars typically have five arms, but some species may have 40 or even 50 arms!

Tipos de equinodermos

Los equinodermos se dividen en varias clases. Las estrellas de mar son los más populares y conforman una clase, pero hay otras clases menos conocidas.

Estrellas de los géneros Ophiothrix y Astrophyton

Tienen brazos largos y delgados y tienden a ser de menor tamaño que las estrellas de mar. La **Figura 34** ilustra una estrella del género Astrophyton.

Erizos de mar y dólares de arena

Los erizos de mar y los dólares de arena son otra clase de equinodermos. Los miembros de esta clase son redondos y sus esqueletos forman una concha interna sólida. No tienen brazos, pero se mueven con sus pies ambulacrales de la misma manera que las estrellas de mar. Algunos erizos de mar caminan con ayuda de sus espinas. Se alimentan de las algas que raspan de la superficie de las rocas y de otros objetos, y mastican con dientes especiales. Los dólares de arena cavan en la arena suave o en el lodo, como se observa en la **Figura 35,** y comen partículas de alimento que encuentran en la arena.

Pepinos de mar

Al igual que los erizos de mar, los pepinos de mar carecen de extremidades. Su cuerpo es blando y áspero; a diferencia de los erizos de mar, son largos y tienen forma de gusano. Se mueven por medio de pies ambulacrales, como los demás equinodermos. La **Figura 36** muestra un pepino de mar.

Figura 34 Astrophyton muricatum

Figura 35 *Los dólares de arena excavan sus guaridas en la arena.*

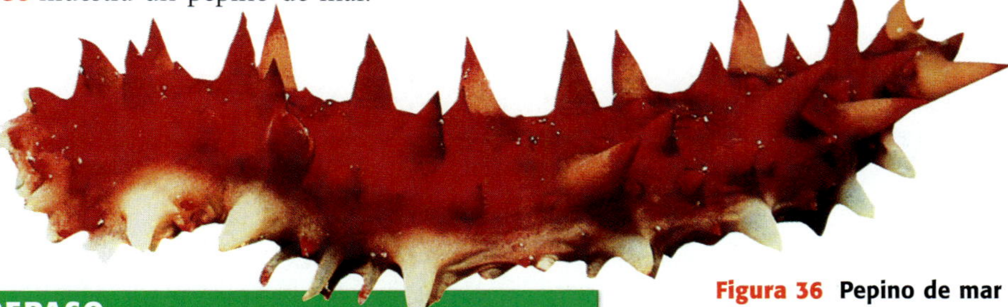

Figura 36 Pepino de mar

REPASO

1. ¿En qué se diferencian los pepinos de mar de los demás equinodermos?

2. ¿Por dónde pasa el agua al fluir por el sistema vascular de agua?

3. **Aplicar conceptos** ¿En qué se diferencian los equinodermos de los demás invertebrados?

347

Chapter Highlights

Section 1

invertebrate an animal without a backbone

bilateral symmetry a body plan in which two halves of an organism's body are mirror images of each other

radial symmetry a body plan in which the parts of a body are arranged in a circle around a central point

asymmetrical without symmetry

ganglia groups of nerve cells

gut the pouch where food is digested in animals

coelom a cavity in the body of some animals where the gut and organs are located

medusa a body form of some cnidarians; resembles a bell with tentacles

polyp a body form of some cnidarians, resembles a vase

parasite an organism that feeds on another living creature, usually without killing it

host an organism on which a parasite lives

Section 2

open circulatory system a circulatory system consisting of a heart that pumps blood through spaces called sinuses

closed circulatory system a circulatory system in which a heart circulates blood through a network of vessels that form a closed loop

segment one of many identical or almost identical repeating body parts

Resumen del capítulo

SECCIÓN 1

Vocabulario

invertebrado (*pág. 326*)
simetría bilateral (*pág. 326*)
simetría radial (*pág. 326*)
asimétrico (*pág. 326*)
ganglios (*pág. 327*)
intestino (*pág. 327*)
celoma (*pág. 327*)
medusa (*pág. 330*)
pólipo (*pág. 330*)
parásito (*pág. 332*)
huésped (*pág. 332*)

Notas de la sección

- Los invertebrados son animales sin columna vertebral.
- La mayoría de los animales tienen simetría radial o bilateral.
- A diferencia de otros animales, las esponjas no tienen simetría.
- El celoma es una cavidad del interior del cuerpo. El intestino está en el celoma.
- Los ganglios son grupos de nervios que controlan las diferentes partes del cuerpo.
- Las esponjas tienen células especiales, llamadas coanocitos, que digieren el alimento.
- Los cnidarios tienen células urticantes para atrapar a sus presas.
- Los cnidarios presentan dos formas corporales, el pólipo y la medusa.
- Las tenias y las duelas son gusanos parásitos planos.

Experimentos

Porosidad de los poríferos (*pág. 608*)

SECCIÓN 2

Vocabulario

sistema circulatorio abierto (*pág. 336*)
sistema circulatorio cerrado (*pág. 336*)
segmento (*pág. 337*)

Notas de la sección

- Todos los moluscos tienen un pie, una masa visceral y un manto. Casi todos también tienen concha.
- Los moluscos y los anélidos tienen celoma y sistema circulatorio.
- En un sistema circulatorio abierto, el corazón bombea sangre a través de los vasos hacia unos espacios llamados senos. En un sistema circulatorio cerrado, la sangre se bombea a través de una red cerrada de vasos.
- Los segmentos son partes corporales, idénticas o casi idénticas, que se repiten.

☑ Comprobar destrezas

Conceptos de matemáticas

VELOCIDAD Y DISTANCIA Si un caracol se mueve a 30 cm/h, ¿qué tan lejos puede llegar en un minuto? En una hora hay 60 minutos:

$$\frac{30 \text{ cm}}{60 \text{ min}} = 0.5 \text{ cm/min}$$

En un minuto el caracol se moverá 0.5 cm.

Comprensión visual

METAMORFOSIS Algunos insectos sufren metamorfosis incompleta, y otros metamorfosis completa. Observa las ilustraciones de las páginas 343 y 344 para identificar la diferencia entre estos dos tipos de metamorfosis.

348

Lab and Activity Highlights

Porifera's Porosity `PG 608`

The Cricket Caper `PG 609`

Datasheets for LabBook (blackline masters for these labs)

SECCIÓN 3

Vocabulario

exoesqueleto *(pág. 340)*

ojo compuesto *(pág. 340)*

antenas *(pág. 340)*

mandíbula *(pág. 340)*

metamorfosis *(pág. 343)*

Notas de la sección

- El setenta y cinco por ciento de todos los animales son artrópodos.

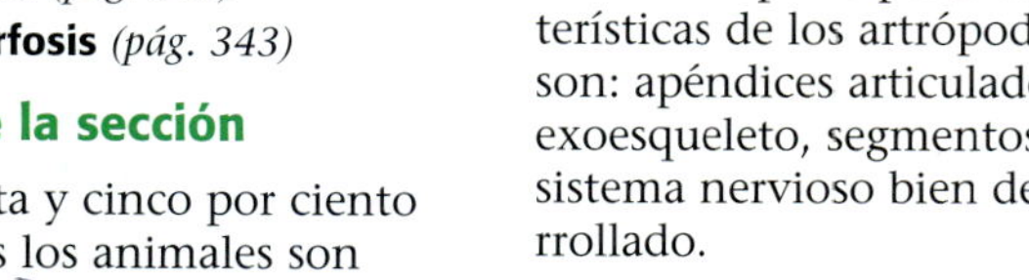

- Las cuatro principales características de los artrópodos son: apéndices articulados, exoesqueleto, segmentos y sistema nervioso bien desarrollado.

- Los artrópodos se clasifican según el tipo de estructura corporal.

- Las cuatro clases de artrópodos son: ciempiés y milpiés, crustáceos, arácnidos e insectos.

- Los insectos pueden sufrir metamorfosis completa o incompleta.

Experimentos

El grillo saltarín *(pág. 609)*

SECCIÓN 4

Vocabulario

endoesqueleto *(pág. 345)*

sistema vascular de agua *(pág. 346)*

Notas de la sección

- Los equinodermos son animales marinos que tienen un endoesqueleto y un sistema vascular de agua.

- La mayoría de los equinodermos presentan simetría bilateral cuando son larvas y radial cuando son adultos.

- El sistema vascular de agua de los equinodermos les permite moverse por medio de pies ambulacrales que funcionan como ventosas.

- Los equinodermos tienen un sistema nervioso simple constituido por un anillo nervioso y nervios radiales.

SECTION 3

exoskeleton an external skeleton made of protein and chitin found on arthropods

compound eye an eye that is made of many identical units, or eyes, that work together

antennae feelers on an arthropod's head that respond to touch or taste

mandible a jaw found on some arthropods

metamorphosis a process in which an insect or other animal changes form as it develops from an embryo or larva to an adult

SECTION 4

endoskeleton an internal skeleton

water vascular system a system of water pumps and canals found in all echinoderms that allows them to move, eat, and breathe

Vocabulary Review Worksheet 14

Blackline masters of these Chapter Highlights can be found in the **Study Guide.**

 **internet**

go.hrw.com

VISITA: go.hrw.com

Visita el sitio web de HRW para encontrar una serie de herramientas de aprendizaje relacionadas con este capítulo. Sólo tienes que escribir la palabra clave:

PALABRA CLAVE: HSTINV

SCI LINKS ₛₘ
NSTA

VISITA: www.scilinks.org

Visita el sitio web de la **Asociación Nacional de Maestros de Ciencias** *(National Science Teachers Association)* para encontrar recursos de Internet relacionados con este capítulo. Sólo escribe el **ENLACE DE CIENCIAS** para obtener más información sobre el tema:

TEMA	ENLACE
TEMA: Esponjas	**ENLACE:** HSTL355
TEMA: Gusanos cilíndricos	**ENLACE:** HSTL360
TEMA: Moluscos	**ENLACE:** HSTL365
TEMA: Artrópodos	**ENLACE:** HSTL370

349

Lab and Activity Highlights

LabBank

Labs You Can Eat, Here's Looking at You, Squid! Lab 7

Inquiry Labs, At a Snail's Pace, Lab 5

Long-Term Projects & Research Ideas, Project 15

Repaso del capítulo

UTILIZAR EL VOCABULARIO

Escoge el término correcto para completar las siguientes oraciones:

1. Los animales sin columna vertebral se llaman _____. *(invertebrados* o *vertebrados)*

2. Mediante el_____ las esponjas absorben agua y la liberan a través de _____. *(ósculo* o *poros)*

3. Los cnidarios tienen simetría_____ y los gusanos planos tienen simetría _____. *(radial* o *bilateral)*

4. La concha de los caracoles es segregada por la/el _____. *(rádula* o *manto)*

5. Los gusanos anélidos tienen _____. *(apéndices articulados* o *segmentos)*

6. La ampolla regula la _____. *(presión del agua en un pie ambulacral* o *presión de la sangre en un sistema circulatorio cerrado)*

COMPRENDER CONCEPTOS

Opción múltiple

7. ¿Qué porcentaje de los animales representan los invertebrados?
   - **a.** 4 por ciento
   - **b.** 50 por ciento
   - **c.** 85 por ciento
   - **d.** 97 por ciento

8. De las siguientes características, ¿cuáles describen a las esponjas?
   - **a.** simetría radial
   - **b.** simetría bilateral
   - **c.** asimetría
   - **d.** simetría parcial

9. ¿Qué células son exclusivas de las esponjas?
   - **a.** glóbulos rojos
   - **b.** células nerviosas
   - **c.** coanocitos
   - **d.** ninguna de las anteriores

10. ¿Qué animal no tiene ganglios?
    - **a.** anélido
    - **b.** cnidario
    - **c.** gusano plano
    - **d.** molusco

11. ¿Qué animal tiene celoma?
    - **a.** esponja
    - **b.** cnidario
    - **c.** gusano plano
    - **d.** molusco

12. Las tenias y las sanguijuelas son
    - **a.** anélidos.
    - **b.** parásitos.
    - **c.** gusanos planos.
    - **d.** depredadores.

13. Algunos artrópodos no tienen
    - **a.** apéndices articulados.
    - **b.** exoesqueleto.
    - **c.** antenas.
    - **d.** segmentos.

14. Los equinodermos viven en
    - **a.** la tierra.
    - **b.** agua dulce.
    - **c.** agua salada.
    - **d.** todas las anteriores

15. *Equinodermo* significa
    - **a.** "apéndice articulado."
    - **b.** "piel con espinas".
    - **c.** "endoesqueleto".
    - **d.** "pie ambulacral"

16. Las larvas de equinodermo
    - **a.** tienen simetría radial.
    - **b.** tienen simetría bilateral.
    - **c.** no tienen simetría.
    - **d.** tienen simetría radial y bilateral.

Respuesta corta

17. ¿Qué es el intestino?

18. ¿Qué diferencias hay entre los arácnidos y los insectos?

19. ¿Qué filo animal agrupa el mayor número de especies?

20. ¿Cómo se mueve un equinodermo?

Organizar conceptos

21. Usa los siguientes términos para crear un mapa de ideas: insecto, esponjas, anémona de mar, invertebrados, arácnido, pepino de mar, crustáceo, ciempiés, cnidarios, artrópodos, equinodermos.

RAZONAMIENTO CRÍTICO Y RESOLUCIÓN DE PROBLEMAS

Escribe una o dos oraciones para responder a las siguientes preguntas:

22. Imagínate que descubres un animal extraño que tiene simetría bilateral, celoma y nervios. ¿Se podría clasificar en el filo de los cnidarios? ¿Por qué?

23. A diferencia de otros moluscos, los cefalópodos se mueven rápidamente. Basándote en lo que sabes sobre las partes del cuerpo de los moluscos, ¿por qué crees que tienen esta capacidad?

24. Los gusanos cilíndricos, los planos y los anélidos pertenecen a distintos filos. ¿Por qué no se agrupan todos en el mismo?

LAS MATEMÁTICAS EN LAS CIENCIAS

25. Si 75 por ciento de los animales son artrópodos y 40 por ciento de los artrópodos son escarabajos, ¿qué porcentaje de todos los animales son escarabajos?

INTERPRETAR GRÁFICAS

El árbol evolutivo siguiente muestra las posibles relaciones entre los distintos filos de animales. El "tronco" del árbol está a la izquierda. Con ayuda de este árbol responde a las siguientes preguntas:

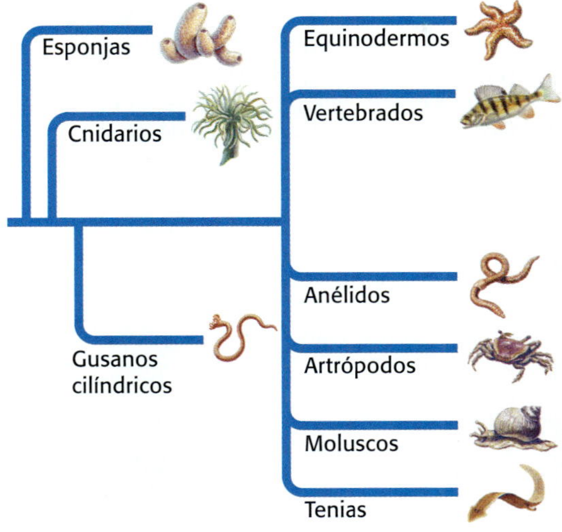

26. ¿Cuál es el filo más antiguo?

27. ¿Con qué están más relacionados los moluscos, con los gusanos cilíndricos o con los planos?

28. ¿Qué filo está más relacionado con los vertebrados?

AHORA, ¿qué piensas?

Revisa tus respuestas a las preguntas de la página 325 que escribiste en el cuaderno de ciencias. ¿Han cambiado tus respuestas? Si es necesario, corrige tus respuestas basándote en lo que has aprendido en este capítulo.

351

NOW WHAT DO YOU THINK?

1. Sponges have collar cells and asymmetry. They have no nerves, no head, and no gut.

2. Humans are vertebrates, and octopuses are invertebrates. Both humans and octopuses have a brain and a closed circulatory system.

Concept Mapping Transparency 14

Blackline masters of this Chapter Review can be found in the **Study Guide.**

Short Answer

17. a pouch lined with cells that release enzymes to digest food

18. Arachnids have two body parts— a cephalothorax and abdomen— and eight legs. Insects have three body parts—a head, a thorax, an abdomen—and six legs. Arachnids have chelicerae. Insects have antennae.

19. arthropods

20. with tube feet and the water vascular system

Concept Mapping

21. An answer to this exercise can be found at the end of this book.

CRITICAL THINKING AND PROBLEM SOLVING

22. No; cnidarians have radial symmetry and do not have coeloms.

23. Because cephalopods do not have a shell, they need to move rapidly to avoid predators. They can move rapidly because they have a large brain.

24. All of the worms are not grouped in the same phylum because they have different characteristics. Annelid worms are segmented; the others are not. Annelid worms also have a coelom and a closed circulatory system. Roundworms have a round cross section, and flatworms do not. Roundworms have a more complex digestive system than flatworms. Flatworms are the only worms with eyespots and sensory lobes.

MATH IN SCIENCE

25. $0.75 \times 0.40 = 0.30 = 30$ percent

INTERPRETING GRAPHICS

26. the sponges

27. flatworms

28. the echinoderms

Background

Tardigrades are also known as moss piglets and can be placed in history by the fact that they were some of the first organisms Antony van Leeuwenhoek examined with his spectacular invention, the microscope.

This is of interest especially because these creatures are not well represented in the fossil record. And after Leeuwenhoek's initial study, tardigrades were largely ignored for the next three centuries, in large part because of their lack of economic importance.

Answer to On Your Own

Architect Eugene Tsui was inspired by the sturdiness of the tardigrade to build what he calls "the world's safest house."

The shape of Ojo de Sol, Tsui's house in Berkeley, California, mimics that of the water bear. He says he designed the house to be oval because "nature doesn't build boxes. They're among the weakest forms because stress is concentrated in corners and along flat surfaces, vulnerable to collapse."

Tsui hoped to design a house to withstand the earthquakes and wildfires, disasters that Northern California is prone to. Fires skirt curved walls instead of penetrating them, and the arches on the upper floor are wood, steel, and concrete. The structure might be safe from some natural events, but in regard to water bears, the question would be, "Can it resist drought?"

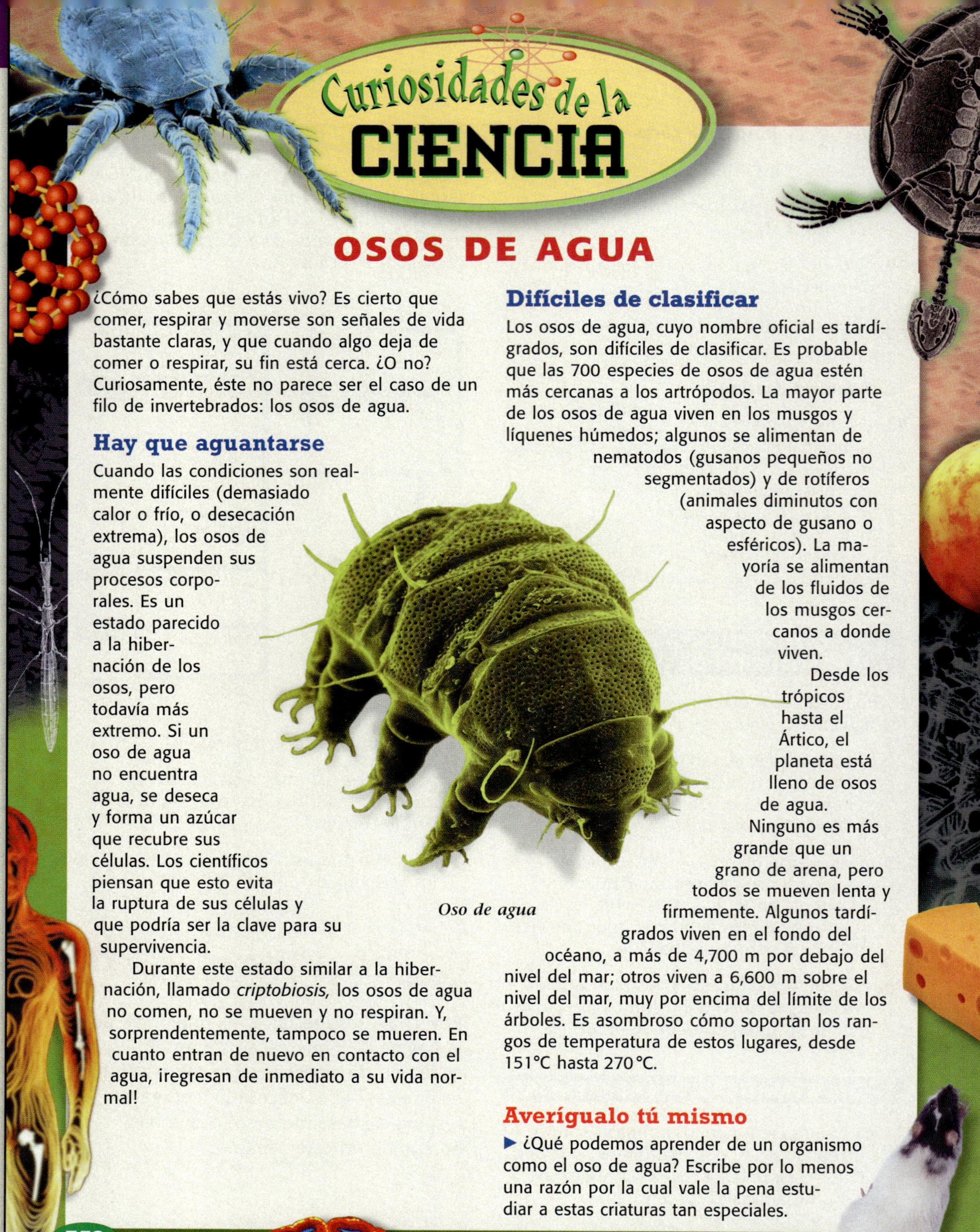

OSOS DE AGUA

¿Cómo sabes que estás vivo? Es cierto que comer, respirar y moverse son señales de vida bastante claras, y que cuando algo deja de comer o respirar, su fin está cerca. ¿O no? Curiosamente, éste no parece ser el caso de un filo de invertebrados: los osos de agua.

Hay que aguantarse

Cuando las condiciones son realmente difíciles (demasiado calor o frío, o desecación extrema), los osos de agua suspenden sus procesos corporales. Es un estado parecido a la hibernación de los osos, pero todavía más extremo. Si un oso de agua no encuentra agua, se deseca y forma un azúcar que recubre sus células. Los científicos piensan que esto evita la ruptura de sus células y que podría ser la clave para su supervivencia.

Durante este estado similar a la hibernación, llamado *criptobiosis,* los osos de agua no comen, no se mueven y no respiran. Y, sorprendentemente, tampoco se mueren. En cuanto entran de nuevo en contacto con el agua, ¡regresan de inmediato a su vida normal!

Oso de agua

Difíciles de clasificar

Los osos de agua, cuyo nombre oficial es tardígrados, son difíciles de clasificar. Es probable que las 700 especies de osos de agua estén más cercanas a los artrópodos. La mayor parte de los osos de agua viven en los musgos y líquenes húmedos; algunos se alimentan de nematodos (gusanos pequeños no segmentados) y de rotíferos (animales diminutos con aspecto de gusano o esféricos). La mayoría se alimentan de los fluidos de los musgos cercanos a donde viven.

Desde los trópicos hasta el Ártico, el planeta está lleno de osos de agua. Ninguno es más grande que un grano de arena, pero todos se mueven lenta y firmemente. Algunos tardígrados viven en el fondo del océano, a más de 4,700 m por debajo del nivel del mar; otros viven a 6,600 m sobre el nivel del mar, muy por encima del límite de los árboles. Es asombroso cómo soportan los rangos de temperatura de estos lugares, desde 151°C hasta 270°C.

Averígualo tú mismo

▶ ¿Qué podemos aprender de un organismo como el oso de agua? Escribe por lo menos una razón por la cual vale la pena estudiar a estas criaturas tan especiales.

352

VENTANA AL MEDIO AMBIENTE

Calamar gigante

"Tenía ante mis ojos un monstruo horrible... Nadó a gran velocidad en la dirección del *Nautilus,* observándonos con sus enormes ojos verdes. La boca del monstruo, parecida al pico de un perico, se abría y cerraba verticalmente". Julio Verne escribió esto en su libro de ciencia ficción *Veinte mil leguas de viaje submarino.* ¿Qué monstruo era el que estaba a punto de atacar al submarino *Nautilus*? Aunque te parezca increíble, era una criatura que en realidad existe: ¡el calamar gigante!

Este calamar gigante ya estaba muerto cuando fue capturado con una red de pesca en la costa de Nueva Zelanda.

Datos sobre los calamares

Los calamares gigantes son los invertebrados más grandes. Su tamaño varía entre 8 y 25 m de largo y llegan a pesar 2,000 kg. Es difícil corroborar estos datos, ya que nadie ha estudiado a un calamar gigante vivo. Sólo se han estudiado calamares gigantes muertos o moribundos que llegan a las costas o que son capturados en redes de pescar.

Estos calamares son muy similares a sus parientes más pequeños. Tienen un cuerpo en forma de torpedo, dos tentáculos, ocho brazos, un manto, un sifón y un pico, pero las partes de su cuerpo son mucho más grandes. Los ojos, por ejemplo, ¡son del tamaño de una pelota de voleibol! Al igual que los calamares adultos de especies de menor tamaño, los gigantes no sólo se alimentan de peces, sino de calamares más pequeños. Dado su tamaño, es difícil imaginar que los calamares gigantes tengan enemigos en el océano, pero así es.

Un enemigo hambriento

Los cachalotes, que pesan unas 20 toneladas, se comen a los calamares gigantes. ¿Cómo se sabe? Se han encontrado hasta 10,000 picos de calamar en el estómago de un solo cachalote; estos picos son demasiado duros para ser digeribles. Los calamares gigantes son un alimento común en su dieta, aun cuando ocasionan algunas cicatrices como resultado de la batalla. Muchos cachalotes tienen marcas en la parte anterior de la cabeza y en las aletas, del tamaño de las ventosas de los calamares gigantes.

¿Realidad o ficción?

▶ Lee el capítulo 18 de *Veinte mil leguas de viaje submarino,* de Julio Verne. Busca otros relatos sobre calamares. Escribe tu propio relato sobre calamares gigantes y coméntalo con tus compañeros y compañeras.

353

Answer to Fact or Fiction?
Answers will vary.

Chapter Organizer

CHAPTER ORGANIZATION	TIME MINUTES	OBJECTIVES	LABS, INVESTIGATIONS, AND DEMONSTRATIONS	
Chapter Opener pp. 354–355	45		**Investigate!** Unscrambling an Egg, p. 355	
Section 1 What Are Vertebrates?	90	▶ List the four characteristics of chordates. ▶ Describe the main characteristics of vertebrates. ▶ Explain the difference between an ectotherm and an endotherm.		
Section 2 Fishes	90	▶ Describe the three classes of living fishes, and give an example of each. ▶ Describe the function of a swim bladder and an oily liver. ▶ Explain the difference between internal fertilization and external fertilization.	**QuickLab,** Oil on Troubled Waters, p. 363 **Making Models,** Floating a Pipe Fish, p. 612 **Datasheets for LabBook,** Floating a Pipe Fish, Datasheet 31 **Whiz-Bang Demonstrations,** The Fish in the Abyss, Demo 10	
Section 3 Amphibians	90	▶ Understand the importance of amphibians in evolution. ▶ Explain how amphibians breathe. ▶ Describe metamorphosis in amphibians.	**Demonstration,** Illustrating Fossilization, p. 364 in ATE **Skill Builder,** A Prince of a Frog, p. 614 **Datasheets for LabBook,** A Prince of a Frog, Datasheet 32	
Section 4 Reptiles	90	▶ Explain the adaptations that allow reptiles to live on land. ▶ Name the three main groups of vertebrates that evolved from reptiles. ▶ Describe the characteristics of an amniotic egg. ▶ Name the three orders of modern reptiles.	**Demonstration,** Exploring Fears, p. 369 in ATE **Demonstration,** Reptile Exhibit, p. 371 in ATE **Long-Term Projects & Research Ideas,** Project 16	

TECHNOLOGY RESOURCES

 **Guided Reading Audio CD**
English or Spanish, Chapter 15

 **One-Stop Planner CD-ROM with Test Generator**

 **Science, Technology & Society,** Salmon Sound Barriers, Segment 25
Eye on the Environment, Fish Farming, Segment 17
What's Slithering in Guam? Segment 19

CLASSROOM WORKSHEETS, TRANSPARENCIES, AND RESOURCES	SCIENCE INTEGRATION AND CONNECTIONS	REVIEW AND ASSESSMENT
Directed Reading Worksheet 15 **Science Puzzlers, Twisters & Teasers,** Worksheet 15		
Directed Reading Worksheet 15, Section 1 **Science Skills Worksheet 26,** Grasping Graphing **Transparency 59,** Chordates	**Multicultural Connection,** p. 356 in ATE **Math and More,** p. 357 in ATE **Real-World Connection,** p. 357 in ATE **Weird Science:** Warm Brains in Cold Water, p. 379	**Review,** p. 358 **Quiz,** p. 358 in ATE **Alternative Assessment,** p. 358 in ATE
Directed Reading Worksheet 15, Section 2 **Transparency 200,** How a Cell Produces an Electric Current	**Physics Connection,** p. 360 **Real-World Connection,** p. 360 in ATE **Connect to Environmental Science,** p. 361 in ATE **MathBreak,** A Lot of Bones, p. 362 **Math and More,** p. 362 in ATE **Connect to Physical Science,** p. 362 in ATE **Across the Sciences:** Robot Fish, p. 378	**Homework,** pp. 359, 361 in ATE **Review,** p. 363 **Quiz,** p. 363 in ATE **Alternative Assessment,** p. 363 in ATE
Directed Reading Worksheet 15, Section 3 **Transparency 60,** Metamorphosis of a Frog	**Connect to Physical Science,** p. 365 in ATE **Apply,** p. 367 **Cross-Disciplinary Focus,** p. 367 in ATE **Multicultural Connection,** p. 367 in ATE	**Homework,** p. 365 in ATE **Self-Check,** p. 366 **Review,** p. 368 **Quiz,** p. 368 in ATE **Alternative Assessment,** p. 368 in ATE
Transparency 61, Reptile History **Directed Reading Worksheet 15,** Section 4 **Transparency 62,** Amniotic Egg **Math Skills for Science Worksheet 35,** Using Temperature Scales **Reinforcement Worksheet 15,** Coldblooded Critters **Critical Thinking Worksheet 15,** Frogs Aren't Breathing Easy	**Math and More,** p. 371 in ATE **Real-World Connection,** p. 372 in ATE **Multicultural Connection,** p. 372 in ATE	**Self-Check,** p. 371 **Homework,** p. 372 in ATE **Review,** p. 373 **Quiz,** p. 373 in ATE **Alternative Assessment,** p. 373 in ATE

Holt, Rinehart and Winston On-line Resources

go.hrw.com

For worksheets and other teaching aids related to this chapter, visit the HRW Web site and type in the keyword: **HSTVR1**

National Science Teachers Association

www.scilinks.org

Encourage students to use the *sci*LINKS numbers listed with the Chapter Highlights to access information and resources on the **NSTA** Web site.

END-OF-CHAPTER REVIEW AND ASSESSMENT

Chapter Review in Study Guide
Vocabulary and Notes in Study Guide
Chapter Tests with Performance-Based Assessment, Chapter 15 Test
Chapter Tests with Performance-Based Assessment, Performance-Based Assessment 15
Concept Mapping Transparency 15

Chapter Resources & Worksheets

Visual Resources

TEACHING TRANSPARENCIES

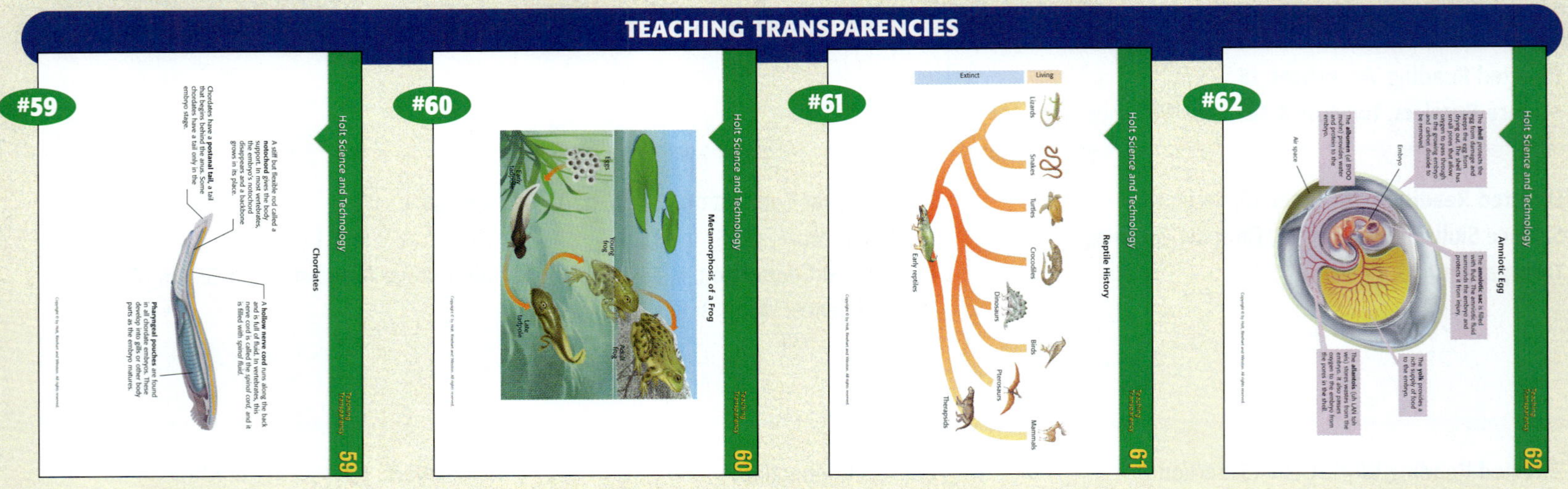

TEACHING TRANSPARENCIES

CONCEPT MAPPING TRANSPARENCY

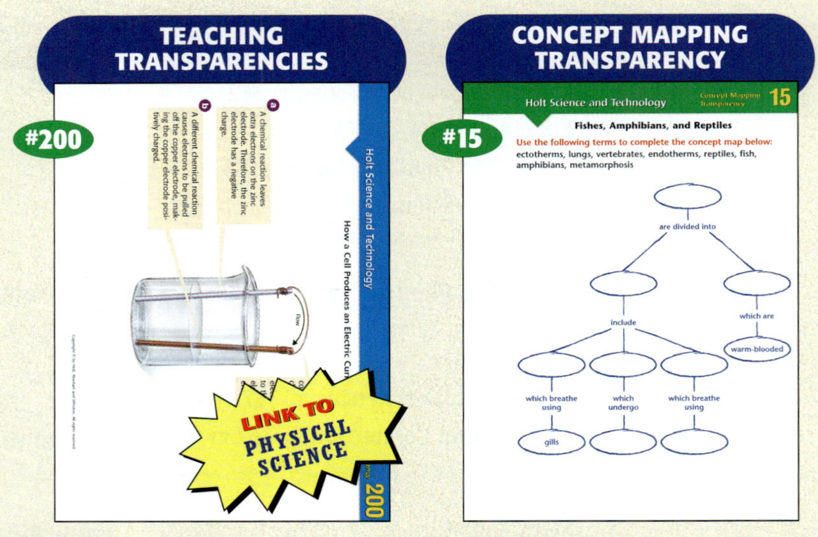

Meeting Individual Needs

DIRECTED READING

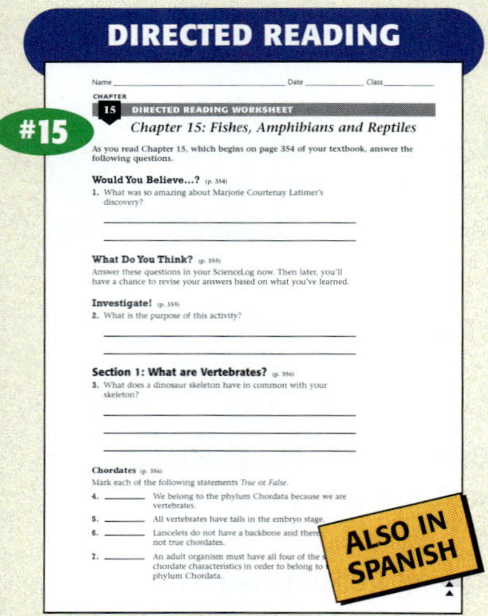

REINFORCEMENT & VOCABULARY REVIEW

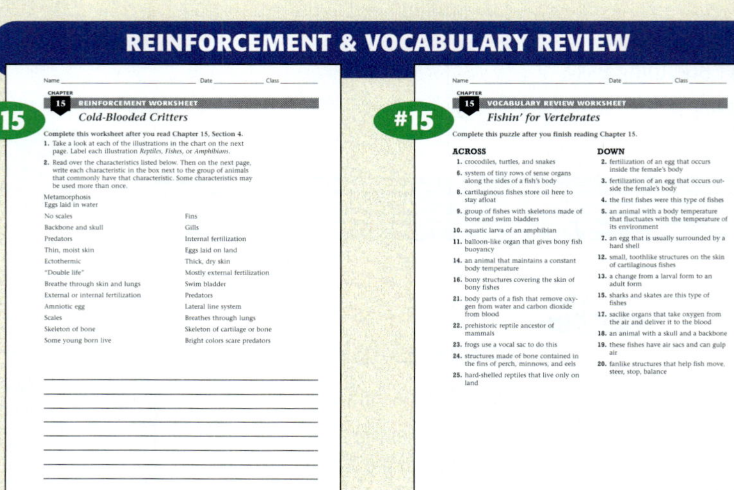

SCIENCE PUZZLERS, TWISTERS & TEASERS

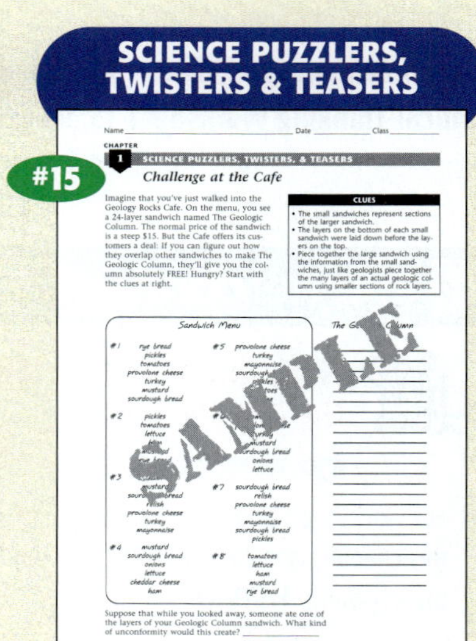

Review & Assessment

STUDY GUIDE

CHAPTER TESTS WITH PERFORMANCE-BASED ASSESSMENT

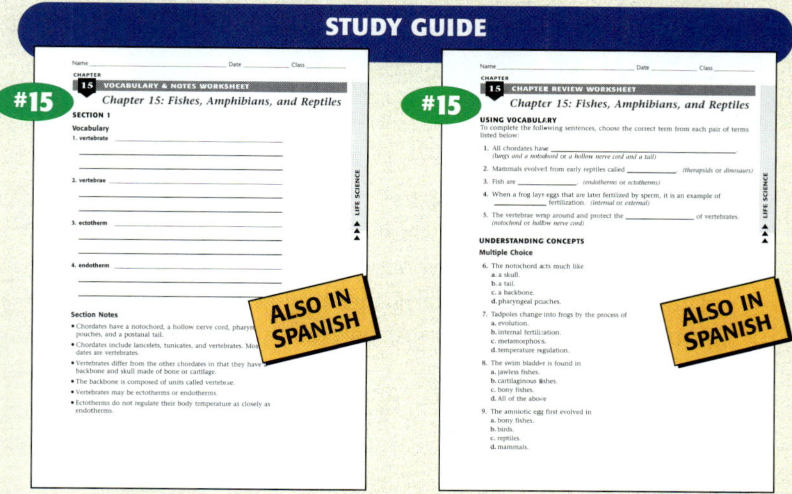

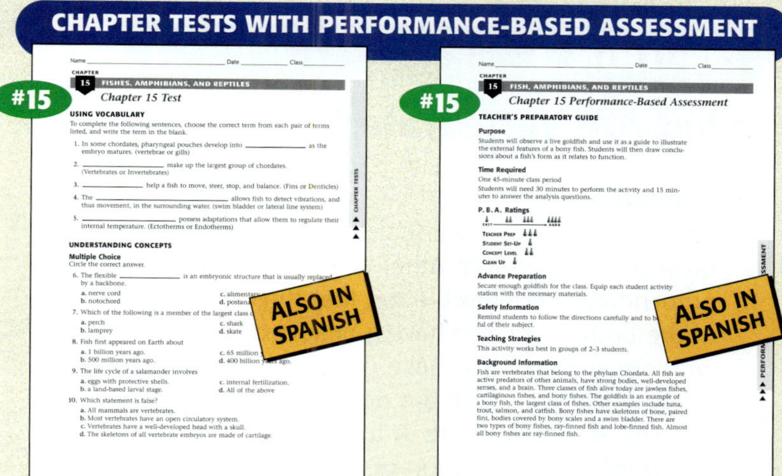

Lab Worksheets

WHIZ-BANG DEMONSTRATIONS

LONG-TERM PROJECTS & RESEARCH IDEAS

DATASHEETS FOR LABBOOK

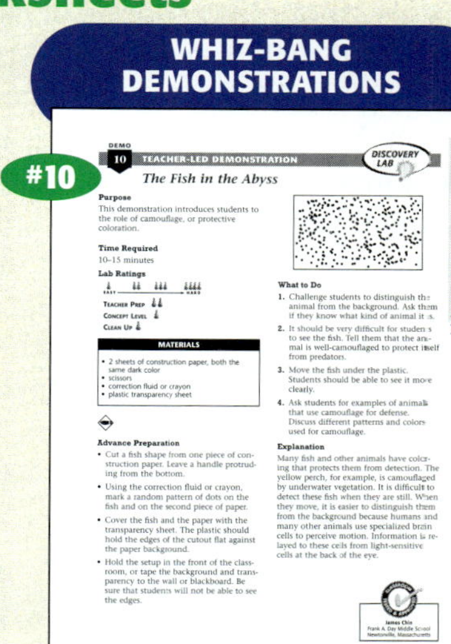

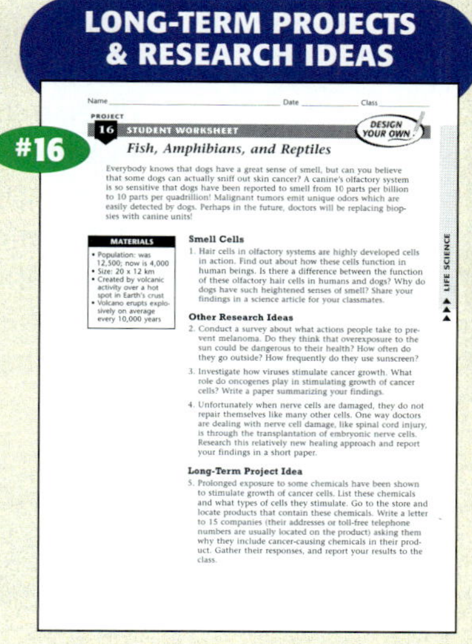

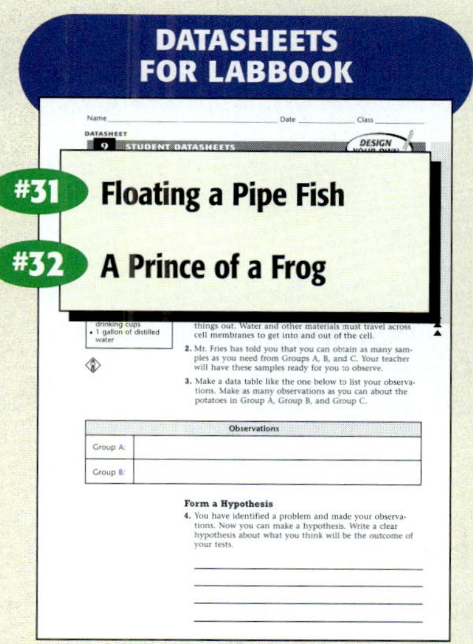

Applications & Extensions

CRITICAL THINKING & PROBLEM SOLVING

SCIENCE TECHNOLOGY

EYE ON THE ENVIRONMENT

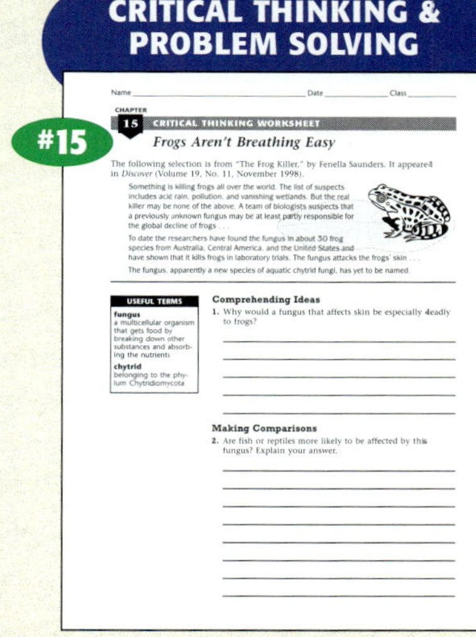

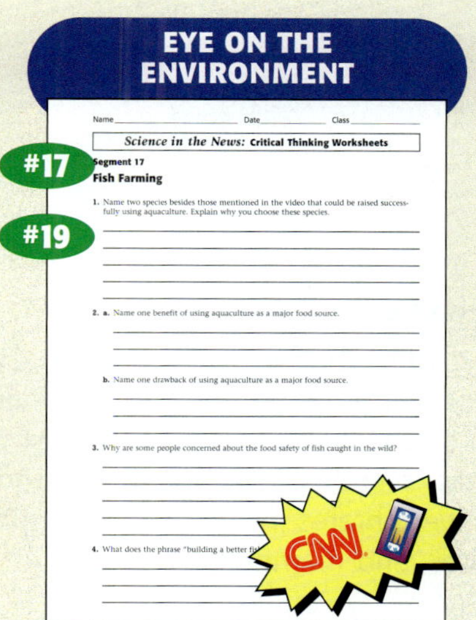

SECTION 1

SECTION 1

What Are Vertebrates?

▶ Chordates

Chordate embryos have four basic features that distinguish them from other animals. The first is the notochord, a semiflexible rod that runs along the length of the body. They also have a hollow dorsal nerve cord, with nerve fibers that make contact with the muscles; pharyngeal slits, which are openings between the pharynx and the outside of the animal; and a postanal tail, which is an extension of the notochord and dorsal nerve cord.

IS THAT A FACT!

■ Cephalochordates (lancelets) have more than 100 pharyngeal slits that are used to strain food particles from water.

▶ Vertebrates

Vertebrates, which belong to the phylum Chordata, first appeared on Earth during the late Cambrian period, more than 500 million years ago. The first vertebrates were jawless creatures; jaws appeared about 100 million years later. Vertebrates include fishes, amphibians, reptiles, birds, and mammals.

SECTION 2

Fishes

▶ Ray-Finned Fishes

The actinopterygians, or ray-finned fishes, are the largest class of fishes. There are about 20,000 species of ray-finned fishes, making them by far the largest group of aquatic vertebrates as well.

• The first ray-finned fish appeared during the Devonian period.

▶ Cartilaginous Fishes

Sharks, skates, and rays are all Chondrichthyes, or cartilaginous fishes. About 800 species of cartilaginous fishes exist, compared with about 20,000 species of bony fishes. Cartilaginous fishes are an amazingly diverse group both behaviorally and physically. For example, sharks range in size from the 14 m (40 ft) whale shark to the 0.6 m (2 ft) dogfish shark. Shark diets differ greatly too. Although most eat other fish, great white sharks have been known to eat seals and other marine mammals. Some sharks, such as the gentle basking shark, eat nothing but plankton.

• Nurse sharks have special sensory organs beneath their nose called barbels that help them locate food on the bottom of the ocean.

IS THAT A FACT!

■ Some ray-finned fishes have evolved special abilities. For example, Siamese fighting fish are capable of breathing air in addition to taking oxygen from water through their gills. Fish such as the walking catfish and mudskipper can even crawl on land!

■ More people are killed in the United States each year by dogs than have been killed by great white sharks in the last century.

SECTION 3

Amphibians

▶ Frogs and Toads

What's the difference between toads and frogs? Toads are types of frogs that belong to the family Bufonidae. These "true toads" generally have stubby bodies, short hind legs, and dry and warty skin, and they tend to lay their eggs in long chains. "True frogs," which are members of the family Ranidae, generally have bulging eyes, strong and long webbed feet, and smooth skin, and they tend to lay their eggs in clusters. There are many exceptions to these rules, however; for example, there are some warty frogs and some smooth-skinned toads.

▶ Salamanders

Salamanders belong to the order Caudata. These carnivorous amphibians are distinguished from other amphibians by the presence of a tail, two pairs of legs of approximately the same size, ribs, teeth on both jaws, and other muscular and skeletal characteristics. Salamanders live in cool, moist habitats in almost all northern temperate areas of the world.

IS THAT A FACT!

- Some salamanders can extend their sticky, mucus-coated tongues as far as half their body length.

- Most salamanders can regenerate lost toes and even entire limbs. As a defense mechanism, they sometimes voluntarily shed their tail. Muscle contractions in the detached tail distract a predator long enough to enable the salamander to make a quick getaway.

SECTION 4

Reptiles

▶ Reptile Characteristics

Most of the 6,000 species of reptiles in existence today live in tropical areas. They include snakes, crocodiles, lizards, turtles, and tuatara, which are lizardlike animals that can be found only on some of the islets of New Zealand. Reptiles range in size from the tiny gecko lizard (3 cm) to the anaconda snake (up to 12 m).

▶ Crocodiles and Alligators

These two closely related carnivorous reptiles belong to the order Crocodylia. How do they differ from each other? Alligators have broader snouts than crocodiles. Also, the fourth tooth on either side of a crocodile's jaw protrudes when the crocodile's mouth is closed.

- Estuarine crocodiles can live to be over 100 years old.

IS THAT A FACT!

- The Chinese alligator, a relatively small animal found in China's Yangtze River region, is an endangered species and may indeed be extinct.

For background information about teaching strategies and issues, refer to the *Professional Reference for Teachers*.

Directed Reading Worksheet 15

Science Puzzlers, Twisters & Teasers Worksheet 15

Guided Reading Audio CD
English or Spanish, Chapter 15

CAPÍTULO

15 Peces, anfibios y reptiles

¿Creerías que...?

En diciembre de 1938, una conservadora de museo que vivía en la costa de Sudáfrica fue, como de costumbre, a visitar el muelle a donde llegaban los pescadores. Marjorie Courtenay Latimer se puso a buscar entre los pescados del día para ver si había alguno interesante, y allí, debajo de un montón de mantarrayas y tiburones, vio la aleta brillante del pez más raro que había visto en su vida. Medía 1.5 m de largo, y cuando Latimer tomó un taxi de regreso al museo, al taxista no le hizo la menor gracia ver a ese monstruo en su asiento trasero. Después de buscar en varios libros, Latimer decidió que se trataba de un celacanto. Los científicos pensaban que estos peces se habían extinguido junto con los dinosaurios, ¡hace 70 millones de años! Fuese lo que fuese, este pez estaba empezando a apestar. Para conservarlo, Latimer lo mandó disecar, pero antes hizo un

rápido boceto y se lo envió a un amigo suyo, J. L. B. Smith, que era experto en peces. Smith se dio cuenta de que el pez era un celacanto. Latimer y Smith se hicieron famosos de la noche a la mañana con su pez disecado. El pez gigantesco que Latimer encontró es uno de los hallazgos más fascinantes del siglo XX. Smith llamó al celacanto *Latimeria*, en honor a su amiga.

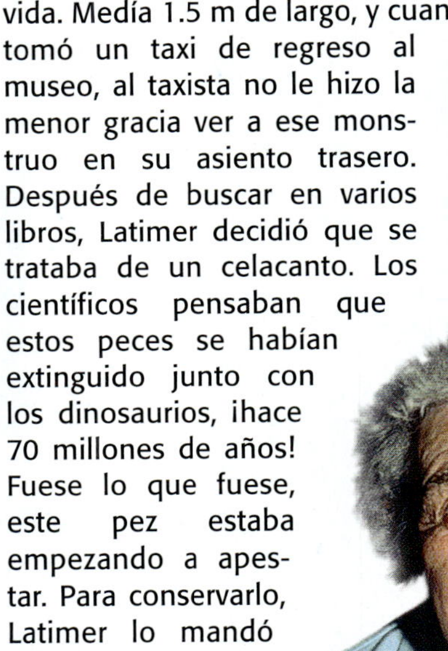

Marjorie Courtenay Latimer descubrió un celacanto viviente.

354

Would You Believe . . . ?

Latimer's decision to have the coelacanth mounted was a good one. But she should have kept better track of the fish's innards! The internal organs were inadvertently discarded during the fish's mounting, leaving many questions unanswered. Fourteen years passed before another coelacanth—internal organs intact— was discovered.

Usa tus conocimientos para responder a las siguientes preguntas en tu cuaderno de ciencias:

1. Cuando decimos que un animal tiene sangre fría, ¿qué quiere decir?

2. ¿Qué diferencia existe entre un reptil y un anfibio?

Descifra el huevo revuelto

Los embriones de peces, anfibios y reptiles se desarrollan dentro de un huevo. Los peces y los anfibios depositan sus huevos en el agua o en lugares muy húmedos, pero la mayoría de los reptiles pone sus huevos en la tierra. Los reptiles modernos tienen huevos con cáscara, que puede ser dura o suave, como de cuero, pero sus huevos son parecidos a los de los pájaros. En esta actividad, vas a abrir un huevo con cáscara para identificar sus partes. Lo que aprenderás se aplica también a los huevos de casi todos los reptiles.

Procedimiento

1. Tu maestro te dará un **huevo de gallina cocido.**

2. Con un **lápiz,** traza la silueta del huevo en una **hoja de papel.**

3. Rompe con cuidado la cáscara y pela el huevo. Busca la membrana delgada que hay entre la clara y la cáscara, y fíjate en la cámara de aire de la parte más ancha del huevo. Dibuja todas las partes del huevo mientras lo vas abriendo.

4. Con un **cuchillo** corta con mucho cuidado el huevo a lo largo, en dos partes iguales. Busca la delgadísima membrana que separa la yema de la clara. Dibuja en tu diagrama las partes que ves.

Análisis

5. Compara tu diagrama con la siguiente ilustración. ¿Encontraste todas las partes? Rotula cada parte del huevo en tu diagrama.

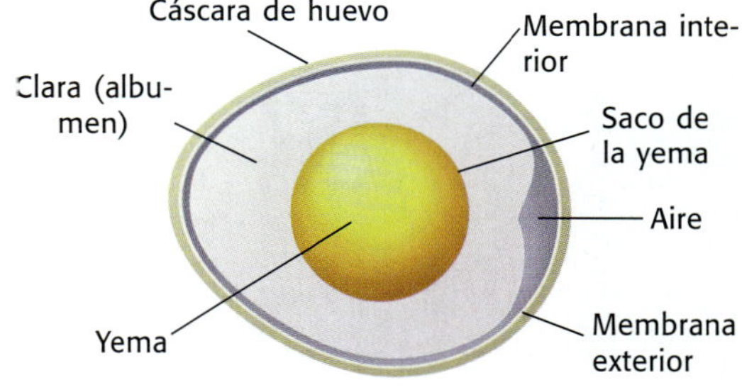

6. ¿Para qué crees que sirve la cámara de aire?

7. ¿Por qué crees que los huevos con cáscara dura les ayudaron a los reptiles a adaptarse a la vida terrestre?

8. ¿Sería más fácil o más difícil identificar las partes del huevo en un huevo crudo? ¿Por qué?

355

Investigate!

MATERIALS

FOR EACH GROUP:
- hard-boiled chicken egg
- pencils
- sheets of paper
- knife

Safety Caution: Remind students to review all safety cautions and icons before beginning this lab activity. Students should be careful when handling knives. To eliminate possible problems, use butter knives. Most butter knives should be sharp enough to slice through an egg. Model for students the best way to cut the egg.

Divide students into small groups. Appoint one student in each group to crack and cut the eggs. Then have students pass the parts around before beginning their diagrams.

Encourage students to discuss the Analysis questions.

Answers to Investigate!

5. Answers will vary.

6. Answers will vary. The air space provides oxygen to the developing embryo.

7. Answers will vary. Students may suggest that hard-shelled eggs allowed animals to move farther from water sources, providing more-diverse food sources.

8. Answers will vary.

Focus

What Are Vertebrates?

This section introduces students to vertebrates and other chordates. Students will learn about the characteristics of vertebrates and the traits that set them apart from other chordates. They will also discover the differences between ectotherms and endotherms.

Bellringer

While you are taking attendance, ask students to ponder this question:

What are some of the physical characteristics shared by dinosaurs and people?

Have each student jot down two or three ideas. Then briefly discuss students' lists before beginning the section.

1) Motivate

ACTIVITY

Assembling Skeletons As an introduction to vertebrates, have students reassemble small mammal skeletons. Obtain owl pellets (the bones of animals eaten by owls) from a nature center or biological supply company. Provide students with a simple illustration of a small mammal skeleton as a reference. In groups of two or three, have students break apart pellets and sort the bones, reconstructing an "exploded view" of whatever small mammal skeletons they find. **Sheltered English**

Directed Reading Worksheet 15 Section 1

VOCABULARIO
vertebrado
vértebra
heterotermo
homotermo

OBJETIVOS
- Enumera las cuatro características de los cordados.
- Describe las principales características de los vertebrados.
- Explica la diferencia entre heterotermo y homotermo.

¿Qué son los vertebrados?

¿Has visto alguna vez el esqueleto de un dinosaurio en un museo? Los huesos fosilizados de los dinosaurios están armados para mostrar cómo era el animal. La mayoría de los esqueletos de dinosaurio son enormes comparados con el esqueleto de la gente que entra a verlos al museo. Pero los seres humanos tenemos muchos huesos parecidos a los de los dinosaurios, aunque más pequeños. Tu columna vertebral es muy parecida a la de un esqueleto de dinosaurio, como puedes ver en la **Figura 1.** Los animales que tienen columna vertebral se llaman **vertebrados.**

Figura 1 *Los seres humanos y los dinosaurios son vertebrados.*

Cordados

Los vertebrados pertenecen al filo Chordata. Los miembros de este filo se llaman *cordados*. Los vertebrados son el grupo más grande dentro de los cordados, pero existen también otros dos grupos de cordados: los anfioxos y los tunicados, que puedes ver en la **Figura 2.** Estos cordados no tienen columna vertebral ni tampoco una cabeza bien desarrollada; comparados con los vertebrados, son mucho más sencillos. Sin embargo, los tres grupos comparten las características de los cordados.

En algún momento de su desarrollo, todos los cordados tienen cuatro partes especiales en el cuerpo: el *notocordio*, un *cordón nervioso hueco*, *hendiduras branquiales faríngeas* y *cola*. Las puedes ver en la **Figura 3,** de la página siguiente.

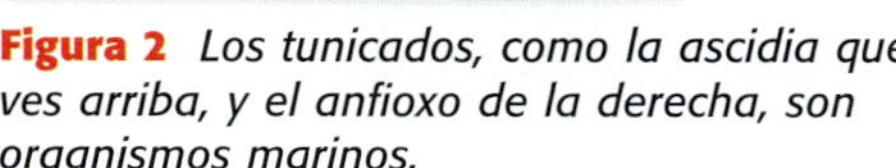

Figura 2 *Los tunicados, como la ascidia que ves arriba, y el anfioxo de la derecha, son organismos marinos.*

356

Multicultural CONNECTION

In many parts of Asia, lancelets are commercially harvested and are an important food source. Have students research the use of lancelets as a food source, examining such things as taste, price per pound, and availability in the United States. Students could present their findings in an oral or written report or use the information they have gathered to develop recipes using lancelets.

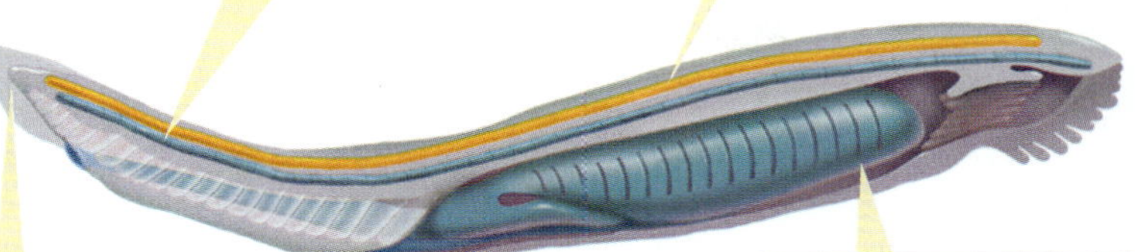

Figura 3 *Las características de los cordados se muestran aquí en un anfioxo. Todos los cordados tienen estas cuatro características en algún momento de su desarrollo.*

La columna vertebral

La mayoría de los cordados son vertebrados. Los vertebrados tienen muchos rasgos que los distinguen de los anfioxos y los tunicados. Por ejemplo, tienen columna vertebral, que es una cadena segmentada formada por huesos llamados **vértebras.** Puedes ver las vértebras de un ser humano en la **Figura 4.** Las vértebras rodean la médula espinal y la protegen. Los vertebrados también tienen una cabeza bien desarrollada, protegida por el cráneo. El cráneo y las vértebras están hechos de cartílago o de hueso. El *cartílago* es el material resistente y flexible del que están hechas tu nariz y tus orejas.

El esqueleto de los embriones de todos los vertebrados está hecho de cartílago. Al crecer, casi todos los vertebrados reemplazan el cartílago por hueso, que es mucho más duro que el cartílago y se fosiliza con facilidad. Se han encontrado muchos fósiles de animales vertebrados que nos dan información valiosa sobre la relación entre distintos organismos.

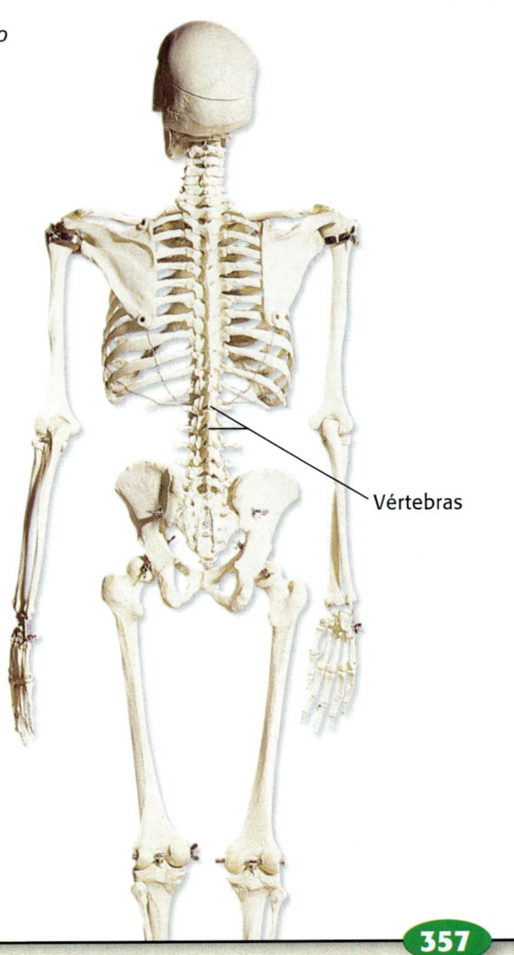

Figura 4 *Las vértebras encajan una en otra para formar una columna de hueso. La columna vertebral protege la médula espinal y sostiene todo el cuerpo.*

357

Tunicates are actually more well-developed as young larvae than they are when they "mature" into adults! As larvae, tunicates have many chordate features, look much like tadpoles, and are able to swim. As they reach adulthood, however, they lose their tail (and therefore their ability to swim), and their nervous system disintegrates.

ACTIVITY

Invite a paleontologist or fossil enthusiast to speak with the class about fossil hunting in your area.

Quiz

1. What is a segmented column of bones that supports the body called? (backbone)

2. Of birds, fishes, mammals, reptiles, and amphibians, which are ectotherms and which are endotherms? (Birds and mammals are endotherms. Fishes, amphibians, and reptiles are ectotherms.)

ALTERNATIVE ASSESSMENT

Writing Have small groups of students compile lists of three or four questions related to the content in this section. Then have students use their questions to quiz each other.

Answer to Explore

Human body temperature will not fluctuate more than a few tenths of a degree, even after exercise. An ectotherm's body temperature fluctuates more dramatically, depending on the temperature of its environment.

Vertebrados... ¿fríos o calientes?

Casi todos los animales necesitan calor, porque las reacciones químicas de sus células sólo ocurren a ciertas temperaturas. La temperatura del cuerpo del animal no debe ser muy alta ni muy baja. Algunos animales controlan su temperatura más que otros.

Conservar el calor Las aves y los mamíferos mantienen la temperatura utilizando el calor de las reacciones químicas de sus células. Su temperatura se mantiene casi constante aunque cambie la temperatura del ambiente. Los animales que mantienen una temperatura constante se llaman **homotermos,** aunque también se les llama animales de *sangre caliente.* Como mantienen constante la temperatura de su cuerpo, los homotermos, como el ave de la **Figura 5,** pueden vivir en climas fríos.

Figura 5 *Los homotermos, como este pájaro piñonero, pueden vivir en climas muy fríos, pues las plumas y otras adaptaciones les ayudan a mantenerse calientes.*

¿Sangre fría? En días soleados, las lagartijas, como la de la **Figura 6,** salen al sol. Al asolearse, la lagartija se vuelve más activa: puede buscar su alimento y escapar de los depredadores. Si la temperatura en el ambiente baja, la lagartija se adormece.

Las lagartijas y otros animales que no pueden controlar su temperatura corporal mediante las reacciones químicas de sus células se llaman **heterotermos.** Su temperatura cambia con la temperatura de su ambiente. Casi todos los peces, anfibios y reptiles son heterotermos. Los heterotermos también se llaman animales de *sangre fría,* pero su sangre puede estar fría o caliente de acuerdo con la temperatura ambiente.

Explora

Usa un **termómetro** que no sea de vidrio para este experimento. Tómate la temperatura cada hora durante por lo menos seis horas. Asegúrate de que hayan pasado por lo menos 20 minutos después de una comida o bebida, antes de ponerte el termómetro. Traza la gráfica de tu temperatura, colocando la hora del día en el eje de las *X* y tu temperatura en el eje de las *Y.* ¿Cambió mucho tu temperatura a lo largo del día? ¿Cuánto? ¿Crees que tu temperatura cambia cuando haces ejercicio? Averígualo. Si fueras un heterotermo, ¿qué tan diferentes serían los resultados?

Figura 6 *Las lagartijas se asolean para absorber calor. Cuando tienen demasiado calor, se ponen a la sombra.*

REPASO

1. ¿Qué tienen en común los vertebrados y los demás cordados? ¿En qué se diferencian?

2. Explica la diferencia entre homotermos y heterotermos.

3. **Aplicar conceptos** Imagínate que tu mascota es una iguana, que es un tipo de lagarto. Crees que está enferma, porque casi no se mueve. El veterinario te recomienda que pongas una lámpara en su jaula. ¿Cómo le puede ayudar esto a tu iguana?

▼ *Answers to Review*

1. Like other chordates, vertebrates have a notochord, a hollow nerve cord, pharyngeal pouches, and a tail at some point in their lives. Unlike other chordates, vertebrates have a backbone and a skull.

2. Endotherms regulate their body temperature by capturing the heat released by the chemical reactions in their cells. Ectotherms do not tightly control their body temperature through the chemical reactions of their cells. Their body temperature depends on the temperature of their environment.

3. An iguana is an ectotherm. The heat lamp will warm the cage and cause the body temperature of the iguana to rise. The lizard will become more active when its temperature rises.

Peces

VOCABULARIO

aletas
escamas
sistema de líneas laterales
branquias
fecundación externa
fecundación interna
dentículos
vejiga natatoria

OBJETIVOS

- Describe las tres clases de peces vivientes y da un ejemplo de cada una.
- Describe la función de la vejiga natatoria y el hígado aceitoso.
- Explica la diferencia entre fecundación interna y externa.

Si encuentras agua, es casi seguro que encuentres también peces. Los peces viven en casi cualquier ambiente acuático, desde las aguas poco profundas de estanques y arroyos hasta las profundidades del océano. Hay peces en las aguas heladas del ártico y en los tibios mares tropicales. Puedes encontrar peces en ríos, lagos, pantanos y hasta en cuevas llenas de agua.

Los peces fueron los primeros vertebrados del planeta; el registro fósil indica que aparecieron por primera vez hace 500 millones de años. Las especies vivientes de peces suman más que todas las otras especies de vertebrados juntas. Hay más de 25,000 y se siguen descubriendo más. La **Figura 7** muestra algunas de ellas.

Figura 7 *Estas son algunas de las muchas especies de peces. ¿Te parecen conocidos?*

a través de las ciencias
CONEXIÓN

Las características de los peces se estudian para construir barcos. Fíjate en el pez robot de la página 378.

Características de los peces

Aunque los peces que ves en esta página son muy diferentes entre sí, también comparten muchas características que les permiten vivir en el agua.

Muchos peces son depredadores, mientras que otros son herbívoros. Como necesitan buscar y encontrar su alimento, los peces tienen un cuerpo fuerte, sentidos bien desarrollados y cerebro.

359

Focus

Fishes

This section introduces students to fishes, the first vertebrates. Students will explore the characteristics of fishes, including their swimming ability and their complex senses of vision, smell, and hearing. Students will also learn about the three classes of fishes.

Bellringer

Have students write a book-title pun on the subject of fishes. Write a few titles on the board to get students' creative juices flowing, such as the following:

I Like Fishes, by Ann Chovie
Life on a Limb, by Anna Perch
Fish Story, by Rod Enreel

1 Motivate

GROUP ACTIVITY

Have students turn the classroom or hallway into a fantasy sea world. Using books, magazines, and other media, students should draw, accurately color, and cut out two or three different fish each. Before posting the fish around the room, have students make a card with the name of the fish, its range, and its size. Sheltered English

Directed Reading Worksheet 15 Section 2

Homework

Writing **Researching Fishes** Have students visit the supermarket and write down each type of canned fish for sale and each type of frozen and/or fresh fish for sale. Students should then choose one of the fish and write a brief, half-page paper describing the size and appearance of the fish in its natural habitat, areas where it can be found, and the amount of the fish harvested each year.

READING STRATEGY

Prediction Guide Before students read the passage about fishes, ask them whether the following statements are true or false.

1. Sharks are considered fishes. (true)

2. Some fish will suffocate if they stop swimming. (true)

3. All fish need to swim to stay alive. (false)

Students will discover the answers as they explore Section 2.

DISCUSSION

Aquarium Presentations
Encourage students who have aquariums to share information about their fish with the class. Discuss the importance of providing oxygen-rich water and other care requirements.

REAL-WORLD CONNECTION

Minnows and carp make up the largest family of fishes, with 1,600 species. They are usually the most abundant freshwater fish in North America, Europe, Asia, and Africa. You may even have one in your own home; goldfish and tetras are carp, and in a large enough setting, they can grow to several inches in length. Under ideal conditions, fish continue to grow for their entire lives.

MISCONCEPTION ALERT

Fish remove oxygen gas that is dissolved in the water. They *do not* use the oxygen that is part of the water molecule itself. Each molecule of water contains one atom of oxygen, but it is unavailable to the fish.

Nacido para nadar El cuerpo del pez tiene fuertes músculos en la espalda que le permiten nadar vigorosamente tras su presa. Los peces se mueven en el agua moviendo sus aletas. Las **aletas** tienen forma de abanico y le permiten al pez desplazarse, mantener la dirección y el equilibrio, y detenerse. Muchos peces tienen el cuerpo cubierto de **escamas,** que lo protegen y reducen la fricción con el agua mientras nadan. La **Figura 8** muestra algunos de los rasgos externos de un pez.

Figura 8 Los peces varían en forma y tamaño, pero todos tienen branquias, aletas y cola.

Sentidos para navegar Los peces tienen los sentidos de la vista, el oído y el olfato bien desarrollados. Muchos también tienen un **sistema de líneas laterales,** formado por una o varias filas de diminutos órganos sensoriales a lo largo de cada lado del cuerpo, y a veces hasta en la cabeza. Este sistema detecta vibraciones en el agua, como las de otro pez nadando cerca. El cerebro del pez, protegido por un cráneo fuerte, ordena toda la información que le llega de los sentidos.

Respiración bajo el agua Los peces respiran a través de las **branquias,** las cuales separan el oxígeno del agua. El oxígeno del agua pasa por la membrana de las branquias a la sangre y ésta lleva el oxígeno por el cuerpo del pez. Las branquias también extraen el dióxido de carbono de la sangre.

Nuevos peces La mayoría de los peces se reproducen por **fecundación externa.** La hembra deposita los huevos en el agua y el macho los cubre con esperma. Otros se reproducen por fecundación interna. En la **fecundación interna,** el macho deposita el esperma dentro de la hembra. Luego, la hembra pone los huevos que contienen los embriones en desarrollo, y después de un tiempo, los huevos se abren para dejar salir a los nuevos pececitos. Existen también algunas especies en las que el embrión se desarrolla dentro de la hembra y los pececitos nacen vivos.

física
CONEXIÓN

Cuando observas un objeto a través de una lupa, tienes que moverla hacia atrás y hacia adelante para enfocar el objeto. Lo mismo ocurre con los ojos de los peces. Los peces tienen músculos especiales para cambiar la posición de la lentes de sus ojos y así enfocar los objetos.

IS THAT A FACT!

Air has 26 times more oxygen than water at the same temperature. Consequently, fish expend much more energy breathing than do most mammals.

Clases de peces

Hay cinco clases muy distintas de peces. Dos de ellas están extintas y las conocemos solamente por sus fósiles. Las tres clases de peces vivientes son: *peces agnatos,* o sin mandíbula, *peces cartilaginosos* y *peces óseos.*

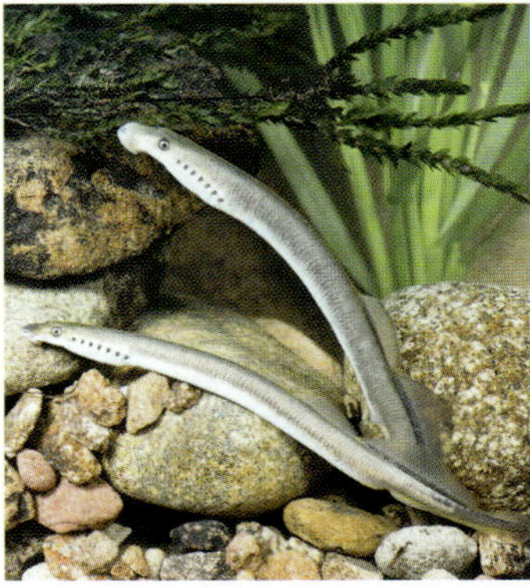

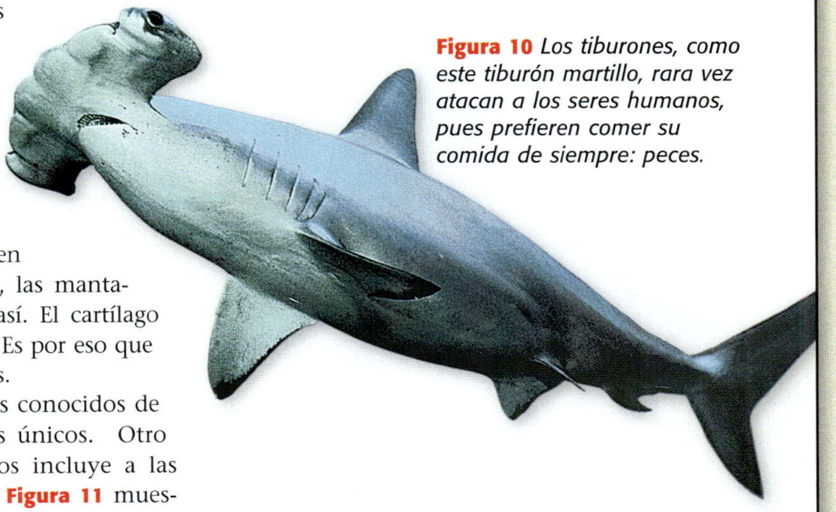

Figura 9 *Las lampreas son parásitos que viven pegados a otros peces.*

Peces agnatos Los primeros peces no tenían mandíbula. ¿Tú crees que se puede comer sin mandíbula? Si no se pudiera, estos peces ya estarían extintos. Pero en cambio, los peces agnatos llevan quinientos millones de años viviendo muy bien. Hoy en día existen alrededor de 60 especies de peces agnatos, entre las cuales se encuentran las lampreas, que puedes ver en la **Figura 9.** Estos peces son parecidos a las anguilas; tienen la piel lisa y resbalosa, y la boca redonda, sin mandíbula. Su esqueleto está hecho de cartílago, y tienen notocordio, pero carecen de columna vertebral. También tienen cráneo, cerebro y ojos.

Figura 10 *Los tiburones, como este tiburón martillo, rara vez atacan a los seres humanos, pues prefieren comer su comida de siempre: peces.*

Peces cartilaginosos ¿Sabías que el tiburón es un pez? Los tiburones, como el de la **Figura 10,** pertenecen a una clase de peces llamados cartilaginosos. En la mayoría de los vertebrados, el cartílago del embrión se convierte poco a poco en hueso. Pero en los tiburones, las mantarrayas y las lisas no sucede así. El cartílago nunca se convierte en hueso. Es por eso que se llaman peces cartilaginosos.

Los tiburones son los más conocidos de estos peces, pero no son los únicos. Otro grupo de peces cartilaginosos incluye a las lisas y a las mantarrayas. La **Figura 11** muestra una mantarraya con espina.

Como todos los admiradores de los tiburones saben, los peces cartilaginosos tienen mandíbulas que funcionan muy bien. También son grandes nadadores y depredadores expertos. Como la mayoría de los depredadores, sus sentidos son muy agudos. Muchos tienen un excelente sentido de la vista y del olfato, y poseen un sistema de líneas laterales.

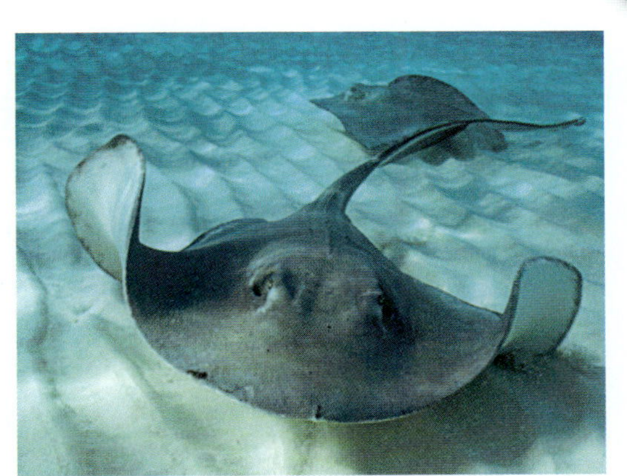

Figura 11 *Las mantarrayas, como esta con espina, se alimentan de crustáceos y moluscos del fondo del mar.*

Homework

Have students indicate which class (or classes) of fishes have the following characteristics; *J* for jawless, *C* for cartilaginous, and *B* for bony.

gills (J, C, B)
denticles (C)
cartilage skeleton (J, C)
swim bladder (B)
scales (B)

READING STRATEGY

Prediction Guide Ask students: What are the three types of fishes alive in the world today?

jawless, cartilaginous, and bony

BRAIN FOOD

Sharks lack some of the features of bony fishes, such as a swim bladder. They also lack movable side fins. In bony fishes, the side fins behind the head can be swiveled so the fish can move backward and forward. Sharks can't do this. How would this affect the shark's movement through the water?

It is difficult or impossible for the shark to stop or to swim backward.

CONNECT TO ENVIRONMENTAL SCIENCE

Once home to 200–500 species of fishes, Lake Victoria, in East Africa, is now home to the Nile perch and little else. Most of the native fishes were small cichlids, with each species unique in appearance and habit. But they were relatively difficult to fish commercially. The Nile perch was being tested in a nearby lake for introduction into Lake Victoria for commercial fishing when it mysteriously appeared in the lake on its own. It has since nearly wiped out the native fishes in this huge lake, to the point that the Nile perch is now largely feeding on its own young and shrimp. As a commercial fishery, the introduction is considered a success. Ecologically, it is a disaster.

MATH and MORE

Carcharodon megalodon shark teeth are as much as 17.5 cm long. Based on its tooth length, scientists estimate the Miocene-era shark was probably 12 m long, which is about twice as long as today's great white shark. Have students imagine that they discovered some shark teeth with the following lengths: 35 cm, 70 cm, and 8.75 cm. Have them use ratios to estimate the sizes of the sharks these teeth came from. (24 m, 48 m, and 6 m)

Answer to MATHBREAK

There are 23,750 species of bony fishes (25,000 × 0.95 = 23,750).

LabBook PG 612
Floating a Pipe Fish

CONNECT TO
PHYSICAL SCIENCE

The electric eel is truly electric— more than half of its body is made up of electricity-producing cells. Found in shallow fresh-water rivers in South America, it can reach 3 m in length and produces a prey-killing jolt of 600 V—enough to kill a person.

You can illustrate an electric current with the following Teaching Transparency.

Teaching Transparency 200
"How a Cell Produces an Electric Current"

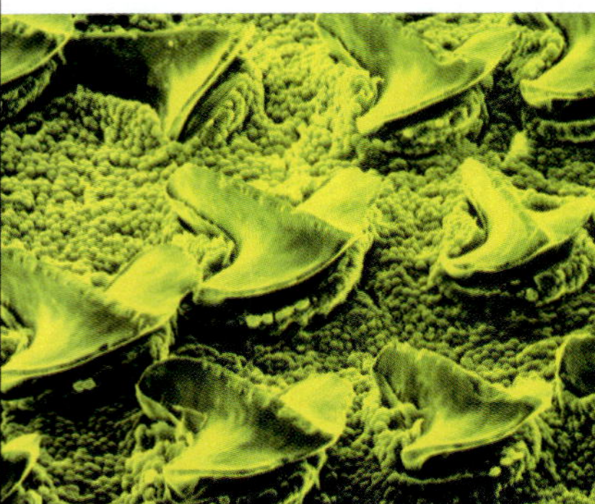
Figura 12 *Los dentículos del tiburón están hechos del mismo material que los dientes humanos.*

Figura 13 *La carpa dorada es un pez óseo.*

¡MATEMÁTICAS!

Un montón de huesos

Si existen 25,000 especies de peces y el 95 por ciento son peces óseos, ¿cuántas especies de peces óseos hay?

La piel de los peces cartilaginosos está cubierta de pequeños **dentículos**, parecidos a dientes diminutos, que le dan una textura como de lija. Si pasaras la mano por el lomo de un tiburón de la cabeza a la cola, se sentiría suave; pero si movieras la mano de la cola a la cabeza, ¡te podría hacer sangrar! Observa una fotografía ampliada de los dentículos en la **Figura 12**.

Para mantenerse a flote, los peces cartilaginosos almacenan mucho aceite en el hígado. ¿Para qué? Averígualo en la sección de Experimentos de la página siguiente. A pesar de su hígado aceitoso, estos peces son más pesados que el agua y tienen que moverse todo el tiempo para no hundirse. Cuando dejan de nadar, se van hundiendo lentamente hasta llegar al fondo.

Los peces cartilaginosos no sólo nadan para mantenerse a flote; algunos necesitan nadar para que el agua entre por sus branquias. Si estos peces dejan de nadar, se asfixian. Pero no todos necesitan nadar continuamente; hay algunos que pueden echarse en el fondo del mar y bombear agua por sus branquias.

Peces óseos Cuando piensas en un pez, seguramente la imagen que te viene a la mente es parecida a la **Figura 13**. Las carpas doradas, el atún, las truchas, el pez gato y el bacalao son peces óseos, que es la clase con más peces en el mundo. El noventa y cinco por ciento de todos los peces son óseos, es decir, su esqueleto está hecho de hueso. Algunos miden sólo 1 cm, y otros tienen más de 6 m de largo. El cuerpo de los peces óseos está cubierto de escamas.

A diferencia de los peces cartilaginosos, los óseos flotan sin tener que nadar todo el tiempo, porque tienen una vejiga natatoria que les da *capacidad de flotación*. La **vejiga natatoria** se parece a un globo y se llena de oxígeno y otros gases que le llegan de la sangre. Puedes ver la vejiga natatoria y otros órganos de los peces óseos en la **Figura 14**.

Figura 14 *Los peces óseos tienen vejiga natatoria, esqueleto óseo y escamas.*

IS THAT A FACT!

Basking sharks, which are the second largest fish, are hunted for the squalene oil in their liver. Although basking-shark livers are extremely large, the Basking Shark Society reports that 2,500–3,000 livers are needed to produce just 1 ton (0.91 metric tons) of squalene oil. Between 1987 and 1994, South Korea alone imported an average of 52 tons (47 metric tons) of the oil from Japan. Squalene oil is used as cooking oil, a skin moisturizer, and a lubricant.

Existen dos clases principales de peces óseos, y de estas dos, la que tiene más especies es la de los *peces con aletas radiadas*. Los peces con aletas radiadas tienen pares de aletas con pequeños rayos de hueso. Entre los peces de aletas radiadas se encuentran muchos peces conocidos, como las anguilas, arenques, truchas, carpas y percas. La **Figura 15** muestra un pez de aletas radiadas.

Los *peces de aletas lobuladas* y los *pulmonados* forman la segunda clase de peces óseos. Los de aletas lobuladas tienen aletas musculosas y gruesas. El celacanto que describimos al principio del capítulo pertenece a este grupo. Existen seis especies de peces pulmonados. Puedes ver un pez pulmonado en la **Figura 16.** Los científicos piensan que los antiguos peces pulmonados son los antepasados de los anfibios de hoy.

Figura 15 *Los peces de aletas radiadas son de los más veloces del mundo. El lucio, como el que ves arriba, puede nadar tan rápidamente como los más veloces corredores humanos, a unos 48 km/h.*

Figura 16 *Los peces pulmonados pueden respirar fuera del agua gracias a sus pulmones, que son una especie de bolsas de aire. Viven en África, Australia y Sudamérica, en aguas poco profundas que suelen secarse en el verano.*

REPASO

1. Nombra las tres clases de peces que existen. ¿A qué clase pertenece el celacanto?

2. La mayoría de los peces óseos se reproducen por fecundación externa. ¿Qué significa esto?

3. ¿Qué es el sistema de líneas laterales y para qué sirve?

4. **Analizar relaciones** Compara lo que los peces cartilaginosos y los peces óseos necesitan para flotar.

Laboratorio

Aceite y agua

¿Qué tiene que ver el hígado aceitoso con la flotación del tiburón? Este experimento te lo explicará. Mide cantidades iguales de **agua** y de **aceite de cocina.** Llena un **globo** con agua y otro con aceite. Amarra los globos de manera que no quede aire adentro. Colócalos en un **recipiente** lleno de agua hasta la mitad. ¿Qué sucede? ¿Cómo impide el hígado aceitoso que los peces cartilaginosos se hundan?

363

▼ **Answers to Review**

1. The three types of fish are jawless fishes, cartilaginous fishes, and bony fishes. The coelacanth is a bony fish.

2. In external fertilization, the female lays unfertilized eggs, and sperm fertilize the egg outside the female's body.

3. The lateral line system is a row of sense organs along the side of a fish's body that detect water vibrations caused by other objects in the water.

4. Cartilaginous fishes have oily livers to help them stay afloat. Bony fishes have a swim bladder that helps them remain buoyant.

---

④ Close

Quiz

1. Of the 25,000 species of fish, about what percent of them are bony fishes? (95 percent)

2. Sharks are a member of what class of fishes? (cartilaginous)

ALTERNATIVE ASSESSMENT

Concept Mapping Create a concept map using the following terms (students must also supply the connecting words linking the terms):

> vertebrate, chordate, notochord, tail, hollow nerve-chord, pharyngeal pouches

*Quick*Lab

MATERIALS

FOR EACH GROUP:
- water
- cooking oil
- balloon
- bowl

Safety Caution: Students should be cautious when working with cooking oil, water, and any other liquids. Oil can stain clothing. Spills should be cleaned up immediately.

Answers to QuickLab

The balloon filled with oil will float, and the balloon with the water will sink. This occurs because the density of oil is lower than that of water. The oily liver of a cartilaginous fish provides buoyancy to a fish and helps it stay afloat.

Focus

Amphibians

In this section, students will discover when and how the first amphibians evolved from fishes. They will find out how amphibians breathe, reproduce, and develop through the process of metamorphosis. Students will also explore the three major groups of amphibians.

🔔 Bellringer

On the board or overhead projector, instruct the students to answer the following questions in their ScienceLog:

What is an advantage to the thin, moist skin of amphibians? What is the primary disadvantage to such skin? (They can absorb oxygen and water right through their skin, but they also lose moisture through skin.)

1) Motivate

DEMONSTRATION

Illustrating Fossilization Show students how fossils are formed. Fill a small bucket halfway with sand. Place a kitchen sponge on top of the sand. Pour sand on top of the sponge until the bucket is nearly filled. Next, pour table salt in a small bucket of warm water, and stir until the salt no longer dissolves. Pour the solution over sand in the bucket, and leave overnight. On the next day, uncover the sponge. The water should have evaporated, and the salt should have hardened the sponge. Explain to students that fossils are formed in a similar manner. (Dissolved minerals enter openings or spaces in plant or animal material and harden.)
Sheltered English

Anfibios

VOCABULARIO
pulmón
renacuajo
metamorfosis

OBJETIVOS
- Comprende la importancia de los anfibios en la evolución.
- Explica cómo respiran los anfibios.
- Describe la metamorfosis de los anfibios.

Al final del período Devónico, hace 350 millones de años, los peces habitaban todas las aguas, pero ninguno de estos vertebrados podía vivir en tierra firme. Este era un lugar maravilloso para los vertebrados: había bosques frondosos, muchos insectos deliciosos y muy pocos depredadores. Pero para adaptarse a la vida terrestre, los vertebrados necesitaban pulmones para respirar y patas para caminar. ¿Cómo ocurrieron estos cambios?

Hacia tierra firme

La mayoría de los anfibios que viven en el planeta hoy en día son ranas o salamandras, como los que ves en la **Figura 17.** Los anfibios primitivos eran muy diferentes; el registro fósil indica que evolucionaron de los antepasados de los peces pulmonados de la actualidad, los cuales desarrollaron pulmones para poder obtener oxígeno del aire. El **pulmón** es un órgano con forma de bolsa que toma oxígeno del aire y lo pasa a la sangre. Las aletas de estos peces primitivos se fortalecieron lo suficiente como para sostener el peso del cuerpo, y con el tiempo se convirtieron en patas.

Los fósiles indican que los primeros anfibios parecían una mezcla entre pez y salamandra, como puedes ver en la **Figura 18.** Fueron los primeros animales vertebrados que pasaban la mayor parte de su vida en tierra firme, y tuvieron mucho éxito. Algunos eran muy grandes (hasta 10 m de largo) y podían permanecer fuera del agua más tiempo que los anfibios del presente, pero tenían que regresar al agua para no deshidratarse, para evitar morir de calor y para poner huevos.

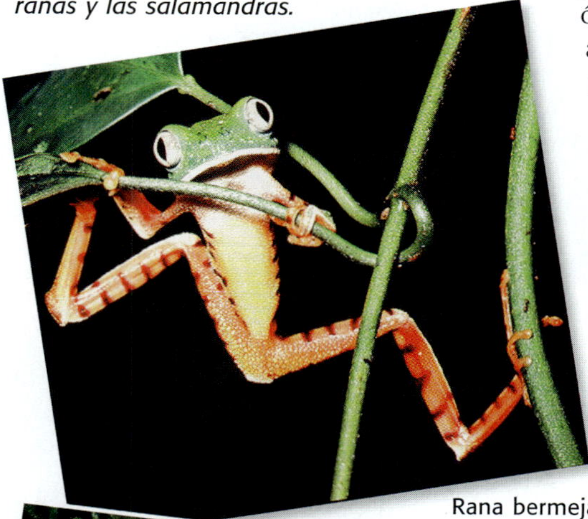

Figura 17 *Entre los anfibios modernos se encuentran las ranas y las salamandras.*

Rana bermeja

Salamandra de la Sierra Nevada

Figura 18 *Los primeros anfibios seguramente tenían un aspecto como este.*

364

IS THAT A FACT!

- A group of frogs is called an army.
- A group of toads is called a knot.
- Fear of frogs is called ranidaphobia.
- Fear of toads is called bufonophobia.

Características de los anfibios

Anfibio quiere decir "doble vida". La vida de la mayoría de los anfibios tiene dos fases; cuando salen del huevo, viven en el agua, como los peces; más tarde se convierten en animales que pueden vivir en tierra firme. Pero incluso los anfibios adultos sólo están adaptados a medias a la vida en la tierra y siempre tienen que estar cerca del agua.

Los anfibios son heterotermos. Como el cuerpo de un pez, el del anfibio cambia de temperatura de acuerdo con la temperatura del medio ambiente.

Piel delicada La mayoría de los anfibios no tienen escamas: su piel es delgada, lisa y húmeda. No beben agua, pues la absorben a través de la piel. Pueden respirar aire gracias a sus pulmones, pero muchos también absorben oxígeno a través de la piel, que está llena de vasos sanguíneos. Algunas salamandras, como la que ves en la **Figura 19,** respiran sólo a través de la piel. La piel de los anfibios es tan delgada y húmeda que estos animales pueden perder agua a través de ella y deshidratarse. Por esta razón, la mayoría de los anfibios viven en el agua o en ambientes húmedos.

La piel de los anfibios suele tener colores muy vivos. Estos colores son una señal para los depredadores, pues muchos anfibios tienen glándulas productoras de veneno en la piel. El veneno puede ser apenas irritante, o puede ser mortal. La piel de la rana venenosa que muestra la **Figura 20** está llena de una de las toxinas más peligrosas que se conocen. Otra rana produce una substancia que hace bostezar a las serpientes. Si la serpiente trata de comerse a la rana, el efecto de la substancia le hace abrir la boca para bostezar, y la rana se escapa de un salto.

La doble vida de los anfibios   Casi todos los anfibios comienzan su vida en el agua. Las ranas se reproducen normalmente por fecundación externa, y las salamandras por fecundación interna. Como los huevos de los anfibios no tienen cáscara ni una membrana especial para evitar la deshidratación, los embriones se tienen que desarrollar en un ambiente muy húmedo.

Figura 19 *La salamandra de cuatro dedos no tiene pulmones, sino que obtiene oxígeno a través de la piel.*

Figura 20 *La piel de esta rana venenosa está llena de glándulas de veneno. En Sudamérica, los cazadores tallan las puntas de sus flechas con esta toxina mortal.*

365

WEIRD SCIENCE

Most toads are well adapted to life on land and leave their fertilized eggs to fend for themselves. The Surinam toad of South America is not like most toads. Highly aquatic, it cannot survive long out of the water. And in a strange mating encounter that involves somersaults, fertilized eggs are attached to the back of the female. The eggs grow into pockets for each developing embryo. Depending on the species, young emerge from the pockets either as tadpoles or as young toads.

Answers to Self-Check

Amphibians use their skin to absorb oxygen from the air. Their skin is thin and moist and full of blood vessels, just like a lung.

ACTIVITY

Observing Development

Obtain frog or salamander eggs either locally or from a biological supply company. Set up an aquarium, and allow students to observe the metamorphosis that follows. Students should sketch the process each step of the way.

Teacher Notes: If the animals were not collected locally, they cannot be released into the wild. Sheltered English

RETEACHING

Poster Project Have students create a poster with captions showing how a tadpole changes into a frog. Sheltered English

Some students may have the mistaken impression that they can get warts from touching frogs and toads. Tell them that warts are caused by human viruses.

Teaching Transparency 60 "Metamorphosis of a Frog"

El embrión del anfibio normalmente se convierte en una larva acuática llamada **renacuajo,** que sólo puede vivir en el agua, obtiene oxígeno a través de branquias y usa su cola para nadar. Más tarde, el renacuajo pierde las branquias y desarrolla pulmones y patas en un proceso llamado metamorfosis. La **Figura 21** muestra la metamorfosis de una rana. La **metamorfosis** es el cambio de forma de larva a adulto. Una vez que el anfibio alcanza la forma adulta, es capaz de vivir en la tierra.

Figura 21
La mayoría de las ranas y salamandras pasan por la metamorfosis. La ilustración muestra la metamorfosis de una rana.

Algunos anfibios se saltan la etapa acuática y sus embriones crecen para convertirse directamente en ranas o salamandras adultas. Por ejemplo, existe una rana sudamericana que pone sus huevos en la tierra húmeda y un grupo de machos los cuida mientras se desarrollan. Cuando un embrión empieza a moverse, una rana macho rápidamente se lo mete en la boca y lo protege en su saco vocal. Cuando el embrión termina su desarrollo, la rana adulta abre la boca y una pequeña ranita salta hacia fuera. Puedes ver esta rana en la **Figura 22.**

Figura 22 *Las ranas de Darwin viven en Chile y Argentina. Un macho puede llevar de 5 a 15 embriones en su saco vocal hasta que las crías miden 1.5 cm de largo.*

366

Coast foam-nest tree frogs mate in trees that overhang ponds and streams. Then the females lay their eggs in large foamy, cocoonlike masses that cake over for protection from the sun. After a week or so of development, the tadpoles are ready to emerge, but only after a rain comes and moistens the foam. Once moistened, the foam drips into the water below, carrying with it the young tadpoles!

Clases de anfibios

Se calcula que existen unas 4,600 especies de anfibios en la actualidad. Estas especies pertenecen a tres grupos: las cecilias, las salamandras, y las ranas y sapos.

Las cecilias Las cecilias no son muy conocidas. Son anfibios sin patas, con forma de gusano o serpiente, como puedes ver en la **Figura 23.** Tienen la piel húmeda y delgada de los anfibios, pero a diferencia del resto, algunas tienen escamas óseas. Muchas tienen ojos muy pequeños debajo la piel y son ciegas. Las cecilias viven en las áreas tropicales de Asia, África y América del Sur; se conocen aproximadamente 160 especies.

Figura 23 *Las cecilias son anfibios sin patas que viven en el suelo húmedo del trópico. Se alimentan de pequeños invertebrados que encuentran en la tierra.*

Salamandras De todos los anfibios que existen en la actualidad, las salamandras son los que más se parecen a los anfibios prehistóricos. Aunque son mucho más pequeñas que sus antepasados, su cuerpo tiene una forma similar, con la cola larga y cuatro patas fuertes. Miden desde unos cuantos centímetros hasta 1.5 m de largo. La **Figura 24** muestra una salamandra.

Se conocen unas 390 especies de salamandras. La mayoría vive bajo troncos y piedras en los húmedos bosques de Norteamérica y se alimentan de pequeños invertebrados. Unas cuantas, como el ajolote **(Figura 25),** no pasan por una metamorfosis, sino que viven toda su vida en el agua.

Figura 24
Las salamandras viven en lugares húmedos, entre los troncos y piedras de los bosques.

Figura 25 *El ajolote es una especie rara de salamandra, que nunca pierde las branquias y vive siempre en el agua.*

Los anfibios son indicadores ecológicos. Cuando muchos de ellos empiezan a morir o a nacer deformes, puede ser señal de que existe un problema ambiental. Algunas veces las deformidades las provocan los parásitos, pero los anfibios también son muy sensibles a los cambios químicos del medio ambiente. Según lo que has aprendido sobre los anfibios, ¿por qué crees que son tan sensibles a la contaminación del agua y aire?

367

IS THAT A FACT!

Although caecilians are more closely related to frogs and salamanders than they are to snakes, the scientific name for caecilians, *Gymnophiona,* means "naked snakes."

Answers to APPLY

Because amphibians' skin is responsible for gas and water exchange, amphibians are highly sensitive to changes in air and water quality.

3 Extend

RESEARCH

Writing Have interested students conduct research about the field of herpetology, the branch of zoology that deals with amphibians and reptiles. Students could prepare reports and present them to the class. As an alternative, invite a herpetologist or other zoologist from the area to speak with the class about herpetology.

4 Close

Quiz

1. Where do most amphibians start their lives? (in water)
2. What is the meaning of amphibian? (double life)
3. What is the largest group of amphibians? (frogs and toads)

ALTERNATIVE ASSESSMENT

Writing Have students compose a song or poem accurately describing the life cycle of an amphibian of their choice.

LabBook | **PG 614**
A Prince of a Frog

Ranas y sapos El noventa por ciento de los anfibios son ranas y sapos. Constantemente se descubren nuevas especies, y habitan en todo el mundo, desde el desierto hasta la selva. Las ranas y los sapos son muy parecidos entre sí, como puedes ver en la **Figura 26.** De hecho, los sapos son un tipo de rana.

Figura 26 *Las ranas tienen la piel lisa y húmeda, y los sapos, que pasan menos tiempo en el agua que las ranas, tienen la piel más seca y llena de protuberancias. También tienen las patas traseras más cortas, y las membranas entre los dedos de sus patas son más pequeñas que en las ranas.*

Los sapos y las ranas están muy bien adaptados a la vida terrestre. Los adultos tienen patas con poderosos músculos para saltar. Tienen el oído muy desarrollado y cuerdas vocales para cantar; además, poseen una pegajosa lengua extensible para atrapar moscas y otros insectos con gran agilidad. La lengua empieza en la parte anterior de la boca, para que la rana pueda sacarla rápidamente para atrapar insectos.

Las ranas son famosas por sus conciertos nocturnos, pero muchas también cantan de día. Como nosotros, empujan aire de los pulmones para que pase a través de las cuerdas vocales en la garganta. Las ranas tienen algo que a nosotros nos falta. Alrededor de las cuerdas vocales tienen una bolsa delgada de piel llamada *saco vocal.* Cuando las ranas vocalizan, el saco se llena de aire como un globo y vibra, como se ve en la **Figura 27.** Las vibraciones del saco vocal aumentan el volumen del canto de las ranas, para que se escuche desde muy lejos.

Figura 27 *La mayoría de las ranas que cantan son machos, y sus cantos tienen distinto significado. Puede ser para defender un territorio o para atraer a las hembras.*

Obseva las características reales de una simpática rana en la página 614.

REPASO

1. Describe la metamorfosis de los anfibios.
2. ¿Por qué necesitan los anfibios vivir cerca del agua o en un ambiente muy húmedo?
3. ¿Qué adaptaciones les permiten a los anfibios vivir en la tierra?
4. Nombra tres clases de anfibios. ¿En qué se parecen? ¿En qué se diferencian?
5. **Analizar relaciones** Describe la relación entre los peces pulmonados y los anfibios. ¿Qué características comparten? ¿En qué se diferencian?

368

Answers to Review

1. After fertilization of an egg, the embryo develops into a tadpole that lives in water. The tadpole loses its gills and develops lungs and legs to become an adult.
2. Amphibians must live near water because they have very thin, moist skin, and they can easily dehydrate.
3. lungs (instead of gills) and legs
4. Frogs and toads, salamanders, and caecilians all have a thin, moist skin; salamanders and frogs and toads have legs, but caecilians do not; caecilians have scales.
5. Both have lungs and can breathe air; amphibians have stronger legs and skin that absorbs oxygen; amphibians go through metamorphosis, but lungfish do not.

Reptiles

Unos treinta y cinco millones de años después de que los primeros anfibios colonizaran la tierra, algunos de ellos desarrollaron rasgos especiales que los prepararon para vivir en ambientes aún más secos. Estos animales desarrollaron una piel gruesa y seca que los protegía de la deshidratación; sus patas se hicieron más fuertes y más verticales, lo que les permitía caminar mejor, y desarrollaron un tipo de huevo que podían poner en la tierra. Estos animales eran los reptiles, los primeros animales que pudieron vivir completamente fuera del agua.

La historia de los reptiles

Los fósiles indican que poco después de su primera aparición, los reptiles se dividieron en grupos. Esta división está ilustrada en forma de árbol genealógico en la **Figura 28.**

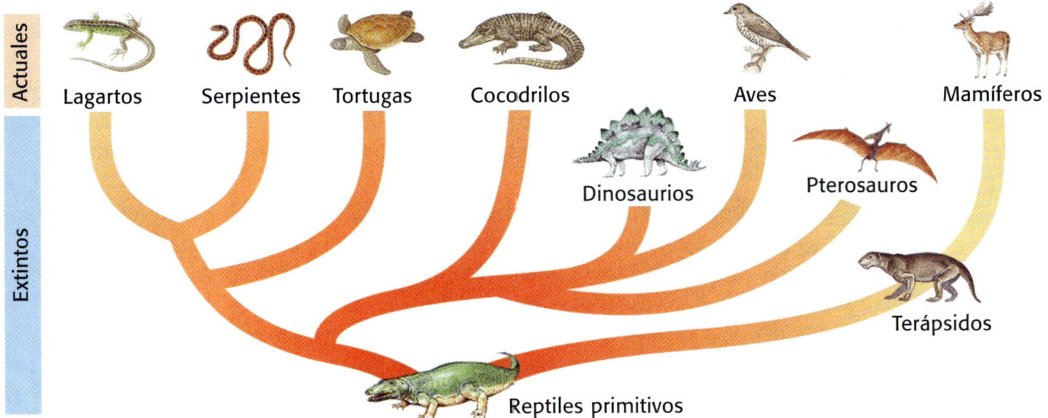

Figura 28 *Los reptiles primitivos son los antepasados de los reptiles, aves y mamíferos actuales.*

La mayoría de los reptiles más fascinantes están extintos. Cuando pensamos en reptiles extintos, casi siempre pensamos en los dinosaurios, pero sólo una fracción de los reptiles prehistóricos eran dinosaurios terrestres; muchos eran reptiles acuáticos y unos cuantos volaban, como los pterosaurus. También había tortugas, lagartos, serpientes y cocodrilos. Además, existía un grupo de reptiles parecidos a los mamíferos, llamados terápsidos. Como puedes ver en la Figura 28, los **terápsidos** son los antepasados de los mamíferos.

369

IS THAT A FACT!

The pteranodon, a particularly large type of pterosaur, had a wingspan of more than 6.5 m (20 ft).

Teaching Transparency 61 "Reptile History"

Directed Reading Worksheet 15 Section 4

2 Teach

ACTIVITY

Organizing Reptiles Before reading about the types of reptiles, take a moment to try to put the following reptiles into their three orders. Match the animal on the left with its closest relative on the right. Check your answers after reading.

turtle	crocodile
snake	tortoise
alligator	lizard

(turtle, tortoise; snake, lizard; alligator, crocodile)

USING THE FIGURE

Encourage students to make comparisons of reptile eggs and amphibian eggs. (Amphibian eggs don't have shells and must be laid in water or moist ground.)

Ask: How might these facts favor reptile reproduction over amphibian reproduction during a drought? (Dehydration would kill off amphibian embryos, but reptile embryos would be more likely to survive.) Sheltered English

GUIDED PRACTICE

Writing Have students make a list of reasons why animals that internally regulate their temperatures have advantages over those that do not. Provide several examples to help students get started.

Teaching Transparency 62
"Amniotic Egg"

Características de los reptiles

Los reptiles están adaptados para la vida en la tierra. Aunque los cocodrilos, las tortugas y algunas especies de serpientes viven en el agua, todos ellos evolucionaron a partir de reptiles terrestres y tienen pulmones para respirar aire, igual que tú.

Para salvar el pellejo Una adaptación muy importante para la vida terrestre es la piel gruesa y seca, impermeable al agua. Esta piel gruesa impide que las células pierdan agua a través de la evaporación. La mayoría de los reptiles no pueden respirar a través de la piel, como lo hacen los anfibios, sino que dependen completamente de sus pulmones para el intercambio de oxígeno y dióxido de carbono. Fíjate en la piel de la serpiente de la **Figura 29.**

A sangre fría Como los peces y los anfibios, los reptiles son heterotermos. Esto quiere decir que, normalmente, los reptiles no pueden mantener una temperatura corporal constante. Mientras hace calor están activos y cuando hace frío se adormecen.

Pocos reptiles pueden generar calor de su propio cuerpo, como algunos lagartos del suroeste de los Estados Unidos, que mantienen una temperatura de alrededor de 34°C incluso cuando hace frío. Aún así, los reptiles modernos sólo viven en climas templados. No pueden soportar el frío de las regiones polares, en las que muchos mamíferos y aves viven bien.

El increíble huevo amniótico Entre las muchas adaptaciones de los reptiles a la vida terrestre, la más importante es el huevo amniótico. El **huevo amniótico** está cubierto por una cáscara, como puedes ver en la **Figura 30.** Ésta protege el embrión en desarrollo para que no se seque. Estos huevos pueden sobrevivir debajo de las rocas, enterrados, en bosques y hasta en el desierto; están tan bien adaptados a la tierra que hasta las tortugas marinas y los cocodrilos salen del agua para poner sus huevos.

Figura 29 *Muchas personas creen que las serpientes son pegajosas, pero en realidad su piel, como la de otros reptiles, es seca y escamosa.*

Figura 30 *Compara los huevos de anfibio de la izquierda con los huevos de reptil de la derecha. ¿Qué diferencias ves?*

WEIRD SCIENCE

The Texas horned toad of southwestern North America is actually a lizard with a bizarre defense mechanism. When threatened, the lizard puffs itself up, increasing its blood pressure until the capillaries around its eyes burst. The predator is stunned when it is then squirted with blood from up to 2.4 m (7 ft) away.

La cáscara es sólo una de las partes importantes del huevo amniótico. Puedes ver el resto en la **Figura 31.** El nombre de huevo amniótico viene del *líquido amniótico* que se encuentra en el saco amniótico dentro del huevo, donde se desarrolla el embrión. El huevo protege el embrión de la acción de otros animales, de infecciones bacterianas y de la deshidratación.

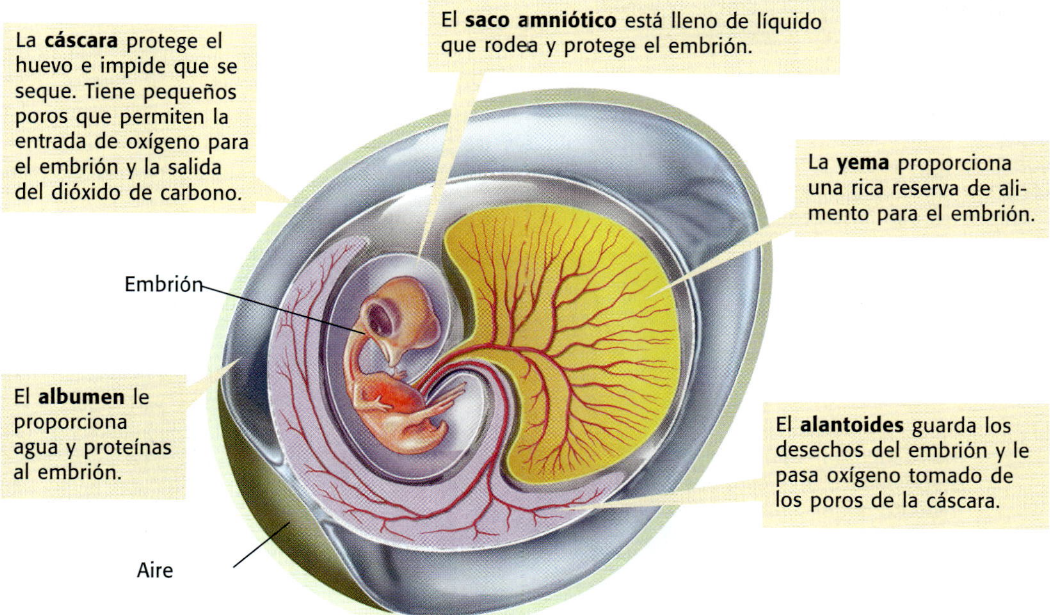

Figura 31 El huevo amniótico

La fecundación del huevo amniótico se lleva a cabo dentro de la hembra. Luego se forma la cáscara y la hembra lo deposita. Como sus huevos tienen cáscara, los reptiles tienen que reproducirse por fecundación interna.

La mayoría de los reptiles ponen sus huevos en la tierra o en la arena. Algunos no ponen huevos, sino que sus embriones se desarrollan en los órganos reproductores de la madre y nacen vivos. En ambos casos, el embrión se convierte directamente en un pequeño reptil, sin pasar por una etapa de larva y sin necesidad de metamorfosis.

Clases de reptiles

En la época de los dinosaurios, de hace 300 millones de años hasta hace más o menos 65 millones de años, la mayor parte de los vertebrados terrestres eran reptiles. Actualmente existen sólo unas 6,000 especies.

Entre los reptiles del presente están las tortugas de tierra y de mar, los cocodrilos, los lagartos y las serpientes.

✔ Autoevaluación

1. ¿Qué adaptaciones de los reptiles son importantes para la vida terrestre?

2. ¿Por qué se tienen que reproducir por fecundación interna los animales que ponen huevos?

(Consulta la página 636 para comprobar tus respuestas.)

371

IS THAT A FACT!

Like modern reptiles and birds, dinosaurs laid their eggs on land. There is a rich fossil record of these eggs. According to the National Geographic Society, eggs (some that contain embryos) and nests have been discovered at over 199 sites. Most of the sites—109—are in Asia. In North America, eggs have been found in 39 locations.

REAL-WORLD CONNECTION

Gerontologists—scientists who study aging in humans—are looking to the animal world to better understand the aging process. Why do some animals, such as turtles, live so long? And not only are turtles long-lived, they show almost no signs of aging until the very end. There is little difference between a 30-year-old tortoise and a 70-year-old tortoise.

Multicultural CONNECTION

People of Indonesia, Sri Lanka, the Philippines, and Southeast Asia have learned that they can use the water monitor—an 8 ft long carnivorous lizard—to find water. If the lizard is seen walking about, it is almost always heading for water.

Homework

Reporting on Reptiles
Have students choose their favorite reptile and write a brief report about the animal. Students should include the following information:

size, appearance, range (where is it found in the world?), habitat, diet, adaptations to its particular way of life (i.e. what tricks or tools does it use to survive), relationship to people, and its status in the world (endangered? rare? common?)

MISCONCEPTION ALERT

Despite lingering urban legends, there has never been a confirmed sighting of an alligator living in the sewers of New York City.

Figura 32 *La parte inferior del caparazón de esta tortuga se dobla en ambos extremos para que la tortuga pueda cerrarla contra la parte de arriba. El resultado es una fortaleza ósea.*

Figura 33 *Las tortugas de mar tienen un caparazón aerodinámico que les ayuda a nadar y girar en el agua con gran rapidez.*

Figure 34 *Los caimanes se diferencian de los cocodrilos por la forma de la cabeza. Los caimanes tienen la cabeza ancha y la punta del hocico redonda, mientras que los cocodrilos tienen la cabeza angosta y el hocico en punta.*

Tortugas Las 250 especies de tortugas de mar y tierra son sólo parientes lejanos del resto de los reptiles.

El rasgo que las distingue del resto de los reptiles es su caparazón. El caparazón hace que la tortuga sea lenta y poco flexible, de modo que es muy poco probable que logre escapar de sus depredadores corriendo. Por otro lado, muchas tortugas pueden meter la cabeza y las patas en la armadura de su caparazón para protegerse, como ves en la **Figura 32.**

La mayoría de las tortugas pasan por lo menos una parte de su vida en el agua. Como ves en la **Figura 33,** las tortugas marinas han desarrollado aletas en sus patas delanteras. La hembra sale a tierra sólo para desovar en la arena de la playa. Las tortugas del desierto son distintas de las demás y viven sólo en la tierra.

Cocodrilos y caimanes Las 22 especies de cocodrilos y caimanes son carnívoras; se alimentan de insectos acuáticos, peces, tortugas, aves y mamíferos. Estos reptiles pasan la mayor parte del tiempo en el agua. Como tienen los ojos y la nariz en la parte superior de la cabeza, pueden ver a su alrededor mientras están escondidos bajo el agua, lo que les da una gran ventaja sobre sus presas. Existen dos tipos de cocodrilos, como muestra la **Figura 34.**

372

SCIENCE HUMOR

Q: Why does a turtle live in a shell?

A: because it can't afford an apartment

Lagartos La mayoría de los reptiles modernos son lagartos y serpientes. Existen cerca de 4,000 especies conocidas de lagartos y alrededor de 1,600 especies de serpientes.

Los lagartos viven en desiertos, bosques, praderas y selvas. Entre ellos se encuentran los camaleones, los gecos, los esquincos y las iguanas. La mayoría comen pequeños invertebrados, pero muchos son hervíboros. El lagarto más grande del mundo, que ves en la **Figura 35,** es el dragón de Komodo, de Indonesia. Mide 3 m de largo y pesa 140 Kg. Sin embargo, casi todos los lagartos miden menos de 30 cm.

Figura 35 *Los dragones de Komodo comen venados, cerdos y cabras. Hasta han llegado a comer seres humanos en raras ocasiones.*

Serpientes Las serpientes no tienen patas. Se mueven contrayendo su cuerpo muscular. En superficies lisas, las escamas del abdomen les ayudan a avanzar.

Todas las serpientes son carnívoras y se alimentan de animales pequeños y huevos. Devoran a sus presas enteras, como puedes ver en la **Figura 36.** Su mandíbula tiene cinco articulaciones, lo que les permite abrir mucho la boca y tragar presas grandes. Los pitones y las boas matan a su presa apretándola hasta asfixiarla. Otras tienen glándulas venenosas y colmillos especiales para inyectar veneno. El veneno mata o inmoviliza a la presa y contiene poderosas enzimas que comienzan la digestión. Las serpientes no tienen muy buena vista ni oído, pero su olfato es muy sensible. Cuando la serpiente agita su lengua bifurcada, las pequeñas partículas y moléculas del aire se pegan a la lengua. Cuando la vuelve a meter en la boca, coloca la punta en dos agujeritos del paladar que sirven para "probar" las moléculas.

Figura 36 *Esta serpiente está devorando el huevo de un ave.*

1. ¿Qué característica distingue a las tortugas del resto de los reptiles?

2. ¿Qué adaptaciones tienen las serpientes para comer?

3. **Aplicar conceptos** Como los reptiles, los mamíferos tienen un huevo amniótico, pero los mamíferos nacen vivos. El embrión se desarrolla a partir de un huevo fecundado dentro del cuerpo de la hembra. ¿Qué partes del huevo amniótico del reptil no le hacen falta a un mamífero? Explica tu respuesta.

373

RESEARCH

Writing Have students investigate and write a report on the differences between the diet of an alligator and that of a similarly sized carnivorous mammal (a lion or tiger). Which of these animals eats more? Why? (The mammal needs to eat more because mammals use a large amount of energy to keep themselves warm.)

Quiz

1. What trait makes turtles stand out from other reptiles? (their shell)

2. What animals are the closest living relatives of dinosaurs? (crocodiles and alligators)

ALTERNATIVE ASSESSMENT

Writing Have students imagine that they are one of the reptiles in this section. They are also restaurateurs trying to attract other reptiles to their new restaurant. Have students prepare accurate menus containing delectables that will surely please their ectothermic thick-skinned customers.

PORTFOLIO

▼ *Answers to Review*

1. Unlike other reptiles, turtles have a shell that they use for protection.

2. Snakes have special jaws so that they can open their mouth wide and swallow their prey whole. Some snakes have venom that begins digesting their prey.

3. Mammals do not need the shell of the amniotic egg because it does not need to be protected from the outside environment. Students might also suggest that mammals do not need the yolk because mammal embryos develop inside the mother and receive nutrients directly from her body.

Reinforcement Worksheet 15 "Coldblooded Critters"

Critical Thinking Worksheet 15 "Frogs Aren't Breathing Easy"

Chapter Highlights

VOCABULARY DEFINITIONS

SECTION 1

vertebrate an animal with a skull and a backbone; includes mammals, birds, reptiles, amphibians, and fish

vertebrae segments of bone or cartilage that interlock to form a backbone

endotherm an animal that maintains a constant body temperature despite temperature changes in its environment

ectotherm an animal whose body temperature fluctuates with the temperature of its environment

SECTION 2

fins fanlike structures that help fish move, turn, stop, and balance

scales bony structures that cover the skin of bony fishes

lateral line system row of tiny sense organs along the sides of a fish's body

gills organs that remove oxygen from the water and carbon dioxide from the blood

external fertilization fertilization of an egg by sperm that occurs outside the body of the female

internal fertilization fertilization of an egg by sperm that occurs inside the body of a female

denticles small, sharp toothlike structures on the skin of cartilaginous fishes

swim bladder balloonlike organ that is filled with oxygen and other gases; gives bony fishes their buoyancy

Resumen del capítulo

SECCIÓN 1

Vocabulario

vertebrado (*pág. 356*)
vértebra (*pág. 357*)
homotermo (*pág. 358*)
heterotermo (*pág. 358*)

Notas de la sección

- En algún momento de su desarrollo, los cordados tienen un notocordio, una médula espinal, hendiduras branquiales faríngeas y cola.

- Entre los cordados se encuentran los anfioxos, los tunicados y los vertebrados. La mayoría de los cordados son vertebrados.

- La columna vertebral y el cráneo, hechos de hueso o cartílago, distinguen a los vertebrados del resto de los cordados.

- La columna vertebral está hecha de unidades llamadas vértebras.

- Los vertebrados pueden ser homotermos o heterotermos.

- Los homotermos pueden controlar la temperatura del cuerpo mediante reacciones químicas en sus células, pero los heterotermos no tienen esta capacidad.

SECCIÓN 2

Vocabulario

aletas (*pág. 360*)
escamas (*pág. 360*)
sistema de líneas laterales (*pág. 360*)
branquias (*pág. 360*)
fecundación externa (*pág. 360*)
fecundación interna (*pág. 360*)
dentículos (*pág. 362*)
vejiga natatoria (*pág. 362*)

Notas de la sección

- Existen tres tipos de peces vivientes: agnatos (sin mandíbula), cartilaginosos y óseos.

- Los peces cartilaginosos tienen el hígado aceitoso, lo que les ayuda a flotar.

☑ Comprobar destrezas

Conceptos de matemáticas

¿CUÁNTAS ESPECIES HAY? Si existen 6,000 especies de reptiles y el 67 por ciento son lagartos, ¿cuántas especies de lagartos hay?

El sesenta y siete por ciento de 6,000 es:

$$6{,}000 \times 0.67 = 4{,}020$$

Hay 4,020 especies de lagartos.

Comprensión visual

METAMORFOSIS La mayoría de los anfibios pasan por una metamorfosis, es decir, cambian cuando se convierten en animales adultos. La Figura 21 de la página 366 ilustra la metamorfosis de una rana. Sigue las flechas para ver cómo pasa de huevo a renacuajo y luego a adulto.

374

Lab and Activity Highlights

Floating a Pipe Fish `PG 612`

A Prince of a Frog `PG 614`

Datasheets for LabBook
(blackline masters for these labs)

SECCIÓN 2

- La mayoría de los peces tienen una vejiga natatoria, que es un órgano parecido a un globo que hace que el pez flote.

- En la fecundación externa, los huevos son fecundados fuera del cuerpo de la hembra. En la fecundación interna, los huevos son fecundados dentro del cuerpo de la hembra.

Experimentos

Manda a nadar a un pez tubo *(pág. 612)*

SECCIÓN 3

Vocabulario

pulmón *(pág. 364)*
renacuajo *(pág. 366)*
metamorfosis *(pág. 366)*

Notas de la sección

- Los anfibios fueron los primeros vertebrados que vivieron en tierra firme.

- Los anfibios respiran tomando aire en los pulmones y absorbiendo oxígeno a través de la piel.

- Los anfibios comienzan la vida en el agua y usan branquias para respirar. En la metamorfosis las pierden, y desarrollan pulmones y patas para vivir en la tierra.

- Entre los anfibios modernos están las cecilias, las salamandras, las ranas y los sapos.

Experimentos

De príncipe a rana *(pág. 614)*

SECCIÓN 4

Vocabulario

terápsido *(pág. 369)*
huevo amniótico *(pág. 370)*

Notas de la sección

- Los reptiles evolucionaron de los anfibios al adaptarse a la vida terrestre.

- Los reptiles tienen una piel gruesa y escamosa que los protege de la deshidratación.

- La cáscara dura del huevo amniótico impide que se seque y protege el embrión.

- Dentro del huevo amniótico, el líquido amniótico rodea y protege el embrión.

- Los vertebrados que evolucionaron de los reptiles primitivos son los reptiles modernos, las aves y los mamíferos.

- Entre los reptiles modernos están las tortugas de tierra y de mar, los lagartos, las serpientes, los cocodrilos y los caimanes.

SECTION 3

lung a saclike organ that takes oxygen from the air and delivers it to the blood

tadpole aquatic larvae of an amphibian

metamorphosis process in which an insect or other animal changes form as it develops from an embryo or larva to an adult

SECTION 4

therapsid prehistoric reptile ancestor of mammals

amniotic egg egg containing amniotic fluid to protect the developing embryo; usually surrounded by a hard shell

Vocabulary Review Worksheet 15

Blackline masters of these Chapter Highlights can be found in the **Study Guide.**

 internet

HRW VISITA: go.hrw.com

Visita el sitio web de HRW para encontrar una serie de herramientas de aprendizaje relacionadas con este capítulo. Sólo tienes que escribir la palabra clave:

PALABRA CLAVE: HSTVR1

SCI LINKS **NSTA** VISITA: www.scilinks.org

Visita el sitio web de la **Asociación Nacional de Maestros de Ciencias** *(National Science Teachers Association)* para encontrar recursos de Internet relacionados con este capítulo. Sólo escribe el **ENLACE DE CIENCIAS** para obtener más información sobre el tema:

TEMA: Vertebrados	ENLACE: HSTL380
TEMA: Peces	ENLACE: HSTL385
TEMA: Anfibios	ENLACE: HSTL390
TEMA: Reptiles	ENLACE: HSTL395

375

Lab and Activity Highlights

LabBank

Whiz-Bang Demonstrations, The Fish in the Abyss, Demo 10

Long-Term Projects & Research Ideas, Project 16

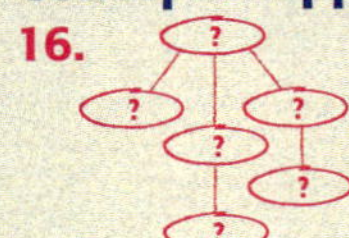

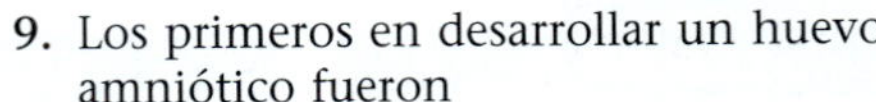

Repaso del capítulo

UTILIZAR EL VOCABULARIO

Escoge el término correcto para completar las siguientes oraciones:

1. En algún momento de su desarrollo, todos los cordados tienen __?__. *(pulmones y notocordio* o *médula espinal y cola)*

2. Los mamíferos evolucionaron a partir de un antepasado primitivo llamado __?__. *(terápsido* o *dinosaurio)*

3. Los peces son __?__. *(homotermos* o *heterotermos)*

4. Cuando una rana pone huevos que más tarde son fecundados por el macho, es un ejemplo de fecundación __?__. *(interna* o *externa)*

5. Las vértebras cubren y protegen __?__ de los vertebrados. *(el notocordio* o *la médula espinal)*

COMPRENDER CONCEPTOS

Opción múltiple

6. ¿Cuál de los siguientes no es un vertebrado?
   a. renacuajo
   b. lagarto
   c. lamprea
   d. tunicado

7. Los renacuajos se convierten en ranas mediante el proceso de
   a. evolución.
   b. fecundación interna.
   c. metamorfosis.
   d. regulación de la temperatura.

8. La vejiga natatoria se encuentra en
   a. los peces agnatos.
   b. los peces cartilaginosos.
   c. los peces óseos.
   d. los anfioxos.

9. Los primeros en desarrollar un huevo amniótico fueron
   c. los peces óseos.
   b. las aves.
   c. los reptiles.
   d. los mamíferos.

10. La yema contiene
    a. alimento para el embrión.
    b. líquido amniótico.
    c. desechos.
    d. oxígeno.

11. Tanto los peces óseos como los cartilaginosos tienen
    a. dentículos.
    b. aletas.
    c. un hígado aceitoso.
    d. vejiga natatoria.

12. Los reptiles están adaptados a la vida terrestre porque
    a. respiran a través de la piel.
    b. son heterotermos.
    c. tienen la piel gruesa y húmeda.
    d. tienen huevos amnióticos.

Respuesta corta

13. ¿Cómo respiran los anfibios?

14. ¿Qué características de los peces les permiten vivir en el agua?

15. ¿Cómo obtiene oxígeno el embrión en el huevo amniótico?

Organizar conceptos

16. Usa los siguientes términos para crear un mapa de ideas: dinosaurio, tortuga, reptiles, anfibios, peces, tiburón, salamandra, vertebrados.

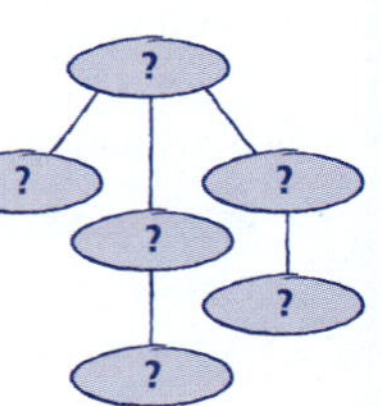

Escribe una o dos oraciones para responder a las siguientes preguntas:

17. Imagínate que encuentras un animal que tiene columna vertebral y branquias, pero no tiene notocordio. ¿Es un cordado? ¿Cómo puedes estar seguro?

18. Imagínate que encuentras un tiburón que no tiene los músculos necesarios para bombear agua por sus branquias. ¿Qué te dice esto sobre el tipo de vida del tiburón?

19. La serpiente de cascabel no tiene muy buena vista, pero puede detectar un cambio de temperatura de hasta tres milésimas de un grado Celsius. ¿Para qué le sirve esta capacidad?

20. Afuera la temperatura es de 43°C, y la temperatura corporal normal del velociraptor es de 38°C. ¿Dónde crees que encontrarías al velociraptor, en el sol o en la sombra? Explica por qué.

21. Una serpiente de Costa Rica puede comerse un ratón cuya masa es un tercio mayor que la de la serpiente. ¿Cuánto comes tú? Apunta tu propia masa en kilogramos. Para convertir tu masa de libras a kilogramos, divídela por 2.2. Si te fueras a comer un platillo cuya masa fuera un tercio mayor que la de tu cuerpo, ¿qué masa tendría el platillo en kilogramos?

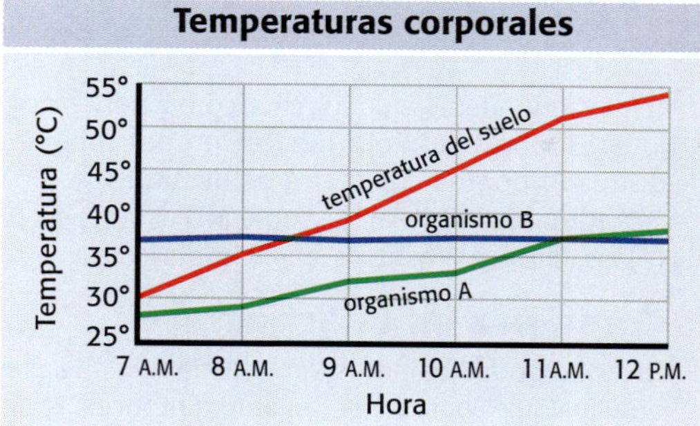

Examina esta gráfica y responde a las siguientes preguntas.

Temperaturas corporales

22. ¿Cómo cambian las temperaturas del organismo A y el organismo B con la temperatura del suelo?

23. ¿Cuál de estos organismos es más probable que sea heterotermo? ¿Por qué?

24. ¿Cuál de estos organismos es más probable que sea homotermo? ¿Por qué?

AHORA, ¿qué piensas?

Revisa tus respuestas a las preguntas de la página 355 que escribiste en el cuaderno de ciencias. ¿Han cambiado tus respuestas? Si es necesario, corrige tus respuestas basándote en lo que has aprendido en este capítulo.

377

CRITICAL THINKING AND PROBLEM SOLVING

17. Yes; animals with a backbone are vertebrates, and all vertebrates are chordates; the notochord is replaced by the backbone as a vertebrate develops.
18. The shark must swim constantly to pass water over its gills.
19. A snake could use such a skill to detect the warm body of a mouse or other prey.
20. The velociraptor would be in the shade. Because it is an ectotherm, its body temperature will respond to the environmental temperature. Since the outside temperature is already higher than the velociraptor's normal temperature, it will try to find a cool spot in the shade to cool its body down.

MATH IN SCIENCE

21. Answers will vary but can be determined by the following: mass (kg) + (mass (kg) × $\frac{1}{3}$) = mass of meal (kg)

INTERPRETING GRAPHICS

22. The body temperature of organism A increases as ground temperature increases. The body temperature of organism B does not change with ground temperature.
23. Organism A is probably an ectotherm. Its body temperature is dependent on the temperature of the environment.
24. Organism B is probably an endotherm. Its body temperature is constant even though the environmental temperature changes.

NOW WHAT DO YOU THINK?

1. Ectotherms are sometimes called cold-blooded. They do not control their body temperature through the chemical reactions of their cells. Their body temperature depends on the temperature of the environment.

2. A reptile has thick, dry skin and lays amniotic eggs. An amphibian has thin, moist skin that it can breathe through, and it must live in or near water.

Blackline masters of this Chapter Review can be found in the **Study Guide.**

Background

Fishes can reverse their direction without slowing down at all. They can also turn with a turning radius as small as 10 to 30 percent of the length of their bodies. Yellowfin tuna have been reported to swim as fast as 73 km/h. Compare all of this with a typical human-made ship. A ship cruises at about 37 km/h, it must reduce its speed by about 50 percent to turn around, and it requires a circle with a radius of 10 times its hull length!

The mobility of a fish and a ship differ in other ways as well. A ship's motion is limited by the rigidity of its hull, while a fish's body is very smooth and flexible. Robotics technology is still too simple to accurately reproduce the flapping of a fish's tail. For this reason, fish-inspired mechanisms had performed rather poorly until the creation of RoboTuna.

Discussion

Lead a classroom discussion about swimming styles. Ask the students to comment on their own swimming technique. Do they swim more like a fish or a rowboat? How do they turn around? What kinds of swimming techniques have other animals adopted? You might also discuss the swimming methods of dogs, jellyfish, frogs, and whales.

Pez robot

¿Qué tiene aletas y cola, nada como pez, y no lo es? ¡RoboTuna! RoboTuna es un pez robot diseñado por los científicos del Instituto Tecnológico de Masachussetts, MIT (Massachusetts Institute of Technology).

Como pez en el agua

No cabe duda: los peces son más rápidos y ágiles que cualquier barco o submarino. Entonces, ¿por qué no se construyen barcos que sean más parecidos a los peces; por ejemplo, con una cola que los impulse? Esto mismo se preguntaron los científicos del MIT, y decidieron construir a RoboTuna, un modelo basado en el atún de aleta azul. El pez robot mide 124 cm de largo y tiene seis motores, piel de espuma de poliuretano y Lycra, y un esqueleto de costillas de aluminio con bisagras conectadas por poleas e hilos.

Fuerza y movimiento

Los científicos del MIT piensan que si el diseño de los barcos fuera más parecido al de los peces, los barcos usarían mucha menos energía y ahorrarían dinero. Al navegar, el barco deja una estela de pequeños remolinos llamados *vórtices* detrás, que aumentan la fricción entre el barco y el agua. En cambio, los peces pueden sentir los vórtices y reaccionan moviendo la cola y creando sus propios vórtices, que contrarrestan el efecto de los vórtices originales, impulsando al pez hacia adelante sin mayor esfuerzo.

RoboTuna tiene sensores especiales que miden el cambio de presión de manera muy parecida a como un atún de verdad siente los vórtices. El pez robot mueve la cola y produce vórtices, y así puede nadar como un pez real. Puede parecer extraño, pero es posible que RoboTuna represente el comienzo de una nueva era en diseño náutico

Observa los vórtices

▶ Llena tres cuartas partes de una fuente para hornear con agua y espera hasta que el agua esté quieta. Amarra un estambre o un listón a la punta de un lápiz y mueve el lápiz por el agua, con el estambre o listón flotando detrás. ¿Cómo reacciona el estambre o listón? ¿Dónde están los vórtices?

Answers to Viewing Vortices

The end of the ribbon or yarn should wiggle back and forth. This is caused in part by the creation of vortices, which trail behind the ribbon or yarn. You might want to put some sand on the bottom of the pan and have students observe how the sand grains move as the pencil passes over them.

CEREBROS CALENTITOS EN AGUA FRÍA

De las 30,000 clases de peces del mundo, sólo unas cuantas están equipadas con sus propios calentadores cerebrales. *¿Calentadores cerebrales?* ¿Para qué necesita un pez calentadores cerebrales? Antes de contestar esta pregunta, tienes que pensar en cómo se mantienen calentitos los peces en las frías aguas del océano.

Es cuestión de calor

La mayoría de los peces y animales marinos son heterotermos. La temperatura del cuerpo de un heterotermo es muy parecida a la de su ambiente. Los homotermos, por su parte, mantienen una temperatura corporal constante sin importar la temperatura de su ambiente. Los seres humanos somos homotermos. Las aves y algunos mamíferos, como los perros, los elefantes y las ballenas son homotermos. Pero sólo unos cuantos peces como el atún, son homotermos. Estos peces son de sangre fría, pero pueden calentar ciertas partes de su cuerpo. Pueden buscar su alimento en aguas extremadamente frías pero pagan un alto precio por la capacidad de nadar en aguas frías, pues consumen mucha energía.

Un organismo homotermo necesita mucha más energía que uno heterotermo. Algunos peces, como el pez espada, el pez aguja y el pez vela tienen adaptaciones que les permiten calentar sólo ciertas partes de su cuerpo: sólo calientan los ojos y el cerebro. Así es: ¡tienen calentadores cerebrales especiales!

▶ *¿Por qué crees que es importante proteger los ojos y el cerebro de temperaturas extrema damente frías?*

Un cerebro tibio

Los peces con calentador cerebral tienen una pequeña masa de músculos adherida a cada ojo, que sirve como termostato, la cual regula la temperatura del cerebro y de los ojos cuando nadan a través de zonas de distinta temperatura. Estos "músculos calentadores" ayudan a mantener las frágiles funciones de los nervios que son importantes para encontrar a sus presas. Por ejemplo, al pez espada le permiten nadar en las tibias aguas de la superficie, y también sumergirse en profundidades de 485 m, donde la temperatura baja a niveles casi helados, permitiendole buscar alimento en una gran variedad de lugares.

Los heterotermos en acción

▶ Pregunta en una tienda de mascotas, en la que vendan peces, cuál es la temperatura ideal de los distintos peces de diferentes partes del mundo. Compara, por ejemplo, la temperatura ideal para las carpas doradas, el pez disco y el pez ángel. ¿Por qué crees que la temperatura de una pecera debe controlarse cuidadosamente?

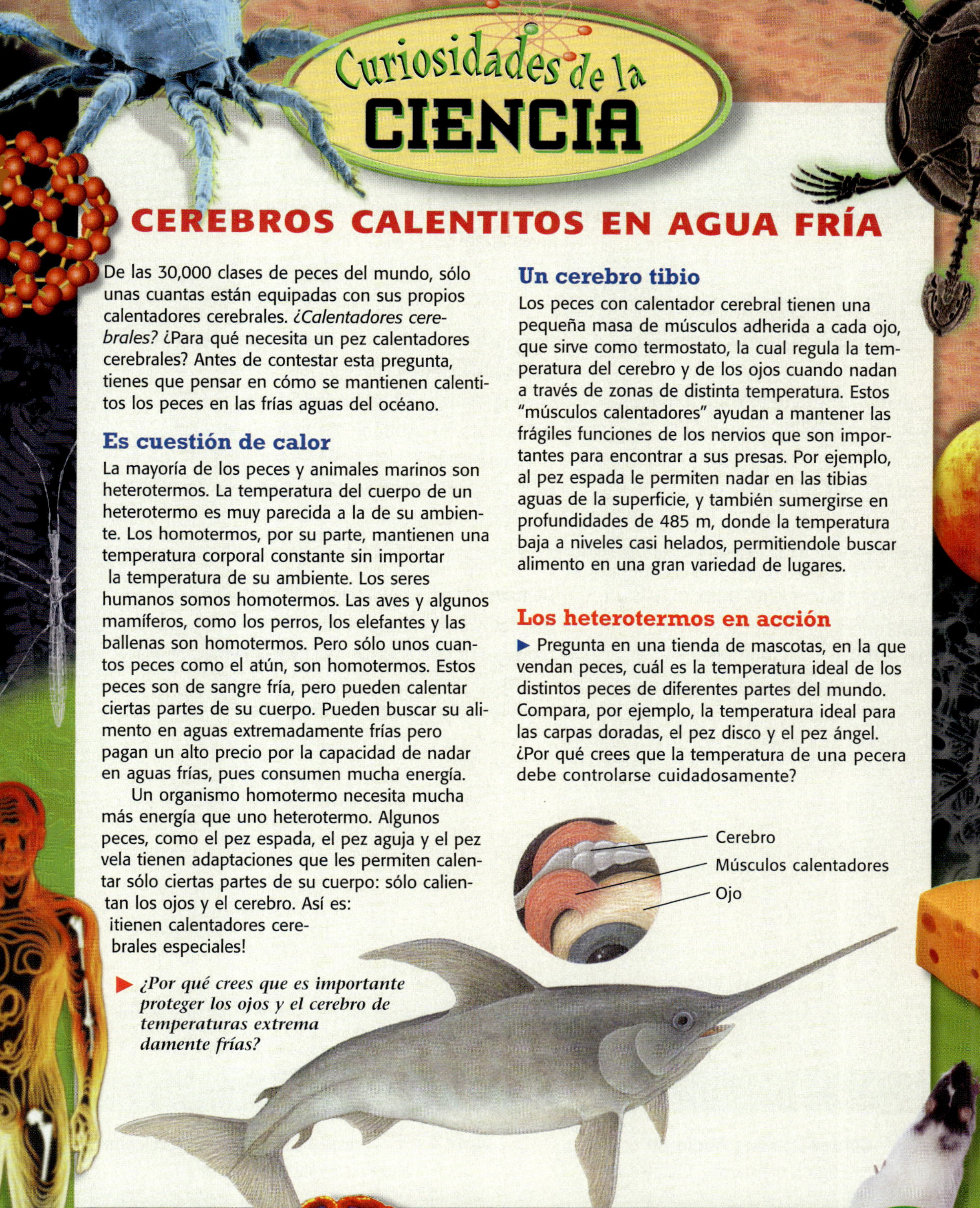

Answer to Ectotherms in Action

Student answers will vary, depending on the fish they select. For instance, a Japanese goldfish is best suited for cool temperatures near 18°C (64°F), a South American discus requires warm temperatures of about 28°C (82°F), and a South American angelfish fares best in a temperature range of 24–26°C (75–79°F). The temperature in fish tanks must be controlled to ensure the fish's survival.

Background

The body temperature of an ectothermic animal, commonly called "coldblooded," is about the same temperature as its surroundings. Amphibians, reptiles, and most fish are ectothermic. When the sun warms the body of an ectothermic animal, its body temperature can rise above that of an endothermic, or "warmblooded" animal.

Endothermic animals include all birds and mammals. The body of an endothermic animal produces most of its heat by metabolizing food. In many endothermic animals, a layer of fat beneath the skin and a covering of feathers, fur, or hair also help the animal maintain a constant body temperature. The principal means of reducing body heat are panting and sweating.

Many ectothermic animals partially control their body temperature through their behavior. For example, ectothermic land animals may bask in the sunlight to become warmer or move to the shade to cool down. Fish may swim closer to the surface of the water to warm themselves. If they become too warm, they may swim to deeper, cooler water.

Chapter Organizer

CHAPTER ORGANIZATION	TIME MINUTES	OBJECTIVES	LABS, INVESTIGATIONS, AND DEMONSTRATIONS
Chapter Opener pp. 380–381	45		**Investigate!** Let's Fly! p. 381
Section 1 Birds	90	▶ Name two characteristics that birds share with reptiles. ▶ Describe the characteristics of birds that make them well suited for flight. ▶ Explain *lift.* ▶ List some advantages of migration.	**Demonstration,** Observing Bird Bones, p. 385 in ATE **QuickLab,** Bernoulli Effect, p. 386 **Making Models,** What? No Dentist Bills? p. 616 **Datasheets for LabBook,** What? No Dentist Bills? Datasheet 33 **Labs You Can Eat,** Why Birds of a Beak Eat Together, Lab 8
Section 2 Mammals	90	▶ Describe common characteristics of mammals. ▶ Explain the differences between monotremes, marsupials, and placental mammals. ▶ Give some examples of each type of mammal.	**Demonstration,** Comparing Skulls, p. 394 in ATE **Discovery Lab,** Wanted: Mammals on Mars, p. 617 **Datasheets for LabBook,** Wanted: Mammals on Mars, Datasheet 34 **Long-Term Projects & Research Ideas,** Project 17

TECHNOLOGY RESOURCES

Guided Reading Audio CD
English or Spanish, Chapter 16

One-Stop Planner CD-ROM with Test Generator

CNN. Eye on the Environment, Development and Preservation, Segment 2

Science Discovery Videodiscs
Science Sleuths: The Blob

CLASSROOM WORKSHEETS, TRANSPARENCIES, AND RESOURCES	SCIENCE INTEGRATION AND CONNECTIONS	REVIEW AND ASSESSMENT
Directed Reading Worksheet 16 **Science Puzzlers, Twisters & Teasers,** Worksheet 16		
Directed Reading Worksheet 16, Section 1 **Transparency 63,** The Digestive System of a Bird **Transparency 64,** Flight Adaptations of Birds **Transparency 182,** Wing Shape Creates Differences in Air Speed **Critical Thinking Worksheet 16,** A Puzzling Piece of Paleontology	**Multicultural Connection,** p. 383 in ATE **Connect to Physical Science,** p. 386 in ATE **Multicultural Connection,** p. 389 in ATE **Across the Sciences:** The Aerodynamics of Flight, p. 410	**Homework,** pp. 382, 385, 387 in ATE **Self-Check,** p. 384 **Review,** p. 388 **Review,** p. 391 **Quiz,** p. 391 in ATE **Alternative Assessment,** p. 391 in ATE
Directed Reading Worksheet 16, Section 2 **Math Skills for Science Worksheet 31,** The Unit Factor and Dimensional Analysis **Reinforcement Worksheet 16,** Mammals Are Us	**Multicultural Connection,** p. 394 in ATE **Connect to Physical Science,** p. 395 in ATE **Cross-Disciplinary Focus,** p. 396 in ATE **Earth Science Connection,** p. 397 **Math and More,** p. 397 in ATE **Connect to Earth Science,** p. 397 in ATE **MathBreak,** Ants for Dinner! p. 398 **Cross-Disciplinary Focus,** p. 398 in ATE **Multicultural Connection,** p. 399 in ATE **Apply,** p. 400 **Multicultural Connection,** p. 402 in ATE **Real-World Connection,** p. 402 in ATE **Environmental Science Connection,** p. 403 **Weird Science:** Naked Mole-Rats, p. 411	**Homework,** pp. 393, 396 in ATE **Review,** p. 395 **Self-Check,** p. 398 **Self-Check,** p. 403 **Review,** p. 405 **Quiz,** p. 405 in ATE **Alternative Assessment,** p. 405 in ATE

Holt, Rinehart and Winston On-line Resources

go.hrw.com

For worksheets and other teaching aids related to this chapter, visit the HRW Web site and type in the keyword: **HSTVR2**

National Science Teachers Association

www.scilinks.org

Encourage students to use the *sci*LINKS numbers listed with the Chapter Highlights to access information and resources on the **NSTA** Web site.

END-OF-CHAPTER REVIEW AND ASSESSMENT

Chapter Review in Study Guide
Vocabulary and Notes in Study Guide
Chapter Tests with Performance-Based Assessment, Chapter 16 Test
Chapter Tests with Performance-Based Assessment, Performance-Based Assessment 16
Concept Mapping Transparency 16

Chapter Resources & Worksheets

Visual Resources

TEACHING TRANSPARENCIES

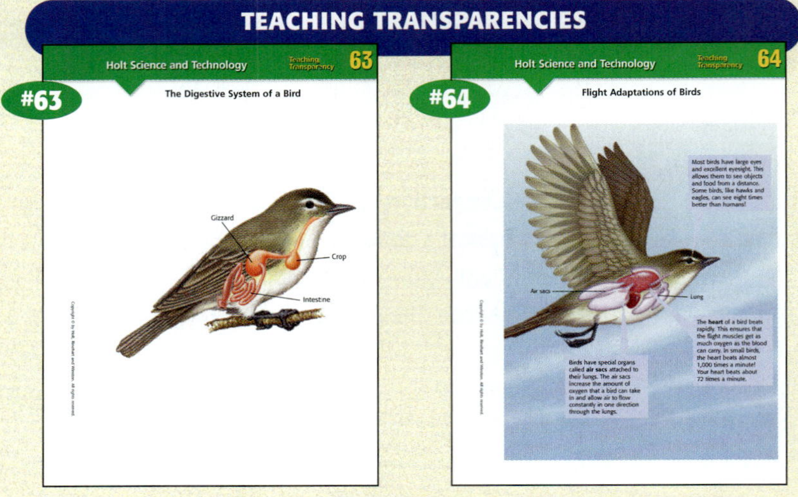

TEACHING TRANSPARENCIES

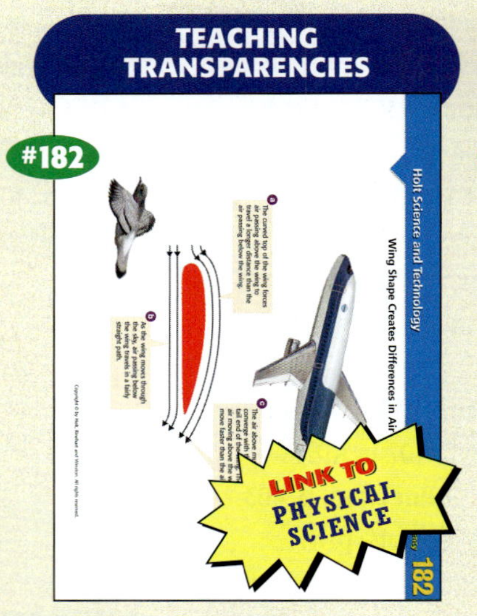

CONCEPT MAPPING TRANSPARENCY

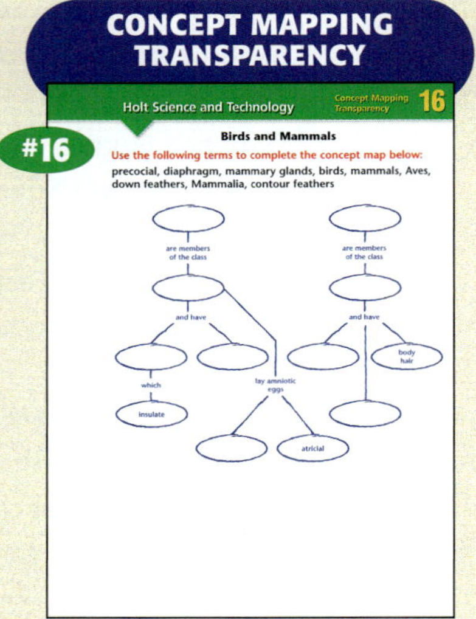

Meeting Individual Needs

DIRECTED READING

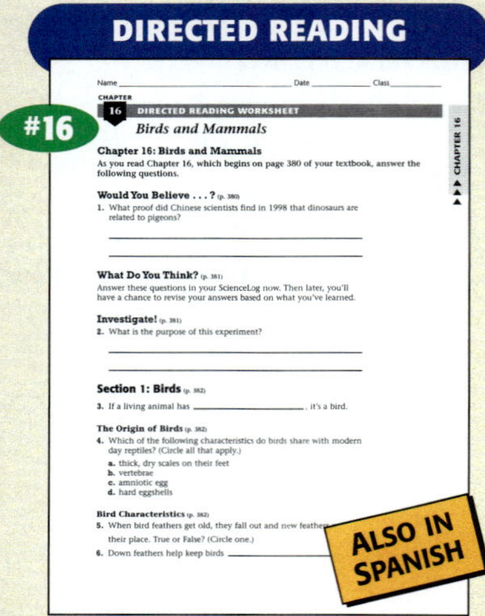

REINFORCEMENT & VOCABULARY REVIEW

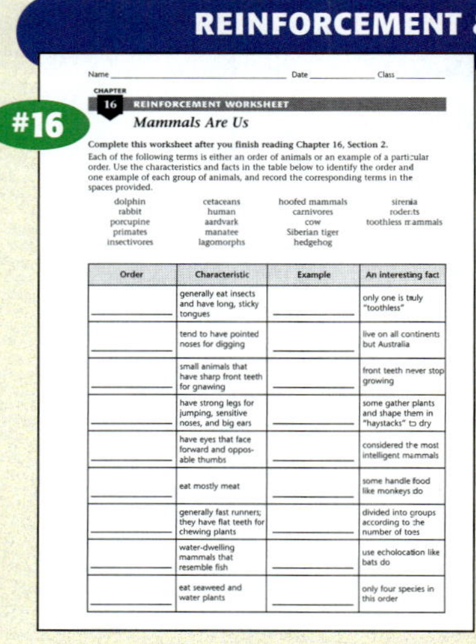

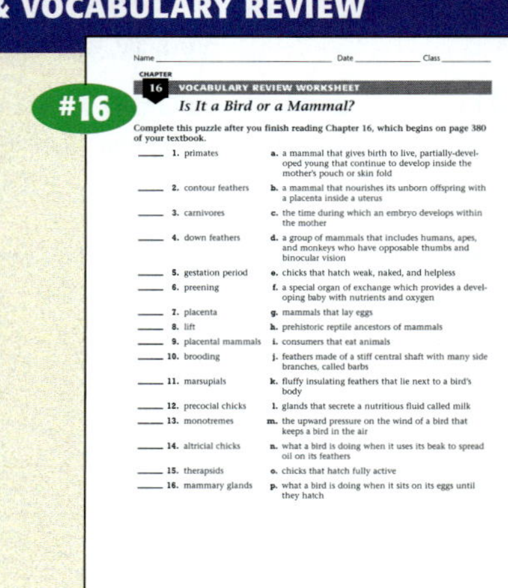

SCIENCE PUZZLERS, TWISTERS & TEASERS

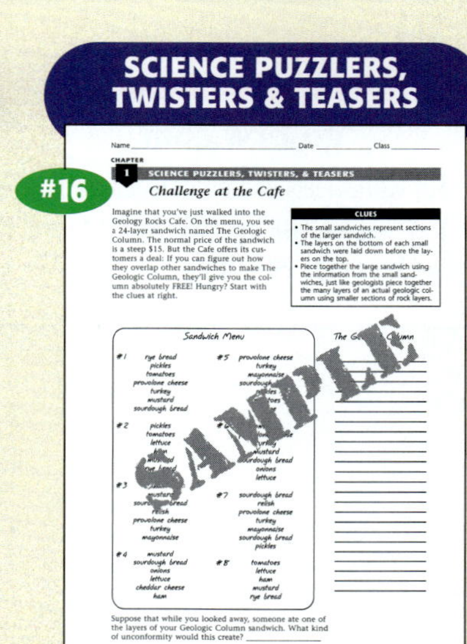

Chapter 16 • Birds and Mammals

Review & Assessment

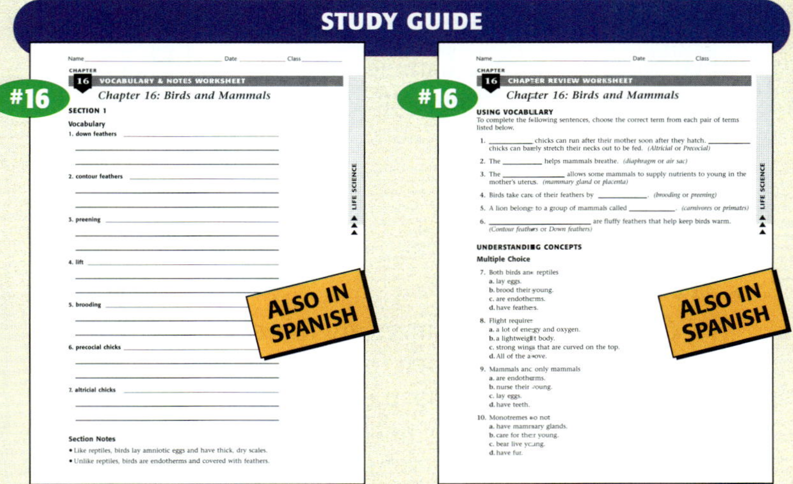

STUDY GUIDE

#16 Chapter 16: Birds and Mammals — Vocabulary & Notes Worksheet

#16 Chapter 16: Birds and Mammals — Chapter Review Worksheet

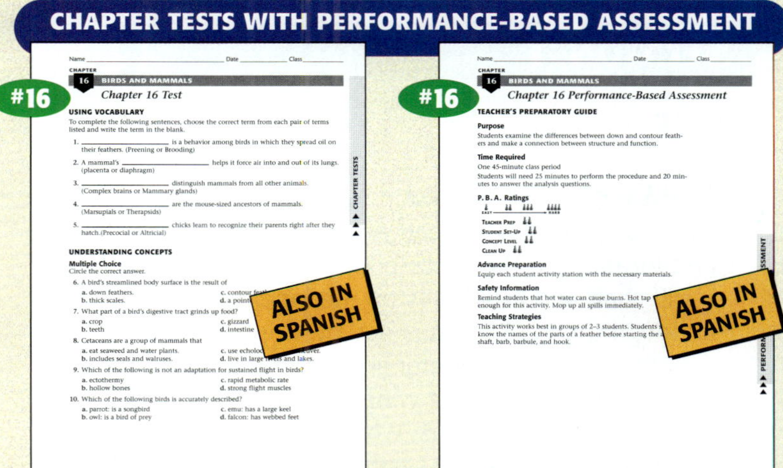

CHAPTER TESTS WITH PERFORMANCE-BASED ASSESSMENT

#16 Chapter 16 Test

#16 Chapter 16 Performance-Based Assessment

Lab Worksheets

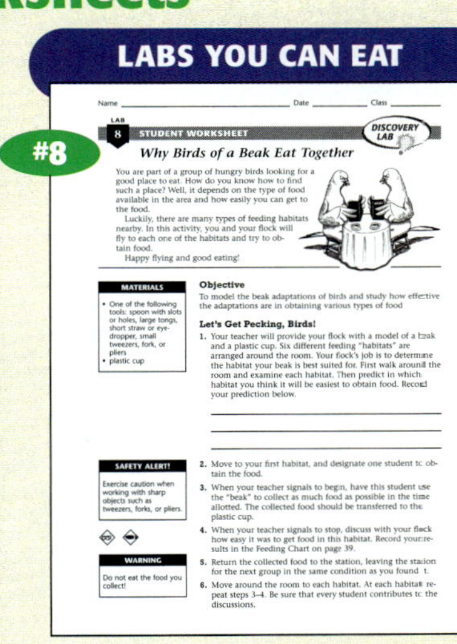

LABS YOU CAN EAT

#8 Why Birds of a Beak Eat Together

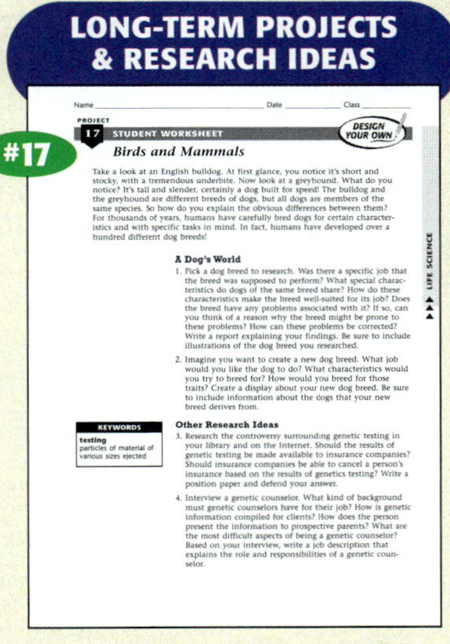

LONG-TERM PROJECTS & RESEARCH IDEAS

#17 Birds and Mammals

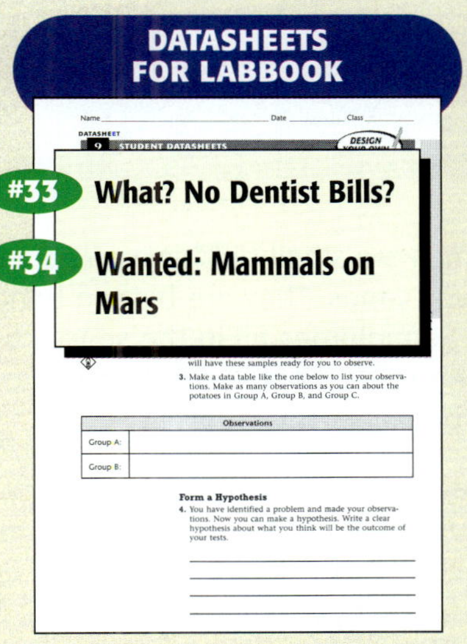

DATASHEETS FOR LABBOOK

#33 What? No Dentist Bills?

#34 Wanted: Mammals on Mars

Applications & Extensions

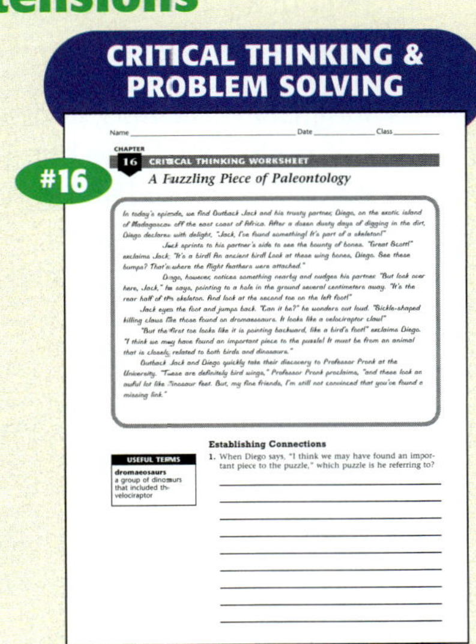

CRITICAL THINKING & PROBLEM SOLVING

#16 A Puzzling Piece of Paleontology

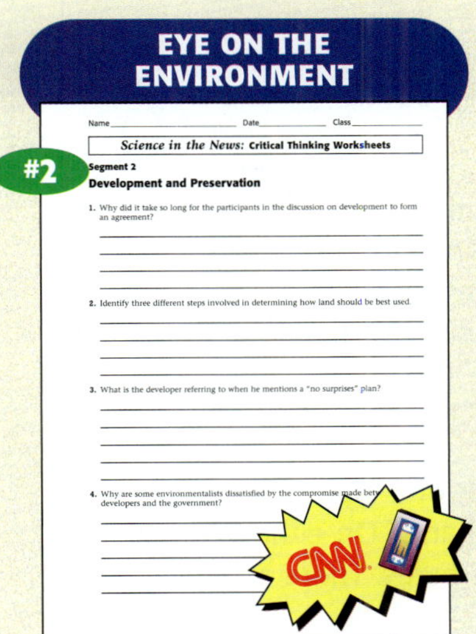

EYE ON THE ENVIRONMENT

#2 Science in the News: Critical Thinking Worksheets — Development and Preservation

Birds

▶ Birds of a Feather

Ornithologists classify feathers based on their function and location on the bird's body. Contour feathers have a stiff shaft and firm barbs on the inner and outer vanes, although the base often is downy. Contour feathers include those on the outer surface of the body and the flight feathers of the wings (remiges) and the tail (rectrices). The auriculars, which cover the ears, are small, modified contour feathers.

- Semiplumes are small but have a relatively large shaft with downy vanes. They are hidden beneath the body feathers. Semiplumes fill in the spaces between the larger contour feathers, provide insulation, and increase buoyancy in water birds.

- Down feathers are tiny, completely fluffy, and have no shaft. In adult birds, down feathers remain hidden beneath contour feathers, but they are the sole body covering for many newly hatched birds. Down is important for insulation and is most abundant in water birds.

IS THAT A FACT!

- The ruby-throated hummingbird has about 940 feathers. The whistling swan has about 25,000. But per unit of body weight, the hummingbird has more feathers than the swan. This is because small birds have a greater need for efficient heat retention.

▶ Stick Together

Branching out from the shaft of a typical contour feather are thin, rather stiff barbs. On either side of the barbs grow even smaller barbules. There are several hundred on each barb. Each barbule has tiny hooks that wrap around the barbules of the adjacent barb so that all the parts of the feather are tightly attached to each other. This construction is what provides firmness to the feathers and increases their ability to insulate and waterproof a bird.

IS THAT A FACT!

- Cormorants and anhingas are fish-eating birds that must dive to catch their food. To decrease their buoyancy and enhance their ability to dive deep, they have unusually heavy bones (for a bird), reduced air sacs, and nonwaterproof plumage. After a dive, they must find a perch and spread their wings to dry off.

▶ Forms of Flight

During flapping flight, a bird's inner wing functions as an anchor for the outer portion and helps provide speed and power. At the bottom of the downstroke, which propels the bird, the "wrist" flexes and begins the upstroke.

- Soaring flight allows a bird to gain and maintain altitude without flapping its wings. Instead, the birds "ride" on warm air currents called thermals. Vultures, hawks, and eagles can use soaring flight because they fulfill the requirements of large size, maneuverability, and light wing load (the ratio of a bird's wing surface area to its weight).

- Kingfishers, kestrals, and some hawks are able to hover, but none match the hummingbird's ability to maintain a position in midair. Hummingbirds move their wings only from the shoulder, which provides unusual flexibility and maneuverability.

IS THAT A FACT!

- Hummingbirds are the only birds that can fly backward.

- A chimney swift that lived 9 years was estimated to have flown 2,000,000 km during its lifetime.

SECTION 2

Mammals

▶ Fur

All mammals have hair, but not all "fur-bearing" mammals have true fur. True fur consists of two layers: a dense undergrowth of short ground hairs, which provide insulation, and longer guard hairs, which protect the skin and the ground hairs by repelling rain or snow.

▶ Rodent Teeth

Rodents, which include rats, mice, beavers, squirrels, guinea pigs, and capybaras, have evolved a unique jaw articulation to compensate for the huge incisors that grow throughout the animal's life. Both the upper and

the lower pair are separated from the cheek teeth by a large gap. When the cheek teeth are engaged in chewing, the lower jaw is pulled back and the incisors do not meet. When the animal is gnawing with its incisors, the lower jaw is pulled forward and downward, so the incisors can meet. Rodents spend a lot of time gnawing with their incisors to keep them worn down to a manageable level.

▶ Beneficial Bats

The insectivorous Mexican free-tailed bats in Texas eat an estimated 20,000 tons of insects each year. They provide a vital, chemical-free system of insect control.

- Guano from insectivorous bats is a valuable fertilizer in many countries because of its high nitrogen and phosphorous content. In some caves, the guano covers and helps preserve archaeologically valuable artifacts and fossils.

- Although some fruit-eating bats can reduce a farmer's harvest, many of them are essential to dispersing seeds that are responsible for new plant growth. Other bats eat only pollen and nectar and are the primary or exclusive pollinators of various plants.

IS THAT A FACT!

- Shrews and other small mammals have a high metabolic rate to compensate for large amounts of heat loss resulting from their large ratio of body surface to volume. A captive shrew (genus *Sorex*) consumed 3.3 times its body weight in a 24-hour period.

- From mid-April to mid-October, hundreds of thousands of Mexican free-tailed bats and thousands of several other species live in Carlsbad Caverns, in New Mexico. At night, the bats fly from the cave in great swarms that resemble clouds of smoke when seen from a distance. It is believed that early explorers of the southwest followed the "smoke" and discovered the cave.

- A three-toed sloth has nine cervical vertebrae and can turn its head 270°. Two-toed sloths and most mammals have seven neck vertebrae.

For background information about teaching strategies and issues, refer to the *Professional Reference for Teachers.*

Directed Reading Worksheet 16

Science Puzzlers, Twisters & Teasers Worksheet 16

Guided Reading Audio CD
English or Spanish, Chapter 16

CAPÍTULO

16 Aves y mamíferos

¿Creerías que . . . ?

Te sorprendería saber que las palomas del patio de tu escuela se relacionan con los velociraptors? El modo de andar gracioso de una paloma no te hace pensar en un feroz animal prehistórico, pero la mayoría de los biólogos coinciden en que las aves descienden de los dinosaurios.

La primera prueba que demostró el vínculo entre reptiles y aves fue el fósil del *Archaeopteryx*, que se descubrió en Alemania hace casi 150 años. Como todo dinosaurio, el *Archaeopteryx* tenía huesos pesados y dientes, pero, como las aves, tenía alas con plumas.

Muchos científicos dudaban sobre la conexión entre los dinosaurios y las aves. Algunas investigaciones incluso sugirieron que el *Archaeopteryx* no era en realidad un dinosaurio.

En 1998, unos científicos chinos descubrieron fósiles de dinosaurios verdaderos que tenían alas y plumas. Estas plumas compartían las mismas características de las plumas de las aves modernas. Hasta ahora, nadie sabe para qué las usaban, pues eran muy cortas para volar. Tal vez servían de abrigo, o quizá los machos las usaban para atraer a las hembras.

El descubrimiento de estos dinosaurios con alas ayudó a convencer a algunos científicos de que las aves descienden de los dinosaurios. Algunos científicos incluso han llegado a decir que los dinosaurios nunca se han extinguido, ya que las aves son dinosaurios.

En este capítulo aprenderás más sobre las aves. También vas a aprender sobre otro grupo de vertebrados, los mamíferos.

Fósil de un *Archaeopteryx*

Fósiles de un dinosaurio con alas

Would You Believe . . . ?

The first fossils of *Archaeopteryx* were found in Germany in 1861. The fossils came from an area that was frequently flooded by water that carried a fine silt, which rapidly covered the animals once they died. The silt preserved the outlines of bones and feathers. The fossils were discovered when limestone was removed from a quarry. Some scientists question whether *Archaeopteryx* could fly because it did not have the characteristic avian breastbone, or keel. But they agree that *Archaeopteryx* was a predator because it had teeth and claws.

¿Tú qué piensas?

Usa tus conocimientos para responder a las siguientes preguntas en tu cuaderno de ciencias:

1. ¿Qué mantiene en el aire a un ave o a un avión cuando vuelan?

2. ¿En qué se diferencian los canguros de la mayoría de los otros mamíferos?

3. ¿Pueden los mamíferos poner huevos? ¿Pueden volar?

¡Investiga!

¡Volemos!

A los humanos siempre les han fascinado las aves y su habilidad para volar. A lo largo de la historia, los seres humanos han tratado de volar. Leonardo da Vinci, en el siglo XV, diseñó máquinas voladoras. Orville y Wilbur Wright, en 1903, finalmente despegaron del suelo cuando hicieron volar el primer avión. ¿Cómo vuelan las aves y los aviones? Esta sección te dará algunas pistas.

Procedimiento  

1. Haz un avión con una **hoja de papel.** Procura que los dobleces sean simétricos y los pliegues estén bien marcados.

2. Lanza el avión con suavidad. ¿Qué sucede?

3. Toma el mismo avión y lánzalo con más fuerza. ¿Hubo algún cambio?

4. Reduce el tamaño de las alas cortándolas con **tijeras.** Asegúrate de que las dos alas tengan el mismo tamaño y que sean simétricas.

5. Lanza el avión otra vez, primero despacio y luego con más fuerza. ¿Qué pasó cada vez que lo lanzaste?

6. Anota todos los resultados en tu cuaderno de ciencias.

Análisis

7. ¿Qué pasó cuando lanzaste el avión de papel original lentamente? ¿Qué pasó cuando lo lanzaste con más fuerza?

8. ¿Qué efectos tiene la velocidad en el vuelo del avión? ¿Piensas que esto sucede en el vuelo de las aves?

9. ¿Qué pasó cuando acortaste las alas? ¿Por qué crees que pasó esto? ¿Crees que el tamaño de las alas influye en la forma de volar de un ave?

10. En base a tus experimentos, ¿cómo diseñarías y lanzarías el avión de papel perfecto? Explica tu respuesta.

What Do You Think?

Accept all reasonable responses.

Students will have a chance to revise their answers in the Chapter Review under NOW What Do You Think?

Investigate!

MATERIALS

FOR EACH STUDENT:
- piece of paper
- scissors

Safety Caution: Remind students to review all safety cautions and icons before beginning this lab activity. Students must be careful with scissors. Instruct them to throw their planes only into areas where other students are not present.

Answers to Investigate!

7. Answers will vary, but forcefully thrown planes probably went farther.

8. Answers will vary. Speed helps keep planes aloft. The same is true for birds.

9. Answers will vary. Smaller wings make the plane more maneuverable but less able to glide; longer wings offer more surface area to keep the plane or bird in the air.

10. Answers will vary, but good design will depend on what kind of flight is desired.

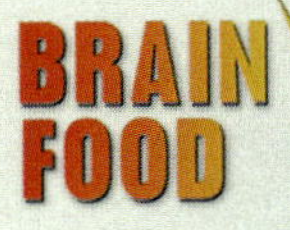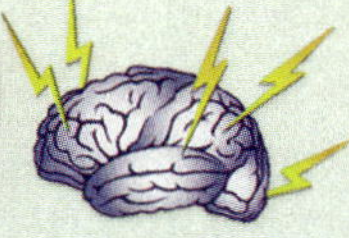

BRAIN FOOD

To prepare for this chapter, encourage students to think about the similarities and differences between birds and mammals. Ask them whether bats are mammals or birds, and challenge them to justify their answer. (Bats are mammals because they give birth to live young and have body hair and because female bats have mammary glands for producing milk.)

Focus

Birds

In this section, students learn which characteristics make birds unique and which characteristics they share with reptiles. Students also learn how birds are adapted for flight. Finally, students learn about bird migration and about how birds raise their young.

Bellringer

Pose this question to students on the board or an overhead projector:

What are some ways that birds are beneficial to people?
(provide meat, eggs, and feathers; offer natural insect and rodent control; serve as pets; pollinate plants; spread plant seeds; consume and eliminate decaying animals)

1) Motivate

ACTIVITY

Light as a Feather Divide the class into small groups. Provide each group with a feather (preferably a wing feather), a paper clip, a small scale, and a meterstick. Tell students to weigh the feather and the paper clip. Next tell the class to let group members take turns dropping the feather and the paper clip from a height of 1 m. Challenge groups to brainstorm about the differences they observe and about the role of feather shape in bird flight.
Sheltered English

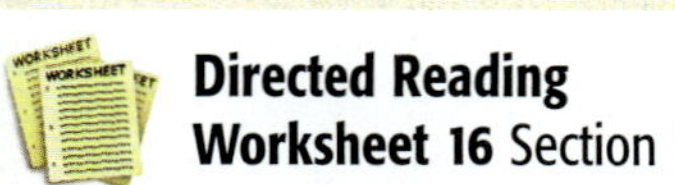
Directed Reading Worksheet 16 Section 1

VOCABULARIO

plumón
plumas de contorno
arreglo de las plumas
flotación

empollar
pollito precoz
pollito altricial

OBJETIVOS

- Menciona dos características comunes de aves y reptiles.
- Describe las características de las aves que las hacen aptas para el vuelo.
- Explica la *flotación*.
- Nombra algunas ventajas de la migración.

Colibrí

Figura 1 *En la actualidad, hay casi 9,000 especies de aves en la Tierra.*

Figura 2 *Las aves tienen un plumón liviano y esponjoso, y plumas de contorno en forma de hoja.*

382

Aves

Gran garza azul

¿Algunas vez has alimentado palomas en algún parque o has observado un halcón volar en círculos en el cielo? ¿Has oído cantar a los pájaros en un hermoso día de primavera? Los seres humanos siempre han observado las aves, quizás porque son más fáciles de reconocer que casi cualquier otro animal. A diferencia de los otros animales, todas las aves tienen plumas, y también se conocen por su habilidad para volar. Los pájaros pertenecen a la clase Aves. La palabra *aves* viene del latín y significa "pájaro". De hecho, la palabra *aviación,* la ciencia de hacer volar aviones, viene de la misma palabra.

Las aves habitan en todo el mundo y son de formas y tamaños diferentes, como se muestra en la **Figura 1.**

Correcaminos

Tucán

Características de las aves

Las primeras aves aparecieron en la Tierra hace 150 millones de años. Como aprendiste al inicio de este capítulo, se cree que las aves descienden de los dinosaurios.

Incluso hoy en día, comparten algunas características con ellos. Igual que los reptiles, son vertebrados. Sus piernas y patas están cubiertas de escamas gruesas y secas, como las de los reptiles. Hasta la piel alrededor de su pico es escamosa. Como los reptiles, las aves tienen *huevos amnióticos,* es decir, huevos con un saco amniótico y una cáscara. Sin embargo, las cáscaras de los huevos de aves son más duras que las de las tortugas y lagartijas. Las aves también tienen muchas características que las apartan del resto del reino animal. Tienen pico en lugar de dientes y mandíbulas, plumas, alas y muchas otras adaptaciones para volar.

Las plumas de las aves Las aves tienen dos tipos principales de plumas: el plumón y las plumas de contorno. Observa los ejemplos de la **Figura 2.** Como las plumas se gastan, las aves mudan las plumas usadas y les crecen unas nuevas.

Homework

Writing

Researching Bird Breeding Tell students that biologists have used their understanding of parent-offspring relationships in birds to breed birds in captivity. Tell students to research how scientists mimic parent birds in order to feed and otherwise support chicks. Encourage them to research the success of these captive-breeding efforts.

PORTFOLIO

El **plumón** está compuesto por plumas esponjosas aislantes que están en contacto directo con el cuerpo del ave. Para conservar el calor de su cuerpo, las aves esponjan el plumón para formar una capa aislante. El aire atrapado en las plumas las mantiene calientes. Las **plumas de contorno** están formadas por una *varita* central dura con muchas ramificaciones, llamadas *barbas*. Las barbas se unen para formar una superficie lisa. En la **Figura 3** puedes ver la estructura de una pluma de contorno. Las plumas de contorno cubren el cuerpo y las alas de las aves para formar una superficie aerodinámica para volar. Las aves cuidan muy bien sus plumas: usan el pico para esparcir aceite en sus plumas en un proceso llamado **arreglo de las plumas.** El aceite, segregado por una glándula cerca de la cola, hace que las plumas repelan el agua y también las mantiene limpias.

Animales de alta energía Las aves son *homotermas,* es decir, mantienen una temperatura corporal constante. Necesitan obtener del alimento mucha energía para volar, lo que se logra por medio de un metabolismo acelerado, el cual genera mucho calor corporal. De hecho, la temperatura corporal promedio de un ave es de 40°C, ¡más alta que la tuya! Cuando las aves tienen calor, extienden sus alas y jadean como los perros. Las aves no producen sudor para enfriar su cuerpo.

¿Comer como un pajarito? Debido a su metabolismo acelerado, las aves comen grandes cantidades de alimento en proporción con el peso de su cuerpo. ¡Algunas aves pequeñas comen casi de manera continua para mantener su energía! La mayoría come una dieta alta en grasas y en proteínas extraída de insectos, nueces, semillas o carne. Este tipo de dieta requiere un aparato digestivo pequeño. Algunas aves, como los gansos, comen hojas de plantas.

Las aves no tienen dientes, así que no pueden masticar su alimento. El alimento va directo de la boca al *buche,* donde se almacena. Las aves también tienen un órgano llamado *molleja,* que a menudo contiene pequeñas piedras. Las piedras de la molleja trituran el alimento para que el intestino lo digiera con facilidad. Las partes del aparato digestivo de un ave se ilustran en la **Figura 4.**

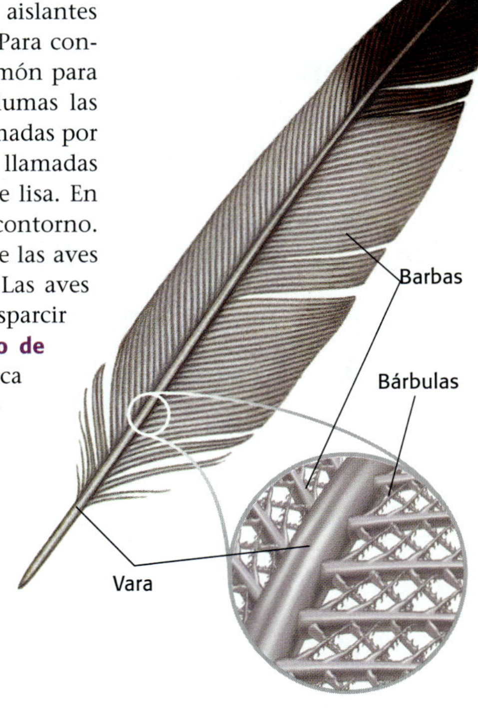

Figura 3 *Las barbas de una pluma de contorno tienen ramificaciones cruzadas llamadas bárbulas. Las barbas y las bárbulas le dan fuerza y forma a la pluma.*

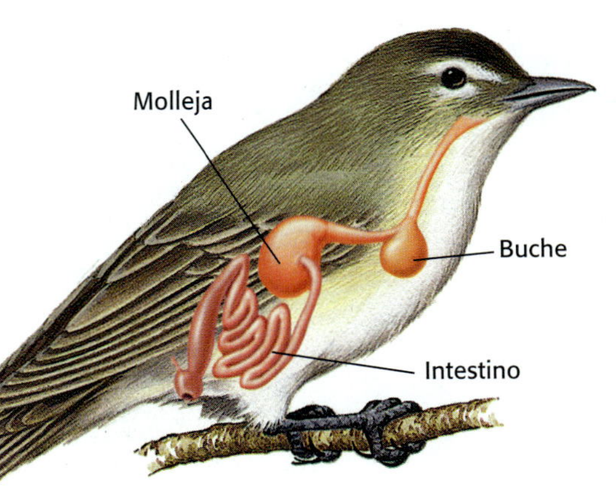

Figura 4 *El aparato digestivo de un ave hace que el alimento se transforme rápidamente en energía aprovechable.*

Q: Why do seagulls fly over the open sea?

A: because if they flew over bays, they'd be bagels (bay-gulls)!

What? No Dentist Bills?
PG 616

Teaching Transparency 63 "The Digestive System of a Bird"

2) Teach

GUIDED PRACTICE

Making Models

> **M A T E R I A L S**
>
> **FOR EACH GROUP:**
> • pipe cleaners
> • straws

Use the board to illustrate how a contour feather is constructed of a main shaft, projecting barbs, barbules, and barbule hooks.

Tell students to insert one or two pipe cleaners into a straw to make it stiff. Then tell them to follow the illustration to construct a model of a feather with at least three barbs and accompanying barbules and hooks.
Sheltered English

Multicultural CONNECTION

Bird myths and mythical birds are common cultural themes around the world. In the Hindu religion, the king of birds is a winged monster called Garuda that feeds on snakes. The fifteenth-century collection of stories *The Thousand and One Nights* presented the mythic roc, or *rukh,* that was so enormous it fed on elephants. And the folklore of the Athapascans, Inuit, and Hopi—three Native American cultures—features a huge eaglelike creature called a thunderbird.

Answers to Self-Check

1. Down feathers are not stiff and smooth and could not give structure to the wings. They are adapted to keep the bird warm.
2. Birds need tremendous amounts of food for fuel because it takes a lot of energy to fly.

READING STRATEGY

Prediction Guide Before students read these pages, ask them:

Why are birds such successful flyers? (Birds possess feathers, a high metabolism, wings, lightweight bones, and air sacs.)

Have them review their answers after they read the text.

USING THE FIGURE

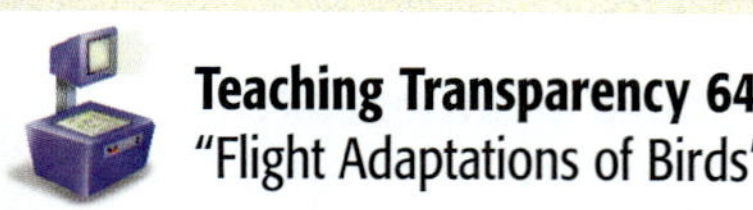
Remind students that several birds, including ostriches, kiwis, and emus, are flightless. Tell students to research the body and wing structure of flightless birds and to write another paragraph for each caption that explains any differences between the characteristics of flightless birds and birds that fly. Sheltered English

Teaching Transparency 64
"Flight Adaptations of Birds"

(Consulta la página 636 para comprobar tus respuestas.)

Alzan el vuelo y desaparecen

La mayoría de las aves vuelan. Hasta las que no lo hacen, como las avestruces, descienden de antepasados que sí volaban.

Las aves cuentan con una larga lista de adaptaciones para volar. Tienen que tomar una gran cantidad de energía de su alimento y una gran cantidad de oxígeno del aire que respiran para poder volar. Las plumas, las alas y los músculos fuertes juegan un papel importante. El cuerpo de las aves es tan liviano que les perimite elevarse. La **Figura 5** de estas dos páginas explica muchas de las características de las aves que son importantes para el vuelo.

Figura 5 **Adaptaciones para el vuelo de las aves**

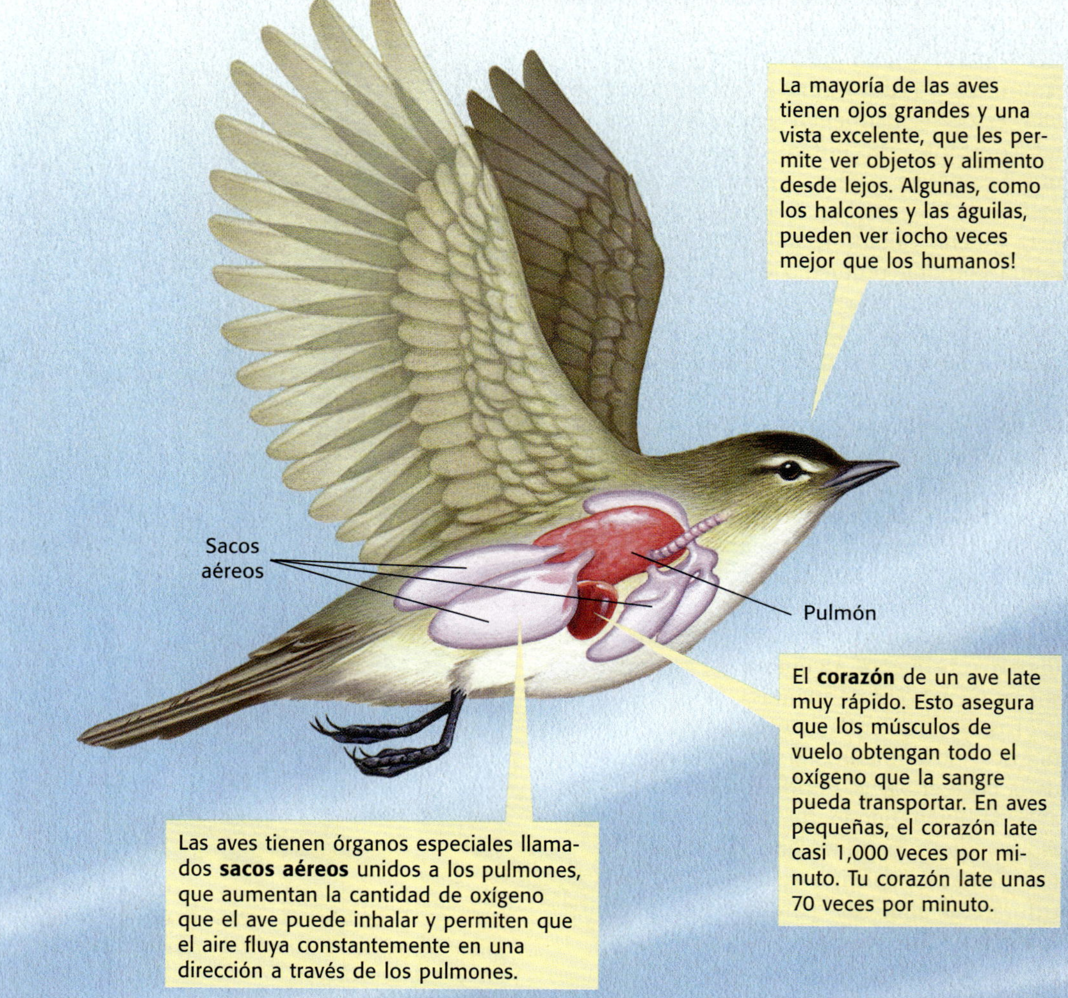

384

WEIRD SCIENCE

Loons are heavy-bodied water birds. Their legs are set far back on their bodies for maximum streamlining while swimming. As a consequence, loons can barely walk on land. And because the ratio of their wing surface area to their body weight is so low, they have difficulty getting airborne. For comparison, think of a big-bodied plane with small wings. To get airborne, loons must "run" on the water while flapping their wings. They may have to taxi for half a kilometer to get aloft!

La forma de las alas de un ave se relaciona con el tipo de vuelo que realiza. Las alas cortas y redondeadas permiten maniobras rápidas, como los movimientos de un avión de combate. Las largas y angostas son mejores para planear, como lo hacen los planeadores.

a través de las ciencias
CONEXIÓN

Estas características ayudan a las aves a volar, pero, ¿cómo vuelan los aviones? Averígualo en la página 410.

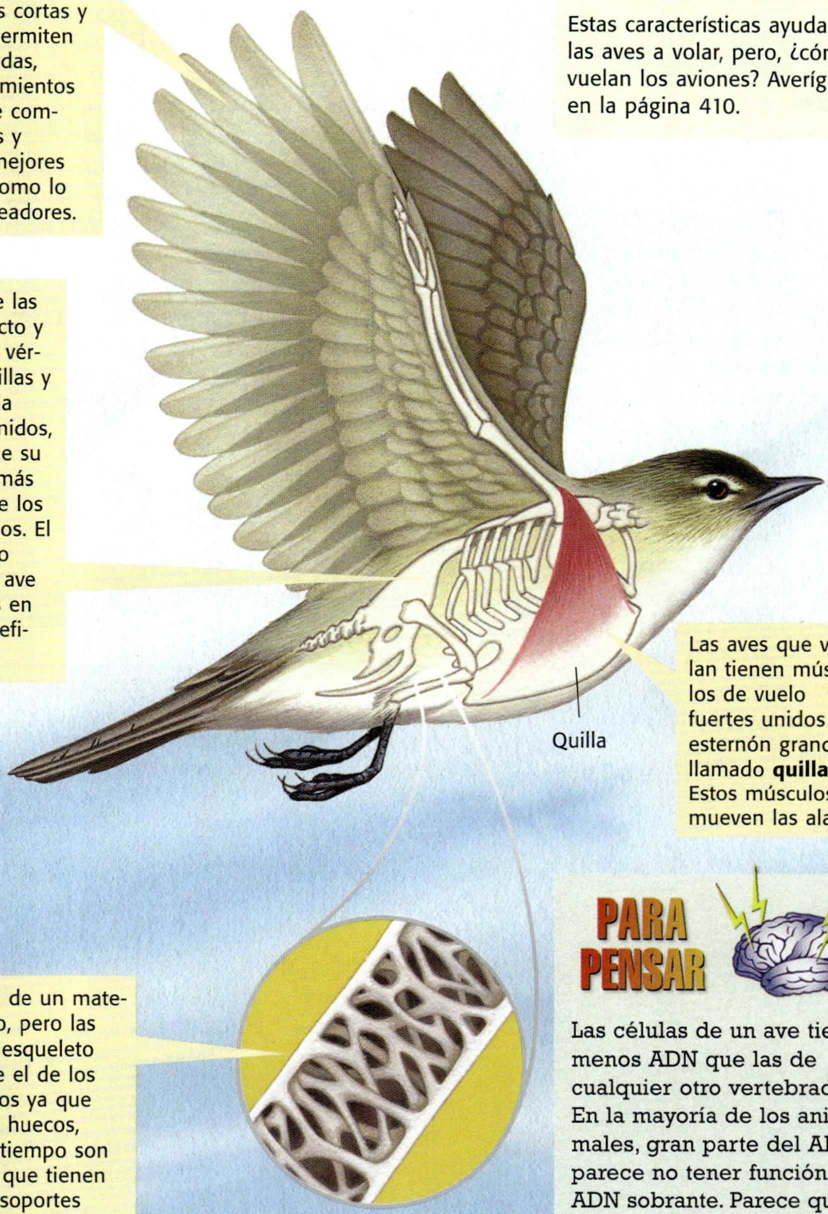

El esqueleto de las aves es compacto y fuerte. Algunas vértebras, las costillas y los huesos de la cadera están unidos, lo que hace que su esqueleto sea más rígido que el de los otros vertebrados. El esqueleto rígido permite que el ave mueva sus alas en forma fuerte y eficiente.

Las aves que vuelan tienen músculos de vuelo fuertes unidos a un esternón grande llamado **quilla.** Estos músculos mueven las alas.

Los huesos son de un material muy pesado, pero las aves tienen un esqueleto más liviano que el de los otros vertebrados ya que sus huesos son huecos, pero al mismo tiempo son muy fuertes ya que tienen unos delgados soportes cruzados que les dan apoyo, igual que las vigas sostienen un puente.

PARA PENSAR

Las células de un ave tienen menos ADN que las de cualquier otro vertebrado. En la mayoría de los animales, gran parte del ADN parece no tener función; es ADN sobrante. Parece que las aves se han deshecho de buena parte de su ADN sobrante, lo que las hace más livianas para el vuelo.

DEMONSTRATION

Observing Bird Bones Obtain several bones from a chicken or a turkey, such as the lower leg bones (drumsticks) and thigh bones, that have been cooked and thoroughly cleaned. Carefully break open the bones so that students can examine the air spaces inside. Sheltered English

Homework

Writing **Researching Falconry** Falconry is an ancient sport that may date back to the eighth century B.C. in Assyria. It was immensely popular among the European upper classes in the Middle Ages. Today, the sport is tightly regulated by the United States Fish and Wildlife Service, which issues permits that allow individuals to trap and train hawks, falcons, and occasionally eagles. Have students research and write a report about the history of falconry. Their report should include information about the role of falconry in raptor conservation.

internetconnect

SCILINKS NSTA
TOPIC: Bird Characteristics
GO TO: www.scilinks.org
***sci*LINKS NUMBER:** HSTL405

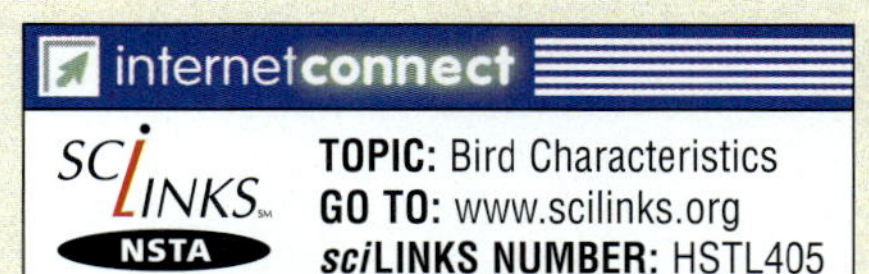

Science Bloopers

The common name of a bird may be an accurate description of what it looks like, but it does not always accurately reflect what kind of bird it actually is. For example, the common nighthawk is not a hawk at all, although it appears hawklike in flight. The nighthawk belongs to the Caprimulgidae family, which are mostly insect eaters. Members of this family were called "goatsuckers" because Caprimulgidae comes from the Latin *caper,* "goat," and *mulgeo,* "to milk." According to legend, the birds sucked milk from goats at night!

MEETING INDIVIDUAL NEEDS

Writing **Advanced Learners** Have students research and prepare a report on the parts and functions of an airplane's wing, comparing the information to the parts and functions of a bird's wing. For example, a bird spreads its primary wing feathers to slow down. Is there a movable piece on the plane's wing that helps slow down the plane when it is approaching the airstrip for a landing?

Quick Lab

MATERIALS

FOR EACH STUDENT:
- drinking straw
- push or straight pin
- tissue paper (3 × 0.5 cm)

Safety Caution: Remind students to review all safety cautions and icons before beginning this lab activity. Students should be especially careful not to stick themselves with the pin when making the hole. They should not use any straw that another student has already used.

CONNECT TO PHYSICAL SCIENCE

In both birds and aircraft, the shape of the wing creates differences in maneuverability and air speed. Designing an airplane wing requires taking into account what kind of flying the plane will need to do. Selective pressures have generated a vast array of wing shapes in birds, enabling different species to fly in very different ways. Use Teaching Transparency 182 to illustrate the effect of wing shape on air speed and lift.

Laboratorio

El efecto Bernoulli

¿Es cierto que el aire que circula rápido crea una presión baja? ¡Por supuesto! Puedes observar este efecto fácilmente con un popote y una hoja de papel. Trabaja en pareja. Con un **alfiler** hazle un hoyo a un **popote**. Corta una pequeña tira de **papel** de unos 3 cm de largo y 0.5 cm de ancho. Sostén la tira de papel tan cerca del hoyo como sea posible, sin que toque el popote. Pídele a tu compañero que sople en el popote. El aire que circula rápido crea una baja presión en el popote. La presión de aire más alta de la habitación empujará el papel contra el hoyo. ¡Haz la prueba!

Teaching Transparency 182 "Wing Shape Creates Differences in Air Speed"

LINK TO PHYSICAL SCIENCE

Levantar el vuelo

Todas estas adaptaciones hacen que las aves estén bien equipadas para el vuelo, pero, ¿cómo superan la gravedad y surcan los aires? Todo lo que vuela necesita alas para despegar del suelo. Las aves sacuden sus alas para atravesar e impulsarse a través del aire. Las alas proporcionan flotación, o la fuerza ascendente sobre el ala que mantiene al ave en el aire.

Cuando el aire fluye por el ala, parte del mismo se fuerza por encima y otra parte por debajo del ala. El ala de un ave tiene la parte superior curva. Como se muestra en la **Figura 6,** el aire de encima tiene que moverse *una mayor distancia* que el de debajo. Como resultado, el aire de arriba se mueve *más rápido* que el aire de la parte inferior. El aire que circula rápido en la parte superior crea baja presión en el aire. Esto se conoce como *efecto Bernoulli*. La presión del aire debajo del ala es mayor y empuja el ala hacia arriba.

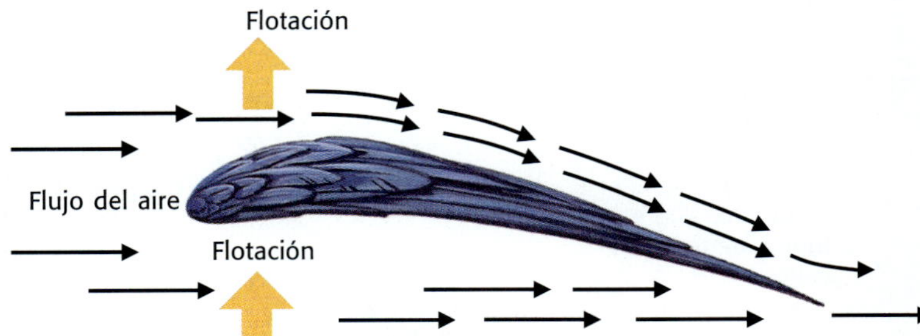

Figura 6 *El ala de un ave está diseñada para producir flotación. El aire que se mueve sobre la parte superior del ala se mueve más rápido que el de la parte inferior. Esto crea una diferencia en la presión del aire que mantiene el ave en el aire.*

Las aves generan flotación extra al agitar las alas. Cuanto más rápido vuela un ave, mayor es la flotación. Otro factor que influye es el tamaño del ala. Cuanto más grande sea, mayor será la flotación. Las aves con alas grandes pueden planear distancias largas sin agitar las alas. Un albatros, como el de la **Figura 7,** puede planear por muchas horas sin agitarlas.

Figura 7 *El albatros errante tiene una envergadura de 3.5 m, la más grande de cualquier ave viviente. Sus alas largas le permiten planear durante largos períodos de tiempo. Un albatros sólo desciende para poner huevos.*

IS THAT A FACT!

The wandering albatross's wings are 3–3.5 m long but barely 23 cm wide. The wings are inefficient for flapping flight, but their unusual shape enables albatrosses to soar for months at a time, alighting only to nest and feed and when winds are too calm for soaring.

Migración

¿Has oído decir que cuando las cosas se ponen difíciles, los más fuertes se las arreglan solos? Las aves deben ser muy fuertes, porque en tiempos difíciles, algunas se van más rápidamente y más lejos que cualquier otro animal. Al poder volar grandes distancias, migran a lugares lejanos.

Algunas aves, como los gansos de Canadá de la **Figura 8,** tienen muy buenas razones para migrar. Cuando migran, pueden encontrar mejores territorios con más alimento. Por ejemplo, en el ártico, durante el verano, hace sol casi las 24 horas del día. Las plantas, los insectos y otros organismos se multiplican de manera explosiva, proporcionando mucha comida. Es un lugar magnífico para que las aves cuiden de sus crías. Sin embargo, el invierno es muy largo y cruel, con muy poca comida. Cuando llega, las aves vuelan al sur en busca de alimento.

Figura 8 *Estos gansos canadienses migran al sur en el invierno.*

El cuidado de las crías

Como los reptiles, las aves se reproducen por fertilización interna y ponen huevos amnióticos que contienen el embrión en desarrollo. Pero mantienen sus huevos cal entitos para que el embrión se desarrolle.

La mayoría de las aves construyen nidos muy elaborados donde ponen los huevos. La **Figura 9** muestra diferentes tipos de nidos. Las aves se sientan sobre sus huevos hasta que nacen los pichones, usando el calor de su cuerpo para mantenerlos abrigados. Esto se llama **empollar.** En algunas aves, como las gaviotas, hembras y machos empollan los huevos. Entre algunas aves cantoras, mientras la hembra empolla los huevos, el macho le trae comida.

Criar aves pequeñas es un trabajo difícil. Algunas aves, como los cuclillos y los garrapateros, han encontrado la forma de que otras aves hagan su trabajo. Un cuclillo deja sus huevos en el nido de otra especie de ave. Cuando el pequeño sale del huevo, sus padres adoptivos lo alimentan y protegen.

Figura 9 *Estos son algunos tipos de nidos de aves. Las aves usan pasto, ramas, musgo, pelo, plumas y muchos otros materiales de construcción.*

IS THAT A FACT!

Migrating takes a lot of energy. Small birds may lose as much as 40 percent of their body weight on a long migratory flight. Even though the costs are high, birds that migrate experience a payoff when they reach areas with good weather and abundant food.

SCIENCE HUMOR

A hen is the only one who can lay down on the job and still get results.

MEETING INDIVIDUAL NEEDS

Writing **Advanced Learners**
Konrad Lorenz (1903–1989), the man in **Figure 10,** was one of the founders of modern ethology, the study of animal behavior by comparative methods. Many of Lorenz's studies involved birds. In 1935, Lorenz observed behavior in young ducks and geese that we now know as *imprinting.* During a critical stage after hatching, young birds will follow the first moving object they see, even if it's not their biological parents. Have students research imprinting and summarize the results of Lorenz's classic bird studies. Have them compose their findings in a short report.

PORTFOLIO

DISCUSSION

Precocial Versus Altrial After students have read this page, ask them to name the advantages and disadvantages of being a precocial chick. (Advantages are that precocial chicks can help find their own food and are not as vulnerable to the elements because their down protects them. A disadvantage is that they are especially vulnerable to predators.)

Ask them to name the advantages and disadvantages of being altrial chicks. (An advantage is that chicks are often safer in a nest. Disadvantages are that parents need to cover the chicks constantly to keep them warm until their feathers grow in, and parents must leave the chicks to find food for themselves and the chicks.)

Figura 10 *Los pollitos precoces aprenden a reconocer a sus padres justo después de salir del cascarón. Pero si sus padres no están ahí, seguirán a cualquier cosa que se mueva, hasta a un ser humano.*

Figura 11 *Los padres de los pollitos altriciales dejan el nido en busca de alimento. Regresan al nido con alimento después de unos minutos y a veces, entre los dos, hacen hasta 1,000 viajes en un día.*

¡Estamos listas! Algunas aves recién nacidas salen del cascarón listas para ser activas y comer bichos. Estos son **pollitos precoces.** Los pollos, los patos y las aves de mar son pollitos precoces. Están cubiertos de plumas suaves y siguen a sus padres en cuanto se levantan. En la **Figura 10**, puedes ver unos pollitos precoces siguiendo a un padre substituto. Los pollitos precoces dependen de su madre para obtener calor y protección, pero caminan, nadan y se alimentan solos.

Se solicita ayuda Los pollitos de halcones y aves cantoras, entre otros, salen del cascarón débiles, sin plumas e indefensos. Estos son **pollitos altriciales**. Nacen con los ojos cerrados y no pueden caminar ni volar. Sus padres deben mantenerlos calientes y alimentarlos durante semanas. La **Figura 11** muestra pollitos altriciales que son alimentados por su madre.

Cuando les crecen las primeras plumas de vuelo, empiezan a aprender a volar. Sin embargo, esto lleva días, y por lo regular los pollitos terminan caminando en el suelo. Los padres deben distraer a los gatos, las comadrejas y otros depredadores para proteger a sus crías.

REPASO

1. Menciona tres características similares y tres diferentes entre las aves y los reptiles.

2. Explica la diferencia entre pollitos precoces y altriciales.

3. ¿Has oído el dicho "come como un pajarito" para decir que alguien come muy poco? ¿Es correcto? ¿Por qué?

4. Menciona algunas de las adaptaciones que hacen que el cuerpo de las aves no pese.

5. **Comprender la tecnología** ¿Puede generar flotación el ala de un avión que no tiene la parte superior curva? Haz un dibujo para ilustrar tu explicación.

388

▼ **Answers to Review**

1. Answers may vary. Both birds and reptiles have scales and amniotic eggs, and both are vertebrates. Unlike reptiles, birds are endotherms, have feathers, and have air sacs.

2. Precocial chicks are fully active after hatching and are covered with downy feathers. Newly hatched altrial chicks cannot walk or fly and do not have feathers.

3. No; birds eat a tremendous amount of food in relation to their mass.

4. Birds' skeletons are compact. Their bones are hollow. Their feathers are light. They even have less DNA than other vertebrates.

5. No; air must move faster over the top of the wing to create lift. A flat wing would cause air to move at the same speed above and below and would not generate lift.

Clases de aves

Hay casi 9,000 especies de aves en la Tierra, que varían en tamaño desde el colibrí de 1.6 g hasta el avestruz del norte de África, de 125 kg. Los cuerpos de las aves también tienen diferentes características, según el lugar donde vivan y lo que coman. Debido a su gran diversidad, las aves se clasifican en 29 órdenes diferentes. Esto puede ser confuso, así que a menudo se agrupan en cuatro categorías no científicas: aves no voladoras, aves acuáticas, aves de rapiña y aves de percha. Estas categorías no comprenden todas las aves, pero muestran qué tan diferentes pueden ser.

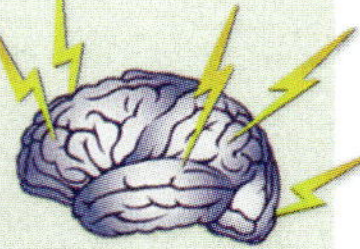

Un huevo de avestruz pesa 1.4 kg. Un solo huevo es tan grande que puede proporcionar los huevos revueltos para cuatro desayunos de toda una familia.

Aves no voladoras

Las avestruces, los kiwis, los emúes y otras aves no voladoras no tienen una quilla grande para el vuelo. Aunque no pueden volar, muchas de estas aves corren con mucha rapidez.

El **kiwi**, de Nueva Zelanda, es un ave de bosque del tamaño de un pollo. Sus plumas son suaves y parecidas al cabello. Duerme durante el día, y por la noche come bayas y caza gusanos y orugas.

Los **pingüinos** son aves no voladoras excepcionales. Tienen una quilla grande y músculos de vuelo muy fuertes, pero sus alas han sido transformadas en aletas. Agitan estas aletas para nadar debajo del agua. Aunque los pingüinos son muy buenos nadadores, en la tierra caminan con torpeza.

El **avestruz** es el ave viviente más grande. Puede alcanzar una altura de 2.5 m y un peso de 125 kg. Sus patas con dos dedos parecen pezuñas. Pueden correr hasta 64 km/h (40 millas/h).

389

Q: What is the difference between *unlawful* and *illegal?*

A: An ill eagle is a sick bird.

Learners Having Difficulty To reinforce the concept that feathers are the defining characteristic of birds, ask students whether or not the following animals are birds: hawk moth, penguin, eagle, bat, ostrich, flying squirrel. Have them justify their answers. Sheltered English

Throughout history eagles have represented power. An eagle was the emblem of the Babylonian god Ashur. An eagle adorned the staff of Zeus. In Norse myths, Odin, the king of the gods, often assumed the shape of an eagle. Native Americans of the plains decorated their war bonnets with eagle feathers. And of course, the bald eagle is the national symbol of the United States.

Be sure students understand that even though some birds look like they are covered with fur, they are actually covered with tiny feathers. The feathers of most birds grow in lines called *tracts.* Between the tracts are gaps. If you blow on a robin's belly, for example, the feathers will separate to reveal patches of bare skin. However, on penguins, ostriches, rheas, cassowaries, and emus, every part of the body is covered with feathers.

③ Extend

GOING FURTHER

Birds of prey such as eagles, owls, falcons and hawks are called *raptors*. There are raptor centers throughout the United States, many of which specialize in the care and rehabilitation of injured birds. The centers frequently serve as wildlife education centers, too. Many house and display for educational purposes birds that are too wounded or too accustomed to people to survive on their own. Have students research and contact these centers for more information about raptors and their care (remind them to include a stamped, self-addressed envelope if they expect a response by mail). If a raptor center is nearby, encourage students to work with the school administration to arrange a visit by a naturalist and a bird or two.

DEBATE

Is It Okay to Destroy Wildlife to Save It? The artist John James Audubon (1785–1851) painted pictures of birds that he shot. This was the only way he could get close enough to see the level of detail he needed in his paintings. Although modern cameras and binoculars provide researchers with excellent detail, scientists still sometimes kill birds for study and museum collections. This allows them to place the birds in proper evolutionary context. The results allow scientists to confirm the discovery of new species and develop specific conservation measures. Ask students to debate the pros and cons of this scientific technique.

Aves acuáticas

En el grupo de las *aves acuáticas* se encuentran las grullas, los patos, los gansos, los cisnes, los pelícanos, los somorgujos y muchas otras especies. Por lo general, estas aves tienen patas palmípedas para nadar, pero también son voladores resistentes.

Este pato tienen un hermoso plumaje que atrae a las hembras. Los patos son resistentes nadadores y voladores.

El **bobo de patas azules** es un ave acuática tropical. Estas aves tienen una danza de cortejo complicada que consiste en levantar las patas una después de la otra.

El **somorgujo común** es el ave más primitiva de las aves modernas. Puede permanecer debajo del agua durante varios minutos en busca de peces.

Aves de rapiña

Las águilas, los halcones y otras aves de rapiña son carnívoras. Pueden comer mamíferos, peces, reptiles, aves u otros animales. Las garras afiladas de sus patas y su pico curvo cortante les ayudan a atrapar y comer su presa. También tienen muy buena vista, y la mayoría cazan durante el día.

Los búhos, como este **búho manchado del norte,** son las únicas aves de rapiña que cazan en la noche. Tienen un agudo sentido del oído que les ayuda a encontrar a su presa.

Las **águilas pescadoras** comen peces. Vuelan sobre el agua y atrapan los peces con las patas.

WEIRD SCIENCE

Some oceanic birds can eject a smelly oil from their stomachs. Fulmars can spew the foul-smelling liquid about 1 m as a defensive weapon. Elimination of the oil can also reduce a bird's weight before flight. The behavior is instinctive; newly hatched fulmars have been observed regurgitating the fluid while still emerging from the shell. The oil is also exchanged by adult fulmars during courtship. Eliminating the rich oil may help birds keep their metabolisms at the proper level.

Aves de percha

Las aves cantoras, como los petirrojos, los reyezuelos y los gorriones son aves de percha. Estas aves tienen adaptaciones especiales para posarse sobre una rama. Cuando un ave de percha aterriza en una rama, sus patas automáticamente se cierran sujetándose a ella. Aunque el ave se quede dormida, no se cae.

Los **loros** no son aves cantoras, pero tienen patas especiales para posarse y trepar a las ramas. Su pico encorvado y fuerte les permite abrir semillas y partir fruta.

Los **paros carboneros** son pequeñas aves que se reúnen con frecuencia en los jardines. Por lo general, se cuelgan de una rama mientras comen insectos, semillas o frutas.

La mayoría de las tángaras son aves tropicales, pero la **tángara escarlata** pasa el verano en Norteamérica. El macho es rojo, pero la hembra es de un color verde amarillento que se camufla entre los árboles.

REPASO

1. ¿Por qué se llama aves de percha a este tipo de ave?

2. Las aves de rapiña tienen muy buena vista. ¿Por qué es tan importante esta característica?

3. **Interpretar ilustraciones** Observa la ilustración de la derecha. ¿Qué pata pertenece a un ave acuática? ¿Cuál pertenece a un ave de percha? Explica tus respuestas.

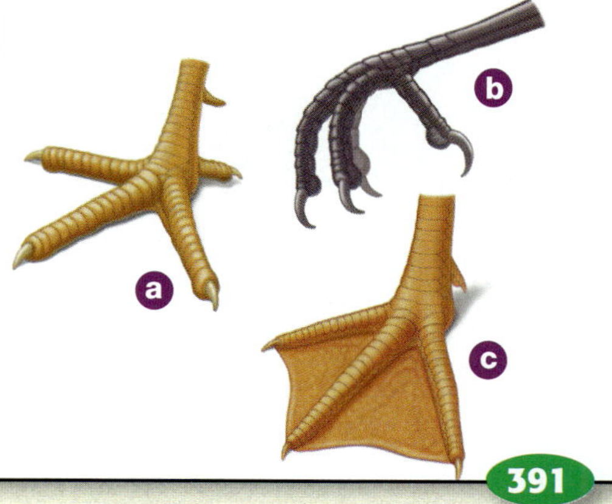

391

Quiz

1. What is the one characteristic that sets living birds apart from the rest of the animal kingdom? (feathers)

2. Why must birds eat a high-energy diet? (They need a lot of energy to maintain a constant body temperature and for flight.)

3. Trace the path of food from a bird's beak through the digestive system. (from the mouth, to the crop, then the gizzard, then the intestine)

4. How do perching birds differ from birds of prey? Give two examples of each kind of bird. (Perching birds, such as robins and sparrows, have special adaptations for perching on a branch; their feet automatically close around a branch when they land. Birds of prey, such as eagles and hawks, have sharp claws; sharp, hooked beaks; and good vision.)

ALTERNATIVE ASSESSMENT

Find pictures of birds in nature magazines, cut them out, number them, and display them for the class. Have students make charts with three columns. Tell them to put the number in the first column, the general type of bird (perching, water, etc.) in the second, and the reasons for their classification in the third.

Sheltered English

Critical Thinking Worksheet 16
"A Puzzling Piece of Paleontology"

Answers to Review

1. Perching birds have special feet that allow them to perch easily on branches.

2. As birds of prey, they are flying high above the ground; they need to be able to spot their prey from great distances.

3. The foot in figure (c) belongs to a water bird. The foot in figure (b) belongs to a perching bird. The water bird has webbed feet. The perching bird has a foot adapted to grasping. (The foot in figure (a) belongs to a pheasant.)

Focus

Mammals

This section introduces the mammals and describes their common characteristics. Students will learn the differences between monotremes, marsupials, and placental mammals and will provide examples of different kinds of mammals.

Bellringer

Write the following on the board or overhead transparency:

In the next 5 minutes, list as many characteristics of mammals as you can think of.

After the 5 minutes are up, ask students what characteristics they have listed, and put their answers on the board. Use their answers as a springboard for discussion about what mammals are and where they live.

1 Motivate

DISCUSSION

Domestication of Animals
Ask students to describe how humans have interacted with wild mammals over time. (Humans have hunted mammals and used their meat for food, their hides for clothing and shelter, and their bones for tools.)

Then ask students how the domestication of mammals, such as horses, cattle, pigs, and dogs, changed the lives of early humans. It became easier to obtain food (eggs, milk), to cultivate crops (oxen pulling plows), to catch wild game (hunting dogs), and to control pests (cats).

VOCABULARIO

terápsido   placenta
glándulas mamarias   gestación
diafragma   carnívoro
monotrema   primate
marsupial
mamífero placentario

OBJETIVOS

- Describe las características comunes de los mamíferos.
- Explica las diferencias entre los monotremas, marsupiales y mamíferos placentarios.
- Da algunos ejemplos de cada tipo de mamífero.

Mandril

Figura 12 *Aunque se ven muy diferentes, todos estos animales son mamíferos.*

Figura 13 *Los terápsidos tenían características de reptiles y de mamíferos, y es posible que se parecieran al que ves en este dibujo.*

392

Mamíferos

Parece ser que, de todos los vertebrados, los que más nos interesan son los mamíferos, quizás porque los seres humanos también lo somos. Pero, con casi 4,500 especies, los mamíferos son en realidad una clase pequeña de animales. Los moluscos, por ejemplo, tienen más de 90,000 especies.

Los mamíferos son de diferentes tamaños, desde los murciélagos más pequeños (que pesan menos que una galleta) hasta las ballenas más grandes. La ballena azul, con una masa de más de 90,000 kg, es el animal más grande que existe. Hay mamíferos en los océanos más fríos, en los desiertos más calientes y en los climas intermedios. En la **Figura 12** ves una pequeña muestra de la gran variedad de mamíferos que existe.

Venado

Rinoceronte

Ballena beluga

El origen de los mamíferos

Las pruebas fósiles sugieren que hace unos 280 millones de años aparecieron reptiles parecidos a los mamíferos, llamados terápsidos. Los **terápsidos** fueron los primeros antepasados de los mamíferos. En la **Figura 13,** puedes ver la versión artística de lo que pudo ser un terápsido.

Hace unos 200 millones de años, los primeros mamíferos aparecieron en el registro fósil. Tenían el tamaño de un ratón y eran homotermos. Como no dependían de su ambiente para calentarse, podían buscar forraje por la noche y evitar a los dinosaurios depredadores durante el día.

Cuando los dinosaurios se extinguieron, había más tierra y alimento para los mamíferos. Estos empezaron a diversificarse y a vivir en muchos ambientes diferentes.

WEIRD SCIENCE

The world's smallest mammal is the Kitti's hog-nosed bat, which weighs only 2 g and is 33 mm long. This tiny bat lives in limestone caves in southwest Thailand.

Características de los mamíferos

Los delfines y los elefantes son mamíferos, así como los monos, los caballos y los conejos. ¡Tú eres uno de ellos! Estos animales son muy diferentes, pero todos los mamíferos comparten muchos rasgos característicos.

Mamas Todos los mamíferos tienen glándulas mamarias, lo que los separa de los demás animales. Las **glándulas mamarias** segregan un fluido nutritivo llamado leche. Todas las hembras mamíferas les dan leche a sus crías. Por lo general, dan a luz a crías vivas y las alimentan, como ves en la **Figura 14.** Aunque sólo las hembras maduras dan leche, los machos también tienen pequeñas glándulas mamarias inactivas.

Figura 14 *Como todos los mamíferos, este becerro obtiene su primer alimento de la leche de su madre.*

La leche está compuesta por agua, proteína, grasa y azúcar. La leche de diferentes mamíferos varía en la cantidad de los nutrientes. Por ejemplo, la humana tiene la mitad de grasa que la de vaca, pero el doble de azúcar. La de las focas tiene casi más de la mitad de grasa. Cuando nacen, los elefantes marinos pesan 45 kg. Después de tomar esta rica leche durante 3 semanas, su peso alcanza los 180 kg.

Cómodo y abrigado Si un perro se ha quedado dormido en tu regazo, ya sabes que los mamíferos tienen una temperatura alta. Todos los mamíferos son homotermos. Como los pájaros, los mamíferos requieren mucha energía del alimento. Descomponen muy rápido el alimento en su cuerpo y usan la energía liberada por sus células para mantener la temperatura. Los mamíferos mantienen constante su temperatura corporal. Ésta cambia solamente cuando hibernan, estivan o tienen fiebre.

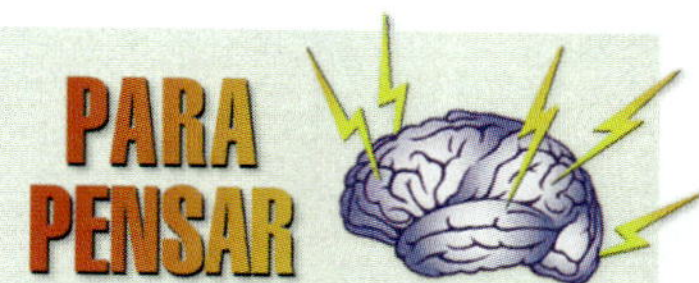

Cuando un canguro trepa por primera vez a la bolsa de su madre, la leche que bebe no tiene grasa. Más tarde, la leche tiene un 20 por ciento de grasa. Una mamá canguro con dos bebés de diferente edad le da leche sin grasa al bebé y leche con grasa al más grande. Cada uno se alimenta de un pezón diferente.

Figura 15 *Los mamíferos se sienten tibios cuando los tocas porque son homotermos.*

393

2) Teach

ACTIVITY

Offspring Number Divide the class into small groups. Tell students to check references to determine the average number of offspring and frequency of births for a field mouse, a pig, a horse, a chimpanzee, and a human. Then tell them to find out how long the babies are dependent on their parents and what the average life span is of each of these mammals. Have students discuss whether they see any pattern. *(Short-lived animals tend to have more offspring and give birth more frequently. Their offspring tend to have a short period of dependence on their parents.)*

The groups should present their findings to the class and compare what they have learned.

Homework

Create a Timeline Have students use library references to create a timeline of mammalian evolution from their first appearance in the Triassic period (Mesozoic era) some 230 million years ago through the Cenozoic era to the present. Encourage students to illustrate their timelines with original drawings or pictures from books or magazines.
Sheltered English

Directed Reading Worksheet 16 Section 2

MEETING INDIVIDUAL NEEDS

Writing **Advanced Learners**
Explain to students that males and females of the same species often vary in size and coloration. Tell them this sex-related difference is called *sexual dimorphism.* Examples in mammals are the male lion's mane, the male deer's antlers, different facial coloration in mandrills (a species of baboon), and larger canine teeth in male baboons. Have students research and write a brief report on possible reasons for sexual dimorphism in mammals.

DEMONSTRATION

Comparing Skulls Display the skulls of several species for students to study. Ideally, include several mammal skulls, a reptile skull, a bird skull, and an amphibian skull. Tell students to carefully examine the skulls, paying close attention to the similarities and differences between classes (major groups) and within the mammal class. Ask them to describe differences in dentition (tooth structure and arrangement) and speculate about advantages and disadvantages of the arrangements they observe.

Sheltered English

---

internet connect

SciLINKS
NSTA

TOPIC: Characteristics of Mammals
GO TO: www.scilinks.org
*sci*LINKS NUMBER: HSTL420

Figura 16 *El pelaje grueso de este zorro ártico mantiene su cuerpo a una buena temperatura hasta en los inviernos más fríos.*

Figura 17 *Los pumas tienen colmillos puntiagudos para despedazar a su presa. Los burros tienen incisivos delanteros afilados para cortar plantas y molares lisos en la parte trasera para triturar.*

Siempre abrigados Los mamíferos tienen adaptaciones que les ayudan a mantenerse a una buena temperatura. Para ello hay que usar un abrigo grueso, y los mamíferos usan unos muy lujosos. Todos, incluso las ballenas, tienen pelo. Éste es otro rasgo distinto de los mamíferos. Los de climas fríos, tienen un pelaje muy grueso, como el zorro de la **Figura 16.** Los mamíferos grandes de climas calientes, como los elefantes, tienen menos pelo. Los gorilas y los seres humanos tienen cantidades similares de pelo, pero el nuestro es más fino y corto.

La mayoría de los mamíferos también tienen una capa de grasa bajo la piel que actúa como aislante. Las ballenas y otros mamíferos que viven en océanos fríos dependen de una capa de grasa que los mantiene calientes.

¡Qué dientes! Otro rasgo que separa a los mamíferos de los otros animales son los dientes. Las aves no tienen dientes y, aunque los peces y los reptiles sí los tienen, todos son iguales. Los dientes de los mamíferos tienen forma y tamaño diferentes, y realizan funciones específicas.

Veamos tus dientes, por ejemplo. Los de adelante son para cortar y se llaman incisivos. La mayoría de las personas tienen cuatro arriba y cuatro abajo. Los siguientes se llaman caninos y sirven para afianzar la comida. Te ayudan a agarrar la comida y mantenerla en la boca. Más atrás están los dientes de superficie lisa llamados molares, que trituran la comida.

Los tipos de dientes reflejan la dieta. Los perros, gatos, lobos y otros carnívoros tienen caninos largos. Los mamíferos que comen plantas tienen molares mejor desarrollados. La **Figura 17** muestra los dientes de algunos mamíferos.

A diferencia de otros vertebrados, los mamíferos tienen dos juegos de dientes. Los primeros se llaman dientes de leche. Éstos son reemplazados por los dientes de adulto permanentes, cuando empiezan a comer alimentos sólidos y su mandíbula se hace más grande.

394

Multicultural CONNECTION

Native peoples of the Arctic region, such as the Inuit, have traditionally hunted marine mammals, such as seals, whales, and walruses. They use the seal's fur for water-repellent clothing and boots. They use seal blubber and whale blubber for lamp oil and for making soaps and lubricants. They even use marine mammal intestines to make waterproof outerwear.

Obtener oxígeno Igual que el fuego necesita oxígeno para arder, todos los animales necesitan oxígeno para "quemar" o descomponer los alimentos. Como las aves y los reptiles, los mamíferos usan los pulmones para obtener oxígeno del aire. Pero además tienen un músculo grande que ayuda a llevar aire a los pulmones. Este músculo se llama **diafragma** y está en la parte inferior de la caja torácica.

Para que pase la mayor cantidad de oxígeno posible de los pulmones a la sangre, los mamíferos tienen un corazón de cuatro cavidades. Esto permite que la sangre con oxígeno esté separada de la sangre sin oxígeno.

Cerebro grande El cerebro de un mamífero es más grande que el de otro animal del mismo tamaño. Esto permite que los mamíferos aprendan, se muevan y piensen rápidamente. El cerebro altamente desarrollado de un mamífero también le ayuda a saber lo que pasa en su ambiente y a responder con rapidez. Los mamíferos se encuentran entre los animales más coordinados. La mayoría de los mamíferos son animales veloces y reaccionan con rapidez, ya sea un conejo que desaparece entre los arbustos o un leopardo que persigue a una gacela.

Los mamíferos dependen de cinco sentidos principales que les proporcionan información sobre su medio ambiente: vista, oído, olfato, tacto y gusto. A menudo, la importancia de cada sentido para cada mamífero depende de su ambiente. Por ejemplo, los mamíferos que están activos de noche dependen más de su habilidad para escuchar que de su habilidad para ver.

Buenos padres Todos los mamíferos se reproducen sexualmente. Como las aves y los reptiles, se reproducen mediante fertilización interna; dan a luz crías vivas y las alimentan. La mayoría nacen indefensos y requieren mucho cuidado. Los padres son muy protectores, y uno de ellos o los dos cuidan a sus crías hasta que crecen. La **Figura 18** muestra un oso pardo cuidando a sus crías.

Figura 18 *Una mamá osa cuida de sus cachorros. Si algo los amenaza, está lista para atacar.*

REPASO

1. Menciona tres características únicas de los mamíferos.

2. ¿Para qué sirve el diafragma?

3. **Hacer deducciones** Imagina que encontraste el cráneo de un mamífero en una excavación arqueológica. ¿Cómo te podrían ayudar sus dientes para saber lo que comía?

395

CROSS-DISCIPLINARY FOCUS

Anthropology The artwork of Australia's Aborigines, an indigenous people, often depicts the country's unique and varied wildlife. Have students research the history of the tribes and the symbolism of the animals in their paintings. Students should bring their information to class to share their findings with their classmates.

READING STRATEGY

Prediction Guide Before students read about marsupials, ask them to respond true or false to the following statement:

Marsupials are native only to Australia. (false)

Have them check their answer after they read the page.

Homework

Writing **Researching Mammals** The platypus is an amphibious mammal. Tell students to find the word *amphibious* in a dictionary if they do not already know what it means. Then have them list examples of other amphibious mammals in their ScienceLog. (Examples include river otters, hippopotamuses, beavers, and muskrats.)

Clases de mamíferos

Los mamíferos se dividen en tres grupos según la forma de desarrollo de las crías: monotremas, marsupiales y mamíferos placentarios.

Monotremas. Los mamíferos que ponen huevos se llaman **monotremas**. Son los únicos mamíferos que ponen huevos y los primeros científicos los llamaron "reptiles peludos". Pero los monotremas no son reptiles; tienen todos los rasgos de los mamíferos: tienen glándulas mamarias, un abrigo grueso de pelo y son homotermos.

Los monotremas sólo se encuentran en Australia y Nueva Guinea. En la actualidad sólo existen tres especies de monotremas. Dos son equidnas, animales cubiertos de espinas con hocicos largos. Los equidnas tienen largas lenguas pegajosas para atrapar hormigas y termitas. En la **Figura 19,** puedes ver un equidna.

El tercer monotrema es el ornitorrinco, que puedes ver en la **Figura 20,** un mamífero nadador que vive y se alimenta en los ríos y los estanques. Tiene patas palmípedas, una cola plana que le ayuda a moverse en el agua, y un hocico plano y elástico que le sirve para cavar en el fango en busca de gusanos, langostinos y otros alimentos. Los ornitorrincos son buenos excavadores y cavan largos túneles en las orillas de los ríos, donde ponen sus huevos.

Un monotrema hembra pone uno o dos huevos con cascarones duros y correosos. Como los huevos de las aves y de los reptiles, los de los monotremas tienen una yema y un albumen para alimentar al embrión en desarrollo. La hembra incuba los huevos con el calor de su cuerpo. Las crías recién salidas del huevo son embriones muy pequeños que no están desarrollados por completo. La madre los protege y los alimenta con leche de sus glándulas mamarias. A diferencia de otros mamíferos, los monotremas no tienen pezones y los bebés no pueden mamar la leche, sino que la lamen de la piel y del pelo alrededor de las glándulas mamarias de su madre.

Figura 19 *Los equidnas son del tamaño de un gato. Tienen garras y hocico largo que les sirve para cavar en busca de hormigas y termitas.*

Figura 20 *Cuando está bajo el agua, el ornitorrinco cierra los ojos y bloquea los oídos. Usa su sensible hocico para encontrar alimento.*

WEIRD SCIENCE

Monotremes lack the glands that produce hydrochloric acid and peptic enzymes that help mammals digest protein. Scientists think that digestion in the echidnas is aided by the grinding action of the dirt they eat.

Marsupiales Probablemente sabes que los canguros, como los de la **Figura 21,** tienen una bolsa. Los canguros son **marsupiales,** o mamíferos con bolsa. Como todos los mamíferos, los marsupiales son homotermos, tienen glándulas mamarias, pelo y dientes. A diferencia de los monotremas, los marsupiales no ponen huevos, sino que dan a luz crías vivas.

Como los monotremas, los bebés marsupiales no están totalmente desarrollados. Al nacer, los embriones de canguro son del tamaño de un abejorro y se arrastran a través del pelo de la madre hasta la bolsa que hay sobre el abdomen. Dentro están las glándulas mamarias. El bebé canguro trepa, se adhiere al pezón y toma leche hasta que es capaz de moverse sin ayuda y dejar la bolsa durante períodos cortos. Estos son los canguros jóvenes.

Hay cerca de 280 especies de marsupiales. El único que habita en el continente americano al norte de México es la zarigüeya, que ves en la **Figura 22.** Los koalas **(Figura 23),** los demonios de Tasmania (dasiuro) y los wallabi también son marsupiales. La mayoría viven en Australia, Nueva Guinea y Sudamérica.

Figura 21 *Después de nacer, el canguro continúa desarrollándose en la bolsa de la madre. Los canguros jóvenes más grandes dejan la bolsa pero regresan si hay cualquier señal de peligro.*

Figura 22 *Cuando está en peligro, la zarigüeya se queda inmóvil, simulando estar muerta para que el depredador la ignore.*

Figura 23 *Los koalas duermen en los árboles durante el día y son activos por la noche. Sólo comen hojas de eucalipto.*

ciencias de la Tierra

CONEXIÓN

¿Por qué la mayoría de los marsupiales viven en Australia? Los fósiles indican que los primeros marsupiales vivieron hace 75 millones de años. Los científicos piensan que en esa época los continentes estaban cerca unos de otros. Los marsupiales primitivos pudieron llegar a Sudamérica, África y Australia. Después de que los continentes se separaran, los marsupiales se extinguieron en África y Europa, pero permanecieron en Sudamérica y prosperaron en Australia. Hoy en día, Australia tiene más marsupiales que cualquier otro lugar de la Tierra.

Answer to MATHBREAK

First calculate the number of ants per day, as follows:

$$\frac{50}{1} = \frac{x}{1,800}$$

$x = 90,000$ ants per day

at a rate of two ants per lick: 45,000 licks.

The time will vary, but if the rate is 60 licks per minute, then

$$\frac{45,000 \text{ licks}}{60 \text{ licks/min}} = 750 \text{ minutes}$$

$$\frac{750 \text{ min}}{60 \text{ min/hour}} = 12.5 \text{ hours per day}$$

Answer to Self-Check

Monotremes are mammals that lay eggs. Marsupials bear live young but carry them in pouches or skin folds before they are able to live independently. Placentals develop inside the mother's body and are nourished through a placenta before birth.

CROSS-DISCIPLINARY FOCUS

Archaeology When archaeologists study the homes of prehistoric people, they usually look for traces of structures or fire circles, and they investigate caves. But scientists studying early South American Indians have learned that they used the shells of early armadillos to build roofs for their homes. These ancestors of modern armadillos had shells up to 3 m long.

INDEPENDENT PRACTICE

Concept Mapping Have students create a concept map that compares the development and life histories of monotremes, marsupials, and placental mammals. Some terms and concepts students can use include:

type of egg with a shell, incubation, gestation, nest, uterus, mammary glands, and method of nursing

---

¡MATEMÁTICAS!

¡Hormigas para cenar!

El oso hormiguero gigante puede sacar la lengua 150 veces por minuto. ¿Cuántas veces puedes sacarla tú? Imagínate que eres un oso hormiguero y necesitas comer 50 hormigas para obtener 1 caloría. ¿Cuántas tendrías que comerte al día? Si atraparas dos hormigas cada vez que sacas la lengua, ¿cuántas veces al día tendrías que sacarla? ¿Cuántas horas por día tendrías que comer?

Mamíferos placentarios La mayoría de los mamíferos son **mamíferos placentarios.** Sus embriones se quedan dentro del cuerpo de la madre y se desarrollan en un órgano llamado útero. Los mamíferos placentarios forman una conexión especial con el útero de la madre llamada **placenta,** que le proporciona alimento y oxígeno de la sangre de la madre al embrión en desarrollo, y desecha los desperdicios del embrión.

El tiempo durante el cual un embrión se desarrolla dentro de la madre se llama gestación. La **gestación** de los animales placentarios va desde unas cuantas semanas en los ratones hasta 23 meses en los elefantes. Los seres humanos tienen una gestación de 9 meses. Después de que los mamíferos placentarios nacen, son alimentados con leche a través de las glándulas mamarias de la madre

✓ Autoevaluación

Explica la diferencia entre un monotrema, un marsupial y un mamífero placentario. *(Consulta la página 636 para comprobar tu respuesta.)*

Clases de mamíferos placentarios

Más del noventa por ciento de todos los mamíferos de la Tierra son placentarios. Los mamíferos placentarios actuales se clasifican en 18 órdenes. Cada orden tiene unas características que ayudan a identificarla. En las siguientes páginas verás las características de las órdenes más comunes.

Desdentados

Este grupo incluye osos hormigueros, armadillos, cerdos hormigueros, pangolines y perezosos. Sólo los osos hormigueros son completamente desdentados. Los demás tienen dientes pequeños. La mayoría de los mamíferos desdentados se alimentan de insectos que atrapan con su larga lengua pegajosa.

No todos los **armadillos** comen hormigas. Algunas especies comen otros insectos, ranas, champiñones y raíces. Cuando un armadillo se siente amenazado, se enrolla sobre sí mismo y se protege con sus placas duras.

El oso hormiguero más grande es el **oso hormiguero gigante** de 40 kg de Sudamérica. Los osos hormigueros nunca destruyen los nidos de sus presas. Los abren, se comen unas cuantas hormigas o termitas y luego pasan a otro nido.

IS THAT A FACT!

Both placental mammals and marsupials have a placenta. So why do we call one group placental? The difference is that in marsupials the placenta does not play a major role in nourishing the fetus, while in placental mammals, it does.

Insectívoros

Los mamíferos que comen insectos, o *insectívoros*, viven en cualquier continente excepto Australia. La mayoría de los insectívoros son pequeños y, por lo general, tienen la nariz puntiaguda para cavar en la tierra en busca de comida. En comparación con los otros mamíferos, tienen el cerebro muy pequeño y unos cuantos dientes especializados. Dentro de los insectívoros se encuentran topos, musarañas y erizos. Los mamíferos primitivos se parecían mucho a los insectívoros actuales.

Los **erizos** viven en Europa, Asia y África. Sus espinas los protegen de la mayoría de los depredadores.

Este **topo** tiene sensores en la nariz que le ayudan a encontrar insectos y a orientarse en los túneles. Aunque tienen ojos muy pequeños, los topos no pueden ver.

Roedores

Más de una tercera parte de todas las especies de mamíferos son roedores y se encuentran en cualquier continente, excepto la Antártida. Entre ellos están la ardilla, el ratón, la rata, el conejillo de Indias, el puerco espín y la chinchilla. La mayoría de los roedores son animales pequeños con bigotes largos y sensibles. Los roedores mastican y roen; todos tienen dientes delanteros filosos para roer. Como mastican tanto, sus dientes se desgastan. Por eso, sus incisivos crecen continuamente, como tus uñas.

Los **capibaras** de Sudamérica son los roedores más grandes del mundo. Una hembra puede pesar hasta 70 kg.

Como todos los roedores, los **castores** tienen dientes para roer, que usan para derribar árboles.

Science Bloopers

Tree shrews are tiny animals that look like squirrels with long, cone-shaped noses. Though they are insectivores and have shrewlike noses, tree shrews are not actually shrews. Until late in the twentieth century, zoologists thought tree shrews were primates. Recently, taxonomists recognized their uniqueness and placed them in their own order, Scandentia.

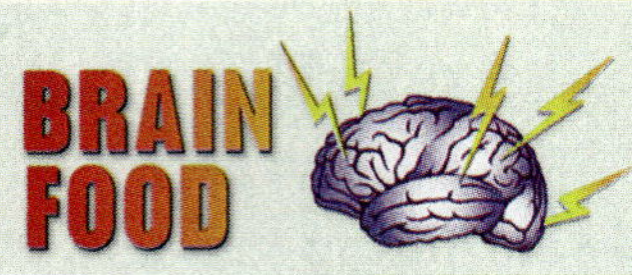

BRAIN FOOD

Tell students that if they were tracking lagomorphs in the snow, they would probably find all four footprints aligned or even prints from the back feet in front of those from the front feet! When moving quickly, lagomorphs rely on their powerful back legs to propel them forward. Their jump distance is increased when they swing their back legs forward until they touch down nearly in line with their front feet and then push off. Lagomorphs are one of the most efficient animals with this type of locomotion. Cheetahs can also do this.

DISCUSSION

Bat Fact and Fiction Ask students if the following statements are true or false:

1. Some bats have wing spans of more than 1.5 m. (True; the flying foxes of Indonesia are the largest bats.)

2. A single little brown bat can catch 1,200 mosquito-size insects in 1 hour. (true)

3. Bats present a serious disease threat to humans. (False; less than one-half of 1 percent of bats carry rabies. Most bite only when threatened.)

Lagomorfos

Los conejos, las liebres y las pikas pertenecen a un grupo de mamíferos placentarios llamados lagomorfos. Como los roedores, tienen dientes filosos para roer, pero a diferencia de ellos, tienen dos juegos de incisivos en la mandíbula superior y una cola corta. Los conejos y las liebres tienen patas traseras largas y fuertes para brincar. Para detectar a sus depredadores, que son muchos, tienen nariz sensible, y orejas y ojos grandes.

Las **pikas** son animales pequeños que viven en lo alto de las montañas de Asia y Europa. Estos animales juntan plantas y las ponen a secar en una especie de "pajares". En el invierno, usan las plantas secas como comida y aislamiento.

Las orejas largas de esta **liebre** le sirven para oír bien.

Mamíferos voladores

Los murciélagos son los únicos mamíferos que pueden volar. Tienen alas hechas de huesos largos y delgados conectados por membranas de piel. Los murciélagos son activos por la noche y duermen en áreas protegidas durante el día. Cuando descansan, se cuelgan de cabeza. La mayoría de los murciélagos comen insectos, pero algunos comen fruta, y tres especies de murciélagos vampiro chupan la sangre de otros animales. Los murciélagos más grandes son los que comen fruta; pesan hasta 1 kg y tienen una envergadura de hasta 2 m. La mayoría de los murciélagos pesan sólo unos cuantos gramos.

Casi todos los murciélagos cazan insectos durante la noche, y se orientan por ecolocación. Emiten ultrasonidos cuando vuelan y el eco de los mis-

En muchos países asiáticos, los **murciélagos** son símbolo de buena suerte, larga vida y felicidad.

mos rebota en los árboles, las rocas, los insectos y otros objetos próximos. Los ecos de un árbol grande y fuerte son diferentes de los que rebotan de una apetitosa y suave polilla. Por lo general, los murciélagos que se guían con la ecolocación tienen enormes orejas para escuchar los ecos de sus propias señales.

APLICA

¿En qué se parecen los murciélagos y los submarinos? Los submarinos usan la ecolocación llamada "sonar" para encontrar y evitar objetos bajo el agua. En base a lo que ya sabes sobre la ecolocación, ¿qué tipo de instrumentos crees que se necesitan para navegar un submarino con "sonar"?

400

Answer to APPLY

A submarine would need instruments to send signals and receive echoes and would need instruments to calculate distances using that information.

Carnívoros

Los **carnívoros** son un grupo de mamíferos que tienen grandes caninos y dientes especiales para cortar carne. El nombre carnívoro significa "que comen carne", así que los mamíferos de este grupo comen principalmente carne. Entre ellos están los leones, los lobos, las comadrejas, las nutrias, los osos, los mapaches y las hienas. Los carnívoros también abarcan un grupo de mamíferos marinos que comen peces llamados pinnípedos. Entre los pinnípedos se encuentran las focas, los leones marinos y las morsas. Algunos carnívoros también comen plantas. Por ejemplo, los osos negros comen pasto, nueces y bayas, y rara vez comen carne. Otros carnívoros no comen otra cosa que animales.

Los **coyotes** son miembros de la familia del perro. Viven en toda Norteamérica y en partes de Centroamérica.

Los gatos se dividen en dos grupos, gatos grandes y gatos pequeños. Todos los gatos grandes rugen. El gato más grande de este grupo es el **tigre siberiano** que pesa hasta 300 kg.

Los **mapaches** tienen patas que parecen manos para atrapar peces y sostener su alimento. Pueden manipular objetos casi tan bien como los monos.

Las **morsas** son pinípedos. A diferencia de otros carnívoros, no usan los caninos para despedazar el alimento sino para defenderse, cavar en busca de alimento y trepar al hielo.

401

Q: Why did the lion cross the savanna?

A: to get to the other pride

Multicultural CONNECTION

Hernando Cortéz reintroduced the horse into North America during the 1500s. Before then, Native Americans had to walk long distances on foot, carrying heavy loads on their backs. The arrival of the horse changed their lives forever. They were able to travel long distances more easily and were able to expand their trade routes. They were also able to hunt bison and other animals more efficiently. For plains tribes, such as the Kiowa, the horse became an important part of practical life and a key part of their culture.

REAL-WORLD CONNECTION

Hoofed mammals are literally the workhorses of the world. South Americans use llamas to carry heavy loads over mountainous terrain. Africans use camels for transportation, and oxen and water buffalo for pulling plows. And Americans have used horses, oxen, and cows for riding, pulling plows and wagons, and providing milk and meat.

Mamíferos con pezuñas

Los caballos, los cerdos, los venados y los rinocerontes son algunos de los mamíferos que tienen pezuñas gruesas. La mayoría de estos animales están adaptados para correr velozmente. Como son herbívoros, tienen molares planos y largos para triturar plantas. Algunos también tienen un aparato digestivo modificado que puede manejar grandes cantidades de celulosa.

Estos mamíferos se dividen en dos grupos, según el número de dedos. Los mamíferos con pezuña impar tienen uno o tres dedos. Los caballos, las cebras, los rinocerontes y los tapires tienen una sola pezuña grande. Los mamíferos con pezuña par tienen dos o cuatro dedos. En este grupo están los cerdos, las vacas, los camellos, los venados, los hipopótamos y las jirafas.

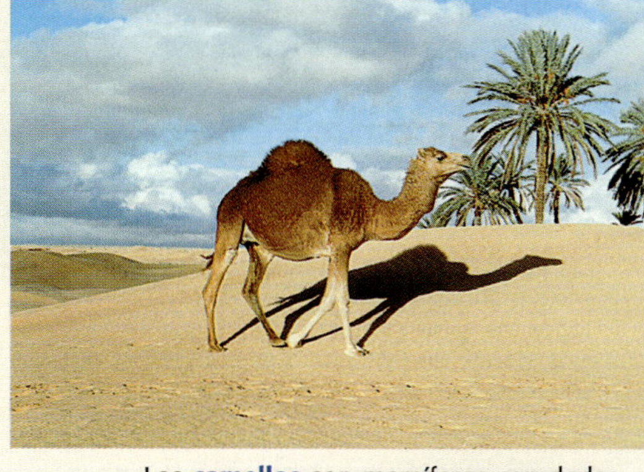

Las **jirafas** son los mamíferos más altos. Tienen el cuello y las patas largas con dedos pares.

Los **rinocerontes** son enormes mamíferos con dedos impares. Sus cuernos continúan creciendo por el resto de su vida.

Los **camellos** son mamíferos con dedos pares. La joroba de un camello es una bolsa de grasa que le proporciona energía cuando escasea la comida.

Los **tapires** son grandes mamíferos con tres dedos, que viven en los bosques. Se encuentran en Centroamérica, Sudamérica y en el sudeste de Asia.

IS THAT A FACT!

The limbs of hoofed mammals are adapted for running and walking long distances over open terrain. These animals have between one and four toes and very long foot bones. In human terms, they are balancing on the tips of their toes.

Mamíferos con trompa

Los elefantes son los únicos mamíferos con trompa. La trompa es una alargada combinación muscular de labio superior y nariz. Los elefantes usan la trompa de la misma forma en que nosotros usamos las manos, los labios y la nariz. La trompa es lo bastante fuerte para levantar un árbol, pero también lo bastante ágil para recoger frutas pequeñas, una por una. Los elefantes usan la trompa para llevarse el alimento a la boca y rociarse la espalda con agua para refrescarse.

Hay dos especies de elefantes: los africanos y los asiáticos. Los africanos son más grandes y tienen orejas y colmillos más grandes que los asiáticos. Ambas especies comen plantas. Como son tan grandes, los elefantes comen hasta 18 horas al día para obtener suficiente alimento.

Los elefantes son los animales terrestres más grandes. Los africanos llegan a pesar hasta 7,500 kg. Los elefantes son muy inteligentes y pueden vivir más de 60 años.

ciencias del medio ambiente
CONEXIÓN

Ambas especies de elefantes están en peligro. Durante siglos, los seres humanos los han cazado para obtener sus colmillos. Los colmillos de los elefantes son de marfil, un material duro que se usa para tallar. Debido a la gran demanda de marfil, gran parte de la población de elefantes ha sido aniquilada. Hoy en día, la caza de elefantes está prohibida.

Los **elefantes** son animales sociales. Viven en manadas que constan de hembras emparentadas, y sus crías.

✔ Autoevaluación

1. ¿Por qué los murciélagos se clasifican como mamíferos y no como aves?
2. ¿En qué se parecen los roedores y los lagomorfos? ¿En qué se diferencian?

(Consulta la página 636 para comprobar tus respuestas.)

403

---

3 Extend

DEBATE

Elephants and People: Room for Both? As the human population in Africa has expanded, villages and ranches have spread into former elephant habitat. Fences prevent elephants from traveling traditional routes to forage and find water. In some areas elephants have raided crops, killed farmers who tried to defend their fields, and even destroyed their natural food sources because too many elephants were trying to live in too small an area. So although elephant populations have declined, current conditions have caused overcrowding among the ones that remain. Should "excess" elephants be killed? The meat from killed elephants is given to the local people, many of whom are in dire need of food. Have students debate the management of elephant populations.

GOING FURTHER

The ears of African and Asian elephants are the most noticeable difference between the two species. Have students explain the evolutionary significance of this difference by exploring how the ears differ and researching the environments in which the animals live. (The larger surface area of the African elephant's ears contains more blood vessels that can be exposed to a cooling breeze passing over the thin skin of the ears; the larger ears serve as "fans" for the animals in the searing African heat. Asian elephants spend a lot of time in the forest, shielded from the sun.)

Answers to Self-Check

1. Bats bear live young, have fur, and do not have feathers.
2. Rodents and lagomorphs are small mammals with long, sensitive whiskers and gnawing teeth. Unlike rodents, lagomorphs have two sets of incisors and a short tail.

RESEARCH

Writing | Have students research and write a short report about one aspect of whale biology. Possibilities include whale behavior, feeding habits, anatomy, and communication. Encourage interested students to do preliminary research to find an interesting subject to study. Have students present their findings to the class.

PORTFOLIO

BRAIN FOOD

Humpback whales have a fascinating fishing strategy. They make their own fishing nets out of bubbles! The bubble net can be up to 98 ft (30 m) in diameter, which is large enough for the whale to get inside. The whale creates this net by swimming around in a spiral directly underneath a school of fish. As the whale swims, it blows air out of its blowhole. This forms a bubble net entirely around the fish, trapping them. The whale then swims up through the net with its mouth wide open, scooping in fish as it goes.

Cetáceos

Las ballenas, los delfines y las marsopas forman un grupo de mamíferos marinos llamados cetáceos. A primera vista, las ballenas y sus parientes pueden parecer más peces que mamíferos. Pero como todos los mamíferos, los cetáceos son homotermos, tienen pulmones y alimentan a sus crías. La mayoría de las ballenas más grandes no tienen dientes y se alimentan de animales marinos diminutos. Pero los delfines, las marsopas, los cachalotes y las ballenas asesinas tienen dientes que usan para comer peces y otros animales.

Los **delfines tornillo** dan vueltas cuando saltan del agua. Como todos los delfines, son inteligentes y muy sociables.

Como los murciélagos, los cetáceos usan la ecolocación para "ver" a su alrededor. Los **cachalotes**, como éste, usan estallidos fuertes de sonido para atolondrar a los peces y poderlos atrapar con más facilidad.

Sirénidos

El grupo más pequeño de mamíferos marinos se llama sirénidos. Abarca sólo cuatro especies: tres tipos de manatíes y el dugongo. Estos mamíferos son completamente acuáticos; viven en las costas y en grandes ríos. Son animales tranquilos que comen algas y plantas marinas.

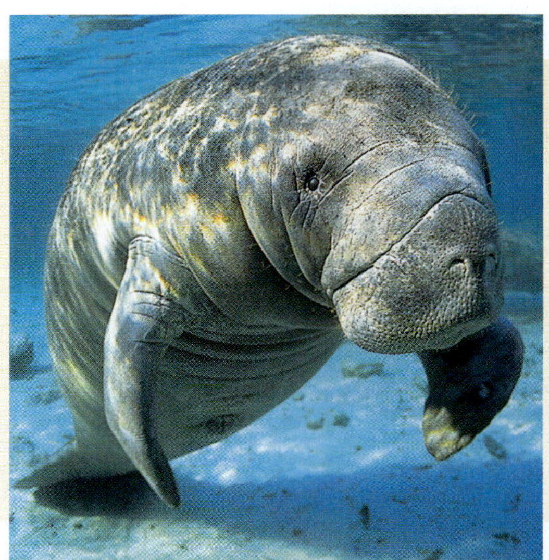

Los **manatíes** también son conocidos como vacas marinas.

WEIRD SCIENCE

Whales and dolphins have lost their body hair to reduce friction when they swim through water. Male competitive swimmers believe that shaving their bodies (even their heads) before they race will help them swim faster, even though Mark Spitz won seven gold medals at the 1972 Olympic Games without resorting to such drastic measures.

Primates

Los prosimios, los monos, los simios y los seres humanos pertenecen a un grupo de mamíferos llamados primates. Hay cerca de 160 especies de **primates.** Todos tienen los ojos hacia el frente, lo que permite que ambos ojos se centren en un solo punto. La mayoría de los primates tienen cinco dedos en cada mano y cinco en cada pie, con uñas planas en lugar de garras. Los dedos de los primates y los pulgares oponibles son prensiles; pueden hacer movimientos complicados, como agarrar objetos. Los primates tienen un cerebro grande en proporción con el tamaño de su cuerpo y se consideran unos de los mamíferos más inteligentes.

Muchos primates viven en árboles. Las articulaciones flexibles de los hombros, y las manos y pies prensiles les permiten trepar a los árboles y pasar de rama en rama. La mayoría de los primates comen hojas y frutas, pero algunos también comen animales.

Los **orangutanes** y otros simios con frecuencia caminan erguidos. Por lo general, el cerebro y el cuerpo de los simios es más grande que el de los monos.

El **mono araña,** como la mayoría de los monos, tiene una cola prensil. Sus brazos largos, piernas y cola le permiten moverse entre los árboles.

REPASO

1. ¿Puedes diferenciar un hipopótamo de un rinoceronte con sólo verles las patas? Explica tu respuesta.

2. ¿En qué se diferencian los monotremas de los demás mamíferos? ¿En qué se parecen?

3. ¿A qué grupo de mamíferos placentarios pertenecen los perros? ¿Cómo lo sabes?

4. **Hacer deducciones** ¿Qué es la gestación? ¿Por qué los elefantes tienen una gestación más larga que los ratones?

Experimentos

Leones, tigres y osos … ¿en Marte? Descubre más en la página 617.

405

---

4) Close

Quiz

1. Why did the mammal population increase when the dinosaurs became extinct? (There was more land and food available for the mammals.)

2. Where do mammals get the energy they need to keep their bodies warm? (Mammals quickly break down food in their bodies and use the energy released from their cells to keep their bodies warm.)

3. What are three characteristics of primates? (Possible answers: eyes that face forward, five fingers on each hand and five toes on each foot, flat fingernails and toenails instead of claws, fingers and thumbs that can make complicated movements, and a large brain)

ALTERNATIVE ASSESSMENT

Have students create a poster that visually summarizes the physiological information they have learned about birds and mammals and that provides examples of each of the animal categories discussed in the chapter. For example, a picture of a bird of prey catching its food would illustrate the grasping talons of raptors.
==Sheltered English==

LabBook PG 617
Wanted: Mammals on Mars

Reinforcement Worksheet 16 "Mammals Are Us"

Vocabulary Definitions

Section 1

down feather a fluffy insulating feather that lies next to a bird's body

contour feather a feather made of a stiff central shaft with many side branches, called barbs

preening activity in which a bird uses its beak to spread oil on its feathers

lift an upward force on an object (such as a wing) caused by differences in pressure above and below the object; opposes the downward pull of gravity

brooding when a bird sits on its eggs until they hatch

precocial chick a chick that leaves the nest immediately after hatching and that is fully active

altricial chick a chick that hatches weak, naked, and helpless

Resumen del capítulo

SECCIÓN 1

Vocabulario
plumón (pág. 383)
plumas de contorno (pág. 383)
arreglo de las plumas (pág. 383)
flotación (pág. 386)
empollar (pág. 387)
pollito precoz (pág. 388)
pollito altricial (pág. 388)

Notas de la sección
- Comos los reptiles, las aves ponen huevos amnióticos y tienen escamas gruesas y secas.
- A diferencia de aquéllos, las aves son homotermas y están cubiertas de plumas.
- Como volar requiere mucha energía, las aves deben comer una dieta energética y respirar eficazmente.

- Sus alas están diseñadas para flotar. La flotación es la presión del aire bajo las alas que mantiene al ave en el aire.
- Las aves pesan poco. Sus plumas son fuertes pero livianas, y su esqueleto es relativamente rígido, compacto y hueco.

- Como pueden volar, migran grandes distancias. Pueden anidar en un hábitat y pasar el invierno en otro. Las que migran pueden aprovechar las fuentes de comida y evitar a los depredadores.

Experimentos
¿No van al dentista? (pág. 616)

☑ Comprobar destrezas

Comprensión visual

FLOTACIÓN El diagrama de la página 386 ayuda a explicar el concepto de flotación. Puedes ver que el aire debe recorrer una distancia mayor sobre el ala curva que la que recorre por debajo. El aire sobre el ala debe moverse más rápido que el aire que circula por debajo para cubrir la mayor distancia en el mismo tiempo. El aire que circula rápido crea baja presión sobre el ala. La presión mayor bajo el ala la empuja hacia arriba, provocando la flotación.

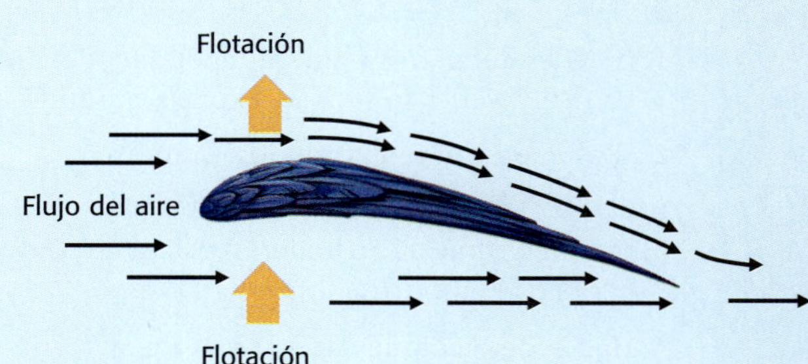

406

Lab and Activity Highlights

What? No Dentist Bills? **PG 616**

Wanted: Mammals on Mars **PG 617**

Datasheets for LabBook
(blackline masters for these labs)

<table><tr><td colspan="3" style="text-align:center">**SECCIÓN 2**</td></tr></table>

Vocabulario

terápsido *(pág. 392)*

glándulas mamarias *(pág. 393)*

diafragma *(pág. 395)*

monotrema *(pág. 396)*

marsupial *(pág. 397)*

mamífero placentario *(pág. 398)*

placenta *(pág. 398)*

gestación *(pág. 398)*

carnívoro *(pág. 401)*

primate *(pág. 405)*

Notas de la sección

- Todos los mamíferos tienen glándulas mamarias; en las hembras, éstas producen leche, que es un fluido altamente nutritivo con el que se alimentan las crías.

- Como las aves, los mamíferos son homotermos.

- Mantienen su rápido metabolismo alimentándose y respirando eficazmente.

- Los mamíferos tienen un diafragma que les ayuda a inhalar aire hacia los pulmones.

- Los mamíferos tienen dientes altamente especializados para masticar diferentes tipos de alimento. Los que comen plantas tienen incisivos y molares para cortar y triturarlas. Los carnívoros tienen caninos para atrapar y despedazar a su presa.

- Los mamíferos son los únicos vertebrados que tienen glándulas mamarias, pelaje y dos grupos de dientes.

- Se dividen en tres grupos: monotremas, marsupiales y mamíferos placentarios.

- Los monotremas ponen huevos en lugar de dar a luz una cría. Producen leche pero no tienen pezones ni placenta.

- Los marsupiales dan a luz a crías vivas, pero éstas nacen como embrión. Los embriones trepan hasta la bolsa de la madre, donde toman leche hasta que se desarrollan.

- Los mamíferos placentarios se desarrollan dentro de la madre durante un período de tiempo llamado gestación. Las madres alimentan a sus crías después de nacer.

Experimentos

Se solicitan mamíferos en Marte
(pág. 617)

internet

 **VISITA:** go.hrw.com

 **VISITA:** www.scilinks.org

Visita el sitio web de HRW para encontrar una serie de herramientas de aprendizaje relacionadas con este capítulo. Sólo tienes que escribir la palabra clave:

PALABRA CLAVE: HSTVR2

Visita el sitio web de la **Asociación Nacional de Maestros de Ciencias** *(National Science Teachers Association)* para encontrar recursos de Internet relacionados con este capítulo. Sólo escribe el **ENLACE DE CIENCIAS** para obtener más información sobre el tema:

TEMA: Características de las aves		**ENLACE:** HSTL405
TEMA: Clases de aves		**ENLACE:** HSTL410
TEMA: El origen de los mamíferos		**ENLACE:** HSTL415
TEMA: Características de los mamíferos		**ENLACE:** HSTL420

407

therapsid a prehistoric reptile ancestor of mammals

mammary glands glands that secrete a nutritious fluid called milk

diaphragm the sheet of muscle underneath the lungs of mammals that helps draw air into the lungs

monotreme a mammal that lays eggs

marsupial a mammal that gives birth to live, partially developed young that continue to develop inside the mother's pouch or skin fold

placental mammal a mammal that nourishes unborn offspring with a placenta inside the uterus

placenta a special organ of exchange that provides a developing baby with nutrients and oxygen

gestation period the time during which an embryo develops within the mother

carnivore a consumer that eats animals

primate a group of mammals that includes humans, apes, and monkeys; have opposable thumbs and binocular vision

 **Vocabulary Review Worksheet 16**

 Blackline masters of these Chapter Highlights can be found in the **Study Guide.**

Lab and Activity Highlights

LabBank

 **Labs You Can Eat,** Why Birds of a Beak Eat Together, Lab 8

Long-Term Projects & Research Ideas, Project 17

Using Vocabulary

1. Precocial, Altricial
2. diaphragm
3. placenta
4. preening
5. carnivores
6. Down feathers

Understanding Concepts

Multiple Choice

7. a
8. d
9. b
10. c
11. c
12. c

Short Answer

13. Marsupials bear live young that continue to develop in their mother's pouches. Like other mammals, they have fur, specialized teeth, and mammary glands.
14. Birds and mammals stay warm by converting food into energy for body heat. Mammals have fur, and birds have feathers, in part, to help retain body heat.
15. The Bernoulli effect is the vacuum created by fast-moving air.
16. Bats need large ears to hear echoes they use in locating food and other objects.

Repaso del capítulo

UTILIZAR EL VOCABULARIO

Escoge el término correcto para completar las siguientes oraciones:

1. Los pollitos __?__ siguen a su mamá apenas salen del cascarón. Los pollitos __?__ apenas pueden estirar el cuello para recibir comida cuando acaban de salir del cascarón. *(altriciales o precoces)*

2. El __?__ les ayuda a los mamíferos a respirar. *(diafragma o saco aéreo)*

3. La __?__ les permite a algunos mamíferos alimentar a las crías dentro del útero de la madre. *(glándula mamaria o placenta)*

4. Las aves se acicalan con un proceso llamado __?__. *(empollar o arreglo de las plumas)*

5. El león pertenece al grupo de mamíferos de los __?__. *(carnívoros o primates)*

6. __?__ son plumas esponjadas que ayudan a calentar a las aves. *(Las plumas de contorno o Los plumones)*

COMPRENDER CONCEPTOS

Opción múltiple

7. Tanto las aves como los reptiles
   a. ponen huevos.
   b. empollan a sus crías.
   c. son homotermos.
   d. tienen plumas.

8. El vuelo requiere
   a. mucha energía y oxígeno.
   b. un cuerpo liviano.
   c. músculos de vuelo fuertes.
   d. Todas las anteriores

9. Sólo los mamíferos
   a. son homotermos.
   b. alimentan a sus crías.
   c. ponen huevos.
   d. tienen dientes.

10. Los monotremas no
    a. tienen glándulas mamarias.
    b. cuidan a sus crías.
    c. tienen bolsas.
    d. tienen pelaje.

11. La flotación
    a. es aire que viaja sobre la parte superior de un ala.
    b. es proporcionada por sacos aéreos.
    c. es la fuerza ascendente que se ejerce sobre un ala para mantener un ave en el aire.
    d. se crea por la presión del diafragma.

12. ¿Cuál de los siguientes animales no es un primate?
    a. un lémur   c. una pika
    b. un ser humano   d. un chimpancé

Respuesta breve

13. ¿En qué se diferencian los marsupiales de los otros mamíferos? ¿En qué se parecen?

14. Tanto las aves como los mamíferos son homotermos. ¿Cómo mantienen su temperatura?

15. ¿Qué es el efecto Bernoulli?

16. ¿Por qué algunos murciélagos tienen orejas grandes?

17. An answer to this exercise can be found at the end of this book.

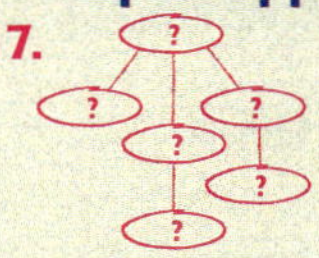

CRITICAL THINKING AND PROBLEM SOLVING

18. Marsupials get their nutrition from their mother's milk. Placental mammals get their nutrition from their mother's body through the placenta.

19. Bats and cetaceans are both active in dark environments, where sound is more helpful than sight.

20. Birds that fly will have a large keel and larger wings. The birds with longer wings probably soared, because longer wings are necessary to provide enough surface area for greater lift.

MATH IN SCIENCE

21. 1,500 Cal

INTERPRETING GRAPHICS

22. The faster the dog goes, the more energy it uses.

23. At 4 km/h, the dog consumes 9 Cal/g/h. At 9 km/h the dog consumes 16.5 Cal/g/h.

24. $\frac{9 \text{ Cal/g}}{h} \times 2{,}000 \text{ g} = 18{,}000 \text{ Cal/h}$

Concept Mapping Transparency 16

Blackline masters of this Chapter Review can be found in the **Study Guide.**

Organizar conceptos

17. Usa los siguientes términos para crear un mapa de ideas: monotremas, homotermos, aves, mamíferos, glándulas mamarias, mamíferos placentarios, marsupiales, plumas, pelo.

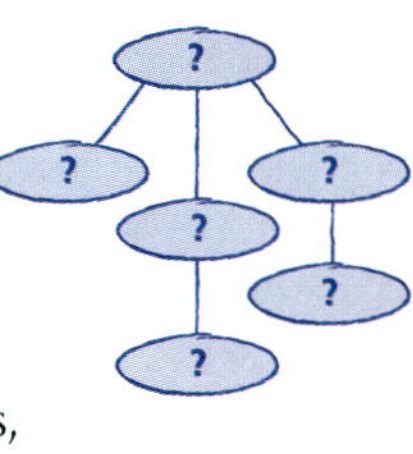

RAZONAMIENTO CRÍTICO Y RESOLUCIÓN DE PROBLEMAS

Escribe una o dos oraciones para responder a las siguientes preguntas:

18. A diferencia de los huevos de las aves y de los monotremas, los de los mamíferos placentarios y los marsupiales no tienen yema. ¿Cómo obtienen los nutrientes los embriones de estos animales?

19. La mayoría de los murciélagos usan la ecolocación. ¿Por qué estos animales no dependen únicamente de la vista para encontrar a su presa y examinar los alrededores?

20. Imagínate que trabajas en un museo y que estás haciendo una exposición del esqueleto de las aves. Desafortunadamente, se perdieron las etiquetas de los esqueletos. ¿Cómo puedes diferenciar los esqueletos de las aves no voladoras de los de las voladoras? ¿Puedes decir qué aves volaban rápidamente y cuáles podían planear? Explica tu respuesta.

LAS MATEMÁTICAS EN LAS CIENCIAS

21. Un ave vuela a 35 km/h. A esta velocidad, su cuerpo consume 60 calorías por gramo de masa corporal por hora. Si el ave pesa 50 g, ¿cuántas calorías usa si vuela durante 30 minutos a esta velocidad?

INTERPRETAR GRÁFICAS

Los homotermos usan mucha energía cuando corren o vuelan. La siguiente gráfica muestra cuántas calorías gasta un perro pequeño cuando corre a diferentes velocidades. Úsala para responder a las siguientes preguntas:

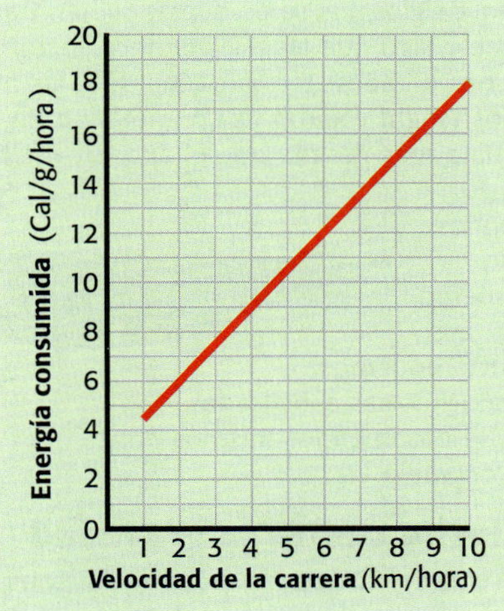

22. A medida que el perro corre más rápido, ¿cómo cambia la cantidad de energía que consume por hora?

23. ¿Cuánta energía por hora consume si corre a 4 km/h? ¿Cuánta energía consume si corre a 9 km/h?

24. La energía consumida se da en calorías por gramo de masa corporal por hora. Si el perro pesa 2 kg y corre a 4 km/h, ¿cuántas calorías usa por hora?

AHORA, ¿qué piensas?

Revisa tus respuestas a las preguntas de la página 381 que escribiste en el cuaderno de ciencias. ¿Han cambiado tus respuestas? Si es necesario, corrige tus respuestas basándote en lo que has aprendido en este capítulo.

409

NOW WHAT DO YOU THINK?

1. Planes and birds stay aloft because air traveling over the top of a wing moves faster than air traveling below it. This generates higher pressure beneath the wing than above it, causing lift.

2. Kangaroos are marsupial. Most mammals are placental.

3. Yes; monotremes are mammals that lay eggs. Bats are mammals that can fly.

The Aerodynamics of Flight

Background

The four aerodynamic forces described in this feature—lift, gravity, drag, and thrust—permit birds and humans to fly. Airplanes are designed differently to take advantage of these forces. Similarly, birds have particular adaptations that make them fast flyers or excellent gliders. Students should recognize how the physical features of airplanes and birds contribute to their ability to fly.

A TRAVÉS DE LAS CIENCIAS

CIENCIAS BIOLÓGICAS • CIENCIAS FÍSICAS

La aerodinámica del vuelo

Durante siglos, los seres humanos han tratado de hacer algo que las aves perfeccionaron hace millones de años: el vuelo. En 1903, los hermanos Wright volaron por primera vez en una máquina más pesada que el aire. Su vuelo duró 12 segundos y sólo recorrieron 37 m. Los aviones modernos son mucho más sofisticados, pero se basan en los mismos principios de vuelo.

La lucha contra la gravedad

El diseño del cuerpo de un jet lucha contra la resistencia aerodinámica, y el de las alas contra la fuerza de gravedad. Para despegar, los aviones deben impulsarse con una fuerza mayor que la fuerza de gravedad, llamada *flotación*. ¿De dónde viene? La parte superior del ala de un avión es curva, y la inferior, plana. El aire debe viajar más lejos y más rápido sobre el ala que debajo de ella. Esta diferencia provoca que la presión sobre el ala sea menor, y hace que el avión se impulse hacia arriba.

Empujar y jalar

La forma del ala no basta para que un avión se eleve. Los aviones también dependen de la *fuerza propulsora,* que permite que el avión avance y que se obtiene de los motores y propulsores. Cuanto más rápido se mueven los aviones, más aire pasa por las alas y aumenta la elevación.

Por lo general, los aviones despegan en contra del viento, el cual choca contra ellos durante el vuelo. Cualquier fuerza que vaya en contra del movimiento de un avión se llama *resistencia aerodinámica* y retarda su movimiento. El cuerpo de un avión tiene curvas lisas para mini-mizar la resistencia aerodinámica. Un viento de cola es una corriente de aire que empuja el avión y acorta el tiempo de viaje. Para aumentar la velocidad, los ingenieros diseñan aviones aerodinámicos para reducir la resistencia. Las alas también se diseñan para aumentar la elevación. Un ala redondeada y más larga da mayor elevación, pero también produce más resistencia aerodinámica. Los atletas también consideran la resistencia aerodinámica cuando escogen su equipo (los corredores y los ciclistas usan ropa muy ajustada para reducirla).

¡Piénsalo!

▶ Hay aviones de muchas formas y tamaños diseñados para distintos fines, como volar rápidamente o transportar cargas pesadas. Investiga y describe cómo difiere la aerodinámica de estas naves.

▲ *El diseño de los aviones está inspirado en las aves.*

410

Answer to Think About It!

Heavy cargo airplanes travel at much slower speeds than fighter jets. Cargo airplanes generally have thick wings to provide enough lift to get their massive contents off the ground. Fighter jets are much lighter and do not require as much lift. They often have thinner wings. Many fighter jets have powerful engines to give them more thrust and speed. The fastest fighter jets can travel over 2,000 mph! Fighter jets also have smooth, aerodynamic surfaces to minimize drag, which would slow them down. Short-distance flyers, such as many personal airplanes, do not have expensive, powerful engines but rely on propellers to provide thrust.

Curiosidades de la CIENCIA

RATAS-TOPO LAMPIÑAS

¿Cómo llamas a un roedor casi ciego que mide 7 cm de largo y que parece una salchicha hervida? Una rata-topo lampiña. Durante más de 150 años, este mamífero de las calientes regiones áridas de Kenya, Etiopía y Somalia, ha dejado perplejos a los científicos por su extraña apariencia y peculiares hábitos.

¿Qué tiene que ver el pelo?

Las ratas-topo lampiñas tienen características tan extrañas que te hacen dudar que sean mamíferos. Su piel rosada grisácea les cuelga libremente, permitiéndoles maniobrar a través de los sistemas de túneles angostos donde habitan. A primera vista, parece que no tienen pelo, una característica clave de los mamíferos. Sin embargo, las ratas-topo lampiñas sí tienen pelo, pero no pelaje. En realidad, tienen bigotes que usan para guiarse a través de los pasajes obscuros, y pelo entre los dedos de sus patas para barrer tierra suelta, como una especie de escobitas. También tienen pelo en los labios para impedir que les entre tierra en la boca cuando cavan nuevos pasajes con sus grandes dientes.

¿Hace frío aquí?

Las ratas-topo lampiñas tienen la peor capacidad homoterma de todos los mamíferos. Su temperatura corporal es casi la misma que la del aire de los túneles, unos 31ºC (casi unos 5ºC más frío que la temperatura corporal de los humanos). Por la noche, estos animales contrarrestan el frío amontonándose unos sobre otros. Por fortuna, la temperatura no cambia mucho en su hábitat natural.

¿Quién está a cargo?

Las ratas-topo lampiñas son los únicos mamíferos que forman comunidades similares a las que forman los insectos sociales, como las abejas melíferas. Una comunidad de ratas-topo lampiñas se compone de 20 a 300 miembros que se dividen las tareas, como lo hacen las abejas, las avispas y las termitas. Cada comunidad tiene una hembra para reproducirse, llamada reina, y hasta tres machos para la reproducción. Todas las hembras son biológicamente capaces de reproducirse, pero sólo una lo hace. Cuando una hembra se vuelve reina, crece más que las otras.

¡Piénsalo!

▶ A primera vista, parece que las ratas-topo lampiñas no tienen algunas de las principales características de los mamíferos. Investiga más y descubre por qué se clasifican como mamíferos.

◀ *Las ratas-topo lampiñas son tan raras que se han vuelto una atracción popular en los zoológicos.*

UNIDAD 6

La ecología

¿ Qué desayunaste esta mañana? Lo que haya sido, es un resultado directo del trabajo conjunto de seres vivos. Por ejemplo, la leche proviene de la vaca, y ésta come plantas para obtener energía. Las bacterias ayudan a las plantas a obtener nutrientes del suelo, que éste contiene porque los hongos descomponen los árboles muertos. Todos los seres vivos de la Tierra están relacionados entre sí. Nuestras acciones tienen un impacto sobre el medio ambiente, que a su vez tiene un impacto sobre nosotros. En esta unidad estudiarás la ecología, que es la interacción entre los seres vivos. Esta cronología muestra cómo los humanos han estudiado e influido en nuestro planeta.

1661

John Evelyn publica un libro que denuncia la contaminación del aire en Londres, Inglaterra.

1771

Joseph Priestley realiza experimentos con plantas y descubre que utilizan dióxido de carbono y liberan oxígeno.

1970

La Agencia de Protección Ambiental, *(Environmental Protection Agency o EPA)* se fundó para establecer y reforzar las normas de control de la contaminación en los Estados Unidos.

1973

El Congreso de los Estados Unidos aprueba la Ley de Protección de Especies en Peligro de Extinción.

1990

Para evitar la captura de delfines por redes de pesca, las empacadoras de atún de los Estados Unidos anuncian que no aceptarán el atún capturado con redes que causen la muerte de delfines.

1852

Los Estados Unidos importan gorriones de Alemania para defenderse de las orugas que destruyen cosechas.

1854

Se publica *Walden*, de Henry David Thoreau, en el que se afirma que debemos vivir en armonía con la naturaleza.

1872

El Congreso establece el primer parque nacional de los Estados Unidos: Yellowstone.

1933

Nace la Organización Civil para la Conservación de la Naturaleza, que siembra árboles, combate incendios forestales y construye presas para controlar inundaciones.

1962

Se publica el libro *Silent Spring*, de Rachel Carson, que describe el uso indiscriminado de pesticidas y la destrucción que ocasionan en el medio ambiente.

1993

Los estadounidenses reciclan 59,500 millones de latas de aluminio (dos de cada tres).

1996

Se abre la presa del cañón Glen para inundar intencionalmente el Gran Cañón. La inundación ayuda a mantener el equilibrio ecológico mediante la recuperación de playas y bancos de arena, así como la renovación de las marismas.

Chapter Organizer

CHAPTER ORGANIZATION	TIME MINUTES	OBJECTIVES	LABS, INVESTIGATIONS, AND DEMONSTRATIONS
Chapter Opener pp. 414–415	45		**Investigate!** A Mini-Ecosystem, p. 415
Section 1 Land Ecosystems	90	▶ Define *biome*. ▶ Describe three different forest biomes. ▶ Distinguish between temperate grasslands and savannas. ▶ Describe the importance of permafrost to the arctic tundra biome.	**Design Your Own,** Life in the Desert, p. 618 **Datasheets for LabBook,** Life in the Desert, Datasheet 35
Section 2 Marine Ecosystems	90	▶ Distinguish between the different areas of the ocean. ▶ Explain the importance of plankton in marine ecosystems. ▶ Describe coral reefs and intertidal areas. ▶ Explain what is unique about polar marine biomes.	**Interactive Explorations CD-ROM,** Sea Sick A **Worksheet** is also available in the **Interactive Explorations Teacher's Edition.** **Design Your Own,** Discovering Mini-Ecosystems, p. 619 **Datasheets for LabBook,** Discovering Mini-Ecosystems, Datasheet 36
Section 3 Freshwater Ecosystems	90	▶ List the characteristics of rivers and streams. ▶ Describe the littoral zone of a pond. ▶ Explain what happens to a pond as the seasons change. ▶ Distinguish between two types of wetlands.	**QuickLab,** Pond Food Connections, p. 429 **Skill Builder,** Too Much of a Good Thing? p. 620 **Datasheets for LabBook,** Too Much of a Good Thing? Datasheet 37 **EcoLabs & Field Activities,** Biome Adventure Travel, EcoLab 7 **Long-Term Projects & Research Ideas,** Project 20

TECHNOLOGY RESOURCES

Guided Reading Audio CD
English or Spanish, Chapter 17

One-Stop Planner CD-ROM with Test Generator

Interactive Explorations CD-ROM
CD 2, Exploration 2, Sea Sick

Multicultural Connections, Saving Pacific Sea Horses, Segment 9

Eye on the Environment, Biosphere Pioneers, Segment 6

Science Discovery Videodiscs
Image and Activity Bank with Lesson Plans: Tragedies in the Commons

CLASSROOM WORKSHEETS, TRANSPARENCIES, AND RESOURCES	SCIENCE INTEGRATION AND CONNECTIONS	REVIEW AND ASSESSMENT
Directed Reading Worksheet 17 **Science Puzzlers, Twisters & Teasers,** Worksheet 17	**Careers:** Ecologist—Alfonso Alonso-Mejía, p. 437	
Transparency 65, Earth's Biomes **Directed Reading Worksheet 17,** Section 1 **Math Skills for Science Worksheet 2,** Subtraction Review **Transparency 66,** Coniferous Forest Biome **Transparency 67,** A Tropical Rain Forest Biome **Math Skills for Science Worksheet 37,** Rain-Forest Math **Transparency 153,** An Example of the Rain Shadow Effect **Reinforcement Worksheet 17,** Know Your Biomes	**Cross-Disciplinary Focus,** p. 417 in ATE **Math and More,** p. 418 in ATE **Real-World Connection,** p. 419 in ATE **Multicultural Connection,** p. 419 in ATE **Multicultural Connection,** p. 420 in ATE **Connect to Earth Science,** p. 420 in ATE **MathBreak,** Rainfall, p. 422	**Self-Check,** p. 421 **Review,** p. 422 **Quiz,** p. 422 in ATE **Alternative Assessment,** p. 422 in ATE
Directed Reading Worksheet 17, Section 2 **Science Skills Worksheet 1,** Being Flexible	**Multicultural Connection,** p. 424 in ATE **Real-World Connection,** p. 425 in ATE **Connect to Physical Science,** p. 425 in ATE **Across the Sciences:** Ocean Vents, p. 436	**Self-Check,** p. 426 **Review,** p. 427 **Quiz,** p. 427 in ATE **Alternative Assessment,** p. 427 in ATE
Transparency 68, River Features **Directed Reading Worksheet 17,** Section 3 **Transparency 69,** Lake Zones **Critical Thinking Worksheet 17,** Risky Development?	**Math and More,** p. 429 in ATE **Connect to Physical Science,** p. 429 in ATE **Apply,** p. 430	**Review,** p. 431 **Quiz,** p. 431 in ATE **Alternative Assessment,** p. 431 in ATE

internet connect

Holt, Rinehart and Winston On-line Resources

go.hrw.com

For worksheets and other teaching aids related to this chapter, visit the HRW Web site and type in the keyword: **HSTECO**

National Science Teachers Association

www.scilinks.org

Encourage students to use the *sci*LINKS numbers listed with the Chapter Highlights to access information and resources on the **NSTA** Web site.

END-OF-CHAPTER REVIEW AND ASSESSMENT

Chapter Review in Study Guide

Vocabulary and Notes in Study Guide

Chapter Tests with Performance-Based Assessment, Chapter 17 Test

Chapter Tests with Performance-Based Assessment, Performance-Based Assessment 17

Concept Mapping Transparency 17

Chapter Resources & Worksheets

Visual Resources

TEACHING TRANSPARENCIES

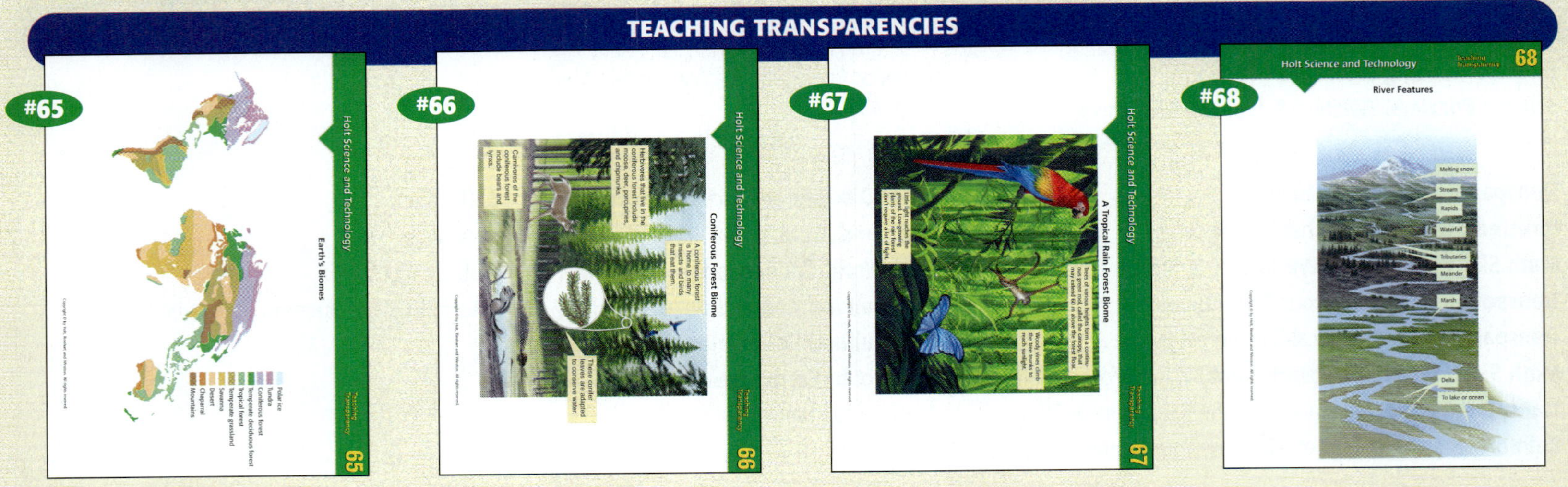

TEACHING TRANSPARENCIES

CONCEPT MAPPING TRANSPARENCY

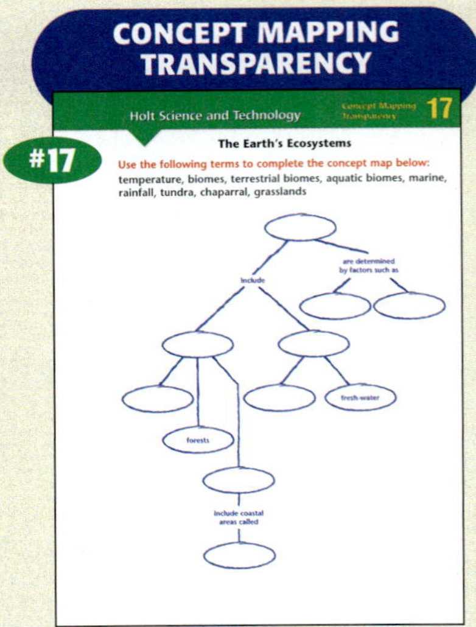

Meeting Individual Needs

DIRECTED READING

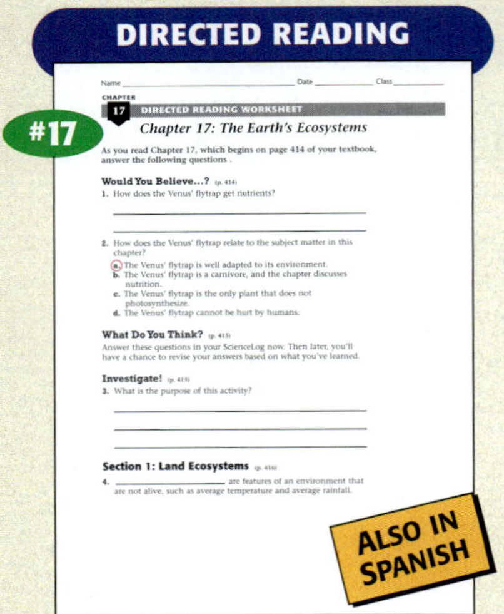

REINFORCEMENT & VOCABULARY REVIEW

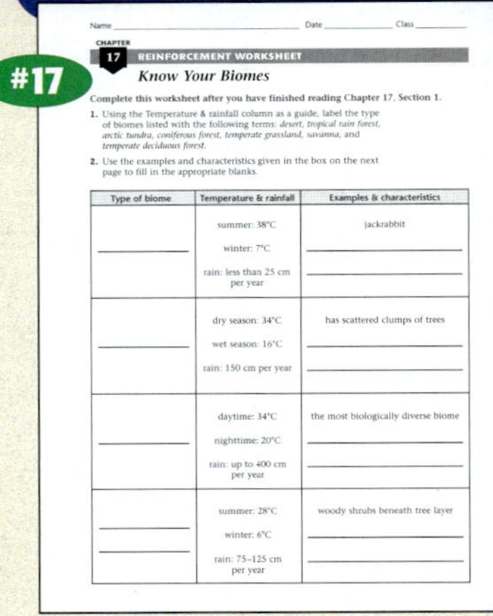

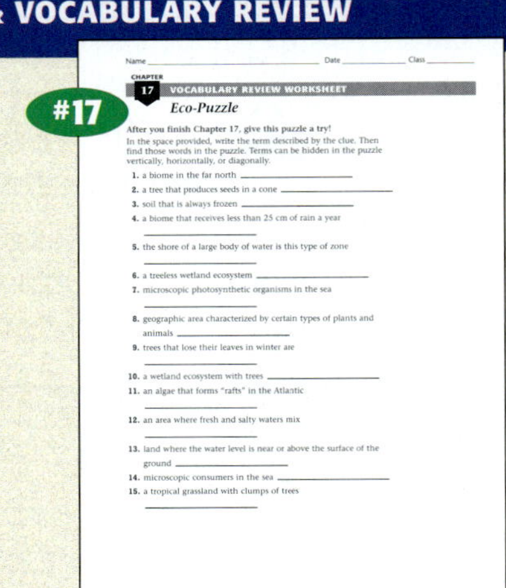

SCIENCE PUZZLERS, TWISTERS & TEASERS

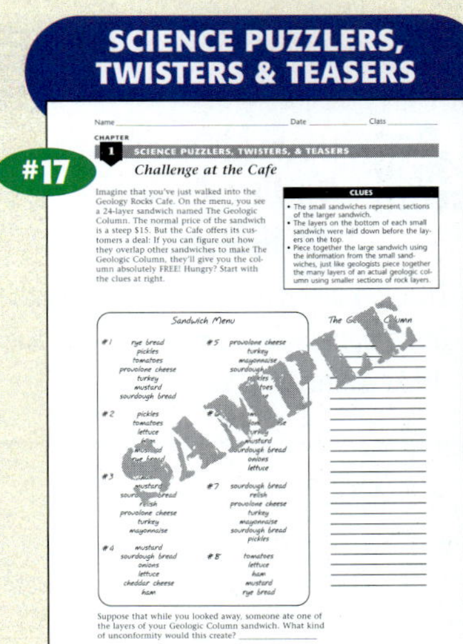

Review & Assessment

STUDY GUIDE

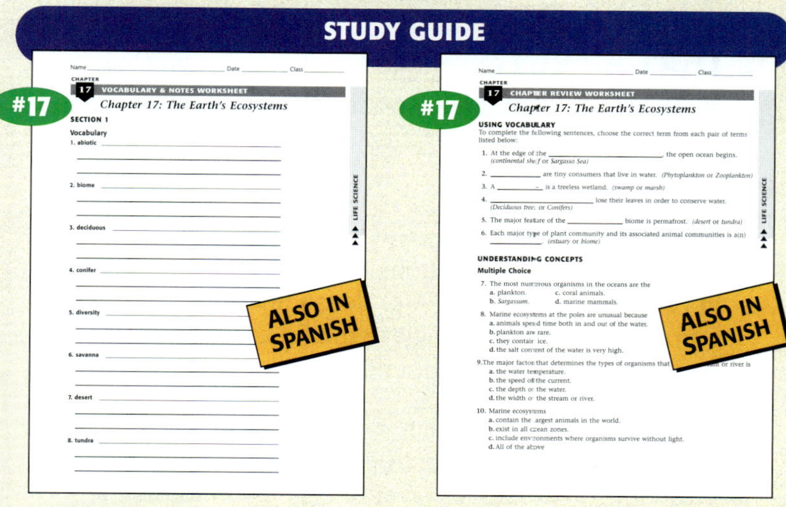

CHAPTER TESTS WITH PERFORMANCE-BASED ASSESSMENT

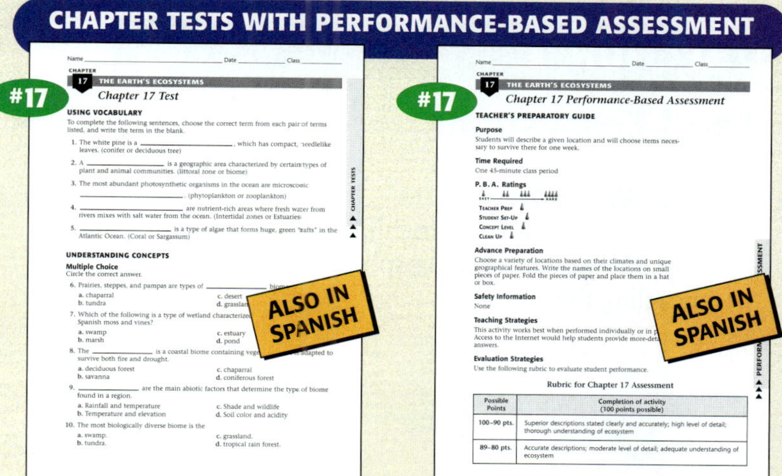

Lab Worksheets

ECOLABS & FIELD ACTIVITIES

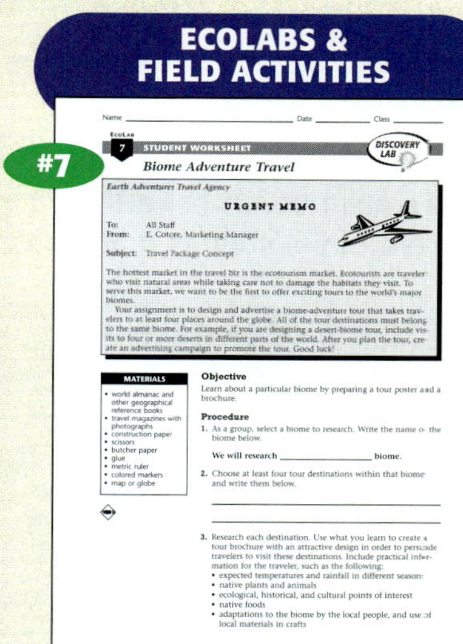

LONG-TERM PROJECTS & RESEARCH IDEAS

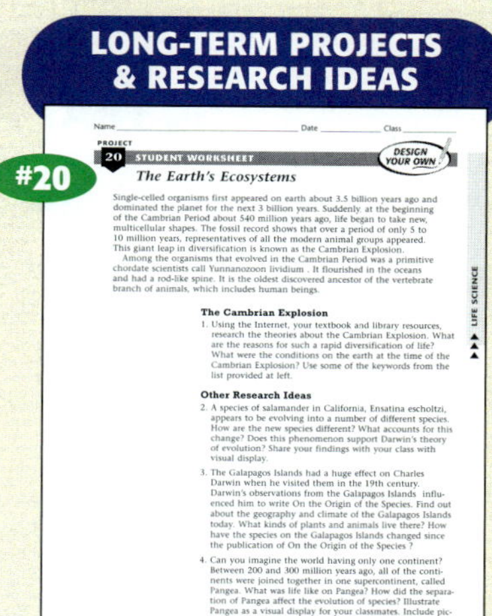

DATASHEETS FOR LABBOOK

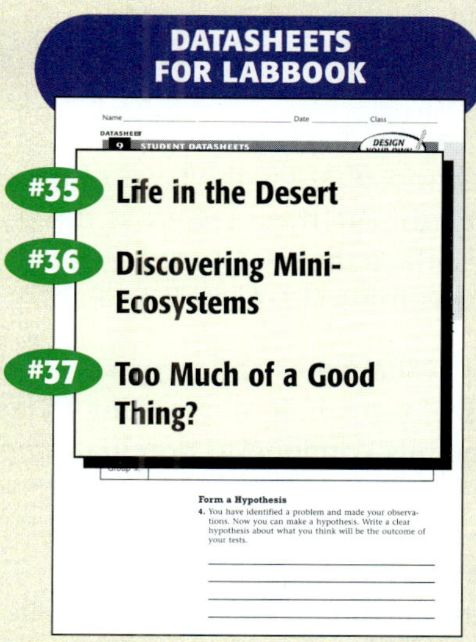

Applications & Extensions

CRITICAL THINKING & PROBLEM SOLVING

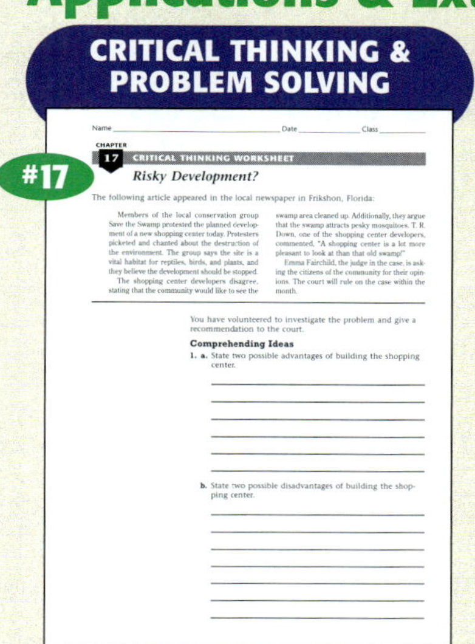

MULTICULTURAL CONNECTIONS

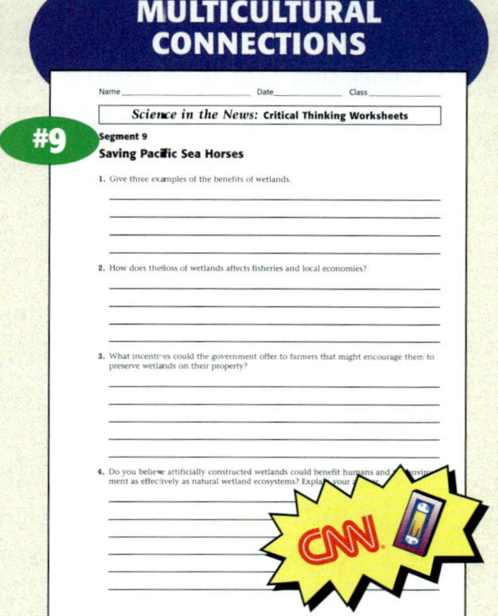

EYE ON THE ENVIRONMENT

INTERACTIVE EXPLORATIONS

Chapter Background

SECTION 1

Land Ecosystems

▶ The Biosphere

All parts of Earth that are inhabited by organisms make up the biosphere. The biosphere is a relatively thin layer encircling the planet.

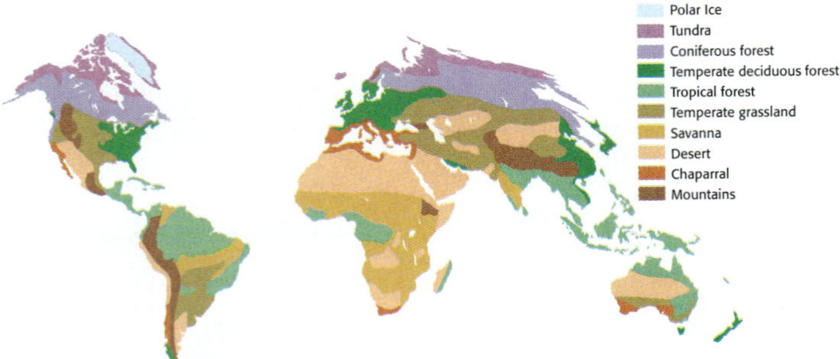

- The Earth can be divided into gas, liquid, and solid parts. The atmosphere is the layer of gases that envelops Earth. The hydrosphere is the portion of Earth's surface that is covered by water. The lithosphere is the soil and rock on Earth's surface.

- Life began in the water and remained in the water for billions of years before migrating to land. Today, aquatic habitats continue to dominate the biosphere.

▶ Biotic and Abiotic Factors

An ecosystem encompasses all of the biotic and abiotic factors in a particular area. Ecology is the scientific study of how organisms interact with one another and with the abiotic factors in their environment. Biotic factors include all organisms—plants, animals, protists, fungi, and bacteria. Abiotic factors include temperature, water, sunlight, wind, rocks, soil, and nutrients.

- Almost all ecosystems are driven by energy from the sun. Thus, the amount of sunlight a region receives has a large effect on the number of producers and consumers that can be supported in the ecosystem.

- Just as chemical reactions are limited by a limiting reagent, populations within ecosystems are limited by limiting factors. The lack of any single abiotic factor can prevent the survival of a population. Limiting factors include the amount of light received, water, temperature, and nutrients in the soil.

IS THAT A FACT!

- One hectare of fertile soil can contain 6 million earthworms and over 2 billion bacteria!

▶ Tropical Rain Forests

Students may think that tropical rain forests are jungles. In fact, a jungle is an area of dense undergrowth within a tropical rain forest. Jungles grow in areas that receive large amounts of sunlight and are near rivers.

- The plants in a tropical rain forest can be divided into three layers. Some trees tower above the forest's canopy. A continuous layer of vegetation forms the upper canopy. Below the canopy is a mid-tree level and a shrubby understory. On the ground are grasses and ferns.

IS THAT A FACT!

- Tropical rain forests cover about 7 percent of Earth's surface. This small portion of Earth contains more than 50 percent of the species that inhabit the planet.

▶ Tundra

A tundra area may receive as little rainfall as a desert. But the soil in a tundra region remains wet due to permafrost, low temperatures, and the low rate of evaporation.

IS THAT A FACT!

- Antarctica has been accumulating ice for more than 25 million years. It contains about 90 percent of Earth's ice and about 70 percent of Earth's fresh water.

- In summer, when ice begins to melt and break off into icebergs, Antarctica shrinks. In winter, Antarctica expands to twice its summer size.

Marine Ecosystems

▶ Oceans

The Earth's oceans include the Pacific Ocean, the Atlantic Ocean, the Indian Ocean, and the Arctic Ocean. Although these oceans have different names, they are all connected.

- Because the oceans are all connected, a change in one marine environment may eventually affect other marine environments.

IS THAT A FACT!

- ◆ Along Australia's northeastern coast is the Great Barrier Reef. It is the largest and most diverse reef system in the world.

▶ Underwater Exploration

New technology for remote-operated vehicles has broadened scientists' ability to explore ocean depths. Using vehicles equipped with cameras, mechanical arms, and remote sensors, scientists have discovered a watery new world that includes deep-sea animals, underwater volcanoes, thermal vents, and entire ecosystems that do not directly depend on light or photosynthesis for energy.

- Early diving suits consisted of a hard helmet, a canvas-and-rubber tunic, leather boots with lead-weighted soles, and additional 13 kg weights. Each boot weighed about 8 kg. Such a suit was called a standard diving suit and was invented by Augustus Siebe in the 1830s.

Freshwater Ecosystems

▶ Wetlands

In the past, wetlands were underappreciated and even considered wastelands. People viewed wetlands as places that should be drained or filled in so that the land could be used for housing or other urban development. Because wetlands are a breeding ground for mosquitoes, they were also considered a health problem.

- Two types of wetlands are introduced in the text: the marsh and the swamp. However, there are many different kinds of wetlands, including inland freshwater wetlands, coastal freshwater wetlands, and coastal saltwater wetlands.

IS THAT A FACT!

- ◆ About 6 percent of Earth's surface is wetlands.

- ◆ Wetland areas in the contiguous United States have shrunk from 81 million hectares to 38 million hectares and continue to shrink each year.

- ◆ Ninety percent of the wetlands in the San Francisco Bay area have disappeared. In their place are airports, houses, industrial parks, and landfills.

Directed Reading Worksheet 17

Science Puzzlers, Twisters & Teasers Worksheet 17

Guided Reading Audio CD
English or Spanish, Chapter 17

¿Creerías que . . . ?

Sin hacer ruido y sin moverse, un atrapamoscas está a la espera. Al poco tiempo, una mosca desprevenida choca con uno de sus sensibles vellos. ¡Zas! La puerta se cierra y su víctima queda atrapada. Cuanto más se resiste, más se cierra la trampa. Su destino está escrito. Cuando el atrapamoscas captura una mosca u otro animal pequeño, sus hojas segregan jugos digestivos y el insecto se digiere lentamente. ¿Existe este extraño animal en la realidad? ¿O será una bestia terrorífica de algún relato de ciencia ficción? Ni lo uno ni lo otro. El atrapamoscas es una planta, aunque poco común. Como sabes, la mayoría de las plantas obtienen del suelo los nutrientes que necesitan para sobrevivir. El suelo de los pantanos contiene pocos nutrientes, pero las plantas como el atrapamoscas son capaces de crecer allí. ¿Cómo?

Toman los nutrientes de una fuente distinta: animales pequeños a los que atrapan y devoran.

A pesar de su impresionante apariencia, el atrapamoscas es muy vulnerable. Su distribución es muy limitada, pues sólo se da en ciertas zonas de Carolina del Norte y de Carolina del Sur. Se cultiva fácilmente en viveros, pero los traficantes furtivos recolectan plantas silvestres ilegalmente. El pasto y los fertilizantes también le ocasionan problemas porque está adaptada a terrenos con pocos nutrientes.

En este capítulo aprenderás que en la Tierra hay muchas zonas diferentes. Hay plantas y animales especialmente adaptados para sobrevivir en cada zona.

414

Would You Believe . . . ?

Plants that feed on animals are called carnivorous plants. Venus' flytraps aren't the only carnivorous plants. A number of plants have evolved insect traps. Most of the traps consist of modified leaves and glands that secrete digestive enzymes.

The pitcher plant has funnel-like leaves that catch water. When an insect goes inside to drink, it is likely to be trapped and drown. The insect's body is then broken down by digestive enzymes in the water.

¡Investiga!

Un miniecosistema

En esta actividad construirás y observarás un ecosistema en miniatura.

Procedimiento

1. Coloca una capa **de grava** en el fondo de un **frasco de boca ancha** de una botella de 2 L de refresco sin cuello. Añade una capa de **tierra.**

2. Agrega diversas **plantas** que requieran condiciones similares para crecer. Escoge plantas pequeñas que no crezcan con mucha rapidez.

3. Rocía **agua** dentro del frasco para humedecer la tierra.

4. Tapa el frasco y colócalo bajo luz indirecta. Describe el aspecto del escosistema en tu cuaderno de ciencias.

5. Obsérvalo cada semana y anota tus observaciones.

Análisis

6. Enumera los factores carentes de vida que conforman el ecosistema que creaste.

7. ¿En qué se parece el miniecosistema a un ecosistema real? ¿En qué se diferencia?

Investigate!

MATERIALS

FOR EACH GROUP:
- gravel
- large widemouthed jar
- soil
- plants
- water

Safety Caution: You may want to have students wear disposable gloves while handling plants. Plants and plant parts should be kept away from the face and eyes because they may scratch or otherwise cause irritation. Have students wash their hands after handling plants, seeds, and soil.

Answers to Investigate!

6. Answers to this question should include representatives of the nonliving factors one would find in nature, such as water, temperature, light, and air.

7. Answers will vary. In general, each mini-ecosystem is similar to a real ecosystem in that it will contain living plants that require similar growing conditions, water, soil, and sunlight. It will differ from a real ecosystem in that it is enclosed, it was artificially assembled, and it did not naturally evolve to its present state.

IS THAT A FACT!

Coral reefs are ancient ecosystems. Most coral reefs are 5,000 to 10,000 years old. Some coral reefs are millions of years old.

Focus

Land Ecosystems

This section introduces the concept of a *biome* and describes several land biomes. Students learn about different types of forest and grassland as well as about deserts and tundra.

Bellringer

Write the following on the board:

cactus	tropical rain forest
tree frog	polar ice
pine tree	desert
polar bear	mountain

Ask students to match the plant or animal in the first column with the environment in the second column where it would most likely be found. They should record their answers in their ScienceLog. (Answer: cactus, desert; tree frog, tropical rain forest; pine tree, mountain; polar bear, polar ice)

1) Motivate

ACTIVITY

Describing Ecosystems Pair up students and have them brainstorm about what distinguishes different land ecosystems—what makes a forest a forest, a desert a desert, and so on. Have the pairs write down their ideas and read them to the class. Write each idea on the board, and place marks next to an idea each time students suggest it. Tell students that they will learn how accurate this popularity contest was as they read this section.

VOCABULARIO

abiótico	sabana
bioma	desierto
árboles de hoja caduca	tundra
conífera	permafrost
diversidad	

OBJETIVOS

- Define el *bioma*.
- Describe tres biomas de bosque.
- Describe las diferencias entre las praderas templadas y las sabanas.
- Describe la importancia del permafrost en el bioma de la tundra ártica.

Ecosistemas terrestres

Imagínate que planeas hacer una excursión y que vas a una agencia de viajes donde hay una máquina de realidad virtual que te permite experimentar distintos lugares antes de partir. Enciendes la máquina y de repente te sientes transportado. Al principio te duelen los ojos debido a la intensa luz del Sol. El viento que te golpea la cara está muy caliente y seco. A medida que tus ojos se acostumbran a la luz, observas un gran cactus a tu derecha y, a lo lejos, pequeños arbustos. Una liebre asustada corre por el suelo seco y polvoriento, y una lagartija difruta del calor del sol sobre una roca. ¿Dónde estás?

Quizá no puedas señalar tu ubicación exacta, pero probablemente te des cuenta de que estás en un desierto, pues la mayoría de los desiertos son calurosos y secos. Estos factores **abióticos,** o carentes de vida, influyen en los tipos de plantas y animales que viven en la zona.

Los biomas de la Tierra

El desierto es uno de los biomas de la Tierra. Un **bioma** es una región geográfica caracterizada por ciertos tipos de comunidades de plantas y animales. Un bioma contiene varios ecosistemas más pequeños, relacionados entre sí. Por ejemplo, el bosque tropical lluvioso es un bioma que contiene ecosistemas de río, de las copas de los árboles, del suelo del bosque y muchos otros. Un bioma no es un lugar específico; por ejemplo, el bioma del desierto no se refiere a un desierto en particular, sino a cualquier ecosistema de desierto de la Tierra. Los principales biomas de la Tierra se muestran en la **Figura 1.**

Figura 1 *La lluvia y la temperatura son los principales factores que determinan el bioma de una región. ¿En qué tipo de bioma vives?*

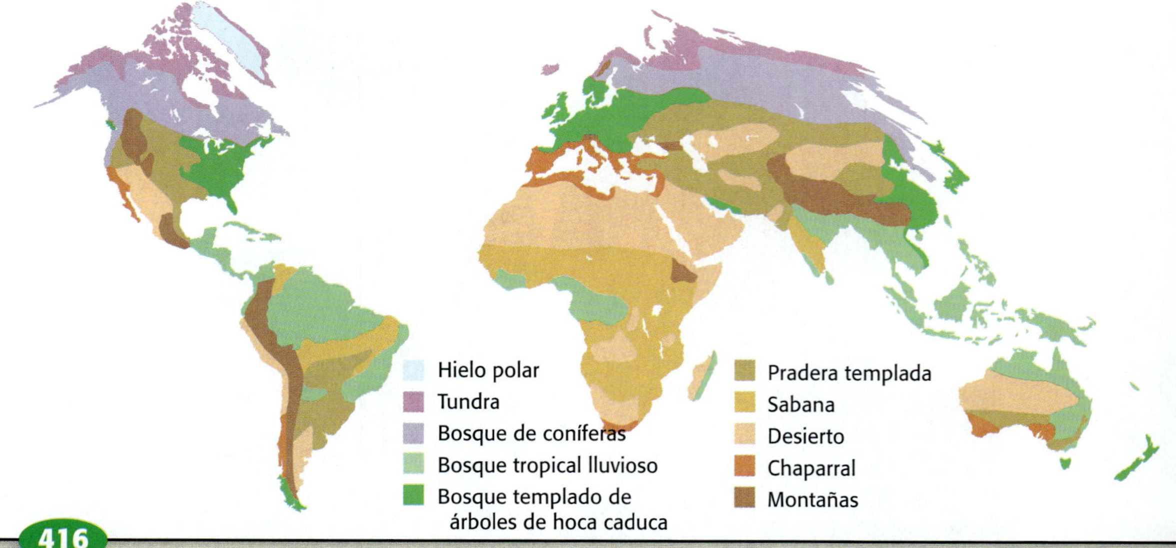

416

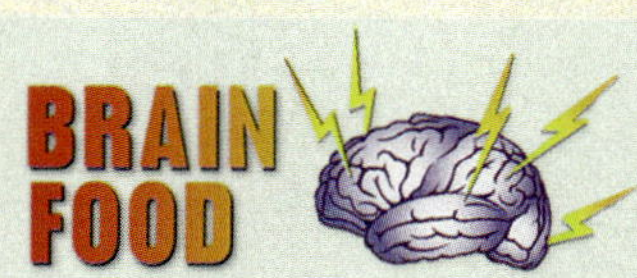

Have students examine the map of biomes in **Figure 1.** Ask, "Does every continent include a biome?" and "Where on the map is Antarctica?" Encourage students to speculate about why Antarctica does not include a biome. Ask them what kinds of plants and animals live in Antarctica. (Students may suggest seals, algae, bacteria, etc.)

Point out that most organisms that inhabit Antarctica are part of an aquatic biome rather than a land biome.

Bosques

Los biomas de bosque se desarrollan donde cae lluvia suficiente y donde la temperatura no es demasiado alta en verano ni demasiado baja en invierno. Hay tres principales tipos de biomas de bosque: bosque templado de árboles de hoja caduca, bosque de coníferas y bosque tropical lluvioso. El tipo de bosque que se desarrolla en una región particular depende de la temperatura y de la precipitación pluvial.

Bosque templado de árboles de hoja caduca ¿Has visto cómo en otoño las hojas cambian de color y se caen de los árboles? Pues ya conoces a los árboles **de hoja caduca.** "Caduco" viene del latín y significa "caer". Al perder sus hojas en el otoño, estos árboles logran conservar el agua durante el invierno. La **Figura 2** ilustra un bosque de este tipo. En la mayoría de estos bosques hay varias especies de árboles y diversos animales, como osos y pájaros carpinteros.

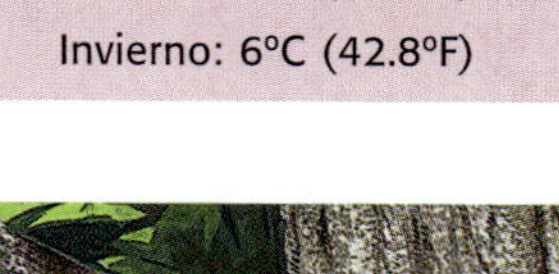

Figura 2 *En un bosque templado de árboles de hoja caduca, los mamíferos, aves y reptiles se desarrollan en medio de una abundancia de hojas, semillas, nueces e insectos.*

MISCONCEPTION ALERT

Students may think that the terms *biome* and *ecosystem* are synonyms. In fact, these terms are not used consistently even by scientists. You may want to point out that both biomes and ecosystems are determined by abiotic and biotic factors. However, a biome is a broader category than an ecosystem. A biome may contain several similar yet separate ecosystems.

COOPERATIVE LEARNING

Provide small groups of students with an almanac and graph paper. Ask groups to find and graph the average monthly rainfall and average monthly temperature for their area. Some students can measure the temperature and rainfall for 1 month. If this is not practical, encourage students to obtain this information from a local weather station or news broadcast. Have other students graph and present the data. You may also want to have students compare the climate of the region in which you live with that of the different biomes discussed in this lesson.

MATH and MORE

Have students compare the amount of rainfall in each of the forest biomes described. Then ask:

Which forest biome receives the most rain? (tropical rain forest)

How much more rain does this biome usually receive than the forest biome that receives the least amount of rain? (The coniferous forest receives a minimum of 35 cm per year. The tropical rain forest receives as much as 400 cm per year. The difference is 365 cm.)

Math Skills Worksheet 2
"Subtraction Review"

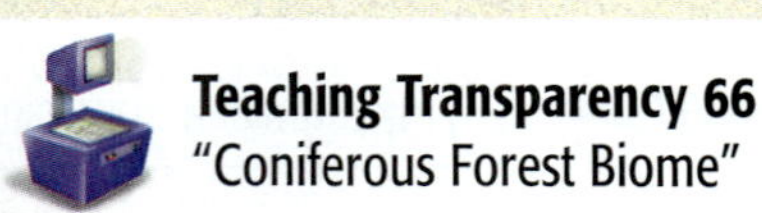

Teaching Transparency 66
"Coniferous Forest Biome"

Precipitación media anual
35-75 cm (14-29.5 pulgadas)

Temperatura promedio
Verano: 14°C (57.2°F)
Invierno: −10°C (14°F)

Figura 3 *Muchos animales del bosque de coníferas sobreviven a los inviernos rigurosos hibernando o migrando a zonas de clima menos frío.*

Bosques de coníferas Los bosques de coníferas no cambian mucho del verano al invierno y se encuentran en zonas con inviernos largos y fríos. Estos bosques están formados principalmente por árboles de hoja perenne, que conservan las hojas durante todo el año. La mayor parte de estos árboles son **coníferas,** es decir, árboles que producen sus semillas en conos. Seguramente has visto un cono de pino porque los pinos son coníferas muy comunes.

La mayoría de las coníferas también pueden identificarse por sus hojas compactas en forma de aguja, cuya cubierta, gruesa y cerosa, evita que se sequen y lesionen durante el invierno.

La **Figura 3** ilustra un bosque de coníferas y algunos de los animales que lo habitan. Observa que no crecen plantas muy grandes debajo de las coníferas, en parte porque llega muy poca luz al suelo.

IS THAT A FACT!

The coniferous forest is also known as a boreal forest or a taiga. More of Earth's land can be categorized under this biome than under any other biome.

Bosque tropical lluvioso El bosque tropical lluvioso posee la mayor **diversidad** biológica de todos los biomas del planeta, pues contiene más especies que los demás biomas. En un área de aproximadamente una cuarta parte del tamaño de una cancha de fútbol puede haber 100 especies de árboles. Aunque algunos animales viven en el suelo, las copas de los árboles, o *bóvedas*, son el sitio preferido para vivir. Si contaras las aves de la bóveda de un bosque tropical, ¡encontrarías hasta 1,400 especies! La **Figura 4** ilustra parte de la diversidad del bioma de bosque tropical lluvioso.

En un bioma de bosque tropical lluvioso, la mayor parte de los nutrientes están en la vegetación. La capa superior del suelo es muy delgada y pobre en nutrientes. Los agricultores que talan el bosque para cultivar sus cosechas deben trasplantarlas a un terreno recién talado después de unos dos años.

Precipitación media anual
Hasta 400 cm (157.5 pulgadas)

Temperatura promedio
De día: 34°C (93°F)
De noche: 20°C (68°F)

Figura 4 Bioma de bosque tropical lluvioso

Multicultural CONNECTION

Scientist and Cornell University professor Eloy Rodriguez studies the chemical properties of plants to discover potentially useful compounds. One aspect of his research is learning about plants used by native cultures of tropical Africa, Asia, and Latin America. Using local knowledge may help Rodriguez find plant compounds with medicinal properties.

Multicultural CONNECTION

Temperate grasslands provide almost ideal growing conditions for grain crops. For this reason, few temperate grasslands remain today. In fact, these areas, such as the American Midwest and Ukraine, are sometimes called the breadbaskets of the world because their temperate grasslands became farmland for grain crops.

CONNECT TO
EARTH SCIENCE

Mountains can influence the climate of surrounding land, resulting in very different ecosystems existing in close proximity. Use the following teaching transparency to illustrate how a mountain can affect climate.

Teaching Transparency 153
"An Example of the Rain Shadow Effect"

LINK TO EARTH SCIENCE

internetconnect

SC*LINKS*
NSTA

TOPIC: Grasslands
GO TO: www.scilinks.org
*sci*LINKS NUMBER: HSTL485

Praderas

Planicies, estepas, sabanas, praderas y pampas son algunos nombres de las regiones donde los pastos son el principal tipo de vegetación. Se encuentran entre los bosques y los desiertos, y existen en todos los continentes. La mayoría son planas o con colinas onduladas.

Praderas templadas La vegetación de las praderas templadas está compuesta principalemente de pastos mezclados con plantas con flores. Hay pocos árboles porque los incendios impiden el crecimiento de las plantas de crecimiento lento. En las praderas templadas del mundo viven mamíferos pequeños que consumen semillas, como los perros de las praderas y los ratones, así como grandes herbívoros, como el bisonte de Norteamérica que aparece en la **Figura 5.**

Precipitación media anual
25-75 cm (10-29.5 pulgadas)

Temperatura promedio
Verano: 30ºC (86ºF)
Invierno: 0ºC (32ºF)

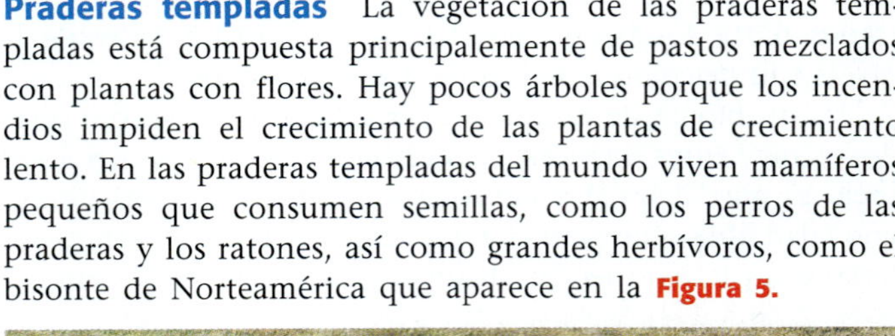

Figura 5 *Los bisontes deambulaban en grandes manadas por las praderas templadas, antes de que los cazadores prácticamente los extinguieran.*

Sabana La **sabana** es una pradera tropical con grupos dispersos de árboles. Durante la temporada de sequía los pastos desaparecen, pero sus profundas raíces sobreviven muchos meses. Durante la temporada de lluvias, la sabana puede recibir hasta 150 cm de lluvia. En las sabanas de África habitan los grupos más abundantes y diversos de herbívoros del mundo, como los que aparecen en la **Figura 6.** Entre ellos están los elefantes, las jirafas, las cebras, las gacelas y los ñúes.

Precipitación media anual
150 cm (59 pulgadas)

Temperatura promedio
Temporada de sequía: 34ºC (93ºF)
Temporada de lluvias: 16ºC (61ºF)

Figura 6 *Los carnívoros, como leones y leopardos, se alimentan de herbívoros, como cebras y ñúes. Por lo general, las hienas y los buitres se comen lo que dejan los carnívoros.*

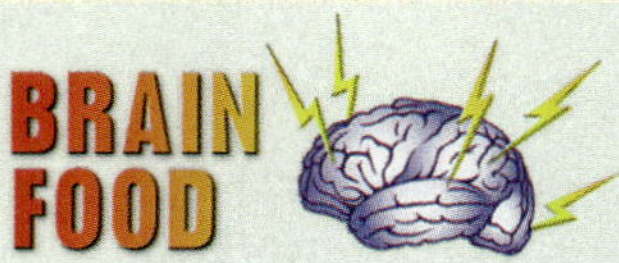
BRAIN FOOD

During the dry season, savanna plants survive as roots. Ask students to discuss how herds of large herbivores might survive when the vegetation dries up.

(Students may suggest that the animals migrate to an area where food and water are more plentiful. Point out that migration is a behavioral adaptation for survival.)

Autoevaluación

Con ayuda del mapa de la Figura 1, compara la ubicación de los bosques de árboles de hoja caduca y la de los bosques de coníferas. Explica las diferencias de ubicación de ambos biomas. *(Consulta la página 636 para comprobar tus respuestas.)*

Desiertos

Los desiertos son regiones calientes y secas donde viven diversas plantas y animales. La mayor parte del agua que cae al suelo se evapora. Los organismos han evolucionado para sobrevivir a temperaturas extremas con muy poca agua. Las plantas crecen muy espaciadas para no competir. Algunas tienen raíces superficiales y extendidas que absorben el agua rápidamente durante una tormenta, mientras que otras tienen raíces muy profundas que llegan al agua del subsuelo.

Los animales también se han adaptado. La mayoría sólo salen de noche, cuando las temperaturas son bajas. Los mosquitos y otros insectos obtienen alimento y agua chupando sangre. Las tortugas se comen las flores u hojas de plantas suculentas y almacenan el agua debajo de sus caparazones durante meses. En la **Figura 7** se observa cómo algunas plantas y animales sobreviven con poca agua en el calor del desierto.

Experimentos

¿Cómo sobreviven los animales en el calor del desierto? ¡Muy bien, gracias! Lee cómo lo hacen en la página 618.

Precipitación media anual
Menos de 25 cm (10 pulgadas)

Temperatura promedio
Verano: 38°C (100°F)
Invierno: 7°C (45°F)

Figura 7 *Hay muchos habitantes del bioma del desierto que están bien adaptados.*

421

GROUP ACTIVITY

Divide the class into groups of four or five students. Then have students play "What Biome Am I?" Each student chooses a land biome. Then students take turns describing their biome to the other members of the group. Members of the group try to guess which biome the student is describing. The successful guesser takes the next turn. Allow students to play the game until everyone has had a chance to describe a biome.

USING THE FIGURE

Have students locate the chaparral biome in **Figure 1,** on page 416. Tell them that the chaparral has a Mediterranean climate. Ask students to hypothesize about what characteristics the chaparral might have. (It is located on the coast and has a mild climate.)

Explain that many chaparral plants are adapted to survive drought and fire. The plants grow back after fires from small bits of surviving tissue.

Reinforcement Worksheet 17
"Know Your Biomes"

SCIENCE HUMOR

Q: Which biome is Elvis's favorite?

A: Grassland

Answer to Self-Check

Deciduous forests tend to exist in mid-latitude, or temperate, regions, while coniferous forests tend to exist in northern areas, closer to the poles.

Answer to MATHBREAK

Biome	Avg. yearly rainfall
Temp. deciduous forest	125 cm
Coniferous forest	75 cm
Tropical rain forest	400 cm
Temperate grassland	75 cm
Savanna	150 cm
Desert	25 cm
Tundra	50 cm

The above rainfall values represent the top end of the average for each biome. Using these values, 375 cm more rain falls on a tropical rain forest than on a desert, and 75 cm more rain falls on a savanna than on a coniferous forest.

Quiz

Ask students whether these statements are true or false. Have students correct any false statements they find.

1. Permafrost thaws only briefly in the summer. (false)
2. Tropical rain forests have more species than any other biome. (true)
3. Grasslands have very poor soil. (false)

ALTERNATIVE ASSESSMENT

Poster Project Have students choose one biome and make a poster to display facts and images of the biome. Posters should provide information about the uniqueness of the biome, indicate where the biome can be found, and compare the biome with other biomes.

Precipitación media anual
30-50 cm (12-20 pulgadas)

Temperatura promedio
Verano: 12°C (53.6°F)
Invierno: −26°C (−14°F)

Tundra

En el extremo norte y en las cimas de las montañas, el clima es tan frío que no crecen árboles. En estos lugares se encuentra el bioma llamado **tundra**.

Tundra ártica La principal característica de la tundra ártica es el permafrost. Durante la corta temporada de crecimiento, sólo la superficie del suelo se descongela. El suelo que está debajo de la superficie, el **permafrost,** permanece congelado todo el tiempo. A pesar de que hay poca precipitación, el suministro de agua no es limitado. Esto se debe a que el permafrost evita que la lluvia se filtre, por lo que el suelo de la superficie se mantiene húmedo. Los lagos y lagunas son, por tanto, comunes.

La capa de suelo descongelado que está encima del permafrost es demasiado superficial para que sobrevivan plantas de raíces profundas. Abundan los pastos, las juncias, los juncos y los pequeños arbustos leñosos. Debajo de estas plantas, en la superficie del suelo, hay una capa de musgos y líquenes. Entre los animales de la tundra **(Figura 8)** hay grandes mamíferos, como caribúes, bueyes almizclados y lobos, así como animales pequeños, como lemmings, musarañas y liebres. Las aves migratorias abundan en verano.

Tundra alpina Por arriba del límite forestal de las montañas muy altas se encuentra otro bioma de tundra. Estas áreas, llamadas tundra alpina, reciben mucha luz solar y precipitación, casi toda en forma de nieve.

Figura 8 *Los caribúes emigran a suelos más ricos durante los largos y fríos inviernos de la tundra.*

¡MATEMÁTICAS!

Precipitación pluvial
¿Dónde llueve más en un año: en un bosque tropical lluvioso o en el desierto? ¿Qué diferencia hay entre las cantidades de precipitación de un bosque de coníferas y una sabana? Para comparar, haz una gráfica de barras de cada bioma con los datos de esta sección.

REPASO

1. ¿En qué se diferencia el clima de las praderas templadas y el de las sabanas?
2. Describe tres adaptaciones de las plantas y animales al clima del desierto.
3. En un bosque tropical lluvioso, ¿dónde se encuentra la mayor parte de los nutrientes?
4. **Aplicar conceptos** ¿Es correcto llamar desierto helado a la tundra ártica? Explica tu respuesta.

422

▼ Answers to Review

1. Savannas are hotter than temperate grasslands and receive twice as much annual rainfall. Also rain falls only during the wet season in the savanna.
2. Answers will vary. Sample response: being active only at night, developing structures that help them get rid of heat, and recycling consumed water.
3. Most of the biome's nutrients are contained in the vegetation.
4. Accept any logical, well-argued response. Sample answer: Yes, it would be accurate because very little rain falls and because liquid water is only available during the short growing season.

Ecosistemas marinos

Cubren casi tres cuartas partes de la superficie de la Tierra y contienen casi el 97 por ciento de las reservas de agua. Los animales más grandes de la Tierra habitan en ellos, junto con miles de millones de criaturas microscópicas, como ves en la **Figura 9.** Sus hábitats van desde las profundidades oscuras, frías y de alta presión a las cálidas playas; desde las aguas heladas de los polos a las costas rocosas. ¿Qué son? Son los océanos y mares de la Tierra. Los ecosistemas marinos se encuentran en cualquier lugar donde haya agua salada. De hecho, la base de los ecosistemas **marinos** es el agua salada. Este factor abiótico tiene una fuerte influencia en los ecosistemas de los océanos y mares.

Regla de los factores abióticos

En la Sección 1 aprendiste que la precipitación pluvial y la temperatura son los principales factores que determinan la clase de bioma terrestre que se encuentra en una región. De la misma manera, los biomas marinos son moldeados por factores abióticos, como la temperatura, la cantidad de luz solar que penetra el agua, la distancia a la tierra y la profundidad del agua. Estos factores sirven para definir ciertas zonas del océano. Como los biomas terrestres, los biomas marinos se encuentran por toda la Tierra y comprenden muchos ecosistemas.

Aguas soleadas El agua absorbe la luz, de modo que la luz solar sólo puede llegar a 200 m de profundidad, incluso en el agua más transparente. Como sabes, la mayoría de los productores elaboran su propio alimento mediante la fotosíntesis y, dado que este proceso requiere luz, casi todos los productores se encuentran únicamente donde llega aquélla.

El conjunto de los productores más abundantes del océano se conoce como **fitoplancton,** compuesto por organismos fotosintéticos microscópicos que flotan cerca de la superficie del agua. Mediante la energía de la luz solar, estos organismos elaboran su propio alimento, tal como lo hacen las plantas terrestres. El **zooplancton** está compuesto por los consumidores que se alimentan del fitoplancton, animales pequeños que, junto con el fitoplancton, constituyen la base de las cadenas alimenticias del océano.

VOCABULARIO

marino
fitoplancton
sargazo
estuario
zooplancton

OBJETIVOS

- Identifica las distintas zonas del océano.
- Explica la importancia del plancton para los ecosistemas marinos.
- Describe los arrecifes de coral y las zonas intermareales.
- Explica qué hace únicos a los biomas marinos polares.

Figura 9 *Los ecosistemas marinos mantienen una gran diversidad de formas de vida. La ballena jorobada es uno de los mamíferos más grandes de la Tierra. El fitoplancton, como el que se observa en la foto pequeña de abajo, está formado por organismos muy pequeños que son la base de las cadenas alimenticias de los océanos.*

423

IS THAT A FACT!

The sea kelp known as *Macrocystis* can grow as much as 60 m in one growing season. This organism can grow lengthwise faster than any other organism known! It grows along the coast of California.

Directed Reading Worksheet 17 Section 2

Focus

Marine Ecosystems

This section introduces different areas of the ocean as ecosystems. Students learn about the importance of abiotic factors in structuring oceanic life zones. The section also focuses on several unique marine ecosystems, including coral reefs and estuaries.

Bellringer

Write the following on the board or an overhead projector, and have students record their answers in their ScienceLog:

How much of the Earth's surface is covered by ocean?

$$\frac{1}{4} \qquad \frac{1}{2} \qquad \frac{3}{4} \qquad \frac{9}{10}$$

(Answer: $\frac{3}{4}$)

What percentage of the Earth's water supply is found in the oceans?

37%   67%   77%   97%

(Answer: 97 percent)

1) Motivate

DISCUSSION

Oceans Students may not realize how connected their lives are to the oceans, especially if they live far from the coast. Ask students:

If all marine life were to die off, how would you be affected? How would others be affected?

Possible answers include: less food available (fish, crab, shrimp, shellfish, etc.), less oxygen in atmosphere (phytoplankton produce one-third to one-half of our oxygen), increased global warming (all that phytoplankton uses up huge amounts of carbon dioxide, the chief greenhouse gas), loss of jobs, and less enjoyment from ocean wildlife.

Multicultural CONNECTION

The Japanese and Koreans use brown seaweed to make *kombu* soup. Red algae is wrapped around rice to make a kind of sushi called *nori* rolls. Seaweed does provide iodine and other minerals, but it is used mostly for the taste and texture it adds to foods.

READING STRATEGY

Prediction Guide Before students read this page, ask them the following questions:

1. How do organisms in the dark depths of the ocean find food?

2. What kinds of organisms live on the ocean floor?

3. What is a thermal vent?

4. Which part of the ocean receives the most sunlight?

Have students evaluate their responses after they read about ocean biomes.

 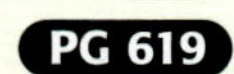 **PG 619**

Discovering Mini-Ecosystems

Biomas de agua maravillosos

En cada parte de los océanos y mares existen biomas únicos y hermosos que son el hogar de muchos organismos excepcionalmente adaptados. Las principales zonas del océano y algunos organismos que habitan en ellas se muestran abajo, en la **Figura 10.**

Figura 10 *La vida en una zona determinada depende de la cantidad de luz que recibe, la distancia a la que está de la tierra y la profundidad a que se encuentra.*

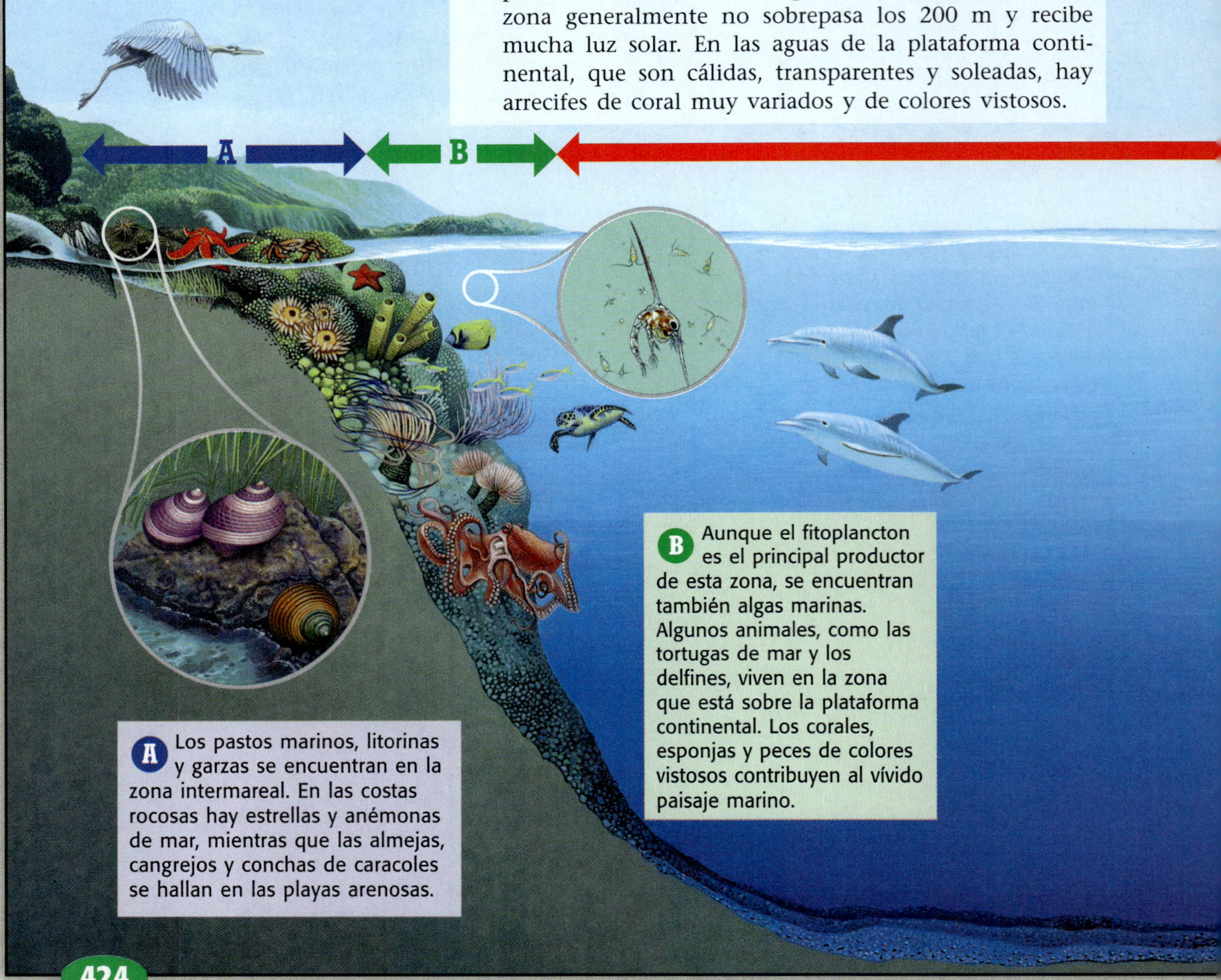

A **Zona intermareal** La zona intermareal es donde el océano se encuentra con la tierra. Esta zona está por encima del agua parte del día, cuando la marea está baja, y con frecuencia recibe el impacto de las olas. Las marismas, costas rocosas y playas arenosas están en esta zona.

B **Zona nerítica** La profundidad del agua aumenta gradualmente a medida que se aproxima el límite de la plataforma continental. La profundidad del agua en esta zona generalmente no sobrepasa los 200 m y recibe mucha luz solar. En las aguas de la plataforma continental, que son cálidas, transparentes y soleadas, hay arrecifes de coral muy variados y de colores vistosos.

B Aunque el fitoplancton es el principal productor de esta zona, se encuentran también algas marinas. Algunos animales, como las tortugas de mar y los delfines, viven en la zona que está sobre la plataforma continental. Los corales, esponjas y peces de colores vistosos contribuyen al vívido paisaje marino.

A Los pastos marinos, litorinas y garzas se encuentran en la zona intermareal. En las costas rocosas hay estrellas y anémonas de mar, mientras que las almejas, cangrejos y conchas de caracoles se hallan en las playas arenosas.

424

WEIRD SCIENCE

Food is scarce in the benthic zone. As a result, organic matter that filters down from the surface is often quickly consumed. Scavengers, including relatives of the freshwater shrimp, will voraciously attack a dead fish that falls to the deep ocean floor, reducing the carcass to bones in just a few hours!

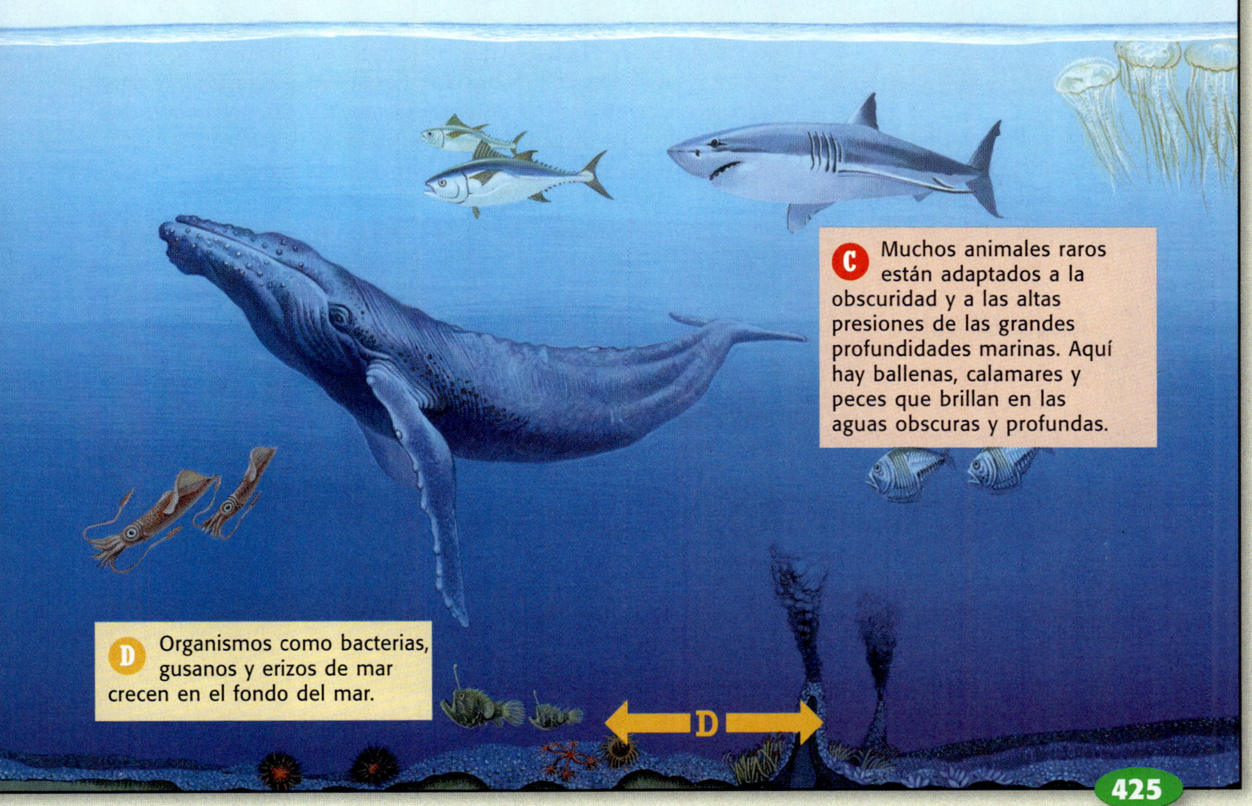

C **Zona oceánica** Más allá de la plataforma continental, el fondo del mar cae abruptamente; son las aguas profundas de alta mar. Hasta una profundidad de 200 m, el fitoplancton es el productor; sin embargo, a mayores profundidades, no llega la luz, por lo que los organismos obtienen su energía consumiendo materia orgánica que cae de la superficie.

D **Zona bentónica** La zona bentónica es el suelo oceánico. Se extiende desde el límite superior de la zona intermareal hasta el fondo de las aguas más profundas del océano. Los organismos que viven en el profundo suelo oceánico se alimentan principalmente de materia que se filtra desde arriba. Algunas bacterias son *quimiosintéticas,* o sea que, utilizan substancias químicas presentes en el agua cercana a las chimeneas hidrotérmicas para elaborar su alimento. Una chimenea hidrotérmica es un lugar del suelo oceánico de donde escapa calor por una grieta de la corteza terrestre.

USING THE FIGURE

Help students understand the information in **Figure 10** by pointing out how it is organized. Make sure that students see the sequence of lettered captions above and below the water and link these letters to the letters indicating parts of the ocean. Also point out how the depth of the ocean changes as you go farther away from the shore. You may also want to help students interpret the magnified views of the ocean floor. Point out that these bubbles represent an up-close view of the area they extend from. Sheltered English

REAL-WORLD CONNECTION

A diving condition called the bends, or decompression sickness, can occur when a person dives very deeply, ascends to the surface too quickly, or stays under the water for a long period of time. The condition occurs when gas forms bubbles in body tissues. Divers suffering from decompression sickness need treatment in a pressure-decompression chamber.

CONNECT TO
PHYSICAL SCIENCE

Red, orange, and yellow wavelengths of light striking the surface of the water are absorbed first. Blue and green wavelengths can penetrate the water more deeply. Thus, producers capable of using blue and green wavelengths of light for photosynthesis can live at greater depths.

IS THAT A FACT!

Deep-sea-vent ecosystems depend on neither sunlight nor photosynthesis for energy.

GROUP ACTIVITY

Have students research and draw a coral reef. Provide reference materials, a large piece of butcher paper, markers, and map pencils. Encourage pairs of students to choose an organism that is part of a coral reef, to research the organism, and to draw it. Then students should place their organism in its habitat on the butcher-paper coral reef. You may want to have one coral reef per class or have all classes contribute to one large coral reef. Sheltered English

GOING FURTHER

Writing

Tell students that Texas has started a "Rigs for Reefs" program in which oil companies are asked to donate oil-drilling equipment that would otherwise be dragged to shore and sold for scrap metal. The old oil rigs and platforms are easily converted to artificial reefs. Have students pretend that they are a member of the board for this or a similar organization. Then have students write a persuasive letter to an oil company explaining why the company should participate in the program.

Science Skills Worksheet 1 "Being Flexible"

Interactive Explorations CD-ROM "Sea Sick"

Un vistazo más de cerca

Los ambientes marinos proveen la mayor parte del agua para la lluvia a través de la evaporación y la precipitación. Las temperaturas de los océanos y las corrientes tienen efectos muy importantes en los climas del mundo y en los patrones de viento. Los seres humanos obtienen enormes cantidades de alimento de los océanos y también vierten enormes cantidades de desechos en ellos. Echemos un vistazo más de cerca a algunos ambientes marinos particulares.

Arrecifes de coral En ciertas aguas tropicales soleadas hay arrecifes de coral en el fondo del mar. Los corales viven en relación cercana con algas unicelulares. Las algas producen nutrientes orgánicos mediante la fotosíntesis y esto proporciona alimento al coral. A su vez, el coral da a las algas un lugar soleado para vivir. La base del arrecife está formada por esqueletos de coral que se han acumulado a lo largo de miles de años. Los arrecifes de coral **(Figura 11)** son el hogar de muchas especies marinas, entre ellas una gran variedad de peces de colores brillantes y organismos como las esponjas y los erizos de mar.

Mar de los Sargazos En medio del océano Atlántico hay un gran ecosistema sin límites terrestres, llamado mar de los Sargazos. El *sargazo* es un tipo de alga que se encuentra pegada a rocas de las costas de Norteamérica, pero que forma enormes balsas flotantes en el mar de los Sargazos. Los animales que están adaptados a este ambiente viven entre las algas y casi todos son del mismo color del sargazo. ¡Algunos hasta se le parecen! ¿A qué crees que se debe esto? ¿Ves algún pez en la **Figura 12**?

Figura 11 *El arrecife de coral es uno de los biomas biológicamente más diversos.*

a través de las ciencias

C O N E X I Ó N

Investiga sobre las chimeneas volcánicas de las profundidades marinas en la página 436.

✓ Autoevaluación

1. Menciona tres factores por los que se caracterizan los biomas marinos.

2. Describe una manera en que los organismos obtienen energía a grandes profundidades en alta mar.

(Consulta la página 636 para comprobar tus respuestas.)

426

Figura 12 *El mar de los Sargazos es un lugar de desove para las anguilas y constituye el hogar de una gran variedad de organismos.*

Answers to Self-Check

1. Accept any three of the following: the amount of sunlight penetrating the water, the distance from land, the depth of the water, the salinity of the water, and the water's temperature.

2. There are several possible answers. Some organisms are adapted for catching prey at great depths, some feed on dead plankton and larger organisms that filter down from above, and some, such as the bacteria around thermal vents, make food from chemicals in the water.

Hielo polar El océano Ártico y las aguas abiertas que rodean a la Antártida constituyen un bioma marino muy poco común. ¿Por qué? ¡Porque tiene hielo!

Las aguas son ricas en nutrientes provenientes de las masas continentales de los alrededores. Estos nutrientes mantienen grandes poblaciones de plancton que, a su vez, mantienen una gran diversidad de peces, aves marinas, pingüinos y mamíferos marinos, como los leones marinos **(Figura 13).**

Estuarios La zona donde el agua dulce de los arroyos y ríos desemboca en el océano se conoce como **estuario.** El agua dulce de los ríos y arroyos se mezcla constantemente con el agua salada del mar. La cantidad de sal de un estuario cambia frecuentemente. Cuando la marea sube, el contenido de sal del agua aumenta y cuando la marea baja, disminuye. El agua dulce que se vierte en el estuario es rica en nutrientes provenientes de la tierra que es arrastrada por el flujo continuo del agua. Dado que los estuarios son tan ricos en nutrientes, mantienen un gran número de organismos. Están repletos de masas de plancton, que suministran alimento a muchos animales más grandes.

Zonas intermareales Los organismos que habitan en las zonas intermareales presentan adaptaciones asombrosas. Las marismas son el hogar de muchos gusanos excavadores y cangrejos, así como de las aves costeras que se alimentan de ellos. Entre los granos de arena de las playas arenosas viven gusanos excavadores, almejas, cangrejos y plancton.

En las costas rocosas, los organismos tienen órganos de sujeción o son capaces de pegarse a las rocas para evitar ser arrastrados por las olas. La **Figura 14** muestra algunos animales capaces de evitar que el agua los arrastre.

Figura 13 *En las costas de la Antártida se reproducen un enorme número de mamíferos y aves.*

Figura 14 *Las anémonas de mar se pegan a las rocas para evitar ser arrastradas mar adentro. Las estrellas de mar pueden arrastrarse bajo las piedras. ¡Los percebes hasta construyen una base de cemento!*

REPASO

1. Explica por qué un arrecife de coral está vivo y muerto al mismo tiempo.

2. ¿Por qué los estuarios tienen vida en abundancia?

3. **Analizar relaciones** Explica cómo la cantidad de luz que recibe una zona determina los tipos de organismos que viven en alta mar.

427

Focus

Freshwater Ecosystems

This section introduces freshwater ecosystems. Students learn about the characteristics of rivers, streams, and ponds. Students learn what happens to a pond as the seasons change. Students also learn how to distinguish between two types of wetlands.

 **Bellringer**

Have students write an answer in their ScienceLog to the following question:

What are four different freshwater ecosystems? (Answers may include the following: stream, river, lake, marsh, pond, swamp, bog, creek.)

1 Motivate

ACTIVITY

Writing Have students write a short description of a personal freshwater experience (something they did at a stream, lake, or wetland). If they have not had such an experience, ask them to use their imagination, or perhaps a field trip can be arranged. Have the students focus on sensations they remember from the experience as well as on what they and others were doing at the time.

VOCABULARIO

afluente   marisma
zona litoral   ciénaga
zona de aguas abiertas
zona de aguas profundas
pantano

OBJETIVOS

- Enumera las características de ríos y arroyos.
- Describe la zona litoral de una laguna.
- Explica qué le sucede a una laguna a medida que cambian las estaciones.
- Distingue entre dos tipos de pantanos.

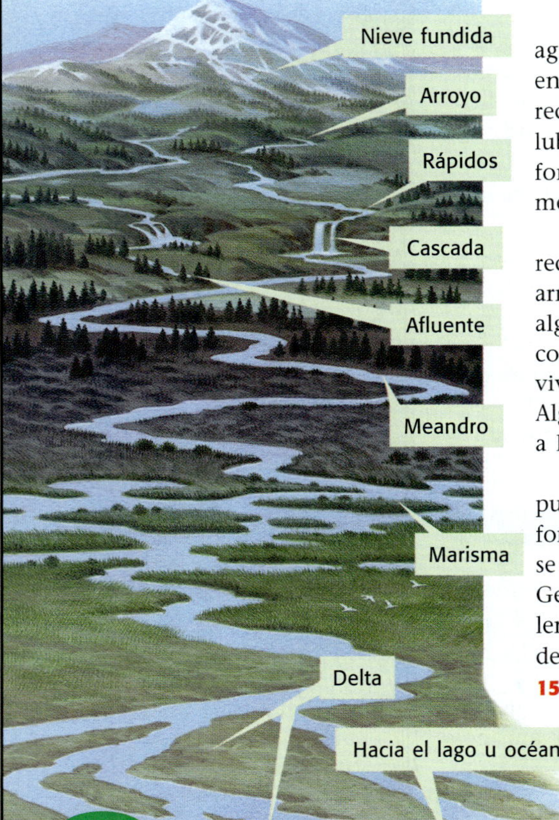

428

Ecosistemas de agua dulce

En su recorrido montaña abajo, se escucha el murmullo de un arroyo. El rumor de un río caudaloso se escucha a su paso por un cañón. Un estanque de jardín está repleto de vida. Un lago casi del tamaño de un mar sacude los barcos durante una fuerte tormenta. En una turbia ciénaga se escucha el eco de ranas y aves.

Estos lugares tienen algo en común: son ecosistemas de agua dulce. Al igual que otros ecosistemas, los de agua dulce están caracterizados por factores abióticos, principalmente la velocidad del agua.

Agua en movimiento

Los arroyos, corrientes de agua y ríos son ecosistemas cuya base es el agua en movimiento. El agua puede comenzar a fluir a partir de hielo o nieve fundidos, o provenir de un manantial, donde el agua fluye hacia la superficie de la Tierra. Un riachuelo o río pequeño que desemboca en uno más grande se llama **afluente.**

A medida que un mayor número de afluentes se agrega a una corriente fluvial, ésta se agranda y ensancha formando un río. Las plantas acuáticas recubren la orilla de los ríos. En las aguas abiertas viven lubinas y percas, entre otros peces, y en el fango del fondo hay animales excavadores, como almejas y mejillones de agua dulce.

Los organismos que viven en aguas en movimiento requieren adaptaciones especiales para evitar ser arrastrados por la corriente. Los productores, como las algas, diatomeas y musgos, se pegan a las rocas; los consumidores, como las formas inmaduras de insectos, viven debajo de las rocas en aguas poco profundas. Algunos consumidores, como los renacuajos, se sostienen a las rocas mediante ventosas.

A medida que un río se hace más ancho y lento, puede formar *meandros* a través del paisaje. Si en el fondo se depositan materiales orgánicos y sedimentos, se forman *deltas*. Las libélulas, los hemípteros, como el *Gerris gibifer*, y otros invertebrados viven en las aguas lentas y sobre ellas. Al final, el agua en movimiento desemboca en un lago u océano. Observa en la **Figura 15** el crecimiento de un río a partir de nieve fundida.

Figura 15 *Observa las características de un río típico. ¿Dónde se mueve rápidamente el agua? ¿Dónde se mueve lentamente?*

🖥 **Teaching Transparency 68**
"River Features"

📄 **Directed Reading Worksheet 17** Section 3

IS THAT A FACT!

The Nile River in Africa is the longest river in the world, at 6,670 km long. The Amazon River in South America comes in a close second, at 6,275 km long.

Aguas estancadas

Los ecosistemas de las lagunas y lagos son diferentes a los de los arroyos y ríos. El lago Superior, que es el más grande del mundo, tiene más cosas en común con una laguna pequeña de castores que con un río. En la **Figura 16** se muestra un corte transversal de un lago típico. Al observar esta ilustración, notarás que el lago ha sido dividido en tres zonas. Más adelante estudiarás estas zonas y los ecosistemas que comprenden.

Donde el agua se encuentra con la tierra Observa de nuevo la Figura 16 y localiza la **zona litoral.** Es la más cercana a la orilla. En ella hay muchos habitantes. Las plantas que crecen en el agua más cercana a la orilla comprenden las aneas y los juncos. Más lejos de la orilla están las plantas flotantes, como los lirios acuáticos, y aún más allá se encuentran las algas que crecen debajo de la superficie del agua.

Las plantas de la zona litoral son el hogar de animales pequeños, como caracoles, pequeños artrópodos y larvas de insectos. Las almejas, gusanos y otros organismos excavan en el lodo. En esta zona también hay salamandras, tortugas de agua, varios tipos de peces y serpientes de agua.

Vida en la parte superior Observa de nuevo la Figura 16. Ubica la **zona de aguas abiertas.** Esta zona se extiende desde la zona litoral hasta la superficie del agua. Sólo abarca hasta donde llega la luz. Este es el hábitat de la lubina, la mojarra de agallas azules, la trucha y otros peces. En la zona de aguas abiertas de un lago, el fitoplacton es el organismo fotosintético más abundante.

La vida en el fondo Ahora observa la Figura 16 y localiza la **zona de aguas profundas.** A esta zona, situada debajo de la zona de aguas abiertas, no llega luz. En ella viven bagres, carpas, gusanos, larvas de insectos, crustáceos, hongos y bacterias. Estos organismos se alimentan de materia orgánica que cae de las capas superiores.

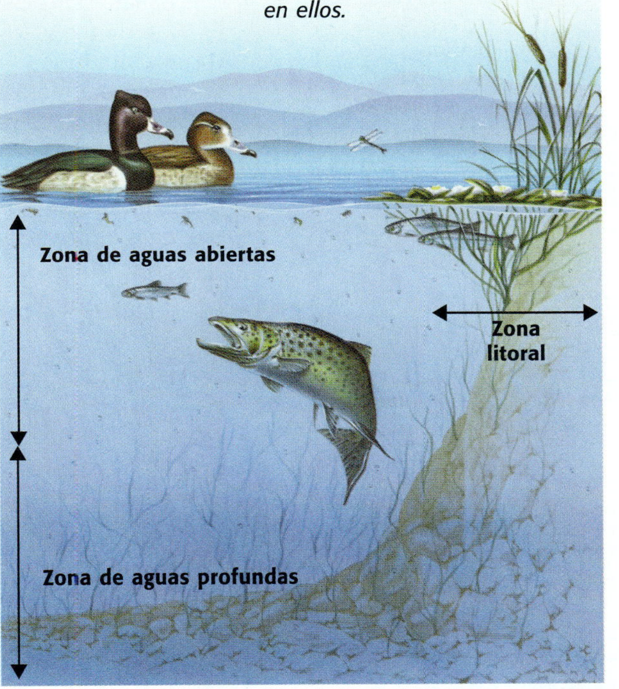

Figura 16 *Al igual que los ecosistemas marinos, los de agua dulce están caracterizados por factores abióticos que determinan qué organismos habitan en ellos.*

Imagínate que eres un biólogo especializado en las plantas que viven en el lago Superior y sus cercanías. Te estás preparando para una expedición de un año a la bahía de Thunder, en la costa canadiense del lago Superior. En lugar de quedarte en un hotel cercano, piensas acampar. Basándote en lo que sabes sobre ecosistemas, responde a las siguientes preguntas: ¿Cómo vivirás en ese lugar? ¿Qué llevarás contigo? ¿A qué problemas te enfrentarás? ¿Cómo los resolverás?

Explora

Mientras explorabas un pantano, descubriste un nuevo organismo. Dibújalo en tu cuaderno de ciencias. Describe su aspecto y cómo está adaptado a su medio. Ahora, intercambia con un compañero o compañera. ¿Te parece posible que un organismo como el suyo exista?

Pantanos

Un **pantano** es un área de terreno donde el nivel del agua está cerca de la superficie del suelo, o por encima de ella, la mayor parte del año. En los pantanos, la vida vegetal y animal es muy variada. Estas áreas juegan un papel importante en el control de las inundaciones. Cuando llueve mucho o se derrite la nieve en la primavera, los pantanos absorben una gran cantidad de agua. El agua de los pantanos también se filtra en el suelo, llenado las reservas de agua subterránea.

Marismas Una **marisma** es un ecosistema de pantano sin árboles, donde crecen aneas y juncos. En la **Figura 17** se ilustra una marisma de agua dulce. Las marismas de agua dulce se encuentran en aguas poco profundas a lo largo de la orilla de lagos, lagunas, ríos y arroyos. Las plantas de las marismas varían de acuerdo con su ubicación y la profundidad del agua. Los pastos, carrizos, aneas y arroz silvestre son plantas comunes de las marismas. En cuanto a los animales, se pueden encontrar ratas almizcleras, tortugas, ranas y tordos charreteros.

Figura 17 *En una marisma, las tortugas encuentran muchos lugares para escapar de los depredadores. Muchas especies cuidan de sus crías en estas áreas protegidas.*

IS THAT A FACT!

Fish living in streams with strong currents face a problem: how do they prevent themselves from being washed downstream? Some have developed suckers that hold them in place. Some stay in relatively calm waters behind rocks, and at least two—a kind of catfish in the Andes and a kind of loach in Borneo—have developed huge lips that they use to clamp on to river debris to hold their position.

Ciénagas Una **ciénaga** es un ecosistema donde crecen árboles y enredaderas. Las ciénagas se encuentran en áreas poco elevadas y junto a ríos lentos. La mayoría de las ciénagas se inundan sólo parte del año, según la precipitación pluvial. Entre los árboles están los sauces, el ciprés de los pantanos, el tupelo acuático *(Nyssa aquatica),* los robles y los olmos. Las enredaderas, como la hiedra venenosa, trepan por los árboles y los musgos cuelgan de las ramas. Los lirios acuáticos y otras plantas de los lagos crecen en las zonas de aguas abiertas. Las ciénagas, como la de la **Figura 18,** son el hogar de diversos peces, serpientes y aves.

Figura 18 *Las bases de los troncos de estos árboles están adaptadas para darles mayor apoyo en el sedimento mojado y blando de esta ciénaga.*

Del lago al bosque

¿Cómo puede desaparecer un lago o laguna, como la de la **Figura 19**? Por lo general, el agua que desemboca en una masa de agua estancada acarrea nutrientes y sedimentos. Estos materiales se van al fondo. Las hojas muertas de los árboles que se inclinan sobre el lago y la vida vegetal y animal en descomposición también se van al fondo. Poco a poco, la laguna o lago se rellena, y las plantas crecen en las zonas recién rellenadas, cada vez más cerca del centro. Con el tiempo, el agua estancada se convierte en marisma y más adelante la marisma puede convertise en un bosque.

Figura 19 *Con el tiempo, la materia orgánica en descomposición y los sedimentos de la tierra rellenarán esta laguna.*

REPASO

1. Describe algunas adaptaciones de los organismos que viven en el agua en movimiento.

2. Compara la zona litoral con la zona de aguas abiertas de una laguna.

3. Describe las diferencias entre una ciénaga y una marisma.

4. **Interpretar ilustraciones** Observa el diagrama de la derecha. Describe los tipos de organismos que pueden vivir en cada zona.

431

4) Close

Quiz

Ask students whether these statements are true or false.

1. Freshwater ecosystems can be grouped as those that are still, those that are flowing, and those that contain salt. (false)

2. Catfish, carp, and other scavengers are likely to be found in the deep-water zone of a lake. (true)

3. Organisms living in fast-moving water do not usually exhibit any special adaptations to their environment. (false)

ALTERNATIVE ASSESSMENT

Writing Have students write a travel brochure for one of the ecosystems in this chapter. Brochures should provide information about climate, recreation, wildlife, and conservation efforts for the area. Encourage students to be informative as well as creative. Sheltered English

LabBook PG 620
Too Much of a Good Thing?

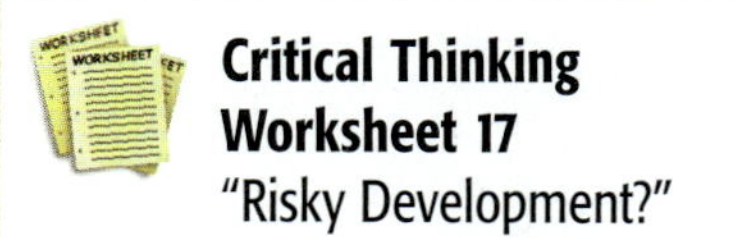

Critical Thinking Worksheet 17 "Risky Development?"

▼ Answers to Review

1. Organisms that live in moving waters tend to use fixed objects, such as rocks, as habitats and/or stabilizers.

2. The littoral zone of a pond is shallower than the open water and contains more nutrients and vegetation. There is more cover and protection from predators for juveniles in the littoral zone.

3. A marsh is a wetland without trees, and a swamp is a wetland with trees.

4. Student answers should reflect the information found on pages 429–431.

Chapter Highlights

Vocabulary Definitions

Section 1

abiotic describes nonliving factors in the environment

biome a large region characterized by a specific type of climate and certain types of plant and animal communities

deciduous describes trees that have leaves that change color in autumn and fall off in winter

conifer a tree that produces seeds in cones

diversity a measure of the number of species an area contains

savanna a tropical grassland biome with scattered clumps of trees

desert a hot, dry biome inhabited by organisms adapted to survive high daytime temperatures and long periods without rain

tundra a far-northern biome characterized by long, cold winters, permafrost, and few trees

permafrost the permanently frozen ground just below the surface of the soil in the arctic tundra

Section 2

marine an ecosystem based on salty water

phytoplankton microscopic photosynthetic organisms that float near the surface of the ocean

zooplankton very small animals that, along with the phytoplankton they consume, form the base of the oceans' food web

Sargassum an alga that forms huge, floating rafts; basis for the name of the Sargasso Sea, an ecosystem of algae in the middle of the Atlantic Ocean

estuary an area where fresh water from streams and rivers spills into the ocean

Resumen del capítulo

SECCIÓN 1

Vocabulario

abiótico (*pág. 416*)
bioma (*pág. 416*)
árboles de hoja caduca (*pág. 417*)
conífera (*pág. 418*)
diversidad (*pág. 419*)
sabana (*pág. 420*)
desierto (*pág. 421*)
tundra (*pág. 422*)
permafrost (*pág. 422*)

Notas de la sección

- La precipitación pluvial y la temperatura son los principales factores que determinan el tipo de bioma.

- Los tres principales biomas del bosque son: bosque templado de árboles de hoja caduca, bosque de coníferas (con veranos cálidos e inviernos fríos) y bosque tropical lluvioso (temperaturas siempre cálidas).

- Las praderas reciben más lluvia que los desiertos y menos que los bosques. Los veranos de las praderas templadas son calurosos, y sus inviernos fríos. En las sabanas hay temporadas de lluvia y de sequía.

- Los desiertos reciben menos de 25 cm de lluvia al año. Las plantas y animales que compiten por las reservas limitadas de agua han desarrollado adaptaciones especiales para sobrevivir.

- El bioma de la tundra se encuentra principalmente en la región ártica. La tundra ártica se caracteriza por el permafrost.

Experimentos

Vida en el desierto (*pág. 618*)

SECCIÓN 2

Vocabulario

marino (*pág. 423*)
fitoplancton (*pág. 423*)
zooplancton (*pág. 423*)
sargazo (*pág. 426*)
estuario (*pág. 427*)

Notas de la sección

- Los tipos de organismos marinos varían según la profundidad del agua, la temperatura, la cantidad de luz y la distancia de la costa.

- En la zona intermareal, el mar se encuentra con la tierra.

- En el fondo del océano hay muchos biomas, tan diversos como los arrecifes de coral y las chimeneas hidrotérmicas.

- El océano abierto comprende biomas únicos, entre ellos, el mar de los Sargazos y los océanos de aguas frías que rodean los polos.

✓ Comprobar destrezas

Conceptos de matemáticas

Precipitación pluvial Con una varilla métrica mide 400 cm del piso del salón de clase. Esta es la precipitación que un bosque tropical lluvioso recibe en un año. Luego, mide 25 cm. Esta es la precipitación que un desierto recibe en un año. Compara ambas cantidades y exprésalo como un cociente.

$$\frac{25}{400} = \frac{1}{16}$$

En un año, un desierto recibe 1/16 de la precipitación de un bosque tropical lluvioso.

Comprensión visual

Bosque tropical lluvioso Observa la Figura 4 de la página 419. Hay tres capas de bosque tropical lluvioso: la superior, la intermedia y la inferior. La superior es la bóveda, donde viven la mayoría de las especies y donde hay más luz solar. La intermedia está debajo de la bóveda y encima del suelo. La inferior es obscura en casi todas las partes del bosque. La mayor parte de las plantas del bosque tropical lluvioso crecen muy alto porque compiten por la luz de la bóveda. El crecimiento de las plantas en la capa inferior no es muy denso por la falta de luz.

432

Lab and Activity Highlights

Life in the Desert PG 618

Discovering Mini-Ecosystems PG 619

Too Much of a Good Thing? PG 620

Datasheets for LabBook
(blackline masters for these labs)

SECCIÓN 2

- Un estuario es una región donde el agua de los ríos desemboca en el océano y en donde el agua dulce y la salada se mezclan debido al movimiento de las mareas.

Experimentos

¡Descubre los miniecosistemas! *(pág. 619)*

SECCIÓN 3

Vocabulario

afluente *(pág. 428)*

zona litoral *(pág. 429)*

zona de aguas abiertas *(pág. 429)*

zona de aguas profundas *(pág. 429)*

pantano *(pág. 430)*

marisma *(pág. 430)*

ciénaga *(pág. 431)*

Notas de la sección

- Los ecosistemas de agua dulce se clasifican de acuerdo con el movimiento del agua. Los arroyos, ríos y riachuelos tienen agua corriente y los lagos y lagunas, agua estancada.

- A medida que los afluentes se agregan a un río, desde la fuente hasta el océano, su volumen y contenido en nutrientes aumentan y su velocidad disminuye.

- La rapidez de la corriente de agua determina los tipos de organismos que hay en un arroyo o río.

- La zona litoral de un lago está habitada por plantas flotantes que constituyen el hogar de una rica diversidad de vida animal.

- Los pantanos comprenden las marismas, donde no hay árboles, y las ciénagas, donde crecen árboles y enredaderas.

Experimentos

¿Demasiada comida? *(pág. 620)*

SECTION 3

tributary a small stream or river that flows into a larger one

littoral zone the zone of a lake or pond closest to the edge of the land

open-water zone the zone of a lake or pond that extends from the littoral zone across the top of the water and that is only as deep as light can reach through the water

deep-water zone the zone of a lake or pond below the open-water zone, where no light reaches

wetland an area of land where the water level is near or above the surface of the ground for most of the year

marsh a treeless wetland ecosystem where plants such as cattails and rushes grow

swamp a wetland ecosystem where trees and vines grow

Vocabulary Review Worksheet 17

Blackline masters of these Chapter Highlights can be found in the **Study Guide.**

 internet

 **VISITA:** go.hrw.com

Visita el sitio web de HRW para encontrar una serie de herramientas de aprendizaje relacionadas con este capítulo. Sólo tienes que escribir la palabra clave:

PALABRA CLAVE: HSTECO

 SCiLINKS. NSTA

VISITA: www.scilinks.org

Visita el sitio web de la **Asociación Nacional de Maestros de Ciencias** *(National Science Teachers Association)* para encontrar recursos de Internet relacionados con este capítulo. Sólo escribe el **ENLACE DE CIENCIAS** para obtener más información sobre el tema:

TEMA: Bosques	ENLACE: HSTL480
TEMA: Pastizales	ENLACE: HSTL485
TEMA: Ecosistemas marinos	ENLACE: HSTL490
TEMA: Ecosistemas de agua dulce	ENLACE: HSTL495

433

Lab and Activity Highlights

LabBank

 **EcoLabs & Field Activities,** Biome Adventure Travel, EcoLab 7

Long-Term Projects & Research Ideas, Project 20

Interactive Explorations CD-ROM

CD 2, Exploration 2, "Sea Sick"

Chapter Review
Answers

USING VOCABULARY

1. continental shelf
2. Zooplankton
3. marsh
4. Deciduous trees
5. tundra
6. biome

UNDERSTANDING CONCEPTS

Multiple Choice

7. a
8. c
9. b
10. d
11. d

Short Answer

12. Tributaries flow into it, increasing the volume of water. The salt and nutrient content of the water increases, as does the cloudiness. The speed of the water decreases.
13. Answers may vary; possible answer: being active only at night, and storing water in body tissue
14. No, wetlands are not *necessarily* always wet. However, wetlands have high water tables most of the year, so wetland soils are wet much of the time.
15. The salt content in the water of an estuary changes constantly as the tides rise and fall, moving water in and out of the estuary. The most important factor in the salinity of an estuary, however, is freshwater runoff from the land. After heavy rainfall, the salinity of the water in an estuary close to land will be very low because the salts will be diluted with additional fresh water.

Repaso del capítulo

UTILIZAR EL VOCABULARIO

Escoge el término correcto para completar las siguientes oraciones:

1. En el borde de ___?___ empieza la alta mar. (*la plataforma continental* o *el mar de los Sargazos*)

2. El ___?___ está formado por diminutos consumidores que viven en el agua. (*fitoplancton* o *zooplancton*)

3. Una ___?___ es un pantano sin árboles. (*ciénaga* o *marisma*)

4. ___?___ pierden sus hojas para conservar el agua. (*Los árboles de hoja caduca* o *Las coníferas*)

5. La principal característica del bioma de ___?___ es el permafrost. (*desierto* o *tundra*)

6. Cada tipo principal de comunidad de plantas y las comunidades de animales asociadas con ella conforman un ___?___. (*estuario* o *bioma*)

COMPRENDER CONCEPTOS

Opción múltiple

7. Los organismos más abundantes del océano son
   a. el plancton.
   b. el sargazo.
   c. los animales de los corales.
   d. los mamíferos marinos.

8. Los ecosistemas marinos de los polos son poco comunes porque
   a. los animales pasan tiempo dentro y fuera del agua.
   b. el plancton es poco común.
   c. contienen hielo.
   d. el contenido de sal en el agua es muy alto.

9. El principal factor que determina los tipos de organismos que viven en un arroyo o río es
   a. la temperatura del agua.
   b. la velocidad de la corriente.
   c. la profundidad del agua.
   d. el ancho del arroyo o río.

10. Los ecosistemas marinos
    a. contienen los animales más grandes del mundo.
    b. existen en todas las zonas oceánicas.
    c. comprenden ambientes donde los organismos sobreviven sin luz.
    d. Todas las anteriores

11. Los dos principales factores que determinan qué tipo de bioma se encuentra en una región son
    a. la cantidad de precipitación pluvial y la temperatura.
    b. la profundidad del agua y la distancia a la que se encuentra de la tierra.
    c. la acción de las olas y el contenido de sal del agua.
    d. Todas las anteriores

Respuesta breve

12. Describe de qué manera una corriente cambia a medida que se acerca al océano.

13. Describe dos adaptaciones de animales al ambiente del desierto.

14. ¿Siempre son húmedos los pantanos? Explica por qué.

15. Explica por qué el contenido de sal en un estuario cambia constantemente.

Organizar conceptos

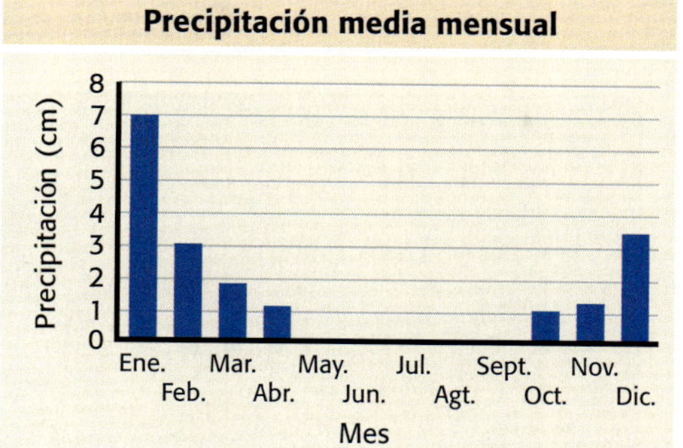

16. Usa los siguientes términos para crear un mapa de ideas: bosque tropical lluvioso, plantas de raíces profundas, arrecife de coral, bóveda, biomas, permafrost, desierto, plataforma continental, tundra, ecosistemas.

RAZONAMIENTO CRÍTICO Y RESOLUCIÓN DE PROBLEMAS

Escribe una o dos oraciones para responder a las siguientes preguntas:

17. Al excavar una región ahora cubierta por pastizales, los paleontólogos descubrieron los restos fósiles de peces y mariscos antiguos. ¿A qué conclusión podrían llegar?

18. Con el fin de construir un nuevo centro comercial, secan un pantano. Posteriormente, las inundaciones se vuelven un problema en esta zona. ¿Cómo puede explicarse esto?

19. Explica por qué la mayoría de las plantas con flores del desierto florecen, producen semillas y mueren al cabo de pocas semanas, mientras que algunas plantas tropicales con flores florecen por mucho más tiempo.

LAS MATEMÁTICAS EN LAS CIENCIAS

20. ¿Cuál es la diferencia media de precipitación pluvial entre un bosque templado de árboles de hoja caduca y un bosque de coníferas?

21. Un área del bosque tropical brasileño recibió 347 cm de lluvia en un año. Con las siguientes fórmulas, calcula la magnitud de la lluvia en pulgadas.

0.394 (pulgadas que hay en un centímetro)

$\times$ 347 cm

_____?_____ pulgadas

INTERPRETAR GRÁFICAS

Las siguientes gráficas representan las temperaturas mensuales y la precipitación pluvial anual en cierta región.

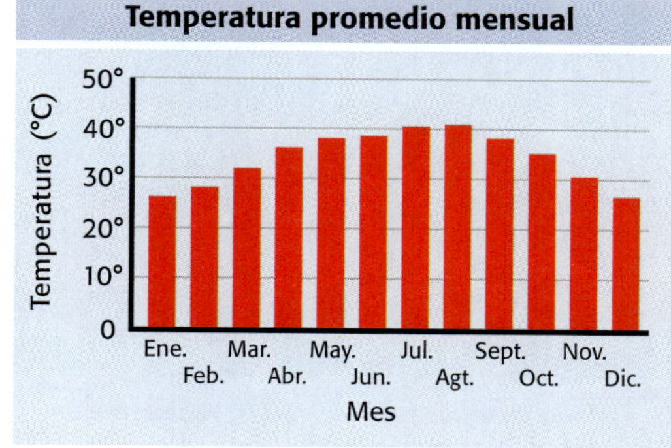

Precipitación media mensual

Temperatura promedio mensual

22. ¿Qué clase de bioma es más probable encontrar en la región representada por estas gráficas?

23. ¿Esperarías encontrar coníferas en ella? Explica tu respuesta.

AHORA, ¿qué piensas?

Revisa tus respuestas a las preguntas de la página 415 que escribiste en el cuaderno de ciencias. ¿Han cambiado tus respuestas? Si es necesario, corrige tus respuestas basándote en lo que has aprendido en este capítulo.

Concept Mapping

16. An answer to this exercise can be found at the end of this book.

CRITICAL THINKING AND PROBLEM SOLVING

17. At one time, this region was covered by water.

18. The wetland had soaked up large amounts of water. Because the water can no longer drain into the wetland, it now runs off the land, causing the flooding.

19. Rainfall is sparse in the desert. When rain does fall, flowering plants of the desert must bloom and bear seeds quickly, while there is still water in the ground.

MATH IN SCIENCE

20. about 45 cm
21. about 137 in.

INTERPRETING GRAPHICS

22. desert
23. No; The graph represents a typical hot, dry desert. Conifers usually grow where the summers are warm and the winters are long, cold, and snowy. Bristlecones and junipers can be found in some mountain deserts.

Concept Mapping Transparency 17

Blackline masters of this Chapter Review can be found in the **Study Guide**.

NOW What Do You Think?

1. There are many differences between a rain forest and a desert, such as water availability and average temperature.

2. Water in a lake is the endpoint of a river. The river may begin as snowmelt or may come from a spring, and it may grow from more and more tributaries joining it as it flows downstream to form a lake.

3. The open ocean contains few species, but some of them are quite large, such as the marine mammals. A swamp has many more species per unit area than does the open ocean.

Background

The first black smoker was found off the Galápagos Islands in 1977, and since then at least four have been found on the bottom of the Pacific Ocean floor 322 km (200 mi) off the coast of Canada. The heat generated by one of them was so intense that the surface of the ocean water was nearly boiling when the chimney was lifted out.

It seems almost impossible that creatures on our planet could survive under such conditions, but one of the species of tubeworms can withstand temperatures up to 80°C (176°F)!

Other recently discovered creatures on the bottom of the ocean thrive on frozen chunks of methane in the same way that these creatures thrive on sulfur. These discoveries of chemosynthetic life have been called a "biological revolution" by one scientist and have sparked debate about the possibilities for life on other planets.

CIENCIAS BIOLÓGICAS • QUÍMICA

Chimeneas oceánicas

▲ *"Son gusanos muy delgados, realmente peludos y aplanados", afirma el científico Bob Feldman acerca de los gusanos tubícolas.*

Imagínate las más remotas profundidades del mar. No hay nada de luz y hace mucho frío. Un hilo de agua de mar penetra hacia las profundidades de la Tierra por las grietas que hay entre las placas del fondo del mar. A su regreso, el agua caliente recoge metales, gases sulfúricos y suficiente calor para aumentar la temperatura del gélido océano a 360°C. ¡A esta temperatura se funde el plomo! El agua caliente de mar es expulsada al océano a través de las chimeneas volcánicas y, cuando este chorro caliente y tóxico entra en contacto con el agua helada, los metales y gases sulfúricos *se precipitan*, o sea, se vuelven sólidos. Estos sólidos forman unos tubos llamados exhalitas, que se prolongan hacia arriba a través del fondo marino. Este ambiente obscuro, frío y tóxico sería mortal para los seres humanos, y sin embargo es el hogar de una comunidad de unas 300 especies, como ciertas bacterias, almejas, mejillones y gusanos tubícolas. Gracias a las exhalitas, estas especies sobreviven.

La vida sin fotosíntesis

Por mucho tiempo, los científicos pensaron que la energía de la luz solar era la base de las cadenas alimenticias de la Tierra y de la vida misma. Sin embargo, en los últimos 15 años los investigadores han descubierto ecosistemas que desafían esta creencia. Ahora sabemos que hay formas de vida alrededor de las exhalitas que pueden sobrevivir sin luz solar. Un tipo de bacteria utiliza los gases tóxicos de las exhalitas de la misma manera que las plantas utilizan la luz solar. En un proceso llamado *quimiosíntesis*, estas bacterias transforman el azufre en energía. Estas bacterias son productores primarios; los mejillones y almejas son consumidores en esta red alimenticia de las profundidades del mar. De hecho, para las bacterias, los mejillones y almejas son un buen lugar para vivir. A su vez, los mejillones y almejas se alimentan de las bacterias. Este tipo de relación entre organismos se llama *simbiosis*. Cuanto más cerca estén los mejillones y almejas de la chimenea, mayor probabilidad tendrán las bacterias de crecer. Por esto, los mejillones y almejas se trasladan frecuentemente para encontrar lugares estratégicos cerca de las exhalitas.

¿Tú qué crees?

▶ Las condiciones alrededor de las exhalitas son similares a las de otros planetas. Investiga sobre estos ambientes extremos, tanto en la Tierra como en otras partes. Comenta con tus compañeros y compañeras dónde y cómo crees que se originó la vida en la Tierra.

436

Answer to What Do You Think?

Answers will vary. Accept any response that is well researched, thoughtful, and well articulated.

PROFESIONES

Casi todos los inviernos, **Alfonso Alonso Mejía** sube a los pocos lugares recónditos del centro de México donde alrededor de 150 millones de mariposas monarca pasan el invierno. Él investiga a las monarca porque quiere contribuir a conservar su hábitat.

Las mariposas monarca son famosas por migrar largas distancias. Las que llegan a México vienen desde el noroeste de los Estados Unidos y el sur de Canadá. Algunas viajan 3,200 km antes de llegar al centro de México.

La amenaza humana para los hábitats

Desafortunadamente, el hábitat de las monarcas está cada vez más amenazado por la tala y otras actividades humanas. Sólo quedan nueve lugares donde puedan pasar el invierno; cinco de ellos son santuarios, pero incluso éstos corren peligro debido a que la gente tala los abetos para leña o con propósitos comerciales.

Investigación al rescate

El trabajo de Alfonso está ayudando a los conservacionistas mexicanos a entender y proteger mejor a las mariposas monarca. Es muy importante su descubrimiento de que las monarca dependen de los arbustos que crecen bajo los abetos, que se conocen como vegetación intermedia y se encuentran entre el suelo y la bóveda del bosque.

La investigación de Alfonso demostró que cuando la temperatura desciende bajo el punto de congelación, como sucede con frecuencia en los lugares altos donde las monarca pasan el invierno, algunas de ellas dependen de la vegetación intermedia para sobrevivir. Esto se debe a que las temperaturas bajas (−1°C a 4°C) limitan su movimiento, de modo que ni siquiera pueden arrastrarse. A temperaturas extremadamente frías (−7°C a −1°C), las monarca que descansan en el suelo del bosque corren peligro de morir congeladas. Pero donde hay vegetación intermedia, pueden trepar lentamente hasta quedar a unos 10 cm por encima del suelo. Esta diferencia de altura proporciona un microclima suficientemente cálido para garantizar su supervivencia.

Antes de los trabajos de Alfonso, no se conocía la importancia de la vegetación intermedia. Ahora, gracias a su trabajo, los conservacionistas mexicanos la protegerán mejor.

¡Participa!

▶ Si te interesa el programa nacional de marcaje para ayudar a los científicos a conocer más acerca de la ruta de migración de las monarca, escribe a Monarch Watch, Department of Entomology, 7005 Howorth Hall, University of Kansas, Lawrence, Kansas 66045.

437

Chapter Organizer

CHAPTER ORGANIZATION	TIME MINUTES	OBJECTIVES	LABS, INVESTIGATIONS, AND DEMONSTRATIONS
Chapter Opener pp. 438–439	45		**Investigate!** Recycling Paper, p. 439
Section 1 First the Bad News	90	▶ Describe the major types of pollution. ▶ Distinguish between renewable and nonrenewable resources. ▶ Explain how habitat destruction affects organisms. ▶ Explain why human population growth is a problem.	
Section 2 The Good News: Solutions	90	▶ Explain the importance of conservation. ▶ Describe the three R's and their importance. ▶ Explain how habitats can be protected. ▶ List ways you can help protect the Earth.	**Demonstration,** Friendly Cleaner, p. 448 in ATE **QuickLab,** Trash Check, p. 452 **Interactive Explorations CD-ROM,** Moose Malady *A **Worksheet** is also available in the **Interactive Explorations Teacher's Edition.*** **Discovery Lab,** Biodiversity–What a Disturbing Thought! p. 622 **Datasheets for LabBook,** Biodiversity–What a Disturbing Thought! Datasheet 38 **Skill Builder,** Deciding About Environmental Issues, p. 624 **Datasheets for LabBook,** Deciding About Environmental Issues, Datasheet 39 **EcoLabs & Field Activities,** A Filter with Culture, EcoLab 8 **Long-Term Projects & Research Ideas,** Project 21

TECHNOLOGY RESOURCES

 **Guided Reading Audio CD**
English or Spanish, Chapter 18

 **One-Stop Planner CD-ROM with Test Generator**

 **Interactive Explorations CD-ROM**
CD 2, Exploration 3, Moose Malady

 **Science Discovery Videodiscs**
Science Sleuths: Dead Fish on Union Lake

  **Multicultural Connections,** Thailand Tire Furniture, Segment 12

Eye on the Environment, Smog Problems in Mexico, Segment 11

Scientists in Action, Forming the Future of Energy Efficiency, Segment 7
Tracking Mercury in the Everglades, Segment 15

CLASSROOM WORKSHEETS, TRANSPARENCIES, AND RESOURCES	SCIENCE INTEGRATION AND CONNECTIONS	REVIEW AND ASSESSMENT
Directed Reading Worksheet 18 **Science Puzzlers, Twisters & Teasers,** Worksheet 18	**Multicultural Connection,** p. 439 in ATE **Careers:** Biologist–Dagmar Werner, p. 458	
Directed Reading Worksheet 18, Section 1 **Transparency 144,** The Greenhouse Effect **Math Skills for Science Worksheet 28,** A Formula for SI Catch-up **Math Skills for Science Worksheet 37,** Rain-Forest Math	**Cross-Disciplinary Focus,** p. 441 in ATE **Earth Science Connection,** p. 442 **Math and More,** p. 442 in ATE **Connect to Earth Science,** p. 442 in ATE **MathBreak,** Water Depletion, p. 443 **Math and More,** p. 444 in ATE **Apply,** p. 446	**Review,** p. 442 **Self-Check,** p. 443 **Review,** p. 446 **Quiz,** p. 446 in ATE **Alternative Assessment,** p. 446 in ATE
Transparency 70, Practicing Conservation **Directed Reading Worksheet 18,** Section 2 **Reinforcement Worksheet 18,** It's "R" Planet! **Critical Thinking Worksheet 18,** Bud Kindfellow Has a Plan	**Math and More,** p. 448 in ATE **Connect to Chemistry,** p. 448 in ATE **Multicultural Connection,** p. 451 in ATE **Cross-Disciplinary Focus,** p. 452 in ATE **Scientific Debate:** Where Should the Wolves Roam? p. 459	**Self-Check,** p. 449 **Review,** p. 450 **Review,** p. 453 **Quiz,** p. 453 in ATE **Alternative Assessment,** p. 453 in ATE

END-OF-CHAPTER REVIEW AND ASSESSMENT

Chapter Review in Study Guide
Vocabulary and Notes in Study Guide
Chapter Tests with Performance-Based Assessment, Chapter 18 Test
Chapter Tests with Performance-Based Assessment, Performance-Based Assessment 18
Concept Mapping Transparency 18

internet connect

Holt, Rinehart and Winston On-line Resources

go.hrw.com

For worksheets and other teaching aids related to this chapter, visit the HRW Web site and type in the keyword: **HSTENV**

National Science Teachers Association

www.scilinks.org

Encourage students to use the *sci*LINKS numbers listed with the Chapter Highlights to access information and resources on the **NSTA** Web site.

Chapter Resources & Worksheets

Visual Resources

TEACHING TRANSPARENCIES

#70

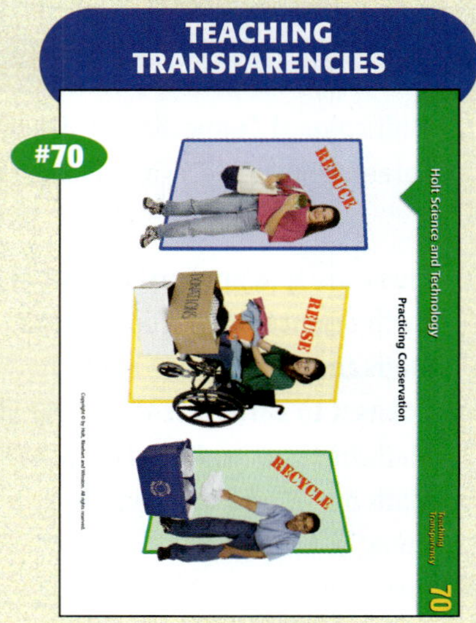

TEACHING TRANSPARENCIES

#144

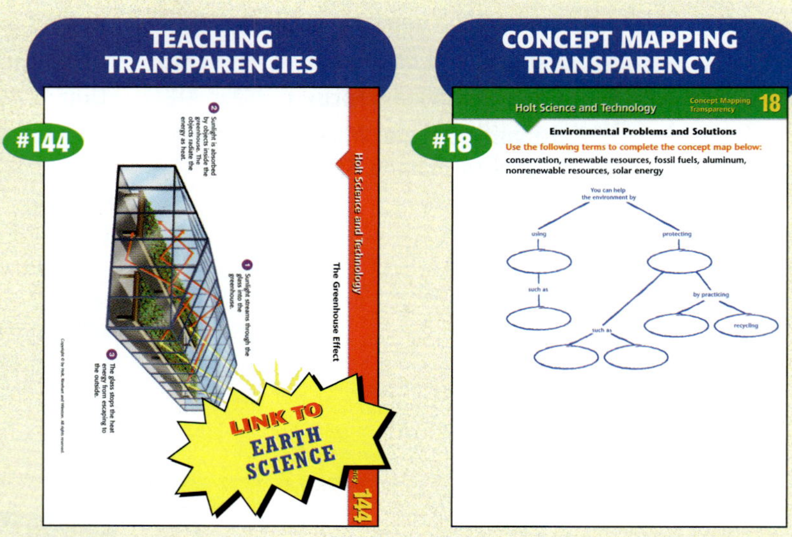

CONCEPT MAPPING TRANSPARENCY

#18

Holt Science and Technology — Concept Mapping Transparency **18**

Environmental Problems and Solutions

Use the following terms to complete the concept map below:
conservation, renewable resources, fossil fuels, aluminum, nonrenewable resources, solar energy

---

Meeting Individual Needs

DIRECTED READING

#18

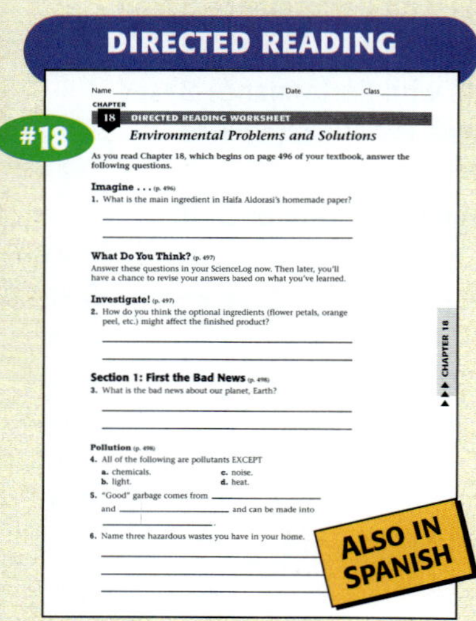

CHAPTER 18 — DIRECTED READING WORKSHEET

Environmental Problems and Solutions

As you read Chapter 18, which begins on page 496 of your textbook, answer the following questions.

Imagine . . . (p. 496)
1. What is the main ingredient in Haifa Aldoransi's homemade paper?

What Do You Think? (p. 497)
Answer these questions in your ScienceLog now. Then later, you'll have a chance to revise your answers based on what you've learned.

Investigate! (p. 497)
2. How do you think the optional ingredients (flower petals, orange peel, etc.) might affect the finished product?

Section 1: First the Bad News (p. 498)
3. What is the bad news about our planet, Earth?

Pollution (p. 498)
4. All of the following are pollutants EXCEPT
   a. chemicals.   c. noise.
   b. light.   d. heat.

5. "Good" garbage comes from _____ and _____ and can be made into _____.

6. Name three hazardous wastes you have in your home.

ALSO IN SPANISH

REINFORCEMENT & VOCABULARY REVIEW

#18

CHAPTER 18 — REINFORCEMENT WORKSHEET

It's "R" Planet!

Complete this worksheet after you finish reading Chapter 18, Section 2.

Use terms and definitions from the following list to design a flier that will encourage students in your school to participate in conservation. Get students' attention by making your flyer colorful, using a catchy phrase, or using any other method you can think of. Your job is to make other students aware of the role they can play in protecting our environment.

- the three Rs: reduce, reuse, recycle
- conservation: preserving resources
- recycling: breaking trash down in order to use it again
- resource recovery: turning garbage into electricity
- plastics
- paper products
- waste wood
- glass
- cardboard
- cans
- to reduce means to use less
- to reuse means to use it again
- to recycle is a type of reuse
- cloth napkins
- paper napkins

VOCABULARY REVIEW WORKSHEET

#18

CHAPTER 18 — VOCABULARY REVIEW WORKSHEET

Solve the Environmental Puzzle

Give this puzzle a try after you finish Chapter 18.

Using each of the clues below, fill in the letters of the term described in the blanks provided on the next page.

1. can be broken down by the environment
2. process of transforming garbage into electricity
3. type of hazardous wastes that take hundreds or thousands of years to become harmless
4. when the number of individuals becomes so large that they can't get all the resources they need
5. the process of making new products from reprocessed used products
6. the clearing of forest lands
7. the world around us
8. an organism that makes a home for itself in a new place
9. harmful substances in the environment
10. a girl who developed a way to make paper without cutting down a tree
11. describes a natural resource that can be used and replaced over a relatively short time
12. describes a natural resource that cannot be replaced or can be replaced only after thousands or millions of years
13. substances used to kill crop-destroying insects
14. the preservation of resources
15. the number and variety of living things
16. poisonous
17. the presence of harmful substances in the environment
18. protective layer of the atmosphere destroyed by CFCs

SCIENCE PUZZLERS, TWISTERS & TEASERS

#18

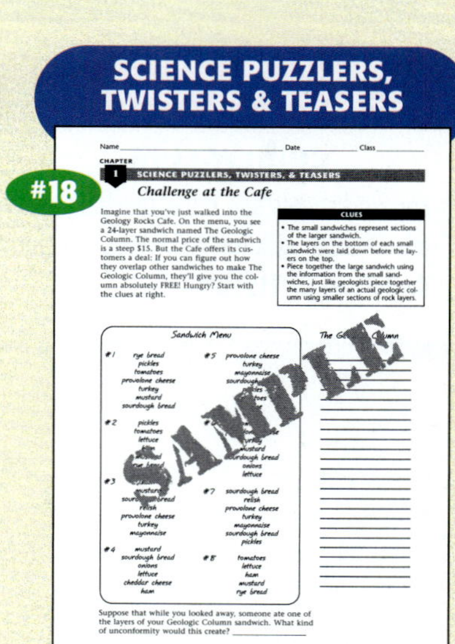

CHAPTER 1 — SCIENCE PUZZLERS, TWISTERS, & TEASERS

Challenge at the Cafe

Imagine that you've just walked into the Geology Rocks Cafe. On the menu, you see a 24-layer sandwich named The Geologic Column. The normal price of the sandwich is a steep $15. But the Cafe offers its customers a deal: If you can figure out how they overlap other sandwiches to make The Geologic Column, they'll give you the column absolutely FREE! Hungry? Start with the clues at right.

CLUES
- The small sandwiches represent sections of the larger sandwich.
- The layers on the bottom of each small sandwich were laid down before the layers on the top.
- Piece together the large sandwich using the information from the small sandwiches, just like geologists piece together the many layers of an actual geologic column using smaller sections of rock layers.

SAMPLE

Suppose that while you looked away, someone ate one of the layers of your Geologic Column sandwich. What kind of unconformity would this create?

Review & Assessment

STUDY GUIDE

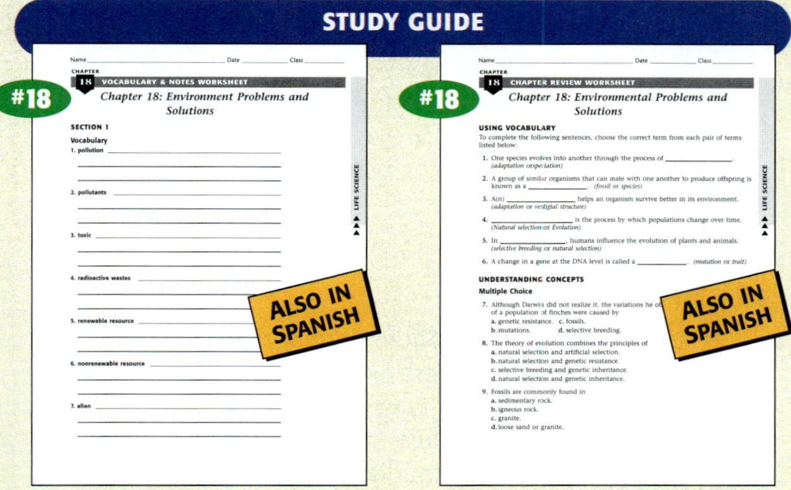

CHAPTER TESTS WITH PERFORMANCE-BASED ASSESSMENT

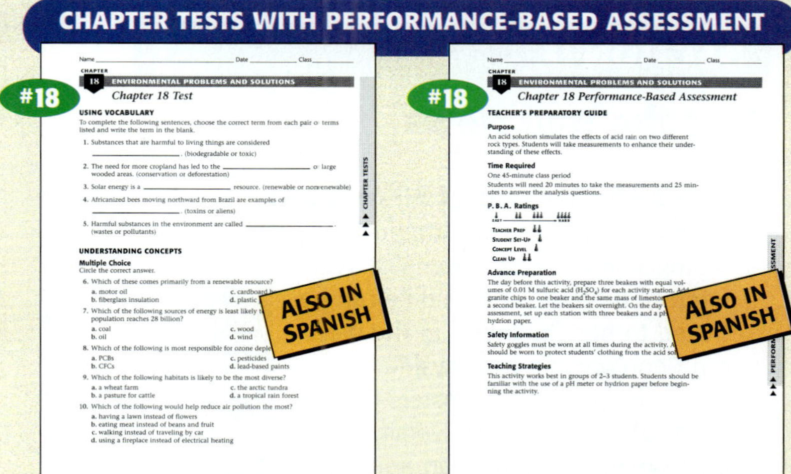

Lab Worksheets

ECOLABS & FIELD ACTIVITIES

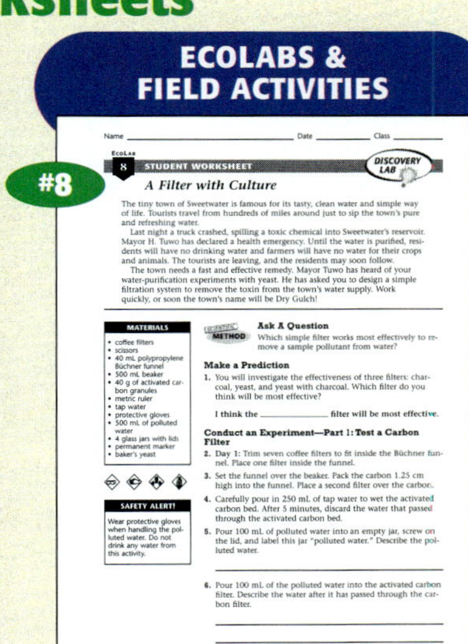

LONG-TERM PROJECTS & RESEARCH IDEAS

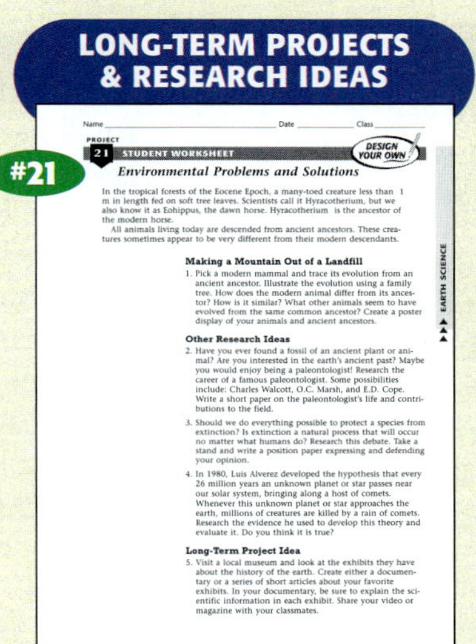

DATASHEETS FOR LABBOOK

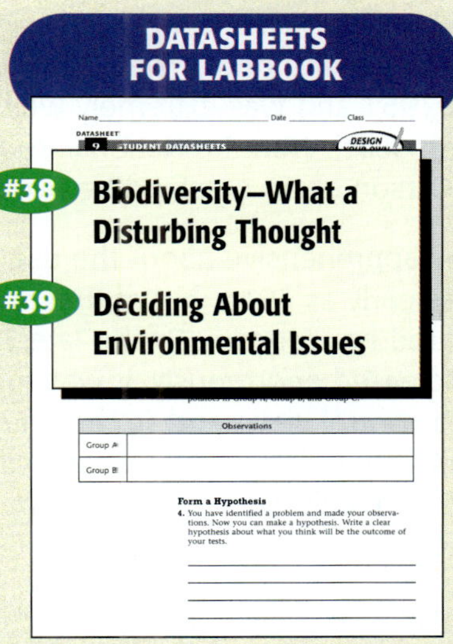

#38 **Biodiversity—What a Disturbing Thought**

#39 **Deciding About Environmental Issues**

Applications & Extensions

CRITICAL THINKING & PROBLEM SOLVING

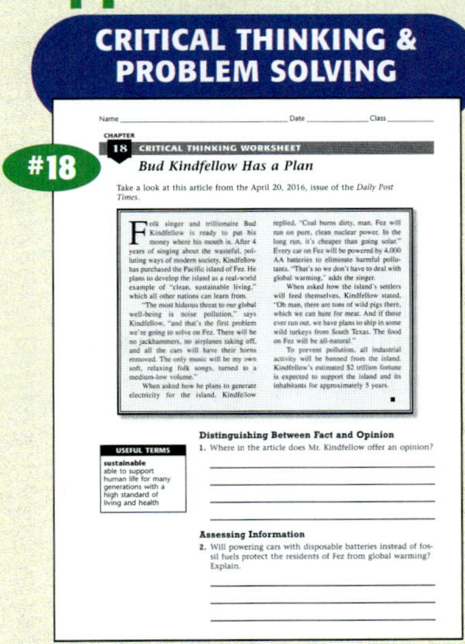

MULTICULTURAL CONNECTIONS

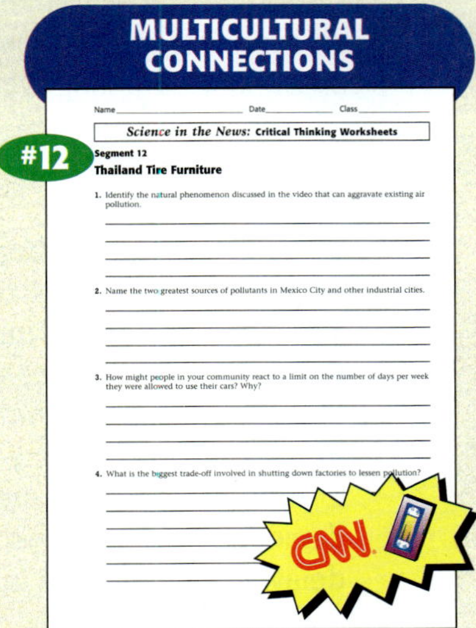

EYE ON THE ENVIRONMENT

SCIENTISTS IN ACTION

INTERACTIVE EXPLORATIONS

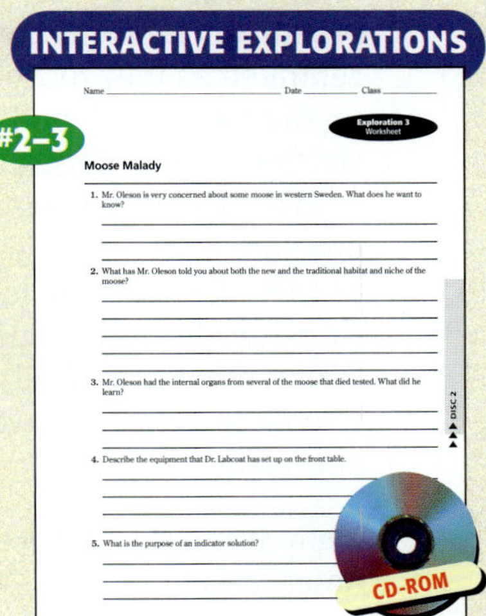

Section 1

First the Bad News

▶ Rachel Carson (1907–1964)

Rachel Carson was an American biologist and environmentalist. Her interest in nature and wildlife can be traced to her childhood in the western Pennsylvania countryside. In 1932, she received a masters degree from Johns Hopkins University and later obtained a job as an aquatic biologist with the U.S Bureau of  Fisheries. She supplemented her income by writing articles about marine life. In 1941, she published her first book, *Under the Sea-Wind*—a vivid depiction of life in the oceans. Her second book, *The Sea Around Us,* was published in 1951 and was extremely well received. It became a best-seller, won the National Book Award, and established Carson as an outstanding science writer.

- Carson was apprehensive about the use of the pesticide DDT as early as 1945, but at that time few other people shared her concern. By the late 1950s, DDT use had increased enormously. It was routinely used on crops to control pests and (rather ineffectively) for mosquito control, but unfortunately it also killed nontargeted animals, such as birds, fish, crabs, grasshoppers, and bees.

- Carson began to research pesticides and turned up alarming evidence that pesticides contaminated soil, water, and air; became increasingly concentrated as they moved  up the food chain; and that such chemicals are so stable that they persist as toxins for a very long time. In the meantime, chemical companies were zealously promoting the "miracles" worked by pesticides such as DDT. Carson compiled her evidence against chemical pesticides in her book *Silent Spring* (1962). The book's release sparked an immediate controversy. Chemical companies were Carson's most vociferous opponents.

- Many scientists, however, praised Carson's book. The American public took Carson's warnings to heart, and by the end of 1962, more than 40 bills concerning pesticide regulations had been introduced in state legislatures. The formation of the Environmental Protection Agency, in 1970, was due partly to Carson's efforts. The United States began to phase out the use of DDT in 1972.

IS THAT A FACT!

- Upon the release of *Silent Spring,* representatives of chemical manufacturers did not shrink from attacking Carson personally. Carson was accused of being a "high priestess of nature," and her book was said to be part of a "communist plot" to destroy the economies of noncommunist countries.

▶ The Problems with Landfills

Landfills take up valuable land, smell bad, and attract pests. They are responsible for a host of less-obvious but equally troublesome problems as well. Decades ago, little thought was given to landfills; they were established wherever cheap or seemingly "useless" land existed. For years, household  trash—some of it containing toxic chemicals, such as those found in household cleaners, paints, bug sprays, plastics, motor oil, and batteries—was plowed into landfills.

- Over time, rainwater percolated through the mountains of garbage, picking up chemical and bacterial wastes and dissolved metals. The rainwater forms a poisonous concoction known as leachate. Leachate can seep into ground water and can eventually contaminate drinking water.

IS THAT A FACT!

- In March 1987, a tugboat hauling a barge loaded with more than 3,000 tons of garbage from New York City headed to North Carolina. Officials in North Carolina turned the boat away because they feared that the trash might be hazardous. The boat traveled from port to port along the Atlantic seaboard for 2 months, and all requests to dump its load were denied. It finally returned to New York. After a 3-month delay, the refuse was incinerated in Brooklyn.

The Good News: Solutions

▶ John Muir (1838–1914)

Scottish-born naturalist John Muir is often considered the father of the conservation movement in the United States. When Muir was 11, he and his family moved to the United States and settled on a farm in Wisconsin. Muir attended the University of Wisconsin for 2.5 years. In 1867, he set out to walk from Indianapolis to the Gulf of Mexico, an experience he described in his classic book, *A Thousand-Mile Walk to the Gulf* (1916). The walk was a turning point in Muir's life; he moved to Yosemite Valley, in California, and devoted himself to the study of the wilderness areas of the West.

• Muir wrote a number of natural-history articles for national magazines that opened the eyes of many to the wonders of nature. In 1876, he proposed to the federal government that it take measures to preserve forests. His efforts played a key role in the establishment of Sequoia and Yosemite National Parks in 1890.

• Two years later, Muir founded the Sierra Club, whose purpose was to "explore, enjoy, and render accessible the mountain regions of the Pacific Coast." Today the Sierra Club is a national organization with a much broader goal that includes the exploration and enjoyment of wilderness as well as the responsible use, protection, and restoration of the world's ecosystems.

▶ The Birth of the "Disposable Society"

Before the 1950s in the United States, recycling and reuse were routine practices. Relatively few disposable items existed, and most products were not wrapped and sealed in multiple layers of packaging. A few items, such as sugar and flour, were packaged, but people either reused the large cotton sacks those items came in or made the sacks into clothing. Some of the sacks had decorative prints, encouraging their use as clothing.

IS THAT A FACT!

• In 1955, *Life* magazine published an article on the new "throwaway" lifestyle and how it liberated Americans from tedious cleanup chores. Among the noteworthy items the article highlighted were disposable curtains, hunting decoys, and barbecue grills.

• The conventional recycling of paper involves the use of harsh de-inking chemicals and requires that the paper be repulped and reprocessed into new paper. Most paper can be recycled only about three times before its cellulose fibers lose their integrity. Some paper is more suitable for recycling than others. In 1996, a mechanical engineer named Sameer Madanshetty devised a very gentle, chemical-free method that uses focused sound waves to "explode" ink off of paper. When paper is placed in water, tiny bubbles form around the inked portions of the paper. Sound waves directed at the bubbles blast the bubbles and remove the ink, which can then be filtered out of the water. The paper itself is undamaged and can be dried and reused repeatedly.

Directed Reading Worksheet 18

Science Puzzlers, Twisters & Teasers Worksheet 18

Guided Reading Audio CD
English or Spanish, Chapter 18

CAPÍTULO

18 Problemas ambientales y soluciones

Imagínate . . .

Haifa Aldorasi tiene quince años, le encanta la ciencia y le preocupa el medio ambiente. Haifa sabe que para fabricar el papel que usamos hay que talar árboles, hacerlos pulpa y añadir substancias químicas. Haifa también está muy interesada en la historia de los Estados Unidos y sabe que los colonizadores hacían papel de trapos de lino y algodón. Este papel no tenía substancias químicas y sigue en buenas condiciones. El papel que usamos hoy no va a durar tanto como el de los colonizadores.

A los 13 años, Haifa reunió sus conocimientos de historia, su amor a la ciencia y su preocupación por el medio ambiente con toda una variedad de verduras, dos años de experimentación y mucha paciencia. ¿El resultado? Haifa fabrica papel de buena calidad sin cortar ni una ramita de árbol ni usar una gota de substancias químicas. Además, su papel va a durar siglos.

Haifa no ha dejado de investigar. Quiere desarrollar un procedimiento eficaz para producir grandes cantidades de papel sin cortar árboles, así que continúa perfeccionando su técnica. Le gustaría llegar a presentar su producto a las grandes compañías papeleras. Mientras tanto, vende un estuche que ella diseñó con todo lo necesario para fabricar papel, para que otras personas puedan hacerlo igual que ella.

En este capítulo estudiarás los problemas del medio ambiente y aprenderás lo que puedes hacer para contribuir a su solución, incluyendo la fabricación de tu propio papel.

438

Imagine . . .

Paper can be made from anything that contains cellulose, an important constituent of all plant cells. Wood contains about 50 percent cellulose, while cotton is about 90 percent cellulose. In fact, wood was not used to make paper in the United States until the 1850s, when a shortage of linen and cotton rags spurred the development of processes for turning wood pulp into paper. At the time, wood seemed a logical choice because the United States had seemingly unlimited forests. Today the need to conserve forests has inspired environmentalists to investigate the use or revival of alternative cellulose sources, such as cotton, kenaf (a fast-growing plant related to the hibiscus plant), hemp, and straw.

¿Tú qué piensas?

Usa tus conocimientos para responder a las siguientes preguntas en tu cuaderno de ciencias:

1. Nombra tres formas en que perjudicamos a la Tierra.

2. Nombra tres formas en las que algunas personas están tratando de impedir que se le haga más daño a la Tierra.

Recicla papel

También puedes hacer papel. Este método no requiere que se corten árboles, sino que utiliza papel que normalmente tiraríamos a la basura.

Procedimiento  

1. Rasga **dos hojas de papel periódico** en pedazos pequeños. Ponlos en una **licuadora** y añade **1 L de agua**. Tápala y licua hasta que obtengas una pulpa aguada. Añade pedacitos de cáscara de naranja o zanahoria, o pétalos de flores. Licua otra vez.

2. Llena un **recipiente grande y cuadrado** con **2 a 3 cm de agua** y coloca un **cuadrado de malla de alambre** adentro. Vacía 250 mL de pulpa de papel sobre la malla, como se muestra abajo. Extiéndela de manera uniforme con una **espátula** o una **cuchara de palo**.

3. Para separar la pulpa del agua, levanta la malla, con la pulpa encima, y deja que el agua escurra sobre el recipiente. Coloca la malla y la pulpa dentro de una sección **de periódico**.

4. Cierra el periódico. Con mucho cuidado, dale la vuelta de manera que la malla quede encima de la pulpa. Tapa todo con una **tabla plana**, y presiona para exprimir el exceso de agua.

5. Abre el periódico y quita con mucho cuidado la malla, de manera que la capa de pulpa quede sobre la mesa. Deja que la pulpa se seque completamente. Finalmente, despega tu nuevo papel del periódico, como se muestra abajo.

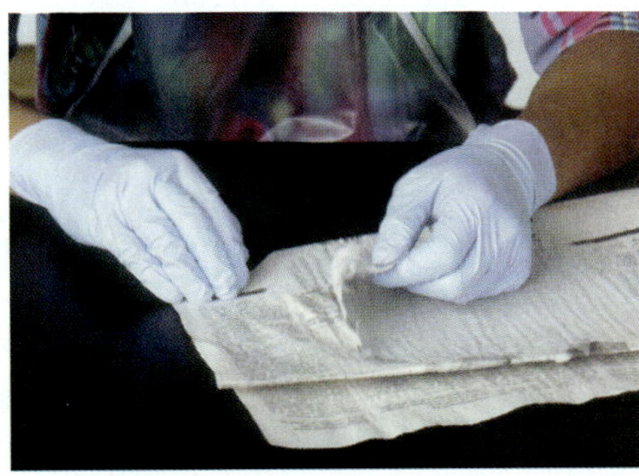

6. En tu papel reciclado escribe una nota a un amigo o amiga, o haz un dibujo.

Análisis

7. En qué se parece y se diferencia tu papel al papel comercial?

8. ¿Cómo podrías mejorar este método de hacer papel?

Multicultural CONNECTION

Making recycled paper is not new. The first "recycled" paper was made nearly 2,000 years ago by a member of the Chinese court named Ts'ai Lun. To make paper, Ts'ai Lun used discarded fish nets, rags, hemp, and grass. The craft of paper-making spread throughout Asia, and by the eleventh century, the Japanese had begun to recycle used paper to make new paper. Today many innovative methods have been developed for recycling various waste materials into paper. In Costa Rica, one company uses banana fiber to make paper, and in Scotland, grass clippings from golf courses have been used to make paper.

Focus

First the Bad News

In this section, students investigate the major types of pollutants and how pollution is generated. Students also distinguish between renewable and nonrenewable resources. Finally, they explore how increases in human population and human activities strain resources and threaten the habitats of other living things.

Bellringer

Ask students:

What is the difference between a renewable resource and a nonrenewable resource? Have them write two examples of each in their ScienceLog. (Renewable resources, such as pine trees, that are used at the proper rate will last forever because they grow back or replenish themselves. Nonrenewable resources, such as oil, do not replenish themselves or grow back, or they would take thousands or millions of years to replenish.)

1) Motivate

GROUP ACTIVITY

Writing In groups of four, have students come up with a list of challenges to the environment. From that list, have each group pick the four that they feel are most important and explain why they should be priorities. Afterward, have groups share their top four challenges, and write them on the board or overhead. This list could be used to generate topics for discussion, research, or action.

VOCABULARIO

contaminación	recursos no renovables
contaminante	biodegradable
tóxico	superpoblación
desechos radioactivos	deforestación
recursos renovables	naturalizado
	biodiversidad

OBJETIVOS

- Describe los principales tipos de contaminación.
- Compara los recursos renovables con los no renovables.
- Explica cómo la destrucción de su hábitat afecta a los organismos.
- Explica por qué el crecimiento de la población humana es un problema.

Figura 1 *El agua que esta fábrica arroja al río está contaminada con substancias químicas y calor. El humo contiene substancias químicas dañinas que contaminan el aire.*

Figura 2 *Cada año tiramos 150 millones de toneladas métricas de basura.*

440

Primero, las malas noticias

Seguramente ya lo has oído o quizá lo has experimentado personalmente: el aire que respiras es malsano, el agua que bebes es perjudicial y la tierra está envenenada. En pocas palabras, nuestro planeta está gravemente enfermo.

La contaminación

La **contaminación** es la presencia de substancias dañinas en el medio ambiente. Estas substancias, o **contaminantes,** aparecen en muchas formas. Pueden ser sólidos, substancias químicas, ruido y hasta calor. Los contaminantes dañan o matan a las plantas y animales que viven en el hábitat afectado, como se ve en la **Figura 1,** y también pueden afectar a las personas.

Montañas de basura Los estadounidenses producen más basura doméstica que ningún otro país del mundo. Si hiciéramos una torre con las latas de bebida que se consumen en un año, la torre llegaría a la Luna... ¡diecisiete veces! El estadounidense promedio desecha 12 kg de basura a la semana, que van a dar a un vertedero como el de la **Figura 2.** Las compañías, industrias y minas también producen mucha basura. Miles de millones de kilogramos de basura se clasifican como *desechos peligrosos,* perjudiciales para las personas y el medio ambiente. Las fábricas de papel, plástico, cemento y plaguicidas producen desechos peligrosos, al igual que centrales nucleares, refinerías de petróleo y plantas procesadoras de metales. Los hospitales, laboratorios y consultorios médicos producen desechos médicos peligrosos. Pero no toda la culpa es de la industria. Los hogares (como el tuyo) también producen desechos peligrosos: coches viejos, pintura, pilas, desechos médicos y detergentes son algunos de ellos.

IS THAT A FACT!

The U.S. Environmental Protection Agency defines hazardous waste as "by-products of society that can pose a substantial or potential hazard to human health or the environment when improperly managed." This includes material that is toxic, corrosive, reactive, or ignitable. Examples include many waste chemicals and radioactive waste from nuclear power plants.

¿Adónde va? ¿Adónde va a dar nuestra basura? La mayor parte va a enormes vertederos de basura. Muchas compañías entierran desechos peligrosos en basureros especialmente diseñados para contener este tipo de basura, pero otras tiran sus desechos peligrosos ilegalmente, arrojándolos en ríos o lagos y contaminando el agua. Otros se queman en incineradores especiales, diseñados para reducir la cantidad de contaminantes que entran a la atmósfera, pero otros se queman de manera inadecuada, aumentando la contaminación del aire.

Las substancias químicas están por todas partes Se usan para curar enfermedades, desde el catarro hasta el cáncer, están presentes en plásticos, termómetros, pinturas, fijadores para el pelo y alimentos enlatados. Están en todas partes; no podemos vivir sin ellas, pero a veces tampoco podemos vivir *con* ellas. Los plaguicidas químicos para matar los insectos que atacan los sembradíos también contaminan el suelo y el agua. Hace más de tres décadas, Rachel Carson, en la **Figura 3,** escribió sobre los peligros de los insecticidas.

Aerosoles, refrigeradores y plásticos usan un tipo de substancias químicas, llamadas clorofluocarbonos (CFC), que ahora están prohibidas. Los CFC se elevan en la atmósfera y destruyen la capa de ozono, una forma de oxígeno que protege la Tierra de los dañinos rayos ultravioleta. Otro tipo de substancias químicas, los policlorobifenilos (PCB), se usaban como aislante, en pinturas, aparatos domésticos y otros productos. Pero los científicos descubrieron que los PCB eran **tóxicos,** o sea, venenosos. Ahora están prohibidos, pero no han desaparecido; se descomponen muy lentamente y siguen contaminando hasta las regiones más remotas, como la de la **Figura 4.**

Basura nuclear Las plantas nucleares producen energía para millones de hogares y negocios, pero también producen **desechos radioactivos,** un tipo diferente de basura, que tarda cientos de miles de años en dejar de ser peligrosa. Estos desechos provocan cáncer, leucemia y defectos congénitos en los seres humanos. Los desechos radioactivos pueden perjudicar a todos los seres vivos.

Figura 3 *El libro de Rachel Carson,* Silent Spring, *se publicó en 1962, y le mostró al público los peligros de los plaguicidas para el medio ambiente, en especial para los pájaros.*

Figura 4 *No hay lugar en la Tierra que sea inmune a la contaminación. Hasta en las más remotas áreas del Ártico se han encontrado contaminantes, como los PCB.*

441

Science Bloopers

Can you imagine children going to school on top of a hazardous waste site? That's what happened in Niagara Falls, New York, in 1954 when a school and neighborhood were built on land covering an old dumping site for a chemical company. Kids played in and around pools of black muck. Parents noticed that their children were burned by the chemicals. Dogs lost their fur. Dark, smelly substances seeped into homes after rainfalls. In 1978, the state ordered the evacuation of 235 families living near the toxic site.

MATH and MORE

On average, for each mile (1.6 km) a car travels, 0.36 kg of carbon dioxide is added to the atmosphere.

For 2 weeks, have students keep a log of how many miles their family travels by car. At the end of 2 weeks, have students figure out how many total kilograms of carbon dioxide their family added to the atmosphere.

CONNECT TO
EARTH SCIENCE

Only 20 percent of the radiation that enters the Earth's atmosphere is absorbed by gases in the atmosphere and transferred in the form of heat. But these gases capture heat in other ways. When land and water absorb radiation, their molecules move faster, increasing their temperature. This energy is transferred to gas molecules in the atmosphere before it can escape into space. As a result, the atmosphere warms up. The Earth's heating process, in which the gases in the atmosphere absorb radiation and transfer the energy in the form of heat, is known as the greenhouse effect. The Earth's atmosphere works much like a greenhouse, as shown in Teaching Transparency 144, "The Greenhouse Effect."

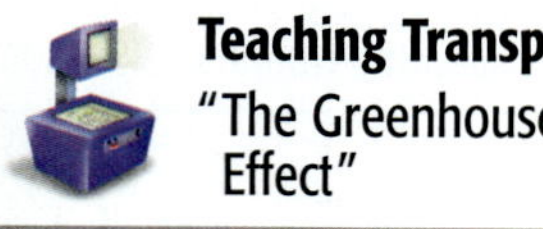
Teaching Transparency 144
"The Greenhouse Effect"

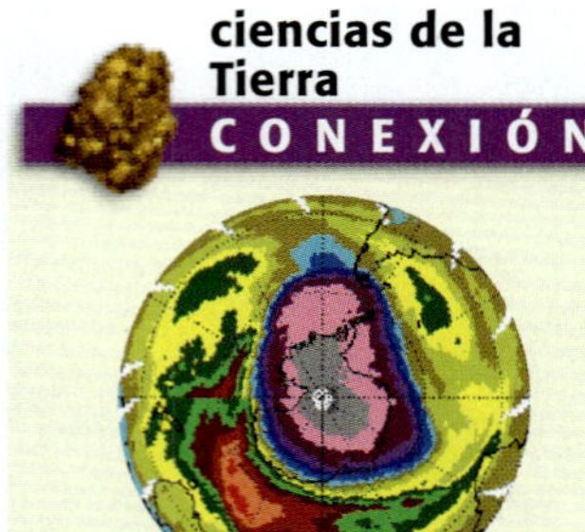

ciencias de la Tierra
C O N E X I Ó N

El ozono de la estratosfera absorbe la mayor parte de la radiación ultravioleta del Sol. Los CFC destruyen la capa de ozono. La imagen del agujero en la capa de ozono (el área gris del centro) se tomó en 1998.

La exposición excesiva a la luz ultravioleta es perjudicial para la salud. Puede causar ceguera, envejecimiento prematuro, cáncer de piel y debilitamiento del sistema inmunológico.

Un alto nivel de luz ultravioleta también contribuye a cosechas más pobres y al desequilibrio de la cadena alimenticia en los océanos.

¡Qué calor! La Tierra está rodeada por la atmósfera, una mezcla de gases entre los que está el dióxido de carbono. La atmósfera funciona como cobija protectora que mantiene al planeta suficientemente caliente para que exista la vida. Sin embargo, desde el siglo XIX, el dióxido de carbono de la atmósfera ha aumentado en un 25 por ciento. El dióxido de carbono y ciertos contaminantes presentes en el aire funcionan como un invernadero. Muchos científicos opinan que su aumento ha provocado una elevación significativa de la tempertatura global. Si ésta siguiera aumentando, se podrían derretir las capas de hielo polar y el nivel de los océanos subiría. Se cree que para el año 2100 el nivel del mar podría subir de 10 cm a 1.2 m. Si el nivel del mar sube 1 m, las costas se inundarían, se contaminarían las reservas subterráneas de agua y desaparecerían las playas actuales.

Demasiado ruido Contaminantes como los malos olores y el ruido afectan a nuestros sentidos. El ruido es molesto e influye en nuestra capacidad de escuchar y de pensar. Si los albañiles y otras personas que trabajan en ambientes ruidosos no protegen sus oídos, pueden quedarse sordos poco a poco. Las estudiantes de la **Figura 5** escuchan música a un volumen prudente para evitar que sus oídos sufran.

Figura 5 *Para prevenir lesiones del oído es mejor escuchar música a un volumen moderado.*

REPASO

1. Describe dos formas en que la contaminación es dañina.

2. Explica por qué el ruido se considera un contaminante.

3. **Aplicar conceptos** Explica cómo cada una de las siguientes cosas que nos son de ayuda perjudican al medio ambiente: hospitales, refrigeradores y construcción de carreteras.

442

▼ Answers to Review

1. Some pollutants are eyesores, such as open landfills. Some pollutants are poisonous to the environment and to living things.

2. Loud noise affects the ability to hear and think and can even permanently damage hearing.

3. hospitals: provide health care and generate medical waste; refrigerators: keep our food fresh and use a polluting chemical as a refrigerant; road construction: helps people travel easily from place to place and destroys habitat

Agotamiento de recursos

Otro problema del medio ambiente es que estamos agotando los recursos naturales. Algunos de los recursos de la Tierra son renovables y otros no renovables. Un **recurso renovable** se puede usar una y otra vez, o existe una fuente ilimitada del mismo. El agua dulce y la energía solar son renovables, al igual que algunos tipos de árboles. Un **recurso no renovable** sólo se puede usar una vez. La mayoría de los minerales son no renovables; también lo son los combustibles fósiles, como el petróleo y el carbón.

Algunos recursos no renovables, como el petróleo, seguramente no se agotarán durante tu vida, pero como usamos tantos al año, no van a durar para siempre. Además, la extracción de estos recursos de la Tierra tiene un alto precio, que se paga en forma de derrames de petróleo, pérdida de hábitats y el daño que ocasiona la explotación minera, como ves en la **Figura 6.**

Figura 6 *Este terreno ha sido explotado para extraer carbón usando una técnica llamada explotación a cielo abierto.*

¿Renovable o no? Algunos recursos que se consideraban renovables se están agotando. Ecosistemas como la selva tropical están siendo contaminados y destruidos, y el resultado es una enorme pérdida de hábitats. En todo el mundo, las capas más ricas del suelo se están perdiendo por la erosión y la contaminación. Lleva miles de años formar unos cuantos centímetros de tierra, que luego pueden perderse en menos de un año. Las reservas subterráneas de agua potable se consumen más rápido de lo que se llenan. Varios centímetros de agua se filtran hacia estas reservas por año, pero, al mismo tiempo, estamos sacando varios *metros* hacia el exterior.

¡MATEMÁTICAS!

Agotamiento del agua

Una reserva subterránea de agua tiene una profundidad de 200 m. El agua se filtra a una velocidad de 4 cm/año, y se bombea a una velocidad de 1 m/año. ¿Cuánto tiempo durará esta reserva?

Para encontrar la pérdida neta de agua de una reserva subterránea, resta la cantidad de agua que entra de la cantidad que se bombea.

¿Cuánto tiempo durará la reserva si el agua entra a una velocidad de 10 cm/año y se bombea a una velocidad de 10cm/año?

Autoevaluación

1. ¿Cómo usas los recursos no renovables?

2. ¿Por qué no es buena idea agotar un recurso no renovable?

(Consulta la página 636 para comprobar tus respuestas).

443

MATH and MORE

While Earth's human population is expected to continue to grow rapidly, not every country in the world is expected to grow at the same rate. The growth rates of some countries, such as Japan and Sweden, are expected to shrink. In 1998, there were 5.9 billion people in the world. If current trends continue, there will be 9.4 billion in 2050, almost double. You can quickly estimate the number of years it will take for a population to double by dividing 70 by the population's current growth rate (GR). Have students find the doubling time for the following regions:

Northern Europe, GR = 0.2
North America, GR = 0.7
Africa, GR = 2.9
Asia, GR = 1.8
Latin America, GR = 2.2

Northern Europe
70 ÷ 0.2 GR = 350 years

North America
70 ÷ 0.7 GR = 100 years

Africa 70 ÷ 2.9 GR = 24 years

Asia 70 ÷ 1.8 GR = 39 years

Latin America
70 ÷ 2.2 GR = 32 years

internetconnect

SCLINKS NSTA

TOPIC: Population Growth
GO TO: www.scilinks.org
*sci*LINKS NUMBER: HSTL515

Especies naturalizadas

Las personas viajan y sin saberlo muchas veces llevan pasajeros consigo. Los barcos, aviones y coches llevan semillas, huevos y organismos vivos de una parte del mundo a otra. Un organismo que llega a un lugar nuevo y se queda a vivir se llama organismo **naturalizado.** Las especies naturalizadas progresan en nuevos lugares porque los depredadores de su hábitat original no existen en el nuevo.

A veces, las especies naturalizadas se convierten en plagas y desplazan a las especies nativas. El mejillón cebra, de la **Figura 7,** llegó como polizón en barcos que viajaban de Europa a Estados Unidos en la década de 1980. La salicaria púrpura llegó de Europa, y ahora le roba espacio a la flora nativa y amenaza a especies raras de plantas en gran parte de Norteamérica. Especies como el diente de león de la **Figura 8,** llevan aquí tanto tiempo que es fácil olvidar que vinieron de otro lugar.

Figura 7 *El mejillón cebra (arriba) es un invasor naturalizado que bloquea las plantas de tratamiento de agua en la zona de los Grandes Lagos. La salicaria púrpura es una planta europea que está desplazando a otras plantas naturales de Norteamérica.*

Figura 8 *El diente de león se ha considerado una planta curativa en China durante más de 1,000 años, pero en Estados Unidos casi toda la gente la ve como maleza.*

El crecimiento de la población humana

En 1800, la Tierra tenía mil millones de habitantes. En 1990, había 5,200 millones. Para el año 2100 puede haber 14,000 millones. Una de cada diez personas se acuesta cada noche con el estómago vacío y millones de personas mueren cada año de enfermedades relacionadas con la desnutrición. Hay quienes opinan que la población mundial es muy grande para que la Tierra pueda alimentarla.

Más personas requieren más recursos y la población de la Tierra está creciendo rápidamente. La **superpoblación** ocurre cuando el número de personas es tan grande que no hay suficiente comida, agua y otros recursos básicos de la vida diaria para todos.

La **Figura 9** muestra que nos tomó la mayor parte de la historia humana llegar a ser mil millones de habitantes. ¿Podrá el planeta mantener a 14,000 millones de habitantes?

Crecimiento de la población humana

Figura 9 *Hoy en día, la población humana de la Tierra se duplica cada pocos años.*

444

Science Bloopers

During the 1930s, sometimes called the Dust Bowl years, soil conservation scientists in the southern United States recommended planting kudzu to hold soil in place. This fast-growing vine is native to China and Japan and can grow 18 m per season. Frosts kill kudzu in Asia, keeping its growth in check. In the South, where heavy frosts are infrequent, kudzu can grow all year. By the 1960s, kudzu had spread over trees and other plants throughout the South. Eradicating the vine is difficult because a single plant may have 50 or more vines attached to a giant root, which can weigh as much as 181 kg!

Destrucción de hábitats

Bio quiere decir "vida", y *diversidad* significa "variedad". La palabra **biodiversidad** quiere decir "variedad de formas de vida", y habla de las muchas y diferentes especies que se encuentran en los hábitats específicos del planeta.

Cada hábitat tiene su propia combinación de ocupantes. Cuando una excavadora abre un agujero o una sierra eléctrica tala un bosque, cuando se desechan substancias peligrosas en el medio ambiente y se construye una presa de agua, un hábitat sufre cambios, queda lastimado, o se destruye. Y cada vez que se destruye un hábitat, hay una pérdida de biodiversidad.

Figura 10 *Los bosques templados se destruyen por las mismas razones que los bosques tropicales.*

Bosques Los árboles nos dan oxígeno, muebles, combustible, frutas y nueces, caucho, alcohol, papel, aguarrás, lápices y postes de teléfono. Hace tiempo, los árboles cubrían el doble del espacio que ocupan hoy. La **deforestación,** que ves en la **Figura 10,** es la tala de bosques. Los bosques se talan para explotar minas, construir presas y hacer carreteras; y para fabricar papel y obtener combustible y madera. Después de que se tala un bosque, casi nada puede crecer ahí, pues el suelo tropical tiene pocos nutrientes. Los nutrientes que tiene están sobre todo en la vegetación.

Pantanos Los pantanos ayudan a controlar las inundaciones al absorber el agua de los ríos desbordados, filtran la contaminación del agua que se desborda y son el lugar ideal para la reproducción de peces, aves acuáticas y otros animales; previenen la erosión y llenan las reservas subterráneas de agua. No obstante, muchos pantanos se secan para construir granjas, casas y centros comerciales, o se dragan para que pasen barcos y lanchas. La contaminación también destruye los hábitats de los pantanos.

Hábitats marinos El petróleo es un gran destructor de hábitats marinos. A veces, el petróleo de fábricas se arroja al mar. Los derrames accidentales y el petróleo que los barcos petroleros echan al mar al enjuagar sus tanques en el agua también contaminan los océanos. El petróleo contamina los hábitats de la costa y los de mar abierto **(Figura 11.)** Como los océanos están conectados, la contaminación de uno puede extenderse por todo el mundo.

Explora

Observa tu salón. ¿Cuántas cosas de madera ves? Haz una lista en tu cuaderno de ciencias. Añade todas las cosas que creas que vienen de los árboles.

Figura 11 *El petróleo del* Exxon Valdez *afectó a más de 2,300 km² de la costa de Alaska.*

445

ACTIVITY

Crude Oil Spills and Seabird Eggs Have groups do the following activity to investigate how oil can seep through eggshells. Give each group four hardboiled eggs and a bowl filled with 250 mL of vegetable oil that has been dyed with oil-soluble red food coloring. Students will place all the eggs in the bowl, remove one egg every 5 minutes, and shell it. Have them record how much time elapses before a shelled egg shows red coloring, indicating that the oil has permeated the shell. Have students draw some conclusions about how crude oil seeping through the shells of developing seabirds might affect the unborn birds and how it might affect future seabird populations. (The oil might kill the developing birds, or it might interfere with their proper development such that if they are born, their chance of survival would be compromised. With fewer birds born or surviving to reproduce, the species' existence could be threatened.) Sheltered English

Math Skills Worksheet 37 "Rain-Forest Math"

Quiz

1. Why is it dangerous to dispose of hazardous waste by burying it? (Buried hazardous waste can exude harmful chemicals that pollute drinking water.)

2. Why is coal a nonrenewable resource? (Coal is nonrenewable because it can be used only once and there is a limited supply of it.)

3. What are two ways pollution can harm people? (Possible answer: Exposure to some chemicals may cause cancer and other health problems; drinking polluted water and breathing polluted air may cause lung damage.)

ALTERNATIVE ASSESSMENT

Have groups of students paint a mural of what the neighborhood near your school might look like in the year 2100 if pollution, resource depletion, and habitat destruction continue at a rapid rate. Sheltered English

Answer to APPLY

It is not a good idea to release 1,000 balloons because the balloons will eventually burst and fall to Earth. The plastic material of the balloons will not degrade and will be a threat to wildlife.

Tu pueblo va a celebrar los 200 años de su fundación. Se ha planeado una gran fiesta y, como parte de la celebración, el municipio soltará al cielo 1,000 globos inflados de helio para cantarle "Las mañanitas" al pueblo. ¿Sabes por qué no es una buena idea? ¿Qué puedes hacer para convencer a los oficiales del municipio de que cambien sus planes?

Figura 12 *Los plásticos perjudican la vida silvestre del mar. Esta gaviota se enredó en un empaque de refrescos.*

Si la raza humana se extingue, otros organismos pueden seguir viviendo. Pero si se extinguen todos los insectos, muchas plantas no podrían reproducirse y muchos animales se quedarían sin comida. Los organismos de los que dependemos y, finalmente todos nosotros, desapareceríamos de la faz de la Tierra.

Muchos plásticos se arrojan en hábitats marinos. Son ligeros, flotan en el agua y, como no son **biodegradables,** es decir, no se descomponen, permanecen en el medio ambiente. Hay animales, como el ave de la **Figura 12,** que tratan de comérselos, se enredan en ellos y mueren. La ley prohíbe arrojar plásticos al agua, pero es difícil capturar a los culpables.

El impacto en los seres humanos

Los árboles y los animales marinos no son los únicos afectados por la contaminación, el calentamiento global y la destrucción de hábitats. El daño que hacemos a la Tierra también nos afecta a nosotros. A veces, es inmediato. Si bebes agua contaminada, te puedes enfermar inmediatamente y hasta morir. Pero a veces el impacto no se siente de inmediato. Algunas substancias químicas causan cáncer 20 ó 30 años después de la exposición. Tus hijos o tus nietos quizá se tengan que enfrentar a un mundo en el que los recursos estén agotados.

Cualquier cosa que amenaza a otros organismos finalmente nos amenaza también a nosotros. Para cuidar el medio ambiente, tenemos que ocuparnos de lo que está sucediendo ahora y al mismo tiempo pensar en el futuro.

REPASO

1. ¿Por qué suelen prosperar las especies naturalizadas?

2. Explica qué tiene que ver el crecimiento de la población humana con la contaminación.

3. **Aplicar conceptos** ¿Cómo nos afecta la destrucción de los hábitats de los pantanos?

446

Answers to Review

1. Alien species often thrive in foreign lands because they have no natural predators.

2. The larger the human population is, the more energy and materials that are needed and the more waste that is produced. Pollution problems happen when there is more waste being produced than we can dispose of without spoiling the environment.

3. Wetlands help control flooding, filter pollutants, prevent erosion, and help restore underground water supplies.

Sección
2

VOCABULARIO

conservación
reciclaje
recuperación de recursos

OBJETIVOS

- Explica la importancia de la conservación.
- Describe "las tres erres" y su importancia.
- Explica cómo se pueden proteger los hábitats.
- Enumera formas en que se puede proteger la Tierra.

Y ahora, las buenas noticias: hay soluciones

Como has visto, el panorama es malo. Pero no del todo: es más, también hay buenas noticias. Hay cosas que podemos hacer (¡y estamos haciendo!) para salvar la Tierra. *Tú mismo* puedes ayudar a salvarla. Si somos responsables del daño que sufre nuestro planeta, también podemos asumir la responsabilidad de curarlo y preservarlo.

Conservación

Una forma importante de salvar la Tierra es la conservación. La **conservación** es el uso responsable y la preservación de los recursos naturales. Si vas a visitar a tus amigos en bicicleta, conservas combustible y, al mismo tiempo, no contaminas el aire. Si usas abono orgánico en tu jardín en vez de fertilizantes químicos, conservas los recursos necesarios para hacer el fertilizante y evitas la contaminación del suelo y el agua.

Conservar es usar menos recursos naturales y reducir la cantidad de basura que producimos. Las "tres erres" de la **Figura 13** describen maneras en que podemos conservar recursos y reducir los daños del planeta. **R**educe, **R**eutiliza y **R**ecicla.

Figura 13 *Estos jóvenes ponen en práctica "las tres erres" usando una bolsa de la compra de tela, regalando ropa que ya no les queda y reciclando plásticos.*

447

Dan and Ed made a great plan,
to recycle all that they can.
They reduced lots of waste
with a bit too much haste,
and recycled their mother's new van.

Teaching Transparency 70 "Practicing Conservation"

Directed Reading Worksheet 18 Section 2

SECTION 2

Focus

The Good News: Solutions

In this section, students discover the importance of conservation and how the strategy "reduce, reuse, and recycle" helps conserve resources. Students also learn about the importance of maintaining biodiversity and protecting habitats. Finally, they explore specific strategies they can use to help protect the environment.

Bellringer

Have students suppose they've finished reading a magazine. In their ScienceLog, have them write down at least two things they might do that would be preferable to throwing the magazine away. (Possibilities include giving it to a friend or relative; donating it to a library or homeless shelter; using it to make a collage; recycling it.)

1) Motivate

COOPERATIVE LEARNING

Reusing Trash Organize students in groups of five or six, and give each group a plastic grocery bag to examine. Have them devise a list of at least five ways the bag can be reused. (It can be reused for groceries; a waterproof covering for books or papers when it rains; cut into strips and used as ribbons; used to line a wastebasket; used to hold wet clothing or shoes; used to wrap sandwiches; used to protect surfaces from spills.)

Have groups share their ideas with the rest of the class. Sheltered English

② Teach

DEMONSTRATION

Friendly Cleaner The following recipe is for safe, all-purpose cleaner. Write it on the chalkboard so that students may copy it if they wish:

    3.78 L (1 gal) hot water

    59 mL ($\frac{1}{4}$ cup) borax

    59 mL ($\frac{1}{4}$ cup) vinegar

    30 mL (2 tbsp) phosphate-free liquid soap

As you prepare the solution, share the following information with students. Borax (sodium tetraborate decahydrate) is a naturally occurring mineral that consists primarily of sodium and boron and works well as a cleanser, a disinfectant, and a deodorizer. Vinegar is a weak acid and is a good degreaser. The liquid soap removes dirt and grease. All the ingredients are nontoxic and are easy on the environment. When the solution is mixed, try it out in the classroom by using it on several surfaces. Sheltered English

MATH and **MORE**

Poster Project A faucet that drips at a rate of one drop per second wastes about 4 L of water per day. Have pairs of students incorporate this fact into a poster reminding people to turn off taps completely.

Reducir

Usar menos recursos naturales es la forma más obvia de conservarlos. Así también reducimos la contaminación y la cantidad de basura. Algunas compañías han empezado a implementar estrategias para conservar recursos y ahorrar dinero.

Reducir la basura y la contaminación La tercera parte de los desperdicios de las ciudades y los pueblos proviene de envolturas. Para conservar recursos y reducir el desperdicio, se pueden envolver productos con menos plástico y menos papel. La comida rápida puede ir envuelta en hojas delgadas de papel en vez de grandes paquetes de plástico no biodegradables. Si no necesitas la bolsa, puedes llevarte tus compras en la mano. Científicos como los de la **Figura 14** trabajan para fabricar plásticos más biodegradables.

Algunas compañías están buscando materiales menos peligrosos para sus productos. Hay agricultores, por ejemplo, que no quieren usar plaguicidas ni fertilizantes químicos. Practican la agricultura orgánica, y usan abono orgánico, estiércol, y fertilizantes y plaguicidas naturales. Los especialistas en agricultura están desarrollando técnicas que son mejores para el medio ambiente.

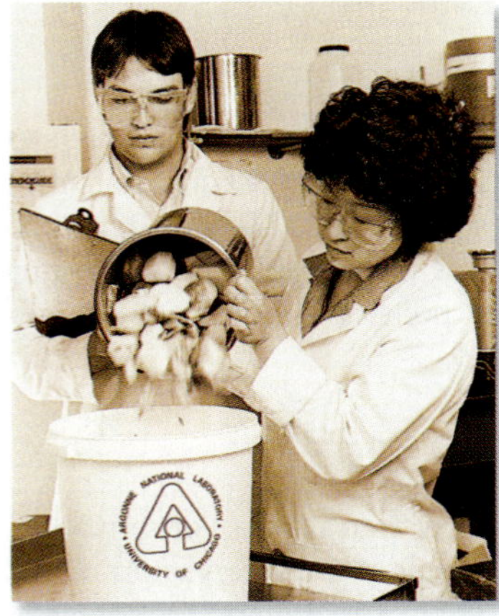

Figura 14 *Estos científicos investigan maneras de usar la basura para hacer plásticos biodegradables.*

Reducir el uso de recursos naturales no renovables Los científicos buscan fuentes alternativas de energía para evitar quemar combustibles y usar energía nuclear. Algunas calculadoras ya usan energía solar y en algunas partes del mundo esta energía calienta agua y produce la electricidad de las casas, como ves en la **Figura 15.** Hay ingenieros que investigan cómo hacer autos que funcionen con energía solar, e investigadores de otras fuentes de energía, como el viento, las mareas, y las cascadas.

Pero reducir el uso de recursos naturales y la cantidad de desperdicios no es sólo tarea de la industria y la agricultura; nosotros utilizamos muchos productos, consumimos mucha energía y producimos grandes cantidades de basura. Cada habitante de los Estados Unidos produce 40 veces más basura que el de un país en desarrollo. ¿Por qué será? ¿Qué podrías hacer para reducir la cantidad de basura que produces? Cada persona debe hacerse responsable de conservar los recursos de la Tierra.

Figura 15 *Los paneles solares en los tejados de estas casas de un barrio de Rotterdam, en Holanda, proporcionan casi toda la energía que necesitan.*

CONNECT TO CHEMISTRY

Biodegradable plastics contain substances that can be broken down by microbes, such as starch. Suggest that students imagine that they work for a company that uses nondegradable plastic in the packaging of their products. Have them write a persuasive letter to the company's directors advocating a changeover to biodegradable plastics.

Reutilizar

¿Usas la ropa que te pasan tus hermanos o hermanas mayores? ¿Parchas un balón desinflado en vez de tirarlo a la basura? Si dijiste que sí, estás ayudando a salvar al planeta al *reutilizar* productos.

Reutilizar productos Cada vez que alguien reutiliza una bolsa de plástico, se necesita fabricar una menos y hay una menos para contaminar la Tierra. Cada vez que alguien usa una pila recargable, se necesita fabricar una menos y hay una menos contaminando el planeta. Reutilizar las cosas es una forma importante de conservar recursos y evitar la contaminación.

Reutilizar el agua Cerca del 85 por ciento del agua usada en los hogares se va por el caño. Las comunidades donde el agua es escasa están experimentando con formas de reutilizarla. Algunas usan plantas verdes o ciertos animales, como las almejas, para limpiarla. El agua no queda bastante limpia para beberla, pero sirve para regar jardines o campos de golf, como el de la **Figura 16.**

Reciclar

Reciclar es una forma de reutilizar procesando la basura para hacer con ella nuevos materiales. A veces se usa material reciclado para hacer el mismo tipo de producto otra vez; otras veces se hace un producto diferente. La banca de la **Figura 17** está hecha de vasos de poliestireno, cajas de hamburguesas y recipientes de plástico que anteriormente sirvieron para guardar detergentes, yogurt y margarina. Todos los recipientes de la **Figura 18** pueden reciclarse fácilmente.

Figura 16 *Este campo de golf se riega con agua reciclada.*

Figura 17 *Esta banca está hecha de plástico reciclado, derretido y moldeado.*

Figura 18 *Estos recipientes son ejemplos de basura doméstica reciclable.*

✓ Autoevaluación

1. ¿Cómo puedes reducir la cantidad de electricidad que usas?
2. Menciona cinco productos que se pueden reutilizar fácilmente.

(Consulta la página 636 para comprobar tus respuestas).

449

Answers to Self-Check

1. Turn off lights, CD players, radios, and computers when leaving a room. Set thermostats a little lower in the winter (wear sweaters). Don't stand in front of an open refrigerator while deciding what you want. (Students will come up with many more.)

2. plastic bags, rechargeable batteries, water, clothing, toys; (Students will name more. Help them understand that the difference between reuse and recycle is that a reused article may be cleaned but is basically unchanged. A recycled article has been broken down and reformed into another useable product.

READING STRATEGY

Prediction guide Ask students: What is "resource recovery"? (Hint: It's not the same as recycling.) Resource recovery is the process of burning garbage to create electricity.

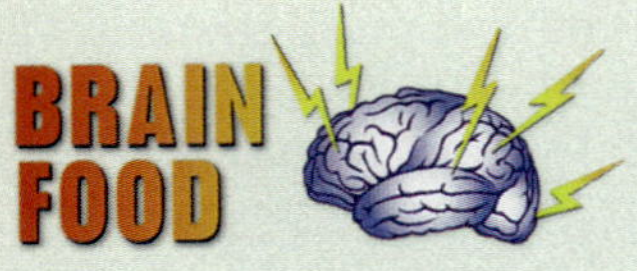

BRAIN FOOD

Provide students with the following data about the time it takes various wastes to break down naturally.

- paper, 2–12 months
- plastic bags, 20–30 years
- aluminum cans, 200–500 years
- plastic rings from soft-drink six-packs, 450 years
- plastic-foam, never

Have students apply the reduce or reuse strategies to the above items. Ask if it is better to reduce the use of some of these items rather than reuse them. Why or why not?

internet connect

SCiLINKS
NSTA

TOPIC: Recycling
GO TO: www.scilinks.org
*sci*LINKS NUMBER: HSTL520

Figura 19 *Cada tipo de material va a su propio contenedor en estos camiones especiales, y la basura para reciclar se lleva a plantas recicladoras para su procesamiento.*

Reciclar la basura El plástico, el papel, las latas de aluminio, la madera, el vidrio y el cartón son algunos ejemplos de materiales reciclables. Cada semana se necesita medio millón de árboles para hacer el papel para imprimir los periódicos de los domingos. Reciclar el periódico puede salvar la vida de muchos árboles. El reciclaje de papel y latas de aluminio ahorra un 95 por ciento de la energía requerida para procesar el mineral crudo. El vidrio contribuye un 8 por ciento al total de basura que producimos, pero se puede volver a derretir para hacer nuevas botellas y frascos. Las pilas de plomo pueden reciclarse para fabricar nuevas pilas.

Los gobiernos de ciudades como Austin, Texas, facilitan el reciclaje de basura. Cada familia recibe contenedores especiales para plástico, aluminio y papel, y cada semana, un camión especial, como el de la **Figura 19,** recoge los elementos reciclables junto con el resto de la basura.

Reciclar recursos La basura que se puede quemar también puede usarse para generar electricidad en plantas procesadoras como la de la **Figura 20.** El proceso de transformar la basura en electricicidad se llama **recuperación de recursos.** La basura que se recoge en las ciudades y pueblos de Estados Unidos podría producir la misma cantidad de electricidad que 15 centrales nucleares grandes. Algunas compañías han empezado a usar así sus propios desechos porque les ahorra dinero y es una forma responsable de manejar los recursos de la empresa.

Reciclar no es difícil, pero en los Estados Unidos sólo se recicla alrededor de un 11 por ciento de la basura, mientras que en Europa se recicla un 30 por ciento y en Japón, un 50.

Figura 20 *Una planta que convierte basura en energía puede proporcionar electricidad a muchas casas y oficinas.*

Explora

Echa un vistazo a tu escuela o a tu casa. ¿Cuántos objetos de madera o de plástico ves? ¿Cuántos pueden reutilizarse o reciclarse? ¿Cuántos están hechos de material reciclado? Haz una lista de estos objetos en tu cuaderno de ciencias.

REPASO

1. Define y explica la *conservación.*

2. Describe las tres principales formas de conservar recursos naturales.

3. **Analizar relaciones** ¿Cómo ayuda la conservación de recursos a proteger a la Tierra y a reducir la contaminación?

▼ Answers to Review

1. Conservation is the wise use and preservation of natural resources. Consuming as few manufactured products as possible and then reusing them in some way is one way to conserve natural resources.

2. The three main ways to conserve natural resources are to reduce, reuse, and recycle.

3. Conservation of resources reduces damage to the Earth by reducing the amount of resources extracted and by reducing the waste and pollution created to alter and package those resources into products people buy.

Mantener la biodiversidad

Imagina un bosque con un solo tipo de árbol. Si una enfermedad atacara a esa especie, podría acabar con todo el bosque. Ahora imagina un bosque con 10 especies distintas. Si un tipo de árbol se enfermara, aún quedarían nueve especies diferentes. Observa la **Figura 21.** El sembradío que ves produce una cosecha importante: algodón. Pero no es un ambiente muy diverso. Para que el algodón crezca bien, el agricultor debe cuidarlo con hierbicidas, plaguicidas y fertilizantes. La biodiversidad mantiene las comunidades estables de manera natural.

Experimentos

¿Qué tanta biodiversidad hay en tu rincón del mundo? Averígualo en la página 622.

Figura 21 *¿Qué pasaría si una enfermedad atacara la plantación de algodón? La biodiversidad es muy baja en sembradíos como este.*

La diversidad de especies es importante porque cada especie hace una contribución única a la comunidad. Además, hay especies muy importantes para los seres humanos. Nos dan alimentos, medicinas, plaguicidas naturales, belleza y compañía.

Protección de especies Una forma de mantener la biodiversidad es protegiendo especies individuales. En los Estados Unidos, la Ley de Protección de las Especies en Peligro de Extinción *(Endangered Species Act)* sirve justamente para esto. Las especies en peligro de extinción están en una lista especial y la ley prohíbe las actividades perjudiciales para los animales o plantas de la lista. La ley también exige que se desarrollen programas para que las poblaciones de especies en peligro se recuperen. Algunas especies en peligro se están recuperando, como el cóndor de California que ves en la **Figura 22.**

Desgraciadamente, el proceso para incluir una especie en la lista de especies en peligro de extinción es muy largo. Hay que añadir muchas especies a la lista, y muchas se extinguen antes de que esto suceda.

Figura 22 *El cóndor de California, que estaba al borde de la extinción, se ha recuperado gracias a su cuidadosa cría en cautiverio. El cóndor es un importante ave carroñera en su medio ambiente.*

451

Multicultural CONNECTION

Although Costa Rica is a small country (51,000 km²), an estimated 505,660 species live there, which is about 4 percent of all living species. In 1989, the Costa Rican government set up the National Institute of Biodiversity (Instituto Nacional de Biodiversidad, or INBio) to catalog native species and to educate residents about the importance of preserving biodiversity. Some of the collecting and cataloging is done by local people who have been trained by scientists. Invite interested students to use the Internet to find out how INBio shares information about beneficial and sustainable uses for some of the unique species found in Costa Rica.

MISCONCEPTION ALERT

In the recent past, humans caused extinction primarily by overhunting and overharvesting. Today extinction of whole groups of species is more likely to result from habitat loss and the introduction of foreign species than from hunting. Saving species today requires protecting not just the animals but the ecosystems that support them.

IS THAT A FACT!

Just outside Detroit, Michigan, looms a 50 m high former landfill once known as Mount Trashmore. From the 1970s to the mid-1990s, Mount Trashmore was used as a ski slope; it even had ski lifts! In addition, methane gas tapped from the decomposing wastes within the landfill was converted to electricity, which fulfilled the energy needs of some 2,000 households. Today the former landfill, renamed Riverview Highlands, is a recreational area offering ice skating, tubing (using recycled tractor tire tubes), and other sporting activities.

MEETING INDIVIDUAL NEEDS

Advanced Learners There are numerous volunteer activities that seek to protect and maintain animal and plant habitats. A few activities are as follows: an organized cleanup of a beach or a riverfront; tree planting; or removing alien species from a forest, prairie, or wetland preserve. Have interested students contact local environmental organizations to find out how they can help organize and lead a group of classmates in volunteer field work.

Sheltered English

CROSS-DISCIPLINARY FOCUS

Art and Design As a habitat disappears, many birds have difficulty obtaining food and shelter. Have students or groups of students research, design, and build a simple birdhouse or bird feeder. Birdhouses and bird feeders are also a good way to make use of scrap lumber.

Sheltered English

INDEPENDENT PRACTICE

Concept Mapping Have students make a concept map with the terms *reduce, reuse,* and *recycle* arranged around the phrase *Things I Can Do to Conserve Natural Resources*. Have them include as many ideas as they can on the map.

PG 624
Deciding About Environmental Issues

Figura 23 *Una forma de proteger los hábitats es reservar para la vida silvestre tierras del gobierno.*

Protección de hábitats Esperar a que una especie esté casi extinta para empezar a protegerla es como esperar a que tus dientes tengan caries para empezar a cepillártelos. Los científicos piensan que es importante proteger a las especies desde antes de que estén en peligro de extinción.

Las plantas, los animales y los microorganismos no viven separados, sino que forman parte de una enorme red de organismos interconectados. Para proteger la red y no interferir con el equilibrio global de la naturaleza, hay que preservar hábitats completos y no sólo especies individuales. Hay que proteger *todas* las especies, no sólo las que están en peligro. Todas las especies que viven en la reserva natural de la **Figura 23** están protegidas, pues su hábitat está protegido.

Estrategias

Hay leyes que sirven para proteger y conservar la naturaleza. Abajo puedes ver los objetivos de estas leyes junto con algunas estrategias que se pueden adoptar para apoyarlas.

Laboratorio

Ojo con la basura

Fíjate en la basura que produces en un día. Clasifícala en grupos. ¿Qué tanto viene de desperdicios de comida? ¿Hay algo que considerarías peligroso? ¿Qué se puede reciclar? ¿Qué se puede reutilizar? ¿Cómo puedes reducir la cantidad de basura que produces?

- **Reducir el uso de plaguicidas.**
  Sólo aplicar plaguicidas que atacan específicamente insectos dañinos. Usar plaguicidas naturales que interfieren con la forma de vida, el crecimiento y el desarrollo de ciertos insectos. Desarrollar más plaguicidas biodegradables que no perjudiquen a los pájaros, los animales o las plantas.

- **Reducir la contaminación.**
  La ley prohíbe el desecho de substancias tóxicas en ríos, arroyos, lagos, mares, tierras de cultivo y bosques.

- **Proteger los hábitats.**
  Conservar los pantanos, Reducir la tala de árboles, usar técnicas de tala respetuosas con el medio ambiente, usar los recursos a un ritmo que permita su crecimiento, y proteger hábitats enteros.

- **Aplicar la Ley de Protección de las Especies en Peligro de Extinción.**
  Acelerar el proceso para añadir especies a la lista.

- **Desarrollar fuentes alternativas de energía.**
  Usar más energía solar, eólica (del viento), y otras fuentes renovables.

452

Q: What are people who damage the habitat of endangered birds engaging in?

A: fowl play

Lo que *tú* puedes hacer.

Reduce, reutiliza y recicla. Protege tu planeta. Todos pueden ayudar; los niños y los adultos pueden ayudar a salvar la Tierra. En esta lista hay sugerencias sobre lo que *tú* puedes hacer. ¿Hay algunas que ya practicas? ¿Se te ocurre algo que añadir?

1. Compra artículos reciclables.
2. Regala tus juguetes viejos.
3. Usa papel reciclado.
4. Escribe por las dos caras de las hojas de papel.
5. Si tienes que usar platos desechables, usa platos y vasos de papel, y no de poliestireno.
6. Cierra la llave del agua mientras te lavas los dientes.
7. No compres artículos hechos con un animal en peligro de extinción.
8. Usa pilas recargables.
9. Apaga la luz, el aparato de sonido y la computadora cuando no los estés usando.
10. Usa ropa que te pasen tus familiares.
11. Usa una bolsa de tela cuando vayas de compras.
12. Usa lonchera o reutiliza tus bolsas del almuerzo.
13. Comparte libros con tus amigos o sácalos de la biblioteca.
14. Usa servilletas y trapos de cocina de tela, no de papel.
15. Recicla vidrio, papel, aluminio y pilas.
16. Haz composta.
17. Compra productos hechos de plástico biodegradable.
18. Camina, monta en bicicleta o usa el transporte público.
19. Compra productos que tengan muy poco o ningún empaque.
20. Arregla las llaves de agua que gotean.

REPASO

1. Describe por qué es importante la biodiversidad.
2. ¿Por qué es importante proteger hábitats enteros?
3. **Aplicar conceptos**   De la lista de arriba, identifica las sugerencias que requieren reducir, reutilizar o reciclar. Algunas abarcan más de una de "las tres erres".

453

---

4) Close

Quiz

1. Give one example each of how you could reduce and reuse natural resources, and one example of how you could recycle a natural resource. (Reduce: Ride a bike to a friend's house instead of getting a ride. Reuse: Buy refillable water jugs, and refill them rather than discarding them when empty; and recycle aluminum cans.)

2. How does the Endangered Species Act protect endangered organisms? (The law forbids activities that would damage any organism that is listed. It also requires the development of programs that help the endangered species recover.)

ALTERNATIVE ASSESSMENT

Writing Organize students into small groups, and have them create a "Test Your Environmental IQ" quiz. Have them use the information they learned in this section to write 10 multiple-choice questions. Have them also devise a scoring system. If possible, have students post their quizzes on a school Web site.

Reinforcement Worksheet 18
"It's 'R' Planet"

Critical Thinking Worksheet 18
"Bud Kindfellow Has a Plan"

Interactive Explorations CD-ROM "Moose Malady"

▼ Answers to Review

1. Biodiversity helps to keep biological communities naturally stable.

2. Entire habitats must be preserved and protected in order to preserve biodiversity because each organism is connected to all others in a huge web. The entire web must be protected to avoid disrupting the worldwide balance of nature.

3. 1—reduce, recycle; 2—reduce, reuse; 3—reduce, recycle; 4—reduce; 5—reduce; 6—reduce; 7—reduce; 8—reduce, reuse; 9—reduce; 10—reduce, reuse; 11—reduce, reuse; 12—reduce, reuse; 13—reduce, reuse; 14—reduce, reuse; 15—recycle; 16—recycle; 17—reduce; 18—reduce; 19—reduce; 20—reduce

VOCABULARY DEFINITIONS

SECTION 1

pollution the presence of harmful substances in the environment

pollutant a harmful substance in the environment

toxic poisonous

radioactive waste hazardous wastes that take hundreds or thousands of years to become harmless

renewable resource a natural resource that can be used and replaced over a relatively short time

nonrenewable resource a natural resource that cannot be replaced or that can be replaced only over thousands or millions of years

alien an organism that makes a home for itself in a new place

overpopulation a condition that occurs when the number of individuals becomes so large that there are not enough resources for all individuals

biodiversity the number and variety of living things

deforestation the clearing of forest lands

biodegradable capable of being broken down by the environment

Resumen del capítulo

SECCIÓN 1

Vocabulario

contaminación (*pág. 440*)
contaminante (*pág. 440*)
tóxico (*pág. 441*)
desechos radioactivos(*pág. 441*)
recurso renovable (*pág. 443*)
recurso no renovable (*pág. 443*)
organismo naturalizado (*pág. 444*)
sobrepoblación(*pág. 444*)
biodiversidad(*pág. 445*)
deforestación (*pág. 445*)
biodegradable (*pág. 446*)

Notas de la sección

- La Tierra está contaminada por desechos sólidos, substancias químicas peligrosas, materiales radioactivos, ruido y calor.

- Algunos recursos naturales de la Tierra son renovables y otros no. Algunos recursos no renovables se están agotando.

- Las especies naturalizadas frecuentemente invaden otras tierras, en las que se extienden, se convierten en plagas y amenazan la existencia de las especies nativas.

- La población humana está en peligro de crecer tanto que los recursos de la Tierra no serán suficientes para mantenerla.

- Varios factores contribuyen a la destrucción de los hábitats de la Tierra, como la deforestación, el relleno de pantanos y la contaminación.

- La deforestación puede provocar la extinción de especies, y casi siempre deja la tierra infértil.

- La contaminación del agua, el aire y el suelo puede dañar o matar animales, plantas y microorganismos.

- Los seres humanos dependen de muchas clases de organismos. La contaminación, el calentamiento global, la destrucción de los hábitats y cualquier cosa que afecte a otros organismos afectará también a los seres humanos.

☑ Comprobar destrezas

Conceptos de matemáticas

PÉRDIDA NETA DE AGUA Supón que el agua se filtra a un yacimiento subterráneo a una velocidad de 10 cm/año. El yacimiento tiene 100 m de profundidad, pero la bomba extrae unos 2 m/año. ¿Cuánto tiempo va a durar el agua?

Primero, convierte todas las medidas a centímetros.

> (100 m = 10,000 cm; 2 m = 200 cm)

Luego, calcula la pérdida neta de agua por año.

> 200 cm 10 cm = 190 cm (pérdida neta anual)

Ahora divide la profundidad del yacimiento entre la pérdida neta anual y averigua cuántos años va a durar el suministro de agua.

> 10,000 cm ÷ 190 cm = 52.6 años

Comprensión visual

QUÉ PUEDES HACER Obviamente, las estrategias para conservar los hábitats del planeta de la página 452 están siendo desarrolladas por científicos y otros profesionales. Si quieres saber que puedes empezar a hacer ahora, repasa la lista de la página 453.

454

Lab and Activity Highlights

Biodiversity—What a Disturbing Thought! PG 622

Deciding About Environmental Issues PG 624

Datasheets for LabBook (blackline masters for these labs)

SECCIÓN 2

Vocabulario

conservación *(pág. 447)*
reciclaje *(pág. 449)*
recuperación de recursos
(pág. 450)

Notas de la sección

- La conservación es el uso consciente y la conservación de los recursos naturales de la Tierra. Con ella podemos ayudar a reducir la contaminación y ahorrar recursos para las generaciones futuras.

- La conservación se puede resumir en "las tres erres": reducir, reutilizar, y reciclar. Reducir quiere decir usar menos cosas desde un principio. Reutilizar es usar productos y materiales una y otra vez. Reciclar es procesar los productos usados para fabricar unos nuevos.

- La biodiversidad es la variedad de formas de vida de la Tierra. Es muy importante para mantener los ecosistemas estables, saludables, y funcionando bien.

- Es posible proteger los hábitats si evitamos su destrucción, usamos menos plaguicidas, contaminamos menos, protegemos las especies y usamos formas de energía renovables.

- Podemos participar en la conservación del planeta si practicamos "las tres erres" en la vida diaria.

Experimentos

Biodiversidad: ¡vaya idea! *(pág. 622)*
Decisiones ambientales *(pág. 624)*

SECTION 2

conservation the wise use of and preservation of natural resources

recycling the process of making new products from reprocessed used products

resource recovery the process of transforming things normally thrown away into electricity

Vocabulary Review Worksheet 18

Blackline masters of these Chapter Highlights can be found in the **Study Guide.**

internet

HRW VISITA: go.hrw.com

Visita el sitio web de HRW para encontrar una serie de herramientas de aprendizaje relacionadas con este capítulo. Sólo tienes que escribir la palabra clave:

PALABRA CLAVE: HSTENV

SCLINKS **NSTA** VISITA: www.scilinks.org

Visita el sitio web de la **Asociación Nacional de Maestros de Ciencias** *(National Science Teachers Association)* para encontrar recursos de Internet relacionados con este capítulo. Sólo escribe el **ENLACE DE CIENCIAS** para obtener más información sobre el tema:

TEMA: Contaminación del aire	ENLACE: HSTL505
TEMA: Agotamiento de recursos	ENLACE: HSTL510
TEMA: Crecimiento de la población	ENLACE: HSTL515
TEMA: Reciclaje	ENLACE: HSTL520
TEMA: Conservación de la biodiversidad	ENLACE: HSTL525

455

Lab and Activity Highlights

LabBank

EcoLabs & Field Activities, A Filter with Culture, EcoLab 8

Long-Term Projects & Research Ideas, Project 21

Interactive Explorations CD-ROM

CD 2, Exploration 3, "Moose Malady"

USING VOCABULARY

1. Pollution
2. Radioactive waste
3. nonrenewable
4. Biodiversity
5. Recycling

UNDERSTANDING CONCEPTS

Multiple Choice

6. d
7. b
8. c
9. b
10. c
11. b

Short Answer

12. ride a bicycle, recycle cans and bottles, wear secondhand clothing, use rechargeable batteries, and others

13. Alien species thrive without their native predators and often crowd out native organisms. Sometimes the habitat of a rare species is taken over by an alien species, causing the native species to become endangered.

Concept Mapping Transparency 18

Blackline masters of this Chapter Review can be found in the **Study Guide.**

Repaso del capítulo

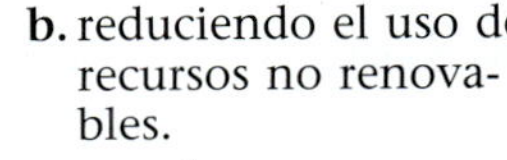

Escoge el término correcto para completar las siguientes oraciones:

1. La ___?___ es la presencia de substancias dañinas en el medio ambiente. *(contaminación* o *biodiversidad)*

2. ___?___ son un tipo de contaminación producida por las centrales de energía nuclear. *(Los CFC* o *Los desechos radioactivos)*

3. Un recurso ___?___ sólo se puede usar una vez. *(nuclear* o *no renovable)*

4. ___?___ es la variedad de formas entre los seres vivientes. *(Biodegradable* o *Biodiversidad)*

5. Cuando ___?___, procesas la basura para hacer un nuevo producto. *(reciclas* o *reutilizas)*

Opción múltiple

6. La protección de los hábitats es importante porque
a. los organismos no viven aislados.
b. es una forma de proteger las especies.
c. si no se practica, se rompe el equilibrio de la naturaleza.
d. todas las anteriores

7. Los recursos de la Tierra se pueden conservar
a. sólo en las actividades industriales.
b. reduciendo el uso de recursos no renovables.
c. si todos hacemos lo que queremos.
d. si se desecha toda la basura.

8. Las especies en peligro
a. son las que están extintas.
b. se encuentran sólo en la selva tropical.
c. pueden recuperarse después de estar casi extintas.
d. están protegidas por la Ley de Protección de las Especies en Peligro de Extinción.

9. El calentamiento global amenaza
a. sólo a gente que vive en climas cálidos.
b. a todos los organismos del planeta.
c. sólo a los organismos de los polos.
d. la cantidad de dióxido de carbono en el aire.

10. La superpoblación
a. no existe en la raza humana.
b. ayuda a disminuir la contaminación.
c. ocurre cuando una especie no puede obtener los alimentos, agua y demás recursos necesarios.
d. sólo existe en las grandes ciudades.

11. La biodiversidad
a. no es importante para la ciencia.
b. mantiene los ecosistemas estables.
c. hace que las enfermedades destruyan poblaciones enteras.
d. sólo se encuentra en el bosque templado.

Respuesta breve

12. Describe cómo ayudarías a conservar recursos. Menciona estrategias que incluyan "las tres erres".

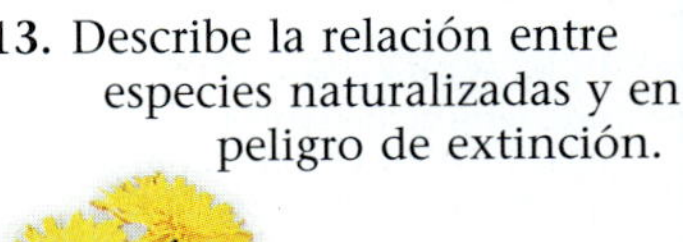

13. Describe la relación entre especies naturalizadas y en peligro de extinción.

Organizar conceptos

14. Usa los siguientes términos para crear un mapa de ideas: contaminación, contaminantes, CFC, cáncer, PCB, tóxico, desechos radioactivos, calentamiento global.

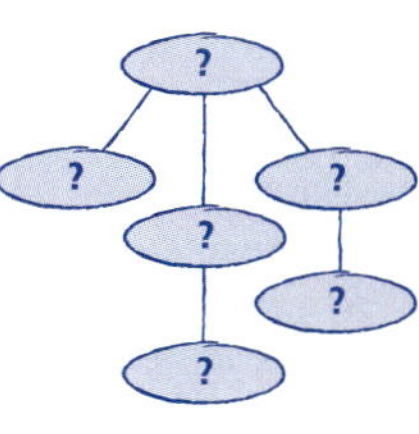

RAZONAMIENTO CRÍTICO Y RESOLUCIÓN DE PROBLEMAS

Escribe una o dos oraciones para responder a las siguientes preguntas:

15. Supón que la reserva de combustibles fósiles se fuera a agotar en 10 años. ¿Y si no estuviéramos preparados? ¿Cómo podríamos prepararnos para algo así?

LAS MATEMÁTICAS EN LAS CIENCIAS

16. Si cada persona de una ciudad con 150,000 habitantes tira 12 Kg de basura a la semana, ¿cuántas toneladas métricas producirá la ciudad al año? (Hay 52 semanas en el año, y 1,000 kg en una tonelada métrica.)

INTERPRETAR GRÁFICAS

La ilustración de arriba muestra cómo los habitantes de una casa usan recursos naturales.

17. Identifica formas en las que están desperdiciando recursos naturales. Da por lo menos tres ejemplos y di qué se podría hacer para conservar recursos.

18. Di cuáles de estos recursos son renovables.

19. Identifica cualquier fuente de desechos tóxicos.

20. Explica por qué la chica con los audífonos contribuye a reducir la contaminacón del aire. ¿Cómo podría hacerle daño esta actividad?

AHORA, ¿qué piensas?

Revisa tus respuestas a las preguntas de la página 439 que escribiste en el cuaderno de ciencias. ¿Han cambiado tus respuestas? Si es necesario, corrige tus respuestas basándote en lo que has aprendido en este capítulo.

457

Concept Mapping

14. An answer to this exercise can be found at the end of this book.

CRITICAL THINKING AND PROBLEM SOLVING

15. We would have severe transportation problems, and most people would not be able to get to work or school. Some places would not have electricity or gas for heating, cooling, or refrigeration. Our survival might even be threatened if needed supplies, such as food and medicine, could not be transported and delivered. We could prepare for such an event by developing alternative sources of fuel, such as solar energy.

MATH IN SCIENCE

16. 93,600 metric tons per year

INTERPRETING GRAPHICS

17. The water is running while the boy is brushing his teeth. He should turn off the water while he brushes. The water sprinklers are running while it is raining. The sprinklers should be turned off; the rain will water the lawn. A boy is throwing away a plastic container that can be recycled.

18. The water is a renewable resource as long as it is replenished faster than it is used.

19. If the girl is not using rechargeable batteries in the radio, the batteries are a source of hazardous waste. The father is pouring oil from the car into the gutter, where it will be a hazardous pollutant.

20. Headphones help reduce noise pollution. If she has the volume too high, it could damage her hearing.

NOW What Do You Think?

1. People damage the Earth by creating harmful pollution and hazardous waste and releasing it into the environment. People reduce biodiversity by clearing forests and filling in wetlands. People also overuse nonrenewable resources and deplete the Earth of certain minerals and water sources.

2. People are trying to prevent further damage to the Earth by reducing our dependence on natural resources, by reusing many of the materials we would otherwise throw away, and by recycling materials that can be reprocessed for other purposes, such as aluminum, glass, plastic, and paper.

Background

The green iguana (*Iguana iguana*) is native to Central America, where it was used by native people as a food source. Iguana meat is tasty, is high in protein, and has a relatively low fat content.

The green iguana population in Costa Rica has declined because people have hunted iguana eggs and adults and because so much of the iguana's habitat has been destroyed. Iguanas are tree-dwelling lizards, so they require a forest habitat to survive.

Teaching Strategies

Ask students why iguana farming might provide Central Americans with environmental benefits. Lead them to conclude that since iguanas are adapted to the forest that is already there, farming them requires preserving areas of forest. Preserving forest and reforesting cleared areas in turn helps preserve fragile tropical forest soils and helps preserve the variety of other creatures that make their home in the forest.

PROFESIONES

BIÓLOGA

Dagmar Werner trabaja en la reserva biológica de Carara, en Costa Rica, protegiendo a la iguana verde. Las iguanas estaban casi extintas debido a la caza, la contaminación y la destrucción de su hábitat. Desde los años 1980, ha criado iguanas en cautiverio para mejorar sus probabilidades de supervivencia y puesto en libertad a miles de iguanas jóvenes. También se dedica a formar a otras personas para realizar esta labor.

En su "rancho de iguanas", Dagmar Werner coloca nidos artificiales para que las hembras pongan sus huevos. Al salir del huevo, las pequeñas iguanas crecen en una incubadora a temperatura y humedad controladas, y reciben una dieta especial. Estas iguanas son más fuertes, crecen más rapido y están mejor protegidas de los depredadores que las salvajes. Normalmente, menos del 2 por ciento de las iguanas alcanzan la edad adulta; las de Werner tienen una tasa de supervivencia del 80 por ciento. Tras liberarlas, Werner mantiene un seguimiento y un registro para determinar si se han adaptado con éxito a su nuevo medio, donde el ambiente no está tan controlado.

Pollo de los árboles

Como las iguanas estaban al borde de la extinción, la estrategia de Werner tenía que ser drástica e inmediata si quería salvarlas, así que combinó su programa de reproducción en cautiverio con un programa educativo para enseñar a los campesinos cómo ganar dinero en la selva. En vez de talar los bosques tropicales para criar ganado, les aconseja que críen iguanas, que pueden ser puestas en libertad o vendidas como alimento. Conocidas como "pollos de los árboles", las iguanas han sido durante miles de años un plato favorito de los habitantes de la selva tropical.

Con este método, los campesinos pueden proteger a las iguanas y ganarse la vida. Pero convencer a los campesinos no ha sido fácil. "Muchos lugareños nunca han pensado en los animales salvajes como seres que necesitan protección para sobrevivir", dice Werner. Como apoyo a su programa, estableció la Fundación Pro Iguana Verde, que organiza festivales y seminarios educativos. Estas actividades hacen propaganda al atractivo tradicional de la iguana, que la gente esté orgullosa del animal y acentúan su importancia económica.

Encuentra otras soluciones

▶ La iguana verde es sólo uno de los animales de la selva tropical en peligro de extinción. Investiga qué otras especies están en peligro, y qué se está haciendo para protegerlas. ¿Te parece que funciona bien?

▶ *Una iguana verde en la reserva biológica de Carara, en Costa Rica*

458

Answer to Find Other Solutions

Students' answers will vary.

¿Dónde vivirá el lobo?

El Servicio de Vida Silvestre y Peces de los Estados Unidos *(U.S. Fish and Wildlife Service)* ha incluido al lobo gris como especie en peligro de extinción en casi toda la nación y tiene un plan para reintroducir lobos en el Parque Nacional de Yellowstone, en la zona central de Idaho y en el noroeste de Montana. El objetivo es establecer una población de al menos 100 lobos en cada uno de estos lugares. Si el proyecto continúa según los planes, los lobos podrían estar fuera de la lista para el año 2002. Pero el plan inquieta a algunos rancheros y cazadores, y les parece insuficiente a varios ecologistas y admiradores de los lobos.

¿Está en peligro el ganado?

A los rancheros les preocupa que los lobos puedan atacar el ganado. Sus pérdidas económicas podrían ser enormes. Actualmente existe un programa de compensación que cubriría las pérdidas si los lobos atacan el ganado. Pero el programa termina cuando los lobos dejen de estar en la lista de especies en peligro de extinción. Los rancheros observan que la amenaza para su ganado no acaba cuando el lobo deja de estar en peligro de extinción; al contrario, la amenaza en ese momento será mayor, pero dejarán de recibir compensaciones.

Por otro lado, algunos biólogos tienen pruebas de que los lobos de áreas con poblaciones suficientemente grandes de venado, uapití, alce y otras presas, no atacan al ganado. De hecho, entre 1995 y 1997, se reportaron menos de cinco ataques de lobos al ganado.

¿Está en peligro la vida silvestre?

Muchos científicos piensan que reintroducir a los lobos estabilizará estas regiones por primera vez en 60 años. En su opinión, eliminarán a los venados, alces y uapitís que ya están muy viejos o débiles y evitarán que sus poblaciones crezcan demasiado.

Los cazadores temen que los lobos puedan acabar con muchos animales de caza en estas regiones y mencionan estudios que indican que las poblaciones grandes de animales no pueden sobrevivir el ataque conjunto de lobos y humanos. La caza es una actividad importante para la economía de estas regiones.

Y nosotros, ¿estamos a salvo?

Algunas personas tienen miedo de que los lobos ataquen a las personas. Sin embargo, en Norteamérica jamás se han documentado ataques de lobos sanos a personas. Quienes apoyan el programa dicen que los lobos son tímidos, que prefieren mantener su distancia de los seres humanos, aunque admiten que hay lugares donde los lobos pueden vivir sin causar problemas y otros en los que su presencia no es buena. Creen que la reintroducción de lobos a estas zonas hará que los lobos vivan bien sin crear problemas.

¿Tú qué crees?

▶ Hay quien cree que la mala fama del lobo viene de los cuentos del "lobo feroz". ¿Crees que estos temores se basan en mitos, o piensas que el lobo puede ser peligroso para las personas y el ganado en las zonas de reintroducción? Investiga el tema y da ejemplos para respaldar tu opinión.

◀ *Un lobo gris en Montana*

Background

In the 1920s, the gray wolf was exterminated from much of the northwestern United States. Ranchers and federal agents killed the animal to protect livestock, which sometimes fell prey to the wolf. Those in favor of wolf reintroduction cite biologists' claims that wolf attacks on livestock are neither as widespread nor as serious as is generally believed. Some opponents of the reintroduction plan argue that wolves should not be classified as endangered at all. According to data from biologists, there are 1,500 to 2,000 wolves in Minnesota, 6,000 to 10,000 in Alaska, and 40,000 to 50,000 in Canada. With such numbers, many people feel that the animal should not receive the special treatment given to endangered species.

Activity

Research Have interested students research the latest news on the wolf reintroduction issue and present a report to the class.

Answer to What Do You Think?
Students' answers might vary, but most of the fears associated with wolves are based on myth.

UNIDAD 7

Los sistemas del cuerpo humano

Tu cuerpo está formado por varios sistemas que trabajan juntos como una máquina bien afinada. Los pulmones toman oxígeno; el corazón bombea sangre que distribuye oxígeno a los tejidos. El cerebro reacciona a cosas que ves, escuchas y hueles, y envía señales a través del sistema nervioso que provocan que reacciones a esas cosas. El aparato digestivo convierte los alimentos en energía que las células del cuerpo utilizan. ¡Estas son sólo unas cuantas cosas que el cuerpo puede hacer! En esta unidad, estudiarás los sistemas del cuerpo. Descubrirás cómo las partes del cuerpo se unen para que puedas realizar todas tus actividades diarias.

3000 a.C.

Los antiguos egipcios son los primeros en estudiar el cuerpo humano de manera científica.

1824

Prevost y Dumas prueban que los espermatozoides son esenciales para la fertilización.

1893

Daniel Hale Williams, un cirujano afro-americano fue el primero en reparar un desgarre en el pericardio, la bolsa que envuelve el corazón.

1930

Karl Landsteiner recibe el Premio Nobel por su descubrimiento de los cuatro tipos de sangre humana.

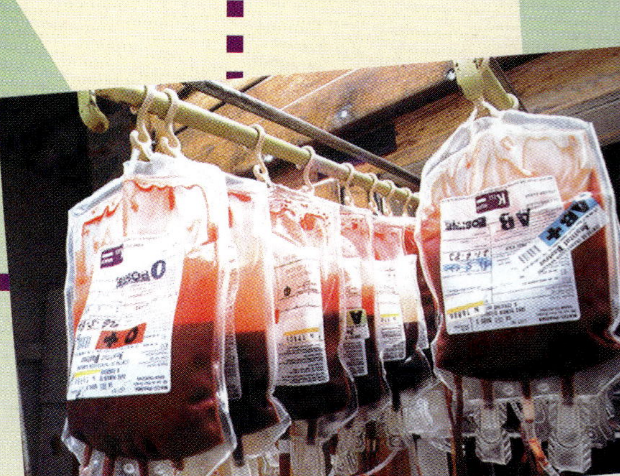

1922

Se descubre la insulina.

500 a.C.
El cirujano indio Susrata realiza operaciones para eliminar las cataratas de los ojos.

1492
Cristóbal Colón llega a las Antillas.

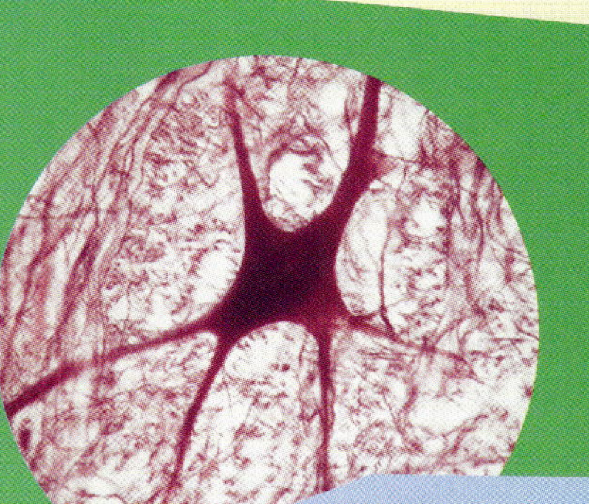

1543
Andreas Versalius publica la primera descripción completa de la estructura del cuerpo humano.

1766
Albrecht von Haller determina que los nervios controlan el movimiento muscular y que todos los nervios están conectados a la médula espinal o al cerebro.

1619
William Harvey descubre que la sangre circula y que el corazón funciona como una bomba.

1982
El doctor Robert Jarvick transplanta un corazón artificial a Barney Clark.

1941
Durante la Segunda Guerra Mundial, en Italia, Rita Levi-Montalcini se ve forzada a dejar su trabajo en el laboratorio de una facultad de medicina por ser judía. Improvisa un laboratorio en su recámara y estudia el desarrollo del sistema nervioso.

1998
El primer transplante de mano se hace en Francia.

Chapter Organizer

CHAPTER ORGANIZATION	TIME MINUTES	OBJECTIVES	LABS, INVESTIGATIONS, AND DEMONSTRATIONS
Chapter Opener pp. 462–463	45		**Investigate!** Too Cold for Comfort, p. 463
Section 1 Body Organization	90	▶ Identify the major tissues found in the body. ▶ Compare an organ with an organ system. ▶ Describe a major function of each organ system.	
Section 2 The Skeletal System	90	▶ Identify the major organs of the skeletal system. ▶ Describe the functions of bones. ▶ Illustrate the internal structure of bones. ▶ Compare three types of joints. ▶ Discuss how bones function as levers.	**QuickLab,** Pickled Bones, p. 469 **Demonstration,** Bone Dissection, p. 469 in ATE
Section 3 The Muscular System	90	▶ List the major parts of the muscular system. ▶ Describe the different types of muscle. ▶ Describe how skeletal muscles move bones. ▶ Compare aerobic exercise with resistance exercise. ▶ Give an example of a muscle injury.	**QuickLab,** Power in Pairs, p. 473 **Design Your Own,** Muscles at Work, p. 626 **Datasheets for LabBook,** Muscles at Work, Datasheet 40 **Inquiry Labs,** On a Wing and a Layer, Lab 6
Section 4 The Integumentary System	90	▶ Describe the major functions of the integumentary system. ▶ List the major parts of the skin, and discuss their functions. ▶ Describe the structure and function of hair and nails. ▶ Describe some common types of damage that can affect skin.	**Skill Builder,** Seeing Is Believing, p. 627 **Datasheets for LabBook,** Seeing Is Believing, Datasheet 41 **Long-Term Projects & Research Ideas,** Project 22

TECHNOLOGY RESOURCES

Guided Reading Audio CD
English or Spanish, Chapter 19

One-Stop Planner CD-ROM with Test Generator

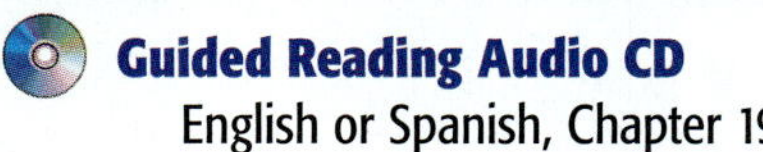

Scientists in Action, Segments 8 and 27
Science, Technology & Society, Manufactured Body Parts, Segment 27
Multicultural Connections, African-American Burial Ground, Segment 5

CLASSROOM WORKSHEETS, TRANSPARENCIES, AND RESOURCES	SCIENCE INTEGRATION AND CONNECTIONS	REVIEW AND ASSESSMENT
Directed Reading Worksheet 19 **Science Puzzlers, Twisters & Teasers,** Worksheet 19		
Directed Reading Worksheet 19, Section 1 **Transparency 71,** Organ Systems	**Cross-Disciplinary Focus,** p. 465 in ATE	**Review,** p. 467 **Quiz,** p. 467 in ATE **Alternative Assessment,** p. 467 in ATE
Directed Reading Worksheet 19, Section 2 **Transparency 72,** What's in a Bone? **Transparency 186,** Machines Change the Size or Direction (or Both) of a Force **Math Skills for Science Worksheet 53,** Mechanical Advantage **Reinforcement Worksheet 19,** The Hipbone's Connected to the… **Critical Thinking Worksheet 19,** The Tissue Engineering Debate	**Multicultural Connection,** p. 469 in ATE **Connect to Physical Science,** p. 470 in ATE	**Review,** p. 471 **Quiz,** p. 471 in ATE **Alternative Assessment,** p. 471 in ATE
Teaching Transparency 73, Types of Muscle **Directed Reading Worksheet 19,** Section 3 **Math Skills for Science Worksheet 31,** The Unit Factor and Dimensional Analysis **Reinforcement Worksheet 19,** Muscle Map	**Chemistry Connection,** p. 474 **Math and More,** p. 474 in ATE **MathBreak,** Runner's Time, p. 475	**Self-Check,** p. 474 **Homework,** p. 474 in ATE **Review,** p. 475 **Quiz,** p. 475 in ATE **Alternative Assessment,** p. 475 in ATE
Teaching Transparency 74, The Skin **Directed Reading Worksheet 19,** Section 4	**Real-World Connection,** p. 478 in ATE **Science, Technology, and Society:** Engineered Skin, p. 484 **Eureka!** Hairy Oil Spills, p. 485	**Self-Check,** p. 477 **Review,** p. 479 **Quiz,** p. 479 in ATE **Alternative Assessment,** p. 479 in ATE

Holt, Rinehart and Winston On-line Resources

go.hrw.com

For worksheets and other teaching aids related to this chapter, visit the HRW Web site and type in the keyword: **HSTBD1**

National Science Teachers Association

www.scilinks.org

Encourage students to use the *sci*LINKS numbers listed with the Chapter Highlights to access information and resources on the **NSTA** Web site.

END-OF-CHAPTER REVIEW AND ASSESSMENT

Chapter Review in Study Guide

Vocabulary and Notes in Study Guide

Chapter Tests with Performance-Based Assessment, Chapter 19 Test

Chapter Tests with Performance-Based Assessment, Performance-Based Assessment 19

Concept Mapping Transparency 19

Chapter Resources & Worksheets

Visual Resources

TEACHING TRANSPARENCIES

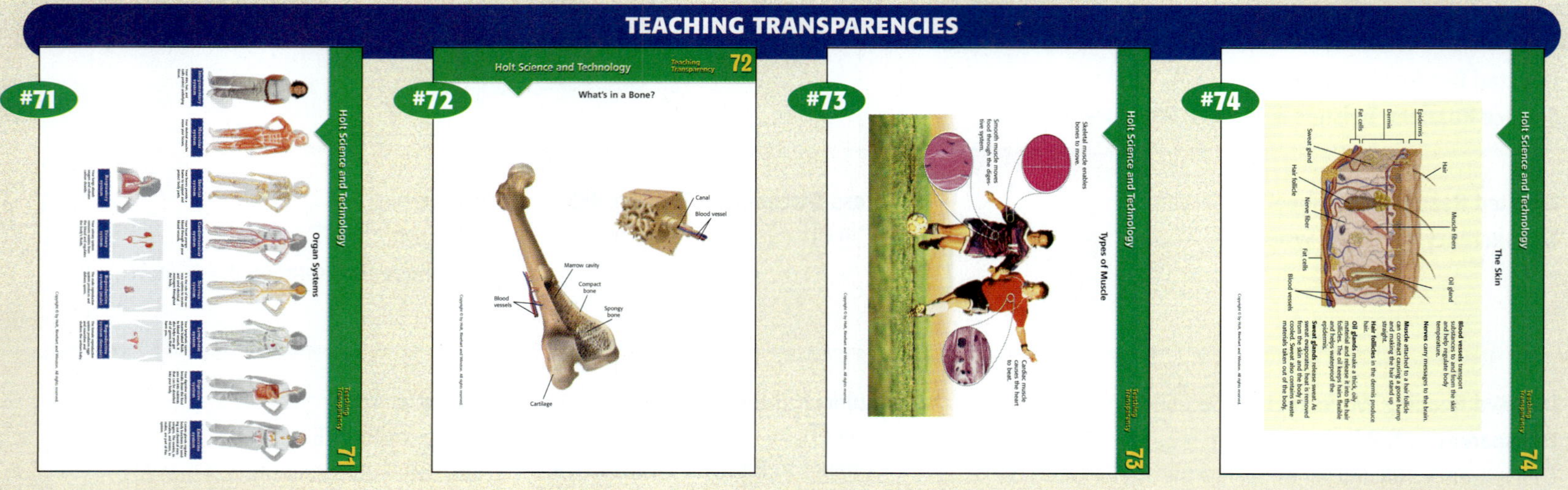

TEACHING TRANSPARENCIES

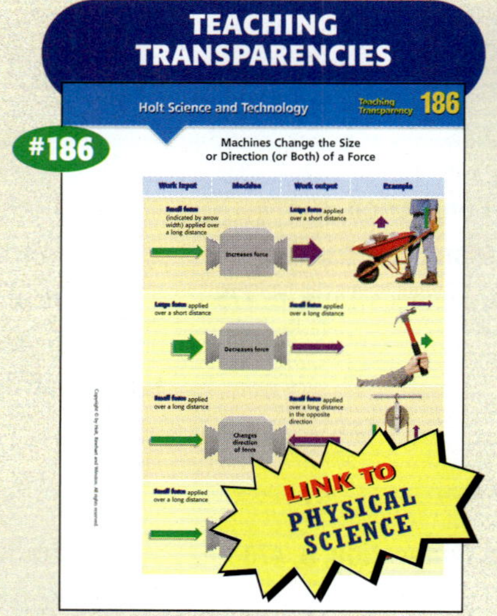

CONCEPT MAPPING TRANSPARENCY

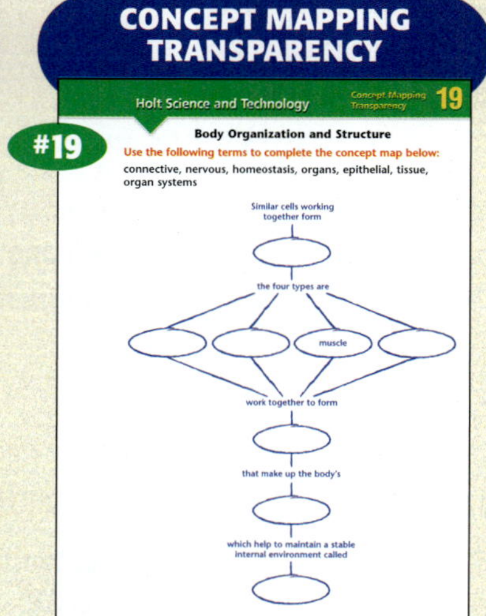

Meeting Individual Needs

DIRECTED READING

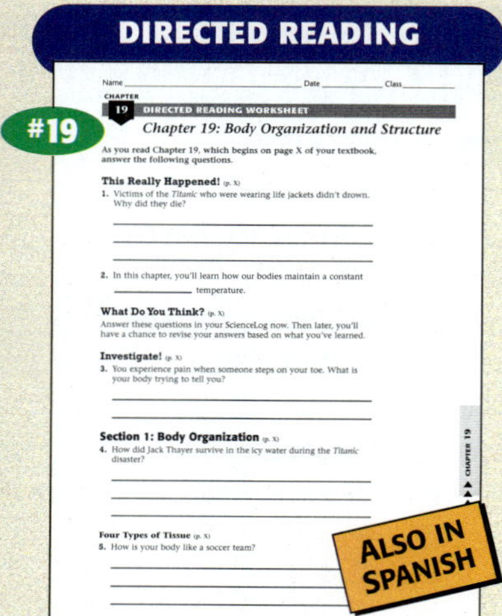

REINFORCEMENT & VOCABULARY REVIEW

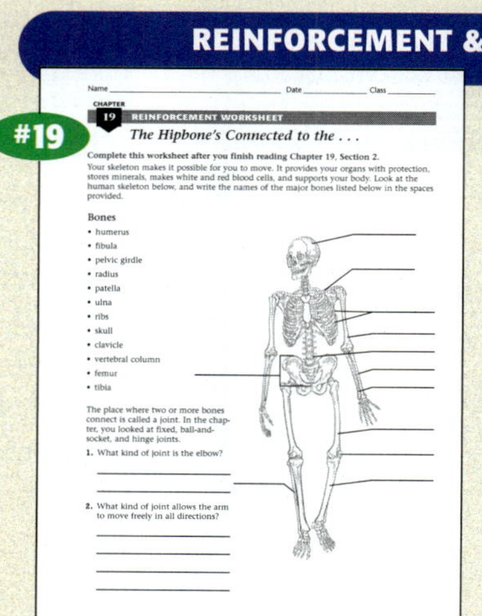

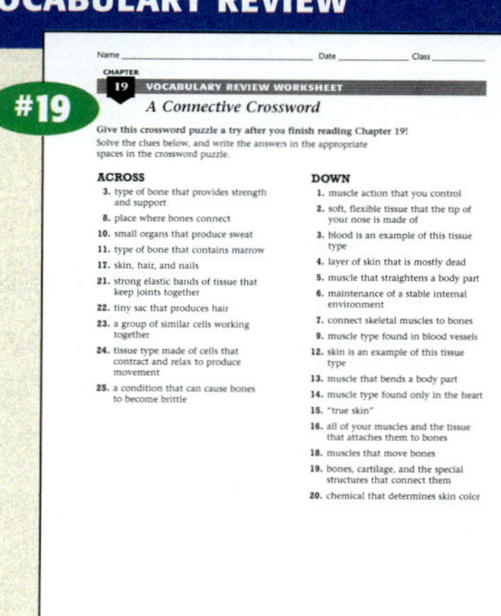

SCIENCE PUZZLERS, TWISTERS & TEASERS

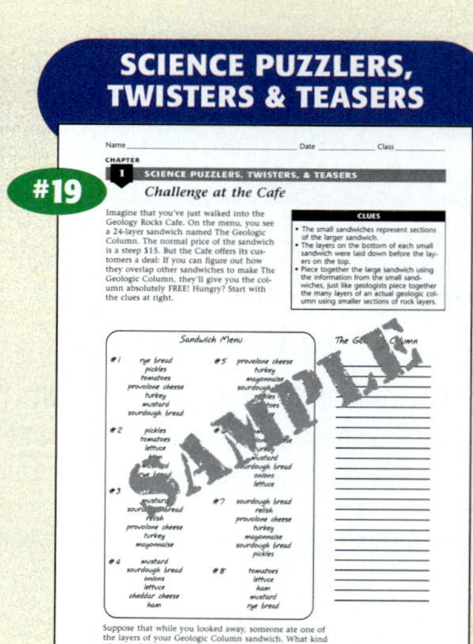

461C   Chapter 19 • Body Organization and Structure

Review & Assessment

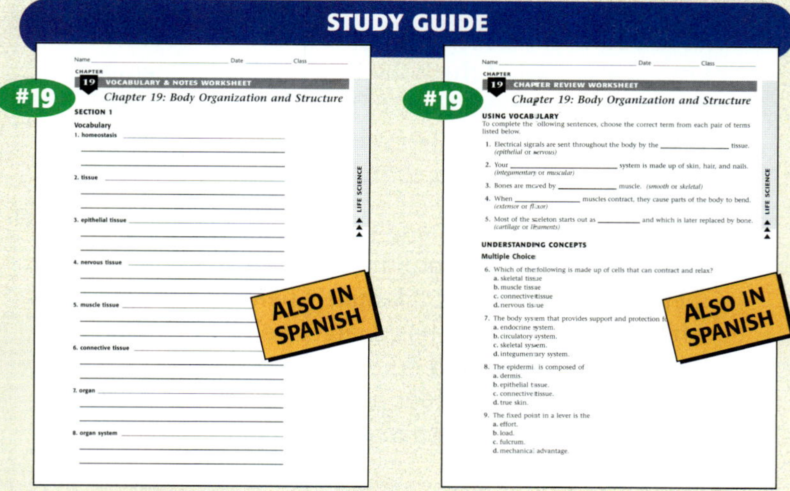

STUDY GUIDE

#19 VOCABULARY & NOTES WORKSHEET
Chapter 19: Body Organization and Structure

#19 CHAPTER REVIEW WORKSHEET
Chapter 19: Body Organization and Structure

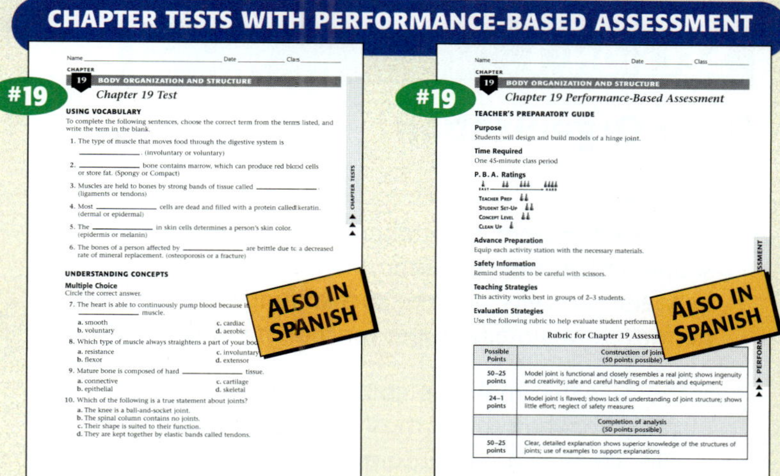

CHAPTER TESTS WITH PERFORMANCE-BASED ASSESSMENT

#19 BODY ORGANIZATION AND STRUCTURE
Chapter 19 Test

#19 BODY ORGANIZATION AND STRUCTURE
Chapter 19 Performance-Based Assessment

Lab Worksheets

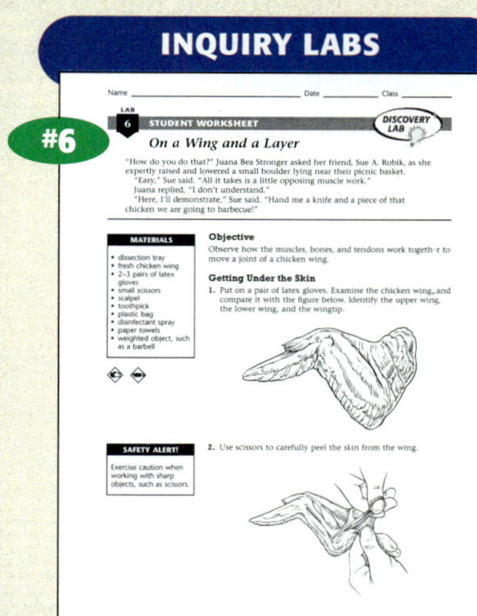

INQUIRY LABS

#6 STUDENT WORKSHEET
On a Wing and a Layer
DISCOVERY LAB

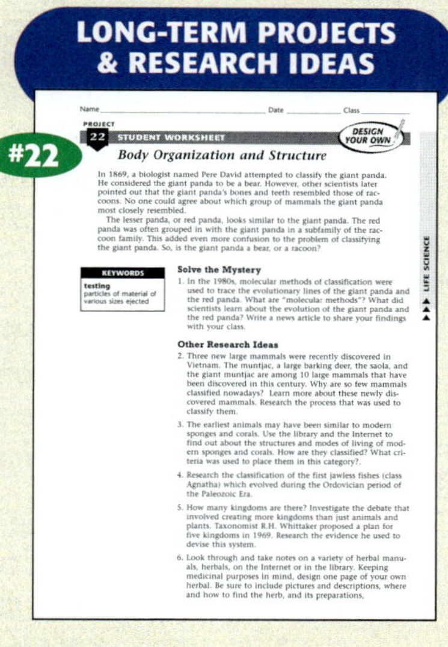

LONG-TERM PROJECTS & RESEARCH IDEAS

#22 STUDENT WORKSHEET
Body Organization and Structure
DESIGN YOUR OWN

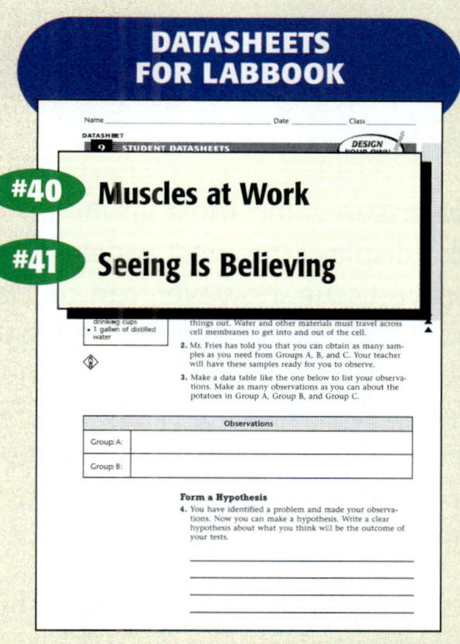

DATASHEETS FOR LABBOOK

STUDENT DATASHEETS
DESIGN YOUR OWN

#40 Muscles at Work

#41 Seeing Is Believing

Applications & Extensions

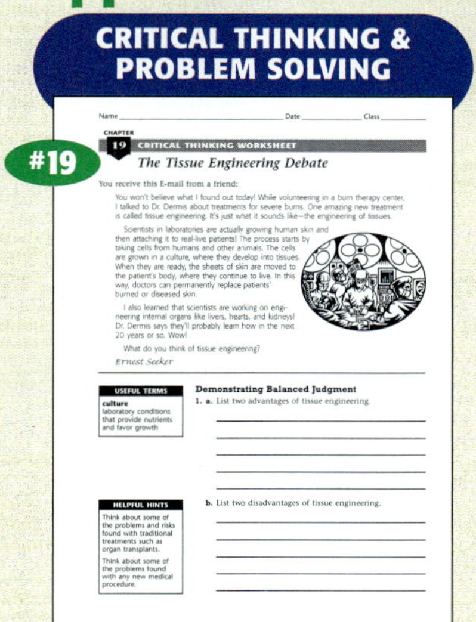

CRITICAL THINKING & PROBLEM SOLVING

#19 CRITICAL THINKING WORKSHEET
The Tissue Engineering Debate

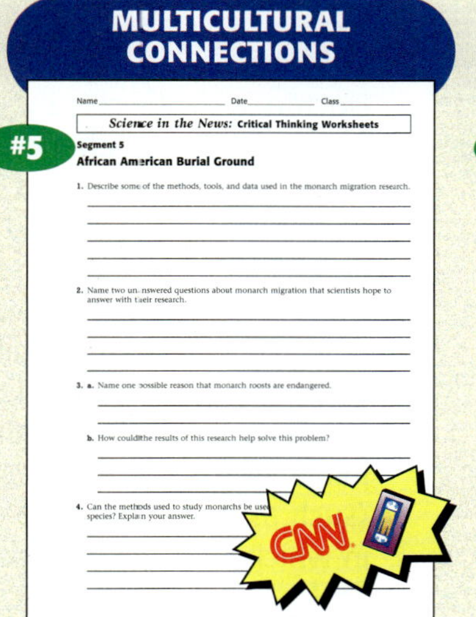

MULTICULTURAL CONNECTIONS

#5 Science in the News: Critical Thinking Worksheets
Segment 5
African American Burial Ground

SCIENCE TECHNOLOGY

#27 Science in the News: Critical Thinking Worksheets
Segment 27
Manufactured Body Parts

SCIENTISTS IN ACTION

#8 Science in the News: Critical Thinking Worksheets
Segment 8
Force in the Circus

#27

Chapter Background

SECTION 1

Body Organization

▶ Tissues

Tissues differ from each other in a number of ways: shape and size of cells, amount and kind of material between the cells, and the special functions they perform to maintain proper functioning of the body.

- The cells that make up epithelial tissue have little space between them; most are welded to adjacent cells, creating a barrier to movement of materials between them; they form continuous sheets of tissue.

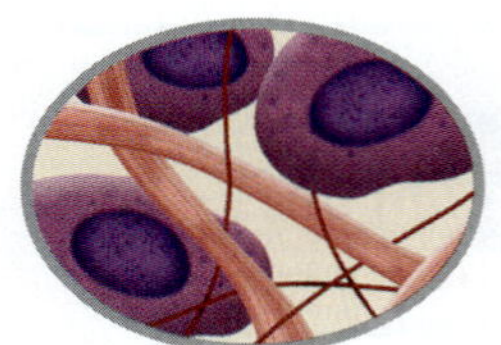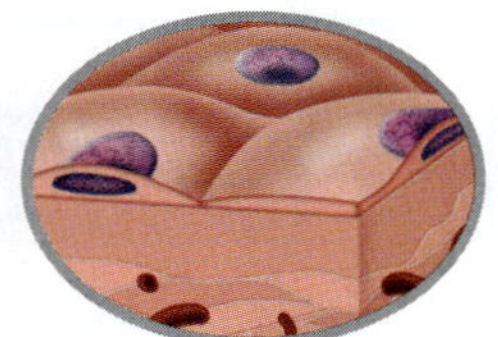

- Connective tissue is the most abundant tissue in the body, and it displays the most variety in form and type. All connective tissue, however, can be classified into one of four types: dense connective tissue—cartilage and bone; loose connective tissue—beneath the skin and around nerves, blood vessels, and body organs; liquid connective tissue—blood and lymph; and fat (adipose) tissue, where the body stores energy as droplets of fat.

- Although bone is considerably harder than other body tissues, it accounts for only about 14 percent of a person's total body weight.

SECTION 2

The Skeletal System

▶ The Human Skeleton

The skeleton provides more than just support for soft tissue. It is key to the regulation of body minerals. It also produces both red and white blood cells. There are officially 206 bones in the adult human, but extra bones, particularly those in the hands and feet, increase that number. The number of bones in children varies with age.

This is because the skeleton forms from over 800 centers of ossification. All of the bony elements are generally not completely united to form an adult skeleton until a person reaches his or her mid-20s. The clavicles are usually the last bones to complete fusion.

- The skeletons of male and female humans are slightly different. The most pronounced differences are in the pelvis. That is because a female's pelvis is adapted for childbearing and thus has a larger pelvic inlet. Women who are malnourished during their childhood typically do not develop the wider pelvis. This can make childbirth dangerous or even fatal for them.

- It is possible to determine an individual's age by looking at the skeleton alone. A younger individual's dentition and bone fusion patterns are indicative of his or her age. In adults, age determination is much more difficult because one must rely solely on signs of skeletal deterioration.

▶ Bones

Each bone is surrounded by a strong fibrous covering called a periosteum. Articular surfaces are also covered in cartilage.

- Bones are made of three types of cells: osteoblasts, osteocytes, and osteoclasts. Osteoblasts are bone-producing cells. Osteocytes are bone-maintaining cells. Osteoclasts are bone-destroying cells.

- For its weight, bone is five times stronger than steel.

▶ Joints

Doctors typically classify joints by structure rather than movement. The three types of joint structures are called fibrous, cartilaginous, and synovial. Fibrous joints are immovable joints (such as those in the skull) in which a fibrous tissue or a hyaline cartilage connects the bones. Cartilaginous joints are slightly moveable joints (such as those in the rib cage) in which cartilage connects the bones. Synovial joints are freely moving joints (such as the knee) in which slick synovial membranes cover the cartilage and ligaments connecting the bone.

SECTION 3

The Muscular System

▶ The Muscular System

There are more than 600 skeletal muscles in the human body. To simplify the study of these muscles, they are organized into the following groups: muscles of the head and the neck, muscles of the trunk, muscles of the upper limbs (arms and hands), and muscles of the lower limbs (legs and feet).

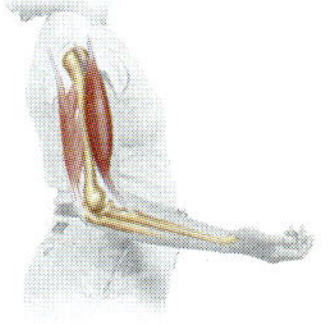 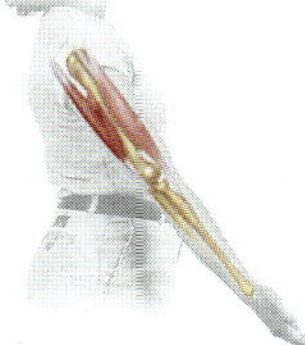

▶ Types of Muscle Cells

When observed through a microscope, the three types of muscles are clearly identifiable. Cells of smooth muscles have a long tapered shape, no clearly defined striations, and a large central nucleus. Skeletal muscle cells are characterized by distinct light- and dark-colored bands across the long tapered cell; each cell has multiple nuclei because several cells merge and the cell membranes become indistinct. The cells of cardiac muscle have one or more nuclei and have an irregular, branched shape.

SECTION 4

The Integumentary System

▶ The Skin

One square inch of the skin can hold as many as 650 sweat glands, 20 blood vessels, and more than 1,000 nerve endings.

- Each person has a unique series of ridges and indentations on the tips of his or her fingers called

fingerprints. No two people have the same fingerprints. Fingerprints help the fingers to grip slippery surfaces. Toes also have a unique pattern on their tips, so along with fingerprints, we all have toe prints as well.

IS THAT A FACT!

- ◤ More than three-quarters of the dust in some homes is made up of dead skin cells!

▶ Hair and Nails

Only mammals have true hair; all mammals have hair somewhere on their bodies.

- The body's most visible signs of aging occur in the integumentary system. Skin becomes thin, dry, wrinkled, and less supple. Dark-colored age spots may develop. Hair turns gray or white and may begin to fall out; hair follicles decrease in number. Sweat glands become less active, causing older people to be less tolerant and adaptable to extremely hot weather. 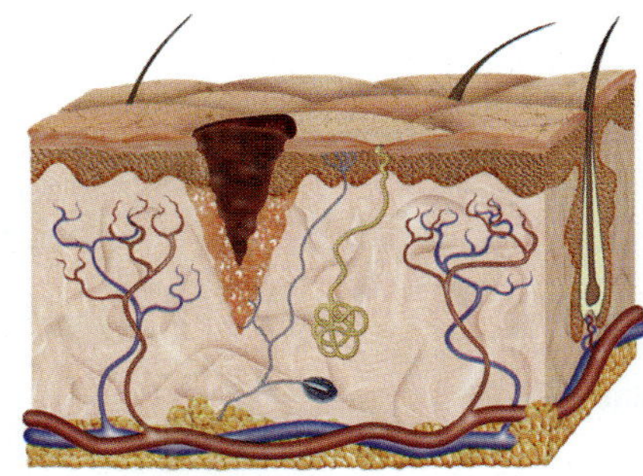

- Hair that is kept short grows an average of 2 cm per month. Growth slows to about 1 cm per month when the hair reaches about 30 cm long. Fingernails grow about 2 cm each year. The fastest-growing nail is on the middle finger. Toenails grow three to four times more slowly than fingernails.

For background information about teaching strategies and issues, refer to the *Professional Reference for Teachers.*

Directed Reading Worksheet 19

Science Puzzlers, Twisters & Teasers Worksheet 19

Guided Reading Audio CD
English or Spanish, Chapter 19

CAPÍTULO

19
La organización y estructura del cuerpo

¡Esto realmente sucedió. . . !

El 14 de abril de 1912, a las 11:40 p.m., el buque de vapor británico *Titanic,* el barco más grande y más lujoso nunca antes construido, chocó con un iceberg. Jack Thayer, un joven de 17 años de Pennsylvania, sintió el impacto en su camarote y fue a cubierta para ver qué había pasado. Para su sorpresa, el barco se hundía y no había suficientes barcos salvavidas para los 2,207 pasajeros y la tripulación. Jack vio como los últimos botes salvavidas, llenos de mujeres y niños, se echaban al agua. Justo antes de que el barco se hundiera en las aguas heladas del mar, Jack saltó. El dolor producido por el frío era como cuchillos clavados en su cuerpo. Nadó hacia un bote salvavidas volcado y se subió a él.

Jack Thayer estaba entre los supervivientes que lograron manterse lo bastante secos para sobrevivir hasta que llegó la ayuda, a las 4 a. m. Unas 1,500 personas murieron. Hasta los que traían salvavidas no pudieron sobrevivir a la temperatura extrema del agua. El agua congelada hizo fallar los sistemas del cuerpo. Apenas podían respirar y los músculos ya no funcionaban. Finalmente, perdieron el conocimiento y su corazón dejó de latir.

El agua alrededor del *Titanic* era demasiado fría para que una persona sobreviviera. En situaciones normales, nuestro cuerpo mantiene una temperatura interna constante. Sigue leyendo para descubrir más detalles interesantes sobre cómo trabaja nuestro cuerpo.

462

This Really Happened...

Though human beings require a fairly constant body temperature to maintain health, the Arctic ground squirrel can survive temperature extremes that would kill a human. Inside their burrows, the winter temperature drops below −17°C. During hibernation, the squirrel's body temperature drops from 37°C to −3°C. An unknown process prevents the squirrel's body fluids from crystallizing and damaging cells.

Usa tus conocimientos para responder a las siguientes preguntas en tu cuaderno de ciencias:

1. ¿Qué relación hay entre células, tejidos y órganos?

2. ¿Cómo ayudan la piel, los músculos y los huesos a que te conserves bien?

Demasiado frío para soportarlo

¿Sabías que el sistema nervioso te envía mensajes continuos sobre las células de tu cuerpo? Por ejemplo, el dolor que sientes cuando alguien te pisa es un mensaje que te dice que necesitas quitar el pie para tu seguridad. Haz lo siguiente para ver qué mensaje te envía el sistema nervioso.

Procedimiento

1. Pon **un poco de hielo** en la palma de tu mano. Deja que el agua derretida caiga en **un plato.** Sostén el hielo hasta que ya no lo aguantes y luego ponlo en el plato.

2. ¿Qué mensaje recibiste del sistema nervioso? ¿Te hicieron esas sensaciones querer soltar el hielo y calentarte la mano?

3. Observa la mano con que sostuviste el hielo y mira luego la otra. ¿Qué cambios observas en la piel? ¿Qué tan rápido se normaliza la mano con que agarraste el hielo?

Análisis

4. ¿Qué sistemas piensas que participaron en restaurar la homeostasis de tu mano?

5. Piensa en un momento de tu vida en que tu sistema nervioso te envió un mensaje, como una sensación intolerable de calor o frío, o de dolor. ¿Cómo reaccionó tu cuerpo? ¿Qué sistemas crees que participaron? Comparte tu historia con una compañera o compañero.

6. Parece que este explorador del ártico tiene mucho frío. Investiga en Internet o en la biblioteca cómo estos exploradores aguantan temperaturas por debajo del punto de congelación.

463

Students will have a chance to revise their answers in the Chapter Review under NOW What Do You Think?

Investigate!

MATERIALS

FOR EACH GROUP:
- ice
- waterproof dish
- paper towel

Safety Caution: Students should clean up the water that results from the melting ice.

Answers to Investigate!

2. Answers will vary, but students should feel some discomfort intense enough to make them want to drop the ice.

3. Students might notice redness when their hand becomes very cold. Individuals will experience warming at different rates.

4. Answers will vary, but students might suggest the cardiovascular system. Point out to students that the redness is an increased blood supply to the cold area. The blood brings warmth to the hand, helping to restore homeostasis.

5. Answers will vary, but students should describe a time when they felt pain, pressure, heat, or some other sensation in response to an outside stimulus.

6. Answers will vary.

Students will likely be surprised to learn in this chapter that the body has only four main types of tissue. As a result, they may think that the body has only a few types of cells. Point out to students that even with a simple classification system, the body still contains more than 200 different types of cells, including red blood cells (erythrocytes), six kinds of white blood cells (neutrophils, eosinophils, basophils, B-lymphocytes, T-lymphocytes, and monocytes), three types of muscle cells (skeletal, smooth, and cardiac), nerve cells, and so forth.

Focus

Body Organization

This section introduces the basic organization of the human body. Students identify the four major tissues of the body and describe how the body's tissues are organized into organs. Students also learn that the body's organs are arranged by function into 11 organ systems. Students will be able to describe the major functions of each body system.

Bellringer

Write the names of the body systems below and their functions on the board or an overhead projector. Scramble the columns so that systems and functions do not match. Ask students to copy both columns in their ScienceLog and draw a line between the body system and its correct function.

- Respiratory system— absorbs oxygen
- Muscular system— moves bones
- Digestive system— breaks down food
- Circulatory system— pumps blood
- Endocrine system—regulates body functions

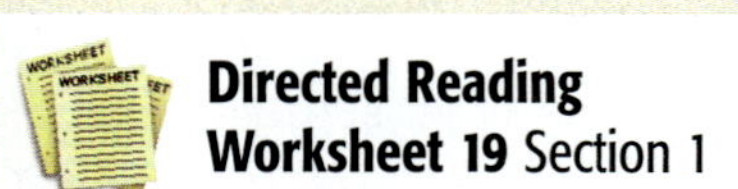

Directed Reading Worksheet 19 Section 1

VOCABULARIO

homeostasis   tejido muscular
tejido   tejido conjuntivo
tejido epitelial   órgano
tejido nervioso   sistema

OBJETIVOS

- Identifica los tejidos principales del cuerpo.
- Compara un órgano con un sistema.
- Describe la función principal de cada sistema.

Figura 1 *El cuerpo tiene cuatro tipos de tejido y cada uno cumple una función específica.*

Organización del cuerpo

El cuerpo tiene una habilidad asombrosa para sobrevivir, incluso en condiciones difíciles. ¿Cómo logró Jack Thayer sobrevivir a una temperatura tan baja? Su cuerpo no permitió que sus condiciones internas cambiaran lo suficiente como para que sus células dejaran de funcionar bien. El mantenimiento de un ambiente interno estable se llama **homeostasis.** Si ésta se interrumpe, las células sufren y a veces mueren.

Cuatro tipos de tejido

Asegurarse de que el ambiente interno permanezca lo suficiente estable como para mantener células sanas no es fácil. Se requieren muchas "tareas" para mantener la homeostasis. Afortunadamente, las células se organizan en equipos para realizarlas. Sí cada miembro de un equipo de fútbol tiene una función especial en el juego, cada célula del cuerpo tiene una tarea específica para mantener la homeostasis. Un grupo de células similares que trabajan juntas forma un **tejido.** El cuerpo tiene cuatro tipos principales de tejido: epitelial, conjuntivo, muscular y nervioso, como ves en la **Figura 1.**

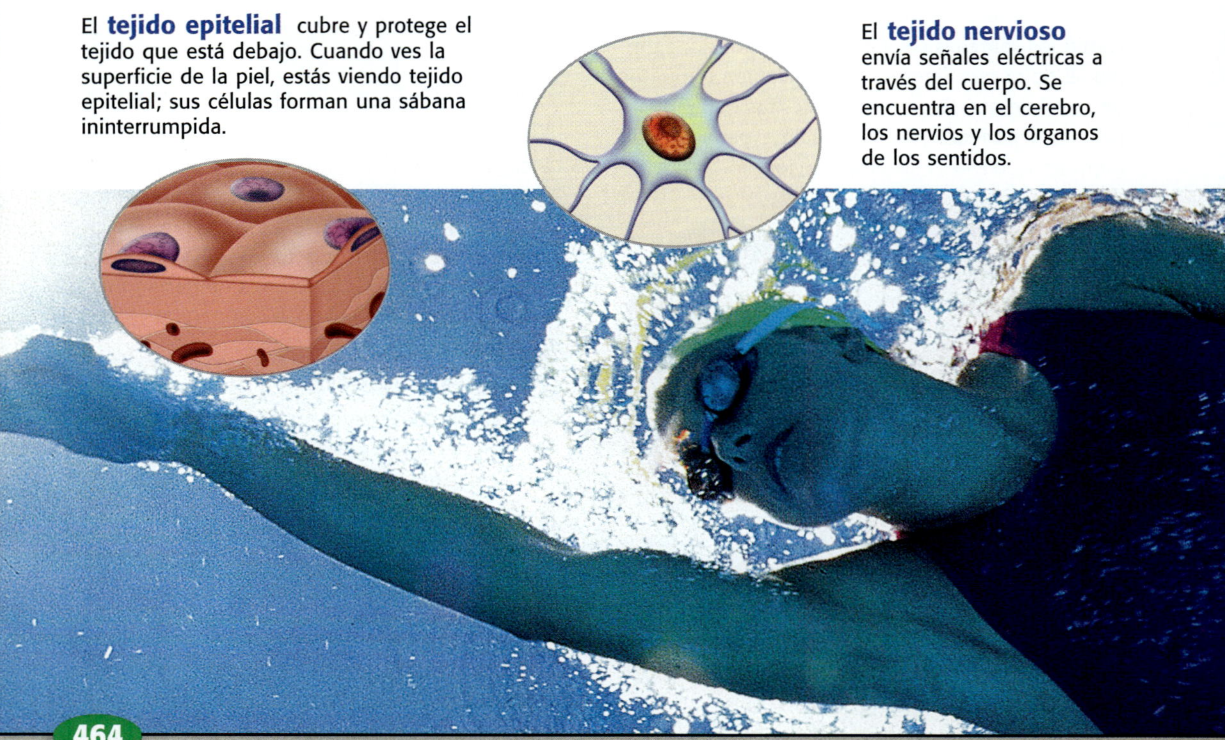

IS THAT A FACT!

The Pompeii worm, *Alvinella pompejana,* can survive a temperature difference of 60°C between its head and its tail! Scientists theorize that a coating of furry bacteria living on the worm's back allow the worm to endure such extreme temperature differences.

Los tejidos forman los órganos

Dos o más tejidos que trabajan juntos forman un **órgano.** Un solo tipo de tejido no puede hacer lo mismo que varios tipos de tejidos que trabajan juntos. El estómago, como ves en la **Figura 2,** usa diferentes tipos de tejido para hacer la digestión.

Los órganos forman sistemas

El estómago hace mucho por ayudarte a digerir los alimentos, pero no todo. Trabaja con otros órganos, como el intestino delgado y el grueso, para digerir los alimentos. Los órganos que trabajan juntos forman un **sistema.** Un fallo en cualquier parte puede afectar a todo el sistema. El cuerpo tiene 11 sistemas principales, que verás en las siguientes dos páginas. ¿Hay alguno que no conocías?

El **tejido muscular** consta de células que se contraen y relajan para producir movimiento.

El estómago es un órgano

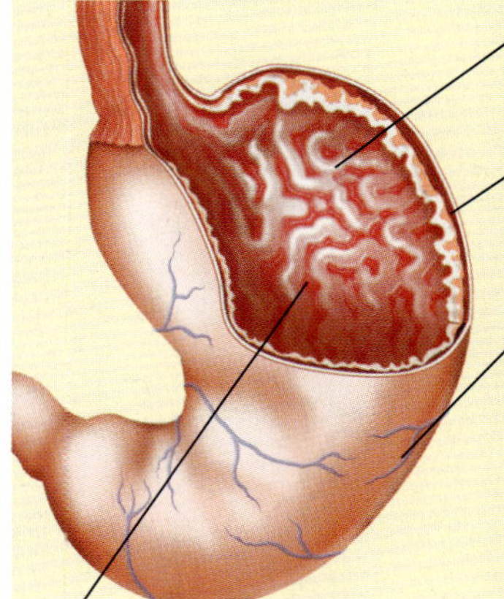

Figura 2 *Los cuatro tipos de tejido trabajan juntos para que el estómago realice la digestión.*

El **tejido conjuntivo** une, sostiene, protege, aisla, alimenta y amortigua los órganos; también evita que los órganos se desbaraten.

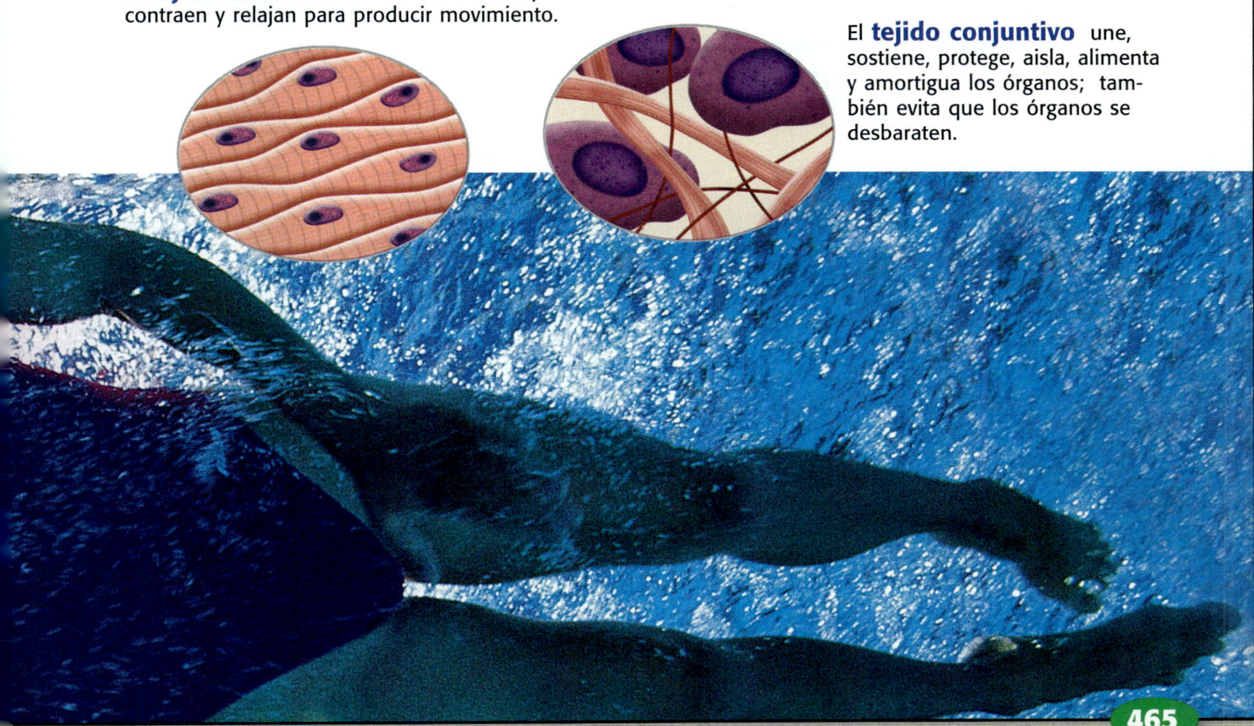

CROSS-DISCIPLINARY FOCUS

History The first transplant of a human heart was performed in Cape Town, South Africa, on December 3, 1967, by Dr. Christiaan Barnard and a team of 30 physicians. In an operation that lasted 5 hours, Barnard removed a heart from the body of a 25-year-old woman and transplanted it into a 55-year-old man named Louis Washkansky. Washkansky lived for 18 days following the transplant before he died of pneumonia. Have interested students conduct research to find out when other organs such as kidneys and livers were first transplanted.

DEBATE

Transplant Ethics There are more than 53,000 people in the United States waiting for organ transplants. There are approximately 20,000 organ transplants performed every year in the United States. The average cost of an organ transplant is $120,000 for a U.S. citizen. Encourage students to research the topic, and have them debate the ethical issues surrounding transplants. Some suggested topics:

- Should transplants happen at all?
- Who should get a transplant?
- Should a child receive a transplant before an older person?

MEETING INDIVIDUAL NEEDS

Learners Having Difficulty
To help students understand and identify the 11 major organ systems of the body, have them make a table with the following headings: Name of Organ System, Function(s), and Main Organs. Have students use the table to organize the information presented on these pages. Encourage students to continue to fill in their tables as they read the remainder of this book.
Sheltered English

Teaching Transparency 71
"Organ Systems"

internetconnect

SciLINKS
TOPIC: Body Systems
GO TO: www.scilinks.org
NSTA
sciLINKS NUMBER: HSTL535

Sistemas

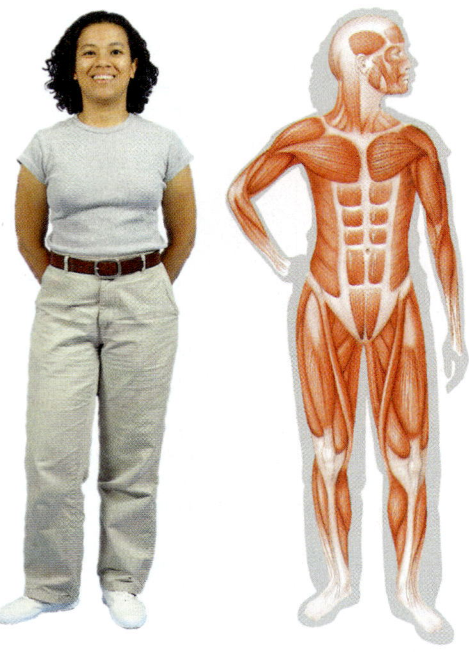

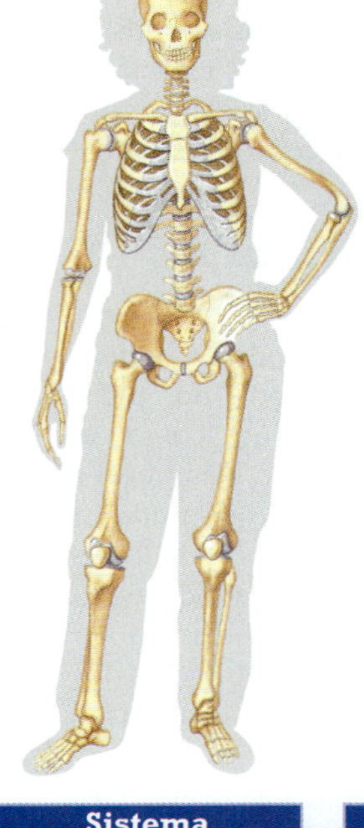

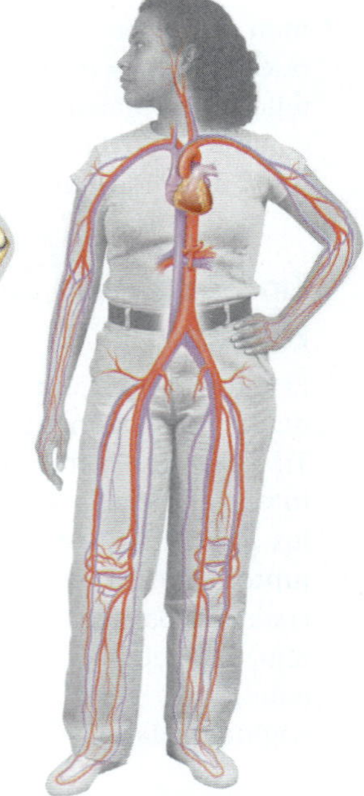

Sistema integumentario
La piel, el cabello y las uñas protegen el tejido que está debajo.

Sistema muscular
Los músculos esqueléticos mueven tus huesos.

Sistema óseo
Los huesos son el armazón que sostiene y protege las distintas partes del cuerpo.

Aparato cardiovascular
El corazón bombea sangre a través de los vasos sanguíneos.

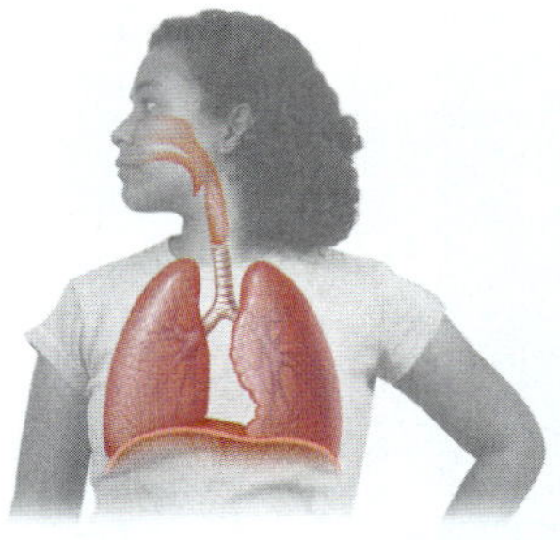

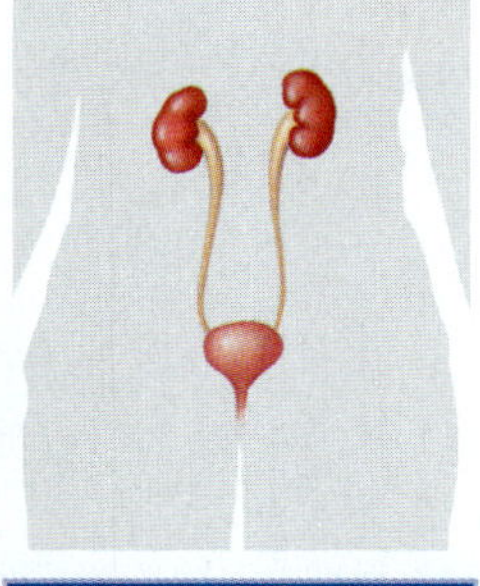

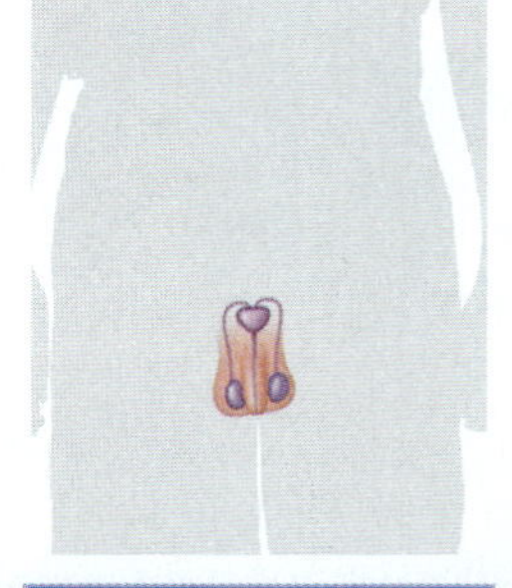

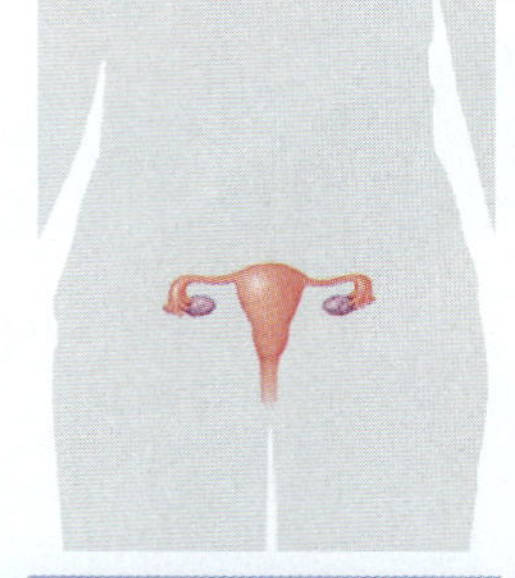

Aparato respiratorio
Los pulmones absorben oxígeno y liberan dióxido de carbono.

Aparato urinario
El aparato urinario elimina desperdicios de la sangre y regula los fluidos del cuerpo.

Aparato reproductor masculino
El aparato reproductor masculino produce y libera espermatozoides.

Aparato reproductor femenino
El aparato reproductor femenino produce óvulos, y alimenta y resguarda al bebé antes del nacimiento.

466

IS THAT A FACT!

Can you learn about the human body by cutting open animals instead of human bodies? In ancient Rome, physician Galen (129–199) did just that! He wrote more than 500 treatises on the human body, and, because of the Roman belief that it was wrong to cut open a human corpse, Galen studied animals and the wounds of fallen gladiators. He never saw inside the human body!

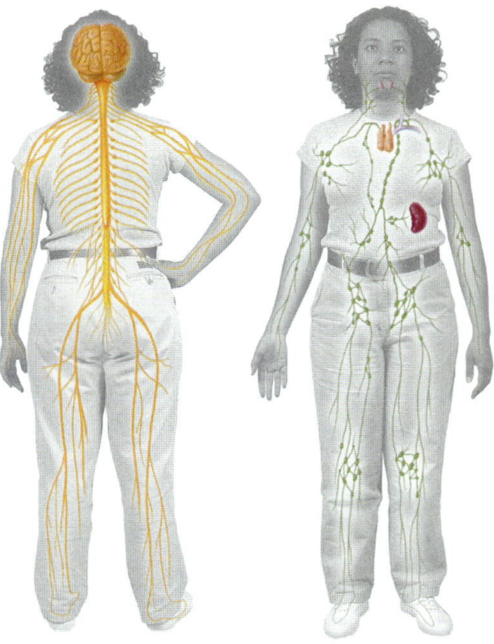

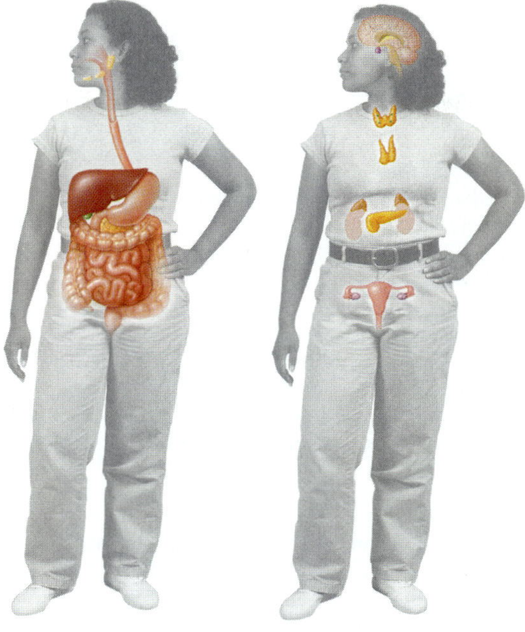

REPASO

1. Explica la relación entre células, tejidos, órganos y sistemas.

2. Compara los cuatro tipos de tejido del cuerpo humano.

3. **Usar gráficas** Haz un diagrama con los principales sistemas y sus funciones.

4. **Relacionar conceptos** Describe una ocasión en que se interrumpiera la homeostasis en tu cuerpo. ¿Qué sistemas del cuerpo piensas que se verían afectados?

Explora

Para recordar los nombres de los sistemas, haz una oración en la que la primera letra de cada palabra represente uno de los once sistemas. Usa las siguientes letras en cualquier orden:

I, M, O, C, R, U, R, E, L, D, N.

467

Focus

The Skeletal System

This section introduces the major organs of the skeletal system and describes the function of bones. The section also illustrates the major internal structure of bones and compares the three major types of joints. The section concludes with a discussion of how bones and joints form levers.

Bellringer

Have students write in their ScienceLog five problems they would have if they lacked bones. (They would have no defined structure, mineral storage, protection, red blood cells, or mobility.)

1 Motivate

ACTIVITY

Locating Bones Review with students that the skeletal system supports the body and protects delicate body parts. Ask students to name the main organ of the skeletal system. (bones)

Encourage students to press the skin in various parts of their body, to feel their bones. Ask students to describe any parts of their body where they cannot feel their bones. (Answers will vary, depending on individual students, but should include the abdomen.)

As you point to various parts of the body, ask students what organ the bones protect. The skull, for example, protects the brain, and the ribs protect the heart and lungs.

Sheltered English

Sección 2

VOCABULARIO

sistema óseo
hueso compacto
hueso esponjoso
cartílago

articulación
ligamento
rendimiento
mecánico

OBJETIVOS

- Identifica los principales órganos del sistema óseo.
- Describe las funciones de los huesos.
- Explica la estructura interna de los huesos.
- Compara los tres tipos de articulaciones.
- Comenta la forma en que los huesos funcionan como palancas.

El sistema óseo

Cuando escuchas la palabra *esqueleto,* quizá te imaginas los restos de algo que ha muerto. Pero tu esqueleto está muy vivo. Tus huesos no son secos ni quebradizos. Están tan vivos y activos como los músculos que están unidos a ellos. Los huesos, los cartílagos y las estructuras especiales que los conectan forman el **sistema óseo.**

La carga de ser un hueso

Los huesos hacen mucho más que sostenerte: realizan varias funciones importantes dentro de tu cuerpo. En la **Figura 3,** puedes ver los nombres de algunos de tus huesos.

Figura 3 *El esqueleto del adulto humano tiene cerca de 206 huesos. En este esqueleto se identifican algunos de los principales.*

Apoyo El sistema óseo te protege y te sostiene. El corazón y los pulmones están cubiertos por las costillas, la médula espinal está protegida por las vértebras, y el cerebro está resguardado por el cráneo.

Almacenamiento Los huesos almacenan minerales como el calcio y los liberan en los tejidos cercanos para que los nervios y los músculos funcionen bien. Las cavidades de los huesos de brazos y piernas también almacenan grasa que se usa como fuente de energía.

Movimiento Los músculos esqueléticos, que están unidos a los huesos por los tendones, los jalan para producir movimiento. Sin los huesos, no podrías sentarte, pararte, caminar ni correr.

Formación de células sanguíneas Algunos huesos están llenos de un material que fabrica los glóbulos rojos y los blancos.

468

Directed Reading Worksheet 19 Section 2

Q: Why didn't the skeleton cross the road?

A: It didn't have the guts.

¿Qué es un hueso?

Un hueso parece no tener vida, pero es un órgano vivo formado por diferentes tejidos. Los huesos están compuestos de tejido conjuntivo y minerales que son depositados por células vivas llamadas *osteoblastos*.

Si miras dentro de un hueso, te darás cuenta de que hay dos tipos diferentes de tejido óseo. Si el tejido no tiene espacios abiertos visibles, se llama **hueso compacto.** Este tejido proporciona al hueso la mayor parte de su fuerza y apoyo. El tejido óseo que tiene muchos espacios abiertos se llama **hueso esponjoso.** Los huesos tienen un tejido suave llamado *médula*. La médula roja, que se encuentra a veces en el hueso esponjoso, produce glóbulos rojos. La médula amarilla, que está en la cavidad central de los huesos largos, almacena grasa. Los canales diminutos dentro del hueso compacto tienen pequeños vasos sanguíneos. La **Figura 4** muestra un corte transversal del hueso más largo del cuerpo.

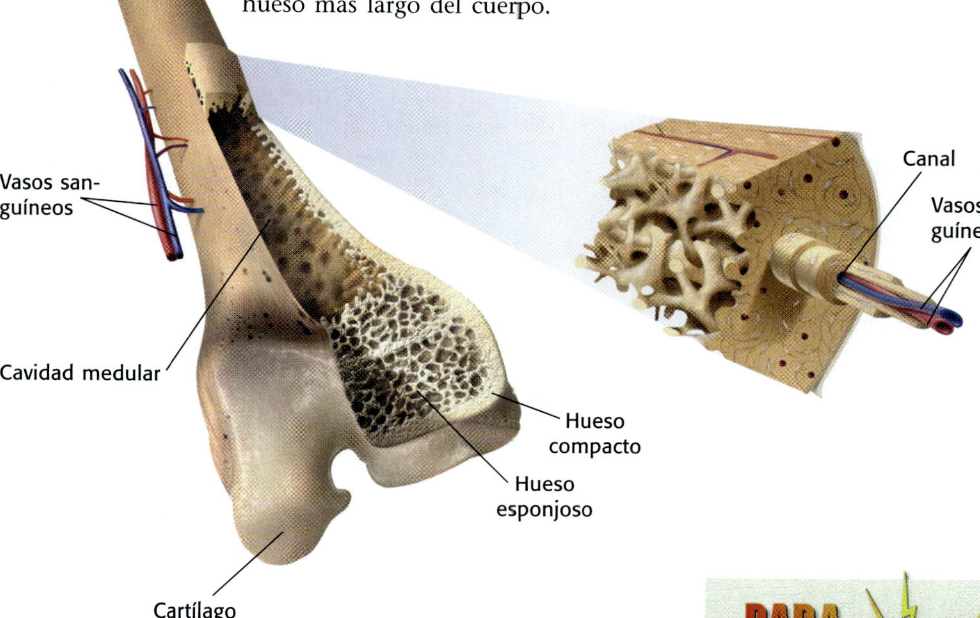

Figura 4 *Esta ilustración de un fémur, el hueso largo del muslo, muestra que se compone de anillos llenos de minerales y de vasos sanguíneos que corren a través de un canal en cada grupo de anillos. La médula rellena el centro de este tipo de hueso, y los vasos sanguíneos entran y salen en ciertos puntos a lo largo del mismo.*

Laboratorio

Huesos en encurtido

Esta actividad te mostrará cómo cambia un hueso cuando se le expone a un ácido, como el vinagre. Pon **un hueso de pollo limpio** en un **frasco con vinagre.** Al cabo de 1 semana, sácalo y enjuágalo con agua. Haz una lista de los cambios que viste o sentiste. ¿Cómo cambió la resistencia del hueso? ¿Qué le quitó el vinagre?

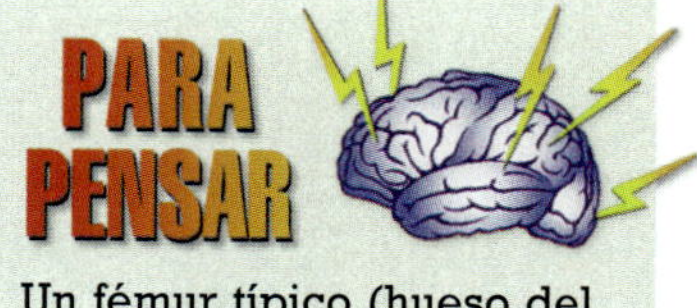

PARA PENSAR

Un fémur típico (hueso del muslo) puede sostener más de 1,000 kg.

469

2 Teach

Quick Lab

MATERIALS

For Each Student:
• clean chicken bone
• jar of vinegar

Safety Caution: Remind students to review all safety cautions and icons before beginning this lab activity.

Answer to QuickLab

The bone becomes more flexible, because the hard minerals in the leg bone dissolve in the vinegar.

Demonstration

Bone Dissection Ask a local butcher to cut a long bone from a pig, cow, or sheep in half lengthwise to expose the bone's internal structure. If a bone is not available from a butcher, obtain a preserved long bone from a biological supply house. Point out to students the differences in structure and location of spongy bone and compact bone. Also point out any cartilage that might remain on the bone. Ask students what makes the bone relatively lightweight for its size. (Students should infer that the air spaces in the spongy bone make the bone relatively lightweight for its size.)
Sheltered English

Teaching Transparency 72 "What's in a Bone?"

Multi**cultural** CONNECTION

Writing

On November 2 each year, people across Mexico celebrate *El Día de los Muertos* (the Day of the Dead). They take food and gifts to the graves of their loved ones and set up elaborate shrines and altars in their homes. These shrines are often decorated with figures of skeletons depicting the person's job when he or she was alive, such as that of a doctor or a teacher. Foods eaten during this holiday are *pan del muerto* (bread of the dead) and candies shaped like skulls. Have interested students write a research paper on this holiday to share with the class.

DISCUSSION

Joint Review After students read about the different kinds of joints, ask them the following questions. You may want to have students move their own bodies before they answer each question.

- What kind of joint do you use when you bend your elbow? (hinge joint) your knee? (hinge joint)
- What kind of joint moves when you swing your arm back and forth? (ball-and-socket joint) Name another location in your body where this type of joint is located. (the hips) Sheltered English

CONNECT TO
PHYSICAL SCIENCE

Students may be surprised to learn that their arms and legs are "machines." Our limbs are levers, and levers are the simplest kind of machine. They allow us to apply, increase, and change the direction of force. Have students think of other levers they may use or know about. (Examples include a seesaw, the claw end of a hammer, a crowbar, and a garlic press.)

Use Teaching Transparency 186 to help students understand how levers make work easier.

Teaching Transparency 186
"Machines Change the Size or Direction (or Both) of a Force"
LINK TO PHYSICAL SCIENCE

Math Skills Worksheet 53
"Mechanical Advantage"

El crecimiento de los huesos

¿Sabías que la mayor parte de tu esqueleto fue una vez suave y elástico? La mayoría de los huesos empiezan como un tejido suave y flexible llamado **cartílago,** que contiene principalmente agua, que le da una textura elástica. Cuando naciste, tenías muy poco hueso de verdad. Pero, al crecer, el cartílago fue reemplazado por hueso. Durante la infancia, hay placas de crecimiento de cartílago en la mayoría de los huesos, creando espacio para que éstos continúen creciendo.

Tócate la punta de la nariz o dobla la parte superior de la oreja. Otras áreas, como éstas, nunca se vuelven hueso. El material flexible bajo la piel en estas áreas es cartílago. Como ves en la **Figura 5,** el cartílago no tiene la densidad mineral para producir una imagen obscura como los huesos del cráneo.

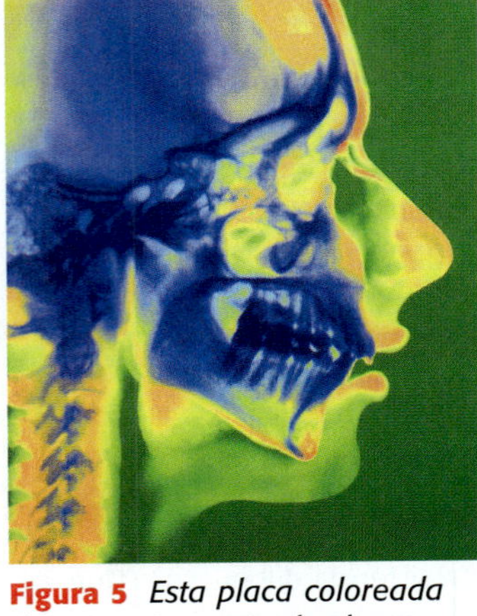

Figura 5 *Esta placa coloreada de rayos X muestra los huesos del cráneo y el cuello.*

¿Para qué sirve una articulación?

El lugar donde dos o más huesos se unen se llama **articulación.** Las articulaciones tienen formas especiales que permiten que te muevas cuando tus músculos se contraen. Algunas articulaciones permiten mucho movimiento; otras están fijas, por lo que permiten muy poco o ningún movimiento. Por ejemplo, las craneales son fijas. Las que tienen mucho movimiento tienden a ser más susceptibles a lesiones que las menos flexibles. Algunos ejemplos de articulaciones móviles se ven en la **Figura 6.**

Figura 6 *La forma de las articulaciones depende de su función.*

Articulación deslizante
Las articulaciones deslizantes permiten que los huesos de las manos se deslicen uno sobre el otro, lo que da cierta flexibilidad al área.

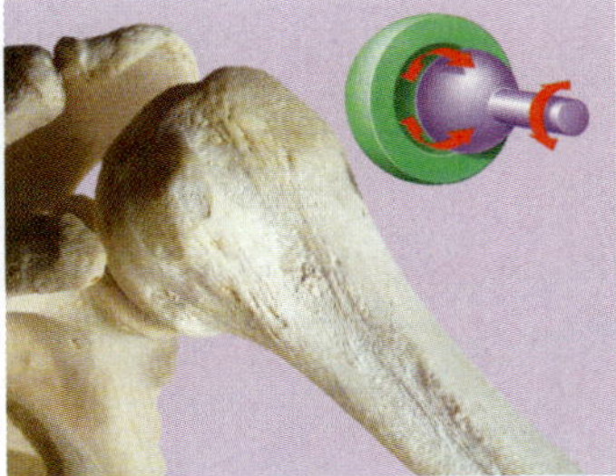

Articulación esférica
Como el control de un juego de computadora, el hombro permite que tu brazo se mueva libremente en todas direcciones.

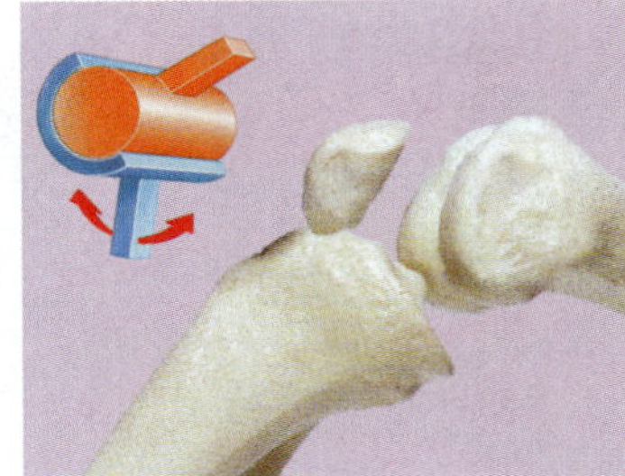

Articulación en bisagra
Como la bisagra de una puerta, la rodilla te permite flexionar y extender la pantorrilla.

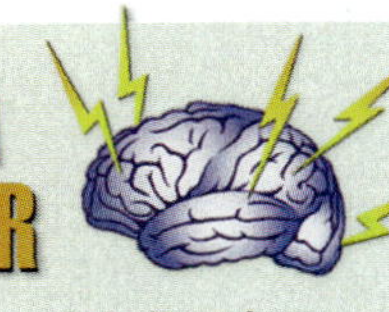

El esqueleto del tiburón no tiene huesos; sólo cartílago.

Hueso a hueso Las articulaciones se mantienen juntas a través de fuertes bandas elásticas de tejido conjuntivo llamadas **ligamentos.** Si un ligamento se estira mucho, se produce una distensión. En general, un ligamento distendido sana con el tiempo, pero uno desgarrado no. Este último tiene que ser reparado quirúrgicamente. El cartílago amortigua el área donde los dos huesos se unen. Si el cartílago se desgasta, la articulación se vuelve artítrica.

IS THAT A FACT!

The only bone in the human body that is not connected to another bone is the hyoid bone. The hyoid bone is found in the throat above the larynx. This bone is easily broken when a person is strangled, and it is often an important piece of evidence in suspected strangulation deaths.

¿Pueden las palancas reducir la carga?

Aunque no lo creas, tus extremidades son máquinas. Un músculo que jala un hueso trabaja como una máquina simple llamada *palanca*. Una palanca es una barra rígida que se mueve en un punto fijo llamado *punto de apoyo*. Cualquier esfuerzo que se aplica a una palanca se llama *fuerza*. La fuerza opuesta que resiste el movimiento de la palanca, como la fuerza descendente ejercida por un peso en la barra, se llama *carga* o *resistencia*. La **Figura 7** muestra cómo se usan en el cuerpo humano los tres tipos de palancas.

En cada tipo de palanca, observa la ubicación del punto de apoyo en relación con la fuerza y la resistencia. En tu cuerpo, los músculos aplican la fuerza sobre la palanca. La resistencia es el peso que resiste el tirón de los músculos. El beneficio obtenido al usar una máquina se mide con un índice llamado **rendimiento mecánico.**

$$RM = \frac{\text{fuerza aplicada a la carga}}{\text{fuerza}}$$

El rendimiento mecánico es la medida de las veces que una máquina simple multiplica una fuerza aplicada a una resistencia. Las palancas de primera y segunda clase pueden tener un rendimiento mecánico importante. Las de tercera clase aumentan la distancia y la velocidad, pero no la fuerza.

REPASO

1. Describe cuatro funciones importantes de los huesos.

2. Dibuja un hueso y señala las estructuras interiores y exteriores. Usa lápices de colores para dibujar y señalar el hueso esponjoso, los vasos sanguíneos, la cavidad medular, el hueso compacto y el cartílago.

3. Describe e ilustra tres tipos de palancas que haya en tu cuerpo.

4. **Interpretar modelos** Estudia los modelos de palancas de la Figura 7. Usa una caja pequeña (resistencia), una regla (palanca) y un lápiz (punto de apoyo) para crear modelos de cada tipo de palanca.

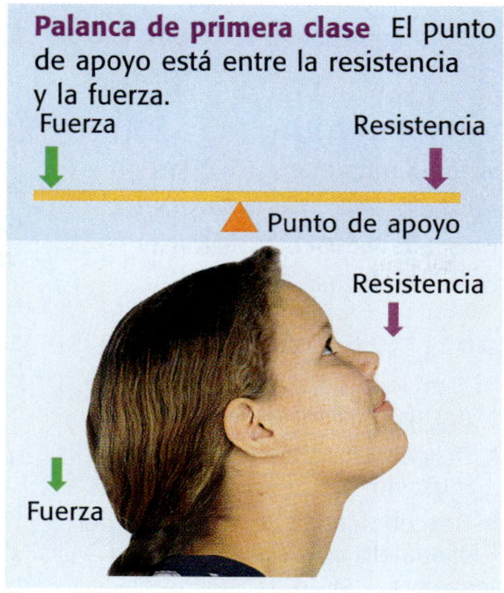

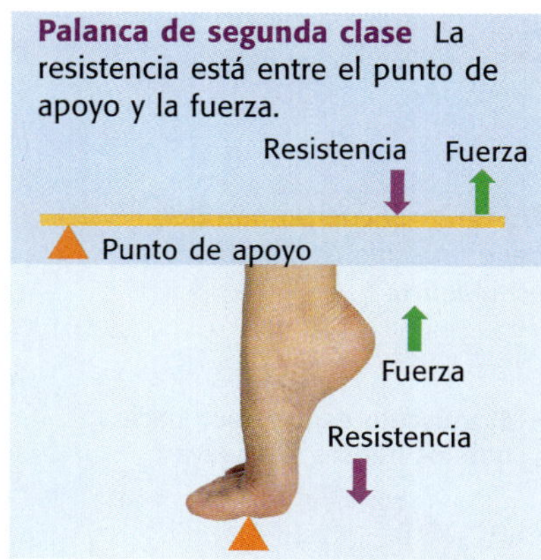

Figura 7 *Hay tres clases de palancas: según la ubicación del punto de apoyo, la resistencia y la fuerza.*

Focus

The Muscular System

This section introduces students to the major parts of the muscular system and describes the different types of muscle. This section also describes how skeletal muscles move bones. Students compare aerobic exercise to resistance exercise. The section concludes with a discussion of typical muscle injuries and how to prevent them.

Bellringer

On the board or an overhead projector, write the following:

In your ScienceLog, list at least five parts of your body that you use to drink a glass of water. (fingers, hands, arm, lips, tongue)

When the students are done, remind them that all the parts needed, including the eyes they used to see the glass, are controlled by muscles.

1 Motivate

ACTIVITY

Poster Project Have students draw an outline of a human body. Have them add smooth muscles, skeletal muscles, and cardiac muscles to their drawings. Point out that certain kinds of muscles are located in certain parts of the body. Ask students to identify the function of each type of muscle. Make sure that students understand that all tissue movement in the body is caused by muscle movement. Sheltered English

---

VOCABULARIO

sistema muscular   tendón
músculo liso   flexor
músculo cardíaco   extensor
músculo esquelético

OBJETIVOS

- Menciona las partes principales del sistema muscular.
- Describe los diferentes tipos de músculo.
- Describe como los músculos del esqueleto mueven los huesos.
- Compara el ejercicio aeróbico con el de resistencia.
- Da un ejemplo de lesión muscular.

Figura 8 *El cuerpo humano tiene músculo liso, cardíaco y esquelético.*

El sistema muscular

¿Alguna vez has intentado quedarte inmóvil por 1 minuto? Trata todo lo que quieras, pero no podrás. En alguna parte de tu cuerpo, algunos músculos siempre están trabajando. Por ejemplo, los músculos impulsan continuamente sangre a través de los vasos sanguíneos. Un músculo hace que respires, y otros te mantienen erguido. Si todos descansaran al mismo tiempo, sufrirías un colapso. Los músculos están hechos de tejido muscular y tejido conjuntivo. Los que están unidos a los huesos y el tejido conjuntivo que los une forman el **sistema muscular.**

Clases de músculo

Hay tres tipos de tejido muscular en el cuerpo. El **músculo liso** se encuentra en el sistema digestivo y los vasos sanguíneos. El **cardíaco** es un tipo especial de músculo que sólo se encuentra en el corazón. Los **esqueléticos** están unidos a los huesos para darles movimiento y protegen los órganos internos. La **Figura 8** muestra las tres clases de músculo.

Cada acción muscular puede ser voluntaria o involuntaria. Las que están bajo tu control son *voluntarias*. Las que no, son *involuntarias*. Las acciones del músculo liso y del cardíaco son involuntarias. Las de los esqueléticos pueden ser voluntarias o involuntarias. Por ejemplo, puedes parpadear cuando quieras, pero los ojos también parpadean automáticamente sin que tú te des cuenta.

Teaching Transparency 73 "Types of Muscle"

IS THAT A FACT!

Horses can sleep standing up. Their legs can support their weight on their bones and tendons without the use of any muscles at all. When they fall asleep and their muscles relax, their leg bones lock in place underneath them, holding them upright for the duration of their nap.

¡Muévete!

Los músculos esqueléticos producen cientos de movimientos voluntarios. Esto lo demuestra una bailarina, una nadadora o incluso alguien que hace una cara chistosa, como en la **Figura 9.** Cuando quieres producir un movimiento, haces que señales eléctricas viajen del cerebro a las células del músculo esquelético. Las células musculares responden a estas señales contrayéndose o acortándose.

De músculos a huesos Fibras resistentes de tejido conjuntivo llamados **tendones** unen los músculos esqueléticos con los huesos. Cuando un músculo se acorta, ocurre un jalón que acerca más los huesos unos a los otros. Por ejemplo, el bíceps de la **Figura 10** está unido por tendones a un hueso del hombro y a otro del antebrazo. Cuando se contrae, el brazo se dobla.

Trabajo en parejas Los músculos esqueléticos trabajan en parejas para producir movimientos suaves y controlados. Muchos movimientos básicos son el resultado de pares de músculos que producen flexión y enderezamiento. Si un músculo dobla una parte del cuerpo, se llama **flexor.** Si hace que una parte del cuerpo se enderece, se llama **extensor.** El músculo flexor del brazo es el bíceps, y el músculo extensor es el tríceps. Descubre algunos músculos flexores y extensores en la sección de Laboratorio de la derecha.

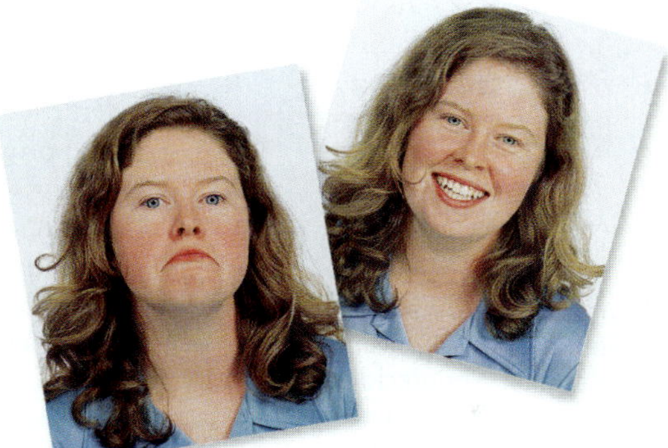

Figura 9 *Para sonreír participan un promedio de 13 músculos, y para fruncir el ceño unos 43.*

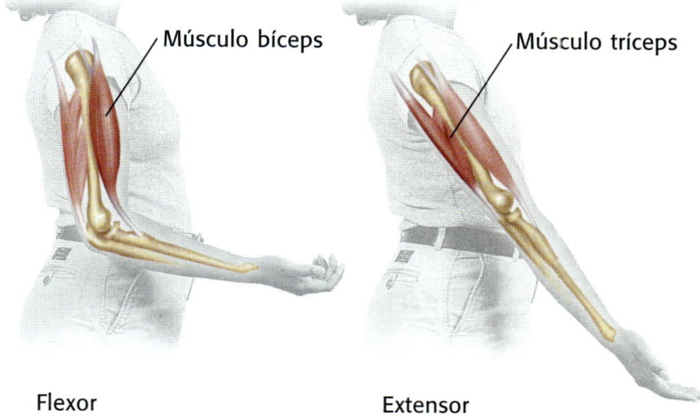

Figura 10 *Los músculos esqueléticos, como el bíceps y el tríceps, están conectados a los huesos por tendones. Cuando el bíceps se contrae, el codo se dobla. Cuando el tríceps se contrae, el codo se endereza.*

Laboratorio

Potencia en pares

1. Siéntate en una silla frente a una mesa y pon una de las manos debajo de la **mesa.** Ejerce una suave presión ascendente, sin esforzarte ni voltear la mesa.

2. Con la otra mano, palpa la parte delantera y trasera de tu brazo.

3. Luego pon la mano sobre la mesa. Ejerce una ligera presión en dirección descendente.

4. Con la otra mano, palpa la parte delantera y posterior de tu brazo.

5. ¿Qué observaste cuando ejerciste la presión ascendente? ¿Qué observaste cuando ejerciste la descendente?

473

2 Teach

ACTIVITY

Muscle Contraction Ask a student volunteer to stand in a doorway with his or her arms and hands relaxed with the palms turned inward. Ask the student to raise his or her hands against the door frame (backs of the hands on the frame) and press steadily against the frame for about 30 to 40 seconds. Then ask the student to relax and step away from the door. Have the rest of the class observe what happens to the student's arms. (The student's arms should rise slowly without obvious effort by the student.)

Explain to students that the arms rise because the muscles that were pushing against the door frame are still shortened, or contracted. Sheltered English

LabBook PG 626
Muscles at Work

Answer to QuickLab

5. Students should feel their biceps tighten when they press up and their triceps tighten when they press down.

Directed Reading Worksheet 19 Section 3

internet**connect**

TOPIC: The Muscular System
GO TO: www.scilinks.org
*sci*LINKS **NUMBER:** HSTL540

SCIENTISTS AT ODDS

In the early part of this century, tens of thousands of people, mostly children, were stricken with polio, a viral disease that paralyzes muscles. Often, polio would leave its victims unable to walk or move. One Australian nurse, Sister Elizabeth Kenny, treated patients using flexible hot wraps and exercise instead of hard, immobilizing casts. Using her treatments, patients avoided paralysis. Largely because Kenny was self-taught and had no formal medical education, the doctors and hospital administrators of the time fought against her practices. Eventually, she and her successes became well known, and her contributions became the beginning of the field of physical therapy. Have students research and report to the class Sister Kenny's story and the opposition she faced in Australia and the United States.

MATH and MORE

Explain to students that a regular program of aerobic exercise enables the heart to work more efficiently, pumping the same volume of blood in fewer beats. Ask students to imagine that their heart normally beats 75 times per minute. After a 3-month program of aerobic exercise, their heart now beats 65 times per minute. What is the difference in heart rate per minute?

(75 beats − 65 beats = 10 beats)

How many fewer times will the heart beat in 1 hour?

(10 beats/minute × 60 minutes/hour = 600 beats/hour)

How many fewer times will the heart beat in 1 day?

(600 beats/hour × 24 hours/day = 14,400 beats/day)

Math Skills Worksheet 31
"The Unit Factor and Dimensional Analysis"

GOING FURTHER

Concept Mapping Divide the class into cooperative groups of three or four students. Have each group create a concept map for the ideas in this lesson. They should connect at least 10 terms and link them with apt phrases. Have students record their finished concept maps in their ScienceLog. Sheltered English

Answers to Self-Check

Curl-ups use flexor muscles; push-ups use extensor muscles.

La química del cuerpo es muy importante para el funcionamiento saludable de los músculos. Si hay un desequilibrio químico en un músculo debido a sudoración excesiva, dieta deficiente, tensión o enfermedad, pueden ocurrir espasmos o calambres. El sodio, el calcio y el potasio, tres elementos químicos llamados *electrolitos,* deben estar equilibrados para evitar calambres y espasmos. En general, la relajación y el masaje ayudan al músculo a recuperar su balance químico.

Úsalo o déjalo

Cuando alguien se rompe un brazo y tiene que usar un yeso, los músculos que rodean el hueso lastimado cambian. Esto pasa porque estos músculos no se ejercitan, y se vuelven más pequeños y débiles. Por otro lado, los músculos que se ejercitan son más fuertes y grandes. Algunos ejercicios dan a los músculos más resistencia, o sea, pueden trabajar más sin cansarse. Los músculos fuertes también benefician a otros sistemas del cuerpo. Cuando un músculo se contrae, sus vasos sanguíneos se aprietan. Esto ayuda a que la sangre circule y aumente el flujo sanguíneo, sin que el corazón trabaje de más.

Ejercicios de resistencia Para desarrollar el tamaño y la fuerza de los músculos esqueléticos, los ejercicios de resistencia son la forma más eficaz. Estos ejercicios requieren que los músculos superen la resistencia (peso) de otro objeto. Algunos, como los abdominales de la **Figura 11,** requieren que superes tu propio peso.

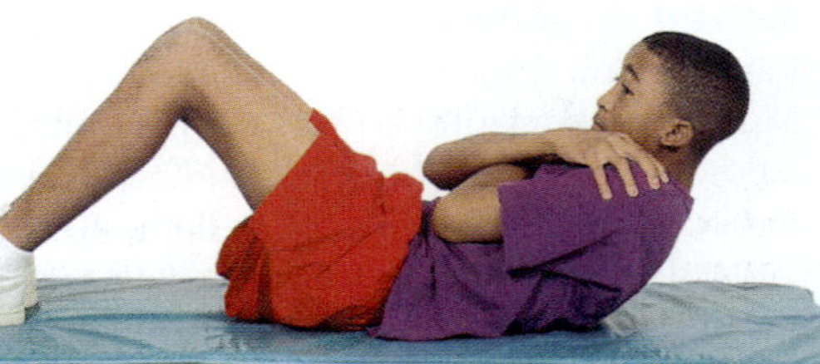

Figura 11 *Los ejercicios de resistencia son pesados, pero realmente ayudan a formar músculos fuertes.*

Ejercicio aeróbico La actividad constante con intensidad moderada, como trotar, montar en bicicleta, patinar, nadar o caminar, se llama ejercicio aeróbico. En cierto modo, el ejercicio aeróbico aumenta el tamaño y la fuerza de los músculos esqueléticos, pero principalmente aumenta su resistenciay fortalece el corazón. Muchas personas, como la chica de la **Figura 12,** disfrutan del ejercicio aeróbico.

Figura 12 *El ejercicio aeróbico es una muy buena forma de divertirse fortaleciendo tu corazón.*

✓ Autoevaluación

¿Qué tipo de músculos esqueléticos ejercitas al hacer abdominales? ¿Qué tipo de músculos ejercitas cuando haces lagartijas? *(Consulta la página 636 para comprobar tus respuestas).*

474

Homework

Researching Injuries For one month, have students read the sports section in the local newspaper or look for articles in sports magazines about injuries to the muscular and skeletal systems sustained by athletes. Have them research and write about these injuries in their ScienceLog. Ask students to identify and count the types of injuries—such as sprained ankles, torn or pulled muscles, bruised ribs, and torn or damaged ligaments. Have students compile their information on bar graphs in which they record the kinds of injuries on the x-axis and the number and frequency of injuries on the y-axis.

Lesión muscular

Cualquier ejercicio se debe empezar de manera gradual para que los músculos adquieran fuerza y resistencia sin lastimarse. Los músculos también deben calentarse gradualmente para reducir el riesgo de una lesión. Sin embargo, como cualquier sistema, el sistema muscular puede lesionarse. Una distensión muscular, o esguince, es la sobretensión o incluso el desgarro de un músculo. A menudo, la distensión muscular ocurre porque el músculo no se preparó adecuadamente para un trabajo determinado. Las distensiones también ocurren por falta de calentamiento o porque no se hace bien un ejercicio.

Como ves en la **Figura 13,** los tendones, como los músculos, pueden lastimarse por excesos. Cuando se lastima un tendón, se calienta o se inflama a medida que el cuerpo trata de repararlo. Esta condición dolorosa se llama tendinitis y, a menudo, se requiere de mucho tiempo para que el tendón sane.

Figura 13 *Un esguince del muslo es un desgarre o distensión de uno de los músculos o tendones de la parte posterior del muslo.*

Los peligros de los esteroides anabólicos Algunas personas tratan de agrandar y fortalecer sus músculos tomando hormonas llamadas *esteroides anabólicos*. Los esteroides anabólicos son substancias químicas poderosas que se parecen a la testosterona, una hormona sexual masculina sobre la que leerás cuando estudies la reproducción humana. No sólo dan a los atletas una ventaja injusta en competiciones, sino que ponen al usuario en riesgo de tener problemas de salud a largo plazo. Su uso afecta al corazón, el hígado y los riñones, y puede producir presión sanguínea alta. Si se toman antes de que el esqueleto esté desarrollado, pueden hacer que los huesos dejen de crecer. Los atletas de competición, como los de la **Figura 14,** se someten continuamente a pruebas para detectar este tipo de droga.

Figura 14 *Con un buen entrenamiento, los atletas pueden desempeñarse muy bien sin el uso de esteroides anabólicos.*

¡MATEMÁTICAS!

El tiempo de un corredor

Rosa, que ha sido corredora desde hace algunos años, ha decidido entrar a una carrera. Actualmente corre 5 km en 30 minutos. Le gustaría disminuir su tiempo en un 15 por ciento antes de la carrera. ¿Cuál será su tiempo cuando cumpla su objetivo?

REPASO

1. Menciona tres tipos de tejido muscular y describe sus funciones en el cuerpo.

2. Compara el ejercicio aeróbico con el de resistencia y da dos ejemplos de cada uno.

3. **Aplicar conceptos** Describe la acción muscular que se requiere para levantar un libro. Haz un diagrama que muestre la acción muscular.

Focus

The Integumentary System

This section introduces students to the major functions of the integumentary system; it also describes the major parts of the skin and discusses their functions. Students learn about the structure and functions of hair and nails and about common injuries to the skin.

Bellringer

Write the following questions on the board or an overhead projector, and ask students to write their answer in their ScienceLog:

When do you see dogs panting? Why do you think they pant? (A dog pants in order to regulate its body temperature. Dogs cannot sweat, and panting cools them down on hot days or after strenuous activity.)

1 Motivate

DISCUSSION

Homeostasis Relay the following story to students:

Dr. Charles Blagden was secretary of the Royal Society of London more than 200 years ago. He wanted to test how mammals regulate their body temperature. He spent 45 minutes in a room with a few friends, a dog, and a steak. The temperature in the room measured 126°C (260°F). Ask students what they think happened to the man, the dog, and the steak after the 45 minutes were up. (Everyone and the dog emerged from the room unharmed, but the steak was cooking! Mammals can regulate their temperature.)

VOCABULARIO

sistema integumentario
glándulas sudoríparas
melanina
epidermis
dermis
folículo piloso

OBJETIVOS

- Describe las funciones principales del sistema integumentario.
- Menciona las partes principales de la piel y explica sus funciones.
- Describe la estructura y función del cabello y las uñas.
- Describe algunos tipos de lesiones de la piel.

El sistema integumentario

Pregunta para ti: ¿qué parte del cuerpo tiene que estar parcialmente muerta para mantenerte con vida? He aquí las pistas: puede ser de diferentes colores, es el órgano más grande del cuerpo y te protege del mundo exterior. ¡Ah! Y está bien visible. ¿Podría ser la piel? Si es así, adivinaste correctamente.

La piel, el cabello y las uñas forman el **sistema integumentario**. (*Integumento* significa "cubierta"). Como los demás sistemas, el integumentario le ayuda al cuerpo a mantener un ambiente interno sano.

La piel: más que un "abrigo"

¿Por qué necesitas piel? He aquí cuatro razones:

- Te protege manteniendo la humedad del cuerpo y evitando que entren en él partículas extrañas.

- Te mantiene "en contacto" con el mundo exterior. Sus terminaciones nerviosas te permiten sentir lo que te rodea.

- Te ayuda a regular la temperatura del cuerpo. Por ejemplo, unos pequeños órganos llamados **glándulas sudoríparas** producen el sudor, un líquido salado que fluye a la superficie de la piel. Cuando el sudor se evapora, la piel se refresca.

- Te ayuda a librarte de desechos. Algunos tipos de desperdicios químicos abandonan el flujo sanguíneo y son eliminados por el sudor.

¿Qué determina el color de la piel? Una substancia química obscura llamada **melanina** determina el color de la piel, como ves en la **Figura15**. Si hay mucha melanina, la piel es muy obscura. Si se produce muy poca, la piel es muy blanca. La melanina de la capa superior de la piel absorbe mucha radiación perjudicial del Sol, lo que reduce el ADN y puede producir cáncer. Sin embargo, *toda* la piel es vulnerable al cáncer y por lo tanto se debe proteger de la exposición solar siempre que sea posible.

Figura 15 *La variedad de color de piel es causada por la melanina. La cantidad de melanina varía entre las personas.*

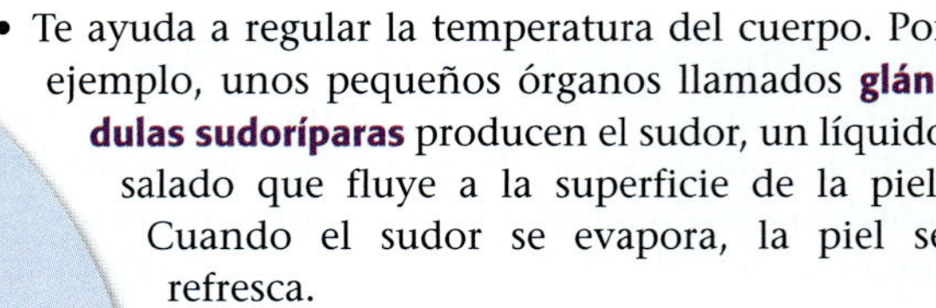

IS THAT A FACT!

In an average adult, the skin has a surface area of about 2 m² and weighs about 4 kg. The skin on the human body varies in thickness from about 5 mm on the soles of the feet to about 0.5 mm on the eyelids.

El cuento de las dos capas

Como sabes, la piel es el órgano más grande del cuerpo. De hecho, la piel de un adulto cubre un área de unos 2 m². Sin embargo, la piel no es tan simple como parece. Tiene dos capas principales: la dermis y la epidermis. La **epidermis** es la más delgada; es lo que ves cuando miras tu piel. (*Epi* significa "sobre"). La capa más gruesa y profunda se conoce como **dermis.**

Epidermis La epidermis se compone de un tipo de tejido epitelial. Aunque tiene muchas capas de células, en la mayor parte del cuerpo sólo tiene el grosor de dos hojas de papel. Es más gruesa en la palma de las manos y en la planta de los pies. La mayoría de las células epidérmicas están muertas y llenas de una proteína llamada queratina, que fortalece la piel.

Dermis La dermis se encuentra debajo de la epidermis. Principalmente es tejido conjuntivo y tiene muchas fibras hechas de una proteína llamada colágeno. Las fibras dan resistencia a la piel y permiten que se doble sin romperse. La dermis también contiene una variedad de pequeñas estructuras, como ves en la **Figura 16.**

Autoevaluación

¿A qué sistema pertenecen los vasos sanguíneos de la piel? *(Consulta la página 636 para comprobar tu respuesta).*

Figura 16 *Bajo la superficie, la piel es un órgano complejo hecho de vasos sanguíneos, nervios, glándulas y músculos.*

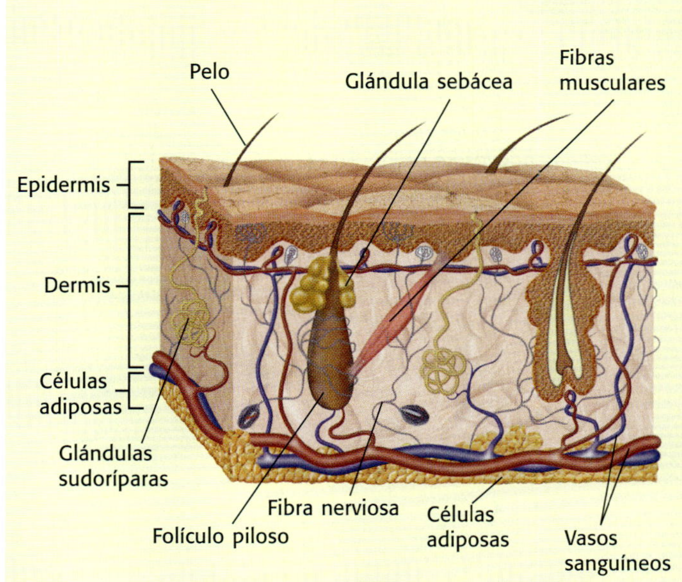

Vasos sanguíneos: transportan substancias y ayudan a regular la temperatura del cuerpo.

Nervios: mandan y reciben mensajes del cerebro.

Fibras musculares: están unidas al folículo piloso y se pueden contraer, lo que hace que el pelo se erice.

Folículos pilosos: (en la dermis) producen pelo.

Glándulas sebáceas: liberan aceite que mantiene el pelo flexible e impermeabiliza la epidermis.

Glándulas sudoríparas: liberan sudor. Cuando éste se evapora, se elimina calor corporal y el cuerpo se refresca. El sudor también tiene materiales de desecho que se expulsan del cuerpo.

A hippo's sweat is a natural sunscreen. Hippopotamuses secrete a pink fluid onto the surface of their skin that helps protect them from the sun's harmful rays.

477

---

2 Teach

ACTIVITY

Measuring Temperature

MATERIALS

FOR EACH GROUP:
- 2 Celsius thermometers
- cotton balls
- water at room temperature
- fan
- clock or watch

Divide class into groups of three to five students. Have students wrap the bulb of each thermometer with a cotton ball and then wet one of the cotton balls. Have students record the beginning temperature of each thermometer and then hold both thermometers in front of a fan. Students should record the temperature of each thermometer every minute for 5 minutes. How do the temperatures of the thermometers differ? Why? (The thermometer with the wet cotton has a lower temperature. Evaporation lowers the temperature.)

How does this relate to what happens when your body sweats? (As sweat evaporates from the body, the skin becomes cooler.)

Answer to Self-Check

Blood vessels belong to the cardiovascular system.

Teaching Transparency 74 "The Skin"

Directed Reading Worksheet 19 Section 4

  **PG 627**

Seeing Is Believing

READING STRATEGY

Prediction Guide Before students read this page, ask them if they agree or disagree with the following statements. Students will discover the answers as they explore Section 4.

- Dark hair gets its color from melanin. (true)
- Hair is made up of both living and dead cells. (true)
- Nails grow longer as new cells form at the nail roots. (true)

REAL-WORLD CONNECTION

Skin cancer is the most common kind of cancer. More than 800,000 new cases of skin cancer are reported each year. You can take several preventive measures to reduce your risk of skin cancer. Avoid being in the sun from 10:00 A.M. to 3:00 P.M., when the sun's rays are most direct. Wear sunglasses, a wide-brimmed hat, and a long-sleeved shirt and long pants. Wear sunscreen with an SPF (sun protection factor) of 15 or higher.

internetconnect

TOPIC: Integumentary System
GO TO: www.scilinks.org
***sci*LINKS NUMBER:** HSTL545

Experimentos

¿Qué tan rápido crecen tus uñas? Averígualo en la página 627.

Pelo y uñas

Un pelo, como el de la **Figura 17,** está formado por varias capas apretadas de células llenas de queratina. Se forma en el fondo de un saco diminuto llamado **folículo piloso.** El pelo crece cuando se agregan células nuevas al folículo y las más viejas se empujan hacia arriba. Las únicas células vivas del pelo son las del folículo piloso, donde éste se produce.

El pelo protege la piel de los rayos ultravioleta y mantiene el polvo y los insectos fuera de los ojos y la nariz. Como la piel, el pelo obtiene su color de la melanina. El cabello obscuro tiene más melanina que el rubio. En la mayoría de los mamíferos, el pelo también ayuda a regular la temperatura del cuerpo. Una contracción de un músculo diminuto unido al folículo piloso hace que éste se doble. En los humanos, cuando los folículos se doblan empujan la epidermis y producen "la carne de gallina". Si el folículo tiene pelo, el pelo se eriza. Cuando los animales peludos hacen esto, se ven más esponjados. Los pelos que se levantan funcionan como un suéter que atrapa el aire tibio alrededor del cuerpo. Los extremos de los dedos de las manos y de los pies están cubiertos por uñas. Las uñas protegen las puntas de los dedos de las manos y de los pies para que puedan permanecer suaves y sensibles. Esto te da un agudo sentido del tacto. Las uñas se forman a partir de la *raíz*, que está bajo la piel en la base y los lados de las uñas. La uña crece al formarse nuevas células. En la **Figura 18,** observa las partes de una uña.

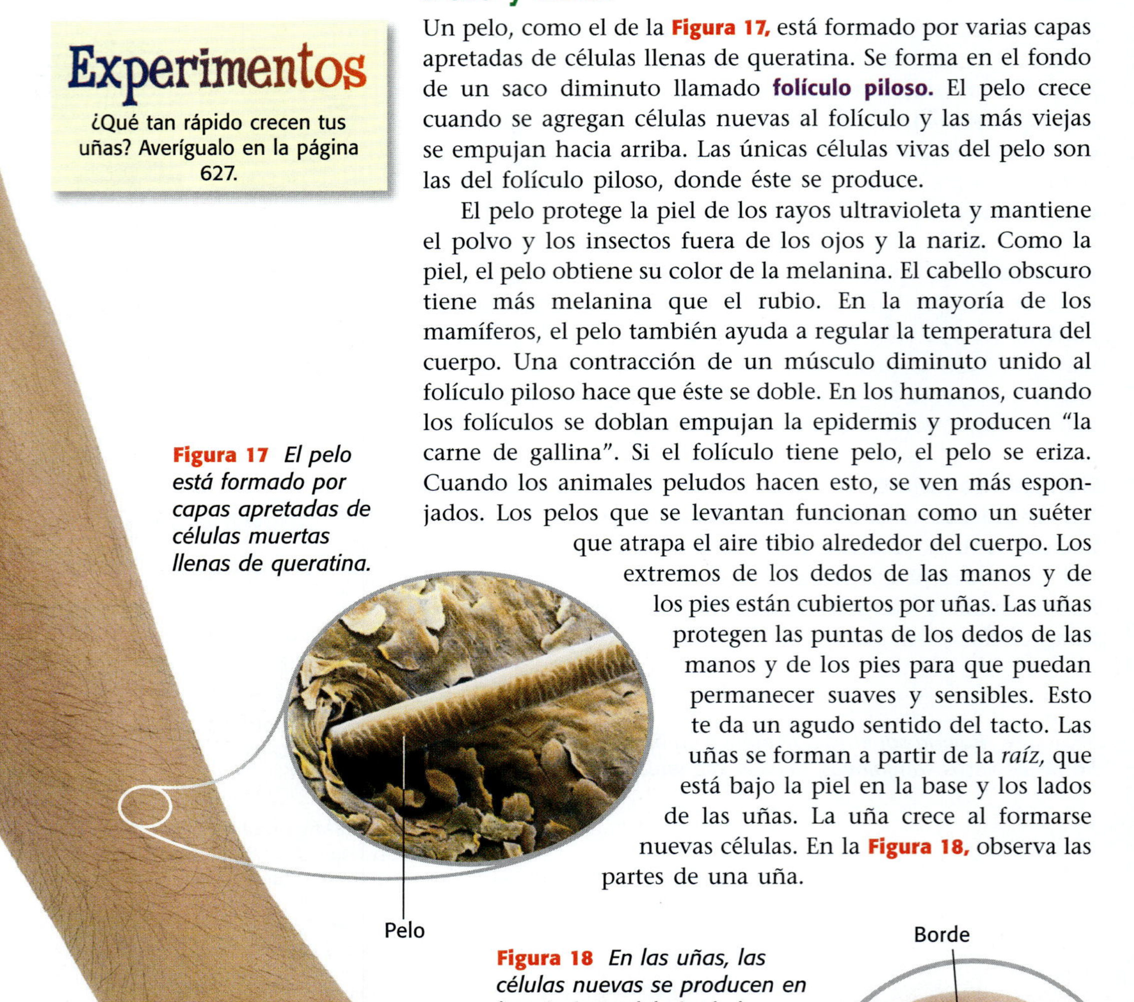

Figura 17 *El pelo está formado por capas apretadas de células muertas llenas de queratina.*

Figura 18 *En las uñas, las células nuevas se producen en la raíz, justo debajo de la lúnula. Las células nuevas empujan a las viejas hacia el borde de la uña.*

SCIENCE HUMOR

Q: Where do sheep get their hair cut?

A: at the baa-baa shop

Difíciles condiciones de vida

La piel es la parte más expuesta del cuerpo. Sirve como barrera protectora pero también se lesiona con frecuencia. Puede ser algo leve, como una ampolla, un piquete o una pequeña cortada. Por suerte, la piel se regenera con sorprendente habilidad, como se ve en la **Figura 19.**

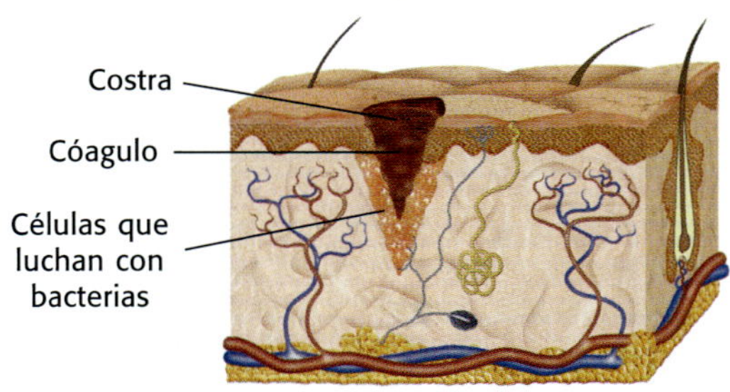

Figura 19 Cómo sana la piel *Para que una herida en la piel sane, se requiere tiempo.*

1 Cuando te cortas, se forma una costra para prevenir que entren bacterias a la herida. Luego, las células encargadas de luchar con las bacterias llegan al área para matarlas.

2 Las células lesionadas se reemplazan por división celular. Al final, sólo queda una cicatriz en la superficie.

Otras lesiones de la piel son más serias. El daño del material genético de las células de la piel puede provocar una división celular sin control, que produce una masa de células llamada tumor. El término *cáncer* se refiere a un tumor que invade otro tejido. Se deben observar con atención las áreas obscuras de la piel, como los lunares, para detectar signos de cáncer. La **Figura 20** muestra un ejemplo de un lunar que posiblemente se ha vuelto canceroso.

La piel también se puede ver afectada por cambios corporales internos. En la adolescencia, las hormonas sexuales hacen que las glándulas sebáceas produzcan un exceso de grasa. Esta grasa se combina con células muertas de piel y bacterias, lo que obstruye los folículos pilosos y causa infecciones. La limpieza adecuada y el cuidado diario de la piel puede disminuir la cantidad de infecciones de la piel.

Figura 20 *Se deben observar los lunares en busca de signos de cáncer. Este lunar tiene dos mitades que no corresponden, una característica llamada asimetría que podría indicar cáncer de piel.*

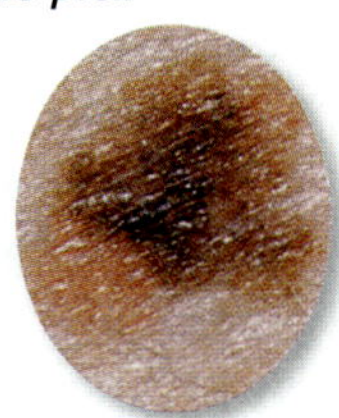

¿Utilizan los médicos vendas hechas de piel verdadera? Averígualo en la página 484.

REPASO

1. ¿Por qué varía el color de la piel entre las personas?

2. Menciona seis estructuras de la dermis y la función de cada una.

3. **Hacer deducciones** ¿Por qué sientes dolor cuando te jalas el cabello o las uñas pero no cuando los cortas?

▼ **Answers to Review**

1. The amount of melanin in the skin determines skin color.

2. nerves, sense; blood vessels, transport; muscle, motion; oil glands, waterproofing; sweat glands, heat regulation; fat cells, insulation

3. The hair and nails don't contain nerves, but the hair follicle and nail root do.

---

3) Extend

DEBATE

Sun Safety Is sun tanning beneficial or dangerous? Lead students in a debate about the benefits of sunlight and the potential dangers and safety issues associated with getting a suntan.

4) Close

Quiz

1. What are five functions of the skin? (keeps moisture in and foreign particles out; provides information about the outside world; helps to regulate body temperature; and removes some wastes)

2. Describe the two layers of the skin. (The epidermis is the thinner outer layer of the skin and contains epithelial tissue and mostly dead cells. The dermis, the deeper, thicker layer of skin, is made up mostly of connective tissue.)

3. How are pimples formed? (Oil glands in the skin produce excess oil that combines with dead skin cells and bacteria to clog hair follicles and cause an infection. The result is a pimple.)

ALTERNATIVE ASSESSMENT

Have students make a colorful drawing of a cross section of skin in their ScienceLog. Have students make their drawings from memory and label the parts and describe the function of each: dermis, epidermis, fat cells, blood vessels, nerves, muscle, hair follicle, oil gland, sweat gland. Sheltered English

Chapter Highlights

VOCABULARY DEFINITIONS

SECTION 1

homeostasis the maintenance of a stable internal environment

tissue a group of similar cells that work together to perform a specific job in the body

epithelial tissue one of the four main types of tissue in the body; the tissue that covers and protects underlying tissue

nervous tissue one of the four main types of tissue in the body; the tissue that sends electrical signals through the body

muscle tissue one of the four main types of tissue in the body; contains cells that contract and relax to produce movement

connective tissue one of the four main types of tissue in the body; functions include support, protection, insulation, and nourishment

organ a combination of two or more tissues that work together to perform a specific function in the body

organ system a group of organs working together to perform body functions

SECTION 2

skeletal system a collection of organs whose primary function is to support and protect the body; the organs in this system include bones, cartilage, ligaments, and tendons

compact bone the type of bone tissue that does not have open spaces; the tissue that gives a bone its strength

spongy bone the type of bone tissue that has many open spaces and contains marrow

cartilage a flexible white tissue that gives support and protection but is not rigid like bone

joint the place where two or more bones connect

Resumen del capítulo

SECCIÓN 1

Vocabulario

homeostasis (*pág. 464*)
tejido (*pág. 464*)
tejido epitelial (*pág. 464*)
tejido nervioso (*pág. 464*)
tejido muscular (*pág. 465*)
tejido conjuntivo (*pág. 465*)
órgano (*pág. 465*)
sistema (*pág. 465*)

Notas de la sección

- Tu cuerpo mantiene un ambiente interno estable llamado homeostasis.

- Cuatro tipos de tejidos trabajan para mantener la homeostasis. Cada tejido tiene una tarea específica.

- Los tejidos trabajan juntos para formar órganos, como el estómago.

- El estómago es parte de un grupo de órganos que transforman el alimento para que pueda nutrir el cuerpo. Un grupo de órganos que trabajan para un propósito común se llama sistema.

- En el cuerpo humano hay 11 sistemas principales.

SECCIÓN 2

Vocabulario

sistema esquelético (*pág. 468*)
hueso compacto (*pág. 469*)
hueso esponjoso (*pág. 469*)
cartílago (*pág. 470*)
articulación (*pág. 470*)
ligamento (*pág. 470*)
rendimiento mecánico (*pág. 471*)

Notas de la sección

- El sistema esquelético comprende huesos, cartílagos y ligamentos.

- Los huesos sostienen y protegen el cuerpo, almacenan minerales y grasas, y producen las células sanguíneas.

- Un hueso típico contiene médula, hueso esponjoso, hueso compacto, vasos sanguíneos y cartílago.

☑ Comprobar destrezas

Conceptos de matemáticas

CALCULAR UN PORCENTAJE En la sección de ¡Matemáticas! de la página 475, se te pidió calcular el porcentaje de un número. Primero expresa el porcentaje como un decimal o una fracción. Luego multiplícalo por el número. Por ejemplo, el 25 por ciento puede expresarse como 0.25 ó 25÷100. Para encontrar el 25 por ciento de 48, multiplica por 0.25 ó 25÷100.

$$0.25 \times 48 = 12$$
$$o$$
$$(25 \div 100) \times 48 = 1{,}200 \div 100 = 12$$

Comprensión visual

MOVIMIENTO DE LAS ARTICULACIONES
Observa los tres tipos de articulaciones de la página 470. Piensa cómo trabajan al lanzar una pelota o subir escaleras. La articulación en bisagra de la rodilla se mueve sólo en dos direcciones. La del hombro, esférica, en muchas. Las de la mano, corredizas, permiten que los huesos se deslicen uno sobre el otro.

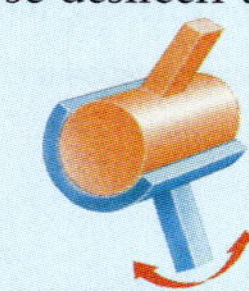 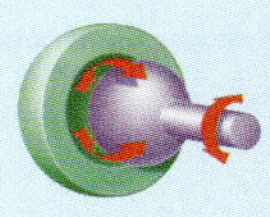 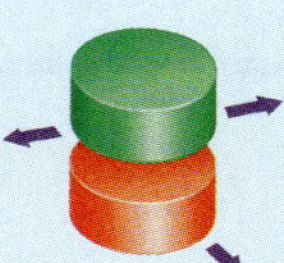

Lab and Activity Highlights

Muscles at Work `PG 626`

Seeing Is Believing `PG 627`

Datasheets for LabBook
(blackline masters for these labs)

Lab and Activity Highlights

LabBank

 **Inquiry Labs,** On a Wing and a Layer, Lab 6

Long-Term Projects & Research Ideas, Project 22

 **Vocabulary Review Worksheet 19**

 Blackline masters of these Chapter Highlights can be found in the **Study Guide.**

Repaso del capítulo

UTILIZAR EL VOCABULARIO

Escoge el término correcto para completar cada una de las siguientes oraciones:

1. Las señales eléctricas son enviadas a través del cuerpo por el tejido __?__. *(epitelial* o *nervioso)*

2. El sistema __?__ está hecho de piel, cabello y uñas. *(integumentario* o *muscular)*

3. Los huesos se mueven gracias al músculo __?__ *(liso* o *esquelético)*

4. Cuando los músculos __?__ se contraen, hacen que algunas partes del cuerpo se doblen. *(extensores* o *flexores)*

5. La mayor parte del esqueleto empieza como __?__, que luego es reemplazado por el hueso. *(cartílago* o *ligamento)*

COMPRENDER CONCEPTOS

Opción múltiple

6. ¿Cuál de los siguientes tejidos contiene células que se contraen y se relajan?
   a. tejido óseo
   b. tejido muscular
   c. tejido conjuntivo
   d. tejido nervioso

7. El sistema que da apoyo y protección a las partes del cuerpo es el
   a. sistema endocrino.
   b. sistema circulatorio.
   c. sistema óseo.
   d. sistema integumentario.

8. La epidermis está compuesta de
   a. dermis.
   b. tejido epitelial.
   c. tejido conjuntivo.
   d. pura piel.

9. El punto fijo de una palanca es
   a. la fuerza.
   b. la resistencia.
   c. el punto de apoyo.
   d. el rendimiento mecánico.

10. Los músculos hacen que los huesos se muevan cuando
    a. los músculos se estiran.
    b. los músculos crecen entre los huesos.
    c. los músculos jalan los huesos.
    d. los músculos empujan los huesos.

11. Los ligamentos son el tejido conjuntivo que conecta los
    a. huesos con los músculos.
    b. huesos entre sí.
    c. músculos entre sí.
    d. músculos con la dermis.

Respuesta breve

12. Resume las funciones de los cuatro tipos de tejidos y haz un esquema de cada tipo.

13. ¿Cómo ayuda la piel a proteger el cuerpo?

14. ¿Qué es la "carne de gallina"?

15. ¿Cuáles son las dos diferencias entre el músculo esquelético y el cardíaco?

16. ¿Cómo se relacionan las funciones del sistema óseo con las del sistema muscular?

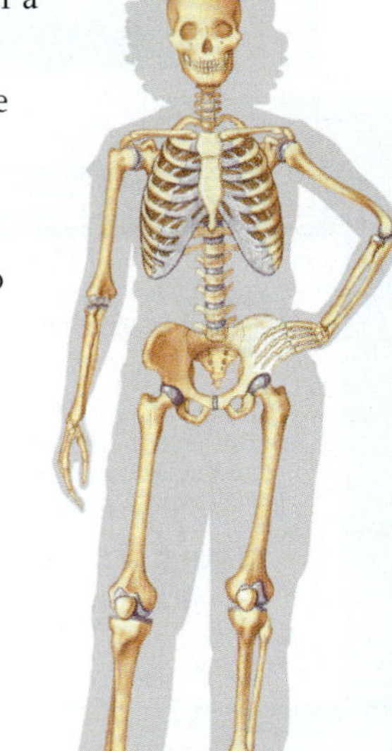

Concept Mapping

17. 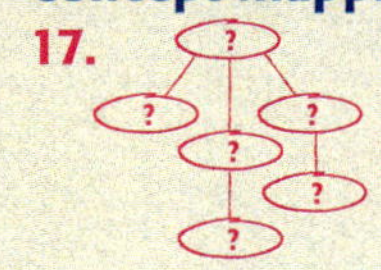 An answer to this exercise can be found at the end of this book.

Concept Mapping Transparency 19

Organizar conceptos

17. Usa los siguientes términos para crear un mapa de ideas: huesos, médula, sistema óseo, hueso esponjoso, hueso compacto, cartílago.

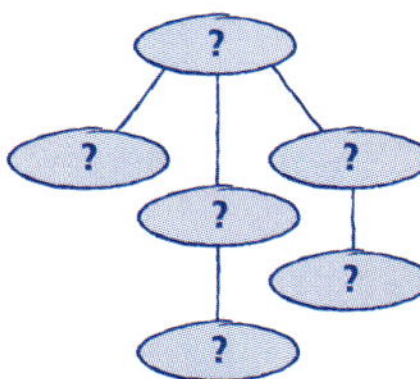

RAZONAMIENTO CRÍTICO Y RESOLUCIÓN DE PROBLEMAS

Escribe una o dos oraciones para responder a las siguientes preguntas:

18. ¿Cómo sería el movimiento del brazo si no tuvieras un extensor y un flexor conectados al antebrazo?

19. A diferencia de los huesos humanos, los de algunas aves tienen cavidades llenas de aire. ¿Qué ventaja les da esto a las aves?

20. Compara la forma de los huesos del cráneo humano con la de los de la pierna humana. ¿Por qué es importante la forma?

21. Compara la textura y sensibilidad de la piel de los codos con la de la piel de las yemas de los dedos. ¿Cómo puedes explicar la diferencia?

LAS MATEMÁTICAS EN LAS CIENCIAS

22. Tus músculos representan cerca del 40 por ciento de todo tu peso. ¿Cuál es la masa muscular de una persona cuya masa corporal total es de 60 kg?

23. Una persona promedio parpadea 700 veces por hora. ¿Cuántas veces parpadearía en una semana si estuviera despierta 16 horas diarias?

INTERPRETAR GRÁFICAS

Observa esta fotografía y responde a las siguientes preguntas:

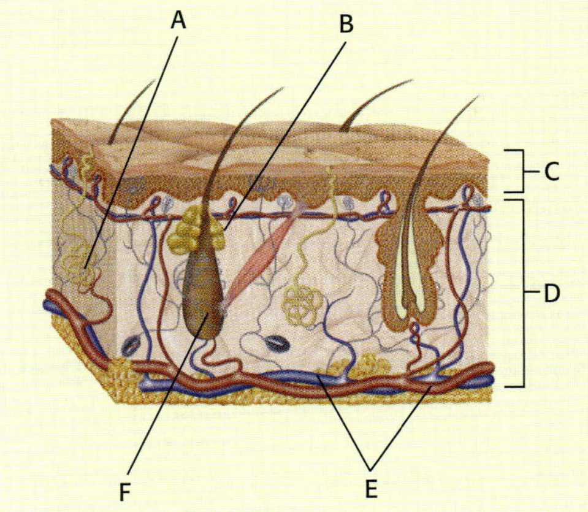

24. ¿Cómo se llama la parte *D*? ¿Qué tipo de tejido abunda más en esta capa?

25. ¿Cuál es el nombre y la función de la parte *A*?

26. ¿Cuál es el nombre y la función de la parte *B*?

27. ¿Qué parte de la piel está hecha de tejido epitelial que contiene células muertas?

28. ¿Cómo ayuda la piel a regular la temperatura del cuerpo?

AHORA, ¿qué piensas?

Repasa tus respuestas a las preguntas de la página 463 que escribiste en tu cuaderno de ciencias. ¿Han cambiado tus respuestas? Si es necesario, corrige tus respuestas basándote en lo que has aprendido en este capítulo.

CRITICAL THINKING AND PROBLEM SOLVING

18. You would not be able to straighten your arm, and you would not be able to bend your arm. You would not be able to pull it back into a different position.

19. Hollow bones are lighter, which makes flight possible.

20. The bones in the skull are shaped to form a strong helmet-like protection around the brain. The leg bone has a design that makes standing upright and moving around possible.

21. Students should notice that the skin on their elbows is hard and rough compared with the delicate, sensitive skin on their fingertips. The skin on an elbow has to be tough and insensitive because we rest our upper body on it.

MATH IN SCIENCE

22. Forty percent of 60 kg is 24 kg.

23. $700 \frac{blinks}{hours} \times 16 \frac{hours}{day} \times 7 \frac{days}{week} = 78{,}400 \frac{blinks}{week}$

INTERPRETING GRAPHICS

24. the dermis, connective tissue

25. Sweat glands produce sweat that helps regulate body temperature.

26. Oil glands produce oil, which waterproofs the skin.

27. C. Epidermis

28. Sweat glands produce sweat, which helps cool the body.

NOW WHAT DO YOU THINK?

1. Organs are made of tissues, and tissues are made of cells.

2. Skin provides protection from the elements; muscles provide movement; and bones provide structure and protection.

Blackline masters of this Chapter Review can be found in the **Study Guide.**

Background

The engineered skin is far better than the scar tissue that would form without it. Scar tissue is weaker and more brittle than the skin it replaces. It does not stretch and grow, making a particularly difficult problem for children suffering from burns. Engineered skin also helps reduce the disfigurement associated with scarring.

One significant limitation of the engineered skin is that when it is new, it lacks sweat glands. Because patients cannot perspire in the areas that have skin grafts, patients with large grafts need to be cautious about overexercising and exposure to the sun.

Piel fabricada

Tu piel es más que un traje bien ajustado, es tu primera línea de defensa contra el mundo exterior. La piel te mantiene a salvo de deshidratación e infecciones, y las glándulas sebáceas funcionan como impermeabilizantes. Pero, ¿qué pasa cuando una porción importante de piel está dañada?

La respuesta es: poner más piel

Algunas veces los doctores hacen un injerto de piel, transfieren piel sana de una persona al área de la piel dañada. Esto es porque la piel es la mejor "venda" para una herida. Protege la herida pero le permite respirar. Y a diferencia de las vendas de tela y de plástico, la piel se regenera a medida que cubre la herida. Aunque, a veces, la piel de una persona está tan seriamente dañada (como pasa con las víctimas de quemaduras) que ya no queda piel para regenerar.

Fabricación de tejido

En los últimos años, los científicos han estudiado la fabricación de tejido para aprender más sobre la forma en que el cuerpo humano sana de manera natural. Con un poco de piel humana joven y sana, y colágeno de vaca, los científicos pueden crear piel humana. Durante el proceso de fabricación, las células forman las capas dérmicas y epidérmicas de la piel tal como si aún estuvieran en el cuerpo. Incluso, la piel humana viva que resulta puede sanar si es cortada antes de usarse para un injerto de piel. Como está viva, la piel debe ser guardada en un medio que le proporcione nutrientes hasta que se coloque en la herida. Con el tiempo, el color de la piel injertada cambia y se vuelve del color de la piel que la rodea.

Una dermis tejida

Los especialistas en los tejidos también han creado otro tipo de piel, pero ésta tiene una

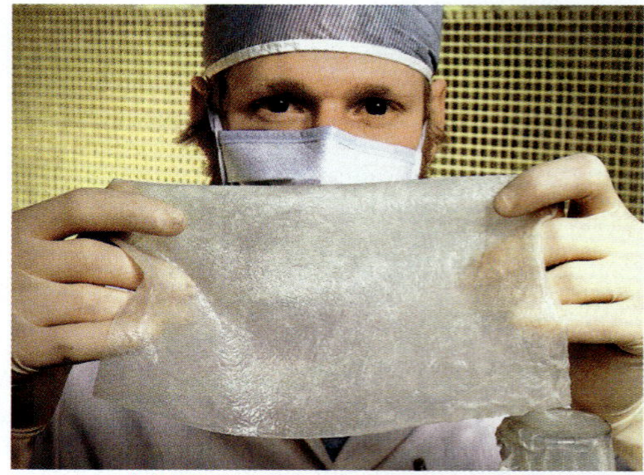
▲ *Esta es una parte de piel fabricada que se usa para injertos.*

capa dérmica y epidérmica inusual. En esta piel, la dermis está hecha de fibras de colágeno tejidas. El área tejida asimila estas fibras y las usa como guía para crear una nueva dermis. La capa epidérmica es una capa temporal de silicón. Protege al cuerpo de infecciones y de deshidratación mientras que la nueva piel se crea.

Después de formarse una nueva capa dérmica bajo la epidermis de silicón protectora, el cuerpo está en mejores condiciones para aceptar un injerto de piel. Los doctores también pueden injertar una porción de piel más delgada. A la larga, un injerto más delgado es mejor para el cuerpo porque es más fácil de sacar de otra parte del cuerpo. La nueva capa dérmica también le da al cuerpo más tiempo para fortalecerse antes del trauma del trasplante de la piel sana a otras áreas.

Averígualo tú mismo

▶ Antes, los doctores recolectaban piel de cuerpos de personas que, antes de morir, decidían donar sus órganos. ¿Qué tipo de problemas podrían surgir si esta piel recolectada se usara en víctimas de quemaduras?

484

Answer to On Your Own

Skin harvested from cadavers can be rejected, and it can also introduce viruses and disease. The bits of skin used to create engineered skin are from the foreskins of babies, and there seems to be a readily available supply. Additionally, youthful skin grows very quickly.

¡Lo encontré!

Derrames de aceite piloso

El aceite y el agua no se mezclan, ¿verdad? El aceite flota en la superficie del agua y se ve a simple vista. Probar esto en tu cocina no es difícil ni peligroso. Pero, ¿qué pasa cuando el agua es el océano y el aceite es petróleo? Se produce un desastre ambiental que cuesta millones de dólares resolver. El peor ejemplo en las aguas del continente americano fue en 1989, cuando el buque petrolero Exxon Valdez derramó 42 millones de litros de petróleo crudo en las aguas de Prince William Sound en la costa de Alaska.

▲ *Esta nutria está empapada de petróleo derramado por el* Exxon Valdez.

Prueba en el patio de atrás

Un estilista de Huntsville, Alabama, hizo una pregunta brillante cuando vio una nutria cuyo pelaje estaba empapado de petróleo derramado por el Valdez. ¿Si el pelaje de la nutria absorbió el aceite, por qué el cabello humano no haría lo mismo? El estilista, Phil McCrory, recogió cabello del piso de su salón para hacer experimentos. Llenó un par de medias con 2.2 kg de cabello y las amarró para formar un bulto en forma de anillo. Después de llenar de agua la alberca, McCrory puso a flotar el bulto. Luego, en el centro del anillo echó aceite usado de motor. Cuando sacó las medias, cerrando el anillo, ¡no había ni una gota de aceite en el agua!

¿Cómo hace esto el cabello?

Lo que McCroy descubrió fue que el cabello

▲ *Phil McCroy entre costales llenos de cabello humano.*

adsorbe el aceite en lugar de **ab**sorberlo. Adsorber significa retener un líquido o gas en capas sobre una superficie. Como cada cabello está cubierto de cutículas diminutas como escamas, el aceite puede pegarse a la superficie del cabello. Compara este proceso con la forma en que trabaja una esponja. Ésta absorbe por completo un líquido, humedeciéndose completamente, no sólo en la superficie.

McCroy presentó su descubrimiento a la Administración Nacional de Aeronáutica y del Espacio, NASA *(National Aeronautics and Space Administration)*. Se demostró que el cabello es el adsorbente más veloz. Un poco más de 1 kg de cabello puede adsorber más de 3.5 L de aceite en sólo 2 minutos.

Se calcula que en una semana, 64 millones de kilogramos de cabello metidos en almohadas de malla pudieron haber embebido todo el petróleo que derramó el *Valdez.* Desgraciadamente, los 2 mil millones de dólares gastados en la limpieza removieron sólo el 12 por ciento del petróleo.

Compara los hechos

▶ Investiga cómo estre descubrimiento se compara con métodos actuales usados para limpiar derrames de aceite. Comparte tus descubrimientos con tu clase.

485

EUREKA!
Hairy Oil Spills

Background

Bioremediation is the use of biological processes, often the action of microorganisms, to eliminate organic contaminants from soils. Hair is a biological waste product that doesn't degrade well in landfills. Using hair for the bioremediation of oil spills would reduce the amount of waste in landfills. Also, since the hair adsorbs the oil, wringing the hair out means the oil can be recovered and the hair can be used again. As a last resort, the oil-saturated mesh pillows can be burned as fuel in order to recover the value of the oil they contain.

Activity

Have a few students work in a group, and ask them to research the current state of Prince William Sound. Has the ecosystem fully recovered? Are there still problems? Ask these students to report their findings to the class.

Answer to Compare the Facts

Current methods of cleanup include skimmers, booms, and dispersing agents. Skimmers recover oil from the water's surface. There are three types of skimmers—weir, oleophilic, and suction. Dispersing agents contain surfactants, which break liquids such as oil into small drops. Booms are used for containment so that other areas are not contaminated. Booms also concentrate the oil into thicker layers on the water's surface, which facilitates recovery. One type of boom has natural oil-eating microbes within it. These microbes biodegrade the contaminants and then the boom biodegrades itself.

McCrory's method has the potential to save millions of dollars. Current remediation procedures cost about $10 per 3.8 L of oil, while his method may cost as little as $2.

Chapter Organizer

CHAPTER ORGANIZATION	TIME MINUTES	OBJECTIVES	LABS, INVESTIGATIONS, AND DEMONSTRATIONS
Chapter Opener pp. 486–487	45		**Investigate!** Exercise and Your Heart, p. 487
Section 1 The Cardiovascular System	90	▶ Describe the functions of the cardiovascular system. ▶ Compare and contrast the three types of blood vessels. ▶ Describe the path that blood travels as it circulates through the body. ▶ Distinguish between blood types.	**Whiz-Bang Demonstrations,** Get the Beat! Demo 12
Section 2 The Lymphatic System	90	▶ Discuss the functions of the lymphatic system. ▶ Identify the relationship between lymph and blood. ▶ Describe the organs of the lymphatic system.	
Section 3 The Respiratory System	90	▶ Describe the flow of air through the respiratory system. ▶ Discuss the relationship between the respiratory system and the circulatory system. ▶ Identify respiratory disorders.	**QuickLab,** Why Do People Snore? p. 501 **Making Models,** Build a Lung, p. 630 **Datasheets for LabBook,** Build a Lung, Datasheet 42 **Skill Builder,** Carbon Dioxide Breath, p. 631 **Datasheets for LabBook,** Carbon Dioxide Breath, Datasheet 43 **EcoLabs & Field Activities,** There's Something in the Air, Field Activity 9 **Whiz-Bang Demonstrations,** Take a Deep Breath, Demo 13 **Long-Term Projects & Research Ideas,** Project 23

TECHNOLOGY RESOURCES

 **Guided Reading Audio CD**
English or Spanish, Chapter 20

 **CNN.** Science, Technology & Society,
Breakthrough Bandage, Segment 30
Modern Acupuncture, Segment 31

 **One-Stop Planner CD-ROM with Test Generator**

CLASSROOM WORKSHEETS, TRANSPARENCIES, AND RESOURCES	SCIENCE INTEGRATION AND CONNECTIONS	REVIEW AND ASSESSMENT
Directed Reading Worksheet 20 **Science Puzzlers, Twisters & Teasers,** Worksheet 20		
Directed Reading Worksheet 20, Section 1 **Math Skills for Science Worksheet 31,** The Unit Factor and Dimensional Analysis **Transparency 75,** The Flow of Blood Through the Heart **Transparency 76,** The Flow of Blood Through the Body **Reinforcement Worksheet 20,** Matchmaker, Matchmaker **Reinforcement Worksheet 20,** Colors of the Heart **Critical Thinking Worksheet 20,** Doctor for a Day	**Math and More,** p. 490 in ATE **Apply,** p. 494 **MathBreak,** The Beat Goes On, p. 495 **Health Watch:** Goats to the Rescue, p. 507	**Homework,** pp. 490, 491 in ATE **Self-Check,** p. 492 **Review,** p. 493 **Review,** p. 495 **Quiz,** p. 495 in ATE **Alternative Assessment,** p. 495 in ATE
Directed Reading Worksheet 20, Section 2		**Self-Check,** p. 496 **Review,** p. 497 **Quiz,** p. 497 in ATE **Alternative Assessment,** p. 497 in ATE
Transparency 77, The Respiratory System **Directed Reading Worksheet 20,** Section 3 **Transparency 180,** Air Pressure and Breathing	**Earth Science Connection,** p. 499 **Multicultural Connection,** p. 499 in ATE **Connect to Physical Science,** p. 500 in ATE **Weird Science,** Catching a Light Sneeze, p. 506	**Review,** p. 501 **Quiz,** p. 501 in ATE **Alternative Assessment,** p. 501 in ATE

Holt, Rinehart and Winston On-line Resources

go.hrw.com

For worksheets and other teaching aids related to this chapter, visit the HRW Web site and type in the keyword: **HSTBD2**

National Science Teachers Association

www.scilinks.org

Encourage students to use the *sci*LINKS numbers listed with the Chapter Highlights to access information and resources on the **NSTA** Web site.

END-OF-CHAPTER REVIEW AND ASSESSMENT

Chapter Review in Study Guide
Vocabulary and Notes in Study Guide
Chapter Tests with Performance-Based Assessment, Chapter 20 Test
Chapter Tests with Performance-Based Assessment, Performance-Based Assessment 20
Concept Mapping Transparency 20

Chapter Resources & Worksheets

Visual Resources

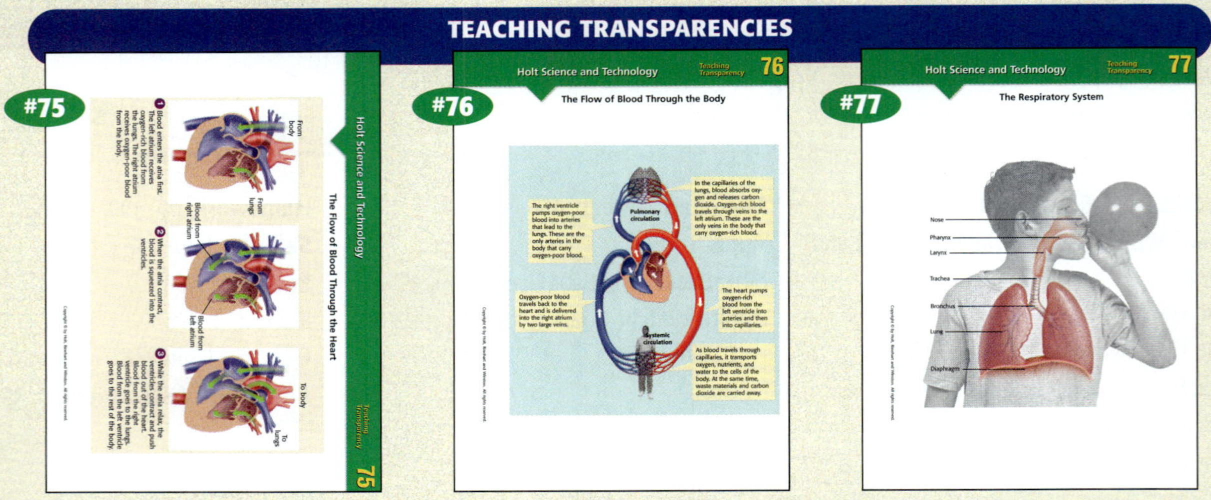

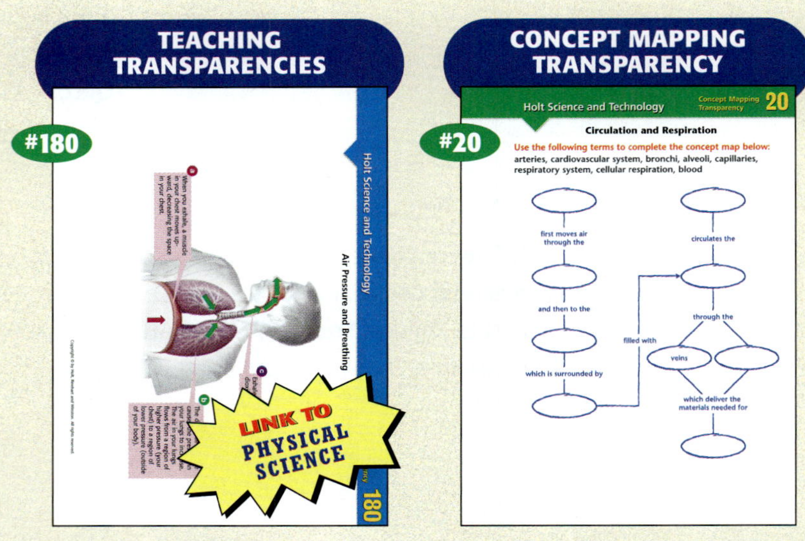

Meeting Individual Needs

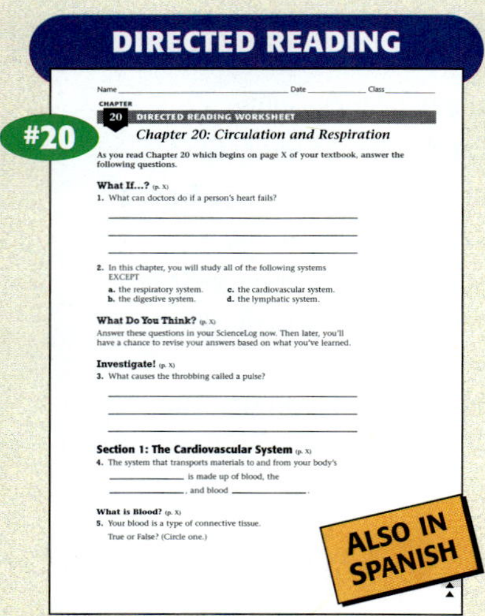

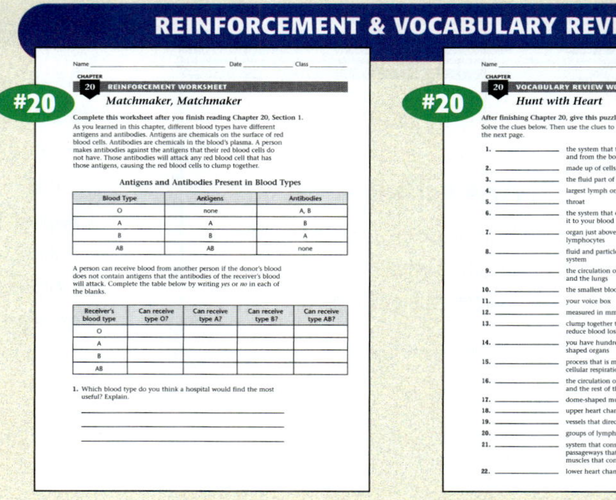

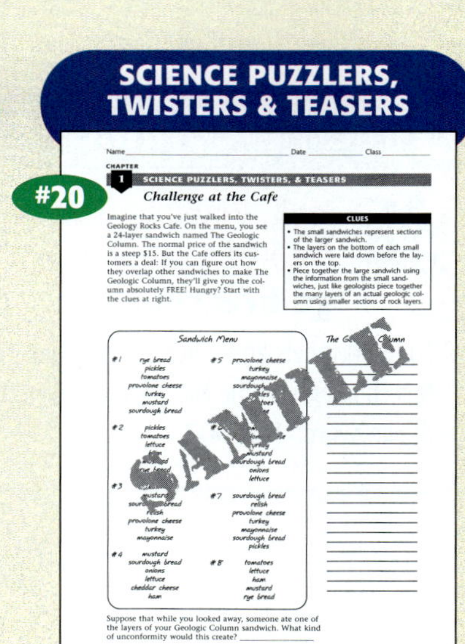

Review & Assessment

STUDY GUIDE

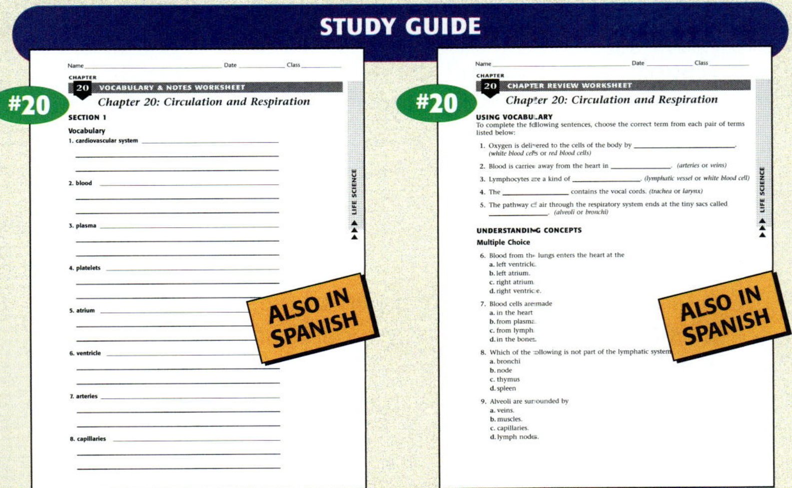

CHAPTER TESTS WITH PERFORMANCE-BASED ASSESSMENT

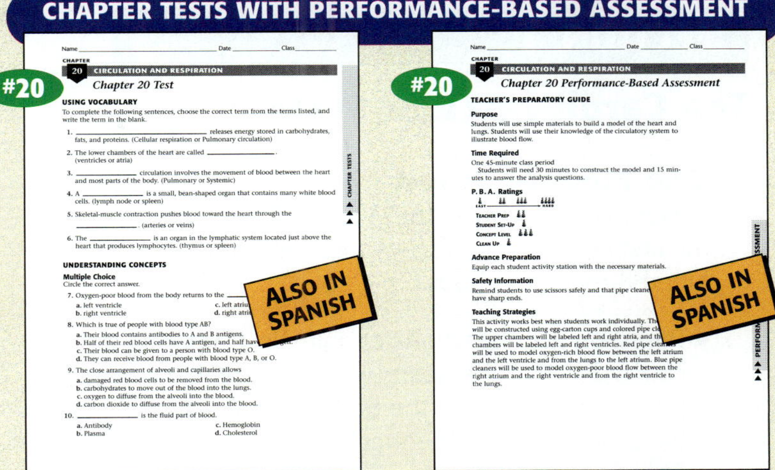

Lab Worksheets

ECOLABS & FIELD ACTIVITIES

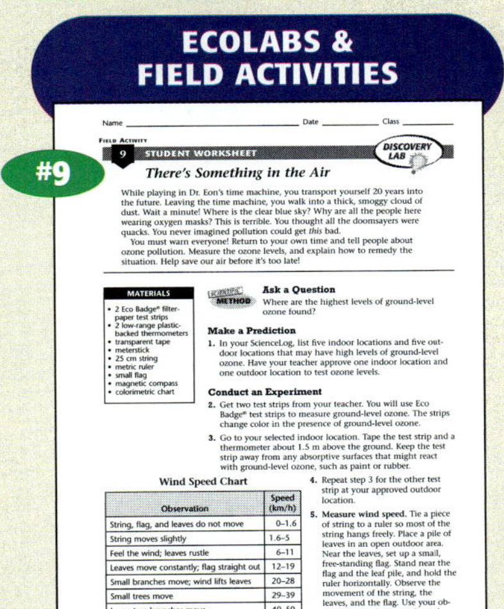

WHIZ-BANG DEMONSTRATIONS

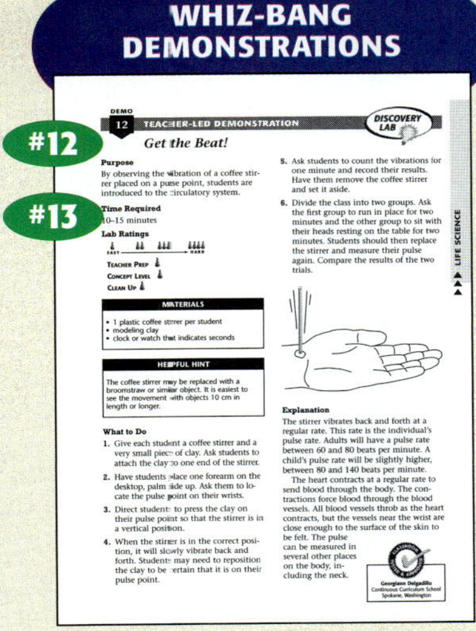

LONG-TERM PROJECTS & RESEARCH IDEAS

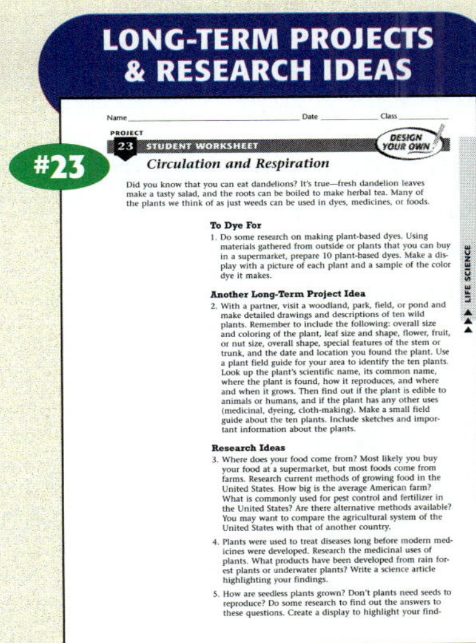

DATASHEETS FOR LABBOOK

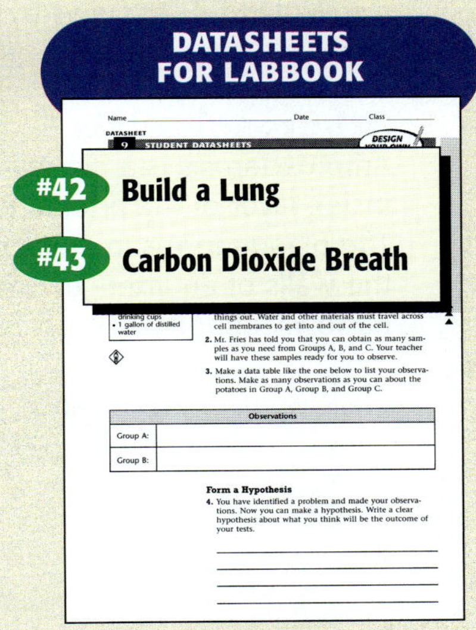

Applications & Extensions

CRITICAL THINKING & PROBLEM SOLVING

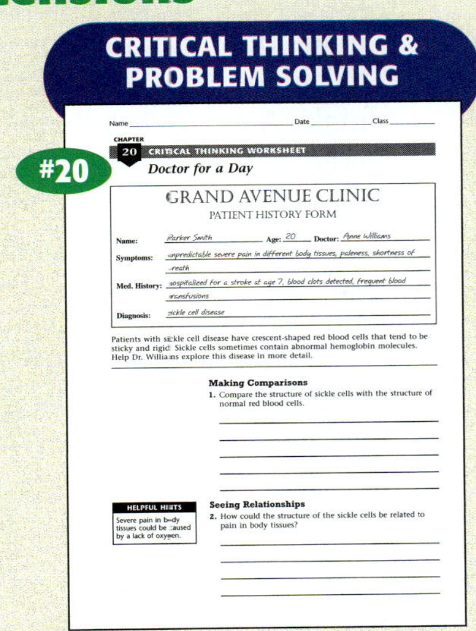

SCIENCE TECHNOLOGY

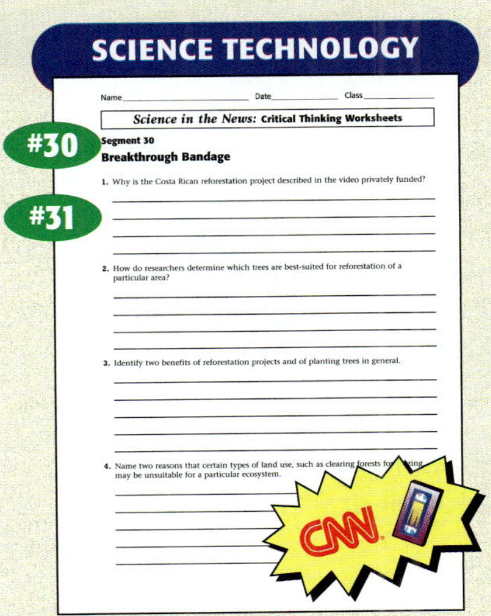

Chapter Background

The Cardiovascular System

▶ The Flow of Blood

William Harvey (1578–1657) is credited with being the first European to discover the circulation of the blood through the body.

- Based on his dissections of animals, Harvey rightly concluded that the heart was a muscle that served to pump blood through the body. In contrast to conventional wisdom and theories of the famous physician Galen (A.D. 129–c. 201), Harvey also correctly maintained that arteries carry blood away from the heart and veins carry blood toward the heart. Harvey was ridiculed for his views by many other physicians.

▶ Atherosclerosis

Atherosclerosis is a disease of the arteries in which the inside layer of the arterial walls thickens with plaque. The plaque forms in areas of turbulent blood flow. As the walls of an artery thicken, the diameter of the vessel narrows, impeding blood flow.

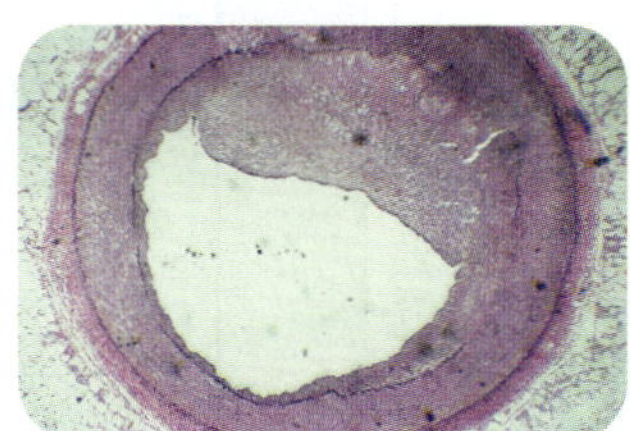

- Cigarette smoking, high blood pressure, obesity, inactivity, high cholesterol level, and a family history of heart disease are all risk factors for atherosclerosis.

▶ Hypertension

Like atherosclerosis, people afflicted with hypertension may not exhibit symptoms of the disease for years. In fact, hypertension is often called "the silent killer." Although blood pressure varies within a wide range across the population, a person whose resting blood pressure is consistently at the high end of that range is said to have hypertension.

- Smoking, obesity, stress, excessive consumption of alcohol, and diabetes mellitus all exacerbate high blood pressure.

▶ Stroke

A stroke occurs when the brain is damaged due to an interruption in the blood flow or leaking of blood from the blood vessels. Atherosclerosis and hypertension are some of the causes of strokes.

▶ Heart Attack

A heart attack occurs when part of the heart muscle dies due to blood and oxygen deprivation. About 1 million people in the United States suffer a heart attack each year.

▶ Blood

If red blood cells (RBCs) have an Rh antigen (a particular kind of protein on the surface of the cell), the blood is positive (+). If an Rh antigen is not present, the blood is negative (−). People have one of the following blood types: A+, A−, B+, B−, AB+, AB−, O+, or O−.

- People make antibodies against the antigens that they do not have on their own RBCs. However, people who have Rh− blood do not always have antibodies against the Rh antigen. These antibodies are made only after the Rh− person has been exposed to blood that is Rh+. For example, type B− people make A antibodies that attack any blood cell with an A antigen on it. They may also make Rh antibodies that will attack any blood cell with an Rh antigen on it. This means that people with type B− blood can't be given A+, A−, B+, AB+, or AB− blood. Type O− blood can be given to anyone because its RBCs have no antigens on their surface that a patient's antibodies could attack. Because of this, a type O− person is said to be a *universal donor.* Type AB+ people are *universal recipients;* they can be given any type of blood because they do not make any antibodies against A, B or Rh antigens.

The Lymphatic System

▶ Lymphocytes

Lymphocytes are white blood cells that specialize in fighting pathogens. The two main kinds of lymphocytes are B cells and T cells.

- About 10 percent of lymphocytes are B cells. When confronted with foreign antigens, B cells produce antibodies that destroy the antigens. This process is called humoral immunity.

- About 90 percent of lymphocytes are T cells, which are formed in the bones and mature in the thymus. Killer T cells locate and attack cells that have foreign antigens on their surface. This type of immunity is called cell-mediated immunity.

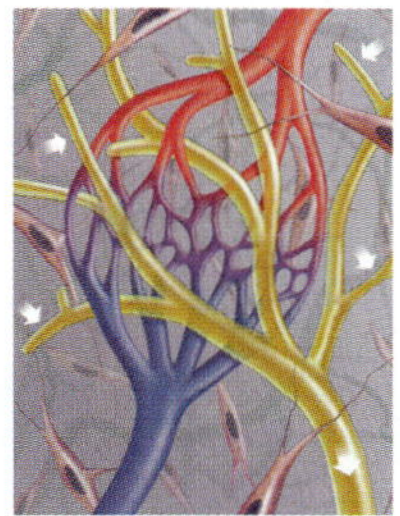

- HIV infects and destroys lymphocytes called helper T cells. When a person's helper T cell count falls below 200 cells per cubic millimeter of blood, the person is diagnosed with AIDS.

IS THAT A FACT!

- The tonsils reach their largest size when a person is about seven years old. Then the tonsils begin to shrink.

The Respiratory System

▶ Control of Breathing

Unless a person consciously holds his or her breath or changes the rate of his or her breathing, breathing is controlled automatically by breathing control centers in the base of the brain (in the medulla oblongata and in the pons).

- Hiccups are caused by a sudden jerky contraction of the diaphragm. When a person eats too much food, the full stomach may irritate the diaphragm muscle, causing it to contract jerkily. Generally, though, the cause of hiccups is unknown.

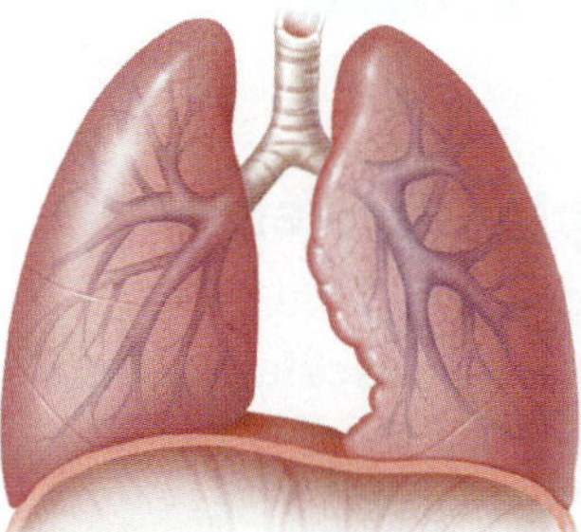

- Yawning is caused by a buildup of carbon dioxide in the lungs. The brain controls breathing by monitoring carbon dioxide, rather than oxygen, levels in the body. Thus, shallow breathing may result in the accumulation of carbon dioxide in the lungs, resulting in a signal from the brain to take a deep breath and exhale the extra carbon dioxide.

▶ Smoking

Tobacco smoking has been implicated in more than 90 percent of lung cancers among men. Among people who do not smoke, 3,000 cases of lung cancer are linked to secondhand cigarette smoke each year.

IS THAT A FACT!

- More than 3,000 adolescents in the United States start to smoke each day. The habit will eventually kill one-third of these children.

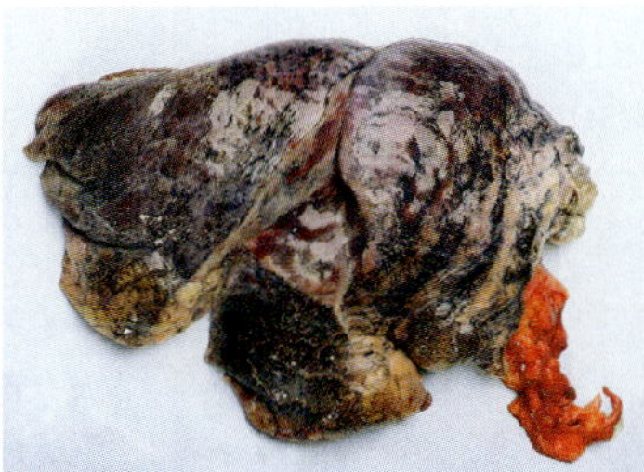

- Each year, parents who smoke at least 10 cigarettes a day cause 8,000 to 26,000 new cases of asthma in children. Moreover, each year between 200,000 and 1 million children who already have asthma have their condition worsened by parental secondhand smoke.

For background information about teaching strategies and issues, refer to the *Professional Reference for Teachers.*

Directed Reading Worksheet 20

Science Puzzlers, Twisters & Teasers Worksheet 20

Guided Reading Audio CD
English or Spanish, Chapter 20

CAPÍTULO 20
Circulación y respiración

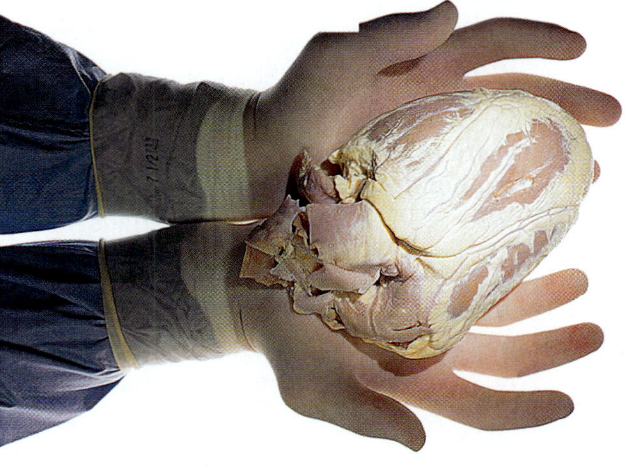

¡Esto realmente sucedió!

El corazón humano es normalmente un órgano muy confiable. Puede latir más de 100,000 veces al día durante toda la vida de una persona bombeando millones de litros de sangre por el cuerpo. Un problema grave del corazón es una amenaza para la vida.

En el pasado, una insuficiencia del corazón causaba la muerte inmediata. En 1969, el Dr. Denton Cooley, del Instituto Cardiológico de Texas *(Texas Heart Institute)*, mantuvo vivo por cinco días a un paciente a quien le falló el corazón, utilizando un corazón artificial diseñado por él mismo.

El diseño de corazones artificiales ha mejorado mucho desde el primer modelo del Dr. Cooley. El corazón eléctrico (arriba a la derecha), es uno de los modelos de prueba más recientes. Pesa sólo 680 g, o sea, es un poco más pesado que un corazón humano de verdad. Los modelos de prueba más recientes son más pequeños y livianos. Estos corazones tienen sensores especiales y microprocesadores que regulan los latidos y responden a los cambios en la presión sanguínea.

Actualmente no hay un corazón artificial que reemplace al corazón humano de manera permanente. El corazón humano es un órgano sofisticado que abastece al aparato cardiovascular. En este capítulo también estudiarás el sistema linfático y el aparato respiratorio, y cómo se relacionan estos sistemas.

486

This Really Happened ...

Each year only about 2,000 human hearts are available for the nearly 45,000 Americans who need a transplant. An experimental and controversial procedure is xenotransplantation, the transplanting of an organ from one species into another species. Hearts, livers, and kidneys from baboons and pigs have been used successfully to extend patients' lives. If the ethical reservations can be resolved, xenotransplantation could alleviate the shortage of human organ donors.

Ejercicio y corazón

¿Cómo responde tu corazón al ejercicio? Esto lo puedes determinar midiendo qué tan rápido palpita tu corazón después de hacer ejercicio. El corazón bombea sangre a través de los vasos sanguíneos y, con cada latido, estos se expanden y luego regresan a su estado original. Esto ocasiona un golpeteo llamado *pulso*. Puedes tomarte el pulso colocando los dedos índice y medio en la parte interna de la muñeca, justo abajo del pulgar.

Procedimiento

1. Tómate el pulso mientras estás inmóvil. Con **un reloj con segundero,** cuenta el número de latidos que hay en 15 segundos. Luego, multiplica este número por cuatro para calcular el número de latidos que hay en un minuto.

$$\text{Número de latidos en 15 segundos} \times 4 = \text{Número de latidos en 1 minuto}$$

2. Trota en el mismo lugar por 30 segundos, detente y calcula de nuevo el ritmo de tu corazón.
   Cuidado: no hagas este ejercicio si tienes problemas respiratorios, presión sanguínea alta o si te mareas fácilmente.

3. Descansa 5 minutos y luego determina tu pulso otra vez.

Análisis

4. ¿Cómo afectó el ejercicio al pulso? ¿Por qué crees que pasó esto?

5. ¿Por qué crees que tu pulso se normalizó después de que descansaste?

487

WEIRD SCIENCE

Leeches make a protein called hirudin that prevents blood from clotting while the leeches drink blood from their victims. Hirudin derived from leeches is useful in preventing dangerous blood clots in patients recovering from surgery.

What Do You Think?

Accept all reasonable responses.

Students will have a chance to revise their answers in the Chapter Review under NOW What Do You Think?

Investigate!

MATERIALS

FOR EACH GROUP
• watch with a second hand

Safety Caution: Have students bring in a signed permission slip for this activity. Any students who have a health problem that is worsened by exercise should be excused from the exercise portion of this activity. Instruct students who feel pain or become dizzy or tired to stop exercising immediately. Some students may feel embarrassed to exercise in front of their peers. You may want to invite a few volunteers to perform this part of the activity instead of having all students exercise.

Answers to Investigate!

4. Students should notice that their heart rate goes up when they are exercising. Explanations will vary. Students might suggest that while exercising, more blood is needed to deliver energy and oxygen to the muscles.

5. Once exercise is complete, less blood is needed, so the heart can slow down.

Focus

The Cardiovascular System

This section introduces the structures and functions of the cardiovascular system. Students compare and contrast three different types of blood vessels and trace the path of blood through these vessels in the body. Students also learn to distinguish between the different blood types.

Bellringer

Ask students to list in their ScienceLog as many song titles, phrases, and slogans that contain the word *heart* as they can in 2–3 minutes. Ask for examples, and list them on the board. Ask for reasons why the word *heart* is the focus of so many songs and slogans.

1 Motivate

DISCUSSION

Invite students to describe a time when their skin was cut. Encourage students to describe not only what happened but also what their blood looked like. Make a list of the words students use to describe their blood. Based on students' experiences with blood and bleeding, lead a discussion about the structure and functions of blood. Then have students read these pages and compare their descriptions of blood with the one in the textbook. Ask:

Can you see individual blood cells when you bleed? Why or why not? How are red blood cells different from other cells in the body? (They lack a nucleus, organelles, and DNA, and they carry oxygen to other cells in the body.)

Sección 1

VOCABULARIO

sistema cardiovascular
sangre
plasma
plaqueta
aurícula
ventrículo
arterias
capilares
venas
circulación pulmonar
circulación sistémica
presión sanguínea

OBJETIVOS

- Describe las funciones del sistema cardiovascular.
- Compara y contrasta los tres tipos de vasos sanguíneos.
- Describe el recorrido de la sangre a través del cuerpo.
- Distingue los diferentes grupos sanguíneos.

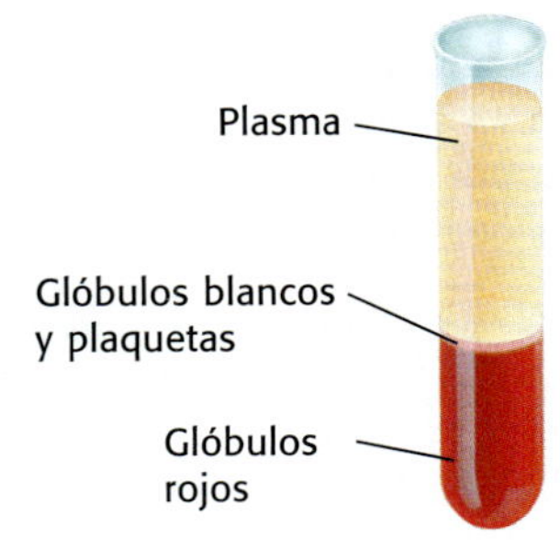

Figura 2 *La sangre está compuesta aproximadamente por un 55 por ciento de plasma y un 45 de glóbulos blancos y plaquetas.*

Figura 3 *Los glóbulos rojos transportan oxígeno.*

488

El sistema cardiovascular

Al escuchar la palabra *corazón*, ¿en qué piensas primero? Muchas personas piensan en el amor, pero el corazón es mucho más que un símbolo del amor. Es la bomba que impulsa el sistema, o aparato cardiovascular. El **sistema cardiovascular** transporta substancias a las células y desde ellas. La palabra *cardio* significa "corazón" y *vascular* significa "vaso." El aparato cardiovascular, que ves en la **Figura 1,** tiene tres componentes: sangre, corazón y vasos sanguíneos.

¿Qué es la sangre?

El cuerpo humano contiene alrededor de 5 L de sangre. La **sangre** es un tejido conjuntivo formado por dos tipos de células, fragmentos celulares y plasma. La **Figura 2** muestra sangre separada en sus cuatro partes principales. El **plasma** es la parte líquida; es una mezcla de agua, minerales, nutrientes, azúcares, proteínas y otras substancias. Los glóbulos rojos, glóbulos blancos y plaquetas flotan en el plasma.

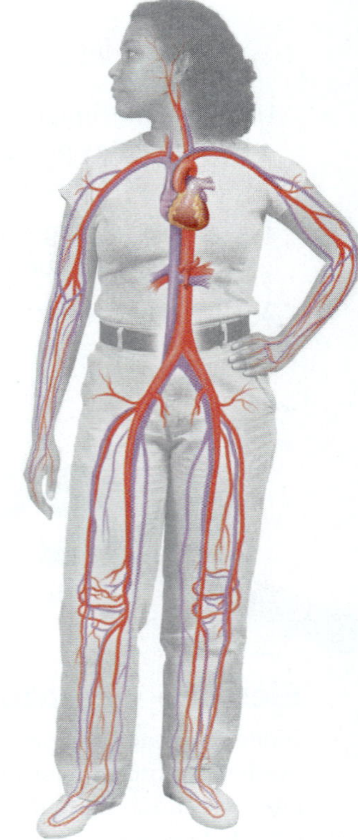

Figura 1 El aparato cardiovascular

Glóbulos rojos Los glóbulos rojos **(Figura 3)** son las células más abundantes de la sangre. Su función es suministrar oxígeno a las células. Las células necesitan oxígeno para llevar a cabo la respiración celular. Cada glóbulo rojo contiene una substancia química llamada *hemoglobina,* a la que se debe su color rojo. La hemoglobina atrapa el oxígeno que inhalas y permite a los glóbulos rojos transportarlo por el cuerpo. Su forma les da una superficie mayor para absorber y liberar oxígeno.

Los glóbulos rojos se producen en la médula ósea. Antes de entrar al torrente sanguíneo, pierden su núcleo y otros organelos. Sin el núcleo, que contiene el ADN, los glóbulos rojos no pueden reemplazar las proteínas desgastadas, y sólo viven unos cuatro meses.

IS THAT A FACT!

About half of the volume of blood is cells. The other half is plasma. Plasma consists of 95 percent water.

Glóbulos blancos Cuando, los agentes *patógenos* (bacterias, virus y otras partículas microscópicas que causan enfermedades) entran a tu cuerpo, casi siempre se encuentran con los glóbulos blancos **(Figura 4)**. Éstos destruyen los patógenos y limpian las heridas para mantenerte sano.

Los glóbulos blancos combaten los agentes patógenos de varias maneras. Unos se salen de los vasos sanguíneos y se mueven por los tejidos buscándolos. Cuando encuentran un patógeno, lo envuelven. Otros liberan substancias químicas llamadas *anticuerpos,* que destruyen a los agentes patógenos. Los glóbulos blancos también te mantienen sano al comerse las células sanguíneas muertas o dañadas. Se producen en la médula ósea, y algunos maduran en los órganos linfáticos.

Plaquetas Entre las células sanguíneas flotan unas partículas diminutas llamadas plaquetas. Las **plaquetas (Figura 5)** son fragmentos de células más grandes que se encuentran en la médula ósea. Estas células permanecen en la médula ósea, pero desprenden fragmentos que entran en la sangre. Aunque las plaquetas viven sólo de 5 a 10 días, son parte importante de la sangre. Cuando te cortas o raspas, sangras porque algunos vasos sanguíneos se han roto; tan pronto sangras, las plaquetas se amontonan en el área lesionada y forman un tapón que reduce la pérdida de sangre **(Figura 6).** Las plaquetas también liberan substancias químicas que reaccionan con las proteínas del plasma para que se originen pequeñas fibras que formarán un coágulo de sangre.

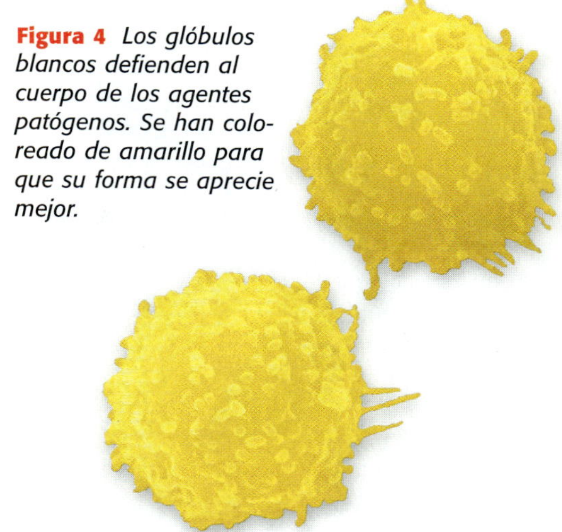

Figura 4 *Los glóbulos blancos defienden al cuerpo de los agentes patógenos. Se han coloreado de amarillo para que su forma se aprecie mejor.*

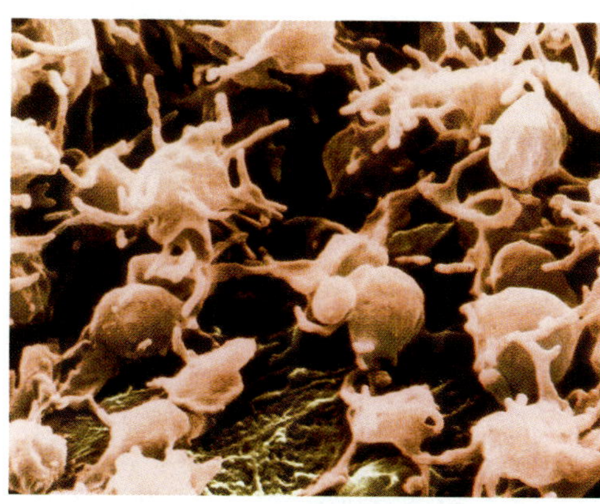

Figura 5 *Las plaquetas reducen la pérdida de sangre.*

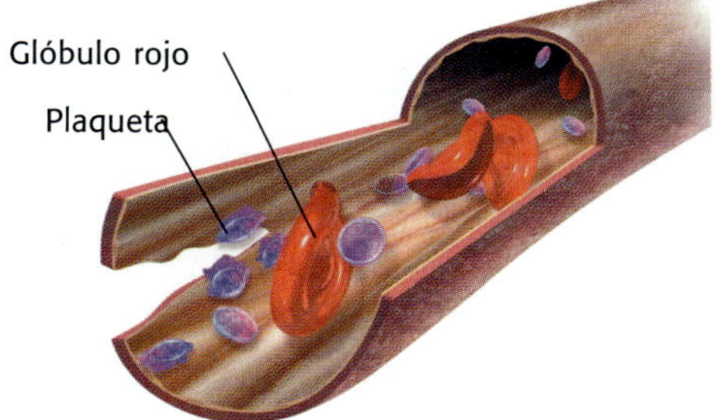

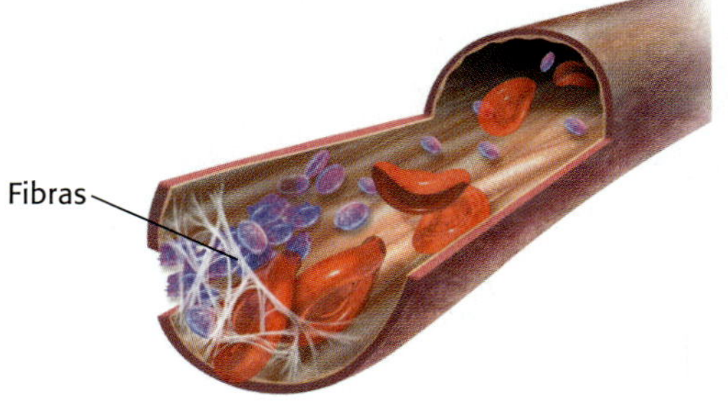

Figura 6 *Las plaquetas liberan substancias químicas en los vasos lesionados y hacen que se formen fibras, que fabrican una "red" que atrapa a las células sanguíneas y detiene el sangrado.*

MISCONCEPTION ALERT

Students may think that deoxygenated blood is blue because it is often depicted that way in illustrations and because blood appears blue in vessels through the skin. The color of blood does change depending on the amount of oxygen it contains; however, human blood is not blue. It varies in color from scarlet to deep red.

Directed Reading Worksheet 20 Section 1

---

2 Teach

READING STRATEGY

Writing **Activity** Before students read the text on this page, have them read the headings aloud. Then ask students to formulate one question that they expect the text under each heading to answer. Have students write the questions in their ScienceLog.

USING THE FIGURE

Have students compare **Figure 3,** which shows red blood cells, with **Figure 4,** which shows white blood cells. Ask students to relate any differences in shape or size to the different functions of these two types of blood cells. Help students understand that the flat shape and small size of red blood cells enable them to carry oxygen through tiny capillaries to cells throughout the body. Help students link the larger size and sticky-looking surface of white bloods cells to their ability to fight pathogens. Sheltered English

MEETING INDIVIDUAL NEEDS

Learners Having Difficulty
To help students visualize the components of blood, have them make a model of blood. Provide students with a variety of materials, such as a resealable plastic sandwich bag, thick corn syrup, plastic-foam balls, and other craft materials. Then have students use the materials to make a model of the parts of blood. Encourage students to be creative in their representations of the parts of blood. Sheltered English

MATH and MORE

Your heart beats about 100,800 times per day. With every beat, about 70 mL of blood is pumped out of your heart. In 1 hour, how much blood does your heart pump out? **(294 L)**

About how much blood does your heart pump out in a day? **(7,056 L)**

Help students visualize this amount by showing them a liter of water.

Math Skills Worksheet 31 "The Unit Factor and Dimensional Analysis"

Homework

Making Models Have students make a model of the human heart to show the path of blood through it. Models can be drawings or three-dimensional constructions. Have students present their completed model to the class. Provide yarn or a pen light so that students can demonstrate to the class the flow of blood through their model heart.

Sheltered English

Teaching Transparency 75 "The Flow of Blood Through the Heart"

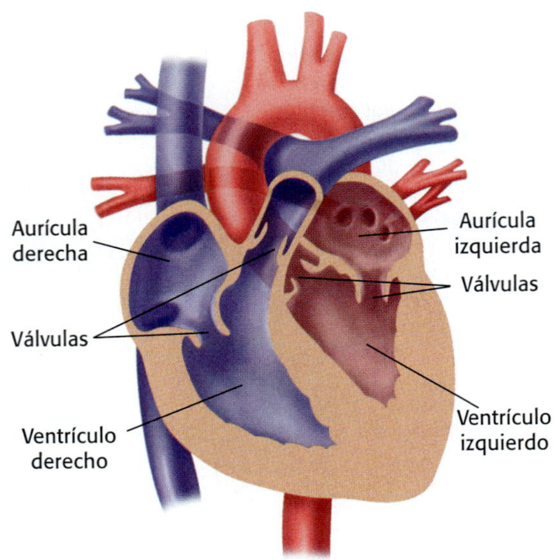

Figura 7 *El corazón es un órgano con cuatro cavidades que bombea la sangre a través de los vasos sanguíneos. Los vasos que transportan la sangre rica en oxígeno están señalados en rojo. Los que transportan la sangre pobre en oxígeno están en azul.*

El corazón

El corazón es un órgano muscular aproximadamente del tamaño del puño. Se encuentra en el centro de la cavidad pectoral y bombea sangre pobre en oxígeno a los pulmones y sangre rica en oxígeno por el cuerpo. Al igual que el de todos los mamíferos, tu corazón tiene un lado izquierdo y un lado derecho, separados por una pared gruesa. Como ves en la **Figura 7,** cada lado tiene una cavidad superior y una inferior. Las cavidades superiores se llaman **aurículas** y las cavidades inferiores **ventrículos.**

Las válvulas son estructuras que parecen tapitas y se localizan entre las aurículas y los ventrículos, así como donde las arterias están sujetas al corazón. Al pasar la sangre por el corazón, las válvulas se cierran para evitar el reflujo de la sangre. El ruido del corazón al palpitar se debe a las válvulas que se cierran. El flujo de la sangre a través del corazón se ilustra en el diagrama de abajo.

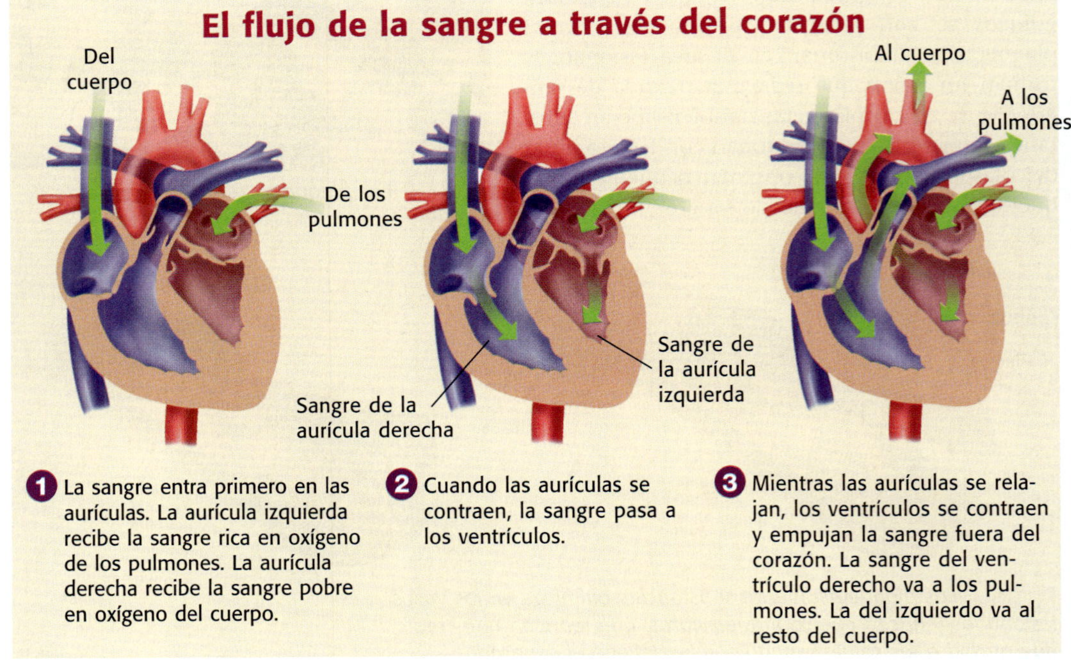

MISCONCEPTION ALERT

The words *left* and *right* as used to diagram the anatomy of the heart might confuse students. When students are looking at a picture of the heart, the left atrium appears on their right, and the right atrium appears on their left. Help students understand anatomical left and right by facing them and asking them to identify your left and right hands in relation to their own.

Vasos sanguíneos

La sangre viaja por el cuerpo a través de los vasos sanguíneos. Un vaso sanguíneo es un tubo hueco que transporta sangre. Hay tres tipos de vasos sanguíneos: arterias, capilares y venas. Sus estructuras y relaciones se ilustran en la **Figura 8.**

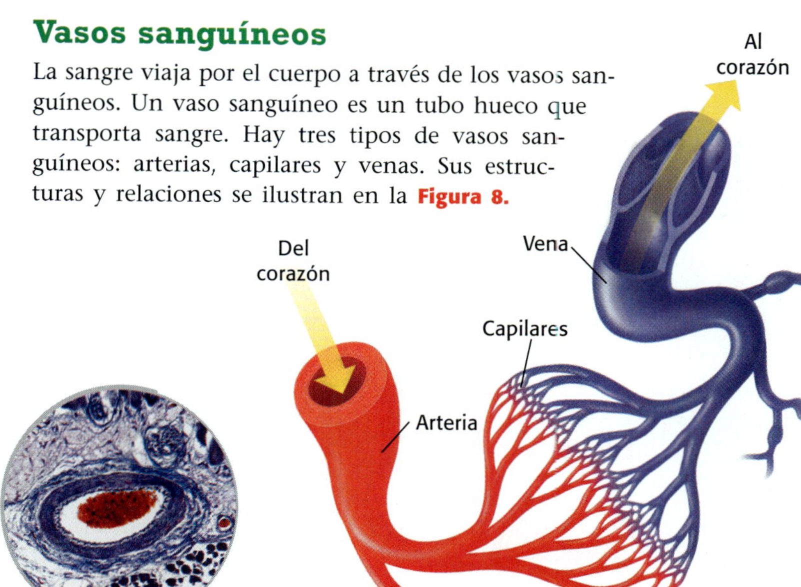

Figure 8 *Las arterias grandes se ramifican en arterias más pequeñas, que a su vez se ramifican en capilares. Los capilares se unen a las venas, que a su vez se unen para formar venas más grandes.*

Arterias Las **arterias** son vasos sanguíneos que transportan la sangre del corazón al resto del cuerpo. Sus paredes están formadas por músculo liso y son gruesas y elásticas. Cada vez que el corazón palpita, la sangre es bombeada fuera del corazón a alta presión. El cambio rítmico en la presión sanguínea se llama *pulso*. El pulso puede sentirse en varias partes del cuerpo, como en la parte interna de la muñeca, bajo el pulgar.

Capilares Un cabello es unas 10 veces más ancho que un capilar. Los **capilares** son los vasos sanguíneos más pequeños del cuerpo. El grosor de sus paredes es solamente de una célula. Como ves en la **Figura 9,** los capilares son tan estrechos que las células sanguíneas deben pasar a través de ellos en una sola fila. La estructura sencilla de los capilares permite que los nutrientes, el oxígeno y muchas otras clases de substancias se difundan fácilmente a través de sus paredes para llegar a otras células del cuerpo. Ninguna célula está a más de tres o cuatro células de distancia de un capilar.

Venas Después de dejar los capilares, la sangre entra en las venas. Las **venas** son vasos sanguíneos que devuelven la sangre al corazón. Las venas pequeñas se unen para formar venas más grandes. A medida que la sangre viaja por las venas grandes, las válvulas evitan el reflujo. Cuando los músculos óseos se contraen, aprietan las venas cercanas y ayudan a impulsar la sangre hacia el corazón.

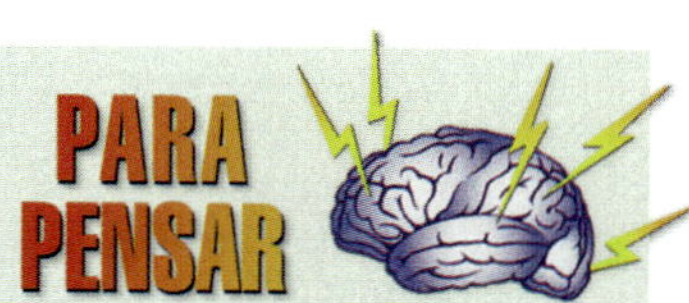

Si se unieran todos los vasos sanguíneos de tu cuerpo, la longitud total sería más del doble de la circunferencia de la Tierra.

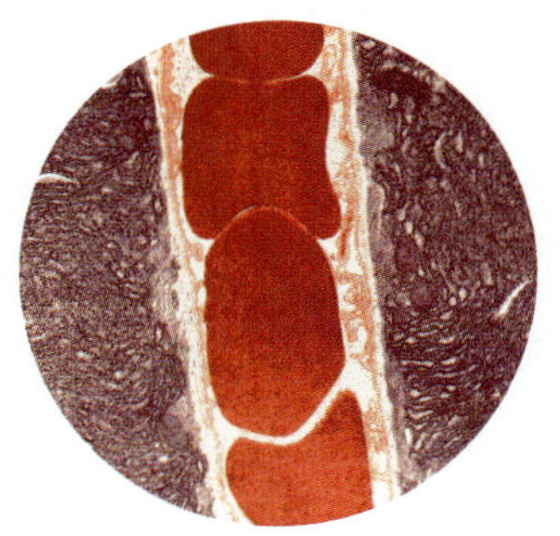

Figura 9 *Estos glóbulos rojos están pasando por un capilar.*

491

WEIRD SCIENCE

Babies born with certain congenital heart defects are blue at birth. The condition, called cyanosis, can be caused by low levels of oxygen in the blood. The low levels of oxygen can be a result of defects in the anatomy of the heart. Deoxygenated blood is pumped to the body before it is pumped to the lungs. This defect can be repaired surgically.

GROUP ACTIVITY

Circulation Relay

MATERIALS

FOR EACH CLASS:
- 5 inflated red balloons (or paper disks)
- 5 inflated blue balloons (or paper disks)
- diagram of The Flow of Blood Through the Body
- 8 stations (cardboard boxes, flags, and so on)

Divide the class into five teams for a relay race.

1. Students begin in the left ventricle carrying a red balloon, which represents an oxygenated blood cell.

2. They travel through the aorta.

3. After passing through the aorta, students carry oxygenated blood to the muscles and exchange the red balloon for a blue one.

4. From the muscles, students carry blood loaded with CO_2 to the right atrium.

5. From the right atrium, students travel into the right ventricle.

6. Students travel through the pulmonary artery.

7. From the pulmonary artery, students travel into the lungs, where they exchange their CO_2 for oxygen (exchange blue balloons for red ones).

8. Carrying oxygenated blood (red balloons), students enter the left atrium and are ready to begin again or hand off their balloons.

Walk one student through the pathway; then have the teams send one student through at a time in a relay race.

Autoevaluación

¿Cómo se relacionan las estructuras de arterias y venas con sus funciones? *(Consulta la página 636 para comprobar tu respuesta).*

Viajar con el flujo

Como ya sabes, una función importante de la sangre es suministrar oxígeno a las células del cuerpo. ¿De dónde toma la sangre este oxígeno? La obtiene de los pulmones, durante la circulación pulmonar. La **circulación pulmonar** es la circulación de la sangre entre el corazón y los pulmones.

Cuando la sangre rica en oxígeno va de los pulmones al corazón, luego debe ser bombeada al resto del cuerpo. La circulación de la sangre entre el corazón y el resto del cuerpo se llama **circulación sistémica.** Ambos tipos de circulación aparecen en el diagrama de abajo.

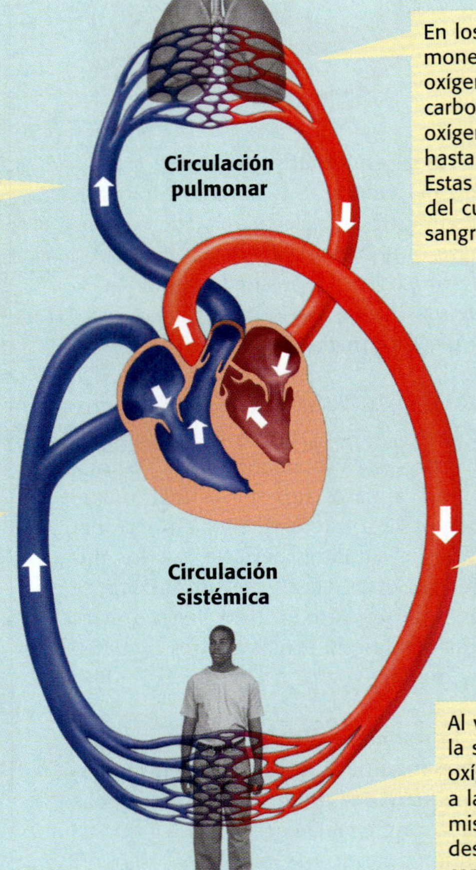

Answers to Self-Check

The hollow tube shape of arteries and veins allows blood to reach all parts of the body. Valves in the veins prevent blood from flowing backward.

Students may think that blood from the heart enters one lung and leaves from the other. Actually, each lung is serviced by vessels carrying blood to and from the heart.

La sangre fluye bajo presión

Cuando riegas con manguera, puedes sentir cómo ésta se endurece a medida que el agua presiona contra sus paredes. La sangre tiene el mismo efecto sobre tus vasos sanguíneos. La fuerza de la sangre sobre las paredes internas de los vasos sanguíneos se llama **presión sanguínea.**

Muchas personas se revisan la presión sanguínea regularmente, como el señor de la **Figura 10.** La presión sanguínea se expresa en milímetros (mm) de mercurio (Hg). Una presión sanguínea de 120 mm Hg significa que la presión sobre las paredes de los vasos basta para empujar una columna angosta de mercurio a 120 mm de altura. En general, la presión sanguínea se mide en las arterias grandes y se expresa con dos cifras. Una presión sanguínea normal es de alrededor de 120/80. El primer número es la presión sistólica. La *presión sistólica* es la presión del interior de las arterias grandes cuando los ventrículos se contraen. Como vimos, la oleada de sangre hace que las arterias se abulten y produzcan el pulso. El segundo número es la presión diastólica. La *presión diastólica* es la presión de las arterias cuando los ventrículos se relajan.

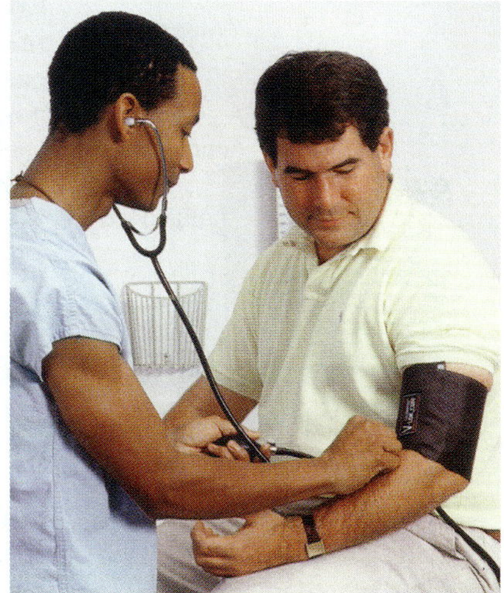

Figura 10 *Un enfermero le toma la presión sanguínea a un paciente. Si la presión sanguínea está constantemente alta o baja, es posible que algo ande mal con el aparato cardiovascular.*

Ejercicio y flujo sanguíneo

Cuando haces ejercicio, tus músculos requieren mucho más oxígeno y nutrientes. Para resolver este problema, el cerebro manda señales al corazón para que los latidos sean más rápidos. La actividad física hace que a los músculos les llegue 10 veces más sangre que cuando el cuerpo está en reposo.

Durante el ejercicio, algunos órganos no necesitan tanta sangre como los músculos esqueléticos. Los riñones y el sistema digestivo reciben menos sangre para que los músculos esqueléticos, cerebro, corazón y pulmones reciban más. Es lo mismo que cerrar en casa algunas llaves del agua para que salga más por otras. Al terminar de hacer ejercicio, los vasos sanguíneos de otras partes del cuerpo se abren y el ritmo del corazón se hace más lento.

REPASO

1. ¿Cuál es la función del aparato cardiovascular?

2. ¿Cuáles son las tres clases de vasos sanguíneos? Compara sus funciones.

3. **Identificar relaciones** ¿Cómo se relaciona la estructura de los capilares con su función?

Explora

Imagina que eres miembro de un equipo científico seleccionado para explorar el aparato cardiovascular. Después de ser reducidos al tamaño de los glóbulos rojos, abordan un submarino diminuto y empiezan su viaje a través del corazón y los vasos sanguíneos. Describe a dónde van, lo que ven y los problemas que surgen durante el viaje.

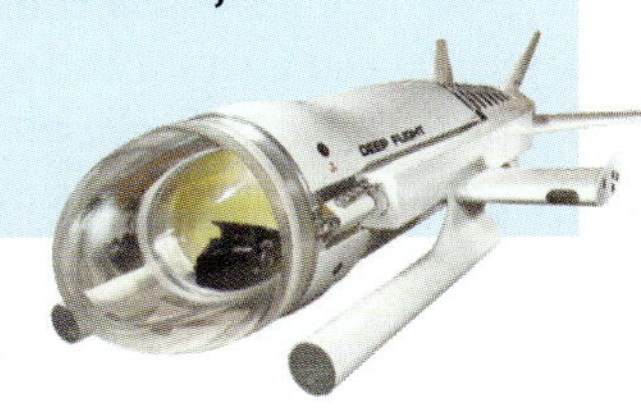

---

③ Extend

GOING FURTHER

Nobel Prize–winning scientist Karl Landsteiner (1868–1943) discovered that some mixtures of blood are compatible and others are not. He found that he could divide the population into different groups based on how their blood reacted with blood from other people. Encourage interested students to research how Landsteiner's experiments led to the discovery of the ABO and Rh blood groups.

USING THE TABLE

Review the material in the table to reinforce the material presented in the text. Help students compare the shapes of the antigens on each red blood cell with the shapes of the antibodies that bind to the antigens. Then ask the following questions:

Which antibodies will a person with type AB blood produce? (none)

Can a person with type O blood receive blood from a person with blood type AB? Why or why not? (No; the type O blood will have antibodies that attack the A and B antigens in the AB blood.)

Can a person with type O blood donate blood to someone with type A blood? Why or why not? (Yes; the type O red blood cells lack antigens.)

Answer to APPLY

You could bring back type A, B, or O.

¿Cuál es tu grupo?

Cuando una persona pierde mucha sangre, se le da sangre donada por otra. A quien la recibe no se le puede dar sangre de cualquier persona, pues hay distintos tipos de sangre. Algunos pueden mezclarse sin peligro, pero la mezcla de otros hace que los glóbulos rojos se aglutinen. Las células aglutinadas pueden formar coágulos que obstruyan los vasos sanguíneos y causar la muerte.

Cada persona tiene uno de los siguientes grupos sanguíneos: A, B, AB u O. El grupo sanguíneo se refiere al tipo de substancias químicas que hay en la superficie de los glóbulos rojos. Estas substancias se llaman *antígenos*. El grupo sanguíneo A tiene antígenos A; el B tiene antígenos B y el AB tiene antígenos A y B. El grupo O no tiene ni antígenos A ni B.

Mezclar o no mezclar Los grupos sanguíneos no sólo tienen substancias químicas diferentes en la superficie de los glóbulos rojos, sino también en el plasma, o parte líquida de la sangre. Estas substancias se llaman *anticuerpos*. Los glóbulos rojos se aglutinan cuando los anticuerpos se unen a ellos.

Como ves en la **Figura 11,** el cuerpo fabrica anticuerpos contra los antígenos no presentes en sus propios glóbulos rojos. Por ejemplo, las personas del grupo B fabrican anticuerpos A, que atacan a las células que tienen un antígeno A. Por lo tanto, las personas de este grupo no pueden recibir sangre de los grupos A o AB. La sangre del grupo O puede darse a cualquier persona porque sus glóbulos rojos no presentan antígenos A ni B en su superficie. Se dice que una persona del grupo sanguíneo O es *donador universal*. Las personas del grupo AB son *receptores universales*, es decir, pueden recibir sangre de cualquier grupo, pues no fabrican anticuerpos contra los antígenos A ni B.

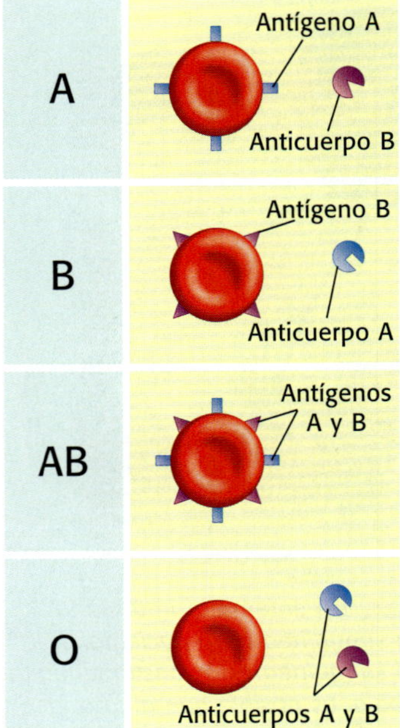

Figura 11 Esta tabla muestra los antígenos y anticuerpos que pueden estar presentes en cada grupo sanguíneo.

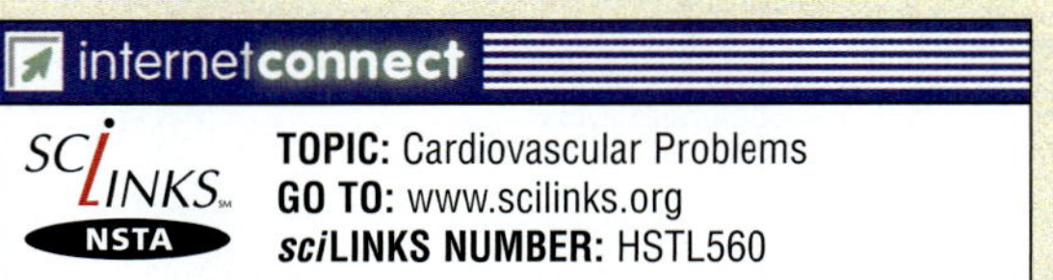

Una mujer joven llega a la sala de emergencias y necesita una transfusión de sangre. Su grupo sanguíneo es AB. Llamas al banco de sangre para pedir sangre AB, pero te dicen que no hay. ¿Qué otro grupo sanguíneo podrían enviarte para la transfusión?

494

IS THAT A FACT!

The first heartbeat occurs in the human embryo at four weeks of age. Although the heart looks like a simple tube at this stage, it begins to beat. This starts blood circulation through the first blood vessels, which eventually grow and develop into the entire circulatory system.

Problemas cardiovasculares

Si algo anda mal con el aparato cardiovascular de una persona, su salud se verá afectada. Unos problemas cardiovasculares involucran al corazón y a los vasos sanguíneos, mientras que otros tienen que ver con la sangre. Cuando hay un problema con el corazón o los vasos sanguíneos, el movimiento de la sangre a través del cuerpo se ve afectado.

La principal causa de muerte en los Estados Unidos es la enfermedad cardiovascular conocida como *aterosclerosis*. La aterosclerosis ocurre cuando substancias grasas, como el colesterol, se acumulan en el interior de los vasos sanguíneos, haciéndolos más estrechos y menos elásticos. La **Figura 12** muestra cómo puede obstruirse un vaso sanguíneo. Si una arteria principal que suministra sangre al corazón se obstruye, la persona sufre un ataque al corazón y parte de ese órgano puede morir. La aterosclerosis también favorece la *hipertensión,* que es una presión sanguínea anormalmente alta. La hipertensión es peligrosa porque hace trabajar en exceso al corazón, debilita los vasos y hace que se rompan. Si un vaso sanguíneo del cerebro se obstruye o se rompe, ciertas partes del cerebro no reciben oxígeno ni nutrientes, y pueden morir. A esto se le conoce como *apoplejía o derrame cerebral.*

Los problemas cardiovasculares se pueden deber al hábito de fumar, los altos niveles de colesterol en la sangre, el estrés y la herencia. Se puede reducir el riesgo de problemas cardiovasculares haciendo lo siguiente: no fumar, tener una dieta baja en sal y en grasas animales, aceites de coco y de palma, comer muchas verduras, frutas y granos enteros, y hacer ejercicio regularmente.

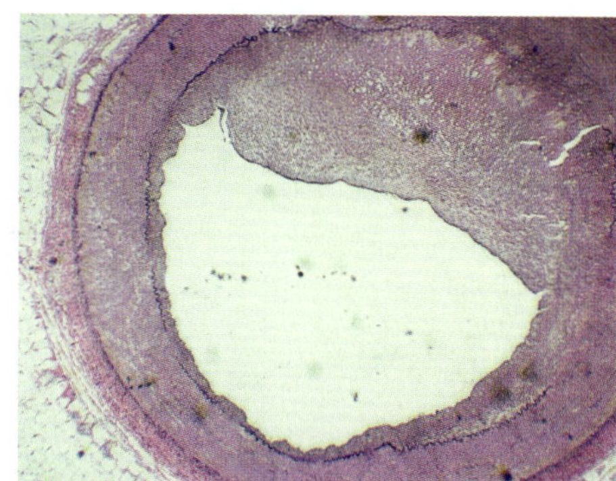

Figura 12 *La aterosclerosis es un problema cardiovascular común. Los depósitos de grasa se acumulan en el interior de los vasos sanguíneos y obstruyen el flujo sanguíneo.*

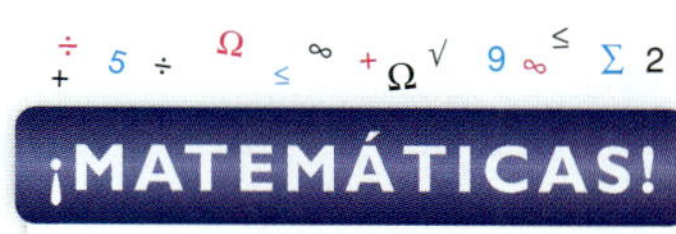

¡MATEMÁTICAS!

¡Que siga latiendo!

Tu corazón palpita alrededor de 100,800 veces al día. ¿Cuántas veces palpita al año?

REPASO

1. ¿Desde dónde y hacia dónde fluye la sangre durante la circulación pulmonar y la sistémica?

2. ¿Qué le pasa al nivel de oxígeno en la sangre al atravesar los pulmones?

3. **Aplicar conceptos** Pedro tiene sangre del grupo A.
   a. ¿Qué tipo de antígenos tiene en sus glóbulos rojos?
   b. ¿Qué anticuerpos puede producir?
   c. ¿Qué grupos sanguíneos podría recibir Pedro si necesitara una transfusión?

Focus

The Lymphatic System

This section introduces the lymphatic system. Students will also learn about the relationship between blood and lymph.

🔔 Bellringer

Draw the following shapes on the board or on an overhead projector:

a circle, a triangle, a straight line, and a cluster of several dots

Ask students to choose the shape(s) that best represents a circulatory system and to explain in their ScienceLog. (The circle and the triangle best represent a circulatory system because they both form a continuous loop.)

1 Motivate

DISCUSSION

Ask students if a doctor has ever felt around their neck when they were sick. Encourage students who have had this experience to share it with the class. Then invite students to explore the purpose of this type of examination. Sheltered English

Answers to Self-Check

Like blood vessels, lymph capillaries receive fluid from the spaces surrounding cells. The fluid absorbed by lymph capillaries flows into lymph vessels. These vessels drain into large neck veins instead of into an organ, such as the heart. Lymph does not deliver oxygen and nutrients.

VOCABULARIO

sistema linfático
capilares linfáticos
linfa
vasos linfáticos
nódulos linfáticos
timo
bazo
amígdalas

OBJETIVOS

- Estudia las funciones del sistema linfático.
- Identifica la relación entre la linfa y la sangre.
- Describe los órganos del sistema linfático.

✓ **Autoevaluación**

¿En qué se parecen el sistema linfático y el aparato cardiovascular? ¿En qué se diferencian? *(Consulta la página 636 para comprobar tu respuesta).*

El sistema linfático

El aparato cardiovascular no es el único sistema circulatorio del cuerpo. Al fluir la sangre por el aparato cardiovascular, de los capilares se escapa líquido que se mezcla con el fluido que baña las células. La mayor parte del líquido es reabsorbido por los capilares, pero no todo. El **sistema linfático** recoge el exceso de líquido y lo devuelve a la sangre.

Al igual que el aparato cardiovascular, el sistema linfático es un sistema circulatorio. Además de recolectar el exceso de líquido que rodea a las células y devolverlo a la sangre, el sistema linfático ayuda al cuerpo a combatir los agentes patógenos.

Vasos del sistema linfático

El líquido recogido por el sistema linfático es transportado a través de vasos. Los vasos más pequeños del sistema linfático se llaman **capilares linfáticos.** De los espacios entre las células, los capilares linfáticos absorben el líquido y las partículas demasiado grandes para entrar en los capilares sanguíneos. Algunas de estas partículas son células muertas que el cuerpo reconoce como extrañas. El líquido y las partículas absorbidas por los capilares linfáticos se llama **linfa.**

Como ves en la **Figura 13,** los capilares linfáticos transportan la linfa a **los vasos linfáticos,** que son vasos más grandes con válvulas. La linfa no es empujada por una bomba; son los músculos esqueléticos los que proporcionan la fuerza para moverla a través de los vasos, y las válvulas evitan el reflujo. La linfa viaja por el sistema linfático y luego se vacía en las venas grandes del aparato cardiovascular que están en el cuello.

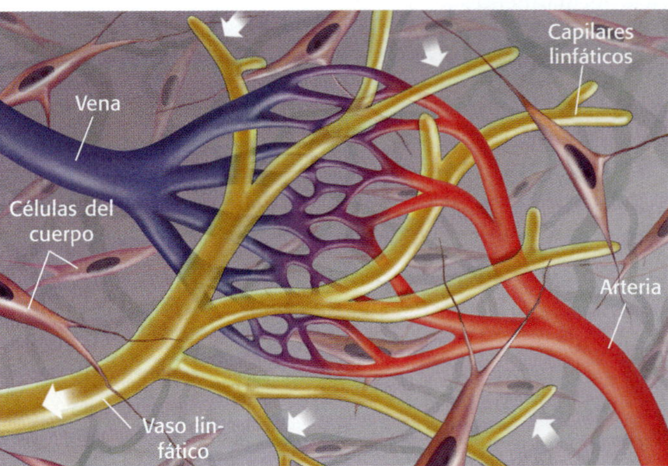

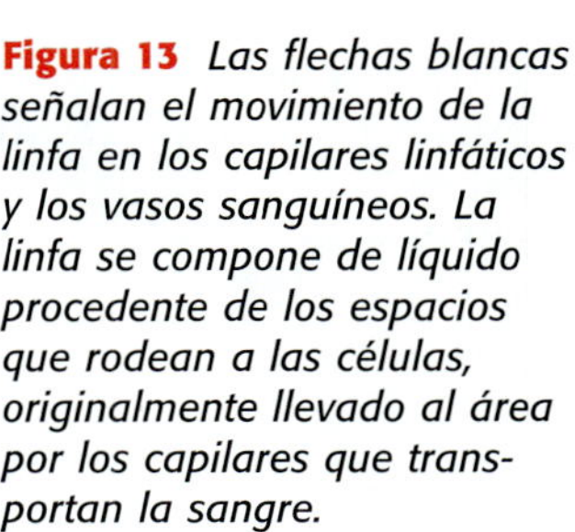

Figura 13 *Las flechas blancas señalan el movimiento de la linfa en los capilares linfáticos y los vasos sanguíneos. La linfa se compone de líquido procedente de los espacios que rodean a las células, originalmente llevado al área por los capilares que transportan la sangre.*

496

SCIENTISTS AT ODDS

The Danish physician Thomas Bartholin (1616–1680), known as the Elder, is credited with being the first person to describe the lymphatic system. His Swedish contemporary Olof Rudbeck (1630–1702) also studied the lymphatic system.

Rudbeck was furious when French anatomist Jean Pecquet and Thomas Bartholin published their findings before he did. Rudbeck spent the rest of his life studying botany.

Órganos linfáticos

Además de vasos y capilares, el sistema linfático está formado por otros órganos, como ves en la **Figura 14.**

Nódulos linfáticos   En su recorrido a través de los vasos linfáticos, la linfa pasa por nódulos linfáticos. Los **nódulos linfáticos** son órganos pequeños con forma de frijol donde ciertas partículas, como agentes patógenos o células muertas, son eliminadas de la linfa. Tenemos cientos de nódulos linfáticos distribuidos por todos los vasos linfáticos.

Los nódulos linfáticos contienen muchos glóbulos blancos, algunos de los cuales se comen a los patógenos. Otros glóbulos blancos, los *linfocitos,* producen substancias químicas que se unen a los agentes patógenos y los marcan para que sean destruidos. Cuando el cuerpo se infecta con bacterias o virus, los glóbulos blancos se multiplican, y por eso a veces los nódulos se hinchan y duelen.

Timo, bazo y amígdalas   El **timo,** que se encuentra justo arriba del corazón, libera linfocitos. Los linfocitos viajan a otras zonas del sistema linfático.

El órgano linfático más grande es el bazo, que se encuentra en la parte superior izquierda del abdomen. El **bazo** filtra la sangre y, como el timo, libera linfocitos. Cuando los glóbulos rojos entran en los capilares del bazo, los más viejos y frágiles se rompen. Los glóbulos blancos grandes del bazo se comen a estas células muertas y las sacan de la sangre. Los glóbulos rojos se digieren y algunos de sus componentes son utilizados de nuevo; por eso el bazo puede considerarse un centro de reciclaje de glóbulos rojos.

Las **amígdalas** están formadas por grupos de tejido linfático localizados en la parte de atrás de la cavidad nasal, en el interior de la garganta y detrás de la lengua. Los linfocitos de las amígdalas defienden al organismo de las infecciones. A veces, las amígdalas se infectan gravemente y hay que extraerlas.

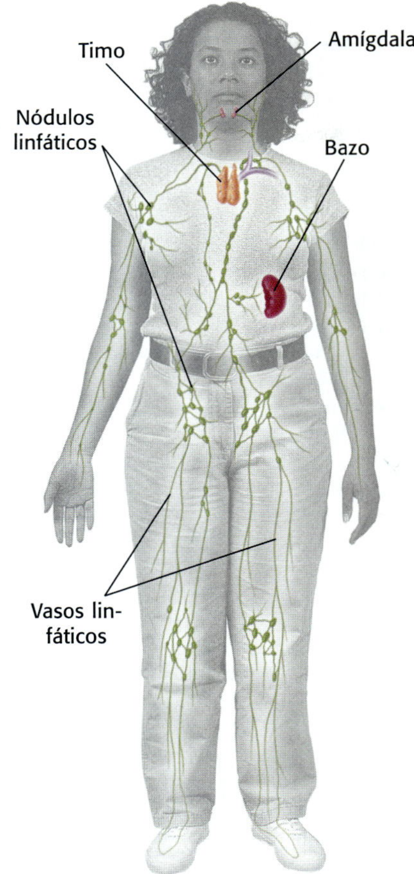

Figura 14 **El sistema linfático**

REPASO

1. ¿Cuáles son las principales funciones del sistema linfático?

2. ¿A dónde va la linfa cuando abandona el sistema linfático?

3. **Identificar relaciones** ¿En qué se parecen los nódulos linfáticos al bazo?

497

Focus

The Respiratory System

This section introduces the structures and functions of the respiratory system. Students will learn to trace the flow of air through the respiratory system and to recognize problems in respiratory function. Students will also find out how the respiratory system and the circulatory system are related.

Bellringer

Write the following on the board or overhead projector:

In your ScienceLog, explain whether the following statements are true or false:

- Breathing and respiration are the same thing. (False; breathing is only one part of respiration.)
- The nose is the primary doorway into and out of the respiratory system. (true)
- The vocal cords are located in the trachea. (False; they are located in the larynx.)

1 Motivate

ACTIVITY

Have students place their hands on either side of their rib cages and breathe deeply several times. Then ask students to describe what they felt while they breathed in and out. (Students should feel their rib cage moving up and expanding during inhalation and moving down and returning to its initial size during exhalation.)

Explain to students that this section focuses on what happens inside the body during breathing.

Sheltered English

VOCABULARIO

respiración
sistema respiratorio
faringe
laringe
tráquea
bronquios
alvéolos
diafragma
respiración celular

OBJETIVOS

- Describe el flujo del aire a través del sistema respiratorio.
- Comenta la relación entre el sistema respiratorio y el circulatorio.
- Identifica las enfermedades respiratorias.

El sistema respiratorio

Respiras todo el tiempo. En este mismo instante lo estás haciendo. Sin embargo, casi nunca lo piensas, a menos que tu capacidad para respirar desaparezca de repente. Es obvio que tienes que respirar para sobrevivir. ¿Por qué es importante respirar?

Afuera el aire malo y adentro el aire bueno

Tu cuerpo necesita un suministro continuo de oxígeno para obtener energía de los alimentos que comes. Aquí es donde respirar es útil. El aire que respiras es una mezcla de varios gases, uno de los cuales es el oxígeno. Cuando respiras, tu cuerpo inhala aire y absorbe el oxígeno; luego, el dióxido de carbono de tu cuerpo se incorpora al aire y éste se exhala.

Sin embargo, el proceso de la respiración no es tan sencillo como suena. La **respiración** es el proceso por el cual el cuerpo obtiene y utiliza oxígeno, y elimina dióxido de carbono y agua. Este proceso se divide en dos partes: por un lado, la inhalación y la exhalación y, por otro, la respiración celular, que implica reacciones químicas que liberan la energía de los alimentos.

La respiración

Gracias al sistema, o aparato respiratorio, podemos respirar. El **sistema respiratorio** consta de pulmones, garganta y conductos que llegan a los pulmones. La **Figura 15** ilustra las partes del aparato respiratorio.

La nariz La nariz es el principal conducto de entrada y salida del aparato respiratorio. El aire es inhalado por la nariz, donde entra en contacto con superficies cálidas y húmedas, aunque también puede entrar y salir por la boca.

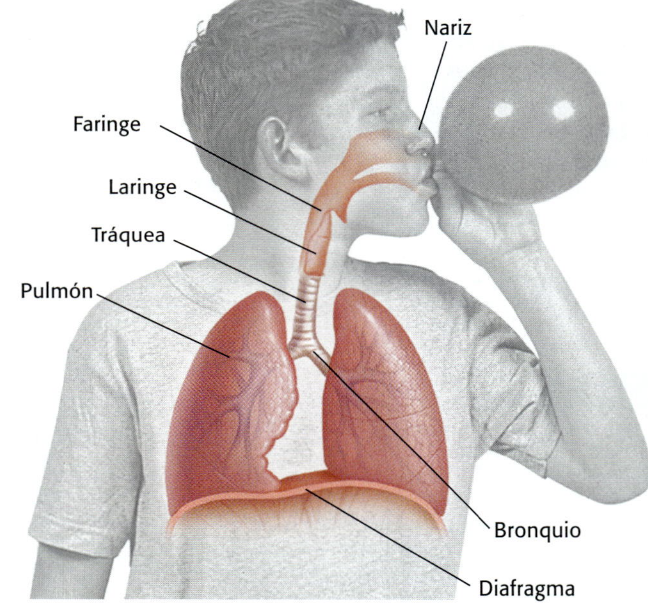

Figura 15 *El aire entra y sale del cuerpo a través del aparato respiratorio.*

498

internet**connect**

SCILINKS NSTA

TOPIC: The Respiratory System
GO TO: www.scilinks.org
*sci*LINKS NUMBER: HSTL570

IS THAT A FACT!

You sneeze when your breathing muscles respond to mucus or dirt that irritates the lining of your nasal cavity. A sneeze consists of a deep breath followed by a 160 km/h surge of air out of the nose!

La faringe El aire fluye de la nariz a la **faringe** o garganta. Con un espejo puedes ver las paredes de la faringe, que está detrás de la lengua. Además de aire, por la faringe también pasan los alimentos y bebidas en camino al estómago. La faringe se ramifica en dos tubos: el esófago, que va al estómago, y la laringe, que va a los pulmones.

La laringe Inclina tu cabeza ligeramente hacia arriba y frótate la parte anterior del cuello con un dedo. Las protuberancias que sientes son la parte exterior de la laringe. La **laringe,** que es el órgano productor de la voz, contiene las cuerdas vocales. Las cuerdas vocales son un par de bandas elásticas que se extienden a través de la abertura de la laringe; los músculos de la laringe controlan su estiramiento. Cuando el aire pasa entre las cuerdas vocales, éstas vibran y emiten sonidos.

La tráquea La laringe protege la entrada de un tubo largo llamado **tráquea**. La tráquea es el conducto por donde el aire va de la laringe hasta los pulmones.

Los bronquios La tráquea se divide en dos tubos llamados **bronquios**. Cada uno va a un pulmón y se ramifica en miles de tubitos llamados *bronquiolos*.

Los pulmones El cuerpo tiene dos pulmones, que tienen aspecto de esponjas. En los pulmones, cada bronquiolo se ramifica para formar miles de saquitos llamados **alvéolos**. Cada alvéolo está rodeado por capilares. La **Figura 16** muestra la disposición de los tubos del sistema respiratorio.

ciencias de la Tierra

C O N E X I Ó N

Cuando las personas que viven en lugares poco elevados viajan a las montañas, generalmente tienen dificultades al hacer esfuerzos. Esto se debe a que, a grandes alturas, la concentración de oxígeno del aire es más baja que la que hay a bajas alturas. Hasta que se acostumbran al cambio, estas personas tienen que inhalar un mayor número de veces para darle a su cuerpo el oxígeno necesario.

Figura 16 *En los pulmones, los bronquios se ramifican en bronquiolos, que llegan a unos saquitos llamados alvéolos.*

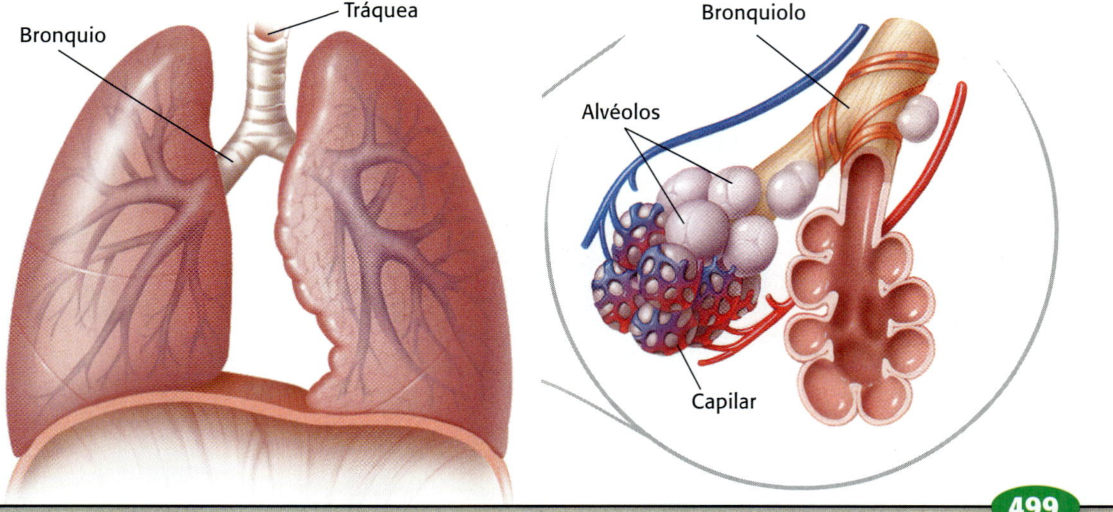

499

2 Teach

ACTIVITY

Investigating Speech Ask students to place their hands lightly on their neck near the larynx and to say, "ah." Have students keep their hands in place and alternate between blowing as they would blow out candles on a birthday cake and saying, "ah." Then ask:

- What happened when you said, "ah"? (The neck vibrated.)
- What happened when you blew without saying anything? (The neck stopped vibrating, but air still rushed out of the mouth.)
- What caused the sound and the vibrations when you said "ah"? (Air rushing past the vocal cord muscles in the larynx caused the muscles to vibrate. This caused the sound.)

Point out that speech is made up of sounds produced both ways, with voicing and without voicing. If students are unconvinced, have them say *sssssssss* (the snake sound) while touching their neck near the larynx. Then have them say *zzzzz* (buzz like a bee) while touching their neck. Students should find that their neck vibrate when they pronounce *z*, which is voiced, but does not vibrate when they pronounce *s*, which is not voiced.

Sheltered English

Teaching Transparency 77 "The Respiratory System"

Directed Reading Worksheet 20 Section 3

Multicultural CONNECTION

Suppose a newcomer to a Peruvian village in the Andes Mountains gets headaches, feels nauseated, and is very short of breath. What is going on? The newcomer is suffering from a lack of oxygen. The villagers don't have these problems; they are extremely well adapted to this high elevation. Because there is very little oxygen in the air, these people have developed lungs and chests that are much larger than those of the newcomer. They also carry more blood in their bloodstream.

ACTIVITY

Concept Mapping Have students create a concept map using the following terms:

respiration, respiratory system, pharynx, larynx, cellular respiration, trachea, bronchi, alveoli, diaphragm

RESEARCH

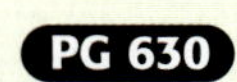

Ask students if they have ever seen a professional football player inhale oxygen using an oxygen mask. Encourage students to find out why football players do this and to write the results of their research in a short report. As a class, discuss whether the practice is useful. (Although panting from physical exertion is a sign of an oxygen deficit, the additional oxygen does not alleviate the problem because the oxygen deficit occurs in the muscles, not in the lungs. In other words, the oxygen deficit is not a result of inadequate respiratory function but of the muscles' inability to take in more oxygen.)

LabBook PG 630

Build a Lung

CONNECT TO PHYSICAL SCIENCE

Lead a discussion of how the body creates changes in air pressure to make breathing possible. The following Teaching Transparency, "Air Pressure and Breathing," is a helpful illustration.

Teaching Transparency 180 "Air Pressure and Breathing"

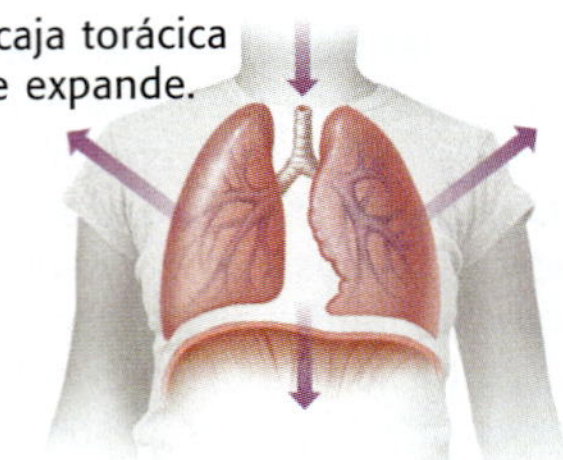

Figura 17 *Cuando el diafragma se contrae, la caja torácica se expande y crea un vacío dentro de los pulmones. El vacío succiona el aire hacia adentro.*

Experimentos

Si quieres saber más sobre los pulmones, pasa a la página 630.

¿Cómo respiras?

Cuando respiras, succionas aire hacia los pulmones o lo expulsas hacia afuera. Sin embargo, los pulmones no tienen músculos para hacer entrar y salir el aire; son los músculos de las costillas y el **diafragma,** un músculo con forma de cúpula debajo de los pulmones, los que llevan a cabo la respiración. Al contraerse, el diafragma aumenta el volumen de la cavidad del pecho; al mismo tiempo, algunos de los músculos de las costillas se contraen y levantan la caja torácica, haciendo que se expanda. Observa este proceso en el diagrama de la **Figura 17.**

¿Qué le pasa al oxígeno? Cuando los glóbulos rojos absorben oxígeno, el aparato cardiovascular lo transporta por todo el cuerpo. El oxígeno pasa al interior de las células, donde participa en una reacción química muy importante: la respiración celular. Durante la **respiración celular,** el oxígeno se usa para liberar la energía almacenada en las moléculas de carbohidratos, grasas y proteínas. Cuando estas moléculas son transformadas, se libera energía junto con dióxido de carbono y agua. El dióxido de carbono y el agua salen de la célula y regresan al torrente sanguíneo. El dióxido de carbono va a los pulmones y se exhala. La **Figura 18** muestra la relación entre la respiración y la circulación de la sangre.

Figura 18 *La sangre es importante para la respiración.*

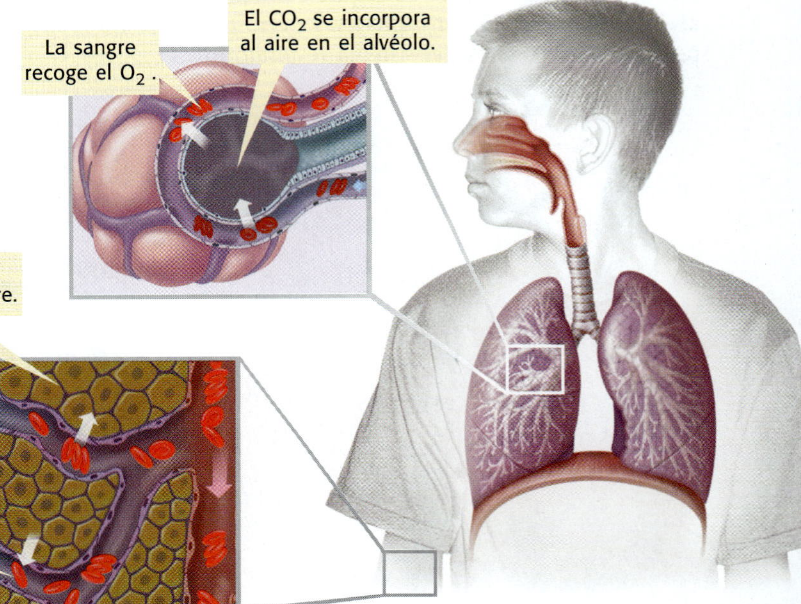

IS THAT A FACT!

The lungs contain about 300 million alveoli. The alveoli provide a tremendous surface area for gas exchange. In fact, a person can breathe easily with only one lung.

Enfermedades respiratorias

Millones de personas sufren enfermedades respiratorias, de las que hay muchos tipos, como asma, bronquitis, enfisema y neumonía.

El *asma* consiste en que el tejido que rodea a los bronquiolos se comprime y secreta grandes cantidades de moco por la presencia de polen, caspa de animales u otros irritantes. A medida que el bronquiolo se estrecha, la persona tiene dificultad para respirar. Cuando una persona tiene bronquitis, neumonía o asma, los conductos de aire que atraviesan los pulmones se obstruyen con moco. La *bronquitis* puede desarrollarse cuando algo irrita el revestimiento de los bronquiolos. La *neumonía* la causan bacterias o virus que crecen dentro de los bronquiolos y alvéolos, haciendo que se inflamen y llenen de líquido. En caso de una disminución significativa del oxígeno que se difunde a la sangre, la persona puede asfixiarse.

Los peligros de fumar

Probablemente ya sabes que fumar cigarrillos es malo para tu salud. De hecho, fumar es la primera causa de las enfermedades cardiovasculares y pulmonares, como *enfisema* y *cáncer pulmonar*. Las personas con enfisema tienen problemas para obtener el oxígeno que necesitan porque sus alvéolos están desgastados, como ves en la **Figura 19.**

El cáncer pulmonar es otra enfermedad respiratoria peligrosa. Las substancias químicas del humo del tabaco pueden hacer que las células pulmonares se vuelvan cancerosas. Las células cancerosas de los pulmones se dividen rápidamente y forman una masa llamada tumor. A medida que el tumor crece, obstruye el flujo del aire e impide el intercambio de oxígeno y dióxido de carbono. Algunas células tumorales pueden separarse y entonces el cáncer es transportado a través de la sangre o la linfa a otras partes del cuerpo, donde las células continúan creciendo y forman otros tumores.

REPASO

1. Describe el recorrido del aire a través del aparato respiratorio.

2. ¿Qué diferencia hay entre la respiración celular y la inhalación y exhalación?

3. ¿Qué enfermedades respiratorias puede desarrollar un fumador?

4. **Identificar relaciones** ¿Cómo se relaciona la función del aparato respiratorio con la del circulatorio?

Laboratorio

¿Por qué ronca la gente?

Pídele a tu maestro o maestra una **hoja de papel encerado de 15 cm².** Tararea tu canción favorita y luego presiona el papel encerado contra tus labios y tararea la canción de nuevo. Cuando termines esta actividad, contesta las siguientes preguntas en tu cuaderno de ciencias:

1. ¿Qué diferencia notaste al tararear con el papel de cera?

2. Con ayuda de tus observaciones adivina la causa del ronquido.

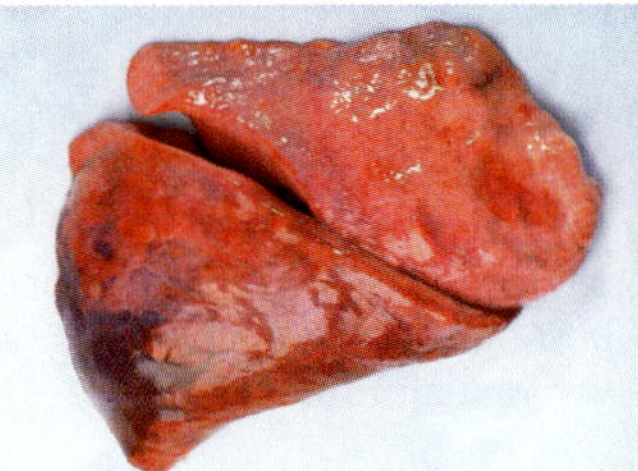

Este es un par de pulmones sanos.

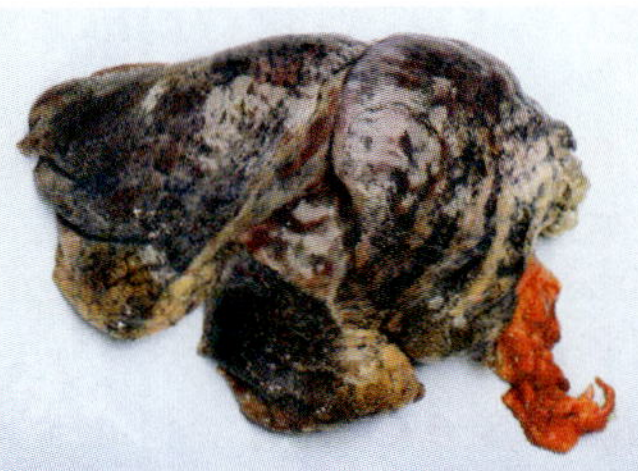

Estos pulmones son de una persona con enfisema.

Figura 19 *El enfisema es una enfermedad respiratoria que puede ser causada por el hábito de fumar.*

501

Chapter Highlights

VOCABULARY DEFINITIONS

SECTION 1

cardiovascular system a collection of organs whose primary function is to transport blood to and from your body's cells; the organs in this system include the heart, the arteries, and the veins

blood a type of connective tissue made up of cells, cell parts, and plasma

plasma the fluid part of blood

platelet a cell fragment that helps clot blood

atrium an upper chamber of the heart

ventricle a lower chamber of the heart

arteries blood vessels that carry blood away from the heart

capillaries the smallest blood vessels in the body

veins blood vessels that direct blood to the heart

pulmonary circulation the circulation of blood between the heart and lungs

systemic circulation the circulation of blood between the heart and the body (excluding the lungs)

blood pressure the amount of force exerted by blood on the inside walls of a blood vessel

SECTION 2

lymphatic system a collection of organs whose primary function is to collect extracellular fluid and return it to the blood. The organs in this system include the lymph nodes and the lymphatic vessels

lymph capillaries the smallest vessels in the lymphatic system

lymph fluid and particles absorbed into lymph capillaries

lymphatic vessels large vessels in the lymphatic system

Resumen del capítulo

SECCIÓN 1

Vocabulario

sistema cardiovascular (*pág. 488*)

sangre (*pág. 488*)

plasma (*pág. 488*)

plaqueta (*pág. 489*)

aurícula (*pág. 490*)

ventrículo (*pág. 490*)

arterias (*pág. 491*)

capilares (*pág. 491*)

venas (*pág. 491*)

circulación pulmonar (*pág. 492*)

circulación sistémica (*pág. 492*)

presión sanguínea (*pág. 493*)

Notas de la sección

- El aparato cardiovascular lleva el oxígeno y los nutrientes a las células y recoge sus productos de desecho para mantener saludable el cuerpo. Está constituido por la sangre, el corazón y los vasos sanguíneos.

- La sangre es un tejido conjuntivo formado por plasma, glóbulos rojos, glóbulos blancos y plaquetas. El corazón es un órgano muscular que bombea la sangre a través de los vasos sanguíneos.

- La sangre abandona el corazón por las arterias y después entra en los capilares. Tras los capilares, la sangre es llevada de regreso al corazón por las venas.

- En la circulación pulmonar, los vasos sanguíneos llevan la sangre del corazón a los pulmones y de regreso al corazón. En la circulación sistémica, la sangre fluye del corazón al resto del cuerpo y luego de regreso al corazón.

- Hay distintos grupos sanguíneos. El grupo sanguíneo de una persona está determinado por ciertas substancias químicas en los glóbulos rojos.

SECTION 2

Vocabulario

sistema linfático (*pág. 496*)

capilares linfáticos (*pág. 496*)

linfa (*pág. 496*)

vasos linfáticos (*pág. 496*)

nódulos linfáticos (*pág. 497*)

timo (*pág. 497*)

bazo (*pág. 497*)

amígdalas (*pág. 497*)

Notas de la sección

- El sistema linfático devuelve el exceso de fluido al aparato cardiovascular y combate las infecciones.

- El sistema linfático comprende la linfa, los capilares linfáticos, los vasos linfáticos, los nódulos linfáticos, el bazo, las amígdalas y el timo.

✓ Comprobar destrezas

Conceptos de matemáticas

LATIDO ININTERRUMPIDO Tu corazón late unas 100,800 veces por día. Esto quiere decir que palpita alrededor de 4,200 veces por hora.

> 100,800 latidos ÷ 24 horas = 4,200 latidos

O sea, palpita alrededor de 70 veces por minuto.

> 4,200 latidos ÷ 60 minutos = 70 latidos

Comprensión visual

CONDUCTOS DE AIRE
Observa de nuevo la Figura 15 de la página 498. Traza con el dedo el camino que el aire recorre para llegar a los pulmones. Al hacerlo, piensa en el papel de la nariz, la faringe, la tráquea, los bronquios, los pulmones y el diafragma en la respiración.

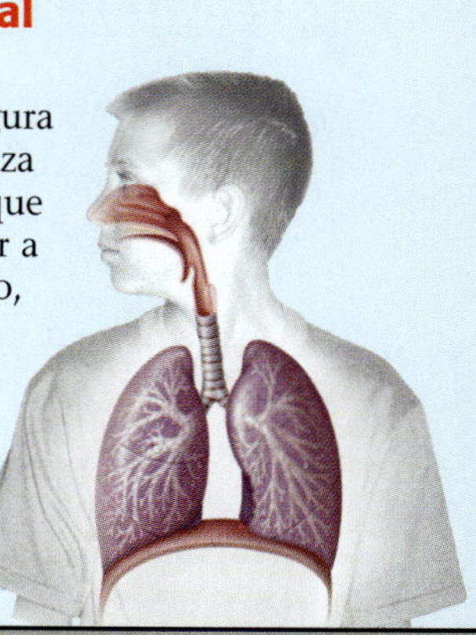

Lab and Activity Highlights

Build a Lung `PG 630`

Carbon Dioxide Breath `PG 631`

Datasheets for LabBook
(blackline masters for these labs)

Vocabulario

respiración *(pág. 498)*
sistema respiratorio *(pág. 498)*
faringe *(pág. 499)*
laringe *(pág. 499)*
tráquea *(pág. 499)*
bronquios *(pág. 499)*
alvéolos *(pág. 499)*
diafragma *(pág. 500)*
respiración celular *(pág. 500)*

Notas de la sección

- El aparato respiratorio introduce y saca aire del cuerpo; comprende la nariz, la boca, la faringe, la laringe, la tráquea y los pulmones.

- El aire entra a los pulmones a través de los bronquios y viaja hasta los alvéolos, saquitos llenos de aire rodeados por capilares del aparato cardiovascular.

- La sangre de los capilares pulmonares absorbe oxígeno y libera dióxido de carbono. El dióxido de carbono se exhala. El oxígeno es llevado por la sangre al corazón y luego a las células del cuerpo.

- Las células del cuerpo necesitan oxígeno para llevar a cabo la respiración celular, un proceso químico que libera la energía contenida en carbohidratos, grasas y proteínas, y la hace accesible a las células.

- La inhalación y la exhalación se deben a la contracción y relajación del diafragma y de los músculos de la caja torácica.

Experimentos

Construye un pulmón *(pág. 630)*
Dióxido de carbono en el aliento *(pág. 631)*

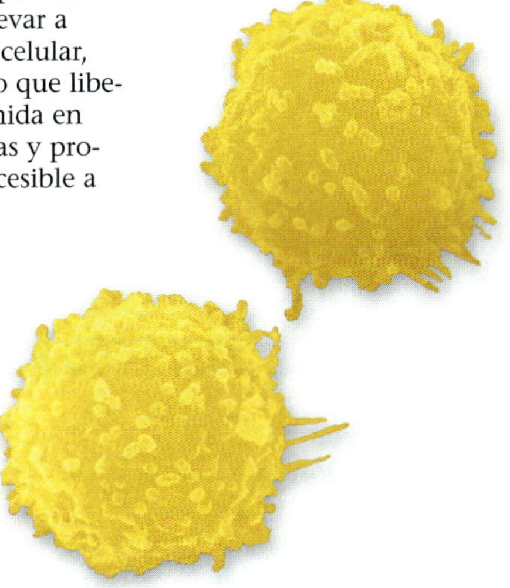

internet

VISITA: go.hrw.com

Visita el sitio web de HRW para encontrar una serie de herramientas de aprendizaje relacionadas con este capítulo. Sólo tienes que escribir la palabra clave:

PALABRA CLAVE: HSTBD2

SCI LINKS
NSTA
VISITA: www.scilinks.org

Visita el sitio web de la **Asociación Nacional de Maestros de Ciencias** *(National Science Teachers Association)* para encontrar recursos de Internet relacionados con este capítulo. Sólo escribe el **ENLACE DE CIENCIAS** para obtener más información sobre el tema:

TEMA	ENLACE
El aparato cardiovascular	HSTL555
Problemas cardiovasculares	HSTL560
El sistema linfático	HSTL565
El aparato respiratorio	HSTL570
Enfermedades respiratorias	HSTL575

503

lymph nodes small, bean-shaped organs that contain small fibers that work like nets to remove particles from the lymph

thymus a lymph organ that produces lymphocytes

spleen an organ that filters blood and produces lymphocytes

tonsils small masses of soft tissue located at the back of the nasal cavity, on the inside of the throat and at the back of the tongue

SECTION 3

respiration the exchange of gases between living cells and their environment; includes breathing and cellular respiration

respiratory system a collection of organs whose primary function is to take in oxygen and expel carbon dioxide; the organs of this system include the lungs, the throat, and the passageways that lead to the lungs

pharynx the upper portion of the throat

larynx the area of the throat that contains the vocal cords

trachea the air passageway from the larynx to the lungs

bronchi the two tubes that connect the lungs with the trachea

alveoli tiny sacs that form the bronchiole branches of the lungs

diaphragm the sheet of muscle underneath the lungs that accomplishes breathing

cellular respiration the process of producing ATP in the cell using oxygen and glucose; releases carbon dioxide and water

Lab and Activity Highlights

LabBank

Whiz-Bang Demonstrations
- Get the Beat! Demo 12
- Take a Deep Breath, Demo 13

EcoLabs & Field Activities, There's Something in the Air, Field Activity 9

Long-Term Projects & Research Ideas, Project 23

Vocabulary Review Worksheet 20

Blackline masters of these Chapter Highlights can be found in the **Study Guide.**

Chapter Review
Answers

Using Vocabulary

1. red blood cells
2. arteries
3. white blood cell
4. larynx
5. alveoli

Understanding Concepts

Multiple Choice

6. b
7. d
8. a
9. c
10. b
11. a

Short Answer

12. Pulmonary circulation carries blood through the lungs and back to the heart. Systemic circulation carries blood from the heart to the rest of the body.
13. The first number is the systolic pressure, or the pressure inside the artery when the ventricles contract. The second number, the diastolic pressure, is the pressure in the arteries when the ventricles relax.
14. Carbon dioxide is a product of cellular respiration which occurs in the body's cells.

Repaso del capítulo

UTILIZAR EL VOCABULARIO

Escoge el término correcto para completar las siguientes oraciones:

1. El oxígeno es llevado a las células del cuerpo por los __?__. *(glóbulos blancos o glóbulos rojos)*

2. La sangre sale del corazón por las __?__. *(arterias o venas)*

3. Los linfocitos son un tipo de __?__. *(vaso linfático o glóbulo blanco)*

4. La __?__ contiene las cuerdas vocales. *(tráquea o laringe)*

5. El recorrido del aire a través del aparato respiratorio termina en unos saquitos llamados __?__. *(alvéolos o bronquios)*

COMPRENDER CONCEPTOS

Opción múltiple

6. La sangre de los pulmones entra al corazón por
   a. el ventrículo izquierdo.
   b. la aurícula izquierda.
   c. la aurícula derecha.
   d. el ventrículo derecho.

7. Las células sanguíneas se producen
   a. en el corazón.
   b. a partir del plasma.
   c. a partir de la linfa.
   d. en los huesos.

8. ¿Cuál de las siguientes estructuras no es parte del sistema linfático?
   a. la tráquea
   b. los nódulos linfáticos
   c. el timo
   d. el bazo

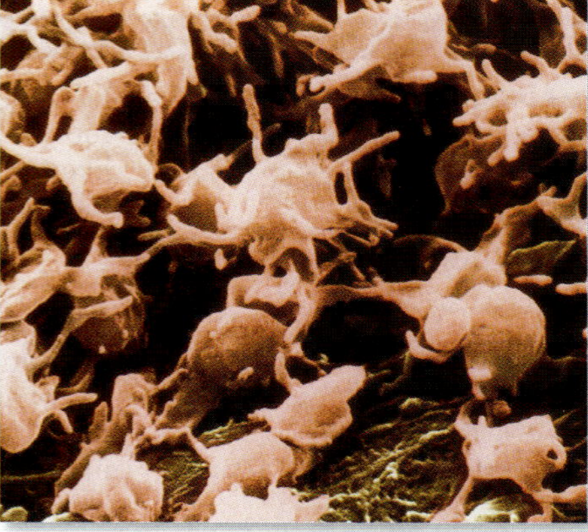

9. Los alvéolos están rodeados de
   a. venas.
   b. músculos.
   c. capilares.
   b. nódulos linfáticos.

10. ¿Qué le impide a la sangre regresarse por las venas?
    a. las plaquetas
    b. las válvulas
    b. los músculos
    d. el cartílago

11. El aire entra en los pulmones cuando el diafragma
    a. se contrae y se mueve hacia abajo.
    a. se contrae y se mueve hacia arriba.
    a. se relaja y se mueve hacia abajo.
    a. se relaja y se mueve hacia arriba.

Respuesta breve

12. ¿Qué diferencia hay entre la circulación pulmonar y la sistémica en el aparato cardiovascular?

13. José tiene una presión sanguínea de 110/65. ¿Qué significan estos números?

14. ¿Qué proceso corporal produce el dióxido de carbono que exhalas?

Organizar conceptos

15. Usa los siguientes términos para crear un mapa de ideas: sangre, oxígeno, alvéolos, capilares, dióxido de carbono.

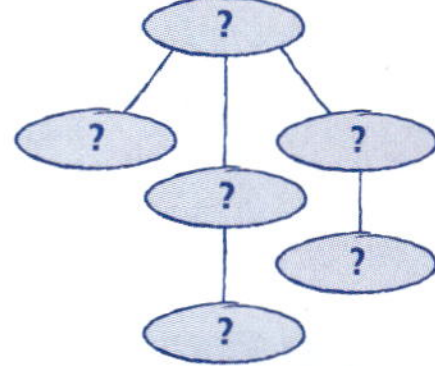

RAZONAMIENTO CRÍTICO Y RESOLUCIÓN DE PROBLEMAS

Escribe una o dos oraciones para responder a las siguientes preguntas:

16. ¿Por qué crees que hay pelos en la nariz?

17. Cuando una persona no se siente bien, a veces el médico analiza una muestra de su sangre para contar los glóbulos blancos que contiene. ¿Para qué sirve esta información?

18. ¿En qué se relacionan la función del sistema linfático y la del aparato cardiovascular?

LAS MATEMÁTICAS EN LAS CIENCIAS

19. Después de que una persona dona sangre, ésta se almacena en bolsas de una pinta hasta que se necesite para una transfusión. Una persona sana tiene, por lo general, 5 millones de glóbulos rojos en cada milímetro cúbico (1 mm^3) de sangre.
    a. ¿Cuántos glóbulos rojos hay en 1 mL de sangre? Un mililitro es igual a 1 cm^3 y a 1,000 mm^3.
    b. ¿Cuántos glóbulos rojos hay en 1 pinta? Una pinta es igual a 473 mL.

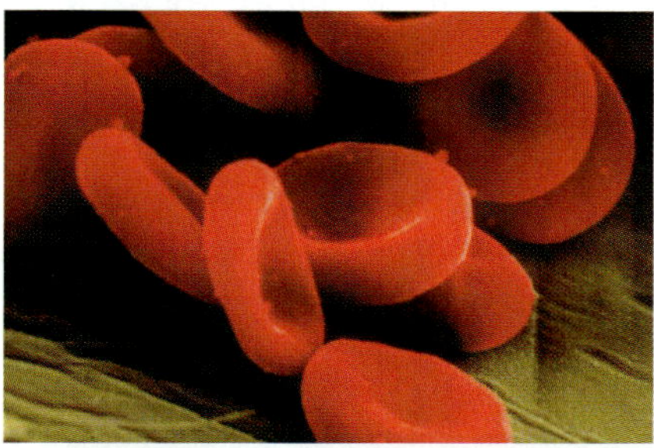

INTERPRETAR GRÁFICAS

El diagrama de abajo muestra un corazón humano. Examínalo y responde a las siguientes preguntas:

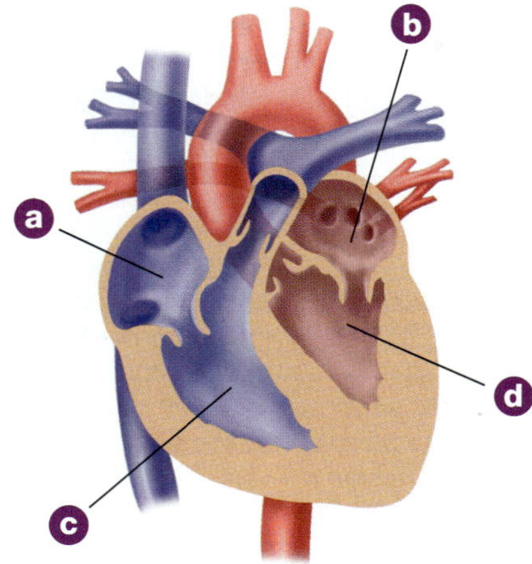

20. ¿Qué letra indica la cavidad que recibe la sangre de la circulación sistémica? ¿Cómo se llama esta cavidad?

21. ¿Qué letra indica la cavidad que recibe la sangre de los pulmones? ¿Cómo se llama esta cavidad?

22. ¿Qué letra indica la cavidad que bombea sangre a los pulmones? ¿Cómo se llama esta cavidad?

AHORA, ¿qué piensas?

Revisa tus respuestas a las preguntas de la página 487 que escribiste en tu cuaderno de ciencias. ¿Han cambiado tus respuestas? Si es necesario, corrige tus respuestas basándote en lo que has aprendido en este capítulo.

Concept Mapping

15. 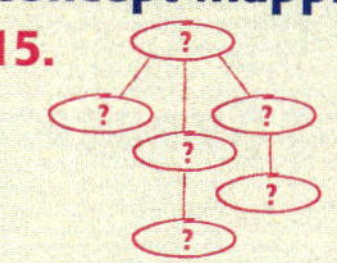 An answer to this exercise can be found at the end of this book.

CRITICAL THINKING AND PROBLEM SOLVING

16. The hairs catch dust and other foreign particles. This helps keep your lungs as clean as possible.

17. Doctors will need to know if the immune system is producing large numbers of white blood cells; this would indicate an infection.

18. The lymphatic system returns leaked fluids back to your blood.

MATH IN SCIENCE

19. a. 5 billion (5,000,000,000) cells
    b. 2.365 trillion (2,365,000,000,000) cells

INTERPRETING GRAPHICS

20. *a*, the right atrium
21. *b*, the left atrium
22. *c*, the right ventricle

  **Concept Mapping Transparency 20**

Blackline masters of this Chapter Review can be found in the **Study Guide.**

NOW WHAT DO YOU THINK?

1. Blood is a type of connective tissue made of cell parts, plasma, and two types of cells. Blood carries nutrients and oxygen to all the cells of the body and carries carbon dioxide and wastes away. White blood cells fight infection, and platelets aid in clotting.

2. Inhaling brings oxygen into the lungs, where it is transferred to the blood and transported throughout the body. Exhaling allows carbon dioxide in the lungs to leave the body.

Activity

Divide the class into pairs. Have students use a flashlight to make their partner's pupils contract. (Be careful not to let them look directly into a bright light for too long.) They should conduct their test for the photic sneeze reflex as a scientific experiment. Have them write a hypothesis, test the hypothesis, and report their results.

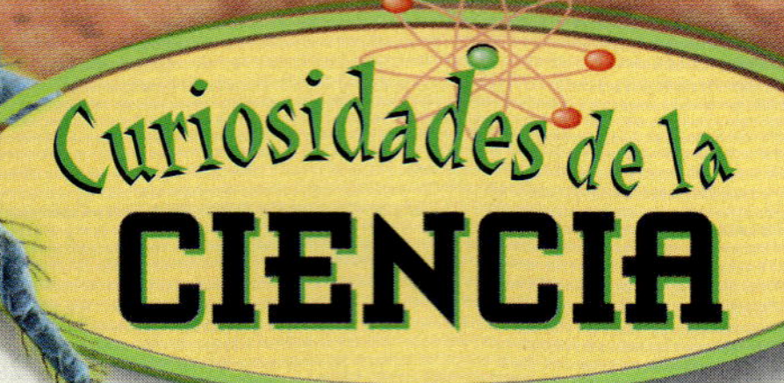

Curiosidades de la CIENCIA

ESTORNUDOS

¿Estornudas cuando sales de una sala de cine obscura a la luz brillante del Sol? Si no lo haces, fíjate en los demás la próxima vez; lo más probable es que varias personas estornuden.

Reflejos equivocados

Por alguna razón, más o menos una de cada cinco personas estornudan cuando salen de un área poco iluminada a una con mucha luz. En realidad, ¡algunas estornudan doce o más veces! Afortunadamente, dejan de estornudar al poco tiempo. La reacción se llama *reflejo de estornudo debido a la luz.* Nadie está seguro de por qué sucede.

El estornudo normal es un reflejo, o sea, se hace sin pensar. La mayoría de las personas estornudan cuando algo les hace cosquillas dentro de la nariz. Al estornudar, el movimiento del aire empuja hacia afuera la partícula extraña. Por ejemplo, si te entra polvo en la nariz, el estornudo lo saca. En el caso de las personas que sufren de estornudos ocasionados por la luz, se trata de un reflejo equivocado.

¡ACHÚ!

Hace unos cuantos años, algunos genetistas estudiaron el reflejo de estornudo debido a la luz; lo llamaron síndrome autosómico dominante de estornudos incontrolables por reacción oftálmica a la luz solar, o ACHOO *(Autosomal Dominant Compelling Helio-ophthalmic Outburst Syndrome)*. Los científicos saben que este síndrome se presenta en familias, por lo que puede transmitirse de padres a hijos. A veces, hasta el número de estornudos por persona es el mismo entre los miembros de una familia.

▲ *¿Estornudas cuando ves luz brillante tras estar en un cuarto obscuro?*

Respuestas posibles

Algunos científicos han dado una posible explicación del síndrome. Las pupilas de las personas se contraen en presencia de luz brillante y los nervios de los ojos están junto a los de la nariz. Por tanto, es posible que las personas con este síndrome tengan los cables ligeramente cruzados: ¡la luz brillante dispara el reflejo de la pupila y también el del estornudo!

Fiesta del estornudo

La luz solar no es la única que desencadena estornudos repentinos. Hay personas que estornudan cuando se frotan el ángulo interno del ojo; otras estornudan cuando se depilan las cejas o se cepillan el cabello. En algunas personas, comer demasiado puede causar estornudos.

Investigación y hechos

▶ Bostezar también es un reflejo. Investiga por qué bostezamos.

506

Answer to Research the Facts

Sample answer: Yawning is a response to a buildup of carbon dioxide or the sight of another person yawning.

Cabras al rescate

Se llaman cabras transgénicas porque sus células contienen un gen humano. Se ven como cualquier otra cabra, pero debido a que tienen un gen humano producen una substancia química que salva vidas.

Genes que salvan vidas

Los ataques al corazón son la causa número uno de muertes en Estados Unidos. Muchos ataques ocurren cuando el flujo de sangre que va al corazón es interrumpido por grandes coágulos de sangre. Las células sanguíneas humanas producen una substancia química llamada *activador del plasminógeno tisular* (APT) que disuelve los coágulos de sangre pequeños. Si se le da APT a una persona que está sufriendo un ataque al corazón, con frecuencia esta substancia es capaz de disolver el coágulo de sangre, detener el ataque y salvar la vida de la persona. Pero es difícil producir grandes cantidades de APT en el laboratorio. Aquí es donde las cabras entran en acción. Un equipo de investigadores de la Universidad de Tufts, en Grafton (Massachusetts), han modificado cabras genéticamente para producir este medicamento salvavidas.

Cabras híbridas

La producción de cabras transgénicas es un proceso complicado. Primero, los óvulos fertilizados se extraen de hembras normales mediante cirugía. Luego, a estos óvulos se les inyectan genes híbridos formados por genes humanos del APT "unidos" a genes de las glándulas mamarias de las cabras. Finalmente, los óvulos modificados se implantan en las cabras mediante cirugía y así empiezan su desarrollo para formar otras cabras. Algunas de las cabras que nacen tienen el gen híbrido y cuando maduran su leche contiene APT. Después, los técnicos separan el APT de la leche para que se utilice en las víctimas de ataques al corazón.

La investigación continúa

Quizá algún día las investigaciones con animales de granja transgénicos, como cabras, borregos, vacas y cerdos, permitan elaborar medicamentos más baratos, en mayores cantidades y con mayor rapidez que los métodos actuales. Esto cambiaría nuestra perspectiva de estos animales.

Descúbrelo tú mismo.

▶ El uso de substancias químicas producidas por animales transgénicos es sólo una de las muchas terapias genéticas. Investiga más sobre la terapia genética; cómo se utiliza y cómo podría utilizarse en el futuro.

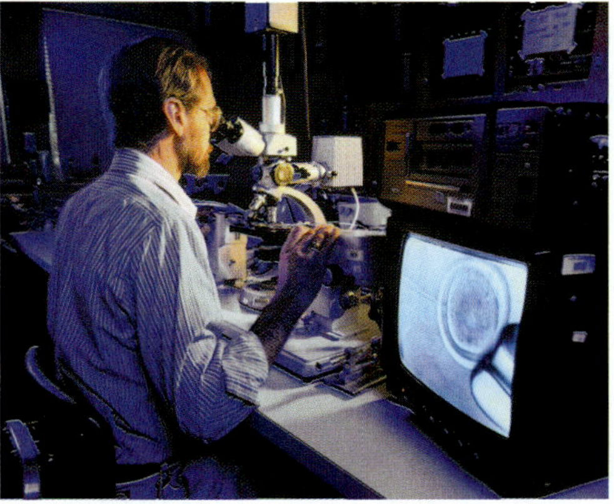

▲ *Un científico de la Universidad de Tufts inyecta genes humanos de APT en los óvulos fertilizados de cabra.*

Answer to Find Out for Yourself

Another form of gene therapy is to obtain healthy genes and insert them into the tissue of a person with a genetic disorder. Such methods are still highly experimental.

HEALTH WATCH
Goats to the Rescue

Background

Many proteins that are produced naturally by the human body could be produced by other animals through genetic engineering. This raises the possibility of turning animals into "drug factories." Genetically altered pigs have been used to produce blood and to grow organs that can be used for human transfusions and transplants. Cows have been altered to produce milk that is more like human milk. A dozen different human proteins have been produced in the milk of various transgenic animals.

However, there are serious safety issues concerning these new animal products. Host animals may carry pathogens that are dangerous to humans. For example, some sheep and goats carry a brain disease known as scrapie. Research companies that offer transgenic drugs must be sure that pathogens are not present in their products. They must ensure that any animal proteins that might trigger allergic reactions or other side effects are removed during the purification process. In addition, these companies must be able to prove that their products will function in the same way as their natural counterparts.

507

Chapter Organizer

CHAPTER ORGANIZATION	TIME MINUTES	OBJECTIVES	LABS, INVESTIGATIONS, AND DEMONSTRATIONS	
Chapter Opener pp. 508–509	45		**Investigate!** How Fast Do You React? p. 509	
Section 1 The Nervous System	90	▶ Explain how neurons in the nervous system work together. ▶ Compare and contrast the central nervous system and the peripheral nervous system. ▶ Describe the major functions of four parts of the brain and the spinal cord.	**Demonstration,** Simulating Neuronal Impulses, p. 512 in ATE **QuickLab,** Knee Jerks, p. 516	
Section 2 Responding to the Environment	90	▶ List four sensations that are detected by receptors in the skin. ▶ Describe how light relates to vision. ▶ Explain the function of rods and cones. ▶ Compare and contrast the functions of photo-receptors, taste buds, and olfactory cells.	**Quick Lab,** Where's the Dot? p. 520 **Skill Builder,** You've Gotta Lotta Nerve! p. 632 **Datasheets for LabBook,** You've Gotta Lotta Nerve! Worksheet 44 **Whiz-Bang Demonstrations,** Now You See It, Now You Don't, Demo 15	
Section 3 The Endocrine System	90	▶ Explain the function of the endocrine system. ▶ List the glands of the endocrine system, and describe some of their functions. ▶ Illustrate the location of some of the endocrine glands in the body. ▶ Describe how feedback controls stop and start hormone release.	**Long-Term Projects & Research Ideas,** Project 25	

TECHNOLOGY RESOURCES

 **Guided Reading Audio CD**
English or Spanish, Chapter 21

 **One-Stop Planner CD-ROM with Test Generator**

 **Science, Technology & Society,** Brain Cell Visuals, Segment 23
Learning from Frog Ears, Segment 26
Easy Touch Toys, Segment 33

CLASSROOM WORKSHEETS, TRANSPARENCIES, AND RESOURCES	SCIENCE INTEGRATION AND CONNECTIONS	REVIEW AND ASSESSMENT
Directed Reading Worksheet 21 **Science Puzzlers, Twisters & Teasers,** Worksheet 21		
Transparency 78, The Neuron **Directed Reading Worksheet 21,** Section 1 **Transparency 79,** What's in a Nerve? **Reinforcement Worksheet 21,** This System Is Just "Two" Nervous! **Transparency 80,** Regions of the Brain **Transparency 81,** The Spinal Cord	**Math Break,** Time to Travel, p. 511 **Real-World Connection,** p. 513 in ATE **Cross-Disciplinary Focus,** p. 515 in ATE **Eureka!** Pathway to a Cure, p. 531	**Homework,** p. 514 in ATE **Self-Check,** p. 515 **Review,** p. 516 **Quiz,** p. 516 in ATE **Alternative Assessment,** p. 516 in ATE
Directed Reading Worksheet 21, Section 2 **Transparency 194,** Wavelength **Reinforcement Worksheet 21,** The Eyes Have It **Critical Thinking Worksheet 21,** There's a Microchip in My Eye!	**Real-World Connection,** p. 518 in ATE **Connect to Physical Science,** p. 518 in ATE **Multicultural Connection,** p. 519 in ATE **Math and More,** p. 520 in ATE **Physical Science Connection,** p. 521 **Science, Technology, and Society:** Light on Lenses, p. 530	**Review,** p. 521 **Quiz,** p. 521 in ATE **Alternative Assessment,** p. 521 in ATE
Directed Reading Worksheet 21, Section 3 **Reinforcement Worksheet 21,** Every Gland Lends a Hand	**Apply,** p. 523 **Math and More,** p. 524 in ATE	**Homework,** p. 524 in ATE **Review,** p. 525 **Quiz,** p. 525 in ATE **Alternative Assessment,** p. 525 in ATE

Holt, Rinehart and Winston On-line Resources

go.hrw.com

For worksheets and other teaching aids related to this chapter, visit the HRW Web site and type in the keyword: **HSTBD4**

National Science Teachers Association

www.scilinks.org

Encourage students to use the *sci*LINKS numbers listed with the Chapter Highlights to access information and resources on the **NSTA** Web site.

END-OF-CHAPTER REVIEW AND ASSESSMENT

Chapter Review in Study Guide

Vocabulary and Notes in Study Guide

Chapter Tests with Performance-Based Assessment, Chapter 21 Test

Chapter Tests with Performance-Based Assessment, Performance-Based Assessment 21

Concept Mapping Transparency 21

Chapter Resources & Worksheets

Visual Resources

TEACHING TRANSPARENCIES

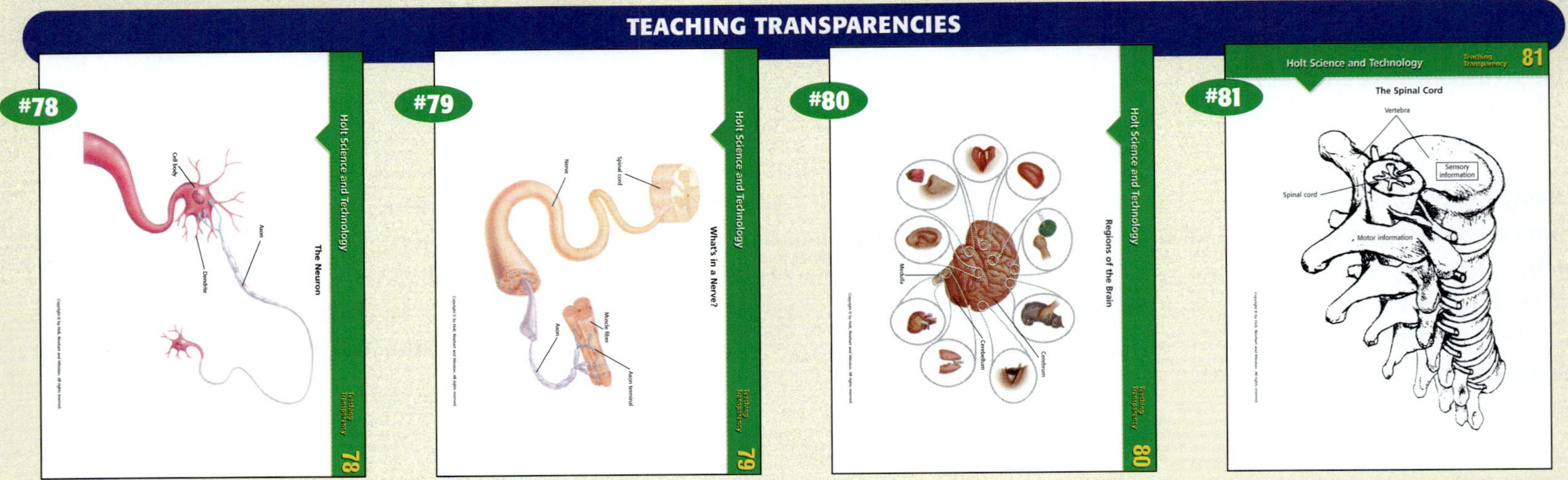

TEACHING TRANSPARENCIES

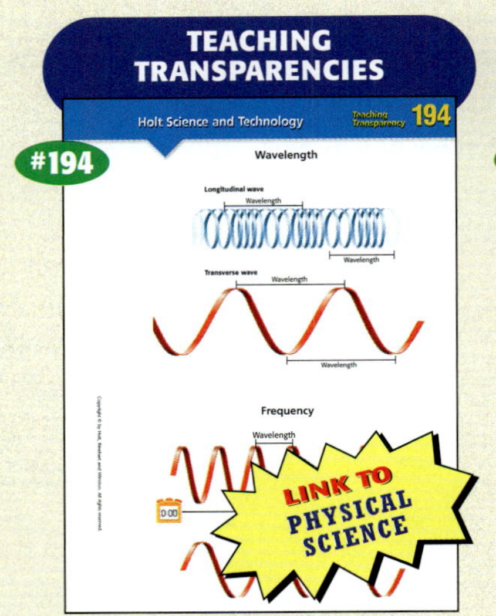

CONCEPT MAPPING TRANSPARENCY

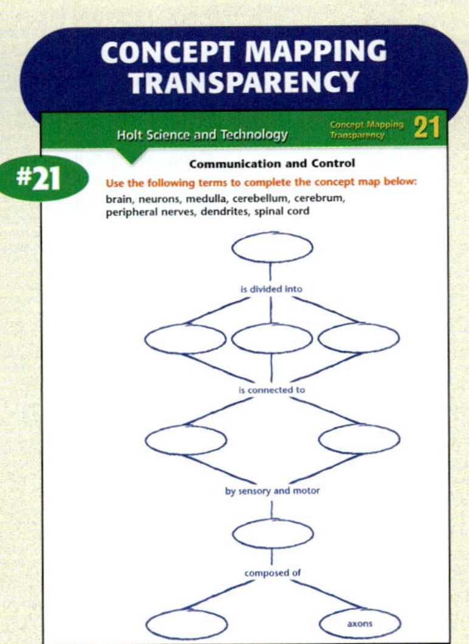

Meeting Individual Needs

DIRECTED READING

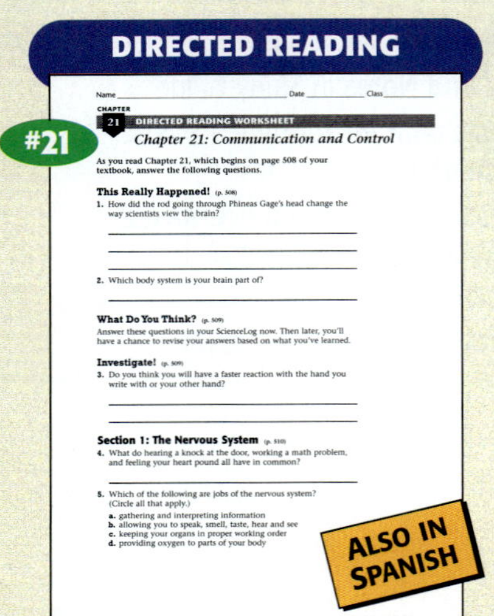

REINFORCEMENT & VOCABULARY REVIEW

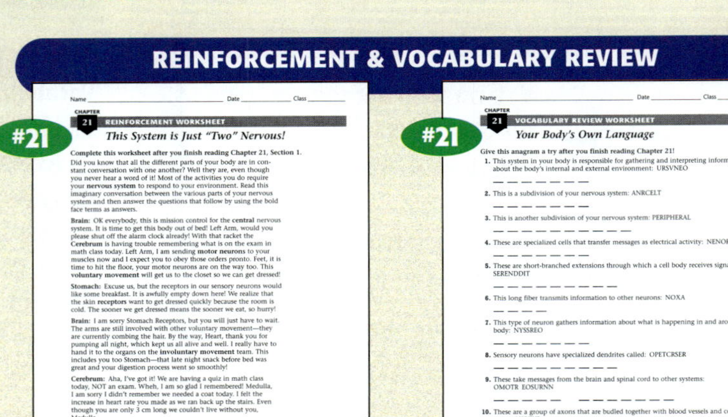

SCIENCE PUZZLERS, TWISTERS & TEASERS

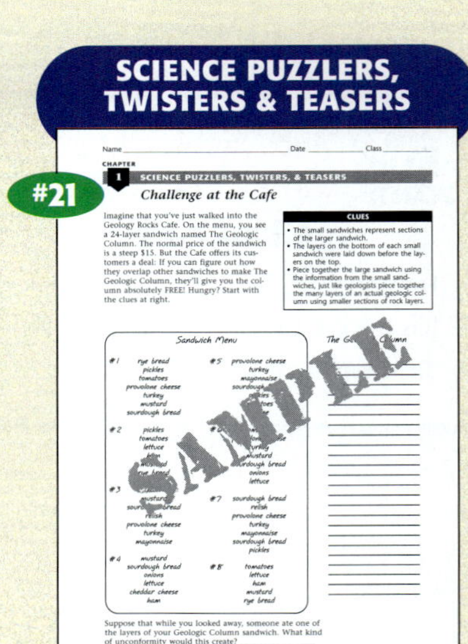

Review & Assessment

STUDY GUIDE

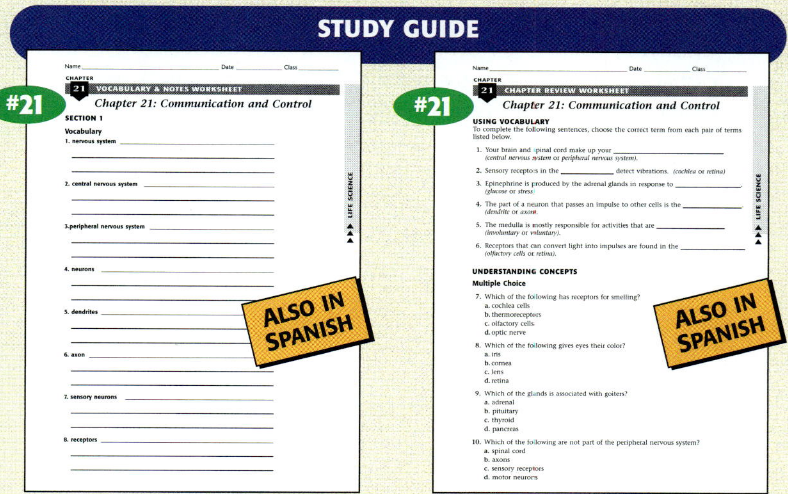

CHAPTER TESTS WITH PERFORMANCE-BASED ASSESSMENT

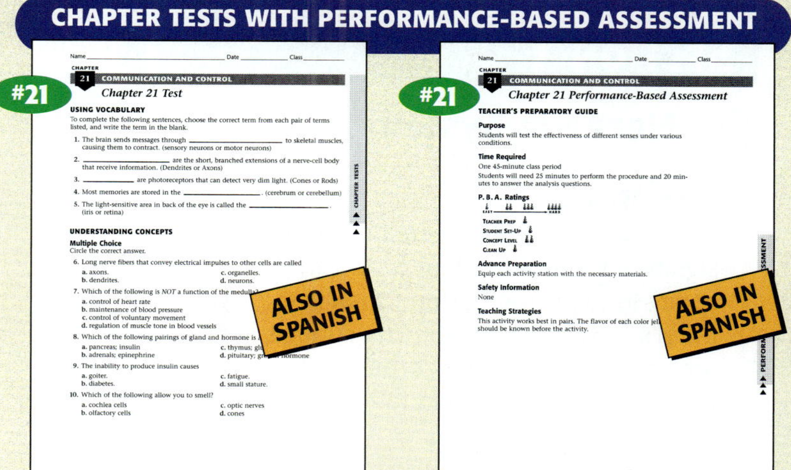

Lab Worksheets

WHIZ-BANG DEMONSTRATIONS

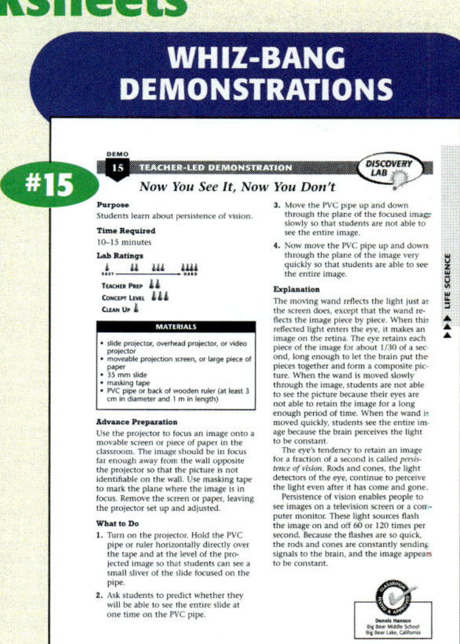

LONG-TERM PROJECTS & RESEARCH IDEAS

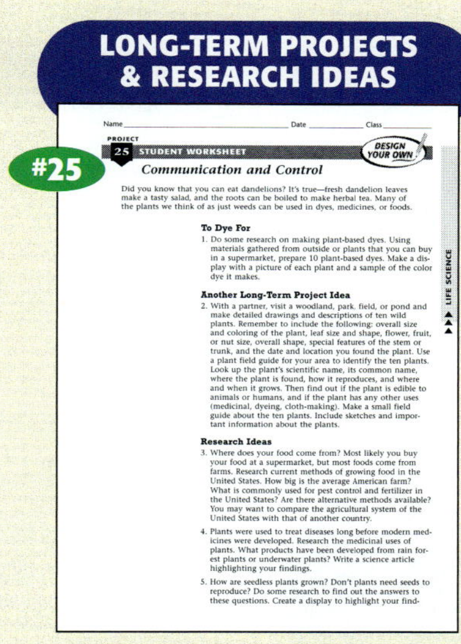

DATASHEETS FOR LABBOOK

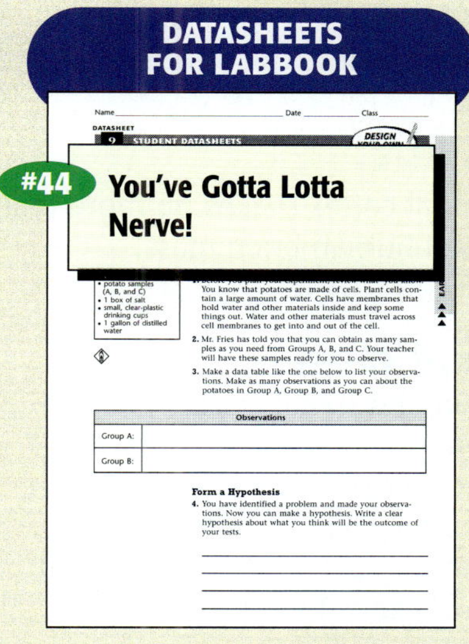

Applications & Extensions

CRITICAL THINKING & PROBLEM SOLVING

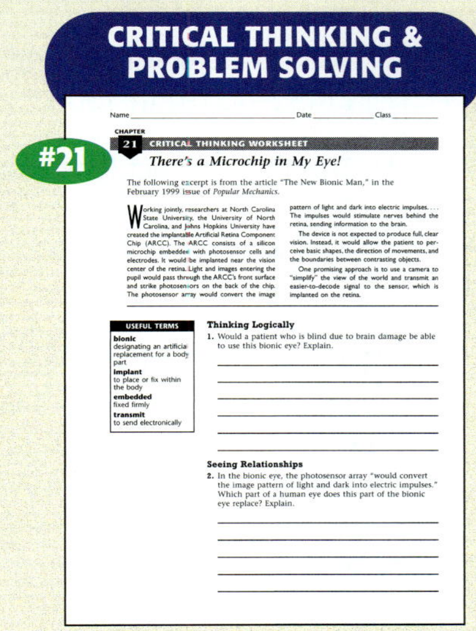

SCIENCE TECHNOLOGY

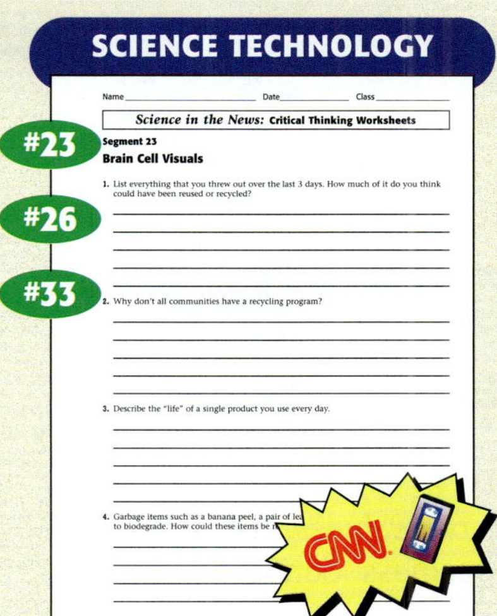

Chapter Background

The Nervous System

▶ The Brain

The cerebrum's surface area is a more significant indicator of an organism's ability to think and perform than is the brain's volume or weight. The surface area of a porpoise's cerebrum relative to its body size is second to that of the human cerebrum.

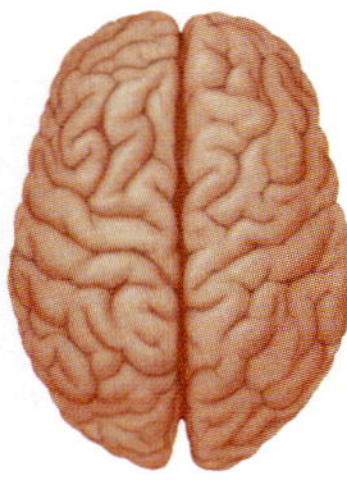

- Computerized scanning techniques allow physicians to take pictures of the brain to detect abnormalities. Scanning techniques include CT scanning, MRI scanning, radionucleotide scanning, ultrasound scanning, and PET scanning.

IS THAT A FACT!

- The cerebral cortex makes up more than 80 percent of the total human brain mass.

▶ The Spinal Cord

Like the brain, the spinal cord contains both gray matter and white matter. The center of the spinal cord is made up of neuron cell bodies and is called gray matter. The outer layer of the spinal cord is made up of axons that traverse the spinal cord. This part of the spinal cord is called white matter.

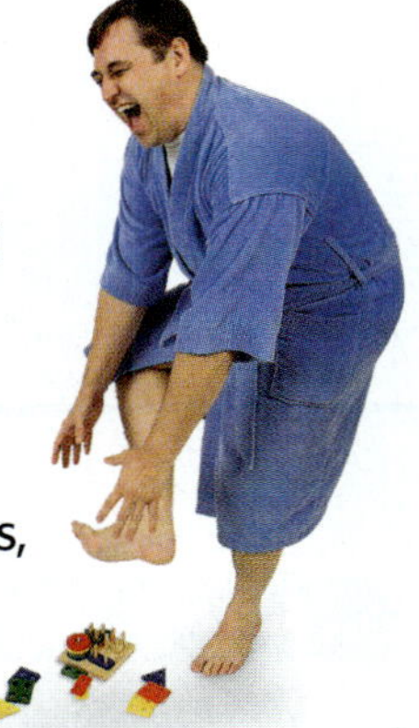

- The spinal cord is protected by 25 bones, the vertebrae and the sacrum. These bones are connected by joints and separated by cartilaginous disks.

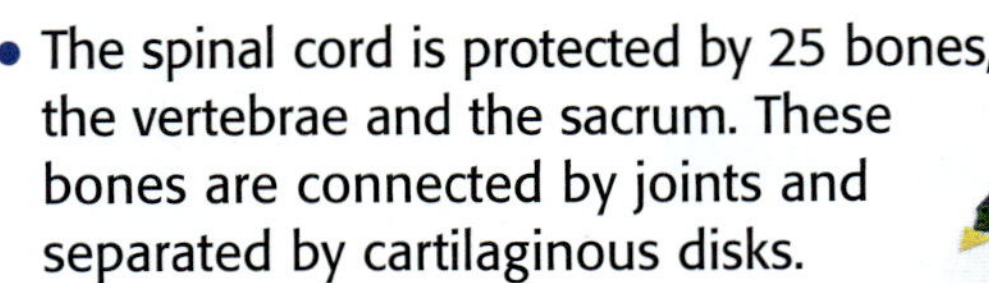

SECTION 2

Responding to the Environment

▶ Hearing Loss

There are two principal kinds of deafness: conductive deafness and sensorineural deafness. Conductive deafness results when transmission of sound from the outer ear to the inner ear fails. It may occur as a result of earwax buildup or damage to the middle ear.

- Sensorineural deafness results when sounds reach the inner ear but are not transmitted to the brain due to either damaged inner ear structures or damaged nerves that carry information from the ear to the brain.

IS THAT A FACT!

- Sensorineural deafness occurs in 1 out of every 1,000 babies.

▶ The Eye Doctor

Ophthalmologists are physicians who specialize in the eyes. An ophthalmologist can examine eyes, prescribe corrective lenses, treat eye disorders, and perform eye surgery.

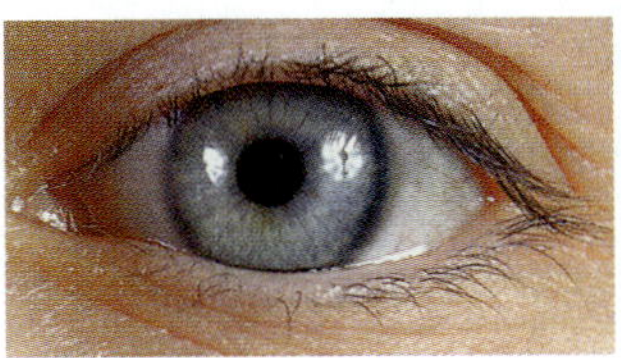

- An optician may only fit and adjust glasses and contact lenses.

- An optometrist can examine and test eyes and prescribe corrective lenses in the form of glasses or contact lenses.

IS THAT A FACT!

- The idea of using contact lenses to correct poor vision was first recorded by Leonardo da Vinci (1452–1519) in 1508.

- The first contact lens was made of glass. It covered the entire frontal surface of the eyeball. This first lens was made by Adolf Fick in 1887.

▶ The Sense of Taste

Saliva dissolves chemicals in the food and drink we consume. After passing through pores in the taste buds, these chemicals stimulate small nerve endings, which send messages to the brain. These messages form our sense of taste.

- People often lose their sense of taste when they lose their sense of smell. This occurs when olfactory bulbs are damaged or when the person has a stuffy nose. It is rare for a person to maintain the sense of smell and to lose the sense of taste.

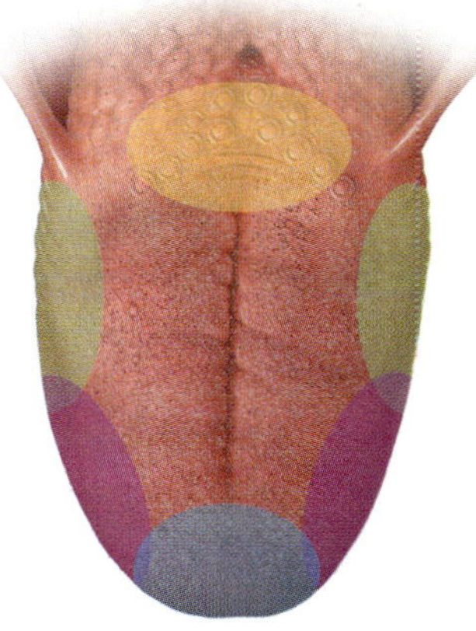

SECTION 3

The Endocrine System

▶ **Endocrine Glands**

There are two main types of glands in the body: exocrine glands and endocrine glands. Exocrine glands, such as sweat glands and salivary glands, secrete substances through ducts to a local area. Unlike exocrine glands, endocrine glands secrete substances directly into the bloodstream (no ducts are involved). The substances secreted by endocrine glands are carried, often to distant parts of the body, by the bloodstream and have effects throughout the body.

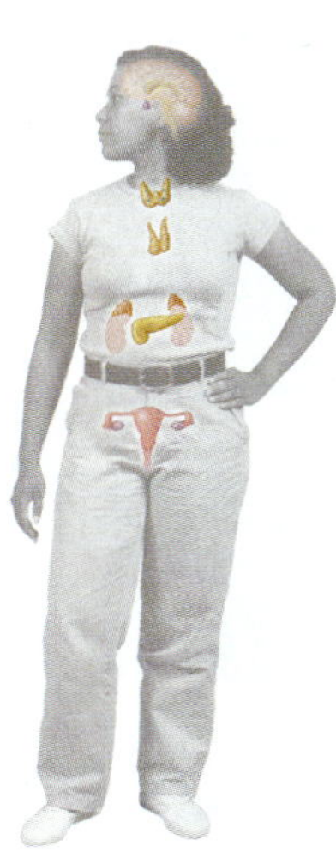

▶ **The Pituitary Gland**

The pituitary gland is often called the master gland because its secretions regulate several other endocrine glands.

- About 10 percent of brain tumors affect the pituitary gland. Although usually benign, these tumors can

have a great effect on the body because they can affect the production of the pituitary hormones.

- Because of the pituitary gland's location in the brain, enlargement of the gland can cause vision disorders by creating pressure on the nearby optic nerve.

▶ **Diabetes**

The term *diabetes* refers to more than one disorder. There is diabetes insipidus (a rare condition) and diabetes mellitus, of which there are two types. Type I requires regular injections of insulin. Type II can often be controlled by changes in diet and exercise, although insulin injections might also be necessary.

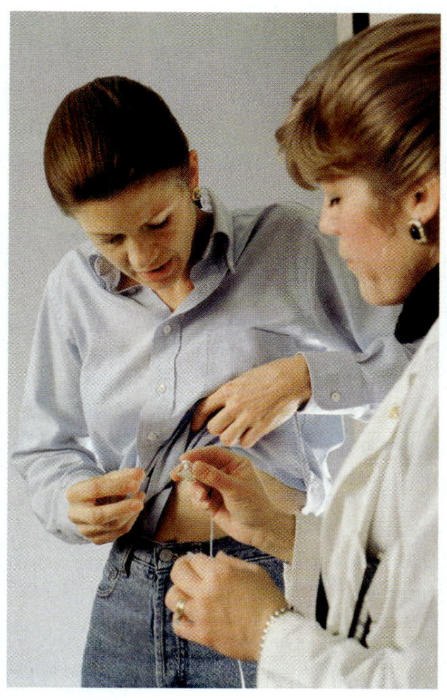

IS THAT A FACT!

- As many as 200 of every 100,000 people in the United States suffer from Type I diabetes.

- About 2,000 of every 100,000 people in the United States have Type II diabetes.

Chapter Preview

Section 1
The Nervous System
- Two Systems Within a System
- The Peripheral Nervous System
- Just a Bundle of Axons
- The Central Nervous System
- The Spinal Cord
- Ouch! That Hurt!

Section 2
Responding to the Environment
- Come to Your Senses
- Something in My Eye
- Did You "Ear" That?
- Does This Suit Your Taste?
- Your Nose Knows

Section 3
The Endocrine System
- Chemical Messengers
- Controlling the Controls
- Hormone Imbalances

Directed Reading Worksheet 21

Science Puzzlers, Twisters & Teasers Worksheet 21

Guided Reading Audio CD
English or Spanish, Chapter 21

CAPÍTULO

21 Comunicación y control

¡Esto realmente sucedió!

El cráneo de Gage muestra el recorrido de la varilla.

Hace unos 150 años, un extraño accidente cambió las ideas científicas sobre la función del encéfalo. El afectado fue un ferrocarrilero llamado Phineas Gage. Un día de 1848, Gage estaba comprimiendo explosivos en un hoyo con una varilla metálica para abrir camino a una vía férrea. Una chispa provocó una explosión que hizo que la varilla saliera disparada y atravesara la cabeza de Gage. La varilla entró por la mejilla izquierda y salió por la parte de arriba de la cabeza. Cualquiera lo hubiera considerado un acci- dente mortal, pero Gage se levantó del suelo, e incluso pudo hablar.

En poco más de 2 meses, sus heridas sanaron, pero él ya no era el mismo. El Gage que todos conocían antes del accidente era un hombre tranquilo, responsable y consi- derado, pero después del accidente se volvió irresponsable e iracundo. Los científicos que estudiaron su cambio de comportamiento descubrieron que el encéfalo tiene otras fun- ciones además de controlar el movimiento y los sentidos. En este capítulo aprenderás más sobre el encéfalo y el resto del sistema nervioso. Verás cómo el sistema nervioso y el endocrino controlan todos los demás sis- temas del cuerpo.

508

This Really Happened . . .

This classic case was reopened in 1994 by two neurobiologists—Hanna and Antonio Damasio, of the University of Iowa. The Damasios reexamined Gage's skull, which the Gage family donated to medical research 13 years after his death and which is now at the Warren Medical Museum, at Harvard University. Using clues in the preserved skull, computer modeling, and neural imaging, Hanna Damasio determined the likely path of the rod through Gage's skull. The rod is thought to have damaged the left frontal lobe, sparing the regions responsible for language and motor control. The Damasios found behavioral changes similar to Gage's in patients who have had this region of their brain damaged.

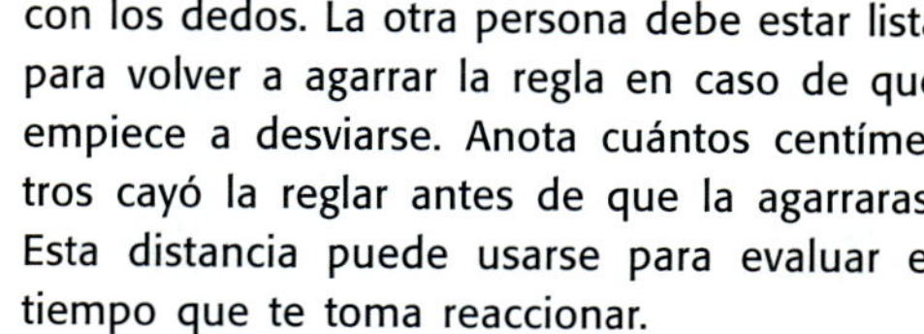

Usa tus conocimientos para responder a las siguientes preguntas en tu cuaderno de ciencias:

1. ¿Qué son los sentidos? ¿Cómo nos ayudan a sobrevivir?

2. ¿Por qué late más rápido tu corazón cuando te asustas?

3. ¿Cómo ayudan los anteojos y lentes de contacto a algunas personas a ver mejor?

¿Reaccionas rápido?

Si quieres mover la mano, el encéfalo tiene que enviar un mensaje hasta los músculos de los brazos. ¿Cuánto tardará en llegar? En este ejercicio, trabajarás con un compañero o compañera para ver con qué rapidez reaccionas.

Procedimiento

1. Siéntate en una silla y pídele a tu compañero o compañera que se pare frente a ti y sostenga una **regla** en posición vertical, como ves a la derecha. Coloca el pulgar a unos 3 cm de distancia de los demás dedos, cerca de la parte más baja de la regla. Esta debe estar colocada para que al caer pase entre el pulgar y los demás dedos.

2. Pídele a la otra persona que suelte la regla sin avisarte. Cuando veas que ya la soltó, trata de atraparla con los dedos. La otra persona debe estar lista para volver a agarrar la regla en caso de que empiece a desviarse. Anota cuántos centímetros cayó la reglar antes de que la agarraras. Esta distancia puede usarse para evaluar el tiempo que te toma reaccionar.

3. Repite el procedimiento varias veces y calcula el número promedio de centímetros. Repite el mismo ejercicio usando la mano que no usas para escribir.

4. Cambia de lugar con tu compañero o compañera, y repitan el procedimiento.

Análisis

5. Compara tus resultados con los de tu compañero o compañera. ¿Qué factores crees que pueden influir en la velocidad de reacción de una persona?

6. ¿Cómo se compara tu tiempo de reacción con una mano y con la otra?

7. Haz una lista de situaciones en las que una reacción rápida sea importante.

509

A typical adult human brain weighs 1.35 kg (almost 3 pounds).

What Do You Think?

Accept all reasonable responses.

Students will have a chance to revise their answers in the Chapter Review under NOW What Do You Think?

Investigate!

MATERIALS

FOR EACH GROUP:
• meterstick

Safety Caution: Remind students to handle the meterstick carefully and to keep it away from their face and far away from the face and eyes of their classmates. Allow students to have a practice trial. Instruct students to look only at the ruler and not at their partner. Point out that looking at their partner could distort the results of the investigation because the partner might give a clue about when he or she will drop the meterstick.

Answers to Investigate!

5. Student answers will vary, but possible suggestions include how rested an individual feels or if distractions occur in the room.

6. Answers will vary, but some students might notice a difference between the reaction time of the left hand and that of the right hand.

7. Answers will vary, but some students might suggest demanding activities, such as driving, sports, or police work.

Focus

The Nervous System

This section introduces the structures and functions of the nervous system. Students will learn the differences between the central nervous system, which includes the brain and spinal cord, and the peripheral nervous system, which consists of nerves that connect every area of the body to the central nervous system.

🔔 Bellringer

Ask students to list as many different functions of the brain as they can in their ScienceLog. After students complete their list, you may want to make a master list on the board or on a transparency. Explain that in this chapter, students will learn how the brain coordinates these many different activities.

1 Motivate

DISCUSSION

Reacting to Stimuli Invite students to describe a time when they reacted quickly. Encourage students to describe not only what happened but also how quickly they were able to react and what they were thinking about as they reacted. Sample experiences include jerking a hand away from a hot object, quickly catching a falling object, and extending one's hand out to brace for a fall. Based on students' experiences, lead a discussion about how quickly the nervous system is able to respond to a stimulus.

Sheltered English

VOCABULARIO

sistema nervioso	receptor
sistema nervioso central	neurona motora
sistema nervioso periférico	nervio
neurona	encéfalo
impulso	cerebro
dendrita	cerebelo
axón	bulbo raquídeo
neurona sensorial	reflejo

OBJETIVOS

- Explica cómo trabajan las neuronas en el sistema nervioso.
- Compara el sistema nervioso central con el sistema nervioso periférico.
- Describe las funciones principales de cuatro partes del encéfalo y de la médula espinal.

El sistema nervioso

Piensa en lo que estas cosas tienen en común: escuchas a alguien llamar a la puerta, escribes el resumen de un libro, sientes que tu corazón se agita después de correr, resuelves un problema de matemáticas, te sobresalta un ruido fuerte y saboreas un mango dulce. Todas éstas son actividades del *sistema nervioso*. El **sistema nervioso** reúne e interpreta información sobre el medio externo e interno del cuerpo, y responde según esta información. El sistema nervioso hace que tus órganos funcionen bien, y te permite hablar, oler, saborear, escuchar, ver, moverte, pensar y experimentar emociones.

Dos sistemas en uno

El sistema nervioso controla y coordina muchas cosas que pasan en el cuerpo. Funciona como un cuartel general; reúne y procesa los datos y se asegura de que a cada parte del cuerpo le llegue la información que le corresponde. Hay dos subdivisiones del sistema nervioso que realizan esta labor: *el sistema nervioso central (SNC)* y el *sistema nervioso periférico (SNP)*.

El **sistema nervioso central** está compuesto por el encéfalo y la médula espinal, que procesan los mensajes que entran y salen del sistema. El **sistema nervioso periférico** consiste en canales de comunicación, llamados *nervios,* que conectan cada área del cuerpo con el sistema nervioso central. La **Figura 1** muestra las principales divisiones del sistema nervioso.

Figura 1 *El sistema nervioso central (color anaranjado) funciona como centro de control del cuerpo. El sistema nervioso periférico (color morado) lleva la información del sistema nervioso central a las partes del cuerpo, y de regreso.*

510

⚛ WEIRD SCIENCE

Human skulls from 20,000 years ago provide evidence that ancient humans cut and drilled holes into each other's heads. This ancient practice is called trephining, and it may have been intended to release evil spirits believed to cause mental problems or illnesses such as migraines or epilepsy. Evidence of this surgery has been found in skulls from Europe, North Africa, parts of Asia, New Zealand, some Pacific Islands, and South America.

El sistema nervioso periférico

¿Cuánto tarda la luz en prenderse cuando enciendes el interruptor? Parece prenderse de inmediato. De manera parecida, unas células especializadas, las **neuronas,** llevan mensajes en forma de veloces impulsos eléctricos por todo el cuerpo. La **Figura 2** muestra una neurona típica. Los mensajes que pasan a través de las neuronas se llaman **impulsos,** y pueden viajar con rapidez, a 150 m/s, o muy lentamente, a 1 m/s.

Estructura de la neurona La neurona consta de un cuerpo celular, dendritas y axones. La parte más gruesa de la neurona contiene el núcleo y los organelos celulares. Observa de nuevo la Figura 2. La neurona normalmente recibe información de otras células a través de ramificaciones cortas, las **dendritas.** Las neuronas pueden tener muchas dendritas, que les permiten recibir impulsos de otros miles de células.

La información se transmite del cuerpo de la célula a otras células a través de una larga fibra, llamada **axón.** Los axones pueden ser largos o cortos; algunos miden casi un metro, y van de la parte baja de la espalda a los dedos de los pies. La punta de los axones muchas veces tiene ramificaciones para comunicarse con más células. La punta de cada ramificación de los axones se llama *terminal*.

$$\div \; 5 \; \div \; \Omega \; \leq \; \infty \; + \; \Omega \; ^\surd \; 9 \; _\infty \; ^\leq \; \Sigma \; 2$$

¡MATEMÁTICAS!

Hora de viajar

Para calcular cuánto tiempo tarda un impulso en viajar cierta distancia, aplica la siguiente fórmula:

$$\text{Tiempo} = \frac{\text{distancia}}{\text{velocidad}}$$

Si un impulso viaja a 100 m/s, ¿cuánto tiempo le tomará viajar 10 m?

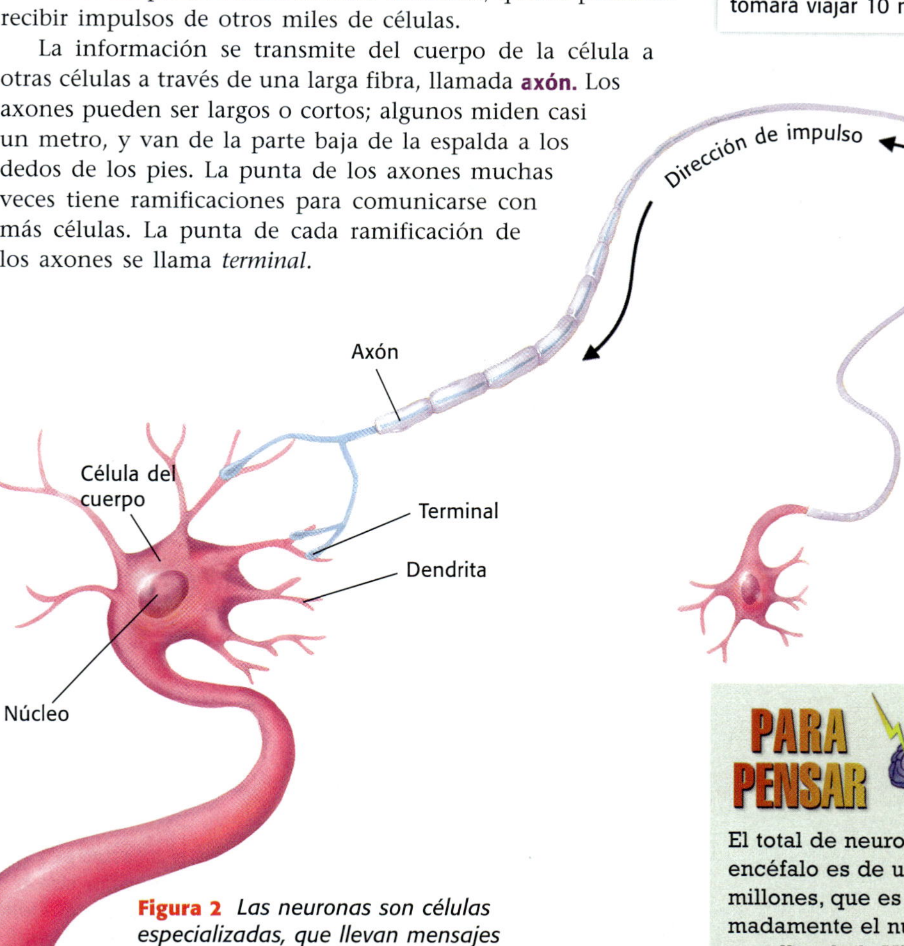

Figura 2 *Las neuronas son células especializadas, que llevan mensajes eléctricos por todo el cuerpo.*

PARA PENSAR

El total de neuronas del encéfalo es de unos 100 mil millones, que es aproximadamente el número de estrellas de la Vía Láctea.

511

Answer to MATHBREAK

$$T = \frac{10m}{100m/s} = \frac{1}{10}\text{s or } 0.1\text{s}$$

ACTIVITY

Writing **Formulating Questions** Before students read the text on these pages, have them identify and read aloud the headings. Then ask students to formulate one question that they expect the text under each heading to answer. Have students write the questions in their ScienceLog. The students can answer their questions as a homework assignment.

USING THE FIGURE

Ask students to imagine that an electrical impulse is being sent by one of the neurons to the other in **Figure 2.** Then have students trace the impulse from the axon of the sending neuron to the dendrites and the cell body of the receiving neuron. Then ask: Where does the impulse go from here? (The receiving neuron sends the impulse to another neuron via its axon.)

MEETING INDIVIDUAL NEEDS

Advanced Learners Provide students with a compound light microscope and prepared slides of different kinds of cells, such as blood cells, liver cells, skin cells, and neurons. Ask students to sketch in their ScienceLog what they see on each slide.

IS THAT A FACT!

Unlike humans, canaries replace old brain cells with new neurons each year. Male canaries even sing a new song every spring. Their brain-cell clusters associated with vocalization grow larger during the spring, when males compose their new melodies and females learn to recognize them. In the fall, the brain clusters shrink, neurons die, and the males forget what they sang. Scientists theorize that this happens so that these birds can acquire new information without having to carry around a large and heavy brain.

Teaching Transparency 78
"The Neuron"

Directed Reading Worksheet 21 Section 1

MEETING INDIVIDUAL NEEDS

Learners Having Difficulty
To help students visualize the central and peripheral nervous systems, have them work in pairs to trace the outline of one student's body on butcher paper. Next have each pair fill in the outline, using different colors for each of the nervous systems. Models should include the brain and the spinal cord (the central nervous system), and sensory and motor neurons throughout the body (the peripheral nervous system). Sheltered English

DEMONSTRATION

Simulating Neuronal Impulses
Ask students to form a circle and hold hands. Explain that each person in the circle represents a neuron. Every left hand represents a dendrite, every body represents a cell body, and every right hand represents an axon. Join the circle, and initiate a nerve impulse by gently squeezing the hand of the student to your right. Instruct students to pass the nerve impulse to the person to their right by gently squeezing his or her hand. Once students understand the mechanics of the activity, have them call out *dendrite, cell body,* and *axon* as the impulse is passed along the circle. Sheltered English

Teaching Transparency 79
"What's in a Nerve?"

Reinforcement Worksheet 21
"This System Is Just 'Two' Nervous!"

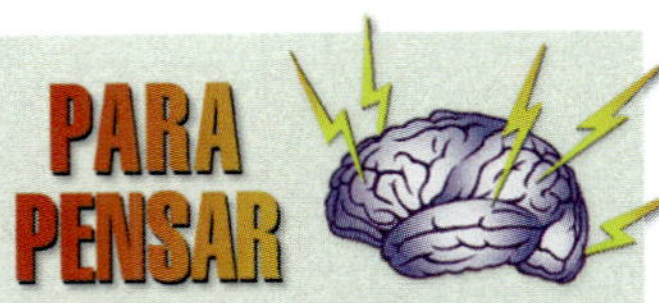

El calamar gigante tiene axones de hasta 2 m de largo, por los que viajan impulsos a una velocidad de 200 m/s.

Recolección de información Las **neuronas sensoriales** recogen información sobre lo que está pasando dentro y alrededor del cuerpo, y la mandan al sistema nervioso central para ser procesada. Las neuronas sensoriales tienen dendritas especiales, llamadas **receptores** que detectan cambios dentro y fuera del cuerpo. Los receptores de tus ojos, por ejemplo, detectan la luz a tu alrededor, y los receptores de tu estómago avisan al encéfalo si tu estómago está lleno o vacío.

¡A la orden! Las neuronas que envían impulsos del encéfalo y la médula espinal a otros sistemas son **neuronas motoras.** *Motor* quiere decir "mover"; cuando los músculos reciben impulsos de las neuronas motoras, se contraen. Las neuronas motoras, por ejemplo, pueden hacer que los músculos alrededor de los ojos se contraigan cuando las neuronas sensoriales detectan luz muy brillante. Este movimiento hace que entrecierres los ojos, y que les entre menos luz. Las neuronas motoras también envían mensajes a las glándulas, por ejemplo, a las sudoríparas. Estos mensajes les hacen producir sudor.

Un manojo de axones

El sistema nervioso central está conectado al resto del cuerpo a través de los nervios. Los **nervios** son axones envueltos en vasos sanguíneos y tejido conectivo. Todo tu cuerpo está lleno de nervios y la mayoría tienen axones de neuronas motoras y sensoriales. La **Figura 3** muestra la estructura de un nervio. El axón de este nervio transmite información de la médula espinal a las fibras musculares.

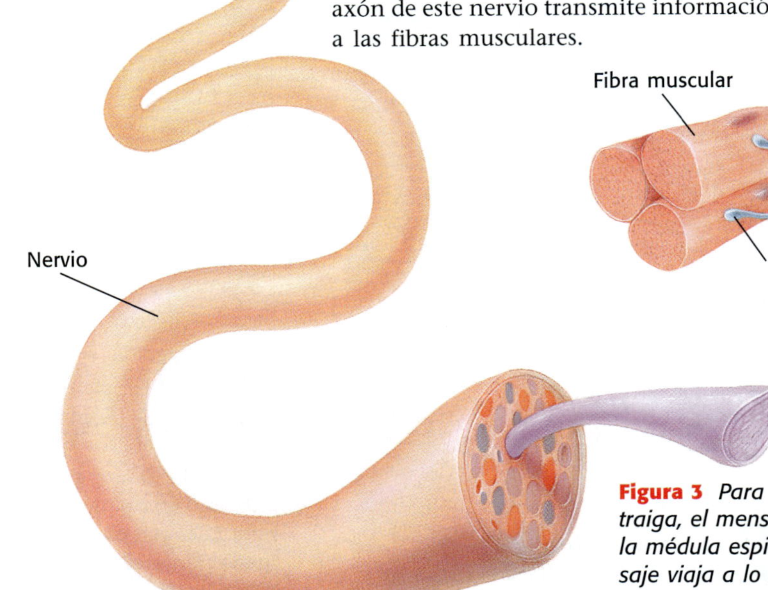

Figura 3 *Para que el músculo se contraiga, el mensaje debe viajar desde la médula espinal al músculo. El mensaje viaja a lo largo del axón de una neurona motora dentro del nervio.*

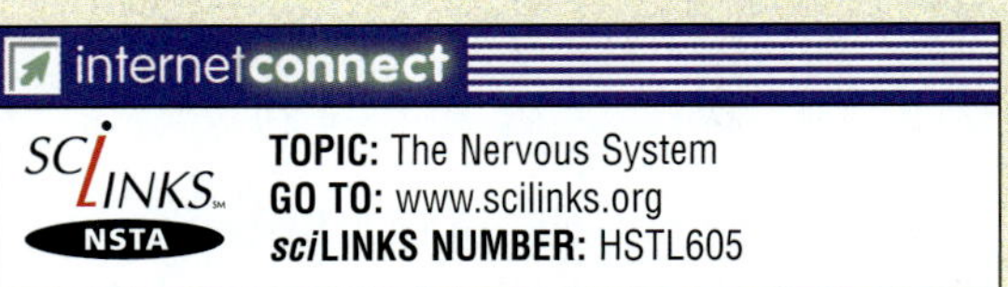

internetconnect

TOPIC: The Nervous System
GO TO: www.scilinks.org
*sci*LINKS **NUMBER:** HSTL605

Q: How do nerves shop?

A: They buy only on impulse.

El sistema nervioso central

El funcionamiento del sistema nervioso central está íntimamente ligado al del periférico. El central recibe información de las neuronas sensoriales y responde con mensajes que envía a las diferentes partes del cuerpo a través de las neuronas motoras.

Misión de control El **encéfalo** es el órgano más grande del sistema nervioso. Cumple cientos de tareas diferentes. Muchos de los procesos que controla son automáticos y se llaman procesos *involuntarios*. Un ejemplo es la digestión. Aunque quieras, no puedes detener la digestión de algo que has comido. Otras actividades que controla el encéfalo son *voluntarias*. Cuando quieres mover un brazo, el encéfalo envía mensajes por las neuronas motoras a los músculos de tu brazo; esto hace que los músculos se contraigan y que tu brazo se mueva. El encéfalo tiene tres partes conectadas entre sí: el cerebro, el cerebelo, y el bulbo raquídeo, cada una con funciones especiales.

¡Qué cerebro! La parte más grande del encéfalo se llama **cerebro.** Tiene la forma de un hongo grande con un largo tallo. Es el área que usas para pensar y donde se guardan la mayoría de tus recuerdos. Controla los movimientos voluntarios y te permite detectar luz, sonido, olores, sabores, sensaciones táctiles, dolor, frío y calor.

El cerebro tiene dos partes, llamadas *hemisferios*. El hemisferio izquierdo maneja la parte derecha del cuerpo y el derecho maneja la parte izquierda, porque los axones de cada hemisferio se cruzan en el lado opuesto del cuerpo en la médula espinal. Pero, en general, los dos hemisferios del cerebro trabajan juntos en la mayoría de los procesos cerebrales. Observa la **Figura 4** para que te des una idea general de las cosas que controla cada hemisferio.

Explora

Cada hemisferio controla diferentes tipos de pensamiento y expresión. Prueba a ver qué hemisferio usas más. Haz una lista de tus actividades favoritas y las materias que más te gustan en la escuela. ¿Qué actividades son del hemisferio derecho y cuáles del izquierdo? Piensa en cosas que haces todos los días, como cepillarte los dientes. ¿Qué mano usas? Cuando tomas una foto, ¿con qué ojo miras a través del objetivo? ¿Qué lado del cerebro crees que usas más?

Figura 4 Los hemisferios cerebrales

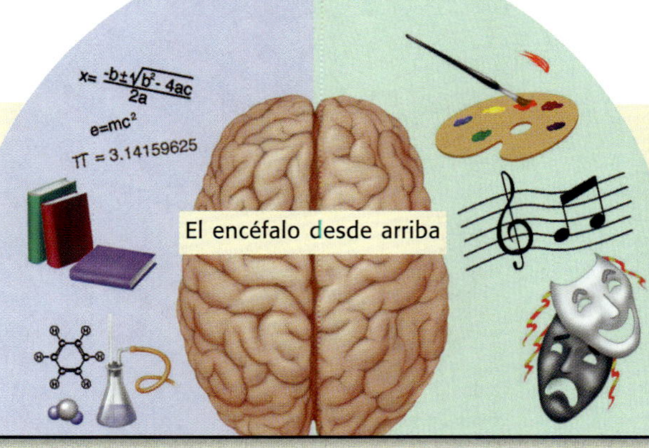

513

IS THAT A FACT!

What is most amazing about the human brain is not its size—a sperm whale's brain is about six times larger. It is the high proportion of brain size to body size as well as the huge surface area. The cerebral hemispheres are folded and wrinkled; but laid out flat, the brain would cover the surface of an office desk. This large surface provides room for highly complex and sophisticated connections.

Answer to Explore

While most activities use both sides of the brain, some are thought of as left-brain or right-brain activities because they are thought to rely on one hemisphere of the brain more than the other. Certain parts of the body are controlled by the hemisphere on the opposite side of the body, this is called contralateral control. For example, the right hand is controlled by the left hemisphere.

READING STRATEGY

Prediction Guide Before students read this section, ask them whether the following statements are true or false.

1. The brain is the body's largest organ. (false)
2. The largest part of the brain is the cerebrum. (true)
3. The medulla is responsible for speech and balance. (false)
4. The spinal cord is about as big around as your thumb. (true)

Have students evaluate their answers after they have read these pages.

REAL-WORLD CONNECTION

Students have probably watched people faint in movies or television shows. Point out that fainting and other forms of unconsciousness, except sleeping, are usually the result of some problem in the brain. Fainting is often caused by suddenly low blood pressure and insufficient blood flow to the cerebrum.

RETEACHING

If students have difficulty distinguishing the structures of the brain, make a life-size model of the brain as a class. Invite two volunteers to make a silhouette of the head; one student should be the model; the other, the tracer. Then draw in a brain. Have students take turns drawing and labeling the parts of the brain on the silhouette. Include the cerebrum, cerebellum, medulla, and the top of the spinal cord. Also have students label the hemisphere that is shown in the diagram. When the life-size model is complete, post it so students can refer to it as they review this section.

Sheltered English

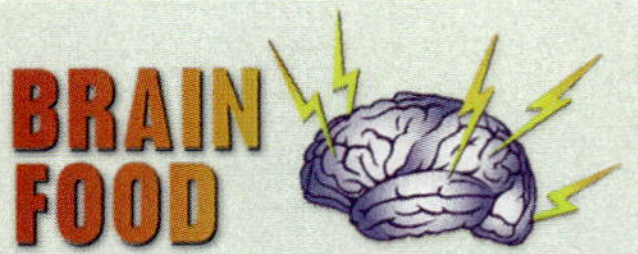

BRAIN FOOD

Early anatomists Herophilus (c.335–c.280 B.C.) and Eristratus (c.276–c.194 B.C.) were experts at dissection, and they produced extensive work on human anatomy and physiology that is now on display at the Museum at Alexandria, in Egypt. Their dissections were discontinued, however, because of the Egyptian belief that the body must be kept intact for the afterlife. Another 15 centuries passed before dissection again was used to study human anatomy. You may want to have students compare the Egyptian views about dissection with modern-day views about organ donation.

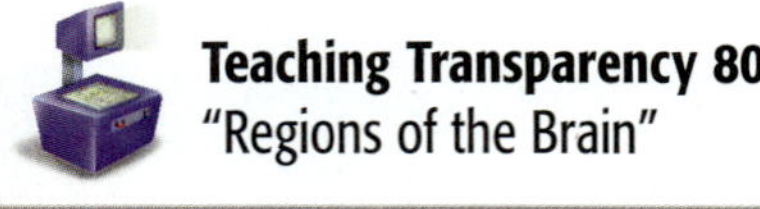

Teaching Transparency 80
"Regions of the Brain"

Figura 5 *El cerebelo hace que se contraigan los músculos esqueléticos para que no pierdas el equilibrio.*

Equilibra equilibrador El siguiente órgano del encéfalo, por tamaño, es el **cerebelo.** Está debajo de la parte posterior del cerebro; recibe los impulsos sensoriales de los músculos esqueléticos y las articulaciones y permite al encéfalo saber la posición de tu cuerpo. Si, por ejemplo, empiezas a perder el equilibrio, como la muchacha de la **Figura 5,** el cerebelo envía impulsos a diferentes músculos esqueléticos para que se contraigan y te mantengan en posición vertical.

Bulbo de bulbos La parte de tu cuerpo que conecta al encéfalo con la médula espinal se llama **bulbo raquídeo.** Mide sólo 3 cm de largo, pero sin él no podrías vivir. El bulbo raquídeo controla la presión sanguínea, la frecuencia cardíaca, la respiración y otras actividades involuntarias.

El bulbo raquídeo recibe constantemente impulsos sensoriales de los receptores de los vasos sanguíneos y usa esta información para controlar la presión sanguínea. Si tu presión sanguínea baja demasiado, el bulbo raquídeo manda impulsos que hacen que los vasos se contraigan para aumentar la presión. También envía impulsos al corazón para hacerlo latir más rápido o más despacio, según sea necesario. La **Figura 6** muestra la ubicación de cada parte del encéfalo y algunas de sus funciones.

Figura 6 *Las diferentes partes del encéfalo controlan diferentes funciones del cuerpo.*

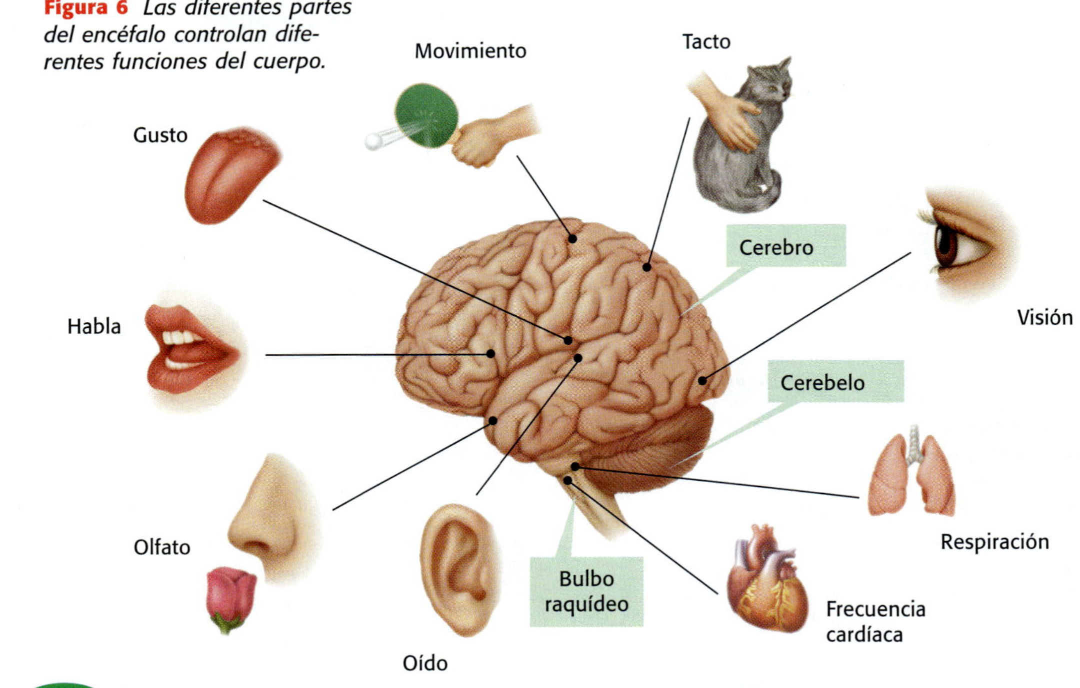

514

Homework

Dream Research Dreaming, sleepwalking, and daydreaming are all phenomena of the brain. Have students research one of these topics and give group presentations before the class. They may use posters, signs, songs, skits, oral reports, and other techniques for their presentations.

IS THAT A FACT!

Synapses form in a human baby's brain at the rate of 3 billion a second. At 8 months, a baby's brain has about 1,000 trillion connections. After that, the number begins to decline. Half the connections die off by the time the child reaches age 10, leaving about 500 trillion.

La médula espinal

La médula espinal forma parte del sistema nervioso central. Tiene más o menos el grosor de tu dedo pulgar y contiene neuronas y grupos de axones que llevan impulsos de ida y vuelta al encéfalo. Como ves en la **Figura 7,** la médula espinal está rodeada por anillos protectores de hueso, llamados *vértebras*.

Las fibras de las neuronas en la médula espinal permiten al encéfalo comunicarse con el sistema nervioso periférico. Las neuronas sensoriales de la piel y los músculos mandan impulsos a través de sus axones a la médula espinal. Esta lleva los impulsos al encéfalo, donde pueden ser interpretados como dolor, frío, calor u otras sensaciones. Los impulsos que van del encéfalo a la médula espinal pasan a las neuronas motoras, que llevan los impulsos a través de sus axones hasta los músculos y las glándulas del cuerpo.

Lesión de la médula espinal Si la médula espinal se lesiona, la información que viene de más abajo de donde se encuentra la parte lesionada quizá no llegue al encéfalo. Del mismo modo, los impulsos motores que el encéfalo envía a las partes que se encuentran más allá de la parte lesionada de la médula tampoco pueden llegar a los nervios periféricos. Miles de personas quedan paralíticas cada año debido a lesiones de la médula espinal. Muchos de estos daños ocurren en accidentes automovilísticos; entre los jóvenes, las lesiones de médula espinal ocurren sobre todo al practicar deportes.

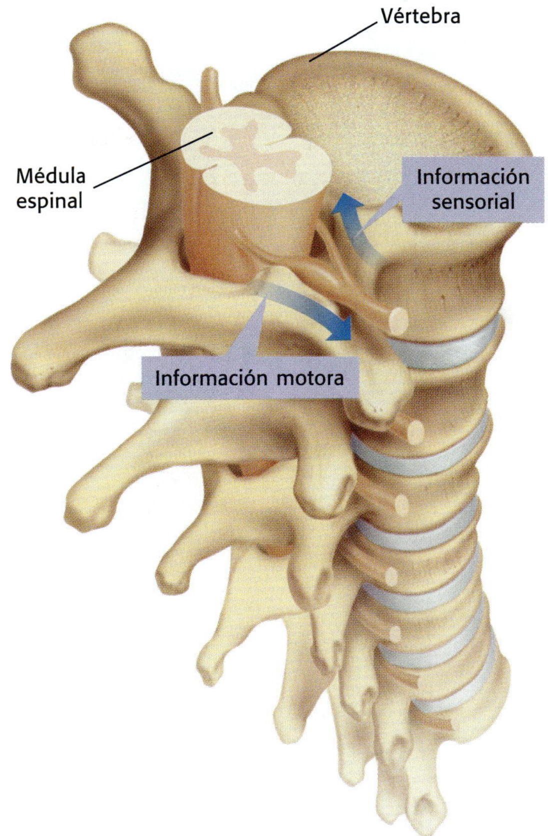

Figura 7 *La médula espinal, que ves aquí en un corte transversal, lleva información del cuerpo al encéfalo y viceversa. Está protegida por una larga columna de vértebras.*

✔ Autoevaluación

1. ¿Qué parte del encéfalo usas cuando haces tu tarea de matemáticas?
2. ¿Qué parte del encéfalo ayuda al gimnasta a equilibrarse en la barra?
3. ¿Para qué sirven las vértebras?

(Consulta la página 636 para comprobar tus respuestas.)

3 Extend

GUIDED PRACTICE

Concept Mapping List the following terms on the board:

brain, hemisphere, cerebrum, cerebellum, medulla, spinal cord, central nervous system, peripheral nervous system, sensory neurons, motor neurons

Have students construct a concept map that uses these terms.

CROSS-DISCIPLINARY FOCUS

Language Arts Long-term memory enables us to recall events that happened to us long ago. These memories can be triggered by a chance stimulus or can be deliberately recalled. Recalling memories helps us to refresh them and makes them last a lifetime. Ask students to write down their earliest memory, their happiest recollection, or their most embarrassing moment in their ScienceLog.

Answer to Self-Check

1. cerebrum
2. cerebellum
3. to protect the spinal cord

Teaching Transparency 81 "The Spinal Cord"

IS THAT A FACT!

Sometimes it is not possible to inject anesthetic into a part of the body that needs to be anesthetized. In these cases, a *nerve block* is performed. In this procedure, anesthetic is injected into or around a nerve that feeds into the part of the body that needs to be anesthetized. Nerve blocks are often performed on nerves that carry messages away from the spine. This blocks the pain impulses before they reach the brain.

Quiz

Ask students whether these statements are true or false.

1. Typically, a dendrite will transmit an electrical impulse to a neighboring axon. (false)

2. Most neurons are made up of either axons or dendrites; few neurons have both. (false)

3. The central nervous system is made up of the brain and the spinal cord. (true)

ALTERNATIVE ASSESSMENT

Writing Have students develop an owner's guide for their central nervous system. The guide should include information about the various components of the central nervous system and a diagram showing their location in the body. Encourage students to share their guide with the class in the form of a presentation or a poster.

Quick Lab

Safety Caution: Students should tap their partner's knee gently. There is no pain involved in a true reflex. Instruct students to tap while standing to the side of their partner to avoid being accidentally kicked.

Answers to QuickLab

2. Students should observe movement in the lower leg; no, an individual has no control over a reflex.

3. Students should indicate that the impulse traveled from the knee to the spinal cord and back to the thigh muscle that moves the lower leg.

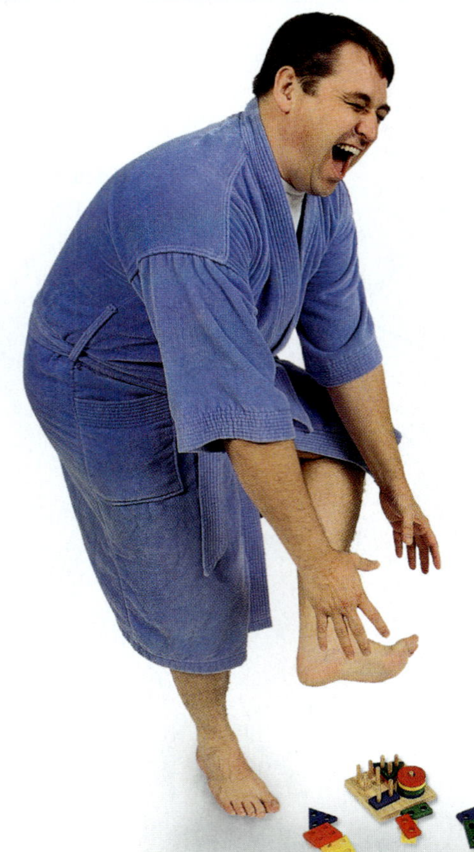

¡Ay! ¡Cómo duele!

¿Alguna vez has pisado algo puntiagudo? Seguramente, levantaste el pie sin pensar. Este movimiento rápido e involuntario se llama **reflejo.** Los reflejos sirven para proteger a tu cuerpo de posibles daños.

Cuando pisas algo puntiagudo, el mensaje "dolor" viaja por la médula espinal y el mensaje "mover el pie" va de regreso a los músculos de tu pierna. Los músculos de la pierna responden antes de que la información haya siquiera llegado al encéfalo. Cuando el encéfalo recibe el mensaje, tu pie ya se ha movido. Si tuvieras que esperar a que el mensaje llegara al encéfalo, tu pie podría lastimarse seriamente. El hombre de la **Figura 8** levantó el pie antes de darse cuenta de que había pisado un juguete.

Figura 8 *Cuando los impulsos de dolor llegan a la médula espinal, se envía de inmediato un mensaje a los músculos de la pierna para que levanten el pie. Los impulsos siguen su camino hasta el encéfalo, pero el grito de dolor probablemente sucede cuando el peligro ya ha pasado.*

Laboratorio

A patadas

1. Siéntate en el borde de un escritorio o una mesa de manera que tus pies no toquen el suelo.

2. Cuando tu pierna esté completamente relajada, pídele a un compañero o compañera que golpee *suavemente* tu rodilla, en la parte de abajo de la rótula, con el canto de la mano. ¿Cómo respondió tu pierna? ¿Pudiste controlar la reacción? Explica por qué.

3. Describe el camino que recorre el impulso que empezó con el golpecito en tu rodilla.

REPASO

1. Haz un diagrama con letreros que muestre el camino de un mensaje eléctrico de una neurona a otra.

2. Explica cómo se conecta el sistema nervioso periférico con el sistema nervioso central.

3. Si una araña peluda está caminando sobre tu brazo izquierdo, ¿qué hemisferio cerebral controla el movimiento que harás para quitártela?

4. Nombra las tres partes principales del encéfalo y describe sus funciones.

5. **Aplicar conceptos** Describe una ocasión en que hayas experimentado un reflejo.

▼ Answers to Review

1. Diagrams will vary but should show the following path: dendrite, cell body, axon, axon terminal, dendrite.

2. The peripheral nervous system consists of communication pathways, or nerves, that connect all areas of your body to your central nervous system.

3. You would use the left hemisphere, since you would have to use your right hand to knock it off.

4. The three major parts of the brain are the cerebrum, cerebellum, and medulla.

5. Answers will vary but should demonstrate an understanding that reflexes involve involuntary muscle actions.

La respuesta al medio ambiente

¿Cómo sabes cuando alguien te toca el hombro o te llama por tu nombre? ¿Cómo sientes el contacto o escuchas el sonido? Los impulsos de los receptores sensoriales en tu hombro y tus oídos viajan hasta el cerebro con la información sobre el ambiente externo. El cerebro necesita esta información para tomar decisiones de las que depende tu supervivencia.

Encuentra el sentido

Los receptores sensoriales detectan la información sobre el ambiente y las condiciones del cuerpo. Esta información se convierte en señales eléctricas y se envía al cerebro para que sea interpretada. Cuando las señales llegan al cerebro, la persona las puede percibir. Está percepción de las señales eléctricas se llama *sensación*. El cerebro tiene pensamientos, sentimientos y recuerdos de las sensaciones.

El cuerpo tiene muchos tipos diferentes de receptores. Los receptores del ojo, por ejemplo, detectan la luz; los del oído, detectan vibraciones llamadas ondas sonoras. Las papilas gustativas de la lengua tienen receptores que detectan las substancias químicas en la comida y la nariz tiene receptores especiales para detectar diminutas partículas en el aire. La piel tiene distintos receptores. Obsérvalos en la **Figura 9.**

VOCABULARIO

retina
fotorreceptores
bastoncitos
conos
nervio óptico
iris
cristalino
cóclea

OBJETIVOS

- Enumera las cuatro sensaciones que detectan los receptores de la piel.
- Describe la relación entre luz y visión.
- Explica la función de conos y bastoncitos.
- Compara y contrasta la función de fotorreceptores, papilas gustativas y células olfatorias.

Experimentos

Tengo que admitirlo: ¡me pones nervioso! Pasa a la página 632 para ver por qué.

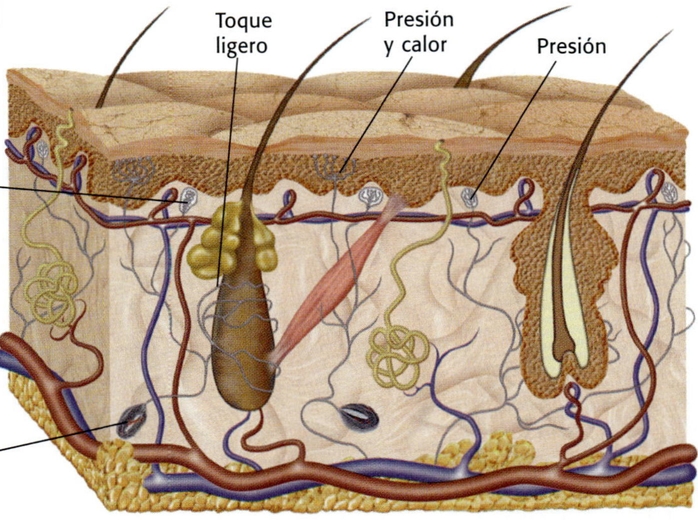

Figura 9 *Este diagrama muestra algunos de los receptores de la piel y lo que detectan.*

517

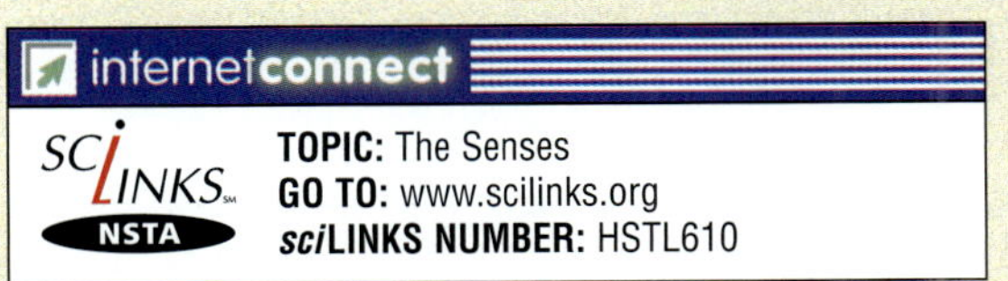

WEIRD SCIENCE

One out of every 25,000 people has *synesthesia*. These people experience a blend of senses. They can hear colors, taste shapes, or other combinations of senses. When one of their senses is stimulated, another sense is also stimulated.

Focus

Responding to the Environment

This section introduces the five senses: touch, sight, hearing, taste, and smell. Students will learn about the specialized cells that are unique to each one of these senses.

Bellringer

Ask students to list the five senses and draw the organ associated with each sense as well as an object detected by each sense. For example, students may draw an ear and a bell to represent the sense of hearing.

1) Motivate

COOPERATIVE LEARNING

Divide the class into groups of three or four students, and assign each group a sense. Each group is to imagine what it would be like to live without the sense that they were assigned. Allow students 10–15 minutes to develop a skit or example of life without the assigned sense.
`Sheltered English`

LabBook **PG 632**
You've Gotta Lotta Nerve!

Directed Reading Worksheet 21 Section 2

REAL-WORLD CONNECTION

Vision is commonly tested by using a Snellen's chart. This familiar chart consists of rows of letters set in decreasing size. A person who can correctly read a line of letters near the bottom of the chart from a distance of 6 m (20 ft) is said to have normal, or 20/20, vision. Sheltered English

MISCONCEPTION ALERT

Students may think that people who are colorblind see in black and white. People who are colorblind can usually perceive colors, but certain colors may appear very similar to one another. This is caused by a lack of at least one of the three cones in the eye. Many people do not know that they are colorblind because they have learned to distinguish other differences in their perception of colors.

CONNECT TO PHYSICAL SCIENCE

In the retina there are three types of cones that only respond to red, blue, or green light. When red cones and green cones are stimulated at the same time, the color yellow is perceived. White is perceived when all three kinds of cones are equally stimulated. Different colors of visible light have different wavelengths. Red has the longest wavelength, and violet has the shortest. Use the Teaching Transparency below to illustrate how wavelengths are measured.

Teaching Transparency 194
"Wavelength"

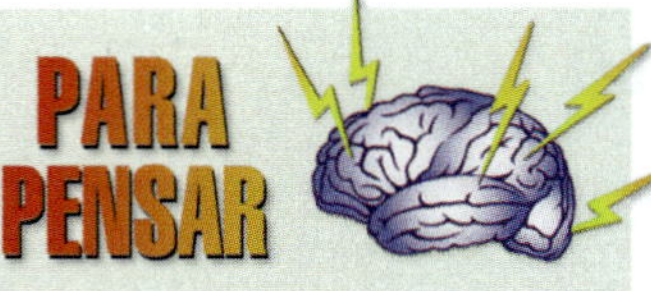

Las zanahorias y otros alimentos ricos en vitamina A pueden mejorar la visión nocturna. La vitamina A es importante para mantener el funcionamiento normal de los bastoncitos de la retina.

Tengo algo en el ojo

Cuando lees esto, estás usando uno de tus sentidos más importantes, la *vista*. La vista es la percepción de la energía en forma de luz. Los ojos tienen receptores especiales que detectan la luz visible, una parte de la energía que llega desde el Sol a la Tierra. La luz visible está hecha de todos los colores del arco iris. Estos colores a veces se pueden ver cuando la luz pasa a través de las gotas de lluvia y forma un arco iris. Los ojos también pueden percibir la luz visible reflejada en los objetos.

El ojo es un órgano sensorial complejo. Examina las partes del ojo ilustradas en la **Figura 10.** La superficie exterior del ojo está cubierta por la córnea, una membrana transparente que protege al ojo pero permite que entre la luz. La luz visible que se refleja en los objetos del ambiente entra por una abertura en la parte anterior del ojo, la *pupila*. Las células de la parte posterior del ojo, de una capa sensible a la luz llamada **retina,** detectan la luz.

La retina está llena de **fotorreceptores** (*foto* quiere decir "luz") que convierten la luz en impulsos eléctricos. Hay dos tipos de fotorreceptores en la retina: los bastoncitos y los conos. Los **bastoncitos** pueden detectar luz muy tenue y son importantes para la visión nocturna. Los impulsos que envían los bastoncitos se perciben como distintos tonos de gris. En la luz brillante, los **conos** permiten ver el mundo a todo color.

La energía en forma de luz produce cambios en los fotorreceptores, que a su vez inician impulsos nerviosos. Estos impulsos viajan por los axones y salen del ojo por el **nervio óptico.**

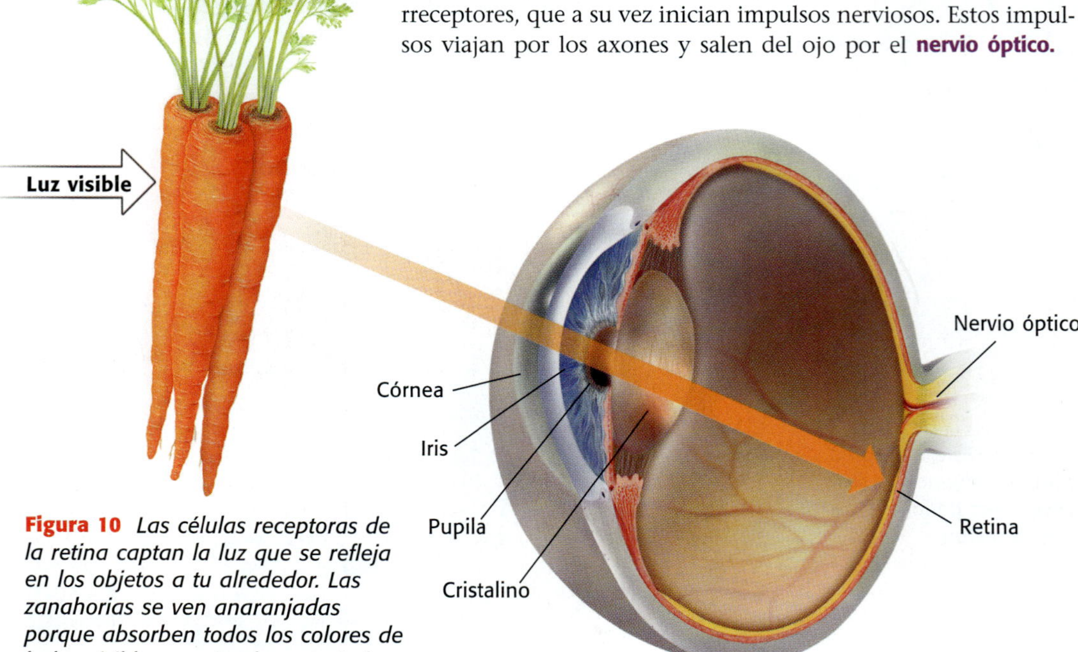

Figura 10 *Las células receptoras de la retina captan la luz que se refleja en los objetos a tu alrededor. Las zanahorias se ven anaranjadas porque absorben todos los colores de la luz visible excepto el anaranjado.*

518

WEIRD SCIENCE

Have students research and report on the tuatara, a reptile that has three eyes.

¡Luz a la vista! Los rayos de luz entran al ojo a través de una abertura llamada *pupila*. La pupila parece un punto negro en el centro del ojo, pero en realidad es una abertura, rodeada por el **iris,** la parte del ojo que tiene color. Un anillo de fibras musculares permite que el iris se abra o se cierre, y que la pupila cambie de tamaño para regular la cantidad de luz que entra por el ojo. Si la luz es fuerte, la pupila se contrae, y si es tenue, la pupila se dilata. Compara el tamaño de las pupilas de la **Figura 11.**

Enfócate La luz viaja en línea recta hasta que pasa por la córnea y el *cristalino*. El **cristalino** es una parte del ojo en forma de lente curva, que se encuentra detrás de la pupila y permite el paso de la luz pero cambia su dirección. El cristalino enfoca la luz que entra por el ojo sobre la retina y cambia de forma para enfocar la luz. Cuando miras algo muy de cerca, el cristalino se curva más y cuando miras algo que está lejos, el cristalino se aplana.

En algunas personas, el cristalino enfoca la luz justo frente a la retina; estas personas tienen miopía. En otras, el cristalino enfoca la luz justo detrás de la retina; estas personas tienen hipermetropía. El uso de anteojos o lentes de contacto casi siempre corrige estos problemas. Enfócate en la **Figura 12** para ver como funcionan las lentes correctoras.

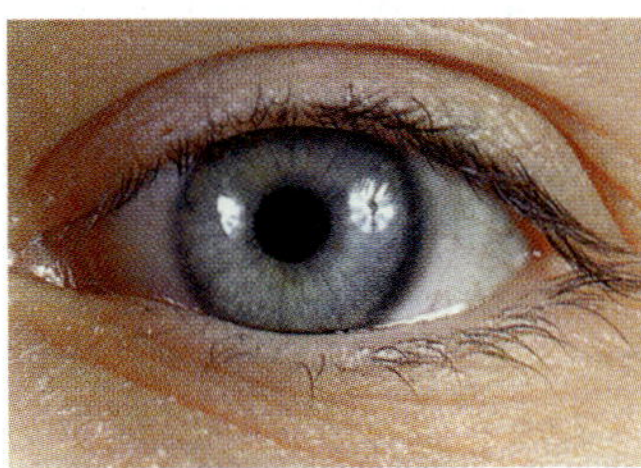

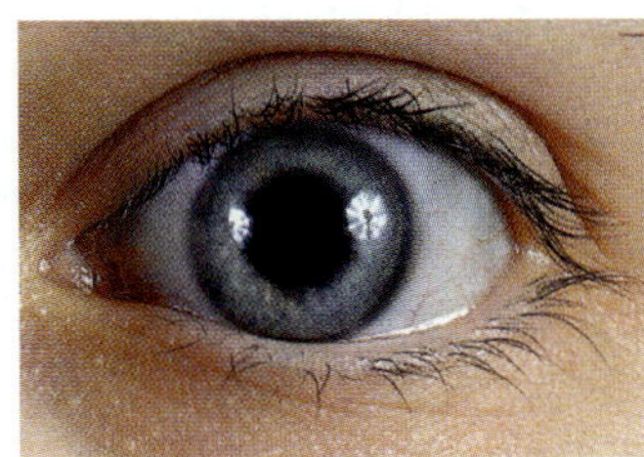

Figura 11 *El iris rodea a la pupila y le da a tus ojos su color. También controla cuanta luz entra a través de la pupila.*

Pupila en luz normal

Pupila en luz tenue

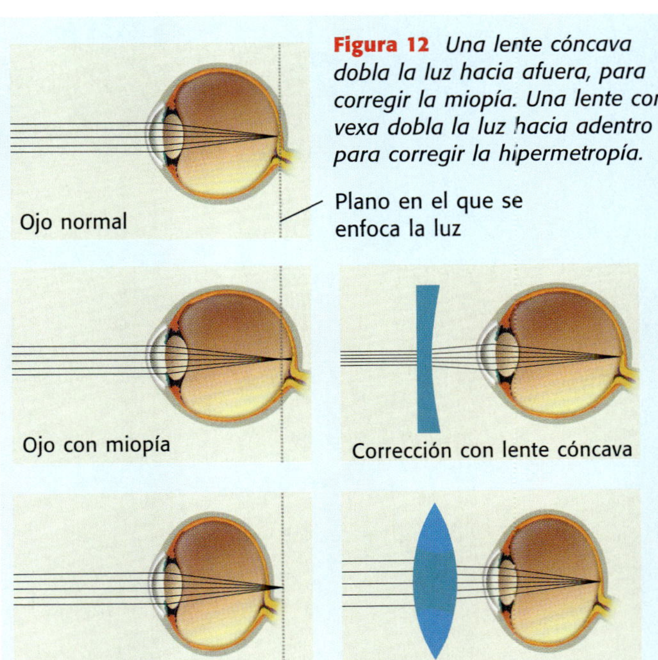

Figura 12 *Una lente cóncava dobla la luz hacia afuera, para corregir la miopía. Una lente convexa dobla la luz hacia adentro para corregir la hipermetropía.*

¿Para qué otras cosas se usan las lentes? Consulta "Luz y lentes" en la página 530 para averiguarlo.

Answers to QuickLab

3. Students should observe that the white dot "disappears" when the image is held about 10 cm from their face. The exact point where this happens will vary from person to person. Encourage the students to try this several times until they get this result.

4. Students should discover that the area where the optic nerve leaves the back of the eyeball does not contain any photoreceptors and is called the blind spot. So when the white dot was focused on the eye's blind spot, it disappeared.

MATH and MORE

Tell students that taste buds are shed approximately every 10 days. Ask students how many times a single taste bud is replaced in a year.

(365 days a year ÷ 10 day cycle = 36.5 replacements in 1 year)

Reinforcement Worksheet 21
"The Eyes Have It"

Critical Thinking Worksheet 21
"There's a Microchip in My Eye!"

Laboratorio

¿Dónde quedó la bolita?

1. Sostén este libro con los brazos estirados frente a ti y cierra el ojo derecho. Enfoca el ojo izquierdo en la bolita negra.

2. Lentamente, acerca el libro a tu cara, sin dejar de enfocar la bolita negra.

3. ¿Qué le pasa a la bolita blanca?

4. Investiga sobre el nervio óptico para explicar lo que sucedió.

Figura 13 *Cuando escuchas un sonido, la onda sonora viaja a través del oído externo, se convierte en vibraciones de huesecillos en el oído medio, en vibraciones líquidas en el interno y, finalmente, en impulsos nerviosos que van al nervio auditivo.*

¡Oye, oye!

Cuando alguien pulsa las cuerdas de una guitarra, ¿qué te permite escuchar el sonido? El sonido comienza cuando algo vibra, por ejemplo, la cuerda de una guitarra. Las vibraciones empujan las partículas de aire alrededor de la fuente de sonido; estas partículas empujan a otras y así se transmite la energía en forma de ondas que se alejan de la fuente de sonido. Lo que escuchamos es la sensación en respuesta a estas ondas sonoras.

El viaje de una onda sonora Tus oídos son órganos especiales para escuchar. Tienen una parte externa, una media y una interna **(Figura 13).** Cuando las ondas sonoras llegan al oído exterior, entran como por un embudo al oído medio, donde hacen vibrar el tímpano. El tímpano, al vibrar, hace vibrar a los huesecillos del oído, uno de los cuales es la **cóclea,** un órgano diminuto del oído medio con forma de espiral. En la cóclea, las vibraciones producen ondas parecidas a las que puedes hacer al golpear suavemente un vaso lleno de agua. Las neuronas de la cóclea convierten estas ondas en impulsos eléctricos y los envían al área del cerebro que interpreta los sonidos.

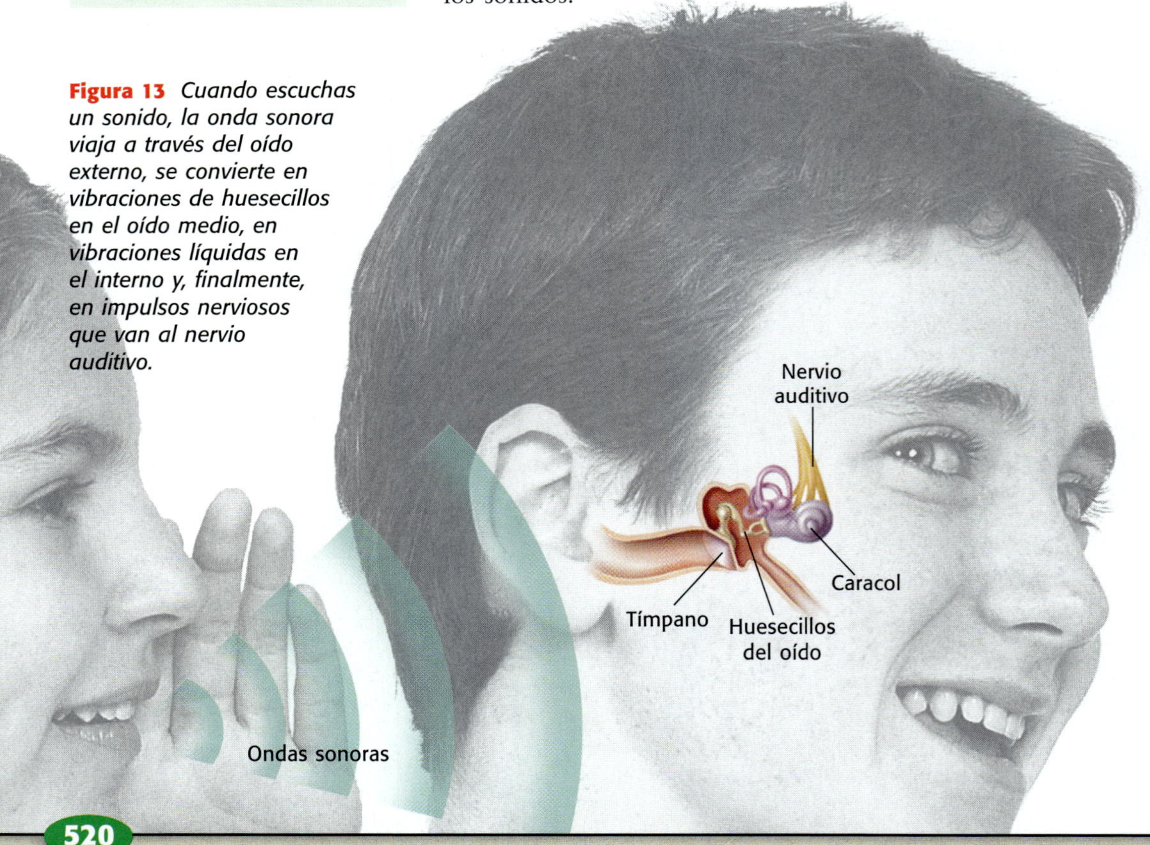

IS THAT A FACT!

Although many people think that the perception of taste occurs primarily in the mouth, 80 percent of the perception of taste is actually the sense of smell.

WEIRD SCIENCE

Some spicy foods, such as chile peppers, actually stimulate pain receptors in the mouth. This is why spicy foods feel like they are "burning" your mouth.

Para todos los gustos

Cuando te llevas un bocado a la boca, el sabor que percibes viene sobre todo de la lengua. El gusto es la sensación que resulta cuando el cerebro percibe ciertas substancias químicas en la boca. Los receptores del sentido del gusto se agrupan en las *papilas gustativas*. La lengua está cubierta de pequeños bultos llamados *papilas*, a cuyos lados se encuentran los receptores de sabor. Como ves en la **Figura 14,** hay cuatro tipos de papilas gustativas. Cada una responde a uno de los cuatro sabores básicos: dulce, agrio, salado y amargo.

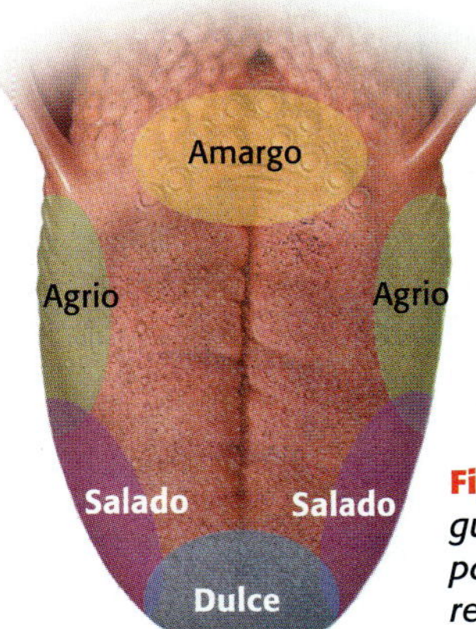

Figura 14 *Las papilas gustativas de diferentes partes de la lengua responden a distintos sabores.*

La nariz también sabe

¿Te has fijado que cuando tienes la nariz tapada no percibes los sabores bien? Prueba un pedacito de menta con la nariz tapada y verás que su sabor no es muy fuerte hasta que tomas aire por la nariz. Esto sucede porque los sentidos del gusto y el olfato están íntimamente ligados. El encéfalo combina la información de las papilas gustativas y de la nariz para dar la sensación de sabor. Los receptores del olfato están situados en las *células olfativas* de la parte superior de la cavidad nasal. Los receptores reaccionan cuando se inhalan substancias químicas y éstas se disuelven en la cubierta húmeda de la cavidad nasal. La mujer de la **Figura 15** está usando su sentido del olfato para comprobar la eficacia de varios desodorantes.

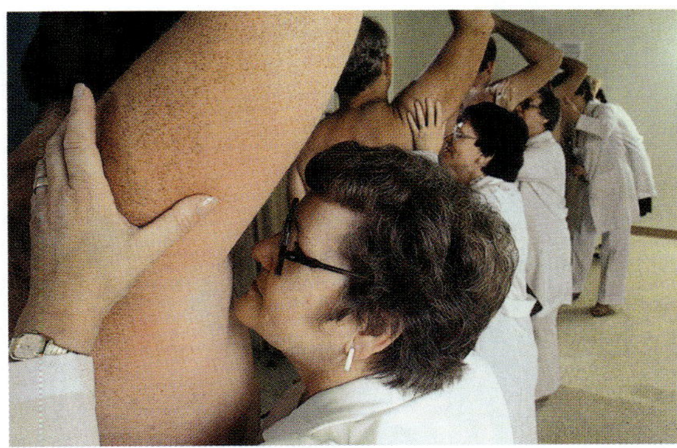

Figura 15 *La nariz de esta mujer detecta las substancias químicas en el sudor y en los desodorantes que usan estos hombres. Su cerebro genera pensamientos y opiniones sobre los olores, que anotará en su informe.*

REPASO

1. Nombra tres sensaciones que puede detectar la piel.

2. Explica por qué sería difícil percibir colores brillantes en una cena con luz de vela.

3. ¿En qué se parece el sentido del gusto al del olfato?

4. **Aplicar conceptos** Si puedes enfocar objetos que están cerca pero los lejanos se ven borrosos, ¿necesitas una lente cóncava o una convexa para corregir tu visión?

a través de las ciencias
CONEXIÓN

El sonido se produce cuando un objeto vibra. Los objetos que vibran más de 20,000 veces por segundo producen sonidos demasiado altos para que los capten los humanos. Los delfines escuchan sonidos de objetos que vibran hasta 150,000 veces por segundo.

521

---

3 Extend

RESEARCH

Writing Prolonged exposure to loud sounds can result in a partial or complete loss of hearing. Have students research at what threshold noises are detrimental to hearing. Encourage students to arrange the information they gather into a report or a visual aid. (Students should find that sounds greater than 90 dB can cause hearing loss over long periods of time. Normal conversations are usually in the 50 dB range. A loud restaurant may be in the 70 dB range. A concert is likely to be in the 120 dB range or higher.)

4 Close

Quiz

Ask students whether these statements are true or false.

1. Rods typically function only in very bright light. (false)

2. Cones provide a colorful view of the world. (true)

3. Different regions of the tongue respond to different chemicals and perceive different tastes. (true)

4. The senses of smell and hearing are closely related. Without one, the other sense is dulled. (false)

ALTERNATIVE ASSESSMENT

Have students choose one of the sense organs and make a model that shows how the organ interacts with the brain. Allow students to choose from an assortment of materials for their model. Ask students to explain their model to the class.
Sheltered English

▼ **Answers to Review**

1. Answers should include three of the following: pain, heat, pressure, cold, and touch.

2. Cones, which perceive color, would not function. Rods function in dim light but only perceive tones of gray.

3. Receptors for smell detect chemicals that are present in air. Receptors for taste detect chemicals that are present in your mouth.

4. concave lens

Focus

The Endocrine System

This section introduces students to the endocrine system. Students will learn how endocrine glands control the body's slower, long-term processes via hormones. They will also learn the location and function of specific endocrine glands. Hormonal imbalances that cause diabetes and goiter are also discussed.

Bellringer

Write the following on the board or overhead projector:

Unscramble the following words, and write them in your ScienceLog:

nalgd *(gland)*
meornoh *(hormone)*
noclotr *(control)*

1 Motivate

DISCUSSION

Endocrine System Ask students how their pulse and breathing rate differed before and after being scared. *(Both should be elevated.)*

Tell students that the endocrine system is responsible for the changes that occurred in their pulse and breathing rate. Ask the students in each pair to tell their partner about a time when they were frightened and to describe the physical responses they had. Make a list on the board of the types of physical responses named. Return to this list when students read about the adrenal glands and the fight-or-flight response in this section.

El sistema endocrino

La función del sistema nervioso es comunicarse con todos los demás sistemas del cuerpo. El sistema nervioso se comunica a través de mensajes eléctricos llamados impulsos cuya función principal es responder a los estímulos. Pero no es el único sistema que tiene esta función. El **sistema endocrino** controla procesos mucho más lentos y de largo plazo, como el equilibrio de líquidos en el cuerpo, el crecimiento y el desarrollo sexual. En vez de mensajes eléctricos, el sistema endocrino envía mensajes con substancias químicas.

Mensajeros químicos

El sistema endocrino controla las funciones del cuerpo con substancias químicas secretadas por las glándulas endocrinas. Una **glándula** es un grupo de células que produce substancias especiales para el cuerpo. Las substancias químicas que producen las glándulas endocrinas se llaman **hormonas.** Las glándulas endocrinas secretan hormonas al torrente sanguíneo, que las lleva a otras partes del cuerpo. Como las hormonas funcionan como mensajeros químicos, una glándula endocrina cerca del encéfalo puede controlar la actividad de un órgano en otra parte del cuerpo.

A menudo las glándulas endocrinas afectan a varios órganos a la vez. Las glándulas suprarrenales, por ejemplo, preparan tus órganos para enfrentarse a la tensión. Estas glándulas producen una hormona, la *epinefrina*, que antes llamábamos *adrenalina*. La epinefrina acelera el pulso del corazón y la velocidad de la respiración para preparar al cuerpo para huir del peligro o luchar por la supervivencia. El efecto de esta hormona se conoce comúnmente como la repuesta de "huir o luchar". Quizá hayas notado estos efectos cuando te asustas o te enojas.

VOCABULARIO
sistema endocrino
glándula
hormona
controles de retroalimentación

OBJETIVOS
- Explica la función del sistema endocrino.
- Nombra las glándulas del sistema endocrino y describe algunas de sus funciones.
- Muestra la posición de algunas de las glándulas del sistema endocrino en el cuerpo.
- Describe cómo los controles de retroalimentación detienen y estimulan la secreción de hormonas.

Figura 16 *Cuando tienes que moverte rápidamente para evitar un peligro, tus glándulas suprarrenales te ayudan al permitir que haya más azúcar en la sangre para usar como energía.*

522

IS THAT A FACT!

Epinephrine occurs naturally in the human body, but it is also administered as a drug by doctors. It can be injected into the heart to help revive a person who has suffered a heart attack. It also dilates the bronchioles of people with asthma.

El cuerpo tiene otras glándulas endocrinas, algunas con muchas funciones diferentes. La hipófisis, por ejemplo, estimula el crecimiento del esqueleto, ayuda a que la tiroides funcione bien, regula la cantidad de agua en la sangre y estimula el parto en la mujer embarazada. La **Figura 17** resume los nombres y algunas de las funciones de las glándulas endocrinas.

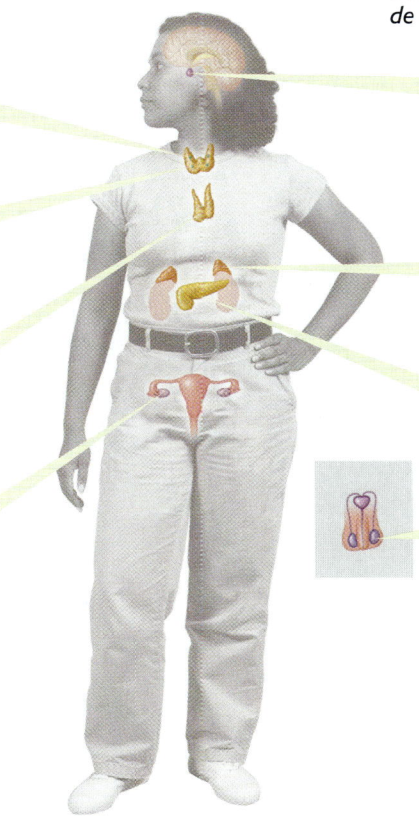

Figura 17 *Las glándulas del sistema endocrino producen substancias químicas llamadas hormonas, que controlan muchas de las funciones del cuerpo.*

APLICA

María se quedó hasta tarde estudiando en la biblioteca. Temía caminar sola a su casa. Echó a andar y, de pronto, notó que una figura obscura caminaba rápidamente detrás de ella. ¡La estaba alcanzando! María sentía que el corazón le brincaba en el pecho. Empezó a correr y, de repente, oyó una voz conocida que la llamaba. Era su papá. Había ido a la biblioteca para asegurarse de que María estuviera bien. ¡Qué alivio!

María tuvo una respuesta de "huir o luchar". Describe en un párrafo una experiencia personal similar. Incluye los términos: *hormonas, "huir o luchar"* y *epinefrina.*

523

IS THAT A FACT!

Since the mid 1980s research at Rutgers University shows that some potent hormones in the last third of pregnancy prepare and motivate mothers to care for their young. The most important of these hormones is oxytocin. It is thought to reach the brain at the same time the mother meets her newborn, helping them to bond.

2 Teach

USING THE FIGURE

Write the following terms on a board or a transparency:

endocrine, pituitary, thyroid, parathyroid, thymus, adrenals, pancreas, ovaries, testes

As a class, pronounce each term. Then locate each gland shown in **Figure 17**. Sheltered English

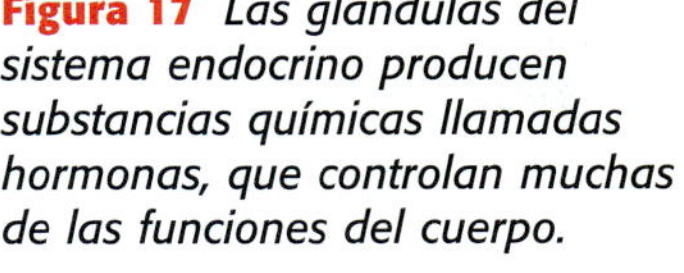

BRAIN FOOD

Some cells found in tumors secrete hormones that are identical to those secreted by endocrine glands. Typically, the tumor cells secrete too much of the hormone and do not respond to feedback control. Ask students to speculate on the effects on the body of these tumor cells and the hormones they produce. (Students should recognize that when hormones are produced in improper amounts or in an uncontrolled way, the body's homeostasis will be disrupted. The specific effects will depend on the body's internal environment and on which hormones the tumor cells secrete.)

Answer to APPLY

Paragraphs will vary. Students should write about a time when they experienced a fight-or-flight response to stress and include the terms *hormone, fight-or-flight,* and *epinephrine.*

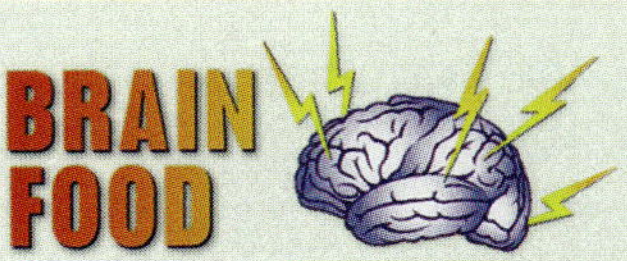

Directed Reading Worksheet 21 Section 3

USING THE FIGURE

Have students relate the events in **Figure 18** to the events that affect a thermostat. Ask: What happens when the temperature becomes too warm? (The air conditioner turns on and cools the room.)

What happens if the temperature becomes too cool? (The air conditioner turns off, and the room warms up.)

What is the overall effect of feedback control? (The level of the variable remains within a narrow range. It never becomes extremely high or extremely low and thus maintains homeostasis.)

MATH and MORE

After students have analyzed **Figure 18,** ask them to draw a graph that shows the changing levels of glucose in the blood over the course of the figure. (Graphs should approximate a sine curve.)

Have students label their graph "Negative feedback." Then draw a graph that shows a line at a nearly 45° angle on the board or a transparency, and label it "Positive feedback." Ask students to compare the graph with the "Negative feedback" curve they drew. Point out that the "Negative feedback" curve shows that the glucose level stays within a narrow range. The "Positive feedback" line shows a variable that constantly increases.

En control del control

¿Cómo saben las glándulas endocrinas cuándo empezar a secretar una hormona y cuándo detenerse? El cuerpo tiene sistemas de control especiales llamados **controles de retroalimentación,** que estimulan o detienen a las glándulas endocrinas. Los controles de retroalimentación funcionan como el termostato del aire acondicionado. Una vez que una habitación alcanza la temperatura deseada, el termostato envía un mensaje para que el aire acondicionado deje de enviar aire frío. De la misma manera, los controles de retroalimentación envían un mensaje a la glándula endocrina para que deje de secretar una hormona específica. La **Figura 18** sigue los pasos de un control de retroalimentación regulador del nivel de azúcar en la sangre.

Figura 18 *En este sistema de control de retroalimentación, el páncreas produce hormonas que ayudan al cuerpo a mantener un nivel adecuado de azúcar en la sangre.*

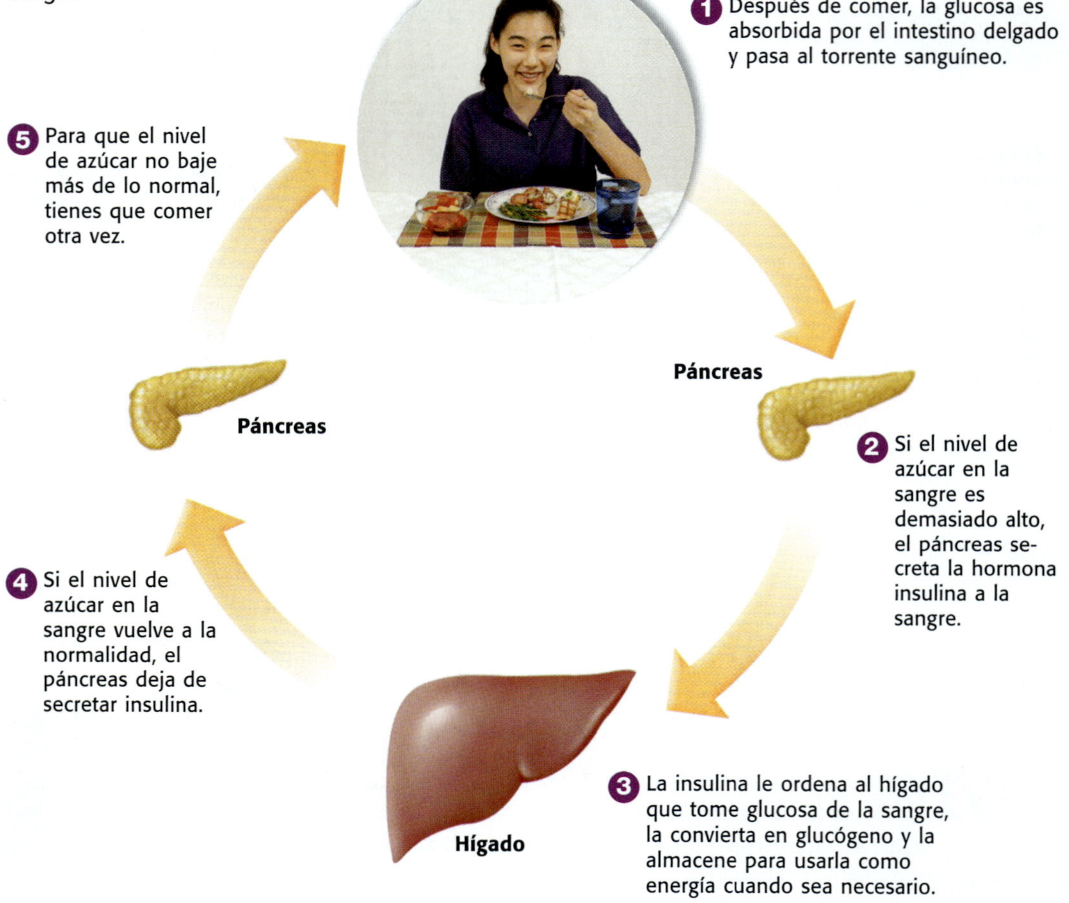

Homework

Writing — **Researching Steroids** Tell students that hormones are used as medicines to treat endocrine disorders. However, other hormones, such as anabolic steroids, are abused to increase athletic prowess. Have students research the effects and dangers of abusing anabolic steroids. Have them write a brief report or prepare an oral presentation to share their findings.

Desequilibrios hormonales

La *insulina* es una hormona producida por el páncreas. Cuando el nivel de glucosa sube después de que la persona ha comido, la insulina hace que las células tomen glucosa y envía un mensaje al hígado para que la almacene. Si el páncreas no produce suficiente insulina, la persona padece *diabetes mellitus* y puede necesitar inyecciones diarias de insulina para mantener el nivel de azúcar en la sangre dentro de los límites normales. Algunos pacientes, como la mujer de la **Figura 19,** reciben la insulina automáticamente de una pequeña máquina que llevan pegada al cuerpo.

Hormona del crecimiento A veces, la hipófisis no produce suficientes hormonas del crecimiento, y el desarrollo del niño o niña puede ser insuficiente. Afortunadamente, si el problema se detecta a tiempo, el médico puede recetar medicamentos con hormonas y vigilar su crecimiento, como ves en la **Figura 20.** Si la hipófisis produce demasiada hormona del crecimiento a una edad temprana, la persona crece mucho más de lo que se esperaba.

Tiroxina Cuando una persona no obtiene suficiente yodo en su alimentación, la glándula tiroides no puede producir suficiente *tiroxina*. Esto hace que la tiroides se inflame y produzca una masa llamada *bocio*. Como la tiroxina acelera el metabolismo, las células están menos activas de lo normal, y la persona se siente fatigada, sube de peso y puede tener otros problemas.

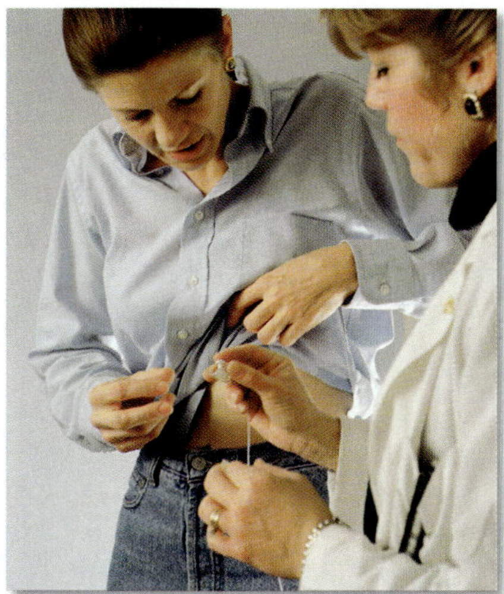
Figura 19 *Esta muchacha tiene diabetes y necesita recibir inyecciones diarias de insulina.*

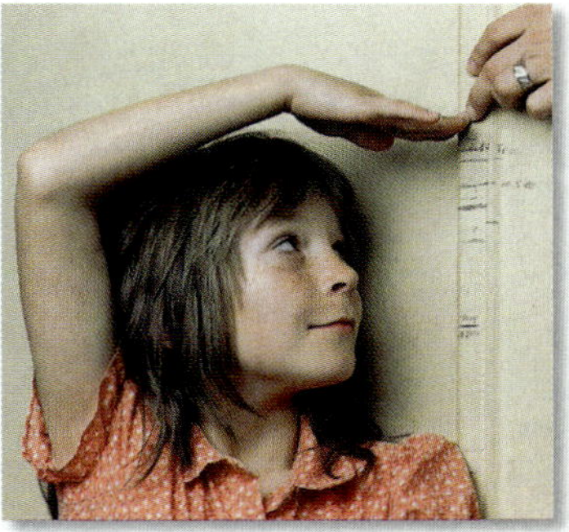
Figura 20 *Esta niña recibe un tratamiento hormonal para el crecimiento. Su mamá está midiendo su estatura.*

> ### REPASO
>
> 1. ¿Qué función tiene el sistema endocrino?
>
> 2. ¿Qué importancia tienen los controles de retroalimentación?
>
> 3. Nombra cuatro glándulas endocrinas y explica su efecto en el cuerpo.
>
> 4. **Aplicar conceptos** La epinefina, la hormona de "huir o luchar", aumenta el nivel de azúcar en la sangre. ¿De qué sirve esto en una situación de peligro o tensión?
>
> 5. **Ilustrar conceptos** Busca en tu casa un ejemplo de control de retroalimentación. Dibuja un diagrama para explicar cómo el control de retroalimentación enciende y apaga una actividad.

525

Quiz

Ask students whether these statements are true or false.

1. The endocrine system is involved with high-speed processes. (false)
2. Hormones are chemicals secreted into the bloodstream. (true)
3. All endocrine glands come in pairs. (false)
4. The body uses feedback control to regulate the secretion of hormones. (true)

ALTERNATIVE ASSESSMENT

Writing Have students write a fictional story about a new hormone that controls a function not discussed in this lesson. Possibilities include a hormone that controls a person's ability to sleep, tell jokes, roll the tongue, or perform some task. Students should describe the feedback control of the hormone and describe the conditions caused by overproduction and underproduction of the fictional hormone. Encourage students to be informative as well as creative. 

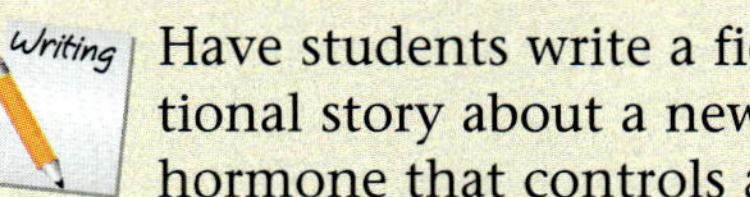

Reinforcement Worksheet 21 "Every Gland Lends a Hand"

Answers to Review

1. Your endocrine system controls body processes such as fluid balance, growth, and sexual development with the use of chemicals called hormones.

2. Feedback controls stop and start hormone release.

3. Answers will vary but should include four of the glands described in **Figure 17.**

4. During stressful situations, your body functions might need to speed up, requiring more energy.

5. Answers will vary, but students might draw a diagram demonstrating how a thermostat keeps water or air within a certain temperature range.

Chapter Highlights

Section 1

nervous system a collection of organs whose primary function is to gather and interpret information about the body's internal and external environment and to respond to that information; includes the brain, nerves, and spinal cord

central nervous system a collection of organs whose primary function is to process all incoming and outgoing messages from the nerves; includes the brain and the spinal cord

peripheral nervous system the nerves, whose primary function is to exchange information between the body and the central nervous system

neuron a specialized cell that transfers messages throughout the body in the form of fast-moving electrical signals

impulse an electrical message that passes along a neuron

dendrite a short, branched extension of a neuron where the neuron receives impulses from other cells

axon long cell fiber in the nervous system that transfers intercellular messages

sensory neuron a special neuron that gathers information about what is happening in and around the body and sends this information on to the central nervous system

receptor a specialized dendrite that detects changes inside or outside the body

motor neuron a neuron that sends impulses from the brain and spinal cord to other systems

nerve an axon through which impulses travel; bundled together with blood vessels and connective tissue

brain mass of nerve tissue that is the main organ of the nervous system

Resumen del capítulo

SECCIÓN 1

Vocabulario

sistema nervioso *(pág. 510)*

sistema nervioso central *(pág. 510)*

sistema nervioso periférico *(pág. 510)*

neurona *(pág. 511)*

impulso *(pág. 511)*

dendrita *(pág. 511)*

axón *(pág. 511)*

neurona sensorial *(pág. 512)*

receptor *(pág. 512)*

neurona motora *(pág. 512)*

nervio *(pág. 512)*

encéfalo *(pág. 513)*

cerebro *(pág. 513)*

cerebelo *(pág. 514)*

bulbo raquídeo *(pág. 514)*

reflejo *(pág. 516)*

Notas de la sección

- El sistema nervioso central se compone de encéfalo y médula espinal. El sistema nervioso periférico se compone de nervios y receptores sensoriales.

- Las neuronas reciben información en unas terminaciones llamadas dendritas y la transmiten a otras células a través de una fibra llamada axón.

- Las neuronas sensoriales detectan información sobre el cuerpo y su ambiente. Las neuronas motoras llevan información del encéfalo y la médula espinal a otras partes del cuerpo.

- El cerebro es el órgano más grande del encéfalo y controla las sensaciones, el pensamiento y el movimiento muscular voluntario.

- El cerebelo es el órgano del encéfalo que le sigue en tamaño al cerebro, y se encarga de registrar la posición del cuerpo y mantener el equilibrio.

- El bulbo raquídeo controla actividades involuntarias, como la frecuencia cardíaca, la presión sanguínea y la respiración.

- Los mensajes de dolor pueden provocar una rápida reacción involuntaria, llamada reflejo, en la que una neurona envía un mensaje al músculo sin esperar a recibir una señal del encéfalo.

☑ Comprobar destrezas

Conceptos de matemáticas

LA VELOCIDAD DE UN IMPULSO Los impulsos viajan muy rápidamente. Como puedes ver en ¡Matemáticas! (página 511), para calcular el tiempo que le toma a un impulso recorrer cierta distancia, primero tienes que saber a qué velocidad viaja. Luego puedes dividir la distancia por la velocidad y obtener el tiempo. Por ejemplo, si un impulso viaja a 150 m/s, le llevaría 0.02 segundos recorrer 3 m.

$$\text{tiempo} = \frac{3 \text{ m (distancia)}}{150 \text{ m/s (velocidad)}} = 0.02 \text{ s}$$

Comprensión visual

EL CAMINO DE LA LUZ Vuelve a observar la Figura 10 de la página 518 para repasar el camino que recorre la luz cuando entra en el ojo. La luz entra por la córnea, que es transparente; luego pasa por la abertura llamada pupila y, finalmente, por el cristalino. En la parte posterior del ojo, los receptores de la retina detectan la luz.

526

Lab and Activity Highlights

You've Gotta Lotta Nerve! **PG 632**

Datasheets for LabBook
(blackline masters for this lab)

SECCIÓN 2

Vocabulario

retina *(pág. 518)*
fotorreceptores *(pág. 518)*
bastoncitos *(pág. 518)*
conos *(pág. 518)*
nervio óptico *(pág. 518)*
iris *(pág. 519)*
cristalino *(pág. 519)*
cóclea *(pág. 520)*

Notas de la sección

- Hay distintos tipos de receptores de la piel responsables de detectar el tacto, la presión, la temperatura y el dolor.

- La retina del ojo contiene fotorreceptores que reaccionan a la luz y hacen que se envíen impulsos al encéfalo.

- El cristalino puede cambiar de forma y ajustar el enfoque para que la imagen que se proyecta en la retina esté enfocada. Los problemas de enfoque normalmente se pueden corregir con anteojos o con lentes de contacto.

- Los receptores especiales dentro de la cóclea reaccionan a las ondas de sonido y envían impulsos al encéfalo.

- Los receptores del gustose encuentran en las papilas gustativas de la lengua.

- Los receptores del olfato están situados en las células olfativas de la parte superior de la cavidad nasal.

Experimentos

¡Me pones nervioso!
(pág. 632)

SECCIÓN 3

Vocabulario

sistema endocrino *(pág. 522)*
glándula *(pág. 522)*
hormonas *(pág. 522)*
controles de retroalimentación
(pág. 524)

Notas de la sección

- El sistema endocrino se comunica con otros sistemas mediante hormonas.

- Las hormonas se producen en las glándulas endocrinas.

- Las glándulas suprarrenales secretan hormonas que ayudan al cuerpo a reaccionar en situaciones de tensión. La epinefrina es una hormona asociada con la respuesta de "huir o luchar".

- Los controles de retroalimentación son mecanismos del cuerpo para hacer que las glándulas secreten hormonas sólo cuando sea necesario.

---

internet

HRW — **VISITA:** go.hrw.com

Visita el sitio web de HRW para encontrar una serie de herramientas de aprendizaje relacionadas con este capítulo. Sólo tienes que escribir la palabra clave:

PALABRA CLAVE: HSTBD4

SCILINKS **NSTA** — **VISITA:** www.scilinks.org

Visita el sitio web de la **Asociación Nacional de Maestros de Ciencias** *(National Science Teachers Association)* para encontrar recursos de Internet relacionados con este capítulo. Sólo escribe el **ENLACE DE CIENCIAS** para obtener más información sobre el tema:

TEMA	ENLACE
El sistema nervioso	HSTL605
Los sentidos	HSTL610
El ojo	HSTL615
Hormonas	HSTL620

527

Lab and Activity Highlights

LabBank

Whiz-Bang Demonstrations, Now You See It, Now You Don't, Demo 15

Long-Term Projects & Research Ideas, Project 25

Vocabulary Review Worksheet 21

Blackline masters of these Chapter Highlights can be found in the **Study Guide.**

VOCABULARY DEFINITIONS, *continued*

cerebrum largest part of the brain; allows detection of touch, sight, sound, odors, taste, pain, heat, and cold

cerebellum second-largest part of the brain; allows the brain to keep track of the body's position

medulla part of the brain that connects to the spinal cord

reflex a quick, involuntary response to a stimulus

SECTION 2

retina layer of light-sensitive cells in the back of the eye

photoreceptors special neurons in the retina that detect light

rods photoreceptors that can detect very dim light

cones photoreceptors that can detect bright light and that help you see colors

optic nerve nerve that transfers electrical impulses from the eye to the brain

iris the colored part of the eye

lens a curved, transparent object that forms an image by refracting light

cochlea an ear organ that converts sound waves into electrical impulses and sends them to the area of the brain that interprets sound

SECTION 3

endocrine system a collection of organs, called glands, whose primary function is to control body fluid balance, growth, and sexual development

gland a group of cells that make special chemicals for the body

hormone a chemical messenger that carries information from one part of an organism to the other; made by the endocrine glands

feedback control the system that turns endocrine glands on or off

Chapter Review
Answers

Using Vocabulary

1. central nervous system
2. cochlea
3. stress
4. axon terminal
5. involuntary
6. retina

Understanding Concepts

Multiple Choice

7. c
8. a
9. c
10. a
11. d
12. c

Short Answer

13. Answers will vary but should describe dangerous or stressful situations that the student has encountered.
14. A ring of muscle fibers causes the iris to open and close.
15. Taste is your awareness of certain dissolved chemicals in your mouth. Smell is your awareness of chemicals in the air.
16. A reflex is a quick, involuntary action. Because the impulse does not need to travel to the brain before a muscle can respond, this reaction can be very quick.

Concept Mapping

17. An answer to this exercise can be found at the end of this book.

Concept Mapping Transparency 21

Repaso del capítulo

UTILIZAR EL VOCABULARIO

Escoge el término correcto para completar las siguientes oraciones:

1. El encéfalo y la médula espinal son parte del __?__. (*sistema nervioso central* o *sistema nervioso periférico*)

2. Los receptores sensoriales de __?__ detectan vibraciones. (*la cóclea* o *el tímpano*)

3. Las glándulas suprarrenales producen epinefrina en respuesta a __?__. (*la glucosa* o *la tensión*)

4. La parte de la neurona que transfiere impulsos a otras células es __?__. (*la dendrita* o *el axón terminal*)

5. El bulbo raquídeo es el principal responsable de las actividades __?__. (*involuntarias* o *voluntarias*)

6. Los receptores que convierten la luz en impulsos se encuentran en __?__. (*las células olfativas* o *la retina*)

COMPRENDER CONCEPTOS

Opción múltiple

7. ¿Cuál de los siguientes tiene receptores del olfato?
   a. las células de la cóclea
   b. los termorreceptores
   c. las células olfativas
   d. el nérvio óptico

8. ¿Cuál de los siguientes da al ojo su color?
   a. el iris
   b. la córnea
   c. el cristalino
   d. la retina

9. ¿Qué glándula está asociada con el bocio?
   a. las suprarrenales
   b. la hipófisis
   c. la tiroides
   d. el páncreas

10. ¿Cuál de los siguientes no forma parte del sistema nervioso periférico?
    a. la médula espinal
    b. los axones
    c. los receptores sensoriales
    d. las neuronas motoras

11. ¿Qué parte del encéfalo regula la presión sanguínea?
    a. el hemisferio cerebral derecho
    b. el hemisferio cerebral izquierdo
    c. el cerebelo
    d. el bulbo raquídeo

12. ¿Cuál de los siguientes tiene relación con el sistema endocrino?
    a. el reflejo
    b. la glándula salival
    c. la respuesta de "huir o luchar"
    d. la reacción voluntaria

Respuesta breve

13. Describe varias situaciones en que tus glándulas suprarrenales puedan secretar epinefrina y provocarte una respuesta de "huir o luchar".

14. ¿Qué hace cambiar el tamaño de las pupilas?

15. ¿En qué se parecen el sentido del gusto y del olfato? ¿En qué se diferencian?

16. ¿Qué es un reflejo? ¿Por qué te permiten los reflejos actuar con rapidez?

528

Organizar conceptos

17. Usa los siguientes términos para crear un mapa de ideas: sistema nervioso, médula espinal, bulbo raquídeo, sistema nervioso periférico, encéfalo, cerebro, sistema nervioso central, cerebelo.

RAZONAMIENTO CRÍTICO Y RESOLUCIÓN DE PROBLEMAS

18. ¿Por qué es importante que el cristalino pueda cambiar de forma dentro del ojo?

19. ¿Para qué sirven los huesecillos del oído medio?

20. ¿Por qué el sistema nervioso tiene un efecto mucho más rápido sobre el cuerpo que el endocrino?

21. ¿Por qué es importante que los reflejos ocurran sin que intervenga el pensamiento?

22. Usa el vocabulario de este capítulo para describir paso a paso el viaje de un impulso que comienza con un dolor en el dedo gordo del pie izquierdo. No olvides mencionar cada tipo de neurona y sus partes, así como los órganos específicos del sistema nervioso.

23. Nombra tres actividades controladas principalmente por el hemisferio derecho del cerebro, y tres por el hemisferio izquierdo.

LAS MATEMÁTICAS EN LAS CIENCIAS

24. El sonido viaja a unos 335 m/s (1 km equivale a 1,000 m). ¿Cuántos kilómetros viajaría un sonido en 1 minuto?

25. Algunos axones pueden enviar un impulso cada 0.4 milisegundos. Un segundo equivale a 1,000 milisegundos. ¿Cuántos impulsos podría mandar uno de estos axones por segundo?

INTERPRETAR GRÁFICAS

Observa la ilustración de abajo para responder a las siguientes preguntas:

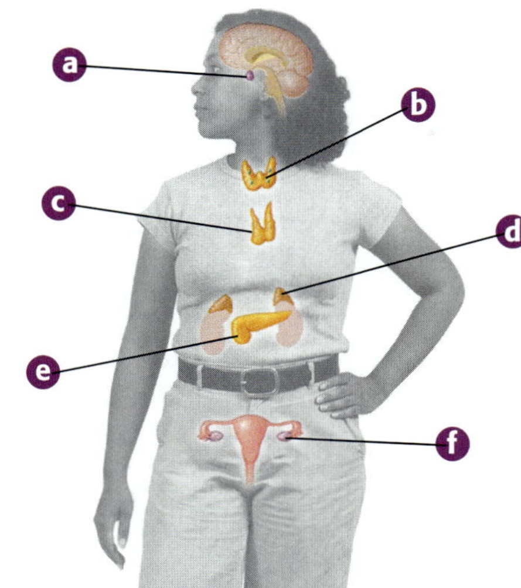

26. ¿Qué letra señala la glándula reguladora del nivel de azúcar en la sangre?

27. ¿Qué letra señala la glándula que secreta la hormona estimuladora del parto en la mujer embarazada?

28. ¿Qué letra señala la glándula que ayuda al cuerpo a defenderse de las enfermedades?

AHORA, ¿qué piensas?

Revisa tus respuestas a las preguntas de la página 509 que escribiste en el cuaderno de ciencias. ¿Han cambiado tus respuestas? Si es necesario, corrige tus respuestas basándote en lo que has aprendido en este capítulo.

CRITICAL THINKING AND PROBLEM SOLVING

18. The lens must change shape to focus objects that are varying distances from the eye.

19. The vibrations of the eardrum make the ear bones vibrate, one of which vibrates against the cochlea, creating waves that are detected by neurons.

20. Electrical messages travel along nerves faster than chemicals can travel in the bloodstream.

21. The speed of a reflex might prevent serious harm to the body.

22. The impulse travels from sensory receptors in the toe along axons leading to the spinal cord. Motor neurons carry a new message to the muscles in the leg telling them to move the foot. The pain message continues along the spinal cord to the cerebrum.

23. Answers will vary. Right cerebral hemisphere: drawing, singing, creating a play. Left cerebral hemisphere: solving a math problem, studying for a social studies test, performing a chemistry experiment.

MATH IN SCIENCE

24. 20.1 km/min
25. 2,500 impulses/second

INTERPRETING GRAPHICS

26. *e* (pancreas)
27. *a* (pituitary)
28. *c* (thymus)

 Blackline masters of this Chapter Review can be found in the **Study Guide.**

NOW WHAT DO YOU THINK?

1. Answers will vary, but some students may suggest that the senses provide the brain with information that can help the individual respond appropriately to environmental changes.

2. Answers will vary, but some students may suggest that during a fight-or-flight response, your body needs extra blood supply to either run or fight for survival.

3. Lenses help bring images into focus. A concave lens corrects nearsightedness, and a convex lens corrects farsightedness.

Background

Many students may be familiar with lenses because they wear eyeglasses. Lenses mounted in frames and used to improve vision were first developed in both Europe and China, probably during the thirteenth century.

Benjamin Franklin invented bifocals in 1784. These were a new type of eyeglasses in which each lens was divided in half. One lens corrected for nearsightedness and the other for farsightedness. Today, single lenses can be ground to include both features. In Franklin's version, each bifocal lens was made of two separate pieces of glass held together by the frame.

Ciencia, Tecnología y Sociedad

Luz y lentes

¿Puedes ver en la obscuridad absoluta? ¡No, claro que no! Para ver, necesitas luz. Pero también necesitas otra cosa: una lente. Una **lente** es un objeto curvo y transparente que *refracta,* o dobla, la luz.

Las lentes son necesarias para enfocar la luz en todo tipo de aplicaciones, desde telescopios, microscopios y binoculares hasta cámaras, lupas, lentes de contacto y anteojos.

La luz rebota

Para entender cómo funcionan las lentes, tienes que saber cómo viaja la luz. Un rayo de luz viaja en línea recta desde su fuente hasta chocar con algún objeto. Cuando la luz choca con un objeto, gran parte rebota, es decir, se refleja. La luz rebota en el mismo ángulo en que chocó con el objeto.

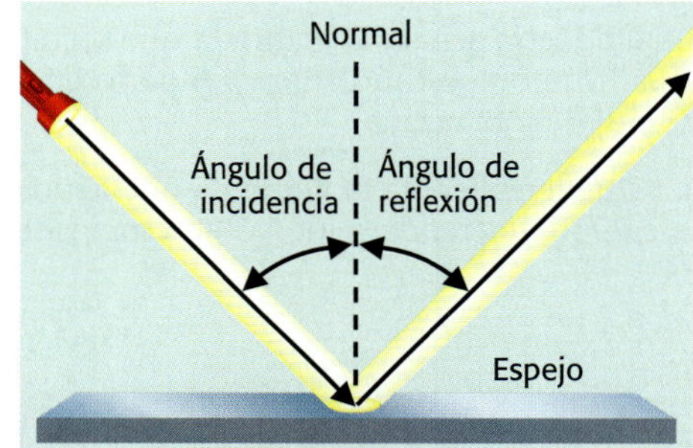

▲ *El ángulo que forma la luz al chocar con un objeto (ángulo de incidencia) siempre es igual al ángulo en que se refleja (ángulo de reflexión).*

Las lentes doblan la luz

La luz puede pasar a través de la lente; sin embargo, una vez que pasa por ella, la luz se refracta. La **refracción** es la forma en que se dobla la luz al pasar de un material transpa-

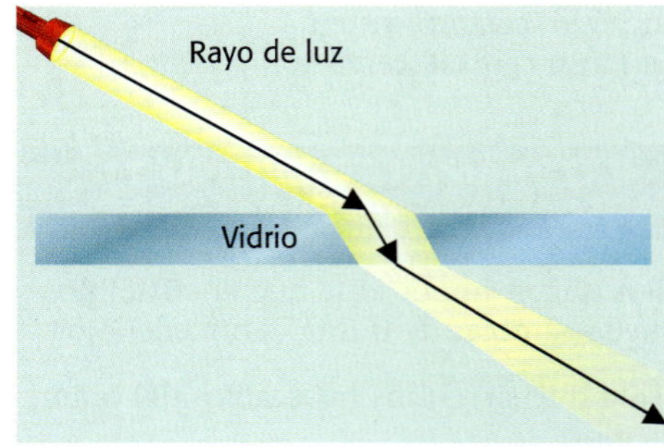

▲ *La luz cambia de velocidad y dirección al pasar a través de un material a otro.*

rente a otro, como cuando la luz que viaja por el aire pasa por una lente de vidrio.

El tipo de lente determina cuánto y cómo se dobla un haz de luz. Cuando una substancia absorbe las ondas de luz, la energía de la luz se transfiere a la substancia. Los electrones de la substancia absorben energía en forma de luz y luego la emiten en forma de calor.

Una lente más delgada en el medio que en los bordes es una lente convexa. Este tipo dobla la luz hacia el centro de la lente. Las lentes convexas se usan en lupas, microscopios y telescopios. El cristalino del ojo es una lente convexa. Una lente más delgada en el medio que en los bordes es una lente cóncava. Este tipo dobla la luz hacia afuera. Ambos tipos de lentes se usan para corregir defectos de la visión. Las convexas se usan en combinación con las cóncavas en las cámaras fotográficas, para enfocar la luz sobre la película.

Luz en tu camino

▶ Investiga qué es una queratectomía fotorrefractiva y cómo corrige la visión de una persona.

530

Answer to Light Your Way

PRK modifies the cornea to correct a person's myopic vision. A surgeon uses an excimer laser to remove a 5–9 mm diameter portion of the cornea. For mild or moderate cases of myopia, the surgeon removes about 5–10 percent of the cornea's thickness, while for extreme cases, up to 30 percent is removed. This procedure has an advantage over RK (radial keratotomy) in that the integrity of the cornea's dome is not compromised, as it is when cuts are made in the cornea during an RK procedure. In the RK procedure, deep incisions are made in a spokelike pattern. The drawbacks of this method are that it is viable only in mild cases of myopia and is ineffective at correcting hyperopia (farsightedness). The incisions in the cornea flatten it, and over time, it continues to flatten, which increases the person's farsightedness.

¡Lo encontré!

El camino a la curación

¿Sabes qué pasaría si tu encéfalo enviara demasiados impulsos a los músculos de tu cuerpo? Primero, el exceso de impulsos aumentaría el número de veces que los músculos se contraen, y sería difícil llevar a cabo movimientos sencillos, como rascarte un brazo o tomar un vaso de agua. Aunque quisieras descansar, los músculos seguirían temblando. Esto les sucede a las personas con la enfermedad de Parkinson, para la que desgraciadamente todavía no existe cura.

La enfermedad

El Parkinson afecta a las células del encéfalo que controlan el movimiento. Estas células necesitan una substancia química llamada dopamina, que hace que la actividad de los nervios sea más lenta, para funcionar bien. Pero si las células que proporcionan dopamina a las que controlan el movimiento de los músculos están dañadas, el encéfalo enviará impulsos constantes a los músculos. El resultado es la enfermedad de Parkinson.

A menudo, esta enfermedad se diagnostica cuando ya está muy avanzada, y la persona ha perdido un 80 por ciento de las células que proporcionan la dopamina. Aunque no existe cura, hay personas que pueden ser tratadas con substancias que imitan la dopamina. Desgraciadamente, estas substancias no funcionan tan bien como la dopamina de verdad. La dopamina no puede administrarse directamente porque no puede pasar del torrente sanguíneo al tejido encefálico.

Un avance

La Dra. Bertha Madras estudia el efecto de la drogadicción en el encéfalo. Cuando estudiaba los efectos de la cocaína sobre el encéfalo, Madras descubrió que una substancia química llamada tropano se adhiere a los mismos nervios que liberan dopamina en el encéfalo. Su descubrimiento puede servir para detectar y diagnosticar el Parkinson lo antes posible y ahorrarle costos médicos al paciente.

Una luz en la obscuridad

Madras y sus colegas pensaban que podían usar tropano para estudiar las células que liberan la dopamina. Le añadieron al tropano un componente radioactivo que lo convierte en una substancia llamada altropano. El altropano también se adhiere a las células que liberan dopamina, pero a diferencia del tropano, el altropano brilla, así que es visible en una tomografía cerebral. Las personas que no padecen de Parkinson tienen áreas grandes a las que se adhiere el altropano. En cambio, en las personas que la padecen, el altropano se adhiere en áreas más pequeñas, debido a la pérdida de nervios. Por tanto, las tomografías de estos pacientes no muestran colecciones tan grandes de altropano. Con este nuevo método para diagnosticar el Parkinson, los médicos pueden detectarla antes de que las neuronas estén demasiado dañadas o se hayan perdido por completo.

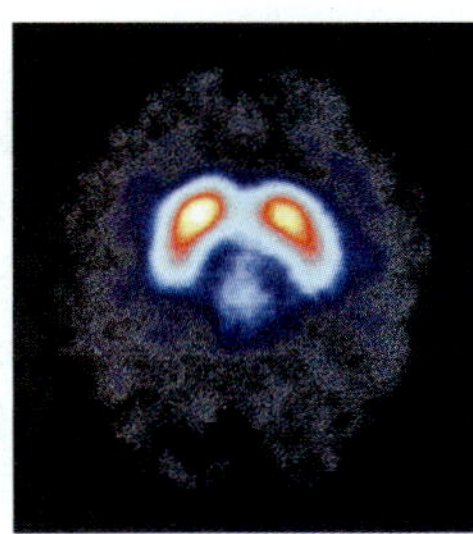 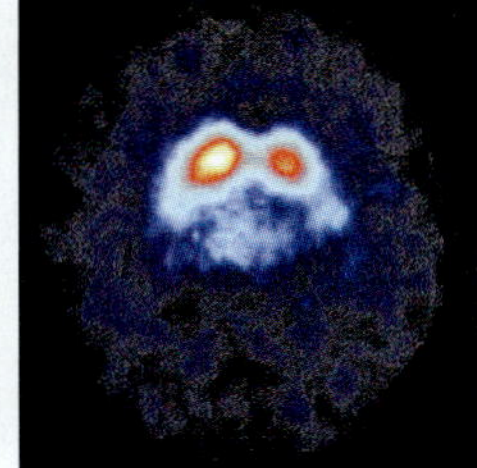

Persona sana   Persona con Parkinson

▲ *Las tomografías del encéfalo, como la de arriba, pueden usarse para diagnosticar el Parkinson.*

Actividad

▶ Investiga qué es una tomografía computarizada por emisión de fotón único, SPECT, *(Single Photon Emission Computed Tomography)* y cómo se utiliza para estudiar el Parkinson.

531

Answer to Activity

The type of brain scan done with altropane is called single photon emission computed tomography (SPECT) imaging. The other imaging method used to diagnose Parkinson's is positron emission tomography (PET). Only seven locations in the United States use the PET method, and it is a complicated procedure. A PET imaging can cost $2,500. SPECT imaging is cheaper than PET, costing about $1,000, and is more widely available at hospitals. SPECT imaging relies on the use of radioactively labeled molecules to create an image.

Chapter Organizer

CHAPTER ORGANIZATION	TIME MINUTES	OBJECTIVES	LABS, INVESTIGATIONS, AND DEMONSTRATIONS
Chapter Opener pp. 532–533	45		**Investigate!** How Grows It? p. 533
Section 1 Animal Reproduction	90	▶ Distinguish between asexual and sexual reproduction. ▶ Explain the difference between external and internal fertilization. ▶ Describe the three different types of mammalian development.	
Section 2 Human Reproduction	90	▶ Describe the functions of the male and female reproductive systems. ▶ Discuss disorders and diseases that are associated with human reproduction.	
Section 3 Growth and Development	90	▶ Summarize the processes of fertilization and implantation. ▶ Describe the course of human development.	**QuickLab,** Life Grows On, p. 547 **Skill Builder,** It's a Comfy, Safe World! p. 634 **Datasheets for LabBook,** It's a Comfy, Safe World! Datasheet 45 **Skill Builder,** My, How You've Grown! p. 635 **Datasheets for LabBook,** My, How You've Grown! Datasheet 46 **Long-Term Projects & Research Ideas,** Project 26

TECHNOLOGY RESOURCES

 **Guided Reading Audio CD**
English or Spanish, Chapter 22

 **CNN Eye on the Environment,** Eagles and DDT, Segment 5

 **One-Stop Planner CD-ROM with Test Generator**

CLASSROOM WORKSHEETS, TRANSPARENCIES, AND RESOURCES	SCIENCE INTEGRATION AND CONNECTIONS	REVIEW AND ASSESSMENT
Directed Reading Worksheet 22 **Science Puzzlers, Twisters & Teasers,** Worksheet 22		
Directed Reading Worksheet 22, Section 1 **Reinforcement Worksheet 22,** Reproduction Review	**MathBreak,** Chromo-Combos, p. 535	**Homework,** pp. 534, 536 in ATE **Self-Check,** p. 535 **Review,** p. 537 **Quiz,** p. 537 in ATE **Alternative Assessment,** p. 537 in ATE
Transparency 82, The Male Reproductive System **Directed Reading Worksheet 22,** Section 2 **Transparency 83,** The Female Reproductive System **Math Skills for Science Worksheet 3,** Multiplying Whole Numbers **Math Skills for Science Worksheet 5,** Dividing Whole Numbers with Long Division	**MathBreak,** Counting Eggs, p. 539 **Cross-Disciplinary Focus,** p. 539 in ATE **Apply,** p. 540 **Math and More,** p. 540 in ATE **Chemistry Connection,** p. 541	**Review,** p. 541 **Quiz,** p. 541 in ATE **Alternative Assessment,** p. 541 in ATE
Directed Reading Worksheet 22, Section 3 **Transparency 133,** How Sonar Works **Transparency 84,** Growth Chart **Reinforcement Worksheet 22,** The Beginning of a Life **Critical Thinking Worksheet 22,** One to Grow On!	**Math and More,** p. 544 in ATE **Multicultural Connection,** p. 545 in ATE **Connect to Earth Science,** p. 545 in ATE **Cross-Disciplinary Focus,** p. 545 in ATE **Across the Sciences:** Acne, p. 552 **Science, Technology, and Society:** Technology in Its Infant Stages, p. 553	**Self-Check,** p. 543 **Review,** p. 547 **Quiz,** p. 547 in ATE **Alternative Assessment,** p. 547 in ATE

Holt, Rinehart and Winston On-line Resources

go.hrw.com

For worksheets and other teaching aids related to this chapter, visit the HRW Web site and type in the keyword: **HSTBD5**

National Science Teachers Association

www.scilinks.org

Encourage students to use the *sci*LINKS numbers listed with the Chapter Highlights to access information and resources on the **NSTA** Web site.

END-OF-CHAPTER REVIEW AND ASSESSMENT

Chapter Review in Study Guide
Vocabulary and Notes in Study Guide
Chapter Tests with Performance-Based Assessment, Chapter 22 Test
Chapter Tests with Performance-Based Assessment, Performance-Based Assessment 22
Concept Mapping Transparency 22

Chapter Resources & Worksheets

Visual Resources

TEACHING TRANSPARENCIES

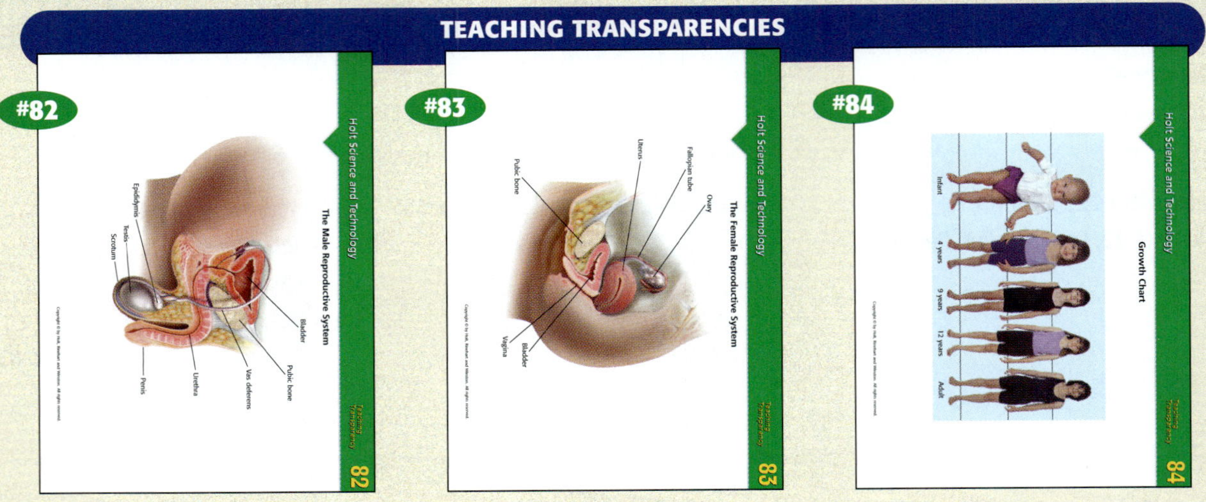

TEACHING TRANSPARENCIES

CONCEPT MAPPING TRANSPARENCY

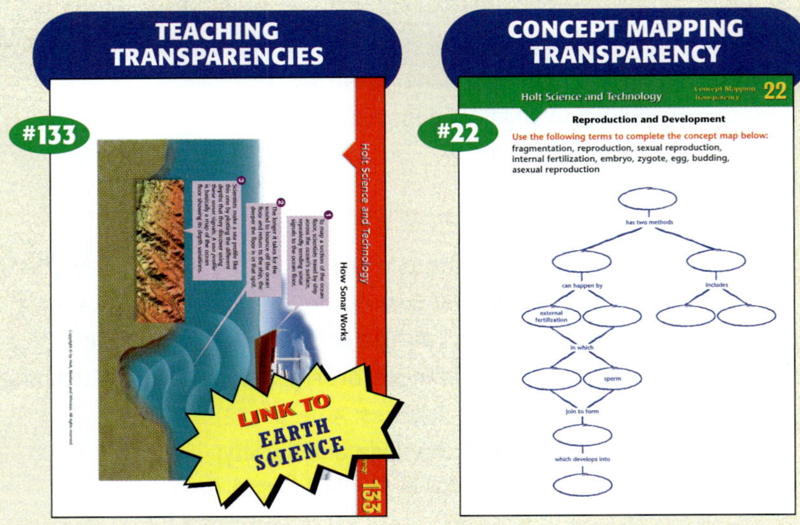

Meeting Individual Needs

DIRECTED READING

REINFORCEMENT & VOCABULARY REVIEW

SCIENCE PUZZLERS, TWISTERS & TEASERS

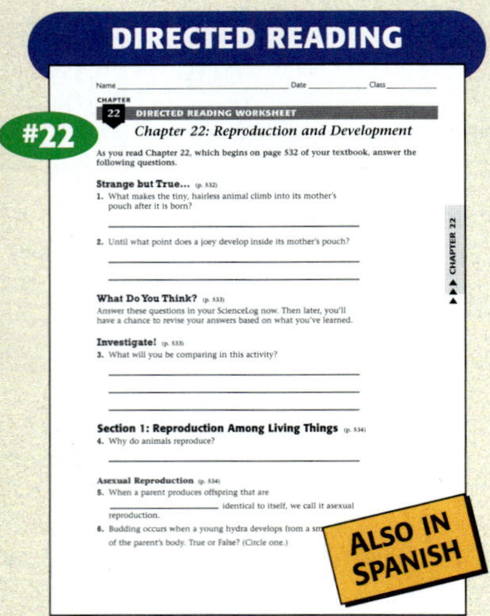

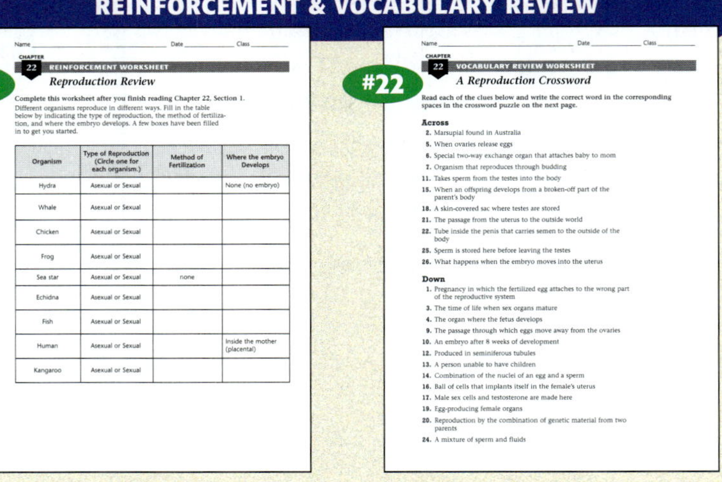

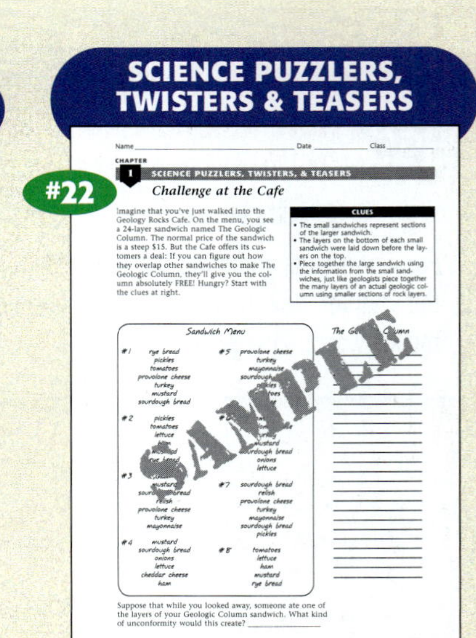

Review & Assessment

STUDY GUIDE

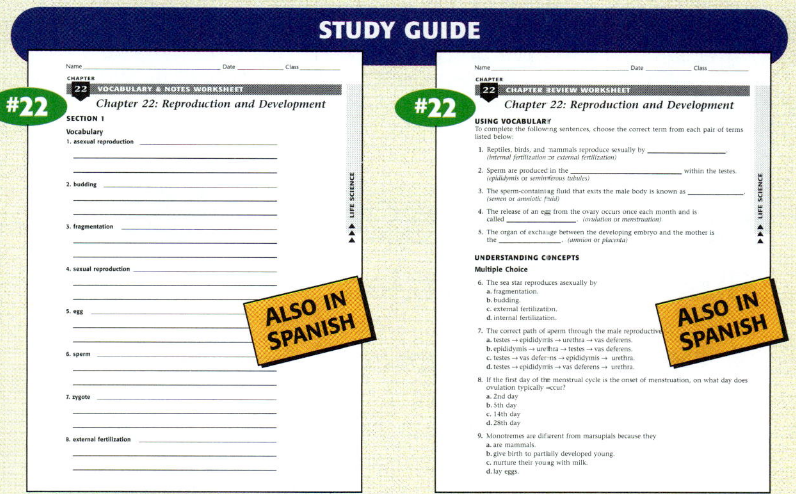

CHAPTER TESTS WITH PERFORMANCE-BASED ASSESSMENT

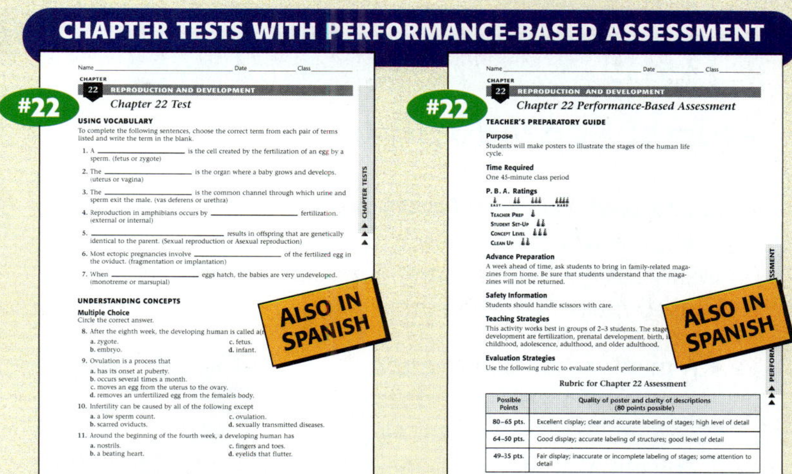

Lab Worksheets

LONG-TERM PROJECTS & RESEARCH IDEAS

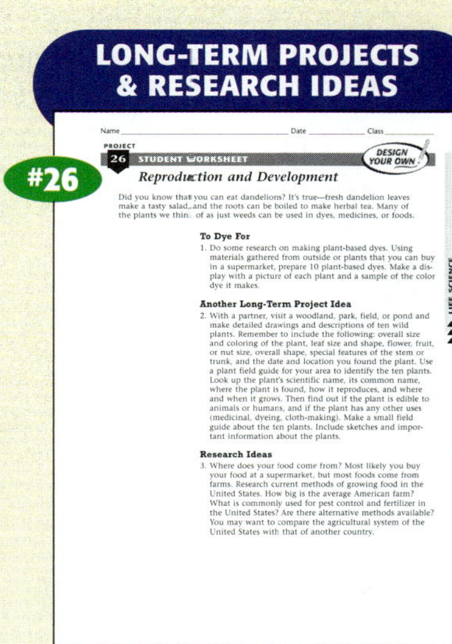

DATASHEETS FOR LABBOOK

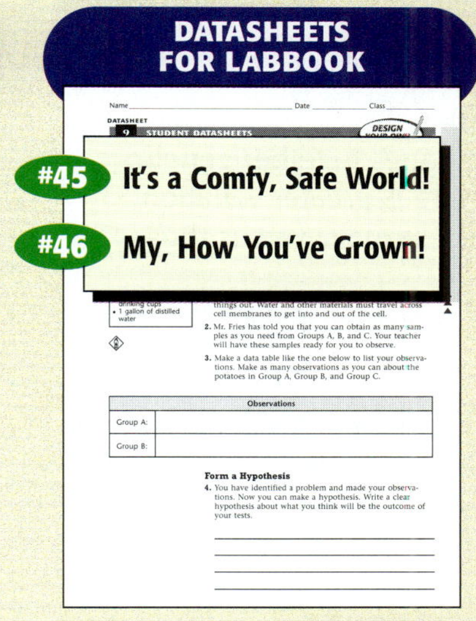

Applications & Extensions

CRITICAL THINKING & PROBLEM SOLVING

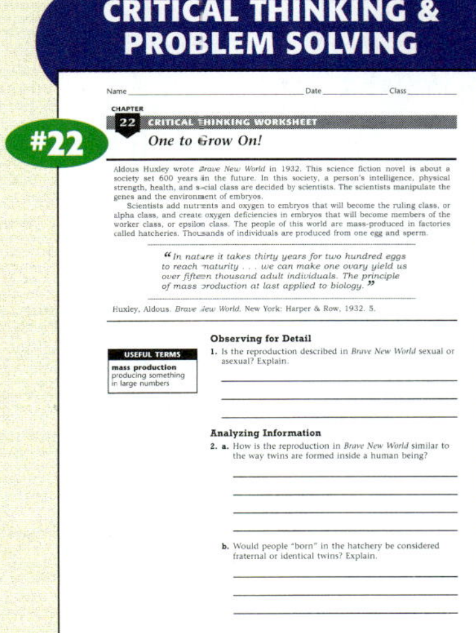

EYE ON THE ENVIRONMENT

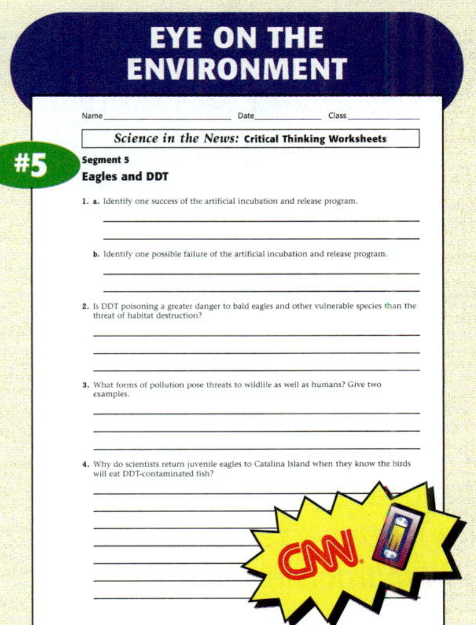

Animal Reproduction

▶ Asexual Reproduction

The most widespread forms of asexual reproduction are binary fission, budding, and fragmentation. Binary fission is used mainly by bacteria.

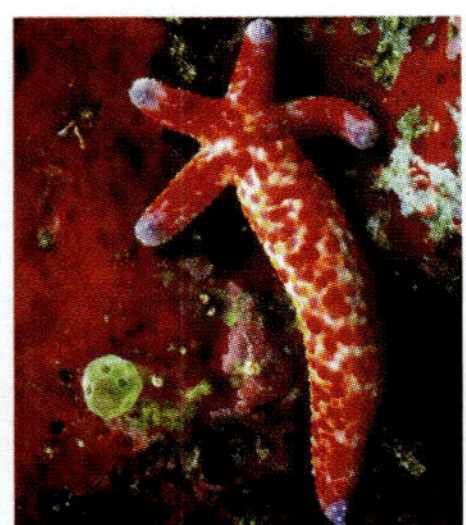

- A major advantage of asexual reproduction is that it does not require a mate. Asexual reproduction also allows animals to produce many offspring in a short period of time. Many animals that do not move around, like sea sponges, reproduce asexually.

▶ Sexual Reproduction

Because sexual reproduction brings together genetic material from two parents, there is greater variation among animals that reproduce sexually versus those that reproduce asexually.

▶ Fertilization

In external fertilization, eggs can be fertilized without physical contact between the parents. Instead, chemical signals coordinate the fertilization process, ensuring that the parents release their sex cells at the appropriate time.

- Internal fertilization requires a more sophisticated reproductive system, including organs for delivering and storing sperm. Fertilized eggs can develop externally, as with birds, or internally, as with placental mammals. Internally protected embryos are more likely to survive, but placental females do not usually produce as many offspring as do egg-laying females.

Is THAT A FACT!

- ◆ Some animal species that reproduce sexually don't have separate sexes. Instead, every individual contains male and female sexual characteristics. This situation is known as *hermaphrodism.*

- ◆ Hermaphrodites can exchange sex cells with one another but can reproduce by themselves as well.

Human Reproduction

▶ The Male Reproductive System

Male reproductive functions mainly concern sperm production. The head of a sperm contains DNA, and the tail region contains mitochondria. The mitochondria are "engines" for the sperm, providing the sperm's tail with the energy to whip back and forth.

- The process of maturation of sperm from germ cells to spermatozoa takes about 74 days. Even then, they cannot yet penetrate an ovum. First they must "ripen" in the epididymis, a process that takes about 10 days. Though the maturation process is lengthy, once sperm are fully mature, they can remain viable for about 6 weeks.

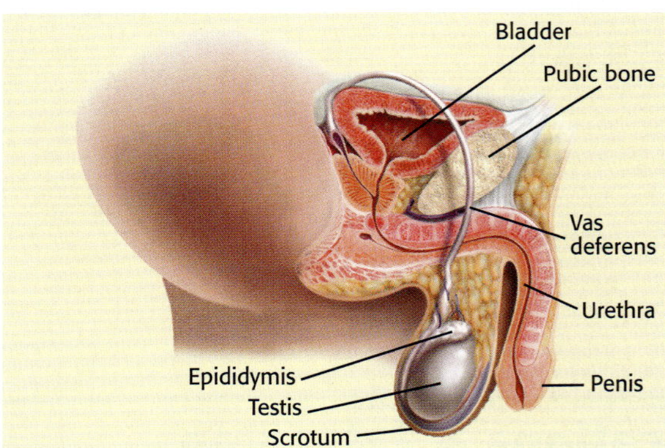

Is THAT A FACT!

- ◆ If the seminiferous tubules—the bundle of tubes that makes up each testicle—were joined together and extended, they would be more than 200 m long!

▶ The Female Reproductive System

The ovaries are the primary female reproductive organs. About the size of large almonds, the ovaries are located on either side of the uterus, each anchored by an ovarian ligament. These tiny organs secrete the hormones largely responsible for development during puberty. They are also responsible for releasing eggs.

- Every menstrual cycle, several ova begin to ripen. In most cases, however, only one egg reaches maturity at a time. This mature, ripened ovum, encased in a Graafian follicle, travels to the surface of the ovary, where it remains until midcycle, when ovulation occurs. Then the Graafian follicle, distended with fluid, ruptures, sending the egg into the abdominal cavity. The fallopian tube then captures the ovum, and it begins the descent to the uterus.

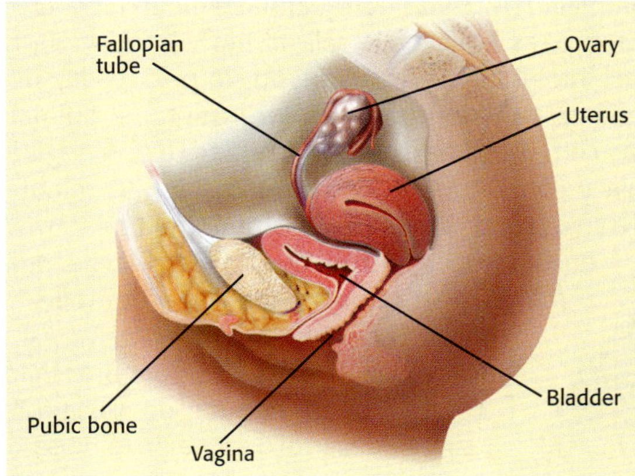

▶ Sexually Transmitted Diseases

Chlamydia is the most prevalent sexually transmitted disease (STD) in North America. Caused by an organism called *Chlamydia trachomatis,* its symptoms include a frequent desire to urinate, pain with urination, and penile or vaginal discharge. 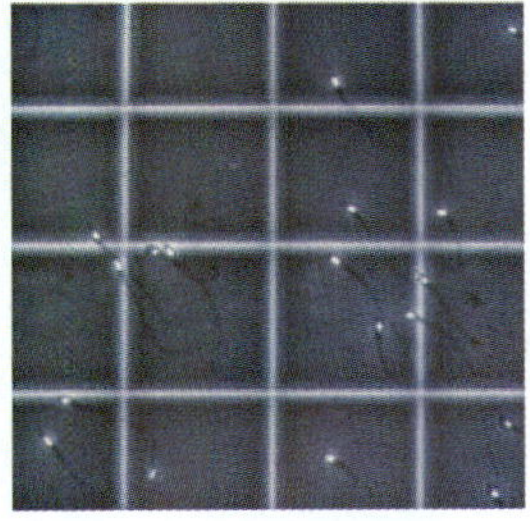 In women, there are often few symptoms in the early stages. Troublesome as the symptoms are, the consequences are worse: chlamydia is a major cause of infertility. If left untreated, it can cause pelvic inflammatory disease and, in women, the subsequent inability to conceive. In men, infection that reaches the testes results in infertility.

- Also known as salpingitis, pelvic inflammatory disease (PID) is an infection of the fallopian tubes, uterus, and cervix. While a number of bacteria can cause PID, the usual culprits are chlamydia and gonorrhea. Though these infections can be completely cured with antibiotics, PID often leaves the fallopian tubes—the conduits between the ovaries and the uterus—scarred, making conception difficult or impossible. Other potential consequences of PID include ectopic pregnancy, peritonitis, and death.

Growth and Development

▶ Sex Determination

One pair of human chromosomes determines the sex of a baby. There are two types of these sex chromosomes: X and Y. Because the egg contains only the X chromosome, the gender of the baby is determined by the father's sperm, which may contain either an X or a Y chromosome. If a sperm containing an X chromosome joins with an egg, the baby will be a girl. If a sperm containing a Y chromosome fertilizes the egg, the baby will be a boy.

▶ Fetal Development

The development of a baby from a single cell progresses at an astounding rate. Early in development, the embryo resembles a tiny tadpole, with a rounded body and tail. By about 7.5 weeks of gestation, however, limb buds—with knee and elbow joints evident—form, and facial features are recognizable. By the  ninth week, nerves and muscles have developed enough that the fetus can move independently. By 12 weeks, the fetus is about 7.6 cm long and has a mass of about 28 g.

For background information about teaching strategies and issues, refer to the *Professional Reference for Teachers.*

Chapter Preview

Section 1
Animal Reproduction
- A Chip off the Old Block
- It Takes Two
- Internal and External Fertilization
- Making Mammals

Section 2
Human Reproduction
- The Male Reproductive System
- The Female Reproductive System
- Irregularities and Disorders

Section 3
Growth and Development
- A New Life
- Before Birth
- Birth
- From Birth to Death

Directed Reading Worksheet 22

Science Puzzlers, Twisters & Teasers Worksheet 22

Guided Reading Audio CD
English or Spanish, Chapter 22

CAPÍTULO

22

Reproducción y desarrollo

Increíble... ¡pero cierto!

Un animalito descansa tranquilmente en un lugar tibio y obscuro. De pronto, es obligado a salir al frío y a la luz. La temperatura es casi intolerable para su cuerpo sin pelo. La luz deslumbrante desorienta su visión borrosa. Por instinto, el animal empieza a trepar con la esperanza de encontrar calor y seguridad. Si resbala y cae al suelo, morirá. Tarda casi 30 minutos en llegar a su destino, una bolsa en el abdomen de su madre. Dentro, encuentra un pezón que le dará leche durante varios meses, hasta que crezca y se desarrolle.

¿Adivinas de qué animal se trata? Así es: un canguro recién nacido. Los canguros nacen en una etapa muy temprana de desarrollo, así que siguen desarrollándose en la bolsa de su madre hasta que pueden comer alimentos sólidos como hierba y otras plantas.

Como todos los seres vivos, los canguros envejecen y mueren, y se deben reproducir para transmitir su herencia genética. Hay muchas formas de reproducción, como descubrirás en este capítulo. Pero la reproducción es sólo una parte de la historia. El desarrollo, o la forma en que los organismos crecen, es también parte importante del ciclo de la vida.

532

Strange but True!

Kangaroos are marsupials. They are pouched mammals that give birth to their offspring at a very early stage in the offspring's development. Even in one of the larger subspecies, such as the red kangaroo, the baby is born after only 33 days of gestation. The newborn, called a joey, looks more like an embryo than a baby—it's about 2 cm long and weighs about $\frac{1}{30,000}$ of the mother's weight. The joeys of the larger kangaroos remain in the pouch for as long as 10 months.

533

¿Tú qué piensas?

Usa tus conocimientos para responder a las siguientes preguntas en tu cuaderno de ciencias:

1. ¿Tienen todos los animales dos progenitores? Explica tu respuesta.

2. ¿Qué te hace físicamente diferente de un adulto?

3. ¿Qué porcentaje de genes heredaste de tu madre? ¿Y de tu padre?

¿Qué tan grande es tu cuerpo?

Como ves, estás envejeciendo. Tu cuerpo crece como el de un adulto. Mira a tus compañeras y compañeros. Sus cuerpos también están creciendo. Pero, ¿tienen tu cuerpo y el de tus compañeros las mismas proporciones que el de un adulto? Haz este experimento para averiguarlo.

Procedimiento

1. Trabaja en parejas. Pídele a un compañero o compañera que te ayude a medir tu altura total, la de tu cabeza y la longitud de una pierna con una **cinta métrica** y una **regla.** Tu maestro o maestra te dirá cómo tomar estas medidas.

2. Calcula la proporción de tu cabeza y la pierna en relación al cuerpo. Aplica las siguientes ecuaciones:

$$\left(\frac{\text{longitud de la cabeza}}{\text{altura del cuerpo}}\right) \times 100 = \frac{\text{proporción}}{\text{de la cabeza}}$$

$$\left(\frac{\text{longitud de la pierna}}{\text{altura del cuerpo}}\right) \times 100 = \frac{\text{proporción}}{\text{de la pierna}}$$

3. Tu maestro o maestra ha escrito en el pizarrón las medidas de su cabeza, cuerpo y piernas, junto con las de otros dos adultos. Calcula sus proporciones. Anota todas las medidas y cálculos en tu cuaderno de ciencias.

Análisis

4. ¿Cómo se compara la proporción de tu cabeza con la de los adultos? ¿Cómo se comparan las proporciones de las piernas?

Investigate!

MATERIALS

FOR EACH PAIR:
- meterstick
- tape measure

Teacher Notes: The head height can be measured by having the student stand next to a sheet of paper with one ear against it. The top of the head and bottom of the chin can be marked with a pencil. That length can be then measured. Total body height can be measured in a similar way. Length should be measured from where the leg bends when the student sits down. The student can hold a meterstick perpendicular to the floor at that level and can mark that height on paper taped to the wall or drop a tape measure down to the floor from the tip of the meterstick. Demonstrate measuring techniques to your students.

Before students come to class, write the measurements of at least three adults on the board. The measurements can be of you and two other teachers.

Answer to Investigate!

4. The student's head height should take up a greater proportion of his or her overall height than the adults' head height. The student's leg length should be about 50 percent of his or her overall height. This should match the leg-length proportion of the adults.

Focus

Animal Reproduction

In this section, students learn about two types of asexual reproduction. Students will review meiosis and learn that sexual reproduction unites an egg and sperm using an internal or external fertilization process. The final section focuses on differences in mammalian reproduction.

Bellringer

Write the following list on the blackboard or on a transparency:

a. bird   **c.** ants

b. human   **d.** sea stars

Ask students to write a paragraph explaining how they think reproduction differs among these four animals.

1) Motivate

DISCUSSION

Reproduction Lead a discussion based on students' answers to the Bellringer. Ask students to think about the similarities and differences among the ways animals reproduce. Birds and ants lay eggs, but humans and sea stars don't. Females and males mate to reproduce in humans, and birds, but not in sea stars. Lead them to conclude that the end result of reproduction is the same for all animal species, but the means differ widely.

Sección 1

Reproducción animal

La vida de algunos seres vivos es muy corta en comparación con la nuestra. Por ejemplo, la mosca de la fruta vive sólo unos 80 días. Otros organismos viven mucho tiempo. Un nogal puede vivir de 2,000 a 6,000 años. Pero, al final, todos los seres vivos mueren. Si una especie logra sobrevivir, sus integrantes deben reproducirse.

VOCABULARIO

reproducción asexual
gemación
fragmentación
reproducción sexual
óvulo
espermatozoide
cigoto
fecundación externa
fecundación interna
monotrema
marsupial
mamífero placentario

OBJETIVOS

- Distingue entre reproducción asexual y sexual.
- Explica la diferencia entre fecundación externa e interna.
- Describe los tres tipos de desarrollo de los mamíferos.

De tal palo, tal astilla

Algunos animales, sobre todo los más simples, se reproducen asexualmente. En la **reproducción asexual** un solo progenitor tiene crías genéticamente idénticas a sí mismo.

Un tipo de reproducción asexual es la **gemación,** que consiste en que una pequeña parte del cuerpo del progenitor se desarrolla como un organismo independiente. La hidra de la **Figura 1** se reproduce asexualmente por gemación. La hidra joven es genéticamente idéntica a su progenitor.

La **fragmentación** es otro tipo de reproducción asexual, en la que un organismo se divide en dos o más partes, cada una de las cuales crece como un individuo independiente. Las estrellas de mar se reproducen por fragmentación. Como las estrellas de mar se alimentan de ostras, hay quienes las cortaban en pedazos y las tiraban al mar con el fin de acabar con ellas, ignorando que cada brazo puede crecer como un organismo independiente. Obsérvalo en la **Figura 2.**

Figura 1 La hidra joven pronto se desprenderá de su progenitor. Sin embargo, los brotes de otros organismos, como el coral, permanecen unidos al progenitor.

Figura 2 El brazo más grande de esta estrella de mar era un fragmento del que creció el resto de la misma. Con el tiempo, todos los brazos tendrán el mismo tamaño.

Homework

Making Tables Ask students to make a table in their ScienceLog. In one column they will list 10 mammals. In the next column they will indicate how each mammal produces young. From the information in this section, students should be able to indicate whether the animal is a monotreme, a marsupial, or a placental. Students should fill in the table to the best of their knowledge, research the correct answers, and put them in an additional column.

Se necesitan dos

La **reproducción sexual** produce crías al combinar el material genético de más de un progenitor. La mayoría de los animales, incluyendo los seres humanos, se reproducen sexualmente. En general, en la reproducción sexual participan dos progenitores: un macho y una hembra. El progenitor femenino produce células sexuales llamadas **óvulos,** y el masculino produce células sexuales llamadas **espermatozoides.** Cuando el núcleo de un óvulo se une con el núcleo de un espermatozoide, se crea una nueva célula, llamada **cigoto.** La unión de un óvulo con un espermatozoide se conoce como *fecundación.*

Repaso de la meiosis El ADN tiene instrucciones conocidas como genes en el núcleo de las células. Los genes se encuentran en los *cromosomas.* Todas las células humanas contienen 46 cromosomas, a excepción de los óvulos y espermatozoides, que tienen 23 y se forman por un proceso llamado *meiosis.*

En los seres humanos, la meiosis implica la división de una célula con 46 cromosomas en cuatro células sexuales, con 23 cromosomas cada una. Durante el proceso de división, los cromosomas se mezclan al azar. Esto significa que cada célula sexual puede tener una combinación genética diferente. Cuando óvulo y espermatozoide se unen para formar un cigoto, se recupera el número original de 46 cromosomas. Esta combinación de genes del padre y de la madre produce un cigoto que crecerá como un individuo único. La **Figura 3** muestra cómo se mezclan los genes a través de tres generaciones.

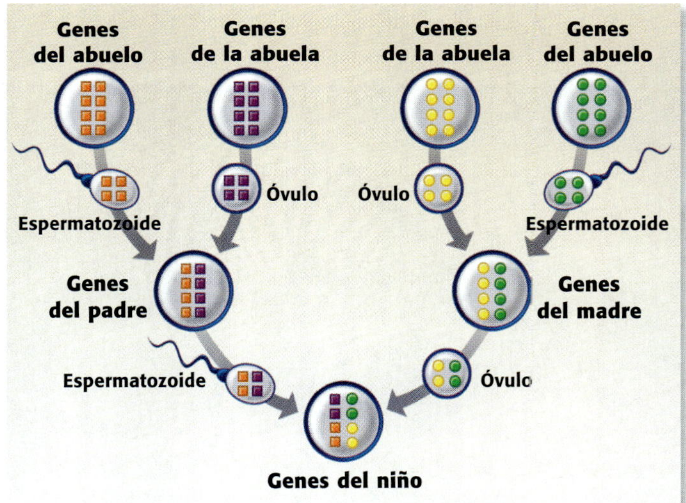

Figura 3 *Los óvulos y los espermatozoides contienen genes. Tú heredaste un número igual de genes de cada uno de tus progenitores. Tus padres heredaron un número igual de genes de sus progenitores.*

Autoevaluación

¿Qué diferencia hay entre la reproducción sexual y la asexual? *(Consulta la página 636 para comprobar tu respuesta.)*

¡MATEMÁTICAS!

Cromo-combinación

Una célula de un organismo está a punto de pasar por la meiosis. La célula tiene 6 cromosomas en 3 pares. ¿Cuántas combinaciones cromosómicas puede haber en las células sexuales que están a punto de formarse? Para descubrirlo, aplica la siguiente fórmula: x^y = posibles variaciones, donde x = número de cromosomas en un par, y = número de pares.

2^3 (ó 2 × 2 × 2) = 8. Por lo tanto, son posibles 8 variaciones.

Una célula humana típica tiene 46 cromosomas en 23 pares. Si la célula pasa por la meiosis, ¿cuántas combinaciones cromosómicas puede haber en las células sexuales creadas?

---

2 Teach

MEETING INDIVIDUAL NEEDS

Learners Having Difficulty
Help students differentiate between sexual and asexual reproduction by having them make diagrams on newsprint or poster board that include all the pertinent information. Students should use a different primary color for sexual and asexual reproduction and include artwork depicting the cells involved. The diagrams should state whether or not the resulting offspring are genetically identical to their parent(s). **Sheltered English**

MISCONCEPTION ALERT

In certain species, a single animal can sexually reproduce alone. For example, in some species of nematodes, sperm are produced and then stored until eggs are produced. This is followed by self-fertilization.

Directed Reading Worksheet 22 Section 1

Answer to MATHBREAK

There are 2^{23}, or 8,388,608, possible chromosomal combinations in a human sex cell.

Answer to Self-Check

In asexual reproduction, one animal produces offspring that are genetically identical to itself. In sexual reproduction, usually the genes of at least two individuals are mixed when sex cells join to form a zygote. This zygote develops into a unique individual.

RETEACHING

Divide the class into groups of four, and provide each group with markers and blank index cards. Each student will write one new term on each card and one fact about each term on a separate card. After shuffling the decks, one student will deal three cards to each member of the group and stack the rest in the center of the table. Students will take turns picking a card from the stack. If it matches one of the cards in their hand, they will keep it and discard another card to the bottom of the stack. The goal is to collect two pairs of matching cards.

3 Extend

GOING FURTHER

Encourage interested students to investigate the role that the endocrine system plays in reproduction. Have them draw diagrams illustrating the interactions among the hypothalamus, pituitary, and ovaries (in females) or testes (in males). Allow time for students to present their findings to the class.

Homework

Encourage students to research an egg-laying mammal. Then have them write a short story about how the egg develops into a new individual.

Fecundación interna y externa

Según el animal, la fecundación puede ocurrir fuera o dentro del cuerpo de la hembra. Algunos peces y anfibios se reproducen por **fecundación externa,** en la que el espermatozoide fecunda el óvulo fuera del cuerpo de la hembra. La fecundación externa debe ocurrir en un ambiente húmedo para que los cigotos no se sequen.

Muchas ranas, como las de la **Figura 4,** se aparean cada primavera. Primero, la rana hembra libera los óvulos. Luego, el macho libera los espermatozoides sobre los óvulos para fecundarlos. Los huevos fecundados se desarrollan solos y, en unas dos semanas, los huevos se transforman en renacuajos.

En la **fecundación interna,** los óvulos y los espermatozoides se unen dentro del cuerpo de la hembra. Reptiles, aves, mamíferos y algunos peces se reproducen así. Muchos animales que presentan fecundación interna ponen huevos fecundados. La hembra pingüino de la **Figura 5,** por ejemplo, pone uno o dos huevos. Los progenitores se turnan para empollarlos, los ponen sobre sus patas y los cubren con su cuerpo para mantenerlos calientes. Cuando los polluelos salen del cascarón, sus progenitores los alimentan hasta que pueden valerse por sí mismos. En la mayoría de los mamíferos, a la fecundación interna la sigue el desarrollo de un óvulo fecundado dentro del cuerpo de la madre. Los canguros y otros marsupiales dan a luz crías vivas en una etapa temprana de desarrollo. Otros mamíferos dan a luz crías bien desarrolladas. Las cebras jóvenes, como las de la **Figura 6,** pueden levantarse y alimentarse casi inmediatamente después de nacer.

Figura 4 Las ranas fecundan los huevos externamente. Algunas especies pueden producir más de 300 crías en una temporada.

Figura 5 Los huevos de los pingüinos se fecundan internamente, pero la mayoría del desarrollo de un bebé pingüino ocurre fuera del cuerpo de la madre.

Figura 6 Esta cebra acaba de nacer, pero ya casi logra ponerse de pie. Dentro de una hora ya podrá correr.

Reinforcement Worksheet 22
"Reproduction Review"

IS THAT A FACT!

There are only 2 or 3 days each month when fertilization can occur in a human female.

El desarrollo de un mamífero

Todos los mamíferos se reproducen sexualmente y alimentan a sus crías con leche. Hay ciertas diferencias en la forma en que los mamíferos producen sus crías, pero cada uno sigue uno de los tres tipos de desarrollo.

Monotremas   Los mamíferos que ponen huevos son **monotremas.** Actualmente, sobreviven dos familias de monotremas, el equidna y el ornitorrinco. Después de que estos animales ponen sus huevos, el período de incubación dura hasta 2 semanas. Al salir del huevo, los bebés están muy poco desarrollados, como ves en la **Figura 7.** Se arrastran hacia un pliegue de piel de la madre y se alimentan de la leche que mana de sus poros.

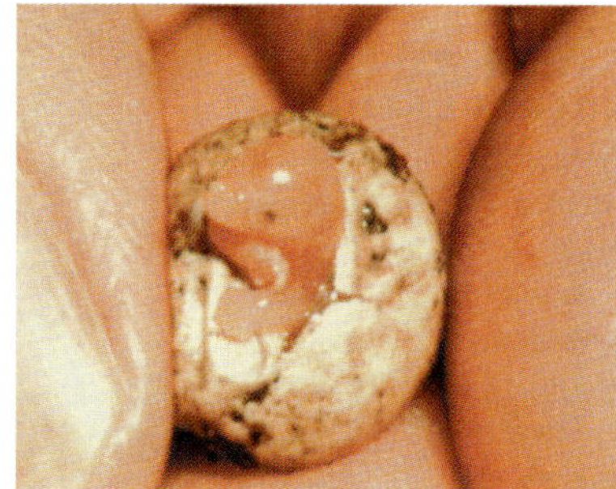

Figura 7 *Unos 10 días después de que un equidna hembra pone un huevo, la cría rompe el cascarón.*

Marsupiales   Los mamíferos que dan a luz a crías vivas parcialmente desarrolladas son los **marsupiales,** de los que hay unas 260 especies. La mayoría tienen bolsas donde sus crías se desarrollan, pero algunas especies sudamericanas no las tienen. Los marsupiales con bolsas tienen huesos adicionales que les ayudan a aguantar el peso de sus crías, como ves en la **Figura 8.** Cuando un bebé marsupial se adhiere al pezón de su madre, el pezón se expande en la boca del bebé para impedir que la cría se separe de su mamá.

Mamíferos placentarios   Hay aproximadamente 4,000 especies de mamíferos placentarios. Entre ellos están las ballenas, los elefantes, los armadillos, los murciélagos, los caballos y los seres humanos. Los **mamíferos placentarios** alimentan a sus crías internamente antes de nacer. Sus recién nacidos están altamente desarrollados en comparación con los marsupiales o monotremas recién nacidos.

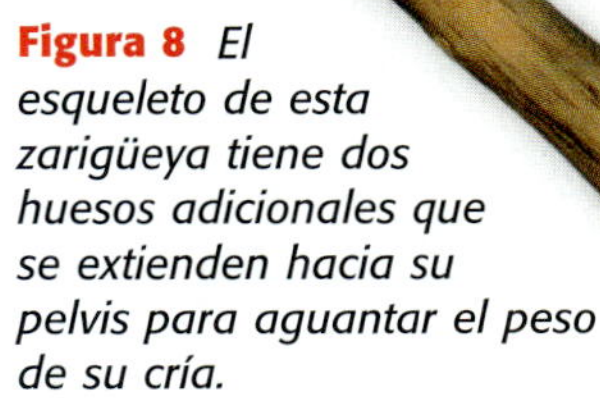

Figura 8 *El esqueleto de esta zarigüeya tiene dos huesos adicionales que se extienden hacia su pelvis para aguantar el peso de su cría.*

REPASO

1. ¿Cuántos progenitores se necesitan en la reproducción asexual?

2. ¿Qué diferencia hay entre monotremas y marsupiales?

3. ¿Cómo se forma un cigoto?

4. **Aplicar conceptos**   Las aves ponen huevos, pero no se consideran monotremas. Explica por qué.

537

Quiz

1. How do the offspring created by asexual reproduction differ from those created by sexual reproduction? (Offspring of asexual reproduction are genetically identical to the parent, while offspring of sexual reproduction have genetic material from two parents.)

2. How do internal and external fertilization differ? How are they alike? (Internal fertilization occurs inside the female's body, while external fertilization occurs outside the female's body. Both result in the creation of single cells with combined genetic information from the parents.)

ALTERNATIVE ASSESSMENT

Concept Mapping   Have students create a concept map using the new terms in this section and the section title.

▼ **Answers to Review**

1. one
2. Monotremes lay eggs and marsupials give birth to live young.
3. A zygote is formed when a sperm enters an egg and the nuclei of the egg and sperm join.
4. Birds are not mammals. They do not nurture their young with milk, and they are not covered with hair.

Focus

Human Reproduction

In this section, students are introduced to the male and female reproductive systems. Students then learn about some of the irregularities and problems that affect the human reproductive system, including multiple births, ectopic births, and sexually transmitted diseases.

Bellringer

Ask students if they think that cloning human beings could be considered reproduction. Why or why not? What kind of reproduction is it? Have students write answers to these questions in their ScienceLog.

1 Motivate

DISCUSSION

Ask students to compare reproduction in birds with reproduction in humans. (Birds lay eggs and must protect and keep the eggs warm while obtaining food for themselves. Human mothers carry their baby inside their body, so the baby is always protected.)

Teaching Transparency 82 "The Male Reproductive System"

Directed Reading Worksheet 22 Section 2

Sección 2

Reproducción humana

Cuando un espermatozoide y un óvulo humanos se unen, empieza a formarse un nuevo ser humano. Unos 9 meses después, la madre da a luz a un bebé. Pero ¿qué pasa antes? ¿De dónde vienen los óvulos y los espermatozoides? Deben ocurrir muchos pasos antes de que los óvulos y los espermatozoides maduren.

El aparato reproductor masculino

El aparato reproductor masculino se ilustra en la **Figura 9.** Produce espermatozoides y los libera en el sistema reproductor femenino. Los **testículos** producen espermatozoides y testosterona, que es la principal hormona sexual masculina y regula la producción de células sexuales y el desarrollo de los rasgos masculinos.

Producción de espermatozoides El cuerpo humano está a una temperatura de unos 37°C, pero los espermatozoides no se desarrollan a esa temperatura tan alta. Por eso los dos testículos permanecen en el **escroto,** una bolsa cubierta de piel que cuelga del cuerpo. El escroto está unos 3 grados más frío. Dentro de cada testículo hay muchos tubos en forma de espiral llamados **tubos seminíferos**, donde se producen los espermatozoides.

Antes de que los espermatozoides salgan de los testículos, se almacenan temporalmente en el **epidídimo.** Los espermatozoides nadan hacia el **conducto deferente,** que va del epidídimo al cuerpo, y se mezclan con fluidos de varias glándulas. La mezcla de espermatozoides y fluidos se llama **semen.** Un varón adulto sano produce millones de espermatozoides al día. Esta producción masiva y continua empieza en la pubertad. La **pubertad** es la etapa de la vida en que los órganos sexuales maduran. Para abandonar el cuerpo, los espermatozoides pasan por el conducto deferente a la **uretra,** que es el tubo que atraviesa el pene. El **pene** transfiere semen al cuerpo de la mujer durante la relación sexual.

VOCABULARIO

testículos	ovarios
escroto	ovulación
tubos seminíferos	trompas de Falopio
epidídimo	útero
conducto deferente	vagina
semen	menstruación
pubertad	infértil
uretra	enfermedad de transmisión sexual
pene	

OBJETIVOS

- Describe las funciones de los aparatos reproductores masculino y femenino.
- Comenta los transtornos y enfermedades asociados con la reproducción humana.

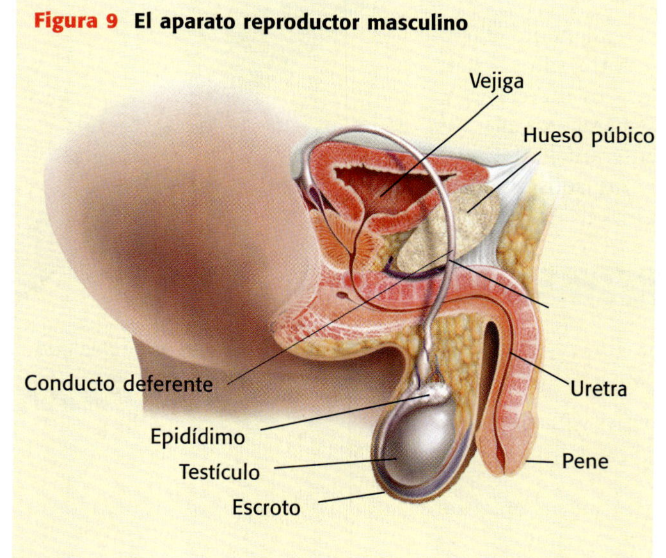

Figura 9   El aparato reproductor masculino

IS THAT A FACT!

Mumps, a common childhood disease, poses a risk to males who contract it during puberty or adulthood. When mumps occurs after childhood, it can cause inflammation of the testes, known as acute orchitis, which in rare cases can result in sterility.

El aparato reproductor femenino

El aparato reproductor femenino se ilustra en la **Figura 10.** Produce óvulos, alimenta los óvulos fecundados y da a luz. En su centro están los **ovarios,** que producen los óvulos y otras hormonas sexuales, como el estrógeno y la progesterona. Estas hormonas regulan la liberación de óvulos y controlan el desarrollo de los rasgos femeninos.

El viaje del óvulo El ovario es del tamaño de una almendra grande y contiene óvulos en varias etapas de desarrollo. Cuando un óvulo madura, se vuelve una célula enorme, unas 200,000 veces más grande que un espermatozide. Durante la **ovulación,** un óvulo desarrollado se expulsa por la pared del ovario y pasa a una de las trompas de Falopio. Las **trompas de Falopio** van de los ovarios al útero. El **útero** es el órgano donde el bebé crece y se desarrolla.

Cuando una mujer no está embarazada, el útero se dobla y queda casi plano. Cada mes, desde la pubertad, el recubrimiento interno del útero se engruesa en preparación para el embarazo. Si hay fecundación, el cigoto diminuto baja de la trompa de Falopio y se instala en el recubrimiento interno del útero. Para nacer, el bebé sale del útero por la vagina. La **vagina** es el pasadizo por donde entró el espermatozoide durante la relación sexual.

Ciclo menstrual En preparación para el embarazo, el sistema reproductor femenino pasa por varios cambios, llamados ciclo menstrual, que en general ocurren cada mes. El primer día del ciclo es el inicio de la **menstruación,** o descarga mensual de sangre y tejido del útero. La menstruación dura unos 5 días. Cuando termina, el recubrimiento interno del útero se empieza a formar de nuevo en preparación para la ovulación. La ovulación ocurre típicamente alrededor del 14º día del ciclo. Si el óvulo no ha sido fecundado cuando llega al útero, se deteriorará. La menstruación lo expulsará y el ciclo volverá a comenzar. El ciclo menstrual de la mujer empieza en la pubertad y termina al final de la edad madura.

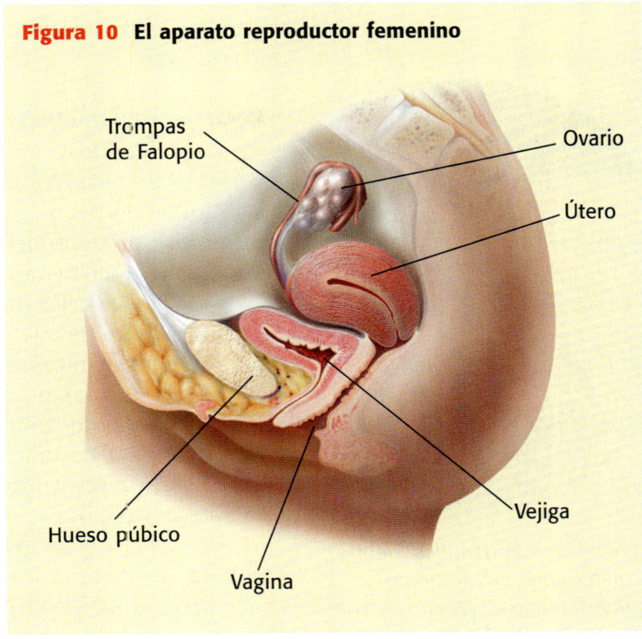

Figura 10 El aparato reproductor femenino

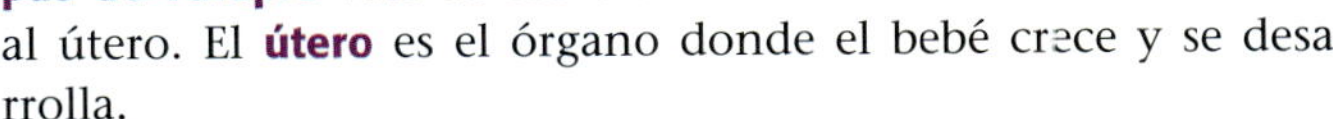

¡MATEMÁTICAS!

Contar óvulos

1. La mujer promedio ovula cada mes desde cerca de los 12 a los 50 años. ¿Cuántos óvulos maduros puede producir durante ese período de tiempo?

2. Los ovarios de la mujer almacenan 2 millones de óvulos inmaduros. Si ovulara regularmente desde los 12 a los 50 años, ¿qué porcentaje de óvulos alcanzaría la madurez?

Teaching Transparency 83
"The Female Reproductive System"

Math Skills Worksheet 3
"Multiplying Whole Numbers"

2 Teach

READING STRATEGY

Activity Draw students' attention to **Figures 9** and **10.** Have them write the new terms in their ScienceLog, leaving a few lines between each term. As students read the section they should fill in definitions that explain the new terms in their own words.

CROSS-DISCIPLINARY FOCUS

Language Arts The words *male* and *female* actually originate from two unrelated Latin words. *Female* comes from *femella,* meaning "girl," while *male* comes from *masculus,* meaning "male."

Answers to MATHBREAK

1. 50 y − 12 y = 38 y
   38 y × 12 eggs/year = 456 eggs. She can produce about 456 mature eggs.

2. $\frac{456 \text{ eggs}}{2,000,000} \times 100 = 0.0228$ percent of her eggs will probably mature.

MATH and MORE

If one set of identical twins is born for every 250 births, and one set of quadruplets is born for every 705,000 births, how many sets of identical twins are born for every quadruplet birth?

$(\frac{705{,}000}{250} = 2{,}820)$

Math Skills Worksheet 5
"Dividing Whole Numbers with Long Division"

MEETING INDIVIDUAL NEEDS

Writing **Advanced Learners**
Encourage interested students to research how fertility drugs cause multiple births. (They will find that fertility drugs stimulate egg ripening and ovulation, which often results in the release of more than one egg at a time.)

Ask them to prepare a brief report and to share their findings with the class.

ACTIVITY

Writing **Puzzle Making** Divide the class into small groups, challenging each to create a crossword puzzle using section vocabulary and concepts. Have students write clues and construct puzzles and then trade with another group. Allow time for students to solve the puzzles.

Irregularidades y transtornos

En la mayoría de casos, el aparato reproductor humano completa sus funciones perfectamente. Sin embargo, como en cualquier otro sistema del cuerpo, a veces hay irregularidades y transtornos.

Nacimientos múltiples ¿Alguna vez has visto un par de gemelos idénticos? A veces son tan parecidos que ni sus padres los distinguen. Por cada 250 alumbramientos, nace alrededor de un par de gemelos idénticos. También es frecuente otro tipo de gemelos, los gemelos fraternos, que pueden ser muy diferentes entre sí.

Los gemelos, como los de la **Figura 11,** son los tipos más comunes de nacimientos múltiples, pero los seres humanos también pueden tener trillizos (3 bebés), cuatrillizos (4 bebés), quintillizos (5 bebés), y así sucesivamente. Los nacimientos múltiples son muy raros. Por ejemplo, de cada 705,000 nacimientos, alrededor de uno es cuádruple. ¿Sabes qué circunstancias provocan un nacimiento múltiple? Para descubrirlo, haz el ejercicio "Aplica" al final de esta página.

Figura 11 *Los gemelos idénticos tienen genes exactamente iguales. Muchos gemelos idénticos que se crían aparte tienen personalidades e intereses similares.*

Embarazo ectópico En un embarazo normal, el óvulo fecundado viaja al útero y se instala en su pared. En un *embarazo ectópico,* el óvulo fecundado se instala en una trompa de Falopio o en otra área del aparato reproductor. Como el cigoto no puede desarrollarse correctamente fuera del útero, un embarazo ectópico puede ser muy peligroso tanto para la madre como para el niño.

APLICA

Raúl y René son gemelos fraternos. Aunque son de la misma edad y son gemelos, no se parecen mucho. Sus amigas Emilia y Carolina son gemelas idénticas. Son tan parecidas que Raúl y René no las pueden distinguir. ¿Por qué unos gemelos son idénticos y otros fraternos? Considera las dos posibilidades de la derecha. En *A,* un solo óvulo fecundado se divide en dos mitades antes del desarrollo y en *B* un ovario libera dos óvulos que son fecundados por dos espermatozoides diferentes. Anota las respuestas a las siguientes preguntas en tu cuaderno de ciencias:

1. ¿Cuál de los dos casos, *A* o *B,* produciría gemelos idénticos? Explica tu respuesta.
2. ¿Podrían los gemelos fraternos ser (a) dos niños, (b) dos niñas, (c) una niña y un niño, o (d) todas las anteriores?

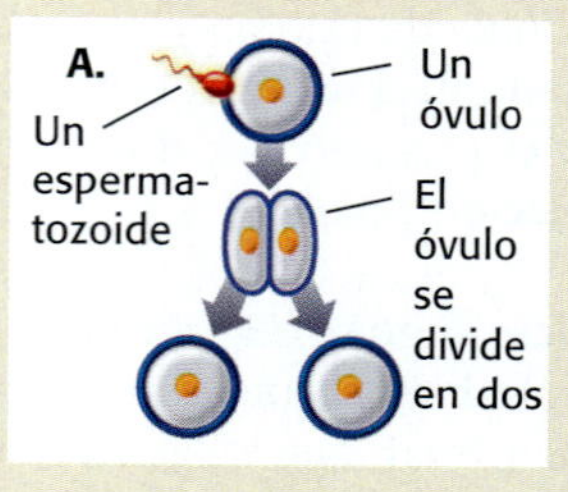

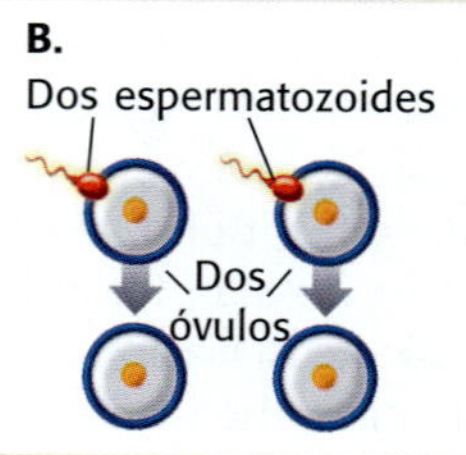

Answers to APPLY

1. *A* would produce identical twins because the two halves of the split egg have identical genetic material.
2. d

Enfermedades del aparato repro-ductor En los Estados Unidos, cerca del 15 por ciento de las parejas tienen problemas para tener hijos. Muchas de estas parejas son **infértiles,** es decir, no pueden procrear. Los hombres pueden ser infértiles porque no producen suficientes espermato-zoides saludables. Esto se llama con-teo bajo de espermatozoides y se observa en la **Figura 12.** Las mujeres pueden ser infértiles porque no ovulan normalmente.

La infertilidad femenina también puede ser provocada por cicatrices en las trompas de Falopio. La cicatriz, generalmente causada por una enfermedad de transmisión se-xual, impide el paso de los óvulos al útero. Las **enfermedades de transmisión sexual** son enfermedades que pasan de una per-sona infectada a otra sana durante el contacto sexual. En los Estados Unidos, las enfermedades de transmisión sexual más comunes son la clamidia, la gonorrea y el herpes genital. El síndrome de inmunodeficiencia adquirida (SIDA) es otra enfer-medad de transmisión sexual muy común. El SIDA se puede transmitir de otras formas, como el uso compartido de agujas, las transfusiones de sangre de donadores infectados y el con-tacto con sangre infectada. No se transmite por contacto casual. El SIDA es causado por el virus de inmunodeficiencia humana (VIH). El cáncer, que es el crecimiento descontrolado de las células, a veces se presenta en los órganos reproductores. Los testículos y la próstata, una glándula que produce el semen, son focos frecuentes de cáncer en los hombres de más de 50 años. En las mujeres, son los ovarios y los senos.

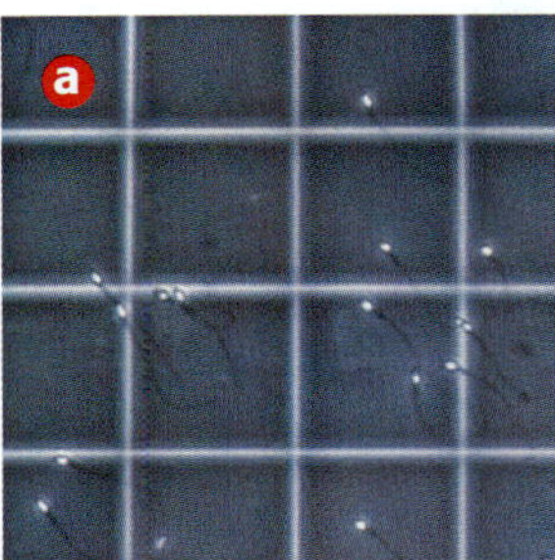
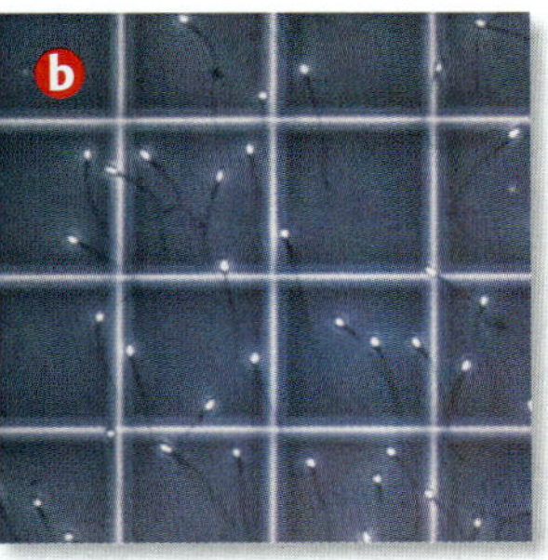

Figura 12 *La microfotografía* (a) *muestra un conteo bajo de espermatozoides. La* (b) *muestra un conteo normal. Un conteo bajo de espermatozoides puede ser causado por muchos factores, como un desequilibrio hormonal o un conducto deferente bloqueado.*

Uno de cada cuatro jóvenes norteamericanos contrae una enfermedad de trans-misión sexual antes de los 21 años.

química

CONEXIÓN

Muchas substancias químicas de agentes contaminantes son similares a las hormonas femeninas. Algunos estudios han comenzado a vincularlas con la menstruación pre-matura y el conteo bajo de espermatozoides.

REPASO

1. ¿Cuál es la diferencia entre espermatozoides y semen?

2. ¿Puede quedar embarazada una mujer en cualquier día del mes? Explica por qué.

3. Define *las enfermedades de transmisión sexual* y da tres ejemplos.

4. **Aplicar conceptos** ¿En qué se parecen los ovarios a los testículos? ¿En qué se diferencian?

541

Focus

Growth and Development

In this section, students learn about egg fertilization and implantation. Students are also introduced to the different stages of growth of a baby *in utero*, culminating in its birth. Finally, they learn about the stages of human development, from birth through adulthood.

🔔 Bellringer

Write this statement on the board or an overhead projector:

Name the stages of physical development you have passed through thus far in your life.

Have students list the stages in their ScienceLog. Remind students that their growth and development began while they were still in the uterus. (Students will likely respond with the following: crawling, walking, talking, growing taller, perhaps developing lower voices for some of the boys.)

1) Motivate

DISCUSSION

Life Stages Ask students to list as many characteristics of each of the following as they can:

> infancy, childhood, adolescence, adulthood

Tell students that while there are individual differences, all people go through these stages.

VOCABULARIO

embrión   cordón umbilical
implantación   feto
placenta

OBJETIVOS

- Resume los procesos de fecundación e implantación.
- Describe los pasos del desarrollo humano.

Crecimiento y desarrollo

Todos empezamos nuestra vida como una simple célula que se convierte en un ser humano. Estamos hechos de millones de células, cada una con una tarea determinada. Tú no eres una excepción. Te has convertido en una persona compleja, capaz de generar miles de pensamientos y acciones. Es difícil creer que una persona tan extraordinaria empezó siendo una simple célula, pero así fue.

Una nueva vida

El proceso natural de la creación de un nuevo ser humano empieza cuando un hombre deposita millones de espermatozoides en la vagina de una mujer durante la relación sexual. La mayoría de los espermatozoides mueren debido al ambiente ácido de la vagina, pero unos cientos de ellos entran a través del útero en las trompas de Falopio, como ves en la **Figura 13.** Los espermatozoides sobrevivientes cubren el óvulo, liberando enzimas que ayudan a disolver su cubierta exterior. En cuanto uno de los espermatozoides logra entrar, una membrana encierra el óvulo fecundado, la cual impide que entre otro espermatozoide.

Implantación El óvulo fecundado baja por la trompa de Falopio hacia el útero. El viaje dura unos 5 días. El cigoto se divide muchas veces durante el viaje. Finalmente llega al útero como una pelotita de células llamada **embrión.** Durante los siguientes días, el embrión debe instalarse en el revestimiento interior, rico en nutrientes, del útero de su madre. Este proceso es la **implantación,** y sólo un 30 por ciento de todos los embriones lo realizan satisfactoriamente. La **Figura 14** muestra un embrión implantado.

Figura 13 **Fecundación e implantación**

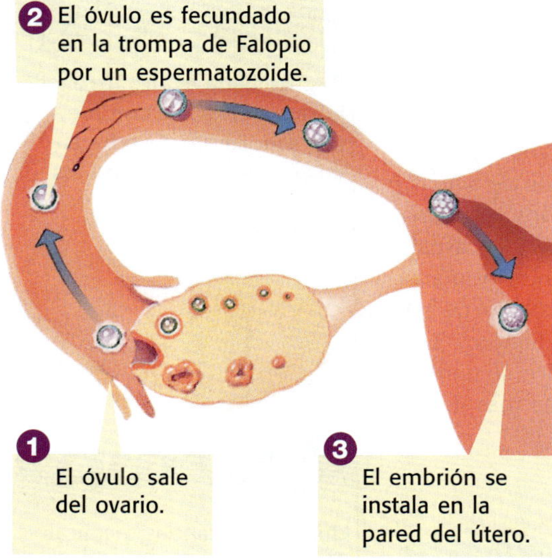

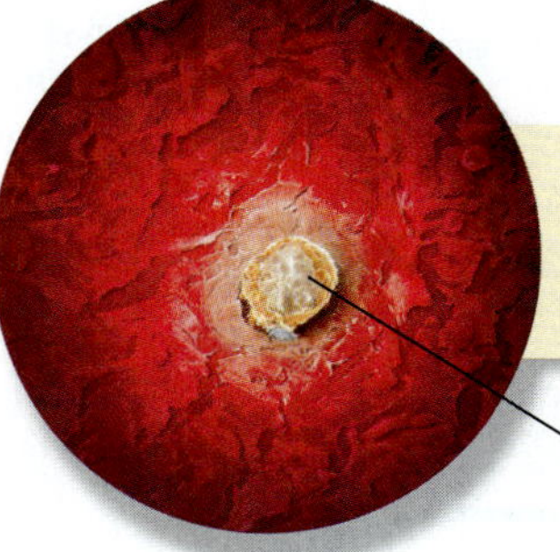

Figura 14 *Este embrión se instaló en la pared del útero de su madre.*

El tamaño real del embrión es ligeramente más pequeño que el punto final de esta oración.

Directed Reading Worksheet 22 Section 3

IS THAT A FACT!

The longest gestation period, 22 months, belongs to the Indian elephant. The shortest, 12 days, belongs to the Virginia opossum.

Antes de nacer

Cuando el embrión se instala en el útero de la madre, se dice que la mujer está embarazada. Para que el embrión sobreviva, un órgano especial de intercambio llamado **placenta** empieza a crecer. La placenta tiene una red de vasos sanguíneos que proporcionan al embrión oxígeno y nutrientes de la sangre de la madre. Los desperdicios que produce son eliminados por la placenta y transportados a la sangre de la madre para que los deseche.

Una semana después de la implantación, se forman las células de la sangre y un tubo cardíaco en el embrión. Luego, el tubo cardíaco empieza a sacudirse bruscamente, iniciando el latido rítmico que continuará durante toda la vida del individuo. A la cuarta semana, el embrión mide casi 2 mm de largo. A su alrededor hay una membrana delgada llena de fluido llamada *saco amniótico,* formada para evitar que el embrión se lastime. El **cordón umbilical** aparece para conectar el embrión con la placenta. Por la placenta, la sangre del embrión y la de la madre establecen un contacto estrecho. Esto permite el intercambio de oxígeno y nutrientes, molécula a molécula. Aunque la sangre del embrión y de la madre fluyen muy cerca una de la otra dentro de la placenta, nunca se mezclan. El cordón umbilical, el saco amniótico y la placenta se ven en la **Figura 15.**

Del primer al segundo mes Cuando el embrión tiene 4 semanas, se empieza a formar parte del cerebro y de la médula espinal. Tiene brotes diminutos a los lados del cuerpo que finalmente se convertirán en brazos y piernas. Entonces se empiezan a formar los orificios nasales, las manos y los pies. Los músculos empiezan a desarrollarse y, por primera vez en su vida, el cerebro empieza a enviar señales a otras partes del cuerpo. A pesar de estas transformaciones, el embrión sigue teniendo el tamaño de un cacahuate. La **Figura 16** muestra un embrión de cinco semanas.

Figura 16 Embrión de 5 semanas

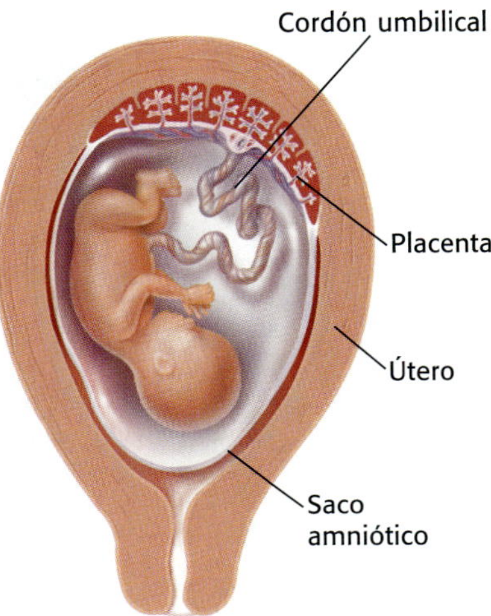

Figura 15 *La placenta, el saco amniótico y el cordón umbilical son el sistema de apoyo de vida del feto.*

✔ **Autoevaluación**

1. ¿Cómo se alimenta un embrión?
2. ¿Por qué es importante que el embrión se instale en el útero y no en cualquier lugar?

(Consulta la página 636 para comprobar tus respuestas.)

IS THAT A FACT!

The European badger and the American marten each have a 250-day gestation period. Due to delayed implantation, the embryo grows during only 50 of those days. The fertilized egg develops for a few days immediately after conception in July or August, remains dormant in the uterus until January, and then completes its growth between January and March, at which time the baby is born.

USING THE FIGURE

Ask students to identify in **Figure 18** the trimester during which the major organs and body structures form. *(the first trimester)*

Ask them to describe the events that occur during the remainder of the pregnancy. *(The structures formed earlier in pregnancy grow and mature during the second and third trimesters.)*

RESEARCH

Writing Tell students that the first few months of pregnancy are crucial for the healthy mental and physical development of the fetus. Drinking alcohol during this period can lead to birth defects and miscarriages. Have students research the causes and consequences of fetal alcohol syndrome (FAS) and present their findings in a short report.

MATH and MORE

Have students calculate the factor of increase in a fetus's body length from the eighth week, when it is about 2.5 cm long, to the fifth month, when it is about 25 cm, to birth, when it is about 50 cm long. *(2.5 cm to 25 cm is an increase by a factor of 10; 25 cm to 50 cm is an increase by a factor of 2; 2.5 cm to 50 cm is an increase by a factor of 200)*

LabBook **PG 635**
My, How You've Grown!

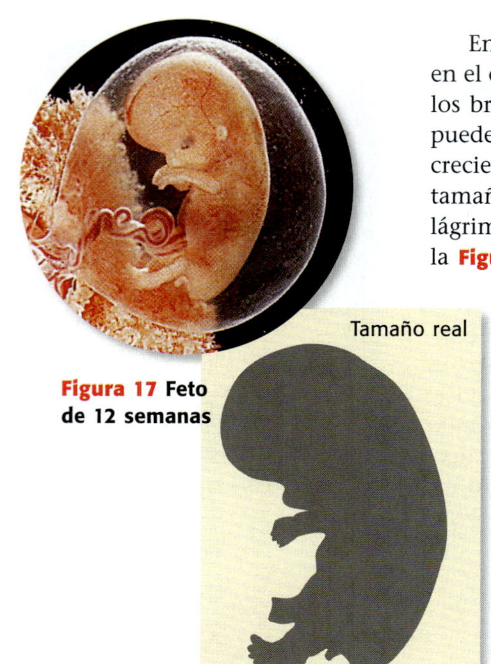
Figura 17 Feto de 12 semanas

En la siguiente etapa se presentan pequeños movimientos en el cuerpo del embrión. El embrión estira las piernas y sacude los brazos. Ahora tiene 8 semanas y está más desarrollado; ya puede llamarse **feto**. Transcurren tres semanas más, y continúa creciendo a gran velocidad; en un mes duplica y triplica su tamaño. Las manos del feto son ahora del tamaño de una lágrima, y su cuerpo tiene el peso de dos hojas de papel. En la **Figura 17** se ve un feto de 12 semanas.

Del tercer mes al sexto Transcurre la 13ª semana de vida del feto y, de pronto, hay nuevos movimientos. Parpadea por primera vez en su vida, pasa saliva, tiene hipo, cierra el puño y mueve los dedos de los pies, que ya tienen uñitas. Para el cuarto mes, el feto empieza a hacer movimientos más grandes. En esta etapa, la madre sabe cuando su bebé estira las piernas o los brazos.

En el quinto mes, el feto mide 20 cm de largo. Se le forman papilas gustativas en la lengua y le crecen cejas. El feto empieza a oír sonidos a través de la pared del útero de la madre. Observa la cronología de la **Figura 18** y revisa los cambios que ocurrieron durante los primeros 6 meses de embarazo.

Figura 18 **Cronología trimestral**

Semanas del primer trimestre	
1 y 2	El óvulo es fecundado por un espermatozoide. El óvulo fecundado se dirige hacia el útero, donde se instala en el revestimiento interior. El óvulo fecundado se llama ahora embrión.
3 y 4	Los principales sistemas han empezado a formarse. El corazón empieza a latir alrededor del día 22. La placenta está completamente formada para la cuarta semana.
5 y 6	Los rasgos faciales empiezan a tomar forma. El esqueleto empieza a desarrollarse.
7 y 8	Empieza el movimiento muscular. El embrión pasa a llamarse feto.
9 y 10	Los brazos, las piernas y los pies se han formado.
11 y 12	Los órganos internos se han desarrollado.

Semanas del segundo trimestre	
13 y 14	El sistema circulatorio está funcionando.
15 y 16	La madre empieza a sentir que el feto se mueve.
17 y 18	El feto responde al sonido.
19 y 20	El feto mide ahora 20 cm de largo.
21 y 22	
23 y 24	Aparecen las pestañas y las cejas.

Semanas del tercer trimestre	
25 y 26	Abre los ojos.
27 y 28	El feto "practica" la respiración.
29 y 30	
31 y 32	
33 y 34	
35 y 36	El feto reacciona a la luz.
Nacimiento	Nace el bebé.

544

internet connect

sciLINKS NSTA
TOPIC: Before Birth
GO TO: www.scilinks.org
***sci*LINKS NUMBER:** HSTL635

IS THAT A FACT!

Between conception and birth, the developing fetus increases in size from a single cell to 6 trillion cells!

Del séptimo al noveno mes En el séptimo mes, los recuerdos del feto empiezan a formarse. Sus pulmones empiezan a "practicar" la respiración, moviéndose hacia arriba y hacia abajo continuamente como si respiraran aire de verdad. Si la madre del feto fuma durante esta etapa, los pulmones del feto se detienen hasta por una hora. Los pulmones del feto de la **Figura 19** empiezan a realizar sus primeros movimientos.

Para el octavo mes, el feto abre los ojos y puede percibir la luz a través de la pared abdominal de la madre y los patrones de sueño empiezan a ser influenciados por la luz del Sol. Cuando el feto está dormido, sueña. ¿Te imaginas qué soñará?

Nacimiento

Por lo general, a los 9 meses el feto está listo para vivir fuera de su madre. La madre pasa por una serie de contracciones musculares llamadas *parto*. Durante esta etapa, el feto sale de cabeza a través de la vagina. Hay muy poco espacio, así que la cabeza del feto se alarga un poco para poder pasar a través de la pelvis de la madre. De pronto, luces brillantes y un aire frío rodean al bebé recién nacido. Abre la boca, llena sus pulmones con aire por primera vez y llora.

El bebé de la **Figura 20** aún está conectado a la placenta por el cordón umbilical. El doctor o la partera que asiste a la madre lo ata y lo corta. El ombligo del bebé es todo lo que queda del punto de unión con el cordón umbilical. Después de que la madre arroja la placenta del cuerpo, el parto ha terminado.

Del nacimiento a la muerte

De todos los animales de este planeta, la duración de la vida de un ser humano es de las más largas. Nuestra infancia dura 2 años, el tiempo del ciclo de vida de la mayoría de los conejos. Nuestra niñez se extiende una década, más de lo que viven muchos gatos o perros. ¡Los seres humanos pueden vivir más de 100 años!

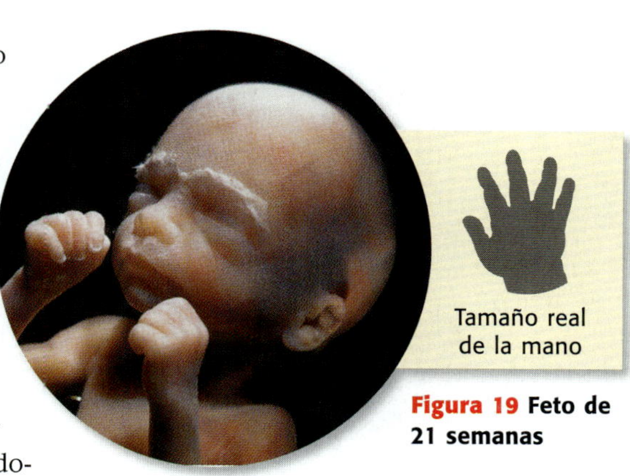

Figura 19 Feto de 21 semanas

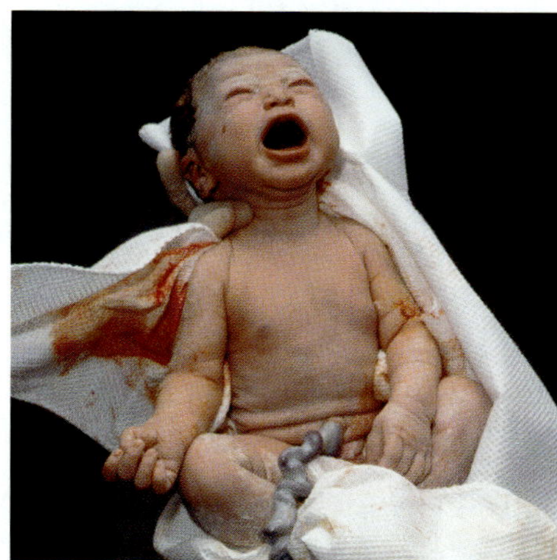

Figura 20 *Este recién nacido todavía está unido al cordón umbilical. La masa promedio de un recién nacido es de 3.3 kg. Su longitud promedio es de 50 cm.*

Experimentos

¿Qué tan bien protege el útero al feto? Sigue las instrucciones de la página 634 para descubrirlo.

545

It's a Comfy, Safe World! PG 634

Fathers-to-be may experience *couvade syndrome,* where they suffer from backaches, nausea, and weight gain. Many cultures have rituals that incorporate the symptoms that expectant fathers may experience. Men do not report couvade syndrome in Western countries as often as they do in other parts of the world. It is not known whether this is because it does not occur as frequently or because men in the West are reluctant to acknowledge the experience.

CONNECT TO

EARTH SCIENCE

It is possible to create pictures using sound waves. Sonograms are pictures obtained by bouncing high-frequency sound waves off of an object. Doctors can use sonograms to "see" a human fetus while it is still in the womb. Sonar, which uses the same principle, is used for navigating and determining an object's position.

Use the following Teaching Transparency to illustrate how sonar works.

Teaching Transparency 133 "How Sonar Works"

CROSS-DISCIPLINARY FOCUS

Language Arts After the umbilical cord is cut and falls off, all that remains is the navel. The word *navel* comes from the Anglo-Saxon word *nafu,* which means "the hub of the wheel." Early Anglo-Saxons believed the navel was the center of the body.

RETEACHING

Divide the class into small groups, and challenge each group to create a board game. Provide each group with a piece of poster board, plain index cards, and markers. Direct them to create a game board that leads players through prenatal development. The first player who is "born" wins the game. Have them use the index cards to write clues that direct players' movements through "gestation." For example, they might write, "Advance to 4 months if you can describe my abilities at 13 weeks." (At 13 weeks, the fetus can blink its eyes, swallow, hiccup, make a fist, and curl its toes.)

Have students create written rules. Then have them exchange games and play.
Sheltered English

Teaching Transparency 84 "Growth Chart"

Reinforcement Worksheet 22 "The Beginning of a Life"

Critical Thinking Worksheet 22 "One to Grow On!"

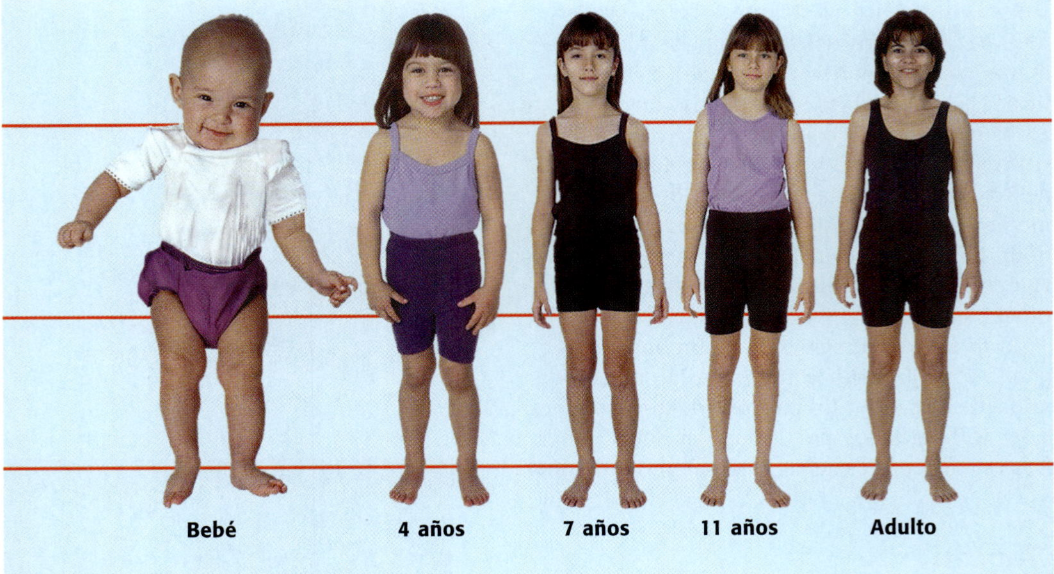

Figura 21 *Este diagrama muestra a una mujer en cinco etapas de desarrollo diferentes. Las etapas se ven en el mismo tamaño para que puedas ver cómo cambian las proporciones del cuerpo a medida que una persona se desarrolla.*

Infancia ¿Por qué etapas has pasado desde que naciste? Probablemente, ya has pasado por la mayoría de las etapas de la **Figura 21.** Fuiste un bebé hasta los 2 años de edad. En este período, creciste rápidamente. Te empezaron a salir los dientes. Adquiriste más coordinación a medida que tu sistema nervioso se desarrolló. Esto te permitió empezar a caminar.

Niñez Tu niñez se extiende de los 2 años a la pubertad. También éste es un período de crecimiento rápido. Tus dientes de leche empezaron a mudar y fueron reemplazados por dientes permanentes. Tus músculos empezaron a estar más coordinados, permitiéndote realizar actividades como andar en bicicleta y saltar a la cuerda. Tus habilidades intelectuales también se desarrollaron durante esta etapa.

Adolescencia Un ser humano se considera adolescente de la pubertad a la edad adulta. Durante la pubertad, los sistemas reproductores masculinos y femeninos alcanzan su madurez. La pubertad ocurre en la mayoría de los varones entre los 11 y los 16 años. El cuerpo del varón se vuelve más musculoso, su voz es más grave y aparece vello en el cuerpo y la cara. El niño de la **Figura 22** pronto experimentará estos cambios. En la mayoría de las niñas, la pubertad ocurre entre los 9 y los 14 años. Durante este período, la cantidad de grasa en las caderas y los muslos aumenta, los senos crecen y aparece vello en áreas como las axilas. A partir de esta etapa, la joven también empieza a menstruar.

Figura 22 *El desarrollo muscular es un cambio de la pubertad en los varones. Este niño pronto se desarrollará físicamente como su padre.*

546

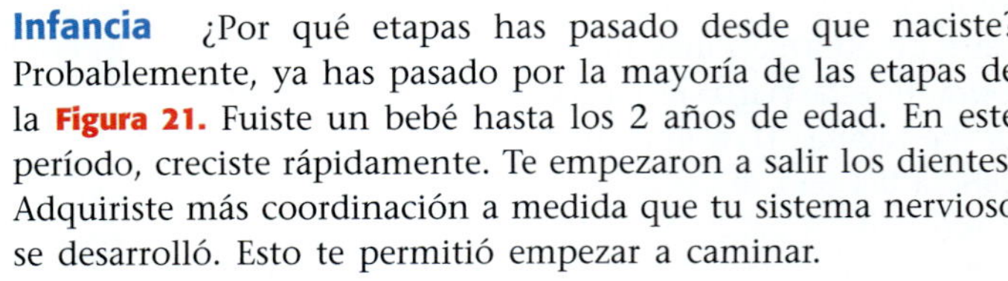
IS THAT A FACT!

Girls attain three-quarters of their adult height by the age of $7\frac{1}{2}$. Boys attain three-quarters of their adult height by the age of 9.

La edad adulta De los 20 a los 40 años, aproximadamente, se considera al ser humano como adulto. En esta etapa se alcanza el máximo desarrollo físico. Alrededor de los 30 años, hay algunos cambios asociados con el envejecimiento. Los cambios son graduales y ligeramente diferentes entre las personas. Generalmente, algunas señales prematuras de envejecimiento son la disminución de la flexibilidad muscular, el deterioro de la vista y del oído, el aumento de grasa corporal y la pérdida del cabello.

El proceso de envejecimiento continúa en el adulto de mediana edad (entre los 40 y 65 años). Durante este período, el cabello se llena de canas, las habilidades atléticas se deterioran y la piel se arruga. Cualquier persona mayor de 65 años se considera un adulto anciano. Aunque el envejecimiento continúa durante este período, los ancianos pueden seguir llevando una vida activa y plena. Los progresos en la ciencia médica han permitido que muchos ancianos sean saludables y productivos por más tiempo que los de generaciones previas. Algunos de los ciudadanos más productivos de este país son adultos ancianos, como ves en la **Figura 23**. Además, muchas personas han podido retrasar el envejecimiento con ejercicio y una dieta nutritiva equilibrada.

Figura 23 *John Glenn, el primer astronauta norteamericano puesto en órbita, regresó al espacio a los 77 años.*

REPASO

1. ¿Qué diferencia hay entre un embrión y un feto?

2. ¿Por qué se forma una membrana alrededor del óvulo una vez que el espermatozoide ha entrado?

3. ¿Qué cambios de desarrollo experimenta el ser humano desde su nacimiento hasta la pubertad?

4. **Aplicar conceptos** Cuando los astronautas trabajan en el espacio, algunas veces están unidos a la nave espacial por una cuerda llamada cordón umbilical. ¿Por qué crees que se llama así?

Laboratorio

La vida crece

Usa la Figura 21 de la página anterior para realizar esta actividad.

1. Con una **regla** mide la longitud de la cabeza del bebé. Luego mide la altura total del cuerpo, incluyendo la cabeza.

2. Calcula el porcentaje de la longitud de la cabeza del bebé en relación con la altura total.

3. Repite estas medidas y cálculos para las otras etapas que de la figura. Responde a la siguiente pregunta en tu cuaderno de ciencias.

Durante el desarrollo del bebé a la edad adulta, ¿le crece más rápida o más lentamente la cabeza que el resto del cuerpo? ¿A qué crees que se debe esto?

Explora

Haz un cartel o una cronología que ilustre las diferentes etapas del crecimiento humano.

¿Te deprime el acné? ¡La página 552 te animará!

*Quick*Lab

MATERIALS

FOR EACH STUDENT:
• ruler
• calculator

Answer to QuickLab

Slower; student explanations should include the following information: babies must be born with large heads to hold large brains, which enable them to learn quickly, and as babies grow older, their bodies begin to catch up in size.

Quiz

1. What is implantation? (It is the process by which an embryo embeds itself in the uterus.)

2. What functions does the placenta serve? (It is a two-way exchange organ that allows oxygen and nutrients to travel to the fetus from the mother and allows wastes to travel from the fetus to the mother.)

ALTERNATIVE ASSESSMENT

Writing Ask students to imagine that they have not yet been born. Have them write first-person stories describing their time in utero. Encourage creativity, but direct students to include the stages of development they went through as a fetus. Allow time for students to share their stories with the class.

▼ **Answers to Review**

1. An embryo is less developed than a fetus.

2. The membrane keeps other sperm from entering. This is important because it helps maintain the characteristic human chromosome number.

3. When a person is first born, he or she is an infant. During infancy, the teeth and nervous system develop. During childhood, beginning at age 2, physical and mental capabilities develop. During adolescence, sexual characteristics develop.

4. The astronaut's umbilical cord supplies life support for the astronaut.

Answer to Explore

Illustrations should be accurate and in the correct chronological order.

VOCABULARY DEFINITIONS

SECTION 1

asexual reproduction reproduction in which a single parent produces offspring that are identical to the parent

budding a type of asexual reproduction in which a small part of the parent's body develops into an independent organism

fragmentation a type of reproduction in which an organism breaks into two or more parts, each of which may grow into a separate individual

sexual reproduction reproduction in which two sex cells join to form a unique individual

egg sex cell produced by a female

sperm sex cell produced by a male

zygote a fertilized egg

external fertilization fertilization of an egg by sperm that occurs outside the body of the female

internal fertilization fertilization of an egg by sperm that occurs inside the body of the female

monotreme a mammal that lays eggs

marsupial a mammal that gives birth to live, partially developed young that continue to develop inside the mother's pouch or skin fold

placental mammal a mammal that nourishes its unborn offspring with a placenta inside the uterus and gives birth to well-developed young

SECTION 2

testes organs in the male reproductive system that make sperm and testosterone

scrotum a skin-covered sac that hangs from the male body and contains the testes

seminiferous tubules coiled tubes inside the testes where sperm cells are produced

epididymis the area of the testes where sperm are stored before they enter the vas deferens

vas deferens the tube in males where sperm is mixed with fluids to make semen

Resumen del capítulo

SECCIÓN 1

Vocabulario

reproducción asexual *(pág. 534)*
gemación *(pág. 534)*
fragmentación *(pág. 534)*
reproducción sexual *(pág. 535)*
óvulo *(pág. 535)*
espermatozoide *(pág. 535)*
cigoto *(pág. 535)*
fecundación externa *(pág. 536)*
fecundación interna *(pág. 536)*
monotrema *(pág. 537)*
marsupial *(pág. 537)*
mamífero placentario *(pág. 537)*

Notas de la sección

- Durante la reproducción asexual, un solo progenitor puede producir crías idénticas a sí mismo. La gemación y la fragmentación son ejemplos de reproducción asexual.

- Durante la reproducción sexual, un óvulo se une a un espermatozoide.

- Los óvulos y espermatozoides son producto de la meiosis y tienen la mitad del número normal de cromosomas. El cigoto recupera el número normal de cromosomas.

- En la fecundación externa, el espermatozoide fecunda los óvulos fuera del cuerpo de la hembra. En la interna, el espermatozoide fecunda los óvulos dentro del cuerpo de la hembra.

- Los monotremas son mamíferos que ponen huevos. Los marsupiales son mamíferos que dan a luz a crías parcialmente desarrolladas. Los placentarios son mamíferos que dan a luz a crías bien desarrolladas.

SECCIÓN 2

Vocabulario

testículos *(pág. 538)*
escroto *(pág. 538)*
tubos seminíferos *(pág. 538)*
epidídimo *(pág. 538)*
conducto deferente *(pág. 538)*
semen *(pág. 538)*
pubertad *(pág. 538)*
uretra *(pág. 538)*
pene *(pág. 538)*
ovarios *(pág. 539)*
ovulación *(pág. 539)*
trompas de Falopio *(pág. 539)*
útero *(pág. 539)*
vagina *(pág. 539)*
menstruación *(pág. 539)*
infértil *(pág. 541)*
enfermedad de transmisión sexual *(pág. 541)*

☑ Comprobar destrezas

Conceptos de matemáticas

ÓVULOS EN EXILIO Una mujer no ovula cuando está embarazada. Por lo tanto, si una mujer tiene tres hijos, liberará por lo menos 27 óvulos menos que si no se hubiera estado embarazada nunca.

> 3 hijos × 9 meses de embarazo = 27 óvulos

Comprensión visual

APARATOS REPRODUCTORES MASCULINO Y FEMENINO Los diagramas de las páginas 538 y 539 muestran los aparatos reproductores masculino y femenino. Obsérvalos otra vez y asegúrate de que reconoces todas las estructuras. También observa las similitudes entre los dos. Por ejemplo, los ovarios tienen una función similar a los testículos, y las trompas de Falopio tienen una función similar a los conductos deferentes.

Lab and Activity **Highlights**

My, How You've Grown! `PG 635`

It's a Comfy, Safe World! `PG 634`

Datasheets for LabBook
(blackline masters for these labs)

semen a mixture of sperm and fluids

puberty the time of life when the sex organs become mature

urethra in males, a slender tube that carries urine and semen through the penis to the outside

penis the male reproductive organ that transfers semen into the female's body during sexual intercourse

ovaries in animals, organs in the female reproductive system that produce eggs

ovulation the process in which a developed egg is ejected through the ovary wall

fallopian tube the tube that leads from an ovary to the uterus

uterus an organ in the female reproductive system where a zygote can grow and develop

vagina the passageway in the female reproductive system that receives sperm during sexual intercourse

menstruation the monthly discharge of blood and tissue from the uterus

infertile the state of being unable to have children

sexually transmitted disease a disease that can pass from an infected person to an uninfected person during sexual contact

Section 3

embryo an organism in the earliest stage of development

implantation the process in which an embryo imbeds itself in the lining of the uterus

placenta an organ that provides a developing baby with nutrients and oxygen from the mother

umbilical cord a cord that connects the embryo to the placenta

fetus an embryo during the later stages of development within the uterus

---

SECCIÓN 2

Notas de la sección

- El aparato reproductor masculino produce espermatozoides y los libera en el femenino. Los espermatozoides se producen en los tubos seminíferos y se almacenan en el epidídimo. Luego abandonan el cuerpo por la uretra.

- El aparato reproductor femenino produce óvulos, alimenta el embrión en desarrollo y da a luz. Cada mes, un óvulo abandona uno de los dos ovarios y pasa al útero. Si no es fecundado, se desintegra y tiene lugar la mestruación.

- Los transtornos del aparato reproductor pueden ser: infertilidad, cáncer y enfermedades de transmisión sexual.

SECCIÓN 3

Vocabulario

embrión *(pág. 542)*
implantación *(pág. 542)*
placenta *(pág. 543)*
cordón umbilical *(pág. 543)*
feto *(pág. 544)*

Notas de la sección

- La fecundación ocurre en una de las trompas de Falopio. De ahí, el cigoto viaja al útero y se implanta en su pared.

- Después de la implantación, se desarrolla la placenta. El cordón umbilical conecta el embrión con la placenta. El saco amniótico rodea y protege el embrión.

- El embrión crece, desarollando extremidades, orificios nasales, párpados y otras características. Alrededor de la octava semana, el embrión está tan desarrollado que recibe el nombre de feto.

- Las etapas de vida del ser humano son: bebé (desde que nace hasta los 2 años), niño (de 2 años a la pubertad), adolescente (de la pubertad a los 20 años), adulto joven (de 20 a 40 años), adulto maduro (de 40 a 65 años) y adulto anciano (más de 65 años).

Experimentos

¡Un mundo seguro y agradable! *(pág. 634)*

¡Oh, cuánto has crecido! *(pág. 635)*

---

internet

VISITA: go.hrw.com

Visita el sitio web de HRW para encontrar una serie de herramientas de aprendizaje relacionadas con este capítulo. Sólo tienes que escribir la palabra clave:

PALABRA CLAVE: HSTBD5

VISITA: www.scilinks.org

Visita el sitio web de la **Asociación Nacional de Maestros de Ciencias** *(National Science Teachers Association)* para encontrar recursos de Internet relacionados con este capítulo. Sólo escribe el **ENLACE DE CIENCIAS** para obtener más información sobre el tema:

TEMA:	ENLACE:
Reproducción	HSTL630
Antes del nacimiento	HSTL635
Irregularidades o transtornos del aparato reproductor	HSTL640
Crecimiento y desarollo	HSTL645

549

Lab and Activity Highlights

LabBank

Long-Term Projects & Research Ideas, Project 26

Vocabulary Review Worksheet 22

Blackline masters of these Chapter Highlights can be found in the **Study Guide.**

Chapter Review Answers

USING VOCABULARY

1. internal fertilization
2. seminiferous tubules
3. semen
4. ovulation
5. placenta

UNDERSTANDING CONCEPTS

Multiple Choice

6. a
7. d
8. c
9. d
10. c
11. d

Short Answer

12. The testes produce sperm, and the ovaries produce eggs.
13. the placenta
14. (must list 4) infancy, childhood, adolescence, young adulthood, middle-aged adulthood, older adulthood
15. an egg and a sperm
16. Budding is a type of reproduction in which a young organism develops off a small part of the parent. Fragmentation is a type of reproduction in which an organism develops from a piece of the parent.

Resumen del capítulo

UTILIZAR EL VOCABULARIO

Escoge el término correcto para completar las siguientes oraciones:

1. Los reptiles, las aves y los mamíferos se reproducen sexualmente por __?__. *(fecundación interna o fecundación externa)*

2. Lose spermatozoides se producen en __?__ dentro de los testículos. *(el epidídimo o los tubos seminíferos)*

3. El fluido del cuerpo humano que contiene los espermatozoides se conoce como __?__. *(semen o fluido amniótico)*

4. La liberación del óvulo desde los ovarios ocurre una vez al mes y se llama __?__. *(ovulación o menstruación)*

5. El órgano de intercambio entre el embrión en desarrollo y la madre es __?__. *(el saco amniótico o la placenta)*

COMPRENDER CONCEPTOS

Opción múltiple

6. La estrella de mar se reproduce asexualmente por
   a. fragmentación.
   b. gemación.
   c. fecundación externa.
   d. fecundación interna.

7. La ruta correcta del espermatozoide a través del aparato reproductor masculino es
   a. testículos → epidídimo → uretra → conducto deferente.
   b. epidídimo → uretra → testículos → conducto deferente.
   c. testículos → conducto deferente → epidídimo → uretra.
   d. testículos → epidídimo → conducto deferente → uretra.

8. Si el primer día del ciclo menstrual es el principio de la menstruación, ¿en qué día ocurre generalmente la ovulación?
   a. 2º día
   b. 5º día
   c. 14º día
   d. 28º día

9. Los monotremas son diferentes de los mamíferos placentarios porque
   a. son mamíferos.
   b. tienen pelo.
   c. alimentan a sus crías con leche.
   d. ponen huevos.

10. Todas las siguientes son enfermedades de transmisión sexual *excepto*
    a. clamidia.
    b. SIDA.
    c. infertilidad.
    d. herpes genital.

11. La fecundación ocurre en _____ y la implantación en _____.
    a. el útero, las trompas de Falopio
    b. las trompas de Falopio, la vagina
    c. el útero, la vagina
    d. las trompas de Falopio, el útero

Respuesta breve

12. ¿Qué órganos reproductores producen espermatozoides y cuáles producen óvulos?

13. ¿A través de qué estructura pasa el oxígeno de la madre al cuerpo del feto?

14. ¿Cuáles son las cuatro etapas de la vida humana tras el nacimiento?

15. ¿Qué células se combinan para crear un cigoto?

16. ¿Qué diferencia existe entre gemación y fragmentación?

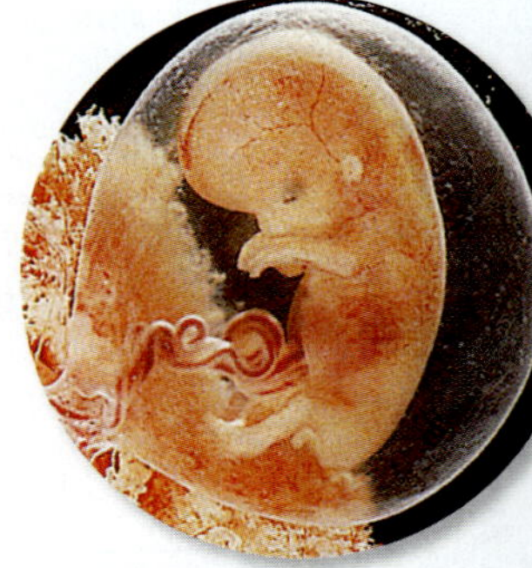

Organizar conceptos

17. Usa los siguientes términos para crear un mapa de ideas: reproducción asexual, gemación, fecundación externa, fragmentación, fecundación interna, reproducción sexual, reproducción.

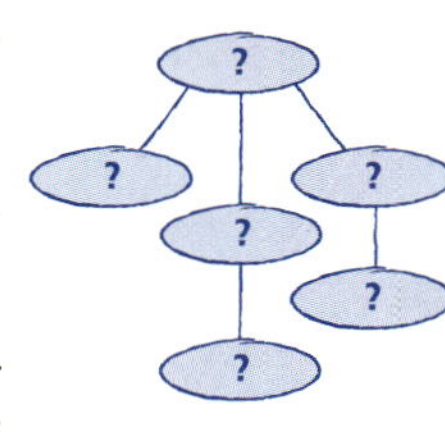

RAZONAMIENTO CRÍTICO Y RESOLUCIÓN DE PROBLEMAS

Escribe una o dos oraciones para responder a las siguientes preguntas:

18. Explica por qué los testículos se encuentran en el escroto y no dentro del cuerpo del varón.

19. ¿Cuál es la función del útero? ¿Cómo se relaciona su función con el ciclo menstrual?

20. ¿Qué importancia tiene la meiosis en la reproducción humana?

LAS MATEMÁTICAS EN LAS CIENCIAS

21. La escuela "El buen porvenir" tiene 2,750 estudiantes. Si 1 par de gemelos idénticos nace por cada 250 alumbramientos, ¿cuántos pares de gemelos idénticos asistirán a la escuela?

22. La señora Rodríguez tuvo un bebé el 30 de abril. Se desarrolló dentro del útero durante 9 meses. ¿En qué mes se fertilizó el óvulo?

23. En Estados Unidos, siete de cada 1,000 bebés mueren antes de cumplir 1 año. Convierte esta cifra a un porcentaje. ¿Es tu respuesta mayor o menor al 1%?

24. En Haití, un pequeño país del Caribe, 74 de cada 1,000 bebés mueren antes de cumplir 1 año. Convierte esta cifra a un porcentaje. ¿Es tu respuesta mayor o menor al 1%? ¿A qué crees que se debe la diferencia entre los Estados Unidos y Haití?

INTERPRETAR GRÁFICAS

La siguiente gráfica ilustra los ciclos de la hormona masculina, testosterona, y de la hormona femenina, o estrógeno. La línea azul muestra el nivel de estrógeno en una mujer durante un período de 28 días. La roja muestra el nivel de testosterona en un hombre durante un período de 28 días.

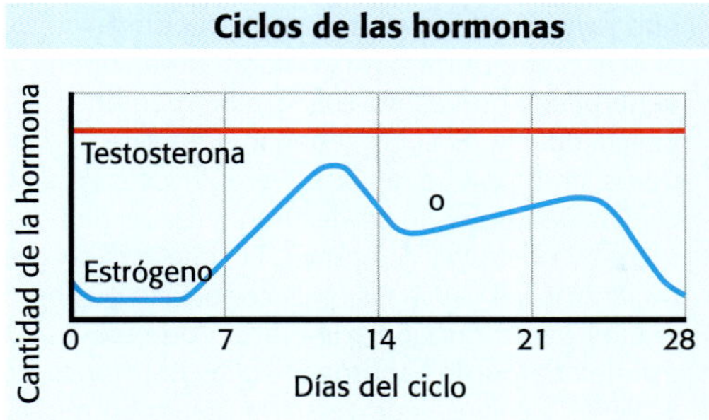

25. ¿Cuál es la principal diferencia entre los dos niveles hormonales durante el período de 28 días?

26. ¿A qué ciclo crees que afecta el estrógeno?

27. ¿Por qué el nivel de testosterona permanece igual?

AHORA, ¿qué piensas?

Revisa tus respuestas a las preguntas de la página 533 que escribiste en tu cuaderno de ciencias. ¿Han cambiado tus respuestas? Si es necesario, corrige tus respuestas basándote en lo que has aprendido en este capítulo.

551

NOW WHAT DO YOU THINK?

1. No; animals that reproduce asexually—such as sea stars—have only one parent.

2. Adults are larger and more sexually developed than children.

3. You inherit an equal number of chromosomes from both your parents.

Concept Mapping Transparency 22

Blackline masters of this Chapter Review can be found in the **Study Guide.**

Concept Mapping

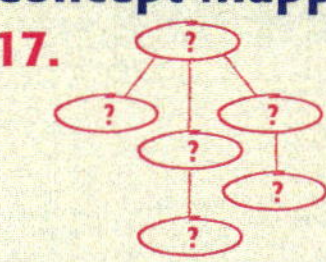

17. An answer to this exercise can be found at the end of this book.

CRITICAL THINKING AND PROBLEM SOLVING

18. Sperm develop best at a temperature three degrees lower than normal body temperature. Therefore, the testes are suspended away from the body to keep them cooler.

19. The uterus is the organ in the female reproductive system in which an embryo can develop into a fetus. Every month it builds up tissue that can help nourish a developing embryo. If no embryo is present, the tissue will discharge and cause menstruation.

20. Meiosis ensures that the characteristic human chromosome number remains the same. Since two cells must combine to form a zygote, each sex cell needs only half the 46 chromosomes that all other human body cells have.

MATH IN SCIENCE

21. 11

22. August

23. Approximately 0.7 percent of American infants die before their first birthday. This is less than 1 percent.

24. Approximately 7.4 percent of Haitian infants die before their first birthday. This is more than 1 percent. This is higher than the American percentage because Americans generally have access to better health care than Haitians.

INTERPRETING GRAPHICS

25. Estrogen levels fluctuate, but testosterone stays at the same level throughout the month.

26. Estrogen affects the menstrual cycle.

27. Testosterone levels stay the same because men continually produce sperm.

Background

This feature will provide students with a look at some of the changes affecting their body. Puberty is a dynamic time because hormone levels fluctuate from day to day. During puberty, acne will be more severe at times. Students should realize that the physical changes that cause acne are a necessary part of maturation and that acne is not necessarily caused by poor hygiene. Students may be reassured to learn that acne generally clears up after puberty, when hormone levels fluctuate less.

Most importantly, students should know that acne is a treatable condition and should understand how common treatments, such as over-the-counter topical creams and antibiotics, work.

Students may be interested to find out that the male sex hormones, called *androgens*, cause acne in both men and women. Indeed, men and women share the same hormones, only in different quantities and levels of activity.

A TRAVÉS DE LAS CIENCIAS

CIENCIAS BIOLÓGICAS • QUÍMICA

Acné

Si eres adolescente, quizá ya tengas cierta experiencia con el acné. Si no, ya la tendrás. Contrariamente a lo que has escuchado, el acné no es causado por los alimentos grasosos ni los dulces, aunque éstos pueden agravar el problema. En general, las fluctuaciones hormonales que ocurren cuando los jóvenes pasan a la edad adulta son las causantes del acné.

¿Qué son los granos?

La piel contiene miles de poros diminutos, cada uno con glándulas sebáceas que producen sebo, el aceite que quizá hayas notado en la superficie de la piel, y que es necesario para mantener la piel saludable. El sebo generalmente sale de los poros sin problema alguno. Pero, a veces, las células de la piel no se desprenden adecuadamente y obstruyen los poros. El sebo que se acumula en ellos causa lesiones conocidas como granos. La producción y liberación de sebo es estimulada por los andrógenos, las hormonas sexuales masculinas, que se activan tanto en mujeres como en hombres durante la pubertad.

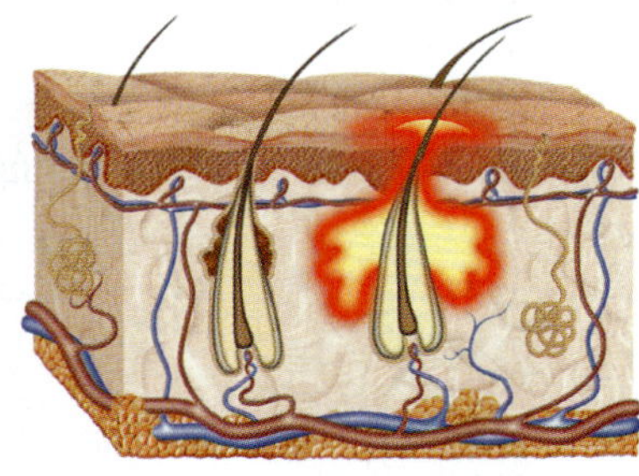

▲ *La acumulación de sebo y células muertas en los poros de la piel causan acné.*

Conoce tus lesiones

Hay dos tipos de lesiones: las no inflamatorias y las inflamatorias. Entre las lesiones no inflamatorias se encuentran las espinillas y los barros. Hay quienes piensan que las espinillas son poros llenos de suciedad. Su color obscuro es, en realidad, el resultado de pigmentos obscuros de la piel o de aceite atrapado en los poros. Los barros son blancos porque su contenido está escondido bajo la superficie de la piel. Las lesiones inflamatorias son causadas por bacterias y, a menudo, son pequeñas protuberancias rojas. Las bacterias viven en poros saludables y, cuando los poros se obstruyen, quedan atrapadas y pueden provocar irritación o infección.

Herencia

La historia familiar parece ser un factor en el desarrollo del acné. Por desgracia, si tus padres o hermanos y hermanas tuvieron acné, eres más propenso a tenerlo. Las causas del acné hereditario siguen siendo imprecisas. La piel puede estar genéticamente programada para producir más sebo del que se produce normalmente en la adolescencia.

¿Hay alguna esperanza?

Algunos productos comerciales pueden retirar las células muertas de la piel y el sebo de los poros. Muchos medicamentos impiden la producción de sebo o estimulan la exfoliación de las células de la piel. Estos tratamientos pueden ayudar a mantener los poros limpios y prevenir el acné. A veces, los médicos recetan antibióticos, como tetraciclina o eritromicina, para tratar casos severos de acné. Los antibióticos son medicamentos que matan las bacterias, como las que irritan las lesiones inflamadas. El lado bueno de esto es que la mayoría de los casos de acné desaparecen cuando se llega a la edad adulta.

Descúbrelo

▶ Descubre qué ingrediente activo se encuentra en los medicamentos comerciales para el acné. Investiga un poco cómo trabaja. Presenta tus descubrimientos al resto de la clase.

552

Answer to On Your Own

The primary ingredient in over-the-counter acne medications is benzoyl peroxide, a strong oxidizing agent that kills bacteria. Once absorbed into the skin, benzoyl peroxide is metabolized into benzoic acid, which then exits the body as benzoate through the urine. Some of the side effects of using medications with this ingredient can include different types of skin irritations, such as burning, blistering, crusting, itching, or severe redness.

Ciencia, Tecnología y Sociedad

La tecnología en sus primeras etapas

Cada año, miles de bebés nacen con enfermedades que amenazan su vida o con defectos graves de nacimiento. ¿Qué pasaría si se les aplicaran tratamientos médicos a estos bebés antes de nacer? Unos médicos de las universidades de San Francisco, Harvard y Vanderbilt practican la cirugía fetal experimental con resultados alentadores.

¿Cuándo puede hacerse cirugía fetal?

Hasta ahora, se han practicado cerca de 100 operaciones fetales en todo el país. Se pueden aplicar tratamientos correctivos entre la 18ª y la 30ª semana de embarazo. Muchos factores determinan si la cirugía fetal es apropiada o no. La cirugía se considera una opción sólo si la condición amenaza la vida del bebé. Sin embargo, los fetos con varios defectos o con anormalidades cromosómicas no son buenos candidatos para a cirugía fetal.

Se han practicado con éxito operaciones en pacientes fetales con espina bífida, hernias diafragmáticas, malformaciones de los pulmones y obstrucciones de las vías urinarias. La espina bífida es un defecto que deja la médula expuesta. La hernia diafragmática es un orificio en el diafragma que causa problemas respiratorios severos.

Cirugía a pequeña escala

La cirugía fetal entra en una de tres categorías. El tipo de tratamiento menos traumático usa un escalpelo con láser o un endoscopio. El escalpelo se usa para eliminar tumores del tórax. Un endoscopio es un dispositivo guiado por video que combina una lente de cámara y unas tijeras que miden menos de 0.2 cm de ancho. El médico guía las tijeras a través de un corte diminuto en las paredes abdominales y uterinas. No puede ver al feto directamente durante la operación porque el corte es muy pequeño. Por lo tanto, debe observar las imágenes de video proporcionadas por el endoscopio durante la operación.

Una opción más traumática es la cirugía fetal abierta. En este tratamiento, se abren el abdomen y el útero de la madre, y el feto queda parcialmente expuesto.

La tercera, y relativamente nueva opción, es el transplante de células tronco (precursoras de las células sanguíneas) al feto. Este tratamiento es esencialmente un transplante de médula ósea para el feto. Se usa para tratar enfermedades genéticas y del sistema inmunológico.

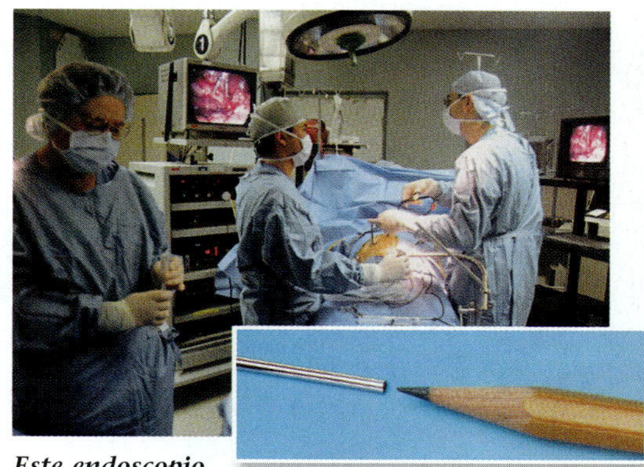

Este endoscopio se usa para practicar una cirugía fetal.

Lo que depara el futuro

Cada cirugía fetal da como resultado la mejora de técnicas y tratamientos, así como la expansión de la lista de defectos y enfermedades tratables. A medida que el número de operaciones aumente, la cirugía fetal será cada vez más común.

Profundizar

▶ Los endoscopios utilizados en la cirugía fetal usan una tecnología llamada fibra óptica. Investiga qué artículos de tu casa usan la fibra óptica.

553

Experimentos

Contenido

La exploración, la invención y la investigación son esenciales para el estudio de la ciencia. Sin embargo, estas actividades pueden ser peligrosas. Para asegurarte de que tus experimentos y exploraciones sean seguros, debes conocer las distintas pautas de seguridad.

Es posible que hayas escuchado el refrán "Más vale prevenir que lamentar". Esto es especialmente cierto en un salón de clases en donde se realizan experimentos y exploraciones. No estar informado o ser descuidado puede dar lugar a graves accidentes. No arriesgues tu seguridad ni la de los demás.

A continuación se presentan importantes pautas de seguridad para el salón de ciencias. Quizás tu maestro o maestra tenga otras pautas y consejos específicos para tu salón de clases y laboratorio. Toma el tiempo de hacer las cosas de manera segura.

¡Reglas de seguridad!

Comienza correctamente

Antes de intentar cualquier experimento en el laboratorio, pídele permiso a tu maestro o maestra. Lee los procedimientos cuidadosamente y ponles especial atención a la información de seguridad y a las notas de precaución. Si no estás seguro de lo que significa un símbolo de seguridad, averígualo o pregúntale a tu maestro o maestra. No importa que exageres cuando se trata de seguridad. Si ocurre un accidente, no importa qué insignificante lo consideres: infórmale a tu maestro o maestra de inmediato.

Símbolos de seguridad

Todos los experimentos e investigaciones que aparecen en este libro y sus hojas de trabajo incluyen importantes símbolos de seguridad para advertirte acerca de los riesgos. Debes familiarizarte con estos símbolos de modo que, cuando los veas, sepas qué significan y qué debes hacer. Es muy importante que leas toda esta sección dedicada a la seguridad para aprender acerca de los peligros que hay en un laboratorio.

Experimentos

Protección de los ojos 

Utiliza gafas de seguridad al trabajar cerca de substancias químicas, ácidos, bases o cualquier tipo de llama o dispositivo calentador. Utiliza gafas de seguridad siempre que exista hasta la más mínima posibilidad de daño a los ojos. Si cualquier substancia entra en tus ojos, avísale a tu maestro o maestra inmediatamente y limpia el ojo con agua de la llave durante al menos 15 minutos. Debes tratar cualquier substancia química desconocida como si fuera una substancia peligrosa. Nunca mires directamente al Sol, pues podrías quedarte ciego permanentemente.

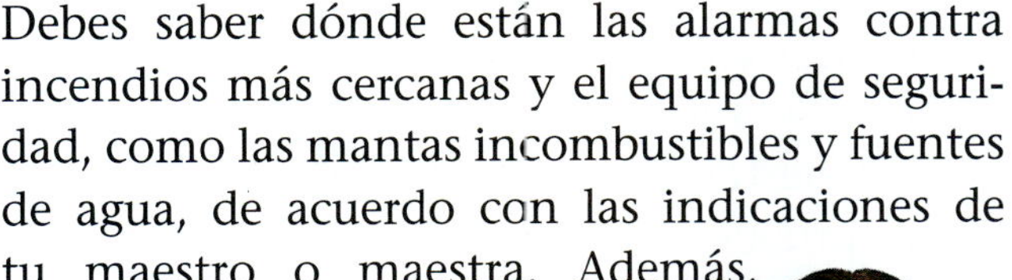

No uses lentes de contacto en un laboratorio, pues las substancias químicas pueden meterse entre los lentes de contacto y los ojos aunque uses gafas de seguridad. Si tu doctor exige que uses lentes de contacto en vez de anteojos, utiliza gafas de seguridad con ojeras en el laboratorio.

Equipo de seguridad

Debes saber dónde están las alarmas contra incendios más cercanas y el equipo de seguridad, como las mantas incombustibles y fuentes de agua, de acuerdo con las indicaciones de tu maestro o maestra. Además, debes saber cómo utilizarlos.

Debes tener especial precaución al utilizar objetos de vidrio. Al echar un objeto pesado a un cilindro graduado, debes inclinar el cilindro de modo que el objeto se deslice lentamente hasta el fondo.

Orden

Debes mantener tu área de trabajo libre de libros y papeles innecesarios. Si tienes el cabello largo, no lo dejes suelto, y fija a tu cuerpo las mangas y otras prendas de ropa sueltas, tales como corbatas o cintas. No uses joyas colgantes ni zapatos abiertos o sandalias. Nunca debes comer, beber o maquillarte en un laboratorio. Los alimentos, bebidas y cosméticos se pueden contaminar fácilmente con materiales peligrosos.

Ciertos productos para el cabello (como el espray en aerosol) son inflamables y, por lo tanto, no se deben usar al trabajar cerca de una llama. No uses espray o gel para el cabello en los días que debes ir al laboratorio.

Objetos punzantes 

Debes utilizar los cuchillos y otros instrumentos afilados con extrema precaución. Nunca cortes objetos mientras los sostienes en tus manos. Debes colocarlos sobre una superficie apropiada para cortarlos.

Experimentos

Calor 

Utiliza gafas de seguridad al usar un dispositivo calentador o una llama. Cuando sea posible, utiliza una placa calentadora eléctrica como fuente de calor en vez de una llama. Al calentar un material en un tubo de ensayo, siempre debes inclinar el tubo lejos de ti y de otras personas. Para evitar quemaduras, utiliza guantes termorresistentes cuando recibas las instrucciones.

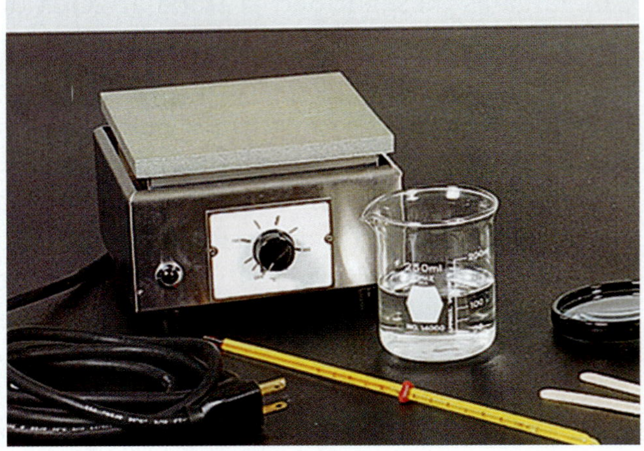

Electricidad

Ten cuidado con los cables eléctricos. Al utilizar un microscopio con lámpara, no debes dejar el cable en un lugar en donde las personas se puedan tropezar con él. No dejes los cables cerca del borde de la mesa de modo que puedan tirar el equipo al suelo accidentalmente. No utilices equipos con cables dañados. Asegúrate de tener las manos secas y que el equipo eléctrico esté apagado antes de enchufarlo. Debes apagar y desenchufar el equipo eléctrico cuando termines.

Substancias químicas    

Utiliza gafas de seguridad al manipular cualquier substancia química potencialmente peligrosa, ya sea un ácido o una base. Si desconoces una substancia química, debes manipularla como si se tratara de una substancia peligrosa. Utiliza un delantal y gafas de seguridad al trabajar con ácidos o bases. Si accidentalmente derramas una substancia sobre tu piel o ropa, lávala de inmediato con agua durante 5 minutos y avísale a tu maestro o maestra.

Nunca mezcles substancias químicas a menos de que tu maestro o maestra te lo pida. Nuca pruebes, toques o huelas las substancias químicas a menos que esto sea parte de las instrucciones. Antes de trabajar con un líquido o gas inflamable, verifica que no haya ninguna fuente de llama, chispa o calor.

Seguridad de los animales 

Siempre pídele permiso a tu maestro o maestra antes de de traer cualquier tipo de animal a la escuela. Debes manipular a los animales de acuerdo a las instrucciones de tu maestro. Siempre trátalos con cuidado y respeto y lávate bien las manos después de manipularlos.

Seguridad de las plantas

No debes comer ninguna parte de una planta o semilla de planta que se utilice en el laboratorio. Lávate bien las manos después de manipular cualquier parte de una planta. Al estar al aire libre, no recojas ningún tipo de planta silvestre a no ser que tu maestro o maestra te lo pida.

Objetos de vidrio

Examina todos los objetos de vidrio antes de usarlos. Asegúrate de que estén limpios y que no tengan roturas o fisuras. Avísale a tu maestro o maestra si encuentras cualquier tipo de daño. Los recipientes de vidrio utilizados para calentar deben ser de vidrio termorresistente.

Does It All Add Up?
Teacher's Notes

Time Required

One 45-minute class period

Lab Ratings

EASY ——→ HARD

TEACHER PREP
STUDENT SET-UP
CONCEPT LEVEL
CLEAN UP

MATERIALS

The materials listed on the student page are enough for 1 student or 1 group of students. Solution A is plain water. Solution B is either isopropyl alcohol or denatured ethyl alcohol. Safety thermometers are recommended for this lab.

Safety Caution

Remind students to review all safety cautions and icons before beginning this lab activity. Tell students that you know the properties of the liquids used in this experiment and that the liquids will not explode or cause harm when mixed. Remind students that they should never taste, touch, or smell *any* unknown chemical.

Caution students to handle mercury thermometers with care. Alcohol is flammable and poisonous. Students should wear goggles and aprons at all times during this lab. A fire extinguisher and fire blanket should be nearby. Know how to use them. The room should be well-ventilated, and students should be familiar with evacuation procedures.

¿Tiene sentido?

Tu maestro o maestra de matemáticas no te dirá esto, pero ¿sabías que a veces 2 + 2 no es igual a 4? (Bueno, en realidad sí, pero a veces no *parece* ser igual a 4.) En este experimento, utilizarás el método científico para predecir, medir y observar lo que ocurre al mezclar dos líquidos desconocidos. Aprenderás que un científico no se propone demostrar una hipótesis, sino más bien ponerla a prueba y, a veces, ¡los resultados no parecen tener sentido!

MÉTODO CIENTÍFICO

Materiales

- 50 mL de líquido A (agua)
- 50 mL de líquido B (alcohol)
- termómetro Celsius
- 5 cilindros graduados idénticos de 50 mL
- marcador para vidrio
- guantes protectores

Observa

1. Ponte los anteojos, los guantes protectores y un delantal. No te los quites durante el experimento. Examina los dos líquidos de los cilindros graduados que te dio tu maestro o maestra. **Cuidado:** No pruebes, toques o huelas ninguna substancia química desconocida.

2. Anota en el cuaderno de ciencias todas las observaciones posibles sobre cada líquido. ¿Tienen burbujas los líquidos? ¿De qué color son? ¿Cuál es el volumen exacto de cada líquido? Toca los cilindros graduados. ¿Están calientes o fríos?

3. Vierte exactamente 25 mL del líquido A en cada uno de los dos cilindros graduados. Junta las muestras en uno de los cilindros graduados y registra el volumen final en el cuaderno de ciencias. Repite el paso con el líquido B.

Formula una hipótesis

4. Basándote en tus observaciones y en tu experiencia previa, formula una hipótesis sobre cómo cambian los volúmenes cuando se combinan los líquidos.

Haz una predicción

5. Haz una predicción basada en tu hipótesis utilizando el formato "Si...entonces...". Explica por qué has hecho esa predicción.

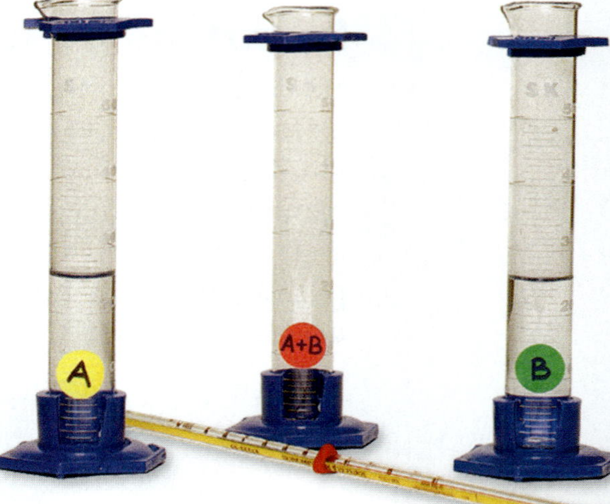

Kevin McCurdy
Elmwood Junior High School
Rogers, Arkansas

Comprueba la hipótesis

6. En tu cuaderno de ciencias, haz una tabla de datos para registrar tus observaciones como la que se muestra a continuación:

	Contenido del cilindro A	Contenido del cilindro B	Resultados de la mezcla: predicciones	Resultados de la mezcla: observaciones
Volumen				
Apariencia				
Temperatura				

7. Con cuidado, vierte exactamente 25 mL del líquido A (agua) en un cilindro de 50 mL. Marca este cilindro con la letra "A". Registra su apariencia, volumen y temperatura en la tabla de datos. Tu maestro o maestra demostrará cómo medir el volumen con un cilindro graduado, y cómo medir la temperatura con un termómetro de laboratorio.

8. Con cuidado, vierte exactamente 25 mL del líquido B (alcohol) en otro cilindro de 50 mL. Marca este cilindro con la letra "B". Registra su apariencia, volumen y temperatura en la tabla de datos.

9. Marca el tercer cilindro, que está vacío, con las letras "A + B".

10. En la columna "Resultados de la mezcla: predicciones" de tu tabla, registra la predicción que hiciste anteriormente. Es posible que cada uno de tus compañeros haya hecho una predicción diferente.

11. Con cuidado, vierte los contenidos de ambos cilindros en el tercero.

12. Observa y registra la apariencia, el volumen total y la temperatura en la columna "Resultados de la mezcla: observaciones" de la tabla.

13. Guarda los materiales como te indique tu maestro.

Analiza los resultados

14. Comenten sus predicciones en clase. ¿Cuántas predicciones diferentes hubo? ¿Qué predicciones fueron corroboradas por las pruebas? ¿Te sorprendió alguna de tus mediciones?

Saca conclusiones

15. ¿Se corroboró tu hipótesis? Si no fue así, ¿puedes explicar los datos?

16. Explica el valor de las predicciones incorrectas.

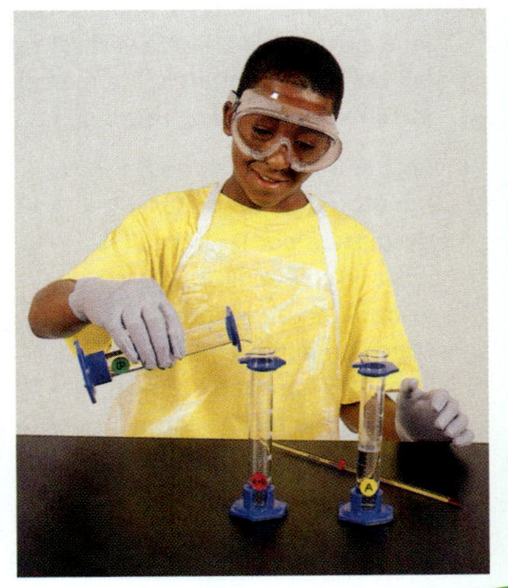

Experimentos

Lab Notes

This lab will be of interest because 25 mL of liquid A + 25 mL of liquid B do not make 50 mL of the mixture! Spaces between molecules of alcohol become filled with water molecules, resulting in less volume. An analogy would be mixing 25 mL of marbles with 10 mL of BBs. The BBs will settle between the marbles, and the result will be less volume. The alcohol-and-water mixture will be cloudy and bubbly for a brief time due to the sudden decrease of volume, leaving tiny bubbles.

Datasheets for LabBook
Datasheet 1

Science Skills
Worksheet 12
"Working with Hypotheses"

Science Skills
Worksheet 15
"Measuring"

561

Answers

3. Students should find that combining two 25 mL samples of liquid A will yield 50 mL of liquid. Combining two 25 mL samples of liquid B will also yield 50 mL of liquid. These steps will provide the controls for the experiment that combines 25 mL of liquid A and 25 mL of liquid B.

All other answers in this lab are based on student observations and will vary. Students may make some unusual predictions. You may want to lead them into questions about volume. Encourage them to think of as many ways to observe and characterize the two liquids as possible.

Time Required

One 45-minute class period

Lab Ratings

EASY ——————→ HARD

TEACHER PREP
STUDENT SET-UP
CONCEPT LEVEL
CLEAN UP

Safety Caution

Remind students to review all safety cautions and icons before beginning this lab activity.

Caution students to exercise proper care when handling the beaker of hot water. Also, caution students to be careful when they are moving around an electrical cord. A clip that will hold the thermometer to the side of the beaker and off the bottom of the beaker while it is heating or cooling is safer and more accurate than a thermometer simply propped up inside the beaker. This is also good scientific practice to model.

Datasheets for LabBook
Datasheet 2

Edith C. McAlanis
Socorro Middle School
El Paso, Texas

Hacer gráficas

Al realizar un experimento, generalmente es necesario recopilar información. Para comprender la información, es bueno organizarla en una gráfica. Las gráficas pueden mostrar tendencias y patrones que quizás no notes en una tabla o lista. En este ejercicio, practicarás la recopilación de información y su organización en una gráfica.

Procedimiento

1. Vierte 200 mL de agua en un vaso de laboratorio de 400 mL. Agrega hielo al vaso de laboratorio hasta que el agua llegue a la marca de 400 mL.

2. Coloca un termómetro Celsius en el vaso de laboratorio. Utiliza un sujetador de termómetro para evitar que éste toque el fondo del vaso. Registra la temperatura del agua con hielo en el cuaderno de ciencias.

3. Coloca el vaso y el termómetro sobre una placa calentadora. Enciende la placa a una temperatura media y, en el cuaderno de ciencias, registra la temperatura cada minuto hasta que el agua alcance los 100°C.

4. Con los guantes termorresistentes, retira el vaso de la placa calentadora y apágala. Sigue registrando la temperatura del agua cada minuto por 10 minutos más.

5. En una hoja de papel cuadriculado, haz una gráfica similar a la que se presenta a continuación. Denomina el eje horizontal (el eje de las *x*) "Tiempo (min)" y márcalo con incrementos de 1 minuto, como se indica. Denomina el eje vertical (el eje de las *y*) "Temperatura (°C)" y márcalo con incrementos de diez, como se indica.

6. Encuentra la marca de 1 minuto en el eje de las *x* y desplázate por la gráfica hasta la temperatura registrada en el minuto 1. Marca un punto en la gráfica justo ahí. Traza cada temperatura de la misma forma. Cuando hayas trazado todos los datos, conecta los puntos con una línea continua.

Materiales

- un vaso de laboratorio de 400 mL
- agua
- hielo
- termómetro Celsius con sujetador
- placa calentadora
- papel cuadriculado
- guantes termorresistentes
- reloj

Análisis

7. Examina la forma de tu gráfica. ¿Crees que el agua se calentó más rápido de lo que se enfrió? Explica.

8. Calcula qué temperatura tenía el agua 2.5 minutos después de poner el vaso sobre la placa calentadora. Explica cómo puedes calcular en forma precisa la temperatura entre las que registraste.

9. Explica por qué una gráfica a menudo proporciona más información que los mismos datos en una lista o tabla.

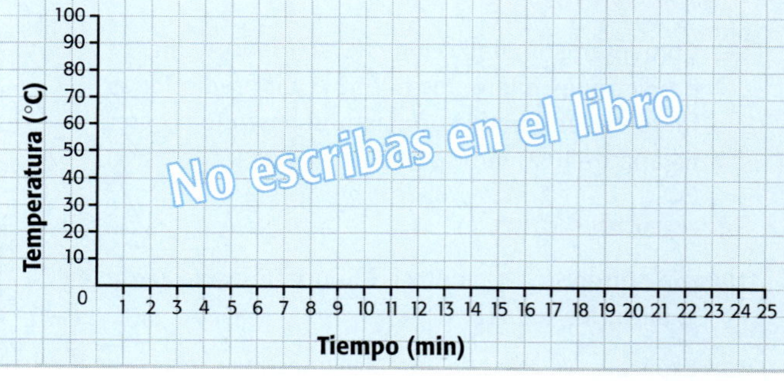

562

Answers

7. Answers will vary according to several factors, including altitude. Students should notice whether there is a gentle slope of the line (indicating gradual heating or cooling) or a steep slope (indicating rapid heating or cooling).

8. Answers will vary, but students should be able to estimate that the temperature was probably reached halfway between 2 minutes and 3 minutes.

9. A list or a chart is organized information, and sometimes it is necessary to put collected data into one of these forms before graphing. Because a graph is like a picture, it can often help students to see what is happening when numbers alone would be confusing. A graph can show a trend or a pattern that may not be readily discernible in a list or chart.

Una ventana a un mundo oculto

¿Te has fijado que los objetos bajo el agua parecen estar más cerca de lo que están en la realidad? Eso se debe a que las ondas de luz cambian de velocidad cuando van del aire al agua. Anton van Leeuwenhoek, un pionero de la microscopía de fines del siglo XVII, utilizó una gota de agua en lugar de un trozo de vidrio para amplificar objetos. Esa gota de agua nos acercó a un mundo oculto. ¿Cómo funcionaba el microscopio de Leeuwenhoek? En esta investigación, construirás un modelo para averiguarlo.

Procedimiento

1. Con una perforadora, haz un orificio en el centro del tablero de anuncios, como se muestra en la ilustración (a).

2. Con la cinta adhesiva pega un pequeño trozo de plástico transparente sobre el orificio, como se muestra en la ilustración (b). Asegúrate de que el plástico sea lo suficientemente grande como para que la cinta adhesiva que uses para pegarlo no cubra el orificio.

3. Usa un cuentagotas para colocar una gota de agua sobre el orificio. Asegúrate de que la gota de agua tenga forma de cúpula (convexa), como se ve en la ilustración (c).

4. Sostén el microscopio cerca de tu ojo y mira a través de la gota. Trata de no mover la gota de agua.

5. Sostén el microscopio sobre un trozo de papel periódico y observa la imagen.

Análisis

6. Describe y dibuja la imagen que ves. ¿Se ve la imagen más grande o de igual tamaño que sin el microscopio? ¿Es la imagen clara o borrosa? ¿Está distorsionada la forma de la imagen?

7. ¿Cómo podrías mejorar tu modelo?

Profundizar
Robert Hooke y Zacharias Janssen contribuyeron considerablemente al campo de la microscopía. Averigua quiénes fueron, en qué época vivieron y qué hicieron.

Materiales
- regla métrica
- un pedazo de un tablero de anuncios de 3 × 10 cm
- perforadora
- cuentagotas
- papel periódico
- cinta adhesiva
- rollo de plástico transparente
- agua

(a)

(b)

(c)

Experimentos

563

A Window to a Hidden World
Teacher's Notes

Time Required
One 45-minute class period

Lab Ratings

TEACHER PREP ▲▲
STUDENT SET-UP ▲▲
CONCEPT LEVEL ▲
CLEAN UP ▲

MATERIALS
The materials listed on the student page are enough for a group of 4–5 students. A 3 × 5 in. index card cut in half lengthwise can substitute for a stiff piece of poster board. It can be difficult to eliminate wrinkles in the plastic over the hole. Some students may need assistance.

Datasheets for LabBook
Datasheet 3

Science Skills Worksheet 24
"Using Models to Communicate"

Answers
6. Answers will vary. Students should see a slightly larger image. It will be blurred, especially around the edges. The image may be distorted.

7. Most students will think their model could be improved by eliminating the wrinkles over the hole.

Going Further
Robert Hooke (1635–1703), one of the world's great inventors, is famous for his discovery of "cells" in cork tissue as seen through his improved microscope. Hooke was also a keen observer with an interest in fossils and geology. Zacharias Janssen, a Dutch lens grinder, mounted two lenses in a tube to produce the first compound microscope in 1590.

Georgiann Delgadillo
East Valley School District
Continuous Curriculum School
Spokane, Washington

Roly-Poly Races
Teacher's Notes

Time Required

One or two 45-minute class periods

Lab Ratings

EASY ——————→ HARD

TEACHER PREP
STUDENT SET-UP
CONCEPT LEVEL
CLEAN UP

The materials listed on the student page are enough for 1–2 students. Remind students that they are handling living things that deserve to be treated with respect. The soil used in this lab should be sterilized potting soil to avoid causing allergic reactions among the students.

Safety Caution

Remind students to review all safety cautions and icons before beginning this lab activity.

Datasheets for LabBook
Datasheet 4

Science Skills
Worksheet 10
"Doing a Lab Write-up"

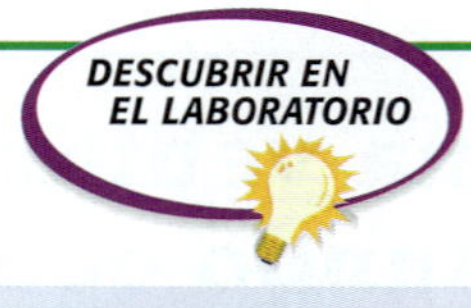

Carreras de cochinillas

¿Alguna vez has visto correr a un insecto? ¿Te preguntaste por qué corría? Probablemente corría en reacción a un estímulo. Algo lo hizo correr. Una de las características de un ser vivo es su habilidad para responder a un estímulo. En esta actividad, estudiarás el movimiento de las cochinillas de humedad, también denominadas *Armadillium vulgare.* En realidad, estos no son insectos sino crustáceos de tierra denominados isópodos. Los isópodos viven en lugares húmedos y obscuros, debajo de piedras o tablas de madera. Debes estimular al isópodo para determinar qué tan rápido se mueve y qué afecta su rapidez y dirección. Recuerda que los isópodos son seres vivos y deben ser tratados con cuidado y respeto.

Materiales

- recipiente pequeño de plástico con tapa
- 1 ó 2 cm de tierra para el recipiente
- rebanada pequeña de papa cruda
- pedazo de tiza
- 4 isópodos
- regla métrica
- cronómetro o reloj con segundero
- guantes protectores

Procedimiento

1. Con un compañero o compañera, decidan cómo llevarán a cabo su carrera de cochinillas. Discutan maneras de estimular a su isópodo para hacerlo mover. Elige cinco o seis cosas para provocar un movimiento, como la temperatura, el sonido, la luz o un leve empujón. Evalúen su decisión con su maestro o maestra.

2. En el cuaderno de ciencias, haz una tabla de datos como la que se muestra a continuación. Anota los tipos de estímulos que quieres usar en la parte superior de las columnas. Marca las filas en la izquierda de la siguiente manera: "Isópodo 1", "Isópodo 2", "Isópodo 3", "Isópodo 4".

Respuestas de los isópodos		
Estímulo 1: ?	**Estímulo 2:** ?	**Estímulo 3:** ?
Isópodo 1		
Isópodo 2		
Isópodo 3		
Isópodo 4		

No escribas en el libro

564

Preparation Notes

Isopods were selected for this lab because they are very common in most areas and can be collected and released. If you choose to use other animals, such as mealworms, that you can obtain at a pet store, be sure to have a plan for appropriate disposal after the lab.

Gladys Cherniak
St. Paul's Episcopal School
Mobile, Alabama

3. Colócate los guantes y úsalos durante el paso 7. Coloca 1 ó 2 cm de tierra en un recipiente pequeño de plástico. Agrega una pequeña rebanada de papa y un pedazo de tiza. Los isópodos utilizarán estas cosas como alimento.

4. Tu maestro o maestra te entregará cuatro isópodos. Colócalos en el recipiente y obsérvalos por unos minutos antes de realizar los experimentos. Escribe tus observaciones en el cuaderno de ciencias.

5. Tu compañero y tú elegirán dos isópodos cada uno para el experimento. Cada uno debe elegir al menos dos de los estímulos del plan que ha sido aprobado por su maestra.

6. Coloca los isópodos cuidadosamente en la "línea de partida". Esta puede ser una línea imaginaria a un lado del recipiente.

7. Prueba los métodos que elegiste para hacer mover los isópodos. Registra las respuestas provocadas por cada estímulo en la tabla de datos. Asegúrate de medir cuidadosamente las distancias que recorrieron los isópodos. Para medir los tiempos usa un cronómetro o un reloj con segundero.

Análisis

8. ¿Cómo se mueven los isópodos? ¿Se mueven sus pies simultáneamente? Describe el patrón del movimiento de los pies.

9. Cuando observaste los isópodos antes de la carrera, ¿viste algún tipo de movimiento? ¿Tenía un propósito el movimiento o era el resultado de un estímulo? Explica.

10. ¿Logró algunos de los estímulos que el isópodo se moviera más rápido o llegara más lejos? ¿Afectó alguna combinación de estímulos el movimiento de los isópodos?

Profundizar

Si bien los insectos no corren por placer como los humanos u otros mamíferos, nosotros, al igual que todos los seres vivos, también reaccionamos a los estímulos. Describe tres estímulos que harán correr a una persona.

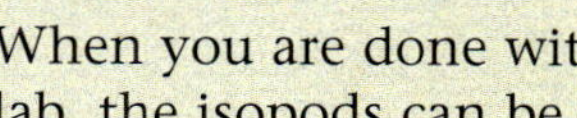

Experimentos

Disposal Information

When you are done with the lab, the isopods can be disposed of according to your policy or released into a natural area.

Answers

8. Isopods move their legs in sequence.

9. Answers will be based on students' observations and will vary.

10. Answers will vary.

Going Further

Answers will vary but might include the stimuli of fear, being late, joy, and a need for exercise.

The Best-Bread Bakery Dilemma
Teacher's Notes

Time Required

Two 45-minute class periods

Lab Ratings

EASY ——————————→ HARD

TEACHER PREP   ▲▲
STUDENT SET-UP   ▲
CONCEPT LEVEL   ▲
CLEAN UP   ▲

MATERIALS

The materials listed on the student page are enough for a group of 3–4 students. Yeast is easily obtained from the local grocery store. The school cafeteria may be willing to donate the amount you need.

You may wish to add other materials in anticipation of students' experimental design. For example, some students may recognize that they could collect CO_2 in a balloon attached to the top of a test tube containing live yeast.

Safety Caution

Remind students to review all safety cautions and icons before beginning this lab activity.

Caution students to be careful of the hot plate and the cord. You should demonstrate the proper laboratory technique for determining the presence of an odor. Hold the container away from your face about 25 cm and just below your nose. Use the other hand to "waft" the odor toward your face. Caution students NEVER to put their noses directly in a container and inhale.

Experimentos

El dilema de la panadería "Rico Pan"

El dueño de la panadería "Rico Pan" cree que la levadura que recibieron está muerta. La levadura es un ingrediente del pan. Se podrían perder miles de dólares si la levadura estuviera muerta. Son organismos vivos, miembros del reino de los Hongos y experimentan los mismos procesos vitales que los demás organismos vivos. Cuando se desarrollan en presencia del oxígeno y otros nutrientes producen dióxido de carbono que se presenta en forma de burbujas y hace que la masa suba.

La panadería "Rico Pan" te ha pedido que compruebes si la levadura está muerta o viva. Has recibido muestras de levadura viva y algunas muestras de la levadura que se sospecha que está muerta. Tu maestro o maestra te dará los instrumentos necesarios para que realices tu prueba.

Materiales

- muestras de levadura (viva, A y B)
- agua
- azúcar
- harina
- lupa
- tubos de ensayo o vasos plásticos
- soporte para tubos de ensayo
- 3 palos de madera para revolver
- cilindro graduado
- placa calentadora
- paleta o cuchara pequeña
- vaso de laboratorio de 250 mL
- termómetro Celsius con sujetador

Procedimiento

1. Haz una tabla de datos similar a la que se presenta a continuación. Deja suficiente espacio para anotar tus observaciones.

2. Ponte anteojos, guantes y un delantal. No te los quites durante el experimento. Examina todas las muestras de levadura bajo una lupa. Quizás quieras oler las muestras para detectar la presencia de algún olor. (Tu maestro te demostrará la manera apropiada para detectar olores en un laboratorio.) Escribe tus observaciones en la tabla de datos bajo la columna denominada "Observaciones".

3. Rotula tres contenedores (tubos de ensayo o vasos plásticos) de la siguiente manera: "Levadura viva", "Muestra de levadura A" y "Muestra de levadura B".

4. Llena un vaso de laboratorio de 250 mL con 125 mL de agua y colócalo sobre una placa calentadora. Utiliza un termómetro para asegurarte que el agua no exceda 32°C. Coloca el termómetro a un lado del vaso con un sujetador de modo que el termómetro no toque el fondo del vaso. Apaga la placa calentadora cuando alcance una temperatura de 32°C.

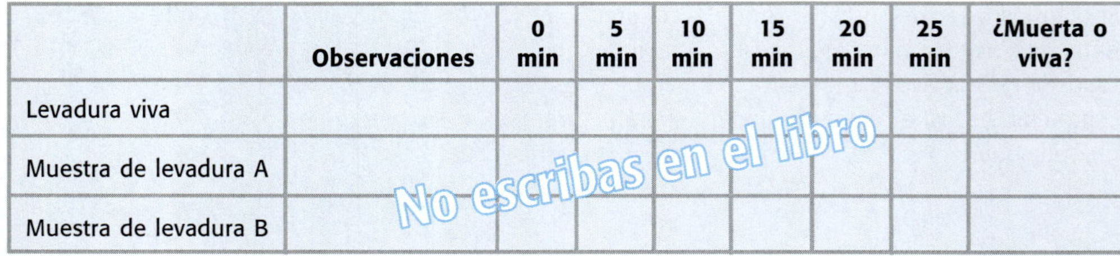

	Observaciones	0 min	5 min	10 min	15 min	20 min	25 min	¿Muerta o viva?
Levadura viva								
Muestra de levadura A								
Muestra de levadura B								

566

Datasheets for LabBook
Datasheet 5

Science Skills
Worksheet 22
"Science Writing"

Susan Gorman
North Ridge Middle School
North Richland Hills, Texas

5. Agrega una cantidad (aproximadamente 1/2 cucharadita) de cada muestra de levadura al recipiente apropiado. Agrega una pequeña cantidad de azúcar a cada recipiente.

6. Agrega 10 mL de agua precalentada a cada recipiente, utilizando un probeta y revuelve con un palo.

7. Agrega una pequeña cantidad de harina a cada recipiente y revuelve.

8. Observa las muestras y la formación de burbujas. Realiza observaciones cada 5 minutos. Escribe tus observaciones en la tabla de datos debajo de la columna de tiempo.

9. En la última columna de la tabla de datos, anota "viva" o "muerta" en función de tus observaciones durante el experimento.

Análisis

10. Enumera y explica cualquier diferencia en las muestras de levadura antes del experimento.

11. Describe el aspecto de las muestras de levadura al finalizar el experimento.

12. ¿Por qué se incluyó una muestra de levadura viva?

13. ¿Por qué se agregó azúcar a la muestra?

14. Según tus observaciones, ¿qué muestra o muestras de levadura están vivas?

15. Escríbele una carta a la panadería "Rico Pan" especificando si pueden o no utilizar las muestras de levadura. Explica tu recomendación.

Profundizar

Diseña un experimento en donde cambies la cantidad de nutrientes o examines distintas fuentes de energía.

Preparation Notes

At least one of the suspect samples should be killed yeast. To do this, place the yeast in an oven at 400°F for 10 minutes or in a microwave oven for a few minutes on high. Do not allow yeast to become moist before use. Toothpicks, coffee stirrers, etc., may be used for stirring. The amounts of each ingredient used are not definite, and you may wish to vary amounts, depending on the results desired.

Lab Notes

To help students prepare for this activity, you may wish to review cellular respiration and fermentation. The equation for respiration is:

$$C_6H_{12}O_6 + 6O_2 \rightarrow 6CO_2 + 6H_2O + energy$$

Answers

10. Answers are based on students' observations and will vary.

11. Answers will vary.

12. Live yeast is included so students can observe bubble formation from the respiration of living organisms.

13. Sugar is added as a nutrient for the living yeast.

14. Answers will vary according to students' experimental protocol.

Mixing Colors
Teacher's Notes

Time Required

One or two 45-minute class periods

Lab Ratings

EASY — HARD

TEACHER PREP — 🧪🧪
STUDENT SET-UP — 🧪🧪
CONCEPT LEVEL — 🧪🧪
CLEAN UP — 🧪🧪

MATERIALS

Materials listed are for each group of 2–4 students. If a sufficient number of flashlights is not available, consider using the spotlights on the school stage or portable floodlight holders instead. Each group should have a set of watercolors that includes red, blue, and green.

Safety Caution

Students should wear aprons when doing Part B of this lab.

Answers

6. Mixing two colors of light together results in a color that is brighter than the original colors.

7. Mixing three colors of light results in a color that is much brighter than the color produced by mixing two colors because more wavelengths are present.

8. The result would be bright, white light because all the wavelengths of light would be combined.

Experimentos

Mezclar colores

Cuando mezclas dos colores, como el rojo y el verde, creas un color diferente. Pero, ¿qué color es? ¿Será un color más claro o más obscuro? El color y la claridad que ves depende de la luz que llega a tus ojos, la cual depende de si estás realizando una adición de color (combinando longitudes de onda al mezclar los colores de la luz) o substracción de color (absorbiendo luz al mezclar colores de los pigmentos). En este experimento tratarás de hacer ambos tipos de formación de colores y verás los resultados ien persona!

Materiales

Parte A

- 3 linternas
- filtros de color rojo, verde y azul
- cinta adhesiva
- papel blanco

Parte B

- cinta adhesiva
- 2 vasos pequeños de papel o plástico
- agua
- brocha
- acuarelas
- papel blanco
- regla métrica

Parte A: Adición de color

Procedimiento

1. Coloca un filtro de color sobre cada lente de linterna. Usa cinta adhesiva para evitar que se muevan los filtros.

2. En un cuarto obscuro, dirige la luz roja sobre una hoja de papel blanco. Luego dirige la luz azul al lado de la luz roja. Verás dos círculos de luces, uno rojo y otro azul, uno al lado del otro.

3. Mueve las linternas para que los círculos se superpongan en la mitad de su diámetro. Examina las tres áreas de color y anota tus observaciones. ¿Qué color se formó en el área mezclada? ¿Es el área mezclada más clara o más obscura que las áreas que tienen un solo color?

4. Repite los pasos 2 y 3 con las luces roja y verde.

5. Ahora dirige las tres luces al mismo punto de la hoja de papel. Examina los resultados y anota tus observaciones.

Análisis

6. En general, cuando mezclaste dos colores, ¿fue el resultado más claro o más obscuro que los colores originales?

7. En el paso 5, mezclaste los tres colores. ¿Fue el color resultante más claro o más obscuro que al mezclar dos colores? Explica tu respuesta en términos de adición de color. (Pista: Lee la definición de adición de color en la introducción.)

8. Según tus resultados, ¿qué piensas que sucedería si mezclaras todos los colores de la luz? Explica.

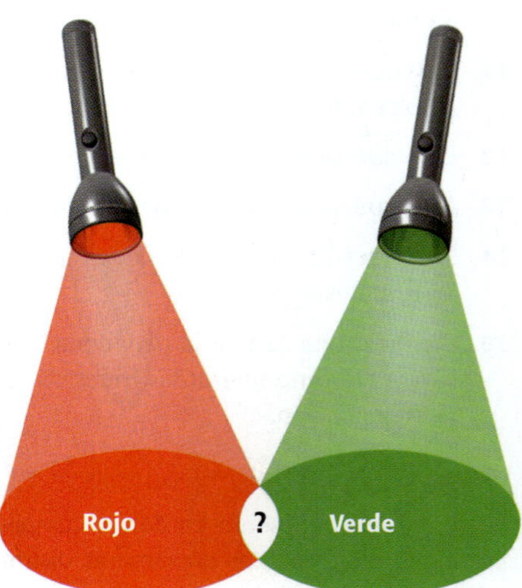

Datasheets for LabBook
Datasheet 6

Barry Bishop
San Rafael Junior High
Ferron, Utah

Parte B: Substracción de color

Procedimiento

9. Coloca un trozo de cinta adhesiva en cada vaso. Rotula un vaso como "Limpio" y el otro como "Sucio". Llena con agua cada vaso aproximadamente hasta la mitad.

10. Humedece bien la brocha en el vaso "Limpio". Con las acuarelas, pinta un círculo rojo con un diámetro aproximado de 4 cm en el papel blanco.

11. Limpia la brocha enjuagándola primero en el vaso "Sucio" y después en el "Limpio".

12. Pinta un círculo azul junto al círculo rojo. Después pinta la mitad del círculo rojo con la pintura azul.

13. Examina las tres áreas: la roja, la azul y la mezclada. ¿De qué color es el área mezclada? ¿Parece más clara u obscura que las áreas roja y azul? Anota tus observaciones en el cuaderno de ciencias.

14. Limpia la brocha. Pinta un círculo verde de 4 cm de diámetro y luego pinta la mitad del círculo azul con pintura verde.

15. Examina las áreas verde, azul y mezclada. Anota tus observaciones.

16. Ahora, agrega pintura verde al área mezclada con rojo y azul, para que tengas un área mezclada con los tres colores. Limpia la brocha.

17. Anota tus observaciones de esta otra área mezclada.

Análisis

18. En general, cuando mezclaste dos colores, ¿fue el resultado más claro o más obscuro que los colores originales?

19. En el paso 16, mezclaste los tres colores. ¿El color resultante fue más claro o más obscuro que al mezclar dos colores? Explica tu respuesta en términos de substracción de color. (Pista: Lee la definición de adición de color en la introducción.)

20. Según tus resultados, ¿qué piensas que sucedería si mezclaras todos los colores de pintura? Explica.

569

Experimentos

Procedure Notes

For further reinforcement in Part B, students can continue to mix colors to confirm their findings about color subtraction (provided their watercolor sets include more than the three required colors).

Answers

18. Mixing two colors of paint together results in a color that is darker than the original colors.

19. Mixing three colors of paint results in a color that is darker than the color that results from mixing two colors. Because each color of paint absorbs some light, colors that have been mixed together absorb even more light.

20. If you mixed all the colors of paint, all colors of light would be absorbed, and a black spot would result.

Disposal Information

Have plenty of paper towels on hand to wipe up water and paint spills. Make sure students clean their brushes thoroughly. Students should use soap and water to clean any water color smudges off their lab tables.

Elephant-Sized Amoebas?
Teacher's Notes

Time Required

Two 45-minute class periods

Lab Ratings

EASY → HARD

TEACHER PREP ▲▲
STUDENT SET-UP ▲▲
CONCEPT LEVEL ▲▲▲
CLEAN UP ▲

Safety Caution

Remind students to review all safety cautions and icons before beginning this lab activity.

Preparation Notes

Some students may find it difficult to work with a nonspecific unit of measurement. If so, the cube models easily convert to centimeters. You may want to add some small items, such as peas, beans, popcorn, or peppercorns, to the sand to represent organelles floating in the cytoplasm. Some students may need to review what a ratio is and how ratios are used.

Experimentos

¿Amibas del tamaño de un elefante?

¿Por qué las amibas no pueden ser tan grandes como un elefante? Porque son organismos unicelulares. Las amibas, como la mayoría de las células, son microscópicas. En realidad, si una amiba alcanzara el tamaño de una moneda de 25 centavos, se moriría de hambre. Para que veas que es cierto, haz un modelo de una célula y compruébalo.

Materiales

- modelos celulares en forma de cubo
- cartulina o cartón
- tijeras
- cinta
- báscula o balanza
- arena fina

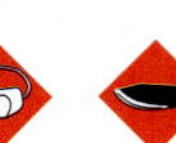
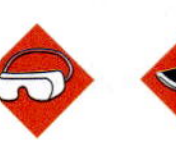

Procedimiento

1. Con cartulina, haz cuatro modelos de célula en forma de cubo con los moldes que te dé tu maestra. Recorta cada modelo de célula, dobla los lados para hacer un cubo y pega las pestañas en los lados. La célula más pequeña tiene lados que equivalen a una unidad de largo, la otra tiene lados de dos unidades, la siguiente tiene lados de tres unidades y la más grande tiene lados de cuatro unidades. Pega la parte de arriba de cada célula de manera que se pueda abrir más tarde. Estos modelos representan la membrana celular, o sea la parte exterior de la célula por la que pasan los alimentos y los desechos.

Modelo celular de dos unidades

2. Haz una tabla en tu cuaderno de ciencias como la Tabla de datos de medidas de la derecha. Utiliza las fórmulas para calcular los datos de tus modelos celulares. En la siguiente página están las equivalencias de los símbolos de las fórmulas. Anota tus cálculos en la tabla. Los cálculos de la célula más pequeña ya están hechos.

Tabla de datos de medidas				
Longitud del lado	Área de un lado ($A = L \times L$)	Área superficial total del cubo ($AT = L \times L \times 6$)	Volumen del cubo ($V = L \times L \times L$)	Masa del cubo
1	1 unidad^2	6 unidades^2	1 unidad^3	
2				
3				
4				

Data Table for Measurements

Length of side S	Area of one side (square unit)	Total surface area of cube cell (square units)	Volume of cube cell (cubic units)	Mass of cube cell (approximate, in grams)
1	1	6	1	4.5
2	4	24	8	30
3	9	54	27	105
4	16	96	64	230

3. Despega la parte de arriba del cubo de cada modelo. Con cuidado, llena por completo los modelos con arena. Averigua la masa de los modelos usando una báscula o una balanza. ¿Qué representa la arena en tu modelo?

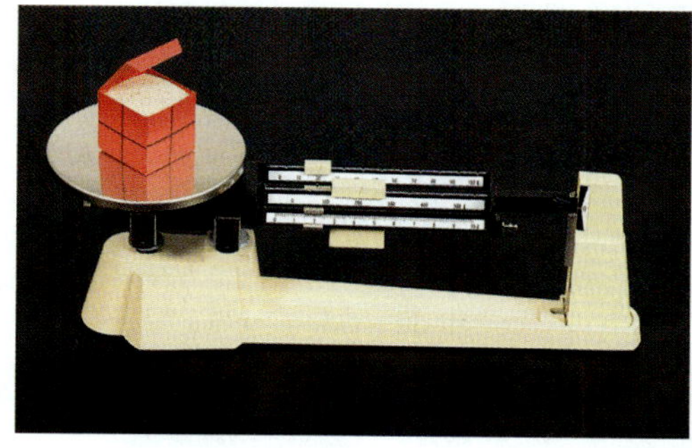

4. Anota la masa de cada modelo en la tabla. (Utiliza las unidades de masa correctas.)

5. Haz una tabla de datos en tu cuaderno de ciencias como ésta.

Tabla de datos de relaciones

Longitud del lado	Relación (razón) del área de superficie al volumen	Relación (razón) del área de superficie a la masa
1		
2		
3		
4		

No escribas en el libro

6. Averigua todas las relaciones (razones) de tus modelos localizandolos datos en las tablas. Por ejemplo, el área de superficie total y el volumen se encuentran en la Tabla de datos de medidas. En la Tabla de datos de relaciones anota las razones de cada modelo.

Análisis

7. A medida que la célula crece, ¿qué sucede con la relación entre el área de superficie total y el volumen? ¿Aumenta, disminuye o queda igual?

8. ¿Qué membrana posee mejor capacidad de suministrar alimento al citoplasma: la de una célula pequeña o la de una grande? Explica tu respuesta.

9. A medida que la célula crece, ¿qué sucede con la relación entre el área de superficie total y la masa? ¿Aumenta, disminuye o queda igual?

10. ¿Qué membrana alimenta mejor al citoplasma: la de una célula con masa alta o la de una con masa baja? Explica tu respuesta frente a tus compañeros o escribe un informe e ilústralo con los dibujos de tus modelos.

Tabla de equivalencia de símbolos

L = la longitud de un lado
A = área
V = volumen
AT = área total

Cell Model Template

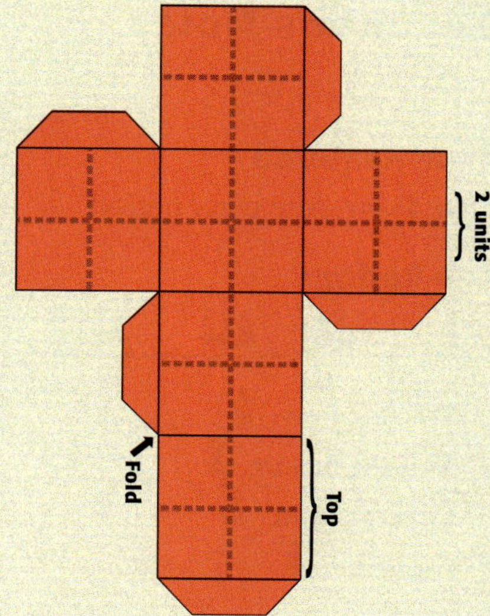

Using the template above, prepare four patterns for students to use to make their cubes. Make one cube 1 unit wide, one cube 2 units wide, one cube 3 units wide, and one cube 4 units wide. The unit can be the size of your choosing.

Answers

3. The sand represents cytoplasm.

4. Masses will vary.

6. See the tables below.

7. decrease

8. A small cell has a higher surface-area-to-volume ratio than a large cell has, allowing more nutrients per cubic unit of volume to enter.

9. decrease

10. low mass

Data Table for Ratios

Length of side S	Total surface area/volume ratio	Total surface area/mass ratio
1	$\frac{6}{1} = 6$	$\frac{6}{4.5} = 1.33$
2	$\frac{24}{8} = 3$	$\frac{24}{30} = 0.80$
3	$\frac{54}{27} = 2$	$\frac{54}{105} = 0.51$
4	$\frac{96}{64} = 1.5$	$\frac{96}{230} = 0.42$

Datasheets for LabBook
Datasheet 7

Cells Alive!
Teacher's Notes

Time Required

One 45-minute class period

Lab Ratings

EASY ———————→ HARD

TEACHER PREP   ♦
STUDENT SET-UP   ♦
CONCEPT LEVEL   ♦
CLEAN UP   ♦

MATERIALS

The materials listed on the student page are enough for a group of 3–4 students. You may wish to collect the algae ahead of time so they are available for students when they begin the lab. Be sure to keep the algae in a warm, damp place out of direct sunlight; a closed plastic bag with water sprayed into it is ideal.

Safety Caution

Remind students to review all safety cautions and icons before beginning this lab activity.

Datasheets for LabBook
Datasheet 8

Terry Rakes
Elmwood Junior High School
Rogers, Arkansas

Experimentos

¡Células vivas!

Seguramente ya has usado un microscopio para observar organismos unicelulares como estos, que por lo general se encuentran en los estanques. En este ejercicio vas a observar algas *Protococcus* que forman una mancha verdosa en troncos de árboles, cercas de madera, macetas y edificios o casas.

Materiales

- algas *Protococcus* (o de otro tipo)
- microscopio
- agua
- cubreobjetos y portaobjetos para el microscopio

Euglena

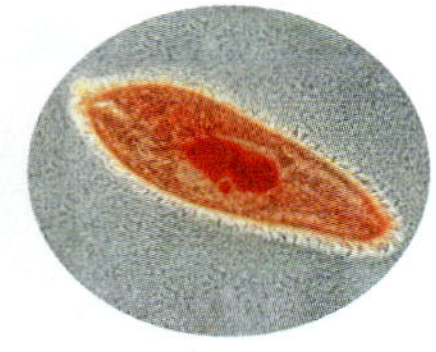
Ameba

Paramecio

Procedimiento

1. Localiza algunas algas *Protococcus*. Coloca una pequeña muestra en un recipiente y tráela al salón de clase. Mezcla la muestra con una gota de agua y coloca un poco en un portaobjetos, como te lo indica la maestra. Si no encuentras un *Protococcus*, busca otro tipo de alga en un acuario; tal vez no sea un *Protococcus* pero será un buen substituto.

2. Utiliza el lente de bajo poder para examinar el alga. Dibuja en tu cuaderno de ciencias las células que ves.

3. Ahora usa el lente de alto poder y examina una sola célula. Dibuja en tu cuaderno de ciencias la célula que ves.

4. Tal vez observes que cada célula contiene varios cloroplastos. Marca un cloroplasto en tu dibujo. ¿Cuál es la función del cloroplasto?

5. En todas las células de algas se debe ver el núcleo. Encuéntralo en una de tus muestras y márcalo en tu dibujo. ¿Cuál es la función del núcleo?

6. ¿Cómo es el citoplasma? Describe cualquier movimiento que veas dentro de las células.

Análisis

7. ¿Son los *Protococcus* organismos unicelulares o multicelulares?

8. ¿Cuál es la diferencia entre los *Protococcus* y las amebas?

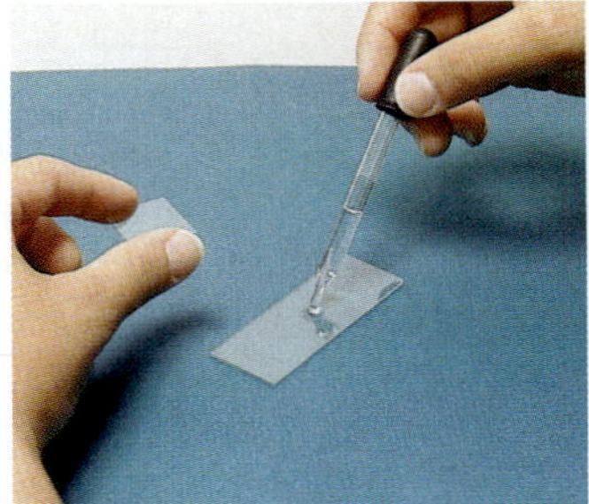

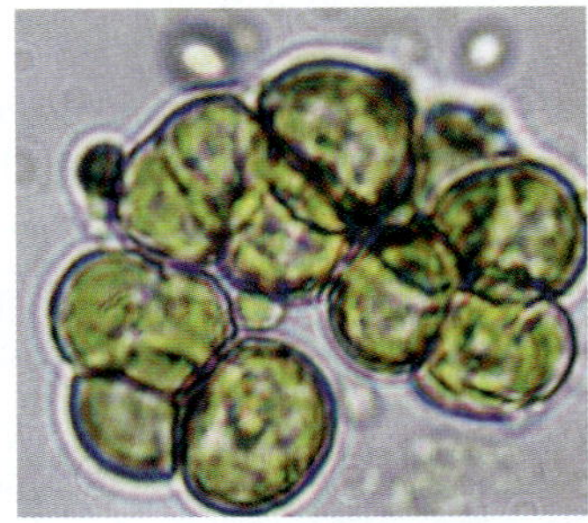
Protococcus

Answers

4. Chloroplasts are the parts of the cell that are responsible for photosynthesis.

5. The nucleus of a cell controls most of the activities that take place in that cell and contains the hereditary information.

6. The cytoplasm is a clear gel-like substance that fills the cell and surrounds the organelles.

7. *Protococcus* is a genus composed of single-celled algae.

8. Many answers are possible, but the following are most likely: *Protococcus* cannot move about like amoebas can; unlike amoebas, they are green and photosynthesize.

¿Cómo se llama esa parte?

Las células animales y las células vegetales tienen en común muchos organelos y otras partes. Por ejemplo, los dos tipos de células tienen un núcleo y mitocondrias; pero también tienen sus diferencias. En este ejercicio, vas a investigar las semejanzas y las diferencias entre las células animales y las células vegetales.

Materiales

- lápices de colores o marcadores
- hojas blancas de papel

Procedimiento

1. Dibuja en una hoja de papel blanco las células vegetal y animal que ves abajo, utilizando lápices de colores o marcadores. Dibuja cada célula en una hoja diferente. Usa diferentes colores para cada organelo.

2. Marca y nombra las partes de las dos células.

3. Debajo de cada dibujo, haz una lista de todas las partes que marcaste y describe su función.

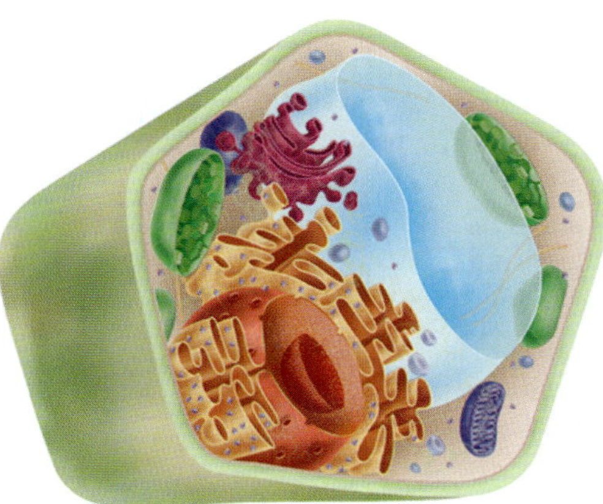

Célula vegetal

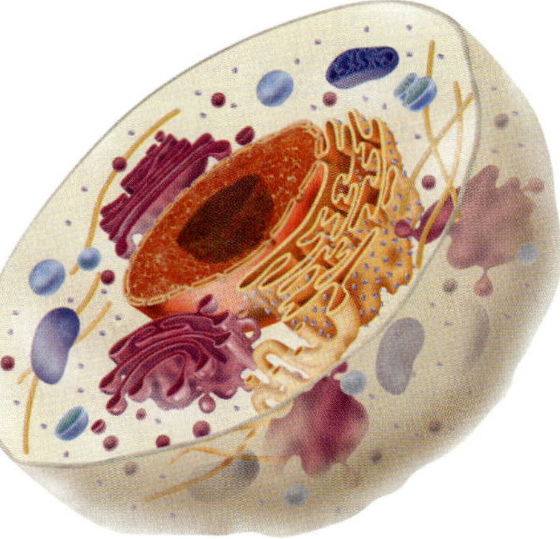

Célula animal

Análisis

4. Menciona por lo menos cuatro estructuras que las células vegetales y las células animales tienen en común.

5. Menciona tres estructuras que posean las células vegetales pero no las animales.

573

Martha Kisiah
Fairview Middle School
Tallahassee, Florida

Datasheets for LabBook
Datasheet 9

Name That Part!
Teacher's Notes

Time Required

One 40-minute class period

Lab Ratings

TEACHER PREP ⚗
STUDENT SET-UP ⚗
CONCEPT LEVEL ⚗
CLEAN UP ⚗

Answers

2. Have students label the following parts: Golgi complex, cytoplasm, nucleus, nucleolus, nuclear membrane, cell membrane, mitochondria, endoplasmic reticulum, vacuole, cell wall, chloroplast, lysosome.

3. **Golgi complex**—area that stores and packages chemicals
   cytoplasm—materials between nucleus and cell membrane
   nucleus—control center containing genetic information
   nucleolus—spherical body in the nucleus
   nuclear membrane—membrane surrounding the nucleus
   cell membrane—membrane surrounding the cytoplasm and the organelles
   mitochondria—releases energy from nutrients
   endoplasmic reticulum—location of ribosomal attachment and part of the cell's internal transport system
   vacuole—bubblelike storage structure
   cell wall—stiff outer covering of a plant cell
   chloroplast—plastid that stores chlorophyll used in photosynthesis
   lysosome—digests large particles

4. Any four structures are correct except the following: centriole, vacuole, cell wall, chloroplast.

5. vacuole, cell wall, chloroplast

The Perfect Taters Mystery
Teacher's Notes

Time Required
Two 45-minute class periods

Lab Ratings

EASY ——————— HARD

TEACHER PREP
STUDENT SET-UP
CONCEPT LEVEL
CLEAN UP

MATERIALS
The materials listed on the student page are enough for 1 class of students. You will need 1 or 2 small potatoes per class. Do not allow students to cut or peel potatoes. You will need to do this ahead of time. Allow students to choose the number of containers they will need for the experiment. They may wish to test several salt concentrations.

Safety Caution
Remind students to review all safety cautions and icons before beginning this lab activity.

Avoid including green or discolored parts of the potato in the pieces students work with. These could cause illness.

Susan Gorman
North Ridge Middle School
North Richland Hills, Texas

Experimentos

El misterio de las papas perfectas

Imagínate que eres el detective en jefe de una compañía de alimentos. El dueño, el Sr. Patatín, te pide que encuentres la manera de mantener las papas frescas y crujientes antes de su cocimiento. Sus empleados ya han probado varios métodos, pero ninguno ha funcionado. Los trabajadores del grupo A pusieron las papas en agua muy salada y sucedió algo inesperado; los del grupo B las pusieron en agua sin sal y ¡pasó otra cosa!; finalmente, el grupo C no puso las papas en agua, pero esto tampoco funcionó. Diseña un experimento para encontrar la manera de que las papas permanezcan frescas y crujientes.

Materiales
Por salón:
- muestras de papa (A, B y C)
- pedazos de papa recién cortados
- 1 caja de sal
- vasos pequeños de plástico transparente
- 4 L de agua destilada

1. Antes de planear el experimento, repasa lo que sabes. Las papas están hechas de células. Las células vegetales contienen una gran cantidad de agua. Las células tienen membranas que mantienen el agua y otras substancias en su interior, y que impiden la entrada de otras. El agua y otras substancias deben pasar a través de la membrana celular para entrar y salir de la célula.

2. El Sr. Patatín te ha dicho que pidieras a los trabajadores de los grupos A, B y C las muestras que necesitaras. Tu maestro o maestra las tienen listas para que las observes.

3. Para anotar tus observaciones, en tu cuaderno de ciencias haz una tabla de datos como la de abajo. Haz todas las observaciones que puedas acerca de las papas que probaron los trabajadores del grupo A, B y C.

Observaciones	
Grupo A:	
Grupo B:	
Grupo C:	

No escribas en el libro

Haz una pregunta

4. Después de hacer tus observaciones, plantea el problema del Sr. Patatín en forma de una pregunta que tu experimento pueda contestar.

574

Lab Notes

Osmosis is often a confusing and misunderstood concept in life science. Quite often, students can repeat the definition of the process but are unable to apply the concept to explain the movement of water in different osmotic environments. In this lab, students will have an opportunity to observe osmosis in a model and obtain measurable results. This lab can be done as a class demonstration if materials and space are limited. The purpose of this lab is to reinforce comprehension of osmosis and to practice the scientific method.

Formula una hipótesis

5. Formula una hipótesis basada en tus observaciones y preguntas. La hipótesis debe mencionar lo que hace que las papas se resequen o se hinchen. Básate en la hipótesis para predecir el resultado de los experimentos. Utiliza el formato "si…, entonces…" para hacer la predicción.

Comprueba la hipótesis

6. Después de hacer la predicción, planea la investigación. Revisa tu plan experimental con tu maestro o maestra antes de empezar. El Sr. Patatín te dará pedazos de papa, agua, sal y un máximo de seis recipientes.

7. Elabora registros tan precisos como te sea posible. Redacta el plan y el procedimiento. Haz tablas de datos. Para estar seguro de los datos, mide cuidadosamente el material y dibuja los pedazos de papa antes y después del experimento.

Saca conclusiones

8. Explica lo que les pasó a las células de papa de los grupos A, B y C de tu experimento. Haz un comentario sobre la membrana celular y la ósmosis.

Comunica los resultados

Escríbele una carta al Sr. Patatín para explicarle tu método experimental, tus resultados y tus conclusiones. Después, haz una recomendación sobre la manera en que sus empleados deben manipular las papas para que se mantengan frescas y crujientes.

Datasheets for LabBook
Datasheet 10

Science Skills
Worksheet 13
"Designing an Experiment"

Experimentos

Answers

8. The potato cells in group A were placed in very salty water. The potatoes shriveled up because water moved out of the cell and into the salty water (from an area of high concentration of water to an area of low concentration of water). This may be confusing to some students, who may think that because the concentration of salt is high outside the potato, the salt should move to the area of lower concentration. Explain that although water can move through a cell membrane by osmosis, salt must be moved across a cell membrane by a process that requires energy.

The potato cells in group B were placed in water with no salt. The potatoes swelled because the concentration of water was lower inside the cell. (The concentration of salt and other molecules was higher inside the potato cell.)

The potato cells in group C turned brown and dried up because the water concentration outside the cell was low. In fact, there wasn't any water at all. The water evaporated as soon as it left the cell membrane. The potato cells turned brown because of chemical reactions with the air.

9. Letters to Mr. Fries will vary according to each student's results. However, all students should explain that through trial and error they found one salt concentration that was closest to the concentration of salt and other molecules inside the potato. This is the concentration that should be used to maintain an osmotic balance in the potato. Furthermore, some students will realize that the potatoes must be kept in water to prevent them from turning brown.

Stayin' Alive!
Teacher's Notes

Time Required

One 45-minute class period

Lab Ratings

EASY ——————————→ HARD

TEACHER PREP
STUDENT SET-UP
CONCEPT LEVEL
CLEAN UP

MATERIALS

The materials listed on the student page are enough for one group of 5–6 students. You may wish to have your students use a calculator to complete this activity.

Preparation Notes

Some students may consider their height and weight to be personal and won't want to weigh and measure themselves with the others in the class. Give these students the option of using the data of a fictional person, such as one of the following:

Jenny	80 lb	4 ft	age 11
Ben	65 lb	3 ft	age 12
Carlos	110 lb	5 ft 2 in.	age 11
Alexa	120 lb	4 ft 6 in.	age 12
Tasheika	90 lb	4 ft 6 in.	age 13

¡Supervivencia!

Cada segundo de vida, las células del cuerpo reciben, utilizan y almacenan energía. Además, se reparan a sí mismas, se reproducen y eliminan sus desechos. En conjunto, estos procesos constituyen el *metabolismo*. El metabolismo del cuerpo es muy similar al de una célula individual. Cada célula necesita energía a una escala pequeña, así que todas las células juntas requieren energía a una escala mayor. Tus células obtienen de los alimentos que comes la energía que necesitas para estar vivo.

El índice metabólico basal mide la energía que el cuerpo necesita para realizar los procesos vitales básicos cuando está en reposo. Entre estos procesos se encuentran los latidos del corazón, la respiración y el mantenimiento de la temperatura del cuerpo. El sexo, la edad y muchas otras cosas influyen en este índice. Es posible que el tuyo sea distinto del de los demás, pero para ti es normal. En esta actividad conocerás la cantidad de energía, medida en calorías, que necesitas cada día para sobrevivir.

Materiales

- báscula
- cinta métrica

Procedimiento

1. Pésate en una báscula. Si la báscula está graduada en libras, convierte tu peso en libras a masa en kilogramos. Para hacerlo, multiplica el número de libras por 0.454.

Ejemplo: Si Carlos pesa 125 lb, su masa en kilogramos es:	125 lb × 0.454 ————— 56.75 kg	

2. Con ayuda de un compañero o compañera y una cinta métrica, mide tu altura. Si la cinta métrica está en pulgadas, convierte la medida de pulgadas a centímetros. Para hacerlo, multiplica el número de pulgadas por 2.54.

Si Carlos mide 62 in, su altura en centímetros es:	62 in × 2.54 ————— 157.48 cm

Datasheets for LabBook
Datasheet 11

Kathy LaRoe
East Valley Middle School
East Helena, Montana

3. Ahora calcula tu índice metabólico basal utilizando la fórmula adecuada. La respuesta te dará el número aproximado de calorías que tu cuerpo necesita cada día para sobrevivir.

Calcular el índice metabólico basal	
Mujeres	**Hombres**
65 + (10 × tu masa en kilogramos)	**66 + (13.5 × tu masa en kilogramos)**
+ (1.8 × tu altura en centímetros)	**+ (5 × tu altura en centímetros)**
− (4.7 × tu edad en años)	**− (6.8 × tu edad en años)**

4. La actividad también influye en el metabolismo. Al hablar, caminar o jugar utilizas más energía que cuando estás en reposo. Para darte una idea de cuántas calorías necesita tu cuerpo por día para mantenerse saludable, elige el estilo de vida más parecido al tuyo de la tabla de la derecha y multiplica tu índice metabólico basal por el factor de actividad.

Análisis

5. ¿Cómo se puede comparar todo tu cuerpo con una sola célula? Explica por qué.

6. Si la actividad aumenta, ¿aumenta también el índice metabólico basal? ¿Provoca el aumento de actividad un aumento en la necesidad de calorías? Explica tus respuestas.

7. Si eres moderadamente inactivo y comienzas a hacer ejercicio todos los días, ¿cuántas calorías adicionales necesitas?

Factores de actividad	
Grado de actividad	**Factor de actividad**
Moderadamente inactivo (actividades normales de todos los días)	1.3
Moderadamente activo (ejercicio 3 a 4 veces por semana)	1.4
Muy activo (ejercicio 4 a 6 veces por semana)	1.6
Extremadamente activo (ejercicio 6 a 7 veces por semana)	1.8

Profundizar

Las mejores fuentes de energía son las que proporcionan la cantidad de calorías adecuada para tu estilo de vida, así como los nutrientes que necesitas. Investiga en la biblioteca o en Internet qué tipos de alimentos son las mejores fuentes de energía para ti. ¿Qué diferencias y semejanzas hay entre tu lista de fuentes óptimas de energía y tu dieta actual?

Haz una lista de todo lo que comes y bebes en un día. Averigua cuántas calorías hay en cada producto que comes y calcula el número total de calorías que consumiste. ¿Qué diferencia hay entre este número de calorías y las calorías que necesitas cada día para realizar todas tus actividades?

Experimentos

577

Lab Notes

Some students will think their basal metabolic rate, or BMR, is impossibly low. Emphasize that the BMR is the number of Calories a body needs just to keep the heart beating, the lungs breathing, and the cells respiring. It is not the number of Calories a person needs for an active lifestyle. Of course, a person can consume fewer than that number of Calories for a day, or even for a few days, without dying. Explain that the Calories required to live during starvation conditions are obtained from stored fat. When there is no more fat, then the energy comes from muscle tissue. Under extreme conditions of starvation, the body even begins to shut down some organ functions that use energy but that are not required for survival, such as the uterine cycle in women. Some students may ask why the BMR numbers are so much higher in males than in females. Explain that before puberty, the numbers are much closer together. But as boys approach puberty, they generally develop a higher muscle-to-fat ratio than girls do. Cellular respiration for muscle tissue requires more energy than for fat tissue.

Answers

5. Just as each cell needs energy on a small scale, your body requires energy on a much larger scale.

6. Technically, the BMR does not change with activity. It is the minimum amount of energy a person needs to stay alive. Activity requires that more energy be added to the BMR, thereby increasing the need for Calories.

7. Students should multiply their own BMR by 1.3, then multiply their BMR by 1.8. Students should subtract the smaller number from the larger number. This represents the additional Calories per day a person would expend shifting from a moderately inactive state to an extremely active one.

Bug Builders, Inc.
Teacher's Notes

Time Required

Two 45-minute class periods

Lab Ratings

TEACHER PREP ▲▲▲
STUDENT SET-UP ▲▲
CONCEPT LEVEL ▲▲▲
CLEAN UP ▲

MATERIALS

You will need 7 small brown-paper bags—one sack for each characteristic listed on page 579. Each bag will contain both dominant and recessive alleles for that characteristic. Cut 1 in. squares of paper to represent alleles. Use a different color paper for each trait (seven colors).

Safety Caution

Remind students to review all safety cautions and icons before beginning this lab activity.

Datasheets for LabBook
Datasheet 12

Kathy LaRoe
East Valley Middle School
East Helena, Montana

Constructora de Bichos, S.A.

Imagínate que trabajas en una compañía que fabrica bichos de juguete. El presidente de la Constructora de Bichos, S.A. te pide que diseñes nuevas versiones de los populares Bichos Galácticos, pero quiere que uses las partes que hay en la bodega. Tu trabajo consiste en crear un nuevo diseño. Ya estudiaste cómo los rasgos pasan de una generación a la siguiente. Ahora vas a usar estos conocimientos para inventar nuevas combinaciones de rasgos y armar las partes del bicho de nuevas maneras. El modelo A servirá como la "mamá" y el B como el "papá". Los dos modelos que ya están a la venta se muestran a continuación.

Modelo A ("mamá")
- antenas rojas
- cuerpo en 3 segmentos
- cola de tirabuzón
- 2 pares de patas
- nariz verde
- patas negras
- tres ojos

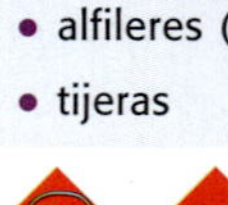

Modelo B ("papá")
- antenas verdes
- cuerpo en 2 segmentos
- cola recta
- 3 pares de patas
- nariz azul
- patas verdes
- 2 ojos

Materiales

- 7 bolsas de alelos (que te dará la maestra o maestro)
- malvaviscos grandes (segmentos del cuerpo y de la cabeza)
- palillos de dientes, rojos y verdes (antenas)
- tachuelas verdes y azules (narices)
- limpia pipas (colas)
- gomitas de dulce, verdes y negras (pies)
- alfileres (ojos)
- tijeras

Haz tu predicción

Si existen dos formas de cada una de las siete características, entonces hay ___?___ combinaciones posibles.

Recopila información

1. Tu maestra te mostrará siete bolsas, una por cada característica. Las bolsas contienen papelitos con letras mayúsculas o minúsculas. Toma dos papelitos de cada bolsa (recuerda que las mayúsculas representan los alelos dominantes y las minúsculas los recesivos). Un alelo es de la mamá y el otro del papá. Apunta los alelos que te tocaron y regresa los papelitos a la bolsa.

578

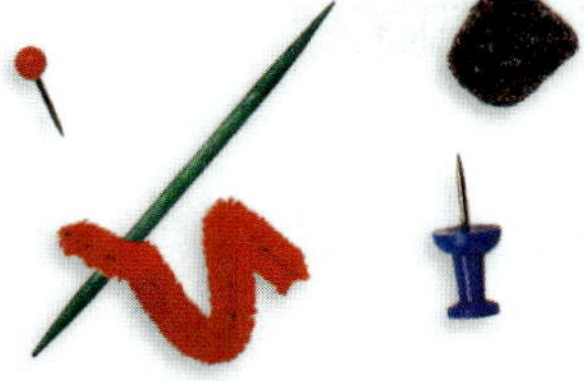

2. En tu cuaderno de ciencias, traza una tabla como la que se muestra abajo. Copia en las primeras dos columnas los alelos que sacaste de las bolsas. Luego llena la tercera columna con el genotipo del nuevo modelo ("bebé").

Rasgos familiares del bicho

Rasgo	Alelos del modelo A "mamá"	Alelos del modelo B "papá"	Genotipo del nuevo modelo "bebé"	Fenotipo del nuevo modelo "bebé"
Color de antenas				
Número de segmentos del cuerpo				
Forma de la cola				
Número de pares de patas				
Color de la nariz				
Color de las patas				
Número de ojos				

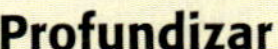

3. Usa la información de la derecha para llenar la última columna de la tabla.

4. Ahora que terminaste de llenar la tabla, puedes escoger las partes que necesitas para armar tu bicho. (Puedes usar palillos para unir la cabeza y los segmentos del cuerpo y para pegar las patas al cuerpo.)

Genotipos y fenotipos

RR o *Rr* = antenas rojas	*rr* = antenas verdes
SS o *Ss* = cuerpo en 3 segmentos	*ss* = cuerpo en 2 segmentos
CC o *Cc* = cola en tirabuzón	*cc* = cola recta
LL o *Ll* = 3 pares de patas	*ll* = 2 pares de patas
BB o *Bb* = nariz azul	*bb* = nariz verde
GG o *Gg* = patas verdes	*gg* = patas negras
EE o *Ee* = 2 ojos	*ee* = 3 ojos

Analiza los resultados

5. Haz una encuesta en tu salón de los rasgos que tienen los nuevos bichos. ¿Qué proporción hay para cada rasgo?

Saca conclusiones

6. ¿Hay algún bicho de la nueva generación que sea igual a los padres? Explica.

7. ¿Cuáles son los posibles genotipos de los padres?

8. ¿Cuántos genotipos distintos puede haber entre los nuevos bichos?

Profundizar

Encuentra una pareja para cruzar a tu bicho "bebé". ¿Cuáles son los genotipos y fenotipos posibles de sus crías?

579

Preparation Notes

This lab will take some time to prepare for, but it is well worth the effort for a genetics-reinforcement lesson.

1. Cut enough squares so that each student will have two alleles for each characteristic. You must have an equal number of both dominant and recessive alleles.

2. Half the squares for each characteristic should be marked with capital letters to indicate dominant alleles. The other half should be marked with lowercase letters to indicate recessive alleles.

3. Label each paper bag with one of the seven characteristics. Place both dominant and recessive alleles in their corresponding paper bag.

4. Have students draw two alleles from each sack. Tell students that the first draw is from "mom" and that the second draw is from "dad."

Going Further

Answers will vary. Have students use Punnett squares to explore possible genotypes.

Answers

Make a Prediction

128

two forms of each of seven characteristics so $2 \times 2 \times 2 \times 2 \times 2 \times 2 \times 2$ (or 2^7) = 128

5. Student ratios should be similar to the ratios determined when the alleles where selected by the teacher.

6. If any students have offspring bugs that look like one of the parents, have them compare the genotype of the offspring with the genotype of the parents. They may look alike but still have different genotypes for some traits.

7. Item 6 will lead students to this answer. Answers will vary.

8. Have students construct Punnett squares using the parental traits. Except for the results obtained by parental genotypes that are all homozygous recessive, the student will see other possibilities for genotypes and phenotypes from the same parents.

Tracing Traits
Teacher's Notes

Time Required

Two 45-minute class periods, separated by several days so students have time to complete their surveys

Lab Ratings

EASY ——————→ HARD

TEACHER PREP ♦
STUDENT SET-UP ♦
CONCEPT LEVEL ♦♦
CLEAN UP ♦

Lab Notes

Family histories will vary. Encourage students to include at least three generations in their histories.

Survey results will vary. Make sure that students actually surveyed each family member who was available. Responses will vary. You may check family members with shaded symbols against the survey results for accuracy.

Percentages will vary. A family member may receive a recessive allele from the father and a recessive allele from the mother. In such a case, this family member will exhibit the recessive form of the trait rather than the dominant form.

Experimentos

Sigue la pista de los rasgos

¿Alguna vez te has puesto a pensar en los rasgos que heredaste de tus padres? Ya sabes que los rasgos se pasan de una generación a la siguiente, pero ¿sabes si tienes algún rasgo que tus padres no tengan? En este proyecto vas a hacer un árbol genealógico, o árbol de familia, como el que se muestra en el diagrama de abajo. Vas a seguirle la pista a un rasgo heredado en tu propia familia o en otra familia que te pueda proporcionar los datos que necesites. Luego, vas a entrevistar a los miembros de la familia para encontrar algunos de los rasgos que comparten. Y, finalmente, vas a investigar cómo un rasgo en particular pasa de generación en generación.

Materiales

- lápiz
- papel

Procedimiento

1. El diagrama de la derecha muestra un árbol genealógico de cuatro generaciones. En un papel aparte, traza un diagrama similar de la familia que escogiste. Trata de incluir tantos parientes como puedas: abuelos, padres, hijos y nietos. Traza círculos para representar a las mujeres y cuadrados para representar a los hombres. Si quieres, puedes añadir más información, como nombres, fechas de nacimiento, o fotografías. Otros miembros de la familia te pueden ayudar a reunir la información que necesites.

2. Haz una tabla como la que se muestra en la página siguiente. Entrevista a cada uno de los parientes que pusiste en el árbol genealógico para determinar quién tiene la forma dominante del rasgo que se describe en la tabla y quién la recesiva. Pregúntales si tienen vello en la falange media de los dedos. Escribe el nombre de cada persona en el cuadro correspondiente. No olvides explicarle a cada uno que es perfectamente normal tener la forma recesiva o la dominante del rasgo.

Árbol genealógico

I
Abuelos

Pedro 1 — Ana 2

II
Padres

Marcela 1 — José 2 · María 3 — Beto 4

III
Hijos

Pablo 1 · María 2 · David 3 — Rosa 4

IV
Nietos

Carlos 1 · Alicia 2 · Tere 3

Because so many children are adopted or live in foster homes or group homes, please emphasize to your students that they may choose any family to study.

Kerry Johnson
Isbell Middle School
Santa Paula, California

Rasgo dominante	Rasgo recesivo	Miembros de la familia con el rasgo dominante	Miembros de la familia con el rasgo recesivo
Vello presente en la falange media de los dedos *(H)*	Vello ausente en la falange media de los dedos *(h)*	*No escribas en el libro*	

3. Pasa la información que escribiste en la tabla al árbol genealógico del paso número 1. Lo puedes lograr coloreando los cuadros o círculos que representen a las personas que tienen la forma dominante de este rasgo. Deja los demás en blanco.

Análisis

4. ¿Qué porcentaje de miembros de la familia tienen la forma dominante del rasgo? Cuenta el número de personas que tiene la forma dominante y divídelo entre el número total de personas que entrevistaste. Multiplica el resultado por 100. A la derecha puedes ver un ejemplo de esta operación.

Ejemplo: Calcular el porcentaje

$$\frac{10 \text{ personas con el rasgo}}{20 \text{ personas entrevistadas}} = \frac{1}{2}$$

$$\frac{1}{2} = 0.50 \times 100 = 50\%$$

5. ¿Qué porcentaje tiene la forma recesiva del rasgo? ¿Por qué no tienen todos la forma dominante?

6. Compara el porcentaje que sacaste para la forma dominante con el de tus compañeros y compañeras. ¿Hay familias que tengan un porcentaje más alto del rasgo dominante?

7. Escoge uno de los miembros de la familia que tenga la forma recesiva del rasgo. ¿Qué genotipo tiene esta persona? ¿De dónde vienen los alelos que forman su genotipo? ¿Cuáles son los genotipos posibles de los padres de esta persona? ¿Tiene hermanos o hermanas? ¿Tienen sus hermanos la forma recesiva o la dominante?

8. Traza una cuadrícula de Punnett como la que está a la derecha. Úsala para demostrar cómo pudo esta persona haber heredado la forma recesiva del rasgo. Dentro de la cuadrícula, apunta el genotipo de la persona que escogiste en la esquina inferior derecha. ¿Puedes determinar el genotipo de los padres? PISTA: puede haber más de un genotipo posible. ¿Qué alelo (dominante o recesivo) tiene que haber pasado de uno de los padres si hay un hermano o hermana que muestra la forma dominante del rasgo?

Answers

7. The genotype of the recessive form of the characteristic must be *hh* (homozygous recessive). Each allele came from one of the individual's parents. Possible genotypes for the parents of the individual expressing the recessive form are *Hh* and *hh*. Does the student know whether either of the parents expresses the recessive form of the trait? Does the student know if the individual chosen has brothers or sisters? Are their genotypes known? If so, have the student decide if each of them has a dominant or recessive genotype. If a dominant genotype is found among the siblings and one of the parents is known to have the recessive form, ask the student what the genotype of the other parent must be *(Hh)*.

8. The Punnett square should show *hh* in the bottom right-hand corner. One of the parents must have the genotype *hh*. The other parent must have either *hh* or *Hh*. If any sibling has the dominant trait, the genotype of the other parent must be *Hh*.

Datasheets for LabBook Datasheet 13

Base Pair Basics
Teacher's Notes

Time Required

One 45-minute class period

Lab Ratings

TEACHER PREP ♦
STUDENT SET-UP ♦
CONCEPT LEVEL ♦
CLEAN UP ♦

MATERIALS

You may want to provide additional materials for the Going Further section.

Safety Caution

Remind students to review all safety cautions and icons before beginning this lab activity.

Students should always exercise care when using scissors.

Lab Notes

You may wish to enlarge the template for your students so the base pairs will be easier to cut out.

Explain to students that the white pieces and the colored pieces indicate an old side and a new side after replication.

Experimentos

Pares de bases

Ya aprendiste que el ADN es una molécula que tiene forma de escalera de caracol. Los pasamanos de la escalera están hechos de moléculas de azúcar y de fosfato. Los lados están sostenidos por moléculas llamadas bases nucleótidas. Estas bases se juntan en pares para formar los peldaños de la escalera. Cada base nucleótida puede formar pares sólo con otra base nucleótida en particular. Cada uno de estos tres pares se llama "par de bases". Cuando el ADN se duplica, las enzimas hacen que los pares de bases se separen. Luego, cada mitad atrae a los nucleótidos necesarios que están disponibles en el núcleo y los reúne para completar una nueva mitad. En esta actividad, vas a crear un modelo de ADN y lo vas a duplicar.

Materiales

- papel o cartulina de color blanco
- papel o cartulina de colores
- tijeras
- lápiz
- bolsa de papel grande

Procedimiento

1. Usa los siguientes dibujos de las cuatro bases nucleótidas para crear una plantilla. Usa las plantillas para trazar las bases en un pedazo de cartulina. Rotula las piezas A (para adenina), T (para timina), C (para citosina) y G (para guanina), como se muestra a continuación. Dibuja otra vez las piezas en cartulina usando colores diferentes para cada base. Dibuja las piezas tan grandes como tú quieras. Haz todas la piezas que puedas.

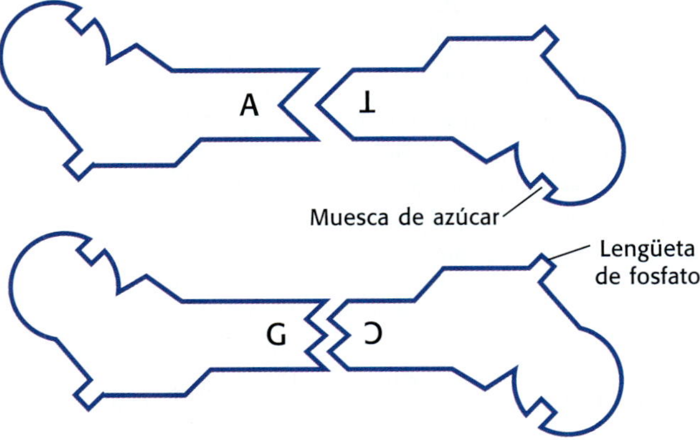

2. Corta con cuidado todas las piezas.

3. Guarda las piezas de colores en una bolsa grande de papel. Esparce todas las piezas en blanco sobre una mesa de trabajo o la mesa del laboratorio del salón de clase.

4. Saca nueve piezas al azar de la bolsa que tiene las piezas de colores. Acomoda las piezas de colores en cualquier orden formando una columna de modo que las letras A, T, C y G estén boca arriba. Asegúrate de unir las muescas de azúcar y las lengüetas de fosfato. Dibuja este esquema en tu cuaderno de ciencias.

Debra Sampson
Booker T. Washington
Middle School
Elgin, Texas

5. Busca las bases nucleótidas blancas que corresponden a las nueve bases de colores que dibujaste. Te será más fácil si recuerdas las reglas para la formación de pares de bases.

6. Acomoda las piezas, uniendo todas las lengüetas con las muescas. Ahora tienes una pieza de ADN que contiene nueve pares de bases. Dibuja los resultados en tu cuaderno de ciencias.

7. Separa los pares de bases, manteniendo unidas las muescas de azúcar y las lengüetas de fosfato. Dibuja este esquema en tu cuaderno de ciencias.

8. Junto a cada base separada que dibujaste en el paso 7, escribe la letra de la base que completa el par.

9. Busca todas las bases que necesitas para completar tu duplicado. Busca las cinco piezas blancas que correspondan a las bases de la izquierda y las piezas de colores que correspondan a las bases de la derecha.

Asegúrate de que todas las muescas y lengüetas se correspondan y que los lados estén derechos. ¡Ya lograste duplicar el ADN! ¿Son los modelos idénticos? Dibuja los resultados en tu cuaderno de ciencias.

Análisis

10. Menciona las reglas para la formación de pares de bases.

11. ¿Qué pasa cuando intentas formar un par con timina y guanina? ¿Se corresponden la una con la otra? ¿Están sus lados derechos? ¿Concuerdan todas las lengüetas y muescas? Explica por qué.

Profundizar

Construye un modelo tridimensional de la molécula de ADN, mostrando su estructura en forma de escalera de caracol. Usa tu imaginación y creatividad para seleccionar los materiales. Tal vez quieras usar agitadores con bolas de goma y palillos, o bien limpiadores de pipa o clips. Haz una exposición de tu modelo para tus compañeros.

Answers

10. G pairs with C, and A pairs with T.

11. The joining areas of guanine and thymine don't match up. They don't fit together.

Datasheets for LabBook
Datasheet 14

Mystery Footprints
Teacher's Notes

Time Required
Two 45-minute class periods

Lab Ratings

EASY ——————————→ HARD

TEACHER PREP ⚗⚗⚗
STUDENT SET-UP ⚗⚗
CONCEPT LEVEL ⚗⚗
CLEAN UP ⚗

Preparation Notes

To set up this lab, you will need to either construct a long, shallow sandbox out of wood or cardboard for use indoors or find (or build) a sandy area outside. Ask a boy and a girl (preferably students who are not in your science class) or two adults, one male and one female, to walk through the sand with their bare feet. The sand should be about 16 cm deep, and the area they walk through should be long enough that three or four footprints can be seen in the sand. Of course, you may want to make the footprints more permanent by using plaster-of-Paris. If you do not have access to sand, soil that will hold a footprint will substitute for sand nicely.

Maurine Marchani
Raymond Park Middle School
Indianapolis, Indiana

Experimentos

Huellas misteriosas

A veces, los científicos encuentran pistas conservadas en rocas que son prueba de las actividades de organismos que vivieron hace miles de años. Pruebas como las huellas conservadas proporcionan información importante sobre un organismo. Imagina que un grupo de paleontólogos pide a los estudiantes de tu clase que analicen unas huellas humanas que se encontraron en las rocas de una zona justo a la salida de la ciudad.

MÉTODO CIENTÍFICO

Formula una hipótesis

1. Tu maestro o maestra les dará unas huellas en arena. Examínenlas y piensen qué información podrían obtener de las personas que caminaron sobre este terreno. Entre todos formulen el mayor número posible de hipótesis acerca de las personas que dejaron las huellas.

2. Formen grupos de tres e investiguen una hipótesis por grupo.

Comprueba la hipótesis

3. En tu cuaderno de ciencias, dibuja una tabla para registrar tus datos. Por ejemplo, si tu hipótesis es que las huellas pertenecen a dos hombres adultos que caminaban de prisa, tu tabla se parecerá a la de abajo.

Materiales

- una caja grande con arena ligeramente húmeda, por lo menos de 1 m² (lo suficientemente grande para que quepan 3 ó 4 huellas)
- regla métrica

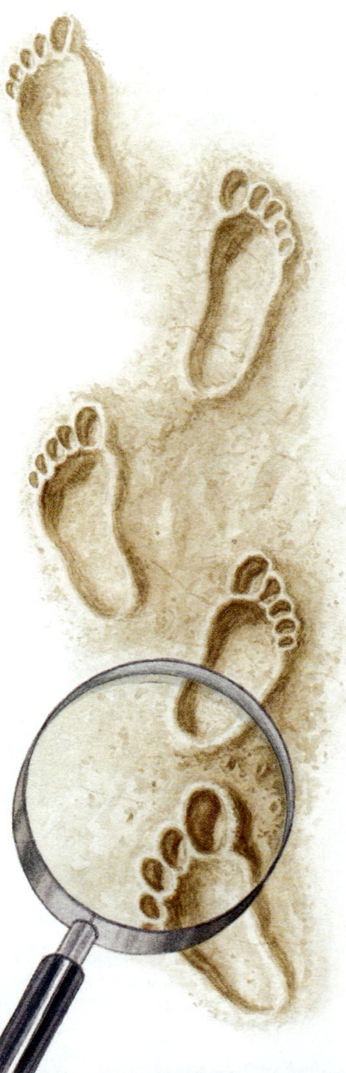

Huellas misteriosas		
	Conjunto de huellas 1	Conjunto de huellas 2
Longitud		
Ancho		
Profundidad de los dedos		
Profundidad del talón		
Longitud del paso		

4. Con ayuda de tu grupo, primero analiza tus propias huellas para sacar conclusiones sobre las huellas misteriosas. Por ejemplo, ¿cuánto mide tu paso cuando corres? ¿Y cuando caminas? ¿Afecta tu peso a la profundidad de la huella? ¿Qué parte de tu pie toca primero el suelo cuando corres? ¿Qué parte lo toca primero cuando caminas? Al correr, ¿qué parte de la huella es más profunda? Haz una lista de los tipos de huellas que se forman con cada actividad. Por ejemplo, podrías escribir algo así: "Al correr, se forma una impresión profunda en el área de los dedos y el paso es largo".

Analiza los resultados

5. Compara los datos de tus huellas con los de las huellas misteriosas. ¿Qué semejanzas y qué diferencias hay?

6. ¿Proceden las huellas misteriosas de una o de más personas? Explica tu respuesta.

7. ¿Hay pruebas suficientes para saber si las huellas misteriosas fueron formadas por hombres, mujeres, niños o una combinación de todos ellos? Explica por qué.

8. Basándote en las observaciones de tus propias huellas, ¿estaban de pie, caminando o corriendo las personas que dejaron las huellas?

Saca conclusiones

9. ¿Tus observaciones corroboran la hipótesis? Explica por qué.

10. ¿Cómo podrías mejorar el experimento?

Comunica los resultados

11. Resume las conclusiones de tu grupo en un documento dirigido a los científicos que solicitaron la ayuda de tu clase. Empieza por enunciar la hipótesis. Después resume el método que utilizaron para recopilar información a partir del estudio de sus propias huellas. Incluye las comparaciones que hiciste entre tus huellas y las huellas misteriosas. Antes de escribir tus conclusiones, haz algunas sugerencias para mejorar tu método de investigación.

12. Haz una presentación de tus descubrimientos frente a tus compañeras y compañeros con ayuda de un cartel, una gráfica o de la computadora.

Experimentos

Lab Notes

Tell the students to imagine that a scientist wishes to analyze footprints found in the rocks near fossilized remains, and he has contacted the class, seeking help in the investigation. The scientist wants to know how the students intend to gather information to make inferences about the humans who left the prints. Explain that a scientist should be able to make the same type of inferences about an organism from fresh tracks as from preserved tracks. Use the mystery footprints in the sand to help students design investigations for gathering data. From that data, students can learn to draw inferences.

Much research in evolution is dependent on scientific inferences. To conclude the laboratory experience, lead the students in a discussion of the importance of large sets of data in helping scientists make inferences.

Datasheets for LabBook
Datasheet 15

Answers

The answers for this activity will depend on the footprints your students observe. They should be able to compare their own activities with variations in the footprints they leave. Then they should be able to apply what they've learned to the mystery footprints.

Out-of-Sight Marshmallows
Teacher's Notes

Time Required

One 45-minute class period

Lab Ratings

EASY————HARD

TEACHER PREP ▲▲
STUDENT SET-UP ▲▲
CONCEPT LEVEL ▲
CLEAN UP ▲

Safety Cautions

Tell students not to eat the marshmallows after the lab. The marshmallows will have been handled thoroughly.

Preparation Notes

Choose several colors of mini-marshmallows and a piece of cloth or a napkin that will make the marshmallows hard to see at first glance.

Datasheets for LabBook
Datasheet 16

Georgiann Delgadillo
East Valley School District
Continuous Curriculum School
Spokane, Washington

Malvaviscos ocultos

Una adaptación es una característica que ayuda a un organismo a sobrevivir en su medio ambiente. En la naturaleza, el camuflaje es una coloración que permite a un organismo confundirse con su entorno. Esta es la hipótesis que comprobarás: los organismos con camuflaje tienen mejores posibilidades de escapar de sus depredadores y, por lo tanto, su posibilidad de sobrevivir es mayor.

Materiales

- 1 pedazo de tela de color (50 cm cuadrados). El color de la tela debe ser parecido al color de uno de los malvaviscos.
- 50 malvaviscos blancos
- 50 malvaviscos de color (preferentemente de un solo color)
- cronómetro o reloj con segundero

Experimentos

Comprueba la hipótesis

1. Se trabajará en parejas.

2. Cuenta 50 malvaviscos blancos y 50 de color. En este experimento los malvaviscos representan las presas (comida)

3. En un pedazo de tela de color, coloca al azar los malvaviscos blancos y los de colores.

4. Un miembro de la pareja será el cazador hambriento (depredador) y el otro registrará los resultados de cada prueba. El cazador hambriento debe mirar la comida unos cuantos segundos y recoger el primer malvavisco que vea. Después debe dejar de mirar.

5. Este proceso continuará sin interrupción por 2 minutos o hasta que el maestro o maestra indiquen que se pare.

Analiza los resultados

6. ¿Cuántos malvaviscos blancos escogió el cazador hambriento?

7. ¿Cuántos malvaviscos de color escogió?

Saca conclusiones

8. ¿Qué representaba la tela del experimento?

9. ¿Influyó el color de la tela en el color de los malvaviscos que el cazador hambriento escogió?

10. ¿Qué color de malvavisco representaba el camuflaje?

11. Describe un organismo que tenga camuflaje como adaptación.

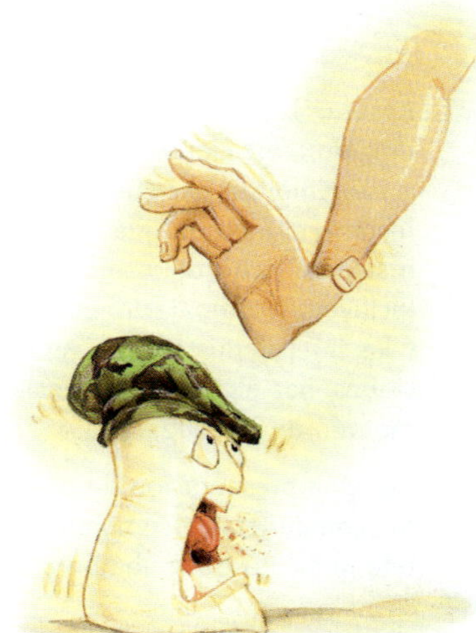

586

Answers

6. Answers will vary according to the students' experiments.

7. Answers will vary, but the marshmallow that most closely resembles the background cloth is expected to be chosen less often.

8. The cloth represents the organism's background, environment, or surroundings.

9. The marshmallow with the color that does not blend in with the color of the cloth will probably be chosen most often because it is easier to see. So the color of the cloth is important to the "organism's" protection.

10. Answers will vary according to the colors used in the lab.

11. Answers will vary, but students should describe an organism that is difficult to see in its natural surroundings.

Chocolates que sobreviven

Imagínate un mundo poblado de dulces y conserva ese delicioso pensamiento por un momento. Ahora, aplica el concepto de selección natural a una población de chocolates cubiertos de dulce. De acuerdo con la teoría de la selección natural, los individuos con adaptaciones favorables tienen mayor probabilidad de sobrevivir. Para la "especie" de dulce de este experimento, la resistencia de la cubierta es una ventaja adaptativa. ¿Qué color de dulce crees que tendrá la tasa más alta de supervivencia? Piensa como un científico de las ciencias de la vida y planea un experimento para descubrir qué color tiene ventaja sobre los otros en términos de resistencia de la cubierta.

Haz una predicción

1. Escribe la siguiente oración en tu cuaderno de ciencias y llena los espacios en blanco. Si la cubierta de color __?__ es la más resistente, entonces un menor número de dulces de este color ___?___ cuando ___________?___________.

Comprueba tu hipótesis

2. Diseña un procedimiento para determinar qué color de dulce es más adecuado para sobrevivir, al no "quebrarse bajo presión". Asegúrate de considerar los materiales y herramientas que puedas necesitar para llevar a cabo el procedimiento. Revisa tu diseño experimental con tu maestro o maestra antes de empezar. Él o ella te dará los dulces que necesites y te ayudará a reunir los materiales y herramientas.

3. En tu cuaderno de ciencias, registra tus resultados en una tabla de datos. Asegúrate de organizarlos de manera clara y comprensible.

Analiza los resultados

4. Escribe un informe en el que describas tu experimento. Explica cómo tus datos confirman o refutan tu predicción. Menciona errores y formas de mejorar el procedimiento.

Profundizar

¿Qué otra característica puede probarse para determinar qué dulce está mejor adaptado para sobrevivir? Explica tu respuesta.

Materiales

- chocolates pequeños cubiertos de dulce y de diversos colores (cada estudiante o grupo decidirá cuántos se necesitan de cada color)
- según el experimento, se pueden necesitar otros materiales

587

Karma Houston-Hughes
Kyrene Middle School
Tempe, Arizona

Datasheets for LabBook
Datasheet 17

Survival of the Chocolates
Teacher's Notes

Time Required

One or two 45-minute class periods

Lab Ratings

Safety Caution

Safety concerns will vary with each design.

Preparation Notes

You will need to be prepared for different experimental designs. Some students may wish to find out if the different colored candy shells differ in hardness by testing which one will crack easiest under physical stress. Others may want to test the colors to see which one will dissolve more readily in water (cold or warm). Encourage different ways of testing. Part of the purpose of this lab is to help students learn how to design an experiment.

Answers

1. The statement made for the student here is for example only. Students may not wish to test for hardness or cracking. Help them make a prediction about their own experiment.

2. Again, students should conduct their own experiment.

Dating the Fossil Record
Teacher's Notes

Time Required

One 45-minute class period

Lab Ratings

EASY ——————————→ HARD

TEACHER PREP 🧪🧪
STUDENT SET-UP 🧪🧪
CONCEPT LEVEL 🧪🧪
CLEAN UP 🧪🧪

Preparation Notes

Think of this lab as a puzzle. Students are told that Sample 2 is the oldest. They must keep in mind that the fossil will not reappear in any sample after it has gone extinct. This should be a challenge, but a fun one. In Part 2 of the procedure, you may want to put the dates given by the geology lab on the board for quick reference. Tell students that *mya* means "million years ago." Tell students that in this exercise they may assume that all samples contain a representative sample of the organisms that lived in each time period. That is to say, the fossil record in this exercise is complete.

James Chin
Frank A. Day Middle School
Newtonville, Massachusetts

Experimentos

La edad del registro fósil

DESARROLLAR DESTREZAS

Te llegaron nueve muestras de rocas de un paleontólogo en California. Tu trabajo es ordenarlas de la más antigua a la más reciente, ségun los fósiles que contienen, y determinar sus edades relativas. Los resultados sobre la edad absoluta de las muestras van a tardar varias semanas, y el paleontólogo necesita la información de inmediato. Ya sabes, por lo que has visto en otras investigaciones, que las rocas de la muestra 2 son las más antiguas.

Procedimiento: Parte 1

1. Forma equipos con tres o cuatro compañeros.

2. Acomoda las tarjetas de los fósiles del más antiguo al más reciente. Comienza con la muestra 2, ya que sabes que es la más antigua. Prueba varias maneras de acomodar las tarjetas hasta que encuentres el orden correcto. **PISTA:** Cuando un organismo se extingue, no vuelve a aparecer en las rocas más recientes.

Materiales

- juego de nueve tarjetas que representan las muestras de roca
- papel y lápiz
- marcadores de colores
- cartulina (61 cm²)

Clave de los fósiles

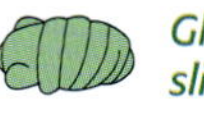 *Globus slimius*
 *Bogus biggus*
 *Circus bozoensis*
 *Microbius hairiensis*
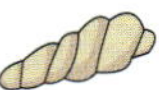 *Fungus amongius*
 *Bananabana bobana*

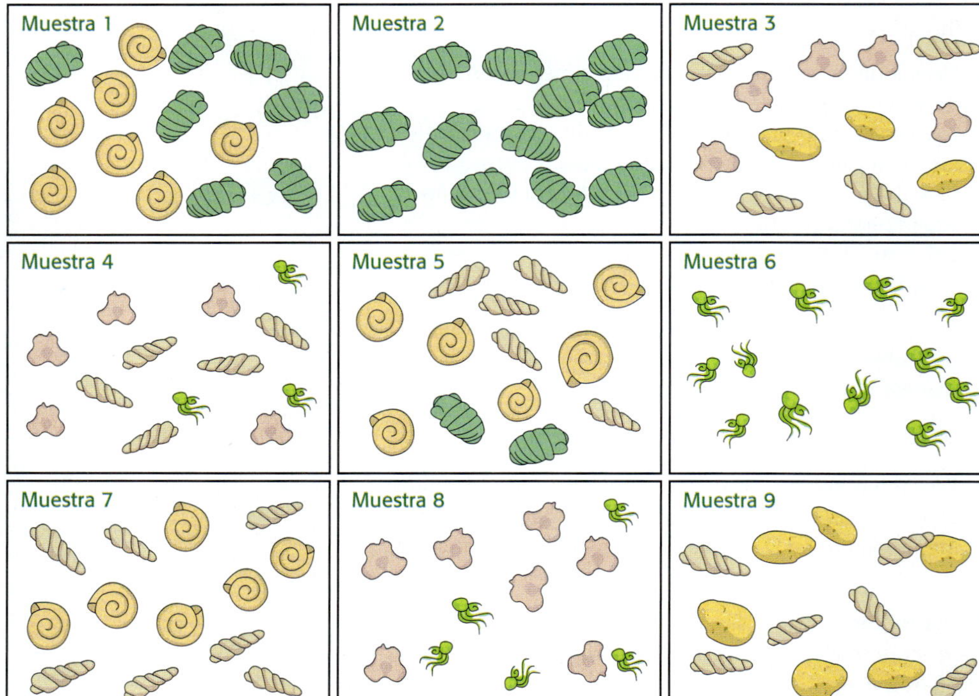

588

3. Haz una tabla de datos como la que se presenta a continuación. En la primera columna apunta las muestras, de la más antigua a la más reciente, de abajo hacia arriba. La muestra 2 ya está anotada.

<table>
<tr><td colspan="6" align="center">Nombre del organismo fósil</td></tr>
<tr><td>Orden de las muestras</td><td>Globus slimius</td><td></td><td></td><td></td><td></td></tr>
<tr><td></td><td></td><td></td><td></td><td></td><td></td></tr>
<tr><td></td><td></td><td></td><td></td><td></td><td></td></tr>
<tr><td></td><td></td><td></td><td></td><td></td><td></td></tr>
<tr><td></td><td></td><td></td><td></td><td></td><td></td></tr>
<tr><td></td><td></td><td></td><td></td><td></td><td></td></tr>
<tr><td></td><td></td><td></td><td></td><td></td><td></td></tr>
<tr><td>Muestra 2</td><td>X</td><td></td><td></td><td></td><td></td></tr>
</table>

4. En la primera fila, de izquierda a derecha y por orden de edad, apunta el nombre de cada fósil. **PISTA:** Examina bien las tarjetas para ver dónde aparece cada fósil en el registro de rocas. Pon una *X* en la columna apropiada para indicar cué fósil o fósiles hay en cada muestra.

Análisis: Parte 1

5. ¿Te parece que las letras forman algún patrón en la tabla? ¿Qué concluirías del patrón?

6. Según la información de la tabla, ¿cuál fósil es el más reciente?

7. Con la información que tienes, ¿puedes dar la edad exacta de alguno de los fósiles? ¿Por qué?

8. ¿Qué información obtienen los paleontólogos de la edad relativa?

Datasheets for LabBook
Datasheet 18

Experimentos

589

Name of Fossil Organism

Order of samples	Globus slimius	Microbius hairiensis	Fungus amongius	Circus bozoensis	Bogus biggus	Bananabana bobana
Sample 6						X
Sample 8					X	X
Sample 4			X		X	X
Sample 3			X	X	X	
Sample 9			X	X		
Sample 7		X	X			
Sample 5	X	X	X			
Sample 1	X	X				
Sample 2	X					

Experimentos

Procedimiento: Parte 2

1. Planeabas preparar una cronología para el paleontólogo de California, pero cuando llegan los resultados del laboratorio de geología, que puedes ver a la derecha, te das cuenta de que las fechas se despegaron de las muestras de roca. Medir la edad absoluta de una roca es muy caro, y no puedes mandar hacer el análisis otra vez. Pero, ¡espera! Ya habías determinado la edad relativa de las muestras. Lo único que tienes que hacer es acomodar las fechas de la más antigua a la más reciente. Anota las fechas en tu tabla de datos.

2. La tabla tiene toda la información que necesitas para desarrollar la cronología para el paleontólogo de California. Usa los marcadores de colores y la cartulina para hacerla. Dibuja una pared de roca donde se vean varias capas. Escribe la edad de cada capa y los nombres de los fósiles que en ella se encuentran. También puedes trazar una línea para marcar las edades de los fósiles y dibujar cada fósil junto a la fecha que le corresponde. ¡Ejercita tu creatividad!

Edades de la fósiles

Los fechamientos que dio el laboratorio de geologia son los siguientes: 28.5 ma, 30.2 ma, 18.3 ma, 17.6 ma, 26.3 ma, 14.2 ma, 23.1 ma, 15.5 ma, and 19.5 ma.

Análisis: Parte 2

3. De acuerdo con la edad absoluta, ¿qué organismo fósil vivió el período más largo de tiempo? ¿Cuál vivió el período más corto de tiempo? Explica tus respuestas.

4. Según la información de la cronología, ¿qué período de vida le asignarías al fósil de *Circus bozoensis*? PISTA: Mide el período entre el año en que el fósil aparece por primera vez en el registro de rocas y el primer año en el que ya no aparece.

5. Determina el período de vida de cada especie fósil en tu lista.

Profundizar

Utiliza la biblioteca o una base de datos de la Internet para investigar si el método que mide la edad absoluta de las rocas que se encuentran alrededor de un fósil es el más confiable para determinar la edad del mismo. Averigua qué circunstancias impiden encontrar la edad absoluta.

590

Age of sample (in millions of years)	Contents of sample
28.5	Sample 1: *Globus slimius, Microbius hairiensis*
30.2	Sample 2: *Globus slimius*
18.3	Sample 3: *Fungus amongius, Bogus biggus, Circus bozoensis*
17.6	Sample 4: *Fungus amongius, Bogus biggus, Bananabana bobana*
26.3	Sample 5: *Globus slimius, Microbius hairiensis, Fungus amongius*
14.2	Sample 6: *Bananabana bobana*
23.1	Sample 7: *Fungus amongius, Microbius hairiensis*
15.5	Sample 8: *Bogus biggus, Bananabana bobana*
19.5	Sample 9: *Fungus amongius, Circus bozoensis*

La vida media de un centavo

Imagínate que has existido por 5,000 años y que tienes otros 5,000 por delante. Eso le sucede al Carbono-14, un elemento inestable que se usa para determinar la edad absoluta de materiales de otros tiempos, como los huesos fósiles. Cada 5,730 años, la mitad del Carbono-14 de una muestra fósil se descompone y produce una forma más estable del elemento. En este experimento, verás cómo los centavos muestran la misma forma de "descomposición".

Materiales

- 100 monedas de un centavo
- bote grande con tapa

Procedimiento

1. Pon 100 monedas de un centavo en un bote grande con tapa. Agítalo varias veces y luego destápalo. Con cuidado, para que las monedas no rueden al suelo, vacía todas las monedas en una superficie plana.

2. Recoge las monedas que hayan caído con el lado "cara" hacia arriba. Apunta el número de monedas que recogiste y el número de las que quedan en una tabla de datos como la de la derecha.

3. Repite el proceso hasta que se te acaben los centavos.

4. En tu cuaderno de ciencias, dibuja una gráfica como la de la derecha. En el eje de las *x* escribe "número de volcadas" y en el eje de las *y* "monedas que quedan". Con la información de la tabla, dibuja la gráfica de las monedas que te quedan en cada volcada del bote.

Análisis

5. Mira la gráfica de la vida media del Carbono-14 de la derecha. Compárala con la que hiciste para las monedas. Explica en qué se parecen.

6. ¿Te acuerdas que la probabilidad de que salga "cara" cuando la lanzas al aire es de $\frac{1}{2}$? Explica por qué el número restante de monedas en cada volcada se reduce a más o menos la mitad.

Número de volcadas	Monedas restantes	Número de monedas recogidas
1		
2	*No escribas en el libro*	
3		

La vida media de un centavo

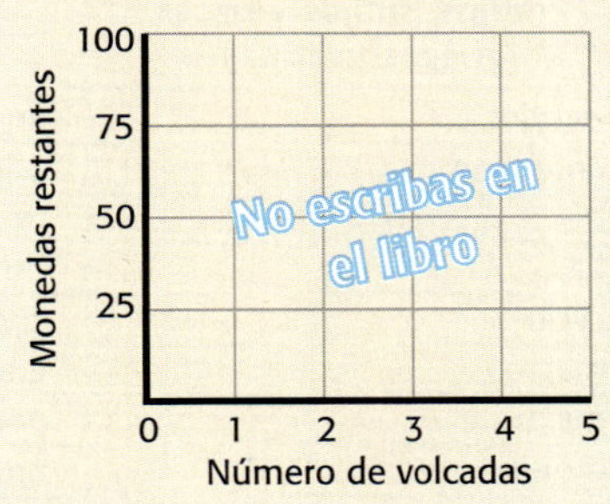

Vida media del Carbono-14

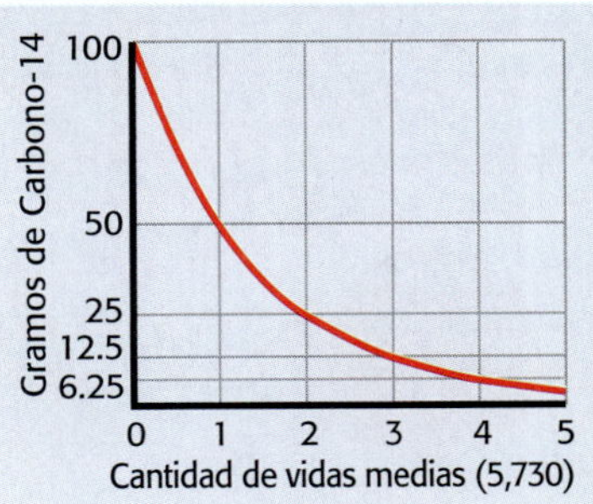

591

Answers

5. The graphs should be very similar in shape. With each half-life and each shake, the number remaining will be reduced by half.

6. The remaining number of pennies is reduced by about half each time the pennies are shaken and tossed because there are only two faces on each coin. The rules of probability dictate that half will land on heads and half will land on tails.

Karma Houston-Hughes
Kyrene Middle School
Tempe, Arizona

The Half-life of Pennies
Teacher's Notes

Time Required

One 45-minute class period

Lab Rating

TEACHER PREP   🍶
STUDENT SET-UP   🍶
CONCEPT LEVEL   🍶🍶
CLEAN UP   🍶

Lab Notes

It is useful to use coin tosses to explain half-life because approximately half the coins will land heads and half will land tails. Therefore, about half the entire quantity of coins tossed will be eliminated with each successive toss. You may want to review definitions of probability with students. Any penny tossed has a 50 percent probability of landing on heads. Similarly, the likelihood of radioactive decay is based on the possibility that in each half-life, each radioactive atom has a 50 percent probability of decay. Because the numbers of atoms in a radioactive sample are normally very large, statistical variation from the expected change is very small. In this experiment, a relatively small number of coins is used so the statistical variation might be larger.

Datasheets for LabBook
Datasheet 19

Shape Island
Teacher's Notes

Time Required
One 45-minute class period

Lab Rating

EASY ——————→ HARD

TEACHER PREP ▲
STUDENT SET-UP ▲
CONCEPT LEVEL ▲▲
CLEAN UP ▲

Lab Notes

This lab will help students demonstrate an understanding of binomial nomenclature by using a key to assign scientific names to fictional organisms. After completing the lab, students should be able to explain the function of the scientific name system.

In the chapter on classification, the vocabulary "two-part scientific name" has been used instead of "binomial nomenclature." You may wish to introduce the latter here.

This activity may be more successful if you review prefixes, suffixes, and root words briefly before beginning. Tell students that the genus name is capitalized but the species name is not and that both words are underlined or italicized.

Experimentos

La isla de las formas

Imagínate que eres un biólogo que explora las regiones desconocidas del mundo en busca de nuevas especies de animales. Navegaste durante muchos días en el océano y por fin encontraste la isla de las formas, cientos de kilómetros al sur de Hawaii. Esta isla tiene algunas formas de vida muy raras. Cada una de ellas tiene una forma geométrica diferente. Has pasado más de un año reuniendo especímenes y clasificándolos según el sistema de Linnaeus. Ya lograste asignarle un nombre científico compuesto a la mayoría de las especies que has reunido. Ahora, debes asignarles los nombres a los últimos 12 especímenes antes de regresar a casa.

Materiales

- lápiz
- papel

Procedimiento

1. En tu cuaderno de ciencias, dibuja cada uno de los organismos de esta página. Junto a cada organismo, dibuja una línea para poner su nombre, como se muestra en la siguiente página. El primero ya tiene nombre; asígnale uno a los otros 12 organismos. Usa este glosario de prefijos, sufijos y raíces griegas y latinas. Observa la tabla y decide cuáles se aplican a los organismos de la ilustración.

Prefijos, sufijos y raíces griegas y latinas	Significado
ankylos	ángulo
antennae	órganos sensoriales externos
tri-	tres
bi-	dos
cyclo-	círculo
macro-	grande
micro-	pequeño
mono-	uno
peri-	alrededor
-plast	cuerpo
-pod	pie
quad-	cuatro
stoma	boca
uro-	cola

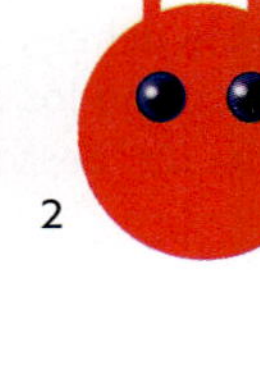

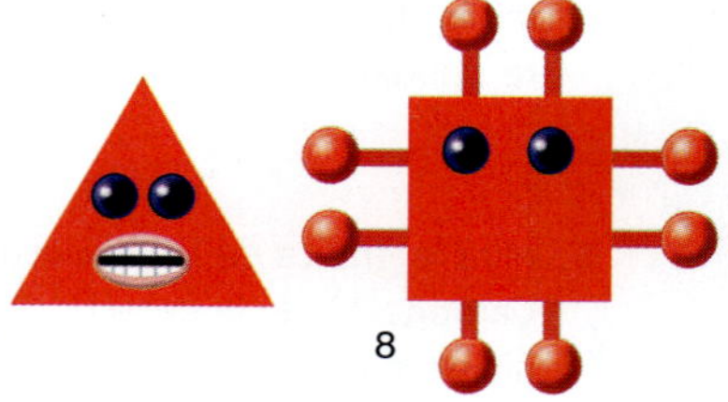

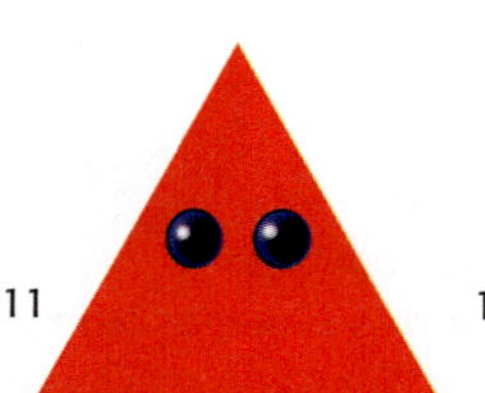

Datasheets for LabBook
Datasheet 20

Maurine Marchani
Raymond Park Middle School
Indianapolis, Indiana

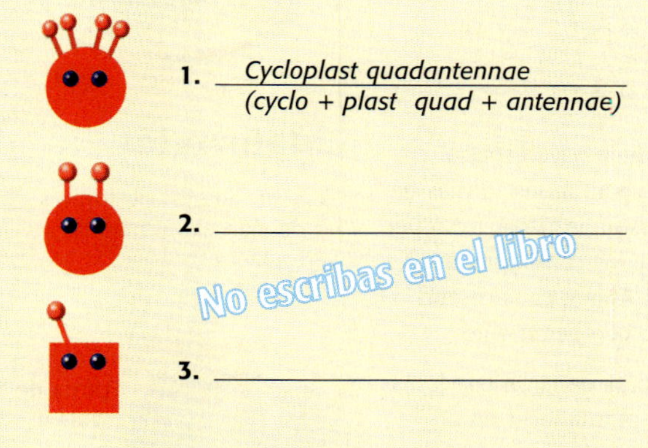

2. Hay otro organismo en la Isla de las formas, pero no puedes capturarlo. Sin embargo, tus provisiones se te están acabando y debes regresar a casa. Has tenido suerte con este animal raro y puedes dibujarlo con detalle. En tu cuaderno de ciencias, dibuja un animal diferente de los demás, y ponle un nombre compuesto.

Análisis

3. Si le das a la especie 1 un nombre común, como cara-redonda-sin-nariz, ¿podría cualquier otro científico saber a qué organismo te refieres? Explica por qué. ¿Qué otros organismos tienen una cara redonda sin nariz?

4. Describe dos características compartidas por todos tus especímenes.

Profundizar

Busca estos nombres científicos. Puedes usar la biblioteca, Internet, un índice taxonómico o unas guías de campo.

Mertensia virginica   Porcellio scaber

Para cada organismo, responde a las siguientes preguntas:

¿Es el organismo una planta o un animal? ¿Cuántos nombres comunes tiene? ¿Cuántos nombres científicos tiene?

Piensa en el nombre de tu fruta o verdura favorita. Averigua si tiene otros nombres comunes e investiga su nombre científico compuesto.

Answers

1. Below are example answers. Student answers may vary, but they should demonstrate an understanding of the key provided. Each name should consist of two words: the first describes the organism generally, and the second describes it more specifically.

   1. *Cycloplast quadantennae*
   2. *Cycloplast biantennae*
   3. *Quadankylosplast monoantennae*
   4. *Quadankylosplast bipod*
   5. *Triankylosplast triantennae*
   6. *Cycloplast stoma*
   7. *Triankylosplast stoma*
   8. *Quadankylosplast periantennae*
   9. *Cycloplast monopod*
   10. *Triankylosplast uromonopod*
   11. *Triankylos macroplast*
   12. *Quadankylos microplast*
   13. *Cycloplast uro*

2. Answers will vary according to student drawings.

3. No. There are five species that have round faces and lack noses.

4. Answers will vary but should indicate that they all have geometric shapes and two eyes. They are all the same color, all animals, and all living.

Going Further

Mertensia virginica, Virginia bluebells, are common wildflowers found in April and May in shady areas, mostly in moist spots near streams. This plant is quite plentiful in some places. Flower buds are pink, turning blue when the flower is fully opened. This wild flower is very common in western Kentucky.

Porcellio scaber is a species of wood louse. Wood lice are crustaceans, related to shrimps, crabs, and lobsters, and they belong to a class of arthropods called Isopoda. They are the only crustaceans that have been able to invade land without the needing to return to water, although they tend to be restricted to fairly damp places.

Voyage of the USS *Adventure*
Teacher's Notes

Time Required
One 45-minute class period

Lab Ratings

Preparation Notes

Some students will find it easier to make the charts on graph paper, so you may wish to add graph paper to the list of materials needed.

Lab Notes

Students should not be allowed to develop a misconception that we are able to travel outside our solar system. This activity should help students categorize organisms or objects by noticing subtle differences. It is a good activity to begin a study of classification of animals, rocks, or plants. This lab may be useful before introducing dichotomous keys, for example.

Experimentos

Viaje de la nave espacial *Aventura*

Imagínate que eres miembro de la tripulación de la nave espacial *Aventura*. La *Aventura* ha estado en una misión para reunir formas de vida fuera del sistema solar. En el viaje de regreso a la Tierra, pasó por una lluvia de meteoritos que arruinó algunos de los compartimentos que contienen las formas de vida extraterrestre. Ahora, es necesario colocar más de una forma de vida en el mismo compartimento.

La nave sólo tiene tres compartimentos que no se dañaron con la lluvia de meteoritos. Tus compañeros y tú deben permanecer en un compartimento, que se debe usar para las formas de vida extraterrestre sólo si es absolutamente necesario. Tienes que decidir qué formas de vida debes colocar juntas. Se cree que las formas de vida similares tienen las mismas necesidades, así que si las agrupan de esta manera, será más fácil cuidarlas. Puedes usar sólo una característica observable para agrupar las formas de vida.

Materiales
- ocho dibujos de formas de vida extraterrestre (opcional)

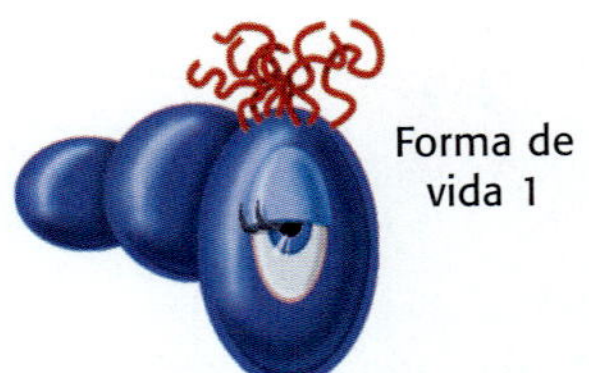
Forma de vida 1

Forma de vida 2

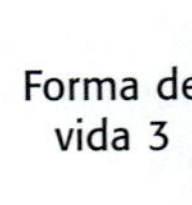

Forma de vida 3

Forma de vida 4

Procedimiento

1. 1. Haz una tabla de datos parecida a ésta. Clasifica cada columna con todas las características posibles de las diferentes formas de vida. En la parte izquierda escribe "Forma de vida 1", "Forma de vida 2", y así sucesivamente, como se muestra en la tabla. Las formas de vida están dibujadas en esta página.

Características de cada forma de vida			
	Color	Forma	Patas	Ojos
Forma de vida 1				
Forma de vida 2				
Forma de vida 3				
Forma de vida 4				

2. Describe cada característica lo más detalladamente posible. Deja suficiente espacio en cada casilla para escribir lo que ves. Según tus observaciones y la tabla de datos, determina qué formas de vida se parecen más entre sí.

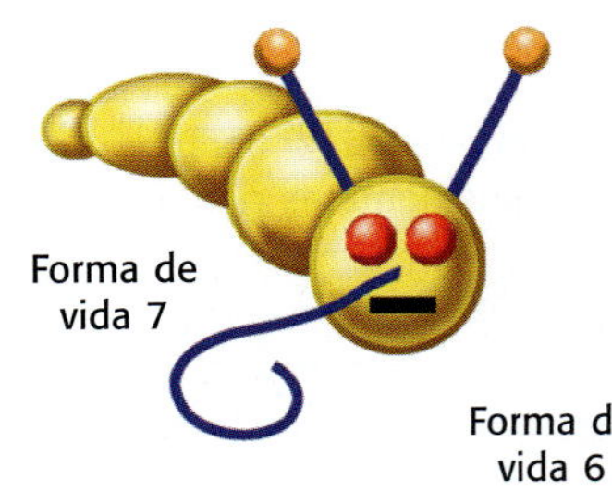
Forma de vida 7

Forma de vida 5

Forma de vida 6

Datasheets for LabBook
Datasheet 21

Georgiann Delgadillo
East Valley School District
Continuous Curriculum School
Spokane, Washington

3. Haz una tabla de datos como ésta. Llena la tabla según las decisiones que tomaste sobre las similitudes entre las formas de vida. Explica por qué las agrupaste de esa manera.

Asignación de compartimientos a las formas de vida		
Compartimiento	Formas de vida	Razones
1		
2		
3		

4. La nave espacial *Aventura* tiene que hacer una parada más antes de regresar a casa. En el planeta X437, descubres la forma de vida más interesante que jamás se haya visto, la CC9, que se ilustra a la derecha. Decide, en base a la agrupación de formas de vida que hiciste, si puedes incluir sin riesgos la CC9 en uno de los compartimentos para el viaje de regreso a la Tierra.

CC9

Análisis

5. Describe las formas de vida del compartimento 1. ¿En qué se parecen? ¿En qué se diferencian?

6. Describe las formas de vida del compartimento 2. ¿En qué se parecen? ¿En qué se diferencian de las del compartimento 1?

7. ¿Hay alguna forma de vida en el compartimento 3? Si es el caso, describe sus similitudes. ¿En qué compartimento permanecerían ustedes para el viaje de regreso a casa?

8. ¿Puedes transportar sin riesgo la forma de vida CC9 de regreso a Tierra? ¿Por qué? Si lo puedes hacer, ¿en qué compartimento la pondrías? ¿Cómo lo decidiste?

Profundizar

En 1831, Charles Darwin partió de Inglaterra en un barco llamado el HMS Beagle. Ya has estudiado los pinzones que Darwin observó en las Islas Galápagos. ¿Qué otros organismos raros encontró en la isla? Por ejemplo, investiga sobre la tortuga de las Galápagos.

Experimentos

595

Answers

There are no right or wrong answers in this activity. The objective is to allow students an opportunity to recognize subtle differences and to recognize that organisms may be more alike than they are different. However, you should make sure the students can give good reasons why they grouped certain life-forms together. There are several ways these seven organisms are similar. For example, four of them are segmented and have no legs. Three of them are geometrically shaped, and three others have mouths. Have students examine them for less observable characteristics, such as what kind of body plan or symmetry they have, how they might obtain food, or whether they might be land dwelling or aquatic.

Going Further

The Galápagos tortoise can have a shell length of 1.3 m, a mass of 180 kg, and live to be 150 years old.

Leaf Me Alone!
Teacher's Notes

Time Required

One 45-minute class period

Lab Ratings

TEACHER PREP
STUDENT SET-UP
CONCEPT LEVEL
CLEAN UP

MATERIALS

The materials on the student page are enough for a group of 4–5 students. Collect plant specimens ahead of time or have students bring in five specimens that they have collected. Students will need to see the leaves as they appear on the stems, so include as much of the plant as possible.

Going Further

This exchange activity can be an effective assessment tool for this lab. Students will enjoy challenging their classmates and this is a good way for students to learn from each other.

Jane Lemons
Western Rockingham
Middle School
Madison, North Carolina

¡Hojas!

Imagínate que eres un naturalista que va solo en una expedición a un bosque tropical. Has encontrado varias plantas que crees que hasta ahora eran desconocidas. Debes llamar a un botánico (científico que estudia las plantas) para confirmar tu sospecha. Como no hay servicio de correo en el bosque, debes dar una descripción completa y precisa por radio. El botánico debe ser capaz de dibujar las hojas de las plantas a partir de tu descripción.

En esta actividad describirás cinco especímenes de plantas utilizando los ejemplos y las listas de vocabulario.

Procedimiento

1. Examina cada ejemplo de hojas para prepararte para describir las hojas de las plantas que te dará tu maestro. Los ejemplos están en la siguiente página. Notarás que se necesita más de un término para dar una descripción completa de una hoja. A la derecha se muestra la descripción una hoja como ejemplo.

2. En tu cuaderno de ciencias, dibuja un diagrama de una hoja de cada planta.

3. Al lado del dibujo descríbela. Incluye características generales, como tamaño y color. Identifica la: forma de la hoja, tipo de tallo, arreglo de las hojas, borde de la hoja, arreglo de las venas y forma de la base de la hoja. Utiliza las listas de vocabulario en las descripciones y en los rótulos de tus dibujos.

Analiza

4. ¿Cuál es la diferencia entre una hoja simple y una hoja compuesta?

5. Describe dos modelos de disposición de las venas de las hojas.

6. Según lo que has aprendido sobre adaptación, explica por qué hay tantas variedades de hojas.

Materiales

- 5 especímenes de hojas
- guía de plantas (opcional)
- guantes protectores

Hoja compuesta

Profundizar

Escoge un compañero o compañera. Con ayuda de las claves y las listas de vocabulario, describe una hoja y comprueba si él o ella puede dibujarla a partir de tu descripción. Después, cambien papeles.

596

Science Skills
Worksheet 2
"Using Your Senses"

Vocabulario de las formas de la hoja

cordada: forma de corazón

lanceolada: forma de lanza

lobulada: con lóbulos

oblonga: hojas con la punta redondeada

orbicular: forma de disco

ovalada: forma de óvalo

peltada: forma de escudo

reniforme: forma de riñón

sagitada: forma de flecha

Vocabulario de los tallos

leñoso: la corteza recubre el tallo

herbáceo: tallos verdes y no leñosos

Vocabulario de los arreglos de las hojas

alternado: las hojas o folíolos alternados a lo largo del tallo o pecíolo

compuesta: hoja dividida en segmentos o varios folíolos en el pecíolo

opuesta: hoja compuesta con varios folíolos unos frente a otros a lo largo del pecíolo

palmeada: una sola hoja con venas dispuestas alrededor de un punto central

palmeada compuesta: varios folíolos dispuestos alrededor de un punto central

pecíolo: tallito que sostiene a la hoja

pinnada compuesta: varios folíolos a cada lado del pecíolo

sencilla: una sola hoja sujeta al tallo por medio del pecíolo

Lab Notes

This activity is designed to help students recognize that leaves have many forms and that the variations of leaf shapes are adaptations for a particular function. This exercise also acquaints students with the identification process used in some field or classification guides. Encourage students to recognize the difference between a stem and a petiole and the difference between a leaf and a leaflet.

Answers

4. A simple leaf is a single leaf at the end of a single petiole. A compound leaf has several leaflets on the end of a petiole in various arrangements.

5. Veins can be arranged in three ways—parallel, extending straight up and down the leaf (which is usually elongated); pinnate, which on a single leaf is a network of veins extending out from a central vein; and palmate, which is an arrangement of veins originating from a single point on a leaf that has several lobes.

6. Answers will vary but should include adaptations for water conservation, light reception, and insect resistance, among others. Explain that leaf shape is only one of the adaptations leaves can have and that the thickness of the leaf and the waxy coating of the leaf are also adaptations. Have them compare an oak leaf with a conifer leaf.

Datasheets for LabBook
Datasheet 22

Travelin' Seeds
Teacher's Notes

Time Required

One 45-minute class period

Lab Ratings

EASY → HARD

TEACHER PREP
STUDENT SET-UP
CONCEPT LEVEL
CLEAN UP

Preparation Notes

Challenge Cards Make up seed-dispersal challenge cards ahead of time. Five basic challenges are listed below. Copy each as needed so that there is one card per student or group.

1. Modify your seed so that it will be carried on an animal's fur as the animal passes by the plant. The seed must travel for at least 1 m. The animal must not be aware that the seed has been attached.

2. Modify your seed so that it will be flung 1 m away from the parent plant (you/group). Be careful that your tests do not fly at others in the class.

3. Modify your seed so that it will glide or float in the air when it falls off the parent plant. The seed must be dropped, not pushed or thrown, and must travel at least 1 m.

4. Modify your seed so that animals will find it desirable to eat and digest.

5. Modify your seed so that it will float on water for at least 1 minute.

Semillas viajeras

Al estudiar las plantas conociste varias adaptaciones muy interesantes y poco comunes. Entre las más interesantes están las modificaciones que permiten la dispersión o distribución de las semillas y frutos lejos de la planta progenitora. La dispersión permite a las plántulas obtener espacio, luz solar y otros recursos sin competir de manera directa con la planta progenitora.

En esta actividad dispersarás una semilla con ayuda de tu creatividad.

Procedimiento

1. Pídele a tu maestra o maestro una semilla y una tarjeta de retos de la dispersión. En tu cuaderno de ciencias, registra el tipo de tarjeta que recibiste.

2. Diseña un plan para utilizar los materiales disponibles para dispersar la semilla como se describe en la tarjeta. Registra el plan en tu cuaderno de ciencias y, antes de comenzar, pídele a tu maestro o maestra su aprobación.

3. Consigue los materiales que necesitas para probar tu método de dispersión de semillas.

4. Con el permiso de tu maestro o maestra haz varias pruebas. Elabora una tabla de datos en tu cuaderno de ciencias y registra los resultados.

Analiza

5. ¿Pudiste completar con éxito el reto de dispersión de semillas? Describe en qué consiste el éxito de tu método.

6. ¿Puedes hacerle modificaciones al método para mejorar la dispersión de las semillas?

7. Describe las plantas que piensas que pueden dispersar sus semillas de manera similar a tu método.

Profundizar

Enséñales a tus compañeros y compañeras tu método de dispersión de semillas.

Experimentos

Materiales

- semilla de frijol
- tarjeta de retos de dispersión de semillas
- diversos artículos del hogar o materiales reciclados (ejemplos: pegamento, cinta adhesiva, papel, clips, ligas, tela, vasos y platos de papel, toallas de papel, cartón)

semilla de mangle

álamo

baya silvestre

abrojo

598

Answers
5.–7. Answers will vary.

Datasheets for LabBook
Datasheet 23

Jane Lemons
Western Rockingham
Middle School
Madison, North Carolina

Construye una flor

Los científicos hacen con frecuencia modelos en el laboratorio para entender procesos o estructuras. Con ayuda de tu creatividad y de tu conocimiento de la estructura de las flores, construirás una maqueta con artículos reciclados y materiales de arte.

Procedimiento

1. En tu cuaderno de ciencias, dibuja una flor similar a la que se muestra abajo, a la derecha. Rotula cada una de las partes. La flor de la ilustración tiene estructuras masculinas y femeninas, pero no todas las flores las tienen.

2. Examina los materiales disponibles para la construcción de tu flor. Decide cuáles son adecuados para cada parte de la flor. Ahora, construye un modelo tridimensional que incluya cada una de las partes de la lista de abajo.

Partes de una flor	
Pétalos	
Sépalos	
Tallo	
Pistilo	(estigma, estilo, ovario)
Estambre	(antera, filamento)

3. En una tarjeta de 3 × 5 pulgadas, dibuja un esquema de tu modelo en el que cada una de las estructuras esté rotulada.

Analiza

4. Haz una lista de las estructuras de las flores y explica la función de cada una.

5. Explica de qué manera la flor que has construido atrae a los polinizadores. ¿Qué modificaciones podrías hacer para aumentar el número de ellos?

6. Describe cómo podría llevarse a cabo la autopolinización en tu modelo.

Profundizar

Haz una exposición de los modelos terminados.

Materiales

- artículos reciclados (ejemplos: platos y vasos de papel, recipientes de yogurt, alambre, cordel, cuentas, botones, cartulina, botellas)

- materiales de arte (ejemplos: pegamento, cinta adhesiva, tijeras, papel de colores, limpiadores de pipa, hilo)

- tarjeta de 3 × 5 pulgadas

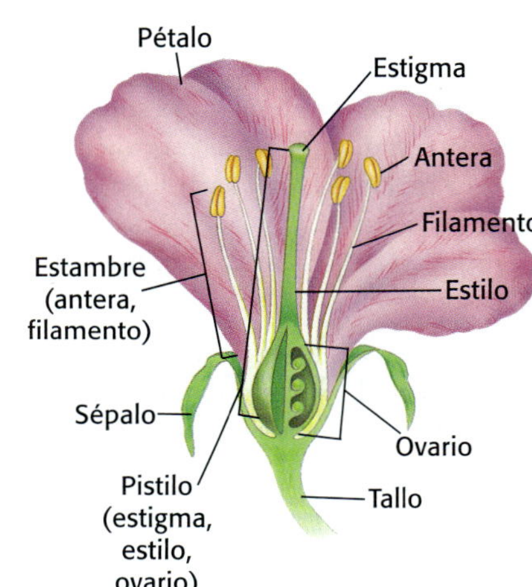

Jane Lemons
Western Rockingham
Middle School
Madison, North Carolina

5. Answers will vary but should include bright flowers, honey guides on the petals, strong scent, and the position and number of pistils and stamens.

6. Answers should explain how the model flower moves pollen from the anther to the stamen.

Build a Flower
Teacher's Notes

Time Required

One 45-minute class period

Lab Ratings

TEACHER PREP
STUDENT SET-UP
CONCEPT LEVEL
CLEAN UP

Answers

4. **petal:** the often colorful crown of a flowering plant that attracts pollinators

   sepal: the leaves that form the base of the flower and that enclose and protect the bud before the flower opens

   stem: the main stalk of the aboveground part of a plant from which leaves, flowers, and fruit develop; water and nutrients move through the stem between the flower and roots.

   pistil: the female reproductive organ of a flower

   stigma: the upper tip of the pistil, which receives pollen

   style: the stalklike part of the pistil between the stigma and ovary that holds the stigma where it can receive pollen

   ovary: the enlarged part of the pistil in which ovules are formed

   stamen: the male reproductive organ in flowers

   anther: the top of the stamen, contains pollen

   filament: the threadlike part of the stamen that holds the anther

Datasheets for LabBook
Datasheet 24

Food Factory Waste
Teacher's Notes

Time Required

One 45-minute class period and about 15 minutes per day for 5 days

Lab Ratings

EASY ———————————→ HARD

TEACHER PREP
STUDENT SET-UP
CONCEPT LEVEL
CLEAN UP

MATERIALS

The materials listed on the student page are enough for 1 student.

- *Elodea* is a common aquarium plant and can be found at some pet stores and most places that sell aquarium fish.
- A 5 percent solution of baking soda and water can be made by dissolving 5 g of baking soda in 95 mL of water.

Safety Caution

Remind students to review all safety cautions and icons before beginning this lab activity.

David Sparks
Redwater Junior High School
Redwater, Texas

Las sobras de la fotosíntesis

Ya sabes que la fotosíntesis de las plantas produce alimentos. Durante la fotosíntesis, también se producen desechos, sin los cuales no podríamos vivir.

En esta actividad vas a observar el proceso de la fotosíntesis y a determinar la tasa diaria de fotosíntesis de dos o tres ramitas de *Elodea*.

Procedimiento

1. Ponte gafas protectoras, guantes y bata de laboratorio.

2. Llena tres cuartas partes de un vaso de precipitados con una solución al 5 por ciento de bicarbonato de sodio y agua.

3. Coloca en el recipiente dos o tres ramitas de *Elodea* de 20 cm de largo. El bicarbonato de sodio le dará a la *Elodea* el dióxido de carbono que necesita para la fotosíntesis.

4. Coloca la parte ancha del embudo sobre las ramitas de *Elodea* de modo que la punta del embudo apunte hacia arriba. La solución debe cubrir completamente la *Elodea* y el embudo, como se ilustra a la derecha.

5. Llena por completo el tubo de ensayo con la solución al 5 por ciento de bicarbonato de sodio y agua. Cubre el tubo de ensayo con el pulgar y dale la vuelta, procurando que no le entre aire. Mete la boca del tubo de ensayo en la solución, y colócalo sobre la punta del embudo, como se muestra a la derecha. Mientras haces esto, trata de que no se salga el agua que está en el tubo.

6. Coloca el vaso de precipitados y su contenido en un área bien iluminada, junto a una lámpara, o si es posible, donde le dé la luz del sol.

7. En tu cuaderno de ciencias, haz una tabla de datos como la que se muestra a continuación.

Materiales

- un vaso de precipitados de 600 mL
- solución al 5 por ciento de bicarbonato de sodio y agua
- embudo de vidrio
- ramitas de 20 cm de largo de *Elodea* (2 o 3)
- tubo de ensayo
- regla métrica
- fuente de luz
- guantes protectores

Cantidad de gas en el tubo de ensayo		
Exposición a la luz (días)	Cantidad total de gas presente (mm)	Cantidad de gas producido por día (mm)
0		
1		
2		
3		
4		
5		

No escribas en el libro

600

Datasheets for LabBook
Datasheet 25

Science Skills
Worksheet 25
"Introduction to Graphs"

8. En tu cuaderno de ciencias, apunta que la cantidad de gas en el tubo de ensayo era 0 en el día 0. (Si no lograste colocar el tubo de ensayo sin que le entrara aire, mide la altura de la columna de aire en el tubo de ensayo, usando una regla métrica y apunta esta medida para el día 0). Mide la cantidad de gas en el tubo de ensayo desde la mitad de la curvatura del fondo del tubo de ensayo, que está apuntando hacia arriba, hasta el nivel de la solución.

9. Mide la cantidad de gas durante cinco días. Apunta tus medidas en el renglón indicado de la tabla de datos, bajo el encabezado "Cantidad de gas presente (mm)."

10. Calcula la cantidad de gas producida cada día restando la cantidad de gas presente el día anterior a la cantidad presente hoy. Apunta estas cantidades en tu tabla de datos en el renglón indicado, bajo el encabezado "Cantidad de gas producido por día (mm)."

11. En tu cuaderno de ciencias, haz una gráfica como la que se muestra a continuación. Traza los datos de la tabla en la gráfica.

Análisis

12. Usa la información de la gráfica para describir lo que pasó con la cantidad de gas en el tubo de ensayo.

13. ¿Qué cantidad de gas se produjo en el tubo de ensayo después del día 5?

14. Escribe la fórmula de la fotosíntesis en tu cuaderno de ciencias. Explica cada parte. Por ejemplo, ¿qué "ingredientes" hacen falta para la fotosíntesis? ¿Qué substancias se producen? ¿Cuál es el gas que las plantas producen como desecho y que nosotros necesitamos para vivir?

15. Escribe un informe en el que describas el experimento, los resultados y tus conclusiones.

Profundizar
La hidroponia es el cultivo de plantas en agua enriquecida con nutrientes, sin usar tierra. Investiga las técnicas de hidroponia y aplícalas al cultivo de una planta.

Lab Notes
You may need to have students practice placing the test tube over the inverted funnel. It may take two or three tries to get the test tube over the funnel stem without letting any air into the tube. First fill the test tube with the solution. Place your thumb over the opening tightly so no air can get in. Submerge your thumb and the top of the test tube underwater. Once the top of the test tube is underwater you can release your thumb to maneuver the test tube over the stem of the funnel. Be sure you have the *Elodea* in place under the funnel before you begin!

You may wish to write the equation for photosynthesis on the board.

Answers

12. Students' graphs should show a gradual increase in the amount of gas in the test tube.

13. Answers will vary according to variables in the classroom, such as the amount of light and temperature.

14. $6CO_2 + 6H_2O$ + light energy $\rightarrow C_6H_{12}O_6 + 6O_2$
CO_2 is carbon dioxide and comes from the baking soda solution. H_2O is the water in the solution. Light energy comes from the sun. $C_6H_{12}O_6$ is sugar (glucose), and O_2 is oxygen. Photosynthesis produces sugar and oxygen. Plants produce oxygen as a byproduct of photosynthesis; oxygen is the gas that we breathe and cannot live without.

Weepy Weeds
Teacher's Notes

Time Required

One or two 45-minute class periods

Lab Ratings

EASY ———————→ HARD

TEACHER PREP
STUDENT SET-UP
CONCEPT LEVEL
CLEAN UP

MATERIALS

The materials listed on the student page are enough for 1 student. The plant used in this lab can be any leafy plant, such as a bean plant or a coleus. The plant shown is a coleus with all but the top four leaves trimmed away.

Safety Caution

Remind students to review all safety cautions and icons before beginning this lab activity.

Lab Notes

If your lab period is short, you may want to eliminate the measurement of the height of water in the test tube at 40 minutes.

Although it is not essential to the activity, you may want to begin with an exact amount of water in each test tube. Students would then know that the difference can be due only to evaporation and transpiration.

Plantas lloronas

Imagínate que eres un científico o científica investigando el modo de drenar un área que está inundada de agua contaminada con fertilizantes. Sabes que las plantas liberan agua a través de los estomas de las hojas, y que cuando el agua se evapora de las hojas, las raíces de la planta absorben más, y la envían hacia los tallos y las hojas. En este proceso, llamado transpiración, la planta absorbe agua y nutrientes de la tierra. De hecho, alrededor de un 90 por ciento del agua que la planta toma a través de sus raíces se libera en la atmósfera como vapor, a través de la transpiración. Se te ocurre una idea: si colocas plantas en el área inundada, éstas transpirarán el agua, y también absorberán el fertilizante por las raíces.

¿Cuánta agua puede absorber y liberar una planta en un período de tiempo dado? En esta actividad, vas a observar el proceso de la transpiración y a determinar la tasa de transpiración del tallo de una planta.

Materiales

- tallo de Coleus o de otra planta
- regla métrica
- 2 tubos de ensayo
- gradilla
- reloj
- papel milimétrico
- agua
- marcador para vidrio

Procedimiento

1. En tu cuaderno de ciencias, haz una tabla de datos como la que se muestra a continuación para registrar medidas.

Altura del agua en los tubos de ensayo		
Tiempo	**Tubo de ensayo con la planta**	**Tubo de ensayo sin la planta**
Inicial		
Después de 10 min		
Después de 20 min		
Después de 30 min		
Después de 40 min		
Al día siguiente		

No escribas en el libro

2. Llena aproximadamente tres cuartas partes de ambos tubos de ensayo con agua y colócalos en la gradilla.

3. Sigue las instrucciones que te dé el maestro para conseguir el tallo de planta. Colócalo de manera que quede parado dentro de uno de los tubos de ensayo. Tus tubos deben quedar como los que ves en la fotografía de la derecha.

4. Con el marcador para vidrio, traza una línea que indique el nivel de agua en cada tubo de ensayo. Asegúrate de que el tallo esté dentro del tubo antes de que lo marques. ¿Por qué crees que esto es necesario?

Datasheets for LabBook
Datasheet 26

Science Skills
Worksheet 27
"Interpreting Your Data"

David Sparks
Redwater Junior High School
Redwater, Texas

5. Con la regla métrica, mide la altura del agua en cada tubo de ensayo. Asegúrate de que el tubo de ensayo esté derecho, y mide desde la superficie del agua hasta la mitad de la curva del fondo del tubo. Apunta estas medidas en el renglón de la tabla de datos que dice "Inicial".

6. Después de diez minutos, mide de nuevo la altura del agua en cada tubo de ensayo. Apunta estas medidas en el renglón de la tabla de datos que dice "Después de 10 min."

7. Repite tres veces el paso 6. En cada ocasión, apunta tus medidas en el renglón indicado.

8. Espera 24 horas y vuelve a medir la altura del agua en cada tubo de ensayo. Apunta estas medidas en el renglón de la tabla de datos que dice "Al día siguiente".

9. En papel milimétrico, traza una gráfica como la que se muestra abajo, usando solamente los resultados del primer día del experimento. Traza dos líneas, una para cada tubo de ensayo. Usa un color diferente para cada línea e indica la clave bajo tu gráfica, como en el ejemplo de abajo.

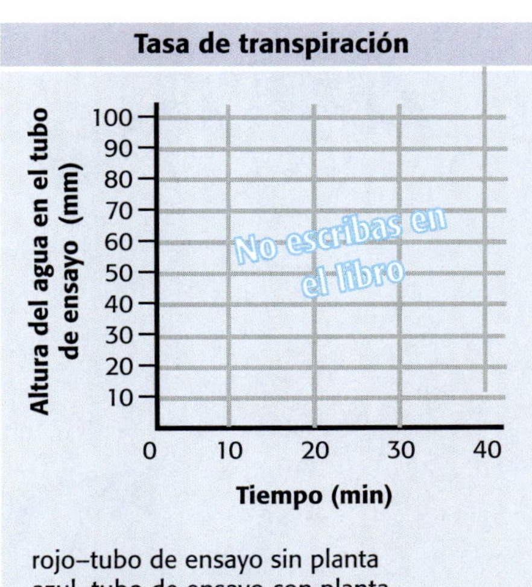

rojo–tubo de ensayo sin planta
azul–tubo de ensayo con planta

10. Calcula la tasa de transpiración de tu planta con las siguientes operaciones:

> **Tubo de ensayo con planta:**
> Altura inicial
> − Altura al día siguiente
> Diferencia en el nivel de agua **(A)**

> **Tubo de ensayo sin planta:**
> Altura inicial
> − Altura al día siguiente
> Diferencia en el nivel de agua **(B)**

> **Diferencia en el nivel de agua debida a la transpiración de la planta:**
> Diferencia **A**
> − Diferencia **B**
> Pérdida de agua (mm) debida a la transpiración de la planta en 24 horas

Análisis

11. ¿Para que sirvió el tubo de ensayo que sólo tenía agua?

12. ¿Qué hizo que el nivel de agua bajara en el tubo de ensayo que contenía el tallo? ¿Sucedió lo mismo en el tubo de ensayo que sólo tenía agua? Explica tu respuesta.

13. ¿Qué tasa de transpiración por día obtuviste con tus cálculos?

14. Usando tu gráfica, compara la tasa de transpiración con la tasa de evaporación de los tubos de ensayo.

15. Prepara una exposición de tu experimento para presentar en clase, usando tus tablas de datos, gráficas, cálculos y ayudas visuales.

Profundizar

¿Cuántas hojas tenía tu tallo? Usa este número para calcular la tasa de transpiración de una planta con 200 hojas. Cuando tengas lista tu respuesta en milímetros de altura en el tubo de ensayo, vacía esta cantidad en un recipiente graduado para medirla en mililitros.

Experimentos

Answers

11. The test tube that holds only water is a control; it will lose water only by evaporation.

12. Water in the test tube containing the plant stem will be lost through evaporation and transpiration. Evaporation is the only means of water loss in the test tube without the plant stem.

13. This answer will vary according to several variables in the classroom, such as the amount of light and the temperature.

14. This answer will vary. Have students compare and contrast the lines on the graph and explain how the graph is easier to interpret than numbers in a data list.

Wet, Wiggly Worms!
Teacher's Notes

Time Required

One 45-minute class period

Lab Ratings

EASY ———————————→ HARD

TEACHER PREP 🔺🔺
STUDENT SET-UP 🔺🔺
CONCEPT LEVEL 🔺🔺
CLEAN UP 🔺🔺

MATERIALS

The materials listed on the student page are enough for a group of 4–5 students.

Safety Caution

Remind students to review all safety cautions and icons before beginning this lab activity. Students may wish to wear protective gloves while handling the worms.

¡Lombrices escurridizas!

Las lombrices de tierra han cavado la Tierra idurante más de 100 millones de años! Viven en cualquier lugar de la Tierra excepto en los desiertos y casquetes polares. Por lo general, no miden más de 30 cm (1 pie) pero pueden llegar a crecer hasta 10 veces más. Las lombrices fertilizan la tierra con sus desechos y la aflojan a su paso. Le sirven de alimento a muchos animales, como pájaros, ranas, roedores y peces. ¡Hay quienes dicen que también son buen alimento para las personas! Si alguien te ofrece plato de *pois de terre*, ten cuidado. Está lleno de lombrices.

En esta actividad, vas a observar el comportamiento de una lombriz viva. Recuerda que las lombrices son animales vivos que deben ser manejados con cuidado. Asegúrate de mantener húmeda la lombriz durante esta actividad. La piel de la lombriz debe permanecer húmeda para que pueda recibir oxígeno. Si la piel se seca, la lombriz se asfixia y muere. Usa un cuentagotas con agua para humedecer la lombriz.

Materiales

- lombriz viva
- linterna
- toallas de papel
- hojas de apio
- charola de disección
- sonda
- tierra
- atomizador
- caja de zapatos
- agua
- cuentagotas
- cartulina
- marcadores de colores
- regla métrica

Procedimiento

1. Pon una toalla de papel sobre una charola de disección. Coloca una lombriz viva sobre la toalla de papel y observa cómo se mueve. Anota tus observaciones en el cuaderno de ciencias.

2. Usa la sonda para tocar con cuidado el extremo anterior (la cabeza) de la lombriz. Con la sonda, toca con cuidado otras áreas del cuerpo de la lombriz. Anota los tipos de respuestas que observes.

3. Coloca unas hojas de apio en un extremo de la charola. Anota cómo reacciona la lombriz ante la presencia de alimento.

4. Con una linterna, ilumina el extremo anterior de la lombriz. Anota la reacción de la lombriz ante la luz.

5. Pon una toalla de papel húmeda en el fondo de la caja de zapatos. Cubre la mitad de la caja con la tapa y deja la otra mitad descubierta.

6. Pon la lombriz en el lado descubierto de la caja, donde le dé la luz. Observa su conducta durante 3 minutos y anota lo que ves.

604

Gladys Cherniak
St. Paul's Episcopal School
Mobile, Alabama

7. Coloca la lombriz en la parte cubierta de la caja. Observa su conducta durante 3 minutos y anota lo que ves. Repite 3 veces los pasos 6 y 7. Si repites estos pasos, puedes darte cuenta si la conducta de la lombriz siempre es la misma.

8. Esparce una capa pareja de tierra, de unos 4 cm de profundidad, en el fondo de la caja. Pon la lombriz sobre la tierra. Observa su conducta durante 3 minutos y anota lo que ves.

9. Humedece la tierra de un lado de la caja y deja el otro lado seco. Coloca la lombriz en el centro de la caja entre la tierra húmeda y la tierra seca. Tapa la caja y espera 3 minutos. Destapa la caja y anota tus observaciones. Repite este procedimiento 3 veces. Recuerda que debes tratar a la lombriz con cuidado. (¡Tal vez tengas que buscarla!)

Análisis

10. ¿Cómo reaccionó la lombriz cuando la tocaste? ¿Tiene unas áreas más sensibles que otras? ¿Cómo influye la luz en la conducta de la lombriz? Con base en tus observaciones, describe cómo puede proteger a un animal su reacción ante un estímulo.

11. ¿Cómo reaccionó la lombriz ante la presencia de alimento?

12. Cuando la lombriz tuvo la opción de estar en la tierra mojada o en la seca, ¿qué hizo? Explica este resultado.

Profundizar

Con base en tus observaciones de la conducta de una lombriz, prepara un cartel mostrando dónde puedes encontrar lombrices. Haz un dibujo con los marcadores o recorta figuras de revistas o de otras publicaciones. Incluye todas las variables que usaste en tu experimento, como la presencia o ausencia de tierra, la tierra húmeda o seca, la luz o la obscuridad, y la comida. Escribe una leyenda en la parte de abajo de tu cartel que describa dónde hay lombrices en la naturaleza.

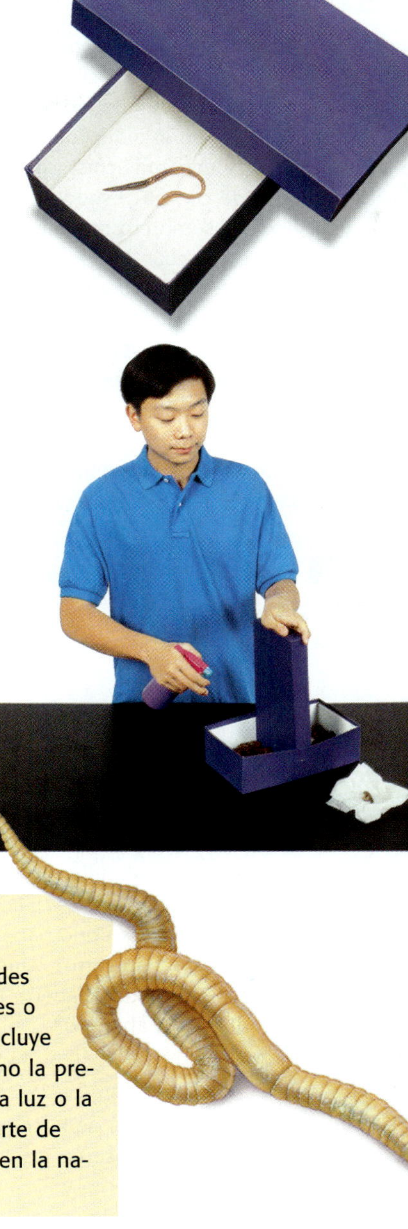

Lab Notes

Earthworms are scientifically classified as animals belonging to the order Oligochaeta, class Chaetopoda, and phylum Annelida. There are about 1,800 species of earthworms. Only two of these are grown commercially. Earthworms have setae, or bristles, located on each segment that help them move. Earthworms have both male and female reproductive organs. They usually do not self-fertilize, but they do exchange sperm as they pass in their burrows. Eggs are deposited in the burrow in a cocoon. The cocoon is manufactured by the clitellum that encircles the body of the worm. Different segments of the earthworm perform different functions, just as each of our body parts do. Earthworms have from 95 to 150 segments, depending on the species.

Datasheets for LabBook
Datasheet 27

Answers

10.–12. Students' answers will vary according to their own observations. They will probably observe that the worm squirms when touched and that some areas are more sensitive than others, such as the clitellum. Students will probably observe that earthworms avoid light. Students should describe the worm's behavior as self-protective.

Going Further

Students' posters should describe warm, wet soil, darkness, and partially decayed organic matter for food.

Aunt Flossie and the Bumblebee
Teacher's Notes

Time Required

One to three 45-minute class periods

Lab Ratings

EASY ———→ HARD

TEACHER PREP △△
STUDENT SET-UP △△△
CONCEPT LEVEL △△
CLEAN UP △△

MATERIALS

The materials listed on the student page are enough for a group of 4–5 students. Materials students may need are construction paper in several bright colors, shoe boxes, scents, honey or some other sweet spread, twine, and binoculars or a hand lens. Encourage students to use recycled materials and to bring in their own supplies.

Safety Caution

Remind students to review all safety cautions and icons before beginning this lab activity.

Tell students to avoid wearing bright floral clothing and perfume or cologne while performing this lab.

All students should be cautious when working with wildlife. Students allergic to insect and bee stings should be excused from this exercise.

La tía Florecita y el abejorro

DISEÑA EL TUYO

Experimentos

La semana pasada, la tía Florecita vino a ver el partido de fútbol, pero todos terminaron viéndola a ella. Un enorme abejorro amarillo con negro la estaba persiguiendo. El abejorro no la picó, pero no dejaba de zumbarle en el oído. Todos trataban de no reírse, pero la tía Florecita se veía muy chistosa. Corría y gritaba, toda perfumada y arreglada con un vestido floreado, joyas extravagantes y un enorme sombrero con un largo listón morado. Corrió hacia las gradas y, cuando pensó que el abejorro se había ido, se sentó, pero éste la volvió a encontrar. Nadie podía entender por qué el abejorro sólo atormentaba a la tía Florecita y a nadie más. La tía dijo que no volvería a ningún otro partido hasta que tú determinaras por qué la perseguía el abejorro.

Tu tarea es diseñar un experimento que determine por qué el abejorro se sintió atraído hacia la tía Florecita. Si quieres, simula la situación usando objetos que tengan las mismas pistas sensoriales que usó la tía Florecita ese día: colores brillantes y fragancias fuertes.

Materiales

- dependerán de cada diseño experimental y de la aprobación del maestro o maestra

MÉTODO CIENTÍFICO

Haz una pregunta

1. Usa la información de la historia para formular tus preguntas. Haz una lista de las características de la tía Florecita el día del partido. ¿Cómo iba vestida? ¿Con qué crees que la confundió el abejorro? ¿Qué fragancia usaba ese día? ¿Cuál de esas características pudo haber afectado la conducta del abejorro? ¿Qué tenía la tía Florecita que pudo haber afectado la conducta del abejorro?

Formula una hipótesis

2. Formula una hipótesis acerca de la conducta de los insectos con base en tus observaciones de la tía Florecita y el abejorro. Una hipótesis posible es: "Los insectos son atraídos por fuertes fragancias florales". Formula tu propia hipótesis.

606

Datasheets for LabBook
Datasheet 28

Barry Bishop
San Rafael Junior High
Ferron, Utah

Comprueba la hipótesis

3. Diseña el procedimiento de tu experimento. Asegúrate de seguir los pasos del método científico. Procura que el procedimiento responda a preguntas específicas. Por ejemplo, si quieres saber si los insectos son atraídos por los colores, tal vez convendría que mostraras recortes de papel de diferentes colores.

4. Haz una lista de los materiales que vas a usar en tu diseño experimental. Puedes usar como carnada objetos de varios colores, hojas de papel de colores, imágenes de revistas o perfumes fuertes. No necesitas usar seres vivos como carnada. Tu maestro o maestra debe aprobar tu diseño experimental antes de que empieces.

5. Determina un lugar y un método para realizar el procedimiento. Por ejemplo, puedes colocar cosas que atraigan a los insectos en una caja en el suelo o colgarlas de la rama de un árbol.
   Cuidado: Asegúrate de no estar muy cerca del área donde pusiste la carnada. No toques los insectos. Pídele a un adulto que te ayude a liberar los insectos que atrapaste o que reuniste.

6. Haz tablas de datos para anotar los resultados de tus pruebas. Por ejemplo, usa una tabla de datos como la de la derecha, y anota los resultados de la prueba de los colores para ver qué insecto fue atraído por ellos. Diseña las tablas de datos que se ajusten a tu investigación.

Analiza los resultados

7. Describe el procedimiento de tu experimento. ¿Apoyaron los resultados tu hipótesis? Si no, ¿cómo cambiarías el procedimiento?

Comunica los resultados

8. Escríbele una carta a la tía Florecita para contarle lo que has aprendido sobre el comportamiento de los insectos. Explícale cuál fue la causa del ataque del abejorro. Invítala a otro partido de fútbol, pero adviértele lo que debe y lo que no debe ponerse.

Efectos del color			
Color	Número de abejas	Número de hormigas	Número de avispas
Rojo			
Azul			
Amarillo			

Experimentos

607

Lab Notes

This lab may need to be done during a certain season in your geographical area. Some students may want to extend their data collection period to several days or weeks.

Answers

7. Students should describe their experimental procedure and include as many steps of the scientific method as possible. All answers will depend on student observations.

8. Letters will vary, but should demonstrate what the students have learned about insect behavior.

Science Skills Worksheet 13
"Designing an Experiment"

Porifera's Porosity
Teacher's Notes

Time Required

One 45-minute class period

Lab Ratings

EASY — HARD

TEACHER PREP ▲▲
STUDENT SET-UP ▲
CONCEPT LEVEL ▲
CLEAN UP ▲

MATERIALS

Students may choose a very light absorbent material, such as a tissue, for the Going Further. You may need to have access to an electronic scale.

Safety Caution

Remind students to review all safety cautions and icons before beginning this lab activity.

Preparation Notes

Many students have never touched a natural sponge. Allow them a few moments to experience the feel of a natural sponge, and help them identify the structures. Tell them that the sponges are animals even though they don't appear animal-like. Review the life cycle of a sponge, and discuss how sponges obtain food.

Answers

5. Answers will vary. Students should subtract the mass of the dry sponge from the mass of the wet sponge.

6. Sponges must hold and filter an enormous amount of water to obtain a small amount of food.

Porosidad de los poríferos

Las esponjas son animales invertebrados acuáticos que constituyen el filo de los poríferos. El nombre *poríferos* viene del latín y significa "que tiene poros". Las esponjas bombean el agua a su interior a través de poros pequeños. De esta manera, la filtran para obtener alimento. El agua sale de la esponja por una abertura que tiene en la parte superior.

Los primeros biólogos pensaban que las esponjas eran plantas, ya que se parecen a ellas de muchas maneras. Por ejemplo, las esponjas adultas se fijan al suelo y no se mueven, no pueden perseguir su alimento y tienen que filtrar mucha agua para obtenerlo. En esta actividad determinarás la cantidad de agua que las esponjas son capaces de absorber.

Experimentos

Materiales

- esponjas naturales
- probeta graduada
- balanza
- calculadora
- agua
- bol (lo suficientemente grande para la esponja y el agua)
- materiales adicionales conforme se requieran

Procedimiento

1. Estima la cantidad de agua que retiene la esponja. Escribe tu predicción en el cuaderno de ciencias.

2. En tu cuaderno de ciencias, elabora las tablas de datos o bosquejos que necesites para registrar tus observaciones.

3. Asegúrate de medir la masa de la esponja antes de agregarle agua y de registrar el resultado. ¿Por qué es éste un paso necesario?

4. Antes de empezar el experimento, pídele a tu maestro o maestra su aprobación. Realiza el experimento y registra tus resultados.

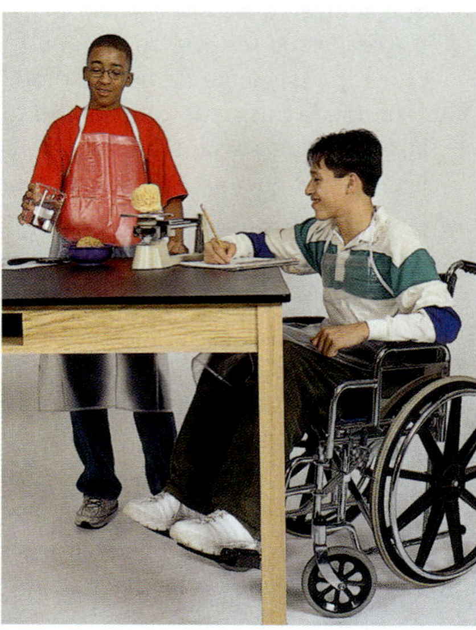

Analiza

5. ¿Cuántos milímetros de agua retiene la esponja por gramo de tejido? Por ejemplo, si la esponja tiene una masa seca de 12 g y retiene 59.1 mL de agua, entonces retiene 4.9 mL de agua por gramo. (59.1 mL ÷ 12 g)

6. ¿Qué ventaja alimenticia tiene la esponja gracias a su capacidad de retener una gran cantidad de agua?

Profundizar

Repite el experimento con otro tipo de material absorbente, como toallas de papel o una esponja artificial. ¿Cuántos mililitros de agua retiene este material por gramo? ¿Cuál es la diferencia con una esponja natural?

608

Going Further

Students will probably find that very few other materials hold as much water as a natural sponge. Students may want to measure the absorbency of disposable diapers. In those, liquid is chemically converted to a gel.

Datasheets for LabBook
Datasheet 29

Kathy LaRoe
East Valley Middle School
East Helena, Montana

Travesuras de grillos

Los insectos constituyen una clase de invertebrados con más de 750,000 especies conocidas y son el grupo animal de mayor éxito sobre la Tierra. En esta actividad observarás la estructura de un grillo y qué comportamientos de adaptación lo hacen tan exitoso. Recuerda que trabajarás con un ser vivo que merece ser tratado con cuidado.

Procedimiento

1. Coloca un grillo en un vaso de precipitados limpio de 600 mL y cúbrelo de inmediato con plástico para envolver. La cantidad de oxígeno que hay en el recipiente es suficiente para que el grillo respire mientras realizas tu trabajo.

2. Mientras el grillo se habitúa al recipiente, en tu cuaderno de ciencias elabora una tabla de datos similar a la que aparece abajo. Asegúrate de dejar espacio suficiente para las descripciones.

Estructuras del cuerpo del grillo	
Número	**Descripción**
Segmentos del cuerpo	
Antenas	
Ojos	
Alas	

No escribas en el libro

3. Sin hacer muchos movimientos, empieza a examinar el grillo. Llena la tabla de datos de las estructuras de su cuerpo.

4. Introduce un pequeño pedazo de manzana en el vaso de precipitados y déjalo sobre una mesa. En silencio, observa al grillo durante varios minutos. Cualquier movimiento hará que el grillo interrumpa su actividad. Registra tus resultados en tu cuaderno de ciencias.

5. Destapa el vaso de precipitados y saca el pedazo de manzana; rápidamente coloca otro vaso de precipitados, de manera que las bocas de ambos recipientes queden juntas; pégalas con cinta adhesiva. Maneja con cuidado los recipientes, recuerda que contienen un animal vivo.

Materiales

- grillos (2)
- vasos de precipitados de 600 ml (2)
- plástico para envolver
- manzana
- lupa (opcional)
- cinta adhesiva
- papel aluminio
- bolsas de plástico con cierre (2)
- lámpara
- hielo
- agua caliente de la llave

Experimentos

609

Alonda Droege
Pioneer Middle School
Steilacom, Washington

Datasheets for LabBook
Datasheet 30

The Cricket Caper
Teacher's Notes

Time Required

One to two 45-minute class periods

Lab Ratings

EASY ———→ HARD

TEACHER PREP   ▲▲▲
STUDENT SET-UP   ▲▲
CONCEPT LEVEL   ▲▲
CLEAN UP   ▲▲

> ### MATERIALS
>
> The materials listed on the student page are enough for a single student or a small group of students. Instead of 600 mL beakers, you may use the bottom halves of 2 clear plastic 2 L bottles. You will need to prepare these ahead of time. The cut on the bottle should be as even as possible to facilitate taping the open ends together in step 5.

Safety Caution

Remind students to review all safety cautions and icons before beginning this lab activity.

Lab Notes

Explain to students that they must move slowly so they won't startle the cricket and alter its behavior. The apple must be removed in step 5 before the containers are taped together. The apple would be an unwanted variable in the tests that follow.

If you decide to extend this activity over two class periods, the cricket will be fine overnight in a covered 500 mL beaker. The cricket will need a slice of potato or apple for food and moisture.

6. Forra con papel aluminio uno de los vasos de precipitados.

7. Si el grillo se esconde en la parte forrada de aluminio, da con cuidado unos golpecitos en el recipiente hasta que lo veas de nuevo. Coloca los recipientes de lado y con una lámpara ilumina la parte que no está cubierta. Observa al grillo y registra su ubicación.

8. Observa y registra la ubicación del grillo después de 5 minutos. Sin molestarlo, cambia con cuidado el papel de aluminio al otro recipiente (esto expone el grillo a la luz). Después de 5 minutos, observa al grillo y registra su ubicación. Repite esta secuencia una vez más para ver si obtienes el mismo resultado.

9. Llena a la mitad una bolsa de plástico con escarcha y llena otra con agua de la llave caliente, también a la mitad. Cierra las bolsas y colócalas sobre la mesa una al lado de la otra.

10. Retira el papel aluminio y mece los recipientes suavemente hasta que el grillo quede en el centro. Coloca los recipientes sobre las bolsas de plástico, tal como se muestra abajo.

11. Observa el comportamiento del grillo por 5 minutos. Anota tus observaciones en tu cuaderno de ciencias.

12. Coloca los vasos de precipitados en posición vertical y espera unos minutos para que vuelvan a la temperatura ambiental. Repite los pasos 10 y 11. (¿Por qué crees que este paso es necesario?) Realiza esta prueba tres veces y anota tus observaciones.

13. Coloca los vasos de precipitados en posición vertical y quita con cuidado la cinta adhesiva. Rápidamente, tapa con plástico el recipiente que contiene al grillo. Déjalo descansar mientras haces dos tablas de datos en tu cuaderno de ciencias similares a las de la derecha.

14. Observa el movimiento del grillo cada 15 segundos por 3 minutos. Llena la tabla de datos titulada "Grillo (solo)" con: 0 = sin movimiento, 1 = movimiento ligero y 2 = movimiento rápido.

15. Pídele a tu maestro o maestra otro grillo y ponlo en el mismo recipiente donde está el primer grillo. En la tabla de datos titulada "Grillo A y Grillo B" registra el movimiento de los grillos cada 15 segundos. Utiliza los códigos del paso 14.

Analizar

16. ¿Los grillos prefieren la luz o la oscuridad? Explica.

17. A partir de tus observaciones, ¿qué deduces acerca de las preferencias del grillo en cuanto a la temperatura?

18. Describe la conducta alimenticia de los grillos. ¿Cómo atrapan el alimento? ¿Lo chupan, lo mastican o lo atrapan con la lengua?

19. Basándote en tus observaciones del grillo A y del grillo B, ¿qué enunciados generales puedes hacer sobre su comportamiento social?

Profundizar

Elabora una tercera tabla que se titule "Grillo y otra especie de insecto". Introduce otro insecto en el vaso de precipitados. Anota tus observaciones durante tres minutos. Haz un resumen sobre la reacción del grillo frente a otro insecto.

Grillo (solo)	
15 s	
30 s	
45 s	
60 s	
75 s	
90 s	
105 s	
120 s	
135 s	
150 s	
165 s	
180 s	

Grillo A y Grillo B		
	A	B
15 s		
30 s		
45 s		
60 s		
75 s		
90 s		
105 s		
120 s		
135 s		
150 s		
165 s		
180 s		

Experimentos

611

Floating a Pipe Fish
Teacher's Notes

Time Required

One 45-minute class period

Lab Ratings

EASY → HARD

TEACHER PREP 🔬🔬
STUDENT SET-UP 🔬🔬
CONCEPT LEVEL 🔬
CLEAN UP 🔬

MATERIALS

The materials listed on the student page are enough for 1–2 students. PVC pipe ($\frac{3}{4}$ in.) is readily available at a hardware store and is relatively inexpensive. The pieces should be cut in advance. If your school has a shop or Industrial Arts classroom, perhaps you could get the pieces cut there. The hardware store may cut them for you. If not, PVC pipe is not very difficult to cut with a hand saw.

For water containers, students may use anything that is at least 15 cm deep. A bowl, plastic basin, or bucket will do fine.

Safety Caution

Remind students to review all safety cautions and icons before beginning this lab activity.

Christopher Wood
Western Rockingham
Middle School
Madison, North Carolina

Manda a nadar a un pez tubo

Los peces óseos pueden controlar la profundidad a la que nadan en un lago, río o en el mar, gracias a un órgano especial llamado vejiga natatoria. Al inflar o desinflar de gases la vejiga natatoria, el pez se eleva o se hunde en el agua. En esta actividad, vas a hacer un modelo de pez con vejiga natatoria. Cuando el pez esté listo, tu tarea es hacerlo flotar en un recipiente de manera que se quede a la mitad entre el fondo y la superficie. Seguramente tendrás que intentarlo varias veces, y poner mucha atención y capacidad de análisis. ¡Buena suerte!

Materiales

- Tubo PVC de 12 cm de largo y 3/4 de pulgada de diámetro
- un globo de forma alargada
- un corcho pequeño
- una liga de caucho
- agua
- recipiente para agua de unos 15 cm de profundidad

Procedimiento

1. Calcula cuánto aire necesitas en la vejiga natatoria (el globo) para que el pez tubo flote en medio del recipiente, entre el fondo y la superficie. ¿Tendrás que inflarlo a la mitad, sólo un poquito o por completo? Recuerda que tiene que caber dentro del tubo, pero también que necesita suficiente aire para aguantar su peso.

2. Infla el globo de acuerdo con tus cálculos. Aprieta el cuello del globo para que no se desinfle y coloca el corcho en la boca del globo. Si está bien cerrado, no deben salir burbujas cuando lo sumerjas en el agua.

Datasheets for LabBook
Datasheet 31

Science Skills
Worksheet 1
"Being Flexible"

3. Coloca la vejiga natatoria, ya tapada con el corcho, dentro del pez (el tubo) y ponle una liga para asegurarla, como ves en la ilustración. La liga impedirá que la vejiga se escape por los extremos del pez.

4. Echa el pez al agua y fíjate dónde se queda flotando. Apunta tus observaciones en tu cuaderno de ciencias.

5. Si el pez no flota en el lugar debido, sácalo del agua, ajusta la cantidad de aire, e inténtalo de nuevo.

6. Puedes dejar salir una cantidad pequeña de aire si alejas del corcho cuidadosamente la boca del globo. Puedes añadir aire inflándolo un poco más. Sigue ajustando la cantidad de aire hasta que el pez tubo flote en medio del recipiente, entre el fondo y la superficie.

Análisis

7. De acuerdo con el cálculo que hiciste en el paso número 1, ¿fue suficiente la cantidad de aire que pusiste en el globo para que el pez flotara? Explica tu respuesta.

8. En relación con el largo y el volumen de tu pez, ¿cuánto aire fue necesario para hacerlo flotar? Expresa tu respuesta en una proporción o un porcentaje.

9. Observando el espacio que el globo inflado ocupa en tu pez, ¿cuánto espacio calculas que ocupa la vejiga natatoria de un pez de verdad? Explica tu respuesta.

Profundizar

Algunos nadadores veloces, como el tiburón, y los mamíferos acuáticos, como las ballenas y los delfines, no tienen vejiga natatoria. Investiga en la biblioteca o en Internet cómo evitan estos animales ir a dar al fondo del mar. Haz un cartel, y explica tus resultados en tarjetas de 3 x 5 pulgadas. En el cartel, incluye dibujos de los peces o mamíferos acuáticos que estudiaste.

Experimentos

Answers

7. Students may tend to inflate the balloon too much, causing the pipe to float on the surface. They may be surprised to find out how little air is needed in the balloon to make it float in the middle of the water column. Remind students that the cork has some buoyancy of its own. You may want to substitute something less buoyant for the cork.

8. Filling less than 10 percent of the internal volume of the pipe with air will make the pipe float in the middle of the water column.

9. Have students give this answer in a proportion or fraction. Less than one-third of the inside of the pipe is taken by the inflated balloon. Explain that the flesh of a fish is not as dense as PVC pipe and the swim bladder does not need to be very large to adjust the floating depth of the fish.

613

Going Further

Sharks and marine mammals, such as dolphins and whales, must swim almost constantly to keep from sinking to the bottom. Dolphins and whales must come to the surface to breathe. A shark must constantly move or lie in moving water in order to keep water moving past its gills. Sharks and marine mammals are able to dive to great depths because they don't have a swim bladder that would rupture under pressure.

A Prince of a Frog
Teacher's Notes

Time Required

One 45-minute class period

Lab Ratings

TEACHER PREP ▲▲▲
STUDENT SET-UP ▲
CONCEPT LEVEL ▲▲
CLEAN UP ▲

Safety Caution

Remind students to review all safety cautions and icons before beginning this lab activity.

You will need to provide protective gloves for the students. Students' hands may make the frog vulnerable to infection. Also, frogs are known to carry salmonella. Students should wash their hands thoroughly with soap and warm water after handling the frog.

Frogs collected in the wild are best for this activity because they are easily released. Frogs from pet stores must NOT be released into the wild.

Kerry A. Johnson
Isbell Middle School
Santa Paula, California

Experimentos

De príncipe a rana

Imagínate que estás haciendo una investigación científica sobre los anfibios. Has escuchado en las noticias que los anfibios de todo el mundo están en peligro de extinción. ¡Qué gran pérdida para el medio ambiente si se extinguieran! Tu trabajo consiste en aprender todo lo posible sobre el comportamiento normal de las ranas para poder ofrecerles esta información a otros científicos que estudian el problema.

En esta actividad, observarás una rana normal en un ambiente seco y en el agua.

Materiales

- recipiente lleno hasta la mitad con agua sin cloro
- roca grande
- rana viva en un recipiente sin agua
- grillos vivos
- cubeta de laboratorio de 600 mL
- guantes protectores

Procedimiento

1. En tu cuaderno de ciencias, haz una tabla parecida a la de abajo para apuntar todas las observaciones que hagas en esta investigación.

Observaciones de una rana viva	
Característica	**Observation**
Respiración	
Ojos	
Patas	
Respuesta al alimento	
Respuesta al ruido	
Textura de la piel	
Comportamiento en el agua	
Color de la piel	

No escribas en el libro

2. Observa la rana en el recipiente sin agua. Dibújala en tu cuaderno de ciencias, y escribe los nombres de las diferentes partes de su cuerpo: ojos, nariz, patas delanteras y patas traseras.

3. Observa los movimientos de la rana mientras respira aire por los pulmones. Describe la respiración de la rana en tu cuaderno de ciencias.

4. Examina con atención los ojos de la rana. Fíjate en su posición en la cabeza del animal. Observa el párpado superior y el inferior, y el tercer párpado transparente. De los tres párpados, ¿cuál se mueve para cubrir el ojo?

5. Estudia las patas de la rana. Escribe en la tabla de datos la diferencia entre las patas delanteras y las traseras.

614

Datasheets for LabBook
Datasheet 32

Science Skills
Worksheet 23
"Science Drawing"

6. Golpea suavemente el lado del recipiente que está más lejos de la rana y observa cómo responde ésta

7. Coloca un insecto vivo, por ejemplo, un grillo, en el recipiente. Observa la respuesta de la rana y toma notas.

8. Ponte guantes y lentes protectores, y levanta cuidadosamente la rana para examinar su piel. ¿Cómo se siente?

Cuidado: Recuerda que la rana es un ser vivo y merece ser tratada con respeto y delicadeza.

9. Coloca la cubeta de laboratorio de 600 mL en el recipiente y pon la rana adentro. Cubre la cubeta con la mano y llévala al recipiente con agua sin cloro. Inclina la cubeta y sumérgela lentamente en el agua hasta que la rana salga nadando.

10. Observa a la rana nadar y flotar en el agua. ¿Cómo usa las patas para nadar? Fíjate en la posición de la cabeza de la rana en el agua.

11. Mientras nada, asómate por un lado del recipiente para ver la parte de abajo de la rana. Luego obsérvala desde arriba. Compara el color de la espalda de la rana y el de su abdomen.

12. Apunta en la tabla de datos tus observaciones sobre el color y textura de la piel de la rana, y su comportamiento en el agua.

Análisis

13. Con las respuestas a las preguntas anteriores, escribe un informe sobre la anatomía y el comportamiento de los anfibios.

14. ¿Qué puedes deducir sobre el campo visual de la rana a partir de la posición de sus ojos?

15. ¿De qué le puede servir la posición de los ojos mientras nada en el agua?

16. ¿Cómo oye una rana?

17. ¿Cómo respira la rana cuando nada en el agua?

18. ¿Qué adaptaciones para la vida terrestre y acuática tienen las patas de la rana?

19. ¿Qué diferencias notaste en la coloración de la parte de arriba y la de abajo de la rana?

20. ¿Cómo come la rana? ¿Qué sentidos usa para atrapar su presa?

Profundizar

Observa otro tipo de anfibio, por ejemplo, una salamandra. Compara las adaptaciones de este anfibio con las que observaste en la rana en esta investigación.

Preparation Notes

If you can divide the class into groups with several observations going on at the same time, you can use a smaller container for each frog. Containers can be a large glass mixing bowl or something similar. Students may bring containers from home as well. Tree frogs are common in pet stores. They are fun to observe, especially if you can find some small crickets to feed them so that students can observe their feeding behavior.

You may substitute another amphibian, such as water doggies, an immature stage of salamanders. Water doggies are especially interesting if they can be kept in the classroom so students can observe their development into salamanders.

Frogs and water doggies may be obtained in pet stores, in the wild, and in bait shops.

Lab Notes

Several years ago, some students who were out collecting frogs for an activity similar to this lab found severe birth defects and mutations among the frogs they found. A good way to introduce this activity may be to find a news clipping from this event or information about frog deformities taken from the Internet. You may also review the material presented in the first chapter of this book.

Answers

13.–20. Have students speculate about the form and function of the frog's structure. Discuss the camouflage coloration of a frog. Ask how the skin of a frog differs from that of a reptile, and how the two different forms have two different functions. Discuss how the frog's skin must stay wet in order for gas exchange to occur.

Going Further

Answers will vary, but students should notice several similar adaptations among amphibians.

615

What? No Dentist Bills?
Teacher's Notes

Time Required

Two 45-minute class periods

Lab Ratings

EASY ——————→ HARD

TEACHER PREP
STUDENT SET-UP
CONCEPT LEVEL
CLEAN UP

MATERIALS

Pea gravel is an acceptable substitute for aquarium gravel. It can be obtained from a local hardware store and is much less expensive.

A 4:1 gravel to birdseed ratio works best.

Safety Caution

Remind students to review all safety cautions and icons before beginning this lab activity.

Answers

4. A bird uses a gizzard instead of teeth.

5. Students should be able to demonstrate how their model gizzard grinds birdseed.

6. Gizzard stones are small pebbles that birds sometimes swallow. They settle in the gizzard and aid in digestion.

7. Model gizzards, no more than three-fourths full, will probably be most effective.

8. Answers may include reducing the amount of food, adding gizzard stones, or adding more liquid to the food.

¿No van al dentista?

Cuando comemos, debemos masticar bien el alimento. Nuestros dientes están hechos para masticar porque la ruptura del alimento en pequeños pedazos es el primer paso del proceso digestivo. Las aves, en cambio, no tienen dientes. ¿Cómo hacen que los grandes trozos de alimento sean lo bastante pequeños para que comience la digestión? En esta actividad vas a diseñar y construir un modelo del aparato digestivo de un ave para averiguar la respuesta a esta pregunta.

Procedimiento

1. Examina el diagrama del aparato digestivo de un ave. Diseña un modelo del aparato digestivo de las aves con los materiales proporcionados por tu maestro o maestra. Incluye en el modelo todos estos órganos: esófago, buche, molleja, intestino y cloaca.

2. Pídele a tu maestro o maestra una bolsa de plástico y los materiales que necesitas, y construye tu modelo.

3. Prueba tu modelo usando el alpiste que te dé el maestro o la maestra.

Análisis

4. ¿Cómo puede un ave descomponer las partículas de alimento sin dientes?

5. ¿Tritura la molleja de tu modelo el alimento?

6. ¿Qué son y qué hacen las piedras de la molleja?

7. ¿Qué tan llena debe estar la molleja para trabajar de manera eficaz?

8. Describe cómo puedes hacer que tu modelo funcione mejor.

Materiales

- bolsas de plástico con cierre, de distintos tamaños
- alpiste
- grava de acuario
- agua
- cuerda
- popote
- cinta adhesiva
- tijeras y otros materiales que sean necesarios

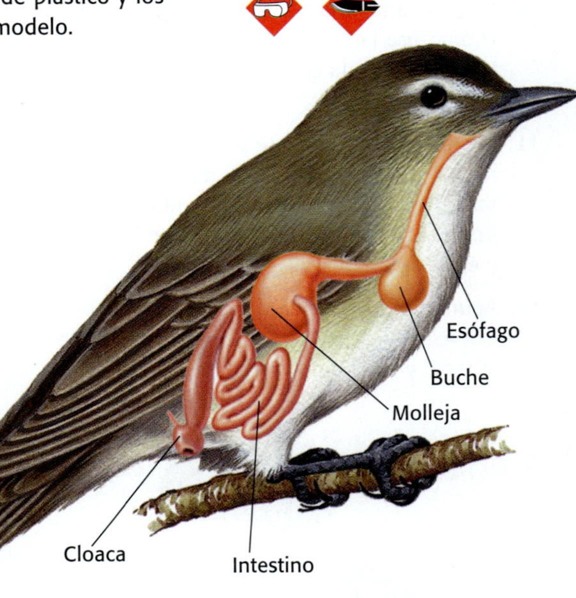

Profundizar

¿Sabías que se han encontrado "piedras de molleja" en algunos fósiles de dinosaurios? Busca en la biblioteca o en Internet información sobre la relación evolutiva entre dinosaurios y aves. Haz una lista de las semejanzas y de las diferencias que encontraste.

616

Going Further

Scientists have long recognized similarities between birds and some dinosaurs, including an S-shaped neck, a unique ankle joint, and hollow bones.

Datasheets for LabBook
Datasheet 33

Randy Christian
Stovall Junior High School
Houston, Texas

Se solicitan mamíferos en Marte

Es el año 2256. Han existido colonias en Marte durante casi 50 años. El agua marciana es escasa pero todavía existe y las temperaturas son extremas pero soportables. El planeta Marte ha desarrollado lentamente una atmósfera donde los seres humanos pueden respirar, gracias a los esfuerzos de muchos científicos durante los últimos 200 años. La Comisión Interplanetaria ha decretado que se envíen algunos mamíferos de diferentes hábitats de la Tierra a Marte. Ahí, se alojarán en un zoológico para que puedan acostumbrarse al clima antes de ser liberados en ambientes naturales.

Tu tarea es preparar una presentación para la Comisión Interplanetaria, que describa por lo menos tres mamíferos que crees que podrían sobrevivir en un zoológico en Marte. Selecciona un mamífero acuático, uno terrestre y uno que viva en el aire parte del tiempo.

Materiales

- marcadores de colores, creyones, bolígrafos u otro equipo de dibujo
- cartulina, papel de periódico u otro papel para dibujar
- otros materiales, según sea necesario

Procedimiento

1. Investiga en la biblioteca o en Internet para obtener información sobre el medio ambiente actual de Marte. ¿Cómo podría la atmósfera cambiar el medio ambiente?

2. Investiga en la biblioteca o en Internet las diferentes especies de mamíferos y su medio ambiente en la Tierra. Usa esta información para seleccionar los mamíferos que piensas que podrían vivir en Marte.

3. Prepara tu presentación para la Comisión Interplanetaria. Puede ser un cartel, un mural, un diorama, una presentación en computadora o cualquier otro formato. Asegúrate de solicitar la aprobación de tu maestro o maestra.

Análisis

4. Menciona y describe los mamíferos que escogiste. Explica por qué crees que tu elección es apropiada.

5. Describe qué adiciones o cambios se deben hacer en el zoológico para alojar a cada uno de los mamíferos.

Profundizar

Si los mamíferos que presentaste se enviaran a Marte y fueran liberados en el planeta, ¿qué otros mamíferos que compitieran con ellos por alimento, resguardo y otros recursos se podrían introducir?

617

Martha Kisiah
Fairview Middle School
Tallahassee, Florida

Datasheets for LabBook
Datasheet 34

Experimentos

Wanted: Mammals on Mars
Teacher's Notes

Time Required

One or two 45-minute class periods

Lab Ratings

TEACHER PREP
STUDENT SET-UP
CONCEPT LEVEL
CLEAN UP

Answers

4. Students should remember that this lab is about mammals. Remind them that they should select an aquatic mammal, which would limit their choices to whales, dolphins, sea lions, beavers, and the like. Similarly, the one that lives in the air part of the time might be a bat, a "flying squirrel," or with a little imagination, a mammal that lives high in a tree, such as a sloth. The reasons for choosing a particular animal should have to do with available resources and the habitat the animal requires.

5. Answers will vary according to the animals chosen by the student. All accommodations should have something to do with mammalian characteristics and habitat resources.

Going Further

Students may have had to make their decisions by the process of elimination. If so, this one will be easy. Just include all the animals here that they couldn't keep in the zoo!

Life in the Desert
Teacher's Notes

Time Required
One 45-minute class period

Lab Ratings

EASY ———————→ HARD

TEACHER PREP ▲

STUDENT SET-UP ▲ ▲

CONCEPT LEVEL ▲ ▲

CLEAN UP ▲

MATERIALS
The sponges used in this lab can be either natural sponges or the synthetic sponges available in grocery stores. Use 3 × 6 in. sponges, 1 per student, cut in half.

Answers

6. Students should describe the kind of covering or protection they provided for their "adapted" sponge. Effectiveness of the adaptation will be measured by the amount of water lost over 24 hours. Students will want their sponges to dry out as little as possible.

7. The unprotected sponge represents the organism that has no adaptation for conserving water. The unprotected sponge should dry out far more than the protected sponge.

James Chin
Frank A. Day Middle School
Newtonville, Massachusetts

Vida en el desierto

Los organismos que viven en el desierto conservan el agua mediante métodos muy poco comunes. La conservación del agua es una función importante para los organismos que viven en tierra firme, pero para los que viven en el desierto es un reto especial. En esta actividad inventarás una "adaptación" para un animal del desierto, representado por un pedazo de esponja, para descubrir cuánta agua puede conservar en un período de 24 horas. Protegerás la esponja mojada para que se seque lo menos posible.

Materiales

- 2 pedazos de esponja seca (8 × 8 × 2 cm)
- agua
- balanza
- ejemplos de descripciones de plantas y animales del desierto
- otros materiales según se requieran

Procedimiento

1. Piensa en un método para evitar que tu "animal del desierto" se seque. El "animal" debe estar al aire libre por lo menos durante 4 horas en un período de 24 horas. Los animales reales del desierto se exponen con frecuencia al calor seco del desierto para ir en busca de comida. Escribe tu plan en tu cuaderno de ciencias, así como tus predicciones acerca del resultado del experimento.

2. Si es necesario, diseña tablas de datos. Antes de empezar, pídele a tu maestro que aprueben tu plan.

3. Sumerge dos pedazos de esponja en agua hasta que empiecen a gotear. Coloca cada pedazo en una balanza y anota su masa en tu cuaderno de ciencias.

4. Inmediatamente empieza a proteger uno de los pedazos conforme a tu plan. Coloca ambos pedazos en un lugar en el que nadie los toque. Saca tu "animal" a comer las veces que quieras, por un total de por lo menos 4 horas.

5. Al término de las 24 horas, coloca otra vez los pedazos de esponja en la balanza y anota su masa.

Análisis

6. Describe la adaptación con la que le ayudaste a tu "animal" a sobrevivir. ¿Fue efectiva? Explica tu respuesta.

7. ¿Cuál fue el objetivo de dejar sin protección una de las esponjas? ¿Qué diferencias hay entre la pérdida de agua de cada esponja?

Profundizar
Comenten en clase las adaptaciones y resultados. ¿Cómo relacionas estas adaptaciones inventadas con las que los organismos reales utilizan para sobrevivir en el desierto?

618

Going Further
Ask students to consider the adaptation they designed for their sponge. Did it compare with a real-life adaptation in a desert organism? In what way? Discuss behavior adaptations for conserving water. Discuss endothermy and ectothermy, and ask students what adaptations conserve water and help regulate body heat.

Datasheets for LabBook
Datasheet 35

¡Descubre los miniecosistemas!

Al estudiar los ecosistemas, aprendiste que un bioma es un ecosistema muy grande que abarca un conjunto de ecosistemas más pequeños relacionados entre sí. Por ejemplo, un bioma de bosque de coníferas puede comprender un ecosistema de río, uno de pantano y otro de lago, cada uno de los cuales puede incluir otros más pequeños relacionados con ellos. ¡Hasta las ciudades tienen miniecosistemas! Puedes encontrar un miniecosistema en un pedazo de acera, en un charco de agua de lluvia, debajo de una llave que gotea, en un área con sombra o debajo de una piedra. En esta actividad diseñarás un método para comparar dos miniecosistemas que se encuentren cerca de tu escuela.

Materiales

- papel y lápiz para tomar notas y hacer observaciones
- otros materiales según la investigación

Procedimiento

1. Examina el terreno que rodea tu escuela y escoge dos zonas distintas para tu investigación. Pídele permiso al maestro o maestra antes de comenzar.

2. Decide lo que quieres averiguar acerca de tus miniecosistemas. Por ejemplo, qué tipo de seres vivos hay en cada zona o cuáles son los factores abióticos de cada miniecosistema.

3. Elabora tablas de datos para registrar tus observaciones. Otra posibilidad es observar los miniecosistemas durante una hora, u observarlos por un período corto, ya sea varias veces al día o varios días a la misma hora. Pídele al maestro o maestra que apruebe tu plan y elabora las tablas de datos apropiadas.

Análisis

4. ¿Qué factores determinan las diferencias entre tus miniecosistemas? Identifica los factores que separan cada miniecosistema del área que los rodea.

5. ¿En qué se parecen y diferencian las poblaciones de tus miniecosistemas?

6. Identifica algunas adaptaciones de los organismos que viven en tus miniecosistemas. Describe de qué manera las adaptaciones facilitan la supervivencia de los organismos en su ambiente.

7. Escribe un informe donde describas y compares tus miniecosistemas con los de tus compañeros y compañeras.

Experimentos

619

Discovering Mini-Ecosystems
Teacher's Notes

Time Required

One to two 45-minute class periods

Lab Ratings

TEACHER PREP
STUDENT SET-UP
CONCEPT LEVEL
CLEAN UP

MATERIALS

Because this is mainly an observation activity, few materials are needed. If they are available, however, binoculars or magnifying lenses may be helpful.

Lab Notes

Even if your school has no area where there is sand, dirt, grass, or trees, ask students to observe puddles, the underside of eaves, the area under drain spouts, and the ground under rocks. Students should observe the areas they have chosen at least twice a day.

Datasheets for LabBook
Datasheet 36

Barry Bishop
San Rafael Junior High
Ferron, Utah

Answers

4. Students may have many answers, but they should include answers such as different vegetation, organisms that live there, the soil type, density of vegetation, and amount of water present.

5. Students should recognize that the populations present in each area are adapted for life in that area.

6. Adaptations that students name will probably include camouflage, deep roots, and burrowing behavior.

7. Answers will vary according to student observations.

Too Much of a Good Thing?
Teacher's Notes

Time Required

One 45-minute class period and one 10-minute observation time every 3 days for 3 weeks

Lab Ratings

EASY ——————————————→ HARD

TEACHER PREP
STUDENT SET-UP
CONCEPT LEVEL
CLEAN UP

Safety Caution

Remind students to review all safety cautions and icons before beginning this lab activity.

Preparation Notes

A review of the causes of eutrophication might be helpful before beginning this lab.

Jason Marsh
Montevideo High and
Country School
Montevideo, Minnesota

Experimentos

¿Demasiada comida?

DESARROLLAR DESTREZAS

Las plantas necesitan nutrientes, como fosfatos y nitratos, para sobrevivir. Los detergentes contienen con frecuencia fosfatos, y es común encontrar nitratos entre los desechos animales y en los fertilizantes. Cuando se arrojan grandes cantidades de estos nutrientes a ríos y lagos, las algas y la vegetación crecen rápidamente y después mueren y se descomponen. Los microorganismos que descomponen las algas y la materia vegetal utilizan el oxígeno del agua, ocasionando la muerte de peces y animales que dependen del mismo para sobrevivir. A menudo, el resultado de este proceso es una laguna o lago que se rellena artificialmente.

En esta actividad observarás el efecto de los fertilizantes en los organismos que habitan en un estanque.

Procedimiento

1. Con un lápiz de cera rotula los frascos así: "Control", "Fertilizante" y "Exceso de fertilizante".

2. Vierte 750 mL de agua destilada en cada frasco. Lee la cantidad de fertilizante recomendada para las plantas en la etiqueta del paquete. En el frasco "Fertilizante" coloca la cantidad de fertilizante recomendada para 1 cuarto de agua, o 1 L si las instrucciones se dan en unidades métricas. En el frasco "Exceso de fertilizante" agrega 10 veces esta cantidad de fertilizante a 1 cuarto de agua. Agita bien cada frasco para disolver el fertilizante.

3. Consigue una muestra de agua de estanque. Mézclala con suavidad para que los organismos que contiene se distribuyan uniformemente. Vierte 100 mL de agua de estanque en cada frasco.

4. Observa al microscopio una gota de agua de estanque de cada frasco. En tu cuaderno de ciencias, dibuja por lo menos cuatro organismos de los que ves. Determina si los organismos que ves son algas (generalmente son verdes) o consumidores (casi siempre se mueven). Describe el número y tipo de organismos que ves.

5. Tapa cada frasco con plástico para envolver (sin sellarlo). Coloca los tres frascos en un lugar bien iluminado, cerca de una ventana. No los expongas a la luz solar.

Materiales

- frascos de 1 cuarto de capacidad (3)
- lápiz de cera
- agua destilada
- fertilizante
- probeta graduada
- varilla agitadora
- agua de estanque con organismos vivos
- cuentagotas
- microscopio
- portaobjetos y cubreobjetos
- plástico transparente
- guantes protectores

Organismos comunes del agua de estanque

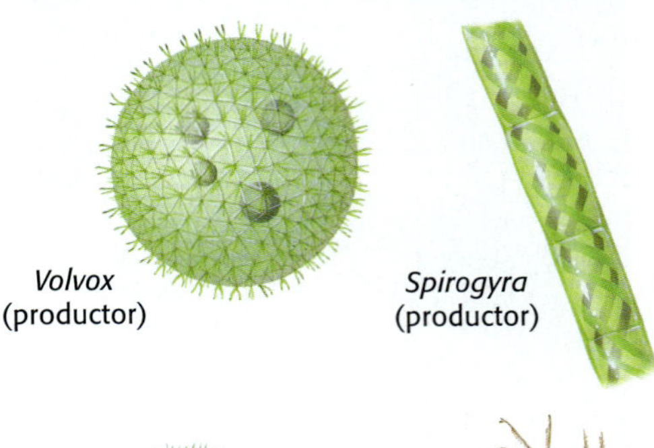

Volvox
(productor)

Spirogyra
(productor)

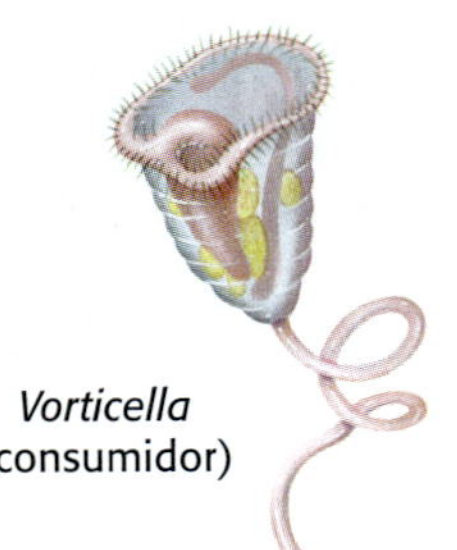

Vorticella
(consumidor)

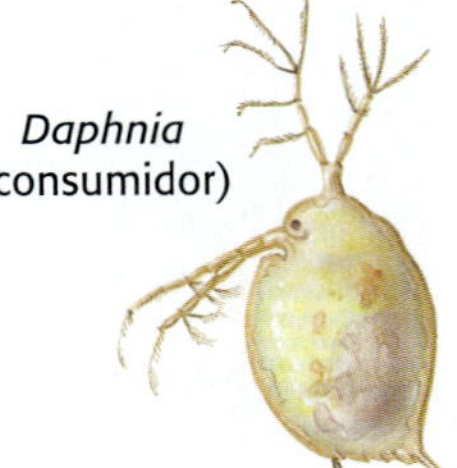

Daphnia
(consumidor)

6. Basándote en tus conocimientos sobre cómo los estanques y lagos llegan a rellenarse y secarse, predice cómo crecerán los organismos de cada frasco.

7. Elabora tres tablas de datos en tu cuaderno de ciencias. Deja suficiente espacio para anotar tus observaciones. Tal como se muestra abajo, una de las tablas debe llamarse "Control", otra "Fertilizante" y otra "Exceso de fertilizante".

Control			
Fecha	**Color**	**Olor**	**Otras observaciones**

8. Observa los frascos al comienzo y por lo menos una vez cada tres días durante tres semanas. Observa el color, el olor y cualquier presencia visible de formas de vida. Anota tus observaciones en las tablas de datos.

9. Cuando las formas de vida comiencen a ser visibles, saca con un cuentagotas una muestra de cada frasco y obsérvala al microscopio. ¿Ha cambiado el número y tipo de organismos desde que observaste por primera vez el agua? En las tablas de datos anota tus observaciones bajo el título de "Otras observaciones".

10. Al final del período de observación de tres semanas, saca de nuevo una muestra de cada frasco y obsérvala al microscopio. En tu cuaderno de ciencias, dibuja por lo menos cuatro de los organismos más abundantes y describe los cambios en el número y tipo de organismos desde tu última observación.

Análisis

11. Después de tres semans, ¿en qué frasco se observa el mayor crecimiento de algas? ¿Qué lo ocasionó?

12. ¿Observaste algún efecto sobre los demás organismos del frasco debido al mayor crecimiento de algas? Explica tu respuesta.

13. Compara tus observaciones con tu predicción. Explica tu respuesta.

14. ¿Cómo puede evitarse o desacelerarse la rápida saturación de las lagunas y lagos naturales?

Experimentos

Datasheets for LabBook
Datasheet 37

Biodiversity—What a Disturbing Thought!
Teacher's Notes

Time Required

One or two 45-minute class periods

Lab Ratings

EASY ———→ HARD

TEACHER PREP

STUDENT SET-UP

CONCEPT LEVEL

CLEAN UP

MATERIALS

Binoculars may not be available in your classroom. Ask students if they might have some at home they could get permission to bring to class for this activity.

Lab Notes

The lab should be reviewed ahead of time. This lab can be extended to a field trip that can get parents involved. Your school may be in a city where there is no suitable undisturbed area. If you are unable to take students on a field trip, help them understand the difference between severely disturbed (a paved parking lot) and an area that is less disturbed, such as an unimproved lot. Some diversity should exist in every area. You may find this lab interesting to repeat during different seasons.

Biodiversidad: ¡vaya idea!

La *biodiversidad* se refiere al número de especies diferentes que viven juntas en una comunidad. La diversidad es importante para la supervivencia de cada organismo de la comunidad. La biodiversidad se refiere también a la diversidad de nichos que los organismos ocupan y a la diversidad de hábitats. Los productores, consumidores y descomponedores tienen cada cual un papel que jugar en el ecosistema, y cada especie tiene su propio nicho. ¿Dónde crees que haya más diversidad: en un bosque o en un terreno que se ha limpiado y en el que crecen cultivos? En esta actividad vas a investigar áreas fuera de tu escuela para determinar cuáles tienen mayor biodiversidad. Vas a usar la información que obtengas en tu investigación para decidir si hay más diversidad en un bosque o en un sembradío.

Procedimiento

1. Elige un área en la que el nivel de intervención es alto (un césped podado, o una jardinera bien cuidada, por ejemplo) y otra en la que haya relativamente poca, o menor, intervención (un lote baldío o una jardinera abandonada, por ejemplo). Pìdele permiso a tu maestro o maestra para explorar estas áreas.

2. Diseña el procedimiento que vas a usar para determinar qué área contiene mayor biodiversidad y enséñaselo a tu maestro o maestra antes de empezar. Para observar los organismos más pequeños, mide un área de 1 m^2, clava estacas en las esquinas, y rodea el perímetro con un cordel. Usa una lupa dentro de esta área para observar todos los bichitos que te encuentres. No es necesario que los identifiques por su nombre científico. Cuando tomes notas en tu cuaderno de ciencias, descríbelos como hormiga A, hormiga B, etc. Observa cada área en silencio, y toma nota de los pájaros y otros animales grandes que las visiten.

Materiales

- materiales y herramientas necesarios para llevar a cabo tu investigación, con la aprobación de tu maestro o maestra. Algunos materiales que te pueden ser útiles son: regla métrica, binoculares, lupa, cordel, pinzas.

Terry Rakes
Elmwood Junior High School
Rogers, Arkansas

3. En tu cuaderno de ciencias, traza las tablas de datos que te hagan falta para clasificar la información. Si observas tus áreas en más de una ocasión, asegúrate de hacer tablas para cada período de observación. Organiza los datos en categorías claras y concisas.

Análisis

4. ¿Confirman tus datos la predicción que hiciste sobre la diferencia en biodiversidad entre un campo sembrado y un bosque? Explica tu respuesta.

5. ¿Qué factores consideraste antes de decidir qué hábitats fuera de tu escuela tenían un nivel de intervención más alto o más bajo? ¿Qué importancia tienen estos factores? Explica tu respuesta.

6. ¿Tuviste problemas para observar y registrar información en cada hábitat? ¿Cuáles fueron los más difíciles? Describe cómo los resolviste.

7. Describe posibles errores en tu método de investigación. Sugiere formas para mejorar el procedimiento y eliminar estos errores.

8. ¿Crees que la biodiversidad fuera de tu escuela ha disminuido desde que se construyó la escuela? ¿Por qué?

9. Las dos áreas que ves en las fotografías tienen plantas hermosas y saludables. Una de ellas, sin embargo, tiene un nivel muy bajo de biodiversidad. Describe lo que ves en cada fotografía y explica la diferencia en biodiversidad.

Profundizar

Se piensa que la selva tropical tiene la mayor biodiversidad de cualquier bioma en el planeta. Investiga este tema en la biblioteca o en Internet. Averigua qué factores existen en la selva tropical que hacen que el bioma sea tan diverso. Compara la biodiversidad de la selva tropical con la de un área boscosa cerca de la comunidad en la que vives.

Pastos y flores silvestres de una pradera

Campo de trigo

623

Datasheets for LabBook
Datasheet 38

Science Skills
Worksheet 19
"Researching on the Web"

Answers

4. Students should explain why they predicted one area would be more diverse than the other. If their predictions were confirmed, have them explain if their reasons were also correct.

5. Generally, any area inside a city or even farmland is considered ecologically disturbed. This is the conclusion you want students to come to. Students should consider human impact when deciding if an area is disturbed or undisturbed.

6. Answers will vary.

7. A possible error might be deciding that an area is undisturbed and finding that it is highly disturbed and has little diversity. A bed of petunias is lovely, but it is disturbed and not very diverse, especially if it is well weeded and insect controlled.

8. Answers will vary. Discuss construction and growth in the neighborhood since the school was built.

9. A wheat field and farmland are less diverse than a tall-grass prairie. The wheat field grows only one plant and provides little habitat for animal organisms. Some animals may visit and feed there, but their nests or dens would be in danger at harvest time.

Going Further

Have students compare a nearby forest and its diversity with a rain forest and the diversity that is found there.

Deciding About Environmental Issues
Teacher's Notes

Time Required

One 45-minute class period

Lab Ratings

EASY ——→ HARD

TEACHER PREP ▲
STUDENT SET-UP ▲▲
CONCEPT LEVEL ▲▲▲
CLEAN UP ▲

Lab Notes

This is a good lab to repeat as environmental issues appear in the news.

You may wish to combine this activity with a video that portrays an international environmental issue. Students can also be encouraged to use the Internet as a source of information.

Debra Sampson
Booker T. Washington
Middle School
Elgin, Texas

Experimentos

DESARROLLAR DESTREZAS

Decisiones ambientales

Cada día tomamos cientos de decisiones; algunas complicadas, pero la mayoría muy simples, como por ejemplo, qué ropa ponerse o qué comer. Las decisiones sobre un problema ambiental pueden ser muy difíciles, porque es necesario considerar tantos factores diferentes. ¿Cómo afectará cierta solución a la vida humana? ¿Qué costo tendrá? ¿Es una solución ética?

En esta actividad, analizarás un problema en cuatro pasos para tomar una decisión al respecto. Averigua cuáles son los problemas ambientales en tu comunidad. Busca en periódicos, revistas, publicaciones de agencias y de otro tipo para enterarte de qué problemas existen, y evalúa uno de ellos. Podrías evaluar, por ejemplo, si tu localidad debería costear los contenedores y camiones de basura especiales para recoger basura reciclable.

Procedimiento

1. En tu cuaderno de ciencias, describe brevemente un problema ambiental.

2. **Reúne información.** Lee sobre este problema en varias publicaciones. Resume los puntos más importantes en tu cuaderno de ciencias.

3. **Considera los valores involucrados,** o sea, lo que tú consideres importante al respecto. Examina el diagrama que se muestra abajo, en él que se consideran varios valores. ¿Cuáles crees que tienen más que ver con el problema que estás evaluando? ¿Hay otros valores que podrían ayudarte a tomar una decisión sobre este problema? Escoge por lo menos cuatro valores que quieras considerar para tu decisión.

Materiales

- periódicos, revistas, otras publicaciones con información sobre problemas ambientales.

Modelo del proceso de decisión en cuatro pasos

Reunir información

↓

Considerar valores

↓

Explorar consecuencias

↓

Tomar una decisión

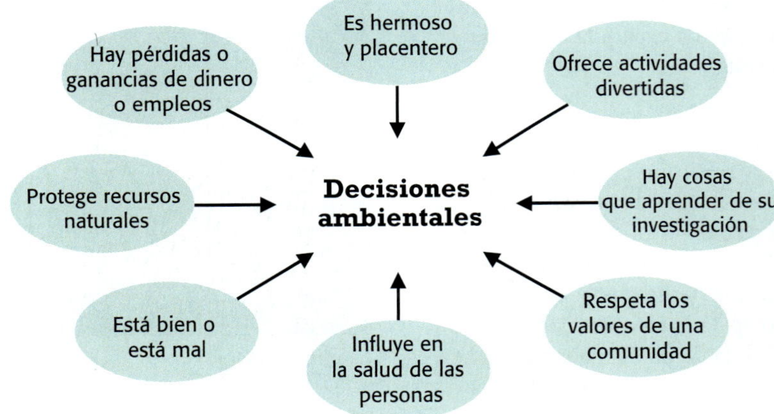

Datasheets for LabBook
Datasheet 39

4. **Explora las consecuencias.** Las consecuencias son el resultado de ciertas acciones. En tu cuaderno de ciencias, haz una tabla como la siguiente para organizar tus ideas sobre las consecuencias derivadas de los valores involucrados en el problema ambiental. Escribe tus valores en la parte de arriba de la tabla. En cada casilla, escribe las consecuencias para cada valor.

Tabla de consecuencias				
Consecuencias	**Valores**			
Consecuencias positivas a corto plazo				
Consecuencias negativas a corto plazo				
Consecuencias positivas a largo plazo				
Consecuencias negativas a largo plazo				

No escribas en el libro

5. **Toma una decisión.** Considera a fondo todas las consecuencias que anotaste y evalúa la importancia de cada una. Decide qué alternativa escogerías.

Análisis

6. En la evaluación, ¿le diste más importancia a las consecuencias a largo o a corto plazo? ¿Por qué?

7. ¿Qué valor o valores influyeron más en tu decisión? Explica tu respuesta.

Profundizar

Compara tu tabla de consecuencias con las de tus compañeros y compañeras. ¿Tomaron las mismas decisiones respecto a problemas parecidos? Si no es así, formen equipos y organicen un debate sobre un problema ambiental específico.

Answers

6. Either short- or long-term consequences can be more relevant, depending on the issue.

7. Answers will vary according to students' perspectives.

Going Further

Encourage students to narrow their topic to a single aspect of an issue, such as the importance of aesthetic value in the preservation of natural areas.

Science Skills Worksheet 3 "Thinking Objectively"

Science Skills Worksheet 12 "Working with Hypotheses"

Muscles at Work
Teacher's Notes

Time Required

One 45-minute class period

Lab Ratings

EASY — HARD

TEACHER PREP
STUDENT SET-UP
CONCEPT LEVEL
CLEAN UP

Safety Caution

Remind students to review all safety cautions before beginning this lab activity.

A digital thermometer that measures temperature from the ear is recommended.

Because of the vigorous nature of the exercise, you may want to ask for volunteers to do the exercising. Also, you should be aware of any health concerns your students have.

Datasheets for LabBook
Datasheet 40

Kathy LaRoe
East Valley Middle School
East Helena, Montana

Los músculos en acción

¿Alguna vez has hecho ejercicio afuera, en un frío día de otoño con ropa ligera? ¿Cómo te mantuviste a una buena temperatura? Las células musculares se contrajeron y, cuando hay contracción, parte de la energía se usa para hacer el trabajo y el resto se libera en forma de calor. Esto te ayuda a mantener una temperatura constante en condiciones frías. Cuando te ejercitas vigorosamente en un caluroso día de verano, tus músculos pueden hacer que tu cuerpo adquiera una temperatura muy alta.

Esta actividad consiste en averiguar cómo la liberación de energía en forma de calor provoca un cambio en la temperatura del cuerpo.

Materiales

- reloj con segundero o cronómetro
- termómetro
- otros materiales aprobados por tu maestro o maestra

Experimentos

Procedimiento

1. En un grupo de cuatro estudiantes, discute sobre varios ejercicios que pueden producir un cambio en la temperatura corporal. Formula una hipótesis y escríbela en tu cuaderno de ciencias.

2. Desarrolla un procedimiento experimental que incluya los pasos necesarios para probar tu hipótesis. Asegúrate de que tu maestro o maestra lo apruebe antes.

3. Asígnales tareas a los miembros del grupo, como tomar notas, registrar datos y contar el tiempo. ¿Qué observaciones y datos vas a registrar? Diseña tablas de datos apropiadas en tu cuaderno de ciencias.

4. Realiza tu experimento según lo planeado en el grupo. Asegúrate de anotar todas las observaciones hechas durante el experimento en tus tablas de datos.

Análisis

5. ¿Cómo determinas si las contracciones musculares provocan la liberación de energía en forma de calor? ¿Sostuvieron los datos tu hipótesis? Explica tus resultados en un informe escrito. Describe cómo mejorar tu método experimental.

Profundizar

¿Por qué tiritamos cuando hace frío? ¿Tiritan de frío todos los animales? Descubre por qué tiritar es uno de los primeros signos de que la temperatura de tu cuerpo está bajando.

626

Answer

5. All answers will depend on the students' observations and their own hypotheses.

Going Further

In a process known as shivering thermogenesis, muscle tone is gradually increased. Shivering increases the workload of the muscles and elevates oxygen and energy consumption. The heat that is produced warms the deep vessels. Shivering can elevate body temperature effectively. It can increase the rate of heat generation by as much as 400 percent. Endothermic animals have the capacity to shiver. Shivering is an automatic response of the body to cold.

Ver para creer

¿Cuántas veces has visto esos anuncios que dicen "Juego de uñas: $25.00", "Puntas: $15.00" o "Relleno: $10.00"? ¿Has pensado por qué la gente paga para que les "pongan" uñas? A pesar de lo que estos anuncios implican, las uñas humanas crecen por sí solas sin necesidad de tratamientos caros. Las uñas son parte del sistema integumentario del cuerpo, que incluye la piel que cubre todo el cuerpo. Las uñas son una modificación de la capa exterior de la piel y crecen continuamente durante toda tu vida.

En esta actividad, vas a medir el tiempo de crecimiento de las uñas de las manos.

Materiales

- marcador permanente
- regla métrica
- papel cuadriculado (opcional)

Procedimiento

1. En tu cuaderno de ciencias, traza con un lápiz tus manos sobre una hoja de papel. Luego ponle algunos de los detalles, como las uñas. Escoge uno de los dedos que dibujaste y marca las partes de la uña, como ves a la derecha. Observa que la matriz de la uña es el área donde la uña está unida al dedo. La ilustración muestra una vista recortada para que puedas ver desde qué tan adentro del dedo empieza tu uña.

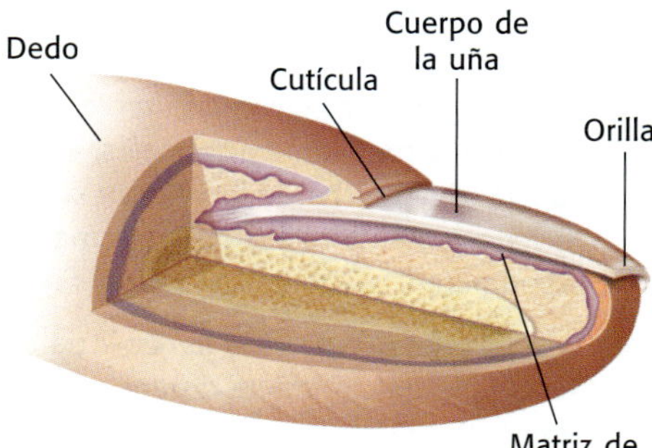

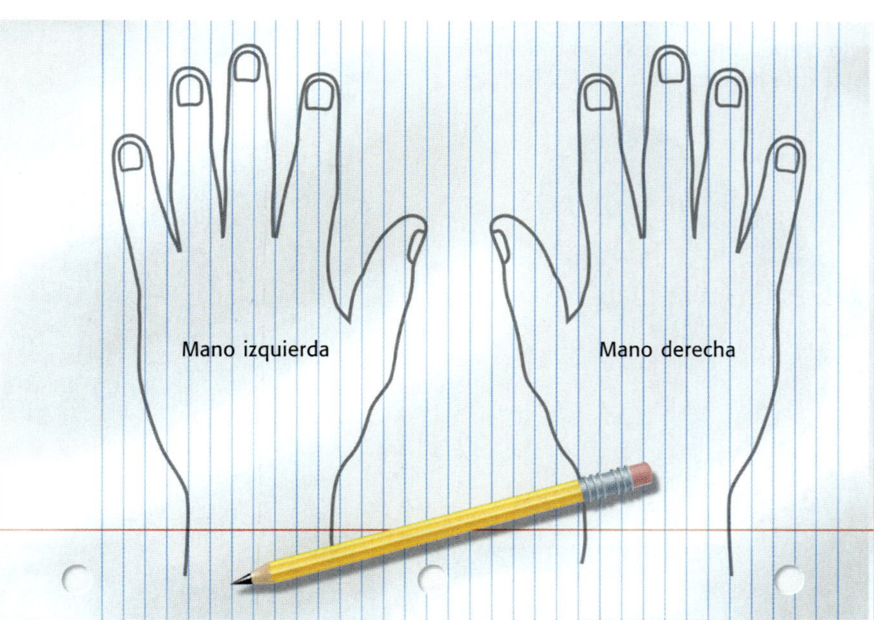

627

Kathy LaRoe
East Valley Middle School
East Helena, Montana

Experimentos

LabBook

Seeing Is Believing
Teacher's Notes

Time Required

One 45-minute class period, and 5 to 10 minutes every other day for 2 weeks

Lab Ratings

EASY ——————→ HARD

TEACHER PREP ♦
STUDENT SET-UP ♦
CONCEPT LEVEL ♦
CLEAN UP ♦

MATERIALS

The materials listed on the student page are enough for 1–2 students. This lab may be done with several different types of marking methods. The fingernail is very hard and not very porous. Marking the nail permanently is a challenge. A permanent marker, such as a laundry-marking pen, may need to be refreshed only once a day. Fingernail polish may be an acceptable alternative. Acrylic paint may also be used.

Safety Caution

Remind students to review all safety cautions and icons before beginning this lab activity.

Datasheets for LabBook
Datasheet 41

Lab Notes

Few topics are as important to students as acquiring knowledge and understanding of their own body. As they develop, students can't help but observe the ways they are changing physically. One part of the body that grows quickly and requires a great deal of their attention is their fingernails. In this lab, students are able to witness the growth of their own body.

Tell students that the graphed data shown in this lab is only an example and will not be the same as their own data. A female adult index fingernail, for example, may be about 12 mm from the cuticle to the beginning of the free edge.

2. Con tu compañero de laboratorio, túrnate para medir la longitud de todas las uñas de las dos manos. Empieza con el pulgar, mide la distancia de la piel a la base de la uña donde empieza la orilla de la uña. Anota la medida de cada uña (en unidades métricas) en el dibujo que hiciste.

3. Encuentra el centro de la matriz de la uña del dedo índice de la derecha (el dedo que le sigue al pulgar). Pon un punto con el marcador permanente en el centro de la matriz de la uña, como se muestra a la derecha. **Cuidado:** Asegúrate de no mancharte la ropa con el marcador. Mide de la marca a la base de la uña. Anota esta medida en el dibujo. Rotúlala como "Día 1".

4. Repite el paso 3 con el dedo índice de la mano izquierda. Luego cambia de lugar con tu compañero de laboratorio.

5. Deja que tus uñas crezcan durante 2 días incluyendo hoy. Las actividades cotidianas y normales, como lavarte las manos, no quitarán la mancha de la uña por completo. Quizás necesites volver a poner la marca periódicamente hasta que acabe el experimento.

6. Después de 2 días, mide la distancia de la marca hasta la base de la uña. Anota la distancia en unidades métricas en tu dibujo. Rotúlala como "Día 3".

7. Continúa midiendo y registrando el crecimiento de tus uñas durante 2 semanas. Vuelve a poner la marca en la uña, según sea necesario. Puedes limarte o arreglarte las uñas como siempre durante este experimento.

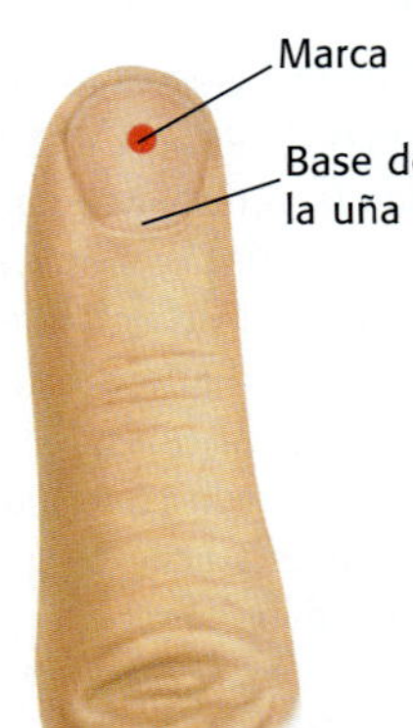

8. Después de completar y registrar tus medidas en el dibujo, prepara una gráfica similar a ésta para mostrar tus descubrimientos. Cada compañero de laboratorio hará una gráfica de sus propias medidas.

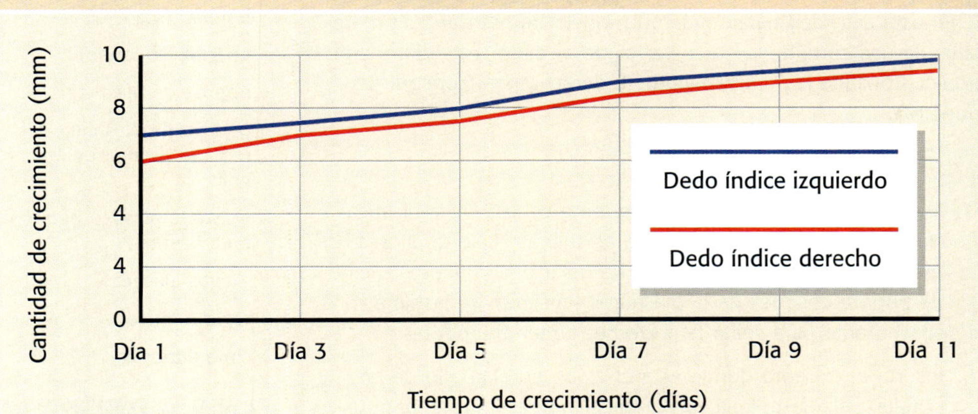

Análisis

9. ¿A qué mano le creció más rápido la uña? Escribe dos explicaciones posibles de por qué una uña podría crecer más rápido que la otra.

10. ¿A qué compañero le crecen más rápido las uñas? ¿A quién le crecen más despacio? ¿Cuál es la diferencia en el crecimiento total de las uñas entre estos dos estudiantes?

11. Entre tus compañeros, ¿hay alguna relación entre la velocidad de crecimiento de las uñas de hombres y de mujeres? ¿Hay una relación entre el crecimiento de las uñas y otras características físicas, como la altura?

Profundizar

Investiga en la biblioteca o en Internet para encontrar las respuestas a las siguientes preguntas:

• ¿Por qué son importantes las uñas? ¿Para qué te sirven? Da por lo menos tres ejemplos que apoyen tus descubrimientos.
• ¿Indican las uñas tu estado de salud o de nutrición?

Diseña un experimento para descubrir la velocidad de crecimiento del cabello. ¿Cómo se compara con la velocidad del crecimiento de las uñas? Repasa tu diseño experimental con tu maestra antes de empezar la investigación.

Experimentos

Answers

All answers to the Analysis questions will depend on student observations and measurements.

Going Further

Nails are extremely versatile. Most of us take for granted the many and unique functions they perform. Fingernails, for example, intensify our tactile sensitivity; they help us when we try to pick up small objects; they protect our fingertips; they serve as tools for scratching; they can be used as weapons. Your students may think of many more ways that toenails and fingernails are important to the body.

Science Skills Worksheet 26 "Grasping Graphing"

Build a Lung
Teacher's Notes

Time Required
One 45-minute class period

Lab Ratings

EASY → HARD

TEACHER PREP
STUDENT SET-UP
CONCEPT LEVEL
CLEAN UP

MATERIALS
You may want to build a model first to use as a reference for students. If so, you may want to substitute a bag smaller than the one that students use to model the diaphragm.

Answers

4. The balloon will inflate when the plastic bag is pulled down.

5. The balloon represents a lung, the plastic wrap represents a diaphragm, and the straw represents a trachea. The bottle represents a body cavity.

6. Air enters the lungs when the diaphragm moves down and creates more space inside the chest cavity. Air is forced out of the lungs when the diaphragm moves up. This should be demonstrated by moving the plastic bag up and down.

Going Further

From the late 1920s to the 1950s, iron lungs were used to treat respiratory paralysis due to poliomyelitis. The patient was encased within an airtight chamber from the neck down. A large set of leather bellows mounted in a separate pumping unit expanded, causing pressure changes inside the chamber. This in turn caused the chest of the patient to expand, drawing fresh air into the lungs through the mouth. Now several models of portable ventilators allow patients much more freedom to move about.

Construye un pulmón

CONSTRUIR MODELOS

Ya sabes que al respirar llevas aire al interior de tus pulmones porque el diafragma hace que tu pecho se expanda. Para comprobar que esto es así, pon las manos sobre las costillas e inhala lentamente. ¿Sentiste que tu pecho se expandía?

En esta actividad construirás un modelo de pulmón con materiales de uso común, y podrás ver cómo el diafragma infla tus pulmones. Consulta los diagramas de la derecha para construir tu modelo.

Procedimiento

1. Sujeta el globo al extremo del popote con una liga elástica. Haz un agujero que atraviese la plastilina e inserta el otro extremo del popote. Asegúrate de dejar por lo menos 8 cm de popote después de la plastilina. Presiona la plastilina suavemente para sellar la parte que rodea al popote.

2. Inserta el extremo donde está el globo en el cuello de la botella. Sella la botella con la plastilina para que el popote y el globo queden adentro.

3. Acuesta la botella con cuidado. Coloca la bolsa de basura sobre el extremo cortado de la botella. Coloca una liga alrededor del fondo de la botella para sujetar el plástico en el extremo. Si lo deseas, refuérzalo con cinta adhesiva. Antes de que el plástico esté completamente sellado, junta lo que sobra de la bolsa en la mano y presiona suavemente hacia el interior de la botella. (Es posible que tengas que hacer un nudo en la mitad de la bolsa para recoger el sobrante.) Con la bolsa en esta posición, termina de sellarla con cinta adhesiva. Esto empujará el exceso de aire fuera de la botella.

Análisis

4. ¿Cómo puedes hacer que el "pulmón" se infle?

5. ¿Qué representan el globo, la envoltura de plástico y el popote de tu modelo?

6. Demuéstra a la clase cómo entra y sale aire del pulmón.

Profundizar
Investiga qué es un "pulmón de acero" y por qué se usaba antes. Averigua qué se usa actualmente para ayudar a las personas con dificultades respiratorias.

Materiales

- la mitad superior de una botella de 2 L
- un globo pequeño
- popote de plástico
- pedazo de plastilina del tamaño de una pelota de golf
- bolsa de plástico pequeña para basura
- 2 ligas elásticas
- regla métrica
- cinta adhesiva

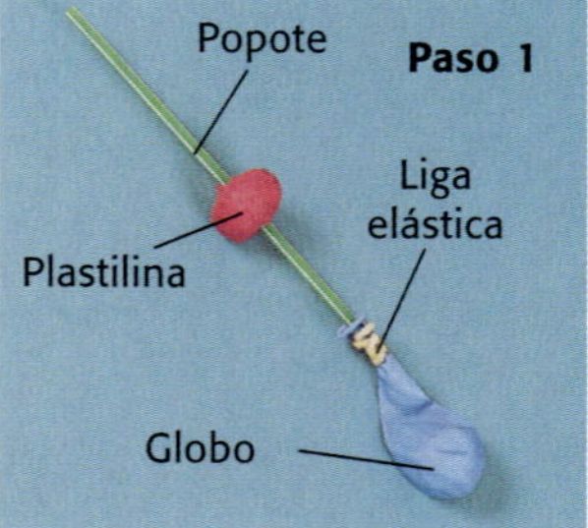

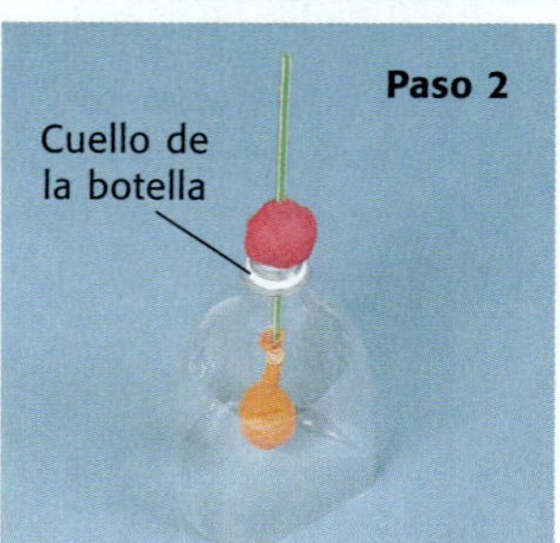

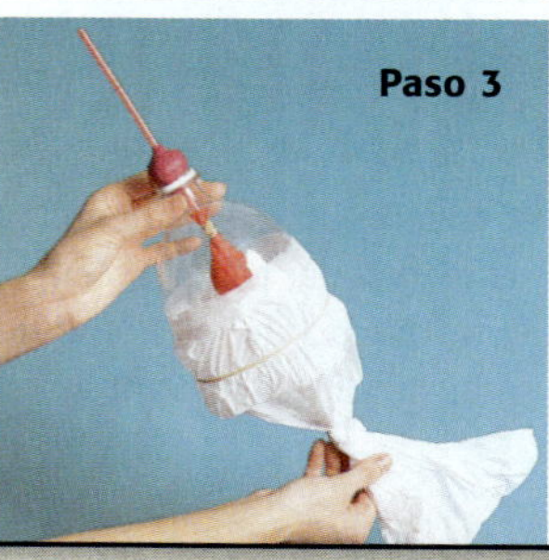

Datasheets for LabBook
Datasheet 42

Yvonne Brannum
Hine Junior High School
Washington, D.C.

Dióxido de carbono en el aliento

Las plantas absorben dióxido de carbono y liberan oxígeno como subproducto de la fotosíntesis. Los animales, incluyéndote a ti, utilizan este oxígeno y liberan dióxido de carbono como subproducto de la respiración.

En esta actividad investigarás el dióxido de carbono que exhalas. La solución con la que trabajarás (rojo de fenol) se vuelve amarilla en presencia de dióxido de carbono. Con esta solución detectarás la presencia de dióxido de carbono en tu aliento.

Materials

- agua
- vaso de precipitados de 150 mL
- probeta de 100 mL
- popote
- solución indicadora de rojo de fenol
- gotero
- reloj con segundero o cronómetro
- guantes protectores

Procedimiento

1. Durante todo el experimento debes tener puestos los guantes, el delantal y los lentes protectores.

2. Coloca 100 mL de agua en un vaso de precipitados de 150 mL. Con un gotero, agrega con cuidado 4 gotas de solución indicadora de rojo de fenol al agua. El rojo de fenol hace que el agua se vuelva naranja.

3. Pon un popote en la solución de agua con rojo de fenol y sopla cuidadosamente. Pon una toalla de papel sobre el vaso de precipitados para evitar salpicaduras. **Cuidado:** No inhales por el popote. No bebas la solución. No compartas el popote.

4. Pídele a tu compañero de laboratorio o a otro compañero o compañera de clase que mida el tiempo requerido para que la solución cambie de color. Debe comenzar a medirlo cuando empieces a soplar. Anota este número en tu cuaderno de ciencias. ¿De qué color se vuelve la solución?

Análisis

5. Compara tus datos con los de tus compañeras y compañeros. ¿Cuál fue el tiempo más largo para ver el cambio de color en la solución de rojo de fenol? ¿Cuál fue el más corto? ¿Cómo explicas la diferencia?

6. ¿Hay relación entre el tiempo que toma el cambio de naranja a amarillo y las características de quien sopla, como su sexo o complexión física?

Profundizar

Haz tijeras o abdominales por 3 minutos y repite el experimento. ¿Cambió el tiempo? Describe y explica cualquier cambio.

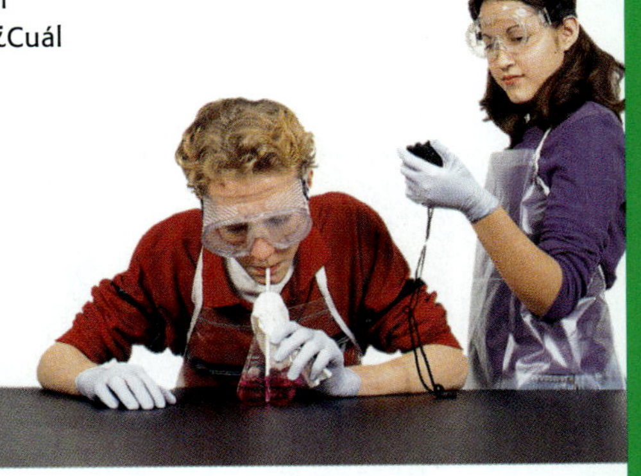

631

Experimentos

Carbon Dioxide Breath
Teacher's Notes

Time Required

One 45-minute class period

Lab Ratings

EASY ———→ HARD

TEACHER PREP   ▲▲
STUDENT SET-UP   ▲▲
CONCEPT LEVEL   ▲▲
CLEAN UP   ▲▲

MATERIALS

You may wish to substitute bromothymol blue indicator solution for the phenol red indicator. The bromothymol blue will turn green in the presence of CO_2. Clear plastic cups (6 oz or 8 oz size) may be used instead of 150 mL beakers if glassware is in short supply or if you have concerns about breakage.

Safety Caution

Remind students to review all safety cautions and icons before beginning this lab activity.

Lab Notes

Tell students that carbon dioxide is in the air of the classroom. They may need to cover their indicator solution to delay the reaction with the air. Tell them not to leave the indicator solution sitting exposed for several minutes before it is used.

Datasheets for LabBook
Datasheet 43

Answers

5. Answers will depend on student observations. It is typical for the solution to change color faster when the student is breathing fast after exercise.

6. In general, an athlete at rest will take the longest time to generate a color change in the indicator solution. There should be little difference observed between genders. There are exceptions, and all answers will depend on students' observations.

You've Gotta Lotta Nerve!
Teacher's Notes

Time Required

One 45-minute class period

Lab Ratings

EASY ——————————→ HARD

TEACHER PREP   ▲

STUDENT SET-UP   ▲

CONCEPT LEVEL   ▲▲

CLEAN UP   ▲

MATERIALS

The materials listed on the student page are enough for 1–2 students. Tell students that they will not be testing for pain. The protective cover on the sharp end of the dissecting pin must remain in place at all times.

Safety Caution

Remind students to review all safety cautions and icons before beginning this lab activity.

Remind students to be safe and gentle with each other in this exercise, respecting the sensitivity and comfort of their peers.

Lab Notes

This activity works best if the student whose hand is being tested looks away or is loosely blindfolded while his or her hand is being tested. Often students will say they feel something when they think they should feel something. Students should be given the choice of being blindfolded or looking away.

¡Me pones nervioso!

La piel tiene miles de receptores nerviosos que detectan sensaciones como calor, frío y presión. El encéfalo está diseñado para filtrar o ignorar la mayoría de la información que recibe a través de estos receptores; de otro modo, sólo con usar ropa se estimularían tantas reacciones que no podríamos funcionar.

Hay partes de la piel, como el dorso de la mano, que son más sensibles que otras. En esta actividad harás un mapa de los receptores de calor, frío y presión en el dorso de tu mano.

Materiales

- marcador lavable de punta fina
- regla métrica
- aguja de disección con un trocito de corcho o un tapón pequeño de goma en la punta
- agua caliente
- agua muy fría
- gotero
- papel milimetrado

Procedimiento

1. Forma grupos de tres. Uno de ustedes será el voluntario o voluntaria que ponga la mano para el experimento; otro llevará a cabo las pruebas y la última persona anotará los resultados. Pregúntale a tu maestro o maestra si pueden turnarse para que cada miembro del grupo pueda participar en cada parte del experimento.

2. Con un marcador delgado de tinta lavable y una regla métrica, traza un cuadrado de 3 × 3 cm en el dorso de la mano de tu compañero o compañera. Dentro del área del cuadrado, traza una cuadrícula, con un espacio entre líneas de unos 0.5 cm. Cuando termines, tendrás una cuadrícula con 36 cuadritos. Fíjate en la fotografía de abajo para asegurarte de que la cuadrícula que trazaste sea correcta.

3. En papel milimetrado, traza tres áreas de 3 × 3 cm. Traza una cuadrícula en cada una, como hiciste en la mano de tu compañero o compañera. Escribe bajo una de las cuadrículas del papel la palabra "Frío", en la siguiente "Calor" y en la tercera "Presión".

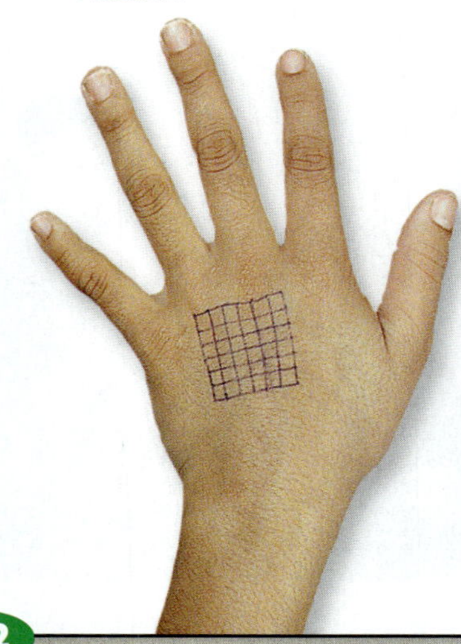

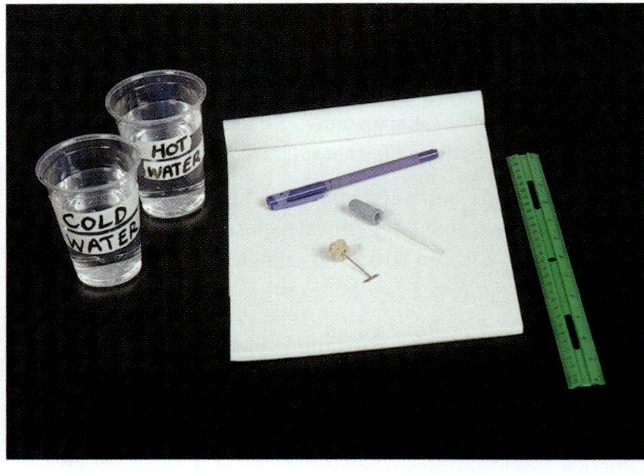

Datasheets for LabBook
Datasheet 44

Science Skills
Worksheet 11
"Understanding Variables"

Christopher Wood
Western Rockingham
Middle School
Madison, North Carolina

---

4. Comienza a ubicar receptores en una esquina de la cuadrícula de la mano. La persona no debe mirarse la mano durante la prueba. ¿Crees que si la persona observara el experimento, esto alteraría los resultados? ¿Cómo? Con el gotero, aplica una gotita de agua fría en cada cuadro de la cuadrícula. Marca con una X en el papel el cuadro que corresponde a la parte de la mano en que la persona sintió frío. Vas a tener que secar cuidadosamente la mano después de unas cuantas gotas.

5. Repite los mismos pasos con el agua caliente. El agua se enfriará lo suficiente al caer del gotero y no te lastimará. De nuevo, marca con una X en el papel milimetrado para indicar la zona en la que se sintió el calor.

6. Repite el procedimiento usando la cabeza (¡no la punta!) de la aguja de disección, con la que tocarás la piel del dorso de la mano para detectar los receptores de presión. Presiona muy suavemente. Marca con una X en el papel para indicar la zona en la que se sintió la presión.

Análisis

7. Cuenta el número de X de cada cuadrícula. ¿Cuántos receptores de calor hay en 3 cm²? ¿Cuántos de frío? ¿Cuántos de presión?

8. ¿Hay áreas del dorso de la mano en las que se superponen los receptores? Explica tu respuesta.

9. ¿Crees que los resultados serían similares o distintos si trazaras una cuadrícula en tu antebrazo?, ¿y en la nuca?, ¿y en la palma de tu mano?

10. Prepara un informe escrito en el que incluyas una descripción de la investigación y un comentario a las preguntas de la sección de Análisis de este experimento.

Profundizar
Averigua en la biblioteca de la escuela o en Internet qué pasa si un receptor recibe estimulación constante. ¿Importa el tipo de receptor? ¿Importa el tipo de estímulo? Explica por qué.

Experimentos

It's a Comfy, Safe World!
Teacher's Notes

Time Required

Two 45-minute class periods

Lab Ratings

EASY ——————→ HARD

TEACHER PREP 🧪🧪
STUDENT SET-UP 🧪
CONCEPT LEVEL 🧪
CLEAN UP 🧪🧪🧪

> **MATERIALS**
>
> This lab may require some larger plastic bags, a meterstick, and various other materials, depending on the students' designs. Soft-boiled eggs will simplify the cleanup. Students may wear gloves.

Safety Caution

Remind students to review all safety cautions and icons before beginning this lab activity. Students should wash their hands after handling eggs.

Lab Notes

This lab should be done over a large plastic sheet. You may also want to do this lab outside.

Answers

6. Answers will vary, but an egg inside a viscous liquid in a plastic bag, wrapped in soft cotton and all inside another bag, should not be damaged when dropped from a height of 1 m. An egg protected only by a shell should break.

7. The answers will vary according to students' observations. Most students will observe that the soft wrapping might be thicker to protect the egg better.

Experimentos

¡Un mundo seguro y agradable!

Antes de nacer, los bebés llevan una vida agradable. Durante el tercer trimestre, se encuentran dentro de su mamita, se chupan el dedo, parpadean y quizá hasta sueñan. Pero, ¿qué tan protegidos están? El pollito, antes de nacer, vive dentro de un cascarón duro y protector hasta que usa todo el suministro de comida. Pero la mayoría de los mamíferos tienen un abdomen suave donde crecen sus bebés, envueltos en un fluido y una placenta dentro del útero, que es un órgano muscular fuerte. ¿Es más seguro el ambiente interno de un mamífero placentario que el del pollito en su huevo? En esta actividad, crearás un modelo del útero de un mamífero placentario y probarás su eficacia en la protección del feto.

Materiales

- bolsas de plástico con cierre
- huevos tibios (pasados por agua), la mitad de ellos sin cascarón
- agua
- aceite mineral, aceite de cocina, jarabe u otro líquido espeso que represente el fluido que rodea al feto
- algodón, tela suave u otros materiales suaves

Procedimiento

1. Piensa en cómo construirás y probarás tu modelo del útero mamífero. Pide a tu maestra o maestro los materiales y construye el modelo. Un huevo tibio sin cascarón representará el feto.

2. Haz una tabla como la de la derecha ("Primera prueba del modelo"). Prueba tu modelo, examina si el huevo sufrió daños y anota los resultados.

3. Modifica el diseño de la tabla, repite la prueba y anota los resultados. Haz la misma prueba del paso 2.

4. Cuando estés de acuerdo con el diseño del modelo, usa otro huevo tibio sin cascarón y uno con cascarón.

5. Haz otra tabla de datos como "Prueba final del modelo" de la derecha. Somete el modelo y el huevo tibio con cascarón a la misma prueba del paso 2. Anota los resultados en la tabla.

Primera prueba del modelo	
Modelo original	Modelo modificado
No escribas en el libro	

Prueba final del modelo	
	Resultados de la prueba
Modelo	*No escribas en el libro*
Huevo con cascarón	

Análisis

6. Explica las diferencias que hayas observado en la habilidad del modelo y del huevo con cascarón para proteger el feto que está en su interior.

7. ¿Qué modificación de las que le hiciste a tu modelo fue la más eficaz en la protección del feto?

Profundizar

Compara el desarrollo de los mamíferos placentarios con el de los mamíferos marsupiales y monotremas.

Going Further

Students may want to research in the library or on the Internet about marsupials and monotremes.

Datasheets for LabBook
Datasheet 45

Randy Christian
Stovall Junior High School
Houston, Texas

¡Oh, cuánto has crecido!

En los bebés humanos, el proceso de desarrollo que tiene lugar entre la fecundación y el nacimiento dura alrededor de 266 días. El nuevo ser crece rápidamente de una simple célula fecundada a un embrión cuyo corazón late y bombea sangre a partir de la 4ª semana. Todos los sistemas y partes del cuerpo del bebé están completamente formados al final del séptimo mes. Durante los últimos 2 meses antes del nacimiento, el bebé crece y sus sistemas maduran. Cuando nace, la masa promedio de un recién nacido es unas 33,000 veces mayor que la de un embrión de dos semanas de desarrollo. En esta actividad descubrirás qué tan rápido crece un feto en poco menos de 9 meses.

Materiales

- papel milimétrico
- lápices de colores

Procedimiento

1. Con hojas de papel milimétrico, haz dos gráficas en tu cuaderno de ciencias, una titulada "Longitud" y la otra "Peso". En la gráfica de longitud, usa intervalos de 25 mm sobre el eje de las *y*. Amplía el eje de las *y* a 500 mm. En la gráfica de peso, usa intervalos de 100 g en el eje de las *y*. Extiende este eje a 3,300 g. Usa intervalos de tiempo de 2 semanas sobre el eje de las *x* para ambas gráficas. Ambos ejes *x* deben ampliarse a 40 semanas.

2. Examina la tabla de datos de la derecha. Marca los datos de la tabla en las gráficas. Usa un lápiz de color para trazar la línea curva que una los puntos de cada gráfica.

Análisis

3. Describe el cambio en la masa de un feto en desarrollo. ¿Cómo puedes explicar este cambio?

4. Describe el cambio en la longitud de un feto en desarrollo. ¿Cómo se compara el cambio de masa con el de longitud?

Profundizar

Con la información de las gráficas, calcula la estatura que un niño o niña de 3 años tendría si continuara creciendo a la velocidad promedio a la que crece un feto.

Aumento de masa y longitud de un feto humano promedio		
Tiempo (semanas)	Peso (g)	Longitud (mm)
2	0.1	1.5
3	0.3	2.3
4	0.5	5.0
5	0.6	10.0
6	0.8	15.0
8	1.0	30.0
13	15.0	90.0
17	115.0	140.0
21	300.0	250.0
26	950.0	320.0
30	1,500.0	400.0
35	2,300.0	450.0
40	3,300.0	500.0

Experimentos

635

My, How You've Grown!

Teacher's Notes

Time Required

One 45-minute class period

Lab Ratings

TEACHER PREP — 1
STUDENT SET-UP — 1
CONCEPT LEVEL — 2
CLEAN UP — 1

Answers

2. Students' graphs should look like those below.

3. The change in mass of a developing fetus is steadily increasing, approximately tripling each month of the first and second trimesters. This is a period of rapid cell division.

4. The change in length steadily increases, doubling and even tripling each month in the first two trimesters. In the third trimester, the rate of lengthening slows.

Going Further

The child would be 2.45 m (8.04 ft) tall!

Datasheets for LabBook
Datasheet 46

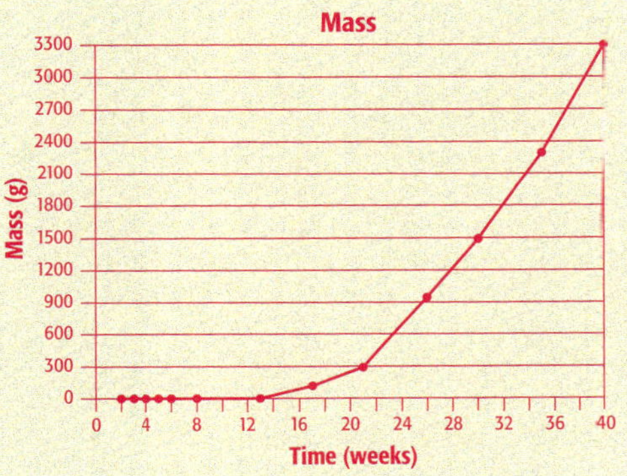

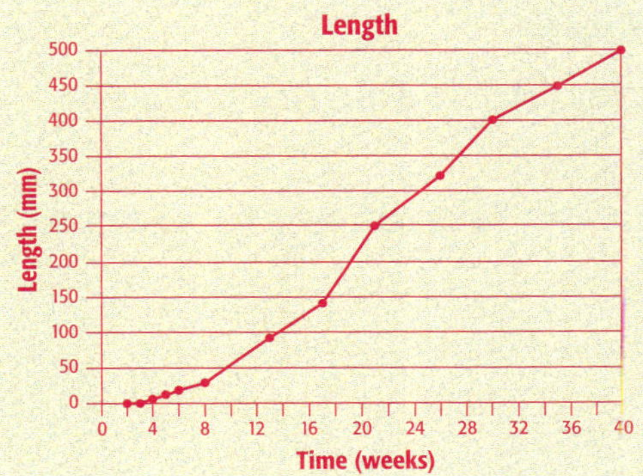

Randy Christian
Stovall Junior High School
Houston, Texas

Autoevaluación: Respuestas

Capítulo 1: El mundo de las ciencias biológicas
Página 12: 2. Los insecticidas y fertilizantes causaron las deformidades.
Página 15: El frasco C es el grupo de control.

Capítulo 2: ¡Está vivo! ¿O no…?
Página 37: Tu despertador es un estímulo. Cuando suena, tu respuesta es apagarlo y levantarte de la cama.

Capítulo 3: La luz y los seres vivos
Página 62: El papel se verá azul porque sólo refleja la luz azul.
Página 68: a. la pupila b. la córnea y el cristalino c. la retina

Capítulo 4: La célula: unidad fundamental de la vida
Página 87: Las células necesitan el ADN para controlar los procesos celulares y para hacer nuevas células.

Página 90: 1. La proporción entre la superficie y el volumen disminuye al crecer la célula. 2. Las células eucariotas tienen núcleo y organelos cubiertos por una membrana celular.

Página 94: Algunas células tienen paredes celulares alrededor de la membrana celular. Todas las células tienen membrana celular, pero no todas tienen paredes celulares. La pared celular les da estructura a algunas células.

Capítulo 5: La célula en acción
Página 109: En agua pura, la uva absorbería agua y se hincharía. En agua mezclada con mucha azúcar, la uva perdería agua y se arrugaría.

Página 117: Después de la duplicación, hay dos cromátides: dos de cada uno de los cromosomas homólogos.

Capítulo 6: Herencia
Página 141: 1. cuatro 2. dos 3. Hacen copias de sí mismos una vez. Se dividen dos veces. 4. Dos, o la mitad de los cromosomas de uno de los padres, están presentes al final de la meiosis. Después de la meiosis, quedarían cuatro cromosomas, el mismo número que en la célula madre. 5. Primero, los cromososmas homólogos se separan.

Capítulo 7: Los genes y la tecnología genética
Página 155: TGGATCAAC

Página 161: 1. 1000 aminoácidos 2. códigos del ADN para formar proteínas. Tu cuerpo está hecho de proteínas, y la forma en que estas proteínas están hechas y combinadas tiene mucha influencia en tu aspecto.

Página 166: El gene humano responsable de una proteína en particular puede insertarse en una bacteria que usa el gene para producir la proteína necesaria con gran rapidez y en cantidades grandes. Por eso se dice que las bacterias son fábricas vivientes.

Capítulo 8: La evolución de los seres vivos
Página 191: 1. b 2. a 3. d 4. c

Capítulo 9: La historia de la vida en la Tierra
Página 205: 5 g, 2.5 g
Página 211: b, c, d, a

Capítulo 10: Clasificación

Página 236: 1. Los dos reinos de las bacterias son diferentes de los demás porque las bacterias son procariotas, es decir, organismos unicelulares que no tienen núcleo. 2. Todos los protistas son eucariotas.

Capítulo 11: Introducción a las plantas

Página 252: Las plantas necesitan una cutícula que proteja a las hojas para que no se sequen. Las algas crecen en un medio acuático y no la necesitan.

Página 256: En las plantas no vasculares, el transporte por medio del agua es muy limitado. Necesitan crecer cerca de la tierra para estar en contacto con la humedad y los nutrientes del suelo. Las plantas vasculares tienen xilema y floema para transportar nutrientes y agua a través de la planta; por lo tanto, pueden crecer más.

Página 268: Los tallos sostienen a las hojas de manera que éstas reciban la luz que necesitan para la fotosíntesis.

Capítulo 12: Procesos de las plantas

Página 281: El fruto se desarrolla a partir del ovario, de modo que sólo puede tener una fruta. Las semillas se desarrollan a partir de los óvulos, de así que debería haber seis semillas.

Página 285: El Sol es el orígen de la energía en el azúcar.

Página 288: 1. (Ver el mapa de ideas a la derecha.) 2. En el fototropismo negativo, la planta crecería en dirección opuesta al estímulo (la luz), de modo que crecería hacia la izquierda.

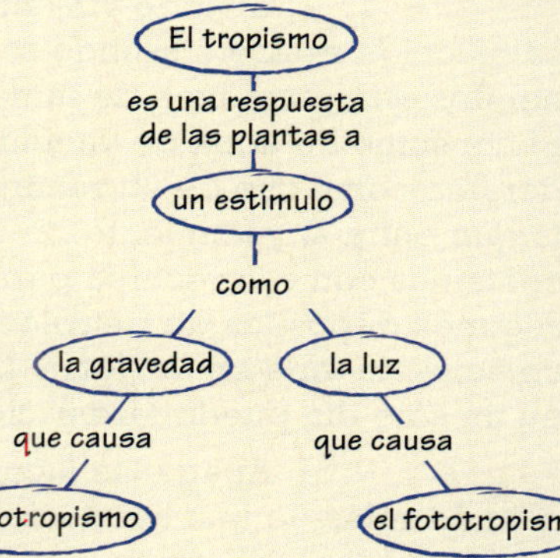

Capítulo 13: Los animales y su conducta

Página 306: Como otros vertebrados, los seres humanos tienen cráneo y columna vertebral.

Página 312: El ritmo circadiano no controla la hibernación. Los ritmos circadianos son ritmos diarios. La hibernación es un comportamiento estacional.

Capítulo 14: Invertebrados

Página 331: Como las medusas contraen su cuerpo para nadar en el agua, deben tener un sistema nervioso que controle esta acción. Los pólipos se mueven muy poco, por lo tanto no necesitan un sistema nervioso complejo.

Página 341: Los gusanos segmentados pertenecen al filo anélidos. Los ciempiés son artrópodos. Los ciempiés tienen patas articuladas, antenas y mandíbulas. Los gusanos segmentados no.

Capítulo 15: Peces, anfibios y reptiles

Página 366: Los anfibios absorben oxígeno a través de la piel. Su piel es delgada, húmeda y está llena de vasos sanguíneos, al igual que un pulmón.

Página 371: 1. La piel gruesa y seca y el huevo amniótico permitieron a los reptiles vivir en la tierra. 2. La cáscara dura del huevo impide la fecundación, así que debe fecundarse antes de que se forme la cáscara.

Capítulo 16: Aves y mamíferos

Página 384: 1. El plumón no es rígido ni liso y, no puede formar la estructura de las alas; está adaptado para abrigar al ave. 2. Las aves necesitan mucho alimento, porque volar requiere mucha energía.

Página 398: Los monotremas son mamíferos que ponen huevos. Los marsupiales dan a luz crías vivas, que llevan en una bolsa o entre los pliegues de la piel antes de que puedan vivir de manera independiente. Los animales placentarios se desarrollan dentro del cuerpo de la madre y reciben nutrientes a través de la placenta.

Página 403: 1. Los murciélagos dan a luz crías vivas, tienen pelo y no tienen plumas. 2. Los roedores y los lagomorfos son pequeños mamíferos que tienen largos y sensibles bigotes, y dientes para roer. Los lagomorfos tienen dos pares de incisivos y cola corta.

Capítulo 17: Los ecosistemas de la Tierra

Página 421: Los bosques de árboles de hoja caduca suelen crecer en latitudes medias o en climas templados, mientras que los bosques de coníferas crecen al Norte, cerca del polo.

Página 426: Las respuestas incluyen: la cantidad de luz que penetra en el agua, su distancia de la tierra, la profundidad del agua, la salinidad del agua, y la temperatura del agua. 2. Hay varias respuestas posibles. Algunos organismos tienen adaptaciones para cazar a grandes profundidades; otros se alimentan de plancton muerto y otros organismos más grandes que se filtran desde arriba, y algunos, como las bacterias que viven alrededor de las fuentes termales, producen alimentos a partir de substancias químicas en el agua.

Capítulo 18: Problemas ambientales y soluciones

Página 443: 1. Cuando viajamos en coche o quemamos carbón en la calefacción estamos usando recursos no renovables. Cuando usamos minerales que provienen de minas, estamos usando recursos no renovables. El consumo de agua de yacimientos subterráneos también puede considerarse como uso de un recurso no renovable, si se bombea más agua de la que entra al yacimiento. 2. Si se agota un recurso no renovable, ya no contamos con ese recurso para satisfacer las necesidades de la Tierra. Algunos depósitos de petróloeo y carbón se han ido formando desde que comenzó la vida en la Tierra. Un bosque maduro que se tala por completo en un sólo día puede tardar cientos de años en crecer de nuevo.

Página 449: 1. Apaga las luces, tocadiscos, radios y computadoras cuando sales de una habitación. Baja un poco el termostato en invierno (y abrígate bien). No dejes la puerta del refrigerador abierta mientras decides qué quieres comer. 2. Bolsas de plástico, pilas recargables, agua, ropa, juguetes, la diferencia entre reutilizar y reciclar es que el artículo que se reutiliza no se transforma, excepto quizás para limpiarlo. Un artículo reciclado se ha procesado para convertirlo en otro producto útil.

Capítulo 19: La organización y estructura del cuerpo

Página 474: Las abdominales ejercitan a los músculos flexores; las lagartijas ejercitan a los músculos extensores.

Página 477: Los vasos sanguíneos pertenecen al sistema cardiovascular.

Capítulo 20: Circulación y respiración

Página 492: La estructura de las venas y arterias, en forma de tubos huecos, permite que la sangre llegue a todas partes del cuerpo. Las válvulas en las venas evitan que la sangre fluya en sentido contrario.

Página 496: Como los vasos sanguíneos, los capilares linfáticos reciben líquidos de los espacios entre las células. Los líquidos que absorben los capilares linfáticos fluyen hacia los vasos capilares. Estos vasos se comunican con las grandes venas del cuello y no con los órganos, como el corazón. El líquido linfático no transporta oxígeno y nutrientes.

Capítulo 21: Comunicación y control

Página 515: 1. cerebro 2. cerebelo 3. para proteger la médula espinal

Capítulo 22: Reproducción y desarrollo

Página 535: En la reproducción asexual, el animal produce crías que son geneticamente idénticas a él. En la reproducción sexual, normalmente se mezclan los genes de por lo menos dos individuos cuando los gametos se unen para formar un cigoto. El cigoto se convierte en un individuo único.

Página 543: 1. El embrión recibe nutrientes de la placenta. Los vasos sanguíneos en el cordón umblical absorben oxígeno y nutrientes de los vasos sanguíneos de la madre. 2. La matriz proporciona los nutrientes y la protección que el embrión necesita para continuar su crecimiento. La matriz es también el único lugar donde se puede formar la placenta.

CONTENIDO

Organizar conceptos: una forma de relacionar ideas

¿Qué es un mapa de ideas?

¿Alguna vez has tratado de contarle a alguien un libro o un capítulo que acabas de leer y te diste cuenta de que sólo puedes recordar unas palabras o ideas aisladas? O quizá has memorizado unos datos para una prueba y semanas después ni siquiera recuerdas los temas de que se trataban.

En ambos casos, puede que hayas entendido las ideas o los conceptos por sí solos, pero no los relacionaste. Si pudieras unir las ideas de alguna manera, podrías entenderlas mejor y recordarlas por más tiempo. Un mapa de ideas te permite hacer esto. Es una manera de ver cómo las ideas o los conceptos se unen. Te ofrecen una "visión global."

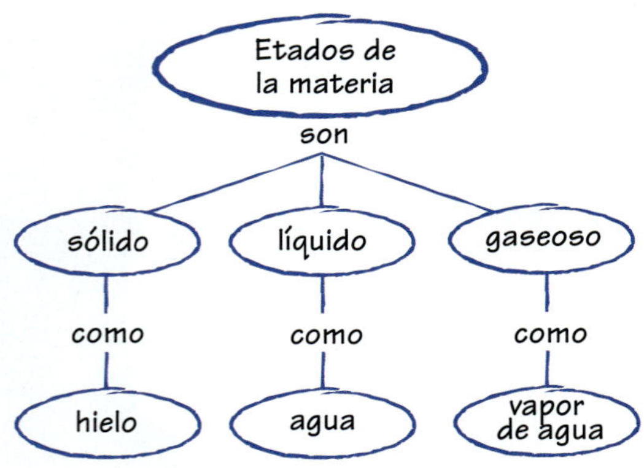

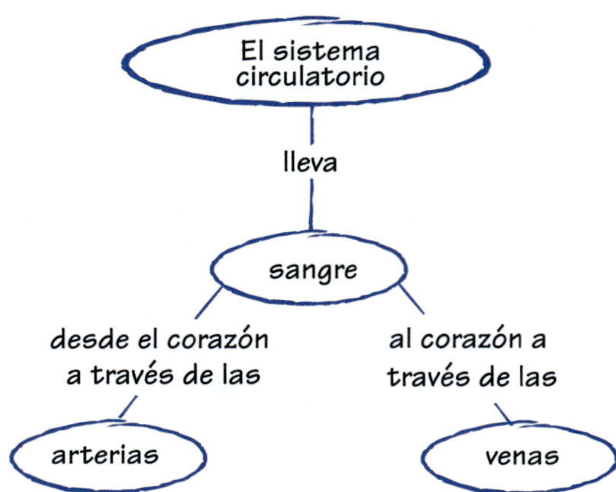

Como hacer un mapa de ideas

1 Haz una lista de las ideas o conceptos principales.

Podría ser útil escribir cada concepto en una hoja de papel aparte. Esto facilitará la reorganización de los conceptos tantas veces como sea necesario para saber cómo se relacionan. Una vez que hayas preparado algunos mapas de ideas de esta forma, puedes pasar directamente de la lista a construir el mapa.

2 Separa las hojas de papel y organiza los conceptos del más general al más específico.

Pon el concepto más general encima y enciérralo en un círculo. Debes preguntarte, "¿Cómo se relaciona este concepto con los demás?" A medida que veas las relaciones, organiza los conceptos del más general al más específico.

3 Une los conceptos relacionados con una línea recta.

4 En cada línea, escribe un verbo o una frase corta para demostrar cómo se relacionan los conceptos. Observa los mapas de ideas de esta página y luego intenta preparar uno para los siguientes términos:

plantas, agua, fotosíntesis, dióxido de carbono, energía solar

Se da una respuesta posible a la derecha, pero no la veas hasta que intentes preparar tu propio mapa.

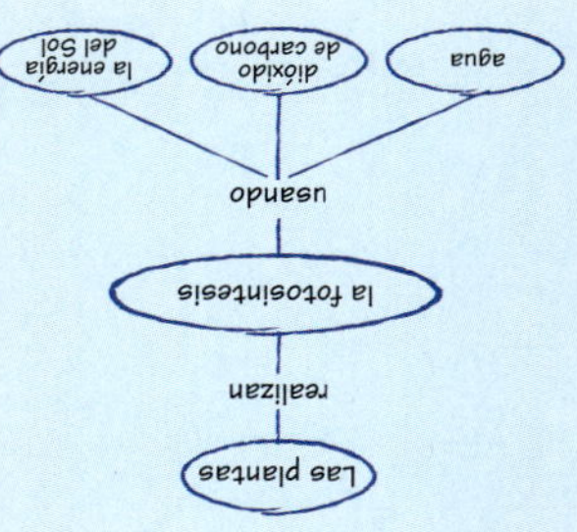

Sistema Internacional de Unidades

El Sistema Internacional de Unidades, o SI, es el sistema de medidas estándar para muchos científicos. El uso de las mismas medidas estándares facilita la comunicación entre ellos.

El SI funciona mediante la combinación de prefijos y unidades básicas. Cada unidad básica puede ser utilizada con distintos prefijos para definir mayor y menor cantidad. La siguiente tabla enumera los prefijos más comunes del SI.

Prefijos SI

Prefijo	Abreviatura	Factor	Ejemplo
kilo-	k	1,000	kilogramo, 1kg = 1,000 g
hecto-	h	100	hectolitro, 1 hL = 100 L
deca-	da	10	decámetro, 1 dam = 10m
		1	metro, litro
deci-	d	0.1	decigramo, 1 dg = 0.1 g
centi-	c	0.01	centímetro, 1 cm = 0.01 m
mili-	m	0.001	mililitro, 1 mL = 0.001 L
micro-	µ	0.000001	micrómetro, 1 µm = 0.000 001 m

Tabla de conversión SI

Unidades del SI	Del SI al Sistema Inglés	Del Sistema Inglés al SI
Longitud		
kilómetro (km) = 1,000 m	1 km = 0.621 mi	1 mi = 1.609 km
metro (m) = 100 cm	1 m = 3.281 pies	1 pie = 0.305 m
centímetro (cm) = 0.01 m	1 cm = 0.394 pulgadas	1 pulgada = 2.540 cm
milímetro (mm) = 0.001 m	1 mm = 0.039 pulgadas	
micrómetro (µm) = 0.000 001 m		
nanómetro (nm) = 0.000 000 001 m		
Área		
kilómetro cuadrado (km^2) = 100 hectáreas	1 km^2 = 0.386 mi^2	1 mi^2 = 2.590 km^2
hectárea (ha) = 10,000 m^2	1 ha = 2.471 acres	1 acre = 0.405 ha
metro cuadrado (m^2) = 10,000 cm^2	1 m^2 = 10.765 $pies^2$	1 pie^2 = 0.093 m^2
centímetro cuadrado (cm^2) = 100 mm^2	1 cm^2 = 0.155 $pulgadas^2$	1 $pulgada^2$ = 6.452 cm^2
Volumen		
litro (L) = 1,000 mL = 1 dm^3	1L = 1.057 fl qt	1 fl qt = 0.946 L
mililitro (mL) = 0.001 L = 1 cm^3	1 mL = 0.034 fl oz	1 fl oz = 29.575 mL
microlitro (µL) = 0.000 001 L		
Masa		
kilogramo (kg) = 1,000 g	1 kg = 2.205 lb	1 lb = 0.454 kg
gramo (g) = 1,000 mg	1 g = 0.035 oz	1 oz = 28.349 g
miligramo (mg) = 0.001 g		
microgramo (µg) = 0.000 001 g		

Escalas de temperatura

La temperatura se puede expresar con tres escalas distintas: Fahrenheit, Celsius y Kelvin. La unidad SI para medir la temperatura es el kelvin (K). A pesar de que 0 K es más frío que 0°C, un cambio de 1 K equivale a un cambio de 1°C.

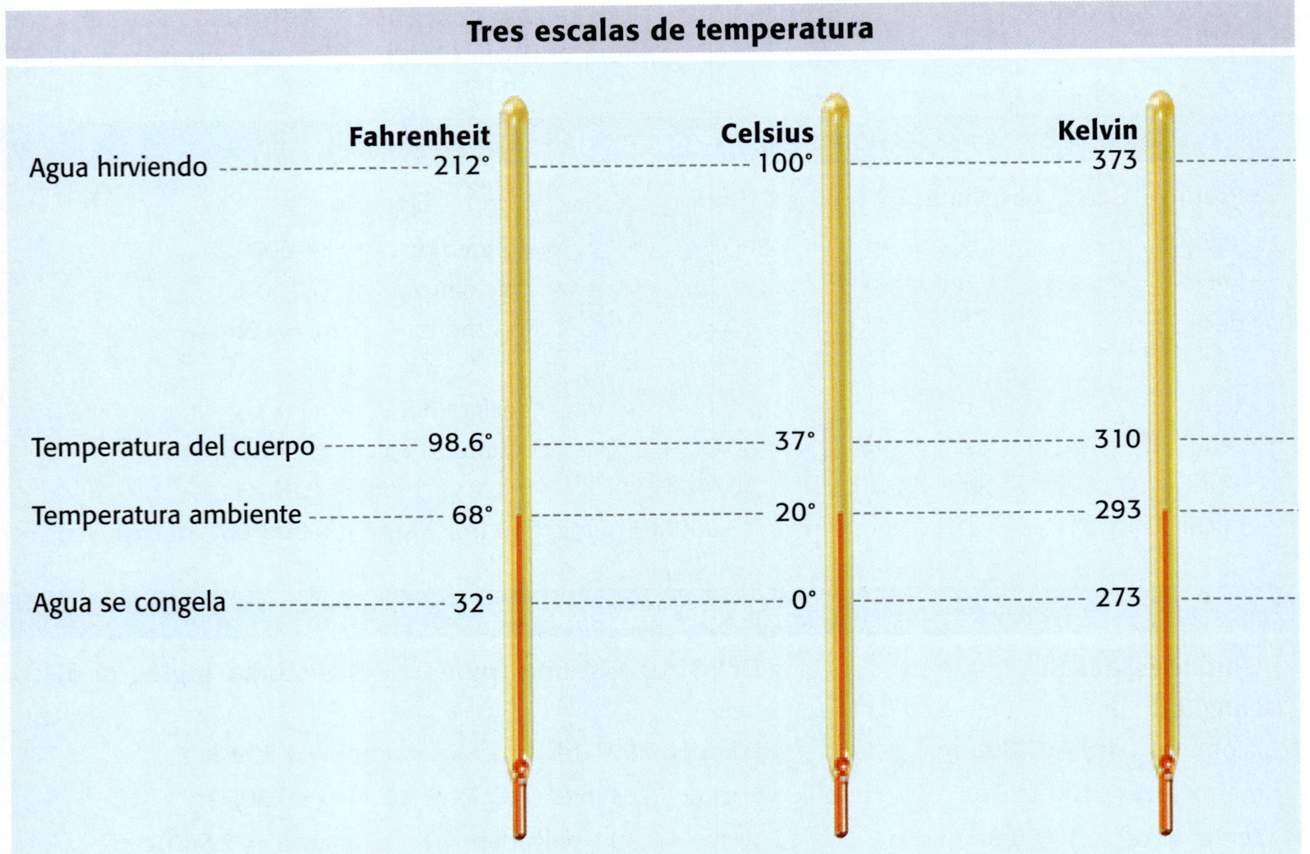

Tabla de conversión de temperaturas		
Para convertir	**Utiliza esta ecuación:**	**Ejemplo**
Celsius a Fahrenheit $°C \longrightarrow °F$	$°F = \left(\dfrac{9}{5} \times °C\right) + 32$	Convertir 45°C a °F $°F = \left(\dfrac{9}{5} \times 45°C\right) + 32 = 113°F$
Fahrenheit a Celsius $°F \longrightarrow °C$	$°C = \dfrac{5}{9} \times (°F - 32)$	Convertir 68°F a °C $°C = \dfrac{5}{9} \times (68°F - 32) = 20°C$
Celsius a Kelvin $°C \longrightarrow K$	$K = °C + 273$	Convertir 45°C a K $K = 45°C + 273 = 318 K$
Kelvin a Celsius $K \longrightarrow °C$	$°C = K - 273$	Convertir 32 K a °C $°C = 32 K - 273 = -241°C$

Técnicas de medición

Cómo usar un cilindro graduado

Cuando utilizas un cilindro graduado para medir el volumen, debes tener en cuenta los siguientes procedimientos:

❶ Asegúrate de que el cilindro esté sobre una superficie plana y nivelada.

❷ Mueve la cabeza de modo que tus ojos estén al mismo nivel que la superficie del líquido.

❸ Lee la marca que se encuentra cerca del nivel del líquido. En los cilindros de vidrio, lee la marca más cercana al centro de la curva.

Cómo usar una vara o regla métrica

Al usar una vara o regla métrica, debes tener en cuenta los siguientes procedimientos:

❶ Coloca la regla contra el objeto que deseas medir.

❷ Debes alinear un borde del objeto con el cero del extremo de la regla.

❸ Observa el otro borde del objeto para ver qué marca de la regla está más cerca de este borde. **Nota:** Cada línea entre los centímetros representa un milímetro, que es un décimo de centímetro.

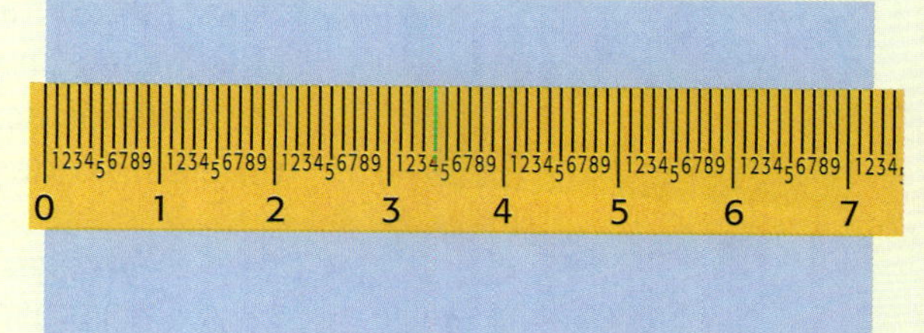

Cómo utilizar una balanza de triple tablón

Al usar una balanza de triple tablón debes tener en cuenta los siguientes procedimientos:

❶ Asegúrate de que la balanza esté sobre una superficie plana y nivelada.

❷ Coloca todas las contramasas en cero. Ajusta la perilla de balance hasta que la manecilla indique cero.

❸ Coloca el objeto que quieres medir sobre el recipiente. **Cuidado:** No pongas objetos calientes o substancias químicas directamente sobre el recipiente de la balanza.

❹ Mueve la contramasa más grande a lo largo del tablón hasta que alcance la última hendidura sin inclinar la balanza. Sigue el mismo procedimiento con la siguiente contramasa.

❺ Luego, mueve la contramasa más pequeña hasta alcanzar el cero. Suma las lecturas obtenidas de los tres tablones para determinar la masa del objeto.

❻ Al determinar la masa de cristales o polvos, debes utilizar un pedazo de papel de filtro. Primero, mide la masa del papel. Luego, agrega los cristales o el polvo sobre el papel y mide la masa nuevamente. La masa real de los cristales o el polvo es la masa total menos la masa del papel. Para encontrar la masa de líquidos, primero mide la masa de sus recipientes vacíos. Luego, mide la masa del líquido y el recipiente. La masa real del líquido es la masa total menos la masa del papel.

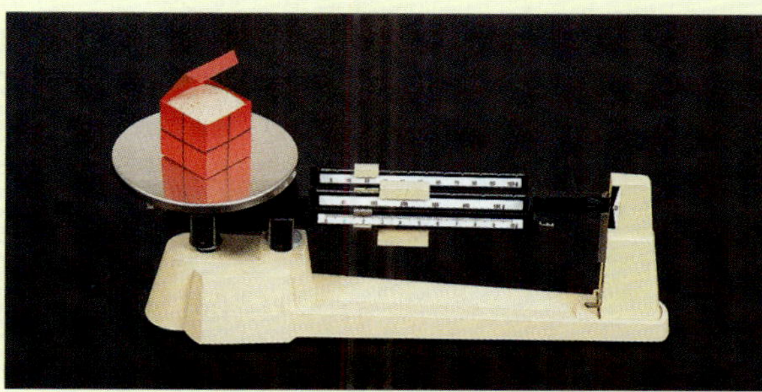

"

Método científico

Los pasos que se siguen para responder preguntas y resolver problemas constituyen el **método científico.** El método científico no es rígido. Se pueden seguir todos los pasos o sólo algunos. Hasta se pueden repetir algunos pasos. La meta del método científico es obtener respuestas y soluciones confiables.

Los seis pasos del método científico

1 Hacer una pregunta Las buenas preguntas son el producto de observaciones **cuidadosas.** Las observaciones se realizan al utilizar los sentidos para recopilar información. A veces puedes usar instrumentos, como microscopios y telescopios, para extender el alcance de tus sentidos. A medida que observas el mundo natural, descubrirás que tienes más preguntas que respuestas. Estas preguntas son el motor que impulsa el método científico.

Las preguntas *qué, por qué, cómo* y *cuándo* son importantes al enfocar una investigación y muchas veces llevan a una hipótesis. (Aprenderás acerca de las hipótesis en el próximo paso.) Aquí tienes un ejemplo de una pregunta que podría llevar a una investigación.

Pregunta: ¿Cómo afecta la precipitación ácida al crecimiento de las plantas?

2 Formular una hipótesis Después de hacer una pregunta, debes convertirla en una **hipótesis.** Una hipótesis es una afirmación clara de lo que tú crees que puede ser la respuesta a tu pregunta. Tu hipótesis representa tu mejor conjetura considerando tus observaciones y los conocimientos de que dispones. Una buena hipótesis debe poderse comprobar. Si no se pueden recopilar observaciones e información, o si no se puede diseñar un experimento para comprobar la hipótesis, ésta no es comprobable y no se puede continuar con la investigación.

He aquí una hipótesis que pudo haber derivado de la pregunta: "¿Cómo afecta la precipitación ácida al crecimiento de las plantas?"

Hipótesis: La precipitación ácida provoca el crecimiento lento de las plantas.

La hipótesis da información que conduce a distintos métodos de comprobación y también puede dar lugar a predicciones. Una **predicción** es lo que tú crees que resultará de tu experimento o recopilación de información. Generalmente se expresan en el formato: "Si…, entonces…". Por ejemplo: **Si** se deja la carne a temperatura ambiente, **entonces** se pudrirá más rápido que la que se guarda en el refrigerador. Es posible hacer más de una predicción para cada hipótesis. A continuación se presenta un ejemplo de predicción para la hipótesis que afirma que la precipitación ácida provoca el crecimiento lento de las plantas.

Predicción: Si una planta se riega solamente con precipitación ácida (que tiene un pH de 4), crecerá a la mitad de su velocidad normal.

3 **Comprobar la hipótesis** Una vez que hayas formulado una hipótesis y una predicción, debes comprobar tu hipótesis. Hay varias maneras de hacerlo. Quizá la más conocida sea la realización de un **experimento controlado.** En un experimento controlado se prueba un solo factor a la vez. Un experimento controlado consta de un **grupo de control** y uno o más **grupos experimentales.** Todos los factores de los grupos (control y experimental) son iguales con excepción de uno, llamado la **variable.** Al cambiar sólo un factor (la variable), se pueden ver los resultados sólo de ese cambio.

A menudo, la naturaleza de una investigación hace imposible un experimento controlado. Por ejemplo, los dinosaurios se extinguieron hace millones de años y el núcleo terrestre está cubierto de miles de metros de roca. Sería difícil, si no imposible, realizar experimentos controlados sobre este tipo de asuntos. En estas circunstancias, una hipótesis se puede comprobar mediante observaciones detalladas. Tomar medidas es una forma de hacer observaciones.

4 **Analizar los resultados** Cuando hayas finalizado tus experimentos y observaciones y recopilado tus datos, analiza toda la información reunida. A menudo se utilizan tablas y gráficas en este paso para organizar los datos.

5 **Sacar conclusiones**
Basándote en el análisis de tus datos, debes concluir si los resultados corroboran tu hipótesis. Si tu hipótesis se ve corroborada, es posible que tú (u otras personas) quieran repetir las observaciones o experimentos para verificar los resultados obtenidos. Si los datos no corroboran tu hipótesis, es posible que tengas que revisar si tu procedimiento tenía errores. Es probable que tengas que rechazar tu hipótesis y formular otra. Si no puedes llegar a una conclusión a partir de tus resultados, quizá tengas que realizar la investigación nuevamente o llevar a cabo otras observaciones o experimentos.

6 **Comunicar los resultados** Después de cualquier investigación científica, comunica los resultados. En un informe escrito u oral, puedes comunicar a los demás lo que has aprendido. Quizá ellos quieran repetir tu investigación para ver si obtienen los mismos resultados. Es posible que tu informe lleve a otra pregunta, que a su vez podría llevar a otra investigación.

El método científico en acción

El método científico no es una sucesión continua de pasos. Hay pasos que se pueden repetir una y otra vez, y otros pueden no ser necesarios. A menudo, los científicos descubren que al comprobar una hipótesis surgen nuevas preguntas e hipótesis que deben ser comprobadas. Otras veces, la comprobación de una hipótesis puede llevar directamente a una conclusión. Además, los pasos del método científico no siempre se siguen en el mismo orden. Sigue los pasos del siguiente diagrama y observa todos los trayectos que puede presentar el método científico.

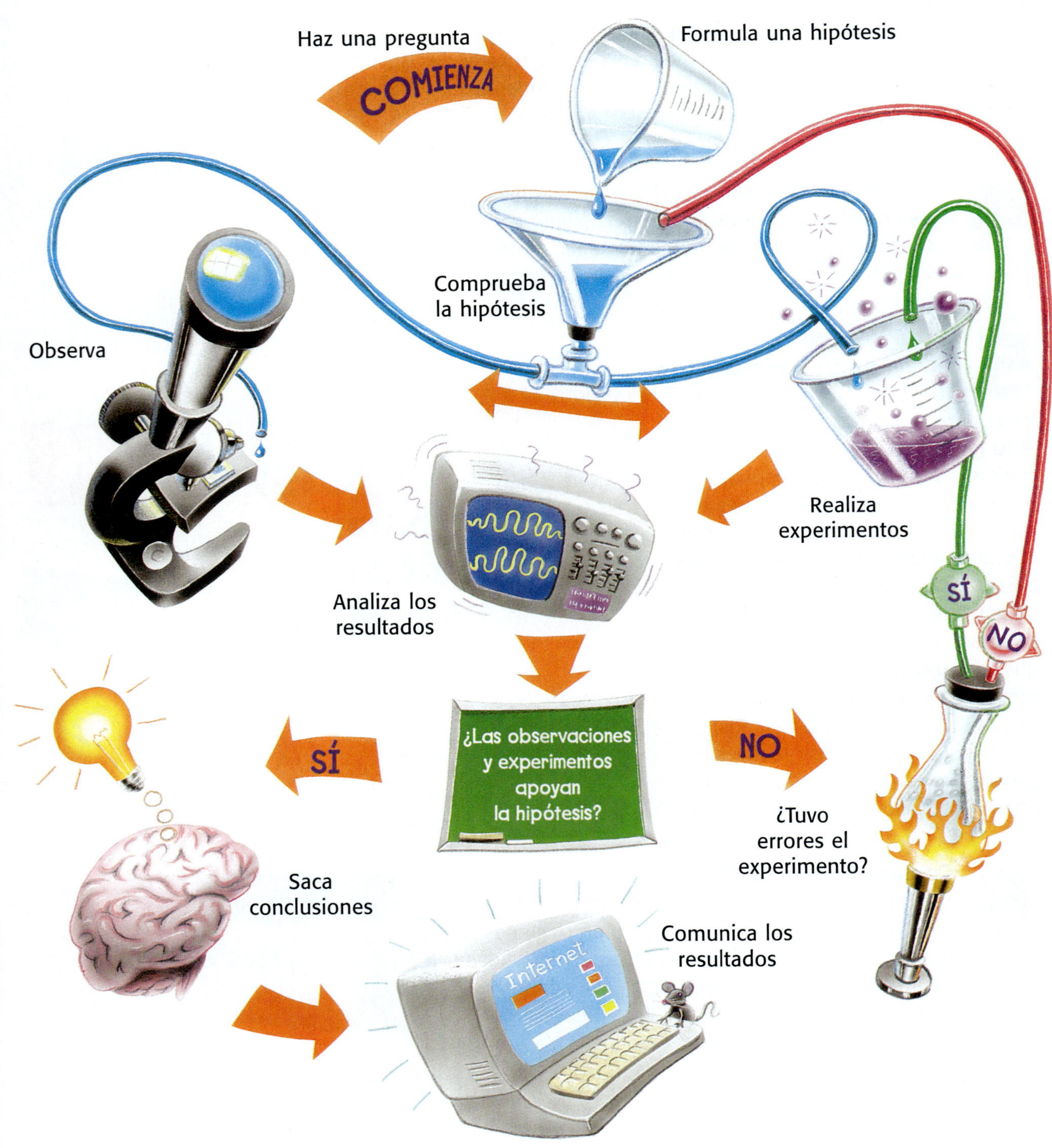

Hacer tablas y gráficas

Diagramas circulares

Un diagrama circular muestra cómo cada grupo de datos se relaciona con la totalidad. Cada parte del círculo representa una categoría de datos. El círculo entero representa todos los datos. Por ejemplo, un biólogo estudia un bosque de árboles de madera dura en Wisconsin y descubre cinco tipos de árboles diferentes. La tabla de datos de la derecha resume sus observaciones.

Árboles de madera dura de Wisconsin	
Tipo de árbol	**Cantidad encontrada**
Roble	600
Arce	750
Haya	300
Abedul	1,200
Nogal americano	150
Total	3,000

Cómo hacer un diagrama circular

1 Para construir un diagrama circular con estos datos, primero debes encontrar el porcentaje de cada tipo de árbol. Para hacer esto, divide el número de árboles individuales entre el número total y multiplica por 100.

$$\frac{600 \text{ robles}}{3{,}000 \text{ árboles}} \times 100 = 20\%$$

$$\frac{750 \text{ arces}}{3{,}000 \text{ árboles}} \times 100 = 25\%$$

$$\frac{300 \text{ hayas}}{3{,}000 \text{ árboles}} \times 100 = 10\%$$

$$\frac{1{,}200 \text{ abedules}}{3{,}000 \text{ árboles}} \times 100 = 40\%$$

$$\frac{600 \text{ nogales americanos}}{3{,}000 \text{ árboles}} \times 100 = 5\%$$

2 Ahora, determina el tamaño de los círculos que constituyen el diagrama. Esto se puede lograr multiplicando cada porcentaje por 360°. Recuerda que un círculo tiene 360°.

$20\% \times 360° = 72°$ $25\% \times 360° = 90°$
$10\% \times 360° = 36°$ $40\% \times 360° = 144°$
$5\% \times 360° = 18°$

3 Luego, revisa que la suma de los porcentajes sea 100 y que la suma de los grados sea 360.

$20\% + 25\% + 10\% + 40\% + 5\% = 100\%$
$72° + 90° + 36° + 144° + 18° = 360°$

4 Con un compás, dibuja un círculo y marca el centro.

5 Luego, utiliza un transportador para dibujar los ángulos de 72°, 92°, 36°, 144° y 18° en el círculo.

6 Finalmente, rotula cada parte del diagrama y elige un título apropiado.

Una comunidad de árboles de madera dura de Wisconsin

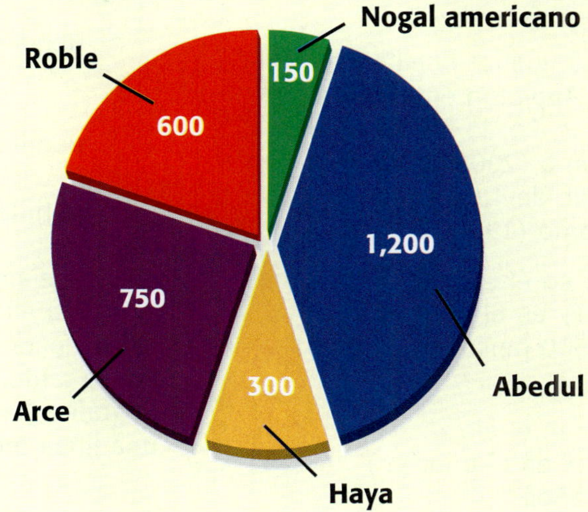

Gráfica lineal

Las gráficas lineales se usan generalmente para representar un cambio continuo. Por ejemplo, la clase de ciencias del Sr. Gómez analizó los registros de población para su ciudad, Appleton, entre 1900 y 2000. Examina los datos de la izquierda.

El año y la población son las *variables* ya que ambas cambian. La población está determinada por el año o depende de él. Así, la población es la **variable dependiente** y el año la **variable independiente.** Cada conjunto de datos se denomina **par de datos.** Para preparar una gráfica lineal, primero se deben organizar los pares de datos en una tabla similar a la de la izquierda.

La población de Appleton, 1900 a 2000	
Año	**Población**
1900	1,800
1920	2,500
1940	3,200
1960	3,900
1980	4,600
2000	5,300

Cómo hacer una gráfica lineal

1 Coloca la variable independiente a lo largo del eje (*x*) horizontal. Coloca la variable dependiente a lo largo del eje (*y*) vertical.

2 Rotula el eje *x* "Año" y el eje *y* "Población" Observa los valores más altos y más bajos de la población. Determina una escala para el eje *y* que provea el espacio suficiente para colocar estos valores. Debes usar la misma escala a lo largo de todo el eje. También debes encontrar una escala apropiada para el eje *x*.

3 Elige puntos de inicio razonables para cada eje.

4 Traza los pares de datos de la forma más precisa posible.

5 Elige un título que represente los datos en forma precisa.

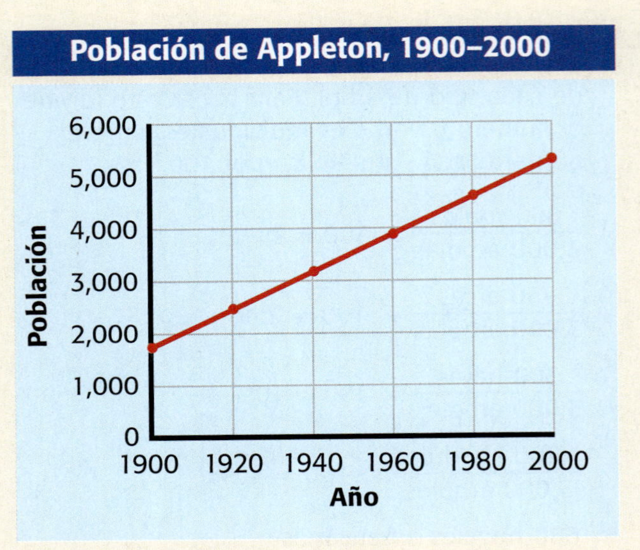

Cómo se determina la pendiente

La pendiente es la relación entre el cambio en el eje *y* el cambio en el eje *x*, o la "elevación sobre el curso".

1 Elige dos puntos en la gráfica lineal. Por ejemplo, la población de Appleton en el año 2000 fue de 5,300 personas. Por lo tanto, puedes definir el punto *a* como (2000, 5,300). En 1900, la población fue de 1,800 personas. Define el punto *b* como (1900, 1,800).

2 Encuentra el cambio en el eje *y*.
(*y* en el punto *a*) − (*y* en el punto *b*)
5,300 personas − 1,800 personas = 3,500 personas.

3 Encuentra el cambio en el eje *x*.
(*x* en el punto *a*) − (*x* en el punto *b*)
2000 − 1900 = 100 años

4 Calcula la pendiente de la gráfica dividiendo el cambio en *y* entre el cambio en *x*.

$$\text{pendiente} = \frac{\text{cambio en } y}{\text{cambio en } x}$$

$$\text{pendiente} = \frac{3,500 \text{ personas}}{100 \text{ años}}$$

pendiente = 35 personas por año.

En este ejemplo, la población de Appleton aumenta en una cantidad fija cada año. La gráfica para estos datos es una línea recta. Por lo tanto, la relación es **lineal.** Cuando la gráfica para un conjunto de datos no es una línea recta, la relación **no es lineal.**

Cómo usar el álgebra para determinar la pendiente

La ecuación del paso 4 también se puede organizar así:

$$y = kx$$

en donde y representa el cambio en el eje y, k representa la pendiente y x representa el cambio en el eje x.

$$\text{pendiente} = \frac{\text{cambio en } y}{\text{cambio en } x}$$

$$k = \frac{y}{x}$$

$$k \times x = \frac{y \times x}{x}$$

$$kx = y$$

Gráfica de barras

Las gráficas de barras se utilizan para demostrar cambios que no son continuos. Dichas gráficas se pueden utilizar para representar tendencias cuando los datos han sido recopilados durante largos períodos de tiempo. Un meteorólogo recopiló los registros de precipitación que aparecen a la derecha para Hartford, Connecticut, entre el 1 y el 15 de abril de 1996 y utilizó una gráfica de barras para representar los datos.

Precipitación en Hartford, Connecticut, del 1 al 15 de abril de 1996

Fecha	Precipitación (cm)	Fecha	Precipitación (cm)
1 de abril	0.5	9 de abril	0.25
2 de abril	1.25	10 de abril	0.0
3 de abril	0.0	11 de abril	1.0
4 de abril	0.0	12 de abril	0.0
5 de abril	0.0	13 de abril	0.25
6 de abril	0.0	14 de abril	0.0
7 de abril	0.0	15 de abril	6.50
8 de abril	1.75		

Cómo hacer una gráfica de barras

1. Utiliza una escala apropiada y un punto de inicio razonable para cada eje.

2. Rotula los ejes y traza los datos.

3. Elige un título que represente los datos en forma precisa.

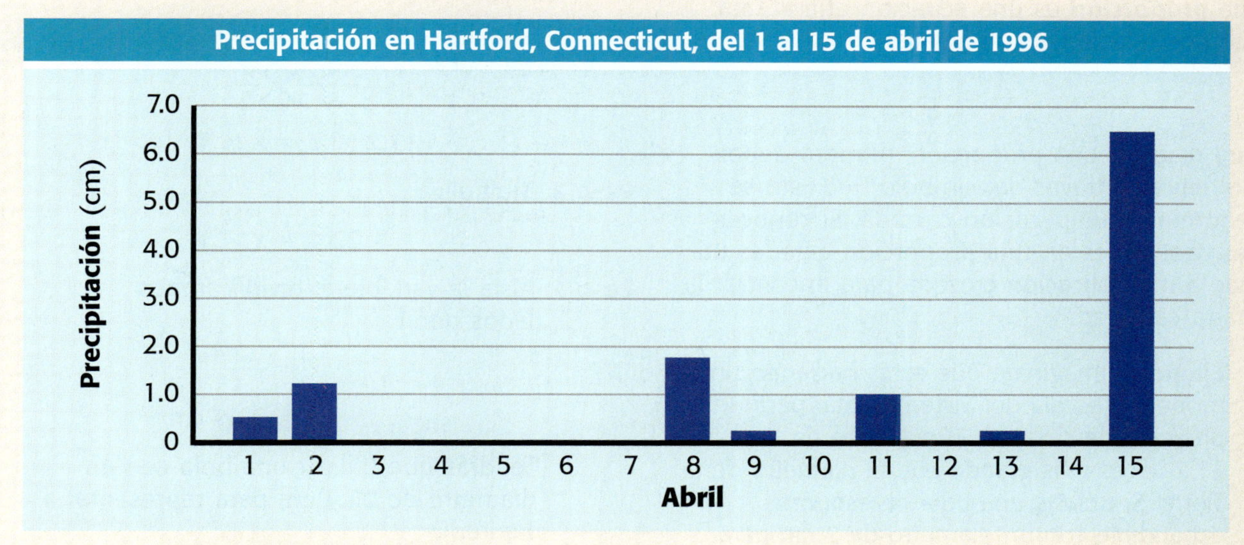

Repaso de matemáticas

Las ciencias requieren el conocimiento de muchos conceptos de matemáticas.
Las siguientes páginas te ayudarán a repasar algunas técnicas matemáticas.

Promedios

El **promedio,** o la **media,** reduce una lista de números a una sola cifra que *aproxima* su valor.

Ejemplo: Encuentra el promedio del siguiente conjunto de números: 5, 4, 7, 8.

Paso 1: Encuentra la suma.

$$5 + 4 + 7 + 8 = 24$$

Paso 2: Divide la suma entre la cantidad de números en el conjunto. Debido a que existen cuatro números en este ejemplo, divide la suma entre 4.

$$\frac{24}{4} = 6$$

El promedio, o la media, es **6.**

Relaciones

Una **relación** es una comparación entre números y generalmente está escrita como fracción.

Ejemplo: Encuentra la relación que existe entre termómetros y estudiantes, si hay 36 termómetros y 48 estudiantes en tu clase.

Paso 1: Haz una relación.

$$\frac{36 \text{ termómetros}}{48 \text{ estudiantes}}$$

Paso 2: Reduce la fracción a su forma más simple.

$$\frac{36}{48} = \frac{36 \div 12}{48 \div 12} = \frac{3}{4}$$

La relación de termómetros en función de los estudiantes es de **3 a 4,** o $\frac{3}{4}$. La relación también se puede expresar como 3:4.

Proporciones

Una **proporción** es una ecuación que afirma que dos relaciones son iguales.

$$\frac{3}{1} = \frac{12}{4}$$

Para resolver una proporción, primero debes multiplicar a través del símbolo "=". Esto se denomina multiplicación cruzada. Si conoces tres cantidades en una proporción, puedes utilizar la multiplicación cruzada para encontrar la cuarta cantidad.

Ejemplo: Imagínate que estás haciendo un modelo a escala del Sistema Solar para un proyecto de ciencias. El diámetro de Júpiter es 11.2 veces más grande que el diámetro de la Tierra. Si utilizas una bola de espuma expandible con un diámetro de 2 cm para representar la Tierra, ¿qué diámetro debe tener la bola que representa a Júpiter?

$$\frac{11.2}{1} = \frac{x}{2 \text{ cm}}$$

Paso 1: Multiplica en forma cruzada.

$$\frac{11.2}{1} \times \frac{x}{2}$$

$$11.2 \times 2 = x \times 1$$

Paso 2: Multiplica.

$$22.4 = x \times 1$$

Paso 3: Aísla la variable al dividir ambos lados por 1.

$$x = \frac{22.4}{1}$$

$$x = 22.4 \text{ cm}$$

Tendrás que utilizar una bola con un diámetro de **22.4 cm** para representar a Júpiter.

Porcentajes

Un **porcentaje** es la relación entre un número determinado y 100.

Ejemplo: ¿Cuál es el 85 por ciento de 40?

Paso 1: Escribe el porcentaje nuevamente moviendo el decimal dos espacios a la izquierda.

$$.85$$

Paso 2: Multiplica el decimal por el número con el cual estás calculando el porcentaje.

$$0.85 \times 40 = 34$$

El 85% de 40 es **34.**

Decimales

Para **sumar** o **restar decimales,** alinea los dígitos en forma vertical de modo que los decimales también estén alineados. Luego, suma o resta las columnas de derecha a izquierda, llevando o tomando prestados números si es necesario.

Ejemplo: Suma los siguientes números: 3.1415 y 2.96.

Paso 1: Alinea los dígitos en forma vertical de modo que los decimales también estén alineados.

$$\begin{aligned} 3.1415 \\ + \ 2.96 \ \ \ \ \end{aligned}$$

Paso 2: Suma las columnas de derecha a izquierda, llevando números cuando sea necesario.

$$\begin{aligned} 1 \ 1 \ \ \ \ \ \\ 3.1415 \\ + \ 2.96 \ \ \ \ \\ \hline 6.1015 \end{aligned}$$

El total es **6.1015.**

Fracciones

Los números te indican cuántos; **las fracciones** te indican *qué parte de un todo.*

Ejemplo: El salón de clases tiene 24 plantas. Tu maestro o maestra te pide que coloques 5 en un lugar con sombra. ¿A qué fracción corresponde esta cifra?

Paso 1: Escribe una fracción con el número total de partes del todo como el denominador.

$$\frac{?}{24}$$

Paso 2: Escribe el número de partes del todo representadas como el numerador.

$$\frac{5}{24}$$

$\frac{5}{24}$ de las plantas estarán en la sombra.

Reducir fracciones

Generalmente es mejor expresar una fracción en su forma más simple. Esto se denomina *reducción* de una fracción.

Ejemplo: Reduce la fracción $\frac{30}{45}$ a su forma más simple.

Paso 1: Encuentra el número entero más grande que puede dividir tanto al numerador como al denominador sin residuo. Este número se denomina el máximo factor común (MFC).

factores del numerador 30: 1, 2, 3, 5, 6, 10, 15, 30

factores del denominador 45: 1, 3, 5, 9, 15, 45

Paso 2: Divide tanto el numerador como el denominador entre el máximo factor común, que en este caso es 15.

$$\frac{30}{45} = \frac{30 \div 15}{45 \div 15} = \frac{2}{3}$$

$\frac{30}{45}$ reducida a su forma más simple es $\frac{2}{3}$.

Sumar y restar fracciones

Para **sumar** o **restar fracciones** que tienen el **mismo denominador,** simplemente suma o resta los numeradores.

> **Ejemplos:**
>
> $$\frac{3}{5} + \frac{1}{5} = ? \quad y \quad \frac{3}{4} - \frac{1}{4} = ?$$

Paso 1: Sumar o restar los numeradores.

$$\frac{3}{5} + \frac{1}{5} = \frac{4}{} \quad y \quad \frac{3}{4} - \frac{1}{4} = \frac{2}{}$$

Paso 2: Escribe la suma o diferencia sobre el denominador.

$$\frac{3}{5} + \frac{1}{5} = \frac{4}{5} \quad y \quad \frac{3}{4} - \frac{1}{4} = \frac{2}{4}$$

Paso 3: Si fuese necesario, reduce la fracción a su forma más simple.

$$\frac{4}{5} \text{ no se puede reducir } y \frac{2}{4} = \frac{1}{2}$$

Para **sumar** o **restar fracciones** que tienen **diferentes denominadores,** primero debes encontrar el denominador común.

> **Ejemplos:**
>
> $$\frac{1}{2} + \frac{1}{6} = ? \quad y \quad \frac{3}{4} - \frac{2}{3} = ?$$

Paso 1: Escribe las fracciones equivalentes con un denominador común.

$$\frac{3}{6} + \frac{1}{6} = ? \quad y \quad \frac{9}{12} - \frac{8}{12} = ?$$

Paso 2: Suma o resta.

$$\frac{3}{6} + \frac{1}{6} = \frac{4}{6} \quad y \quad \frac{9}{12} - \frac{8}{12} = \frac{1}{12}$$

Paso 3: Si fuese necesario, reduce la fracción a su forma más simple.

$$\frac{4}{6} = \frac{2}{3}, \text{ y } \frac{1}{12} \text{ no se puede reducir.}$$

Multiplicar fracciones

Para **multiplicar fracciones,** debes multiplicar los numeradores y los denominadores y luego reducir la fracción a su forma más simple.

> **Ejemplo:**
>
> $$\frac{5}{9} \times \frac{7}{10} = ?$$

Paso 1: Multiplica los numeradores y los denominadores.

$$\frac{5}{9} \times \frac{7}{10} = \frac{5 \times 7}{9 \times 10} = \frac{35}{90}$$

Paso 2: Reduce.

$$\frac{35}{90} = \frac{35 \div 5}{90 \div 5} = \frac{7}{18}$$

Dividir fracciones

Para **dividir fracciones**, primero debes escribir nuevamente el divisor (el número con el cual *se divide*) en forma invertida. Esto se denomina el recíproco del divisor. Luego puedes multiplicar y reducir si fuese necesario.

> **Ejemplo:**
>
> $$\frac{5}{8} \div \frac{3}{2} = ?$$

Paso 1: Escribe el divisor nuevamente como su recíproco.

$$\frac{3}{2} \rightarrow \frac{2}{3}$$

Paso 2: Multiplica.

$$\frac{5}{8} \times \frac{2}{3} = \frac{5 \times 2}{8 \times 3} = \frac{10}{24}$$

Paso 3: Reduce.

$$\frac{10}{24} = \frac{10 \div 2}{24 \div 2} = \frac{5}{12}$$

Notación científica

La **notación científica** es una forma abreviada de representar números muy grandes o muy pequeños sin necesidad de agregar todos los ceros.

> **Ejemplo:** Escribe 653,000,000 en notación científica.

Paso 1: Escribe el número sin los ceros.

653

Paso 2: Coloca el decimal después del primer dígito.

6.53

Paso 3: Encuentra el exponente contando el número de espacios que tuviste que correr el decimal.

6.53000000

El decimal se corrió ocho espacios hacia la izquierda. Por lo tanto, el exponente de 10 es 8 positivo. Recuerda que si el decimal se hubiese corrido a la derecha, el exponente sería negativo.

Paso 4: Escribe el número en notación científica.

$$6.53 \times 10^8$$

Área

El **área** es el número de unidades cuadradas que se requieren para cubrir la superficie de un objeto.

> **Fórmulas:**
>
> Cuadrado = lado $\times$ lado
> Rectángulo = longitud $\times$ ancho
> Triángulo = $\frac{1}{2}$ base $\times$ altura
>
> **Ejemplos:** Encuentra las áreas.

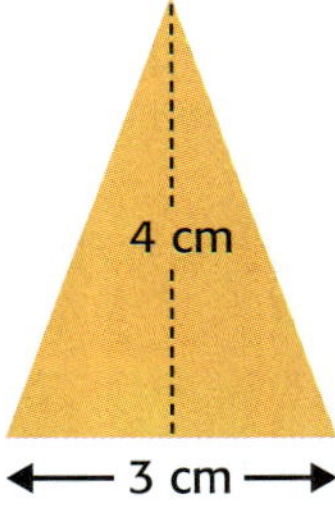

Triángulo
Área = $\frac{1}{2} \times$ base $\times$ altura
Área = $\frac{1}{2} \times 3$ cm $\times 4$ cm
Área = **6 cm^2**

Rectángulo
Área = longitud $\times$ ancho
Área = 6 cm $\times$ 3 cm
Área = **18 cm^2**

Cuadrado
Área = lado $\times$ lado
Área = 3 cm $\times$ 3 cm
Área = **9 cm^2**

Volumen

El **volumen** es la cantidad de espacio que ocupa un objeto.

> **Fórmulas:**
> Cubo =
> lado $\times$ lado $\times$ lado
>
> Prisma =
> área de la base $\times$ altura
>
> **Ejemplos:**
> Encuentra el volumen
> de los sólidos.

Cubo
Volumen = lado $\times$ lado $\times$ lado
Volumen = 4 cm $\times$ 4 cm $\times$ 4 cm
Volumen = **64 cm^3**

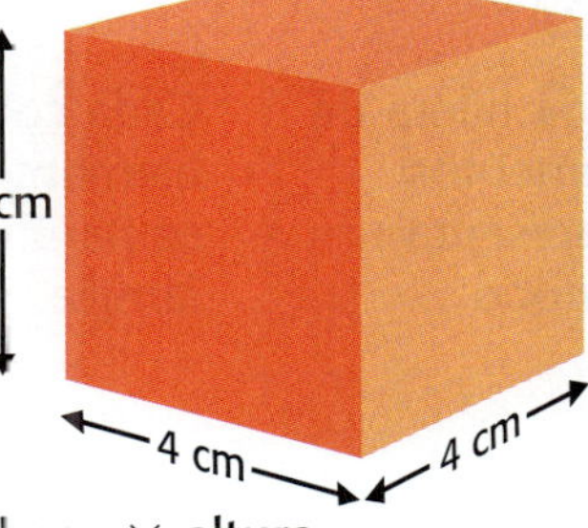

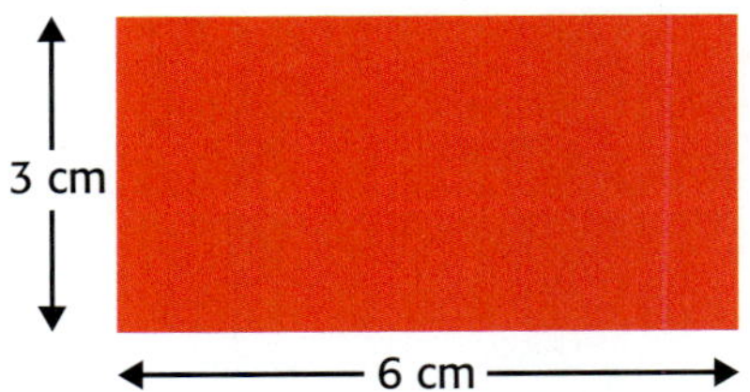

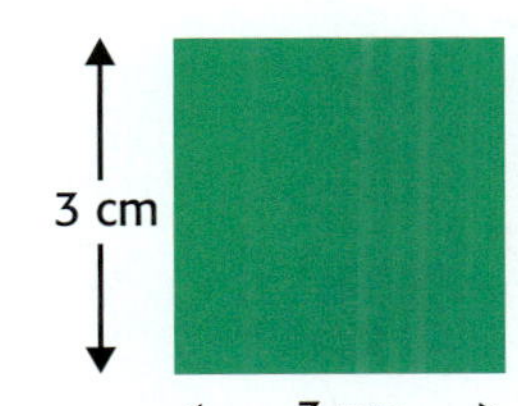

Prisma
Volumen = área de la base $\times$ altura
Volumen = (área de un triángulo) $\times$ altura
Volumen = $\left(\frac{1}{2} \times 3 \text{ cm} \times 4 \text{ cm} \right) \times 5$ cm
Volumen = 6 cm$^2 \times 5$ cm
Volumen = **30 cm^3**

Tabla periódica de los elementos

Cada cuadro de la tabla contiene el nombre del elemento y su símbolo químico, número atómico y masa atómica.

Número atómico — 6
Símbolo químico — C
Nombre del elemento — Carbono
Masa Atómica — 12.0

El color de fondo indica el tipo de elemento. El carbono es un no metal.

El color del símbolo químico indica el estado físico del elemento a temperatura ambiente. El carbono es un sólido.

Fondo
- Metales
- Metaloides
- No metales

Símbolo químico
- Sólido
- Líquido
- Gas

Período 1

1
H
Hidrógeno
1.0

	Grupo 1	Grupo 2	Grupo 3	Grupo 4	Grupo 5	Grupo 6	Grupo 7	Grupo 8	Grupo 9
Período 2	3 **Li** Litio 6.9	4 **Be** Berilio 9.0							
Período 3	11 **Na** Sodio 23.0	12 **Mg** Magnesio 24.3							
Período 4	19 **K** Potasio 39.1	20 **Ca** Calcio 40.1	21 **Sc** Escandio 45.0	22 **Ti** Titanio 47.9	23 **V** Vanadio 50.9	24 **Cr** Cromo 52.0	25 **Mn** Manganeso 54.9	26 **Fe** Hierro 55.8	27 **Co** Cobalto 58.9
Período 5	37 **Rb** Rubidio 85.5	38 **Sr** Estroncio 87.6	39 **Y** Itrio 88.9	40 **Zr** Circonio 91.2	41 **Nb** Niobio 92.9	42 **Mo** Molibdeno 95.9	43 **Tc** Tecnecio (97.9)	44 **Ru** Rutenio 101.1	45 **Rh** Rodio 102.9
Período 6	55 **Cs** Cesio 132.9	56 **Ba** Bario 137.3	57 **La** Lantano 138.9	72 **Hf** Hafnio 178.5	73 **Ta** Tántalo 180.9	74 **W** Wolframio 183.8	75 **Re** Renio 186.2	76 **Os** Osmio 190.2	77 **Ir** Iridio 192.2
Período 7	87 **Fr** Francio (223.0)	88 **Ra** Radio (226.0)	89 **Ac** Actinio (227.0)	104 **Rf** Ruterfordio (261.1)	105 **Db** Dubnio (262.1)	106 **Sg** Seaborgio (263.1)	107 **Bh** Bohrio (262.1)	108 **Hs** Hassio (265)	109 **Mt** Meitnerio (266)

Cada hilera de elementos representa un período.

Cada columna de elementos representa un grupo o familia.

Lantánidos	58 **Ce** Cerio 140.1	59 **Pr** Prosedimio 140.9	60 **Nd** Neodimio 144.2	61 **Pm** Promecio (144.9)	62 **Sm** Samario 150.4
Actínidos	90 **Th** Torio 232.0	91 **Pa** Protacnídio 231.0	92 **U** Uranio 238.0	93 **Np** Neptunio (237.0)	94 **Pu** Plutonio 244.1

Estos elementos se escriben debajo de la tabla para que ésta no se extienda.

Las líneas en zigzag nos recuerdan dónde se encuentran los metales, los no metales y los metaloides.

Grupo 18

2		
He		
Helio		
4.0		

Grupo 13	Grupo 14	Grupo 15	Grupo 16	Grupo17	
5	6	7	8	9	10
B	**C**	**N**	**O**	**F**	**Ne**
Boro	Carbono	Nitrógeno	Oxígeno	Flúor	Neón
10.8	12.0	14.0	16.0	19.0	20.2
13	14	15	16	17	18
Al	**Si**	**P**	**S**	**Cl**	**Ar**
Aluminio	Silicio	Fósforo	Azufre	Cloro	Argón
27.0	28.1	31.0	32.1	35.5	39.9

Grupo 10	Grupo 11	Grupo 12							
28	29	30	31	32	33	34	35	36
Ni	**Cu**	**Zn**	**Ga**	**Ge**	**As**	**Se**	**Br**	**Kr**
Níquel	Cobre	Zinc	Galio	Germanio	Arsénico	Selenio	Bromo	Criptón
58.7	63.5	65.4	69.7	72.6	74.9	79.0	79.9	83.8
46	47	48	49	50	51	52	53	54
Pd	**Ag**	**Cd**	**In**	**Sn**	**Sb**	**Te**	**I**	**Xe**
Paladio	Plata	Cadmio	Indio	Estaño	Antimonio	Telurio	Yodo	Xenón
106.4	107.9	112.4	114.8	118.7	121.8	127.6	126.9	131.3
78	79	80	81	82	83	84	85	86
Pt	**Au**	**Hg**	**Tl**	**Pb**	**Bi**	**Po**	**At**	**Rn**
Platino	Oro	Mercurio	Talio	Plomo	Bismuto	Polonio	Astato	Radón
195.1	197.0	200.6	204.4	207.2	209.0	(209.0)	(210.0)	(222.0)
110	111	112						
Uun	**Uuu**	**Uub**						
Ununnilium	Unununium	Ununbium						
(271)	(272)	(277)						

Los nombres y símbolos de los elementos 110 – 112 son temporales. Están basados en el número atómico del elemento. Los nombres y los símbolos oficiales serán aprobados por un comité internacional de científicos.

63	64	65	66	67	68	69	70	71
Eu	**Gd**	**Tb**	**Dy**	**Ho**	**Er**	**Tm**	**Yb**	**Lu**
Europio	Gadolíneo	Terbio	Disprosio	Holmio	Erbio	Tulio	Iterbio	Lutecio
152.0	157.3	158.9	162.5	164.9	167.3	168.9	173.0	175.0
95	96	97	98	99	100	101	102	103
Am	**Cm**	**Bk**	**Cf**	**Es**	**Fm**	**Md**	**No**	**Lr**
Americio	Curio	Berkelio	Californio	Einstenio	Fermio	Mendelevio	Nobelio	Lawrencio
(243.1)	(247.1)	(247.1)	(251.1)	(252.1)	(257.1)	(258.1)	(259.1)	(262.1)

El número en paréntesis es la masa del isótopo más estable del elemento.

Repaso de ciencias físicas

Átomos y elementos

Todos los objetos en el universo están compuestos de partículas de algún tipo de materia. La **materia** es todo lo que ocupa un espacio y tiene masa. Toda materia está compuesta de elementos. Un **elemento** es una substancia que no puede ser dividida en componentes más simples por medios químicos comunes. Esto se debe a que cada elemento sólo está compuesto de un tipo de átomo. Un **átomo** es la unidad más pequeña en la que se puede dividir un elemento sin que pierda sus propiedades.

Estructura atómica

Los atómos están compuestos de pequeñas partículas denominadas partículas subatómicas. Los tres tipos principales de partículas subatómicas son **electrones, protones** y **neutrones.** Los electrones tienen cargas eléctricas negativas mientras que los protones tienen cargas positivas y los neutrones no tienen carga eléctrica. Los protones y los neutrones se encuentran muy unidos para formar el **núcleo.** Los protones le dan al núcleo una carga positiva. Los electrones de un átomo se mueven en una región alrededor del núcleo conocida como **nube de electrones.** Los electrones de carga negativa son atraídos a los núcleos de carga positiva. Un atómo puede tener muchos niveles de energía en donde se pueden ubicar los electrones.

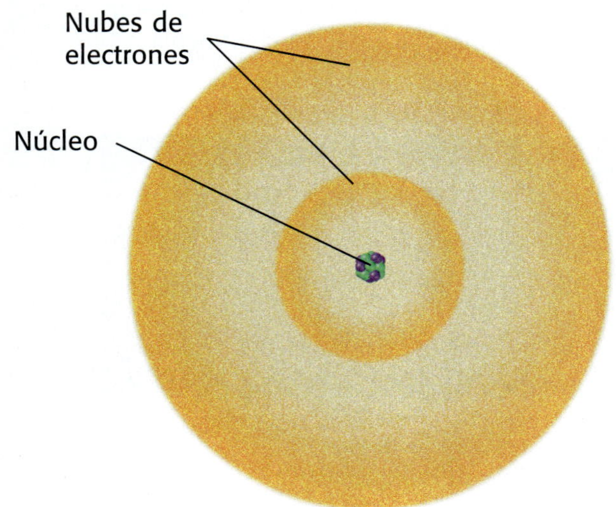

Número atómico

Para facilitar la identificación de los elementos, los científicos le han asignado un **número atómico** a cada tipo de átomo. El número atómico es equivalente al número de protones en un átomo. Los átomos que presentan el mismo número de protones son el mismo tipo de elemento. En un átomo no cargado, o neutro, existe un número equivalente de protones y electrones. Por lo tanto, el número atómico también equivale al número de electrones en un átomo no cargado. El número de neutrones, sin embargo, puede variar en un elemento determinado. Los átomos de un mismo elemento que tienen distintos números de neutrones se denominan **isótopos.**

Tabla periódica de los elementos

En la tabla periódica, los elementos están dispuestos de izquierda a derecha en orden de menor a mayor número atómico. Cada elemento en la tabla se encuentra en un cuadro separado. Cada elemento tiene uno o más electrones y uno o más protones que el elemento que se encuentra a su izquierda. Cada hilera horizontal de la tabla se denomina un **período.** Los cambios en las propiedades químicas a través de un período corresponden a los cambios en las disposiciones de los electrones de un elemento. Cada columna vertical de la tabla, llamada **grupo,** enumera los elementos con propiedades similares. Los elementos de un grupo tienen propiedades químicas similares porque tienen el mismo número de electrones en su nivel exterior de energía. Por ejemplo, los elementos helio, neón, argón, criptón, xenón y radón tienen propiedades similares y se conocen como gases nobles.

Moléculas y compuestos

Cuando los átomos de dos o más elementos se unen químicamente, la substancia resultante se denomina **compuesto.** Un compuesto es una substancia nueva con propiedades distintas a las de los elementos que lo componen. Por ejemplo, el agua (H_2O) es un compuesto que se forma cuando se unen los átomos de hidrógeno (H) y oxígeno (O). La unidad más pequeña de un compuesto que contiene todas sus propiedades se denomina **molécula.** Una fórmula química indica los elementos de un compuesto. También indica el número relativo de átomos de cada elemento presente. La fórmula química del agua es H_2O, lo cual indica que cada molécula de agua contiene dos elementos de hidrógeno y uno de oxígeno. El subíndice se usa después de un símbolo para indicar cuántos átomos de ese elemento se encuentran en una molécula del compuesto.

Ácidos, bases y pH

Un ion es un átomo o grupo de átomos que tiene una carga eléctrica debido a la pérdida o ganancia de uno o más electrones. Cuando un ácido como el ácido hidroclórico (HC1), se mezcla con agua, se divide en iones. Un **ácido** es un compuesto que produce iones de hidrógeno (H^+) en el agua. Luego, los iones de hidrógeno se mezclan con una molécula de agua para formar un ion hidronio (H_3O^+). Una **base,** sin embargo, es una substancia que produce iones hidróxido (OH^-) en el agua.

Para determinar si una solución es ácida o básica, los científicos utilizan el pH. El **pH** es la medida de la concentración de iones hidronio en una solución. La escala de pH fluctúa entre 0 y 14. El punto medio, pH = 7, es neutral, es decir ni ácido ni básico. Los ácidos tiene un pH de menos 7, las bases tienen un pH mayor de 7. Mientras menor sea el número, más ácida es la solución. Mientras más elevado sea el número, más básica será la solución.

Ecuaciones químicas

Una reacción química ocurre cuando sucede un cambio químico. (En un cambio químico, se forman nuevas substancias con nuevas propiedades.) Una ecuación química es una manera útil de describir una reacción química por medio de fórmulas químicas. La ecuación indica las substancias que reaccionan y los productos que se forman. Por ejemplo, cuando el carbono y el oxígeno se mezclan forman dióxido de carbono. A continuación se presenta la ecuación de la reacción: $C + O_2 \rightarrow CO_2$.

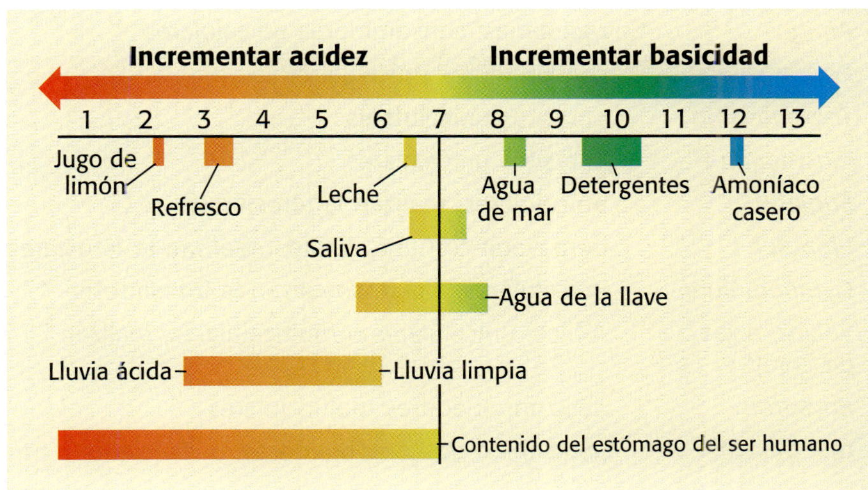

Los seis reinos

Reino de las arqueobacterias

Los organismos de este reino son procariotas unicelulares.

Arqueobacterias		
Groupo	**Ejemplos**	**Características**
Metanógenos	*Methanococcus*	se encuentran en la tierra, pantanos, el aparato digestivo de los mamíferos; producen gas metano; no pueden vivir donde hay oxígeno
Termófilos	*Sulpholobus*	se encuentran en ambientes demasiado calientes; requieren azufre, no pueden vivir donde hay oxígeno
Halófilos	*Halococcus*	se encuentran en ambientes con alto contenido de sal, como el mar Muerto; casi todos pueden vivir donde hay oxígeno

Reino de las eubacterias

En este reino se conocen más de 4,000 especies de procariotas unicelulares.

Eubacterias		
Groupo	**Ejemplos**	**Características**
Bacilos	*Escherichia coli*	forma de bastones; viven libres, simbióticos o parásitos; algunos pueden fijar nitrógeno; algunos causan enfermedades
Cocos	*Streptococcus*	en forma esférica, causan enfermedades; forman esporas para resistir ambientes desfavorables
Espirilos	*Treponema*	en forma de espiral; responsables de varias enfermedades serias, como la sífilis y la enfermedad de Lyme

Reino de los protistas

Los organismos de este reino son eucariotas. Hay organismos unicelulares y multicelulares.

Protistas		
Groupo	**Ejemplos**	**Características**
Sarcodinas	*Amiba*	radiolarias; consumidores unicelulares
Ciliados	*Paramecium*	consumidores unicelulares
Flagelados	*Trypanosoma*	parásitos unicelulares
Esporozoarios	*Plasmodium*	parásitos unicelulares
Euglenas	*Euglena*	unicelulares; realizan la fotosíntesis
Diatomeas	*Pinnularia*	la mayoría son unicelulares; realizan la fotosíntesis
Dinoflagelados	*Gymnodinium*	unicelulares; algunos realizan la fotosíntesis
Algas	*Volvox*, algas de coral	4 filos; unicelulares o multicelulares; realizan la fotosíntesis
Moho viscoso	*Physarum*	descomponedores multicelulares
Moho de agua	mal blanco	parásitos o descomponedores, unicelulares o multicelulares

Reino de los hongos

En este reino hay eucariotas unicelulares y multicelulares.
Como las plantas, los hongos se clasifican en divisiones y
no en filos. Hay cuatro tipos principales de hongos.

Hongos		
Groupo	**Ejemplos**	**Características**
Hongos filiformes	moho del pan	esféricos; descomponedores
Hongos de sacos	levadura, hierba mora	en forma de saco; parásitos y descomponedores
Setas	champiñones, tizones, royas	en forma de sombrilla; parásitos y descomponedores
Líquenes	musgo de los renos	simbióticos con algas fotosintéticas

Reino vegetal

Los organismos de este reino son eucariotas multicelulares.
Tienen sistemas especializados para procesos de vida diferentes. Se clasifican en divisiones y no en filos.

Plantas		
Groupo	**Ejemplos**	**Características**
Briofitas	musgos, hepáticas	se reproducen por esporas
Musgos	*Lycopodium,* pinillo	se reproducen por esporas
Colas de caballo	juncos	se reproducen por esporas
Helechos	lenguas de ciervo, helecho sensorial	se reproducen por esporas
Coníferas	pinos, abetos, cedros	se reproducen por semillas; conos
Cicadáceas	*Zamiase*	reproducen por semillas
Gnetáceas	*Welwitschiase*	reproducen por semillas
Ginkgos	*Ginkgos*	reproducen por semillas
Angiospermas	todas las plantas con flores	se reproducen por semillas; flores

Reino animal

Este reino contiene eucariotas multicelulares. Tienen tejidos
especializados y sistemas complejos.

Animales		
Groupo	**Ejemplos**	**Características**
Esponjas	esponjas vítreas	sin simetría ni segmentación; acuáticas
Cnidarios	medusa, coral	simetría radial; acuáticos
Tenias	planaria, solitaria, trematodo	simetría bilateral; sistemas
Gusanos cilíndricos	*Trichina,* anquilostomas	simetría bilateral; sistemas
Anélidos	lombrices, sanguijuelas	simetría bilateral; sistemas
Moluscos	caracoles, pulpos	simetría bilateral; sistemas
Equinodermos	estrella de mar, erizo de mar aplanado	simetría radial; sistemas
Artrópodos	insectos, arañas, langostas	simetría bilateral; sistemas
Cordados	peces, anfibios, reptiles, aves, mamíferos	simetría bilateral; sistemas complejos

Cómo usar el microscopio

Partes del microscopio compuesto

- El **ocular** amplía la imagen 10×.

- El **objetivo de poco aumento** amplía la imagen 10×.

- El **objetivo de gran aumento** amplía la imagen ya sea 40× ó 43×.

- El **dispositivo de revólver** sostiene los objetivos y se puede girar para cambiar de un aumento a otro.

- El **tubo** mantiene la distancia correcta entre el ocular y los objetivos.

- El **tornillo macrométrico** mueve el tubo hacia arriba y hacia abajo para permitir enfocar la imagen.

- El **tornillo micrométrico** mueve el tubo ligeramente para agudizar el enfoque de la imagen.

- La **platina** sostiene la laminilla.

- Las **pinzas de la platina** sujetan la laminilla que se va a observar.

- El **condensador** controla la cantidad de luz que entra a través del portaobjetos.

- La **fuente de luz** proporciona luz para observar la laminilla.

- La **columna** sostiene el tubo.

- La **base** sostiene el microscopio.

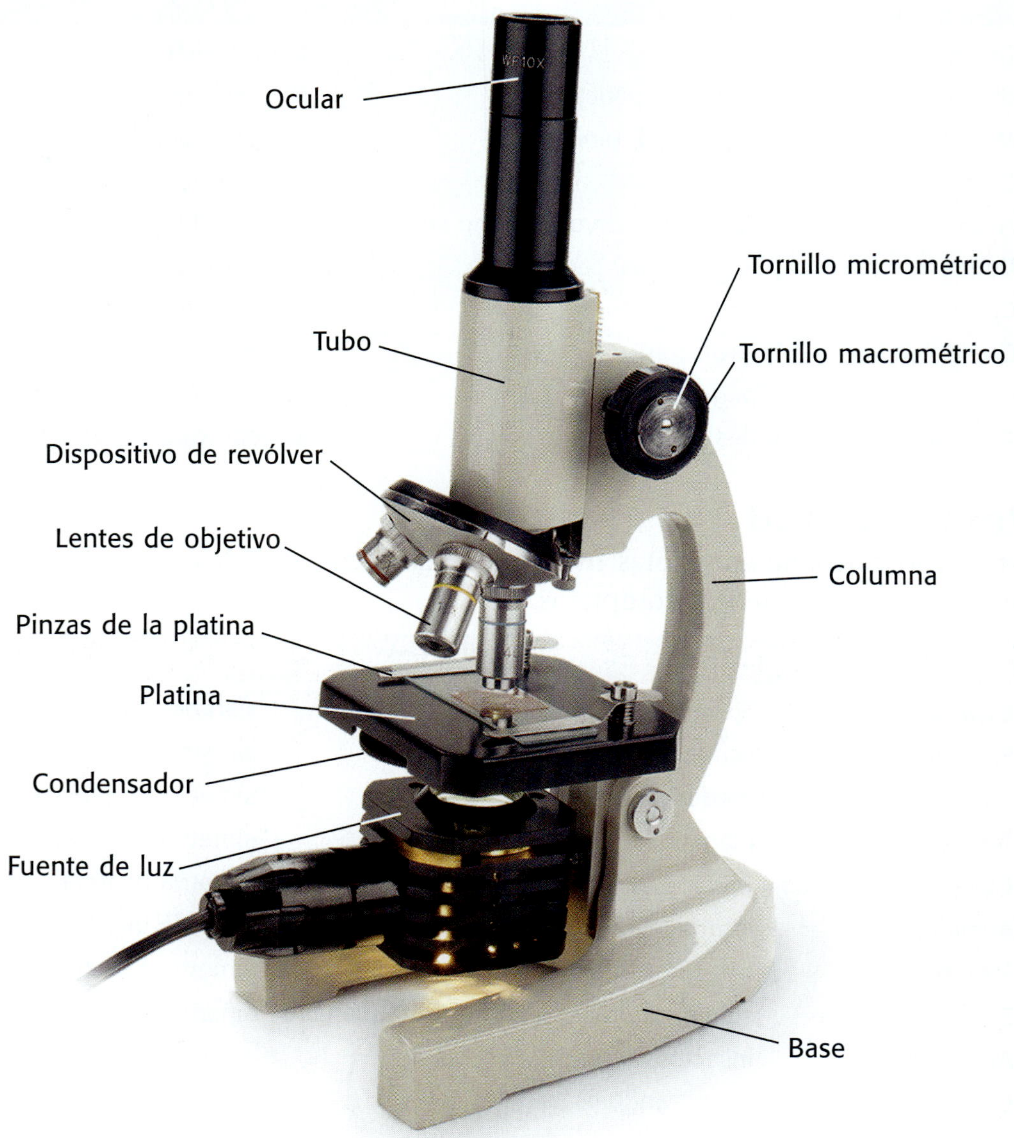

Uso adecuado del microscopio

1 Lleva el microscopio a la mesa del laboratorio; sujétalo con las dos manos. Con una mano sostén la base y con la otra agarra la columna del microscopio. Trata de que el microscopio esté cerca de tu cuerpo mientras lo llevas a la mesa.

2 Coloca el microscopio sobre la mesa, asegúrate de que quede a unos 5 cm de la orilla.

3 Revisa qué tipo de fuente de luz usa el microscopio. Si usa una lámpara, conéctala y asegúrate de que el cable no te estorbe. Si el microscopio tiene un espejo, ajústalo para reflejar la luz a través del orificio de la platina.
Cuidado: Si el microscopio tiene un espejo, no uses luz solar directa como fuente de luz. Esta luz te puede lastimar los ojos.

4 Siempre empieza a trabajar con el objetivo de bajo aumento alineado con el tubo. Ajusta el dispositivo de revólver.

5 Coloca una laminilla sobre el orificio de la platina y asegúrala con las pinzas.

6 Mira a través del ocular. Mueve el condensador para ajustar la cantidad de luz que entra a través de la platina.

7 Observa la platina al nivel de tus ojos. Gira lentamente el tornillo macrométrico para bajar el objetivo hasta que casi toque la laminilla. No permitas que el objetivo la toque.

8 Mira a través del ocular. Gira el tornillo macrométrico para levantar el objetivo de bajo aumento hasta que se enfoque la imagen. Siempre enfoca levantando el objetivo a partir de la laminilla. *Nunca enfoques hacia abajo.* Con el tornillo micrométrico, perfecciona el enfoque. Mantén los ojos abiertos mientras observas la laminilla.

9 Asegúrate de que la imagen esté exactamente en el centro de tu campo de visión. Luego cambia al objetivo de gran aumento. Enfoca la imagen sólo con el tornillo micrométrico. *Nunca uses el tornillo macrométrico con el objetivo de gran aumento.*

10 Cuando acabes de usar el microscopio, quita la laminilla. Limpia el ocular y los objetivos con papel especial para lentes. Guarda el microscopio en su lugar. Recuerda que debes cargar el microscopio con las dos manos.

Examinar una preparación húmeda

1 Limpia una laminilla y un cubreobjetos con papel especial para lentes.

2 Coloca la muestra que deseas observar en el centro de la laminilla.

3 Con un gotero, pon una gota de agua sobre la muestra.

4 Mantén el cubreobjetos al filo del agua y a un ángulo de 45° de la laminilla. Colócalo de manera que roce con la gota de agua. Asegúrate de que el agua fluya por la orilla del cubreobjetos.

5 Baja el cubreobjetos lentamente para evitar que se hagan burbujas de aire.

6 El agua se puede evaporar de la laminilla a medida que trabajas. Agrega más agua para que la muestra se mantenga fresca. Coloca la punta del gotero en la orilla del cubreobjetos. Agrega una gota de agua. (También puedes usar este método para agregar colorantes o soluciones a una preparación húmeda.) Elimina el exceso de agua de la laminilla con la punta de una toalla de papel. No levantes el cubreobjetos para agregar o eliminar agua.

Glosario

A

abdomen parte del cuerpo de un animal que contiene los intestinos y otros órganos del aparato digestivo (339)

abiótico se refiere a los factores del medio ambiente que no forman parte de los seres vivos (416)

absorción transporte de energía hacia las partículas que forman la materia, realizado por ondas luminosas (60)

ácido nucleico compuesto bioquímico, formado por subunidades llamadas nucleótidos, que almacena la información que se necesita para fabricar proteínas y otros ácidos nucleicos (45)

adaptación característica que le ayuda a un organismo a sobrevivir en su medio ambiente (176)

adenina una de las cuatro bases que se combinan con azúcares y fosfatos para formar una subunidad de nucleótidos del ADN; la adenina forma un par con la timina (152)

ADN abreviatura del ácido desoxirribonucleico, que es el material hereditario que controla todas las actividades de una célula, contiene la información para la creación de nuevas células y da las instrucciones para la síntesis de proteínas (38, 93, 152)

afluente arroyo o río pequeño que desemboca en uno más grande (428)

alelos formas alternativas de un gene que regulan las mismas características (135)

aletas estructuras en forma de abanico que les ayudan a los peces a moverse, dar la vuelta, detenerse y mantener el equilibrio (360)

alvéolos pequeños sacos que forman las ramificaciones de los bronquiolos de los pulmones (499)

amígdalas masas pequeñas de tejido blando situadas en la parte de atrás de la cavidad nasal, dentro de la garganta, detrás de la lengua (497)

anaeróbico que no necesita oxígeno (209)

anfibio tipo de vertebrado heterotérmico que por lo general comienza su vida en el agua con branquias y más tarde desarrolla pulmones (364)

angiosperma planta cuyas flores producen semillas (253)

antenas apéndices de la cabeza de los artrópodos que responden al tacto o al gusto (340)

aparato de Golgi organelo celular que modifica, prepara y transporta materiales al exterior de la célula (96)

árbol genealógico diagrama de la historia de una familia que se utiliza para seguir la trayectoria de un rasgo a través de varias generaciones (164)

área medida de la superficie de un objeto (24)

arqueobacterias bacterias que prosperan en un medio ambiente con condiciones extremas (235)

arreglo de las plumas actividad en la que un ave utiliza el pico para esparcir aceite en sus plumas (383)

arterias vasos sanguíneos que transportan la sangre del corazón al resto el cuerpo (491)

articulación parte donde se unen dos o más huesos (470)

asimétrico sin simetría (326)

ATP abreviatura del adenosin trifosfato que es la molécula que le suministra energía a la célula para realizar sus actividades (45)

aurícula una de las cámaras superiores del corazón (490)

australopitco homínido primitivo que evolucionó hace más de 3.6 millones de años (216)

autopolinizante planta que contiene las estructuras reproductoras femeninas y masculinas (131)

axón fibra celular larga del sistema nervioso que transfiere mensajes de una célula a otra (511)

B

bacterias organismos unicelulares muy pequeños que carecen de núcleo; se llaman también células procariotas (90, 235)

bastoncitos fotorreceptores que detectan la luz tenue (518)

bazo órgano que filtra la sangre y produce los linfocitos (497)

biodegradable que se puede descomponer por la acción del medio ambiente (446)

biodiversidad número y variedad de los seres vivos (445)

bioma región de grandes dimensiones caracterizada por un tipo de clima específico y ciertos tipos de comunidades animales y vegetales (416)

branquias órganos que extraen el oxígeno del agua y sacan el dióxido de carbono de la sangre (360)

bronquios los dos tubos que conectan los pulmones a la tráquea (499)

bulbo raquídeo parte del cerebro que se une a la médula espinal (514)

C

cabeza parte del cuerpo de los animales en donde se encuentra el cerebro (339)

camuflaje coloración o textura que le permite a un animal pasar desapercibido en su medio ambiente (309)

capilares los vasos sanguíneos más pequeños (491)

capilares linfáticos los vasos más pequeños del sistema linfático (496)

carbohidrato substancia bioquímica formada por dos o más azúcares enlazados que se usa para suministrar y almacenar energía (43)

carnívoro que come animales (401)

cartílago tejido flexible que da apoyo y protección sin tener la rigidez de los huesos (470)

cefalotórax parte del cuerpo de los arácnidos formado por la cabeza y el tórax y que tiene, por lo general, cuatro pares de patas (341)

celoma cavidad del cuerpo de algunos animales en donde están localizados los intestinos y los órganos (327)

célula estructura cubierta por una membrana que contiene todos los materiales necesarios para la vida (36)

célula eucariota célula que contiene un núcleo central y una estructura interna compleja (91, 210)

célula procariota célula que no posee núcleo ni ningún otro organelo cubierto por membranas; también conocida como bacteria (90, 209)

centrómero región que mantiene unidas las cromátidas cuando se duplica un cromosoma (117)

cerebelo parte del cerebro que controla la posición del cuerpo (514)

cerebro parte del encéfalo que detecta el tacto, la visión, el sonido, el olor, el gusto, el dolor, el calor y el frío (513)

ciclo celular ciclo de vida de una célula; en las eucariotas, consiste en la duplicación de los cromosomas, la mitosis y citoquinesis (116)

ciclo de las rocas proceso continuo mediante el cual una clase de roca se transforma en otra (202)

ciénaga ecosistema pantanoso en el que crecen árboles y plantas trepadoras (430)

ciencias naturales estudio de los seres vivos

cigoto óvulo fertilizado (535)

circulación pulmonar circulación de la sangre entre el corazón y los pulmones (492)

circulación sistémica circulación de sangre entre el corazón y el cuerpo que no incluye los pulmones (492)

citoplasma líquido que rodea los organelos de una célula (87)

citoquinesis proceso de división del citoplasma después de la mitosis (119)

citosina una de las cuatro bases que se combina con azúcares y fosfatos para formar una subunidad de nucleótidos del ADN; la citosina forma un par con la guanina (152)

clase nivel de clasificación que le sigue al filo; los organismos de un filo están organizados en clases (229)

clasificación organización de los seres vivos en grupos según sus similitudes y presuntas relaciones de evolución (228)

clave dicotómica sistema para identificar organismos desconocidos que consiste en un par de oraciones descriptivas, de las cuales sólo una se aplica al organismo desconocido y conduce a otra serie de oraciones, hasta que se identifica el organismo desconocido (232)

clorofila pigmento verde de los cloroplastos que absorbe la energía de la luz para realizar la fotosíntesis (284)

cloroplasto organelo de las células de plantas y algas en el cual tiene lugar la fotosíntesis (95)

cóclea órgano del oído que convierte las ondas sonoras en impulsos eléctricos y los envía a la parte del cerebro que procesa el sonido (520)

comportamiento aprendido comportamiento que se ha aprendido en base a la experiencia u observación (310)

comportamiento innato comportamiento que no responde a la influencia de los genes ni depende del aprendizaje (310)

comportamiento social interacción entre animales de la misma especie (314)

comunicación transferencia de una señal de un animal a otro, que provoca un tipo de respuesta (314)

comunidad conjunto formado por todas las poblaciones de especies diferentes que viven e interactúan en un área determinada (84)

conducto deferente tubo del sistema reproductor masculino en donde los espermatozoides se mezclan con algunos líquidos para formar el semen (539)

coníferas árboles que producen las semillas en frutos de forma cónica (418)

conos fotorreceptores que pueden detectar la luz brillante y contribuyen a que podamos ver los colores (518)

conservación el uso y la preservación inteligentes de los recursos naturales (447)

consumidor organismo que obtiene su energía comiéndose a los productores o a otra clase de organismos (40, 307)

contaminación presencia de substancias perjudiciales en el medio ambiente (440)

contaminante substancia perjudicial presente en el medio ambiente (440)

controles de retroalimentación sistemas que hacen que las glándulas endocrinas comiencen a funcionar o dejen de hacerlo (524)

cordón umbilical cordón que une el embrión con la placenta (543)

cotiledón hoja de una plántula dentro de una semilla (263)

Cro-Magnon tipo de seres humanos con rasgos modernos que probablemente migraron de África hace unos 100,000 años y eventualmente llegaron a todos los continentes (218)

cromátidas copias idénticas de los cromosomas (117)

cromosoma estructura en forma de espiral formada por ADN y proteínas que se produce en el núcleo de la célula durante la división celular (116)

cromosomas homólogos cromosomas cuya información se corresponde entre sí (117)

cromosomas sexuales cromosomas que contienen los genes que determinan el sexo de las crías (143)

cruza real planta que siempre produce otras que poseen los mismos rasgos (132)

cruza selectiva cruza de organismos que tienen rasgo determinado (186)

cuadrícula de Punnett tabla que se utiliza para representar todas las posibles combinaciones de los alelos provenientes de los padres (135)

cutícula capa cerosa que cubre la superficie de los tallos, las hojas y otras partes de la planta que se encuentran expuestas al aire (251)

D

deforestación tala y limpieza de los bosques (445)

dendrita prolongación corta y ramificada de una neurona que le sirve para recibir las señales de otras células (511)

dentículos estructuras pequeñas y afiladas, parecidas a los dientes, que se encuentran en la piel de los peces cartilaginosos (361)

depredador animal que se come a otros animales (308)

dermis capa de piel que se encuentra bajo la epidermis (477)

descomponedor organismo que para obtener energía descompone los restos de organismos muertos y consume y absorbe los nutrientes (40)

desechos radioactivos desechos peligrosos que tardan cientos o miles de años en volverse inofensivos (441)

desierto bioma caluroso y seco habitado por organismos que se han adaptado a sobrevivir en temperaturas muy altas durante el día y largos períodos sin lluvia (421)

diafragma músculo que se encuentra por debajo de los pulmones de los mamíferos y cuya contracción ayuda a inhalar el aire dentro de los pulmones (395, 500)

difusión movimiento de partículas de un área de alta concentración a otra de baja concentración (108)

dispersión liberación de luz producida por partículas de materia que han absorbido un exceso de energía (61)

diversidad medida del número de especies contenida en un área (419)

E

ecosistema comunidad de organismos y su medio ambiente (84)

edad absoluta cálculo de la edad de un objeto o suceso que se hace midiendo la cantidad de átomos inestables en las rocas que rodean la muestra (203)

edad relativa cálculo de la anterioridad o posterioridad de un suceso u objeto, por ejemplo un fósil, con relación a otros sucesos u objetos (203)

embrión organismo en su más temprana etapa de desarrollo (306, 542)

empollar sentarse un ave sobre sus huevos hasta que nacen los pichones (387)

encéfalo el órgano más importante del sistema nervioso (513)

endocitosis proceso en el que la membrana celular rodea una partícula y la encierra en una vesícula para llevarla al interior de la célula (111)

endoesqueleto esqueleto interno (345)

enfermedades de transmisión sexual enfermedades que se transmiten durante el contacto sexual (541)

enzima proteína que permite que ciertas reacciones químicas ocurran rápidamente (42)

epidermis capa exterior de la piel (477); capa de células que cubre el exterior de las raíces, los tallos, las hojas y algunas partes de las flores (265)

epidídimo lugar de los testículos en que se almacena el esperma antes de pasar al conducto deferente (538)

época precámbrica período en la escala de tiempo geológico que comienza con el origen de la Tierra, hace 4,600 millones de años, y termina en el momento en que aparecieron los organismos complejos, hace unos 540 millones de años (208)

era cenozoica período en la escala del tiempo geológico que comenzó hace unos 65 millones de años y se extiende hasta el presente (213)

era mesozoica período de la escala de tiempo geológico que comenzó hace unos 248 millones de años y que duró aproximadamente 183 millones de años (212)

era paleozoica período en la escala de tiempo geológico que comenzó hace unos 570 millones de años y terminó hace aproximadamente 248 millones de años (211)

escala de tiempo geológico división de la historia de la Tierra en intervalos diferenciados de tiempo (204)

escamas estructuras óseas que cubren la piel de los peces (362)

escroto saco cubierto de piel que cuelga del cuerpo de los machos y que contiene los testículos (538)

especiación proceso por el que dos poblaciones de la misma especie se vuelven tan diferentes que ya no pueden cruzarse (192)

especie el más específico de los siete niveles de clasificación, caracterizado por organismos que pueden aparearse para producir crías fértiles (176, 229)

espectro electromagnético rango completo de las ondas electromagnéticas (56)

espermatozoide gameto producido por el macho (535)

esporofito etapa del ciclo de vida de una planta durante el cual se producen las esporas (251)

estambre órgano sexual masculino de la flor, que consta de un filamento que termina en una antera productora de polen (271)

estigma parte de la flor situada en la punta del pistilo (271)

estímulo cualquier cosa que afecta la actividad de un organismo, órgano o tejido (37)

estivación período de poca actividad que algunos animales experimentan en el verano (311)

estomas aberturas de la epidermis de la hoja que permiten la entrada del dióxido de carbono y la salida del agua y del oxígeno (269)

estuario lugar donde el agua dulce de los arroyos y ríos desemboca en el océano (427)

eubacterias clasificación que contiene la mayoría de las bacterias de vida independiente que se encuentran en diferentes hábitats (235)

evolución proceso por el cual las poblaciones acumulan los cambios heredados a través del tiempo (177)

exocitosis proceso por el que la célula extrae partículas grandes de su interior; durante la exocitosis, una vesícula que contiene las partículas se funde con la membrana celular (111)

exoesqueleto esqueleto externo de los artrópodos, compuesto por proteína y quitina (340)

experimento controlado experimento que ensaya un solo factor a la vez (14)

extensor músculo que hace que una parte del cuerpo se enderece (473)

extinción masiva período en que un gran número de especies desaparece al mismo tiempo (205)

extinto especie de organismo que ha desaparecido por completo (205)

F

factor cualquier elemento que puede afectar el resultado de un experimento (14)

familia nivel de clasificación que viene después del orden; los organismos de un orden están organizados en familias (229)

faringe porción superior de la garganta (499)

fenotipo caracteres visibles que un organismo ha heredado de sus padres (135)

fermentación descomposición de azúcares para producir ATP en ausencia de oxígeno (113)

feromona substancia química producida por los animales para comunicarse entre ellos mismos (315)

fertilización externa fertilización de un óvulo que ocurre fuera del cuerpo de la hembra (360, 536)

fertilización interna fertilización del óvulo que ocurre dentro del cuerpo de la hembra (360, 536)

feto embrión en las últimas etapas de desarrollo dentro del útero (544)

filo nivel de clasificación que viene después de reino; los organismos de un reino están organizados en filos (229)

fisión binaria método de división simple de las células utilizado por las bacterias, en el que una célula se divide en dos (116)

fitoplancton organismo microscópico capaz de realizar fotosíntesis que flota cerca de la superficie del océano (423)

flexor músculo que permite que una parte del cuerpo se doble (473)

floema tejido vegetal especializado que transporta moléculas de azúcar de una parte a otra de la planta (264)

flotación fuerza ascendente sobre un objeto causada por las diferencias de presión arriba y abajo del mismo; la flotación se opone a la fuerza de la gravedad (386)

folículo piloso órgano pequeño de la capa dérmica de la piel que produce pelo (478)

fosfolípido tipo de lípido que constituye gran parte de la membrana celular (44)

fósil resto o impresión solidificada de un organismo que vivió hace mucho tiempo (178, 202)

fotorreceptores neuronas especializadas de la retina que detectan la luz (518)

fotosíntesis proceso por el cual las plantas capturan la energía de la luz del Sol y la convierten en azúcar (112, 210)

fototropismo cambio en el crecimiento de una planta como respuesta a la luz (287)

fragmentación tipo de reproducción en la que un organismo se divide en dos o más partes, cada una de las cuales puede llegar a ser un individuo independiente (534)

G

gameto óvulo o espermatozoide; un gameto tiene la mitad del número de cromosomas que se encuentra en las demás células del cuerpo (138)

gametofito período del ciclo de vida de una planta durante el cual se producen los gametos de ambos sexos (251)

ganglios grupos de células nerviosas (327)

gemación tipo de reproducción asexual en la que una pequeña parte del cuerpo del progenitor se desarrolla como un organismo independiente (534)

género nivel de clasificación que viene después de familia; los organismos de una familia están organizados en géneros

genes segmentos de ADN localizados en los cromosomas que contienen las instrucciones de la herencia y pasan de padres a hijos (135)

genotipo combinación de alelos que se hereda de los padres (135)

gestación tiempo durante el cual un embrión se desarrolla dentro de la madre (398)

gimnosperma planta que produce semillas pero no produce flores (253)

glándula grupo de células que fabrica substancias químicas especiales para el cuerpo (522)

glándulas mamarias glándulas que secretan un líquido alimenticio llamado leche (393)

glándulas sudoríparas órganos pequeños de la dermis que producen el sudor (476)

glóbulos blancos células sanguíneas que protegen el cuerpo contra los agentes patógenos (489)

glóbulos rojos células que llevan el oxígeno de los pulmones a todas las células del cuerpo y que devuelven el dióxido de carbono a los pulmones para que éstos lo eliminen (488)

gravitropismo cambio en el crecimiento de una planta como respuesta a la fuerza de gravedad (288)

guanina una de las cuatro bases que se combina con azúcares y fosfatos para formar una subunidad de nucleótidos del ADN; la guanina forma un par con la citosina (152)

H

herencia la transferencia de rasgos de padres a hijos (38, 130)

heterotermo animal cuya temperatura corporal varía con la temperatura del medio ambiente (358)

hibernación período de inactividad que experimentan algunos animales en invierno, durante el cual sobreviven consumiendo la grasa que han acumulado en el cuerpo (311)

hipótesis explicación o respuesta posible a una pregunta (12)

homeostasis mantenimiento de un medio ambiente interno estable (37, 464)

homínido familia a la que pertenecen los seres humanos y algunas especies similares que se han extinguido, algunas de los cuales fueron antepasados de los seres humanos (215)

homotermo animal que mantiene una temperatura corporal constante a pesar de los cambios de temperatura de su medio ambiente (358)

hongos grupo de organismos complejos que obtienen alimento al descomponer y absorber los nutrientes de otras substancias de su medio ambiente (238)

hormona mensajero químico que lleva la información de una parte a otra del organismo; en los mamíferos, las hormonas se producen en las glándulas endocrinas (293, 522)

hueso compacto tipo de tejido óseo que no tiene espacios vacíos (469)

hueso esponjoso clase de tejido óseo que tiene muchos espacios libres y que contiene médula (469)

huésped organismo en el que vive un parásito (332)

huevo amniótico huevo que contiene líquido amniótico para proteger al embrión en desarrollo; por lo general está rodeado de una cáscara dura (370)

I

implantación proceso por el cual un embrión se fija en el interior del útero (542)

impresiones de ADN análisis de fragmentos de ADN con fines de identificación (166)

impulso mensaje eléctrico que pasa por una neurona (511)

infértil condición de no poder tener hijos (541)

ingeniería genética manipulación de los genes que les permite a los científicos poner genes de un organismo en otro (165)

intestino órgano donde se digiere la comida en el cuerpo de los animales (327)

invertebrado animal que carece de columna vertebral (305, 326)

involuntario movimiento muscular que no está bajo control consciente (472)

iris parte pigmentada del ojo (519)

L

laringe parte de la garganta que contiene las cuerdas vocales (499)

latente estado inactivo de una semilla (282)

lente cóncavo lente que es más delgado en el centro que en los bordes (67)

lente convexo lente que es más grueso en el centro que en los bordes (67)

lente objeto curvo y transparente que forma una imagen al refractar la luz (67, 519)

ligamento banda fuerte de tejido que conecta a los huesos entre sí (470)

linfa líquido y partículas absorbidas por los capilares linfáticos (496)

linfocito leucocito que destruye algunos organismos patógenos (497)

lípidos compuestos bioquímicos, entre los que se encuentran las grasas y los aceites, que no se disuelven en agua; los lípidos almacenan la energía y forman parte de la membrana celular (44)

lisosoma vesícula especial de la célula que digiere las partículas de alimento, los desechos y los invasores que provienen del exterior (98)

litoral área de un lago o laguna que está más cerca de tierra firme (429)

longitud de onda distancia entre un punto determinado de una onda y el punto correspondiente que se encuentra en una onda adyacente (56)

M

mamífero placentario mamífero que posee una placenta para alimentar al feto dentro del útero y cuyas crías están bien desarrolladas al nacer (398, 537)

mandíbula maxilar inferior de algunos artrópodos (341)

marino ecosistema de agua salada (423)

marisma ecosistema pantanoso y sin árboles, donde crecen plantas como las aneas y los juncos (430)

marsupial mamífero que da a luz crías vivas, parcialmente desarrolladas, que continúan desarrollándose dentro de una bolsa o pliegue de la piel de la madre (397, 537)

masa cantidad de materia que forma parte de un objeto; su valor es el mismo, independientemente de la posición del objeto (26)

medusa forma del cuerpo de algunos cnidarios parecida a un hongo con tentáculos (330)

meiosis división celular que produce células sexuales (139)

melanina substancia química que determina el color de la piel (476)

membrana celular capa de fosfolípidos que cubre la superficie de la célula y actúa como barrera entre el interior de la célula y su medio ambiente (87)

menstruación pérdida mensual de sangre y tejidos del útero (539)

metabolismo procesos químicos combinados que se efectúan en una célula o en un organismo vivo (38)

metamorfosis proceso por el cual un insecto u otro animal cambia de forma al desarrollarse de embrión o larva a adulto (343, 366)

método científico serie de pasos que utilizan los científicos para responder preguntas y resolver problemas (10)

metro unidad básica de longitud del sistema métrico decimal (23)

microscopio compuesto miscroscopio formado por un tubo con lentes, una platina y una fuente de luz (19)

microscopio electrónico microscopio que usa pequeñas partículas de materia para producir imágenes aumentadas (20)

migrar viajar de un lugar a otro como respuesta a las estaciones o a las condiciones ambientales (311)

mitocondrias organelos celulares rodeados una membrana doble que descomponen las moléculas de alimento para producir ATP (95)

mitosis división nuclear de las células eucariotas en la que cada célula recibe una copia de los cromosomas originales (117)

monotrema mamífero que pone huevos (396, 537)

multicelular formado por muchas células (83, 305)

músculo cardíaco el tipo de músculo que se encuentra en el corazón (472)

músculo esquelético clase de músculo que mueve los huesos y protege los órganos internos (472)

músculo liso clase de músculo que se encuentra en los vasos sanguíneos y el sistema digestivo (472)

mutación cambio en el orden de las bases del ADN de un organismo; eliminación, inserción, o sustitución (162, 189)

mutágeno que puede alterar o causar cambios en el ADN (162)

N

naturalizado organismo que establece su hogar en un lugar diferente de donde vive normalmente (444)

Neandertal especie de homínido que vivió en Europa y en el oeste de Asia de 230,000 a 30,000 años atrás (218)

nervio axón a través del cual viajan los impulsos nerviosos; los nervios están agrupados en haces con vasos sanguíneos y tejido conjuntivo (512)

nervio óptico nervio que transfiere impulsos eléctricos del ojo al cerebro (518)

neurona célula especializada que lleva mensajes a través del cuerpo en forma de energía eléctrica que se traslada con rapidez (511)

neurona motora neurona que envía impulsos del cerebro y de la médula espinal a otros sistemas (512)

neurona sensorial neurona especial que reúne información sobre lo que está pasando dentro y alrededor del cuerpo y la envía al sistema nervioso central (512)

nódulos linfáticos órganos pequeños en forma de frijol en cuyo interior hay fibras minúsculas que funcionan como redes para retirar partículas de la linfa (497)

núcleo organelo de las células eucariotas cubierto por una membrana que contiene el ADN y que sirve como centro de control de la célula (90)

nucleótido subunidad de ADN formada por un azúcar, un fosfato y una base nitrogenada (152)

 O

ojo compuesto ojo formado por muchas células idénticas que trabajan en conjunto (340)

onda perturbación que transmite energía a través de la materia o el espacio (54)

onda electromagnética onda que no necesita un medio para desplazarse (55)

orden nivel de clasificación que le sigue a clase; los organismos en una clase están organizados en órdenes (229)

organelo estructura interna de una célula que en ocasiones está rodeada por una membrana (87)

organismo ser que puede realizar independientemente los procesos necesarios para la vida (83)

órgano combinación de dos o más tejidos que trabajan juntos para llevar a cabo una función específica en el cuerpo (81, 306, 465)

orientarse determinar la dirección que debe seguirse para ir de un lugar a otro (312)

ósmosis difusión de agua a través de la membrana celular (109)

ovario órgano del sistema reproductor femenino de los animales que produce óvulos (539); estructura de las flores que contiene los óvulos y que se va a convertir en fruta después de la fertilización (271)

ovulación proceso por el cual un huevo se expulsa a través de la pared del ovario (539)

óvulo gameto producido por la hembra (535)

ozono molécula de gas formada por tres átomos de oxígeno que absorbe los rayos ultravioleta del Sol (210)

 P

palanca máquina simple que consiste en una barra rígida que puede girar sobre un punto fijo denominado punto de apoyo; existen tres clases de palancas según la ubicación de la fuerza aplicada, la fuerza producida y el punto de apoyo en relación con la resistencia: palancas de primera clase, de segunda clase y de tercera clase (471)

paleontólogo científico que estudia los fósiles para reconstruir la historia de la vida durante los millones de años previos a la aparición de los seres humanos (202)

Pangea masa de tierra que hace 200 millones de años contenía todos los continentes que existen en la actualidad (206)

pantano área de terreno donde el nivel del agua está cerca o por encima de la superficie del suelo durante la mayor parte del año (431)

parásito organismo que se alimenta de otro ser vivo sin matarlo (332)

pared celular estructura que rodea y le suministra resistencia y apoyo a la membrana celular de algunas células (93)

pene órgano reproductor masculino que transfiere el semen al cuerpo de la hembra durante las relaciones sexuales (538)

permafrost parte profunda del suelo de la tundra ártica que está permanentemente congelada (422)

pétalos estructuras de la flor que participan generalmente en la atracción de los polinizadores y que con frecuencia son de colores brillantes (270)

pistilos estructuras reproductoras femeninas de la flor, formadas por el estigma, el estilo, y el ovario (271)

placenta órgano especial de intercambio que le suministra al feto en desarrollo nutrientes y oxígeno (398, 543)

plancton organismos muy pequeños que flotan en la superficie del mar o cerca de ella y constituyen la base de la red alimenticia del mismo (423)

planta no vascular planta que depende de los procesos de difusión y de ósmosis para el traslado de materiales de un lugar a otro (252)

planta vascular planta que posee tejidos especializados llamados xilema y floema, que transportan materiales de una parte de la planta a otra (253)

plantas de hoja caduca árboles cuyas hojas cambian de color en otoño y se caen en invierno (290, 417)

plantas de hoja perenne árboles que conservan las hojas durante todo el año (290)

plaquetas fragmentos de células que ayudan a coagular la sangre (489)

plasma parte líquida de la sangre (488)

plumas de contorno plumas formadas por una varita central dura, con muchas ramificaciones en los costados, llamadas barbas (383)

plumón plumas livianas y aislantes que están en contacto directo con el cuerpo de un ave (383)

población grupo de individuos de la misma especie que conviven al mismo tiempo en un área determinada (83)

polen partículas con apariencia de un polvo fino que contienen los gametofitos masculinos de las plantas productoras de semillas (258)

polinización transferencia de polen al fruto femenino de las coníferas o al estigma de las angiospermas (261)

pólipo, la forma de florero del cuerpo de algunos cnidarios (330)

pollito altricial pollito que al salir del huevo se encuentra débil, sin plumas e indefenso (388)

pollito precoz pollito totalmente activo que sale del nido inmediatamente después de nacer (388)

presa organismo que le sirve de alimento a otro organismo (308)

presión sanguínea fuerza que la sangre ejerce sobre las paredes interiores de un vaso sanguíneo (493)

primate grupo de mamífero al que pertenecen los seres humanos, los simios y los monos y que se caracteriza por la presencia de pulgares oponibles y visión binocular (214, 405)

probabilidad posibilidad matemática de que ocurra un evento (136)

productor organismo que utiliza la energía solar para producir azúcares (40)

prosimio primeros antepasados de los primates; también se conoce como prosimios a un grupo de primates que existe en la actualidad, entre los que se encuentran los lemures y los lemures africanos (216)

proteínas compuestos bioquímicos formados por aminoácidos que regulan las reacciones químicas, transportan y almacenan materiales y proporcionan apoyo en distintos lugares del organismo (42)

Proyecto del Genoma Humano esfuerzo conjunto de científicos de todo el mundo para descubrir la ubicación de cada gene y crear un mapa de la totalidad del genoma humano (167)

pubertad momento de la vida en que los órganos sexuales alcanzan la madurez (538)

pulmón órgano parecido a una bolsa que toma el oxígeno del aire y lo envía a la sangre (364)

punto de referencia objeto fijo que se usa para determinar la posición durante la navegación (313)

pupila abertura del iris en la parte anterior del ojo (519)

R

raíz central raíz principal que crece hacia abajo de la cual salen muchas ramificaciones pequeñas (265)

raíz fibrosa tipo de raíz en la que hay varias raíces del mismo tamaño que se separan de la base del tallo (265)

rasgo dominante se observa cuando se hereda por lo menos un alelo dominante para una característica determinada (133)

rasgo recesivo rasgo que se manifiesta solamente cuando dos alelos recesivos se heredan para la misma característica (133)

rasgo cualidad distintiva que puede pasar de una generación a otra (186)

receptor célula especializada, como las dendritas, que detecta cambios dentro o fuera del cuerpo (512)

reciclaje técnica que consiste en volver a procesar productos usados para obtener productos nuevos (449)

recuperación de recursos transformar en electricidad las cosas que normalmente se desechan (450)

recurso no renovable recurso natural que no se puede remplazar o que sólo se puede remplazar en miles o millones de años (443)

recurso renovable recurso natural que se puede usar y remplazar en un período de tiempo relativamente corto (443)

reflejo respuesta rápida e involuntaria a un estímulo (59)

reflexión cambio de dirección que sufre una onda al chocar con una barrera u objeto (59)

registro fósil secuencia histórica de la vida indicada por los fósiles que se han encontrado en las capas de la corteza terrestre (178)

reino el más general de los siete niveles de clasificación (229)

reino animal grupo de organismos complejos, multicelulares, formados por células que no tienen pared celular; estos organismos pueden desplazarse de un lado a otro y poseen un sistema nervioso que les permite percibir y reaccionar a lo que pasa a su alrededor (239)

reino de los protistas reino de organismos eucariotas unicelulares o multicelulares simples que contiene todos los eucariotas que no son plantas, animales ni hongos (236)

reino vegetal reino que contiene las plantas, que son organismos complejos y multicelulares que utilizan la energía solar para fabricar azúcares por medio de la fotosíntesis (237)

relación de superficie a volumen superficie exterior de una célula en relación con su volumen (88)

reloj biológico control interno de los ciclos naturales (312)

renacuajo larva acuática de un anfibio (366)

rendimiento mecánico medida de las veces que una máquina multiplica la fuerza que se aplica sobre una resistencia (471)

reproducción asexual reproducción en la cual un solo progenitor produce crías genéticamente idénticas al progenitor (38, 534)

reproducción sexual reproducción en la que dos gametos se unen para forman un cigoto (38, 535)

reptil vertebrado heterotérmico que se desarrolla a partir de un huevo amniótico y se caracteriza por tener la piel gruesa y seca (369)

respiración intercambio de gases entre las células vivas y su medio ambiente; incluye la respiración y la respiración celular (498); ver *respiración celular*

respiración celular proceso durante el cual la célula produce ATP a partir de oxígeno y glucosa, liberando dióxido de carbono y agua (113, 285, 500)

retículo endoplásmico organelo celular cubierto por membranas que produce lípidos, descompone medicamentos y otras substancias y prepara las proteínas para sacarlas de la célula (94)

retina capa de células sensibles a la luz que se encuentra en la parte posterior del ojo (518)

ribosoma organelo de la célula en donde se realiza la síntesis de proteínas a partir de aminoácidos (94, 161)

ritmo circadiano ciclo diario (312)

rizoides pequeños filamentos que mantienen las plantas no vasculares en su lugar (254)

rizoma tallo subterráneo de un helecho (256)

S

sabana bioma tropical cubierto de hierba con algunos grupos de árboles aislados (420)

saco amniótico membrana delgada, llena de líquido, que rodea al feto de un mamífero (543)

sangre tejido conjuntivo formado por glóbulos rojos, glóbulos blancos, plaquetas y plasma (488)

sargazo alga que forma plataformas flotantes enormes; le da su nombre al mar de los Sargazos, un ecosistema de algas del océano Atlántico (426)

sedimento partículas finas de arena, polvo o barro que el viento o el agua depositan a través del tiempo (202)

segmento parte del cuerpo que se repite varias veces (337)

selección natural proceso por el cual los organismos que poseen un rasgo favorable sobreviven y se reproducen más rápidamente que los organismos que carecen del mismo (188)

semen líquido que contiene los espermatozoides (538)

sépalos estructuras parecidas a hojas que cubren y protegen una flor que no ha madurado (270)

simetría bilateral propiedad de un organismo que consiste en que sus dos mitades son imágenes idénticas la una de la otra (326)

simetría radial propiedad de un organismo en el que las partes del cuerpo están organizadas en un círculo alrededor de un punto central (327)

sistema grupo de órganos que trabajan juntos para llevar a cabo funciones determinadas dentro del cuerpo (82, 465)

sistema cardiovascular conjunto de órganos que transporta la sangre, formado por el corazón, las arterias y las venas (488)

sistema circulatorio abierto sistema circulatorio formado por un corazón que bombea la sangre a través de espacios llamados senos (336)

sistema circulatorio cerrado sistema circulatorio en que la sangre circula a través de una red de vasos sanguíneos que forman un sistema cerrado (336)

sistema de líneas laterales hilera o hileras de pequeños órganos de los sentidos localizadas en los costados del cuerpo de un pez (360)

sistema endocrino conjunto de glándulas que controlan el equilibrio de los fluidos del cuerpo, el crecimiento y el desarrollo sexual (522)

sistema esquelético conjunto de órganos cuya función primaria es darle apoyo y protección al cuerpo; entre ellos están los huesos, los cartílagos, los ligamentos y los tendones (468)

sistema integumentario conjunto de órganos que le ayudan al cuerpo a mantener un medio ambiente interno estable y sano; entre ellos están la piel, el pelo y las uñas (476)

sistema linfático conjunto de órganos que reúnen los líquidos extracelulares y los devuelven a la sangre; entre ellos están los nódulos y vasos linfáticos (496)

sistema muscular conjunto de órganos cuya función principal es el movimiento; entre ellos están los músculos y el tejido conjuntivo que los une a los huesos (472)

sistema nervioso conjunto de órganos que recopilan e interpretan información sobre el medio ambiente interno y externo del cuerpo y responden a esa información; entre ellos están el cerebro, los nervios y la médula espinal (510)

sistema nervioso central conjunto de órganos que procesa todos los mensajes que pasan por los nervios; el cerebro y la médula espinal son dos órganos que forman parte de este sistema (510)

sistema nervioso periférico conjunto de nervios cuya función principal es llevar información de todas las áreas del cuerpo y el medio ambiente exterior al sistema nervioso central y del sistema nervioso central al resto del cuerpo (510)

sistema respiratorio conjunto de órganos cuya función principal es tomar el oxígeno y expeler el dióxido de carbono; entre estos órganos están los pulmones, la garganta y las vías que llevan a los pulmones (498)

sistema vascular de agua sistema de bombas de agua y canales de los equinodermos que les permite moverse, comer y respirar (346)

superpoblación condición que se presenta cuando el número de individuos de un medio ambiente crece tanto que no hay suficientes recursos para todos (444)

T

taxonomía ciencia de identificar, clasificar y darle un nombre a los seres vivos (230)

tecnología aplicación del conocimiento, los instrumentos y los materiales necesarios para resolver problemas y realizar tareas; objetos que se utilizan para realizar determinadas tareas (18)

tectónica de placas estudio de las fuerzas que producen el movimiento de los trozos de la corteza terrestre en la superficie del planeta (207)

tejido conjuntivo uno de los cuatro principales grupos de tejido del cuerpo; entre sus funciones están el apoyo, la protección, el aislamiento y la alimentación (465)

tejido epitelial uno de los cuatro tipos de tejido del cuerpo; cubre y protege los tejidos que están debajo (464)

tejido muscular uno de los cuatro tipos principales de tejido del cuerpo; contiene células que se contraen y relajan para producir movimiento (465)

tejido nervioso uno de los cuatro tipos principales de tejido del cuerpo, cuya función consiste en enviar señales eléctricas a través del cuerpo (464)

tejido grupo de células similares que se unen para llevar a cabo una función específica en el cuerpo (306)

temperatura medida de lo caliente o frío que es algo (26)

tendón tejido conjuntivo fuerte que une los músculos esqueléticos a los huesos (473)

teoría celular teoría que afirma que: (1) todos los organismos están compuestos de una o más células, (2) la célula es la unidad fundamental de la vida en todos los seres vivos, y (3) todas las células provienen de otras ya existentes (86)

teoría explicación que unifica una gran variedad de hipótesis y observaciones que se han sometido a verificación a través de experimentos (18)

terápsido reptil prehistórico, antepasado de los mamíferos (369, 392)

territorio área ocupada por un animal o un grupo de animales, de la cual se excluyen otros miembros de la especie (314)

testículos órganos del sistema reproductor masculino que producen los espermatozoides y la testosterona (538)

tiamina una de las cuatro bases que se combinan con azúcares y fosfatos para formar una subunidad de nucleótidos del ADN; la tiamina forma un par con la adenina (152)

tiempo de generación período entre el nacimiento de una generación y el de la siguiente (190)

timo órgano linfático que libera linfocitos (497)

tórax parte central del cuerpo de un artrópodo o de otro animal que contiene el corazón y los pulmones (339)

tóxico venenoso (441)

transpiración pérdida de agua de las hojas de una planta a través de los estomas (286)

transporte activo movimiento de partículas a través de las proteínas de la membrana celular en sentido opuesto a la difusión; requiere que las células utilicen energía (110)

transporte pasivo difusión de partículas a través de las proteínas de la membrana celular de las áreas de alta concentración a las de baja concentración (110)

tráquea tubo por donde pasa el aire de la laringe a los pulmones (499)

trompas de Falopio tubos que van de los ovarios al útero (539)

tropismo cambio en el crecimiento de una planta como respuesta a un estímulo (287)

tubos seminíferos tubos en forma de espiral del interior de los testículos en donde se producen los espermatozoides (538)

tundra bioma del extremo norte de la Tierra, caracterizado por inviernos largos y fríos, permafrost, y pocos árboles (422)

U

unicelular formado por una sola célula (83)

uretra tubo delgado que lleva la orina y el semen al exterior a través del pene de los machos (538)

útero órgano del sistema reproductor femenino donde crece y se desarrolla un cigoto (539)

V

vacuola estructura grande del interior de las células vegetales, encerrada por una membrana, que sirve para almacenar agua y otros líquidos (97)

vagina órgano del sistema reproductor femenino que recibe el semen durante las relaciones sexuales (539)

variable factor que cambia en un experimento controlado (14)

vasos linfáticos vasos grandes del sistema linfático (496)

vejiga natatoria órgano similar a un globo, lleno de oxígeno y otros gases, que les da a los peces óseos la capacidad de flotar (363)

venas vasos sanguíneos que llevan la sangre de los diferentes sitios del cuerpo al corazón (491)

ventrículos cámaras inferiores del corazón (490)

vertebrados animales que poseen cráneo y columna vertebral, como los mamíferos, las aves, los reptiles, los anfibios y los peces (304, 356)

vértebras segmentos de hueso o cartílago que forman la columna vertebral (357)

vesícula compartimento cubierto por una membrana que se forma cuando parte de la membrana celular de una célula eucariota rodea un objeto y se separa del cuerpo de la célula (97)

vestigio resto de una estructura anatómica que cumplió una función en algún momento (179)

vida media tiempo que tarda en desintegrarse la mitad de una muestra radioactiva determinada (203)

volumen cantidad de espacio que ocupa o que contiene un cuerpo (24)

voluntario movimiento muscular que está bajo el control del individuo (472)

X

xilema tejido vegetal especializado que transporta el agua y los minerales de una parte de la planta a otra (264)

Z

zona de aguas abiertas zona de un lago o laguna que se extiende por toda la superficie del agua desde el litoral y cuya profundidad está determinada por la distancia que la luz penetra dentro del agua (429)

zona de aguas profundas área de un lago o laguna donde no llega la luz (429)

zooplancton animales diminutos que, junto con el fitoplancton que consumen, forman la base de la red alimenticia de los océanos (423)

Índice

Los números en **negrita** se refieren a una ilustación en dicha página.

A

676 Índice

H

I

tabla de conversiones, 642
unidades del SI, 22, 642
temperatura corporal, 358, 462
de las aves, 383
de los humanos, **26**
de los mamíferos, 393-394
de los reptiles, 370
regulación de la, 37, 476, 478
tendón, 473, **473**
daño al, 475
tenias, 332-333, **333**
teoría, 18
teoría celular, 86
teoría endosimbiótica, 78, 96, **96**
terápsido, 369, **369**, 392, **392**
terminaciones nerviosas
en la piel, 476, **477**
termómetro, **11**, 26, **26**
territorio, 314, **314**
testa, 259, **259**
testículos, **523**, 538, **538**
testosterona, 475, 538
Thayer, Jack, 462
tiburón, **306, 361,** 361-362, **470**
dentículos del, 362, **362**
esqueleto del, 470
Tierra
atmósfera temprana, 209
cambios en la vida sobre la, 177, **177**
capas de la, 178
edad de la, 177, 187
historia de la, 204
tigre, 9, **9,** 317
color del pelaje, 158
tigre siberiano, 9, **9,** 401
timina, 152-153, **152**
timo, 497, **497, 523**
tipo sanguíneo, 43, 494, **494**
tiroides, **523**, 525
tomate, 267, 282, **282**
tonelada métrica, 26
topo, 399, **399**
tórax
de los artrópodos, 339, **339**
de los insectos, 343, **343**
tortuga, **239**, 371-372
tortuga de mar, 372, **372,** 424, **424**
tortugas, 370-372, **372,** 429-430, **430**
tóxico, 441
transfusión, 494
transpiración, 286, **286**
transporte
activo, 110, **110**

pasivo, 110, **110**
tráquea, **498,** 499, **499**
trilobites, **178,** 211
trompa de falopio, 539, **539**
tropismo, 287, **287-288**
tubérculo, **283**
tuberculosis, 8, 190
tubo polínico, 280, **280**
túbulos seminíferos, 538, **538**
Tumlinson, Jim, 278
tumor, 501
tundra, **416,** 422
alpina, 422
ártica, 422, **422**
Tyrannosaurus rex, 231, **231**

U

unicelular
organismo, 83, **83**
unidades del SI, 22, **22,** 641
prefijos, 641
tabla de conversión, 641
uñas, 478, **478**
uretra, 538, **538**
útero, 398, 539, **539,** 542, **542**

V

vacuolas, 97-99, **97, 98-99**
vagina, 539, **539**
válvula
del corazón, 490, **490**
venosa, 491
vara métrica, 643
variable, 14, 645
variación genética, 188, **188**
vaso de precipitados graduado, 24
vasos sanguíneos, 491, **491**
de la piel, **477**
vegetal (reino), 237, 659
vejiga natatoria, 362, **362**
vena
de la hoja, 269, **269**
sistema circulatorio, 491, **491-492**
vertebrado, 304, 356-358
vértebras, 357, **357**, 515, **515**
vesículas, 97, 111, **111**
vestigios, 179, **179**
vida media del, 203, **203**
Virchow, Rudolf, 86
visión, 65, 395, 518-519, **518-519**
binocular, 214, **214**
vitamina A, 518

vitamina D, 58
vitamina K, 235
vivir en grupos
beneficios de, 317
desventaja de, 317
volumen, 22, 24-25, **25,** 88, **88** 653, **653**
cálculo del, 653
de objetos irregulares, 25, **25**
de objetos rectangulares, 25, **25**
unidades del SI, 22, 641
vuelo
de las aves, 384-385, **384-386,** 386-387
mecánica del, 386, **410**
músculos de, **385**

W

Wallace, Alfred Russel, 188
Watson, James, 153, **153**
Wegener, Alfred, 206
Wilkins, Maurice, 153

X

xilema, 264, **265,** 266-267, **267**

Y

yema, 80, **371**

Z

zarigüeya, 397, **397, 537**
zona bentónica, 425, **425**
zona de aguas abiertas, 429, **429**
zona de aguas profundas, 429, **429**
zona intermareal 424, **424,** 427, **427**
zona nerítica, 424, **424**
zona oceánica, 425, **425**
zooplancton, 423
zorrillo, 226, 232, 309

Credits

Abbreviations used: (t) top, (c) center, (b) bottom, (l) left, (r) right, (bkgd) background

ILLUSTRATIONS

All illustrations, unless otherwise noted below, by Holt, Rinehart and Winston.

Table of Contents Page viii(tr), Morgan Cain & Associates; ix(tl), Marty Roper/Planet Rep; (bl), Frank Ordaz/Dimension; (br), John White/The Neis Group; xv(cl), Christy Krames; xix(tl), Carlyn Iverson.

Unit One Page 2(bl), Kip Carter.

Chapter One Page 12(cr), Michael Morrow; 13(all), Michael Morrow; 15(cr), Ralph Garafola; 16(c), Ross, Culbert and Lavery/The Mazer Corporation Corporation; 18(bl), The Mazer Corporation; 20(tl), Blake Thornton/Rita Marie; 22(tl), Blake Thornton/Rita Marie; (cl, bl), Stephen Durke/Washington Artists/Washington Artists; (b), The Mazer Corporation; 23(cl), Susan Johnston Carlson; (c), Frank Ordaz/Dimension; (c), Steve Roberts; (c), Morgan Cain & Associates; (cl), Ross, Culbert and Lavery; 24(bl), Terry Kovalcik; 26(bl), Stephen Durke/Washington Artists; 31(cr), Annie Bissett; (bl), Mark Heine.

Chapter Two Page 39(tl), Will Nelson/Sweet Reps; (br), Terry Kovalcik; 42(cl), Morgan Cain & Associates; 43(all), Morgan Cain & Associates; 44(cl), Blake Thornton/Rita Marie; (bl, br), Morgan Cain & Associates; 45(tr), David Merrill/Suzanne Craig; (cr), The Mazer Corporation; (cr), John White/The Neis Group; (cr, br), Morgan Cain & Associates; 47(tr), Morgan Cain & Associates; 48(bl), Morgan Cain & Associates.

Chapter Three Page 55(tl), Will Nelson/Sweet Reps; (br), Blake Thornton/Rita Marie; 56(cl, b), Sidney Jablonski; (cl), Mike Carroll; 57(b), Sidney Jablonski; 58(cl), Blake Thornton/Rita Marie; 59(bl), Dan Stuckenschneider/Uhl Studios/Preface; 60(all), Dan Stuckenschneider/Uhl Studios/Preface; 62(cl), Gary Ferster; 65(tl), Stephen Durke/Washington Artists; 66(t), Stephen Durke/Washington Artists; 67(cr), Keith Kasnot; 68(c), Dan Stuckenschneider/Uhl Studios; 69(all), Dan Stuckenschneider/Uhl Studios; 70(c), Blake Thornton/Rita Marie; 71(cr), Dan Stuckenschneider/Uhl Studios; 72(tr), Stephen Durke/Washington Artists.

Unit Two Chapter Four Page 82(c), Michael Woods; (br), Christy Krames; 83(c, cl), Morgan Cain & Associates; (cr), Christy Krames; 84(bl), Yuan Lee; 86(bl), David Merrill/Suzanne Craig; 88(tl), Terry Kovalcik; (c, b), Morgan Cain & Associates; 89(cr), Morgan Cain & Associates; (bc), Terry Kovalcik; 90(br), Morgan Cain & Associates; 91(cr), Morgan Cain & Associates; 92(all), Morgan Cain & Associates; 93(all), Morgan Cain & Associates; 94(all), Morgan Cain & Associates; 95(all), Morgan Cain & Associates; 96(all), Morgan Cain & Associates; 97(br), Morgan Cain & Associates; 98(all), Morgan Cain & Associates; 99(c), Morgan Cain & Associates; (br), Blake Thornton/Rita Marie; 103(all), Morgan Cain & Associates.

Chapter Five Page 107(all), Mark Heine; 109(tl), Stephen Durke/Washington Artists; 110(tl), Terry Kovalcik; (bl, br), Morgan Cain and Associates; 111(all), Morgan Cain and Associates; 112(bl), Morgan Cain and Associates; 113(bc), Morgan Cain and Associates; 114(tl), Robin Carter; (tl) The Mazer Corporation; (tr, cl, cr, bl, br), Morgan Cain and Associates; 118, 119(all), Alexander and Turner; 123(cr), Morgan Cain and Associates.

Unit Three Page 127(all), John White/The Neis Group.

Chapter Six Page 131(b), Mike Wepplo/Das Group; 132(tl), Michael Woods; (tl, cl, b), John White/The Neis Group; 133(c), Michael Woods; (cr, bc, br, r), John White/The Neis Group; 134(tl, cl, cr, bl, br), John White/The Neis Group; (tl, tr, cl), Michael Woods; (cr), The Mazer Corporation; 135(tr, cr), John White/The Neis Group; (br), The Mazer Corporation; 136(tl), John White/The Neis Group; 139(r), Alexander and Turner; 140,141(all), Alexander and Turner; 142(cr), Alexander and Turner; 143(bc), Alexander and Turner; (br), Rob Schuster/Hankins and Tegenborg; (br), Blake Thornton/Rita Marie; 144(br), The Mazer Corporation; 145(br), Blake Thornton/Rita Marie; 146(tr), Rob Schuster/Hankins and Tegenborg; 146(tr), John White/The Neis Group; 147(bl), John White/The Neis Group.

Chapter Seven Page 150(t), The Mazer Corporation; (c), Stephen Durke/Washington Artists; 152(b), Rob Schuster/Hankins and Tegenborg; 154(tl), Marty Roper/Planet Rep; (c), Alexander and Turner; 155(cl), Alexander and Turner; 156(c), Morgan Cain & Associates; 157(br) Alexander and Turner; 158(cl), John White/The Neis Group; 160,161(c), Rob Schuster/Hankins and Tegenborg; 162(tl), Rob Schuster/Hankins and Tegenborg; 163(cl), Rob Schuster/Hankins and Tegenborg; 166(cl), The Mazer Corporation; 167(cr), The Mazer Corporation; 168(cl) Marty Roper/Planet Rep.

Chapter Eight Page 174(tl), Michael Morrow; (br, l), Will Nelson/Sweet Reps; 177(b), Steve Roberts; 179(tr, cr), Ross, Culbert and Lavery; (b, br), Rob Wood/Wood, Ronsaville, Harlin; 180(all), Rob Wood/Wood, Ronsaville, Harlin; 181(all), Rob Wood/Wood, Ronsaville, Harlin; 182(cl), Christy Krames; 183(tc, c), Sarah Woods; (tr, cr), David Beck; (cr), Frank Ordaz/Dimension; 185(tr, c), Tony Morse/Ivy Glick; (bl, bc, br), John White/The Neis Group; 186(tl), Carlyn Iverson; 187(bl), Ross, Culbert and Lavery; 188(all), Will Nelson/Sweet Reps; 190(cl), Frank Ordaz/Dimension; (bl, bc, br), Carlyn Iverson; 192(c, cl, cr), Mike Wepplo/Das Group; (bl, br), Will Nelson/Sweet Reps; 193(all), Carlyn Iverson; 194(tc), Rob Wood/Wood, Ronsaville, Harlin; 195(all), Carlyn Iverson; 197(all), Ross, Culbert and Lavery.

Chapter Nine Page 201(c), Michael Morrow; 202(all), Mike Wepplo/Das Group; 203(br), Mike Wepplo/Das Group; (br), The Mazer Corporation; 204(tc, c, bc), Barbara Hoopes-Ambler; (cl), The Mazer Corporation; 205(t), John White/The Neis Group; 206(cl), MapQuest.com; 207(tr), MapQuest.com; (br), Walter Stuart; 209(c), John White/The Neis Group; 210(cr), Craig Attebery/Jeff Lavaty; 211(all), Barbara Hoopes-Ambler; 212(all), Barbara Hoopes-Ambler; 213(tr, cr), John White/The Neis Group; (br), Terry Kovalcik; 214(bl), Todd Buck; (br), Will Nelson/Sweet Reps; 215(all), Christy Krames; 219(cr), The Mazer Corporation; (cr), Christy Krames; 220(cr), Barbara Hoopes-Ambler; 223(br), The Mazer Corporation; (tr), John White/The Neis Group; 224(br), Greg Harris.

Chapter Ten Page 227(bl), Terry Kovalcik; 229(c), Michael Woods; (c), David Ashby; (c), Ponde and Giles; (c), Frank Ordaz/Dimension; (c), Will Nelson/Sweet Reps; (c), Graham Allen; (c), Chris Forsey; (c), The Mazer Corporation; 230(tl, cl), Will Nelson/Sweet Reps; (cl), Michael Woods; (cl), Ponde and Giles; (bl), The Mazer Corporation; 231(tr), Blake Thornton/Rita Marie; (b), John White/The Neis Group; 232(c), Marty Roper/Planet Rep; (bl, tl), John White/The Neis Group; 233(tr), John White/The Neis Group; (bl), Cy Baker/WAA; (br), The Mazer Corporation; 239(br), Will Nelson/Sweet Reps; (br), Chris Forsey; (br), Michael Woods; (br), Frank Ordaz/Dimension; (br), The Mazer Corporation; 240(cr), John White/The Neis Group; 242(cr), Marty Roper/Planet Rep; 243(tr), Cy Baker/WAA; (tr), The Mazer Corporation; 244(tr), Barbara Hoopes-Ambler; (cl), John White/The Neis Group.

Unit Four Chapter Eleven Page 248(tc), Marty Roper/Planet Rep; 251(tr), Morgan Cain and Associates; (bl), The Mazer Corporation; 253(c), The Mazer Corporation; (c, cl, cr), John White/The Neis Group; 254(bc), The Mazer Corporation; (bc), Ponde and Giles; 256(bc), The Mazer Corporation; (bc), Ponde and Giles; 259(tl), Sarah Woods; (tr), Keith Locke/Suzanne Craig; (br), James Gritz/Photonica; 261(bc), The Mazer Corporation; (bc), Will Nelson/Sweet Reps; 262(tl), Marty Roper/Planet Rep; 263(tr), The Mazer Corporation; (tc, tr, c, cr), John White/The Neis Group; 264(bl), Will Nelson/Sweet Reps; 265(tc), John White/The Neis Group; 267(all), Will Nelson/Sweet Reps; 269(all), Will Nelson/Sweet Reps; 270(c), Will Nelson/Sweet Reps; 272(br), Sarah Woods; 274(br), John White/The Neis Group; 275(all), Will Nelson/Sweet Reps.

Chapter Twelve Page 278(cl), Dan McGeehan/Koralick Associates; 280(all), Will Nelson/Sweet Reps; 281(all), Will Nelson/Sweet Reps; 282(br), Will Nelson/Sweet Reps; 284(all), Stephen Durke/Washington Artists; 285(bl), Ponde and Giles; 286(cr), Ponde and Giles; 287(bl), Carlyn Iverson; 289(cl, bl), Rob Schuster/Hankins and Tegenborg; (cr), Stephen Durke/Washington Artists; 291(all), Rob Schuster/Hankins and Tegenborg; 294(cl), Will Nelson/Sweet Reps; 296(tr), Will Nelson/Sweet Reps; 297(all), Carlyn Iverson.

Unit Five Chapter Thirteen Page 302(tr), Tony Morse/Ivy Glick; 305(cr), Sidney Jablonski; (c), Barbara Hoopes-Ambler; (c), Sarah Woods; (c), Steve Roberts; (c), Bridgette James; (c), Michael Woods; 306(bl), Kip Carter; 310(cl), Keith Locke/Suzanne Craig; 312(tr), Gary Locke/Suzanne Craig; (bl), Tony Morse/Ivy Glick; 316(tl), John White/The Neis Group; 318(br), Sidney Jablonski; 319(br), Sidney Jablonski; 321(bl), Sidney Jablonski.

Chapter Fourteen Page 327(all), Barbara Hoopes-Ambler; (tc), Sarah Woodward; (tr, cr, br), Alexander and Turner; 329(all), Alexander and Turner; 330(all), John White/The Neis Group; 331(all), Morgan Cain & Associates; 332(cl), Alexander and Turner; 335(tr), The Mazer Corporation; (c, cl, cr), Alexander and Turner; 339(br), Felipe Passalacqua; 341(c), John White/The Neis Group; (cr), Will Nelson/Sweet Reps; 343(bl), Steve Roberts; (r), Marty Roper/Planet Rep; 344(all), Bridgette James; 346(all), Alexander and Turner; 351(cr), The Mazer Corporation; (cr), Barbara Hoopes-Ambler.

Chapter Fifteen Page 355(cr), Kip Carter; 357(tc), Alexander and Turner; 360(cl), Will Nelson/Sweet Reps; 362(br), Kip Carter; 364(br), Peg Gerrity; 366(all), Will Nelson/Sweet Reps; 368(bl), Marty Roper/Planet Rep; 369(c), Barbara Hoopes-Ambler; (c), Chris Forsey; (c), Ponde and Giles; (c), Morgan Cain & Associates; 371(c), Kip Carter; 374(br), Will Nelson/Sweet Reps; 376(bl), Will Nelson/Sweet Reps; 377(tr), Rob Schuster/Hankins and Tegenborg; (bl), Marty Roper/Planet Rep; 378(bl), Ron Kimball; 379(bc), Ka Botz.

Chapter Sixteen Page 381(tr), John White/The Neis Group; 383(tr), Will Nelson/Sweet Reps; (cr, br), Kip Carter; (br), Will Nelson/Sweet Reps; 384(all), Will Nelson/Sweet Reps; 385(all), Will Nelson/Sweet Reps; 386(cr), Will Nelson/Sweet Reps; 391(all), Kip Carter; 392(tl), Howard Freidman; 406(br), Will Nelson/Sweet Reps; 409(cr), Sidney Jablonski.

Unit Six Chapter Seventeen Page 416(b), MapQuest.com; 417(b), Will Nelson/Sweet Reps; 418(all), Will Nelson/Sweet Reps; 419(all), Will Nelson/Sweet Reps; 421(tr), Marty Roper/Planet Rep; 423(tc), Will Nelson/Sweet Reps; 424(all), Yuan Lee; 425(all), Yuan Lee; 428(all), Will Nelson/Sweet Reps; 429(br), Mark Heine; 431(br), Rob Schuster/Hankins and Tegenborg; 432(tc), Will Nelson/Sweet Reps; 433(cl) Mark Heine; 434(bc), Will Nelson/Sweet Reps; 435(all), Rob Schuster/Hankins and Tegenborg.

Chapter Eighteen Page 444(bl), Will Nelson/Sweet Reps; 457(t), John White/The Neis Group.

Unit Seven Chapter Nineteen Page 464(all), Morgan Cain & Associates; 465(all), Morgan Cain & Associates; 466(all), Christy Krames; 467(all), Christy Krames; 469(bc), Keith Kasnot; 470(all), John Huxtable/Black Creative; 471(all), Annie Bissett; 473(tc), Christy Krames; 477(tr), Marty Roper/Planet Rep; (bl), Morgan Cain & Associates; 479(all), Morgan Cain & Associates; 480(all), John Huxtable/Black Creative; 482(br), Christy Krames; 483(tr), Morgan Cain & Associates.

Chapter Twenty Page 488(tr), Christy Krames; (cl), Todd Buck; 489(b), Keith Kasnot; 490(all), Kip Carter; 491(tlc), Kip Carter; 492(bc), Kip Carter; 494(tl), Jared Schneidman/Wilkinson Studios; (br), Marty Roper/Planet Rep; 496(cr), Kip Carter; 497(tr), Christy Krames; 498(bl), Christy Krames; (br), Christy Krames; (br), John Karapelou; 500(tl, br), Christy Krames; (c, bc), Kip Carter; (cl), Eyewire, Inc.; 502(br), Kip Carter; 505(tr), Kip Carter.

Chapter Twenty-One Page 510(bl), Christy Krames; 511(bc), Scott Barrows/The Neis Group; 512(bc), Scott Barrows/The Neis Group; 513(all), Brian Evans; 514(all), Brian Evans; 515(tr), Christy Krames; 517(br), Morgan Cain and Associates; 518(cl), Carlyn Iverson; (br), Keith Kasnot; 519(all), Keith Kasnot; 520(cl), Christy Krames; 521(tr), Keith Kasnot; 522(b), Dan McGeehan/Koralick Associates; 523(all), Christy Krames; 524(all), Keith Kasnot; 526(br), Keith Kasnot; 527(c), Dan McGeehan/Koralick Associates; 528(tr), Christy Krames; 529(tr), Christy Krames.

Chapter Twenty-Two Page 535(bl), Rob Schuster/Hankins and Tegenborg; 538(bl), Keith Kasnot; 539(tr), Keith Kasnot; 540(br), Rob Schuster/Hankins and Tegenborg; 542(cl), David Fischer; 543(cr), Christy Krames; (bc), Mary Kate Denny; 551(cr), Sidney Jablonski; 552(bl), Morgan Cain and Associates.

LabBook Page 554 (tl), Stephen Durke/Washington Artists; 562(br), The Mazer Corporation; 565(cr), Keith Locke/Suzanne Craig; 567(cr), Blake Thornton/Rita Marie; 568(br), Stephen Durke/Washington Artists; 570(tl), David Merrill/Suzanne Craig; (cr), Rob Schuster/Hankins and Tegenborg; 573(all), Morgan Cain & Associates; 580(cr), The Mazer Corporation; (br), Kip Carter; 582(cl), Rob Schuster/Hankins and Tegenborg; 584(cr), Frank Ordaz/Dimension; 585(all), John White/The Neis Group; 586(br), Keith Locke/Suzanne Craig; 587(cr), Keith Locke/Suzanne Craig; 591(cr), The Mazer Corporation; (cr, br), Rob Schuster/Hankins and Tegenborg; 592(all), Rob Schuster/Hankins and Tegenborg; 593(tl), The Mazer Corporation; (cr), Keith Locke/Suzanne Craig; 594(cr), Rob Schuster/Hankins and Tegenborg; 595(cr), Rob Schuster/Hankins and Tegenborg; 596(cr), Will Nelson/Sweet Reps; 597(cr), Will Nelson/Sweet Reps; 599(cr), Sarah Woodward; 601(br), John White/The Neis Group; 605(br), Carlyn Iverson; 606(br), Keith Locke/Suzanne Craig; 607(tr), John White/The Neis Group; (cr), The Mazer Corporation; 609(br), Marty Roper/Planet Rep; 611(br), Keith Locke/Suzanne Craig; 613(tr), John Huxtable/Black Creative; (br), David Merrill/Suzanne Craig; 616(cr), Will Nelson/Sweet Reps; 617(b), Blake Thornton/Rita Marie; 618(cr), John White/The Neis Group; 620(all), Carlyn Iverson; 627(cr), Morgan Cain & Associates; (b), Rob Schuster/Hankins and Tegenborg; 628(tr), Kip Carter; 629(t), Rob Schuster/Hankins and Tegenborg.

Appendix Page 639(cl), Blake Thornton/Rita Marie; 642(t), Terry Guyer; 646(all), Mark Mille/Sharon Langley; 654, 655(all) Kristy Sprott; 656(bl), Stephen Durke/Washington Artists; 657(b), Bruce Burdick.

PHOTOGRAPHY

Cover and Title Page (cl), Frans Lanting/Minden Pictures; (c), Peter Peterson/Tony Stone Images; (cr), Chris Jaffe; (bl), Carr Clifton; (br), Gerry Ellis/ENP Images; owl: (cover, spine, back, title page), Kim Taylor/Bruce Coleman.

Table of Contents Page v(br), Uniphoto; vi(cl), Leonard Lessin/Photo Researchers; vii(tl), Dr. Jeremy Burgess/Science Photo Library/Photo Researchers; (tr), E.R. Degginger/Color-Pic; (br), Robert Brons/BPS/Tony Stone Images; viii(tl), Frans Lanting/Minden Pictures; ix(tr), Biophoto Associates/Photo Researchers; x(tl), Centre National de Prehistoire, Perigueux, France; (bl), David B. Fleetham/Tom Stack & Associates; xi(tl), SuperStock; (tr), Runk/Schoenberger/Grant Heilman Photography; xii(tl), Richard R. Hansen/Photo Researchers; (bl), Darryl Torckler/Tony Stone Images; xiii(tl), Brian Parker/Tom Stack & Associates; (tr), James Beveridge/Visuals Unlimited; (br), Tui De Roy/Minden Pictures; (br), Konrad Wothe/Westlight; xiv(tl), Rob & Ann Simpson/Visuals Unlimited; (bl), David Young/Tony Stone Images; xv(tr), Dr. Dennis Kunkel/Phototake; (cr) Enrico Ferorelli; (bl), Lennart Nilsson/Albert Bonniers Forlag AB, A CHILD IS BORN; xxi(bl), S.C. Bisserot/Bruce Coleman.

Feature Borders Unless otherwise noted below, all images copyright ©2001 PhotoDisc/HRW. Pages (104, 124, 224, 378, 410, 436, 552), *Across the Sciences*: all images by HRW. Pages (32, 225, 277, 437, 458), *Careers*: sand bkgd and saturn, Corbis Images; DNA, Morgan Cain & Associates; scuba gear, ©1997 Radlund & Associates for Artville. Pages (485, 531), *Eureka*. Pages (198, 299, 322, 353), *Eye on the Environment*: clouds and sea in bkgd, HRW; bkgd grass and red eyed frog, Corbis Images; hawks and pelican, Animals Animals/Earth Scenes; rat, John Grelach/Visuals Unlimited; endangered flower, Dan Suzio/PhotoResearchers, Inc. Pages (105, 149, 507), *Health Watch*: dumbbell, Sam Dudgeon/HRW Photo; aloe vera and EKG, Victoria Smith/HRW Photo; basketball, ©1997 Radlund & Associates for Artville; shoes and Bubbles, Greg Geisler. Pages (50, 244, 459), *Scientific Debate*: Sam Dudgeon/HRW Photo. Pages (33, 125, 199), *Science Fiction*: saucers, Ian Christopher/Greg Geisler; book, HRW; bkgd, Stock Illustration Source. Pages (148, 276, 484, 530, 553), *Science Technology and Society*: robot, Greg Geisler. Pages (245, 298, 323, 352, 379, 411, 506), *Weird Science*: mite, David Burder/Tony Stone; atom balls, J/B Woolsey Associates; walking stick and turtle, EclectiCollection.

Unit One Page 2(tc bkg), O.S.F./Animals Animals Earth Scenes; (br), University of Pennsylvania/Hulton Getty Images/Liaison International; 3(tl), Hulton Getty Images/Liaison International; (tr), National Portrait Gallery, Smithsonian Institution/Art Resource, NY; (bl), Peter Veit/DRK Photo; (br), O. Louis Mazzatentangs/National Geographic Image Collection.

Chapter One Page 4(br), Minnesota Pollution Control Agency; 7(tl), NASA; (tc), Gerry Gropp; (tr), Chip Simons Photography; (br), Charles C. Place/Image Bank; 8(tl), Hank Morgan/Photo Researchers; (bl), Mark Lennihan/AP Wide World Photos; 9(cr), Dale Miquelle/National Geographic Image Collection; (br), George Holton/Photo Researchers; 12(tl), Fernando Bueno/Image Bank; 13(tr), Mark Gibson; 14(tl), (br), John Mitchell/Photo Researchers; 18(br), John Reader/Photo Researchers; (cr), Greg Greico/PENN State; 19(bl), CENCO; (bc), Robert Brons/Tony Stone Images; 20(cl), Sinclair Stammers/Science Photo Library/Photo Researchers; (cr), Personal SEM/RJ Lee Instruments Ltd.; (bl), Microworks/Phototake; (br), Karl Aufderheide/Visuals Unlimited; 21(tr), Scott Camazine/Photo Researchers; (cr), Alfred Pasieka/Photo Researchers; (br), Howard Sochurek/Stock Market; 23(tr), David Austen/Publisher's Network; 25(tr), Science Kit & Boreal Laboratories; 27(br), Dr. Jeremy Burgess/Science Photo Library/Photo Researchers; 29(cr), CENCO; 30(tr), Charles C. Place/Image Bank; 32(tl), Eric Pianka/University of Texas; (br), Charles C. Place/Image Bank.

Chapter Two Page 34(tr), Patrick Landmann/Liasion International; (cr), Chris Landmann/Liasion International; (br), Chris Landmann/Liaison International; 35(bl), Steve Dunwell/Image Bank; 36(tl), Cabisco/Visuals Unlimited; (cl), VU Science/Visuals Unlimited; (br), Wolfgang Kaehler/Liaison International; 37(cl, cr), David M. Dennis/Tom Stack & Associates; (br), Fred Rhode/Visuals Unlimited; 38(tl), Stanley Flegler/Visuals Unlimited; (cr), James M. McCann/Photo Researchers;

(bl), Lawrence Migdale/Photo Researchers; 40(cl), Robert Dunne/Photo Researchers; 41(tr), Wolfgang Bayer/Bruce Coleman; (cr), Rob & Ann Simpson/Visuals Unlimited; 42(cl), Hans Reinhard/Bruce Coleman; (bl), L. West/Photo Researchers; 4E(cr), Stanley Flegler/Visuals Unlimited; 47(tl), Wolfgang Bayer/Bruce Coleman; 49(bl), Dede Gilman/Unicorn Stock Photos; 50(tc), NASA.

Chapter Three Page 52(t), Visuals Unlimited; (bc), Cindy Foesinger/Photo Researchers; 54(c), Michael Fogden and Patricia Fogden/Corbis; (cl, cr), Leonard Lessin/Photo Researchers; 56(bl), Robert Wolf/HRW Photo; 57(tr), Cameron Davidson/Tony Stone Images; (bl), Leonide Principe/Photo Researchers; (bc), Hugh Turvey/Science Photo Library/Photo Researchers; (br), Blair Seitz/Photo Researchers; 62(c), ©2001 PhotoDisc; (bl), Renee Lynn/Davis/Lynn Images; 63(cl), Leonard Lessin/Peter Arnold; (bc), Daniel Schaefer/HRW Photo; 64(c), Index Stock Imagery; 66(tl), Richard Megna/Fundamental Photographs; 67(all), E.R. Degginger/Color-Pic; 74(all), E.R. Degginger/Color-Pic; 75(tr), NASA; (tr), SuperStock.

Unit Two Page 76(tr), Kevin Collins/Visuals Unlimited; (c), Ed Reschke/Peter Arnold; (c), Glen Allison/Tony Stone Images; (br), Cold Spring Harbor Laboratory; 77(tc), Ed Reschke/Peter Arnold; (cl), Matthew Brady/National Archives/Images; (cr), Keith Porter/Photo Researchers; (bc), Dan McCoy/Rainbow; (br), Dr. Ian Wilmut/Liaison International.

Chapter Four Page 78(cr), Biology Media/Photo Researchers; 80(cl), ©2001 PhotoDisc; (bl, bcl, bc), Dr. Yorgos Nikas/Science Photo Library/Photo Researchers; (br), Lennart Nilsson/Albert Bonniers Forlag AB, A CHILD IS BORN; 81(tc), Fred Hossler/Visuals Unlimited; (tr), G.W. Willis/BPS/Tony Stone Images; (tl), National Cancer Institute/Science Photo Library/Photo Researchers; (br), G. Shih-R. Kessel/Visuals Unlimited; 83(tr), Robert Brons/BPS/Tony Stone Images; (bc), Michael Abbey/Visuals Unlimited; (tl), David M. Phillips/Photo Researchers; (br), Edward S. Ross; 84(cl), Joe McDonald/DRK Photo; 85(c, cl), C.C. Lockwood/DRK Photo; (bc), Kevin Collins/Visuals Unlimited; (br), Leonard Lessin/Peter Arnold; 86(tl) Doug Sokell/Visuals Unlimited; (inset, b), K.G. Murti/Visuals Unlimited; (inset, r), D.M. Phillips/Visuals Unlimited; (cl), Dr. Jeremy Burgess/Photo Researchers; 87(tr), Dr. Petit/Rapho/Gamma Liasion International; (br), Biophoto Associates/Science Source/Photo Researchers; 89(tr), AP/Wide World Photos; 97(tr), Photo Researchers; 100(tr), Michael Abbey/Visuals Unlimited; (c), Joe McDonald/DRK Photo; 102(br), Biophoto Associates/Science Source/ Photo Researchers; 104(cl), Hans Reinhard/Bruce Coleman; (bkgd), Andrew Syred/Tony Stone Images; 105(tc), Dr. Smith/University of Akron.

Chapter Five Page 107(cl), David M. Phillips/Visuals Unlimited; 109(cr), Stanley Flegler/Visuals Unlimited; (bl), David M. Phillips/Visuals Unlimited; 111(tr), Michael Abbey/Science Source/Photo Researchers; (cr), Dr. Birgit H. Satir; 112(cl), Runk/Schoenberger from Grant Heilman Photography; 113 (br), E.R. Degginger/Color-Pic; 115(tr), Clive Brunskill/ALLSPORT; 116(bl), CNRI/Science Photo Library/Photo Researchers; 117(tr), L. Willatt, East Anglian Regional Genetics Service/Science Photo Library/Photo Researchers; (br), Biophoto Associates/Photo Researchers; 118(all), Ed Reschke/Peter Arnold; 119(tr, cr), Biology Media/Photo Researchers; (br), R. Calentine/Visuals Unlimited; 120(cl), Stanley Flegler/Visuals Unlimited; 121(tr), Ed Reschke/Peter Arnold; 122(br), CNRI/Science Photo Library/Photo Researchers; 123(all), Biophoto Associates/Science Source/Photo Researchers; 124(tr), Lee D. Simon/Photo Researchers.

Unit Three Page 126(tc), Library of Congress/Corbis; (c), Kenneth Eward/Science Source/ Photo Researchers; (bl), John Reader/Science Photo Library/Photo Researchers; (br), NASA; 127(c), Marine Biological Laboratory Archives; (cl), John Reader/Science Photo Library/Photo Researchers; (cr), Ted Thai/Time Magazine; (bc), Biophoto Associates/Science Source/Photo Researchers.

Chapter Six Page 128(tl), Gerard Lacz/Peter Arnold; (tr), Dr. Paul A. Zahl/Photo Researchers; (bl), Runk/Schoenberger/ Grant Heilman Photography; (br), Frans Lanting/Minden Pictures; (br), Corbis; 131(tr), Runk/Schoenberger/Grant Heilman Photography; 135(tl), Archive Photos; 137(tr), Gerard Lacz/Animals Animals Earth Scenes; (br), ©2001 Photodisc; 138(cl), Phototake/CNRI/Phototake; (bc), Biophoto Associates/Photo Researchers; 142(tl), Dr. F. R. Turner, Biology Dept., Indiana University; 143(tr), CNRI/Phototake; 144(tr), Frans Lanting/Minden Pictures; 148(tc), Hank Morgan/Rainbow; 149(tc, c), Dr. F. R. Turner, Biology Dept., Indiana University; Dr. F. R. Turner, Biology Dept., Indiana University.

Chapter Seven Page 151(cr, whorl), Leonard Lessin/Peter Arnold; (cr, arch), Reprinted from "The Science of Fingerprints" courtesy of the FBI; (br, loop) Archive Photos; 153(tr), Science Photo Library/Photo Researchers; (cr), Science Source/Photo Researchers; (br), Archive Photos; 155(c), Dr. Gopal Murti/Science Photo Library/Photo Researchers; 156(tl), Phil Jude/Science Photo Library/Photo Researchers; 157(tc, tr), U.K. Laemmli/Universite de Geneve; (c), Biophoto Associates/Photo Researchers; (br), Dan McCoy/Rainbow; 158(br), Lawrence Migdale/Photo Researchers; 159(br), Sara Krulwich/New York Times Permissions; 163(cr, br), Jackie Lewin/Royal Free Hospital/Science Photo Library/Photo Researchers; 165(c), Dr. Chris R. Somerville/Science Photo Library/Photo Researchers; (cl), Remi Benali & Stephen Ferry/Liaison International; (cr), Science VU/Monsanto/Visuals Unlimited; (br), Science VU/Keith Wood/Visuals Unlimited; 166(tl), Biophoto Associates/Science Source/Photo Researchers; (cl), R.Kessel-G.Shih/Visuals Unlimited; (cr), see ON PAGE credit; (bl), SIU/Visuals Unlimited; 167(c), Biophoto Associates/Photo Researchers; (cl), Science Photo Library/Custom Medical Stock Photo; 169(tr, cr), Science VU/Monsanto/Visuals Unlimited; (c), Dr. Chris R. Somerville/Science Photo Library/Photo Researchers; (cl), Remi Benali & Stephen Ferry/Liaison International; 170(bc), Kenneth Eward/Science Source/Photo Researchers; 172(tc), Volker Steger/Peter Arnold.

Chapter Eight Page 176(c), Doug Wechsler/Animals Animals Earth Scenes; (cl), James Beveridge/Visuals Unlimited; (cr), Gail Shumway/FPG International; 178(cr), Ken Lucas/Visuals Unlimited; (cr), John Cancalosi/Tom Stack & Associates; 183(br), H.W. Robinson/Visuals Unlimited; 184(cl), Jonathan S. Blair/National Geographic Image Collection; (br), William E. Ferguson; (b), Christopher Ralling; 186(tl), Stephanie Hedgepath/Jimanie; (r), Jeanne White/Photo Researchers; (cr), Fritz Prenzel/Animals Animals Earth Scenes; (c), John Daniels/Bruce Coleman; (bc), Perry Phillips/Dennis & Catherine Quinn/Brigadier Bulldogs; (b), Yann Arthus-Bertrand/Corbis; (br), Robert Pearcy/Animals Animals Earth Scenes; 187(t), Library of Congress/Corbis; 189(cr), ©2001 Photodisc; 191(t, c), M.W.F. Tweedie/Photo Researchers; 193(br), Susan Van Etten/PhotoEdit; 197(bl), Breck P. Kent/Animals Animals Earth Scenes; (br), Pat & Tom Leeson/Photo Researchers; 198(tr), Doug Wilson/Photolibrary; (c), Thomas W. Martin/Photo Researchers.

Chapter Nine Page 200(t), Centre National de Prehistoire, Perigueux, France; (br), Jerome Chatin/Liaison International; 202(cr), Louis Psihoyos/Matrix International; 208(cr), SuperStock; (bl), Ken Lucas/Visuals Unlimited; 209(tr), Science VU/NVSM/Visuals Unlimited; 210(tl), M. Abbey/Photo Researchers; (bl), Andrew H. Knoll/Harvard University; 214(c), Art Wolfe/Tony Stone Images; (cl), Daniel J. Cox/Tony Stone Images; (cr), Renee Lynn/Photo Researchers; 216(tl), Daniel J. Cox/Liaison International; (bl), John Reader/Science Photo Library/Photo Researchers; 217(tr), John Reader/Science Photo Library/Photo Researchers; (cr), David L. Brill; (bl), John Gurche; 218(tl), Neanderthal Museum/Mettman, Germany; (bl), John Reader/Science Photo Library/Photo Researchers; 219(tr), David L. Brill; 221(tr), Daniel J. Cox/Tony Stone Images; 222(br), John Reader/Science Photo Library/Photo Researchers; 225(all), Bonnie Jacobs/Southern Methodist University.

Chapter Ten Page 226(t), Jeff Lepore/Photo Researchers; 228(bl), Ethnobotany of the Chacobo Indians, Beni, Bolivia, Advances in Economic Botany/The New York Botanical Gardens; 230(tl), Library of Congress/Corbis; 234(cl), Biophoto Associates/Photo Researchers; 235(tr), Sherrie Jones/Photo Researchers; 235(bl, bc), Dr. Tony Brian & David Parker/Science Photo Library/Photo Researchers; 236(tl), M. Abbey/Visuals Unlimited; 236(cl), Stanley Flegler/Visuals Unlimited; 236(bl), Chuck Davis/Tony Stone Images; 237(cl), Corbis; 237(bc), Art Wolfe/Tony Stone Images; 238(tl), Robert Maier/Animals Animals Earth Scenes; 238(cr), Sherman Thomson/Visuals Unlimited; 238(br), Richard Thom/Visuals Unlimited; 239(tr), SuperStock; 239(cl), G. Randall/FPG International; 239(cr), FPG International; 241(tc), Robert Maier/Animals Animals Earth Sciences; 245(cl), Peter Funch.

Unit Four Page 246(t), David L. Brown/Tom Stack & Associates; (c), AP/Photos; (bc), Dr. Jeremy Burgess/Science Photo Library/Photo Researchers; 247(tc), Larry Ulrich/DRK Photo; (tr), National Graphic Center; (cr), Runk/Schoenberger/Grant Heilman Photography; (bl), Greg Vaughn; (br), Debra Ferguson/AG Stock USA.

Chapter Eleven Page 250(t), Robert Schafer/Tony Stone Images; (bl, br), SuperStock; (bc), Tom & Michelle Grimm/Tony Stone Images; 251(br), Peter Guttman/Corbis; 252(tl), Roland Birke/Peter Arnold; (tc), Runk/Schoenberger/Grant Heilman Photography; (bl), John Gerlach/Animals Animals Earth Scenes; 254(cl), Doug Sokell/Tom Stack & Associates; 255(tr), Runk/Schoenberger/Grant Heilman Photography; (br), John Weinstein/The Field Museum, Chicago, IL/GEO85637c; 256(tl), Larry Ulrich/DRK Photo; (c), SuperStock; 257(tr), Ed Reschke/Peter Arnold; (br), Runk/Schoenberger/Grant Heilman Photography; 258(tr), Robert Barclay/Grant Heilman Photography; (bl), Heather Angel; (br), Phil Degginger/Color-Pic; 260(tl), Tom Bean; (cr), Jim Strawer/Grant Heilman Photography; (bl), Walter H. Hodge/Peter Arnold; (br), John D. Cunningham/Visuals Unlimited; 261(tr), Patti Murray/Animals Animals Earth Scenes; 262(cl), William E. Ferguson; (bl), Werner H. Muller; (bc), Grant Heilman/Grant Heilman Photography;

(br), SuperStock; 265(tr), Runk/Rannels/Grant Heilman Photography; (cr), Ed Reschke/Peter Arnold; (bl), Dwight R. Kuhn; (bc), Runk/Schoenberger/Grant Heilman Photography; (br), Nigel Cattlin; 266(tc), Harry Smith Collection; (cl), Larry Ulrich/DRK Photo; (bc), Albert Visage; (br), Dale E. Boyer; 267(tr), Stephen J. Krasemann/Photo Researchers; (cr), Tom Bean; 268(tl), E. R. Degginger/Color-Pic; (tr), Index Stock Imagery; (c), E. R. Degginger/Color-Pic; (cr), Gary A. Braasch/Braasch Photos; (cr), William E. Ferguson; 269(br), Ken W. Davis/Tom Stack & Associates; 270(tl), SuperStock; (bl), Kevin Adams; 271(tr), George Bernard/Science Photo Library/Photo Researchers; (bl), Galen Rowell/Corbis; (br), Patrick Jones/Corbis; 272(c), The Field Museum, Chicago, IL/GEO85637c; 273(tr), SuperStock; 276(bl), Sanford Scientific, Waterloo, NY; 277(tl), Mark Philbrick/Brigham Young University); (br), Phillip-Lorca DiCorcia.

Chapter Twelve Page 282(tl), W. Ormerod/Visuals Unlimited; (tc), George Bernard/Animals Animals Earth Scenes; (tr), ©2001 Photodisc; 283(tr), Paul Hein/Unicorn Stock Photos; (cl), George Bernard/Animals Animals Earth Scenes; (cr), Jerome Wexler/Photo Researchers; 284(br), Gregg Hadel/Tony Stone Images; (c), David Parker/Science Photo Library/Photo Researchers; 286(tl), Dr. Jeremy Burgess/Science Photo Library/Photo Researchers; 287(br), Cathlyn Melloan/Tony Stone Images; 288(all), R.F. Evert/University of Wisconsin; 289(c), Dick Keen/Unicorn Stock Photos; (bc), E. Webber/Visuals Unlimited; 290(tr), W. Cody/WestLight; (bl, bc, br), Rich Iwasaki/Tony Stone Images; 291(all), Bill Beatty/Visuals Unlimited; 292(bl), R.F. Evert/University of Wisconsin; 293(br), Sylvan Wittwer/Visuals Unlimited; 295(c), R.F. Evert/University of Wisconsin; 297(bl), W. Cody/WestLight; 298(all), David Liittschwager & Susan Middleton/Discover Magazine; 299(cr), Cary S. Wolinsky.

Unit Five Page 300(tr), Runk/Schoenberger/Grant Heilman Photography; (c), M. Gunther Bios/Peter Arnold; (cr), Grant Heilman/Grant Heilman Photography; 301(tl), Johnny Johnson/DRK Photo; (tr), Gail Shumway/FPG International; (cl), M. Corsetti/FPG International; (cr), John James Audubon/Collection of the New-York Historical Society; (bl), Art Wolfe/Tony Stone Images; (br), Susan Erstgaard.

Chapter Thirteen Page 302(b), Doug Wechsler/VIREO; (bc), G. Bilyk/ANSP/VIREO; 303(t), James L. Amos/Peter Arnold; 304(br), David Fleetham/FPG International; 306(tl), David M. Phillips/Photo Researchers; (cr), Fred Hossler/Visuals Unlimited; 307(tr), Stephen Dalton/Photo Researchers; (c), Gerard Lacz/Peter Arnold; (cr), Manoj Shah/Tony Stone Images; (crb), Keren Su/Tony Stone Images; (b), Stephen Dalton/Photo Researchers; 308(bl), Tim Davis/Tony Stone Images; 309(tr), J.H. Robinson/Photo Researchers; (bl), W. Peckover/Academy of Natural Sciences/VIREO; (cr), Leroy Simon/Visuals Unlimited; 310(tl), Breck Kent/Animals Animals Earth Scenes; (bl), A.J. Copley/Visuals Unlimited; 311(tr), George D. Lepp/Tony Stone Images; (bl), Michio Hoshino/Minden Pictures; 313(tr), FPG International; (br), Breck P. Kent; 314(cl), Fernandez & Peck/Adventure Photo & Film; (bl), Peter Weimann/Animals Animals Earth Scenes; 315(tr), Lee F. Snyder/Photo Researchers; (br), Johnny Johnson/Animals Animals Earth Scenes; 316(tl), Ron Kimball/Ron Kimball Photography; 317(tr), Matthews & Purdy/Planet Earth Pictures; (cr), Richard R. Hansen/Photo Researchers; 318(tr), Stephen Dalton/Photo Researchers; (c), Keren Su/Tony Stone Images; 319(cr), Lee F. Snyder/Photo Researchers; 320(bc), Leroy Simon/Visuals Unlimited; 322(tr), Wayne Lankinen/DRK Photo; 323(tc), Wayne Lawler/Auscape.

Chapter Fourteen Page 324(cl), Tim Branning/Topic Productions; 325(tl), Barbara Gerlach/Visuals Unlimited; (tc), Norbert Wu/Peter Arnold; (tr, cr), Larry West/FPG International; (cl), Tom Corner/Unicorn Stock; (bl), James H. Carmichael, Jr./Image Bank; (bc), E. R. Degginger/Color-Pic; (br), G.K. & Vikki Hart/Image Bank; 326(tr), SuperStock; (c), J. Carmichael/Image Bank; (c), Carl Roessler/FPG International; (bl), David B. Fleetham/Tom Stack & Associates; 328(tl), Jeffrey L. Rotman/Peter Arnold; (bl), Keith Philpott/Image Bank; (br), E.R. Degginger/Color-Pic; 329(br), Nigel Cattlin/Holt Studios/Photo Researchers; 330(tl), Lee Foster/FPG International; (cl), Biophoto Associates/Science Source/Photo Researchers; (bl), Randy Morse/Tom Stack & Associates; 331(cr), Charles Seaborn/Woodfin Camp & Associates; 332(tl), T.E. Adams/Visuals Unlimited; (br), CNRI/Science Photo Library/Photo Researchers; 333(tr), R. Calentine/Visuals Unlimited; (br), A.M. Siegelman/Visuals Unlimited; 334(tr), Holt Studios International/Photo Researchers; (c), E. R. Degginger/Color-Pic; (cl), SuperStock; (cr), Stephen Frink/Corbis; 335(br), David M. Phillips/Visuals Unlimited; 336(cl), David Fleetham/FPG International; (bl), North Wind Picture Archives; 337(tr), Milton Rand/Tom Stack & Associates; (br), Daniel Schaefer/HRW Photo; 338(tl), Mary Beth Angelo/Photo Researchers; (cr), St Bartholomew's Hospital/Science Photo Library/Photo Researchers; 339(tr), SuperStock; (cl), Will Crocker/Image Bank; (bl), Sergio Purcell/FOCA/HRW Photo; 340(tl), CNRI/Science Photo Library/Photo Researchers; (cl), A. Kerstitich/Visuals Unlimited; (bl), E.R. Degginger/Color-Pic; 341(bl), David Scharf/Peter Arnold; 342(tl), R. Calentine/Visuals Unlimited; (c), Stephen Dalton/NHPA; (tc), Dwight Kuhn; (tr), SuperStock; (bc), Oliver Meckes/Photo Researchers; (bl), Gail Shumway/FPG International; (br), Uniphoto; 343(cr), Joe McDonald; 345(tr), Robert Dunne/Photo Researchers; (c), Chesher/Photo Researchers; (cl), Darryl Torckler/Tony Stone Images; (blt), Paul McCormick/The Image Bank; (bl), Cabisco/Visuals Unlimited; 347(tr), Andrew J. Martinez/Photo Researchers; (cr), Marty Snyderman/Visuals Unlimited; (bc), Daniel W. Gotshall/Visuals Unlimited; 349(tc), SuperStock; (cl), Uniphoto; 350(br), Keith Philpott/Image Bank; 352(cl), Diane R. Nelson/Visuals Unlimited; 353(bl), Dr. Mark Norman.

Chapter Fifteen Page 354(tr), Visuals Unlimited; (bc, br), K.Hissmann & H. Fricke/Max-Planck-Institute; 355(tr), Dale Jackson/Visuals Unlimited; 356(cr), Louis Psihoyos/Matrix International; (bl), Randy Morse/Tom Stack & Associates; (bc), Norbert Wu/Peter Arnold; 357(br), Grant Heilman Photography; 358(tl), Stephen J. Krasemann/Photo Researchers; (cr), Uniphoto; 359(c), Doug Perrine/DRK Photo; (c), Steven David Miller/Animals Animals Earth Scenes; (tl), Jane Burton/Bruce Coleman; (tr), Brian Parker/Tom Stack & Associates; (cr), Ken Lucas/Visuals Unlimited; 361(tr), Hans Reinhard/Bruce Coleman; (c), Index Stock Imagery; (br), Martin Barraud/Tony Stone Images; 362(tl), Science VU/Visuals Unlimited; (c), Navaswan/FPG International; 363(tr), Bruce Coleman; (cl), Steinhart Aquarium/Photo Researchers; 364(cl), Michael Fogden/DRK Photos; (bl), Nathan W. Cohen/Visuals Unlimited; 365(tr), David M. Dennis/Tom Stack & Associates; (br), C. K. Lorenz/Photo Researchers; 366(bl), Michael & Patricia Fogden; 367(tr), Bruce Coleman; (c), Stephen Dalton/NHPA; (cr), Zig Leszczynski/Animals Animals Earth Scenes; 368(tc), Leonard Lee Rue, III/Photo Researchers; (cl), FPG International; 368(tr), Breck P. Kent; 369(tr), Rob & Ann Simpson/Visuals Unlimited; 370(tl, cl), Gail Shumway/FPG International; (bc), Stanley Breeden/DRK Photo; (br), Joe McDonald/Visuals Unlimited; 372(tl), Leonard Lee Rue, III/Bruce Coleman; (cr), Mike Severns/Tony Stone Images; (bl), Kevin Schafer/Peter Arnold; (br), Wayne Lynch/DRK Photo; 373(tc), Wolfgang Kaehler Photography; (cr), Michael Fogden/DRK Photo; 374(cl), Uniphoto; 375(tr), Brian Parker/Tom Stack & Associates; (cr), Michael Fogden/DRK Photos; 376(tr), Steven David Miller/Animals Animals Earth Scenes.

Chapter Sixteen Page 380(bl), James L. Amos/Photo Researchers; (bc), O. Louis Mazzatenta/National Geographic Image Collection; 382(tr), Stan Osolinski/FPG International; (c), James Brandenberg/Minden Pictures; (cl), Anthony Mercieca/Photo Researchers; (cr), Gail Shumway/FPG International; (bl-inset), Runk/Schoenberger/Grant Heilman Photography; (bl), Douglas Faulkner/Photo Researchers; 386(br), Ben Osborne/Tony Stone Images; 387(tr), George H. Harrison/Grant Heilman Photography; (cr), Frans Lanting/Minden Pictures; (c), D. Cavagnaro/DRK Photo; (bc), Joe McDonald/DRK Photo; 388(tl), Thomas McAvoy/Time Life Syndication; (bl), Hal H. Harrison/Grant Heilman Photography; 389(cl), Kevin Schafer/Tony Stone Images; (cr), APL/J. Carnemolla/Westlight; (br), Gavriel Jecan/Tony Stone Images; 390(tr), S. Nielsen/DRK Photo; (cl), Tui De Roy/Minden Pictures; (c), Wayne Lankinen/Bruce Coleman; (bl), Greg Vaughn/Tony Stone Images; (b), Fritz Polking/Bruce Coleman; 391(tr), Frans Lanting/Minden Pictures; (cl), Stephen J. Krasemann/DRK Photo; (c), S. Maslowski/Visuals Unlimited; 392(tr), ©2001 Photodisc; (c), Nigel Dennis/Photo Researchers; (cl), Gerard Lacz/Animals Animals Earth Scenes; (c), Tim Davis/Photo Researchers; 393(cl), Hans Reinhard/Bruce Coleman; 394(tl), David E. Myers/Tony Stone Images; (c), Tom Tietz/Tony Stone Images; (bc), Konrad Wothe/Westlight; 395(cr), Kathy Bushue/Tony Stone Images; 396(cl), Erwin & Peggy Bauer/Bruce Coleman; (bl), Dave Watts/Tom Stack & Associates; 397(tr), Jean-Paul Ferrero/AUSCAPE International; (c), Hans Reinhard/Bruce Coleman; (bl), Art Wolfe/Tony Stone Images; 398(c), John D. Cunningham/Visuals Unlimited; (bl), Wayne Lynch/DRK Photo; 399(tr), Gail Shumway/FPG International; (cl), D.R. Kuhn/Bruce Coleman; (b), Frans Lanting/Minden Pictures; (br), Lynda Richardson/Peter Arnold; 400(tr), David Cavagnaro/Peter Arnold; (c), John Cancalosi; (cr), S.C. Bisserot/Bruce Coleman; 401(tr), Gail Shumway/FPG International; (cr), Uniphoto; (cr), Joe McDonald/Bruce Coleman; (bl), Arthur C. Smith III/Grant Heilman Photography; 402(tr), Scott Daniel Peterson/Liaison International; (cl), S. R. Maglione/Photo Researchers; (cr), Roberto Arakaki/International Stock; (br), Gail Shumway/FPG International; 403(c), Art Wolfe/Tony Stone Images; 404(tr), Flip Nicklin/Minden Pictures; (cl), Francois Gohier; (cr), Tom & Therisa Stack/Tom Stack & Associates; 405(tr), Inga Spence/Tom Stack & Associates; (cl), J. & P. Wegner/Animals Animals Earth Scenes; (br), World Perspective/Tony Stone Images; 406(cr), Frans Lanting/Minden Pictures; 407(cl), Gerard Lacz/Animals Animals Earth Scenes; 408(br), S.C. Bisserot/Bruce Coleman; 410(tc), Tom & Pat Leeson/Photo Researchers; (br), Will & Deni McIntyre/Tony Stone Images; 411(bl), Raymond A. Mendez/Animals Animals Earth Scenes.

Unit Six Page 412(tr), Carr Clifton/Minden Pictures; (cr), SuperStock; (bc), Tom Blakefield/Corbis; 413(tl), SuperStock; (tr), St. Meyers/Okapia/Photo Researchers; (c), Keystone View Company/FPG International; (cl), Erich Hartmann/Magnum Photos; (bc), David Young/Tony Stone Images; (br), Tom Smart/Liaison International.

Chapter Seventeen Page 414(t), David Sieren/Visuals Unlimited; (cl), J. H. Robinson/Photo Researchers; (br), Runk/Schoenberger/Grant Heilman Photography; 420(cr), Grant Heilman/Grant Heilman Photography; (br), Tom Brakefield/Bruce Coleman; 422(cl), Kathy Bushue/Tony Stone Images; 423(bl-inset), Manfred Kage/Peter Arnold; (bl), Stuart Westmorland/Tony Stone Images; 426(tl), Jeff Hunter/Image Bank; (br), Zig Leszczynski/ Animals Animals Earth Scenes; 427(tr), Johnny Johnson/DRK Photo; (br), H. Richard Johnston/Tony Stone Images; 430(tr), Phyllis Ked/Unicorn Stock; (br), Dwight R. Kuhn; 431(tr), Hardie Truesdale/International Stock; (cr), Don & Pat Valenti/DRK Photo; 433(tl), Jeff Hunter/Image Bank; 436(tl), Dr. Verena Tunnicliffe; 437(all), Lincoln P. Brower.

Chapter Eighteen Page 438(b), Richard Aldorasi; 440(cr), Grant Heilman/Grant Heilman Photography; (bl), Arthur Tilley/Tony Stone Images; 441(tl), National Wildlife Magazine; (br), Ken Griffiths/Tony Stone Images; 442(tl), 1999 NASA GSFC 916; (cr), Roy Morsch/Stock Market; 443(cl), Jacques Jangoux/Tony Stone Images; 444(tl, cl), Runk/Schoenberger/Grant Heilman Photography; (tc), John Eastcott/Woodfin Camp & Associates; 445(tr), Rex Ziak/Tony Stone Images; (br), Martin Rogers/Uniphoto; 446(cl), Fred Bavendam/Peter Arnold; 448(tl), Argonne National Laboratory; (bl), Emile Luider/Rapho/Liaison International; 449(tl), Jeff Greenberg/PhotoEdit; (cr), Kay Park-Rec Corp.; (br), J. Conteras Chacel/International Stock; 450(cl), Martin Bond/Science Photo Library/Photo Researchers; 451(tr), Toyohiro Yamada/FPG International; (cl), Uniphoto; (br), K.W. Fink/Bruce Coleman; 452(tl), Stephen J. Krasemann/DRK Photo; 453(tr), Will & Deni McIntyre/Tony Stone Images; (cr), Stephen J. Krasemann/DRK Photo; 454(cl), Arthur Tilley/Tony Stone Images; (cr), K.W. Fink/Bruce Coleman; 456(bc), Runk/Schoenberger/Grant Heilman Photography; 458(all), Karen M. Allen; 459(bc), Art Wolfe.

Unit Seven Page 460(tc), Geoffrey Clifford/Woodfin Camp & Associates; (cr), CNRI/Science Photo Library/Photo Researchers; (bl), Brown Brothers; (br), SuperStock; 461(t), J & L Weber/Peter Arnold; (tr), Liaison International; (bl), AP/Photos; (br), Enrico Ferorelli.

Chapter Nineteen Page 462(tr), New York Times/Corbis; 462(b), C. J. Ashford/Denis Cochrane Collection/e.t. archive; 463(tr), Simon Fraser/Science Photo Library/Photo Researchers; 464, 465(b), David Madison/Tony Stone Images; 470(tl), Peter Dazeley/Tony Stone Images; (c, cl, cr), Sergio Purcell/FOCA/HRW Photo; 472(cl), G.W. Willis/Tony Stone Images; (b), Bob Torrez/Tony Stone Images; (bl), E. R. Degginger/Color-Pic; (br), Manfred Kage/Peter Arnold; 474(bl), Chris Hamilton; 475(tr), Shelby Thorner/David Madison; (cr), Wally McNamee/Corbis; 478(cl), Robert Becker/Custom Medical Stock Photo; 479(cr), Dr. P Marazzi/Science Photo Library/Photo Researchers; 481(cl), Peter Dazeley/Tony Stone Images; 484(tr), Dan McCoy/Rainbow; 485(tr), Huntsville Times; (cl), Liaison International.

Chapter Twenty Page 486(t), Enrico Ferorelli; 488(bl), Dr. Dennis Kunkel/Phototake; 489(tr), Don Fawcett/Photo Researchers; (cr), Custom Medical Stock Photo; 491(tl, tr), Meckes/Nicole Ottawa/Photo Researchers; (br), David Phillips/Science Source/Photo Researchers; 493(tr), Custom Medical Stock Photo; (br), James Wilson/Woodfin Camp & Associates; 495(tr), Ken Wagner/Phototake; (br), Russell Dian/HRW Photo; 501(cr, br), Matt Meadows/Peter Arnold; 502(tr), Dr. Dennis Kunkel/Phototake; 503(cr), Don Fawcett/Photo Researchers; 504(tr), Custom Medical Stock; 505(bl), Dr. Dennis Kunkel/Phototake; 506(tr), Index Stock Imagery; 507(tr), Russell Dian/HRW Photo; (bl), Jim Gipe.

Chapter Twenty-One Page 508(tr), Warren Anatomical Museum/Harvard Medical School; (b), Vermont Historical Society Library; 519(br), Bruno Joachim/Liaison International; 521(cr), Louis Psihoyos/Matrix International; 525(tr), Will & Deni McIntyre/Photo Researchers; (br), Ted Spiegel/National Geographic Image Collection; 531(all), Journal of Nuclear Medicine.

Chapter Twenty-Two Page 532(cl), SuperStock; 534(cl), Cabisco/Visuals Unlimited; (br), Innerspace Visions; 536(tl), Michael Fogden/Animals Animals Earth Scenes; (cl), Guy Mannering/Bruce Coleman; (br), Clem Haagner/Photo Researchers; 537(tr), CSIRO Wildlife & Ecology; (cr), E. R. Degginger/Bruce Coleman; 540(tl), Chip Henderson/Tony Stone Images; 541(all), James King-Holmes/Science Photo Library/Photo Researchers; 542(tr), Lennart Nilsson/Albert Bonniers Forlag AB, A CHILD IS BORN; 543(bc), Petit Format/Nestle/Science Source/Photo Researchers; 544(tl), Lennart Nilsson/Albert Bonniers Forlag AB, BEING BORN; 545(tr), Lennart Nilsson/Albert Bonniers Forlag AB, A CHILD IS BORN; (cr), Keith/Custom Medical Stock Photo; (br), Sergio Purcell/FOCA/HRW Photo; 547(cl), NASA/Liaison International; (br), ©2001 Photodisc; 548(cr), Guy Mannering/Bruce Coleman; 550(br), Lennart Nilsson/Albert Bonniers Forlag AB, BEING BORN; 553(cr), Tom McCarthy/Rainbow; (inset), Vince Viverito, Jr./Richard Wolf Medical Instruments Corp., Vernon Hills, IL.

LabBook "LabBook Header": "L", Corbis Images, "a", Letraset Phototone, "b" and "B", HRW, "o" and "k",images copyright ©2001 PhotoDisc/HRW Page 544(tc), Scott Van Osdol/HRW Photo; 557(cl), Michelle Bridwell./HRW Photo; (br), ©2001 PhotoDisc; 559(tr), Jane Birchum/HRW Photo; 572(tl), Runk/Schoenberger/Grant Heilman Photography; (tc), Runk/Schoenberger/Grant Heilman Photography; (tr), Michael Abbey/Photo Researchers; (br), Runk/Schoenberger/Grant Heilman Photography; 598(tr), Runk/Schoenberger/Grant Heilman Photography; (cr), R. Calentine/Visuals Unlimited; (bc), Breck P. Kent; (br), Stephen J. Krasemann/Photo Researchers; 612(cl), Navaswan/FPG International; 614(br), Rod Planck/Photo Researchers; 621(br), David Hoffman/Tony Stone Images; 623(tr), Tom Bean/DRK Photo; (br), Darrell Gulin/DRK Photo; 634(br), E.R. Degginger/Color-Pic.

Appendix Page 660(c), CENCO

Sam Dudgeon/HRW Photos v(bl), vii(bl), viii(bl), xvi(tl, br), xvii(bl), xviii(br), xix(br), 5(bl), 11(b), 12(bl), 17(tr), 26(c), 35(tr), 42(bc), 43(cr), 44(tl), 49(tr, cr, br), 53(bl), 56(bc), 64(bl), 72(bl), 73(bl), 79(all), 81(bl), 106(t), 108(bc), 115(cr), 134(tl), 150(cl), 151(tr), 156(bl), 158(tl), 159(tr), 249(tr), 279(all), 284(bl), 285(tr), 415(b), 446(tr), 453(cr, br), 466(tl), 468(bc), 473(all), 474(cr), 479(br), 487(all), 509(br), 514(tl), 516(tl), 517(cl), 519(tr, cr), 523(br), 524(all), 526(cl), 533(br), 544(tl, tr, br), 546(bl), 549(cr), 556(bl), 557(b), 558(tr, bl), 559(tl), 560(br), 561(br), 563(all), 564(c), 567(cl), 569(tr), 571(tr), 572(cr), 574(cr), 575(br), 576(all), 577(br), 578(all), 579(br), 586(cr), 589(b), 590(b), 591(bkgd), 600(c), 602(br), 605(all), 608(cr), 610(all), 626(br), 628(b), 630(all), 631(br), 632(all), 633(b), 640(all), 643(b) Systems of the Body background photos by Sam Dudgeon/HRW Photos: 82, 83, 466, 467, 473, 488, 497, 498, 510, 520, 523, 529, 500

Peter Van Steen/HRW Photos v(tr, bl), xx(cr), 4(t), 5(br), 6(bl), 9(tr), 11(tr), 25(bl, br), 28(c), 129(tr, br), 175(br), 303(br), 304(cl), 381(all), 393(br), 439(all), 447(bl, bc, br), 450(tl), 455(tr), 463(bc), 471(all), 476(all), 478(bl, br), 546(t), 559(b), 612(br), 615(b), 619(bl), 622(b), 625(br), 643(t)

John Langford/HRW Photos xi(bl), 56(bl), 60(bl), 91(br), 113(bl), 120(cr), 557(tr), 604(all)

Stephanie Morris/HRW Photos 61(br), 70(bl), 558(bl)

Victoria Smith/HRW Photos 150(c), 249(br), 453(cr), 583(b), 558(cl)

Annotated Teacher's Edition Credits

TE Frontmatter: Page T7(cl), 2001 Photodisc; T8(cl), Lawrence Migdale/Photo Researchers; T9(bkgd), Digital Stock; T10(tl), 2001 Photodisc; T11(tr), 2001 Photodisc; T12(tl), Sam Dudgeon/HRW Photo; T13(tr), 2001 Photodisc; T14(bl), Randy Morse/Tom Stack & Associates; T15(tr), 2001 Photodisc; T16(tl), Dale Jackson/Visuals Unlimited; T17(br), 2001 Photodisc; T18(tr), Frans Lanting/Minden Pictures; T18(bl), Gerald & Buff Corsi/Visuals Unlimited; T19(tr), 2001 Photodisc; T20(cl), 2001 Photodisc; T21(tr), Sam Dudgeon/HRW Photo; T22(cl), Gay Bumgarner/Tony Stone Images; T23(br), 2001 Photodisc.

Master Materials List: Unless otherwise noted all images: Image Copyright ©2001 PhotoDisc/HRW; Page xxiv (CD), HRW Photo; xxv (marshmallows), Sam Dudgeon/HRW Photo; xxvi (graph paper), Sam Dudgeon/HRW Photo; xxvii (salt, lima bean seeds), Digital Stock Corp.; xxvii (toothpicks), Sam Dudgeon/HRW Photo; xxvii (sugar), Digital Stock Corp.

Lab Approval Portraits: All photos courtesy of the reviewers

TE Background Illustrations: Page 33F(bl), Morgan Cain & Associates; 33F(br), Morgan Cain & Associates; 51E(cr), Dan Stuckenschneider and Preface; 51F(tr), Keith Kasnot; 51F(br), Dan Stuckenschneider; 77E(tl), Morgan Cain & Associates; 77E(bl), Christy Krames; 77F(tl), Morgan Cain & Associates; 77F(tr), Morgan Cain & Associates; 77F(bl), Morgan Cain & Associates; 77F(br), Blake Thornton/Rita Marie; 105E(tr), Morgan Cain and Associates; 105E(bl), Morgan Cain and Associates; 105F(tr), Alexander and Turner; 127E(tl), Alexander and Turner; 127E(cr), Mike Wepplo/Das Group; 127E(bl), John White/The Neis Group/The Neis Group; 127F(bl), Rob Schuster/Hankins and Tegenborg; 127F(bl), Blake Thornton/Rita Marie; 149E(tl), Alexander and Turner; 149F(cl), Rob Schuster/Hankins and Tegenborg; 173E(tl), Christy Krames; 173E(cl), Mike Wepplo/Das Group; 173E(bl), Rob Wood/Wood, Ronsaville and Harlin; 173E(br), John White/The Neis Group; 173F(tl), John White/The Neis Group; 173F(br), Tony Morse/Ivy Glick; 199E(tl), Mike Wepplo/Das Group; 199E(tr), Barbara Hoopes-Ambler; 199E(br), Barbara Hoopes-Ambler; 199F(tl), Barbara Hoopes-Ambler; 199F(tr), Christy Krames; 225E(tl), Michael Woods; 225E(tr), Mazer; 225E(tr), Cy Baker/WAA; 225E(cl), Graham Allen; 225E(bl), John White/The Neis Group; 225E(br), John White/The Neis Group; 247F(tl), Sarah Woods; 277E(br), Pond and Giles; 301E(c,cr,br) Barbara Hoopes-Ambler; 301F(tl), Tony Morse/Ivy Glick; 323E(bl), John White/The Neis Group; 413E(tl), GeoSystems Global Corporation; 413E(cr), Will Nelson/Sweet Representatives; 461E(cl), Morgan Cain & Associates; 461F(br), Morgan Cain & Associates; 485F(tr), Christy Krames; 485F(cl), Kip Carter; 507E(tl), Brian Evans; 507F(tl), Keith Kasnot; 507F(bl), Christy Krames; 531E(cr), Rob Schuster/Hankins and Tegenborg.

TE Background Photography: Page 3E(tl), Minnesota Pollution Control Agency; 3E(tr), Chip Simons Photography; 3E(br), Dale Miquelle/National Geographic; 3F(tl), John Mitchell/Photo Researchers; 3F(tr), Dr. Jeremy Burgess/Science Photo Library/Photo Researchers; 3F(bl), John Reader/Photo Researchers; 3F(br), Howard Sochurek/The Stock Market; 33E(tl), VU/Science VU/Visuals Unlimited; 33E(bl), Wolfgang Kaehler/Liaison International; 33E(br), Fred Rohde/Visuals Unlimited; 51E(tl), Leonard Lessin/Photo Researchers; 51E(bl), John Langford/HRW Photo; 51F(bl), Renee Lynn/Davis/Lynn Images; 51F(bl), PhotoDisc; 77E(tr), Photo Researchers; 77E(br), Kevin Collins/Visuals Unlimited ; 105E(tl), Michael Abbey/Science Source/Photo Researchers; 105E(br), Biophoto Assoicates/Science Source/Photo Researchers; 105F(cl), L. Willatt, East Anglian Regional Genetics Service/Science Photo Library/Photo Researchers; 105F(br), Ed Reschke/Peter Arnold; 127E(tl), Corbis; 127F(cr), Runk/Schoenberger/Grant Heilman Photography; 149E(cr), Jackie Lewin/Royal Free Hospital/Science Photo Library/Photo Researchers; 149E(bl), Sam Dudgeon/HRW Photo; 149F(tr), Remi Benali & Stephen Ferry/Gamma Liaison; 149F(br), Cellmark Diagnostics; 199E(bl), Louis Psihoyos/Matrix International; 199F(bl), Neanderthal Museum/Mettman, Germany; 199F(br), David L. Brill ; 225F(tl), Biophoto Associates/Photo Researchers; 225F(tr), Dr. Tony Brian & David Parker/Science Photo Library/Photo Researchers; 225F(bl), Sherrie Jones/Photo Researchers; 247E(tl), Peter Guttman/Corbis; 247E(tr), Larry Ulrich/DRK Photo; 247E(bl), John Gerlach/Animals Animals Earth Scenes; 247F(cr), Phil Degginger/Color-Pic; 247F(bl), Robert Barclay/Grant Heilman Photography; 277E(tl), Paul Hein/Unicorn Stock Photos; 277E(cl), George Bernard/Animals Animals Earth Scenes; 277F(tl), Rich Iwasaki/Tony Stone Images; 277F(bl), E. Webber/Visuals Unlimited; 301E(tl), Manoj Shah/Tony Stone Images; 301E(cr), Peter Weimann/Animals Animals Earth Scenes; 301F(bl), Gerard Lacz/Peter Arnold; 302F(cr), Sylvan Wittwer/Visuals Unlimited; 323E(tr), E. R. Degginger/Color-Pic; 323E(cl), Keith Philpott/Image Bank; 323E(br), Milton Rand/Tom Stack & Associates; 323F(tl), SuperStock; 323F(cr), Chesher/Photo Researchers; 323F(bl), Sergio Purcell/FOCA/HRW Photo; 353E(tl), Norbert Wu/Peter Arnold; 353E(cr), Index Stock Photography; 353E(bl), Randy Morse/Tom Stack & Associates; 353F(tl), FPG International; 353F(cr), Wayne Lynch/DRK Photo; 353F(bl), Stephen Dalton/NHPA; 379E(tl), Ben Osborne/Tony Stone Images; 379E(cr), Fritz Polking/Bruce Coleman; 379E(bl), Anthony Mercieca Photo/Photo Researchers; 379F(tr), D.R. Kuhn/Bruce Coleman; 379F(cl), Lynda Richardson/Peter Arnold; 379F(br), Wayne Lynch/DRK Photo; 413F(tl), Stuart Westmorland/Tony Stone Images; 413F(cr), Don & Pat Valenti/DRK Photo; 413F(bl), Jeff Hunter/Image Bank; 437E(tl), National Wildlife Magazine; 437E(cr), Arthur Tilley/Tony Stone Images; 437E(bl), Grant Heilman/Grant Heilman Photography; 437F(tr), J. Conteras Chacel/International Stock; 437F(cl), K.W. Fink/Bruce Coleman; 437F(br), Peter Van Steen/HRW Photo; 461E(tr), Sam Dudgeon/HRW Photo; 461E(br), 461F(tl), Bob Torrez/Tony Stone Images; 461F(tr), Robert Becker/Custom Medical Stock Photo; 485E(cl), Ken Wagner/Phototake; 485E(cr), Dr. Dennis Kunkel/Phototake; 485F(br), Matt Meadows/Peter Arnold; 507E(cr), Sam Dudgeon/HRW Photo; 507E(bl), Sam Dudgeon/HRW Photo; 507F(cr), Will & Deni McIntyre/Photo Researchers; 531E(tl), Innerspace Visions; 531E(bl), Guy Mannering/Bruce Coleman Inc.; 531F(tr), Lennart Nilsson/Albert Bonniers Forlag AB, A CHILD IS BORN; 531F(bl), Chip Henderson/Tony Stone Images; 531F(br), Peter Van Steen/HRW Photo.

Bert J. Sherwood
Science Teacher
Socorrow Middle School
El Paso, Texas

Patricia McFarlane Soto
Science Teacher and Dept. Chair
G. W. Carver Middle School
Miami, Florida

David Sparks
Science Teacher
Redwater Junior High School
Redwater, Texas

Elizabeth Truax
Science Teacher
Lewiston-Porter Central School
Lewiston, New York

Ivora Washington
Science Teacher
Hyattsville Middle School
Hyattsville, Maryland

Elsie N. Waynes
Science Teacher and Dept. Chair
R. H. Terrell Junior High School
Washington, D.C.

Nancy Wesorick
Science Teacher
Sunset Middle School
Longmont, Colorado

Christopher Wood
Science Teacher
Western Rockingham Middle School
Madison, North Carolina

Alexis S. Wright
Middle School Science Coordinator
Rye Country Day School
Rye, New York

John Zambo
Science Teacher
E. Ustach Middle School
Modesto, California

Gordon Zibelman
Science Teacher
Drexel Hill Middle School
Drexell Hill, Pennsylvania

Answers to Concept Mapping Questions

The following pages contain sample answers to all of the concept mapping questions that appear in the Chapter Reviews. Because there is more than one way to do a concept map, your students' answers may vary.

CHAPTER 1 Life and Living Things

15.

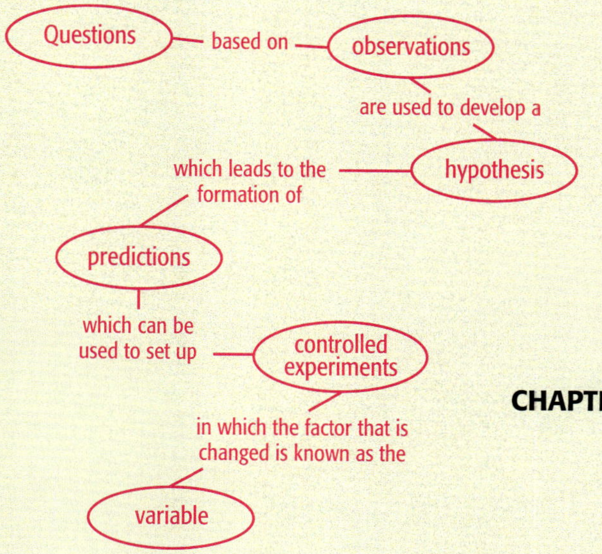

Questions — based on → observations

observations → are used to develop a → hypothesis

which leads to the formation of → predictions

predictions → which can be used to set up → controlled experiments

in which the factor that is changed is known as the → variable

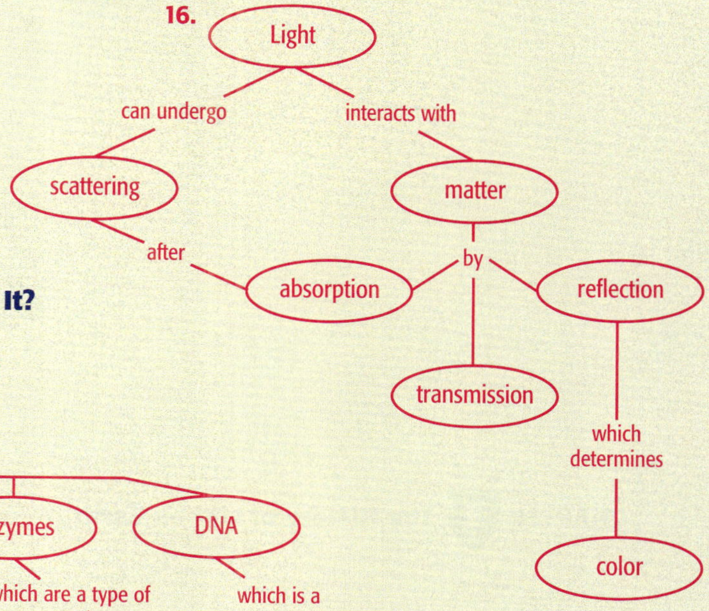

CHAPTER 2 It's Alive!! Or, Is It?

16.

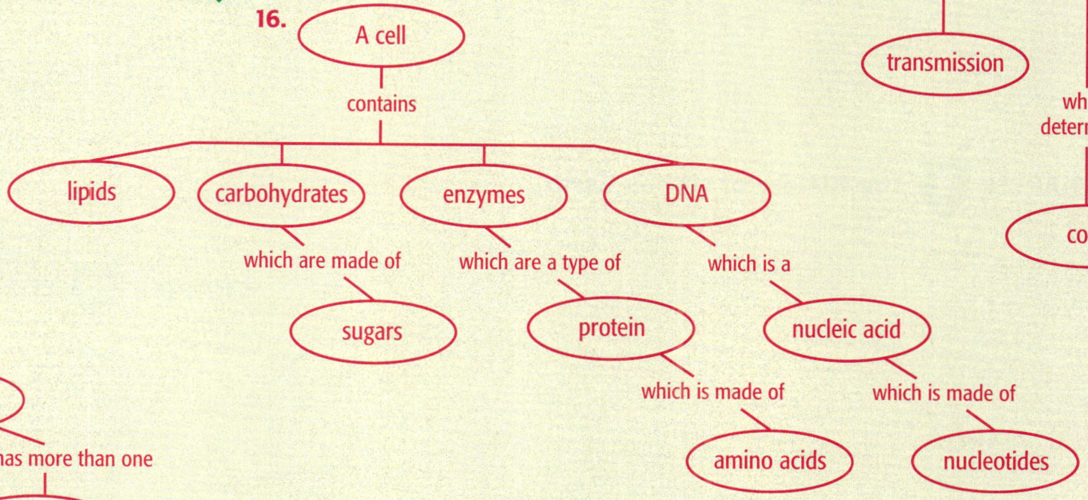

A cell → contains → lipids, carbohydrates, enzymes, DNA

carbohydrates → which are made of → sugars

enzymes → which are a type of → protein

DNA → which is a → nucleic acid

protein → which is made of → amino acids

nucleic acid → which is made of → nucleotides

CHAPTER 3 Light and Living Things

16.

Light → can undergo → scattering

Light → interacts with → matter

scattering → after → absorption

matter → by → absorption, reflection, transmission

reflection → which determines → color

CHAPTER 4 Cells: The Basic Unit of Life

15.

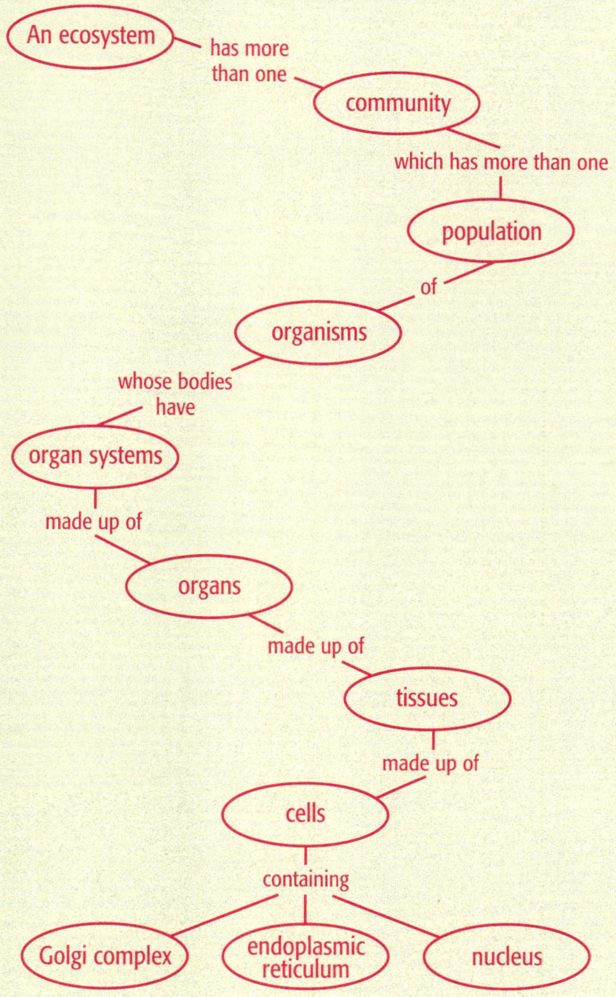

An ecosystem → has more than one → community

community → which has more than one → population

population → of → organisms

organisms → whose bodies have → organ systems

organ systems → made up of → organs

organs → made up of → tissues

tissues → made up of → cells

cells → containing → Golgi complex, endoplasmic reticulum, nucleus

CHAPTER 5 The Cell in Action

15.

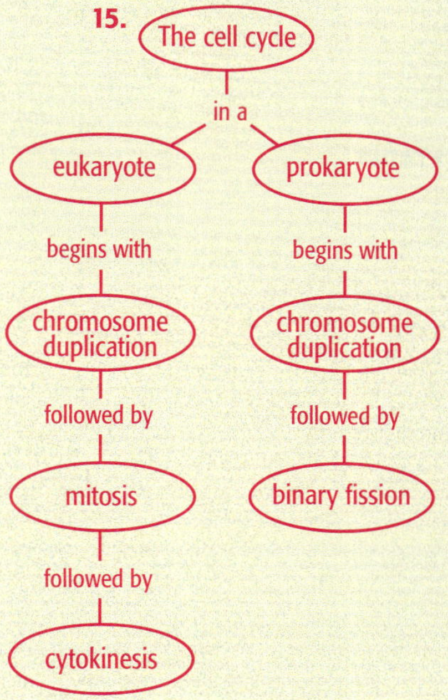

The cell cycle → in a → eukaryote, prokaryote

eukaryote → begins with → chromosome duplication → followed by → mitosis → followed by → cytokinesis

prokaryote → begins with → chromosome duplication → followed by → binary fission

CHAPTER 6 Heredity

16.

Cell division → occurs following → meiosis, mitosis

meiosis → which produces → sex cells

sex cells → which can be either → eggs, sperm

eggs → which always contain an → X chromosome

sperm → which contain either a(n) → X chromosome, Y chromosome

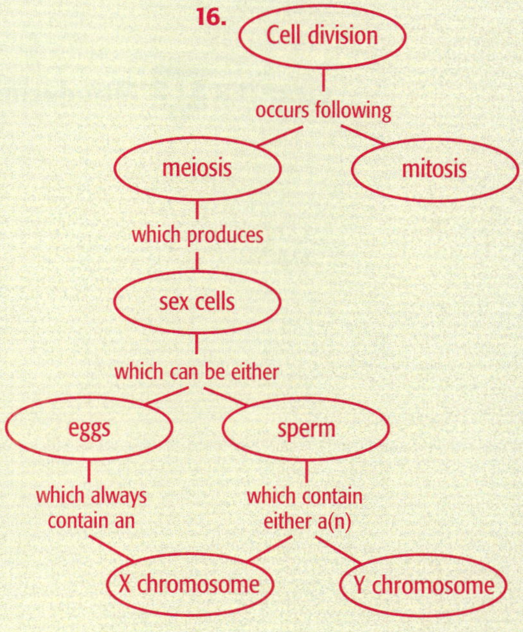

15.

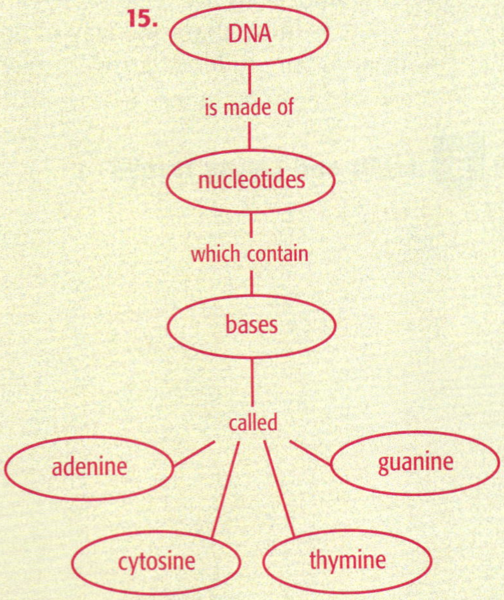

DNA

is made of

nucleotides

which contain

bases

called

adenine

guanine

cytosine

thymine

17.

Darwin

developed a theory of

natural selection

which includes the steps

struggle to survive

genetic variation

overproduction

successful reproduction

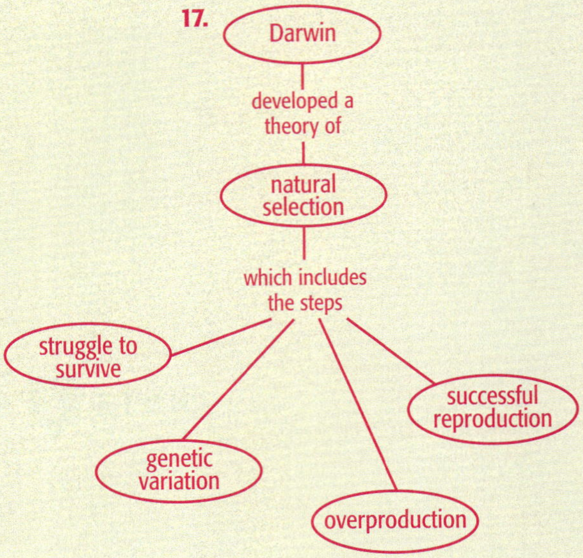

CHAPTER 9 The History of Life on Earth

14.

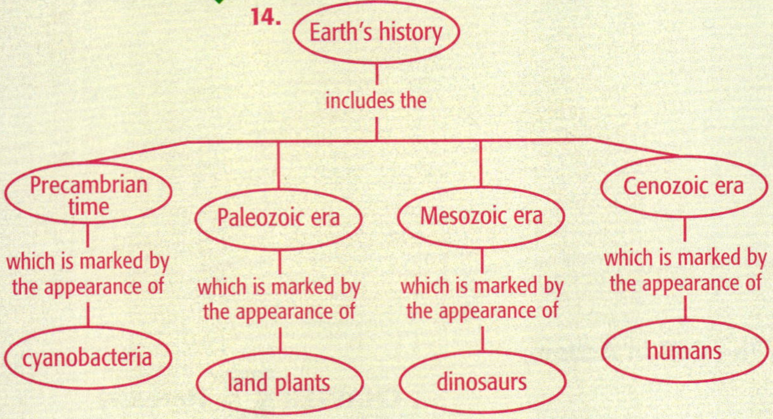

Earth's history

includes the

Precambrian time

which is marked by the appearance of

cyanobacteria

Paleozoic era

which is marked by the appearance of

land plants

Mesozoic era

which is marked by the appearance of

dinosaurs

Cenozoic era

which is marked by the appearance of

humans

CHAPTER 10 Classification

15.

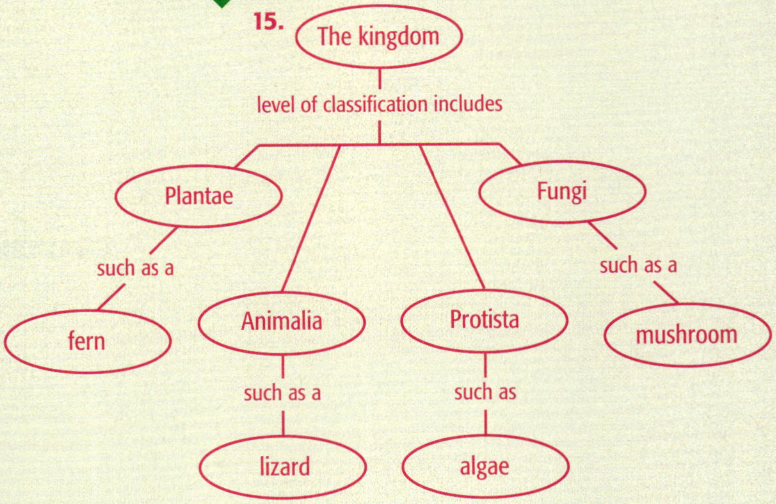

The kingdom

level of classification includes

Plantae

such as a

fern

Animalia

such as a

lizard

Protista

such as

algae

Fungi

such as a

mushroom

CHAPTER 11 Introduction to Plants

18.

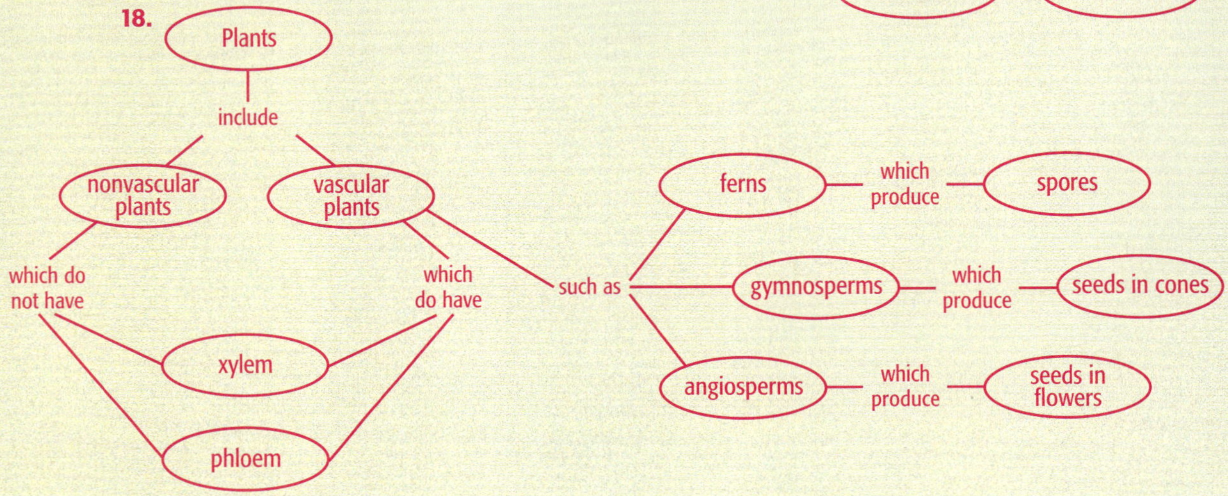

Plants

include

nonvascular plants

which do not have

vascular plants

which do have

such as

xylem

phloem

ferns

which produce

spores

gymnosperms

which produce

seeds in cones

angiosperms

which produce

seeds in flowers

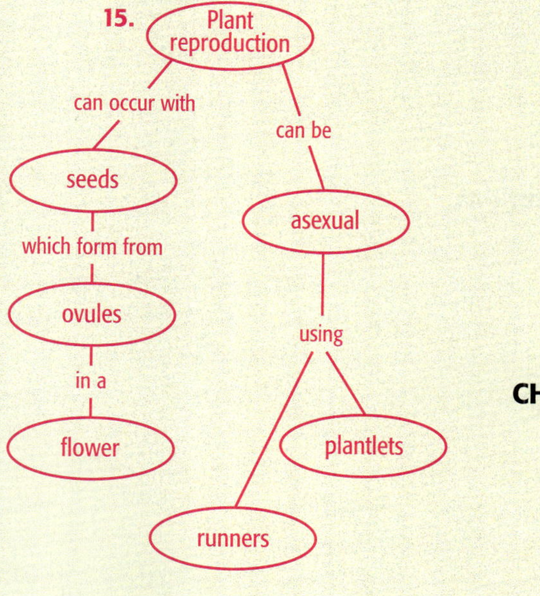
CHAPTER 12 Plant Processes
15. Plant reproduction
can occur with
seeds
which form from
ovules
in a
flower
can be
asexual
using
runners
plantlets

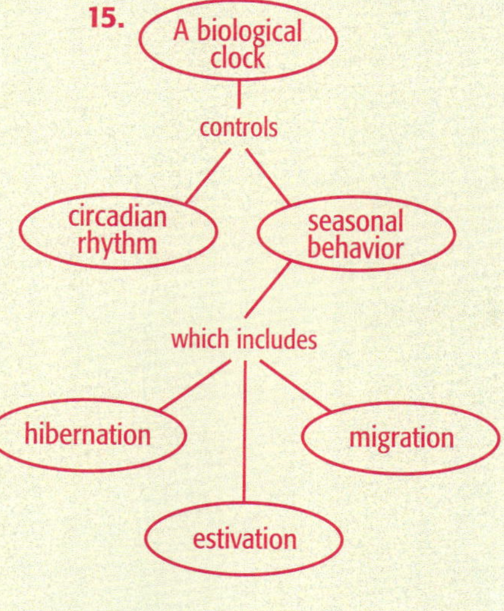
CHAPTER 13 Animals and Behavior
15. A biological clock
controls
circadian rhythm
seasonal behavior
which includes
hibernation
estivation
migration

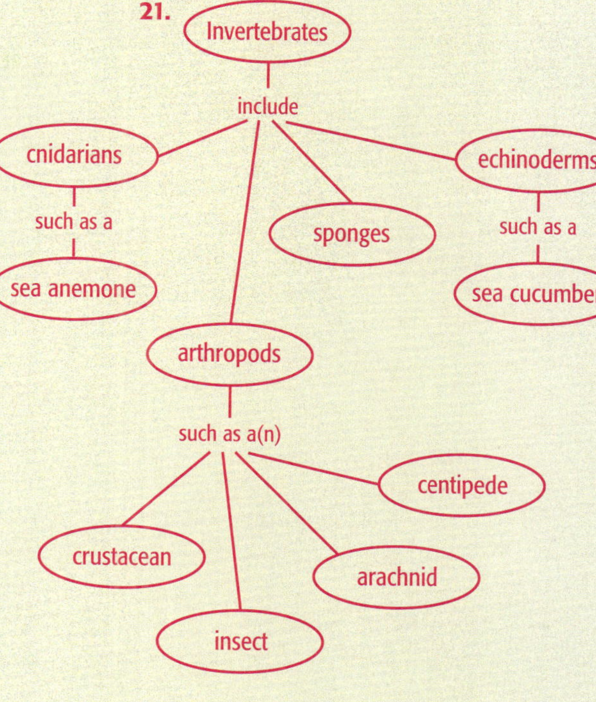
CHAPTER 14 Invertebrates
21. Invertebrates
include
cnidarians
such as a
sea anemone
sponges
echinoderms
such as a
sea cucumber
arthropods
such as a(n)
crustacean
insect
arachnid
centipede

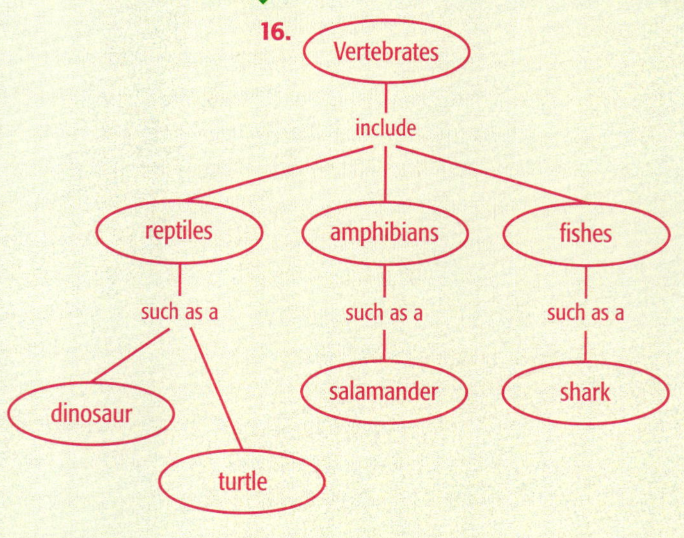
CHAPTER 15 Fishes, Amphibians, and Reptiles
16. Vertebrates
include
reptiles
such as a
dinosaur
turtle
amphibians
such as a
salamander
fishes
such as a
shark

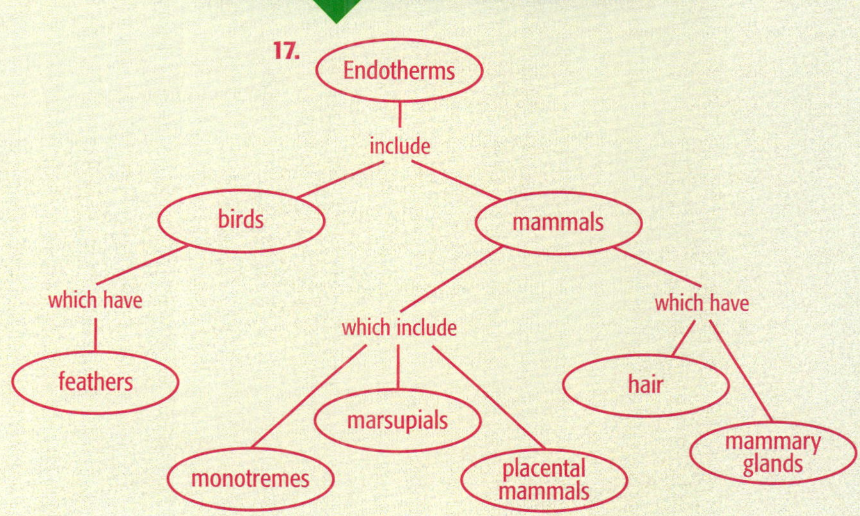
CHAPTER 16 Birds and Mammals
17. Endotherms
include
birds
which have
feathers
mammals
which include
monotremes
marsupials
placental mammals
which have
hair
mammary glands

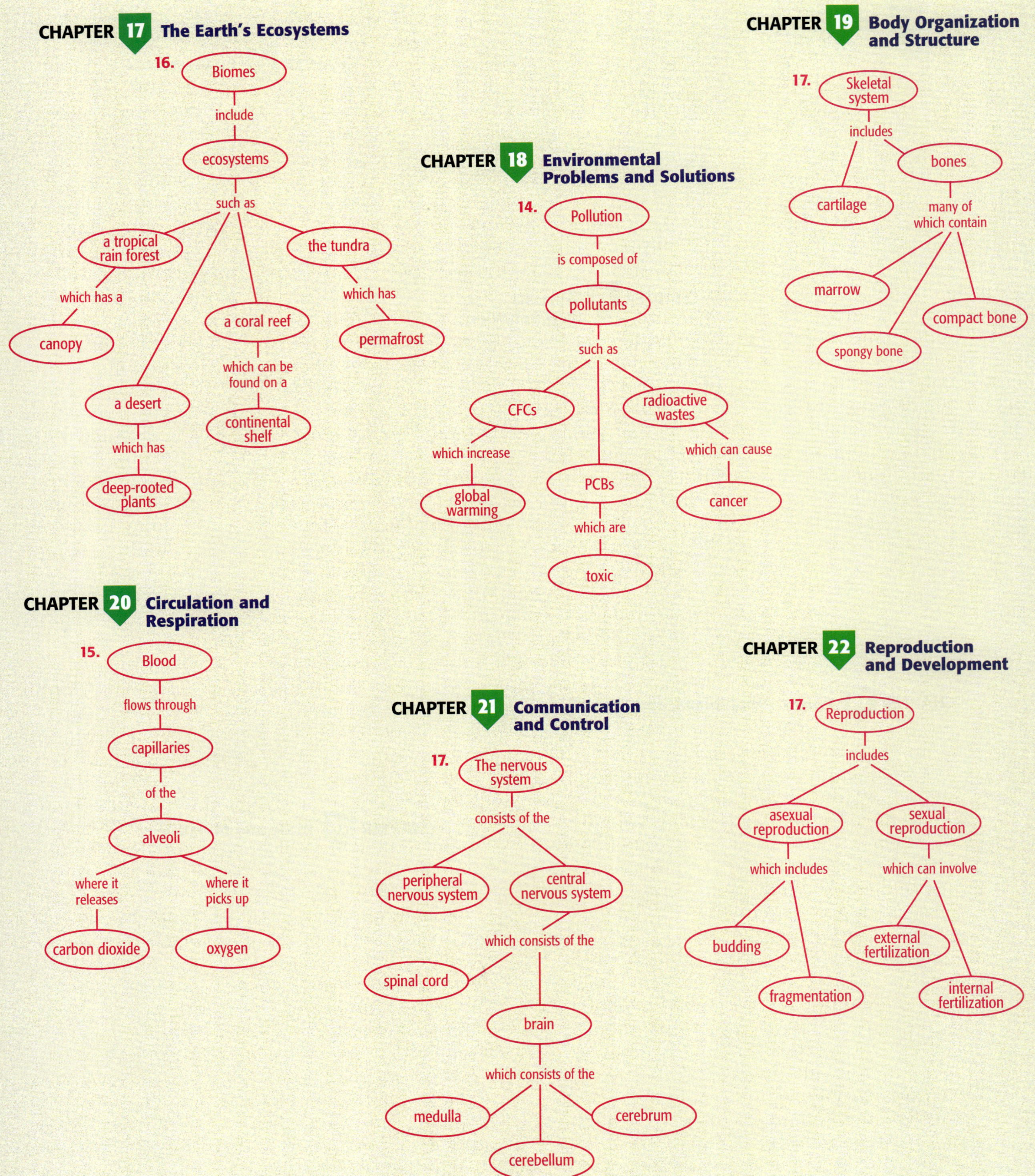

CHAPTER 17 The Earth's Ecosystems
16. Biomes
include
ecosystems
such as
a tropical rain forest
which has a
canopy
a desert
which has
deep-rooted plants
a coral reef
which can be found on a
continental shelf
the tundra
which has
permafrost

CHAPTER 18 Environmental Problems and Solutions
14. Pollution
is composed of
pollutants
such as
CFCs
which increase
global warming
PCBs
which are
toxic
radioactive wastes
which can cause
cancer

CHAPTER 19 Body Organization and Structure
17. Skeletal system
includes
cartilage
bones
many of which contain
marrow
spongy bone
compact bone

CHAPTER 20 Circulation and Respiration
15. Blood
flows through
capillaries
of the
alveoli
where it releases
carbon dioxide
where it picks up
oxygen

CHAPTER 21 Communication and Control
17. The nervous system
consists of the
peripheral nervous system
central nervous system
which consists of the
spinal cord
brain
which consists of the
medulla
cerebellum
cerebrum

CHAPTER 22 Reproduction and Development
17. Reproduction
includes
asexual reproduction
which includes
budding
fragmentation
sexual reproduction
which can involve
external fertilization
internal fertilization